Commission of the European Communities

1993 European Community Wind Energy Conference

Proceedings of the International Conference held at Lübeck-Travemünde, Germany

8-12 March 1993

Edited by:

A.D. GARRAD
Garrad Hassan & Partners Limited, Bristol, United Kingdom

W. PALZ
Commission of the European Communities, Brussels, E.C.

S. SCHELLER
Bundesministerium für Forschung und Technologie, Bonn, Germany

Publisher: H.S. Stephens & Associates

The Conference was organised by the Commission of the European Communities

ISBN 0-9521452-0-0

British Library Cataloguing in Publication Data

European Community Wind Energy Conference:
Proceedings of an International
Conference Held at Travemünde, Germany,
8-12 March 1993. — 1993
I. Garrad, A.
621.45

Publication Nº EUR 15083 EN of the
Commission of the European Communities,
Scientific and Technical Communication Unit
Directorate-General Telecommunications, Information
Industries and Innovation, Luxembourg
and H.S. Stephens & Associates.

LEGAL NOTICE

Neither the Commission of the European Communities nor any person acting on behalf of the Commission is responsible for the use which might be made of the following information.

Published and available from:
H.S. Stephens & Associates
Pavenham Road
Felmersham
Bedford MK43 7EX
United Kingdom

Printed by Uniskill Limited, Witney, Oxfordshire, U.K.

PREFACE

I am happy to introduce the proceedings of the European Community Wind Energy Conference 1993 held in Lübeck-Travemünde in Northern Germany. This conference was sponsored and organised by the Commission of the European Communities and was the major European event of the year in Wind Energy. It was attended by more than 600 delegates from all the European Community member states and 30 other countries around the world. Moreover, there were 150 industrial and commercial participants in the exhibition which presented a complete cross section of the European wind industry and was another highlight of the conference.

I take this opportunity to express my gratitude to the Bundesministerium für Forschung und Technologie in Bonn for co-sponsoring the conference and supporting it in many ways. It was a particular honour that State Secretary Hermann Neumann gave an important opening speech putting wind energy into the perspective of German energy policy. I am also grateful to the government of the host Land Schleswig-Holstein for its great interest and support for this event.

The conference marked an important milestone in wind energy achievement in Europe, namely the installation of the first 1000 MW of wind power in the European Community countries. The many thousands of operational turbines demonstrate the efficiency and vigour of Europe's industry which is leading the world in this emerging sector.

The main subject of discussion during the five days of the conference was the state of the art in wind technology and the current problems of wind energy implementation. During the conference, 90 papers were presented orally and 167 others as specialist papers.

The conference confirmed that wind technology is becoming increasingly competitive with conventional power sources for large-scale electricity production. The potential for further technological innovation was highlighted by the Commission's programme for the development of a new generation of wind turbines in the megawatt size as well as a whole new range of technical concepts.

At the closing session, Dr. Hermann Scheer, member of the Bundestag and President of Eurosolar, introduced the newly created Poul la Cour prize, named after the Danish pioneer who, 100 years ago, developed the world's first wind turbine for electricity production and was a great promoter of the social and technical implementation of wind turbines in rural areas. The prize was awarded by Dr. Andrew Garrad to Mr. Grove Nielsen for his important contribution to the modern development of wind turbines which started in the middle of the 1970s in Denmark.

I hope that these proceedings will become a reference work for the current state of wind technology and will provide practical support for all those working for the further development and promotion of electricity from the wind.

Brussels, 7 April 1993

Dr. Wolfgang Palz
Conference Director
Commission of the European Communities

1993 European Community Wind Energy Conference & Exhibition

SPONSORED BY:

Commission of the European Communities
— Directorate General XII for Science, Research & Development
— Directorate General XVII for Energy
Bundesministerium für Forschung und Technologie
Land Schleswig-Holstein
Stadt Lübeck

Conference preparation in co-operation with the
European Wind Energy Association

STEERING COMMITTEE

A.D. Garrad *Chairman*	Garrad Hassan & Partners	UK
H.J.M. Beurskens	ECN	Netherlands
G. Caratti	CEC, DG XII	E.C.
E. Hau	ETA PLAN	Germany
P. Helm	WIP-Munich	Germany
H. Klinger	Min. für Soziales, Gesundheit und Energie des Landes Schleswig-Holstein	Germany
P.H. Madsen	Risø National Laboratory	Denmark
E. Millich	CEC, DG XVII	E.C.
W. Palz	CEC, DG XII	E.C
W. Rieß	University of Hannover	Germany
F. Sanchez Sudon	CIEMAT – IER	Spain
E. Sesto	ENEL and EWEA	Italy
H.S. Stephens	H.S. Stephens & Assocs.	UK
H. Wollmerath	Engineering Office	Germany
A. Zervos	CEC, DG XII	E.C.

GERMAN NATIONAL COMMITTEE

Prof. W. Rieß *Chairman*	University of Hannover
E. Hau *Vice Chairman*	ETA PLAN
P. Helm *Secretariat*	WIP-Munich
H. Wollmerath *Secretariat*	Engineering Office
D. Austerman	Member of German Parliament
K.H. Buhse	Schleswag AG
P.H. Carstensen	Member of German Parliament
R. Kottkamp	Niedersächsisches Min. für Wirtschaft, Technologie und Verkehr
B. Krahl	Ministerium für Wirtschaft, Mittelstand u. Technologie des Landes Brandenburg
J.P. Molly	DEWI
M. Opel	Member of German Parliament
W. Palz	CEC, DG XII
K. Rave	Min. für Soziales, Gesundheit und Energie des Landes Schleswig-Holstein
W. Reichelt	Stadtwerke Lübeck
G. Reiners	EWE
B. Richter	Windtest Kaiser-Wilhelm Koog GmbH
H. Rudolph	Min. für Wirtschaft, Technik, Energie, Verkehr und Tourismus des Landes Mecklenburg-Vorpommen
H. Scheer	Member of German Parliament
S. Scheller	BMFT
G. Schmitz	Preussen Elektra
I. Weiß	HEVAG Rostock

SCIENTIFIC COMMITTEE

G. Ambrosini	ENEA-CRE Casaccia	Italy
F. Avia	CIEMAT-IER	Spain
H.J.M. Beurskens	ECN	NL
A. Crespo	Universidad Politécnica de Madrid	Spain
R. Derdelinckx	Vrije Universiteit Brussel	Belgium
K. Diamantaras	CEC, DG XVII	E.C.
V.A. Dobrovolski	Moscow Aviation Institut	CIS
J.B. Dragt	Technical Univ. of Delft	NL
G. Elliot	National Wind Turbine Centre	UK
A.D. Garrad	Garrad Hassan & Partners	UK
E. Hau	ETA PLAN	Germany
N. Jenkins	U.M.I.S.T.	UK
D.P. Lalas	Lamda Technical Ltd	Greece
E. Luken	NOVEM	NL
B.T. Madsen	BTM Consultants	Denmark
P.H. Madsen	Risø National Laboratory	Denmark
B. Maribo Pedersen	Technical University of Kopenhagen	Denmark
H. Matsumiya	Mechanical Eng'g. Lab.	Japan
H.G. Matthies	Germanischer Lloyd	Germany
D.G. Milborrow	National Wind Power	UK
J.P. Molly	DEWI	Germany
D.I. Page	ETSU	UK
E. Peltola	VTT	Finland
E.L. Petersen	Risø National Laboratory	Denmark
D.C. Quarton	Garrad Hassan & Partners	UK
A. Restivo	Universidade do Porto	Portugal
W. Rieß	University of Hannover	Germany
J. Schmid	Fraunhofer Gesellschaft	Germany
E. Sesto	ENEL-CREI	Italy
P. Shi	Chinese Wind Energy Development	China
G. Simonnot	ADEME	France
P. Smulders	Eindhoven University of Technology	NL
H. Snel	ECN	NL
E. Spinadel	University of Buenos Aires	Argentina
E. Szechenyi	ONERA	France
S.-E. Thor	FFA	Sweden
R. Thresher	NREL	USA
C. Tantareanu	ICEMBERG	Romania
J. van Leuven	University of Antwerp	Belgium
S. Wagner	University of Stuttgart	Germany
R. Watson	University College Dublin	Eire
A. Zervos	CEC, DG XII	E.C.

CONFERENCE ORGANISER ON BEHALF OF THE CEC:

WIP-Munich,
Sylvensteinstr. 2,
D-8000 München 70,
Germany

CONFERENCE ORGANISATION

H. Ehmann	WIP-Munich
A.Zepf	WIP-Munich

CONFERENCE SECRETARIAT

A. Baubin	CEC, DG XII

Die Bedeutung der erneuerbaren Energlen für Energieversorgung und Umwelt

Eröffnungsansprache des Parlamentarischen Staatssekretärs
beim Bundesminister für Forschung und Technologie
Bernd Neumann, MdB
anläßlich der
Europäische Windenergie-Konferenz ECWEC '93
Lübeck-Travemünde, 08.03.1993

(Anrede)

Ihrer Einladung zur Teilnahme an der Eröffnungsveranstaltung der Europäischen Windenergiekonferenz 1993 bin ich gerne und mit Interesse gefolgt. Ich freue mich, daß die Konferenz in diesem Jahr in Deutschland stattfindet und die neuen Bundesländer besonders berücksichtigt werden sollen. Dies begrüße ich ausdrücklich und wünsche Ihnen allen in den nächsten Tagen interessante Fachvorträge und anregende Diskussionen und dazu darf ich Ihnen auch herzliche Grüße von Bundesforschungsminister Matthias Wissmann übermittein.

Meine Damen und Herren!

Es besteht keine Aussicht, daß das drastische Wachstum der Erdbevölkerung in den nächsten Jahrzehnten gestoppt werden kann. Die Vereinten Nationen haben abgeschätzt, daß im Jahr 2000 etwa 6,3 Milliarden Menschen die Erde bevölkern werden, für das Jahr 2025 wird eine weltere Steigerung auf etwa 8,5 Milliarden erwartet. All diese Menschen wollen leben und benötigen dazu Energie. Es ist daher weltweit von einem steigenden Energiebedarf insbesondere fossiler Brennstoffe auszugehen, die heute weltweit nach wie vor 96% des Primärenergieverbrauchs decken.

Steigender Verbrauch fossiler Brennstoffe bedeutet weiter anwachsende Kohlendioxid-Emissionen und damit möglicherweise eine weitere Verschärfung des Klimaproblems. Was kann, was muß getan werden, um den CO_2-Ausstoß zu verringern und damit, wie manche sagen den Weg in die Klimakatastrophe zu bremsen?

Wie Sie wissen, hat sich Deutschland, ebanso wie mehrere andere Nationen, darauf festgelegt, hiergegen entschiedene Schritte zu unternehmen. Das Bundeskabinett hatte deshalb bereits im Dezember 1990 beschlossen, für das Jahr 2005 eine Reduktion der CO_2-Emissionen in Deutschland um 25-30% anzustreben, bezogen auf das Jahr 1987.

Wie können wir dieses Ziel erreichen?
Vor 20 Jahren, zur Zeit der ersten Ölkrise, war die Antwort der damaligen Bundesregierung unter Kanzler Willy Brandt eindeutig und klar. Sie lautete Kernenergie.

Eine so schlichte Antwort ist heute nicht mehr möglich. Dies liegt aber nicht daran, well die Kernenergie seither unsicherer geworden wäre oder technisch und wirtschaftlich nicht das bringt, was damals erwartet worden war. Es liegt vielmehr daran, daß seit dieser Zelt eine allgemeine Akzeptanzkrise bei vielen Großtechnologien eingetreten ist. Und die Kernenergie ist hierfür ein Beispiel, an dem sich trefflich streiten laßt. Ich will hier die Argumente dafür und dagegen gar nicht wiederholen. Das wäre völlig müßig. Was für mich allerdings zählt, ist die reale Welt, so wie sie heute ist. Und da sehe ich auch auf absehbare Zelt in unserem Land keine andere Quelle, die soviel CO_2-freie Energie liefern kann, wie die Kernenergie. Deshalb glaube ich Herr Kollege Möller, wird es keinen Konsens geben, wenn vor der Klammer steht: völliger Ausstieg.

Natürlich läßt sich auch diese Naturkraft nur dann verantwortlich nutzen, wenn die größtmögliche Sicherheit gegeben ist. Diese Maßgabe gilt für die Kernenergie genauso wie für alle anderen Energiequellen, die ja ebenfalls sicherheitstechnische Vorkehrungen erfordern. Auch die Nutzung etwa der Wasserkraft mit Hilfe von Staudämmen hat ihre Risiken für Sicherheit und Umwelt. Es ist unsere Aufgabe, Energieerzeugung in jeder Form verantwortbar und so sicher wie möglich zu gestalten.

Ich sehe in der realen Welt auch in der Zukunft eine Vielzahl von Kernreaktoren in den Ländern um uns herum. Ein Ausstieg Deutschlands aus der Kernenergie würde dort sicher nicht zur Nachahmung Anlaß geben. Es würde uns aber die Möglichkeit nehmen, mit unserer Sicherheitstechnik positiven Einfluß auf andere auszuüben. Auch dies gehört zu einer Energiepolitik. Und wir können in unserem Land auf den Ergebnissen der Reaktorsicherheitsforschung aufbauen, für die allein das BMFT von 1974 bis heute ca. 1,2 Mrd. DM aufgewandt hat.

Die Überlegungen, die ich hier als kurze Gedankenskizze zur Kernenergie vorgetragen habe, sind nicht überall populär, dessen bin ich mir bewußt. Dennoch müssen sie dargelegt werden, um vor Wunschdenken zu warnen, das in diesem Zusammenhang oftmals gehegt wird, um die zukünftige Rolle erneuerbarer Energien politisch zu überhöhen. Damit wird meiner Meinung nach auch den erneuerbaren Energien ein schlechter Dienst erwiesen. Es gibt kein entweder — oder ! keinen alleinseligmachenden Weg in der Energiepolitik! Vielmehr müssen wir unter den Gesichtspunkten der Umwelt, der globalen Verantwortung, der Wirtschaftlichkeit und Nachhaltigkeit zu einem sinnvollen Mix der Energiequellen kommen.

Meine Damen und Herren!

Wir haben in Deutschland mehrere Möglichkeiten, erneuerbare Energien zu nutzen. Am wichtigsten ist zur Zeit die Wasserkraft. Sie macht derzeit etwa 3% der Stromerzeugung aus und sie ist — wie auch die Windenergie — regional ungleich verteilt.

Selbst bei einer — optimistischen — Verdoppelung des Anteils der erneuerbaren Energien an der Primärenergieerzeugung bis zum Jahr 2005 ergäbe sich in Deutschland erst ein Antell von etwa 5%. Um dies zu erreichen, müßten z. Bsp. stilliegende Wasserkraftwerke reaktiviert und die Windenergie massiv ausgebaut werden; auch Biomasse und Müllverbrennung müßten verstärkt dazu beitragen. Der Anteil der Photovoltaik wird dagegen nur bescheiden bleiben. Ein derartiger Ausbau ist nicht ohne erhebliche Anstrengungen möglich, Anstrengungen finanzieller und administrativer Art. Auch Eingriffe in die Umwelt sind dazu nötig.

Lassen Sie mich am Beispiel der Windenergie aufzeigen, was wir hier bisher schon erreicht haben.
Von den weltweit über 20,000 stromerzeugenden Windanlagen stehen in Europa etwa 5,000 mit einer Gesamtnennleistung von rund 1,000 Megawatt. Dieser Aufschwung war bei uns in Deutschlnd besonders deutlich zu erkennen: Vor etwa fünf Jahren gab es vermutlich mehr alte Windmühlen als moderne Windkraftanlagen; heute sind dagegen schon mehr als 1,200 moderne Anlagen mit einer Gesamtnennleistung von über 180 MW in Betrieb. Jährlich kommen weitere 50-70 MW hinzu, vorwiegend an windreichen Küstenstandorten, aber auch zunehmend im Binnenland.

Diesen eindrucksvollen Aufschwung hat insbesondere das "250 MW Wind"-Programm der Bundesregierung bewirkt, das seit 1989 besteht und über das auf dieser Konferenz ja noch im Einzelnen berichtet werden wird. Die Küstenländer Niedersachsen und Schleswig-Holstein setzen die Förderung der Windenergie mit eigenen Programmen fort. Wenn es gelänge, die jetzigen Planzahlen zu realisieren, würde die Windenergie jährlich über 4 Mrd. Kilowattstunden Strom erzeugen und läge damit nicht im Promille- sondern schon im Prozentbereich unserer gesamten Primärenergieerzeugung.

Wenn dies auch im Hinblick auf den Gesamtbedarf ein kleiner Beitrag ist und die Forschungs- und

Entwicklungsaufwendungen dafür relativ groß sind, so will die Bundesregierung dennoch Fakten setzen und deutlich machen, daß wir die erneuerbaren Energien ernst nehmen als ein Mittel, auch zur CO_2-Reduktion beizutragen. Lassen Sie mich hinzufügen — und da denke ich über meine Eigenschaft ais Staatssekretär im BMFT hinaus — daß über die Förderung von Forschung und Entwicklung eine weitergehende Demonstrations- und Anwendungsförderung wünschenswert wäre.

Bei der Realisierung dieses geschliderten Ausbaus der Windenergie in Deutschland müssen noch einige Hürden überwunden werden. Für technische Probleme lassen sich früher oder später Lösungen finden — zu denen wir mit dem "250 MW Wind"-Programm und mit FuE-Fördermaßnahmen unseren Beitrag leisten werden. Auch die Finanzierung dieses Ausbaus dürfte möglich sein, zumal die Kosten für Windkraftanlagen sinken werden, so daß zum Teil auch ohne Förderung ein wirtschaftlicher Betrieb möglich sein wird. Wenigstens an günstigen Standorten sind wir nicht mehr weit davon entfernt.

Als ein größeres Problem sehe ich vielmehr Standort- und Genehmigungsfragen. Schon heute ist erkennbar, daß nicht alle technisch möglichen Standorte auch realisiert werden können Akzeptanzfragen oder solche des Landschafts- und Naturschutzes spielen da auch bei der Windenergie eine wesentliche Rolle. Hier müssen die vor Ort zuständigen Behörden einschließlich der Politik noch viel Überzeugungsarbeit leisten — und manchmal auch über ihren eigenen bürokratischen Schatten springen!

Das Bundesministerium für Forschung und Technologie hat für die Förderung der Windenergie im Rahmen seines Energieforschungsprogramms seit 1974 insgesamt etwa 300 Mio DM zur Verfügung gestellt. Diese Investition auf die Zukunft zahlt sich jetzt aus.

Rechnet man die Aufwendungen für die anderen Bereiche der erneuerbaren Energien und der rationellen Energieverwendung sowie die Mittel der Bundesländer hinzu, dann gehört Deutschland mit den USA und Japan zur Spitzengruppen in diesem Bereich.

Abschließend die Frage und die Beantwortung:
Wie geht es weiter?

Trotz der bisher schon erkennbaren positiven Entwicklung der Windenergie in Deutschland ist eine Fortsetzung dieses Trends ohne staatliche Förderung bisher noch nicht möglich. Deshalb müssen Bund und Länder ihre Unterstützungsmaßnahmen fortsetzen.

Von der Bundesregierung wird daher das "250 MW Wind"-Programm unverändert bis 1996 fortgeführt; die notwendigen Fördermittel sind eingeplant. Daneben werden weiterhin Forschungs- und Entwicklungsarbeiten mit den vorher genannten Zielen gefördert. Zusätzlich unterstützt die Bundesregierung die Anwendung der Windenergie in Ländern der Dritten Welt im Rahmen des ELDORADO-Programms, das jetzt angelaufen und auf eine große Resonanz gestoßen ist. Darüber hinaus wird sich die Bundesregierung an der Entwicklung von Großwindanlagen der 1MW-Klasse beteiligen, die zusammen mit der Kommission der Europäischen Gemeinschaften im Rahmen des JOULE II-Programms gefördert wird. Großwindanlagen können besser als kleine Anlagen die bei uns nur begrenzt vorhandenen windgünstigen Standorte ausnutzen.

All diese Maßnahmen haben ein gemeinsames Ziel; Sie sollen der Windenergie dazu verhelfen, in wenigen Jahren auch ohne staatliche Unterstützung wettbewerbsfähig zu sein.
Ich bin zuversichtlich, daß wir gemeinsam dies erreichen können, damit auch die Windenergie den von ihr erwarteten Beitrag zu Energieversorgung und Umwelt leisten kann.

Ich danke Ihnen für Ihre Aufmerksamkeit.

THE CHAIRMAN'S OPENING SPEECH

On behalf of the Steering Committee, the Commission of the European Communities and our other sponsors, I would like to welcome you to this conference — the biggest ever wind energy event to be held on European soil.

During the week which we spend here European wind energy will pass a major milestone. The one thousandth MW of wind energy will be installed within the Community boundaries. Europe is now the most rapidly expanding wind energy market in the world and, within the next couple of years, is expected to exceed the installed capacity in the USA. Many of the people who are responsible for that achievement are delegates who can, quite properly, use this week as an opportunity for celebration.

However, we should not be complacent. In our terms 1000 MW is a lot of power, and it is certainly a lot of windmills, but it is still minuscule compared with the conventional sources installed throughout the Community. It is just the beginning.

The main obstacles to wind energy in Europe are not now technical. Modern wind turbines are returning availability figures in excess of 98% — a dramatic change from even 5 years ago. There are, of course, technical improvements which can still be made, and it is a measure of the resolve to make those improvements, that there are so many people here with technical contributions to make.

There will always be public debate about the acceptability of different forms of energy. The debate over wind energy is now important in countries where there is more than a minimal level of capacity. There is no glib answer to the objections raised to the appearance of wind turbines on the horizon. To me they symbolise clean energy, to others they are an eyesore. This debate cannot be ignored and, indeed, the environmental issues which affect wind energy have been placed in a prominent position within the conference programme.

Economic arguments about energy sources are quickly revealed to be essentially political and the real obstacle is political commitment. Political commitment can be shown in several ways. In many countries it is shown through the attitude of the utilities. The documented performance and reliability of wind turbines show that they can no longer be brushed aside as a technical indulgence or environmental pipe dream. They are a serious means of energy production. That opinion is not, however, shared by many utilities throughout the Community and we would do well to ask "Why?" A firm commitment by them, not only to support generation from wind energy, but also to allow renewable energy sources fair and sensible access to the grid, is a vital part of assuring the future for this industry, and for cleaner power.

Political commitment can also be shown through the market incentives and investment provided for wind energy by government agencies. Significant interest in wind energy has always been produced by market incentive programmes.

Political commitment has been shown by the German Federal Government through its 250 MW programme and by the Schleswig Holstein Land, through its support and encouragement of wind energy here. It is therefore no coincidence that this conference is being held in Germany, which is presently the most active of all European countries. Within Germany some 95% of the national capacity is installed in Schleswig Holstein. It is entirely fitting that the European Wind Energy Community should be gathered here in Schleswig Holstein at this propitious moment.

The location of Travemünde, on the very edge of old East Germany, is also a poignant reminder that the countries to the east of us face the joint challenge of major environmental problems and a chronic shortage of energy. Wind energy can form part of the solution to those two problems and it would be fitting for the delegates here at this conference to make a pledge that the next 1000 MW should be installed, not to the west of Travemünde, but to the east.

I hope that you all spend an interesting, demanding and rewarding week here, and have the opportunity to discuss and resolve some of these important issues.

Dr. Andrew Garrad
Conference Chairman

Rede von Dr.W. Palz, EG Kommission, anläßlich der Eröffnung der EG-Wind Konferenz in Travemünde am 8. März 1993

Ich heiße Sie alle sehr herzlich im Namen der Kommission der Europäischen Gemeinschaft zu dieser wichtigen Windkonferenz hier in Travemünde, im Norden Deutschlands willkommen.

Die Konferenz, die wir heute eröffnen, fällt in eine Zeit großer Veränderungen in der energiepolitischen Entwicklung in der Welt. Die 80er Jahre waren markiert, unter Leitung der Führungsmacht USA, durch ein kontinuierlich abnehmendes staatliches Interesse an der Entwicklung neuer Energiekonzepte und gleichzeitig damit eine substantielle Regression der staatlichen Förderprogramme für Energie insgesamt und in ganz besonderem Maße der Fördermittel für die erneuerbaren Energien. Nur wenige Staaten wie etwa die Bundesrepublik Deutschland, widerstanden diesem Trend.

Die energiepolitische Landschaft begann sich erst zu verändern zum Ende des letzten Jahrzehnts, als die Diskussion des Treibhauseffekts das konventionelle Energiedenken erschütterte und schließlich anläßlich der Tagung in Rio im letzten Jahr zu der weltweiten Bewußtseinsbildung führte, daß die konventionellen Denkschemen in der Energiepolitik neu überdacht werden müssen.

Seit Anfang dieses Jahres sehen wir, wieder eingeleitet durch die USA und den dortigen Regierungswechsel, neue Anzeichen dafür, daß den erneuerbaren Energien endlich die ihnen angemessene Bedeutung zuerkannt wird. Auch in der Europäischen Gemeinschaft ist das politische Klima dabei, sich mehr zu Gunsten der erneuerbaren Energien zu verändern. So hat das Europäische Parlament in Straßburg am 19. Januar 93 einstimmig eine Resolution verabschiedet, in der für die Förderung der erneuerbaren Energien in der EG verstärkte Anstrengungen verlangt werden. Der EG- Ministerrat für Forschung seinerseits ist dabei, für die Energieforschung im JOULE II Programm und dabei insbesondere für die erneuerbaren Energien, erhebliche Sondermittel für die Jahre 1993-94 zu beschließen. Wir rechnen damit, daß uns dies seitens der Kommission ermöglichen wird, noch in diesem Jahr im Rahmen einer allgemeinen Ausschreibung, die demnächst veröffentlicht werden soll, neue Fördermittel auch der Entwicklung der Windtechnologie zukommen zu lassen.

Das neue politische Klima stimmt uns für die zukünftige Entwicklung der Windenergie in der EG zuversichtlicher als bisher. An dieser Stelle ist aber hervorzuheben, daß die europäische Windindustrie und Windgemeinschaft insgesamt, in den letzten Jahren, trotz widriger Konjunktur, ganz Erstaunliches geleistet hat. Nicht zuletzt ist dieser Kongress in Travemünde der geeignete Rahmen, die soeben geschafften 1000 MW installierter Windturbinenkapazität in der EG gebührend zu feiern. Qualitativ und quantitativ ist dies eine bemerkenswerte Etappe und es ist hier angebracht, das Erreichte kurz zu beleuchten.

Die Energieerzeugung pro Rotorfläche konnte in den letzten Jahren verdreifacht werden und zwar von 400 kWh/m^2 noch vor 10 Jahren auf mehr als 1200 kWh/m^2 pro Jahr für gute Standorte.

Die mittlere Dauerleistung konnte allgemein auf über 30% gesteigert werden. Immer häufiger findet man Maschinen, die bis zu 48% erreichen.

Die Verfügbarkeit heutiger kommerzieller Maschinen liegt über 90% und erreicht in vielen Fällen 98%.

Die Standardturbine von 55kW von vor 10 Jahren wurde abgelöst von weit größeren Maschinen; 500 kW Einheiten sind heute in der Kommerzialisierung begriffen. Der Landbedarf konnte durch die Nutzung der größeren und effizienteren Turbinen enorm gesenkt werden.

Die Kosten des Stroms aus Wind konnten kontinuierlich reduziert werden und liegen heute für gute Standorte bei 4 bis 7 ECU-cents pro kWh.

Von 1991 nach 92 war die Marktentwicklung besonders rasant. Während 1991 noch 175 MW in der EG installiert wurden, waren es im letzten Jahr 340 MW. Die mittlere Turbinenleistung stieg von 160 kW auf 270 kW. Die

Marktentwicklung kommt vor allem der kleinen und mittleren Industrie zugute. Seit Anfang der 80er Jahre wurden im Windbereich 3400 neue Arbeitsplätze geschaffen. Industriepolitisch ist eine Rationalisierung eingetreten: von über 200 Turbinenherstellern Mitte der 80er Jahre sind heute 30 übrig.

Insgesamt hat die installierte Leistung in der EG 1000 MW erreicht, davon etwa die Hälfte in Dänemark. Bemerkenswert ist, daß die Windtechnologie in allen anderen EG- Ländern ebenfalls Fuß fassen konnte. Dies gilt auch für Frankreich, das lange Zeit von dieser Entwicklung ausgeklammert war.

Bisher wurden in der EG insgesamt 6,3 TWh Elektrizität durch Wind erzeugt. Dadurch wurde u.a. die Emission von 3.5 Millionen Tonnen CO_2 und vielen tausend Tonnen SO_2 und anderer Schadstoffe vermieden.

Gestatten Sie mir bei dieser Gelegenheit noch einige Worte über den Beitrag der EG- Kommission zu dieser Entwicklung. Die Kommission unterstützt den Bau und die Entwicklung innovativer Turbinen und hat so gut wie alle in den letzten Jahren neu entwickelten Maschinen finanziell direkt gefördert, sei es durch die Programme der General-Direktion XVII (Thermie), die Programme der General-Direktion XII (Joule) oder auch indirekt durch die regionalen Entwicklungsprogramme der EG in Irland, Portugal, Spanien, Italien und Griechenland. Schließlich werden zur Zeit in Brüssel Überlegungen angestellt, den Betrieb von Windanlagen in der Zukunft auch in landwirtschaftlichen Betrieben im Rahmen des EG-Agrarmarkts gezielt zu fördern.

Die technologische Herausforderung der Zukunft ist die Entwicklung der nächsten Generation von Windturbinen im Megawatt-Bereich. Es sind heute in der EG 17 Pilotmaschinen von mehr als 750 kW Leistung im Betrieb. Alle wurden von der Kommission gefördert.

Das technologische Innovationspotential für große Maschinen ist aber bei weitem noch nicht ausgeschöpft: zur Zeit sind in der EG 12 verschiedene neue Turbinentypen im Entwurf oder im Bau. Auch diese werden von der EG-Kommission gefördert. Hervorzuheben ist hier das sogenannte 'Windenergie Großanlagen Programm (WEGA II)' der GD XII, in dem eine Palette von 7 verschiedenartigen Technologien in grenzüberschreitender Zusammenarbeit entwickelt werden. Unter den neuen Konzepten, die dabei zum Zuge kommen, sollen hier nur die beiden vielversprechenden getriebelosen Anlagen erwähnt werden, die in Zusammenarbeit zwischen der EG und dem BMFT in Deutschland entwickelt werden.

Lassen Sie mich schließen mit dem Wunsch, daß diese Konferenz dazu beitragen möge, die heute besonders drängenden Probleme am Markt in gemeinsamer Arbeit anzugehen. Es ist vor allem notwendig, sehr rasch das Problem der Baugenehmigungen und der allgemeinen Standortplannung zu rationalisieren. Dazu gehört auch die bessere Standardisierung der Turbinentypen und noch weitere Verbesserungen in der Geräuschemission.

Hier direkt an der alten, nun Gott sei Dank verschwundenen Grenze zum nicht mehr existierenden Ostblock darf nicht unerwähnt bleiben, daß es unsere Aufgabe ist, in fairer Zusammenarbeit aller Völker die neue Windtechnologie, die wir in Europa geschaffen haben, auch den Menschen in Ost-Europa und der Dritten Welt zugute kommen zu lassen.

Ich wünsche uns allen, die wir hier aus allen Kontinenten für die gute Sache der Windenergie zusammengekommen sind, daß wir fruchtbare und angenehme Tage des konstruktiven Gesprächs in Travemünde erleben werden.

Ich danke Ihnen.

Dr. Andrew Garrad,
Chairman of the Conference.

The launch of this outstanding international event.

The Press Conference:
A highly interesting debate about the future of wind energy converters and strategies on their application.

Dr. Andrew Garrad awarding the Poul la Cour prize to Erik Grove-Nielsen from Denmark for outstanding work in wind energy technology.

CONTENTS

---

* Paper numbers are identical to those of the conference papers. Missing numbers indicate that papers have since been withdrawn.

Mechanical Testing

Machine Design

Noise

Dynamics and Control

Large Scale Applications

WIND ENERGY - WHERE ARE WE NOW?

Dr.David Lindley
National Wind Power Ltd.
Riverside House, Meadowbank
Bourne End, Buckinghamshire, U.K.

ABSTRACT: By the end of 1993 over 1100 MW of wind turbines will be operational in the Member States of the European Community. The rate of installation of these turbines has been dramatic with over half of all turbines installed in 1991, 1992 and 1993. These installations, with a turnkey cost of approximately 1.7 billion ECU (£1.4 billion), have been financed by over ten major European Banks, about twenty electricity generating utilities and many hundreds of individuals. This development in Europe has been able to take advantage of the Californian market of the 1980s, where the majority of turbines were supplied by European manufacturers. The turbines now being installed in Europe have lower costs, are more reliable and are larger, with installations in 1993 averaging 300 kW per turbine. At the current rate of installation Europe will have a larger installed wind turbine capacity than the USA by the end of 1995.

Most Member States now recognise the need to encourage the development of Renewables and a combination of legislation and grant aid to support and encourage windfarm development is in place in over half of all European countries.

1. INTRODUCTION

When the European Wind Energy Associations' Strategy document, 'A Plan of Action' (1) was published in October 1991, 510 MW of wind turbines were recorded as being installed by July of that year. By the end of the year, 643 MW had been installed and this grew to 891 MW by the end of 1992. Table I shows that by the end of 1993, installed capacity will have increased to 1123 MW.

Energy production within Denmark, Germany and the Netherlands alone will reach 1.25 billion kWh this year, more than will be generated by wind in either California's Altamont or Tehachipi Passes, and more than twice the amount projected from turbines in the San Gorgonio Pass. The rapid growth in the UK will result in energy production from wind of close to 0.5 billion kWh/yr from the end of 1993 onwards. Worldwide wind generation is likely to reach 4.3 billion kWh in 1992.

Table I: European Installed Wind Turbine Capacity (MW)

INSTALLED MW			
Country	1991	1992	1993
Denmark	418	470	520
Netherlands	83	116	120
Germany	90	170	220
United Kingdom	10	30	131
Spain	15	45	57
Belgium	6	6	6
Italy	5	6	20
Greece	5	26	26
Portugal	2	2	2
Sweden	8	12	12
France	1	1	1
Ireland	-	7	8
TOTAL	643	891	1123
MW installed/yr.	166	247	232

Table I also shows that over half the total capacity in Europe has been installed in 1991, 1992 and 1993. At the present rate of installation of over 200 MW per annum, Europe will overtake the United States by the end of 1995 as the world's leading user of wind energy. Indeed, at the current rate of development, Europe could have almost 3000 MW of installed capacity by the year 2000. Turbine manufacturing capacity in Europe is large enough to sustain an annual implementation rate of at least twice this amount, thus enabling the installation of closer to 5000 MW by the year 2000.

A closer look at installations in 1992 and 1993 also shows that average turbine size is about 300 kW and it is interesting to note that if all turbines now installed in Denmark were replaced with turbines of this size, they would in total contribute about 10% of all Danish electricity needs. Other European countries will benefit from the increasing size of the latest turbines as these will enable large energy yields from a given land area.

2. CREDIBILITY

It does not seem that long ago that Wind Energy was referred to, especially by some opponents in the nuclear industry, as a 'punk' technology. In their eyes, the technology lacked credibility and it seemed hard to imagine that Banks and Electric Utilities would ever invest in Wind Energy. The developments in California in the middle years of the 1980s demonstrated how the Banking System could respond and close to US$3 billion has already been invested in windfarms in the USA. It is now very encouraging to see the large number of windfarms in Europe which have been financed with a combination of equity and non-recourse project finance provided by the European Banking Community. Table II lists some of the banks involved in European windfarm deals in the last three years.

Clearly, after the Banks and their advisors had completed their due diligence procedures, they had concluded that the technology was credible, that it worked and that it was profitable.

Table II: European Banks involved in Windfarm project Finance.

Bank	Country
o Barclays Bank plc	U.K.
o County NatWest	U.K.
o Westpac Bank	U.K.
o Hambros Bank	U.K.
o Triodos Bank	Netherlands
o Bayerische Vereins Bank	Germany
o Danske Bank	Denmark
o ABN Amro Bank	Netherlands
o ING Bank	Netherlands

It is also most encouraging that so many European electrical utilities have invested in Wind Energy. Only 3 or 4 years ago this did not seem possible and many manufacturers will have their stories of how difficult it was to sell turbines to an electrical utility.

Now, over twenty European electrical utilities have invested in windfarms and they include those shown in Table III.

Table III: European Electrical Utility Windfarm Investors.

Utility	Country
o National Power plc	U.K.
o PowerGen plc	U.K.
o East Midlands Electricity plc	U.K.
o Yorkshire Electricity plc	U.K.
o S.W.E.B. plc	U.K.
o MANWEB plc	U.K.
o South Wales Electricity plc	U.K.
o Yorkshire Water plc	U.K.
o ENDESA	Spain
o Sevillana	Spain
o PEN	Netherlands
o PNEN	Netherlands
o PGEN	Netherlands
o PPC	Greece
o ENEL	Italy
o ELSAM	Denmark
o Elkraft	Denmark
o Vestkraft	Denmark
o Norrkraft	Denmark

3. GLOBAL WARMING

Much has been written about the subject so that all that needs to be said here is that there is no question that the earths atmosphere is being seriously impacted by the continued combustion of fossil fuels (2, 3). In the UK for example, a Government Committee, (the so called Renewable Energy Advisory Group, REAG) has recently published its report (4). It refers to recent international negotiations on environmental issues which have focused increasingly on the concept of sustainable development. This is most simply defined in the words of the World Commission on Environment and Development as a process "which meets the needs of the present without jeopardising the ability of future generations to meet their own needs". The report goes on to refer to Agenda 21 which was adopted by the United Nations Conference on the Environment and Development (UNCED, or "Earth Summit") in Rio de Janeiro in June 1992. It re-emphasises that "much of the world's energy .... is currently produced and consumed in ways that could not be sustained if technology were to remain constant and if overall quantities were to increase substantially". Hence it says "the need to control atmospheric emissions of greenhouse and other gases and substances will increasingly need to be based on efficiency in energy production, transmission, distribution and consumption and a growing reliance on environmentally sound energy systems, particularly new and renewable sources of energy".

The successful completion of regulations on a framework Convention on Climate change which was signed by over 150 nations at UNCED is likely to provide further impetus for Renewables. The Convention emphasises the need for developed countries "to take the lead in combatting climate change and the adverse effect thereof", endorses the cautionary principle (on which to base environmental policy) and commits parties to take action on greenhouse gas emissions. The Convention enters into force when 50 countries have ratified it, which will probably be towards the end of 1993. Under the Convention, developed countries which are party to it are committed to reporting to other signatories their programme for returning their emissions of CO_2 and other greenhouse gases to 1990 levels by the year 2000. The ultimate objective of the Convention is the stabilisation of greenhouse gas concentrations in the atmosphere at a level that would prevent dangerous anthropogenic interference with the climate system.

The European Commission has given expression to its general policy towards renewable energy sources in several Regulations and Recommendations. A number of recent proposals from the Commission have also addressed or have implications for renewable energy. These include:-

a) The ALTENER proposal for the promotion of renewable energy including an objective of an increase in the contribution of renewable energy sources to total energy demand in the Community from the present 4 per cent to 8 per cent by 2005.

b) Proposals for a carbon/energy tax starting at a rate equivalent to US$3/barrel in 1993 rising through annual increments of $1 to $10 after seven years from which renewables (except large scale hydro) would be exempted. This is one way to help ensure that the environmental consequences of fossil fuel burning are reflected in energy prices.

c) Proposals for Common rules for the internal market in electricity which would involve priority being given within transmission systems to renewable generating systems.

d) The recommendations and motions contained within the CEC so called 'Bettani' report (5) of the Committee on Energy, Research and Technology on the Promotion of Renewable Forms of Energy. The motions included:
   - o call for the financial support of renewable energies to be equal to that of thermo nuclear fusion
   - o a call for the creation of an independent European Association for the Promotion of Renewable Energies
   - o Calls on the Commission to draw up a proposal for the use of revenues raised by levying a tax on CO_2 emissions to develop and introduce renewable forms of energy.

The report says that a possible Community target for wind energy is 100,000 MW.

In the CEC as a whole, the emission of 15 million tonnes of CO_2 would be avoided each year by replacing 1 per cent of the CEC's fossil burning electrical generating capacity by wind energy (1). For its contribution to the Rio targets, the UK must reduce its annual CO_2 emissions by 10 million tonnes before the year 2000. REAG has recommended that the Government support the construction of 1500 MW (Declared Net Capacity) of renewables by the year 2000 (4). If this is implemented, it would have the effect of saving 10 million tonnes of CO_2 thus meeting the UK obligation. If wind turbines were installed in the UK at a rate of 150 MW/annum (only slightly more than the rate achieved in the last 12 months), wind energy would save about 4 million tonnes/annum by the year 2000.

It is also interesting to note the number of individuals and companies who want to make a contribution at a practical level. In the UK for example, National Wind Power has signed an agreement with Body Shop International plc (an international cosmetics company) to build a windfarm large enough to provide all the companies electricity needs for its 250 UK shops and two factories. The companies objective is to leave no 'footprint'.

Elsewhere in Europe, others are looking to provide the means by which individuals can invest in a windfarm at a level sufficient to provide them with all their electricity needs. A very large proportion of the turbines in Denmark have been financed in this way.

4. ECONOMICS

The cost of generating electricity from a windfarm is dominated by the initial turnkey capital cost of the windfarm and by the terms governing the repayment of capital. The cost of the turbines represent about 60 per cent to 70 per cent of the total turnkey cost, though the exact percentage will depend on the site and other factors. Other costs involved include:

- 'Balance of Plant' costs (which include roads, foundations, electrical work and grid connection)
- Development costs (including meteorological work, legal, planning and environmental work)
- interest charges during construction
- financing costs

After commissioning the windfarm, the operational costs include:

- maintenance and repair
- land lease costs, rates, taxes and insurances
- ongoing management costs
- provision for decommissioning.

These operational costs are modest compared with the interest and principal repayments payable on the capital employed.

Table IV illustrates the effect of these parameters on the cost of producing electricity from a notional windfarm of say 10 MW capacity on a 7.5 m/s site in the UK, producing 3000 MWh per MW of capacity and selling all of it.

The top section of the table shows a range of initial costs per kW of capacity. The middle section shows possible running costs per kWh produced. The bottom section adds in the effect of interest and capital repayment for different borrowing scenarios. (Note: The 9, 10, 12 and 13 per cent figures are the pre-tax real rates of return on the project). It can be seen that the final cost per unit generated is dominated by the cost of capital. It is clear (and obvious) that long term loans and low interest rates make a very big difference to the cost of generation.

The left hand column depicts the recent past as far as capital costs are concerned. If the farm had been paid for with an eight year loan at 12 per cent (pre-tax project real rate of return), corresponding to the underlined figure, the capital charges alone would be over 8p (9.6 ECU cents) per unit of electricity produced - bringing the total generating cost to 10.1p/unit - far in excess of current electricity prices. It is interesting to note here that the windfarms built, or under construction in the UK, are being built with six or seven year loans because the power supply contracts that each project has been awarded expire at the end of December 1998. This explains why the windfarms in the UK receive 11p/kWh (index

Table IV: Windfarm Generation Costs

Initial Costs £/kW				
Turnkey construction	1,100	1,000	800	600
Interest during construction	55	50	40	30
Prospecting, fees, etc	100	90	80	70
	1,255	1,140	920	700
Running costs p/kWh				
O & M	0.8	0.7	0.6	0.5
Rent, insurance, grid, fees, etc.	0.9	0.8	0.7	0.6
Total cost p/kWh				
25 years at 9%	5.9	5.3	4.4	3.4
20 years at 10%	6.7	6.1	5.0	3.9
15 years at 11%	7.6	6.8	5.6	4.4
10 years at 11%	8.8	8.0	6.5	5.1
8 years at 12%	10.1	9.1	7.4	5.8
6 years at 13%	12.2	11.0	9.0	6.9

linked to inflation) until 31st December 1998. They need this tarriff in order to repay the loans so quickly.

If we now contemplate machine and construction costs falling throughout the rest of this decade and we assume that turbines become more reliable so that operations and maintenance costs fall as a result of experience, then we can construct the last two columns as representative of steps in the process. It is then necessary to consider the point in the evolution of the technology when Banks and other investors are prepared to provide longer term loans and accept rates of return compatible with other low risk energy projects.

Thus from the table it is assumed that the turnkey capital costs fall to 720 ECU/kW (£600/kW) and finance can be arranged with a 20 year loan with a pre-tax real return of 10 per cent on the project, then generation costs will fall to 4.7 ECU cents/kWh (3.9p/kWh).

These are not unrealistic goals for an industry still in its infancy. They will be assisted in so far as 'conventional' electricity generating costs rise to:

a) pay for tighter environmental constraints on conventional plant
b) recognise the 'social' costs and hidden subsidies for conventional electricity generation, and
c) reflect carbon taxes in one form or another to encourage a switch to 'clean' energy sources so that Europe can meet its commitments made at Rio de Janeiro in June 1992.

Clearly, a reduction in real interest rates paid for capital also has a very beneficial effect.

The conclusion is therefore that if windfarms can be financed on the same terms applicable to conventional power plant (i.e. loans repayable over 15, 20 or more years) they will be able to compete with 'clean' coal and nuclear electricity generation.

5. MARKETS

There are now some signs that the rate of market growth is slowing down in Denmark, Germany and the Netherlands. This may be only temporary as planners wait to gain experience and take account of public opinion before cooperating with further development in those countries. Notwithstanding this observation, Table I shows that over 100 MW of new plant will have been built in these countries in 1993.

The most dramatic growth in the next few years seems likely to take place in the UK, Spain, Portugal and possibly in Italy and Greece. The UK for example, has increased capacity from 10 MW in 1991 to over 130 MW by the end of 1993. By then it will have 19 windfarms, with 369 turbines with an average rating of 320 kW as detailed in Table V. It is expected that a new Non Fossil Fuel Obligation (NFFO) will be announced by the UK Government in mid 1993 with contracts announced in Spring 1994. It is widely expected that there will be an annual order made then for the next several years. This is likely to mean that very little new capacity will be built in 1994 but that at least 100 MW/year will be built until the end of the decade.

Beyond Europe there is still a potential market in the USA amounting to some tens of thousands of Megawatts, whilst 'deals' in the Ukraine have been reported which amount to in excess of 1000 MW. India also continues to demonstrate interest in increasing the number of windfarms it already has in operation.

6. CONCLUSION

Though the development of modern wind turbines is in its technological infancy compared, for example, with the almost 50 years that nuclear power has been under development, much has already been achieved and Europe seems set to become the leader in both technology development and implementation. The factor that most severely limits

Table V(a): U.K. WINDFARMS

Operational (Total 76.5 MW).

Site	Developer	Turbine Type	No.of Turbines	Installed MW	Commissioning Date
Delabole, Cornwall	Windelectric	Vestas (DK)	10	4	Nov. 1991
Carland Cross, Cornwall	Renewable Energy Systems	Vestas	15	6	Aug. 1992
Cemmaes, Powys	National Wind Power	WEG (UK)	24	7.2	Nov. 1992
Chelker, Yorkshire	Yorkshire Water	WEG	4	1.2	Dec. 1992
Haverigg, Cumbria	Windcluster	Vestas	5	1.25	Oct. 1992
Rhyd-y-Groes	Ecogen	Bonus (DK)	24	8.4	Dec. 1992
Blyth Harbour, Northum.	Blyth Harbour	HMZ (BE)	9	2.7	Jan. 1993
Winterton-on-Sea, Norf.	Euros Power	Vestas	10	2.25	Dec. 1992
Great Orton, Cumbria	Carter Wind Turbines	Carter (USA)	10	3	Jan. 1993
Coal Clough, Lancs.	Renewable Energy Systems	Vestas	24	9.6	Feb. 1993
Penrhyddlan, Powys	Ecogen	Mitsubishi (JP)	43	12.9	Jan. 1993
Llidiartywaun, Powys	Ecogen	Mitsubishi	60	18	Jan. 1993

Table V(b): U.K. Windfarms.

Under Construction (Total 41.8 MW)

Site	Developer	Turbine Type	No. of Tur- bines	Installed MW	Commission- ing Date
Cold Northcott, Cornwall	National Wind Power	WEG	22	6.7	April 1993
Llangwyryfon, Dyfed	National Wind Power	WEG	20	6.0	May 1993
Kirkby Moor, Cumbria	National Wind Power	Vestas	12	4.8	Aug. 1993
Werfa, Glamorgan	Windstar Turbines	Windstar (Ch)	20	0.5	May 1993
Ovenden Moor, Yorkshire	Yorkshire Windpower	Vestas	23	9.2	Summer '93
Taff Ely, Wales	Perma Energy	Nordtank	20	9.0	Oct. 1993
Goonhilly, Cornwall	Cornwall Light & Power	Vestas	14	5.6	July 1993

further development is nothing to do with technology, or the commitment of industry or the banking system, or indeed with the number of available sites or the size of the resource. It is entirely to do with POLITICAL WILL. The Governments of Europe, and for that matter of all other countries in the Developed World need to translate their good intentions into actions.

Wind Energy is already set to save the emission of about 2.5 million tonnes of CO_2 annually within the CEC by the end of 1993 and about 5 million tonnes worldwide. With Political Will, it could easily meet at least half of the CEC member states Rio Commitments to reducing CO_2 emissions. Several Member States have already shown that there are several practical ways to encourage the sustained development and implementation of the technology. It now requires their continued commitment to ensure the necessary growth in the market which will allow further cost reductions to be made.

REFERENCES

[1] EWEA, Wind Energy in Europe, 'Time for Action' and 'A Plan of Action', published by The European Wind Energy Association, October 1991.

[2] Houghton, J T, et al, Climate Change: The IPCC Scientific Assessment, published by Cambridge University Press, 1991, ISBN 0 521 40360 X.

[3] Houghton, J T, et al, Climate Change : The Supplementary Report to the IPCC Scientific Assessment, published by Cambridge University Press, 1992. ISBN 0 521 43829 2.

[4] Renewable Energy Advisory Group : Report to the President of the Board of Trade - Energy Paper No.60, published by H.M.S.O., ISBN 0-11-414287-4.

[5] Bettini, V, Report of the Committee on Energy, Research and Technology on the promotion of 'renewable' forms of energy. Nov. 1992, published as a session document of the European Parliament, DOC.EN/RR/218/218450.

STUDY OF ENVIRONMENTAL ASPECTS OF THE WIND PARKS IN SPAIN

C. Lago *, A. Prades **, E. Soria *,A. Diaz **.

* CIEMAT-IER
22, Avenida complutense, 28040 Madrid, Spain
Tel. 34-1 3466534, Fax. 34-1 3466037
** CIEMAT-ITN
22, Avenida complutense, 28040 Madrid, Spain
Tel. 34-1 3466356, Fax. 34-1 3466037

ABSTRACT: The environmental aspects of Cabo Villano and Tarifa wind farms have been studied in this paper together with a study on public opinion of the affected populations. The nuisance on birds, the visual impact and the noise have been evaluated before the installation of the wind farms. The favourable response of the people regarding the installation of these wind farms has been one of the most relevant findings of this study. Now, we will carry out a second wave of questionnaires and interwieus to see the posible evolution of public opinion once with the wind parks are operating, as well as their effects on the environment foccussing on birds, landscape and noise.

1. INTRODUCTION

The Spanish New General Planning in Energy (PEN) foresees a installed rated power of 174.2 MW by the 2000.This development has increased the total installed power from 7 MW in 1991 to 53.1 MW at the end of 1992. Within this development framework the study of environmental aspects in Tarifa and Cabo Villano wind parks has been incorporated.

Cabo Villano is located at the northwest edge of the Iberian Peninsula, in the Death Coast, municipality of Camarinas. It is integrated by 22 wind mills (1 VESTAS V20-100 Kw, 1 V25-200 Kw and 20 MADE AE-20M-180 Kw). The installed power is 3.9 MW, but adding the 1.2 Mw of the experimental wind turbine AWEC-60, the total power is 5.1 MW. The participants in the project are: UNION FENOSA, ENDESA and VALOREN PROGRAM.

PESUR and EEE are located at Tarifa municipality, in the south of Spain. PESUR, located at Cerro del Tesoro, is integrated by 155 wind turbines AWP-56-100 kW and 34 wind mills MADE AE-20M-180 kW. The total power installed is 21.6 MW. The participants are COMPANIA SEVILLANA DE ELECTRICIDAD, ENDESA Y ABENGOA-WIND POWER (AWP). EEE located at Sierra de Enmedio, is integrated by 66 wind turbines, 50 ECOTECNIA 20/150 kW and 16 MADE AE-20M- 180 kW. The total power installed is 10.3 MW. The participants are: COMPANIA SEVILLANA DE ELECTRICIDAD, INSTITUTO DE FOMENTO ANDALUZ, IDAE and ECOTECNIA.

Obviously the environmental problems of wind energy applications are of small importance compared with the problems originated from conventional energies (greenhouse effect, water pollution, soil and air pollution, health effects, ...), but they must be taken into account to be able to minimize their possible negative effects over the environment and the neighbour populations.

European experiences have demonstrated a higher acceptability towards wind energy projects if a previous appropriate information on the environmental aspects, the specific project and wind technology state of art are performed. The degree of acceptance is also increased if there exists a close participation of the local administration from the beginning of the project planning.

2. STUDY OF THE ENVIRONMENTAL ASPECTS AND PUBLIC OPINION

The environmental aspects together with a deficient information of the technology, performance and project characteristics are the aspects that may generate a bad public opinion of wind energy, restricting therefore its development. To avoid this, as far as possible, there are two ways. On the one hand, to take care of those environmental aspects, trying to minimize their effects. On the other hand, to give information to those populations that will be affected by the wind parks, as long as desinformation itself may generate rejection. Therefore it is necessary to carry out a detailed study of the affected populations.

Experience obtained through wind parks developments has foccussed the study of the environmental aspects on three of them : visual impact, noise and birds disturbance.

At we said it is essential to consider the sociocultural peculiarities, idiosyncrasy and worries, everydaylife characteristics, etc, of those populations, if it is intended to inform properly.

The methodological approach for analysing these "Attitudes, values and expectancies about wind parks and wind energy", has combined both quantitative (questionaire) and qualitative (in-depth interwies) techniques. Those methods were applied to the so-called "experimental groups" (directly affected by the phenomenon under study), and to the "control groups" (with similar socio-cultural characteristics but not directly affected by the wind parks).

At Cabo Vilan wind park, the experimental group is Camarinas where the wind park will be located; (n=116) and Muxia affected by visual impact; (n= 116). The control group is Fisterra (n=115).

At Tarifa wind parks, the experimental group is the village of Tarifa (n=117), and the control one is Barbate (n=121).

It can be said that we are dealing with low developed populations in both cases. Their main worries are determined by the subsistence economy, the emigration processes (specially among young people due to the absence of opportunities), and the lack of adecuate infrastructures: roads, hospitals, etc. Those populations are very distant of the "environmental awareness", the energetic problems, etc, typical of highly developed societies. The level of progress required for those considerations to arise has not been reached. The expectancies are focussed in the acquisition of higher consumption levels, as it is considered the main indicator of "quality of life".

Very important levels of unknowledge regarding any energy source as well as wind energy, were detected. At galician townships 48% of the population declared they "don't know anything", and 49% "know something". The only better known energy sources were those of traditional use (coal, gas, etc).

At Andalucia, a higher level of knowledge was detected (56% of the sample knows "something"). But this "something" reflects an approximate, rough, not accurate, kind of knowledge. In fact, a significative part of the population, at first, did not discriminate between "eolica"(wind energy) and "hidraulica" (hydraulic), and they showed hard difficulties to understand the concept "energy source".

Regarding the levels of specific information about wind parks received by the populations, Camarinas and Muxia, 84% of the population declared they had not received any information, while at Tarifa this percentage was of 58% of the sample.

Once we have the results in our hands, we decided to carry out some informative sessions. At Camarinas the informative sessions were held on June 1992, while at Tarifa the date will be confirmed promptly.

Wind Projects are not submitted to Environmental Impact Assessment (EIA), given the slight impact they have on the environment. The impact of an activity on the envoriment will have more or less incidence depending on the following aspects :

1. Character of the action itself
2. Ecological fragility of the territory where the action will take place.
3. Ecological quality of the place where the project will be developed.

The more intensive the action is, the more fragile and more quality the territory has, the impact will be greater. Wind projects' character generates little impact, therefore to evaluate it we should foccuss on the other two aspects (ecological fragility and quality).

The present vegetation in Cabo Villano is integrated by brooms, furzes and heathers, with a great covering of soil, without any erosion problems and with some pines reafforestation. This vegetation is climatic in the most coastal border, because the continuous winds do not allow the development of arboreal stratum. The wind park is situated in this border; its environmental quality is good and it will not be disturbed by the wind project, because only between 1-5% of the total surface of the park has been occupied by wind turbines and the wind park's development has no influence on the ecosystem processes.

The landscape's quality will be affected with the turbines installation decreasing its value due to visual impact. At the ecological level there are no fragility problems but at landscape's level there are some, as it is an open landscape from and toward the sea.

PESUR and EEE projects, are located inside the Natural Park of Los Alcornocales, and both projects have received the corresponding licences for their installation inside Natural Park. The wind projects are situated in the South limit of the Park, in areas that due to their proximity to Tarifa and Facinas, have been exploted in the traditional way, eliminating the climatic vegetation. Even so the environmental quality is acceptable with a good covering of soil without big erosion and a landcape distributed according to its use. The ecological quality will not be affected by the wind parks installation, but regarding fragility two aspects must be considered: visual impact of wind turbines with rotating blades, and possible influences in the very important birds migrations ruotes through Gibraltar Strait, mainly of the electrical grids inside the wind parks.

One of the most important characteristics is that wind energy impacts are reversible, thus the dismantling of a wind installation leaves the environment practically in the same way as before its execution.

2.1. BIRDS' EFFECTS

Wind turbines can be a problem for the birds. Obviously depending of the height and speed of blades rotation there exists a potential danger to the flying birds. Birds collision with a blade provokes some damage or death, but so far, there has not been any registration of death at spanish wind parks.

Bird census availability, as well as migratory routs and residents movement's data, are the main prerequisites for a proper evaluation of wind park's impact on birds behaviour. At Spain, unfortunately extensive studies of this kind have not been accomplished at national level, although some have been carried out at some specific places. So far, available literature is very scarce, and that's why consistent studies of wind energy impacts on birds are so hard and difficult to carry out.

In many cases, especially when the wind park is installed in a protected area, it has been suggested the necessity of a subterranean electrical grid if not possible along the whole way at least in the most difficult areas. The cost will be increased therefore wind parks promoters criticize this measurement. This requirement is usually proposed for the wind parks but not for the national grid, which volume and extension cause almost the totality of the electrocution and collision problems.

Until now, the three wind turbines installed in Cabo Villano, have not occasioned any collision neither death to the resident or migratory birds in the area. Resident birds seem to become accustomed to wind turbines, in the same way than to another obstacle or building existing in the territory.

Breeding birds in Vilan Da Fora Island, some of very great importance as Uria aalge, do not seem negatively affected by wind turbines. The migration routes of marine birds and Limicolas (verified by naturalist groups as "Habitat" and "Sociedad Galega de Historia Natural"), have been located nearer the sea than wind parks. There are no registrations of any collision or death of this bird's kind. In any case, when the new wind park is installed, a study will be make to check if any bird is affected by the wind park.

The Camarinas and Muxia opinion demonstrates a majority consensus on the fact that wind turbines do not represent any danger to the birds: 79% of the people think "Don't affect anything".

Gibraltar Strait together with Bosforo are the two most important migration areas in Europe. It has not been verified any bird's collision or death in the small wind turbines installed at Tarifa wind farm neither Monte Ahumada wind park (seven wind turbines of 150 Kw), but it is possible that when the wind parks become bigger (as PESUR and EEE), some problems with birds appear, specilly considering the importance of Tarifa's migration routes.

2.2. VISUAL IMPACT

Visual intrusion on landscape is the most frequent objection of windturbines, as well as the main factor in the configuration of public attitudes towards wind energy developments. Therefore it can not be ignored. This environmental impact is the hardest to quantify and the least investigated, considering other kind of environmental disturbances. The reason is its subjective character, its difficulty to estimate and quantify.

The visual effects of a wind park depend on:

- The ability of the observer to register visual impressions according to his perception process.
- The landscape: topography, buildings, vegetation and weather.
- The wind park: size and height, material, colour.

Although everyone has its own perception, actually there is a general consensus about landscape evaluations. This does not mean that most people have the same opinion about landscape or aesthetic, but there are some regular opinions in their judgements. Anyway, it shoud be remarked that the data gathered by means of questionaires are not extrapolable in an universal way.

Cabo Vilano wind park can not be seen from Camarinas; the nearest village is located at 3 Km away. It can only be seen from the sea and from the village of Muxia, located on the other side of Camarinas fiord.

The reasons for visual intrusion of wind parks on landscape can be detailled as follows :

- Wind turbines in both cases (from the sea and from Muxia) are outlined above the sky, standing out with more spotlessness than if outlined above vegetation.
- Wind turbines are vertical estructures, which outlines them in a landscape of horizontal components.
- Wind turbines are artificial, isolated estructures, therefore their visual impact is lower than for lineal estructures occupating big extensions.
- Moving objects draw observer's attention, thus the rotating blades become dominant points on landscape.
- Visual intrusion decreases with distance, so the vision from Muxia (located more than 5 Km away) become quite minimum.

Results of the public opinion study reveals that at Camarinas and Muxia 86% of population considers that wind turbines do not disturb landscape, declaring in a 80% that they are "very nice" or "quite nice", while only a 10% thinks they "are not nice at all".

Tarifa wind parks can not be seen from the nearest villages (Facinas and Tarifa), but can be seen from a stretch of the road. Visual dominance is lower because wind turbines are diseminated on a mountainous orography. The pertinent photograpic report has not been done yet.

Anyway, Tarifa population consider (90%) that wind turbines do not spoil the landscape, and only a 10% declares that "they are not nice at all".

2.3. NOISE

Wind turbines in operation make some noise. The disturbance degree depends on four factors: itself operation wind turbines's noise, the wind turbines' position, distance between the wind turbines and inhabitants, and the background noise of the area.

Camarinas, the nearest village from Cabo Villano's wind park, is located at 3 Km., thus the noise is not a problem for people. Anyway only at low wind speed, the wind turbine's noise is higher than the one of the wind, but at higher wind speed the background noise level is superior than the noise produced by wind turbines.

Public opinion's results, agrees with that because 92% of the interviewed people in Camarinas and Muxia consider that wind turbines "Don't make noise".

At PESUR and EEE wind parks noise problems are not foressen because they are far from Tarifa and Facinas villages. There can only be disturbances at the houses disseminated in the mountains and located nearer the wind parks. 82% of the people from Tarifa considers that wind turbines "don't make noise", and the 10% thinks that the wind turbines make a "little" noise.

The noise from AWEC-60 has been measured during last February, according to Recommended Pratices for Wind Turbines Testing and Evaluation Acoustics Measurement of Noise Emission from Wind Turbines. All the measurements are within the limits established in the AWEC-60 contract signed with the CE. The maximum level was 60.1 dB at a wind speed of 10.8 m/s with 1200 Kw of operating power. This value es lower than the maximum permited values.

3. CONCLUSIONS

The most relevant environmental aspects of wind energy development have been described. We conclude with its main advantages:

- Reduction of the environmental problems.
- Decrease of energetic dependence.
- Save of currencies.
- Generation of employment.
- Increase of energetic supply safety, through local generation.
- Absence of contamination.
- Small exclusive occupation of the territory.
- The possible environmental effects are, anyhow, reversible and low.

Wind energy can improve local conditions, increasing its incomes, estimulating the economic activity in the area. For the analysed spanish population, the most appreciated aspects were the expected infrastructure improvements (60%) and the increase of tourism (75%). Regarding the possibility of employment generation, expectancies were less favourable, specially at the experimental groups, where 50% of the population declares that wind parks "will not generate any".

Anyway, the favourable reaction towards the wind parks was surprising, although also previssible after the first approach and the pre-test. Only 1.3% at Camarinas and Muxia declares that the wind park is prejudicial, being this percentage of 1% at Tarifa (At Camarinas and Muxia 52% considers it advantageous, and 64% at Tarifa).

In principle protected natural areas remain out of wind energy developments. Legislative instruments to exclude enviromental protected areas of possible wind energy projects development have to be defined. It is also necessary to define on which of those areas it would be feasible to carry out such projects, considering all the necessary restrictions (Law for the preservation of natural areas and wild life, Law for Coasts and Law for the ground regime and urban order, at a national level besides legislations from the Autonomous Communities).

Wind energy is actually an economically competitive energy, even without integrating the external environmental costs of the traditional energies. It has a clear technological development (mature technology for the 250-5000 Kw wind turbines), and costs under continuous decrease.

Today wind policies are centralized in most of the countries, and the population is often against this. From some experiences in other countries it is suggested that for wind energy, a decentralized approach with some local involvement could get a better acceptance.

Therefore, from different points of view, promoting the development of wind energy should be a logical option. A clear political decision is the requisite to obtain a reasonable encouragement of its wide posibilities.

REFERENCES

1- "Localizacion e Inventario de las Colonias de Arao Comun (Uria aalge, Pontoppidan) en las Costas de Galicia; Determinacion de las Posibles Causas de su Disminucion.
F. Barcena.
Boletin de la Estacion Central de Ecologia. Ano XIV. Numero 28. 1985

2- "La Migracion de las Aves en el Estrecho de Gibraltar. Volumen I: Aves Planeadoras".
F. Bernis.
U.C.M. Catedra de Zoologia de Vertebrados. 1980.

3- "A multifaceted study: Environmental issues impacting on wind energy developments".
P. Bosley and K. Bosley.
Communication Department, Townson State University. European Community Wind Energy Conference.
September, 1990. Madrid.

4- "Wind Energy in Europe: Time for Action. A strategy for Europe to realise its enormous wind potential". EWEA Strategy Document. Third Draft.
European Comission DG XVIII.
1990.

5- "Inventario Abierto de Espacios Naturales de Proteccion Especial. Cadiz".
ICONA.
1990.

6- "The Contribution of the Social Sciences to Wind Power Research Environmental Ideology and Public Perception".
H. Martin Edge.
Robert Gordons Institute of Technology. Scotland. European Community Wind Energy Conference.
September, 1990. Madrid.

7- "La Migracion de las Aves en el Estrecho de Gibraltar. Volumen II: Aves no Planedoras".
Jose Luis Telleria.
U.C.M. Catedra de Zoologia de Vertebrados. Madrid. 1981.

8- "Parque Eolico de Cabo Villano. Primera Ampliacion de 3 MW".
Union Fenosa.
1990.

9- "Attitudes and Expentancies about Wind Turbines and Wind Farms".
M. Wolsink.
Department of Environmental Sciences. Amsterdam University.
Environmental Impact Assesment Review 1988. Elsevier Science Publishing Co.

10- "Propuesta de: Normas Complementarias y Subsidiarias de Planeamiento de las Provincias de La Coruna, Lugo, Orense y Pontevedra".
Xunta de Galicia. Conselleria de Ordenacion do Territorio e Obras Publicas.
1990.

FINAL RESULTS ON THE RESEARCH PROGRAMME CONCERNING THE SOCIAL AND ENVIRONMENTAL ASPECTS RELATED TO THE WINDFARM PROJECT OF THE DUTCH ELECTRICITY GENERATING BOARD

drs. F. Lubbers and drs. L.R. Pheifer

Dutch Electricity Generating Board (N.V. Sep)
P.O. Box 575
6800 ET Arnhem
The Netherlands

SUMMARY

This paper deals with the activities of the Advisory Committee for Social and Environmental Aspects (AMMA) related to the windfarm project at Oosterbierum of the Dutch Electricity Generating Board (Sep). In this paper the results of the research programme of AMMA will be summarized.

1 INTRODUCTION

Mid 1982, the Ministry of Economic Affairs and the Dutch Electricity Generating Board (Sep) developed a joint scheme for the installation of an experimental windfarm with a total capacity of 5.4 MW. The experimental windfarm comprises 18 wind turbines with a capacity of 300 kW each, an axial height of 35 metres and a rotor diameter of 30 metres. One of the objectives of the experimental windfarm is gaining experience in physical planning aspects and environmental aspects as well as social acceptance related to the experimental windfarm. To achieve this objective a advisory committee was set up (advisory committee for social and environmental aspects, AMMA for short). AMMA adressed itself to seven sub-surveys, namely:

1 **Bird survey** (contractor: Rijksinstituut voor Natuurbeheer en N.V. KEMA)
The bird survey consists of:
- bird behaviour survey and bird collision survey
- bird disturbance survey
- methodological survey.

2 **Traffic counts** (contractor: University for Agriculture Wageningen)
Traffic counts are part of this survey, in order to obtain some idea of the extent to which a possible disturbance to meadow birds, as well as the effect on agricultural management, is caused by traffic attracted by the experimental windfarm.

3 **Agricultural survey** (Contractor: University for Agriculture Wageningen)
Research into the effects of a windfarm on farming practices. As there are no mixed farming methods employed in the vicinity of the windfarm and no influence on pasture is expected as a result of the experimental windfarms, the research therefore is aimed solely at arable farming.

4 **Landscape analysis** (Contractor: Buro Maas)
By means of a detailed landscape analysis one can get insight in the effects of the windfarm on the landscape. This is essential to the assessment of how well a windfarm fits into the landscape.

5 **Research into the visual experience of windfarms** (Contractor: Technical University Delft)
This research is aimed at the question of how people rate a windfarm as a new visual-spatial element in the landscape and the arguments they put forward in that regard.

6 **Socio-psychological survey** (Contractor: State University Groningen)
Within this survey, data about social experience and social acceptance are collated to gain insight into the reactions of the population to this new large-scale element in the landscape.

7 **Research into undisturbed wind catch** (Contractor: Bureau van Heesewijk B.V./Centrum voor Energiebesparing)
The undisturbed wind catch is of great importance, because the output of a windturbine depends on the supply of wind. The research focussed on the influence of changes in the surface-roughness on the production output, the legal possibilities of a constantly good and undisturbed wind catch as well as the consequences with regard to purpose and use of land in the vicinity of wind turbines.

The total budget for the AMMA research runs to approx. Dfl. 1,500,000; a relatively large amount when compared to similar research projects abroad. The research program started during the second half of 1984 and was completed in 1992. In this paper the results of the sub-surveys; bird survey, landscape analysis, survey concerning the visual experience of windfarms, the socio-psychological survey concerning the social experience and acceptance of wind energy and the survey into undisturbed wind catch will be discussed in more detail as well as the evaluation of the results of the surveys in the light of the ultimate prospective possibility of the wind energy option in the Dutch system of electricity supply.

2 RESULTS OF THE AMMA-SURVEY

2.1 Bird survey

The survey conducted on the flying behaviour of birds showed that during the daytime and also by night, birds generally detect the wind turbines in time to adapt their flying routes and avoid going too near the moving rotor blades. However, this does not mean that there have been no bird fatalities on account of the experimental windfarm. During the survey period the windfarm frequently caused one or several fatalities. Extrapolation of the results of the bird survey near the experimental windfarm and of a bird survey conducted near the windfarm at Urk to a value of 1,000 MW wind energy capacity leads to the conclusion that the numbers of bird fatalities per linear kilometre of windfarm never exceed those per kilometre of motorway or high-voltage overhead line; per kilometre of windfarm they are even less than in the vicinity of lighthouses or high transmission towers in coastal areas with large bird populations.

Lighting of wind turbines will cause the number of bird fatalities to rise rather than fall. Another option to reduce the number of bird fatalities, viz. stopping the wind turbines in certain weather conditions, is hardly feasible in practice. The configuration of a windfarm can however contribute to the reduction of nuisance to birds. The preference for a certain windfarm configuration from a viewpoint of bird nuisance depends on local bird life and has to be considered for each project separately.

The disturbance survey does not indicate any disturbance of summer birds around the experimental windfarm within the survey period. However, a negative effect on two out of four summer birds included in the survey cannot be ruled out. This result is in general confirmed by limited results of small surveys conducted abroad on this subject. For foraging and also for migratory birds, however, a considerable disturbance can be demonstrated from time to time. Therefore, the conclusion is drawn from the disturbance survey that the disturbance aspect is to be considered as more negative to birds than the fatality aspect due to collisions as indicated above.

In the bird survey near the experimental windfarms particularly radar, heat image cameras and residual light amplifiers proved to be effective in observing bird movements during the night. Residual light amplifiers can be used to observe birds during the night at short distances (about 20 to 30 metres) in dry weather and with reasonable visibility. The heat image camera can be used to observe birds during the daytime as well as by night, and to a certain extent also during spells of fog and rain, up to a distance of at least several hundreds of metres. The survey radar proved useful only for qualitative determination of migrations at greater altitudes (higher than 50-100 metres) in the general area surrounding the windfarm.

Especially the heat image camera supplied an abundance of information on the flying behaviour of birds at wind turbine altitude.

The fatality survey is extremely time-consuming. However, without the determination of detection probability rates and if searches are carried out less frequently, the results of fatality surveys will be rather insignificant.

With the completion of the survey on the flying behaviour of birds near the experimental windfarm and the survey conducted near small wind turbines, at the Kreekraksluizen and at Urk, the conclusion may be drawn that most questions in this field have been answered. The advisory committee "Bird nuisance due to wind turbines", which was set up by NOVEM in February 1991, recently arrived at a similar conclusion. The committee holds the opinion that only a limited further research effort is required for a survey on bird fatalities. This concerns in fact a location-dependent survey to determine the bird supply at different altitudes, enabling an estimate of the probability of bird fatalities in the event that wind turbines are installed.

2.2 Agricultural survey

There were few agricultural objections with regard to the establishment of the experimental windfarm. This can be partly accounted for by the fact that due to land allotment at the windfarm site agricultural utilisation has greatly improved. The turning of rotor blades generally causes no problems for agricultural utilization of the lands between or near the turbines.

2.3 Landscape analysis

With regard to integration into the landscape there are four aspects of a windfarm playing a role. These are:

- number of wind turbines
- height of the wind turbines
- configuration shape
- design of the wind turbines.

The surveys have shown that smaller groups of wind turbines are considered to be preferable to a large windfarm with a great many turbines. The number of 18 turbines in the experimental windfarm is already viewed as fairly ample. The most preferable size of a group of wind turbines will be different in each individual situation.

With respect to the visual-landscape effect the height of the wind turbines will play a less prominent role than their number. Smaller groups of heigher wind turbines will be easier to integrate into the landscape than a large group of smaller wind turbines.

Both the line-shaped windfarm as well as the solitary and park configuration are options for the configuration shape. The configuration should be based primarily on the production output.

For the effect on the landscape the following aspects play a significant role:

- styling of windturbines
- colour (in each individual situation it should be assessed whether assimilation with or variation from the surrounding landscape should be preferred)
- turning speed (the turning of the rotor blades causes a troubled image, which is aggravated by dissimilar turning speeds and by a high turning speed; therefore a low or intermediate turning speed is preferred)
- noise production.

Landscape experts assess the visibility of a windfarm as a positive characteristic, for instance from a point of view of symbolism and clarity.

2.4 Research into the visual experience of wind turbines

Visual motives are important, but not eventually decisive in the consideration whether or not to go ahead with the building of a windfarm. As is also the case with landscape aspects, visual-spatial aspects will mainly play a role in the elaboration of the configuration shape and design. Sites located in industrial estates, on sea embankments or at sea will be more eligible based on visual-spatial considerations. In other locations the configuration shape and design of the turbines will have to take better account of the qualities of the surrounding landscape. Generally speaking, where windfarm configuration is concerned smaller groups consisting of a smaller number of larger wind turbines are assessed more positively than a large number of smaller wind turbines in such other locations.

Unlike the landscape experts, people living in the neighbouring area of a windfarm are likely to prefer lower visibility of the windfarm, for instance by hiding the windfarm, at least to some extent by planting trees and shrubs. However, planting will affect the energy output of a windfarm.

2.5 Socio-psychological survey

Wind energy and wind power stations are assessed highly positively in comparison with nuclear, coal-fired or gas-fired electricity generation. This public opinion is very strong indeed and hardly subject to change. The output of windfarms and the ratio of operating hours vs. standstill periods clearly influences the general opinion on windfarms. If the output is low and the turbines are idle for long periods of time, this will also tend to reduce social acceptance and also visual experiences. Surveys indicate that people, when asked for their opinion on the establishment of a wind power station in their own neighbourhood, react more positively than if this concerned a nuclear, coal-fired or gas-fired power station. However, once they are actually confronted with a plan to establish a wind power station in their neighbouring area, its integration will be experienced as negative. Particularly visual nuisance, deterioration of the landscape and noise are key factors in this. Locations of windfarms in industrial estates or in linear configurations along large infrastructural constructions will be designated as more suitable.

Once a windfarm has been established, with the passage of time, habituation and a certain degree of social acceptance will occur. The time it takes before this happens will depend on the number and nature of the complaints filed against the plan and the degree to which possible nuisance can be minimized.

2.6 Research into undisturbed wind catch

In general, windfarms will occupy a large indirect surface area, which may amount to an area with a 6 km radius around the windfarm, as is the case with the experimental windfarm. In determining the extent to which claims should be laid upon this extensive surrounding area, the most important factor is essentially the balancing of the loss of income with a windfarm against the expenses to be incurred to prevent this loss of income.

With taller wind turbines (>40 m rotor height) changes in terrain roughness cannot affect wind turbine output to a similar degree, so that the claim upon the surrounding area can be smaller.

3 EVALUATION OF THE RESULTS IN THE LIGHT OF THE ULTIMATE PROSPECTIVE POSSIBILITY OF THE WIND ENERGY OPTION IN THE DUTCH SYSTEM OF ELECTRICITY SUPPLY

Calculations carried out from the Sep anemometer network show that for a windcapacity set up for 1000 MW, a required production capacity which is approximately 165 MW less is required in power stations (saving on production capacity). This means that if wind energy is to contribute substantially to the national power supply, a large number of windturbines will have to be erected.
The results of the AMMA study show that in principle there are good opportunities for windfarm sites in the Netherlands. The only areas which the AMMA study excludes for the installation of windfarms are some parts of the coastal area with important bird-migration, bird sanctuaries and stopovers as well as areas which are still considered unspoiled or where people go to rest and relax.
The energy yield and the economics of a windfarm project must be considered as crucial prior conditions in the search for locations. As the productivity of these projects increases, their assesment will improve, not only from the financial and economic standpoint, but also from the point of view of social acceptance, how windfarms fit in the landscape and how they are visually perceived in their environment. In view of wind energy application on a larger scale, it is important that the results of the study concerning landscape suitability, how the windfarms are visually perceived in the environment and the demands that are necessarily made on the space surrounding them, indicate that windfarms of the future will be feasible using windturbines with a larger capacity than the ones used untill now. The AMMA study in fact shows that smaller clusters of large windturbines are more suitable from the point of view of landscape and visual perception than a large windfarm with relatively smaller windturbines, such as the ones used in the experimental windfarm. Larger windturbines will also be less demanding on the space surrounding them, since topography has less effect on their energy yield.

LOCALIZATION STUDY - 300 MW WIND POWER

Gunnar Grusell
Vattenfall Energisystem AB
Jämtlandsgatan 99
Box 528
16215 Vällingby
Sverige

Before starting to talk about Vattenfall's Localization Study, I would like to make a brief summery of the Government survey entitled "Status for Wind Power", SOU 1988:32.

In June 1985, the Swedish Parliament decided that continued efforts should be directed towards developing wind power. During the 1990s, a decision may need to be taken regarding the development of energy production pending the phasing out of nuclear power.

The continued development of wind power should be carried out in several stages, in which further generations of large wind power plants are successively built and evaluated. In parallel with the continued technical development of wind power, further consideration should be given to the preconditions governing the localization of wind power plants.

A special survey should be set up to propose where in Sweden the most suitable sites are for large wind power plants in groups. One basic guideline should be that the proposed locations permit a total production of approx. 10 TWh/y on land and approx. 20 TWh/y at sea.

The survey should also determine conflicting interests when it comes to land and water use and the environment, which may exist in those areas that are conceivable localization for the plants.

The report first of all presents the ongoing development of wind power in Sweden and then gives an overall description of the work connected with the localization of suitable wind power areas. A study has only been made of suitable wind power areas in the south of Sweden, where electricity consumption is highest. Two alternative safety distances to buildings have been studied: 300 and 500 m.

The Swedish planning system is based on close cooperation between authorities at a central, regional and local level. The study group has constantly worked together with representatives from the county and local authorities. In addition, two formal drafts have been distributed for comment.

After studying a large number of sites both on land and at sea, the study group has proposed that development should in the first instance be centred on 37 sites on land and 17 at sea.

The study gives a technical description of wind power plants and of the preconditions for localization. This is followed by a description of the environmental impact. Wind power plants give rise to noise, and create interference for TV reception and other telecommunications systems. A possible accident could lead to a turbine rotor, or a part of a rotor, breaking off and being hurled away.

The study describes the conceivable environmental impact from wind power plants, such as noise, safety, and ratio and TV interference. An account is also given of conservation interests - such as nature conservation, the preservation of cultural interests, bird sanctuaries, etc. - and of other development interests - the Armed Forces, telecommunications, aviation, sea transport, roads and railways, fishing, agriculture and raw materials handling.

1. VATTENFALL's LOCALIZATION STUDY: 300 MW WIND POWER – 100 3 MW PLANTS

1.1 Background

On the basis of the wind power areas proposed in the study entitled "Status for Wind Power", Vattenfall's localization study for 300 MW wind power, also known as the "300 MW Study", shall conduct a survey of the localization potential, energy production and costs for 300 MW of wind power within the area in which Vattenfall is the supplier of prime power.

The aim of the study is to obtain a clearer idea of the potential and opportunities offered by wind power, and to have the areas proposed for wind power included in the synoptic planning of the local authorities concerned.

The "300 MW Study" shall be based on and study more deeply the basic data presented in "Status for Wind Power". Material for the study shall also be gathered from county administrations, local authorities, local energy companies, the Swedish Meteorological and Hydrological Institute and other interested parties. First an inventory will be made of conceivable areas and this will be followed by an assessment of different factors before a choice is made of the best areas with respect to wind supply and suitable preconditions for a wind power development. The greatest possible care shall be given to public and private interests within the areas in question.

1.2 Historical background

The study gives an account of the developments in wind power in Sweden since the end of the 70s. 1977 saw the completion of the first pilot plant, with an output of approximately 60 kW. Since then, another ten wind power plants have been built as part of a research and development programme with the Government and power companies as the principal interested parties.

Discussions on the design and economy of wind power plants have often concerned the question of optimum size. The small and medium-sized wind power plants with their simple technology are currently reasonably well-developed, but if wind power is ever to play a significant role in the national electricity system, development will have to move towards larger units in the MW class, and preferably with an output of 2-4 MW.

In summer 1991, The Government decided to allocate a 25% investment grant for wind power plants of more than 60 kW. The total grant amounts to SEK 250 million over a 5-year period. From January 1993 onwards, the grant was raised to 35%.

At the end of 1992, there were a total of 90 wind power plants with a total potential output of 20 MW. Later on in the Conference you will receive more information on the Swedish wind power programme and on three Swedish development projects – the 400 kW unit in Lysekil on the Swedish West Coast and the 3 MW unit at Näsudden on Gotland, as well as a 1 MW unit which is in the process of being built.

1.3 Localization criteria

When selecting a site for 100 wind power plants each 3 MW in size, carefully considered assessments and evaluations will have to be made with regard to wind supply, construction and economic factors, conflicting interests and landscape aspects.

The map of southern Sweden below shows the gross onshore and offshore energy areas presented in "Status for Wind Power", as well as the sites in Bohuslän and the northern part of Halland on the Swedish West Coast and Gotland that have been studied in the "300 MW Study".

MAP OF SOUTH SWEDEN

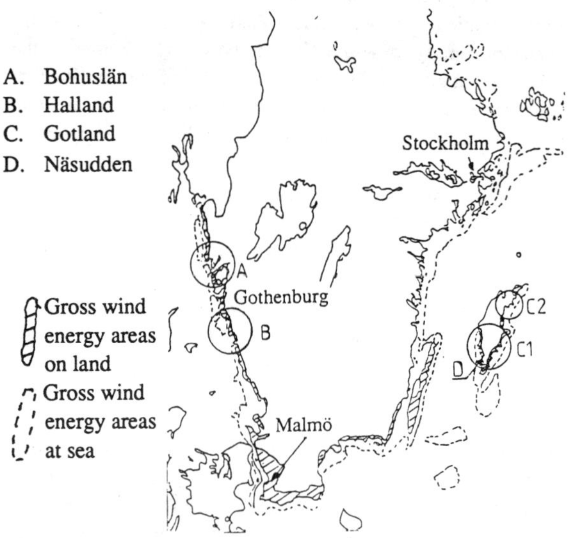

The technical structure and cost estimates are based on experience gained in connection with the two Swedish MW wind power units and assessments made by suppliers. An outline survey has been made of the local conditions, and the arrangement of the individual wind power sites has been adapted to the availability of wind, topography, ground and foundation conditions, existing road systems and power and telecommunications networks, etc.

Information on conflicting interests within the areas studied has in the first instance been obtained from local authorities and county administrations. These interests can be divided into the following categories: preservation interests, such as nature conservation, recreation and the preservation of areas of cultural value, and development interests, such as industry and housing, agriculture and forestry, aviation and military interests, and radio and telecommunications interests.

The new wind power interest does not necessarily have to be a threat against the already established. With sound planning there is every likelihood that the new interest can be integrated with established interests without too much conflict.

The establishment of wind power on a large scale will change the landscape of the area in question. The areas concerned are often stretches of coastline with an attractive landscape, such as the varied cliff beaches in Bohuslän and the flat coastal landscape of Halland and Gotland.

Wind power plants cannot be hidden and will inevitably become a prominent feature of the landscape. Whether or not people will regard them in a positive or a negative way depends on a number of factors. The attitude people take to different types of energy and their impact on the environment, as well as to the encroachment or extra value they represent for the individual, is difficult to judge. The findings of this study suggest that wind power could well become an environmentally accepted and attractive feature of our landscape, and a valuable addition to our future energy supply.

1.4 Wind power sites studied

"Status for Wind Power" specifies that the gross area on land in Sweden could in theory accommodate 70 and 35 TWh/year at protective distances from buildings of 300 and 500 m respectively. Following a review of other land-use interests – both conservation and development interests – the study group proposes that continued expansion of wind power should in the first instance take place in 37 onshore and 17 offshore sites in the southern part of Sweden.

If use is made of all the onshore sites, the energy production would be 2.9 TWh/y (490 wind power plants) with a 500 m protective distance to buildings and 6.7 TWh/y (1 150 wind power plants) with a 300 m protective distance.

Within Vattenfall's distribution area – Areas A, B and C – the energy production according to "Status for Wind Power" will be 1.2 TWh/y (190 wind power plants) with a 500 m protective distance to buildings and 2.3 TWh/y (380 wind power plants) with a 300 m protective distance.

"Status for Wind Power" also presents a development of offshore wind power, in total 22.5 TWh/y (2 740 wind power plants) of which 12.3 TWh/y (1 530 wind power plants) falls within Vattenfall's distribution area.

In the "300 MW Study", seven areas have been studied on the Swedish West Coast and eight on the island of Gotland. The areas have been studied on economic data maps on a scale of 1:10 000. The wind supply must be greater than 4.5 MWh/m2 and year, which corresponds to a median wind of 7.2 m/s at a height of 100 m. The wind power plants have been arranged at the corners of equilateral triangles with sides of 7 rotor diameters or 560 m. This arrangement has also been used to calculate energy production and costs.

The results of this first attempt were that the Vattenfall distribution area would accommodate 1.5 TWh/y (250 wind power plants) with a 300 m protective distance and 0.9 TWh/y (150 wind power plats) with a 500 m protective distance . This is 35% and 20% respectively less than the figures given in "Status for Wind Power". The reasons for this are: smaller wind areas since the lowest wind supply is 4.5 MWh/m2 and year in the "300 MW Study" and 4.0 MWh/m2 and year in "Status for Wind Power", the map studies have been carried out on more detailed maps and certain areas with other strong land use interests have been excluded.

An inspection has been made of all the wind power sites. It could be concluded that the locations of the foot points indicated on the map are relatively good from the wind supply point of view. It was also concluded that it should be possible in a later design stage to adapt the unit locations to existing buildings, local wind conditions, economic entities, ground conditions, topography, existing roads and electrical power network, etc.

On Gotland, practically all the sites can be reached via the existing road network. On the other hand, many of the sites on the West Coast are difficult to reach via the road network owing to the topography, viaducts and bridges. Therefore, the study group has investigated whether materials and equipment can be transported by boat and helicopter during both the construction and operation periods.

The local authorities and county administrations concerned have been informed of the proposal. Central authorities, including the Ministry of Natural Resources and the Environment, the Swedish National Board for Industrial and Technical Development (NUTEK), the Environment Protection Board, the Armed Forces and the Civil Aviation Administration have also been informed. All of these organizations have in principle shown a positive attitude towards wind power. However, in order to be able to achieve positive results with a large-scale wind power development and gain acceptance for extensive wind power expansion, careful planning and coordination with established land use interests is essential.

1.5 Wind energy

With the aid of the calculation model known as WAsP (Wind Analysis and Application Programme), comparative calculations have been made of the median wind (in m/s) and wind power input (W/m2) between a number of weather stations in respective regions. These calculations show the standard of measurement accuracy and at the same time indicate which measurement stations are most representative for each region.

The calculations are conducted in such a way that the terrain at the measuring point is described with regard to topography, surface properties and buildings, etc, or in other words factors that affect the local wind velocity. After this, a similar description is made of the evaluation site.

The program subsequently converts the wind velocity into undisturbed wind at the measuring point and the wind velocity is then calculated at the evaluation site by taking into consideration the local disturbances that are present at the site in question.

If the wind supply at the evaluation site is then calculated in this way with the help of two closely positioned wind gauges, it should be possible to achieve good correlation. This is frequently not the case, and where this happens uncertainties have been stated. These uncertainties are proportional to the deviations.

Use has been made of 10 minute mean values from wind measurements at a height of 10 m every third hour over a 15 – 20 year period. The following graph shows the Weibull distribution at Näsudden at 96 m on Gotland during 1980- 88.

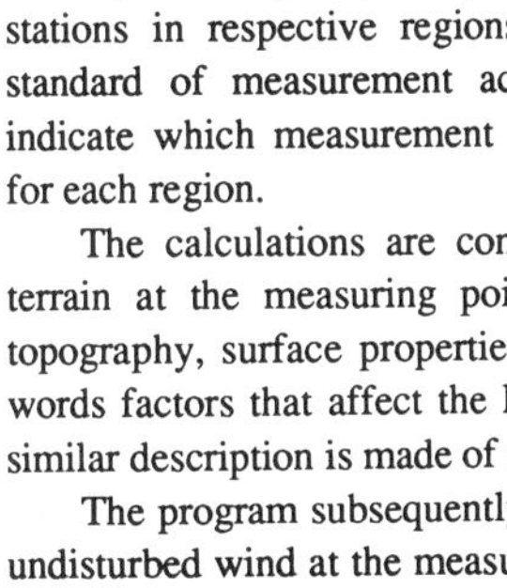

Median wind 8.0 m/s
Wind power 627 W/m2
Weibull factors
- scale factor A 9.5 m/s
- form factor C 2.23

% of year´s hours

The gross annual production of the wind power plant can be calculated with the help of the unit's wind to power output ratio, see the graph below and the site wind supply and wind velocity distribution.

Losses in energy production as a result of the units being arranged in groups have been calculated with the help of the Danish program referred to as "PARK". An average group of between 10 and 20 units is subject to group station losses of approximately 5%. The following table specifies the uncertainties in wind energy evaluations for two areas

	Mean wind velocity	Energy production
West coast of Gotland	± 1- 3%	± 2-6 %
Bohuslän	± 5-10%	± 25 %

To this shall be added uncertainties in group station losses that may be estimated to be± 30% of the group station loss, or in other words ± 1-2% of the energy production.

1.6 Technical description

The present land use and access to roads, power lines, transformer stations, etc. are factors that may have an influence on the final choice and arrangement of wind power sites. When planning and designing wind power groups, it is important to restrict land claims by giving the greatest possible consideration to ongoing land use, existing road and cable runs and in all other respects make use of the potential offered by the the area in the best and most favourable way.

An outline calculation for 100 pcs 3 MW wind power plants shows that on average per wind power plant:

- the net area for the unit location, roads, cables and other electrical equipment is 14 000 m2 of which 1 300 m2 is for the unit site, and
- the development area, or in other words the area of land within which a given number of wind power plants can be positioned, is 17 ha.

The electrical interconnection of wind power plants comprises a smallest unit or group of 2-6 wind power plants. The wind power plants in a group of this kind are connected together separately to a common transformer station. Several units can be interlinked to the regional network. Individual groups or several groups make up a group station. The following diagram shows the design of a group station with two smaller groups with 3 and 4 wind power plants respectively.

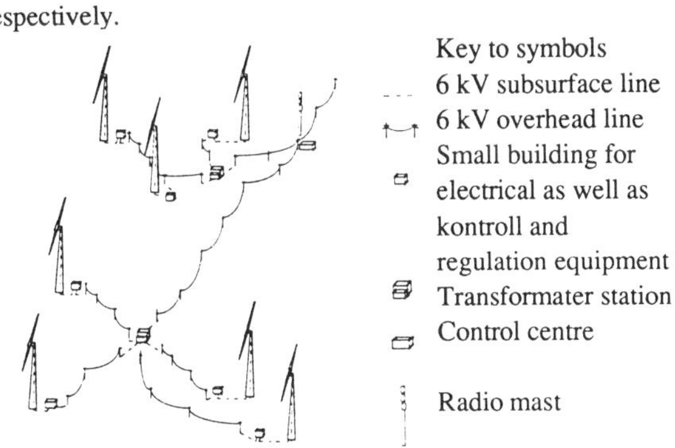

From the units a 6 kV overhead line (or alternatively subsurface line) is laid to a 6/40 or 6/70 kV transformer station. From the wind power group, a connection cable is laid to the existing power network.

Prototypes of the second generation of 3 MW wind power plants studied are about to become operational on the German North Sea coast and on Gotland in Sweden. Both projects are being conducted under the framework of Swedish/German cooperation. Preussen Elektra is the client for the German plant and Vattenfall AB for the Swedish one. The generator housing has been designed and manufactured by Kvaerner Turbin AB of Kristinehamn/Trollhättan and the turbine blades by Messerschmidt-Bölkow-Blohm (MBB) of Bremen. The table below presents the basic technical data on the plant and the illustration shows a view of the plant with a turbine diameter of 80 m and hub height of 90 m. (Aeolus II in Germany)

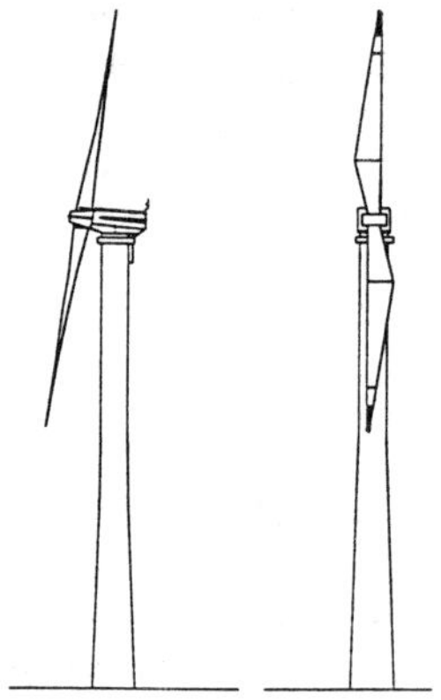

VIEWS OF NÄSUDDEN II

Turbine diameter	80 m
Speed	21 och 14 rpm
Tower heigh	79 m
Output	3 och 1 MW
Energy	6,9 GWh/år
(with 90 % availability)	
Starting wind	6 m/s
Rated wind	14 resp 9 m/s
Shutdown wind	25 m/s
Weigh of rotor	9 ton
Weigh of hub	23 ton
Weigh of machinery	129 ton

Our experience of the two pilot units is that the construction work is both costly and time-consuming. Consequently, complete and tested units such as the generator housing and minor buildings for control and regulation equipment etc should be manufactured in assembly shops so that the amount of work that has to be done on the construction site can be minimized. The transport and assembly of units as bulky and heavy as this requires careful planning.

With the aid of modern communications equipment, a large number of wind power plants in several groups can be monitored and remote-controlled from a joint regional operation and maintenance centre. With equipment of this kind, the availability of wind power can be increased and production costs lowered.

The period required for the construction of 100 wind power units is entirely dependent on the manufacturing capacity and number of suppliers. If a supplier is assumed to manufacture 15 units per year, it is estimated that it will take 12 – 14 years including pilot study, design and all necessary permits before all 100 wind power plants are in operation.

1.7 Electricity production and production costs

With the aid of the model shown below, an average production cost has been calculated for some 250 wind power plants distributed over 15 sites.

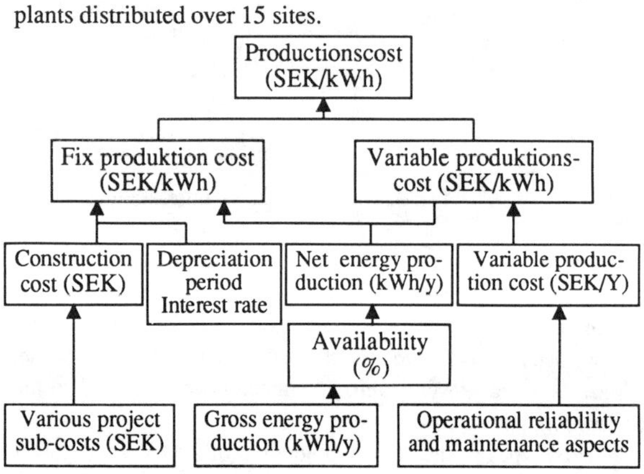

The supply of wind on the Swedish West Coast is higher than on Gotland, and the following table shows the energy production for the average unit in the fifteen different sites.

	Gross	Net	
Bohuslän	8.0 – 8.9	6.9 – 7.4	GWh/y
Halland (northern part)	8.8 – 9.4	7.4 – 7.7	GWh/y
Gotland (west coast)	7.0 – 7.7	5.9 – 6.2	GWh/y
Gotland (east coast)	5.6 – 6.3	4.7 – 5.4	GWh/y

The total energy production has been calculated to be 1.5 TWh/y for 250 wind power plants and 0.7 – 0.8 TWh/y for 100 plants.

The construction costs are higher for all areas on the Swedish West Coast than for the areas on Gotland. This relative cost varies between 0.93 and 1.0 units on the West Coast and between 0.87 and 0.89 units on Gotland, and is divided into 70% for wind power plants, 15% for towers with a lift and 15% for other facilities, electricity, telecommunications equipment and installation.

Assuming a 25-year depreciation period, 5% real interest and operation and maintenance costs of SEK 0.5 million/unit, the production cost has been put at between 50 and 75 öre/kWh for the sites on the West Coast and the west coast of Gotland. An additional 1 öre per kWh will be charged for laying branch lines and strengthening the existing network. Development efforts will be directed from now on at bringing these costs down to a competitive level compared with other other power generation techniques with similar environmental data.

Näsudden on Gotland is the largest land-based wind power site presented in "Status for Wind Power" and in the "300 MW Study". The potential has been estimated at 300 GWh/y, which is equivalent to 40% of Gotland's energy consumption.

It is proposed that 26 of the 100 wind power plants with a 3 MW output should be sited at Näsudden. It is calculated that these plants will be capable of producing 180 GWh/y.

1.8 Conclusions

Within the 15 sites studied, the "Status for Wind Power" study assessed the potential to be 380 wind power plants whereas the "300 MW Study" judged the potential to be 150 plants. The main reason for the difference is that the areas in the "300 MW Study" are smaller and have better wind supplies.

Approximately half the 100 wind power plants have been localized to the Swedish West Coast and the western coast of Gotland, and the total energy production has been put at 0.7 – 0.8 TWh/y.

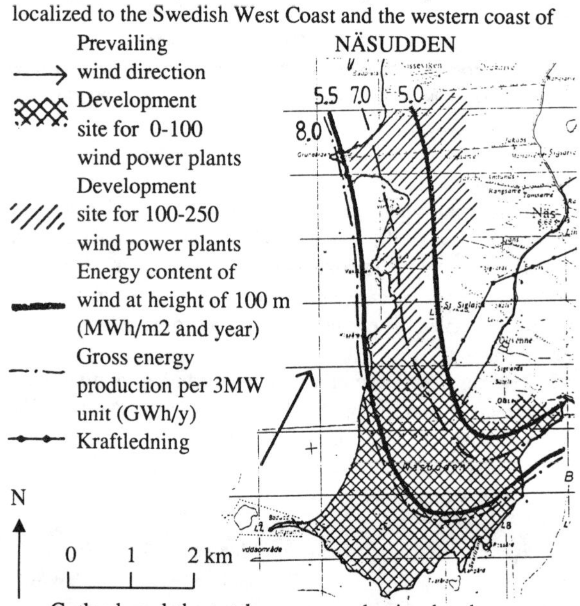

The production cost has been estimated to be between 50 and 75 öre/kWh. This presupposes serial production of the Näsudden II unit. Development work must from now on be directed at lowering the costs to a competitive level compared with other techniques with the same kind of environmental data.

In order to facilitate large-scale wind power development in the future, the best areas for wind energy should be included as areas of interest in the synoptic planning of the local authorities concerned. (1 ECU= 9,09 SEK 4 March 93)

A COMPARATIVE ANALYSIS OF PLANNING CONTROL AND ENVIRONMENTAL ASSESSMENT FOR WIND FARM DEVELOPMENTS IN THE UNITED KINGDOM, GERMANY, DENMARK AND THE NETHERLANDS

Marcus Trinick Solicitor and Partner
Bond Pearce, Solicitors, 1 The Crescent,
Plymouth PL1 3AE, United Kingdom

1. Introduction

In this contribution I set out to compare the basis of physical planning control in some of the EC countries which have been seen intensive wind farm development either over a number of years, or, as a result of new national programmes, over the last two or three years.

For reasons of time I have had to limit my overview to Germany, Holland, Denmark and the United Kingdom. This is in no way a comment on the quality or quantity of wind energy development in other EC countries.

By way of further introduction I mean by the United Kingdom only England and Wales, partly because that is where wind energy development has taken place and partly because the planning systems of Scotland and Northern Ireland differ in a number of ways from those of England and Wales.

In the second part of my talk I discuss the emergence of formal environmental assessment as part of the fabric of planning control.

The evolution of any national system of planning control reflects a mosaic of constitutional and political history, a country's physical characteristics, historic land use patterns and, largely a product of these factors, social perceptions.

What I have found is, in the light of these influences, no surprise. The planning systems of Germany, Holland and Denmark are very similar while that of the UK exhibits some important and unique characteristics. However, what is striking, and I think encouraging, is the extent to which the UK. system is tending to converge with those of the other countries surveyed.

2. The Development Plan

A feature known to all the planning systems studied is what can be called generically the "development plan". This companies a series of structured documents which contain, depending upon the level in the system at which they are produced, either general strategic policies or specific detail on such topics as development infrastructure, housing, employment, recreation, countryside protection, and of course, energy. Except in Denmark mature policies on wind farm development are only now beginning to emerge, but in the U.K. local authorities in high resource areas are now concentrating on the relevant issues, and are producing draft policies of a wide (and worrying) variety.

The role of the development plan has historically distinguished the UK planning system from those of the other countries studied. The development plan in the UK has always been part of a system which allows great administrative discretion, and was just one of a variety of considerations to be borne in mind when a local planning authority decided whether or not to grant planning permission. On the other hand, the other countries have used the development plan as a legal definition of acceptable future development.

This plan-led approach of countries except the UK has emerged largely as a product of political systems. The Dutch *Bestemmingsplan*, the German *Bebauungsplan* and the Danish *Lokalplan* are all derived from written constitutions which guarantee the property rights of individuals and their relations with the State. They are detailed plans which legally bind each citizen and the administrative decision body to a predetermined approach to development activity.

More recently, the UK has made a significant move towards the plan-led approach. In addition to an obligation to have regard to the provision of the development plan when considering a development proposal, it is the law as from September 1991 that a decision must be made according to the policies of the development plan either where no other factors exist or where other considerations which do exist are of less importance then the plan. Only where it is decided that other considerations are of greater importance than the policies in the development plan may a decision be made against the advice of those policies. The importance of these other considerations in comparison with policies in the plan is, not surprisingly, a subject for much debate.

3. Front Line Development Control

The lowest level of the administrative structure has primary decison making responsibility throughout the countries studied. In the UK that level is the District Council which, outside the larger towns, comprises an area with an average population of 100,000. As an exception, applications for electricity generation projects exceeding 50 MW must be made to the Secretary of State for Trade and Industry.

The District Council is also charged with the job of preparing the most detailed development plan available within the system, the Local Plan. The similarities between the UK and the other countries stop there. The zoning plans of Dutch, German and Danish systems (the *Bestemmingsplan*, the *Bebauugsplan* and the *Lokalplan*) are exactly what the UK Minister for Housing and Planning has recently indicated their UK equivalent (the *Local Plan*) should not become prescriptive [1]. In other words, while UK planning applications should normally be decided according to the development plan, the plan itself is flexible and zoning policies are rare outside town and cities.

The consequence of the legally binding effect of zoning plans in countries except the UK is that building or construction permits for proposals which do not conform to approved plans will not generally be granted. Logically, proposals which do accord with the approved plans are rarely refused. This rigidity is diluted to some extent by the fact that zoning plans may be altered to fit the nature of the development. However, in such a case the plan must be altered. The tendency to compromise and improvise with decision making procedures in the UK has, for better or worse, avoided rigidity. I am sure that it is also the case that at a local level there is more flexibility in Germany and Denmark (and perhaps Holland) than in law is apparently the case.

4. The Advantages and Disadvantages of Rigid Plans

Prescriptive zoning in plans may give a large measure of certainty and speed to a planning system while the requirements of developers remain constant and predictable. However, it may not be well equipped to cope with new forms of development which confront the local decision making bodies (*the District Councils, Gedeputeerde Staten, Landkreis*[2] *and Kommune*) with the need for an alternative and creative approach. An obvious example of a new form of development is wind energy. The general response to unwelcome rigidity has been to formulate more flexible, general plans such as the Dutch *Hoofdlijnen Bestemmingsplan,* the German plans under section 35 of the *Bundesbaugesetz,* and the Danish *Kommuneplan*.

The flexibility inherent in the UK system does I believe have advantages, especially in relation to the often complex planning issues which arise out of wind farm developments. In particular, it is very difficult for any public authority to do more than look at the protection of the countryside, nature conservation and potential noise emissions. No public authority can carry out the anemometry necessary to properly measure mean wind speeds, while in remoter areas the cost of grid connection will be a serious issue. A reasonable degree of flexibility is essential in any development which requires a natural and variable resource.

5. Subsidiarity

The proposed new Article 130s(2) of the ECC Treaty as amended by the Treaty on European Union gives jurisdiction to the Council of Ministers to adopt measures concerning town and country planning. Article 130s is part of the Environment Chapter of the Treaty[3]. The principle of subsidiarity[4] only applies in areas which do not fall within the Community's exclusive competence. Subsidiarity's effect is that the Council of Ministers can only exercise its jurisdiction to adopt measures in the field of town and country planning where the objectives of the proposed action cannot be sufficiently achieved by the member States of the Community. I have already described how the historic differences between the UK planning system and the other systems studied have been removed to some extent. This is an ongoing process prompted by the member States themselves. Whether in fact action to harmonise planning systems at the Community level is needed is still a matter for debate.

6. Environmental Impact Assessment for Wind Farm Developments

A fundamental stage in the approach to developing a wind farm is likely to be the provision of an environmental statement, within which an assessment of the likely significant effects of the development will be assessed. Environmental impact assessment (EIA) has its origins in the US National Environmental Policy Act 1969 (NEPA). This imposed a broad obligation upon the Federal administration to prepare an Environmental Impact Statement (EIS) where development proposals involved "major Federal actions significantly affecting the quality of the human environment".

An EIS incorporates a statement describing the project and an assessment of its anticipated effects. The character of NEPA was illustrated by the requirement that the likely consequences of a proposal upon the environment had to be analysed not only scientifically, but also in the wider social context. The EIS also had to discuss alternative measures which might achieve the same or similar objectives to those of the proposal.

The European Community's First Action Programme on the Environmental[5] in 1973, and subsequent Action Programmes, all embody the international environmental law precautionary principle: the best environmental policy consists in preventing the creation of pollution or nuisances at source, rather than subsequently trying to counteract their effects. With this in mind, the EC Commission published in 1980 its proposal for a Directive on the assessment of the effects of certain public and private projects on the environment.[6] Five years later the proposal emerged as Council Directive 85/337/EEC[7] ("the EIA Directive").

The preamble to the EIA Directive noted the disparities between the existing laws in force in each of the member States concerning EIA. Member States' systems for assessing the impact a project would have upon the environment were still at an early stage. Eight years later national laws on EIA across the EC are still inconsistent and subject to different interpretations.

United Kingdom

The 1985 EIA Directive is implemented in the UK by the Town and Country Planning (Assessment of Environmental Effects) Regulations 1988. The 1988 Regulations follow the format of the EIA Directive by making an EIA compulsory for certain types of development. In addition the local authority may require an EIA for other specified types of development where an assessment will be justified by the nature, size and location of the proposal. If a developer disagrees with the local authority's view that an EIA is required he can refer the matter to the Secretary of State for the Environment for a final binding decision.

An EIA must contain information on all the likely environmental effects of a development, and in fact these are specified in detail within the 1988 Regulations.

At present an EIA cannot be required for a wind farm, but new legislation has already been drafted to change this position. It seems likely that EIAs will only be required for wind farms comprising more than 10 machines or for 4 MW unless within a short distance of a National Park or AONB.

Germany

Specific EIA legislation exists in Germany. In 1975 the Government issued Principles for the Assessment of the Environmental Impact of Public Actions by the Federal Government[8]. In 1977 the Administrative Procedure Act

1977[9] gave the public a legal right to make representations on development proposals to the *Landkreise* or *Kreisefreie Städte*. The most recent legislation, the Act of 12 February 1990[10] (UVPG) implements the EIA Directive and follows its format in requiring an EIA for the projects listed in its Annex.

The 1975 Principles apply only to public actions (including the preparation of legislation and the various plans used within the development control system) by the federal authorities[11]. They have no legal status and cannot be viewed as a legal precedent for ensuring environmental protection.

In addition the *Bundesnaturschutzgesetz* (Federal Nature Protection Act) of 1976 contains a requirement that if effects upon the countryside and nature are not assessed prior to the issue of a development plan or a building permit then such an assessment must take place before any work can commence. The effects with which the 1976 Act is particularly concerned are those which could harm the natural ecosystem, or seriously or permanently affect the landscape. If adverse impacts are identified, and no mitigating measures can be taken to minimise or avoid their effects, then the development cannot proceed.

Denmark

While there is no specific Danish legislation on EIA, the National and Regional Planning Act 1973 and Municipal Plan Act 1975 require that the preparation of regional and municipal plans (Landsplandirektiver and Lokalplan) includes consideration of alternatives to all forms of development anticipated by the plans. Whilst this consideration is not intended to be specific to individual projects, it does go some way in achieving the aims of the NEPA EIA model.

The Municipal Plan Act 1975 has a significant influence in terms of EIA because it requires the preparation of a *Lokalplan* where major building works are proposed[12]. It is open to the *Kommune* to insist that the developer participates in the preparation of the *Lokalplan*, and in this way the cost of such assessments is apportioned between the state and the developer.

Netherlands

IN 1981 the Dutch Government was considering legislation requiring EIA for specific types of project. Legislation, delayed by the emergence of the EIA Directive, has since extended the General Environmental Hygiene Provisions Act 1980 and defined the scope of the EIA procedure itself. My understanding is that an EIA is required for, amongst other developments, a wind farm.

From this short overview it can be seen that member States have not adopted a uniform approach to EIA. This has in part resulted from the wording of the Directive. The European Parliament indicated its concern in this respect in 1987 when it proposed supplementary measures to tighten up policy towards EIA and remove the inconsistencies with the EIA Directive[13].

7. Conclusions

The written constitutions of Germany, Denmark and Holland have led naturally to a system of planning control based on plans which have a binding effect. In comparison the UK has no written constitution, and the planning system has evolved on the basis of compromise and improvisation, and with a common belief that flexibility in the application of policies is desirable.

The UK system is converging with that of the other countries studied in that the development plan has acquired a much greater significance. This is partly in response to the enormous development pressures of the 1980s. Nevertheless, policies within UK development plans still guide rather than prescribe, and this may remain the position for the foreseeable future.

In the field of environmental assessment there remain inconsistencies of approach towards the implementation of Council Directive 85/337/EEC. It would seem very desirable that other countries should achieve a higher degree of consistency, especially where environmental emissions may travel considerable distances. However, such is not the case with wind energy, and consistency of approach is desirable rather than essential.

For a developer visiting Earth from another planet the ability of human beings to complicate physical planning with layers of decision making and policy control would be very obvious. However, for those of us who increasingly operate in the European rather than a single national context I am relieved that the differences which I have isolated in this talk are matters of emphasis rather than fundamental principle. Certainly in talking to those working in the planning systems of Germany, Holland and Denmark there is any easy recognition of common difficulties and reactions to development. This is after all to be expected since the common denominators of all political systems are the politicians themselves.

1. "...[Local] plans contain too much detail. Many contain non land-use policies ... Others seek to determine every planning application before it is received ... Plans are not prescriptive ..." [1992] JPL Occasional Papers No.19 para. 44.
2. Or the *Kreisefreie Städte* which combines the roles of the *Gemeinden* and *Landkreise*.
3. The proposed new Title XVI.
4. Contained in the proposed new Article 3b of the EEC Treaty as amended by the Treaty on European Union.
5. OJ [1973] C 112/2 (20.12.1973).
6. OJ [1980] C 169/14 (09.07.1980).
7. Council Directive 85/337/EEC of 27 June 1985 on the assessment of the effects of certain public and private objects on the environment OJ [1985] L 175/70.
8. Public announcement of the Federal Minister of the Interior, 12 Sept. 1975.
9. *Verwaltungsverfahrensgesetz* (VwVfG) of 1 January 1977; in addition, the consultation process afforded by the *Bundesbaugesetz* (Federal Building Law of 1960) and *Baugsesetzbuch* (Federal Building Law Code of 1987).
10. *Gesetz zur Umsetzung der Richtlinie des Rates vom 27 Juni 1985 die Umweltverträglichkeitsprüfung bei bestimmen öffentlichen und privaten Projekten.*
11. By reason of the fact that the 1975 Principles only apply to federal actions, some of the *Länder* have introduced their own regulations:
Baden-Württemburg — *Joint enactment by the Ministries for the Economy and for Transport and the Ministries for Food, Agriculture and the Environment of 1 March 1979.*
Bavaria — *Principles of the Bavarian Government for the assessment of the environment impact of public actions in Bavaris of 12 September 1978.*
Berlin — *Circular concerning EIA in planning procedures of 5 Sept. 1978.*
Saarland — *principles for the assessment of the environmental impact of public actions in the Saarland of 26 October 1976.*
12. A similar requirement is incorporated within the Environmental Protection Act (as amended) whereby the Minister for the Environment may issue regulations requiring an ***amter*** (county) to prepare "preliminary surveys and evaluations of environmental consequences in connection with the establishment of a large, separate installation".
13. European Parliament, Interim Report of the Committee, Public Health and Consumer Protection on supplementary measures in respect of environmental impact studies (Doc. A 2-18/87).

IMPACT OF WIND TURBINES ON BIRDLIFE
an overview of existing data and lacks in knowledge

drs. ir. J.H.B. Benner, Consultants on Energy & the Environment (CEA)
P.O. Box 21421, 3001 AK Rotterdam, Netherlands

SUMMARY

Wind energy generally is considered to be one of world's best options for clean production of electric energy. Actual application in Europe however turns out to be limited by all kinds of economical and planological factors. One of the bottle necks for wind energy is its impact on birdlife. In order of the European Community CEA, Consultants on energy and the environment and Environmental Biology Leiden University have made a status report on this impact. The report presents an overview of wind energy programs, an analysis of recent studies on the impact of wind turbines on birdlife and data on the attitude of nature conservation organisations in Europe and the USA. The studies were analysed on legitimacy of assumptions and validity of conclusions.
Most of the wind energy programs and all european studies deal with bird kills and disturbance in coastal areas. The number of victims per turbine per year in this type of area appears to be acceptable. The disturbing effect of wind turbines on breeding birds appears to be negligible. The disturbance of resting and migrating birds by wind turbines however is reasonably clear. Negative attitudes towards wind energy projects are not particularly due to avian considerations, but rather to a general objective to protect landscapes and habitats, undisturbed by human infrastructures and disturbance.
The report concludes that all new locations for wind energy projects should be weighed on the disturbance aspect. Reference data for such a weigh are available for coastal areas, although the impact on local migration between feeding grounds and high water refuge areas needs further research. Future research also is needed for application of wind energy both on offshore locations and grass-/farmlands. Wind energy developers and conservationists should have a close contact in order to establish consensus on how to deal with remaining uncertainties.

Introduction

Actual application of wind energy in Europe turns out to be limited by several kinds of economical and planological factors, of which the impact on birdlife is an important one. Birdlife can be influenced either by direct collision of birds with wind turbines or by disturbance of their breeding, resting or foraging in the vicinity of these turbines. This subject was studied by Consultants on Energy and the Environment (CEA) in Rotterdam, together with the department for Environmental Biology of the Leiden University. The study fits in a larger project (part of the european Joule-programme) on the environmental impact of wind energy installations, coordinated by the University of Amsterdam, Department of Environmental Science. In this project CEA was associated contractor for the bird item.
The study of CEA and Leiden University had the following goals:
- to inventorize existing wind programs and bird areas,
- to select potential conflicting location types,
- to inventorize attitudes of nature conservation organizations,
- to gather and analyse conducted studies on effects for birds,
- to indicate lacks in knowledge.

The study concerned all European countries with published wind energy programmes. Most attention was paid to the leading wind energy countries: Denmark, Germany and The Netherlands. Other countries involved in the study are Greece, Italy, Spain, Sweden and the United Kingdom. Some remaining countries, like France, do have an interesting wind supply, but were not included due to lack of actual plans. Results of conducted field studies in the United States of America and the attitude of american nature conservation organizations were taken into account as well.

Framework of the study

Figure 1 gives a schematic overview of the study.
In the first step the actual problem was specified. The next step was to collect and inventorize data on wind energy programs, important areas for birds and already conducted field studies. Based on this information typologies were defined concerning biogeographical zones, habitats and the impact and effects of wind turbines on birdlife. These typologies were used in the analysis of wind programs, bird areas and the available field studies.
Wind energy programmes in Europe were analysed to determine which types of locations were most frequently selected for wind energy. Using this information the locations for wind programs were confronted with important areas for birds to indicate possible conflicts. The results of this confrontation were used in the detail analysis of field study results and in inventorizing the attitudes of nature conservation organizations and lacks in knowledge. To get information about these attitudes important nature conservation organizations in four countries (Sweden, Denmark, the Netherlands and the U.S.A.) were interviewed.

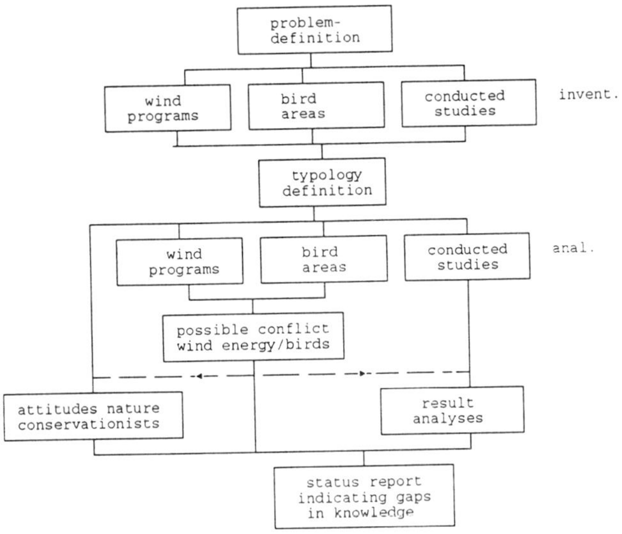

Figure 1: Schematic overview of the study

Wind energy programmes

After an areatypology was defined in the study actual wind energy programmes were analysed to determine which types of area were most frequently selected for developing wind energy projects. Table 1 shows the selected types of areas for wind energy in the western part of Europe. The division in areatypes is according to the definitions in the study.

Table 1: Distribution of wind energy plans over the types of area in the countries included in the study.

Country	Areatypes							
	Off-shore	Coast	Inland wetland	Forest/ heath	Grass-/ farmland	Bogs	High-lands	Industry/ Villages
Denmark	+	++	-	-	+	-	-	-
Germany	-	+	-	-	+/-	-	-	+/-
Greece	-	++	-	-	-	-	-	-
Italy	-	+	-	-	-	-	-	-
Netherlands	+/-	++	-	-	+	-	-	+
Spain	-	+	-	-	-	-	++	-
Sweden	++	++	-	-	-	-	-	-
U.K.	+	+	-	-	+	-	+	-

++ : many plans occur
\+ : plans occur
+/- : some plans occur
\- : plans occur hardly or not at all.

No wind energy program concerns inland wetlands, forest/heath or bogs. These area types were therefore excluded from further analyses.

Analysis of existing knowledge

Recent studies on the impact of wind turbines on birdlife both from the USA and Europe were analysed on legitimacy of assumptions and validity of conclusions. In the study only final results were analysed.

Bird victims

The number of birds found needs to be converted in order to get an estimate of the real number of victims. For this reason the cause of death, the removal by scavengers, the observer efficiency and the wounded victims need to be taken into account. In all the studies that are analysed the cause of death was determined, so no overestimation because of this point has taken place. Studies in which no scavenger removal or observer error surveys were carried out are numerous. All of these studies underestimate the number of victims caused by wind turbines.
In studies in which finding probability was not or just partly determined, no estimate of the number of victims per turbine per year could be made. The numbers mentioned in the studies analysed are estimates on the basis of correction factors.
The number ranges from a minimum of 0.1 to a maximum of 309 per turbine per year. The number of victims in the study of Karlsson (1983) is higher, but this is a result of mass collisions in just one night. This kind of 'disaster-nights' are not likely to occur very often and therefore the number of Karlsson is not representative for other years.
In none of the studies wounded victims were taken into account, except for that of Winkelman (1992 a). She assumed that the total emigration of wounded victims from the area of research equals the immigration into the area. This kind of assumption only seems legitimate for areas with a considerable number of turbines, in which victims of one turbine can shift to another turbine, and the length of the outer border of the area is relatively small compared to the size of the area. And even in this case emigration exceeds immigration. Therefore, all of the analysed studies underestimate the number of victims to some extent due to this factor.

Disturbance

In Europe several studies to the disturbance of birds by wind turbines have been carried out. The disturbing effect of wind turbines on breeding birds seems negligible. The disturbance of resting and migrating birds by wind turbines, however, is reasonably clear, although dependent on the species. No data are available on the impact of wind turbines on local migration between areas for foraging and high water refuge areas.
Whether or not disturbance of migrating birds is undesirable is disputable. This disturbance nearly always results in effective avoidance of the turbine by birds, often by minor adaptations in course of migration.
Studies from the USA all concern wind power locations in highland areas. Aim of these studies was to investigate the effect of wind turbines on birds of prey. Whether or not bird populations are negatively effected by wind turbines is unclear, but the California Energy Commission stated that efforts should be made to mitigate losses, regardless of whether bird populations are significantly affected by turbine-related mortality.
The existing knowledge is summarized in table 3. The matrix contains a great number of question marks. Nevertheless not every lack in knowledge has to be translated to the same extent in a need for further research.

Resulting recommendations and lacks in knowledge

On the basis of the analysed studies, recommendations can only be given for future wind power locations in coastal areas. The most important concern is the disturbance of resting birds, so, in planning new locations, areas that are rich in resting birds, such as wetlands, should be avoided. More data are desired on the impact of wind turbines on local migration between areas for foraging and high water refuge areas.

Although no actual research data are available it is probable that few problems will arise at wind power locations in grassland areas. Breeding birds do not seem to be significantly affected by turbines. The number of resting and migrating birds is lower than in coastal areas, so disturbance of resting and migrating birds as well as the number of victims will be smaller. However, to really verify this statement complementary research is needed.

Points of concern for future offshore turbine locations, as planned in Sweden and, to a smaller extent in Denmark and the Netherlands, cannot be given. To assess the problems in these areas research is necessary.

Additional effects caused by wind turbines in built-up areas are negligible. Already existing edifices have disturbing effects as well, and the impact of the wind turbines used in built-up areas is often comparative small.

Table 3: Matrix indicating the existing knowledge on impact to birds

type of landscape	impact				
	collisions		disturbance		
	local migration	seasonal migration	breeding birds	resting birds	migrating birds
offshore:	?	?	?	?	?
coastal zone:	+/-	+/-	-	+	+
grass-/farmlands:	?	?	?	?	?
highlands:	+/-	?	?	?	?
industry/ villages:	?	?	?	?	?

? : no research results
- : no impact
\+ : impact
+/- : impact exist, but seems to be small

Attitude of nature conservation organizations

Whether or not wind energy developers consider the bird problem as a real issue, it is important to realize that if individuals or interest groups fear for an unacceptable impact, the possibility to realize a wind project can be seriously influenced. Therefore in a wind energy project it is necessary to investigate the attitude of important nature conservationists in the area and ask for and listen to their opinion on the bird issue in a given situation.

The attitude of nature conservation organizations towards wind energy differs enormously between countries. The main trend in the development of environmental awareness in all north-western European countries however is the same.

Wind energy gradually became subject to an aggregated environmental awareness. Discussions on the environmental conveniences and inconveniences of wind power generation intensified with the increasing number of wind turbines in the country side. Unequal levels of documentation on the environmental convenience (e.g. estimated decrease in CO_2 pollution) and inconvenience (e.g. unknown impact to birds, insects and bats) have caused many distractions and misunderstandings. This is reflected in the dilemma many conservations have shared - how to measure uneven environmental pros and contras of wind power generation.

Following prevailing governmental planning of the late eighties, areas have been designated in which siting of wind turbines is prohibited. Among those areas are:

- Wetlands of international importance (Ramsar-convention),
- EC Special Protection Areas (Directive 79/409 on the Conservation of Wild Birds),
- Areas with other more expedient utilization (e.g. raw materials reclamation, forestry).

General conclusions

Actual application of wind energy in Europe turns out to be limited by several kinds of economical and planological factors. One of the bottle necks for wind energy is its impact on birdlife.

An analysis of wind energy programs shows a general interest in coastal areas due to the occurrence of a good wind regime. The type of projects in this area type concerns mainly large scale windfarms. Wind programs in almost all leading wind energy countries give attention to future development of locations offshore. In flat countries like Denmark, the northern part of Germany and The Netherlands many solitary wind turbines are placed on farmlands near farms.

Most of the European studies deal with bird kills and disturbance in coastal areas. The number of victims per turbine per year in this type of area seems acceptable. The disturbing effect of wind turbines on breeding birds seems negligible. The disturbance of resting and migrating birds by wind turbines, however, is reasonably clear, although dependent on the species. The disturbance nearly always results in effective avoidance of the turbine by birds, often by minor adaptations in course of migration.

Based on the studies analysed, recommendations can be given for future wind power locations in coastal areas. The most important concern is the disturbance of resting birds. Therefore in the planning stage of new windfarms areas which are rich in resting birds, such as wetlands, should be avoided. More data are required on the impact of wind turbines on local migration between areas for foraging and high water refuge areas.

During the last 15-20 years wind energy gradually became subject to an aggregated environmental awareness. Unequal levels of documentation on the environmental benefits and disadvantages have caused many distractions and misunderstandings. This is reflected in the dilemma many conservationists have shared - how to measure uneven environmental pros and contras of wind power generation. Since environmental issues generally do not know national frontiers, the problem of bird kills and disturbance should be handled on an international scale. As a result of the international concern several areas have been designated where siting of wind turbines is prohibited, e.g.:

- Wetlands of international importance (Ramsar-convention),
- EC Special Protection Areas (Directive 79/409 on the Conservation of Wild Birds).

At present one of the major concerns among conservationists is that many valuable and vulnerable environmental resources are located outside the protected areas while gaps exist in the knowledge about possible impact of wind turbines on these resources. Negative attitudes towards location of wind turbines are not particularly due to avian considerations, but rather to a general objective to protect landscapes and habitats, undisturbed by human infra-structures and disturbance.

Wind energy developers and conservationists should have a close contact in order to establish consensus on how to deal with remaining uncertainties. When no further investigations on the magnitude of the effects of wind turbines take place the bird issue might restrict the number of areas available for the use of windpower.

Although inevitable with loss of nuance the general results of the study are summarized in table 4.

Table 4: Summary of study results

type of landscape	results				
	existence of wind energy programs	size of expected impact on birdlife	extent of existing knowledge	attitude of nature conserva-tionists	general conclusion
offshore:	future trend	unknown	unknown	conservative	further research required
coastal zone:	many programs	significant impact expected	several studies available	conservative	knowledge mostly satisfactory
grass-/farmlands:	several programs	some impact expected	unknown	conservative	further research required
highlands:	some programs	some impact expected	unknown	varying	research low priority
industry/ villages:	some programs	no impact expected	unknown	positive	research low priority

MAIN LITERATURE

- Bennett G., "Towards a European Ecological Network", Inst. for Eur. Env. Policy (1991).
- Berkhuizen J.C. and Postma A.D., "Impact of Windturbines on Birdlife", Proc. EWEC'91, CEA (1991).
- Grimmett R.F.A. and Jones T.A., "Important Bird Areas in Europe", ICBP Technical Publication 9, Int. Counsil for Bird Preservation, Cambridge (1989).

REPORTS ANALYSED

Estep, J.A. *et al.*, 1989. Avian mortality at large wind energy facilities in California: Identification of a problem. Staff report California Energy Commission.

Hartmann, G. & G. Vauk, 1990. Biologisch-ökologische Begleituntersuchungen zum Bau und Betrieb von Windkraftanlagen: Endbericht. NNA-Berichte-3/Sonderheft, 1990.

Karlsson, J., 1987. Fåglar och vindkraft: Kompletterande resultatredovisning. In: Vindkraft Fåglar, Vindkraftsutredningens Betänkande sou 1988: 32.

Møller, N.W. & E. Poulsen, 1984. Vindmøller og fugle. Vildtbiologisk Station, Kalø.

Musters, C.J.M., G.J.C. van Zuylen & W.J. ter Keurs, 1991. Vogels en windmolens bij de Kreekraksluizen. Report Milieubiologie Rijksuniversiteit Leiden.

Orloff, S. & A. Flannery, 1992. Wind turbine effects on avian activity, habitat use, and mortality in Altamont Pass and Solano County wind resource areas, 1989-1991. Final report, California Energy Commission & BioSystems Analysis, Tiburon, CA.

Pedersen, M.B. & E. Poulsen, 1991. En 90m/2MW vindmølles indvirkning på fuglelivet. Fuglenes reaktioner på opførelsen og idriftssættelsen af Tjæreborgmøllen ved Det Danske Vadehav. Danske vildundersøgelser, hæfte 47. Miljøministeriet & Danmarks miljøundersøgelser.

Petersen, B.S. & H. Nøhr, 1989. Konsekvenser for fuglelivet ved etableringen af mindre vindmøller. Ornis Consult, Kopenhagen.

Swelm, N. van, 1988. Vogels en de multi-windturbine op de Maasvlakte. Stichting Ornithologisch Station Voorne, Oostvoorne.

Winkelman, J.E., 1984. Vogelhinder door middelgrote windturbines. Een verkennend onderzoek naar vliegedrag, slachtoffers en verstoring. RIN-report 84/7, IBN-DLO, Arnhem.

--, 1989. Vogels en het windpark nabij Urk (NOP); aanvaringsslachtoffers en verstoring van pleisterende eenden, ganzen en zwanen. RIN-report 89/15, IBN-DLO, Arnhem.

--, 1992 a. De invloed van de SEP-proefwindcentrale te Oosterbierum (Fr.) op vogels. 1. Aanvaringsslachtoffers. RIN-report 92/2, IBN-DLO, Arnhem.

--, 1992 b. De invloed van de SEP-proefwindcentrale te Oosterbierum (Fr.) op vogels. 2. Nachtelijke aanvaringskansen. RIN-report 92/3, IBN-DLO, Arnhem.

--, 1992 c. De invloed van de SEP-proefwindcentrale te Oosterbierum (Fr.) op vogels. 3. Aanvlieggedrag overdag. RIN-report 92/4, IBN-DLO, Arnhem.

--, 1992 d. De invloed van de SEP-proefwindcentrale te Oosterbierum (Fr.) op vogels. 4. Verstoring. RIN-report 92/5, IBN-DLO, Arnhem.

STUDY OF MOBILE WIND POWER COMPLEX (WPC) 'ZHAVORONOK' EFFICIENCY ON THE TERRITORY OF CIS

V.G. Nikolaev, Y.A. Grintsevich, V.N. Kiselev.

State Scientific Research Center TSAGI
17, Radio Str., Moscow, 107005, USSR

ABSTRACT

Atlas of generation of electricity by the self-contained wind-power complex (WPC) 'ZHAVORONOK' on the territory of CIS and of several foreign regions, is intended to reveal the regions with the wind resources, sufficient for WPC effective use; the Atlas also contains several recommendations on achievement of maximum power and economic efficiency of WPC use. In addition to the maps, reflecting of annual and daily generation of electricity in prescribed regions of the Globe, this Atlas contains the data on WPC 'ZHAVORONOK' specifications, and the estimations of economic effect of its introduction and use.

During this Atlas development, the technique of definition of generation of electricity by wind-power sets on the territory of CIS was used; this technique was developed by the specialists of N. Zhukovsky Central Institute of Aero-hydrodynamics on the basis of the specialized computer system 'FLUGER'using data on the wind parameters, prepared by the Main Geophysical observatory (MGO) specialists.

The authors hope that the materials and recommendations, contained in the ATLAS, will promote the production and deliveries of WPC, and will favour WPC effective on the territory of CIS and abroad.

INTRODUCTION

Power, ecologic and economic problems aggravation makes the mankind to search for the new ways of power supply, and to develope them. Such non-traditional ways are as follows: solar, wave, geothermal, tidal, bio-organic and the other types of energy use.

It is known long since that one of the most accessible and perspective power sources is the wind. The wind-power sets (WPS) have been used by a man for at least many centuries. During this period such sets developed from the simple windmills into the complex wind-power stations of the great capacity. Nowadays, the wind-power engineering became the powerful, high-tech industry, characterized by the wide geographic and functional spread.

The leaders of the world wind-power engineering are such countries as Denmark, USA, Italy, Germany, Japan, where with the help of WPSs a great deal of ecologically clean power is generated, the structure of power consumption is being improved, and considerable effect of power supply is being achieved.

In the former USSR, the wind-power engineering developed rather intensively up to 1950-s. Then there was a period of stagnation as a result of the shift of domastic interests to hydro-, heat- and nuclear power sources. However, today in the countries of CIS a considerable increase of interests to the wind-power engineering is observed again. It is accompanied by many developments of WPS of various types, capacity and purpose, and the transition to their pilot and quantity production is observed.

Up-to-date WPSs differ in their functional purposes (electrical, water eifting and power WPS), in lay-out (self-contained WPSs or the sets operating together with the other sources, for example, with diesel or solar ones), in desing (for example, orthogonal or axial sets), and in capacity, which defines their use possibilities more exactly.

WPSs of small and mean capacity are intended to provide the farms,country-cottages, villages, factories and the other power users, located in the regios with the favourable wind situation, with electricity and thermall power WPSs of mean capacity are often used as the base modules for formation of power complexes including one or several WPS, operating together with diesel or hydroulic sets, with the network, or for independent user.

The powerful megaWatt WPSs are used as additional power source for the available power systems, or as seft-contained power supply for large residential or indus tital objects in the regions with stable and powerful wind flows.

Among numerous domestic and foreing WPSs of traditional type, there is rather remarkable and even unique WPC "ZHAVORONOK", the self-contained mobile WPC, which may be rapidly restationed and started. The independence and guaranteed power productivity of the complex is ensured by parallel work with the 30 kW WPS with the Diesel-electric set of the same capacity. The WPC mobility is achieved by its placing within the 12 m container, which is used as the storage and transportation facility, and as a platform for power complex, provided with necessary gangets for its mounting, adjusting and rapid starting.

The review of up-to-date WPSs, which are being produced and developed now, shows that nowadays there are WPSs of WPS type neither in our country no abroad. At the same time, owing to the advantages, connected with its mobility there is a wide field of WPC application both in CIS and abroad.

However, the possibility and efficiency of use of WPC, like of any other WPS, are defined not only by its technical perfection, but also by the wind use as the power source with optimum specifications for the given WPS type. Thus, the estimation of wind-power potential of atmospheric circumterrestrial layer, and of WPC power productivity in the regios of its possible use is the necessary stage of the work during both its development and realization.

This Atlas is intended for revealing the regions, appropriate for WPC high effective use. The first part of the Atlas containes the data on WPC "ZHAVORONOK" specifications. The second part is devoted to development of the maps representing the annual and dayly generation of electricity in the prescribed regions of the Globe. The last part of the Atlas represent the results of analysis of WPC power and economic efficiency on the territory of CIS and of foreing countries, and also several recommendations on achievement of WPC use maximum efficiency.

During the Atlas development, the computer information technique of definition of WPSs productivity on the territory of CIS was used; this technique was developed by the specialists of the State Research Centre of N.Zhukovsky Central Institute of Aerohydrodynamics.

Among numerous developments of WPS of traditional stationary type, there is remarkable SCWPS "ZHAVORONOK", which is self-contained mobile wind-power complex designed for parallel work of 30 W WPS and automatic Disel set of the same power.

During this work a study of wind-power resources of CIS was carried out, the most perspective regions for SCWPS "ZHAVORONOK" placing were revealed, and the ways of this WPS most effective use were offered. The estimates of the economic effect after this WPS introduction were given as well.

The main stages of the works were as follows:

1. The estimate of the wind resources in prescribed regions of CIS and the Globe. During this stage the authors formed the archieve of data concerning meteorologic and aerologic sounding of the atmosphere boundary layer in the prescribed regions. This archive may be read by computer. On the basis of the long-term meteoobservations, the local statistic models of the most probable characteristics of the wind were developed. These characteristics are: annual, daily and long-term change of the wind velocity and direction, velocity high-altitude distribution, probability of calms and the wind velocities within the prescribed limits defined regarding altitude and relief peculiarities of the region. Using statistic and hudrodynamic modelling techniques, wind-power potential of the atmosphere near-land layer, and the other characteristics were studied, regarding the topographic peculiarities of the region.
2. Maximum reliable estimate of power efficiency of SCWPS "ZHAVORONOK" in prescribed regions of CIS (mean annual and mean monthly elaborations, coefficient of the rated power use, probability of the calm and storm downtime and SCWPS operation at varions power regimes).
3. Preparation and analysis of the maps of mean monthly generation of electricity by SCWPS "ZHAVORONOK" for prescribed regions.
4. Working out the recommendations on optimum placing and effective use of SCWPS "ZHAVORONOK" in prescribed regions on the territory of CIS.
5. Preparation of make-up page of the Atlas of SCWPS "ZHAVORONOK" power productivity on the territory of CIS and some other regions of the Globe.

The results obtained make it possible to affirm that several regions of CIS have sufficient wind resources, which enable to set up a relatively inexpensive (compared to the traditional sources) power supply for factories and agricultural objects, for residential districts and special objects on the basis of SCWPS "ZHAVORONOK" in the nearest 2-3 years.

The most effective these variants might be for far regions and regions difficult of access, where it is difficult to set up power lines or fuel supply.

In our opinion, the results obtained are of practical interest during SCWPS power and economic effeciency, during the solution of the problems of SCWPS ratronal placing and use on the territory of our country and abroad, and also during the choice of optimum variants of power supply based on SCWPS basis.

The benefit of the work results at the combined analysis of the sets generation and power efficiency, and of the regimes of needed nerwork power load, is undoubtful. Analysis of use of the possible fields of SCWPS of "ZHAVORONOK" type demonstrates high efficiency of their use for power supply dirrected to agricultural production, farms, private possessions, geologic parties, drilling rigs arctic and antarctic expeditions, temporary settlements, zones of natural calamities and industrial catastrophes, regions with instable social and political situation leading to power supply stoppage, etc.

The most important consumer properties of SCWPS are as follows: their transportation and keeping convenience, drive in setting up at minimum time and means speding for the place preparation and for installation, reliability and reservation of the power source at the wind lack and at the accidents, and automation level together with the use convenience.

Review of the serial and being developed now WPSs showed that neither in our country nor abroad such wind-power complexes, which meet the above mentioned requirements, are not designed yet.

The world practice demonstrated high efficiency of container transportation of various cargos. Nowadays the technical means necessary for such container transportation (container stock, hoist and transportation means, storage facilities, railway and motor-car stock, river and sea vessels, aviation means, service points, etc) have been provided. Geometry and rigging-fastening units are standardized by ISO. SCWPS placing in 12-meter container, corresponding to the universal container 1A (State Standard 18477-79) regarding its dimesions and rigging-fastening units, will make it possible to solve rationally all the problems of its transportation.

WPS composition.

The self-contained mobile wind-power complex "ZHAVORONOK" has two power sets with nominal power 30 kw each to provide uninterrupted power supply for the uses:

- wind-power set (WPS);
- automated Diesel set (ADS).

The main regime of operation is the WPS generator work. ADS is a reserve source. If there is such a need, the WPS and ADS parallel operation is foreseen. WPS is constructively arranged in the container, the dimensions and transport units of which correspond to the requirements of ISO and to the State Standart 18477-79.

WPS consists of:

- rotor (windwheel), including blades and bush;
- nacelle, arranged on the turning platform and including step-up gear, induction generator of acceleratated current, high-speed connecting shaft brake, actuators and sensor;
- folding tower.

ADS consists of:

- Diesel engine;
- accelarated current generator;
- equpment and controll desk.

To provide the operative preparation of the complex for the work and to provide the complex functioning, the WPS container has a system of deployment and levelling, including outriggers with hanging out and levelling jacks, hydrocylinders for the tower lifting, for container doors opening, for extending the landing gears, for outriggers output, and also it has hydroequpment and pumping station.

For convenience of WPS setting up, the equipment includes the control desk for out-of-door operatioin.

The inner and outer loads power supply is provided by the direct current source arrangement in the container; there are also arrange the cable systems and electro-technical equipment including charging unit, batteries, protective units, ect.

All the systems of power supply and consumption are combained in current distribution unit, which includes the joints of the outer loads. The sets and systems control of the complex according to infomational sensors data are provided by the control system of WPS, consisting of: change unit, central computer, amplifier-transducer, control panel and control. The general block diagram of WPS composition is given.

WPS main characterisics.

Destination: Power supply for small agricultural production, private buildings, polar stations, drilling rigs, and for technical means during rescue works, ect.

The peculiarities:

- Keeping and transportation like the "container transportation";
- minimum time for WPS preparation for operation;
- no need in capital outlays for preparation of the WPS installation place;
- WPS high automation.

Maximum power, kW:

WPS	30
ADS	30
during WPS and ADS parallel operation	60
Current	
accelerated, voltage, V	380/220
direct, 2 kW, voltage, V	28.5
The wind vilocites operating range, m/s	4.9 - 25
The wind rated velocity, m/s	9
Windwheel: diameter, m	15
the nunber of blades	3
rotation frequency, rpm	80
Generator:	
type	induction
rotation frequency, rpm	300
Equipped WPS weight, t	14
WPS dimensions, m	
in transportation state	12.192x2.438x2.438 (contained of A type)
windwheel axis height above the ground, m	12
Time of setting up from trasportation state to operational one, min no more than	10
Resourse, hours	15000
Use duration, yeas	15

Independence:
during operation between maintenance periods, with refuelling, 1 year
-with Diesel generator fuel reserve 10 days.
WPS restationing: by helicopter MI-26, by container truck, by aeroplane IL-76, by river or sea vessels, by railway.
WPS price, mln roubles (1992) 14
Beginning of serial production 1994 - 95.

The complex dimensions correspondence to the 12 m universal container provides its trasportation convenience, and protection from external influence. High level of automation of wind-power complex, achieved by up-to-data technologies used by domestic industry, provides its fitting up efficiency at minimum preparation of the place; convenient use and service together witn relative inexpensiveness and high reliability of the complex are its peculiarities.

The SCWPS advantages compared to stationary WPS

1. Fast and inexpensive fitting up and dismantling.
2. Use simplicity.
3. Protection from external influence.
4. Mobility, possibility of fast change of the location.
5. Possibility of use in regions difficult of access.

The set advantages may be realized by the following ways:

1. Use of the annual and seasonal wind motion, and the set mounting at the places with maximum wind activity (the passes and hills windward sites, reservoirs coasts, etc.).
2. Dismounting and evacuation in case of the natural calamities (storms, floods, fires, and erthquatres).
3. Power supply for national economic objects (nomadic cattle- brecding, remote construction works, fishing seasons and lumberings, natural resources extractions and laying in, geological prospecting, and elecrto-capacious watch works).

STROMERZEUGUNG AUS WINDENERGIE IN DEUTSCHLAND - POTENTIALE, KOSTEN UND IMPLEMENTIERUNGSSTRATEGIEN -

Andreas Wiese*, Martin Kaltschmitt**
*Institut für Energiewirtschaft und Rationelle Energieanwendung (IER), Universität Stuttgart
Heßbrühlstr. 49a, Tel.: 0711/78061-0, W-7000 Stuttgart 80
**Kuratorium für Technik und Bauwesen in der Landwirtschaft (KTBL),
Bartningstr. 49, Tel.: 06151/7001-0, W-6100 Darmstadt

Aufgrund ihres hohen Potentials und aufgrund der Kostengünstigkeit bei gleichzeitiger geringer Umweltbelastung könnte die windtechnische Stromerzeugung in ein bis zwei Jahrzehnten in Deutschland einen ähnlichen Beitrag zur Stromerzeugung liefern wie derzeit die Wasserkraft. Vor diesem Hintergrund werden bei dieser Untersuchung zunächst die technischen Potentiale einer Stromerzeugung aus Windenergie bestimmt. Dazu werden die für eine Windkraftnutzung besonders prädestinierten Gebiete identifiziert und um die technisch nicht nutzbaren Flächenanteile vermindert. Ausgehend von diesen für eine Windkonverterinstallation geeigneten Flächen errechnen sich unter Berücksichtigung des lokal unterschiedlichen Windenergieangebots die Stromerzeugungspotentiale. Werden sie zusätzlich um die Netz- und Speicherverluste vermindert, ergibt sich daraus letztlich das korrespondierende Endenergiepotential. Anschließend werden die Kosten analysiert, mit denen diese Windstromerzeugung verbunden ist. Darauf aufbauend wird unter Berücksichtigung moderater Kostendegressionen eine Implementierungsstrategie der Windkraftnutzung bis zum Jahr 2020 untersucht; sie ist bis zum Ende des Betrachtungszeitraumes durch eine installierte Leistung von 3 300 MW, eine Stromerzeugung von rund 8 600 GWh/a und mittlere Stromgestehungskosten von 11 Pf/kWh gekennzeichnet.

1 EINLEITUNG

Aufgrund der zunehmenden Umweltverschmutzung und wegen der Gefahr einer globalen Klimaveränderung, verursacht durch die Emissionen klimarelevanter Spurengase, wird verstärkt die Nutzung umweltfreundlicher und klimaneutraler Techniken insbesondere zur Stromerzeugung gefordert. Eine solche Möglichkeit stellt die Nutzung der Windenergie dar, die - außer durch vergleichsweise hohe technische Potentiale - an Standorten mit hohem Windenergieangebot auch durch relativ niedrige Stromgestehungskosten gekennzeichnet ist. Deshalb wächst - nicht nur aufgrund der staatlichen Förderung (u. a. 250 MW-Windprogramm) - die jährlich installierte Anlagenzahl bzw. die installierte windtechnische Leistung sowohl weltweit als auch in Deutschland relativ schnell. In diesem Kontext ist es das Ziel dieses Beitrages, die grundsätzlichen Möglichkeiten und Grenzen einer großtechnischen Nutzung der Windenergie innerhalb der Bundesrepublik Deutschland zu analysieren.

2 POTENTIALE

Aufgrund des regional sehr unterschiedlichen Windenergieangebotes ist zur Bestimmung der windtechnischen Stromerzeugungspotentiale eine örtliche hoch aufgelöste methodische Vorgehensweise erforderlich, die zunächst die Analyse der windtechnisch nutzbaren Flächenpotentiale für jeden Kreis in Deutschland ermöglicht. Daraus können dann die korrespondierenden Stromerzeugungspotentiale auf der Grundlage gemessener Ganglinien der stundenmittleren Windgeschwindigkeit und den Leistungskennlinien für typische Referenzanlagen bestimmt werden. Unter Berücksichtigung von Netz- und Speicherverlusten errechnen sich daraus die Endenergiepotentiale.

2.1 Flächenpotentiale

Zur Bestimmung der Flächenpotentiale wird ausgegangen von der Windgeschwindigkeitsverteilung innerhalb Deutschlands, bei der einzelnen Gebieten die dort vorherrschenden jahresmittleren Windgeschwindigkeitsniveaus zugeordnet werden. Demnach kommen vor allem die küstennahen Zonen in Norddeutschland für eine Windkraftnutzung in Frage [3]. Zusätzlich existieren im Binnenland in den Mittelgebirgen (z. B. Schwäbische Alb, Erzgebirge, Harz) Gebiete mit hohen und damit windtechnisch nutzbaren Windgeschwindigkeiten. Insgesamt sind in Deutschland eine Gebietsfläche von ca. 21,1 Mio. ha durch jahresmittlere Windgeschwindigkeiten zwischen 3 und 4 m/s, ca. 4,7 Mio. ha durch mittlere Geschwindigkeiten zwischen 4 und 5 m/s, ca. 0,7 Mio. ha durch 5 bis 6 m/s und knapp 0,1 Mio. ha durch mehr als 6 m/s gekennzeichnet.

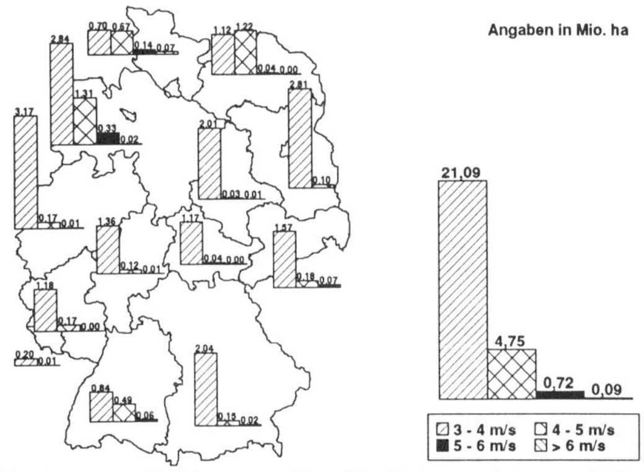

Abb. 1: Gebiete mit ähnlichen jahresmittleren Windgeschwindigkeiten in Deutschland /1/

Diese Flächenpotentiale, d. h. die Flächen, auf denen im Jahresdurchschnitt ein gewisses Windgeschwindigkeitsmittel gegeben ist, werden für jeden Stadt- und Landkreis in der Bundesrepublik Deutschland bestimmt. Die für die einzelnen Bundesländer aggregierten Werte zeigt Abb. 1 [1].

Die in Abb. 1 dargestellten Gebiete sind allerdings aufgrund einer Vielzahl von Restriktionen nicht vollständig für eine Windkonverterinstallation nutzbar. Werden diese entsprechenden Restriktionen (vgl. [1]) - wiederum unter Beachtung der regionalen Unterschiede - berücksichtigt, kann ebenfalls für jeden Kreis das für eine Windkraftanlagenaufstellung technisch geeignete Flächenpotential ermittelt werden. Werden diese auf Kreisebene erhobenen Werte für die Gebietsfläche der Bundesrepublik Deutschland aggregiert und wird davon ausgegangen, daß nur Flächen mit jahresmittleren Windgeschwindigkeiten von mehr als 4 m/s (bezogen auf 10 m Höhe über Grund) für eine Konverterinstallation in Frage kommen, ergibt sich eine Fläche von ca. 5,6 Mio. ha. Davon sind der weitaus größte Teil (ca. 85 %) durch jahresmittlere Windgeschwindigkeiten zwischen 4 und 5 m/s, etwa 13 % durch jahresmittlere Geschwindigkeiten zwischen 5 und 6 m/s und lediglich ca. 2 % durch mittlere Windgeschwindigkeiten von mehr als 6 m/s gekennzeichnet.

2.2 Stromerzeugungspotentiale

Aus den dargestellten Flächenpotentialen können die korrespondierenden Stromerzeugungspotentiale abgeleitet werden. Dazu sind zunächst die auf den technisch verfügbaren Flächen installierbaren

Anlagenanzahlen unter Berücksichtigung eines bestimmten Abstandes zur Minimierung der Abschattungsverluste zwischen einzelnen Konvertern ermittelt. Werden zusätzlich als Referenztechnologien drei derzeit markttypische Konverter (Kleinanlagen mit 80 kW, mittlere Konverter mit 300 kW und Großanlagen mit 1 200 kW) zugrundegelegt, kann ausgehend von der installierbaren Anlagenanzahl die für jeden Kreis installierbare Windkraftleistung bestimmt werden.

Zur Bestimmung der diesen installierbaren Anlagen korrespondierenden Stromerzeugungspotentiale stehen an 39 Standorten gemessene Ganglinien der stundenmittleren Windgeschwindigkeiten zur Verfügung. Aufbauend auf dieser Datenbasis wurde eine methodische Vorgehensweise entwickelt [1], die es ermöglicht, für jeden Kreis und jedes darin vorkommende Windgeschwindigkeitsmittel eine Ganglinie der stundenmittleren Windgeschwindigkeit im Jahresverlauf zu simulieren. Mit den technischen Daten bzw. dem systemtechnischen Verhalten der Referenzanlage kann dann die korrespondierende Stromerzeugung für jeden Stadt- und Landkreis in der Bundesrepublik Deutschland bestimmt werden.

Abb. 2 zeigt die sich auf der Basis dieser Vorgehensweise ergebenden technischen Stromerzeugungspotentiale in einzelnen Bundesländern und in der Bundesrepublik Deutschland für jahresmittlere Windgeschwindigkeiten mit mehr als 4 m/s, bezogen auf eine Meßhöhe von 10 m über Grund. Je nach Anlagentechnologie liegt das so berechnete Stromerzeugungspotential zwischen 104 und 128 TWh/a. Besonders deutlich wird in der Darstellung auch die regionale Verteilung dieses Windpotentials in Deutschland. Je nach Konvertertechnologie beträgt das Stromerzeugungspotential in den drei Küstenländern zwischen 85 und 90 % des Gesamtpotentials in Deutschland. Von den Binnenländern weisen Baden-Württemberg (3,1 bis 3,8 TWh/a) und Sachsen (2,0 bis 2,6 TWh/a) die größten Potentiale auf.

2.3 Endenergiepotentiale

Die in Abschnitt 2.2 dargestellten Stromerzeugungspotentiale ergeben sich unter der Annahme, daß auf sämtlichen windtechnisch nutzbaren Flächenpotentialen Windkraftanlagen installiert werden. Die daraus

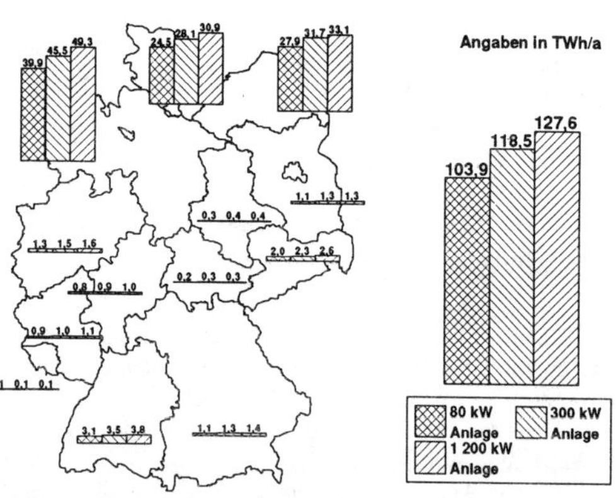

Abb. 2: Technische Erzeugungspotentiale einer windtechnischen Stromerzeugung in Deutschland (jahresmittlere Windgeschwindigkeit mehr als 4 m/s) /1/

resultierenden Endenergiepotentiale ergeben sich unter Berücksichtigung netz- und bedarfsseitiger Restriktionen. Dazu werden im folgenden vier verschiedene Ansätze diskutiert, die zu jeweils unterschiedlichen Endenergiepotentialen führen, die in Abb. 3 dargestellt sind. Innerhalb der dadurch aufgespannten Bandbreite bewegen sich letztlich die tatsächlichen Endenergiepotentiale.

Ansatz I. Bei dem Ansatz "Maximale technische Nutzung" wird davon ausgegangen, daß die zur Verfügung stehenden Flächenpotentiale vollständig mit Windkraftanlagen belegt werden. Dann ist aber - zumindest in einigen Küstenregionen - die Durchdringung des Netzes mit fluktuierender Energie so groß, daß bei einem hohen Windenergieangebot ein Teil der erzeugten elektrischen Energie zwischengespeichert werden muß.

Unter diesen Randbedingungen entsprechen die technischen Erzeugungspotentiale in der Bundesrepublik Deutschland einem

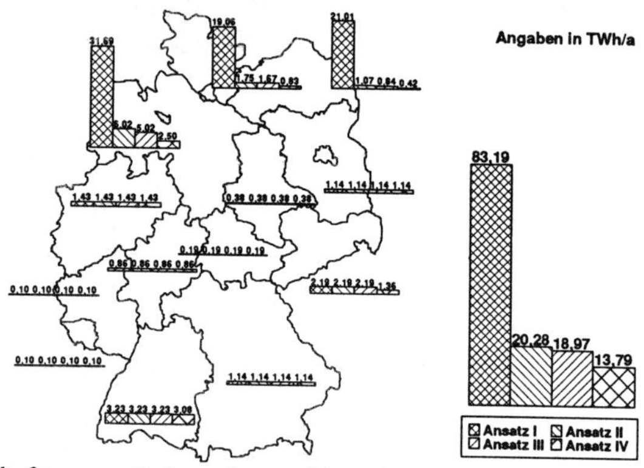

Abb. 3: Endenergiepotentiale einer Stromerzeugung aus Windkraft in Deutschland /1/

technischen Endenergiepotential von rund 83 TWh/a (vgl. Abb. 3). Dazu müßten ca. 61 GW an Windkraftleistung vornehmlich an der deutschen Nord- und Ostseeküste installiert werden. Dies entspricht - nimmt man die zugehörigen Erzeugungspotentiale als Ausgangspunkt - einer mittleren Vollaststundenzahl von rund 1 900 h/a. Ursache für diese für die Windkraftnutzung vergleichsweise geringe Vollaststundenzahl ist die Tatsache, daß ein Großteil des Potentials an Standorten mit jahresmittleren Windgeschwindigkeiten zwischen 4 und 5 m/s verwirklicht wird. In den Bundesländern, in denen das Windenergieangebot und damit die technischen Erzeugungspotentiale besonders groß sind, ist die Elektrizitätsnachfrage vergleichsweise gering; dies gilt im wesentlichen für Schleswig-Holstein, Niedersachsen und Mecklenburg-Vorpommern. Hier reduzieren sich die aus den technischen Erzeugungspotentialen berechneten Endenergiepotentiale besonders deutlich. Dies liegt darin begründet, daß es unter den hier gesetzten Randbedingungen in diesen Regionen zu sehr hohen Durchdringungen des Netzes mit fluktuierender elektrischer Energie kommt; dadurch tritt hier regional ein sehr hoher Speicherbedarf auf.

Ansatz II. Beim Ansatz "Kein regenerativ bedingter Speicherbedarf" wird kein zusätzlicher Speicherbedarf zugelassen. Grundsätzlich kann dann nur von einer 10 bis 20 %igen Durchdringung des Netzes mit fluktuierender Energie ausgegangen werden [6]; hier wird deshalb von einer mittleren und regional begrenzten Durchdringung von 15 % ausgegangen. Unter diesen Rahmenannahmen ergibt sich für Deutschland ein gesamtes technisches Endenergiepotential von ca. 20 TWh/a (vgl. Abb. 3); zu seiner Erschließung müßte eine windtechnische Leistung von knapp 10 GW installiert werden. Die mittlere Vollaststundenzahl liegt bei rund 2 100 h/a; sie ist damit höher als bei Ansatz I, da hier nur die Teile der technischen Erzeugungspotentiale ausgenutzt werden, die durch das höchste jahresmittlere Windangebot gekennzeichnet sind.

Ansatz III. Bei dem Ansatz "Minimale Last" wird davon ausgegangen, daß bei einer bestimmten Anzahl installierter Anlagen die größte momentane windtechnische Leistung gerade die Summe der Nennleistungen der Windkraftanlagen ist. Obwohl kurzfristig durchaus über die Nennleistung hinausgehende Leistungen von den Windkraftanlagen abgegeben werden können, kann bei einer großräumigen Verteilung der Konverter und der dadurch bedingten Vergleichmäßigung als maximale theoretisch ins Netz eingespeiste Leistung die Summe der Nennleistungen aller Windkraftanlagen unterstellt werden. Damit in diesem Fall kein Speicherbedarf auftritt, muß die ins Netz eingespeiste Leistung maximal gleich der zu diesem Zeitpunkt auftretenden elektrischen Last sein. Wird von der minimalen jährlichen Last ausgegangen und unterstellt, daß es nur innerhalb eines Bundeslandes zum Ausgleich der windtechnischen Stromerzeugung kommen kann, läßt sich die installierbare Windkraftanlagenleistung bestimmen, bei der auf keinen Fall Speicherbedarf auftritt.

Diesen Überlegungen zufolge würde sich in Deutschland ein Endenergiepotential von ca. 19 TWh/a ergeben (vgl. Abb. 3); dazu müßten ca. 9,3 GW an windtechnischer Leistung installiert werden. Da die installierte Leistung bzw. der Potentialausnutzungsgrad bei diesem Ansatz in etwa im Bereich der vergleichbaren Werte des Ansatzes II liegt, bewegen sich auch die mittleren Vollaststundenzahl in einer vergleichbaren Größenordnung.

Ansatz IV. Im Gegensatz zu dem Ansatz "Minimale Last und konventionelle Grundlast" wurde bei dem zuvor diskutierten Ansatz unterstellt, daß zum Zeitpunkt des Auftretens der minimalen Last sämtliche konventionellen Kraftwerke abgeschaltet werden können. Dies ist jedoch im Hinblick auf die derzeitige Stromversorgungsstruktur wenig realistisch, da unter diesen Bedingungen die Spannungs- und Frequenzstabilität des Netzes in der gegenwärtigen Form u. U. nicht sichergestellt werden kann. Deshalb wird hier unterstellt, daß ständig rund 50 % der nachgefragten elektrischen Energie durch den konventionellen Kraftwerkspark gedeckt werden müssen.

Unter diesen Randbedinguingen liegt das gesamte Endenergiepotential einer windtechnischen Stromerzeugung bei ca. 14 TWh/a (vgl. Abb. 3); die korrespondierende in Windkraftanlagen installierte elektrische Leistung bewegt sich dann bei etwas mehr als 7 GW. Mit ca. 2 020 h/a liegt hier die mittlere Vollaststundenzahl unterhalb der Vollaststundenzahl der Ansätze II und III, da ausschließlich die Potentiale in den Küstenländern nur z. T. genutzt werden können, während sich die konventionellen Grundlast von 50 % der minimalen Last in den Binnenländern aufgrund der dort gegebenen geringen Potentiale kaum potentialmindernd auswirkt. Damit verringert sich gegenüber Ansatz II in diesem Ansatz der Anteil der windtechnischen Leistung, der in Windgebieten mit hohen jahresmittleren Windgeschwindigkeiten installiert werden kann; dies führt zu einer Verringerung der mittleren Vollaststundenzahl.

Vergleich. Die vier diskutierten Ansätze stellen denkbare technische Endenergiepotentiale einer windtechnischen Stromerzeugung in der Bundesrepublik Deutschland dar. Dabei ist Ansatz I durch das mit Abstand höchste Endenergiepotential gekennzeichnet. Bei diesem Ansatz werden - zumindest regional - zusätzliche Speicher in zum Teil erheblichen Umfang benötigt. Auch die Struktur des gegenwärtigen Netzes der öffentlichen Versorgung selber bedarf unter den diesem Ansatz zugrunde liegenden Randbedingungen erheblicher Änderungen (Netzverstärkungen, eigene "Windnetze", zusätzliche Umspannstationen etc.). Diese Überlegungen sind daher eher als eine theoretische Variante zu betrachten, die das windtechnisch gewinnbare Endenergiepotential unter ausschließlicher Berücksichtigung der Flächenverfügbarkeitsrestriktion angibt. Bei diesem Ansatz könnten ca. 19 %, bezogen auf den elektrischen Endenergieverbrauch in der Bundesrepublik Deutschland im Jahr 1991, durch Windenergie gedeckt werden. Deutlich geringer, untereinander aber sehr ähnlich, sind die möglichen Beiträge zur Deckung des elektrischen Endenergiebedarfs bei den Ansätzen II, III und IV. Die aus diesen Überlegungen resultierenden Endenergiepotentiale entsprechen einem Anteil von ca. 3 (Ansatz IV) bis 5 % (Ansatz II) am gegenwärtigen Endenergieverbrauch an Strom. Die diesen Ansätzen zugrundeliegende Leistung von ca. 8 GW in Deutschland ließe sich aber ohne erhebliche Umstrukturierungen im konventionellen Kraftwerkspark und im Netz in die öffentliche Versorgung integrieren.

3 KOSTEN

Die für die Berechnung der spezifischen Stromgestehungskosten einer Windkraftnutzung relevanten Gesamtinvestitionen setzen sich zusammen aus den Anlagenaufwendungen ab Werk (2 200 bis 7 900 DM/kW bei Anlagen zwischen 20 und 90 kW und 1 600 bis 3 000 DM/kW bei Konvertern mit 90 bis 500 kW), aus den Kosten für Transport und Montage (ca. 5 bis 6 % der Anlageninvestitionen ab Werk), den Kosten für das Fundament (ca. 6 % der Kosten ab Werk bei Standardfundamenten) und die Netzanbindung (10 bis 25 % der Investitionen ab Werk) sowie den sonstigen Kosten (z. B. Planungskosten, Wegekosten; ca. 15 % der Investitionen ab Werk). Die jährlich anfallenden Betriebskosten ergeben sich aus den Wartungs- und Instandhaltungsaufwendungen, den Versicherungskosten und u. U. den Pachtkosten für das Aufstellungsgelände (ca. 1 bis 3 % der gesamten Investitionen pro Jahr) [2].

Aus den Investitionen und Betriebskosten können mit Hilfe der Annuitätenmethode die über die Abschreibungsdauer der Windkraftanlage konstanten mittleren barwertigen Kosten berechnet werden (Abschreibungsdauer entsprechend der technischen

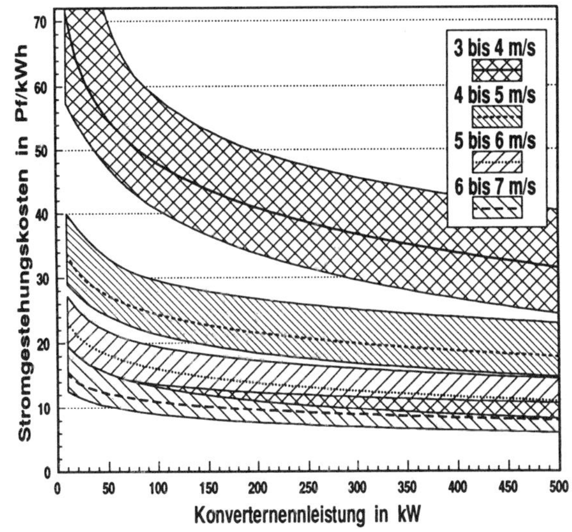

Abb. 4: Spezifische Stromgestehungskosten einer Stromerzeugung aus Windenergie

Lebensdauer 20 Jahre, realer Zinssatz 4 %). Abb. 4 zeigt die resultierenden Stromerzeugungskosten in Abhängigkeit der installierten Leistung im Bereich von 50 bis 500 kW für unterschiedliche Windgeschwindigkeiten. Demnach sind die spezifischen Kosten mittelgroßer Windkraftanlagen mit einer installierten Leistung zwischen 100 und 500 kW deutlich geringer als die kleinerer Anlagen. Ursache sind zum einen die niedrigeren spezifischen Investitionen, zum anderen die meist größeren Turmhöhen, die am gleichen Standort eine höhere jahresmittlere Windgeschwindigkeit in Nabenhöhe und damit eine höhere Energieausbeute bedingen.

Verglichen mit dem Aufwand für elektrische Energie, die der Letztverbraucher im Durchschnitt zu zahlen hat, ist somit die Windkraftnutzung an günstigen Standorten kostengünstiger als der Tarif. Aufgrund des unregelmäßigen und wenig vorhersehbaren Windenergieangebots muß jedoch im Regelfall ein Tarifkunde eine Mischkalkulation zwischen dem Einspeisetarif und den vermiedenen Strombezugskosten zugrunde legen. Unter diesen Bedingungen ist die Windkraftnutzung in deutlich weniger Fällen und damit nur noch an Standorten mit hohen mittleren Windgeschwindigkeiten und sonstigen günstigen Rahmenbedingungen kostengünstiger. Werden die Stromgestehungskosten aus Windenergie demgegenüber aus der Sicht eines Versorgungsunternehmens gesehen, müssen sie im wesentlichen mit den variablen Kosten der Mittellastkraftwerke konkurrieren, da die Stromerzeugung aus Windkraft nur durch einen geringen Kapazitätseffekt gekennzeichnet ist (vgl. [7]) und im wesentlichen Mittellast verdrängt (vgl. [8]). Damit ist unter diesen Bedingungen eine Stromerzeugung aus Windkraft nur noch in deutlich weniger Fällen kostenkünstiger.

4 IMPLEMENTIERUNGSSTRATEGIEN

Aufgrund der aufgezeigten Potentiale und vor allem wegen der vergleichsweise geringen Stromgestehungskosten stellt die Windkraft eine Option dar, umweltverträglich und ressourcenschonend elektrische Energie in einer energiewirtschaftlich relevanten Größenordnung bereitzustellen. Vor diesem Hintergrund soll im folgenden eine mögliche Implementierungsstrategie einer Windkraftnutzung in der Bundesrepublik Deutschland definiert und analysiert werden.

Werden die zukünftigen Möglichkeiten der Windkraftnutzung untersucht, ist davon auszugehen, daß sich die windtechnischen Potentiale in Zukunft nicht substantiell verändern werden; trotz des weiterhin geringfügig steigenden Flächenverbrauchs ist aufgrund einer zukünftig besseren Kenntnis und Erschließung von einer näherungsweisen Konstanz der technischen Potentiale auszugehen. Dies gilt aber nicht für die Kosten, die für die Stromerzeugung aus

Windenergie aufzubringen sind; wegen der technologischen Weiterentwicklungen kann es hier zu erheblichen Veränderungen der Investitionen in der Zukunft kommen. Deshalb wird hier davon ausgegangen, daß sich auch mittelfristig der Trend zu Horizontalachsenkonvertern nicht ändern wird. Technische Weiterentwicklungen sind im wesentlichen nur bei den Windkraftanlagen mit einer installierten elektrischen Leistung von einem Megawatt und mehr zu erwarten. Insgesamt ist aber nicht davon auszugehen, daß nennenswerte Wirkungsgradverbesserungen bei den Windturbinen erreichbar sind; lediglich die technischen Verfügbarkeiten der Großwindkraftanlagen werden deutlich zunehmen. Auch kann nur bei den Großwindanlagen von größeren Kostendegressionen ausgegangen werden [4, 5].

Bei dieser Implementierungsstrategie wird unterstellt, daß - ausgehend von einer als realistisch einzuschätzenden Windkraftleistung von 300 MW im Jahr 1995 - im Zeitraum von 1995 bis 2000 jeweils 100 MW/a, von 2001 bis 2010 jährlich 150 MW/a und von 2010 bis 2020 je 100 MW/a zusätzliche Leistung installiert werden. Im Jahr 2020 wären dann innerhalb der Gebietsgrenzen der Bundesrepublik Deutschland rund 3 300 MW an windtechnischer Leistung installiert. Weiterhin wird davon ausgegangen, daß zunächst die Potentialanteile ausgenutzt werden, die durch das höchste mittlere Windenergieangebot gekennzeichnet sind. Mit zunehmender Erschöpfung dieser optimalen Standorte erfolgt - aufgrund der weiteren Anlageninstallation - ein allmählicher Übergang auch auf weniger windstarke Standorte.

Tab. 1 zeigt die für die Bundesrepublik aggregierten Ergebnisse dieser untersuchten Implementierungsstrategie; für die unterstellte installierte Gesamtleistung sind die korrespondierende jährliche Stromerzeugung und die zugehörigen mittleren spezifischen Stromgestehungskosten dargestellt. Demnach würde sich am Ende des Betrachtungszeitraumes bei einer dann installierten Leistung von 3 300 MW eine jährliche Stromerzeugung von ca. 8 569 GWh/a bei mittleren spezifischen Stromgestehungskosten von rund 11 Pf/kWh ergeben. Unter diesen Rahmenannahmen wäre das Flächenpotential mit jahresmittleren Windgeschwindigkeiten von mehr als 6 m/s

Tab. 1: Implementierungsstrategie einer windtechnischen Stromerzeugung in Deutschland bis zum Jahr 2020

Jahr[1]	gesamte installierte Leistung	Stromerzeugung	Stromgestehungskosten[2]	vermiedene CO_2-Emissionen[3]
1995	300 MW	906 GWh/a	11 Pf/kWh	0,519 Mio. t
2000	800 MW	2 405 GWh/a	10 Pf/kWh	1,377 Mio. t
2005	1 550 MW	4 630 GWh/a	10 Pf/kWh	2,651 Mio. t
2010	2 300 MW	6 381 GWh/a	10 Pf/kWh	3,654 Mio. t
2015	2 800 MW	7 475 GWh/a	10 Pf/kWh	4,280 Mio. t
2020	3 300 MW	8 569 GWh/a	11 Pf/kWh	4,907 Mio. t

1 jeweils am Jahresende.
2 Abschreibungsdauer 20 Jahre, Zinssatz 4 %.
3 bezogen auf den gegenwärtigen Kraftwerksmix in der Bundesrepublik Deutschland.

vollständig und das Potential mit jahresmittleren Windgeschwindigkeiten zwischen 5 und 6 m/s zu ca. 20 % ausgeschöpft. Dies würde - wird der gegenwärtige Kraftwerksmix zugrunde gelegt - einer Reduktion der energiebedingten Kohlendioxidemissionen aus der Strombereitstellung von rund 50,5 Mio. t im Jahr 1995 und etwa 4,9 Mio. t im Jahr 2020 entsprechen.

5 SCHLUSSBETRACHTUNG

Ziel der vorliegenden Untersuchung ist eine Analyse der Potentiale und Kosten sowie der mittelfristigen Möglichkeiten und Grenzen der Windkraftnutzung in der Bundesrepublik Deutschland. Dabei hat sich gezeigt,

- daß auf den windtechnisch nutzbaren Flächen bei einer ausschließlichen Installation von 80 kW-Anlagen ein Stromerzeugungspotential von rund 104 TWh/a, von 300 kW-Konvertern von ca. 119 TWh/a und von 1 200 kW-Anlagen von etwa 128 TWh/a erreicht werden kann; damit sind - wird als oberstes Ziel eine Maximierung des Energieertrags angestrebt - Großanlagen zu bevorzugen;
- daß unter zusätzlicher Berücksichtigung netz- und bedarfsseitiger Restriktionen sich technische Endenergiepotentiale zwischen ca. 14 und 83 TWh/a ergeben; die untere Grenze ergibt sich unter der Annahme, daß ständig eine konventionelle Grundlaststromerzeugung von der Hälfte der minimalen Last zur Aufrechterhaltung der Netzstabilität gegeben sein muß und keine zusätzliche Speicherkapazität geschaffen wird; die obere Bandbreite errechnet sich bei einer vollständigen Integration der technischen Erzeugungspotentiale unter Berücksichtigung der dann gegebenen Speicher- und Netzverluste;
- daß die Stromgestehungskosten gegenwärtig bei mittleren Windgeschwindigkeiten zwischen 4 bis 5 m/s bei ca. 17 bis 30 Pf/kWh und bei mehr als 5 m/s bei ca. 8 bis 20 Pf/kWh liegen; sie sind damit - verglichen mit den Gestehungskosten der konventionellen und anderen regenerativen Energieträger - relativ kostengünstig;
- daß - und das hat die untersuchte Implementierungsstrategie gezeigt - damit die Windenergie eine vielversprechende Option darstellt, kostengünstig einen doch merklichen umweltverträglichen und ressourcenschonenden Beitrag zur Strombereitstellung in der Bundesrepublik Deutschland zu leisten.

LITERATUR

[1] Kaltschmitt, M.; Wiese, A. (Hrsg.)
Erneuerbare Energieträger in Deutschland - Potentiale und Kosten
Springer-Verlag, Berlin, Heidelberg, New York, Frühsommer 1993

[2] Wiese, A. u. a.
Vergleichende Kostenanalyse einer wind- und solartechnischen Stromerzeugung
Elektrizitätswirtschaft 91 (1992), 6, S. 292 - 299

[3] Deutscher Wetterdienst (Hrsg.)
Karte der Windgeschwindigkeitsverteilung in der Bundesrepublik Deutschland
Deutscher Wetterdienst, Offenbach a. M., 1991

[4] Molly, J. P.
Windenergie - Theorie, Anwendungen, Messung
Verlag C. F. Müller, Karlsruhe, 1990, 2. Auflage

[5] Deutsches Institut für Wirtschaftsforschung, Institut für Systemtechnik und Innovationsforschung (Hrsg.)
Kostenaspekte erneuerbarer Energiequellen in Deutschland und auf Exportmärkten
Deutsches Institut für Wirtschaftsforschung und Institut für Systemtechnik und Innovationsforschung, Berlin und Karlsruhe, 1992

[6] Nitsch, J.; Luther, J.
Energieversorgung der Zukunft
Springer-Verlag, Berlin, Heidelberg, New York 1990

[7] Kaltschmitt, M.; Voß, A.
Kapazitätseffekte einer Stromerzeugung aus Windkraft und Solarstrahlung
Elektrizitätswirtschaft 90(1991), 8, S. 365 - 371

[8] Fischedick, M.; Kaltschmitt, M.
Integration einer windtechnischen und photovoltaischen Stromerzeugung in den konventionellen Kraftwerkspark
8. Internationales Sonnenforum, Berlin, 1992

A SYSTEM FOR REGIONAL WIND RESOURCE ANALYSIS APPLIED TO SOUTH WEST ENGLAND

H Scholes
CSM Associates Ltd
Rosemanowes, Herniss, Penryn, Cornwall, UK

ABSTRACT: As part of a recent renewable energy resource study of the South West of England undertaken by the regional electricity company, SWEB Plc, and the Energy Technology Support Unit (ETSU) under contract to the UK Department of Trade and Industry, CSM Associates Ltd have developed a coherent suite of software to undertake onshore wind resource analysis on a regional basis. The software described here can be considered as a specialised form of geographical information system. It includes resource/constraint analysis, intelligent identification of potential wind farms, assessment of their impact on the distribution system, and their probable generation costs under a variety of financial scenarios.

1. BACKGROUND

Of various renewable energy resources investigated in the joint SWEB/ETSU regional study, onshore wind has probably the most immediate potential and is possibly the most significant. The study defined four levels of resource.

Technical Resource.
The energy resource which could be exploited if technical feasibility is the only limitation. All environmental, planning and other practical considerations are ignored.

Realisable Resource.
The realisable resource is the technical resource limited by basic physical constraints, such as roads, rivers, topology, settlements, suitability of specific sites for development, etc.

Accessible Resource.
The accessible resource is the resource which could be exploited if institutional constraints such as environmental and planning considerations are imposed in addition to the physical constraints. This resource definition must be used with caution as planning decisions and attitudes to development can be reversed. Development of some technologies may also be permitted in close proximity to restricted areas such as Sites of Special Scientific Interest.

Economic Resource.
The economic resource is the extent of the resource if economic constraints are placed on the resource in addition to all the previously defined constraints. Clearly, pertinent economic criteria must be defined in advance, i.e. the methods and parameters used in the evaluation. Unlike the other resource levels the economic resource is multi-valued, i.e. the resource size varies with generation cost.

A suit of analysis software was developed by CSM Associates to assess these different resource levels for the onshore wind resource. There are four stages in the assessment process; a resource/constraint analysis, a cluster analysis to identify potential wind farms, an electricity distribution grid impact analysis and finally an economic analysis of the remaining potential wind farms.

2.0 RESOURCE/CONSTRAINT ANALYSIS

2.1 Inputs

For an area the size of the SWEB region it was deemed most appropriate to compute the first three resource levels using a Geographical Information System (GIS) on a square kilometre grid basis. A GIS code called SITE was used for this purpose. SITE was originally developed by the Camborne School of Mines to estimate the size of Hot Dry Rock geothermal energy resources and was adapted to perform the onshore wind analysis. Ancillary codes for digitising, rasterisation and data format conversion were developed as required.

The starting point of the resource analysis is an estimate of the annual mean wind speed (AMWS) for each square kilometre in the region. This data is normally derived from computer models. In the case of the SWEB region it was provided by ETSU who had already undertaken wind speed modelling exercises using the NOABL computer model. The second major input is the type of wind turbine to be employed. In reality many types of turbine will be utilised but to simplify the problem a ''typical'' wind turbine was defined with characteristics averaged from a range of commercially available machines with the following specification:-

Installed capacity (C)	400kW
Rotor diameter (d)	34 m
Minimum spacing	333.3m (10d)
Availability (a)	95%
Average array efficiency (e)	90%
Ex-factory cost	£275/m^2 of swept area
Installation/commissioning cost	15% ex-factory cost
Site infrastructure cost	30% ex-factory cost
Grid connection cost	Site specific
Operating & Maintenance costs	1.5% capital cost
Reactive power/meterage charges	0.15 p/kWh
Land rental	2.5% gross revenue
Rates	£2.50/installed kW
Insurance	1% capital cost

The third major input to the resource analysis is the number of turbines that can be sited in each square kilometre. This is a function of some initial population and any constraints to siting. Each constraint reduces the number of turbines by the fraction of the square kilometre covered by that constraint. The constraints are organised to be mutually exclusive. This implies a constraint hierarchy. In general nationally designated constraints were assumed to take precedence over local ones. Each level of resource is the result of the application of an appropriate set of constraints to the previous level.

Not all the square kilometres in which wind turbines can be sited will be sufficiently productive, there will be some with wind speeds that are too low to generate power economically. In order to put the resource estimates into an appropriate context, two wind speed thresholds were adopted. These can be interpreted as "probably" economic (7.0 m/s and above) and possibly economic (6.5 m/s and above).

2.2 Technical Resource Estimation

The technical resource calculation assumes that the only constraint to siting is a minimum spacing criterion. The number of turbines that can be sited in any square kilometre (N) is limited only by wake interaction effects. The separation between machines is expressed in terms of the rotor diameter (d). For sites with no dominant wind direction a spacing of 10d is usual. For sites with highly directional wind fields spacings as low as 5d have been used. As the analysis can not be site specific at this stage, the 10d spacing was adopted, i.e. for a 33m diameter machine, N = 9.

The power from a single wind turbine is computed as follows:-

$$P = C\ L(V)\ e \quad \text{kW}$$

Where C is the installed capacity, e is the array efficiency (including other losses), and L(V) is the load factor for a specific annual mean wind speed V.

The power available from any single square kilometre

$$P = C\ L(V)\ e\ N/1000 \quad \text{MW}$$

and the annual energy output is

$$E = C\ L(V)\ e\ N\ a\ 8760/1000 \quad \text{MWh}$$

where a is the machine availability factor. For the technical resource N = 9 for all squares. The total technical resource is the sum of the resource in the individual squares.

2.3 Realisable Resource Estimation

The realisable resource is the technical resource limited by physical siting constraints. Two methods for defining more realistic initial populations of turbines were available at this point.

The first method was based on the technical maximum, 9 per sq km reduced by the following constraints, A roads, B roads, towns, villages, woodland, rivers, railways, quarries, airfields, electricity distribution grids, radio/TV transmitters and repeaters. Isolated dwellings are not taken into account with this method hence it overestimates the realisable resource.

The second method was derived from a turbine siting study undertaken by the Institute of Terrestrial Ecology (ITE) for ETSU. The study examined 32 different land classes taking random samples of each class throughout the UK landmass (12 per class) and assessed the number of turbines that could be sited on each sample. It took into account isolated dwellings and some linear features such as roads and railways. ITE land class data was available for the SWEB region and hence it was possible to define an initial turbine population in each square kilometre. The following additional constraints were then applied to estimate the realisable resource, towns, villages, woodland, quarries, airfields, electricity distribution grids, radio/TV transmitters and repeaters.

It should be noted that neither approach accounts for microwave communication links which are thought to be a significant problem to wind farm placement in the South West. It had proved impossible to acquire information about this constraint within the time scale of the study. The resource could therefore be over-estimated, however the effect of this constraint may be ameliorated by judicial positioning of wind turbines within any one farm.

2.4 Accessible Resource Estimation

The accessible resource is the realisable resource limited by institutional constraints. Social, environmental and aesthetic considerations will limit wind farm development in certain areas. Unlike the physical constraints institutional constraints are "soft" and therefore possibly amenable to change. Again the effect of each constraint is to factor the number of turbines that are present in each square kilometre. The following institutional constraints were considered:-

- Areas of Outstanding Natural Beauty (AONBs)
- Areas of Great Landscape Value (AGLVs)
- National and Regional Parks
- Heritage Coast
- National Trust land
- National and Local Nature Reserves (NNRs, LNRs)
- Sites of (Special) Scientific Interest (SSIs, SSSIs)
- Green belts
- Environmental Sensitive Areas (ESAs)
- Metalliferous mining, China clay and Quarrying consultation zones
- National, County and District planning considerations
- CEC Special Protected Areas (e.g. Ramsars)

Historic monuments were ignored in the analysis, as the data consisted primarily of cairns, standing stones and small scale features which were unlikely to directly impact turbine siting. Certain conservation areas belonging to voluntary bodies were ignored as the sites were very small and usually covered by SSSIs, NNRs, & ESAs. Military zones were also ignored as in many parts of the South West they coincide with other constraints. The remaining areas with the exception of target ranges could in any case be used for wind turbine siting.

The one remaining constraint to wind turbine siting is planning consent. A point of debate in this area is whether wind farms would be allowed in AGLVs. The first commercial wind farm operating in the South West is sited in an AGLV, however applications in other AGLVs have been rejected. Resource estimates were therefore computed both excluding and including ALGV. The inclusion of AGLVs resulted in an approximately 50% increase in the resource size.

2.5 Economic Resource Estimation

The total economic resource is the accessible resource, however it may not be cost effective to exploit all of it. To compute the cost of generating that energy it was necessary to consider complete wind farms. A means of identifying potential wind farms in the mass of available resource was therefore required. The site specific costs associated with connecting these wind farms to the electrical distribution system also needed to be assessed. In addition there was a question whether the distribution system could accept all the wind farms identified.

3.0 IDENTIFYING POTENTIAL WIND FARM SITES

3.1 K-Mean Clustering

The method selected to identify individual wind farms, K-mean clustering, attempts to assign a large number of data points (the square kilometres) to a fixed, and smaller, number of clusters (the wind farms). The method is implemented in a computer program in three distinct phases. In the first phase (thinning) outlying and isolated data points are eliminated. The second phase (clustering) then groups the remaining data points into the most favourable clusters. This second phase is iterative. The third phase allows the user to modify the solution found in the second phase using some simple wind farm design rules.

3.2 Thinning

The thinning phase is conducted as follows. All the squares are examined and those containing less than a specified minimum are discarded. Next the number of turbines within a user supplied distance of the square currently under consideration is determined. If the number of turbines found is greater than a user supplied threshold then the square is retained, otherwise it is dropped from the analysis. This procedure is performed for each square in turn.

3.3 Clustering

The clustering approach is based on a non-hierarchical K-mean clustering algorithm due to Anderberg [1]. The actual method used is an iterative version of the method of MacQueen [2]. The process consists of two steps. The first is to assign the starting positions of the cluster centres. Then each square is added to the cluster which is least adversely affected by having that square added to it. This is done until all the squares have been assigned. This defines MacQueen's method. Measures are taken to ensure that the initial cluster centres are well dispersed throughout the data set, even if the data is supplied in a well ordered form. The second step consists of going through the data considering each square. If it proves advantageous, in terms of the defined cost function, to swap a square from one cluster to another then this is done. This process continues until a complete pass is made through the data without a single swap taking place. This is the iterative modification to MacQueen's method.

3.4 Modification

The wind farm identification problem requires additional constraints to be placed on the clustering. The first constraint is that no wind farm should exceed a maximum number of turbines. The second limit is that very spread out wind farms will not be economic and so should be split into smaller denser farms. Both of these constraints are catered for by the program. After the initial clustering has been performed the user is informed which is the largest farm produced and also the sparsest. The sparsest is defined in terms of the length of connections between squares and the total number of turbines in the farm. The user can then set limits for these two constraints. The data is then re-examined and the largest and sparsest farms split up if necessary.

In each case a new cluster centre is introduced as the most isolated point in each farm. The data is then re-clustered until no further swaps take place. A compromise is needed between only splitting the very biggest farm and splitting all the farms which are too big. The former would take too long and the latter would introduce too many new farms. The compromise is that a farm can only be split if it is more than a defined distance away from any previously split farms in this iteration. Once this splitting and re clustering has taken place the user is informed of the current state of the largest and sparsest farms and how many exceed the requested limits. This continues until the user terminates the process or all the farms are within the originally requested limits.

4.0 WIND FARM INTERACTION WITH THE DISTRIBUTION SYSTEM

4.1 Objectives

Once the potential wind farms had been identified their interaction with the distribution system was examined. The objectives of this analysis were:-

(a) To identify the points on the existing 33kV/132kV grids and substations where wind farms could be connected.
(b) To identify those connections that could be accepted in capacity and fault level terms without reinforcement of the distribution system.
(c) To identify the amount of reinforcement required to accept the remaining wind farms (or some proportion of them)
(d) To assess the cost of connection in each case.

4.2 The Distribution System

The distribution system in the SWEB region is very complex both in topological terms and in its ability to accept new generators. To make the problem tenable some simplifying assumptions were applied. The first of these was that no connections at less than 33kV were allowed. Ignoring the 11kV network significantly reduces the complexity of the problem. The second was that there were unlikely to be any wind farms large enough to require direct connection to the 132kV grid. The problem was therefore effectively reduced to the 33kV distribution system.

The following basic data about the distribution system was supplied by SWEB, the topology of the 33kV, 132kV and 400kV networks, the location of sub-stations, bulk supply points, junctions and breakpoints, transformer capacities and fault levels of each component of the network. This information was input into a purpose written distribution system analysis program, TOPOL. The program builds, links and checks the distribution system topology and capacity, then allows the most favourable connection for a wind farm to be identified and costed.

4.3 Connecting Wind Farms to the Distribution System

TOPOL can be interrogated about connection options and costs for specific wind farm sites simply by giving the site location and installed capacity. The reduction in available capacity at substation and bulk supply point (BSP) transformers and on the 33kV lines is accumulated so that further wind farms use the newly adjusted capacities in determining their connectivity. The following parameters were used in the grid connection algorithm:-

Search radius for a 11kV substation	= 10km
Search radius for a 33kV line segment	= 30km
Search radius for a BSP	= 30km
Power output of single wind turbine	= 0.41MVA
Fault level induced by the wind farm	= 48.0MVA
Line rating for 0.1" cables	= 12.4MVA
Line rating for 0.15" cables	= 15.9MVA
Line rating for 11kV cables	= 5.3MVA

There are six different ways in which a wind farm may be connected:-

A single dedicated 11kV circuit
Two dedicated 11kV circuits in parallel
Connection to existing 33kV circuit including 0.1" dia line
Connection to existing 33kV circuit excluding 0.1" dia line
Dedicated 0.15" dia 33kV line to a BSP
Two dedicated 0.15" dia 33kV lines to a BSP in parallel

4.4 Connection Costs

The cost is computed for each different method and the cheapest available option is used. Costs were based on the following formula which varies for each connection type. An allowance for a proportion of underground cabling and an indirect connection path is made in the cost/km values.

COST (£k) = fixed charges + cost/km x distance to wind farm

The order the farms are presented in can have a profound effect upon connectivity and the capacity accepted by the distribution system. Initial tests with the data sorted by both increasing and decreasing installed capacity gave results that were not in line with expectations. Finally it was decided to order the wind farms by lowest generation cost based on a connection cost of 15% of capital cost.

The results of the above analysis is a set of connection costs for each of the constraint scenarios. Wind farms that could not be connected are indicated by a zero connection cost. From this the total connectable resource is computed along with that which is unconnectable, i.e. the amount of grid reinforcement required. These more accurate connection costs were then used in the following economic analysis.

5.0 ECONOMIC ANALYSIS

5.1 Inputs

The primary inputs to the economic analysis are the capital cost and annual running costs of each wind farm. These consisted of the ex-factory capital cost of the "typical" wind turbine in terms of cost/m2 of swept rotor area, plus other costs expressed as percentages of the capital cost, installed capacity, power produced or the gross revenue. To these the site specific site connection costs derived in the grid connection analysis were added. The annual output of each wind farm can be computed from the output of the individual square kilometres that make them up.

5.2 ETSU's Discounted Cash Flow Methodology

The one remaining and possibly most important factor is the method of assessing generating costs. Normal practice within ETSU is to use a simple discounted cash flow (DCF) approach which assumes that the project is fully equity funded. The DCF is carried out over the full operational lifetime of the plant, typically 25 years for a wind farm. Although many more approaches to investment appraisal are available [3] and are more sophisticated than DCF analysis this simplistic approach allows a reasonable comparison to be made between different renewable energy schemes with very different economic characteristics.

5.3 SWEB's Discounted Cash Flow Methodology

SWEB's approach is more complicated as projects are typically highly geared. Although the exact method of investment appraisal is undertaken very much on a project by project basis, for the purposes of the study a standard "commercial" DCF methodology was specified by SWEB consisting of the following elements. Projects would be undertaken with a 20% equity, 80% loan funding split. These loans are paid off during the first 10 years of the project. The interest rate for the loan is a function of its size, i.e. base rate+5% up to £5M, base rate +4% up to £10M and base rate +2% over £10M. Also the annual loan repayments are deflated by 5% per annum to reflect the effect of real world inflation. SWEB's DCF is carried out over the "commercial" lifetime of the project which in the case of a wind farm is 15 years.

5.4 Implementation

The above differences apart the two DCFs are substantially the same. A one year build period is assumed. Annual costs and revenues begin in the year following. The revenue stream is modified by varying the unit sale price of electricity (p/kWh) until the net present value (NPV) of the project's DCF is zero or slightly positive. The unit price of electricity at this point is the breakeven generation cost at a particular discount rate. It should be noted that this is an iterative process. The analysis was undertaken using Microsoft's Excel spreadsheet. Both types of DCF were implemented in a single Excel macro function, (Estimate). This function is called repeatedly by another macro (Breakeven) which implements the iterative solution. DCFs were undertaken for lifetimes of 15 and 25 years and discount rates of 8%, 15% and 25%.

6.0 WIND FARM PROXIMITY CONSTRAINTS

Finally to understand the possible sensitivity of the wind energy resource to planning considerations a five kilometre proximity constraint was applied to the wind farms identified in the previous analysis. The premise was that planners are unlikely to allow wind farms to dominate a landscape or vista and therefore will demand some spatial separation between farms. This had little effect in the case of AGLVs excluded and reduced the resource by approximately 10% when AGLVs were included.

7.0 SUMMARY

A coherent suite of software for regional onshore wind resource analysis has been developed. It has been successfully applied in a regional renewable energy study of South West England allowing the technical, realisable, accessible and economic resource levels to be comprehensively estimated. The software has subsequently been used in a similar regional study of North Wales for Manweb Plc.

ACKNOWLEDGEMENTS

The work described in this paper was carried out under contract as part of the UK Department of Trade and Industry's (DTI) Renewable Energy Research and Development programme, managed by the Energy Technology Support Unit (ETSU). The views and judgements expressed in the paper are those of the author and do not necessarily reflect those of ETSU or the DTI.

The Author wishes to thank Dr J Craig of ETSU and Mr T Masood of SWEB Plc for their support during the development of the above software and the preparation of this paper.

REFERENCES

[1] Anderberg M.R., 1973, Clustering analysis for applications. pp359, Academic Press Inc. (London) Ltd.

[2] MacQueen J.B., 1967, Some methods for Classification and Analysis of Multivariate Observations. Proc. Symp. math. Stastist. and Probability, 5th, Berkely,vol 1,281-297, AD 669871, Univ. of California Press Berkley.

[3] Lumby S. 1988, Investment appraisal & financial decisions, Chapman and Hall

Allocation of areas for wind farms in Schleswig-Holstein and Hamburg

M. Scherweit
WINDTEST KWK GmbH
Sommerdeich 14 b
2222 Kaiser-Wilhelm-Koog
Germany

S. Glocker
WINDTEST KWK GmbH
Sommerdeich 14 b
2222 Kaiser-Wilhelm-Koog
Germany

ABSTRACT: Increasingly critical discussion by the public take place regarding the installation of wind energy converting systems (WECS). Therefore it is essential to allocate suitable areas for the installation of turbines and in particular wind farms together with all decision-makers involved. For the Schleswig-Holstein districts of Dithmarschen (1991), Ostholstein (1991/92) and Steinburg (1992) and for the Federal State of Hamburg the WINDTEST KWK has prepared an expertise on the connection of wind farms, to serve as a basis for discussions in connection with approval by the competent authorities. From this situation an explanation of the procedure for the determination of wind potential and suitable areas is presented.

1 INTRODUCTION

The offered grants by the government of the Republic and of the states have lead to an increasingly visible interest in the economical use of wind energy, especially in the northern states. More and more conflicts arouse because of the increasing number of istalled turbines. In this connection the determination of suitable areas for the realisation of wind farm projects becomes very important.

For the situation is very complex, the local authorities are often expected to much of approving the installation of WECS. Therefore some of the districts take charge of drawing up general and standardised recommendations for the determination of suitable areas (wind potential, nature and landscape preservation, network connection, directional radio routes and partly monument protection) to serve as a basis for planning and discussion for the local authorities concerns. The recommendations include considerations of wind potential, nature and landscape preservation, network connection, directional radio routes and partly monument protection. The local authorities decide weather to do special plannings under the regulations of the construction law or to set up a plan for land utilisation.

2 Procedures

To take the criteria of the above-mentioned representatives into account teams on district level have to be formed. WINDTEST assumes coordinational functions and summarizing of the results. Therefore several restricting criteria imposed by the construction law, nature and landscape preservation and energy supply are considered and the areas to be examined are analysed as to their suitability.

3 Restricting criteria required of the representatives

3.1 Wind potential

One essential part of the expertise relates to an investigation on the wind potential prevailing in defined areas. To this effect the European Wind Atlas is employed, taking into account surface roughness and local orography. For demonstrating the calculated wind potential, the distribution of lines of identical wind velocities (isovents) on reasonable scaled maps is suitable. These maps offer an idea of the suitability of locations for the installation of wind farms and estimates on possible quantities of energy generated.

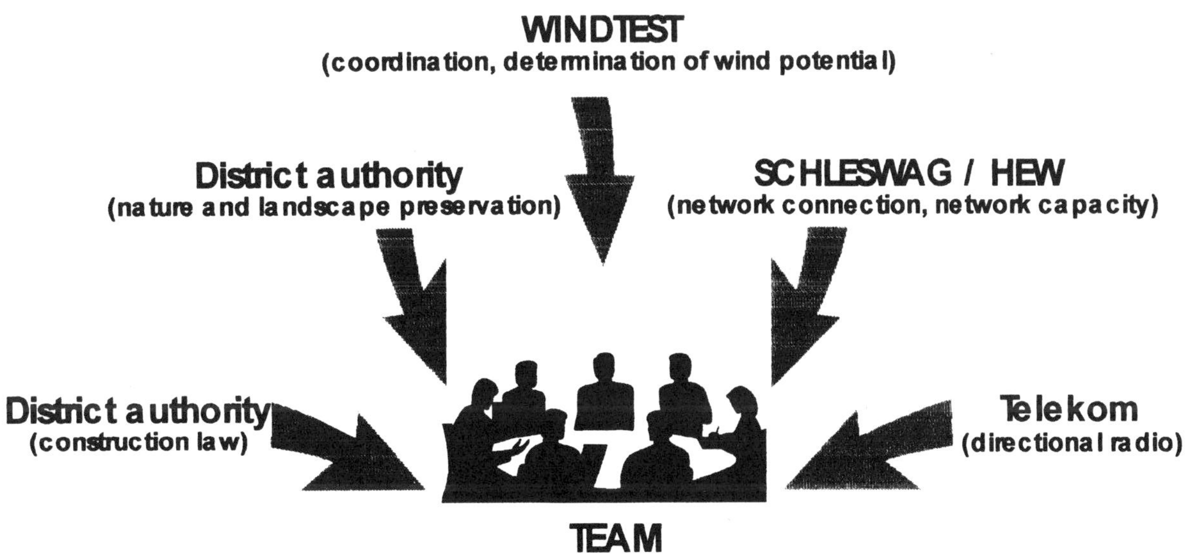

Figure 1: Accound team on district level

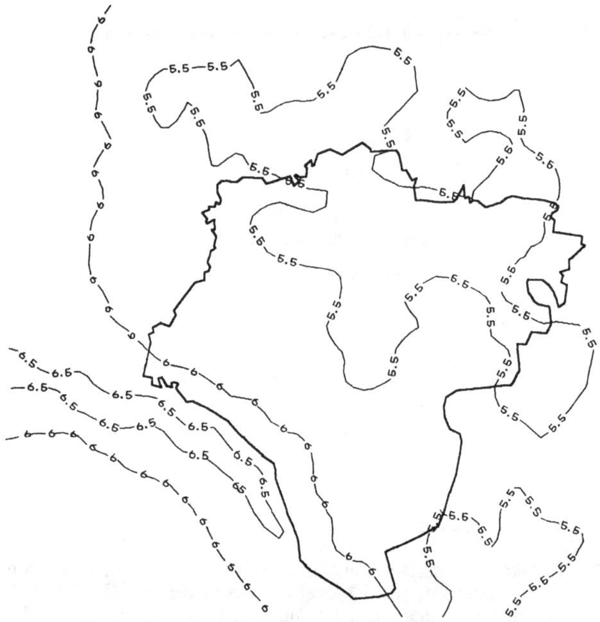

Figure 2: Isovents over the district of Steinburg

3.2 Matters of construction law

The requirements of the government of Schleswig-Holstein for the planning of WECS is considered to fulfil the standards of the construction law [1]. The section about general principals of planning gives details about minimum distances to be observed. This is to avoid negative effects on living conditions (e.g. shadows of blades and noise-emission) and safety (e.g. pieces of ice can be thrown off the blades in winter).

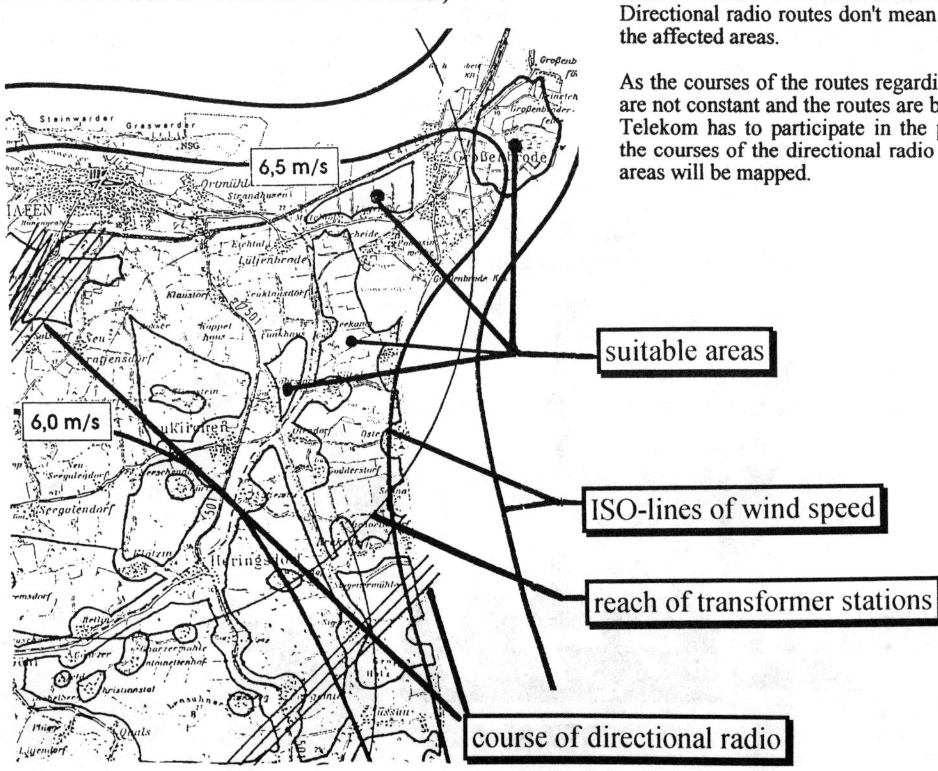

Figure 3: Summary of the restricting criteria imposed by the involved parties

3.3 Nature and landscape preservation

For the installation of WECS means a changing to the landscape it is necessary to do careful plannings in the preliminary stage. Regions which are already legally protected or which are under the subject of the regulation to guarantee a nature reserve are taken out of the further procedure.

Existing nature reserves are not necessarily to be excluded, but have to be taken in consideration as areas subdivided into different category Areas which can be classified as very sensitive are not suitable for the installation of wind farms.

3.4 Possibility of network connection

Feeding wind energy into the local grid is current technical standard. Because of the existing electrical infrastructure to supply the state Schleswig-Holstein and Hamburg with energy the considerations of the utilities are hardy concerned with operating an independent stand alone network. Therefore a potential operator of a WECS has to rely on the possibility to be connected to the network.

Usually the connection occurs at a 20kV transformer station so that at this level an input/output-junction has to be available.

As the laying of suitable cables bears high expenses the distance between wind farm and a possible junction is an important criterion. For instance, for the Schleswig-Holstein district the local utility SCHLESWAG AG requires economical network connection in an 8 km area around the closest 20kV transformer station. Besides possible capacities for the current 20 kV-grid will be considered.

3.5 Directional radio routes

To avoid interferences on directional radio routes caused by WECS the inclusion of the Telekom in the procedure of finding suitable areas for the installation of wind farms is essential. Directional radio routes don't mean necessarily an exclusion of the affected areas.

As the courses of the routes regarding the width and the height are not constant and the routes are being expanded permanently Telekom has to participate in the plannings. For this purpose the courses of the directional radio routes through the affected areas will be mapped.

4 Conclusion

To decide the further proceedings the criteria of all parties involved are summarized presented on a map showing the areas suitable for the installation of wind farms. This is shown in figure 3.

With the wind available constituting as an essential category, a weighting and grading of the areas determined is made (figure 4).This is of great importance for the discussions with the local authorities concerned.

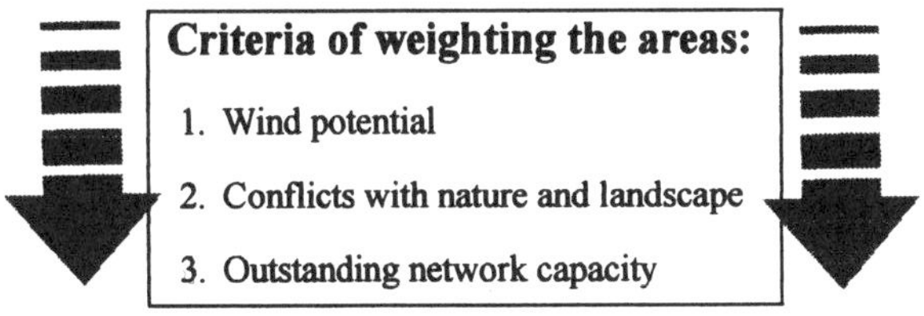

figure 4: Weigting the suitable areas

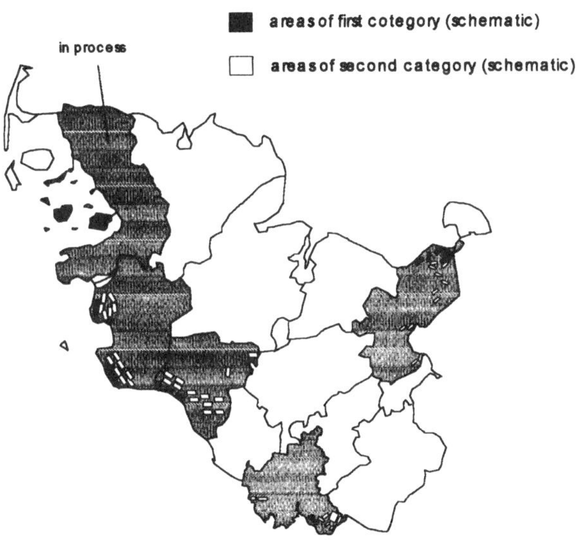

Figure 5: Suitable areas over Schleswig-Holstein (first and second category)

One further part of the expertise is to estimate possible quantities of energy generated on the areas determined. For this the upper limit of the possible installed power is defined as a certain concentration (distance between rows = 5 - 10 rotor diameter) of single WECS. A possible lower limit is given by the consideration of the acoustically demanded area of the wind farm, with keeping the respective immission limits (figure 6).

For both calculations of the possible potential WECS of the todays technical standard (P=300 kW/sound power level = 100_d B(A)) are used. These determinations are to estimate the prospects of wind energy and to point out the local emphasis which have to be considered in the planning stage (e.g. network connection).

Figure 7 shows an illustration of the possible potentials of wind energy (V_{wind} > 6 m/s) for the regions Dithmarschen, Ostholstein (without Fehmarn), Steinburg and Hamburg. For the upper level an installation of high density and for the lower level an exploitation of the acoustic possibilities is taken into account.

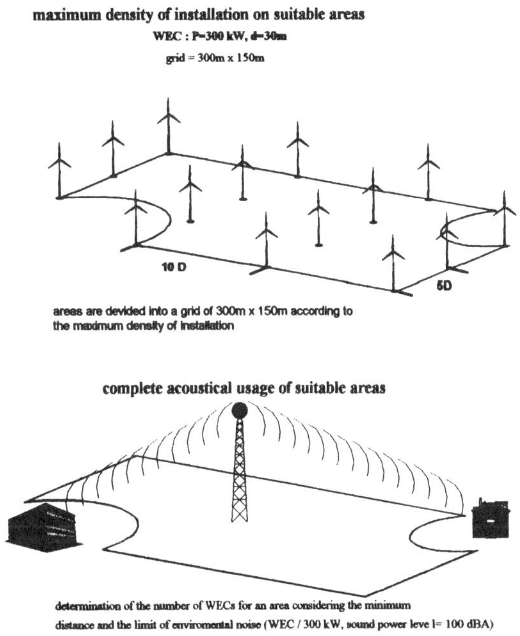

Figure 6: Possibilities to estimate possible quantities of energie generatet on the suitable areas

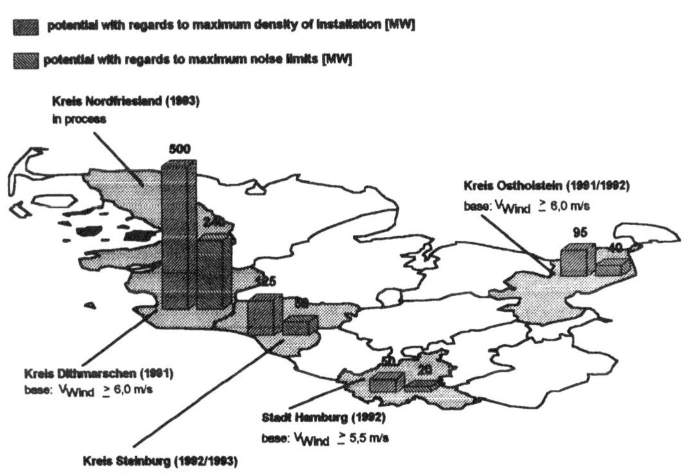

Figure 7:Possible potentials of wind energie in Schleswig-Holstein and Hamburg

For the mentioned regions the **calculated potential is 350-770 MW.**

But for the next year a **realistic quote will be 250 MW**. This means ca. **0,5 TWha year,** which is as much as **5 % of the energy consumption of Schleswig-Holstein.**

5 References

[1] Runderlaß der Landesregierung von Schleswig-Holstein "Grundsätze zur Planung von Windenergieanlagen"
IV 8, IX 3, XI 8
Amtsbl. Sch.-H. 1991 S. 560

VERSICHERUNGSCHUTZ FÜR WKA HEUTE UND IN DER ZUKUNFT

Peter Paulowski

Versicherungsbüro
für Umweltprojekte
Peter Paulowski GmbH
Bahnhofstr. 11
2085 Quickborn

Seit 1989 ist die Realisierung von Windparkprojektennach anfänglichen Zögerungen geradezu boomartig, vorangegangen. Nicht zuletzt die kurzfristigen Erhöhungen Einspeisevergütungen von DM 0,09 auf 0,1657 heute, plus der jeweiligen Bundes- und Landesförderungen, haben dazu beigetragen.

Trotz der zahlreichen Hürden im Genehmigungsverfahren, haben sich sehr viele Interessierte auf das Abenteuer Windkraft eingelassen, da selbst bei vorsichtiger Wirtschaftlichkeitsberechnung, Renditen zustande kommen, die andernorts nur mit erheblicher Risikobereitschaft zu erzielen sind.

An dieser interessanten und kaum risikobehafteten Geldanlage, sind auch die Versicherer nicht ganz schuldlos. Warum ist das so ?
Wie wir alle wissen und vielleicht auch aus eigener Erfahrung kennen, sind Kredit institute hier zulande immer weniger bereit, ein Risiko zu tragen, daß ihren Erträgen gerecht wird.

In dem Maße, in dem die Kreditgeber ihr Risiko mindern wollen, wird es den Versicherern aufgebürdet. Denken wir hier nur an den Punkt "Kündigungsverzicht im Schadenfall". Es sollte eigentlich jedem bekannt sein, daß diese Klausel zum Schutz der Versichertengemeinschaft besteht, um sich schneller von "Schwarzen Schafen" trennen zu können, die das Schadenportefeuille durch zweifelhafte Schäden unnötig in die Höhe treiben. Darüber hinaus, wollten sich Banken schon bestätigen lassen, daß im Falle einer Leistungsverweigerung aufgrund einer Obliegenheits verletzung, trotzdem gezahlt wird.
Zur Abrundung dieses Themas möchte ich nicht unerwähnt lassen, daß mir schadenbedingte Kündigungen von seriösen Versicherern, ohne sachlichen Hintergrund, nicht bekannt sind.

Die für dieses Risiko kompetenten Gesellschaften sind ohnehin an einer Hand abzuzählen, sie sehen die langfristige Zusammenarbeit, das heißt die dauernde Erfüllbarkeit der Verträge. Das kann nur klappen, wenn der Versicherer eine Chance sieht in die Gewinnzone zu kommen.

Der zur Zeit angebotene Versicherungsschutz greift stark in den Bereich der kaufmännischen Risiken ein. Das Betreiber Risiko beschränkt sich hingegen beinahe ausschließlich auf die Windverhältnisse am Standort und die **Überlebensdauer des Herstellers**, nicht der Anlage. Das ist auf die Dauer ein ungesundes Verhältnis.

Die derzeitige Einstellung, daß bei jedem Ausfall einer Anlage, gleichwohl aus welchem Grund, der Versicherer entschädigen soll, ist unseres Erachtens äußerst kurzsichtig und zeigt die momentane Grundeinstellung. Verneint der Versicherer nun auch noch die Leistung, stößt er vollends auf Unverständnis. Eigentlich fehlt nur noch der Anspruchswunsch bei Stillstand der Anlagen durch Wartung.

Schon heute hat die Versicherungswirtschaft mehr als doppelt so viel an Schaden aufwendungen gehabt, als an Prämien eingenommen wurden. Jeder ordentliche Kaufmann weiß, daß auf Dauer keiner von tiefroten Zahlen leben kann. Das diese Tendenz künftig noch steigt, da die ersten Anlagengenerationen aus der Garantie gelaufen sind oder jetzt laufen, ist jedem klar, der sich mit dieser Materie beschäftigt.

Was passiertin der Zukunft?
Die Versicherer sind gezwungen zu handeln, das heißt man wird darüber nachdenken ob der Deckungsumfang verändert werden muß. Es könnte z. B. das Blitzschlag-Risiko nur noch separat, gegen eine risikogerechte Prämie, versichert werden.
Die Grundprämien würden entsprechend der Schadensituation steigen, eine Veränderung der Selbstbehalte ginge damit Hand in Hand. Der schadenverlaufsabhängige Rabatt könnte fortfallen, es wäre allenfalls über eine Gewinnbeteiligung zu diskutieren.

Um dieses Schlimme Szenario nicht ausufern zu lassen sind alle gefordert: Hersteller, Betreiber und Versicherer.
Wichtig ist es das Blitzschlag-Risiko zu mindern. Nach unseren Informationen soll der Germanische Lloyd dieses Problem untersuchen.
Vielleicht ist es auch möglich, Reparaturen durch erworbene Erfahrungen, ind der Zukunft preiswerter durchzuführen?

Die Betreiber müssen einsehen, daß bei zu erwartenden hohen Renditen auch kaufmännische Risiken zu tragen sind, dazu gehören auch vorbeugende Investitionen, die die Ertrags- und Schadensituation nachhaltig verbessern, ohne zu überlegen ob nicht noch ein Dritter, i. d. R. der Versicherer an den Kosten beteiligt werden kann.

Es wäre zu begrüßen, wenn die Aufklärungs arbeit der Verbände und Beratungsfirmen auch in diese Richtung gehen würden.

Sollten diese Probleme nicht in den Griff zu bekommen sein, hätte das äußerst negative Auswirkungen auf die Windkraft.

Alle wissen, daß der Zweit- und Drittgrös ste Technische Versicherer, die Albingia und der Gerling, bei dem derzeitigen Prämienniveau, nicht bereit sind Windkraft-Risiken zu zeichnen. Damit sollte auch jedem klar sein, daß die wenigen Versicherer die sich heute qualifiziert mit dem Versicherungsschutz für WKA beschäftigen, zurückziehen könnten.
Ein Ausweichen auf andere Gesellschaften ist dann aus rückversicherungstechnischen Gründen kaum vorstellbar.
Wenn nun kein geeigneter Versicherungsschutz mehr zu bekommen ist, wäre auch keine Bank mehr bereit eine Finanzierung vorzunehmen, daß aber ist das Ende der Windkraft. So weit sollten wir es nicht kommen lassen.

Wir alle sind uns darüber im klaren, daß die Anpassung von Prämien und Selbstbehalten nicht das Allheilmittel sind.
Wichtig sind für die Gesellschaften auch Ausgleichsprämien aus anderen Risikobereichen. Wir als Spezialvermittler, sind daher bereits seit längerer Zeit bestrebt, den Gesellschaften Ausgleichsgeschäft zukommen zu lassen, daß für Kompensation sorgt.

Aufgrund der aufgezeigten Problematik sind wir der Überzeugung, daß die Windkraft nur Zukunft hat, wenn das Miteinander erkannt wird, denn:

WIR SITZEN ALLE IN EINEM BOOT

Hersteller
Banken
Versicherer/Vermittler
Betreiber

RAISING FINANCE: A CHALLENGE FOR WIND FARM DEVELOPERS

Melissa R. Taylor
14 Alwyne Square
London N1 2JX U.K.
tel +44 71-359-2916 fax +44 71-226-5466

ABSTRACT

Raising finance is a major challenge in developing a wind farm. There are 3 reasons why wind farm finance is difficult.

Project finance is complex, but it brings many benefits, such as greater debt:equity leverage than a corporate loan, and hence a lower cost of capital. But project finance also has its costs: complexity, and higher fees and interest rates.

Everything else must be ready first. The lender's chief motivation is to minimize that risk. As a result, all aspects of the project must be in place or near completion before one can raise debt financing.

Banks are hard to persuade. Not all banks are able to do project financing. Even those that have project finance skills are often unwilling to consider wind farms, because they are perceived as very risky.

Though raising finance for a wind farm is difficult, there are several things a developer can do to make it easier.

- Keep accurate records of all related transactions.
- Get professional help early on in the development stage.
- Be realistic about your approach, your assumptions and your timing.

Ladies and gentlemen, I come before you today as a lender, investor and advisor who has seen worthwhile and technically excellent projects fail because their owners fail to understand how hard it is to get other people to hand over their money. This challenge is particularly serious for wind farm devleopers, as I believe it is the case that nobody has ever acquired a signed contract guaranteeing wind supply. And this only adds to the financing difficulties. Today I will discuss project finance, the method used to raise funds for many wind energy projects.

Much of what I say today also directly applies to small projects. Most of the stringent lending requirements for project financing are the exact things about which the developer should be cautious. In the end, a developer of any sized project wants his money back and a profit. Small or large, a strong project has a better chance of succeeding.

Before closing I will mention a few tips on managing your lenders. Let's start by defining what we mean by project finance.

> Project finance is any financing of a venture where the financiers look to the cash flow and earnings of that particular project as the principal source of repayment for the borrowed funds, and to the assets of the venture as the key collateral.

Project finance relates to the structure of the transaction - the limited recourse nature where the lender looks specifically to the project and its assets for repayment. While limited recourse means there is little other source of repayment, this does not mean that the lenders will bear the project risk, as lenders will also expect to have bonds, construction and performance warranties, insurance and some guarantees for the project.

I. **Project finance is complex.** But it also brings many benefits.

A. Project finance is complex because of the large number of participants - and the related contracts. The major players in a wind farm are the investor, lender, contractor, developer, electric company and operator.

Wind is easy in that to date, no one has signed a fuel supply agreement, so there is 1 less person and 1 less document to deal with. Also note that to make things more complex, there is often more than one investor, lender, or contractor. Many contracts make up the project documents. And remember that each party will have attorneys and solicitors that will have to agree on their documents, and then the lender's counsel will have to review and agree on everything that the others have done.

B. As I said, project finance does bring certain advantages and disadvantages.

1. Advantages

¶ Properly structured, project finance can limit the risk borne by the sponsors and the developer.
- the developer is protected when the contractor bears responsibility for delivering a fully operating wind turbine through his completion guarantees and operational warranties.
- If he fails to do so, the contractor must pay damages sufficient to compensate the owners for the lost revenue

¶ Project finance can be kept off the balance sheet of sponsoring companies. The ability to stay off balance sheet is based on
- actual ownership percentage
- control of the project
- guarantees made to the lenders in support of the project

As requirements for off balance sheet lending vary by country, a developer must get expert tax, accounting and legal advice.

¶ Greater debt:equity leverage is possible in project finance than a corporate loan. A project might be able to maintain an 80% debt / 20% equity capital structure. The higher levels of security, as well as the fact that a project does not have other on going corporate obligations, allow it to maintain this higher gearing - or leverage.

¶ Higher leverage is a benefit because the interest expense on the debt is tax deductible, resulting in a lower cost of capital and an enhanced return on investment.

2. Disadvantages

¶ Complex structure and heavy documentation is a disadvantage. Many contracts and permits must be in place before any money will be lent against the project.

¶ This leads to another disadvantage: far more time is needed to arrange a transaction, largely due to the documentation requirements.

¶ Finally, because of the complex, limited recourse nature of project finance, banking and arrangement fees, as well as interest rates are higher than traditional corporate lending. Legal fees and fees for outside experts like engineering and wind consultants will be higher too.

2%-5% of the capital costs as your financing, expert and legal fees. This percentage will of course be a larger proportion of a small project as the same work must done no matter what size the project.

C. Considering all this time and effort, when does someone use project finance? It may be used when
- the project is economically viable in its own right
- ordinary corporate finance is not practical
- the developer and corporate sponsors want the project off balance sheet
- the project is being developed as a joint venture and the sponsors want to limit their risk
- political or country risks bar other structures
- tax or accounting considerations make project finance attractive

II. Everything else must be prepared before you contact your lender.

A. A project lender has limited upside return from a project and immense downside risk. If the project is the most successful one in the country, the lender will never receive more than the "spread" that is paid over the base interest rate.

B. The lender's chief motivation, therefore, is to contain that risk. As a result, all aspects of the project must be in place or near completion before one can raise debt financing.

¶ Adequate equity finance must be arranged. The amount of equity required will be influenced by the structure of the financing, the recourse that the lender has to sponsors, and the credit environment.

¶ Site selection must be complete and the project must have a clean leasehold or title to the site.

¶ Planning permission should be well underway; full planning permission must be granted before any funds will be disbursed from a bank.

Planning has increasingly become a project killer. Good projects are simply not getting approval. Furthermore, because planning is such a sensitive issue, many banks will not even hold a serious discussion until planning permission is received

¶ Project contracts should be in final stages.

C. Here there is always the problem of putting the cart before the horse. The project contracts should not be fully negotiated and signed before the developer seeks financing because the lender will usually require changes. Nevertheless, a developer needs to have very well developed draft contracts in order to convince the lender that this is a real project under development.

III. Investors and banks are hard to persuade. Not all banks are able to do project financing. Even those that have project finance skills are often unwilling to consider wind farms, because they are perceived as very risky.

A. First, wind farms are often too small for expensive project finance methods. The same work goes into a financing whether it is for £5 million or £50 million. Obviously, the lender will receive much greater income from a larger project so the incentive is to ignore the smaller ones.

For most banks in the UK and the US the threshold is around £10 million. Some banks will consider £5 million. The smallest project financing that I know is of £3 million.

B. Wind data and history present a big problem. This scares off many lenders as wind energy is seen as avant garde, "going where no man has gone before."

In many cases, there is limited wind speed history. In addition to on-site data, you will need a comparable, nearby site where the wind history can be fully documented. We all know that the wind blows, but where and how much for how long?.

No one has yet to sign a long term fuel supply agreement for wind. But this creates a very real downside for the lender: the lack of traceable, dependable "fuel supply." Just how does one analyze the wind? Is it creditworthy? Are supplies sustainable? Are there delivery problems?

¶ Actual site readings covering one year are the absolute minimum. These must be very well correlated with another local site which has a much more extensive history The correlation period must be fairly long to cover more than one weather or drought cycle.

¶ Your wind analysis program must be thoroughly tested. You may have to run your program on data from an existing, operating windfarm to prove that your predicted results match the actual results.

¶ You must have all of your methodology and output figures vetted by a qualified third party. This point alone can cause troubles, for in this specialized industry, there may be no one else in your area or country that has these skills. Importing expertise is not always seen as a positive because local knowledge can be deemed to be of critical importance.

C. Finally, despite facts to the contrary, wind energy technology is considered new and risky. Lenders tend to lend to what they know - like tried and true gas or coal turbines that have been around for donkeys years.

Though raising finance for a wind farm is difficult, there are several things a developer can do to encourage the lender to see your project in a positive light.

I. **Keep accurate records of all related transactions**, including real estate or property leasing activities for the site, development, scientific and analytical work, and all corporate documents related to the authorized existence of your company. Proper records help you appear more organized and professional to the lender

II. **Get professional help early on in the development stage**. Expert assistance will save you time and money. The most common outside experts are financial advisors and legal counsel. Before financing is complete, you will often work with insurance advisors, engineers, environmental engineers and wind experts.

A. The best favor you can do for yourself is to be sure that your advisor and your legal counsel are experienced in project finance. The legal documentation for project finance is very different from corporate work. Your interests will not be as well served (or protected) using only your general corporate counsel.

B. Select your banker with care. If she or he does not exhibit a healthy - I'd say a nosey - interest in your work, background, technology and project, then this is not the bank for you. For the banker to get approval to lend you money she must make a thorough analysis of your project. If she is not bombarding you with questions, she is not doing her job.

Besides wondering whether or not the wind will blow, here are a few of the key pieces of information that you should have to hand for your banker

¶ Developer information: ownership, management and structure, experience, financial resources

¶ Project: business rational, key players, contractual arrangements including key terms, financial profile, project technology

¶ Business plan: cash flow projections, cash flow sensitivity analysis, forecast balance sheet and profit and loss, forecast equity returns

This is also a good source of discipline for the developer as these questions should have thought through and sorted out prior to seeing the bank. So be prepared to answer these points and many more.

C. Banks are not the only source of funds. Others include leasing companies or other non-bank financial institutions, some insurance companies, and occasionally international agencies and government export support agencies.

III. Be realistic about your approach, your assumptions and your timing.

A. You must project professionalism and trustworthiness to your lender. Once you lose your credibility, it is almost impossible to regain it.

B. Developing a wind farm always takes longer than you think. From initial contact to funding of a loan, a 6-month period is not at all unlikely.

C. It always costs more than you think. You will remember my earlier estimates of 2%-5% of the capital costs for financing, expert and legal fees.

D. Bankers feel that developers are notoriously optimistic about the viability of their projects. Therefore potential lenders will typically take a far more hard headed view. Sensitive areas include equity contributions and optimistic projections, unreliable technology (is this a new turbine or more tried and true?), debt repayment period is too long, base case projections are too speculative

Summary

Today I have tried to explain to you first why project finance can be such a challenge for wind farms, and then give you a few pointers on how to improve your odds of success. What you should remember is that - everything you do to make your project stronger, benefits you ultimately. Whether or not these actions were at the request of a lender, a strong project should be your goal. This is key because as the equity holder, your money goes into the project first; but you are the <u>last</u> in line to get your money back.

FINANCING COMERCIAL WINDFARMS IN SPAIN

FERNANDO R. VIZCAINO
PLANTA EOLICA DEL SUR, S.A.
Autovía Sevilla-Coria, km. 3,5
41920 SAN JUAN DE AZNALFARACHE (Sevilla) - ESPAÑA

1.- INTRODUCTION

The objective of this report is to give an understanding of the different steps taken to obtain the necessary outside financial backing for the building of the largest commercial wind farm in Europe. These steps will include not only the measures taken and the results but also the initial negotiations.

2.- INITIAL PROPOSAL

The initial financial proposal for our commercial wind farms that was made public was as follows in million $(US):

FINANCIAL NEEDS		FINANCIAL RESOURCES		%
INVESTMENT	36	EQUITY	5	13
		SUBSIDIES	9	23
TAXES (V.A.T.)	3	LOAN	25	64
	39		39	100

This proposal is founded on the importance of the three contracts which guarantee:
1) the turnkey project of the wind farm,
2) its maintenance and operation, and
3) the sale of energy.

These contracts are between PESUR and its partners. Apart from having common interests as owners, each sustains its own interests as equipment suppliers, utility company or the promoter of regional activity and growth.

This first proposal, based on the financial and technical stability of the partner firms, would:
1.- Supply the contract of a wind farm that would guarantee the energy production in relationships to the investment.
2.- Guarantee a kw/h price previously fixed by law for the electric energy sales.

Both of these guarantees would give security to the foreseeable figures in the feasibility plan.

3.- RESULT

After a first presentation of the feasibility plan to banks, consultors, etc. the following results were obtained:

The security margin or the ratio between the Present Net Value (VAN) of the cash flow and the size of the loan would be located around 1.5.

This simple calculation gave a loan without partner guarantee of $20 Million (U.S.).

The financial leverage or the proportion between our own and outside financial sources would then be about 1, given the characteristics of the project and its originality. This was a common demand of the different financial institutions with whom we negotiated with in order to obtain the necessary outside finances. It also gave more balance to the proportion of the financial sources.

Based on the previous proposal, a second and final financial proposal was establishes as follows in million $(US):

FINANCIAL NEEDS		FINANCIAL RESOURCES		%
		EQUITY	5	13
		SUBSIDIES	9	24
		PARTNER LOANS	5	13
INVESTMENT	36			50
TAXES (V.A.T.)	3	LOAN	20	50
	39		39	100

4.- CONCLUSION

This change of strategy in the proportion of the different financial sources is caused by the search for the firm's financial relief. Returning the loan to the partners and the payment of its interests would be handled with flexibility after attaining the security level. Thus with more participation of the partners in the project, a decrease of economical risk of the financial institution is obtained.

A twenty million dollar loan was granted by CAJA MADRID without partner guarantee. Throughout ten years, nine equal payments would be made. The trimestrial MIBOR interest was with a 0.65% spread an payable on a trimestrial basis.

MARKET VALUE OF GENERATED ELECTRICITY DETERMINED BY MARGINAL COST PRINCIPLE

B. Deppe and T. Morovic
Forschungsgesellschaft für umweltschonende
Energieumwandlung und -nutzung mbH
Kronshagener Weg 130 a
2300 Kiel 1
Germany

Abstract: Presently in the Federal Republic of Germany the proceeds for the electricity generated by wind power plant is regulated by the "Stromeinspeisungsgesetz". According to this regulation, the price for wind generated electricity is calculated on a simple average cost pricing basis. The latter principle, however, shows no relation between the proposed price and the displacement of long run production costs which occurs by using wind energy for the electric utility and hence no relation to market pricing principles. Consequently, economic efficient pricing for wind generated electricity means a responsibility cost related payment based on the marginal cost pricing principle. Very preliminary results show that it is more profitable to produce electricity by wind power plants under marginal cost pricing conditions than under average cost pricing conditions. This due to the fact that wind power plants rather supply electricity in peak load periods than in off peak load periods.

1. GENERAL PROBLEM AND PURPOSE OF THE WORK

It is well known that the world wide amount of fossils fuels is limited and their use leads to steadily increasing environmental problems. Thus, the environmentally-sound energy conversion and efficient use of energy is an inevitable precondition for a sustainable future. On the other hand, an economically efficient use of limited ressources is only to be reached by an instrument which allows for the correct measure of scarcity. In competitve markets this instrument will be the price. As an instrument of coordination for market activities the price induces highest efficiency in allocation of ressources and therefore the welfare optimum. In a competitive market the price would reflect the correct value of electricity produced, too. The user or consumer of electricity would have to pay the costs induced by this additional demand. This general economic principle is called marginal cost related or responsibility cost related payment for energy supply. The marginal production costs of electricity are related to the running periods of different types of plants, characterized by different cost structures. The decision on the running time itself is dependent on grid load characteristics (demand curve). It is therefore necessary to evaluate the price for energy produced by wind power plants according to the marginal cost principle, as only this principle considers the displacement of long run production costs and demand curve at the same time and hence indicates the real costs occurring for the electric utility by using wind energy. The main objective of the work is to develop and apply a computer-aided tool for the evaluaton of price for the wind generated electricity based on the marginal cost pricing principle. The results will allow for identification of economic threshold and recommendation for an increased use of wind power plants.

2. THE COMPUTER-AIDED SIMULATION MODEL

2.1 Theoretical Basic

Due to technical peculiarities in the generation and distribution of power by the electrical utilities, basic problems are encountered in fixing electricity prices and the economically correct evaluation of the power unit (kWh) supplied. In economic theory this is called the peak load problem and the price-fixing is termed peak load pricing.

The basis of the simulation model described below is simplifying energy price model complying with economic theory [1]. Despite some restrictions such models are suitable for providing general directions for business-economically and overall-economically correct pricing.

2.2 The price of energy produced by wind power plants according to the marginal cost price principle

The following simulation model compares the proceeds based on a simple average cost pricing basis with the proceeds based on marginal cost pricing. Figure No. 1 shows a conceivable result.

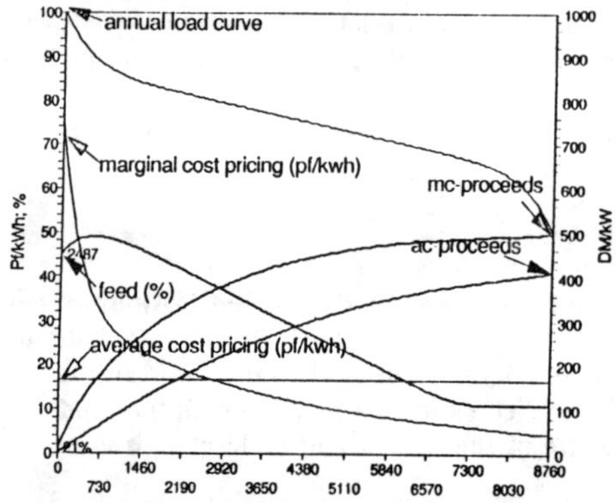

Figure 1: Comparison of the proceeds based on a simple average cost pricing and the proceeds based on marginal cost pricing [2].

On the X axis, the hours of the year are shown. On the left Y axis, electricity price in Pf/kWh is plotted as well as the grid load of an electric utility in 1/100 and the supply probability in 1/100 of the wind power plant to be examined. On the right Y axis the annual proceeds of a wind power plant are given in DM/kW.

The curve entitled "annual load curve" reflects the respective grid load of the electric utility. It describes the relation between load period and the extent of the supply required. It can be distinguished between peak load, medium load and base load. The maximum grid load (100 %) is only achieved for one hour per year, whereby a defined base load has to be provided all year long, i.e. 8.760 hours.

The marginal cost payment curve (marginal cost pricing) in Pf/kWh describes the supply payment at any point in time, which the utility would have to pay to the wind power plant. This curve results directly from the use of power stations mix in dependence of the annual load curve. Due to their varying cost structure, different types of power station are used, depending on the duration of supply required. In base load periods, power stations with low variable costs and high fixed costs (base load power plants) are preferred. During medium and peak load periods, power stations with higher variable and lower fixed costs (medium and peak load power plants) are employed. This process results in minimum cost power generation (optimum power station mix), which is directly coupled to the respective grid load, presenting an economically desirable state in the current circumstances of scarcity.

Assuming a given power station mix and given electricity prices to be paid by the electric utility, a minimum cost overall function for the entire power station park can be approximated (system overall cost function). The marginal cost oriented price, the system marginal cost function, is the first derivative of this overall cost function. Here, the marginal costs at a defined time of the year reflect the variable costs of the respective marginal supplier, i.e. costs of the power station which was used last in order to meet the demand.

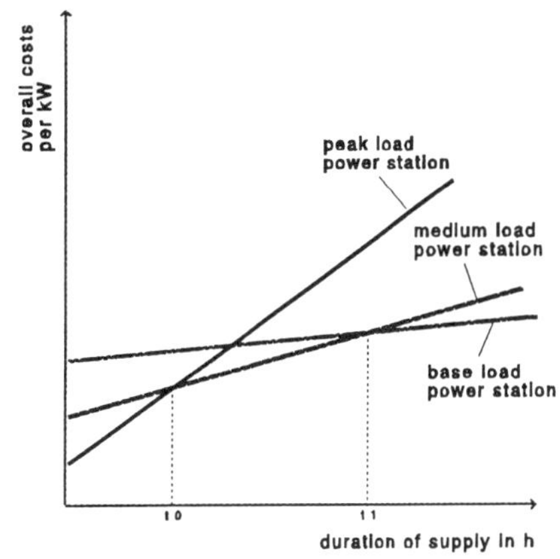

Figure 2a): Overall costs per kW of the base load power station, the medium and peak load power station depending on the duration of supply.

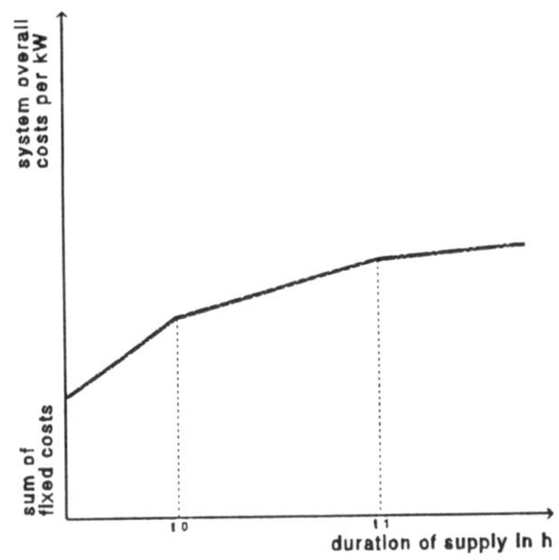

Figure 2b): Overall costs function per kW determined for the power station mix; the optimized system overall cost function

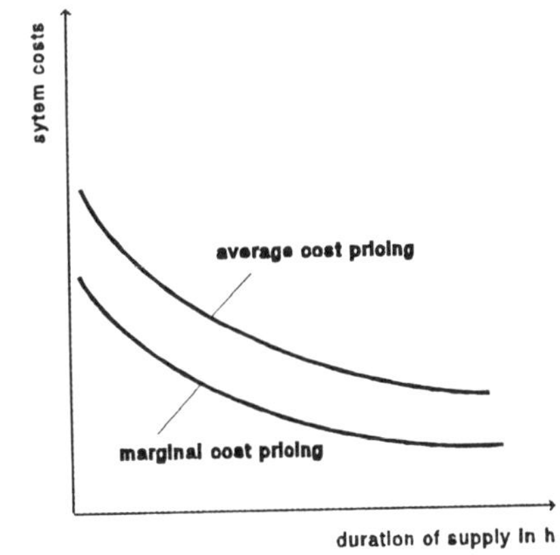

Figure 2c): The system average cost curve and the system marginal cost curve derived from the system overall cost function.

Figs. 2 a) to c) display this relation. Fig. 2 a) shows the overall costs per kW of the base load power station and the medium and peak load power stations depending on the duration of supply, measured in kW. Assuming an optimum, and thus minimum cost, use of the individual types of power plants, an overall cost function per kW can be determined for the power station mix (Fig. 2 b)). In Fig. 2 c) the system average cost curve and the system marginal cost curve have been approximated from the system overall cost function. For an existing power station mix, a decrease in the system marginal costs is dependent on the duration of supply (in kW) required. For additional units of kW, which are required all year by the consumers, lower marginal costs are incurred than for additional units of kW, which are only provided for a single hour of operation. The actual marginal cost curve results from the actual annual load curve of an entire electric utility.

In low load periods, the electricity price corresponds directly to the variable costs, mainly fuel costs, of a base load power station, in medium load periods it corresponds to the variable costs of a medium load power station, and in peak load periods it corresponds to the overall costs of the peak load power station. Thus, the marginal costs are dependent on the respective grid load, i.e. they are time dependent. Higher the grid load, higher the electricity price. This reflects not only the varying avoided costs of the utility (production side) but also the differing appraisal, in terms of time, of the commodity electricity on the consumers side (annual load curve as a demand curve).

The average cost curve (average cost pricing) represents a payment which is time independent. In accordance with current regulations, the operators of wind power plants receive at least 90 % of the average electricty per kWh payed by the consumer.

The supply curve shows the supply probability, measured by the capacity installed of the wind power plant. If the supply probability curve is integrated over the entire year, the sum of the full load hours results.

The curves labelled ac-proceeds and mc-proceeds describe the relation between the extent of the proceeds over the year taking into account the two pricing principles.

The average cost proceeds curve is determined by adding up the products of the supply probability and the average price over the entire year. The marginal cost proceeds curve, in contrast, is determined by adding up the products of the supply probability and the price derived by the marginal cost principle over the entire year.

Assuming that the supply probability of wind power plants at peak load periods is higher than during base load periods (decreasing run of the curve), i.e. peak load periods coinciding with the supply of electricity by wind power plants, the annual marginal cost proceeds are higher than the annual average cost proceeds. For a slope of the supply probability corresponding to the load duration curve, the average cost proceeds per kWh and the marginal cost proceeds per kWh are identical.

Marginal cost payment, i.e. payment based on micro-economical considerations in accordance with free market regulations, results in wind power plants ceteris paribus being already more economical than for payment based on average cost pricing. The objective of the empirical studies is to establish the important correlation between load duration curve and supply probability on the basis of actually measured data.

3. CONCLUSIONS

If every electricity unit supplied by wind power plants is evaluated differently in terms of time, i.e. taking into account the costs avoided by the utility, a wind power plant could already be economical, depending on its supply behaviour. Even without subsidies, the use of wind power plants as an alternative source of energy is advanced just by properly conveyed price signals. This result is also supported by a possible internalization of external effects such as environmental damage and so-called user costs for non-renewable sources of energy, as extraction of these raw materials today results in the abandoning of utilization possibilities for future generations. Considered in perspective to other power plants, the production costs of an electricity unit generated by wind power are lower and the use of wind power plants more economical.

In addition, by marginal cost oriented supply payment, competitive elements are implemented. This might result in decentralized electricity generation being more advantageous.

If, in addition, the consumer is informed about how much the electricity unit required actually costs by introducing a load-variable charge, i.e. the conferring of the time-dependent supply payment onto the market, capacity savings might result due to shifting electricity or even absolute electricity savings due to an increased price awareness on the part of the consumers.

4. EMPIRICAL RESULTS

In fig. 3, first empirical results referring to the coincidence between an annual load curve of an electric utility and the wind power availability (supply probability curve) of a wind power plant are shown for the year 1992. It can be recognized that the wind power is well distributed during the year and thus contributes for the base load. On the other hand, the main contribution occurs in peak load periods rather than in off peak load periods. The later result is not as pronounced as expected. The curves shown represent the result of only one year; to achieve more reliable results the data of the period 1989 - 1992 will be accumulated and interpretated in the next step.

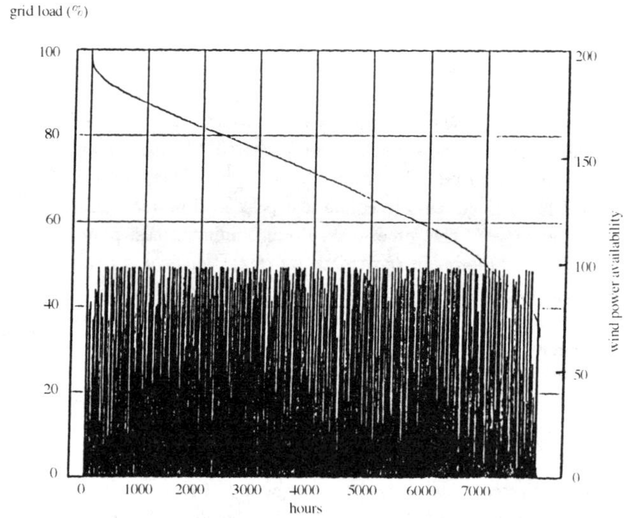

Figure 3: Annual load curve of an electric utility compared with power supply probability of a wind power plant for the year 1992.

[1]Crew, M.A./Kleindorfer,P.R. (1975) Optimal Plant Mix in Peak Load Pricing, in: Scottish Journal of Political Economy, Vol.XXII, No.3, S.277-291.

Hoven, I./Schulz, W. (1988) Kostenorientierte Stromtarife, in: Elektrizitätswirtschaft, 12. Jg., Heft 4, S.221-239.

Turvey, R. (1968) Peak-Load Pricing, in: Journal of Political Economy, Bd. 76, S. 101-113.

Waldmann, R. (1981) Die Theorie des peak-load pricing, eine Modellanalyse von Struktur und Ergebnissen früherer Ansätze zur Spitzenlastpreisbildung und deren Weiter-entwicklung, Dissertation, Giessen.

Wenders, J.T. (1976) Peak load pricing in the electric utility industry, in: The Bell Journal of Economics; Bd. 7, S.232-241.

Wiilliamson, O.E. (1966) Peak-load Pricing and Optimal Capacity under Indivisibility Constraints, in: The American Economic Review, Vol.56, S.810-827.

[2] Tönnies, J. (1992) Fremdeinspeisung - Der Markt würde mehr zahlen!, in: Energiedepesche, Nr. IV-Dez. 1992, S.40/41.

[3] Authors own research

[4] Stromeinspeisungsgesetz §3(2)

THE IMPORTANCE OF A GOOD WIND YEAR TO START ON WHEN BUILDING A WIND FARM

Sten Frandsen and Erik Lundtang Petersen

Department of Meteorology and Wind Energy
Risø National Laboratory
DK-4000 Roskilde, Denmark

Abstract: *An analysis of the influence of long-term variations in annual mean wind speed on the economy of a wind power plant has be carried out, using one British and one Danish time series of wind speed extending 90 years back. The analysis shows that even if the long-term average wind speed is known precisely, natural variations in wind speed makes the final wind power plant economy a function of the year in which the project is initialized. Such variations cannot be predicted and must be treated as an uncertainty.*
Employing the two series of annual mean wind speeds, cost of energy was found to vary as must as ±13 % from an arbitrarily chosen mean value of 0.40 DKK/kWh.

INTRODUCTION

An analysis of the uncertainty of the annual energy output from - and thus project economy of - a wind power plant exposed to the natural wind will include a large number of individual components. The following 5 categories, Ref. [1], could be used as basis for a general analysis of the uncertainties on annual energy output/overall economy of a wind energy conversion system:

1) Power curve uncertainty.
2) Wind turbine performance variation with time and specimen.
3) Availability.
4) Wind turbine array efficiency.
5) Wind resources.
6) Construction costs, O&M, lifetime.
7) Financing and interest rates.

Identification of uncertainties on each of the seven components is not an easy task. However, here we will deal with only item 5) wind resources, and then only a fraction of that source of uncertainty. It is well known that the mean wind speed over a year period is not constant from one year to another. Considering the accumulated production over e.g. 20 years from a wind turbine, such fluctuations in the annual production are of little consequence to the economy - **if** the variations are frequent compared to the lifetime of 20 years. On the other hand it is not clear whether long-term variations in annual mean wind speed occur that could influence the integrated economy of a wind power project. In other words, could a row of especially good "wind years" in the beginning of the lifetime of a wind farm significantly enhance the resulting project economy? For the purpose of investigating the matter two long-term recordings/ observations of wind speed have been analyzed and a year-by-year calculation of the *Internal Rate of Return (IRR)* has been performed.

LONG TERM CLIMATE VARIATIONS

Observation of wind speed over long periods, say more than 50 years, are rare. Even more seldom are wind speed series generated on basis of instrument recordings. Since obviously the uncertainty of those rare series must be expected to be considerable, we have chosen to select two series recorded at widely separated sites and employing different methods. One series is recorded in Fornaes, Denmark, and the other in Southport, UK. The series (annual averages) are plotted in figure 1.

Fornaes is the most eastern point of Jutland, a light house overlooking Kattegat. The data sets cover the period 1872-1988 and is detailed described in Ref. [2].

The data are based on a visual inspection mainly of the state of the sea but also notice of the behaviour of trees etc. is made. The strength of the wind was noted 5 times a day following the well-known Beaufort scale and later transformed to m/s (not a trivial procedure: more than 30 methods exists!).

Southport is located at a well exposed site on the west coast of Britain at about 54° N. The record extends from 1898 to 1954, it is measured with the same instrument - an anemometer - and it is presented and analyzed in Ref. [3].

The two series of annual mean wind speeds are plotted together on Fig. 1. For obvious seasons the absolute level could not be expected to be the same (different reference/measurements levels, different topographic settings, difference wind sequences). However the relative variations can be compared and despite some discrepancies the overall tendencies seem to be reasonable equal especially in the period from 1915 to 1950.

The wind speeds before 1895 at Fornaes seems unrealistic and are therefore not used in this study.

Since variations in annual mean wind speed extending over longer periods are of more consequence to the economy than variations with shorter periods, we have computed the power spectra for the series in search of obvious energetic frequencies, figure 2 and 3. The over-all significance of the individual spectral estimates are for several reasons not impressing. However, both spectra seem to have significant peaks around

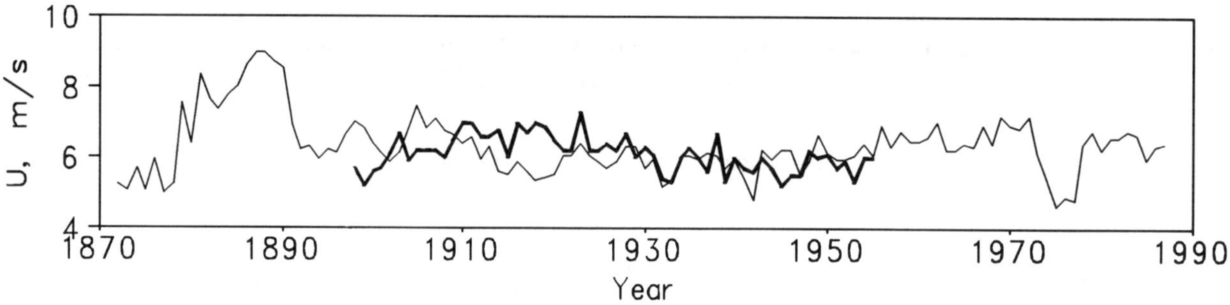

Figure 1. Annual mean wind speeds for one UK met station in Southport (heavy line) and one Danish met station in Fornaes. The British wind speed series is based on instrument recording, and the Danish series has been derived from visual observations.

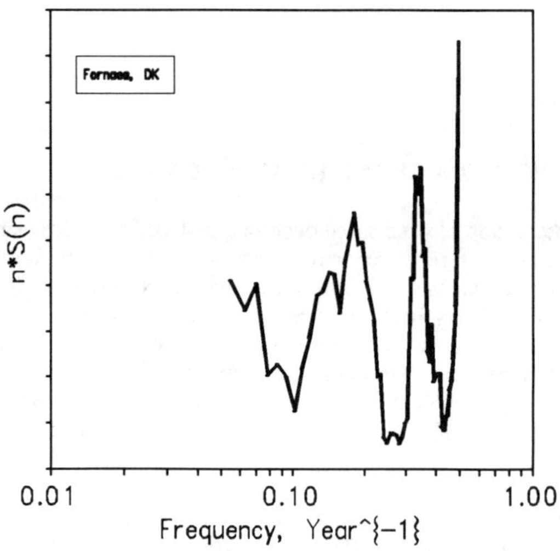

Figure 2. Power spectrum of the Fornaes wind speed series.

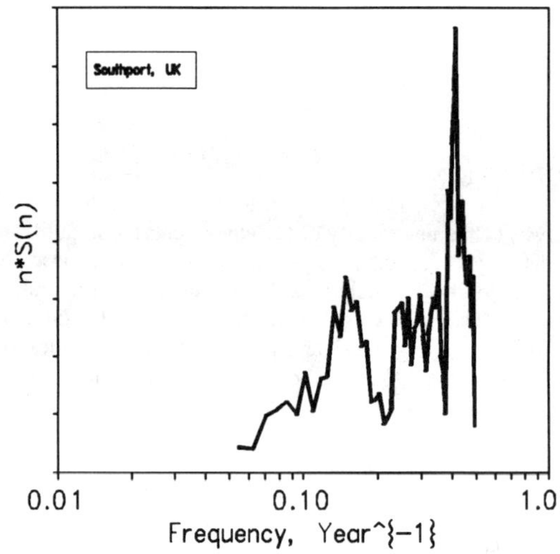

Figure 3. Power spectrum of the Southport wind speed series.

1/6 to 1/8 year^{-1} corresponding to variation periods of 6 to 8 years. Unfortunately - as it will be seen below - the low part of the spectrum is not covered by the present data series.

INTERNAL RATE OF RETURN

The internal rate of return is a measure of how well an investment pays off. Disregarding inflation, the initial - and possible running - investments in a project are compared with the net income generated by the project over its (technical) lifetime. The IRR of the project is the constant bank-interest rate in real terms (inflation subtracted) that would be needed for the investment to produce the same resulting surplus or deficit as the project if the investment money were placed on a bank account. It follows that an income generated during the fist years of a project is worth more than the same income obtained at the end of the project. For that reason it is good for the economy to have higher than average income (i.e. higher wind speed) early in the wind turbine's lifetime.

Background

As indicated, we have discarded the first, most probably erroneous part of the Fornaes-series. The statistics of the remaining data are given in table 1. It is seen that the two series are similar in statistical properties, which to some extent support the credibility of the present analysis. The maximum deviation of the annual mean wind speed from the long term mean is of the order ± 25 %, and the standard deviation 8-10 %.

SITE	FORNAES, UK	SOUTHPORT, UK
Mean wind speed	6.2 m/s	6.1 m/s
RMS, σ_u	0.5 m/s	0.5 m/s
Max{u}/Min{u}	7.5/4.6 m/s	7.3/5.2 m/s
WT-investment	1,6 Mio. DKK	1.6 Mio. DKK

Table 1. Key input data for the computation of running value of the IRR for the two locations.

Employing the power curve from a standard grid-connected wind turbine, the energy production has been calculated year-by-year for the two sites. Arbitrarily, the investment has been set to 1.6 million DKK and the selling price of energy minus operation & maintenance costs to 0.40 DKK/Kwh. The technical lifetime of the machine is assumed to be 20 years.

Each year the IRR has been calculated by including the

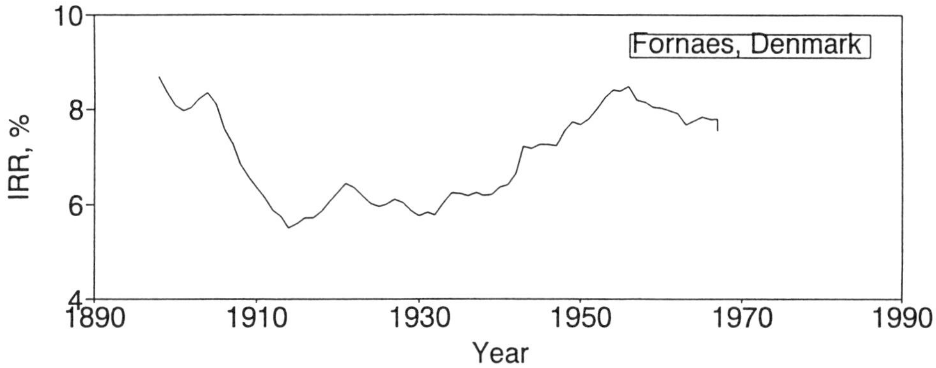

Figure 4. Graphical illustration of the variation of the IRR for the Fornaes site, Denmark. The average level is chosen arbitrarily.

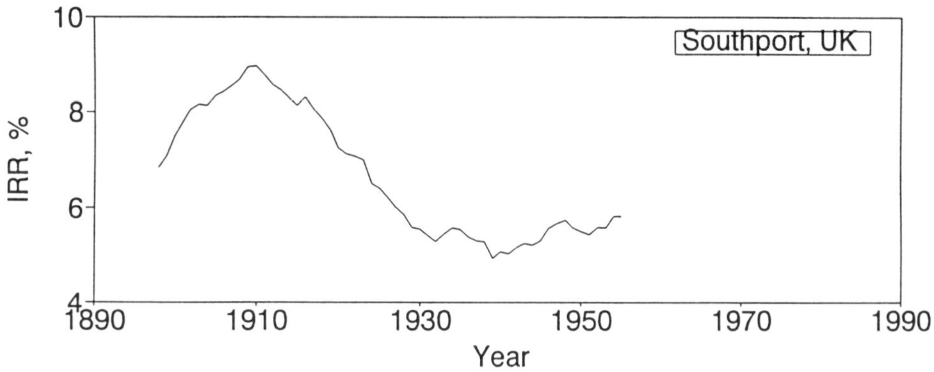

Figure 5. Graphical illustration of the variation of the IRR for the Southport site, UK. The average level is chosen arbitrarily.

following 20 years of income from selling electricity. The Southport series is extending over only 57 years so we have added the first part of the series to the end of the series to allow computation of the IRR up till 1955.

The year-by-year values of the IRR are shown in figure 4 and 5. Effectively, the running calculation of the IRR acts like a filter on the wind speed series, and though non-linear, the filter basically eliminates short-term variations. The figures indicate that the predominant IRR variation is found in periods of typically 60-80 years, not in phase, but with similar characteristics.

SITE	FORNAES, UK	SOUTHPORT, UK
Mean{IRR}	7.0 %	7.0 %
RMS, σ_{IRR}	1.0 %	1.2 %
Max{IRR}/Min{IRR}	8.7/5.2 %	9.0/4.9 %
Mean{Product}	380 MWh/year	370 Mwh/year

Table 2. Key output data for the computation of running value of the IRR for the two locations.

The statistics of the IRR for the two hypothetical wind turbine installations are given in table 2. The mean value is 7 %, and for the time period considered the IRR varies between extremes of approx. 5 and 9 %.

Reversing the approach, the results given in terms of IRR can be translated to *Levelized Production Costs*: Assume that the investment is unchanged and that the discount rate (typical interest rate excluding inflation) is 7 %, what is the variation in cost of energy? Of course, in average the production costs are 0.40 DKK/kWh. The standard deviation turns out to be $\sigma_{cost} \approx 0.03$ DKK/kWh while minimum and maximum costs over the time span considered are approx. 0.35 and 0.45 DKK/kWh, respectively.

CONCLUSION

An analysis of the influence of long-term variations in annual mean wind speed on the economy of a wind power plant has be carried out. The analysis shows that even if the long-term average wind speed is known precisely, natural variations in wind speed makes the final wind power plant economy a function of the year in which the project is initialized. Such variations cannot be predicted and must be treated as an uncertainty.

Employing two series of annual mean wind speeds, cost of energy was found to vary as must as approx. ± 13 % from an arbitrarily chosen mean value of 0.40 DKK/kWh.

REFERENCES

[1] Frandsen, S. and C.J. Christensen; Accuracy of Estimation of Energy Production from Wind Power Plants, Accepted for publ. in J. of Wind Engineering 1993.

[2] Kristensen, L. and Knud Frydendahl; Danmarks Vindklima fra 1870 til Nutiden (In Danish, Denmark's Wind Climate from 1870 until Present Days), Environmental Agency, Report nr. 2 1991.

[3] Palutikof, J.P. et al; Wind speed Variation and Climate Change, J. of Wind Engineering, vol.10, No.4, 1986.

WIND POWER IN GERMANY - MARKET ANALYSIS AND DEVELOPMENT

Dr. Norbert Allnoch

Institute of Geography
University Münster
Robert-Koch-Str. 26
4400 Münster
Germany

ABSTRACT: For three years now, due to supporting programs of the Federal Government and the Laender, this sector has been booming. Based on an opinion poll carried out among producers of wind power plants, the author analyses the market for mains-connected wind converters (≥ 10 kW) and describes the chances for future development of the use of wind energy in Germany.

The market for wind power plants in the Federal Republic of Germany continues to expand. Up to now an ever increasing demand for wind power plants can be stated. Reasons for this are on the one hand the fact that the Federal Research Ministry (BMFT) has increased the maximum permissible level of wind generated energy fed into the grid in a large scale test program from 100 to 250 MW and on the other hand the subsidies of the Laender and also the legal payment increase of currently 16.5 Pfg for each kWh of power fed into the mains supply. As a consequence, the amount of wind power in use in Germany has risen from 109 MW in 1991 up to 174 MW at the end of 1992. This represents an increase of about 60% (Fig. 1).

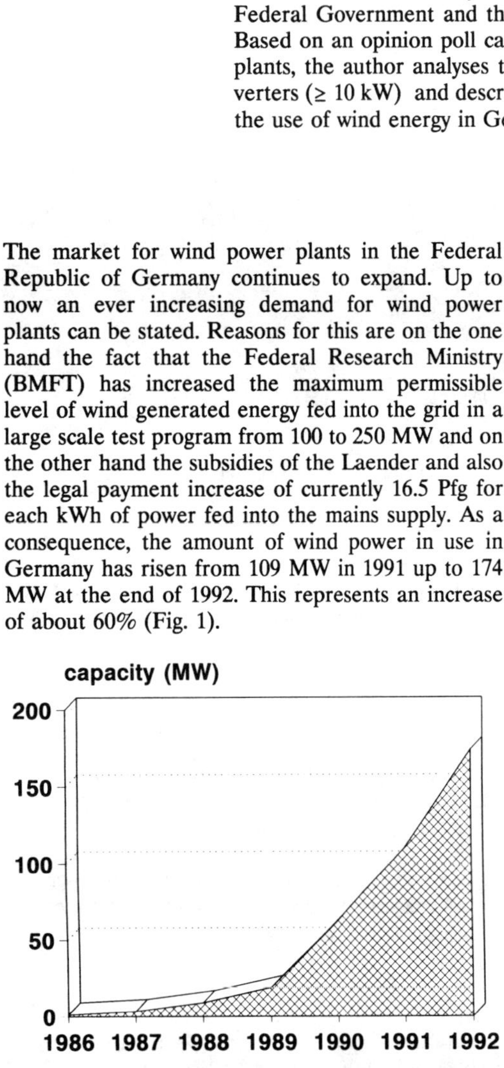

Fig. 1: Development of the total capacity of wind turbines in Germany

The aim of this presentation is to analyze the market situation for wind power plants linked with the main grid in Germany. Questionning of wind power plants suppliers showed that in the period from 1982 to 1992 1,133 turbines were installed in Germany altogether. The output of the turbines in 1992 of 64.5 MW is so far the highest annual output to date and compared to 1991 (47 MW) represents an increase of nearly 37%. In terms of the quantity of wind converters the figure is increasing by about 30% to 364. As a consequence of the high demand more and more suppliers (+15%) are forcing themselves into the wind power plants market. Measured in terms of turnover its market volume rose about 30%, i.e. from 210 to 220 million German marks. Including extras, buyers made an average investment of about 3.3 million DM per MW installed capacity in 1992.

Wind power plants tend to be built near the coastline rather than inland, reflecting the varying potentials according to locality. In comparison to the previous year a shift can be seen on the subject of wind power plants being installed, i.e. the output of turbines in coastal regions and inland. While the number of new turbines which were constructed in coastal areas rose in 1992 by more than 35% to 231 (1991: 171), the corresponding quantity inland rose only about 21% to 133 (1991: 110) turbines. With reference to the overall wind power capacity of 64.5 MW 1992 in Germany, about 81% (51.9 MW) were installed in coastal regions and about 19% (12.6 MW) in the inland (Fig. 2).

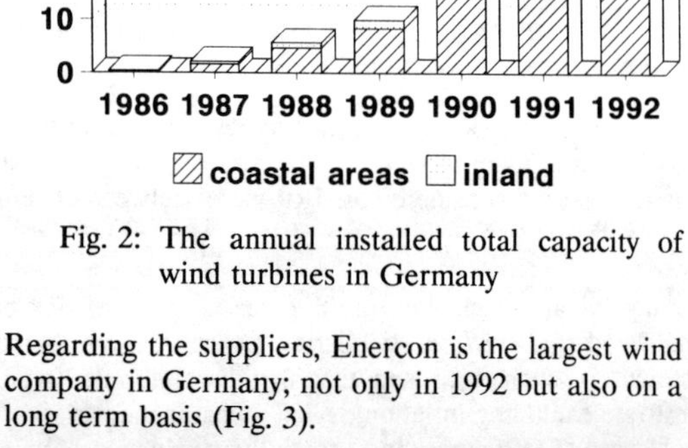

Fig. 2: The annual installed total capacity of wind turbines in Germany

Regarding the suppliers, Enercon is the largest wind company in Germany; not only in 1992 but also on a long term basis (Fig. 3).

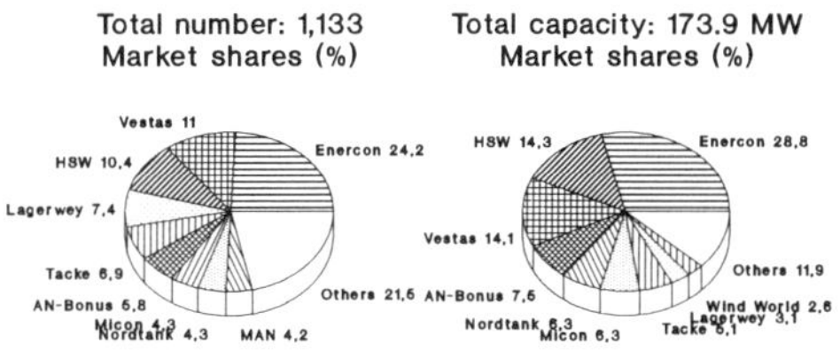

Fig. 3: The German market according to number and capacity (1982 - 1992)

With an capacity installed rate of 17.6 MW in 1992, Enercon is leading the "Top 5", followed by Vestas (7.7 MW), Nordtank (7.5 MW), AN-Bonus (7.1 MW) and Micon (5.4 MW).

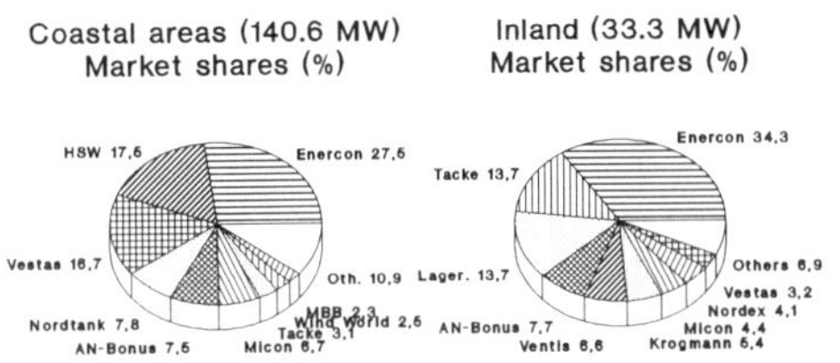

Fig. 4: The German market according to installed capacity for coastal areas and inland

With regard to the regional activities of the suppliers concerning coastal and inland areas, the structure of the wind power market is rather revealing. In the period from 1982 to 1992 Enercon, HSW, and Vestas have achieved a market share of about 62% (86,7 MW) of the overall wind power capacity installed in the coastal regions (140,6 MW), these manufacturers suffered a decline to 47% (24,6 MW) in these regions in 1992. The year 1992 saw some striking shifts of market shares in inland areas. Regarding the wind power capacity installed (12,6 MW), Enercon (29,6%) is now followed by the suppliers Tacke Windtechnik (14,7%), and Nordex (8,4%).

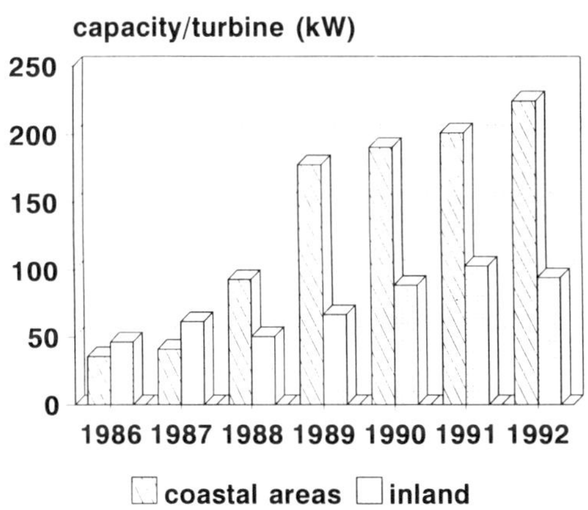

Fig. 5: Development of the annual on an average installed capacity per wind turbine in Germany

Although the average capacity of wind power plants has increased from 167 kW in 1991 up to 177.2 kW in 1992, differences can be noticed for the development in coastal regions from inland areas (Fig. 5). While in coastal regions the average figure rose from 205 kW up to 225 kW, this number decline from 107 kW (1991) to 94,5 kW inland.

Altogether, the expected trend to larger wind turbines has also been verified in 1992, although the 500 kW-class does not yet effect the figure. Nearly all bigger producers work on a production-type or rather on the improvement of therir large plants in this capacity-catagory; the new impulses will first effect the German market in terms of figure in 1993/94 [2].

The prices for turbines will increase in 1993 on an average of 4,2% compared to 1992, whereat the range - depending on the type of the plant - extends from 0% to over 10%.

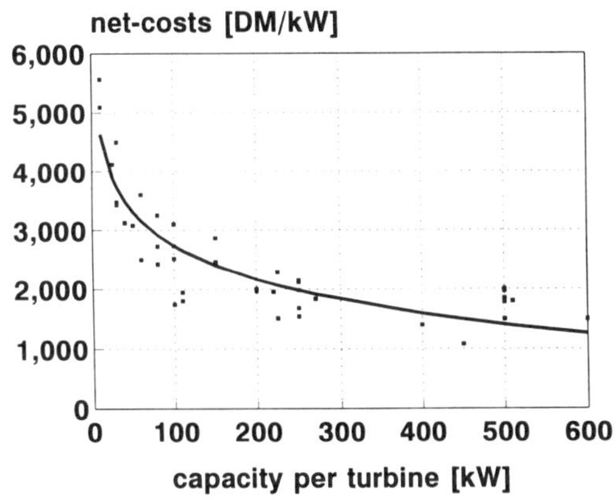

Fig. 6: The specific net-costs per kW depending on the installed capacity (dating: Jan. 1993)

Fig. 6 shows the specific net-costs, whereat for the price (including delivery and setting up) those prices were considered which are related to a turbines height of 30 m. Regarding the course of the function it becomes obvious, that the specific costs (DM per kW) - relatively - decrease with the increasing of installed capacity.

Concerning the subsidies of wind energy in Germany it only seems to be obvious, that the financial ressources are going ro run short and the average support for use of wind energy will be reduced [2]. The new quiding rules for subsidies being expected for the northern Laender of Germany (Schleswig-Holstein, Niedersachsen und Mecklenburg-Vorommern) have led to reinforced endeavour of the producers to improve the quality of their plants concerning the energy output and the noise emission.

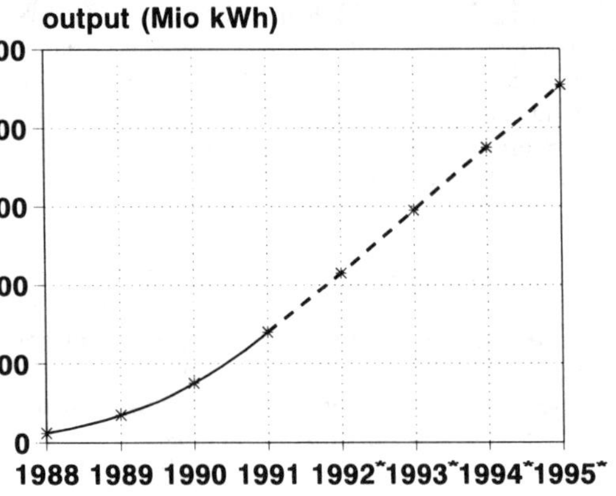

Fig. 7: Development of the electricity output of wind turbines in Germany (* expected)

The output of wind produced power has risen to the level of about 140 million kWh in 1991 and is expected to reach 450 million kWh by the end of 1995 under stable conditions of support (Fig. 7). By that time 1,800 turbines will have been installed in Germany with a total capacity of about 300 MW [2]. Since there is no follow-up support programme in sight after the BMFT-programme will have finished, the suppliers of wind power plants in Germany have to face fierce competition. Before the 250 MW programme was started there were only about ten worth mentoining manufacturers on the German market, whereas now there are over 30. From the economic point of view this number is too high considering a market volume of 210-220 million German marks. Without a follow-up support programme, in a few years time the number of suppliers could possibly fall under ten owing to company fusions and bankruptcies.

Concerning the subsidies, 1993 will be again a good wind year, although signs going along with the first announced lowering of the prices of an aggravating competition can be seen. Many producers try to oppose this by being present with their turbines in as many capacity categorys as possible or rather by looking for new markets through export.

REFERENCES

[1] Allnoch, N. (1992): Expansionskurs hält noch an. Der Markt für netzgekoppelte Windkraftanlagen in Deutschland. In: Energie 44, 10, S. 24-28.

[2] Carstensen, U. (1993): Förderung im Tief. In: Wind Energie Aktuell 3,1, S. 8.

NEWS FROM THE EUROPEAN WIND TURBINE DATABASE EUROWIN

Klein, Hans-Peter; Schmid, Jürgen
Fraunhofer-Institut für Solare Energiesysteme (FhG-ISE), Oltmannsstrasse 22, 7800 Freiburg
Tel.: 0761/4014-179/216
Fax : 0761/4014-100/217

Abstract :

We are approaching rapidly the 1000 MW threshold of total installed wind jenergy capacity in Europe. There is now a total installed wind energy capacity of about 822 MW registered and it can be expected that in summer 1993 we exceed the 1000 MW margin in the EUROWIN database. Ambitious targets concerning the development of wind energy technology lead in nearly all of the member countries of the European Community to a remarkable dynamic in the installation of new wind energy capacities. The EUROWIN project is financed by the Commission of the European Communities DG XII whithin the JOULE II - programme. In a network of wind energy experts coming from 10 member countries of the Community the development of European wind energy technology is investigated and evaluated by different criteria. The latest publication is a report comprising the results until september 1992, see [1]. Beside the statistical evaluation of performance data until end of 1991 , national reports from 11 authors give an actual survey of the present wind energy situation until september 1992. The contribution to the ECWEC '93 shows an actual update of the EUROWIN database and some examples of the mentioned report "EUROWIN - Annual Reports 1990+1991".

1 Introduction :

The following update and the mentioned report, see [1], summarize the most important results obtained by the evaluation of performance data coming from European wind turbines in the years 1990 and 1991 and 1992.

The evaluations show that besides the traditionally leading countries such as Denmark, The Netherlands and Germany, also Spain, Great Britain, Greece and Portugal are increasing its activities based on favourable wind conditions which can be found in their coastal areas.

The following Table 1 is an update of the database from 04.03.1993 which surveys the installed and registered wind energy capacities and some preliminary results of performance data. The different column shows the actual number of registerd **units**, the actual registered **capacity** in [MW], the reported capacity in 1991 and 1992 in [MW], likewise the reported cumulative **energy generation** in 1991 and 1992 in [GWh]. The numbers in brackets ( ) indicate the preliminary results for 1992, final results shall be given after finishing the evaluations for 1992 in a consecutive report by the end of 1993.

Table 1: *Update of the EUROWIN database on March 4th 1993 for participating European countries in the EUROWIN project*

country	units registered 1993	capacity registered 1993 [MW]	capacity reported 1991 [MW]	reported 1992 [MW]	energy generation reported 1991 [GWh]	reported 1992 [GWh]	status
EC member countries :							
B	23	4.6	2.4	(2.4)	5.3	(3.6)	30.09.92
D	922	140.8	40.1	(112.6)	52.0	(194.2)	31.12.92
DK	2750	372.7	289.7	340.9	510.2	553.8	31.12.92
E	301	31.1	3.1	9.2	3.25	(6.1)	31.01.93
F	10	0.58	0	( )	0	( )	31.08.92
GB	273	87.64	.46	( )	.56	( )	25.02.93
GR	134	24.4	0	( )	0	( )	02.03.93
I	48	9.96	0	(1.45)	0	(0.936)	04.03.93
IRL	21	6.45	0	( )	0	( )	31.12.92
L	0	0	0	0	0	0	31.12.92
NL	554	116	44.5	68	57.1	(99.9)	26.02.93
P	70	8.57	0	( )	0	( )	13.10.92
SUM*EC*	**5106**	**802.48**	**380.2**	**(533.0)**	**628.7**	**(857.9)**	**04.03.93**
Not-EC countries :							
CH	( )						
N	( )						
S	88	19.0	6.39	(19.0)	13.29	32.49	15.02.93
SF	7	1.39	0	(1.39)	0	2.37	26.02.93
SUM*Not-EC*	**95**	**20.39**	**6.39**	**(20.39)**	**0**	**34.9**	**04.03.93**
Total Europe :							
SUM*Total*	**5201**	**822.9**	**386.41**	**(553.4)**	**642.2**	**(892.8)**	**04.03.93**

update : 04.03.1993

The different columns in **Table 1** mean :

country

B	Belgium
CH	Switzerland
D	Federal Republic of Germany
DK	Denmark
E	Spain
F	France
GB	Great Britain
GR	Greece
I	Italy
IRL	Ireland
L	Luxembourg
N	Norway
NL	The Netherlands
P	Portugal
S	Sweden
SF	Finland

units		number of registered WEC-units at the date of status
capacity		
	registered	total registered wind energy capacity of WECs installed and in operation in [MW]
	reported	reported wind energy capacity with performance data in [MW]
energy generation		
	reported	reported cumulative wind energy generation in [GWh]
status		date of last input to the database
(#.#)		preliminary results final results shall be given in a consecutive report in autumn 1993

The following figures survey some significant trends in the development of European wind energy technolgy.

2 Trends in the development

Figure 1 shows the distribution of wind energy capacity in Europe and its increase between August 1992 and March 1993.

Significant are the increase rates especially in Great Britain (now 87.64 MW), Spain (now 31.1 MW) and Greece (now 24.4 MW). The Netherlands have exceeded the 100 MW margin and have an installed capacity now of 116 MW.

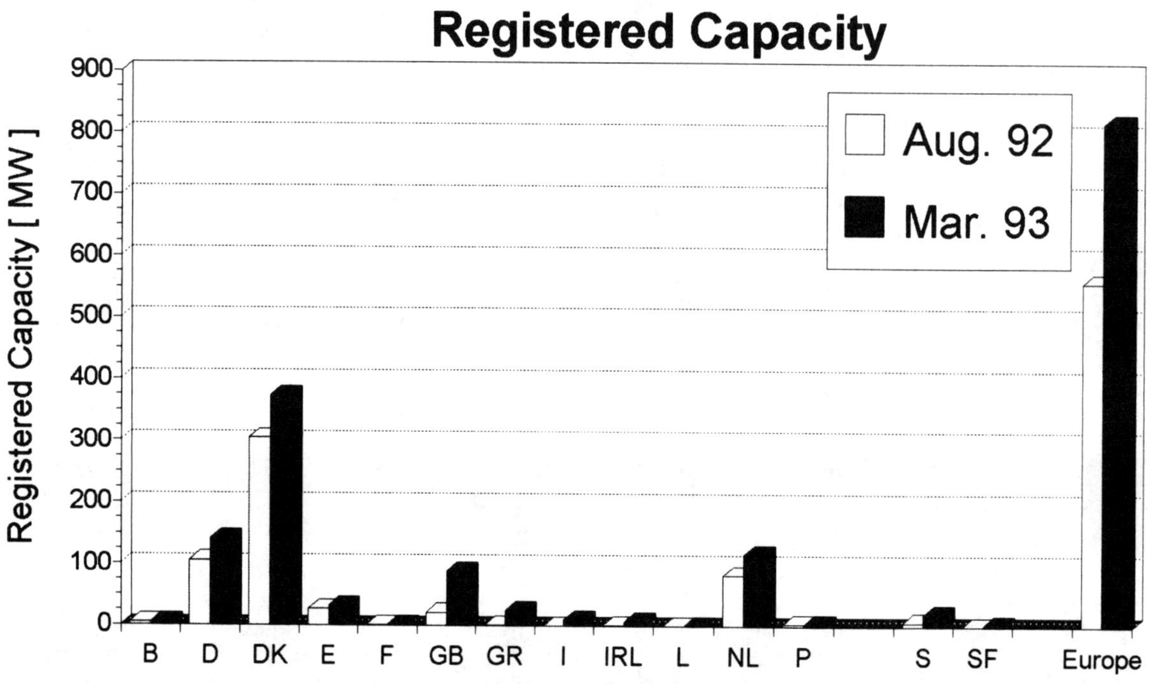

Figure 1 : Distribution of wind energy capacity in Europe and its increase between September 1992 and March 1993

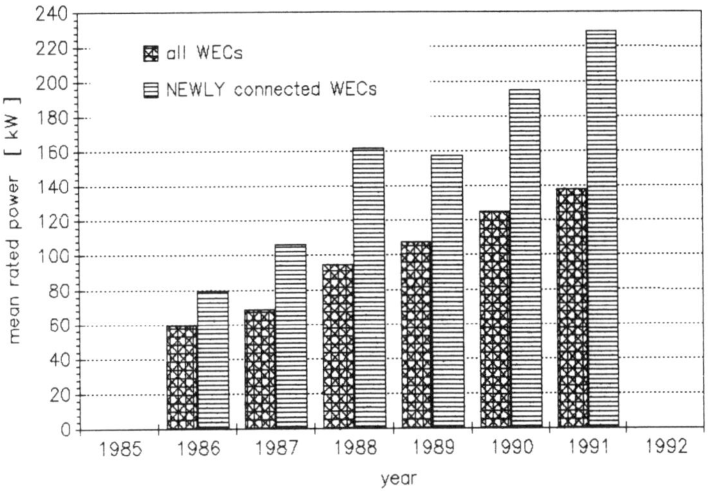

Figure 2 : *Development of the mean rated power for all and newly connected machines in Europe between 1986 and 1991*

In 1991 the average rated power was about 130 kW for registered WECs and about 230 kW for the newly connected machines.

2 Energy generation

While the evaluations from 1992 are not yet completed the following Figure 3 shows an example from the mentioned report, see [1], concerning the annual profile of mean specific energy in [kWh/m² month] and corresponding capacity factor [%] which indicates the mean percentage rated power as an average from all evaluated machines.

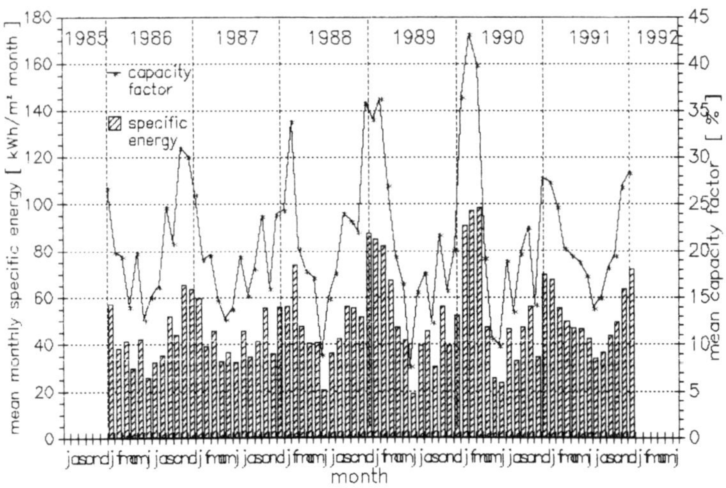

Figure 3 : *annual profile of mean specific energy and corresponding avg. capacity factor of European wind turbines from January 1986 to december 1991, from [1]*

The following Table 2 shows the avg. monthly specific energy in [kWh/m² month] for different member countries of the EC between 1991 and 1991.

Table 2 : *Average specific monthly energy of operational wind turbines in Belgium, Denmark, The Netherlands, Germany, Spain and total Europe from [1]*

	Belgium [kWh/m²month]	**Denmark** [kWh/m²month]	**The Netherlands** [kWh/m²month]	**Germany** [kWh/m²month]	**Spain** [kWh/m²month]	**Europe** [kWh/m²month]
1986	xxx	45.3	31.5	xxx	xxx	44.5
1987	xxx	42.6	24.4	xxx	xxx	41.9
1988	xxx	51.5	50.3	32.5	xxx	51.1
1989	xxx	51.9	37.5	19.4	xxx	50.2
1990	100,9	55,7	51,2	47,4	45,9	55.3
1991	94,1	52,6	52,8	36,9	40,7	50.8
avg.	**97,5**	**51,02**	**46,45**	**36,43**	**43,99**	**50.14**

The averages are 'weighted' with the number of evaluated units respectively the corresponding reported capacity. For example Belgian machines (they are all from the Zeebrugge windfarm) achieved in 1991 an avg. of 94.1 [kWh/m2 month] at a reported capacity of 2.4 [MW], see Table 1. In comparison Dutch machines have an average in the same year of 52.8 [kWh/m2 month] at a reported capacity of 44.5 [MW], see Table 1.

Table 3 : *List of Top 10 rated WECs with highest specific energies per year between 1986 and 1991*

ranking order	**1986** [kWh/m²year] [kW]/[m]/c/y	**1987** [kWh/m²year] [kW]/[m]/c/y	**1988** [kWh/m²year] [kW]/[m]/c/y	**1989** [kWh/m²year] [kW]/[m]/c/y	**1990** [kWh/m²year] [kW]/[m]/c/y	**1991** [kWh/m²year] [kW]/[m]/c/y
1	1120 75/17/DK/85	1177 95/19.4/DK/86	1343 200/25/DK/?	1579 200/25/DK/88	1662 200/25/DK/88	1554 50/20/E/89
2	1096 75/17/DK/85	1099 95/19.5/DK/86	1271 95/19.4/DK/86	1418 200/25/DK/88	1505 130/19.4/DK/88	1474 200/25/DK/88
3	1083 55/15/DK/84	1041 75/17/DK/85	1193 150/23/DK/87	1334 95/19.6/DK/86	1487 200/25/DK/88	1424 200/25/E/90
4	1081 45/12/DK/80	1020 75/15/DK/86	1171 130/20.5/DK/87	1329 150/23/DK/88	1457 200/25/DK/?	1378 225/27/DK/90
5	1036 75/17/DK/85	995 95/19.4/DK/87	1163 150/23.5/DK/?	1319 200/25/DK/88	1408 150/20/E/89	1378 200/25/DK/90
6	1026 75/16/NL/86	953 99/20.5/DK/86	1149 95/19.4/DK/86	1308 150/23/DK/88	1404 150/23/DK/88	1378 225/27/DK/90
7	1021 75/17/DK/85	947 55/15/DK/84	1137 95/19.5/DK/?	1305 200/25/DK/88	1362 200/25/DK/88	1378 225/27/DK/88
8	1017 55/11/DK/84	940 95/19.4/DK/87	1127 80/17/DK/86	1285 200/25/DK/?	1354 150/23/DK/88	1344 200/25/DK/?
9	993 75/17/DK/85	939 80/17/DK/86	1114 75/17/DK/85	1282 130/19.4/DK/88	1311 200/22.5/B/85	1341 250/26/NL/89
10	964 55/16/DK/83	933 75/17/DK/85	1106 95/19.5/DK/?	1190 130/20.5/DK/88	1303 250/24/DK/88	1332 130/19.4/DK/88

example: Nr. 6 1986 1026 [kWh/m² year]/ 75 kW nominal power/ 16 m rotor diameter sited in The Netherlands/ starting year 1986

References :

[1] Klein, H.P.; Schmid, J.; October 1992
"EUROWIN-Annual Reports 1990+1991"
Fraunhofer-Institut für Solare Energie Systeme
Oltmannsstrasse 22, D-7800 Freiburg

[2] Schmid, J.; Klein, H.P.; 1991
"Performance of European Wind Turbines, A Statistical Evaluation from the European Wind Turbine Database EUROWIN"
Elsevier Science Publishers Ltd.
Crown House, Linton Road, Barking, Essex IG11 8JU, England

STATISTICAL WEC DESIGN AND COST TRENDS

J. P. Molly, A. Keuper, M. Veltrup
Deutsches Windenergie-Institut (DEWI)
D - 2940 Wilhelmshaven, Ebertstr. 96
Germany

During the Glasgow EWEA Wind Energy Conference in 1989 a presentation on statistical design and cost parameters of wind energy converters combined with a prognosis concerning the economics of WECs with growing sizes was given by the author. Today WECs of up to almost 40 m rotor diameter are commercially available on the market. This means a doubling of the installed unit-power from 250 kW to 500 kW and a diameter increase of 15 m within four years. This paper compares the prognosis of 1989 with the statistically evaluated technical and economical development reached today in order to show, if the design changes and economic progress happend as expected.

Focussing on the specific tower head mass per rotor disc area unit a clear tendency to light weight machines can be stated. Three years ago the average maximum of the mass specific rotor torque was about 5.3 Nm/kg. Today the leading machines reach values between 6 and 7 Nm/kg. These technical advantages must be combined with the cost development per kg machined tower head material which is between 22 DM/kg to 40 DM/kg for rotor sizes from 35 to 40 m. In general high mass costs are associated with low tower head mass and vice versa.

Finally the energy costs are calculated for the energy production of one year at a reference site. The result shows that the trend predicted in 1989 is still correct, as well as the absolute value calculated for the energy production costs which diminish with growing size. Assuming mass specific torque values of up to 10 Nm/kg cost competitive WECs of the one megawatt class can be expected within the next years.

1. INTRODUCTION

Since the end of the seventies several megawatt class wind energy converters (WEC) have been built and operated worldwide. The first generation, with rotor diameters of 75 to 100 m and rated power of two to three megawatts, was operated under more or less severe restrictions due to various technical problems. After this first courageous step full of technical innovations, but little success, the engineers stuck to a more conservative and smaller design for the second generation of large WECs. With about 60 m diameter, three blades and close to one megawatt installed power the designers hoped to have found the solution for cost competitive large WECs. But the raised hopes could not be fulfilled, in spite of the fact that in some cases the same development teams were involved.

These facts gave reason for studies on the expected cost development with growing WEC sizes. Hau (1) predicted size independent energy generation costs between 30 m and 100 m rotor diameter, whereas Molly (2) expected a slight decrease until about 60 m, followed by slight increase until about 90 m. So far the two results showed very similar trends.

To underline this trend calculated and statistical data had been compared with the theoretical result. Today, about four years later, it seems to be interesting to compare the expected trend with the actually obtained technical and economical standards of the WECs. All costs of the year 1992 are corrected to 1989 currency values in order to make the data comparable.

2. TOWER HEAD MASSES

The important parameter 'tower head mass' has a clear tendency towards light weight machines. As Fig. 1 shows, all specific masses (tower head mass divided by the rotor disc area) of actually available medium size WECs between 25 and 40 m rotor diameter are belonging to the light class. The connection lines between the dots represent the size development within the product line of a single manufacturer. These product trends, extrapolated to larger diameters, let grow the expectation to find future cost competitive large WECs (45 m and more) in the light weight class.

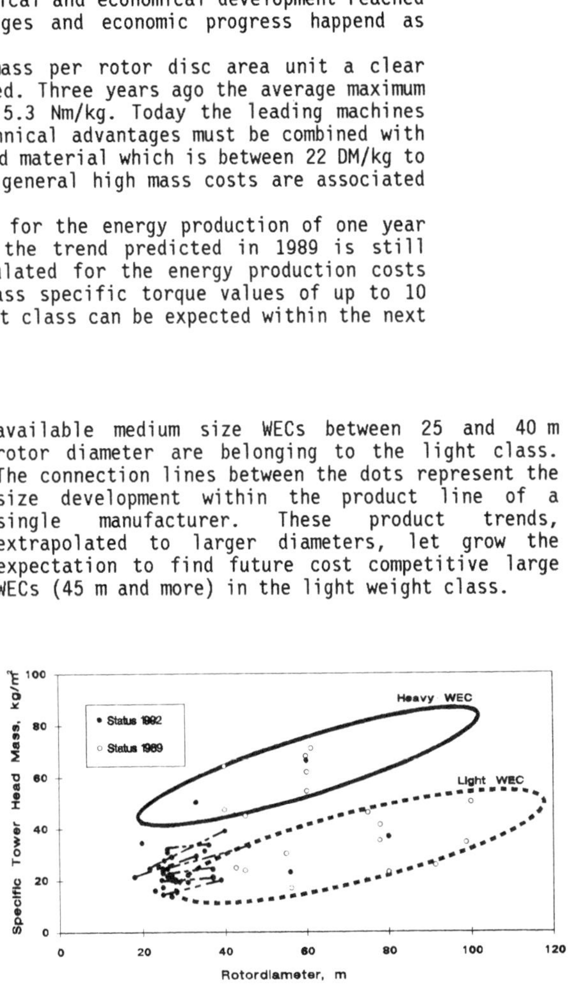

Fig. 1 Tower head mass development with rotor diameter. Connection lines show the in-house product line development with size of some manufacturers.

Nevertheless, a weight difference of up to 100 % between the lightest and heaviest WECs of same size can be found. This indicates, at least for part of the todays commercially available WECs, a considerable mass and therefore also cost reduction potential. The relation between mass and cost reduction can not be proportional, because the cost intensive material machining time will not change significantly. Of course a mass reduction and thus making a better use of the material is only possible, if load cases and material properties are well known. In 1989 the top value of a 25 m WEC for the mass specific torque (rotor torque divided by tower head mass) was about 5.2 Nm/kg (Fig. 2).

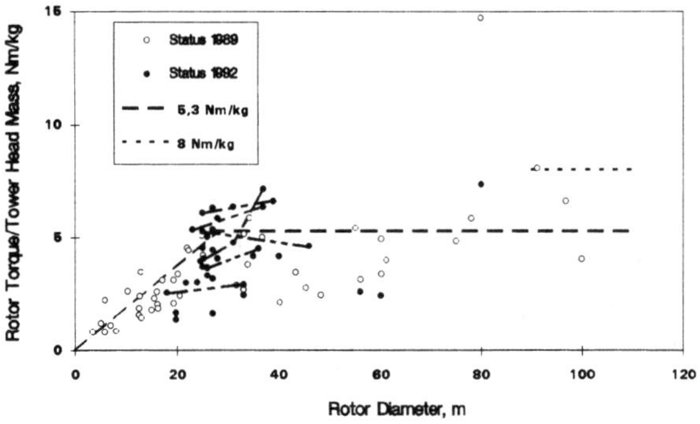

Fig. 2 Mass specific rotor torque development with size. Connection lines show the in-house product line development with size of some manufacturers.

Today WECs of more than 7 Nm/kg are produced in series, an improvement of nearly 40 %. Also the company specific product lines show an increase in the material utilization factor, but in most cases with a lower gradient. The average gradient for all WECs indicates a potential for further increase of the mass specific torque up to 7.5 Nm/kg for WECs of 40 m rotor diameter and up to 10 Nm/kg for WECs of 60 m diameter. That this prediction is not

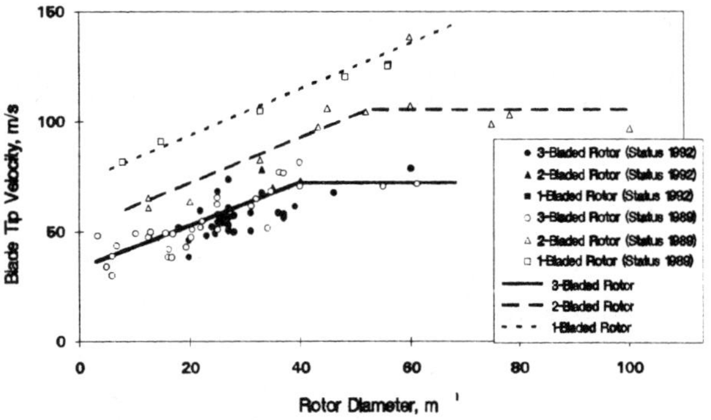

Fig. 3 Blade tip velocity development with size and blade number.

totally impossible is corroborated by the initial design value of the AEOLUS II with 15 Nm/kg at 80 m rotor diameter, a value which corresponds very well to the assumed gradient. For unknown reasons however, this value decrease to 7.3 Nm/kg in the version finally built.

The rotor torque for a given power output can be diminished by an increase of the rotor speed. In Fig. 3 the change of the rotor tip speed with size is shown. If the today realized tip speeds are compared with those four years ago, no increase can be stated. In contrary, a little tendency towards lower tip speeds is visible. Certainly the reason can be found in the rotor noise emission which grows overproportionally with the increase of the tip speed. In this case the existing noise protection laws and not load considerations determine the technical feasibilities. Therefore research in the field of aerodynamic noise reduction needs to be done urgently in order to shift the noise relavant tip speed to higher values.

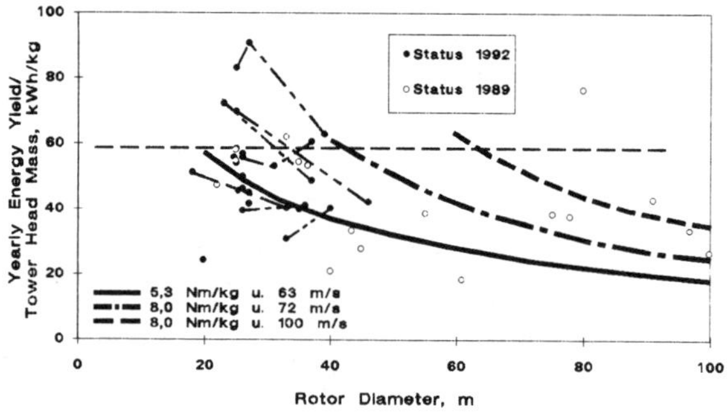

Fig. 4 Theoretical up-scaling function for the mass specific yearly energy yield compared with some real WEC data. Connection lines show the in-house product line development with size of some manufacturers.

A good measure for the economy of a WEC is given by the yearly energy yield by unit mass of the tower head. It had been shown (2) that this value is invers proportional to the rotor size. The three curved lines in Fig. 4 show the theoretical up-scaling curve for three different WEC design parameter sets. Some company product lines for growing WECs follow the theoretical tendency thus underlining the correctness of the up-scaling formular. Others, who have increasing values with size obviously changed their design, in such a way reaching better mass specific energy yields. In 1989 the best WEC produced about 58 kWh/kg and had 25 m rotor diameter, in 1992 the best value found was 90 kWh/kg with 27 m size. But other designs are also above the old best value, specially some with diameters near to 40 m.

3. COST DEVELOPMENT

Of course the economy of a WEC will be improved only if the price per kilogramme machined material grows less than the mass specific yearly energy production. In Germany in 1992 WECs may have been more expensive than necessary due to the different subsidies available. Fig. 5 shows the 1992 situation (corrected to 1989 currency values) compared with data from 1987 to 1989 which at that time were derived mostly from Danish WEC prices. The prices are related to the rotor disc area. Remarkable is the slight increase from 25 m on

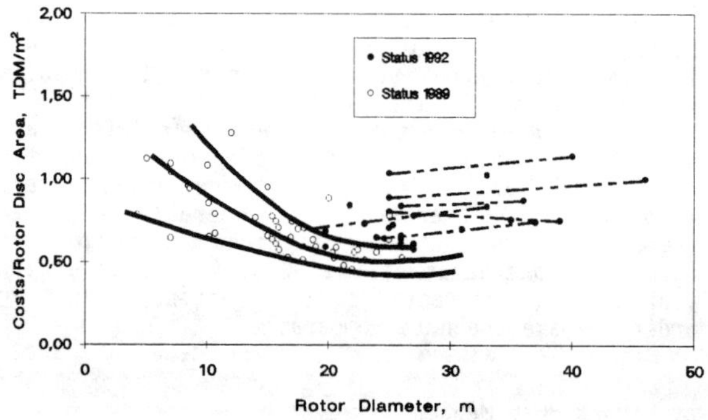

Fig. 5 Rotor disc area related WEC prices of today, corrected to 1989 currency values.

towards larger rotor diameters within the in-house product lines. This tendency results in an increase of the energy cost, if the additional gain caused by the higher hub heights of larger rotors does not compensate the area specific cost growth. As known from the market development, larger WECs are more economic than smaller ones and therefore are prefered by the customers, i.e. the area specific costs rise less than the gradient given by the energy increase with height.

An other interesting measure is given by the cost per machined unit of the tower head mass. For simplicity reasons the price used in the evaluation includes the tower, because it is quite difficult to get the prices for the towers seperately. Fig. 6 shows at 25 m rotor diameter a large scatter of mass specific prices from 23 DM/kg to 62 DM/kg. Above 35 m rotor diameter the scatter reduces to 26 DM/kg to 41 DM/kg. It is not clear, if the lower boundary really increases with size or if this is

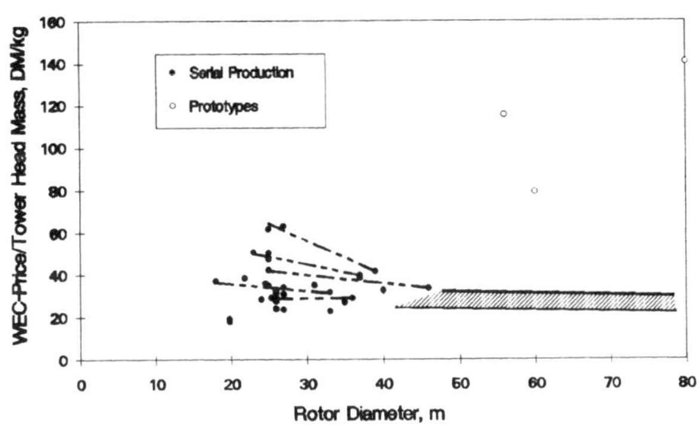

Fig. 6 Cost development per unit tower head mass with size. Connection lines show the in-house product line development with size of some manufacturers.

only caused by the lack of data. As a rule, high mass specific costs go along with low tower head masses and vice versa. In-house product lines in general show a cost reduction with size which all together seem to concentrate in a band of 25 DM/kg to 30 DM/kg for WECs larger 50 m diameter (Fig. 6).

For a quick comparison of the achieved economics of different WECs the costs ex works devided by the yearly energy yield give an appropriate measure. All costs were again corrected to 1989 currency values. The energy is calculated for a site with an annual wind speed of 6 m/s at 10 m height using a Rayleigh distribution and a vertical wind speed according to the one-seventh power law. Together with the power curve of the WEC the cost development with size of Fig. 7 is found. The data for the real WECs as well as their tendency towards larger diameters are still within the prognosis of 1989. Some of the manufacturers have growing costs with size for their in-house product lines. Others show decreasing specific energy costs, which coincide with the prognosis. Taking all data together, an average cost decrease can be stated, also following the predicted line. Which future cost development will occure is impossible to say at the moment. Four years ago an optimistic and a conservative assumption for the future energy cost development had been established. Some product lines which reached the threshold of 40 m size seem to follow the conservative forecast, others have the tendency to fall further.

The described analysis still allows to assume an energy cost reduction for the now coming 1 MW class of some European manufacturers. These WECs will have rotor diameters of 50 to 55 m according to the development subsidized by the Commission of the European Communities. Their market introduction can be expected for 1995/96 which means that commercially available WECs at this point will have reached twenty times the installed power of those built 10 years earlier.

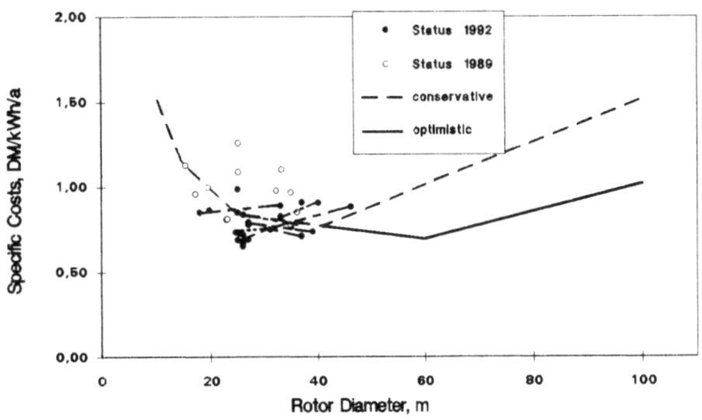

Fig. 7 Electricity generation costs related to the yearly energy yield. Connection lines show the in-house product line development with size of some manufacturers.

4. LITERATURE

(1) **Hau, E.**, et al. Next Generation of Large Wind Turbines. Final Report, München, Mai 1991

(2) **Molly, J. P.**, Maximum Economic Size of Wind Energy Converters. EWEC'89, 10.-13. Juli 1989, Glasgow

(3) **Klein, H.**, EUROWIN, Report on Operational Data 1986/87

REALIZATION OF WINDPARK PROJECTS BY MULTIPLE OWNED OPERATING COMPANIES

Dipl. Ing. Henning Holst
WINKRA GmbH, Zingel 5, W-2250 Husum
Germany
Tel: +49 4841 6044
Fax: +49 4841 81886

ABSTRACT: In Germany in the last 2 years private operating companies, mostly of farmers and locals have been established to realize large windfarm projects. This paper points out the approach and the advantages of projects handled by these companies.

1. INTRODUCTION

The large number of single applications for the federal 250 MW wind programme has led the provincial governments in Germany to release planning issues for the determination of windfarm areas.

The applicants, mostly farmers with relatively small sites, planned to erect single turbines in most cases. They were faced with numerous difficulties.

2. DISADVANTAGES OF SINGLE OPERATORS

Some of the reasons to alter the planning approach are listed as follows:

2.1 The visual impact of many single machines in a rural area leads to the resistance of the local population.
2.2 The capacity of a soft grid is limited. The potential operators then started to struggle for these capacities.
2.3 According to different financial means of these investors there were applications for turbines varying in rated power, hub height, appearance etc.
2.4 Some of the sites were not approvable.
2.5 Depending on the site the energy yield can vary up to 25%.
2.6 The financial risk for single operators is very high.
2.7 There are no cost saving effects because every machine would have been installed separately.
2.8 A fair distribution of limited subsidies is not possible.

3. CONSEQUENCES

In order to establish feasible projects with political support, WINKRA is working as a consultant especially on wind projects to be installed by multiple owned operating companies. Unlike the Danish model for cooperatives, these companies are neither limited in the size of the shares nor in any other way. In the presented case WINKRA is doing the planning and consulting work for 2 windfarm areas on the island of Fehmarn. The former single operators grouped together and established 2 large operating companies.

4. ADVANTAGES OF THE ACHIEVED SOLUTION

All involved parties benefit from the results of coordinated planning and a jointly owned and developed project:

4.1 The visual impact of 2 large windfarms is concentrated on 2 limited areas.
4.2 The windfarm areas have been chosen in an optimal size and shape and on the most suitable site.
4.3 The connection into the 30 kV grid allows for a maximum capacity.
4.4 The most suitable machines can be chosen.
4.5 The investors can purchase shares of the project according to their financial means.
4.6 Subsidies are given to one big project, all investors are benefiting equally.
4.7 Significant price reductions are possible for purchasing the machines, external costs, maintainance and insurance.
4.8 Public acceptance rises.
4.9 Political support is given preferably to big projects and joint efforts.

By shaping these operating companies, the originally planned capacity of various single machines could be raised significantly by the use of 500 kW turbines. This enabled the operating companies to sell additional shares and thus raise the acceptance. The locals who have to accept the turbines in their surrounding should benefit in the first place from renewable energy projects. Multiple owned operating companies of local people are a good basis for a popular implementation of wind energy.

5. CONCLUSION

By shaping multiple owned operating companies many local people can benefit from harnessing wind energy. The planning issues are much easier to meet by a jointly owned project. Public acceptance is no issue if locals have the choice to be financially involved in wind energy installations. We are convinced that this model is applicable in other countries of the community and allows small investors to benefit from the implementation of renewable energy sources.

The European Community Wind Energy R&D Programme within JOULE II

by

W. Palz, G. Caratti, K. Diamantaras, A. Zervos

Commission of the European Communities
Directorate-General XII for Science, Research and Development
200 Rue de la Loi
1049, Brussels, Belgium

Abstract

Within the 12 countries of the European Community, the total installed wind capacity has reached today about 1000 MW. This expansion has been brought about with the aid of large programmes of research, development and demonstration on the part of the Member States and the Commission of the European Communities.

Following formal assessment procedures, the Commission of the European Communities has put together the JOULE II programme from the proposals submitted in response to a general call (deadline February 1992). This paper describes the wind energy segment of the programme which runs from 1992 to 1994.

The section of Renewable Power Plants in the wind sector of the JOULE II programme is divided as follows - showing also the funding for the different elements which reflect the Commission's priorities based on the assessment of the proposals received rather than regulatory requirements which are not laid down in advance at this detailed level:

- Large Wind Turbines 7 MECU
- Wind Energy Generic R&D 4 MECU
- Utility Issues 1.5 MECU

a) Large Wind Turbines: while the WEGA turbines were fully experimental machines, the machines of its follow-up programme (WEGA II) have been more oriented towards the market. This action, which is one of the main aims of the JOULE II programme, benefits from a Commission contribution of 7 MECU and involves a total investment of about 25 MECU. The programme compares the "best examples" of several concepts such as stall control, partial pitch, 2 or 3 blades, vertical or horizontal axis etc. All the proposed machines are light weight, seen as the key to low cost, and most have an advanced electrical conversion system.

b) In the Wind Energy Generic R&D sector of the JOULE programme, 15 projects were supported. Support is given to projects dealing with eight different facets of generic R&D which are: rotor design, development and testing; electric conversion system; aerodynamics; design tools; materials and fatigue; structural loads and standards; wind farm integration; and operational experience and environmental aspects.

The projects were chosen with a view to the establishment of a clear link to further development and application on full size wind turbines.

c) Utility Issues: in this section the programme has been constructed in three areas:
- the variability of supply from wind power when integrated into weak utility grids
- integration to strong, mainland grids, and
- wind forecasting for operational control in despatching centres.

The paper describes in detail these activities and while JOULE II is ongoing, details of a supplementary JOULE II call for proposals to be announced within March '93 will be presented.

1. Introduction

Within its pluri-annual and pluri-national research and development programmes in the field of Non Nuclear Energy, the Commission of the European Communities (CEC), Directorate-General XII for Science, Research and Development, has been supporting, for more than a decade, shared-cost projects in the field of renewable energies among which wind energy is one of the priorities.

The last such programmes, named JOULE II (Joint Opportunities for Unconventional and Long-term Energy options) runs from 1991 to 1994 with a total budget for research contracts of 138 MECU. JOULE includes four main energy-related research areas, namely:
i) energy modelling and strategy, ii) fossil fuels, iii) renewable energies (divided into area 3A: solar-derived renewables and area 3B: geothermal energy and deep reservoir geology and iv) energy conservation and rational use of energy.

The budget for research contracts allocated to area 3A in the first round amounted to about 40 MECU, out of which 11 MECU were explicitly reserved for wind energy R&D projects. It should be pointed out that the Commission provides financial support to a project's total cost up to 50% of the eligible costs. Hence, the total budget volume involved in the programme is higher than the double of the support from CEC.

In the frame of a financial supplement of the Third Framework Programme for Research and Technological Development, the Council of the European Communities has recognized the importance of JOULE II by allocating to it additional 104 MECU as a supplement for the years 1993 and 1994 [1]. This was the highest relative budget increase of all the Commission's R&D programmes. As a result, the total JOULE II budget for research contracts was raised to 242 MECU (approximately 60 MECU/year) out of which 90 MECU are earmarked for solar-derived renewables.

While JOULE represents only about 5% of the total public spending in the Community on non nuclear energy [2], it plays indeed an essential role for the development of renewable energy technologies in the Community.

Among its benefits are:

* it creates European networks around key R&D topics

The added value of collaborative research, with the set up of large project addressing the whole chain from basic research to market demand and avoiding duplication of research work, is perhaps more important than the financial support itself provided by CEC.

* it promotes new important initiatives by providing catalytic funding

Very often, the initiative to embark on important strategic actions is first taken by CEC in consultation with the delegations of EC Member States. In other words, the Commission provides "seed money" which then triggers complementary funds from national delegations. From the side of the latters, there is a better comfort in supporting a national project when this is part of larger one internationally acknowledged.

* it balances national efforts

There presently exists a large disparity between Member States on the level of effort and thematic contents of the national programmes in renewable energies. Given the mid-/long-term nature of this research, there are in fact more opportunities for researchers from wealthier EC countries in the North to receive national support than for those from poorer areas in the South. Some countries or regions have indeed hardly any fund available for research in specific technologies. The JOULE programme represents the instrument to re-balance these difference and to enable anybody to contribute to a wide range of technological options, with a view to improve the Community's long-term social and economic cohesion.

The JOULE programme, which is open to anybody in the Community, brings together universities, research centres industries and consultants on specific projects under ad-hoc consortia. The selection is made through open calls for proposals published in the EC Official Journal with the assistance of independent experts and in consultation with the delegations of the EC member States. In the context of the European Enlarged Area agreement, undertakings from EFTA countries may participate fully in the Community framework R&D programmes. Switzerland's recent withdrawal from this agreement has delayed its negotiations, but it is expected that it will effectively come into force as of 1 January 1994. Some Community R&D programmes have also recently been opened up to participants from Central and Eastern European countries and special programme has been created to support the costs of research organisations therein which are wishing to associate.

Since the start of the Community's R&D programme, progress in the wind energy technology has been tremendous. Just ten years ago, the average power rating of commercial wind turbines was 55 kW and now this is 500 kW. The installed cost at good windy sites has practically halved from 0.63 ECU/kWh/year to 0.32 ECU/kWh/y (which means, with 8% interest rate, 20 years economic lifetime and 2% O&M cost, a production cost of less than 0.04 ECU/kWh).
The attractiveness of this figure, when compared to conventional electricity, is pushing wind energy towards self supportance. A Further stimulus will come once the hidden costs of electricity productions -i.e. the externalities such as the effects on environment, security of supply, job creation, subsidies, etc. - will fully be accounted for in these comparisons.

In Denmark, wind energy already supplies about 3 % of the total yearly electricity consumption, but it was calculated that, by replacing existing turbines for today's best commercial ones rated at 500 kW, this share would rise to 10% [3]. Although this is theoretical consideration, as spacing between turbines depends on the size, it nevertheless demonstrates the progress of the technology.

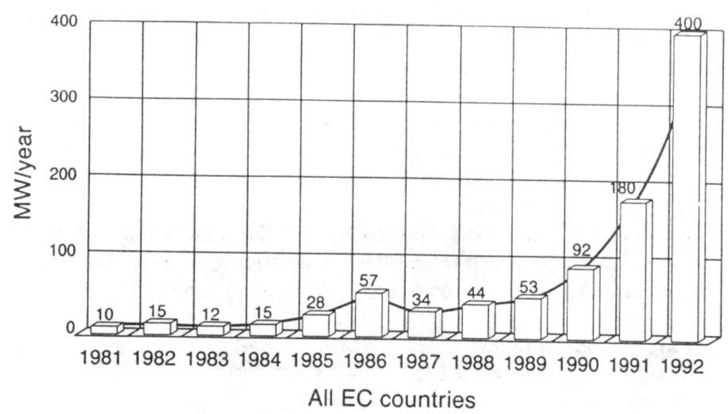

Figure 1. Evolution of yearly wind energy market in the EC

The evolution in wind power in the Community in the last 12 years is shown in figure 1. The yearly market in 1992 was 400 MW, i.e. 40 times as much as in 1981. The total installed wind power has reached the symbolic landmark of 1000 MW and is approaching California, where 1600 MW are installed producing about 3 TWh/y. Figure 2 illustrates the breakdown of wind power among the EC member countries. Denmark has the lyon's share with half of the total EC installed power, but other countries such as Germany, UK and Spain are increasing very rapidly thanks to important programmes which have been set up therein to stimulate private initiative.

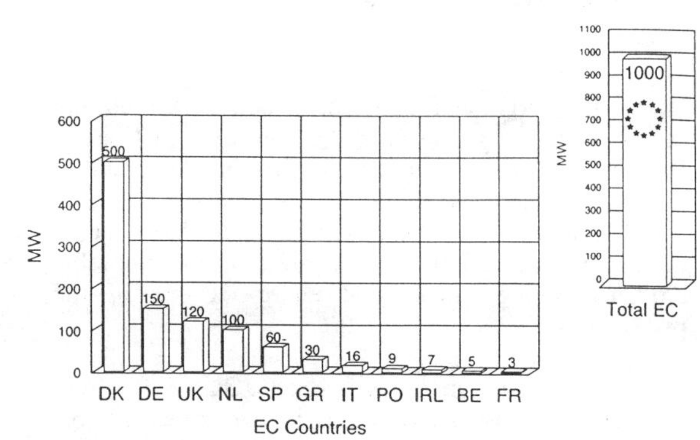

Figure 2. Wind power installed in the EC member countries (end 1992)

Together with the other renewables, wind power will have an important role in the developing world, where 4/5 of the world population presently lives, to provide electricity for rural development and for basic household needs. In China, for instance, the rural private industry communities have passed from 1 million in 1978 to 18 millions in 1993 with an increase in production of 30 %, but this growth is threatened by the lack of energy.

There is a large wind energy potential in India, whose GNP is progressing at a 10% yearly rate. The potential in some of Eastern and Central European countries is considerable, particularly in Russia, Ukraine, The Baltic States and Rumania. Wind turbines can be manufactured locally under licensing agreements and do not require highly specialised maintenance crews. The challenge of providing clean, safe and decentralised energy to developing countries will be an imperative to reduce CO_2 emission in the next century.

Another positive consequence of developing wind power, when compared with imported fuels, is the aspect of job creation. It has been estimated that to produce wind turbines about 4 jobs/$MW_{manufactured}$ are needed [4] and that to operate and maintain wind power about 1 job/$MW_{installed}$ [5]. If wind power supplies 10% of the total electricity in Europe, corresponding to about 100.000 MW [5], then this will bring about 120.000 long-term jobs (100.000 for O&M and 20.000 for production and replacement of wind turbines). This, without taking into account the effects on the induced markets and the other associated benefits such as the effects on rural areas.

It is essential to tackle renewable energies in aggregate and not as individual technologies, they should form part of a dynamic, decentralised and diversified energy supply system. Provided appropriate technology-pushed and market-pull actions are implemented, the share of renewables shall gradually increase to replace conventional and more polluting fuels. The UNCED has recently published the results of a major study developing a world energy supply scenario which predicts that renewable energies would account, by the year 2050, for over 3/5 of the world's electricity market, half of which would be by intermittent renewables, the other half being biomass and hydro [6].

To this end, the action of JOULE II's renewable energy programme is focused towards three application-oriented sectors, i.e. the solar house, renewable power plants, renewables for rural development. Wind energy is concerned with the second and third sectors and the whole of the programme have been re-organised under these headings. For the definition and implementation of the various actions, the Commission is advised by specific groups of operators, such as utilities for the second sector and national research centres and municipal and regional authorities for the third sector.

2. The Wind Energy Programme

The three general objectives of JOULE's wind energy programme are the following:

* to develop advanced large wind turbines

While the size of today's best commercial wind turbines is in the 400-500 kW range, larger wind turbines of advanced and lightweight concept have the potential to become cost-effective in the mid-term and will use the available land more efficiently.

* to support generic wind energy technology

There is potential simultaneously to reduce cost and increase efficiency and availability of existing wind turbines by supporting generic R&D on specific scientific and technical topics yet unresolved.

* to find alternative sites for wind turbines

As the availability of good sites for wind turbines in many European regions is limited, there is a need to identify unconventional sites such as in complex terrain, offshore or having lower wind regimes.

Altogether the programme includes 30 coordinated actions involving many research institutes from the Community and the EFTA countries.

COORDINATOR	PARTNERS	type	diam. (m)	P MW	blades	control	rot. speed	Innovating elements
Nordic WP (SE) LM (DK)	Flender (DE); Vattenfall (SE)	H	52	1.0	2	stall	variable	combin. of stall/ var. speed/teeter hub/2 blades/soft yaw
Enercon (DE)	Aerpac (NL)	H	55	1.0	3	pitch	variable	no gear-box, new generator type, pitch adjustable separately for each blade
Nedwind (NL)	G.H. (GB); Stork (NL); ECN (NL)	H	52,6	1.0	2	stall	constant 2 speeds	optim. of stall charact. via pitch, 4+2 generators, no teeter
WEG (GB)	Pfleiderer (DE)	H	50	1.0	2	pitch	variable	combin. of 2 blade-teetered hub variable speed-pitch control
Heidelberg Motors (DE)	Union Fenosa (ES); RWE (DE); WIP (DE); NTUA (GR)	V	64 (X42)	1.2	2	stall	variable	no gear box, new generator - electrical system
Bonus (DK)	TUD (DK); LM (DK); Flender (DE); Anmu (DE); Risoe (DK); G.H. (GB); Elsam PR (DK)	H	50	0.75/ 1.0	3	stall	constant	adv. nacelle, new blades, tips and airfoils optimised for noise reduct.
Vestas (DK)	Siemens (DE); Hansen (BE)	H	55	1.0	3	pitch	possible variable	new airfoils, new pitch control

Table 1 : Projects supported in the WEGA II programme

2.1 Large Wind Turbines

While the WEGA turbines supported by the Commission in the framework of the JOULE I programme were fully experimental machines, the machines of the present follow-up programme (WEGA II) have been orientated more towards the market. A call was prepared having as a strategy to include an ample variety of different concepts which are promising both from the technical and economic standpoints and to consider the outlook of cost effectiveness, expressed in terms of parameters such as total cost per unit of annual output at a reference site, specific weight of major components and ease of transport and erection.

This action, which is one of the main results of the JOULE II programme, benefits alone of a Commission contribution of 7 MECU (including a general coordination contract) and involves a total investment of about 25 MECU. In table 1, the selected projects are listed along with the main features of the associated machines. Major European turbine manufacturers together with blade manufacturers have replied to this call. In table 2 the main characteristics of the machines in the WEGA and WEGA II programmes are compared. An important progress was achieved in the last five years as demonstrated by the tremendous reduction in the weights in both the blades and the nacelle. The cost of the electricity produced by the new machines, when installed at good windy sites and for series production, can clearly compete with that from conventional power plants.

This action will undoubtedly provide a stimulus to the European manufacturers and keep Europe in the forefront of research in the next future.

The strong commitment of the utilities is a confirmation that a market for this size is opening up in Europe.

	WEGA (1986-1991)	WEGA II (1991-1996)
Rotor diameter (m)	55-60	50-55
Generator rating (MW)	1-2	0.75-1
Blade weight (ton)	6-9	2-4
Nacelle weight (ton)	80-200	29-70
Specific weight (kg/MWh/y)[1]	39-57	12-21
Specific electr. gen. (kWh/m^2)[1]	900-1 300	1 200-1 350
Specific el. cost (ECU/kWh/y)[2]	n.a.	0.3-0.5

1 At reference site, w_s = 7.5m/s at 50m hub, 100% array efficiency

2 for series production (50 unit/year) including installation but no O&M and under the assumption of one year of operation only

Table 2 : Comparison between 1st and 2nd generation of large wind turbines

2.2 Generic R&D

There are still many open scientific and technical issues which need to be resolved. This programme looks at several different aspects of wind energy technology. Particular importance is attached to R&D for future cost reduction, in particular new components - such as carbon-fibre blades, gearless generators to reduce weight and cost of wind turbines- development of standards, simplified and integrated design tools, research on dynamic inflow, operational experience and the reduction of noise. The list of the projects supported by this programme along with the partners and EC contribution is given in Table 3.

Area	Scope	Partners	Grant (kECU)
Wind	Common approach to how to get permission to put up wind turbines in European countries	Elkraft (DK) Preussenel (D) PPC (GR) Iberdrola (E) Vattenfall (S)	139
Wind	Common specifications for the supply of medium-sized wind turbines	ENEL(I) Scottish H (UK) Vattenfall (S) PPC (GR) SEP (NL)	98
Wind	Wind power production tools in central dispatch centres	ELSAM (DK) SEP (NL)	200
Wind	Technical and economic effects of connecting increasing numbers of large wind turbines to weak public utility networks	Garrad-H. (UK) ENDESA (E) South West El (UK) Un Comillas (E) Un Manchester (UK)	250
Wind	Planning for large wind energy penetration on a power system	EDP (P) ESB (IRL) (observer)	100
Wind	Technical and economic aspects of assimilation of wind farms on utility networks	NWP (UK) Walker & Ass. ELSAM (DK) RAL (UK)	245
Photovoltaics	Development of a stand-alone PV system of 3 to 6 kW for utility service options	Iberdrola (E) Bayernwerk (D) WIP (D) ENEL (I) Siemens (D)	81
Biomass	Decentralised production of electricity of biomass and institutional barriers to the European bio-energy market * this project is formally part of the biomass programme	NRI (UK) ENEL (IT) East Midlands (UK) Green Land (UK) IVA (SF)	310 *
All renewables	Develop and implement an advance control system for the optimal operation and management of medium size power systems with a large penetration from renewables	Armines (F) NTUA (GR) Amber (GR) INESC (P) RAL (UK) Conphoebus (I)	300
All renewables	Network on monitoring, data reduction, data banks, modelling and simulation	EUREC (B) NRMC (IRL) FhG (D) Conphoebus (I)	100

Table 3. Wind Energy Generic R&D projects

Area	Partners	Coun try	(kEC U)	Scope
Rotor design, development and testing	ONERA HMZ A Tout Vent	FR BE FR	380	Develop design method for aeroelastic tailoring of blades. With special reference to graphite fibre, design a set of blades for an HMZ machine
Electric conversion system	ECN Eur.Or.Energy Group	NL UK	270	Develop variable reluctance generator for variable speed machines.
	Centre for Ren.Energy Magnacut Berendsen	DK UK DK	140	Develop multipole permanent magnet generators
Aerodynamics	ECN, Techn.Univ.Delft Tech.Univ.Denmk Univ.Stuttgart Garrad Hassan NTUA Teknikgruppen	NL DK DE UK GR SE	230	Extend dynamic inflow studies to yawed conditions and partial pitch span
	ECN Tech.Univ.Delft Aer.Res.Inst.	NL NL SE	39	Study to investigate use of wind tunnels to analyse unsteady stall phenomena on full size blades
Design tools	NTUA Lamda Tech. CRES Garrad Hassan	GR GR GR UK	206	Develop and validate design tools to produce methods of direct value to industry, certifiers and standards authorities.
	Windmaster Garrad Hassan Germanischer Lloyd	NL UK DE	126	Design tools specific for wind loading and fatigue
Materials and fatigue	Risø, LM Glasfiber Sciotech, Bristol Polyt. Aerpac, ECN, NLR DLR WEG & Univ. Bath	DK UK NL DE UK	500	Fatigue properties of coupons, blade components and sections, research in the development of blades of advanced type
	FFA AS Veritas RAL, City & Notting. Univ. Aerpac, Delft Tech.Uni. LM Glasfiber	SE NO UK NL DK	300	Condition monitoring of wood/epoxy and glass/ polyester blades
Structural loads and standards	ECN, Stork Product Elsamprojekt	NL DK	100	System and structural reliability analysis based on the 1 MW Newecs 45
	Wind Energy Institute CRES FFA	DE GR SE	100	Develop methodology for predicting long term load spectra from short term test
	ECN, CIWI, Stork Risø CIEMAT Germanischer Lloyd	NL DK ES DE	300	Devise fatigue test procedures for rotors
	Garrad Hassan TNO, ECN, KEMA Nat.Wind Power CRES FFA	UK NL UK GR SE	360	Converge the work of several JOULE projects to produce guidelines for loads in wind farms on simple terrain and to begin to analyse loads in complex terrain
Operational experience and noise reduction	ECN DK-Teknik&Tech. Univ. Danish Acoustic Inst. NEL Wind Inst. Stutt.Univ. Univ. of Oldenburg NTUA Univ. du Havre Inst. for Computer App. DLR ECN & NLR	NL DK DK UK DE DE GR FR DE DE NL	1000 (*)	Noise measurement and development of predicition methods and investigation of blade tip modifications to reduce noise and improve performance
	Fraunhofer Inst. Various subcontractors	DE	520 (*)	Maintain Eurowin database, include local wind speed data and extend treatment of failures and incidents and accidents

NB The first named proposer in each group is the project co-ordinator

(*) including an extension to be negotiated

Table 4. Utility issues programme

Based on a study for the Commission on renewable energies performed by EUREC, future cost reductions in wind systems are likely to results from falling machine costs due to: series production, larger machines on taller towers, improvements in the performance of larger machines resulting from advanced generator technology and gearless design concepts and optimisation. This study states that by the year 2000, there should be an overall wind generation cost reduction of up to 25% and further reduction may derive by flexible machines and lighter designs.

2.3 Utility issues

As the follow-up to a series of studies on the penetration of wind energy in the Community electricity production systems, the Commission has set up within JOULE II a new action whereby all the major European utilities involved in renewables energies participate. Not only is this programme a stimulus to initiate important collaborative actions between utilities -which will no doubts be prime actors in the future deployment of renewable energies-, but also is a forum for them to discuss and exchange experience in such novel fields. In the particular case of wind energy, there proved to be a transfer of technological knowledge from countries, such as Denmark and the Netherlands, where the European wind power development commenced to the benefit of those countries which followed some time later. The presence of research organisations and consultants associated to specific studies has enlarged views and improve the effectiveness of research. The group deals with a variety of electricity producing renewables in possibly integrated system with a view to exploit their complementarities and synergies. The list of the projects in this action is given in Table 4.

3. Future plans

The bulk of the wind energy programme has been set up through the first call for proposals which closed in February 1992. However, some topics of JOULE II's work programme have not been launched in this call for budget reasons and a new call is under preparation with a provisional deadline on 25 June 1993. The main subject of this new call which is of interest to wind energy concern wind measurement and modelling in complex terrains and offshore. However, three specific topics which were not treated satisfactory in the first call have been added, i.e. aerodynamics of wind turbine rotors, development of standards, wind park public acceptability. Furthermore, it is planned to launch a call for proposals restricted to wind turbine manufacturers on the development of advanced components for large wind turbines along with a new action for utilities.

For the future a strong involvement of the European industry also in the generic R&D projects will be sought. It is also felt that greater importance should be attached to research on non-technical issues, such as planning, environmental issues, public acceptability in close cooperation with the Directorate-General for Energy (DG XVII).

The Commission has recently drafted a proposal for a 4th Framework Programme for Research and Technological Development which will cover the period 1994-1998. This proposal includes R&D on wind energy so that continuity of support to wind energy will be ensured. A Council decision may be expected by the end of 1993.

4. Conclusions

The 1990s is a decade of "strong wind" for the European wind energy community when a tremendous progress has been achieved as compared to the 1980's and when 1000 MW of wind power have been reached. New market opportunities for wind power have arisen, particularly in the Eastern European and in emerging developing countries such as China and India. By means of programmes specifically devised to support wind energy, e.g. JOULE, Thermie and Valoren, the Commission of the European Community has largely contributed to this success. The future of wind energy will however depend on many factors which are not necessarily of a technological nature, among which the availability of suitable sites for installation, public acceptability, planning, and last but not least, the price of conventional fuels. The Commission has been, and will in the next future continue to support wind energy R&D in consideration of its important role for energy supply scenarios, the improvement of the environment and creation of new jobs.

References

[1] Council Decision of 15.3.1993

[2] GMV Conseil, "The European scenery of the JOULE programme", Paris, 7.12.1992

[3] E. Hau, Lesson learned from WEGA-1 and other large machine programmes, Travemünde, March 1993

[4] Private conversation with VESTAS, Denmark

[5] EWEA, "Wind energy in Europe", October 1991

[6] B. Johanson et al. "Renewable for fuels and electricity" Island Press, 1993

[7] G.T. Wrixon, A.M. Rooney "Renewable energy 2000: a report prepared for EUREC", January 1993, (soon to be published)

THERMIE PROGRAMME AND ENERGY POLICY

H. NACFAIRE, Head of Sector, and P. NAGHTEN, Expert
NRES – New and Renewable Energy Sources
Commission of the European Communities – Directorate General for Energy (DG XVII)
Energy Technology – Strategy, Dissemination, Evaluation
B – 1049 Brussels

EC energy policy aims to ensure security of supply, backed by an efficient internal energy market and cooperation in energy matters with third countries. In addition, the environmentally – sound production and consumption of energy is of vital importance for the achievement of sustainable growth.

Energy technology is an integral part of this policy, and the Thermie programme administered by DG XVII of the Commission is the main vehicle for the promotion of innovative and efficient European energy technology. The first part of this paper describes how Thermie works, and how the Commission has engaged a network of 40 OPETs (Organisations for the Promotion of Energy Technology) to assist it with the so-called Associated Measures: essentially market–orientated activities of promotion, analysis and information.

In the second part of the paper, the significant part played by wind energy technology – from previous demonstration programmes on up to Thermie – is considered. The evolution of wind energy is described, with an overview of the available statistics. Finally, some current and future OPET activities in the field are outlined.

We would like to put into the context of this Conference the efforts being made by the Community executive, i.e. the Commission of the European Communities, and in particular the Directorate - General for Energy, DG XVII, in relation to the common energy policy and in the technological field.

1. Common Energy Policy

As with others areas of Community interest (e.g. trade, social affairs) the EC has a common policy on energy.

Energy consumption in the European Community represents 14% of world consumption and 8% of production; the Community accounts for 38% of the world energy trade, and about half of its energy supplies comes from third countries.

That is why a big effort is now being made in the energy sector, because although the European Community's energy supplies are at present satisfactory from the point of view of quantities and prices, security of supplies remains, and will continue to be, of great importance to its energy policy.

This policy can be said to have two broad objectives:

1.1. The first objective is to complete the internal energy market as part of the programme to establish a single market within the twelve Member States. A single market will certainly make an important contribution to the security of supply in all Member States.

Our objective of completing the internal market necessitates a number of flanking measures, and we are currently addressing the following priorities:

* stepping up our efforts to promote energy efficiency and conservation and renewable energies, in particular in order to improve our environment, and especially to combat the greenhouse effect and reduce carbon dioxide emissions (SAVE and ALTENER programmes);

– SAVE (Specific Actions for Vigorous Energy efficiency), of four years' duration, is focused on encouraging the initiation of different "policies" in the field of the rational use of energy in order to improve that overall energy intensity and efficiency. The main axes are:

- Regulatory actions
- Exchange of information at international and Community level

– ALTENER, of five years' duration, must develop specific actions for greater penetration for renewable energy sources in the following areas:

- harmonisation of legislation and formulation of common technical standards
- promoting third party financing
- training ...

* With a view to stabilising emissions of carbon dioxide at 1990 levels by the year 2000, the Commission has proposed the introduction of an energy/CO_2 tax equivalent to $3 per barrel of oil running up to $10 per barrel in the year 2000. However, this proposal will enter into force only when similar measures are adopted by other industrialised countries, to avoid handicapping European industries. This tax will apply to all sources of energy with the exception of Renewable Energy sources

* development and promotion of new energy technologies (THERMIE programme), and

* strengthening of energy infrastructures throughout the Community (REGEN programme)

1.2. The second objective concerns the establishment of cooperation relations with third countries.

* The international dimension of the energy market remains a decisive factor for the European Community, which is dependent on the outside world for nearly half of its energy supplies. An expansion of trade with third countries, and especially our neighbours in Eastern Europe and on the Mediterranean, is essential to safeguard the security and reliability of our supply system.

* Turning to the countries of Central and Eastern Europe and the former Soviet Union, I should mention the European Energy Charter which was signed in December 1991 at The Hague by 44 States: all the European countries, including the former Soviet Union and the non-European OECD countries.

The aim of the Charter was to trigger economic recovery, to support the implementation of a market economy system and to help to ensure political stability in the Republics of the former USSR and Eastern European countries by joint efforts to stabilise the region's production of energy resources; to contribute to the coverage of its energy needs while modernising the local energy industries in order to better protect our common environment.

* The opening up of the European Community to neighbouring countries was given practical expression with the adoption of:

- the PHARE programme for the countries of Central and Eastern Europe,

- the programme of technical assistance for the former Soviet Union and,

- the programme for the new Mediterranean policy.

2. Energy Technologies

Switching over to the technical field, I shall devote the second part of my address to the THERMIE programme which is the flagship of the Community's promotional programme in the field of energy technology. It answers the double imperative of improving security of energy supplies and the preservation of the environment.

THERMIE covers a period of five years starting in 1990. The budget allocated is 700 MECU. I do wish to draw your attention to the fact that the efforts of the Commission in the field of innovative energy technologies go back to 1975 and therefore over the whole period between 1975 and 1990.

The table (Fig. 1) gives the distribution of the total amount of support granted under THERMIE and the former demonstration programme between 1975 and 1992.

During that period the Community has assisted over 2885 projects for a total amount of 1.76 billion ECU of which 141 were wind energy projects, attracting total support of 48,9 MECU (the figures for wind projects exclude 1992 projects, as data are not yet available on accepted support, i.e. contracts actually signed).

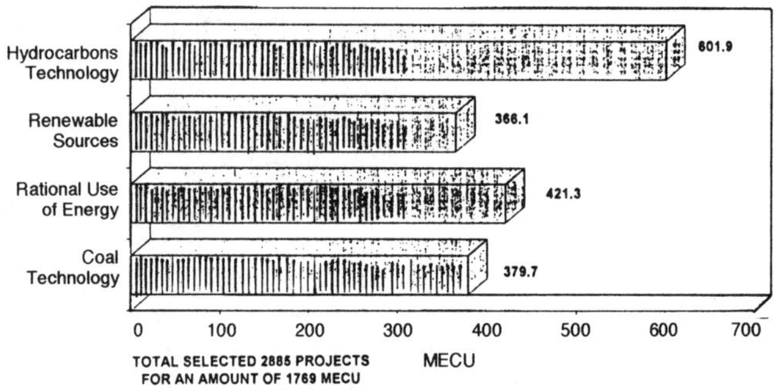

Fig 1. EC Support for Energy Technology

2.1. Areas

THERMIE covers the following areas:

- The rational use of energy in all economic sectors (industry, building, transport and urban infrastructure, electricity and heat)

- Renewable energies which include solar energy, both thermal and photovoltaic, biomass and waste, wind energy, hydroelectric and geothermal energies

- Solid fuel which includes the clean use of coal

- Hydrocarbons namely exploration, production, transport and storage activities.

2.2. Categories

In the fields stated above, the projects financed by the Commission cover three categories:

- Innovative projects which concern a first application of a market size process or new product: financial support is maximum of 40%

- Dissemination projects which consist in a second demonstration of innovative projects, provided that the economical and geographical context is different to the original one. These projects are generally carried out in the less developed areas of the Community. This is the reason why I did stress that THERMIE counts as one of its objectives that of helping the development of such regions. Subsidies are a maximum of 35%.

- Targeted projects which are encouraged or promoted by the Commission whenever a possibility exists for a breakthrough of a significant technological nature in a given field.

One of the basic stipulations for a project to be eligible for the THERMIE programme is that its realisation must present important technical and financial risks which act as a deterrent to investment according to normal commercial criteria; however, it is necessary that the project offers prospects of technical and financial viability since it must be borne in mind that the aim of the programme is the commercial exploitation of the technology concerned.

2.3. Dissemination

In addition to the assistance given to projects, the Commission undertakes, in the framework of THERMIE, the funding of a number of activities called associated measures. This is due to the fact that the help given to investment projects is not sufficient to promote innovative energy technologies and to see to their market penetration. It is therefore of importance to make these innovative technologies known and to increase their potential through a vigorous policy of information dissemination.

For this, the programme has expressly set aside 10 to 15% of the THERMIE funds to finance these supporting actions, and the amount available to be used in this respects stands between 70 and 100 million ECU for 5 years.

The said actions cover mainly the analysis of the characteristics and market potential evaluation, evaluation of projects financed by the Community, dissemination of information of energy technologies and project results.

Such actions of dissemination are carried out through the organisation of technical seminars and forums of technological cooperation, participation in technical fairs, the production of brochures and the development and exploitation of databases.

To achieve these supportive and dissemination actions, the Commission has set up a network of 40 organisations called OPET (Organisation for the Promotion of Energy Technologies). In each and every Member State, one or more OPET's acting on behalf of the Community, disseminate the results of projects and assist in their market penetration(See Fig.2).

The results of this network proved very positive since they are anchored locally. Added to the impetus which the Commission is able to give because of its size, is the fact that the organisations know their ground which leads to greater efficiency. It has been estimated that the OPET's place at the disposal of the Commission more than 3000 experts in the various energetic sectors.

Fig 2. The OPET Network

2.4. Cooperation with third countries

Within the framework of the THERMIE programme a new channel of cooperation with third countries has been opened for promoting the development and use of energy technologies offering the possibility of combined actions, always with the main objective of encouraging the use of efficient energy technologies in those third countries through technology transfer.

In this way, third countries can be included in the CEC's dissemination strategy in such a way that all the above mentioned actions could be developed in those countries though independent agreements.

In fact, the success achieved by the OPET's has already led to an extension of the OPET network outside the European Community, with a number of Energy Centres in Eastern Europe (Prague, Warsaw, Budapest, Sofia), in some previous Soviet republics (Moscow, St. Petersburg, Minsk, Kiev) in the Baltic states and two more are planned in the Mediterranean Region: one for the Maghreb and one for the Mashrek. Some information activity takes also place in Scandinavia, Austria and some other third countries.

3. The Specific Case of Wind Energy

Support for energy technology – formerly by means of "pure" demonstration, and now through Thermie – has been described in general terms as an instrument of Community energy policy. But what of the specific case of wind energy? A brief look at the statistics serves to illustrate the very creditable performance by the wind energy sector in attracting EC support.

3.1. A Brief Review

Since its introduction, in 1983, as an area of application for energy demonstration projects, the wind energy sector has been particularly successful in attracting EC support.

Fig. 3 (Ref. 1) shows that over the period 1983–1992 the Community has selected 186 out of 564 proposals for support, an average offer rate of 33%. The actual uptake of support has in fact been a bit less (about 150 projects) due to projects not proceeding, or contracts for support not signed. In terms of support granted, wind energy projects account for almost 20% of support allocated to all renewable energy projects, i.e. 67.5 MECU out of a total of about 360 MECU over the period.

It is calculated that wind energy projects supported under the demonstration/Thermie programmes represent a total investment of more than 300 MECU, with its attendant job creation benefits both in direct equipment manufacture and in sub-supply businesses such as electrical switchgear, siteworks etc.

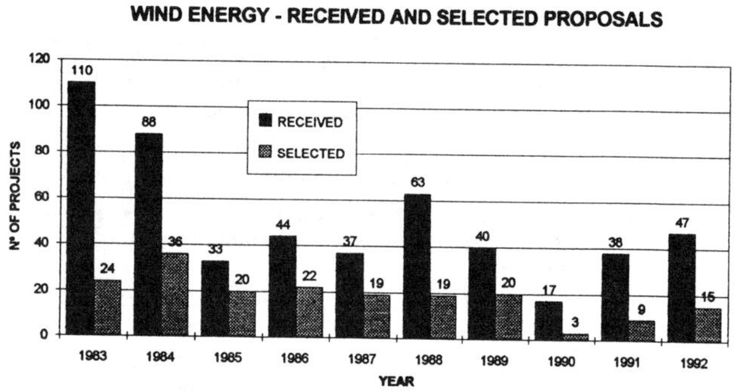

Fig 3. Wind Energy Projects

The significant positive environmental impact of wind energy is shown in Fig. 4 (1) which gives the cumulative avoided emissions compared to both oil- and gas-fired generation.

AVOIDED EMISSIONS DUE TO USE OF WIND ENERGY
(CUMULATIVE FIGURES. PERIOD JAN 1981 - APRIL 1992)

EMISSIONS (tons)	COAL STATIONS	OIL STATIONS
CO2	3532050	2089796
SO2	19426	40619
NOX	4238	2884
PARTICULATES	2678	706
V.O.C.	795	795
SLAG AND ASH	206036	117735

Fig 4. Wind Energy – Environmental Benefits

Fig. 5 shows the general evolution in the European Community of installed wind power capacity from 1981 to 1992, including of course locally financed commercial wind farm developments as well as projects supported by means other than demonstration/Thermie (e.g. VALOREN). In fact the 'milestone' of 1000 MW is expected to be passed by the first quarter of 1993.

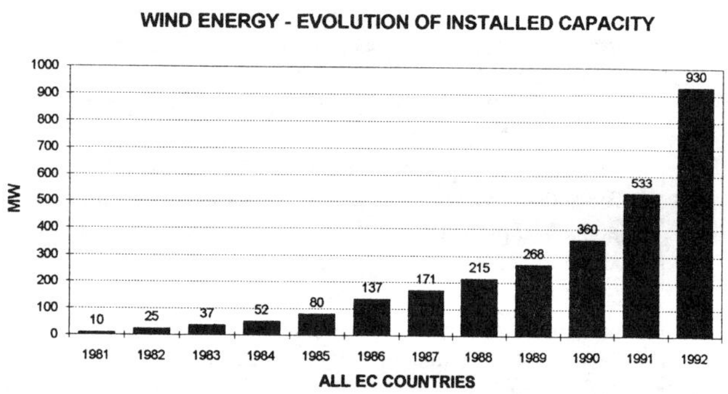

Fig 5. Wind Energy Growth, 1981 – 92

3.2. The Marketing Challenge

With the emphasis in Thermie on promotion and dissemination, the aim is now to better identify the key technologies best suited to specific market segments. Thus in the current programme of Associated Measures, the OPET network is helping to promote wind energy technology to specific target groups, whether these be geographical areas (e.g. island regions) or user groups (local authorities, maritime communities etc.).

Listed below are some of the main activities undertaken by the Commission and the OPETs as part of the Thermie 'Associated Measures':

1. Seminars organised by OPETs, e.g.:
Rhodes (GR), June 1992– "Mediterranean countries"
Herning (DK), Sept. 1992– "Potential for Wind Parks"

2. Market Studies (Ongoing):
Wind energy applications in island regions.
Wind energy potential in:
Eastern Europe; Baltic States; Ukraine; Bulgaria (four separate studies)

3. Publications:
Guidelines for Wind Energy Plants (in German)
Maxibrochure on Wind Energy applications
Guide on Wind Energy for Island Regions

In addition, two detailed reviews of EC-supported wind energy projects have been carried out, with a view to assessing the effectiveness of support funding in stimulating development, and as an aid to the planning of future strategies.

Furthermore, the Associated Measures part of Thermie gives the Commission the possibility to support major events such as this conference, and to work more closely with the professional associations, and other key actors, with a view to motivating investors and decision-makers.

With regard to the technology itself, the trend in Thermie projects is towards increasingly larger machines (400 kW and more), mainly in dissemination type applications, with parallel improvements in control and safety systems. The advent of commercial Megawatt scale generators is also awaited with interest.

However, there is still considerable scope for the deployment of smaller machines for specialised applications, whether as stand alone units or as part of integrated energy supply networks. Furthermore, the development of harmonised standards for wind energy generators (as called for in the ALTENER proposals) can be expected to contribute to improving the economic viability of future generations of machines.

In conclusion then, we can say that the main objectives of EC energy policy are well served by the promotion of innovative European technology. Wind energy in particular represents a proven resource which can contribute, by means of import substitution, to security of supply. At the same time, wind energy technology offers an opportunity for European industry to develop new markets at home and abroad. The Thermie programme, and the OPET network, are there to help 'seize the moment'.

References:
(1) *Review of EC–Supported Wind Energy Projects Final Report, 1992; CRES (OPET)*

PROMOTING WIND ENERGY IN GERMANY

G. Hauerstein, H. Räde*, W. Sandtner, N. Stump*, R. Windheim*
Federal Ministry for Research and Technology, P.O. Box 20 02 40, D-5300 Bonn 2
*Project Management Organization Biology, Energy, Ecology (BEO), P.O. Box 19 13, D-5170 Jülich

1) Introduction

The Federal Ministry for Research and Technology (BMFT) supports the use of wind energy since 1974. In particular, energy economy goals are being aimed at, i.e. contributing perceptibly to our electricity supply at acceptable costs in the longer run, reducing our energy imports and broadening the basis of our energy supply. The goals also include considerable contributions towards securing the energy supply of a growing number of people in the Third World. Moreover, particular significance is attached to reducing loads on our environment and climate [1].

The goals in energy economy are to be achieved by a further development of wind energy conversion technology. The costs of electricity from wind energy converters (WECs) are to be reduced to such a level that these machines may be competitive by the end of the century. This goal is apparently not easy to achieve, but considerable progress has been made since 1974.

After a phase of applied basic research and the testing of first generation prototypes, initial demonstration programmes were launched to test technical reliability on a larger scale. A large variety of WEC types rated between 5 and 300 kW were installed under seven major demonstration projects which also included four different wind farms. A total of 214 WECs were thus supported with a total rated power of 14.5 MW (Table 1).

Table 1: Demonstration projects small and medium WECs

Project	Period	Number of turbines	Total rated power [kW]
Application of small WECs in Germany (MAN)	83-86	20	400
Test of WECs in Developing countries (KfW)	84-93	49	1.470
DEMO Program 250 kW (13 manufactures)	86-87	48	2.943
DEMO Program 80-800 kW (5 manufactures)	87-90	25	3.650
Windfarm Westküste (Windf. Westk. GmbH)	86-89	32	1.330
Windfarm Cuxhaven (ÜNH)	87-88	25	1.000
Windfarm Husum (City of Husum)	90-92	15	3.750
		214	14.543

These demonstration projects formed an adequate basis for the "250 MW Wind" programme [2] offered since 1989. This broad-based test programme is being supported by the German Federal States too. On the income side, the test programme provides for additional earnings exceeding the legally stipulated payment of currently around 0.17 DM per kilowatt hour of electricity fed into the grid. In the German coastal region, medium-sized WEC can now already achieve an annual electricity production at a cost of approx. 1 DM/kWh relative to the plant price ex works. This is a clear improvement in comparison to the results from earlier demonstration programmes with greatly scattering figures around 2 DM/kWh and year.

A further support programme ELDORADO [3] offers German manufacturers the possibility of implementing WECs in countries with different climatic conditions.

In general, a trend towards wind energy converters of increasing rated power is to be observed, which is also reflected in the support given to various research and development projects for large WECs. It should be pointed out in mind that, in accordance with all predictions, the existing potential in our relatively densely populated country can be best exploited by large WECs. Experts of the advisory panel the "Ad-Hoc Committee on Large Wind Energy Converters" were of the opinion that the electricity production costs of future WECs in the MW range may already be comparable with those of current commercial WECs on a medium-term basis, although considerable expenditure on R+D and testing must be involved.

2) "250 MW Wind"

Interest is currently focused on the "250 MW Wind" programme for wind energy utilization. In conjunction with a scientific measuring and evaluation programme WMEP [4], the data from supported WECs will be recorded and evaluated over a period of 10 years. As an incentive and compensation for their participation in the test, the operators receive financial support from the BMFT, which normally depends on the amount of electricity produced. (In general 0,06 DM/kWh). This also motivates operators to keep their machines as far as possible in well operating conditions. In September 1990, i.e. slightly more than 12 months after the directive originally conceived as "100 MW Wind" was announced in May 1989, support applications were filed for as many as 1000 machines. To date, applications have been filed for a total of more than 7000 machines. This is much more than can be approved under the directive. In the meantime, almost half of the power available within the programme has been approved and 30 % (about 700 WECs) has been installed. The averaged rated power was 214 kW per machine in 1992 (Fig. 1). The total generated power in 1992 was 227 million kilowatthours.

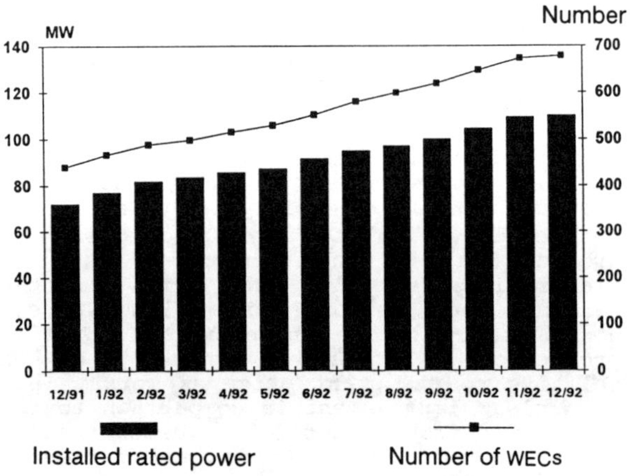

Fig. 1 : "250 MW Wind"-programme. Installed rated and number of WECs for 1992

3) Use of Wind Energy Under Different Climatic Conditions

According to a study by the World Bank, almost 50 % of the inhabitants in developing and threshold countries do not have access to central supply grids (electricity, oil, gas etc.). They may be assisted by decentralized concepts, and renewable energies are considered to be one option for decentralized energy supply.

Against this background, BMFT projects are targeted at the following major applications

- WECs grouped in wind farms and connected to a grid: either in weak diesel-based grid for fuel conservation or in extended grids for additional electricity production,

- WECs in stand-alone operation or in combination with diesel units for the supply of individual consumers such as water pumps, drives for agricultural equipment and village power supply.

Especially worth mentioning is a project operated by the Kreditanstalt für Wiederaufbau (KfW) on behalf of the BMFT. 49 WECs each with a rated power of 30 kW are being operated in seven wind farms connected to small isolated grids (see Table 2).

Table 2: Wind farm project of KfW

Country / Locality	Number of WECs à 30 kW	Wind velocity (m/s)	In operation since
Madeira / Porto Santo	8	7.0	5/88
Azores / Santa Maria	8	6.8	8/88
Haiti / Port-de-Paix	5	6.5	10/88
Mauritius/ Rodrigues	4	8.5	12/88
Cape Verde / Mindelo	10	8.9	11/89
Argentina / Rio Mayo	4	7.1	2/90
China / Sijiao Island	10	7.2	6/91

The experience acquired under this programme confirms that

- parallel operation of small WECs with diesel units ensures reliable electricity supply at reasonable costs,
- WECs should be grouped in wind farms which have a certain minimum size to ensure economic operation and to minimize servicing costs.
- The WECs must have been tested in our region and proven to be reliable and require minimum maintenance.

Recently BMFT launched the indirect-specific support programme ELDORADO Wind in 1991 which is being jointly carried out with several partner countries. The type of support selected by the BMFT is intended to motivate a larger number of users in southern climatic zones to construct and operate WECs in cooperation with German partners. The aim is to achieve an installed total power of 20 MW in these countries within the next five years on the basis of findings derived from domestic measures.

Response to the ELDORADO programme is surprisingly positive. Several projects have already been started (see Table 3), others are in preparation.

Table 3: Projects already approved within ELDORADO programme

Country/Locality	Application	Total rated power	Type of WECs
Egypt	Wind farm	1 MW	Ventis 100
Brazil	Wind farm	1 MW	Tacke 250
China	Wind / Diesel	150 kW	Aeroman 33
China / Dalian	Wind farm	1 MW	HSW 250
China / Zhurihe	Wind farm	1 MW	HSW 250
Russia	Wind farm	300 kW	HSW 30

4) Research and Development Projects [5]

In parallel with the demonstration programmes the BMFT has concentrated on the further development of WECs in the MW range. This development to MW units started in the second half of the eighties and has already led to the second generation of large WECs (Fig. 2). They comprise the 1.2 MW WEC 60 (60 m, 3-blade on Heligoland, WEC 60 II of the same size in Kaiser-Wilhelm Koog and the 640 kW Monopteros (50 m, 1-blade) in Wilhelmshaven. This second generation also includes AELOUS II (80 m, 2-blade) with 3 MW rated power erected last year in Wilhelmshaven. None of these machines has so far been operated other than as a prototype.

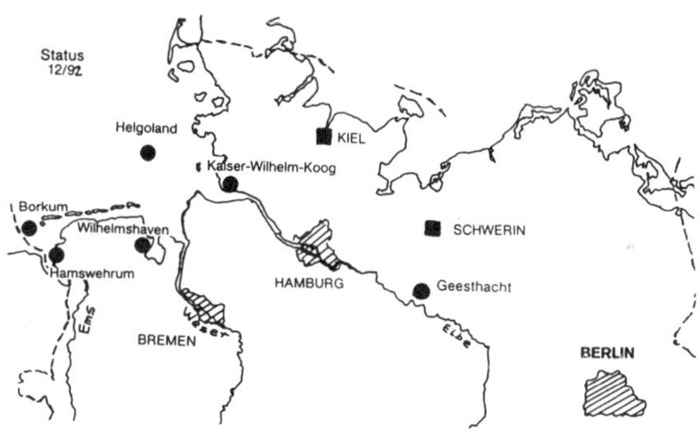

Type	Location ●	Power	Realization
MON 50	Wilhelmshaven	3 à 640 kW	Summer 89
WKA 60 AWEC 60	Helgoland Cabo Villano (Spain)	1.2 MW 1.2 MW	Summer 90 Summer 89
AEOLUS II WKA 60 II	Wilhelmshaven Kaiser Wilhelm-Koog	3 MW 1.2 MW	Spring 93 Autumn 91
HSW 750	Kaiser Wilhelm-Koog	750 kW	93
E-36	Hamswehrum	400 kW	Spring 92
TW 500	Borkum	500 kW	Summer 92
E-40	Geesthacht	500 kW	93

Fig. 2 : Prototypes with a rated power more than 400 kW in Northern Germany

Two medium-sized German companies started constructing 1MW WECs using a different approach in order to obtain the economy of commercial medium-sized WECs on a relatively short-term basis. This was to be achieved incorporating innovative concepts and components. One manufacturer intends to use a design with vertical rotation axis for wind energy utilization (H rotor). This WECs works with a special generator without gear box. A similar gearless concept, combined with an advanced electronic control system, is envisaged by the other manufacturer for a horizontal axis converter. Both manufacturers are currently developing and constructing 1-MW prototypes, iniciated by the JOULE II programme of the C.E.C..

5) Funds and Outlook

The BMFT has so far spent roughly DM 300 million on the promotion of wind energy [1]. Not included is the expenditure on cooperation with countries of southern climate zones for which several ten million DM have been additionally spent to date. Approximately DM 120 million of the total amount of roughly DM 300 million was spent on large WECs, roughly DM 100 million on the promotion of small and medium-sized machines and the remainder on other R+D projects including, since 1989 and with a clearly rising trend, the "250 MW Wind" programme. The expenditure for the "250 MW Wind" programme was about DM 40 Mio up to-date.

The "250 MW Wind" programme is a long-term programme, since every plant will be scientifically evaluated over a period of 10 years, including plants only put into operation in 1996 or 1997. When this programme ends, roughly DM 350 million will probably have been spent on support and on the measuring programme WMEP.

BMFT funds for separate R+D projects should also be available in the future. For small and medium-sized WECs, however, the level of development is already very advanced so that further developments will be generally financed by the manufacturers themselves. Interest will probably focus on large units with the goal of further reducing the costs of wind power, e.g. by increased service life, improved technical reliability and reduced plant costs as well as by optimum integration into the electricity grid.

REFERENCES

[1] Renewable Energy, The Federal Minister for Research and Technology. Address see title.

[2] Full guideline published by Bundesanzeiger No 37, Febr. 22, 1991 p. 950f.

[3] Full guideline published by Bundesanzeiger No 198, Oct. 23, 1991. English Translation avail. from BEO, Dept. 13, Address see title.

[4] Annual report and monthly reports avail. from ISET, Königstor 59, D-3500 Kassel.

[5] Annual Report "Energieforschung und Energietechnologie". Fachinformations-Zentrum Karlsruhe, D-7515 Eggenstein Leopoldshafen 2.

THE UNITED STATES WIND ENERGY PROGRAM IN A DECADE OF CHANGE

D. Ancona, R. Loose, J. Cadogan
U.S. Department of Energy
1000 Independence Ave, S.W.
Washington, D.C., 20585, U.S.A.

Use of wind energy as an electric power generating option is increasing throughout the U.S. and internationally. This process is expected to accelerate over the next decade as a result of several factors. Technological advances, reduced system costs, and new government financial incentives are all contributing to the technology's increasing prominence. Inclusion of external costs into electricity rates is being discussed or examined in most states. Substantial wind energy installations have either been negotiated or are being seriously considered in at least eight states. At the Federal level, new energy legislation has been signed that authorizes production incentives of 1.5 cents per kilowatt-hour for wind plants installed after 1993. The economic package of the new Clinton administration includes substantial increases in federally-funded conservation and renewable energy research and development (R&D) through 1998. At the same time, wind energy technology has progressed to the point that U.S. Department of Energy-supported advanced turbine prototypes are being fabricated and installed, and are expected to be able to produce energy at five cents per kilowatt-hour (at 6 m/sec. sites, measured at 10 meters). The wind industry is poised at the confluence of these developments, preparing for strong growth in the last half of the last decade of the century.

This paper will discuss some of these new initiatives in legislation and utility planning, and the role that the Federal government has in supporting wind energy research and development. The U.S. Department of Energy Wind Energy Program is strengthening joint efforts with industry in a program to maintain and improve existing wind installations, develop advanced wind turbines for the near- and mid-term, and upgrade the technology base through applied research. Three national laboratories are working with industry and universities to effect near-term improvements in the efficiency and reliability of wind farms, and develop a new generation of wind turbines for the next century, while continuing the fundamental research that supports these efforts.

The Impact of New U.S. Energy Policy

The tone of the new Administration's energy policy has been set with a proposal for a broad-based energy tax, based on the BTU content of the energy source. In the electric utility sector, this tax would provide a substantial stimulus to greater use of renewables and implementation of conservation measures. Hazel O'Leary, the newly-appointed Secretary of Energy, has been clear in her support of alternative energy. Prior to her appointment, she worked with Northern States Power (NSP), a utility active in wind energy development. NSP recently announced their intention to install 100 megawatts of wind turbines in the state of Minnesota by the end of the decade.

The President's economic plan also includes proposals for major revisions to the funding structure of the U.S. Department of Energy (DOE). Significant increases in funding are proposed in the areas of Federal energy management, technology transfer, renewables and gas technologies, and alternative-fueled vehicles, as well as fundamental high-technology projects such as the Superconducting Super Collider. A substantial de-emphasis within DOE in the defense and nuclear areas will, in part, provide the savings necessary to fund these efforts. Enhanced funding for energy management will take the form of grants to improve energy efficiency in public housing, schools, and hospitals, and investments in energy efficient technologies in Federal buildings. In the short term, the DOE Wind Energy Program may be affected by the element of the short-term economic stimulus package that would increase funding for cooperative research partnerships between national laboratories, industry, and universities. The DOE Wind Energy Program is already committed and involved in this approach to advancing the commercial state of wind generators, and this proposal is a welcome boost to ongoing efforts. In the longer term, (FY 1994-98), increases are proposed for both cooperative research partnerships and funding for Conservation and Renewable Energy R&D. Anticipated increases in funding for the Wind Energy Program will serve to enhance the current cooperative programs with utilities and the wind industry.

Another major step forward for wind and other renewable energy systems was the passage of the Energy Policy Act of 1992[1] (EPAct) by the U.S. Congress in October. This new law provides wind energy production incentives, access to transmission lines, and other programs helpful to wind energy development. The new Law provides a tax credit of $0.015/kWh for wind and "closed loop" biomass facilities for a 10 year period for installations placed in service between 1 January 1994 (January 1993 for closed-loop biomass) and 1 July 1999. The credit is phased down as a national average price (called the reference price) from the eligible source in the previous year exceeds 8.0 cents per kWh and eliminated as the reference price reaches 11 cents per kWh. Contrary to previous incentives, investor owned utilities are eligible for the production incentive. In addition, renewable energy production incentive payments of \$0.015/kWh are provided to publicly owned (e.g., municipal, cooperatives, and public utility districts) and qualified renewable energy facilities, although funds for these payments are subject to annual appropriations. A qualified facility is one which generates electricity using solar, wind, biomass or geothermal energy. Other new EPAct procedures provide for transmission access (including any enlargement of transmission capacity necessary) for wholesale transactions that would not unreasonably impair the continued reliability of the affected electric systems and that would in other respects be in the public interest. EPAct also implements programs designed to encourage the use of wind systems through training and education. A wide range of other energy issues are also covered by the new law, including the creation of a class of unregulated generation called exempt wholesale generators (EWGs).

The State of the U.S. Wind Industry

Although landmark progress has been made on the federal level in legislation and research, new installations and actual energy production have been modest in the last year. By the end of 1992, new wind capacity was less than 15 MW. Total wind turbine installations remain approximately the same as 1991, about 1600 MW. Generation from the approximately 16,000 turbines is estimated to be slightly higher than prior years. (See Figure 1).

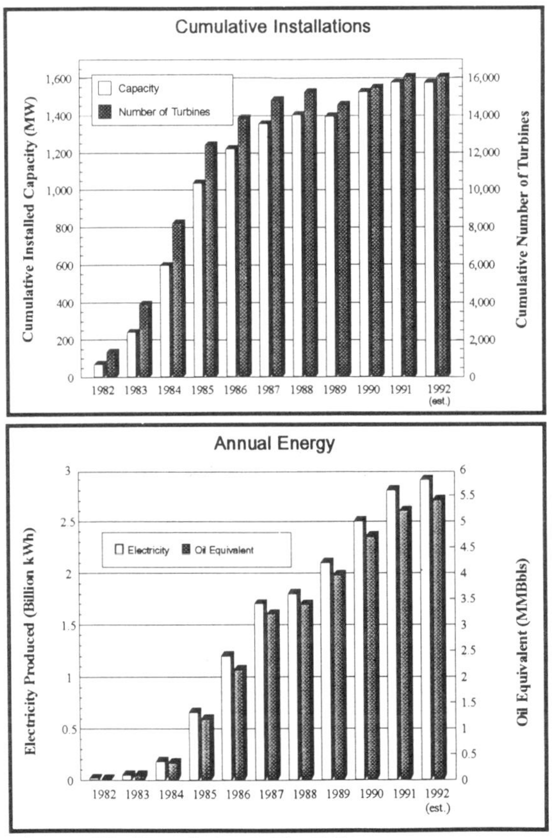

Figure 1. Wind Installation In The U.S.

Although projected new installations for 1993 are also small, the industry mood is upbeat. A number of the new U.S. Windpower 33M-VS turbines have been installed in demonstrations in several regions of the country, and designs spawned by the first phases of the DOE Advanced Wind Turbine program will soon be commercially available. The way has been paved among utilities, public utility commissions, and state energy offices for the introduction of wind generators onto utility systems. Finally, it is hoped that January 1st of 1994 will effectively serve as the launch date for this new generation of turbines, as the recently-passed energy production tax credits become effective. The principal market inhibitor for new wind generation has been the low spot market price of natural gas, typically below \$2.00 per million Btu. Considering the sum of all these factors, the projected installed capacity of wind turbines in the U.S. is expected to grow to 6.3 GW by the year 2010.[2]

At the grass roots, industry news has been mixed. In February of this year, the state of Iowa dedicated its first commercial wind farm. Small but symbolic, the five turbine farm is seen as another important step in the spread of utility-integrated wind power eastward. The Bonneville Power Administration (BPA), a federal power authority for the northwestern portion of the U.S., has committed to an initial 50 MW of wind to be on-line by 1995. On a discouraging note, Hawaiian Electric Industries announced in October that it had decided to shut down its windfarms at Kahuka due to losses from operations and continuing mechanical problems.[3] The company's main windfarm at Kahuka on the island of Oahu consists of two projects, the fourteen Westinghouse wind turbines rated at 600 kW each, and the 3.2 megawatt M0D5-B wind turbine. The decision to shut down the wind farm came after one of the 600 kW turbines was destroyed following a hydraulic pitch change system failure, resulting in an overspeed runaway. Operation of the Mod-5B turbine is continuing until a disposition decision is made. As of October, 1992, the turbine had operated over 20,561 hours and produced 26,776 MWh. In California, short term avoided energy costs are now below 3 cents per kWh, raising concern about the viability of some windfarm projects after the ten year period of the relatively high Interim Standard Offer #4 contracts. Discussions are underway between purchasing utilities and Qualifying Facilities (QFs) about alternative arrangements such as lower payments over a longer term which would leave ratepayers unaffected.

Federally-Funded Research and Development

This past year has seen the start-up of several new initiatives intended to complement the U.S. Department of Energy's existing program, among them projects to address utility integration issues through demonstrations involving installations of state-of-the-art turbines and efforts to re-engineer existing turbines for near-term cost and efficiency improvements. After a near doubling of funding from \$11.1 million in Fiscal Year (FY) 1991 to \$21.4 in FY 1992, the funding level has risen again this year to \$24.0 million.

DOE recently announced the establishment of a National Wind Technology Center, intended to be the new base of operation for the National Renewable Energy Laboratory's Wind Technology Division, and to include full turbine performance verification and other testing and research facilities. The center, located at Rocky Flats, Colorado on 280 acres of land, will provide the U.S. with a world-class wind energy research and development center. It will provide specialized component test facilities, computer modeling facilities, and a turbine test capability for use by industry on a user facility basis. The center will support a wide variety of activities, including the development of performance standards.

Advanced Wind Turbines

Industry needs for research to advance mid-term and long-term wind turbine design are addressed through the Advanced Wind Turbine (AWT) program. The AWT program is a collaborative venture with utilities and industry with the goal of bringing wind-generated electricity costs to levels competitive with those of fossil fuel generators in most regions of the U.S. Figure 2 illustrates the fundamental program structure.

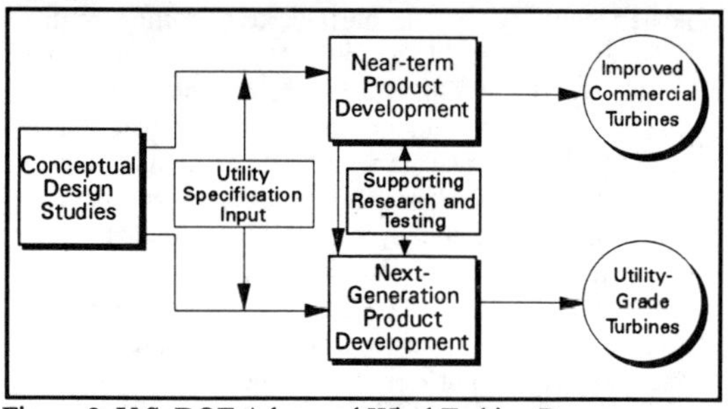

Figure 2. U.S. DOE Advanced Wind Turbine Program

The AWT program has two major thrusts: the development of improved turbines incorporating incremental refinements into existing designs by the mid-1990s and the development of next-generation technology by the turn of the century. Both of these elements began with conceptual design studies in late 1990 with the award of three subcontracts, and were concluded in 1992. The Near-term Product Development phase of the program is currently being funded, with three contracts awarded so far in 1992, and two more being negotiated. The specific goal of the Near-term Product Development activity is to develop wind turbine systems to be commercially available in the 1994-95 time frame that produce electricity for $0.05 per kilowatt-hour at 13 mph wind sites.

Both near-term and next generation technologies will incorporate multiple advanced concepts. For instance, the advanced turbines will incorporate blades with advanced airfoils designed specifically for the turbines, based on aerodynamics research performed at the DOE's National Renewable Energy Laboratory (NREL). In field testing, these airfoil families have demonstrated energy capture 30% greater than that of commonly used blades. For one near-term design, a single piece generator housing is being used, to reduce long term manufacturing costs. In another near-term design, ailerons are being used for rotor speed control. Power electronics, with associated variable speed operation, is being considered for the next generation turbines.

Utility Integration Program

The wind program has initiated a series of activities to help address the concerns that electric utilities have expressed about integrating wind systems into their normal operations. Two of these programs, both of which were formally initiated in 1992, are the Utility Performance Verification activity and the Power Marketing Administration initiative.

The Utility Performance Verification activity is being performed in conjunction with the Electric Power Research Institute (EPRI). This goal of this major new activity will be to deploy and evaluate commercial/prototype wind turbines in typical utility operating environments. This technology demonstration is a necessary bridge between current advanced turbine development activities to a time when utilities will begin large-scale purchases of wind turbines. The activity seeks to establish deployment and testing projects with at least two host utilities by mid-1993 and will document the experience from these tests over the period 1993-1997. The tests will involve a number of turbines (nominally 20 at each host site or approximately 6 MW). Funding for these projects will come primarily from the host utilities, with additional funds from EPRI and DOE. In addition, NREL, DOE, EPRI and vendors will provide technical support as required. A second phase is also being contemplated which would evaluate the Next Generation Technology from the AWT Program in the late 1990s.

The Power Marketing Administration initiative is a program of testing and validation of wind turbines as a source of electricity supply that is being initiated in parallel with the EPRI project. The Bonneville Power Administration, the Western Area Power Administration and possibly other power marketing administrations are engaging in this cooperative undertaking. The geographical areas of authority for these organizations coincide in a number of instances with regions of the country having a large percentage of the national wind resource. The 50 MW program planned by BPA for the Northwest region (mentioned earlier in this paper) is the first step in this program.

Collaborative Wind Technology Program

Activities under the Collaborative Wind Technology Program respond to the needs of the industry for assistance in research to maintain current competitiveness, as well as helping address aspects of system development not addressed in other programs. These include the need for things as diverse and practical as the redesign of existing turbine components and access to turbine testing facilities.

The Cooperative Wind Technology Applications activity assists the wind industry in seeking near-term commercial opportunities, both domestic and international, for wind energy. Seven companies have received awards under the initial round of this activity. These cooperative cost-shared research activities are assisting the industry in the design, development, testing, and analysis of solutions to current operational problems. Specific problems being addressed include, but are not limited to, verification testing of existing, new, or improved wind turbines; design, development and testing to verify new technology components or subsystems; site engineering optimization; control and utility interconnection optimization; and application demonstrations in new operational environments (for example, hybrid systems using small machines).

A Value Engineered Turbine (VET) effort was initiated in 1992 to assist the U.S. wind industry in developing near-term commercial wind turbines for both the domestic and international marketplace. The focus of this activity is on re-engineering and/or remanufacturing conventional wind turbine configurations. Many of these designs have proven to be reliable suppliers of electricity in California and Europe. However, the cost of energy achieved by these machines has been above the goals that have been set for wind energy. The VET will support, on a cost-shared basis, value analyses, manufacturing, and commercialization of these machines. Awards under this activity are expected in June 1993.

Applied Research Program

Applied research is necessary to ensure that fundamental design tools will be available for performing detailed designs of future advanced wind turbines. For example, the program has developed new computer models for aerodynamic performance, structural dynamics, and yaw dynamics that run on personal computers. These codes are being used by many wind turbine companies and university research programs. However, enhancements to the models are required to predict the behavior of advanced rotor configurations that will result from advanced conceptual design studies.

Aerodynamics research currently underway involves exploring the characteristics of the wind at actual wind turbine sites (atmospheric fluid dynamics) and investigates the aerodynamic phenomena experienced and created by wind turbines during operation. One of the efforts has been the analysis of Combined Experiment pressure measurements to determine the effects of delayed stall on wind turbine steady and dynamic loads. Testing is done on the heavily instrumented research turbine operated by NREL and located at their test facility at Rocky Flats, Colorado (Figure 3). Aerodynamic investigations on vertical axis turbines are being undertaken at Sandia National Laboratories (SNL). Much of SNL's efforts, in support of all areas of applied research, have been conducted on the 500 kW, 34 meter vertical axis wind turbine test bed located in Bushland, Texas.

Figure 3. National Renewable Energy Laboratory Combined Experiment Turbine

A new Structural Test Facility, located at NREL, has been developed to perform static and dynamic tests of wind turbine components, such as blades. Static tests will determine ultimate strength limits and weak points in the component designs for later detailed investigation. Dynamic tests using hydraulic actuators will submit components to cyclical fatigue loads similar to those experienced during actual operation.

Research is continuing into the effect of wind turbines on operational control and power quality in utility grids, electrical utility interconnection, and wind/hybrid systems. Because wind turbines usually operate in clusters, research on controls and adaptive controls will take a systems approach to all operational control issues. The goal of activities in this area is to define electrical characteristics and controls that will minimize undesirable power fluctuations and maximize the energy contribution that wind turbines can make to a utility grid.

Work is also continuing on development of advanced airfoils and blade designs, variable speed generator and advanced drive train research, advanced tower concept development, and the evaluation of aerodynamic rotor brakes.

Summary

During the next decade, broad-based development of the vast wind resources in the United States will begin. A financial incentive of $0.015/kWh was recently enacted for wind and certain biofuels. More legislation is planned that will affect the economics of fossil fuels, and thereby benefit renewables.

The Federal wind technology program is increasing its emphasis on working closely with utilities and industry. Funding levels are increasing as new cooperative efforts are initiated. In the near-term, activities will be focussed on assisting industry in solving current problems and on identifying advanced designs that will be more broadly competitive in the mid-1990s. In the longer term, an advanced turbine activity targeted at the design of next generation wind turbines should ensure wind technology's continued success into the next century.

As a result of improving technology, new financial incentives and increasing concerns about the environment and dependence on fossil fuels, the use of wind energy to generate electricity is expected to grow dramatically. By 2010, the installed capacity of wind systems in the United States is expected to grow four-fold from today, to 6.3 GW. This world-leading deployment of wind systems will be accompanied by expected rapid expansion of wind system exports.

References

1. Energy Policy Act of 1992; Public Law 102-486; October 24, 1992.

2. Annual Energy Outlook 1993 with Projections to 2010; U.S. Energy Information Administration; Publication #DOE/IEA-0383(93); January, 1993.

3. Windpower '92 Proceedings; American Wind Energy Association; Seattle, Washington; October 19-23, 1992.

THE WIND ENERGY PROGRAMME OF THE INTERNATIONAL ENERGY AGENCY

W G Stevenson
Scottish Hydro-Electric plc, 16 Rothesay Terrace
Edinburgh EH3 7SE, United Kingdom

B Pershagen
NUWEC energikonsult, Svanvagen 46
S-611 62 Bykoping, Sweden

ABSTRACT

Fourteen parties from twelve countries collaborate in wind energy under the auspices of the International Energy Agency. The activity includes co-operative research, recommendations for wind turbine testing, and information exchange with emphasis on the development of large-scale systems. Nine projects have been successfully completed and three projects are on-going. A joint experimental study of stall aerodynamics by full-scale atmospheric field testing is presently being set up. The overall programme is controlled by as Executive Committee in which each of the participating countries is represented. The Executive Committee publishes a semi-annual Wind Energy Newsletter and an Annual Report, highlighting the progress of the joint tasks and the national activities in the participating countries.

1 THE LEGAL FRAMEWORK

The International Energy Agency (IEA) is the energy forum for 23 industrialised countries, all of which are members of the Organisation for Economic Co-Operation and Development (OECD). IEA was created in 1974 to ensure the collective energy security of its members. One of the objectives is to encourage collaboration in energy technology research, development and demonstration (R,D&D). One such ares is wind energy. Co-operative projects in wind energy R,D&D have been organised since 1977 with participation from fifteen of the IEA member countries.

The IEA projects are set up under Implementing Agreements which define the commitments of the participants and establish a management structure; they also provide for the protection of intellectual property supplied to or arising from the projects. The contracting parties are either governments or parties designated by their governments. The IEA has no central funds to finance projects. All resources must therefore be furnished by the contracting parties by either costsharing or tasksharing.

The wind energy collaboration is effected under the Implementing Agreement for Co-Operation in the Research and Development of Wind Turbine Systems, or IEA R&D Wind for short. IEA R&D Wind has presently 14 contracting parties from 12 countries. Some of the countries are also members of the Commission of the European Communities (CEC), see Figure 1. The agreement is of the umbrella type where individual projects, called Tasks, are defined in Annexes to the Implementing Agreement. The overall programme is controlled by an Executive Committee in which each contracting party is represented. Each Task is managed by an Operating Agent, designated by the Task participants.

So far, thirteen Tasks have been initiated, ten of which have been successfully completed, see Table 1. The detailed results of the Tasks have been documented in more than 200 technical reports. Overviews have been presented at earlier wind energy conferences [1]. This paper concentrates on recently completed and ongoing activities.

IEA Wind	IEA + CEC	CEC
Austria	Denmark	Belgium
Canada	Germany	France
Japan	Italy	Greece
Norway	Netherlands	Ireland
Sweden	Spain	Luxemburg
USA	UK	Portugal

Figure 1: IEA R&D Wind participation and CEC member countries

2. RESEARCH PROJECTS

2.1 Wind-Diesel Systems

This Task was set up in 1985 and has involved about thirty experts from ten countries in a task-sharing arrangement, coordinated by the UK National Engineering Laboratory. The overall objectives were:

- To define cost effective models and techniques suitable for obtaining wind and load data necessary for planning and specifying decentralised wind systems;
- To apply and further develop models suitable for analysing the performance of winddiesel systems.

At an early stage it was decided to produce a work of reference which would convey to a wider community the potential difficulties and stage of development of wind-diesel technology. The final report of the Task has therefore taken the form of a handbook on the siting and implementation of wind-diesel systems. The authors comprise the foremost experts from the participating countries who by discussion and information exchange have agreed upon the contents which include:

1 Wind-diesel options and their applicability
2 Matching the wind-diesel system to the community
3 Assessing the wind resource
4 Designing a system
5 Wind-diesel case studies
6 Modelling techniques
7 Installation and setting up wind-diesel systems

Table 1: IEA R&D Wind Tasks

No.	Title	Operating Agent	Duration
I	Environmental and Meteorological Aspects of Wind Energy Conversion Systems	National Swedish Board for Energy Source Development	1978-81
II	Evaluation of Models for Wind Energy Siting	Battelle Pacific Northwest Laboratories, USA	1978-83
III	Integration of Wind Power into National Electricity Supply Systems	Kernforschungsanlage Julich GmbH Germany	1977-83
IV	Investigation of Rotor Stressing and Smoothness of Operation of Large-Scale Wind Systems	Kernforschungsanlage Julich GmbH Germany	1977-80
V	Study of Wake Effects behind Single Turbines in Wind Turbine Parks	Netherlands Energy Research Foundation	1980-84
VI	Study of Local Flow at Potential WECS Hill Sites	National Research Council of Canada	1982-86
VII	Study of Offshore WECS	Central Electricity Gener. Board, UK	1983-88
VIII	Study of Decentralised Applications for Wind Energy	National Engineering Laboratory, UK	1985-90
IX	Intensified Study of Wind Turbine Wake Effects	National Power plc, UK	1985-91
X	Systems Interaction	US Department of Energy	Postponed
XI	Base Technology Information Exchange	Department of Fluid Mechanics, Technical Univ. of Denmark	1988-
XII	Universal Wind Turbine for Experiments (UNIWEX)	Institute for Computer Applications, Univ. of Stuttgart, Germany	1988-
XIII	Co-operation in the Development of Large-Scale Wind Energy Systems	National Renewable Energy Laboratory, USA	1991-

8 Assessing the economics

A contract has been negotiated with Cambridge University Press who will publish and distribute the book on the open marked in 1993 [2].

2.2 Wake Effect Studies

Wind tunnel tests of wind turbine wake effects were pioneered in the late 1970's as a subtask of Task 1 (see Table 1). The work was extended as Task V, in which field measurements and the development of predictive models were included in addition to laboratory-scale tests. At the close of Task V a reasonable understanding had been gained of single wake behaviour. Estimates could be made of the array losses in large wind farms, but validation against full-scale data was needed. A continued study was therefore initiated in 1985 as Task IX.

Task IX has involved eight countries in a task-sharing arrangements, coordinated by National Power plc of the UK. The Task has seen the development of more reliable wake models and the aquisition of considerable amounts of data from single turbines, pairs of interacting turbines, and from full size wind farms. The experimental data and theoretical predictions were brought together in a benchmark exercise, based around Nasudden (Sweden) for the evaluation of single wakes and Taendpipe (Denmark) for wind farm models. The principal conclusions are:

- Site- and machine-specific factors have a very strong influence on array behaviour;
- Array energy losses are real and significant at spacings below seven diameters, but less than predicted by earlier wind tunnel tests;
- Turbulence is enhanced in arrays, again less than earlier studies indicated, but sufficient to cause measurable increases in fatigue damage rates.

The final report was issued in August 1992 [3]. The report outlines the national contributions to the Task and summarises the work completed. It then highlights the principal results and conclusions from the studies and assess those areas where further work may be required.

2.3 The UNIWEX Project

UNIWEX is a computer-controlled, two-bladed experimental wind turbine of 16m rotor diameter, installed at the Ulrich Hutter wind test field near Schnittlingen, Germany. The main goals of the project are the experimental study of aerodynamics, loads and control for different hub concepts, and the validation of wind turbine dynamic design codes. Seven organisations from three countries are participating in the project, which was initiated in 1988 and is operated by the Institute for Computer Applications at the University of Stuttgart, Germany.

The programme of work has comprised:

- Recommissioning of the wind turbine, including instrumentation, trial runs and development of evaluation software;
- Investigation of operational behaviour and load spectra for different rotor types (rigid hub, teetering hub, individually flapping blades, individually lead-lagging blades, combination of flap and lead-lag, teetering hub with flap-pitch coupling, and individual flapping with flap-pitch coupling) for a variety of control strategies within a wide range of values for

parameters like wind speed, rotational speed and power output;

- Verification of numerical simulation codes, including identification and adaptation of a numerical model for the blade, the nacelle and the total turbine, and simulation and comparison with experiments;
- Study of special aerodynamic effects, such as stall control, rain effects and icing.

In all, some 1000 measurement campaigns have been carried out. The work has been documented in a large number of scientific reports [4]. The final Task report is nearing completion and will be published in 1993. The simulation software has been a unique data base for further comparison with numerical predictions.

2.4 Field Rotor Aerodynamics

Several countries are investigating aerodynamic phenomena experienced and created by wind turbines during operation. One of the efforts is to determine the effects of delayed stall on wind turbine steady and dynamic loads. A cooperative project is being set up to measure the aerodynamic profile characteristics under three-dimensional, rotating atmospheric turbulent conditions. The method can be based either on pressure tap measurements at different radial blade positions or on the direct measurement of aerodynamic forces.

Prospective participants are the National Renewable Energy Laboratory, USA; Riso National Laboratory, Denmark; ECN and Delft Technical University, the Netherlands; and ETSU, UK. A joint measuring programme will be defined, including a critical evaluation of the available instrumentation and the development of data processing requirements and data base format. The measurements will be performed at the participants' field activities in a task-shared effort. Coordination of planning, performance and evaluation will be made the Operating Agent on a cost-shared basis.

3. RECOMMENDED PRACTICES FOR WIND TURBINE TESTING AND EVALUATION

The evaluation of wind turbines must include many aspects, such as power performance, reliability, durability, cost of energy, and environmental impact. The development of internationally agreed evaluation procedures is needed for establishing a sound market. IEA R&D Wind has long since addressed this issue by having experts from the participating countries agree on recommended practices for wind turbine testing and evaluation. To date, recommendations have been issued in eight areas, see Table 2.

Table 2 Documents in the series of Recommended Practices for Wind Turbine Testing and Evaluation. Available from the Department of Fluid Mechanics of the Technical University of Denmark, Lundtoftevej 100, DK-2800 Lyngby, Denmark.

This activity is one of the subtasks of Task XI Base Technology Information Exchange and is directed by a Standing Committee (SC), consisting of the Operating Agent and four other members appointed by the Executive Committee. The SC meets once or twice a year in order to discuss the need for updating the existing recommendations and the need for new recommendations in view of the development of the technology and the implementation of wind energy systems. The updating or preparation of new recommendations is carried out by ad hoc expert groups. Recently an ad hoc group on Wind Speed Measurements/Anemometry was established. The work of this group may lead to a new document in the series of Recommended Practices to be issued in 1993.

When the first recommendations were issued about ten years ago, there were practically no technical guidelines for the evaluation of wind turbines. The work is facilitated by the fact that the recommendations are not to be regarded as standards. They are serving a useful purpose until standards are established, which is normally a very time-consuming procedure. The national and international standards now being developed are in many cases inspired by or based on the original IEA recommendations.

4. JOINT ACTIONS

The activities of Task XI include Joint Actions, which represents a looser form of cooperation than a Task but may develop into a Task if necessary. Specific important areas can be selected for Joint Actions by the Executive Committee. As a minimum, workshops or symposia are arranged once a year where current issues of important are discussed, recent and ongoing research is presented and discussed, and the need for future research defined. So far, four Joint Actions have been initiated in the areas of aerodynamics, fatigue, offshore installations, and wind-diesel systems.

In the Joint Action on Aerodynamics, symposia are arranged with participation of invited experts from the member countries. To date, six symposia have been held with presentations in topics of current interest, such as new airfoil sections, theoretical and experimental studies of rotors in yaw, and three-dimensional flow through rotors. Proceedings are published by the Operating Agent, which is the Department of Fluid Mechanics of the Technical University of Denmark.

In the Joint Action on Fatigue, a group of experts has agreed on a reference load spectrum with variable amplitude for wind turbine blade fatigue testing. The spectrum is known as WISPER (Wind Turbine Load Spectrum Reference). The spectrum has been frequently used in blade fatigue testing, among others in a programme co-sponsored by the Commission of the European Communities.

5. EXPERT MEETINGS

Besides the activities covered under the Joint Actions within specific areas, the Operating Agent of Task XI arranges Topical Expert Meetings, normally twice a year, where the subject to be discussed is chosen by the Executive Committee. Up to 1992, twenty-two such meetings have been

Vol	Title	1st Edition	2nd Edition	3rd Edition
1	Power Performance Testing	1982	1990	
2	Cost of Energy from Wind Energy Conversion Systems	1983	In prep	
3	Fatifue Characteristics	1984	1989	
4	Measurement of Noise Emission from Wind Turbines	1984	1988	In prep
5	Electromagnetic Interference	1986		
6	Structural Safety	1988		
7	Quality of Power, Single Grid-Connected Wind Turbines	1984		
8	Glossary of terms	1987	In prep	

held. The subjects reflect the progress of wind technology during the past fifteen years. The meetings are similar to the symposia mentioned above with respect to selection and number of participants. Proceedings are edited and published by the German contracting party, the Research Centre Julich. The meetings have proved to be a very effective means of communicating research results and discussing programme plans.

6. COOPERATION ON LARGE WIND TURBINE SYSTEMS

At the end of 1992, about 22,000 wind turbines were installed in the member countries, representing a rated capacity of around 2,400 MW. The rate of increase is about 250 MW per year at the present. Most of the capacity is found in wind farms of small and medium-sized wind turbines, owned by private investors or cooperatives, although utilities are taking an increasing interest in the deployment of large wind systems. The average unit size in present wind farms is about 110 kW, while that of new wind farms is increasing and in 1992 was about 200 kW. Units of 500 kW are now on the market, and the leading manufacturers are developing megawatt-scale machines.

The IEA set up an Implementing Agreement for Co-operation in the Development of Large-Scale Wind Energy Conversion Systems (LS WECS) in 1977, in addition to the R&D Agreement, described above. The condition for joining the LS Agreement was that the country be involved in "national projects for the design, construction or operation of at least one LS WECS with a rated capacity of approximately 1 MW or more". The original participants were Denmark, Germany, Sweden and United States, who all had projects for the development of megawatt-sized wind turbines at the end of the 1970's. Later participants were Canada, Italy, The Netherlands, Spain and the United Kingdom.

The development of large wind turbines proved to be more difficult than expected, and the commercial breakthrough, stimulated by government-sponsored marked incentives, cam instead with small and medium-sized machines utilised in wind farms. As a result of the changed direction of development, the character of the IEA activities also changed gradually. There was no longer a need for two Implementing Agreements, and in 1991 the two Agreements were merged. The large-scale WECS activities were incorporated in the R&D Agreement as a separate Task XIII with the emphasis still on the development of large wind turbines but also including issues pertaining to large wind farms. Eleven parties from Canada, Denmark, Germany, Italy, Japan, Norway, the Netherlands, Spain, Sweden, United Kingdom and United States are now participating in Task XIII, which is operated by the US National Renewable Energy Laboratory.

Task XIII is essentially an agreement for the exchange of information on large wind systems. Input to a computerised information system is provided by a network of contact persons in the participating countries. Consensus has been achieved on the form and content of the material to be collected and disseminated. The information includes national wind energy installation and generation summaries, design and operational data on large size prototype wind turbines, and summary data for large wind farms. Incidents and accidents involving large wind systems are reported on a voluntary basis. Information on national wind energy programme funding levels is also compiled by the Operating Agent.

7. EXECUTIVE COMMITTEE ACTIVITIES

Much of the information exchange takes place at the Executive Committee meetings twice a year when the activities in the participating countries are reviewed with emphasis on large scale developments. The highlights of the national R,D&D programmes and the progress of the cooperative Tasks are summarised in the Wind Energy Newsletter, which is issued twice a year and is widely distributed in the wind energy community. The Executive Committee also publishes an Annual Report which gives more detail of the joint projects and the national activities in the member countries. The Annual Report is distributed to all IEA countries and is also available from the Swedish contracting party to the Agreement, the Swedish National Board for Industrial and Technical Development.

As part of the current restructuring of IEA's collaboration programme in energy technology and R&D, the Executive Committee is developing a strategy document for a five-year continued cooperation in wind energy. The document will include a state-of-the-art summary of wind technology and markets, and the identification of specific technical and environmental issues for cooperation. Other important issues are the participation of industry, the dissemination of information, the widening of the cooperation by invitation of new contracting parties and association of non-IEA countries.

8. CONCLUSIONS

The IEA cooperation has provided a useful framework for joint international efforts in wind energy during more than fifteen years. Nine comprehensive research projects have been successfully completed, three projects are currently in progress, and new cooperative action is being planned. The sharing of research and information has clearly stimulated the development of wind technology, has eliminated unnecessary redundancy in the national programmes, and has encouraged efficient approaches to solve common problems. Last but not least, it has created a cooperative spirit among the participating professional groups that seems to be unique.

REFERENCES

[1] H J M Beurskens, B Pershagen, Proceedings of the European Wind Energy Conference (EWEC '91)

[2] G Elliot, R J Hunter (Editors), Wind-Diesel Systems, Cambridge University Press (In print)

[3] D J Millborrow, J F Ainslie, Intensified Study of Wake Effects behind Single Turbines and in Wind Turbine Parks, Final report of Task IX, National Wind Power, Leatherhead, UK (1992)

[4] IEA Wind Energy Annual Report 1992 Swedish National Board for Industrial and Technical Development (1993)

REVIEW OF EUROPEAN WIND ENERGY PROGRAMMES

H.J.M. Beurskens
Netherlands Energy Research Foundation, ECN
P.O. Box 1
1755 ZG Petten
The Netherlands

D. Lalas
Lamda Technical Ltd.
64 L Riancour Street
GR-115.23 Athens
Greece

Abstract

This paper gives a general review of European national programmes.
Papers from Norway, Sweden, Spain and Denmark submitted to this conference (session C) and the draft annual report (1992) of the IEA R&D Wind Programme were used as sources of information.
The paper first describes in general terms tendencies of the past wind energy programmes and than links them to the present developments. Not only the individual aspects are reviewed (R&D, wind turbine development, market stimulation, utility involvement, regulatory issues and operational experiences), but also the synergetic aspects of their integration is addressed.
The paper summarizes the tendences in a number of conclusions which follow the aspects mentioned before. The main conclusion, however, is that the integration of R&D, industrial development and market stimulation works.

1. PAST DEVELOPMENTS; LESSONS LEARNT

In the past most of the national wind energy programmes were characterized by R&D activities directly related to resource assessment and to the development of experimental, mostly large size, wind turbines. Generic research (aerodynamics, aeroelastics, structural mechanics, materials, design criteria and methods, electrical conversion and control technology and safety requirements) was defined by the scientific community with little reference to problems arising from the design process in the factories.

This is clearly demonstrated by the large wind turbine design and development projects which in the early eighties were carried out by high tech industries, mostly with their roots in the aircraft technology, in countries like the USA (MOD series), Germany (GROWIAN), Sweden (Marglarp, Näsudden). These industries were interested in the development of advanced concepts as such and not necessarily in commercializing their developments: Boeing, Hamilton Standard, MBB, MAN, Fokker, Stork, all stepped out, the moment the 100%-paid government development contracts were not available anymore and they were expected to continue wind energy activities partly on their own.

As a result the know-how and experience, built up within the framework of large wind turbine development programmes, which in some countries (USA, D, GB, S) took the major part of the available funds, became only partially available to the smaller companies developing wind turbines to be marketed immediately.

Know-how from the generic R&D projects, carried out by research establishments (Risø, SERI (at present NREL), Sandia, ECN, TNO, NLR, FFA, D(FV)LR, CIEMAT, ENEA, etc.) and universities, belonged to the public domain and thus was less interesting to companies because results in general could not be used on an exclusive basis. Other reasons why the results were not intensitively used, included:

- R&D (with possible exceptions in Denmark) was not directly addressing problems the industrial designers were confronting with on the work floor.

  Examples : flexible rotor concepts were investigated while industry was considering different concepts.
- Results were difficult to transfer because software was not compatible with the capabilities of the industry.
- Because the number of sales of machines were small, the human resources of the industry in a lot of cases, were insufficient to absorb know-how from the scientific community.

The manufacturers though of small (before 1990) and medium size machines (after 1990) managed to convince R&D policy makers and planners of the commercial viability of a large market niche for their machines, to generate a genuine interest among private investors and utilities and to direct the aims of R&D programmes towards problems related to the category of machines the industry was marketing.

As a result the notion grew that a new technology cannot be successfully developed and implemented by R&D efforts only on a market place that is dominated by government and utility initiatives. Besides R&D efforts, demonstration and market stimulation schemes came to be considered necessary and were occasionally implemented. In the USA a market stimulation programme (PURPA) through tax incentives probably came too early. Initially the tension between demand and supply of machines caused products of poor quality to be installed. Further the subsidy level was not in accordance with the non economic part of the generation cost of wind energy. As the result the annual sales of installations dropped to less than $^1/_3$ of its maximum value within one year time after the tax credits were abandoned (1986).

Other countries were confronted with a "vicious circle", trapping further commercial development (NL, D). Utilities, considered as the most important buyer of turbines, postponed purchases for different reasons, the most important one being prohibitively high cost of wind turbines and unknown reliability of the technology offered.
Manufacturers were not able to decrease manufacturing cost and to improve reliability because the number of machines

being sold was too small resulting in margins too low to enable financing of development on their own. Furthermore they were not willing to invest in production facilities because a firm market perspective was missing. The only player capable of breaking this closed circle was the government by providing a market by a long term (> 5 years) stimulation programme (NL) or large scale test/demo programmes (D). Later these schemes were followed by initiatives in other countries like the UK (Non Fossil Fuel Obligation - NFFO), Sweden, Italy and Norway.
The Danish example shows that potentially the present success of the Dutch, German, British, Swedish and Spanish programmes could be made permanent essentially without subsidies if the subsidies are gradually (and in time) adopted to actual cost developments.

Until now we mainly addressed the role of the industry and the R&D community in the development process of the wind energy industry. A third essential element in this process is the demand side of the market. Here private initiatives and utilities play an equally important role. Private initiatives very often acted as a catalyst for utilities and the private sector in Europe continues to be responsible for the major part of the sales volume.

The role of the utilities needs to be addressed separately, because of its diversity, see table 1. In Italy and Greece, the utility essentially controls all wind energy developments; in some other countries utilities are fully integrated in the development process (S, DK) and in some they play a vital role in offering a market perspective to the industry by incorporating wind energy in their energy planning and by purchasing wind turbines on a compatitive basis. Specific conclusions are difficult to draw, their role, however, is essential as a market driver!

Table 1: Utility involvement in wind energy development.

Utilities in:	R&D resource assess-ment	Wind turbine develop-ment	Test/demo projects	Purchasing w.t.'s on compatitive basis
DK	+	+	+	+
S	+	+	+	+
I	+	+	+	0
GB	+	0	+	+
N	+	0	+	+
NL	0	0	+	+
GR	+	0	+	+
D	0	0	+	+
E	+	+	+	0
EIR	0	0	0	+
P	0	0	0	+
F	0	0	0	0
B	0	0	0	0
L	0	0	0	0

Last but not least the legal and regulatory aspects have to be mentioned. The fact that they are mentioned in the last instance by no means implies that they are least important. In practice they have been overlooked too often resulting in long preparation times for some projects and even cancellation of others. Densely populated countries like the Netherlands were the first to be confronted by these problems. The question however has always been how to attack the problems. Only a few aspects such as acoustic noise emission and immission, public acceptance, effect on birds life can be addressed by research. Regulations and physical planning problems have to be solved rather as part of the normal planning procedures.

2. PRESENT DEVELOPMENTS; DID WE LEARN FROM THE PAST?

2.1 Introduction

Besides the four papers submitted to this conference from countries with most exciting programmes (Norway, Sweden, Spain and Denmark), to provide a more complete spectrum of developments, we also took into consideration the draft IEA WIND R&D programme's 1992 Annual Report and appropriate US data. We will describe developments in the following areas:
a: R&D; b: wind turbine development; c: market stimulation; d: utility involvement; e: regulatory issues; and f: operational experience.

2.2 R&D

Although the total budgets for wind energy development are increasing, the budgets for generic R&D tend to decrease, most significantly in Germany where a 50% co-financing is required and to a lesser extent in the Netherlands and Sweden. Areas of high priorities remain the "conventional" ones of:

- aerodynamics (3-dimensional effects, dynamic stall phenomena, dynamic inflow, dedicated aerofoils),.
- fatigue characteristics of (blade) materials and components including the development of reference testing methods for large blades,
- the development of standards on load spectra, load cases, safety systems, structural integrity, etc.

Subjects which have been recently added to the list of priorities are:

- the development of low and variable speed electrical conversion systems,
- availability technology (probabilistic reliability and safety analysis, life time and load monitoring),
- acoustic noise emission of rotor blades,
- advanced components (passive control concepts, flexible hub configurations, aerodynamic control methods).

Spain, Great Britain, Greece and to a lesser extent also Sweden adopted the policy to primarely let their programmes follow priorities set by the European Commission for the JOULE programme. Participation in European programmes is actively stimulated. Spains' R&D programme almost completely consists of JOULE activities.
Norway's R&D activities mainly consist of resource assessment and the development of autonomous wind diesel systems (AWDS). This is due to the particular situation in Norway where the best wind resources are located at the low islands in Mid-Norway (Sør Trøndelag). Norway has achieved a prominent place in AWDS development.
Germany has put a heavy emphasis on the collection of operational experience of its 250 MW experimental programme (collection of data for 10 years is obligatory!).

Many analytical and empirical descriptions of aspects of the behaviour of wind turbine installations have been incorporated in computer programs. Offering these packages has become a commercial activity and the market is very competative. This leads to the conclusion that only the development of specific models such as the description of aerodynamic phenomena (3-dimensional flow, dynamic stall and inflow) and specific mechanical response models, (hub configurations with complex flexibility and passive control systems) can be considered as part of the R&D domain.
The incompatibility of software modules developed by different institutes and engineering consultants seriously prevents an effective transfer of knowhow to manufacturers. Also an effective use of these modules for design and optimisation iterations is time (and money) consuming. For

this reason standardization through shell- or file management programmes deserves priority as well.

The integral approach (R&D, industrial development, market stimulation) apparently is considered a necessity.
Depending on the development stage of the technologies, priorities have to be shifted. In the time perspective, this means more emphasis on industrial development and market needs. In doing so one should be careful that areas which become relatively less important do not get neglected entirely due to lack of critical mass.

2.3 Wind turbine development

The development activities can be divided into optimization of the present commercial category of medium size turbines and development of a third generation of MW-size machines.
A separate category is the "stretching" and "repowering" of medium size turbines.

Medium size machines:
Examples of upscaling from the 100-200 kW range to the 400-700 kW range are:

Examples: Spain	:	MADE: 300 kW/25m and 500 kW Ecotécnia: 150 kW/24m to 400 kW
Sweden	:	NORDIC: 400 kW/35m Zephyr: 250 kW
Denmark	:	all Danish manufacturers (Vestas, Nordtank, Bonus) are working on turbines in the 400 to 500 kW range
Germany	:	HSW: 750 kW/40m ENERCON: 400-500 kW/40m Tacke: 500 kW/36m
Netherlands	:	NEDWIND: 500 kW/40m WINDMASTER: 750 kW/40.2m Lagerwey: 225 kW/27m
Italy	:	Alenia/WEST: 320 kW/33m Riva Calzone: 350 kW/30m
GB	:	WEG: 400-450 kW/37m Markham: 600 kW (Villas license)

The development activities are financed by the manufactures themselves, sometimes with assistance from national programmes (Sweden, Denmark, Netherlands) and/or from utilities (Italy, Sweden, Germany, Spain).

It is these machines and their predecessors where the major increase of the European installed power by about 200 MW in 1992 came from and where in the medium term the extension of the power is expected from. The total European wind power is (see table 2) close to 1000 MW, more than half the installed power in the US. At present growth rates Europe will likely surpass the US in installed wind power in 1995. This may not come to pass, if new wind energy sites being developed in the mid west materialize and if the utilities' interest grows.

A critical factor in further development of the European wind energy market is the successful removal of bottle necks like public acceptance, acoustic noise emission, physical planning aspects, cost and safety. The new R&D issues of availability technology, acoustic noise emission, life time monitoring all are directly related to removing these obstacles.

Table 2: Installed wind power in Europe

Country	Total installed power, in MW, end 1992	Added in 1992 MW (units)
DK	418 (3218)	75 (393)
D	142 (964)	80
NL	100	
E	50	40 (233)
GB	76	
GR	27	
S	20 (95)	
N	3	
EIR	6	
I	15	
B	5	
P	8	
Total	≈ 900	≈ 200

Another remarkable development is the implementation of so-called advanced concepts in these medium sized machines. Concepts which initially were considered the exclusive domain for MW size machines trickle down to the medium size machines. In scaling up machines the increased loading of components makes it necessary to apply materials with a better specific strength-mass ratio, to apply flexible elements for the relief of dynamic loads and to apply passive elements. All these measures lead to lower weights and to lower cost as well. Apparently these concepts are already now considered advantageous for commercial machines.

Examples: Nordic (S) 400 kW	:	teetering hub, stall control variable speed;
Zephyr (S) 250 kW	:	partial individually passive pitch, 2 speeds;
Enercon (D) 500 kW	:	slow running synchronous generator, direct drive (!), variable speed, fast active blade pitch control.

MBB stopped its wind energy activities which were based on the one-blade concept. These activities still are pursued by Riva Calzone of Italy.

There has always been a dispute concerning the viability of MW-size turbines as the results of the first generation of MW-size wind turbines were not encouraging, especially regarding cost and siting. A number of machines though, especially the one developed through the WEGA project of the CEC JOULE-I programme, have shown that commercial machines of MW-size can be produced.
The wind turbines realised within the CEC-JOULE-I programme (WEGA-project), the Danish Tjaereborg turbine (2MW, 60m) the Spanish WKA60 (1.2MW, 60m) and the British Richborough (1MW, 55m) are under test and especially the Tjaereborg turbine forms a valuable source of measuring data for verifying design tools (JOULE I and II) and gaining operational experience. The technical concepts were not taken up for future developments. The manufacturer of the WKA turbine, MAN has resigned from the wind energy business.

At the same time, medium size machines are being scale up simultaneously incorporating advanced concepts.
The development of MW-size machines has been taken up in Sweden, Germany, Netherlands, Denmark. The machines being developed now are smaller than the first generation experimental units: 1 MW or slightly more.

The new developments of MW size machines are characterized by considerable lower specific weights compared to the previous generation.
(The WEGA tubrines '86-'91 : 39-57 kg/MWh (per year);
WEGA-II proposals '91-'96 : 12-21 kg/MWh (per year)).
It is surprising to see other actors appearing on the supplyer side of the market. Most of the established manufacturers of medium size machines made proposals to the European Commission to develop a 1 MW turbine. New manufacturers committed themselves to actually commercialize their machines. All proposals contain innovative developments. Some of these developments will be tried out on medium size machines, such as NORDIC and ENERCON.

Some MW-size machines:

Näsudden-II (S)	: 3 MW/80m; full pitch, stiff hub, 2 speeds
Aeolus II (D)	: See Näsudden II, difference: variable speed
WEG* (GB)	: 1 MW/50m; teetering hub, active blade pitch, variable speed
BONUS* (DK)	: 0.75-1MW/50m; stall, flexible hub construction, fixed speed
VESTAS* (DK)	: 1MW/55m; rigid hub,
Union Fenosa* (E)	: 1.2 MW/58m; vertical axis, H-type rotor, direct drive, permanent magnet generator, variable speed
Avedore (ELKRAFT)* (DK)	: 1MW/50m, both pitch and stall, variable speed
ENERCON*(D)	: 1MW/55m; full fast pitch, stiff hub, variable speed, direct drive
NORDIC*(S)	: 1MW/50m, teetering hub, yaw control, variable speed
NEDWIND*(NL)	: 1.1 MW/50m, 2 speeds, stall control
GAMMA 60*(I)	: 1.5 MW/60m, variable speed, yaw power control
HMZ*(B)	: 1.0 MW/45,5m

(The development of the Wind turbines labelled with * are cofinanced by the European programmes JOULE-II or THERMIE).

Including the machines from the list above, at present 20 wind turbines with an installed power of 1 MW or more are being tested or developed in Europe and one in the USA. Almost all of them have different main features. From a technical scientific point of view this is an extremely favourable situation to compare different concepts.
From a commercial point of view these are too many to find a place on the market. As the European Commission plays an important catalytic role in these developments, it is an excellent position to assess the results and distill as many lessons as possible from this excellent "laboratory" it helped to realize. With the assistance of CEC programmes, directed towards the evaluation of the turbines (both design and operational) in combination with the selection process on the market place, the European industry could build a unique position on the world market.

In general the off shore application of wind turbines is considered far away. It will only be considered seriously if land locations are depleted. Wind energy from off shore turbines is more expensive than wind energy from land based machines (numbers: 50% higher cost). Nevertheless Sweden and Denmark have started test projects on the basis of commercial land machines to gain experience. Their projects are realized in shallow coastal waters.

Sweden : Nøgersund : 1 * 220 kW
Denmark : Vindeby : 11 * 450 kW

Denmark plans another 5 MW off-shore wind park between Ebeltoft and Samsø.

2.4 Market stimulation

Earlier we noticed that integration of R&D, industrial developments and market stimulation is a prerequisite for a successful large scale introduction of the wind energy technology. Almost all countries which have significant wind energy programmes have introduced some sort of market incentive. The mechanisms to stimulate the demand side of the market however vary widely as can be seen from Table 3. It seems that in the initial phase, when the difference between actual generation cost of wind energy and acceptable cost is large, investment subsidies work satisfactory. In a later stage, subsidies of the tariffs of the generated electricity seem to be a more practical method. Further analysis is needed. The subsidy level in Denmark is at present the lowest. The German programme is presented as a large experimental programme with the aim to gain operational experience (10 years measuring programme obligatory). In practice this programme also appears to be an effective stimulus of the market.

Table 3. Market stimulation and incentives

Country	Investment subsidies	Tariff	Others
DK	-	70% to 85% selling price, excl. taxes	guarantee for repayment of long term financing of large projects outside EC
D	Some local investment subsidies	+ DM 0,06/kWh during 10 years	
NL	35% of total investments; bound to maximum	Some utilities pay extra high feed back tariffs	
GB	-	"bid-in" system NFFO (at present £ 0,11/kWh)	-
S	25% ('92), 35% ('93), of total investments (>60 kW)	-	-
N	50% of total investments	-	-
I	30-80% of total investment	ECU 0.11/kWh during 5 years	-

2.5 Utility involvement

Utilities are involved in wind energy developments in various ways:

- passively . implementing technical regulations which facilitate connection of decentralized power units to the grid;
  . offering tariffs reflecting at least avoided fuel cost, the value of capacity credit and avoided environmental cost.

- actively
  . carrying out own R&D projects;
  . supporting the development of wind turbine technology by:
    * designing, building and testing own wind turbines [(DK: Nibe, Tjaereborg, Avedore);
    * joining industries in designing wind turbines (ENEL (I), Union Fenosa and MADE (E), EPRI (USA), Vattenval and Südkraft (S)].
    * acting as launching costumer [(PEN (NL), PEB and EGD (NL), Preussen Electra (D), National Power (GB), etc.)].
  . developing own wind turbine plants on the basis of competative bids.

All sorts of support mentioned here can be found in Europe. It is also noticed that the number of utilities involved in wind energy in Europe has been increasing and that this apparently is crucial in establishing a firm market perspective from which industries can derive confidence to intensify developments, further optimize their manufacturing facilities and convince banks to provide the necessary capital.
The question which remains however is, what way of involvement is most effective. We think that the utilities role of launching customer, developer and operator of their own wind plants with commercial wind turbines and stimulating private unitiatives by setting favourable regulations and tariffs is the most natural role fitted to utilities, which implies the least conflicts of interest.

2.6. Regulatory issues

In the past these have been underestimated. However these will remain important in the future as well.
The best sites will be used up increasingly. As a consequence, less suitable sites will be considered which means that environmental and economic restrictions will cause project preparation activities to become costlier and the economics less favourable. In this respect, it is good to notice that in different national programmes (NL, D, S, GB) and the JOULE-II programme, the issue of reducing acoustic noise emissions from wind turbines has been included and given a high priority. Another important item for the R&D domain is the public acceptance. From R&D in this field (NL, DK) lessons can be learned directly on how to involve the public in the preparations.
The development of standards and certification systems will facilitate establishing rules for realising wind farms because it alleviates local authorities from the difficult job of developing their own. The sharing of experiences in this field is not the subject of any programme known to us, but should be considered as a high priority activity (Task CEC?).

2.7. Operational experience

In different countries systematic data bases exist for operational data of commercial wind turbine projects (DK, NL, D, USA (California)). The most systematic approach to gather data is the 10 years monitoring requirement of the 250 MW German experimental programme. The EUROWIN data base, co-financed by the European Commission also provides valuable data. It is a well known fact that it is extremely difficult and costly to collect detailed data such as number and kind of outages related to the **total** wind turbine park of Europe. Even a less ambitious goal of collecting data on a representative, non-selective, population of turbines producing general parameters is very useful for utilities, private developers, financers and policy makers. The following general parameters would form a useful set:

* technical availability, with some relation to the total time the project is in operation;
* capacity factors assuming 100% technical availability;
* annual energy outputs (in kWh per square meter swept rotor area);
* turn key cost, possibly split up into machine cost, civil works including installation and grid connection cost and project preparation cost.

Available values, which are being reported by the different country representatives indicate:

* technical availabilities of over 95% for newly installed projects,
* a wide range of capacity factors, from 0.05 (mostly old projects) to close to 0.45. The majority of newly installed parks show values close to 0.25.
  Here a warning has to be made. Of course high capacity factors are preferred by utilities. However, the wind is a highly varying resource. Too high capacity factors (> 0.3) therefore lead to decreased energy outputs (low generator capacity compared to the rotor swept area), which has a negative influence on the economic performance.
* annual energy output values, varying from 500 kWh/m^2 to about 1200 kW/m^2 for the best locations. Some indicative correlations with annual averaged wind speeds would help in better interpreting these data.
* total investment costs varying a lot with local circumstances and the commercial "deals". Wind turbines typically cost ECU 300 to 500 per m^2 rotor area. Total investment cost may be a factor 1.25 to 3(!) higher because of civil works, installation cost and last but not least: connection cost to the grid.
* more and more evidence coming available that maintenance of larger wind farm projects can be realised at annual cost of about 1% of the initial investments cost.

2.8. Conclusions

The following conclusions can be drawn:

* Successful wind energy development programmes combine the R&D, industrial development and market stimulation aspects.
* As generic R&D becomes relatively less important, the size of activity has to be maintained above critical mass level in order to be effective in producing meaningful results.
* Concepts of medium sized machines (as a result of upscaling) and MW size machines converge.
* A whole series of innovations are being implemented. These innovations tend in the direction of flexible components, variable speed generator systems, direct drives.
* There are 20 wind turbines of 1 MW or more being tested and designed. All have different technical features. This puts Europe in a unique situation on the world market.
* A careful evaluation of these technologies will be necessary to control the inevitable future process of natural selection on the market place.
* Market stimulation programmes have been proven to be successful. An analysis of the effectiveness of instruments (investment subsidies, tariffs) in relation to the need of market incentives is necessary.
* Utilities are getting more and more involved in wind energy developments. Their natural role seems to be as a
  - stimulator of private power generation by appropriate regulations and tariffs;
  - launching costumer of prototype machines;
  - purchaser of commercial wind farms.
* regulatory issues become more important in the future as the best wind turbine sites get depleted;
* systematic collection of operational data provides very useful information for utility planners and policy makers.

THE POTENTIAL OF WIND ENERGY IN SØR-TRØNDELAG, NORWAY

L. Tallhaug and T.A. Nygaard
Institutt for energiteknikk
Box 40
N-2007 Kjeller
NORWAY

ABSTRACT: The aim of the study has been to estimate the potential for wind energy in the county of Sør-Trøndelag, and to provide a detailed ranking of projects, sorted by increasing cost of energy (COE). The data base with all projects includes all necessary investments, the wind resource at the site, environmental issues, O & M costs and more. The study includes a sensitivity analyses where the connection between some main parameters and the potential has been established.

1. INTRODUCTION

1.1 The Norwegian energy system

The Norwegian power production system has an installed capacity of 26.000 MW, serving a population of about 4 million people within an area of 324.000 km^2. Approximately 99 % of the electricity production is hydropower and approx. 0.01 % is windpower (10 turbines and 3 MW total installed capacity). The annual energy production varies from 80 TWh to 125 TWh due to variation in precipitation. The production and end-use of energy in 1990 is shown in table 1.1.

	Production	End-use
Electricity	121	97
Oil and gas	1430	70
Biomass	11	11
Coal. other	3	12

Table 1.1. Energy production and end use in 1990 (TWh). Note that the precipitation in 1990 was higher than average.

Due to lower growth in electricity demand than expected, Norway has now become a net exporter of electricity. But the fact that more than 20 TWh of oil is burned for heating purposes makes the electricity demand sensitive for increases in carbon taxes.

1.2 Norwegian wind energy activities

Since the mid 70's the department of trade and energy has supported some wind energy activities in Norway. The main activities through the 80's was development of a wind diesel system and testing of danish wind turbines in the norwegian climate. As a result of this activity there has been erected eight turbines with a total rating of 2755 kW, in addition to two privatly owned (55 kW and 75 kW).

During the last year the priority has changed towards more industrial development and less testing of foreign technology. The total annual governmental funded wind energy activity has been about 1.500.000 ECU/year. Now when the testing of foreign technology is nearly withdrawn the budget is reduced to about the half in 1993. Next year the funding system is going to be changed, and the activity level is so far uncertain.

1.3 Previous studies of the potential of wind energy in Norway

In 1981 a study of the potential for wind energy in Norway was finished. The aim of this study was to estimate the number of turbines rated 2750 kW that could be located along the Norwegian cost under certain restrictions.

In that study the potential was estimated to be about 11 TWh/year avoiding areas protected for reasons of nature care. The cost of energy was not calculated in the project.

A simplified revision of the potential was made in 1990 because the technology development had resulted in smaller commercial turbines.

2. AIM OF THE PROJECT

The aim of this study has been to estimate the potential of wind energy in the county of Sør-Trøndelag, and to provide a detailed ranking list of projects, sorted by increasing cost of energy (COE).

Sør-Trøndelag is a county in the middle part of Norway. This part of Norway is characterised by flat islands outside the coastline, well suited for wind energy conversion.

In the study from 1981, the contribution of Sør-Trøndelag was estimated to be approximately 25% of the national wind energy potential.

3. METHOD OF APPROACH

The wind conditions recorded at the weather stations indicated that the most interesting areas for wind energy are located along the coast.

The rough mountainous terrain inside the coastline generally have less favourable wind conditions. Furthermore have concern about possible icing problems and visual impact reduced the attention paid to these areas.

The first site survey was based on maps in scale 1:50.000 and contour interval 20 meter. To ensure that important items not were left out on the maps most of the interesting areas were visited. With this basis detailed maps (scale 1:25.000 and contour interval 5 meter) for each site could be ordered.

On the detailed maps each and every turbine was plotted and the cost for road construction and grid connection was assessed.

All choices of technology and costs were made on the basis of today's level. This was done to make the plan as realistic as possible in contrast to some studies on renewables in the past, where considerable cost reductions were included.

3.1 Choices of technology

The turbine used in the study is the largest commercial Danish turbine on the market. This means a rotordiameter of 39 m and a generator rating 500 kW. All the assumptions are listed in table 3.1.

Turbine rating	500 kW
Rotordiameter (D)	39 m
Hub height	40,5 m
Spacing in windfarms	6 x D across prevailing wind direction 8 x D parallel to the prevailing wind direction
Farm efficiency	n is the number of turbines
• n < 6 and line formed	η = 98 %
• 6 < n < 15	η = 96 %
• 15 < n	η = 94 %
Distance to houses	Limited by noise (see separate figure), minimum 300 m

Table 3.1 Main assumptions

3.2 Upgrading the public grid

All necessary investments related to transmission of the energy are incorporated.

The public grid is designed for transmission of energy from the inland to the coastal areas. Some capacity is therefore free in the existing grid. To make use of this capacity, small windfarms are incorporated in all areas. These projects are not given any penalty for their use of the grid.

The approach for the distribution of the costs for upgrading the grid on the rest of the projects is simply to assume that all projects are realised the same year. The problem with a more realistic approach is that the ranking is not independent of the speed of the development.

The windfarms are mainly placed at a long distance from the main consumers. This means that the grid needed for transmission of the energy forms a tree with branches outwards to the outermost islands. The costs for each gridpart are distributed on the different projects that use the part proportional to installed capacity.

As a rule of thumb, it is assumed that the local grid can handle wind power up to 50% of its peak capacity.

3.3 Investments

All investments for every windfarm are calculated. The parts of the investments that are independent of site are given in table 3.2 as a function of number of turbines in the farm. Other necessary investments are calculated separately for each farm. The items are given in table 3.3.

	1000 NOK	ECU
Windturbine	2950	355.000
Spare parts	50	6000
Foundation	250-n	30.000 - 120n
Transportation	100-n/2	12.000 - 60n
Unexpected	200	24.000
Installation	150	18.000
Service building	50	6.000
Administration	150-n/2	18.000 - 60n
Roads	150	18.000
Land lease	50	6.000
Cables and transformers inside the windfarm	250-n	30.000 - 120n

Table 3.2 Site independent investments in 1000 NOK and ECU. n = number of turbines in the farm. 1000 NOK ≈ 120 ECU.

- expropriation: where one or two buildings obstruct a large windfarm, costs are added for expropriation. This possibility is used only a few times. For houses 1 mill NOK are added, for cabins 0,5 mill.
- roads: the costs for roads in table 3.2 includes the roads between the turbines and 300 m towards the existing road. The costs for additional roads varies between 500 NOK pr. m to 1000 NOK pr. m depending on the terrain.
- Bridges: for islands near the main land extra costs for bridge construction are added. The cost is set to 7000 NOK/m^2.
- upgrading the main grid: costs are added as explained in a previous chapter. The costs includes lines, subseacables and transformers.

Table 3.3 Site specific costs. 1000 NOK ≈ 120 ECU.

3.4 Estimating the energy capture.

Wind monitoring was carried out at eight locations in the most interesting areas for wind energy. To estimate the long term wind conditions at these locations, comparisons were made with simultaneous measurements at a reference station with known long term wind data [1]. In this project, the method was modified to take differences between wind direction at reference station and site into account.

The relationships between wind speed and direction at reference station (V_R ,D_R) and site (V_S ,D_S) are established using standard regression technique for averages of some hours. For the different direction classes at the reference station, the mean values (μ_{Ri}, μ_{Si}), standard deviation (σ_{Ri}, σ_{Si} and cross correlation coefficients ρ_i are calculated. The long term mean at the site for the actual season given the direction class d_{Ri} at the reference station is estimated by:

$$E\,(V_S|d_{Ri}) = \mu_{Si} + \frac{\rho_i \sigma_{Si}}{\sigma_{Ri}}\,(E\,(V_R|d_{Ri}) - \mu_{Ri})$$

$E(V_R|d_{Ri})$ is the weighted average of monthly long term means, where the weight factors reflects the number of samples within each month in the correlation period.

The long term directional distribution on the site is estimated by the long term directional distribution at the reference station, and the conditional directional probability at the site observed during the short term measurements.

$$p\,(d_{Sj}) = \sum_{i=1}^{12} p(d_{Ri}) \cdot p(d_{Sj}|d_{Ri})$$

Finally, the long term means at each direction class at the site can be estimated.

$$E(V_S|d_{Sj}) = \sum_{i=1}^{12} E(V_S|d_{Ri}) \cdot p(d_{Sj}|d_{Ri}) \cdot p(d_{Ri}) / p(d_{Sj})$$

In this project, the long term distributions of the reference station was extrapolated to the site simply by adjusting the scale factor in the Weibull distribution with the ratio of long term means at site and reference station, while keeping the shape factors unchanged.

Table 3.4 shows the estimated long term means, the distance to the reference station (Titran) and the cross-correlation coefficient.

Station	E(V)	Distance to ref [km]	Cross correlation
Bremvåg 30m	6.0	20	0.89
Hjertåsen 30 m	6.7	21	0.91
Andersskog 25 m	6.6	24	0.90
Grande 30 m	6.9	64	0.68
Ulvøya 30 m	7.4	38	0.91
Føllingen 30 m	7.9	23	0.96
Vallersund 30 m	7.3	77	0.75
Linesøy 30 m	7.7	90	0.76

Table 3.4 Main results from the regression calculation

Fig. 3.5 shows the directional distribution of the reference station, the estimated directional distribution at Grande, and the observed long term distribution at Ørland weather station 5 km from Grande. It can be seen that significant direction changes are present , and that the estimated directional distribution at Grande coincide fairly well with the measured distribution at Ørland. The local terrain is fairly flat for at least ten km in all directions at Titran, and completely flat for at least five km at Grande. The shift to direction 5 (southeast) is probably caused by the fiord Trondheimsfjorden. The fiord seem to channel the wind, especially during the winter, when cold air is draining out from the mainland . Simple models as WASP [2]is considered to be a valuable tool for estimation of local terrain effects, but predictions without measurements seem to be difficult in these areas.

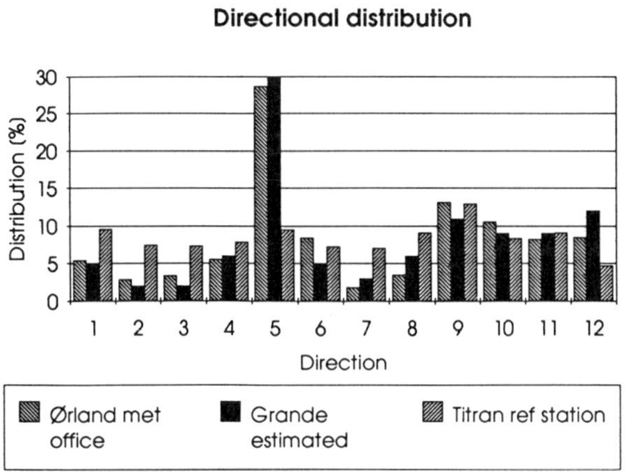

Figure 3.5 the directional distribution of the reference station, the estimated directional distribution at Grande, and the observed long term distribution at Ørland weather station 5 km from Grande.

The wind conditions at each wind farm were estimated subjectively based on:

- the estimated long term wind conditions at each monitoring station.
- the long term wind data from the weather stations.
- estimation of local terrain effects at the monitoring station and some of the sites with the model WASP.
- general knowledge about the wind conditions from earlier projects.

3.5 Economy

As mentioned above the main goal of the project was to provide a ranking list of projects, sorted by increasing cost of energy (COE). The key parameters for calculating the COE are given in table 3.6. All the way through the project the policy was to be as realistic as possible.

The economy for each project is calculated within the database. As an average the O & M costs are 0,085 kr/kWh or 0,01 ECU/kWh.

Real rate of interest	7 %
Economic lifetime	20 years
O & M	2 % of total investments pr. annum
Availability	95 %

Table 3.6: Key parameters for the economic calculations

4. RESULTS

The resulting list consists of 68 windfarms. Four windfarms have been left out because of conflict with areas of environmental protection. The most favourable project produces energy for 0,042 ECU/kWh, and approximately 700 GWh/year can be produced for less than 0,06 ECU/kWh. The total amount of energy is 1300 GWh/year, and the least favourable project costs about 0,085 ECU/kWh. A little less than 200 GWh/year comes directly in conflict with environmentally protected areas and are not included in the figures. These numbers are also illustrated in figure 4.1.

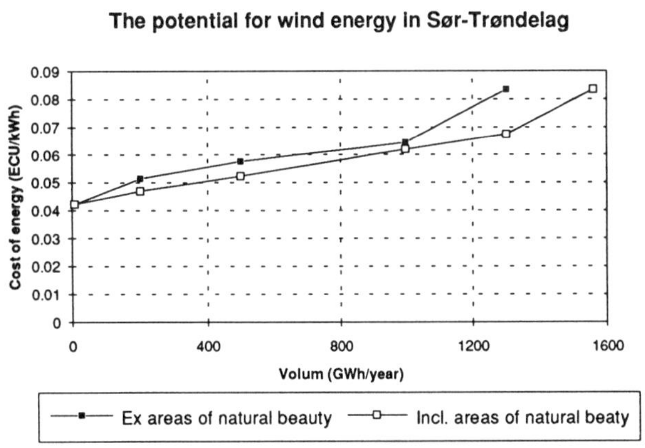

Figure 4.1 The potential for wind energy in Sør-Trøndelag with and without taking account for environmentally protected areas.

There is a strong correlation between the assumed windforce and the COE. The average windspeed varies from 6,3 m/s (40 meters above ground level) to 8,5 m/s, and the specific investment varies from 1120 ECU/kWh to 1550 ECU/kWh with an average at about 1270 ECU/kWh. The best projects are those which do not result in an upgrading of the grid and at the same time are exposed to good wind conditions.

The area in question includes approximately 25% of the Norwegian potential for wind energy. The Norwegian consumption of electricity about 100.000 GWh/year.

When (if) the need for more energy arises there excist a lot of options. The COE for these options varies from option to option and from case to case within each option. The list in table 4.1 gives an illustration of the possibilities and the costs. To give a more complete picture of the economy, curves like the one in figure 4.1 should be given for each technology option.

	Total potential (TWh/year)	Potential with a COE < 0,06 ECU/kWh (TWh/year)
Hydro power	≈40	≈40
Energy conservation	≈18	≈18
Gas fired power plants	----	dep. on gas price
Biomass	≈15	≈13
Solar energy	≈8	≈2

Table 4.1 Options to meet an increasing demand.

4.1 Sensitivity study

Both the costs and the size of the wind energy potential are highly sensitive to a lot of parameters. The most important parameters are:

- estimation of the wind resources
- distance between the turbines within a windfarm
- size of the turbines
- size of the generator compared to the swept area
- degree of detail in the topographic information

In this study only the variation in the size of the potential is quantified.

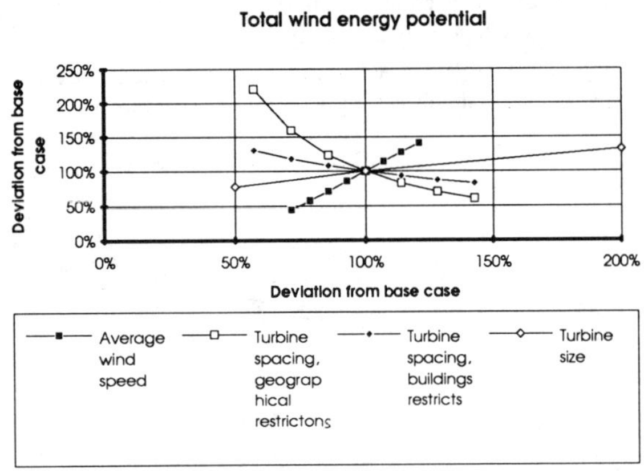

Figure 4.2 Size of the potential as a function of different parameters. The two lines for turbine spacing are the upper and lower limit.

4.1.1 Estimating the wind resources

The average wind speed is a very important parameter both for the size of the potential and the costs. At the same time it is the most difficult parameter to quantify. Figure 4.2 illustrates the sensitivity of mean windspeed on energy capture.

4.1.2 Distance between the turbines

Figure 4.2 illustrates two curves for a variation of the turbine spacing. The steepest curve illustrates the sensitivity to the turbine spacing if one assumes that each windfarm is limited by geographical objects only. The lower curve gives the opposite, if every windfarm is limited by distance to houses. The truth is somewhere in between. In the scattered populated Norway the truth is probably closer to the upper curve.

The reason why these two curves differ is that if one assumes that every windfarm is limited by geographical objects, there will not be a restriction zone due to noise. If one assumes that all the farms has an restriction zone, the size of the zone is dependent of the spacing and the size of the turbines.

Also the cost is sensitive to the spacing. If the spacing is too large the best sites will be wasted, and the sites with poor windconditions must be realized earlier. Better knowledge about the loads on turbines in wakes, and turbines that can withstand the turbulence better will have a substantial effect on the costs of a realization of a part of the potential.

4.1.3 Turbine size

Large turbines will imply a better exploitation of the land because they use a thicker part of the atmospheric boundary layer, and because of a larger hub height. This effect is also illustrated in figure 4.2 and shows that a 1 MW 56 m turbine (200 %) will increase the potential (in our case) with approximately 30%.

4.1.4 Size of the generator

Because the diameter of the rotor determines the spacing, a large generator will extract more energy out of a good site than a small. The same arguments as in chapter 4.1.2 can then be used here.

4.1.5 Degree of detail in the topographical information

In the study from 1990 which included a part of the same area it was used rather coarse maps in scale 1:50.000 and a contour interval of 20 m. Using the coarse maps a lot of turbines were placed between small hills. The effect was that the potential was overestimated by more than 30%. This illustrates the importance of detailed geographical information, especially when one deals with a country with an undulating ground.

5. CONCLUSION

The study shows that in the county of Sør-Trøndelag, about 1,3 TWh/year of wind energy can be produced at a COE less than 0,09 ECU/kWh. In a previous study the contribution of Sør-Trøndelag was estimated to be about 25% of the total national potential.

The study also shows that high degree of detail in the information is very important for the size of the potential. In the previous study the potential was overestimated with more than 30% due to lack of detail in the topographic information.

Also the high sensitivity of the size of the potential for parameters like turbine spacing in the windfarms and turbine size is shown.

The public acceptance is not directly taken into account in this study. Even if every windfarm is designed with visual impact and noise in mind, some areas will be dominated by wind turbines. The limit will be governed by subjective considerations, and is difficult to assess theoretically.

[1] Nygaard,T.A., Estimating expected energy capture at potential wind turbine sites in Norway, Journal of Wind Engineering and Industrial Aerodynamics, 39 (1992).

[2] Troen, Ib and Lundtang Petersen, Erik, Siting of wind turbines, ECWEC 1988, Herning Denmark.

DEVELOPMENT OF WIND ENERGY IN DENMARK

Poul Nielsen
DEFU, Lundtoftevej 100
DK-2800 Lyngby, Denmark

In recent years, there has been a rapid growth in the number of wind turbines connected to the Danish electricity supply system. As of 1 January 1993, the total number of grid-connected turbines was approximately 3400 units with a total output capacity of approximately 450 MW. In 1993 the total power production by wind turbines was approximately 800 GWh corresponding to 2.75% of the Danish electricity consumption.

The first 100 MW utility programme was fully implemented by the end of 1992. It was expected that a second 100 MW programme would be completed by the end of 1993. This programme is, however, delayed due to serious planning difficulties.

Refurbishment of the 630 kW Nibe A wind turbine with a new 40 m stall-controlled rotor is completed, and the machine is now in test operation. The wind farm on Masnedø south of Zealand consisting of five 750 kW machines and operated by ELKRAFT is in normal operation. In Jutland, ELSAM has finished the test operation of the 2 MW Tjæreborg wind turbine south of Esbjerg. A 1 MW prototype machine for Avedøre is in the manufacturing phase.

1. LARGE-SCALE WIND TURBINES

The Danish wind power programme for the development of large-scale wind turbines is sponsored jointly by the national government, the Commission of the European Communities and the Danish electric utilities. The main achievement of the programme has been the installation of large-scale prototypes at Nibe, Masnedø and Tjæreborg. A prototype machine for Avedøre is in the manufacturing phase. The work at these locations has been carried out by Danish utilities, public institutions and industry.

1.1 The Nibe Wind Turbines

The two 40 m/630 kW wind turbines at Nibe in northern Jutland, called Nibe A and Nibe B, were commissioned in 1979/80. Both turbines have 3-bladed, upwind rotors with active yawing. The A machine is stall-controlled; the B machine has full-span pitch control.

As of 31 December 1992, the operating statistics were as follows:

Nibe A:	Running time	: 7,495
	Energy output	: 1,624
Nibe B:	Running time	: 28,752
	Energy output	: 8,041

Nibe A is being recommissioned after refurbishment; Nibe B has been in normal operation since 1984.

Due to blade malfunctions, Nibe A has been operating only a few hours since 1983. The first blades on Nibe A were of an iron/-fibreglass construction supported by nine bracing rods. These blades were damaged by metal fatigue soon after installation. In 1991 the A machine was provided with cantilevered, laminated wooden blades. Power limitation is still achieved by stall-control. The design work was funded by the Danish government and the refurbishment by ELSAM.

Originally, all blades on the B machine were fabricated with an outer section made of fibreglass and an inner one of fibreglass shells attached to a tapered steel spar. In February 1984 the blades of Nibe B were replaced by three new 20 m laminated wooden blades. Since then this machine has operated satisfactorily.

It is intended to prolong the lifetime of both turbines by 7-10 years in order to test their operation with the wooden blade technology. The average wind speed at the site is 7.1 m/s at hub height (45 m). Assuming 100% availability, the annual energy production per turbine will be about 1300 MWh.

1.2 The Masnedø Wind Farm

ELKRAFT Power Company Ltd., the regional power company of Zealand, operates a wind farm comprising 5 units of WINDANE 40 manufactured by Danish Wind Technology Ltd. The site is on Masnedø, a small island close to the southern coast of Zealand. The main specifications for these machines are similar to those for Nibe B.

As of 1 December 1992, the operating statistics were as follows:

	MAV 81	MAV 83	MAV 84	MAV 85
Hours	18,656	21,454	22,696	22,912
MWh	4,926	4,792	5,989	6,707
%	38	89	94	91

The availability is average for 1992.

The five Masnedø turbines were installed late in 1986, and commissioning progressed well until October 1987, when a fire completely destroyed one of the turbines (MAV 82). It has now been rebuilt and placed back into operation. The design of the new wind turbine is in many ways an improvement over the previous one. The same turbine experienced in spring 1992 an overspeed accident but no serious damage has been observed.

Most of the turbines have had gearbox failures. As a result of this, they have all been modified. As a safety precaution, the maximum power output of all turbines was for a certain period limited to 450 kW, until the power control system was able to prevent excessive power peaks by wind gusts. The pitch control system now seems to operate well.

Surface cracks have been observed in a number of turbine blades. Until recently provisional repairs were made, either at the site or in the workshop. Now two of the turbines are equipped with new wooden blades, and the remaining three turbines are provided with new fibreglass blades.

1.3 The Tjæreborg Wind Turbine

In 1988 ELSAM, the regional power company of Jutland and Funen, commissioned a 60 m/2 MW wind turbine at Tjæreborg south of Esbjerg on the west coast of Jutland. This machine has a 60 m upwind rotor with three pitch-controlled, cantilevered 30 m fibreglass blades and a 1:68 epicyclic gearbox. The 2 MW generator is of the induction type. The 60 m tower is made of concrete.

As of 31 December 1992, the machine has operated for 11,698 hours and generated 8,121 MWh of electricity. The cumulative hours of availability was 17,032 hours. The average availability for 1991 and 1992 was for both years approximately 75%, ref. figure 1.

In August 1989, one year after the turbine was commissioned, its gearbox failed by fracture of the main gear wheel. The damaged parts were quickly replaced by the manufacturer, but then high temperature was found at the two bearings of the output shaft. This has further reduced the number of operational hours, but the installation of an improved bearing designed with better lubrication and cooling has now solved the problem. Later, a 1P sound from the first main bearing was observed. This is caused by friction between the bearing and the main shaft, but only for a short time after a start from cold condition.

1.4 The Avedøre Wind Turbine.

The ELKRAFT Power Company Ltd. has designed a 50 m/1 MW wind turbine with blades that can be used in both pitch and stall control. The site is at the Avedøre power station 10 km south of Copenhagen. The project is financially supported by the CEC.

The machine is now in the manufacturing phase, and commissioning is scheduled for October 1993. The turbine will first, for a period of about one year, be operated in the stall mode. Then some minor mechanical modifications will be made in order to change the operational mode to pitch control. A measurement system will collect all relevant data during both periods.

2. SMALL-SCALE WIND TURBINES

As of 1 January 1993, the total number of grid-connected wind turbines in Denmark was about 3400 units with a total electrical capacity of about 450 MW. In 1992 the total power production was about 800 GWh corresponding to 2.75% of the Danish electricity consumption. The total increase in 1992 was approximately 180 turbines and 35 MW capacity. This is a considerable slowdown compared to previous years and caused by lack of local acceptance of wind turbines, ref. figure 2.

2.1 Technology

To a great extent the well-known "Danish design" is based on the proven Gedser design: 3-bladed, upwind rotors, active yawing, stall-regulation of power output and induction generators connected to the low voltage consumer level at 380 V in a delta-coupling.

The latest development trend shows that this concept is unchanged. Stall-regulation is still the most common means of power limitation, although pitch-control is in progress. Generally, two types of towers are used: steel lattice towers or steel tubular towers. In both cases they are 30 to 35 m high. Induction generators are still used but they are now often designed to have 660 V between the phases in a star-coupling.

The power range is rapidly expanding. In 1992, machines with ratings of 400 to 500 kW have been sold by manufacturers. The utilities have recently also asked for tenders of 1 MW machines, so this machine size is expected to be included in the range of commercially available machines rather soon.

2.2 Accounting Rules

The development of private, small-scale wind turbines is no longer promoted by public installation subsidies. Favourable pay-back rates and exemption from electricity taxation is now the main economic driver. Windfarms installed by the Danish electric utilities are not subsidized, although the utilities receive 0.1 DKK as reimbursement for the carbondioxide taxation.

By 1 November 1992, new accounting rules for wind-generated electricity came into force. From that date on private wind turbine operators receive 85% of the sales price of electricity to general consumers plus a subsidy of 0.27 DKK per kWh delivered to the grid. This gives 0.85 x 0.45 + 0.27 = 0.65 DKK per kWh (≈ 0.068 ECU per kWh). The consumer price of DKK 0.45 per kWh is excluding taxation. The full price for a Danish low-voltage customer consuming about 3000 kWh per year is about 1.00 DKK per kWh (≈ 0.133 ECU per kWh).

Note: All calculations are performed in Danish kroners = 100 øre. At the time of writing, the exchange rate is about 1 ECU ~ 7.50 DKK.

2.3 Utility Windfarms

Traditionally, Danish wind turbines were erected as individual units. Since 1984, however, many wind turbines have been installed in windfarms. These windfarms are distributed all over the country, and the main part of them is now owned by the electric utilities.

In 1985, an agreement was reached between the Danish government and the electric utilities committing them to install 100 MW in windfarms over the next five years. There has been a delay caused by difficulties encountered in getting planning permissions. The agreement was, however, fully implemented by the end of 1992.

In 1990, a new agreement was reached on another 100 MW to be implemented before the end of 1993. As a first step, a joint government and utility committee in cooperation with local authorities has recommended a number of suitable sites. Despite this, the implementation of the 1990-agreement is at least one year delayed due to local opposition to wind turbines.

The experience and economy for Danish utility windfarms is described in details in the conference proceedings from the EWEA windfarm conference in Herning, Denmark, September 1992. More recent data are not available.

2.4 Offshore Windfarm

The first Danish offshore windfarm was commissioned in mid-91. The site is in the ELKRAFT area at Vindeby, northwest of Lolland in the Baltic Sea. The windfarm consists of 11 wind turbines in two rows. The water depth varies between 2 and 6 m. The distance from the shoreline ranges from 1200 to 2400 m, and the distance between turbines is approximately 300 m. Each turbine is rated at 450 kW, and the hub height and rotor diameter are respectively 37.5 and 35 m.

Although the specific energy output is expected to be about 60% higher than for average onshore sites, or about 1130 versus 700 kWh per m^2 of swept rotor area, the cost of energy produced by this windfarm (0.084 ECU per kWh) is estimated to be 50% higher than for average inland sites. The total costs of the project, including a two year measurement programme, were 11 million ECU. The reported availability is higher than 96%.

3. APPROVAL AND CERTIFICATION SYSTEM

In May 1991, a new approval and certification system for Danish wind turbines was introduced as the result of two years of preparation by the Danish Energy Agency, the Test Station for Wind Turbines at Risø, the Danish manufacturers, the Danish Wind Power Association and various insurance and certification companies. The previous system of type approvals consisted mainly of a general design review and a review of the load and strength calculations. The general purpose of the new approval and certification system is to improve the quality level of Danish wind turbines.

The new approval system for wind turbines specifies requirements for documentation of all design criteria such as load cases and loads, fatigue evaluation, safety levels, power curves, noise emission, quality procedure for manufacturing, transporting, installing and subsequent servicing of the wind turbines. The manufacturers must also have a fully introduced and certified quality assurance system according to ISO 9002.

All wind turbines to be installed in Denmark after 1 July 1992 must be approved and certified according to the new system. Furthermore, it is a condition for obtaining export guarantees from the Danish Wind Turbine Guarantee A/S that wind turbines must be approved and certified according to the new system.

The new approval and certification system is managed by the Danish Energy Agency assisted by the certification secretariat at Risø.

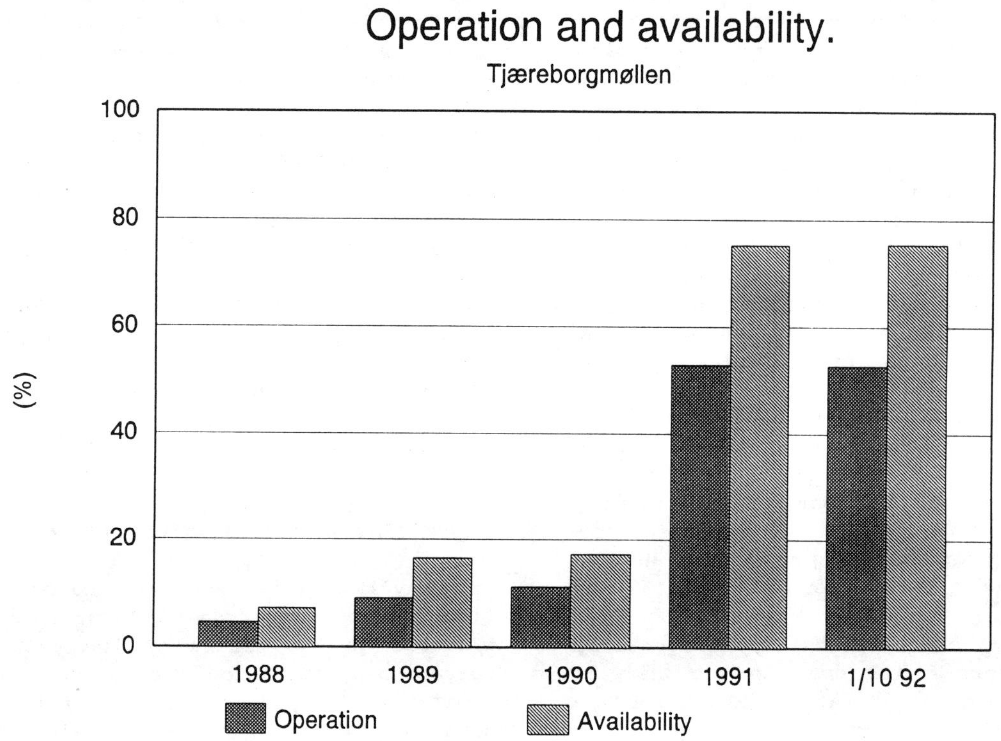

Figure 1. The Tjæreborg Wind Turbine: Operation and Availability.

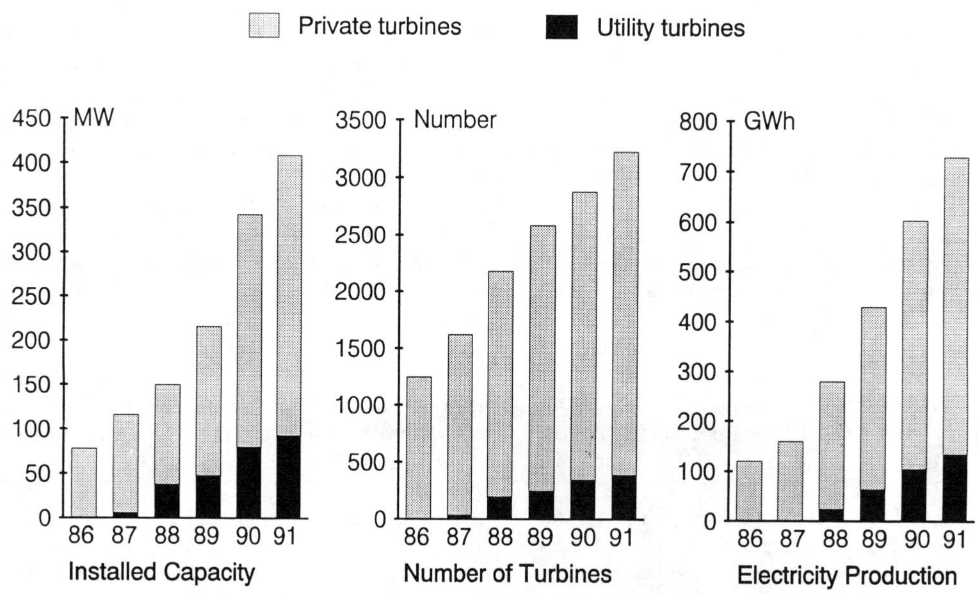

Figure 2. Accumulated data for Wind Turbines in Denmark.

WIND ENERGY DEVELOPMENT IN SWEDEN

Kenneth Averstad, Vattenfall AB, S-162 87 Vällingby, Sweden
Olef Sandberg, NUTEK, S-117 86 Stockholm, Sweden

NUTEK is the Swedish National Board for Industrial and Technical Development. Vattenfall is the biggest Swedish utility, producing about 70 TWh (billion kWh) of electricity per year, which is 50% of the cosumption in Sweden.
This paper very shortly will describe the present status of the Swedish research and development of wind energy.

1. ABSTRACT

In Sweden a wind energy programme has been operated since 1975. In the 70's and in the early 80's the resources were concentrated on development of large wind turbines as the task was to investigate possibilities to replace a part of our nuclear power production system. During the 1980's the emphasis of the programme was on the erection, operation and evaluation of the two megawatt size machines, WTS 3 of 3 MW (Maglarp) and WTS-75 of 2 MW (Näsudden I). During the last 4 years the programme has diversified to technical development of medium size wind turbines and since July 1991 it also includes a market stimulation programme.

The annual research budget for wind in Sweden is about 13 MSEK. The research programme covers aerodynamics, structural dynamics and computer calculation methods, materials, noise, electrical and control systems and meteorological research.

Three different prototype turbines are now being developed in Sweden by the industry and utilities:

1) The Näsudden II turbine has a rated power of 3 MW. It was installed on its tower on October 13, 1992. After commissioning a two year evaluation programme will be carried out. The project is a cooperation with Germany. A sister turbine was manufactured by Kvaerner and MBB in parallel with the Swedish machine. The German turbine is equipped with a variable speed system designed in Germany.

2) The second technical development is a 400 kW advanced turbine by Nordic Windpower AB, also under contract with Vattenfall. The machine is a stall control, variable speed design with a teeter hub. The weight of the machine has been reduced about 1/3 compared to an equivalent three-bladed stiff design. First operation was on September 15, 1992. After the now ongoing test period Vattenfall will run an almost two year evaluation programme.

3) The third prototype is a 250 kW wind turbine developed by Zephyr Energy AB. This design is fixed speed, partial passive pitch design with individually flapping blades. The second machine of this design is now in operation.

Since July 1, 1991, there is a market stimulation programme of 25% on investment costs for wind projects. Over a five year period 250 MSEK are allocated and this is estimated to give around 110 MW of installed new capacity. The subsidy was increased to 35% from January 1, 1993.

For the coming years the research will concentrate on aspects of importance for simple, flexible designs, mainly in aerodynamics and structure. An increased importance will be put on environmental aspects, primarily noise and visual impact. Wind resources will be further evaluated, particularly off-shore. The evaluation programmes will play an important role in this development.

2. INTRODUCTION

The Swedish wind energy programme has been operated since 1975. In the 70's and in the early 80's the resources were concentrated on development of large wind turbines, 2 - 3 MW per unit, as the task then was to investigate the wind energys possibilities to replace a part of our nuclear power production system. Two machines were designed, manufactured and put into operation in 1982 and 1983. The first turbine was the WTS 3 of 3 MW at Maglarp and the second was the WTS-75 of 2 MW at Näsudden. During the 80's the emphasis of the programme was on the operation and evaluation of the two megawatt prototypes. The experiences from technical solutions, materials, costs etc are very well documented. The direction towards large turbines in megawatt size was based on the knowledge that a considerable energy contribution can only be produced with large machines on high towers.

During the last years the basis for the Swedish energy politics has been that the energy system shall emanate from what nature and environment can carry and that the system shall, to as large extent as possible, be grounded on enduring, renewable and preferably domestic energy sources. Wind energy fits well into such an energy system. The wind programme has thus diversified to technical development of advanced medium size wind turbines and since July 1991 it also includes a market stimulation programme. A Swedish wind energy industry has begun to establish.

The overall electricity situation is now such that the nuclear and the hydro plants are running very well with high availability and low production costs. Nuclear produces almost 50% and hydro almost 50%. The rest is mainly peak production with different fuels. The expected increase in electricity demand is low and the need for new power plants of any kind is therefore small in Sweden. The parliament have decided not to shut off any nuclear units during the 90's and how to do with the nuclear, when we reach the year 2010, is still to be decided about in the Parliament in the future.

3. WIND ENERGY IN THE ENERGY SYSTEM

Sweden has a good wind resource. It is mainly located at the coasts, the large lakes, in the northern fjelds and off shore. The wind regime in the southern parts of Sweden is comparable with that in Denmark and in the Netherlands.

The pattern of energy consumption in Sweden accords well with the energy production from wind turbines. The consumtion curve and the wind energy production curve follows each other over the year.

The large part of hydro power in our energy system gives a great advantage when integrating wind power into the system. Water can be stored in the reservoirs during windy days and it can be used again when the wind is not producing.

4. HOW MUCH CAN BE BUILT OR WHAT IS THE REAL RESOURCE

In 1988 a governmental survey presented results on how much wind energy there are possibilities to exploit in Sweden. The survey found that it is possible to develope between 3 and 7 TWh on land and about 20 TWh off shore. The study was based on production from 3 MW turbines.

The total electricity consumption in Sweden was in 1992 about 140 TWh. The wind resource corresponds accordingly to 2 - 5% on land and about 15% off shore. The siting of the turbines have to carefully consider a lot of other competing interests such as housings, leisure time activities, military interests, etc.

The lower figure (3 TWh/år) corresponds to the electricity consumption in 600'000 households.

5. TECHNOLOGY DEVELOPMENT PROJECTS

5.1 Large megawatt turbines, Maglarp and Näsudden I

The two large prototypes were both made two-bladed but represent different turbine technologies.

The Maglarp turbine has a rotor diameter of 78 meters and a rated output of 3 MW. It is downwind, have a teeter hub, a steel tube tower that is rather soft and blades of glassfibre reinforced plastic material. Maglarp is still operating and is the wind turbine in the world that has produced most electricity or more than 34 GWh (million kWh). It is operated by the utility Sydkraft AB in southern Sweden.

Näsudden I had a full blade pitch regulating system, a stiff hub, a stiff concrete tower and blades with load-carrying beams of welded steel. It was upwind and had a rotor diameter of 75 meters and a rated output of 2 MW. Näsudden I was dismantled in 1990 after having received a severe crack in one of the blade beams. It was decided to dismantle the nacelle but to keep the tower. The tower will be utilized by the Näsudden II unit.

Näsudden II

The cooperation between Sweden and Germany, which was started in the Näsudden I project, was enlarged in the development of the second generation of megawatt units. Two parallel projects were started, one German and one Swedish. The German utility Preussen Elektra ordered one 3 MW machine and Vattenfall ordered another 3 MW machine. Turbine manufacturer is Kvaerner Turbin AB of Sweden. The blades are manufactured by MBB in Germany. The main technical difference between the turbines is that the Swedish one is two-speed (14 / 21 rpm) while the German has a variable speed system (14 - 21 rpm).

The rotor diameter is 80 meters and the blades are made of a composit of carbonfibre and glassfibre reinforced plastic. The turbine technology is based on Näsudden I with an upwind, full pitch, stiff hub and two-speed design.

The two projects imply a great step ahead technically and substantially lower costs than the first generation.

When all tests are approved including a 500 hours test operation period, Vattenfall will take over the responsibility for the wind turbine from Kvaerner Turbin AB. The guarantee period will be 2 years after taking over.

Näsudden II is sited on a Swedish island in the Baltic Sea called Gotland and it will be operated by Vattenfall.

The German turbine Aeolus II, is sited at Jade Windpark near the town Wilhemshafen on the German North Sea coast. It will be operated by the German utility Preussen Elektra.

Both units were shipped on barges through the Trollhätte Canal in western Sweden and then over the sea to Germany and to Gotland respectively. Aeolus II were erected in August and Näsudden II were erected onto the old tower October 13, 1992. The total lifting weight was about 195 tons including the lifting sledge.

Evaluation Programme Näsudden II and Aeolus II

Vattenfall will run a monitoring and evaluation project which will be finally reported in 1995/96. The evaluation is financed 50% by NUTEK and 50% by Vattenfall. The evaluation will also include a comparison with the German 3 MW wind turbine.

3rd generation of 3 MW wind converters

Vattenfall, Kvaerner Turbin and MBB have made a study for defining the basis for the third generation of 3 MW wind power plants. The study was reported internally in December 1992.

The involved companies in Germany and Sweden have also started the discussions on how to organize and finance the third generation project in the following stages. Cooperation is also under discussion on a governmental level between BMFT (Bundesministerium für Forschung und Technik) and NUTEK.

For a future development of wind power in off shore sites, large turbines have to be of a cost effective design.

Risholmen 750 kW

The 750 kW Howden wind turbine outside Gothenburg was repaired during January 1992 and did since beginning of February until mid of June operate with a time availability of 92 - 99% per month. In end of June it lost one of its blade tips during starting operation. The accident has been investigated in cooperation with the manufacturer. A descision has been made to dismantle the machine and replace it with another wind turbine.

5.2 Advanced medium size turbines.

NWP 400 (Nordic Windpower 400 kW)

The technology represented by the Maglarp turbine with a teeter hub has been further developed in a project called Lyse Wind Power Station. Lyse is situated on the Swedish west coast closed to the town Lysekil. In the project we intend to compare the standard three-bladed technology and the new flexible two-bladed technology in turbines of about the same size and the same location.

The new advanced turbine has been developed by the Swedish company Nordic Windpower AB. Manufacturing was made in Hägglunds workshops in Örnsköldsvik in the northern part of Sweden.

A standard Bonus 450 kW machine was erected and put into operation in June 1992 and the Nordic Windpower 400 kW turbine was erected in August 1992. NWP 400 went into test operation on September 15. A testing programme will run until May 1993. If all tests are completed satisfactorily, Vattenfall will take over the responsibility for the operation and maintenance.

The two machines are sited on two small islands, called Basteviksholmarna, near the town Lysekil. An information building has been built on the site. It will contain a small wind energy exhibition and also information about the involved companies and the town Lysekil.

NWP 400 is designed with the intention to minimize loads due to both fatigue and due to extreme wind speeds at stand-still. The small projected area of the unit minimizes parking loads. The properties of the teetering hub and of the yaw control means that short term wind speed variations due to wind shear and turbulence are de-coupled from the machine. This is especially important for turbines installed in wind farms, where ordinary rigid hub units suffer from fatigue.

The turbine is upwind, has stall control and the blades are of glassfibre reinforced polyester.

The Swedish wind research programme has resulted in some very useful computor tools for wind turbine design, which have been further developed in a pre-project (Hönö 2) and in this project.

Technical data from NWP 400 are as follows:

Hub height	40 m
Rotor diameter	35 m
Rotor speed	20 - 41 rpm
Nacelle with rotor weight	13,7 tons
Tower weight	14,8 tons

One very big difference between the two machines is the weight. The Danish 46 tons stands against 28 tons for NWP 400. This will hopefully lead to that NWP 400 will be cost effective and competitive in a serial production.

NWP 400 has been certified by Det norske Veritas.

An evaluation programme will be run by Vattenfall for two years after taking over the NWP 400 turbine in 1993.

Zephyr 250 kW

A third technology for horizontal axle wind turbines has been designed and tested by the Swedish manufacturing company Zephyr Energy AB.

The company has now manufactured two 250 kW wind turbines with a self regulating turbine. They are two bladed and the diameter is 28 meters. The machine type is called Zephyr WTS 28.250. It has two rotation speeds 55 and 37 rpm. The power output is 250 kW at the higher speed and 75 kW at the lower speed. The tower is a conical tube tower. The hub height is 32.8 meters.

The turbine is self regulating through a passive pitch regulation. Power regulation by blade brakes which are controlled by a aerodynamic moment. Each blade consists of two halves. The inner half is fixed and the outer half is for pitching. This technical solution also maximizes the overspeed to about 10%.

Another feature with this turbine is the flapping hub. The blades move individually which decreases the forces substantially. At lower wind speeds the slower rotation speed is used, which gives lower noise.

The first Zephyr turbine went into operation in 1990, and though it is a prototype in this size, it has been producing very well. The capacity factor has been among the very best even in comparison with the new Danish turbines operating in Sweden. The second Zephyr 250 kW turbine went into operation in autumn 1992. A third turbine is ordered by the community of Falkenberg on the Swedish west coast.

6. WIND ENERGY OFF SHORE

A great part of the wind resource in Sweden is off shore. A group of private companies therefore carried out a feasability study during 1990 on design and construction for a large windfarm outside the Swedish south-east coast. The study was made for a farm with 98 turbines, each with a rated output of 3 MW. The study was financed to 50% by NUTEK.

The result was, that the production cost off shore is significantly higher compared with the same turbines on land, though the average wind speed is higher off shore. More research efforts have to be allocated to the technology for off shore wind turbines to make them cost-effective. This is true also for foundation technology, transport and erection of off shore wind farms.

The off shore wind resource is also insufficiently charted and the site, which was chosen for the feasibility study, turned out not to be the very best.

To procure practical experiences from off shore siting and operation the world's first wind turbine was built in the Baltic, in 1990. It is situated outside Nogersund on the Swedish south coast. A 220 kW turbine from WindWorld, Denmark, was placed on a three-legged steel and concrete foundation. The wa ter depth is 6 meters. Owner of the plant is the utility Sydkraft and contractor was BPA, a Swedish construction company.

In the project evaluation programme, which is continued until 1994, studies are beeing made of how fish and birds are influenced by the turbine and how noise from the turbine is transferred to the water. The influence from iceforces and salt water corrosion are other tasks in the evaluation programme.

7. MARKET STIMULATION

From July 1, 1992, there was an investment cost support of 25% when building wind power plants in Sweden, which from January 1, 1993, is increased to 35%. The turbines shall have a rated output higher than 60 kW. The support was introduced to get an incentive to faster development of wind power.

The support is administrated by NUTEK and it amounts to 250 MSEK during a five year period until June 30, 1996. It can be given with 35% of approved investment costs. The average cost for a wind energy project is about 9'000 SEK/kW which means that the decided support will be sufficient for a total of about 80 MW new rated output.

To qualify for support, the wind turbines must be certified. A certification procedure has been set up. Another rule is that the project must have all local permits before receiving any support money. The owners who have received a subsidy are then taking part in an evaluation programme of wind turbine performance.

The investment support has implied that the number of wind turbines in Sweden has grown three times in 18 months. On December 31, 1992, we had 89 turbines with a total rated output of 20 MW. The large prototypes are included in this figure. During the first year decisions were taken on 22 MSEK in support. This is conciderably less than the 50 millions which was planned per year. The reason is lack of profit for projects that are not sited in extremely windy areas. Wind produced electricity is in Sweden in average paid with 0.25 SEK/kWh. If the wind turbine owner is a private company, a farmer or a for the purpose grounded wind energy association and the produced electricity can be used by the owner himself, the economic situation gets another. Then he will be spared from paying energy tax on the used electricity, corresponding to 0,11 SEK/kWh. The cost for electricity to households and heating of houses is about 0,50 SEK/kWh.

8. WIND RESEARCH

As a support for the development of wind energy in Sweden there are a research programme since 1975. The research programme is to 100% financed by the government through NUTEK. For 1992 the budget was 13 MSEK.

The aeronautical Research Institute, FFA, is performing studies in aerodynamics, structural mechanics, the rain-flow-count method, materials, acoustics, control technology, safety and standards.

Electric machinery for wind systems is studied at the Chalmers University of Technology in Gothenburg, who also operate the Wind Turbine Test Station on the island of Hönö off the west coast of Sweden.

Basic wind research is carried out at the Institution of Meteorology at the University of Uppsala, MIUU, for example on the arise of wakes behind turbines and meteorological phenomenon in the atmospher on lower heights where the wind turbines take its energy.

Research is also made at the Meteorological Institute, SMHI, mainly on wind resources, and at KTH, the Royal Institute of Technology, Institution for Electrical Energy Systems on integration of wind power into the electricity power system.

9. FUTURE DEVELOPMENT

9.1 Technology development and demonstration projects

The development goes towards larger turbines. The European Community has in its frame programme on non-nuclear energy research, JOULE II, decided to support the development of turbines in the size of 1 MW. Seven projects in 5 different countries was selected for support. Swedish companies is a part in that programme with a project grounded on Nordic Windpowers advanced 400 kW turbine. Nordic Windpower will develop its new 1 MW turbine together with some European manufacturing companies within the EC countries. Vattenfall AB will be the operator of the machine and run the evaluation programme.

The goal is that NWP 1000 will be of a simple, light and cost-effective design, which in a later step can be commersially competitive.

To continue the technical development of large turbines in the size of 3 MW there is, from a Swedish point of view, a need for a continued international cooperation.

9.2 Market stimulation

With the increased investment support level, more turbines will be built. This will then increase the need of planning efforts in communities and counties. Areas especially suited for wind power have to be planned for that purpose. Several County administrations now make closer investigations of the wind resources in their county and try to find areas where a minimum of conflicts is probable to arise.

9.3 Future Research

The research programme will probably continue at a some lower cost level compared with 1992. Some tasks will have higher priority. One is noise. The increased number of projects have shown that noise is one of the most severe obstacles for siting wind turbines in Sweden. Procedures and tools for effective siting of wind turbines in the landscape have to be developed. Together with noise, the impact on the visual landscape is the biggest hindrance for establishing wind turbines and wind farms. Impact on birds is another area that demand more research, preferably as international joint research projects.

A major part of the wind resource is off shore. This need to be better charted, in the Baltic and in other sea areas. Computer programmes for wind energy predictions have to be compared to real turbine productions. Some research is intended to be realized in multinational cooperation projects.

10. EVALUATION OF SMALL AND MEDIUM SIZE WIND TURBINES IN SWEDEN 1989 - 92

A project for reporting of operation experiences and also economics started in 1989. Monthly and annual reports have been published and a summary for 1989-91 was published in September 1992. In February 1993 the annual report from 1992 was published.

The Swedish wind turbines produced last year 33 GWh. The number of small and medium size turbines increased from 50 to 87 during 1992. Total installed output was 20 MW in end of 1992.

Total investment costs including permission procedures, engineering, roads, foundation, grid connection are in average 8'200 SEK/kW.

Will the wind turbines in Sweden give a profit? At our present situation with an average selling price of 25 öre/kWh you need a very good windy site to get your money back within 20 years, even after having received the 35% investment support.

11. VATTENFALL AB's SITING STUDY, LOK 300

Vattenfall's siting survey was completed in November 1991. The goal was to show an optimal siting of 300 MW rated output. We found that if we site the first 300 MW in an optimal way they can produce 750 GWh per year. To reach the double figure 1.5 TWh/year, we need to site 750 MW rated output.

12. COMMUNITY SURVEY PLANNING

In Sweden the different communities have a monopoly on planning within their borders. It is the local community's duty, to make a future development of the wind resource possible, and the tool is the community's survey planning, where they define all known both preservation interests and exploitation interests.

The different communities have treated the wind resource very differently in their planning. Some are positively active and have established specific areas for wind energy use.

13. WIND ENERGY COOPERATION BETWEEN UTILITIES

The major Swedish utilities cooperate in wind energy issues through a separate company established for the purpose. The company is called Kraftföretagens Vindkraft AB (KVAB), which means Utilities' Wind Power Ltd. KVAB is the owner of the 750 kW Howden turbine at Risholmen. KVAB is now completing a feasibility study for a 5 MW-farm to be sited in the southern part of Sweden. It is meant to contain about 10 turbines with an output of 400 to 500 kW each. The aim is to find a site for a wind farm project with low production costs and high availability.

14. CONCLUSION

Wind energy development in Sweden has during the 70´s and 80´s mostly contained technology research, demonstration and evaluation of large scale wind turbines. Now in the 90´s the market stimulation programme and the efforts to take the Swedish wind research experiences into a commersial market, with the new advanced turbines, show the way to a very challenging future for both the utilities, other investors and the industry. Sweden is planning to join the European Community and this will give a greater market for new turbines and for wind energy.

"WIND ENERGY IN SPAIN"

F. Sanchez/E. Soria
CIEMAT-IER
Avda. Complutense, 22
28040 Madrid (SPAIN)

ABSTRACT: During the year 1992, a great number of projects for wind farms were carried out, implying around 50 MW of installed power. The use of wind power has experienced an important qualitative jump, going from small installations which still keep their experimental character, to wind farms which constitute commercial exploitations. Most of the installed windturbines have been manufactured in Spain, either by own developments of the spanish windturbines manufacturers, or by means of agreements or joint ventures with other foreign companies. The models that have been installed correspond to machines between 100 and 180 Kw of rated power, with regard to the measurement of wind energy resources, the works for the better knowledge of the windy areas, have continued, getting a deeper characterization of the interesting places in order to optimize the windturbines installation. The most interesting areas in the country are: The Gibraltar Strait zone, where most of the projects for wind parks are actually carried out; Canary Islands, with good wind conditions in the regime of trade winds, and with some other favorable conditions for the wind energy development, by the fact of their insularity and also the NW of the Peninsula (Galician corner). In the near future, the wind energy installations are going on, and in this way, either the Public Adminstration or the electrical companies, have some plans for stablishing the horizon for the year 2000, of several hundreds of MW.

1. INTRODUCTION

1986, can be considered the beginning of a new phase for the use of Wind Energy in Spain.

Throughout several initiatives from the Public Administration, together with the collaboration of Autonomous Communities and Electrical Companies, the first projects of wind farms were undertaken, still with a reduced dimension, 300 Kw of total installed power, with windturbines of 30 Kw of national development. The most significative results of these first actuations, can be estimated as:

- Improvement of the technology and increase of the administrative experience in Wind Parks.
- Determination of the areas with the best wind potential within the country.

In 1989, the second Renewable Energy Plan (PER-89) began, which showed a more important presence of wind energy, if compared with other renewable energies sources. In 1990, because of different support measurements, there were in Spain eight wind parks with a total installed power of about 8 MW.

The new installations represented a considerable improvement in the windturbines performances, because of a bigger madurity degree of the technology, increasing the size of the machines to be installed and getting windturbines in the range of 100 Kw of nominal power. In this stage, within the program of big machines of the DG XII, CEE, the installation in Cabo Villano (La Coruga) of the windturbine AWEC-60 of 1,2 MW of nominal power was carried out. The prototype of three blades with a diameter of 60 m. incorporated some new developments in the concept of big windconverters, like the speed system with variable rotation thoughout the hypersynchronous cascade and a new development of the blades, carried out with national technology.

The objective established by the Renewable Energy Plan, of reaching 100 MW of installed power for the end of 1995, seemed to be overcome at the view of the projects started in the zone of the Gibraltar Strait. Two wind parks, with a total power of 30 MW, were put in operation in Tarifa at the end of 1992.

At Figure 1, we can see the evolution of the power installed in Spain from 1986 and the estimated foresight until 1995.

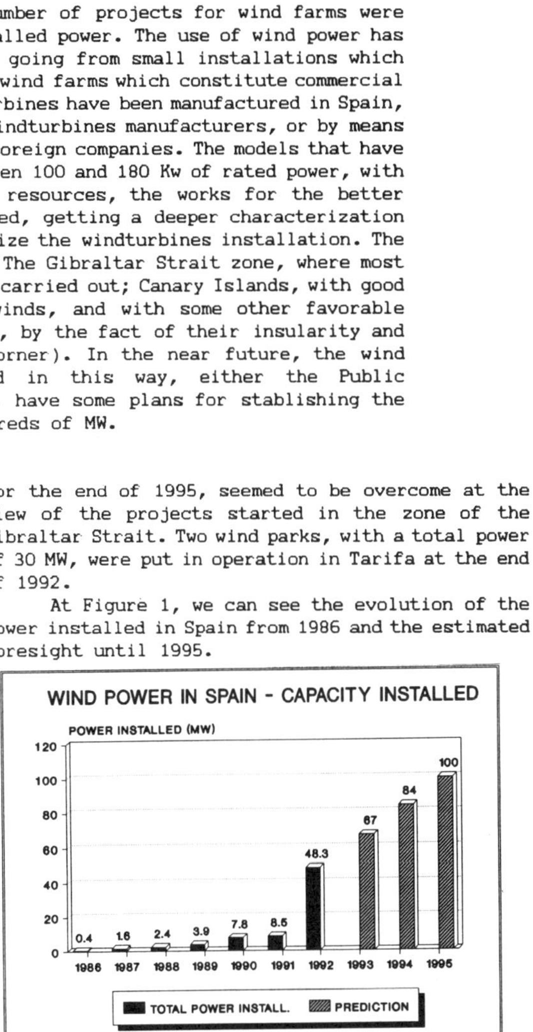

CIEMAT-IER.FIG.1

Many projects are in advanced definition phase, either at the Iberian Peninsula or in Canary Islands.

An indicator of the increasing interest of Wind Energy in Spain is shown by the recent indications of Empresa Nacional de Electricidad (ENDESA) to install 750 MW before the year 2000.

2. INSTALLATIONS

At the enclosed tables, current information about the wind farms in Spain is included. Table number 1 shows the first wind farms built in Spain, mainly with little windturbines of 30 Kw of rated power.

TABLE 1

WINDFARMS OF RATED POWER UNDER 1 MW IN OPERATION IN SPAIN UPDATE JAN 93

WINDFARM		WINDTURBINES				
NAME	TOTAL POWER (MW)	MANUF.	MODEL	RATED POWER (kW)	No	YEAR
LA MUELA	0,55	GESA	10/3	30	12	1987
		GESA	AE 15	75	1	1988
			AE 18	110	1	1988
ESTACA DE BARES	0,36	GESA	10/3	30	12	1987
ONTALAFIA	0,30	ECOTEC.	12/30	30	10	1988
TARIFA	0,65	ECOTEC.	12/30	30	10	1988
		ECOTEC.	20/150	150	1	1989
		ECOTEC.	24/200	200	1	1992
CABO ROSES	0,59	GESA	AE 15	75	2	1990
		GESA	AE 18	110	4	

At table number 2, the wind farms over 1 MW of rated power are included.

One of the most important concentration of windconverters is located at the Tarifa area on the Southern of Spain by the Gibraltar Strait, with a total amount of 272 wind turbines of 100, 150 and 180 Kw of rated power and a total capacity of 33 MW. The wind data at the Tarifa area are very promising; the results obtained at the Experimental Wind Park, supported by CIEMAT, show an annual average wind speed of 8.7 m/s at 30 m height, with prevailing winds blowing from the East. The total investment for the wind farms is about 5.400 Millions of Pts with an external supporting of a 14 % of private funds and a 25 % of official funds throughout the FEDER (VALOREN Program) and the Spanish Ministry of Industry and Energy and the Autonomous Government; the remaining 61 % of the total cost is obtained from long term loans. The estimation of the wind potential for the whole area, gives a value near 250 MW of wind capacity and several projects are considered, in order to increase the current installations.

In the North-West coast, at the Galician Region, the wind farm of Cabo Villano is located, which includes the AWEC-60 windturbine (prototype of 1.2 MW of rated power and 60 m rotor diameter, that is a part of the WEGA Program of the CE), and 20 windturbines of 180 Kw developed by a Spanish Company. A new 15 MW wind farm is being considered close to the wind farm. The most conservative estimations for the area, show an amount of 300 MW of wind capacity for the Galician Region in a terrain with no other commercial values, but with the added difficulty of the very complex terrain.

The Canary Islands, located in the trade winds regime, is the other interesting area of Spain. Nowdays, there are 5 MW of wind energy machines spread in several wind farms along the islands. Several approved plans trend to install 15 MW more at the islands. The last estimations for the wind capacity show a value of 100 MW, but the grid

TABLE 2

WINDFARMS OF RATED POWER OVER 1 MW IN OPERATION IN SPAIN UPDATE JAN 93

WINDFARM		WINDTURBINES				
NAME	TOTAL POWER (MW)	MANUF.	MODEL	RATED POWER (kW)	No	YEAR
GRANA-DILLA	1,68	MADE	AE-20	150	1	1990
		ECOTEC.	20/150	150	1	1990
		W.E.G.	25/250	250	1	1990
		CENEM.	FL-19	300	1	1990
		VESTAS	V 25	200	1	1990
		ENERC.	34	330	1	1991
		MADE	AE-26	300	1	1992
MONTE AHUMADA	1,05	MADE	AE-20	150	7	1989
CABO VILLANO	3,9	VESTAS	V 25	200	1	1990
		VESTAS	V 20	100	1	
		MADE	AE-20M	180	20	1992
PESUR	21.1	MADE	AE-23	180	34	1992
		A.W.P.	AWP-100	100	150	
E.E.E.	10.3	ECOTEC.	20/150	150	50	1992
		MADE	AE-20	180	16	
COSTA CALMA	1.1	ACSA-VESTAS	V 27	225	5	1992
TENEFE	1.1	ACSA-VESTAS	V 27	225	5	1992
MONTAÑA MINA	1.1	ACSA-VESTAS	V 27	225	5	1992

connection with the small insular grid could require certain studies about the wind energy penetration in the grid.

An interesting example of wind farms at the Canary Islands, is the Granadilla Wind Farm. This park has been conceived like an experimental plant, joining several types of wind turbines, between 100 and 300 Kw of rated power, including a vertical axis wind turbine in operation from 1991.

At the figure n' 3, the statistics for the year 1992 are shown for the different wind machines included at the park.

At the table number 3, the windfarms under planification are described.

The activities for the installations of new windturbines are foreseen to continue during the next years. With all this, the foresights of the installed power could be near the proposed objective of 100 MW for the year 1995.

3. INDUSTRIAL ACTIVITIES

The industrial activities in Spain show a considerable growing from the first developments carried out during the year 1991. Actually, there are two main manufacturers of windturbines who have developed models of wind generators with 300 Kw of rated power and they are involved in projects for spreading the wind capacity.

One of the companies is the public utility ENDESA, that is included at the major project of wind energy in Spain, through the subsidiary company

TABLE 3

WINDFARMS AT PLANNING STAGE
UPDATE JAN 93

WINDFARM		WINDTURBINES			Nº	YEAR
NAME	TOTAL POWER (MW)	MANUF.	MODEL	RATED POWER (kW)		
JANDIA	10	ACSA-VESTAS	V 27	225	45	1992
BAJO EBRO	4	ECOTEC. MADE	20/150 AE-20M	150 180	16 9	1993
CABO VILLANO (Phase 3rd)	15	MADE	AE-26	300	50	1993
INALSA	5.6	MADE AWP	AE-20 AWP-100	180 100	9 40	1993
BARRANCO TIRAJANA	1.3	MADE	AE-23	180	7	1993

MADE, that has developed several types of wind converters from the first model of 24 Kw. The most used unit, performed by MADE is the 180 Kw wind turbine, three bladed and stall regulated machine with 23 m. rotor diameter.
A new model of 300 Kw is now under testing at the Granadilla wind farm. Three new prototypes of the 300 Kw will be erected at the Tarifa region. The new commercial plants developed by ENDESA will be based at this model.
The basic objective of the machine is the cost reduction, with regard to the previous 130 Kw model. The design concept is similar to the other models (three blades and stall regulation) and the responsible entities hope that the new product will permit more economical projects and a better use of the terrain.

The next step in the program of this manufacturer is the development of a wind turbine of 500 Kw, rated power; the project trends to achieve a national development to be in operation at the end of 1994.

The other spanish manufacturer is the company ECOTECNIA; this is a little cooperative of specialists in renewable energies that has centered its main activity on wind energy. Its works have been oriented towards the development of its own development of its own technology. In 1983, it developed a 30 Kw wind turbine and after that a new model of 130 Kw, 20 m. diameter was manufactured and tested; more than 75 units of this model equivalent to 11,75 MW are now in operation in several wind farms. Now a new 24/200 Kw wind turbine of 24 m. diameter is in operation in Tarifa. Other activities of the company have focussed in the isolated applications and the wind potential studies, developing new site evaluation models for complex terrains, which are usual in Tarifa and Tortosa (Ebro Valley) regions.

Two other companies are also involved in the manufacturing of wind turbines, but with joint ventures with foreigner manufacturers. One of this is located at the Canary Islands, and have an agreement with Vestas Company; this is Aerogeneradores Canarios (ACSA), that is involved in the wind farms of the islands.

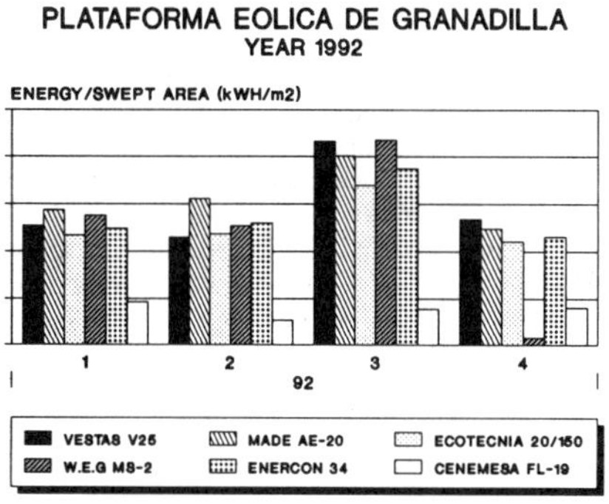

CIEMAT-IER. FIG.2

The other company is AWP that through a relation with U.S. Wind Power, built in Spain the 100 Kw wind turbine.

The recent developments of wind farms during the year 1992, represent an important amount of activities for the companies involved and they are creating the appropriate infrastructure to attack more ambitious projects.

4. R & D PROJECTS

The Spanish Renewable Energy Plan, only considered actions for Demonstration and Application of the renewable resources, R & D aspects were induced in the Plan of Energy Research (Plan de Investigacion Electrotecnica PIE).

There are several ongoing projects, carried out in public and private centers. Several Departments at Universities (Univ. Politécnica de Madrid, Univ. Politécnica de las Palmas,...) are working in projects covering different fields as wake effects of wind turbines, wind-diesel systems, component material properties for utilization in blades of wind turbines, assessment of wind characteristics, grid penetration studies.

The utilities have several projects in course, in a wide range of activity from experiences from a 1 MW class wind turbine to studies of grid interferring and cost evaluation.

At the large scale wind turbine, the most relevant experience is the AWEC-60 project, conceived by a Spanish German consortium and with funds of the E.C. and OCIDE (Spain) to develop, manufacture and test a 1.2 MW wind turbine. The machine put in operation in 1990, is now under supervision of UNION ELECTRICA FENOSA and it is involved in a test program. In the future and in the frame of the WEGA-II program, the utility UNION FENOSA will develop a new 1.2 MW vertical axis wind turbine without gear-box that will be placed at the Cabo Villano site.

In the medium power range, the efforts are conducted in two projects, carried out by the national utility ENDESA; one of this is the test program for a prototype of 300 Kw based at the design concept of the previous designs widely spread at the spanish wind farms. The prototype MADE AE- is a three bladed stall regulated wind turbine, now being tested at the Granadilla wind farm. The program will include measurement of power curve, noise emission, loads over the wind turbine,...; this prototype was followed by three more at the Tarifa area, the machine that ENDESA group applied to its wind farms projects. The other activity is a new design of medium size wind turbine of 500 Kw rated power, the project that will involve the spanish groups working in wind energy, and that will be supported by the Official Organism of Electrical Research of the utilities and trend towards a commercial wind machine that reduces cost and implies a better usage of the terrain; the project with two years of duration is now under discussion.

The main center for R & D is the Instituto de Energias Renovables (IER) from CIEMAT, that has been working since 1985 in the field of wind energy. Its wind department is actually involved in several projects, some of them in the framework of the JOULE I and II Programs of the DG XII of the CE, and in cooperation with other R & D centers from Spain and EC countries.

The main areas of activity are: Resource Assessment, that includes besides wind measurement, projects about short-term prediction models of energy production, mainly applied at the Tarifa area.

Environmental Issues, that consist of some studies about the influence of the wind farms on the nature and terrain classifications for wind energy uses. The development of composite blades is being tested in cooperation with the manufacturers, also design and testing prototypes like a new 300 Kw development and start-alone, and wind diesel applications, for instance a new 100 Kw wind turbine with a diesel engine for supplying electricity to a fisherman village at the Canary Islands.

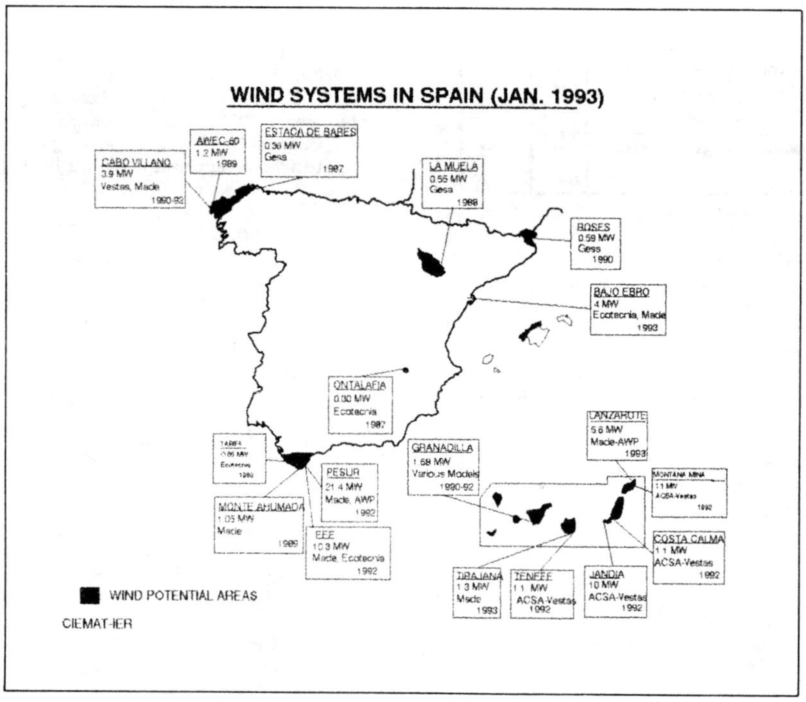

WIND ENERGY ACTIVITIES IN DEVELOPING COUNTRIES

R.K. Pachauri and Jami Hossain
Tata Energy Research Institute
9 Jor Bagh, New Delhi 110 003

ABSTRACT: Wind energy resources and current status of wind energy programmes in developing countries are reviewed. Problems and constraints to wind energy programmes are identified and some recommendations are made for future strategies. Observations are made on the basis of detailed examination of hypothetical high wind penetration in the Tamil Nadu electricity utility.

1.0 INTRODUCTION

Increasing population, aspirations for improved living standards and general economic and industrial growth in most of the developing countries result in an increasing energy demand. However, due to the environmental constraints and the risks associated with high dependence on fossil fuel based energy generation, further capacity addition with conventional centralized methods would be constrained. As a result of increased attention towards alternative forms of energy and methods of conversion and their growing economic viability, the significance of renewable energy devices operating in dispersed and decentralized mode as serious components of modern energy supply has been growing. Wind energy is one of the clean and renewable energy sources which holds the promise of meeting a significant portion of energy demand in developing countries in both, direct grid connected mode and the remote stand alone 'niche' applications.

Large windfarms in California and Europe and to a significant extent in India as well as demonstration projects in several developing countries have established beyond doubt the viability of this technology in potentially windy regions. In this paper, we review the wind energy resources in developing countries and current status of programs on wind energy utilization in China, India and Egypt. Problems and constraints in wind energy programs are identified and recommendations are made for future strategi-s.

2.0 WIND RESOURCE IN DEVELOPING COUNTRIES

It is estimated that the extractable energy in the surface winds, keeping in view all possible constraints and limitations is of the order of 10^3 GW worldwide[**1**]. The world-wide wind energy resource distribution shows that typically wind speeds are high in higher latitudes. Following these average distributions, one would expect generally high wind speeds in southern parts of South America, southern Africa and northern China. In the tropical and sub-tropical regions of the developing world comprising the Indian subcontinent, middle east, Brazil and northern Africa, one would expect generally lower wind speeds. However, wind is modified locally by differences in the heat capacities of water and land, topographic features and vegetation. These smaller scale features can cause considerable variations in wind energy over relatively short distances, particularly over coastal or mountainous areas. The examples are the Aralvoimozhy gap in South Tamil Nadu, India , where average annual wind speeds of about 8 m/s are experienced due to the 'tunnel effect' of the gap and where already 8 MW of windfarm capacity is operational. Similarly Fujian province in south eastern coastal area of China which is bounded by Dai Yun mountain range and the central mountain range of Taiwan experiences average annual wind speeds of 6.5 m/s to 8.5 m/s[**2**]. Although the area under such regions is small on a global scale, it is significant in terms of potential for wind powered generation. A broad country wise description on wind resources is given below.

China: The country is known to experience high wind speeds along stretches of its eastern and south eastern coastlines extending from Jilin province, east L nning province, Zhejiang province, parts of Shandong province, Fujian province and islands off Guangdong. Several sites with average annual wind speeds higher than 6 m/s have been identified along this coastal stretch. Wind speeds as high as 11.7 m/s are experienced at Tian Pool in Jilin district. High wind speeds are also experienced in Inner Mongolia Autonomous region, Northern Gansu Province and Tibet[**3**].

India: India has been quick to realise the potential and significance of wind power. Till recently, the country was regarded as a region with low or mediocre wind speeds. However, wind mapping and the wind monitoring program initiated by the Ministry of Non-conventional Sources has resulted in identification of several sites with wind speeds higher than 6 m/s. Some of the sites, due to the complex terrain effects, experience wind speeds in the range of 8.5 - 9.5 m/s. The potential regions are the coastal areas, inland gaps and ridges in Tamil Nadu, Gujarat, Maharashtra, Andhra Pradesh, Karnataka, parts of Orissa, parts of Kerala and some regions in Himalayas like the Ladakh region.

Egypt: Wind energy resource assessment carried out in 1984 has revealed that regions near the Red Sea and on the Mediterranean coast experience average wind speeds from 6-10 m/s.

Other Countries: The other countries with potential regions are Taiwan (>5.6 m/s), Lebanon (Mediterranean Coast and Mountains, >5.6 m/s), Algeria (Atlas Mountains, Interior Desert, >5.6 m/s), Madagascar (Coastal and Mountains, >5.6 m/s), Argentina (Southeast Coast, Andes Mountains, >5.6 m/s) etc..

3.0 POWER AND ENERGY SECTOR IN DEVELOPING COUNTRIES:

All developing countries are characterized by power and energy deficits. The energy sector in India has been considered a priority area in development planning. With the growth in the economy and increased mechanisation of agriculture and allied activities, the commercial energy consumption including electricity has gone up. Power generating capacity which is owned and operated by utilities, has grown at the rate of over 9.7% per annum since 1950. Of the 64818 MW of capacity commissioned until March 31, 1991, more than 50% had been added during the past ten years alone. In the hydro-thermal mix the share of hydro has declined from 80% of thermal to 41%. The power supply situation in India is characterized by peak and energy shortages which are likely to continue over the next several years. Despite an increased investment in and outlay for Power, in 1990 there was a peaking deficit of over 9300 MW and an energy shortage of nearly 15,000 GWh. Due to the limited resources available for new investments, the central government, state governments and the utilities are facing new challenges in implementing capacity expansion plans. On the other hand the productivity in the industrial sector is adversely affected by these shortages.

China's total installed generating capacity is over 100 GW with electricity generation of 500 TWh/yr. Urban or developed areas with 20% of the population account for 80% of electricity consumption. Consequently, one third of the population lives in non grid connected areas. Serious electricity shortages exist.

Egypt has proven resources of oil, gas, hydropower and to some extent coal. In 1987/88 proven reserves of oil and natural gas were 480 million metric tonnes and 12 trillion cubic feet respectively. At present the country exports energy. But demand forecasts indicate that beyond the year 2000, the country may have to import energy. In the last ten years (1982-1991) domestic consumption of energy grew @ 5.5% annually while energy exports of crude oil and petroleum products grew at a lower annual rate of about 2%.

In general all developing countries are either facing energy and power deficit or like in the case of Egypt may have to face crises in the near future. Therefore, in the context of developing countries, sustenance of developmental activities would in the long run depend upon locally available alternative forms of energy. In fact, the existing ambitious government programs on alternative and renewable energy form... in India, China and Egypt demonstrate a significant change in attitudes towards favouring renewables within a framework of sustainability and environmental concern.

4.0 STATUS OF WIND ENERGY PROGRAMS:

The potential for installation of windfarms is related to the characteristics of the conventional power system (grid) on the one hand and availability of the regions with sufficient wind resource on the other. The potential computed merely on the basis of land and wind speeds is fairly large and the limitations are imposed due to characteristics of the grid like the availability of feeders, substations, nature of generation mix and the response of different components in it etc. Studies have shown that 25 - 35% of the generating capacity can be based on windfarms[**4**]. In India the technically realisable potential for windfarms computed as a percentage of the generating capacity turns out to be about 45,000 MW by the year 2010.

The first grid connected windfarms in India were installed in 1986 in the coastal areas of Tamil Nadu, Gujarat, Maharashtra and Orissa with 55 kW individual machines. The performance of these windfarms was encouraging and the capacity of windfarms at these sites was doubled. As a result of extensive wind mapping and wind monitoring programs some very good sites have been identified. Some of these sites like Muppandal and Kayathar in Tamil Nadu, Kaptagudda and Jogimatti in Karnataka and Dhank in Gujarat are comparable to the best sites in the world. Wind speed data collected from 21 wind monitoring stations in 4 states has been analyzed and published[**5**]. Initial developments in India took place under government sponsored demonstration projects. Convincing demonstration and incentives provided by the government has generated much interest in the private sector. As of December 1992 total of about 50 MW windfarms had been installed including private windfarms/turbines of about 5.00 MW (table 1).

The government of India has offered several incentives to encourage private sector participation in setting up power projects and there are additional incentives for setting up windfarms. These are summarised as follows:

- ❖ The private sector is allowed to set up power projects.
- ❖ NRI (Non Resident Indian) participation in power projects is allowed and encouraged.
- ❖ Wind Electric Generators and components can be imported under Open General Licence.
- ❖ All Wind Electric Systems are exempt from central excise.
- ❖ An import duty of 35% and additional duty of 5% has been imposed on WEG. However, subsequent amendments allow certain components listed below to be imported duty free.

❶ Blades
❷ Gear Box
❸ Brake Assembly

❹ Yaw Components
❺ Sensors
❻ Special Bearings
❼ Hubs
❽ Flexible Couplings
❾ Brake Hydraulics
❿ Controller Parts

❖ Under the Income Tax Act, the Government of India is allowing 100% depreciation in the first year of commissioning of the WEG.

Several Indian states with potential have come forward with windfarm promotional polices and incentives. In brief, these measures include permission to wheel and bank energy generated from windfarms, help in site selection and in providing detailed information on wind resources and the grid. The measures undertaken by different states are summarized in table 2. Monetary aid and loans from several sources (primarily World Bank and DANIDA) are now available through IREDA for promotion, commercialisation and installation of renewable energy systems. The largest program among all renewable sources is for windfarms with the objective of private sector participation to install 100 MW of windfarms.

The program for stand alone and autonomous systems has not been as successful as that of grid connected windfarms. This is largely due to institutional problems.

In China also attention has been focused on these two types of systems. Unlike in India the program on stand alone systems has been more successful. China developed indigenous capability to manufacture these systems. The western parts of China, Quighai, Xinjiang and Inner Mongolia is rich in wind energy resource but is sparsely populated with dispersed load centres. It is not economical to electrify settlements in these regions with centralised power grid. Integration of small autonomous wind electric systems (10-100 kW) with local grids and stand alone systems (upto 1 kW) is one of the ways of meeting electricity demand in this region. At present 100,000 of such systems are operational in China and the manufacturing capability approaches an impressive 20,000 systems annually. China has set up six demonstration windfarms totalling a capacity of 4.3 MW. Two 20 kW systems are working in parallel with 280 kW diesel generator at Dachen island, Zhejiang province[**6**]. In Egypt, wind electric pumping systems (from 10 kW to 50 kW capacity) have been installed mainly in the north-western part of the country while two windfarms of 400 kW each are connected to the local grids on the Red Sea.

Indigenous capability to manufacture grid connected WEG has been partly developed in India and at present it is assessed that except the rotor all other components can be indigenously procured. The public sector organisation BHEL is implementing a windfarm project with 200 kW WEG in Andhra Pradesh. NEPC, a Madras based company in collaboration with Micon of Denmark is offering 80% indigenised WEG. In China emphasis seems to be on small systems which are already under production while in Egypt it seems much needs to be done towards developing indigenous capability.

4.0 TECHNICAL AND ECONOMIC PERFORMANCE OF WEG:

Windfarms are reported to be doing well in China while little is known about there performance in Egypt. However, in the case of India data on performance of all the windfarms is being obtained by the Ministry of Non-conventional Energy Sources and can be accessed. Some windfarms in India have been operating since 1986 and it is possible to get a realistic indication of the expected performance during their life time. The technical performance can be evaluated on the basis of a number of parameters discussed below:

Capacity Utilization: Capacity Utilization Factor (CUF)[**7**] is the index of overall performance of WEG. The windfarms at potential sites in India have generated electricity at an average CUF Factor of 15% to 30%.

Availability: Availability of a machine in good working condition demonstrates the technical viability on the one hand and the effective maintenance and service on the other. Analysis of the performance of data on windfarms in India has shown that the low wind is the major contributor to machine downtime, and the faults in machine and grid cause relatively small downtimes. For the windfarms operational in Tamil Nadu, the technical availability is observed to be higher than 95% and at some locations is as high as 98%.

System reliability: High technical availability is itself an indicator of high machine reliability. The reliability of any system in integrated mode is related to the probability of the failure of the system to meet peek demand on it. In the case of a windfarm there are a number of wind turbines. Since the availability of independent WEG is high, the probability of the failure of windfarm as a whole is very small. For large penetrations there will be a number of clusters with thousands of machines. Therefore, the overall reliability of the wind powered generating capacity is expected to be high and the probability of a significant capacity on outage is expected to be negligible. One of the indicators of system reliability is the Loss of Load Expectation (LOLE)[**8**]. In a simulation exercise on TNEB power system LOLE was not found to change adversely for large wind penetrations. The factors that affect system reliability are: reliability of individual units in the system, manner of system operation and the temporal matching of wind speed time series with system demand.

Repair and Maintenance: High occurrences of major component/sub system failures was not observed during the operation of windfarms. There were very few instances of major component replacements like generators or blades. Generally the O&M costs were about 2-3% of the capital costs.

Machine Life Time: The lifetime of grid connected WEG is taken to be about 20 years for the purpose of project accounting. Some suppliers of equipment claim a lifetime of 30 years.

In view of the few component failures in Tamil Nadu and Gujarat in the last five years it can be assumed that the machines will operate for 20 years without being completely replaced.

Economics of Operation: Several studies carried out earlier, have established the economic viability of windfarms at the potential sites[**9**].

The levelized annual cost of Rs. 1.6/kWh (5 cents/kWh) to Rs. 3.5/kWh (11 cents/kWh) for wind generated electricity is comparable or marginally higher than the cost of generation of Rs.1.65/kWh (5 cents/kWh) from coal based plants[**10**].

5.0 RECOMMENDATIONS FOR FUTURE STRATEGY:

The potential for wind energy utilisation in different modes (grid connected, autonomous etc.) in developing countries is large. The techno-economic viability is established and to a large extent the systems have demonstrated the performance convincingly. However, efforts at large scale integration of WEG with the conventional systems (grid or diesel generator) and commercialisation of the technology will face several managerial and logistic hindrances. In the Indian context, the implications of several private organisations availing the facility of wheeling and banking with the grid and dispersed windfarms feeding electricity intermittently are neither analyzed in depth nor clearly understood. While there are limits imposed by the considerations of penetration, energy banking and wheeling would further limit the extent to which windfarms can be installed due to the system constraints.

There are some underlying similarities in system characteristics, the mode of operation and the nature of generation mix in most of the developing countries. Therefore, the basic issues related to grid integration of WEG and the measures required to achieve/ensure higher levels of penetration would be somewhat similar. One of the major handicaps associated with all planning exercises in these countries is the lack of reliable or authentic data. In India as mentioned a program was initiated to monitor the wind speeds in potentially windy areas in the last few years and the data thus collected is being processed at present. Generally, however, sufficient information on the resource availability of wind is not available. In the developing countries, therefore, the first crucial step towards the wind option would be the collection and development of a strong and reliable database on the renewable energy resources.

In most of the developing countries, indigenous development/production has not taken place and at least in the initial stages the generation devices would have to be imported from developed countries. The technology in developed countries, however, is basically designed for resource and ambient conditions in high latitude areas. For example, wind electric generators have to be designed for optimal energy capture from wind on the one hand and the extreme structural loads corresponding to high wind speeds on the other. Therefore, a very cautious effort has to be made to supply/install those systems in developing countries which are optimally designed for local conditions. In the developing countries, indigenisation/adaptation should be taken up as a parallel activity in the overall strategy of wind power integration. Initially all projects were implemented with imported technology, but subsequently one expected R&D on WEG in both govt. and private institutions to be stepped up. However, this did not happen. There has been no effort to promote R&D with industry participation. Some projects on rotor development have been undertaken with 100% capital funding in the government organisations. Taking lessons from the European experience with capital and market led funding, one would not expect the capital funded projects in India and rest of the developing world to achieve the desired goals.

Some of the windfarms installed since 1986 have performed rather poorly. This is due to both, unstable grid conditions and inadequate wind speeds. It appears that these windfarms were implemented hastily in which prior and sufficient examination of the factors affecting capacity utilisation could not be carried out. A similar trend is observed in California where the first windfarms were installed in a haste to avail the tax benefits etc.. In all future projects, it is very important to allow sufficient time for assessing the feasibility of windfarm in a given region or at a given site. In order to encourage participation of private sector in installing wind turbines, the government allows 100% depreciation in the first year of the commissioning of the windfarms. An organisation can avail of this facility if the windfarm is commissioned before September of the year. While this measure is a sincere effort to encourage windfarming the time dependent element in this regulation may result in a compromise on site, machine quality and the grid quality.

With several organisations indigenously manufacturing wind turbine components, there is a need for quality control, testing and certification. At present developing countries are lacking in such facilities and this is an area where international cooperation is possible and desirable.

References

[1] **Wind Energy Systems: Export Market Potential, A SERI report, Jan. 1984, ISBN: 0-88016-011-X.**

[2] **C. Zhinan & C. Xiayao; "The Wind Power Resource and its Development Potential in Fujian Province"; Wind Engineering, Vol 12, No. 2, 1988; pp. 118-124.**

[3] **D. Huizhu & J. Twidell; "A brief review of wind energy exploitation in China"; Wind Engineering Vol. 12, No.5, 1988.**

[4] **J. Hossain; "High Penetrations of renewables in the power grid of Tamil Nadu"; Wind Engineering Vol. XV, No.6, 1991. J. Hossain; "Grid Integration of Renewables in Developing Countries: A case study of high wind penetration in the Tamil Nadu Electricity Utility"; Energy Policy (Forthcoming).**

[5] **A. Mani; "Wind Energy: Resources Survey for India-I" 1990, Allied Publishers Ltd., ISBN 81-7023-297-X.**

Table 1 : List of Demonstration Windfarms in India

Site	Date of commissioning	Capacity (MW)
Mandvi I	16.01.86	1.10
Mandvi II	29.09.89	.50
Tuticorin I	18.01.86	.55
Okha	08.03.86	.55
Puri I	01.05.86	.55
Deogar I	23.05.86	.55
Tuti II	09.09.86	.55
Kayathar IA	07.04.88	1.35
Kayathar IB	12.06.89/31/03.90/	1.40
Puri II	01.01.88/01.03.88	.55
Deogad II	December 1988	.55
Tirumalla	January 1989	.55
Okha Madhi	25.07.89	1.80
Kheda	18.02.90	.50
Lamba	30.04.90	10.00
Kayathar II	02.05.90	6.00
Muppandal	31.03.90	4.00

[6] C. Yingrong; "Renewables in China: Case Study"; Energy Policy, Vol. 19, No.9, November 1991; pp. 892-896.

[7] CUF is defined as (Annual Energy Generation from NET)/(8760 x Rated Capacity).

[8] For detailed explanation of reliability concepts in power systems see, R. Billinton & R. N. Allan, 'Reliability evaluation of power systems', Pitman publishing Ltd., 1984, ISBN 0-273-08485-2 and G.J. Anders, 'Probability concepts in electric power systems', A Wiley-Interscience publication, 1989, ISBN 0-471-50229-4.

[9] J. Hossain et. al.; 'Analysis and recommendations for wind electric systems', A study conducted by TERI for Ministry of Non-conventional Energy Sources, December 1986. J. Hossain et. al., 'Wind energy: Some economic and policy implications,' Financial Express, March 1987, New Delhi. J. Hossain, K. Thukral & V.V.N. Kishore,' Windfarms in India: Minimizing the costs of generation', Journal of Wind Engineering and Industrial Aerodynamics, 27 (1988). R. Bhatia,' Wind energy utilisation in India: Analysis of economic and financial aspects', Proceedings of the 12th Annual international conference of the International Association for Energy Economics, 4-6 January 1990, New Delhi, pp 326-337. K.S. Rao,' Performance of Okha and Mandvi windfarm', Seminar organised by Hindustan Brown Boveri, 7th feb 1987, New Delhi.

[10] J. Hossain; 'Strategies for indigenous development of grid connected wind electric generators in India', A report submitted to PACER secretariat of USAID, ICICI, Bombay, January 1992.

Table 2 Summary of Incentives in Different States.

SNo.	Item	Andhra Pradesh	Tamil Nadu	Karnataka	Gujarat
1.	Power Wheeling & Banking.	Introduced	Introduced	Introduced	Introduced
2.	Buy back of generated/ surplus power by SEB.	Permitted	Permitted	Permitted	Permitted
3.	Selling and sharing of generated power to other industries.	Yet to be	Yet to be	Permitted (subject to a fee)	Permitted (Any industrial unit can be declared beneficiary of a given wfarm).
4.	Charges for Power Wheeling.	2% of energy generated	2% of energy generated	2% of energy generated	2% of energy generated
5.	Charges for Power Banking.	Included in above.	-do-	Included in above	Included in above
6.	Buy back by SEB	Negotiable	@ Re. 1.00/kWh	@ Rs 1.25/kWh	@ Rs. 1.75/kWh
7.	Capital Subsidy	Upto Rs. 15 lac per project.	10% of the project cost	As extended to other industries.	

COMPARISON OF THE FINANCING ARRANGEMENTS AND TARIFF STRUCTURES FOR WIND ENERGY IN EUROPEAN COMMUNITY COUNTRIES

E.J. van Zuylen and A.J.M. van Wijk
Ecofys, Research and Consultancy, Biltstraat 110, NL-3572 BJ The Netherlands
tel +31 30 732144; fax +31 30 731424

C. Mitchell
University of Sussex, Science Policy Research Unit, Mantell Building, Brighton BN1 9RF, Great Brittain;
tel +44 273 686758; fax +44 273 685865

ABSTRACT: Various mechanisms intended to stimulate wind energy are in place in European countries: investment subsidies (the Netherlands), tariff support (Denmark and the UK), combinations of these two (Germany) and tax measures (Denmark). Financing of wind energy shows a similar variation, with different required rates of return for each country. A steady growth of installed capacity can be seen in Denmark (70 MW/yr) and, to a lesser degree, in the Netherlands (35 MW/yr). An acceleration of growth can be observed in Germany (60 MW/yr) and, more recently, in the UK (66 MW under construction).

1 INTRODUCTION

A number of European countries have mechanisms in place, such as tariffs, subsidies or tax measures, which are intended to stimulate wind energy. These measures and the means of financing wind energy projects are reviewed. Denmark, the Netherlands, Germany and the United Kingdom are considered and some remarks on the situation in Italy, Spain and the United States are made.

2 SUPPORT MECHANISMS BY COUNTRY

2.1 Denmark

In Denmark there are no investment subsidies. Wind energy is supported by an electricity tariff paid for partly by the utilities and by the Government in the form of restitution of CO_2-tax and an energy tax (including VAT) and by tax measures.

Tariffs: The utilities are required to pay 85% of the consumer price without taxes, currently 0.045 ECU/kWh. In addition, the CO_2-tax (without VAT), corresponding to 0.015 ECU/kWh, and the general energy tax (including VAT) on fossil fuels, corresponding to 0.025 ECU/kWh, are refunded to co-operatively and single owned turbine owners. Utilities have only the CO_2 tax refunded.

Other arrangements: Profit gained from wind turbine investment is tax free for co-operatively owned turbines to a maximum of 150% of individual investors consumption and for utilities. Single owned turbines can be depreciated against tax. Support for wind energy is part of an integrated plan which includes a fixed price for, and a clear division of responsibility between utilities and developers for, grid connection.

Finance: Projects of co-operatively owned single turbines (or small groups) is the most common way of financing and producing wind energy. The investors must live in the municipality (or neighbouring) in which the turbine is sited in and each investor may only buy shares up to an equivalent of 150% of the investors electricity consumption. Either the investor puts up their own money (usually a total of around 4000 to 5000 ECU) or borrows it from a local bank at around 12% (at current levels). Usually, the bank loan is paid off first and only then the investor receives any return on their investment. The incentive, in a country with minimum to maximum tax of 50-68%, is that the investment is tax free. Single owners are generally able to raise 100% of the total capital costs (usually 60% from building societies and 40% of banks). Building societies generally require about 2% lower returns than banks. The utilities are also expected to erect wind farms. The finance is raised from consumer electricity bills and notionally a return of around 7% on investment is expected.

Conclusion: Despite the relatively low tariff (see Table 1), investments in wind energy is more or less common practise. Important ingredients of this are favourable tax arrangements, the coordinated development of the sites (including grid connection) and proven reliability to give comfort to investors and banks. This results in a more or less steady increase in the installed capacity (about 70 MW per year). About 450 MW is installed of which less than 25% is utility owned.

2.2 The Netherlands

In the Netherlands, wind energy is stimulated by Government investment subsidies (TWIN-program). However, in addition a number of utilities support wind energy through an increased electricity tariff and/or an investment subsidy.

Tariffs: Utilities are required to buy all wind energy offered to them. The standard tariff consists of a fuel component and a capacity component and is about 0.035 ECU/kWh. Utilities may choose to pay more. For instance PEB pay 0.055 ECU/kWh while PEN and PGEM pay 0.069 ECU/kWh. Negotiations are underway at the moment between the Dutch turbine owners association (PAWEX) and the electricity utilities (EnergieNed) for a fixed price to be paid by all utilities.

Subsidies: The Ministry of Economic Affairs gives investment subsidies up to 35% of the total capital costs. The amount is determined by a formula using the rotor surface and the rated power of the specific turbine. A further subsidy for silent turbines was stopped in 1993.

Finance: Utilities are allowed to add up to 2% to domestic electricity bills in order to pay for their obligations to reduce CO_2 emissions under the National Environmental Plan. This so-called MAP payment may be disposed of as the utility likes for energy saving or for example to support wind energy projects. In general, utilities have financed wind farms with Government investment subsidies (the TWIN program) and MAP subsidies which together total about 50% of capital costs. The other 50% is raised from corporate loans which are repaid with about 9% interest over 10 years. Some utilities prefer not to invest their own money and have either put wind farms out to turnkey tender or have launched wind funds, such as the successful EGD fund, which gives average returns of about 8% over 15 years when EGD will buy back the shares.

Several wind projects have been financed via joint ventures between utilities and private developers. The utility share has been financed as described above while private developers have been able to bring cheaper, long term fixed finance to the table. Further, farmers are able to obtain finance (85–100%, depending on location) from their local bank, usually based on the collateral of their farm at normal bank rates of return over for instance 10 years. Co-operatives, now responsible for around 30 turbines of 80 to 250 kW, are financed from individuals either bringing in their own capital or borrowing a certain amount from banks at normal bank rates. Rates of return have so far averaged between 2–8% over 10 years. Banks are prepared to loan money to the co-operatives using the wind turbines as collateral rather than the homes or possessions of the chosen representatives of the co-operatives.

Conclusions: The majority of 100 MW wind turbine capacity has been installed by utilities or utilities in partnership with private developers (about 80%). Single turbines and (increasingly) small wind farms have been erected by farmers and by co-operatives run on the Danish lines. In 1992 private investors have claimed the largest part of the TWIN budget. The growth rate of last years is about 30–35 MW per year, with a Government goal of 400 MW in 1995 and 1000 MW in 2000.

2.3 Germany

Wind energy in Germany benefits from 3 main supports: a law that requires utilities to pay a fixed, long term, high tariff for electricity generated by renewables; an additional Government tariff or subsidy; and finally extra subsidies can be obtained in a number of "länder".

Tariffs: The standard tariff paid by the utilities for wind electricity is 90% of the small consumers price currently 0.08 ECU/kWh. This tariff is financed by a levy on the small consumers price. In addition, 0.03 ECU/kWh is paid by the Government within the BMFT 250 MW program if this support is not taken up as an investment subsidy. The BMFT tariff subsidy stops when the developer has regained the investment 2 times or after 10 years of turbine operation.

Subsidies: It is possible that developers take an investment subsidy from the BMFT, based on hubheight and rotor diameter with a maximum of 60% or 46.000 ECU, instead of the extra BMFT tariff.

Other arrangements: There are various regulations in addition to the standard tariff and the BMFT arrangements, mostly concerning investment subsidies ranging from 20 to 45% of the total turbine costs, in a number of "länder". In Nieder Sachsen (Lower Saxony) for example, the support is depending on the electricity production of the turbine in standard conditions and on the noise level. Wind turbine projects that are not included in the BMFT program can be granted even more subsidy as well as turbines at inland locations.

Finance: The majority of wind projects in Germany obtain finance through the Deutsche Ausgleichbank (DAB), which is subsidized by the Government. In general, provided a company has an annual turnover less than 250 MECU, it is possible to obtain a loan of a maximum 0.5 MECU or 50% of the total capital costs and in addition, a top-up loan of a maximum 5 MECU or 25% of the total capital costs via a normal bank. The DAB lends to the bank at 6% over up to 15 years and the bank lends to the developer at 7% over up to 15 years. Banks are willing to top-up DAB loans to make up 85–100% of total capital costs (depending on the size and the location of the project) at around 1% over normal bank rates of return. A number of companies exist as gatherers of equity investors for example Beratung Verwaltung Treuhand which gives average returns to investors, post tax until costs are paid off, of around 8–10%. In addition, a number of banks, for example the GLS bank, arrange wind funds for particular projects or give low interest equity loans.

In 1992, around 50 MECU was loaned by DAB to wind energy projects in former West Germany and 7.5 MECU in former East Germany. For utilities, not eligible for DAB loans, investment in wind energy comes from corporate funds and is expected to be repaid at the nominal company rate of return.

Conclusion: Less than 20% of the installed capacity is utility owned. The annual increase in capacity is estimated to be about 60 MW, with about 170 MW installed and a government goal of 250 MW installed in 1994. The rate of growth increased substantially in 1990 following the setting of the electricity tariff for renewables and the implementation of the BMFT program.

2.4 The United Kingdom

The electricity act of 1989 established the Non-Fossil Fuel Obligation which obliged regional electricity companies to pay different premium prices until 1998 for different electricity technologies in England and Wales. Renewable electricity in Scotland and N. Ireland is paid either the pool price or, occasionally, a higher especially negotiated price. The NFFO is levied on all electricity bills (11% in 1992) although only 2% of the total goes to renewable electricity. The rest is going to pay for nuclear power. So far, there have been tranches in 1990 and 1991 and a third tranche is expected mid year 1993.

Tariff: The 1991 order set a tariff of 0.14 ECU/kWh to be paid until 1998. The Department of Energy (as was) decided on a capacity to be contracted for from the wind band. The lowest bid per kWh was given the first contract to generate within the wind band of the NFFO. The second lowest bid was given the next contract and so on. The price paid to all the capacity was determined by the highest bid price that completed the total capacity.

Finance: The majority of projects in the UK have been project financed and have total capital costs above 6 MECU. On the whole projects have obtained loan finance at about 1–2% over the base rate until 1998 for about 80% of the total capital costs. The other 20% equity capital has been obtained from regional electricity companies, water companies, manufacturers, parent companies and the developers themselves. The equity returns have been in the order of 15–20% real until 1998.

In a few isolated cases, projects which have lower capital costs than 6 MECU have been project financed in a similar way as described above. Another project found an alternative way to be financed via the Business Expansion Scheme and low interest return equity capital from the Mercury Provident Bank. However, on the whole, financing small projects at sites with windspeeds below 7 m/s has proved difficult. There are few alternative financing mechanisms, such as co-operatives, utility bonds and equity gathering companies available. This is in large part due to the newness of investment in wind energy and probably these mechanisms will develop over time.

Conclusion: The 2 NFFO orders were for 28 MW in 1990 and for 192 MW in 1991, although only about half of this is expected to come on line due to problems with planning permissions. The Governments goal is currently for 1000 MW of renewable energy by 2000. However, it is unclear whether this is installed or "declared net" capacity (1 MW installed = 0.43 MW declared net capacity for wind). The 1998 end-date has been a serious drawback for investors. Delays to commissioning, such as obtaining planning permission, directly influences the profitability of projects. In 1992 29 MW was installed and at the moment 67 MW is under construction with another 15 MW with financing but not under construction.

2.5 Other countries

In Italy, a tariff of about 0.11 ECU/kWh has been established for wind energy for 5 years. Furthermore, in this country investment subsidies (20 – 40%) can be obtained. So far, only utilities have significant plans for Finally, in Spain 0.08 ECU/kWh is paid to private investors, based on the Energy Conservation Act.

In other European countries (e.g Sweden or Belgium) the tariffs are about 0.03 ECU/kWh, which is related to directly avoided fuel and capacity costs. In addition, Sweden has investment subsidies up to about 25%. In the United States an energy bill is in effect that includes a 0.012 ECU/kWh production tax credit or production "payment" for non profit companies. Further, only Oklahoma has an investment tax credit up to 8000 ECU. Finally, in the US there is a tendency to internalize environmental costs, which may be favourable for renewable energy sources.

3 INFLUENCE OF THE EC

The European Community can wield direct influence on the arrangements of a member state. One of the criteria is that the measures do not discriminate between (products of) member states. Furthermore, market protection is not allowed. On the other hand the stimulation of wind energy (and other renewables) is considered positively, because of its importance for the environment and because it not yet a proven, profitable technology. The NFFO is an example of an arrangement that had to be adjusted (given a 1998 end-date) after EC involvement.

In addition, the EC directly supports wind energy. The JOULE program supports R&D projects, while the THERMIE demonstration program offers 40% subsidy of total capital costs of setting up 1 turbine of an innovative type and 35% for dissemination projects.

4 OVERVIEW

Tariffs: In Table 1 an overview is given of the tariffs for wind energy in the various countries together with the existing small consumer electricity prices. The prices for small consumer electricity are given including standing charges and taxes (exclusive VAT) for a usage of 3.300 kWh/yr.

Table 1 Overview of the tariffs (in ECUct/kWh) for wind energy and the small consumer electricity prices in various European countries.

Country	Wind energy tariff	Small consumer prices	Remarks
Denmark	8.3	11.2	Up to a share of 9.000 kWh/yr per participant
The Netherlands	3.5 – 6.9	8.4	Depending on fuel price
Germany	11.8	13.3	Including 3.2 ECUct/kWh extra <u>or</u> extra investment subsidy (Table 2)
United Kingdom	2^{e} order 13.6	10.9	Tariffs valid until 1998. Then ± 4.6 ECUct/kWh
Italy	8.7	11.3	Fixed until 1995
Spain	7.7	13.0	
Sweden	3.4	7.8	

Subsidies: In Table 2 an overview of the various investment subsidies is presented.

Table 2 Investment subsidies in the various European countries.

Country	Subsidy	Remarks
Denmark	none	
The Netherlands	maximum 35% (1993)	Total costs; budget 1993: 16 MECU)
Germany	maximum 60% or 46.000 ECU	Total costs without foundation (see remark in Table 1)
United Kingdom	none	
Italy	20% – 40%	
Spain	none	
Sweden	25%	

Other arrangements: In the various countries extra arrangements to ease and stimulate the implementation of wind energy are in use. An overview is presented in Table 3.

Table 3 Overview of supplementary arrangements in the various countries. Further, in most countries R&D budgets are available.

Country	Arrangement
Denmark	Fixed grid connection costs; profit gained from wind energy is (partly) tax free
The Netherlands	Limited investment subsidies in various provinces
Germany	Considerable subsidies for several "bundesländer"
United Kingdom	none

Finance: The way wind projects are financed varies from one country to another, which has implications for the profitability and the implementation speed in the present situation. A summary is given in Table 4.

Table 4 Overview of financial aspects, specific for the various countries.

Country	Finance
Denmark	Primarily co-ops; utilities via consumer bills; acceptance of wind energy by banks.
The Netherlands	Primarily utilities via consumer bills; utility bonds; co-ops; private developers with low interest long life loan.
Germany	Based on DAB grants, some project financing, equity gathering companies, co-ops, low interest wind fund.
United Kingdom	Primarily project financing, high interest equity finance, problems for projects under 6 MECU and windspeed below 7 m/s

5 CONCLUSIONS

There is a wide variety of mechanisms in place intended to stimulate wind energy. in the countries that we looked at in detail: The Netherlands, Denmark, Germany and the United Kingdom. The accent in the Netherlands is on investment subsidies; in the UK and Denmark there is tariff support only; and in Germany a complex set of both subsidies and tariffs. A steady growth of installed capacity can be seen in Denmark and to a lesser degree, in the Netherlands. An acceleration of capacity growth can be observed in Germany and, more recently, in the UK.

Despite higher wind speeds than on continental Europe, financing of wind energy is most difficult in the UK and differs in three main ways from the continent. The UK is primarily project financed; there are few alternative mechanisms of finance; and utility investors expect to make higher returns on their investment (to be paid to their shareholders) than, for example, the 5–9% required to be re-invested by Danish and Dutch utilities. Utility investment is an important financing mechanism im the other countries examined, but central in the Netherlands. Both Denmark and Germany have additional financing mechanisms based on tax rebate and government supported loans

APPENDIX

The various national currencies in this paper are converted to ECU's using the exchange rates given in Table 5.

Table 5 Exchange courses used for comparison purposes (valid: end 1992)

Country	National currency	ECU
Denmark	100 DKK	13.53
The Netherlands	1 fl	0.46
Germany	1 DM	0.51
United Kingdom	1 £	1.24
Italy	10.000 Lit	5.85
Spain	100 Peseta	0.72
Sweden	100 SEK	14.22

REFERENCES

- Förderung der windenergie in Nieder Sachsen, Klug, H. and Molly, J.P., DEWI Magazine Nr 1, August, 1992.
- Privatejede vindmøllers økonomi, RISØ, Energityrelsen, june, 1991.
- The Non-Fossil Fuel Obligation, 1991 Renewables Tranche, Office of Electricity Regulation, Notes for Generators, 1991.
- Nationaal ondersteuningsprogramma toepassing windenergie in Nederland (TWIN), 1991-1995, NOVEM,.
- Prospectus Molens voor Milieu B.V., EGD, Groningen, 1991
- Förderfibel Energie, 1992, Forum Fur Zukunftenergien EV, Bonn, Germany.
- GKG-Windkraftfonds 1 (2 and 3), GLS bank, Bochum, Germany.
- UK Wind Update, 1993 Safe Energy, Feb/Mar No.93, Edinburgh, Scotland.
- Loans for Energy Savings (016 GER:ERP), 1993, Deutsche Ausgleich Bank, Bonn, Germany.

ACKNOWLEDGEMENTS

The comparison of subsidies and tariffs was financially supported by NOVEM. The comparison of financial arrangements was supported by SERC. A great number of people in various countries has been very helpfull in gathering the information.

NEW SUBSIDY METHOD STIMULATING THE DEVELOPMENT OF MORE EFFICIENT WECs

Helmut Klug, Jens Peter Molly, Joachim Gabriel, Dieter Westermann

DEWI

(Deutsches Windenergie-Institut),

Ebertstr. 96, 2940 Wilhelmshaven, Germany

The new wind turbine subsidies in the German State Lower Saxony now depend on measured power curves and sound power levels encouraging the installation of quiet and efficient turbines. The measurements have to be carried out according to the IEA recommendations Power Performance Testing and Acoustics-Measurement of Noise Emission from Wind Turbines by recognised independant institutions. The subsidy is calculated by the following formula developed by DEWI on behalf of the State Lower Saxony:

$$S = E^2(A/F + B/(8760P))$$

with

E: annual energy production (MWh)
F: acoustically affected area (hectares, 45 dB, see Fig. 1)
P: measured power curve maximum (MW, see Fig. 2)
A: acoustic evaluation factor (e.g. 1)
B: power evaluation factor (e.g. 300).

The annual energy production is calculated from the measured power curve for a reference location with an average wind speed of 5.5 m/s at ten meters height. A circle around the turbine at a distance from the wind turbine at which the sound pressure level is 45 dB(A) gives the area F (see Fig. 1). Penalties for tonality (e.g. gear box noise, see Fig. 3) are determined according to the german guideline DIN 45681 and added to the sound power level. The other States of Northern Germany will accept the formula for their financial support programme in the year 1993.

Measurements from different institutions are only comparable when the same measurement and anemometer calibration procedure is used. The anemometers from other institutions will be calibrated by DEWI in a 1 m² low turbulence wind tunnel equipped with a LASER-Doppler-Anemometer (see Fig. 5). The importance of the anemometer calibration is shown in Fig 6.

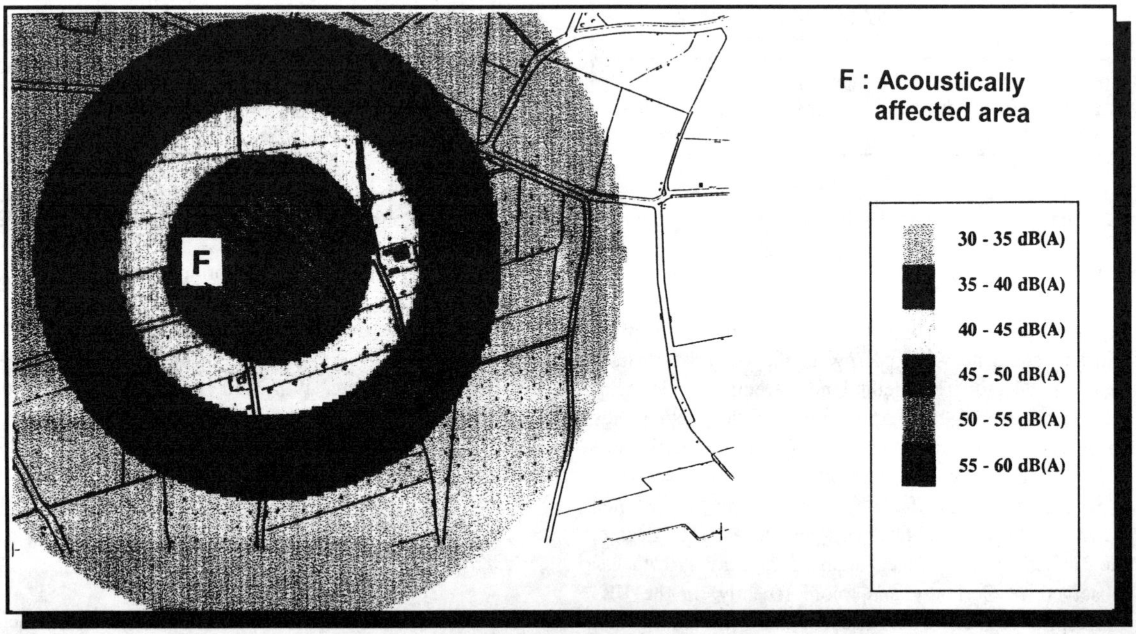

Fig. 1: Acoustically affected area (F) according to the German Guideline VDI 2714 (Outdoor sound propagation). The measured sound power level at a wind speed of 8 m/s at 10 meters height is used for the calculation.

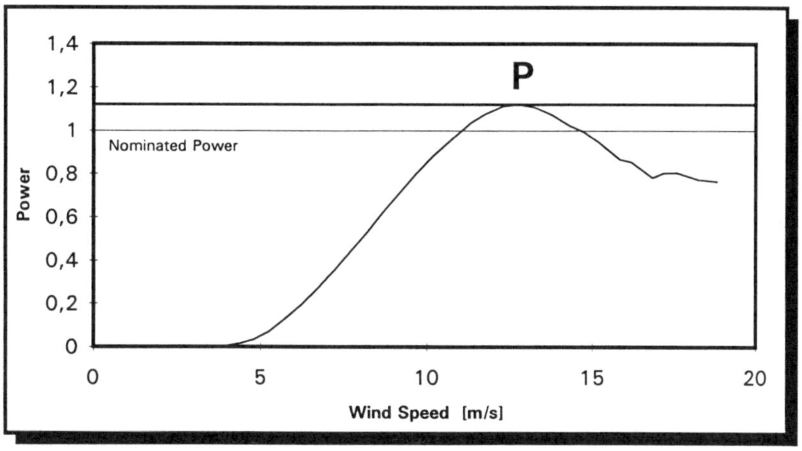

Fig. 2: Measured power curve showing maximum P and nominated power.

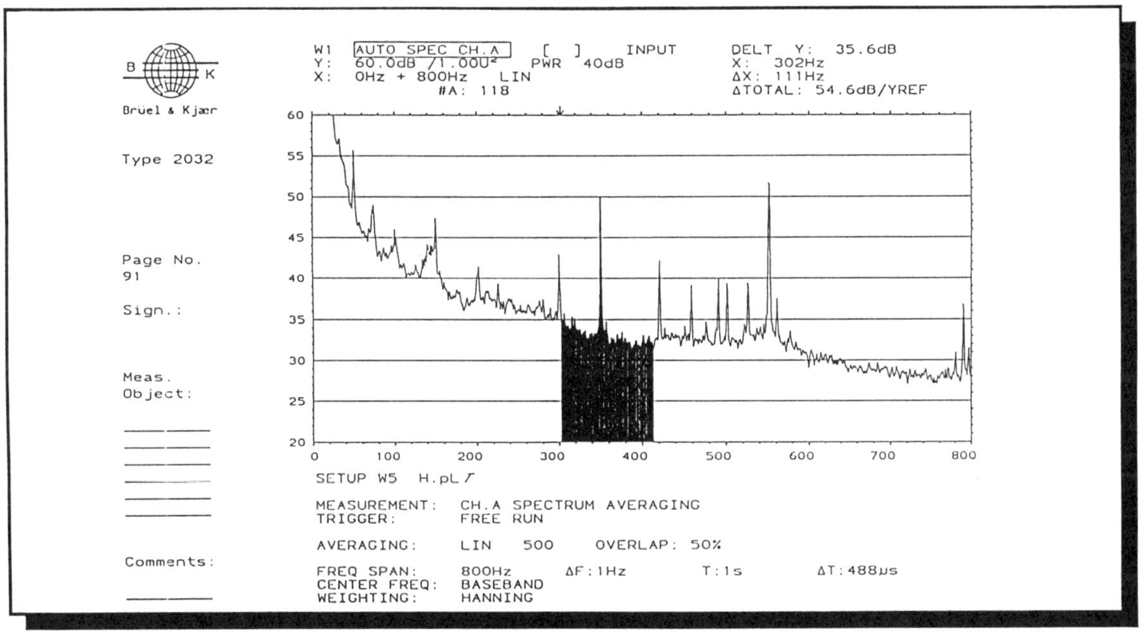

Fig. 3: Narrow band (1Hz) frequency-analysis of the wind turbine noise. The pure tone at 350 Hz (gear box noise) results in a penalty of K_T=3 dB added to the sound power level of the wind turbine.

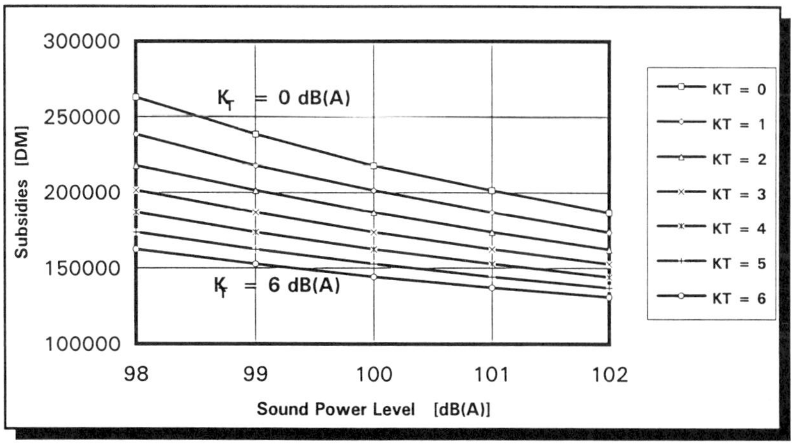

Fig. 4: Dependance of the subsidies on the sound power level and the pure tone components (P=500 kW, A=1, B=300, E=1200 MWh, penalties between 1 and 6 dB)

Fig. 5: LASER-Doppler-System
used for the anemometer calibration. The accuracy is much
better than other methods because no pressure and temperature
correction is needed and there is no influence of the LASER beam on the flow field.

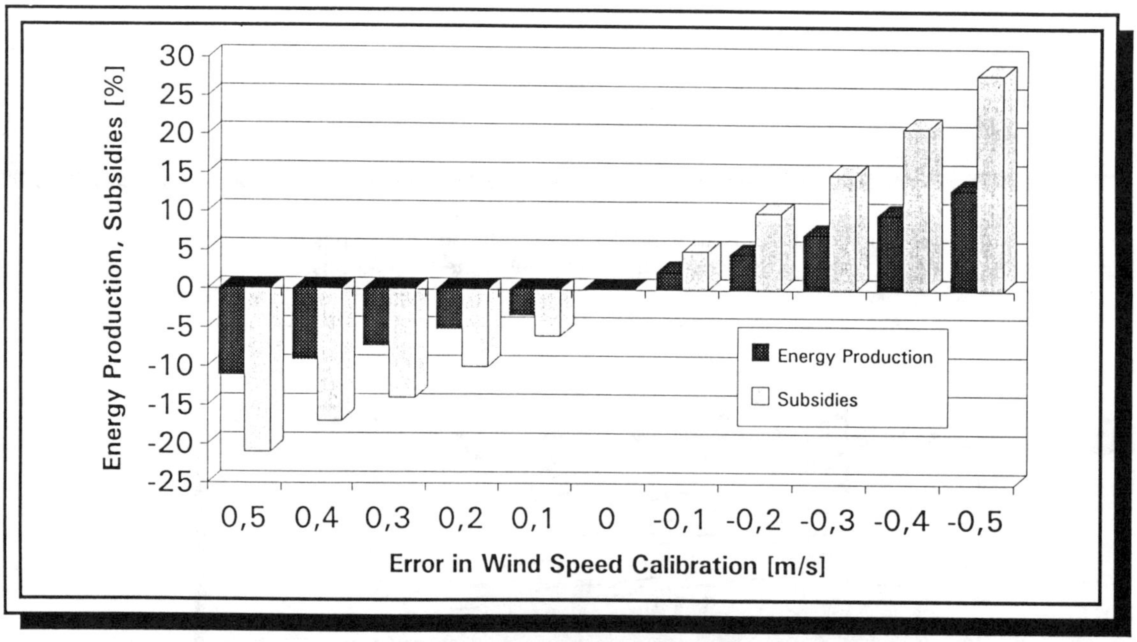

Fig. 6: Influence of anemometer calibration
on the calculation of the annual energy production E and the subsidies. A deviation from
the real wind speed in the order of 0.3 m/s results in 8 % error in E and 15 % error in the subsidies.

CANARIES WIND 2000 PROYECT

R. Calero, J. A. Carta

Departamento de Ingeniería Mecánica
Universidad de Las Palmas de G.C.
Edificio de Ingenierías. Campus de Tafira
35017 Las Palmas
SPAIN

ABSTRACT: The present article explains, in a schematic way, the several research projects carried out in the Canarian Archipelago, under the generic title of "Canaries Wind 2000 Project", with the purpose of knowing the archipelago´s wind potential, and in particular, the posibilities of applying it to the sea water desalination.
The different objectives above defined, are pointed out as well as the works developed to achieve them: methodologies, human an economic resources, find results, etc.

INTRODUCTION

The Canarian Archipelago is composed of 7 islands, with a total surface of 7.500 Km^2, a native population of 1,5 million people, and a foreing (tourist) population over 6,5 million visitors per year.
Due to the shortage of rain and the overexploitation of aquiferous layers, most of consumed water in obtained from sea water desalination, by different procedures, what makes water cost very high.
Canary Islands are immersed in the Trade Winds what makes the wind potential very high in certain areas of the islands. Because of these 2 factors, the possibility of resorting to wind energy came out not just as a way of producing electrical energy but also as a way of desalinating sea water on a large scale.
This desalinated water, of high cost, could be a perfect "storage" of variable wind energy.
In this situation came out the project Canaries Wind 2000, supported by the Canarian Government Industry Office within the framework of the general objetives proposed in the Energy Plan of Canaries (PECAN).
In the present stage, the research phase is almost ended, with some projects still runnning, in the area of sea water desalination with wind energy, directed to validate in an experimental way the theoretical results obtained, as well as to develope the necessary equipment to enable the general application of this technology.
At the same time and as a consequence of the knowledge of the archipelago´s high wind potential, important projects for the use of this energy have started, being a total of 6 Wind Parks, in 4 islands, connected to the electrical network with a total power of 23 Mw.

OBJECTIVES IN THE RESEARCH PHASE:

In Canaries Wind2000 Project, 7 main objectives were established to be developed sequentially:

- Objective 1: Making of every islands´ wind park.
- Objective 2: Localization of possible "wind parks" in each island, evaluation of usefal surface and valuation of their wind potential.
- Objective 3: Study on the maximum electrical production that could be obtained in each island from the wind and its temporal evolution (per hour an per month).
- Objective 4: Study on the technical and economic viability of specific applications of the wind energy and in particular, of the sea water desalination.
- Objective 5: In terms of the future electrical power evolution, to find the wind power adaptable to each island electrical network in every moment.
- Objective6: To plan the future development of wind energy in Canaries, for the electrical energy production and the desalinated water production (including, in this las case, the direct connection between the wind parks and the desalination plants, with no connection to the electrical networks).
- Objective7: To analyse the economic, legal and environmental aspects, that allow the application of wind energy, on a large scale, in the Canarian Archipelago.

<u>Objective1:</u>
To obtain the wind maps of each island, a total of 31 register stations (Second Wind AL2000) were installed in a first phase for almost 2 years, and at the present time, in a second phase, a total of 36 stations are still working (including AL2000 and NOMAD).
At the same time as the recording of data, some necessary programs for their processing were developed, including wind speed estimations for a long term, as well as numerical models based on finite elements (FEA) which simulaate the wind behaviour as it flows over an island.
About 7 research workers have took part in this phase of the work (belonging to the Mechanical and Civil Engineerin Departments of Las Palmas de Gran Canaria University) and about 164.000 ECUS have been invested.

<u>Objective2:</u>
The selection and evaluation of wind parks is finished in Fuerteventura and Gran Canaria islands, being in process the evaluation of possible parks in the other islands.
For this study, networks of anemometers were installed in each zone and specifc computer programs were developed to simulate the "wind map of each park" (numerical models based on the method of finite elements FEA).
In Fuerteventura island 2 possible parks were determined:
-the one in "Cañada del Río", with an approximated surface of 4,7 Km^2 and a annual average speed of 7,3 m/s and
-the one in "Punta de Jandía" with a total surface of 3,6 Km^2 and an average speed of 7 m/s.
In Gran Canaria island 3 possible parks were determined:
-the one in "Arinaga", with 501 Ha and 7 m/s of annual average speed.
-the one in "Punta de Gaviota", with 892 Ha and 7,5 m/s of average speed

-and the one in "Tenefé" with 686 Ha and 7 m/s of annual average speed.
About 9 research workers have took part in this phase (from the Mechanical Engineering and Mathematics Departments of Las Palmas de Gran Canaria University), working for 2 years, with an investment of about 50.000 ECUS.

Objective3

This objetive, directed to know the maximum generation of electrical energy in the islands, is finished in Fuerteventura and Gran Canaria, being in process the study on possibilities in the other islands. The results achieved are shown in the following box:

ISLAND	GRAN CANARIA	FUERTEVENTURA
PARK	- ARINAGA - PUNTA GAVIOTA - TENEFE	- CAÑADA DEL RIO - PUNTA DE JANDIA
WIND POWER (MW)	69 122 94	73 37
GENERATED ELECTRIC ENERGY BY YEAR (kWh/year)	193.615 337.953 144.451	190.000 110.000

The data above represent average values that can be obtained among several types off wind turbines and several arrangements of their positions.
However, it has been developed a computer program which optimizes the type, number and position of the wind turbines in a park, to achieve the maximum electrical energy during one year.

Objective 4:

The purpose of studying the technical and economic viability of accioning desalination plants with wind energy has been fully accomplished at a theoretical level.
Since that, computer programs have been developed to simulate the behaviour of a wind park connected to a desalination plant (both isolated and connected to de electrical network).
Different desalination systems have been studied (reverse osmosis, vapour compression) as well as several design and functioning possibilities for such desalination plants ("module plants", with different module sizes and working in systems of constant and variable parameters).
The results for Gran Canaria island are shown in the following box, pointing out that they have been obtained with wind turbines Villas of 600 Kw, connected to reverse osmosis plants of 50 membranes each.

PARK	- ARINAGA - PUNTA GAVIOTA - TENEFE	TOTAL
Nº WIND TURBINE	115 204 157	
Nº MODULE	212 371 269	
UNSALTED WATER PRODUCTION (m^3/year)	28.400.000 49.700.000 21.000.000	**99.100.000**

Such results mean that the wind desalinating potential in Gran Canaria island represent more than 2,5 times the desalinated water demand in the year 2000, and equals the capacity of all the reservoirs in the island.
From an economic point of view, the results indicate that there is an optimum combination between the wind park size and the connected desalination plant size and type, which provides a maximum production at a minimum cost.
In the case of Gran Canaria, the calculated costs for one cubic meter of produced water only by wind energy, are slightly smaller than the present costs for water obtained from desalination plants connected to the electrical network supplied by conventional thermical plants.
This part of the project has been entirely designed in the Mechanical Engineering Department of Las Palmas de Gran Canaria University, and 4 research works have took part in it, with an investment of about 30.000 ECUS.

Objective5:

This objective, directed to know how much wind energy of that avaible in each island cuould be used, and its future evolution, has been only partially achieved, since there are not just theoretical aspects involved but also, in a major way, practical and management aspects.
In fact, the islands´ electrical systems weakness and the great wind potentials in some islands, introduce important practical uncertainties in the connectation of these to the present electrical network, what makes the level of penetration in the network very difficult to define.
At the present time there is an important project going on in Fuerteventura island, where a wind park of 11 Mw is being installed, connected to a network with a thermic power of 50 Mw (with a vally power of 15 Mw and a consumption in several desalination plants of almost 40 % of the total in the island).
The practical consequences to obtain from this system, which will tend to optimize the wind penetration by acting on the generation system (park and thermic plants) and on the charges (mainly on the group of desalination plants, both present and future ones), will settle a sound base to know the real potential of wind energy in Canary Islands (including the possible electrical systems remakes to allow the maximum wind penetration).
Besides of that, in the Electrical Engineering Department of Las Palmas de Gran Canaria University, electrical systems models for the islands are being developing to analyse the maximum penetration potentials; and in the Mechanical Engineering Department, studies directed to know the future electrical energy demand evolution (both general and desalination uses) are also being developing.
With both studies, it´s intended to know the possible evolution of the future use of wind potential en each island.
About 7 research workers have been involved in there studies, with an investment of 55.000 ECUS.

Objective 6:

As accordingly to the indefinite in objective 5, this new objective can´t be reached in total plenitude.
However, we have to point that one shows two slope: the one, to planning the wind energy investment that had to be connected to the insular network, the other one, investment planning when wind park can be directly connected at desalination plants, in this case "penetration" problems dissapear (and it is minimized to the rest of available wind energy connected to a network).
Even though the theorical levels, this direct connecting is possible technically and economically, it´s not possible to define an invest plan without checking the theorical results with an experimental phase (at the present time it is at the beginning).
As reference at the present time and admiting a penetration degree of 18 % from installed power, the wind power values that can be connected to a network in the 4 island with major potential are shown in the following box:

ISLA	POTENCIA TOTAL INSTALADA	POTENCIA EOLICA
LANZAROTE	80	14,4
FUERTEVENTURA	54	11,5
GRAN CANARIA	480	86,4
TENERIFE	370	66,6
TOTAL		**178,9**

<u>Objective 7:</u>
As accordingly to above mentioned in the objectives 5 and 6, the objective 7 isn´t carried out, even though in our opinion, wind energy importance is known by canarian population, so the firsts steps are being carried out in different meanings: in the first place, the necessity to actualize the Canaries Energetic Plan (PECAN), in order to arrange this important energetic source, not evaluated at the moment of its editing. In the second place, to consider water-energy as all in (especially, when, more and more, Canaries water is from desalination sea water),
In the third place, the necessity to define the economic and enviroment repercussions so that the masive implantation of wind energy not suppose a demage to the insular ecosystem, and fundamentally, it don´t produce a negative impact in the tourist sector, essential in the islands economy.
In the fourth place, real expectatives are being developing about the possibilty to introduce a new industrial sector in the region based on the application of wind energy to desalination sea water.

CONCLUSIONS

In general, Canaries Wind 2000 Project let to know the very important role that wind energy means in Canarian Archipelago, and the following conclusions are reached:
1.- In all and each of the Canary Islands, the electric energy from wind origen suppose, as a minimum, a 30 % predicted demand in the year 2000.
2.- From electric energy from wind origen and without connecting to the insular network, it is possible to produce all the unsalted water demanded in the year 2000.
3.- In both cases, the Kwh and one cubic meter of water costs are competitives with present cost.

WHY DO I SAY "YES" NOW?
Wind farms in Patagonia.

E. Spinadel, C. Luna Pont, G. Dutt.
GENCo group - University of Buenos Aires
Paseo Colón 850
1063 Buenos Aires - Argentina - tel/fax (54-1) 795 32 46

At ECWEC '90, I presented the view that wind energy parks were not needed in Argentina over the next decade. Argentina's electrical requirements for the nineties were fully covered, partially by the existing utilities and partially by several new hydroelectric or nuclear power stations now under construction. Rather, wind diesel generators were seen to be most relevant, for providing electricity to isolated communities. Spread all over Patagonia and some other areas with good wind conditions, there are numerous small towns and other human settlements not served by any electrical grid or at least by a weak local grid and localised in places that will never be served by any net.

This point of view remains absolutely correct today.

Our GENCo group goes on working in this line. Our new test station at the La Plata river front in the University of Buenos Aires City is been equipped mainly for this purpose. The Chilean University of Magallanes in Punta Arenas will too install a test station in the near future. Both Universities, in close collaboration will certify windmills to be used in this part of the world.

Now, it is well known that engineers are expected to build for today, to planify for tomorrow and to investigate for the day after tomorrow. In our field, today means the year running, tomorrow stands for the next 5 to 10 years and the day after tomorrow means the following 25 to 30 years. In the GENCo group, we devote part of our today and our tomorrow to the development of wind-diesel systems for isolated communities. But certainly, we also investigate for the day after.

This 'day after' is strongly related to the electrical requirements of the southern end of our continent, once the next century has begun.

It is fairly licit and reasonable to suppose that the large region beginning in the industrial area of Sao Paulo, Brasil, and going down to the extreme south of continental Argentina and Chile, will be a strongly integrated productive area by the year 2020. This year is considered as a reference by many energetic as well as ecological studies. All the activities related to the so called Mercosur-program points also in this direction. As it is well known, Mercosur covers the regional economies of Argentina, Brasil, Paraguay nd Uruguay, and Table 1 shows the present energy requirements of this region. These values should be increased in not less than 8% considering the additional requirements of Bolivia and at least of the central and south areas of Chile.

The expected population growth of this extended region permits the reasonable assumption of some 70x10^6 inhabitants with access to the by then existing interconnected electrical net in the year 2020. An extremely conservative point of view assigns to these inhabitants an annual mean electrical energy requirement value of 1800kW per capita.

So, the electric energy demand for this region will be not less than 126.000GWh/year by 2020.

That means that the total generation park must have a peak value of 25GW. This figure certainly must be increased by the necessary reserves (considering an availability of 5800 hours per year and exploitation losses of assumed 12%, realistic value as for the transmission distances involved).

TAB.1 **ENERGY REQUIREMENTS IN MERCOSUR AREA****

COUNTRY	**A) 10^3 inhabitants**	**B) GIP 10^6 1988 U$S**	**C) 10^3 tep final requirement**	**D) GWh electricity req.**
Argentina	31,929	85,730	32,224	41,200
Brasil	147,404	336,072	109,793	213,181
Paraguay	4,158	6,207	2,814	1,709
Uruguay	3,077	8,947	2,034	3,720
Total	**186,568**	**486,956**	**146,865**	**259,810**
	B/A) GIP/inhab 1988 U$S / inhab	**C/A) fin.req./inh. tep / inhab.**	**C/B) energy dens. tep / U$S**	**D/A) elect./inhab kWh / inhab.**
Argentina	2685	1.0	0.4	1290
Brasil	2280	0.7	0.3	1446
Paraguay	1493	0.7	0.5	411
Uruguay	2908	0.7	0.2	1209
Mean Mercosur	**2342**	**0.8**	**0.4**	**1089**

TAB 2 **INSTALLED CAPACITY IN MERCOSUR (MW)****

	1989	%	1999	%
Hydro	57,529	82.5	89,592	84.0
Nuclear	1,675	2.4	3,665	3.4
Oil	6,240	9.0	5,837	5.4
Gas	2,877	4.1	4,557	4.2
Coal	1,405	2.0	3,205	3.0
Total	**69,726**	**100.0**	**107,357**	**100.0**
Peak load	49,100	70.4	80,291	74,8
Mean capacity reserve	46.1%		37.3%	

Table 2 shows values of installed capacities in 1989 and assumed for 1999, for the Mercosur area.

TAB. 3 **CONFIRMED RESERVES IN MERCOSUR****

Country	Oil 10^3 bar	Nat.gas 10^9 m3	Coal 10^6 ton	Hydro MW
Argentina	2.168	744	684	44,500
Brasil	2,760	116	10,215	106,500
Paraguay	-------	-----	--------	25,000
Uruguay	-------	-----	--------	2,700
Total	4,928	860	10,899	178,700

Table 3 shows the energy reserves as known by 1989 for the Mercosur area.

Having analysed these values, the possible solutions for the electric energy requirements by 2020 seems to be very simple. Nevertheless taking into consideration some environmental aspects, severe doubts immediately arise.

Nowadays, the southern cone of the American continent is a very important 02 regeneration area, as the balance between the emitted CO2 by all processes and the regeneration by green areas and plankton in the adjacent oceans is strongly positive. The world community can not permit that this condition should be inverted. The world needs that this vast region remains being a sort of lung for mankind. Seen from this point of view, an increase of CO2 emission due to the burning of more fossils or a disminution of the O2 recuperation as a result of covering green areas in the environmentally fragile Amazonas and La Plata basins caused by new dams and water reservoirs, would be equivalent to suicide and therefore could not be accepted.

Fig. 1 gives a resume of possibilities due to wind energy in Patagonia.

The emplacement of a small wind farm in Río Mayo town can be seen in this figure. In 1990 this farm, integrated by 4 Aeroman 12.5 (30kW each) totalling 120kW, started generating electricity into the local grid served by 1.3MW diesel groups. Table 4 resumes the values obtained during the first year of operation.

TAB. 4 **RIO MAYO WIND FARM****

Capacity:	**Isolated Local Grid**
4 x Aeroman 12.5: 30 kW each	1 x Preexisting Diesel: 1,3 MW

Mean values for period February 1990 - February 1991

Capacity Factor	32.2%
Total Production	332.1 MWh
Availability	87,5%
mean wind speed	8.0 m/s resulting penetration 15.5% of total energy demand
Max. mean wind speed (Nov.1990)	10,9 m/s.resulting penetration 30,0% of total energy demand

Fig.1 **SOUTHERN MERCOSUR AREA**
HYDRO AND WIND POWER IN PATAGONIA

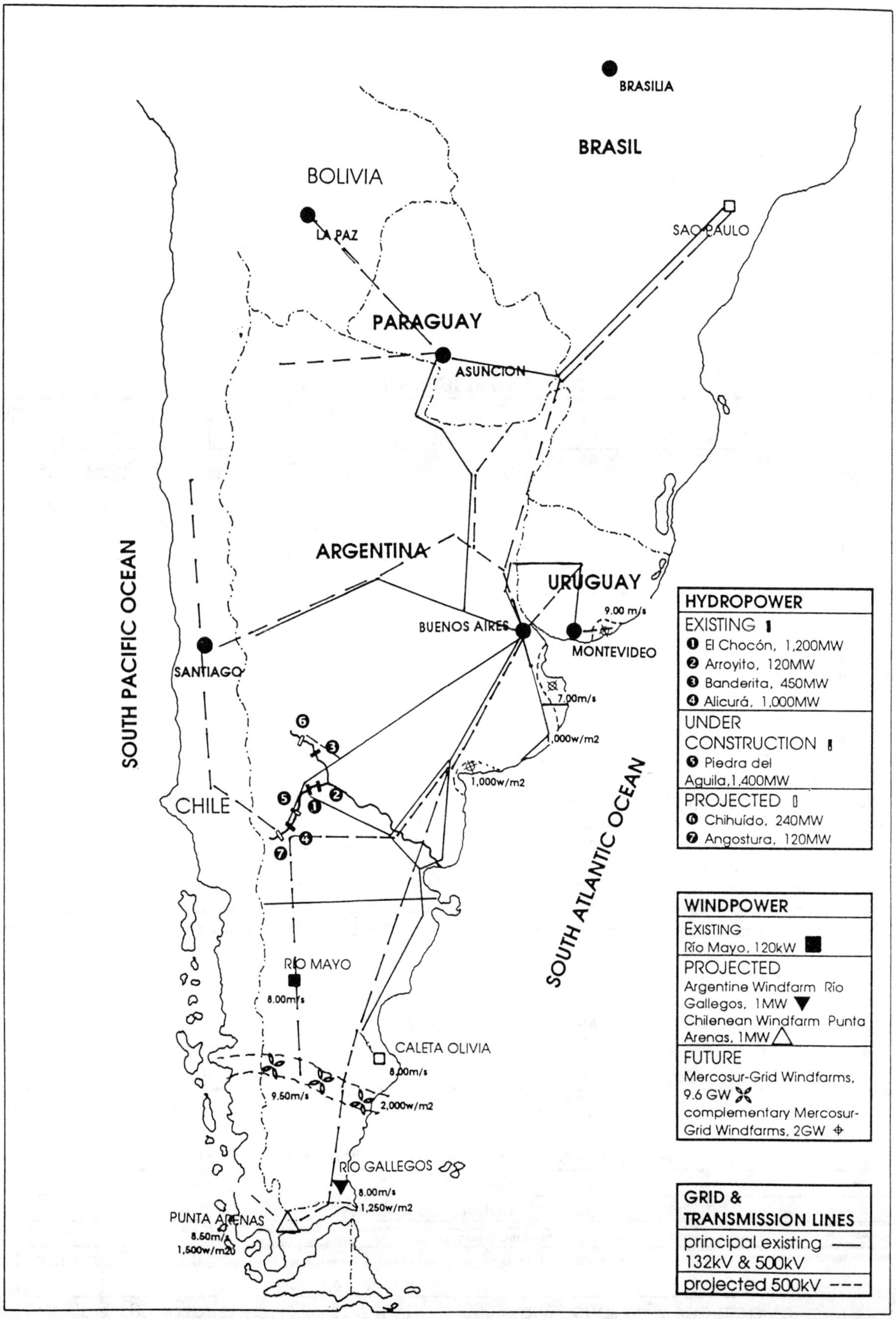

As by these figures, the amortisation of the inverted capital within 10 to 12 years seems to be realistic. Unfortunately and due to a lack of logistical support 2 of the 4 engines are out of service so far, but that is just another aspect that must be (and is already) taken into consideration by planning wind farms in Patagonia.

Coming back to Fig. 1 it can be observed that in a band located south of Río Mayo, the available wind energy is quite superior to the existing in that emplacement. It is in this band where we hope to install the large wind farms of the next twenties.

Let us assume three conservative values:

Capacity factor	25%
Availability	90%
Penetration in the grid	15%

We then can affirm that by these figures, the 15% penetration of the total of 126.000GWh per year, requires from the wind farms an electric generation of 18.900GWh per year. With a capacity factor of 25% and an availability of 90%, that means an installed capacity of 9.6GW. All these units working in Patagonia will be complemented by other 2GW located at the Atlantic coast line in Argentina and in Uruguay (see Fig. 1) to compensate losses in all the lines coming up from Patagonia. It must be said that in the present approach, these lines will come up from the south in coincidence with the existing lines transmitting hydropower from central west Patagonia, entering into these lines through existing 132kV lines that will be convertd to 500kV, shown in the same figure, to which the wind farms will be connected following the rules of the art. Together with the Buenos Aires University, several other Argentine Universities are already studying this aspect.

Even if this project cannot be considered nowadays as more than a preliminary study, it appears as being really seducing, because once carried out, we may foresee that in this part of the world, in the year 2020, electricity will be provided 65% by renewable sources (hydro, wind and thermal) and only by 20% fossil and 15% nuclear. These values are much better than the goal proposed by the Basel Manifesto of 33.3% for each.

It also must be pointed out, that the Patagonian market is really promising in the near future for European windmill manufacturers. We hope to establish fluent relations with some of the, in order to permit us to prosecute with our project and make it profitable for all of the parts involved. First step for this will probably be the installation of two planned pilot wind farms, each one with a capacity of 1MW, serving one the local grid of Río Gallegos in Argentina and the other the local grid of Punta Arenas, Chile, after essays in our mentioned test stations.

** Ref: World Bank, OLADE, Mercado Eléctrico, GENCo Series.

APPENDIX

The previous pages showed that with proper technology transfer, our countries would be able to install wind parks in southern Patagonia, with a total power output of some 9.6GW with an additional 2GW in the La Plata estuary. This market is indeed a very promising one for European wind turbine manufacturers, as we cannot realistically hope to be able to build all the required components within our region, even including significant technology transfer.

But we might go still further with our thoughts. In several possible European sites, wind speeds of about 5 m/s are considered to be fair enough. In our proposed Patagonian sites, the mean annual wind speed is over 9 m/s. That is, the same wind turbine in Patagonia would generate six times as much electricity as in Europe. As there are no frontiers for CO2 pollution, let us assume a European company installing wind farms for our grid, and that due to distances and other local conditions, each installed kW costs 50% more than it could cost installed in Europe. Then, for each ECU invested, four times as much electricity would be generated in Patagonia compared to the same investment in Europe. And then, if the EC accredits to the supposed company for its own country only 50% of the avoided CO2 emissions, it would still be twice as much as they could obtain by installing that wind farm in their own country. So, we may conclude that it would be very cost effective for them to join us in our project and start working together right now.

A PROPOSAL TO ASSESSMENT FOR DEVELOPING WIND ENERGY RESOURCES IN HOKKAIDO, JAPAN

K. Okuda
Sapporo Science and Technology College
1-22, W 17, Ohdohri, Chuhohku,
Sapporo, 060, JAPAN

ABSTRACT: Although wind energy resources have been developed in Europe, America and the other countries with wind parks, Japan has not yet developed his wind energy resources extensively as for contributing to "alternative, renewable and clean energy developing", excluding his pilot plants of wind turbines with capacity under 300 kW in Japan [1]. The author proposes the plans to erect many wind parks of small, propeller type wind turbines with capacity under 85 kW in order to contribute to developing "alternative, renewable and clean energy resources" in Hokkaido, Japan. Hokkaido is the northernmost island and one of the districts with prevailing winds in Japan.

1. INTRODUCTION

The author proposes a trial estimation method for developing wind energy resouces to make use of small, propeller type wind turbines in Hokkaido.

The quantity of electric power demand for some cities or towns in Hokkaido might be supplied if a certain number of small wind turbines would be erected in wind parks.

The author will estimate the wind energy by referring the results of the turbine experiments in fluctuating winds in Hokkaido [2].

2. THE REASONS WHY THE AUTHOR SELECT HOKKAIDO AS THE WIND PARK SITES, AND SMALL WIND TURBINES AS THE TURBINE UNITS.

2.1 For wind park sites:

Hokkaido has a broad land, so wind park sites fit for turbines will be found there easily, along the seashore or at the hills a little distant from a seashore.

2.2 For the Typhoon:

The Typhoon (very strong storm) will attack Hokkaido with few probability a year in Japan.

2.3 For the design and production for small type of wind turbines:

The blade root stresses for bending, twisting and vibration of a small turbine blade will be less than that of a larger turbine blade. The turbine blade design for a smaller turbine is easier than that of a larger one.

It is useful that the production cost of one turbine unit will be made lower than that of a unit at present, if the units will be produced by mass production system, and then the number of turbine units in a wind park will be increased or decreased easier according to electric power demand in the wind park district.

The smaller wind turbine unit may have simpler automatic regulation system than that of larger unit for fluctuating winds.

3. THE PROCEDURE OF WIND ENERGY ESTIMATION FOR WIND PARKS

3.1 Selection for wind park sites:

The author will select wind park sites at 18 districts in Hokkaido, differing from prevailing wind force classes: 9.0, 8.3, 5.9, 4.9, 3.9 m/s, mean wind speed a year, as A, B, C, D and E district, with referring to Fig.1, Fig.2 and Fig. 3. The last Fig. will show 22 districts fit for wind park sites, in dotted lines[3].

3.2 Estimation for the turbine output:

According to the various districts, turbine output, blade diameter, and hub height should be calculated. The wind turbine output will be estimated as the following formula in SI units:

$$P = C_P(1/2)\,\rho\,\pi R^2 V^3 \times 10^{-3} \quad [kW] \qquad (1)$$

where P: wind turbine output, [kW]
C_P: power factor, C_P is supposed as 0.3 from the results of experiments [2].
ρ : air density, ρ = 1.25 [kg/m³], at 10°C, R: rotor radius [m],
V: wind speed [m/s],

Putting constants into factors, the above formula will be as follows:

$$P = 0.5890 \times 10^{-3} \times R^2 V^3 \quad [kW] \qquad (2)$$

3.3 The results of estimation

Table 1 shows PLAN I, the first step for erecting pilot wind parks. The outputs of a turbine unit are equal to 25 kW or less than the former, to aim to have more number of units for smaller turbine units.

Table 2 shows PLAN II, the second step for erecting pilot wind parks. The outputs of a unit are the same as those of PLAN I, to aim to have more number of units for larger turbine units.

Table 3 shows PLAN III, the full-scale step for erecting wind parks, with unit capacity under 85 kW. The site districts should be restricted A, B and C, the numbers of districts are doubled those of PLAN II.

3.4 Total available capacity of the turbines:

According PLAN I, -II and -III,the author supposes number of units per a district, number of districts, total wind turbine units and the unit capacities. Mutiplied them, total capacity A [kW] of the wind parks will be derived as in Table 1 -2 and -3. Wind park efficiency (η) will be taken as follows [2]:

$$\eta = \alpha \times \beta \qquad (3)$$

where α : availability of a wind turbine
β : capacity factor of a wind turbine

Then, available capacity $A\eta$ [kW] and available capacity a year $B = A\eta \times 8760$ [kWh] will be derived as in the Tables.

Table 1. PLAN I, the first step for erecting pilot wind parks.

District	A	B	C	D	E	Sum Total
Rated wind speed [m/s]	9.0	8.3	5.9	4.9	3.9	
Turbine diameter [m]	15	15	10	10	10	
Hub height [m]	20	20	15	15	15	
Capacity of a unit [kW]	25	20	3	2	1	
No. of units per a district	20	20	40	100	100	
No. of districts	2	3	4	4	5	18
Total wind turbine units	40	60	160	400	500	1,160
Total capacity A [kW]	1,000	1,200	480	800	500	3,980
Wind park efficiency η	0.4	0.4	0.3	0.3	0.2	
Available capacity Aη	400	480	144	240	100	1,364
Available cap. a year B [kWh]	$3,504\times10^3$	$4,205\times10^3$	$1,261\times10^3$	$2,102\times10^3$	876×10^3	$11,948\times10^3$

Table 2. PALN II, the second step for erecting pilot wind parks.

District	A	B	C	D	E	Sum Total
Rated wind speed [m/s]	9.0	8.3	5.9	4.9	3.9	
Turbine diameter [m]	15	15	10	10	10	
Hub height [m]	20	20	15	15	15	
Capacity of a unit [kW]	25	20	3	2	1	
No. of units per a district	250	200	125	100	60	
No. of districts	2	3	4	4	5	18
Total wind turbine units	500	600	500	400	300	2,300
Total capacity A [kW]	12,500	12,000	1,500	800	300	27,100
Wind park efficiency η	0.4	0.4	0.3	0.3	0.2	
Available capacity Aη	5,000	4,800	450	240	60	10,550
Available cap. a year B [kWh]	43800×10^3	42048×10^3	$3,942\times10^3$	$2,102\times10^3$	526×10^3	$92,418\times10^3$

Table 3. PLAN III, the full scale step for erecting wind parks.

District	A	B	C	Sum Total
Rated wind speed [m/s]	9.0	8.3	5.9	
Turbine diameter [m]	28	25	25	
Hub height [m]	30	30	30	
Capacity of a unit [kW]	84	53	19	
No. of units per a district	100	80	80	
No. of districts	4	6	8	18
Total wind turbine units	400	480	640	1,520
Total capacity A [kW]	33,600	25,440	12,160	71,200
Wind park efficiency η	0.4	0.4	0.3	
Available capacity Aη	13,440	10,176	3,648	27,264
Available capacity a year B [kWh]	$117,734\times10^3$	$89,142\times10^3$	$31,956\times10^3$	$238,832\times10^3$

4. DISCUSSION ON THE ESTIMATED RESULTS

Picking up the the data from the above Tables (total turbine units and total capacity a year B [kWh]), Table 4 will be obtained. B divided by total electric power demand of a common dwelling house a year in Japan will be reduced the number of dwelling houses C and populations D of a city or a town.

From the above Table, PLAN II and PLAN III shows that the electric power demand of a city with population of 130 and 348 thousands will be supplied with the wind parks of 2,300 and 1,520 units.

FROM THESE FIGURES, THE DEGREE OF CONTRIBUTION FROM WIND ENERGY RESOURCES TO RURAL ELECTRIC POWER SUPPLY WILL BE CONCEIVED CLEARLY. THIS IS CONCLUSION OF THE PAPER.

Table 4. Total electric power supply to rural cities or towns from wind energy resources.

PLAN	No. of units	Available capacity a year B [kWh]	No. of dwelling houses C	Population D
I	1, 160	$11,948\times10^3$	5, 000	17, 500
II	2, 300	$92,418\times10^3$	38, 500	130, 000
III	1, 520	$238,832\times10^3$	99, 510	348, 000

5. BACKGROUND FOR CONSIDERING THE ASSESSMENTS

5.1 The alternative energy to the fossil fuels:
The wind energy may be one of the natural energy, of course, that will be a part of alternative energy to the fossil fuels, should be researched and developed for descendants of our human beings. The fossil fuels must be spent with savings at present, because those will be limited in few decades, they said.

5.2 The national energy supply problem:
It is hopeful for national energy supply problem, if the wind energy resources could be researched and developed with some contributions (though they would be with few percentage) to national energy demands, although there are many and much hardships, obstacles etc. to erect wind parks in Japan as follows:

(a) Problems of the acquisition or rent for wind turbine sites; the relaxation of national park laws, because there will be many national parks in the districts suitable for wind turbine sites in Japan.

(b) Problems of the acceptance or consensus of rural inhabitants those will be in wind parks.

(c) Problems of the acceptance of national government and rural governments for the wind park erection.

(d) Problems of ecology or social circumstances of the districts : whether the wind parks will hurt or break the district environments or not ? ; the effects of noises, disturbances for electromagnetic wave ; whether birds or animals will be hurt or not ? , and so on.

REFERENCES

[1] H. Matsumiya et al, The research and development of wind turbine generator systems in Japan, Proceedings of ECWEC '90, R 6, Spain, 1990.
[2] K. Okuda et al, Estimation methods for characteristics of a small type of wind turbine, operating in fluctuating winds, Proceedings of ECWEC'88, p. 160, Denmark, 1988.
[3] "Utilization of wind energy", Institute for Material and Energy Resources, Science and Technology Organization; JWEA,1978. (in Japanese.)

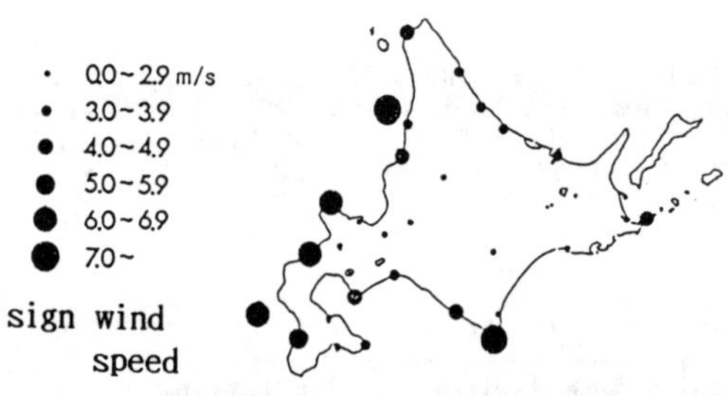

Fig. 1 One example of mean wind speed observed during a year in Hokkaido, Japan.

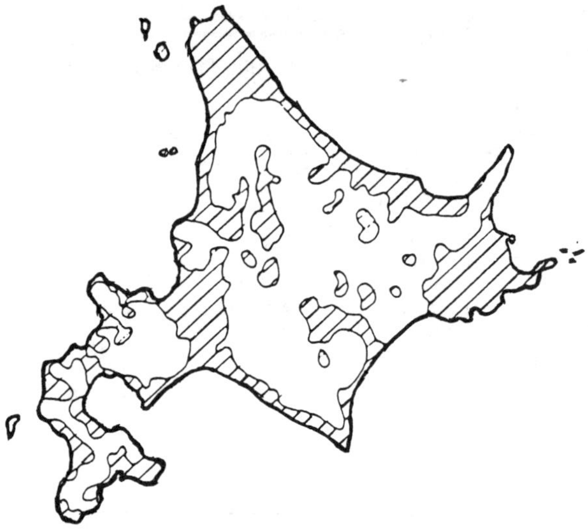

Fig. 2 Prevailing wind districts in Hokkaido

sign ⁄⁄⁄ : districts with mean wind speed a year equal to or over 4 m/s.

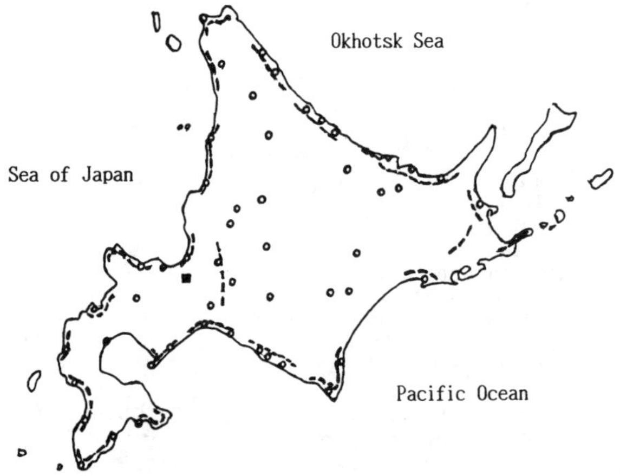

Fig. 3 Wind park districts those will be prearranged in Hokkaido,
sign ■, o : cities and towns.

WIND POTENTIAL AND ECONOMIC EVALUATION OF SOME WIND PROJECTS IN VIETNAM

Prof. Nguyen duc Loc
Hanoi National Institute of Technology
Dai co Viet Road, Hanoi, Vietnam

ABSTRACT

Wind resource had been accessed by using wind data of 206 Meteorological stations in the whole country. The results showed that mean wind speed lie along the coast-line of more than 3000 km long bordered by Pacific Ocean and the Gulf of Thailand. Small scale wind turbines ranging from 0.04 to 0.5 KW for electricity generation and wind pumps for both high and low water heads had been designed, constructed, tested and evaluated by various Institutions, Universities and factories. Much experience in design of windmills have been gained due to severe regime in northern and central parts of the country caused by many strong typhoons coming from the Pacific Ocean.
A standardised methodology of wind assessment is also proposed for establishing wind map or wind Atlas for the countries of Asia and the Pacific.
The tendency of sharp increasing of commercial energy prices in Vietnam and the demand of clean energy for improving environment and living standard of the people in the coastal and island areas make wind energy more and more attractive. With the particularity of low income of the users (up to 200 USD/year), economic evaluation of wind projects application is also introduced for guiding users to choose right scale and type of windmills.

Assessment of Wind Potential

The assessment of wind potential of Vietnam has been carried out since 1981 when a National Programme on New and Renewable Sources of Energy (NRSE) was set up. Wind data have been measured by 206 meteorological stations and processed during the period of 1956-1985. A national study coded 52C.01.01 for wind potential assessing was finished in 1985. Then another study for establishing Wind Energy Handbook of Vietnam was published in 1990. (Including average wind speeds, maximum wind speeds and their frequency, wind density, energy pattern factor, wind flowers, yearly total wind energy, wind mapping, etc...)

The figures gathered show that:

- The average wind speed of Vietnam is not very high but with the coastline of more than 3000 Kms at the Eastern and southern sides bordered by the Pacific Ocean the Gulf of Thailand, these places and islands are considered as good places to lay windmills in. It is not because there is high wind speed there but this due to the higher number of wind hours in a day and higher wind days in a month. As the result, wind energy gathered in these areas is much higher than that in the Northern part (see table 1 and wind map).
- The territory of Vietnam (mainly the North and the Centre) is much affected by annual strong typhoons from the Pacific Ocean. The maximum wind speed typhoons sometimes rises to 40-50 m/sec. The about-mentioned characteristic has caused so many difficulties for vietnamese wind designer. On the other hand they have got much experience which could be exchanged with experts from other countries in the world.

In the future, technical exchange between developed and developing countries on assessment of developable potential of wind energy resources using standardized methodology and wind mapping for the Region should be promoted. To this end a proposal of the methodology of wind power assessment is mentioned in the appendix 2.

In addition, it is a fact that so far in Vietnam there is a shortage of experimental data of wind speeds are very old and obsolete. So the data collected from certain meteorological stations are not reliable and correct enough. The process of treatment of wind data was carried out with rejection of errors and ihomogeneitics caused by subjective and objective reasons.

The Research and Development of Wind Energy Technology

It is noted that concern of vietnamese Government for renewable energy systems in general and wind energy application in particular can be traced back some ten year ago when NRSE National Programme was set up. The Programme had a network of 5 NRSE research Centres (in which 4 Centres of the Ministry of Education and Training and 1 Centre of the Ministry of Energy). Many Universities, research Institutes of 5 Ministries (including Institute of Meteo-Hydrology) were involved in the activities, projects of the Programme.The programme has also a NRSE technology dissemination network at 22 provincial Boards of Science throughout the country.

On wind energy technology development, before the years of 1974 Vietnam has imported three types of windmills: wind turbines WINDCHARGER (200 watts) from USA and SAHOR and OASIS wind pumps. OASIS and WINDCHARGER are considered reliable in the southern part of the country (no typhoons) but due to low average wind speed of the areas, the power of WINDCHARGE is lower than that design power expected. The year from 1978 to 1985 were the time of attempting to design and manufacture locally first models of vietnamese windmills (Salvonius, Sail type DH-6, wind pumps DN-4,B5-10,wind turbine D-6, FD-4). All of them are small scale. So far only a small number of these windmills are still operating. The majority of them were out of work. The reasons were many:

- Poor design.
- Strong typhoons.
- Wind application without wind potential assessment or without any economic analyzes.
- Poor maintenance or without training.
- Bad affect by the hot, salty and humid climate.
- Investment cost (price) is too high compare with low income of the users.
- Subsidy price of electricity 5 years ago (nearly free).
- Lack of know-how and technology transfer.
- Other social factors.

From these above mentioned failures, vietnamese wind experts have got much experience. In additional, through international and regional conferences, seminars, training, courses, post graduated education and information from outside, vietnamese wind experts have learned many useful things and a lot of success has been made.

Wind turbine PH-500 (500 W) of Hanoi Institute of Technology (North), Wind charger PD 170-6 (120W) of Ho chi Minh City Institute of Technology (South), Wind pump D3.2 (D=3.2 m) for high head of water and wind charger 100W of the Institute of Energy (North). Wind pump OASIS of the Collective named "2 nd September", wind pumps BP 380-10, BP 380-8 of the Ho chi Minh City Institute of Technology (South) are considered of good quality and reliability.

From 1993 a new wind project named VIWA funded by GEF, Switzerland-Vietnam Association and ASA Programme of Germany Carl- Dwisberg Company was set up. This is cooperation project between Germany, Switzerland and Vietnam. The project has 3 objectives:

- Improving the technical characteristics of 3 models of windmills: wind turbine 200 KW, wind turbine 150 W and wind pump 2 KW.
- Establishing a demonstration point of wind energy technologies combined with environment(in the South).
- Establishing a production workshop for manufacturing locally small scale windmills (in the North).

It is necessary to point out that Vietnam has passed the most difficulty stage of wind energy technology development. That was the stage of many attempting and failures. But now it is not the time when the mechanism of the technology dissemination works automatically. So financial support from Government and other Internation Organizations

and NGO's is needed for decreasing the price of wind energy units, for demonstration programme, training courses to convince the users for implementing credit and subsidy programme and for promoting local industry to manufacture some windmills components.

International Co-operation

The International Cooperation for Wind Energy Development can be done under the form of technology transfer from developed countries to developing countries as well as under the form of technical and economical exchanges among developing countries (TCDC/ECDC exchanges). These activities could be financially supported by the United Nations.

The process of technology transfer has many stages:

- Feasibility studies: Feasibility studies should be undertaken for assessing for scope of wind power, priority areas and scope for local manufacturing of windmills. These studies could be supported by the United Nations and Government Funds.
- Demonstration Programme: The programme is very important and necessary to persuade and convince the users, to create the market. These activities could be funded through Government Aids, the EC or World Bank.
- Technology transfer: The transfer of wind energy technology can take place through training courses, consultant services, licensed production or joint-ventures with public and private companies in developing countries

Economic Evaluation of Wind Projects

Before 1984 decisions on the installation of windmills in Vietnam were made on the basic of agreements between Institutions and local Authorities without considering carefully wind potential and economic efficiency. As a result, the energy gathered was small, the investment cost was high and economic benefit was negligible. From this fact the NRSE National Programme of Vietnam decided to fund two studies focusing on:

- Economic and Social constraints of the wide-spread utilization of NRSE technologies in general and wind energy in particular.
- To work out a wind energy handbook to help wind experts choose the right areas for windmills installation.

One of the important data of the handbook is the yearly total wind energy over Vietnam territory (KWh/m2). A computer programme for calculating the cost per Kilowatt Hour of the electricity generated by wind turbines was considered as one of the most important parameters for windmills installation decisions. The computer programme shows what the prices of electricity generated by windmills are they are to be installed in different areas of the country. Then we compare the cost of wind energy with the energy generated by diesel of gasoline engines or with the price of electricity of national grid (including transmission and distribution cost) and choose the right appropriate technology. We can see the result on table and figure 2.

Another computer calculating programme has been established to analyze the benefit and cost of wind projects. Three critters are chosen: Net present value (NPV), Internal Rate of Return (IRR) and Payback Period (PBP). Due to the great change of commercial fuel prices, prices of energy equipment and sometimes also due to the discount rate in Vietnam, these figures were given in computer programme as dynamic data and we can input or update them before making decision.

Conclusion

1. The shortage of conventional energy in remote areas, the trend of sharp increasing of commercial fuel price and low living standard in an agricultural country as Vietnam make wind energy more and more attractive.

2. Vietnam's coastline of about 3000 kms and islands are considered good places to lay windmills in. The central and southern parts of the country have better wind potential than that of the North. In areas rich in wind resource, the cost per kilowatt hour of small scale wind turbine is equal to that of diesel generator and is half of the cost of the gasoline one. Of course, the price of electricity generated by medium or large scale wind turbines will be lower and competitive with the commercial electricity price. From this conclusion, medium and large scale turbines should be imported.

3. Due to the low income of individual users in the remote areas, coastal areas and islands, small scale (family size) windmills should be manufactured by vietnamese central or local industries. This is because these units must be of high quality, reliable performance, simple maintenance and must be cheap. In addition, as batteries take account for a large percentage of cost in the components of small scale generators, the quality batteries should be taken into consideration (long life time, high efficiency, good protection by charging controller).

4. For domestic use, the social benefit of wind-generators is considered more important than that the economic one. Social efficiency should be understood as the influence of telecommunication (TV, Radio, Cassette, Video...) on people's life, the effect of electric lighting on their children education and their population planning, the presentation of urbanization etc .... To fill in the gap between the high price of wind sets and the low income of the users, subsidy and credits should be introduced. The cost of these programmes should be financially supported by their Government and other International Organizations. Without these supports low income users cannot afford to have wind sets installed in their homes.

5. Research and development of wind energy technology and cost reductions of small scale windmills should be carried out and promoted through bilateral cooperation between developing and industrialized countries. Feasibility studies (assessment of wind power scope, priority areas and scope for local manufacturing), TCDC exchanges and Wind Energy Demonstration Programme should be supported by the European Community, the United Nations and other International, Governmental or Non-Governmental Organizations.

Reference

[1] WIND ENERGY, *Proceedings of the International New and Renewable Energy Conference INREC'90*, Bejing, China, June 14-18, 1990.

[2] PHAN MY TIEN: *Assessment of wind energy availability in Vietnam*, National study coded 52C.01.01a, Hanoi, Vietnam, 1990.

[3] NGUYEN DUC LOC: *Development of wind energy in Vietnam, Proceedings of ESCAP-Regional Expert Group Meeting on Wind Energy Technology* Nov 28- Dec 2, 1990, Shantou, China.

[4] NGUYEN DUC LOC: *Wind Energy Application in Vietnam, Expert Group Meeting on the Application of Wind Energy for Electricity Generation*, Yogyakarta, Indonesia, 2-6 Dec, 1991

[5] NGUYEN DUC LOC, PHAM VAN HOA: *Economic and Social problems of the widespread application of New and Renewable Sources of Energy*, National study coded 52C.01.02, Hanoi, Vietnam.

Table 1

WIND ENERGY POTENTIAL OF VIETNAM

Stations	I(v = m/s)	II(k)	III(w/m²)	IV(E = KWh/m²)
Lai chau	2.0	3.0	22.5	131.8
Pha din	3.2	3.2	-	751.1
Thai nguyen	2.3	2.5	-	154.3
Lang son	2.7	3.6	-	379.2
Co to	4.4	2.9	-	1,317.9
Bai chay	3.3	2.9	64.0	562.0
Ha noi	2.5	2.5	24.2	212.4
Bach long vi	7.3	2.2	119.0	4,487.0
Nam dinh	3.6	2.5	72.0	631.0
Van ly	4.3	2.3	72.0	933.5
Thanh hoa	2.6	2.9	29.5	259.0
Dong hoi	3.9	3.1	108.6	952.0
Quy nhon	4.1	3.1	106.6	935.0
Cam ranh	4.2	2.7	124.3	1,065.7
Phu quy	6.8	2.1	108.0	3,554.2
Play ku	3.1	4.1	69.6	610.0
Da lat	3.0	4.5	66.2	580.0
Tay ninh	2.4	2.3	-	179.3
Tan son Nhat	3.2	2.9	56.1	492.0
Vung tau	3.9	3.0	101.1	886.0
Soc trang	2.7	4.2	49.2	431.0
Phu quoc	3.7	3.3	97.5	855.0
Rach gia	3.2	2.8	47.7	476.0
Bac lieu	2.8	3.5	-	383.5
Truong sa	6.3	2.1	307.1	2,692.0

Note : I . Yearly average wind speed (m/s).
II . Energy pattern factor.
III. Energy density.
IV . Yearly total wind energy (KWh/m2).

Distribution of yearly total Wind Energy
over Vietnam territory (Kwh/m2)

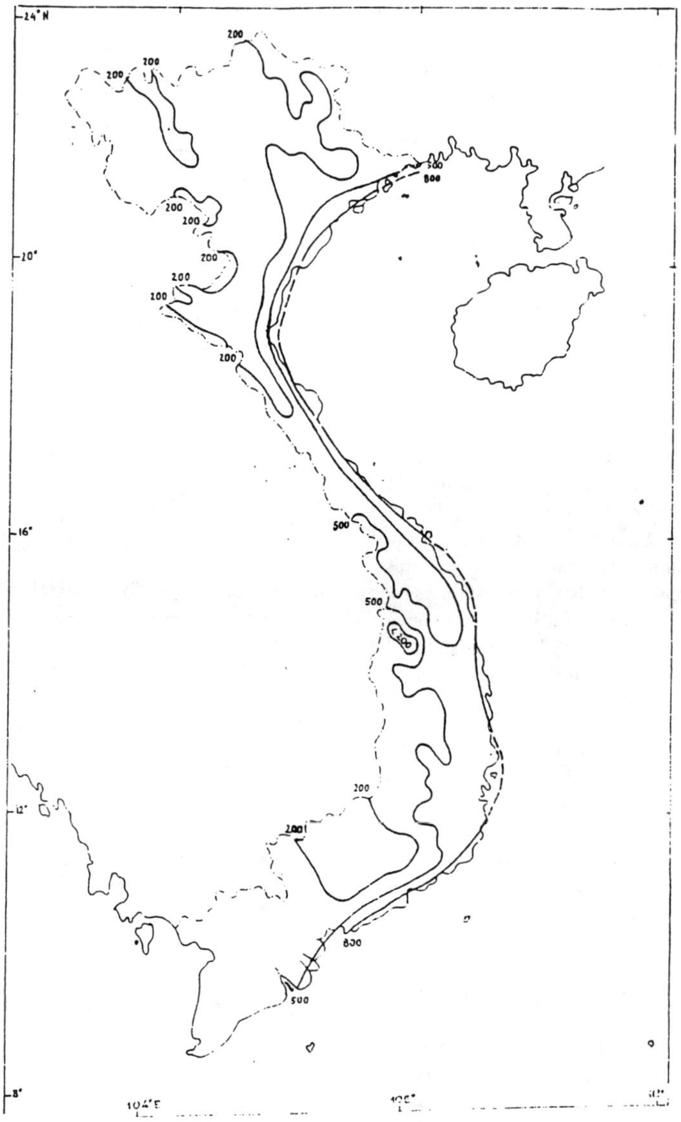

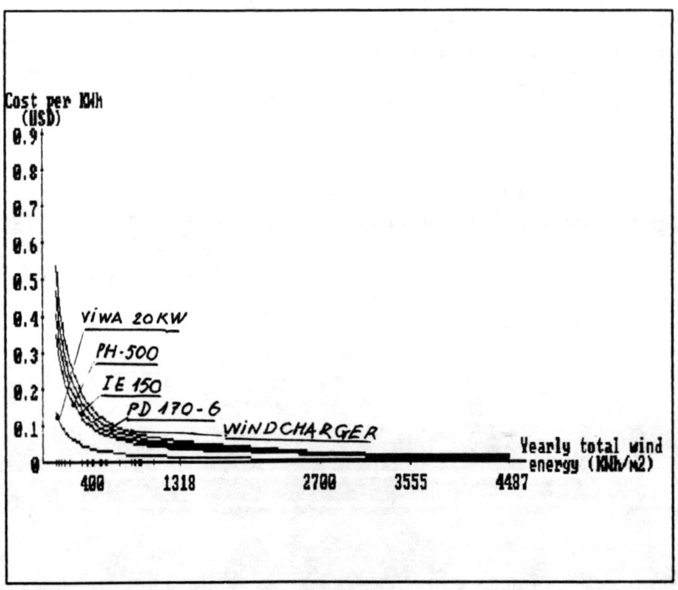

Figure 2

DESIGN, ANALYSIS AND TESTING: METHODS AND PROCEDURES FOR LARGE MACHINES ILLUSTRATED BY THE GAMMA 60.

S.Avolio, A.Aliberti, I.Sebastianelli(*)
G.S.Doman(**), S.Ferrari(***), C.Golia(****)

(*)Alenia - Energy and Environment Division
Via Vitorchiano, 81 - 00189 Roma.
(**)Alenia- WEST consultant - Wind Energy System
Taranto, Via Archimede Zona Industriale 74100 Taranto.
(***)ENEA - Viale Regina Margherita, 125 - 00198 Roma.
(****)University of Naples - 81031 Aversa.

ABSTRACT: The design of Gamma 60 HAWT prototype has been carried out on the basis of clearly defined, aeroelastically determined loads, to assure a useful life to be 30 years and making use of:

- finite elements computing methods for designing the more critical components;
- suitable materials and machining processes;
- laboratory and field tests for experimental checking.

The design was subjected to a continous process of the Design Review; the entire process was carried out during the design, production, assembly, and installation phases of the project. Detailed analytical investigations have been carried on with the support of the main suppliers and developping new calcolations in house too. A specific programme of experimental work for checking the design analysis has been drawn up in order to ascertain that the various sub-systems and components operate correctly and effectively, as well as, to obtain the necessary and/or useful feedback to improve both design and product for the future machines.

GENERAL INTRODUCTION

Technical features of Gamma 60 can be broken down into "basic" characteristics (upwind rotor and teetered hub), which were chosen from among well-proven solutions, and "advanced" characteristics (variable speed and power regulation through yaw control), which represent the innovative aspect of Gamma 60 as compared with other large wind turbine generators [1]. These technical features have been implemented in order to reach two basic objectives: reduction of initial and maintenance costs and increasing of WTG performances (energy capture and fatigue life). Operating speed varies over a broad range because rotor performance can thereby be kept at near-peak conditions, and beacuse more power can be transmitted through the gearbox and generator without increasing torque or generator current. It also allows to increase the fatigue life.

Controlled yaw is used instead of pitch control for aerodynamic torque modulation. As compared with pitch control, this feature makes for a much simpler machine for purposes of both manufacture and maintenance, eliminating all control components from the rotating system, and thus facilitating access for inspection.

The structural design philosophy is that the structural components should be made relatively flexible. The "softness" of the rotor, tower, and drive system has the effect of attenuating both the harmonic and the transient stresses in the machine. This makes for reduction in weight, while retaining long, conservative fatigue life.

1. INTRODUCTION

Gamma 60 design is made up of concepts, construction methods and materials all of which have been favourably utilized in prior projects by Hamilton Standard, or have been evaluated in special tests. While the power electronic frequency converter system is new to large wind turbines, it is actually state of the art production hardware that is well proven in the form of variable speed industrial drive motor systems. The result is a highly advanced system that has been analyzed with great throughness in very complete computer simulations. The analytical tools and methods used were very successfull; also they are well proven and among the very best in existance.

2. DESIGN METHODOLOGY

Gamma 60 is designed by respecting the concept to realize an intrinsecally safe machine; this means that in each of the load cases any failure doesn't have any catastrofic effects towards the external environment. That objective was achieved by doing the following stages:

- definition of the modes /phases of operating of the system (e.g., start up, stopping, emergency, etc.);
- identification for each mode/phase of critical components and parameters for purpose of the correct operation and control of the unit;
- identification of the critical areas using F.M.E.A. (Failure Mode Effects Analysis) methodologies;
- comparative checking by estimating the reliability of the various solutions proposed and by pinpointing redundancy needs.

The following safety features were adopted (Fig.1):

- an emergency system completely independent of the control logic,
- redundancy of the actuating oleodynamic system and of the relevant critical components,
- redundancy of the sensors dedicated to the recording of the quantities used by the regulator.

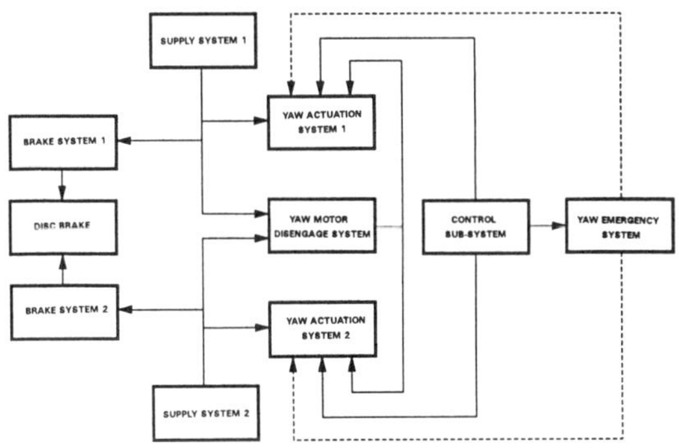

Fig.1 - Blocks diagram of the safety features

The rotor contributes to the safety and structural integrity of the entire wind turbine generator thanks to the philosophy that, in rotor structure and drive system, "softness in goodness". That concept has been applied with limits dictated by knowledge of the sources of instability that can occur. Thus the blades retain high torsional stiffness and high in-plane bending stiffness to guard against flutter and mechanical instability of the coupled rotor and flexible tower system. Except for those two aspects of rotor stiffness, the soft system concept was applied encouraging flexibility and allow modal natural frequencies to fall where they may and assuming that any instabilities can be corrected by the introduction of damping (i.e. mechanical damping of nacelle yaw motion and electrical damping of shaft torsional activity).

The implementation of the above statements together with the target to get high availability, involved management and technical resources in the Design Review process. The methodology employed included, briefly, the following:

- careful evaluation of design specifications,
- specific attention to interface problems (both physical and operating) as between the various sub-systems,
- careful selection of suppliers, who were evaluated in terms of both designing and manufactoring capabilities.
- definition and implementation of a specific quality assurance plan,
- definition of a series of basic criteria for maintenance work and checking observance of the same.

The review process involved a heavy presence of internal and external experts from the major italian universities.

3. ANALYSIS

The system has been analyzed in the normal and extreme operating conditions, considering the fatigue driving loads and the limit loads, respectively (Fig.2).

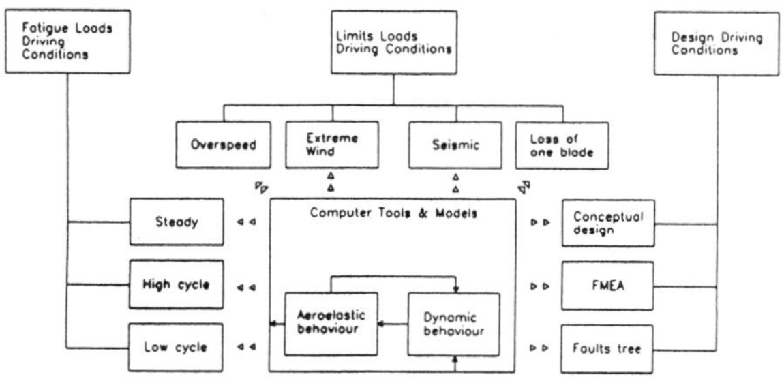

Fig.2 - Checklist for loads and operating conditions

The final design loads have been established for the main "design driving load cases", and the response of the simulations to the envelope of external and internal conditions was examined. Structural loads were determined by considering the activity of the control system. The stress due to the limit loads does not exceed the yield point for the materials so the "no damage" criterion is always satisfied.

Fatigue loads driving conditions was established by taking a very conservative 2 sigma Fichtl up gust; it creates the greatest response because gusts of large magnitude are of longer period than 15 seconds, so they are easily attenuated by the control system.

The Gamma 60 system with the physical properties of the above features concerning the subsystem components was modeled in a complete system dynamic simulation model using the F 762 code qualified by H.S. both in its analytical roots and by experimental verification in WTS-4. Simulation was then run to explore both stability and loads. The modes of instability have been investigated by carrying on an important and extensive qualification activity of GAROS code which is based on a combination of the finite element method with a component mode coupling technique of the rotor and tower substructures. Instabilities were eliminated by suitable adjustment of the natural frequency of the nacelle yaw mode, and by eliminating a delta three angle in the teeter hinge.

The Campbell plot of Figure 3 shows the predicted natural frequencies for tower and rotor uncoupled.

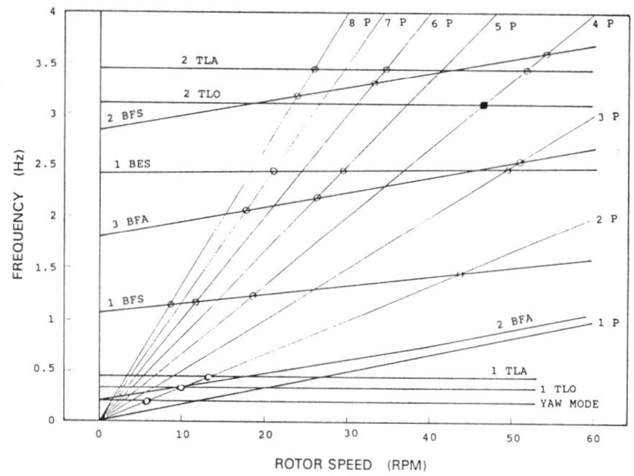

Fig.3 - Blade and tower uncoupled frequencies

It is notable that the first symmetrical out-of-plane mode (1BFS) of the rotor is resonant at 45 rpm at 2P and at 18 rpm at 4P. That condition does not create important stress because, at small amplitudes, the aerodynamic damping arrests the growth of vibratory bending.

Blade inserts are one of the most important components of the blade attachment; the step-tapered, hollow, steel components are bonded to the fiberglass blade root and fastened to the hub with 43 studs. In order to achieve maximum uniformity of the shear stress distribution in the bond joint between the steel insert and the fiberglass, the insert is designed with incremental reduction in outside diameter and wall thickness. The bonded joint has been analyzed by 3D finite element programmme, already tested in prior projects (WTS-3 and 4). Allowable stress in the adhesive bond joint beetwen the inserts and the fiberglass buildup are obtained after fatigue tests performed on several 1/2 scale models. That investigation confirmed the importance of minimizing manufacturing tolerances. The stress results are plotted on Goodman diagram (Fig.4) in which a substantial design margins exist for 2 sigma gust high cycle fatigue endurance case, as well as, for the 85 rpm overspeed load case.

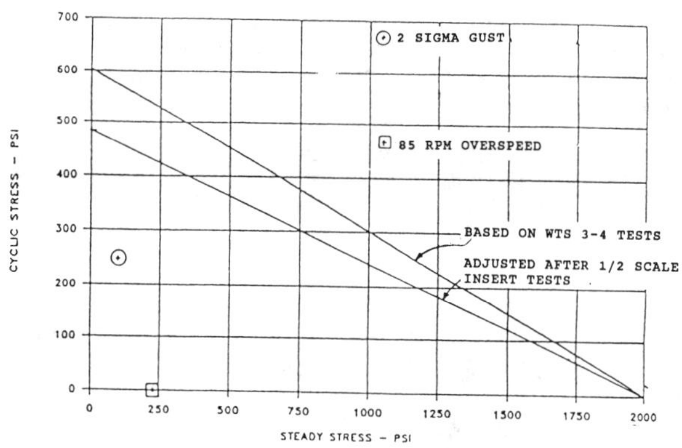

Fig.4 - Blade insert bond stress results

Figures 5 and 6 present the most significant results of the fatigue analysis on large 3D finite element models of hub and shaft with reference to the part that interfaces the hub (hammerhead). Furthrmore the portion of the shaft beetwen the axial bearing and the disc brake have been modelled with NASTRAN using the solid elements, Hexa an Penta type.

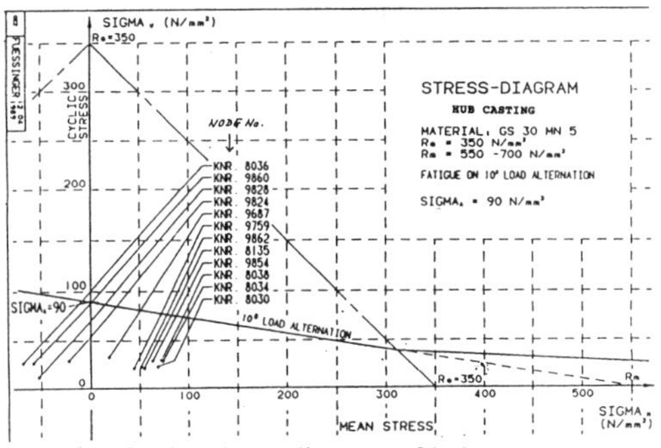

Fig. 5 - Goodman diagram of hub

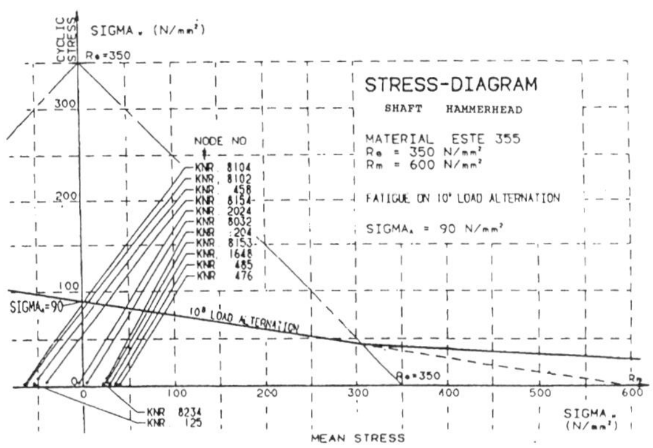

Fig. 6 - Goodman diagram of shaft hammerhead

4. EXPERIMENTAL OPERATION

While not all strain gages have as yet been read and evaluated, the experimental campaign has permitted to evaluate the following loads and stresses [2]:

- drive system torque,
- rotor blade bending,
- tower bending,
- hydraulic system pressure.

The drive system torque is held constant at or below the nominal rated (main rotor shaft) value of 380 KNm by the action of the regulator. Superimposed upon that value is a torque variation proportional to the acceleration and decelleration of the rotor inertia in the generator. Both of these values are firmly determinable and confirmed by test data.

4.1 Rotor blade stresses

Many sensors have been installed on the Gamma 60 rotor blades; calibration of the bridges was made by using the blade's own weight for chordwise and thrust loading for flatwise. The operating load conditions can be clearly viewed as compared to the design limit conditions in Fig.7 for the root blade (R=3.5m).

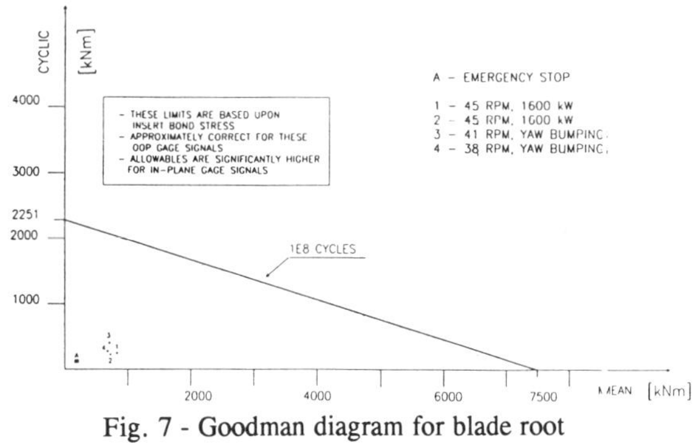

Fig. 7 - Goodman diagram for blade root

The hub design is extremely conservative and loads would have to be many times higher than design values to become in any way critical. Hub loads are reactions to blade loads so that the hub is secure if the blade loads at the design analysis locations on the blade are as predicted. Out of plane bending is of most interest because the strengths are lower than in plane for both the blade and for the attachment bolts. Also there are flatwise natural frequency crossings at both 2P and 4P that occur in the operating range. The allowable loads shown in Fig. 7 are for the insert bonds of the blade root attachment bolt system; the blade spar and load allowables are much higher. Although rotor stress will continue to be recorded and studied it is very clear that rotor fatigue life will be theoretically infinite. The most important resonance crossing occur at about 44 rpm so that its effects are included in Fig.7.

4.2 Tower stresses

The main section of the tower welded on site has been instrumented for bending measurements and for fatigue cycles evaluation. Tower stresses are not greatly affected by the relatively high frequency loads that come from rotor dynamic. The first bending modes of the tower (1TLO and 1TLA) are damped by the rotor aerodynamics and by the viscous damping in the yaw action system. Only the second longitudinal mode is significant as a resonance crossing because it is resonant at 4P about 46 rpm. To date however the 4P vibratory stress levels at this crossing have only been +/-2 N/mm² suggesting that a frequency avoidance control feature can be eliminated. Plotting the continous operating stresses as seen in Fig. 8 shows large margins of safety in fatigue. In effect these large margins are the allowance for large, long term cycles that will be examined through rainflow techniques during the post commissioning test program.

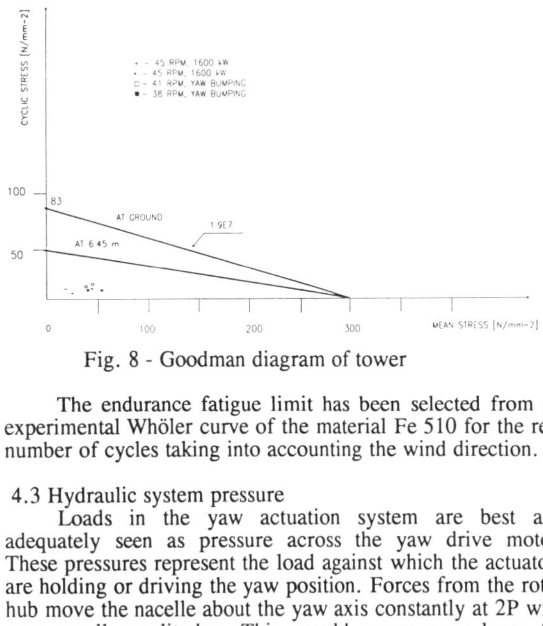

Fig. 8 - Goodman diagram of tower

The endurance fatigue limit has been selected from an experimental Whöler curve of the material Fe 510 for the real number of cycles taking into accounting the wind direction.

4.3 Hydraulic system pressure

Loads in the yaw actuation system are best and adequately seen as pressure across the yaw drive motor. These pressures represent the load against which the actuators are holding or driving the yaw position. Forces from the rotor hub move the nacelle about the yaw axis constantly at 2P with very small amplitudes. This would cause very large 2P torques in the yaw drive and large 2P pressure variation except that the stiffness of the system is greatly reduced by accumulators at each side of each motor. The resulting very small pressure cycles at 2P are then seen as delta pressure across the motor. A small part of the yaw actuation loads comes from the shaft torque component, the nacelle inertia and yaw bearing friction; the maximum required yaw torque per motor would be about 25 KNm which is 19 Bar across the motor. A major contributor is due to the teeter bumpers that are now set to contact with increasing force as teeter amplitude goes above +/-1 Deg. The design torque capacity of the system is 200 KNm per motor. It is clear that the size of the actuation system components can be reduced.

Based upon the available data and with the knowledge of the system behaviour from the studies in system simulation it can concluded that the loads and stress behaviour is excellent and very much as expected. The margins of safety in fatigue are very large as was predicted in the design analysis and the conservatism applied to blade attachment design.

5. CONCLUSION

Test results have confirmed that the control, loads and stability analysis of the Gamma 60 system were correct and successful. Therefor it can be concluded that the system has the large margin of safety that were intended. The extensive analytical effort has been rewarded in that there have been no "surprises" in the test program.

The ongoing experimental program will explore the prospect of future refinement for cost reduction purposes.

6. ACKNOWLEDGEMENTS

The Gamma 60 Project has seen the efforts of various bodies: Alenia-West and Hamilton Standard for System activities and blades; major suppliers included: Escher Wyss (Ravensburg-D and Schio-I) for design and manufacturing of hub, mechanics and actuation; Rexroth (D) for hydraulics; Thyssen (D) for the gearbox; Twiflex (GB) for the brake; Galtarossa (I) for steel structures and civil works; Ansaldo (I) for the electric subsystem; Gavazzi Space (I) for electronic apparata.

The internal team was made by: A.Aliberti, F.Anaclerio, S.Avolio, C.Calò, G.Caturano, F.Cecilia, G.Doman, S.Fugaro, S.Golinelli, P.Lecce, G.Moleti, F.Morabito, M.Musolino, G.Nettis, M. Orlando, C.Palmari, C.Pierdominici, M.Rotondi, I.Sebastianelli, L.Serafini, A.Smarchi.

Thanks go also to all the other people that in different stages have contributed to the success of the program.

7. REFERENCE

[1] S.Avolio, S.Golinelli, G.Moleti, F.Morabito, C.Pierdominici, V.Arcidiacono, S.Corsi, M.Pozzi, G.Botta -"Dynamic analysis of Gamma 60 WTG control system: design and theoretical and experimental validation", 1993 ECWEC Proceedings.

[2] S.Avolio, G.Moleti, F.Morabito - " Supervisory, control and data acquisition: when a prototype needs to have prototype components", 1993 ECWEC Proceedings.

FATIGUE BEHAVIOUR OF FIBRE GLASS WIND TURBINE BLADE MATERIAL

P.A. Joosse
Stork Product Engineering B.V.
P.O. Box 379, 1000 AJ Amsterdam, the Netherlands
D.R.V. van Delft
Delft University of Technology, Stevin laboratory
P.O. Box 5048, 2600 GA Delft, the Netherlands

The Dutch research program on fatigue of glass fibre reinforced polyester (GRP) is near completion. A review of the generated test data is given. An introduction to present work, both tests and evaluation is presented. The main task now is to establish a master curve for the fatigue behaviour of GRP. This master curve should give an accurate, general purpose formulation to be used for wind turbine rotor blades. The requirements of the formulation are discussed and two formulations are compared with available data. The formulation based on a normalized stress range gives a safe representation of the available test data.

1 INTRODUCTION

Fatigue is a major design driver for wind turbines. Nowadays wind turbines are becoming more cost-competitive, with increasing rated power and diameter and therefore increasing investment costs. This calls for realistic fatigue design lines based on thoroughly tested fatigue behaviour.

Most of the wind turbines have rotor blades made of glass fibre reinforced plastics. The fatigue behaviour of these materials has been the subject of research for several years, carried out in a number of European countries. Although a lot of test data is available, the debate about the fatigue formulation itself has started only recently. The consequences of the formulation can be large: different interpretations of the same set of fatigue data can have an effect of 30% on blade mass [2].

In the design-process of the first Dutch MegaWatt wind turbine, it became clear that a realistic fatigue formulation was not available. Therefore a national fatigue research program was started in the Netherlands in 1984 to achieve a general purpose fatigue curve for the materials and design details as used by industry.

In this paper we will focus on the formulations for fatigue life prediction and compare these to the available data.

The research is carried out by a working group consisting of research institutes and manufacturers. The institutes are ECN, NLR, Delft University of Technology and N.V. KEMA. The manufacturers/designers are Aerpac, Polymarin, NedWind and SPE. During the last years there has been an interaction with the EC-program, leading to combined research and presentations (e.g. [4]).

2 REVIEW OF THE PROGRAM

2.1 Purpose of the program

The purpose of the program is to give a master curve for the fatigue behaviour of glass reinforced polyester (GRP) in laboratory conditions. If possible the effect of parameters like humidity, temperature, matrix (epoxy) and fibre material (carbon) and clamping pressure (on bolted joints) will be given separately. As this curve will be used in design and certification, only materials and details as used by the manufacturers are tested.

For the present phase the main objective is to establish this curve. For this purpose available data was reviewed, new tests were defined and a statistical interpretation of the test data was started.

The second topic of the present phase is to find a cost effective blade root connection. In the past attention was paid to several joint types, of which bolted joints proved to be the best. In a new research project this joint will be compared to alternatives like studs etc. After reviewing the potential joints, the 'most promising' joint will be selected and tested. The new research into alternative blade root connections (part of an EC project) will not be discussed here.

2.2 Subjects of the program

Within the research program flat coupons and joints are tested. The flat coupons are used to find basic material data for laminates, typical for rotor blades. The laminates are symmetrical and laminated of non-woven material. Most of the laminates are made of glass reinforced polyester, a limited number of carbon/glass hybrids is tested.

A significant part of the research deals with joining glass/polyester to steel. Different joining techniques were evaluated, e.g. bonding, bolting and a Hütter-type joint. To compare the small joint specimens with a real product a few full-scale blade root connections are tested.

Since the loading of a wind turbine rotor is variable in nature, both constant amplitude (c.a.) and variable amplitude (WISPER) tests are carried out. The WISPER spectrum gives an indication of the effect of varying loads on the degradation.

2.3 Summary of the test results

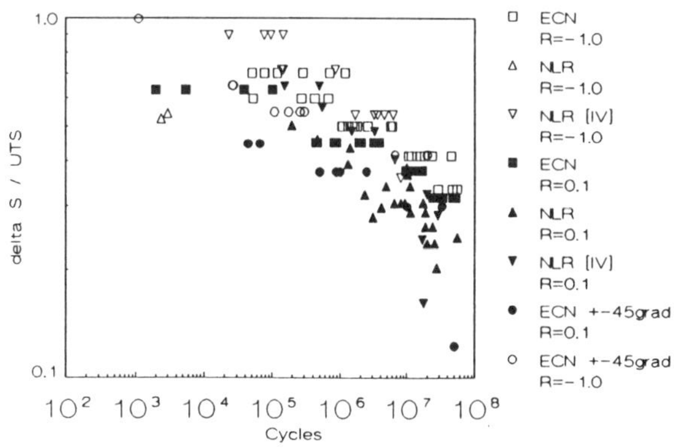

Figure 1: Normalized stress range plotted against the number of cycles to failure, for 0° and ±45°.

As a result of the fatigue research programs a vast amount of fatigue test data is available. Test results are presented in fig. 1 for glass-polyester laminates with a lay-up of

0° or (±45°)$_s$, for R=0.1 and R=-1 (R being the ratio between the minimum and maximum stress in the fatigue cycle). Test results are normalized with the static tensile strength.

Most rotor blades have a lay-up using a combination of 0° and ±45° layers. It is expected that the fatigue behaviour of this combination can be calculated from the fatigue curves of the base materials, using some sort of rule-of-mixtures.

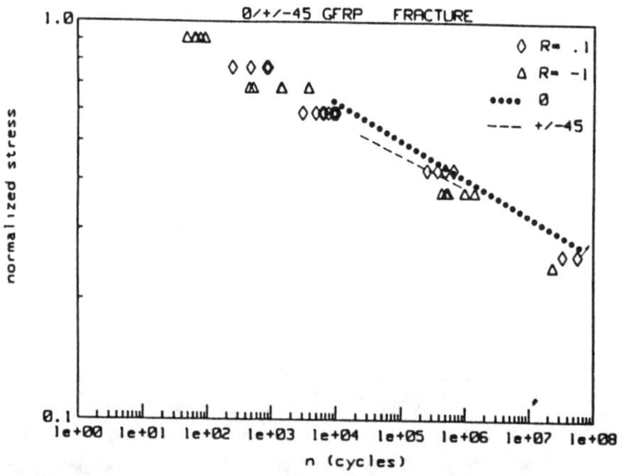

Figure 2: Results for three glass/polyester laminates: 0°, 45° and 0°/±45° lay-ups [1].

Results as given in fig. 2 of c.a. fatigue tests confirm this prediction for a symmetric lay-up consisting of 50% 0° and 50% ±45°. WISPER results of the 0/±45 laminate are a different story (fig. 3). One WISPER sequence is roughly the equivalent of 2 months, so we are interested in the region from 100 to 200 WISPER passes. In this region the 0° laminate has a higher allowable stress, which seems to contradict the constant amplitude results.

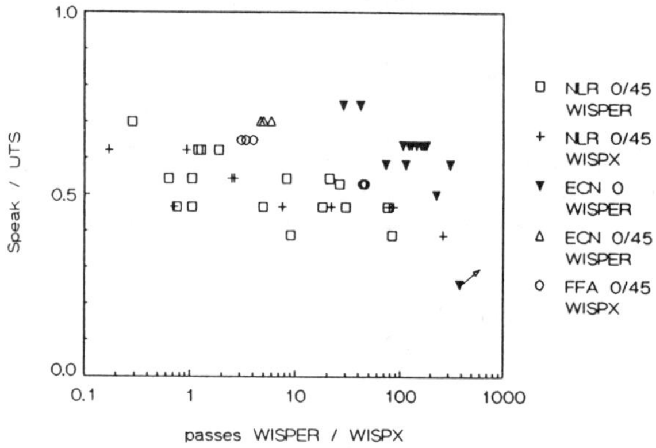

Figure 3: Comparison of WISPER results for 0° and 0°/±45° laminates.

The influence of the matrix material seems to be marginal: fatigue results of glass/epoxy and glass/polyester fit the same curve. The fatigue behaviour of a hybrid material (glass/car-bon on a 50-50 basis) with polyester was not significantly better than glass with polyester [1].

When the fatigue results of the lateral constrained joints are compared to the results of the flat coupons they coincide very well for ordinary c.a. tests.

The effect of moisture is expected to be more severe for bolted specimens than for flat coupons. To quantify this degradation bolted specimens were pre-conditioned in water. Lifetimes for c.a. loading are typically one order shorter than 'dry' specimens. At present only 3 specimens are tested with WISPER, but the lifetimes are similar to the dry specimens. Moisture possibly has less effect on the low-load region of the spectrum. Additional tests will be carried out in the future.

The limited number of tests on pre-stressed bolted joints proved this joint to be very efficient [6].

2.4 Present work

The large scatter in fatigue test results allows for a wide variety of empirical fatigue curves. To gain information in the high-cycle region the so-called Ultra-High Cycle fatigue tests (500 million cycles) are carried out. These UHC-tests started in 1991 and have a typical testing time of 1.5 year. The tests are carried out on 0/±45° laminates for R=-1 and R=0.1. For the UHC-tests two or four specimens are tested simultaneously in each dedicated test system to reduce the test period. At this moment only run-out results (depicted in figure 5 as 'TUD 0/45/-45') are available. For R=0.1 four run-out results are presented at $1*10^8$ cycles and four for R=-1 with cycles up to $3*10^8$. The stiffness reduction for all specimens is very small, indicating the specimens are still far from failure.

The influence of the mean stress level is known for only a few R-values. Tests will be carried out for three more R-values in the near future, especially in the compression-compression region.

Present WISPER-results show a rapid decrease in life cycles at lower WISPER peak values. This could mean that the well-known Palmgren-Miner damage accumulation law is too optimistic for the low-load/high-cycle area. Since rotor blades are designed for this low-load region it is of prime importance to get more information. For such low-load tests a long testing period (several months) is required. These tests will start mid 1993.

The most important activity is to achieve a master curve for glass/polyester. As a first step the data available from previous research (national or EC-projects) will be evaluated statistically. Different fatigue formulations will be compared with the data. The fatigue formulation which reduces the scatter most and gives the best fit will be selected. The statistical analysis and the most-promising fatigue formulation will be discussed with the parties concerned: rotor blade manufacturers, certification institute etc.

Newly generated data from the R-value tests and UHC-tests will be incorporated in the final evaluation, which is expected mid 1994.

3 FATIGUE LIFE FORMULATIONS

3.1 Introduction

The fatigue life (or S-N) formulation gives a relation between an imposed fatigue cycle, defined by a stress or strain and a ratio, and the number of fatigue cycles.

For composite materials theory has not produced fatigue formulations which are generally applicable, despite all the research into damage growth and fracture mechanics. As a consequence the S-N formulations are empirical: interpretations of various test results, sometimes with a micro-mechanical model.

In literature different ways of presenting fatigue data can be found, for example: tension-tension fatigue is commonly described by the maximum stress or strain. The following presentations are possible: stress versus cycles or strain versus cycles, maximum values or ranges, with linear or logarithmic axes. Since in most analyses linear regression to the data is applied to obtain the S-N curves, the different presentations result in different algorithms.

3.2 Requirements to formulations

The fatigue formulation will be used in the design of a wind turbine rotor blade. An accurate, safe description of the fatigue life has to be given for the following conditions:

- life cycles up to 10^8
- varying temperature and moisture
- different combinations of the mean and range of the stress (or strain)

The fatigue formulation has to be valid for materials and design details as used by rotor blade manufacturers. For glass/matrix rotor blades this means:

- low volume fraction
- different lay-ups
- different batches of fibre and matrix
- coupons and details (joints, ply drop-off)

As we are looking for a general formulation it is necessary to normalize the data or to present the data on a strain-basis. At present we use normalization by the static strength, because strain-life curves do not have a good fit.

3.3 Overview of the formulations

At the start of the Dutch fatigue program very few formulations were available, of which the formulation given by Mandell [3] seemed the most promising. Mandell's degradation model gives an adequate approximation for a variety of fibre textiles and matrix materials in tension-tension fatigue. The Mandell equation relates the maximum allowable stress in the fatigue cycle to the lifetime; about 10% of the initial strength is lost every decade. By this it gives a model with which different types of glass/matrix materials and lay-ups can be described.

A draw back of the Mandell formulation is, that the effect of the stress ratio is not accounted for. A modification was therefore proposed, called the Appel-Olthoff formulation (see [4]), to take this ratio into account. The tension-compression data did, however, not follow the Appel-Olthoff curve.

An alternative way to describe the fatigue behaviour is a range versus cycle formulation. This formulation, being common for steel and aluminum, was used for glass/epoxy in the helicopter industry. As was shown in [2] the scatter in the test results was much smaller, for different R values, with this presentation.

From the test data available it was not clear whether the formulation should be based on a linear vs. logarithmic or a double logarithmic scale. The double logarithmic formulation was adopted because it represented the data marginally better and it is commonly used for other materials. (At the same time a test program was started to make verification possible: the UHC-tests.)

It was felt wise to take into account the effect of the cyclic mean stress, with a Goodman-line: a straight line between the ultimate strength and the stress amplitude at a fully reversed stress cycle.

In the present Dutch certification criteria [5] this range-cycle formulation is used. Because of the log-log nature the allowable amplitude for high cycle fatigue is higher than for the Appel-Olthoff formulation. The maximum amplitude is taken at a mean stress equal to the mid of the ultimate stresses.

3.4 Comparison with test results

The present run-out results of the UHC tests are shown in figure 4. Lines are drawn for the log-log and the lin-log case from [6], giving a scatter band of plus and minus twice the standard deviation. The run-out results are in the scatter band of the log-log algorithm suggesting this algorithm to be correct.

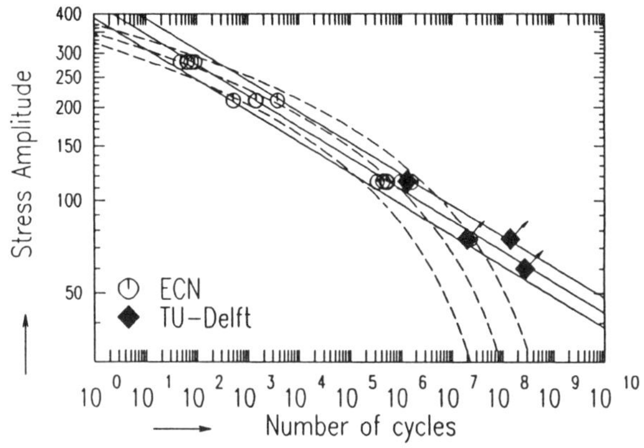

Figure 4: Results of 0/±45° laminate, R=-1, compared to lin-log and log-log algorithm.

The formulations mentioned above, normalized by the tension strength, are compared with the test results for R=.1 and R=-1 in figure 5. For both R-values the results fall within a narrow band, except for a NLR-tested laminate at R=-1. This laminate gives a lower curve with a flat slope. The reason for this is not quite clear, it may be caused by the anti-buckling device.

For tension-tension fatigue (R=0.1) both curves represent the test data well up, to 10^8 cycles. For fully reversed fatigue (tension-compression, R=-1) the Appel-Olthoff formulation is too optimistic for the lower cycle region and too conservative for the high cycle region. The range-type curve gives a kind of lower bound to the results.

Compared to the present (run-out) results of the UHC tests the range-Goodman line seems to be conservative for R=-1 and not too optimistic for R=0.1.

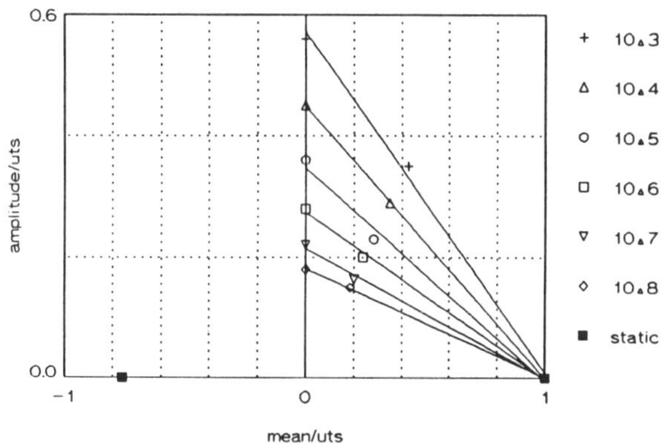

Figure 6: Goodman diagram for the 0°/±45° laminate.

Presenting the results of the 0/±45 laminate in a Goodman diagram, a linear fit proves to be an adequate approximation (fig. 6). At the moment no data is available in the compression-compression part of the diagram. Constant life lines of the 'certification' formulation fall below the 'test lines' of figure 6.

For the available test results the range-Goodman formulation seems to give a better representation than the Appel-Olthoff formulation. As is shown in [2] the range-Goodman formulation allows for a (relative) low-weight design.

4 CONCLUSIONS

For economical design there is a need for an accurate formulation of the fatigue behaviour. Over the past years a vast amount of fatigue test data has been generated on glass/matrix material. Additional tests are still necessary to evaluate the validity of the present fatigue life formulation and to gain more insight in the environmental influence such as moisture. Tests will be carried out for more R-values and in the ultra-high cycle region.

The main activity at present is to establish an accurate, general purpose fatigue life formulation to be used in the design of a wind turbine rotor blade. Different formulations will be compared to the statistically evaluated test data.

The present, provisional, fatigue life formulation is based on the stress range, with a Goodman-type correction for the cyclic mean stress. The stresses are normalized by the static strength. This range-Goodman formulation gives a safe representation of the available test data.

ACKNOWLEDGEMENT

The Dutch national program is sponsored by NOVEM, the Dutch agency for energy and environment, and works in close cooperation with a European program, sponsored by the European Community, DG XII.

REFERENCES

[1] P.W. Bach
Fatigue properties of glass- and glass/carbon-polyester composites for wind turbines
Report ECN-C-92-072, November 1992

[2] D.R.V. van Delft, F. Hagg, P.A. Joosse
The influence of fatigue design line criteria on the rotor blade design
Proceedings ECWEC '90, Madrid, 10-14 Sept. 1990

[3] J.F. Mandell, D.D. Huang, F.J. McGarry
Tensile fatigue performance of glass fiber dominated composites
M.I.T. research report R80-4, Nov. 1980

[4] M. Poppen, P.W. Bach
Fatigue testing using the WISPER sequence
Proceedings EWEC '91, Amsterdam, 14-18 Oct. 1991

[5] L.W.M.M. Rademakers e.a.
Proposal to change the technical criteria with respect to the fatigue properties of fibre reinforced composites (in dutch)
ECN, DE-Memo-91-58, 10 Dec. 1991

[6] R.H.F. Tikkemeijer, P.A. Joosse
Summary of the research program "Fatigue of wind turbine materials"
SPE, Amsterdam, 6 June 1991

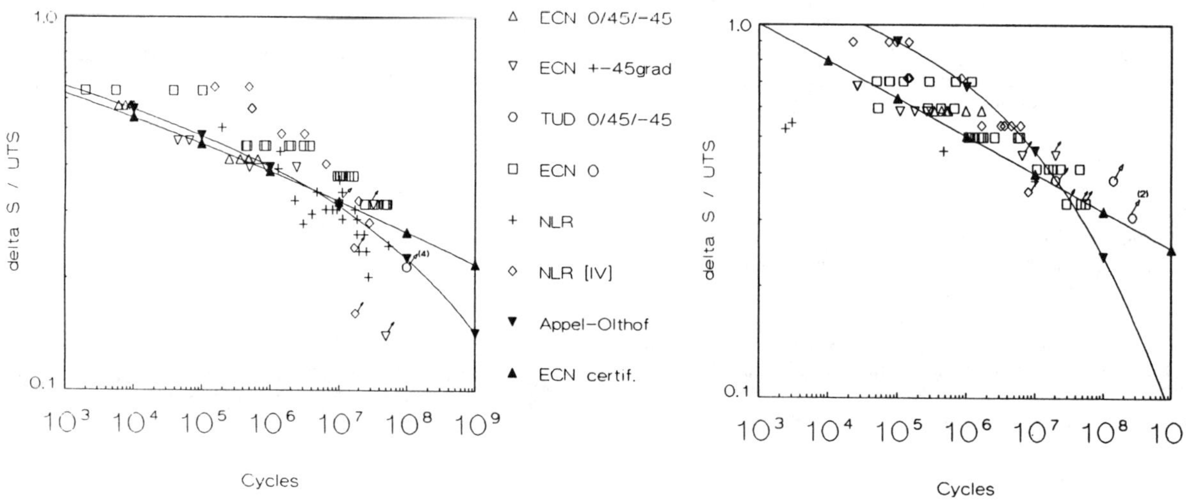

a) tension-tension (R=0.1)

b) tension-compression (R=-1)

Figure 5: Comparison of test results with two fatigue formulations: Appel-Olthoff (Mandell-type) and the present Dutch certification (range-type).

FATIGUE TESTING OF WOOD COMPOSITES FOR AEROGENERATOR BLADES. PART VIII. STATISTICAL TREATMENT OF CONSTANT LIFE DATA FOR DESIGN OPTIMIZATION.

I P Bond, M P Ansell, C L Hacker
School of Materials Science
University of Bath
Bath, Avon
BA2 7AY, UK

ABSTRACT

The safe operation of dynamically loaded components is of vital importance in the design of wind turbine generators (WTGs) and certification procedures and standards are being developed to define safe fatigue loads for components including blades. S-N data has been analysed in the form of N-S data in order to define materials tolerance limits. Constant life diagrams have been created from UCS, UTS and the R=-1 N-S curve to characterise the fatigue performance of Khaya, poplar and beech wood composites and the statistical treatment has been applied to the diagram. Based on this information new wood species, joint configurations or resin systems can be rapidly assessed in fatigue for life prediction purposes.

1 INTRODUCTION

The commercial realisation of wind energy has meant that research now needs to concentrate on optimizing and standardizing the criteria by which wind turbine blades are designed and manufactured.

Critical to the success of a wind turbine is the fatigue design and the resultant performance of rotor blades. The achievement of a good design relies on the use of fatigue data at all R ratios (σ_{min} / σ_{max}). This data can be obtained from a constant life diagram. These diagrams are plots of mean stress versus alternating stress for a whole spectrum of different R ratios and they compare families of curves each representing a constant life. The generation of such diagrams presently requires a large amount of costly fatigue testing of the representative blade material at numerous R ratios to provide S-N plots from which constant life data can be extracted.

If one performs only minimal static and fatigue testing and statistically treats the data to improve its reliability, simplified constant life diagrams can be produced which offer a safe estimate of fatigue life. These diagrams can be subsequently used in the fatigue design and life prediction of blades/blade materials.

The generation of these diagrams is relatively straightforward requiring only monotonic tensile and compressive testing and fatigue testing at R=-1 (reversed loading). This allows the rapid assessment of the effects of changes in blade construction on static and fatigue performance. Likely changes are the use of a new joint geometry, alternative veneer type, alternative veneer thickness or new resin system.

2 EXPERIMENTAL PROCEDURE

Monotonic tensile and compressive data and fatigue data at R=-1 (reversed loading) were gathered by appropriate axial tension/compression testing of representative wood/epoxy/GRP composite samples using a 200kN Mayes servohydraulic test machine. Each sample incorporated factory standard scarf joints in typical blade configurations and were all manufactured at Wind Energy Group's blade production facility in Southampton. Several species of wood were tested, Beech (*Fagus sylvatica*), Poplar (*Populus canadensis* or *Populus serotina*) and Khaya (*Khaya ivorensis*) as part of a parallel initiative to evaluate various wood species as alternatives to the currently used Khaya. (1).

Log cycles to failure data gathered at three or four stress levels (17.5, 20, 22.5 & 25 MPa) were chosen to be statistically treated. N-S plots and simplified, triangulated constant life diagrams could then be derived from the statistically treated data and used as a basis for design.

3 STATISTICAL ANALYSIS

R=-1 fatigue data and ultimate compressive strength data were plotted on the same graph, in the mathematically correct form of 'peak stress magnitude' as abscissa (x) or independent variable and 'log cycles to failure' as ordinate (y) or dependent variable. This was because a linear regression analysis using the '*method of least squares*' was required for log cycles (y) with respect to peak stress (x). This mathematical analysis of the data gives an estimate of the equation that will best predict one variable (y) from another (x). (7, 12).

A histogram of the residuals i.e. difference between actual data point 'log cycles to failure' values and those predicted by the regression line, gave a roughly normal distribution, figure 1. The variance in 'log cycles to failure' values at each 'peak stress level' was found to be an approximately constant value. Both these factors justified this type of regression fit. (12).

The 95% tolerance constraint is effectively a lower limit to the data, above which one can say with 95% confidence that 95% of all future data points will lie. (4, 5, 6). Alternatively it can be described as a limit above which one can say with 95% confidence that there is a 95% survival-probability of any future sample. It takes the form of;

$$y = x - z.\sigma$$

where y is the 95% lower tolerance limit,
x is the mean value of the data set,
σ is the standard deviation of the data set,
and z is a tabulated factor for a one-sided tolerance limit for a normal distribution and is dependent upon the probability constraints applied and the number of data points in the set.

These probability constraints can obviously be chosen to reflect the degree of certainty that is required to be enforced upon the data. The background theory and necessary constants are distributed throughout the literature. (5, 6).

Figure 1 - Distribution of Residuals for Poplar, Khaya and Beech

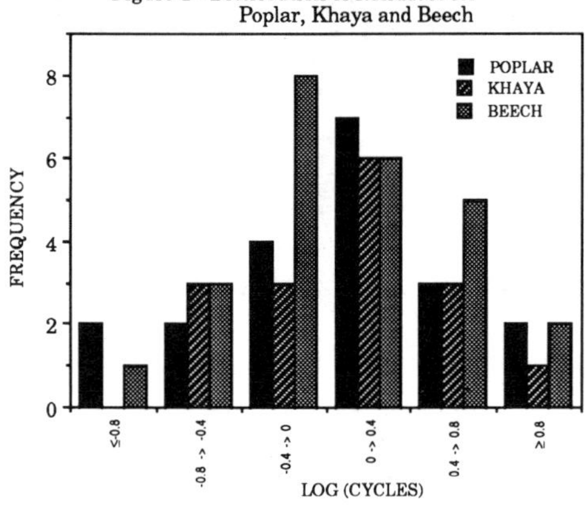

Each set of 'log cycles to failure' data at the various fatigue stress levels were treated to obtain the 95% lower tolerance values. These were also plotted on the S-N graphs described in section 4 and linear regression analysis applied to these points to obtain a 95% lower tolerance boundary. It must be noted that the relatively few data points at each fatigue test stress level, typically five, is the minimum that can be treated using statistical methods. More testing is required and will be carried out to further enhance this work and increase the sample size.

4 RESULTS AND DISCUSSION

The triangulation of constant life lines can be considered a safe and reliable method of simplification by studying figure 2 which shows a comprehensive constant life diagram for unjointed Khaya obtained after extensive axial fatigue testing at numerous R ratios. (2,3). As can be seen the triangulation of the diagram by including only R=-1 and static tensile and compressive data does not cause the simpler shape to pass outside the more comprehensive performance envelope. The application of 95% tolerance constraints further reduces the size of the performance envelope of the simplified diagram, giving an even greater margin of safety for use in design.

Figure 2 - Comprehensive constant life diagram for unjointed Khaya with the imposition of simplified constant life lines.

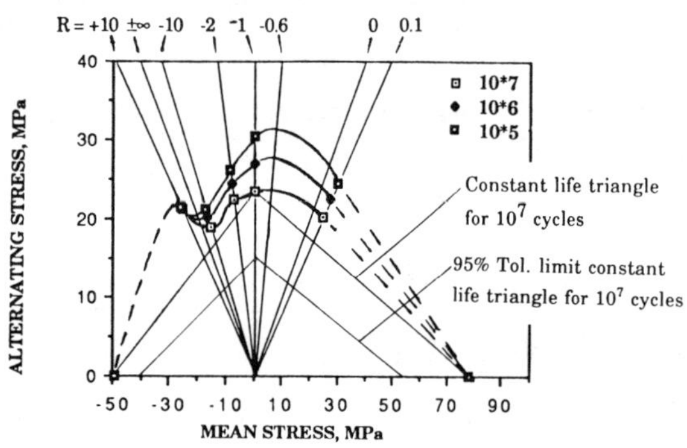

Figures 3, 4 and 5 are N-S plots for the three species tested, Khaya, poplar and beech respectively, and they include fatigue data and monotonic compressive strength data. The compressive data was included because wood, being weaker in compression, suffers most damage under the compressive half-cycle of the fatigue loading, therefore the compressive strength is a limiting factor in the fatigue performance of the wood. Tolerance constraints of 95% were applied to the log cycle and compressive strength data sets to give a lower bound on each of the N-S diagrams above which, with 95% confidence, samples had a survival probability of 95%.

Figures 4 and 5 show the poor fit of the regression line to the 95% tolerance values. The large scatter in the fatigue data (up to 2 orders of magnitude) and the relatively few data points at each stress level makes statistical inference precarious and often causes low tolerance values. i.e. log cycles = 0.96 at 25 MPa in figure 4. This ambiguity will be reduced upon the completion of further testing.

Figures 3 and 4 show a remarkable similarity, implying that there is very little difference in performance between Khaya (density 550-650 kg/m^3) and poplar (density 380-530 kg/m^3) in both static and fatigue, see figure 6. Both gradients and intercepts of the regression equations are in close proximity, the poplar giving a slightly poorer performance, as would be expected of this lower density wood.

Figure 3 - N-S Plot of All scarf Khaya at R=-1

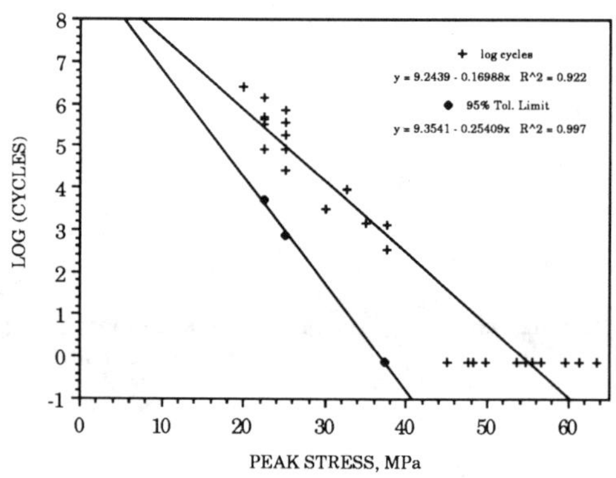

Figure 4 - N-S Diagram for All Scarf Poplar at R=-1

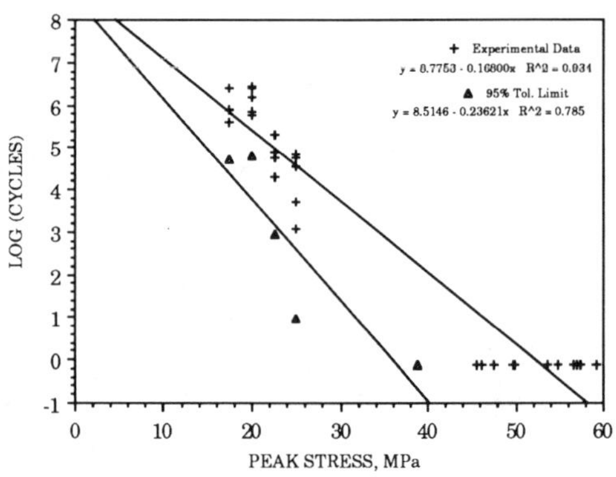

Figure 5 - N-S Plot for All Scarf Beech at R=-1

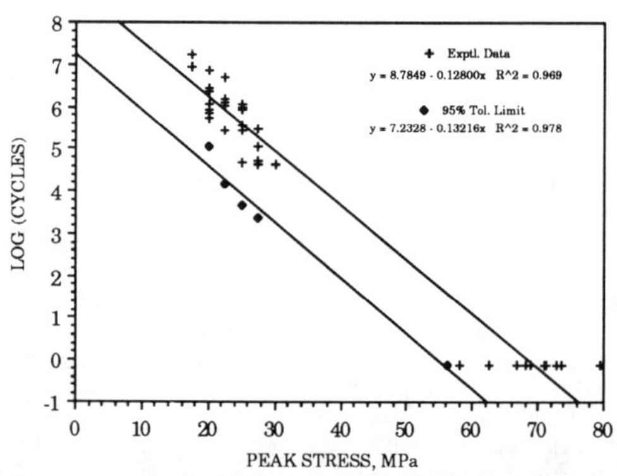

Peak stress values can be obtained for lives of 10^5, 10^6 and 10^7 cycles from the various N-S regression equations. From these values the corresponding mean and alternating stress values can be deduced via the relationship below;

$$R = \sigma_{min} / \sigma_{max} = \sigma_{mean} - \sigma_{alt} / \sigma_{mean} + \sigma_{alt}$$

where R is the R ratio,
σ_{mean} is mean stress,
and σ_{alt} is alternating stress.

This allows the construction of triangulated constant life diagrams for the three types of sample tested, figures 7, 8 and 9. Each diagram includes two sets of constant life lines, those derived from the regression fit to the experimental data and those derived from the regression fit to the 95% lower tolerance boundary. Few data points at each test stress level along with a relatively large degree of scatter in the data has resulted in low 95% tolerance boundaries predicting low stress parameters at lives of 10^5, 10^6 and 10^7 cycles. This is particularly apparent for the scarfed beech (figure 5) where the high cycle constant life lines indicate an exceptionally poor fatigue performance, figure 9.

Figure 6 - N-S Plot of All Scarf Poplar and Khaya at R=-1

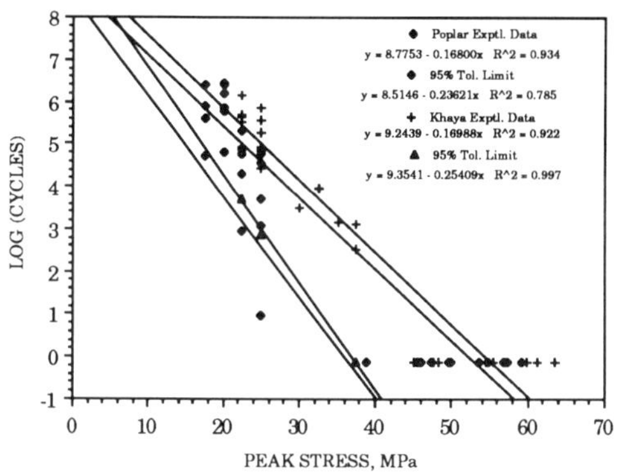

Figure 7 - Constant Life Diagram for All Scarf Khaya

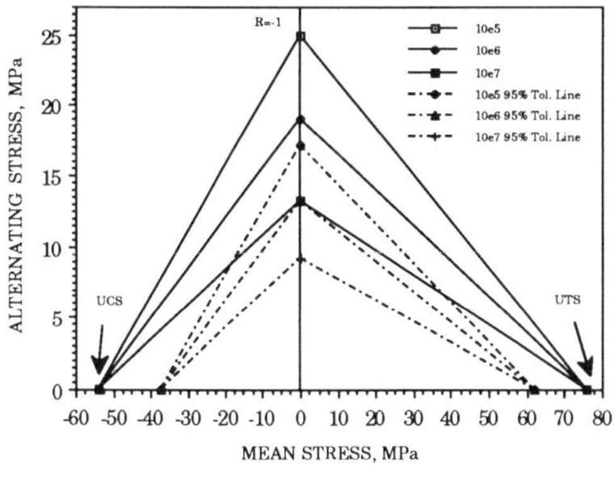

Figure 8 - Constant Life Diagram for All Scarf Poplar

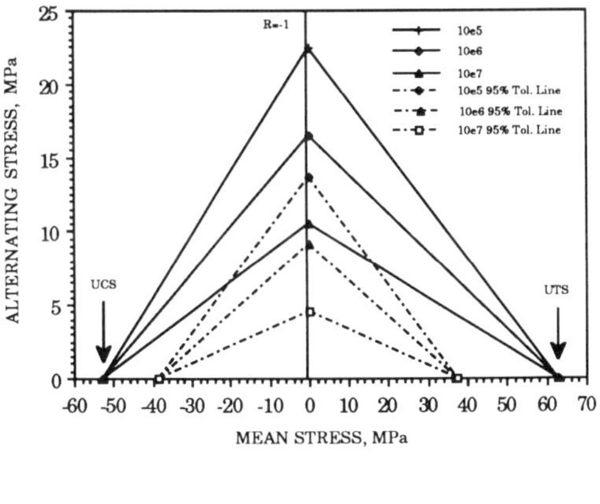

Figure 9 - Constant Life Diagram for All Scarf Beech

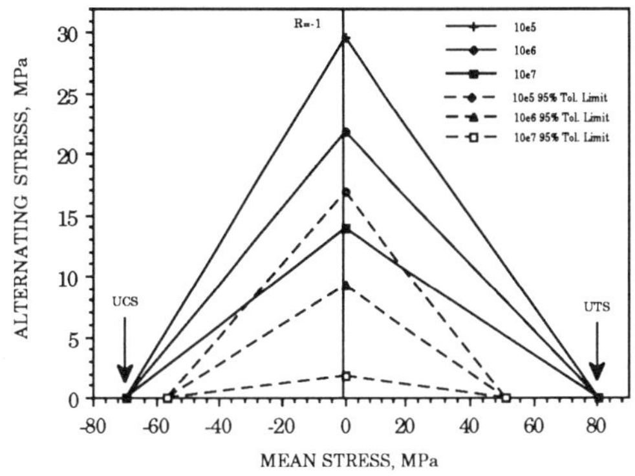

Figure 10 - Constant Life Diagram for All Scarf Khaya and Poplar

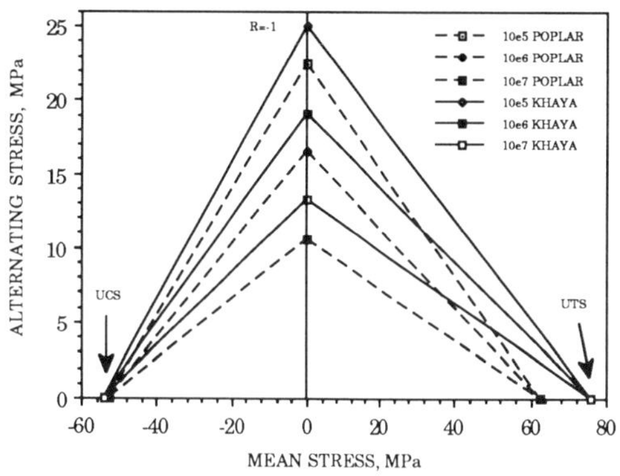

Figure 10 is a comparison of two constant life diagrams for Khaya and poplar. The poplar, although a less dense and cheaper wood, still attains an overall fatigue performance comparable to that of Khaya. This ability to compare various material and design changes after only minimal testing is of great use in the continued development and optimization of blade design.

The fatigue design of wood composite blades in the UK has, in recent years, been based on data and experience gained with *Khaya ivorensis*, a tropical hardwood. Recently, both economic and environmental concerns have led to the need for an alternative temperate species as a substitute. Figure 10 implies that substituting poplar for Khaya, would not be overtly detrimental to blade performance and would considerably reduce cost, both financial and environmental. In fact, recent full scale blade testing has proved this implication to be valid.

5 CONCLUSIONS

a). A test and analysis methodology has been demonstrated that allows a rapid assessment of the fatigue performance of new blade materials. By performing only static tensile and compressive tests and fatigue testing at an R ratio of -1 (reversed loading), a simplified, triangulated constant life diagram can be constructed and used in the fatigue design or life prediction of blades.

b). The requirement for data at only one R ratio and static tensile and compressive data means that a useful constant life diagram can be generated without a comprehensive test programme. This has the two-fold advantage of improving statistical accuracy because many samples can be tested at R=-1, and reduces the time and cost needed to obtain a valuable indication of fatigue performance of new species and the effects of changes in manufacturing methods.

c). The application of 95% lower tolerance limits to N-S curves and their subsequent use in simplified constant life diagrams improves safety margins over and above the simple triangulated procedure and gives a reliable estimate of a material's minimum performance level.

6 ACKNOWLEDGEMENTS

The authors are very grateful to the Department of Trade and Industry for funding this work under Contract No. E/5A/6111/2818. The work has been carried out in collaboration with Wind Energy Group Ltd, (Blade Plant), Southampton, who supplied all sample material. Special thanks to Mark Hancock of WEG for his many thoughts and ideas.

7 REFERENCES

1. Bonfield, P.W., Bond, I.P., Hacker, C.L., Ansell, M.P., "Fatigue testing of wood composites for aerogenerator blades. Part VII. Alternative wood species and joints", *Proc. of the 14th BWEA Wind Energy Conference - Wind Energy Conversion 1992.*, Ed. Clayton, B., Mechanical Engineering Publications., pp.243-250.

2. Bonfield, P.W., Ansell, M.P., "Fatigue properties of wood in tension, compression and shear", *J. Mat. Sci.*, Vol. 26, No. 17, 1991, pp. 4765-4773.

3. Bonfield, P.W., "Fatigue evaluation of wood laminates for the design of wind turbine blades", *PhD Thesis*, University of Bath, Bath. 1991.

4. Little, R.E., "Review of statistical analyses of fatigue life data using one-sided lower statistical tolerance limits", *Statistical Analysis of Fatigue Data. ASTM STP 744.* R.E. Little and J.C. Ekvall, Eds., American Society for Testing and Materials, 1981, pp. 3-23.

5. Natrella, M.G., *Experimental Statistics. Handbook 91*, U.S. Department of Commerce, National Bureau of Standards, U.S. Government Printing Office, Washington, D.C., 1963.

6. Hahn, G.J., "Understanding statistical intervals", *Industrial Engineering*, Vol. 2, Dec. 1970, pp. 45-48.

7. McNatt, J.D., "Linear regression of fatigue data", *Wood Science*, Vol. 11, No. 1, July 1978, pp. 39-41.

8. Wirsching, P.H., Wu, Y.T., "Probabilistic and statistical methods of fatigue analysis and design", *Pressure Vessel and Piping Technology 1985 - A decade of Progress, ASME*, 1985, pp. 793-819.

9. Hwang, W., Han, K.S., "Statistical study of strength and fatigue life of composite materials", *Composites*, Vol.18, No.1, Jan. 1987, pp. 47-53

10. Weibull, W., "A statistical distribution function of wide applicability", *J. Appl. Mech.*, 1951, pp. 293-297.

11. Little, R.E., Jebe, E.H., "Statistical design of fatigue experiments", Applied Science Publishers, 1975.

12. Chatfield, C., "Statistics for technology - A course in applied statistics", 2nd Edn., Chapman and Hall, 1978.

RELIABILITY ANALYSIS IN WIND ENGINEERING

L.W.M.M. Rademakers, A.J. Seebregts and B.A. van den Horn
Netherlands Energy Research Foundation, ECN,
P.O. Box 1, 1755 ZG Petten
The Netherlands

ABSTRACT: Presently, wind turbines are being designed in accordance with deterministic design rules. These rules concern the design of main components and the safety systems. They neither facilitate the quantification of the degree of conservatism in the applied safety margins nor explicitly address the reliability and availability of the wind turbine. By a case study of the Lagerwey LW 15/75 wind turbine design, it has been demonstrated that the wind turbine industry can take advantage of the safety and reliability practices which are frequently applied in e.g. aerospace, nuclear and offshore industry.

1. INTRODUCTION

1.1 Background

To minimize the likelihood of a structural failure of a wind turbine, the deterministic design rules as published for instance in the documents 'Regulations for the Type-certification of Wind Turbines: Technical Criteria' [1] and the IEC TC88 Draft Document 'Safety of Wind Turbine Generator Systems' [2] concern the design of main components, such as the rotor and the tower, and the safety systems. As wind turbines increase in size and power, these rules may not be adequate to ascertain a safe and well-balanced design. The rules neither facilitate the quantification of the degree of conservatism in the applied safety margins nor explicitly address the reliability and availability of the wind turbine. The lessons learnt from operating experience are not systematically used during all phases of the lifetime of a wind turbine, viz. during design, manufacturing, construction, and operation.

The safety and reliability of wind turbines can be enhanced by adoption of techniques from other branches of industry. Therefore, the project 'Probabilistic Safety Assessment (PSA) of Wind Turbines' has been carried out to determine a methodology that takes advantage of existing probabilistic reliability techniques. The project has been carried out under a joint contract of the Commission of the European Communities DG-XII for Science, Research and Development, and NOVEM, the Netherlands Agency for Energy and Environment.

1.2 Project Objectives

The objective of the project is to develop a methodology that can be applied complementary to existing rules. This methodology should enable:

- the wind turbine industry to make a well-balanced design, and
- certifying bodies to assess the safety and to quantify the availability of wind turbines, with due attention to the influence of quality assurance, inspection and maintenance procedures, and the human factor.

The feasibility of applying reliability-based methods in conjunction with existing design and certification rules should be investigated. The extent to which the PSA methods could be used in quantifying the degree of conservatism of the current design regulations and of the prescribed safety margins should be assessed. Considering the state of the art of wind turbine engineering, the shortcomings will be indicated and recommendations for further development will be given. Although the emphasis of the project has been laid on safety aspects, much attention has also been paid to the availability of wind turbines and their sub-systems.

1.3 Working Method

The project has been performed by ECN in co-operation with Lagerwey Windturbine, KEMA, TNO Building and Construction, and Stork Product Engineering. A review committee has been consulted to obtain relevant comment from the scientific and technical community. The project has been divided into three subsequent phases:

Pase 1. Survey and Comparison of Methods for Safety and Reliability Analysis

This survey provides the rationale for selecting a suitable set of methods for demonstrating on a case study. The results of this first phase are reported in [3]. Methods for reliability analysis have been divided into two categories viz. *System Reliability* and *Structural Reliability*.

System reliability uses more or less standardized techniques, which are applied succesfully in other branches of industry. They are often used for the evaluation and assessment of complex control and safety systems. Qualitative methods from system reliability like Failure Mode, Effect, and Criticality Analysis (FMECA), Event Tree Analysis (ETA), and Fault Tree Analysis (FTA) are therefore expected to be beneficial for the wind turbine industry on the short term. Quantitative methods are also expected to be useful on the short term, although appropriate reliability data might cause a problem.

Techniques for structural reliability analysis are frequently applied for the tuning of design rules for structural components. In this respect, especially the fields of offshore industry, flood protection and bridge and building design should be mentioned. The direct application in the design of particular structures is limited to very large projects, as for instance nuclear installations, storm surge barriers and aerospace devices. Also in case of wind turbine engineering, an indirect use of structural reliability theory as a support for developing new design rules seems to be most appropriate.

Phase 2. Case Study of the Lagerwey LW 15/75 Wind Turbine

The case study of the Lagerwey LW 15/75 provided a realistic case for applying and developing a methodology for probabilistic safety assessement. The demonstration of the methodology and the results of the case study are summarized in Chapter 2.

Phase 3. Development and Assessment of the Methodology

Based on the experience and insights gained in the case study, a the applied method to enhance the safety, reliability and availability of wind turbines has been assessed. The applicability, the benefits and the limitations of the methodology are summarized in Chapter 3, together with recommendations for improvement. The results of the project phases 2 and 3 are reported in [4].

2. CASE STUDY OF THE LW 15/75

For the safety and reliability analysis of the Lagerwey LW 15/75 design, seven tasks have been execucuted. These subsequent tasks and the results of the case study are briefly discussed below.

Task 1: Description of System Design and Operation

Procedure. Before starting a detailed analysis of a wind turbine, it is necessary for the reliability analyst(s) to become familiar with the specific wind turbine. This should result in a detailed description of the system design and operation, including maintenance, test and inspection procedures, and design modifications [5], [6], [7], [8]. State Diagrams and Event Trees have been used to explain the control and safety actions.

Wind Turbine Design. The Lagerwey LW 15/75 wind turbine, see Fig. 2.1, is equipped with an upwind, two bladed rotor that has a diameter of 15.6 m. The rated electrical power of 75 kW is produced with an asynchronous generator at wind speeds between 3 and 30 m/s. The power control is realized by pitch control of the blades towards feather position in combination with a variable rotor speed. This power control method requires a fairly complex electrical system, which is tuned with the passive blade pitch control mechanism. It is recommended by the manufacturer to shut-down the turbine manually at wind speeds over 30 m/s by yawing the nacelle such that the rotor is at 150 degrees towards the wind direction. The nacelle is placed on a steel tower of a length that can be either 24, 30 or 36 meter. In order to ensure safe operation the wind turbine is equipped with two independent and diverse safety systems, viz. a passive blade pitching mechanism, and a yaw mechanism.

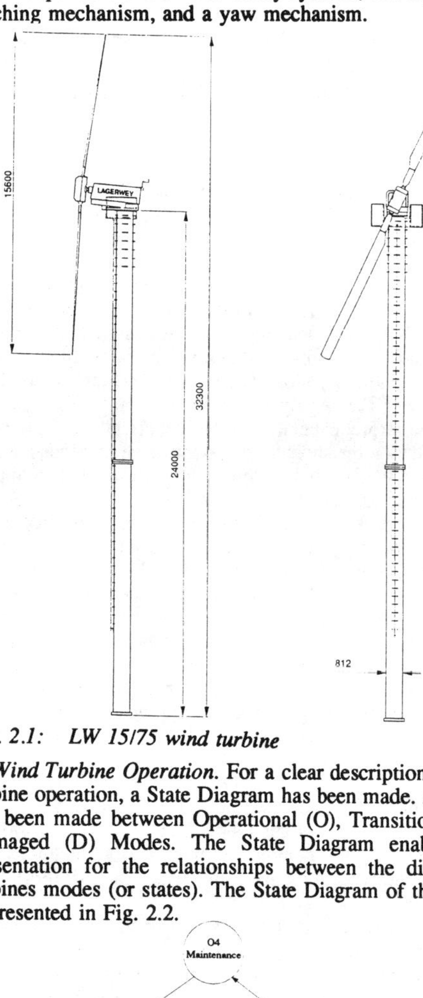

Fig. 2.1: LW 15/75 wind turbine

Wind Turbine Operation. For a clear description of the wind turbine operation, a State Diagram has been made. A distinction has been made between Operational (O), Transitional (T), and Damaged (D) Modes. The State Diagram enables a clear presentation for the relationships between the different wind turbines modes (or states). The State Diagram of the LW 15/75 is presented in Fig. 2.2.

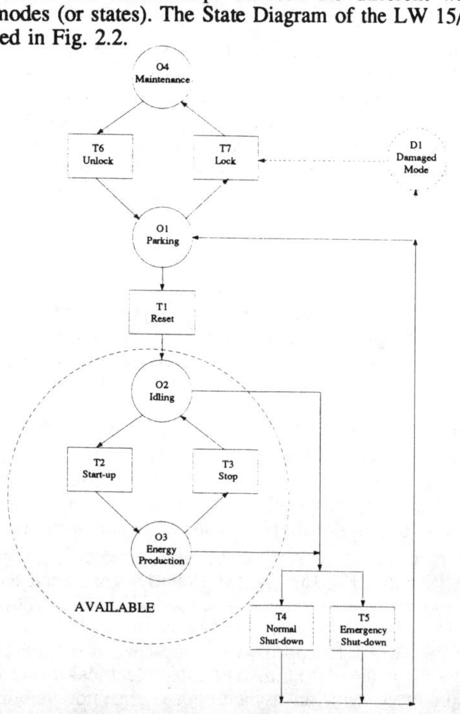

Fig. 2.2: State Diagram of the LW 15/75

Control and Safety Actions. Generally, the control systems of the LW 15/75 have to keep the wind turbine within its design limits during normal operation and to ensure electrical energy is generated and supplied to the grid in an efficient manner. In case of a failure or in case of abnormal external conditions, the control systems in combination with the safety systems have also to keep the turbine within its design envelope to prevent damage or unsafe operation.

The safety systems of the LW 15/75 will be activated in case of the following events: an overspeed situation, excessive vibration, overload of the generator, loss of load, too high inverter temperature, and abnormal cable twist which have all been considered in the case study. In case of a normal shut-down (T4), the nacelle will be yawed out of the wind; the yaw motor is activated by the grid. In case of an emergency shut-down (T5), the yaw motor is powered directly by the generator. This happens for instance during grid failure (S5). A survey of control and safety actions is given in Table 2.1.

Table 2.1: Survey of events, resulting in control or safety actions.

CONTROL ACTION			
Code	Event	Control Action	Remarks
C1	Exceeding rated rotor speed (125 rpm)	Pitch regulation	Backup safety action S1
C2	High inverter temperature (40 °C)	Cooling	Backup safety action S3
C3	Maximum cable twist	Yawing, cable twist protection	No backup action
C4	Exceeding cut-out wind speed (30 m/s)	Normal shut-down	Not applicable for all wind turbines
'C5'	Exceeding survival wind speed (54 m/s)	None	Wind turbine is presumably in parked situation
SAFETY ACTION			
Code	Event	Safety Action	Remarks
S1	Overspeed (150 rpm)	Normal shut-down	Pitch regulation has failed
S2	Excessive vibration	Normal shut-down	Various causes
S3	Very high inverter temp. (90 °C)	Normal shut-down	Cooling has failed
S4	Failure of inverter control	Emergency shut-down	
S5	Grid Failures	Emergency shut-down	
S6	Generator overload	Emergency shut-down	
S7	Manual shut-down	Normal shut-down	Various causes, also high wind speed (C4)

To describe, explain and visualize the sequence of the control and safety actions resulting from an initiating event, *Event Trees*, [7], [8], [12] have been developed. These event trees provide a manner to show the different consequences of the succesful or unsuccessful operation of the control and safety systems. The resulting situations are called the *End States*. The event tree technique is illustrated by a situation with a 'High inverter temperature' C2. (See Fig. 2.3.) As can be seen in the figure, four control and safety systems have to fail before a dangerous situation occurs.

Others. For the performance of a complete safety and reliability analysis, it is not sufficient to consider only the turbine itself, but also aspects like operating procedures, assembly and construction, maintenance, inspection and repair, and design modifications. Especially operating procedures appeared to be of great influence on the availability of the wind turbine.

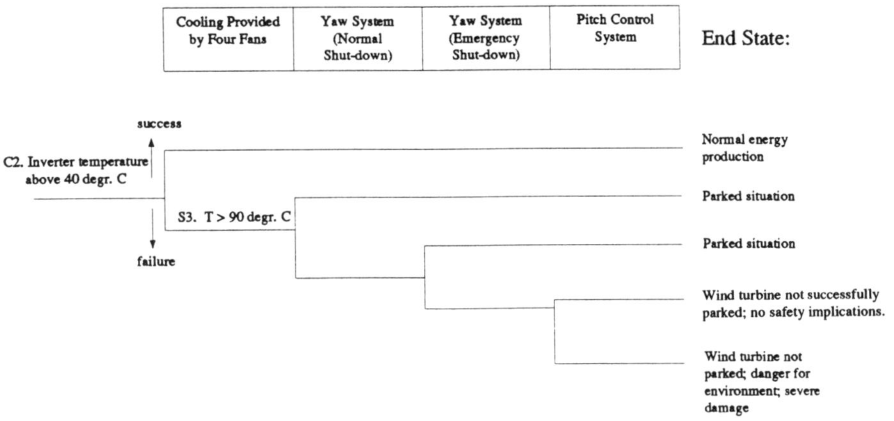

Fig. 2.3: Event tree of a situation with a high inverter temperature (C2)

Task 2. FMECA

Procedure. For the completion of the list of undesired events and for the identification of critical components in the wind turbine, a Failure Mode, Effect, and Criticality Analysis (FMECA), [5], [6], [9], [10], has been performed. FMECA is a systematic procedure for identifying the modes of failure and for evaluating their consequences. The essential function of an FMECA is to consider each major part of a system, how it may fail and what the effects of this failure on the system could be. Furthermore an FMECA can be used to collect historical information and to assist in the analysis of field experience. The FMECA should be an integral part of the conceptual design process and should be periodically updated to reflect changes in design or application. An updated FMECA should be a major consideration in design reviews, inspections, or other major system review points in the program. Moreover, an updated FMECA can be useful for scaling up existing types. A close co-operation with the designers and maintenance engineers is essential for the reliability engineer to determine all possible functions and failure modes of a sub-system or component.

Structural Breakdown. The structural breakdown is necessary to identify the sub-systems of a wind turbine and to define the system boundaries.

Identification and Definition of Items. The items to be analyzed, such as failure mode, failure effect, and criticality, depend strongly on the turbine to be analyzed and on the interests of the manufacturer. In Fig. 2.4, an example is given of the yaw drive, which is part of the yaw mechanism.

Selection Criterion for Critical Modes. The items L, S, A represent the Likelihood of a failure, it's Severity and the required Action. The likelihood has been derived from reliability data or engineering judgement. The severity has been based on the consequences of a failure and refer to the following aspects: the wind turbine operation, the economic aspects and personnel injury. The assessment whether an identified failure mode requires additional action is based on both the likelihood and severity. The more likely a failure is to occur and the more severe the consequences, the higher the risk.

Results. From the FMECA, a set of critical components and initiating events has been derived. This survey has been used to check the measures the manufacturer has taken to prevent these failures and events from happening. Since Lagerwey has a long term experience with this type of turbine, most of the preventive measures conceived during the FMECA have already been realized, either as a design modification, or by changing the maintenance procedures.

SYSTEM N-YS Yaw System		SUB-SYTEM or COMPONENT N-YS-YAD Yaw Drive		SHEET: 1 of 1 REVISION: 1					DATE: 24/11/1992 NAME: LR	
FUNCTION:	1. Yawing and carrying out safety actions									
FAILURE MODE	**EFFECT ON SYSTEM**	**EFFECT ON WTB**	**FAILURE CAUSE**	**DETECTION METHOD**	**PREVENTIVE ACTION**	**RESPONSI-BILITY**	**L**	**S**	**A**	**LOAD SITUATION**
1. Yaw drive shaft fails to rotate ('Free yawing')	1. No yaw 2. Failure of safety system	1. Decrease of efficiency 2. No safety action possible	1. Motor not electrically powered (further analyzed with FTA) 2. Motor burned after emergency stop 3. Yaw drive disconnected from frame due to high mechanical loads 4. Mechanical damage (fracture of shaft or tooth) 5. Fatigue fracture; oscillations of nacelle, due to failure of yaw brakes 6. Capacitors burned during emergency stop	1. Visual, inspection	1. FTA to analyze electrical part 2. Regular inspection, monitoring of status signal 3. Installation of encoder 4. Robust construction	1. Designers 2. Service 3. Assembly	2.2E-5	3	yes	All
2. Yaw drive shaft blocked ('No yaw possible')	1. No yaw 2. Failure of safety system	1. Decrease of efficiency 2. No safety action possible	1. Blocking of gearbox, no lubricant 2. Too high friction of yaw brakes 3. Contamination or corrosion of yaw bearing	1. Visual, inspection	1. Maintenance, testing	1. Service 2. Assembly	1.3E-5	3	yes	All

Fig. 2.4: Example of a filled out FMECA sheet.

Task 3. Event Sequence Analysis and System Modelling

Procedure. The event sequence analysis involved safety actions leading to unavailability of the wind turbine. The analysis has been focused on the operational modes 'Idling', O2, and 'Energy Production', O3, and on the transitional modes 'Normal Shutdown', T4, and 'Emergency Shutdown', T5. These modes are the most important with respect to the unavailability and safety of the wind turbines. For the event sequence analysis and system modelling, techniques have been used like event trees (see Fig. 2.3) and fault trees respectively.

Example of Fault Tree During Overspeed. System modelling will be illustrated here with the overspeed situation S2. It is assumed that the pitch mechanism (control action C1) has failed already and that the rotor rotational speed has exceeded the value of 150 rpm. The overspeed event should be detected by a tachometer which should activate the yaw mechanism. The yaw mechanism can be powered by either the grid (normal shut-down) or by the generator (emergency shut-down). The probability of a successful yaw action has been investigated with a fault tree analysis, [8], [9], [11], [12]. The top part ot the fault tree of the yaw mechanism is presented in Fig. 2.5.

Results. Simular to the overspeed situation, the initiating events S2 to S7 and their resulting safety actions have been analyzed. From the qualitative analyses, unbalances were found in the design of safety systems, depending on the initiating event. Excessive vibration had the lowest level of defence, viz. only the yaw system.

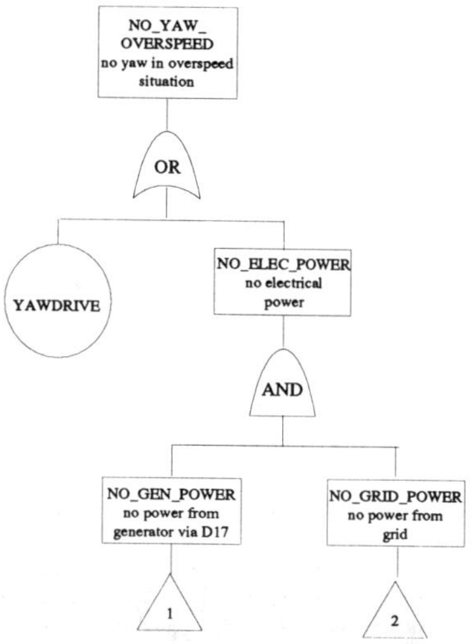

Fig. 2.5 Fault tree of the yaw system during overspeed.

Task 4. Data Collection and Parameter Estimation

Procedure. To get additional information on the actual behaviour of the wind turbine design, specific operational experience has been collected and analyzed, [12], [13], [14]. This information is not only meant for quantifying the unavailability and safety of the wind turbine but also for qualitative insights. The following types of data have been collected and analyzed:

- maintenance and failure data,
- operational data, consisting of:
  - wind turbine mode's data (e.g. starts, stops, transitions and durations), and
  - external conditions (mainly wind and grid conditions).

Sources of Information. Maintenance and failure data have been derived partially from maintenance sheets, audits, logbooks and [16]. In addition, generic data contained in reliability data handbooks, [14], [15], European, [17], [18], [20], and American, [19], [20] wind turbine databases, and engineering judgment have been used to finally arrive at the parameters required for the quantification process and to compare the specific LW 15/75 failure data with. The operational data have been derived from logbooks, measurements, and function tests.

Data Collection and Analysis. For the collection of the specific data, 11 turbines have been analyzed, representing operational experience of about 26 years. Amongst others, the following parameters have been derived for reliability quantification purposes:

- failure rates or failure frequencies of components and sub-systems,
- frequencies of initiating events that require safety action, and
- downtimes.

Results. The collection of data for these eleven LW 15/75 wind turbines has shown some limitations in the current pratices of data collection. Not all types of data required for reliability purposes have been recorded. Despite the shortcomings of data collection, both qualitative and quantitative information, has been derived from the available operational experience. In general, the component failure frequencies of the LW 15/75 compared favourably to the reported generic wind turbine data.

Task 5. System Reliability Quantification

Procedure. The fault trees and event trees developed in Task 3 have been quantified with the failure rates, initiating event frequencies, and downtimes, estimated in Task 4, [7], [11], [12], [13], [21]. Estimates have been obtained of safety system unavailabilities and wind turbine unavailability. Components contributing to the safety system unavailabilities have been ranked according to their importance in order to identify possibly weak points in the LW 15/75 design, to highlight unbalances, and to indicate effective areas for improvement. The resulting absolute values should be considered as rough estimates because of data uncertainties and modelling assumptions. When drawing conclusions from these values, these limitations should be considered. Sensitivity analyses have been performed to show the robustness of the conclusions.

Safety System Unavailabilities. The unavailability of the yaw system has been derived by quantification of the fault tree. The unavailability of the pitch mechanism has been estimated somewhat conservatively from the number of overspeed alarms. The number of alarms has been corrected for spurious alarms. The end states of the event trees have been quantified by multiplying the availabilities or unavailabilities of the successive events.

The results show an estimated severe damage frequency of 1 per 35 wind turbine operating years. Partial damage with no safety implications occurs approximately once in 8 years. Successful shut-downs occur 5 times per year. These values are based on the above mentioned operational experience and the assumptions that: 1) the wind turbine structure will not fail if the safety systems properly function (too optimistic), and 2) it will fail in case of unavailability of the safety systems (conservative). Especially the detection units (e.g. tachometer and vibration detection) contribute to the damage frequencies. This is because there is no redundancy in these components.

Wind Turbine Unavailability. As can be seen in the state diagram in Fig. 2.2, there are three principal wind turbine modes that contribute to the wind turbine unavailability, viz. Maintenance (O4), Parking (O1), and the Damaged Mode (D1). The outcomes of the event sequences result in either Parking (O1) or the Damaged Mode (D1). The contribution from maintenance comes from planned, periodic maintenance and from unplanned maintenance due to entering the damaged mode or, as a separate category, as a result of direct critical component failures for which no safety actions are designed. Each of these contributions has been quantified.

Successful shut-downs contribute the most to the wind turbine unavailability, especially the activation of the 'grid failure' alarm.

Task 6. Structural Reliability

Procedure. The unsafe situation near the site of a wind turbine is often caused by failures of structural parts like the blades, the hub and the tower. The failure rates of these components should be very low and usually cannot be determined from operational experience. To determine the failure probabilities of these components, additional anayses should be performed. In civil engineering structural reliability appeared to be appropriate [22], [23], [24], [25], [28]. With structural reliability, the system behaviour of a load carrying structure can be modelled, taking into account the different failure mechanisms (e.g. yield or fatigue) of the components. Since only little experience is available about structural reliability in wind turbine engineering [30], a limited method called Mean Value Approach (MVA) [25] has been used to make a probabilistic fatigue analysis of the tower foot. As a first attempt, the analysis has been performed in order to investigate the applicability and feasability of structural reliability in wind turbine engineering, much more than to determine the accurate failure probability. More information about the probabililistic fatigue analysis of the tower foot can be found in Appendix A.

Results. Although the probabilistic model was limited, insight was gained in how the failure probability is influenced by uncertainties in the parameters. The uncertainties in the S-N curve or the welding and in the annual average wind speed appeared to be dominant. Model uncertainties appeared to be of less importance than as might be expected, and also the turbulence intensity appeared to be of little influence.

Task 7. Interpretation and Discussions of the Results

The results and recommendations of the LW 15/75 case study are summarized as follows:

- *Results*

  Depending on the initiating event, a safety action must be performed using one or two of the safety systems. It has been concluded from the qualitative analyses with event trees and fault trees, that there are unbalances in the safety systems design. Excessive vibration of the nacelle had the lowest level of defence (only the yaw system). Quantitative results showed an estimated severe failure frequency of 1 per 35 wind turbine operating years. Partial damage with no safety implications are expected to occur approximately once in 8 years. Successful shut-downs occur 5 times per year. If safety is to be improved, modifications can best be focused on the improvement of the detection units. A strong improvement of the wind turbine availability can be made by automatic resetting, which causes the downtimes to decrease.

- *Interpretation of Results*
The system reliability analysis results are valid for the so-called 'useful life' period where failures are random and component failure rates are more or less constant. Reported gross failures (from other than the 11 units) which have resulted in hazardous situations appeared to be caused not by random failures during the useful life period, but were mostly the result of unexpected situations not foreseen in the design stage. Especially activation of the yaw mechanism at high wind speeds was found to be critical. In early designs of the LW 15/75, this situation led to excessive vibrations and high loads. The phenomena causing these effects were not known in the design phase. A load and operation monitoring programme together with the analysis of more load situations are strongly recommended to gain more insights in the dynamic behaviour of the turbine and to reduce the occurrence of these failures. To improve the safety at high wind speeds, it is recommended to enhance the yaw mechanism or to add an extra safety device. Otherwise, it is recommended to install an anemometer that causes automatic shut-down at lower wind speed.

3. CONCLUSIONS AND RECOMMENDATIONS ON THE METHODOLOGY

The results of this project are summarized as follows:

- *Feasability*
The short term benefits of system reliability analysis (Thorough System Description, State Diagrams, FMECA, Event Tree and Fault Tree Analysis) have been proven. Probabilistic analyses to assess the safety and reliability of wind turbines constitute a necessary supplement to the existing deterministic studies and design rules, to achieve a more balanced design. Probabilistic analyses provide a systematic approach in the design and operation of a wind turbine, a systematic feedback from operational experience into the design process, and can form a basis for discussions. Qualitative analysis methods can be applied immediately.
- *Limitations*
In addition to deterministic studies, reliability analysis techniques are not the only means to further enhance safety and reliability. There is still a fundamental lack of knowledge of physical phenomena in wind turbine engineering that may cause hazardous situations. Even when these phenomena are known, quite often they cannot be adequately modelled with the current computer codes. It is therefore highly recommended to perform analyses of additional load situations during the design stage and to use measurement, test and monitoring programmes during the early operational phase of a wind turbine. The results of these programmes can be used for updating the design process.
A survey of methods to be applied during the lifetime of a wind turbine is given in Table 3.1.

Table 3.1: Reliability analysis techniques in the different operating stages.

Operating Stage	Reliability Analysis Technique	Objective
Design and Development	State Diagram (and System Description)	1. Familiarization with turbine design and operation. 2. Enabling a clear presentation and transfer of knowledge. 3. Definition of a clear set of terms.
	FMECA	1. Identification of all the ways a piece of equipment can fail, trace it's cause and determine it's effect. 2. Give recommendations to reduce the likelihood of a failure, and to minimize the hazardous effects, e.g. maintenance and testing procedures, design modifications, or adjustment of the manufacturing process. 3. Design of checklists.
	Event Tree Analysis (ETA)	1. Analyze possible hazardous events and designed safety actions; show chronological relationships between initiating event and subsequent events. 2. Completion of list with load situations. 3. Ranking of events contributing to the different end states (quantitative!).
	Fault Tree Analysis (FTA)	1. Identification of events and combinations of events leading to an undesired event. 2. Identification of weak spots and unbalances in system design. 3. Ranking of events contributing to the safety system unavailability (quantitative!).
	Structural Reliability	1. Search for unbalances in load carrying structure. 2. Determination of dominating factors contributing to uncertainties in limit states. 3. Underpin safety factors (longer term).
	Quality Assurance	1. Enabling a systematic design approach. 2. Minimization of design and production errors.
Burn-in	Operation and Load Monitoring	1. Verification of design assumptions, obtain data for structural reliability analysis. 2. Completion of list with load situations. 3. Determination of loads during rare events which have not been foreseen in the design phase.
	Maintenance, Inspection and Testing	1. Determination of early failures. 2. Verification of maintenance, inspection and testing procedures.
	Data Collection and Failure Reporting	1. Systematical approach for reporting incidents and accidents and feedback to the design.
	FMECA, ETA, FTA	1. Updating and completion of analyses performed in design stage. 2. Updating of procedures and checklists.
	Quality Assurance	1. Minimization of early failures. 2. Minimize operating and maintenance failures.
Useful Life	Data Collection and Analysis	1. Statistical analyses, trend analysis, parameter estimation. 2. Periodic feedback to designers, owners and certification institutes. 3. Input for cost-benefit anlyses.
	Operation and Load Monitoring	1. Collection of operational data. 2. Determine lifetime consumption.
	Maintenance and Inspection	1. Periodic: minimization of random failures. 2. Corrective: restoration of failures. 3. Collection of failure data
	FMECA, ETA, FTA	1. Update earlier analyses with specific data and experience.
	Quality Assurance	1. Minimization of operating and maintenance errors.
Wearout	Maintenance and Inspection	1. Search for cracks and wear in main components. 2. Minimization of random failures.
	Operation and Load Monitoring	1. Search for wear and changes in behaviour, e.g. changes in eigenfrequencies or noise production.
	Data Coll. and Analysis	1. Statistical analyses to determine trends and to assess wear.

- *Data Collection and Analysis*

In view of limitations in the current practices of data collection, priority must be given to a systematic collection of data, not only concerning maintenance and failure data (component failures, downtimes, failure causes, etc.) but also concerning operational data (external conditions, starts and stops, etc.). Computerized data collection and analysis systems are strongly recommended. Detailed recommendations can be found in [4].

- *Certification*

On a short term, the certification institutes can benefit from especially qualitative analysis methods. The systematic approach and presentation, e.g. state diagrams and event trees enable a direct insight in the operation of the considered wind turbine. FMECAs performed by the designers provide a survey of critical events which the certification institutes can review during the certification process. At this moment, the critical events are often conceived by the certification institutes themselves in a more or less unstructured way. They should therefore ask at least for the results of an FMECA. In addition, quantification of the safety system unavailabilities will be necessary for evaluating safety concepts

- *Structural Reliability*

Although the demonstrated MVA for the structural reliability analysis of mechanical components appeared to have limitations, a further application of probabilistic design techniques is strongly recommended. The application of the MVA in addition to the currently used deterministic design rules gives qualitative insights in the uncertainties of the fatigue damage calculations. Fruitful discussions in wind turbine industry which must lead to a wider acceptance and introduction of probabilistic design techniques can be generated only if these techniques are wider applied, parallel to the improvement of the methodology and the identification and quantification of uncertain parameters. In the near future, these techniques must be used to further underpin or adjust the currently used safety factors.

REFERENCES:

[1] Stam, W.J. et al: 'Regulations for the Type-Certification of Wind Turbines: Technical Criteria', ECN-91--001, February, 1991, The Netherlands.

[2] IEC Standard: 'Safety of Wind Turbine Generator Systems' (Draft document), International Electrotechnical Commission, Technical Committee 88, 1992.

[3] Rademakers, L.W.M.M. et al: 'Reliability Analysis Methods for Wind Turbines; Task 1 of the Project: Probabilistic Safety Assessment for Wind Turbines', ECN-C--92-018, Petten, June 1992.

[4] Rademakers, L.W.M.M. et al: 'Methodology for Probabilistic Safety Assessment of Wind Turbines; Illustrated by a Case Study of the Lagerwey LW 15/75 design', ECN-C--93-010, Petten, March 1993.

[5] ANSI/IEEE Std 352-1987: 'IEEE Guide for General Principles of Reliability Analysis of Nuclear Power Generating Station Safety Systems', New York, NY 10017 USA, 1987, ISBN 0-471-60205-1.

[6] 'Hazard and Operability Study; Why? When? How?' Report of the Directorate-General of Labour, Ministry of Social Affairs, First edition 1979.

[7] 'Procedures for Conducting Probabilistic Safety Assessments of Nuclear Power Plants (Level 1)', IAEA, Safety Series No. 50-P-4, Vienna, 1992.

[8] NUREG/CR-2300: 'PRA Procedures Guide', USNRC, January 1983.

[9] ANSI/IEEE: 'IEEE Guide for General Principles of Reliability Analysis of Nuclear Power Generating Station Safety Systems', Std 352-1987, New York, 1987.

[10] 'MOD-2 Failure Mode and Effect Analysis', USDOE, July 1979, NASA CR-159632.

[11] NUREG-0492: 'Fault Tree Handbook', Systems and Reliability Research Office of Nuclear Regulatory Research, U.S. Nuclear Regulatory Commission, Washington D.C., January 1986.

[12] Commissie Preventie van Rampen door Gevaarlijke Stoffen: 'Methoden voor het Bepalen en Verwerken van Kansen' (Methods for the Determination and Processing of Probabilities), CPR12, Directorate General of Labour, Voorburg, The Netherlands, May 1985.

[13] Terpstra, K. et. al.: 'PHAMISS; Reliability Computer Program for Phased Mission Analysis and Risk Analysis: User's Manual', Volume 1 and 2, ECN-183, Petten, The Netherlands, April 1986.

[14] IEEE Std 500-1984, 'IEEE Guide to the Collection and Presentation of Electrical, Electronic, Sensing Component, and Mechanical Equipment Reliability Data for Nuclear-Power Generating Stations', IEEE, Inc., New York, 1983.

[15] 'Nonelectronic Parts Reliability Data 1991', Reliability Analysis Center, Rome, NY 13440-8200, 1991.

[16] Curvers, A. and Schuurman, J.J.: 'Bedrijfsgegevens van IPW-windturbineprojecten: eerste fase' (Operating experience of IPW wind turbine projects: first phase)', ECN-C--91-041, June 1991.

[17] Schmid, J., and Klein, H.P: 'Performance of European Wind Turbines - A Statistical Evaluation from the European Wind Turbine Database EUROWIN', Fraunhofer Institute for Solar Energy Systems, Freiburg, July 1991.

[18] W.J. Stam, et al.: 'Collection of Data on Accidents and Incidents of Wind Turbines for Updating of Safety Criteria', ECN-C--90-017, May 1990.

[19] Electric Power Research Institute: 'Assessment of Wind Power Station Performance and Reliability', EPRI GS-6256, Project 1590-10, Final Report, March 1989.

[20] WindStats Newsletter: Vrinners Hoved, Dk-8420 Knebel, Denmark, or: PO Box 496007, Ste 2170, Redding, California 96099-6007, USA.

[21] Seebregts, A.J.: 'Short Manual CESAR-I: Code for Evaluating System Availability and Reliability', ECN-I--90-027, June 1990.

[22] Ang, A.H.S and Tang, W.H.: 'Probability Concepts in Engineering Planning and Design, Vol. I: Basic Principles', J. Wiley & Sons, 1975.

[23] Ang, A.H.S and Tang, W.H.: 'Probability Concepts in Engineering Planning and Design, Vol. II: Decision, Risk and Reliability', J. Wiley & Sons, 1984.

[24] Karadeniz, H., van Manen, S., Vrouwenvelder, A.: 'Probabilistic Reliability Analysis for the Fatigue Limit State of Off-shore Structures', Bulletin Technique du Bureau Veritas, July 1984.

[25] Rackwitz, R.: 'Principles and Methods for a Practical Probabilistic Approach to Structural Safety' Subcommittee for First Order Reliability Concepts for Design Codes of the Joint CEB-CECM-CIB-FIP-IABSE Committee on Structural Safety, Published by CEB, 1976.

[26] Winkelaar, D.: 'SWIFT: Program for Three-Dimensional Wind Simulation; Part 1: Model Description and Program Verification', ECN-R--92-013, December 1992, The Netherlands.

[27] Snel, H. and Lindenburg, C.: 'PHATAS-II, Program for Horizontal Axis Wind Turbine Analysis and Simulation version II, Aeroelastic Modelling', ECN-C-92-027, May 1992, The Netherlands.

[28] Vrouwenvelder, A., Ligteringen, H., and Moan, T.: 'Reliability Analysis for Offshore Structures', Engineering Committee on Oceanic Resources (ECOR), May 1991, London.

[29] Grol, H., Snel, H., and Schepers, J.G.: 'Wind Turbine Benchmark Exercise on Mechanical Loads; A State of the Art Report, Volume I (Part A) Main Body of the Report', ECN-C--92-030, May 1991, The Netherlands.

[30] Veers, P.S.: 'Fatigue Reliability of Wind Turbine Components', SAND--90-2538C, Sandia National Laboratories, Albuquerque, NM 87185.

APPENDIX A: Probabilistic Fatigue Analysis of the Tower Foot

A-1. General

As part of the case study of the LW 15/75, the possible application of reliability techniques for the assessment of the structural integrity of wind turbine components has been investigated. These techniques have been denoted as *Structural Reliability*. For the probabilistic reliability analysis of mechanical components, four levels can be identified:

Level 0: This level is a deterministic analysis based on fixed data; a deterministic (high) load, a deterministic (low) strength and an 'overall- safety-factor'.

Level I: This level is intended for routine design use. For each of the stochastic variables a certain unfavourable value is chosen, the so-called characteristic value, usually based on a 5% confidence limit. In addition, a set of partial safety factors is applied; these establish the margin between load and strength for the various cases to be considered. This approach is yet commonly used in wind turbine engineering.

Level II: The analysis at Level II is an approximative analysis based on the so-called "first order/second moment" principle.

Briefly, what it amounts to is that only the mean value and the standard deviation of each stochastic variable are taken into account, linearization being applied if necessary. Level II approximations are used to provide a basis for Level I procedures. By means of Level II studies it is objected to underpin the partial safety factors of a Level I analysis as soundly as possible. Moreover, Level II analyses are used for special and important structures for which an above-normal standard of reliability is considered essential, e. g. offshore structures.

Level III: This level comprises complete and exact analyses performed with the aid of analytical or numerical integration procedures or Monte Carlo simulations. Level III analyses generally demand too much computational effort to be practicable for dealing with major problems, due to the large number of stochastic variables to be considered in most cases. The scope for applying them exists mainly in verifying and complementing Level II approximations.

A-2 Mean Value Approach

As a first attempt for the probabilistic fatigue analysis of the tower foot, a limited Level II method called Mean Value Approach (MVA) has been applied. First, a selection has been made of wind parameters, wind turbine parameters and fatigue parameters which have been considered stochastically instead of deterministically. In Table A-1 the mean values and standard deviations of these parameters are listed. Some values of these means and standard deviations have been derived from measured data but most have been based on engineering judgement. The MVA assumes that these parameters are normally distributed.

Table A1: Stochastic variables in the fatigue analysis of the tower foot.

i	Designation	x_i	μ_i	σ_i	$(\sigma/\mu)_i$
1	Turbulence intensity	σ/U_{10}	19.28 %	1 %	0.052
2	1st tower eigenfrequency	f_{tower}	0.97 Hz	0.049 Hz	0.050
3	Maximum lift coefficient	$C_{l\ max}$	1.60	0.04	0.024
4	Torsion spring	$M_{pitch\ max}$	730 Nm	50 Nm	0.068
5	Uncertainties in load level	Model uncertainties	1.0	0.125	0.125
6	Wall thickness	t_{wall}	10 mm	0.5 mm	0.050
7	Constant of S-N curve	log$C$	12.511	0.207	0.017
8	Annual average wind speed	U_{ann}	7.28 m/s	0.728 m/s	0.100

With the mean values of the stochastic variables the fatigue damage of the tower foot has been calculated deterministically. The premisses of this calculation e.g. the turbulence intensity, the required load cases, and the site, have been chosen in accordance with the Dutch certification criteria [1]. The wind has been modelled with SWIFT [26] and the loads in the tower foot have been determined with PHATAS-II [27]. The loads have been transformed to stresses and with the Palmgren-Miner rule the damage, D_{mean}, has been calculated.

Subsequently, all stochastical variables have been varied independently from μ_i to $\mu_i + \sigma_i$. For each new value the corresponding fatigue damages D_i has been determined for i = 1...8.

A reliability function Z has been selected in a deterministic way such that Z < 0 corresponds with fatigue failure and Z ≥ 0 with non-failure. Of course Z is related to the miner sum and

$$Z = 1 - D \tag{1}$$

has been chosen (with D being the fatigue damage). When Z is linearized in the mean values of the stochastic variables, the mean value μ_Z and the standard deviation σ_Z of Z can then be approximated by the expressions

$$\begin{aligned} \mu_Z &= Z(\mu_1,\mu_2,\ldots,\mu_n) = 1 - D(\mu_1,\mu_2,\ldots,\mu_n) \\ &= 1 - D_{mean} \end{aligned} \tag{2}$$

$$\sigma_Z = \sqrt{\sum_{i=1}^{n} \left[\frac{\partial Z}{\partial X_i} (\mu_1,\mu_2,\ldots,\mu_n) \sigma_i \right]^2} \tag{3}$$

The partial derivatives in (8) cannot be determined analytically and are therefore numerically approached

$$\begin{aligned} \frac{\partial Z}{\partial X_i} = \frac{\partial D}{\partial X_i} &\approx \frac{D(\mu_1,\mu_2,\ldots,\mu_i + \sigma_i,\ldots,\mu_n) - D_{mean}}{\sigma_i} \\ &= \frac{\Delta D_i}{\sigma_i} \end{aligned} \tag{4}$$

From (3) and (4) it appears that the standard deviation of Z is given by

$$\sigma_Z = \sqrt{\sum_{i=1}^{n} (\Delta D_i)^2} \tag{5}$$

and the reliability index β can be written as

$$\beta = \frac{\mu_Z}{\sigma_Z} = \frac{1 - D_{mean}}{\sqrt{\sum_{i=1}^{n} (\Delta D_i)^2}} \tag{6}$$

Now, the failure probability of the tower can be derived

$$P_F = \Phi_N(-\beta) \tag{7}$$

where Φ_N is the standard normal cumulative distribution function. For positive β it appears that the larger β, the larger the safety. The importance factors α_i^2 are defined by the squares of the direction cosines of β and can be expressed as:

$$\alpha_i^2 = \frac{(\Delta D_i)^2}{\mathrm{var}\ Z} \tag{8}$$

They are a measure for the percentage of variability due to each random variable.

A-3 Numerical Results

The numerical results are presented in Table A-2 and A-3.

Table A-2: Fatigue damage.

i	$D_{mean}+\Delta D_i$	ΔD_i^2
1	0.2225	0.0009
2	0.1818	0.0001
3	0.1770	0.0003
4	0.2081	0.0002
5	0.2748	0.0067
6	0.2235	0.0009
7	0.3109	0.0139
8	0.3124	0.0142

Table A-3: Results.

Parameter	Results
μ_Z	0.8070
σ_Z	0.1929
β	4.18
P_F	$1.4\ 10^{-5}$

The importance factors α_i^2, obtained as a byproduct of the calculation of β are displayed in Fig. A-1. They must sum to unity and are a measure of the percentage of variability due to each stochastic variable. It appeared that the fatigue coefficient log$C$ (37.3%) and the average annual wind speed U_{ann} (38.1%) are by far the most important sources of variability. The modelling uncertainties contributed 18% to varZ. The remaining 8.6 % is divided among the other stochastic variables.

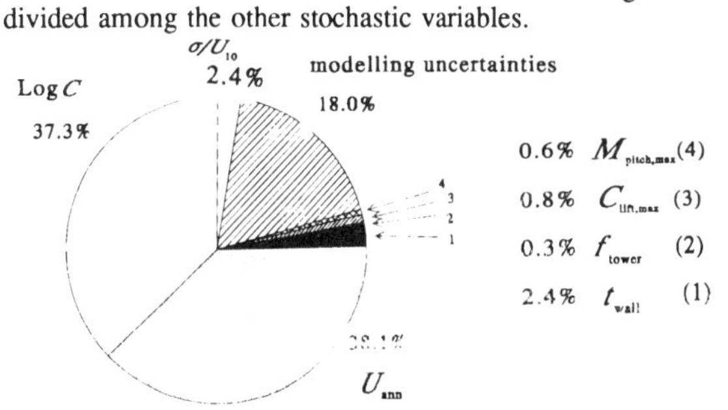

Fig A-1: Importance factors obtained with MVA.

THE DANISH CODE OF PRACTICE FOR LOADS AND RELIABILITY OF WIND TURBINES.

by Carl Jørgen Christensen and Peter Hjuler Jensen,
The Test Station for Wind Turbines,
Dept of Meteorology and Wind Energy, Risø National Laboratory,
DK-4000, Roskilde, Denmark

ABSTRACT

The Danish Code of practice for Wind Turbines, DS-472, is described. The code is build on limit state design and uses the partial coefficient method. The level of safety (reliability) of the code is compared to the newest draft of the forthcoming IEC standard, that is under preparation.

1. INTRODUCTION.

DS-472, a new Danish code of practice for wind turbines, was issued in the summer of 1992 [1].

Also a new international standard is under development under the auspices of the International Electrotechnical Commission (IEC/TC88). This IEC standard is expected to be finished in the not too distant future.

In this paper the Danish code is described. On relevant points, the Danish code is compared to the newest IEC draft. Of particular interest is the comparison of the level of safety build into the two codes.

2. THE DANISH CODE, DS-472.

2.1. The Basis for the Code.

The Danish code uses the method of limit state design including the partial coefficients method. This method is the most commonly used for codes, in particular in Scandinavian codes. The basis for the method is given in DS-409, Code of practice for the Safety of Structures [2], which again rests on the Nordic NKB-55 [3].

The IEC standard is build on a very similar basis, except that in this case the background standard is ISO-2394 [4], which is equivalent to the above mentioned Danish (Nordic) documents.

Strictly speaking the Danish code is only valid in Denmark itself, but fortunately the climate in many other parts of the world is not worse than in Denmark. The Danish code covers loads and mechanical safety including impact of the electrical system on the mechanical safety. It is only valid for larger machines (rotor diameter > 5m).

In contrast the IEC code also talks about electrical safety in general. And IEC covers safety in connection with the design and installation of the turbine sites.

2.2. The Limit States.

A limit state is a state of a structure and the loads imposed upon it, beyond which the structure no longer satisfies the design requirements.

The purpose of design calculations is to keep the probability of a limit state from being reached below a certain value prescribed for the WTGS. The design requirement for the limit state can in very simple cases be expressed by the equation

$$R_d - R_k/\gamma_R \geq L_k\gamma_L - L_d \qquad (1)$$

R and L designate values of resistance and load, index $_k$ indicates characteristic values and index $_d$ design values, The γ's are partial coefficients for load (index $_L$) and resistence (index $_R$). Very simply stated the equation (1) expresses, that the design value of the resistance function of a component should be stronger than that of the load, it is expected to have to meet.

The danish code operates with four kinds of limit states:

- ultimate limit states
- fatigue limit states
- accidental limit state
- serviceability limit state

It is not common to define a separate fatigue limit state. Both the Eurocode 1 and the equivalent ISO-2394 treat ultimate and fatigue limit states as one type, but does require fatigue state checking in cases, where it is needed. The Danish code working group decided, that fatigue is so important for wind turbines, and that the calculation methods for fatigue checking are so different from ultimate state calculations, that the rational solution was to define two independent limit states. ISO 2394 specifically allows this subdivision of the ultimate states, as does the general Danish code DS-409 (Reliability of Structures).

The IEC code will not make this clear separation, but will require fatigue checking.

2.3 Control and Safety Systems.

The requirements put on the control and safety systems are rather elaborate in the Danish code. The control system is defined as a system watching essential function parameters and taking care of normal operations of the system. It shall do that in such a way, that the turbine is kept within a well defined operational region. This region is defined by

- a max 10min wind speed (V_{max}) for allowed operation
- a max power on the defined power curve (long term average)
- a max 10min Power for wind speeds up to V_{max}.
- a max speed of rotation for normal operation.
- a wind speed, under which the turbine can be completely stopped, eg for maintenance.

The safety system shall go into operation when detecting a fault or breakdown in the wind turbine or its control system. It shall be able to keep the turbine within a well defined extreme operational region. This region is in turn defined by

- an extreme speed of rotation
- a max transient speed of rotation.

The rationale behind these strict definitions is of course, that the different limiting values for the normal and extreme operational limits are essential as input parameters to calculation of the limit state conditions for the various essential load cases.

The regulation of the safety system is rather detailed. The essential requirement is, that the safety system shall be tolerant to any single fault in the turbine and shall have precedence over the control system. One controversial detail, that will interest most people, is the compulsory double brake system, where at least one brake must be aerodynamic. We still don't think that anybody have designed a single mechanical brake for the larger machines, that we are talking about here. Such brakes will sooner or later cause a fatal accident.

The IEC code is not too different in its requirements to the control and safety system. The main difference is the above mentioned Danish requirements on the brakes. Both codes have the requirement of single fault tolerance and requests the safety system to have the highest priority. IEC requests a fail-safe construction on the control system in general. Denmark only requires that against faults in energy supply. The Danish, on the other hand recommends the full fail-safe construction as a possibility to fulfil all requirements.

2.4. Load Cases.

The wind turbine has many operational modes - production, start, stop, stopped with a fault, etc. In connection with standards, each of these modes are actually called design situations and can each contain critical load cases.

The Danish code defines groups of load cases as shown in Table I

Table I: Combining external conditions with operational mode to find load cases.

Load Case	Design Situation	External Conditions
Normal	Normal	Normal
Extraord	Normal	Extreme
	Fault *	Normal
Accidental	Multiple Faults	Normal

* one fault or common mode failures

We can not in this paper go into the details of pinpointing load cases. But the following can be said about the relation to the IEC standard.

The framework for selecting design situations and combining them with external conditions is largely identical to what has been done in the Danish code. When it comes to pinpointing load cases, there are differences. Roughly speaking, the Danish code is more detailed in its description of emergency and fault situations, whereas the IEC standard is more detailed in the production design situation.

It is common practice in Denmark, and it is stated in the Danish code, that it is the responsibility of the designer to identify the worst load cases and to demonstrate the safety of the construction in these cases. Therefore the Danish load case definitions are not quite as strict and detailed as somebody could wish. Actually the designer is required to think. It will of course be up to the users (eg possible certifying authorities) to develop a reasonable interpretation of the load cases and make sure, that they do cover the actual design and are used optimally and safely.

2.5 External Conditions and Load Calculations

Wind conditions in Danish codes of practice have for 10 years been described using a stochastic approach. The present wind turbine code is no exception. A few characteristics of the wind conditions are the following.

The normal wind climate is ruled by:

- a Weibull distribution of the wind speed
- the surrounding terrain is classified according to roughness into one of 4 classes with each their characteristic roughness
- Weibull parameters can be calculated once the roughness is known.
- a logarithmic wind speed profile is prescribed. Wind speed in height h is

$$V_{10\text{min}} - V_{10\text{min},hub} \cdot \frac{\ln(h/z_0)}{\ln(h_{hub}/z_0)} \tag{2}$$

- a von Karman turbulence model is assumed with a spectrum defined (using the turbulence length scale l_1 and the frequency n) as

$$\frac{nS_1(n)}{(I_1 V_{10\text{min}})} - \frac{nl_1/V_{10\text{min}}}{(1 + 1.5\, nl_1/V_{10\text{min}})^{5/3}} \qquad I_1 - 1/\ln(h/z_0) \tag{3}$$

- extreme wind speeds are defined as a function of height over ground

$$v_{2s} - v_b \cdot k_t \cdot (\ln(h/z_0) + 3) \qquad v_{10\text{min}} - v_b \cdot k_t \cdot \ln(h/z_0) \tag{4}$$

where v_b is a basis wind speed of 27 m/s as determined for the Danish climate. The constants z_0 and k_t are given in the code.

As a possible method for calculating extreme loads with rotor parked is specified a Davenport type calculation. The following wind load distribution on the blade element of chord $c(r)$ can be assumed

$$p(r) - 0.5\rho v^2 c(r) \cdot \phi \tag{5}$$

If the eigenfrequency of the dominating resonance is lower than a 'typical wind excitation frequency', v_{10min}/R,the equation is used with a 10min extreme wind speed and an amplification factor ϕ, that could be as high as 3 as in this case the rotor (or the blade) eigenfrequency is assumed to be excited. In the opposite case, the 2s extreme wind speed shall be used, but with a amplification factor of unity.

The IEC code deviates from the Danish code in its extensive use of cosine-bell deterministic gusts (Figure 1). This does make it a bit difficult to compare to the Danish code. But actually the IEC code also contains provisions for doing the Davenport calculations. Now, the gust shapes has to be considered also by Danish designers. Here a couple of facts comes to mind. The cos shapes generally have rise times of 5-10 sec.

This means that direct excitation of components in the turbine will be very weak. The gust will, however, have a

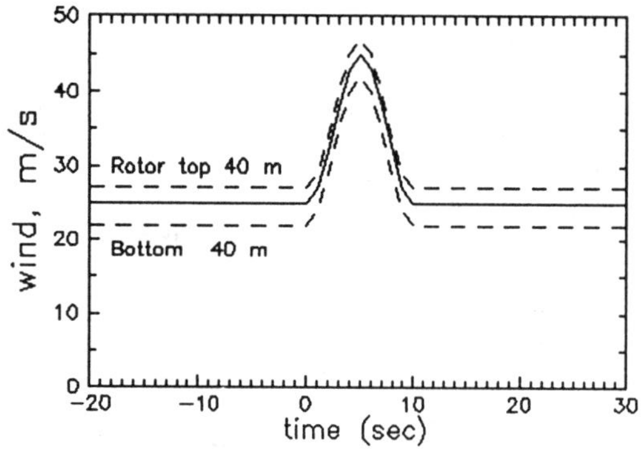

Figure 1: IEC extreme operating gust at 3 levels: rotor top, hub, bottom.

possibly strong influence via the control system. Eg an extreme coherent gust with direction change is defined. This will sometimes cause the turbine to operate in back wind. And quite likely an active yawing system is not fast enough to avoid the back wind. We think, that it is realistic, that such a situation should be examined, and as a matter of fact the Danish code has a load case of a very similar nature.

Apart from this control system influence, of course one has to look at the peak values of the cos gusts as extreme wind situations. These peak values are really not too different from the values used in the Danish codes. It seems, that they can all be accommodated with the limiting values of an IEC class II turbine. The IEC classes are as shown in Table II.

Table II: IEC turbine classes and equivalent values for Danish roughness class 0

Class	V_{ref}	V_{ave}	I_{Ave}
I	50	10	.17
II	42.5	8.5	.17
III	37.5	7.5	.17
IV	30	6	.17
	V_{e50}	V_{ave}	I_{ave}
IEC-II	59.5	8.50	.17
DK-0	52.8	8.17	.11

It is seen, that the IEC class II is a bit harder than the worst Danish climate. As will be obvious later, the partial coefficients for materials decreases the safety level of the IEC code somewhat compared to the Danish code. This will be remedied by the higher climatic values chosen for the class II turbine. It is also clear, that this class II will be the interesting one in Europe, where only a few locations will have worse conditions than the Danish roughness class 0. Class I will be of interest mainly on tops of mountains or in some locations in the northwesterly part of Europe close to the Atlantic.

2.6 Fatigue Calculations.

No compulsory method to be used for fatigue calculations is spelled out. But a possible, quite simple method is outlined for those who wish. The method can only be recommended, if a series of restrictions is fulfilled in the design. Stated shortly, it is only useful for the classical Danish concept (stall regulated, 3-bladed etc.). This concept is the one, we know well enough and where sufficient measurement data exist.

In this method the fatigue inducing part of a dynamic force is described in terms of an accumulated load width distribution.

Figure 2 illustrates a typical blade root lead/lag moment width distribution. The distribution is calculated as the sum of two distributions.

One is a deterministic, constant width (eg gravity) load acting on a rotating blade. The corresponding accumulated

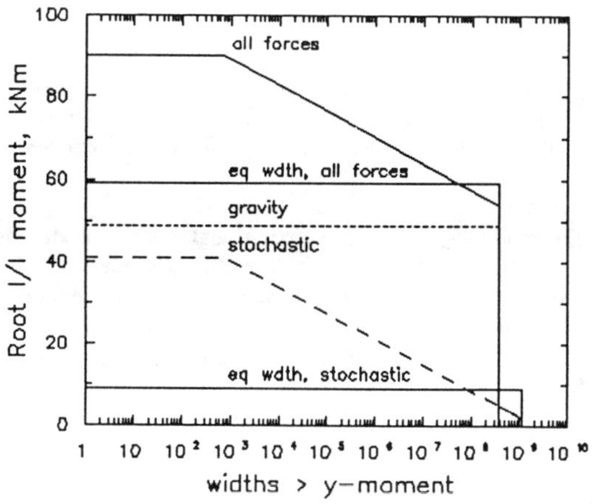

Figure 2. Blade Root M_y accumulated width distribution according to DS-472.

load width distribution is a simple step function, in Figure 2 the curve marked 'gravity'. It contains N_r = 3.4*10^8 widths (ie 1P cycles during a 20 yr lifetime of the wind turbine). The other distribution is a standardized shape multiplied with a constant depending on the turbine construction and containing $3N_r$ widths (Figure 2, 'stochastic'). This signifies a stochastic load pattern. The summed distribution in Figure 2 is marked 'all forces'. The resulting width distribution is a blade root lead/lag moment distribution, where the deterministic gravity load dominates. The discontinuity at N_r cycles again corresponds to the number of gravity cycles in the rotor. These N_r largest widths consist of deterministic gravity induced widths superimposed with stochastic widths induced by air forces.

The standard distribution used as a basis for defining the stochastic distribution can be expressed by the following two equations with I for the relative turbulence, A for the Weibull form factor:

$$\hat{F}_{\Delta}(N_v) = \beta\ (\log(N_f) - \log(N_v)) + 0.18 \quad (6)$$

$$\beta = 0.11\ (I + 0.1)\ (A + 4.4)$$

The distribution is cut off at F_{Δ}=2. The code gives rules for constructing distributions for loads and moments in various parts of the turbine. As exemplified in Figure 2, the rules use a simple conservative summation of the deterministic (gravity induced) and the stochastic parts of the loads:

$$S_\Delta(N_\nu) = C_F F_\Delta(N_\nu) + C_G G_\Delta(N_\nu) \quad (7)$$

Formulas for the scaling factors on the two terms in the summation are given in the code for many forces and moments. The factors depend only on turbine parameters. This straight addition of the two accumulated definitions conforms to the IEA recommended practices for fatigue loads.

The form of these distributions comes from many measurements of load width distributions by the Test Station for Wind Turbines. This distribution has shown up to work quite well as illustrated in Figure 3, where the standardized

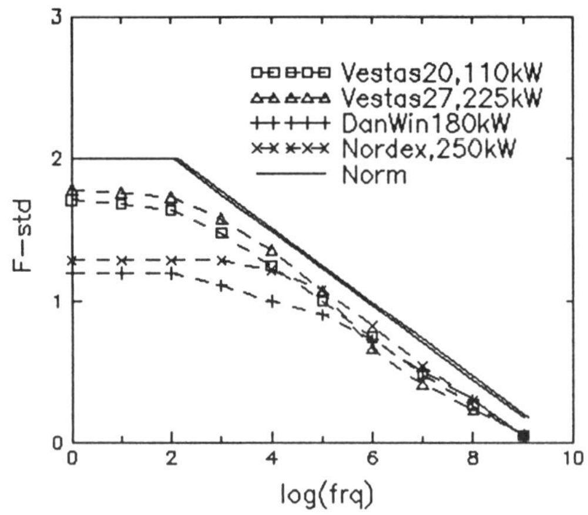

Figure 3 Comparing the danish code standard spectrum derived from 4 different machine with the code definition.

distribution has been derived from blade root flat wise moments on 4 different turbines of different sizes.

2.7 Level of Safety and Partial Coefficients.

The level of safety in a code is determined by the combined choices of all four factors in (1). This means, that measures of load and material resistance, and the partial coefficients for loads and materials shall be defined.

The load values are defined similarly in the Danish code and in the IEC draft, which both use 50 yr recurrence interval for wind load extrema.

The partial coefficients for loads are somewhat different in the Danish code and in the IEC draft. For ultimate limit states the two codes give partial coefficients for load as shown in the Table III. For the fatigue and for the serviceability limit states the Danish and IEC codes agree on 1.0.

Table III: Partial coefficients for ultimate loads, DS versus IEC

Load Type	DS 472	IEC
Aerodynamic	1.3	1.3
Functional	1.3	1.4
Gravitational	1.0	1.1
Inertial	1.0	1.2

It is seen, that the IEC partial coefficients for loads tend to be somewhat larger than the Danish ones.

For materials the Danish code relies on existing Danish materials codes. Only for GRP and for wood, DS-472 has prescribed a special value for fatigue of 1.7 for the normal safety class.

The IEC committee had difficulties with the lack of good international materials codes, and have therefore chosen to allow the use of national codes for materials. It does request the user of national codes to make sure, that the codes used gives a 'level of safety' corresponding to the level the IEC code is aiming at. The IEC Code establishes a minimum level of safety. It does require the materials resistance values to correspond to at least 95% survivability. And it does not allow materials partial coefficients in ultimate limit states under 1.15. It will mean, that most national codes which will be used are tougher. For steel (ultimate) we use 1.27 in Denmark. It is seen, that this, compared to the minimum of 1.15, together with the larger load factors of the IEC code leads us to a comparable safety level. For fatigue the IEC prescribes 1.25 for non-fail-safe components. This is to compare with a danish steel fatigue number of 1.56, which means, that here the Danish code is two safety classes stronger than the minimum IEC requirement. (One safety class is approximately a factor 1.1). For GRP and wood (fatigue) the danish code is even more conservative. This presumably reflects the fact, that these materials are new and/or not too well known.

3. CONCLUSIONS.

The Danish code of practice, DS-472, have been available less than a year, but is coming in extensive use. Among others, the certification office of the Test Station for Wind Turbines the code use as a part of their certification rules. Already earlier drafts of the Danish codes have been in use in the certification process. This means, that by now a considerable experience with the code has been obtained. The experience so far shows, that the code has its powers and only few and small shortcomings. The comparison with the IEC Standard seems to indicate, that the safety levels of the two codes are quite comparable. Therefore we are confident, that it will be useful and without major problems. We do regret, however, that the strict danish requirement about aerodynamic brake systems have not been adapted, as we still consider it of significant importance for the safety. We do, however, appreciate the huge advantage in having an international code, that will make the documentation of machines for use everywhere much easier for all manufacturers, Danish or others.

REFERENCES:

[1] DS-472: last og sikkerhed for vindmøllekonstruktioner, (in Danish), Copenhagen, May 1992.

[2] DS-409, Code of practice for the Safety of Structure. Dansk Ingeniørforening, Copenhagen 1983

[3] NKB 55: Retningslinier for Last og Sikkerhed for Bærende Konstruktioner. Nordic Committee for building regulations. 1987

[4] ISO 2394: General principles on reliability of structures, International Organization for Standardization, Geneva, 1986.

MONITORING FATIGUE LOADS USING CYCLE COUNTING FIELD COMPUTER

H. Seifert, H. Söker
Deutsches Windenergie-Institut (DEWI)
D-2940 Wilhelmshaven, Ebertstr. 96
Germany

ABSTRACT: Fatigue loads on components of wind turbines are commonly determined on the basis of short time sequences. However, recording raw data is time consuming and expensive. As an alternative, the presented project suggests long term fatigue load monitoring using on-line cycle counting methods. In parallel meteorological data of the site and operational data of the wind turbines have to be recorded in order to be correlated with the obtained fatigue load spectra.

In order to determine the rotor blade load spectra of wind turbines operating in a wind farm, on-line cycle counting data acquisition systems (DAS) were installed on the hubs of two 300kW pitch controlled wind turbines. Under the prevailing wind direction one of the turbines operates in undisturbed flow, whereas the other turbine operates in wake flow. An independent data logger for meteorological and operational data assessment was installed in the utilities substation. Despite some problems during commissioning of the DAS this paper presents first results. It is planned to monitor the fatigue loads for at least one year.

1. INTRODUCTION

As in any industrial application, the duration of a wind turbine's life is a key parameter for the evaluation of its economic potential. Assuming a life span of 20 years, components of the turbine have to withstand a number of load cycles of up to 10^8. Such high numbers of load cycles pose high demands on the fatigue characteristics of both, the used materials and the construction. However, the specific features of these fatigue loads are mostly unknown yet and can be expected to vary with the type of turbine and the site of operation. Measurements have to be carried out in order to adopt and verify load assumptions for different wind turbine configurations and different sites. Improved load assumptions will be of great importance for the development of the next generation of larger scale wind turbines.

2. METHOD OF APPROACH

Fatigue loads on components of wind turbines are commonly determined by applying cycle counting methods on short time sequences of the respective loads [1]. The resulting frequency distributions of the load cycles are extrapolated and combined according to long term operation statistics in order to obtain an amplitude spectrum. This fatigue load spectrum forms the input data for life time calculations. However, recording raw data requires data acquisition systems with high expenditure of costs and man power. Only a limited number of time sequences may be evaluated and rare events like extreme wind speeds, high gusts and fast changing wind directions may not be recorded at all. As an alternative, the presented project suggests long term fatigue load monitoring using on-line cycle counting methods. Thus compilation of a cost effective, easy to handle and reliable data acquisiton system (DAS) capable of on-line data reduction is the primary task when following this approach. The second objective is to evaluate the fatigue loads on the blade roots of two wind turbines operating in a wind farm. Under the prevailing wind direction one of the turbines operates in undisturbed flow, whereas the other turbine operates in wake flow. On each turbine strain gauges are mounted in half bridge configuration to measure bending moments at the blade root in edgewise and flapwise direction. The signals are sampled and reduced on line employing the Rainflow counting algorithm. The data are stored as frequency distributions of the counted load cycles in form of matrices.

In parallel, meteorological data of the site and operational data of the wind turbine have to be recorded in order to be correlated with the derived fatigue load spectra. For this purpose an independent data logger and a meteorological tower were installed in the wind farm.

Both data acquisition systems are read out once a month. It is planned to obtain load spectra of edgewise and flapwise blade root bending moments as well as the corresponding meteorological and operational statistics for a time span of at least one year.

Two 300 kW wind turbines in the windfarm of Hamswehrum, a small village in the northern part of Lower Saxony, are equipped with data acquisition systems. The wind farm is commercially operated by the local utility EWE (Energieversorgung Weser Ems) and consists of 11 ENERCON 33 wind turbines with 33m rotor diameter, variable speed and pitch control. The turbines are positioned in rows of five and six, respectively. Turbine number twelve is the prototype of a new 36m diameter turbine (Figure 1). Under the prevailing wind direction one of the turbines (#7) operates in undisturbed flow whereas the other turbine (#3) operates in wake flow. The prevailing wind direction is depicted on the top of Figure 1.

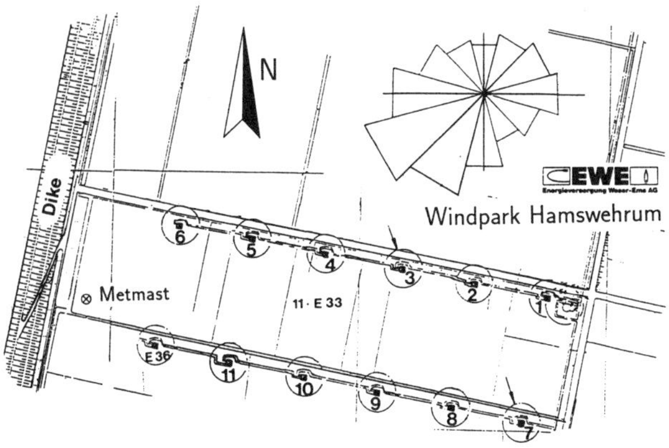

Figure 1: Wind Farm Hamswehrum

3. DATA ACQUISITION SYSTEM

The DAS is made up by two independent subsystems as shown in Figure 2.

Figure 2: Data Acquisition System

3.1 Field Computer

A commercially available field computer (FC), protected against moisture and rain, is mounted on the rotating hub of each of the observed wind turbines. Each system is equipped with two strain gauge modules. These modules supply and sample two strain gauge brigde circuits and are calibrated to measure bending moments in edgewise and flapwise direction. The sampling rate is set to 100 Hz. On each channel's signal the FC performs Peak-Valley and Rainflow counting and stores the results as frequency distributions of the counted load cycles in form of matrices. The FC are read out via RS232 communication link to a laptop computer. The data are checked directly on the turbine, making it possible to detect faults immediately. A modem connection to the telephone line may be an option in the future.

The FC's external energy supply system is battery buffered and is fed from the grid through slip rings. The slip rings are divided into two parts for assembly on already operating turbines and transmit safety extra-low voltage.

3.2 Ground data logger

The meteorological data i.e. wind speed at heights of 18m and 42m, wind direction at 42m, temperature, and air pressure as well as the operational data such as power output, rotor speed and status signals are recorded by the ground data logger (GDL). They are sampled at a rate of 1Hz. The samples are then stored as 10-min-averages and corresponding standard deviations. They are evaluated with commercially available software which is developed and maintained by DEWI. Figure 3 shows an example for the evaluation of the wind speed at 42m height. The percentage of in wake operation of the two turbines is determined from the wind direction evaluation.

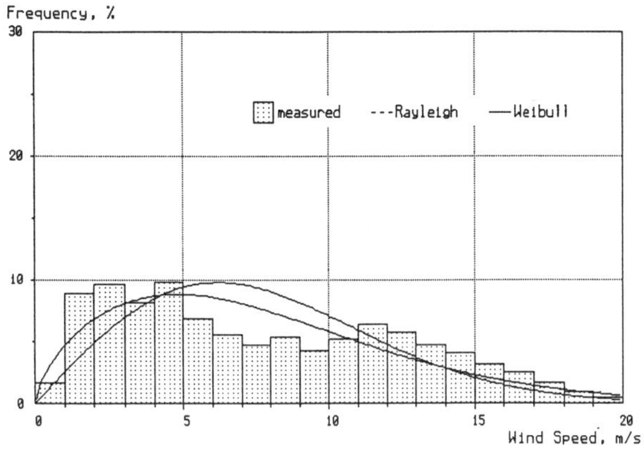

Figure 3: Frequency Distribution of Wind Speed at a Height of 42m

The advantages of the presented DAS are comparably low costs of the hardware and automatic operation combined with a rather quick evaluation of the data.

3.3 Calibrations

The measured bending moments are calibrated with the moment calculated from the blade mass and its centre of gravity. In order to ensure reliability of data, time sequences of the signals are recorded while the turbine is rotating slowly. The range between the extrema of that time sequence can be related to the blade bending moments and a possible zero drift can be detected. Recording these time sequences is part of the procedure when reading out data wich is done once a month.

3.4 First experiences

Laboratory tests of the field computers led to satisfying results. However, after the first in field test they were found damaged, obviously due to electrostatic discharge. In cooperation with the FC's manufacturer an improved, additional electrostatic shielding concept was developed and a special "bleed-off" circuitry is now used when establishing the communication link. In addition a special proceedure was defined for data read out.

The following in field test started promising. Reading out the data from turbine #7 after six days gave the expected results. At this point the second FC was installed on turbine #3 and the test was started. Three days later the data were read out again. This time the FC of turbine 7 presented good results on both channels whereas the FC on turbine #3 delivered good results on the flapwise channel only.

A month later, the next attempt to retreive data from the two FC turned out to be rather disappointing: The system of turbine #7 was damaged again, presumably due to a lightning stroke and the system on turbine #3 offered bad results on the edgewise channel, because its plug was corroded. Thus, the most pressing work in the near future is to improve the reliability of the DAS.

On the other hand, the GDL proved to be more reliable. It operated since installation without any failure.

4. FIRST RESULTS

A preliminary evaluation was performed on the recorded data in order to give an idea of the nature of the results that can be expected. From turbine #7 data were taken after six days and again after nine days of typical operation i.e. the turbine was operating without failure and the wind speed stayed in the turbine's range of operation. The data of the six days measurement were extrapolated to be equivalent to a period of time of nine days and compared to the nine days measurement. Figure 4 and Figure 5 show the Rainflow matrices of the nine days measurement.

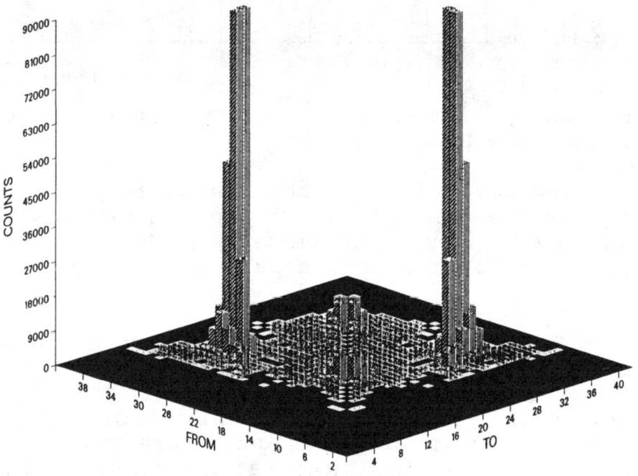

Figure 4: Rainflow Matrix of Edgewise Bending Moment (Turbine #7)

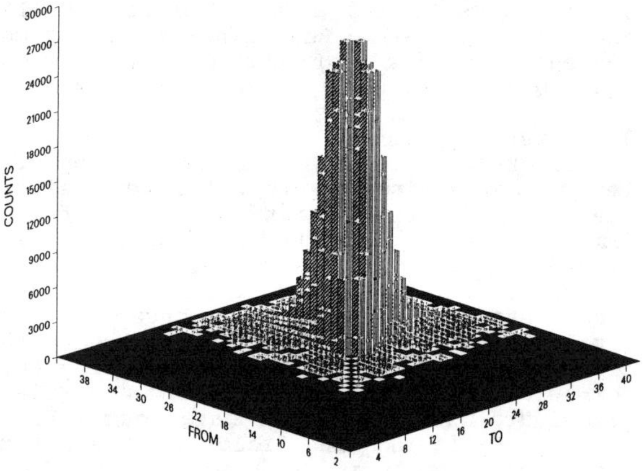

Figure 5: Rainflow Matrix of Flapwise Bending Moment (Turbine #7)

The edgewise bending moment is dominated by the deterministic gravity load of the rotating blade having a relatively high mass. On the other hand the flapwise bending moment shows the expected stochastic behaviour due to the wind speed fluctuations. This is clearly confirmed by the obtained amplitude spectra which were choosen in order to facilitate the comparison (see Figure 6).

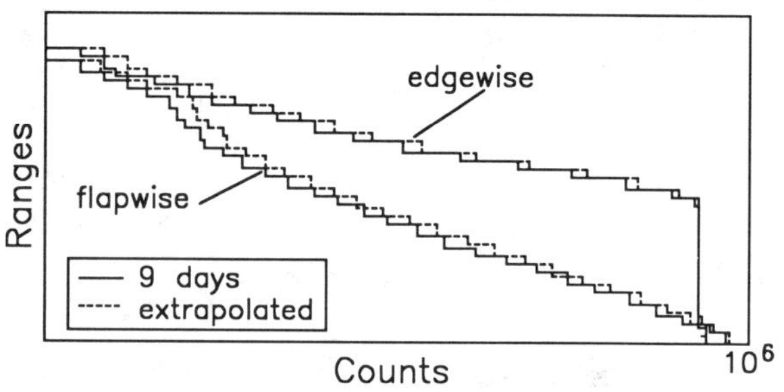

Figure 6: Amplitude Spectra (Turbine #7)

Each pair of curves depicts the extrapolated spectrum and the spectrum of nine days continuous measurement. The differences between the extrapolated curves and the measured curves are of minor influence and can be ignored.

The measurements of the flapwise bending moment of turbine #3 (not calibrated in the same way as on turbine #7 see Figures 7 and 8) over periods of time of three days and 34 days were evaluated in the same way as the data of turbine #7. However, the amplitude spectra now show a tremendous difference between the shapes of the extrapolated and the measured matrix (Figure 9). In that 34 days period the high wind speeds of the January storms were recorded, leading to 10-min-averages of up to 31 m/s.

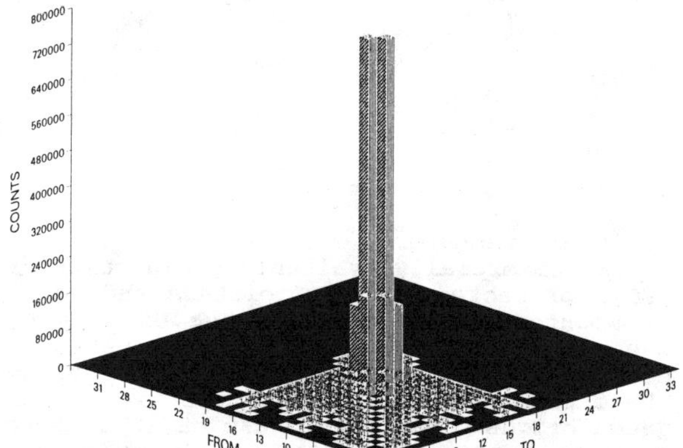

Figure 7: 3 Days Rainflow Matrix of Flapwise Bending Moment (Turbine #3)

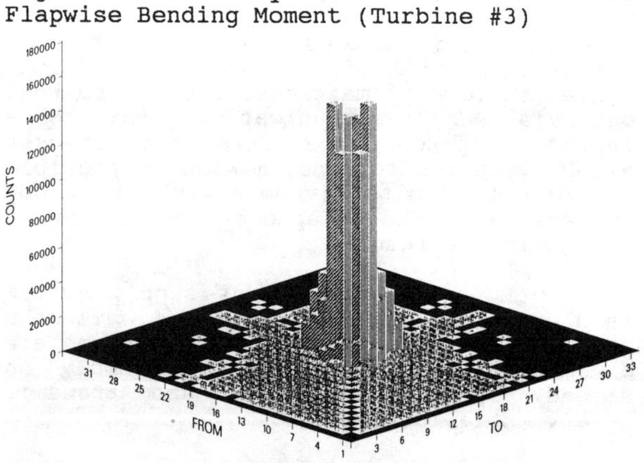

Figure 8: 34 Days Rainflow Matrix of Flapwise Bending Moment (Turbine #3)

The evaluation of the amplitude spectrum reflects this in showing higher amplitudes at lower cycle numbers compared to the extrapolated curve. Also, in the interval between 10^4 to 10^6 the loads are underpredicted by the extrapolated curve.

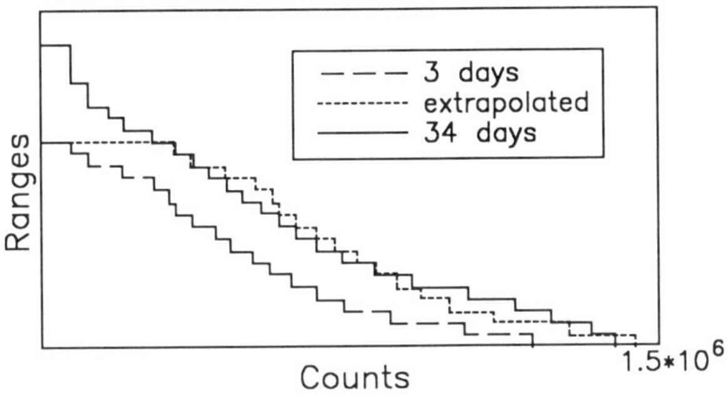

Figure 9: Amplitude Spectra Flapwise Bending Moment (Turbine #3)

Looking at the iceberg model of the lifetime spectrum [2] the stated results seem to be correct.

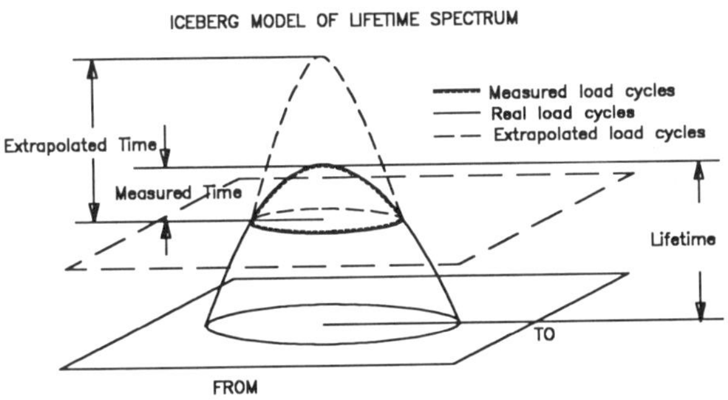

Figure 10: Iceberg Model of Lifetime Spectrum

4. CONCLUSIONS

This first evaluation of the on-line counted Rainflow matrices shows that long time monitoring of fatigue loads can lead to more reliable results at low and high numbers of load cycles. This is especially important for the rotor blades of the next generation of larger wind turbines. It is important to perform such measurements at different turbines in order to improve the load assumptions. Compared to the costs of large scale wind turbines the costs of the proposed monitoring method are marginal, whereas the results can help to save material without diminishing safety.

The next step in the project will be the comparison of load spectra measured at wind turbines of the same type operating in the wind farm. Load spectra are also gathered from a turbine operating at a site with varying turbulence intensity in Greece and at a stall-controlled turbine operating at Alsvik wind farm in Sweden. This is part of a Joule II programme DEWI carries out in cooperation with the Center for Renewable Energy Sources (C.R.E.S.) and the Aeronautical Research Institute of Sweden (FFA).

5. ACKNOWLEDGEMENTS

Although measuring in a commercial wind farm, owned by the local utility EWE, DEWI is allowed to stop the turbines for installation and calibration of the DAS. The authors wish to thank EWE and especially Mr Mentjes who always assists the DEWI staff during their work on the field.

The described measurements are part of a research programme supported by the Commission of the European Communities. The partners are C.R.E.S. in Greece and FFA in Sweden.

REFERENCES

[1] P.H.Madsen et al.(Ed): Recommended Practices for Windturbine Testing and Evaluation: 3.Fatigue Loads, International Energy Agency Programme for Reasearch and Development on Wind Energy Conversion Systems 2nd Ed. Paris (1990).

[2] T.Haberle, T.Kramkowski, H.Söker: Ermittlung von Ermüdungslasten: Rohdatenauswertung oder direkte Datenreduktion?, Proceedings DEWEK'92 Conference, Deutsches Windenergie-Institut Ed. Wilhelmshaven (1992) 71.

DEVELOPMENT AND TEST OF AN EXTRUDED ALUMINIUM ROTOR BLADE FOR HORIZONTAL AXIS WIND TURBINES AND COST MINIMIZING SERIES PRODUCTION

Axel Andreä
aerodyn Energiesysteme GmbH
Redderbek 1
D-2333 Damendorf

ABSTRACT: The economic efficiency of wind turbines is essentially characterised by their production costs. In this connection it is important to exploit to the full the opportunities for cost savings on a large number of system components. The rotor blades of a wind turbine in particular offer substantial possibilities of reducing costs, as rotor blades are currently manufactured mainly from fibre composites in a cost-intensive and largely craft process. Extruded aluminium rotor blades seem to be virtually predestined for series production, and their potential for horizontal-axis wind turbines is being investigated as part of a research project subsidised by the Federal Ministry for Research and Technology (BMFT). The production and use of these rotor blades for turbines of up to 200 kW is considered to be feasible. With a view to scaling up, prototype rotor blades are being developed and produced initially for trials on a 30 kW pitch-regulated wind turbine.
A blade flange manufactured by the aluminium vacuum-casting process is provided as the blade connection component; it is bolted transversely to the load-bearing cross-section of the extruded section.
The critical requirements for the structure as a whole are determined by fatigue loads. The fatigue strength of the structure will be investigated in detail in laboratory. At the same time field tests are being planned which are intended to demonstrate the suitability of rotor blades of this kind for horizontal-axis wind turbines.
An initial calculation of the costs of rotor blades manufactured individually and in series shows that a reduction in manufacturing costs of up to 50% can be achieved with extruded aluminium rotor blades compared with fibre composite rotor blades manufactured by the conventional process.

1. INTRODUCTION

Economic aspects are becoming increasingly important in the use of wind energy. The justification of the use of this form of energy to supplement the forms of energy traditionally used requires cost-effective production of the energy unit.The reduction of the manufacturing costs of a wind turbine constitutes the precondition for the achievement of economic effectiveness. The full potential for reducing costs in the production process is a long way from being achieved in many areas. Many manufacturers are endeavouring to obtain economies in the areas of system components or manufacturing processes without reducing product quality. In this connection it may be appropriate in certain circumstances to refrain from striving for a maximum degree of efficiency in energy conversion and to allow for small trade-offs. As long ago as 1986, aerodyn tackled the problem of possible cost reductions in the production process in the context of the development of a 30 kW wind turbine (Heubel and Siegfriedsen [1]). Considerable potential for cost reduction was seen in the manufacture of rotor blades by the aluminium extrusion process.

The use of the extruded aluminium rotor blades on a two-blade wind turbine with a rated output of 30 kW, however, did not lead to a success. After only a short period of operation cracks appeared in the vicinity of the blade roots, and these ultimately led to the destruction of the rotor blade.

2. DAMAGE ANALYSIS

Following the receipt of a promise of support from the Federal Ministry for Research and Technology (BMFT) at the end of 1991 for the continuance of the project, work on it was taken up again.

Prior to the revision of the design, a detailed analysis of the damage to the defective rotor blade was carried out.

2.1 Design of the damaged rotor blade

Because of the size of the available extruder, the extruded part of the rotor blade, which had an overall length of 6.25 m and a blade chord of 0.4 m, was manufactured as two part- sections in the longitudinal direction (Fig. 1). The material used was AlMgSi1. The nose and rear sections were joined by a permanent clip connection, which was additionally bonded with special adhesive. The blade tip was a component produced by the GRP negative process, which was bonded to the jointed part-sections of the blade.

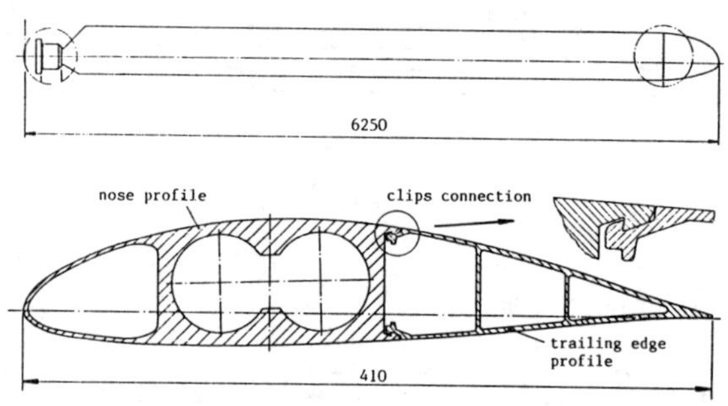

Fig. 1: Blade dimension and cross section of the extruded part sections of the damaged rotor blade, [1]

The characteristic inner contour of the load-bearing cross-section serves as the connecting structure to the blade flange. The basic unit of the blade flange consisted of a cylindrical injection-moulded component made from solid material, out of which the external contour of the load-bearing cross-section of the nose section was milled so that the latter could be fitted exactly into the flange. The necessary connection between the two parts was ensured by the use of clamping sleeves with tapering tensioning bolts to provide a frictional link (Fig. 2). In order to prevent the penetration of moisture into the hollow sections of the blade, cover plates were fitted and welded on in the transitional area between the blade section and the flange.

Following the issue of the type test for this blade design, a start was made on the manufacture of the rotor blades, followed by their fitment on the turbine.

2.2 Course of the damage and consequences

After only a short period of operation hairline cracks appeared in the area of the weld seam of the attached cover plate, "Washing out" of the adhesive in the area of the clip connection was also observed.

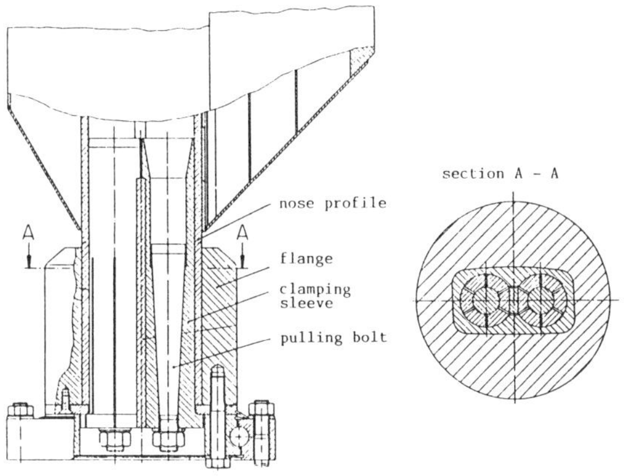

Fig. 2: Joint of extruded nose section and blade flange using clamping elements

Finally, after some months of operation, the first blades on a wind turbine broke in the transitional area between the blade section and the flange. According to a thorough analysis, the cracks started to form in the weld seam of the attached cover plate and spread in the area of the rigidity transition of the nose section - due to the inserted clamping sleeves - and into the cross-section, ending up with a overload break (Fig. 3). The formation of the cross-sectional surfaces of the break indicates, from the lines of rest, a typical break caused by material fatigue (Meyer [2]).

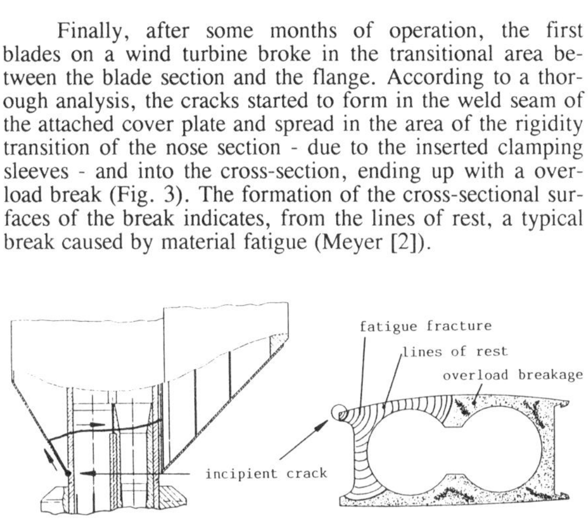

Fig. 3: Crack propagation on the damaged rotor blade and principle view of the cross section of rupture

2.3 Examine the first design

First of all it was necessary to check the operating load assumptions which had formed the basis for the dimensioning according to (Matthies H. et. al. [3]). To this end comprehensive measurements were carried out on an undamaged rotor blade fitted to a turbine. These showed that the load alternation figures in some instances exceeded the assumed values by 25%. On the basis of the measured loadings, a simplified single-stage loading spectrum was compiled and a deterioration calculation carried out for the cross-section at the break. The Wöhler stress- cycle lines required for this were taken from the data in [2]. The results show that there is a 97.7% probability that the material will fail after approximately 1.8E+6 load changes (Fig. 4). This matches well with the numbers of load changes of 2.0E+6 occurring during the operating time.

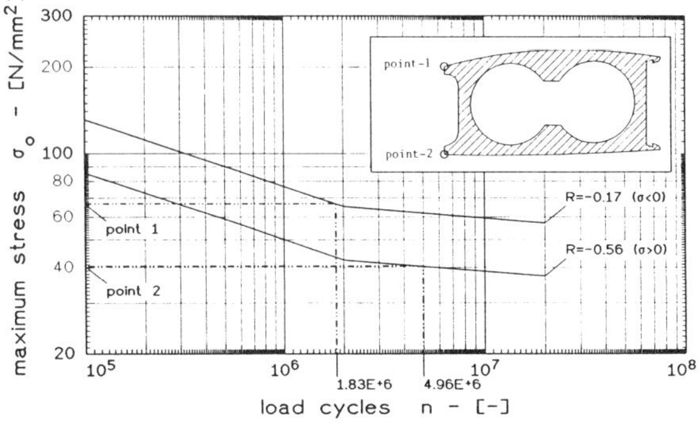

Fig. 4: Damage analyses of the defective blade cross section based on simplified fatigue loads derived from measurements

3. REDESIGN OF THE EXTRUDED ALUMINIUM ROTOR BLADE

3.1 Choice of aerodynamic profile

A large number of boundary conditions limits the choice of aerodynamic profile to a small range. Thus the demand for maximum energy yield and low aerodynamic drag coefficients of an extruded aluminium rotor blade conflicts with the rectangular contour of the rotor blade dictated by production considerations and with the demand for the maximum section modulus in the direction of flapwise bending combined with the use of the minimum extended mass.

The design of the prototype rotor blade was undertaken for a 30 kW two-blade pitch-regulated wind turbine. The experience gained from the use of the FX84-W175 section with the damaged blade was very good with regard to control behaviour, sensitivity to dirt and noise generation [1]. Thus parameter studies were first carried out on the basis of a selected section chord of 400 mm.

The partly concave contour pattern of the section gave rise to no problems in manufacture by the extrusion process, whereas other manufacturing processes prohibited the use of such section shapes on grounds of cost.

Regarding the extreme thin trailing edge which is possible to realise by the extrusion process and the relatively low aerodynamic lift coefficient at the outer rotor radius, due to const. cross section in longitudinal direction, a remarkable reduction in aerodynamic noise generation is possible.

3.2 Design of the rotor blade

3.2.1 Constructional criteria for decision

In the design an effort was made to use a section as thin as possible from the aerodynamic point of view. To what extent this could be achieved with the necessary strength was the subject of investigation. 18%, 21% and 24% thicknesses were therefore studied for the FX84-W section. The basis for the choice of dimensions was the static load from the century gust of wind as given in [4]. As the section is manufactured from part-sections, as described, the position of the join in the section chord must also be determined, taking into account the extruder available.

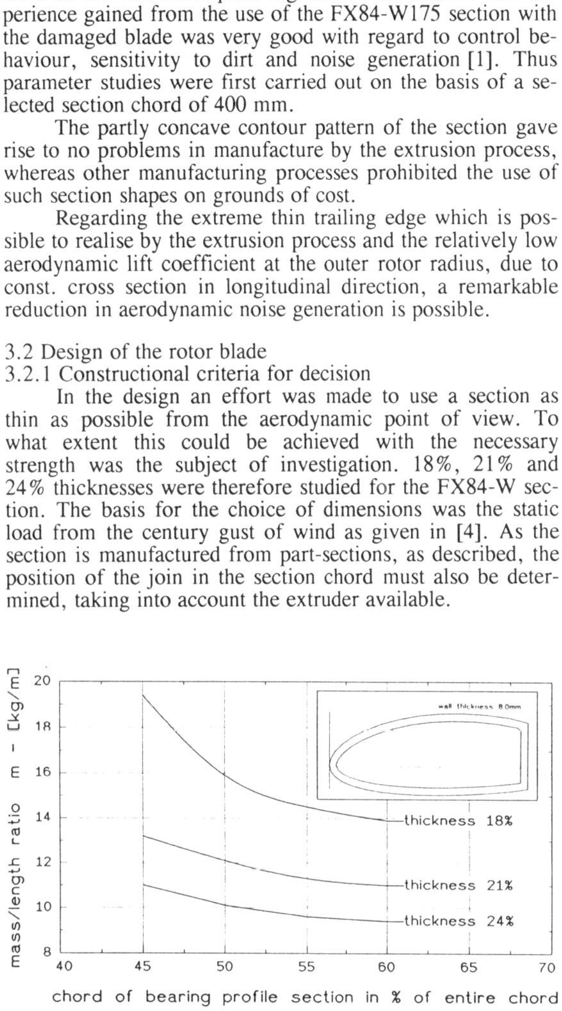

Fig. 5: Required extended mass of different thicknesses using FX84-W profile section under extreme load depending on the chord of the load bearing part section

Using the parameters of the material AlMgSi1, a remarkable saving in weight, relatively independent of the section chord of the load-bearing cross-section, is recorded when changing from the 18% to the 21% section (Fig. 5). A further saving in weight appears to be possible if a higher tensile-strength material is selected (Fig. 6), but at this point other decision criteria come to the fore. The differences in the static strengths of the materials are extreme, while the difference in the fatigue strength is very small, with the result that with materials of increasing strength a change takes place from dimensioning based on extreme loads to designing based on fatigue strength. This effect is further intensified by the higher rate of utilisation of the material, which results in reduction in wall thickness as the alternation of tension due to the fatigue loads increases (Fig. 7). The proportion of fluctuating inertia loads is certainly lower as a result of the reduction in weight, but this reduction is hardly significant, as the fluctuation proportions due to aerodynamic loads predominate with the blade dimensions used here.

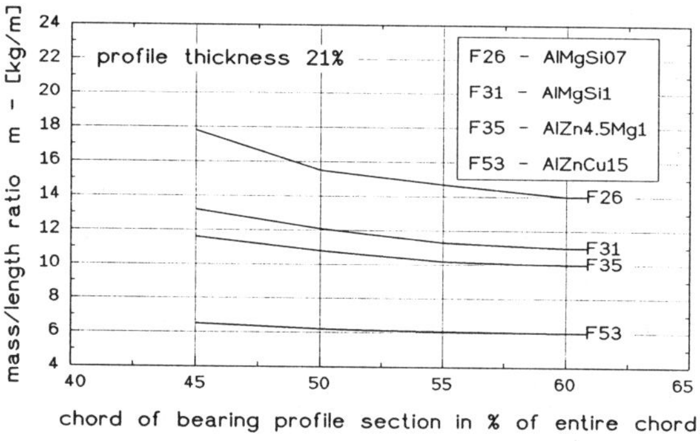

Fig. 6: Additional possibilies of mass reduction using high strength material

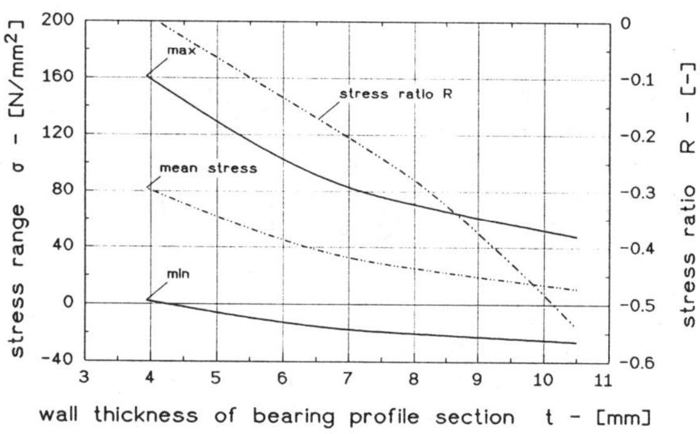

Fig. 7: Stress range due to variation in wall thickness using a rectangular fatigue load spectrum

Another aspect is the suitability of a high-strength aluminium alloy for use at wind turbines in marine atmospheric conditions. The resistance to corrosion decreases as the strength increases, notably due to the presence of Cu as an alloying element. If a material of this kind is used, higher material and manufacturing costs will also be incurred, as well as expensive and complicated corrosion-protection measures.

Detailed studies of the interrelationships described above led to the choice of the aerodynamic section FX84-W218 and the material AlMgSi1 for the load-bearing cross-section, which represents 65% of the optimised total section chord of 410 mm of the section.

3.2.2 Dimensioning

Since the beginning of 1992 a set of rules has been available for the dimensioning of aluminium components based on Eurocode 3 for steel structures [5]; these rules were used in the design of the rotor blade cross-section. The assumed loads for demonstrating fatigue strength were calculated on the basis of a simplified deterioration-equivalent trapezoidal spectrum. The stress ranges derived from this form the basis for the Palmgren-Miner deterioration calculation, using the associated Wöhler stress-cycle curve corresponding to the notch classification of the load-bearing cross-section. By clever shaping of the inner contour it was possible to construct the cross-section in such a way that both the fatigue strength in relation to the specified service life of 20 years and the requirements with regard to extreme stresses can be maximised.

3.2.3 Construction of the join between the part-sections

As the rear section does not perform any load-bearing function but merely determines the contour, the cross-section can be made extremely light. The material used is AlMgSi05.

As a result of the damage analysis particular attention has been paid to the joint. The critical problem is the transfer of shear flow from the nose section to the rear section. A rigid connection by bonding or by bonding in conjunction with riveting is ruled out because of the experience in the past and also because of possible contact or frictional corrosion. The holes drilled for the rivets also produce a notch effect in the load-bearing cross-section, which must be avoided.

Analysis of these problems led to the choice of a flexible bonded joint with an adhesive layer thickness of 2-3 mm between the nose section and the rear section (Fig. 8). The FEM (finite element method) calculation shows that because of the extremely low modulus of elasticity ($E = 5 N/mm^2$) shear flow into the rear section is very small (Fig. 9).

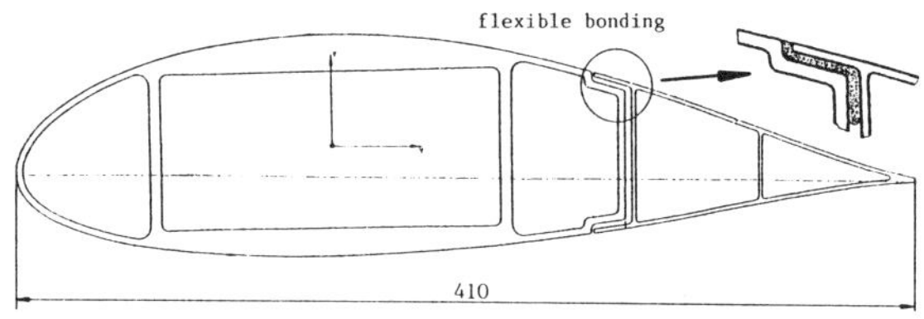

Fig. 8: Cross section of the devellopped aluminium rotor blade with joint of part profiles by flexible bonding

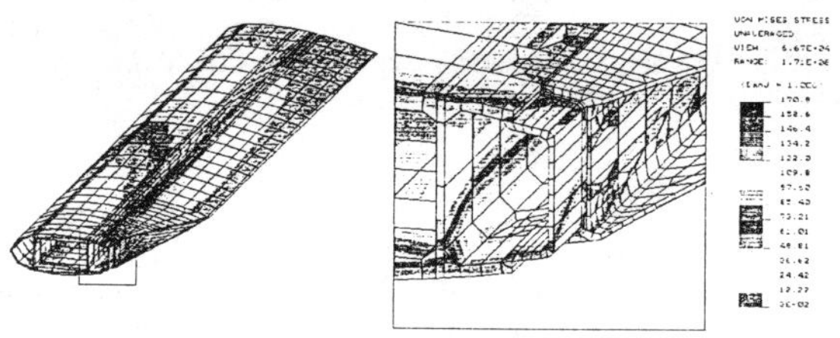

Fig.9: FEM-calculation of the aluminium rotor blade with respects to the flexible bonding under extreme load condition

The joining process during manufacture is carried out with the aid of a mechanism which ensures that the previously applied adhesive is distributed in the bond joint in the most effective way.

3.2.4 Construction of the blade connection

The transfer of the force flow from the rotor blade to the blade flange and ultimately to the hub sets the most severe demands on the design. On the basis of a concept study that was carried out, a solution was found involving a link between the rotor blade and the flange using a high-strength friction grip (HSFG) joint (Fig. 10). The advantage of this construction lies in the induction of compressive stresses in the area of the drilled holes, as a result of which the notch effect at this point can be almost completely eliminated (Warren [6]).

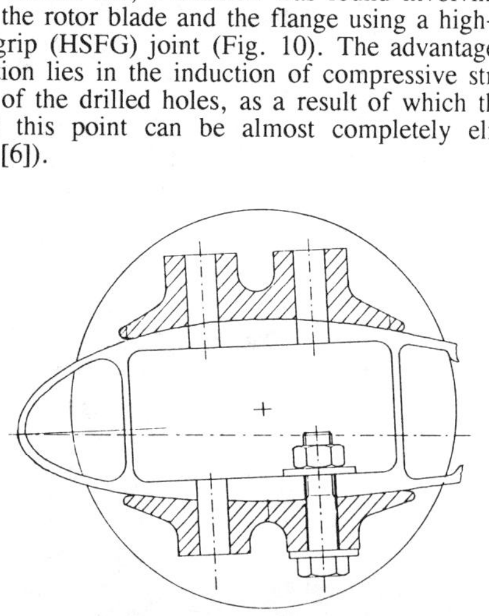

Fig. 10: Cross section with bolted joint of load bearing section and flange using HSFG-bolts

The blade flange is manufactured by the vacuum casting process, using AlSi7Mg as the material. It consists of two parts in the form of jaws, bolted to the top and bottom of the profile and form one unit after assembly (Fig. 11).

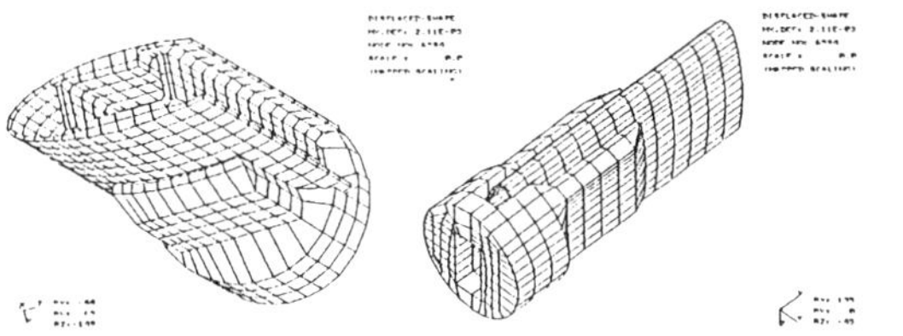

Fig. 11: Grid model as principle 3D-view of the blade-flange connection

4. COST COMPARISON

Following the finalisation of the design, a cost estimate was prepared for the developed rotor blade in the 30 kW output class, together with a comparison with a rotor blade produced in the conventional manner from fibre composites. According to this, taking all costs into consideration, the labour costs are 17% less, whereas the material costs are 14% greater. The share of the blade flange in the total cost of the blade increases by a factor of 2 (Fig. 12). In this example the total costs of the aluminium rotor blade are about 51% of those of the GRP rotor blade. The relationship of the individual cost items of the aluminium blade to those of the GRP blade can be seen in Fig. 13. In addition to the low labour costs, amounting to only about 35%, even the material costs are worthy of note, as they represent only about 70% of those of a GRP rotor blade. The costs of the blade flange are almost the same.

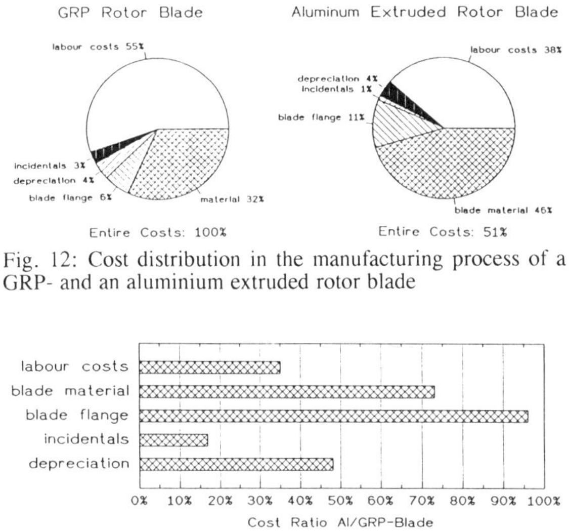

Fig. 12: Cost distribution in the manufacturing process of a GRP- and an aluminium extruded rotor blade

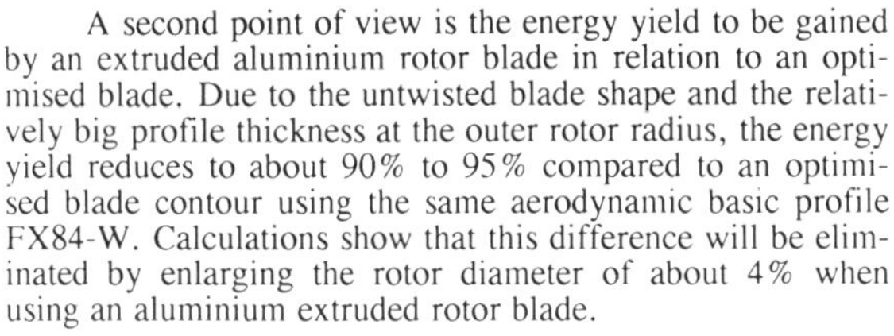

Fig. 13: Cost relation of an aluminium extruded rotor blade compared to a GRP blade

A second point of view is the energy yield to be gained by an extruded aluminium rotor blade in relation to an optimised blade. Due to the untwisted blade shape and the relatively big profile thickness at the outer rotor radius, the energy yield reduces to about 90% to 95% compared to an optimised blade contour using the same aerodynamic basic profile FX84-W. Calculations show that this difference will be eliminated by enlarging the rotor diameter of about 4% when using an aluminium extruded rotor blade.

5. PROSPECTS

The fatigue strength of the designed structure will be investigated in detail in the laboratory. At the same time field tests are being planned which are intended to demonstrate the suitability of rotor blades of this kind for horizontal-axis wind turbines.

On the basis of the input data gained under Section 3, preliminary calculations have been made of the possibilities of scaling up the rotor blades for wind turbines in higher output classes. Fictitious turbines were used in a first approximation. Once the blade dimensions have been fixed and the section chord optimised with a view to maximising the energy output and the manufacturing tolerances have been estimated, the principle of the design can be regarded as transferable. Only the choice of the location of the joins in the section still requires special investigation. Last but not least, consideration must be given to the consumption of material and the optimum utilisation of the extruders. It is considered possible and economically interesting to work towards an output class of about 200 kW.

6. SUMMARY

The possibility of designing an aluminium rotor blade to be manufactured by the extrusion process for wind turbines with a horizontal rotor axis was investigated, and a prototype is beeing produced for a turbine with a rated output of 30 kW. The crucial basis of the design work was the analysis of the damage to a rotor blade which failed in 1987 because of a fatigue fracture. The knowledge from the damage analysis enabled a design to be worked out that matched the demands made on it, with particular attention given to fatigue strength. The rotor blade is manufactured from extruded part sections, which are bonded together with a flexible adhesive. Extended test activities shall prove the validity of such rotor blades.

Detailed studies of this design principle from the "scaling up" aspect show that the production of such rotor blades for output classes of up to 200 kW is both technically possible and of economic interest. A detailed comparison between the manufacturing costs of an extruded aluminium rotor blade and those of a GRP blade shows that a cost reduction of about 50% can be achieved in production.

REFERENCES

[1] Heubel, G. / Siegfriedsen, S.
Demonstration der Einsatzmöglichkeiten des Windenergiekonverters aeolus 12.5 mit 5 Anlagen im Netzparallelbetrieb
BMFT-Forschungsbericht, Nr. 03-28-726 B
Damendorf, Mai 1988

[2] Meyer, J.
Gutachten zum Schaden an der Windkraftanlage HSW 30 der Husumer Schiffswerft
Bericht MB 88.080
Germanischer Lloyd Hamburg
Hamburg, Juni 1988

[3] Matthies, H. / Meyer, J. / Nath, C. / Richter, B.
Richtlinien für die Prüfung, Abnahme und Überwachung von Windkraftanlagen
BMFT-Forschungsbericht
Nr. 03 E-8384-A
Hamburg, September 1986

[4] N. N.
Vorläufige Richtlinie für die Zertifizierung von Windkraftanlagen
Fassung Juni 1989, Nachdruck Oktober 1989
Germanischer Lloyd Hamburg
Hamburg, Oktober 1989

[5] N. N.
European Recommendations for Aluminium Alloy Structures Fatigue Design
ECCS-Technical Committee 2, Aluminium Alloy Structures
First Edition 1992

[6] Warren, A. S.
Fatigue Behaviour of 6063 Aluminum Alloy Extrusions for Wind Turbine Applications
Massachusetts Institute of Technology
Cambridge, Massachusetts 1990

FULL SCALE TEST ON THE GRP ROTOR BLADE OF THE 500 kW NEDWIND-35 WIND TURBINE

D.R.V. van Delft [1)]
L. Drost [2)]
J.L. van Leeuwen [1)]
J. Olthoff [3)]

[1)] Delft University of Technology
STEVIN Laboratory; Steel & timber structures
Stevinweg 1
P.O. Box 5048
2600 GA Delft ; NETHERLANDS

[2)] NedWind Rhenen bv
Remmerden 9
P.O. Box 118
3910 AC Rhenen ; NETHERLANDS

[3)] NedWind Heerhugowaard bv
Stevinstraat 6
P.O. Box 1126
1700 BC Heerhugowaard ; NETHERLANDS

ABSTRACT: A full scale fatigue test has been carried out on the GRP rotor blade of the 500 kW NedWind-35 wind turbine. This fatigue test was a final verification of the fatigue resistance of the blade and the connection to the hub. Before fatigue testing stress measurements were carried out.

The blade has been subjected to test loadings (comprising 12 millions of cycles) which were based on design load cases, the properties of the GRP blade material and the Dutch design criteria. The severity of the test load is equivalent to the design load of 20 years.

Improvements in the design and production were made based upon these observations during the test.

1 INTRODUCTION

The 500 kW NedWind-35 wind turbine is provided with rotor blades made of GRP. This is a two bladed (stall regulated) wind turbine with a rotor diameter of 35 meters.

As a final design verification a test program on a blade was carried out. During this a static load test to measure the blade stiffness, natural frequencies and stresses was carried out. These measured values where checked against the design values.

After this static measurements a fatigue test was undertaken to check the fatigue resistance of the blade. During the fatigue tests the preload in the bolts connecting the blade to the hub section was monitored as well.

Having finished the fatigue test after 12 million cycles the stiffness and frequencies were measured again. Finally the blade was subjected to a ultimate load test until static failure.

The results of the test program are reported in ref [1] and [2]. In this paper the fatigue test phase of the test program of the blade will be discussed.

2 THE ROTOR BLADE CONSTRUCTION

2.1 The GRP blade construction

The rotor radius of the 500 kW NedWind-35 windturbine is 17.5 m. The GRP blade starts at a rotor radius of 2000 mm which results in a length of the GRP part of 15500 mm. The 2000 mm long tip section acts as a control tip and is connected to the blade by means of a steel tube which has its bearings attached in the 'main' blade.

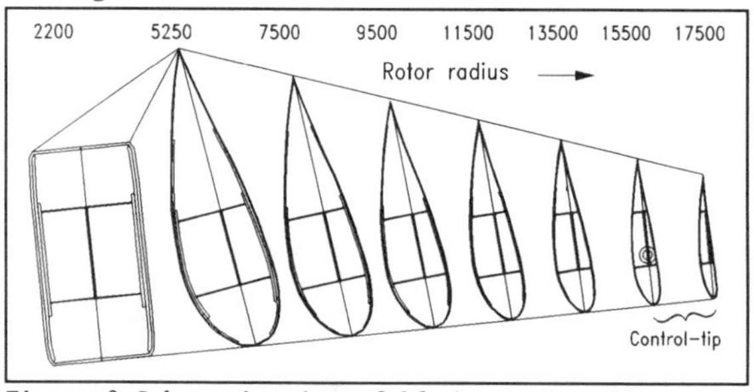

Figure 2 Schematic view of blade construction

An overview of the rotor blade in the test rig is given the picture in figure 1 while figure 2 a schematic impression of the blade construction is given. As shown figure 2 the blade profile has a

Figure 1 Overview of rotor blade in test rig

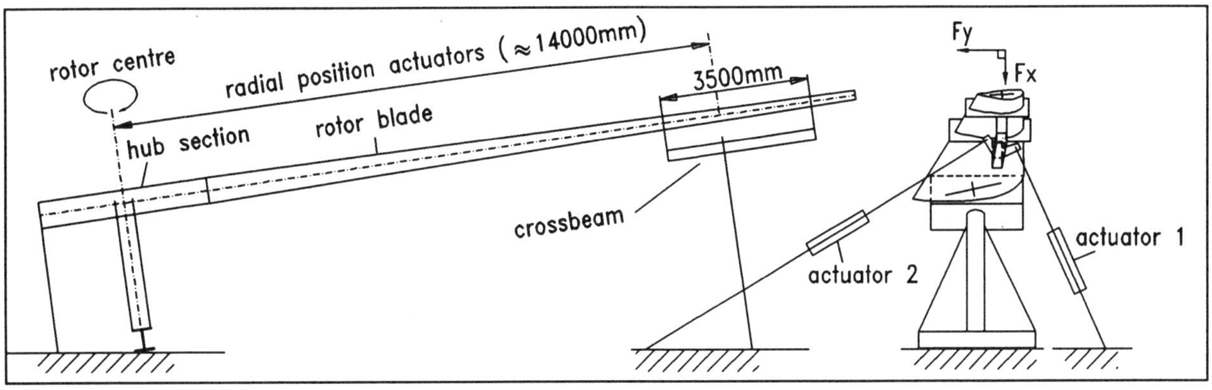

Figure 3 Schematic overview of test rig

box-shaped end for the connection with the hub. The blade has been build up with a laminated GRP skin and provided with an internal 'H'-beam acting as a stiffener. An overview of the material properties is given in table I.

The total mass of the blade as tested in the laboratory is 1755 kg with the mass center at a rotor radius of R = 6980 mm.

2.2 The blade-to-hub connection

The blade-to-hub connection consist of single strap bolted joints on both sides (upwind- and downwind side) of the blade.
During mounting of the blade to the hub the high strength steel bolts were tightened according to a tightening procedure to assure the prescribed preload of the m27 bolts.

Twelve bolts (see figure 5) were provided with strain gauges and calibrated. For these bolts the preload after tightening with the applied procedure was checked.

3 THE TEST SET UP FOR THE FATIGUE TEST

3.1 General

In the test rig set up the box shaped hub section and the main shaft were included. A photo of the test rig set up is given in figure 1 and in figure 3 a schematic overview.

The test set up was similar to full scale tests on rotor blades carried out earlier in the Stevin laboratory [3] en [4].

3.2 The load introduction

The loads were applied to the blade by two hydraulic actuators. Both actuators were positioned at the same rotor radius being 14150 (later this was changed to 14112 and 13662 mm). By these two actuators a force varying in magnitude as well as varying in direction could be applied. The actuators were connected to a crossbeam (having a length of 3.5m) which divided the loads over two cross sections of the blade as shown in figure 3.

On both ends the actuators were provided with elastic hinges [3].
The actuators had a capacity of 50 kN and 30 kN and a maximum stroke of 750 mm and 400 mm for actuator 1 and actuator 2 respectively.

3.3 The clamping of the hub to the floor

The main shaft and hub section were mounted to the floor of the laboratory in such a way that the shaft was almost in vertical position. This shaft was provided with triangular stiffener plates in one plane to prevent movement in that direction. Shaft and plates were welded to a H-beam which was clamped to the floor. The end of the hub without a blade was fixed in horizontal and vertical direction by tension bars.

3.4 Measuring and control system

The fatigue loading on the blade and the measurements of the transducers connected was controlled by the computer program FATEP running on a PC 386 machine ("PC 1" in figure 4). This program FATEP has been developed by the Stevin laboratory.

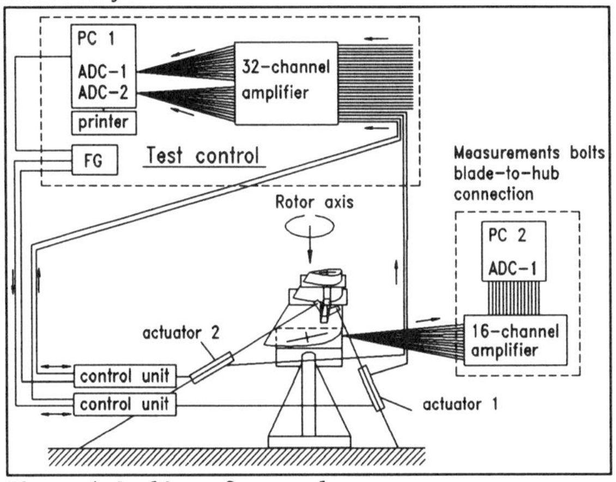

Figure 4 Outline of control system

The 386 computer was equipped with ADC cards for data acquisition and connected via a programmable function generator ("FG" in figure 4) to the actuator control units for the load excitation.

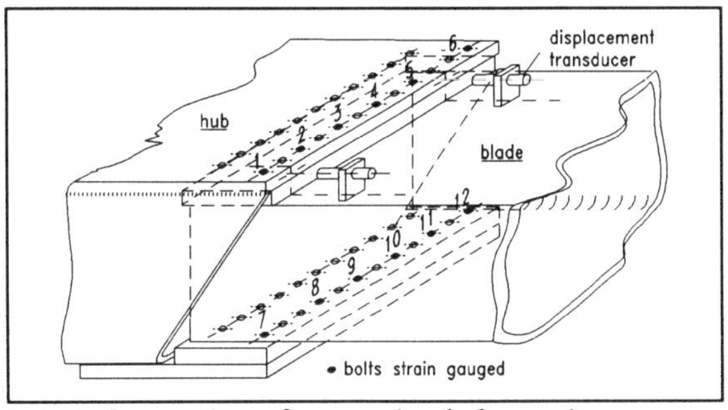

Figure 5 Location of measuring bolts and displacement transducers at blade-to-hub connection

During the fatigue test and several static tests the following values were continuously monitored:

- the applied load by means of load cells at the force excitation
- strains in the blade from strain gauges at different radial positions
- displacements of both the actuator pistons
- preloads from twelve strain gauged bolts in the blade-to-hub connection (located as shown in figure 5). This twelve bolts were monitored continuously by an independent measuring system as indicated in figure 4.
- displacements of the blade-to-hub connection,

location as indicated in figure 5

4 THE FATIGUE TEST

4.1 Fatigue loading

The starting point for the fatigue loading during the test was that in about 10 millions of cycles the fatigue loading at the blade (between radial positions R=2.2m and R=15.5m) was at least as severe as the design load.

From the original set of 83 load cases 6 typical load cases were used for the test loading. They were modified and increased by an additional load factor. As a result the original 500 millions of cycles reduced to 10 millions of cycles for the test load.

Two examples of test load cases which were applied are plotted in figure 6, where the forces in X-, and Y-direction at the cross section where the actuators are positioned are given. The X-, and Y-directions coincide with the flapwise, and edgewise directions respectively (as indicated in figure 3).

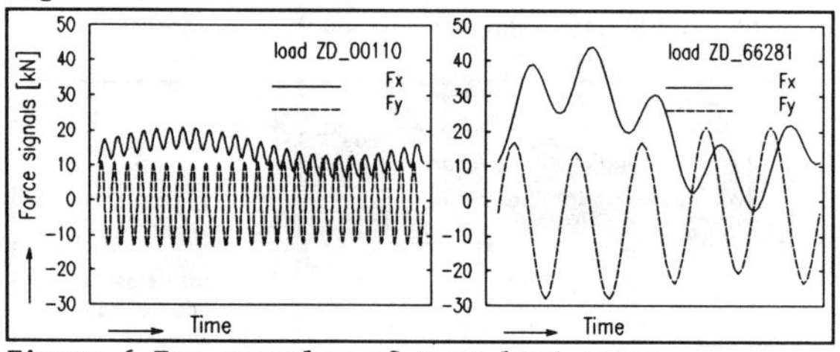

Figure 6 Two examples of test load cases

For the evaluation of the fatigue test loading these loads are compared with the latest design load cases as supplied by NedWind bv.

4.2 Observations during the fatigue test

In an early stage a crack was observed in the web section near the blade root at 0.65 million cycles. The blade design was modified and the crack repaired accordingly.

After 10.87 million cycles, which was beyond the equivalent life of 20 years, a failure occurred at the section were the tip was connected to the blade. This crack was repaired either so that the test on the blade could be continued. (blade section at rotor radius higher than R=13662 mm not loaded after this event because of removal of crossbeam).

During the stages in between these two events no changes in the response of the structure and no damage could be observed.

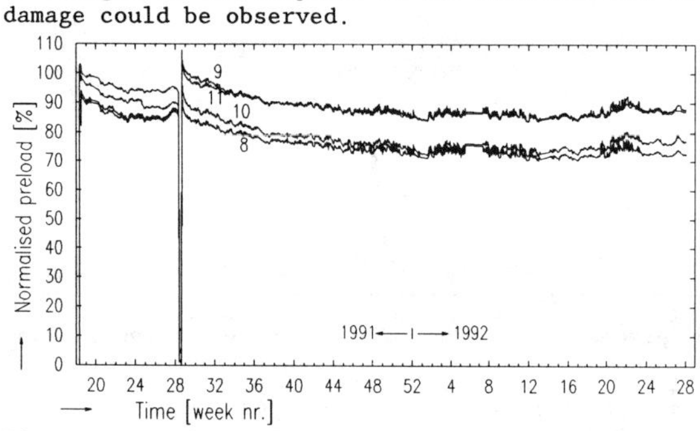

Figure 7 Preload in bolts 8 to 11 blade-to-hub connection during test

Monitorings of the twelve strain gauged bolts showed that an average decrease in preload of the bolts in the blade-to-hub connection of 20 % developed during the test period as shown in figure 7 for bolts 8 to 11. Despite of this decrease of preload no increase of displacements at this connection could be observed [1].

5 FATIGUE CALCULATIONS

5.1 General

In order to arrive at the test load and to evaluate the severity of test load with respect to the design load, fatigue calculations based on the design load as well as based on the test loads have been done.

5.2 Starting points of the fatigue design

For the fatigue design of the blade as tested the following starting points were used:

- A required design life of 20 years
- An allowable damage resulting from the Miner summation of 1.0
- A load factor of 1.35 according to the Dutch design criteria [5]
- Fatigue design procedure according to the Dutch design criteria [5]
- Stress concentration factors resulting from FEM analysis
- Timeseries of the cross section forces at 10 cross section in the blade for 83 load cases which were supplied by ECN.

5.3 Material fatigue design data

The design values for the UTS (Ultimate Tensile Strength), UCS (Ultimate Compressive Strength) and Young's modulus, used in the fatigue and stress calculation, is given in table I.

Table I Design values GRP material for fatigue calculations

component	UTS [MPa]	UCS [MPa]	E [MPa]
SKIN	260	167	16350
STIFFENER	69	129	8200
SKIN at Bolted connection	260	167	16350

These values of the UTS and UCS are the design values obtained according to the procedure given in Dutch design criteria [5].

The S-N curve used in the fatigue calculation according to [5] is expressed as:

$$N = \left(\frac{\frac{(UTS+|UCS|)}{2*C_1*C_2*C_3} - \left| S_{mean} - \frac{(UTS-|UCS|)}{2*C_1*C_2*C_3} \right|}{S_{amp}} \right)^{10} \quad (1)$$

wherein:

N = number of allowed stress cycles
$Smean$ = average stress
$Samp$ = stress amplitude
C_1 to C_3 = partial material factors

5.4 Procedure of fatigue calculations

Based on the starting points fatigue calculations have been carried out for 7 cross sections of the blade at about 20 locations at the circumference of each of these cross sections.

The calculations were carried out with the computer code FAROB. For each location the time series of the cross section forces were compiled into time series of the stresses. The damage at each location was calculated according to Dutch design criteria [5] using the S-N formulation according to equation (1). In this procedure the stress amplitudes are established by rainflow counting on the stress histories for each considered location and the damage is expressed by the Miner summation.

The fatigue damage was calculated for the design loads as well as for the test loads.

Furthermore for a better comparison of the test load against the fatigue load on a the basis of stress and strength, the minimum required fatigue strength to survive the loads for each location have been calculated on the following basis. For the applied S-N curve according to equation (1) the fatigue strength is characterized by the Ultimate Tensile Strength (UTS) and the

Ultimate Compressive Strength (UCS) of the material. Assuming a constant ratio between the UCS and UTS and substituting the partial material factors equation (2) can be rewritten as:

$$N=\left(\frac{\frac{UTS(1+RUS)}{2*C_1*C_2*C_3}-\left|S_{mean}-\frac{UTS(1-RUS)}{2*C_1*C_2*C_3}\right|}{S_{amp}}\right)^{10} \quad (2)$$

wherein:

$$RUS=\frac{|UCS|}{UTS}$$

The minimum required fatigue strength to survive a load history is the value of the UTS for which the calculated damage is equal to one. This minimum required UTS value for which the detail will survive the load history was obtained by varying the UTS value in the S-N curve (2) until the Miner summation gave a damage 1.0. This minimum required UTS value is referred to as UTS^1.

6 COMPARISON OF TEST LOAD WITH DESIGN LOAD

The severity of the test load can be evaluated by comparing this UTS^1-value for the test load with the UTS^1-value resulting from design loads. The ratio of these UTS^1-values, which is referred to as the "UTS-test-factor" from now on, expresses the severity of the test loading with respect to the design loads on the basis of strength. A value of the UTS-test-factor larger than unity means that the test load is more severe than the design load. This ratio depends on the assumed S-N formula but is not very sensitive to the actual UTS and UCS values. It is for this S-N formulation also independent on the assumed stress concentration factors.

The UTS-test-factors at the locations of maximum UTS^1-values, which are the most critical locations, are listed in table II for the SKIN material (total 8 cross sections have been considered). The values for the stiffener material are similar [1].

Table II UTS-test-factors SKIN material

radial position of cross section	UTS^1 test		UTS^1 design		UTS-test-factor	
	up-wind	down-wind	up-wind	down-wind	up-wind	down-wind
15510	76	149	65	117	1.17	1.27
14495	103	190	140	149	0.74	1.28
13501	119	215	182	191	0.65	1.13
11501	174	309	129	200	1.35	1.55
9501	202	356	133	218	1.52	1.63
7501	199	345	140	234	1.42	1.47
5251	298	271	202	202	1.48	1.34
2219	241	500	211	412	1.14	1.21

Similar to GRP material also a comparison of the design and test load for the steel tube connecting the tip section was made. For the cross section at a rotor radius of 15550 the maximum value of the minimum required fatigue strength (expressed by the stress range at 10^7 cycles) for the steel tube was 83 MPa and 66 MPa for the test and design load respectively.

7 DISCUSSION OF THE RESULTS

The UTS-test-factor is for the highest UTS^1 at the cross sections a factor of 1.2 to 1.7. For other locations at the relevant cross section this can differ but for relatively high stressed regions in a cross section the figures are similar. This design load included a factor 1.35 so that in total the fatigue test load was a factor of 1.6 more severe than the fatigue load to be expected in operation.

Assuming that a factor of 1.2 compensates the effect that the fatigue loading does not include the possible effect of humidity and UV radiation it can be concluded that at the end of test the blade had been subjected to a test load which was up to a radius of 15510 similar or more severe than the design load (including the load factor of 1.35).

At the blade root connection the highest UTS^1-value to survive the test load was 500 MPa. This is about a factor of 2 higher than the fatigue design UTS-value of 260 MPa. This indicates that the design calculation procedure for this detail is conservative. This is most probably caused by a too conservative assumption of the SCF values as used originally. This is supported also by the fact that during the test the mesurements of displacement transducers at the connection (see figure 5) showed no change in displacements.

Although failure at the blade tip connection (between rotor radius R=15500 mm and R=14400 mm) occurred after it had been subjected to a fatigue load equivalent to a design load of 20 years, the fatigue strength seems to be less than expected. At failure the highest calculated UTS^1-value is 190 MPa whereas the fatigue design UTS-value is 260 MPa. Some undisturbed sections having the same UTS-value were tested to a UTS^1-value up to 356 MPa without failure. Based on the examination of the failure the production procedure was improved for the detail where the failure occurred.

After the fatigue test an utlimate strength test was performed by flapwise loading. The blade collapsed at a section located at a radius of 7.5m. At collapse the bending moment was 3 times higher than the maximum moment following from the ultimate design analysis.

8 CONCLUSIONS

- The blade has been subjected to a fatigue load equivalent to the design load of 20 years. The latter including a load factor of 1.35 and an extra factor of 1.2 to compensate for humidity, UV radiation etc. During continuation of the testing beyond the 20 years equivalent loading, failure occured at the section where the tip was connected to the blade.
- At all other sections neither fatigue damage nor reduction of stiffness could be detected.
- The result of the test indicates that the fatigue calculation at the blade root connection may be too conservative.
- Although an average decrease in preload force of the bolts in the blade-to-hub connection of 20 % developed during the test period no increase of displacements at this connection could be observed.

9 REFERENCES

[1] J.L. van Leeuwen and D.R.V. van Delft
Full scale fatigue test on the GRP rotor blade of the NedWind-35 wind turbine, TU-Delft, Stevin rep. 06-92-21 (confidential)

[2] J.L. van Leeuwen and D.R.V. van Delft
Research on the fatigue performance of the rotor blade of the NedWind-35 wind turbine, TU-Delft, Stevin rep. 06-92-20 (confidential)

[3] D.R.V. van Delft et.al
Fatigue testing of a full scale steel rotor blade of the WPS-30 wind turbine.
Journal of Wind Engineering and Industrial Aerodynamics; 27(1988) 1-13

[4] D.R.V.van Delft, D.C.Corbett, J.L.van Leeuwen
Full scale fatigue tests of wood-epoxy blades Amsterdam EWEC '91, 256-260

[5] "Regulations for type-certification of wind turbines: Technical criteria"; ECN-91-001; february 1991, *including the amendment on fatigue of GRP:* Memo-91-58 ,ECN, Petten, 10 Dec. 1991

CONFORMITY ASSESSMENT - CURRENT LR THINKING

Dr. J.R. Maguire
Lloyd's Register Industrial Division
29 Wellesley Road, Croydon, Surrey, CRO 2AJ, UK

1. INTRODUCTION

Over the last decade Lloyd's Register (LR) has been involved in a number of wind turbine projects (Ref.1) and over the last 2 years has been developing a set of rules for classification and certification. However, during these last 2 years the wind turbine market has evolved rapidly on both a European and international level, particularly with regard to the development of standards. LR's original aim of its own comprehensive, stand-alone,set of rules (standards) is no longer consistent with European and international thinking. The purpose of this paper is to outline LR's current thinking in relation to rules, standards, certification and classification.

2. BASIC ASSUMPTIONS

2.1 Approach

LR will set safety (and quality) standards for different stages of the wind turbine life cycle (Figure 1) and invite a "safety case" (Reference 2) to be put forward, detailing the safety (and quality) approaches adopted. At each stage LR will examine the approach, to ensure it meets the requested standard, and if acceptable will issue an appropriate certificate.

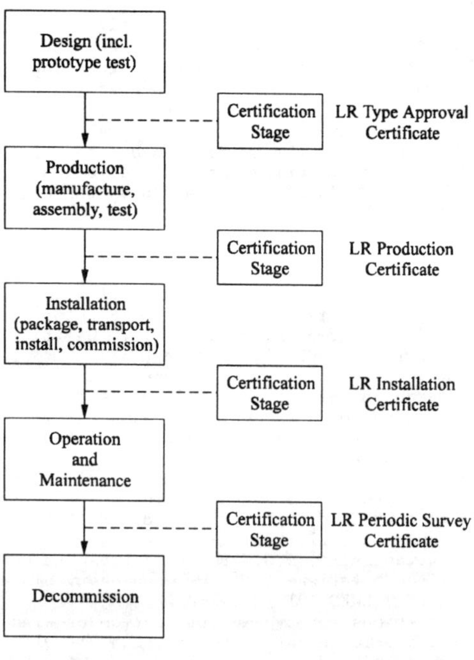

Figure 1: Idealised Wind Turbine Life Cycle and Relation to LR Certification

2.2 LR Certification/LR Classification

Both LR Certification and LR Classification are forms of certification, as illustrated in Figure 2. Classification can be seen to be a "cradle to grave" concept, covering all of the wind turbine life cycle, and importantly covering feedback of in-use experience. Classification has traditionally (within LR) been carried out with reference to "in-house" (and national and international) standards, and has traditionally been of interest to all members of an industry (particularly insurers).

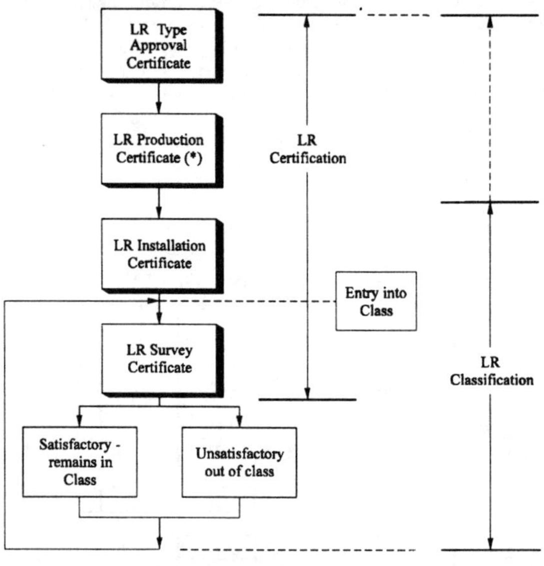

(* = not always requested - for major items only. May accept certificates from others.)

Figure 2: Relation of LR Certification to LR Classification for Wind Turbines

3. TERMINOLOGY

Certain "concepts" come up regularly in relation to wind turbine certification, but these concepts are labelled differently depending on whether one is British, European or international. The sections below detail some of the major concepts and their possible labels.

3.1 The Global Approach

The CEC recognise a global approach to conformity assessment (Figure 3) which is modular in nature (Table B1). With respect to wind turbines this may need interpretation in light of the Machinery Safety Directive (Directive 89/392/EEC). In the UK, DTI guidance is awaited on this matter.

3.2 LR Type Approval

LR Type Approval is broadly equivalent to EC Type Examination. A manufacturer will submit to LR technical documentation related to a particular type of wind turbine. LR will ascertain conformity with the essential requirements (see later), carrying out (or witnessing) type tests of the example wind turbine. If satisfactory, a LR Type Approval Certificate (EC Type Examination Certificate) will be issued. Figure 4 relates the wind turbine life cycle to conformity assessment.

3.3 LR Production & Installation Certificates

These certificates can be achieved in any of four ways - refer to Table B1. The four ways are:

(1) "Conformity to type" (product checks);
(2) "Production QA" (quality system checks);
(3) "Product QA" (quality system checks);
(4) "Verification of Products" (product checks).

The Machinery Safety Directive does not indicate which of these 4 routes should be chosen. Various European/international standards make the choice. This may yet change but in the final event the modules permitted will be restricted for any artefact.

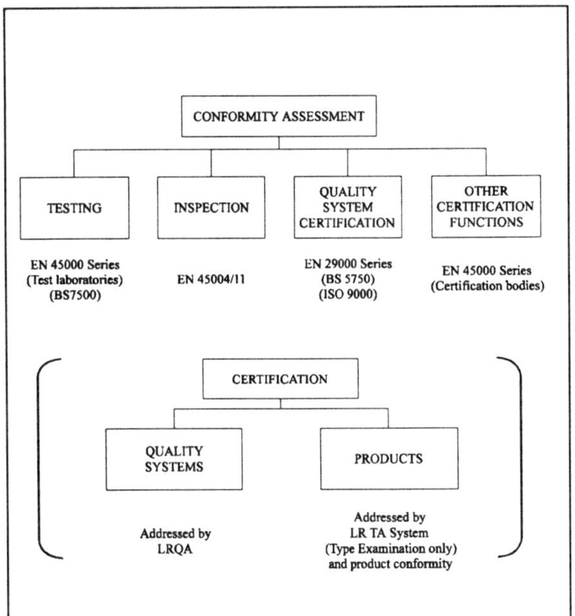

Figure 3: European Conformity Assessment
(Source : DTI, April 1992)

	A. Internal Manufacturing Check	B. (Type examination)				G. (Unit Verification)	H. (Full QA)
DESIGN	**Manufacturer** Keeps technical documentation at the disposal of national authorities. **Aa.** Intervention by the notified body.	**Manufacturer** submits to notified body – Technical documentation. – Type. **Notified body** – Ascertains conformity with essential requirements. – Carries out tests, if necessary. – Issues EC type examination certificate.				**Manufacturer** – Submits technical documentation.	EN 29001 **Manufacturer** – Operates an approved Quality System (QS) for design. **Notified body** – Carries out surveillance of the QS. – Verifies conformity of the design. (1) – Issues EC design examination certificate. (1)
	A.	C. (Conformity to type)	D. (Production QA) EN 29002	E. (Product QA) EN 29003	F. (Verification of Products)		
PRODUCTIONS	**Manufacturer** – Declares conformity with essential requirements. – Affixes the CE mark. **Aa. Notified body** – Tests on specific aspects of the product. (1) – Product checks at regular intervals. (1)	**Manufacturer** – Declares conformity with approved type. – Affixes the CE mark. **Notified body** – Tests on specific aspects of the product. (1) – Product checks at regular intervals.	**Manufacturer** – Operates an approved Quality System (QS) for production and testing. – Declares conformity with approved type. – Affixes the CE mark. **Notified body** – Approves the QS. – Carries out surveillance of the QS.	**Manufacturer** – Operates an approved Quality System (QS) for inspection and testing. – Declares conformity with approved type, or essential requirements. – Affixes the CE mark. **Notified body** – Approves the QS. – Carries out surveillance of the QS.	**Manufacturer** – Ensures conformity with approved type, or to essential requirements. **Notified body** – Verifies conformity. – Issues certificate of conformity. – Affixes the CE mark.	**Manufacturer** – Submits product. **Notified body** – Verifies conformity with essential requirements. – Issues certificate of conformity. – Affixes the CE mark.	**Manufacturer** – Operates an approved QS for production and testing. – Declares conformity. – Affixes the CE mark. **Notified body** – Carries out surveillance of the QS.

(1) Supplementary requirements which may be used in specific directives

Table B1 – Conformity Assessment Procedure in EC Legislation

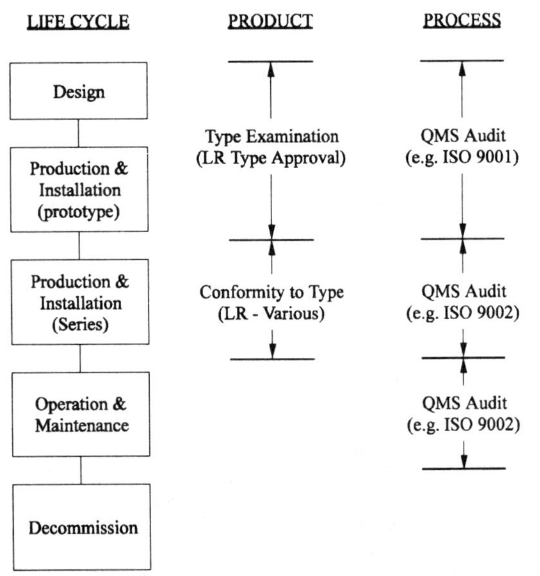

Figure 4: Relation between Life Cycle and Conformity Assessment for Wind Turbines

4. KEY DOCUMENTS

There are a number of standards/certification documents which LR consider to be "key" as regards formulating the essential health and safety requirements (EHSR's) referred to in the Machinery Safety Directive. These are briefly reviewed below.

4.1 International Energy Agency (IEA) Recommended Practices

There are currently 8 practice notes in print, including No. 6 - Structural Safety and No. 8 - Glossary of Terms. A ninth is being written on anemometry. Two (No. 1 - Performance and No. 4 - Noise) are being rewritten. The IEA notes that standards can never be the same from country to country but that it is desirable that the framework for standards be as common as possible. As far as safety philosophy, safety systems, load cases and a safety factors the IEA suggests adopting either Dutch or a mixture of Dutch/Danish practice. LR's current thinking is to broadly aim to follow Danish (certification) practice at present.

4.2 (CEC) Machinery Safety Directive

This entered into force on 31 December 1992. It does not explicitly reference wind turbines yet but the DTI have suggested this is likely to be the most relevant directive. Under the directive most machines must:

(1) satisfy the EHSR's (assume IEA/Danish);
(2) in some cases be type-examined depending on supporting standards;
(3) carry a CE mark and information (not done at present).

The manufacturer must draw up and keep a technical file. The notified body may need to be involved in the process. In view of the relevance of the CEC documents to a large number of machines installed (and planned) in Europe it is thought that LR should take on board the principles of this directive as closely as possible.

4.3 (CEC) Proposal for harmonisation

With a desire to create a common European certification procedure a group of 4 wind turbine test stations (CRES-Greece, ECN-Netherlands, NEL-UK and RISO - Denmark) proposed a harmonisation of procedures and technical regulations in Feb. 1991. The proposal was based around Dutch/Danish certification systems but has never apparently been taken up due to lack of CEC funds. It is understood that the Danes, with UK support, have put forward their certification system as a CEC role model. LR's current thinking is to align with this.

4.4 International Electrotechnical Commission (IEC)-Safety Standard TC88

The IEC have produced a draft standard on safety of wind turbine systems, which went for third draft discussion in Rotterdam in October 1992. Originally envisaged for completion in 1995 it is now hoped that this draft will be finalised substantially earlier (1993). This standard on safety deals with:

(1) safety philosophy;
(2) QA;
(3) engineering integrity;

The draft covers design, installation, maintenance and operation, with QA being integral at all stages.

There has been a close liaison between the IEC, CENELEC and national (eg. BSI) standard bodies, resulting in very similar technical contents in national, European and international standards. As the IEC draft is the "highest level" standard LR's current thinking is to align with this as closely as possible. The question of whether formal QA systems are mandatory or advisory has yet to be answered.

4.5 (Danish) Technical Basis for TA & Certification of Windmills

This document was published by the Danish National Energy Board (DNEB) in May 1991. The DNEB approval system, self-administered, covers type approval, certification of QC systems for production (to EN29002), and certification of QC systems for installation (to EN29002). The technical basis for the system is split into:

(1) principles of approval;
(2) procedures for applicants;
(3) technical requirements.

It assumes the existence of uniform standards, specifications and other technical rules with respect to design, production and operation of wind power installations. Where standards do not exist technical requirements are to be drawn up.

It is interesting to note that the Danish system emulates on the LR "class" philosophy i.e. the Danish system envisages ongoing maintenance of Type Approval by ongoing submissions of information. (This ongoing feedback of information is also a feature of the Dutch certification system).

Comparing the Danish system with LR's system the following broad evolution is under way:

(1) LR has developed a document entitled "Conformity Assessment of Wind Turbines" which embraces/explains the LR Type Approval/QA/QC system, and interprets it for wind turbines. This emulates the Danish "Technical Basis" document, and includes principles of approval and procedures for applicants.

In turn, the "Conformity Assessment" document refers to another supporting document entitled "Interpretation of LR's Type Approval System for Wind Turbines". This can be used for "stand-alone" Type Approval.

(2) LR needs to further develop/modify its provisional rules for certification into a document similar to the Danish "Technical Requirements". This should call up all the relevant standards by reference, and only aim to "plug missing gaps" where necessary. It should therefore be a relatively short document compared to the current provisional rules.

(3) The "Conformity Assessment" document has been structured to be country independent (somewhat similar to CEC Eurocodes). In this way the LR system can become applicable world-wide, with only local conditions (e.g. local environment) to be inserted as appropriate.

5. CONCLUSIONS

A significant change in LR thinking has taken place over the last 2 years. LR's provisional rules have been changed and restructured in line with current (and emerging) European and international standards. The (proven) LR Type Approval system is now forming the focus for dealing with standards and other technical requirements, as part of an integrated approach to conformity assessment.

It is now clear that LR must more actively support and align itself with co-operative international standards activity. It is also clear that all certification bodies involved with wind turbines should now harmonise their (broadly similar) certification and classification approaches, for the benefit of this industry.

REFERENCES

(1) Lloyd's Register Technical Association, "An Introduction to Modern Wind Energy Technology", D.C. Surplus, Paper No. 3, Session 1991-92.

(2) The Institution of Electrical Engineers, "Safety-Related Systems", Professional Brief, Sept. 1992.

(3) Lloyd's Register Industrial Division, Report CIV/9238/R001/jrm, "A Conformity Assessment System for Wind Turbines".

(4) Lloyd's Register Type Approval System, "Procedure", 1990.

(5) Lloyd's Register Industrial Division, Report CIV/9235/R005/jrm, "Interpretation of LR Type Approval System for Wind Turbines".

INFRA RED THERMOGRAPHY AS A TOOL FOR IDENTIFYING FLAWS AND STRESS CONCENTRATIONS IN COMPOSITE WIND TURBINE BLADES

A G DUTTON, A D IRVING
Energy Research Unit
Rutherford Appleton Laboratory
Chilton
Didcot
Oxon OX11 0QX
UK

G M SMITH, B R CLAYTON
Department of Mechanical Engineering
University of Nottingham
University Park
Nottingham NG7 2RD
UK

ABSTRACT

The theoretical basis of flaw detection using infra red thermography (IRT) is considered in comparison with competing condition monitoring techniques. Particular attention is directed to the mode where heat is generated during some fatigue process when it is shown that IRT indicates some "first order" response to damage. In typical wind turbine blade materials such as wood/epoxy and glass/polyester, this response has both transitory (thermoelastic) and cumulative (thermo-viscoelastic) components. Data are presented showing the thermal response of wood/epoxy and glass/polyester under tensile fatigue loading. Intensified heating effects are seen around stress concentrations. The thermo-viscoelastic response is found to give a measure of the recent load history of the object. A third heating mechanism due to frictional rubbing between free crack surfaces is indicated by specimens loaded in bending.

1. INTRODUCTION

Proving the structural integrity of novel wind turbine blade designs or certificating the safety of an in-service blade with a history of severe loads poses potential future problems for the wind turbine industry. Costly design failures could be avoided by the type-testing of new designs provided that suitable full-field condition monitoring techniques were available to indicate areas of high stress concentration. Such stress concentrations could arise due to inadequate design codes or a failure to meet manufacturing tolerances and must be identified as early as possible in the design process. The problem is that full-field monitoring techniques are not currently available for immediate application to composite materials.

However, the authors have recently had some success with the application of infra red thermography (IRT) for this purpose. During a full-scale wood/epoxy blade fatigue test a hot spot was identified, at only a third of the final lifetime, which increased in temperature relative to the surrounding blade material during subsequent testing and which corresponded to the final point of failure of the blade [1]. During a second test a similar hot spot was identified at the very start of loading [1]. However, this hot spot was of lower relative temperature and did not increase during the subsequent loading before the test was terminated without failure [2]. The implication of these results was that IRT could identify points of incipient failure and that both critical and non-critical conditions could be detected. More fundamental work was required to determine whether this critical boundary could be easily defined. This paper will address some aspects of this work.

2. THEORY

Wood/epoxy and glass/polyester composites are visco-elastic materials. This means that, due to forces of internal friction, there is a phase lag between the stress and strain vectors, resulting in the material having some finite 'memory' of its load history. During cyclic fatigue loading this phase lag results in work being done on the composite, which is then manifested (mainly) as heat.

A second, concurrent mechanism is that of the thermoelastic effect, described by Lord Kelvin in the 19th century, by which a change in the sum of the principal stresses at a point in a body results in a change in the temperature at that point. In elastic materials undergoing fatigue these variations are assumed to be adiabatic, resulting in a technique (SPATE) to measure the sum of the principal stresses at the surface of a body. Typical temperature differences can be an order of magnitude less than those arising from internal friction effects and accurate stress readings are only obtained after close correlation with the load frequency. Irreversibilities due to visco-elastic material properties are not well understood.

The effects of both these mechanisms are intensified around stress concentrations. It is the aim of current work to determine whether and in what way the heating characteristics change as damage accumulates in a body. It is well known that the elastic modulus of composites decreases as damage accumulates; the effect on the visco-elastic phase lag is less clear. However, changes in the stress-strain hysteresis curves for wood/epoxy composites as damage accumulates, reported by Bonfield and Ansell [3], appear to show evidence of both these phenomena.

3. LIFE CYCLE TESTING OF WOOD/EPOXY COUPONS

A series of four life-cycle tests were carried out at the University of Bath on wood/epoxy coupons with internal flaws in the form of joints. The basic specimen geometry comprised a five veneer thick laminate, with unjointed central ply and various joint geometries in the two pairs of outer plies capped with an outside layer of woven glass fibre reinforced epoxy. The dogbone-shaped coupons had a central working section 16mm wide by 20mm thick. The four joint geometries were as follows:

(i) butt-jointed on the inner layers/scarf-jointed on the outer layers,

(ii) all scarf-jointed,

(iii) all underscarf-jointed (ie scarf-jointed with excess epoxy in the joint),

(iv) all overscarf-jointed (ie scarf-jointed with an overlap of wood causing geometric distortion of the test coupon).

Each coupon was tested at constant loading rate in constant amplitude, reversed loading fatigue for a range of stress levels aimed at producing failure within about half a day's testing. Full-field thermograms of the temperature field on the side of the specimen were recorded throughout the load history of each test coupon.

The loading rates were sufficiently high to cause a general increase in temperature of the central working area of each specimen of the order of 3-5°C during the full test. However, superimposed on this general increase in temperature were significant local 'hot spots' generated around the joints. These were more intense in the underscarf- and butt-jointed cases. In the latter case (figure 1) both butt-joints were elevated above the surrounding temperature field and a further difference of about 2°C existed between them for most of the test. The temperature at one joint then increased sharply to more than 320K (47°C) immediately prior to failure at that joint. The temperature rise was then very rapid and the peak temperature was in excess of the instrument settings. About 40 seconds after the actual failure a temperature of almost 340K (67°C) was recorded at the failed joint. The loading regime for the butt-jointed specimen is shown in table I. Note the rapid

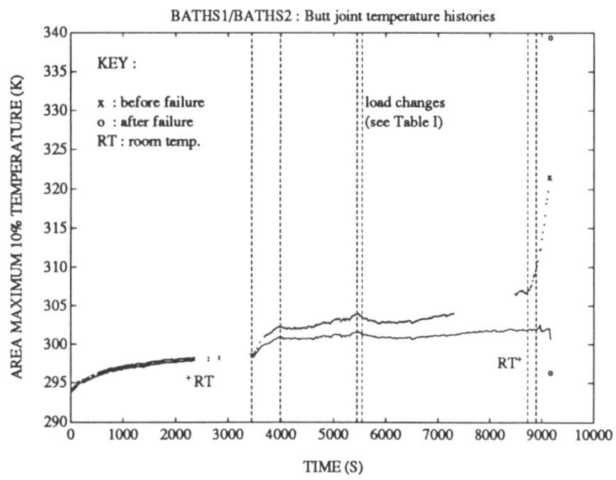

Figure 1 : Khaya/epoxy outer scarf/inner butt jointed coupon fatigue test - temperature histories at the two butt joints.

Table I : Test parameters for butt-jointed coupon in tensile fatigue

Time (s)	Stress level (MPa)	Frequency (Hz)	Comments
0	20	5.0	
3500	25	4.0	Load increased first
5500	20	5.0	
8750	25	4.0	Frequency reduced first
9130			Failure

rise in the temperature of both joints following the first load change at 3500s. This was because the load increase was applied before the frequency decrease (required to maintain a constant loading rate) and vividly demonstrates the effect of loading rate in this kind of test. Figures 2 and 3 show thermograms of the critical joint at the onset of failure and 20 seconds later. Note that in figure 2 the shape of the isotherms indicates the direction in which the crack subsequently propagated across the coupon. Each frame comprises two

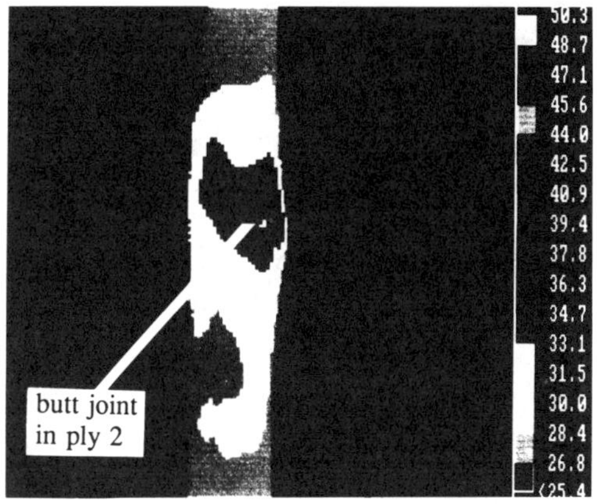

Figure 2 : Khaya/epoxy outer scarf/inner butt jointed coupon fatigue test - thermogram (°C) at failure initiation

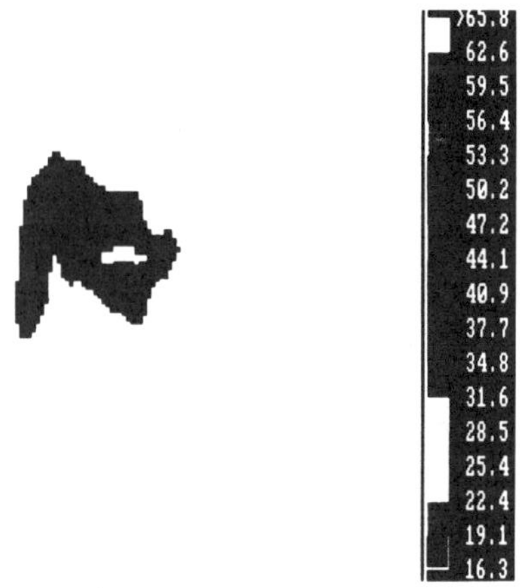

Figure 3 : Khaya/epoxy outer scarf/inner butt jointed coupon fatigue test - thermogram (°C) approximately 20s after failure initiated

interlaced frames, so figure 3 actually shows the deformation caused by failure, with the glass skin bulging outwards on the right hand side and the left hand outer ply splitting away from the main coupon on the other face.

Figure 4 shows a composite thermogram of the underscarf-jointed coupon at about two-thirds final lifetime. Note that all four joints are clearly visible. Failure of this coupon initiated at the lower right hand joint and proceeded diagonally upwards and across the specimen without involving any of the other joints. A peak temperature in excess of 320K (47°C) was recorded at the critical joint during failure.

4. FATIGUE TESTING OF GLASS/POLYESTER COUPONS

A series of tensile (pull-pull) and 3-point bending fatigue tests have been carried out on glass/polyester coupons at the University of Nottingham. These tests, on rectangular coupons with cross-section 25mm by 10mm, also investigated the stress-raising effects of small holes and in-plane delaminations.

Figure 5 shows the effect of load range on the overall mean temperature rise during a tensile test. The mean temperature is evaluated over a large representative area excluding the specimen edges and is compared with the

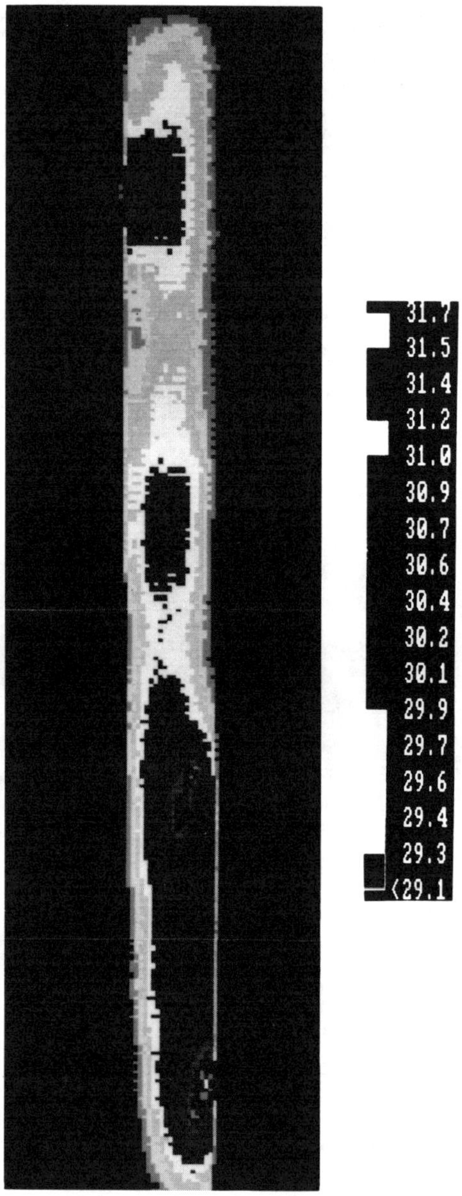

Figure 4 : Khaya/epoxy underscarf-jointed coupon fatigue test - composite thermogram (°C) at two-thirds of final lifetime

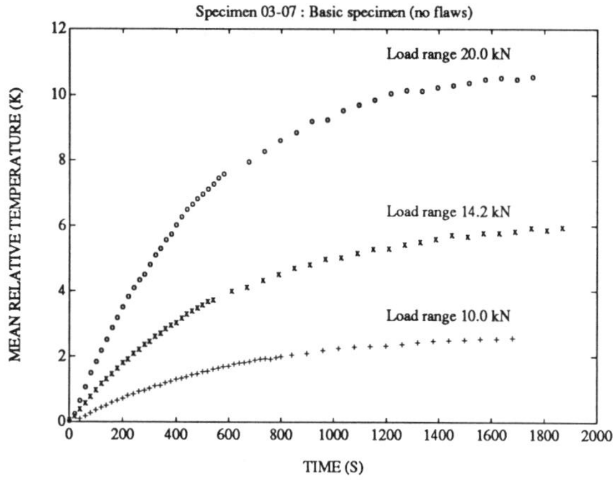

Figure 5 : Glass/polyester tensile fatigue test - temperature histories relative to control specimen at three load ranges.

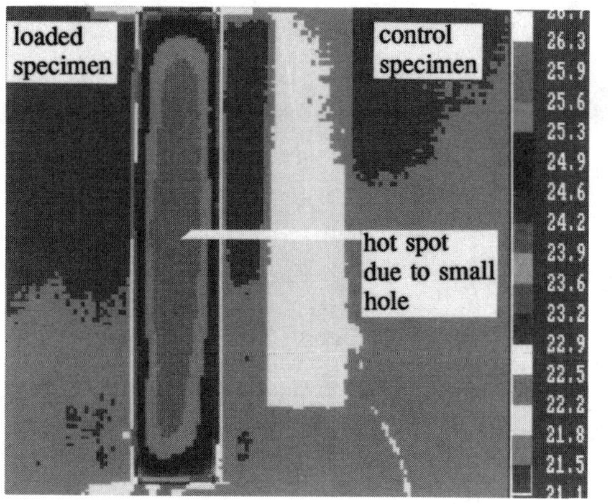

Figure 6 : Glass/polyester coupon fatigue test - thermogram (°C) showing specimen with 1 mm diameter hole

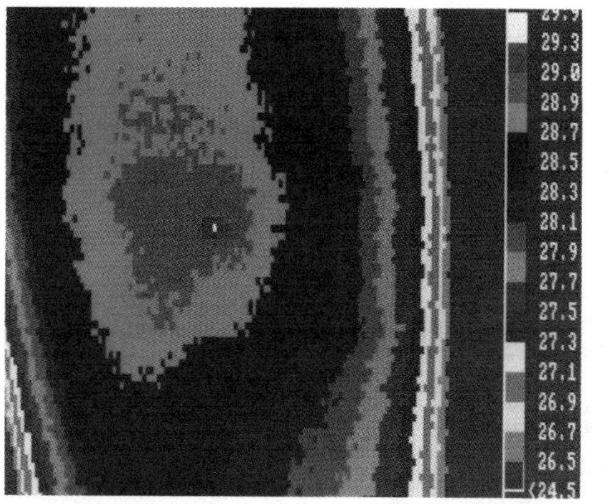

Figure 7 : Glass/polyester coupon fatigue test - thermogram (°C) showing specimen with 1 mm diameter hole (close-up)

temperature of a similar area on a control specimen to eliminate gross heating effects within the laboratory. The temperature rise at the specimen surface was very uniform. The upper load range in figure 5 corresponds to a peak stress of 80 MPa, the lower load range to a peak stress of 40 MPa, which are low by comparison with the statically determined ultimate tensile strength of over 200 MPa for this material.

The effect of drilling small holes through the centre of the specimen was also investigated. Figure 6 shows the small local temperature rise due to the stress concentration effects of a 1 mm diameter hole. The hole itself is only 1 pixel wide in figure 6. A close-up of the affected region (figure 7) shows a steep temperature gradient at the edges of the hole, which appears to be the hottest point. This is because the stress concentration is highest at the edges of the hole, giving higher temperatures, and because the hole approximates to a black-body cavity whose emissivity, 1.0, is higher than the surrounding material. (Since the infra red camera measures thermal emittance rather than temperature, a variation of emissivity looks the same as a variation of temperature.)

In bending of an unflawed specimen hot spots appeared symmetrically on either side of the central roller. The central portion of the specimen appeared cooler, presumably due to the high rate of conduction into the central (steel) roller. (Note here the effect of the large relative difference in thermal conductivity between steel and glass/polyester composite which results in hot spots persisting for a considerable time in the composite, thereby increasing the chance of detection.) In a partially delaminated specimen (figure 8), made by inserting a strip of melinex at mid-section during the lay-up and later removing it, the hot spots were concentrated in the region of the delamination tip and above the side roller (on the split side of the specimen). For a central deflection of 2.4 ±1.4 mm the peak temperature rise was about 4.0°C. Further work is required to discover the extent to which the more central hot spot moves with the crack tip and whether it is caused by stress concentration effects or by external friction. Comparison with the unflawed specimen bend test suggests that the hot spot above the side roller is due to friction at the free crack surfaces.

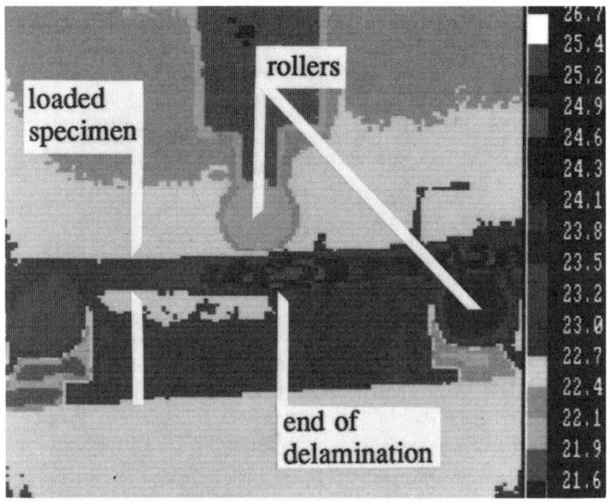

Figure 8 : Glass/polyester coupon bending fatigue test - thermogram (°C) showing temperature distribution in a specimen with central delamination

From thermograms captured at the maximum sampling rate of 25Hz it was possible to determine the temperature variations due to the thermoelastic effect in the bend specimen. The neutral axis of the unflawed specimen was clearly delineated as a region with no temperature variation. The peak to peak (ie maximum to minimum stress) cyclic temperature variation then rose to ±0.2°C at the specimen edges. In the immediate vicinity of the central roller the cyclic temperature variation was as high as 0.5°C, indicating a high stress concentration beneath the roller. Thus temperature variations due to the thermoelastic effect are much smaller and harder to detect than the cumulative temperature increase arising from external or internal friction. They do, however, carry different information about the stress field.

5. DISCUSSION

As a group, the scarf-jointed coupons resulted in lower peak temperature differences for similar loading regimes, even when the joints were poorly made (eg underscarf-jointed). The higher temperatures attained in the test on the butt-jointed wood/epoxy coupon suggests that high stress concentrations occurred round the joints, implying that the joints themselves carried little or no load. Until shortly before failure, the peak temperature of a hot spot occurring at any of the joint types tested was typically located in the solid ply to one side of the joint. Further work is required to demonstrate whether this peak temperature measurement corresponds with the point of maximum stress. Immediately before and during failure there was a further large burst of heat generation that obscured the finer detail of the fracture process. The reason for this rapid increase in heating is not yet clear. Some proportion of the increase can be attributed to the very high stress level, most, probably, to friction as the surrounding wood splinters, and possibly even some to the fracture process itself.

Similar high temperature rises due to internal friction effects alone were demonstrated by the glass/polyester coupon tests, even at stress levels well short of the material's ultimate tensile strength. These temperature rises were intensified around stress concentrations, such as small holes. Further work is required to study how these effects develop as failure is approached.

6. CONCLUSIONS

Large temperature rises have been measured in glass/polyester and wood/epoxy composites undergoing coupon fatigue tests. These temperature rises are related to the load range and frequency and arise from the fundamental viscoelastic properties of these materials. Additional temperature increases due to the same mechanism have been measured at stress concentrations such as joints in wood/epoxy coupons and small holes drilled in glass/polyester coupons.

External friction due to the rubbing of free crack surfaces has also been shown to cause a considerable local temperature rise in delaminated glass/polyester composites under bending loads. The proportion to which external friction is involved in the rapid heating effects observed at and around joints in wood/epoxy laminates during failure is yet to be resolved.

These results elucidate the temperature rises previously measured during blade fatigue tests and will be input to an ongoing JOULE-II project to look at the application of NDT techniques to wind turbine blades.

ACKNOWLEDGEMENTS

The life-cycle fatigue tests on wood/epoxy laminates were carried out by Ian Bond and Clare Hacker of the Department of Materials Science, University of Bath, under the supervision of Dr Martin Ansell. The authors are grateful to Martin Ansell and his team for the opportunity to monitor these tests.

REFERENCES

[1] Dutton, A.G., Irving, A.D., Lipman, N.H., Clayton, B.R., Aftab, N., Bond, L.J., 'Infra red condition monitoring of wind turbine blade fatigue tests', Proc. of the 14th British Wind Energy Association Conference, ed. B.R. Clayton, MEP, 1992

[2] Hancock, M., Richmond, B.N., Morris, M.J., Jones, W.A.R., 'Fatigue testing of 12m wind turbine blades', Proc. of the 14th British Wind Energy Association Conference, ed. B.R. Clayton, MEP, 1992

[3] Bonfield, P.W., Ansell, M.P., 'Fatigue testing of wood composites for aerogenerator rotor blades. Part VI: spectrum fatigue loading, life prediction, and damage rates', Proc. of the 13th British Wind Energy Association Conference, ed. D.C. Quarton, V.C. Fenton, MEP, 1991

CERTIFICATION OF WIND ENERGY CONVERSION SYSTEMS

D. Frey
Germanischer Lloyd
Vorsetzen 32
D-W-2000 Hamburg 11

For the past 125 years Germanischer Lloyd have offered to interested circles their services as an independent organization of technical experts. For a period exceeding ten years several staff members have been concerned with the certification of wind energy conversion systems.

By certification a wind energy converter is verified to meet specified requirements. The requirements to be specified shall have to make sense technically and suit the intended purpose; for instance, wind energy converters should in general have a useful life of 20 years.

The five standards listed below contain definitions of requirements to be met by wind energy conversion systems.

- IEC Standard (at present in the drafting stage)
- Riso Regulations (Denmark)
- ECN Regulations (Netherlands
- IfBT Regulations for Tower and Foundation (Federal Republic of Germany)
- GL Regulations (Federal Republic of Germany)

Normally, these standards are more or less binding in the 'native countries' of the organizations concerned.

For manufacturers of wind energy conversion systems it is desirable that these standards comply with each other, so that the work involved in the certification need be performed only once and distorted competition is avoided.

The IEC aim at establishing such an internationally binding standard, which will be recognized and declared to be binding by as many countries as possible. It must be the objective of expert organizations (such as Germanischer Lloyd) to achieve optimum compliance between their own standards and this generally accepted standard, without losing right of their safety target, in order to be able to confirm to their clients that the requirements imposed by the IEC Standard have also been met.

In 1989 Germanischer Lloyd published Preliminary Regulations for the Certification of Wind Energy Conversion Systems, approx. 500 copies of which in total have been distributed in the German and English languages. A final version of these regulations has been worked on for some time. However, efforts at harmonizing it with other rules and regulations have led to considerable delays. The final version was finished just before this conference.

The certificates of Germanischer Lloyd cover the whole turbine. The following certificates are available:

* DESIGN APPROVAL

Designapproval is made up entirely of 'paper work', covering mainly:

— **Design loads**

The load cases to be examined are described in the Regulations for Certification and to be determined either by computation or by reasoning. 'By reasoning' is to be understood to mean that load cases not resulting in any new knowledge or information will not be examined by computation. The examination of design loads serves as a basis for further plan approval.

— **Safety concept and safety system**

The safety concept is to ensure that the system can be safely operated and/or be caused to assure a safe condition under all operating conditions to be considered.

— **Rotor blades**

On the basis of the design loads proof is to be furnished that the envisaged useful life will be achieved and extreme loads will not cause destruction.

— **Machinery**

See rotor blades

— **Tower and foundation**

See rotor blades

— **High-voltage electrical equipment**

This equipment is to be designed such that, taking into account the special boundary conditions of wind energy converters, safe operation will be ensured.

* PROTOTYPE TEST

In line with the motto "Confidence is good, control is better" it is advisable to test the prototype, in order to establish the extent to which the design assumptions reflect actual facts. Also. the prototype test serves the purpose of optimizing the mode of operation. Normally, the instrumentation required (measuring equipment and evaluation aids) is costly and requires highly qualified specialists. For many manufacturers it is useful to avail themselves of the services for assistance offered by the Measurement Department of Germanischer Lloyd.

The prototype test will, as a rule, result in revision of the design, such as to be able to produce the series as economically as possible.

After having arrived at the optimum mode of operation or adjustment, power curve and acoustic power level measurements can be carried out at the prototype.

* SUPERVISION OF PRODUCTION

The design defines the required quality characteristics of the individual system components. Supervision of production is aimed at ensuring that these quality characteristics are in fact achieved. Since the individual components employed are normally extremely expensive, supervision of production has to include the discussion of possible repairs of manufacturing defects, so that the product quality required will nevertheless ultimately be ensured.

The scope of production supervision will be determined on the basis of the design particulars.

Since the individual components will usually be produced by different manufacturers, it will be of advantage if the supervisory organization maintains an extensive inspection network.

* SUPERVISION OF TRANSPORT

Particularly for shipment by sea the individual components are required to be adequately packed. This means that on the one hand packing must be appropriate, resistant and sufficiently tight, while on the other hand relevant corrosion protection has to be provided.

In some sea ports approved cargo gear only is admissible.

Also, within the scope of transport supervision the condition in which wind energy converters have reached defined destinations is certified.

* SUPERVISION OF SYSTEM INSTALLATION

Proper construction of the foundations for the mains supply system and erection and final assembly of the tower and nacelle are verified and construction progress is certified.

* START-UP

Following start-up certification, the systems are to start operating.

All systems will separately be subjected to thorough visual checking. The tightening torques of the screws, for instance, will be spot checked.

Large-scale operational tests of the safety systems will be carried out.

If the wind conditions permit, the mode of operation specified by the plant management will be checked.

Normally, start-up certification constitutes completion of a project. The start-up certificate serves as proof of completion to be presented by manufacturers to financing companies.

* PERIODICAL INSPECTION

Since wind energy conversion systems are also subject to natural wear and tear, they have to be serviced at regular intervals. Such maintenance is to ensure not only output, but also system safety. By periodical inspection at regular intervals of two to four years the system is checked as to whether it is in good condition and working order.

* DIVISION OF FUNCTIONS AT GERMANISCHER LLOYD

If requested by clients, the different certificates may be issued. Thus, for instance, clients fairly frequently wish to obtain certificates combining design approval, production supervision and start-up supervision.

For the individual sectors covered by certification specialist knowledge is required. Different Germanischer Lloyd departments and Inspection Offices normally provide their services quite unbureaucratically.

Activities are coordinated by Germanischer Lloyd Head Office in Hamburg.

Design approval is carried out at Head Office by staff specially qualified in the different subject matters to be handled.

At present two systems are undergoing prototype testing. For this purpose the Measurement Department provides the qualified staff, appropriate equipment and proven software for evaluation.

Supervision of production, transport and system installation are activities traditionally performed by the Inspection Offices. Services are rendered worldwide in close cooperation with Head Office.

Start-up certification is at present attended to by Head Office.

Periodical inspections currently have the highest growth rate and are, if distances to be travelled permit, performed by Head Office. Periodical inspections provide valuable information for design approval.

STRENGTH VERIFICATION OF A GRP ROTOR BLADE FLANGE BY ANISOTROPIC FINITE ELEMENTS, CARRIED OUT ON THE BLADE OF THE WIND POWER GENERATOR HSW 250

G.U. Steinhagen

INNOBRAIN
Beratung für Innovative Technologie
Brahmsallee 26 . D-2000 Hamburg 13
Tel.: 040-451084 - Fax.: 040-456904

It is known that GRP structures can be considered a continuum when analysed by the Finite-Element-Method. When the model of three-dimensional anisotropic brick elements is used, the continuum theory requires knowledge of the materials elastic behaviour in all three dimensions. The two-dimensional strength and distorsion characteristics of plane structures are well known. This paper adds the unknown characteristics of the third dimension. As thin slices are discretized by finite elements the plane/plate problem can be simplified by "decoupling" since the bending and torsion moments are negligible. Each slice can be considered a stack of GRP layers. For each slice, the three-dimensional elasticity matrix is calculated on the computer. Consequently, using the prepared three-dimensional brick elements allows us to evaluate the strain of the GRP rotor blade flange which is important to obtain an operating licence.

1. ELEMENT ELASTICITY MATRIX

1.1 Advantages of the Continuum Concept

The continuum theory, in contrast to the mesh theory takes into account the resin load capacity which is quite realistic to describe the load bearing behaviour. On the other hand the continuum theory can be described with translucent mathematics provided that the individual elements from thin slices so that the bending and torsion moments become negligible compared to the basic stresses. These stress equations contain no Young's-moduli for bending which considerably reduces the equations.

As to using the theory of plates it would require the knowledge of all stiffness values for bending and torsion stresses. This information is widely unknown to GRP manufacturers, and the extra effort for generating it will not necessarily result in better data.

The individual layers of the slice (= multilayered composite) are considered homogenous by the continuum theory. This concept, however, is possible only before failure occurs. As a stress limit criterion for the GRP rotor blade flange we use a maximum permissible strain which is safely below.

1.2 Mathematics

The basics of the continuum theory have been discussed elsewhere [1, 2, 3, 4]. If we consider the load condition in the laminated plane for the special case of orthotropy [1, 2], the mean stresses of the slice using

$$[\sigma] = [E] \cdot [\varepsilon]$$

are (Fig. 1)

$$\sigma_X = c_{11}\,\varepsilon_X + c_{12}\,\varepsilon_Y$$
$$\sigma_Y = c_{12}\,\varepsilon_X + c_{22}\,\varepsilon_Y$$
$$\tau_{XY} = c_{33}\,\varepsilon_{XY}\,.$$

Having a slice with a glass layer of thickness t_K in it the glass volume fraction φ can be derived from the glass mass fraction ψ_G (k= layer index, G= glas, H= resin) as

$$\varphi_K = \frac{1}{1 + \dfrac{(1-\psi_G)\,\varrho_G}{\psi_G\,\varrho_H}}$$

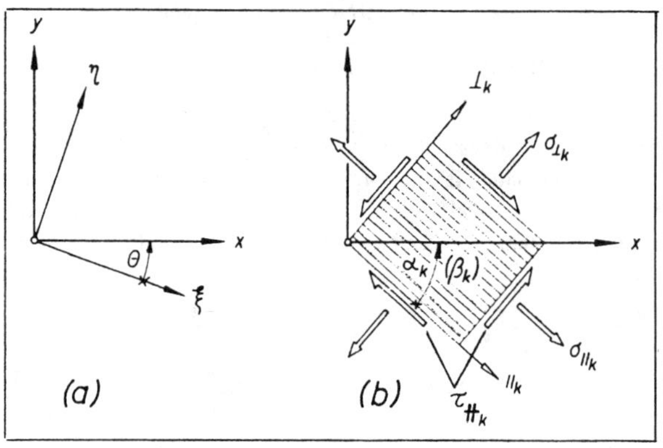

Fig. 1: Geometry of laminate consisting of n unidirectional layers (k= 1 .. n).
a) The X and Y directions symbolize the orthotropic axes of the laminate. The ξ and η directions symbolize the mean stress axes of the laminate.
b) Stresses in the k-th layer. The angle α_K (= β_K when X and Y are the directions of the main normal force) is postive in case you move counter-clockwise from the $\parallel_K$ direction to the X direction.

With the shear moduli for glass and resin being

$$G_G = \frac{E_G}{2\,(1+\upsilon_G)} \qquad G_H = \frac{E_H}{2\,(1+\upsilon_H)}$$

the values for the k-th single layer and the directions of the main normal force, we have

$$E_{\parallel} = \varphi_K\,E_G + (1-\varphi_K)\,E_H$$
$$\upsilon_{\perp\parallel} = \varphi_K\,\upsilon_G + (1-\varphi_K)\,\upsilon_H$$

Perpendicular to the direction of the glass layers we have

$$E_\perp = \frac{E_H}{1-\upsilon_H^2}\;\frac{1+0{,}85\,\varphi_K^2}{\varphi_K \dfrac{E_H}{E_G(1-\upsilon_H^2)} + (1-\varphi_K)\sqrt[4]{1-\varphi_K}}$$

$$\upsilon_{\parallel\perp} = \upsilon_{\perp\parallel}\,\frac{E_\perp}{E_\parallel}$$

$$G_{\#} = G_H\,\frac{1+0{,}60\sqrt{\varphi_K}}{\varphi_K \dfrac{G_H}{G_G} + (1-\varphi_K)\sqrt[4]{1-\varphi_K}} \quad .$$

The stiffnesses of the unidirectional layer k with the orthotropic axes $\parallel$ and $\perp$ are .

$$C_\parallel = \frac{E_\parallel}{1-\upsilon_{\perp\parallel}\upsilon_{\parallel\perp}} \qquad C_\perp = \frac{E_\perp}{1-\upsilon_{\perp\parallel}\upsilon_{\parallel\perp}}$$

$$C_{\perp\parallel} = \upsilon_{\perp\parallel}\,C_\perp \qquad C_{\#} = G_{\#} \quad .$$

Using the matrix of elasticity of the k-th layer which is transformed onto the orthotropic direction, and utilizing the notation of the tree-dimensional matrix of the vector

$[\sigma]^T = [\sigma_X\ \sigma_Y\ \sigma_Z\ \tau_{XY}\ \tau_{YZ}\ \tau_{XZ}]$,

we obtain

$$C_{11K} = C_\parallel\cos^4\alpha + C_\perp\sin^4\alpha + (\frac{C_{\perp\parallel}}{2} + C_{\#})\sin^2 2\alpha$$

$$C_{22K} = C_\perp\cos^4\alpha + C_\parallel\sin^4\alpha + (\frac{C_{\perp\parallel}}{2} + C_{\#})\sin^2 2\alpha$$

$$C_{44K} = C_{\#} + \frac{C_\parallel}{4} + \frac{C_\perp}{4} - (\frac{C_{\perp\parallel}}{2} + C_{\#})\sin^2 2\alpha$$

$$C_{12K} = C_{21K} = C_{\perp\parallel} + \frac{C_\parallel}{4} + \frac{C_\perp}{4} - (\frac{C_{\perp\parallel}}{2} + C_{\#})\sin^2 2\alpha$$

The stiffnesses of the multilayered composite (index S: slice) can be calculated according to the law of superposition for parallel acting stiffnesses

$$C_{11} = \sum_{K=1}^{n}\frac{t_K}{t_S}C_{11K} \qquad C_{22} = \sum_{K=1}^{n}\frac{t_K}{t_S}C_{22K}$$

$$C_{44} = \sum_{K=1}^{n}\frac{t_K}{t_S}C_{44K} \qquad C_{12} = C_{21} = \sum_{K=1}^{n}\frac{t_K}{t_S}C_{12K}$$

The analog structures shown in Fig. 2 and 3 illustrate the development of the stiffnesses for the unidirectional single layer in the Z direction. If the coordinates $\parallel_K = X$, $\perp_{(1)K} = Y$, $\perp_{(2)K} = Z$ are used, and as well $\tau_{XY} = \tau_{YZ}$, we may write for the vertical direction

$G_{XY\,K} = G_{YZ\,K}$

If more than one layer is concerned, the principle of sequential stiffness effects of the individual layers (rovings, mesh, fabric, resin), perpendicular to the plane direction, is valid. Also, if the layers are twisted, transformation of the

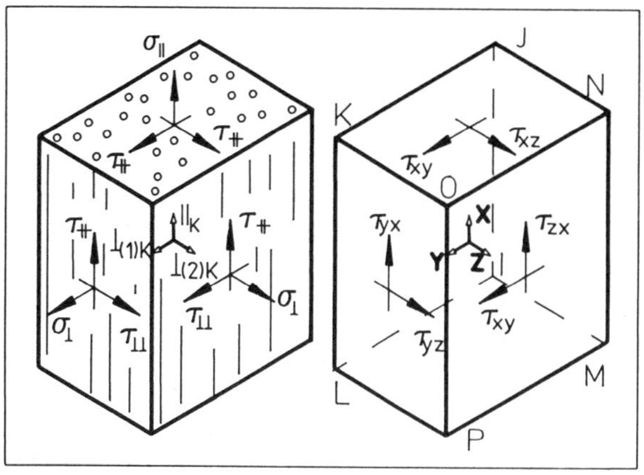

Fig. 2: Types of stress in a unidirectional GRP: axial stress $\sigma_\parallel$, transverse stress $\sigma_\perp$, axial/transverse shearing stress $\tau_{\parallel\perp} = \tau_{\#}$, transverse/transverse shearing stress $\tau_{\perp\perp}$.

Fig. 3: 3D element, showing nodes and coordinate system

equations is necessary utilizing twisting angle α_K or $90\text{-}\alpha_K$. In addition, perpendicular to the layering, the ratio of the area of layer $A_{XY\,K}$ and the mean area of the slice A_{mS} , must be considered in the stiffness calculations [4]

$$\chi_K = \frac{A_{XYK}}{A_{mS}} \quad .$$

Knowing the glass volume of the slice, i.e.

$$\varphi_S = \frac{\sum_{K=1}^{n}\varphi_K t_K}{\sum_{K=1}^{n} t_K}$$

and the quantities $E_{\parallel S}$ and $\upsilon_{\parallel S}$ which are calculated with the φ_S value, we can now calculate the heretofore missing matrix stiffnesses (note that α_K has no effect on C_{33})

$$\frac{1}{C_{33}} = \sum_{K=1}^{n}\frac{1}{\chi_K C_{22K(\alpha=0)}}$$

$$\frac{1}{C_{55}} = \sum_{K=1}^{n}\frac{1}{\chi_K C_{44K(90-\alpha)}}$$

$$\frac{1}{C_{66}} = \sum_{K=1}^{n}\frac{1}{\chi_K C_{44K(\alpha)}}$$

$$\upsilon_{\parallel\perp S} = \upsilon_{\perp\parallel S}\,\frac{C_{33}}{C_{11}}$$

$$C_{13} = C_{31} = \upsilon_{\perp\parallel S}\,C_{33}$$

$$C_{23} = C_{32} = \upsilon_{\parallel\perp S}\,C_{33} \quad .$$

The fabric and resin layers are considered isotropic structures. The approximate equations available for fabrics have been verified by experiments [3]. These are

$$E_M \approx 4000 + 30180\,\varphi - 4840\,\varphi^2$$

$$\upsilon_M = 0{,}34 - 0{,}075\,\varphi$$

$$G_M = 1430 + 11000\,\varphi \quad .$$

According to [1], results will be more realistic when secants moduli are used instead of initial Young's-moduli in the direction of $\perp_K$. Therfore, our computer programme [5] utilizes

$$E_{SM\perp} = 0{,}9\,E_{\perp} \qquad G_{SM\#} = 0{,}8\,G_{\#} \quad .$$

Creep, which is time-dependent, is thus well traced.

This allows to determine the 6x6 matrix [E] of the elastic constants for anisotropic three-dimensional elements, however, limited to the special case of orthotropy. These constants are used in the stress-strain relationship

$$[\sigma] = [E] \cdot [\varepsilon] \quad .$$

2. ROTOR BLADE FLANGE OF THE HSW 250

2.1 General Considerations

The governmental strength requirements for GRP power generator rotor blades [6] are based on static rather than dynamic load conditions. The load limit is given by a maximum permissible strain of 0,2% in compression and 0,3% in tension, using a security factor of 1,5.

We used the century wind gust and the blade at 6^{00} position and wind coming in at a 15^0 angle as the maximum load condition for the stall controlled wind power generator.

The calculations were carried out via the Finite-Element-System ANSYS, under license no. 08460. The slected anisotropic element STIF64 allows for the input of HOOKE's matrix [E] as well as its inverse matrix $[E]^{-1}$.

2.2 Modelling Concept of the System

The rotor blade flange is designed as a cylindrical GRP shaft and a flange at right angle. The flange and part of the cylinder (approx. flange height) are interlocked by the teeth of a cast-iron support ring with fingers reaching into the flange (Fig. 4).

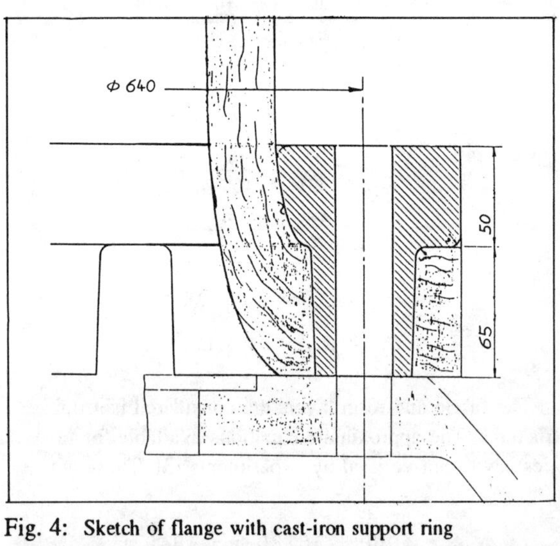

Fig. 4: Sketch of flange with cast-iron support ring

The ring carries the bolt joints and prevents the crushing of the soft GRP material. In addition, the support ring reduces the unavoidable secondary bending of the structure.

The GRP structure is an alternating layering of rovings, mesh, fabric and resin. The load-bearing roving layers are surrounding the teeth of the interlocking iron support ring. In the shaft the rovings are placed at an angle of $\alpha_K = -45^0$ to $+45^0$.

Finite elements were defined as slices 1 mm to 4,5 mm thick. We made sure that the layers of a slice, as a whole, do satisfy orthotropic requirements. This means that in a slice the main load direction coincides with the orthotropic direction X. All element coordinates X are located parallel to an imaginary line connecting the nodes I and J. Fig. 5 shows how this was done in the flange.

Part of a 10^0 flange is shown in Fig. 6. The model of 180^0 was actualized by mirror effects and duplication.

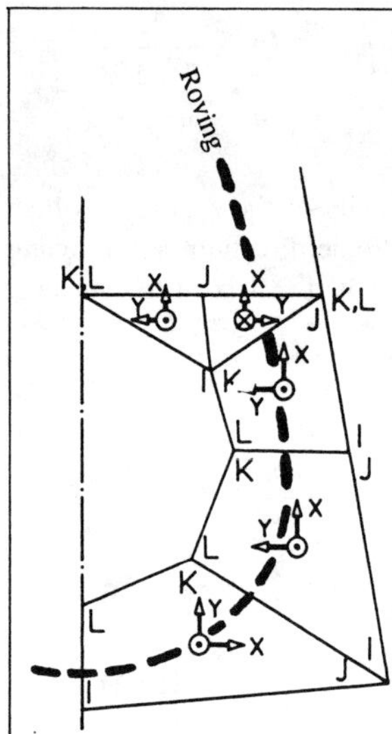

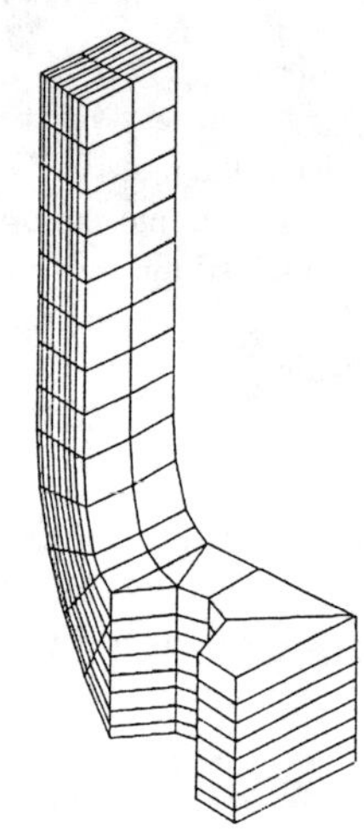

Fig. 5: Element coordinates in the flange (STIF64)

Bild 6: Element structure, selected over 10^0

2.3 Boundary Conditions

The two determining load factors at the rotor blade root are the moments of impact and torsion (which is 10 times smaller). Due to the fact that the geometry is rotationally symmetric it was possible to achieve both mechanical and geometrical symmetry by placing the geometrical cutting plane of the 180^0 model in the mechanical symmetry plane.

The only load is bending due to the resulting moment. The other stresses, which exist due to axial forces and shearing, are by comparison 200 times smaller and are thus negligible (Fig. 7).

Deformation suppressions were given in the cutting plane. Fixing all nodes in the holes in three coordinate directions simulated the influence of the rigid ring elements. In the transition area flange/cylinder the support of the rigid ring up to a height of 50 mm was simulated by permitting node displacements only tangentially to the surface. In this area maximum strains were expected.

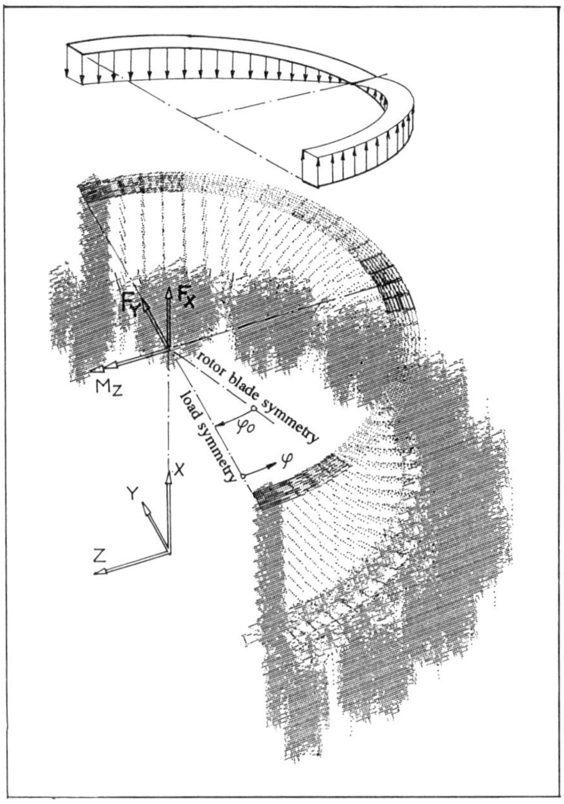

Fig. 7: Nodes with geometrical and mechanical boundary conditions

2.4 Results and Discussion

We found maximum strains to occur in the transitional zone between flange and cylinder (Fig. 8: black area).

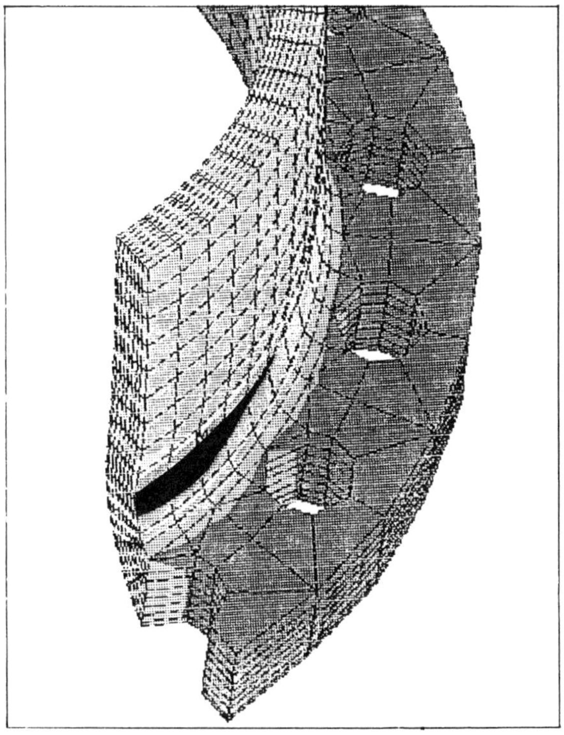

Fig. 8: Strain concentration in the transitional zone from the flange to the cylinder

The stress pattern closely resembles the strain pattern, as the stiffness among the slices (having basically the same layering pattern throughout) is similar.

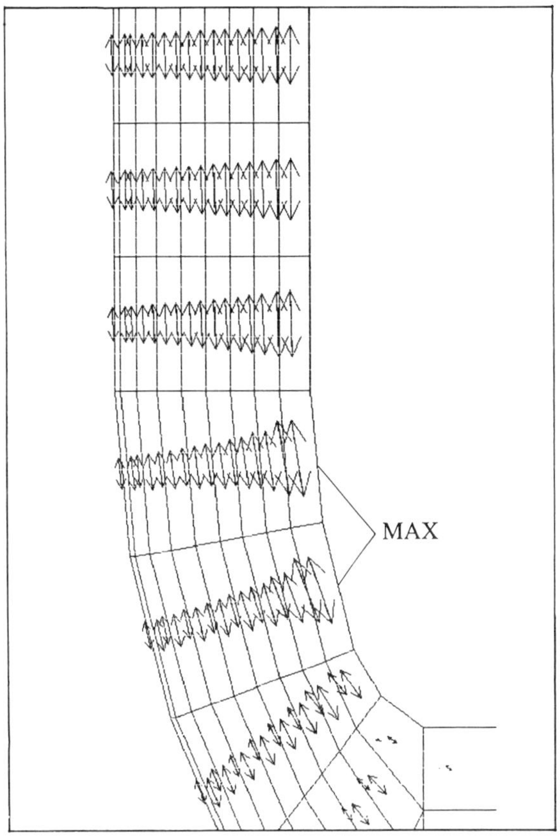

Fig. 9: Cylinder showing most of the strain (MAX) occuring at the outer surface of the curvature

Fig. 9 shows the zone of maximum stress and strain occurrance. It appears important to put the load-bearing roving layers at the outer surface of the cylinder curvature. Also, the iron support ring appears to significantly help limit the strain.

REFERENCES

[1] A. Puck: Zur Beanspruchung und Verformung von GFK-Mehrschichtenverbund-Bauelementen, Kunststoffe Bd. 57 (1967) und Bd. 58 (1969).

[2] J. Wiedemann: Elastizität und Festigkeit von Bauelementen aus GFK, Beiheft zum 'Kunststoffberater', Umschau Verlag 1969.

[3] A. Matting, G. Niederstadt: Kunststoffe Bd. 57 (1967).

[4] K.K. Chawla: Composite Materials, Science and Engineering, Springer Verlag 1988.

[5] G.U. Steinhagen: Rechenprogramm 'ANSOTROP' 1990. E-Moduln und Matrix-Steifigkeiten eines 2-dim. oder 3-dimensionalen anisotropen GFK-Mehrschichten-Verbundes (als Kontinuum).

[6] Institut für Bautechnik, Berlin: Vorläufige Richtlinien für statische Nachweise von Windkraftanlagen, 1988.

RECOMMENDATIONS FOR THE DESIGN OF WING BLADES FOR WIND TURBINES

Rayner M Mayer

Sciotech, 9 Heathwood Close, Yateley, Hants, GU17 7TP, UK

ABSTRACT The work of the Joule-funded program which has been carried out between 1990 & 1993 is assessed. The material and structural response to fatigue loading is outlined for those composite materials commonly used in wing blades. The origins of a set of recommendations, which could form part of a future structral code of practice is described against the need to comply with the essential requirement of the new EC approach directives.

1 INTRODUCTION

Good design requires the wind turbine to be designed around the wing blade. The blade's aerodynamic performance will determine the loading on the turbine components whilst its ability to withstand structural loading will determine its design life.

This collaborative program investigated between 1990 & 1993 specific aspects of the design of wing blades for which our knowledge was incomplete [1]. It has concentrated on the response of materials and structures to repeated loading and how structural design concepts can be verified.

Much knowledge has been developed of which 5 papers are being given at this conference. The entire work is being written up for publication [2] whilst a comprehensive summary is given in [3].

The Joule partners are listed in table 1 and various facets of the project are described briefly in the succeeding sections.

Table 1 Joule partners

Riso National Laboratory	Riso
DLR, Stuttgart	DLR
NLR, Emmeloord	NLR
ECN, Petten	ECN
RUG, Gent	RUG
DNV, Hovik	DNV
FFA, Stockholm	FFA
University West of England	UWE
Bath University	Bath
Sciotech +	Scio

+ technical coordinator

2 MATERIALS

Wing blades are made from composite materials being either fibre-reinforced plastics or wood-veneer laminated plastic. In the blade root area, metallic materials are generally incorporated to permit easy load transfer to the turbine hub.

We have considered the following aspects in some depth -

* alignment of fibres
* influence of matrix and fabric
* use of hybrid fabrics
* properties of wood composites
* influence of moisture and impact damage
* influence of spectral loading

Measurements have been made on coupons fabricated by hand lay-up or filament winding and these include the following types of loading

* tensile
* flexural
* compression
* reverse loading (tension/compression)

Design information has been generated about working stresses and strains up to 10^7 cycles.

3 COMPARISONS

There are a multiplicity of test methods, but few standards. So we have conducted a benchmark test amongst 7 of the partners, each using a portion of a large laminate plate (2.4 x 1.2m). The average of the means from each test are listed in table 2.

Table 2 Averages of the mean values of the tensile and flexural properties as measured by various institutes

	E (GPa)	σ (MPa)	ε (%)
flexure	22.9+-2.1	618+-26	2.8+-0.2
tension	24.5+-1.0	514+-58	2.1+-0.2

It is clear that the scatter is small and so it can be assumed that the results are self-consistent. The effect of cyclic loading will have a smaller scatter as most results are measured relative to a monotonically loaded sample.

The effect of scale has been considered by comparing the structural response of coupons laminated in the laboratory with those cut from spars and a complete spar.

4 STRUCTURES

The major problem with testing is that of introducing the load and monitoring the accumulation of the damage. The former generally requires special adaptor pieces whilst for the latter strain gauges, thermocouples and visual inspection have been found useful.

The damage generally tends to manifest itself locally rather than globally and an initial strain survey is recommended to determine the positions of high strain.

A variety of blade roots have been tested, some production and some experimental, and valuable information has been gained on the load-carrying capability of the designs.

5 GOOD WORKING PRACTICES

We have developed a set of good working practices from our investigations. These cover the following [2] -

* material selection
* material characterisation
* manufacture
* coupon testing
* component testing

They are more comprehensive than those currently required by existing rules and regulations [4], and also serve as a check-list to the designer, who may wish to add his own requirements.

One example is given from each section to illustrate the proposed approach -

r1) identify fully fibres and fabrics

* fabric manufacturers are generally vague about the source of fibres and the fibre size may be indeterminate. We recommend that the fibres should be specified fully.

r4) specify reinforcement unambiguously

* there is currently no agreed terminology or classification of fabrics and an ever widening set of processes has become available. We recommend that a systematic classification be adopted such as that given in [5].

r7) check degree of cure using barcol hardness

* for prolonged exposure to cyclic loading, the component needs to be fully cured all over. Barcol hardness being a hand held indentor is capable of checking the degree of the cure by measuring the microhardness immediately below the indentor and comparing with the manufacturer's recommended value.

r10) validate test methods

* in the absence of agreed methods, it is essential to use valid test techniques. This is not easy to achieve and requires some careful thought as to how to establish the value of a test technique. We would recommend using initially a material combination or component whose response is known.

r17) evaluate the accumulation of damage during structural testing

* the way in which a component may fail is as important as its failure life. Monitoring the accumulation of damage should allow one to follow the progression of damage from one stage to the next. It is the transition from steady-state to runaway that needs to be detected if one wishes to determine the safe-life condition.

6 EC DIRECTIVES

There are some five new approach directives, which are applicable to wing blades (table 3).

Table 3 Relevant EC directives

product liability
product safety
construction products
machinery
electro-magnetic compatibility

The demonstration of compliance with the essential requirements of these directives should be agreed at EC level and a code of practice formulated. The recommendations described above could then form part of this code.

In this way, the CE mark could be affixed to wing blades in their own right.

7 CONCLUSIONS

A comprehensive set of design information has been derived within this Joule-funded project. The collaboration between the various partners has been fruitful and has enabled our understanding of the response of composite materials and structures to cyclic loading to be increased.

The information is applicable to many types of components and structures and this will hopefully lead to new applications of these attractive materials

ACKNOWLEDGEMENTS

* the Renewable Energy Unit, DG XII for a funding grant under its Joule program
* Giancarlo Caratti and Jens-Peter Molly for their encouragement and advice
* various national funding bodies for grants

REFERENCES

[1] R M Mayer Fatigue and design of wing blades for wind turbines in Wind Energy: Technology & Implementation EWEC '91 Amsterdam Elsevier 267

[2] Design of Composite Structures: fatigue considerations and applications to wind turbines; to be published

[3] Proc Wind Energy Contractors Meeting Alghero June 1992 EUR 14607 EN (1992) 241

[4] The classification societies like DNV and Germanisher Lloyd have produced their own rules; so have the Danish Energy Authority and ECN

[5] R M Mayer Design with reinforced plastics Design Council (1993)

THE EFFECT OF MOISTURE ON THE FATIGUE PERFORMANCE OF GLASS FIBRE REINFORCED POLYESTER IN BOLTED JOINTS

P.W. Bach

Netherlands Energy Research Foundation, ECN
P.O.Box 1
NL-1755ZG Petten
The Netherlands

ABSTRACT: Although GRFP's generally are found to have good weathering properties, the effects of moisture on the fatigue properties are not negligible. In order to investigate the influence of moisture, fatigue tests on preconditioned GFRP bolted joints were performed at 100% R.H. and compared to the results of dry GFRP bolted joints from a previous fatigue study. At the highest stress level a reduction in lifetime of about one order of magnitude is observed for the humid specimens. The reduction in fatigue strength by the moisture is somewhat decreasing with stress level. This effect is more pronounced for the stress ratio 0.1. The results of the WISPER tests are in accordance with the diminishing effect of moisture in the low stress range as found in a literature survey and observed in the R= 0.1 constant amplitude tests.

1. INTRODUCTION

The most fatigue critical part of a wind turbine is the rotor blade structure and its connection to the hub. In a design life extending over more than 20 years it has to endure high cycle (> 10^8) fatigue loading of variable amplitude due to gusts, windshear, start and stop procedures etc. For a mechanical construction the fatigue of a wind turbine blade is a quite unusual matter due to the large number of cycles and the typical load spectrum.

Very few data are available in this high cycle fatigue range for the materials and substructures used in wind turbines, especially in combination with environmental effects. One of the materials that is extensively used in wind turbine blades is glass fibre reinforced polyester (GFRP).

Probably the most critical part of a wind turbine is the rotor blade connection to the hub, for which often a bolted joint is applied. Although the rotor blade is usually well protected against weathering by a gel-coat, often moisture can find its path to the joint. The sensitivity of GFRP bolted joints to moisture is investigated.

The design strength and stiffness of rotor blades have to be maintained during the lifetime of a wind turbine. Especially the stiffness is of importance because stiffness degradation of the materials has an effect on the aerodynamics and the resonance frequency of the rotor blade. In the fatigue investigation special attention is paid to monitoring the stiffness degradation of the specimens.

2. EFFECT OF MOISTURE ON FRP LAMINATES

Although GRFP's generally are found to have good weathering properties, the effects of moisture on the fatigue properties are not negligible [3].

The water absorption process can be described as follows:

- Diffusion of water molecules through the matrix to the glass/resin interface;
- Weakening of the interfacial fibre matrix bond through chemical and physical attack on the coupling agent;
- Exposure of the glass fibre surface following partial dissolution of the coupling agent;
- Chemical attack on the glass surfaces from which alkali silicates and silica molecules are leached out.

The major effects of absorbed moisture on the laminates are:

- Lower glass transition temperature of the matrix resin;
- Damage of the interface between the fibres and the resin;
- Reduction of the cure induced residual stresses through swelling which may retard failure.

Hence the effects of moisture on the properties are not all negative.

Especially when loaded in a humid environment glass fibres are susceptable to subcritical crack growth, a phenomenon which is often referred as static fatigue. This will influence the time-under-load dependent properties such as creep and cyclic fatigue.

Also some synergy with moisture and fatigue can be expected. A composite subjected to water and fatigue loads will show increased matrix cracking which will accelarate the uptake of water.

From a literature survey it is found that the effects of moisture absorption on the mechanical properties of GFRP's are not known accurately.[3]. This is mainly due to the large scatter in results which are sometimes even inconsistent. The effect of moisture on the fatigue properties for both glass/polyester and glass/epoxy laminates diminishes in the very high cycle and low stress range. For wind turbine applications this is a positive observation. In table 2.1 a rough indication of the effects, as estimated from the survey, of moisture on the mechanical properties of fibre reinforced plastics are listed. From the viewpoint of moisture effects the combination of glass/epoxy or carbon/epoxy gives the best fatigue properties.

material	UTS	fatigue 1E3 cycles	fatigue 1E7 cycles
glass/polyester	-15	-35	-10
glass/epoxy	-10	-20	-5
carbon/polyester	-5		
carbon/epoxy	-2	±0	±0

Table 2.1 Indication of the effect of moisture on the mechanical strength of FRP. Data in % reduction [3].

3. MATERIALS AND METHODS

3.1 Glass-polyester bolted joints

In order to investigate the influence of moisture on the fatigue performance of glass-polyester structural details bolted specimens were fabricated.

Specimens were cut on-axis from a 0°and +/-45° reinforced glass fibre polyester plate (15 mm thick). The plate contains 9 layers of 700 g/m^2 unidirectional fibres plus 100g/m^2 chopped fibre mat (Ahlstrom 700/100) and 8 layers of stitched 700 g/m^2 + or - 45° (Ahlstrom R-24-700-45) glass fibres in a symmetrical lay-up. The plate was produced under industrial conditions by hand lay-up and with a glass/polyester ratio of 53 weight% (~ 33 vol.%). The resin is the iso-polyester Synolite 593-E-1.

In the specimens 2X2 holes were drilled with a diameter of 16 mm and with an H8 fit. It is well known that full bearing strength in bolted composites is only attained when the width/diameter and the end/diameter ratios are above a minimum value.[1]. The ratios used (w/d=3.5 and e/d=4) are above this minimum value.

The performance of bolted joints can be improved when lateral constraint is applied as is usual in the connection to the hub. Therefore the specimens were thickened around the holes to a total thickness of 20 mm with Synolite TP 347 VG epoxy(UTS=77 MPa), in order to fit precisely in the steel flange. In figure 3.1 the details of the bolted specimens are depicted.

The Ultimate Tensile Strength UTS of dry specimens was measured in a previous project with a loading rate of about 10 kN/sec. The UTS is 118 kN per hole , which means a bearing stress of 490 MPa.

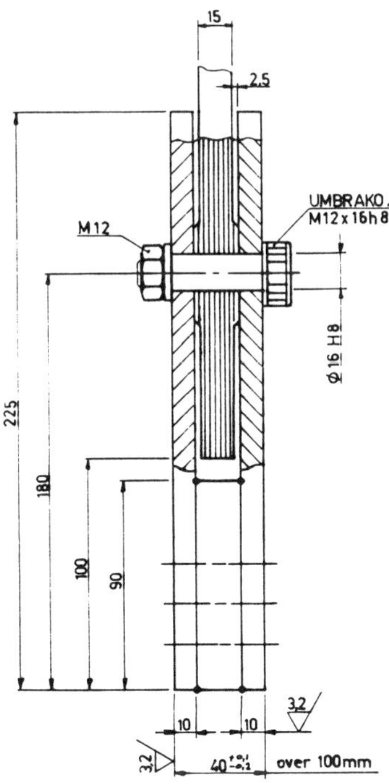

Fig. 3.1. The FRP bolted joint specimen.

3.2. Constant amplitude tests

The constant amplitude tests were performed on load controlled servo-hydraulic testing machines (Instron 1343) with tension-tension or tension-compression loading (stress ratio R = $\sigma_{min}/\sigma_{max}$ = 0.1, -1 or -0.5).

During fatigue tests the fibre reinforced plastic specimens will experience hysteristic heating dependent on the stress amplitude and the frequency.

Due to the expected high cycle fatigue life the tests of the coupon specimens were performed with a frequency upto 5 Hz. The temperature of the specimen was kept low with a cooled air flow of 15- 18 °C.

After measurement of the initial stiffness, the stiffness degradation was determined on-line during the test at regular intervals (100-10.000 cycles). The stiffness was calculated as the delta load divided by the delta displacement of the specimen as measured by the displacement transducer of the actuator. The stiffness degradation of the specimens will influence the load performance in the closed-loop servo-hydraulic system. In order to have a constant load until failure the load was corrected on-line during the fatigue life of a specimen.

3.3. Variable amplitude tests

Current design procedures for wind turbines are primarily based on constant amplitude data. However, this simple kind of loading is usually insufficient to represent the interaction effects of high and low cycles that occur in a realistic type of loading.

Especially in the aerospace field variable amplitude fatigue testing is widely used and standard loading sequences have been developed in order to predict the material fatigue behaviour under realistic conditions. The main applications of standard load sequences are for ranking materials, structural design alternatives, fabrication techniques etc.

A variant of the WISPER spectrum has been derived (WisperX), in which the ranges below the size 17 have been omitted. The number of cycles in this spectrum is reduced by almost a factor 10, 12,831 versus 132,711. If the damage development caused by these spectra does not differ too much the testing time can be significantly reduced.

The WISPER spectrum represents a mean 2-month period of the flapwise loading of an imaginary wind turbine blade. As a result WISPER does not simulate sufficiently well the typical loading in a blade of a particular turbine in a particular wind regime. This means that it can never be used instead of a design spectrum.

The WISPER load sequence is intended for comparative purposes only, i.e. to evaluate materials, structural details, dimensions, design alternatives and lifetime prediction methods.

A lifetime prediction can be based on the rainflowed WISPER or WisperX 64x64 from-to matrix. A damage 64x64 matrix can be obtained by dividing the number of cycles in each matrix element with the number of cycles to failure calculated with the relevant S-N equation or constant life diagram.

The relative damage of one WISPER sequence is determined by summation of the damage data of the elements.

The available amplitude tests were performed on load controlled servo-hydraulic testing machines (Instron 1342 and Instron 1343) equipped with the Harrier computer interface and a Hewlett-Packard 9816 micro-computer.

In order to control the fatigue machines in the WISPER sequence tests a special software programme was developed. After measurement of the initial stiffness, the stiffness degradation is determined on-line at regular intervals. Variable amplitude testing is possible in two modes and with two wave shapes.

a. Constant frequency, only the amplitudes are variable.
b. Constant loading rate. For each individual load cycle the frequency must be calculated in order

to obtain a constant loading rate. In this mode both the frequency and amplitude are variable.

c. Tests can be performed with a saw tooth or a sine as the wave shape.

It is known that for glass fibre reinforced plastics the loading rate has an effect on the (fatigue) strength due to the visco-elastic behaviour and subcritical crack growth. Further the frequency and the load amplitude will affect the hysteretic heating of the specimens. So there can be some uncertainty in performing WISPER tests.

In a previous investigation the influence of the loading parameters on the WISPER fatigue lifetime was studied [2]. It was observed that there was no difference in fatigue life within the tested conditions.

The WISPER fatigue tests were performed with a constant loading rate and a sinusoidal wave shape, because these are the most realistic for the stress cycles in a turbine blade.

4. RESULTS

Although the laminate of a rotor blade is usually well protected against weathering by a gelcoat, often moisture can find its path to the joint. The sensitivity of GFRP bolted joints to moisture is investigated by preconditioning the laminate in water of 50 °C for 14 days and testing under 100% relative humidity. The initial condition of the specimens is from stack for several months at about 22 °C and an R.H.of 40%.

The uptake of water by the preconditioning is determined by weighing the specimens from the glass/polyester laminate. In figure 4.1 the curves of the change in moisture content are shown. Curves were obtained for absorption in water of 25 °C and of 50 °C and for desorption in an exsiccator at 25 °C. During the preconditioning the moisture content in the laminate increased with 0.13 weight %. Even after 100 days in water of 50 °C the ingress of moisture still continues.

The desorption of the laminate shows the same kinetics as the absorption.

The bolted joint is especially sensitive to the moisture content near the hole where an increased diffusion can take place due to edge effects and the damage caused by the drilling of the holes (capillary action). To simulate this phenomenon a strip with a width of 10 mm was grinded and preconditioned. The rate of moisture uptake is almost doubled. So the change in moisture content near the holes in the laminate is about 0.26 %.

The diffusion of moisture into composite materials can usually be described by the Fick law. The curves are plotted with the square root of the time on the abscissa. For Fickian absorption a straight line can be expected when the moisture content is far from saturation. The slope of the curves decreases somewhat so there is probably some deviation of the Fick's law by edge effects or the moisture content is approaching the saturation level.

In order to investigate the influence of moisture, fatigue tests on preconditioned GFRP bolted joints were performed at 100% R.H and compared to the results of dry GFRP bolted joints from a previous fatigue study [4].

The joints failed by the degradation of the bearing capability of the hole. Halfway the lifetime of the specimen also some damage started to develop in the 45° layers of the laminate in the middle part of the specimen.

The damaged zone near the hole in diagonal direction is about 2 x the diameter after failure and is in accordance with the stress distribution as predicted with a Final Element Method calculation [5]. There is some difference between the dry and preconditioned specimens in the magnitude of the damaged zone. This zone is about 20% smaller for the humid specimens.

In the R= 0.1 humid tests at the lower stress level the bolts failed due to corrosion assisted fatigue. In a later stage zinc coated bolts were applied.

In Figure 4.2 the number of cycles until frac-

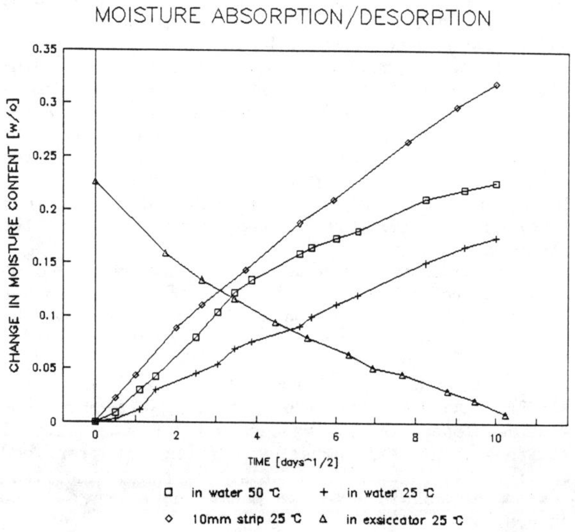

Fig. 4.1. Absorption and desorption of water in the GFRP bolted joint laminate.

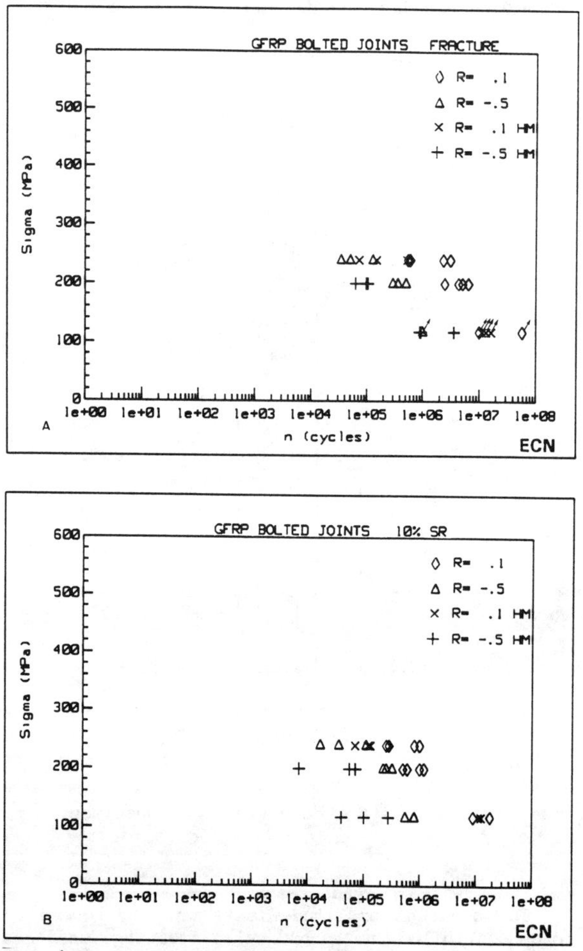

Fig. 4.2. Fatigue results of the dry and humid GFRP bolted joints.
A: Fracture B: 10% stiffness reduction.

ture of the dry and humid specimens versus the maximum initial bearing stress level is shown for R = 0.1 and R = -0.5. Some specimens did not fail during the planned fatigue test. Such a so-called Run-Out is indicated with an arrow. At the highest stress level a reduction in lifetime of about one order of magnitude is observed for the humid specimens. At the lower stress level there is no indication of the effect of moisture due to run-out of the specimens. However, an indication of the effect can be found by the results for 10% stiffness degradation of the specimens as shown in figure 4.2.B. The reduction in fatigue strength by the moisture is somewhat decreasing with stress level. This effect is more pronounced for the stress ratio 0.1.

The results for the WISPER tests are illustrated in figure 4.3 for the failure criteria fracture and 10% stiffness reduction. The tests with the moisture condition were run at a rather low stress level, which has a better predicting value for the stress levels used in practice. Within the limited test matrix there is no difference in the behaviour

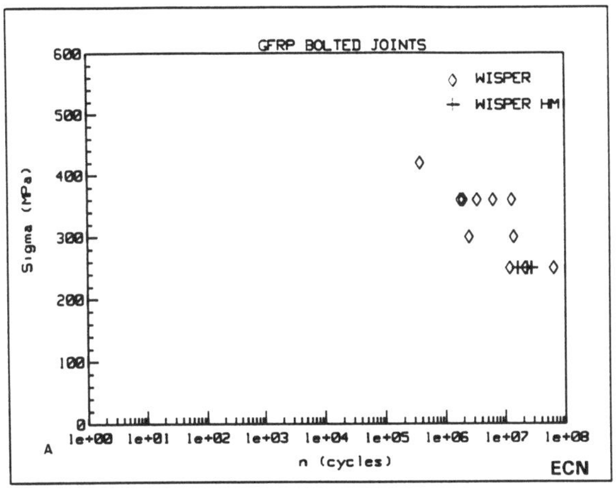

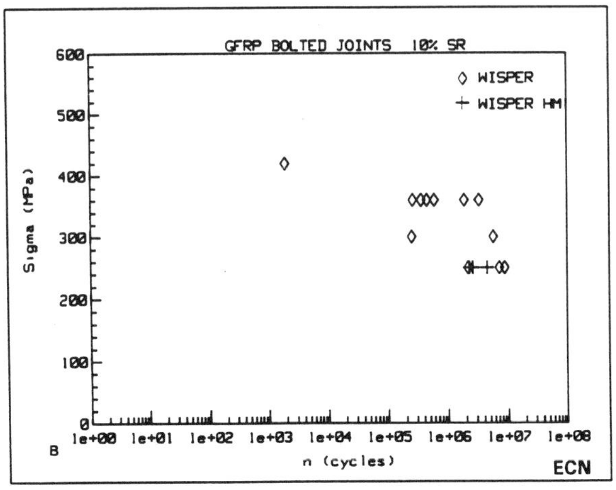

Fig. 4.3. WISPER fatigue results of the dry and humid GFRP bolted joints.
A. Fracture B. 10% stiffness reduction.

of the dry and humid specimens.

The WISPER spectrum contains only a small minority of through zero cycles and the fatigue behaviour will be largely governed by stress ratios >0. So the results are in accordance with the diminishing effect of moisture in the low stress range as found in the literature survey and observed in the R= 0.1 constant amplitude tests.

5. CONCLUSIONS

Although GRFP's generally are found to have good weathering properties,the effects of moisture on the fatigue properties are not negligible.

Fatigue tests on wet preconditioned GFRP bolted joints performed at 100% R.H show at the highest stress level a reduction in lifetime of about one order of magnitude. The reduction in fatigue strength by the moisture is somewhat decreasing with stress level. This effect is more pronounced for the stress ratio R=0.1.

The results of the WISPER tests are in accordance with the diminishing effect of moisture in the low stress range as found in the literature survey and observed in the R= 0.1 constant amplitude tests

ACKNOWLEDGEMENTS

The author wishes to thank mrs. M. Jong and mr. B.A.J. Schaap of ECN for preparing and performing the fatigue tests, and NOVEM and the EC-JOULE programme for sponsoring this work.

REFERENCES

[1] Godwin, E.W., Matthews, F.L. (1980), A Review of the Strength of Joints, In Fibre Reinforced Plastics, Composites, July 1980, p. 155.
[2] Bulder, B.H., Bach, P.W.(1991), A Literature Survey on the Effects of Moisture on the Mechanical Properties of Glass and Carbon Fibre Reinforced Laminates, ECN-C-91-033.
[3] Bach, P.W. (1991), High Cycle Fatigue Testing of Glass Fibre Reinforced Polyester and Welded Structural Details, ECN-C-91-010.
[4] Bach, P.W., Schaap, B.A.J. (1990), Vermoeiingsonderzoek aan windturbinematerialen en bladverbindingen, ECN-C-90-012.
[5] Roorda, B. (1988), Analyse pen/gat verbindingen fase 2 vermoeiingsonderzoek, Polymarin rapport R&D-880622.

ENVIRONMENTAL EFFECTS ON GL-EP ROTOR BLADES

Ch. W. Kensche

German Aerospace Research Establishment
Institute of Structures and Design
D-7000 Stuttgart 80, Pfaffenwaldring 38-40

ABSTRACT: The effects of moisture and hailstone impact were investigated on Gl-Ep (Glass-Epoxy) material, typically for rotor blades of wind turbines. The laminates of 2 mm thickness were exposed to an artificial climate of 90 per cent relative humidity and 45°C up to saturation. The plates and also those in a dry or respectively *as-received* status were impacted by iceballs simulating the hailstones. Static properties of the material like tension, compression and interlaminar shear were tested as well as the fatigue properties at R=-1 (tension-compression). Due to the hailstone impact there could not be observed a decrease of the quality, however moisture degrades the static properties significantly. Concerning the fatigue properties at strains of ±0.8 per cent and more the load cycles show a decrease of order of magnitude. However, at lower applied strain levels which are interesting for the structural design no moisture influence could be stated.

1. INTRODUCTION

The service life of wind turbine rotor blades made of GFRP (Glass Fibre Reinforced Plastics) may be reduced by hostile surrounding conditions as humidity and hailstone impact. Thus, the high cycle fatigue data available e.g. for Gl-Ep (Glass-Epoxy) [1] cannot be used satisfactorily. Improved knowledge is necessary for design and certification whether the material properties will be diminished or not. therefore the effects of moisture and hailstone impact were studied solely and in combination and compared with data of unaffected material.

2. MATERIAL

The material used was Interglas fabric with the lay-up (1×92125/±45° / 2×92146/0°)$_{sym}$ and the finish I550. It was laminated with the epoxy resin system L20/SL from Bakelite. The curing and postcuring conditions were 24h at room temperature (RT), 8h/60°C and 15h/80°C. Several plates of the size 260×500 mm^2 and a mean thickness of 2.2 mm were manufactured with a fibre volume fraction of about 40 %. The plates were cut into two parts, one of them taken for exposure to humidity, the other one stored in laboratory climate to avoid differences in properties due to the influence of fabrication.

3. EXPOSURE TO HUMIDITY

The exposure of the material took place in a tropical chamber at 90 % relative humidity. To accelerate the penetration velocity of the moisture, the surrounding temperature was elevated up to 45°C. The increase of the moisture content was measured on traveller specimens. Figure 1 shows that saturation was achieved after about 3 months of exposure at a humidity level of 0.4 to 0.5 %. The plates were stored in the humid climate without further weight gain up to the time they were prepared for testing.

To know the absolute moisture content, traveller specimens were also desiccated (RT). The loss of weight was between 0.1 and 0.2 % and corresponds with the moisture content in the *as-received* status (furtheron called *dry*).

Similar results were obtained in a FEM simulation of a 30 years' moisture absorption for a rotor blade of 10 mm thickness under the assumption of cyclic cold-wet and dry-warm weather conditions related to a one-year cycle measured at the DLR wind turbine testfield [2]. The calculation showed that a 20 years natural weathering resulted in a constant moisture content of 0.4 to 0.5 %. This justifies a relation between this and the artificially produced moisture absorption. It should be mentioned that the slowing effect of the moisture penetration due to the gel coat protection was not considered in the simulation.

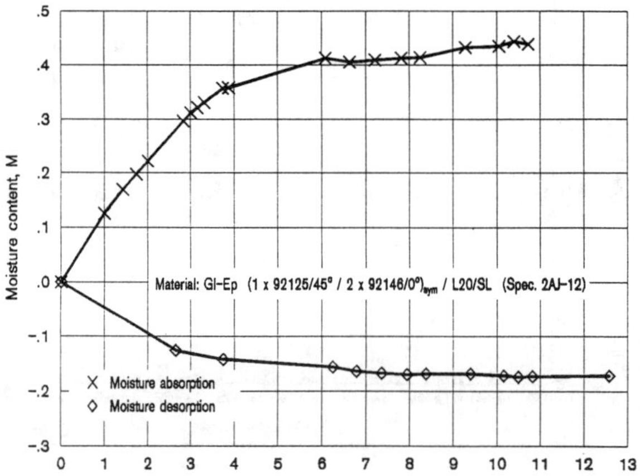

Figure 1. Moisture absorption and desorption of traveller specimen at 45°C/90%RF

4. HAILSTONE SIMULATION

Literature differenciates up to 7 forms of hailstones depending on the time of growth and the conditions in thunderstorm clouds [3]. The mean density is about 0.9 g/cm^3. The frequently milky appearance is due to voids the volume of which defines the density.

The terminal velocity of hailstones can be described by the empirical formula

$$V_\infty \approx 9 \times D^{0.8}$$

with V_∞ in m/s and D in cm. For definition of the impact velocity of a hailstone onto a rotor blade, the terminal velocity of the hailstones was superposed with maximum velocity components of a rotorblade of the type DEBRA 25 with diameter 25 m and 50 min^{-1} r.p.m. This resulted in a hailstone velocity of more than 100 m/s.

In laboratory Gl-Ep plates were impacted during preinvestigations with artificially manufactured iceballs up to diameters of 37 mm to choose a suitable size of the projectiles and to improve the experience in impacting. In Figure 2 the principle of the impact testing rig is shown.

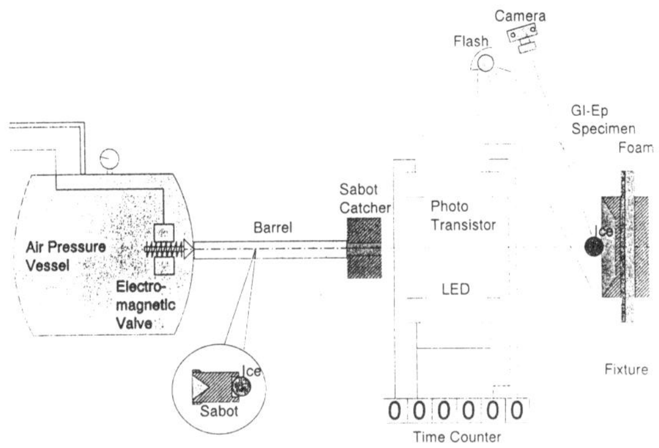

Figure 2. Impact testing rig

Airpressure controlled, the iceball is accelerated in the tip of a sabot up to the necessary velocity which is measured by two light barriers just before striking the specimen. This triggers a camera with a time-justable flash light to take a photograph either before or during the impact, shown in Figure 3 and Figure 4 by the example of 37 mm iceballs. Behind the circularly clamped GFRP plate a hardfoam disk is mounted to prevent it from failure in the rim. After impacting the whole plate it was cut into pieces of the necessary size of test specimens.

Figure 3. Iceball of ø37 mm before impact

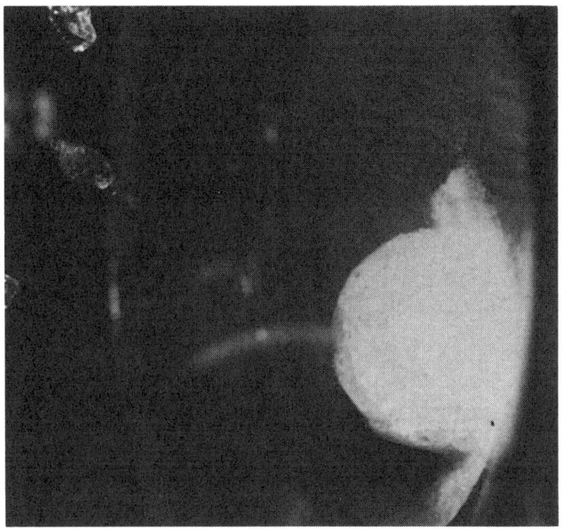

Figure 4. Impact of a ø37 mm iceball

The damages by the hailstone impact on the front side of the specimen are barely visible, whereas the rear side shows a whitened area of a size corresponding to the diameter of the iceball. The impact tests were continued with a diameter of 20 mm. Thus, the 20 mm wide test specimens got a damage over the whole cross section, and as many specimens as possible could be cut out of the plates. Considering the chosen velocity of about 105 m/s the hailstone impact energy resulted in about 27 Joule.

The whitening on the rear side of the 2.2 mm thick Gl-Ep plate after hailstone impact is not due to delaminations but to microcracks between matrix and fibres in the outer ±45° fabric, see the photomicrograph in Figure 5. Figure 6 shows a photomicrograph of the whole cross section with microcracks and voids.

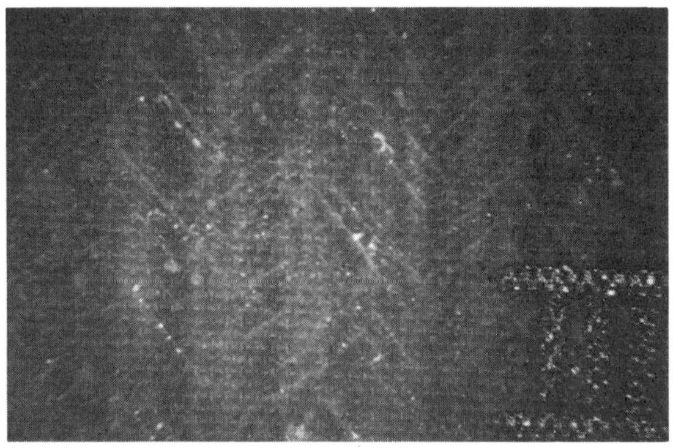

Figure 5. Damage by hailstone impact on rear side of Gl-Ep plate

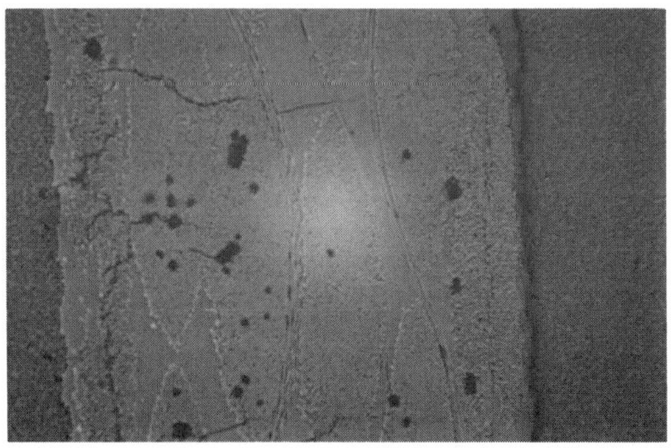

Figure 6. Cross section micrograph of an impact damaged specimen

5. MATERIAL INVESTIGATIONS

5.1 Static properties, tension

Static tension tests were carried out on six specimens each for the status *dry, undamaged* and *damaged* as well as *wet, undamaged* and *damaged* of the size 250×20×2 mm^3. The results are shown in Figure 7 for the E-Modulus and in Figure 8 for the strength. Only a rather limited tendency of property decrease due to humidity is detectable. It is interesting to note that the damaged specimens achieve a bit higher strength than the undamaged ones.

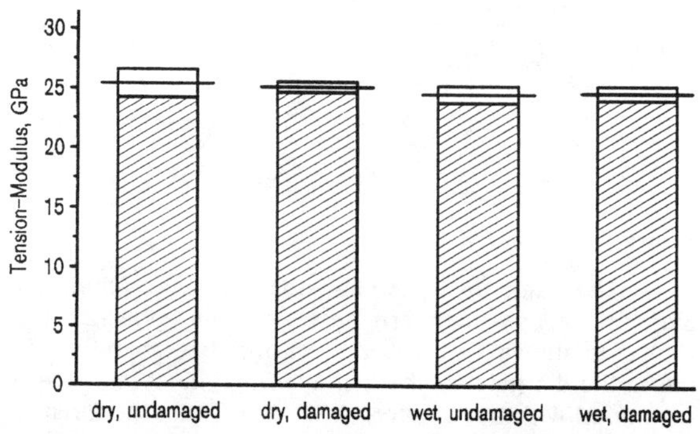

Figure 7. Tension Modulus of differently conditioned Gl-Ep

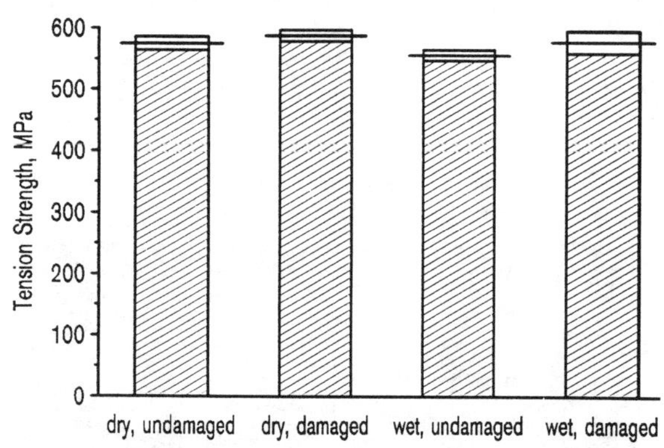

Figure 8. Tension strength of differently conditioned Gl-Ep

5.2 Static properties, compression

The problem in compression tests is that the free compression length must not be too large to avoid instability failure. In our investigation the *Celanese-testing rig* was used which has a normed free length of 8 mm. This appeared too small for testing the effects of impact and humidity on compression strength. Thus, to get detailed information about the beginning of the instability, at first some parameter tests were made by a variation of the free length between 8 and 32 mm, see Figure 9. The diagram shows that instability begins at a free length of 12 mm. Therefore, the tests were continued with this size of free length for the compression tests.

The tests were carried out with six specimens each for the four different combinations of conditioning. Figure 10 shows the results for the compression modulus, Figure 11 for the compression strength. While the stiffness of the humid specimens in compression is about 5 to 10 % lower than for the dry ones, the 30 % decrease in strength due to humidity is detrimental. No significant differences can be recognized due to the hailstone impact.

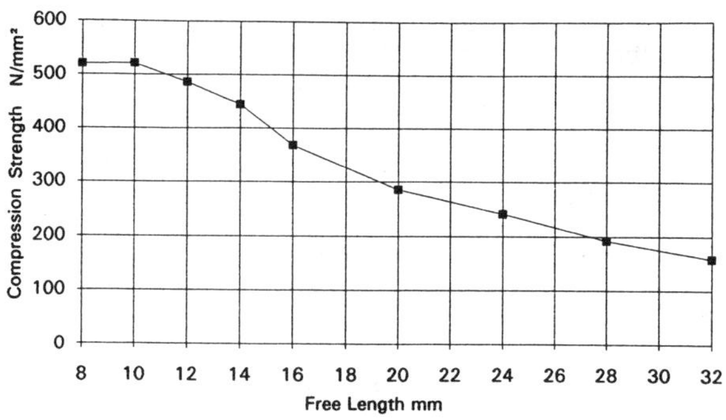

Figure 9. Compression strength versus free length

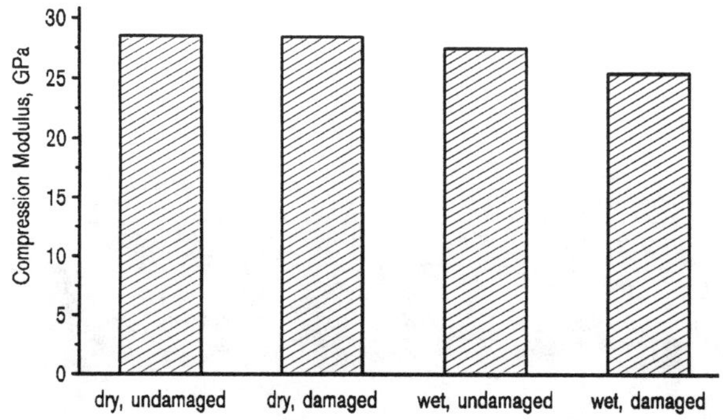

Figure 10. Compression modulus of differently conditioned Gl-Ep

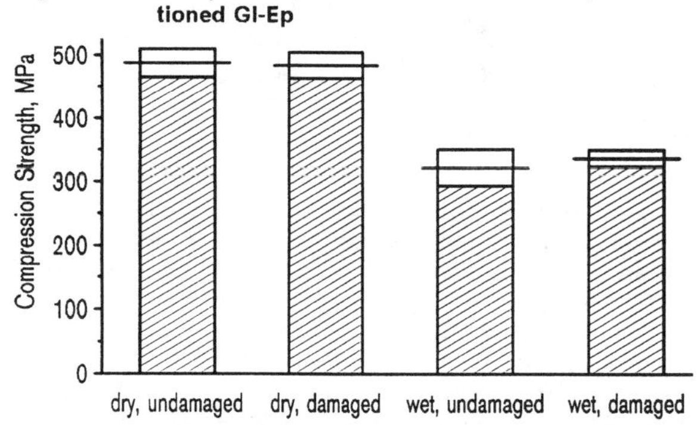

Figure 11. Compression strength of differently conditioned Gl-Ep

5.3 Static properties, ILS (interlaminar shear strenth)

ILS in structures occurs especially in bondings between spar caps and shear webs as well as in areas of load introduction. Thus, some ILS-tests were performed on short-beam specimens to get also information about the conditioning effects on that loading. Figure 12 shows no influence by the hailstone impact, however, an about 15 % degradation due to humidity.

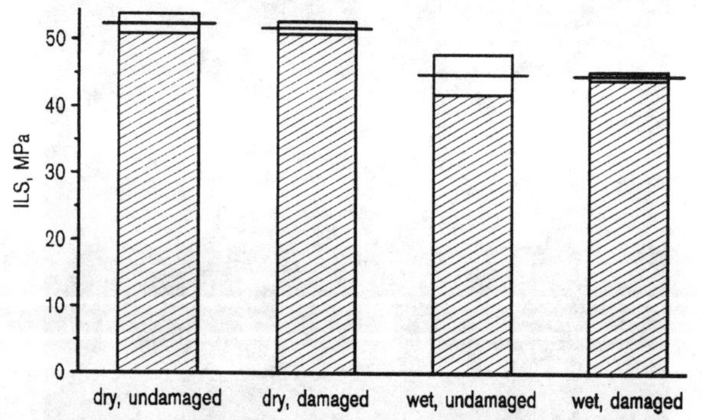

Figure 12. Interlaminar shear strength of differently conditioned Gl-Ep

5.4 Fatigue investigations

The fatigue properties of Gl-Ep material in the different modes of conditioning are of special interest since they can give informations about the life time of the rotor blades. A basis for comparison with non-damaged Gl-Ep of the same material is available from investigations described in [1].

For the fatigue tests which were carried out at an R-ratio of -1 (tension-compression), the application of an antibuckling guide was necessary to avoid instability of the specimens (dimensions 250×20×2.2 mm^3). For observation purposes and also to admit failure mechanisms of the conditioned specimens probably differing from the dry/undamaged ones, the antibuckling guide was supplied with a window of the size 16×16 mm^2. The long-time tests with humid specimens (at strain levels lower $\pm$0.8 %) were carried out in humid atmosphere to prevent the material from dry-out effects.

As a first step comparison tests were performed with *dry undamaged* specimens to compare the results with those from [1]. Figure 13 shows that the new data points are within the scatter of the previous investigation and beyond the curve of 95% survivability from [1].

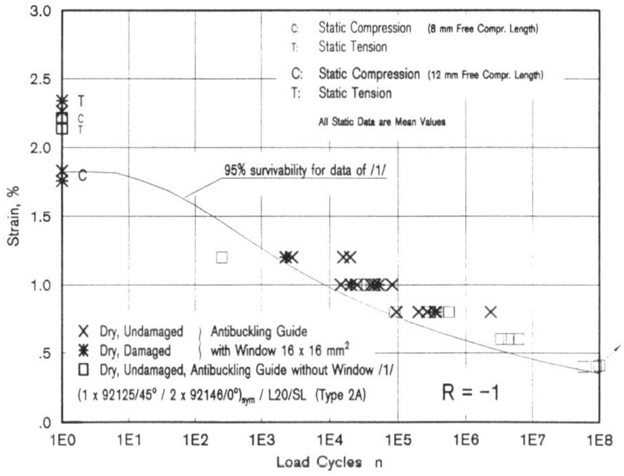

Figure 13. ε-n data of unconditioned Gl-Ep at R=-1

The tests were continued by fatigue testing of the *dry damaged*, *wet undamaged* and *wet damaged* specimens. The stiffness was measured by clip gauge extensometer in a first static tension test immediately before starting the fatigue loading and at 10^2 and 10^3 load cycles. There is a tendency of stiffness degradation from the condition *dry undamaged* down to *wet damaged*. However, the largest decrease of stiffness is only about 4 % and it should not be concluded from the behaviour of flat specimens to that of total structures because these do not show a decrease of the global stiffness, e.g. deflection at a defined load, also shortly before failure [1].

Figure 14 shows the ε-n data for the conditioned material together with the *dry undamaged* one. No influence of the hailstone impact induced damages is visible. But by the influence of the humidity on the fatigue at strain levels of $\pm$0.8 % and more the load cycle number is reduced significantly to more than an order of magnitude. All the data points are beneath the above mentioned curve of 95% survivability. However, at a strain level of $\pm$0.6 % load cycle numbers were achieved for the wet material in a similar level as for the dry one. Thus, another failure mechanism seems to occur at this lowest tested strain level. Since, however, rotor blade design allowables will - due to reasons of stability - not exceed strain levels of 0.6 %,

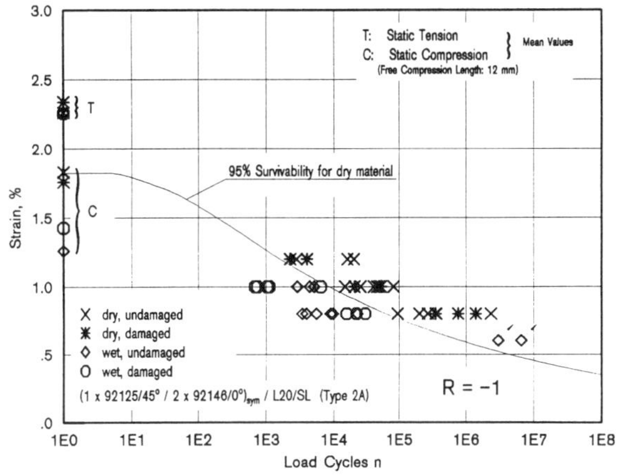

Figure 14. ε-n data of differently conditioned Gl-Ep at R=-1

fatigue effects may not be expected by humidity and also by hailstone impact.

6. CONCLUSIONS

Static and fatigue investigations were carried out on rotor blade specific Gl-Ep laminates. The material was conditioned and tested in the status *dry undamaged*, *dry damaged*, *wet undamaged*, and *wet damaged*.

It was shown that the hailstone simulating iceball impact does not harm the properties of the Gl-Ep investigated, whereas by the influence of humidity a significant decrease of the static properties and also of the fatigue properties at R=-1 and strain levels of $\pm$0.8 % and more must be stated. However, at lower strain levels no effects of humidity could be found in fatigue compared to the unconditioned material. I.e. that on Gl-Ep rotor blades the service life will not be affected by humidity, since the design strains presumably will not be higher than about 0.6 %. Nevertheless, it should be taken care of a good, well inspected gel coat protection to slow the penetration of moisture.

ACKNOWLEDGEMENT

The investigation was carried out within the JOULE programme which was 50%-financed by the CEC. The laboratory work was done by Mr. Reiter and Mr. Jägg who was mainly responsible for the data evaluation [4]. The statistical evaluation of the fatigue data in [1] (Figure 13 and 14) was performed by W. Sys (RUG/Belgium) within the preparation of a book (*in statu nascendi*) about the previous CEC programme *Material Testing*.

7. BIBLIOGRAPHY

[1] Ch. W. Kensche *High Cycle Fatigue of Glass Fibre Reinforced Materials for Wind Turbines.* DLR Forschungsbericht 92-17 (1992)

[2] A. Röhm *Berechnung der Feuchteaufnahme einer GFK-Struktur und der daraus resultierenden Spannungen mit der Methode der Finiten Elemente.* DLR, IB 435-09/89 (1989)

[3] Laube, Höller *in Landolt-Börnstein: Numerical Data and Functional Relationship in Science and Technology, Volume 4, Meteorology.* Springer Verlag

[4] S. Jägg *Einfluß von Hagelschlag und Feuchte auf das Festigkeits und Steifigkeitsverhalten von GFK-Laminaten.* DLR, IB435 - 17/92 (1992)

FATIGUE INVESTIGATION ON MATERIALS FOR WINDTURBINE BLADES

A. Silvera (**), F. Arias (*) , J. Vázquez (**) , E. Soria (*)

(*) C.I.E.M.A.T. Avda.Complutense 22, 28040 Madrid (Spain)
Phone 34-1-3466039 Fax 34-1-3466037

(**) Universidad Politécnica de Madrid. Dept.Ingeniería Rural. Avda.Complutense s/n ,
28040 Madrid (Spain)
Phone 34-1-5432600 Fax 34-1-5434879

ABSTRACT: We present in this work some results obtained by means of mechanical tests (statics and dynamics) of an assigned composite material,made in fiber glass reinforced polyester (GFRP) with a specific laminated secuency,convenient to windturbine blades. In fact,with a perfec set up material (weight and volume respecting to fiber glass) statics tests were carried out in order to determine the Young modulus and the breaking limit.

On the basis of this data,dynamic fatigue tests were carried out,for R = 0,1 - 0,2 - 0,3 - 0,4 and load levels of 50 % , 40 % , 30 % , 20 % and 10 %.

We have studied the influence of the ultraviolet radiation on the fatigue strength,collating the extracted results with artifically aged specimens,for R = 0,3 and 30 % load level in the Wöhler diagram.

Finally,it has been checked the influence of humidity on the material endurance to fatigue through the ageing of specimen from dipping in water at ambient temperature.

The results showed in this paper take part of a Research program under way in order to get a lot of data,which will allow a safe and economical design of wind turbine blades,made with these materials,in Spain.

1. INTRODUCTION

The tested material is suitable for the construction of several parts of the windturbine blades,which work with a high structural responsibility and whose efforts are mainly orientated according to the predominant direction.

This material is specially suitable to absorb the predominant efforts,which are the traction-compression,and also the cutting efforts,that is it,those whose direction is perpendicular to the formers and parallel to the surface of the material.

Nevertheless,the tests are exclusively made on traction,because these efforts are the most characteristics of the design.

The main characteristics of the material are:

* volume of the fibre 55 % aprox
* secuency $((\pm 45)/0_5/(\pm 45)/0_2/\bar{0})_S$

The specimen are 5mm of thickness,enough for the characteristics of the test.

The dimensions have been taken according to technical reason to allow both,reliable results and a good fastening to the grips of the machine.

This dimensions are shown in fig Nº1.

In fig.Nº2 and Nº3 are shown the different aspects of the test machine.

Once the specimens have been made we have proceed with the static tests.

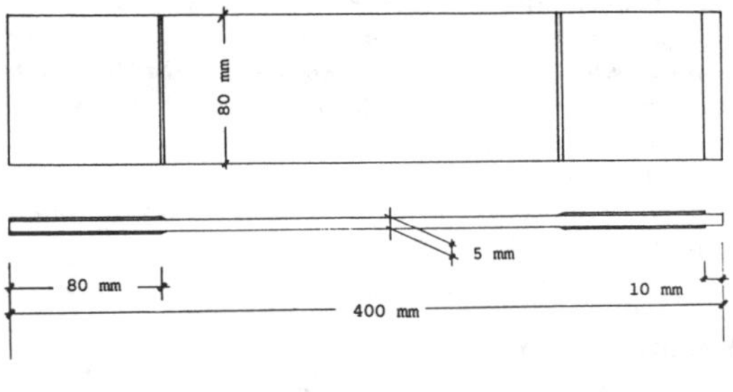

Figure Nº1 : Dimensions of the specimen

2. STATICS TESTS

From the static break of a certain number of coupons we have obtained it.s tension-deformation diagram.

39 specimens have been broken to obtain the break limit. We have arrived to the following final data:

* N = 39
* mean tension σ_{med} = 659,07 MPa
* maximum " σ_{max} = 714,31 MPa
* minimum " σ_{min} = 562,12 MPa

From the breaking of 20 coupons we have statistically deduced the characteristic values for the Young modulus and for the deformation of the specimen:

* N = 20
* Young modulus = 38 441 MPa
* elongation (L_o = 100 mm) = 2,02 mm
* " of the specimen = 0,71 %

All these data are according to Weibull distribution.

Figure Nº2 : Aspects of machine of tests

3. DYNAMICS TESTS

Once the value of the media tension of break has been known,we proceed to evaluate the material performance in the states of loading by R = 0,1 - 0,2 - 0,3 - 0,4 , with the media tension of 50 % , 40 % , 30 % , 20 % , 10 % of the breaking load.

We pretend to make a Wöhler diagram for the coupons made with the material as it has been delivered,that.s it,without suffering from any kind of process.

The dispersion found in the tests is very high,becoming higher when σ_{med} is smaller.

In every point tested we have looked for the distribution function most according to the experimental ressults,and we have found out that the one closest to the mayority ofe de cases is the Lognormal and the Weibull the most fitted for the rest of the cases,in one of the cases the difference of adjustment between both functions is very slight.

According to the s/n diagram we think the data σ_{med} = 66 MPa is good for the quick design,with no need of previous studies of accumulation of damage.

Figure Nº3 : Aspects of machine of tests

4. INFLUENCE OF THE ULTRAVIOLET RADIATION

We have radiated 18 coupons with UV rays in the solar spectrum gama,using a bench of 6 lights type Osram Ultravitalux. These lights give specially high intensities of ultraviolet radiation in the gama of 280 - 325 y 315 - 400 nm (UV-B y UV-A),apart from the visible spectrum.

The specific radiation received by the specimens has been 40 Mjul/m^2.

The tests have been made in turns of 6 coupons each time.The specimens have been disposed over an expositor,as it is shown in picture Nº 4 and have received radiation during a period of 216 hours for omly one of their sides.

We also measure the superficial temperatures in the specimens for the different daily hours,obtaining the values shown in fig.Nº 5 as an example of a typical day.

According to the tests,we have not notice,by simple visual inspection,any kind of considerable degradation or change in the color of the coupons. We also haven.t appreciate any evidente change in the surface of the specimens under the microscope.

To test if the UV radiation affects the fatigue endurance in the material,we have tested 18 specimens (without having been treated before) and some other aged in the load level R = 0,3 al 30 % .

Once we have proceeded to the statistic test of comparation of means,whose study we include,we are in conditions to state that there¨s no significant difference between the medium values of life of the radiated samples and the ones which have received no treatment.

Figure Nº4 : Installation for UV radiation

4.1 Comparation of results

Not treated coupons	Radiated coupons
R = 0,3	R = 0,3
30 % s/ load level	30 % s/ load level
Nº of tests = 18	Nº of tests = 18
C.V. = 43,94 %	C.V. = 43,51 %
distr.LOGNORMAL	distr.LOGNORMAL
fitting K-S 0,12	fitting K-S 0,90
shape p. 0,358	shape p. 0,483
cycles 8 090,16	cycles 7 178,40

Test of comparation of two samples

Number of observations = 36
Number of variables = 1

Comparation of two indepentent samples

	without treat.	**radiated**
mean	8 725,00000	7 933,33333
s.dev.	3 937,0880	3 452,427
N	18	18
observations	1 a 18	19 a 36

Difference = 791,6665
s.dev. of the difference = 1 234,2422 (S_D)

t = 0,6414 (D.D.L. = 34) VARIABLE NvsUV

Unilateral test , probability = 26,63 %
Bylateral test , probability = 53,26 %

The minimum significant = S_D x t_λ , in that t_λ correspond to the "t" of Student for different intervals of confidence.

For 10 % the minimum significative difference is 1 234,24 x 1,69 = 2 085

On the other hand the bylateral test shows the probability that the difference between means due to hazard is 53,26 % of the cases.

From all this,we deduce that the differences found between the cycles of the specimens treated and without having been treated don.t reach a significant value so as to state categorically that the UV radiation applied affects to the endurance of the coupons.

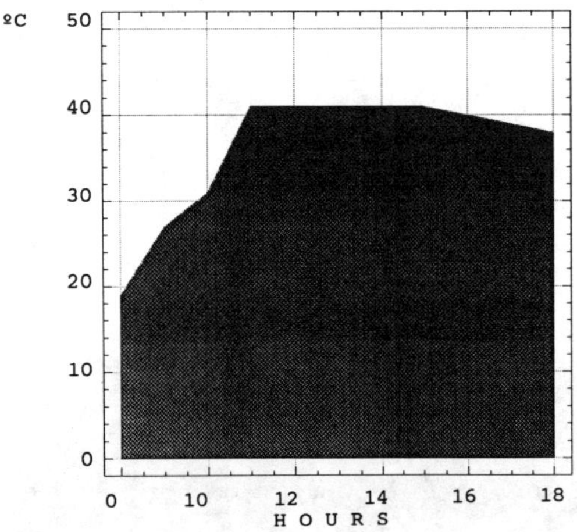

Figure Nº5 : Superficial temperatures in the specimens for a typical day

5. MOIST INFLUENCE

We have submerged 14 specimens in water at ambient temperature during 1,5 months.

To test if the humidity conditions affects the fatigue endurance in the material,we have tested 14 coupons (without having been dipped in water before) and some other treated,in the same load contitions.

Difference between means = 614,286

Conf.interv.for diff.in means 95 %

26 D.F.

Computed "t" statistics = 0,206369

S.dev.of the diff. = 2 977,6

The minimum significative difference is 2 977,6 x 1,706 = 5 079,78

Consecuently,we are in conditions to state that there.s no significant difference between the medium values of life of the treated samples and the ones which have received no treatment.

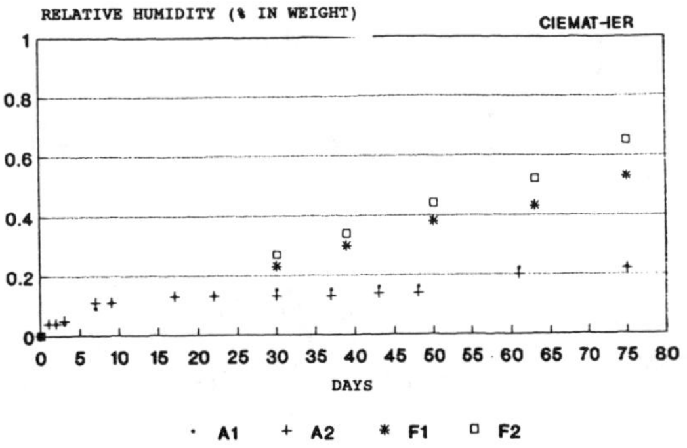

Figure Nº6 : Water absorption in specimen

6. CONCLUSIONS

The tested material (with the lamination secuence before indicated) is fitted for the construction of the structural parts of the windturbine blades. With this specified lamination secuence and thanks to these tests we dispose of a material whose elastic and fatigue characteristics are determined ,and then,it.s apt for making designs.

The ultraviolet radiation practically affects nothing to the fatigue endurance of the material.

Humidity does not affects the fatigue endurance of the material either under the ambiental conditions referred previously.

7. REFERENCES

1 - Research program for fatigue-critical structures of small and medium sized wind turbines.
German Aeroespace Research Establishment
Dipl.Ing.Ch.Kensche.Institute for Structural Research and Design.
Wind Energy - 1.Proceedings of the first Contactors Meeting.Brussels 5-6 May 1.986

2 - Investigation of the fatigue properties of aerogenerator structural materials
University of Ghent
Prof.A.Vinckier . Laboratory for Strength of Materials and Welding Technology.
Wind Energy - 1 . Proceedings of the first Contractors' Meeting. Brussels 5-6 May 1.986

3 - Materials data for WT design basis:fatigue properties of glass fibre reinforced polyester.
Metallurgy Department.Riso National Laboratory. Roskilde . Denmark
H.Lilholt
Wind Energy - 1.Proceedings of the first Contactors' Meeting. Brussels 5-6 May 1.986

4 - High cycle fatigue investigation of windturbine materials
Netherlands Energy Research Foundation,ECN
P.W.Bach
Proceedings of an International Conference held at Herning Congress Centre.
Denmark 6-10 June 1.988

5 - Fatigue of glass polyester composite materials for wingblades.
Metallurgy Department.Riso National Laboratory. Roskilde. Denmark
S.I.Andersen and H.Lilhot

6 - Investigation of the fatigue properties of aero-generator structural materials
Laboratory for Strength of Materials and Welding Technology.
Saint Pieters,niewstraat 41,B-9000,Ghent,Belgium.
Ir.W.Sys.Prof.ir.A.Vinckier

FATIGUE PROPERTIES OF WINGBLADE MATERIALS AND COMPONENTS MANUFACTURED BY THE TRANSVERSE FILAMENT WINDING PROCESS

G A Lowe and N D Satterly,
Faculty of Engineering
University of the West of England, Bristol
Coldharbour Lane
Frenchay
Bristol BS16 1QY

ABSTRACT: The main task of the University of the West of England (formerly Bristol Polytechnic) within the EC JOULE Programme has been to test glass reinforced polyester coupons and D-spars manufactured using the transverse filament winding process. Associated research work funded by the UK Department of Energy has investigated the effect of changing the manufacturing technique in order to improve the design and production process particularly with regard to large wind turbine blades.

Apart from standard static and fatigue testing in order to characterise the material, the work has also concentrated on examining the relationship between the results obtained from tests of spar samples, produced by typical Blade Manufacturers, with those from tests on coupon specimens extracted from spars and flat plates.

In order to compare the effect of cyclic loading on small specimens and large structural components manufactured by the same method, various damage accumulation techniques have been incorporated into the test programme. The measurement of damage using flexural reduction in stiffness, hysteresis area, surface temperature and visual inspection are all techniques used or examined in the work.

There is a clear indication that the fatigue life of spars made using the tape winding process is considerably lower than the life obtained from testing small coupon specimens used to characterise the material. The production quality of coupons and components should be similar in order to compile accurate data for design purposes.

1. INTRODUCTION

The rotor blades of wind turbine generators can expect to be subjected to over 100 million fatigue cycles within their lifetime. Thus the choice of materials and an adequate knowledge of their static and fatigue properties are essential elements of rotor design.

The University has been involved in the EC Joule programme and associated work funded by the UK Department of Energy (ETSU), investigating the fatigue behaviour of typical wind turbine blade spars manufactured from transverse filament tape.

The method of manufacturing by tape winding has been used since the blades made for the turbines at Nibe in 1978 [1]. It allows automated manufacture of the spar, thus leading to a reliable and reproducible component.

This investigation consists of two distinct aspects:

(i) Base material properties obtained from test plates produced under laboratory conditions.

(ii) Blade spars and coupons produced from the same materials using bulk manufacturing techniques, characteristic of volume production. The spar is considered as the primary load carrying element of the chosen blade design and hence the aerodynamic shells do not form part of this investigation.

The objectives of the work were to establish the following:

(i) How long such material and components last under a given loading cycle.

(ii) How damage accumulates in such materials and structures. Using various damage parameters monitored throughout the tests to establish whether their rate of change can be used to estimate the severity of the damage and hence assess structural integrity at any point in the life of the component.

The above is of direct interest to both designers and operators alike. Such information is vital to produce and operate efficient and cost-effective machines.

2. MATERIALS AND MANUFACTURE

2.1 Materials

Glass reinforced isophatalic polyester resin has been used for the production of flat plates and spar sections. This material is typical of that used for production of wind turbines.

Two types of glass tape reinforcement were used, the traditional woven and the more recent knitted construction. Both materials are in the form of a 210mm wide tape, with polyester yarns in the longitudinal (warp) direction and glass rovings in the transverse (weft) direction. The polyester in the woven construction passes over and under the glass fibres causing them to kink. This is believed to give rise to stress concentration, with a consequential reduction in fatigue performance. However, the knitted construction keeps the fibres straight, eliminating the kinks and associated stresses. When winding takes place and tension is applied to the polyester yarns the above problem is extenuated.

Additionally the knitted construction allows more stretch and hence is more compliant to the tapered shape of the mandrel used for manufacture of the spar.

2.2 Manufacture

Flat Plates were manufactured at the UWE using the leaky mould technique, a closed mould method with a nominal pressure of 0.1 bar applied to consolidate the laminate. Post curing for 3 hours at 80°C was carried out. These were made to provide a benchmark for comparison with the tape wound items.

Spar sections were produced by two manufacturers using their own versions of the tape winding process. Glass tape is wound on to a shaped rotating mandrel with the resin applied automatically or manually by a pumped roller. An overlap of about 40% and a 0° laminate is maintained by careful control of the angle θ to account for the taper and change in geometry of the mandrel (see Fig.1). This 0°

laminate is essential for maximum strength and stiffness [2]. Following winding the mandrel is allowed to stand for 24 hours before the component is pressed off.

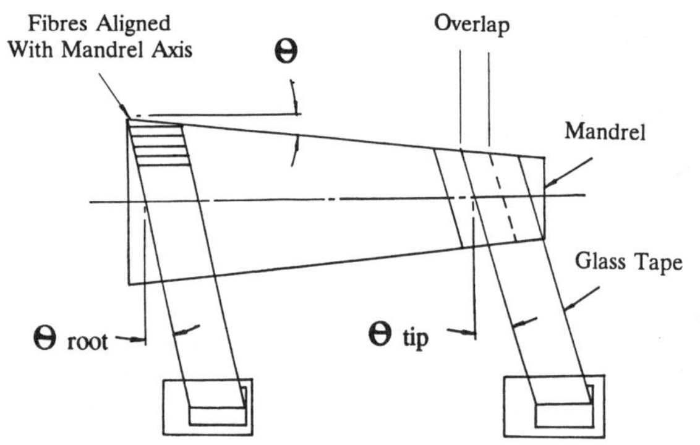

Figure 1. Transverse Filament Tape being wound on to a rotating mandrel

3. MATERIAL CHARACTERISATION

Flat plates and spar sections have been characterised by a series of tests.

Flexural testing has established the flexural modulus, strength and failure strain using standard three point bend tests [3]. This type of testing has been chosen because the tape wound material is unidirectional and of varying thickness. Some tensile testing has been carried out but valid failures have been impossible to obtain.

Fibre volume has been found from resin burn off tests [4].

Interlaminar shear strength has been measured [5]. This test is a good quality check and the type of failure is independent of features such as joints that exist in the tape wound sections.

The results of some of the tests are shown in figure 2.

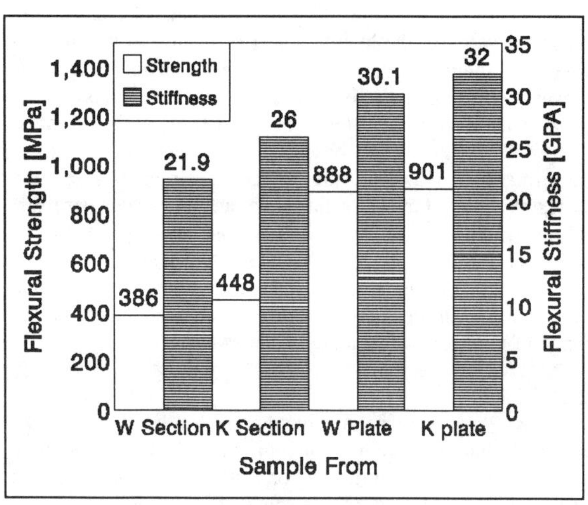

Figure 2 Flexural Strength and Stiffness

The results show that the flat plates are of better quality than the spar sections and that the knitted material is better than the woven. The plate results show similar strengths and stiffness. It was noted that the mode of failure for knitted and woven samples was tension. It is therefore believed that the small difference in properties is due to the kinks in the woven construction.

The greater difference in properties between the two types of spar material is explained by the failure mode. Whilst the knitted samples always failed in tension the woven samples exhibited significant signs of compression damage. The latter is almost certainly due to the kinks permitting buckling on a macroscopic scale.

The large difference between the properties of the plate and spar coupons can be explained by a general reduction in quality arising from the production method, including high voidage and the effect of the tape joints.

4. FATIGUE TESTING OF COUPONS

Testing was carried out in flexure using a similar arrangement as used for the static testing with some modifications [6]. The test machine was a Schenck Pulsermat operating at approximately 8 H_z under load control. An 'R' ratio of 0.25 and a span to depth ratio of 16:1 was used.

The stiffness of the samples was used as the failure criterion, a 10% reduction of stiffness from the original meaning failure (100% life) [6].

The results are displayed in the S-N curves shown in figures 3 and 4.

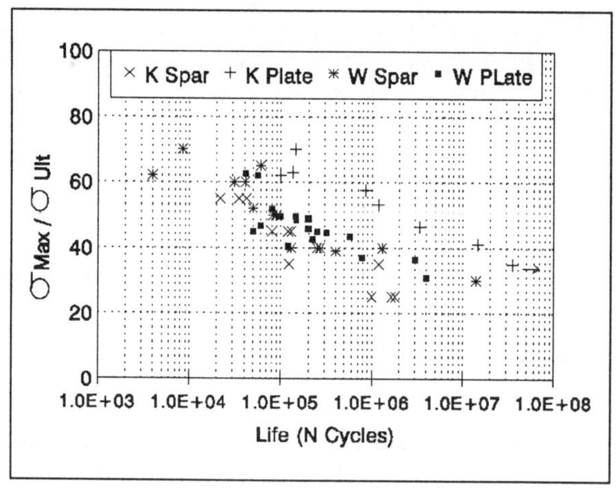

Figure 3 Normalised S-N Curve

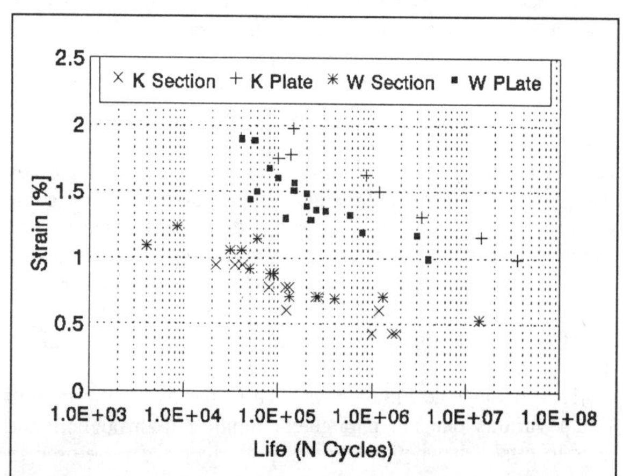

Figure 4 Fatigue Curve : Strain - Life

The first figure shows normalised data, enabling a comparison of the materials fatigue characteristics, regardless of its ultimate strength [7]. The plate data indicates that there is some feature of the woven material that significantly affects the fatigue performance. This is again the kink stresses which are more serious in fatigue than in the static situation. This effect is not evident in the spar coupon results indicating that there is a more dominant feature affecting the material. It is suggested that the poor quality and the overlap joints cause the lower values, emphasising the need to optimise the production process to ensure that the material is used to its full capacity.

The second curve, relating test strain to life, obviously shows that the tape winding technique has not produced material of as good quality as the plates. This highlights the need for manufacturers to ensure that fatigue data used for design purposes originates from representative coupon tests.

Additionally this curve shows that the data converges in the low stress, high cycle tests. This could be considered an advantage for components such as wind turbine blades that are subjected to many millions of low stress cycles, as differences in quality become less important.

5. METHODS OF MONITORING

Throughout the fatigue testing programme various parameters were monitored on both coupons and spars, using a computer controlled data logging system developed at the University. The parameters recorded were flexural stiffness, hysteresis area, surface temperature, local strain (spar only) and visual inspection.

The time history of these parameters is used to correlate rate of change of the parameter with amount of damage with regard to the life of the sample.

Figures 5 and 6 show how the stiffness, hysteresis area and surface temperature change during the normalised life [8] of typical coupons under high and low stress (short and long life) respectively.

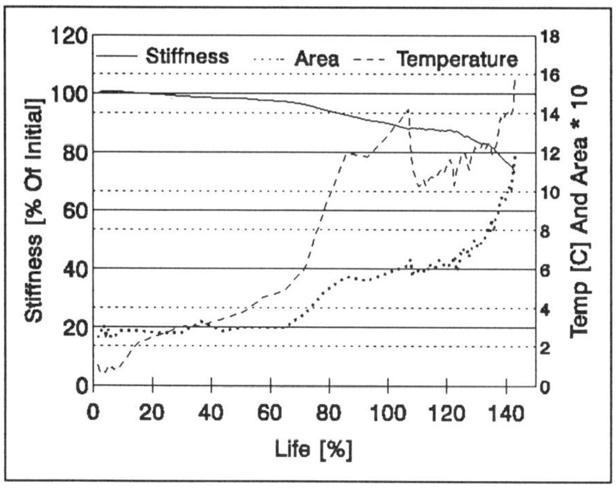

Figure 5 Damage Monitoring - High Stress

The shape of the stiffness curves is similar, indicating little dependence on stress. However, the other two parameters are very different, the high stress producing large changes. There is a good correlation between the hysteresis area and temperature as would be expected from a viscoelastic material such as GRP, both showing an accelerating trend towards the end of the sample's life. The dependence of these parameters on high stress and hence high damage rate is obvious.

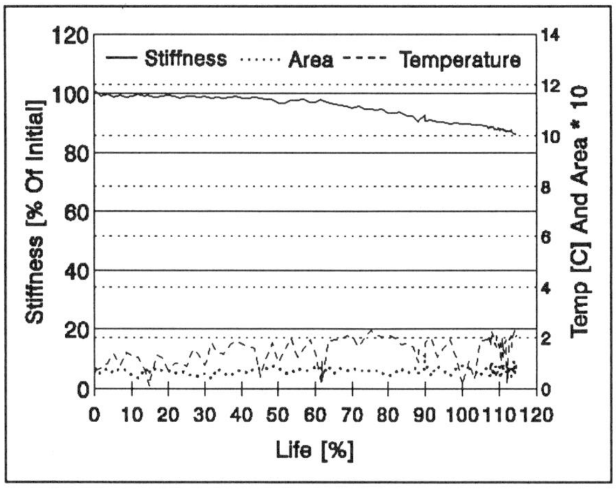

Figure 6 Damage Monitoring - Low Stress

6. COMPONENT TESTING

Three spars of two different designs have been tested at between 3000 and 6000 microstrain in the University structures testing rig as shown in figure 7. It has been found that their fatigue lives are significantly shorter than predicted from the coupon data, by a factor of approximately 10. In this context it should be noted that only a small number of components have been tested and the data is subject to scatter. Damage initiated in all cases from the tape joints, the spars manufactured from the heavier knitted tape (less layers) clearly failed in this way, possibly due to load transfer in shear across the lap joint. This did not appear to be the case with the lighter woven tape (more layers) which suffered more fibre breakage, probably because of the lower load transfer between layers.

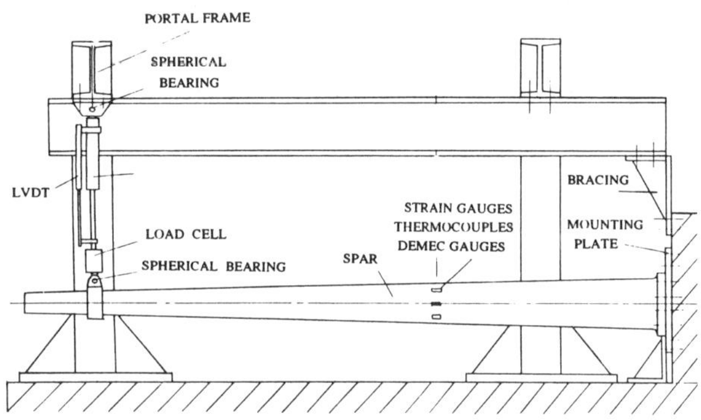

Figure 7 Spar Test Arrangement

The damage monitoring techniques were applied to the spars but the results were less informative than those obtained from the coupons. The component size makes the positioning of local techniques difficult and the damage, when it does occur, remains confined to a small area. Even when instrumentation is correctly positioned sensing the damage was very difficult with the methods used. Any changes in the global parameters that have been recorded i.e stiffness and hysteresis area, have been small. After losing only 5-8% of global stiffness spars failed catastrophically whereas coupons would hardly have shown any signs of damage.

All techniques were subject to scatter which could mask real effects when changes are small. The scatter can be induced by various factors such as experimental error, change of ambient temperature, and non permanent creep.

The best indicator of the spar condition proved to be visual inspection; monitoring the growth rate of an area of damage was found to illustrate the various phases of damage accumulation [8] and displayed the same characteristics as that shown by the coupons in figures 5 and 6.

7. CONCLUSIONS

- The research has showed that the knitted construction is significantly better in fatigue than the woven material, although static properties are similar. This is believed to be due to the kink stresses induced in the woven construction when loaded both statically and dynamically.

- It has been found that the fatigue lives of components are significantly shorter than that predicted from coupon data, typically by a factor of 10.

- Whilst differences in quality do have an effect on fatigue performance, for typical wind turbine applications i.e high cycle - low stress conditions, these effects are less significant indicating that GRP appears to be tolerant to manufacturing errors and to be a good material for turbine blades. The results also show that quality control of the production method is important in achieving the best performance.

- Coupons should be representative of the lay-up of the final component to achieve true data.

- Failure of the spar component could initiate from overlaps in the tape. Reduction in the amount of load transfer per layer e.g by decreasing the thickness of the tape, may help to alleviate this problem.

- Damage monitoring methods have been seen to work effectively on coupons but were less convincing when used on components. Changes in parameters have been found to be small, the data needs to be carefully analysed to discriminate between environmental factors and experimentally induced scatter to see any real variation. Care should be taken in assessing the results.

- It is believed that a solution to effective monitoring of damage in operational turbines will require much further research,possibly into completely new and novel methods.

REFERENCES

[1] Wingblades of glass fibre reinforced polyester for a 630 KW wind turbine, H Lilholt et al, Proc ICCM-3 (Elsevier, Amsterdam 1980).

[2] Composites - A Design Guide, T Richardson (Industrial Press NY, 1987).

[3] BS2782: Part 10: Method 1005, Determination of Flexural Properties - Three Point Method.

[4] BS 2782: Part 1: Method 107K: 1970, Determination of Fibre Content of Resin MAtrix Composites by REsin Burn Off.

[5] BS 2782: Part 3: Method 341A: 1977, Determination of Interlaminar shear Strength of Reinforced Plastics.

[6] A Vamas Round Robin on Fatigue Test Methods for Polymer Matrix Composites, Part 1 - Tensile and Flexural Tests of Unidirectional Material, G D Sims (NPL Report DMA(A)180, April 1989).

[7] Testing of Wood in Fatigue, M P Ansell et al (proceedings of an IEA Workshop on Fatigue in Wind Turbines, 21 - 22 March 1988).

[8] Damage - Tolerant Polymer Composite Systems, K L Reifsnider (Journal of Metals, Nov 88).

TEST PROCEDURES FOR PROTOTYPES AND MACHINERY OF WIND ENERGY CONVERTERS PRODUCED IN SERIES

S. Glocker, V. Köhne, A. Petersen
WINDTEST KWK GmbH
Sommerdeich 14 b
D-2222 Kaiser-Wilhelm-Koog
Germany

R. Rohden
Germanischer Lloyd
Vorsetzen 32
2000 Hamburg 11
Germany

ABSTRACT: Tests and experiments are of fundamental significance for Wind Energy Converters (WECs). This technology is still in the developmental stage, further essential progress is expected as a result of practical research. Appropriate test procedures and measurement campaigns are important for prototypes as well as for series produced WECs. The range of tests is including measurements of power performance, loads and vibrations, grid effects and noise. In addition the governments of the coastal states of northern Germany demand for results of standardized measurement campaigns to give financial support to the operators. Common test procedures are described in detail and selected results are presented.

1. INTRODUCTION

Wind energy has become an increasing commercial factor in Germany during recent years. Therefore the significance of measurement campaigns on research and development has increased. On request of manufacturers and operators Germanischer Lloyd (GL) and WINDTEST Kaiser-Wilhelm-Koog GmbH (WINDTEST) carry out measurements on WECs with regard to efficiency, safety, static and dynamic strength, as well as noise emission and electrical characteristics. The results either confirm with the assumptions being made in the design calculations or lead to a modification of the design. This in turn has effects upon the cost and the quality of the product (e. g. service life, efficiency).

Some of these measurement campaigns are required for the application for subsidies from the governments of the coastal states of northern Germany. Power characteristics and sound pressure levels have a direct effect on the total amount of the financial support.

GL examines and certifies WECs according to the Society rules and/or standards and regulations at the site of operation. The Society supervises the manufacture, transport and construction commissioning of WECs as well as the testing of materials and components and operational testing. The complete certification procedure includes tests and periodical surveys of WECs.

A further application of measurements is to serve as an instrument for damage analysis for manufacturers and insurance companies.

Measurements are not only used for verification, but do also furnish a proof of quality thereby serving as a sales argument.

2. SCOPE OF TESTING FOR PROTOTYPES AND COMMERCIAL WECS

A useful scope of testing covers the following special measurement campaigns (see figure 1):

- Power measurements including recording of service data,
- Measurements of loads and vibrations,
- Noise measurements and
- Measurements of grid effects.

During the period of concept realisation and design material and component tests are an important instrument for the verification of the design parameters. New concepts often are based on newly developed components such as rotor blades, generator systems or gearboxes. Some of the component tests can be carried out on test beds, but there is no guarantee that the test conditions are realistic in all details. As a consequence component tests should be verified by measurements on prototypes.

During prototype testing one of the most important tasks is testing and evaluation of fatigue loads to confirm design assumptions. The reduction of noise emission and the optimization of power performance are main subjects of WEC-tests and special measurement campaigns are required. Noise and grid effect measurements are carried out after all revisions and improvements of the prototype have been made.

For commercial and series produced WECs it is recommended to perform the complete test procedure, because each certification, subsidizing and authorisation require parts of the measurements. In most cases changes of the WEC's concept or design lead to new examinations and approvals.

In Germany the northern coastal states Schleswig-Holstein, Lower Saxony, Mecklenburg-Vorpommern and Hamburg have defined special test procedures to be fulfilled before granting subsidies [4]. Power performance and noise measurements according to IEA-recommendations [1] [6] are necessary, including measurements of effects upon grid. The subsidies directly depend on the quality of the

- Power characteristics represented by the calculated energy output at a standardized site (5,5 m/s in a height of 10 m) and the
- Noise emission represented by the area taken by the WEC where a sound pressure level of 45 dB(A) is being exceeded.

In addition measurements of the effects upon the grid will be

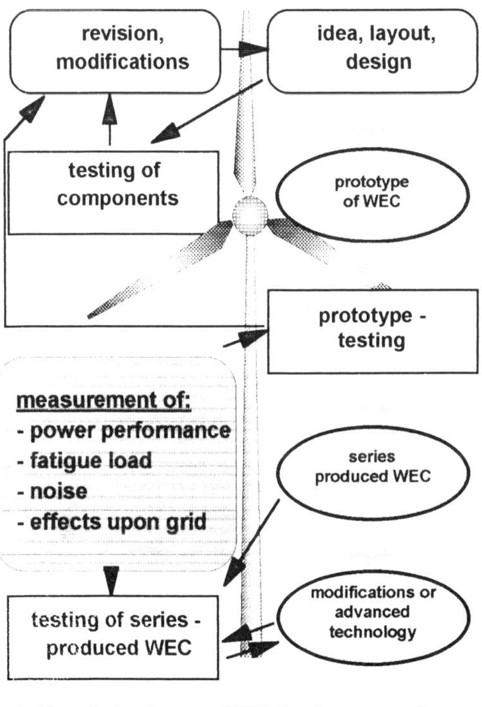

Fig. 1: Correlation between WEC-development and tests

obligatory, but will not have influence on the subsidies.

3. SPECIFIC MEASUREMENT CAMPAIGNS

3.1 Power performance measurements

Figure 2 is an overall view of the power performance measurement procedure including data collection, averaging and data analysis.

The procedure aims at measurement and evaluation of power curves for energy production calculation. The power curve is a graph which despicts the net power of a WEC as a function of the wind speed. The wind speed, measured at hub height, and other meteorological parameters are refered to the centre of the WEC-rotor assuming they are not disturbed by the presence of the turbine. This reference can only be taken at reasonable homogenious terrains having a minimum of influence on the wind sheer. Site calibration by flow modelling or experimental investigations should be handled with care, because the required accuracy of the measurements hardly can be achived. All sensors should be of high accuracy. Especially anemometers have to be calibrated before and after the measurement campaign.

Depending on the sampling rate the time histories of the data can include informations about dynamic effects such as start-up and shut-down or the power control bahaviour. Time histories can be triggered by different events.

Data are reduced to 1 min or 10 min-values and stored as pre-averaged times series.

The collected data have to be processed using particular methods, e.g. the determination of the power curve according to different recommendations [1][2][3]. In case of sensor or system failures, grid variations and special meteorological conditions (e.g. wake effects on measurement tower or WEC, icing, high turbulence intensities and in some cases precipitation) the respective data sets cannot be used for the determination of the power curve. The selected data should be corrected to standard air density (1.225 kg/m³) according to the recommendations. They are sorted using the "method of Bins". Depending on the recommendations a minimum of data sets for each bin is required. The data around cut-in and cut-out suffer from high uncertainties because of dynamic effects. These have a high impact upon the statistical determination of the power curve.

The measuring time required for a power curve depends largely on the environmental conditions of the site. Different campaigns have shown that it usually takes about three to eight weeks to complete a power curve.

It is important to present the results immediately, preferably during data collection. WINDTEST is working on a new measurement system which will show specific evaluations online.

The power performance measurement procedure of WINDTEST is accredited according to DIN EN 45 001 since 1992.

3.2 Load and vibration measurements

In cooperation with WINDTEST GL carries out load and vibration measurements on the test station in Kaiser-Wilhelm-Koog during prototype testing.

As it is shown on the flow chart of the data aquisition system (see figure 3) the PCM (Pulse Code Modulation) method is used within these measurements. This method enables a safe data transfer from the rotating part of the WEC as well as a simultaneous data aquisition of several channels with high sampling rates.

For the data aquisition a software was developed by GL, which carries out measurements according to the IEA-Recommendations [5] in automatic operation.

The measured data are transfered to the GL-mainframe computer, where software for an efficient evaluation of extensive data files is available.

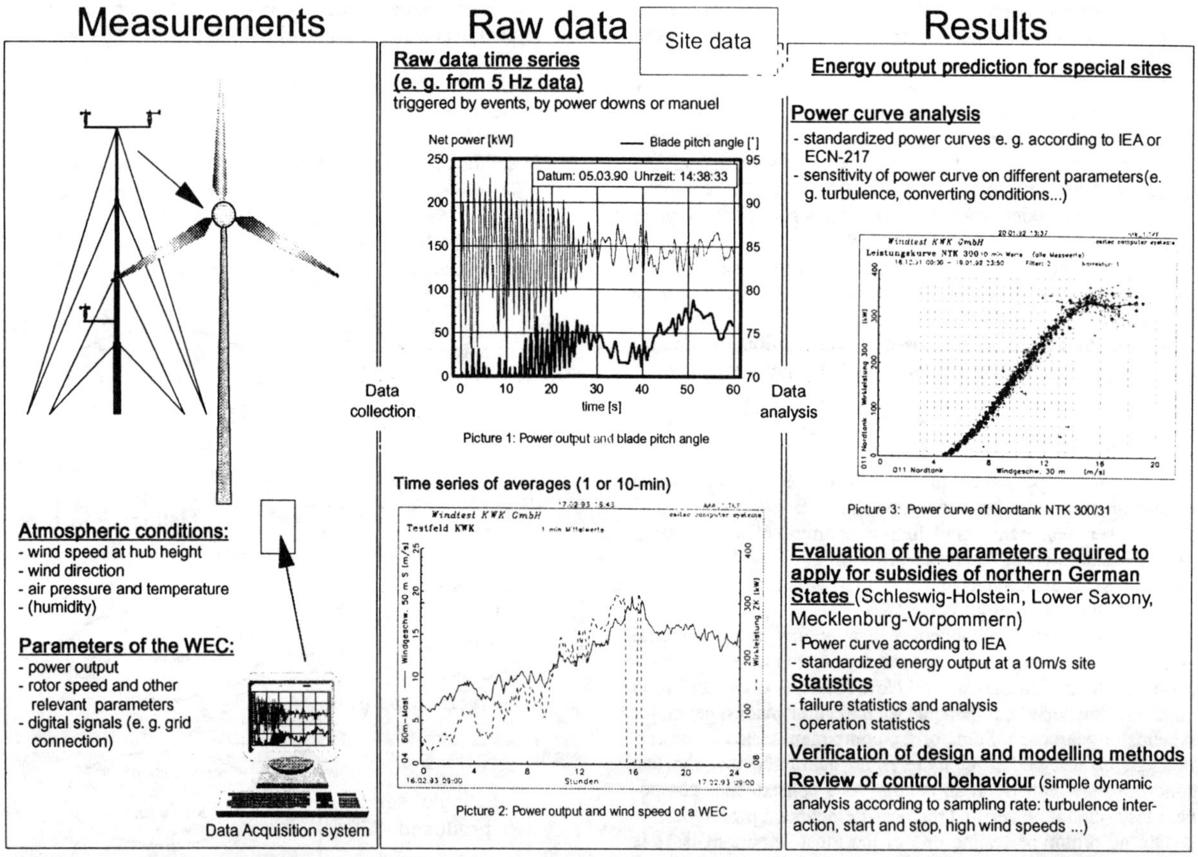

Fig. 2: Power performance measurement procedure

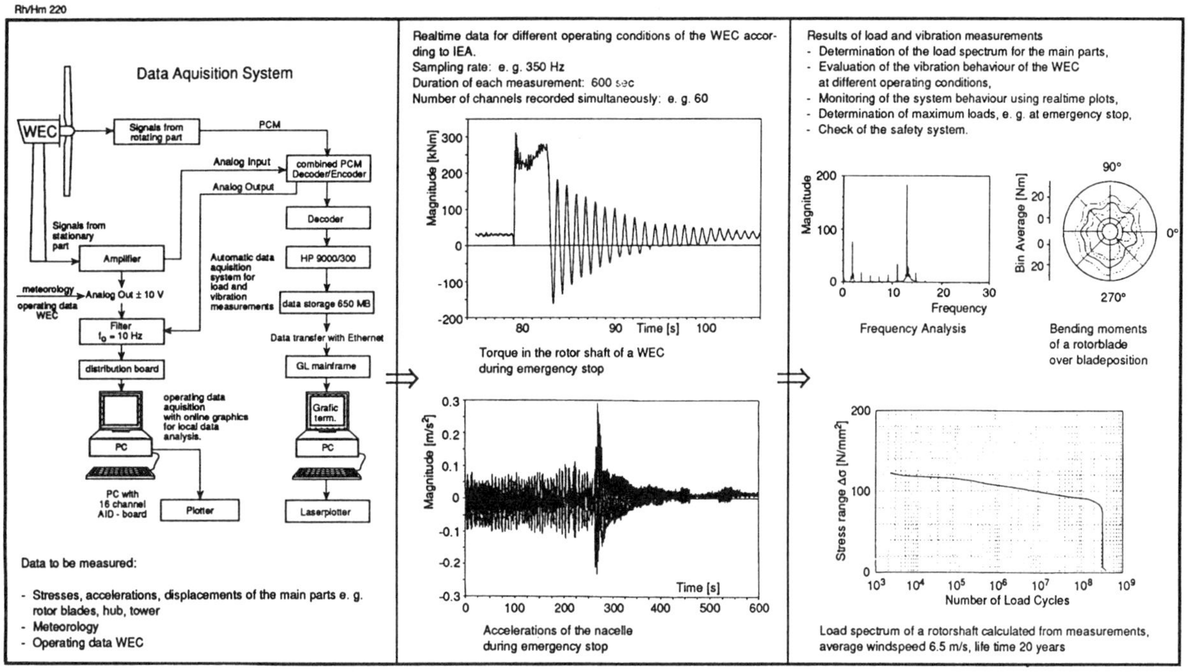

Fig. 3: Load and vibration measurements

The reasons and results of the load and vibration measurements (short time measurements) can be summarized as follows:

- Determination of the load spectrum for the main components in order to carry out calculation of expected lifetime, check of the design and comparison with the calculations.
- Vibrational behaviour of the WEC at different operating conditions in order to find resonant frequencies if present.
- Monitoring of the system behaviour using realtime plots, e. g. to optimize the control system.
- Determination of the maximum loads for the main components, e. g. at emergency stop.
- Check of the safety system.

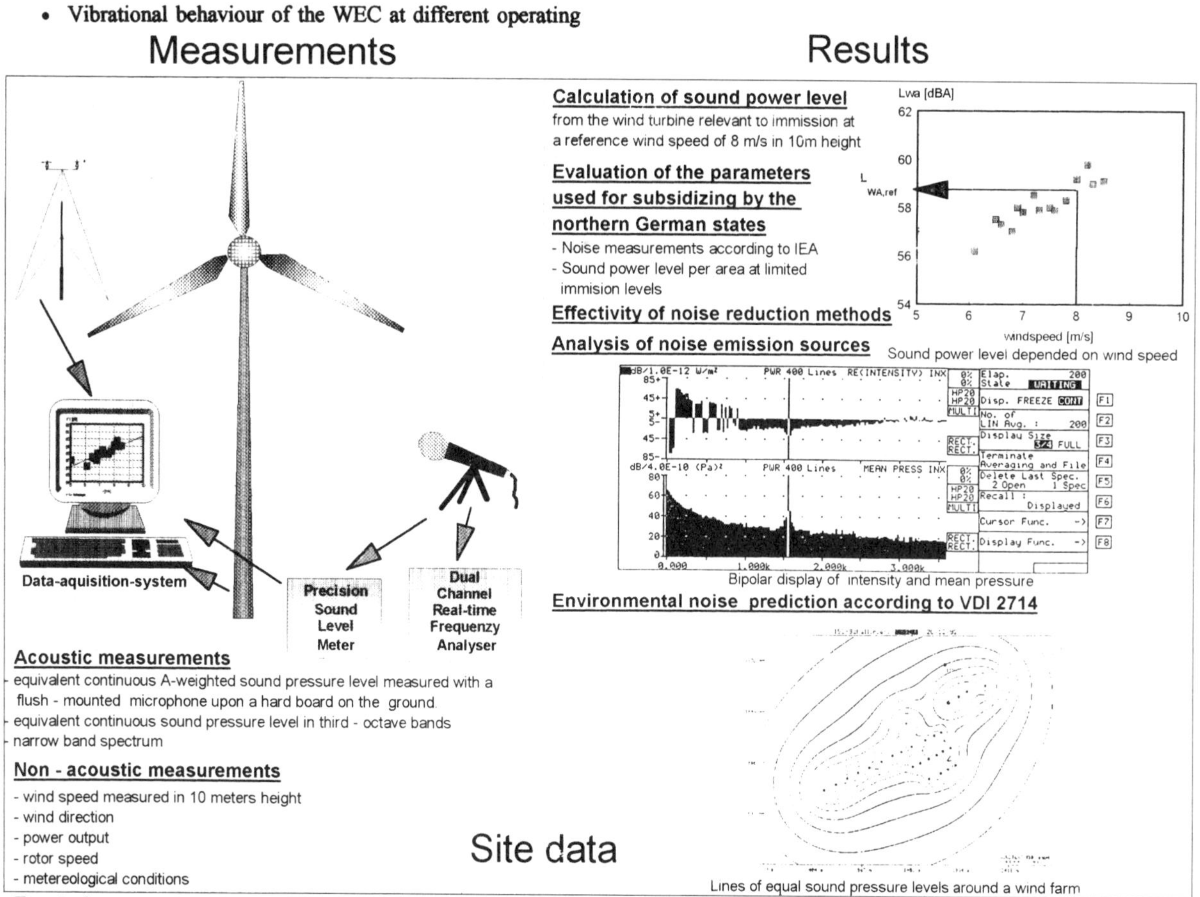

Fig. 4: Acoustic measurements

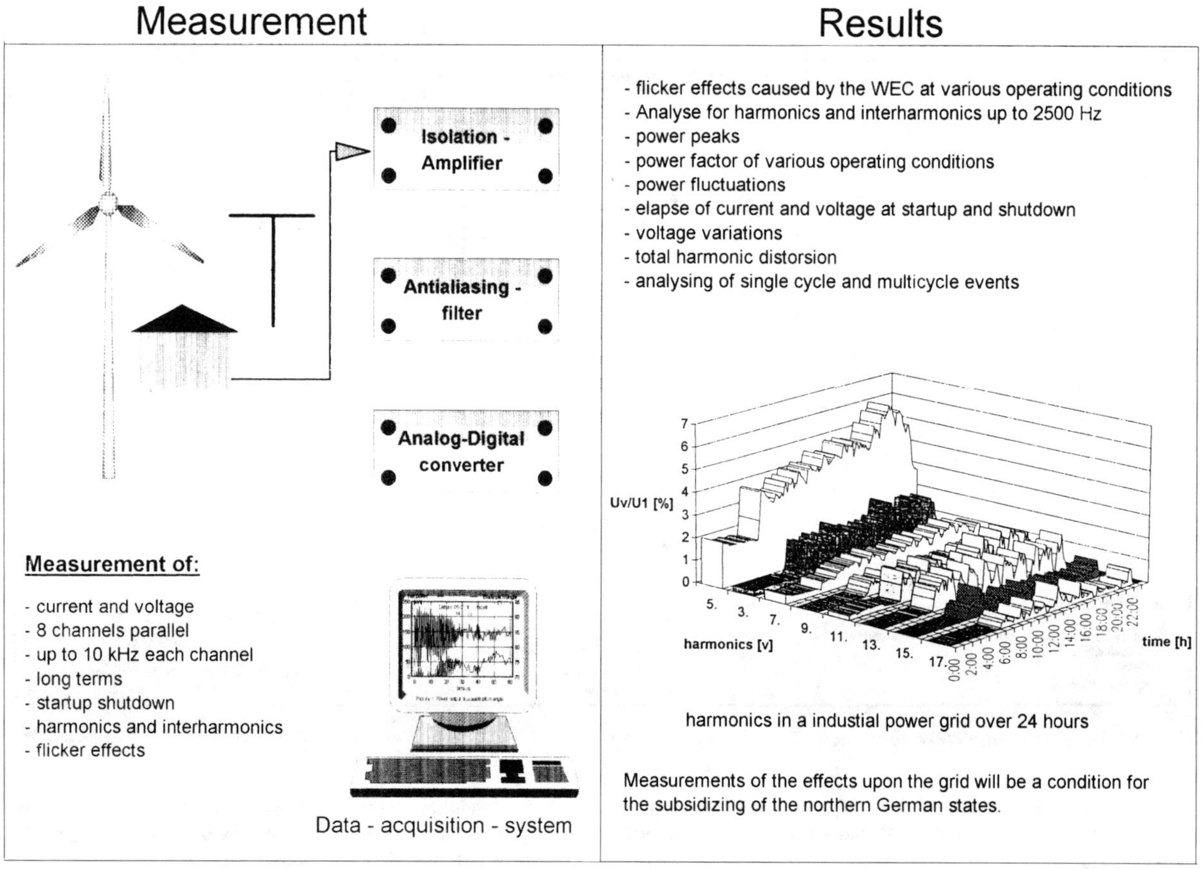

Fig. 5: Measurements of effects upon the electrical grid

3.3 Noise measurements

Noise emission of WECs is an essential topic for the development of wind energy. The noise emission of WECs can be determined according to the IEA method [6]. The continuous A-weighted sound pressure level (L_{Aeq}) has to be measured in five fixed positions around the WEC.

All measurements have to be carried out with a microphone which is flush-mounted on a hard board placed to the ground. The wind speed during the measurements should be in the interval of 5 to 10 m/s in a height of 10 m. The results in downwind direction are used for the determination of the relevant sound power level (see figure 4). A picture in figure 4 shows the relation between the noise emission and the wind speed measured 10 m above ground together with a narrow-band frequency spectrum.

Based on measured sound power levels; WINDTEST evaluates predictions of the environmental noise for planed WECs which can be compared to the permissible values fo the location. After the erection, measurements are carried out to verify the predictions.

3.4 Measurements of effects upon electrical grid

Usually WECs are connected to the grid of the local utility. In some circumstances this can affect applications connected to the grid.

The main affects are flicker and harmonics, added by disturbances caused by start-up and shut-down. To examine a WEC, measurements will be set up at the connection point to the grid. With the measurements information will be obtained to decide whether a WEC has neglectable effects upon the grid or not.

Currents and voltages at the power connection point have to be measured. Depending on the measured data the different effects can be analysed. The measurement of harmonic distorsion and flicker effects demands long-time measurements, while the measurement of start-up and shut-down requires short-time measurements with a high sampling rate. An overview about the measuring procedure is shown in figure 5.

4. CONCLUSION

Measurement campaigns are a very important instrument for the development of WEC. A useful scope of testing procedures for prototypes includes measurements of loads, vibrations, power performance, noise and the effects upon electrical grid. Series produced WECs can profit from tests carried out on the prototype.

For successful WEC-development it is essential to carry out the tests directly and in close cooperation with the manufacturer. Furthermore the range of measurements should be flexible and adapted to the specific task. It is necessary to immediately evaluate the results and to hand out the reports as soon as possible to give fast feedback to the developing engineers.

5. REFERENCES

[1] IEA Recommended Practices for Wind Turbine Testing and Evaluation, 1. Power Performance Testing, 2. Edition 1990.

[2] Recommendation for Wind Turbine Power Curve Measurements, Technical Requirements for Type Approval and Certification of Wind Turbines in Denmark.

[3] ECN-Recommendations for a European Wind Turbine Standard on Performance Determination (ECN-217).

[4] Technische Randbedingungen für die Landesförderung der Windenergie, Deutsches Windenergie-Institut GmbH, WINDTEST Kaiser-Wilhelm-Koog GmbH, Entwurf vom 03.11.92.

[5] IEA Recommended Practices for Wind Turbine Testing and Evaluation, 3. Fatigue Loads, 2. Edition 1990.

[6] IEA Recommended Practices fo Wind Turbine Testing and Evaluation, 4. Acoustic Measurement of Noise Emission from Wind Turbines, 2. Edition 1988.

SOME EXPERIENCES OF FATIGUE AND STATIC TESTING OF FULL SCALE WINDMILL BLADES UP TO 17.0M LONG

Dr L F Boswell, Dr W J Morris, Mr W A R Jones, Dr C D'Mello
Wind Energy Unit, Civil Engineering Department
City University, Northampton Sq, London EC1V OHB

ABSTRACT: The Wind Energy Unit at City University have been involved in the fatigue and static testing of full scale windmill blades over the last 2 years. The blades, made of laminated timber with an epoxy coating made by the Wind Energy Group Ltd (WEG) have varied in length from 12.0m to 17.5m.

The tests have highlighted areas where modifications were needed to improve the strength of the blade and the programme has been one of continued improvement not only in the fatigue life of the blade but also in the economy of blade production.

The paper describes the methods of fatigue testing with both hydraulic and rotating mass actuators and also the basic laboratory facilities and equipment that are required for such tests.

The method of static testing and the equipment required is outlined together with the advantages of performing such a test.

The results obtained during the performance of the fatigue tests are described together with the failure modes of blades at the termination of a test.

1. Introduction

The Wind Energy Unit at City University has been involved over the last two years in a contract with the Wind Energy Group Ltd (WEG) a subsidiary of Taylor Woodrow for the fatigue and static testing of their M.S.2 and M.S.3 blades. This was to fulfil contracts with both WEG and D.T.I. to research into alternative woods and production methods, for wood epoxy composite blades. The blades have varied in length from 12.0m to 17.5m and were constructed from Khaya, Beech and Poplar laminates.

The methods of testing using hydraulic and rotary mass actuators and the laboratory facilities and equipment required is described together with the method of static testing. The relative advantages of the test methods are outlined and some typical test results are described.

2. Methods of dynamic fatigue testing and laboratory requirements

Initially the fatigue tests on the blades were carried out using a hydraulic actuator (Fig. 1). Four blades of 12.0m in length were tested using this method, but for the longer M.S.3 blades (16.0m) a rotary mass actuator (Fig. 2) had to be constructed to cope with the larger forces and deflections required. Apart from the method of actuation the laboratory requirements were the same for both methods.

The basic equipment/facilities required consisted of:

(a) A 400m thick reinforced concrete strong floor capable of resisting direct forces of at least 50 tons.

(b) A reaction frame consisting of 250 x 250mm steel members to which was bolted a 100mm thick baseplate. The blade to be tested was bolted horizontally to the baseplate with 36 no. 20mm diameter steel studs.

(c) An overhead crane with a minimum capacity of 10.0 tonnes. The crane was required for lifting the blade and placing it in position on the base plate and also for lifting the actuating equipment. It was also used during the static tests.

(d) The strain measurement and recording equipment. The strains at critical locations on the surface of the blade were measured with either strain gauges or demec gauges. The strain gauge measurements were monitored and collected at intervals on a computer whilst the demec gauge measurements - both static and dynamic were measured manually.

(e) Deflection measurement equipment at locations of interest along the blade. This consisted of:
   (i) a staff gauge at the point of load application The maximum and minimum deflection were noted at intervals during a fatigue test on this gauge and hence the range, mean deflection and any creep could be measured.
   (ii) an ultrasonic gauge at the point of load application for continuous recording of the deflection which was then recorded on the computer disc or displayed on the monitor.

(f) Control equipment for both the hydraulic and rotary mass actuators. For a full description see section 3.

(g) Temperature and humidity measurements. The temperature was kept constant at 22°c but the humidity varied according to external conditions.

(h) Specialised equipment for occasional additional tests consisted of:
   (i) thermographic measurement equipment for determining the location and developments of "hot-spots" on the blade. This test was useful in identifying potential areas of failure and was carried out by the staff of Rutherford Appleton Laboratory using specialised equipment.
   (ii) accelerometer for measuring the natural vibration and damping characteristics of the blade in both the loaded and unloaded condition. It was necessary to determine the natural vibration of the blade accurately so that the fatigue test could be conducted at a frequency below this value where much better deflection control and stability could be obtained.

3. Actuators

(a) Hydraulic Actuator (Fig. 1 and Plate 1)

The servo hydraulic actuator used for the initial fatigue tests on the Wind Energy Group's 12m blades consisted of a 2 tonne load capacity jack with a ± 250mm maximum stroke. The alternating movement of the jack was controlled by an electronically actuated Moog valve.

The initial mean deflection was produced by a combination of hydraulic jack load and weights attached to the underside of the blade opposite the jack. The use of the added weights resulted in a reduction in blade frequency with a consequent increase in stroke thus enabling some flexibility in the method of testing.

The hydraulic operating pressure of 21MPa ($21MN/m^2$) and flow rate of 45l/min for the actuator was produced by a $27MN/m^2$ pump located in a chamber below the strong floor. The frequency and stroke of the actuator was set and monitored on a control panel which also had facilities for the continuous recording of the load cell reading attached to the jack together with the jack deflection.

(b) Rotary Mass Actuator (Fig.2 and Plate 2)

The hydraulic actuator performed satisfactorily for the range of 12.0m long blades that were tested but during the latter stages of the tests the load and deflections that were required were near the operating limit of the actuator.

In view of the larger loads and deflections that were required on the 17.5m long W.E.G. blades to be tested it was decided to design and construct a rotary mass actuator that would be suitable for testing these larger blades.

The rotary mass system consisted of an electric motor which drove two rotating buckets containing lead through a reduction gear box. The rotational speed of the motor and hence the rotation of the buckets was monitored by an encoder which maintained a constant speed of rotation as set on the control panel. This enabled very accurate deflection control and gave an accuracy of less than + 1.0mm at the lower end of the frequency curve.

The mean deflection of the blade was obtained by placing lead weights in the container adjacent to the electric motor and gearbox.

The deflection range was controlled by the rotational frequency of the buckets (for a given weight of lead in the buckets) and this was determined by the strain range that was required on the target gauge to give the correct R - ratio. ($R = \delta min/\delta max$) with the mean deflection equal to the range.

The target gauge was selected as the gauge which gave the largest tensile strain on a longitudinal section of the blade which for the 17.0m blades ocurred at about 6.0m from the root.

The strain range and R - ratio required were provided by the client (WEG). A typical blade fatigue test consisted of a number of continuous test sequences lasting for about $0.25.10^6$ cycles. The mean and alternating strains were increased at the end of each test sequence until failure. Usually fatigue failure could not be obtained in a suitable time scale when resort was then made to a static test to failure.

Following each test sequence incremental load/deflection/strain readings were taken together with a natural vibration test so that any change in the blade structural characteristics could be monitored. All the strain and deflection measurements together with the rotational speed of the actuator were monitored, sampled and collected on a computer disc . Real time sampling enabled the location of the operating point on the frequency curve for the blade to be displayed on the monitor.

4. Static Tests (Plates 3 & 4)

The static tests were performed (a) with the blades anchored horizontally in the reaction frame with the pressure side of the blade facing downwards and the load applied upwards from underneath the blade with the aid of hydraulic jacks or (b) by applying the load downwards onto the pressure side of the blade with the aid of a known weight suspended from the crane. Edgewise tests were also performed using this method.

For method (a) the load on the blade was measured either with the aid of a load cell or by recording the hydraulic pressure acting on the piston of the hydraulic jack. The load was transferred from the jack to the blade through a ball joint attached to the cradle on the blade thus ensuring that the resultant load was applied through the centre of gravity of the jack and at right angles to the blade.

For method (b) the load on the blade was measured with the aid of a direct reading load measuring gauge suspended from the crane hook. The load was applied to the blade via the cradle that was attached to the blade at the point of load application.

Single point loading provided an economical method for fatigue testing the critical inner 50% of the blade. Static tests were performed with both single and two point loading - the latter being representative of the inner 80%.

For each test and for each load increment the deflections and strains along the length of the blade were measured with the aid of dial gauges and demec gauges respectively. The blade angle at the point of load application was also noted so that the load normal to the blade could be obtained.

5. Comparison of Test Methods

The hydraulic actuator worked well for the initial tests on the 12.0m long blades but for some of the tests the actuator was working near its operating limit even with additional weight suspended from underside the blade. There wás evidence, as recorded on a load cell, of irregular load transfer between the actuator and the blade due to the different frequency characteristics of the two systems.

The rotary mass actuator was designed and constructed in the Civil Engineering laboratory and has proved to be an elegant and economical solution for fatigue tests on blades of any size. The actuating system and the blade move together at the same frequency producing a smooth transfer of load between the actuator and the blade, which is likely to be more representative of actual wind loading conditions.

The static tests on the blades were relatively simple to perform and were usually carried out if the blade had not failed following the sequence of fatigue tests. They provided considerable information on the blade strength characteristics in a relatively short time, but the results of course were representative only of extreme load conditions.

6. Test Results (Plates 5 & 6)

The results obtained from the tests to date have highlighted the likely modes of failure of the blades both statically and dynamically. The failures can be classified as:

(a) Tension failures - mainly at joints in the timber laminates (early blades).

(b) Compression failures in the wood.

(c) Shear joint failures.

(d) Anchorage failures due to fatigue of the retaining steel studs (early blades).

The results from early blades have already been published [1].

Analysis of these failures has led to to a continued improvement in the design and manufacture of the blades and a considerable increase in strength. Although coupon tests are useful for basic design purposes they are limited by their size and boundary conditions and are not always able to reproduce the type of failures that have been shown to occur on full scale blades. Often blade failures have been shown to be due to the integrated effects of forces acting on the blade in three dimensions.

7. Conclusions

Our experience so far in the fatigue and static testing of windmill blades has shown that the accelerated testing techniques employed can lead to a better understanding of blade structural behaviour and provide useful information about the lifetime of the blades.

By using the most damaging part of the fatigue spectrum where the mean was equal to the range it was possible to represent a large part of the fatigue life of the blade in about 2 weeks thus enabling a turn around of about 1 blade/month.

The test programme has been closely integrated with W.E.G. (the blade manufacturer) who have used the test results to develop their blade structural design and method of manufacture for the new blades to be tested and in a short time a wealth of useful data on the blades structural characteristics was obtained.

As a result of information gathered from previous tests the Wind Energy Group have been able to optimise their blade design to such an extent that in a recent static test the failure load was more than three times the working load despite the previous cycling of the blade. (Plate 3).

REFERENCES

[1] Hancock M., Richmond B.N., Morris W.J., Jones W.A.R. Fatigue Testing of 12m Wind Turbine Blades. 1992 B.W.E.A. Conference, Nottingham.

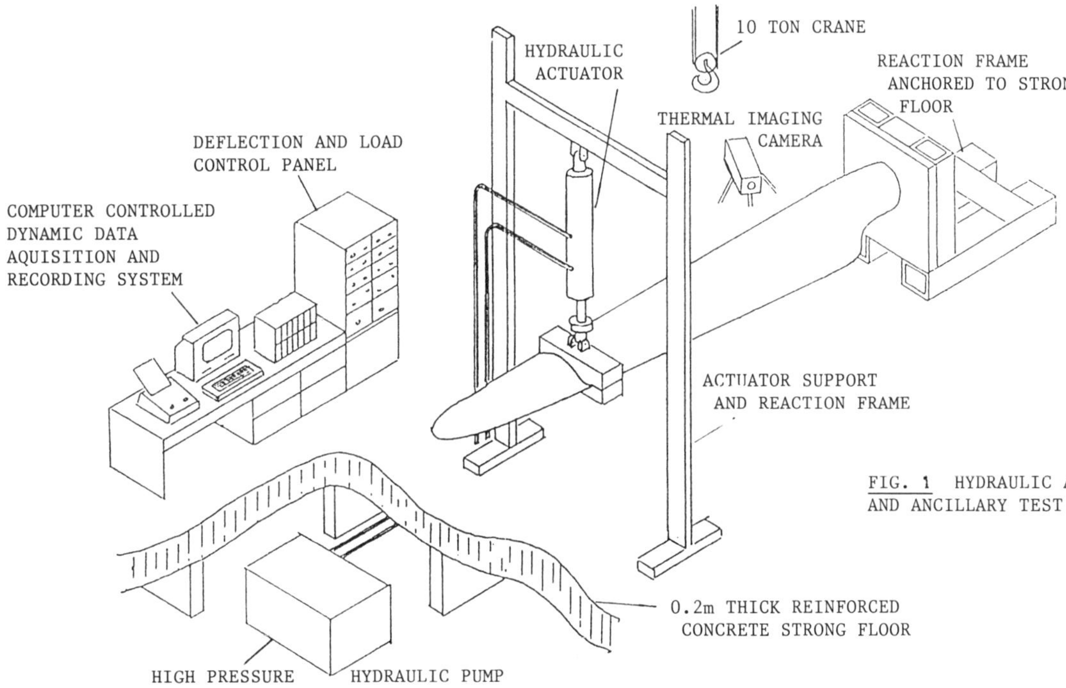

FIG. 1 HYDRAULIC ACTUATOR AND ANCILLARY TEST FACILITIES

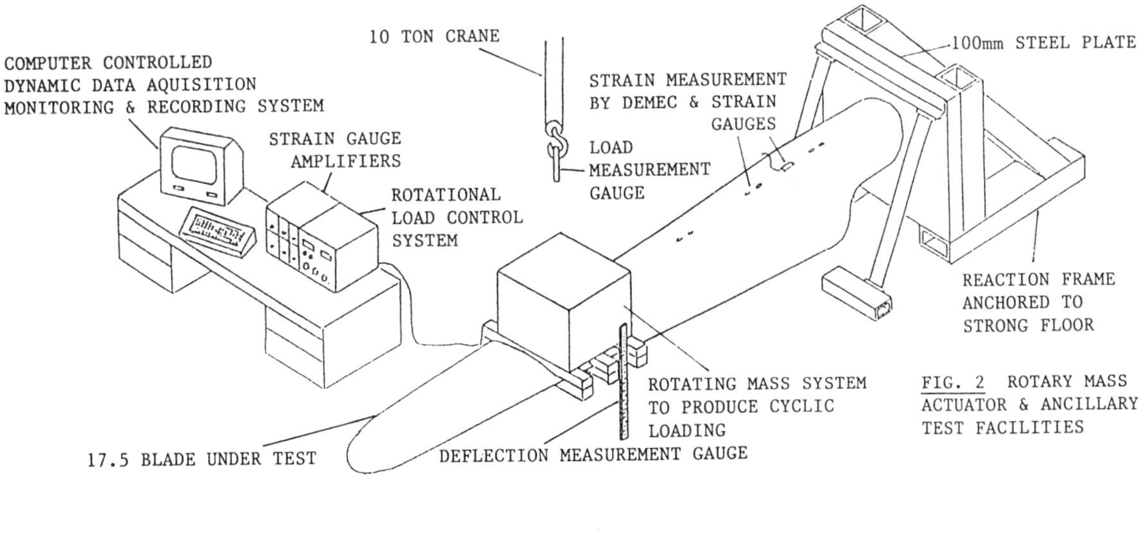

FIG. 2 ROTARY MASS ACTUATOR & ANCILLARY TEST FACILITIES

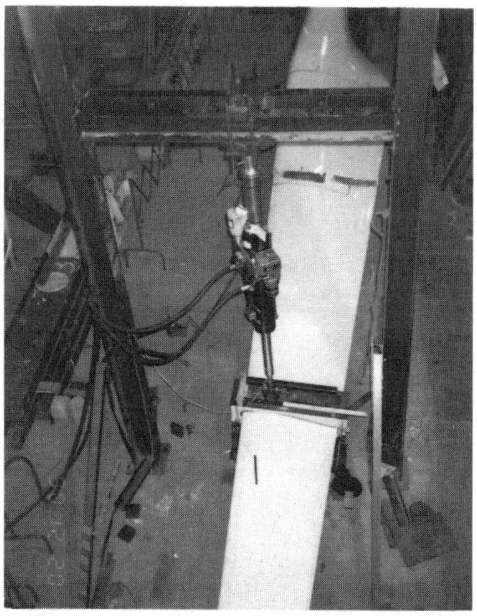

PLATE 1

PLATE 2

PLATE 3

PLATE 4

PLATE 5

PLATE 6

PLATE 1. HYDRAULIC ACTUTOR SYSTEM

PLATE 2. ROTATING MASS SYSTEM

PLATE 3. 2 POINT STATIC TEST ON WEG 400 BLADE

PLATE 4. STATIC TEST WITH OVERHEAD LOAD

PLATE 5. TYPICAL BLADE TENSION FAILURE

PLATE 6. TYPICAL BLADE COMPRESSION FAILURE

MECHANICAL CHARACTERISTICS OF BLADE DETERMINATION FROM AEROELECTRICAL TURBINE WITH HORIZONTAL AXIS

F.Gyulai x), I.Santău x), T.Cioară xx), A.Botici xxx)
Technical University of Timişoara
x) Center Research of Aeroenergetical Energy
xx) Laboratory Vibrations of Mechanical Department
xxx) Laboratory of Steel Construction Department
B-dul M.Viteazu 1
TIMISOARA ROMANIA

ABSTRACT: Within the frame work of research developed at the Technical University of Timişoara was paied attention to the verification of mechanical resistence of wind turbines'blades as a preventive measure against the potential demages. In this way was realised three testing stations which are shown in this paper.

INTRODUCTION

An important objective in the begining stage of research work for the wind energy promotion, developed in Technical University of Timisoara, it was represented by the avoidance for a certainly of possible demages on the placements. With this end in view was imposeing an rigorous individually test of the blades. More than usual controls (technological and dimensional) from factory, we have imposed three tests for a global control. Each of the blades which are realised or imported in Romania are submited for these tests. On the basis of these three global control tests, the Wind Energy Research Center from Techical University of Timişoara approve the assembling permit on the wind turbines placement.

THE TESTING STATIONS FOR THE MECHANICAL CONTROL

The three testing stations which were realised at the Technical University of Timişoara are useful for two distinct activities :
- the control and the approveing for assembling of the blades
- scientific research concerning the development of designing metodologies.

The three testing station erected with this end in view are :
- The station for static Testings, which is placed in Laboratory of Steel Construction Department from Faculty of Constructions ;
- The Station for Vibration Testings, which is placed in the Vibrations Laboratory of Mechanical Department from Faculty of Mechanics ;
- The Overspeed Blades Testing Station, erected over a special field and used by the Wind Energy Research Center.

On these stations have been tested since present four blades families which were realised within frame work of the research programme. The mechanical caracteristics of these can be seen in table no. 1.

Table no 1

Name of blade	Rotor diameter	Blade length	Length of airfoil chord		Mass (Weight)
			hub	extremity	
	m	m	m	m	kg/pcs
SK 1	1o	3,7oo	8oo	4oo	9o
OPS 7	1o	3,8	55o	25o	85
OPS 7-o1	3o	12,4	165o	7oo	126o
OPAFS-o1	3o	12,4	165o	7oo	13oo

The first two blades have been assembled and tested on a 3o kW laboratory wind unit, which is erected upon one of the Technical University's building.

The 3-th set blades is assambled on a 3oo kW experimental wind unit placed in a mountain placement from western zone of Romania, in mountain climatics and it is in operation from some years.

The 4-th set blades is a technological attempt where the steel strength structure is covered with polyester reinforced with fibre glass.

If we take into consideration the technological conditions existing since present in Romania, a strength structure like this is a matter of interest for moment.

The blades have been designed for the following conditions :
- operational wind speed : 4 - 3o m/s
- survival speed : 8o m/s
- maximal operational peripherical speed: 8o m/s
- maximal testing peripherical speed : 12o m/s
- extreme calculation peripherical speed (brake protection) : 16o m/s.

The aerodynamical calculation have been made on the basis of lifting method and singularities method, and the strength calculation on the basis of finite elements method.

THE STATION FOR STATIC TESTINGS

This station it's compoused of a supporting device which permite to fixe the blade by her flage. The blade it's fixed with screws in horizontal position haveing the subpressure side downwards. On the blade are placed 14 strain gauges (fig. 1).

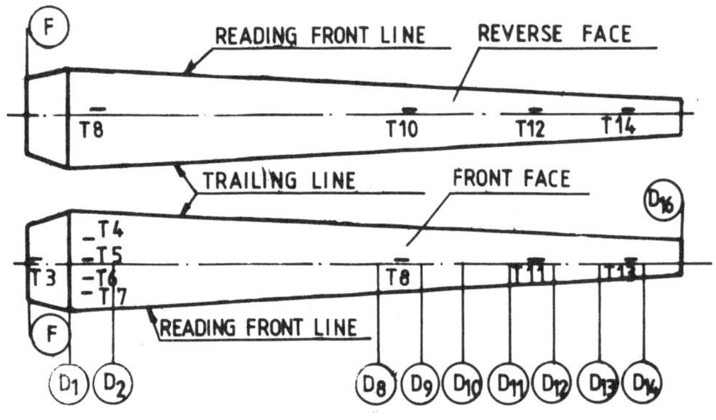

Figure 1. Gauge points position.

The testings of blade it's made useing bags containing sand with a mass that it have been determinated before hand (fig. 2).

Figure 2. Station for Static Testings

The load was realised successively according to wind speed of (6-32)m/s ; 6o m/s and 8o m/s. According testings were written 1,21 Δ p; 1,74 Δp for W = = (6 - 32)m/s respective 147o Pa and 384o Pa as uniform distributed pressure for 6o m/s and 8o m/s.

The testings and the pressure repartition on the blades can be seen in detail in [1] . At the same time in [1] are exposed the testing combinations that we have taken in consideration to estimate the blades.

For each testing stage were measured the punctual novement on the lenght of leading edge and trailing edge of blade.

Unitary stress found on the basis of strain measurements were calculated on those points that we supoused critical on the pressure side and subpressure side of the blades. The points were determinated through the analysis of - strain state.

On the basis of static testings will be authorised the blades assembling on wind unit if the results of teoretical investigation are iuto the admisible error limit with the experimental results and if the structure haven't behaviour anomalies.

The testings of OPS 7 -ol blades with a steel structure have given the following conclusion: a very good concordance between the calculated and measured punctual movements into a error limit of maximum 5 % - fig. 3 a, a good concordance between the teoretical and measured unitary stress into a error limit of maximum lo % - fig. 3 b, c, d, e.

For a clear representation in fig.3 can be seen, comparatively, the experimental and theoretical results concerning the rated load on the blade for a normal operation 1,21 Δp. Details about the results of the other calculation stages it's shown in [1] and [2] .

Withint the frame of static testings have been determinated the supporting capacity.

The diagrams of maximal and effective bending moment, comparative with bending moments which can be tolerate by the blade until tearing-strength, can be seen in [2] .

On the basis of these testing results were calculated the safety coefficient for each transverse sections of blade which are 1,52 - 1,78.

THE STATION FOR VIBRATION TESTINGS

The main aims of the vibration testing are:
- selection of the most appropriate structural couple of the blades, from a manufactured set, to be mounted on a given wind turbine. As selection criterion is stated the deviation values of the natural frequencies between the same mods of the blades, the relative percentage deviation limit being imposed at 1 % ;
- detection of the weakness sides of the blade by resonant testing. If there are bad manufactured side of the blade the vibration signals, measured the blade structure, are harmonically disturbed, by nonlinear effects in bad joints ;
- validation of the dynamic model of the blade structure.

The vibration testing stand is consisting of one rigid body 1, (Figure 4) to which the blade 2 is connected in the same structural mode like of its installation on the rotor of the wind turbine: clamped in the flange 3 of a tubular axle 4, which is axed by two rolling bearing 5, mounted into a rigid metalic support 6, fixed to the massive rigid body 1.

The whole ensemble, massive rigid body 1 - metallic support 3 - blade 2, is bears by four air cushions connected to massive rigid body 1. The vibration system of the elastic supported below of 1,5 Hz. In this condition for frequency range of elastic blade natural frequencies, of testing interest, 3,5 - 5o Hz, the ensemble can be considered as a free - free system, being practically isolated.

The excitation force is applied in a single point Pe on the massive rigid body 1, by a shaker 8 and a force control transducer 9. For this excitation procedure it is demonstrate, [1] , that the all mods can be ease excited.

For modal parameters estimation (natural frequencies, modal ratios and modal shape) is applied the procedures and data processing developed in the Vibration Testing and Research Laboratory 4 - 7 .

The experimental results on a couple of three metallic blades which were mounted on a 3oo kW wind turbine are given in table No.2. There are printed the values for, natural frecvencies, modal damping ratios, and modal frequency relative deviations to mean value of three blades, corresponding to first six natural modes of which. Also are printed the procentage deviations in frequency. The modal shapes for bending mods four from six are draws in figure 5.

The three blade of couple was admixed after vibration testing: the maximum value of frequency deviation .683 for the second mode of the blade 3/1 and weakness structure effects not were observed during resonant testing.

Table No. 2 The values of modal parameter for a couple of three metallic blades

	blade 1/1	blade 2/1	blade 3/1
	natural frequency Hz	/ modal damping	ratio /
		1st MODE	
mean values	5.5271 / .219	5.5oo3 / .218	5.5496 / .223
freq. dev. mH	1.43	-25.4	23.9
deviation %	.259	-4.61	4.31
		2nd MODE	
mean values	11.426 / .3o3	11.485 / .314	11.574 / .37o
freq. dev. mH	-69	-1o	79
deviation %	-.6o4	-.o87	.683
		3th MODE	
mean values	2o.884 / .239	2o.929 / .246	21.o43 / .397
freq. dev. mH	-68	-23	91
deviation %	-.326	-.11	.432
		4th MODE	
mean values	28.625 / .2o1	28.761 / .213	28.932 / .2o9
freq. dev. mH	-13o	-2o.7	15o
deviation %	-.43	-.o72	.52
		5th MODE	
mean values	36.158 / .3o3	36.124 / .341	36.174 / .443
freq. dev. mH	6	-28	22
deviation %	.o16	-.o77	.o61
		6th MODE	
mean values	46.726 / .145	46.614 / .125	46.73o / .154
freq. dev. mH	36	-76	4o
deviation %	.o77	-.163	.o86

THE OVERSPEED BLADES TESTING STATION

This plant offer the possibility to test each functional blades, haveing in view :

- the blades quality and behaviour in operation for nominal rotation speed and even more than that, till maximal tolerated rotation speed obtained in operation.
- to find the deflection of blade.

The Overspeed Blades Testing Station is a complex plant which have included : observations platform (A), electrical supply unit (B), installations for protection (C) and the platform for mounting operations (D) fig. 6.

On the observation platform are placed the driving system of blade, containing : electric D.C. motor with P = 8oo kW install power at n = 1ooo rot/min, 3 gear boxes (one of them conical) and the radial-axial bearing of main shaft (E) fig. 6.

The possibility to couple the driving system permite to drive the main shaft and consequently the blade (F) fig. 6 in revolutions speed $n = [o...4oo]$ rot/min. The driving system permite to change continue the blades rotation speed, or to maintain rigorous constant for purpouse values of rotation speed. The plant (fig. 6,7) permite testings of blades for wind turbines with max. diameter D_{max} = 6o m.

Untill to start the testings, the blade having horizontal axe it's fixed on the main shaft of driving system, through the flange, which assured at the same time to install the blade on the hub of wind turbine. The installing angle of blade it established, so that, to permite at the nominal rotation speed, that the aerodynamical forces to balance the forces of weight.

At the same time the installing angle can be a parameter of testings, it can be gradually changed. The overspeed testing, made for each blade, driving succesively for some purpouse values of rotation speed which are selected depending on nominal value of rotation speed and rotor diameter of wind turbine.

For each constant value the blade it turned 1o seconds, and measured the rotation speed (rot/min) and the deflection at the extremity of blade.

The OPS 7-o1 blades, designed for an aeroelectrical unit with nominal rotatio speed n = 5o rot/min, were tested at overspeed for the following values : 3o, 4o, 5o, 6o, 7o and 75 rot/min.

The testing results of 3 blade can be seen in fig. 8. Result from that, the

Figure 7. Overspeed Blades Testing Station

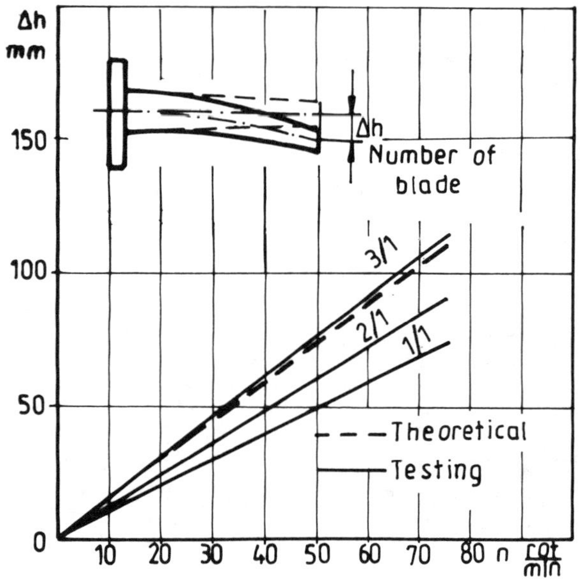

Figure 8. Comparison between the analytical and experimental results obtained in the overspeed testing.

deflections h at the extremity of blades are a litle changed, into a limit of 13,5 %, for the 3 blade. That difference can be provided by variation, in tolerated limits, of installing angle of blades or of position angle of airfoils from blade sections.

Comparing deflection values determinated on the experimental way and the values determinated by analytical method, result that the blades 1/1 and 2/1 have deflections shorter than calculated deflection, and the blade 3/1 have a deflection longer, than calculated value, with aprox. o,8 %.

On the basis of testings, result conclusions concerning the permision to instalate the blade having in consideration the following problems:
- the maximal deflection at the extremity of blades must be smaller then the admissible value obtained by designing calculation for a max. operation speed;
- the relative error, defined by

$$\frac{h_{max} - h_i}{h_{max}} \cdot 1oo,$$

between the maximal deflection (h_{max}) acording to maximal rotation speed and the deflection h_i of one blade of a set blades which are destinated for a wind unit, can't be more than 2o %;
- the outside surface of blade must not to have macroscopic fissures.

Concerning the scientific results that can be found through overspeed testings, we specify:
- verifications for calculation hypothesis that have been taken to design the blade;
- studies about the influence of parameter position angle;
- determination of pressure distribution on the blade;
- determination of stress distribution in blades strength.

On the basis of static, vibration and overspeed testings that have been made for blades on the mentioned stations, will be obtaine the quarantee of blades operation, 8 which have tendency to the international standards.

These testing stations and methods permite on the other hand to obtain scientific results necesary to improve the methods of blades designing.

CONCLUSIONS

- The testing stations and methods for wind units, that have been shown in this paper, permite to determine the static, vibration and overspeed characteristics of blade.
- On the basis of the obtained results through static, vibration and overspeed testings will be assure: the assembling permit on the wind turbines placement and on the other hand the scientific results wich will facilitate to improve the designing methods of blade.

REFERENCES

1 A.Botici - Contribuţii la calculul şi alcătuirea paletelor din metal pentru aerogeneratoare cu ax orizontal -Teză de doctorat,Universitatea Tehnică Timişoara,1977
2 A.Botici - Wind Energy catching research on the Wind Turbines Rotor Blades,Symposium and Demonstration of the use of Wind Energy in Romania (2) Timişoara, 3-6 May 1992
3 I.Santău,F.Gyulai,- Experimental researches on the Dynamic characteristics of the Blades of Wind Turbines with Horizontal Spindle,Symposium and demonstration of the use of Wind Energy in Romania (1),Timişoara,3-6 May, 1992
4 T.Ghe.Cioară - Modal testing for the Blades of one Horizontal Axis Wind Turbine.Proceedings of 17th Interna-

tional Seminar on Modal Analysis - Katholike University Leuven,23-25 Sept.,1992,p. 1331-1342
5 x x x Theoretical and Experimental Research on Dynamics of a Horizontal Axis Wind Turbine,Annual Technical Reports of VTRL of Technical University of Timişoara (in Roumanian)
6 T.Ghe.Cioară - On a Curve-Fitting Algorithm for Modal Paraameter Estimation, Proceedings of 6th international Modal Analysis Conference, February 1-4,1988,Orlando,Florida, p. 717-723
7 T.Ghe.Cioară - Consideration on Graphical Representation of Modal Shape and Vibration Motion Distribution of a Complex Structure (to be published
8 P.H.Jensen - Recommendation for an European wind turbine safety standard and load cases standard made by the cooperation of the European wind energy test sites,European Wind Energy Conference and Exhibition EWEC'89, Iuly 1989.

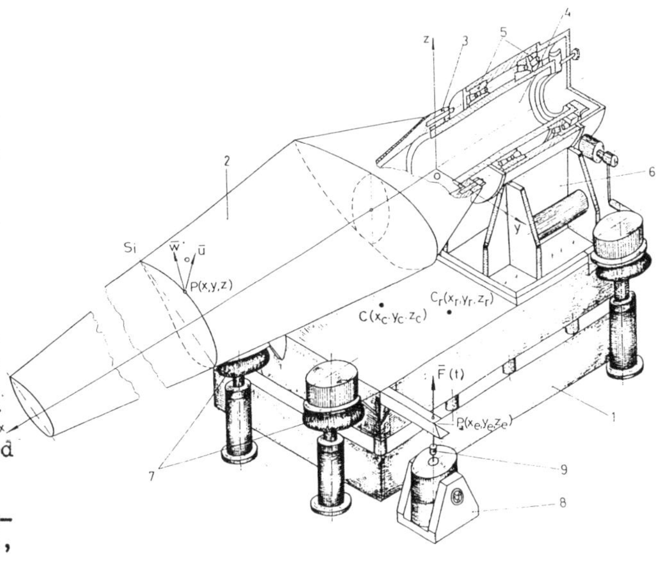

Figure 4. Vibration testing stand.

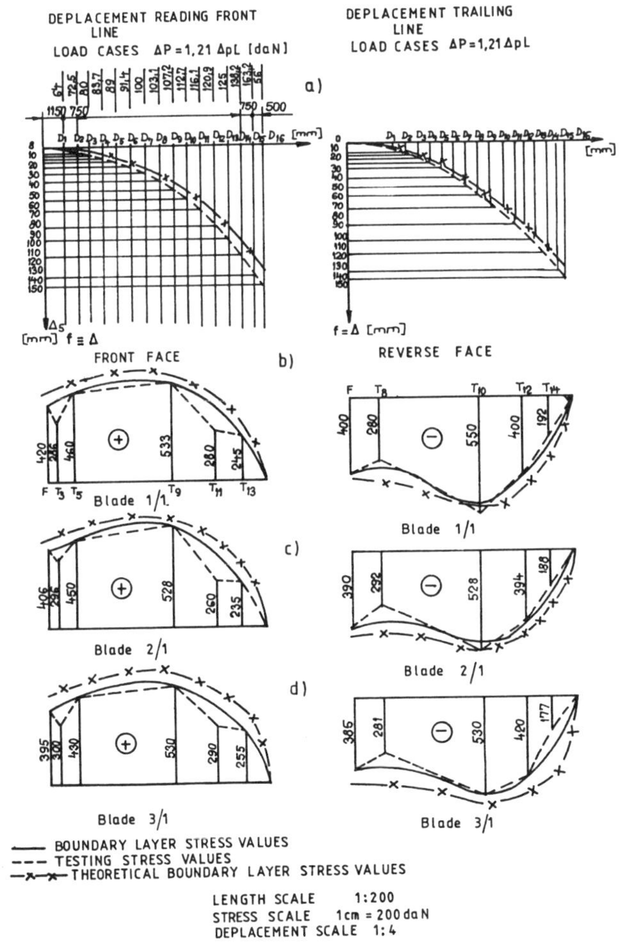

Figure 3. Comparison between the analytical and experimental results obtained in the static testing.

Figure 5. Blade modal shapes.

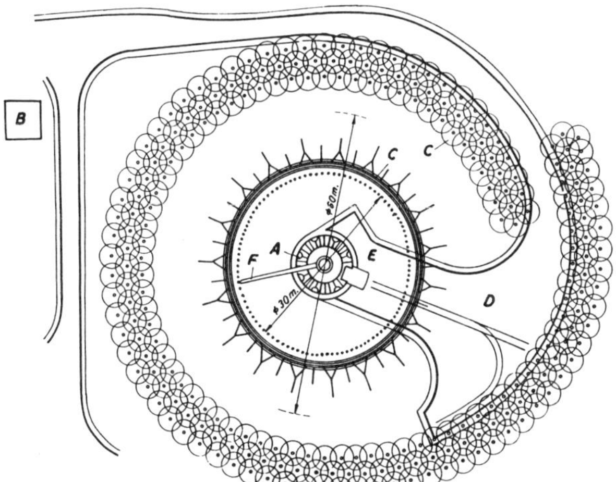

Figure 6. Station for overspeed blades of aeroelectric units.

ACCURATE GEOMETRICAL CALCULATION FOR ROTOR BLADE SURFACES OF A HORIZONTAL AXIS WIND TURBINE

Z.Q.Ye
Scientific Research Department
Shantou University, Shantou 515063, China
Z.X.Cheng
Department of mechanical Engineering
Gansu University of Technology, Lanzhou 730050, China

ABSTRACT: A principle and program for accurately calculating the coordinates of rotor blade surfaces of a fast running horizontal axis wind turbine according to the geometrical parameters obtained from the aerodynamic design is given in this paper. It makes the geometrical errors of the rotor blades minimum during the design stage. The calculation method is based on the principle of coordinate transformation. The method and its program, in conjunction with the aerodynamic optimum design program[1],has been used for designing and manufacturing a series of small-size rotor blades which are of high efficiency and good performance. Practical example is also given in the paper.

1. INTRODUCTION

In order to ensure the aerodynamic characteristics of the design rotor, it is necessary to make the rotor blade surfaces manufactured according to the geometry (e.g. Chords, Setting angles, and Airfoil Sections) obtained from the aerodynamic design. Under the same manufacturing technology and equipment conditions,whether the geometrical shape of the rotor blade is consistent with the original design or not will depend on how accurately determine the space coordinates of the points on upper(section) and lower(pressure) surfaces of the rotor blade. According the authors, in the design of the rotors of wind turbine or other turbomachinery,the drawing method is commonly used ,therefore geometrical error is caused in the design.A calculating principle and computer program for determining the coordinates of the points on upper and lower surfaces is given to ensure the geometry of the rotor blade from any deviation which could be prevented.

2. THE ORIGINAL PARAMETERS FOR DETERMINING THE GEOMETRY OF THE ROTOR BLADE.

Acording to the aerodynamic calculation of the rotor blade, the following geometrical parameters can be obtained.(as shown in Fig.1)

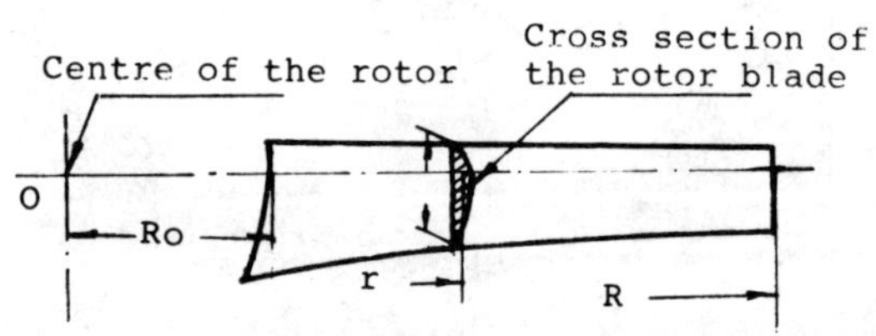

Figure 1: Geometry shapes of rotor blade plane.

2.1 Radius R and root radius R_0 of the rotor;

2.2 Chord b of airfoil section at different radius r of the rotor blade;Because the flow around the rotor blade is axisymmetric, the flow surfaces are concentric cylindrical surfaces with different radii. The airfoil section is also in a cylindrical surface, and the chord b is an arc with radius r.

2.3 Airfoil geometries at different radii ; They include: (Fig.2) [2]

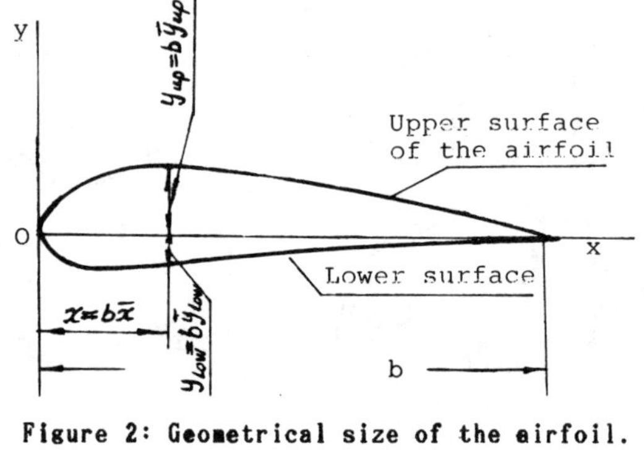

Figure 2: Geometrical size of the airfoil.

(a) Relative value of the coordinate in the OXY coordinate system for points on upper surface of the airfoil

$$\bar{x} = \frac{x}{b} \qquad \bar{Y}_{upper} = \frac{Y_{upper}}{b}$$

(b) Relative value of the coordinate in the OXY coordinate system for points on lower surface of the airfoil

$$\bar{x} = \frac{x}{b} \qquad \bar{Y}_{lower} = \frac{Y_{lower}}{b}$$

(c) The radius of the leading edge and the slope of the airfoil

2.4 Setting angles ψ of the airfoil section at different radius (Fig.3). It is the included angle between the geometrical chord and the rotating surface of the rotor blade.

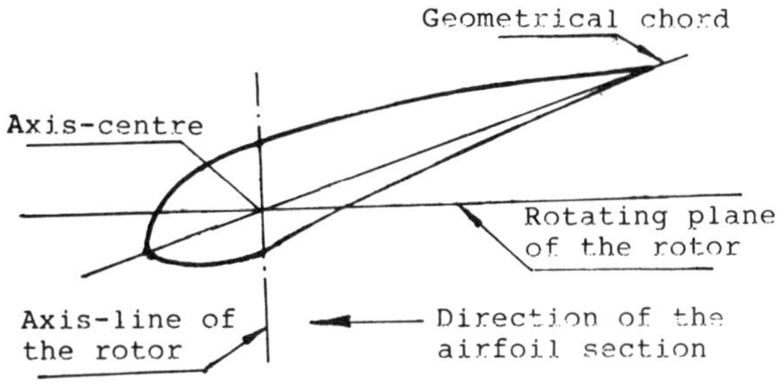

Figure 3: Setting angle of the airfoil section.

3. DETERMINATION OF THE SPACE COORDINATES OF THE POINTS ON ROTOR BLADE SURFACES

The space coordinates of the points on rotor blade surfaces can be determined according to the original parameters mentioned above. For calculation, it is essential to set up a left-handed space coordinate system OXYZ with the OXZ plane being the rotating plane of the roter shown in Fig.4.

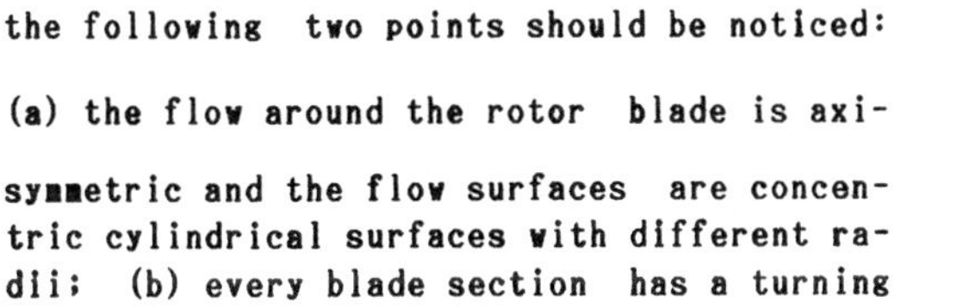

Figure 4: Coordinate system for determining the rotor blade geometry

the following two points should be noticed:

(a) the flow around the rotor blade is axisymmetric and the flow surfaces are concentric cylindrical surfaces with different radii; (b) every blade section has a turning angle called setting angle relating to the rotating plane OXZ. Therefore, in the OXYZ coordinate system a wind turbine blade can be regarded as a two dimensional blade through twisting transformation and transforming the coordinate system from cylindrical coordinates into OXYZ coordinates. Therefore, the determination of the coordinates of the points on rotor blade surface can be carried out in two steps.

3.1 Twisting transformation.

An airfoil section of a two dimensional rotor blade in radius r is shown in Fig.5 . The airfoil section twisting a setting angle ψ rounding its pressure centre D and relating to the rotating plane OXZ can be regarded as the transformation between the coordinate system OXY and the coordinate system O'X'Y' (Fig.5). The axis X' in the new coordinate system O'X'Y'

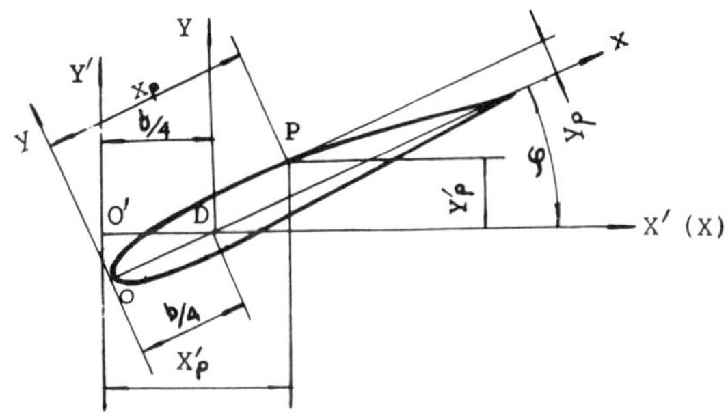

Figure 5: Twisting transformation of the airfoil section

and the axis x in the coordinate system OXY、y' and y are in the same direction respectively. The orignal point of the coordinate moves b/4 to the left (the distance between the pressure centre D and the leading edge of the airfoil section). When the coordinates of a point of the airfoil surface in the coordinate OXY are known, the coordinates in the coordinate system O'X'Y' can be obtained through transformtion and rotation of the coordinate system OXY.

$$X_P' = X_P \cos\psi - Y_P \sin\psi + \frac{b}{4} - \frac{b}{4}\cos\psi$$

$$= [\bar{X}_P \cos\psi - \bar{Y}_P \sin\psi + \frac{1}{4}(1-\cos\psi)]\cdot b \quad (1)$$

$$Y_P = X_P \sin\psi + Y_P \cos\psi - \frac{b}{4}\sin\psi$$

$$= [\bar{X}_P \sin\psi + \bar{Y}_P \cos\psi - \frac{1}{4}\sin\psi]\cdot b \quad (2)$$

3.2 Transformation between the planar flow surface and the cylindrical flow surface

Because the flow surface which the airfoil section is in is a cylindrical surface with the radius r, the coordinates of a point on the surface in the coordinate system OXYZ will change when the coordinate system O'X'Y' is bent to a cylindrical surface. Fig.6 shows the situation of the airfoil section on the cylindrical flow surface viewing from the direction of the axis OY, suppose point P is a point on the airfoil

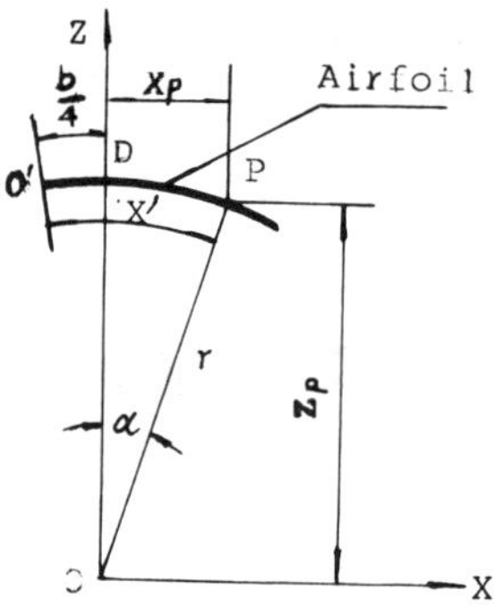

Figure 6: Transfomation between the planar flow surface and the cylindrical flow surface

surface. Its coordinate on the planar flow surface are X_P', Y_P', Z_P' (or r), while its coordinates in the coordinate system OXYZ are X_P, Y_P, Z_P, But $Y_P=Y_P'$. Let's find out X_P and Z_P: Assume the included angle between the radius vector OP and the coordinate axis OZ to be α , thus

$$\alpha = \frac{1}{r}\left(X_P' - \frac{b}{4} \right) \tag{3}$$

$$X_P = r \cdot \sin\alpha = r \sin\left[\frac{1}{r}\left(X_P' - \frac{b}{4}\right)\right] \tag{4}$$

$$Z_P = r \cdot \cos\alpha = r \cos\left[\frac{1}{r}\left(X_P' - \frac{b}{4}\right)\right] \tag{5}$$

substitute equation (1) into (4) and (5), then the coordinates of the point P on the airfoil section surface can be expressed.

$$X_P = r \cdot \sin\left\{\frac{b}{r}\left[\bar{X}_P\cos\psi - \bar{Y}_P\sin\psi + \frac{1}{4}(1-\cos\alpha) - \frac{1}{4}\right]\right\}$$

$$= r\sin\left\{\frac{b}{r}\left[\left(\bar{X}_P - \frac{1}{4}\right)\cos\psi - \bar{Y}_P\sin\psi\right]\right\} \tag{6}$$

$$Z_P = r\cos\left\{\frac{b}{r}\left[\left(\bar{X}_P - \frac{1}{4}\right)\cos\psi - \bar{Y}_P\sin\psi\right]\right\} \tag{7}$$

from equation (2)

$$Y_P = Y_P' = \left[\bar{X}_P\sin\psi + \bar{Y}_P\cos\psi - \frac{1}{4}\sin\psi\right] \cdot b \tag{8}$$

According to the equation (6), (7) and (8), the coordinates of the point P in the coordinate system OXYZ can be accurately calculated when the chord b of the airfoil section, setting angle ψ and the coordinates of the point on airfoil surface $\bar{X}$ and $\bar{Y}$ (including section surface and pressure surface) is given.

4. PRACTICAL EXAMPLE OF GEOMETRICAL CALCULATION FOR ROTOR BLADE

The calculation method and its program, inconjunction with the aerodynamic optimum design program (which was presented by authors in AMSTERDAM EWEC'91), has been used for designing and manufacturing a series of small-size rotor blades from 200w to 20kw wind turbines which are listed in the key scientific projects of the Country's Seventh Five-Year Plan. A smallist rotor blade as a practical example is shown to explain the calculation procedure.

4.1 Original parameters

Diameter of the rotor:

D = 2R = 1.65 (m)

None working radius of the rotor :

R = 0.2R = 0.165 (m)

Airfoil and its chords and setting angles in different radii :

relative radius $\bar{r} = \frac{r}{R}$	airfoil	chord b(m)	setting angle ψ (rad)
0.2		0.125	-0.276
0.3		0.105	0.194
0.4	NACA 6412	0.094	-0.157
0.5		0.086	-0.137
0.6		0.079	-0.124
0.7		0.074	-0.114
0.8		0.068	-0.105
0.9		0.064	-0.099
1.0		0.060	-0.094

Geometrical parameters of the airfoil [2] :

$\bar{X}$	$\bar{Y}_{upper}$	$\bar{Y}_{lower}$
0.0000	0.0000	0.0000
0.0125	0.0273	-0.0123
0.0250	0.0380	-0.0164
0.0500	0.0536	-0.0199
0.0750	0.0657	-0.0205
0.1000	0.0758	-0.0199
0.1500	0.0918	-0.0167
0.2000	0.1034	-0.0125
0.2500	0.1114	-0.0078
0.3000	0.1165	-0.0038
0.4000	0.1180	0.0020
0.5000	0.1116	0.0055
0.6000	0.0995	0.0078
0.7000	0.0823	0.0085
0.8000	0.0603	0.0073
0.9000	0.0333	0.0039
0.9500	0.0179	0.0016
1.0000	0.0000	0.0000

4.2 Computer program for the calculation of space coordinates of the points on rotor blade surfaces.

A computer program has been designed according to the original parameters and the equation (6) (7) (8). The outline is shown in Fig. 7. The calculation method is based on the principle of coordinate transformation as mentioned above. Besides, to make the processing handy, the arc sections are changed to straight sections using the linear insert-value method in the calculation of the coordinates of the points. After the program has been executed, the results are printed in the form of a data file containing the coordinates of 540 points for the upper and lower surfaces.

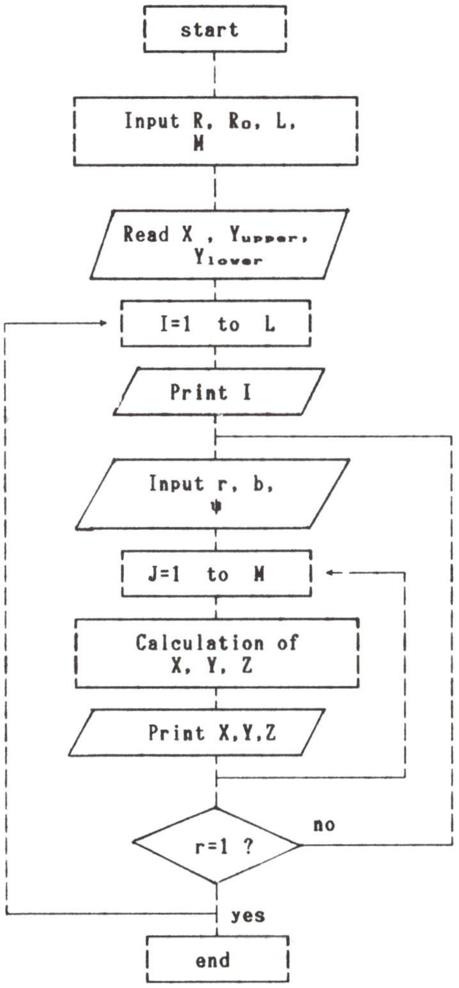

Figure 7: Outline of the Computer Program

For accurately measuring the aerodynamic characteristics of the design rotors and making the comparison of their power coefficients between the design and the experiment, the rotors were tested in the low speed wind tunnel or on testing sites. The results have shown that their power coefficients C_p are all above 0.4 and that the C_p curves turn out to be even and wide for their high-efficiency zones.

5. CONCLUSION

Calculating method for the geometry of the rotor blade surfaces has been developed to minimize its geometrucal errors. The calculation method is based on the principle of coordinate transformation.

Computer program for accurately calculating the space coordinates of the rotor blade surfaces is quite efficient and capable of ensuring the processed precision and aerodynamic characteristics of the rotor.

The results of the design rotors have shown that they have high efficiency and good performance.

REFERENCES

[1] Z.Q.Ye, Z.X.Cheng, J.G.Chen, and S.B.Bai Aerodynamic optimum design procedure and program for the rotor of a horizontal axis wind turbine. Journal of Wind Engineering and Industrial Aerodynamics 39(1992)179-186

[2] Airfoil Data for Wind Machine, Vol.1 China wind Committee and CARDE (1984)

[3] И.Н.Бронштейн И К.А.Семендяев: Справочник по математике, 3 изд, Москва, (1953)

ARCTIC WIND TURBINE TEST STATION

E. Peltola
VTT Technical Research Centre of Finland
Laboratory of electrical and automation engineering
PO Box 34
SF 02151 Espoo FINLAND

B. Tammelin
Finnish Meteorological Institute
Technical climate research
PO Box 503
SF 00101 Helsinki FINLAND

ABSTRACT A test wind turbine for developing the wind technology for cold and icing conditions will be constructed during summer 1993. The turbine is to become an as continuously as possible working prototype. Its operational experience should help in avoiding problems in future wind farm projects under similar or almost similar conditions. The machine will be equipped with such means for avoiding the icing and low temperature problems that have been theoretically or in small-scale and laboratory experiments shown to be most promising.

1 INTRODUCTION

The newly found wind climate of the arctic mountains (fells) in the Finnish Lapland provides good possibilities for large scale wind energy production. The annual mean wind speed upon the fells higher than 500 m above sea level is about 7.5-8.5 m/s. This is significantly more than the synoptic mean wind speed, 3-4 m/s, in Lapland. The calculated wind energy potential in Lapland is about 14 TWh/a.

The main problem for wind energy production on the fell top areas is severe icing or rime accretion during the winter months from November to March. However, it seems that even if the wings were heated during the time of rime formation the net production of a 300-500 kW wind turbine would be about the same as at the best sites on the coastal areas.

The icing problem has been earlier investigated by using a small battery charging turbines with rated capacities of 1,5 kW and 2,5 kW. The results of the latter experiment carried in 1991 in Pyhätunturi fell have been reported e.g. in [1], [2] and [3]. The turbines met very heavy icing conditions and were regularly heavily iced.

A larger grid connected wind turbine has been operating in Enontekiö, 68,5 °N, since February 1991. This turbine has given useful experience on the behaviour of the turbine in low temperature. Also some icing events have occurred.

This paper presents the backgrounds, goals and present status of an effort to put up a large wind turbine on the Pyhätunturi fell as the next step on the development of wind turbines suitable for the climate conditions of the fells in Lapland. The project is carried out as a part of the Finnish national research programme NEMO on Advanced energy systems by Technical Research Centre of Finland VTT, Finnish Meteorological Institute IL and power company Kemijoki Oy.

2 THE GOALS

The main goal of the project is in first phase to specify and construct a 150 kW wind turbine on the top of the Pyhätunturi fell. Former experience gained during the measurements on the fell and in running the wind turbine in Enontekiö has been used. The turbine is due to be erected in summer 1993. Thus the winter 1993-94 will be the first season for experiments.

The turbine is to become a continously (or as continuously as possible) working prototype. Its operational experience should help in avoiding problems in future wind farm projects under similar or almost similar conditions. The machine will be equipped with such means for avoiding the icing and low temperature problems that have been theoretically or in small-scale and laboratory experiments shown to be most promising.

3 THE TEST SITE

The test turbine will be erected on the southernmost top area of Pyhätunturi ridge of fells. The fell ridge is located in eastern Lapland, 67 °N. The highest tops of the ridge are about 540 m a.s.l. The site of the turbine has an elevation of 470 m a.s.l., which is about 30 m less than the height of the closest top (see Fig. 1). Except parts of this southern top the Pyhätunturi ridge of fells belong totally to a national park.

The access to the site is relatively easy since there are service roads upto the top due to the existing broadcasting and telecommunication stations and the skilifts. The local 20 kV grid is also near for the same reason.

The site is a top plateau area, which makes the construction and erection work and later modification, which require mobile cranes, possible. The plateau is, however, behind a 30 m higher top, which is about 500 m towards W-NW from the site. Although this location is not the best with regard to maximize the energy production, it gives a possibility to study the flow over and around the hill and it's effect on the turbine.

4 THE WIND CLIMATE ON THE LAPLANDS FELLS AND AT THE SITE

The wind measurements at the tops of the fells were started in 1986 to study the building climatology as well as the climatic limitations for the skiing resorts [4]. It can be said that the wind energy resources were found when the measurements were made at the Pyhäkero fell (700 m as), located in the western Lapland.

The annual mean wind speed at the synoptic satations, located on the low lands, is only 2.8-3.5 m/s [5]. The annual variation is very small. During the winter months ground level inversions are very numerous and quite strong

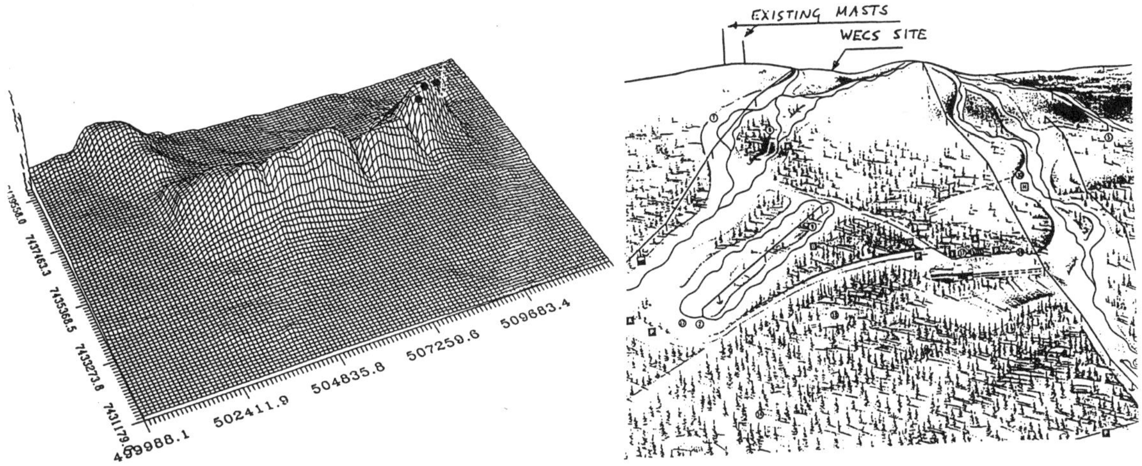

Figure 1. The topography of the Pyhätunturi ridge of fells (left) and the site of the arctic test turbine and the skiing resort seen from east (right).

[6],[7],[8]. Especially during the winter months the vertical increase of the wind speed is very strong [9]. The tops of the fells reaching 500 m above sea level are most of the time above the groud inversion and thus swept by the winds of the free atmosphere, where the geostrophic wind is about 9.5-10 m/s (see Fig. 2, [5]).

The annual mean wind speed on Pyhätunturi is 8,5 m/s. The maximum 10 minute average measured during the last three years has been 43 m/s (December 1992).

The annual variation of wind speed at heights above 500 m is very strong, very similar to the coastal areas of Finland (Fig. 3). This gives very high monthly wind speeds for the mountain areas. It is also very typical for the winter months that very high wind speeds, more than 25 m/s, are quite frequent. For example the mean wind speed during december 1992 at Pyhätunturi fell was 15 m/s.

5 RIME ACCRETION

Rime accretion takes place when supercooled droplets woving with the wind strike the surface (wings, tower etc.) and freeze. The rate of rime accretion is affected by water content and size of supercooled cloud droplets, wind velocity vector respect to the vertical surface of the object and collection efficiency of the object [10].

The formation of different types of rime increases with inreasing height. At the top of high fells (>700 m) rime formation is almost a daily problem during January-February. The number of rime days decreases significantly below 400 m above sea level, but unfortunately the wind potential also declines rapidly at these heights [9],[11],[12]. At Pyhätunturi the monthly number of rime days is 12-17 during November-March and the monthly cumulative rime accretion in January-March even as high as 250-300 kg/m^2.

Rime acretion specially on wings is a severe problem to the wind energy production at the fells. One possibility to avoid or at least to minimize this problem is to heat the wings. This method has been used succesfully to protect the anemometers against rime accretion [8],[13].

It is estimated that a heating power of 1.2 kW/m^2 could be enough even in Lapland [1]. If this power was kept on only during the time of rime formation the loss of the wind

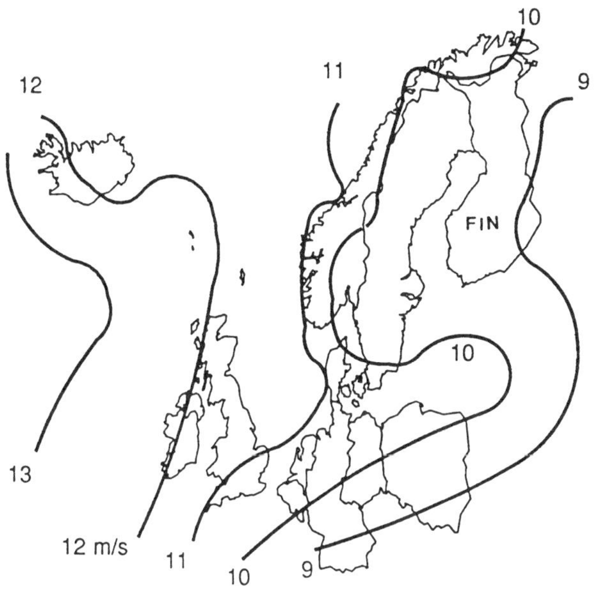

Figure 2. The annual mean of geostrophic wind speeds in northern Europe. [5]

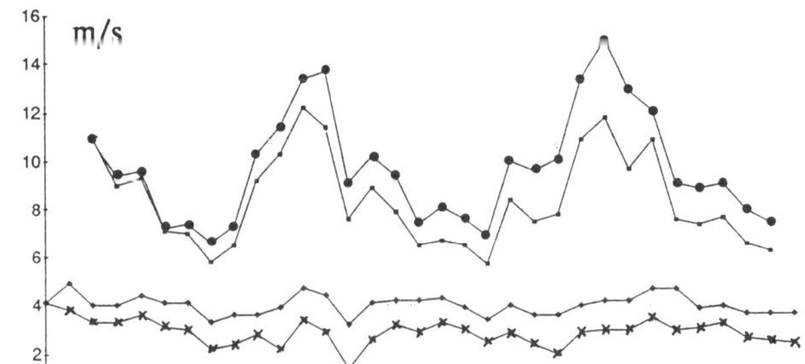

Figure 3. The monthly mean wind speeds during 1990-92 measured at Pyhätunturi and at nearby synoptic stations (Sodankylä observatory and Rovaniemi airport).

energy production would be only 10 % annually, if large WECS (200-500 kW) were used [12]. The loss of energy is relatively largest at the lower altitudes where the wind speed is smallest and the loss decreases with the altitude as the wind speed increases. For smaller WECS which are not using the high wind speeds the effect of altitude is somewhat different.

6 THE TEST TURBINE AND EXPERIMENTS

The test turbine will be a modified 150 kW turbine of Danish design. At first stage it will be equipped with blades with special coating and partially with heating elements. Some parts of the tower will also be surface treated. Other measures to ensure the safe use of the turbine include heating elements for some key components.

The measurements comprise of meteorological measurements and the measurements of turbine. The meteorological measurements will include

- wind and turbulence measurements in a 50 m mast using ice-free gauges
- visibility and weather measurements
- weather and cloud height monitoring with a video
- icing measurements using RMC-method (Rotating Multi-Cylinder)

The main elements (wind speed and direction, temperature) of the meteorological measurements have been carried out on the top of the fell since 1989. Later measurement points on the slopes have been added. The site will be equipped with a new wind measurement mast while the measurements on the top are maintained.

The turbine itself will be equipped with strain gauges on the blades, a video camera for monitoring the ice build-up and slip rings for data transmission from the rotor. The power production and blade heating demand are measured, too.

7 CONCLUSIONS AND FURTHER WORK

The arctic test wind turbine shall be operational in summer 1993. The regular operations and research tasks shall start immediately after that. During the winter 1993-94 we hope to get answers to the following questions:

- technical possibilities to wind energy production in arctic windy areas
- effectiveness of blade heating system and special blade surface treatment, alone and together
- control limits for the heating to keep the blades free of ice
- operations of the safety systems.
- more knowledge on the conditions when the measures against icing are needed.

For example the effect of ageing on the ice adhesion properties of the coating materials remains a longer term task.

The test turbine together with the supporting facilities are to be used as a research object for five years after which it will, hopefully, remain in normal operation.

An international meeting, BOREAS II, will be held at the site in spring 1994.

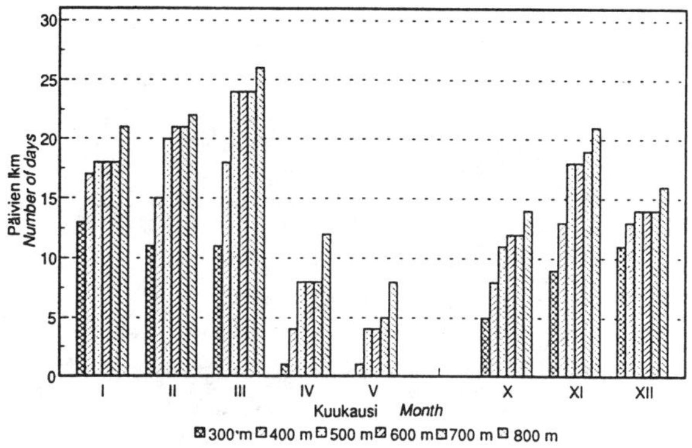

Figure 4. Calculated mean number of rime days M_d at different heights in Sodankylä in 1987-89.

REFERENCES

[1] Böhmeke, G. Wind power plants in the weather conditions of Northern Finland, VTT Research Notes 1352, 86 p + 8 app.

[2] Böhmeke, G. The development of wind technology for Finnish climate, Wind Energy Technology and Implementation. Proceedings of the EWEC '91. Amsterdam, The Netherlands, October 14-18, Elsevier Science Publisher, 1991, pp. 114-117.

[3] Böhmeke, G and Peltola, E. Icing test and rework of WECS for use on Laplands fells. Proceedings of BOREAS, Hetta 10.-13.2.1992, Helsinki, NEMO, FMI, VTT, 1992. pp. 219-232.

[4] Tammeling B. Orografian aiheuttamista ilmastollisista eroista Saariselän alueela (Microclimatological differences due to orography in Saariselkä). Meteorological publications, Finnish Meteorological Institute No 7, 1988, 67 p.

[5] Tammelin, B. Suomen tuuliatlas - Vindatlas för Finland - The Finnish wind atlas. The Finnish Meteorological Institute, Helsinki, 199, 300 p.

[6] Huovila, S. and Tuominen, A. Some features of ground inversion in Finnish Lapland. Proceedings of BOREAS Hetta 10.-13.2.1992, pp. 144-152.

[7] Tammelin, B. Väderläksförhållanden vintertid i Laplands fjällområden. 17. Nordiska Meteorologmötet, Helsinki, 1990, s. 281-293.

[8] Tammelin, B., 1992. The climate of the fells. Proceedings of BOREAS Hetta 10.-13.2.1992, p. 284-290.

[9] Tammelin, B. Wind energy in the northern Finland. European Community Wind Energy Conference. Proceedings of an International Conference held at Madrid, Spain 10-14 September 1990, p. 101-105.

[10] Makkonen, L. and Autti, M. The effects of icing on the wind turbines. Wind Energy Technology and Implementation. Proceedings of the EWEC '91. Amsterdam, The Netherlands. October 14-18, Elsevier Science Publisher, 1991, p. 575-580.

[11] Tammelin, B. and Säntti, K. Huurrekertymät tunturien lakialueilla - Rime accretion on the top areas of the fells.Meteorological Publications No. 19, Finnish Meteorological Institute, 1992.

[12] Tammelin, B. and Säntti, K., 1993. Huurteen vaikutus tuulienergiatuotantoon tunturien lakialueilla. (The effect of rime accretion upon wind energy production at the fell tops). Meteorological publications No 23, Finnish Meteorological Institute, Helsinki. (in print).

[13] Tammelin, B. Experiences of wind gauges used for measurements in cold regions. Proc. of BOREAS Hetta 10.-13.2.1992, pp. 241-261.

NÄSUDDEN — A 3 MW WIND POWER PROJECT IN SWEDEN

Jan Blix, Vattenfall Ab, S-162 87 Vällingby, Sweden
Anders Wickström, Kvaerner Turbin AB, P.O. Box 1005. S-681 01 Kristinehamn, Sweden

ABSTRACT: Since 1989 the construction and manufacturing of two 3 MW wind turbines has been going on in the premises of Kvaerner Turbin AB in Sweden. One unit has during August 1992 been erected in Wilhelmshaven in Germany and the other was placed on top of its tower in Näsudden on the island of Gotland in Sweden in Ocotber, 1992.
This development project has illustrated the importance of having well defined load cases and other criteria at the startup of the construction work. Changes at a late stage of the project should by all means be avoided. In this project discussions and calculations of relevant load cases have continued despite the fact that the construction work was started. The brake had to be redesigned from a parking brake to an emergency brake. There has been late changes in the lightning protection system.
An extensive evaluation and test programme is foreseen for the Näsudden II Wind Turbine. This will also include a comparison with Aeolus II, the German sister machine, which has a different control system and operates with variable speed by means of a thyristor converter. Parallel to the completion of the Näsudden II-project a development study for the third generation of MW-sized Wind Turbines has been initiated. It aims at further simplifying and customizing the present design for series production and thus bringing down the costs substantially. The study is subdivided into three phases with intermediate decision points whether to carry on or not.

1. BACKGROUND

During the winter 1988/1989 decisions were taken in Germany and Sweden to construct and build two MW-prototypes, one in each country. The construction should be based on a development study, which had been carried out in cooperation between German and Swedish interests. The contracts, one for each prototype, were signed between the same parties, that had participated in the study, i.e. Preussen Elektra and MBB for the German prototype Aeolus II and Vattenfall and Kvaerner Turbin AB for the Swedish prototype Näsudden II. MBB and Kvaerner Turbin AB have been acting as subcontractors to each other in the respective projects. Financial support has been given to both projects, in Germany by the BMFT, the state of Niedersachsen and the EEC and in Sweden by the Swedish Naitonal Energy Administration and the Swedish Energy Development Corporation. The latter organisation is owned by utilities and power producers in Sweden.
In Sweden a 2 MW - prototype, Näsudden I, was erected in the beginning of the 80:ies. It was in operation for more than 11000 hours and produced 13 million kWh before it had to be taken out of service in 1988. The reason for that was the detection of a major fatigue crack in the load carrying steel spar in one of the blades. After it was taken out of service the Näsudden I machine has been dismantled and the same tower is now used for Näsudden II.

2. TECHNICAL DESCRIPTION

2.1 General

In the construction of the new prototypes a lot of experience has been drawn from the construction and operation of Näsudden I. Certain design features are retained while others have been considerably changed. The rated output, 3 MW, is 50% higher than for Näsudden I, while the weight of the complete rotor, nacelle and machinery has been reduced with almost 20%.

2.2 Blades

The two bladed rotor has a diameter of 80 m and the blades are full span pitch controlled. The blades are built in carbon fibre and glass fibre reinforced plastics and the weight of each blade has been reduced to 9.5 tons, compared to 21 tons for the Näsudden I blades.

2.3 Nacelle

The nacelle consists of a foundation ring and a beambed built up from welded steel, over which a partly sound-absorbing cover has been placed. The beambed supports a two-step planetary gear with a bevelled endstage for connection to the vertically placed generator. The primary shaft and the turbine bearings are integrated with the gearbox.
An emergency brake capable of braking the rotor from 30% overspeed is placed on the primary shaft.

2.4 Electrical system

The main components for Aeolus II and Näsudden II are essentially the same but the electrical system and the control system differ. The Aeolus II will be operated with variable speed by means of power electronics while Näsudden II will be operated with two fixed rotor speeds, 14 and 21 rpm respectively. For Näsudden II a double wound asynchronous generator with increased slip is used. The generator is mounted vertically at the top of the tower with the stator fixed to the tower, while the rotor participates in the yawing motion of the nacelle. In this way the cables for the produced power can be fixed to the tower all the way up and sliprings are unnecessary.

2.5 Control system

The microprocessor based control system consists of standard industrial programmable controllers (PLC). In order to simplify operation and maintenance a Man-Machine-Interface system is used. This system includes terminals with colour screens both in the nacelle, control room and the remote control centre.

2.6. Noise

The operation with a lower rotor speed at low wind speeds reduces the noise impact on the surroundings.

COMPARISON BETWEEN NÄSUDDEN I AND NÄSUDDEN II

Näsudden I		Näsudden II
2 MW	Rated Power	3/1 MW
75 m	Rotor diameter	80 m
25 rmp	Rotor speed	21/14 rpm
21 tons	Weight of one blade	9.5 tons
78 tons	Weight of rotor	40 tons
210 tons	Total weight above tower	170 tons
5.0 GWh	Annual energy production (Availability 90%)	7.3 GWh

3. ASSEMBLY AND ERECTION

3.1 Workshop assembly of nacelle

The workshop assembly took mainly place during the winter 1991/92. The complete nacelle was mounted and the activity was followed by the workshop testing. The only remaining assembly on site was the mounting of the blades.

3.2 Transportation to site

The nacelle was transported to site in May 1992. The transportation was made by trailer a short distance form the workshop to a nearby harbour in Trollhättan. A barge carried the machine to the island of Gotland and the last distance to site was once again executed by trailer.
The blades arrived to site from MBB in Germany in the end of September 1992.

3.3 Uplift of the nacelle

The uplift was performed by means of a special lifting sledge, constructed to slide on rails on the concrete tower. After the attachment of the sledge to the tower, the nacelle was lifted, by cranes, from the trailer to the sledge. In this position, the blades were mounted.
The sledge, holding the complete unit, was then hoisted to the top of the tower by means of hydraulic jacks climbing on hanging wires. This operation took about 8 hours. Being at the correct height, the complete nacelle was tilted about 90 degrees to a horizontal position and adjusted to the screw joint on top of the tower. The tilting operation took about half an hour.
The same lifting sledge was used for lifting of Aeolus II in Wilhelmshaven.

3.4 Present status

The first rotation is expected to take place in the beginning of March 1993 and the first connection to the electrical grid about a fortnight later. A 500 hour test operation period should take place in April/May and the date for take-over is set to 1993-07-15. This implies that the time delay compared to the original time schedule is somewhat more than one year.

4. EVENTS AND EXPERIENCES IN THE PROJECT WHICH HAVE CAUSED DELAYS

4.1 Organisation

The organisation and formal procedures around the project have been complex and time consuming.
There has been two main contractors, one for the Näsudden II machine and an other for the Aeolus II. The contractors have been responsible for differents parts of the machines. The customers for each machine has referred to different regulations and organised different approval authorities.
For all major decisions in the project, these participants must be united for the approval.

4.2 Wind and load situation

The blades, made of carbon- and glassfibre reinforced epoxy, are not dimensioned by fatigue but from buckling at emergency stop. The loads from that occurrence were easily determined. Other load cases for the static dimensioning were also well defined and given in the contractual Technical Specification.
For the fatigue loads, MBB had used a simple wind assumption for the design of their Monopteros machines which would cover all other cases:
Wind shear 0.25 m/s/m at rated wind speed for 2.0e+8cycles.
This assumption was accepted by Germanischer Lloyd.
For a 2-bladed rotor with stiff hub and a diameter of 80 meters, this wind assumption is considered to be too rough and severe. Especially for subsystems which experience the sum of the loads from each blade, such as the beambed of the nacelle, the assumption leads to unrealistic load variations. Thrust and drive torque variations are to some extent neutralized whereas other load components are amplified. An improved model for the fatigue loads was needed. Additionally, any fatigue load assumption had to be verified against the Technical Specification and approved by Vattenfall and Germanischer Lloyd.
From the wind assumption point of view, the Technical Specification was an inheritance from the Näsudden I project. During the project it was found that it did not give all data needed for the fatigue load calculations. Additional research papers were studied and used in the verification of the wind shear and gust spectrum which was finally decided and agreed upon.
As these problems were observed during the design phase of the project, a lot of time and effort had to be put into this activity. This affected the time schedule of the whole project negatively as the design and manufacturing had started before the final loads were established. Redesigns in a late stage had to be done for some systems.

4.3 Emergency stop philosophy

From a concept study carried out prior to the project, it was concluded that the full span blade pitch system was fulfilling the requirements for emergency braking. The redundance in the hydraulic system was taken into account and accepted.
During the design phase, the redundance in the mechanical system was questioned and not accepted by all parties.
Due to this, the mechanical brake on the primary shaft, originally designed as a parking brake, had to be redesigned to an emergency brake. As this change of philosophy wasn't decided until the design work had progressed for some time, it had a substantial influence on the time schedule and the cost of the project.
The brake has the capacity of braking the turbine at simultaneous pitch blocking and loss of grid at 25 m/s.

4.4 Rotor speed philosophy

From the concept study of the Näsudden II project, it was decided that the machine should operate at two constant rotor speeds.
In a late stage in the design phase, questions of the rotor speed concept started. Reinvestigations in this issue were carried out and many discussions were held. This also affected the original time schedule.
Based on these investigations and discussions, it was decided to implement a variable rotor speed concept for the Aeolus II machine whereas the two constant rotor speeds concept remained for Näsudden II.

4.5 Lessons to be learned

The events and experiences mentioned above are not unique for this project but can be seen in a number of other large development projects, even apart from wind power technology.
Based on the experience from Näsudden II, we have for the work with the next generation of large WECS in Sweden learned the importance to:
- Define wind conditions, loads and caculation algorithms before start of the design.
- Collect views and acceptance from all parties involved in the design certification before start.
- Maintain continuity in the developing process.

5. EVALUATION AND TEST PROGRAMME

5.1 Production Results and Operating Experiences

Concurrently with the operation the Näsudden II wind turbine will be evaluated and assessed by Vattenfall with respect to its qualities as a power plant. For this purpose a continuous recording of production data and of measures taken to operate and maintain the wind turbine will be made. The measurements, recording of data and generation of reports are made both by the control system and by a separate measuring and data acquisition system, which has been installed specifically for the evaluation.

5.2 Technical evaluation

One objective of the technical evaluation is to investigate the power performance by measuring and calculating power curves, efficiency curves and losses in different operating modes. Another objective is to evaluate loads, dynamic behaviour and expected life. The results of this might initiate a revision of the relevant load specification.
Measurement of the electrical characteristics of the delivered power will take place to assess the quality as described in the IEA Recommended Practices: "7 :Quality of Power Single Grid-Connected WECS".
The control system performance will be tested at for instance starts and stops, changes between 1 and 3 MW generator winding, and at turbulent winds.

5.3 Environmental impact

Measurements of electromagnetic interference will take place and comparisons will be made with results from earlier measurements at Näsudden.
Noise measurements will be made in accordance with the IEA Recommended Practices for Wind Turbine Testing and Evaluation: "4. Acoustics Measurement of Noise Emissions from Wind Turbines".
The Näsudden II wind turbine will not be equipped with any ice detectors. The only way to note any formation of ice is if it occurs when there is personnel at the site, or if the operation is disturbed in any way. One indication of icing conditions might be differences in the signals from the two wind speed sensors.

5.4 Comparison with results from the evaluation of Aeolus II

Aeolus II will be operated by Preussen Elektra. The wind turbine is going to be evaluated with respect to production results and operating experiences by Preussen Elektra and with respect to loads and stresses by MBB. A common programme for comparisons of the results from the evaluation programmes for the two prototypes has been discussed but so far no decisions have been taken. In order to facilitate comparisons and the exchange of measurement data from the measuring systems of the two wind turbines, there has however been an ambition to make them as similar as possible with respect to sensors and sensor locations.

6. DEVELOPMENT STUDY III

Parallel to the completion of the Näsudden II-project a new project "Development Study of a large Wind Turbine System, Generation III". (Development Study III) has been initiated. Development Study III is performed in collaboration between Kvaerner Turbine AB and Vattenfall AB with participation from specialists within the Swedish wind energy research programme.

The study has been divided into three phases with gradually increasing accuracy, working time needed and costs. Each phase shall be finished and evaluated before the next phase is begun. In this way there are two check points where the project can be interrupted if the work so far shows that the aim of the development study can not be fulfilled.

* PHASE 1, GENERAL CONDITIONS

During phase 1 the general conditions for the whole study shall be agreed upon. The main items are:

- Wind and load description.
- Design requirements.
- Allowed costs.

* PHASE 2, DESIGN STUDY

A design study, a preliminary design and cost estimations shall be carried out in order to be able to choose the most competitive design.

* PHASE 3, DETAILED DESIGN

The design work on the chosen design shall be refined, resulting in design solutions for all sub systems. The manufacturing costs per unit shall be calculated for a small wind farm and with the assumption of series production.
The aim for the complete Development Study III is to reach a design for the third generation of large wind turbines which can produce electricity at a competitive cost.

6.1 Results from phase 1

Phase 1 was completed in December 1992 and created a well founded platform for the following work in phase 2 and 3. Among the most important results the following can be mentioned:

6.1.1 Loads and technical requirements

A document "External Conditions and Load Cases" based on the IEC/TC 88 paper "Safety of Wind Turbine Generator Systems" was developed. The document is adapted for Swedish conditions and supplemented with some additional conditions, not specified by the IEC. Operation in a wake is accounted for through a wake deficit function were the wake wind speed field is determined by the distance between WECS in a farm, the position of the wake centre line and empirical values of the wind speed reduction in a wake. Thus, the most important loads during wake operation are generated when the studied turbine is operating in a partial wake and not by the increased turbulence level in the wake.
In the study a Technical Specification is made up. In this specification it is stated that the fatigue load calculations shall be made according to a method developed by Teknikgruppen AB. The fundamentals for this method are the time simulation programme for wind turbines VIDYN (ref. [3]) and the wind matrix based on long term wind measurements giving a RFC (Rain Flow Count) evaluated description of the wind speed variations in time and space (ref.[4]).

6.1.2 Economy

A target cost for the WECS in Development Study III, 3,00 SEK/kWh/year, has been established. Based on this cost, cost goals for the different subsystems in a WECS have been calculated.

6.2 Phase 2 and 3

Discussions are presently taking place about the continuation of Development Study III with phase 2. The current plans are to start the work in phase 2 during April 1993 and finish in the autumn 1994, whereafter a decision for phase 3 should be taken.

REFERENCES

[1] "Experience from the Swedish 2MW wind turbine at Näsudden and development of the next generation of LWTS." Paper presented at the EWEC '89 conference in Glasgow, 1989.

[2] "The realization of the 3MW wind power project Näsudden II in Sweden." Paper presented at the ECWEC '90 conference in Madrid 1990.

[3] Ganander H. Fatigue loads and fatigue spectrum due to wind, based on a wind matrix and dynamic calculations. IEA R&D WECSAnnex XI "Workshop on Wind Turbine Fatigue" at Risø Danish Nat. Lab., April 1991.

[4] Bergström H., Ganander H., Johansson H. Wind description for designing WECS, based on RFC evaluated wind measurements, European Wind Energy Conference, Amsterdam EWEC '91, 1991.

THE DESIGN OF THE NEDWIND 50

L.F. Drost, J.N. de Jong, J. Olthoff
NedWind Rhenen bv
P.O. Box 118, 3910 AC Rhenen
the Netherlands

G.A.M. van Kuik
Stork Product Engineering
P.O. Box 379, 1000 AJ Amsterdam
the Netherlands

ABSTRACT: The NedWind 50 is the first stall regulated one-megawatt wind turbine, to be erected in autumn 1993. The turbine is the prototype of a commercial series: it is not an innovative test turbine with promising but experimental systems. In the design, the state of the art of the 500 kW class of turbines is applied on a megawatt scale. The paper describes the important design features, and some special aspects of the rotor and nacelle construction.

1. Introduction

In 1990 NedWind started a discussion with the E-utilities EGD and PEB about designing and installing a 1 MW wind-turbine. Together with NOVEM and the European Commission DG XVII NedWind investigated the funding of this project. This resulted in the start of the NedWind 50 project in 1991.
Two prototypes will be built. A special design team was formed. Members of the Energy Research Centre ECN, Stork Product Engineering and Technical University of Delft assisted the NedWind designers and engineers.

2. Design history

In 1990 preliminary calculations showed that the 1 MW size wind turbines on the long run should be competitive with the existing medium size wind turbines.
The target for the team was to design a commercial wind turbine, to manufacture and to install the two turbines in 1993 / 1994.
Phases of the development programme:

- start design: March '91
  based on:
  . functional specifications
  . certification criteria
  . low development risk
  . NedWind 35 state of the art
  . standard components
  . low noise level
- start engineering: July '92
- start manufacturing: March '93
- installation first turbine: October '93
- demo: '94
- measurement programme: '93 / '94

In a parallel traject a new 500 kW wind turbine, the NedWind 40, was developed based on the design experiences of the NedWind 50 and the operational data of the NedWind 35.
The first NedWind 40 has been installed in December 1992 The operational results are very satisfying. A measurement programme has been started in January 1993 and the first results affirm the design data.
During the design phase it appeared that upscaling of the NedWind 35 was not possible in all aspects. Major changes are the full span pitch instead of tip pitch, the bedplate - and the drive train construction, as described in the next sections.

3. General description of the NedWind 50

3.1. Main characteristics

Figure 1 shows a sketch of the turbine.
It is a classical stall turbine, except for the number of blades (2 instead of 3). Furthermore the drive train construction is unusual. A survey of the main characteristics is given in table 1.

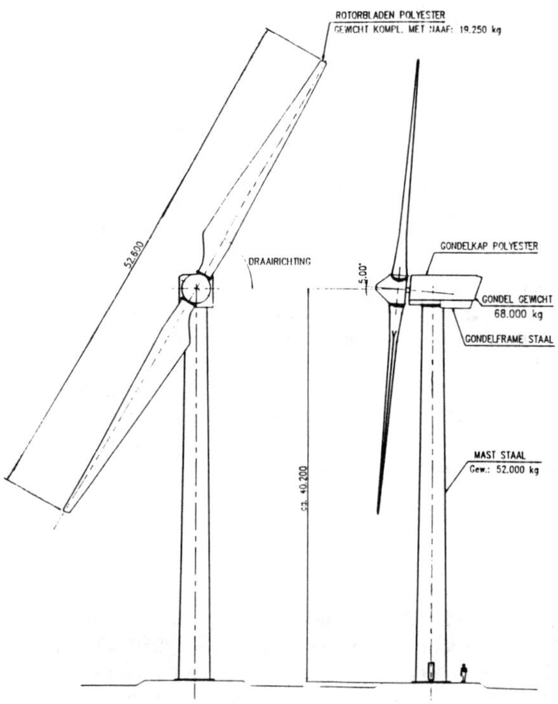

Figure 1: sketch of the NedWind 50 (general layout)

Two prototypes are under construction. The first one is scheduled to be erected in the autumn of 1993 near the Eemshaven in the province of Groningen. The second one will be commissioned a few months later at the location of the SEP wind farm in Friesland. At the turbine near the Eemshaven a measurement program will be undertaken, see section 5.

Performance characteristics	
Rated power	1 MW
Cut-in wind speed	5 m/s
Cut-out wind speed	20 m/s
annual yield at Eemshaven	2.3 GWh/y
Rotor	
Number of blades	2
Diameter	52.6 m
Rotation speed	24.65 rpm
Low rpm (option)	16.3 rpm
Design lambda	8
Twist	10 degrees
Rotor position	upwind
Tilt angle	5 degrees
Cone angle	0 degrees
Hub height	40.2 m
Hub	cast cylinder
Blade material	GRP
Blade section	NACA 636 and 646
Target source level at low rpm	96 dB(A)
Nacelle and drive train	
Transmission	parallel gearbox with 4 fast shafts
Generators	4*250 kW induction
Brakes	4 on high speed shafts
Construction	brakes & generators mounted on gearbox
Suspension	3 point elastomeric: 2 for gearbox, 1 for main shaft
Top mass incl. rotor	68 tonnes
Yaw system	
Type	plastic slide bearing activated by cable & intermittently operating hydr. cylinders
Tower	
Type	conical, steel
Height	38.5 m
Diameter foot	3.325 m
Diameter top	2.2 m
Tower mass	52 tonnes
Safety system	
First system	blade pitch to stall
Second system	4 mechanical brakes on high speed shafts

Table 1: Main characteristics of the NedWind 50

3.2. Rotor design

The NedWind 50 rotor design is very similar to the rotor design of the 500 kW turbines NedWind 35 and 40, a two bladed rotor with a relatively high design tip speed ratio of 8. The tip speed is equal to the NedWind 35 for reason of noise emission. The choice of the diameter is a matter of an optimization of the cost-/benefit ratio. The choice of the aerofoils is conservative: the use of specially designed aerofoils on a megawatt turbine, introduces too much risk. The 250 kW NedWind turbines use NACA 64-6xx sections with very good experience. These are used for the NedWind 50 on the outboard.

On the inboard section the NACA 63-6xx is used, because a better behaviour in deep stall. Furthermore the maximum thickness at 30% of the chord fits very well on the circular blade root construction. The most important reason for choosing full span pitch control instead of tip control is the possibility to adjust the maximum power and to optimize the power curve.

3.3. Safety system

The full span pitch control is used to slow down the rotor for operational as well as emergency stops. The direction of the pitching is towards stall. When the rotor speed is very low, the mechanical brakes bring the rotor to a complete stop. In case of grid failure, the blade pitch mechanism and the mechanical brakes are activated simultaneously. The four mechanical brakes are capable of stopping the rotor up to 38 m/sec.

3.4. Performance

The hub height of the NedWind 50 is chosen as low as possible, the increase of the annual yield for higher hub heights is balanced by the increase of turbine costs. For both prototype locations a hub height of 40 m appeared to be enough.
Figure 2a shows the performance for a coastal location with (Weibull) k = 2.06. The PV curve shown is calculated with an assumed drive train efficiency of 90%, and with the blade pitch fixed. Using the blade pitch to optimize the power curve, meanwhile maintaining stall as the power control mechanism, increases the annual yield significantly. This is shown in figure 2b. This performance will be tested.

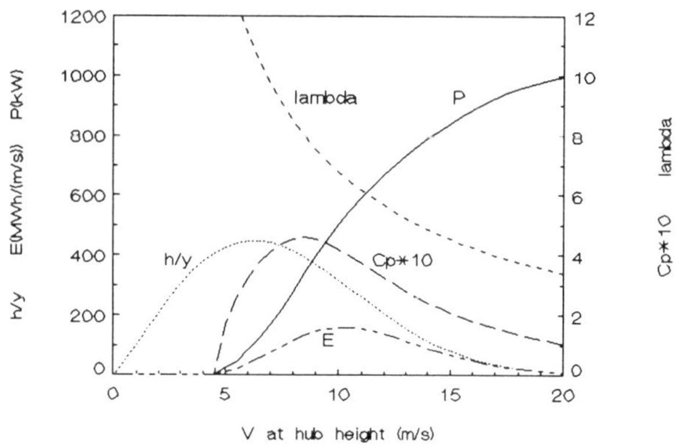

Figure 2a: Performance of the NedWind 50 with 1 pitch angle

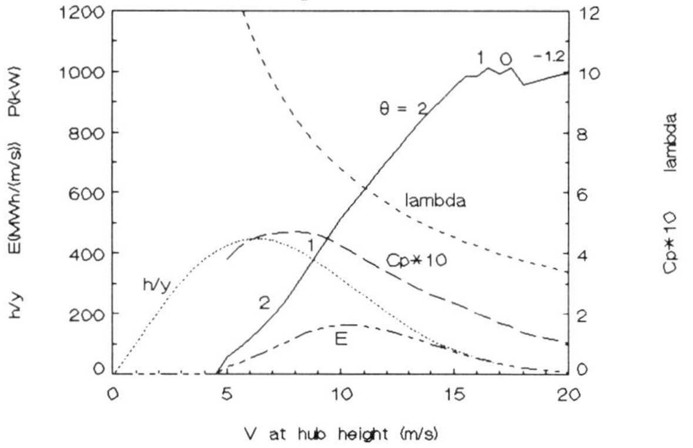

Figure 2b: As figure 2a, for optimized tip-pitch angles

3.5. Comparison with other turbines

Although the NedWind 50 is the first stall regulated one-megawatt turbine, it is intended to be a state of the art turbine. A comparison with other large turbines is shown in figure 3, taken from the CEC Post WEGA study, where the tower top mass is plotted against the rotor diameter. The NedWind 50 tower top mass fits very well.

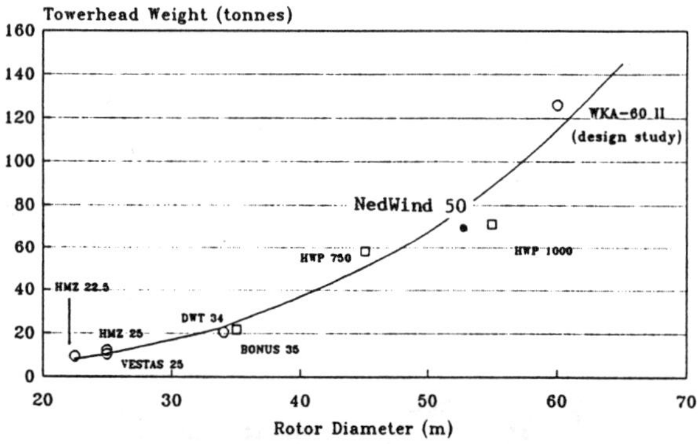

Figure 3: Comparison of tower top mass with other large turbines.

4. DESIGN FEATURES

4.1. The aerodynamic design

Two problems in designing a stall rotor will be highlighted: the required 3-D correction of the aerofoil characteristics, and the avoidance of stall induced vibrations. It is well known that the 2-D performance data of aerofoils are inadequate when the flow has a strong 3-D character. This is the case at the blade root where the stall is most pronounced and the flow has a radial component, and at the blade tips where the flow is essentially three dimensional.

In both regions quasi 3-D aerofoil characteristics have been derived, based on experimental data available in the literature.

Stall induced vibrations have been observed on some stall turbines. This phenomenon has been analyzed, and is believed to originate as follows. The frequency of the rotationally sampled turbulence is concentrated at integers of the rotor frequency (nP). If the rotating flap-eigenfrequency of the blade coincides which such an integer, the excitation by the turbulence can induce large vibrations, specially when the aerodynamic damping of the flap movement is very low due to stall. Simulations for the described situations indeed reproduced the stall induced vibrations. This leads to a design criterium for the rotating flap frequency of the blade: this should be far enough from nP. For the NedWind 50 blade, this frequency becomes 3.5P.

4.2 Rotor blade construction

The rotorblade construction is defined by the desired manufacturing method at NedWind Heerhugowaard, which is automated impregnation and machine lay-up in open mould halves. This means that the structural material will be situated in the skins which forms the blade halves.

To achieve the desired fatigue strength and static strength a non woven multi axial glass fibre reinforcing fabric was chosen. This material combines high fatigue strength with good handling characteristics. With the given fatigue spectrum the desired thickness of the skin was determined.

Finite strip calculations were carried out to predict the buckling stability of the skin under compressive loads.

As a result of these calculations two shear webs were considered to be necessary. These shear webs act as stiffeners against buckling, meanwhile carrying the crushing and shear loads.

To prevent buckling as well as aerodynamic deformation the trailing edge section of the blade was constructed as a sandwich panel.

The blade root was designed as a pre-stressed bolted polyester steel connection.

The allowable design criteria for this type of connection were established in a rotorblade material test program, were extensive fatigue testing was performed on coupons as well as on bolted connections.

The bolted connection was chosen for its controlability and reliability during production and lifetime.

The resulting total polyester weight of the blade is about 3000 kg (figure 5). As the NedWind 50 is the largest blade so far to be used in stall, a flutter divergence calculation seemed appropriate. The torsional rigidity of the blade itself however appeared to be less significant than the stiffness of the pitch mechanism Flutter appeared not to be a design driving parameter.

4.3. The drive train

The rotor is carried by a 5.6 m long main shaft, supported at the front side by a mainshaft bearing and at the rearside by the gearbox. The length of the shaft is such that the gearbox is located behind the towercentre, seen in the wind direction.

The gearbox is suspended on two points.

The whole drive train including the rotor is thus supported on three points, giving a kinematically determined suspension, figure 4. The mainshaft bearing is mounted on four rubber Spherolastic bearings, the gearbox on two. This way the drive train has been acoustically isolated from the bedplate and the tower, limiting the contact noise from the transmission. The Spherolastic bearings are expected to have a lifetime of at least 20 years, but they can be exchanged if necessary without dismantling the turbine.

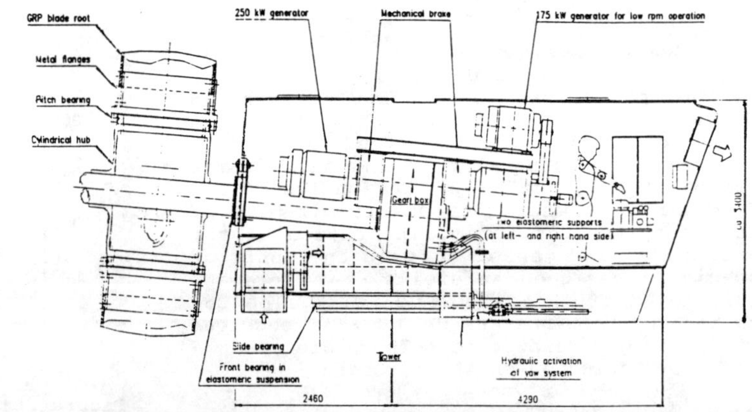

Figure 4: sketch of the nacelle, drive train and hub

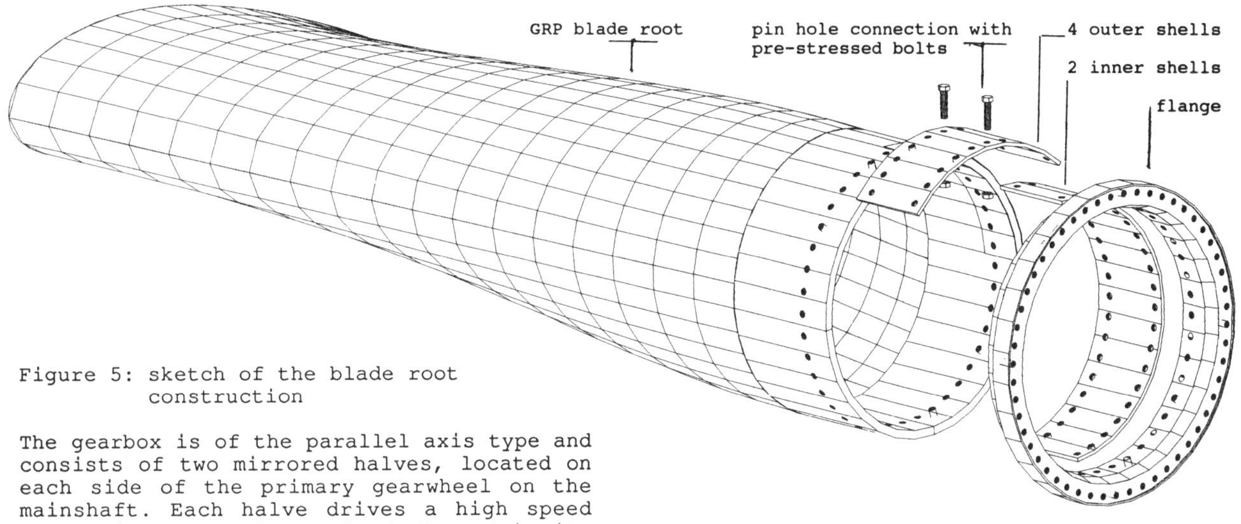

Figure 5: sketch of the blade root construction

The gearbox is of the parallel axis type and consists of two mirrored halves, located on each side of the primary gearwheel on the mainshaft. Each halve drives a high speed shaft with two ends. Each shaft end in its turn drives a 250 KW generator, one at the front and one at the rear of the nacelle, giving a total of 4 x 250 KW installed generator power. The low speed step of the gearbox in this way has become more or less similar to that of a planetary gearbox and could be designed considerably cheaper because the forces on opposite sides of the wheel are counteracting each other. The generators are mounted on coupling housings, which on their turn are mounted on the gearbox. This assures a simple and accurate aligning during assembly of the drive train. In the coupling housing disc brakes and a flexible coupling are mounted. The disc is located on the high speed shaft, driving a flexible coupling on the generator shaft. The brake is of the spring powered, hydraulically released type. The four brakes together have 16 callipers, mounted on yokes, giving a symmetric distribution of brake forces on the shaft ends. The unit of generator, coupling housing and brake is exactly identical to those used on the 500 KW NW40 turbine. A single generator and brake are used on the NW25 turbine which has an installed power of 250 KW, so the whole line of NedWind turbine models is built with identical generator/brake modules. With this modular set-up significant cost reductions in the purchasing of the generators and in the production of the turbines have been achieved. The generators are coupled to the grid by means of a thyristor switch of the 250 KW class. Once a generator is coupled to the grid the softswitch is switched through to the next generator. If the wind is strong enough this is switched on too and the softswitch is coupled to the third generator, etcetera. Because of this only one small softswitch and a number of relay switches are necessary. The 250 KW generator is a type that is produced in large quantities. Electrically there is no difference with the 500 KW model, but a 1000 KW model will be of the high voltage type. An advantage of one 1 MW generator would be the slightly higher machine efficiency. A disadvantage is that it is running more often in partial load, especially at low wind regimes. The bedplate is a welded construction of steel plate. It is basically an A shaped wedge with the mainshaft front bearing at the top of the A and the tower opening between the legs.

A closed box at the top assures the required stiffness against torsion. The legs of the A carry the gearbox. The rotor thrust is lead through the gearbox to the bedplate.
The hydraulic cilinder driving the pitch mechanism is located on the rear of the gearbox. A push-pull bar goes through the hollow main shaft and acts as a synchronising mechanism between the two blades as well. The rotation of the shaft is uncoupled by a thrust bearing in the nacelle, between the hydraulic cilinder and the push-pull bar.
The machine layout has been analysed in a number of ways. At first a drive train modal analysis has been made. Furthermore a Finite Element Analysis has been made of the bedplate. For a number of unit loads in the rotor centre stresses (so called unit stresses) and displacements have been calculated in the whole structure. A number of potentially critical spots have been selected and for these spots fatigue calculations have been made on the basis of linear combinations of unit stresses with the 130 load cases of the fatigue spectrum. The results of these calculations have been fed back in the design of the bedplate, finally resulting in a bedplate weight of 7.5 tonnes.

5. **The measurement program**

The program, which is part of JOULE-II, is aimed at improving the understanding of the behaviour of a large, stall regulated wind turbine. Four activities are distinguished:
1 installation of monitoring system
2 collection of detailed dynamic load and response data,
3 evaluation of the current design methods against measured data in order to improve the predictive tools,
4 verification of the NedWind 50 design.

The results of the activities 2 and 3 will be available for the wind energy community. The measurements will be done at the first prototype in Groningen, and will start immediately after the commissioning of the turbine. The partners involved are: NedWind (all activities), Garrad & Hassan (1,2,3), SPE and ECN (3,4).

THE FOLKECENTER 525 KW WINDTURBINE: DESIGN, MEASUREMENT SYSTEM, FIRST RESULTS

N. Vilsboell, P. Kunwald
Folkecenter for Renewable Energy
P.O. Box 208
DK-7760 Hurup Thy
Denmark

C. Tantareanu
Folkecenter for Renewable Energy
Denmark, and
Power Research and Modernization
Institute, Romania

1. BACKGROUND AND AIM OF THE MEASUREMENT PROGRAMME

In 1986 Folkecenter started the conception work for a 525 kW windturbine. The project was intended to demonstrate the industrial feasibility of a Danish standard WT concept (three-bladed rotor up-wind, stall regulation etc) on the 525 kW power level.

The project was jointly funded by the Commission of the European Communities, the Danish Energy Ministry and Folkecenter. The cost shares are:
CEC-DG XII..............2.00 mil DKK
Danish Energy Ministry...1.20 mil DKK
Folkecenter............ 2.25 mil DKK

The WT is part of the CEC measurement programmes on innovative WT concepts. Therefore the Denmark 36 windturbine was prepared and instrumented for a comprehensive test programme.

After the CEC funding approval the detailed design was performed. At the end of 1990 the siting was decided near the Hanstholm harbour.

On 18th September the 525 kW WT was connected to the grid. In October it has reached the nominal regime point expected for the actual set-up blade pitch angle (500 kW for -2.7 degrees angle) and has operated in the stalled regime (over the nominal wind speed) too.

As well as being a research pilot, the WT is supplying electricity to the grid in commercial conditions. The energy production in the October month, despite quite often stops required by the research work was on a satisfactory level: 79,000 kWh. In November 1992 the windmill produced 150,000 kWh.

The research measurement system is completely separate from the measurement system included in the WT control. It is built according to the European research standards in the field. During the spring and summer of the 1992 the measurement system (the instrumentation and its associated logging system) was assembled and tested in laboratory and since October was transferred to the real WT.

The results of the measurement project aim to validate and improve the calculation code of WT blades and WT structure. A particular attention will be given to the stresses on the yaw motors.

Because the site topography could be considered inhomogeneous as there is an important hill to the South direction, another objective is to investigate the dynamic loads and the WT performances according to different wind turbulence regimes.

2. BASIC FEATURES OF THE DANmark 36 WINDTURBINE AND OF THE HANSTHOLM SITE

The Denmark 36 windturbine is fairly representative in its main characteristics as a Danish WT. It is stall regulated, three bladed rotor with rigid hub, the rotor oriented up-wind, induction generator and a stiff tower. To mention the integrated gear box, no other exterior bearings being necessary in the machinery lay-out (see fig.1). A view of the WT is shown in figure 2.

Table no. 1 presents a technical basic description.

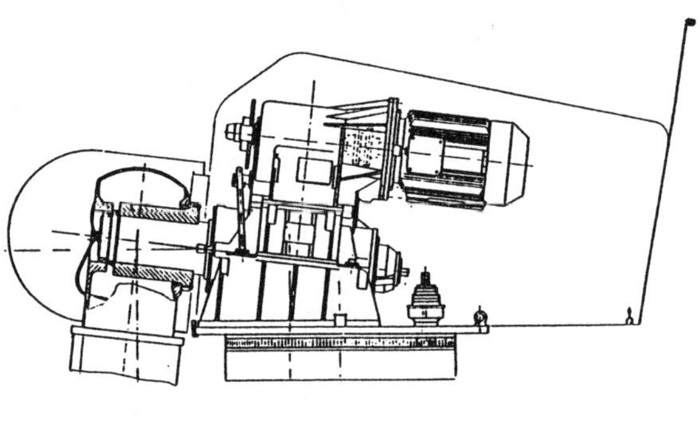

Fig. 1. The lay-out of the machinery

Fig.2 The 525 Kw prototype is situated near the Harbour of Hanstholm.

Table 1. Basic technical description of DANmark 36 wind turbine.

Characteristic wind speeds
- cut-in: 4 m/s
- rated: 16 m/s
- cut out: 25 m/s

Rotor
- 36 m diameter
- rotational speed 31 rpm
- NACA 63-2 blades profile with rotating tips (manufacturer LM)

Transmission (manufacturer Tacke)
- 3 staged gear box
- ratio 48.

Generator (manufacturer ABB)
- induction
- 400 V

Tower
- conical with a diameter from 3 m at base to 1.8 m at top
- height 38.8 m

Weights
- rotor (blade with hub) 9 tones
- nacelle (without rotor) 13 tones
- tower 26 tones

Safety systems
- mechanical disk brake at high speed shaft
- aerodynamic brake using blades tips pivotable at 85 degrees

Control system

The windturbine has a highly advanced control system and a new thyristor system allows, not only a soft cut-in to the grid, but a better generator efficiency at low loads through the control of the generator voltage.

The site

The WT is located on the Danish North West coast near the Hanstholm city and harbour, away from residential areas. The foundation is on the beach on some 200 m far from the sea. The surrounding terrain is flat (beach and sea) on the NW-NE sector. On the East and West sectors there is a beach with small irregularities. On the SW-SE sector is located an important 20-25 m height hill at about 300 m distance from the windturbine. This hill gives an irregular topography to the site and creates different operation conditions for the WT.

The measurement house has the dimensions 3 m x 5 m x 2 m and is located at 5 m East from the WT. It contains the electrical boards and the data acquisition equipment.

3. GENERAL DESCRIPTION OF THE MEASURING SYSTEM.

As basis for our measurings is used: ECN 217 (Recommendations for a European windturbine standard on performance determination).

All the measurements are collected with a sampling frequency of 25 Hz except the rotating speed of the generator shaft which is calculated 1 time per revolution.

The data are collected in bins of 10 minutes duration.

After this the measurements are stopped temporarily for calculation in order to calculate minimum, maximum, mean value and variance of all data which are saved. On basis of the mean wind speed and the wind turbulence it is examined if the bin contains new data, if not, they are cancelled, otherwise they are saved in a so-called "Capture matrix", see figure

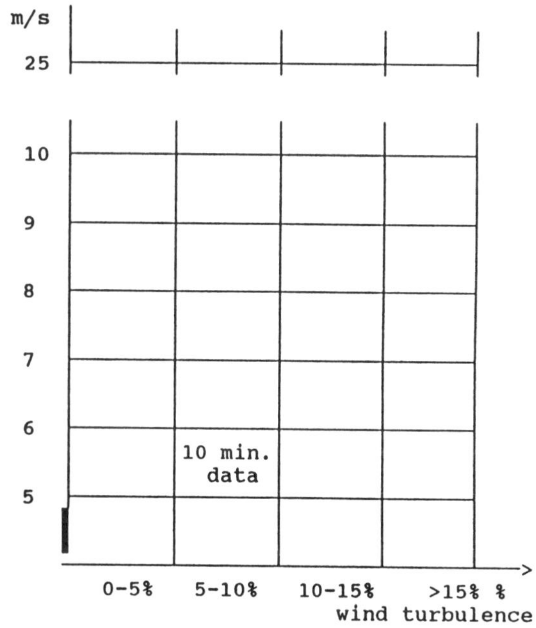

Furthermore rain flow counting for all the strain gauge measurements are carried out.

Hardware

As measuring computer is used an "IBM" compatible PC with various data collecting boards.

Measuring on electric parameters from the windturbine

As converter for measuring on the electric capacity of the generator is used a standard DEIF converter with a class accuracy of 0.5, except for the main effect converter which is a DEIF converter rebuilt to a response time which has been changed from 80 Ms till 20 Ms. This change has been done in order to be able to make FFT analysis of the effect signal of frequencies <10 Hz. However, this means some 50 Hz noise on the measuring signal, which is filtered away in a 5th order low pass filter.

The following converters have been used
- **Main effect : 0-3,000 W with current transformer 1,000/5**
- **Reactive effect : 0-3,000 Am with current transformer 1,000/5**
- **Aux effect : 0-2,500 W with current transformer 60/5**
- **Phase - phase voltage : 320-480 VAC**
- **Phase current : 0-5 A with current transformer 1,000/5**
- **Frequency : IMCON-FC5 45-55 Hz**

Measuring the meteorology

To measure the meteorological data a measuring tower has been situated 2 rotor diameters west of the WT, equipped with the following instruments:

In hub height (40 meters) windspeed, wind direction and temperature are measured.

Ten meters height (meteorological height) windspeed and wind direction are measured.

Two meters height (meteorological height) the temperature is measured.

Besides measuring on atmospheric pressure and digital measuring of fall of rain are carried out.

Calibration
- Cup anemometer is calibrated every 3 months
- Wind wane is controlled every 3 months
- The temperature is controlled by measuring with the other pt100 Resistance in the double sensor.

Measuring of stress
- On every blade root the bending is measured flap wise and chord wise together with the torsion,
- on the main shaft the bending is measured in the X and Y direction together with the torsion,
- the measuring signals are lead through a collector ring as a current signal to the main cross-field,
- on the generator shaft is measured the torsion, the measuring signal is lead via a capacitive telemetry system to the main cross-field,
- bending in X and Y direction is measured, together with torsion on the tower at the foot of the WT.

Calibration

All strain gauges measurements are being calibrated.

Other measurements

The blade and the nacelle position are absolute encoders carried out by a 10 bit gray coded each.

The speed of the generator shaft is measured by frequency measuring with input from an inductive sensor giving a pulse per revolution.

4. FIRST MEASUREMENTS

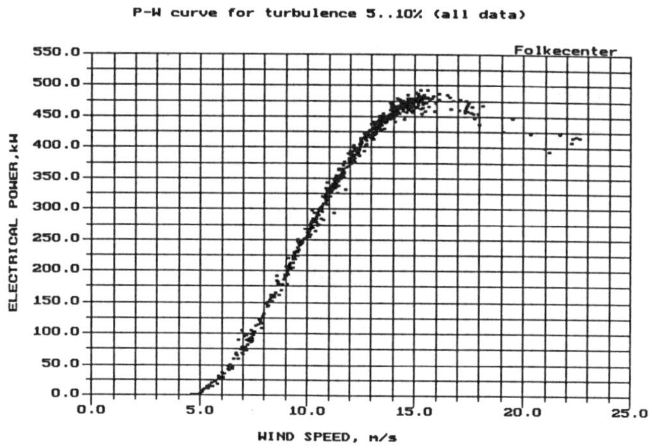

Above is the power curve, taken with a wind turbulence in the area of 5% to 10%. The curve is made with a tip angle set at -2.7 degrees, in the mean time we have changed this angle to -2.0 degrees which produced a power curve at a little over 500 kW.

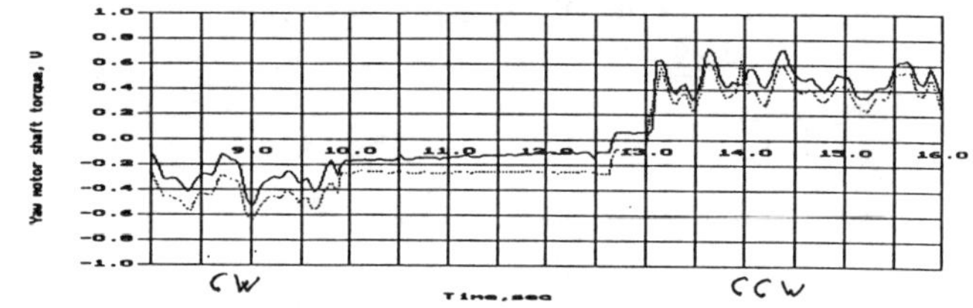

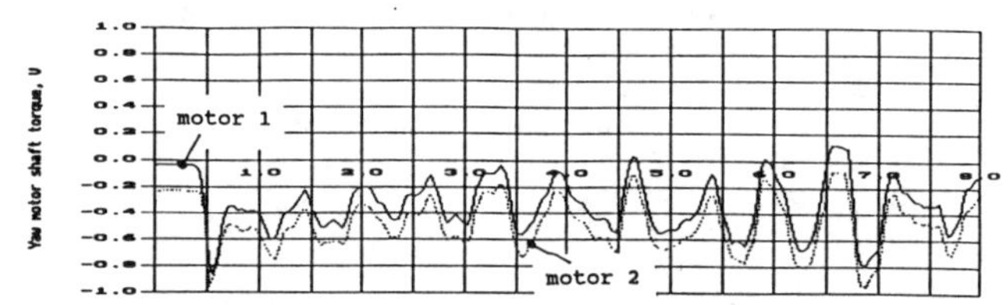

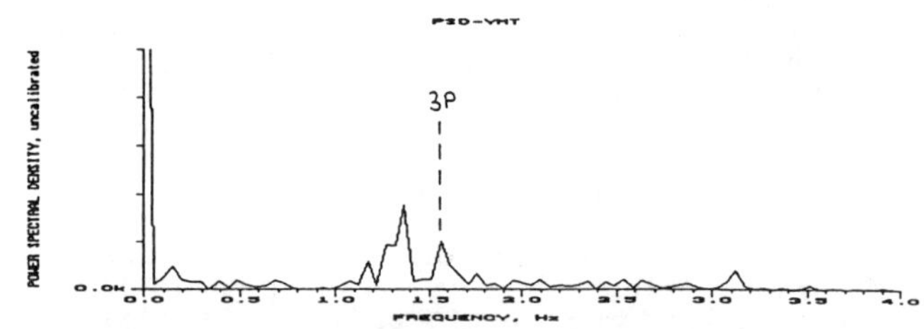

The torsion torque on the main shaft of the yawing system can be seen in the upper figure. First curve shows clockwise and counterclockwise yawing. At low windspeed a signal that resembles a unit step with a long settling time will occur, at stalling no settling will occur -see curve nr. 2.

FFT -analysis of curue 2 shows the frequncy spectrum. Observe that the shaft is strained with varying frequencies, with the frequency of the blade passing the tower (3p) pronounced.

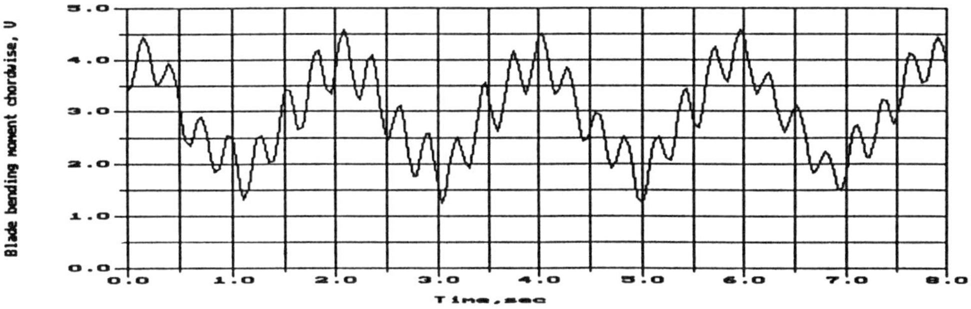

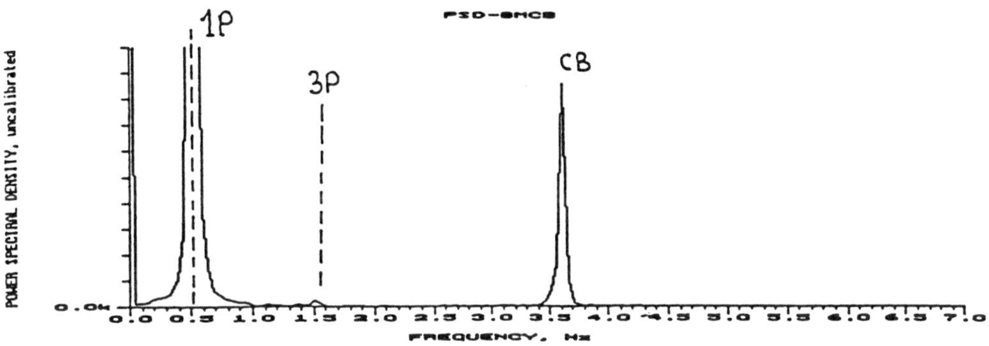

The above curve shows the cordwise bending moment of the blade, both in time and frequency domains. One can see clearly the frequency of the rotor (1P) and also the frequency of the tower bending

References

1. Kunwald P, Tantareanu C, THE RESEARCH MEASUREMENT ON THE DANMARK 36-525 Kw WIND TURBINE , Folkecenter November 1992.

2. Vilsbøll N, Kunwald P, Tantareanu C, FIRST MEASUREMENTS ON DANMARK 36 - 525 Kw WINDTURBINE, Folkecenter December 1992.

3. POWER PERFORMANCE TESTING , IEA , 2nd edition 1990.

4. Pedersen T F a.o., LOADS FOR WIND TURBINE IN INHOMOGENEOUS TERRAIN. Risø-M-2922,1991.

TEST RESULTS FOR FLEXTEETER
A FLEXIBLE ROTOR WITH FAST PASSIVE PITCH CONTROL

J.W.M. Dekker, C.M. de Groot
Netherlands Energy Research Foundation ECN, P.O. Box 1,
1755 ZG Petten, The Netherlands

F. Hagg, P.A. Joosse, G.A.M. van Kuik
Stork Product Engineering B.V., P.O. Box 379,
1000 AJ Amsterdam, The Netherlands

H.J. Heerkes
AERPAC B.V., P.O. Box 573,
7550 AN Hengelo,The Netherlands

ABSTRACT: A rotor with teeter, flexbeam and passive pitch control has been tested successfully. An instability which occurred during the test, was suppressed by installing a damper between the passive control tip and the fixed inner part of the blade. Afterwards no more instabilities were met. As a pre-requisite for cost reduction, appreciable load reductions were demonstrated. Yawing has been tested extensively. It does not constitute a severe load case for a flexible rotor, so that it can be used as an auxiliary safety system.

1. INTRODUCTION

A rotor called FLEXTEETER, equipped with an elastomeric teeter, a flexible beam and passive control tips has been tested at the '25m HAWT' rotor test facility of ECN (see fig. 1). This test was the final experiment of the FLEXHAT programme. FLEXHAT is a generic name for a cluster of research projects, with the objective to develop technology for the next generation of windturbines. It is funded by the Netherlands Energy Research Foundation (NOVEM) and the European Commission. SPE was responsible for the design, AERPAC for the manufacturing and ECN for the testing. At previous ECWEC and EWEC conferences [1],[2] the results of the FLEXHAT programme as a whole and test results of a rotor with passive control tips (20-WPX-PRT1) in combination with variable speed operation were discussed. The promising power control capabilities of the passive pitch rotor were shown at that time. In the present paper test results of the FLEXTEETER rotor will be discussed. This rotor has the same outer dimensions as an AERPAC 20-WPX-PRT1 rotor, but it is equipped with a fully flexible connection with the hub.

2. THE FLEXTEETER EXPERIMENT

2.1 Description of the FLEXTEETER rotor

The FLEXTEETER rotor is the last rotor to be tested within the FLEXHAT programme. The new element in this test is the flexible hub connection: an elastomeric teeter with passive teeter limiters and a flexbeam. There is no active control of the teeter motion. A complete description is presented in [1],[2]. Fig. 1 shows the rotor at the 25m HAWT rotor test installation. In order to facilitate easy adjustments of the limiter position and stiffness, a balcony has been mounted to the hub. Fig. 2 shows a schematic lay-out of the flexible hub construction. Geometrical data of the rotor, the natural frequencies and some information on the measurement equipment are collected in Appendix A.

The FLEXTEETER rotor operates normally at variable rotor speed. At a certain rotor speed, NLIM, the passive tips become active. Above NLIM the generator power is kept constant at the power limit, PGMAX. Below NLIM constant tip speed ratio is maintained by controlling the generator current. Occasionally experiments were made at constant rotor speed in order to compare the measured loads with data from other configurations of AERPAC 20-WPX blades tested earlier.

2.2 The aim of the experiment

A teetering rotor is not new, a teetering rotor with a flexbeam is not very usual, but a teetering rotor with a flexbeam, variable speed and passive control tips is unique. The dynamics of such a rotor are complex, which justifies extensive testing. In normal operation the limiters are not necessary: the aerodynamic rotor damping is enough.

Fig. 1 FLEXTEETER rotor mounted at the ECN '25m HAWT' rotor test facility.

However when (a part of) the rotor is in stall the damping is small, leading to larger teeter excursions. In these circumstances the elastomeric limiters are required. By modifying the distance between the limiters and the flexbeam (see fig. 2), the range of "free" teeter excursions can be changed. An increased range is favourable for load reduction, but unfavourable for damping of the teeter motion in stalled conditions. The first aim of the experiment was to find an optimum position of the limiters. A second important aim is the verification of the dynamic modelling in the design tools.

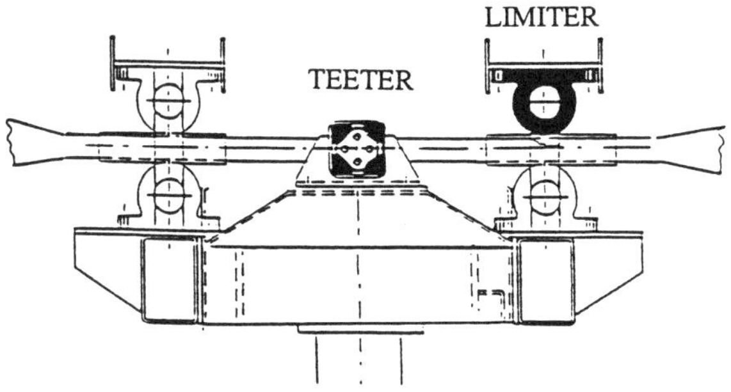

Fig. 2 Schematic view of the FLEXTEETER rotor hub.

2.3 Survey of the measurements

The following conditions were tested:

- Normal operation conditions, including starts and stops.
- Yawing. The passive tips safeguard the rotor against overspeeding (see discussion in [1],[2]), but cannot bring the rotor to standstill. This is done by slowly yawing the turbine out of the wind. Therefore yawed conditions have been analyzed as well.
- Off-design conditions: blocking of one tip.
- Power control by stall: this is done by applying a fixed rotor speed with the control tips not active.

3. STABILITY

During the first phase of the FLEXTEETER measurements a severe stability problem occurred: in several circumstances both tips oscillated violently near the natural frequency of the tip torsion mechanism, combined with an excitation of the second flap mode of the blades. By analyzing the rotating video recordings, it appeared that the tip rotation was about twice the rotation measured at the screw cylinder, which is the heart of the pitching mechanism. This implies that the rod, connecting the tip and the screw cylinder was distorted by high torsional loads, close to the strength limits. These problems occured mostly at higher windspeeds (>15 m/s) during start-up, emergency shut down and also at free run conditions. The position of the teeter limiters did have some influence on the occurrence and intensity of the instabilities. The common factor in the observed instabilities at higher windspeed is stall at the tips. During start-up this is the usual stall on the profile suction side (positive stall). In free-run at high wind speed conditions, the tips brake the rotor by producing negative lift, accompagnied by negative stall. The aerodynamic instability generating the heavy oscillations is somehow related with stall flutter. However this is a one-degree-of-freedom mechanism. Since the tip-torsion- and the second flap eigenfrequency are very close, the combined eigenfrequency was excited, leading to oscillations in flap and torsion. The flap eigenfrequency, and thus also the combined eigenfrequency, depends of course on the stiffness and position of the elastomeric teeter limiters. This explains the influence of these limiters on the occurence of the instability.

During one measurement run the limiters were effectively removed by mounting them with 4 cm free space at both sides of the flexbeam. This measurement was made at a moderate windspeed around 11 m/s with a power limit of 25 kW. Right after the start of this run a severe instability occurred, lasting about 3 minutes. Afterwards in the same measurement run of one hour no more instabilities were observed, with all the external conditions virtually unchanged. In these conditions however the blades were not in stall condition. In fig. 3 measured signals during the transition from instable to stable operation are depicted.

In order to suppress the observed instabilities the rotor was taken down for an overhaul. Two measures were taken:

a. a damper between the tip and the upper blade part was installed.

b. the connection rods between tip and the screw cilinder were replaced by new ones.

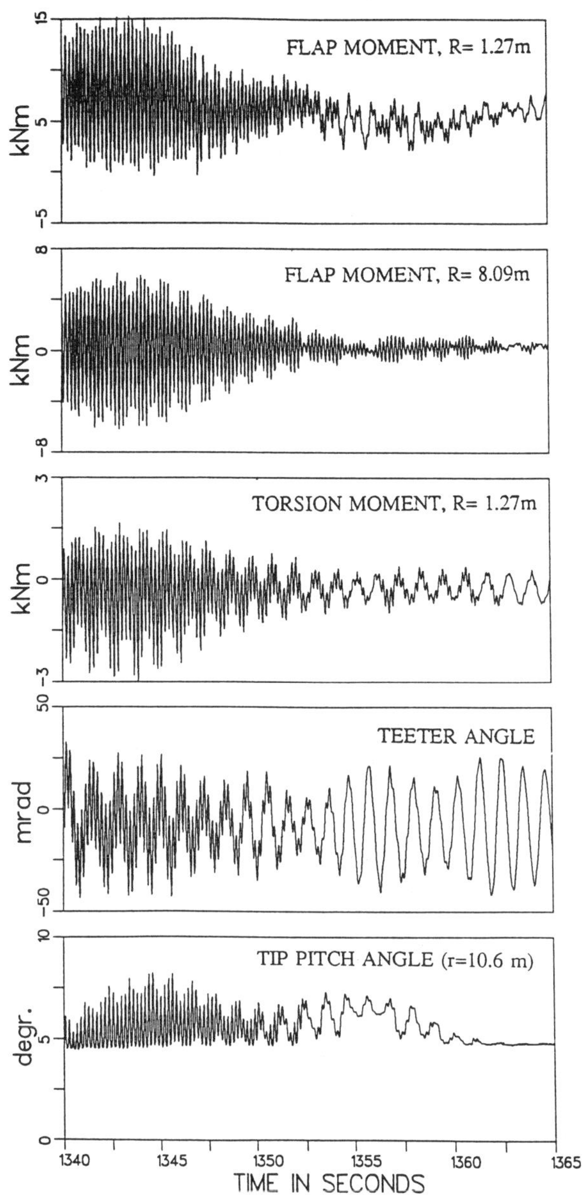

Fig. 3 Rotor signals of blade B, during the transition from unstable operation to stable operation.

The new rods could be made only slightly thicker, for lack of space between the connection rod and bearing pipe. Thus only a small increase (0.5 Hz) of the tip-torsion eigenfrequency could be realized. Nevertheless replacement was necessary because of the high torsional loads at the rods during instabilities, close to the strength limits. During the subsequent test of the rotor in a windspeed range from 5 to 20 m/s no instabilities were observed anymore.

4. LOAD REDUCTION AND TEETER MOTION

4.1 Load reduction

Load reduction at the blade root region and consequently also at the hub and the tower is the most important justification for introducing flexible windturbine blades. The quantification of this load reduction will be made by comparing the flap moment at the blade root (r=1.27m) of the FLEXTEETER rotor with AERPAC-20-WPX blades tested earlier at the ECN rotor test facility. Five cases are distinguished:

Table 1. Average values, standard deviations and 1P amplitudes for FLEXTEETER and other versions of AERPAC 20-WPX rotor blades in various configurations.

case *		a	b	c	d	e
type rotor speed power control maximum el. power flexibility		20-WPX-PRT1 variable passsive pitch 10 kW rigid hub	FLEXTEETER variable passive pitch 25 kW teeter flexbeams	FLEXTEETER variable passive pitch 100 kW teeter flexbeams	FLEXTEETER 43 rpm stall 100 kW teeter flexbeams	20-WPX-THR 48 rpm stall 120 kW rigid hub
windspeed m/s	av	16.1	16.8	16.3	16.7	16.7
	σ	1.5	2.5	2.6	2.5	1.6
	1P	-	-	-	-	-
rotor speed rad/s	av	6.40	6.44	6.12	4.52	5.03
	σ	0.15	0.14	0.19	0.02	0.02
	1P	-	-	-	-	-
flapmoment kNm	av	-6.82	1.83	6.17	15.29	18.60
	σ	9.58	2.42	2.94	4.08	7.92
	1P	6.94	2.91	2.82	4.56	5.54
teeter angle mrad	av	-	2.3	4.1	2.2	-
	σ	-	13.6	12.1	33.2	-
	1P	-	19.2	16.9	46.7	-

* see more detailed description in section 4.1

a. the 20-WPX-PRT1 configuration [4],[5],[6] with passive control tips, operated at variable speed and a low power limit, but with the blades fixed rigidly at a cone angle of 0°.
b. the FLEXTEETER configuration, with passive control tips, teeter and flexbeam operated at variable speed and low power limit (PGMAX= 25 kW).
c. as b. but now with PGMAX =100 kW.
d. as b. but operated at constant rotorspeed, n=43 rpm, that is below the limit speed at which the tips start pitching. Thus rotor power is controlled by stall also at about 100 kW. In fig. 4 power curves for case c and d are compared.
e. the 20-WPX-THR configuration [7] without passive control tips, operated at constant rotor speed, n=48 rpm, maximum power in stall 130 kW, cone angle 5°.

The position, stiffness and pre-stress of the limiters after the overhaul in the cases b, c and d are identical and described in appendix A.

Average values and standard deviations at five measurement runs (duration 0.5 hr), corresponding with the cases a,b,c,d,e are collected in table 1. In fig. 5 standard deviations of the flapmoment (r=1.27m) as a function of average windspeed at the rotor hub are given for case a and b. These standard deviations and averages are obtained over 600 s time intervals. It follows from the figure, that the expected load reduction necessary for a substantial cost reduction has been achieved indeed.

Spectral analysis of the dynamic loads, corresponding with the data in table 1, is depicted in fig. 6, 7 and 8 so that the load reduction due to the teeter and flexbeam is shown in more detail.

In fig. 6 normal variable speed operation of FLEXTEETER (case b) and 20-WPX-PRT1 (case a) are compared. The flexbeam reduces the 1P symmetric eigen-freqency from 2.4 Hz (20-WPX-PRT1) to 1.3 Hz (FLEXTEETER). Dynamic loads at 2P and the 1st symmetric flap eigenfrequency (F in fig. 6, 7 and 8), dominant for 20-WPX-PRT1, are virtually vanished for FLEXTEETER. Due to the teeter the 1P intensity is also reduced appreciably (6 times).

Normal operation of FLEXTEETER at low power (case b) and at high power (case c) are compared also in fig. 6. The 1P flap loads for 100 kW are somewhat lower than for 25 kW. The '0P' loads however are higher. This may be connected with tip pitch activity. At the 100 kW run, the tip pitch is active in the full range, while in the 25 kW case the tip activity is more limited to the end of the pitching range.

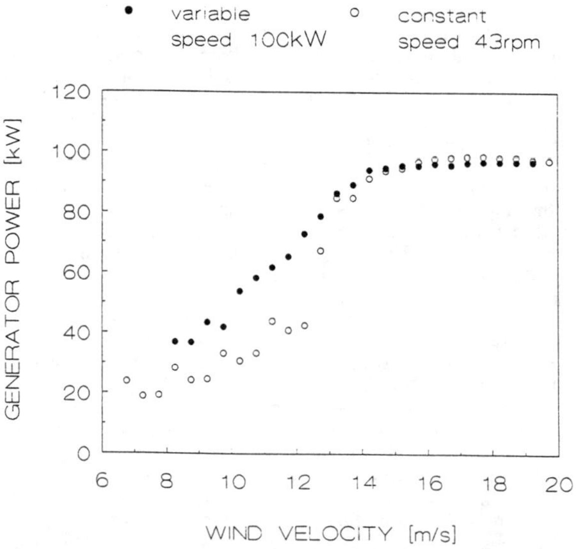

Fig. 4 Generator power curves for FLEXTEETER at variable rotor speed (PGMAX=100 kW) and at fixed rotor speed (n=43 rpm).

Normal operation of FLEXTEETER at PGMAX=100 kW (case c) is compared in fig. 7 with operation of FLEXTEETER at a constant rotorspeed of 43 rpm (case d). This constant rotorspeed generates a power curve with a flat power level of 100 kW from 14m/s windspeed (see fig. 4). The tip pitching influence on the flap moment vanishes in stall regulation, having no tip pitch activity at all. The 1P contribution at stall (43 rpm =0.72 Hz) is much larger than with variable speed, passive pitch operation at 100 kW. Since the load reduction at 1P by the teeter is very sensitve for damping, it is supposed that in the stall case the aerodynamic damping is much lower. The necessary teeter damping is delivered in case d by the elastomeric limiters situated at r=0.70m, between the flexbeam and the teeter hinge. The flap loads are measured on the flexbeam at r=1.27m, outside of the limiters. Thus larger reaction forces (corresponding with larger teeter amplitudes) generated by the limiters are felt at the load measuring radius on the flexbeam. The intensity at the 1st symmetric flap eigenfrequency is also less reduced in case d for the same reason: lack of aerodynamic damping.

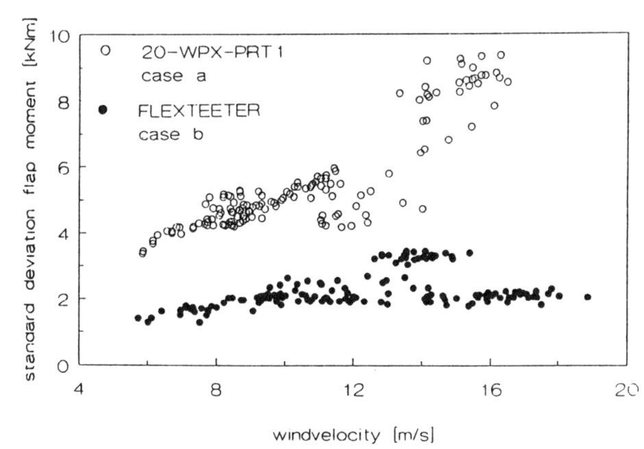

Fig. 5 Standard deviations of blade root flap moment (r=1.27m) for the FLEXTEETER rotor compared to the 20-WPX-PRT1 rotor.

Finally a comparison between FLEXTEETER and the firstly tested rigid rotor (20-WPX-THR, [7]), both in stall at fixed rotor speed, is made in fig. 8. It follows (see also table 1) that in stall still an appreciable dynamic load reduction can be obtained by introducing flexibility (teeter and flexbeam). The larger part of the dynamic load reduction takes place at the 1st symmetric flap eigenfrequency, at 3P and at 2P. At 1P the reduction is smaller, for reasons mentioned allready in the discussion of fig. 7.

4.2 Teeter motion

It follows from fig. 6 that for FLEXTEETER the remaining flap moment fluctuations are mainly at 1P. Therefore they are accompanied by a 1P fluctuation of the teeter angle. 1P amplitudes of teeter motion and flap moment at the flexbeam (r=1.27m) are collected in table 1. The data in the table show that in the stall case d, the teeter motion is about 2.5 times larger than at variable speed. The corresponding 1P amplitude of the flap moment is about 1.6 times larger in the stall case. The apparent stiffness (flap moment/teeter motion) in case d (100 kNm/rad) is lower than

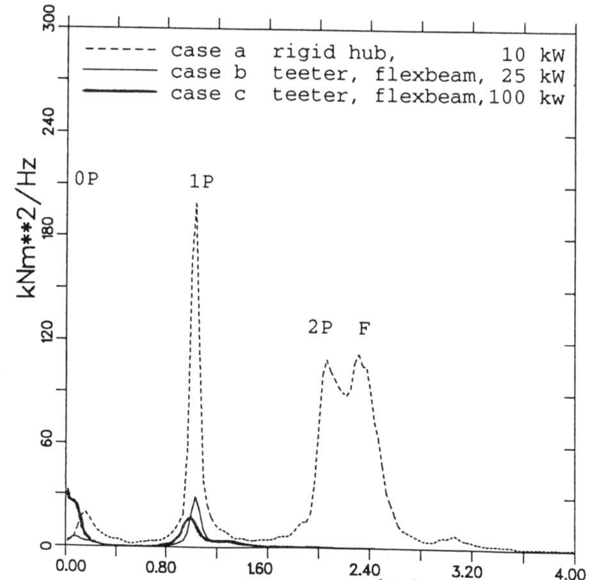

Fig. 6 Power spectral densities of flap bending moment in the blade root (r=1.27m) at a wind speed of 16 m/s for a rigid rotor with passive tip pitch (AERPAC 20-WPX-PRT1) and a fully flexible rotor (FLEXTEETER) also with passive pitch control. Outer dimensions of the compared rotors are the same.

in case b and c (160 kNm/rad). This is related to the limiter pre-stress. In case b and c the flexbeam motion at the limiters (r=0.70m) is about 1.3 cm, while in case c this is about 3.3 cm. In the last case the motion is much larger than the pre-stress range (1 cm). Modelling the limiters as ideal springs leads to a stiffness 4 times the stiffness of one limiter within the pre-stress range and 2 times outside this range. This means that without limiter pre-stress, lower flap moment 1P amplitudes are to be expected in the cases b and c, while in case d almost no change is expected. Even smaller 1P flap moments can be realized when the limiters are mounted with sufficient free space between flexbeam and limiter. In such a case the flap moment is determined by the stiffness of the teeter hinge (about 40 kNm/rad) only. Stability may be endangered however in this case, as was seen before the overhaul of the rotor.

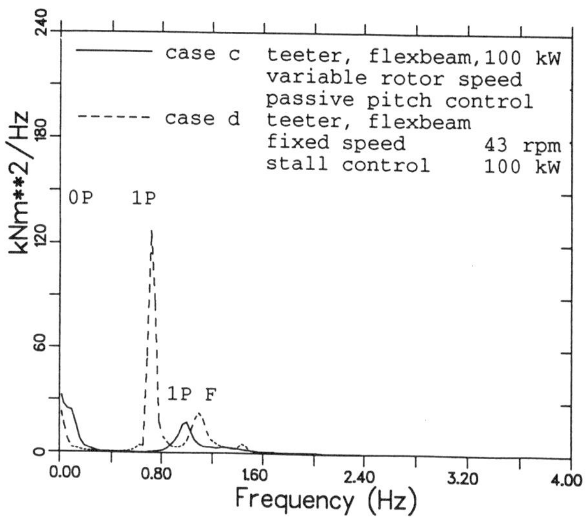

Fig. 7 Power spectral densities of flap bending moment in the blade root (r=1.27m) at a wind speed of 16 m/s for FLEXTEETER at normal operation (maximum generator power level: 100 kW) and for FLEXTEETER at constant rotor speed (43 rpm) with the rotor in stall at generator power also 100 kW.

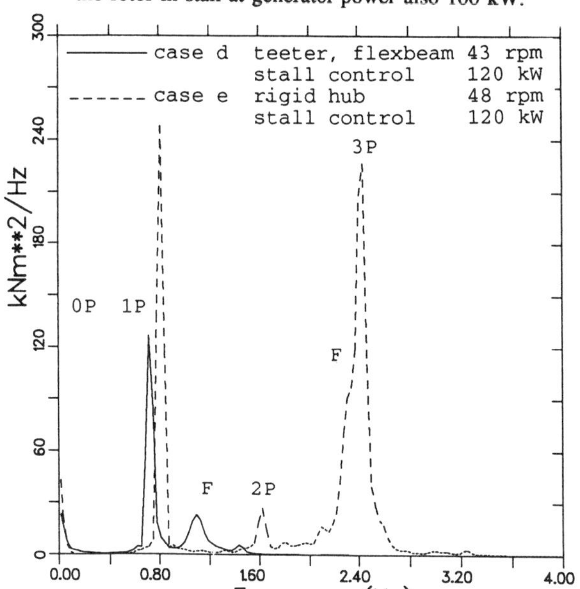

Fig. 8 Power spectral densities of flap bending moment in the blade root (r=1.27m) at a wind speed of 16 m/s for FLEXTEETER at constant rotor speed (43 rpm) with the rotor in stall at generator power level 100 kW and for a rigid rotor (AERPAC 20-WPX-THR) at constant rotor speed (48 rpm) with the rotor in stall at generator power 120 kW. Outer dimensions of the compared rotors are the same.

5. YAWING

In the FLEXHAT safety design, the two passive control tips form the first safety system, which has to be designed in such a way that they will keep the rotor speed allways below a safe limit, even in the case of a generator breakdown at high windspeed. Yawing in combination with a small parking brake is envisaged as an auxiliary, secondary safety system for bringing the rotor to a standstill in case of maintenance. For this reason yawing (angular velocity 1.2°/s) has been programmed at the test of FLEXTEETER as the normal stop procedure. Moreover a special test run at a windspeed of 14.5 m/s was made, in which stepwise yawing has been realized with a yaw angle step of 15°, each yaw step lasting for about 20 minutes. Operational conditions during this test were as described in case b in section 4. The main results are given in fig. 9.,which shows the 1P amplitudes of the flap moment (r=1.27m) together with the 1P teeter amplitude The 1P component is the remaining component in the flap moment of FLEXTEETER, when operated as described in case b (see section 4). It follows from fig. 9, that this main component does not depend strongly on the yaw angle. The extra 1P flap loads caused by yaw misalignment are not very large, as follows from the teeter amplitudes, which do not increase much, when going to larger yaw mis-alignment. Thus yawing is not a severe load case for FLEXTEETER, when operated with variable speed. The use of yawing as an auxiliary safety system for stopping the rotor is therewith completely feasible. The question, whether this is true also for FLEXTEETER in stall at constant rotor speed (case d) is scheduled for testing in the near future.

6. CONCLUSIONS

A flexible rotor, called FLEXTEETER, has been tested succesfully at the ECN test facility.

An instability observed at the first part of the test was suppressed by the installation of a damper between the pitchable tip and the fixed part of the blade.

Substantial reduction of dynamic loads (3 to 4 times) at the blade root region due to flexbeam and teeter has been demonstrated.

In case of constant speed operation and stall control a still appreciable (2 times) but smaller load reduction was obtained by the introduction of the same kind of flexibility.

Passive power control by means of stall introduces more dynamic flap loads (40%) in the blade root than passive pitch control in case of a flexible rotor like FLEXTEETER.

Yawing a flexible rotor can be used as an auxiliary safety system for stopping the rotor without introducing severe dynamic loads.

Optimalisation of the teeter damping and stiffness can lead to still smaller dynamic loads in the blade root of a flexible rotor.

REFERENCES

[1] G.A.M. van Kuik, R. H. F. Tikkemeier (SPE) and J.W.M. Dekker (ECN). The development of advanced rotor systems. A survey of the FLEXHAT program, including full scale results. Proceedings of the European Community Wind Energy Conference, Madrid, Spain, September, 1990.

[2] G.A.M. van Kuik (SPE), J.W.M. Dekker (ECN) The FLEXHAT programme: Technology development and testing of flexible rotor systems with fast passive pitch systems. Proceedings of the European Wind Energy Conference, EWEC'91, Amsterdam, The Netherlands, Oktober, 1991.

[3] F. Hagg, G.A.M. van Kuik (SPE) and J.W.M. Dekker (ECN). Design of FLEXTEETER. SPE report 91-057, October 1991.

[4] J.W.M. Dekker, C.M. de Groot and M. Späth. Results from a test of rotor blades with a passively controlled tip. Calibration and test under motor control. ECN-C--90-019, June 1990.

[5] J.W.M. Dekker, C.M. de Groot, H.Ch. Rieffe and M. Späth. Test of rotor blades with passively controlled tips: Normal and abnormal operation. ECN-C--91-026, May 1991.

[6] S. Cervenka, J.W.M. Dekker, and H. Ch. Rieffe. Test of rotor blades with passively controlled tips. Phase 3: Endurance test. ECN-C--91-053, August 1991.

[7] J.W.M. DEKKER, C.M. de Groot and M. Späth. Mechanical measurements on VSH 20-WPX-THR rotor blades at the '25 m HAWT rotor test facility, Part 1 and 2. ECN-89-35 and ECN-89-36, April 1989.

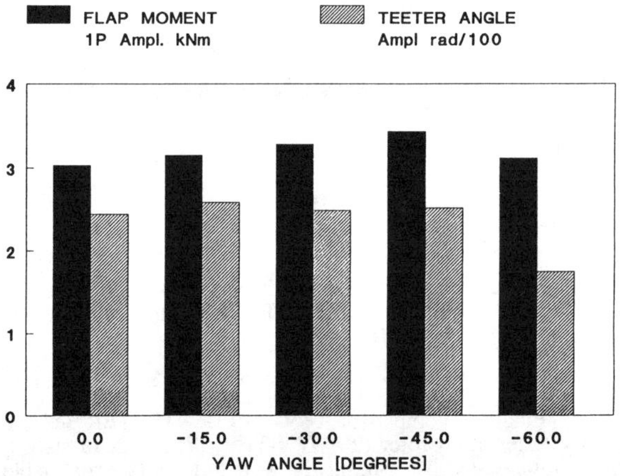

Fig. 9 1P amplitudes of blade root flap moment (r=1.27m) and teeter angle as a function of yaw angle for FLEXTEETER, during operation with various yaw angles at a wind speed of 14.4 ± 2.5 m/s.

APPENDIX A Rotor data for FLEXTEETER

Geometry, see also [3]	
two blades, upwind, diameter:	21.2 m
cone angle	0°
tilt angle	5°
teeter range	±10°
blade mass, including tip	606 kg
centre of gravity	4.99 m
tip length	2.50 m
flexbeam between	0.60 and 1.60 m
limiters on each side of the flexbeam	r= 0.70m
stiffness one limiter	75kNm/rad
idem before overhaul	150kNm/rad
pre-stress	1 cm
variable speed from	40, 80rpm
tip pitch action n > NLIM	55 rpm
normal operation: constant tipspeed ratio below NLIM, constant power above it	
Natural frequencies at standstill	
1st antisymmetric flap	0.2 Hz
1st symmetric flap	0.77 Hz
1st symmetric lead-lag	3.5 Hz
2nd antisymmetric flap	5.5 Hz
2nd symmetric flap	5.7 Hz
tip torsion	6.2 Hz
Rotor sensors installed at blades A and B, see also [4]	
flap, lead-lag and torsion moment on the flexbeam at	r=1.27m
idem at the blade division	r=8.09m
blade displacement in flap direction relative to the teetering rotor plane	r=1.59m
tip displacement in radial direction at the screw cilinder	r=6.6 m
teeter angle (only one side)	r=0.462m
Moreover rotor angular position,rotor angular speed and mechanical torque at the main shaft were measured.	
During operation blade B was monitored with a video camera fixed to the hub.	

DEVELOPMENT OF AN ADVANCED TWO BLADED HAWT

P.B. Simpson, R.S. Haines, R.H. Sauven
Wind Energy Group Limited
Greenford House
309 Ruislip Road East
Greenford
Middlesex UB6 9BL
ENGLAND

ABSTRACT: Continuing a line of development pursued over a period of eight years, the Wind Energy Group has further optimised the design of wind turbines at the most suitable scale for installation in windfarms. This paper describes the process of machine optimisation, based on six years of commercial windfarm operating and maintenance experience, the testing of two earlier prototypes and the installation and operation of machines in six countries, resulting in the 36.4 m diameter WEG 400. The design concept of the WEG 400 incorporates a lightweight, two-bladed, teetered hub rotor, in combination with an integrated drive train and 'soft' tower, providing the basis for commercial production of a very cost competitive machine. Optimisation of performance has been achieved by adjustments to rotor diameter, rotational speed and rating. Noise emission has been reduced by reduction in blade tip speed, refinement of gearbox design and vibration isolation measures. Value engineering techniques have been applied to various elements of the design. Details of the results of design optimisation and of operational trials at the UK National Wind Turbine Test Centre are presented and the design improvements carried forward into the production model are described.

1. INTRODUCTION

Wind Energy Group (WEG) Ltd is a wholly owned subsidiary of Taylor Woodrow Construction Holdings plc, which is one of the largest UK construction companies, active in a variety of fields, ranging from civil engineering construction to energy related services.

WEG Ltd has developed a line of commercial horizontal axis wind turbine generators and has been a major participant in the UK Department of Energy (now Department of Trade and Industry) programme of investigation of wind energy technology for over 15 years.

The first medium scale wind turbine designed and built by WEG was the MS1, a 250 kW, 20 m diameter machine that was installed in 1983 on Burgar Hill, Orkney (Fig. 1) [1]. This machine served as a one third scale prototype of the 3 MW, 60m machine that WEG built on the same site four years later, in the context of the UK Department of Energy development programme [2].

Following demonstration of the satisfactory performance of a two-bladed teetered hub rotor with the MS1 turbine, a necessary preliminary to the construction of the 60m diameter machine, the next turbine designed and built by WEG was the MS2, a 20m diameter, three-bladed, 200 kW machine. The prototype MS2 was installed in December 1984 at Ilfracombe, Devon (Fig. 1) and the very satisfactory operation of this machine early in 1985 was followed by the installation later that year of a further 20 MS2's in a windfarm in Altamont Pass, California, which was fully operational in December 1986 [3].

Recognising the potential of the two-bladed teetered hub concept for achieving economies in machine design at medium scale, as were evident at large scale, in 1987 WEG adapted the design of the MS2 to incorporate the teetered hub principle. Taking advantage of advances in the laminated wood blade construction, the lighter rotor arrangement and built in reserves in the drive train, the diameter of the turbine and the rating were increased to 33 m and 300 kW respectively. A prototype MS3, as this turbine was named, was installed in Altamont Pass in late 1988, followed by a second machine at a test centre at Carmarthen Bay in South Wales (Fig. 1) [4].

In 1990 MS3's were sold to Aeritalia for ENEL and to Northern Ireland Electricity. Two machines were also built for use in the UK Department of Energy R & D programme, for installation at the National Wind Turbine Test Centre (NWTC), on Myres Hill, near East Kilbride in Scotland (Fig. 1) [5].

In 1992 WEG Ltd sold sixty-five MS3 turbines to a UK windfarm developer, First Windfarm Holdings Ltd., for installation in three windfarms at sites near Cemmaes and Llangwyryfon in mid Wales and Cold Northcott in Cornwall [6]. Four MS3s were also sold to Yorkshire Water Services for installation at a site adjacent to Chelker reservoir (Fig. 1).

The further process of optimisation, based on the design, installation, operating and maintenance experience of the MS1, MS2 and MS3 turbines, as well as rigorous testing of the prototype installations, resulting in the WEG 400 turbine, is the subject of this paper.

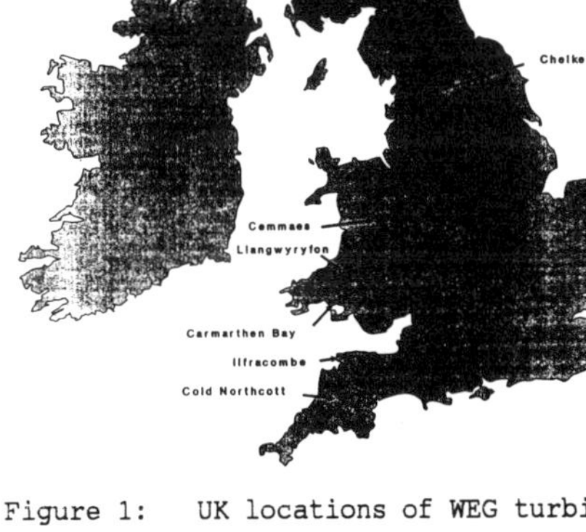

Figure 1: UK locations of WEG turbines

2. DESIGN CONCEPT

The adoption of a teetered hub with a two-bladed rotor provides the basis for taking maximum advantage of the economy in use of materials offered by the two-bladed configuration, in comparison with a three-bladed configuration. Uncoupling the out of plane blade moments from the rotor shaft relieves both the blade root sections and the pitch bearings (in the case of full span pitch control) of a considerable proportion of the loading they would otherwise have to carry, permitting lighter blade and bearing construction. Although the additional complexity in hub and pitch control mechanism design somewhat erodes the advantage, overall there is a very significant cost saving when the further benefits accruing to drive train, nacelle bed plate (or equivalent), yaw mechanism and drive are taken into account. In particular, when the low speed shaft bearings are integrated with the gearbox and, also, when the gearbox casing is integrated with the nacelle support structure to achieve further economies, the advantage offered by the teetered hub concept increases. As described below, the WEG 400 incorporates both these principles of integrated arrangement, and therefore derives maximum benefit from the teetered hub concept.

3. DESIGN DEVELOPMENT

3.1 Drive Train Arrangement and Support

The general arrangements of the MS2 and MS3 nacelles, showing the drive train arrangements and bed plate structures, are shown in Figs. 2 and 3. The integration of the low speed shaft bearings with the gearbox is an important feature, contributing to the relatively low total tower head masses of 8,500 kg and 10,750 kg respectively.

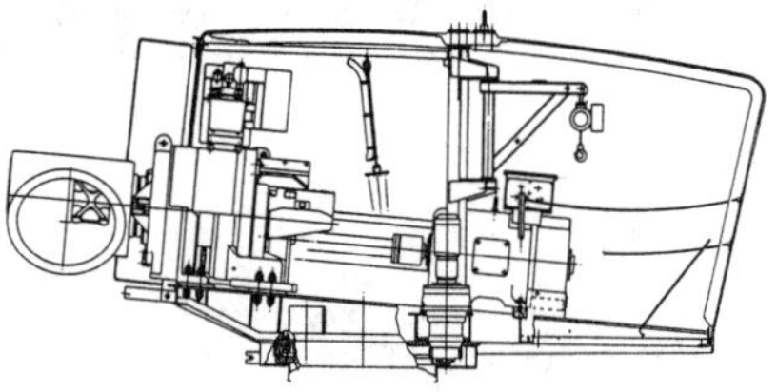

Figure 2: MS2 - General Arrangement of Nacelle

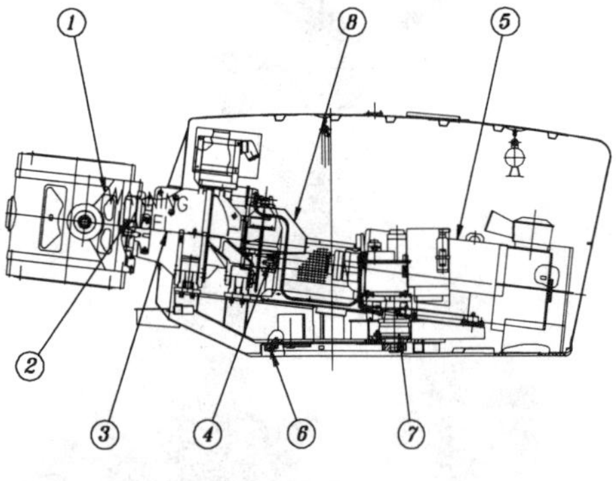

1. Rotor Hub. 2. Teeter Damper. 3. Gearbox. 4. Pitch Servo Actuator. 5. Generator. 6. Yaw Bearing. 7. Yaw Drive. 8. Pitch Sensor.

Figure 3: MS3 - General Arrangement of Nacelle

The central feature of the development of the WEG 400 general arrangement, shown in Fig. 4, is the use of the gearbox casing to carry load horizontally back towards the tower head, not just vertically as before. This eliminates the duplication of load paths formerly provided by the gearbox casing and nacelle bed plate, resulting in integration of the gearbox casing with the nacelle structure, with corresponding reduction in use of materials, weight and cost. The tower head mass is 11,000 kg, only a marginal increase over that of the lower rated MS3.

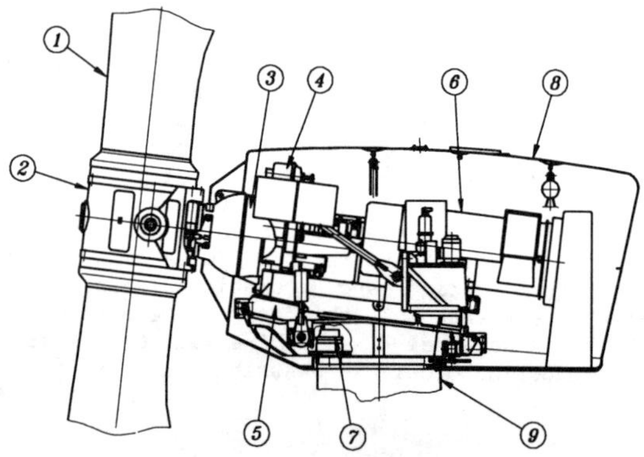

1. Blades. 2. Hub. 3. Gearbox. 4. Mechanical Brake. 5. Compliant Mounted Transmission. 6. Generator. 7. Yaw Drive. 8. Nacelle. 9. Tower.

Figure 4: WEG 400 - General Arrangement of Nacelle

Some form of compliance is required in the drive train of a wind turbine of this class, in order to smooth fluctuations in aerodynamic torque and limit dynamic magnification of some elements of system response, particularly at blade passing frequency. In the WEG MS3 this was achieved by means of an induction generator with high slip. With increased rating, this method incurs an increased cost in terms of the greater generator frame size required, particularly with the move in the industry to totally enclosed generators. The cost of the energy lost in rotor slip must also be taken into account. In view of these factors it was decided to investigate alternative forms of compliance for the WEG 400 in conjunction with a less costly, low slip generator.

Well-known devices such as fluid and elastic couplings were considered, but it was concluded that mounting the entire drive train flexibly was a superior solution in this application. The elements of this solution comprise a relatively light framework supporting the gearbox and generator, and a system of elastomeric mountings, with very high flexibility about the roll axis, connecting this framework to the nacelle bed. A major advantage of this approach is that the flexible mountings work as anti-vibration mountings as well as effectively reducing drive train stiffness, so that structural transmission of noise is strongly reduced. Moreover, the mountings concerned provide an economic form of flexibility that is well-tested in other industries. With the flexible mountings providing compliance, by virtue of the reduced requirement for slip the WEG 400 uses a generator with essentially the same frame size as that of the MS3.

The principle was tested by modifying an existing fabricated pallet to form the bed member and fabricating a new framework to support the gearbox and generator. The dynamic behaviour of the system was satisfactorily tested during both transient and normal running conditions, above and below rated power. In particular, the normal running behaviour was characterised by open-loop System Identification tests. A production version of the system was then developed, using the same concept and torsional characteristics but with components optimised for the purpose.

Like the generator, the gearbox internals are also unchanged in most respects from those in current production for the MS3, the initial design having intentionally included a margin for uprating. Two specially modified gearboxes have been tested in a back-to-back configuration at the new torque level of the uprated machine. One change since the initial design, an increase in the size and span of the main bearings to cater for increase in rotor diameter, has already been incorporated in the current MS3, so as to increase the degree of commonality.

3.2 Blades

As a result of a programme of blade development which WEG has undertaken with UK Department of Trade and Industry (DTI) support over the last two years, the rotor design has been improved in terms of weight, energy capture and quietness.

Part of the programme has included material testing at Bath University and fatigue testing at City University, North London (both of these activities are reported separately in this conference), which have together provided a very sound data base on which to assess the strength of the blade in service. As was anticipated, the strength reserves of the existing root connection and inner part of the original blade have been found to be well able to withstand the loads from a tip extended radially by over 10%.

3.3 Noise Reduction

Noise reduction has been a major objective in the design of the WEG 400. The incorporation of vibration isolation in the drive train mounting has already been described. At a more detailed level, the design of the gearbox housing has been modified, without affecting the moving parts, to avoid features likely to serve as efficient noise radiators. The design has a number of other features tending to reduce noise transmission, such as the wrap-around, flexibly mounted cladding and the use of SG cast iron for the nacelle bed. Additional barriers of a more superficial nature can easily be introduced, depending on the outcome of investigations currently in hand.

The prototype turbine on Myres Hill has the same tip speed as the MS3, but aerodynamic noise has been reduced by changes in detail. These include sharper trailing edges to reduce boundary layer and a tapered tip to reduce vortex shedding. Test results demonstrating this are reported below.

3.4 Value Engineering

Apart from the development of the integrated gearbox nacelle structure arrangement and flexible drive train support, the significant improvement made in moving from the MS3 to the WEG 400 has been in blade design.

Part of the DTI blade programme with WEG has focussed on the improvement of production methods. The blades tested at City University have progressively incorporated the new production features, which have not only dramatically reduced cycle times but have also resulted in increase in strength.

4. PROTOTYPE TESTS

Testing of the dynamic behaviour of the modified pallet and flexibly mounted framework supporting the gearbox and generator was undertaken on one of the two MS3's installed at the NWTC on Myres Hill. Longer, modified blades, with the improved tip geometry described earlier, were also fitted to this machine, increasing rotor diameter to 35.4 m. At the rotational speed of 45 rpm chosen for the prototype gearbox blade tip speed is no greater than that of the MS3, whilst rated output of the turbine is increased to 450 kW. Operational trials were conducted throughout 1992 and the performance of the machine was found to be very satisfactory. In particular, the dynamic behaviour was as predicted and the improvement in aerodynamic sound power level achieved with the modified blade tip geometry was confirmed.

4.1 Power Quality

As described above, the WEG 400 uses a low-slip generator, achieving the higher power rating at the same frame size as the high slip MS3 generator. The compliant drive train support system was designed to compensate for the consequent reduction in drive train compliance and hence maintain acceptable power quality. The prototype tests have demonstrated that power quality has in fact improved relative to the MS3.

Fig. 5 shows a 20 sec period of operation of the prototype at 450 kW rating. The period starts with power just below rated level. The following can be seen:

i) Just below rated wind speed, the control system is already acting to limit possible exceedances of 450 kW.

ii) Above rated wind speed, the pallet and control system are working together to limit power fluctuations to less than 20%, with a standard deviation of up to 35 kW.

iii) Although the principal driving frequency for power is 2P, i.e. 1.5 Hz, the response is at 0.9 Hz, which is the drive train natural frequency, showing how effectively the torque fluctuations are being absorbed.

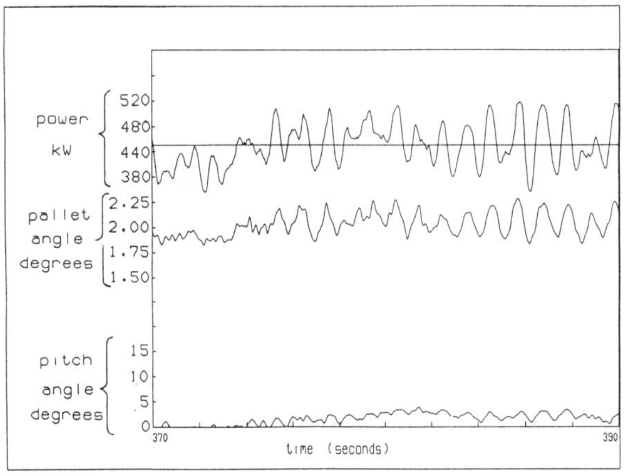

Figure 5: Example of WEG 400 Prototype Operation at 450 kW

4.2 Noise Emission

Measurements of noise spectra at the NWTC using standard IEA techniques clearly show that MS3 aerodynamic noise is principally due to a peak centred on 1 kHz. Theory suggests that this is due to coincidence of the peak in the trailing edge vortex noise with that of the turbulent boundary layer noise. Directly comparable measurements of the prototype WEG 400 spectrum shows that this peak has been effectively removed, albeit with insignificant increase in noise at lower and higher frequencies. The small increase at the lower frequencies is due to the slower moving inboard blade section, whilst the increase at higher frequencies is due to the sharper trailing edge and the narrow chord tip. The results demonstrate that an overall noise reduction of about 2 dB(A) has been achieved.

A further reduction of 1 dB(A) in aerodynamic sound power level has been designed into the production turbines by reducing tip speed.

5 PERFORMANCE

Initially, for the prototype turbine on Myres Hill, increase in diameter from 33 m to 35.4 m with reduction in rotational speed to hold tip speed constant resulted in 27% increase in predicted energy capture. For the production turbines, in order to gain further reduction in noise emission and taking fuller advantage of the reserves in blade strength, rotational speed has been reduced to 40 rpm and rotor diameter increased to 36.4 m. With rating held at 400 kW, this results in some reduction in energy capture, but preserves 24% increase relative to the performance of the 33 m diameter machine.

The power-windspeed curve of the WEG 400 is shown in Fig. 6. Variation of predicted annual energy yield with annual mean wind speed at hub height is detailed, assuming 100% availability and without allowance for cluster losses.

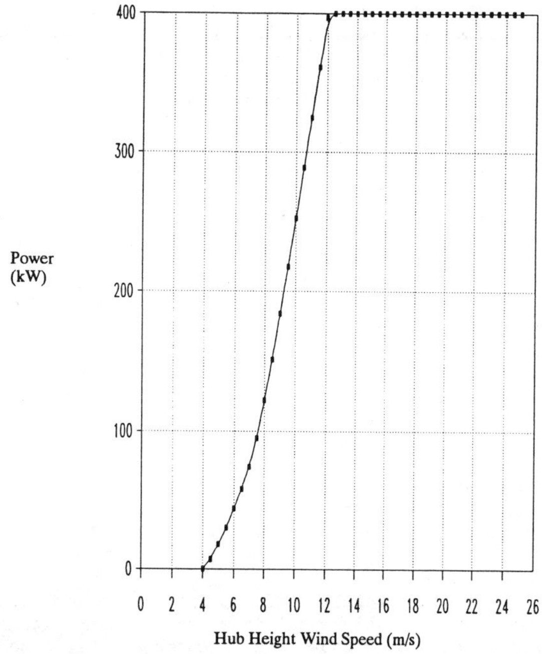

m/s (hub)	MWh
5.0	433
5.5	580
6.0	737
6.5	897
7.0	1057
7.5	1213
8.0	1361
8.5	1501
9.0	1630
9.5	1747
10.0	1852

ASSUMPTIONS

- **Rayleigh distribution of wind speed**
- **Measured at the generator**
- **No turbulence**
- **Two speed model**
- **Air density 1.225 kg/m^3**

Figure 6: WEG 400 Power-Windspeed Curve

REFERENCES

[1] J R C Armstrong, G R Ketley, B J Cooper, "The 20 m diameter wind turbine for Orkney", Proceedings Third BWEA Wind Energy Conference (1981) pp 54 - 62.

[2] P B Simpson, A L Burton, D Lindley, "The LSI, 3 MW Horizontal Axis Wind Turbine", Fifth International Conference on 'Energy Options - the Role of Alternatives in the World Energy Scene' (1987).

[3] P M Elliott, C R Gamble, D Lindley, "WEG's California Windfarm", Proceedings Twelfth BWEA Wind Energy Conference (1990).

[4] Dr D Lindley, Dr C R Gamble, Dr J G Warren, "Performance of the WEG MS3 300 kW Wind Turbine", Conference Publication EWEC Part Two (1989) pp 623-630.

[5] C R Gamble, N Jenkins, R H Sauven, A L Burton "Installation and operation of MS3 wind turbines in Ireland, Scotland, and Sardinia". Proceedings Thirteenth BWEA Wind Energy Conference (1991) pp 33 - 38.

[6] D Lindley, P Musgrove, J R C Armstrong, M L Hitner, "Early experience in UK Windfarms", European Community Wind Energy Conference (1993).

LESSONS LEARNED FROM WEGA I AND OTHER LARGE MACHINE PROGRAMMES

Erich Hau
ETAPLAN GmbH, 8000 München 81, Oberföhringer Str. 168b

The development of megawatt wind turbines was initiated mainly by governmental R&D organisations with the intention of providing an additional renewable energy technology which could be used by public utilities. In the United States the first experimental units committed to this aim were erected in 1978. Within a few years a multi-megawatt prototype, the MOD-2, achieved operational status. Five units of this type were built and tested from 1979-82. (Fig. 1)

In Europe, Sweden and Germany also, showed interest in this technology with the building of experimental machines. The German project, GROWIAN, so far the largest wind turbine ever been built, unfortunately only achieved some hundreds operational hours, but the large turbines built in Sweden (WTS-3 and WTS-4) have operated for some twenty and ten thousand hours respectively.

Experiments in Canada have had a special focus on vertical axis wind turbines. The world largest Darrieus-rotor, with a capacity of 4MW, has been developed and has been under test since 1987.

Building upon some of the experience gained with this first generation of multi-megawatt experimental wind turbines, a second generation of machines was launched in the mid Eighties. These turbines were somewhat smaller in size - one lesson which was learnt from the first multi-megawatt turbines.

In Britain, Holland and Italy advanced two blade concepts, the LS-1, the NEWECS-45 and the GAMMA 60, have been erected and are being tested.

The Danish utilities, which had collected positive experiences with the Nibe experimental turbines, decided to built a 61m diameter, 2 megawatt turbine. This project, later called Tjaereborg Wind Turbine, together with a Spanish-German project AWEC-60 and a British prototype, built by James Howden and Company formed the WEGA large wind turbine programme, which attracted substantial financial support from the Commission of the European Communities.

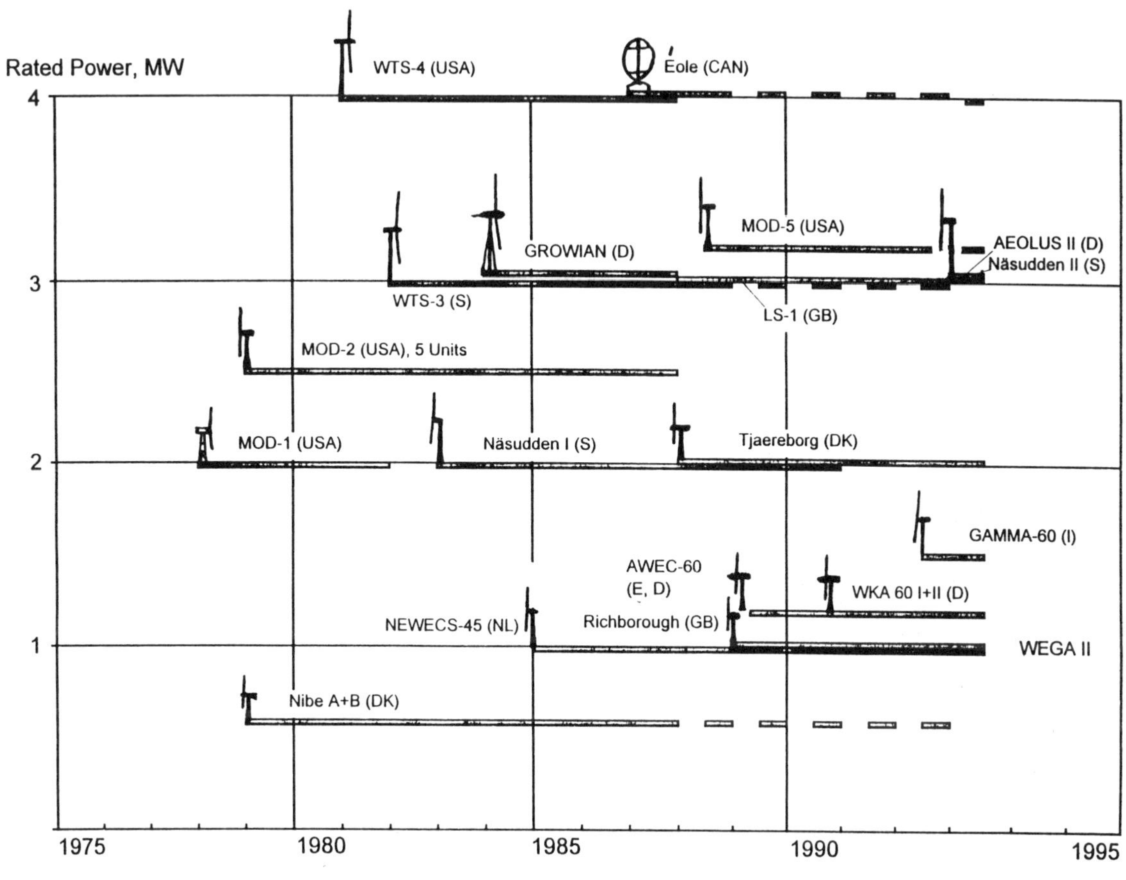

Figure 1: Large Experimental Wind Turbines

In total about 25 large wind turbines with power outputs of around one megawatt and more have been erected and evaluated in the period from 1978 to 1993. In general terms it can be concluded, that nearly all of these turbines have successful demonstrated their functional integrity from the technical point of view. All projects have proved to have had a larger R&D component side than it was generally expected in the beginning of the development and there have been a number of technical problems and failures.

At the present time, the commercial use of wind energy for electricity production is based on series produced wind energy turbines with power ratings up to approximately 500kW. In a good wind regime turbines of this kind can achieve energy production costs near to conventional generating costs. Within the last ten years the size of the commercial wind turbines has been increased as well as the cost effectiveness. (Fig. 2)

The technical problems of large wind turbines have either been solved or are capable of solution within a short time scale, but the main obstacle to the commercial use of the large wind turbines remains inferior economics. The specific installation costs (ECU/kW or ECU/m²) for the machines are up to four times higher than those for small and medium sized turbines. (Fig. 3) The higher energy captures resulting from the utilisation of the higher wind speeds obtained at the higher rotor hubs is not sufficient to compensate for this effect. As a result the economics of the large wind turbines are remarkable inferior to the smaller commercial WEC's.

The main reason for the high production costs of the megawatt wind turbines lies in the high component weights. A comparison with the today's smaller commercial turbines highlights the unfavourable specific towerhead weights of the larger machines (Fig. 4). Although there are intrinsic physical reasons for an increase of the specific masses per swept rotor area, a significant potential for weight and cost reduction can still be identified.

	1983	1993	Improvement Factors
Rated Power	55 kW	500 kW	9,1
Rotor Diameter	15 m	40 m	2,7
Swept. Rotor Area	177 m²	1.256 m²	7
Hub Height	20 m	40 m	2
Energy Yield			
at V_{w10m} =6m/s	110.000 kWh	1.400.000 kWh	12,7
specifically	621 kWh/m²	1.115 kWh/m²	1,8
Rotorblade Weight	285 kg	1.100 kg	
specifically	1,6 kg/m²	0,85 kg/m²	1,9
Towerhead Weight	5.500 kg	23.000 kg	
specifically	31 kg/m²	18 kg/m²	1,7
Total Weight	12.000 kg	51.000 kg	
specifically	68 kg/m²	40 kg/m²	1,7
Ex Works Price	70.000 ECU	450.000 ECU	
(Basis 1993)	1.272 ECU/kW	900 ECU/kW	1,4
specifically	395 ECU/m²	358 ECU/m²	0,9
	0,636 ECU/kWh*a	0,321 ECU/kWh*a	2

Figure 2: Commercial Wind Turbines

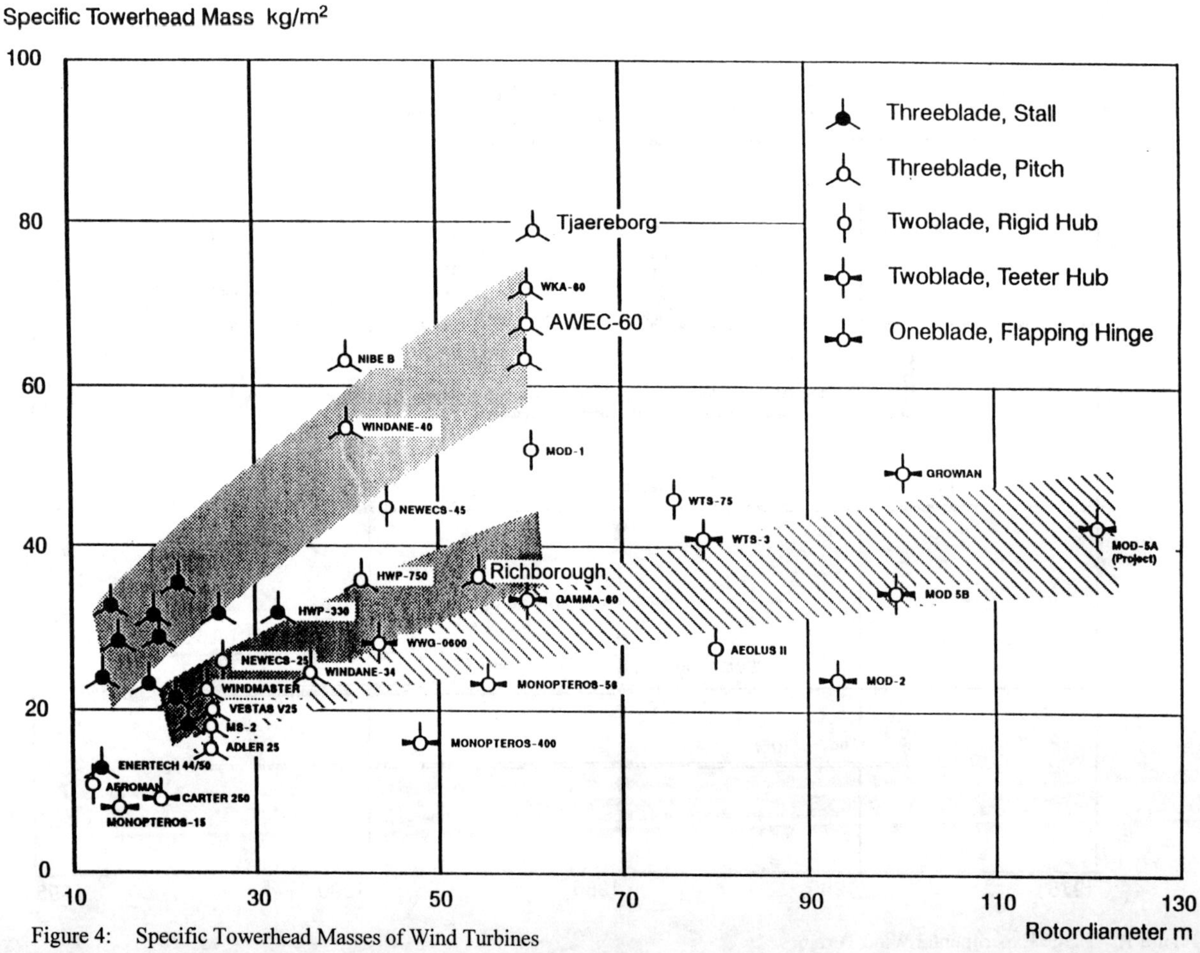

Figure 4: Specific Towerhead Masses of Wind Turbines

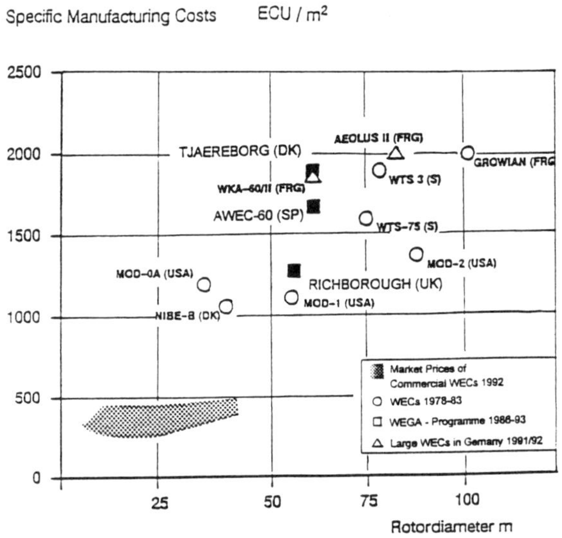

Figure 3: Specific Manufacturing Costs of Wind Turbines

Within the context of the continuing development of large wind turbine technology the WEGA turbines can be regarded as the 'second generation' of large experimental machines. Designed in the years 1984 to 1986, they built upon experience gained from the 'first generation' of large experimental machines which were erected in the early eighties.

Although the WEGA programme has not yet come to an end one important conclusion can already be drawn. Large wind turbines cannot become commercially viable using the technology upon which the current WEGA machines are based. This is clearly true for the Tjaereborg wind turbine and the AWEC-60 if less so for the Howden Richborough machine. The main reason lies in the high, uncompetitive costs of the towerhead machinery which, in turn, are due to the high component weights.

In the commercial wind turbine industry, on the other hand, a continuous growth in the size of series produced commercial machines can be observed. These machines are becoming more lightweight in construction and, also, have gradually increased in size up to the 500kW; the level which is the industry standard today. The technological gap between the commercially driven developments of the small to medium sized machines and R+D concepts of the large experimental wind turbines is obvious and in need of investigation.

In order to address this problem DG XII of the CEC initiated a strategic study entitled - **Study on the Next Generation of Large Wind Turbines**. The study is being carried out by an international group of experts with the aims of identifying the optimum size range of wind turbines and also of comparing the cost effectiveness of different wind turbine concepts including commercial and R+D designs. The cost calculations include all factors in a turn-key installation together with operation and maintenance costs over the lifetime of the plant. The results of the study should assist with the next step in the development of large wind turbines. (WEGA II)

The main economic conclusions of the strategic study are that there is no clear optimum size in the size range with rotor diameters from 30 to 80m (Fig. 5). Unit costs of electricity production slowly increase with size but the increase is so slow that it can be offset by improvements in design and by economies of scale in other cost areas such as site and operation costs. Interim results from the study have also shown that energy generation costs from large wind turbines could be reduced to the same levels as today's commercial wind turbines if lightweight design principles are adopted.

- No Physically Determined Optimal Size
  (Study on the Next Generation of Large Wind Turbines, EC, DG-XII)
- "Optimal Size"
  Determined by the Next Reasonable Step from the Commercial Available Wind Turbines to the Next Generation
- WEGA II (and Others)
  Rotor Diameter 50 - 55 m
  Rated Power 0,75 - 1,25 MW

Figure 5: Optimum Size of Wind Turbines

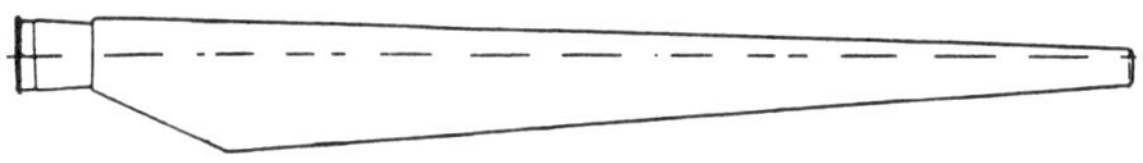

WEGA I
- Glas/Polyester
- Hub Attachment by Heavy Steel Flanges
- Classic Aircraft Airfoils (NACA 44xx, 230xx)
- Weight (60m ø) : 6500 - 9000 kg
- specifically: 2,3 - 3,2 kg/m²

WEGA II
- Glas/Epoxy, Wood/Epoxy, Carbon reinforced
- Hub Attachment by Bonded Alu-Rings and Bolts
- Specially Designed Airfoils (LS-1, FX ...)
- Weight (50m ø) : 2800 - 3000 kg
- specifically: 1,4 - 1,5 kg/m²

Figure 6: Progress of Rotorblade Technology

The essential progress towards light weight design principles is associated to rotor blades (Fig. 6) and drive train design (Fig.7).

The forthcoming generation of large wind turbines, WEGA II and others, also show a larger variety of the overall concept features compared to the WEGA I wind turbines.

The WEGA II programme includes wind turbines with three and two blade rotors. It is a remarkable fact that due to the availability of lightweight rotor blades the three blade rotor remains an attractive option in the size range of 50-55m diameter (Fig. 8).
Regarding power control optimised stall controlled rotors seem to feasible also for rotors of this size (Fig.9). This lesson has been learned mainly from the successful commercial wind turbines, which use this technology up to rotors of 40m diameter and 500 kW rated power.

One of the mot exciting innovative elements in the WEGA II programme is the development of gearless wind turbine concepts (Fig. 10). Direct rotor driven electric generators in combination with a frequency converters do not show the extreme unfavourable increase of the generator diameter and

WEGA I

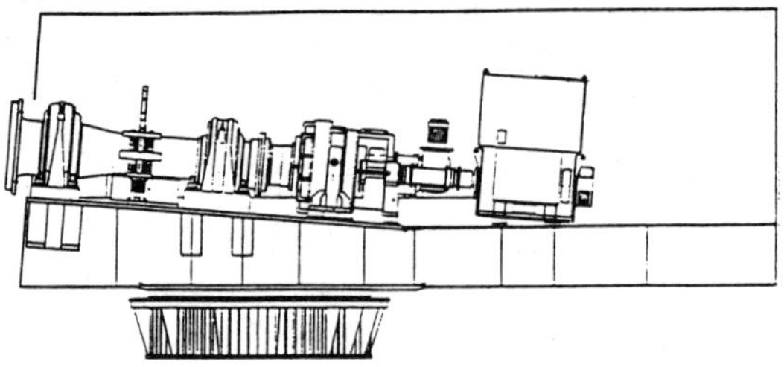

- Modular Design, Long, Separated Rotor Shaft
- Heavy Bedplate
- Large Nacelle

WEGA II

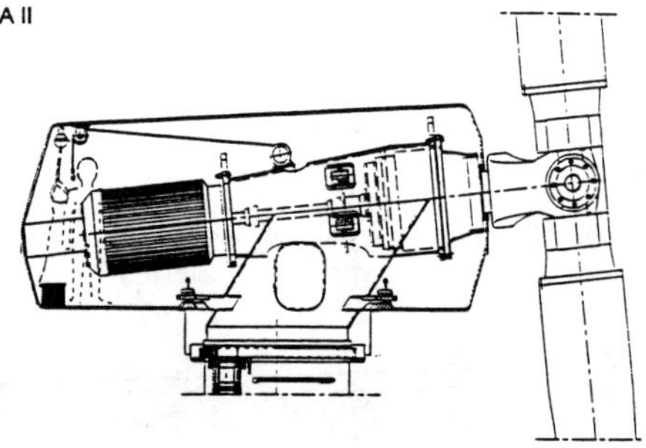

- Integrated Design, Short, Optimized Load Path from Rotor to Tower
- No, or Small Bedplate
- Compact Nacelle

Figure 7: Progress in the Conceptional Design of Drive Train and Nacelle

WEGA I

Three Blade Rotors

⇓

WEGA II

Three Blades	Two Blades
• Still Attractive at 50-60 m Ø due to the Availability of Lightweight Blades	• Two Blade Rotors with Very Low Weight • Noise Reduction by Variable Speed or Two Speed Operation

Figure 8: Three and Two Blade Rotors in the Forthcoming WEGA II Programme

weight as experienced in the past. Using this technology also a vertical axis concept could be a cost competitive alternative to the horizontal axis wind turbines.

A variety of large wind turbines in the megawatt range will be developed and tested during the next three years. It can be expected that at least some of the prototypes will establish the technical basis for an economically viable megawatt wind turbine. Certainly substantial additional work will be necessary to put these machines into series production, but as of today there are good prospects that large wind turbines, in the megawatt power range, will eventually achieve a commercial application in the late nineties. The utilisation of wind energy will then enter into a new phase of importance.

WEGA I

Blade Pitch Control

- Full or Partial Span
- Heavy and Complex Hydraulic Powered Systems

⇓

WEGA II

Pitch Control	Stall
Advanced Systems, i.e. individual electrical blade pitching motors	Optimized Stall Characteristics airfoils, rotor speed, better theoretical models, etc.

Figure 9: Power Control by Blade Pitch and Stall

WEGA I

Conventional Transmission with Mechanical Step-up Gearbox

⇓

WEGA II

Conventional Transmission and: Gearless Concepts with Direct Driven Multi Pole Electrical Generators and Frequency Converters

Horizontal Axis

Vertical Axis

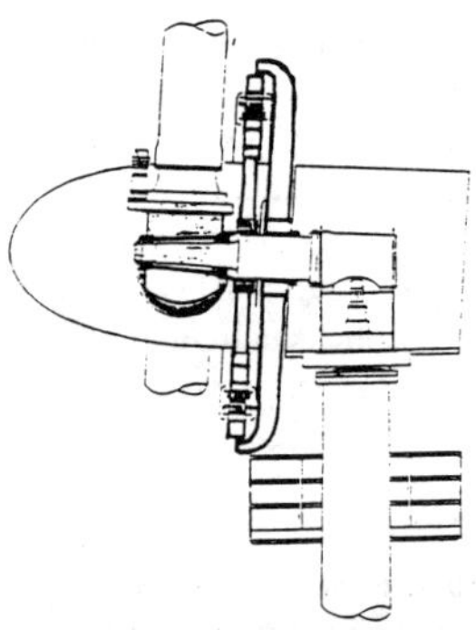

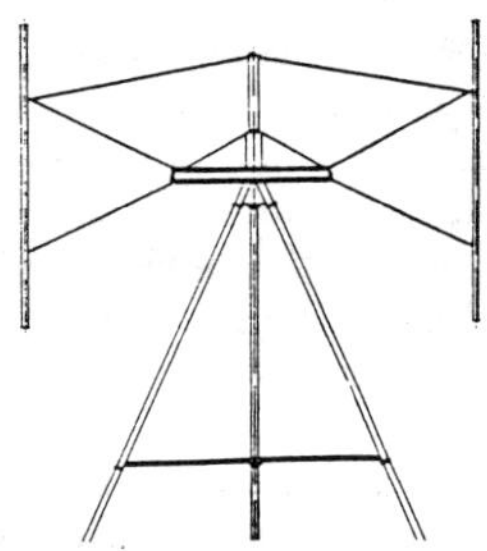

• Heavy Towerheads • Manufacturing Costs of Generators	• Longer Lifetime • Lower Maintenance Costs

Figure 10: Innovative, Gearless Drive Train Concepts at Horizontal and Vertical Axis Wind Turbines

REFERENCES

E. Hau, J. Langenbrinck, W. Palz
WEGA-Large Wind Turbines. Springer Verlag, 1993

E. Hau et. al. Study on the Next generation of Large Wind Turbines. Commission of the European Communities, DG-XII.

CEC, DG XII, WEGA Follow-up Actions.

Peter Christiansen B.Sc., Elec.eng.
ELSAMPROJEKT A/S
Kraftværksvej 53
DK-7000 Fredericia
Denmark.
Tel: +45 75 56 44 11
Fax: +45 75 56 44 77

ABSTRACT: Since the mid-eighties, CEC, DG XII has contributed to designing and developing three MW size wind turbines, known as the WEGA machines. In the same period, the LS1 Wind Turbine has been built and operated on Orkney Islands. As these projects represent a unique possibility of comparing different wind turbines with regard to design loads, operation strategies and actual measured loads, two follow-up projects The WEGA MEASUREMENT PROJECT and The WEGA DESIGN REVIEW AND EVALUATION PROJECT, have been initiated.
The purpose of the projects is to ensure that the maximum amount of know-how is gained, by performing a design review for the four turbines in order to identify their cost reduction potential and to evaluate the technological potential of these large turbines. The design reviews will be based on a comparison between measured loads and design loads together with an evaluation of the safety margins of the designs. The coordinator ELSAMPROJEKT A/S will finally make a comparison between the turbines at a technical and economic level to reach valid conclusions on a common basis.
Finally, in order to give an overview of the essential results of design, construction and the first period of operation of the three WEGA turbines a WEGA synopsis report has been prepared by ETA Gmbh. and published by DG XII.

1. BACKGROUND.

Since the mid-eighties, CEC, DG XII, has contributed to the design and the development of three large scale wind turbines known as the WEGA turbines. The turbines are:

* AWEC-60, operated by UNION FENOSA S.A.
* Richborough Wind Turbine, operated by PowerGen Plc.
* Tjæreborg Wind Turbine, operat. by I/S VESTKRAFT.

In the same period the LS1 Wind Turbine, operated by Scottish Hydro-Electric, has been built and operated on the Orkney Islands.

As these projects represent a unique possibility of comparing different wind turbines with regard to design loads, operation strategies and actual measured loads, two follow-up projects, the WEGA MEASUREMENT PROJECT and the WEGA DESIGN REVIEW AND EVALUATION PROJECT have been initiated. The purpose of the projects is to ensure that the maximum amount of know-how is gained, by performing a design review on the four turbines in order to identify their cost reduction potential and to evaluate the technological potential of these large turbines.
To ensure that similar measurements are performed and reported in a coordinated way the project coordinator, ELSAMPROJEKT A/S, has prepared 10 report templates outlining the extent of measurements and reporting formats for the three WEGA turbines. The LS1 turbine is associated to THE WEGA DESIGN REVIEW AND EVALUATION PROJECT and essential measurement results will be available for this turbine as well.
The strategy for the WEGA DESIGN REVIEW AND EVALUATION PROJECT is that each of the contractors will arrange for a design review to be performed on the main components of their respective turbines. The design reviews will be based on a comparison between measured loads and design loads together with an evaluation of the safety margins of the designs. On the basis of technical summaries provided by the contractors, the coordinator will finally make a comparison between the turbines at a technical and economic level to reach valid conclusions on a common basis.

Finally, in order to give an overview of the essential results of design, construction and the first period of operation of the turbines a WEGA synopsis report has been prepared by ETA Gmbh. and published by DG XII just prior to this conference (ECWEC'93).

2. STATUS FOR THE TURBINES.

2.1. AWEC-60.

At the end of November 1992, the accumulated statistics on operation for AWEC-60 were as follows:

Available hours	: 3,500 h
Operating hours	: 1,700 h
Production	: 700 Mwh

A repaired crack in one of the blades has caused the power set point to be reduced to 300 kW during "normal" operation. It is, however, raised to rated power (1200 kW) during measuring campaigns.
The organization responsible for performing the design review is in place. A number of Spanish companies involved in wind energy activities have each been contracted to perform the review on one or more of the components (systems).

2.2. Richborough Wind Turbine.
At the end of November 1992, the accumulated statistics for Richborough Wind Turbine were as follows:

Available hours	: 7,124 h
Operating hours	: 4,933 h
Production	: 1,830 Mwh

The turbine has been out of operation for more than half of 1992 due to an electrical fault in the generator. The turbine is back in operation but with reduced operating range pending a solution (repair) of a problem with the yaw brakes. It is anticipated that the measurement period will be finished in the third quarter of 1993.
The design review on the turbine will be performed by Renewable Energy Systems.

2.3. Tjæreborg Wind Turbine.
At the end of November 1992, the accumulated statistics for Tjæreborg Wind Turbine were as follows:

Available hours	: 16,353 h
Operating hours	: 11,234 h
Production	: 7,462 Mwh

The turbine is performing well after initial problems with the gearbox and has been in regular operation for most of the period. The measurement programme has been completed and the measurement system stopped at the end of 1992.
The design review on the turbine will be performed by ELSAMPROJEKT A/S. The reports are expected to be finished before the summer of 1993.

2.4. WEG, LS1.
At the end of November 1992, the accumulated statistics for WEG, LS1 are as follows:

Available hours	: - h
Operating hours	: 6,846 h
Production	: 9,204 Mwh

A 12-month monitoring period was started on February 28, 1992. The turbine is performing well.
The design review on the turbine will be performed by the supplier and designer of the turbine, Wind Energy Group. The reports will be finished by the end of May 1993.

2.5. WEGA Synopsis Report.
The WEGA synopsis report has, as one of the first results of the projects been published just prior to this conference.

3. THE WEGA MEASUREMENT PROJECT.
3.1 Introduction.
The purpose of the study is to organize, carry out and analyze the results from the measurement programs for the three wind turbines in the WEGA program in such a way, that a maximum of information can be obtained in coordinated action between the WEGA projects.

The short term objective of the project, i.e. those to be obtained within the time frame of the project are:

* To obtain better and unified technical reports from the projects and ensure that the same weight is attached to essential issues in all three projects.

* To create a comprehensive and well documented database consisting of time series describing the aerodynamics as well as the mechanical behaviour of the turbines.

The long-term objective is that the data base with time series shall be available to researchers in the wind energy and aerodynamics fields and in this way help to introduce wind energy in the European electricity supply system in an economic and safe way.
The strategy for the project is that the reports made by the contractors shall be structured according to common template reports, one for each subject. The templates are be prepared by the coordinator in cooperation with the contractors.

3.1. Report Templates.
The following nine templates have been prepared:

* **Measurement System Calibration Report.**
This report describes the data acquisitions system, the sensors and the their calibration. The report is fundamental for the understanding of all reports based on results from the measurement system and for the subsequent use of the data base. It also describes the site and presents the coordinate systems to which the data are referred.

* **Duty Cycle Measurement.**
This report presents the duty cycle for the turbine i.e. it basically reports the number of hours for which the turbine will be in each characteristic fatigue load situation according to the design basis and how many events it will experience. The report is the basis for fatigue life calculations.

* **Loads During Normal Operation Mode.**
The report presents the measured wind turbine loads obtained during the normal operation mode, such as start-up, energy production, stopping and standstill/idling. The wind turbine loads are summarized in tables and graphs together with eigen-frequencies.

* **Extreme Loads and System Events.**
The report identifies extreme loads and system events based on measured time series obtained during extreme meteorological conditions, during extreme mechanical loads or during turbine events.

* **Fatigue Load Characteristics.**
The report on the fatigue load story of the turbine. The loads are presented as range/mean matrices from rainflow counts on signals in all major sections of the turbine. The report follows IEA recommendation #3 on fatigue loads, closely.

* **Aerodynamic Behaviour of the Rotor.**
The report covers the investigation of the aerodynamic behaviour of the rotor. It presents the concept of the rotor transfer function (characterized by amplitude and phase plots) obtained by varying the pitch angle of the blades.

* **Control quality analyses.**
The report presents the control system of the wind turbine. It reports on the control quality in terms of time history plots, basic statistics, amplitude spectra, control accuracy, control activity on basis of time series for wind

accuracy, control activity on basis of time series for wind speed, power and pitch angle. It also presents the rotor damping function on turbulence.

*** Power Curve Sensitivity analysis.**
The reports describes the S-curve determination and the investigation of power curve sensitivity due to wind direction, yaw error, wind shear, turbulence and other factors relevant to wind turbine energy production calculations.

*** Cost of Operation and Maintenance.**
The report is the template to be followed when reporting on the costs of operation and maintenance. The report distinguishes between non commercial (prototype) costs and commercial costs, the latter being the most interesting in the evaluation of the energy costs from large wind turbines.

3.3 Data Standardization and Terminology.
The report **"Data Exchange Formats and Terminology"** describing the common data formats to be used when submitting data to the commission is finished and made available to the contractors.
The report defines three types of time series, from capture matrices, from cyclic buffers and from campaign measurements.
For these three types of time series, file formats and terminology are defined in the report. The report also suggest a library system to be used when cataloguing the time series for later retrieval and use.

4. THE WEGA DESIGN REVIEW AND EVALUATION PROJECT.

4.1 Strategy.
The strategy for the project is as follows:

Loads are measured under the WEGA MEASUREMENT PROJECT and reported according to common report templates prepared as part of the project in order to facilitate subsequent analyses. Similar measurements funded by the UK Dept. of Energy have been made at LS1 and will be used in this project.
On the basis of a comparison between measured loads and design loads, a design review will be performed for the four machines with the aim of finding the cost reduction potential of the designs through component and/or sub-system design improvements. The improved design shall be of the same basic concept as the turbines.
The contractors will arrange for the design review of their respective turbines to be performed, whereafter the coordinator will make a comparison at a technical level to reach conclusions, where possible, on a common basis.

4.2 Tasks.
*** WEGA synopsis report.**
A WEGA synopsis report, with the following aims, has been produced and published by DG XII:

* To give an overview of essential results of design, construction and the first period of operation of the three WEGA turbines.
* To disseminate the first R&D-results to the public but also to ensure that gained experience and learned lessons are utilized in the next generation of large wind turbines.

*** Design review of the three WEGA turbines.**
A design review shall be performed for the following three WEGA turbines:

* Richborough Wind Turbine operated by PowerGen Plc.
* AWEC-60 operated by Union Electrical Fenosa S.A.
* Tjæreborg Wind Turbine operated by I/S VESTKRAFT.

The review should comprise the following components and sub systems:

* Rotor blades.
* Blade bearings.
* Hub.
* Pitch mechanism.
* Main shaft, hub flange and bearings.
* Gearbox.
* Rotor brake.
* Electrical power system.
* Bedplate and yaw-bearing.
* Yaw drive system.
* Nacelle housing.
* Tower.

The design review shall comprise the following 5 subtasks:

*** Comparison of load cases.**
Each party will provide a description of essential load cases used in the turbine design.

*** Load assumptions and actual loads.**
Each party will provide information on assumed extreme loads and load spectra (range/mean matrices) associated with load cases used in the design.
Actual loads will be reported in the form of measured extreme loads and fatigue load spectra derived by rain flow counting.

*** Identification of safety margins.**
For components and sub-systems included in the design review, safety margins used in the mechanical design will be reported together with the influence of the national standards on allowable stress levels and presumed lifetime of the components.

*** Design evaluation.**
On the basis of the design review each party will present recommendations for an improved design of components and sub- systems included in the study.
The recommendations should be based on the precondition that the requirements of the national standards are fulfilled and the purpose should be to identify the cost reduction potential of the designs.

*** Summaries.**
A summary of design review and cost evaluation shall be prepared by each party. The summary should include an estimate of the manufacturing cost of the improved turbine.

4.3. LS1 association.
As all three WEGA turbines are of the same basic concept there is a considerable interest in making a comparison between those and a wind turbine of similar size but of a different technical concept such as the LS1 operated by Scottish Hydro-Electric. LS1 is a two-bladed turbine with teetered hub and narrow band variable-speed.

To make this comparison possible information on essential results concerning the following subjects will be made available.

* Load cases.
* Design loads and measured loads.
* Safety margins.
* General design improvement ideas.

The information will be presented in a form that makes it possible to compare it with the summaries described above.

4.4. Coordination.

Within the project framework the coordinator has a special technical task to perform in order to fulfil the objective of the project. On the basis of the summaries provided by the other 4 parties in this project, the coordinator will make a comparison between the turbines at a technical and economic level to reach conclusions, where possible, on a common basis.

5. RESULTS.

5.1. Introduction.

Few results from the design review are available yet due to delays with obtaining the necessary measurements. However the work is in progress and as one of the first results a comparison of the loadcases used in the 4 projects can be presented here. In the following the four wind turbines are identified by the use of the following abbreviations:

A:	AWEC-60
L:	LS-1
R:	Howden HWP 1000/55, Richborough
T:	Tjæreborg Wind Turbine

5.1. Basic design principles.

The design of the wind turbines A, L and T were based on fairly similar principles: An estimations of the loads, especially the wind, defining of load cases divided into one or more groups related to e.g. the condition of the turbine or conditions of the wind and finally a calculation of the load cases resulting in the design loads of the wind turbine.
The design of R was handled a little different as the load cases from the beginning were individually assigned to the individual components of the wind turbine and not to the whole turbine as such. It can be noted that due to the withdrawal of Howden, the supplier of R, from the wind turbine industry, the amount of documentation available is fairly restricted.

5.2 Design wind conditions

The design extreme wind speed for the 4 turbines lies in the range 60 to 67 m/s.

AWEC-60	: 64.5 m/s
LS-1	: 67.0 m/s
Howden HWP 1000/55, Richborough	: 60.0 m/s
Tjæreborg Wind Turbine	: 61.0 m/s

L, placed on the Orkney Islands, has, not surprisingly, the highest and R in Kent the lowest.

5.3. Load cases

Though that the design load cases of each wind turbine have not been defined or characterized in the same manner, an attempt has been made to make a common classification of the load cases with the purpose to compare the main conditions of each load case. The following table presents a brief overview of the load cases used in the design of the turbines.

WEGA Wind turbines. Load cases				
Wind Turbine	**A**	**L**	**R**	**T**
WIND TURBINE OPERATIONABLE				
Normal wind, production	x		x	x
Gusts, production	x	x	x	x
Normal wind, start & stop				x
Normal wind, idling/parked	x			x
Extreme wind, idling/parked	x	x	x	x
Extreme wind shear	x			x
Extreme yaw	x	x		x
WIND TURBINE IN FAILURE MODE				
Overspeed	x	x	x	x
Emergency stop		x	x	x
Yaw error				
Pitch fail		x	x	x
Blade position fail		x		
Loss of blade / mass imbalance		x	x	x

OPERATION AND EVALUATION OF YAW TEST RESULTS THE TJÆREBORG WIND TURBINE

Peggy Friis, Marianne Knudsen
ELSAMPROJEKT A/S, 53, Kraftværksvej, DK-7000 Fredericia, Denmark

Kurt S. Hansen
Department of Fluid Mechanics, Technical University of Denmark, DK-2800 Lyngby, Denmark

Abstract: The Tjæreborg wind turbine is a prototype of the MW class, built for demonstration and R&D purposes within the framework of the Danish wind energy programme. In January 1993 the measuring programme was completed and the turbine performed well. Our experience is based on more than 12.500 hours of operation. The turbine is an up-wind machine and thus the interesting feature of yaw misalignment has been investigated and related to main loads. The measurements show that it is possible to keep the average yaw misalignment of the turbine within ±5° and that operation is allowed within an average yaw misalignments band of ±20°, taking loads and energy production into consideration.

Background

The Tjæreborg Wind Turbine is a three-bladed up-wind turbine with a rotor diameter of 61m. The rated power is 2 MW. It was built in 1987 by the Danish electric utility, ELSAM. As part of the WEGA programme [1] it was granted subsidy from the CEC, DG XII. The Danish Ministry of Energy has also supported this project.

Operational results

The turbine was commissioned in 1988. After two and a half years with various problems, the turbine has now been in regular operation for almost two and a half years. As regards operation and availability, the results have been very satisfying, as is evident from figure 1.

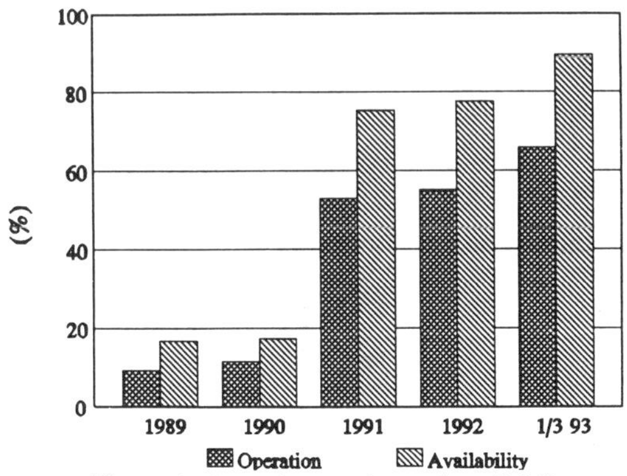

Figure 1. Annual operation and availability

The accumulated operational status as of March 1, 1993 is that the turbine has been in operation for 12,582 hours and available for operation for 18,233 hours and the generator has produced 8,780 MWh.

The test operation was completed in January 1993. Both operation and availability are expected to be increased in the future when the turbine has been put into commercial operation.

This assumption is based on the fact that, in order to gain experience with respect to operation and loads, the turbine has been deliberately operated under various conditions which have not always been optimum as regards operation, availability, loads etc. Furthermore, a lot of stops have been provoked in order to maintain the sensors of the measuring programme.

The annual increase in the number of operating hours has been about 5000 hours/year over the period with regular operation. Figure 2 illustrates the corresponding annual increase in the energy production.

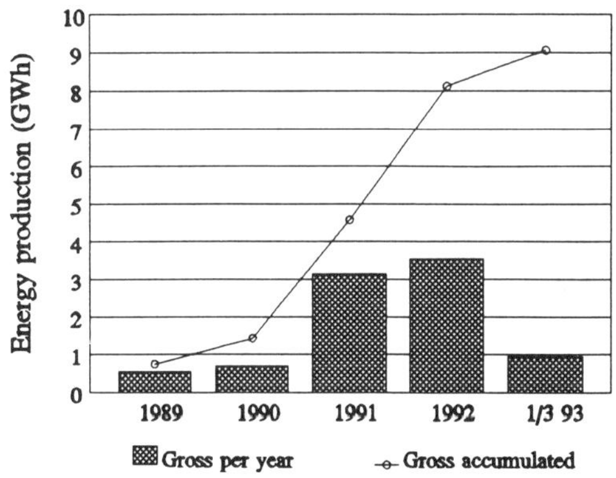

Figure 2. Annual energy production

The average production in 1991 and 1992 was 3.28 GWh/ year exactly corresponding to the expected annual energy production at 80% availability.

It is encouraging to see that the windy period in January and February 1993 has already brought the energy production up to 25% of the expected annual production.

Measuring programme

The scientific measuring programme ran for about 5 years and was completed in January 1993. During this period a methodical gathering of data was carried out. Measurements have been performed with a view to collecting data of various load situations and events at all wind speed regimes, in the form of both statistics and time histories. Some of the results are presented in reference [2].

Yaw test results

Operational performance and optimization of the yaw system have been interesting features and the yaw quality has been followed attentively, because a perfect yaw performance maximizes the energy production and minimizes the loads.

The main questions have been: How to achieve a perfect yaw performance, how perfect it should be, and in which yaw misalignments band operation is allowed?

Yaw misalignment is defined as the nacelle position minus wind direction, e.g. if the yaw misalignment is positive then the rotor is to the left in relation to the wind direction when looking down wind.

Adjusting the yaw operation and control strategy to the measured relative wind direction (WDN) by means of a wind vane, located behind the rotor on the top of the nacelle, proved to be a challenge.

The signal from the wind vane is the only information about yaw misalignment used by the control system. The wind vane measures the wind direction relative to the up-wind nacelle position.

The measured signal from the wind vane is affected by the nacelle casing and the turbulence in the rotor wake.

When the turbine is in operation the measured signal from the wind vane varies with the yaw misalignment and the wind speed.

Figure 3 shows the ratio between the measured signal from the wind vane and the yaw misalignment. The plotted data are derived from the extreme yaw misalignment campaign measurements. The set data points in figure 3 are based on 1 min. average values.

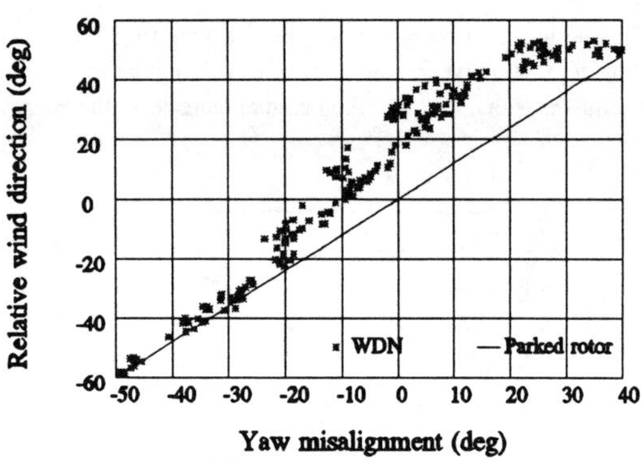

Figure 3. Absolute and relative yaw misalignment

At a yaw misalignment between ±20° the relation between yaw misalignment and measured wind direction at the wind vane is YM = 0.543 * WDN + 20.8°, the gain being 1.8.

When the rotor system is parked, the relative yaw misalignment is found to be increased by 1.2 as compared to the absolute value.

The changes in wind direction due to rotor wake is handled by adding a constant offset to the measured values. The current offset parameter during operation is 15°. The current gain used is 1.6.

The influence of the wind speed on the wind vane is reflected in the yaw quality during operation as shown in figure 4. The data used are 10 min. average values based on the latest set of control parameters.

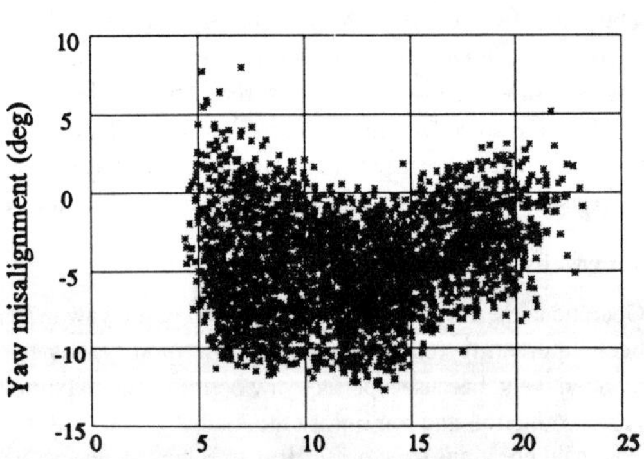

Figure 4. Yaw quality during operation

Figure 4 shows that, in an average band of ±5°, the average yaw misalignment is -5° over the whole wind speed range. With an offset of 20° it will be possible to keep the average yaw misalignment at 0° ±5°.

Figure 5 shows the sensitivity of the power curve to yaw misalignments. The data are classified according to the following options for yaw misalignment: YM = 0° ± 2.5°, YM = -10° ±5° and YM = 30° ±15°.

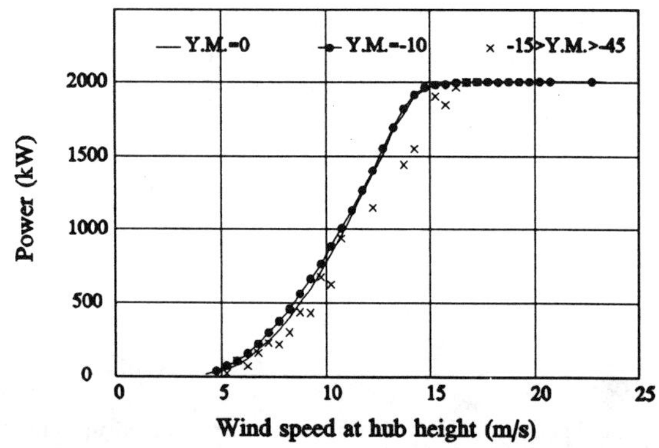

Figure 5. Power curve sensitivity to yaw misalignment

The influence exercised by an average yaw misalignment is found to be insignificant within a mean value of ±15°. The power curve is based on 10 min. average values.

The turbine has been deliberately operated under extreme yaw misalignment with the purpose of performing structural load measurements under special conditions. Among other things, the results are used for the operation and the control strategy of the yaw system.

There are some difficulties in calculating the mean bending moment measured by the strain gauges at the blades as well as at the main shaft, because the 0-point drifts continuously during time and the yaw misalignment campaign measurements have on several occasions been running for more than 5 hours without any operational stops. This should be borne in mind when evaluating the results.

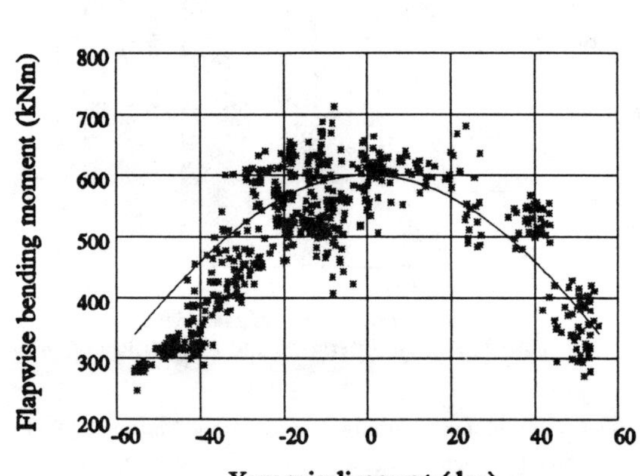

Figure 6. Mean Flapwise Bending Moment.

Figure 6 shows the mean flapwise bending moment of blade number 1 as a function of the yaw misalignment. The sets of data are based on 30 second average values, and the measurements are plotted when a linear wind speed adjustment to a wind speed reference of 8 m/s has been carried out.

The mean flapwise bending moment of the blade is not particularly influenced by a yaw misalignment in the range of ±20°, and it decreases at even larger misalignments.

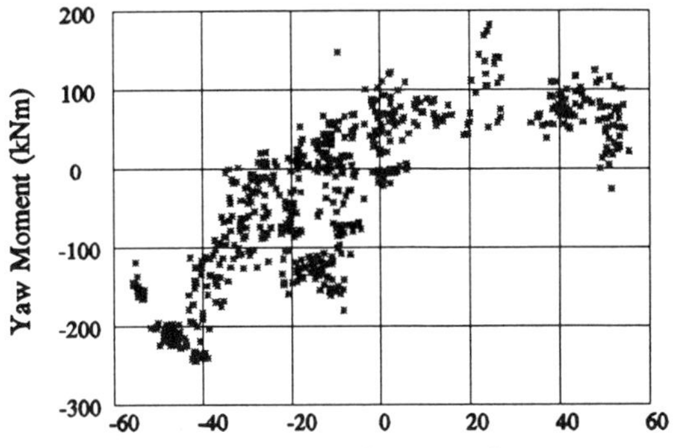

Figure 7. Yaw moment

Figure 7 shows the mean yaw moment as a function of the yaw misalignment. The data are based on 30 second average values, adjusted to a wind speed reference of 8 m/s.

The moment of yaw is derived from the 2 bending moments measured on the main shaft and transferred to a non-rotating system of coordinates and as such it is relative to the tower.

The moment of yaw helps bringing the turbine into an up-wind position, though there is an affinity to the left. The experience gained from the yaw misalignment campaign measurements was that it is difficult to obtain an extreme misalignment to the right of above 45° - 50°.

Conclusions

Concluding from the results it is possible to keep the turbine in an up-wind position at an average yaw misalignment of 0° ±5°. Operation can be allowed within an average yaw misalignments band of ±20°, taking loads and energy production into consideration.

Regarding the blade loads, the yaw misalignment is not all that important and a wide band is allowable. Regarding loads on the yaw system drive train, the best solution is a band of ±15° but on a short time basis higher values are allowed. The energy production is optimized in the band of ±15°, however, there is no reason to limit the operation because of such circumstances.

Acknowledgement

The authors want to thank the staff of the power plant I/S Vestkraft, responsible for the daily operation of the wind turbine. They have been most helpful and have shown a lot of patience during the campaign measurement where the interests of the two teams were of somewhat conflicting character.

References

[1] Peter Christiansen.
CEC, DG XII, WEGA Follow-up Actions.
Report no.: EP93/271
Presented at ECWEC 1993 in Travemünde.

[2] Kurt S. Hansen
An evaluation of measured and predicted fatigue loads for the Tjæreborg Wind Turbine.
Afm-note: VK-248
Presented at ECWEC 1993 in Travemünde.

[3] B. Maribo Pedersen, et al.
Some experimental investigations on the influence of the mounting arrangements on the accuracy of cup-anemometer measurements.
Presented at EWEC 1991 in Amsterdam.

[4] Stig Øye
Simulation of loads on a pitch controlled wind turbine during an emergency stop.
Afm-note: VK-246
Presented at Fifth IEA-Symposium on Aerodynamics, Stuttgart 1991.

[5] Peggy Friis & Kurt Hansen
Operating experience with Tjæreborg Wind Turbine, 1989 - 1990.
Report no.: EP93/271
Presented at ECWEC 1990 in Madrid.

[6] Peter Christiansen.
Design, construction and first operating results from the 60 m/2 MW Tjæreborg Wind turbine.
Report no.: EP90/678
Presented at WEC 1989 in San Francisco.

[7] Peggy Friis & Kurt Hansen, et al.
Evaluation of test results and operation experience with Tjæreborg Wind Turbine.
Report no.: EP90/676
Presented at EWEC 1989 in Glasgow.

[8] Peggy Friis
Commissioning of the 2 MW Wind Turbine, Esbjerg 1.
Report no.: EP90/673
Presented at ECWEC 1988 in Herning.

[9] Kurt Hansen, B. Maribo-Pedersen, et al
Preliminary test results from Esbjerg 1 -
60 m/ 2 MW wind turbine.
Report no.: EP90/672
Presented at ECWEC 1988 in Herning.

EXPERIENCES AND TEST RESULTS WITH THE AWEC-60 PROJECT

F. AVIA , E. SORIA
C.I.E.M.A.T.
Avda. Complutense 22
28040 Madrid (Spain)
Phone: 34-1-3466039
Fax: 34-1-3466037

ABSTRACT: The AWEC-60 is one of the large scale wind turbine generators supported by the Commission of the European Communities in the frame of the CEC WEGA Programme. The wind turbine was put into operation at the beginning of 1990 and was transferred to the utility UNION FENOSA which operates the wind plant. A large number of parameters, covering meteorological conditions, machine status, performance, electrical variables, etc, are being logged regularly and special campaigns of high frequency data collection have been undertaken. This report describes the main results obtained from the analysis of the recorded data, and the experience gained during the operation of the prototype, making emphasis in operation data, power curves, behaviour of the control system, statistics of the maintenance work performed in the different subsystems, etc... Comparison between theoretical and measured data, and some conclusions and guidelines for future developments of large scale wind turbines are presented.

1.INTRODUCTION.

The AWEC-60 project was conceived in the spring of 1985 by a spanish-german consortium with the aim of developing, manufacturing, test and operating a large scale windturbine of 1.2 MW rated power.The technical characteristics of the project are described in previous reports [1,2].The project is supported by the DGXII of the ECC, the BMFT (Germany) and OCIDE (Spain).The prototype was installed in Cabo Villano (La Coruna) Spain, and the acceptance test was completed in March 1990. The windturbine is in operation under the supervision of the utility Union Fenosa S.A., and at March 1993 has accumulated more than 2000 hours connected to the grid.During this period, the monitoring and evaluation of the windturbine has been performed inside the scope of the JOULE project "Follow-up WEGA Projects" that pursue to analyze the results from the measurement programs for the three windturbines included in the WEGA program, in such a way, that a maximum of information can be obtained in this coordinated action.

2.OPERATION AND MAINTENANCE EXPERIENCES.

The wind turbine was commissioned at the beginning of 1990, and during the first year of operation several problems appeared, mainly in the yaw system (hydraulic motors, and brake system) and control system. The windturbine was put in regular operation in December 1990, and was stopped in May 1991 , when during a programmed revision a crack in the main spar of blade number two was detected. The blade was dismounted to carry out the reparation works and was mounted again during the first week of October 1991. Since them, the windturbine has been operating regularly, accumulating a total of near 2000 hour grid connected at the end of 1992. During this period, the wind turbine is operating most of the time at reduced set-power, that will be maintained until a new blade will be available. Only for the realization of some specific tests (like noise measurements campaigns performed in February 1992), the wind turbine operates at design rated power 1.2 MW.
Graph 1 shows the statistics of operation.

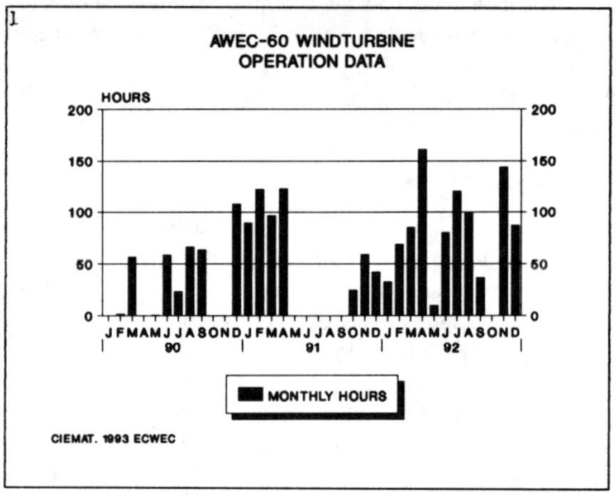

Graph 2 shows the man-hours spent in operation and maintenance works for the whole wind turbine from the beginning of the operation and Graph 3 is the break-down for the different subsystems in percentage of the total hours for the year 1992. Neither the hours dedicated for the maintenance of the measurement and data acquisition system, nor the hours spent in for the reparation of the blade have been included in the graph.

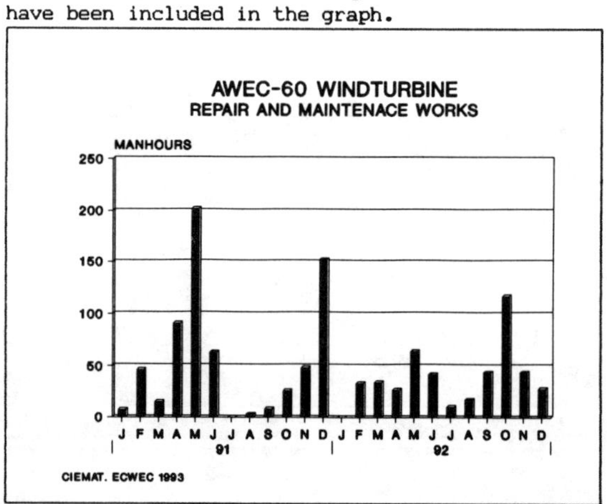

The yaw system and hydraulic unit have consumed the large number of hours of the repairing and maintenance works, being these systems the most critical part of the wind turbine.The work performed included change of one of the hydraulic motors of the yaw, change of the pinions, adjustment of the yaw brake, repairing of several linkages, etc ...
Most of the time included in the auxiliary system have been consumed for the auxiliary diesel group and cooling system.
The hours included for the blades are only for checking, cleaning and painting, and not the consumed for the reparation of blade 2.

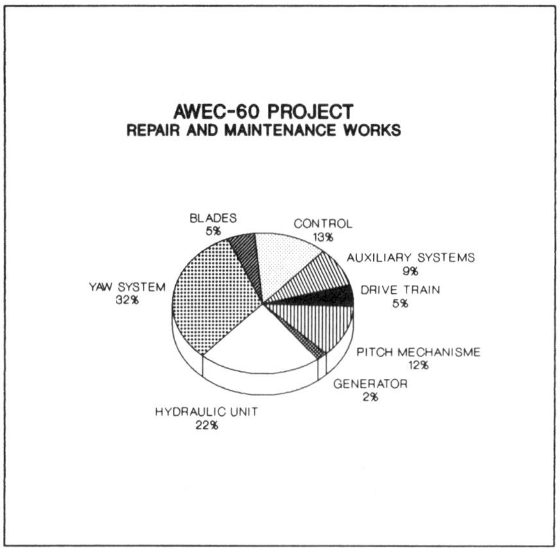

3. INCIDENTS.

The main incidents registered during the operation of the wind turbine deal with the yaw system and hydraulic unit, and with the blades. Also some incidents experimented in the electrical system, pitch mechanism and control systems have been reported. The design has to be improved in order to increase the reliability of the windturbine, but no unresolved appears from the technical point of view. A malfunctioning of the yaw operation system was detected in the early time. A failure in one of the two hydraulic motors was detected, and several damages were found in the pinion and the shaft of the yaw motor. A modification in the yaw sequence was performed in order not to release completely the yaw breakers during the yaw sequence.
Also a modification in the control strategy when the windturbine is in operation in oversynchronous cascade mode [Ref 5] was performed. When operating with the resistors connected to the rotor of the generator, the heat dissipated produces a high increase of the temperature in the tower base (were the resistors are located), that makes necessary to limit the set-power of the machine.

After the detection of a crack in the main spar of the blade number two, the blade was dismounted to analyze the problem and to carry out the reparation work. The crack appeared at the main spar in the 8.0 m. blade section and implies a serious damage in the structural part. Graph 6 shows the location of the crack. The reparation procedure was defined after a carefully inspection for the CIEMAT technicians, and performed "on site". The cause of the crack was a manufacturing defect produced during the process of situating the laminated inside the molds, that produced the wrinkling of some layers of the laminate. The defect was not detected in the inspection performed, and two main conclusions should be realized, first that the manufacturing process used for manufacturing the main spar has to be improved in order to avoid the possibilities of defects like delaminations, porosity-voids, resin concentration, and layer wrinkling, and on the other hand that new more reliable quality control procedures for non destructive evaluation has to be developed. The methods used in the final blade inspection [Ref. 4] were hammering techniques and visual inspection through the thickness, due to the transparency characteristics of the laminate , appear not to be adequate because the defect was not detected.
For the reparation of the blade the procedure was defined according with the particular characteristics of the laminate. The following sequence was preparation of the "on site building", removing the secondary structure of the blade, preparation of the area to be repaired (terraces realization) lamination process , assembly and finishing. It took about three weeks to make the work.

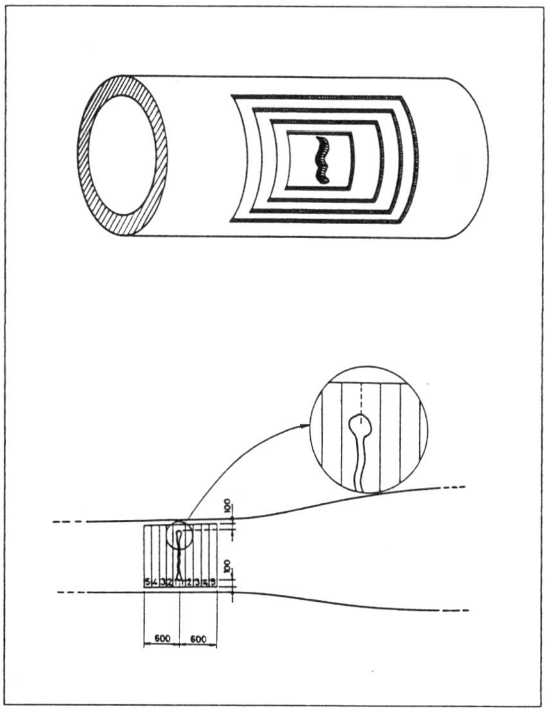

4.POWER CURVE.

The power curve of the wind turbine was obtained for the two prevailing wind directions. Description of the data process, methodology for the correction of the data and analysis is widely described in the document "AWEC-60 power curve" [Ref 6]. In graph 5, the power curves obtained in relation to the two meteorological mast, and the theoretical power curve are presented.

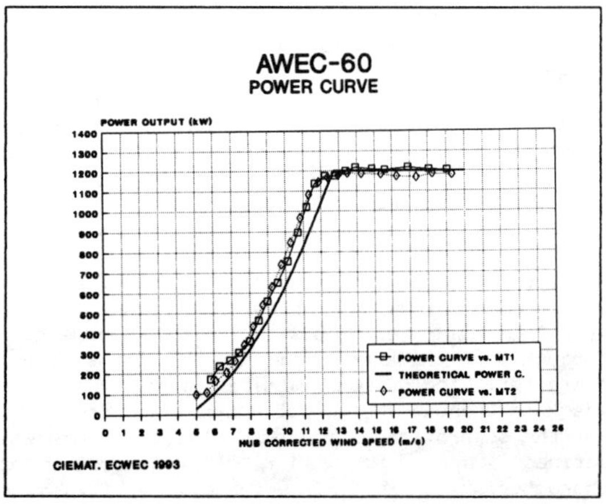

The discrepancy with the theoretical value is very low being the measured values lightly higher than the theoretically calculated for the partial load operation mode. That is due to the fact already known, that for the calculation the coefficients used were lower than the actual coefficients.

5.NOISE MEASUREMENTS.

The sound power level from the windturbine was measured according to the specification of the recommended practices of the AIE, at the distance of 76 m. (H + D/2) downwind the tower centre. The results obtained are presented in Table I.

Operation conditions	Wind speed	Up-wind (76 m)	Down-wind (76 m)	Down-wind (152 m)
Stopped	6,0 m/s	40 dB<	40 dB<	40 dB<
Cut-In	5,5 m/s	49,3 dB	48,3 dB	40 dB<
Partial load	7,0 m/s	52,5 dB	51,2 dB	40 dB<
Rated power	11,7 m/s	57,6 dB	58,3 dB	47,3 dB

6. ELECTRICAL SYSTEM.

Three different modes of operation have been considered in the project:

- -Operation with oversynchronous cascade
- -Operation with rotor resistors
- -Operation with short-circuit rotor

The three operational modes have been tested. The windturbine runs regularly with the oversynchronous cascade and only for some specific test the other two modes of operation are used. The oversynchronous cascade has demonstrated to be a very suitable method to introduce the variable speed concept in the operation of the windturbine. A comparative study to assess the influence of the variable speed system on the loads over the windturbine is pending of realization. However, from the experience gained from the operation it is possible to conclude that the behaviour of the machine is more elastic and smoothly, running in the oversynchronous cascade mode. Graphs 6 is a print-out of a one minute period in a gusty wind of about 15 m/s, showing that the rotor speed is following the power variations while the shaft torque is more or less constant.
The generator slip reaches values of near 10 %

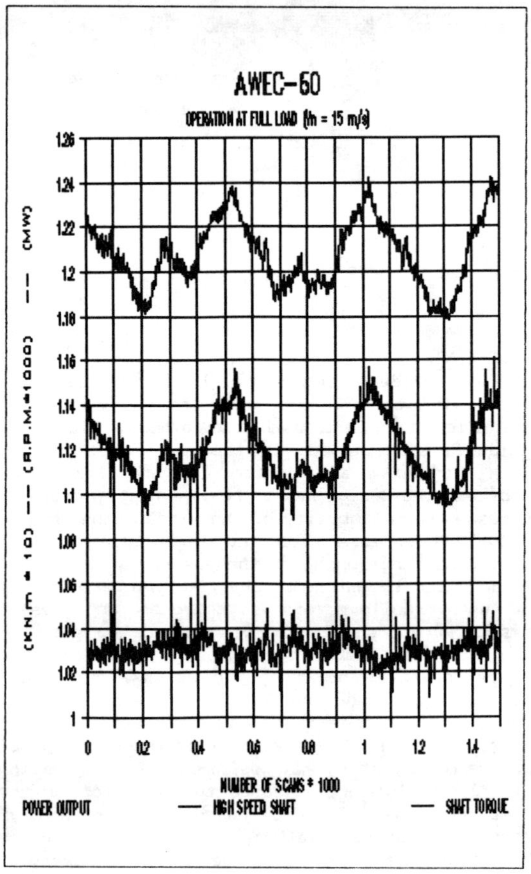

7. CONTROL SYSTEM.

The control system has resulted to be one of the most well designed systems of the windturbine, being very flexible and easy to modify, and very reliable and secure.

During the operation of the wind turbine several modifications were realized in the control system of the machine.

As it was stated in previous paragraph, the yaw sequence was modified. Also the wind vane sensor on the nacelle of the windturbine was modified to prevent the frequent stop of the windturbine due to exceed the maximum misalignment permitted by the control.

The pitch control maintains the power output for the full load operation mode (over rated wind speed) without power excursions due to the combination of the variable speed and the pitch angle variation.

In graph 7 the main functional parameters for a typical start-up and grid connection at a wind speed near the cut-out is presented.

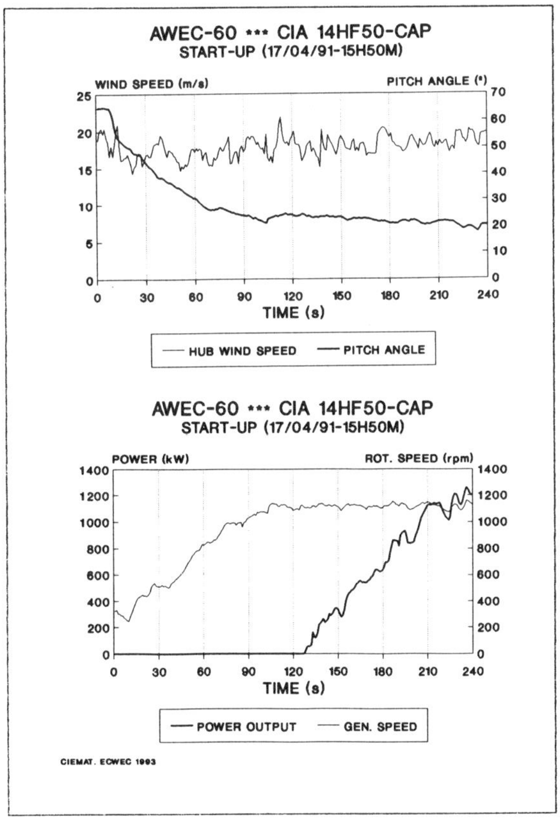

8.CONCLUSIONS.

The experimental test program is following its realization with particular emphasis in the measurements campaign of loads. The initial stress measurement system has been replaced by a new one, and the strain gages have been also substituted.On the other hand, and inside the scope of the JOULE project "WEGA Design Review and Evaluation Project", a study of the redesign of the windturbine is carried out with the objective of assessing the possibilities to reduce the cost on a new windturbine designed under the same technical concept. In particular, the study is focused in the nacelle, drive train, yaw system and blades, that have been identified as the subsystems with higher potentiality of cost/weight reduction. The technical concept of variable speed and full-span pitch control has shown an excellent behaviour that allows to predict a promising future for its utilization in the new generation of large scale windturbines. The cost increase due to the implementation of the oversynchronous cascade versus the cost of the normal short-circuit rotor of the induction generator is lower than the 2 % of the windturbine cost. The company Bihar that supplied the oversynchronous cascade is performing an optimization study for Union Fenosa. The experience accumulated will be applied in the new designs of spanish wind turbine, and in particular in the new 500 kW windturbine of the utility ENDESA.

REFERENCES.

1.- "Development of an advanced grid connected WEC in the 1 MW class (AWEC-60)" . A. Cano et alt. EWEC 1988. Herning . Dk.

2.- "Operation and maintenance experiences and test results from the AWEC-60 project". E. Soria et alt. ECWEC 1990. Madrid. Spain.

3.- "Development of composite blades for megawatt size wind turbines". F. Arias et alt. ECWEC 1990. Madrid . Spain.

4.- "Quality control in windturbine blades". M. Esteban et alt. ECWEC 1990. Madrid. Spain.

5.- "Electrical system of wind generator AWEC-60". A. Matas. ECWEC 1990. Madrid . Spain.

6.- "The AWEC-60 Power Curve". F. Avia. CIEMAT. (In spanish)

7.- "The AWEC-60 Start-Up Sequence". F. Avia. CIEMAT (In spanish)

Entwicklung und Betrieb der Großwindanlagen AEOLUS II und WKA 60

D. Knünz*, G. Schmitz**

*MBB Förder- und Hebesysteme GmbH
Postfach 1840, 2870 Delmenhorst

**PreussenElektra Windkraft Niedersachsen GmbH
Postfach 4849, 3000 Hannover 91

Um mit Windkraftanlagen einen signifikanten Beitrag zur CO2-freien Stromproduktion leisten zu können und die Primärenergieresourcen zu schonen, sind Großwindanlagen erforderlich. PreussenElektra Niedersachsen GmbH hat mehrere MW-Anlagen, WKA 60 und AEOLUS II, installiert. Damit werden die technischen und kommerziellen Rahmenbedingungen für diese Art der Stromproduktion ermittelt.

1. Einführung

Das Einsatzgebiet von Windkraftanlagen ist zum einen an Standorten zu suchen, wo Wind als einzige Primärenergiequelle vorhanden ist. Der Einsatz und die Größe der Anlage dort richtet sich nach dem örtlichen Bedarf und den wirtschaftlichen Möglichkeiten. Zum anderen kann man Windkraftanlagen in industrialisierten Ländern aus ökologischen Gründen fordern, um neben Wasserkraft und Kernenergie Quellen für CO2-freie Stromerzeugung zu besitzen. Für diese Anwendungsfälle sind Großwindanlagen hervorragend geeignet, um auf begrenztem Raum mit einem bestehenden Stromnetz das Windangebot optimal zu nutzen und dadurch einen signifikanten Anteil und damit einen energiewirtschaftlich meßbaren Beitrag zur CO2-freien Stromproduktion zu erreichen.

Für die Bundesrepublik Deutschland bedeutet dies z.B.:

Um nur 1% des jährlichen deutschen Stromverbrauchs (rund 460 Mrd kWh) mit 3 MW-Windkonvertern zu erzeugen, müßten hiervon etwa 650 Anlagen installiert werden, was eine Doppelreihe von Anlagen entlang der Nordseeküste von den Niederlanden bis Dänemark erforderte. Es leuchtet sofort ein, daß ein solcher Anspruch mit kleinen Anlagen gar nicht befriedigt werden kann, so daß es sich lohnt, über Großwindanlagen nachzudenken und ihre Entwicklung zu fördern.

Der ad-hoc-Ausschuß "Großwindanlagen" beim BMFT gibt in seinem Abschlußbericht vom April 1992 folgende Empfehlungen (auszugsweise)(1):

"Aufgrund der ermittelten hohen Standortpotentiale und der gegenüber mittelgroßen Anlagen verminderten Umwelteinflüsse wird die verstärkte Entwicklung und Nutzung von Großwindenergieanlagen empfohlen.
Durch Großwindanlagen kann sich das Potential der Windenergienutzung in Deutschland gegenüber kleineren Anlagen durch höhere Windenergienutzung etwa um den Faktor 2 vergrößern. Unter diesen Voraussetzungen ist am ehesten ein energiewirtschaftlich nennenswerter Windenergiebeitrag zur Stromerzeugung erreichbar. Hierzu kommt, daß mit Großanlagen die Windenergie landschaftsschonender genutzt werden kann."

Die von PreussenElektra und Schleswag in Zusammenarbeit mit dem BMFT und der EG, Generaldirektion Energie und den Ländern Schleswig-Holstein und Niedersachsen geplanten, geförderten und durchgeführten Projekte haben zum Ziel, die Effektivität und Verfügbarkeit der Windkraftanlagen im MW-Bereich zu ermitteln. Durch die Einspeisung der elektrischen Energie in unterschiedlich konfigurierte Netze eröffnet sich darüber hinaus die Möglichkeit, Netzrückwirkungen zu analysieren. Mit diesen Grundlagen sind die EVU in der Lage, bei künftigen Entscheidungen in Gremien des BMFT u.a. die Erfordernisse fundiert mitgestalten zu können.

2. Ausgewählte Großwindanlagen

2.1 Anlagenbeschreibung WKA 60

Die WKA 60 der Firma MAN Technologie GmbH ist eine inzwischen in drei nahezu baugleichen Exemplaren existierende Großwindanlage mit einer installierten Leistung von 1.2 MW.

Die drei Maschinen sind in Nordwestspanien (AWEC 60), auf der Insel Helgoland (WKA 60-1) sowie im Kaiser-Wilhelm-Koog (WKA 60-2) aufgestellt. Sie unterscheiden sich - z.T. wegen der örtliche Gegebenheiten - in der Turmkonstruktion und im elektrischen System. Die AWEC 60 hat zudem in Spanien entwickelte und gefertigte Rotorblätter.

Die beiden deutschen Anlagen besitzen einen Dreiblattrotor von 60 m Durchmesser mit starrer Nabe. Die Rotorblätter sind in Glasfaserverbundbauweise gefertigt. Es sind die gleichen, wie sie auch bei der Elsam-2000-Maschine eingesetzt sind. Sie sind aus drei Bauteilen, dem über einen Dorn gewickelten Holm und der Ober- und der Unterseite des Flügelprofils gefertigt, welche miteinander verklebt werden.

Die Rotorblätter sind über vorgespannte Vierpunktlager mit der Nabe drehbar verbunden. Die Nabe ist als Präzisionsstahlgußteil hoher Festigkeit ausgeführt. Sie wiegt etwa 15 to.
Die Rotorblätter werden hydraulisch-mechanisch einzeln angesteuert. Das Regelungssystem synchronisiert die Blattverstellung. Das hydraulische System zur Versorgung der Steuerung ist im Maschinenhaus untergebracht, wohingegen die Blattverstellzylinder und die hydraulischen Akkumulatoren zur Schnellabschaltung in die Nabe eingebaut sind, was die Zugänglichkeit für Wartung und Reparatur wesentlich erschwert und die Verfügbarkeit der Maschine bisher negativ beeinflußte.

Das maschinentechnische Konzept der Anlage beruht auf dem konsequenten Einsatz konventioneller Bauteile des Maschinenbaus. Die langsam laufende Rotorwelle ist auf zwei Pendelrollenlager gelagert, zwischen denen die Rotorbremse - als ausschließliche Haltebremse - angebracht ist. Das Getriebe besteht aus zwei Planeten- und einer Stirnradstufe auf der schnellen Seite. Zwischen Getriebe und Generator ist eine Bogenzahnkupplung mit einer Sollbruchstelle als Überlastschutz angeordnet. Am Getriebe ist ein elektrisch angetriebener, aber hydraulisch einkuppelbarer Rotor- positionierantrieb für Montage- und Reparaturarbeiten ange- flanscht. Diese Bauweise erlaubt zwar einen relativ guten Zugang zu den einzelnen Komponenten, baut aber groß und schwer (siehe Bild 1).

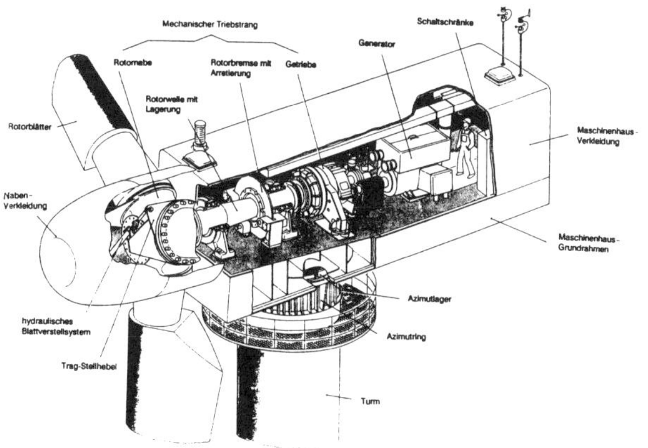

Bild 1: WKA 60 Maschinenhaus

Die WKA 60 ist mit einem Synchrongenerator mit Gleichstrom- zwischenkreis ausgerüstet. Dadurch ist die Rotor-/ Generatordrehzahl unabhängig von der Netzfrequenz. Die Beanspruchung der Maschine wird durch dieses "weiche" Drehzahlkonzept insgesamt geringer. Außerdem bringt der Betrieb bei niedriger Windgeschwindigkeit mit abgesenkter Drehzahl deutlich niedrigere Geräuschemissionen.

Der Turm der WKA-60-1 ist eine auf den Felsen der Insel Helgoland gegründete Spannbetonkonstruktion aus Ortbeton.

Der Turm der WKA-60-2 ist aus vorgefertigten Betonschüssen von je 2,5m Höhe zusammengesetzt. Dabei dient das Fundament des demontierten Growian als Fundament für die WKA 60-2.

Beide Maschinen werden fernüberwacht. Die WKA-60-1 als einer der Energieerzeuger des Kraftwerkes Helgoland kann vom etwa 800m entfernten Kraftwerk aus den Lastanforderungen entsprechend gesteuert werden.

Die WKA-60-2 soll mit maximaler Ausnutzung automatisch laufen. Deswegen wird in der ca. 30 km entfernten Warte nur der Betriebszustand der Anlage angezeigt. Dazu gehören auch auch Störmeldungen - insbesondere solche, die ein sofortiges Eingreifen erfordern wie z.B. "Ausfall des Hydrauliksystems " und die Brandmeldungen. Vom Hersteller wird zudem eine aufwendige tägliche Begehung der Anlage gefordert.

2.2 Bau und Betriebserfahrungen WKA 60

WKA 60-1 auf Helgoland

Während der Planungsphase wurde festgestellt, daß es bei bestimmten Wetterlagen (schwacher Nordostwind) am Standort der Anlage zu plötzlichen Windrichtungsänderungen von bis zu 30 Grad kommt. Der daraus resultierende Lastfall führte zu langwierigen Diskussionen im sog. projektbegleitenden Ausschuß und zu ebensolchen neuerlichen Auslegungsrechnungen.

Während der Fertigung der Stahlblechtragkonstruktion des Maschinenhauses führten schweißbedingte Verformungen zu erheblichen Schwierigkeiten bei der anschließenden mechanischen Bearbeitung (Überschreiten der Fertigungstoleranzen).
Wetterbedingte Verzögerungen traten bei der Montage von Maschinenhaus und Flügel auf.

Während des ersten Betriebsjahres blieb die technische Verfügbarkeit erheblich unter den Erwartungen. Die Gründe dafür waren:

- wiederholte Leckagen an den Hydrauliksystemen; notwendiger Austausch der Blattverstellzylinder in der Nabe

- wiederholte Störabschaltungen durch falsche Stellung des Rotors zum Wind (s.o.: Lastfall , plötzliche Windrichtungsänderungen)

- elektrische Störungen im Inselnetz durch Oberschwingungen bei bestimmten Leistungsstufen der WKA

- Leistungsschwingungen im Inselnetz zwischen WKA und Kraftwerk; dadurch bedingt notwendige Begrenzung der WKA auf etwa 67% der Nennleistung über längere Zeit hinweg

- Blitzschlag mit Zerstörung eines Flügels

Die Betriebsergebnisse der ersten Jahre sind in der folgenden Tabelle zusammengefaßt:

Jahr	1992	1991
Betriebsbereitschaft	4.750 h	6.150 h
Betrieb am Netz	3.030 h	3.360 h
Wartestellung	1.670 h	2.760 h
Störabschaltungen	149	186
Notabschaltungen	32	33
Zeiten mit externer Leistungsbegrenzung	1.366 h	1.566 h
Energieerzeugung	1.203 MWh	1.266 MWh

Tabelle 1: Betriebsstatistik WKA 60 I/ Helgoland

WKA 60-1 auf dem Kaiser Wilhelm Koog

Da die beiden Maschinen nahezu baugleich sind, gab es bei Bau und Montage keine nennenswerten Schwierigkeiten. Die Montage mittels Autokränen ist im Bild 2 und 3 gezeigt.

Bild 2: WKA 60-2 Maschinenhaus-Montage

Allerdings entstanden während der Inbetriebsetzung und des Leistungsbetriebes bisher die folgenden Probleme:

- Leckagen im Hydrauliksystem, die zwangsläufig tägliche Kontrollgänge notwendig machen

- Oberschwingungen im elektrischen Netz mit unmittelbaren Auswir kungen auf andere Windenergieanlagen und Überla stung /teilweise Zerstörung der Kompensationsaggregate. Die Nachbesserung der elektrischen Filteranlage verursachte einen zusätzlichen Kostenaufwand von 150 TDM.

- häufiges Störabfahren durch Ansprechen der Turmschwingungsüberwachung aufgrund von Defekten

Als nachteilig erwies sich die lange Beschaffungszeit für verschiedene Komponenten, im Extremfall bis zu 9 Monaten. Dies läßt sich mit den Erfordernissen einer kommerziellen und rationellen Stromerzeugung nicht vereinbaren.

Bild 3: WKA 60-2 Rotorblatt-Montage

2.3 Anlagenbeschreibung AEOLUS II/ Näsudden II

Die AEOLUS II Windkraftanlage ist eine Weiterentwicklung der 1983 auf der schwedischen Insel Gotland installierten Anlage Näsudden I mit 75 m Rotordurchmesser, 2 MW Nennleistung und 80 m Turmhöhe.

Diese Entwicklung wurde, wie bereits bei der Näsudden I, gemeinsam von den Firmen MBB und Kvaerner Turbin AB (KTAB) im Zuge eines deutsch-schwedischen Kooperationsprojektes durchgeführt. Es werden gleichzeitig 2 Anlagen des Typs AEOLUS II gebaut. Die deutsche Anlage wird in Wilhelmshaven, die schwedische auf der Insel Gotland installiert.

Diese beiden Anlagen unterscheiden sich in den Punkten Turm und Generatorsystem. Die schwedische Anlage wird auf den von der ersten Anlage vorhandenen Turm gesetzt und der Frequenzabgleich über eine Turmverlängerung mittels einer Schweißkonstruktion bewerkstelligt, während für den AEOLUS II das dynamische Turmkonzept neu erarbeitet wurde.

Beim AEOLUS II liegt die 1. Biegeeigenfrequenz des Turmes mit installierter Gondel zwischen der 1. und der 2. Rotordrehfrequenz. Bei Näsudden II liegt die 1. Turm EF über der 1. und 2. Rotordrehfrequenz, so daß ein Durchfahren der Turm EF nicht notwendig ist.

Die Generatorsysteme unterscheiden sich dadurch, daß für den AEOLUS II ein drehzahlweiches System mit Sychrongenerator mit nachgeschaltetem Gleichrichter und netzgeführtem Wechselrichter verwendet wird, während bei Näsudden II ein polumschaltbarer Asynchrongenerator mit 2 Drehzahlen, je nach Windenergieangebot, starr am Netz betrieben wird.

Der AEOLUS II ist vom Grundkonzept so wie Näsudden I aufgebaut mit 2 Blattrotor, starrer Nabe, Planetangetriebe mit integrierter Rotorwellenlagerung und angebautem Winkelgetriebe. Der Generator ist zur Vermeidung der Leistungsübertragung durch Schleifringe turmfest angebracht. Die Konzeption des Maschinenhauses ist in Bild 4 dargestellt.

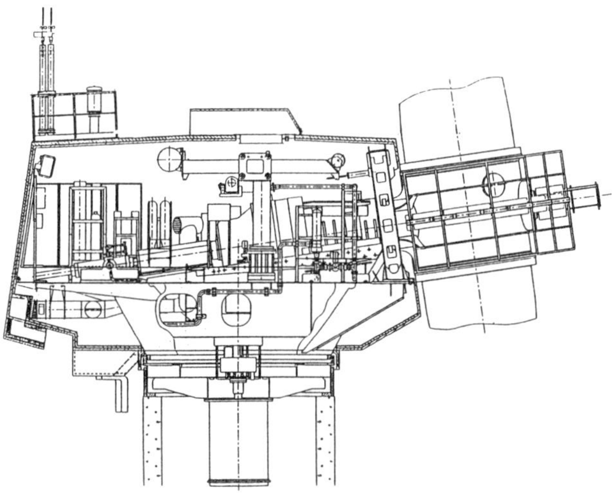

Bild 4: AEOLUS II Maschinenhaus

Die Rotorblätter werden über eine zentral im Maschinenhaus angebrachte Verstellhydraulik synchron verstellt. Das zweite unabhängige Bremssystem wurde durch eine auf der langsam laufenden Getriebeseite angeordnete Not-Aus-Scheibenbremse verwirklicht.

2.4 <u>Projektdurchführung AEOLUS II</u>

Das Projekt AEOLUS II wurde in deutsch-schwedischer Kooperation durchgeführt. Es wurde folgende Arbeitsteilung vereinbart:

MBB: Lastendefinition und Berechnung
Aerodynamische Auslegung
Dynamische Gesamtsystemauslegung
Rotorblattentwicklung

KTAB: Maschinenhauskonstruktion und Berechnung
Nabe mit Blattverstellung
Entwicklung der Subsysteme

Die Komponenten Turm mit Gründung, die elektrischen Systeme und die Betriebsführung entwickelte und baute jeder für seine Anlage selbständig, während die o.g. Entwicklungen gem. Arbeitsteilung auch für die Maschine des Partners durchgeführt und für beide Anlagen gefertigt wurden.

Die Näsudden I-Rotorblätter bestanden aus einem lasttragenden Stahlholm und einer GFK-Schale zur Profilierung, während die AEOLUS II-Blätter aus faserverstärktem Kunststoff gefertigt wurden. Dies eröffnete die Möglichkeit, den Rotordurchmesser von 75m auf 80m zu vergrößern und die Nennleistung von 2 auf 3 MW zu erhöhen. Trotz des größeren Durchmessers der AEOLUS II-Anlage konnte das Blattgewicht von 22 to auf 9 to gesenkt werden.

Die neuen Rotorblätter sind aus glas- und kohlefaserverstärktem Kunststoff mit ober- und Unterseite und an der Blattvorder- und Hinterkante. Die Schubübertragung von Ober- zu Untergurt erfolgt über PVC-Schaum. Die Blattaußenhaut besteht aus einer Glas- sandwichhaut, die gegen Beulen und Knittern durch Stringer und Rippen gestützt wird (siehe Bild 5). Es wurde die Profilfamilie FX84W eingesetzt. Pro Blatt wurden im wesentlichen 3.100 kg CFK, 1500 kg GFK und 4.200 kg PVC-Schaum eingesetzt.

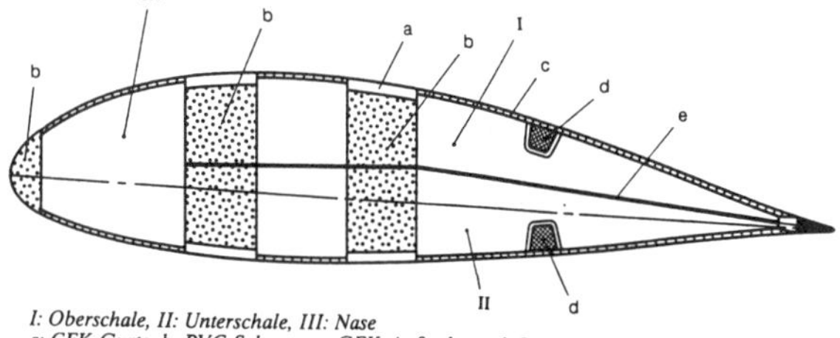

I: Oberschale, II: Unterschale, III: Nase
a: CFK-Gurte, b: PVC-Schaum, c: GFK-Außenhaut, d: Stringer (Schaum, GFK), e: Trennebene zwischen Ober- und Unterschale

Bild 5: AEOLUS II Rotorblattschnitt

Die Rotorblattfertigung erfolgte in Lemwerder bei A & R Faserverbundtechnik. da es sich bei diesen Rotorblättern um 40 m lange Teile handelt, mussten neue Verfahren für Fertigung und Handling geschaffen werden. Die Blätter werden in den drei Hauptkomponenten Oberschale, Unterschale und Nase in den entsprechenden Negativ-Formen gefertigt, für die Verklebung vorbereitet und dann zusammengebaut (Bild 6 und 7).

Bild 6: Blatthälfte entformen

Bild 7: AEOLUS II Fertig geklebtes Rotorblatt

Der Blattanschluß erfolgt durch axial im Anschlußquerschnitt eingesetzte Dehnschrauben, welche über Querbolzen im Blatt befestigt sind (siehe Bild 8).

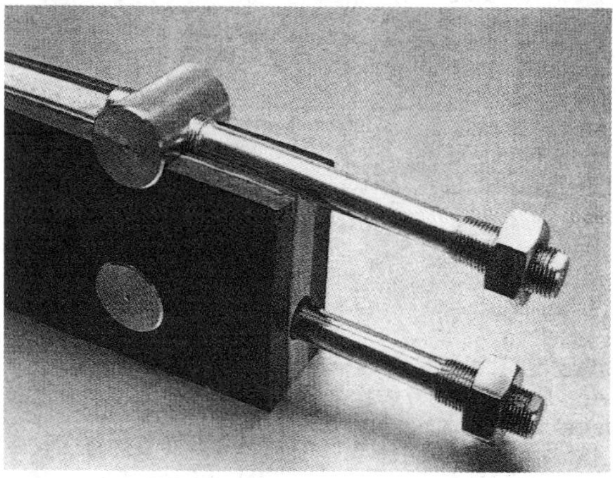

Bild 8: Rotorblattanschluß

Besonderes Augenmerk wurde dem Blitzschutzsystem (2), (3), gewidmet. Die am meisten gefährdete Blattspitze wurde in nichtrostendem Stahlblech ausgeführt. Daran angenietet sind Kupfermeshbahnen, welche alle in Kohlefaser ausgeführten Blattbereiche abdecken. Sie leiten die Blitzeinschläge in die Nabe ab. Von dort aus wurde eine direkte Erdverbindung über das Maschinenhaus und den Turm in die Fundamenterde hergestellt.
Die bei diesen großen Blättern notwendigen technologischen Untersuchungsprogramme wurden projektbegleitend durchgeführt, wobei partiell eng mit Instituten und anderen Einrichtungen zusammengearbeitet wurde (4). Jedes Rotorblatt wurde anschließend einem Belastungstest unterzogen (siehe Bild 9).

Der Turmbau mit Fundamentierung wurde in Wilhelmshaven mit Ortsbeton durchgeführt. Da die Anforderung an die Turmbiegeeigenfrequenzen sehr hoch sind, wurde die Pfahlgründung sehr sorgfältig geplant und durchgeführt (siehe Bild 10).

Bild 9: AEOLUS II Rotorblatt im Test

Bild 10: AEOLUS II Fundament

Die 36 Pfähle mit einer durchschnittlichen Länge von 18 m tragen die 16 m im Durchmesser und 4 m in der Höhe messende Fundamentplatte. Darauf wurde im 24-Stunden-Betrieb der Turm in Gleitbauweise kontinuierlich hochgezogen. Nach 28 Tagen waren die 88 m Endhöhe des Turmes erreicht.

In Trollhättan fertigte und montierte KTAB das Maschinenhaus mit allen Inneneinrichtungen und der Nabe mit Blattverstellmechanik. Die 160 to schwere Rotorgondel wurde über Schifftransport nach Wilhelmshaven gebracht und auf einem Spezialanhänger vom Hafen auf die Baustelle gefahren.

Das Maschinenhaus der AEOLUS II-Anlage wiegt 160 to und die beiden Rotorblätter zusammen 18 to.

Da diese enormen Gewichte nur noch unter größtem Aufwand mit Kranhilfe auf den 88 m hohen Turm montiert werden könnten, wurde ein neues Montage-Verfahren entwickelt und angewendet.

Auf einem Schlitten, welcher auf 2 an der Turmaußenfläche angebrachten Schienen glitt, wurde die Rotorgondel montiert und an die Nabe die Rotorblätter angebaut (siehe Bild 11).

Der Schlitten wurde über hydraulische Heber an Stahllitzen, welche an einer turmkopfseitig angebrachten Konsole befestigt waren, mitsamt dem Maschinenhaus zur Turmspitze gefahren. Dort wurde ein Teil des Schlittens mit der daran befestigten Rotorgondel auf das Turmoberteil mittels hydraulischen Zylindern geschwenkt und mit Fundamentschrauben befestigt (siehe Bild 12).

Bild 11: Maschinenhaus montiert am Schlitten

Nach Installation der elektrischen Kabel, des Generators an das Winkelgetriebe und der Installation der Feuerlöscheinrichtung und der Flugwarnleuchten konnte mit der Inbetriebnahme begonnen werden.

Derzeit erfolgen die Inbetriebnahmearbeiten, die voraussichtlich bis April/Mai - einschließlich des Probebetriebes - abgeschlossen werden.

Bild 12: AEOLUS II, Kippen des Maschinenhauses auf den Turm

3. Bau- und Betriebsgenehmigungsaspekte am Beispiel des AEOLUS II

Windkraftanlagen mit einer elektrischen Nennleistung von mehr als 300 kW bedürfen einer Genehmigung nach dem Bundesimmissionsschutzgesetz in Verbindung mit der 4. Verordnung zur Durchführung des Bundesimmissionsschutzgesetzes. Diese ist von den zuständigen staatlichen Gewerbeaufsichtsämtern nach dem vereinfachten Verfahren zu erteilen. Sie schließt die gesondert zu beurteilenden Baugenehmigungen nach den Bauordnungen der einzelnen Bundesländer einvernehmlich mit ein.

Für die Genehmigung der Windkraftanlage AEOLUS II im Jade-Windpark ist das Staatliche Gewerbeaufsichtsamt in Oldenburg zuständig. Da hauptsächlich auch der Standort der Windkraftanlage und das Turmbauwerk mit seiner Gründung zu begutachten und zu genehmigen waren und dabei entscheidend Belange der Stadt Wilhelmshaven betroffen waren wurden wesentliche , zum gesamten Genehmigungsverfahren gehörende Prüfungen und Gutachten durch das Bauordnungamt der Stadt Wilhelmshaven beauftragt und betreut. Folgende weitere Behörden wurden hinzugezogen:

- Bezirksregierung Weser Ems gemäß dem Bundesbaugesetz und zur Flugsicherung
- Umweltdezernat der Stadt Wilhelmshaven einschließlich der Unteren Naturschutzbehörde
- Liegenschaftsamt, Stadtplanungsamt und Referat für Wirtschaft und Recht der Stadt Wilhelmshaven
- Feuerwehr
- Tiefbauamt der Stadt Wilhelmshaven einschließlich der Unteren Wasserbehörde

Die Eignung des Standortes ist durch gutachterliche Stellungnahmen nachgewiesen worden hinsichtlich

- der Schallemission und -immission
- der sicherheitstechnischen Machbarkeit
- der Baudynamik und Statik der Gründung und des Bauwerkes und
- durch ein Boden- und Gründungsgutachten

Die maschinentechnischen Anlagenteile und besonders die Rotorblätter wurden vorab begutachtet und ihre Herstellung baubegleitend durch einen Gutachter der Behörde kontrolliert. Als Gutachter wurde der Germanische Lloyd eingesetzt. Ein ökologisches Gutachten steht noch aus.

Um bereits während des laufenden Genehmigungsverfahren mit der Errichtung beginnen bzw. einzelne Einrichtungen bereits nutzen zu können, wurde die gesamte Genehmigung in mehreren Schritten beantragt. Bis einschließlich Anlageninbetriebnahme und Probebetrieb wurden die Genehmigungen erteilt.

Eine weitere Genehmigung für den Dauerbetrieb steht noch aus. Sie soll bereits die Erfahrungen aus dem Probebetrieb berücksichtigen.

Die Verwaltungsgebühren einschließlich der Kosten für die Gutachten und die begleitende Kontrolle betrugen bisher etwa 430.000,- DM. Weitere Kosten von etwa 50.000,- DM sind noch bis Ende des Genehmigungsverfahrens zu erwarten.

4. Resümee

Die Stromerzeugung in einer nennenswerten Größe erfordert in Mitteleuropa zwingend den Einsatz von Großwindenergieanlagen. Eine vorläufige Bewertung nach der Montage beider Anlagentypen zeigt, daß auf den Leichtbau des Maschinenhauses und der Rotorblätter besonderen Wert gelegt werden muß. Das WKA-60-Maschinenhaus mit Rotorblättern (60m Rotordurchmesser) wiegt 214 to während das 3MW-Maschinenhaus des AEOLUS II mit Blättern nur 178 to wiegt bei 80 m Rotordurchmesser. Zudem erlaubt das Montagekonzept der AEOLUS II-Anlage eine kostengünstige Montage von großen Windkraftanlagen.

Wie in der "Ad-hoc-Ausschuß Großwindanlage"-Studie des BMFT ausgeführt ist die Förderung des Baues von Großwindanlagen sinnvoll. Es ist jedoch nur mit Hilfe von Förderungen der Prototypentwicklung und des Nullserienbaus ein Engagement von Herstellern machbar, wobei der zukünftige Absatzmarkt planbar werden muß.

Die grundsätzliche Bereitschaft der EVU, solche von der Wirtschaft hergestellten Systeme einzusetzen ist vorhanden. Es müssen jedoch vertretbare kommerzielle Randbedingungen geschaffen werden.

(1) Ad-hoc-Ausschuß Großwindanlagen des Bundesministeriums für Forschung und Technologie; Bonn April 1992

(2) Rotorblätter für Großwindenergieanlage in Faserverbundbauweise, Kunststoffe 82 (1992) 5, D. Knünz und H.R. Rühmann

(3) Blitzschutz Rotorblatt, Dok Nr SP 000.1866 AEOLUS1-6320, MBB, Okt. 1991

(4) Rotorblatt in Faserverbundbauweise für die Windkraftanlage AEOLUS II, MBB-UE-0013-92-PUB R. Wachsmuth, D. Knünz, S. Habib

SUPERVISORY, CONTROL AND DATA ACQUISITION: WHEN A PROTOTYPE NEEDS TO HAVE PROTOTYPE COMPONENTS

S. Avolio, G. Moleti, F. Morabito

Alenia - Environment and Energy Division
Via Vitorchiano 81
ROMA (Italy)

ABSTRACT: GAMMA 60 is an advanced two-bladed, 60 m-rotor, upwind Wind Turbine Generator, with a rated power of 1500 kW, developed by Alenia-WEST (formerly Aeritalia). The prototype has been developed with funds from ENEA, ENEL, and the DG XVII (DIRECTORATE GENERAL FOR ENERGY) of the European Community, as well as from Alenia-WEST. ENEL placed an order for the supply of the prototype, plus two further units, to be defined on the basis of experience gained on the first machine. The GAMMA 60 prototype was installed in the first half of 1992 at ENEL's Alta Nurra test site in Sardinia and, after the end of the ongoing commissioning activity, is now being tested jointly by ENEA, ENEL, and the manufacturers.
The paper presents the description of the hardware and software architecture of the Supervisory, Control And Data Acquisition System of Gamma 60 prototype WTGS. In this respect the architecture of the system, the major requirements in terms of hardware and software and the major problems encountered are presented. Attention is also given to the industrialization of the Supervisory and Control features.

0. GENERAL INTRODUCTION

Technical features of GAMMA 60 can be broken down into "basic" characteristics (upwind rotor and teetered hub), which were chosen from among well-proven solutions, and "advanced" characteristics (variable speed and power regulation through yaw control), which represent the innovative aspect of GAMMA 60 as compared with other large wind turbine generators. These technical features have been implemented in order to reach two basic objectives: reduction of initial and maintenance costs and increasing of WTG performances (energy capture and fatigue life). Operating speed varies over a broad range because rotor performance can thereby be kept at near-peak conditions, and beacuse more power can be transmitted through the gearbox and generator without increasing torque or generator current. It also allows to increase the fatigue life. Controlled yaw is used instead of pitch control for aerodynamic torque modulation. As compared with pitch control, this feature makes for a much simpler machine for purposes of both manufacture and maintenance, eliminating all control components from the rotating system, and thus facilitating access for inspection.

The structural design philosophy is that the structural components should be made relatively flexible [2]. The "softness" of the rotor, tower, and drive system has the effect of attenuating both the harmonic and the transient stresses in the machine. This makes for reduction in weight, while retaining long, conservative fatigue life.

1. FOREWARD

The Control System of WTGS is according to its characteristics closely connected with "WTGS System"; therefore, the development of the Gamma 60 control system was carried out in a way different from the other sub-systems.

In practice, we can distinguish three different areas:

- Control algorithms and models analysis;
- Supervision and management Hardware and Software·
- Acquisition and data processing Hardware and Software;

This paper deals with the hardware and software of the prototype; another paper [1] deals with algorythmes definition and validation .

1. THE STORY OF GAMMA 60 CONTROL SYSTEM

1.1 The Controller

The Control System of Gamma 60 WTGS was developped starting from a simulation program named F097, by Hamilton Standard. After the porting realized by Alenia in VMS (Digital proprietary operative system), this program was named CODE W (COde Dynamic Evaluation Wind turbine). It consists of a main program which simulates the machine environment and structure and of a subroutine that performs the control loops.

After the dynamic validation, this architecture made it possible to take out the controller software and then to implement it on a microprocessor board.

A board 68020 microprocessor based was selected, then interfaced with a data acquisition unit 80C86 based.

1.2 The Supervisor:

The requirements of the supervision unit regard the management of the machine operation states and the acquisition of about 300 digital and analogical channels, at 25 Hz prevalently (the real quantum time is 40 msec). This system is justified by the necessity of to conduct conduct an adeguate series of exprerimental tests, since the machine is a prototype.

Therefore, in order to avoid a duplication of signals or sensors, one thought of coming up with a single machine serving both proper supervision and acquisition. Nevertheless, this choice clashed with the future machines already put into plan by the industry.

Thus, one decided to use a system having a scaling down architecture wich would allow a safeguard the on investments of the next machines.

We selected a Digital VMS system, an operative system rather then a dedicated hardware, one in order to have, in the long run, an adequate warranty of continuity and the possibility to scale down the hardware for a prototype or an industrial machine.

2. THE ARCHITECTURE OF CONTROL AND SUPERVISION SYSTEM

2.1 Hardware

The Control and Supervision System of Gamma 60 WTGS presents a distributed architecture (Fig. 1).

We can identify six areas of acquisition and command:

- Hub
- Nacelle
- Box
- Tower
- Wind tower
- Supervisor

The Hub is connected with the nacelle by radio. The acquired data regard the strain gauges; therefore, they are necessary for the prototype machine only. A dedicated line made of optical fiber allows this data to be transmitted to the supervisor.

In the nacelle, there are the regulator, the data

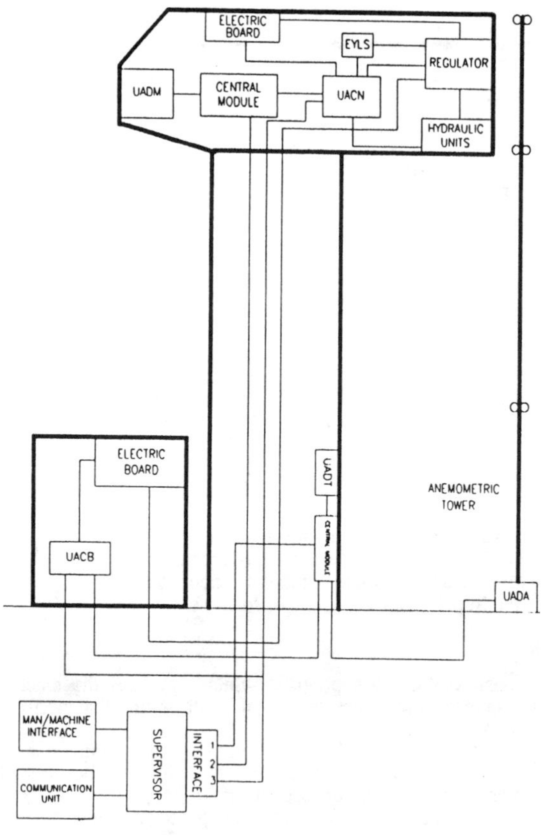

Fig. 1

acquisition units and an emergency unit called Emergency Yaw Logic System (EYLS). This unit, based on very simple electronic components and relays, monitors the efficient running of the controller and the supervisor.

Furthermore, some on/off signals are present related to a overspeed breaker and to a panic button situated in the control room

If an anomaly occurs, EYLS starts the hydromechanical backup that stops the WTGS by mere mechanical and hydraulic means.

The Controller is interfaced with the external world by a RS232 asyncronous line, towards both its data acquisition unit and the Supervisor.

This fact permitted the Controller to be tested through some time histories originated by CODE W, so that the implementation correctness was validated.

The Controller Software is self-contained; there are some logical flags controlled by the supervisor, but they only have the purpose to start/stop the control loops only. Therefore, the controller is not obliged to talk continuously with the Supervisor.

In the box situated near the tower there is the power electrical converter and the respective acquisition and command units.

Some acquisition units are allocated in the tower (strain gauges measures) and also on the two wind towers (wind and weather measures). These units, made for the prototype machine, will not be present in the industrial machine.

The Supervisor is situated in the control room, that is about 700 m. from WTGS. The connection between Supervisor and acquisition units is realized through optical fibers, except for some initial problems caused by defective plugs, this kind of connection proved to be excellent in a site that, unfortunately, shows a high quotient of lightning.

Every unit sends data to the Supervisor in PCM telemetry. The Supervisor has the function of acquiring data, to manage the operative conditions (supervision and command), to record events (monitoring), to execute pre-elaborations, to display on/off line acquired data.

Data flow is continuous and appropriate controls are executed in order to verify the reception congruence (synch lost, frame and format counter, transmission redundancy). During the period of development in the workshop frequent synchronism losses have been noticed, owing to electrical noise in the frame syncronizer rack; however, one must point out that parameters corntrolling the transmission have always proved to be adequate and avoided the acquisition of incorrect data.

The interface between Supervisor and acquisition system has been realized through the DMA (direct memory access) boards.

The Supervisor consists of a DEC μVax 3400 cluster connected with 3100 (graphic workstation); furthermore, there also exists a dedicated electronics board wich functions as a standard synoptical.

2.2 Software

Three software areas can be distinguished:
- Controller
- Data acquisition and PCM transmission
- Supervisor

Controller software consists of a part developed on 80C86 and of a part worked out on 68020 (Fig. 2).

80C86 has the function of interfacing the controller with sensors and actuators, while 68020 attends the regulation loops. Double processor solution allows to relieve 68020 of acquisition tasks and permit loops implementation with a high level language (pascal). 80C86 operates as a system clock also; therefore, 68020 performs the regulation single cycle only after a communication interrupt sent by 80C86. Moreover, 68020 is connected with the Supervisor to receive the commands that concern controlling flags (start, stop, emergency, etc.).

A configuration file EEPROM located, permits controller working parameters to be modified.

Acquisition software deals with tables located on the EEPROM that are present in the acquisition units.

Therefore, the program produces channel scanning and joins transmission controls. Parallel tables are present at the supervisor level, so as to address and control received data.

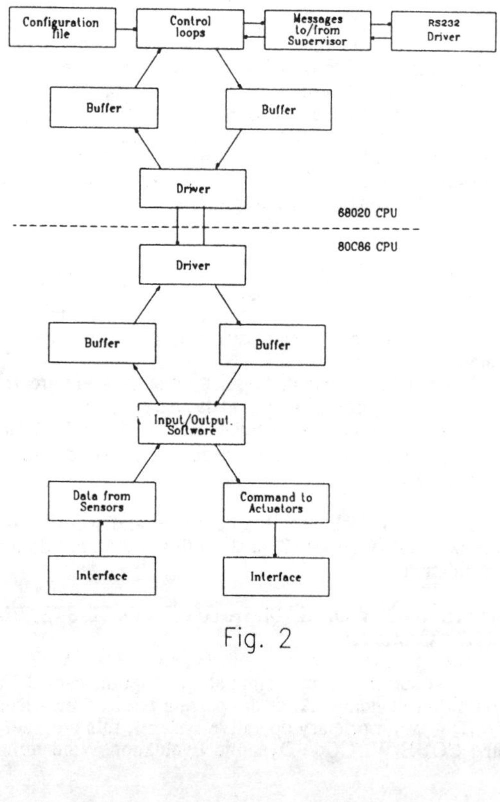

Fig. 2

Supervision software is located on μVax 3400 that is situated in the control room. It is possible to distinguish a "kernel" software and a software that concerns prototype machine; "kernel" software regards machine working while the second type regards particular acquisition data, pre-elaboration and graphic display.

Data coming from peripheral units are inserted in a common memory area that is refreshed continuosly (Fig. 3).

This Common Area, together with virtual devices (mailboxes) represents the logical interface for all the supervisor system processes.

The Common Area is a buffer that consists of several vectors on which acquisition software makes the signals already available. System status is updated by acquisition software and by supervision software.

Appropiate configuration files allow the management of the single channel to be easily modified, for instance by establishing whether the channel is able to produce an alarm or it would be better to submit it to a digital filtering and, in general, to define the channel configuratione.

The choice of a common memory area has permitted the management of (in a multitask environment such as VMS) different supervision processes. There are several tasks attending, for example, to communication with the field, to the recording of historical (logbook) and temporary memories (black-box) or to emergency stops management.

In all, there are 25 processes that interact through both the common area and the mailboxes. Such a high number of processes is justified by the prototype character of the machine.

As all the prototype programs are supported by the common area and don't interact with the "kernel", they can be eliminated in industrial machine.

In any case software architecture and operative system permit the task scheduled with dynamic priority by putting the processes which are waiting for a timer or a device not available in hybernation state.

In this way, the processing times have been optimized and as a result, the whole system works virtually in background at the telemetry handler process.

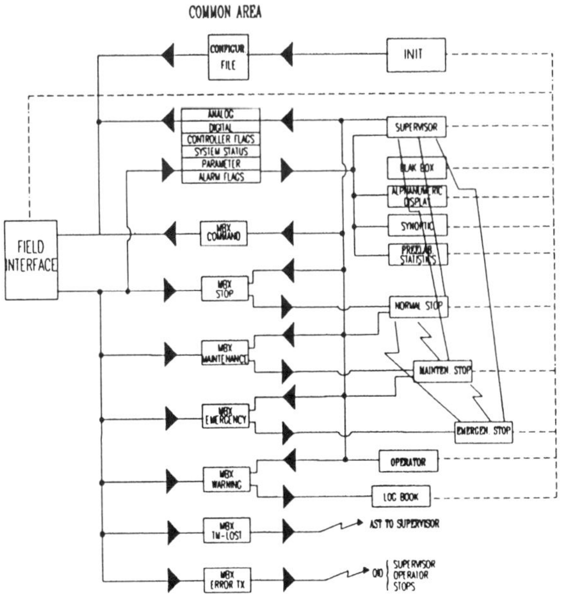

Fig. 3

Regarding the man-machine interface software, it must be pointed out that the interface consists of a task which attends the alphanumeric display and another task provides for the graphic display in X-window environment.

These programs work on a cluster connected workstation through a local area network (LAN) and not on the principal computer.

Therefore, an extent of the common area concept also exists also in the LAN and not only in the single computer.

By using a workstation it is possible to relieve the Supervisor of the display tasks and also to consent the industrialization of system with optional modules.

3. CONCLUSIONS

3.1 Utility in prototype situation

Gamma 60 WTGS Supervision and Control system has been conceived to provide the features that are needed in a prototype machine. It is based on two DEC minicomputers and a dedicated electronic board and has acquisition, display and pre-elaboration on-line abilities. Within the limits of an experimental campaign, acquired and recorded data can be made available on geographic network by extending the common area further on. By using a modem a specific software also makes remote communication possible - but only with an alphanumerical display - with a personal computer.

3.2 For serial machines

Perhaps, the most important characteristic of the system is representeted by the possibility of its re-utilization at the time the product is in serial production.

The use of the operative system VMS and the pulling down of the hardware costs permit the software developement expenses to be recovered. When the Supervisor system was conceived, its cost was about 25000 \$/Mips. Considering the operative system continuity warranty, it was possible to go in two different ways, by making use of medium-large computers adequate for wind farm (multitask) or, alternatively, by using little computers for each machine having no longer having the prototype necessities.

As hardware prices have dropped and performance has increased accompanied by a falling down of the costs (2500 \$/Mips), we think that the second way gives better assurance of a higher reliability and profit level. Therefore, one can reasonably conceive that the control and supervision system development and industrialization will go towards an architecture in which each machine is locally controlled and connected to a network where a separate unit attends to the functions of measure, monitoring, and control of the wind farm.

4. REFERENCES

[1] S. Avolio, S. Golinelli, G. Moleti, F. Morabito, C. Pierdominici, V. Arcidiacono, S. Corsi, M. Pozzi, G. Botta.
Dynamic analysis of GAMMA 60 wind turbine generator and control system: design and theorethical and experimental validation.
1993 European Community Wind Energy Conference Proceedings.

[2] S. Avolio, A. Aliberti, I. Sebastianelli, G.S. Doman, S. Ferrari, C. Golia.
Design, analysis and testing: methods and procedure for large machines illustrated by the GAMMA 60.
1993 European Community Wind Energy Conference Proceedings.

ACKNOWLEDGEMENTS

The Gamma 60 Project has seen the efforts of various bodies: Alenia-West and Hamilton Standard for System activities and blades; major suppliers included: Escher Wyss (Ravensburg-D and Schio-I) for design and manufacturing of hub, mechanics and actuation; Rexroth (D) for hydraulics; Thyssen (D) for the gearbox; Twiflex (GB) for the brake; Galtarossa (I) for steel structures and civil works; Ansaldo (I) for the electric subsystem; Gavazzi Space (I) for electronic apparata.

The internal team was made by: A.Aliberti, F.Anaclerio, S.Avolio, C.Calò, G.Caturano, F. Cecilia, G.Doman, S.Fugaro, S.Golinelli, P.Lecce, G.Moleti, F.Morabito, M.Musolino, G.Nettis, M. Orlando C.Palmari, C.Pierdominici, M.Rotondi, I.Sebastianelli, L.Serafini, A.Smarchi.

Thanks go also to all the other people that in different stages have contributed to the success of the program.

DOE/NREL ADVANCED WIND TURBINE DEVELOPMENT PROGRAM

C. P. Butterfield
B. Smith
A. Laxson
B. Thresher
National Renewable Energy Laboratory
1617 Cole Blvd.
Golden, Colorado 80401
U.S.A.

P. Goldman
U.S. Department of Energy
Wind/Hydro/Ocean Division
1000 Independence Avenue, S.W.
Washington, DC 20585
U.S.A.

ABSTRACT: The development of technologically advanced, high-efficiency wind turbines continues to be a high-priority activity of the U.S. wind industry. The National Renewable Energy Laboratory (formerly the Solar Energy Research Institute), sponsored by the U.S. Department of Energy (DOE), has initiated the Advanced Wind Turbine Program to assist the wind industry in the development of a new class of advanced wind turbines. The initial phase of the program focused on developing conceptual designs for near-term and advanced turbines. The goal of the second phase of this program is to use the experience gained over the last decade of turbine design and operation combined with the latest existing design tools to develop a turbine that will produce energy at $0.05 per kilowatt-hour (kWh) in a 5.8-m/s (13-mph) wind site. Three contracts have been awarded, and two more are under negotiation in the second phase. The third phase of the program will use new innovations and state-of-the-art wind turbine design technology to produce a turbine that will generate energy at $0.04/kWh in a 5.8-m/s wind site. Details of the third phase will be announced in early 1993.

1. INTRODUCTION

The Advanced Wind Turbine Program is currently in its second phase. In Phase II, each of the contractors is required to evaluate a promising existing turbine, propose design improvements for the same turbine, fabricate two prototypes of the improved turbine design, and test these prototypes in a demanding wind site. The contractors must demonstrate that their designs have met the cost of energy (COE) and reliability goals during these tests. Later, under Phase III, contractors•will be required to start their design process with a completely new approach that must use innovative design concepts capable of meeting the much more challenging COE goal. The development of this class of turbine will require the latest design tools, extensive component development testing, and the use of advanced materials.

This paper will describe the complete program, the turbines currently being designed, the innovative design techniques, and the evaluation testing planned for the prototypes. It will also contain the results of preliminary tests that have already been conducted on "preprototype" turbines.

2. BACKGROUND

Wind generation of electricity for delivery to the electric grid must compete with other utility sources of electricity in a complex world of financing, tax credits and penalties, competing cost formulas, and set-asides. Despite the actual complexity of the energy market, relatively new technologies such as wind turbines must still compete on a pure COE basis with the well-developed fossil fuel infrastructures for the production of electricity. Wind systems can only grow by eliminating perceptions of poor reliability and availability through technological developments, innovation, and improvement. Wind turbine operators have made great strides to eliminate these perceptions using current technology, but that is not enough. Existing wind turbines produce electricity at $0.07–$0.09/kWh. Decreases on the order of 50% are needed to help make wind energy competitive. New advances in aerodynamics, materials, generators, controls, and design tools must be incorporated into turbine designs to make wind-generated electricity competitive with conventional energy sources.

3. ADVANCED WIND TURBINE PROGRAM

The next 5 years are critical for the U.S. wind industry. Market opportunities are multiplying as utilities, state and local governments, and the public are becoming more aware of the potential for wind power. The wind industry must place itself in a position to take advantage of these opportunities. The DOE programs are designed to assist industry activities in meeting these challenges. First, existing wind plant operators and developers must remain healthy and competitive. Second, new products that combine the best of existing designs with new research and technology must be ready for deployment in the near term. Third, a next generation of utility-grade wind turbines must be developed for the year 2000 and beyond.

The goal of the Advanced Wind Turbine Program is to assist U.S. industry in incorporating advanced wind turbine technology into utility-grade wind turbines for the near term (1993–1995) (Near Term Product Development [NTPD] Program), and to provide the next generation of utility-grade wind turbines for the mid-term (late 1990s) (Next Generation Product Development [NGPD] Program). The NTPD Program is in progress with three contractors developing prototype turbines. The technical advances incorporated in these turbines will be the main topic of this report.

The NGPD Program is split into two parts. The first part, entitled the Innovative Subsystems Project, will focus on development of innovative subsystems or components that can be designed separately from the full turbine, bench tested, and proof-of-concept tested on a full-scale turbine. The second part, entitled Turbine Development Project, will focus on the development of utility-scale turbines and produce prototype machines for field testing.

These programs will help introduce new products into the marketplace, support the development of new jobs domestically, help create greater stability and strength in the wind industry by attracting utility participation, broaden the national use of wind energy, and increase competitiveness in the international marketplace.

4. NEAR-TERM PROTOTYPE DEVELOPMENT

Under the NTPD Program, three contractors are developing prototype wind turbines. R. Lynette and Associates has already deployed the WC-86 machine, which incorporates several design improvements from the earlier ESI-80. Atlantic Orient Corporation is developing the 15/50, a dramatic improvement over the Enertech 44/60 turbine after which it is designed. And, Northern Power Systems is developing the NPS 250, which sports a redesigned rotor.

4.1 WC-86

R. Lynette and Associates (RLA), located in Redmond, Washington, U.S.A., is developing a 275-kW, 26.3-m, downwind, teetered rotor turbine based on the ESI-80 design. The ESI-80 design was developed as the next generation of the ESI-54 in the early 1980s and enjoyed partial success. Most of the failings of the ESI-80 were caused by simple engineering oversights, poor attention to design details, or a poor knowledge of extreme loads. With a thorough re-engineering effort, a better knowledge of the design loads, and an optimum blade redesign, RLA believes this design will be a robust and reliable turbine with improved energy production, suitable for utility-scale power generation. Redesigning the ESI-80 has allowed RLA to take advantage of the substantial operating history of this machine, retain component designs that proved reliable, and focus redesign efforts on those areas of the turbine that were unreliable [1].

Simplicity and Performance Improvements: Figure 1 illustrates RLA's WC-86 turbine along with the basic design specifications. Simplicity has been maintained throughout the machine with the fixed pitch and free yaw configuration. The rotor diameter has been increased from 24.5 m (80 ft) to 26.3 m (86 ft). Proven wood composite material has been used to design a more aerodynamically efficient blade that incorporates the new, NREL-developed S8XX thick family airfoils (Figure 2). Use of the new airfoils plus an optimized planform has resulted in a substantial increase in performance. Figure 3 compares the original ESI-80 power curve with the measured

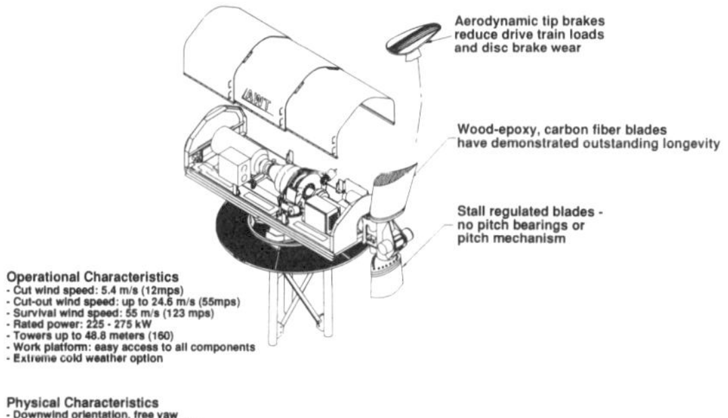

Figure 1: R. Lynette and Associates WC-86 Turbine

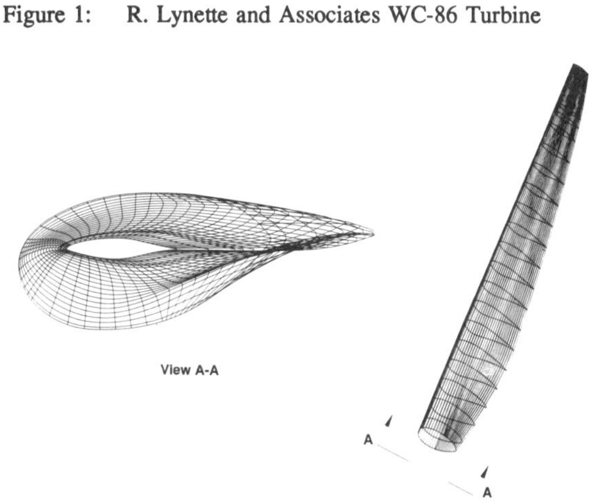

Figure 2: RLA WC-86 Wood Epoxy Blade

power curve of the WC-86. A percent improvement curve is shown in the same figure and illustrates a dramatic 70% improvement at 8 m/s.

Tip Vanes: One key to the success of any stall-regulated turbine is a reliable aerodynamic brake. RLA has retained the tip vane concept and uses it both as a fail-safe brake and as an active brake for normal shut-down operations. The ESI-80 tip vane proved to be very effective when used in both these braking modes. The vane is mounted at the tip of the blade and allowed to pivot about a hinge line that is perpendicular to the chord line located at the leading edge. The vane is held in a low-drag orientation with an electro-magnet and released either by a control system command or centrifugal forces. When released, the vane is forced to rotate into a high-drag orientation (parallel to the leading edge) by centrifugal forces and aerodynamic forces.

The ESI-80's tip vanes principal failings were insufficiently sized hinge pins, overweight fiberglass vanes, and unreliable slip rings. The slip rings are used to transmit electrical power from the nacelle to the rotor. RLA re-engineered all these problem parts while retaining the successful features of the system. Slip rings have been replaced with a split rotating transformer that transmits electrical power to the rotor using magnet coupling. The split feature allows wind turbine technicians to easily remove the transformer coils in the unlikely event of failure. The split, rotating transformer was supplied by Atlantic Orient Corporation for the same application in its new turbine design, which is discussed later.

The tip vane has been redesigned into a low-drag, lightweight, composite structure. The new vane has been tested in a wind tunnel to confirm its low drag in the undeployed orientation (parallel to the tip path) and to determine its maximum drag in the deployed orientation. These force measurements were also helpful in accurately simulating the deployment transient loads that were needed to design the hinge structure. RLA has conducted rotor performance tests that prove that the tip vanes actually enhance production at low to moderate wind speeds and reduce peak power at cut-out wind speeds (23 m/s). Figure 4 shows the tip vanes during wind tunnel testing.

Low-Speed Shaft Brake: A low-speed shaft brake was incorporated into the WC-86 to eliminate braking loads in the gearbox and improve the reliability of the brake. The new system is a more traditional multiple caliper disk brake, which is spring applied and released by air pressure. The brake will experience light duty because the tip brakes will provide the primary braking energy for all normal high-wind shut-downs, but it is sized to be capable of stopping the machine in all normal and abnormal operating conditions.

Prototype Tests: RLA will initially deploy at least two prototypes. Both of these turbines will be located in a very turbulent site in the Tehachapi, California, wind farms. The first one was installed in February 1993. The next prototype will incorporate all the experience gained during assembly and

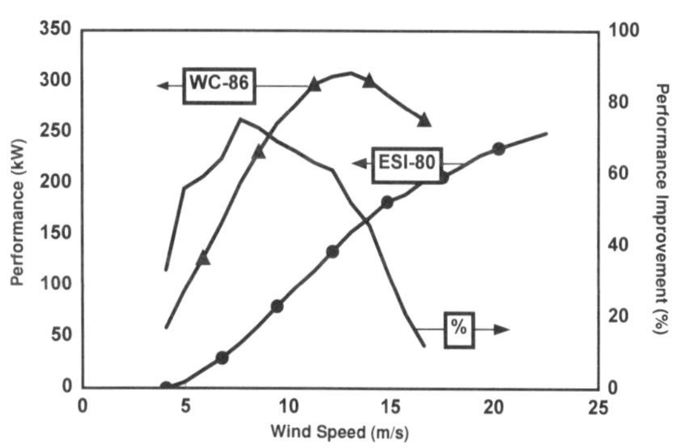

Figure 3: RLA WC-86 Performance Improvement

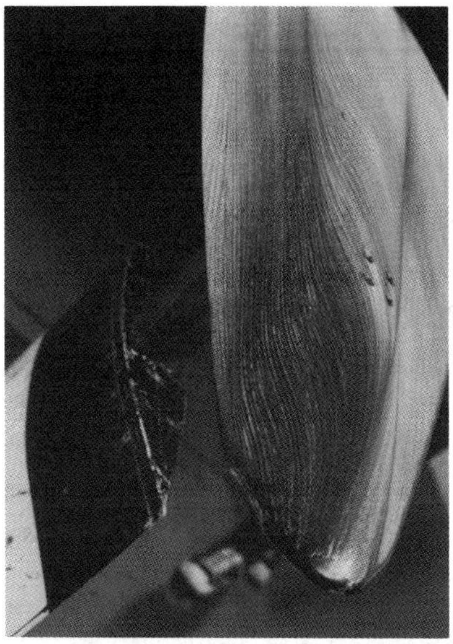

Figure 4: RLA WC-86 Tip Vane Wind Tunnel Test

testing of the first. The second prototype is expected to be installed in the late fall of 1993.

4.2 AOC 15/50

The Atlantic Orient Corporation (AOC) is located in Norwich, Vermont, U.S.A. The AOC has developed an improved and simplified version of the Enertech 44/60 called the AOC 15/50 (15-m diameter with a 50-kW rating) [2]. Figure 5 illustrates this new design and lists some of the major specifications. This down-wind, three-bladed turbine design is suited for remote applications, small wind farms, and even large wind farms where sufficient land is available. Its integrated main frame/gearbox/generator eliminates many of the critical bolted joints found in conventional turbine designs and creates a very efficient load path from the rotor to the tower top. The new configuration has allowed AOC to reduce the material needed for a safe design. The integrated drive train has also eliminated maintenance-prone couplings between the gearbox and the generator.

Several other features have been added to the machine without increasing the complexity. An optional yaw damper has been added for turbines located in turbulent sites. The damper is a passive hydraulic system that limits yaw rates and hence gyroscopic loads. In most sites, with average turbulence, this device will not be needed. A cast tower top plate has been added to further improve the efficiency of the load path. A fail-safe electrodynamic brake was developed using passive electrical components. The tip vanes are electromagnetically retained, much like the RLA tip brakes. As mentioned earlier, AOC developed a split rotating transformer that is used to transmit tip brake control power from the nacelle to the rotor. Finally, AOC has developed a new wood epoxy composite blade using the new NREL S809 thick family airfoils. Like RLA, AOC has dramatically improved its performance by increasing its rotor diameter while retaining a modest peak power. This approach limits the size of the generator and the gearbox required, yet improves the performance in the most probable energy-producing wind speed range. Figure 6 shows this new blade, and Figure 7 shows the improved power curve.

Component Fatigue Testing: The new wood composite blade was supplied by Gougeon Brothers Inc. Static and fatigue tests were conducted at the NREL fatigue test facility [3]. Fatigue loads were applied at a spanwise location, which exercised the majority of the inboard portion of the blade. Load sequences (blocks) were applied repeatedly while the cyclic amplitude was increased. Failure occurred at 55,733 cycles for load amplitude ratios of R = 0.1, with maximum loads of 2/3 (68%) of design failure strength. Analysis shows that this blade would probably exceed the Danish Standard and has allowed AOC to quantify the blade design limits.

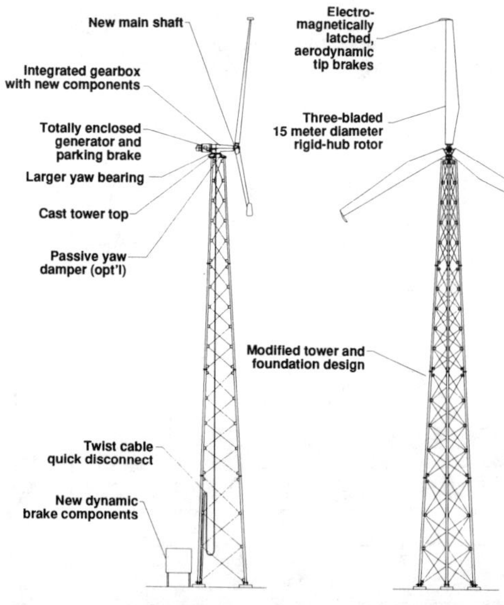

Figure 5: Atlantic Orient Corp. 15/50 Turbine

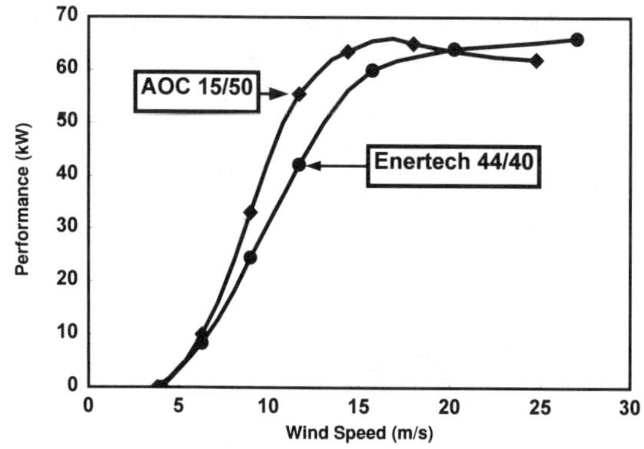

Figure 6: AOC 15/50 Blade Layout

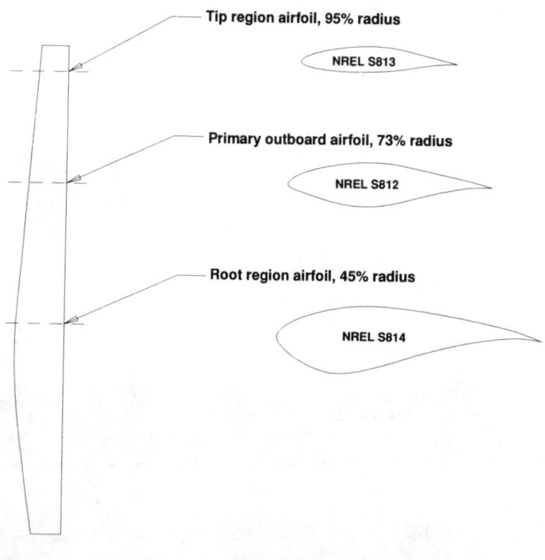

Figure 7: AOC 15/50 Performance Improvements

Because the tip brakes are so important in the safe and reliable operation of the machine, AOC decided to fatigue test them under simulated conditions. Figure 8 shows a spin test stand that is designed to cycle the tip brakes through a sequence of start-ups, deployments, and shut-downs. This system will realistically simulate operating loads and allow AOC to subject the system (including the rotating transformer) to a lifetime of deployments before the prototype has been field tested. This test stand may also be used to test cold weather icing conditions.

A test stand has been designed to simulate high yaw rate loading cycles on the entire load path from the hub joint through the yaw bearing to the tower top. The generator will be used as a motor to drive the gearbox and main shaft while simulated thrust and gyroscopic loads are applied using pneumatic cylinders. These extreme load conditions will be simulated until AOC design engineers are satisfied that there are no design weakness.

Prototype Tests: At least three prototypes will be deployed. The first will be deployed at the U.S. Department of Agriculture (USDA) Agricultural Research Station in Bushland, Texas, U.S.A. The second will be installed at a severe cold weather site on top of Mount Equinox, Vermont, where winter icing is common. This site is operated by Green Mountain Power, a Vermont utility. The last will be installed in an extremely turbulent site in the Palm Springs, California, wind farms.

4.3 NPS 250

Northern Power Systems (NPS) is located in Moretown, Vermont, U.S.A. Northern's wind turbine design experience dates back to the mid-1970s when its researchers developed the High Reliability 2-kW (HR-2) turbine [4]. Northern's last design was the North Wind 100-kW machine (NW-100) [5]. The 250-kW, two-bladed, teetered, up-wind machine that NPS is developing under the current DOE contract is a scaled-up version of the NW-100. Figure 9 illustrates this turbine and lists some of its key features.

Most of the NPS work has been focused on the rotor design. Learning from the successes and failures of the NW-100, Northern researchers have eliminated the full span pitch control system used in that design and substituted aileron controls. They have designed a more robust hub that incorporates teeter dampers and an active teeter brake. The blade is a hybrid composite that combines skins and spars, fabricated using filament winding techniques, with a flow-through hub section. Figure 10 shows an exploded view of how the various components will be assembled into an efficient low-cost blade. Figure 11 illustrates the hub saddle, which captures the blade center section in a wrap-around elastomeric blanket. The flow-through hub will eliminate all costly and inefficient blade root joints, which have been the site of past blade failures in some turbines. With this hub design, all the critical loads, such as flapwise bending, in-plane gravity, and centrifugal loads, are isolated in the rotor structure. Only torque, thrust, and unbalanced loads are transmitted through the hub to the nacelle structure.

Ailerons: The most important new development featured in this turbine is the ailerons. These devices deflect in the downwind direction and act as spoilers or drag brakes when aerodynamic brakes are needed. When power modulation is needed, above rated wind speed, the ailerons can be deflected through small angles to increase or decrease power. Small

Figure 8: AOC 15/50 Tip Brake Test Stand

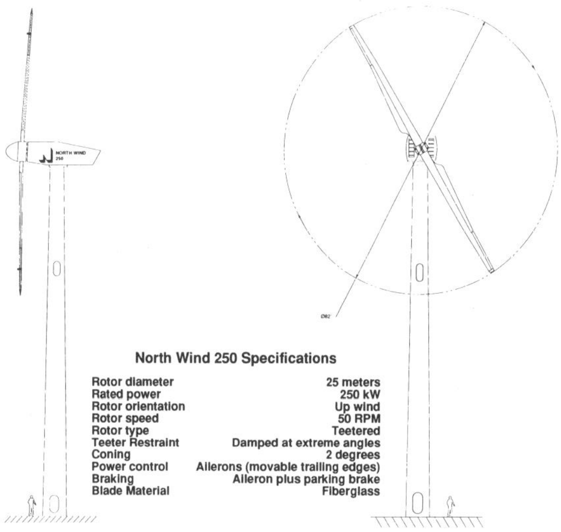

Figure 9: Northern Power Systems 250-kW Turbine

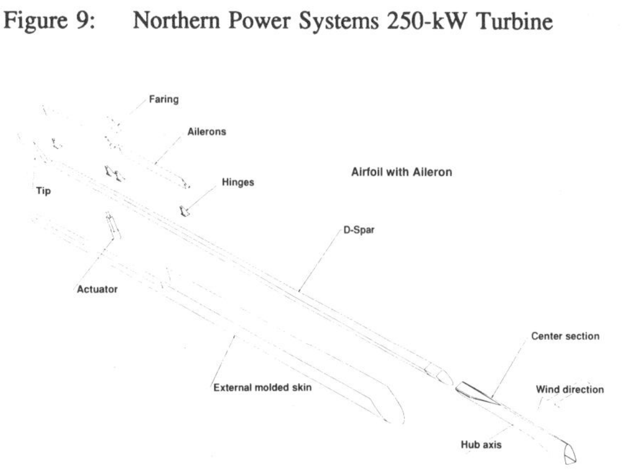

Figure 10: NPS 250 Blade Assembly

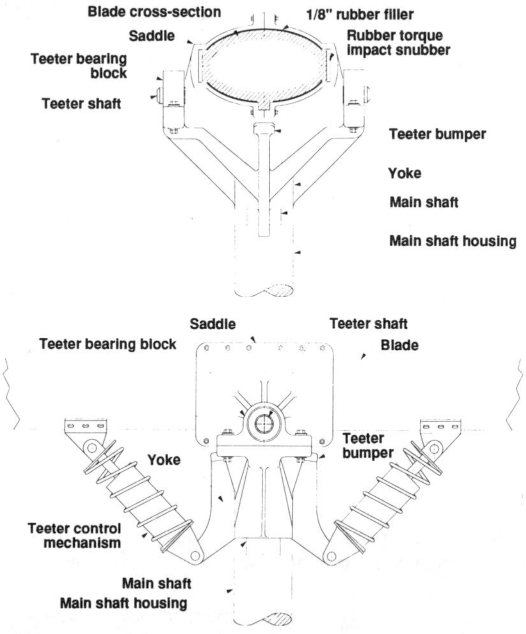

Figure 11: NPS 250 Teetered Hub Assembly

positive deflections (upwind) will increase camber, which will increase lift (and drag). Small negative deflections (downwind) will decrease camber and decrease lift. Because only 30% of chord and span is used to accomplish the needed control, the actuation loads and actuators are small. This approach does not require costly joints in the load-bearing spar of the blade.

Figure 12 shows tests being conducted at the Wichita State University wind tunnel on an airfoil with an aileron. These tests were run to measure the airfoil performance under a wide range of angles of attack and aileron deflection angles. Results have helped NPS aerodynamicists optimize the aileron configuration. Ailerons such as these have been tested on NASA's 40-m MOD-0 test bed and three different NPS test turbines ranging in size from 2 to 20 m [6]. Each of these tests has shown that ailerons can be used effectively in controlling rotor speed. A parking brake will still be required for maintenance and bringing the rotor to a complete stop, but, as in the case of RLA and AOC, the aerodynamic brake will be used as the primary braking device.

Prototype Tests: NPS has tested the aileron concept on one three-bladed turbine and two two-bladed, teetered turbines. Tests for a prototype turbine will begin in the fall of 1993. The test site will be located on a wind farm in Altamont Pass, California.

5. CONCLUSIONS

The DOE/NREL Advanced Wind Turbine Program has initiated the development of three promising new turbine designs. Each of the designs has taken advantage of years of operating experience with similar turbines. Using current technology and more advanced design tools, the three companies have improved on designs that were deployed in the mid-1980s. The new designs all show dramatic performance, reliability, and COE improvements. They are all being subjected to comprehensive component test programs and multiple prototype field test programs. When these efforts are complete, these turbines will represent some of the most extensively engineered wind turbines in the world.

Figure 12: Aileron Blade Wind Tunnel Tests

6. REFERENCES

[1] J Rogers,"WC-86 Advanced Wind Turbine Development," SED—Vol. 12, 11th ASME Wind Energy Symposium, ASME (1992).

[2] Z R Zickefoose, "Enertech 15 kW Wind System Development, Phase II - Fabrication and Test," RFP-3515-UC-60 (December 1982).

[3] M D Zuteck, W D Musial, and B Johnson, "The Design and Fatigue Testing of the AOC 15/50 Wind Turbine Blade," presented at Windpower '93, San Francisco, California, American Wind Energy Association (July 1993).

[4] D J Mayer and J H Norton, "North Wind Power Company 2-kW High Reliability Wind System, Phase I - Design and Analysis Technical Report," RFP-3310/2-UC-60 (July 1981).

[5] C Coleman and B McNiff, "Dynamic Response Testing of the North Wind 100 Wind Turbine," SERI/TP-257-3424, Sloar Energy Research Institute (May 1991).

[6] D R Miller, and R L Putoff, "Aileron Controls for Wind Turbine Applications," NASA TM-86867, IECEC, San Francisco, California (August 1984).

TESTINGS ON THE 300 KW WIND UNIT IN A MOUNTAIN PLACEMENT

C.Albu x), S.Lázárescu x), A.Bej x), F.Novac x),
M.Nicodim xx)
x) Technical University of Timişoara
Wind Energy Research Center
Bd. M. Viteazu no. 1
TIMISOARA ROMANIA
xx) Hidrotim S.A. Timişoara
Calea Martirilor no. 1
TIMISOARA ROMANIA

ABSTRACT: In Romania, in the conditions of mountain placement were erected a wind unit model with a 300 kW install power. The testings that have been made during one working year stay at the basis of evalution of technical conditions imposed by the mountain placement, and as basis to achive the design of an industrial wind unit that its now in execution and assembling stage.

1. INTRODUCTION

The design and the execution of the 300 kW wind unit are included into the research program of wind energy, financed from budget by the Department of Science.

Technical University of Timişoara, as coordinator of this program, since 1990 cooperated with specialized companies: in design with Hidrotim S.A. Timişoara, in execution with. C.M.Bocşa, in erection with Electromontaj S.A. Timişoara and with RENEL the company for electric energy distribution and service.

The design of 300 kW wind unit executed in 1986-1987 meant an attempt of beginning in the field of using wind energy, in the easteuropean countries. The wind unit were executed and erected in 1989-1990 and in December 1991 were put into service. This unit with a 300 kW install power, is a research station of Wind Energy Research Center from Technical University of Timişoara.

2. THE PLACEMENT

The wind unit is installed in a mountain placement at 1400 m altitude in the Semenic Mountains from wester zone of Romania. The climatics determinate severe operational conditions for the unit.

Energetical availability of placement assure the necessary wind speed of 14-15 m/s for working approx. 6250 h/year as part of total 8760 h/year, and during the year 2160 hours the wind speed value is 4 m/s and 350 hours over 25 m/s.

3. WIND UNIT CONCEPTION

The wind unit is an experimental model conceived to assure the possibility of a complex research concerning the foundation, the tower, the conversion system of energy, the control and driving system, the joint system with the network and the wind unit operation in the energetical system.

A general view of wind unit can be seen in fig. 1. The wind unit is equiped with a machines line which has two generators with a max. install power of 300 kW. There are the possibility to couple an electrical resistances group or a suprasynchronous cascade in the circuite of generators rotor, aiming the operation at the variable rotation. On the other hand is possible to put into short-circuit the rotor circuite, in this case the unit operation is realised on the natural characteristics of electric machines. In all situations the power is limited with blades adjustment (electromagnetical).

The position of rotor is down wind, it is a rotor with 30 m diameter equiped with 3 blades (steel construction) each of them with a length of 13 m and a mass of 1340 kg.

The nacelle conception can be seen in fig. 2.

4. CONTROL AND MEASURING SYSTEMS

Aiming to measure the meteorogical parameters: wind speed, wind direction, humidity, atmospherical pressure and temperature, exist a transducers set mounted on a tower at 10 m high. The wind speed and wind position are measured on a telescopic arm of nacelle at 30 m high, too. The unit is equiped with transducers for electric power, stator current, generators speed rotation and turbine shaft moment.

Inside the rotor is placed a transducer for blades position.

The transducers signals converted in unified signal 4 - 20 mA can be use both on the analogic instruments from nacelle and the display.

Vibrations are measured with accelerometer which is placed inside the nacelle, and with the measuring bridges from tower installed on three levels of tower.

The wind unit can be driven through the manual operations from nacelle and from bellow, too. The acquisition equipment for operating parameters and meteorological data make the same time the automatic control through the soft.

The joint between P.C. and process is made with specialized distinct interfaces for data acquisition and for process control. The principle scheme is shown in fig. 3.

Data are stored on the hard-disk, in real time, aiming so, to enlarge storing capacity. Acumulated data make up a library which have contained the primary operation data of wind unit.

5. TEST PROGRAM

In the stage of puting into service and during the 1992 the research work aimed some distinct stages :

The first stage included the verification testings of unit operation, of protection systems, of control and measuring systems and of joint equipments at the network.

In this stage the wind unit operated with both generators off and on, aiming to verify the following aspects :
- unit orientation by wind and stability on the wind direction during the operation - fig. 4.
- starting/stoping at 25 r.p.m.
- starting/stoping at 33 r.p.m.
- braking without load, using the both braking systems : hydraulic brake on the turbine shaft and the mechanical brake on the gear-box output shaft.
- coupling/uncoupling in electrical network
- overspeed at 55 r.p.m.

In the second stage the generators were coupled at the synchronism:
- rotation speed of generator G.1 = 5oo r.p.m.
- rotation speed of generator G 2 = 1ooo r.p.m.

After some defects at the generator G 1 transmision (5oo r.p.m.) the testings has continued only with generator G 2. During this stage the wind unit has operationed on the natural characteristic of generator in whole wind speed field till max. speed 25 m/s and till powers of 3oo kW.
- it have been made braking testings for normal operation with aerodynamical braking and braking with service brake.
- it have been made testings of throw of electrical load (separation from discharge line) with emergency braking which consist in puting in work simultaneous the both braking systems and to cut out the generator.

After a general revision and after then some technical solutions were optimized (transmision, collector for signals and control, and power collector, transducers), in 1993 will pass on the stage of operating on variable speed rotation with max. 5o r.p.m.

6. RESULTS OF FIRST TESTINGS

The first testing result processing certify that the energetical turbine performances that have been estimated in designing stage are practical come true.

On the whole, the unit behaviour, during these testings, emphasized the correctness of principle solutions adopted for all subansambles, as well the necessity to optimize some mechanical and electrical elements.

An example of parameters registration during the operation, can be seen in fig. 4.

Testings have been made in severe climatic conditions with temperature variation between -2o°C and +25°C, relative humidity between 6o and 96 %, deposition of white frost and ice, and snow, as well frequently storms with electrical discharges.

The adopted solution for foundation of this experimental model wind unit is technological easy to make but she involve a complication for the tower.

Althrough the foundation solution offered the possibility to achive an extended earthen electric contact, in the placement conditions can't be achived an earthing resistance smaller then 8 Ω comparatively with a value of 1 Ω proposed in designing stage.

The downwind rotor, with continue orientation, unlimited, assure a good stability in operation - fig. 5

The solution involve collectors necessary to transmit the energy, the signals and the commands. In placement conditions have appeared electrical piercements in insulating materials even at reduced tensions of 24 V.c.c.

The adjustment system of blades being placed inside the rotor create major dificulties when is necessary to intervene in this zone and determinate to grow the probability of defects apparition.

Althrough the gear-box (made by Renk-Reşiţa) has a faultless operation, the machines line lenght involve a rigid frame therefore very heavy, in contrary case appear systematicall machines unbalance of cinematic line.

The both braking systems have a very good efficiency. The braking process with disk on the turbine shaft is uniform, silenced and did not introduce shocks and vibrations. It have the disavantage of large gauges and masses.

The braking process behind gear-box, load very much shes gearings and it's made with considerable shocks.

The white frost depositions on the blade have as effect the change of aerodynamics and the unbalances appearence which determinate to grow the vibrations level and evean to make more difficult to start the turbine.

Turbine rotation and proper blades vibration don't determinate always the whole white frost to get out of blades, but a few hours of sun radiation has a good effect in geting out the white frost and the ice.

Concerning the wind speed transducer (anemometer) and wind direction transducer, the white frost deposition has as effect the futility of these transducers obstructing so the unit operation, and a few hours of sun radiation has not significant effects in this case.

The conditions of excesively humidity and the low atmospherical pressure of the mountain placement, where is erected the wind unit, determinate a high wears of electrical coupling elements and the diminution of fiability.

An analysis about defects source can be made with fig. 8.

7. CONCLUSIONS

The experience that we have accumulated since now,through the testings on the model wind unit, permite to us to get conclusions for an industrial prototype wind unit, designed to work in mountain condition,that is now in execution - anssambling stage.

It have been improved the fundation solution achiving an anchored foundation in rock, this solution aiming a more simplified tower.

The integrated gear-box, made by Renk-Reşiţa, has as effect the machines line shortening.

The fractioned blades (having flaps) by polyester reinforced with fibreglass are designed to achive the limitation of power, in normal operation, and to achive the aerodynamical braking through the flaps in cases of emergency.

The aerodynamical brake is supplemented, in the emergency situation, by the mechanical brake.

Fig.1. General view of 300 kW experimental wind unit in a mountain placement.

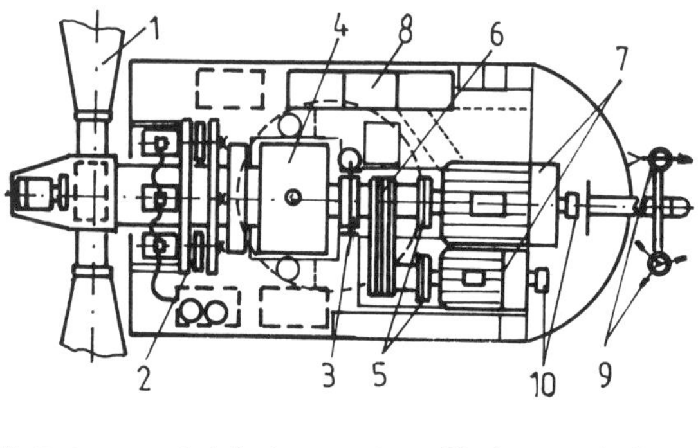

1.Rotor and blades
2.Emergency brake
3.Working brake
4.Gearbox
5.Electromagnetic coupling
6.Belt transmision
7.Generators
8.Instrument cubicles
9.Wind transducers
10.Generators speed transducers

Fig.2. Nacelle of wind unit.

8. REFERENCES

1 F.Gyulai, C.Albu, S.Lăzărescu - Comparision between computed and Measured performances at the 300 kW wind unit from Semenic location,"Energetics National Conference" Present Tendinces and direction in romanian energetics", Neptun 15-18 iun.1992, Romania

2 F.Gyulai, Gârbacea A., Wind Energy Research in Romania, Winddirections, Vol.XII nr.1/summer 1992.

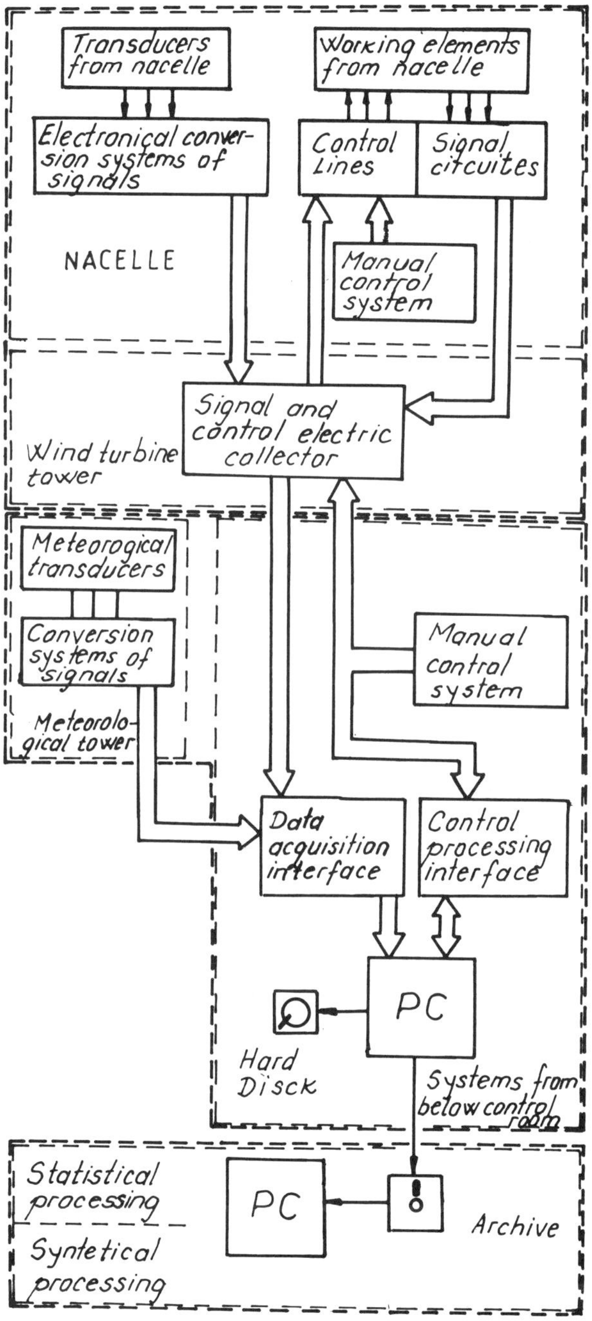

Fig.3. Principle scheme to draiving and data aquisition.

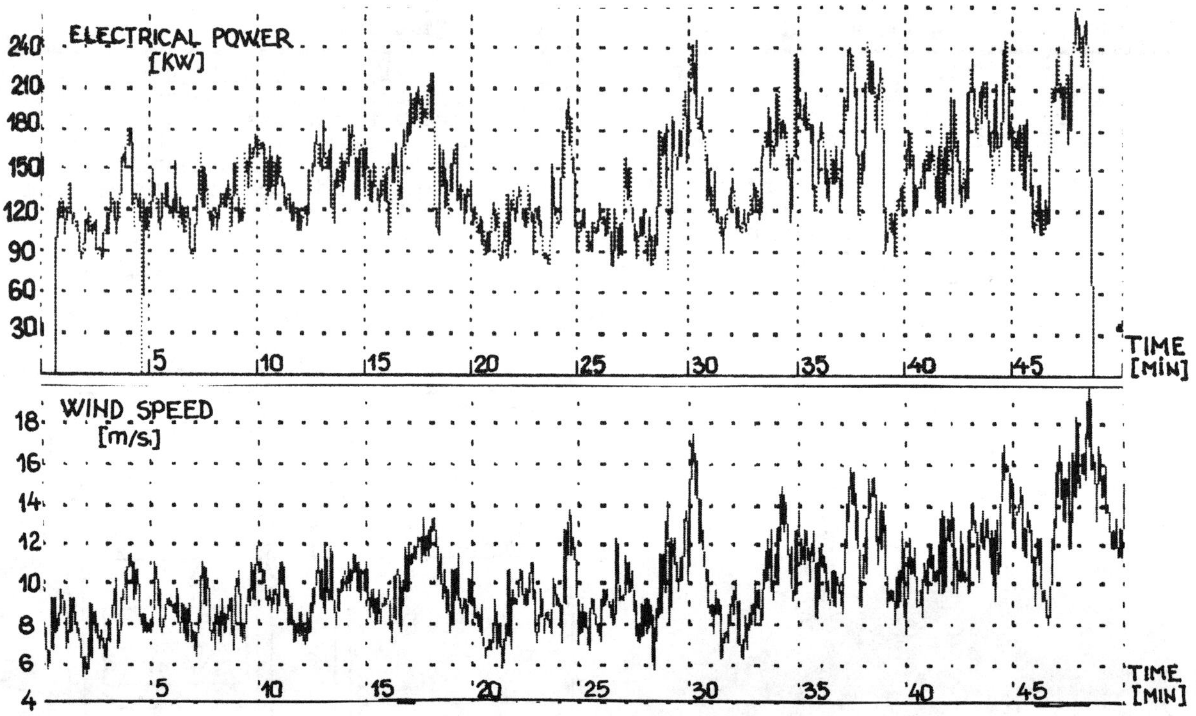

Fig.4. Parameters registration during the operation.

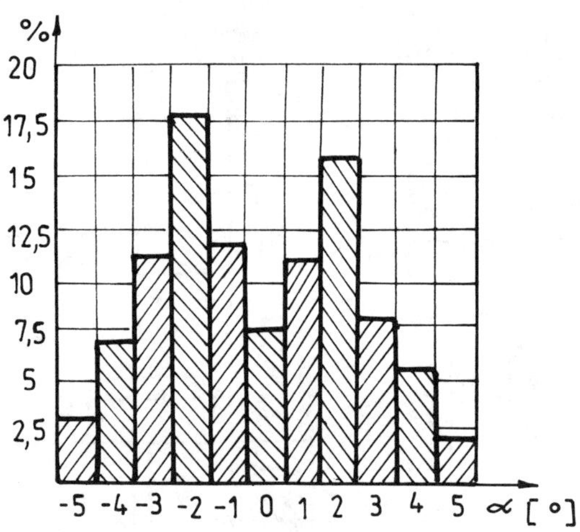

Fig.5. The deviations of the nacelle axis from the wind direction by statistic evaluation.

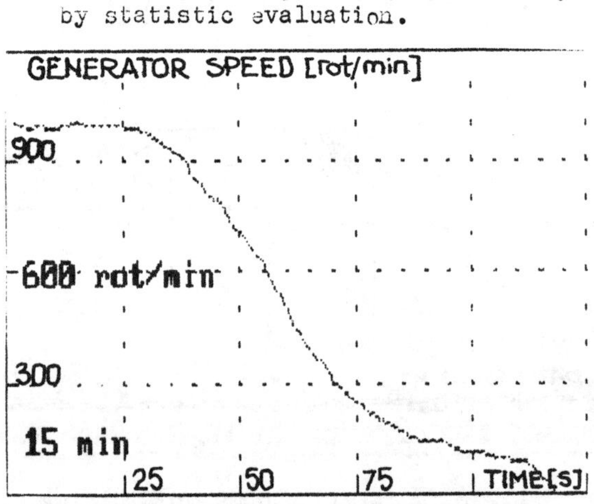

Fig.6. Braking with hydraulic brake.

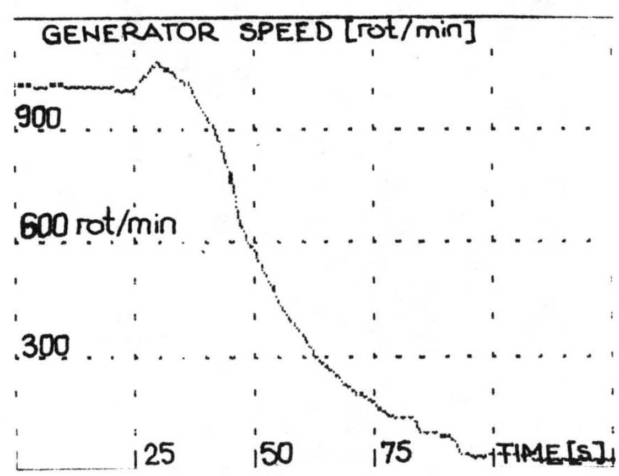

Fig.7. Emergency braking with hydraulic and mechanical brake.

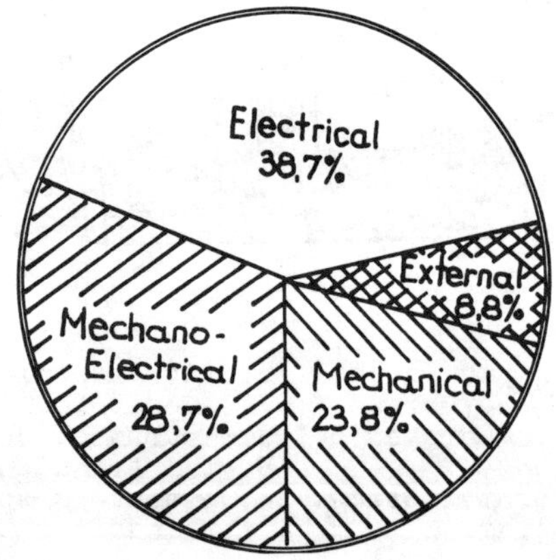

Fig.8. Statistic evaluation of the causes of malfunctions.

THE IMPROVEMENT OF THE COST/BENEFIT RATIO BY APPLYING FLEXHAT TECHNOLOGY

F. Hagg, Th. van Holten, P.A. Joosse, G.A.M. van Kuik
Stork Product Engineering
P.O. Box 379, 1000 AJ Amsterdam
The Netherlands

ABSTRACT: The Dutch Flexhat program is almost finished, resulting in a number of full-scale tested technology improvements with respect to the present state of the art. By a conceptual design study, it is shown that applying this technology in an existing commercial wind turbine leads to a 30% reduction of the kWh price.

1. INTRODUCTION

The Dutch Flexhat program started in 1984 aiming at the development the technology for advanced wind turbines. The progress and results of this program have been reported at several wind energy conferences, [1],[2] and [3]. The main features of the Flexhat technology are: a flexible rotor with an elastomeric teeter, power control by centrifugally activated tips, sophisticated power electronics for the variable rpm conversion system, standard components for the mechanical parts.

The application of these components results in a decrease of rotor and drive train loads, and an improvement of operational aspects. A survey is presented in table 1. A more comprehensive description can be found in the mentioned conference papers. All rotor components and the power electronics for the variable speed operation have been tested at the 25mHAT of ECN. The electronics for the damping of drive train oscillations has been tested at the ECN laboratory.

During the program, the potential reduction of the kWh price by fully employing this technology was estimated at 30%. Now the program is almost finished and the full-scale tests at the ECN test turbine confirm the expected decrease of the load spectra on the blades and drive train, this estimation has been worked out in much more detail in the Veriflex study [4] (VERIfication of the FLEXhat technology). The approach in this study was to compare the performance and costs of an existing, commercially available turbine with the performance and costs of a redesigned version of it. The redesign was restricted to the rotor and the conversion system, keeping all other components the same.

For the state of the art turbine a three bladed, stall regulated turbine with a diameter of 25 m, and 250 kW installed power was chosen. This turbine is denominated further on as the Original 25/250. A sketch of the Veriflex turbine is shown in figure 1.

2. METHODOLOGY OF THE VERIFLEX STUDY

The basic idea is to stretch the rotor diameter and to increase the installed power. This is possible as the specific rotor loads, i.e. loads per square meter of rotor area, decrease when the original rotor is replaced by a Flexhat rotor. The limits of the stretching are dictated by the strength limits of the unmodified components: the drive train (except the generator), nacelle bed plate, tower etc..

Two methods for establishing the cost/-benefit ratio have been applied:

- a global, simple method, indicating trends and orders of magnitudes based on scaling laws. As the reference turbine a generalized 3-blade, 250 kW, 25 m diameter stall turbine is defined: the Original 25/250.
- a qualitative optimization method called Optihat, [5], starting from an existing, commercially available turbine which resembles the Original 25/250. The optimization has been carried out as a preliminary design study.

The qualitative method combines the calculation of the load spectra and extreme loads with predictions of the relevant construction details such as thickness of the laminate. Combined with the manufacturers cost data and with general cost algorithms, an optimization of the cost/benefit ratio is possible by varying relevant design drivers such as diameter and design lambda. The costs for transport and assembling are included in the comparison, as are the typical costs for grid connection within a wind farm. These costs have been supplied by a Dutch utility for a specific site.

The first method will be used here for explanation, and the results of the qualitative method for the final comparison.

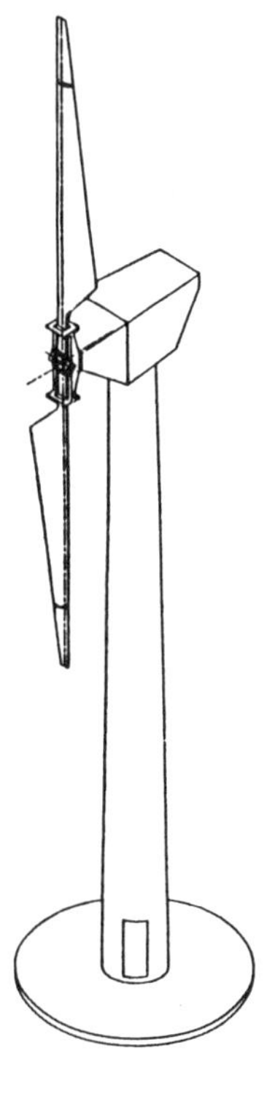

Figure 1: A sketch of the Veriflex turbine

load reduction in:	by application of:
rotorblades transmission conversion system nacelle bed plate yaw mechanism	flexbeam and teeter peakshaving of torque by rpm control system and (centrifugally activated) peakshaving of power by control system and tips teeter teeter
improved reliability	by:
aerodynamic control teeter operation of turbine	passive system standard elastomeric components all aspects (peak shaving of torque, power control, noise reduction, damping of drive train oscillations by power electronics
operational advantages	by:
control of acoustics constant power above V_{rated} power control of turbine or wind farm	variable rpm set point characteristics of rpm control variable power set point

Table 1: Survey of Flexhat components and design features.

3. STRETCHING OF THE DIAMETER AND INSTALLED POWER, USING A GLOBAL ANALYSIS

3.1 Cost/benefit comparison for equal diameter and power

As an intermediate step in stretching the rotor diameter and the generator power, the consequences of replacing the stiff three bladed rotor with a two bladed flexible rotor of the same diameter and power is studied: the Veriflex 25/250. For the Original turbine the following cost breakdown is assumed in the global method:

- rotor costs including hub: 27%
- conversion system: 10%
- other turbine costs: 63%
  complete turbine: 100%
- farm costs: 44%
  total costs: 144%

These estimates of the turbine costs breakdown are confirmed by the general cost analysis in [6]. Compared with the original rotor, the costs of the new rotor, having the same tip speed ratio, will be approximately 2/3rd. This is made plausible in the table below.

number of blades	3	2	
rotor type	stiff	stiff	flex
rpm	fixed		variab
1) external loads on each blade	1	3/2	3/2
2) chord and thickness	1	3/2	1
3) efficiency of construction	1	3/2	3/2
4) costs for each blade 1):3)	1	1	1
5) rotor costs	1	2/3	2/3
6) fatigue of nacelle bed plate	1	>> 4/3	< 1
7) extreme tower-loads at stand still	1	1	2/3

Table 2: Comparison of stiff 2- and 3-bladed rotors with a flexible 2-bladed rotor, having the same rpm at low wind speeds.

It is clear that changing from 3 to 2 blades is the most important step for the rotor costs. The disadvantage of the stiff two-bladed rotor, the high fatigue loads on the low speed axis, nacelle, bed plate, yaw system and tower by windshear and non-coherent gusts, is removed by introducing flexibility: the teeter and flexbeam eliminate almost all loads with a frequency of 2,3,4P (P is rotor frequency) and the 1P yawing and tilting moments.

Due to the variable rpm the blade area of the flexible rotor can be smaller than the area of the stiff rotors: the design tip speed ratio is higher. This results in a higher (than original) tip speed at V_{rated}, but the same tip speed at low winds where noise emission is critical. This increased design tip speed ratio explains the value "1" for the chord and thickness of the flexible blade.

The efficiency of the construction is the capability to carry the loads: for the two-bladed rotor the loads are 3/2 times larger than for the three-bladed rotor, but the chords and thickness are increased equally. For the flexible rotor the external loads result in much lower internal loads compared with the stiff rotors, because of the rotor dynamics. This explains the increased efficiency of the construction despite the small chord and thickness.

The decrease in rotor costs is canceled by the increased costs of the variable rpm conversion system. Such a system is 2 to 3 times more expensive than the fixed rpm induction generator. Here the factor 2.5 is used. However, the benefit is an increased annual yield as the power curve can be optimized for each wind speed. The difference in the annual yield for a stall rotor and an optimized rotor is known to be 10% approximately.

The comparison yields the following results:

component	Orig. 25/250	V-flex25/250
	% related to	orig.turbine
rotor incl.hub & tip control	27 %	18 %
E-system	10 %	25 %
other components, transp. & assembling	63 %	63 %
subtotal	100 %	106 %
farm costs	44 %	44 %
total costs	144 %	150 %
annual yield	100 %	110 %
costs/benefit	100 %	95 %

Table 3: Cost/benefit comparison for equal power and diameter.

The final difference in the cost/benefit ratio is very small. This is not too surprising: no use has been made of the surplus strength in the drive train, nacelle, yawing system and tower.

3.2 Cost/benefit comparison for enlarged diameter and power.

If the rated wind speed is kept constant, some simplified rules describe the stretching of the rotor diameter D and installed power P, and the increase of costs and energy yield. For constant V_{rated} P and D are related by:

$$P \div D^2$$

The rotor costs then follow the square cube law, with a slightly modified component (see [6]):

$$\text{Rotor costs} \div D^{2.7}$$

For the conversion system the costs are, in first order, assumed to be proportional to the installed power:

$$\text{Costs of conversion system} \div P$$

Finally for constant V_{rated} the energy yield E is proportional with the rotor area:

$$E \div D^2$$

Applying these rules to a target diameter of 31 meter, an installed power of 380 kW results. The following table shows the cost breakdown. Possible strength limits in the unmodified components of the Original 25/-250 are not taken into account yet.

component	Original 25/250	V-flex 25/250	V-flex 31/380
	% related to orig.turbine		
rotor incl.hub & tip control	27 %	18 %	32 %
E-system	10 %	30 %	38 %
other components, transp. & assembling	63 %	63 %	63 %
subtotal	100 %	111 %	133 %
farm costs	44 %	44 %	44 %
total costs	144 %	155 %	177 %
annual yield	100 %	110 %	169%
costs/benefit	100 %	98 %	73 %

Table 4: Cost/benefit comparison for enlarged power and diameter.

3.3 Limitations by the strength of unmodified components and noise emission

The available strength of the original transmission, nacelle bed plate, yaw system and tower may put limits on the stretching of the rotor diameter and installed power. Furthermore, the noise emission at low wind speeds is a limiting factor.

- The strength of the transmission is not a problem: the nominal torque Q of the Veriflex-25/250 is 2/3rdf the nominal torque of the Original turbine, since the rpm above V_{rated} is 1.5 times higher. Furthermore, the usual margin in the dimensioning of drive trains of stall turbines is 90%, while for the flexible rotor only 25% is needed. This is due to the peak shaving of the torque by the variable speed system and the passive control tips. This leads to a margin of 1.5*(1.9/1.25)= 2.3 for the Veriflex with the same diameter and power. This margin is used for the stretching. Keeping V_{rated} and the tip speed the same, the following relations apply:

$$\Omega \div D^{-1}\ ,\quad P \div D^2 \quad \Rightarrow \quad Q \div D^3$$

According to this, the margin of 2.3 allows a diameter of 33 meter.

- The nacelle, bed plate, yaw system and drive train bearings are not critical, since the teeter and flexbeam reduce the loads very efficiently.
- The dimensioning of the tower of the Original turbine is assumed to be dictated by stiffness demands. This gives a margin w.r.t. strength and fatigue demands. Global calculations show that this margin allows a diameter of 32 meter. The extreme loads on the tower during standstill are determined by the blade area. For the same diameter, the Veriflex area is 2/3 times the Original area (see table 2), so the rotor areas are equal for $D = 25*\sqrt{1.5} = 30.6$ m.
- Noise emission is mainly determined by the tip speed. At low wind speeds the Veriflex rpm is approximately 60% of the rated rpm, and equal to the fixed rpm of the Original turbine. The tip speeds are constant during scaling of the diameter, so the noise emission is expected to remain the same.

3.4 Conclusions of the global method

The global analysis shows that the cost benefit ratio improves by almost 30% when the original, 25 m ∅, three bladed stiff rotor is replaced by a 31 m ∅, two bladed Flexhat rotor, meanwhile installing a 380 kW variable speed conversion system instead of the fixed rpm induction generator.

This conclusion has to be confirmed by detailed optimization of the important design drivers (diameter, tip speed ratio, installed power, etc.), and by detailed load calculations, dimensioning and cost analysis. A survey of these calculations is shown in the following paragraph.

4. DETAILED OPTIMIZATION AND DIMENSIONING

The detailed analysis has been reported in [4]. An existing, commercially available turbine is used as reference turbine. The design calculations of the manufacturer have been used to establish the stiffness and strength limits of the unmodified components. The cost data of the manufacturer have been used as input for the optimization method [5].

The optimization of the enlarged Veriflex turbine with the code Optihat, revealed a preference for 360 kW instead of 380 kW installed power, with a diameter of 32 m. The improvement of the cost/benefit ratio for this Veriflex-32/360 is 30% compared with the Original-25/250. In this analysis the wind farm costs depend on the installed power, instead of being constant as assumed in the global method. Figure 2 shows the sensitivity of the cost/benefit ratio for a 10% increase of the important design parameters w.r.t. the final values. In figure 3

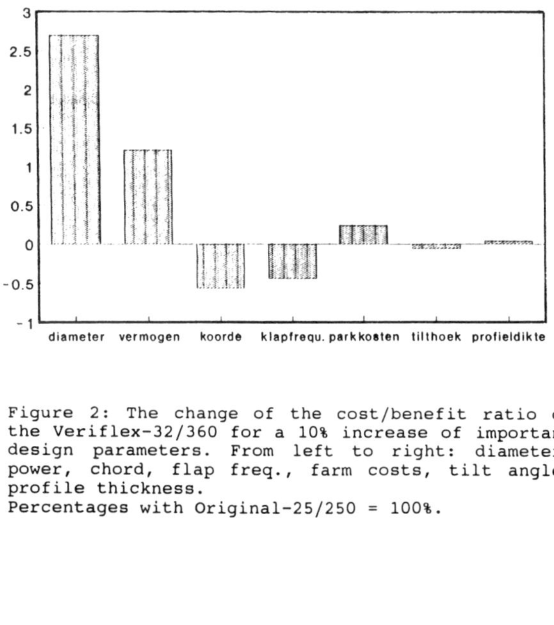

Figure 2: The change of the cost/benefit ratio of the Veriflex-32/360 for a 10% increase of important design parameters. From left to right: diameter, power, chord, flap freq., farm costs, tilt angle, profile thickness.
Percentages with Original-25/250 = 100%.

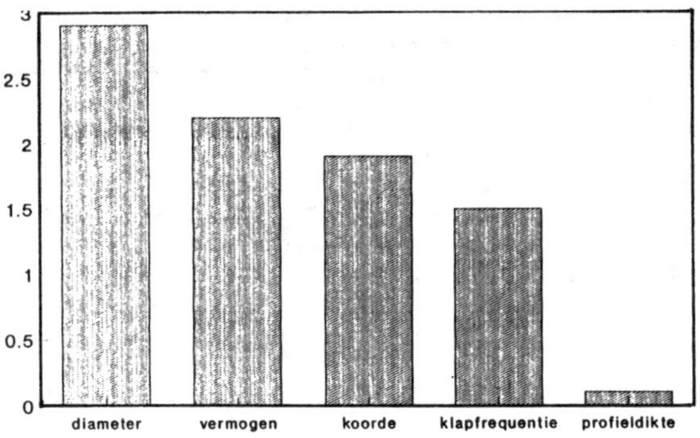

Figure 3: As figure 2, but for optimum values of the individual parameters. From left to right: diameter, power, chord, flap freq., profile thickness.

the same sensitivity is shown for the optimum values of each individual parameter. Both figures show that the cost/benefit ratio is not very sensitive.

With respect to the strength and dimensioning analysis with the computer code Flexlast, the most significant results are the following:

- The only assumption in the global method is confirmed: the loads on the nacelle bed plate and the yawing system, using the Veriflex rotor, are lower than the design loads.
- The tower strength is enough for stretching of the diameter to 33.5 m, if the tower base access opening is strengthened locally.
- The tilt angle of the enlarged rotor has to increase to 15°. This is necessary to accommodate for the large tip deflections connected with the teeter movement. The construction of the turbine allows such an increased tilt angle without changing the entire nacelle construction.
- The limiting factor for the stretching is the gravitational load on the main shaft.

The preliminary design of the rotor hub with the teeter and flexbeam is shown in figure 4.

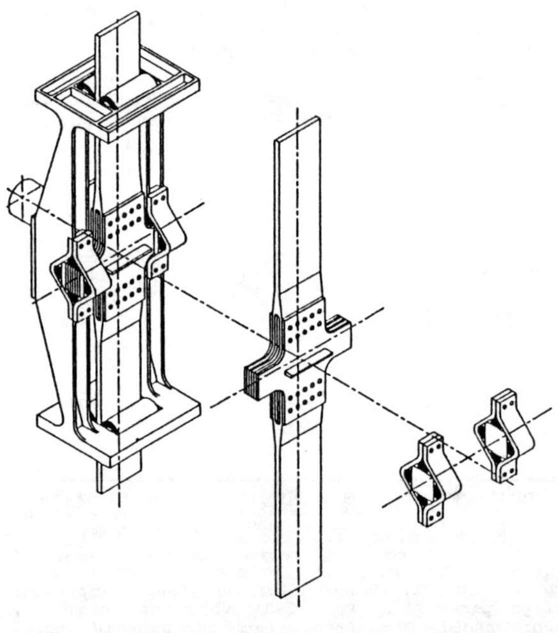

Figure 4: Sketch of the Veriflex hub with the teeter and flexbeam

5. CONCLUSIONS

It is possible to improve the cost/benefit ratio of an existing state-of-the-art turbine by applying Flexhat technology. This is demonstrated by a preliminary design study in which the three bladed, 25m Ø, stiff rotor of a commercially available turbine, is replaced by a flexible, two bladed rotor with 32m Ø, and at the same time the 250 KW induction generator by a 360 kW variable speed generator. This is possible without exceeding the strength limits of the other, unmodified components, since the specific rotor loads (loads per m^2) for a Flexhat rotor are significantly lower than for a conventional one. The result is a 30% better cost/benefit ratio.

REFERENCES

[1] G.A.M. van Kuik, R.H.F. Tikkemeier (SPE), J.W.M. Dekker (ECN): The development of advanced rotor systems - a survey of the FLEXHAT program, including full scale results. Proceedings of ECWEC'90, Madrid, september 1990.

[2] G.A.M. van Kuik (SPE), J.W.M. Dekker (ECN): The FLEXHAT program - technology development and testing of flexible rotor systems with fast passive pitch systems. Proceedings of EWEC'91, Amsterdam, october 1991.

[3] J.W.M. Dekker, C.M. de Groot (ECN), F. Hagg, P.A. Joosse, G.A.M. van Kuik (SPE), H.J. Heerkens (Aerpac): Test results for the Flexteeter - a flexible rotor with fast passive pitch control. Proceedings of EC-WEC'93, Travemünde, march 1993.

[4] Th. v. Holten, F. Hagg, P.A. Joosse, G.A.M. van Kuik (SPE), J.Th.G. Pierik (ECN): VERIFLEX, a summary of the concept- and feasability studies (in Dutch). SPE report 92-011, februari 1992

[5] F. Hagg, N. Dames: Description of Optihat, in Dutch. SPE report 87-007.

[6] E. Hau e.a.: Study on the next generation of large wind turbine, final report for CEC-DGXII, ETA GmbH, Munich, may 1991.

ANNOYANCE FROM WINDTURBINE NOISE ON SIXTEEN SITES IN THREE COUNTRIES.

M. Wolsink*, M. Sprengers*,
A. Keuper+, T.H. Pedersen#, C.A. Westra*

* Department of Environmental Science University of Amsterdam
Nieuwe Prinsengracht 130; 1018 VZ Amsterdam (NL)
+ Deutsches Windenergie Institut, Wilhelmshaven (D)
# Lydteknisk Institut, Lyngby (DK)

SUMMARY: On locations in Denmark, the Netherlands and Germany the actual annoyance caused by wind turbine noise was investigated. Residents of communities living around existing turbines were interviewed and emitted noise levels were computed based on site measurements. The number of people actually indicating annoyance by wind turbine noise is fairly small. But at the same time the amount of annoyance is hardly related to the objective sound level.

1. INTRODUCTION.

1.2 Wind turbine noise.

Noise has been recognized as one of the main factors in potential environmental impact of wind energy application. Nevertheless much is still unknown about the actual annoyance caused by noise produced by wind turbines.

The European Community partially financed a research project on the annoyance from wind turbine noise. The most important question to be answered was the actually existing annoyance caused by operational wind turbines. On several locations in Denmark, the Netherlands and Germany, residents of communities living around existing wind turbines were interviewed. At the same time the emitted noise levels were computed based on sound measurements on the sites. The main objectives of the study were to establish the level of annoyance, and the amount of annoyance that can be attributed to possible other causes than the objective sound levels.

1.2 Other causes of annoyance.

The establishment of annoyance from noise is not easy, because simple dose-effect models appear to be inadequate (1). Apart from the source (the windturbines) many other causes for negative feelings may exist. These are interfering the direct relation between the level of noise and the amount of annoyance. In order to estimate the annoyance caused by wind turbines these intervening variables must be included in the research.

The appraisal of noise is a highly subjective matter and therefore several factors of individual nature will affect it, beside turbine noise itself. Psychological research on reactions to environmental noise fully agrees on the influence of the following attitudinal factors (2):

* The awareness of non-noise problems increases annoyance.
* Fear of the noise source increases annoyance.
* The belief that the noise source is important decreases annoyance.
* The belief that the noise could be prevented increases annoyance.

In the case of wind turbines the first factor will be mainly visual intrusion, because it is the strongest determinant in public acceptance of wind power application. The second factor will be of very little weight. The third and fourth factor may be affected by the decision making process in the phase of planning the wind turbines. We will refer to them in the section were the results are discussed.

Beside these attitudinal factors some situational factors may also moderate annoyance. The literature on this subject is ambiguous. For instance the hypothesis that annoyance decreases with time, is sometimes supported. In about the same number of studies the opposite hypothesis is accepted (2).

The only clear situational factor is the impact of non-noise intrusion caused by the same source: it increases noise annoyance. The most important factor of wind turbines in the view of the public, is the visual impact. The characterization of the visual impact as 'spoiling the scenery' is only one side of an evaluative scale of turbines in the landscape. The other side of it is a positive scenic evaluation of wind turbines, which is made by a substantial part of the population, mainly dependent of the type of landscape on the site of the turbines. Therefore this situational factor is basically the same as the first attitudinal factor, the awareness of non-noise problems.

Even more complicated is the relation between noise and psycho-social stress or health effects (4). Annoyance caused by wind turbine noise may cause stress, but at the same time many other intervening variables exist. Stress due to noise may be caused by other sources of noise, but also by factors that are not related to either wind turbines or environmental noise. In general residential dissatisfaction and daily hassles will be the main causes for stress. High levels of stress from situational and personal factors (and therefore that are not linked in any way to wind turbine noise) may also cause an increase in the perceived turbine noise and the level of annoyance. Actually wind turbine noise is only one of many factors that may decrease residential satisfaction.

2. RESEARCH DESIGN.

2.1 Sites and respondents.

As the installed capacity grows the noise problem will become more significant in siting decisions of wind turbines and wind farms. At the moment the number of sites of wind farms in Europe is still considerably small. Sites were easily selected without annoying anyone, although some planning mistakes have been made (6).

One of the major problems in this research was the selection of operating turbines with people living near them. The research was carried out in the three European countries with the largest installed capacity: Denmark, the Netherlands and Germany. Nevertheless the actual number of persons that objectively can be annoyed by noise from operating turbines in these countries, does not exceed a few hundred. In these countries sixteen sites were selected with residents living within the 35 dB(A) region. These sites are marked in the figures 1A-C.

Because a certain variance had to be included in the study, some residents below the sound level of 35 dB(A) were included in the survey. The total number of valid interviews in the study was 574 (see table 1). In Denmark

Table 1: Total sample: country and number of sites.

country	N of sites	N of respond.	% of total	residents (%) longer than 5 yrs
	---	---	----	----
Netherlands	4	159	27.7	75
Germany	3	216	37.6	62
Denmark	9	199	34.7	75
Total	16	574	100	70

and the Netherlands the average time people were living in their homes was about the same. In Germany the time was considerably shorter. This was caused by the site of West-Fehmarn, were the type of housing was different. Many cottages are sited around the wind turbines there and a different type of residents resulted. One out of four respondents in Germany was living less than two years in the house were they were interviewed.

All other demographic variables as sex, age, etc. were evenly distributed. One third of the subjects (32%) were living in a household with persons below 14 years of age.

2.2 Questionnaires.

For the design of the questionnaire a research model was constructed. Attitudinal and situational variables as outlined in section 1.2 were included in the questionnaire. It was constructed in Dutch and then translated into Danish, German and English. In the research model the following construct-variables were included.

For the appraisal of wind turbine noise three variables were constructed: annoyance, perceived loudness and interference. These are the central variables of which the dependency of the noise-stratum has to be established.

Attitudes to wind power application were measured and a list of attributed aspects of wind power application were judged by the respondents, as used in wind energy research (7,8).

The perceived residential quality and satisfaction was measured by a set of nine statements ("we enjoy living in our home", "it's a pleasant area to live in", etc.). Annoyance of noise from other sources (traffic and work activities) is a part of residential quality and it was included in the questionnaire.

Stress due to turbine noise was measured with items from a scale developed by King et.al. (5). It contained reported states of mind such as 'angry', 'nervous', 'irritated' and 'anxious'.

The social and psychological data were merged with data on sound levels and environmental conditions, such as surrounding trees and obstacles between turbines and homes. These are based on measurements on the locations themselves.

3. NOISE AND ANNOYANCE.

3.1 Objective sound levels.

Sound levels have been measured on sites, and sound level strata have been calculated with 5 dB(A) intervals. The sound contour strata plots were combined with the actual distance between a subjects' house and the wind turbine location to obtain the objective sound level. Most of the subjects (55%) live in the region surrounded by the 35 dB(A) sound level, one of the sampling criteria.

The average sound level that subjects are exposed to is approximately 35 dB(A) (sd. = 5 dB(A)) from an almost normal distribution (table 2).

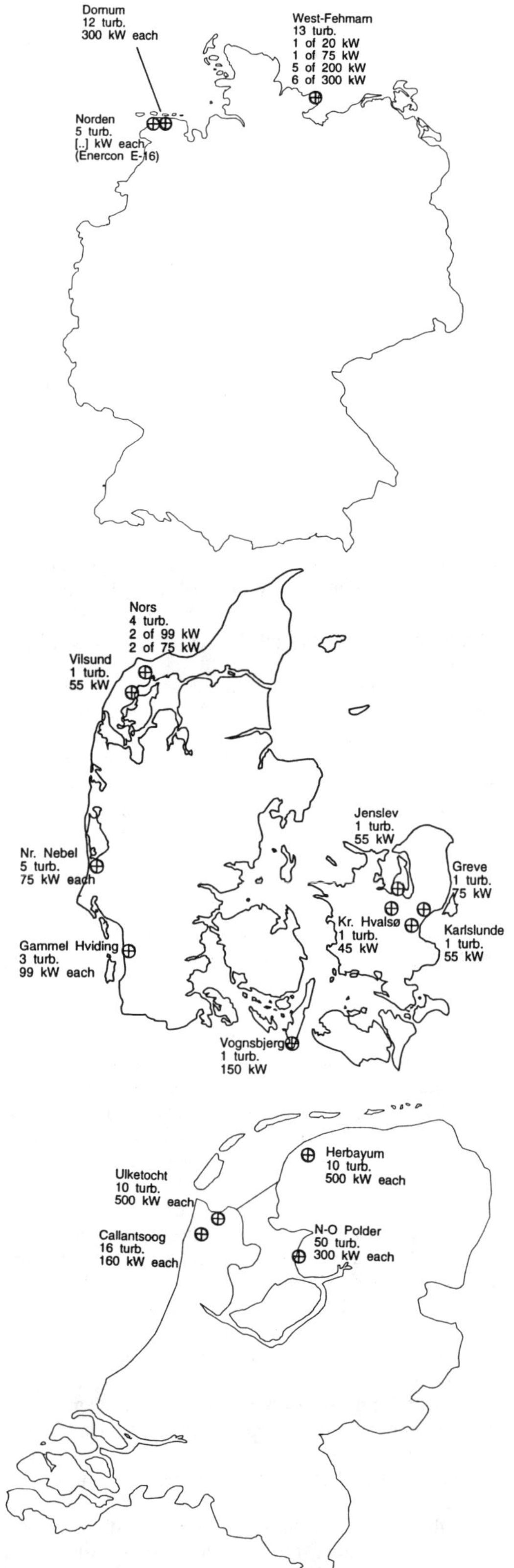

Fig.1: A-C: The sixteen sites in Germany (above), Denmark (middle) and the Netherlands (below).

Table 2: Measured sound levels at residents dwellings.

sound level	N of resp.	% of total
< 25 dB(A)	13	2.3
25-30 dB(A)	64	11.2
30-35 dB(A)	177	30.8
35-40 dB(A)	225	39.2
40-45 dB(A)	69	12.0
45-50 dB(A)	18	3.1
50-60 dB(A)	8	1.4
Total	574	100.0

3.2 Perceived noise annoyance attributed to turbines.

In order to assess *noise annoyance* the respondents were first asked whether they had complaints about noise from various sources in the neighbourhood. Half of the complaints about noise annoyance are attributed to noise from wind turbines (table 3).

Table 3: Resident reporting (any) annoyance caused by wind turbine noise.

annoyance	N or resp.	% of total
not at all annoyed	537	93.6
(some) annoyance	37	6.4
total	574	100.0

The first obvious hypothesis is about the dose-effect relation. According to the dose-effect model there should be a clear and direct relationship between the complaints about noise annoyance from wind turbines and the objective measured sound level. However, we already stated that many intervening variables exist. Analyses shows that the dose-effect model is not an appropriate description of reality. There is only a very weak correlation between sound level and complaints about noise annoyance caused by wind turbines (Kendall's coefficient for correlating rank order variables τ =.09; p<.05).

A second instrument for measurement of annoyance refers to the occurrence of certain characteristics of the noise emitted by wind turbines: loudness, duration, high-low, strong-weak, and intensity. According to Van Kamp (4) the judgments about these sound characteristics were combined to a uni-dimensional index of *perceived loudness*. In general, the perceived loudness is very low with an average of .86 in an empirical range of 0 to 5 (sd=1.00). About one third of the subjects (35%) do not perceive any of the sound characteristics. Residents complaining about wind turbine noise also perceive more sound characteristics (τ =.38; p<.001).

A third measurement of noise annoyance relates to the intensity of the *interference* of noise with various *daily activities*. Subjects were asked to indicate to what extend noise interfered with activities like resting, leisuring, sleeping, talking, reading, thinking, listening to the radio, and watching television. There are few complaints about noise interfering with daily activities: the average score on the resulting indicator is only 1.40 (sd=2.47) on a range of 0 to 40. Respondents with complaints about turbine noise reported more interferences of daily activities (τ =.56; p<.001).

As established in research on the annoyance of air traffic noise and road traffic noise (4), the three indicators of annoyance show a reasonable correlation. So the three indicators were combined in one uni-dimensional indicator for the level of noise annoyance, based on their relative contribution a single determining factor.

The origin of the noise that is complained most about is the sound produced by the blades, followed by unspecified sounds during operation. Most of the annoyance is experienced between 16.00 pm and midnight. When people are reporting annoyance, in most cases they had that feeling when they were outdoors.

4. INTERVENING VARIABLES.

4.1 Noise annoyance and situational factors

First we will direct our attention to situational factors to determine the level of annoyance among residents in the direct vicinity of wind turbine sites. The distinguished situational factors are:

* the distance between a subjects residence and the wind turbine site;
* the location of the residence relative to the location of the site with regard to the dominant wind direction;
* other buildings between a subjects house and the wind turbine site;
* natural barriers, like trees etc., between a subjects house and the wind turbine site;

In order to estimate the contribution of these factors we used regression analysis with the indicator for noise annoyance as dependent variable.

The result of this analysis is that the amount of noise annoyance is lower when buildings are standing between the respondents home and the wind turbine site. Furthermore none of the situational factors are related to the level of noise annoyance. Even the objectively measured sound level is not significantly related to noise annoyance in this analysis.

4.2 Noise annoyance and characteristics of residents.

The disappointing results with regard to the relation of situational factors and noise annoyance is all the more reason to pay attention to attitudinal and other characteristics of subjects in order to determine the amount of annoyance. We analysed the possible effects of the general attitude towards wind energy, of the visual impact, residential satisfaction, other sources of general dissatisfaction, and the estimated pros and cons of large scale application of wind power.

The analyses of annoyance causes major problems due to the very skewed distributions of the scores on the dependent and some of the independent variables. With regard to the dependent variable, the level of annoyance, most of the residents that were interviewed (93%) were not annoyed by wind turbine noise. They hardly ever heard the wind turbines. They perceived no special sounds in the wind turbine noise (65%); nor did the noise interfere with their daily activities (85%).

With regard to the independent variables satisfaction with housing and neighbourhood and irritation or stress due to wind turbine noise, the situation was very much alike. Most of the subjects were very satisfied with their housing and neighbourhood (about 70%) and a very small number of residents (< 15%) were reporting irritation or stress due to wind turbine noise. The general attitude towards wind power application and the appraisal of the visual impact of wind turbines on the landscape had less skewed distributions.

We have tested the hypotheses which directly relate the level of annoyance with:

(a) the general attitude towards large scale use of wind power energy;
(b) appraisal of the impact of turbines in the landscape;
(c) satisfaction with housing and neighbourhood;
(d) an index for other sources of dissatisfaction;
(e) stress caused by wind turbine noise.
(f) objective factors like sound level and buildings between tubines and homes.

Because these factors are related to one another, and may also interfere with some of the situational factors, we carried out a multivariate analyses of the level of annoyance with characteristics of subjects. It shows that stress due to noise of wind turbines is the major explanatory effect of the level of annoyance. Most of the subjects who tend to get stressed by the wind turbines have a high level of annoyance.

The perceived effects of wind turbines in the landscape (i.e., visual intrusion) plays an intermediate role: in general, noise annoyance decreases the less intruding wind turbines are judged in the landscape. This factor shows some remarkable differences between the the three countries when the results are compared. For these differences we did not find an explaination.

No effects were found for any of the situational aspects like the actual objectively measures level of noise, buildings, trees and fences between the wind turbine site and the subjects house. Additional to the subjective characteristics there is however still one intermediate direct and multivariate effect. The longer a turbine site is operational, the less the level of annoyance, even if we control for subjective characteristics. In an equation with standardized effects:

$$\text{ANN} = .69\,\text{SN} + .11\,\text{DH} + .10\,\text{LS} - .06\,\text{TIME} + -.58 + e$$

The coefficients give an estimation of the relative weight of each aspect. (ANN = annoyance; SN = stress due to noise; DH = daily hassles; LS = landscape; TIME = time of turbines are operational). The prediction results in an overall explained variance of 53%.

5. CONCLUSION.

The results presented here should be treated with caution. There are a number of methodological problems involved in a project in which subjects from different countries have been interviewed using the "same" questionnaire. Translating questionnaires without loss of information is extremely difficult, and interview effects may interfere differently between countries. It may result in incompatibilities in the data.

Second, there are analytical problems involved due to extremely skewed data. The fact that the actual level of annoyance among the large majority of the subjects is extremely low, leaves very little variation in the major response variables. Extremely little variation means that there is hardly anything left to explain. The prediction that somebody is not annoyed, irrespective of any other effect, is in most cases already the right prediction. Moreover, caution must be taken in order not to ascribe possible effects to outliers, which would make the conclusions not very robust.

The first main result of the study is fact that the number of people actually indicating annoyance by wind turbine noise is fairly small. Of all persons interviewed only 6.4 % is reporting any noise annoyance caused by wind turbines.

The second result is that the amount of annoyance is hardly related to the objective sound level. As a consequence we have to conclude, and that may be disappointing for technical oriented scientists and decision makers, that a reduction in the emissions and a decrease in the sound levels will not result a the disappearance of annoyance and complaints about the noise. The fact that someone is complaining is mainly determined by the personality of the individual. Personal characteristics and circumstances determine the tendency to complain. This conclusion must not be misunderstood. The fact that sound level is not predicting annoyance *does not mean* that people are *not really annoyed* when they are reporting it.

6. DISCUSSION

The fact that only a very small number of resident around operational wind turbines is annoyed by noise can be interpreted as a positive result. It means noise is not a major problem in the case of operational wind turbines. Nevertheless expectations about annoyance in the future caused by wind turbines that will be built, is a crucial factor in the planning phase of wind power projects. From attitude studies we know that the public does not think noise is a problem that is generally troubling the application wind power (7,9). Noise is considered to be a problem that can be prevented, but because of this fact wind power developers must be prudent. All research on noise annoyance indicates that the belief that the noise could be prevented increases the annoyance.

Noise will mostly be an issue when it comes to siting turbines. Utilities and authorities often fail to present projects that sufficiently deal with all locational aspects, such as scenic values and future noise levels. Particularly in this phase of planning wind power projects people want to be taken serious when they have certain doubts.

In judging the way they are treated by the authorities and planners one of the criteria for the community is the way planning agencies use the existing noise regulations and standards. Specific noise standards for wind turbines only exist in Denmark (3). In Germany and the Netherlands the standards are derived from general recommendations. From case studies on siting decisions we know that people feel that noise annoyance is not adequately considered when these standards are use in a flexible way, only to make the siting of turbines possible (9).

The solution is to be found in careful selection of sites and adaptation of the layout of wind farms, not in adjusting noise standards. Most neighbouring residents do not appreciate flexible standards. They feel the standards are manipulated and exactly that will lead to distrust and to the feeling something might be very wrong with the assessments of noise emissions.

References.

(1) Baum, A., J.E.Singer and C.S.Baum (1981) Stress and the environment. *J. of Social Issues* 37 (1) 4-31.

(2) Fields, J.M. (1990) A quantitative summary of non-acoustical variables' effects on reactions to environmental noise. *Noise-Con 90.* University of Texas, Austin. p.303-308.

(3) Hayes, M.D. (1992) Noise impact assessment of windfarms. *Wind Engineering* 16 (3) 141-157.

(4) Kamp, I. van (1990) *Coping with noise and its health consequences.* Styx & PP Publications, Groningen, NL.

(5) King, M.C., G.D. Burrows and A.V. Stanley (1983) Measurement of stress and arousal. Validation of the stress and arousal check-list. *British J. of Psychology* 74 (3) 285-279.

(6) Westra, C.A., M. Wolsink and D. IJtsma (1990) *Besluitvorming windturbineparken.* (transl. Wind farm decision-making) Department of Env. Sciences, University of Amsterdam.

(7) Wolsink, M. (1990a) *Maatschappelijke acceptatie van windenergie.* (transl: Public acceptance of wind energy) Thesis Publishers, Amsterdam.

(8) Wolsink, M. (1990b) Wind power as a social dilemma: the siting problem. *European Community Wind Energy Conference Madrid.* p.725-729.

(9) Wolsink, M. (1991) The noise issue in siting decisions. in: *Wind Energy: Technology and Implementation. European Wind Energy Conference 1991* Vol.1. Elsevier, Amsterdam p.747-751.

STRUCTURE BORNE NOISE FROM WIND TURBINE ROTOR BLADES

Bent Andersen

dk-TEKNIK
Gladsaxe Møllevej 15
DK-2860 Søborg
Denmark

Phone: +45 3969 6511
Fax: +45 3969 6002

Jørgen Jakobsen

Danish Acoustical Institute
Akademivej, Building 356
DK-2800 Lyngby
Denmark

Phone: +45 4593 1211
Fax: +45 4593 1990

ABSTRACT: The purpose of this study was to determine whether typical wind turbine rotors were likely to emit structure-borne mechanical noise, transmitted through the main shaft. This was investigated by measuring the radiation properties of a typical rotor blade and then carrying out in-situ measurements of the vibration of rotor blades during normal operation. The radiation efficiency of different blade sections was determined by measuring the vibration velocity as well as the sound intensity of the radiated noise created by mechanically exciting the root of the blade. The radiation efficiency turned out to be greater than -10 dB in a large frequency range - consequently rotor blades are potential noise radiators. Then vibration velocity measurements were carried out for three wind turbines, two of which were equipped with the earlier investigated type of blades. During the in-situ measurements the vibration velocity signals were transmitted to an analyzer near the turbine by means of small radio transmitters attached to the root of the blade. Simultaneously with these measurements, the total noise emission and the vibration velocity of the main bearing were measured. The structure-borne rotor noise proved to be insignificant for two of the turbines. This contribution was, however, significant in a narrow frequency band for the third turbine, even though extensive measures were taken to reduce all mechanical noise sources.

1. INTRODUCTION

Noise from wind turbines is often separated into aerodynamic noise from the rotor and mechanical noise from the machinery (gearbox, generator, cooling fans, hydraulic power packs etc). It is now well established that for a properly designed wind turbine the aerodynamic noise will normally be the most important contribution to the total A-weighted noise level. The mechanical noise is, however, often the most annoying part of the noise due to characteristics of the noise. Usually the gearbox is the main source giving rise to both tonal characteristics and a special time variation (modulation) of the noise level.

The mechanical noise is radiated partly as airborne noise directly through openings and slits, or through the nacelle walls. The major part of the mechanical noise is, however, transmitted as structure-borne noise to the radiating surface elements. Earlier investigations have shown [1] that the nacelle walls and roof as well as the tower may be important noise radiators in a limited frequency range critical to gear noise. It has also been suggested that the rotor blades might emit structure-borne mechanical noise. The blades have a significant surface area, and are rather stiff, lightweight constructed with low internal damping. Due to their stiff coupling to the gearbox they might contribute significantly to the total noise emission.

2. METHOD

The structure-borne noise emitted from the rotor of a wind turbine was determined in two stages:

- First the radiation efficiency was determined for a rotorblade - that is the relation between the emitted acoustical power and the structural vibrations
- Then the vibrations of the rotor surface was measured for the operating wind turbine.

3. PROCEDURE AND RESULTS

3.1 Measurement of radiation efficiency for rotor blades.

The radiation efficiency is defined as:

$$\sigma = P/(\varrho c S \langle v^2 \rangle) = I/(\varrho c \langle v^2 \rangle)$$

where
- P is the emitted acoustical power [W]
- ϱ is the density of the air [kg/m^3]
- c is the sound speed in air [m/s]
- S is the surface area [m^2]
- $\langle v^2 \rangle$ is the mean square vibration velocity of the surface [m^2/s^2]
- I is the emitted acoustical intensity [W/m^2]

In logarithmic form the radiation efficiency is:

$$10 \log \sigma = L_I - L_v + 33.8 \text{ dB}$$

where L_I is the emitted sound intensity level [dB re 1 pW/m^2] and L_v is the vibration velocity level [dB re 1 nm/s].

The investigation was performed for a self-supporting glassfiber reinforced polyester rotor blade made by LM. The blade length was approximately 12 m, and it was equipped with a front spoiler at the outer 2.75 m.

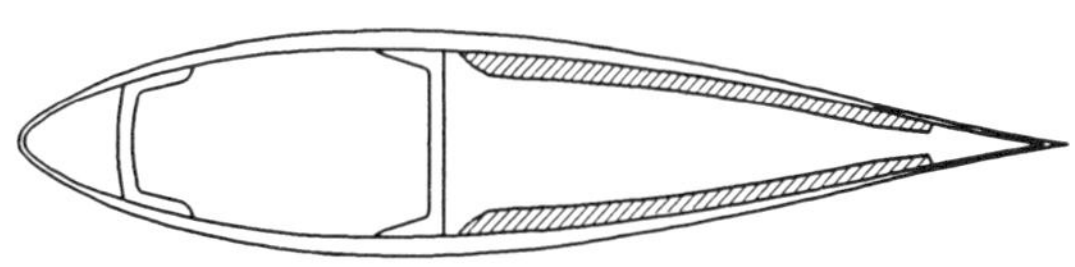

Figure 1: Sketch showing a cross section of the blade with integral, load bearing box girder and sandwich layers.

As shown in the sketch in Figure 1 the blade is designed with an integral box girder and two sandwich layers towards the trailing edge. The radiating surface is thus very inhomogeneous, and for the measurements each side of the blade was divided into 7 sections as shown in Figure 2.

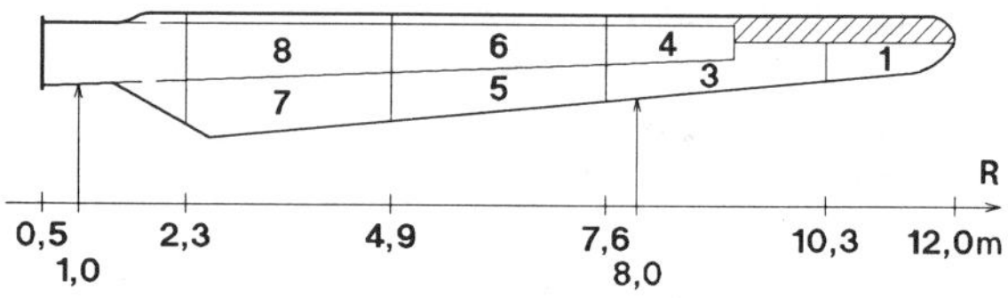

Figure 2: Sketch showing the sectioning and supporting of the blade in relation to the radiation efficiency measurements.

The measurements were performed in a testing hall, and the blade was mechanically excited by a limited band of pink noise (150 Hz - 7000 Hz) from an electromagnetic vibration exciter. Two types of excitation were used:

- longitudinally (that is radially to the rotor) in an eccentric point at the root flange
- penpendicular to the plane of the blade near the edge of the root flange (simulating torsional vibrations of the main shaft).

The mean vibration velocity level was measured in 8 positions in each section using 2 grams accelerometers (Brüel and Kjær 4344). The accelerometers were attached to the surface by means of bees' wax.

For each type of excitation the emitted sound intensity level was measured for each section of the blade by the scanning method. The sound intensity probe was a Brüel & Kjær 3545, and the low frequency range (200 Hz - 630 Hz) was measured using a 50 mm spacer, while a 12 mm spacer was used for the frequency range 800 Hz - 5000 Hz.

All measurements were performed in 1/3-octave bands using a Brüel & Kjær 2133 real time analyzer.

3.2 Resulting radiation efficiency

The radiation efficiency was determined for each section and for each type of excitation. No systematic differences were found for the results from each side of the blade. Consequently, results from sections opposite to each other were averaged.

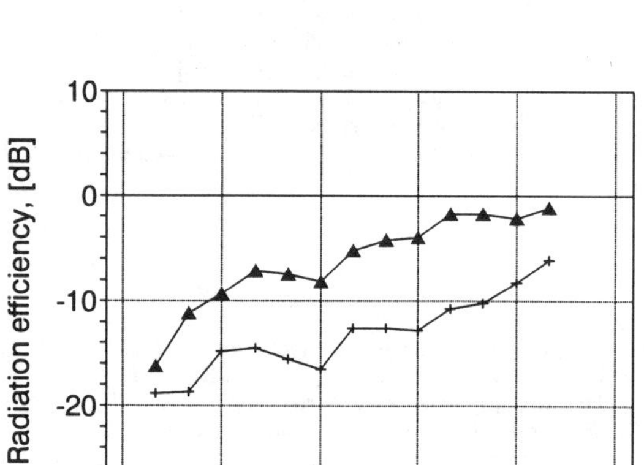

Figure 3: Radiation efficiency in dB for the stiffened, box girder, and the unstiffened, sandwich layer part of the rotor blade.

The radiation efficiency found for longitudinal (radial) excitation was approximately 5 dB lower than the value determined with perpendicular (torsional) excitation. In order to simulate real excitation these results were averaged to give the radiation efficiency for each section. It appeared that the results could be divided into two groups for which the radiation efficiencies were similar. One group consisted of the sections over the box girder (No. 4, 6 and 8) - the other of the unstiffened sandwich-layer sections (No. 1, 3, 5 and 7). Figure 3 gives the radiation efficiency averaged for sections in each group.

3.3 Measurements of vibration velocity during operation.

The vibration velocity of the rotor blade was measured for two wind turbines equipped with the type of blades for which the radiation efficiency was determined (LM 12 HF). Different excitation spectra were aimed at by testing the rotors of two different types of wind turbines mounted with different gearboxes (three stages respectively two stages helical gears): a Danwin 27/225 kW and a Micon 250 kW.

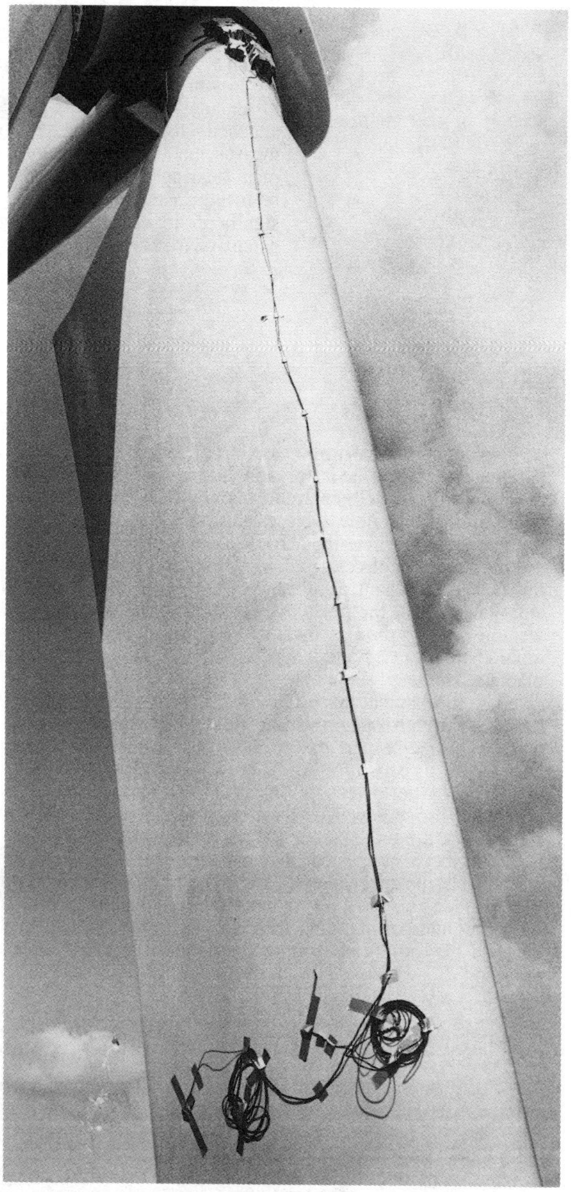

Figure 4: Photograph showing the fitting of accelerometers, charge amplifiers and radio transmitters.

The blades were sectioned in the same way as during the radiation efficiency measurements, and vibration velocity levels were averaged for three positions in each section. The vibration measuring equipment was identical to the equipment used for the radiation efficiency measurements. The charge amplifiers were fastened to the root of the blade together with FM radio transmitters as shown in the photograph in Figure 4. The signals from the corresponding receivers were analyzed with a two-channel 1/3-octave band, real time analyzer. Due to limitations in the radio link only the frequency range 50 Hz - 2500 Hz could be measured.

During the measurements the produced electrical power was approximately 80 kW corresponding to a wind speed at hub height of approximately 8 m/s. The range of variation for the produced electrical power was approximately 60 kW - 130 kW for both wind turbines. As the measurements were performed sequentially, the vibration velocity level at the main bearing of the rotor shaft was continuously recorded in order to allow for a possible load correction. It appeared, however, that the variation of the total velocity level at the main bearing was within $\pm$ 1.5 dB for both measurement series. Consequently no load correction was applied.

3.4 Results

The A-weighted sound power level (L_{WA}) of the structure-borne noise emitted from each section of the blade was calculated according to the formula given in section 3.1. The total structure-borne noise from the rotor was calculated by adding the contributions from each section and assuming that contributions from both sides of the blade and from all three blades were similar. The A-weighted 1/3-octave band spectra of the structure-borne noise from the rotor are given in Figure 5 for each of the turbines.

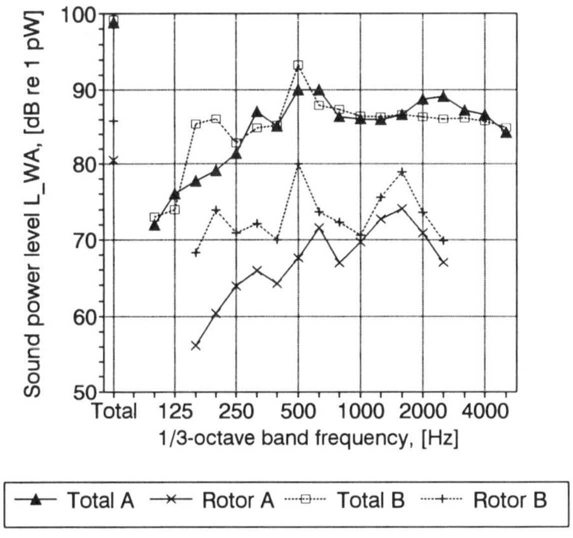

Figure 5: A-weighted sound power level per 1/3-octave band (dB re 1 pW per 1/3-octave) of the total noise and the structure borne rotor noise.
Turbine A: Danwin 27/225 kW, B: Micon 250 kW.

Figure 5 also displays the total A-weighted sound power level per 1/3-octave band emitted from each wind turbine (determined from measurements upon a ground board 40 m downwind).

4. DISCUSSION

The radiation efficiencies determined for a typical, self-supporting GRP rotor blade proved to be greater than -10 dB for frequencies above 200 Hz for the load bearing part of the blade (see Figure 3). The radiation efficiency for the unstiffened sandwich-layer part of the blade was 5 - 10 dB smaller. This is, however, to some extent defeated by a proportional higher vibration velocity level for this part of the blade. Thus the blades really proved to be potential radiators of gearbox noise.

Figure 5 shows, however, that for both turbines the contribution from structure-borne rotor noise to the total sound power level of the turbines was insignificant. It appeared from the measurements that the vibration velocity level of the blades for one of the turbines was approximately 5 - 10 dB higher than for the other. Even then the structure-borne rotor noise was 8 - 15 dB smaller than the total sound power in each 1/3-octave band.

The structure-borne rotor noise also has been measured for another wind turbine equipped with longer blades, but of the same make and general design. This turbine was specially designed for low mechanical noise - even the rotor was vibrationally isolated by a flexible coupling to the main shaft.

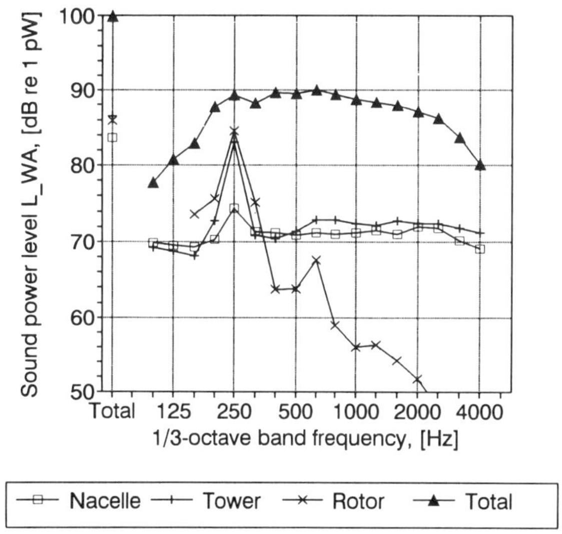

Figure 6: Contributions to the A-weighted sound power level of a 300 kW wind turbine with 17 m rotor blades.

At the most prominent tooth meshing frequency of the gearbox (230 Hz) the contribution to the total sound power level of the structure-borne noise emitted from the rotor was almost equal to the noise from the tower (see Figure 6). Each contribution is only 5 dB below the total sound power level, and is thus dominating the total sound power level at the 1/3-octave band in question.

REFERENCES

[1] Andersen, B., Jakobsen, J. and Kristensen, J.B.: "Noise from small wind turbines. Determination of partial noise sources, and summary of noise measurements during 1982 - 1984" (in Danish).
Report LI 109/87 or L/70.84.782 from Danish Acoustical Institute or dk-TEKNIK, 1987.

[2]: Jakobsen, J. and Andersen, B.: "Structure-borne noise from wind turbine rotors" (in Danish).
Report LI 195/91 or F/70.90.732 from Danish Acoustical Institute or dk-TEKNIK, 1991.

A THEORETICAL AND EXPERIMENTAL INVESTIGATION OF NEW TIP SHAPES

Ioannis Antoniou, Helge Aagaard Madsen and Uwe Schmidt Paulsen

The Test Station for Wind Turbines

Risø National Laboratory, DK-4000 Roskilde

ABSTRACT: Aerodynamic noise has become one of the main issues which the wind turbine industry must deal with at the present time. One of the most important sources of aerodynamic noise from wind turbines is the blade tip region, which also affects the aerodynamic characteristics of the rotor. This paper presents preliminary results from ongoing tests conducted for a series of different tip shapes. The selection of the different shapes has mainly been based on a literature survey and results from helicopter rotors. The tests comprise load and inflow measurements, noise measurements and flow visualization recordings. A 100 kW wind turbine Tellus T-1995 is used as a test bed. The results from the different tips are to be compared to those of the rotor's standard tip. The experimental setup is presented together with load and preliminary noise measurement results. The merits and problems of the noise measurement technique adopted are also discussed.

1. INTRODUCTION

The tip area affects both the aerodynamic and the acoustic performance of the blade due to the high speeds encountered and the formation of the tip vortex. Detailed experimental load data from this region has until recently not been available and this lack has hampered the development of better prediction codes. The usual blade element momentum and vortex codes are known to underpredict the power in the stall regulation region [1] and to overpredict the loads in the tip area. With the objective to study the behavior of a series of different tip shapes a new program has been initiated. During this program load and local inflow measurements, noise measurements and flow visualization recordings being combined to give a better understanding of the flow at the tip region. Along with the experimental program, a theoretical one has been initiated which aims at to useing N-S codes to predict the flow field around the tip region and thereby to develop the field of computational aeroacoustics. Concerning the noise generated at the tip region, George et al [2] identified it to be caused by the highly turbulent separated flow involved in the formation of the tip vortex, which is radiated as trailing edge noise. George and Chou [3] and Brooks et al. [4] have proposed prediction models for airfoil noise based on experimental data. The model of Brooks et al [4] is based on tests conducted on stationary 2D and 3D airfoils. While these results lead to the conclusion that tip noise is of secondary importance to the overall broadband noise spectrum, they state that this does not hold for rotating airfoils due to the higher velocities at the tip area.

Regarding data from different tip shapes, most of the experimental results come from research on helicopters e.g. [5-7]. The effort has been concentrated in modifying the tip section so that improved rotor blade efficiency would be realized. First the modifications aim to influence the rollup process of the tip vortex so that circulation i.e. lift distribution is changed and the tip vortex itself becomes more diffuse. Thus interaction with the following blade (BVI) becomes less with a positive result on both performance and rotor acoustics. Second tailoring of the tip area aims at reducing the profile drag which causes savings i.e. due to reduced, control loads.

2. THE EXPERIMENTAL SETUP AND INSTRUMENTATION

The three bladed stall regulated HAWT has a diameter of 19m. The blades used are commercial LM 8.2m of the 63_n-2nn profile series and have both twist and taper. One of them is modified so that three blade segments, 0.5m in span, are suspended each on a 3 component balance attached to the main spar of the blade (Fig. 1). One balance is situated at the tip

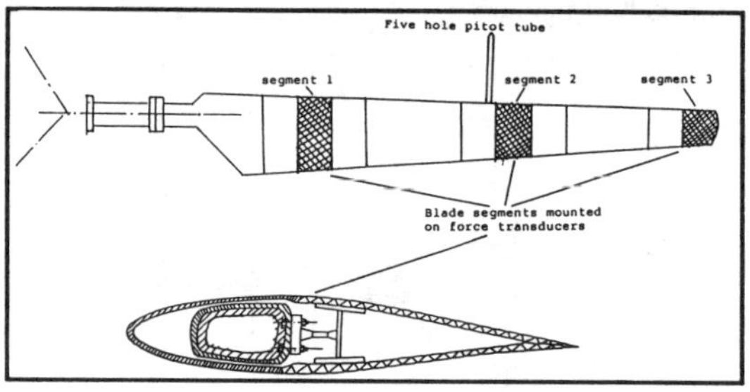

Fig. 1. The Test Blade Layout

area and thus the blade can easily accommodate new tips which are fastened directly on the balance. The quantities measured are the normal force F_N, the chordwise force F_T and the pitching moment C_M. An accelerometer is mounted close to the tip in order to enable the subtraction of the inertia force component from the measured balance force. A total of 60 channels are measured on the turbine and the met. towers. Of these 16 are rotor measurements are transmitted to the ground via a PCM telemetry system. Strain gauges are also placed in the roots of the blades and the flapwise blade moments are measured.

To measure the local inflow velocity vector a five hole pitot is mounted just outside the mid segment on a tube bent to the pressure side of the blade and the corresponding differential pressure transducers are embedded in the blade close to the pitot tube. For the flow visualization recordings, a camera mounted on a rotating arm at the back of the nacelle is used. The rotating arm can follow the rotors rotation. The data acquisition system samples at 25Hz and allows identification of every record acquired to a camera frame. A detailed description of the experimental setup and of the wind turbine can be found in [8]. For acoustic tests a 1.27cm diameter free field microphone (type 4165, Brüel and Kjær) was placed on the tower at tip height on the suction side of the rotor. The distance of the microphone from the rotor was varied during the tests both in the horizontal as well as in the vertical direction.

Noise signal processing: The noise signal is recorded simultaneously on a DAT recorder and on a real time spectrum analyzer that gives the noise spectrum on line. Prior to each measurement campaign a calibration signal is written on the DAT recorder and the same signal is used for calibrating the spectrum analyzer. The spectra produced by the analyzer are A-weighted and a gated multispectrum (i.e. a spectrum consisting of consecutive spectra averaged over a certain time period) is obtained. The time window and the time delay were adjusted freely while the negative slope of a rectangular pulse (rotor status) was used to initiate a triggering cycle on the analyzer (Fig. 2) every second rotor revolution and when the instrumented blade was vertical.

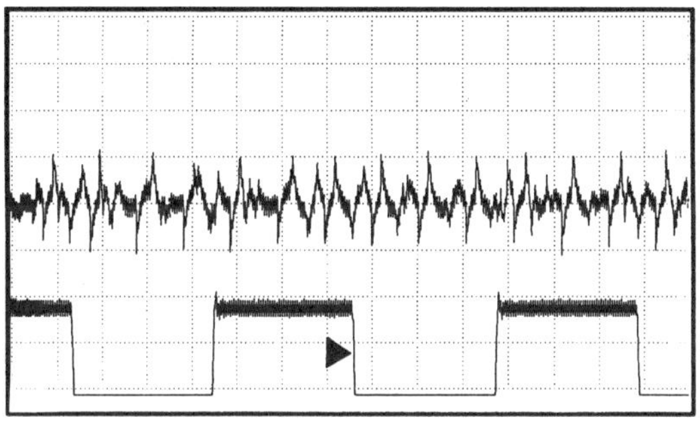

Fig. 2. The Rotor Position and Noise Signals

This setup allowed spectra for the same rotor azimuthal positions to be averaged. The user during the recordings followed the noise events by using earphones.

Near field noise tests have been performed at Risø in the past [9]. Results seemed promising and this motivated us to try refine and develop further this technique.

3. CHOICE OF TIP SHAPES

The following four tips :
1) Ogee
2) Tapered (1:4)
3) Swept-Tapered (1:4, 33° rel. to the leading edge)
4) Elliptic

shown in (Fig. 3), along with the standard blade tip, have been chosen for this study. All tips have the same area as the standard tip. This is the result of a compromise. Had the constant radius approach been chosen the result would have been an area decrease of up to 50% (for the Ogee tip) and serious technical problems in mounting the tips on the balance (due to their reduced size). The constant area approach results in a difference of 1% up to 3% in blade length and tip speed (Elliptic, Ogee). The percentage of the tip area outside the radius of 9.5 m is 6% for the Elliptic, 10% for the Ogee and around 15% for the Swept and Tapered. The geometry of the test blade i.e. profile, thickness distribution, and twist is retained also in the tip area. A modification introduced is that the first three (Ogee, Tapered, Swept) were given a sharp TE.

The Ogee tip has been studied in detail in helicopter rotor tests [10]. Results from full scale as well as model tests show that the vortex produced by it is more diffused relative to a square tip. This along with the fact that the tip vortex is expected to form due to the elongated leading edge, in the outer short chord part of tip, will prevent it to interact with the inboard trailing edge flow. The Tapered and Swept Tapered tips were chosen as a set in order to enable the study of the effects of sweep. Tapered and Swept Tapered tips have

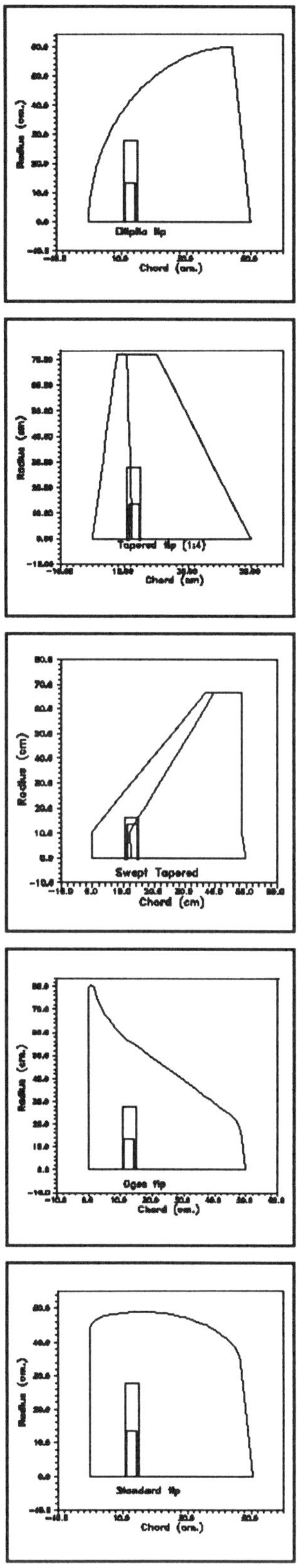

Fig. 3. The Tips chosen for this study

also been studied in helicopter tests [11]. Noise reduction is expected due to the gradual decrease in chord length and thickness which should influence in a positive way the turbulent boundary layer-trailing edge noise, the separation stall noise and the tip vortex formation noise. The Elliptic tip due to the gradual reduction of its chord to zero is actually equivalent to continuously tapered tip and does not suffer by any step transition. Thus for the same reason as above the Elliptic tip is expected to produce less noise. In Figs. 4 and 5 calculations

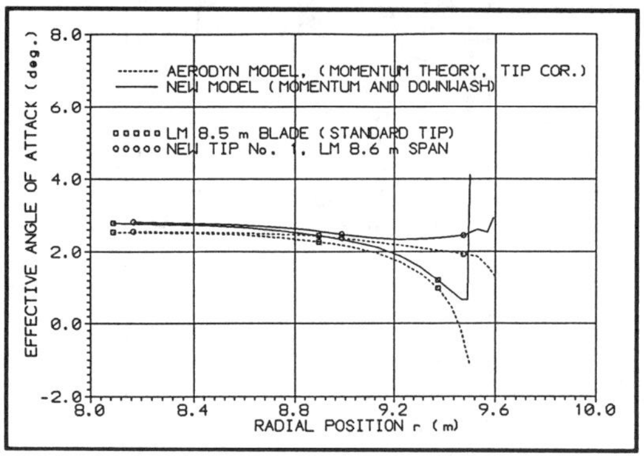

Fig. 4. Calculation of the effective aoa with the two methods

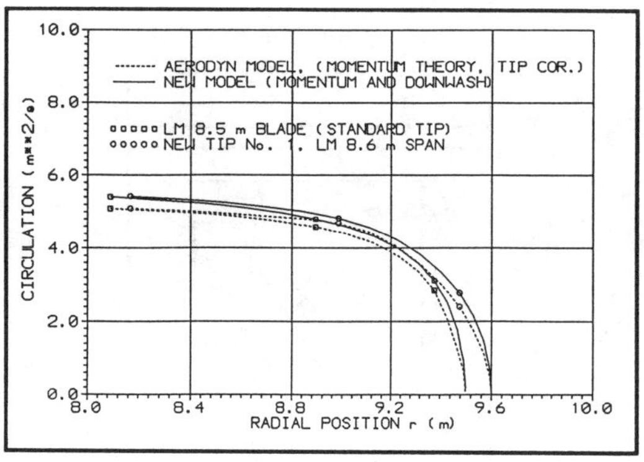

Fig. 5. Calculation of the circulation distribution with the two methods

of the effective angle of attack (aoa) and the circulation distribution are shown for the Standard and the Elliptic tip (under the name NEW TIP No.1). They have been done both with a blade element momentum theory code and a combined vortex-blade element code. It is seen that Tapering of the tip has resulted in a more uniform distribution of the effective aoa and a more gradual reduction of the circulation distribution near the tip. Thus these calculations are in line with the above findings, but they also indicate that for a more detailed prediction of the tip flow, more advanced codes should be applied. Recently an effort to apply the N-S code FIDAP in such problems has been initiated at the Test Station for Wind Turbines along with research in Computational Aeroacoustics.

Wagner [12] has studied the steady and unsteady pressure distribution and the tip vortex qualities for the Ogee, Tapered Swept (without taper) and Rectangular Tips (non-rotating). The maximum velocities in the vortices decrease as the outer chord decreases and the vortex formed by the Ogee is seen to be the weakest and to have the biggest diameter followed by the Tapered tip. For the case of the Rectangular tip two vortices were observed, one of which is the strongest of all. Furthermore this vortex is seen to be the one that moves the longest over the chord.

4. RESULTS

4.1 Load-Inflow Measurements

In what follows, results are presented vs. the experimental aoa as measured by the Pitot tube. The reason for selecting the aoa was its high correlation to the other parameters presented. This is important at least for two reasons: 1) a low standard deviation on the different parameters is obtained which is necessary in order to see for example the small differences in power for the different tips , and 2) only a small number of measurements are needed to cover the operating interval (typically 3 x 10 minutes). The aoa has been corrected in order for the linear part of the experimentally determined power curve to match the theoretical one as this has been derived by a simple blade element momentum code. This is done in order to be able to analyze the measured local airfoil characteristics in relation to the theoretical ones. Adjusting the aoa means that a constant of 4.7 deg. must be subtracted from the measured aoa.

In Fig. 6 results for the lift C_L and drag C_D coefficients for the Standard and Ogee tip at high rpm are presented. Prior to the measurements during a day with light wind the wind turbine was rotated slowly and the offsets of the different parameters were fitted relative to their azimuthal position, via a 5th degree polynomial. This procedure was repeated when the Ogee tip was mounted. In the measurements presented the inertia force component, measured by the accelerometer, has been subtracted from the normal force measured by the balance. Furthermore the minimum C_T coefficient has arbitrarily been set to -0.01 in the range around zero aoa, as it was not possible to get a satisfying zeroing during the tests.

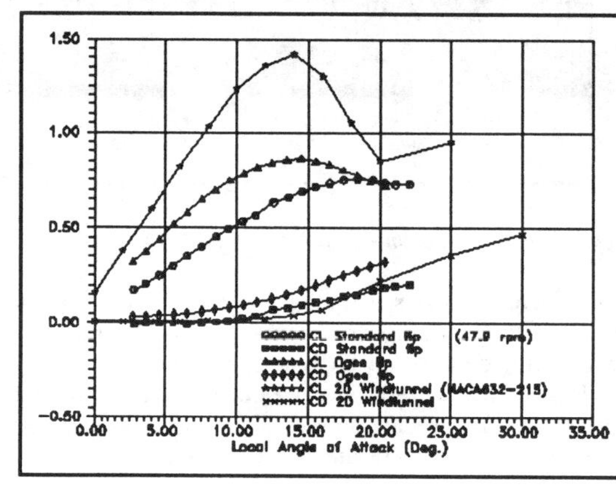

Fig. 6. The Lift and Drag Coefficients for the two tips

In the same figure, 2D data for a similar airfoil are also presented. These data were derived for comparison purposes [13] and were used in a benchmark test of rotor models where the LM blade was a test case. Clearly the behavior of both tips has strong 3D characteristics. The C_L slope is for both tips lower and the maximum value is well below C_{Lmax} for the 2D case. The slope of the Ogee is seen to be higher than the slope of the standard tip. There is also a difference in their maximum value which partly should be attributed to the different planform. No attempt was made to introduce corrections for that purpose (it should be kept in mind that the C_L measurement is actually the result of a bending moment measurement). The maximum value of C_L occurs at smaller aoa for the Ogee tip relative to the standard one and afterwards its slope becomes negative whereas this is not the case for the standard tip (flat stall) neither from the present, nor from previous measurements [8] to which the present data compare well. Thus the two curves show that the Ogee tip stalls at lower aoa and that its post-stall behavior is different than that of the standard tip which has a more smooth behavior at least for the aoa encountered. Referring to Figs. 4 and 5 again we see that this behavior is well explained since the Ogee tip due to its reduced chord will also experience a more uniform distribution of the aoa and a reduced circulation at the tip area relative to the standard one.

The C_D curves are also shown for both tips and the 2D case. The three sets are seen to differ considerably between them, but is interesting that the slopes of the Ogee and that of the 2D data are alike in the post-stall area whereas the slope of the Standard tip is lower. Partly the higher drag values for the Ogee tip are due to its longer profile for the same planform area as the Standard tip.

In Figs. 7 and 8 the measured electrical power curves before and after the replacement of the Standard by the Ogee tip are shown both at high and low rpm. The data are averaged down to 1s and each power curve is based on 4 to 5 sets of 10 min. long time series. The power curve at low rpm for the Standard tip is given only for high aoa due to lack of data. The peak power is found at the same angle for both tips at high rpm. As shown in Fig. 7 this angle is beyond the point of maximum C_L for the Ogee and before the corresponding one for the Standard tip, which means that in the case of the Ogee, stall is deeper at maximum power. The power curve for the Ogee tip is seen to be systematically higher compared to the power curve of the Standard tip. This difference reaches

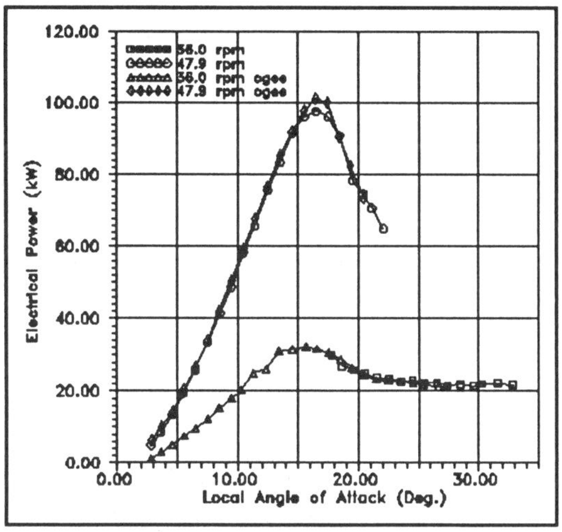

Fig. 7. The Electrical Power vs. the aoa for the two tips

a maximum at peak power. The difference between the two power curves is seen in Fig. 8. It is interesting that at maximum turbine efficiency (around 11°-12°) it would result in an increase in production by around 5 kW if all three tips were changed.

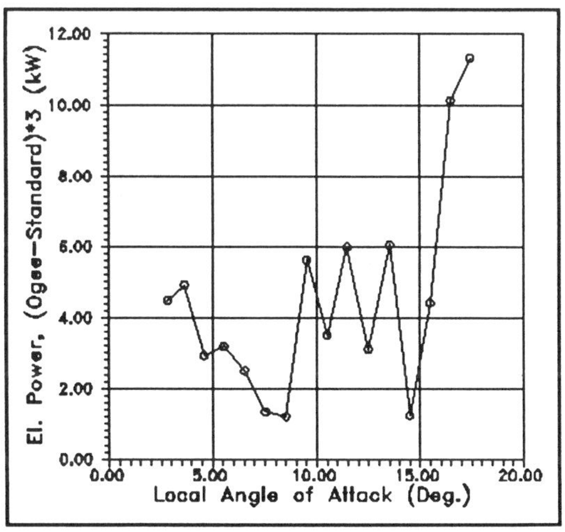

Fig. 8. Difference in the Electrical Power vs. aoa for the two tips

4.2 Acoustic Noise Measurements

In what follows raw noise signal recordings and spectra from the tips are presented. No weighting has been applied to the recordings and comparisons between them will take place in relation to their characteristics but not their magnitude.

Let us first look at the noise signal in Fig. 9 (rotor equipped with the Standard tip) which was recorded when wind was 6-8 m/s from a southwest direction and the microphone was placed behind the rotor at tip height and 2-3 m away from it. The acoustic signal is dominated by a low frequency pressure variation dictated by the blade passage which causes a negative pressure peak (NPP) followed by a

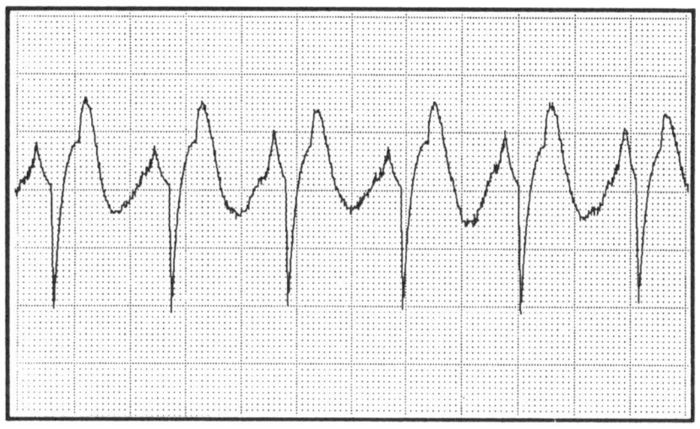

Fig. 9. Microphone at the tip (six blade passages)

smaller positive one (PP). Following these recordings by earphones a very regular noise pattern typical of the blade passage as experienced by an observer on the ground, could be heard. Occasionally during recordings from SW directions, which allowed similar distances from the rotor a noise of hammering (slapping) nature occurred in the earphones which distorted and masked the regular signal and which could not be identified as mechanical or heard from the ground. The situation became worse during recordings from NW directions when the microphone had to be situated closer to the rotor. An example of such a recording is seen in Fig. 10 (rotor equipped with the Ogee tip, wind 8-11m/s, mic. at tip height 1-1.5m from rotor). In Fig. 10 the secondary NPP due to the passage of the Ogee tip is marked with a triangle. The hammering noise was identified with the occurrence of the strong secondary NPP following the primary one. Finally by moving the microphone around the tip area, studying the recordings and monitoring the signal with earphones the conclusion was reached that the secondary NPP is caused due to the passage of the tip vortex over the microphone and that it is due to the lower pressures encountered within the vortex. This motivated a more detailed

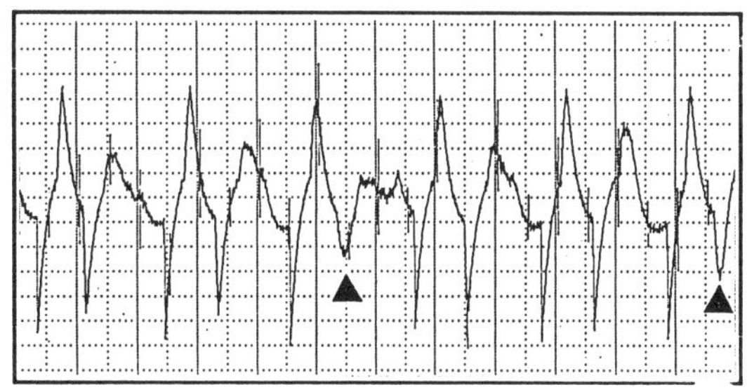

Fig. 10. Microphone at the tip (the triangles mark the secondary NPP of the Ogee tip)

look in the signals. To identify the blade passage with certainty the rotor position square pulse was used extensively (Fig. 2) in combination with the noise signal. From the recordings the secondary NPP was found systematically smaller for the Ogee tip. These differences can be explained in two ways. Either the vortex of the Ogee tip is weaker than the vortex of the Standard tip (assuming the depth of the peak is a measure of the pressures in the vortex) or simply the centers of the vortices are found at different radial positions. The latter seems however less possible since a deeper NPP for the Ogee was never observed and the Ogee tip is known to produce a weaker vortex relative to a square one. At this point it should be mentioned that when due to rotation of the turbine, the microphone was found even closer to the rotor disk the secondary NPP signals became extremely strong. The hammering noise dominated over the rest, and no differences could be observed in the signals. For the same wind conditions (NW wind, 8-11 m/s) the microphone was moved inboard (30-40 cm) and outboard (around 1m) of the rotor disk. The corresponding noise recordings are shown in Fig. 11 and 12. The secondary NPP of the Ogee tip in Fig. 11 is again marked with a black triangle. Now the secondary NPP relative to the ones found at the tips are seen to be reduced indicating a smaller influence of the tip vortex (apparently due to a larger distances from its center) which is however still present. By monitoring the signal via the earphones the reduction of the level of hammering noise could also be certifyed which actually became now intermittent. Briefly, it seems possible that one can estimate the size of the tip vortex by monitoring the noise signal. Quite different are the recordings in Fig. 12. Now the differences between the passages of the three tips are neither visible nor can be heard. The NPP typical of the impulsive nature of the noise is still present whereas the positive one has now been reduced very much and the secondary NPP is not existing. Looking at the

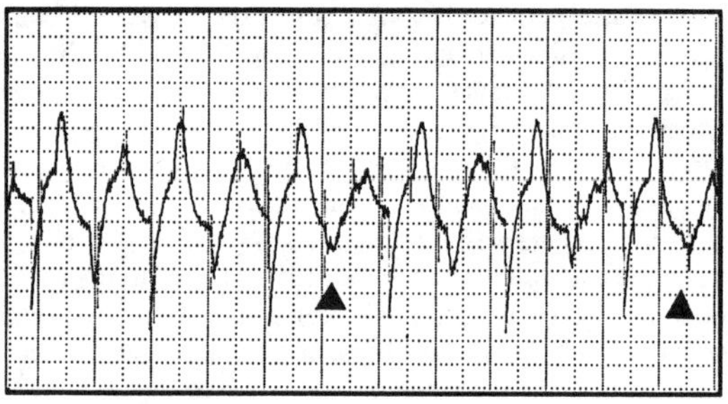

Fig. 11. Microphone inboard (the triangles mark the secondary NPP of the Ogee tip)

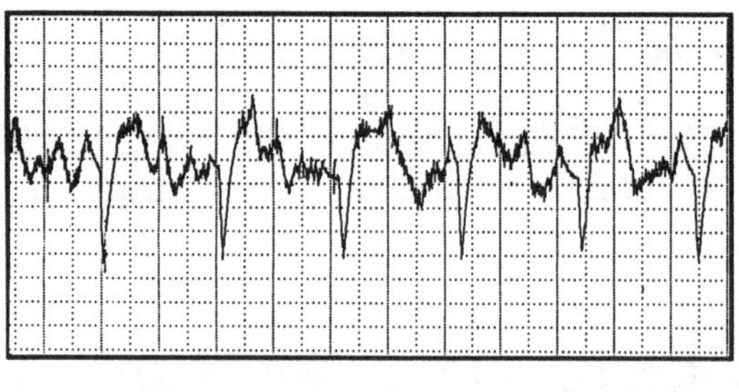

Fig. 12. Microphone outboard

spectra (not shown here) the SPL was seen to have suffered a pronounced reduction whereas by monitoring the signal with earphones we could confirm that the hammering noise was totally eliminated.

Spectra have been obtained simultaneously with the recordings of the above signals. Such a sequence covering 240^0 of rotor revolution is shown in Fig. 13. To obtain the spectra a delay time of either 0.487sec. (corresponding to a period equal to 140^0) or 0.522 sec. (150^0) and a gate width of 0.139 sec. (40^0) were used. The averaging time was 5 sec.which corresponded to 35 triggered blade passages, this time gave reproducible results and could actually had been lowered without loss of certainty.The spectra are A-weighted and the column to the right side gives their sound pressure level (SPL). The top spectrum corresponds to the passage of the Ogee tip and the fourth from top corresponds to the passage of the following blade whereas the rest of the spectrum cover the subsequent intervals in the order of recording. There exist marked differences between the spectra in relation to the blade's azimuthal position and the results confirm the potential of the method to evaluate in detail the noise emitted with respect to its phase. Unfortunately the above results makes it

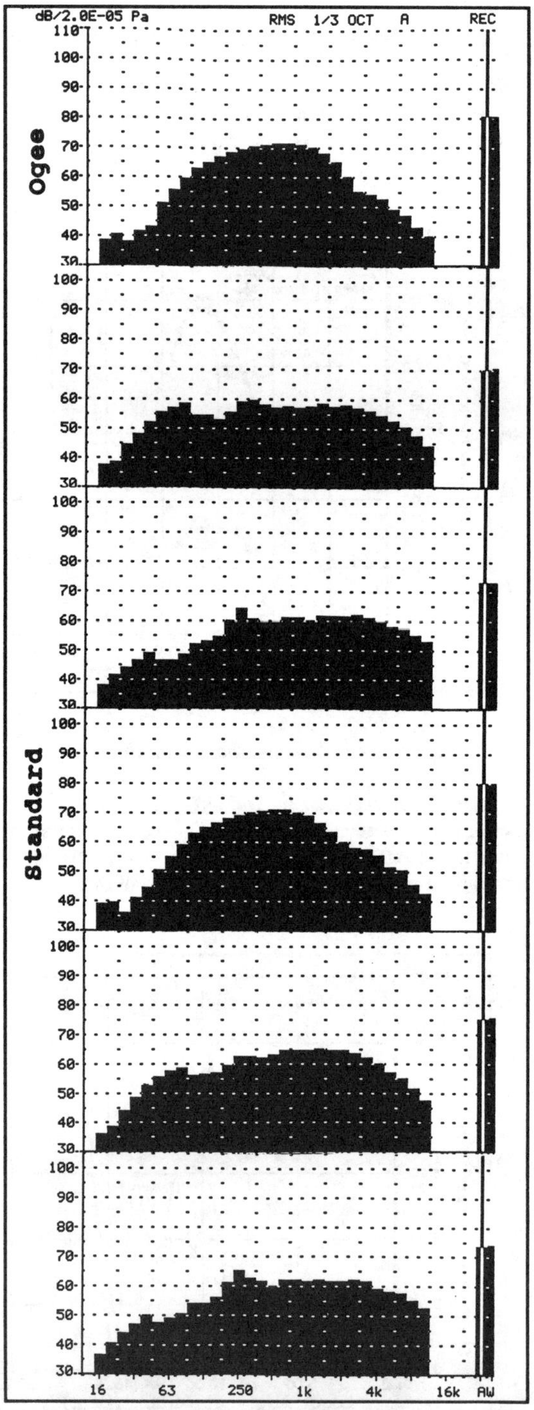

Fig. 13. A sequences of spectra that covers 240º of rotor rotation (40º window per spectrum)

difficult to reach a clear conclusion as to which tip is more noisy for two reasons. First the signals have been contaminated by the passing of the tip vortex over the microphone and second because at a later stage it was realized that the window of 40^0 was not adequate for the calculation of the SPL. By looking in Fig. 9 it is obvious that a gate width corresponding to a period of at least 60^0 should have been used if the whole pulse (negative plus positive peak) should be taken into consideration. The use of a 40^0 width where the blade passage occurred after a period of 20^0 prohibited the recording of the whole pulse and the extraction of valid results.

5. CONCLUSIONS

An experimental setup which allows measurements of loads and noise on tips has been presented, along with preliminary results from an Ogee and a Standard tip. Preliminary analysis of the load measurements shows that the Ogee tip has certain performance advantages compared to the Standard one. The technique employed for noise measurements has potentials both in measuring the SPL of a certain tip but also in providing information on the phase and strength of the emitted noise for different rotor azimuthal positions, by choice of a suitable gate widths. Care should be taken to avoid the obstruction of the results from the tip vortex. Signals at the tip or just inboard are alike but signals outboards of the tip are quite different. The results show also that the vortex formed by the Ogee when compared to the Standard tip is weaker. A more detailed analysis (not presented here) of the load measurements showed the complex turbulent nature of the flow field in the tip region. Therefore application of more advanced computational codes (e.g. N-S codes) is necessary for better predictions and this work has been initiated at Risø.

6. ACKNOWLEDGEMENTS

One of the authors (I.A.) acknowledges with thanks the support received from the CEC/DGXII in the form of a bursary which made possible his participation in this project.

REFERENCES

[1] Rasmussen, F., Petersen, S.M., Larsen, G., Kretz, A. and Andersen, P.D. "Investigation of Aerodynamics, Structural Dynamics and Fatigue on Danwin 180 kW", Risø M-2727 Report (1988).

[2] George, A.R., Najjour, F.E., and Kim, Y.N., "Noise Due to Tip Vortex Formation on Lifting Rotors". AIAA-80-1010. (1980).

[3] George, A. R., and Chou, S.I., "Broadband Rotor Noise Analysis", NASA CR-3797, (1984).

[4] Brooks, F.I., Pope, D.S., and Marcolini, M.A., "Airfoil self Noise Prediction", NASA Ref. Pub. 1218 (1989).

[5] Prieur, J., Lafon, P., Caplet, M., and Desopper, A., "Aerodynamics and Acoustics of Rectangular and Swept Rotor Blade tips". J.Am. Helicopter Society, January (1989), 42.

[6] McVeigh, M.A. and McHugh, F.J. "Influence of Tip Shape, Blade Number and Airfoil on Advanced Rotor Performance" J.Am. Helicopter Society, October (1984), 55.

[7] Stroub, R.H., Rabbolt, J.P. Jr, and Niebanck, C.F., "Rotor Blade Tip Shape Effects on Performance and Control Loads from Full Scale Wind Tunnel Testing", J.Am. Helicopter Society, October (1979), 28.

[8] Madsen, H.A., "Aerodynamics of a Horizontal-Axis Wind Turbine in Natural Conditions", Risø M-2903 (1991).

[9] Dunbabin, P., "Analysis of Noise Emission from the BONUS KOMBI Wind Turbine", Wind Engineering, V.16, No. 3.

[10] Mantay, W.R., "Full Scale Testing of an Ogee Tip Rotor", NASA-CP-2052 pt. 1 (1978).

[11] Lee, A., and Masher, M., "A Study of the Noise Radiation from Four Helicopter Rotor Blades" NASA-CP-2052 pt. 1 (1978).

[12] Wagner, W.J., "Comparative Measurements of the Unsteady Pressures and the Tip Vortex Parameters on Four Oscillating Wind Tip Models", 10th European Rotorcraft Forum, Hauge (1986).

[13] Petersen, H., "Benchmark Test on Power Curve Computations on Wind Turbines - A Compendium", Risø (1986) DRAFT.

NOISE CONTROL ON THE BONUS 450 kW Mk III TURBINE

K. Ø. Petersen
H. Stiesdal
T. Rønnow

BONUS Energy A/S
Fabriksvej 4,
DK-7330 Brande
Denmark

S. Tenbrock

A. Fr. Flender AG
Alfred-Flender-Str. 77
D-4290 Bocholt
Germany

K. Medelbye

ABB Motors A/S
Petersmindevej 1
DK-500 Odense C
Denmark

ABSTRACT: Noise will be one of the most important limiting factors in the dissemination of wind power in populated areas of Europe. To reach even moderately ambitious goals for installed power it is necessary to limit the noise output of commercially attractive turbines.

The purpose of the work described in the present paper has been to reduce the noise output from both aerodynamic and mechanical sources of a commercial turbine, the BONUS 450 kW Mk III. Three important sources for the noise generation are the rotor, the gearbox, and the generator. Each of these sources has been addressed in the noise reduction project.

The aerodynamic rotor noise has been reduced with attachments at the blade tips. The gearbox grinding is modified for reduced noise generation. The generator has a special forced-flow cooling arrangement with noise reduction features.

The work has been successful, with a significant noise reduction from the BONUS 450 kW Mk I turbine. The BONUS 450 kW Mk III turbine has a source intensity according to the IEA recommendation of less than 97 dB(A) at 8 m/s in 10 m height with no significant pure tone content.

1. INTRODUCTION

Noise will obviously be one of the most important limiting factors in the dissemination of wind power in the populated areas of Europe, and to reach even moderately ambitious goals for installed power it is necessary to reduce the noise output of commercially attractive wind turbines to the largest extent possible.

BONUS Energy A/S has for several years assigned much development effort to the noise reduction of the company's wind turbines. The present paper describes the methods and results of the very dedicated noise control design work undertaken on the BONUS 450 kW Mk III turbine.

Three important sources for the noise generation are the rotor, the gearbox, and the generator. Each of these sources has been addressed in the noise reduction project.

The LM 17.0 m rotor blades have been equipped with attachments at the blade tip, shaping the tip vortex in a controlled manner.

The drivetrain arrangement design permits the gearbox to be completely suspended in rubber, thereby limiting the structural noise transfer. The gearbox grinding is modified for reduced noise generation. The canopy is tightly sealed to prevent noise leakage.

The generator is mounted on rubber bushings and has a special forced-flow cooling arrangement with noise reduction features.

The cooling arrangement prevents the temperature problems often associated with tightly sealed machine canopies. In the development process the standard BONUS 450 kW Mk I turbine has been used as reference.

2. TEST TURBINE

The BONUS 450 kW turbine is a proven specimen of the well-known Danish wind turbine concept, with an upwind, three-bladed, fixed pitch rotor, stall regulation, active yaw, and a grid-connected asynchronous generator. The prototype was erected in 1987 and the production version has been marketed since 1989. At the beginning of 1993 approximately 60 BONUS 450 kW turbines had been installed.

In the MK III version the BONUS 450 kW turbine has the following main data:

Rated power	450 kW
Rotor diameter	37 m
Rotor speed	29 rpm
Hub height	35 m
Blade type	LM 17.0

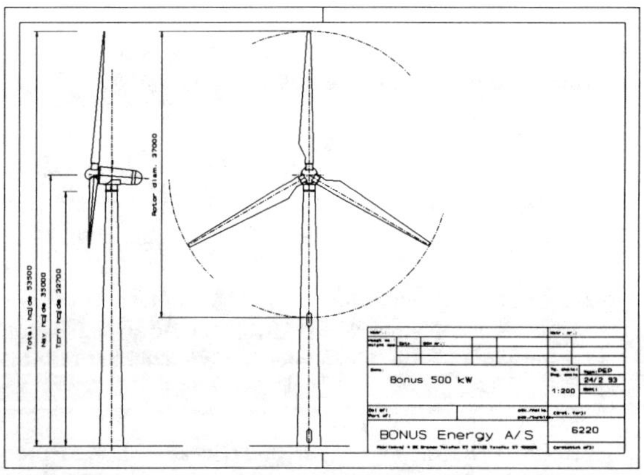

Fig.1. BONUS 450 kW Mk III turbine

3. AERODYNAMIC NOISE

Aerodynamic noise from wind turbine rotors is influenced by a number of factors. The noise from the trailing edge and from the blade tip give important contributions, and there is a more diffuse effect of the incoming turbulence itself. The new development reported here has primarily focused on the reduction of the tip noise.

The BONUS 450 kW Mk III turbine is fitted with LM 17.0 m blades. The LM 17.0 m blade has a high-efficiency planform of the main blade with convex leading and trailing edges. The planform of the tip itself is characteristic by a longer trailing edge than leading edge, stretching the tip into a sharp point at the trailing edge. The tip planform generates a rather concentrated tip vortex, and this results in moderate tip losses and a large efficient radius of the blade. The planforms of both the main blade and the tip contribute to the excellent performance of the LM 17.0 m blade. However, the downside of the optimized tip planform can in BONUS' opinion be the noise from the tip where the high air speed of the concentrated tip vortex may generate a higher source noise intensity and may be more sensitive to minor imperfections than with an ordinary tip planform.

Through a long development process on both turbines fitted with the LM 17.0 m blade, and on turbines fitted with other blades of similar tip planforms, BONUS has developed an attachment to the tip which results in a lower source noise intensity from the tip. The attachment is in the form of a small torpedo-shaped body.

Fig.2. Tip torpedo

Flow studies with a rotating video camera indicate that the noise-reduction of the tip attachment is caused by the vortex-shaping effect of the torpedo, but it is difficult to conclude firmly as regards the specific mechanisms in detail.

4. TRANSMISSION NOISE

The transmission noise reduction is the result of two separate development processes - a reduction of the gearbox source noise, and a reduction of the noise transfer through the nacelle construction.

4.1. Gearbox source noise

The gearbox is one of the prominent sources for tonal noise, and from a nuisance consideration it is particularly important to control the gearbox noise. The aerodynamic noise has to a certain extent a white noise character and may therefore be masked by blending into the background sound impression, but the tonal character of a badly damped wind turbine gearbox leaves little chance of this masking.

The BONUS 450 kW Mk III turbine gearbox is manufactured by Flender A.G. and is of the type PEAK 4265, a combined planetary and helical design. The first, high-torque stage is planetary, and the two next lower-torque stages are helical. By this hybrid design the gearbox benefits from the best characters of the two basic types of gearing, resulting in a compact and low-noise unit.

In the development of the low-noise design the following methods have been applied:

- The corrective grinding, which account for all elastic deformations of shaft, bearing and gears, has been optimized for the partial load range as the nuisance effect of the tonal noise is highest at low ambient noise levels.
- With the special shape of this corrective grinding the load distribution is even at peak loads within acceptable limits and ensures a high gear safety.
- The modification of the profile angle combined with a small tip and root relief reduces the gear mesh shock without noticeable reduction of the overlapping in the partial load range.
- The grinding of the high speed stage is performed on selected types of gear grinding machines to avoid resonances with machine frequencies of the grinding machine.

The improvements in noise performance have been substantial. Figures 3 and 4 show the 1/3-octave spectra of the gearbox type before and after modification. The sound pressure level improvement is from 80 to 72 dB(A).

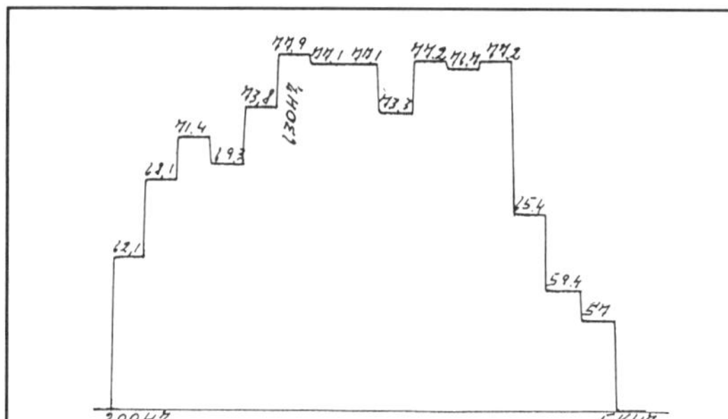

Fig.3. Gearbox noise before modification

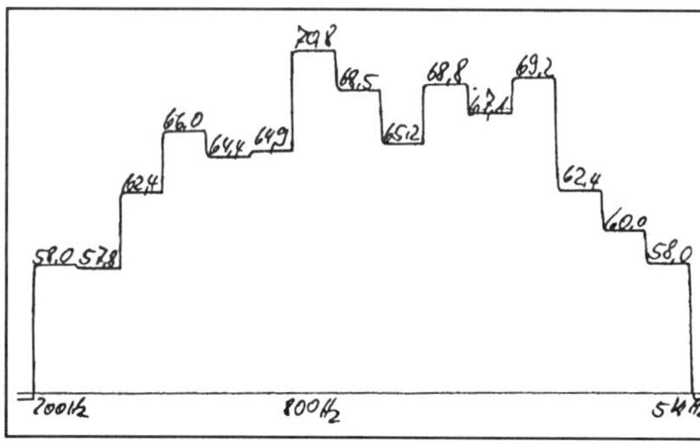

Fig.4. Gearbox noise after modification

4.2. Noise transfer

The nacelle design of the BONUS 450 kW Mk III turbine is particularly well suited for the limitation of noise transfer.

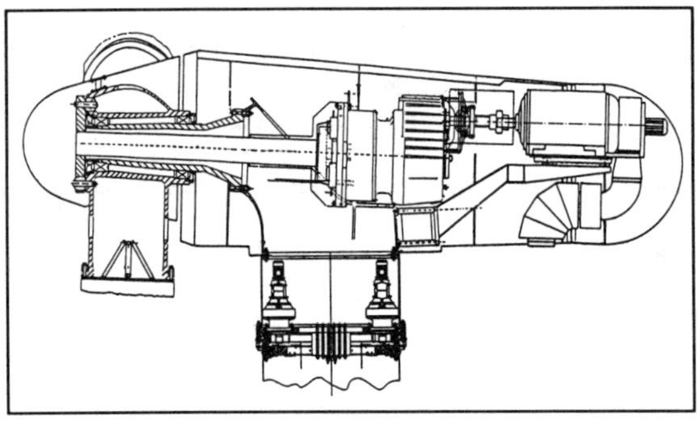

Fig.5. Nacelle layout

The wind turbine rotor hub is suspended on a fixed hollow shaft. The torque is transferred to the transmission shaft with a large Flender Rupex rubber coupling. The transmission shaft is connected to the gearbox with a Stüwe shrink disc, and the gearbox itself is suspended by a symmetric torque arm mounted in rubber bushings. The high speed shaft is connected to the generator with a double rubber coupling. By this arrangement the gearbox is completely suspended in rubber; it has no metallic contact to the nacelle bedplate. The bedplate itself is made from nodular cast iron which has excellent damping properties.

The canopy is a steel construction, clad on parts of the inside with noise absorbing material. The canopy is completely sealed so noise leakage is avoided.

5. GENERATOR

The basic design criteria for modern wind turbine generators include requirements for high efficiencies at low loads, low reactive power consumptions, controlled temperature rises, and a low noise emission.

The two most critical generator noise sources are the cooling system and the electromagnetic noise produced in the stator/rotor construction.

5.1. Cooling system

The standard cooling system for a totally enclosed generator is usually based on a fan fixed to the generator shaft. This type of cooling system can only be modified slightly for noise reduction. ABB and BONUS have therefore decided to develop a special cooling arrangement for the BONUS 450 kW Mk III turbine generator. The arrangement is based on a cap around the generator, a cooling fan with a two-speed motor, and an outlet tube system including a noise absorbing element.

The two speed Dahlander fan motor (4/2-pole) has been selected to obtain low ventilation power consumption and a low noise emission at moderate generator power levels (up to 10 m/s wind speed at hub height). At low speed the cooling fan consumes approximately 300 W and at high speed approximately 2400 W. The narrow-band frequency analyses have generally shown good noise performance of the generator/fan combination, but a pure tone at 447 Hz has required the addition of an additional noise absorbing element in the fan outlet.

5.2. Electromagnetic noise

Reduction of the electromagnetic noise is quite a complex process, as the primary performance requirements tend to lead to conflicting requirements. For the present development designers from ABB have studied different combination of materials, slot designs, and numbers of poles. The pole-numbers in the stator and rotor systems are important from resonance considerations. Three different rotors were designed and produced with different number of pole-pairs.

The first rotor performed very well in the generator regime, with high efficiency and low temperatures, but unfortunately also with a prominent pure tone at 1725 Hz, so this design was discarded. The second rotor performed wery well with respect to the noise, but the efficiency was low and the reactive consumption was excessive, so this design was discarded too.

The third rotor performed well on both efficiency and on noise. However, the efficiency is marginally lower than the efficiency of the optimum design with no noise considerations. The narrow band frequency analysis shows a number of peaks, but these do not exceed the pure tone release levels, so this rotor design was chosen as the optimum compromize for the BONUS 450 kW Mk III turbine.

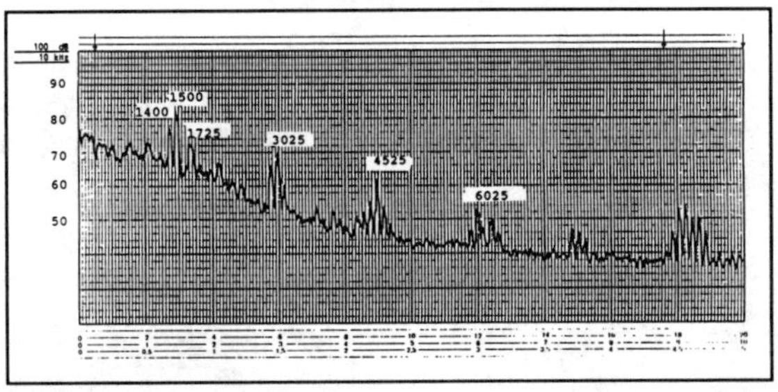

Fig.6. Narrow band analysis of test rotor 3

At the ABB Strömberg test laboratory all relevant measurements were performed for the selected generator concept.

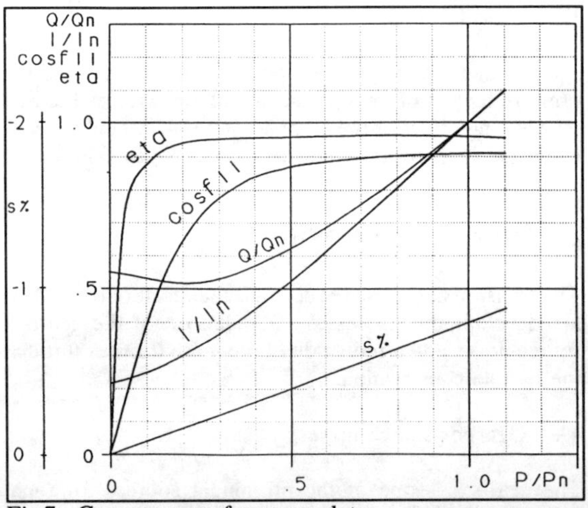

Fig.7. Generator performance data

Full load tests with 500 kW power output and with the original fixed fan showed a temperature rise of 92.8 °C by the resistance method according to IEC 34. With the special BONUS cooling system the temperature rise was measured to 89.5 °C by the resistance method. Compared with the original ABB design the reactive consumption at no load was reduced for 187 to 125 kVAr, and the efficiency at 10 percent loading was increased from 78 to 87 percent.

The prototype generator was installed on a BONUS 450 kW Mk III turbine in the spring of 1992 and the generator performance have been verified by field measurements.

6. MEASUREMENT RESULTS

As an important part of the development process several noise measurements have been performed at the BONUS 450 kW MK III prototype situated in Dybe in the north western part of Jutland, Denmark.

According to [1] the total sound power level have been determined to less than 97 dB(A) at 8 m/s in 10 m height.

Furthermore, analysis of frequency spectra from different power ranges have been performed.

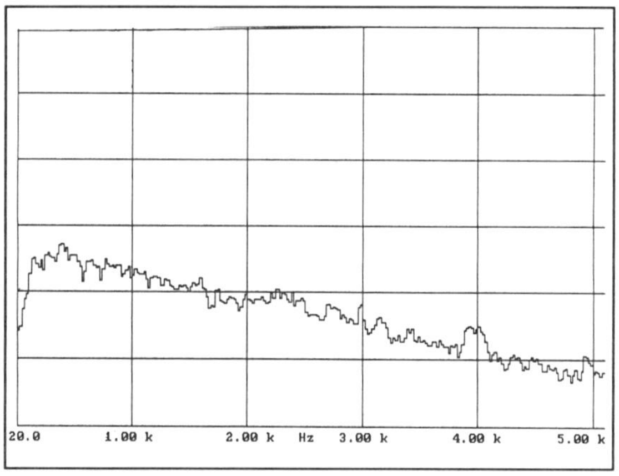

Fig.8. Frequency spectrum at 63 kW power output

In the low power range no single peak tones appear. The noise emission from the rotor blades is also kept within acceptable limits, which is visible in the higher frequency range, above 1 kHz.

Some minor pure tones appear at high power range.

In the low frequency range (below approximately 400 Hz) peaks from the harmonics of the teeth mesh at the planetary gear stage appear.

At 480 Hz a small peak from by the generator fan blades is visible.

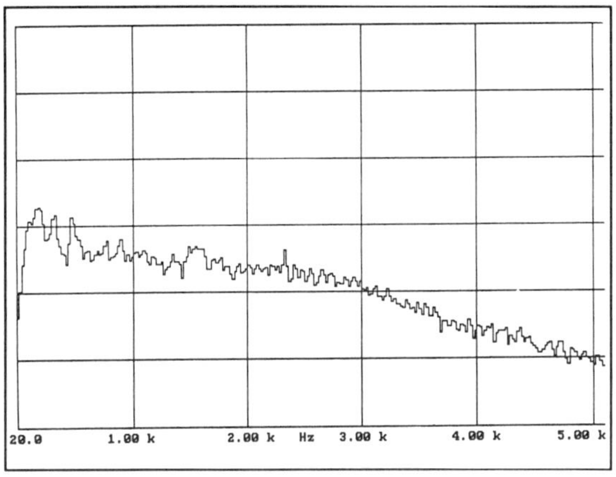

Fig.9. Frequency spectrum at 314 kW power output

7. CONCLUSION

Special work have been carried out in order to reduce the noise emission from the BONUS 450 kW Mk III wind turbine. The three main noise sources the rotor, the transmission system, and the generator, have been evaluated extensively.

To reduce the aerodynamic noise BONUS has decided to use a sharp trailing edge and a special BONUS designed tip torpedo.

The drivetrain arrangement of the turbine suspends the main gearbox completely in rubber, and consequently the transferred structural noise is reduced. Special gear geometries and corrective grindings have been developed to reduce the gear mesh shock in the high speed stage at partial loads. Measurements have proven that no single gear tone occurs over a wide power range.

The generator is mounted on rubber bushings and has a special force-flow cooling arrangement with noise reduction features. The cooling fan is driven by a two-speed Dhalander motor, and the fan is only running with 2880 rpm at high power range. Measurements have shown that a small peak in the frequency spectrum occurs at high power.

Even though the machine canopy is tightly sealed temperature problems have been prevented by the special forced-flow cooling system.

With a measured total sound power level of 96.7 dB(A) the BONUS 450 kW MK III is today one of the most quiet turbines in the 400-500 kW range.

The goal of the noise reduction project has been achieved by a very good cooperation with the sub-supplier of the main components.

Fianlly, it must be mentioned that some of the noise reduction design features are protected by Danish and international patents.

REFERENCES

[1]: "Documentation of Noise Emission from 450 kW Mk III BONUS Wind Turbine, IEA-method", 260 2 5038/STP 925, Department of acoustics, DTI.

NATIONAL AERO-ACOUSTIC RESEARCH ON WIND TURBINES IN THE NETHERLANDS

F.Hagg, Stork Product Engineering (SPE), Amsterdam
J.C Bruggeman, Institute of Applied Physics (TNO-TPD) Delft
A.G.M Dassen, National Aerospace Laboratory (NLR), Emmeloord

ABSTRACT: An aero acoustic wind turbine research program for noise reduction is defined, based on a literature search and a study on proper facilities to examine the models, which has the best opportunities for noise reduction. Found is a model based on an irregular shaped trailing edge and the vagueness of the mechanisms for low frequency noise. As a first step an irregular shaped trailing edge and inflow turbulence will be tested in a wind tunnel.

1. INTRODUCTION

During the design of a wind turbine and the definition of a location, noise is an important restriction. Machine noise can be avoided by isolation measures, which will lead to reduction in the near future.

Another important source, the aerodynamic noise of the rotor, cannot be isolated and will then be the most important noise source. For this reason a research project, funded by the Netherlands Agency for Energy and the Environment (NOVEM) has been started, in the Netherlands, with the objective to reduce the aerodynamic noise of a wind turbine.

The first phase, definition of the project [1] was executed in 1992 and is the subject of this paper. The next phase, research in the wind tunnel, has just started in January 1993, The work is done by a team of specialists from Netherlands Energy Research Foundation (ECN), NLR, SPE, TNO-TPD (full names see above) and the wind institute of the Technical University Delft (TUD).

The definition phase was split into different tasks:

- Literature search and survey of industry and institutes.
- Brainstorm and selection of the most promising measures.
- Selection of the model development and experimental equipment needed.
- Selection of the specific facilities and measuring techniques.
- Gobal qualification of the selected facilities.
- Set up of a research program.

2. LITERATURE SEARCH

In literature seven aerodynamic noise generating mechanisms were distinguished:

- Non uniform inflow.
- Inflow turbulence.
- Trailing edge noise.
- Blunt trailing edge noise.
- Laminar-turbulent vortex shedding.
- Tip noise.
- Flow over holes, slits and non-streamlined bodies.

Non-uniform inflow is seen as the cause of noise up to a frequency of 100 Hz, due to unsteady forces on the rotor blades [2]. Non-uniform inflow is caused by wind shear, skew wind and tower influence.

Frequencies of noise caused by non-uniform inflow are expected to be 10-100 times the rotor frequency for 'down wind' and 1-10 times for 'up-wind' rotors.

Interaction of inflow turbulence with the rotor blades is seen as the main low frequency source mechanism for frequencies up to 2000 Hz. The models which have been developed to date are not clearly understood or accurate, i.e. the most frequently used Grosfeld model [3], is only correct for frequencies lower than 200 Hz.

The leading edge interaction model could be a possible starting point for further model development [4].

However the atmospheric turbulence intensity is very low in the 1000 Hz range, so the low frequency noise is not clearly understood.

Trailing edge noise is the main source in the high frequency range, due to interaction of the turbulent boundary layer with the trailing edge. The model is clearly understood and validated in different experiments [5].

Howe [6] has found a theoretical model to reduce trailing edge noise, based on the effect that an edge does not interact efficiently if the rotation of an eddy is perpendicular to the edge. Therefore an irregular 'saw tooth' shaped edge will reduce the trailing edge noise. An algorithm based on the boundary layer thickness and the 'saw tooth' dimensions is given.

Blunt trailing edge noise is a tonal source, due to vortex shedding behind the trailing edge [7]. The frequency depends on the thickness of the trailing edge, the displacement thickness and the air speed. This noise can be avoided by choosing a thin edge which shifting the frequency to the ultrasonic range.

Another tonal source is due to a resonant interaction of trailing edge noise with the unstable laminar-turbulent transition [8] ($10^5 < Re < 10^6$, based on the blade cord). This noise can be avoided by choosing an operation $Re > 10^6$ or by placing a turbulence strip near the leading edge.

Tip noise is caused by interaction of the tip vortex with the lateral and trailing edge [9]. The mechanism is similar to trailing edge noise. The noise can be avoided by choosing tip plan forms, which do not make contact with the tip vortex. Another possibility is making the 'trailing edge' noise of the tip less efficient, for instance with an irregular edge.

However, 'real' tip noise (caused by the tip vortex) is seen as a minor source. The difference between the most ideal tip and a tip with a smooth elliptical plan form can only be 3 dB.

Noise due to slits, holes and other irregularities, not particular to wind turbines, is well documented. Prediction algorithms are available.

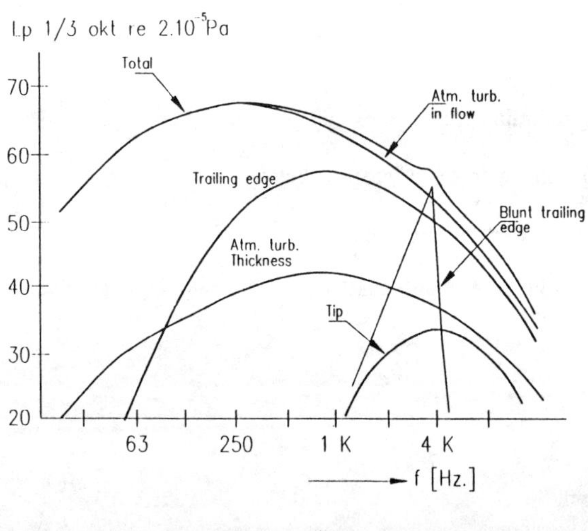

figure 1: Overview of several aerodynamic noise sources [3]

3. SELECTION OF MEASURES

Table 1 shows the trade-off between possible test facilities and the different models, which were distinguished during the literature search.

research	test facility, method in the field (car, glider	existing turbine	acoustic wind tunnel	rotor test facility	FEM
inflow turbulence					
< 10 Hz	+	+	-	++	-
> 10 Hz	+	+	++	+	-
rot. sampling	-	+	-	++	-
airfoil	+/-	-	++	+/-	-
trailing edge	-	+/-	++	+	-
irregular	+/-	+/-	++	+	-
porous	+/-	+/-	++	+	-
damping material	+/-	+/-	++	+	-
blunt	+/-	+/-	++	+	-
other blade wake	-	+	+	++	-
vibrating blade	+/-	+/-	++	++	+
laminar-turbulent vortex shedding	+/-	+/-	++	+	-
tip vortex	+	+	+	++	-
stall	+/-	+	++	++	-
tolerances	+/-	+/-	++	++	+/
slits/holes	+	+	++	+	-

table 1: necessity and capability of facilities for different research

Two main models and/or measures are found, which appear to be most promising for noise reduction:

- Optimizing the trailing edge shape, possibly by application of irregular edges (sinus or saw tooth or porous), in order to bring down the high frequency noise level.
- Increasing the understanding of the low frequency (LF) noise mechanism.

The first measure is found to be the most promising from the literature search. However as no clear model of this is available at the moment, the latter is chosen.

As mentioned in the literature search, the only LF model available is based on the inflow of atmospheric turbulence and this is not clearly understood, while the peak frequency of the atmospheric turbulence encountered by the rotor blades is typical 0.1 Hz. Even rotational sampling cannot produce the peak frequencies of 50 to 2000 Hz, which are present in the aerodynamic noise spectrum. Of course high frequency turbulence is present but the intensity is very low. Inflow turbulence noise of aeroplanes encountering the atmospheric boundary layer is hardly audible.

4. MODEL DEVELOPMENT, VALIDATION, FACILITIES

The reduction of the trailing edge noise by an irregular edge is a theoretical model, which must be validated experimentally.

Inflow turbulence noise will be studied experimentally in the wind tunnel, although with high frequency turbulence up to 8000 Hz. The measured noise and turbulence levels will be compared with the normal atmospheric turbulence spectrum.

After the 'validation' of this model the understanding of the inflow turbulence will be greater and may eventually lead to a better LF model. If inflow turbulence is not responsible for the LF noise another model must be found. This new model and its validation will be part of the study.

The expectation is that a better insight of the LF noise will lead to reduction of the LF noise.

In the trade-off mentioned above, a wind tunnel and a rotor test facility are found to be the best facilities for experimental research.

As a first step detailed experiments with extended parameter variation are needed to establish the valid parameter ranges and the sensitivity of the models. For this study an acoustic wind tunnel will be the best option. For further 3-D validation, fine tuning and finding design measures a rotor test facility will be more convenient.

4.1 Selection of the facilities

In another study different wind tunnels and rotor test facilities were compared with regard to costs, testability and availability.

In this study, the best wind tunnel is the DNW (German Dutch Wind tunnel) with very good noise specification. However, due to the budget available, the KAT (small anechoic tunnel) of NLR is selected. The small dimensions and velocity, limiting the maximum attainable Reynold number to $3\ 10^6$, could be a problem. If no turbulent boundary layer occurs a turbulence strip will be used on the test model.

The best 3-D facility would be a special acoustic rotor test facility in a large 'windspeed' open tunnel (for example the large open tunnel at Stuttgart, but than with special noise reducing measures).

However, the variable speed, variable pitch rotor test facility of TUD, which can even rotate vertically at ground level, is again selected for budgetary reasons. The proximity of a busy highway could be a problem, although the noise of this highway can be filtered out with special filtering techniques.

4.2 Qualification of the selected facilities

In the definition phase some global qualification tests are performed in order to make sure that the facilities can meet the research objectives. The main issues are the background noise and the availability of the facility, measuring equipment and personnel.

For both facilities the background noise is found to be critical, so much attention must be paid to special filtering techniques. Elliptic or parabolic mirrors or higher order correlation techniques with synthetic acoustic antennas (SYNTAKAN of TUD/TPD) must be considered.

An overview of the tunnel background noise is given in appendix A.

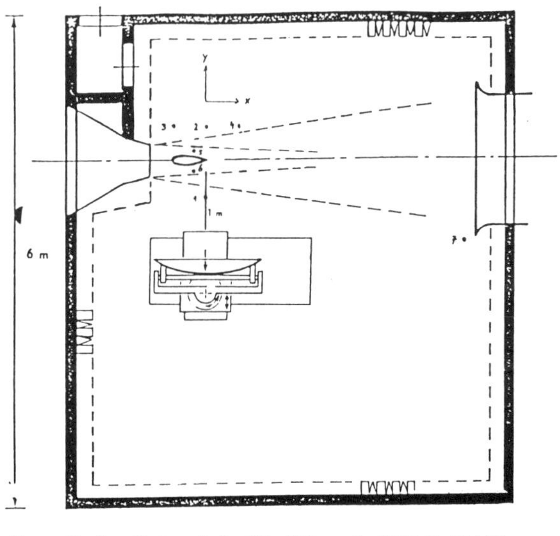

figure 2: Small Anechoic Wind Tunnel of NLR (KAT)

5. TEST PROGRAM

The test program has a modular set up, so it can be terminated at certain points (see flow sheet). It also allows for involvement other parties and/or cooperation with other projects.

After the definition phase, already carried out, the first step is the preparation of the KAT and the validation of the theoretical models in the tunnel. Six test models will be made and tested under different conditions:

- A reference test model
- Five test models with different irregular shaped trailing edges (see figure 3. and table 2 below)

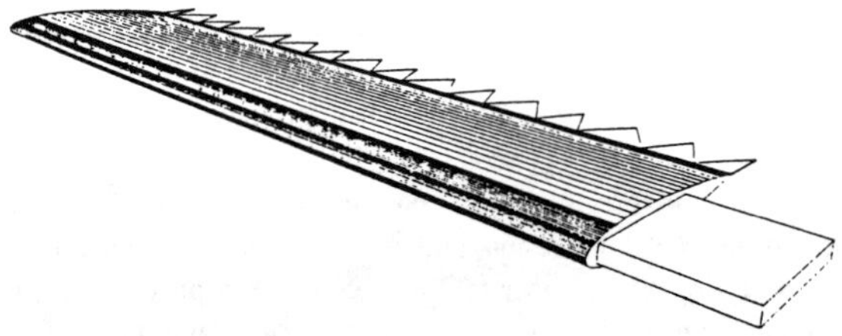

figure 3: test model

tooth height mm	tooth module mm 0	5	10	25	50
0	x				
10		x			
20			x		
50		x		x	x

table 2: test model variations

The following parameters will be varied:

- airspeed to 80 m/s.
- angle of attack to 30°.
- inflow turbulence with frequency peaks up to 8000 Hz and intensities up to 10%.

Turbulence is defined as the velocity of the flow disturbance divided by the characteristic length scale.

The reference test model (see table 2, tooth height and module are zero) will be used for comparison and for the inflow turbulence test. The turbulence will be generated with different grids. The intensities are in the same range as the intensities encountered by a rotor blade rotating in atmospheric turbulence with intensities up to 100% (λ=6-10).

The test models and grids will be designed and tested. During and after the tests the theoretical models will be evaluated and recommendations for the next stage will be made.

The phase described above is contracted by the NOVEM and started in January 1993. The first wind tunnel tests will be executed in April 1993.

After this phase the following steps are foreseen:

- fine tuning of the models, possibly including new ideas, models and measures for the LF noise.
- 3-D tests on the rotor test facility of TUD
- Generation of design measures and tests on a prototype.

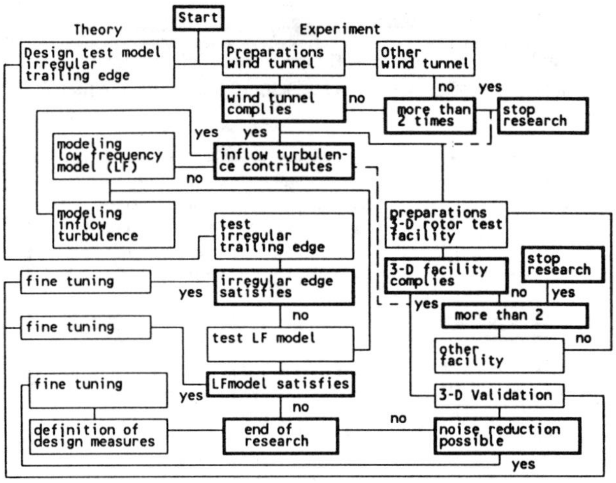

figure 4: Flow sheet of test program

6 CONCLUSIONS

Two important aerodynamic noise sources, which can lead to noise reduction are distinguished:

- trailing edge noise, responsible for the high frequency part.
- inflow turbulence noise, now thought to be responsible for the low frequency part.

Other noise sources like blunt trailing edge noise, laminar-turbulent vortex shedding and noise of holes, slits and other irregularities can be avoided by known measures.

Tip noise is seen as a minor noise source and is mainly caused by the same mechanism as trailing edge noise, which is part of the study. Reduction of tip noise is the topic of other projects [9].

For the reduction of trailing edge noise a theoretical model is available and will be verified by experiments.

The inflow turbulence noise generating mechanisms are not clearly understood and will be tested by experiment to improve the understanding and generate better models, which may lead to noise reduction.

Nevertheless, there are enough indications to believe that the insight gained from the research discussed in this paper will lead to noise reduction in the future.

REFERENCES

[1] F. Hagg, N.J.C.M. v.d. Borg, J.C. Bruggeman, G.J.W. v. Bussel, A.G.M. Dassen, Definition of an aero-acoustic research program, SPE 92-025, April 1992 (Dutch).

[2] H.H. Hubbart, K.P. Shepherd, Aeroacoustics of large wind turbines, J. Acoust. Soc. Am. Vol. 89(6), 1991.

[3] F.W. Grosfeld, Prediction of broadband noise from horizontal axis wind turbines, AIAA J. Propulsion Vol 1(4), 1985.

[4] M.S. Howe, Contributions to the theory of sound production by vortex-airfoil interaction, Proc. R. Soc. Lond A 420, 1988.

[5] A.G.M. Dassen, Tip noise of wind turbines, results of a literature study and proposals for 'silent' brake tips, NLR-CR-92-407-L, dec 1992.

[6] M.S. Howe, Noise produced by a sawtooth trailing edge, J. Acoust. Soc. Am. Vol. 90(1), 1991.

[7] W.K. Blake, Mechanisms of flow induced sound and vibration, Vol I and II, Academic Press 1986.

[8] T.F. Brooks, M.M. Marcolini, Scaling of airfoil self-noise using measured flow parameters, AIAA J. Vol. 23(4), 1985.

[9] A.G.M. Dassen, Experimental validation of a wind turbine aerodynamic noise prediction code, ECWEC 1993 Travemünde.

[10] Brooks T.F., Hodgson T.H., Trailing edge noise prediction from measured surface pressures, J. Sound and Vibration, 78(1), 1981.

APPENDIX A: WINDTUNNEL MEASURING TECHNIQUES

The quality of airfoil noise measurements in a wind tunnel depends greatly on the level of the tunnel background noise. Therefore the application of a technique capable of separating the airfoil noise from the tunnel background noise was included in the program.

At the moment, the opportunities of an acoustic mirror, in-flow measurements and a multiple microphone setup are investigated.

In the past, a 1 m diameter acoustic mirror, having a ellipsoidal surface, was used for noise measurements in the Small Anechoic Tunnel (KAT). This mirror focuses the high-frequency (> 2 kHz) airfoil noise on a microphone which is placed out of the flow. In this way the ratio of airfoil noise and tunnel background noise can be significantly improved.

The same may be expected from the application of in-flow microphones due to the microphones being relatively close to the noise source to be measured. Since the levels are obtained in the acoustic near field these measurement can only be used for comparison of different airfoils.

Finally, the application of an array of microphones can be considered. Several processing techniques are described in literature, mostly using the perceived phase differences and providing the opportunity to separate (the origin of) different noise sources. One of the most straight-forward techniques will be discussed further from the results obtained from some simple measurements of 'airfoil' noise in the KAT. These measurements were recently performed during the measurements of the background noise of the KAT when the idea arose to place an old strut, very much resembling a 'real' airfoil into the 0.5 m diameter flow. The span length of the 'airfoil' was 0.6 m, its chord length 0.5 m. The flow was tripped by fixing a roughness strip to the airfoil surface at 25% of the chord.

Several microphones were positioned around the blade, all out of the flow. A symmetric pair of microphones was placed on opposite sides of the airfoil, each exactly 1 m from the airfoil trailing edge. The noise at these positions was measured with, and without, airfoil in the flow. Assuming that any difference appearing is due to noise emanating from the airfoil, its contribution can be determined by subtracting both spectra.

It is explained in detail in [10] that due to the influence of extraneous noise sources, the use of (single microphone) autopower spectra will lead to poor results. The application of crosspower spectra, obtained from the symmetric microphone pair offers the opportunity to almost completely eliminate the influence of tunnel background noise. For a detailed description of the data-processing refer to [10].

The results are presented in fig. 5; G_{M1M1} and G_{M2M2} represent the autopower spectra, G_{M1M2} the crosspower spectra. For frequencies between approximately 700 Hz and 2 kHz an increase in the difference between the autopower levels, obtained with (w airfoil) and without (w/o airfoil) was found to appear. Since the levels are mainly determined by the tunnel background noise over the largest part of the frequency range, substraction of both levels leads to poor results for the airfoil noise spectrum.

However, the crosspower spectra indicate a very low level of coherent background noise for frequencies higher than approximately 1 kHz. It is important to note that if the airfoil is the only source of coherent noise perceived by both microphones, the level of the crosspower spectra is equal to the level of airfoil noise. This clearly indicates the importance of increasing the frequency of the airfoil noise to a range where the coherent background noise level is insignificant (> 2 kHz).

At this moment an optimization of the experimental setup is executed. It has already been demonstrated that airfoils having a reduced chord length are favourable. In these cases thinner boundary layers can be obtained, leading to a higher frequency of the broadband trailing edge noise.

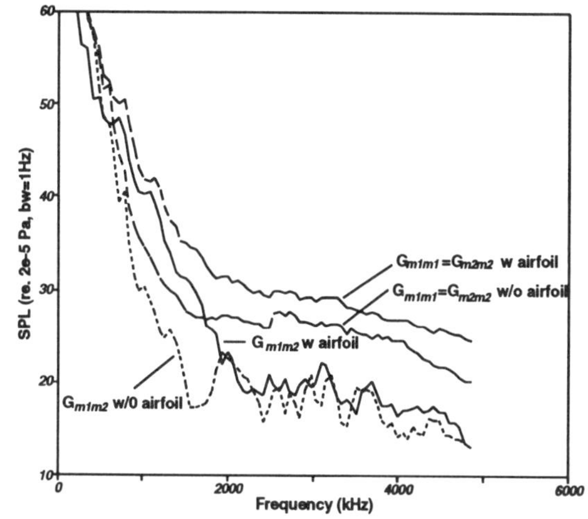

Figure 5: Autospectra and cross spectra of microphones on opposite sides of an 'airfoil' placed in the open jet of the Small Anechoic Tunnel of the National Aerospace Laboratory for a flow of 80 m/s.

TONAL GEAR NOISE FROM WIND TURBINES

Mr. A. Crone and Mr. T. Astrup
Ødegaard & Danneskiold-Samsøe ApS
Kroghsgade 1
DK-2100 Copenhagen Ø

ABSTRACT: This paper deals with different aspects of tonal gear noise, radiated from wind turbines. The impact of tonal noise on human beings is discussed, and it is shown that the tonal noise is likely to be more annoying than normally determined by use of the standard procedures for measuring and evaluating noise from wind turbines. The paper also discusses the generation, transmission and radiation of tonal gear noise, and several methods for describing the potential noisiness of gearboxes (source strengths) are presented and evaluated. Finally, are described methods and special considerations to be made when the effort is to reduce the tonal gear noise radiated from wind turbines.

1. INTRODUCTION

Noise radiation from wind turbines is an essential limiting factor for the siting of wind power plants in populated areas.

Basically wind turbine noise can be divided into two main components:

- Aero-dynamic noise generated around the moving rotor blades.
- Machinery noise.

In general, both components have to be considered when evaluating potential sitings for wind turbines.

In most modern wind turbines, the air-borne noise from the nacelle interior is well-damped. Therefore, the major part of the machinery noise from the wind turbine is caused by structure-borne noise which is transmitted as vibrations through the structure to the radiating external surfaces.

The radiated machinery noise is in general dominated by noise generated in the gearbox. This is partly because the gearbox is the component where the biggest oscillating forces act, and partly because the gearbox is the component which is the most difficult, effectively, to vibration isolate.

The other machinery components e.g. the generator and hydraulic pumps, are normally effectively isolated by means of resiliently mounts and couplings.

2. EVALUATION OF TONAL NOISE

The noise from the rotor and from the machinery is of different kind and its impact on human beings is different. The aero-dynamic noise appears as broad-band noise, whereas the machinery noise acts mainly as a tonal noise.

As an example Figure 1 shows a narrow-band frequency spectrum of the sound power emitted from a wind turbine. Both types of noise are present in the spectrum; the broad-band noise as the flat, smooth decreasing part of the spectrum, and the tonal noise as the distinct peaks.

2.1 Human Perception of Tonal Noise

Broad-band noise and tonal noise are perceived differently by the human hearing.

The broad-band noise is of the same character as the natural ambient noise e.g. the noise arising from the wind blowing in vegetation or in the ear. Human beings have a natural instinct for "ignoring" this type of noise.

Contrary, tonal noise is perceived clearly by human beings, especially if the level of the noise is varying as machinery noise from wind turbines often does due to the fluctuating wind. The human hearing focuses naturally on the varying tonal noise in contrast to the broad-band noise, as a part of identifying speech.

Tonal noise is therefore perceived as more annoying than broad-band noise of equal sound pressure level, even though the broad-band noise will have a reducing, 'masking' effect on the level of the tonal noise perceived by human beings, when the two types of noise occur simultaneously.

The increased annoyance of tonal components compared with broad-band noises of the same sound pressure level is recognized by many national and international standards for the evaluation of environmental noise, cf ISO 1996, ISO 3891, BS 4142. A "tone adjustment" of 2-7 dB to be added to the measured, total A-weighted sound pressure level is prescribed as a compensation for the excess annoyance not presented by the directly measured sound level.

The magnitude of this penalty makes the elimination of tonal noise the primary task of all noise control efforts, since it will generally be difficult to achieve a similar reduction of the annoyance by reducing broad-band noise components alone

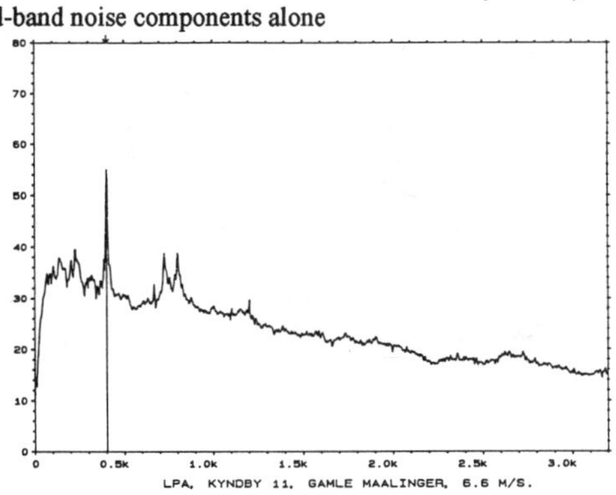

Figure 1: Narrow-band frequency spectrum of the sound pressure from an approx. 200kW wind turbine, placed in *Kyndby Windturbine Farm*. L_{pA} in dB re 1 µPA. Wind speed 8 m/s, 10 m above ground.

2.2 Objective Evaluation

In the *Joint Nordic Method* [1], tonal noise components are evaluated by comparing their sound pressure levels with the level in a standarized, critical frequency band around the tonal frequencies. If the levels of the tones exceed the level in the critical band (minus the tonal components) with a certain amount, usually 4 dB in the frequency range relevant for the gear tones, the tones are judged as 'clearly audible' and a 5 dB penalty has to be added to the measured level of the tones.

2.3 Tonal Noise as Function of Wind Speed

The level of the noise radiated from wind turbines depends on the actual wind speed at which it is evaluated.

The relation between the sound pressure of the aero-dynamic noise generated at the rotor or in the vegetation, and the wind speed, has been found to be 2 - 3 dB/ms^{-1}.

The tonal gear noise from wind turbines is found to have a relatively weaker relationship to the wind speed, normally 0.5 - 1 dB/ms^{-1}.

The different relationship between the two types of noise and the wind speed implies that the tonal gear noise will be most dominating in the total noise at moderate wind speeds, say 4-6 m/s whereas it in many countries are common to evaluate the noise from wind turbines at 8 m/s.

The potential annoyance arising from the tonal noise, will therefore often be underestimated when following the common, present procedures for evaluating noise from wind turbines.

3. TONAL NOISE FROM GEARBOX

The gearbox is the most important contributor to the tonal noise radiated from wind turbines.

3.1 Generation of Gear Noise

Different noise sources exist in the gearbox. The most important ones are related to the tooth mesh and generate dynamic forces with the frequency of the tooth mesh and its harmonics. Tonal components above the 5th harmonic of the tooth mesh have often been observed to be significant in the total noise from wind turbines. The tonal component of highest noise level in the total wind turbine noise, can normally be related to the tooth mesh of the fastest running gear stage.

One of the two most important sources, is the *tooth impact*, occurring when a couple of teeth collide in an non-ideal engagement due to the deformation of the gear wheels from the contact forces.

The other important source is the forces arising due to the *parametric excitation* from the on-going change in the contact stiffness, when the number of tooth pairs in contact changes with the turning of the gear wheels .

3.2 Transmission of Gear Noise

The forces arising from internal sources in the gearbox will transmit as vibrations through the bearings or through other gear stages to the gearbox housing and connected shafts. The further transmission paths of the noise generated in the gearbox are illustrated in Figure 2.

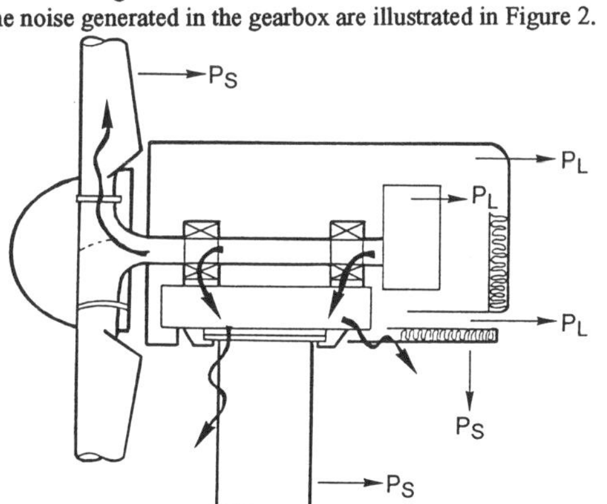

Figure 2: Sketch of possible transmission paths and emitting surfaces for the gearbox noise. P_S = Structure-borne noise, P_A = Air-borne noise.

As stated earlier, the contribution from the air-borne gear noise is negligible in most modern wind turbines, yielding the structure-borne noise as the important contributor.

The specific paths through which the structure-borne noise is transmitted further out in the wind turbine structure may depend on the actual design of the gearbox and its foundation.

In Danish mid-range wind turbines of the size 100 - 400 KW, a hollow shaft gearbox is commonly used. This type of gearbox is mounted directly on the rotor shaft, which then, through its bearings, acts as the main support of the gearbox. The resulting torque on the gearbox is absorbed in one or two rods, linked between the gearbox and the foundation.

For this concept, the significant part of the structure-borne noise is usually transmitted through the input rotor shaft, out to the radiating surfaces.

Experiments, where flexible elements have been introduced in the supporting rods, have showed no difference in the tonal gear noise emitted from the wind turbines. This implies that no significant gear noise is transmitted through these supporting rods.

In bigger turbines, 400 - 1000 KW, it seems that a variant of the above-mentioned arrangement will become common in the future. Here, the backmost rotor shaft bearing is integrated in the gearbox. The support of the rotor shaft system and the gearbox is obtained through the frontmost bearing and the rods connected to the gearbox.

Experience with the influence of this concept on the transmission of structure-borne noise from the wind turbine, is rather sparse. Significant parts of the structure-borne noise may be transmitted to the foundation structure, through the supporting rods as well as through the rotor shaft bearing.

3.3 Radiation of Gear Noise

The structure-borne noise generated in the gearbox, can in principle be transmitted to, and radiated from the surface of the nacelle, tube tower and rotor blades, as shown in Figure 2.

The contribution from the individual parts to the total radiated gear noise, may depend on the specific design of the wind turbine.

The gear noise radiated from the exterior of the nacelle will in most cases be negligible, counting newer Danish wind turbines. This, because the nacelle is normally designed to fully enclose the machinery and its foundation, thereby shielding against noise radiated from these parts of the structure. Mostly, these nacelles are also connected to the structure by flexible elements which will reduce the transmission of structure-borne noise to the nacelle.

The major part of the tonal gear noise will then usually be radiated from the tube tower and the rotor blades. In general, it is not possible a priori to rank the contribution from these two radiators.

Measurements on several wind turbines show that this relation may vary with the specific construction and the frequency of the tones. In one wind turbine type, where several tonal components between 500 Hz and 700 Hz gave troubles, measurements indicated that some of the components were mainly radiated from the tower while others were mainly radiated from the rotor.

For most of the wind turbines of the 'Danish concept', the frequency of the significant tonal components in the noise, will be in between 250 Hz and 800 Hz. Within this "Radiation Window" the tonal components generated in the gearbox seem to be radiated especially well from the wind turbines. On several types of wind turbines, the maximum tonal radiation have been found within this frequency range, although the maximum vibration level measured on the gearboxes was found at other frequencies. This may be due to some specially well-suited properties of the transmission and radiation characteristics of the structure, within this frequency range.

4. SOURCE DESCRIPTION OF GEAR NOISE

Besides the effectiveness with which the wind turbine structure transmits and radiates the vibrations from the gear, the amount of gear noise from the turbine is determined by the source strength of the gearbox, in respect to the structure-borne noise.

4.1 Sound Power Level

A common practice at German gear manufacturers for determining the noise source strength, is to measure the sound power emission from the gearbox when driven in a test bed. The obtained level is then related to some emperic data given in *VDI 2159*.

In Figure 3 is shown a diagram from this recommendation, valid for helical gears. The diagram is obtained by statistical analysis of measurements made on a very large number of different gearboxes. It shows, as a function of the rated mechanical power of the gearbox, the sound power level at different fractiles.

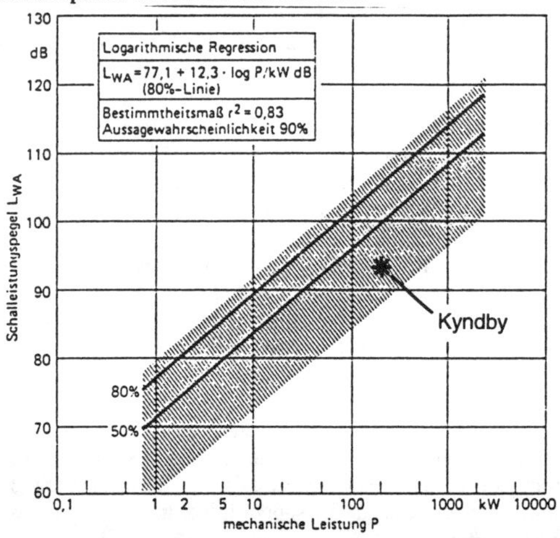

Figure 3: Diagram from VDI 2159 showing the relationship between sound power level at different fractiles, as function of rated mechanical power of the gearbox, helical gears.

Although useful for evaluation of the over-all quality of a given gearbox, this procedure is not usable for evaluating the amount of gear noise which may be radiated from a wind turbine when the gearbox is mounted. The reason is that there is no general relationship between the noise emitted from the gearbox and the vibrations generated.

As an example, one of the Danish wind turbine farms, *Kyndby Vindmøllepark,* suffered in its start from a very high level of tonal noise, which could be related to the gear mesh in the fastest running stage of the gearboxes. A typical sound power level, which was measured on the gearboxes prior to delivery from the manufacturer, is shown as an asteric in Figure 3, implying that relatively quiet gearboxes may cause high gear noise radiation from wind turbines when mounted.

4.2 Velocity Ratings

Another possibility for evaluating the noise source strength of gearboxes could be to use vibration measurements according to *ISO/DIS 8579-2*, although still in draft. Vibration measurements made on the bearing housings of the gearbox when driving in a testbed, serve as basis when using this standard.

The source strength of a given gearbox is obtained by comparing the frequency spectra of the measured vibration velocities with some 'Velocity Rating' curves (VR), illustrated in Figure 4. The gearbox gets its rating from the curve which just envelopes the spectra of all the measurements.

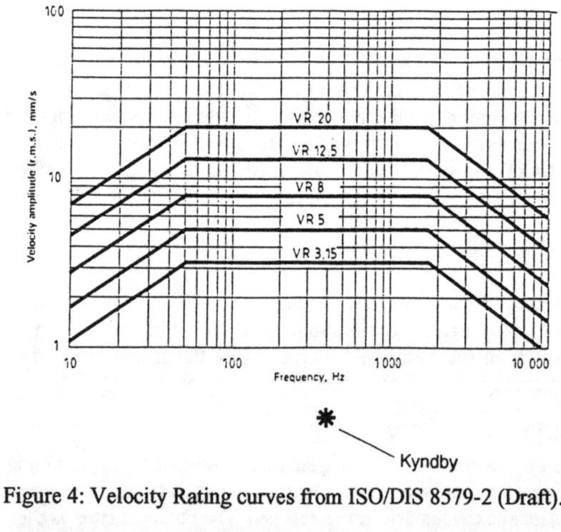

Figure 4: Velocity Rating curves from ISO/DIS 8579-2 (Draft).

To evaluate the quality of the given gearbox, the standard operates with four classes of quality, 'A'-'D', where 'A' counts gearboxes with the highest quality, for navy purposes etc. As function of the rated power of the gearboxes, the standard gives recommended maximum Velocity Ratings which may be full-filled by gearboxes classified in the four categories.

By using the standard, the recommended VR for the *Kyndby* gearboxes, discussed above, is 4 mm/s if the gearboxes are classified in the best quality category. For comparison, is the velocity (0.6 mm/s) which is measured at the problematic gear tone frequency on one of the most noisy gearboxes shown in Figure 4 as an asteric. As it is seen, these recommended velocity levels are of little use when evaluating wind turbine gearboxes with respect of the noise they will generate when mounted in a wind turbine.

4.3 Transmission Error

A third method for determining the noise source strength of wind turbine gearboxes is measurement of the *Transmission Error* (TE) of one or more of the gear stages. Considering a single gear stage, the TE describes the difference between the actual, angular position of the output shaft and the ideal position it may have possessed if the geometry of the gear wheels was perfect and the teeth were inflexible. The TE may be expressed in radians or in micro meters related to the contact line.

The close relationship between the TE and the vibrations generated in the gear has been documented in several reports, e.g. [2].

TE's are standarized in *DIN 3963* where the measures are directly related to the DIN quality grade.

Until now, no explicit data exist on the relations between TE and gear noise radiation from wind turbines. The method is nevertheless expected to have a big potential for noise evaluation of wind turbine gearboxes, but measurement practice still has to be developed. In a project run by ØDS, for the Danish Energy Agency, TE measurements will be made on a 1 MW gearbox for a wind turbine developed by a Danish utility company, *Elkraft.*

4.4 Velocity Levels of Gear Tonal Components

Vibration measurements made on gearboxes when driven in a test bed, can be used for evaluating the noise potential of the gearboxes when mounted in wind turbines. The governing quantities are the velocity levels of the tonal components, averaged over several measurement positions. The measurements have to be made under operational conditions which are similar to the conditions where the noise emission is to be evaluated.

Figure 5 shows the results of such measurements made on three different types of wind turbines and gearboxes. The figure shows the measured sound power levels of the tonal component at the gear mesh frequency of the fastest gear stage, radiated from the completed wind turbine, versus the average velocity level of the same component, measured on the gearboxes in a test bed.

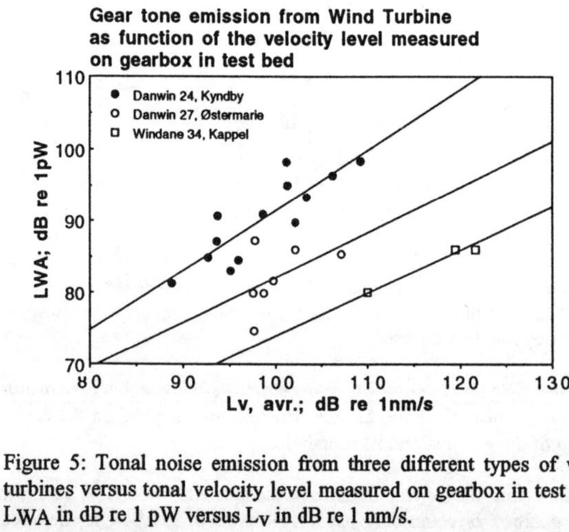

Figure 5: Tonal noise emission from three different types of wind turbines versus tonal velocity level measured on gearbox in test bed; L_{WA} in dB re 1 pW versus L_V in dB re 1 nm/s.

The diagram in Figure 5 shows the rather good relationship between the velocity level on the gearbox and the tonal emission from the wind turbine. It also shows that this relationship is "biased" between different types of wind turbines, although of same concept. Measurements on gearboxes of different make mounted in turbines of same type, indicate that the relationship between the velocity level and the noise emission is more dependent on the specific wind turbine construction than on the gearbox construction.

Given a tonal emission level which can be accepted, this procedure may be used for determining acceptance criteria on vibration measurements made as a factory test of the gearbox.

The procedure has been used in the *Kyndby* case, described earlier, where the gearboxes were dismantled and sent to the manufacturer for modification. From noise measurements made at neighbours' to the wind turbine farm, the acceptable tonal emission was determined. Relating this level to the velocity level on the gearbox, by means of the diagram shown in Figure 5, the maximum acceptable velocity level was found. Several phases of modification of the gearboxes were needed before acceptable velocity levels were reached and the gearboxes were accepted for re-mounting in the wind turbines.

5. REDUCTION OF GEAR NOISE

Troublesome tonal gear noise emission may be altered in different ways.

5.1 Alternation of Transmission Properties

One way is to reduce the transmission of the structure-borne noise from the gearbox when these paths are known.

Gear noise emission from the nacelle and tower may be altered by introducing flexible elements in the supporting rods, considering hollow shaft gearboxes of the type with one integrated main bearing. For the other type of hollow shaft gearbox, where the turbine has two rotor shaft bearings, the flexible rods will probably not alter the gear noise.

Gear noise emission from the tube tower may be reduced by mounting a concentric ring chamber filled with sand, in the upper part of the tower. Reduction of the emitted tonal noise of 12 dB or more has been observed for tonal components at frequencies above 500 Hz.

5.2 Alternation of Source Strength

If a significant part of the gear noise is radiated from the rotor, which may be very possible, it is difficult to alter the transmission path of the structure-borne noise without a radical change in the design of the rotor shaft and gearbox system. In this case the source strength of the gearbox may be altered.

The first thing to do is to make sure that the gearbox is not running in a resonance condition at its normally fixed, operating speed. In several cases it has been observed that gearboxes were operating in conditions where the gear mesh frequencies were close to torsional natural frequencies of the shaft system. This was, amongst other cases, actual in the *Kyndby* case.

Figure 6 shows the velocity level at the gear mesh frequency of the fastest running gear stage, at different operating speeds, measured on two of the original gearboxes from the Kyndby turbines. The expected response is also shown in the figure.

The figure shows a very obvious peak where the gear mesh frequency coincides with the torsional natural frequency, at approx. 480 Hz. The gearboxes were operating at 1000 RPM, implying an original gear mesh frequency of approx. 410 Hz.

From the curves in Figure 6, it was decided to change the number of teeth on the pinion of the output gear stage, and thereby to change the gear mesh frequency to approx. 350 Hz, close to where the response shows a minimum in the figure.

After this modification, the measurements of the velocity response at different speeds were repeated.

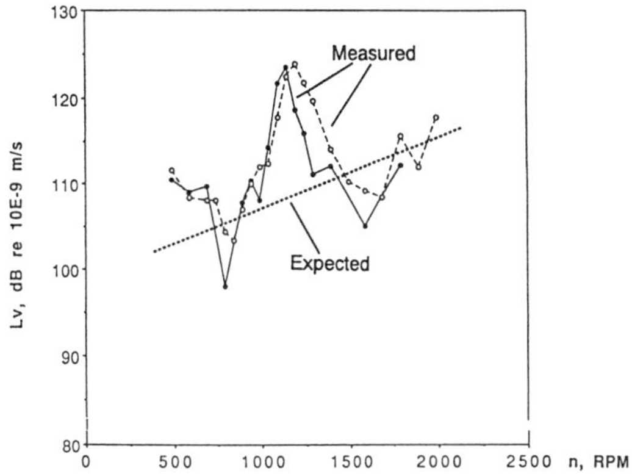

Figure 6: Velocity level of tonal component at gear mesh frequency of fastest running gear stage, as function of the operating speed. Measured in a testbed, on a gearbox from *Kyndby Wind Turbine Farm.*

The response showed the same relation to the excitation at the gear mesh frequency, but the general velocity level had decreased by 8 to 10 dB. The velocity level of the troublesome tonal component showed, at the operating speed of 1000 RPM, to have decreased by approx. 15 dB by this modification.

Experiments made after the changing of the tooth profile showed that the general reduction in the tonal velocity level of 8 to 10 dB, was caused by an optimal tooth profile, mainly determined by the size of the tip reliefs on the pinion and gear wheel, although the backlash of the gear stage also showed to have an influence on the velocity level generated.

The remaining 5 to 7 dB was caused by removing the frequency of the excitation away from the natural frequency of the shaft system.

CONCLUSION

Mechanical noise, especially originating in the gearbox, may very well be a limiting factor when siting wind turbines.

Normal procedures for evaluating noise from wind turbines may underestimate the annoyance of tonal gear noise components.

The gear noise is primarily transmitted as structure-borne noise (vibrations) from the gearbox to the radiating surfaces.

Reducing the gear noise emission by means of flexible elements, e.g. rubber mounts, is difficult to do effectively due to the big forces acting on the gearbox. Experiences from constructions where such measures have been tried, show only gains between none and moderate.

The noise potential, and thereby the gear noise emission, may be minimized or reduced considerably, by ensuring a proper margin between the tooth mesh frequencies of the gearbox, and the torsional natural frequencies of the wind turbine shaft system, and by using specially optimized tooth profiles.

REFERENCES

[1] T H Petersen: Methods for Evaluating the Prominence of Audible Tones in Noise. Lydteknisk Institut, Denmark, 1988.

[2] G Niemann, J Baethge: Drehwegfehler, Zahnfederhärte und Geräusch bei Stirnrädern. VDI-Z 112 (1970) Nr. 4- Februar (II)

NOISE REDUCTION OF HORIZONTAL AXIS WIND TURBINE DUE TO TIMING BELT

Y. Shimizu, T. Maeda, E. Mizuno and H. Fujiwara
Department of Mechanical Engineering
Mie University
Kamihama-cho 1515, Tsu, Mie 514, JAPAN

ABSTRACT: In wind turbine operation, a rotor generally drives a generator via gears. The gear drive, however, causes the noise. The paper describes that the timing belt drive instead of gear is very useful for the reduction of the noise from the wind turbine.

1. INTRODUCTION

The wind turbine is the best generation system for the natural environment, because the wind turbine transforms the wind energy into the electric power without the pollusion. The wind turbine, however, radiates the noise such as aerodynamic noise of the blades and meshing noise of gears. Since the wind turbine is operated in the open air, the noise is radiated to the surroundings. When the wind turbine may be constructed near the place of residence, the noise radiation is very important. The gear drive, however, causes the strong noise. The timing belt drive for the reduction of the noise from the wind turbine shows good result.

2. METHOD FOR NOISE MEASUREMENT

The measurement of the noise from the wind turbine is carried out according to the IEA (the International Energy Agency) recommendation [1].

The noises from the wind turbine are measured by the microphones of 13 mm in diameter. The microphones are mounted on the boards at the ground level. The distance between the microphone and the leading edge of the board is set for 1 m. The board has a length of 1.8 m, a width of 1.8 m and a thickness of 16 mm. The board is covered with the aclylic board of 1 mm in thickness owing to reflection of the sound. The microphone is covered with the wind screen with a diameter of 90 mm owing to reduction of the back ground noise (see Fig. 1).

The setting position of the microphones is shown in Fig. 2. In Figure, the nomenclature mean as follows.

R_1:distance between wind turbine and microphone at reference position $= H+D/2 = 12.1$ [m]
H :hub height $= 8.0$ [m]
D :diameter of rotor $= 8.2$ [m]
R_3: $= 3R_1 = 36.3$ [m]

The noise obtained by a microphone are, simultaneously, recorded with the noise of the other microphone. The comparison of the simulatneous recorded data of R_1 and the other position are discussed.

Figure 3 shows the data processing diagram of the measurement. The noises sensed by the microphones are transmitted to the voltage signals, and they are amplified due to the precision sound-level meters. The signals are recorded on the audiotape by means of the digital recording device. The time of the measurement is also recorded on the audiotape. The wind velocity, the rotor speed and the time are recorded on the magnetic disk with a small computer. The sound pressure signals recorded on the tape are reproduced by the recorder, and processed with the use of a FFT analyzer. The results of the measurement are examined with the instantaneous wind velocity and the instantaneous rotor speed recorded by the computer.

The noise are recorded by digital recording with the use of 8mm-videotape recorder. The recorder has constant frequency response over the range of hearing for the person, 20 Hz to 20 kHz.

Figure 4 shows the sketch of the measurement site. There are the buildings of lectures at the position of 72 m southwest from the wind turbine, our laboratory at 40 m northwest, the handball court in east neighborhood and the road along seaside at 83 m east. The background noise is measured in the daytime as the equivalent continuous A-weighted sound pressure level, 60 dB, and the equivalent continuous unweighted sound pressure level, 45 dB. The grass is cut about 15 cm over the measurement area. The irregular noise is generated such as sound from birds, voice of human, noise of cars etc. The results of the measurement are processed except for the data of these irregular noise. Equivalent continuous sound pressure level is determined with the use of sound data for steady turbine-rotation over a time interval of 20 seconds.

A test wind turbine has a diameter of 8.2 m. It is the upwind type with three blades. The height of the tower is 8 m.

The experiment is carried out under the conditions of the wind velocity 4.0 - 8.0 m/s and the rotor speed 30 - 90 rpm. The pitch angle of blades is set at 0 degree from rotational plane. The nacelle turns in a parallel direction with the wind.

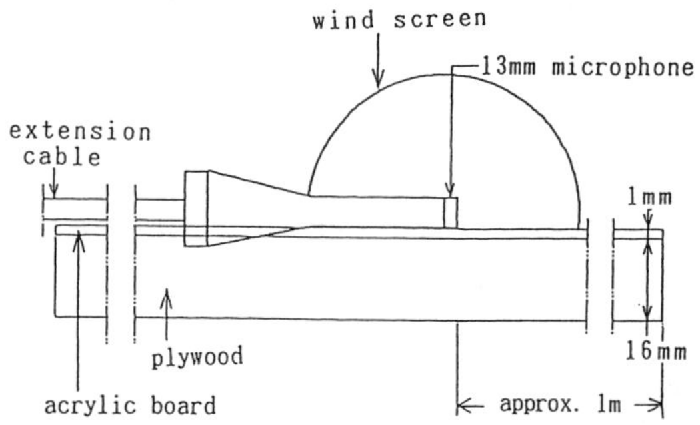

Figure 1: Mounting of microphone.

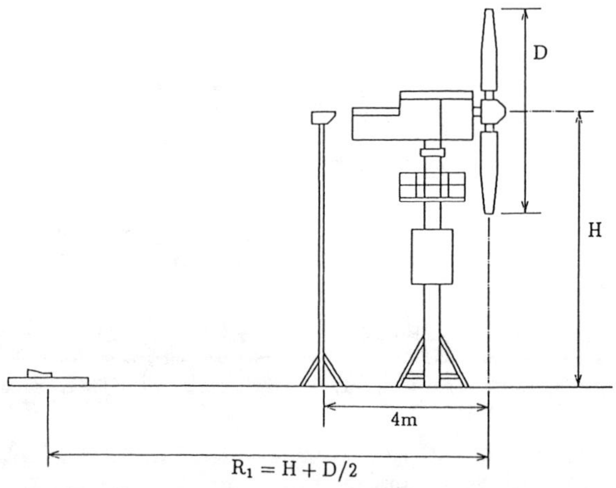

Figure 2: Definitions of R_1.

The number of the pulley teeth on the rotor shaft is 120. The transmission shaft with the pulley of 32 teeth is driven via timing belt. Another pulley, 96 teeth, is set on the transmission shaft, and the generator shaft with 40 teeth pulley is driven via another timing belt. The timing belts change the rotational speed ratio 4:15 on the first stage and 5:12 on the second stage. Finally, the generator is driven by nine times of the rotor speed. The generator of rated output at 8 kW is mounted on the test wind turbine.

3. EXPERIMENTAL RESULTS AND DISCUSSION

3.1 Noise of Test Wind Turbine with Timing Belts

The results of the noise measurements for the test wind turbine with timing belts are presented in this section. Table 1 shows the frequency of mechanical noise sources against the rotor speed. The mechanical noise emission from test wind turbine is dominated by the rotation of the rotor shaft and the generator shaft, and the meshing of the pulley and the timing belts.

Figure 5 shows the A-weighted sound pressure spectra at a height of 8.0 m downwind of wind turbine, $R_1/3 = 4.0$ m. The measurements of noise are carried out by the use of a microphone which is set at the top of the pole of 8.0 m high. The spectra is analized from the data of rotor revolution of 70 rpm. Since this measuring position is just downwind of the nacelle, the mechanical noise is dominant compared to the aerodynamic noise level. The strong spectra are seen at the frequencies 280, 840, 980 and 1680 Hz, which agree with the harmonics of the mechanical noise from the nacelle as shown in Table 1.

Figure 6 shows the A-weighted sound pressure spectra on the ground level downwind of wind turbine, R_1. The spectra arised by the mechanical noise, as shown in Fig.5, becomes relatively small. The spectral distribution over the whole frequency range is also different from those in Fig. 5.

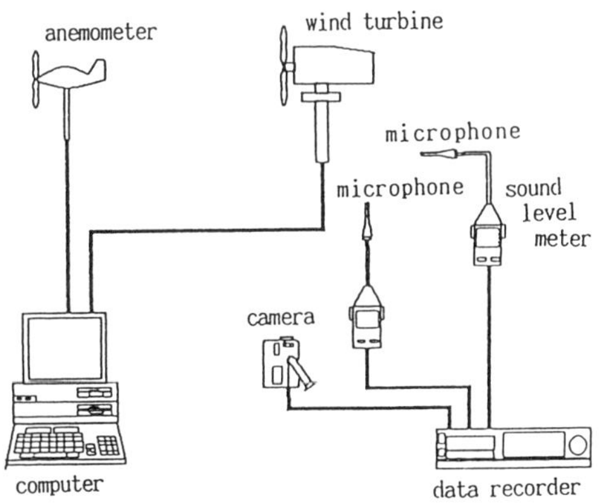

Figure 3: Data processing diagram.

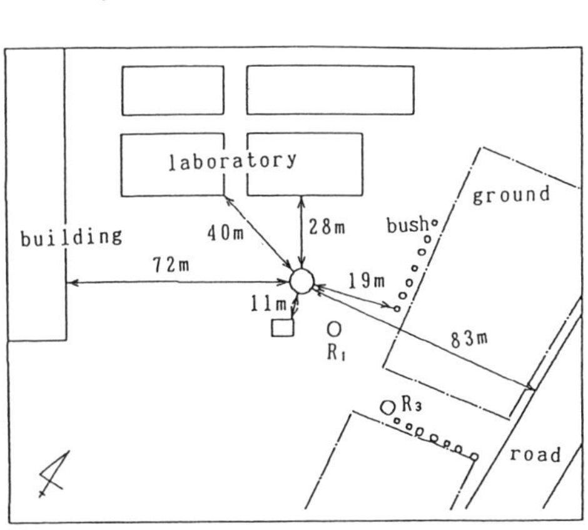

Figure 4: Acoustic environment near wind turbine.

Figure 7 shows the A-weighted sound pressure level in third-octave bands at the downwind positions of turbine, R_1 and R_3, under the rotor speed of 90 rpm. The sound pressure level at R_1 takes the values from 40 dB to 50 dB in the frequency range of 125 to 4000 Hz. The sound pressure level at R_1 decreases in the higher frequency range, and finally becomes the same level of the background noise at 5000 Hz. The strong sound level can be seen at the frequency 200, 400 and 1000 Hz, which agree well with the harmonics of the mechanical noise frequency as shown in Table 1. However, as the results of the spectral analysis, the dominant frequencies are different from those of mechanical noise in third-octave bands. The detailed discussion is found later in Figs. 9 and 10.

The distance from the turbine to R_3 is three times as far as R_1. The sound power P (W) is defined as the energy per second emitted from noise sources. It is assumed that the noise of the wind turbine generates at the point of rotor center. The sound intensities on the spherical surface at the radius r_1 and r_2 from the rotor center become $P/4\pi r_1{}^2$ and $P/4\pi r_2{}^2$ (W/m^2), respectively. The sound pressure levels L_{r1} and L_{r2} are calculated for r_1 and r_2, respectively,

$$L_{r1} = 10\log\frac{P/4\pi r_1{}^2}{I_0} \qquad (1)$$

$$L_{r2} = 10\log\frac{P/4\pi r_2{}^2}{I_0} \qquad (2)$$

Table 1: Frequency of mechanical noise sources.

Roter Speed (rpm)	Rotating Frequency of Roter Shaft (Hz)	Belt Mesh Frequency of Roter Shaft (Hz)	Belt Mesh Frequency of Generator Shaft (Hz)	Rotating Frequency of Generator Shaft (Hz)	Generator Speed (rpm)
30	0.50	60.0	180.0	4.5	270
40	0.67	80.4	241.2	6.0	360
50	0.83	99.6	298.8	7.5	450
60	1.00	120.0	360.0	9.0	540
70	1.20	140.0	420.0	10.5	630
80	1.33	159.6	478.8	12.0	720
90	1.50	180.0	540.0	13.5	810

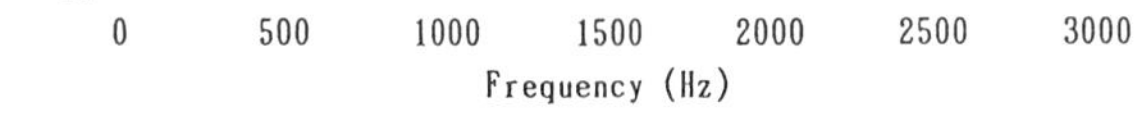

Figure 5: A-weighted sound pressure spectra just downwind of nacelle at $R_1/3$, height of 8.0 m.

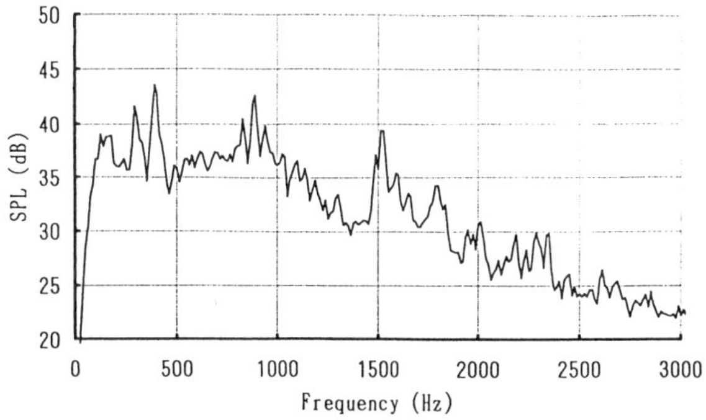

Figure 6: A-weighted sound pressure spectra downwind position R_1.

where I_0 is the reference sound intensity ($= 10^{-12}$ W/m^2). The difference between L_{r1} and L_{r2} is

$$L_{r1} - L_{r2} = 20 \log \frac{r_2}{r_1}. \qquad (3)$$

The ratio of the radius $r_1 = \sqrt{R_1^2 + H^2} = 14.5$ (m) at R_1 to $r_2 = \sqrt{R_3^2 + H^2} = 37.2$ (m) at R_3 is determined as $r_2/r_1 = 2.56$. The sound reduction between R_1 and R_3 is 8.2 dB by calculation with substituting $r_2/r_1 = 2.56$ to Eq.(3). Since the measured sound level at R_1 is 58.6 dB in Fig. 7, the sound level at R_3 calculated from Eq.(3) becomes 50.4 dB, which agrees well with the experimental value of 50.2 dB at R_3. In the case of timing belt drive, both measured and calculated sound pressure level at R_3 decrease to almost the same level as the background noise 50.3 dB as shown in Fig. 7. It is also seen that the spectra over the whole frequency range decrease to the background level as shown in Fig. 7. The sound level at R_2, 53.7 dB, is obtained in the same manner as R_3 using Eqs.(1)-(3).

Figure 8 shows the A-weighted sound pressure level in third-octave bands at upwind and downwind positions R_1 with rotor speed of 90 rpm. The sound level at downwind position R_1 takes larger values in 1 - 3dB at 160, 200, 400, 500 and 1600 Hz than those at upwind position R_1. The differences of sound pressure level are, however, hardly seen between upwind and downwind positions. The sound pressure level generally reduces in upwind direction compared with those in downwind direction. In this experiment, the sound pressure level is hardly independent of the wind direction, because the wind velocity is low, 4.0 - 8.0 m/s, and the tower is high enough for the distance of turbine to position R_1.

Figure 9 shows the A-weighted sound spectra at R_1 for rotor speed from 30 rpm to 90 rpm. The strong spectra are obtained at 250, 400, 800, 1500 and 1800 Hz, which are independent of rotor speed.

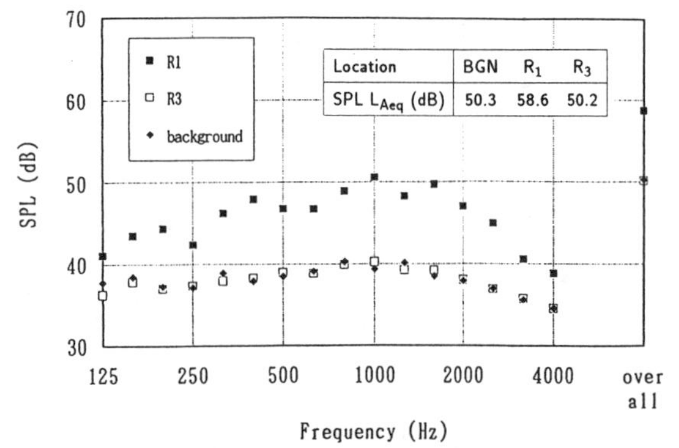

Figure 7: A-weighted sound pressure level in third-octave bands downwind position R_1 and R_3.

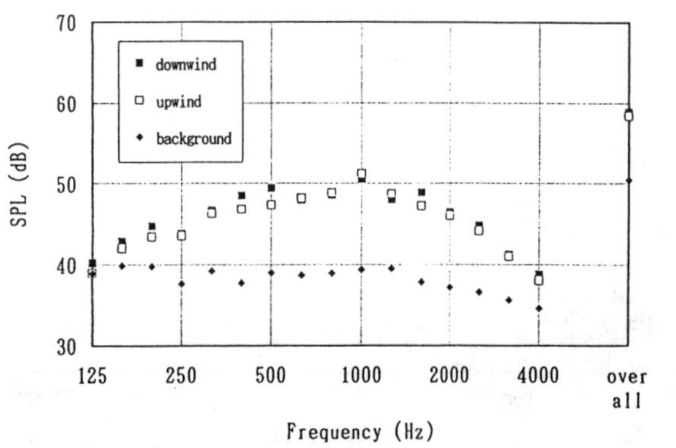

Figure 8: A-weighted sound pressure level in third-octave bands at upwind and downwind position R_1.

Figure 10 shows the relationship between the sound intensity and the rotor speed. The sound intensity at 30 rpm and 40 rpm is almost the same level as the background noise, while the sound intensity increases considerably over 50 rpm. Lowson[3] suggests that the wind turbine noise is classified into the discrete frequency noise, the self induced noise and the inflow noise. The discrete frequency noise is generally dominated in the noise from turbomachinery. The discrete frequency noise has a strong dependency on blade speed and number of blades. The sound intensity of the discrete frequency noise increases as high as the sixth power of the rotor speed (a broken line in Fig. 10). However it can be seen that the sound intensity of the test wind turbine increases in the ratio of fifth power of rotor speed as shown in Fig.10. The noise of the test wind turbine is mainly affected by the self induced noise due to blade geometry or aeroacoustic instabilities. It seems that the flow around the blade root is unstable because the root of test wind turbine blade is not twisted. The effects of self induced noise from test wind turbine is obvious by the independency between the frequency of strong spectra and the rotor speed as shown in Fig. 9. The sound intensity from test wind turbine is strongly affected by the blade geometry since the dominant frequency and the rotor speed are independently in Fig. 9. The inflow noise is hardly observed in this experiment.

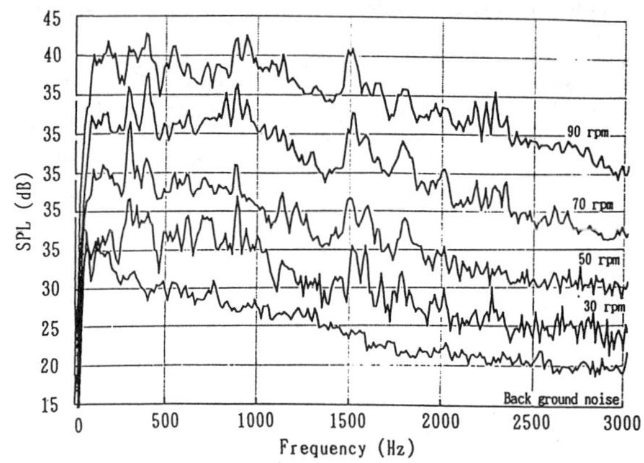

Figure 9: A-weighted sound pressure spectra at R_1 for various rotor speed.

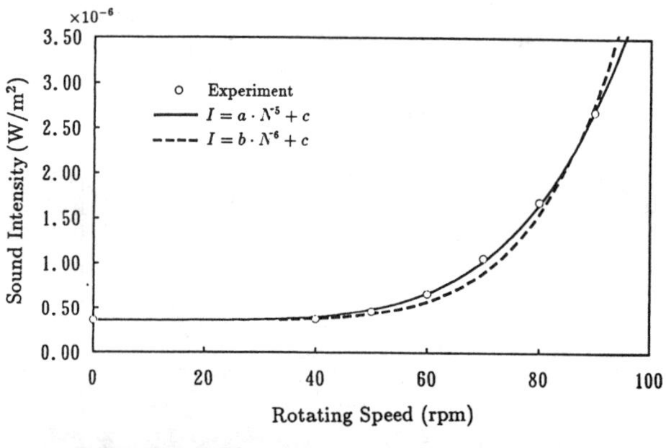

Rotating Speed (rpm)	BGN	30	40	50	60	70	80	90
SPL L_{Aeq} (dB)	49.8	50.9	50.9	52.6	54.0	56.1	57.1	59.4
Sound Intensity ($\times 10^{-6}$ W/m^2)	0.36	-	0.37	0.46	0.66	1.06	1.69	2.69

Figure 10: Relationship between sound pressure intensity and rotor speed.

3.2 Comparison of Wind Turbine Noise with Timing Belts Drive and Gear Drive

Figure 11 shows the comparison of the A-weighted sound pressure spectra between the timing belts drive and the gear drive at R_1. The noise from WINDMEL, which is a horizontal axis wind turbine, is referred to the data of the gear-driven wind turbine [2]. The noise from WINDMEL is measured at the height of 7.5 m, and 11m away from the tower. Comparison between the test wind turbine and WINDMEL is shown in Table 2. The reference distance R_1 for WINDMEL is estimated as 14.7 m, which is refered to the distance between the microphone and rotor center. The sound reduction of the ideal reference position to the real measuring position is 0.95 dB. In case of the gear drive, the sound pressure level reaches the maximum value of 65 dB at the meshing frequency 300 Hz. On the contrary, the quickly projecting frequencies of the sound pressure level can not be seen with timing belts, and the level is almost kept under 40 dB. As the results of these analyses, it is investigated that the noise of the wind turbine is considerably reduced due to timing belts without gears.

Table 2: Comparison between test wind turbine and WINDMEL.

	WINDMEL	MIE
Rated Wind Speed	8 m/s	8m/s
Rated Output	15 kW	8kW
Hub Height	14.8m	8m
Roter Location	Upwind	Upwind
Diameter	15m	8m
Number of Blades	2	3
Airfoil	FX-77-W Series	NACA4415 NACA4412

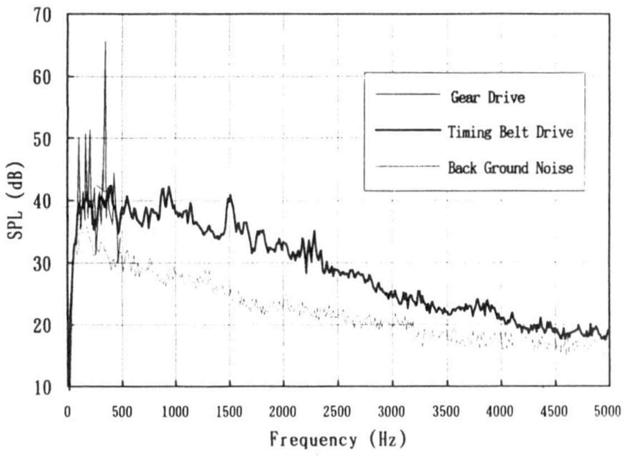

Figure 11: Comparison of A-weighted sound pressure spectra between timing belts drive and gear drive.

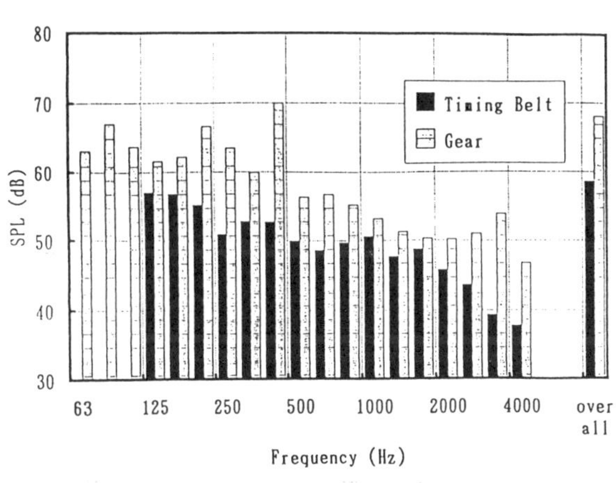

Figure 12: Sound pressure level in third-octave bands for timing belts drive and gear drive.

Figure 12 shows the comparison of the unweighted sound pressure level in third-ocrave bands for the test wind turbine and WINDMEL. The equivalent continuous sound pressure level over whole frequency range is kept in low level, 58.6 dB, for timing belt drive, while the sound pressure level for gear drive takes much higher value, 68.1 dB.

Table 3 shows the characteristics for the various driven system. Since timing belts drive is favorable to the noise and the maintenance, timing belt is the best driven system for the middle class wind turbine. A timing belt's life is about 3×10^8 bending times. As the rated speed of the generator on the test wind turbine is 900 rpm, the timing belt's life is estimated as four months in the case of continuous rated operation. In case of the test turbine, twin timing belts are stretched between every pulleys, so that the timing belt's life becomes eight months. The wind velocity, however, changes unsteadily, so that the total averaged-output of the wind turbine is almost 1/8 compared with the rated output even if the wind turbine is operated continuously. Therefore, the timing belt's life on the test wind turbine is estimated about five years, which becomes longer at more weak wind area or becomes shorter at more strong wind area.

4. CONCLUSIONS

(1) The noise emitted from the wind turbine is considerably reduced due to timing belts without gears.

(2) In the case of timing belt drive, the sound pressure level at R_3 decreases to almost the same level as the background noise.

(3) The sound intensity of the test wind turbine increases in the ratio of fifth power of rotor speed.

ACKNOWLEDGEMENTS

The authors would like to thank Mr. Y. Nii, Mr. T. Gotanda, Mr. T. Yamane, Mr. H. Mizutani, Mr. H. Matsumiya and Mr. S. Kawamura for offering their experimental data of WINDMEL.

REFERENCES

[1] Ljunggren, S. and Gustafsson A. Ed., "Recommended Practices for Wind Turbine Testing and Evaluation - 4.Acoustics.Measurement of Noise Emission from Wind Turbines 2nd Ed.", IEA Programme for Research and Development on Wind Energy Conversion Systems, (1988).

[2] Nii, Y., Gotanda, T., Yamane, T., Mizutani, H., Matsumiya, H., and Kawamura, S., "Mechanical Noise Sourc of A 15m Diameter Variable Speed HAWT", Wind Energy, (1991), 732-736.

[3] Lowson, M.V., "Application of Aero-Acoustic Analysis to Wind Turbine Noise Control", Wind Engineering, Vol.16, No.3, (1992), 126-140.

Table 3: Caracteristics for various driven system.

	Gear	Chaine	V-Belt	Timing Belt
Toruque	◎	○	△	○
Efficiency	◎	○	○	◎
Speed Contorol	Const.	Const.	Variable	Const.
Size	S	M	L	M
Life	L	M	S	M
Maintenance	○	○	△	○
Reliability	◎	○	○	◎
Price	×	△	○	△
Noise	△	△	◎	◎

EXPERIMENTAL INVESTIGATION TO REDUCE THE AERODYNAMIC NOISE OF HAWTS

Volker Cornelius
Deutsches Windenergie Institut, Ebertstr. 96, Wilhelmshaven
Institut für Flugmechanik, Rebenring, TU Braunschweig
Hänselmannstr. 6, D-3300 Braunschweig

1. Introduction

Power augmentation of HAWTs is achieved amongst others by using the station at higher tipspeed ratios. Thus results an increase of aerodynamic sound which limits the choice of the location. The profile and the tip vortex were within this study especially investigated of how to diminish aerodynamic wing noise. Therefore the acoustic wind tunnel of the University of Oldenburg was used for intensive tests as well as an Aerowatt windturbine (D = 10m).

2. Blade Section Noise

In the laminar boundary layer instabilities occur called Tollmien — Schlichting Waves. They cause in interaction with the trailing edge loud and tonal sounds. These sounds occur on the suction side of the blade using narrow angles of attack ($a < 3^\circ$) and on the pressure side using wide ones ($3^\circ < a < 15^\circ$).

To reduce this type of noise, a turbulator has been glued to the blade section. The turbulator causes the tripping of the laminar boundary layer into a turbulent one. Thus the resonances between the instabilities of the boundary layer and the trailing edge are disturbed. Figure (2) shows the spectral noise analysis of a 70cm blade with and without a turbulator in the wind tunnel.

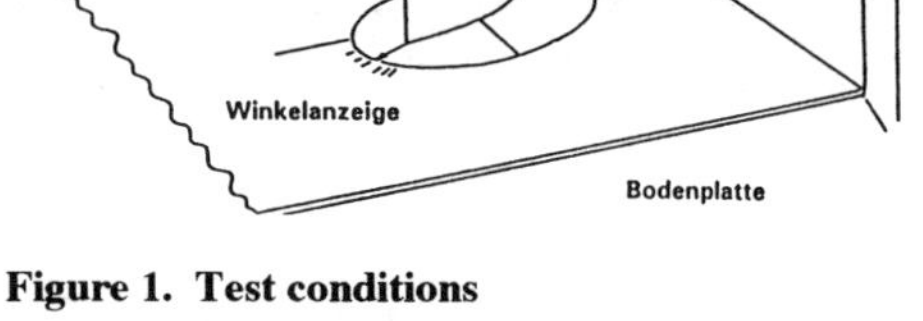

Figure 1. Test conditions

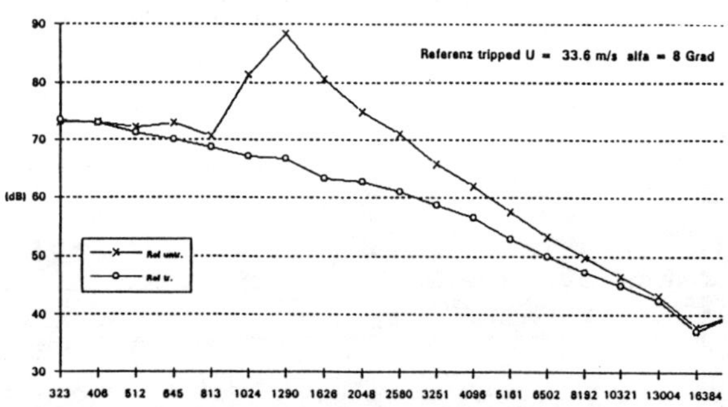

Figure 2. 70cm blade in the windtunnel with/ without turbulator

Being very important for the noise reduction, the position of the turbulator was determined by tests in the wind tunnel with different angles of attack. The turbulator has to be placed on the pressure side of the profile under 56% of the chord. Without the turbulator there are significant peaks in the spectral analysis, which are up to 20 dB higher than the average noise. At increasing windspeed the peaks turn to higher frequencies. At the rotation blade with its different flow conditions at every segment these particular peaks sum up to broadband noise. Using the turbulator, those peaks are banished **and** the overall noise reduced.

The aerodynamic properties of the turbulator could not be tested.

3. Tip Vortex Noise

Tip vortexes occur because of pressure-levelling between upper and lower blade section and cause a hissing sound. This sound increases at higher tipspeed and wider angles of attack.

This study investigates five different tip-shapes in order to determine their alternative acoustic properties. All tips were tested at 4 different windspeeds and 6 angles of attack.

The following shapes were tested:

1. cut-off tip
2. rounded tip
3. triangular tip
4. Mie Vane (flat winglet)
5. winglet

3.1 Cut Off Tip

The cut off tip has been the reference for all other tips. It has good noise properties and is a simple shape.

3.2 Rounded Tip

As a result of this study the rounded tips gave the highest noise emission. The tip vortex separates at the edge of the tip. But the separation point is unstable due to the rounded contour. Noise increases by the unstable flow conditions.

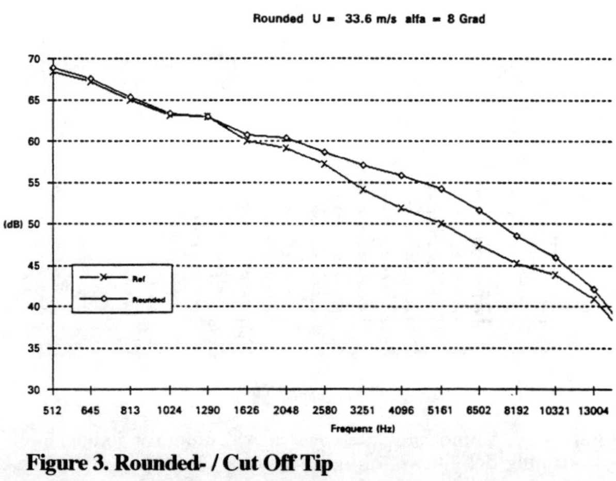

Figure 3. Rounded- / Cut Off Tip

3.3 Triangular Tip (TNT)

This tip type did not show any remarkable noise reduction compared to the reference tip at fixed angle of attack.

As an advantage of the TNT the noise emission at various angles the did not change. This could be an important property in respect to the flow disturbation of the tower. With this tip shape the periodic noise of windturbines may be diminished.

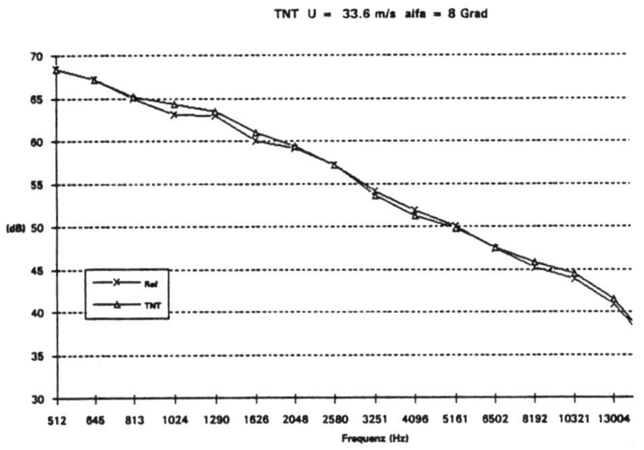

Figure 4. Triangular- / Cut Off Tip

3.4 Mie-Vane

No noise improvements could be measured by using a Mie Vane.

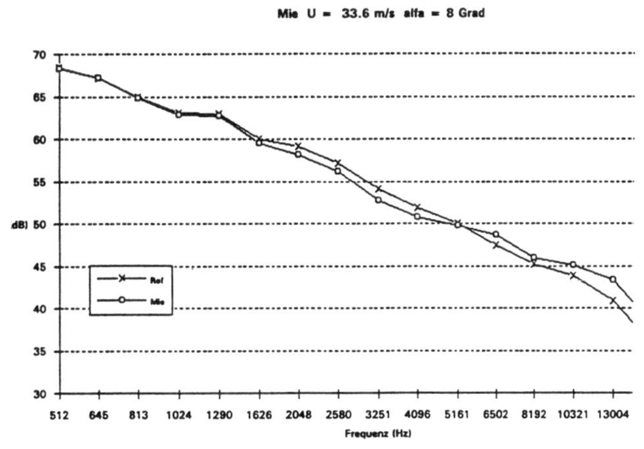

Figure 5. Mie Vane / Cut Off Tip

3.5 Winglet

The winglet gave the lowest noise emission of all tested tip shapes. It also showed a constant noise behaviour at different angles of attack.

If a winglet may contribute to power augmentation is not known, but it might be the future tip type of wind turbines.

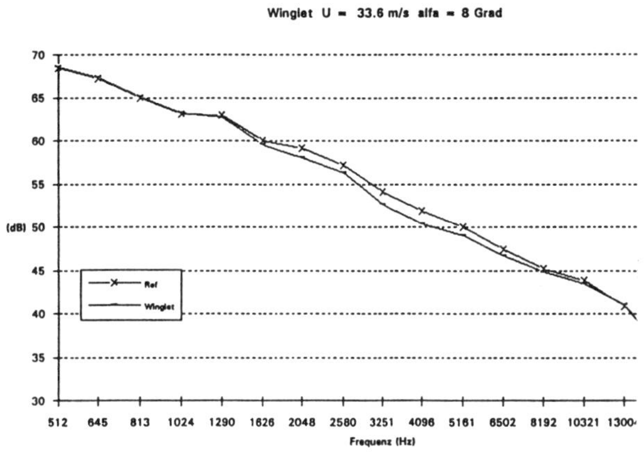

Figure 6. Winglet / Cut Off Tip

4. Conclusion

The aerodynamic noise of wind turbines increases at higher tip speed. The tower causes a periodic noise due to sudden change of the flow conditions.

Airfoil noise is the significant noise type. A turbulator can reduce the blade section noise up to 20dB. A remarkable rotor noise reduction was achieved at the Aerowatt wind turbine with a simple turbulator.

In order to diminish the tip vortex noise the cut off tip and the specially profiled winglet turned to be the shapes with the smallest noise emission. Compared to the conventional rounded tip shape the noise emission of the winglet was about 5dB lower. In respect to various angles of attack the triangular tip had the best acoustic properties.

Only windtunnel measurements were successfuly done. Further wind turbine tests are required to get the best tip shape.

EXPERIMENTAL VALIDATION OF A WIND TURBINE AERODYNAMIC NOISE PREDICTION CODE

A.G.M. DASSEN
NATIONAL AEROSPACE LABORATORY
P.O. BOX 90502
1006 BM AMSTERDAM
THE NETHERLANDS

ABSTRACT: For the purpose of validating a wind turbine aerodynamic noise prediction code, an extensive series of measurements was carried out. It was shown that the application of selective techniques provided accurate spectra of the rotor noise. Since the agreement between the results of the measurements and the predictions were not satisfactory, the source terms in the code had to be adjusted. As a result, the code can now be expected to reliably predict the rotor noise spectra of medium-sized (200 kW - 500 kW) to large wind turbines (> 500 kW) as a function of the turbine dimensions and operating conditions. In the present paper, the course of the validation and its main results are discussed.

NOMENCLATURE

B	= number of blades
C	= chord length (m)
D_R	= rotor diameter (m)
f	= frequency (Hz)
h_0	= site roughness (m)
h_{HUB}	= hub height (m)
P_{MAX}	= maximum electrical power output (MW)
U_{REL}	= relative flow speed at the blade tip (m/s)
Re_C	= Reynolds number based on chord length ($U_{REL}C/\nu$)
Str	= Strouhal number ($f\delta^*/U_{REL}$)
α_{85}	= relative angle of attack at 85% span (degrees)
δ	= boundary layer thickness (m)
δ^*	= boundary layer displacement thickness (m)
λ	= tip speed ratio
ν	= kinematic viscosity (m^2/s)
ϑ_P	= blade pitch angle (degrees)

1. INTRODUCTION

In order to realize governmental target rates of installed wind power in The Netherlands, the erection of a considerable number of large-size wind turbines (> 500 kW) and wind farms is pursued. It became clear quite soon that these objectives pose problems in a densely populated country as The Netherlands because of the allocation of sites being severely hindered by the expected noise nuisance experienced by neighbouring residents. This problem is enhanced by the scaling up, leading to the application of larger wind turbines having higher tip speeds. As a consequence a significant increase in the level of the flow-related noise is experienced.
In 1984 the modelling of the flow-related noise of wind turbines was taken in hand at the National Aerospace Laboratory [1]. It was intended to create a design tool, capable of reliably predicting the influence of the turbine dimensions and its operating conditions on the noise emission. The first version of the prediction code was completed by 1986. This code, called RHOAK, is mainly based on the Grosveld [2] description of the source mechanisms. It was developed for predicting the noise spectra (25 Hz - 10 kHz) at any position in the geometrical near-field (within several rotor diameters) of a wind turbine. Besides the site roughness, the operating conditions and the dimensions of the wind turbine, accurate aerodynamic data for the boundary layer thicknesses and the lift and drag coefficients are required as input.
A first validation of this code took place in 1987 by spectra obtained from measurements at the 25 m HAWT of the Energy Research Centre (ECN), leading to a second, adjusted, version of the code. Since there was still doubt about the general applicability of the code it was decided to define a supplementary program. The goal of this program was to obtain rotor noise spectra of three different turbines, within a large range of operating conditions. The measurements were performed during the last months of 1990 and the first half of 1991, leading to an improved version of the code in 1992.

2. SELECTION AND DESCRIPTION OF THE TURBINES

For the selection of the wind turbines and the sites a large number of criteria were defined. Most of all, the turbines should offer a wide range of operating conditions and differ as much as possible in rotor dimensions. Further, much attention was paid to the opportunities to record all conditions.
Finally, the NEWECS-45 in Medemblik (in the province of Noord-Holland), the WPS-30 in the Oosterbierum wind farm and the PI-250 in the Herbayum wind farm (both in the province of Friesland) were selected. All turbines are pitch-controlled and of the up-wind type. Table 1 summarizes most geometrical data of the turbines and the sites.

Table 1: Survey of turbine and site data.

WT type	B (-)	h_0 (m)	D_R (m)	h_{HUB} (m)	P_{MAX} (MW)
NEWECS-45	2	0.01	45	60	1.0
WPS-30	3	0.01	30	35	0.35
PI-250	3	0.1	23	30	0.2

3. DESCRIPTION OF THE TEST PROGRAM

3.1 Equipment

Since the RHOAK-code is developed for the prediction

of rotor noise, equipment offering the opportunity to distinguish this noise from the noise emanating from the nacelle was preferred. The high resolution synthetic acoustic antenna (SYNTACAN) developed and owned by the Delft University of Technology [3], is a very suitable tool for directional noise immission measurements. This antenna consists of a sparse array of 32 microphones. It covers the range from 90 to 1400 Hz with a resolving power (based on the -3 dB points) of 1.5°. The noise sources are presented in a polar diagram. By integrating the levels over the angle of noise incidence, the immission levels of different noise sources can be determined accurately.
An acoustic parabola offers the opportunity to discriminate noise sources at higher frequencies. A 1.5 m diameter acoustic parabola, owned by STORK FDO B.V., provides the opportunity to measure the contribution of the nacelle for frequencies higher than 800 Hz. The potential (angular resolution and gain) is highly dependent on the frequency of the noise to be measured. The parabola is therefore used in combination with a ground board microphone serving as a reference. The application of ground boards is very usual when performing noise measurements under windy conditions. The Institute of Applied Physics (TNO-TPD) owns four ground boards, each board containing two flush-mounted microphones. This provides the opportunity to suppress the influence of the wind-induced noise by application of a correlation technique when post-processing the recorded signals.

3.2 Measuring method

After the wind direction was determined, the SYNTACAN and the parabola were placed at an exactly up-wind position of the turbine tower. Two ground boards were placed at opposite sides, in the rotor plane, and two boards were placed right up-wind and right down-wind of the tower.
A sensor was mounted on the turbine brake disc providing a trigger signal. This enabled the SYNTACAN to obtain data within a short time gate in which the rotor blades were acoustically observed in a nearly fixed position.
For calibrating the parabola, a noise source radiating pink noise was hoisted up as close as possible to the nacelle. The noise of this source was measured with the acoustic parabola and the ground board, placed as close as possible to each other. During all measurements the parabola was optically directed towards the middle of the nacelle.
In the case of measurements being taken from a wind turbine in a wind farm (Oosterbierum and Herbayum) all other turbines were stopped by turning their rotors away from the wind direction.
Measurements were taken during a one hour period. The signals of the eight ground board microphones and the parabola microphone were recorded continuously. Within this time period, SYNTACAN was able to perform about 20 acquisitions. During one day, recordings representing 4 to 6 series, each lasting one hour, were taken.
In all cases, the wind speed and wind direction, the RPM, the yaw angle and the blade angle could be monitored continuously. These conditions were recorded every 20 s.

3.3 Operating conditions

In all cases it was tried to select a dry day in which the wind speeds were forecasted to be between 7 and 12 m/s. During two days the wind speed was fairly high causing the turbine to operate at the maximum rated power output. In these cases the blade pitch angle is increased (relative to the rotor plane) leading to a lower torque and a so to a limited output. A survey of the conditions is given in table 2a.

Table 2a: Survey of wind and turbine conditions.

WT type	wind speed (m/s)	RPM (1/min)	ϑ_P (degrees)
NEWECS-45	6.2 - 14.6	32 - 44	2.3 - 10.0
WPS-30	6.2 - 13.3	29 - 44	2.0 - 9.0
PI-250	5.5 - 13.0	42.3	2.0

The ranges of some derived conditions are shown in table 2b. It will be shown that the application of these conditions facilitates the determination of the relations between the measured noise spectra and the flow parameters.

Table 2b: Summary of derived conditions.

WT type	λ (-)	U_{REL} (m/s)	α_{85} (degrees)
NEWECS-45	6.8 - 12.4	77 - 104	2.0 - 6.1
WPS-30	4.9 - 8.0	45 - 69	4.5 - 10.0
PI-250	3.8 - 9.1	51 - 53	10.8 - 15.5

4. DATA ANALYSIS

The absolute levels of the parabola nacelle noise measurements were determined using the transfer function obtained from the pink noise source measurements. Because almost no rotor noise emanates from the inner parts of the rotor, the spectra determined were considered to represent mechanical (nacelle) noise only. These levels were subtracted from the total noise levels obtained from the ground board measurements in order to get the spectra for the rotor noise.
The SYNTACAN spectra were determined by integrating the levels over the angle of rotor (or nacelle) noise incidence.
In the frequency range where both SYNTACAN and the parabola provided levels for the nacelle noise, good agreement (within 3 dB) was found.
All recorded microphone signals were averaged over a time period of 20 s leading to more than 1800 spectra per hour of measurement. For the determination of the rotor noise levels at the ground board positions in the rotor plane, it was assumed that the radiation of the nacelle is omni-directional.
After the spectra of the rotor noise at all positions were determined, the averaging of all reliable spectra was carried out. Therefore, a limited number of intervals, representing equal conditions and mostly containing a large number of spectra, were selected. The width of these intervals is determined by the accuracy of the measurements and the recorded conditions. For example, all spectra determined for $98 \leq U_{REL} < 108$, $6.9 \leq \lambda < 7.6$ and $4.3 \leq \alpha_{85} < 5.0$ were averaged, giving one NEWECS-45 spectrum. This procedure provided 17 intervals, giving the same number of averaged spectra (for one position). In this way it turned out to be possible to use all spectra for comparison with the predictions as will be described in 5.4.

5. RESULTS

5.1 Rotational speed

In the RHOAK code it is assumed that the emitted sound energy levels are strongly related to the flow speed relative to the blade. For the in-flow turbulence noise, mostly contributing to the lower frequency part of the spectrum, a fourth power velocity dependency is used. The energy of the (high frequency) noise, caused by the interaction between the boundary layer and the trailing edge, is taken to be proportional to the fifth power of the relative flow speed. The dependency on the flow speed of all turbines was investigated for several frequencies. In fig. 1 the measured rotor noise levels as a function of the relative tip speed are shown for two frequencies. Although a significant scatter can be observed, the general trend shows that the assumed relations are confirmed to a reasonable extent.

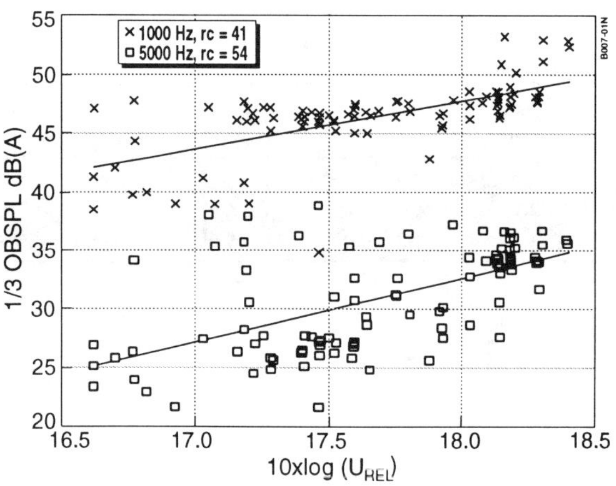

Figure 1: Measured rotor noise 1/3-octave band levels as a function of the blade tip speed

5.2 Angle of attack

The angle of attack is not directly modeled in the source terms of the code noise mechanisms. As a start of every computation the induced velocities are calculated by using blade element theory. The effective angle of attack, being defined as the angle between the chord line and the sum vector of the relative velocity and the induced velocity, is subsequently calculated. The boundary layer thickness, δ, depending on the instantaneous Reynolds number (based on the chord length), Re_C, and effective angle of attack, is used for the computation of the level and the peak-frequency of the trailing edge noise.

The frequencies giving the highest levels are modeled in the second version of the RHOAK code by a Strouhal number (Str) based on the boundary layer thickness and the relative velocity. The value for the Strouhal number giving the maximum levels was taken to be 0.4. From the results of the measurements it was however shown that for large angles of attack ($> 10°$) a better agreement is found for Strouhal numbers based on the displacement thickness of the boundary layer (δ^*). This can possibly be explained by boundary layer separation at the trailing edge, leading to a deflected ratio between the boundary layer thickness (thickness defined by the distance to the blade surface where 99% of the main flow velocity is attained) and the boundary layer displacement thickness. The results are presented in fig. 2.

For all conditions, the Strouhal numbers for the measured peak-frequencies of the NEWECS-45 and the WPS-30 were found to be concentrated between 0.1 and 0.2. These values are quite consistent with values resulting from previous studies of broadband trailing edge noise [4]. The spectra of

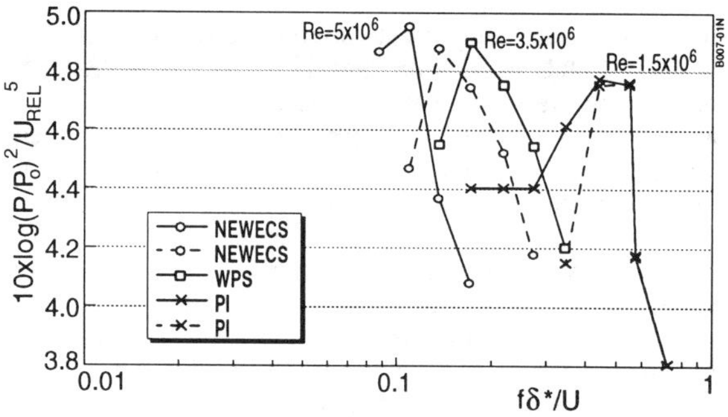

Figure 2: Measured maximum levels of trailing edge noise

the PI-250 showed severe deviations for almost all conditions. This is ascribed to the boundary layer at the suction side of the blade being partially laminar. It is outlined by Blake [5, pp. 762-767] that, in case $Re_C < 2x10^6$, the development of an unstable boundary layer and coherent vortex shedding at the trailing edge, finally leading to tonal noise, is very likely to occur. Therefore, the highest level in the PI-spectra is assumed to originate from coherent vortex shedding on some parts of the blade rather than from turbulent boundary layer trailing edge noise.

5.3 Directivity

It was demonstrated by the directivity measurements that almost no difference exists between the levels at positions right in front of (up-wind) or right behind (down-wind) the turbine. In the rotor plane significant differences (up to 10 dB) between the side of the descending blades and the side of the ascending blades were found.

These differences were measured for the higher frequencies, only. It is described in [6] that most of the trailing edge noise is radiated towards the nose of the blade in case the wave length is much smaller than the chord of the blade. It is indeed shown by the measurements that a significant increase in the levels at the descending side occurs in case the ratio of chord length (at 85% span) and the wave length is over 4.

5.4 Comparison with predictions

As expected from the agreement observed between the modeled and the measured dependency on the flow, a suitable agreement for the height of the levels was found. However, the prediction of the spectrum shape yielded some unsatisfactory results. This is particularly valid for the extent to which the level rate increases at the lower frequencies and decreases at the higher frequencies. As expected, the agreement for the high frequency part of the PI-250 spectra was in almost all cases very poor.

5.5 Adjustments

After a detailed comparison for all conditions was made, it became clear that several adjustments were necessary to obtain a code capable of reliably predicting the rotor noise spectra of a wind turbine. Because of the good consistency between the measured deviations and the physical phenomena explaining these deviations, it was expected that the adjustments would provide a general applicable code.

The adjustments made are purely based on the spectra and the relations obtained from the described validation measurements. The spectra from the previous validation (at the 25 m HAWT of the Energy Research Centre) were considered as well.

The most important adjustment consisted of the Strouhal number for broadband trailing edge noise now being based

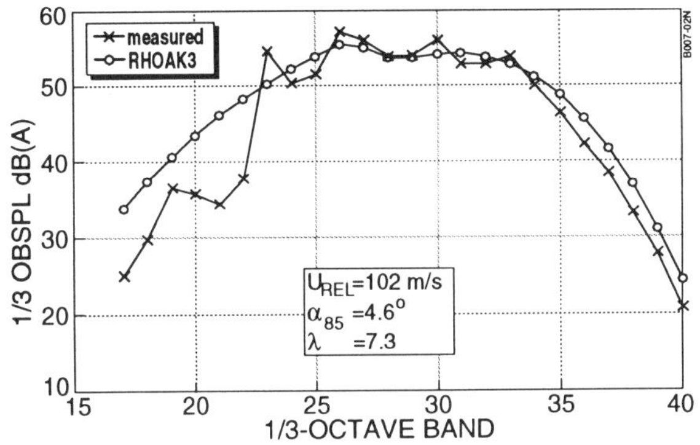

Figure 3: Comparison of predicted and measured NEWECS-45 rotor noise spectra, using RHOAK3

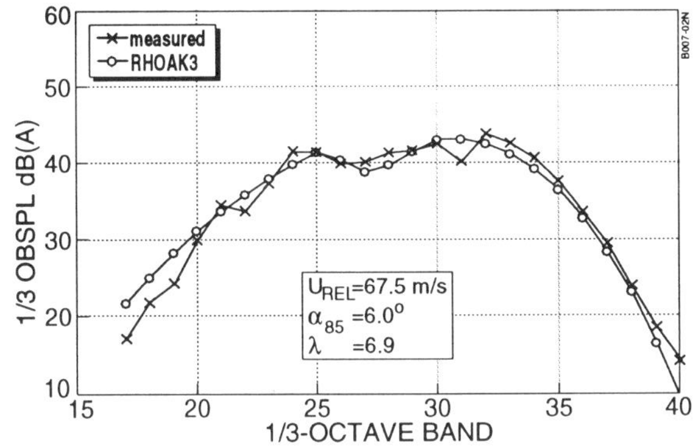

Figure 4: Comparison of predicted and measured WPS-30 rotor noise spectra, using RHOAK3

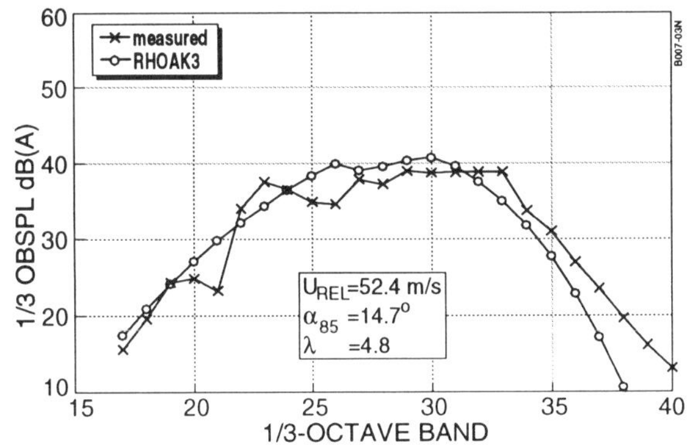

Figure 5: Comparison of predicted and measured PI-250 rotor noise spectra, using RHOAK3

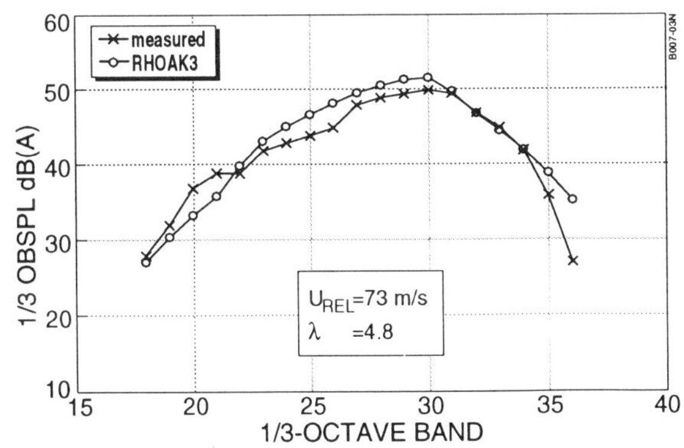

Figure 6: Comparison of predicted and measured 25 m HAWT rotor noise spectra, using RHOAK3

on the displacement thickness of the boundary layer and using a value of 0.15. Further the rate of level increase for the low frequency part and the rate of decrease for the high frequency part of the spectrum was brought in accordance with the rates observed.
Finally, all results obtained from the directivity measurements which showed consistency for most conditions, were included. In fig. 3 to fig. 6 the results of the adjusted RHOAK code (RHOAK3) are presented. For all turbines, the deviations of the predicted levels are mainly within a few dB's of the measured levels, except for the higher frequency part of the PI-250 spectrum where the agreement is rather poor.

6. RESULTS AND CONCLUSIONS

It is shown that the rotor noise of three different wind turbines could be accurately measured, separately from the mechanical noise emanating from the nacelle, by application of several selective techniques.

The assumed fourth to fifth power dependency of the flow related noise on the blade rotational speed was demonstrated by the results of the measurements over a large range of conditions.

The peak-frequencies of the broadband trailing edge noise are proved to occur within a Strouhal number range of 0.1 to 0.2. In this case the Strouhal number has to be based on the displacement thickness of the boundary layer.

In the case of the Reynolds number (based on the chord) smaller than $2x10^6$ on large parts (of the outer section) of the blades, tonal noise is very likely to occur, its frequency being related to the thickness of the trailing edge.

In case the wave length of the radiated rotor noise is at least four times smaller than the chord length of the blade (at 85% span), a directivity of the noise is observed.

As a consequence of the deviations between the predicted levels and the measured levels, the code had to be adjusted. It is now felt that the adjusted RHOAK3 code will reliably predict the rotor noise spectra of medium-sized (> 200 kW) to large (> 500 kW) wind turbines for a large range of conditions. The deviation in the predicted over-all levels is expected to be within 3 dB(A) of the measured levels.

7. ACKNOWLEDGEMENT

This work was supported by the Netherlands Agency for Energy and Environment (NOVEM).

8. REFERENCES

[1] Wolf, W.B. de, Aërodynamisch geluid van windturbines (in Dutch), NLR MP 87004 U, 1987.
[2] Grosveld F.W., Prediction of broadband noise from horizontal axis wind turbines, J. Propulsion, 1(4), 1985, pp. 292-299.
[3] Boone, M.M., Design and development of a synthetic acoustic antenna for highly directional sound measurements (Thesis, Delft University of Technology), 1987.
[4] Brooks, T.F., Hodgson T.H., Trailing edge noise prediction from measured surface pressures, J. Sound & Vibration, 78(1), 1981, pp. 69-117.
[5] Blake, W.K., Mechanics of flow-induced sound and vibration, Academic Press, Londen, 1986.
[6] Amiet, R.K., Noise due to turbulent flow past a trailing edge, J. Sound & Vibration, 47(1), 1976, pp. 387-393.

Aerodynamic Noise from Wind Turbines

P.A. Dunbabin
University of Edinburgh, Department of Physics
Mayfield Road
Edinburgh EH9 3JZ
United Kingdom

January 28, 1993

1 Abstract

The aim of this work was threefold:

- to examine and compare noise data from a variety of machines,
- to extract from these comparisons information about the relative importance of tip speed and angle of attack in determining noise emission,
- to develop a noise prediction code which can be used for any machine geometry.

Sound pressure level data from 12 wind turbines of different designs have been examined and compared to isolate the effects of two parameters, namely:

1. Air speed at the tip
2. Angle of attack at the tip

Tip air speed was found to have a strong influence on the intensity of low frequency sound emission (at or below 200 Hz). Comparisons of data indicate that Sound Pressure Level (SPL) is proportional to V^{5-7} in this range. SPL's of frequencies between 250 and 500 Hz seem to be proportional to V^{4-5}, while SPL's of high frequencies seem to be almost independent of tip air speed, the deciding factors being angle of attack and tip shape.

The angle of attack in the tip region was found to exert an influence over the intensity of medium to high frequency sound. However, the effect of angle of attack is much less clear than that of air speed at the tip, and we plan to carry out further controlled experiments to clarify this point.

An Aerodynamic Noise Prediction Code has been written which takes into account the detailed design of the rotor. It gives a good first estimate of noise emission for a variety of wind turbine designs, and we believe it has good potential as a design tool.

Comparisons of the Edinburgh and ISVR codes are given. The version of the ISVR code used the 1987 version, which may have been updated since. It is clear that the ISVR code is better for small, stall regulated machines, (less than about 100 kW), while the Edinburgh code is generally better for larger ones.

1 Comparison of Experimental Sound Spectra from Twelve Wind Turbines

In order to facilitate comparison of data, all sound data were normalised:

1. All data were expressed as third octave, linearly weighted sound pressure levels (eg. including spectra recorded in full octave bands).
2. Where appropriate, 6 dB was subtracted to account for reflection off a hard board.
3. Where necessary, background noise was subtracted logarithmically.
4. Data were normalised to a standard distance of 1.5 rotor diameters as measured from the hub.
5. Atmospheric absorption was calculated using the Glegg program, and found to be minimal at all frequencies, except for the largest machine, A.
6. For several of the machines, the wind speeds were measured at a height of 10 m, rather than at hub height. These values were extrapolated to hub height.
7. Turbines G and L are of interest, as mechanical noise was minimised by the use of anechoic lining. These turbines are quieter than the others at frequencies of 4 kHz and above.

1.1 Effect of Tip Air Speed

Data from machines with similar tip angles of attack but different tip speeds were compared, in order to assess the influence of tip speed in determining noise emission. In the first instance, the effects of chord were ignored. Angles of incidence were calculated using a standard blade element code, taking pitch schedules into account where necessary.

Tip air speed is the dominant factor at low frequencies (less than 630 Hz). Sound Pressure level at low frequencies around 100 Hz is found to scale with V^{5-7}. At higher frequencies, around 250 Hz Sound Pressure Level scales with V^{4-5}.

1.2 Effect of Blade Incidence

To isolate the dependence on incidence α, normalised spectra from a single machine taken at different windspeeds were compared. The average value of the angle of attack over the outermost 10% of the blade was calculated.

Sound Pressure Levels of intermediate frequencies (1–2 kHz) appear to increase with increasing angle of attack. However, measurement errors are proportionately much higher, for two reasons:

1. the wind noise on the microphone varies strongly with wind speed
2. in general, turbulence also increases with increasing wind speed

Thus this result is uncertain. One experiment which could clarify this point would be measure noise emission from a single machine operating successively under stall and pitch control.

2 The Edinburgh Noise Prediction Code

This code is designed to predict the intensity and spectrum of aerodynamic sound emitted from any given design of horizontal axis wind turbine under any reasonable operating conditions. The output of the program is a noise spectrum at a given observer location, expressed as Sound Pressure Level in third octaves.

2.1 Description

The prediction code consists of two parts:

1. An aerodynamic performance prediction code, which divides the blade into a number of radial elements (typically 30), and calculates the angle of attack, local air velocity, Reynolds number, and lift and drag coefficients at each.

2. A sound prediction code, which is applied at each blade element, and includes calculations for both aerofoil self-noise and noise due to interaction of the blade with inflowing turbulence. This part of the program directly uses the results of the aerodynamic performance code described above.

The basic information required to run the program is:

1. Blade plan geometry and twist distribution.

2. Aerofoil sectional data: lift, drag and moment coefficients tabulated against incidence.

3. Wind speed.

4. Rotor rotational speed.

Figure 5: Turbine L
Effect of Blade Incidence

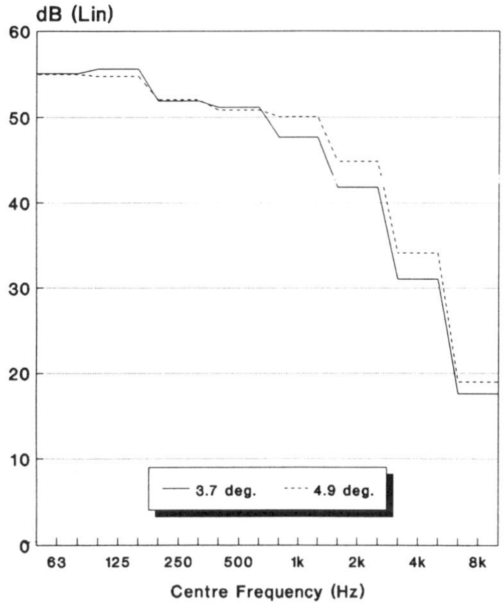

Other variables may be included, for example, yaw angle, rotor coning angle, wind shear and tower shadow, and pitch control schedule.

2.2 Aerofoil Noise

Five noise mechanisms are modelled:

1. Inflow turbulence noise

2. Noise from turbulence in the boundary layer as it passes over the trailing edge. Suction and pressure boundary layers were considered separately.

3. Noise from separated flow.

4. Noise from vortex shedding at the tip (both round and flat tips were tested).

5. Noise due to vortex shedding at a blunt trailing edge.

The inflow turbulence noise calculation is based on work by Grosveld and Glegg *et al.*, with three important differences:

1. noise is assumed to be due to atmospheric, not blade, turbulence

2. noise is calculated from aerofoil pressure data instead of assuming a 2π lift slope

3. Amiet's noise cancellation factor is not included as it is not appropriate to rotating blades.

The remaining noise sources were calculated using an empirical model by Marcolini *et al.* of NASA. This model was chosen, as many theoretical models appear to overestimate the influence of tip speed on high frequency noise emission.

Figure 1: Machines A and B

Comparison of tip speed at similar tip incidence (A=5.3, B=5.3)

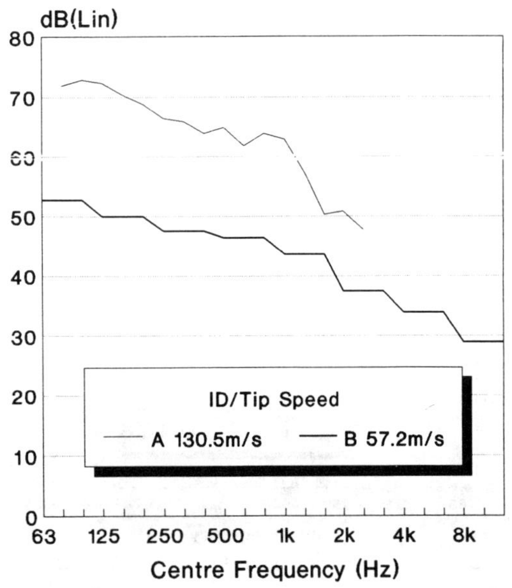

Figure 2: MACHINES A, C, L

Comparison of tip speed at similar tip incidence (A=4.3, C=4.1, L=3.7)

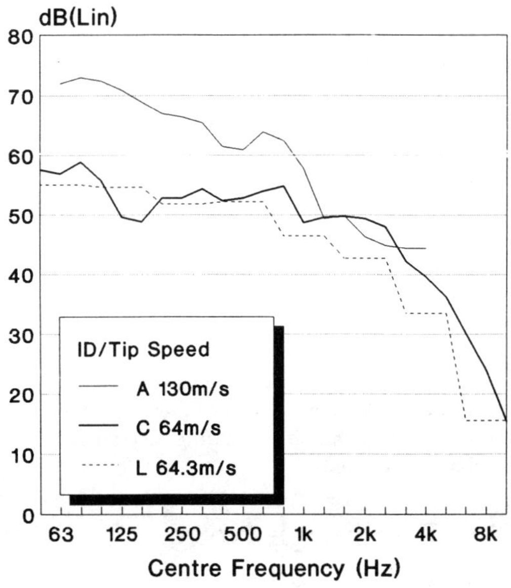

Comparisons of Experimental Data to Illustrate the Influence of Tip Air Speed on Noise Emission

Figure 3: MACHINES D, H

Comparison of tip speed at similar tip incidence (D=6.1, H=6.1)

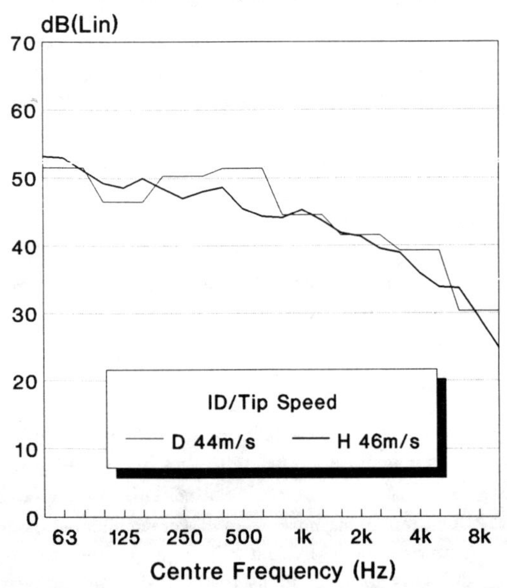

Figure 4: MACHINES E, F

Comparison of tip speed at similar tip incidence (E=6.7, F=6.6)

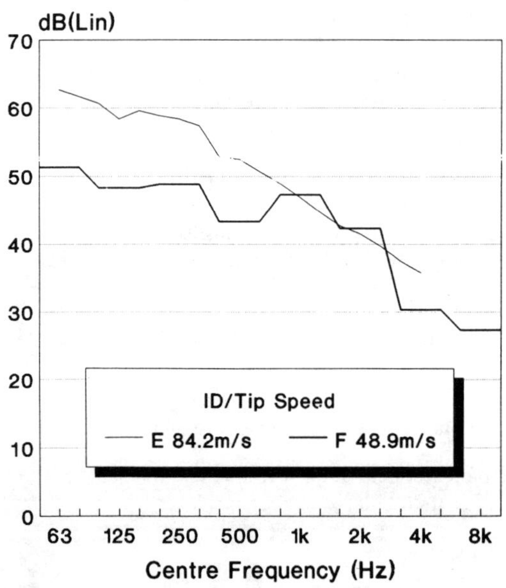

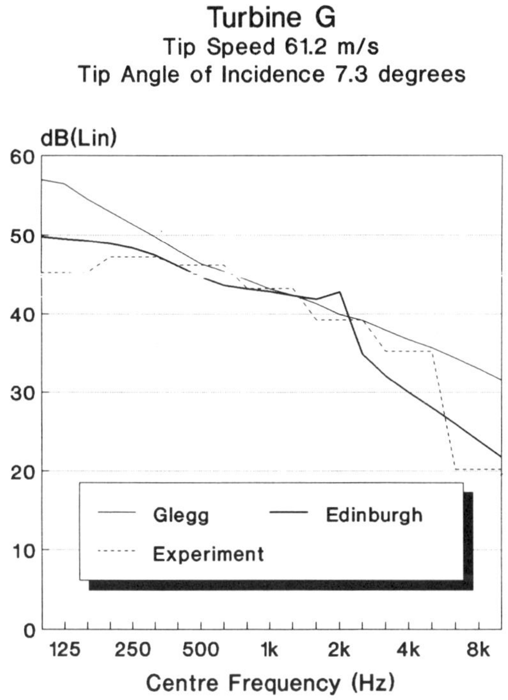

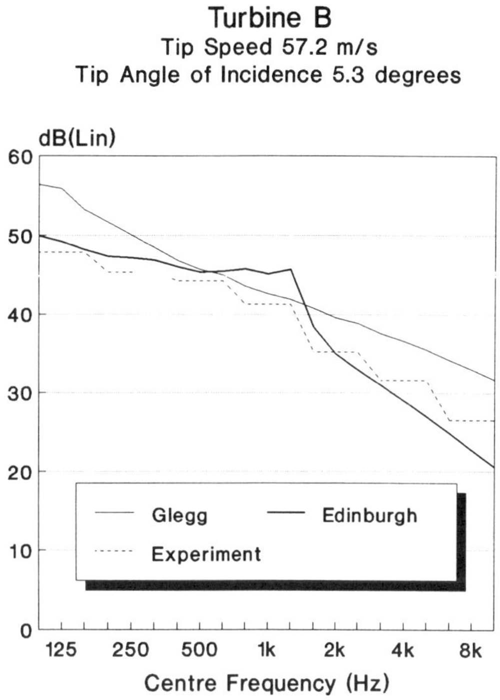

Figures 6-9

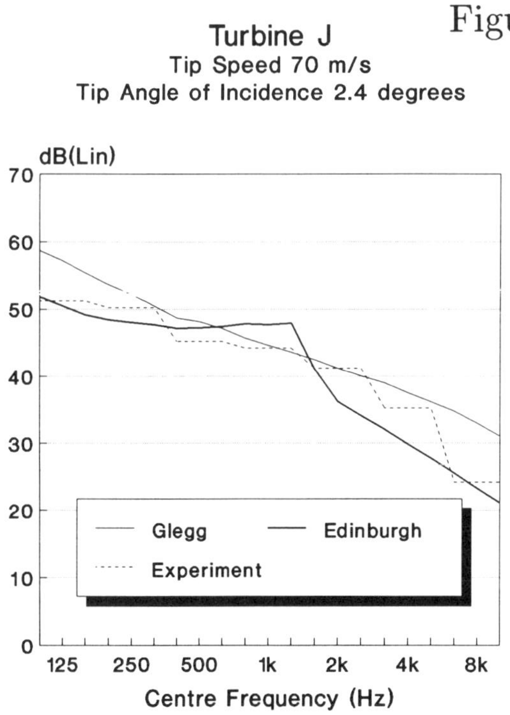

3 Results

Comparisons of experimental data with predictions made by both the Edinburgh and Glegg codes are shown above. It appears that the Edinburgh code is more accurate for medium to high power machines, while the Glegg code is better for smaller stall regulated ones. The peak due to trailing edge noise is not always predicted at the frequency. This may be due to an inaccurate estimate of the trailing edge bluntness, as some turbine manufacturers were unable to provide this data.

References

[1] Ljunggren, S., "Acoustic Measurement of Noise Emission from Wind Energy Conversion Systems", International Energy Agency Programme, 1984.

[2] Glegg, Baxter, and Glendinning, "Prediction of Broadband Noise From Wind Turbines", *Journal of Sound and Vibration*, 1987, pp. 217–239.

[3] Brooks, Pope, and Marcolini, "Airfoil Self Noise Prediction", *NASA Rpt. 1218*, July 1989.

AERODYNAMIC NOISE FROM WIND TURBINES

J. Jakobsen
Danish Acoustical Institute
Akademivej, Building 356
DK-2800 Lyngby, DENMARK
Phone: +45 45 93 12 11
Fax: +45 45 93 19 90

B. Andersen
dk-TEKNIK
Gladsaxe Møllevej 15
DK-2860 Søborg, DENMARK
Phone: +45 39 69 65 11
Fax: +45 39 69 60 02

ABSTRACT: The aerodynamic noise from wind turbine generators has been investigated in this project which was mutually supported by EEC (JOUR-CT 90-0107) and the Danish Ministry of Energy. A measurement series with focus on tip noise was carried out, using a specially developed measurement technique. It was found that certain tip shapes caused an increase of the total noise, and one shape, resulting in a significant decrease of the total rotor noise, was identified. A number of supplementary results and observations have been made, and the outlines of a general qualitative rotor noise prediction method are given.

1 INTRODUCTION

1.1 Noise from wind turbines

The environmental noise is at present the most important hindrance to the exploitation of wind power. The development of low noise wind turbines is neccesary to utilize many sites in densely populated areas. At the same time there is an increasing public demand for quiet wind power.

The noise from wind turbines is for convenience regarded as two components: machinery noise and aerodynamic noise. The machinery noise originates from the machinery components, usually the main part is gear box noise, and is radiated through the nacelle walls (and in particular through openings) or - as structure borne noise - from the tower, the nacelle or even from the rotor blades. The generation and control of machinery noise is quite well understood. It is technically possible to manufacture wind turbines without significant contributions from machinery noise by consistent application of usual noise control strategies.

1.2 Aerodynamic noise

The aerodynamic noise is responsible for the main part of the total noise from most wind turbines. Even for wind turbines with noticeable machinery noise, the aerodynamic noise contributes to most of the A-weighted sound power, but due to its broad-band character it is evaluated less loud than the machinery noise containing tone components. The knowledge about aerodynamic noise from wind turbines is incomplete and rather simple. Emperical and semi-empirical models exist [1,2], using only few parameters and giving results which generally deviate considerably from corresponding measurement results from other wind turbines than those used for development of the models. The more general models available for aerodynamic noise from propellers and helicopter rotors have not yet been well adapted to the range of parameters relevant for wind turbine rotors, but work is in progress [3].

The scope of the present project was to investigate more deeply the aerodynamic noise from wind turbines and its relation to different design and operational parameters. Specifically the tip noise has been examined by an extensive measurement series.

2. TIP NOISE MEASUREMENTS

A series of measurements was made where the tips of a 300 kW Bonus wind turbine were removed and 6 different sets of test tips were mounted in turn by implanted mounting adaptors. Supplementary noise contol had been carried out to reduce the machinery noise of the test wind turbine. Following an inconclusive series of pilot measurements, a measurement series with better manufactured tips and using a newly developed measurement technique was performed.

2.1 Test tips

The tip designs chosen for the experiments had been discussed in a project group. Tip shapes giving very different aerodynamic conditions were chosen, as is illustrated in Figure 1. Tip M ("tip torpedo") was chosen to verify very promising results from a pilot study [4], and tip O ("Ogee") was inspired from tests with quiet tips for helicopter rotors. Tip E was designed with an uniform rounding of the pressure side, wich should give much the same flow conditions as the "torpedo", and tips I, J and R were different designs aiming at combining the leading edge wake and the tip wake like with a delta wing.

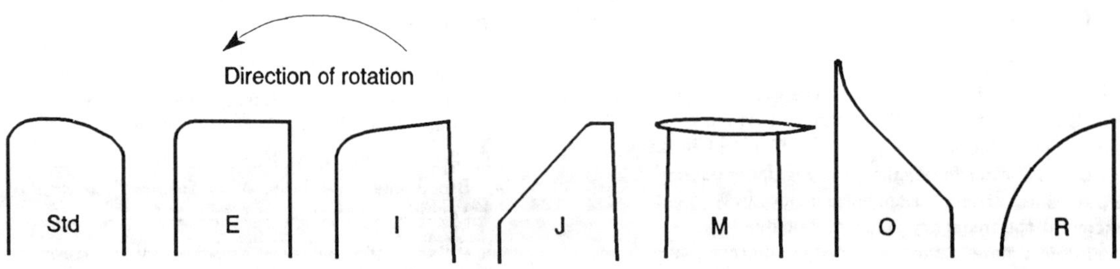

Figure 1 Sketch of the shape of the test tips and the standard tips of the LM 17 m rotor.

2.2 Measurement technique

For the main part of the experiments a newly developed measurement technique was applied. The total noise from the wind turbine was measured with a microphone on a ground board, placed in the downwind direction of the wind turbine in a distance approximately equal to the hub height. Simultaneously with the noise measurements the wind speed was measured at 10 m height in the same distance directly upwind. The noise was analyzed in 1/3-octave bands in the range 50 Hz - 10 kHz in 1-minute periods, and the results stored on a PC together with the wind data from the same periods. A regression analyses of noise level vs. wind speed was made in each frequency band, and the result at the reference speed (v_{10} = 8 m/s) was determined. At least 30 data pairs were recorded. Figure 2 shows a raw plot of the A-weighted sound power level vs. wind speed, and Figure 3 illustrates that the change of the noise per 1 m/s is not the same in the entire frequency range.

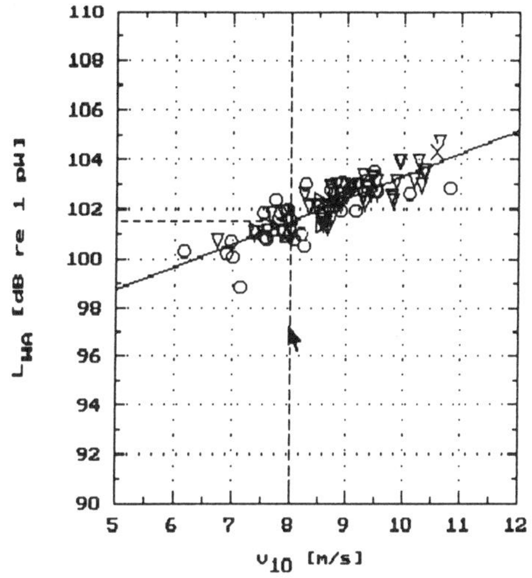

Figure 2 Example of plot of A-weighted sound power level vs. vind speed. Measurement of test tip M.

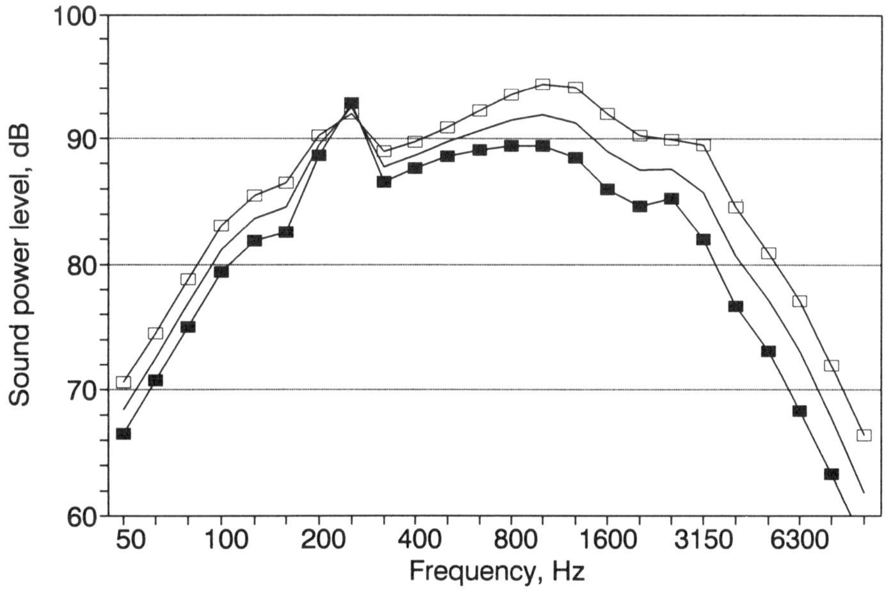

Figure 3 Spectra of A-weighted sound power level at different wind speeds, calculated from regression analysis. Measurement of test tip M.

2.3 Measurement results

The results of the tip noise measurements are shown in Table 1 as the total A-weighted sound power levels at 8 m/s. These have been corrected for the influence from machinery noise, which caused a prominent tone at 230 Hz with some influence (1-2 dB) on the total level, and correction has also been made for the slightly different rotor radii. The latter correction is found from -60 · log (R_{test}/R_{std}) and is between 0 and -1 dB. Figure 4 and 5 shows the A-weighted spectra of the sound power level at 8 m/s for the different test tips. In the figures the average of the two measurements of similar wind turbines with standard tips have been included. These spectra have not been corrected neither for tone component nor for different rotor diameters.

Test tip	Wind speed [m/s]	Tip pitch [°]	L_{WA} corr. tone	L_{WA} corr.diam.
J	6.9-10.5	0	100.3	99.7
M	6.1-10.8	0.3	100.9	100.4
I	5.5- 8.7	0.3	104.1	103.6
R	7.7-10.1	0.3	99.6	99.1
O	4.0- 7.1	0.3	96.6	95.6
O	5.6- 9.6	1.0	97.5	96.5
Std.1	6.4- 9.5	1.0	99.1	99.1
Std.2	5.4- 8.6	1.0	98.1	98.1

Table 1 Main result of tip noise measurements, indicating range of wind speeds during measurement, tip pitch and A-weighted sound power level at v_{10} = 8 m/s [dB re 1 pW] disregarding the machinery noise.

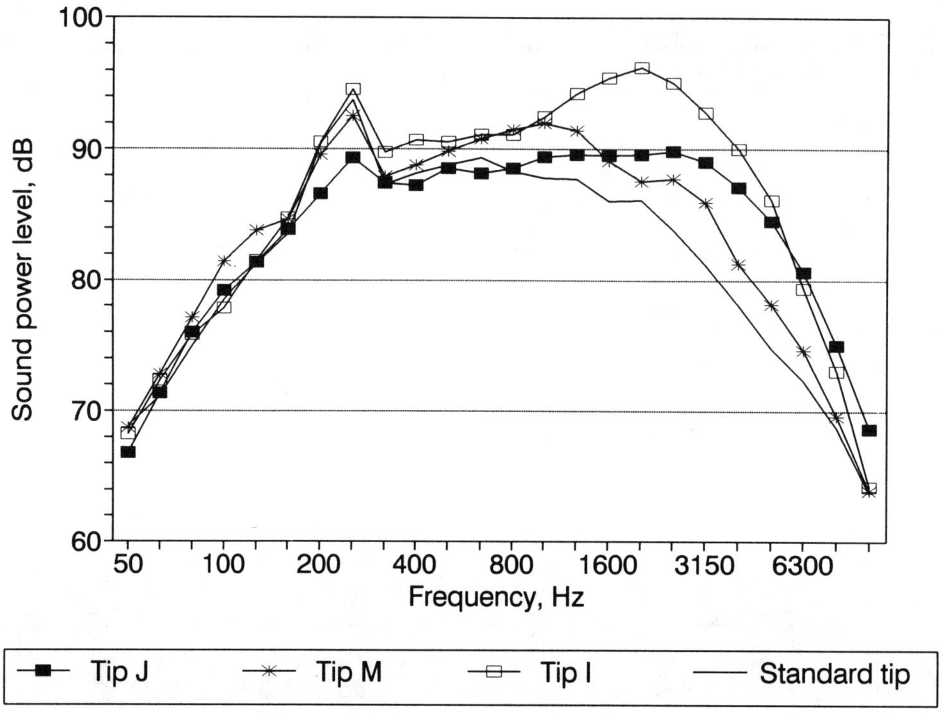

Figure 4 A-weighted sound power level of the test wind turbine at v_{10} = 8 m/s with test tips J, M and I [dB re 1 pW].

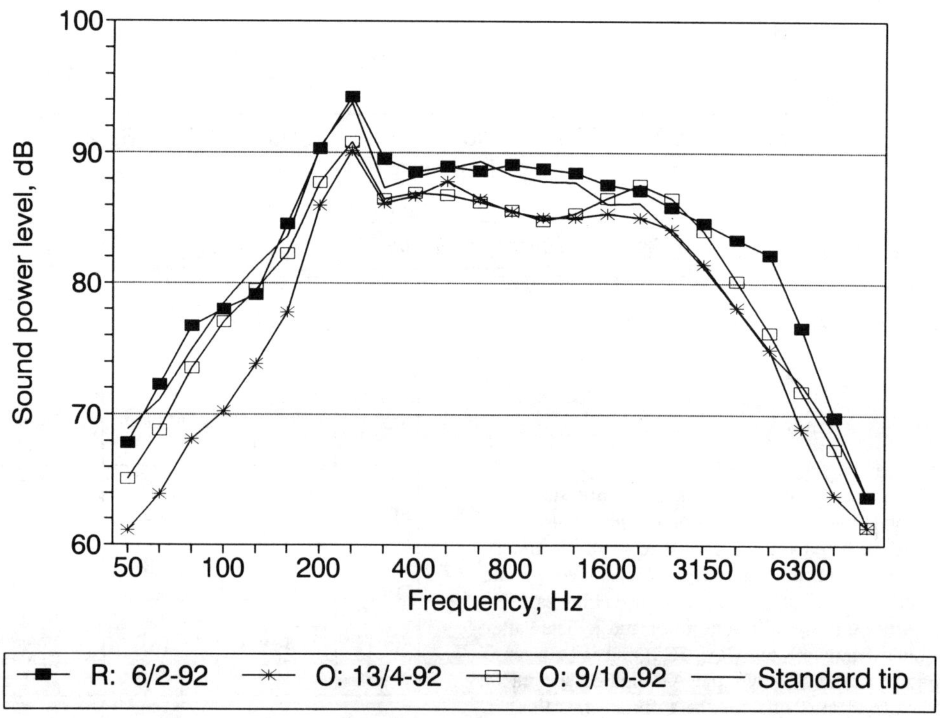

Figure 5 A-weighted sound power level of the test wind turbine at v_{10} = 8 m/s with test tips R and O, before and after final pitch adjustment [dB re 1 pW].

2.4 Summary of tip experiments

It has been found that occasionally tip noise has a significant influence on the total noise from a wind turbine. One of the test tips ("Ogee") resulted in a reduction of the total A-weighted noise of 2-3 dB, and another set of test tips caused an increase of about 5 dB. These differences are significant because the uncertainty using the described improved measurement technique is only about 1 dB on the total A-weighted level. The remaining tips resulted in more or less the same noise as the standard tips regardless of the very different shapes and flow conditions. Differences can be clearly seen on the spectra, however.

It can be concluded that the tip shape can have a marked influence on the total noise from wind turbines. The noise generating mechanisms have not been identified, and the project's results do not allow for a tip noise model to be set up.

3. OTHER RESULTS FROM THE PROJECT

Within the frames of the project numerous measurements have been made, of which only a few have been mentioned here. Much experience has been gained, and some of it is mentioned in this section.

3.1 Near field measurements

During some of the measurements another measurement technique was tested where the microphone was placed very close to the tips. As the sound is reflected from the tube tower, causing a complex interference pattern, the microphone was placed close to the tower's surface to obtain a coherent reflection which can be corrected for.

This measurement technique would allow for individual measurement of the noise from the three tips assuming a reliable triggering signal, and would thus speed up the measurements and reduce the costs considerably.

The noise levels measured at tip height had no obvious connection with the noise levels measured at ground level at a distance. Further it was found that the divergence between the noise levels from the different test tips showed very differently at tip height from the corresponding divergence found at a distance. Thus the near field measurements gave results different from those at ground level at a distance, and it was decided to abandon the near field measurements.

The cause for this divergency is not obvious. The Doppler-shift of the fast moving sound source close to the microphone cannot fully explain the differences. A more probable explanation involves an expected pronounced directivity in the near field of the aerodynamic noise sources responsible for the tip noise.

3.2 Slits, holes and imperfections

On several occasions it was found that even very small imperfections of the blade shape near the tip caused a pronounced change of the aerodynamic noise. An entire pilot measurement series showed inconclusive mainly due to insufficient fitting between the test tips and the rest of the rotor blades. Discontinuities about the size of 1 mm caused different whisteling sounds from the three blades.

The rotor was supplied with an aerodynamic brake operating by turning of the tips. At times the pivotable tips did not engage fully, lieving a small leak through the blade about 1.5 m from the tip. Also this slit has resulted in a significant noise.

The test tips were fastened through mounting holes approx. 50 mm which were subsequently sealed with tape. When the tape had fallen off due to age, an intense whisteling noise occurred.

The first measurement series with the O-tips was made in light wind only, and the results showed some differing tendencies. Following the measurements were repeated during stronger winds, and it showed that the tip pitch had been adjusted in the meantime. These results seem to contradict a more general finding that the overall noise level increases about 1 dB for each 1 degree increase of pitch.

4. ROTOR NOISE MODEL

The results from the tip measurement series together with experience from several other investigations of noise from wind turbines has led to outline a very qualitative rotor noise model. It must be emphasized that this model is based only on systematical observations, and as so may reflect different operating conditions of different wind turbines. The purpose of the model is mainly to serve as a reference for further observations, and it is in no way a substitution for the recent efforts to establish a theoretically based rotor noise model.

Following parameters have been considered:

- Tip speed: Vast importance, probably about $60 \cdot \log(v)$
- Trailing edge thickness: Very important. Good results have been obtained by shapening of trailing edges in many cases. Probably an optimum thickness is in the range 1-2 mm.
- Trailing edge shape: Unknown
- Cord: Uncertain, not very large influence
- Tip pitch: probably about 1 dB/degree
- Tip shape: Important, but unpredictable. Up to 5 dB higher or lower than "usual" tip shapes may be expected.
- Wind speed: About 1 dB per m/s, at least at moderate wind speeds where deep stall does not occur.
- Imperfections: Very important, if they occur in the tip region where the speed is high. The shape must be smooth and even; absolutely no holes or slits through the blade. Moveable parts must be fully withdrawn when the turbine operates.

REFERENCES

[1] F.V. Grosveld, "Prediction of Broadband Noise from Horizontal Axis Wind Turbines", Journal of Propulsion 1 (1985), p. 292-299.

[2] U. Michel, "Berechnung des Aerodynamichen Lärms von Windkraftanlagen", proceeedings DAGA 1984.

[3] M.V. Lowson, "Assessment and Prediction af Wind Turbine Noise", Flow Solutions Report 92/20, Bristol 1992.

[4] B. Andersen, "Noise Sources on Typical Danish Wind Turbines", IEA 18th meeting of experts, Petten 1989.

The project is reported in:

J. Jakobsen, B. Andersen, "Aerodynamical Noise from Wind Turbine Generators. Experiments with Modification of Full Scale Rotors", which has not yet been published.

ENGINEERING DESIGN TOOLS FOR WIND DIESEL SYSTEMS: PRESENTATION AND VALIDATION OF THE LOGISTIC MODELLING PACKAGE

D.G. Infield, A. Scotney, ERU, RAL, Oxon, UK
P. Lundsager, H. Bindner, RISO, Denmark
K. Uhlen, T. Toftevaag, EFI, Norway
J.T. Pierik, ECN, The Netherlands
L. Manninen, TKK, Finland
M. Falchetta, ENEA, Italy
V. van Dijk, Uni. Utrecht, The Netherlands

ABSTRACT: A design tools software package for wind diesel systems has been developed with support from the CEC JOULE programme. Work has been undertaken cooperatively by several European countries. The complete package includes modelling of both logistic and dynamic aspects of system operation. It has been designed to be user-friendly and runs on a PC. This paper describes the logistic modelling capabilities and validation against a range of different system configurations.

1. INTRODUCTION

The potential market for wind diesel technology is very large. Effective exploitation of this market requires the production of reliable and cost effective systems, possible only through the replication of proven design concepts. Due partially to a lack of suitable design tools, systems have been developed and installed without accurate expectations of their performance. Some early demonstration projects failed to be cost effective or were not technically viable. More attention therefore needs to be placed on performance prediction and economic assessment of systems prior to application.

The engineering design tools package, developed with support from the CEC JOULE programme, aids design and assessment of wind diesel systems. It has capabilities for overall performance prediction and dynamic and transient modelling. This paper is devoted to the performance modelling. A parallel paper [1] describes the modular dynamic model.

Wind diesel has been an active research area in a number of European countries over an extended period. System models have been developed, usually with the aim of representing the particular hardware under experiment at a given research centre. Infield et al [2] provides a review of the research undertaken and the models developed. It was not considered feasible, to develop one comprehensive model capable of adequately coping with this diversity. Instead, it was decided to create a common environment from which the various national models could be run and results interpreted. Some adaptation of the individual models was required, but the main effort was concentrated on the specification and programming of the user interface.

2. LOGISTIC MODELLING TOOLS

Each of the participants in this project supports a logistic model, table I. Individually none of the models constitute a comprehensive logistic simulation tool, however collectively they account for most system scenarios currently considered. A new global Shell acts as the model user interface, replacing the original individual simulation package user interfaces, where these existed. All pre and post simulation activity has been transferred to the Shell with the models adapted accordingly.

The Shell has been developed by RAL in consultation with the other participants. Shell specification was of central importance as this defined the parameterisation of all aspects of the modelling. The Shell is written in Turbo Pascal. Supported models have been written in a range of languages and are run as independent codes from the Shell.

Table I: Supported models

Participant	Model	References
RAL, UK	RALMOD	[3]
ECN & Utrecht Uni, The Netherlands	SOMES	[4]
RISO, Denmark	WDILOG	[5]
EFI, Norway	VINDEC	[6]
ENEA, Italy	E_WISDA	
TKK, Finland	TKKMOD	[7]

2.1. The Program Suite

A number of separate programs units collectively constitute the integrated package, figure 1. Overall control of the suite of programs is handled by the Executive, separate programs handle the pre and post simulation activity (Input and Output) and the six models are retained as individual codes. Communication between the separate program units is undertaken using standard files.

2.2. Input

A wide range of pre-simulation functions are supported, including processing of data files and specification of component and system files.

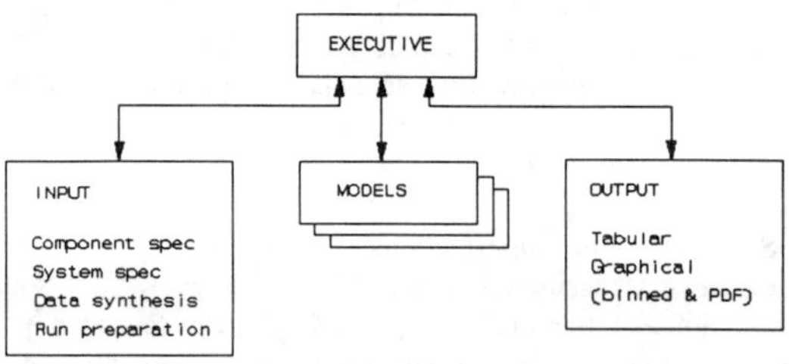

Figure 1. Logistic model - structure

2.2.1 Data

A number of functions for the processing of wind and load data files are catered for by the Shell. Three data types are identified: 'real'; actual collected time series data, 'pseudo'; time series data synthesised within the Shell, and 'stats'; statistical parameters without time series data. A standard data file format is used (one for wind and one for load) consisting of a data header, followed by the time series data (where applicable). The data header consists of overall statistical data. The format of the header is derived from the collective requirements of the six models. Data processing functions are as follows:

Convert: Using this option the user may convert a wind or load data file (time series data only - any format) to the standard Shell format. Convert creates a new file in the 'wind' or 'load' directory devoted to Shell formatted data files.

Scale: All wind and load time series data files (Shell formatted) can be scaled with a new mean and standard deviation, using this function. The user may overwrite the existing file or save the data to a new named file.

Create: This option enables the user to create new wind or load data files, of types 'stats' or 'pseudo'. In the first case the user simply enters key statistical data, which is subsequently saved in the standard header file format. The Pseudo option enables the synthesis of both wind and load data from within the shell. Basic statistical information entered by the user is used to generate a time series data file of up to 100,000 data points.

2.2.2 Components

The shell incorporates a component data base. Seven component templates are specified;

* wind turbine * diesel-generator set * flywheel
* battery * rectifier * inverter * dump load

Each component template consists of a global specification derived from the collective requirements of the six simulation models. An on-screen checking facility indicates to the user which parameters are required by each model. This assists the user in choosing data fields to be specified.

Diesel-generator or wind turbine components, may be of type 'real' or 'simplified'. In the first case the user has to make a full specification. In the second case routines are invoked to automatically calculate some of the component parameters. For the diesel-generator, the user is required to specify the diesel type (turbo-charged or naturally aspirated) and the power rating. Pre-defined characteristics are then used to calculate the fuel consumption parameters. For the wind turbine, the user must specify the power rating, cut-in, rated, and furling wind speeds (defaults are provided for the last three fields), from which the power curve is automatically calculated.

2.2.3. Systems

Once the user has defined the components required for their prospective system, they can proceed to system specification. This specification comprises the following elements:

Configuration: Eight system configurations are available; simple wind diesel, wind diesel with asynchronous flywheel, wind diesel with battery, wind diesel with DC link, wind diesel with DC link and asynchronous flywheel, wind diesel with DC link and battery, DC system and simple wind diesel with mechanically coupled flywheel.

Control strategy: Five strategies are available; continuous diesel, intermittent diesel (normal mode), intermittent diesel (cycle charge mode), diesel only (normal) and diesel only (cycle charge).

Control parameters: Depending on the strategy chosen, up to four control parameters must be specified; power hysteresis level, diesel minimum load, diesel minimum run time and exponential smoothing factor.

Components: The user selects named components via a component template list the content of which is determined by the chosen configuration.

The listed configurations and control strategies are global, representing the collective capability of the model suite. At each stage in the system specification process the on-screen checking facility indicates to the user which models will run with the selected option. System specifications are saved to named files.

2.2.4. Initiating a Simulation

With appropriate wind and load data files, component specifications and a system specification file, the user is able to initiate a simulation run. A sequence of menus enable selection of wind, load and system files, along with optional additional parameters (including: wind and load data scaling parameters and environmental data). Once again, at each stage in this process the on-screen checking facility indicates to the user which models will run with the selected option. Finally the user is presented with a list of the models compatible with the chosen simulation specification. Once the model has been chosen the simulation can be initiated.

Two additional simulation options are included within the package. Inevitably certain idiosyncratic features of specific models have not been included in the new Shell. Scaled down model specific environments have been retained to facilitate access to such features, where appropriate. Also, for those models which were developed with their own user interface, the option to run the model independently of the Shell facilities has been retained, Here software control is handed over to the model specific user interface (and the user must consult the independent model documentation to proceed).

2.3. Output

2.3.1. Output Files

Upon completion of a simulation run, software control is handed to the Output program. A new named output file is created with content derived from the current standard output and input files. The new output file is named after the system file and appended with a unique numbered file extension (the extension is incremented for each subsequent run of the same system).

The Output program may also be accessed directly from the main Input menu. In this case the user must first retrieve an existing output file.

2.3.2. Tabular Output

Three tabular summaries are available; summary of simulation inputs, short summary of simulation results and extended summary of simulation results. Options included enable the user to view or print any of these tables.

2.3.3. Graphical Output

The Shell supports a comprehensive graphing facility. Graphical output presently includes six binned data functions (wind turbine power, diesel-generator power, fuel consumption, run time, diesel starts and dump load power) and probability density functions for wind speed and load power. Graphing options provided include; re-scaling, griding, cursor (for taking spot measurements), overlays, printing and plotting.

2.4. Help Facility

The modelling software package supports a fairly comprehensive context sensitive, on-line help facility. This function operates at three levels:

Option check prompt: Use of this function depends on the context as described elsewhere in the paper.

Menu information: A hot-key opens a window containing information on the overall function of the current menu.

Option information: A hot-key opens a window containing information on the specific menu option currently highlighted.

3. MODEL VALIDATION

3.1 Purpose and Scope of Validation

Validation plays two important roles with regard to the modelling package. Firstly, it gives a level of confidence to the design engineer in its application. Secondly, it should highlight the strengths and weaknesses of the individual models available through the Shell. These have generally been developed with particular systems in mind and will not all be equally applicable to a given situation. The user will need to know which models perform best for which applications. Ideally the validation would have covered a wider range of wind diesel system designs and control algorithms. Unfortunately limitations of resource precluded study of more than two systems. Previous model validation work has demonstrated that the modelling of systems in which the diesel runs continuously is not problematic; reliable predictions can be obtained from even simple models. Difficulties arise with intermittent diesel engine operation where accurate modelling of the energy storage and controls is the main challenge.

For these reasons two systems allowing intermittent diesel operation, but with radically different approaches to energy storage were selected. One, based at RAL, utilises short term flywheel energy storage, mechanically coupled to the diesel engine via a clutch [8]. The other, operated by EFI on the Norwegian Island of Froya [9], includes longer term battery storage and a power electronic interface.

Detailed data sets were collected from the systems. Data provided by EFI covered 392 hours of uninterrupted system operation at ten minute intervals. For the RAL system, data was collected at 30 second intervals for 144 hours. The differences in data collection rates reflect the nature of these two very different systems. Measured data was supplied to all modellers together with an appropriate standard input file for each system. The results presented here are the first results of simulations made by individual modellers. The results might be improved with discussion of the suitability of the input parameters chosen.

3.2 Results

3.2.1 Validation Against EFI Data

Early on in the validation exercise ECN, together with the Utrecht University, identified some minor anomalies with the data set. It was realised that during the first 40 hours of measurement there were occasions when the wind turbine was switched off, despite wind speeds above the cut in speed. In addition, comparison of overall energy flows indicated an underestimate of average system losses. To compensate for these two effects a modified data set was created. Anomalies in the wind turbine output were dealt with by setting the wind speed to zero during these 'off' periods; this resulted in a reduction of the mean wind speed for the data set from 6.2 to 5.6 m/s. The additional system losses were compensated by adding a constant 0.858 kW to the load. A number of the participants used this modified data set (EFI-MOD), in addition to the original set supplied by EFI.

Table II compares predictions from the different models using original data as input, with the measured results. Percentage errors are indicated in brackets. Model results using the EFI-MOD data set are shown in Table III.

All models were run against the EFI data despite difficulties of representation in some cases. TKKMOD was developed for lead-acid batteries rather than the Ni-Cad ones used. In addition the control strategies available were not really appropriate. Finally, RALMOD was not developed for long term storage and so no meaningful calculation of diesel starts or storage energy flows is provided in this instance.

3.2.2 Validation Against RAL Data

Initially the supplied wind turbine characteristic was based on historic data. After the collection of the validation data set this was recognised to be no longer appropriate; significant changes had occurred to the performance of the passive pitching mechanism in the interim. A revised power curve was supplied based on the validation data. A key difficulty created by this was that the control strategy implemented was based on the previous understanding of the wind turbine performance. With the revised performance curve, the criteria for switching off the diesel would never, in an averaged sense, be satisfied. This is because the rated power of the wind turbine was less than the applied constant load plus standing losses. In practice the diesel does cycle on and off, reflecting poor control by the passive pitch mechanism of short-term variations in power.

Difficulties were experienced with the fuel flow measurement. As a result, emphasis has been put on prediction of the diesel load. Past experience has shown that, if this has been modelled accurately, calculating the correct diesel fuel consumption follows straightforwardly. Not all models were considered suitable for running on this data set.

Table IV summarises the validation results.

3.3 Discussion of validation

Examination of the results based on the original EFI data (Table II) shows a high degree of consistency between the models but poor agreement with the measured results. That comparisons are much improved when using the modified input (Table III) confirms the decision to adapt the input but indicates just how demanding the collection of validation data is in the case of complex systems, in particular when control strategies dependent on a number of different system components are involved. Diesel engine loading and hence fuel consumption is proportionally easier to calculate than dump load activity. Diesel on/off cycling, critically dependent as it is on controls and battery modelling, is hard to predict accurately. RALMOD is not intended to model battery storage and consequently makes a significant error in the calculation of the dump load. Overall, the models have performed well (generally within a few percent) for the most important quantity: diesel fuel consumption. This should give confidence to future users of the package, for systems such as this.

Interpretation of the results of validation against the RAL system is more problematic. This is mainly for the reasons discussed above. However the issues raised are of general importance. Difficulties in representing wind turbines such as the one used in this experiment should be acknowledged. Of the five models assessed in this case, four, based on the steady state wind turbine curve, calculated that the diesel would run continuously (zero or one start depending on initialisation), whilst WDILOG determined a number of on/off cycles. In this case a control parameter (minimum diesel run time) was adjusted to compensate for non-ideal modelling of the flywheel energy store, and this resulted in approximately the correct number of diesel starts. Differences are also believed to relate to the way losses in the system are represented. Diesel on/off cycling is highly sensitive to the energy balance for systems with limited energy storage (like the RAL system) and losses can be a key factor in determining this balance. Losses associated with the flywheel and the electrical generator of the diesel set are dealt with differently by the models. Cycling is

Table II: Validation Results using original EFI input data

	Measured	SOMES	VINDEC	WISDA	WDILOG	RALMOD
Wind turbine output (kWh)	4801	5785(+20)	5778(+20)	5783(+20)	5693(+19)	6331(+32)
Diesel set output (kWh)	4656	4111(-12)	3977(-15)	4510(-3)	4236(-9)	4069(-13)
Diesel run time (hours)	284	241(-15)	258(-9)	256(-10)	71(-75)	263(-7)
Dumped energy (kWh)	1261	1919(+52)	1734(+38)	2021(+60)	2021(-60)	2567(+104)
Number of diesel starts	48	12(-75)	20(-58)	14(-71)	49(+2)	-
Storage - energy in	223	338(+52)	257(+15)	253(+13)	124(-44)	-
Storage - energy out	141	255(+81)	182(+29)	206(+46)	88(-38)	-
Total fuel consumption (*l*)	1812	1566(-14)	1603(-12)	1645(-9)	1766(-3)	1582(-13)

Table III: Validation Results using EFI-MOD input data

	Measured	SOMES	VINDEC	TKKMOD	WDILOG	RALMOD
Wind turbine output (kWh)	4801	4942(+3)	4991(+4)	4684(-2)	4897(+2)	5317(+11)
Diesel set output (kWh)	4656	4524(-3)	4503(-3)	5015(+8)	4757(+2)	4732(+2)
Diesel run time (hours)	284	262(-8)	277(-2)	263(-7)	76(-73)	290(+2)
Dumped energy (kWh)	1261	1160(-8)	1135(-10)	1088(-14)	1412(+12)	1886(+50)
Number of diesel starts	48	20(-58)	23(-52)	28(-42)	45(-6)	-
Storage - energy in	223	385(+73)	279(+25)	488(+119)	151(-32)	-
Storage - energy out	141	308(+118)	197(+40)	424(+200)	107(-24)	-
Total fuel consumption (*l*)	1812	1701(-6)	1752(-3)	1765(-3)	1919(+6)	1794(-1)

Table IV: Validation results using RAL data

	Measured	SOMES	VINDEC	WDILOG	RALMOD	E-WISDA
Wind turbine output (kWh)	165	152(-8)	215(+30)	157(-5)	166(+1)	162(-2)
Diesel set output (kWh)	414	308(-26)	393(-5)	380(-8)	432(+4)	420(+1)
Diesel run time (hours)	132	144(+3)	144(+9)	93(-30)	144(+9)	144(+9)
Dumped energy (kWh)	103	7(-93)[1]	140(+36)	69(-33)	130(+26)	118(+15)
Number of diesel starts	66	1(-98)	1(-98)	63(-5)	0(-100)	0(-100)
Total fuel consumption (*l*)	-	229	214	211	223	220

[1] Dump load energy calculated by SOMES is very low because the model is unable to account for the minimum load on the diesel generator.

also sensitive to control parameter values, the dynamic representation of the wind turbine and turbulence levels. It should be stressed though that the main aim of a logistic modelling package is to predict energy flows and that such matters as diesel cycling are of secondary importance. Despite reservations there is a high level of agreement over the total diesel output and the predicted fuel consumption.

4. CONCLUSIONS AND SCOPE FOR FURTHER WORK

The Shell is substantially complete and all models have been successfully integrated with it. The package is currently undergoing testing. Comprehensive user / program documentation is in the process of being written. A general release of the package is planned for the summer, 1993.

This project has focused on the integration of existing models within a single user environment. In this way a diversity of wind diesel modelling expertise has been made readily accessible to the design engineer. The package does however have a number of significant limitations, which would be addressed in further work. A flexible approach to the specification of system architecture and control would be introduced. Wind turbine modelling would be extended to include clusters of machines. The capacity to model several diesel generators / diesel scheduling would be included. The range of component templates could be extended substantially. Extensive component libraries would be helpful. Routines for the economic analysis of systems could be considered and ideally the user interface should be standardised as a Windows application.

REFERENCES

[1] K Uhlen, et al, 'Engineering Design Tools for Wind Diesel Systems: Presentation and Validation of the Modular Dynamic model', ECWEC93, 1993.

[2] D G Infield, et al, 'Wind Diesel System Modelling and Design', Proc ECWEC90, 1990.

[3] D G Infield, et al, 'Current Progress in the Development of a Wind Diesel System for Autonomous Electricity Generation', Proc, 'Alternative Energy Systems: Electrical Integration and Utilisation', Pergamon 1985.

[4] V A P van Dijk, et al, 'SOMES A Simulation and Optimisation Model for Autonomous Energy Systems', Manual Version 2.0, (1990).

[5] P Lundsager, 'WDILOG, A Logistic Simulation Model for Wind Diesel Systems', RISO-M-2822, 1991.

[6] O Skarstein, K Uhlen, 'Design Considerations with respect to long term Diesel Savings in Wind Diesel Plants', Wind Engineering, Vol 13, No 2, 1989.

[7] L M Manninen, et al, 'Wind Diesel system analysis for northern locations'. Proc EWEC91, 1991.

[8] D G Infield, 'An Assessment of Flywheel Energy Storage as Applied to Wind diesel Systems', Wind Engineering, Vol 14, No 2, 1990.

[9] K Uhlen, et al, 'Design and Operation of the Full Scale Norwegian Wind Diesel Laboratory Model', Proc EWEC89, 1989.

AUTONOMOUS WIND/DIESEL/BATTERY-SYSTEM A TECHNOLOGY TRANSFER TO THE PEOPLE'S REPUBLIC OF CHINA

G. Cramer[1], R. Kniehl[2]

[1] SMA Regelsysteme GmbH
Hannoversche Straße 1 - 5, 3501 Niestetal/Germany
[2] CES - Consulting & Engineering Service
Kleinschmidtstraße 35, 6900 Heidelberg/Germany

ABSTRACT: The rapid growth of the Chinese economy has resulted in a great demand for electrification even of remote regions. The backlog is high, so that for ecological reasons alone the use of renewable energies is a must. The high wind potential in the regions of great demand shows how important a role the use of wind energy can play in China, which is also shown in the public programs for the electrification of rural regions. The domestic industry already supplies the local market with own products for base supply, namely with large-scale manufactured wind energy converters (up to 200 W power) for battery charging. Converters with higher power and the system technology necessary for an autonomous electricity supply for villages do not as yet exist in China.

The demonstrations of wind/diesel/battery systems in the frame of the German-Chinese cooperation supported by the German government was a basis for the industrial cooperation between SMA and an industrial enterprise in Inner Mongolia, P.R. China, established in the meantime.

The modular wind/diesel/battery systems developed by SMA allow for an operation with minimized diesel running hours as the diesel can be switched off temporarily and load peaks can be balanced by the battery storage in case of insufficient wind supply. A wind/diesel/battery system mainly consists of conventional components. Apart from the microprocessor-equipped operation control and the wind energy plant all components are available on the local market of an industrial threshold country like the P.R. China. This fact enables a local production share of approx. 50% to be reached in China right from the very beginning.

A stage plan was developed for the transfer of system technology, with the intention, that SMA together with their Chinese partner would increase the local production share to reduce the import of components successively. This conception allows for a local share of 70 % without the partner having to pay licence fees for the system technology. The step towards a complete take-over will then be taken within a licence agreement or a joint venture company.

1. INTRODUCTION

The People's Republic of China, with a population of approx. 1.2 billion shows an enormous economical growth with two-figure increase rates in the last year. At the same time, the conditions for the population and of the economy in the rural regions of China are rather typical examples for a developing country. Nearly 50 % of the population live in the wide rural regions, in which complete electrification in accordance with the standards of industrialized nations is not possible because the population density is too low and the electricity demand not great enough to be able to realize an interconnected network technically and economically.

Figure 1: Containers of the wind/diesel/battery system in Inner Mongolia (left container: battery and switch cabinet; right container: diesel set) with AEROMAN wind turbine (33 kW)

Therefore there exists in China an urgent demand for decentralized power supply. About 30000 villages are not electrified presently. Thousands of villages are partially supplied by means of small centralized diesel power sets, which can work only for limited periods of time due to the high diesel prices and the uncertain availability.

From an ecological point of view alone all available ecologically beneficial resources must be exhausted for the supply of the increasing power demand, which no doubt will also happen in the rural regions because of the universal economic growth.

Presently, this aim points to the increased use of wind energy in China due to the great potential. In the northern and north-western steppe and desert regions, and on the thousands of islands along the Chinese coast, high wind energy potential and great demand for decentralized electrification coincide in an ideal way.

Figure 2: Wind charger, 200 W

The potentials of the use of wind energy have been technically and economically proved today. This is also valid for hybrid systems like the wind/diesel/battery systems intended for China, which guarantee a continuous supply of electric power due to an optimal use of wind energy and a minimization of diesel running hours. With these systems developed by SMA, which have been in the production stage since 1990, up to 75 % of diesel fuel can be saved on a good wind location as compared with a mere diesel power set supply. Thus the example of China clearly shows the extent of the positive ecological effects such a technology can have.

The Chinese government realized in time that the electrification of the rural regions will only be ecologically and economically possible if the available renewable energies are used extensively. Therefore ambitious and long-term programs exist in the meantime to achieve this aim. China undertakes great efforts for this purpose, but also concedes that this will not be possible in the medium-term without technology transfer and the cooperation with industrialized countries. Presently, the most important tasksare to strengthen the developing industry, and to produce the necessary great number of decentralized power supply systems as far and as fast as possible locally, supported by the co-operation of suitable foreign partners. The great political importance of these measures and the subsequent political support makes the realization of such cooperations easier.

The development and optimization of the wind/diesel/-battery systems by SMA since 1987 led to the production stage in 1990. SMA systems presently work in Ireland, Northern Ireland, Jordan, and Australia, and are satisfactory technically and economically.

Two wind/diesel/battery systems have been in operation successfully for demonstration purposes in Inner Mongolia, P.R. of China, since 1991 within the frame of the German-Chinese cooperation under orders from the Federal Ministry for Research and Technology (BMFT). Along with this, wind measurements and analyses of the demand structure of typical villages are carried out as part of the special energy program of the Federal Ministry for Economic Cooperation (BMZ), which will result in conceptions for local supply. The Chinese partner organization, the Ministry for Machinery and Electronic Industires with its local representations, the planning boards and local industry are convinced of the ecological and economical benefits of the wind/diesel/battery systems and aim for a local production of the systems in cooperation with SMA. SMA is also interested in this cooperation, and is willing to transfer the 'know-how' to China. The Commission of the European Community presently supports the elaboration of a cooperation strategy for the technology transfer with a Chinese industry partner.

How serious the Chinese partners are about realizng this cooperation is shown in the accelerated administrative support of the project. A first cooperation agreement between SMA and an Inner Mongolian industry enterprise could be signed by the end of 1992. Parts of the first systems with a share of about 50 % will be produced locally in China already in the base stage of the transfer. This local share will be increased in future stages. Parallel with the technical support the partner will also be advised in the marketing of the wind/diesel/battery systems.

2. TECHNOLOGY AND PERFORMANCE

To supply remote villages and settlements not connected to the public grid continuously with electric power is an ambitious technical task considering ecological and economical points of view. SMA has developed a modular supply system as a solution for this task. Meanwhile these wind/diesel/battery systems have proved their technical reliability and economical competitiveness worldwide under different extreme environmental conditions partially for more than six years.

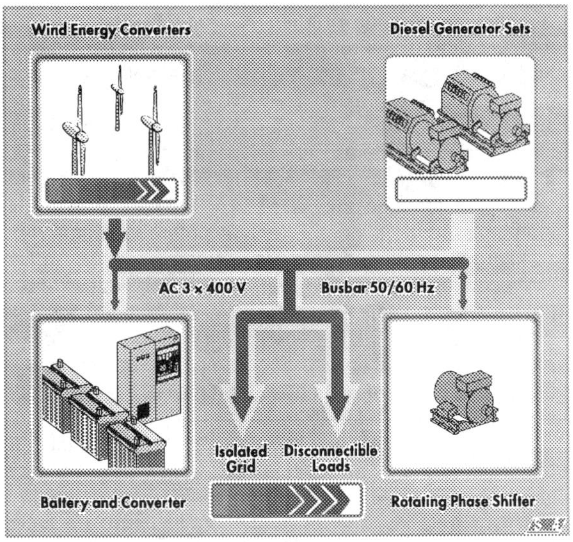

Figure 3: Good wind conditions

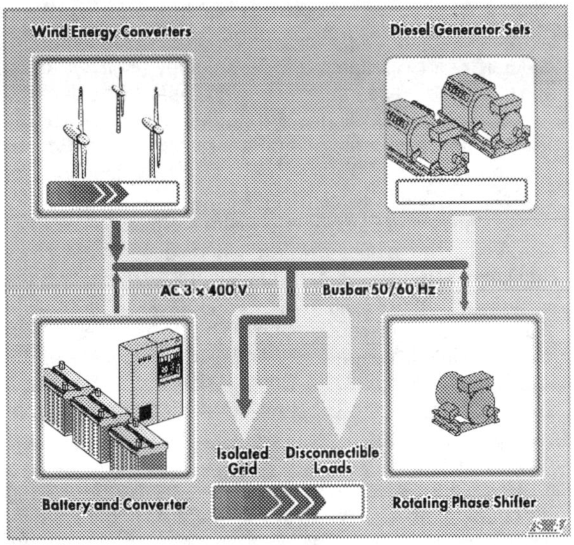

Figure 4: Medium wind conditions

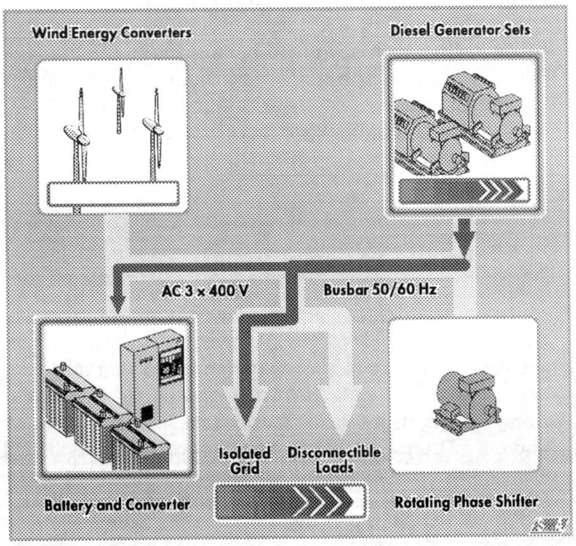

Figure 5: Weak wind conditions

The modular autonomous electric supply systems realized by SMA basically consist of one or more diesel power sets, a battery storage with converter, a rotating phase shifter, an optional number of wind energy converters, and eventually a photovoltaic generator. All modules are coupled on the 3-phase AC side, i.e. on grid side (isolated grid). In case of sufficient wind energy offer and low consumer demand the wind energy converters supply enough electric power to prevent the diesel generators from running.

Before the diesel power set can be switched off the rotating phase shifter is to be connected. Now this synchronous generator takes over voltage control and reactive power supply.

Due to the fact that diesel power sets have a very poor efficiency in the lower partial load range, the nominal power of the diesel is usually dimensioned for about 2/3 of the peak consumer power. The high short-time demand peaks are getting balanced by the battery storage in case the wind energy converters do not supply it.

To avoid frequent starting of the diesel due to changing wind velocity resp. varying consumer demand the system is equipped with a battery storage. In this way short-time load peaks in the isolated grid can be balanced by the battery system to unload the diesel engine. It is sufficient to design the battery system as a short-time storage. The supply to the isolated grid or the battery charging is carried out via a line-controlled inverter. This thyristor inverter is a standard product from drive engineering and stands out for its robustness and its good price/performance ratio.

The careful operation of the battery due to an intelligent charging supervision and control and an acid circulation during charging results in an efficiency of more than 80 % and a battery life of more than 8 years.

The use of two differently sized diesel power sets results in a further increase of fuel saving and economic benefits of the system. The dimensioning of the diesels should be carried out in a way that the peak consumer power can be supplied by the two engines together, and the smaller one should have half the output power of the greater one. The diesel power sets will be connected in a way that they can be operated as close to their nominal power as possible, i.e. with the highest efficiency.

Furthermore, this makes possible to reduce the battery capacity, because now the battery storage has to balance the starting periods of the diesel power sets only, when power variations occur due to load resp. wind velocity changes.

Wind turbines with a good speed/power control via a fast blade pitch setting device are preferably used as wind energy converters. The turbines are equipped with an asynchronous generator, and designed for grid-parallel operation. They can feed the isolated grid at every possible location, because they don't need any additional control lines besides the power connection. Even stall-controlled wind energy converters can be integrated in such a grid on principle.

The microprocessor-equipped operation control selects the most economic and secure operation mode depending on the present power output of the wind energy plant, the charging state of the battery, the load conditions, and the consumer power to be expected.

The operation of these plants could be significantly simplified using a powerful microprocessor system (80186-based) and a menu-driven graphical operating interface. The new systems contain a remote control and operating data output via Modem and telephone line besides its operating data acquisition and output and an automatic

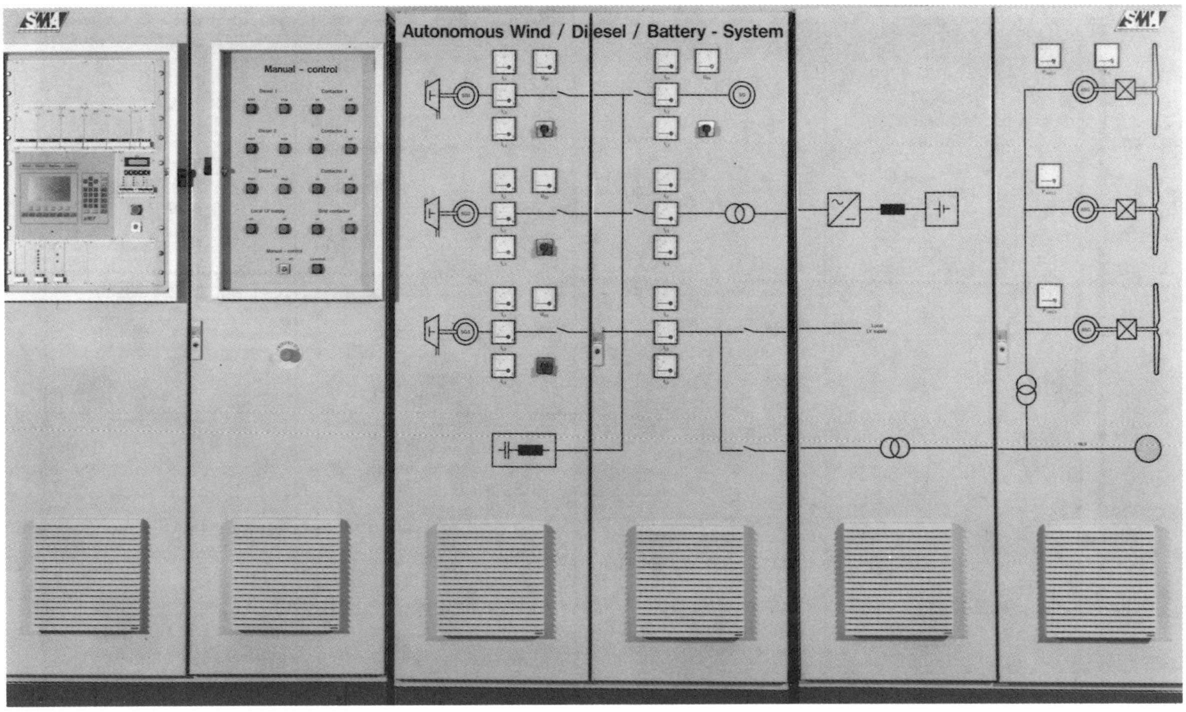

Figure 6: Switch cabinet with block diagram of the wind/ diesel/battery system on the Irish island of Rathlin

servicing demand for the several partial systems depending on their operating times.

A typical problem with isolated low-power grids is the very low electric power demand during the night and the early morning hours as compared with the daytime. As a rule, the wind energy converters have to be driven down during these times to a degree that less than half the wind energy offer is used. Thus the economy of the plant is reduced respectively.

Remedial measures may be the controlled connection and disconnection of electric power consumers with storages, like water pumps, storage heaters, boilers, etc..

The system conception and the control and supervisory technique are modularly designed not only making possible different system configurations (e.g. diesel/battery systems), but also the realization especially of plants in the power range from 300 kW to 2 MW using greater wind energy converters (e.g. 250 kW per turbine). Later enlargements or e.g. the installation of PV generators in some years, when photovoltaic will become less expensive, can easily be carried out.

One system constellation in particular shall be mentioned here, which consists of several diesel power sets with different power and a battery storage, but no wind or photovoltaic devices. The fuel saving effects resulting from the optimized operation of the diesel engines, the automatic operation, and the remote control via telephone modem make the commercial use of these systems possible in Australia. More than 10 of these plants in the power range from 100 kW to 700 kW with system technique from SMA are successfully in operation for the supply of remote villages at present.

3. TECHNOLOGY TRANSFER

In China and especially in Inner Mongolia a local production of small wind energy plants up to 200 W power for the basic electrification (battery charging) has been built up during the last years as an achievement unequalled by other developing countries. Today more than 100,000 of these very small plants work in the Inner Mongolia.

The most successful manufacturer of this plant size is the cooperation partner for the initiated local production of the wind/diesel/battery systems. Furthermore, this company produces a 5 kW plant in series.

The basis module of the SMA wind/diesel/battery system consists besides others of a 33 kW wind turbine AEROMAN. This size of plant has not existed in China until now; generally the development of bigger plants has only been sporadic up to now, but shall be strongly accelerated in the near future with the support of foreign cooperation partners according to the plan targets.

There are no principal limits for the power of the wind/ diesel/battery systems. As a matter of fact, conceptions for some Megawatts of electric power can be realized, though the autonomous system should not be smaller than 30 kW due to the system expenditure. There is no binding to a specific wind energy converter type.

The starting point for the technology transfer is the demonstration of two wind/diesel/battery systems in villages in Inner Mongolia supported whithin the frame of the scientific/technical cooperation with China by the BMFT. Aims of the demonstration are on one hand to prove the technical reliability under extreme climatic conditions in Inner Mongolia, and on the other the economy of the continuous village power supply, i.e. that the agricultural production can be improved (e.g. via directed irrigation), and the pre-manufacturing of products and the local crafts can be strengthened in this way. The continuous supply on the a.m. islands, on which "Seafood-Farming" is carried out, is of essential economic signifigance, where cooling and pre-production play a crucial role.

In the frame of the demonstration project the Chinese partners and operators were made familiar with the technology by means of theoretical and practical training. The local industry was also integrated early on information and demonstration events.

The local production of the wind/diesel/ battery system is made easier by the fact that a great number of parts are conventional components available on the local market or

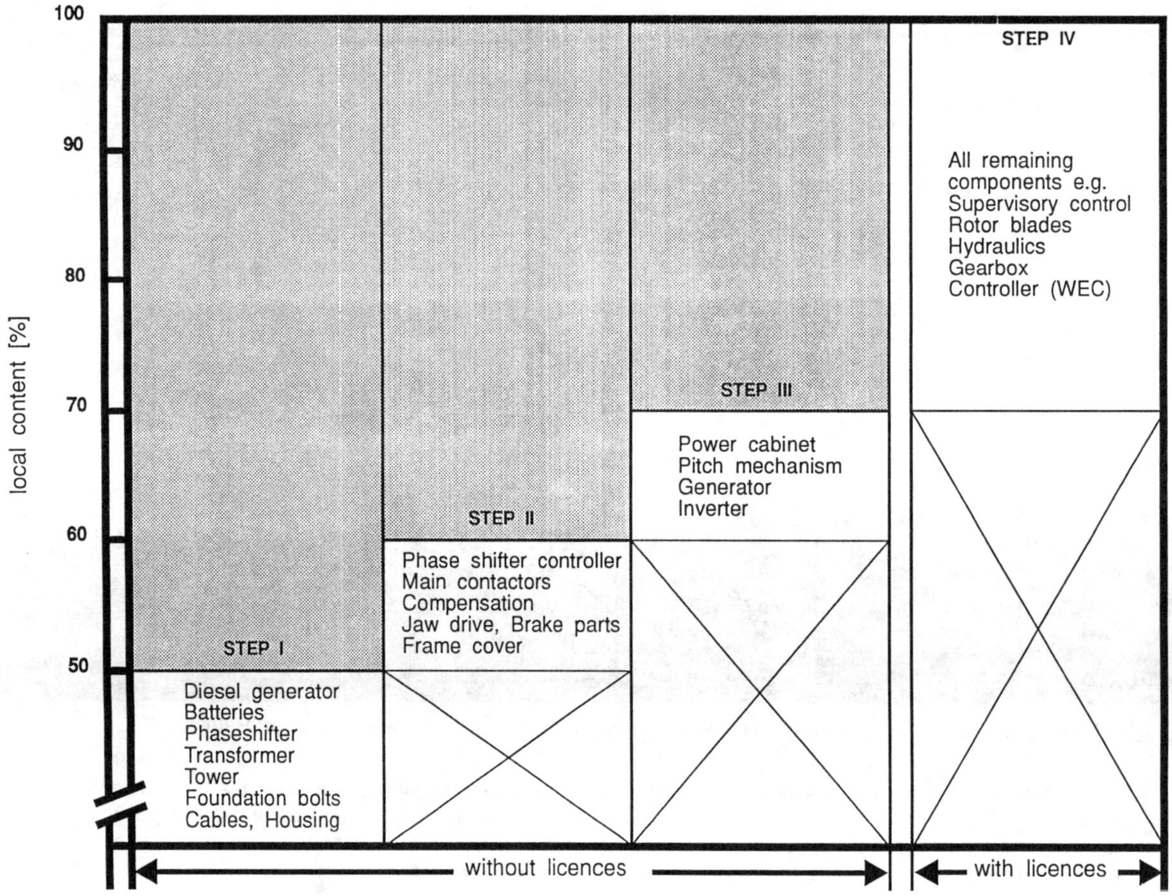

Figure 7: Steps of technology transfer of wind/diesel/battery systems

capable of being produced locally. This makes the high initial local production share of about 50 % of the system price possible at importation. Four stages were defined for the transfer and the appertaining increase of the local share. The most remarkable thing about this is that during stages I to III all specifications and documents necessary for the local production will be made available to the Chinese cooperation partner without charges. The German partner will supply the other components and with his know-how be responsible for the solution of problems. And so a local production share of about 70 % will be obtained without charging licence fees. In stage IV it is intended that a licence agreement will be signed concerning the remaining components, and the know-how be brought into a joint-venture company.

This procedure enables an instantaneous cooperation start to be made without the delay of long licence negotiations. Furthermore, in this way the German partner will still be interested in the functioning of the system as long as he may supply components. This pragmatic approach has been met with great interest on the Chinese side.

Figure 7 shows the four transfer stages schematically

Stage I intends that diesel power set, phase shifter, batteries, transformers, wind converter tower, foundation parts, all cables, and containers are taken from Chinese production.

Stage II additionally intends electric high-power components, the phase shifter control, generator, wind direction follow-up, and brake components from Chinese production.

Stage III additionally contains the switch cabinet and the pitch-control mechanism. It is under discussion to start the rotor blade production already in stage III, because just this work-intensive component would reduce the cost due to the low Chinese wages.

Stage IV consists of all remaining components and makes the complete local production possible after the respective transfer of know-how.

The transfer's time-schedule depends essentially on the input of the cooperation partners. It can be seen today, that the local production share can be increased up to 100% within four years, whereby it is to be considered that the gradient of the increase will become lower with the project progress due to technological reasons.

4 ECONOMY

The distribution and use of wind/diesel/battery systems in a relevant degree concerning the power industry in developing countries today depends on three basic preconditions:

a) The systems must be ecologically efficient.
This requirement is fulfilled by the substitution of diesel fuel by wind energy.

b) It must be possible to produce the systems at least to a great extent in the user land.
This local production can be obtained through industry cooperation, intensive, long-term technology transfer, and political support.

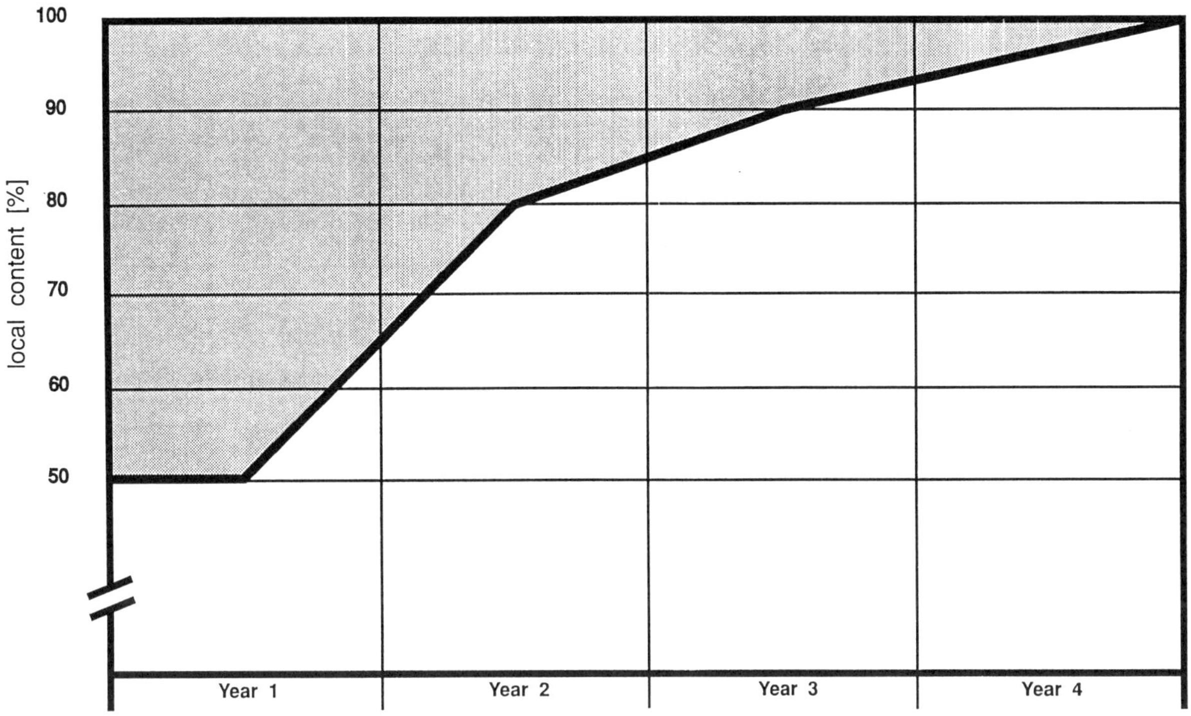

Figure 8: Time schedule for the technology transfer

c) The systems must be capable to compete economically with a simple diesel power set supply.

The proof of the economy of wind/diesel/battery systems depends on a great number of parameters and must therefore be carried out for each specific plant.

The economy of a wind/diesel/battery system essentially depends on two parameters. On one hand the wind velocity conditions at the respective site are decisive for the diesel fuel saving by the installed wind energy plant, on the other hand the local diesel price determines whether it is possible to generate electric power using a wind/diesel/battery system at a lower price than using a simple diesel power set in spite of the higher investments.

Actually, the investments are also of extreme relevance, and make local production necessary.

Furthermore, some other financial and technical parameters have to be considered when calculating the economies, e.g. the rate of inflation, the increase rate of the fuel prices, the interest rates, the component life, servicing costs, etc..

To be able to carry out these complex calculations for different system configurations and sites, SMA developed a program five years ago, which calculates the kWh cost for a wind/diesel/battery system and a conventional diesel power set using the annuity calculation. To be able to use these calculations for the special situation of developing countries and multiple diesel systems, a number of extensions had to be implemented. An example is the difficult calculation of the servicing costs for diesel power sets, which can now be carried out variably and in detail.

The program package is PC compatible and so the costs of the systems can be analyzed with the prospective customer directly on site.

Cost calculations specifically for the Chinese situation and a local production share of 50 % have shown that, at yearly average wind velocities above 6 m/s, and at fuel costs of more than 0.60 DM/liter the generation of electric power using wind/diesel/battery systems is less expensive than using simple diesel power sets.

HYBRID WIND DIESEL SYSTEM FOR A VILLAGE IN THE CANARY ISLANDS. DEVELOPMENT AND COMMISSIONING.

A. Gonzalez (1), R. Calero (2), J.A. Carta (2), L. Ojeda (2)
J. Gonzalez (3), A. Perez (3), J. Torres (3), R. Vega (3)

(1) CIEMAT-IER
Avda.Complutense, 22
28040 MADRID, SPAIN

(2) E.T.S. Ingenieros Ind.
Dpto. de Ingenieria Mecanica
Tafira Baja 35017
GRAN CANARIAS - SPAIN

(3) E.T.S. Ingenieros de Telecomunicacion
Dpto. de Electronica y Comunicaciones
Tafira Baja 35017
GRAN CANARIAS - SPAIN

ABSTRACT: The project "Hybrid Wind/Diesel system to supply electricity to a fishermen village in Fuerteventura, Canary Islands", was funded by the VALOREN Program, the Council of Pajara and the "Consorcio de Abastecimiento de Agua", of Fuerteventura and the Renewable Energies' Institute (IER) from the CIEMAT, in Madrid.
The whole system has already been installed and consists of a 225 kW Wind Energy Converter and two 60 kvA diesel engines with flywheel and synchronous generator of 75 kvA each one, to produce electricity for all the services of the village: 56 m^3/day R.O. seawater desalination plant, cold chamber and ice production for fishing conservation, 25 m3/day sewage water treatment plant, public and private lighting, etc., for an average of 300 people, the most of them fishermen families.
The final configuration and installation of all the equipment including the above mentioned loads, designed from the consumption and wind data will be shown.
Besides, the control system: Hardware, software, sensors, power supply, etc., implemented from the control requirements study: Definition of the different operation modes and identification of all the measurement and protection points as well as the devices used will be shown and/or described, including the system commissioning trial.

1. INTRODUCTION.

In August 1988 and after some meetings maintained from January 1988 with responsible people from the Local Government of Fuerteventura, one of the Canary Islands, in Spain, the project: "Hybrid Wind/Diesel system for the supply of electrical energy, water, and other services in P. de la Cruz, in Fuerteventura was presented to the VALOREN Program, in Spain. Before that, the initial idea about this project and its carrying-out was reported in the 1988 E.C. Wind Energy Conference, in Denmark. The project finally consists of a 225 kW Wind Energy Converter, two diesel engines of 60 kvA each with flywheel, a 56 m^3/day R.O. seawater desalination plant with a freshwater storage tank of 1000 m^3, a cold storage chamber for fish conservation with an ice production system of total power of 8 kW, a 25 m^3/day sewage water treatment plant and public and private lighting for the narrow streets and dwellings of the village.

In January 1989, the VALOREN Program granted almost the 50% of the foreseen budget: 160 millions of pesetas, civil works included, and from then up to date the project has been developed and installed, being now in the first steps of the commissioning.

Besides this project has been financed by: The Council of Pajara which is the Local Government of the region where is located the project, the Consorcio de Abastecimiento de Agua de Fuerteventura, which is the organism depending on the Autonomous Government responsible of the production and distribution of the fresh water in this island where there is no natural fresh water almost from several centuries ago, and the CIEMAT through the Renewable Energies' Institute (IER). Some representative people of these organisms have worked during all this time taking the first decisions of purchasing after the technical reports shown periodically for a technical committee composed of several engineers and architects, coming from the polytechnical school of Las Palmas and coordinated by the renewable energie's systems Department of the IER.

2. PROJECT DESCRIPTION.

2.1. Previous studies. Input data.

In reference with the wind data used in the project design, a measurement campaign was carried out in Fuerteventura from June 1987 until November 1989 in different sites. Wind speeds and directions, calms, etc., were registered during the whole period in "P. de la Cruz" in the "Peninsula de Jandia", the location of the project among others. All these parameters were monthly collected and a Weibull distribution for every month was adjusted calculating the K and C factors through the momentum method -see figure 1-. After a statistical treatment the historical data were correlated with the actual ones. Also, the long term band of monthly average wind speeds for this location was calculated using the Corotis method -see figure 2-.

The wind data (hourly values of the monthly average days) of the whole 1988 and their Weibull distribution -calculated by the momentum method- were used as input data since were in the middle of the mentioned band and they had a strong correlation with historical ones.

To calculate the input load data four different periods in the year were defined: Summertime, Holy week, week end and labour days, in which the envisaged population were 425, 285, 145 and 60 people, respectively.

The needs of fresh water was calculated in 125 l/person.day for all the uses, without saving but in the case of Fuerteventura the actual consumption is around 80 l/person. day which in the case of using toilets of low water consumption -recommended- this amount fell up to 60 l/person.day, which has been the hypothesis of the design.

Besides, the combination between the RO seawater desalination plant and the fresh water storage will allow to produce all the water needed in 4 months, managing the performance of the mentioned RO plant in such a way that only when there is enough wind potential the RO plant works, avoiding the production of fresh water with the diesel(s) engine(s).

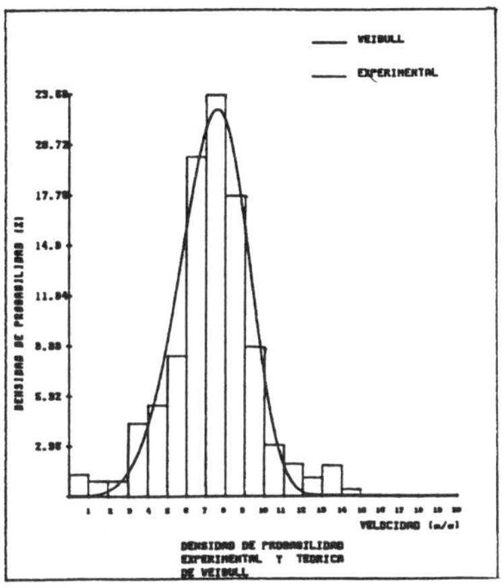

Fig. 1. Wind speed density function. Measured data (Sept. 88) and Weibull distribution.

For the calculation of electrical power and energy demand required for this use and for the other ones: Cold chamber and ice production, sewage water treatment plant, public and private

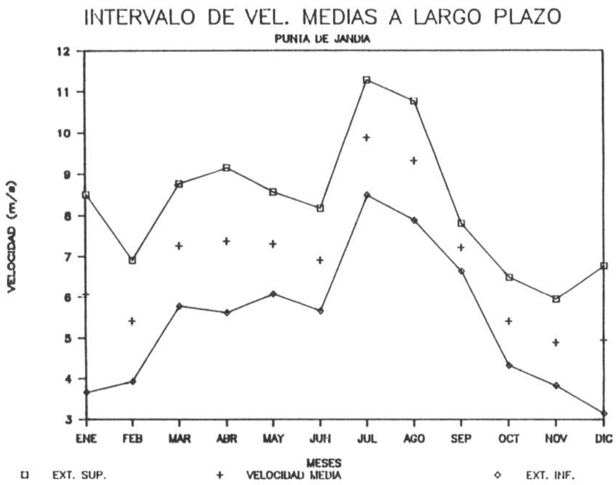

Fig. 2. Long term band of monthly ave. wind speeds.

lighting, etc., different criteria and standards have been taking into account. The total power demand for each one of the periods considered is shown in the figure 3.

From the detailed data of electrical needs and from the wind data already mentioned it was calculated the size of the wind energy converter and diesel(s), once chosen the basical features of the hybrid system design. The hourly average electrical powers for all the monthly average days were calculated using the power curve of a 200 kW wind energy converter, very similar to the installed one. These values for the month of August 1988 are shown in the figure 4.

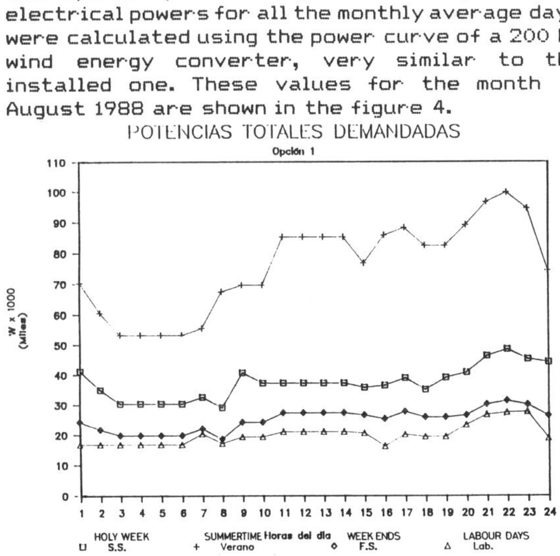

Fig. 3. Total power demand for all services in every period considered.

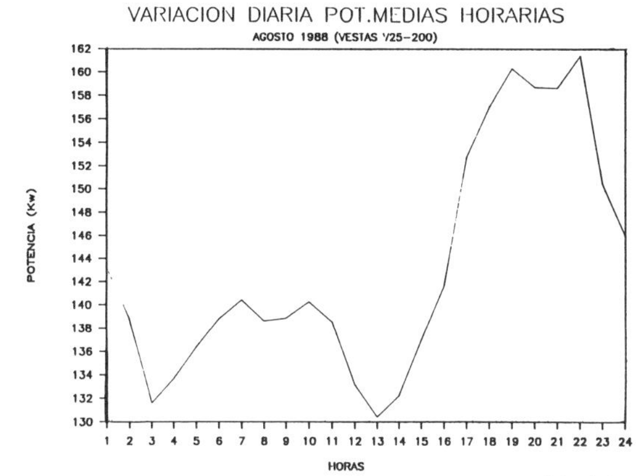

Fig. 4. Foreseen ave. elect. power supplied by a 200KW WEC for a monthly ave. day of August 88.

2.2. Description of the hybrid system and equipment installed.

2.2.1. General installations.

The wind energy converter (WEC) is a VESTAS V-27: The first technical features of this machine are:

- Rotor
  - Diameter 27 m
  - Rotor speed 43/32 rpm
  - Power regulation pitch
  - Blades Number 3
  - Material Fibreglass & polyester

- Generator AC

Type	6 pole	8 pole
	asynchronous	
Output(kW)	225	50
Current(A)	400	100
Voltage(V)	400	400
Cos φ	0.31	0.72
Inst.react (KVAr)	87.5	25
Reslt.cos φ	0.94	0.91
Reslt.curr	345 A	80 A

- Operational data
  - Start-up wind speed 3.5 m/sec.
  - Cut-out wind speed 25.0 m/sec
  - Nominal output at 13.5 m/sec.

- Tower
  - Type Cone-shaped tubular
  - Height 30 m
  - Height 12 tons

- Yaw system
  - Type Active yawing
  - Control Windvane
  - Speed 0.6 degr./sec.

- Control unit
  - Vestas multi processor

All the elements and materials of the wind energy converter installed, have been assembled by the official supplier of the manufacturer in Gran Canaria island, in the Canary Islands.

The WEC already installed as well as the building containing all the first equipment installed is shown in the figure 5.

Fig. 5. View of the WEC installed in P. de la Cruz.

Each one of the two diesel engines installed has a power of 60 kvA and is mechanically connected through an electromagnetic clutch to the flywheel and directly to the 75 kvA synchronous generator. Each one of the flywheel could deliver a half of the envisaged peak power demand -100 kW - during 30 sec. aprox. without the frequency decreases below the minimum level allowed: 50 Hz - 4%. They have been wholly developed and manufactured in Spain. The figure 6 shows this equipment during the assembling phase.

Fig. 6. Diesel generators & flywheels during the assembling phase.

One of first services supplied for this project, explicitly given in its title is the fresh water production, as it has been said before.

The seawater desalination plant used is a 56 m^3/day reverse osmosis plant in which the average electrical consumption will be of 7 kw.h/m^3 with a total power of 16.5 kW. To storage the fresh water for the consumption a storage tank of 1000 m^3 has been built being of use the different level between the ground and the ramp where fisher boats are stranded. See figure 7 and 8.

To optimize the fishing exploitation a cold chamber and an ice production system were installed. The capacity of fishing storage is aprox. 1200 Kg and the total electrical power required for both systems is 8 Kw. The building where these two services are contained is the nearest one to the sea, to facilitate to the fishermen their work.

In every grid of electrical supply there are two essential parameters to maintain their stability: Voltage and frequency. Through the different sensors, current and voltage transformers, multiconverters, etc., all the measures needed for the control process are acquired and via the corresponding electronic cards shape the input of the PC whose CPU calculate all the values of the control variables which through the output ports will be sent to the different system actuators.

The basic operation modes of the hybrid system are the following: Only diesel(s), diesel(s) and WEC and only WEC. In the first case, when the wind speed is below the cut-in value of the WEC the frequency and the voltage are ensured by the diesel engine(s) which are self-regulated in speed and by the excitation current of the synchronous generator(s), giving the active and reactive power needed.

When diesel(s) and WEC are working together, the active power is delivered by both systems and the reactive power needed will be given to the whole system from the synchronous generator(s) over-exciting it (them).

In the case in which the WEC works alone, all the active power will be delivered by itself and the synchronousgenerator(s)/flywheel(s)disconnected of the stopped diesel engine(s) through the electromagnetic clutch(es) would deliver the reactive power demanded for the whole system: WEC and loads, increasing or decreasing the synchronous field (rotor) current(s). Besides there is a capacitor bank to give more reactive power (in steps) if it is required.

For the diesel(s) start-up and to avoid a high rate of clutch(es) reposition it exists a pneumatic system which will start up the diesel engine(s) unloaded before the connection already at nominal speed and through the clutch to the flywheel and synchronous generator.

Also it exists a 100 kW dump load to allow the dissipation of the excess of active power generated from the WEC when be necessary.

Fig. 7. Reverse osmosis plant during the assembling phase.

Fig. 8. Fresh water storage tank built of 1000 m_3.

Other important service given to the village is the public and private lighting. The power demanded in the case of public lighting is around 4 Kw and for the private lighting a consumption between 2.5 and 4 Kw.h per day and dwelling has been foreseen. The envisaged peak power demand for this use according to the four different periods of occupation: Labour days, weekends, holy week and summertime is 15.5, 19, 32.5 and 67.5 Kw, respectively and the average power demand along the day for the same periods: 10, 13.5, 23.5 and 45.5 Kw. All the domestic uses have been considered with the exception of the air conditioning.

The rest of services: the water pumping through the installed piping, the fuel storage tank and pumping, the drainage system with a sewage water treatment plant, etc. are included in the project. The unifilar diagram of all the installations is shown in the figure 9.

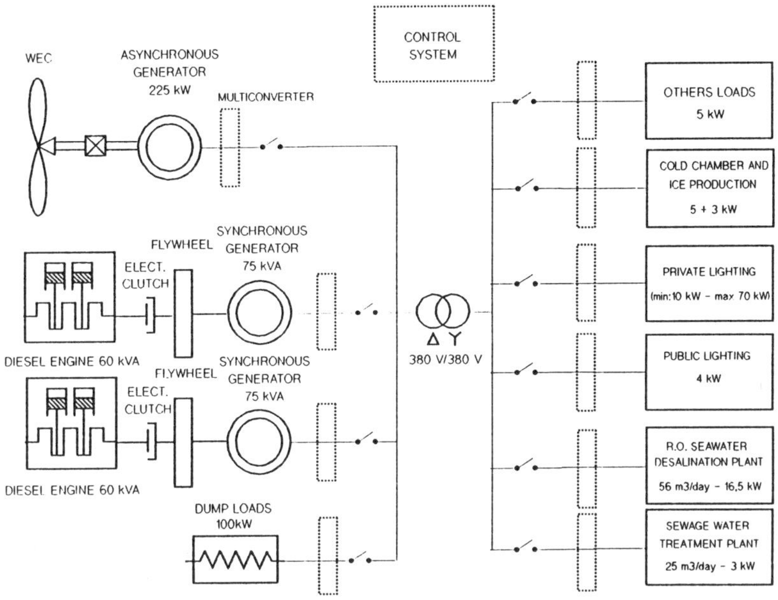

Fig. 9. Unifilar diagram of the whole system & loads.

2.2.2.Control system.

From several years ago it is very well known that to optimize the diesel fuel consumption in the case of hybrid wind diesel systems a certain amount of energy storage is required. In this project the choice of energy storage has been carried out using inertial storage through a flywheel, directly connected to the shaft of each one of the synchronous generators.

The hardware included in the control system consists of sensors, voltage and current transformers, converters and multiconverters, integrators for energy acquisition data, analogical and/or digital input and output cards, industrial PC, protection devices, etc. The range of analogical outputs is 4-20 mA and in the case of bidirectional parameters from +10 to -10 mA. The range of digital parameters is 0-24 V DC, or in the case of contactors and other electrical actuators, 0-48 V DC. In short there are 67 analogical inputs, 39 digital inputs, 10 counter inputs, 65 digital outputs and 2 analogical outputs.

In the figure 9, all the real points of measurement, protection and control are displayed in dotted line.

The software implemented in the industrial PC (microprocessor 80386 at 33 MHz) consists of a main program which after the inicialization process gives the order of diesel(s) start(s), synchronizing them. Immediately after, a subroutine of checking and acquisition data will save in the data memory the value of every parameter to be used for the main program and subroutines. From this point, the system is asked. Depending on it one of the three main subroutines: Only diesel, only WEC and Diesel-WEC in parallel, will be run. After that, the system runs the following subroutines: Temperature control, system surveillance, voltage control and control of the reverse osmosis plant, returning to the mentioned subroutine of checking and acquisition data and closing the loop.

Also a screenware package has been implemented which allows the continuous monitoring of the system in real time.

3. COMMISSIONING.

A commissioning trial program has been developed to test every one of the different equipment installed: wind energy converter, diesel engines, R.O. desalination plant, electrical boards and protections, control system, etc. measuring and recording the first parameters of their performance as well as ensuring the proper operation of the control, communication and alarm systems. Some steps of this commissioning trial were carried out with the control system, testing its good performance in combination with the first sensors, converters and multiconverters used in this project, in the laboratory of electrical machines in the School of industrial engineers of Las Palmas, in Gran Canaria.

Also a transient stability analysis of the hybrid wind/diesel system, using the EMTP package, was carried out for the worst conditions of operation: instantaneous wind gusts, idem for disconnection of important loads, etc. and the results obtained showed the good performance of the system from the point of view of the transient stability due to its high mechanical inertia.

The start-up of the whole system will be carried out in April 93.

REFERENCES

1. A. Gonzalez, E. Soria: Wind and Diesel power system for supplying electricity to a fishermen village. E.C. Wind Energy Conf. in Herning, Denmark. June 1988.

2. R. Calero, J.A. Carta, L. Ojeda, A. Lopez: Estudio y Evaluacion de los recursos y necesidades. Definicion de las bases del proyecto. Internal report. Las Palmas. December 1989.

3. G. Elliot and R. Hunter. International Energy Agency Programme for Research and Development of Wind Energy Conversion Systems. Annex VIII. Decentralised Applications for Wind Energy. July 1991.

4. J. Abel Perez: Diseno del control de un sistema aerogenerador/grupodiesel/Plantadesaladora por osmosis inversa. Universidad Politecnica de Las Palmas. E.T.S.I.I. Marzo 1993.

SIMENERG: A NOVEL TOOL FOR DESIGNING AUTONOMOUS ELECTRICITY SYSTEMS.

R. Chaer, R. Zeballos, G. Casaravilla and W. Uturbey

Instituto de Ingeniería Eléctrica, Facultad de Ingeniería J. Herrera y Reissig 565.
CP 11.300 Montevido-URUGUAY. Fax:(598-2) 71.54.46, Tel.(598-2) 71.09.74, E-mail: rch@iie.edu.uy

ABSTRACT: This work describes the methodology and point of view adopted when developing SimEnerg. SimEnerg is a software tool-kit to help the analysis of autonomous electricity systems. The developed tools are briefly described. The output of a simple application is presented.

1 Introduction.

SimEnerg is a set of software tools specially designed to help the evaluation and design of autonomous electricity systems. It was developed in the frame of two research projects about energy-generating stations from wind and solar sources, as an alternative to the grid connection for users placed far from the national distribution grid. The two projects are being carried out by the Institute of Electrical Engineering of Uruguay for, and under the financial support of, the State Power Board of Uruguay (U.T.E.).

Having a system to study, SimEnerg allows the easy building of an efficient simulator for it. One can evaluate the service quality and other aspects of the system by simulating many years of its life. Also, by leaving free some of the design parameters and choosing a cost function, one can carry out an optimization on the unconstrained parameters, in order to minimize the cost function guaranteeing a given energy availability of the system with a given confidence.

SimEnerg is an open set of tools, so once the user has understood its general philosophy, he can modify and add other more refined tools without limit. It has been designed using Object Oriented Programming (OOP) Methodology, resulting in an easy to understand and easy to extend tool by users, even if they are not experienced programmers. For the implementation it was used Turbo Pascal 6.0, a fully OOP language for PCs under DOS.

In this paper a metaphoric language is intentionally used, placing the simulation of a system on a Theatre Stage on which the different Actors play their roles directed by a Director. If possible, no programming-words are used. This was useful when developing different pieces of SimEnerg by different application specialists without bothering them with the later software implementation, and will be good for the potential users helping them to have a mental model of the tools and their possibilities. Carrying out a simulation with this model is really simple; it would be enough to choose the suitable Actors, teach details to the Director and, taking a comfortable place watch the mise en scene.

Who are the Actors? Any Object, identifiable entity of the system, who exchanges energy with some other is an Actor. For example, in a system built of an array of solar panels, a battery bank, a water pump and a lamp, the solar panels can be considered as one actor, the battery bank as another, the water pump another and the lamp another one. In order to define a Class of Actor, it is necessary to describe the aspects of its behavior which are relevant for its role.

Once a class of Actor is defined, it is available for future simulations. SimEnerg has the set of the most frequently used Classes of Actor already defined, which are enumerated below in section III.

The Director has the responsibility of making things work well. We must teach him the details of the Script of each theatrical work (simulation).

Which is the Script of the Theatrical work? Surprisingly, the script is always almost the same, being the main variations the elapsed time (years of evaluation of the system), the Actors who play, and the details the Audience wants to pay attention to (simulation results).

In application-domain words, the story is that the actors exchange energy among them. Each Actor can be of the type: Energy Source, Energy Sink or Energy Accumulator, (e.g., Wind-Generator, Lamp for lighting and battery bank, respectively).

2 More About The Stage Model.

In this section more details about the model used for the system under evaluation are given. As mentioned before, the system will be modelled as a Stage, a set of Actors and a Director. The life of the system will be the theatrical work played by the Actors under the management of the Director.

The Script will be played act by act, each of which takes a time (virtual time, i.e., time in the simulated world) called Time-Step or Beat.

At each Beat, the Actors offer and demand Energy Packets (EP). An EP has the amount of energy that the Actor wants to exchange and a Priority Number (PN). The PN is like a price that the Actor puts to the amount of energy in the EP.

With all the Offers and Demands in hand, the Director decides which transactions (exchanges of energy between actors) actually take place.

The criterion used by the Director to dispatch the EPs at each beat is the following: he begins satisfying Demand with the biggest PN using for that the Offer that has the smallest PN, and continues doing that while there are Offers with PNs less or equal than the PN of the Demand with greater PN that has not been satisfied yet.

The Dispatch Criterion of the real system is modelled setting the PNs with which the Actors offer and demand. For example, if in an autonomous wind-diesel station, with a battery bank (on a DC link) the energy-dispatch criterion is that the energy must be taken first from the Wind Generator, and only if it is not enough, from the battery bank, then, the PN of the Wind Generator must be smaller than the PN of the Battery Bank when offering.

The Actor can (by himself) change his PN based on his own history. This is used in one of the diesel-generator models to simulate the behaviour of a human operator who has a criterion to turn the machine On and Off.

loads used for lighting purposes. In this class we can set two daily working hour-ranges and these ranges can be different for Winter, Spring-Autumn and Summer. Other parameters of this Class are Power and PN.

3.2.5 Tdemgran.

This is the most complex, thus the most realistic model of loads. The idea behind this model is that there are groups of apparatus that have similar behaviour, and then can be modelled as one Actor of this class. The parameters are: Maximum Possible Power, Probability of turn-on of an entity of the group given that it was at the Off-State, Probability of turn-off of an entity of the group given that it was at the On-State, Number of entities in the group and the PN. All these parameters are given at the Actor's birth as a list of records of parameters, each of which has an activating-hour.

3.3 The Offerers.

3.3.1 Tpanel.

Tpanel is a Class of Actors that models a Solar Array Panel. The parameters are: Panel Area, Conversion-Performance-Index and Incidence Angle. At the birth of an Actor of this class a "Source of Sun" must be set. This Source of Sun will be giving to the Tpanel Actor the sun radiation at the geographical point where it was placed. The Conversion-Performance-Index is affected on each change of season to take into account the variation of the average temperature.

3.3.2 Taerogen.

This Class models a Generic Wind Generator. The actual machine is chosen at the Actor's birth from a pre-defined set. Adding a new machine is very easy: you only need to supply a text file with a two-column table representing the Power-Output versus Wind-Speed and the name assigned. One must also connect the Taerogen with a "Source of Wind" that will play the role of the wind at the real geographical place.

3.3.3 Tdiesel01.

Tdiesel01 is a very simple Diesel-Generator model. It allows the user to define the Power versus Fuel-Oil-consuming characteristic, and also Turn-On and Turn-Off hours.

3.3.4 Tdiesel02.

This Class models a Diesel-Generator-Station with an operator that can turn off/on motors taking into account if the alternative source is enough to supply all the load plus a user-defined surplus. An Actor of this class registers the following quantities for each motor: time of uninterrupted-operation histogram, fuel-oil and oil consumed, history of Turn-On Turn-Off events.

4 Modelling of the resources.

In this section the methodologies used to modelling the wind and solar resources as random sources are commented. A complete description of methods and application results for Uruguay is available in the SimEnerg documentation. The models implemented until now are valid only inside Uruguay, but they can be fitted to other areas when data is available. It is possible to change the models for more complex or more simple ones depending on the data available or the kind of simulation one is building. The possibility of an easy-check of the influence on desired outputs due to model refinement is an important aid in SimEnerg. One can begin a system simulation with a very simple model of a source and then refine it until the variations of the output are not relevant. The models that are described below were refined until the histogram of the year-deficit of a testing-system showed no significant changes.

4.1 Wind Resource.

The Source of Wind built supplies the average of the wind-speed every hour. In order to tune the source to a geographical place, a historical hourly series of the wind-speed for that place is needed, (one year minimum). The auxiliary variable z corresponding to the square-root of the wind-speed to the power three is defined; and then, the power spectrum of z (in each season) is analyzed, separating a signal that we call the deterministic one, and the noise. To model the noise a recursive digital filter is estimated, and a histogram for the filter-input-random-source of noise is computed. So from the initial data one has a model of z for each season, that together with the transform from z to the wind-speed completes the wind source.

In order to introduce speed variation of frequencies greater than an hour, we are working on a model based on a definition of the "characteristic length of obstacles" for a place.

4.2 The Solar Energy Resource.

From eight years of measures of the daily average Global Radiation at one geographical point of Uruguay a random-deterministic model has been built. The random component multiplies the deterministic signal and has a memory of three days. With the monthly average Global Radiation at many points of the country, a conversion map has been created that gives the month-conversion-coefficients to fit the model to any place in the country. Random models are applied to the average Global Radiation in order to get Direct and Diffuse Radiation on an hourly base. We have mounted a test station to check these models having very good results.

5 A Simple Example.

Until now SimEnerg has been used in the analysis of two cases. One was the optimal design of the solar-panel area and the battery-bank capacity for a rural school. The other was the analysis of the influence of adding a wind-generator to a little town in the North of Uruguay called Vichadero. Nowadays Vichadero is power-fed from three diesel-generator machines of 160, 170 and 380 kW (220V/50Hz).

A typical demand curve for a day in January is shown in fig.1. Two Actors of the Class TDemGRan were used to model the system's load, and the Diesel-Generators were modelled using TDiesel01 Actors.

In order to evaluate the system's Service Quality an histogram of the energy deficit per week was used. This histogram is shown in fig.2 for the system without the Wind-Generator. Each bar in the histogram corresponds to the energy equivalent to five minutes of black-out in a week. One can see from fig.2. that people at Vichadero can expect to have more than five minutes of black-out (or the equivalent amount of energy deficit) in the 75% of the weeks.

Fig.3. shows the same histogram when a BONUS-55 Wind-Generator was used. As you can see, now the expectation of a black-out of more than five minutes (in a week) is less than 11%.

The fuel-oil consumption was 15800 kg/week in the first case and 14000 kg/week in the second.

The time when simulating 7 years of the system's life (using a time step of 1 hour) was 12 and 14 minutes respectively running in a PC with a 486DX of 33MHz.

The simulation of a system progresses from its beginning beat by beat, until an Evaluation Period (specified at the beginning) is completed. Then, the user-defined "Catching Task" is run in order to gather results. Next, actors are notified that a new Evaluation Period is beginning and then the simulation is continued. This cycle is repeated until the Number of Evaluation Periods reaches the value specified at the beginning.

Simulations are carried out taking Actors from an existing Library, and if needed, new classes of Actor can be defined. To implement a new Class, one must add to a Basic Class of Actor only the new characteristics, teaching it specially how to do the following actions: Offer-Energy, Demand-Energy, Receive notification of an accepted Offer, Receive notification of an accomplished Demand, Birth, Life and Death. This method for the definition of a Class of Actor (specialization of a parent one), is possible due to the Object Oriented approach used for the design and implementation of SimEnerg.

The Basic Class of Actor (from which all others are derived) has a default behaviour for the set of principal actions already defined. For instance, a pre-defined action is to take account of the following Energy quantities in each of the possible PNs: Energy Delivered, Energy Received, Energy Offered but not delivered (Excess) and Energy Demanded but not received (Deficit). These accounts are available at the end of each evaluation period and can be used in the user-defined "Catching Task", and are cleared to zero at the beginning of the next evaluation period.

The Director knows how to do the following actions: Start the simulation, Get the Offers, Get the Demands, Decide Eps' exchanges, Notify acceptation of Offers, Notify accomplishment of Demands, Look if it is time for an evaluation, Look if it is time to finish the simulation. He does these actions beat by beat.

Another important component of the model is the Time Kernel (TK) which is a generator of time-events. The beats that give life to the simulation are one among all the time-events that the TK is continuously sending to the Stage. The basic idea behind the TK is to have an easy way to specify things such as: "Plot this value every week" (to have an on-line plot of a value during the simulation), or "On every month look at the temperature and correct your performance index" (when teaching an actor the role of a simple battery model).

The pre-defined events generated by the TK are the changes of: Minute, Hour, Day, Week, Month, Season and Year.

Each kind of event has a queue of waiting tasks, and when an event takes place the tasks from its waiting queue are fired.

Frequently, the simulated system has a random component, such as the wind or the load. Due to this, it is more interesting to have a probabilistic measure of the quantities, than a unique value of them. For example, the histogram of the Energy Deficit during each evaluation period of the Actors playing as loads, can be used as a system's Quality Index. In order to facilitate this kind of measures the Class of Evaluators is defined in SimEnerg. The user teaches to an Evaluator, at its birth, on which energy account and on what PN-Range it must work. An Evaluator plays in the stage making a histogram of the energy account that the user has specified.

3 The Actors.

In this section the Classes of Actor already implemented for SimEnerg and their principal characteristics are briefly enumerated. The names of the Classes of Actor are in Spanish, and are the same as those in the software implementation.
First we define the Basic Class of Actor called Tactor, and then the other Classes of Actor are described.

Tactor defines the general characteristics of all objects that we call Actors. It is the parent class of any other class of Actors that actually models a real component. It gives support for the accounting of the energy quantities described before. Thus, the user does not need to bother himself teaching this behavior to his own Classes of Actor. This property is naturally inherited from Tactor.

3.1 Accumulators.

Accumulators are Actors who can play as a source and a sink of energy.

3.1.1 Tbateria01.

The class Tbateria01 implements a simple model for a lead-acid battery type. The parameters the user can set at the birth of an Actor of this class are: Rated Capacity, Rated Temperature, Temperature Coefficient of the Capacity, Charging-Performance-Index, Discharging-Performance-Index, Maximum-Charging-Current, Maximum-Discharging-Current and Percentage of Discharge Allowed.

An Actor of this class offers either, the amount of energy which (if accepted by the director) leaves himself at the discharging limit, or the amount of energy corresponding to the maximum of the discharging current multiplied by the time-step, the smaller of them, multiplied by the Discharging-Performance-Index. On the other hand, he demands either, the amount of energy necessary to fully charge the battery, or the amount of energy corresponding to the maximum-charging-current multiplied by the time-step, the smaller of them, divided by the Charging-Performance-Index.

The Capacity is affected with temperature changes only on seasons' changes.

3.2 The Demanders.

Demanders are Actors who play as Energy Sinks. Five classes of demanding Actors are described below, in order of complexity. The user can use as many Actors of this kind as needed to model the Real Load of the system close enough. Even if the pre-defined Demanding Classes of Actor are not well suited for a given system load, it is possible and easy to derive a new class of demanding Actor specialized in the new requirements.

3.2.1 Tdemandante01.

This class models the most simple load type, corresponding to a constant one. The parameters of an Actor of this class are: Power and PN.

3.2.2 Tdemandante02.

The parameters of this Class are: Power, PN, Turn-On Hour and Turn-Off Hour. Tdemandante02 is intended to model those pieces of the system-load that have a daily cycle. A good example of this type of load is the one presented by the set of televisions in the system.

3.2.3 Tdemandanteestacional.

This Class, like Tdemandante02, allows the definition of Turn-On and Turn-Off hours, and in addition, a Turn-On Month, following which the Turn-On and Turn-Off hours are valid and a Turn-Off Month after which the actor goes to sleep until the next year, when the Turn-On Month comes again.

This Class is intended to model apparatus that are used seasonally such as electric heaters, air-conditioners, etc.

3.2.4 Tiluminacion.

Tiluminacion (Lighting), like its name suggests, is the class to model the behaviour of the

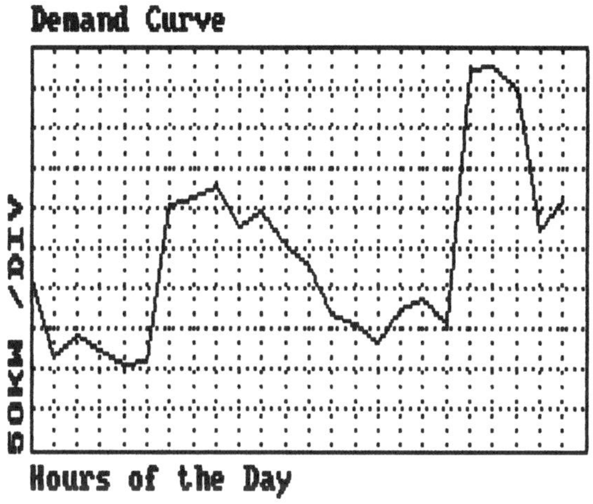

Fig. 1

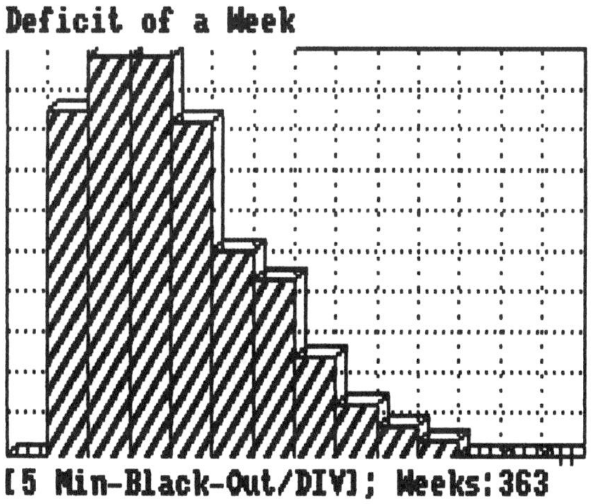

Fig. 2

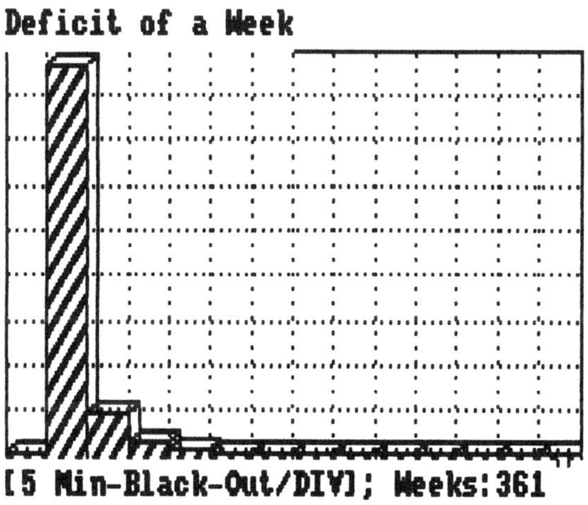

Fig. 3

6 Final Words.

This work, besides being a presentation of SimEnerg, is a suggestion of a methodology for the development of software tools for the analysis of autonomous-systems. Autonomous systems are not big enough to justify the development of a software program to face their analysis, so an approach like the one exposed here, in which incremental development of the software is possible, is good. New models are developed and programed only when they are needed and then added to a Library. After a Critical Mass (as the one included in SimEnerg) has been reached, the increment needed to face new problems is little enough to justify its development.

HYDROGEN STORAGE IN ISOLATED GRIDS WITH WIND ENERGY - ELECTROLYZER TECHNOLOGY AND ECONOMICS

H.Dienhart, G.Hille, W.Hug* T.Kürsten, F.Steinborn+
The German Aerospace Research Establishment
ITT-Dep. for System Analysis and Technology Assessment
*ITT-Hydrogen Energy Technology Group
+ZSW-Center for Solar an Hydrogen Energy Research
Pfaffenwaldring 38-40
D-7000 Stuttgart 80
Tel: +49-711-6862-674 Fax: -349

ABSTRACT: Hydrogen for seasonal storage - made up of electrolyzer, pressure storage and gas engine - is an attractive option for early hydrogen application. Special effects of electrolyzer operation with fluctuating wind like dynamical input, frequent shut down periods or frequent operation at partial load for different types of electrolyzers are investigated by experiment and simulation.

Furthermore economics of hydrogen storage on the basis of state of state art components in a small isolated grid (100 kW) made up of wind converters, battery, hydrogen storage and diesel is investigated and compared with hybrid systems with battery storage only. Economically hydrogen storage has substantial advantages compared with hybrid systems with battery at high renewable fractions. On the other hand hybrid systems with hydrogen storage still have to proof working reliable in big power systems.

1. INTRODUCTION

Within the Saudi-German project HYSOLAR one major task is the investigation of operation and improvement of efficiency of electrolysis operated with fluctuating renewable energy converters like photovoltaic (PV) or wind energy converters (WEC).

Furthermore systems studies are performed to identify early applications of solar hydrogen (H_2). In contrast to central energy systems where hydrogen is seen as long term option [1] isolated systems with seasonal storage demand have better conditions for early H_2 application like electricity storage or/and fuel for heating.

Isolated grids with H_2 storage are investigated here. This application does not require any infrastructure for distribution. Additionally H_2 storage does not suffer from self discharge and has economic advantages in big storage systems. The relatively cheap storage tank can be designed very big whereas the expensive charging system (electrolyzer) and decharging unit (e.g. gas engine) do not have to increase.

Precondition for H_2 utilisation is the economic competiveness and reliable operation compared with conventional diesel or wind/diesel systems. On the other hand tremendous external effects of fossil energy and depletion of fossil resources stresses the necessity of increasing the renewable fraction (share of renewable energy at total load) to achieve a sustainable energy system.

In the first part of this paper technical aspects of wind electrolyzer operation is described in detail. Differences between steady state and highly fluctuating conditions on efficiency, gas purity and operation strategy are described. In principle different types of electrolyzer (EL) - in terms of operating pressure and activation of electrodes - will differently react on fluctuating input. This requires a careful selection of EL depending on the application.
In the next section a techno-economic evaluation of small isolated grids with hydrogen storage is done by simulation to assess the role of H_2 in isolated grids with wind energy and high renewable fractions. Wind energy has low electricity costs at good sites which makes these systems suitable for high renewable shares. The hybrid system investigated is made up of state of the art technology with gas engine instead of fuel cell and EL with non activated electrodes resulting in relatively high specific energy demand.

Furthermore electricity costs of pure diesel supply and wind/diesel/battery systems are calculated which enables an economic assessment of H_2 storage compared with competing technologies.

2. TECHNICAL ASPECTS OF WIND ELECTROLYZER OPERATION

2.1 Inherent Particularity of WEC Coupled Alkaline Electrolyzers

For the performance of alkaline electrolyzers that are operated at constant power, the I-V characteristic normally gives sufficient information. The effective surface area and optional the catalytic activity of electrodes, the ohmic losses in the electrolyte and in the diaphragm as well as pressure and temperature during operation are the main parameters that influence this characteristic. Operation at elevated temperature and pressure lowers the specific power consumption for H_2 production.

Intermittent operation of PV and WEC coupled alkaline electrolyzers is characterised by distinct daily and seasonal power fluctuations with varying overload, partial load, shut-off and dynamic periods. Fig. 1 shows a typical WEC power profile over a period of 4 days. This power profile (Profile A) and the mean value (Profile B) represent an total of about 287 kWh DC energy .

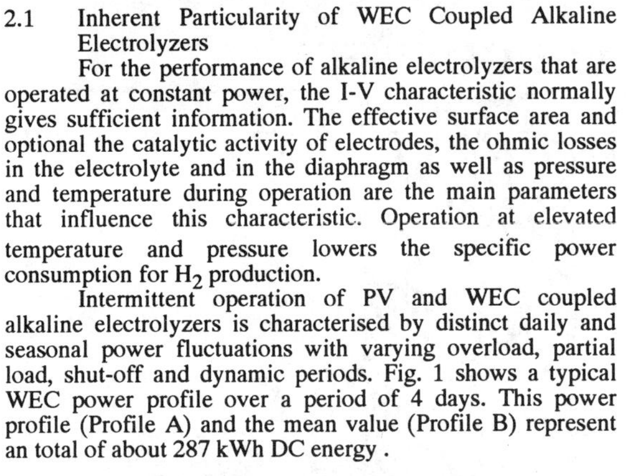

Figure 1: WEC power profile (10 minute mean values)

The inherent particularity of WEC coupled electrolyzers is discussed in the following chapters.

2.1.1 Partial Load Efficiency

The efficiency of an electrolyzer is determined by the thermodynamic efficiency, the current or Faraday efficiency and the gas efficiency. Gas efficiency considers the amount of H_2, which is not available for the user due to gas purities lower 100Vol%.

The thermodynamic or voltage efficiency is defined as the ratio of the voltage, equivalent to the lower (LHV) or higher heating value (HHV) of H_2 and the cell voltage of the system. Decreasing cell voltage with decreasing current density is the reason for higher thermodynamic efficiencies at partial load.

Faraday losses occur by parasitic currents along the gas and electrolyte ducts in the cell block and depend on current density, pressure, temperature and geometric configurations. The decrease of the Faraday efficiency down to 40% and less with decreasing current density is caused by the simultaneous increase of the parasitic current share along the gas ducts due to the increasing amount of electrolyte and therefore lower resistance. Higher temperature and pressure and thus lower electrical resistance along the gas and electrolyte ducts cause worse efficiency.

Gas impurities - apart from those that are caused by leakages - are due to inherent diffusion processes through the diaphragms. Pressure differences during intermittent operation, electrolyte temperature, efficiency of gas separators and again the current density are factors that influence the purity of the product gases. Gas purity becomes worse with decreasing current density. According to safety standards the operational limit for an electrolyzer is 2 Vol% (H_2 in O_2 and O_2 in H_2). Compared to the safety limit consumer sometimes require better H_2 purity. It depends on the storage and utilisation technology.

2.1.2 Potential Stabilization during Shut-off Periods

To prevent degradation of activated electrodes due to corrosion mechanisms, potential stabilization is necessary after power shut-off. Therefore a current density in the range between 5 and 10 mA/cm has to be supplied to WEC coupled electrolyzers during dead calms to compensate parasitic currents and to stabilize the potential of the electrodes. As a result, gases with a high amount of gas impurities are produced. Compared to PV coupling with periodical day and night cycles the accurate forecast of shut-off periods of WEC coupled electrolyzers caused by dead calms is not possible.

If the electrolyzer operates at ambient pressure, product gases during shut-off times are vent to atmosphere as long as the amount of impurities exceeds the safety limit. Ongoing operation at higher current density purges the system with gases of better purity and the product gases can be stored again if the required gas purity is reached.

2.1.3 Overload

As discussed in the previous chapters operation at partial load can reduce the current efficiency as well as the purity of the product gases remarkably. Therefore it would be an advantage especially from the safety point of view to lower nominal load of the electrolyzer to about 80% of the maximum power output of the WEC by reducing the surface area of the electrodes. This improves partial load behaviour and reduces the investment costs of the electrolyzer.

On the other hand the specific energy consumption at nominal load increases slightly and a heat exchanger for removal of excess heat during operation periods with peak wind velocities can be necessary.

2.1.4 Dynamics

One of the main criteria of the electrolyzer control unit is the dynamic response to level and pressure changes in the periphery and the cell block of the electrolyzer due to rapid load fluctuations in order to avoid system instability.

Recent experiments with one of the HYSOLAR 10 kW electrolyzers showed, that changes up to 200 mA/cm /s, corresponding to 40% of nominal load per second did not cause any problems. Experiments with frequencies higher 1 Hz failed by the dynamic response of the used rectifier. Short term measurements with the pitch controlled 100 kW WEC DEBRA [2] indicated no higher dynamics even at partial load, so that from the present point of view dynamics does not cause problems in WEC operation.

Level changes in the gas separators as well as different pressure drops in the anodic and cathodic gas and electrolyte channels can promote diffusion processes inside the cells, resulting in worse purity of the product gases. A systematic investigation of these phenomena will be part of future works.

2.1.5 Particular Features of Pressure Electrolyzers

Apart from reduced energy consumption for H_2 compression to conditions of pressure storage the thermal behaviour of an electrolyzer can be improved by operation under pressure. Lower heat losses at higher system pressure are due to a decrease of the amount of water vapour in the product gases with increasing system pressure.

In comparison the influence of system pressure on cell voltage and specific energy consumption of advanced electrolyzers is small. The disadvantage of a higher reversible potential at elevated pressure is compensated by the reduced amount of gas in the cell and the smaller covering of the electrodes with gas bubbles.

Current efficiency as well as gas purity becomes worse with increasing system pressure. The reduced gas volume and the higher diffusion rate are responsible for this.

Pressure electrolyzers normally keep the system pressure after shut-off to avoid energy losses when starting operation again. Gas production due to potential stabilization of activated electrodes during shut-off periods has low gas purity, so that the safety limits can be exceeded. This requires either expansion of the gases to atmospheric pressure or purging of the system with nitrogen, resulting in energy losses or additional operating costs.

2.2 Simulation Runs

The presented results are performed by the simulation model SIMELINT (SIMulation of ELectrolyzer in INTermittent operation), calculating thermal behaviour, performance, gas purities and current efficiency of a 10 kW electrolyzer for any given power or current profile. The results are in very good correspondence with measured constant and intermittent operation data, gained at the HYSOLAR 10 kW test and research facility in Stuttgart [3].

Input for the presented simulation runs are power profiles, recorded between 6.12.90 and 9.12.90 at the site of the 100 kW WEC DEBRA on the basis of 10 minutes mean values. With respect to the nominal load of the regarded HYSOLAR 10 kW electrolyzer, an advanced alkaline prototype with nominal current density of 500 mA/cm [4], the WEC profile is scaled down to 10% (Fig. 1).

Four different cases of constant and intermittent electrolyzer operation have been evaluated. These are:

1. intermittent operation of the 10 kW electrolyzer, the minimum current density is fixed to 2 mA/cm , power profile A
2. intermittent operation of the 10 kW electrolyzer, the minimum current density is fixed to 30 mA/cm , power profile A
3. steady state operation of the same 10 kW electrolyzer at 30% of nominal load (143 mA/cm), power profile B and
4. constant operation of a 3 kW electrolyzer at 102% of nominal load (462 mA/cm) and 78.3°C, power profile B.

Auxiliary power consumption of 200 W, system pressure of 5 bar absolutely and ambient temperature of 21°C was considered in the calculation. Starting temperature for simulation run 1, 2 and 3 was 45.7°C, corresponding to the steady state temperature of the electrolyzer in case 3. In simulation run 4 the performance data of the 10 kW electrolyzer are transferred to a fictive 3 kW electrolyzer, which is operated at is steady state temperature of 78.3°C. The results of the four simulation runs are summarized in Table 1.

Periods with gas impurities, exceeding the safety limits of 2 Vol%, are not considered in cumulative DC energy and H_2 amount. In the simulation runs the minimum and the maximum values for current efficiencies are limited to 20 and 98%.

Table 1. shows, that in intermittent operation (case 1) the total efficiency of 69.44 related to the lower heating value of H_2 is reached. The current density for potential stabilization was considered to be 2 mA/cm , a value being rather too low, according to the present state of the art of electrode

Simulation run No.		1	2	3	4
Power profile No.		A	A	B	B
Nominal load	kW	10	10	10	3
Minimum current density	mA/cm^2	2	30	143	462.4
DC energy input	kWh	286.2	299.8	287.9	287
Produced hydrogen	Nm^3	66.25	69.13	67.5	66.26
Spec. energy consumpt.	kWh/Nm^3	4.32	4.337	4.265	4.331
Current efficiency	%	20 - 97.6	67.7 - 97.6	94.6	97.6
Thermodyn. efficiency.	%	67 - 100	66.4 - 88.4	74.2	70.8
H_2 in O_2	Vol%	0.06 - 3.1	0.06 - 1.75	0.18	0.11
Total efficiency	%	69.44	69.17	70.33	69.26
Exceeding safety limit		yes, 3x	no	no	no

Table 1: Summarized results of simulated electrolyzer operation

development. Fig. 2 shows the electrolyzer response to a section of the used intermittent DC power input (Fig. 1, Profile A) in detail. The resulting profiles for current density, temperature and gas impurity H_2 in oxygen are given for the section of the 2nd day in the considered operation period.

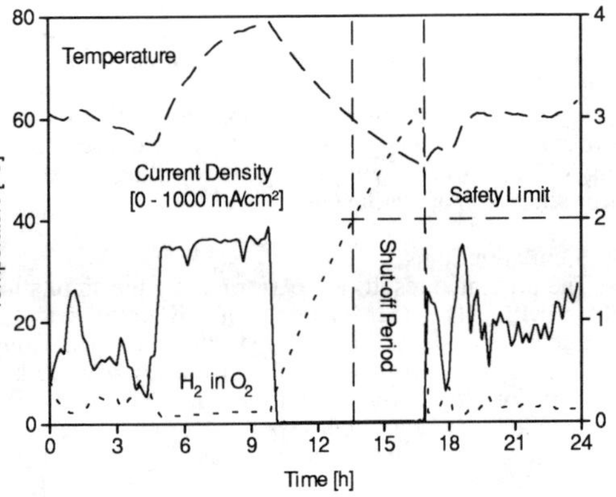

Figure 2: Electrolyzer response to power profile A (2nd day, current density for potential stabilization 2 mA/cm)

As an disadvantage, three times the safety limit was exceeded. The delay time between ongoing current flow and H_2 storage, caused by the gas capacity of the separators (geometric volume x pressure), was the reason for additional losses. The course of the efficiencies for the same day are given in Fig. 3.

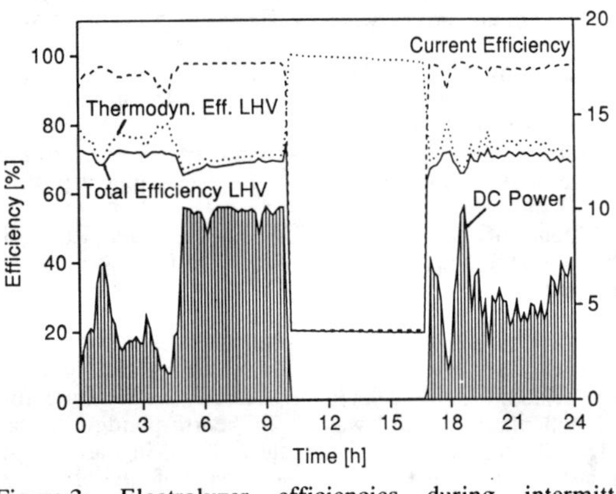

Figure 3: Electrolyzer efficiencies during intermittent operation (Profile A, 2nd day, current density for potential stabilization 2 mA/cm)

The influence of the low current efficiency during potential stabilization on the total efficiency is quite negligible.

To avoid the disadvantage of exceeding the safety limit during periods of potential stabilization, the electrolyzer can be operated at an elevated current density level as grid or battery supplied ground load, so that even under unfavourable conditions (shut-off immediate after overload operation with high system temperature) a safe and reliable operation is guaranteed. This can be an additional advantage, if the produced H_2 is also used for heating purposes and not only for seasonal storage of electrical energy. This was the basis for simulation run 2. Compared to the corresponding simulation run without ground load, Table 1 shows as well an higher required DC energy as well as the benefit of more produced H_2 and a better gas purity. The effect on the total efficiency depends on the power profile an the frequency of shut-off periods. In the considered case a slightly lower total efficiency can be observed.

Constant operation at 30% of nominal load (case 3) results in the best total efficiency with 70.33%. The main reason for this fact is the high thermodynamic efficiency.

Last not least in case 4 constant operation at nominal load of a down scaled electrolyzer is simulated, resulting in an total efficiency in the range between case 1 and 2. The main benefit of this case are the reduced investment costs of the electrolyzer.

3. ECONOMIC ASPECTS OF HYDROGEN STORAGE IN ISOLATED GRIDS

3.1 Boundary Conditions

The hybrid system investigated (Fig. 4) is made up of wind energy converters (WEC), battery for short term storage and H_2 for seasonal storage. Furthermore a back up engine can be regarded (e.g. diesel). The concept of the hybrid system is derived from the well established wind/diesel/battery systems Cape Clear [5] and extended by state of the art components for H_2 storage, a low pressure alkaline electrolyzer with non activated electrodes (spec. energy consumption=4.8 kWh/m^3), single stage H_2 compressor (80 bar), pressure storage and H_2 gas engine. This conventional technology leads to a mean efficiency of only 18% whereas battery has an overall efficiency of about 70% (including inverter. Important technical data of all components are shown in Tab. 2.

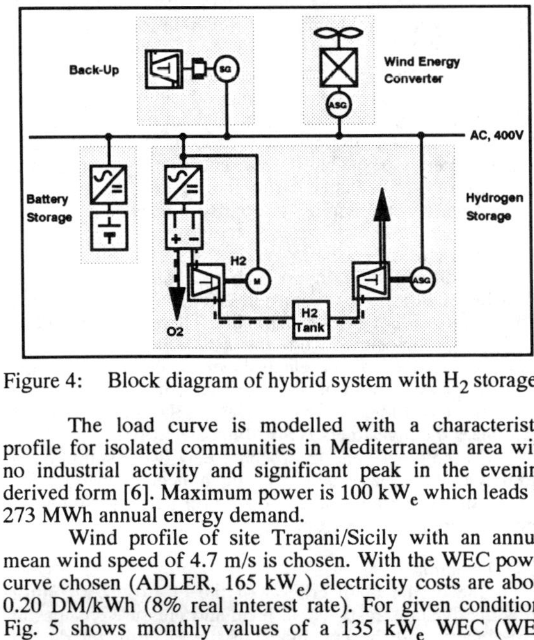

Figure 4: Block diagram of hybrid system with H_2 storage

The load curve is modelled with a characteristic profile for isolated communities in Mediterranean area with no industrial activity and significant peak in the evening derived form [6]. Maximum power is 100 kW_e which leads to 273 MWh annual energy demand.

Wind profile of site Trapani/Sicily with an annual mean wind speed of 4.7 m/s is chosen. With the WEC power curve chosen (ADLER, 165 kW_e) electricity costs are about 0.20 DM/kWh (8% real interest rate). For given conditions Fig. 5 shows monthly values of a 135 kW_e WEC (WEC annual output=annual load) and load. Seasonal storage has to transfer wind electricity accumulated during periods of excess (e.g February-May) to periods of lack of supply (e.g. June, July). This stresses the importance of seasonal storage for this example.

Important financial data on the basis of year 1992 are summarized in Table 2. The cost data are derived from manufacturers survey in Germany and do not include taxes, custom duties and transportation costs to remote sites. Costs of EL are based on a market survey [7] and depend strongly on the chosen power range. Costs of H_2 gas engines are derived from [7] where cogeneration units are documented.

The Economic lifetime of the components is either 20 a (5 a for battery) or 30000 h of operation in case of diesel and gas engine with maximum of 20 a. Real interest rate is 8% which is usual for federal development banks.

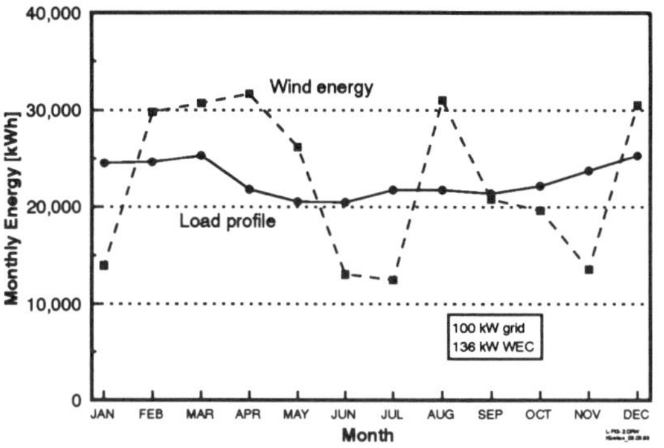

Figure 5: Seasonal profiles of load and wind energy production (annual WEC output = annual load)

System	Nominal Power	Invest. Costs	O & M	outp. related Costs	Economic lifetime	Notes
Unit	kW_e	DM/kW_e	% of IC 3)	DM/kWh		
WEC	165	3300	2.5	0	20 a	characteristic ADLER_165
Diesel	100	1100	8	0	30 000 h	eta = 30%
Battery	100 1)	400 2)	2	0	5 a	eta = 69%
- Periphery		600 2)			20 a	max. DOD=30%
- Inverter	100	1400			20 a	
Electrolyzer +Compressor	100	4450	3	0.066	20 a	eta = 59% consumpt. compressor = 0.091 kWe/kWth (LHV)
H2 Tank	20000 1)	30 2)	2	0	20 a	80 bar
H2 Gas Engine	100	2700	3.5	0.025	30 000 h	eta = 30%

1) Unit-kWh 2) Unit-DM/kWh 3) = Investment Costs

Table 2: Technical and economic parameters for hybrid systems (status 1992)

Cost minimal hybrid systems are derived by simulation with the program INSEL-NETZ [9] (calculates energy flow on hourly basis) and the program HYCOST [10] (cost calculation on the basis of the annuity method). Variable optimization parameters are WEC (300 - 600 kW_e in steps of 100 kW), EL (100, 200, 300 kW_e), battery (0.5, 1, 3 days of mean load) and H_2 tank (steps of 2000 kWh). H_2 gas engine is constant at 100 kW to match the load.

3.3 Results of Autonomous H_2 Hybrid System Design

Electricity generation costs for H_2 hybrid systems without diesel (renewable fraction 100%) are shown in Fig. 6. Storage size is expressed in "days of mean load" which accounts for about 750 kWh for the load profile regarded.

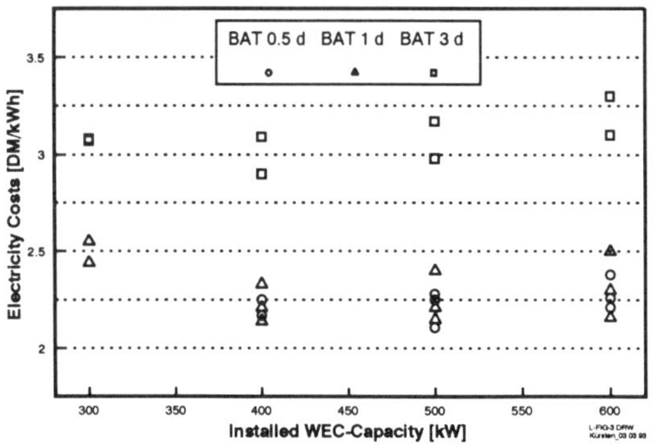

Figure 6: Electricity generating costs for autonomous systems with H_2 storage (no back up); differnt EL (100-300 kW_e sizes for 3 batteries investigated

The least cost option has the following data:

- WEC = 500 kW_e
- Battery = 0.5 days of mean load
- EL = 200 kW_e
- H2 tank = 48 days of mean load

This system has electricity generating costs of 2.11 DM/kWh. One also realizes that some other hybrid systems - with different component sizes - have only about 2% higher costs. This means that additional optimization criteria like minimal capital requirement or maximal excess electricity (in case of additional consumer like hydrogen production for heating) could be relevant.

The partition of WEC production - which is about 3.6 times annual load - is plotted in Fig. 7. About 60% of load is supplied directly by WEC, 20% by hydrogen and battery each. Because wind energy is fairly cheap economic optimisation favours dumping wind electricity (about 2 times annual load) over big storage. This has to be investigated in more detail as utilizing this energy could be an additional source of income.

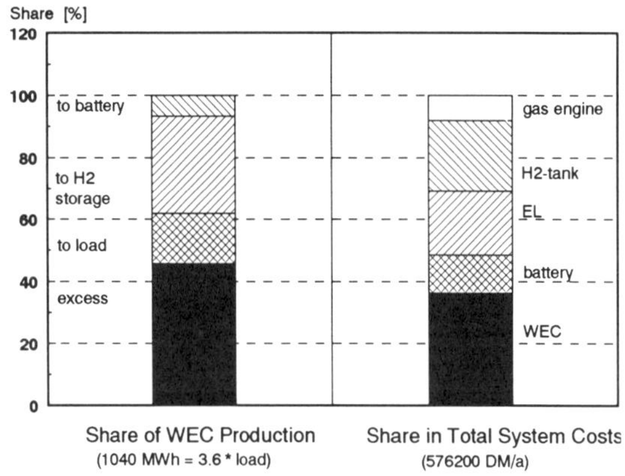

Figure 7: Partitition of WEC production and share in annual costs for least cost H_2 system (load=273 MWh/a)

For the least cost system examined Fig. 7 also shows share in annual costs. Battery for example contributes about 12% to total costs. For the next best systems seen in Fig. 6 the share of components will change. Only gas engine has a constant cost share of about 8% because this component has constant size.

3.4 Sensitivity Analysis

A variation of parameters can be done to assess their impacts on the costs of hybrid systems compared to the reference system. EL investment costs for example depend strongly on nominal power which means a smaller grid has higher and a bigger grid has lower costs. Calculation with 50% lower EL costs (EL in the 1MW range) reveal only 8% lower electricity costs. This is due to the small share in total costs of EL (21% in reference case).

On the other hand an increase of investment costs of all components - possibly due to transportation to remote location - almost proportionally affects the electricity costs.

A more detailed sensitivity analysis is planned for future work.

3.5 Comparison with other Systems

For assessing the results of systems with H_2 storage Fig. 8 shows electricity costs for different hybrid systems with battery storage (systems 1-7). A H_2 system with renewable fraction 95% (system no. 8) was also calculated which incorporates a diesel back up of 100 kW (lifetime 20 a). This significantly reduces electricity costs (by 18%) compared with the autonomous system No. 9. The impact of even smaller renewable fraction on H_2 hybrid system has to be investigated.

One realizes that even for unfavourable technical and economic conditions of present status H_2 storage is an economical must at high renewable fractions. Otherwise big and expensive batteries are necessary.

Considering the hybrid systems with 95% renewable fraction H_2 storage offers a cost reduction of 32% compared with battery only.

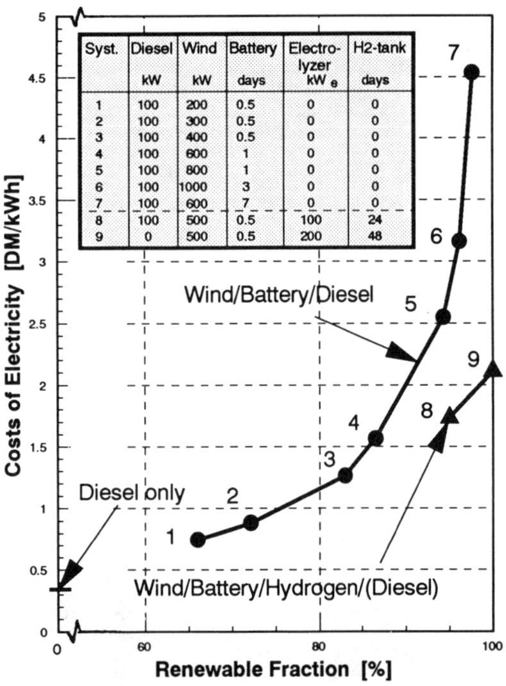

Figure 8: Comparison of electricity generation costs for various optimized systems (fuel price = 0.05 DM/kWh = 0.5 DM/l diesel, storage content in days of mean load)

3.6 CO_2 Avoidance costs

A pure economic evaluation without considering external effects of energy supply is not appropriate for solving current environmental problems like greenhouse effect. Reducing CO_2 emission in comparison to pure fossil supply can results in extra costs, depending on system configuration and fuel price. For example CO_2 avoidance costs for hybrid H_2 system (95% renewable fraction, low diesel price of 0.5 DM/l) are about 1560 DM/t. This is in the range of combined cycle coal plants compared with conventional coal fired plants as reference (based on US-American boundary conditions). Furthermore electricity costs with H_2 are significantly reduced for high renewable fractions which results in a higher CO_2 reduction for given amount of specific CO_2 avoidance costs available.

4. CONCLUSIONS

Following conclusion are derived form this work. The special boundary conditions for economic evaluation have to be considered.

- in principal WEC electrolyzer operation is possible; specific operation strategy has to be applied depending on type of EL
- dynamics of WEC output will not seriously affect EL operation
- in systems with frequent partial load nominal power of EL should be smaller than max. load; this has to be optimized for the special application
- for EL with activated electrodes energy for potential stabilization is required (about 1-2% of nominal power); an optimization between additional energy consumption versus improvement of efficiency is necessary for any special case
- using pressurized EL in systems with frequent shut-off periods an optimization between reduced energy demand for compression of H_2 (for storage) versus additional energy for building up pressure after expansion during shut off is necessary
- economically H_2 is very attractive for high renewable fractions in wind systems, even for present unfavourable technical conditions (low efficiency of 18%) and high costs of H_2 technology
- back up in H_2 storage systems improves the cost situation considerably; this holds even for an increase of fuel price as back-up fuel has only a small share on total costs
- utilisation of excess electricity could improve economic situation (e.g. production of hydrogen gas)
- H_2 storage has not proven working in power systems yet, especially operation strategies for these complex systems have to be tested in reality

REFERENCES

[1] J. Nitsch, Strategies for a Solar Hydrogen Economy (Published in German), VDI-Report No. 912, 1992

[2] H.Böhnisch, A.Kussmann, K.Reininger, R.Thiesen, Type-specific Dynamic Measurements on Wind Turbines at the Test Site Schnittlingen (Published in German), The German Aerospace Research Establishment Reports, DFVLR-Mitteilungen 88-31, 1988

[3] W.Hug, H.Bussmann, A.Brinner, Intermittent Operation and Modelling of an Alkaline Electrolyzer, Proceedings of the 9th World Hydrogen Energy Conf. Paris, 1992

[4] A.Brinner, H.Bussmann, W.Hug, W.Seeger, Test Results of the HYSOLAR 10kW PV-Electrolysis Facility, Int. J. Hydrogen Energy, Vol. 17, No. 3, pp. 187-197, 1992

[5] G.Cramer, W.Kleinkauf, T.Shott, Wind/Diesel/Battery Systems - Application, Experiences, Economy of Different System Configuration and Sizes, Proceedings of ECWEC, Madrid 1990

[6] H.Klaiß, F.Staiß (Eds.), Solar Thermal Power Plants in the Mediterranean Area, (in German), Springer Verlag, 1992

[7] S.Klein, A.Siegel, T.Schott, H.Klaiss, Economic Optimization of Photovoltaic-Electolyzer Systems, DLR-German Aerospace Research Establishment, Internal Report (in German) IB-444 003/88, Stuttgart 1988

[8] Technologies for Hydrogen Applications: Significance, End-Use Technologies and Technical and Economic Development Potentials, Centre for Solar and Hydrogen Technology (ZSW), The German Aerospace Research Establishment (DLR), Ludwig Bölkow Systemtechnik (LBST) (in German), March 1992

[9] H.Dienhart (DLR), F.Steinborn (ZSW), Program Documentation Simulation Code INSEL-NETZ, DLR-Internal Report (in German), Stuttgart 1993

[10] H.Dienhart, Program Documentation of Hybrid System Cost Calculation Management Program, DLR-Internal Report (in German), Stuttgart 1993

Experimental Wind-Diesel System at the German Institute for Wind Energy - Layout of the Storage Unit and First Operational Experience

H. G. Beyer *, T. Degner, G. Gerdes, F. Santjer

Deutsches Windenergie-Institut, D-2940 Wilhelmshaven, Ebertstraße 96
* Fachbereich Physik, Carl von Ossietzky Universität Oldenburg, D-2900 Oldenburg, Postfach 2503

Abstract: At the test site of the Deutsches Windenergie-Institut a wind-diesel test station is operated. In the final stage the test station will offer the opportunity for the comparision of various system configurations and system controll strategies. The layout of the test station is described and first results from system operation are reported.
The analysis of the system performance is accompanied by simulation calculations. Simulations are also used to size the components of the system. The paper reports the results of calculations concerning the layout of the battery storage that will be implemented in the system.

1. Introduction

In 1992 the first basic components of a wind-diesel test station were installed at the wind turbine test site of the Deutsches Windenergie-Institut (DEWI). The test station will give the opportunity to compare the performance of wind-diesel systems with various configurations. Special emphasis will be put on the comparision of complex state of the art designs (including battery storage and inverters) with systems of simpler structure. For both types of concepts advanced control strategies should be applied. Wind turbines with different design concepts (fixed speed pitch controlled, fixed speed stall controlled, variable speed) may be used to identify the best match of turbine and system design.

The evaluation of the performance data gained with the system is backed up by simulation calculations. The simulations will be used to extrapolate short term experimental data to annual performance figures under different climatic conditions. During the planning phase simulations are used for the sizing of the system components. Especially the layout of the battery storage is under consideration.

2. System configuration

The wind-diesel test station is located about 20 km north of the city Wilhelmshaven and 4 km from the North Sea coast. The terrain is flat, the average annual wind speed is 5.1 m/s at 10 m height above ground level.

The final layout of the system is shown in figure 1. Different types of turbines (in the range up to 50 kW) can be connected to the system. Up to now a fixed speed, pitch controlled turbine is installed (HSW 30, manufacturer Husumer Schiffswerft). The turbine is a two bladed machine with 2 stage induction generator (8.5 kW resp. 30 kW rated power). Rotor diameter is 12.5 m and hub height 22.4 m.

The diesel generator set consists of a diesel engine (46 kW), an electromagnetic clutch, and a synchronous generator (30 kW).

The wind turbine simulator consists of a computer controlled DC-motor (50 kW) and a generator. The motor is able to reproduce the torque/rotational speed characteristics of a wind turbine rotor implemented in the computer control. The generator is interchangeable. At the moment the mounted generator is identical to the one used in the HSW 30 turbine.

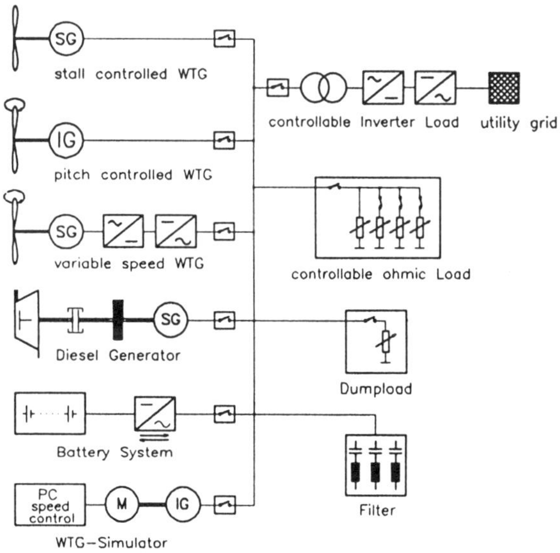

Figure 1. *Final layout of the experimental wind-diesel test station. The test station will give the opportunity to compare the performance of wind-diesel systems with various configurations.*

Two types of load simulators are implemented. The controllable inverter load can be continuously regulated to consume electric power up to 75 kW. This power is fed into the utility grid. The controllable ohmic resistors can be regulated to a total load of 127 kW in steps of 1 kW. Both loads are able to simulate various electrical consumers independently.

The battery storage and the other two wind turbines are considered to be installed this year.

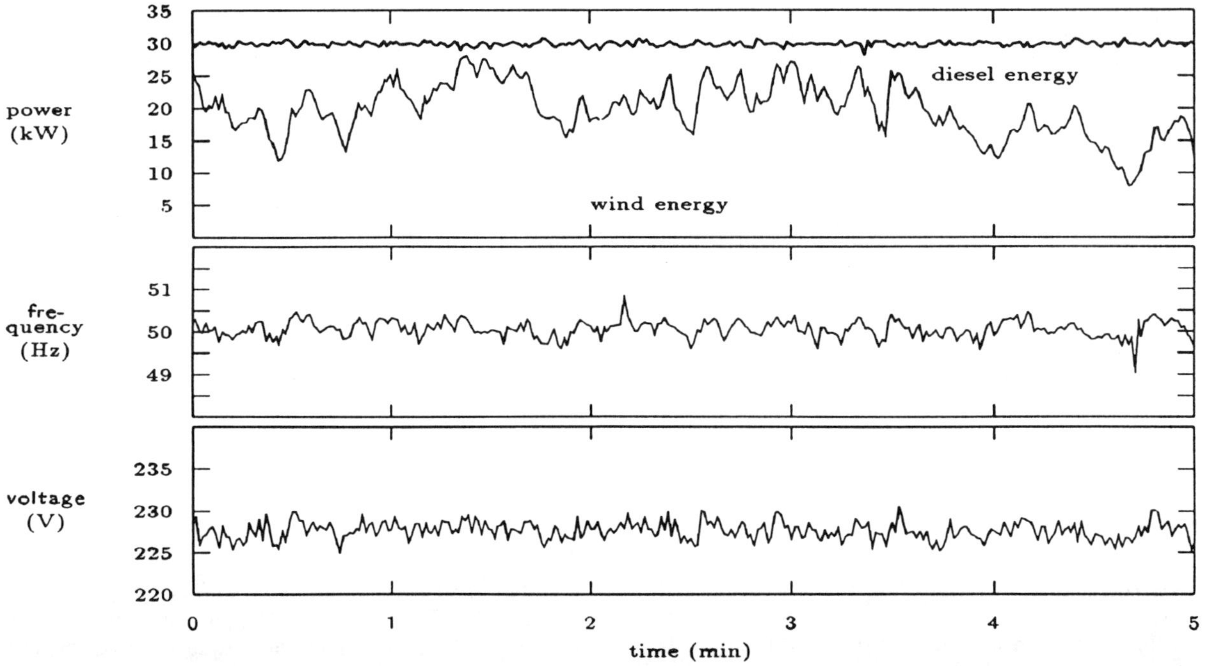

Figure 3. *A five minute section of the parallel operation of wind turbine generator and diesel generator. Shown is the power output of the wind turbine (top, light line). The difference to the load demand (top, bold line) is delivered from the diesel generator set. In addition system frequency (middle) and system voltage (bottom) are shown.*

The single components of the wind-diesel system are linked by a standard AC-grid (400 V, three phases, frequency 50 Hz). The layout of the system allows the connection of additional components (wind turbines, diesel generators, etc.).

3. Measurements at the system

The introduction of wind energy in a diesel electric grid aims at the reduction of the diesel fuel consumption. To analyse the fuel saving effect of the wind energy the knowledge of the fuel consumption/electric power relation is required. Therefore first measurements of the fuel consumption of the diesel at different load levels are performed. To gain first experience with the complete system the system behaviour under parallel operation of wind turbine and diesel generator is monitored.

3.1. Fuel consumption of the diesel

Figure 2 shows the fuel consumption of the diesel as a function of the electrical power output of the synchronous generator. The squares represent measured values while the solid line is the approximation used in the system simulation (see next chapter). The fuel consumption was measured under constant load conditions. For the evaluation of negative load conditions the diesel was powered by the wind turbine simulator. In the range from -10 kW to 25 kW the fuel consumption of the diesel is a quasi linear function of the electrical power output. At high load levels the fuel consumption increases more than linear due to the lower efficiency of the synchronous generator when operated above rated power (30 kW).

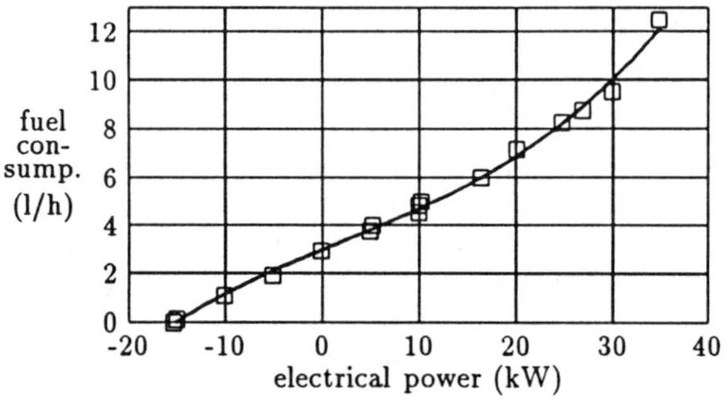

Figure 2. *Specific fuel consumption of the diesel at the DEWI test station. Rated power of the diesel generator is 30 kW. Squares show measured values while the solid line gives the approximation used in the system simulation*

3.2. Parallel operation of wind turbine and diesel generator

We analyse a 23.5 minute period with parallel operation of the diesel generator and the wind turbine. Figure 3 shows a five minute clip of this period. The thin line shows the electrical power output of the wind turbine. The difference to the load (bold line) is supplied by the diesel generator. In addition grid frequency and voltage are shown.

For this measurement the controllable inverter load was used to simulate a constant load of 30 kW. The plot shows that the load is almost but not exactly constant. The diesel control keeps the frequency constant within 49 Hz and 51 Hz (± 2 %) and the voltage between 225 V and 230 V (± 1 %).

Average power output for this time period is 8.3 kW from the diesel generator and 21.7 kW from the wind turbine generator, that is 72 % of the load demand.

The measured fuel consumption of the diesel was 1.79 l, i.e. 4.55 l/h. Compared to the consumption at 30 kW load (9.5 l/h) the fuel saving is 5 l/h or 52 %.

The average fuel consumption can be estimated from the electrical power output of the diesel generator using the relation fuel consumption versus power output (figure 2). This gives an average fuel consumption of 4.6 l/h which is close to the measured values. Further measurements, carried out over longer time spans, have shown that the fuel consumption of the diesel can in general be obtained from the relation given in figure 2 with high accuracy.

4. Layout of the storage unit

The wind-diesel test station should be equipped with a battery storage unit. Simulation calculations are performed to investigate the effect of the storage size on the wind-diesel system. We analyse the diesel fuel consumption, the diesel start/stop cycling frequency and the energy flow to the storage. We present a sizing of the storage for two different battery types and calculate the expected battery lifetime.

4.1. System model

In the simulations the power flows in the system are modelled. The system simulation is done in the time domain for a time span of one year. As typical time scales in the variations of the wind speed are in the order of seconds, a time step of approximately one second was chosen. The wind turbine model represents fixed pitch three blade wind turbines. The rotor is characterised by a $c_p(\lambda)$ curve. The generator is of induction type with a maximum slip of 8% and power rating of 30 kW. Rotor diameter is 11.9 m and hub height 22.4 m. The diesel generator has 30 kW rated power, the specific fuel consumption is derived from measurements at the diesel of the test station (see fig. 2). The minimum diesel generator load requirement is 20% of the rated power. The storage unit is not specified technically. It is modelled by an energy balance with charging efficiency of 0.6. The storage size is given in units of the time the storage can cover the average load demand. Storage sizes from 0.4 minutes up to 9.4 days of load demand are analysed. The power to charge or discharge the storage is limited by the power ratings of the rectifier and inverter (here: 30 kW). For the consumer load we assume a daily load pattern. The load varies between 6.6 kW (during night time) and 20 kW, average load is 8.9 kW. The load is constant within one hour.

The system control strategy ensures that the load demand is covered allways. The diesel is stopped if (i) the power output of the wind turbine is greater than the load demand (ii) the five minute average of the wind turbine power output is 1.05 times greater than the load demand (iii) the storage is fully charged.

The Results are compared to two wind-diesel systems without storage:

- System 1: the diesel is running continuously
- System 2: the diesel is stopped if (i) the power of the wind turbine generator is greater than the load demand and (ii) the five minute average of the wind turbine generated power is 1.05 times greater than the load and (iii) the diesel is operated at least one minute.

4.2. Wind speed time series

Input for the model of the wind turbine generator are hybrid wind speed time series. The low frequency part of the time series is based on measured hourly wind speed time series while the high frequency part consists of synthesized wind speed fluctuations.

During 1.10.91 to 30.9.92 wind speed was measured at the test station in 10 m and 34 m height a.g.l.. Under the assumption of a logarithmic wind profile these data were used to calculate hourly mean speed at 22.4 m height. Annual wind speed is 5.4 m/s in 10 m height and 6.2 m/s in 22.4 m height. The short time fluctuations of the wind speed are synthesized taking into account the turbulence characteristics of the wind speed in the atmospheric boundary layer. A turbulence intensity of 0.16 was assumed (see /1/ for further details).

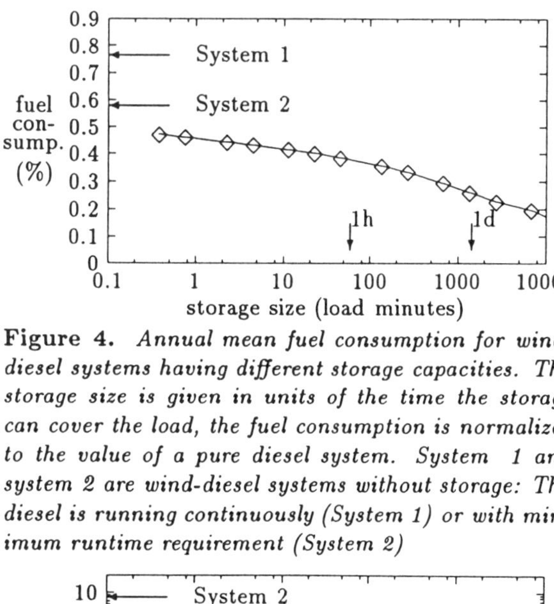

Figure 4. *Annual mean fuel consumption for wind-diesel systems having different storage capacities. The storage size is given in units of the time the storage can cover the load, the fuel consumption is normalized to the value of a pure diesel system. System 1 and system 2 are wind-diesel systems without storage: The diesel is running continuously (System 1) or with minimum runtime requirement (System 2)*

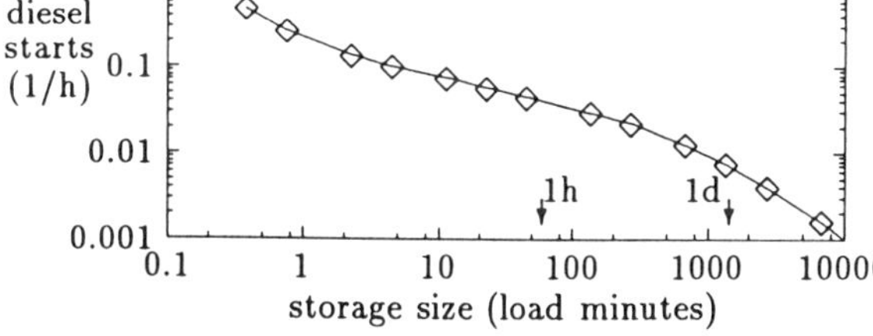

Figure 5. *Annual average diesel cycling frequency as a function of the storage size. System 2 is a wind-diesel system without storage.*

4.3. Results

Under the climatic conditions mentioned above the wind turbine generator produces about 8.7 kW in the annual mean, which is nearly the same as the the average load demand. Due to the mismatch between power productions and load demand the produced power can be used in the system partly only.

Fig. 4 and 5 show the system performance with respect to different storage capacities. Compared to the system with no storage even a marginal storage capacity leads to a significant reduction in the start/stop cycling frequency. The expected fuel savings are given compared to a system without wind turbine and without storage.

E.g. the system with a 11 minute storage needs 45 % less fuel than the system 1 (diesel running continuously). On the other hand the fuel saving obtained with a 100 minute storage is only by a factor of about 1.2 greater than the one obtained with a 10 minute storage. Substantially higher fuel savings can be obtained with large storage sizes only (one day up) which are out of economic scale.

4.4. Constraints in the application of batteries

When sizing a battery storage one must consider, that the amount of energy which can be drawn from a battery depends strongly on the ratio of discharge power to storage capacity. Thus the nominal battery size can not directly be linked to the storage size as defined above.

We evaluate data from a manufacturer for two battery types suited for high current applications: a Pb-battery with bar-shape plates and a NiCd battery with sintered plates /2/. We assume a maximum discharge power of 20 kW (equivalent to the maximum load) and a cutoff voltage of 1.75 V/cell for the Pb resp. 1.0 V/cell for the NiCd-battery.

Under these assumptions figure 6 shows the required nominal capacity as function of the storage size. To realize a 10 minute storage a Pb-battery must have a nominal capacity of 70 minutes and a NiCd-battery a nominal capacity of 23 minutes. With a nominal battery bank voltage of 400 V the Pb-battery consists of 200 cells 30 Ah each and the NiCd battery of 333 cells 10 Ah each.

The calculations show that it is reasonable to choose a nominal battery capacity which is equivalent to a 10 minute storage. Smaller storage sizes will not result in much smaller nominal battery capacities. An increase of the storage size will not have a reasonable pay off as seen in figures 4 and 5.

Regarding the battery storages it must however be taken into account that each battery can handle only a limited amount of total energy flow. Battery lifetimes are given from the manufacturer as 1000 charge/discharge cycles for the Pb and 3000 cycles for the NiCd battery. We take one cycle as the amount of energy drawn from a battery during discharging from 100 % to 20 % state of charge. As the NiCd battery is about three times more expensive in costs per nominal capacity, the costs per delivered energy unit are comparable for both types of batteries.

On the basis of the power flow to/from the storage the expected battery lifetimes for both battery types was evaluated. The result is shown in figure 7. Up to a storage size of 20 minutes lifetime of NiCd and Pb batteries are comparable. For greater storage sizes the lifetime of the NiCd battery exceeds the lifetime of the Pb battery. However as mentioned above there is no economic advantage.

We like to point out that the absolute lifetimes for the batteries given here will depend strongly on various factors which among others are the characteristics of the fluctuations of the wind speed, the consumer load properties (e.g. the ratio of maximum load to mean load) and the system control strategy.

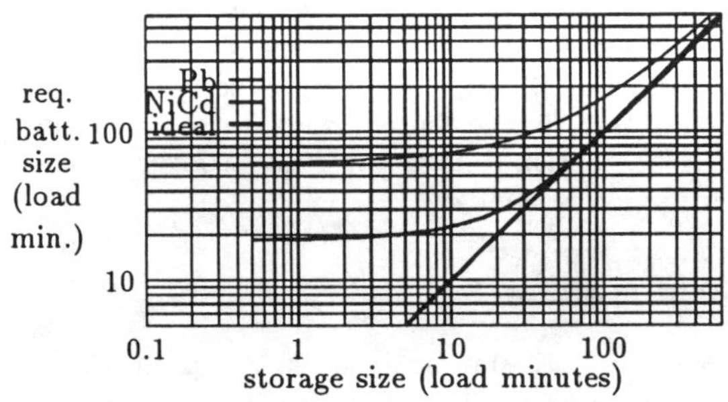

Figure 6. *The figure shows the required nominal battery size/storage size relation. The values are calculated for a ratio maximum discharge power to average load of 2.3. Given is the relation for a Pb-battery with bar-shape plates (light line), for a NiCd-battery with sintered plates (medium line) and for an 'ideal battery' (bold line).*

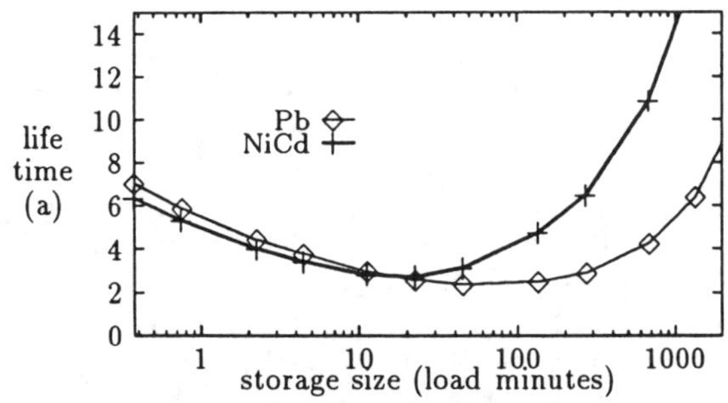

Figure 7. *Expected battery lifetime of Pb and NiCd batteries operated in wind-diesel systems. Lifetime is calculated on the basis of the power flow through the storage.*

References:

/1/ T. Degner, *Synthese von Zeitreihen der Windgeschwindigkeit*, internal report, Carl von Ossietzky Universität Oldenburg, 1992

/2/ VARTA Batterie AG, *VARTA-Grundsatzbroschüre, Staionäre Batterien, Systemmerkmale*

ENGINEERING DESIGN TOOLS FOR WIND-DIESEL SYSTEMS: PRESENTATION AND VALIDATION OF THE MODULAR DYNAMIC MODEL.

K. Uhlen, T. Toftevaag,Norwegian Electric Power Research Institute (EFI), Trondheim, Norway
H. Bindner, P. Lundsager, Risø National Laboratory, Roskilde, Denmark
J.T.G. Pierik, Netherlands Energy Research Foundation (ECN), Petten, Netherlands
D. G. Infield, A. Scotney, Rutherford Appleton Laboratory (RAL), Didcot, UK,
M. Falchetta, ENEA Casaccia, Rome, Italy,

ABSTRACT: A design tools software package for wind diesel systems has been developed with support from the CEC JOULE programme. The package has been developed cooperatively by several EEC and EFTA countries. The complete package includes modelling of both dynamic and logistic aspects of system operation and runs on a PC. This paper describes the dynamic model JODYMOD and its validation against measured results from two different wind diesel systems.

1. INTRODUCTION

The CEC JOULE project " Engineering Design Tools for Wind-Diesel Systems" (Jour-0078) has resulted in two different modelling packages: A logistic one for analysis of long term system performance and a dynamic one called JODYMOD, developed for stability analysis and controller design [1], [2]. The modular layout of the model allows the user to specify a wide range of system designs by combining component models.

This paper deals with the dynamic model and emphasis is put on the validation of the model. The wind-diesel systems at Risø [3] and Froya [4], working in different operational modes, are used for the validation thus demonstrating the flexibility of the model. Identification of model parameters is discussed and results from the model validation work are presented.

The modular dynamic model developed in this project will soon be available to users.

2. THE DYNAMIC MODEL JODYMOD

The dynamic model is based on a concept similar to the finite element (FEM) technique in structural mechanics, where the equations of a model are built by the program according to the user's system specification [1]. The general simulation program Cypros/ESIM [5] has been chosen as environment for the model.

The model is modular in the sense that it gives the user the opportunity to build a system model by combining a set of available component models. The component model library is given in table 1. Each component has inputs and outputs, and a connectivity vector is used to connect the outputs to the inputs in order to build the system equations. The electrical and mechanical component equations are outlined in [1].

The electrical machine models allow the machine equations to be solved in a matrix structure suited for automatic assembly. Synchronous and induction machines are represented in a combination of the representations of [6] and [7], and saturation of the electrical machines are dealt with in an iterative process operating on the entire electrical system equations.

Mechanical components such as wind turbine drive trains, diesel generator assemblies etc. are represented by FEM type equations. A transformation [8] extracts the rigid body motion of each component and connects it to the corresponding electrical machine through the air gap torque. A routine is included that allows the user to define models of rotational systems with gears with an option to represent a suspended gearbox and a synchronous flywheel. The user may also input the structural matrices directly.

A prototype version of JODYMOD has already been operative for some time as part of a validation exercise. This paper presents results from this ongoing exercise.

Table I: Component library available in JODYMOD. The table includes component name and max number of same component type in a system model. TS is a Cypros file format [5]

Type	Compo nent	Max #	Component description
Elec-trical	BBL	1	Busbar/Trans. line/Consumer load
	DPL	1	Dump load
	CAP	1	Capacitor bank
	BAS	1	Battery storage w/ converter
Electro Mech.	SGE	2	Synch machine, electrical model
	AGE	2	Induc machine, electrical model
Mech-anical	DEL	4	Diesel engine, combustion model
	WTL	4	Wind turbine, aerodynamic model
	STR	4	Structural mechanic system
		10	Max # DOF's per structure
Sensors	SEN	10	Sensor model, filter & noise
Con-trollers	VRG	4	Simple voltage regulator
	AVR	4	Advanced voltage regulator
	GOM	4	Mechanic diesel governor
	GOE	4	Electronic diesel governor
	PID	5	Generic PID controller
	GDC	1	MIMO dynamic compensator
Signals, Input	SIG	20	Signal source (set points etc)
	ETS	10	External source (TS file input)

3. SCOPE OF MODEL VALIDATION

The model is validated experimentally against measured results from two actual systems. The scope of the validation is primarily to verify the model's ability to predict system stability and power quality. Both component models and system models are verified. Emphasis is put on electrical machine models and voltage regulators, including an assessment of the problems encountered in the determination of machine parameter values. The validation includes both short term step transients and long term steady operation.

Step transients are interesting in the context of detailed component and controller behavior. Step transients are evaluated by 'visual' means, i.e. by inspection and reasoning. The long term results are interesting in relation to the expected output from a system in terms of system stability and power quality. Such results are evaluated by 'statistical' measures such as binning and density functions. In this validation exercise rain flow counting is tentatively applied to represent power quality.

4. RESULTS FROM THE FROYA SYSTEM

4.1 Description of the system

The Norwegian system is developed jointly by Norwegian Electric Power Research Institute and ABB Energy, and the system is installed in the island Froya. This is a modular design consisting of a 55 kW stall regulated wind turbine, a 50 kW diesel generator set (with clutch, but no flywheel), battery storage unit with line-frequency converter and a switched resistor type dump load. All components are connected to a common AC busbar.

The only mode of operation investigated in this paper is represented by stand alone wind turbine controlled by the dump load and the synchronous generator (diesel engine is stopped). Excitation of the SG is controlled by a PI type controller while the dump load controller is P-type only. The converter is active only to maintain a very low charging current. The parameters of the system are well known from previous measurements.

4.2 Response to step load change

The figures in this section show results from a test where a 9.5 kW resistive load is switched on.The initial load is 22 kW and the wind speed during the transient is between 10 and 11 m/s. The corresponding step load change is simulated with a constant wind speed input of 10.5 m/s.

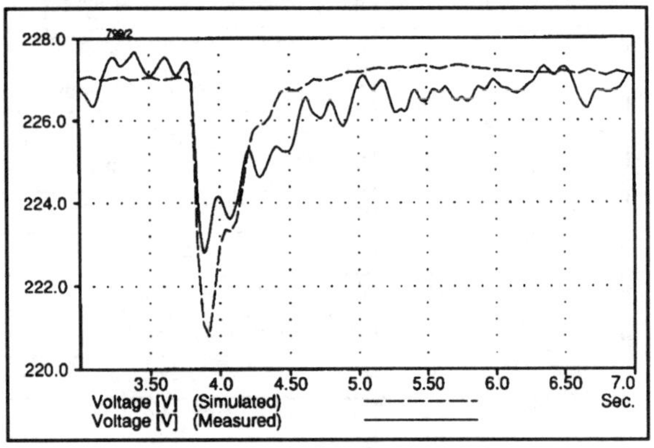

Figure 1: Voltage during 9.5 kW step load change (Froya)

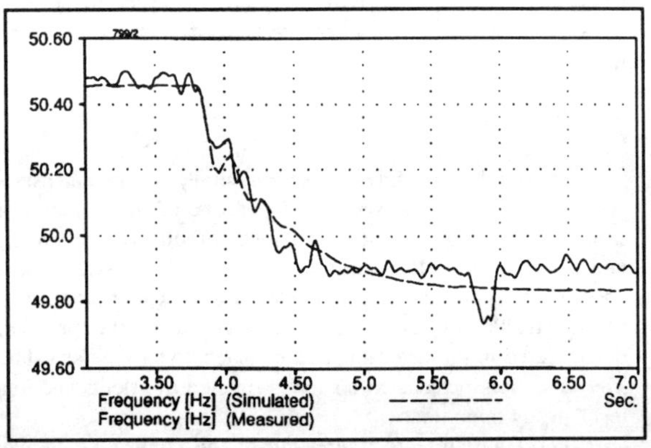

Figure 2: Frequency during 9.5 kW step load change (Froya)

The curves show measured and simulated responses in busbar voltage and frequency and the electrical power flows during the transient.The curves show that the main system dynamics is very well represented by the model. Note the oscillatory mode at approx 5 Hz in the response of the wind turbine generator (fig. 3) and the synchronous generator (fig 4). The oscillations are less pronounced in the simulated results. This may be explained by the fact that the wind turbine in the simulation is excited by a constant wind speed only. The main damping modes are also seen to correspond very well. The busbar voltage deviation fig 1 is a little overestimated by the model.

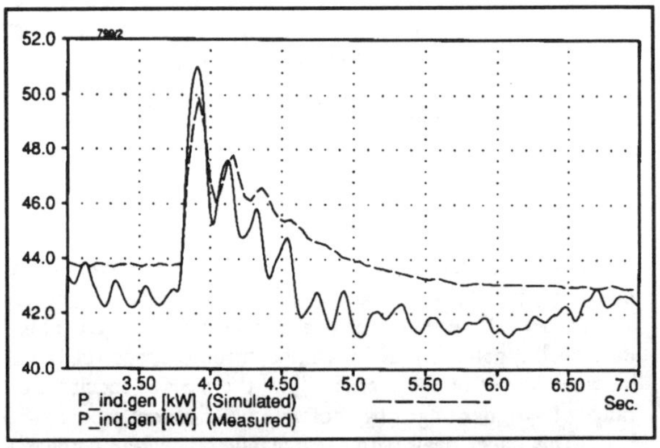

Figure 3: IG power during 9.5 kW step load change (Froya)

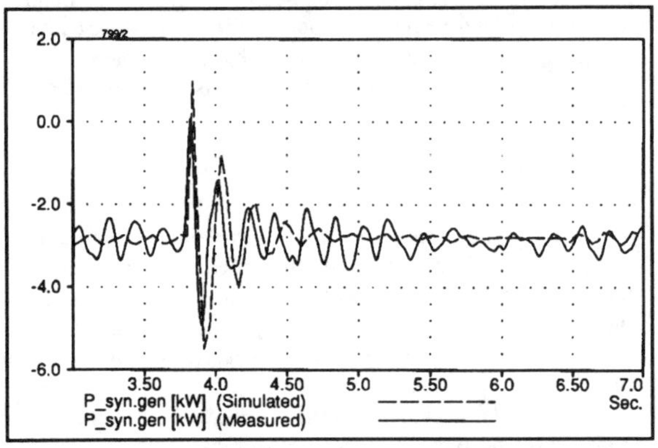

Figure 4: SG power during 9.5 kW step load change (Froya)

The system behavior is highly nonlinear mainly because of variable saturation in both machines. Therefore small differences between measured and simulated results are to be expected.

4.3 Long term results from the Froya System

A constant active consumer load of 19 kW is applied and the system performance in terms of power, voltage and frequency responses is recorded for 30 minutes at a sampling rate of 10 Hz (0.1 sec). The wind speed varies between 8 and 14 m/s during the experiment. The wind power distribution is shown in fig 5, where the lack of measured power above 55 kW is due to a limitation in the power transducer.

The resulting power quality is illustrated in fig 6 and fig 7 comparing simulated and measured busbar frequency and voltage distributions, respectively.

The main conclusions from these plots are that the power quality in terms of distribution functions is predicted very well by the model. The trend is also here that the model shows somewhat larger deviations than the measurements for the busbar voltage. The simulated wind power turbulence is also seen to be a little overestimated.

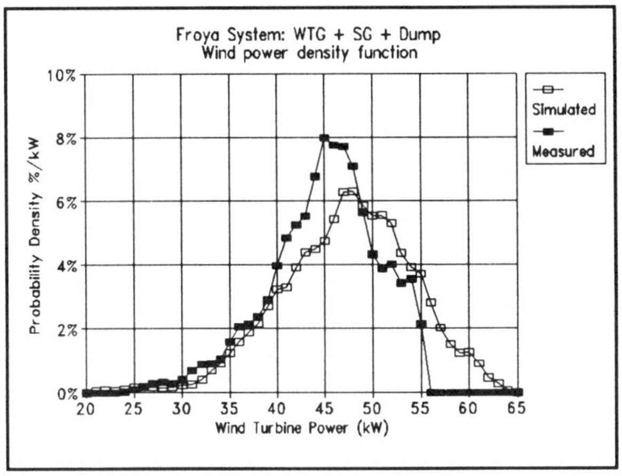

Figure 5; WTG power distribution (Froya long term run)

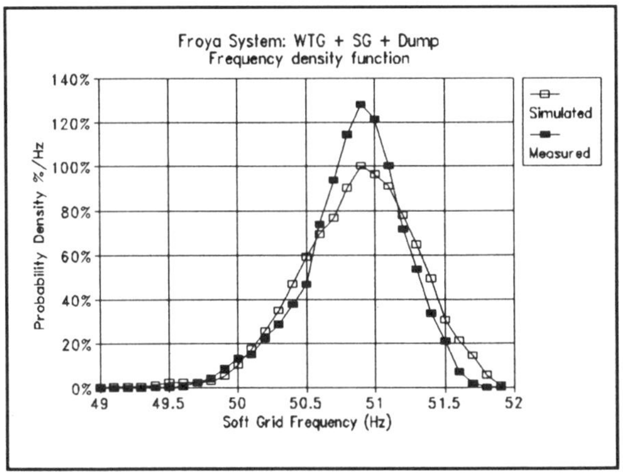

Figure 6: Frequency distribution (Froya long term run)

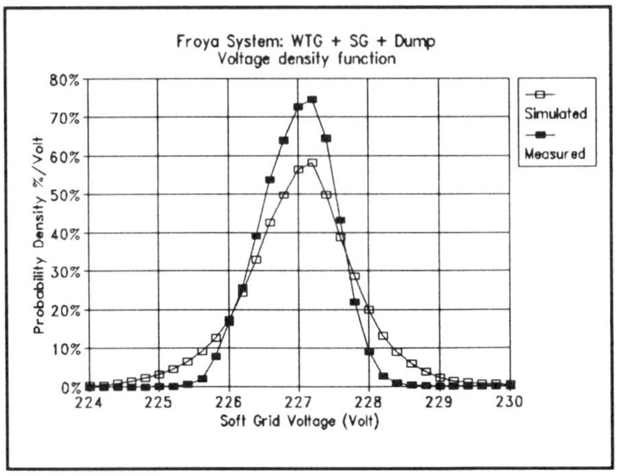

Figure 7: Voltage distribution (Froya long term run)

5. RESULTS FROM THE RISØ SYSTEM

5.1 Description of the system

The Danish wind diesel system is developed and installed at Risø for R&D and demonstration purposes. The system consists of a 55 kW WTG, a 30 kW DGS and a 75 kW switched dump load, all connected to a common AC busbar with chopper controlled resistive and switched capacitive load simulators. The DGS has a clutch and a small (3 kgm^2) flywheel. The SG has a standard compound regulator relying heavily on saturation.

The only mode of operation investigated in this paper is stand alone wind turbine controlled by the dump load and the synchronous generator (diesel engine is stopped). Excitation of the SG is controlled by a compound controller while the dump load controller is a P-type controller. Parameters for the system are well known from previous measurements, except that the determination of parameters for the SG and especially the compound controller has been difficult. The SG is a 4 pole machine rewound to 2 poles.

The voltage control modelling will be very sensitive to the modelling of the saturation and compound control parameters, and the controller has no integral action. Setting the compound parameters to get the correct steady state value in one operating point will not necessarily produce the correct dynamic response.

Therefore only long term results are included to show that statistical simulation results may be useful even if single events (voltage transients in this case) may not be very well modelled.

5.2 Long term results from the Risø system

The wind turbine operates in turbulent wind with output 20 to 60 kW. Consumer load is 0 kW, the SG loss is approx 6 kW (represented by absolute viscous damping in the structural model) and excess wind power is taken by the dump load. The results below are from a 10 minute simulated run compared to the same 10 min of a two hour run of stable operation, scanned at an 8 Hz rate.

The simulated system became unstable after some time due to problems with the voltage regulation. Fig 8 shows why: In the real system the voltage increases with the frequency. This is desirable because increased frequency corresponds to increasing power into the system. The simulated system has the opposite characteristics which may lead to instability. An increase in power input leads to higher frequency (the dump load control) which leads to decreasing voltage. This reduces the dump load power and the system will increase frequency again and the voltage will drop further. During this the SG operates in a (highly) saturated state, which tends to strengthen this effect.

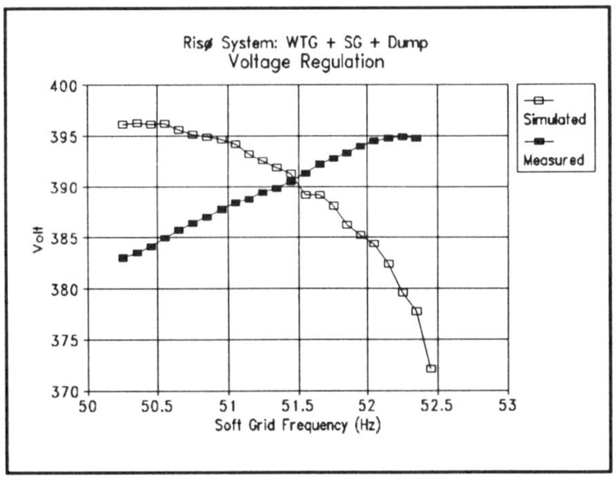

Figure 8: Voltage regulation (Risø long term run)

Fig 10 reflects this problem in the way the simulated and measured voltage probability density functions deviate. In the simulation the range is wider. In spite of a peak at higher voltage the simulated voltage has in fact more frequent occurrences at lower voltage.

In spite of this the simulated and measured frequency probability density functions fig 9 compare rather well. The peaks in the frequency probability densities reflect the differences in voltages as only the dump load controller is active.

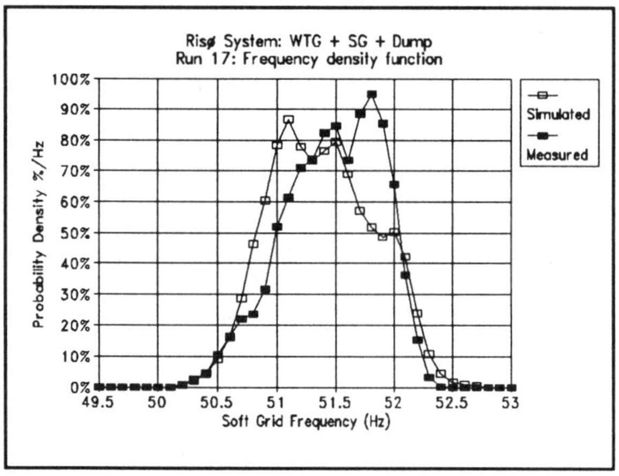

Figure 9: Frequency distribution (Risø long term run)

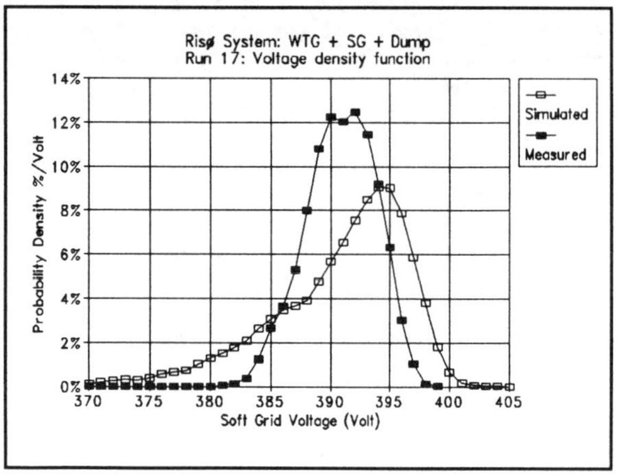

Figure 10: Voltage distribution (Risø long term run)

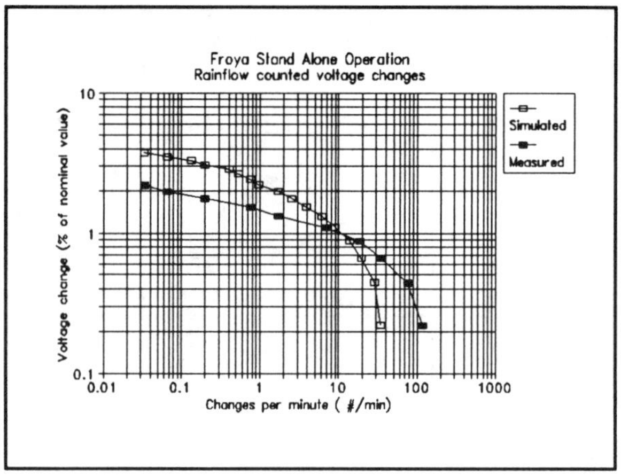

Figure 11: Rainflow counted voltage amplitudes (Froya)

6 DISCUSSION OF RESULTS

For a full description of power quality in a weak grid, frequency fluctuations should be included in addition to voltage fluctuations. Also, changes in mean value as well as fast changes should be accounted for. Fast changes in voltage are accounted for in the usual measures for power quality. In weak grids fast frequency fluctuations should be included to account for e.g. power fluctuations in electric machines.

The probability density functions shown above describe the variations in 'mean' values, but they do not include information of amplitude and occurrence of fast changes. Such phenomena could be quantified by rainflow counting.

Fig 11 shows the rain flow counted voltage changes from the long term Froya run. This measure may represent a valuable basis for assessing power quality in weak grids. The corresponding rainflow count of the Froya frequency agree somewhat better than the voltage. The rain flow counts for the Risø run deviate more because of the problems with the voltage regulation. Again the frequency fits better than the voltage.

7 CONCLUSIONS

Main dynamic modes (oscillatory modes and damping) are represented very well. Steady state active and reactive power distributions agree well. Deviations during transient response are generally represented with acceptable accuracy, with the exception discussed below.

For the two cases investigated the main trend seems to be that the model predicts larger deviations than reality. This may be at least partly due to the fact that the wind speed is input to the WTG load model without spatial or time filtering.

A major problem is the modelling of synchronous generators with compound type voltage regulators or low gain regulators in general. This is, however, not a specific weakness of this model but rather a general modelling problem of identifying correct sets of parameters describing the nonlinear nature of this system. Acceptable performance depends both on the description of magnetic saturation of the synchronous machine and the compound regulator modelling.

The steady state frequency and voltage distributions agree well, with the exception of the voltage from the compound regulated Risø system. This means that except for compound regulated SG's, i.e. simplest SG's, power quality is well predicted.

REFERENCES

[1] P.Lundsager et. al: "Progress with the European Wind-Diesel Modelling Software Package". Proc 1991 European Wind Energy Conference EWEC'91, Amsterdam, the Netherlands, 1991.

[2] D.G.Infield et. al: "Engineering Tools for Wind Diesel Systems". Proc. CEC Wind Energy R&D Contractor Meeting, Alghero, June 1992.

[3] P.Lundsager, P.Nørgård: "The 55/30 kW Experimental Wind/Diesel System at Risø National Laboratory". Proc. ECWEC'88, Herning, 1988.

[4] Kj.Uhlen, Ø.Skarstein, T.Toftevaag, J.O.G.Tande: "Design and Operation of the Full Scale Norwegian Wind/Diesel Laboratory Model". Proc. EWEC'89, Glasgow, 1989.

[5] CYPROS (CYbernetic PROgram Systems), Computer Aided Modelling Design and Engineering Tools. CAMO A/S, Jarleveien 4, N-7041 Trondheim, Norway.

[6] J.Th.G. Pierik: "AWIDIMOD: A Computer Program for Dynamic Analysis of Autonomous Wind-Diesel Systems". ECN report ECN-89-84, Petten, 1989.

[7] Kj. Uhlen: "VINDI - A dynamic simulation program for wind-diesel power plants" (in Norwegian). EFI Trondheim 1988.

[8] P. Lundsager and P. Hauge Madsen: "Theoretical and Experimental Results from the Operation of Risø's Wind/Diesel System". Proc EWEC'86, Rome, Italy, 1986.

ADVANCED MODELLING AND IDENTIFICATION OF AUTONOMOUS WIND-DIESEL POWER SYSTEMS. APPLICATION ON THE FRENCH ISLAND DESIRADE.

G. Kariniotakis, G. Stavrakakis
Technical University of Crete
Dept. of Electronic & Computer Engineering
73133 Chania, Crete, Greece

E. Nogaret, M. Bordier
Ecole des Mines de Paris
Centre d'Energetique
06904 Sophia Antipolis, France

ABSTRACT : Transient stability of isolated Wind-Diesel power systems must be examined in order to determine the optimum penetration level at which the power system remains stable. This level assures also the optimum profit from the wind generators production. In order to analyse the power system transient phenomena, detailed models for the main power system components, like the Diesel engine, synchronous generator, speed and voltage regulators, wind turbine, asynchronous generator, load and transmission network are presented. These models have been integrated in a simulation algorithm able to analyse networks of different complexity level, to consider wind parks of different size and to take into account various types of load. A software has been developed with the possibility to simulate operating scenarios and disturbances. The case-study of the small Wind-Diesel autonomous power system of the Desirade island, where a wind park of 12 wind turbines has been installed and represents a large share of the demand, is presented and analysed. Several typical perturbations are examined, like the loss of the whole wind power or the operation of the wind turbines in turbulent wind. The results obtained demonstrate that the perturbations caused to the voltage and frequency are within acceptable limits.

1. INTRODUCTION

The introduction of a relatively small amount of wind derived electrical power into a utility grid does not normally present any interfacing or operational problems. The situation is completely different when substantial amount of power is penetrating a small or medium isolated utility system.

The assessment of the effects of the wind power penetration to isolated power systems necessitates the calculation of the disturbances (voltage and frequency deviations as well as instabilities) that may be provoked by the wind turbines under several abnormal operating conditions e.g. start-up or sudden disconnection of the wind farm or part of the wind farm, wind gusts, short circuits on the network etc.

In order to evaluate these disturbancies, models of the Diesel engine, Diesel engine governor, synchronous generator, automatic voltage regulator, wind turbine, induction generator, transmission network and the load have been elaborated.

The above models are integrated in a simulation algorithm for the entire isolated power system. This algorithm gives the possibility to analyse networks of different complexity level, to consider various power stations and wind parks, to take into account various types of load etc. Various operation modes and perturbation types can be simulated.

Applications of the computer program are given for the case of the small size isolated power system of the French island La Desirade in the Carribean (installed Diesel capacity : 880 kW, wind park : 144 kW).

The unknown parameters of the wind generators installed in the island have been identified from the machine characteristic curves. The nonlinear optimization Simplex method of Box, under constraints assuring the physical meaning of the results, has been applied for the identification.

The simulations and the measurements, which have been performed in the Desirade, demonstrate the good performance of the wind park even in the case of high penetration conditions and also during power system perturbations.

2. MODELLING OF THE WIND-DIESEL AUTONOMOUS POWER SYSTEM

In order to develop a simulation algorithm able to perform transient stability analysis of an autonomous Wind-Diesel power system, appropriate models for all the main power system components are given in this section.

2.1 The wind turbine model

The mechanical power P_{Am} that is produced by a wind turbine (WT) is given by:

$$P_{A_m} = 0.5\, \rho\, C_p\, A\, U_w^3 \qquad \text{(watts)} \tag{1}$$

where U_w is the instantaneous wind speed, ρ is the air density, A is the rotor swept area and C_p is the power coefficient. For analytical expressions of C_p, the reader is referred to [1].

The mechanical rotational speed ω_A of the wind turbine is given by the well-known swing equation:

$$\frac{d\omega_A}{dt} = \frac{\omega_o}{2H_A}\left(T_{A_m} - T_{A_e} - \frac{D_A}{\omega_o}\omega_A\right) \tag{2}$$

where $T_{Am} = P_{Am}/S_n(\omega_o/\omega_A)$. In equation (2), angular velocities are in rad/sec, while the other quantities are given in the per-unit (p.u.) system where the machine kVA rating (S_n) is chosen as basis. For a more detailed model which describes accurately the dynamics of the various parts of the WT the reader is referred to [2].

2.2. Asynchronous generator model

Wind turbines are equiped with asynchronous generators which in the simulations are considered as "negative loads" that is, as asynchronous motors in generating operation. Thus, the asynchronous machine equations are the following in p.u. [3]:

$$V_d = E_d' + r_s I_d - X' I_q\ , \quad V_q = E_q' + r_s I_q + X' I_d \tag{3}$$

$$\frac{dE_d'}{dt} = \frac{1}{T_o'}\left[-E_d' - (X - X')I_q\right] + s\omega E_q' \tag{4}$$

$$\frac{dE_q'}{dt} = \frac{1}{T_o'}\left[-E_q' + (X - X')I_d\right] - s\omega E_d' \tag{5}$$

$$T_{A_e} = E_d' I_d + E_q' I_q \tag{6}$$

Output active P_{Ae} and reactive Q_{Ae} power, voltage V_t and current I_A are given by the following relations (where * denotes the conjugate quantity):

$$P_{A_e} = \mathrm{Re}\{V_t I_A^*\},\quad Q_{A_e} = \mathrm{Im}\{V_t I_A^*\},\quad V_t = \sqrt{V_d^2 + V_q^2},\quad I_A = \sqrt{I_d^2 + I_q^2} \tag{7}$$

2.3 Diesel engine - sped regulator model

The Diesel engine model gives a description of the fuel consumption rate as a function of speed and mechanical power at the output of the engine - Figure 1.

The differential equations of the state variables [δ, ω, m_B, m'_B, P_c], describing the dynamics of the Diesel engine and the speed regulator, can be easily extracted from the block diagram of Figure 1.

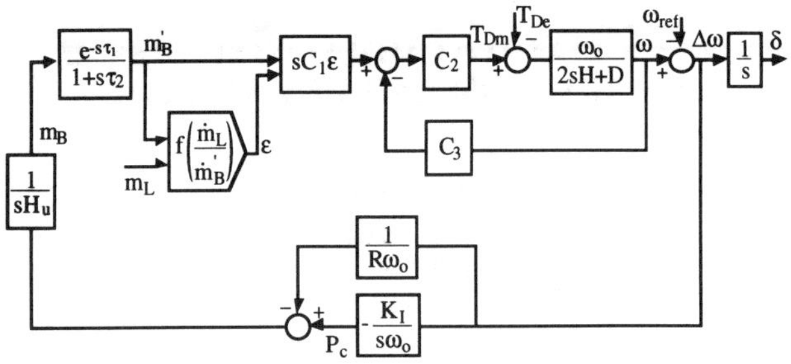

Fig. 1: Diesel engine - Governor model.

In some practical cases the necessary parameters for the above detailed model are not available. Then simplified models, like the one of reference [4], has to be used.

2.4 Synchronous generator

The synchronous generator model permits to examine both transient and subtransient phenomena. This is important for the accurate assessment of the Diesel generators-wind turbines interaction, especially in the case of wind turbines having small time constants.

The equations describing the stator and rotor dynamics and the electrical torque of the synchronous generator are in p.u. [5]:

$$V_d = E_d'' - r_s I_d + X_q'' I_q \,, \qquad V_q = E_q'' - r_s I_q - X_d'' I_d \tag{8}$$

$$\frac{dE_d''}{dt} = -\frac{1}{T_{qo}''}\left[E_d'' - (X_q - X_q'')I_q\right] \tag{9}$$

$$\frac{dE_q'}{dt} = \frac{1}{T_{do}'}\left[E_{fd} - \frac{X_d - X_d''}{X_d' - X_d''}E_q' + \frac{X_d - X_d'}{X_d' - X_d''}E_q''\right] \tag{10}$$

$$\frac{dE_q''}{dt} = -\frac{1}{T_{do}''}\left[E_q'' - E_q' + (X_d' - X_d'')I_d\right] \tag{11}$$

$$T_{D_e} = E_d'' I_d + E_q'' I_q - (X_d'' - X_q'')I_d I_q \tag{12}$$

The machine current, output voltage and power, given by (7) for the asynchronous generator, are also valid here.

2.5 Automatic voltage regulator (AVR)

The most commonly used AVR general models are those defined by the IEEE and especially Type 1 model [4]. The differential equations of the state variables [V_R, V_A, V_F, V_r] of the IEEE Type 1 AVR model can be easily extracted from the following block diagram:

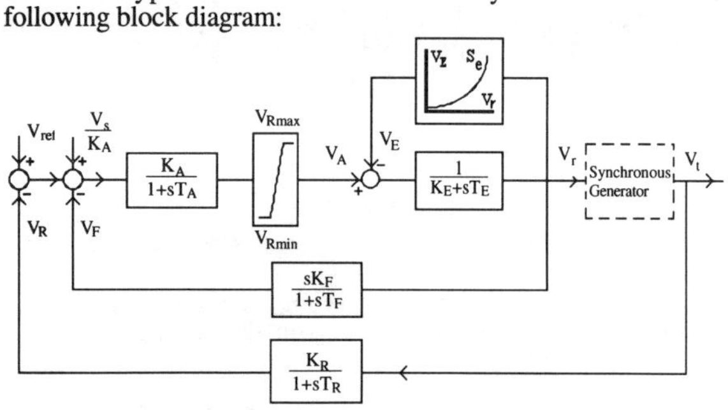

Fig. 2: IEEE Type 1 Automatic Voltage Regulation (AVR) System.

where $S_e = f(V_r)$ is the saturation function of the AVR.

The synchronous generator field voltage E_{fd} is related to the excitation voltage V_r by the relation $E_{fd} = K_f V_r$ where K_f is characteristic parameter of the synchronous generator [4].

2.6 Load modelling

When isolated small grids are examined, the whole load or a part of this can be modelled as an impedance $Z_L = R_L + j\,\omega L_L$.

Motor load is modeled by the induction machine equations (3) to (6) with the appropriate parameters. The mechanical load torque can be represented by a polynomial function of the motor speed [6].

For details on models of other types of load like lighting load and also for agreggated load models, the reader is referred to [7].

2.7. Electrical network

The more general case of a power system, where the transmission network is not neglected and an arbitrary number of synchronous generators and/or asynchronous machines is connected at each bus-bar, is considered.

The bus-bar voltages V_N and the corresponding injected currents I_N are related through the network impedance matrix Z_N by the relation $V_N = Z_N\ I_N$ which is written in the synchronous reference frame D-Q.

This relation is combined with the synchronous generators and asynchronous machines stator equations and the following equation is obtained in a matrix form :

$$I_{dq} = \left[T^{-1}\, Z_N'\, T + Z_m'\right]^{-1} E_{dq} \tag{13}$$

where T is the transformation matrix which rotates the machine quantities from the synchronous reference frame to the individual reference frame of each machine according to it's angle δ, while Z_N' and Z_m' are the appropriate matrices including the network and the machines impedances respectively. Relation (13) provides the current coordinates of each machine (synchronous or asynchronous generator or motor) connected to the network, in the individual machine reference frame.

3. SIMULATION OF A SMALL SIZE ISOLATED POWER SYSTEM.

In this paper it is examined the case of the small Wind-Diesel isolated power system of the french island La Desirade, Figure 3.

The power station of the island is equiped with one Diesel-generator of 160 kW and three identical Diesel-generators of 240 kW.

The wind park consists of 12 identical wind turbines (WTs) of 12 kW each. One capacitor bank (CB) of 10 kVAr is connected to each WT and provides it with the 75 % of the necessary reactive power for its excitation.

The load demand in the power system of Desirade is between 150 and 350 kW.

In the following paragraph, it is presented the methodology to identify the unknown wind generator electrical parameters, which are necessary for the simulation model.

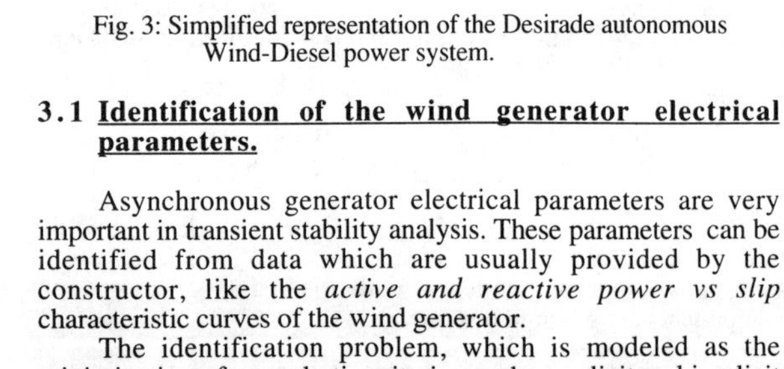

Fig. 3: Simplified representation of the Desirade autonomous Wind-Diesel power system.

3.1 Identification of the wind generator electrical parameters.

Asynchronous generator electrical parameters are very important in transient stability analysis. These parameters can be identified from data which are usually provided by the constructor, like the *active and reactive power vs slip* characteristic curves of the wind generator.

The identification problem, which is modeled as the minimization of a quadratic criterion under explicit and implicit constraints, is solved using the nonlinear Simplex method of Box [8] (Complex method).

The minimization criterion is the sum of the squared differences of the actual data P_{ACi}, Q_{ACj} in i=1,..n and j=1,..n points of the *active and reactive power vs slip* characteristic curves respectively and the estimated P_{IDi}, Q_{IDj} obtained using the unknown parameters:

$$min\ (J) = \sum_{i=1}^{m} (P_{ACi}-P_{IDi})^2 + \sum_{j=1}^{n} (Q_{ACj}-Q_{IDj})^2 \quad (14)$$

P_{IDi} and Q_{IDj} are calculated from relations (3-5) by setting derivatives to zero and equation (7). The unknown parameters of the optimization problem (14) are: the stator and rotor resistances r_s and r_r, the stator and rotor reactances X_s and X_r and the magnetizing reactance X_m. The above parameters are related to the parameters of the asynchronous generator model of Paragraph 2.2 by the following relations [3]:

$$X = X_s + X_m, \quad X' = X_s + \frac{X_r X_m}{X_r + X_m}, \quad T_o' = \frac{X_r + X_m}{\omega_b r_r} \quad (15)$$

The following constraints are considered: (i) each one of the five parameters takes only positive values, (ii) $X_m > X_r$ and $X_m > X_s$ since always the magnetizing reactance has a greater value, (iii) $0 < T_o' \leq \alpha$, with α to be an appropriate upper limit for the rotor open circuit transient time constant, depending on the type and the size of the machine. In the case examined here T_o' is in the order of some tenths of a second.

The estimated parameters of the wind generators in the Desirade island are given in the Appendix.

4. CASE-STUDIES

4.1 Abrupt change of penetration level

In this case-study it is analyzed the effect of the loss of the wind park when 10 WTs are connected producing 120 kW (penetration level 60 %). The Diesel unit G_2 is operating at the lowest acceptable level (1/3 of its nominal power) while the load is taken as 200 kW / 150 kVAr. Capacitor banks (CB) provide the 75 % of the wind park reactive demand. It is considered that at instant t_o a permanent fault happens on the line connecting the wind park. Then the wind park with the capacitor banks are disconnected.

Results: The loss of the wind park results an increase of 120 kW and a decrease of 22 kVAr of the Diesel active and reactive power respectively. Due to the above difference between active and reactive power, higher frequency than voltage oscillations are observed.

The bus-bars voltage variations shown in Figure 4 are between $-0.1 \leq \Delta V \leq 0$ %, while the frequency variations shown in Figure 5 are between $-0.65 \leq \Delta f \leq 0$ Hz. The above variations are dependent on the Diesel engine parameters used. If different but possible mechanical parameters are used the above frequency variation can reach the value of -1.3 Hz. Similar conclusions can be made for the voltage variations according to the electrical parameters used.

4.2 Operation under turbulent wind

The objective of this case-study is to examine the behaviour of the power system when the wind turbines operate in turbulent wind. It is considered that at $t<t_o$ 8 WTs are connected providing the system with 36 kW (penetration 15%). CBs are also connected. The load of the power system is taken as 260 kW / 182 kVAr.

The two Diesel-generators G_2 and G_3 operate with power factor and spinning reserve close to 76 % and 53 % respectively.

The wind-speed during the time period examined is shown in Figure 6.

Results: The power output of each WT varies between 4.5 kW and the maximum value of 13.2 kW. This corresponds to a penetration level variation between 14% and 40%. The Diesel spinning reserve of 53% is necessary when WTs production decreases suddenly. Then, Diesel engines are not overloaded and the load is satisfied. In the contrary case, when penetration increases, the Diesel production does not fall under undesirable operation limits (e.g. 1/3 of rated power).

The bus-bar voltage variations, see Figure 7, are between: $-3.9 \leq \Delta V \leq 2.1$ %. Apart from the first abrupt voltage drop (-3.9 %, 0.4 sec) the other variations are between ± 2 %.

The power system frequency variations are between: $49.94 \leq f \leq 50.19$ Hz, see Figure 8. These limits are within the acceptable voltage and frequency variations, determined by the electric utility of the island, which are ± 5 % and ± 1 Hz respectively.

Althougth the variations of the system variables were in acceptable limits, this is not valid for the rate of change of these variables. This rate of change is an additional criterion for the acceptable operation of the Wind-Diesel power system.

5. CONCLUSIONS

In this paper models for the main Wind-Diesel power system components were presented. These models have been integrated in a simulation algorithm able to describe accurately the transient phenomena induced by the wind park operation and by various power system perturbations.

The input of this algorithm can be any scenario on the power system operation conditions. A wide range of perturbations can be examined like balanced faults, machine switchings, operation under turbulent wind etc. By choosing the appropriate operation scenarios, studies can be performed for the penetration level and security limits, the stability properties of the autonomous power system, the wind generators power quality etc. The case-studies presented concern the autonomous power system of the French island Desirade. Various perturbations were presented, while the variations of the most important variables were analyzed (frequency, voltage etc.). From these case studies, useful results can be extracted for the transient behavior of a small Wind-Diesel isolated power system.

REFERENCES

[1] P. M. Anderson, A. Bose, "Stability Simulation of Wind Turbine Systems", *IEEE Transactions on Power Apparatus and Systems*, vol. PAS-102, No. 12, December 1983.

[2] A. M. de Paor, "Aerodynamic design of optimum wind turbines", *Applied Energy*, vol. 12, pp. 221-228, 1982.

[3] Paul C. Krause, "Analysis of Electric Machinery", McGraw Hill, 1987.

[4] P. M. Anderson, A. A. Fuad, "Power System Control & Stability", The Iowa State University Press, 1977.

[5] D. W. Olive, "Digital Simulation of Synchronous Machine Transients", *IEEE Transactions on Power Apparatus and Systems*, vol. PAS-87, No. 8, pp. 1669-1674, August 1968.

[6] W. W. Price et al, "Load Modelling for Power Flow and Transient Stability Computer Studies", *IEEE Transactions on Power Systems*, vol. 3, No. 1, February 1983.

[7] F. Iliceto, A. Ceyhan, G. Ruckstuhl, "Behaviour of loads during voltage dips encountered in stability studies. Field and Laboratory tests", *IEEE Winter Meeting*, Winter 1972.

[8] S. S. Rao, "Optimization theory and applications", 2nd edition, Willey Eastern Ltd, 1984.

LIST OF SYMBOLS

T_{Am}	Mechanical torque developed by a wind turbine (p.u.).
H_A	Total wind turbine and induction generator inertia time constant.
ω_o	Synchronous angular velocity [=2 π 50 Hz] (rad/sec).
ω_e	Angular velocity of the synchronous reference frame.
s	Slip of induction machine defined as the ratio $s=\frac{\omega_e - \omega_A}{\omega_e}$.
V_d, V_q	Coordinates of terminal voltage V_t: $V_t = V_d + jV_q$ (p.u.).
E_d', E_q'	Induction machine voltages behind transient reactance X' (pu.)
I_d, I_q	Coordinates of induction machine current I_A: $I_A=I_d+jI_q$ (p.u.)
r_s	Induction or synchronous generator stator resistance (p.u.)
X	Induction generator open circuit reactance (p.u.).
X'	Induction generator open circuit transient reactance (p.u.).
T_o'	Induction generator rotor open circuit transient time constant (sec).
δ, ω	Synchronous generator angle and angular velocity respectively.
$\dot{m}_B$	Diesel engine consumption rate (kg/sec).
$\dot{m}_B'$	Diesel engine combusted fuel rate (kg/sec).
$\dot{m}_L$	Diesel engine air supply rate (kg/sec).
ε	Diesel engine efficiency of the combustion.
τ_1	Diesel engine time delay in the combustion process (sec).
τ_2	Diesel engine time constant of the combustion process (sec).
C_2	Diesel engine stroke volume proportionality constant.
C_3	Diesel engine constant describing mechanical losses.
H_u	Diesel engine heat value of the fuel (kJ / kg).
K_I	Governor summing loop amplification factor (p.u.kW/pu.Hz)
R	Diesel engine permanent speed droop (p.u. Hz / p.u. kW).
D	Load damping coefficient (p.u. kW / p. u. Hz).
H	Synchronous generator inertia time constant (sec).

$X_{i=d,q}$ Synchronous generator reactance of direct (i=d) or quadrature (i=q) axis (p.u.).

$X'_{i=d,q}$ Synchronous generator transient reactance of direct (i=d) or quadrature (i=q) axis (p.u.).

$X''_{i=d,q}$ Synchronous generator subtransient reactance of direct (i=d) or quadrature (i=q) (p.u.).

T'_{do}, T''_{do} Synchronous generator open circuit transient and subtransient respectively time constants of direct axis (sec)

T''_{qo} Synchronous generator open circuit subtransient time constant of quadrature axis (sec).

E'_d Synchr. generator voltage behind the transient reactance X'_d (p.u.).

E''_d, E''_q Synchronous generator voltages behind the subtransient reactances X''_d and X''_q respectively (p.u.).

R, A, F, E AVR subsystems indices: Regulator input filter (R), Amplifier (A), Stabilizer (S), Exciter (E).

$K_{i=R,A,F,E}$ Gain constant of the i subsystem of AVR.

$T_{i=R,A,F,E}$ Time constant of the i subsystem of AVR (sec).

APPENDIX: Power system parameters

Diesel engines - Synchronous generators.

	S_n	P_n	V_n	r_s	X_d	X_q	X'_d	X''_d	X''_q	T'_{do}	T''_{do}	T'_{qo}	T_D	H
$G_{2,3,4}$	300	240	380	0.5	120	100	32	16	15	4.5	0.035	3.5	0.03	3
	(kVA)	(kW)	(V)	(			%		)	(	sec			).

Speed and voltage regulators : R = 3%, K_I = 16.5 p.u.
IEEE 1st Type AVR model :
Input filter : K_R=1, T_R=0.05, Amplifier : K_A=200, T_A=0.1
Stabilizer : T_F=0.028, K_F=0.5, Exciter : K_E=-0.17, T_E=0.5

Transmission lines: Resist.: R_{TL}=0.14 Ω/km. React.: X_{TL}=0.25 Ω/km

Wind generators : Nominal power : P_n = 12 kW, Power factor at nominal power : p.f. = 0.83, Nominal voltage : V_n = 380 V, Blades length R = 5 m, H_A = 0.259 sec.

The asynchronous generator emlectical parameters as estimated by the identification process are (in the p.u. system on the generator KVA base): r_s = 0.027, r_r = 0.026, X_s = 019, X_r = 0.19, X_m = 4.90

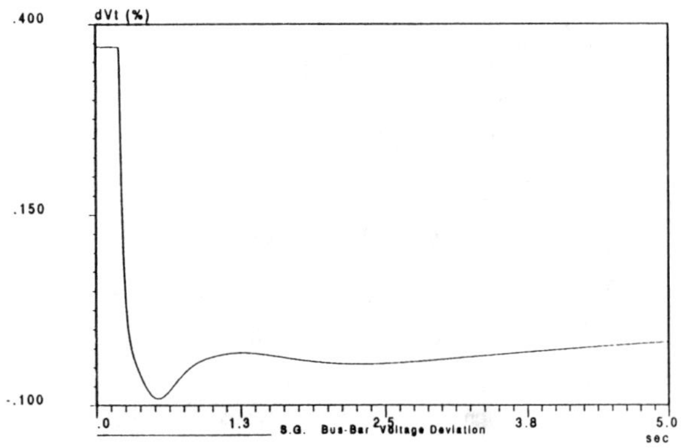

Fig. 4 : Bus-bar voltage variations

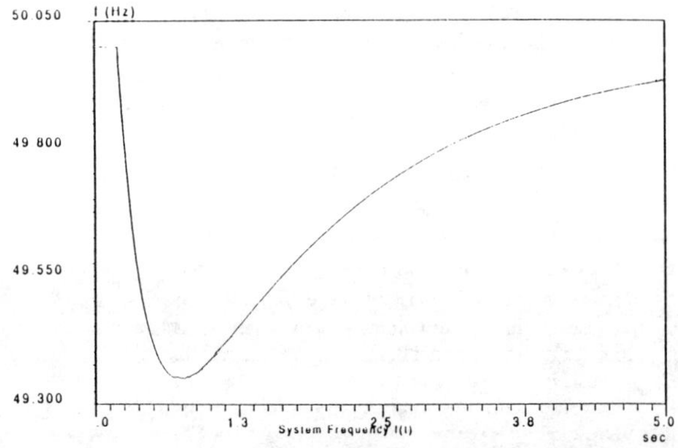

Fig. 5 : Power system frequency variations

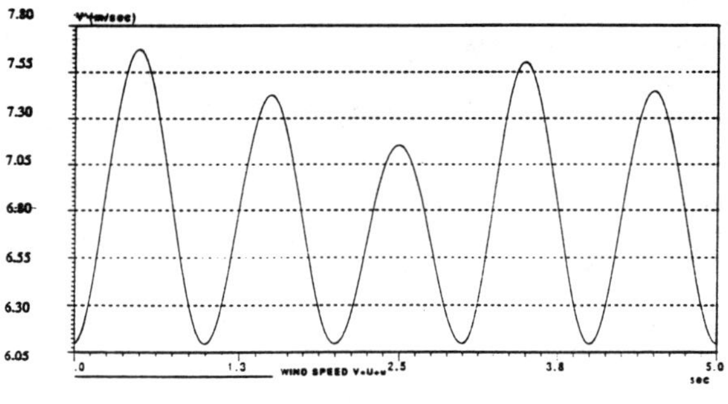

Fig. 6 : Wind speed variations

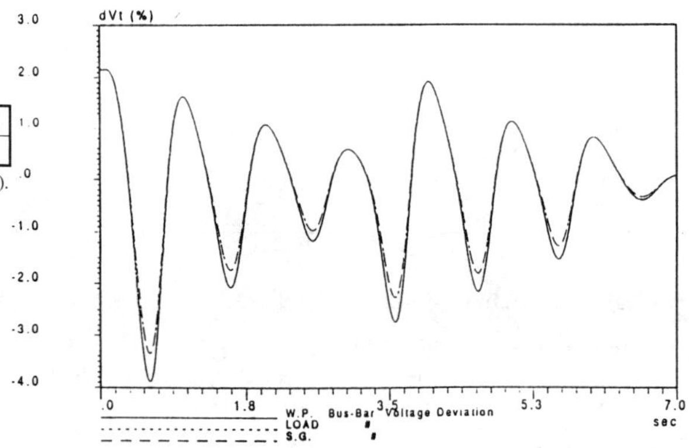

Fig. 7 : Bus-bar voltage variations

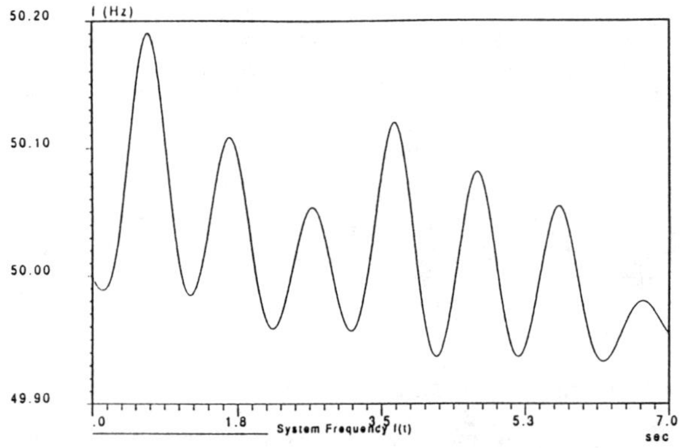

Fig. 8 : Power system frequency variations

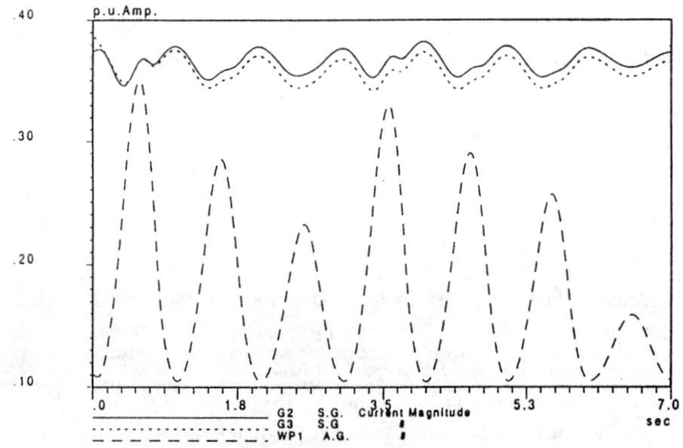

Fig. 9 : Current variations

IMPROVEMENTS OF OPERATING STRATEGIES FOR THE ANHOLT WIND-DIESEL SYSTEM.

Kristiansen, J.

ELSAMPROJEKT A/S
Kraftværksvej 53
DK-7000 Fredericia
Denmark
Fax: + 45 75 56 44 77
Tel: + 45 75 56 44 11

Summary.

Since 1989, a wind-diesel system has been in operation on Anholt Island. At the same time the measurement programme was started. In the period since, many experience have been gained with this practical application and have served as basis for improvements in the operation strategies for the system as well as for demands for the hardware in the system.
Anholt is a small island with no electrical connection to the power system on the mainland. Previously, the electricity supply on Anholt was based solely on diesel generating equipment, but to reduce the fuel consumption and thereby energy costs, windpower was introduced in the existing system in 1989. At the time of planning and implemention of the project, only few results from other public full scale systems existed. Therefore, a simple system was chosen in order to make the system cost-effective and easy to maintain.
This paper presents results of the operation during the measuring period and deals with problems concerning minimum loads on the diesel engines and electrical stability, in periods with low power consumption and high wind power production.
Furthermore, the measurement programme has shown, that changing the control strategy for the diesel generators gives a considerably decrease in fuel consumption. These changes are implemented in early 1993.

1. INTRODUCTION.

Anholt is an island with approx. 70 households. It is situated in the middle of Kattegat about 50 km north east of the town Grenaa. In the summer period, the electricity consumption grows considerably due to the many tourists visiting the island. Hence the load pattern varies considerably over the year. The maximum consumption is approx. 500 kW and the minimum is 50 kW. There is no electrical connection to the electric system on the mainland.
The results of a feasibility study carried out were positive and resulted in erection of a Nordtank 55 kW stall controlled wind turbine and a central control system for the combined wind-diesel system. In 1989, the turbine and the control system were installed and connected to the grid. The diesel generator sets at the power station are controlled and operated as before the erection of the wind turbine and is separated from the wind-diesel control system.
The purpose of the project is to demonstrate that a series-produced standard wind turbine is able to operate together with standard diesel generators in an isolated electricity supply system and to contribute to the energy supply of the area. Further the connection of the wind turbine should have no significant impact on the power quality.
The first measuring programme was started in 1989 and ended in 1991. An extension was decided to improve the operation conditions for the wind turbine and the diesel generator sets. This will be ended in mid 1993.

2. THE POWER SYSTEM.

The power system on the island is small and totally isolated as there is no electrical connection to the mainland.

The power station comprises 1 suction and 3 turbo charged diesel generating sets with a total installed power capacity of 990 kW.
The diesel generated power production is at 0.4 kV level and is distributed directly to the Anholt town from the power station.

Further, there is a 15 kV cable connection via the marine station to the harbour, where the distribution is at 0.4 kV level as well.

The wind turbine, a 55 kW standard stall-controlled turbine, is connected to the grid at the harbour via a 150 m LV-cable. The connection is at 0,4 kV level.

The power system on the island is shown in figure 1.

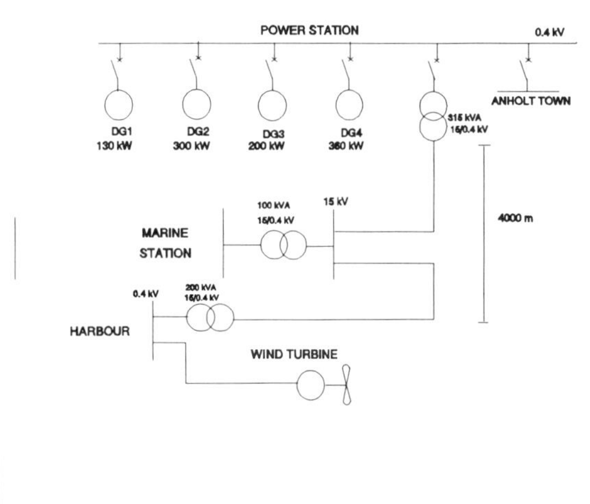

Figure 1. Power system on Anholt.

3. CONTROL SYSTEM.

The control system for supervising the wind-diesel system is a combined data acquisition and control system.

The data acquisition system collects and presents the data measured from the wind turbine, the power station and a meteorological tower at the wind turbine.

The control system contains software for control logic. On the basis of measured data from the turbine, the power station and the meteorological tower, the control logic starts and stops the wind turbine. The control logic is based on the operational ranges of the diesel engines and wind power penetration.

The configuration of the control system is shown in figure 2.

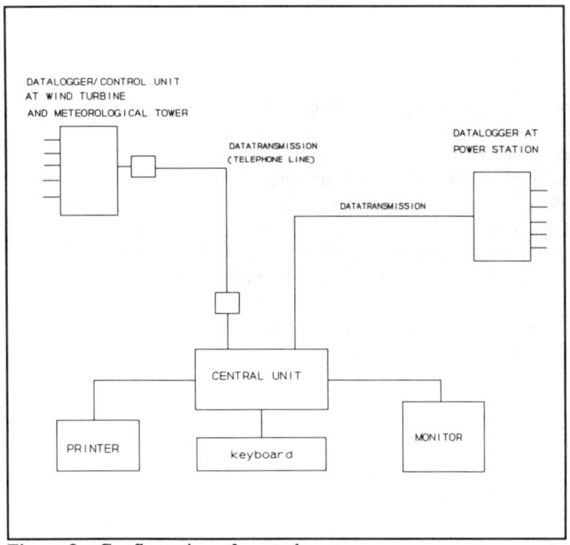

Figure 2. Configuration of control system.

4. METHOD OF OPERATION.

The control logic for the operation of the wind turbine has been designed according to the following constraints given by the existing system:

- The introduction of the wind turbine must not affect the control logic for the existing diesel generator sets. Integration of the diesel control logic in the central control system for the wind diesel system is not allowed.

- The minimum load for the turbo charged diesel generator sets at the power station is 40%. This limit is prescribed by the diesel engine manufacturer. As the power station already existed before the erection of the wind turbine, the diesel engines are not optimized according to a wind diesel system, therefore the wind diesel ratio and the operational range of the diesel engines are not optimum.

- The introduction of wind power must not decrease the power quality significantly. As the system is no experimental system, the consumer's demands for the electricity quality and secure supply are high.

The above mentioned restrictions and the necessity to keep the turbo charged diesel engines within their operating range has been fundamental for the control strategy.

In periods with high level of wind power and low load, power control of the turbine is done by starting and stopping the wind turbine as there are no dumpload facilities. The simple control strategy ensures, that the wind turbine operates only when the above restrictions are fullfilled and when the electricity consumption is large enough to ensure the diesel generator sets to control the grid with a high stability in voltage and frequency.

4.1 Changed operation strategy for Diesel generator sets.

Based on evaluations of the operation of the diesel generator sets a new control strategy for the power station has been developed. This is made to increase the mean load level of the diesel generator sets and thereby the efficiency of the engines. Furthermore, the changes minimize the numbers of starts and stops of the diesel generator sets as well as minimizing the time with stopped wind turbine. Investigations made show that the changes gives a decrease in oil consumption of approx. 10%. The changed control strategy is implemented in the beginning of 1993

5. RESULTS OF OPERATION AND MEASUREMENT.

The measurement period was started on May 17, 1989 and is still ongoing. In this period many experiences with the practical application have been gained and they serve as basis for improvements in the operation strategy for the system as well as for demands for the hardware in the system.

Lightning strikes have caused severe damage to many of the hardware components. Overvoltage and transients were introduced through the power supply and through the data transmission line (telephone line) from the turbine. The problems were solved by introducing overvoltage protection in the power supply and in selected points in the system. Since these modifications were carried out, there has been no damage of components in the system, caused by lightning and there has been only short periods with the system stopped due to faults.

Table 1 illustrates statistics for availability for the system and for the turbine during the measurement period.

[Hours]	1989	1990	1991	1992
Total Hours in period	5466	8760	8760	8760
Measurement system active	2529	4375	6350	8069
Turbine producing	762	1744	3391	5589
Turbine not producing	1767	2631	2857	2480
WT stopped, low load on DG	364	902	825	711
WT stopped, high load on DG	9	167	51	11
WT stopped, low wind	1267	1161	1847	1644
WT stopped due to fault	127	401	134	114
Availability Meas.system [%]	46.3	50.0	71.3	92.1
Availability WT [%]	80.2	66.4	83.8	89.8

Table 1. Availability for the system and the wind turbine.

The availability of the turbine is calculated for the periods with active measurement/control system. The main reason for the reduction in availability is caused by the constraints in the control logic for operation of the turbine.

Stop of the turbine due to a high load level on the diesel generator sets is almost eliminated by modifications in the control logic.
Time with stopped turbine due to low load is reduced by lowering the permissible minimum load on the diesel generator sets. This has been done in the end of 1992.
The system availability has been improved during the measuring period. The improvements are primarily caused by the above mentioned modifications and protection against lightning.
The relatively low availability for the turbine is mainly caused by the control logic for the wind diesel system.

Production [MWh]	1989	1990	1991	1992
Diesel generators total	399	713	1,063	1,289
Wind turbine	18	47	73	148
Potential extra prod. of WT.	13	48	31	30
Capacity factor WT.	0.13	0.20	0.22	0.33

Table 2. Energy production on diesel generator sets and wind turbine.

The capacity factor for the turbine is calculated for the periods with active measurement/control system. The main reason for the low value is caused by the limits in the control logic for the system.
The availability improvements and changes in the control logic limits have increased the wind power produced share of the energy production. Figure 3 shows the trend in the wind energy production. The wind energy produced makes out an increasing part of the potential wind energy contribution.

The capacity factor for the wind turbine has clearly increased, even in 1991 which was a poor wind year. Figure 3 illustrates the wind power produced part of the total production, the potential wind energy and the capacity factor for the wind turbine.

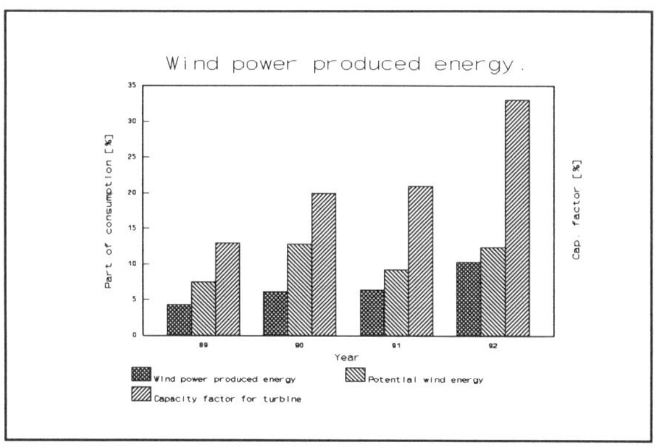

Fig. 3 Energy produced by the wind turbine compared with the potential wind energy and capacity factor for the wind turbine.

5.1 Operation situations.

During the measurement period, modifications of the control logic have been made to increase the amount of wind energy produced and to reduce unnecessary start and stop of the wind turbine and the diesel generator sets. The changes have been made on basis of the experiences and measurements gained during the period.
During periods with low load and potentially high wind power production, the turbine may have to be stopped due to low load on the diesel engines. Figure 4 indicates a period with a load of approx. 90 - 170 kW and with good wind conditions. The turbine is stopped due to low load (P_{DG2} < 100 kW ≈ 50%) on diesel generator set No. 2.

During the period, a large amount of wind power energy is lost due to the low load limit for the diesel engines. A large part of this energy is now utilized by lowering the limit to 40% load on the diesel generator sets. This has been done at the end of 1992.
Operation of the wind turbine could be further improved by letting the diesel engines run occasionally at a level below 40%. This is only allowed if no excessive wear or deteoriation of the diesel engines occurs. Further investigations of operating the diesel engines in the low load range have to be carried out in cooperation with the manufacturer.

5.2 Voltage quality.

Earlier the island was supplied only by diesel generator sets. The demands for the voltage quality and secure electricity supply is therefore high.
These demands and the necessity of keeping the turbo diesel engines within their operating ranges have been fundamental for the design of the control strategy for the wind-diesel operation.

Restrictions caused by the load range on the diesel engines are dominating. Therefore, the demands for high power quality (frequency and voltage stability) are of secondary importance regarding the control logic.

Figure 5 indicate data measured for frequency and voltage in the grid. The minimum and maximum values for frequency and voltage are shown as well.

The frequency is stable and only minor distorbances occur. The distorbances are caused by start and stop of the turbine as well as the diesel generator sets.
Measurements from the period 1989 through 1992 show frequency variations of up to +1/-2 Hz caused by start and stop of the turbine only.
Whether the wind turbine is connected or not has no stationary influence on the voltage at the power station. At the wind turbine and the harbour, start and stop of the turbine cause voltage fluctuations of up to + 2/-3.5% and stationary variations of up to ± 2%.

In summer period, a starting order is used where DG4 is base load unit and DG3 is second diesel generator to be started.
Figure 13 shows a situation where the turbine is stopped due to low load during night time. By lowering the minimum load limit for the DG4 to 40% these periods are reduced considerably.

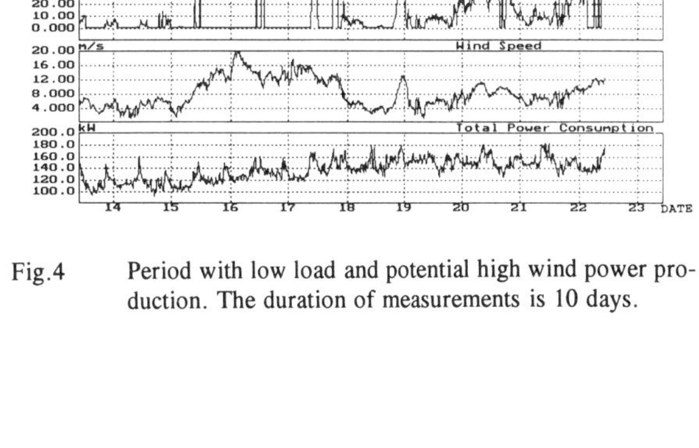

Fig.4 Period with low load and potential high wind power production. The duration of measurements is 10 days.

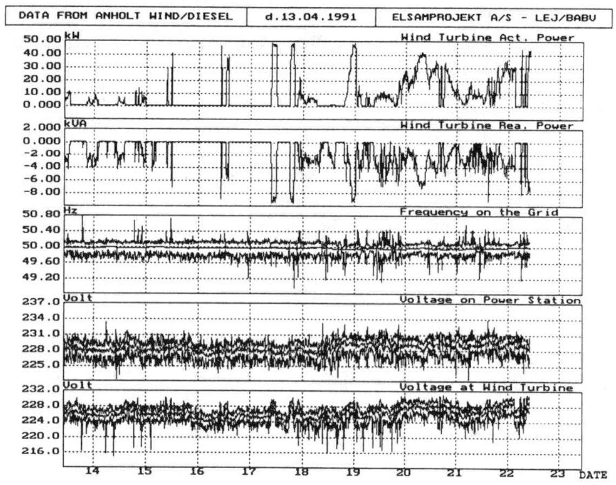

Fig. 5 Measurements in the grid during a period with low load. The duration of measurements is 10 days.

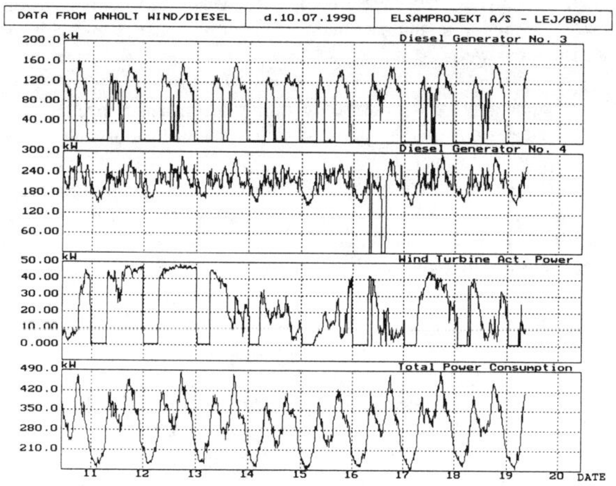

Fig. 6 the turbine is stopped due to low load during night time. By lowering the minimum load limit for the DG4 to 40% these periods are reduced conciderably.

6. CONCLUSIONS AND FUTURE WORK.

This project has proven that it is feasible to implement wind energy in a small public diesel electric grid in a very simple way without decreasing the reliability of electricity supply and without significant impact on the electricity quality.

To improve the wind energy production, some modifications in the control system for the wind diesel system has been necessary. The hardware has been modified to prevent long periods with the system stopped due to faults and the control logic for operating the wind turbine has been improved on basis of the experiences gained through the measurement period. Based on the investigations made on basis of the measurement programme, the overall efficiency of the system has been increased by means of the abovementioned improvements and changes in control logic for the diesel generator sets.

The measurement period continues to mid 1994 and in this period work will be done to improve the efficiency of the system.

Operating the diesel engines at low load will be investigated in close cooperation with the diesel engine manufacturers. The aim is to reduce the minimum load of the diesel engines to a lower limit, without significant increase of wear on the engines and decrease of the electricity quality.

The changed operating strategy for the diesel generator sets will be evaluated with regard to increased efficiency and the impact on lifetime and wear.

REFERENCES.

[1] Lundsager, P., Sherwin, R. Using simple wind-diesel systems without energy storage to obtain high penetration and market acceptance in the near future. Windpower'90 Conference, 1990.

[2] Rasmussen, A., Kristiansen, J. Anholt 55 kW wind turbine. Operation strategy, experiences and results. Proceedings of 1990 European Community Wind Energy Conference, Madrid.

[3] Rasmussen, A., Kristiansen, J. Final report for the 55 kW wind turbine on anholt. Final report Commission of the European Communities Directorate General for Energy, WE/619/84.

[4] Kristiansen, J. Operating experiences with the Anholt wind-diesel system. CEC Thermie Seminar, Rhodos 1992.

A STATISTICAL INVESTIGATION OF THE BEHAVIOUR OF AN ISOLATED WIND TURBINE

A. Hoque*, Q. Ahsan** and W.C. Beattie*

*Department of Elect and Electronic Engg
The Queen's University of Belfast
Ashby Building, Stranmillis Road
Belfast, BT9 5AH, N. Ireland, UK

**Department of Elect Engg
University of Bahrain
PO Box: 32038, Isa Town
Bahrain

ABSTRACT: The paper studies the application of a wind turbine generator (WTG) to an isolated island community in Bangladesh and examines the statistical probability of being able to supply the electrical load demand. For a reliable electrical energy system from intermittent sources namely wind turbine, a probabilistic model has been considered. The probabilistic model is applied to a realistic system. In the numerical evaluation the segmentation method is applied to determine the reliability of the system. The probabilistic model of a WTG is applied in the same manner as a conventional generating unit in the probabilistic simulation. The model is based on the concept of the multi-state representation of a generating unit. The probability density function (PDF) of the capacity output from a WTG is derived from the bivariate function of wind velocity and temperature. To justify the capacity generation model of a WTG, involving the bivariate function of wind velocity and temperature, a capacity generation model of a WTG is also considered which is a univariate function of wind velocity. Based on simulation this paper presents a comprehensive comparison between the WTG and the thermal (diesel) generator (THG). The results clearly reveal a new dimension for generation planners regarding alternative sources.

1. INTRODUCTION

Wind power has been harnessed by man for many centuries both on land and sea. The increasing awareness on toxic waste removal, ecological damage control, acid rain and most importantly, the irreversible damage to the earth's atmosphere in the developing and the developed countries of the world together with the limited number of new hydro sites, has refocused attention on WTG as a possible alternative. The increasing interest in WTGs is also due to its low operating cost and high availability as one of the most simple ways of converting a natural resource to electric power in remote and isolated areas.

There are numerous papers that discuss WTGs. Most of these papers deal with the characteristics or the model of WTGs and many of these models are developed in a deterministic fashion [1-3]. However, the main input variables to a WTG, wind velocity and temperature, are random in nature. They vary from place to place, season to season, day to day, hour to hour or even for a smaller interval of time. The domain of variation for temperature and that of wind velocity are different and there exists a mapping between the two domains. That is, the two random variables are correlated and it requires a joint PDF to describe the nature of variations. Bakirtzis et al. [4] described probabilistic method of predicting the customer owned grid-connected wind arrays based on the statistical data of customer load and wind velocity.

To the authors' knowledge, the models reported so far in the literature are developed considering the temperature as a constant quantity. In nature, the temperature is not constant and for an 11.1°C change the output of the WTG varies by 3.5% [5].

Unlike many attempts in the past, this paper considered a model of a WTG developed [6] from the bivariate distribution function of temperature and wind velocity and it is compared with the model developed from the univariate function of wind velocity only. The two multi-state models of a WTG considered are applied separately to evaluate the expected energy generation (EEG), the expected energy not served (E(ENS)) and the LOLP of the system.

The paper studies the application of a WTG, as a stand-alone source of energy to an isolated community in Bangladesh and examines the statistical probability of being able to supply the electrical load demand of such an area. This paper also presents a comprehensive comparison between the WTG and the THG in terms of reliability and cost. This paper addresses the question of achieving a high reliability to international standards. In practice, the energy supply to the island only exists for 5-7 hours in every day and clearly any improvement on this would be deemed acceptable. There is clearly a need to define an acceptable level of availability in such communities, but this has not been considered in this paper.

2. CASE STUDY

An isolated area has been considered to study the application of a WTG. The considered isolated area is Kutubdia, an island of Bangladesh. This island is located in the southern end of Bangladesh and it is an island in the Bay of Bengal. The island has an area of 70 square kilometers and a population of 72,000.

3. NUMERICAL EVALUATION

For a reliability study of an electrical energy system

from an intermittent source, namely WTGs, a probabilistic model [6] has been considered. In the numerical evaluation, the segmentation method [7] is used to determine the reliability of the system.

3.1 Basic Data

The wind velocities of the island during June 10 to 17 and their corresponding temperatures are considered in this study. The three hourly wind velocities and the corresponding temperatures are shown in Figure 1. The wind velocities are modified to take into consideration the difference between the heights of an anemometer and the hub of a WTG.

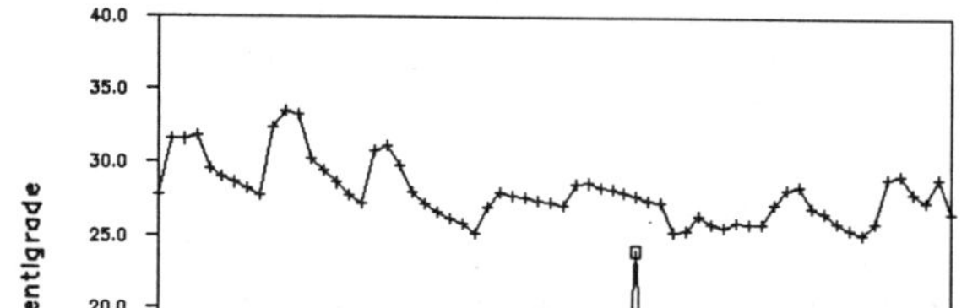

Figure 1: Three hourly wind velocity and temperature

For this research, a WTG of the Energy Development Co., USA, model (4-45) is considered [5]. The cut-out and cut-in velocities of this WTG are 34.7 and 4.3 knots respectively. In this study, a modified rated output of 47 kW is considered for this WTG assuming a higher blade efficiency. The forced outage rate (FOR) is considered to be 0.2. It extracts a constant output power for any wind velocity that lies between 23.4 to 34.7 knots.

The load data used [6] in the simulation is the load of the island during the study period. The hourly average load of the study period is shown in Figure 2.

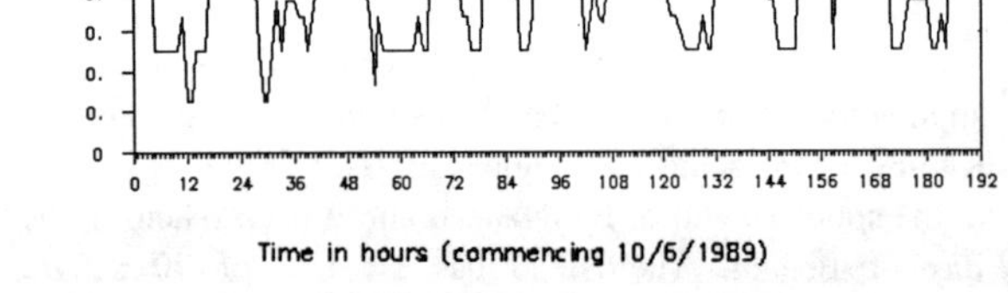

Figure 2: Hourly average load of the typical island

3.2 Electrical Power Generation from a WTG

Considering the data of the Figure 1, the chronological variation of temperature and wind velocity are sampled at equal intervals of time. Assigning equal probability of occurrence to each sample, a joint distribution of temperature and wind velocity is obtained. In this case, temperature and wind velocity values are assumed to be correlated and the sampling interval is considered to be three hours. Following the procedure described in the methodology [6], a probable value of power output corresponding to each pair of temperature and wind velocity values of the joint PDF of temperature and wind velocity is translated into the PDF of equivalent electrical power generation. This distribution is shown in Figure 3. The distribution assumes that the WTG is 100% reliable.

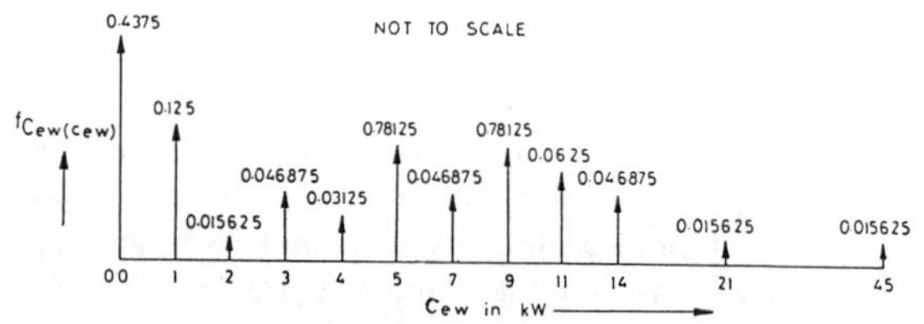

Figure 3 PDF of generation with bivariate function

For comparison, the PDF of equivalent electrical power generation is also derived assuming that the output of a WTG is a function of wind velocity only. A constant value of temperature of 15.15°C [8] is considered in developing this PDF. This is shown in Figure 4. Then the random

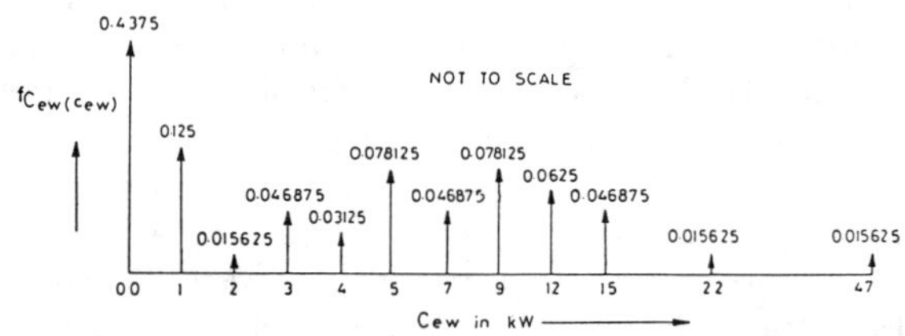

Figure 4 PDF of generation with univariate function

outage of the WTG is incorporated in the PDFs of equivalent electrical power generation by considering the FOR of the WTG. The resulting distributions for the above considered models [6], one derived from bivariate function of temperature and wind velocity and the other derived from univariate function of wind velocity only, are already shown in the Figures 3 and 4 respectively.

3.3 Simulation Results

The model is applied to evaluate the EEG, the E(ENS) and the LOLP corresponding to different number of WTGs. In simulation, it is considered that the WTG is the only source of electrical energy to meet the electrical load demand. The simulation results are presented in Table 1. For comparison, the results obtained using the model from the univariate function of wind velocity only are also included.

3.4 Observations

The comparison of the PDFs of equivalent capacity

generation of Figures 3 and 4 and their probabilities, clearly reveals that there is a distinct difference between the two considered models.

It is observed from Table 1, that with the increase of number of WTGs in the system, the EEG increases while the LOLP decreases, as expected. A difference between the results of two models is clearly observed from this Table. The application of the model reveals that the results of generation expansion analysis will be different from the reality especially for a large variation of temperature.

Table 1: Comparison of the simulated results of the bivariate (BIVA) and the univariate (UNIV) models.

No of WTGs	Installed Capacity (MW)		EEG (MWh)		LOLP (10^{-2})	
	BIVA	UNIA	BIVA	UNIV	BIVA	UNIV
30	1.35	1.41	19.44	20.16	99.99	99.99
35	1.58	1.65	22.68	23.52	99.99	99.99
40	1.80	1.88	25.92	26.88	99.98	99.97
45	2.03	2.16	29.16	30.24	99.95	99.93
50	2.25	2.35	32.40	33.59	99.90	99.86
56	2.52	2.63	36.28	37.61	99.78	99.70
60	2.70	2.82	38.86	40.29	99.66	99.56
65	2.93	3.06	42.08	43.63	99.46	99.34
70	3.15	3.29	45.30	46.96	99.22	99.08
75	3.38	3.53	45.51	50.28	98.95	98.79
80	3.60	3.76	51.70	53.59	98.67	98.51
85	3.83	4.00	54.89	56.89	98.40	98.23
90	4.05	4.23	58.07	60.19	98.13	97.94
95	4.28	4.47	61.25	63.47	97.86	97.64
100	4.50	4.70	64.41	66.74	97.60	97.27
112	5.04	5.26	71.95	74.51	96.53	95.84
160	7.20	7.52	99.76	102.72	80.51	78.08

4. GENERATION SYSTEMS

4.1 Wind Turbine Generator (WTG)

The model of a WTG used to evaluate the reliability and cost of the potential power system of the typical island is developed [6], utilizing the wind velocity and the temperature of Kutubdia. The developed multi-state model has been shown in Figure 3.

Note that the maximum possible output of this WTG is 45 kW. For the purpose of sample cost analysis, data available from published literature [5,9,10] and local authorities [11] are utilized. The annual operation and maintenance cost of a WTG is of fixed type.

4.2 Thermal (Diesel) Generator (THG)

The THGs of 100 kW and 200 kW with FOR of 0.04 and 0.1 are considered [9,12,13]in order to find out a most desirable generation mix. The operation and maintenance costs of these two thermal units depend on the amount of energy produced and the major portion of the cost is the fuel cost.

4.3 Simulation Results

The power system of the area is evaluated considering each of the above two different types of units separately.

In the case of a WTG; the EEG, the E(ENS) and the LOLPs are evaluated for WTG units ranging in number from 15 up to 160, each of capacity 45 kW. The THGs are also considered as the only source of energy to meet the electrical load demand of the system. The system is evaluated assuming that the THGs of two different capacities, 100 kW and 200 kW, with different numbers comprise the generating system.

The cost of the generating system is also evaluated for WTGs as well as THGs, depreciation not being considered.

The results obtained considering two types of generating units as stand-alone sources are compared in the Figures 5 to 7.

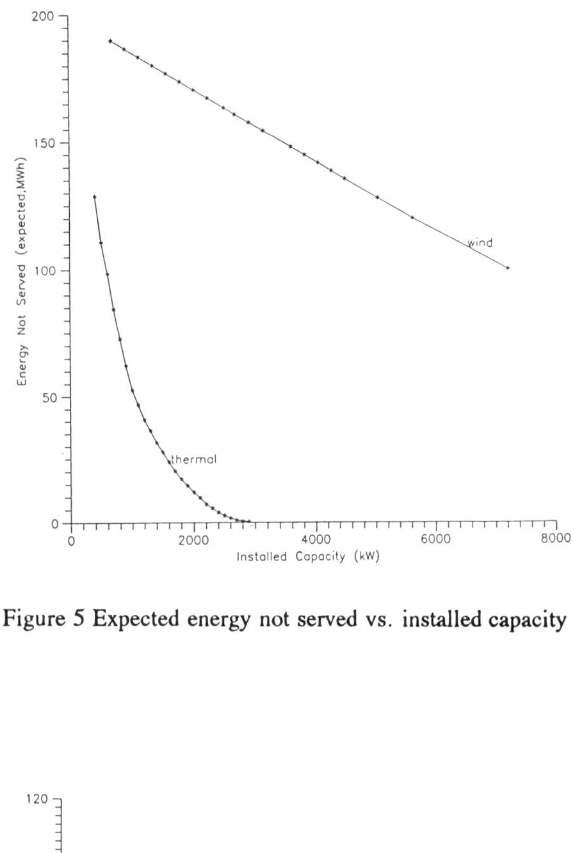

Figure 5 Expected energy not served vs. installed capacity

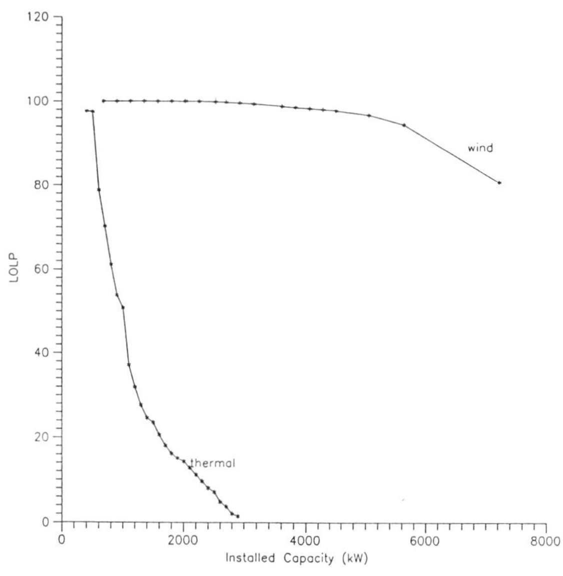

Figure 6 LOLP vs. installed capacity

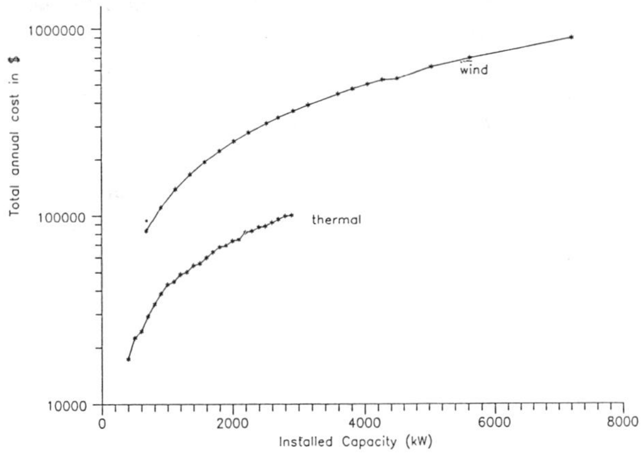

Figure 7 Total annual cost vs. installed capacity

5. DISCUSSION

In what follows, the salient points observed from the computer simulation of different types of units are discussed.

Figure 5 shows, the E(ENS) is very high when WTGs are considered as the sources of power supply, while the E(ENS) of the system becomes very low when THGs are considered as the sources of power supply. For example, at 2500 kW installed capacity, the values of expected energy not served are 1.38% and 81.7% of the total energy demand for THGs and WTGs respectively. This is in a system with a peak demand of 2520 kW.

Figure 6 shows, the rate of improvement of the reliability of the system is very slow in case of WTGs, as a stand-alone source. The reason behind the poor performance WTGs in improving the system reliability is the low probability values of output capacities. Although the rated capacity of a WTG is 45 kW, the expected capacity output of a WTG is only 3.37 kW.

Figure 7 shows, the THGs are also superior to WTGs regarding cost also. For an example, for an installed capacity of 2000 kW the THGs require only 0.073 M$ to meet the total annual cost while WTGs require about 0.24 M$.

6. CONCLUSION

The comparison of different types of unit clearly reveals that the addition of THGs at any level of installed capacity, provides better improvement of the system reliability as well as involves lower additional cost compared to the addition of WTGs. For example, at an installed capacity of 1500 kW comprising of THGs the system LOLP and the total annual cost are 23.55% and 0.056 M$ respectively. Now an addition of 500 kW thermal units increases the annual cost by 30.36% and decreases the LOLP by 39.19%. However, if 500 kW of WTGs are added, the annual cost increases by about 98.21% (with respect to THGs cost) and the LOLP decreases only by 0.04% . Therefore, the THGs installation in the system to meet the demand is the most desirable solution for the considered isolated area from the cost as well as reliability point of view.

7. ACKNOWLEDGEMENT

The authors would like to acknowledge the support of the Commission of the European Communities in providing post doctoral fellowship to Dr Aminul Hoque. The authors also wish to thank Bangladesh University of Engineering and Technology (BUET), Dhaka, Bangladesh and The Queen's University of Belfast (QUB), UK for the provision of the necessary research facilities.

8. REFERENCES

[1] Saifur Rahman and B.H. Chowdhury, "Effects of Clusters on the Electric Power from Windfarms", IEEE Trans. on PAS, Vol. PAS-103, No. 8, pp. 2158-2165, August, 1984.

[2] P.C. Krause and D.T. Man, "Transient Behaviour of a Class of Wind Turbine Generators During Electrical Disturbances", IEEE Trans., 1981, PAS-100(5).

[3] G. Thomann and M. Barfield, "The Time Variation of Wind Speeds and Windfarm Power Output in Kansas", IBID, 1988, 3, (1), pp. 44-49.

[4] A.G. Bakirtzis et al., "A Probabilistic Costing Method for the Evaluation of the Performance of Grid-Connected Wind Arrays", IBID, 1989, 4, (1), pp. 43-40.

[5] V. Daniel Hunt, "Wind Power: A Handbook on Wind Energy Conversion System", Van Nostrand Reinhod Company, New York.

[6] A. Hoque, "A Methodology for Analyzing Electrical Energy Options for Isolated Areas", a Ph.D Thesis, 1990, Bangladesh University of Engineering and Technology, Dhaka, Bangladesh.

[7] Q. Ahsan and K.F. Schenk, "Two Area Production Cost Evaluation by the Segmentation Method", IEEE Trans., 1985, PAS-104 (8), pp. 2140-2146.

[8] K.L. Kumer, "Engineering Fluid Mechanics", Revised 3rd. Edition, 1984, pp.759-760.

[9] Bakirtzis et al., "Short Term Generation Scheduling in a Small Autonomous System with Unconventional Energy Sources", IEEE Trans. on Power Systems, Vol.3, pp. 1230-1236, August, 1988.

[10] G.T. Chinery et al., "Break-even Cost for Incorporating Photovoltaic Generation into the TVA Power System", IEEE Trans. on Power Systems, Vol. PWRS-2, No.3, August, 1987.

[11] Power Engineering Department, Siemens Bangladesh Ltd.,Bangladesh, 1990.

[12] Xifan Wang et al., "Reliability Modeling of Large Wind Farms and Associated Electric Utility InterFace Systems", IEEE Trans. on PAS, Vol. PAS-103, No. 3, March, 1984.

[13] M.A. Khallat and S. Rahman, "A Probabilistic Approach to Photovoltaic Generator Performance Prediction", IEEE Trans. on EC, Vol. EC-1, No. 3, pp. 34-40, September, 1986.

ARE FEASIBILITY STUDIES RELIABLE AT HIGH WIND ENERGY PENETRATION LEVELS

Jens Carsten Hansen **John Olav Giæver Tande**

Dept. of Meteorology and Wind Energy
Risø National Laboratory
DK-4000 Roskilde, Denmark
(Phone: +45 42371212 - Fax: +45 46755619)

ABSTRACT. Many power systems around the world show economic prospects for introducing wind energy to high penetration levels.
Few actual high wind energy penetration systems have been built and decisions to fund such projects have to be taken based on feasibility studies including power system modelling.
This paper compares the validity of feasibility studies of high wind energy penetration projects with studies for wind energy penetration at the level of "traditional" grid connected wind farms. Differences in study techniques and possible reasons for investors scepticism towards high wind energy penetration systems are identified. By dividing uncertainties in a) those always present in wind power projects and b) those special for high wind energy penetration systems, real differences in project risks and uncertainties appear.
The case of the capital of the Republic of Cape Verde is considered. Feasibility study uncertainties elements for high and low penetration levels are identified and quantified for the simple case excluding capacity value of wind power. Sensitivity analyses are performed.
For the case studied, the uncertainty in the cost of wind energy is practically independent of the wind energy penetration level and dominated by uncertainties in wind resource predictions.

1 Introduction

Can investors and funding agencies trust results of feasibility studies for introduction of wind energy to high penetration levels in electric power systems?

Obviously, they often don't think so. Many isolated small to medium scale power systems exist around the world with good economic prospects for introducing wind energy to penetration levels well above what today normally is considered the technologically proven level, but the high wind energy penetration projects are still very few. Why?

Attraction of funds for building of high wind energy penetration systems relies on feasibility studies. Work is therefore needed regarding validation of different study and modelling techniques. This paper suggests and discusses validation of a simple system modelling approach for small to medium scale isolated power systems for which high wind energy penetration seems realistic.

2 Study and modelling techniques

Feasibility studies are multidisciplinary undertakings composed of a large number of substudies, including

- national and local development planning
- climate, wind resources and other physical and infrastructural conditions
- physical planning
- electricity demand and load forecasting
- site selection
- electric grid design
- control systems design
- wind farm layout, optimization, equipment requirements, design and lifetime evaluation
- power system dynamics
- power station electricity generation and performance modelling - conventional and wind
- environmental effect
- institutional aspects and staffing
- project implementation planning and logistics
- operation and maintenance
- dismantling or rebuilding
- financial implications
- economics
- uncertainties

Most of the analyses and modelling techniques applicable for the many substudies are of course the same for traditional grid connected wind farms and for high wind energy penetration. These parts of the feasibility studies therefore generally are considered acceptable if performed according to the established standards and guidelines.

The remaining substudies where analysis and modelling techniques from traditional grid connected wind farms may not be applicable are primarily related to the power system performance, controls and dynamics. These are the possible additional sources of feasibility study uncertainties as compared to traditional grid connected wind farms. The study and modelling techniques for these three key substudy elements should therefore be chosen preferably so that investor risks due to the additional uncertainties are eliminated or at least reduced as much as possible.

Firstly, investor risks related to introduction of special power system controls and dynamics due to the high wind energy penetration level can be eleminated. This can be achieved by assuming power system investments in grid and controls in the high end of realistic costs and using rude design assumptions to facilitate true conclusions of modelling approximations. Using minimization of investor risks as the overall priority for design of the electric grid and control system, the chosen designs may not necessarily end up being technical or economic optimal. Furthermore, investments may be overestimated and certain projects may not be realized, though feasible. However, at the present level of the technological development of high wind energy penetration systems these uncertainties must be dealt with in a convincing way, and initial projects will and should be built at the most favourable locations anyway.

Secondly, modelling uncertainties of electric energy generation by conventional power stations and wind farms as input to the economic analysis depends on the models used and its compatibility with the power system design, in particular the ability to model the control system and operator behaviour for the combined operation of wind power and conventional plants. The simpler the control system the easier understood and operated by the operator and the easier accurately modelled by the feasibility study. The determination of the capacity value of wind power is determined through the supply reliability contribution from wind using statistical methods independent of the wind energy penetration level.

Provided that appropriate rude system design and modelling assumptions are made, the only special modelling requirements for wind farms at high penetration levels may be limited to calculating the annual energy production of wind farms and conventional generating sets including

- variations in time of wind power output and consumer loads
- operational requirements and strategies of the conventional power station, including spinning reserve and technical minimum loads on the individual generating sets.

3 Validation approach

The validation of feasibility study models for high wind energy penetration of course ultimately should be done by comparing results with actual data from a representative set of projects. Until enough projects and data exist real validation can not be made satisfactorily, but system design and modelling techniques can be chosen so that investor risks and uncertainties are minimized as described above. The increased uncertainty of entering into a high wind energy penetration project can then be illustrated by analysing the sensitivity of the energy production modelling to the key variables in the modelling - wind, electricity consumption, technical minimum load and spinning reserve.

4 Case

The case of Praia, the capital of the Republic of Cape Verde, is considered. Using one and the same set of modelling techniques, sensitivity and uncertainty analyses of the economic feasibility of low and high wind energy penetration levels are compared. The uncertainty elements and sensitivity factors are identified and quantified for low as well as high wind energy penetration, 2.5% and 25% respectively.

4.1 Assumptions

Table 1 lists some of the key assumptions and power system data.

Table 1: *Outline of key assumptions and power system data.*

economic	
Economic lifetime	20 years
Real discount rate	5%
Annuity factor	18.05
wind farm	
WT capacity (kW)	N*190
Total avail+loss+red factors	0.85
Investment (DKK/kW)	9000
Annual O&M (% of WT inv)	2.5
Retrofit costs (% of WT inv)	20
Retrofit year	10
Social costs	0
Salvage value	0
conventional	
conv. capacity (kW)	9974
Annual load (GWh)	27.7
Fuel costs (DKK/kWh)	0.28
O&M costs (DKK/kWh)	0.09
Social costs	0

In 1992 a load forecast for 1995 - the possible year for implementation of a high penetration wind power project - was prepared. The diurnal variations of the load was found by scaling the 1992 loads. Seasonal load variations are practically non-existing. Figure 1 shows the forecasted loads and load patterns assumed by the study, different for week-days and weekend-days.

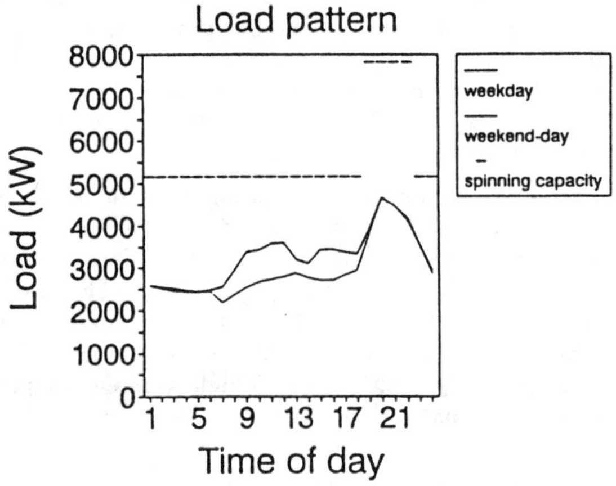

Figure 1: *Forecast of load patterns for weekdays and weekend-days with indication of the associated spinning capacity at the chosen operational strategy for 1995 at Praia.*

Wind conditions at the candidate site at Praia, Mte. S. Filipe, at 30 m height above ground fits a Weibull distribution with scale parameter $A = 10.3 m/s$ and shape $k = 3.3$, annual mean wind speed $\bar{U} = 9.3 m/s$ and a theoretical maximum production of 812.5 MWh/year for an average 190 kW wind turbine as used by the study.

The diurnal variation of the wind has been found by making WA^{S}P analyses of 3 years of data for each hour of the day. The result is illustrated by the wind power production factor for one 190 kW wind turbine in figure 2.

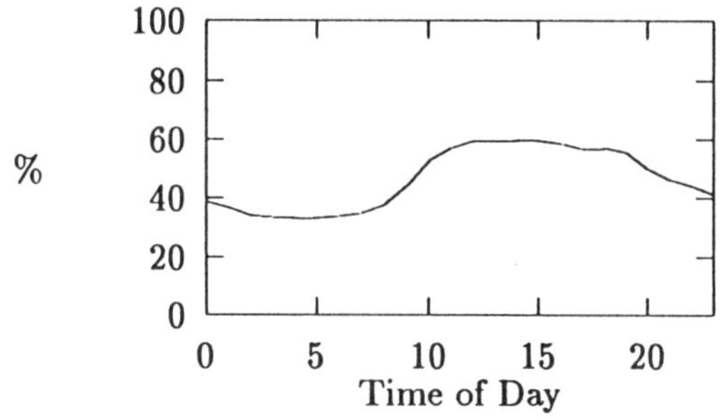

Figure 2: *Hour by hour wind power production factor for one 190 kW wind turbine at Mt. S. Filipe, Praia.*

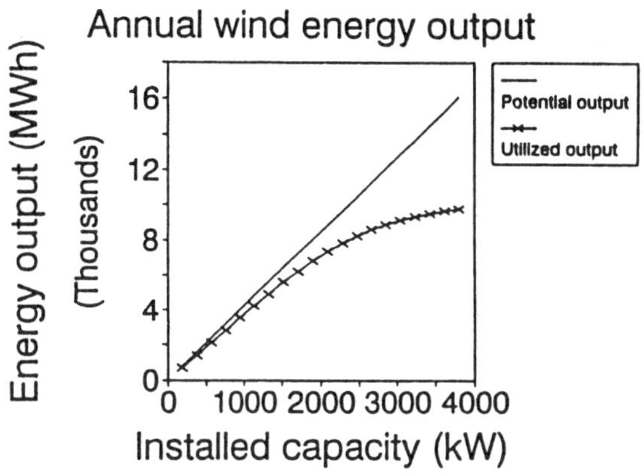

Figure 3: *Total and net output from Mte. S. Filipe wind power plant as a function of the installed wind power capacity.*

4.2 Modelling

The model uses a combination of time series simulation and a statistical analysis approach, assuming power system regulation to be taken care of by the conventional power station and a fast dump load, theoretically of the size of the wind farm. In practice it will be smaller, but the investment for the large size is included.

The total potential energy output and the net output excluding the energy dumped is estimated for different sizes of the wind farm. The result is shown in figure 3.

The total output is calculated from assuming the wind turbine power curve, the wind speed distributions hour by hour, wind turbine availability, wake effects, losses etc. as common for such analysis.

The net output is calculated from considering the operation of the total power system. Thus, the annual net wind energy output is:

$E_{w,n} = E_w - E_d$

where E_w is the total wind energy output and E_d is the excess energy dissipated. The dissipated energy is found from a power system analysis determining the net load distribution, $\Pr(P_n)$, i.e. the distribution of $P_n(t) = P_l(t) - P_w(t)$. $P_l(t)$ is the load without wind power. The dissipated energy is:

$$E_d = \sum_{i=1}^{24} T_i \cdot \int_{P_{l,min}-P_{w,max}}^{P_{g,min,i}} \Pr(P_n) \cdot (P_{g,min} - P_n) dP_n$$

for $P_n < P_{g,min}$ and else zero.

$P_{g,min}$ is the technical minimum load for the diesel power station. The spinning capacity must be determined respecting $P_{g,min}$. In this the required spinning capacity is conservatively estimated neglecting $P_w(t)$.

Cost of Wind Energy

The cost of energy is expressed as the levelised production cost (LPC) distributing costs and energy output over the lifetime, balancing costs and income. The result is suited for comparison with other wind turbines or other sources with similar function and operation characteristics, provided similar cost computation methodologies are applied, e.g. with respect to environmental impact and other social costs.

At the above assumptions, the levelised cost of energy from a wind farm at Mte. S. Filipe as a function of the number of wind turbines is shown in figure 4.

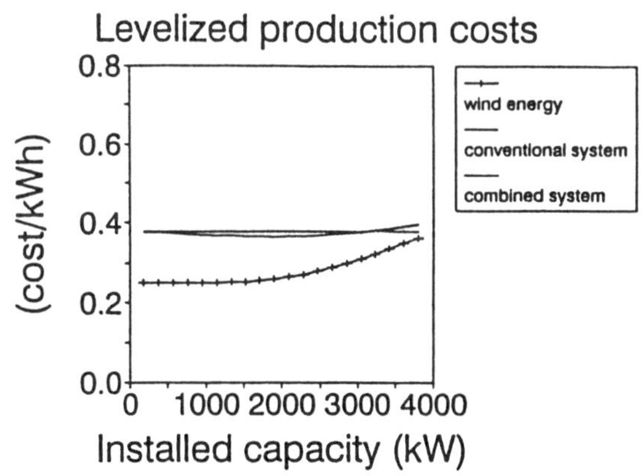

Figure 4: *Levelised cost of energy from the assumed wind power plant at Mte. S. Filipe as a function of the installed wind power capacity.*

Economic analysis excluding capacity value

An economic analysis is made in order to be able to determine the optimal amount of wind power capacity in the system and the economic consequencies for the investor. For simplicity this paper only considers the situation that the conventional generating capacity in the power system is sufficient to cover the electricity demand throughout the 20 year lifetime of a wind farm. The capacity value of wind power is thus not included, and economic benefits of wind power considered in this analysis are exclusively the savings achieved for the existing power system in fuel and varying O&M costs. Savings in outage costs and social costs, i.e. environmental benefits, are not included.

The net profit and the internal rate of return of the investment, IRR, are shown in figure 5. The result for this case study shows maximum net profit at a wind power capacity of approximately 2MW at the assumptions mentioned, excluding the environmental benefits and capacity value of wind power.

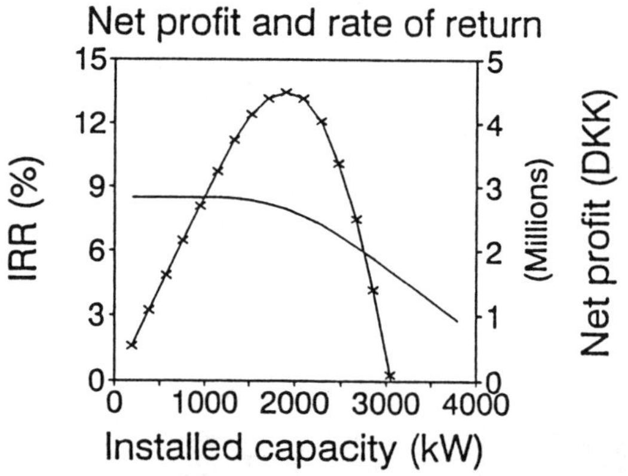

Figure 5: *Calculated net profit and internal rate of return (excluding the environmental benefits and capacity value of wind power) versus the installed wind power capacity.*

4.3 Sensitivity and uncertainty analysis

Table 2 shows the sensitivity of the levelized production cost (LPC) to variations in 15 key parameters in the analysis for both low and high wind energy penetration. Estimates of the total uncertainties of the cost of wind energy are found treating the parameters as independent variables.

Table 2: *The sensitivity of the levelized production cost (LPC) of energy from wind power at low and high wind energy penetration. The sensitivity factor to variations in 15 key parameters is given for 1 and 10 wind turbines, 2.5% and 25% penetration respectively. Uncertainties for the 15 parameters assumed by the uncertainty estimate is shown.*

parameter	sensitivity factor		uncertainty assumed (%)	
	1 WT	10 WT	1 WT	10 WT
Investment	0.704	0.704	5%	
O&M	0.212	0.212	5%	
disc rate	0.071	0.071	0%	
lifetime	-0.720	-0.720	0%	
air density	-1.010	-0.845	1%	
power curve	-1.010	-0.845	5%	
A(Weibull)	-1.891	-1.7142	10%	
k(Weibull)	-0.119	-0.176	10%	
availability	-1.010	-1.045	2%	
losses	-1.010	-1.045	3%	
reductions	-1.010	-1.045	5%	
array eff	-1.010	-1.045	5%	
tech min load	0	0.163	50%	
spinn reserve	0	0.018	50%	
load	0	-0.350	20%	
Total uncertainty			21.7%	22.7%

The result of the sensitivity analysis of the cost of wind energy for 1 and 10 · 190 kW wind turbines at a ~ $10MW$ power station is illustrated for a selection of 6 out of the 15 parameters in figures 6 and 7. It is seen that for both low and high wind energy penetration, feasibility study results are most sensitive to variations in the annual average wind speed. The three parameters - Technical minimum load, spinning reserve and load - to which there is no sensitivity at low wind energy penetrations do not increase uncertainties in determination of the cost of wind energy significantly at the system design and modelling assumptions made.

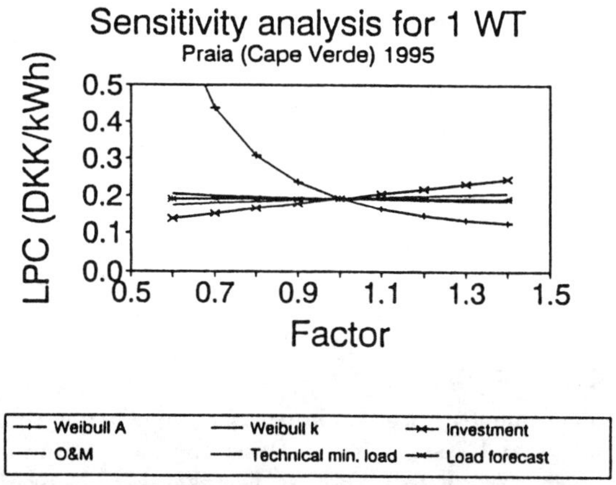

Figure 6: *Sensitivity of cost of wind energy for 1·190 kW wind turbines at Mte S. Filipe.*

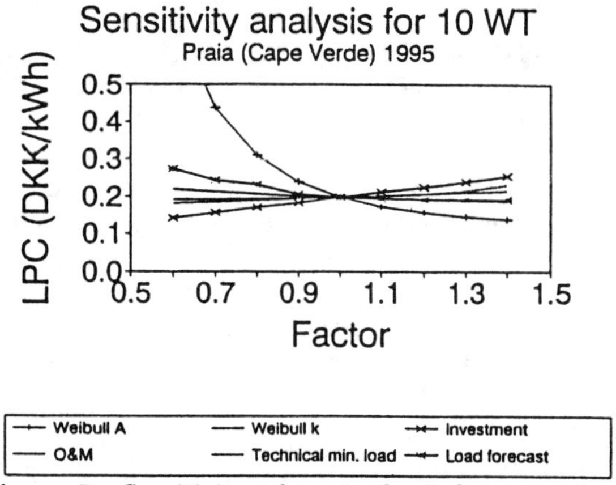

Figure 7: *Sensitivity of cost of wind energy for 10·190 kW wind turbines at Mte S. Filipe.*

5 Conclusion

Uncertainties in feasibility studies for high wind energy penetration systems may be limited to being only insignificantly higher than for traditional grid connected wind farms, provided that appropriate design and modelling assumptions are made. The technical minimum load of the conventional gensets, the required spinning reserve and the consumer loads are the three extra parameters at high penetrations, but cost of wind energy is for both high and low wind energy penetration most sensitive to uncertainties in wind speed.

This study shows that cases exist for which feasibility studies are just as reliable for high wind energy penetrations as they are for low penetration traditional grid connected wind farms.

ON THE DESIGN OF OPTIMIZED PROFILES FOR STALL REGULATED HAWTs

P. Chaviaropoulos, B. Bouras,
G. Leoutsakos, K.D. Papailiou
National Technical University of Athens
Laboratory of Thermal Turbomachines
P.O. Box 64069, 157 10 Athens, Greece

ABSTRACT: The paper presents a methodology developed for the design of optimized airfoils for stall regulated HAWTS. Given optimized lift and drag characteristic curves for maximum energy capture, the method focuses on the design of the corresponding profile shape. Although the problem is a multi-disciplinary one, the present design philosophy is based on a single-point approach, which, through an iterative procedure aims to achieve the best possible compromise of the imposed requirements. The adopted methodology is based on a combination of an inverse viscous and an inverse inviscid counterparts. The viscous part provides an optimized suction side velocity distribution for a given Reynolds number. A compatible pressure side velocity distribution is then constructed for the given design lift coefficient. At a final stage the profile shape is provided by the inverse inviscid solver using an iterative procedure to effect profile closure automatically. The resulting profile performance is fully assessed using a direct viscous inviscid interaction method. Numerical results are presented for two designed profiles corresponding to tip and hub sections of an optimized rotor. For the tip case the design philosophy is verified against experimental measurements.

1. INTRODUCTION

The paper presents the methodology developed at NTUA for the design of optimized airfoils for stall regulated constant speed horizontal axis wind turbines. This methodology was developed under the JOUR-0079-C(SMA) contract in cooperation with RISOE and the University of Southampton. Using a blade element code to feed back an optimization Kernel RISOE [1] proved that a significant increase of energy capture may be obtained for a classical stall regulated HAWT (the tests have been performed using a Danwin machine) if its rotor is suitably redesigned. In this framework RISOE provided the optimized lift and drag characteristic curves for the tip, mid and hub sections of the rotor, while NTUA focused to the definition of the corresponding profiles shape.

The definition of a profile with desirable lift and drag curves is not a straight-forward task. Evidently, the design problem as posed above is a multi-disciplinary one and, thus, it cannot be tackled with a "one-shot" method. In any case, however, one should satisfy the essential items of the design requirements which are the maximum C_l value (which is crucial for stall-regulated machines), the smooth transition of the C_l-curve from the pre-stall to the post-stall region and the drag reduction in the linear pre-stall zone. In the range of Reynolds numbers that wind turbine blades are operating drag may be reduced by designing laminar airfoils instead of using the classical NACA series which are characterized by early transition from laminar to turbulent flow. Monitoring of the suction-side transition point in the positive incidence off-design region is one of the key features of the NTUA philosophy for both drag reduction and C_{Lmax} control.

The present design methodology is based on an inverse viscous and an inverse inviscid method. Once the design point has been fixed, the inverse viscous counterpart [2],[3] provides an optimized suction side velocity distribution for a given overall Reynolds number. A compatible pressure side velocity distribution is then constructed for the design lift coefficient. At a final stage the profile shape is provided by the inverse inviscid counterpart [4] using an iterative procedure to effect profile closure automatically. The resulting profile performance is fully assessed in both design and off-design conditions using a direct viscous-inviscid interaction method [5],[6].

The steps followed in the present design methodology are fully discussed in the paper. Some brief description of the numerical tools involved is also given. Computational results are presented for two designed profiles which correspond to the tip and hub sections of an optimized HAWT rotor. The tip-section has been constructed and tested in the Southampton University wind-tunnel. For this case the obtained numerical results are compared with measurements. Comparisons indicate very satisfactory airfoil performance in both pre- and post-stall regions and verify the validity of the design philosophy.

2. THE DESIGN PHILOSOPHY

The wind turbine blade section design and optimization problem faced in the present work may be posed as follows:

Design a blade profile subject to:

(a) Given Reynolds no of operation (based on chord)
(b) Given C_{Lmax}
(c) Operation within a specified range of incidence angles.
(d) C_D with minimum possible values and as long a flat C_D-curve as possible.
(e) Maximum blade thickness meeting prescribed minimum for structural-mechanical considerations.
(f) A maximum inviscid flow velocity with as low a value as possible, for low noise emission reasons.
(g) Although not required, it is desired to have a reasonably flat C_L distribution in the post-stall region.

Although the problem is defined as a multi-point design one we shall be obliged to face it (with our current means) as a single point design problem. This means that only an approximate satisfaction of the overall constraints may be finally expected. The NTUA design procedure may be summarized in the following steps.

STEP 1. Define the design point. The design point which is currently selected is the one where the suction side velocity distribution is at the limit of separation. The desired C_{lmax} is, then, providing a basis for estimating the value of the inviscid C_l of the design point. The choice of the first design point is performed empirically, on the basis of previous experience. The difference in incidence between the operating point with maximum C_L and the point at which separation first appears at the trailing edge is used. Checking with the direct (analysis) method, when an airfoil shape has been designed, provides the basis for a better estimation of the above mentioned incidence difference.

STEP 2. For the $C_{L\text{-}inviscid}$ compute the corresponding mean values of the suction ($\bar{u}_s$) and pressure side ($\bar{u}_p$) velocity distributions. Their difference weighted by the arc length of each side provides the circulation (and the lift), while their sum must be very nearly equal to 2. For the given Reynolds and $\bar{u}_s$ compute with the inverse shear layer solver the corresponding optimized suction side velocity distribution with the following features

a) It is produced according to the inverse viscous flow theory and optimization procedure described in [2],[3] and which is based on minimizing the viscous losses of an airfoil.
b) The boundary layer approaches the limit of separation at the trailing edge region.
c) The boundary layer possesses and extended laminar part. Transition takes place between 40% and 60% of the chord on the suction side and possibly more on the pressure side.
e) The suction side maximum velocity is low, so as to reduce the noise levels as much as possible. It is to be noted, in this respect, that the noise level of an airfoil is proportional to the 5th-7th power of the velocity level.

STEP 3. Construct the corresponding pressure side velocity distribution. Using the predefined ($\bar{u}_p$) mean value, the pressure side velocity is constructed analytically using simple curves (straight line for the mid-part and parabolas for the near-leading and trailing-edge regions). The proper selection of the pressure side characteristics is critical since it effects directly important features of the profile, like the thickness, the camber and the negative off-design behaviour. Care should also be taken to maintain the velocity mean value close to the desired value. On the other hand it should be noted that no unique solution exists and several options have to be exploited with the design specification compromizes in mind.

STEP 4. Use the inverse inviscid solver to achieve a closed blade. In general the initial suction and pressure side velocity distributions do not provide a completely closed profile. An iterative procedure is then followed using two independent control variables for the closure, namely the ratio of the suction and pressure side arc-lengths and the C_L of the design. This procedure is rapid, needing 5 to 10 iterations without affecting much the desired C_L.

STEP 5. Use the direct analysis method to validate the designed blade. This is done initially at the all important design point, but then the complete desired range of incidences is examined. Particular emphasis is placed on identifying the value of C_{Lmax} and the corresponding incidence angle. On the other hand, it is important that the pressure side velocity peak at negative incidences is controlled, so that reasonable velocity distribution values are obtained, which do not produce excessive C_D values.

Rapid inviscid calculations demonstrate how the velocity distribution on the suction side behaved with incidence. It was observed that, for a considerably wide range of incidences the suction side velocity distribution did not appear to change considerably, at least for tip section airfoils. It was decided, then, to employ this characteristic in order to regulate stall by creating, for the stall incidence (and C_{Lmax} value) a sufficiently pronounced peak near the leading edge, in order to enforce transition there. Consequently, through the jump in the transition point, a large separation region appears in the last part of the airfoil, which ensures stall. If stall appears at the required incidence and produces the desired C_{Lmax}, then the design procedure has succeeded in this respect.

In fact, two typically different cases may be distinguished in the design of the wind turbine blades
a) The tip blade section case for which an extended laminar part is adopted. In this case the leading edge geometry is shaped in such a way so as to produce a leading edge peak near the point of desired maximum lift. Past that point, the leading edge peak results in shifting transition back to the leading edge. As arranged, this ensures extended turbulent separation and hence stall. In this way, practically up to the stall incidence, the drag coefficient remains low. Care is taken to ensure low velocity peaks at high negative incidence for the pressure side velocity distribution.
b) The hub blade section case, where the leading edge radius as well as the blade thickness as a chord percentage are both increased. The laminar part of the airfoil is reduced and the previously applied design approach, cannot be applied. Stall appears gradually, when the maximum deceleration attains a critical value, while the decelerating part of the suction side velocity distribution increases also gradually with incidence.

Once the design is completed, extensive and systematic calculations are performed on the blade using the direct method.

The iterative design procedure during which the design control parameters are either radically changed or just trimmed - depending on the deviation from the target specifications -stops when a reasonable compromise is achieved between the required conditions.

3. BRIEF DESCRIPTION OF THE COMPUTATIONAL TOOLS INVOLVED

3.1 Inverse viscous method

The aim is to establish an optimum (minimum losses) suction side velocity distribution for a specified incidence. Following the methodology described in [2],[3], this translates in defining an optimum L_k-X image curve, given a Reynolds number and desired airfoil circulation (design point and incidence). The image curve is defined based on minimizing laminar and mainly turbulent boundary layer viscous dissipation and extending the laminar region for as long as the free stream turbulence level allows. Once this is done the corresponding optimized velocity is obtained by solving the transformed integral shear layer equations and since the abscissa X is directly related to the flow losses, this inverse procedure controls the drag contribution of the suction side velocity at a desired level. It should be pointed out that the choice of an optimum suction side shear layer does not necessarily give the optimum blade shape, since the contribution of the pressure side velocity - satisfying the closure conditions -, will finally define blade circulation and total drag.

3.2 Inverse inviscid method

The NTUA inverse inviscid method is based on an alternative definition of the inviscid flow equations using as "physical" coordinates the potential function and the streamfunction (Φ-Ψ plane). On the (Φ-Ψ) plane the mass-conservation and irrotationality condition provide a second order p.d.e for the velocity modulus, which may be solved with a unique solution if the velocity boundary conditions are known. These boundary conditions, as stated before, are provided by the inverse shear layer solver (suction side) and a corresponding pressure side distribution which mean level is fixed from the desired C_L of the design. If the suction and pressure side velocities are compatible in terms of blade closure, this procedure ends with the definition of the corresponding blade shape. To increase the accuracy of the computation of rounded leading edge profiles some improvements have been performed in the original NTUA inverse inviscid method using a C-type grid generator for the discretization on the (Φ-Ψ) plane. More details can be found in [4].

3.3 Direct viscous-inviscid interaction method

Airfoil performance was assessed using a viscous inviscid interaction procedure involving a inviscid potential compressible flow solver and a shear layer solver coupled strongly in a direct scheme, through appropriate boundary conditions applied at the common surface shared by the inviscid and viscous flows [5],[6]. The two-zone model underlying the interaction is that of Lock and Firmin [7]. The number of iterations necessary for convergence is of the order of 10 for attached flows and 30 or more for separated flows, where the shear layer is solved directly up to near separation and inversely from there onwards.

The effect of the viscous displacement thickness of the inviscid flow is represented by mass transpiration through the surface, formulated as a transpiration velocity.

4. APPLICATION OF THE DESIGN METHOD. RESULTS AND DISCUSSION

The above design procedure was applied to two different kinds of design.
a) A tip blade section design with $Re=10^6$ and level of external turbulence $T_u=0.2\%$, suitable for wind tunnel testing,

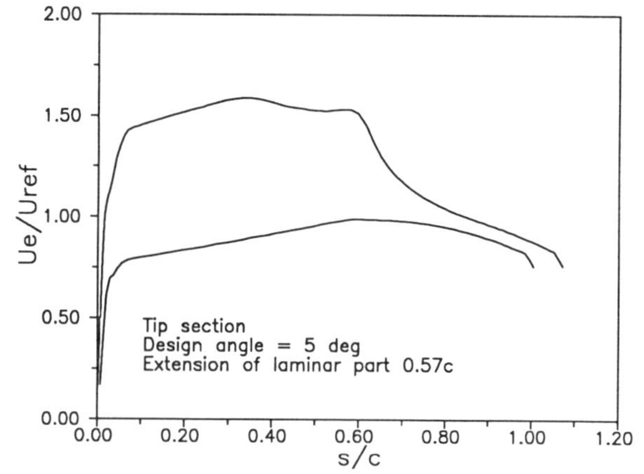

Fig. 1 Optimized suction and pressure side velocities for the tip section.

which took place at the Southampton University facilities, aimed principally at code and methodology validation. The optimized suction and pressure side velocity distributions are shown in figure 1. The resulting profile shape which is 18% thick is shown in figure 2. The above airfoil was manifactured at Southampton

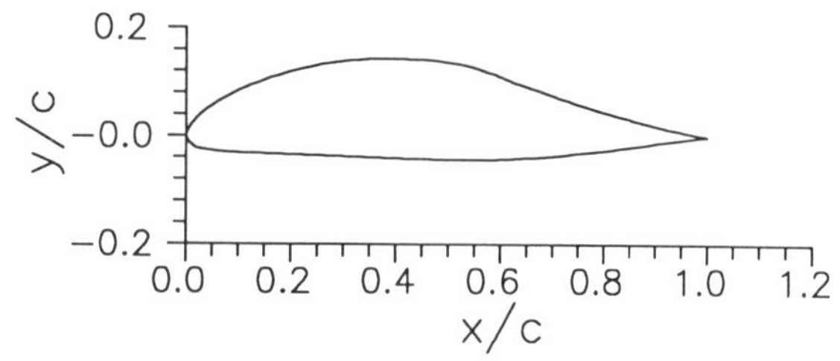

Fig. 2 Tip-section profile shape.

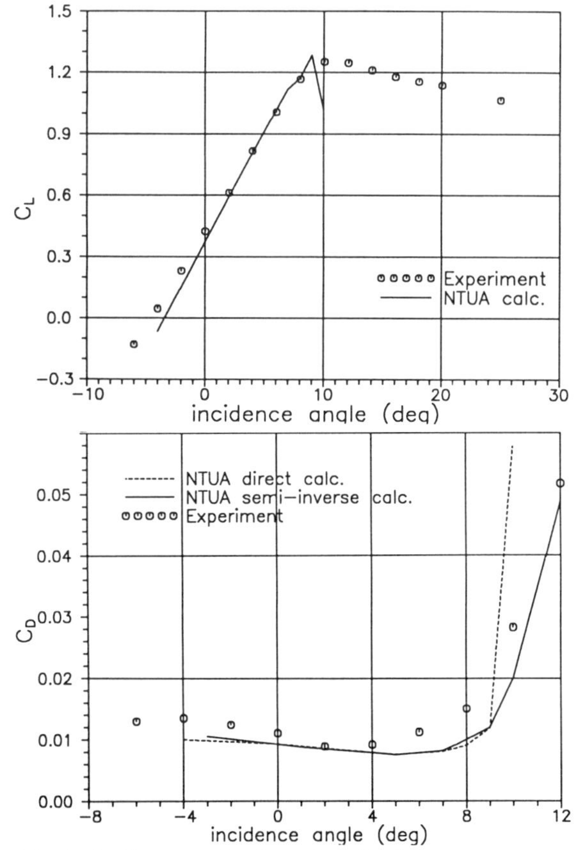

Fig. 3 C_L, C_D curves for tip-section airfoil.

University and tested at the design Reynolds number and turbulence level. The characteristic C_L and C_D distributions versus the incidence angle are shown in figure 3. The experimental results validate the design philosophy in the respects (a) C_{Lmax} close to the desired one (b) Smooth and relatively flat post-stall lift characteristics. The NTUA C_L calculations using the viscous-inviscid interaction procedure are also shown on the above graphs. The C_{Lmax} and the corresponding incidence angle are reasonably well predicted but the post-stall prediction capabilities are poor which is due to convergence problems of the direct method in that region. The C_D prediction appears to be reliable in both linear and post-stall regions with improved results when the direct method is run on the semi-inverse mode. Furthermore the direct method accuracy was cross-checked using a full Navier-Stokes solver [8]. Figure 4 shows the C_p distribution at 6°, with both viscous-inviscid interaction and Navier-Stokes solutions as well as experimental measurements from two different span-wise locations (rows A,B) on the airfoil. The latter indicates the two-dimensional nature of the flow. The two prediction methods produce very comparable results on both pressure and suction sides, the Navier-Stokes calculation being more accurate in the trailing edge region.

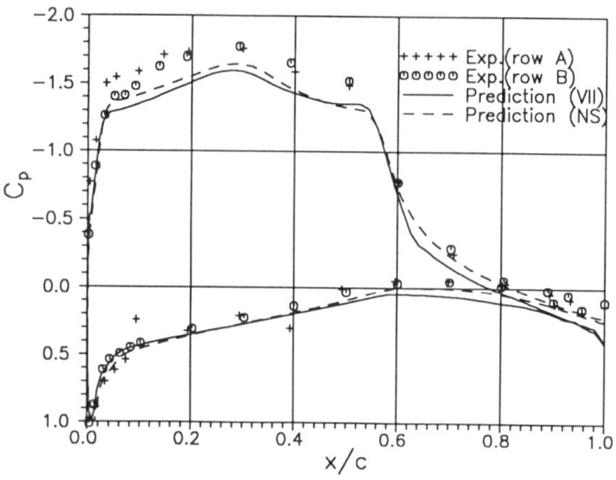

Fig. 4 C_p distribution at $a=6^o$ for the tip section.

b) Three blade sections (hub, mid, tip) according to RISO's specifications for optimum wind turbine operation. These designs were aimed at improving the performance of an existing Danwin Wind Turbine. The Re numbers were $1.8x10^6$ for the hub and $2.35x10^6$ for the mid and tip sections. Turbulence level was taken as $T_u=2\%$, value, considered realistic for wind turbine operating conditions, although quite a few doubts exist as to the real turbulence level value, which must be used for wind turbine design. Due to space limitations results will be presented for the hub section only. Figures 5, 6 and 7 show the optimized pressure

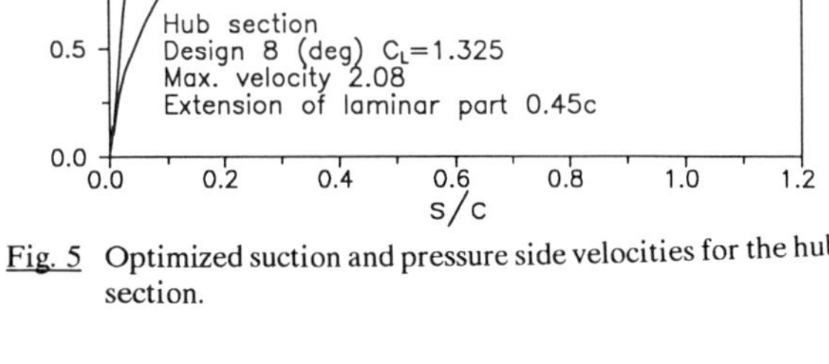

Fig. 5 Optimized suction and pressure side velocities for the hub section.

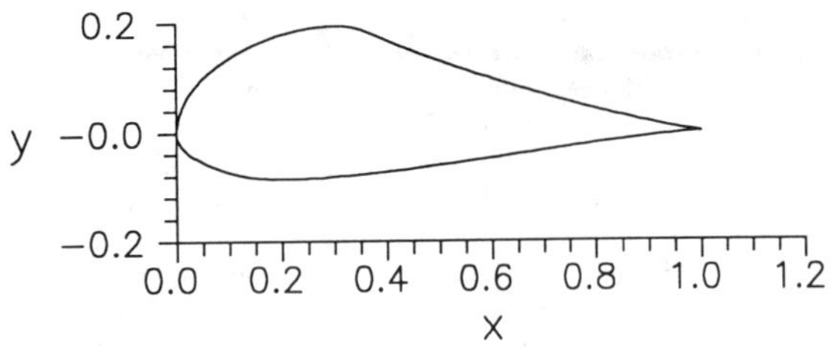

Fig. 6 Hub section profile shape.

and suction velocity distributions, the profile shape and the C_L, C_D curves for the hub section. It should be noted that both the C_L and C_D distributions in the post-stall region are obtained by extrapolating the tip-section airfoil post-stall behaviour to the hub case.

Following the application of our optimization method to the application field of wind turbines, the following comments are in order.

1. The design principles used for obtaining stall appear to be functioning properly. However one sensitive point that ought to be investigated is the validity of the transition criteria in the operating conditions of a wind turbine owing to the nature of atmospheric turbulence (high amplitude, low frequency).

2. Although an absolute optimum could not be reached, the available tools gave a considerable level of control in the design procedure. Viewing critically what was done, it seems that one should further investigate whether the optimal velocity distribution family provides the best compromise for this particular application, or some of its features could be selected differently, for increasing blade circulation.

3. During the application of the optimization procedure, it was realized that flexibility and rapidity of the tools was very critical. The inverse (design) part of the method was found to be rapid, as far as both the viscous and inviscid counterparts were concerned. Besides profile closure and thickness, which were automatically satisfied, off design requirements (as far as the inviscid part of the method was concerned) could be treated successfully as well. The difficulty came with the direct method, which encountered complications in predicting post stall behaviour and was computationally expensive, especially when separation was present.

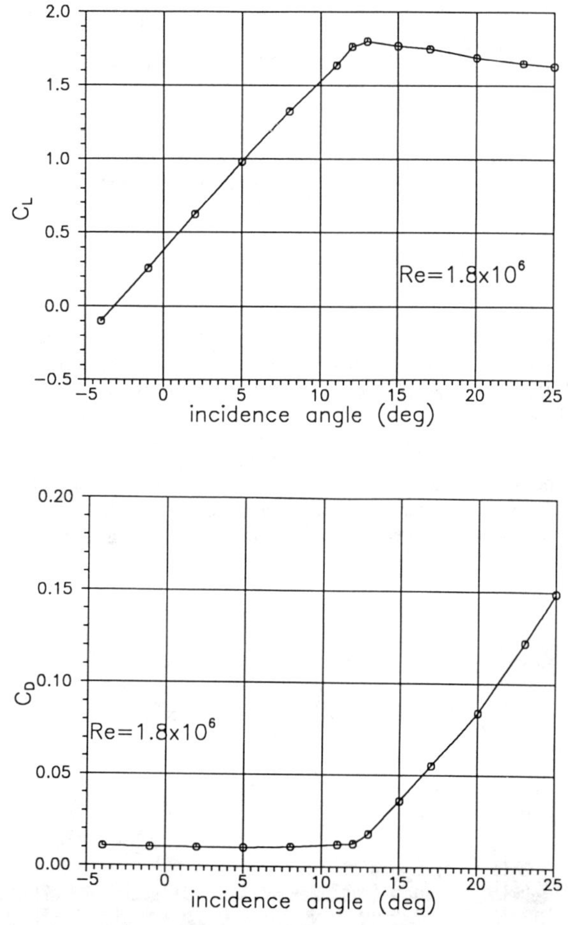

Fig. 7 C_L, C_D curves for hub-section airfoil.

5. CONCLUSIONS

A methodology is presented for the design of optimized airfoils for stall regulated HAWTS. Given optimized lift and drag characteristic curves the method provides the corresponding profile shape, combining an inverse viscous and an inverse inviscid counterpart. The selection of the design point as well as the specification of the corresponding design "target pressure" distribution are discussed in the paper. The resulting profile performance is fully assessed using a direct viscous inviscid interaction method. Numerical results are presented for two designed profiles corresponding to the tip and hub sections of an optimized rotor. For the tip case the predictions are compared against experimental results which validate the design philosophy.

REFERENCES

1. "Aerofoil Section Design and Assessment", JOUR-0079-C(SMA) EEC Project, Final Report, To appear in 1993.

2. Papailiou, K.D., "Le Foll's Method and the Calculation of Attached and Separated Two-Dimensional Boundary Layers", VKI LS 1981-1.

3. Bouras, B., Karagiannis, F., Leoutsakos, G., Giannakoglou, K., Papailiou, K.D., "Arbitrary Blade Section Design Based on Viscous Considerations. Background Information", Paper presented at the Int. Symp. on Num. Sim. in Turbomachinery, Joint ASME-JSME Fluids Engg. Meeting, Portland, Oregon, June 23-26, 1991.

4. Chaviaropoulos, P., Dedoussis, V., Papailiou, K.D., "A Robust Inverse Inviscid Method for Airfoil Design", Proceedings of the First European Computational Fluid Dynamics Conference, Brussels, 7-11 September, 1992.

5. Bouras, B., Karagiannis, F., Chaviaropoulos, P., Papailiou, K.D., "Arbitrary Blade Section Design Based on Viscous Considerations. Optimization Procedure", Paper presented at the Int. Symp. on Num. Sim. in Turbomachinery, Joint ASME-JSME Fluids Engg. Meeting, Portland, Oregon, June 23-26, 1991.

6. Bouras, B., Ph.D. Thesis under preparation, NTUA 1993.

7. Lock, R.C., Firmin, M.C.P., "Survey of Techniques for Estimating Viscous Effects in External Aerodynamics", IMA Conference, London, 1981.

8. Giannakoglou, K., Lymberopoulos, N., Tourlidakis, A., Nikolaou, E., Elder, R.L., Papailiou, K.D., "Adaptation of a 3-D Pressure Correction N-S Solver for the Accurate Modelling of Tip Clearance Flows", Revue Francaise de Mechanique, No 1992-93.

THE MEASUREMENT OF FLOW AROUND AN OPERATIONAL WIND TURBINE USING PARTICLE IMAGE VELOCIMETRY

D G Infield
Energy Research Unit
Rutherford Appleton Laboratory
Chilton, Didcot
Oxfordshire, OX11 0QX, UK.

G H Smith and I Grant
Fluid Loading and Instrumentation Centre
Department of Civil and Offshore Engineering
Heriot-Watt University
Edinburgh, EH14 AS, UK.

ABSTRACT: An introduction is provided to Particle Image Velocimetry (PIV), a non-invasive, flow field measurement technique. Results from its application to the measurement of the flow field around a wind turbine operating in a wind tunnel are presented. The calculation of derived quantities, such as angle of attack and circulation, is discussed. Some initial attempts at full-scale outdoor experiments are also described.

1. BACKGROUND

Future developments in the understanding and computation of wind turbine aerodynamics require an extensive data base of fundamental measurements. An important component of such a data base would relate to the quantification of the flow field around operational wind turbines. Although established point measurement techniques such as hot-wire anemometry have much to offer, [1], a major role is seen for whole field techniques such as Particle Image Velocimetry. Developed mainly over the last ten years for general fluid flow measurement (mostly of liquids) the technique has recently been extended and refined for aerodynamic application by the authors.

2. PARTICLE IMAGE VELOCIMETRY

PIV involves the imaging of particles introduced into the flow over an extended region. Velocities are calculated from the displacement between successive images of individual particles as they translate with the flow. References [2], [3] and [4] provide a detailed description of the technique.

A high intensity multiple pulse light sheet is used to illuminate the particles. A photograph is taken of the region of interest with an exposure time which results in multiple images of the seeding particles. Figure 1 depicts the overall arrangement. From the optical magnification of the camera optics, the time between successive light pulses and the measured displacement on the photographic image, the particle and hence flow velocity is straightforwardly calculated.

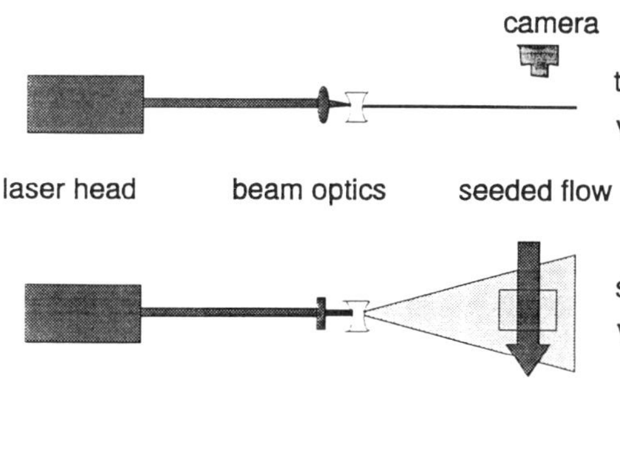

Figure 1 : Overall experimental arrangement for PIV

$$\overline{V} = M\overline{d}/t$$

where $\overline{V}$ is the in-plane velocity, $\overline{d}$ is the measured in-plane displacement vector, t is the time between light pulses, and M is the optical magnification factor

2.1 Laser Illumination

Choice of the light source is influenced by several factors, the most important being the requirement for high intensity and a pulse length and rate compatible with the fluid velocity. The ease with which the light sheet can be formed and manipulated is also relevant.

Since PIV requires that the seeding particles must be accurately located from the images, the light pulse needs to be short in relation to the time for the particle to travel a significant fraction of its diameter. For the flow velocities associated with wind turbines this results in a very short pulse duration. Viewing of the particles is at right angles to the beam propagation direction and as 'side' scatter is particularly inefficient, high beam intensity is required. Together these requirements have led to lasers being the preferred form of illumination for PIV work, especially for aerodynamic applications. For the work reported here a frequency doubled Yag laser with a wavelength of 532 nm (green) has been used with an output energy of about 180 mJ.

The pulse characteristics of the Yag laser are summarised in Table I, below. The very short pulse width of about 6 ns results in very sharp particle images. For example, a 60 micron seed particle at 10 m/s will travel less than 0.1% of its diameter during the pulse. In double pulse mode the energy associated with each pulse is less than for single pulses. The pulse energy from the laser peaks at a repetition rate of around 10 Hz. Single shot application results in significant loss of energy. These considerations need to be taken into account when designing the experiments.

2.2 Image Recording and Analysis

The image is photographed onto 35 mm film using a standard single lens reflex camera. For the wind tunnel work a Nikon 110 mm macro lens at an approximate magnification of 0.25 was used. Kodak T Max film was used for the wind tunnel experiments, developed in T Max developer at a dilution of 1:4 for 15 minutes. For outdoor work, Ilford HP5 film (400

Table I: Yag laser pulse characteristics

Output wavelength	532 nm (doubled)
Pulse energy	180 mJ
Number of pulses	2
Pulse width	6 - 7 ns
Pulse separation	0.04 - 0.2 ms
Maximum repetition rate	16 Hz

ASA) was generally used with the Yag laser, being a good compromise between grain size (and hence resolution) and sensitivity. A shutter speed is selected to encompass the pulse separation time.

Two distinct methods for extracting information from the photographic image are available. The first approach, described by Pickering and Halliwell [5], involves illuminating the negative with coherent light and analysing the resulting fringe pattern. The second approach, used throughout the study presented here, utilises incoherent white light to view the negative. The image is magnified and scanned by a high resolution video camera. A 512*512 pixel frame board is used to digitise the image which is then transferred to a PC for analysis. Sophisticated routines are available for identification of particle image pairs and the calculation of flow velocities. Additional software exists for the averaging and interpolation of flow fields and the calculation of derived quantities such as vorticity and circulation.

3. WIND TUNNEL EXPERIMENTAL ARRANGEMENT

To date most applications of PIV have involved liquid flows where fluid velocities are generally low and seeding of the flow is straightforward. In the work described here considerable effort was expended to develop an experimental method appropriate to air flow measurement covering sizable regions of flow.

Experimental work has been undertaken at Heriot-Watt in a closed return wind tunnel, some results from which are presented in [5]. The wind turbine was placed in the open working section to minimise blockage effects.

3.1 Wind Turbine

A small commercial wind turbine manufactured by Marlec was used in this study. Overall rotor diameter is 0.9 m with a hub diameter of 0.24 m. It has six moulded plastic blades, non-tapered, with a simple cambered profile. The turbine was designed for battery charging at up to about 50 Watts. For experimental purposes the wind vane of the turbine was removed and the rotor locked in the upwind position. The generator was electrically loaded by a 12 volt battery with solid state by-pass to prevent overcharging. An opto pick-up was fitted to the hub. This enabled rotation rate to be measured, and was also used for triggering of the laser and the camera.

3.2 Experimental Geometry

The Yag laser was located on the wind tunnel roof with the beam being projected down into the working section. Sheet forming was achieved using a glass rod of 5 mm diameter to diverge the beam. Figure 2 shows the experimental arrangement.

For the examination of flow around an individual blade the sheet was positioned to intercept a horizontal blade at the selected blade radius whereas a position along the rotor centre-line was adopted for study of the near wake region. This is shown together with the blade numbering in Figure 3.

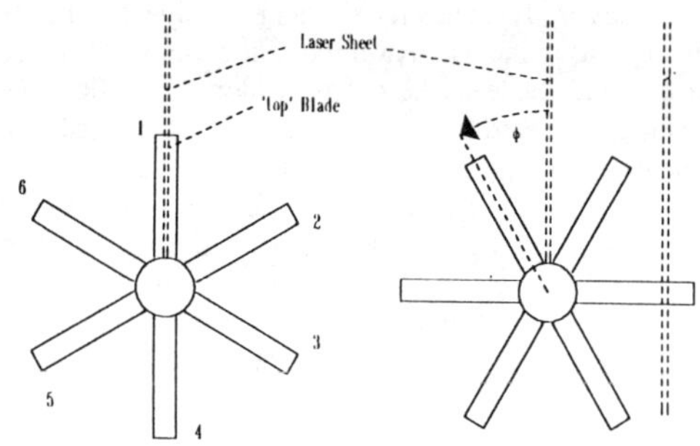

Figure 3 : Laser sheet positioning

3.3 Flow Seeding

Flow volume rates and hence seeding rates are much higher than for hydrodynamic work. In addition, the flow field covering a relatively large area is to be investigated. At a typical wind speed of 10 m/s about 14.4 m3 of air pass through the working section each second. It was found that seeding could be introduced to the tunnel downstream of the turbine, causing no disturbance to the flow incident on the rotor. This was a clear advantage of a recirculating tunnel. After some use

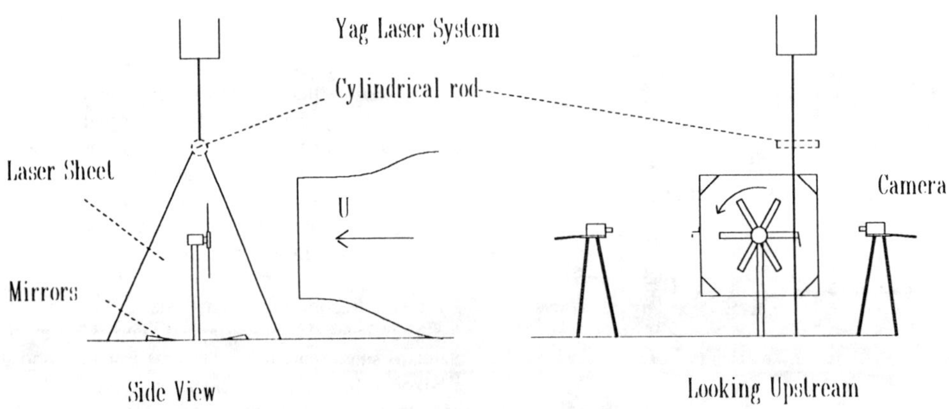

Figure 2: Wind Generator and Laser in Wind Tunnel.

the tunnel accumulated enough seeding to seed the flow automatically.

Seeding particles of polycrystalline form, with a mean diameter of about 30 microns, were used. These particles were found to have high scattering efficiencies.

3.4 Laser Control

The signal from the opto pick-up on the hub of the wind turbine is used to trigger the laser. The rotational speed of the turbine was found to be close to 10 Hz which meant that the repetition rate was advantageous from the point of view of laser energy. A variable delay was used so that the blade position during the pulse could be adjusted. The double pulse separation was controlled by a separate time delay.

4. RESULTS FROM WIND TUNNEL WORK

4.1 Flow Around an Individual Blade

Figure 4a is an example of a photographic image taken with the blade frozen in the horizontal position, the leading edge is towards the top of the picture and the flow is moving left to right. Particle pairs are identifiable. As flow conditions in the tunnel are steady, flow fields calculated from a number of images can be averaged. Ten images have been averaged in this way to give Figure 4b. The length of the vector gives the magnitude of the in-plane component. The disturbance caused by the blade is visible but it is more informative to transform the frame of reference to either that of the blade or that of the fluid.

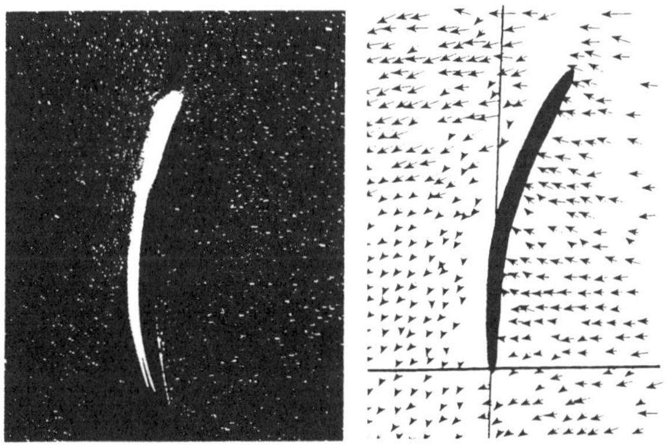

Figure 4 : a) Photographic PIV image b) Averaged flow field

From the measured blade rotation rate, and radial position the local blade velocity is straightforwardly calculated. Changing coordinates accordingly, the flow seen by the blade is produced as shown in Figure 5a and, depending on the definition adopted, the angle of attack can be derived immediately.

Alternatively by transforming to the frame of reference of the fluid by removing the mean measured flow velocity at the rotor disc, the circulation around the blade becomes apparent, Figure 5b.

Contour integration provides the numerical value of the blade bound circulation. This has been calculated a number of blade stations and the results are plotted in Figure 6. Some inconsistency is evident. This arose because data was collected from arbitrary blades at that stage of the experimental work. Since then significant variations between the performance of different blades have been identified. This matter is discussed in more detail in relation to tip vortices in the next sub-section.

4.2 Investigation of the near wake

Figure 7 shows the photographic image of a section through the near wake. To the right, the rear of the rotor is visible and the trailing tip vortices are very clearly visualised

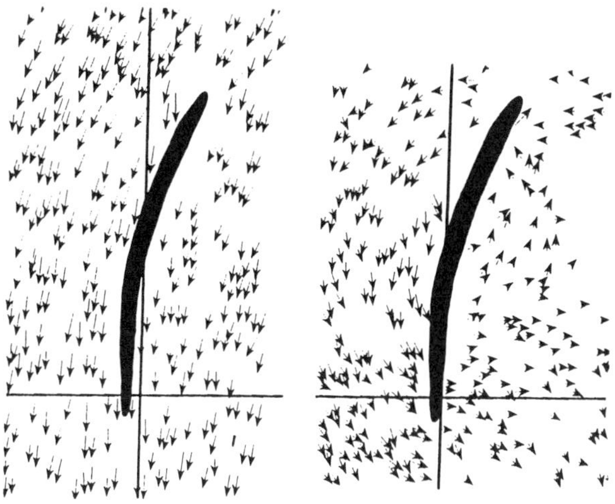

Figure 5 : a) Blade reference frame b) Fluid reference frame

in the wake. From the calculated velocity field vorticity contours can be produced as shown in Figure 8 with the structure of tip vortices highlighted. Contour integration around individual tip vortices gives the circulation. This has been calculated for a series of concentric paths so as to give the dependence on the distance from the vortex core. Results, normalised by the maximum measured blade bound circulation, for all six blades, with the vortex in position 1 as given by Figure 8, are shown in Figure 9. It is clear that the tip vortex from blade 3, and to a lesser extent blade 5, is weaker than for the other blades. If the vortex from blade three is examined at different positions (ages) it appears, see Figure 10, that the problem could be associated with an abnormally long role up process.

5. FIELD APPLICATION TO FULL SIZE ROTORS

Preliminary work is currently being undertaken to assess whether the PIV technique can be extended to the measurement of flow around full size turbines. Early outside work in Scotland established that seeded ambient air flow, at a height of about 10 m could be successfully imaged. Introduction of seeding into the flow at the appropriate height required the development of specific apparatus, and great care was needed to ensure focused images and adequate illumination levels.

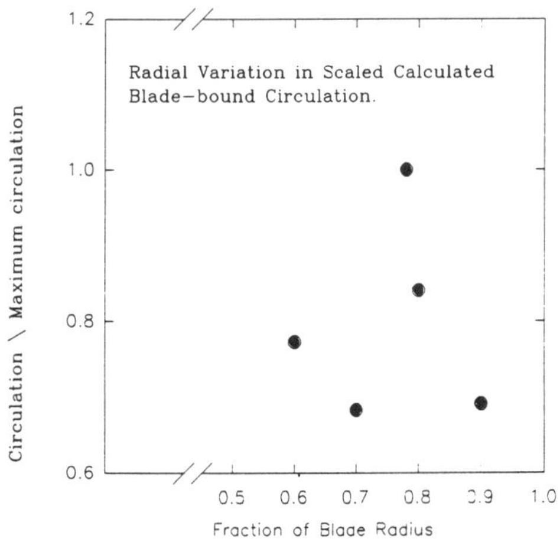

Figure 6 : Variation of normalised circulation along blade

More recent work outdoors at RAL has provided images of flow around the blade of a stationary rotor. The turbine in question has a hub height of 15.5 m and a rotor diameter of 17 m. A laser and camera arrangement has been devised to produce images at 80% of the radius, equivalent to the wind tunnel image of Figure 4.

As for the wind tunnel work an opto pick-up has been fitted to the rotor for triggering purposes. This time however the laser triggering is more complex as the blade passing frequency is not compatible with high laser pulse energy. A repetition rate of about 10 Hz is required and a designated double pulse must be timed to coincide with the required blade position. In these circumstances the camera too must be automatically triggered. A PC based control system has been developed for these purposes.

6. CONCLUSIONS AND POTENTIAL FUTURE WORK

To date most effort has been concentrated on the development of experimental technique suited to aerodynamic application. Results from the wind tunnel work have established that PIV has much to offer the wind turbine aerodynamicist. Further refinements in technique are expected to give detailed information on the tip vortex roll up process and flow separation from the blade at high angles of attack. It may be that light can be shed on the complex aerodynamics towards the blade root that are associated with the abnormally high lift coefficients identified there.

Outside work is more problematic although the early indications are promising.

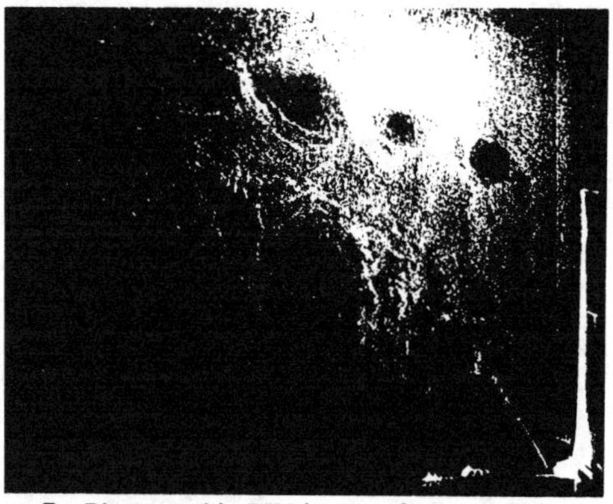

Figure 7 : Photographic PIV image of near wake region

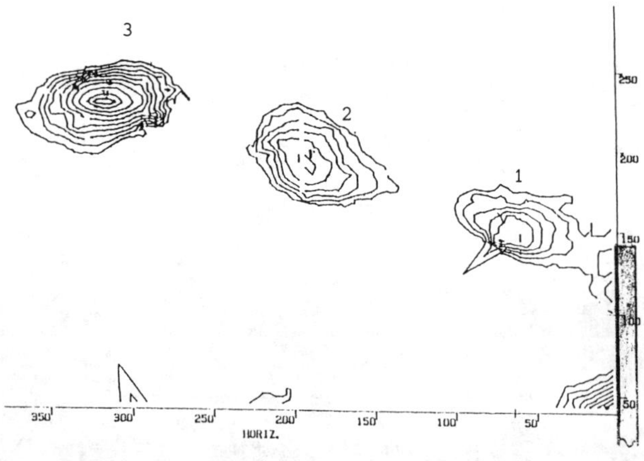

Figure 8 : Vorticity contours for near wake region

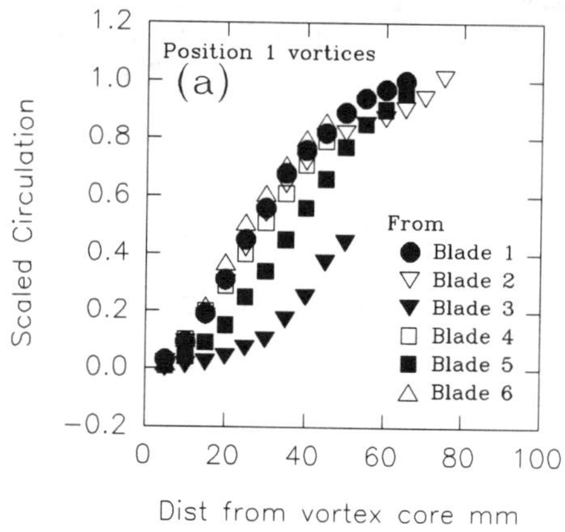

Figure 9 : Tip vortex circulation in first position

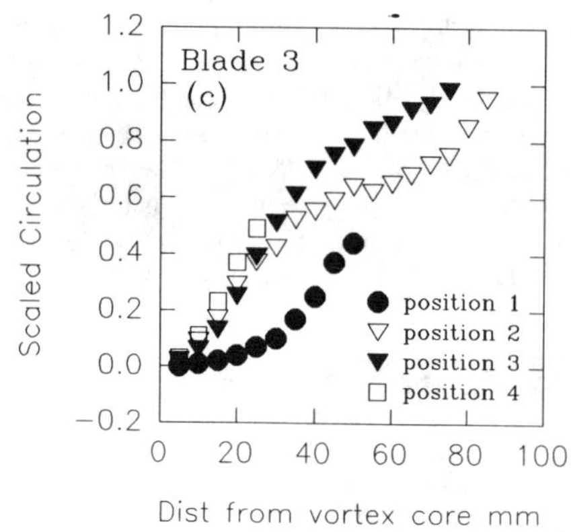

Figure 10 : Tip vortex circulation at different positions

REFERENCES

[1] N-J Vermeer, Local circulation on rotating wind turbine blades from velocity measurements in the wake of a model rotor. Proc. BWEA 14, MEP (1992).

[2] T D Dudderer and P G Simpkins, Laser speckle photography in a fluid medium. Nature, 270 (1977).

[3] R Meynart, Instantaneous velocity field measurements in unsteady flow by speckle velocimetry. Applied Optics, 22 (4) pp 535-40 (1983).

[4] I Grant and G H Smith, Modern developments in Particle Image Velocimetry. Optics and Lasers in Engineering, 9 pp 245-64 (1988).

[5] G H Smith, I Grant, A Liu, D Infield and T Eich, The wind tunnel application of Particle Image Velocimetry to the measurement of flow over a wind turbine. Wind Engineering, Vol 15 No 6 (1991).

INVESTIGATION AND MODELLING OF DYNAMIC INFLOW EFFECTS.

H. Snel and J. G. Schepers (editors),
Netherlands Energy Research Foundation, ECN,
P.O. Box, 1, 1755 ZG Petten

Summary

This paper describes the progress in the CEC JOULE project Dynamic inflow. The following organizations (and persons) co-operate in the project:

- Netherlands Energy Research Foundation, (ECN) (coordinator), together with Stork Product Engineering (SPE), both NL: H. Snel, J.G. Schepers, Th. van Holten.
- University of Stuttgart, (UniSt, D) : S. Wagner, R. Bareiß, K. Braun.
- Garrad Hassan and Partners, (GH, UK): R. Rawlinson-Smith.
- Technical Univerisity of Denmark (TUDk, Dk): S. Øye.
- Delft University of Technology (DUT, NL): G.W. van Bussel, N.J. Vermeer.
- National Technical University of Athens (NTUA, Gr), together with Universitė du le Havre (UdH, Fr): S.Voutsinas, S. Huberson.
- Teknikgruppen AB (TA) together with FFA, both S: H. Ganander, A. Björck.
- KEMA, NL: H. Hutting.

The aim of the project is to develop one or more engineering models for the wake induced instationarity and nonuniformity in the rotor inflow. The models are being implemented in state of the art computer codes for dynamic load predictions. They are validated by comparing calculated results with measurements, and with results from more sophisticated vortex wake models. The Danish 2MW Tjaereborg machine has been used quite succesfully for experimental data, while the first results of tests in the open jet wind tunnel of the Delft University of Technology are becoming available.

Section 1 looks at the different situations in which dynamic inflow can be of importance and how they are, or will be, covered in the project. Axial symmetric flow situations, such as in coherent windgusts and in (idealized) fast blade pitch angle changes, have first been adressed. This part of the project has now been concluded. The main results are summarized in section 2. Section 3 discusses the first results for yawed operation, both from experiments and the preliminary modelling, while section 4 shows the further validations and verifications that will done within the project.

The project is sponsored by the CEC under the Joule-I and Joule-II programmes and by the various National Agencies or participants.

1. IMPORTANCE OF DYNAMIC INFLOW MODELLING

The term dynamic inflow is used to indicate the dynamic response of the inflow velocities in the rotorplane, to changes in the load conditions on the rotor. Figure 1 shows a simple example, in which the blade pitch angle undergoes a rapid change from an intial value θ_1 to a new value θ_2. Blade element-momentum (bem) theory gives two different equilibrium values of the induced velocity pertaining to the two pitch angles, viz. $u_{i,1}$ and $u_{i,2}$. In most aero-elastic response calculations, it is assumed that the inflow is always in equilibrium with the loads. In reality, however there is time needed to accelerate the air from velocity $U\text{-}u_{i,1}$ to $U\text{-}u_{i,2}$. This means that when the change in pitch angle is sufficiently fast, the inflow velocity will essentially still be equal to $U\text{-}u_{i,1}$ and only gradually change to the new equilibrium value. The consequence of this for the angle of attack α is also shown in figure 1. Instead of the instant change from α_1 to α_2 as suggested by equilibrium theory, there is an important overshoot in the angle of attack, indicated by the heavy line. It will be clear that for a finite rate of change of the pitch angle, the actual 'overshoot' will depend on the time scale involved in the adjustment of the inflow.

The blade loads will exhibit an 'overshoot' (as compared to equilibrium values) in accordance with the overshoot in the angle of attack. In fact there is another time dependent effect here, as loads do not instantaneously respond to angles of attack either. There is also time needed for the accomodation of the flow around the section. This effect will be referred to by instationary profile aerodynamics (ipa). One can intuit that this involves the acceleration of a volume on the scale of the sectional chord, whereas the inflow involves masses on the scale of the rotordiameter. Indeed the 'ipa' has a much shorter time scale (see also [1] for a more exact discussion) than the dynamic inflow phenomena. It can be argued that in wind turbine aerodynamics, its main importance is in stall. The present project chiefly concerns the instationarity due to dynamic inflow, although the effects of ipa will be touched upon.

Similar phenomena as those occuring for a rapid pitch change will occur for any change in conditions that modify the induced velocity in the rotor plane, e.g wind gusts. In the case of yawed flow operation, the loading of the turbine blades changes periodically, and the inflow will respond with time delay. In the Joule-I project, full span pitch, wind gusts (axially symmetric) and yawed flow are considered. In a Joule-II continuation, also partial span pitch will be addressed.

Dynamic inflow can be modelled by realizing that the distribution of vortices in the wake is responsible for the induced velocity in the rotor plane. By following the creation and transport of wake vorticity in time, and calculating the velocity induced by it in the rotor plane, a so called free vortex wake model is obtained. From a fundamental point of view, these methods describe the main characteristics well, although neglecting viscous effects such as vorticity diffusion. However, they need a large amount of computer time.

The present project concentrates on methods that can be used by designers in PC or workstation based aero-elastic analysis codes, referred to as 'engineering methods'. The models proposed by the various participants are in fact adaptations of the bem method, transforming the algebraic equilibrium equations into differential equations reflecting the dynamics of the process. In the construction of the models, the basic concepts of wake vorticity are used, but translated to time delays in different parts of the rotor plane. Vortex wake models of various degrees of complexity are used in the project to enhance the detailed understanding of the phenomena and with that validate the engineering methods. At the same time, field measurements and wind tunnel measurements are used for the quantitative validation.

2. AXIAL SYMMETRIC CASES

2.1.Pitch steps

Four pitching transients (i.e. pitching steps as sketched in fig. 1) at different loading conditions were measured at the Danish Tjaereborg turbine and compared with calculations. The computational models, developed and implemented within the present project, were described in some detail in [1].

The models can be divided into differential equation models (ECN,de, GH and TUDk), prescribed wake models (ECN,iw), and free wake models (DUT, NTUA, UniSt). The first type of models are relatively simple modifications of the standard bem methods, where a time derivative of the induced velocity is added to the equation of the axial induced velocity. Time delay functions (and their dependence on radial position) were obtained by considering the velocities induced by simplified wake geometries.

The measured values for blade and shaft loads were averaged over a number of realizations, and over the three blades in order to filter out structural effects, stochastic wind influences and deterministic effects as (average) windshear and tower shadow. The averaging process is described in [2]. Typical results for the calculated and measured flatwise blade bending moment and rotor shaft torque are shown in figures 2 and 3.

A clear overshoot in the loads can be seen. The measured range between maximum and minimum flatwise moment is about 375 kNm where the range between the equilibrium values is about 210 kNm. For the rotorshaft torque, these numbers are 210 kNm and 45 kNm. In general the calculated ranges are in good agreement with the measured ones. Table 1 shows the measured overshoots in flatwise moment for all the cases which have been investigated. Index 1 refers to the initial and final (equilibrium) values, index 2 refers to the intermediate one. The value of $(d\theta/dt)_1$ gives the pitching rate for the first step, and $(d\theta/dt)_2$ gives the value for the second, returning step.

λ	θ_1	θ_2	$d\theta/dt_1$	$d\theta/dt_2$	M_1	M_2	M_{max}	M_{mi}	1-2	max - min
9.5	1.0	3.0	1.0	1.0	530	420	590	380	110	210
8.1	0.1	3.7	7.2	7.2	670	460	750	375	210	375
7.8	0.2	3.4	slow	5.3	675	470	735	500	175	265
5.6	1.2	3.2	1.5	1.5	920	810	930	770	110	160

Table 1. Comparison of equilibrium and dynamic load variations.

The indication 'slow' for dθ/dt means that for this case the pitch change was done at a slow, exponential rate. From the table it can be seen that the overshoots depend on the loading (the induction factor) and the pitching rates. The overshoots become less apparent for lower induction factors, i.e. higher wind speeds and for lower pitching rates. Since pitch control normally becomes active at high wind speeds, this implies that the effect of dynamic inflow on the fatigue life may be limited.

2.2 Time scales

The pitching steps are performed with a finite pitching rate. It is clear that the overshoot will depend on the change which has already occurred in the wake at the end of the pitching movement, i.e. on the time scale of the process in comparison to the time scale of the pitch change. Information about the time scale has been obtained by determing the relevant values from the measurements and the model results.

Thereto an exponential behaviour of the load F is assumed according to:

$$F(t) = F_1 + (F_2 - F_1)*[1 - \exp\{-(t - t_1)/f(t)\}]$$

with f(t) the time scale.

It must be emphasized that the determination of the time scale from the measured results is very difficult due to the fluctuations in the measured signal, which results in an uncertainty in the equilibrium values. The time scale appeared to be strongly dependant on the exact values of the equilibrium values. Nevertheless some agreement between calculated and measured time scales for time steps not too long after t_1 (say $t<t_1+D/U$) was found. In this period the most relevant change in load takes place.

At time $t=t_1+0.5D/U$, both the measured and calculated time scales have values between 0.3 and 0.5D/U for most cases. The spread in the calculated time scales was about 0.15D/U. A more reliable validation of the time scale is expected from wind tunnel measurements.

2.3 Coherent wind gusts

Calculations reported in [1], showed little effect of dynamic inflow modelling on the calculated load fluctuations due to coherent wind gusts. This was explained by the relatively small change in induced velocities that result from the gusts which are considered. To investigate this in a detailed matter, an expression is derived from the blade element impulse relations which gives the change in the equilibrium values of the induced velocity, as function of the change in wind speed and the design parameters.

The expression takes the following form:

$$\Delta u = H\,\Delta U$$

in which Δu represents the change in induced velocity and ΔU gives the change in wind speed while:

$$H = \sigma\,\lambda_r^2\,dc_l/d\alpha\,(\theta + \alpha_0)\,/\,4(1 - a)$$

The main assumptions used in the derivation of this equation are:

neglect of a', c_d, tip losses and higher order terms;
linearized theory;
small inflow angle;

It appears that the value of H is very small for the Tjaereborg turbine. For other turbines the factor H may be larger. In effect, for the WPS-30 turbine, evaluation gives a higher value. Therefore the dynamic inflow effect may be more significant for other turbines.

So far, analysis of cases of steps on the wind speed has been based on calculations only. Field measurements for well defined gusts are very rare. It is expected that wind tunnel measurements with the use of a gust generator, will give more information about this subject. Such measurements are planned within the project, to be performed at the free jet wind tunnel facilities of the DUT.

3. YAWED OPERATION

Extension of the models to yawed operation is done to improve the predictive capabilities of the models for asymmetric conditions. Apart from blade and shaft loads, the asymmetric hub loads (yawing and tilting moments) form the main interest. From the practical point of view yaw misaligment is a fact of life, and is of importance to the load spectra of yaw bearing, tower and shaft.

Field measurements of loads correlated with yaw misalignment are difficult in the sense that the actual (time dependent) misalignment angle is very hard to measure, while (unknown) windshear also interacts with the misalignment effect. It was decided to use field measurements to determine general trends and order of magnitude, while wind tunnel measurements will be used to study the effects in a controlled environment. In the Joule-I project, field measurements are obtained from the Tjaereborg turbine.

3.1.Tjaereborg measurements

The measurement campaigns on the Tjaereborg turbine have been finalized and the results are being processed. Figure 5 shows some preliminary results for the measured yaw moment, averaged over 30 seconds. Figure 4 defines the yaw related quantities as used in the study.

Notwithstanding the large amount of scatter, it is clear that a reasonably high restoring yaw moment exists. For a yaw misalignment angle of around 45 degrees the scatter is less and the measurements consistently show a yaw moment of some -170 kNm.

The 'standard' bem methods do not yield an appreciable yaw moment. In fact, for a stiff rotor without tilt or coning, the aerodynamic situation is symmetric around the 0 (and 180) azimuth position. The yaw moment is obtained by integrating the loads multiplied by $\sin(\Phi_r)$, and gives an average of zero for each blade. In order to obtain a nonzero contribution, the loads should contain 'sine' like terms.

Application of a standard model to the Esbjerg turbine shows the following interesting results:

- Without blade flexibility (hence blade loads in phase with aerodynamics) there is a very small destabilizing average moment of about 10 kNm;
- Including effects of blade flexibility (there is a small effect of blade coning, but more importantly, blade loads lag behind in phase with respect to aerodynamic loads) there is a restoring yaw moment of approximately -30 kNm (20% of the observed value).

It is clear that sufficient accuracy cannot be obtained with equilibrium wake modelling. Dynamic inflow modelling gives a further phase shift of the blade loads, related to the wake geometry, see section 3.3. This is assumed to give a more realistic result for the calculation of the asymmetric loads.

3.2 Wind tunnel measurements

The wind tunnel environment makes it possible to do measurements of the inflow itself, at different positions directly behind the rotor plane. Information at this level is very important for the correct modelling, as the engineering models work on basis of inflow equations. The first results of the measurements at the DUT free jet tunnel are now becoming available. Important data of the tunnel, the model and the measurement setup are shown in figure 4.

Figure 6 shows the axial induction factor derived from the measurements for a number of radial stations, as function of the azimuth, for a yaw misalignment of 30 degrees. The measurements have been averaged over one revolution of the rotor. Apart from the tip sections (90% and 95% radial position), the induced velocity is maximum in the region between 120 and 180 azimuth. For the tip sections, a phase shift of approximately 60 degrees is measured.

Velocity measurements have been done for 10 and 20 degrees misalignment as well. The measurements will be complemented with blade load measurements.

3.3 First modelling efforts

For the modelling of yaw misalignment effects, a number of aspects are of importance:

a) In the upper half plane the blade will be retreating and in the lower half plane advancing with respect to the inplane wind velocity component. This tends to increase the angle of attack of the blade in the lower halfplane and therefore increase the vorticity shed there. This effect is symmetric around zero azimuth and hence of 'cosine' type.

b) The skewed wake geometry, or the proximity to the rotorplane of the vortices in the wake strongly influences the inflow. The trailing tip vortices are closer to the 'upwind' side of the rotor plane, while the root vortices pass closer to the downwind side. The velocities induced by the trailing tip vortices will be symmetric about 90-270 degrees azimuth line, and hence be of 'sine' type which does contribute to the yaw moment. This effect was first modelled by Glauert [3], who proposed the following distribution of induced velocity:

$$u_i = u_{i,0} (1 - r/R\ K_c \sin \Phi_r)$$

Later researchers derived values for K_c on the basis of the vortex wake geometry. A slightly higher loading on the upwind part of the rotor results and hence a restoring yaw moment. If the wake vorticity distribution due to the effects of advancing and retreating blade (described under a) is accounted for, a cosine term has to be added to the above expression.

The root vortex effects will be anti-symmetric around the 270 degrees azimuth but of much higher harmonic content. Its effects will be mainly of importance in the region close to 270 degrees, where the distance to this vortex is small. Finally, as a result of the periodicity of the blade loading, there will be shed vorticity in the wake whith both sine and cosine like contributions to the induced velocity.

Both effects should be present in the modelling. The 'standard' bem method only contains part a), while the classical expressions by Glauert for the induced velocity in helicopter forward flight aerodynamics, contain only the sine like part of b). Free vortex wake models, as used by NTUA and UniSt, accelaration potential model of DUT and the prescribed wake model of ECN, inherently account for both effects.

Figure 7 shows preliminary results of the first efforts to obtain engineering models. The geometry choosen here was that of the Esbjerg turbine, whithout tower shadow, tilt or windshear effects. The wind speed was taken uniform at 9 m/s, and the misalignment angle was set at 30 degrees. The figure shows calculated induced veloties at r/R = .5 in function of azimuthal position, following a blade.

The models are too much in a development phase to describe the details here, but a few remarks can be made. All models must be consistent with the earlier derived dynamic wake models for axial symmetry. The model used by TUDk is of the Glauert type, the Pitt and Peters [4] model used by GH gives rise to likewise terms. They do show the same general behaviour as the NTUA free vortex wake model. The ECN integral wake (a very simple prescribed, cylindrical wake model) shows the local influence of the root vortex, maybe in exagarated form because all root vorticity has been concentrated in a vortex eminating at r=0. The importance of the root vortex, producing rapid load variations around the 270 degrees azimuth position, was also noted in [5]. The ECN d.e.

(differential equation) model does not yet reflect the wake proximity term. Only the advancing and retreating blade features were implemented in a dynamic manner, the wake geometry terms will be added.

Qualitative disagreement with the wind tunnel measurements, (a maximum of the induced velocity at about 150 degrees azimuth) is evident. This means also that the Delft measurements are in disagreement with the Glauert type of modelling. The measured velocity distribution in the rotorplane, with slightly lower velocities in the upwind part, suggest a destabilizing yawing moment. Force measurements still have to be done and analysed to arrive at a conlusion. The scene seems set for a lot of qualitative flow description and interpretation in order to arrive at more satisfactory modelling.

5. FURTHER VALIDATION

Further validation of the models (once established) and the calculated loads is necessary. In the context of the Joule-II program, measurement programs will be done on the 30 m diameter Holec WPS-30 turbine and the 16 m diameter UNIWEX turbine, both for axial symmetry (pitching steps) and for yaw misalignment. Together with the Tjaereborg measurements, a good range of diameters will be covered.

Finally, the effects of windgusts, as analysed in section 2, will be validated by measurements in the Delft wind tunnel, where a gust generator will be installed.

6. CONCLUSIONS

The engineering models for dynamic inflow, developed in the Joule I project, were validated against measurements at the Tjaereborg turbine. The measured load overshoots due to pitching transients can be well reproduced by the models. Also the predicted time scales of the wake inertia phenomena compare reasonably well to the measurements, although a definite conclusion cannot be drawn, due to the difficulty of the exact determination of time scales from field measurements.

For coherent wind gusts the dynamic inflow effect is relatively small, according to the models. Verification of this must await wind tunnel measurements.

It is recommended that the equilibrium wake methods which are used in the present state of the art aeroelastic codes should be replaced by above models.

For yawed flow conditions, the first measurements from the Tjaereborg turbine and from wind tunnel tests were presented. The most important aspects of the flow which must be reflected in the models were identified, and it was shown that present equilibrium models do not account for all of these. Preliminary results of different models used by the participants were shown and discussed. There is still some way to go to arrive at a satisfactory model.

REFERENCES

[1] H. Snel and J. G. Schepers: Engineering models for dynamic inflow phenomena. Journal of Wind Eng. and Ind. Aerod., 39 (1992), 267-281.

[2] S. Øye: Tjaereborg wind turbine, First dynamic inflow measurements, Technical University of Denmark, AFM Notat Vk-189, April 1991.

[3] H. Glauert: Miscellaneous Airscrew Problems, in: Aerodynamic Theory (ed. W.F. Durand), Div. L, Chapter XII. Dover Publications Inc, N.Y. 1934.

[4] D.M. Pitt and D.A. Peters: Theoretical Prediction of Dynamic Inflow Derivates. Vertica, Vol.5 pp 21-34, 1981.

[5] W.Z. Stepniewsky and C.N. Keys: Rotary Wing Aerodynamics. Dover Publications Inc. N.Y. 1984.

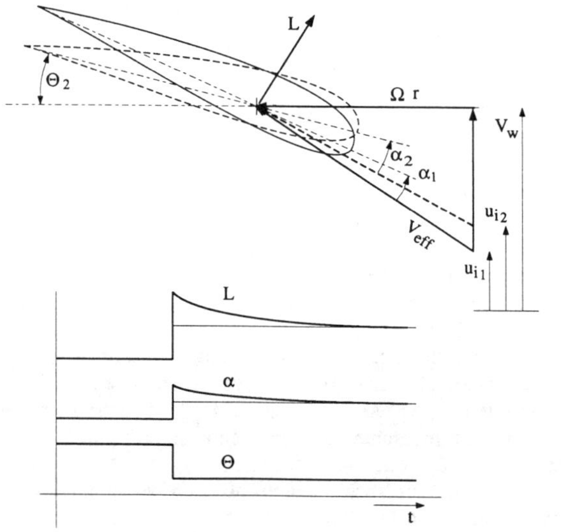

Figure 1. Influence of wake inertia (dynamic inflow) on loads after a pitch step.

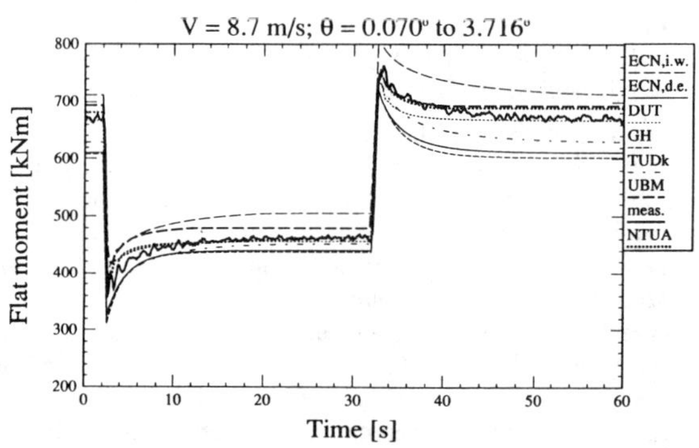

Figure 2. Measured and calculated flapping moments as response to pitch steps.

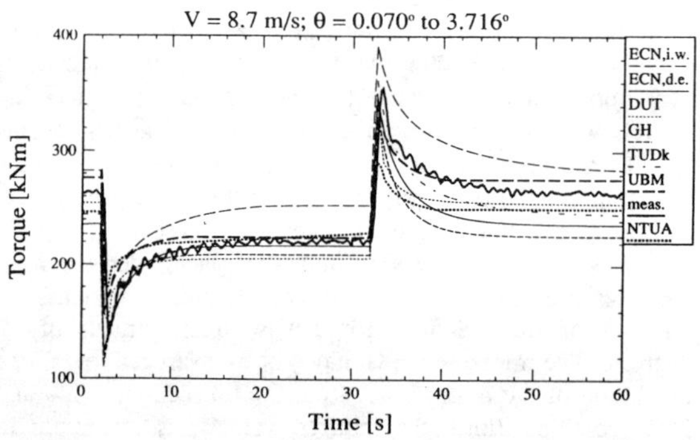

Figure 3. Measured and calculated shaft torque as response to pitch steps.

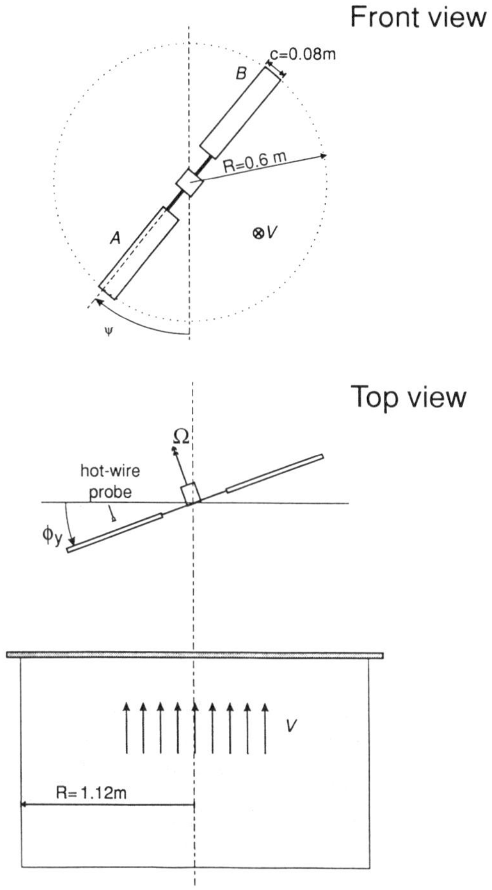

Figure 4. Definition of rotor azimuth and yaw misalignment angle, and of wind tunnel setup.

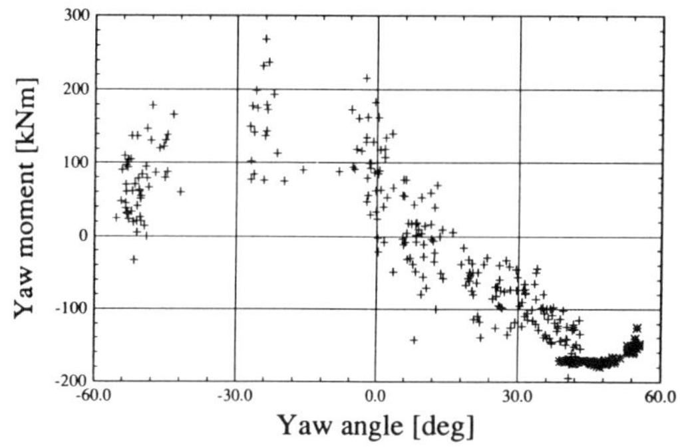

Figure 5. Measured yawing moment as function of yaw misalignment angle, Tjaereborg turbine.

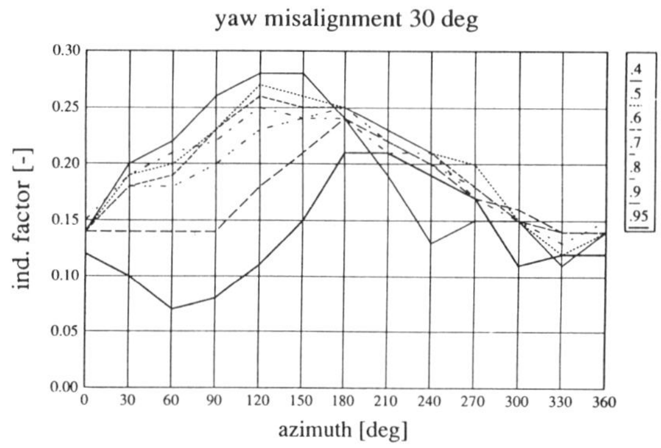

Figure 6. Measured axial induction distribution, DUT wind tunnel.

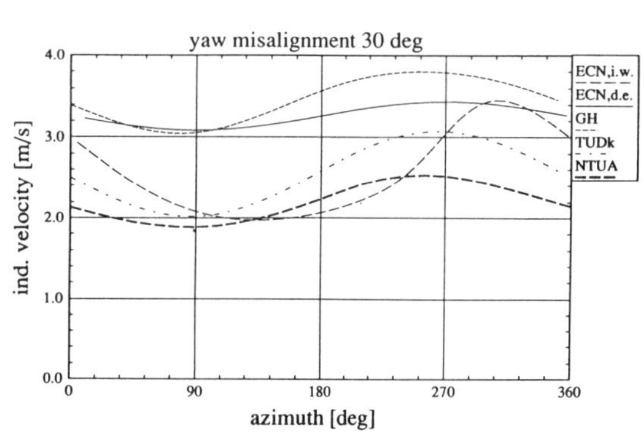

Figure 7. Calculated induced velocity distribution at r=15m, 30 degrees yaw, Tjaereborg geometry.

PRESSURE MEASUREMENTS ON A ROTATING WIND TURBINE BLADE ON THE OPEN AIR ROTOR RESEARCH FACILITY OF THE DELFT UNIVERSITY OF TECHNOLOGY.

A. Bruining.

Institute for Wind Energy, Delft University of Technology.
Stevinweg 1, 2628 CN Delft, The Netherlands.
tel. +31 15 785167, fax: +31 15 611465.

ABSTRACT.

This paper gives a short explanation of the experimental arrangement for measuring pressure distributions on a 10m. diameter rotating wind turbine blade. Results of the 30% span and 70% span sections are compared with two-dimensional wind tunnel pressure data on the 70% span section.

For the 70% span section the maximum normal force coefficients, calculated from the rotating pressure distributions display the same level as the wind tunnel tests. The rotating 30% span data tend to show higher levels of the normal force coefficient, indicating a possible stall delay. The shape of the pressure distributions, especially the 70% span section looked similar to the wind tunnel data.

1. INTRODUCTION.

To investigate the rotational (3D) effects on a wind turbine rotor blade, a full-scale experiment was set up with a two bladed 10m. diameter upwind rotor. See also [ref.1]. The experiments were carried out on the Rotor Research Facility of the Delft University of Technology. On this Open Air Research Facility most variables can be controlled except of course the wind conditions. The tower is tiltable for easy access to the nacelle and rotor. Compared with wind tunnel experiments, the tests are more complicated due to the atmospheric unsteady wind conditions, but on the other hand they can also give information about unsteady effects.

2. MODEL.

The rotor blade incorporates the NLF(1)-0416 airfoil section and has no taper or twist. The blade chord is 0.5m. and the blade length is 4.4m. The rotor blade has 59 high quality pressure taps at each of the 4 span wise locations at 30%, 50%, 70% and 90% of the radius. The diameter of each pressure tap is 0.4mm. The blade pitch can be manually adjusted between ±10° in intervals of 1°. See also [ref. 2] and fig. 1 for a drawing of the blade layout.

The planform without taper and twist gives the opportunity to test the same blade in the wind tunnel at two-dimensional stationary conditions. Wind tunnel tests on the 70% section were carried out in the Low Speed Wind Tunnel of the Faculty of Aerospace Engineering. In the comparison of the wind tunnel data and rotating data uncertainties due to differences in model geometry for this section can thus be avoided.

3. INSTRUMENTATION.

3.1 Connection of the pressure taps.

The pressures were recorded with two PSI high speed differential electronic transducers. Each transducer has 32 ports and a built-in calibration facility. The transducers are located at approximately 80% of the radius. The total pressure was measured with a Kiel pitot tube at 76% radius during the measurements of the 70% section, and at 36% radius during the measurements of the 30% section. The reference port of the transducers was connected to a Kiel pitot tube on the hub. The Kiel pitot tube indicates a correct total pressure within ± 45° of the incoming flow. To avoid different behaviour in dynamic response the pressure taps of the 70% section were connected to identical tubes of 70 cm. length and an inside diameter of 2 mm.. Unfortunately for blade construction reasons it appeared to be impossible to locate the pressure transducers close to the 30% span section. Consequently it was necessary to connect the taps of the 30% section to 2 m. long tubes with different dynamic characteristics than those of the 70% section.

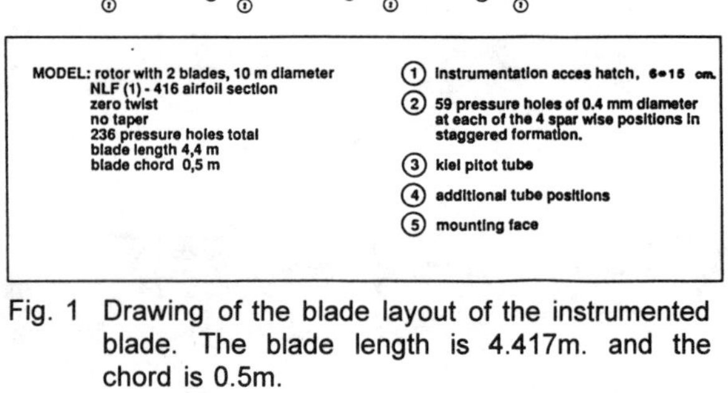

Fig. 1 Drawing of the blade layout of the instrumented blade. The blade length is 4.417m. and the chord is 0.5m.

The signals of the pressure ports were numerically corrected with the dynamic pressure of the incoming flow ($\frac{1}{2}\rho U^2$), to compensate for the reference port being connected to the total pressure instead of the standard atmosheric pressure. This layout for the tubing system was chosen to avoid the problem of measuring the static atmospheric pressure in unsteady wind conditions. The wind speed U was recorded with a cup anemometer on a moveable tower positioned one rotor diameter in front of the rotor at rotor axis height level.

Additional collected signals were: the rotation speed, the wind speed, the rotor torque, the temperature of the transducers, and strain gages for the blade flapping moments and the tower bending moments. A video camera recorded the flow pattern on the suction side of the rotor blade. This pattern was visualized with tufts. All electrical signals to and from the rotor were transferred over a slip ring set with 30 rings. Special attention was paid to eliminate possible disturbing effects due to the slip rings and the long electrical wires during the transmission of the electrical signals.

3.2 Measurement system and control.

A PC based software system controlled the measurering equipment and stored the rough data on disk. To present on line the data in scope graphs, for following the signals in time and pressure distributions charts, also the system averaged and corrected the data real time.

The measurements were controlled by three triggering systems: one trigger at every 360° rotation when the instrumented blade is in top position to define the azimuth position, an internal timer trigger in the measurement software and a trigger at every 6° rotor rotation. For analysis the PC system offers a data replay function at measurement speed, or at an user defined speed. In post processing mode the program can generate standard Ascii files in which the data is binned on a arbitrary signal, for instance the local calculated speed ratio or a local incoming flow angle, or time series of all the signals with or without averaging.

3.3 Measurement speed.

During the tests all pressure ports and additional signals were read at a 3 msec. internal time trigger, and stored in the computer memory. After every 6° rotation of the rotor these values were averaged to a measurement value in order to reduce the electronic scatter. This means that the sampling rate of one 32 port transducer was in the order of 10 kHz. Consequently at high rotation speed a measurement value consists of less averaged readings than at low rotation speed. At 60 rpm. a value of a one channel consists of 5 averaged readings.

4. TESTS.

The 70% section and consecutively the 30% section were measured which a blade pitch setting of -2°, with rotational speeds from 8 rpm. to 95 rpm. and wind speeds below 12 m./sec. to avoid excessive blade loads. At each measurement run the rotation speed of the rotor was kept constant. The generator therefore switched automatically to motor operation if necessary.

During the initial tests on the 70% section a harmonic ripple was discovered on all ports with a frequency equal to the rotation speed and an amplitude in the order of 60 Pa. to 100 Pa. Additional tests were carried out to understand these effects. For this purpose pressure taps or the reference port were systematically blocked. Also the rotor was quickly stopped with the emergency brake at different azimuth positions. The harmonic ripple also occurred at zero wind conditions. See fig. 2 for an example with a quick stop from 8 rpm. The tests showed that the harmonic ripple was probably caused by the changing altitude of the transducers and taps during rotation, combined with viscous effects in the tubes. See also [ref. 3] for graphs of screen dumps of replay sessions during analysing of this phenomenon. The 30% section measurements demonstrated some different behaviour due to the different tubing and are not yet fully understood.

5. DATA REDUCTION.

5.1 Data correction.

Three corrections on the pressure signals were used:

- a correction on the reference channel with the incoming wind dynamic pressure $\frac{1}{2}\rho U^2$
- an individual centrifugal force correction on each pressure port and reference port
- a harmonic ripple correction.

These corrections are calculated on line during the data collection and the data replay.

The centrifugal force correction was calculated with:

$$P_{transducer} = P_{tap} + \frac{1}{2}\rho\omega^2\left(r^2_{transducer} - r^2_{tap}\right) \qquad (1)$$

The harmonic ripple was reduced with an numerically generated signal with an amplitude and phase shift that was based on the behaviour of the pressure distributions during data replay exercises. On all pressure signals the same anti ripple correction was applied simultaneously.

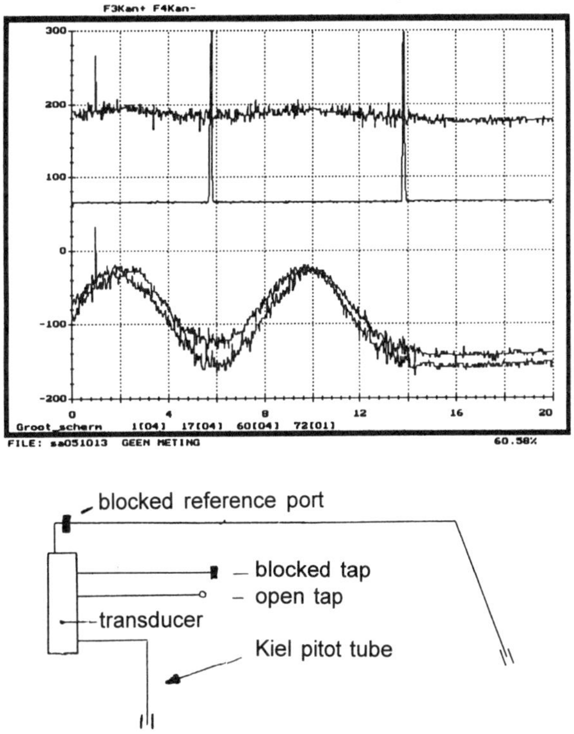

Fig. 2 Example of the variation of some signals of the 70% section at nearly 0 wind condition and 8 rpm. with a blocked reference port. The lower signal is from the pitot tube located at the leading edge of the rotor blade. The blocked tap is nearly constant. The rotor is stopped with the instrumented blade in top position after 13 sec. replay of the measurement run. The pulse corresponds with the top position.

5.2 Calculation of C_p.

The pressure coefficient C_p is defined as:

$$C_p = \frac{P_{tap}-P_{atm}}{q} = \frac{P_{tap}-P_{atm}}{P_{pitot\ blade}-P_{atm}} \qquad (2)$$

C_p was on line calculated with:

$$C_p = \frac{\Delta p_{tap} + \frac{1}{2}\omega^2 r^2_{tab} + \frac{1}{2}\rho U^2 + \delta p_{ripple}\sin(\omega t+\psi)}{\Delta p_{pitot\ blade} + \frac{1}{2}\omega^2 r^2_{tap} + \frac{1}{2}\rho U^2 + \delta p_{ripple}\sin(\omega t+\psi)} \qquad (3)$$

where Δp indicates the recorded differential pressure at a port, ω the rotational frequency, t the time, ψ the phase shift and δp_{ripple} the amplitude of the ripple correction. [Ref. 3] gives a more detailed description of the derivation of C_P. The calculation of C_P was carried out on line during the data acquisition.

6. RESULTS.

The pressure distributions were sorted on local inflow angle φ, calculated from $U/\omega r$, in intervals of 1 degr. and averaged in each interval followed by an integration to yield C_n. Fig. 3 and fig. 4 give the results of these calculations plotted against φ and the comparison with the wind tunnel results. The wind tunnel data, however, are plotted against α, giving an offset to the rotating data plotted against φ. Fig. 5 through fig. 7 contain a three-dimensional impression of some examples of measured C_P distributions of the 30% span and 70% span section. Fig. 8 and fig. 9 show curves of the 30% span section combined with wind tunnel data for a Reynolds Number of $0.5*10^6$. Fig. 10 finally relates data of the 70% span section with the wind tunnel measurements of the same section at $1.0*10^6$.

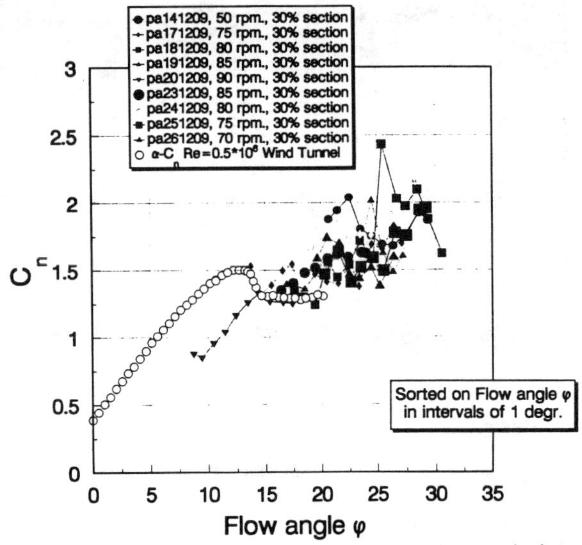

Fig. 3 Comparison of C_n from the wind tunnel data (versus α, open symbols) with the rotational data from the 30% span section (versus φ).

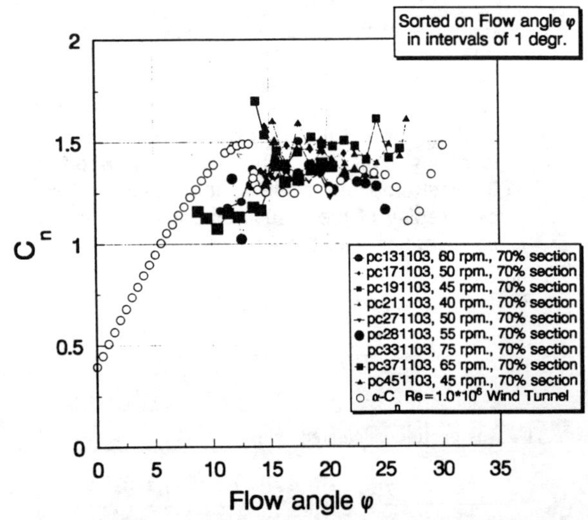

Fig. 4 Same as fig. 4, but now with rotational data from the 70% span section.

7. CONCLUSIONS.

1. The shape of the pressure distributions of the 70% section compares very well with the two-dimensional wind tunnel measurements.
2. The shape of the pressure distributions of the 30% span also rather well matches the wind tunnel measurements of the 70% section. At the higher inflow angles the pressure distributions of the first 50% of the chord showed an unstalled distribution while the aft part looked more or less stalled.
3. The maximum normal force coefficient C_n, calculated from the C_P's, of the 70% span section is of the same magnitude as the two-dimensional wind tunnel data. C_n of the 30% can exceed the maximum values of the wind tunnel data, hinting to a possible stall delay in the rotational situation.
4. When the "altitude ripple" is not removed the pressure distributions showed more cyclic variation, the suction peek becomes higher and the shape of pressure distributions changes. Without this correction the unsteady pressure distributions can easily be interpreted as full wind shear or yawed flow effects.
5. The pressure signal variations of the 30% span section are not yet fully understood. Additional tests are necessary.
6. The atmospheric reference pressure can not be measured easily in the unsteady open air conditions. The total pressure corrected with the dynamic pressure from the wind was applied to avoid this difficulty.
7. It is difficult to define, and hence to calculate, a local angle of attack that can be used to compare the open air pressure data with the two-dimensional wind tunnel data. The local inflow angle, calculated from $U/\omega r$ was used to sort the pressure data as starting point. The induced velocities, due to the finite blade span and the energy extracting or consuming process of the rotor, were neglected.

   In future the analysis model will be improved with the induced velocities and corrections for tip and turbulent wake effects. Arrangements will be made to measure the local incoming flow angle.

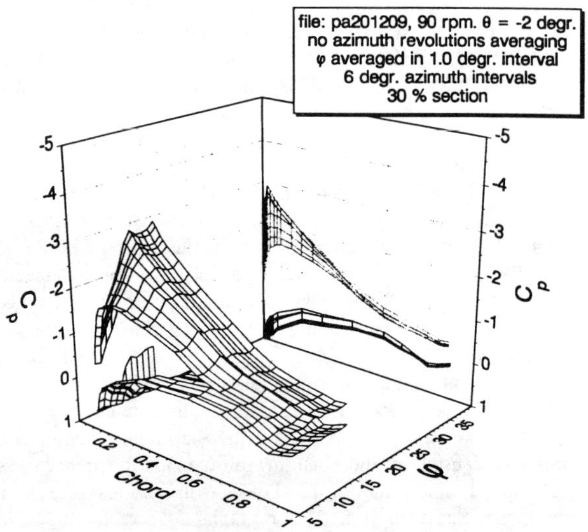

Fig. 5 A three-dimensional example of measured C_P distributions of the 30% span section, binned on flow angle φ in 1 degr. intervals. The back plane contains the projection of the data.

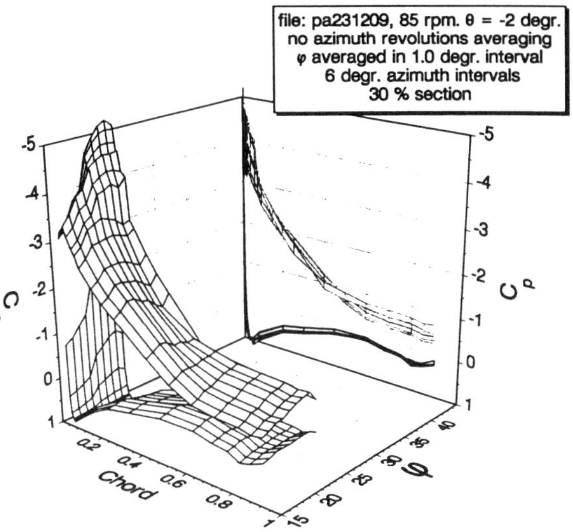

Fig. 6 Same as fig. 5, with C_p distributions selected of higher flow angles φ.

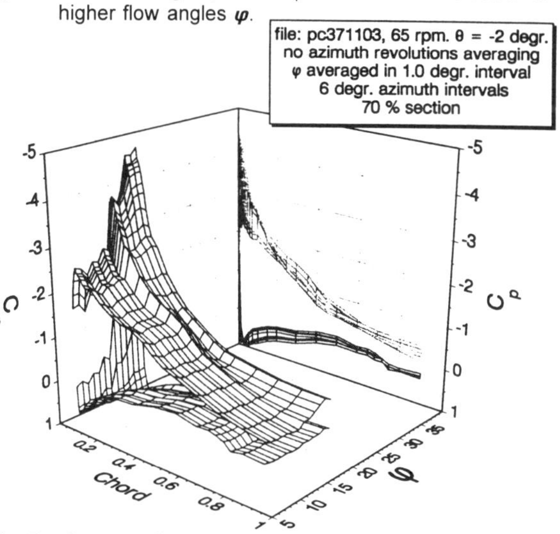

Fig. 7 Same as fig. 6, with data of the 70% span section.

8. ACKNOWLEDGEMENTS.

I would like to thank Eric van der Pol and Cees van Beek for their unwearing helpful assistance during the setup, the measurements; and the development of the measurement software with the additional hardware.

9. REFERENCES.

[1] A. Bruining, W.A. Timmer, Airfoil characteristics of rotating wind turbine blades, Journal of Wind Engineering and Industrial Aerodynamics, 39 (1992) pag. 35-39, Elsevier Science Publishers B.V., Amsterdam.

[2] A. Bruining, W.A. Timmer, Airfoil characteristics of rotating wind turbine blades, Proceedings of the Fifth IEA Symposium on the Aerodynamics of wind turbines (Stuttgart, 1991) ETSU.

[3] A. Bruining, Airfoil characteristics from a rotating wind turbine blade. Proceedings IEA meeting Dec. 1992, held in Petten and Delft. To be published.

[4] C.P. Butterfield, S. Huyer, D. Simms, Recent Results from Data Analysis of Dynamic Stall on Wind Turbine Blades. Proceedings of the Fifth IEA Symposium on the Aerodynamics of wind turbines, Stuttgart, 1991, ETSU.

[5] H. Snel, R. Houwink, W. J. Piers, Sectional prediction of 3D effects for separeted flow on rotating blades, NLR TP 92409L National Aerospace Laboratory NLR.

[6] G. Ronsten, Static pressure measurements on a rotating and non-rotating 2.375m. wind turbine blade. Comparison with 2d calculations. FFA. Sweden.

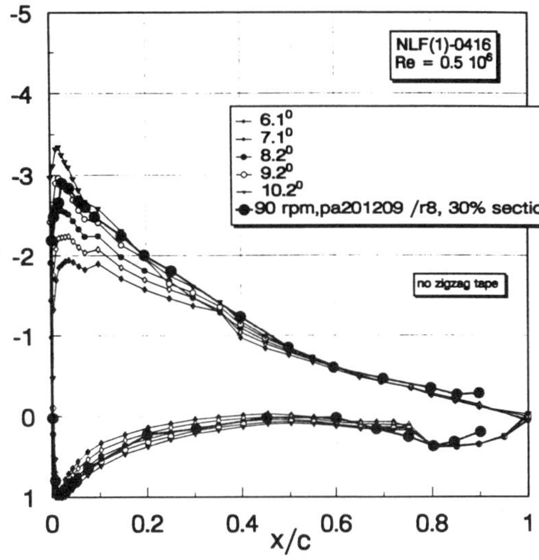

Fig. 8 Comparison of two-dimensional C_p distributions measured in the wind tunnel at the 70% section and a Reynolds number of $0.5*10^6$ (small symbols) with a C_p distribution measured at 90 rpm. (large circle symbols). The rotating test example was selected of data that was binned on flow angle intervals of 1^0.

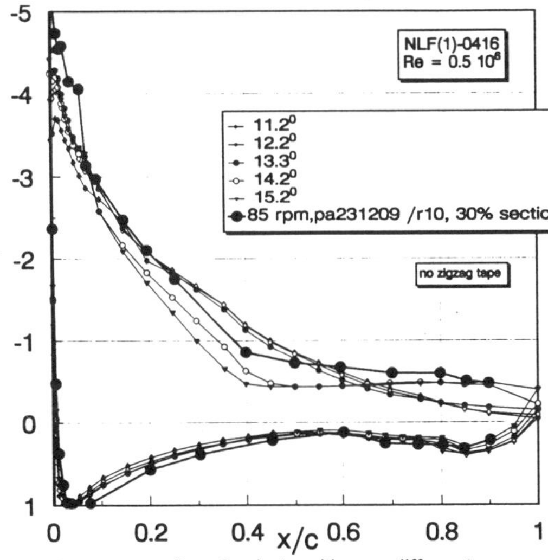

Fig. 9 Same as fig. 8, but with a different example selected of the binned data, measured at 85 rpm.

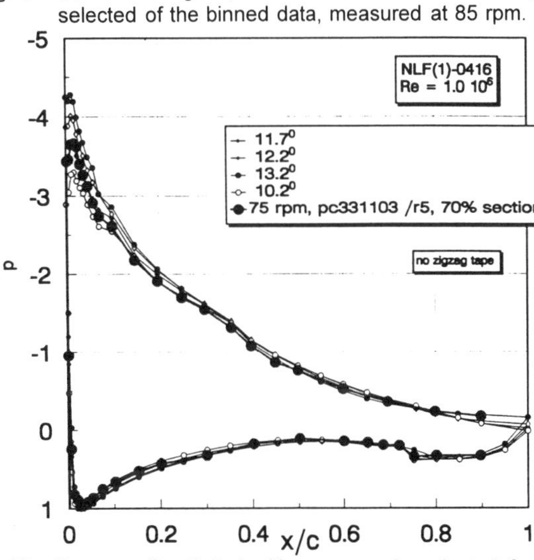

Fig.10 Same as fig. 9, but with an example selected from the binned data of the 70% span section, measured at 75 rpm., compared with wind tunnel data at a Reynolds number of $1.0*10^6$.

RESULTS FROM A JOINT WAKE INTERFERENCE RESEARCH PROGRAM

Axel Albers*, Hans Georg Beyer, Theo Kramkowski*, Marius Schild,
Annette Schomburg, Wolfgang Schlez, Hans-Peter Waldl, Ubbo de Witt*

Carl von Ossietzky Universität Oldenburg, Dept. of Physics, Renewable Energy Group, D-2900 Oldenburg, Germany
FAX ++ 49 441 798 – 3326

* *Deutsches Windenergie Institut, D-2940 Wilhelmshaven, Germany FAX ++ 49 4421 4808 – 43*

We report on the results and current work of a measuring program concerning wake effects. Analyses of farm efficiencies are made in two wind farms. Measurements of the mean velocity field and the turbulence characteristics in the wake flowfield of single wind turbines are performed. The results of the single wake measurements of mean wind speed are compared to the Ainslie model. Turbulence measurements are presented.

Introduction

At the Deutsches Windenergie Institut (DEWI) and the Carl von Ossietzky Universität Oldenburg, a comprehensive measurement program is carried out in cooperation to improve the database of full scale wind farm effects.

Analyses of farm efficiencies are performed in two wind farms. Measurements of the mean velocity field and the turbulence characteristics in the wake flowfield of single wind turbines are performed. Two different wind turbine types, a *ENERCON 16* (rotor diameter (D) 16m) and a *ENERCON 33* (rotor diameter 33m) are analyzed. Both turbines are operated with variable rotational speed.

In one of the monitored farms, investigations of the blade bending moments are in progress (/Seifert/) and will in future be evaluated using the results of the single wake turbulence measurements.

The goal of these investigations is to gain more detailed information on the performance of standard models for the prediction of wake velocity and turbulence. The experience gained with the models will be used for a computer code for calculating the farm efficiency and for optimizing wind farms.

Wind Farm Measurements

As reported in /Beyer89/, former work had shown that the total performance of a small straight line wind farm (5*55kW) can be modelled quite well with the PARK code (/Jensen/). Measurements of farm efficiencies are performed now at two wind farms with larger turbines but with a similar simple geometry. The first one is a windfarm consisting of two straight lines operated by the *Stadtwerke Norden* (array configuration illustrated in Fig. 1) with a total of ten turbines. The performance of five turbines (1-5) ENERCON E-32 with a rotor diameter (D) of 32.0 m and one turbine (6) ENERCON E-33 with a rotor diameter of 33.0 m (variable rotor speed, pitch controlled, rated power 300 kW) is analyzed. The data are monitored in the framework of the German WMEP measuring program. A meteorological mast is installed at a distance of two rotor diameters from the array to registrate wind velocity and direction at hub height.

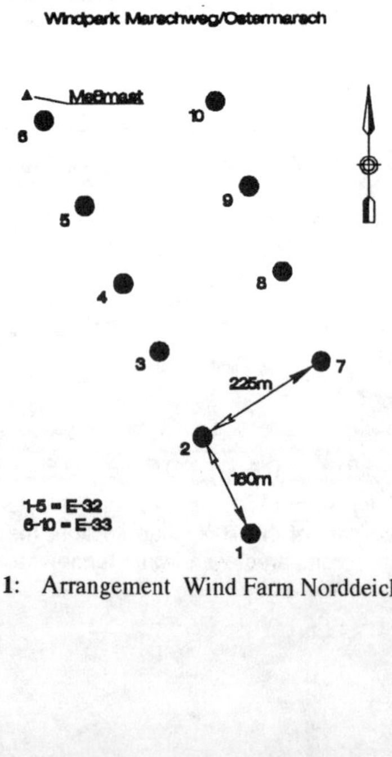

Figure 1: Arrangement Wind Farm Norddeich/Marschweg

An example for the performance of this park during wake immersion is given in Fig. 2. The power output of the turbines operating in the wake is reduced to approx. 50% of the output of the turbine operating in the free flow. The detailed evaluation of the measurements is difficult because the power characteristics of the turbines are not identical.

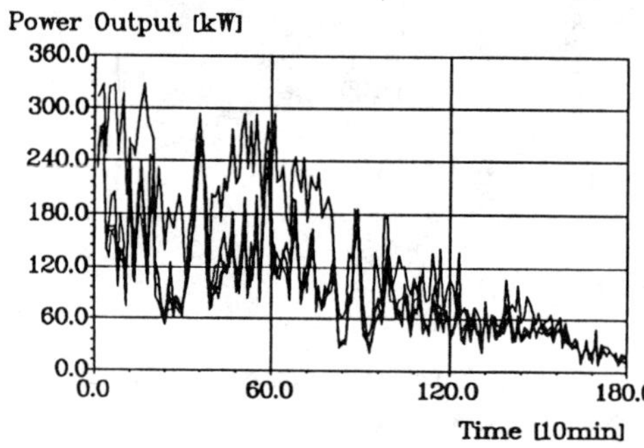

Figure 2: Time series of turbines one to six during wake immersion

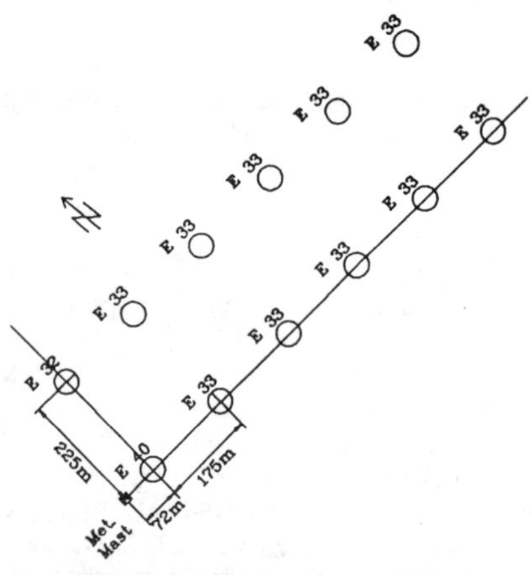

Figure 3: Arrangement Wind Farm Hamswehrum

The second farm under investigation is built up in two rows with a spacing of 175 m between the turbines and 225 m between the lines (see Fig. 3). In this farm the monitoring equipment has been installed recently. Both long term performance data and short term power fluctuations of the turbines will be registrated.

Norddeich Fledderweg, *ENERCON 16*

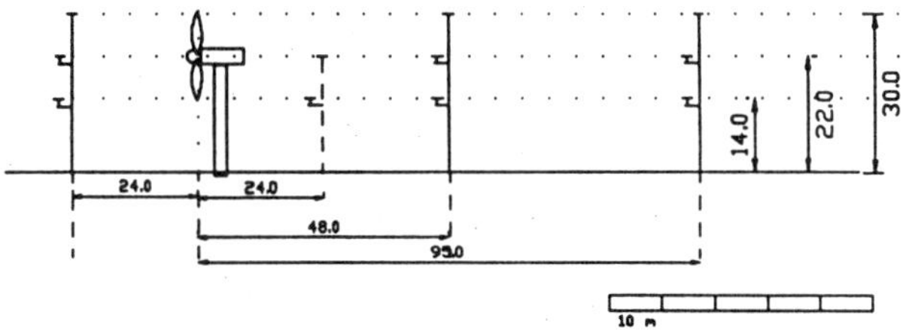

Figure 4: Arrangement of turbine *ENERCON 16* and masts

Single Wake Measurements

Wake velocity measurements are performed at a 55 kW *ENERCON 16* wind turbine (diameter 16.2 m, hubheight 22.4 m) which is stall regulated and operating at variable rotational speed. The turbine is sited in a small straight line wind farm Norddeich/ Fledderweg (5x 55 kW, sited at a distance of approximately four rotor diameters) at the North Sea Coast.

The wind velocity was measured at in total three meteorological masts (Fig. 4). One is sited at two rotor diameters upwind from the turbine in the prevailing wind direction, measuring the undisturbed wind velocity and direction at hub height and at the upper and lower edge (hubheight ± rotor radius) of the rotor disk.

For one measuring campaign, an additional mast was sited at a distance of two rotor diameters behind the turbine. It was equipped with anemometers at hub height and at the height of the lower edge of the rotor disk.

In a second period, two masts were used for registrating the wake wind speed at a distance of four and six rotor diameters from the wind turbine. These masts have the same equipment as the free mast.

For all wind measurements, standard cup anemometers are used. The time resolution of the data recording was one minute respectively one second for selected time periods.

Mean Wake Flow

The data sets gained with this setup are used for an extended test of the Ainslie wake model (/Ainslie/). Fig. 5 shows the mean wind speed (base : one minute averages) measured at hub height in the wake flow at the three masts as a function of wind direction (□). All values are normalized to the free wind speed and classified with respect to wind direction bins. Mean free wind speed during this measuring period was about 8 *m/s*. The wind speed is reduced to about 40% at two diameters behind the rotor and recovers with increasing distance to the rotor to 70% (4 D) and 80% (6 D).

At one edge of the wake, the wind speed is higher than the undisturbed velocity for a certain range of the wind direction. This effect is supported by former measurements of the farm efficiency (/Beyer89/) which shows the same phenomenon. Unfortunately the respective wind directions are not covered by the data gained with mast *2D*.

In the same figure, calculations with the Ainslie model / Ainslie/ are presented (Δ). The model is based on the solution of the axial-symmetric reynolds equation. Using the boundary condition proposed by Ainslie the equations are solved applying the /PDE/ PROTRAN/ code. The calculations predict a slightly higher wind speed over the whole range of wind directions but are in total in a reasonable accordance with the measured wind speed values.

It should be noted that the model calculations are based on a c_t*(windspeed)* — characteristics of the wind turbine calculated by the (measured) rotational speed and the thrust coefficient $c_t(\lambda)$ characteristics supported by the blade manufacturer. A measurement of the rotor thrust by registrating the wind turbine tower root bending moment is in progress but not finished yet. So, a more certain comparison between calculated and measured wind speed values can be expected in the near future.

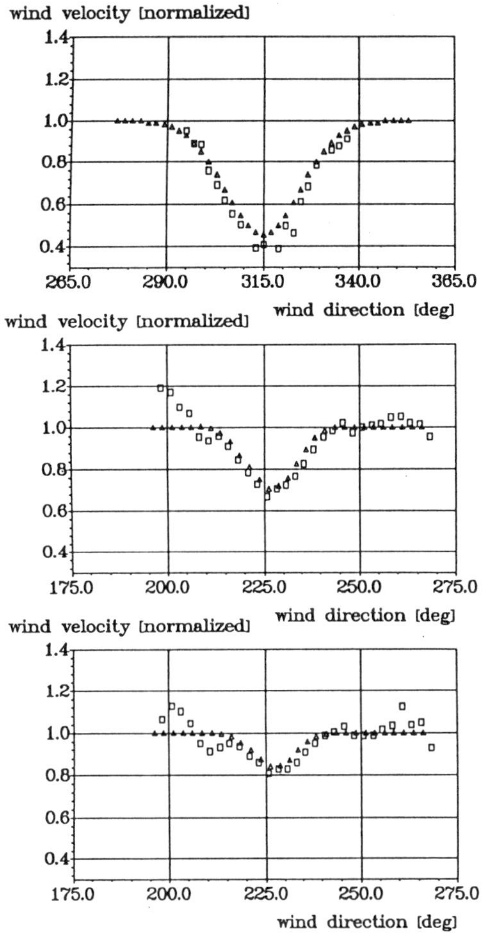

Figure 5: Horizontal profile of the wake velocity (normalized to the free wind speed) at hub height versus wind direction. Distance to the rotor: 2 *D* (upper), 4 *D* (middle), 6 *D* (lower figure). Measured (□) and calculated (Δ) values

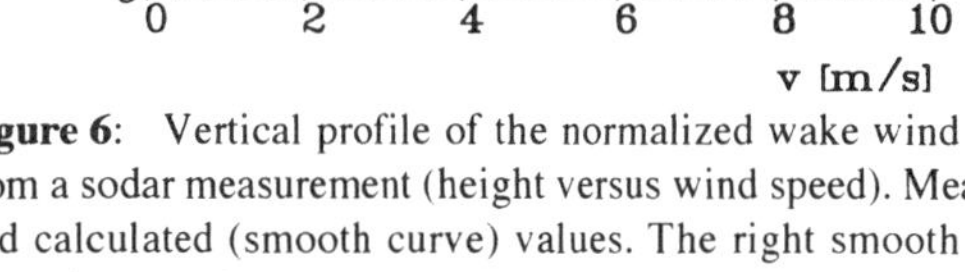

Figure 6: Vertical profile of the normalized wake wind speed from a sodar measurement (height versus wind speed). Measured and calculated (smooth curve) values. The right smooth curve gives the used free stream logarithmic wind profile.

The Ainslie model produces a axisymmetric wake profile. To take into account the vertical wind profile in the boundary layer, it is suggested to superimpose a logarithmic profile to the wake model calculations. Here the velocity deficit of the wake flow and the wind speed change in the profile relative to the reference height of the rotor hub are added. Fig. 6 shows a measurement in the wake flow with a mini sodar (see /Schomburg/) which gives a vertical profile of the wake flow (distance to the turbine approx. 3 rotor diameters). The comparison with the model calculations (/Ainslie/ + logarithmic profile) supports this quite simple assumption.

Norddeich Marschweg, *ENERCON 33*

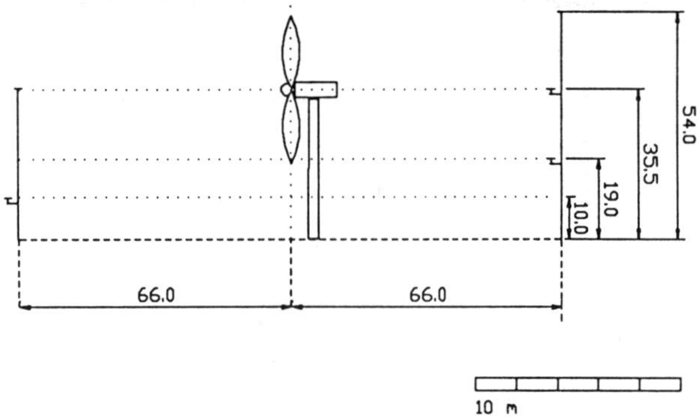

Figure 7: Arrangement of turbine *ENERCON 33* and masts

Measurements of the near wake flowfield of a larger turbine are performed at a single wind turbine of type ENERCON E-33. (see Fig. 7)

The upwind mast is equipped with two cup-anemometers, mounted at hubheight and at 10 meters height. The downwind mast is installed at a distance of 2 rotor diameters. It is equipped with three cup-anemometers, mounted at hubheight and at height of the upper and lower rotor tip. In addition the wind vector is registered at hubheight with one ultrasonic anemometer with a high time resolution. The measurements of the wind speeds and characteristic turbine quantities are recorded with a time resolution of 1 second.

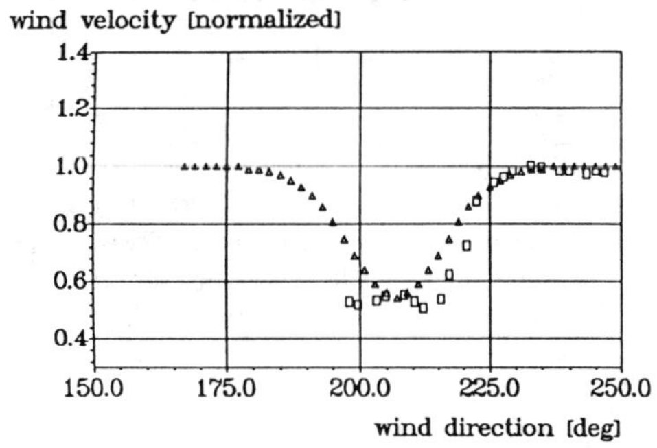

Figure 8: Horizontal profile of the wake velocity of a *ENERCON 33* (normalized to the free wind speed) at hub height versus wind direction. Distance to the rotor: 2 *D* Measured (□) and calculated (Δ) values.

As for the smaller turbine the wind speed time series will be analyzed with respect to the wind direction and compared to the predictions of the Ainlsie model. Figure 8 shows a first result of the comparison. Here the measured data cover only one half of the wake. The maximal velocity deficit is represented quite well by the model.

Turbulence in the Wake

On the basis of wind speed data registered with a time resolution of one second the standard deviation of the wind speed fluctuations in the free and the wake flow are analyzed. To allow a detailed investigation of the cross flow structure of the turbulence the standard deviations are calculated for one minute time intervals. These calculations are performed for the measurements at the *ENERCON 16* turbine. Figure 9 shows the standard deviation in the wake flow normalized to the free flow value as a function of wind direction. The results for the 4D and 6D masts are given. In addition the pro-file for the normalized mean wind speed is given for orientation in the upper plot. Unfortunately only one half of the wake is covered by the wind direction in this measuring period.

The turbulence in the wake flow at 4 *D* is increased up to 150% compared to the undisturbed flow. At 6 *D*, the turbulence is lower, but it a increase to 130% still remains at this distance from the rotor.

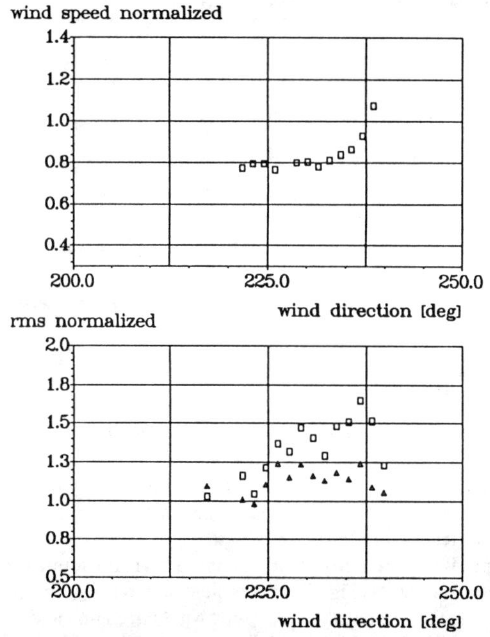

Figure 9: Horizontal profile of the standard deviation in a part of the wake flow of the *ENERCON 16*- turbine (lower plot). The values are given for a distance of 4 *D* (□) resp. 6 *D* (Δ). The center of the wake is at approx. 225°.

Acknowledgement

Part of this work is funded by the German BMFT and the utilities Energieversorgung Weser-Ems AG, Hamburger Electricitts-Werken AG, and Schleswag AG. The WMEP-Data for one farm is supported by ISET. We are grateful to the co-workers of the Stadtwerke Norden and to the farmers Mr. Martens and Mr. Wübbens for supporting our work.

Literature

/Ainslie/ J.F. Ainslie, Calculating the flowfield in the wake of wind turbines, Journal of Wind Engineering and Industrial Aerodynamics, 27, 1988, 213-224

/Jensen/ N.O. Jensen, A Note on Wind Generator Interaction, Risø National Laboratory, Risø, 1983

/Beyer89/ Beyer et al, Operational Behaviour of a small straight Line Wind Farm, EWEC '89, Glasgow 1989

/Seifert/ H. Seifert, Monitoring Fatigue Loads using cycle counting Field Computer, this conference

/Schomburg/ A. Schomburg, Use of a Mini-SODAR for Investigation of Turbulence in a Wind Farm, this conference

/PDE/PROTRAN/IMSL, 1986

TURBULENCE IN WINDFARMS

Jørgen Højstrup and Michael S. Courtney
Department of Meteorology and Wind Energy
Risø National Laboratory, Roskilde, DENMARK

ABSTRACT. Turbulence data from three windfarms in flat and mountaneous terrain types are presented. Turbulence properties are presented in the form of covariances, spectra and probability distributions. Standard deviations and shear stresses in the windfarm are far from their wellknown equilibrium values. Velocity spectra are strongly influenced by scales of sizes comparable to the diameter of the wake. Probability distributions do not differ much from turbulence in general, they exhibit the same non-Gaussian behaviour as the undisturbed free stream turbulence.

1. INTRODUCTION

Atmospheric turbulence in general is quite well understood, both in homogeneous terrain and for relatively simple terrain inhomogeneities, but until recently we did not know very much about turbulence in windfarms, except that turbulence intensities were high, and that windturbines situated in arrays were more prone to mechanical failure than were turbines in the free flow.

Knowledge on the detailed behaviour of turbulence in a windfarm is interesting mainly for two reasons:

- For prediction of mechanical loads on the turbines
- For modelling of the flow inside the windfarm and downstream.

Risø National Laboratory has performed several large field programmes for a variety of terrain conditions aimed at obtaining sufficient information on this important aspect of windfarms. An overview of the programme can be seen in fig.1. The most recent set of measurements are being reported in [11], earlier presentations of data from Nørrekær Enge can be found in [5], [6], [10] and [9]. The Tændpibe measurements were reported in [1], [2], [3] and [4]. The Alta Mesa measurements were described in [8].

This paper is an attempt to summarize a few of the main results from the vast amount of data that has been collected. The three windfarms have quite different characteristics with respect to terrain, size and turbine sizes:

- TÆNDPIBE. Flat coastal terrain, 35 * 75 kW VESTAS.

- NØRREKÆR ENGE II. Flat terrain close to a coastline, surrounded by slightly more complicated terrain. 42*300kW NORDTANK.

- ALTA MESA. Very rough, complicated mountaneous terrain. 140*180 kW DANWIN.

2. MECHANICALLY GENERATED WAKE TURBULENCE

In order to get a qualitative feel for the turbulence to expect in a wake let us consider only vertical gradients in windspeed, then the amount of turbulent kinetic energy generated per time unit by mechanical forces in a shear layer can be written:

$$\frac{de}{dt} \sim -\langle uw \rangle \frac{\partial U}{\partial z} \qquad (1)$$

In the horizontally homogeneous case this expression is very simple with a constant <uw> and the windspeed increasing logarithmically with height. In the wake case the situation is much more complicated with both the correlation and the gradient assuming much different values.

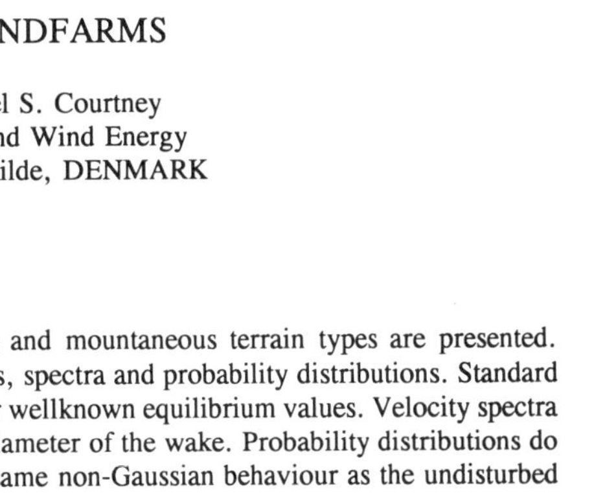

Figure 1 Overview of the Risø wake and windfarm program.

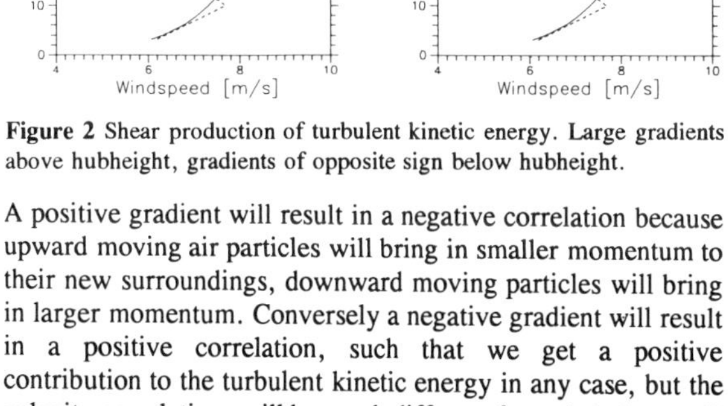

Figure 2 Shear production of turbulent kinetic energy. Large gradients above hubheight, gradients of opposite sign below hubheight.

A positive gradient will result in a negative correlation because upward moving air particles will bring in smaller momentum to their new surroundings, downward moving particles will bring in larger momentum. Conversely a negative gradient will result in a positive correlation, such that we get a positive contribution to the turbulent kinetic energy in any case, but the velocity correlations will be much different from their "normal" values. This is illustrated in fig.2 where a typical (2D-wake) profile is shown together with a free field profile for neutral stability. We assume that the relevant gradient for a certain vertical excursion is the average gradient over the height range in question. We look at hubheight, 31m (leftmost figure) and 23m and 58m. At hubheight we see an indication of large gradients at small excursions ("1") decreasing at intermediate scales ("2" and "3"). Below the wake we see gradients with the opposite sign, indicating a sign change for the correlations. Above the wake we see much increased gradients indicating much larger production of energy.

The profile of windspeed standard deviations from Nørrekær Enge (fig.3) does indeed confirm these assumptions, i.e. excessive levels of turbulence in the lower and upper parts of the wake, with highest levels in the upper part. If we derive the equilibrium energy levels from the actually measured profiles and velocity correlations, then we arrive at values that are far from those measured, more energy than expected in the center of the wake, less energy in the outer portions of the wake. The reason is of course that it takes time to change the energy of turbulence, and the flow moves too fast past the disturbed area to reach equilibrium except for the shortest wavelengths. Also there must be quite vigourously mixing of turbulence across the wake.

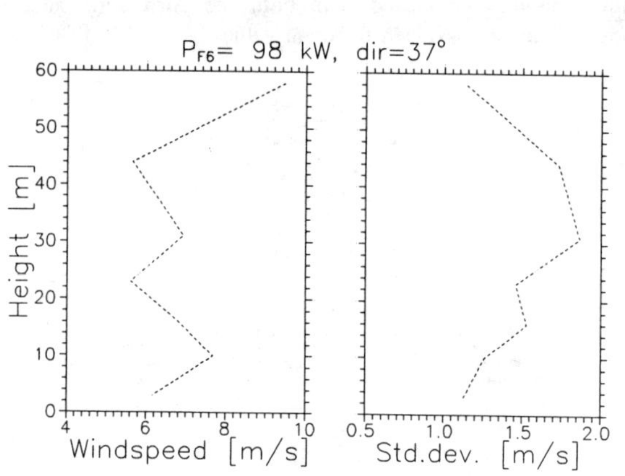

Figure 3 2D wake profiles from Nørrekær Enge. Note the two peaked structure of both speed and turbulence profiles and highest turbulence levels above the wake.

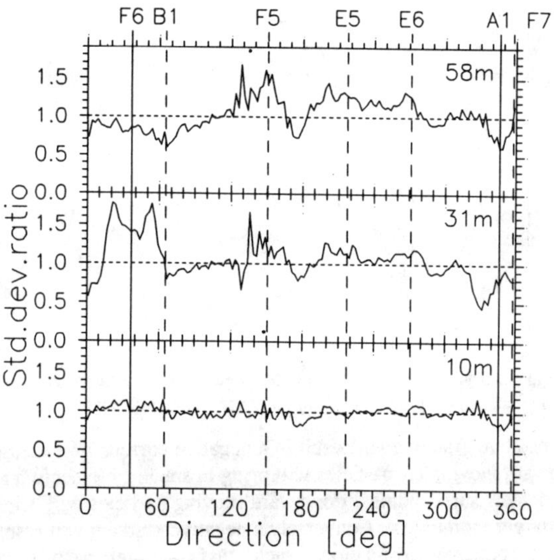

Figure 4 Ratios of standard deviations measured at two masts in Nørrekær Enge. The full vertical lines denote 2D wake situations, the dashed lines are 5D wakes.

Fig.4 shows a horizontal scan of the ratio of turbulence measured at two masts σ_{M2}/σ_{M1} from Nørrekær Enge. The vertical lines show winddirections where one of the two masts are in the wake of a turbine. Full lines denote 2D-wakes, dashed lines are 5-8D wakes. Mast M1 is placed close to turbine A1, mast M2 is placed close to F6. In the 2D wakes we see increases in turbulence level of about a factor of 2, and for the 5-8D wakes we see increases of 30-60%, with the largest increases seen above the wake. Some typical values for second order moments are shown in table 1 from three 30-60minute runs from the three windfarms. The data was derived from measurements by sonic anemometers.

Table 1. Second order normalized moments from different farms. In the table is specified hubheight, rotor diameter, downstream distance, and height of measurement in rotor diameters above hub. The Tændpibe measurements are an average of 5 runs, Nørrekær and Alta Mesa are one run (1 hour) each.

Site	Hub [m]	D [m]	x [D]	z-H [D]	σ_U/σ_W	σ_V/σ_W	σ_W/u_*
Equilibrium					1.91	1.54	1.25
Tændpibe	23.5	17	>5	-.3	1.7	1.4	-
				.5	1.6	1.2	1.3
Nørrekær	31	28	5	-.3	1.4	1.1	2.3
				1	1.9	1.3	1.0
			2	-.3	1.5	1.0	-
				1	1.6	1.3	1.7
Alta Mesa	23	23	2	.4	1.2	1.1	2.4

The standard deviations were normalized by σ_W because the usual normalization with u_* is very much influenced by the wakes, and in some cases we get positive <uw> correlations, meaning that we do not have a meaningful u_* measurement.

General conclusions: Nothing seems to be in equilibrium, but in the 5D wakes the vertical component seems close to equilibrium values except for the highest level. The horizontal values are less than expected, i.e. the horizontal components with their large content of low frequency energy have not yet adjusted to the increase in local production. In the 2D wake all components seem far from equilibrium. The general tendency is towards the turbulent kinetic energy more evenly distributed on the three components than seen in equilibrium turbulence.

3. VELOCITY SPECTRA

The PSD for the sonic anemometers are shown in figs. 5-7. The cospectra of uw are severely disturbed in all cases, just as expected. We see positive deviations in the lower levels at length scales comparable to the wake width (in frequency about 0.1 Hz), leading to decreases in shear stress, and similarly large negative deviations above the wake center leading to increased shear stress values. In the Alta Mesa data another mechanism seems to be removing energy from the cospectrum at lower frequencies, leading to a net loss in shear stress. This site is situated on top of a ridge running perpendicular to the winddirection, such that the accelerated flow on top of the hill will cause diminished shear stress.

The power spectra of the velocity fluctuations have all had their dominant length scale move towards much shorter values than expected (see fig.7) due to the input of excess energy at scales comparable to the wake width, The length scales of the u-component seem to be changed by an order of magnitude, such that the dominant scale length is 2*height instead of the expected 20*height or more. Also the scales of the lateral and the vertical component seem closer together than expected.

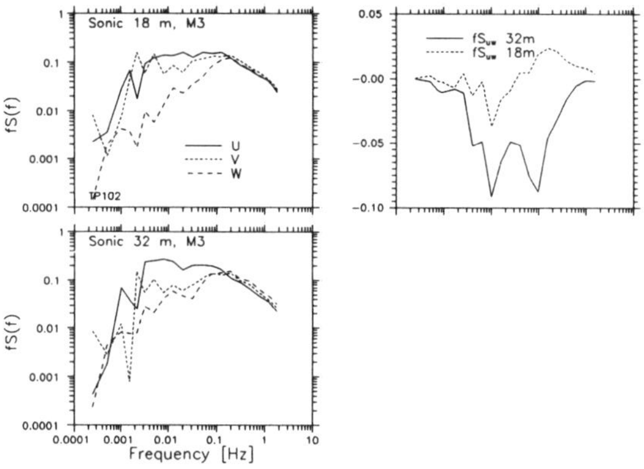

Figure 5 Power spectra and cospectra of velocities from Tændpibe.

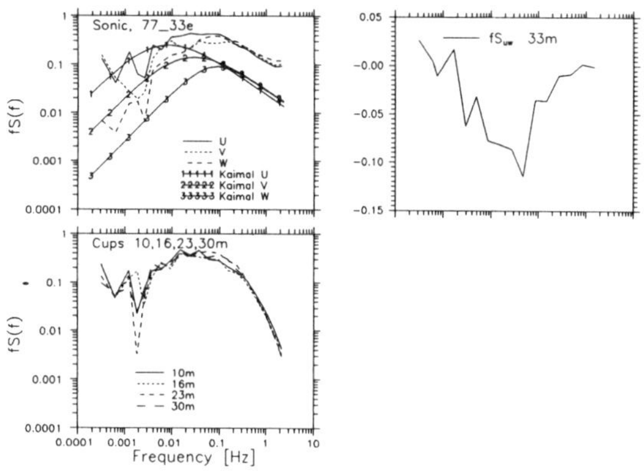

Figure 7 Alta Mesa. Power spectra of u,v,w and cospectrum of uw. For comparison is shown a neutral model spectrum plotted using the measured windspeed and u_*.

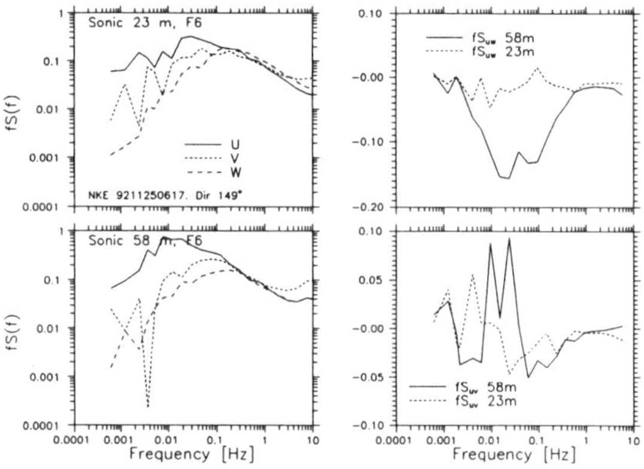

Figure 6 Power spectra of all three velocity components and cospectra of uw and uv from Nørrekær Enge.

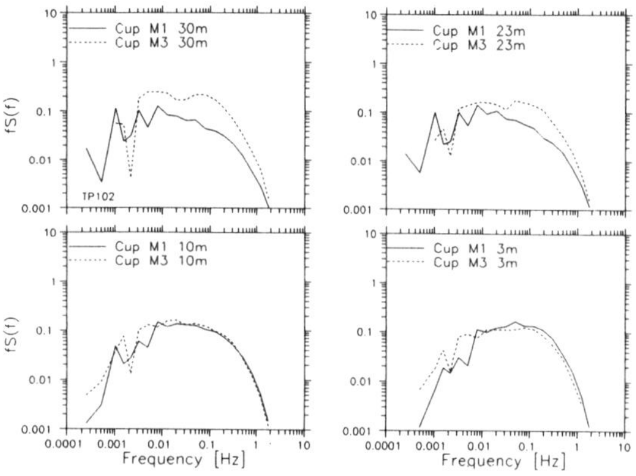

Figure 8 Tændpibe. Cupanemometer spectra. Upstream in full line, 5-6D wake in dashed line.

For two of the farms we can make direct comparisons between upstream and wake spectra (measured by fast response cupanemometers), here shown in fig. 8 for Tændpibe, where we see large increases in spectral levels at scales comparable to the wake width.

4. PROBABILITY DISTRIBUTIONS

Probability distributions are very similar from all three farms and we show here just one example (fig.9). The top set of distributions are the distributions of velocities and winddirections, whereas the bottom ones are the distribution of velocity differences, i.e. an approximation to the time derivatives of the signals. In general we see exactly the same behaviour as we do in free flow turbulence, velocities close to normal distributions, although there is a tendency towards more narrow distributions of the velocities inside the wake [10], but the derivatives have much longer "tails" in their distributions. This of course means that turbulence is not the result of a Gaussian process, and the fatigue damage resulting from such turbulence must be assumed to be more severe than the damage resulting from Gaussian turbulence. Another way to illustrate the non-Gaussian behaviour of turbulence is to look at the joint probability distribution of windspeeds and windspeeds lagged a certain amount of time [12]. An example is shown in fig.10 where the actually measured distribution is compared with a Gaussian joint distribution with the same second order statistics. This type of statistics contains information on both the probability distributions we have seen above, and we can see that the measured distribution has a somewhat wider extent than the Gaussian model.

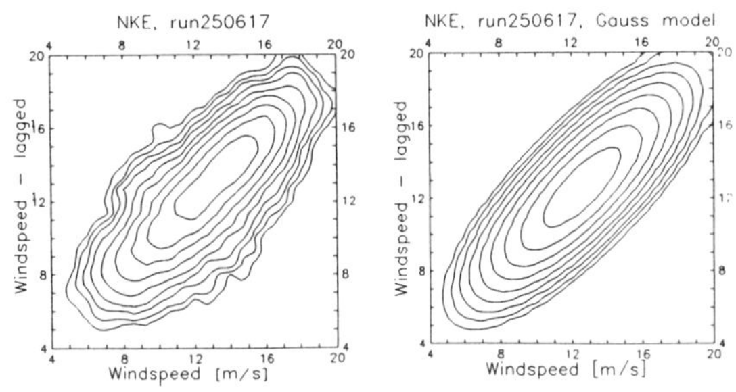

Figure 10 Joint probability distributions of velocities and velocities one second later. Fo comparison is shown a Gaussian model with the same correlation.

5. CONCLUSIONS

Wake turbulence is characterized by

- high turbulence intensities
- non-equilibrium second order statistics
- spectral distributions strongly controlled by the wake dimensions with length scales an order of magnitude shorter (for the horizontal component).
- Non-Gaussian. Velocities are approximately normally distributed, derivatives have much wider tails.

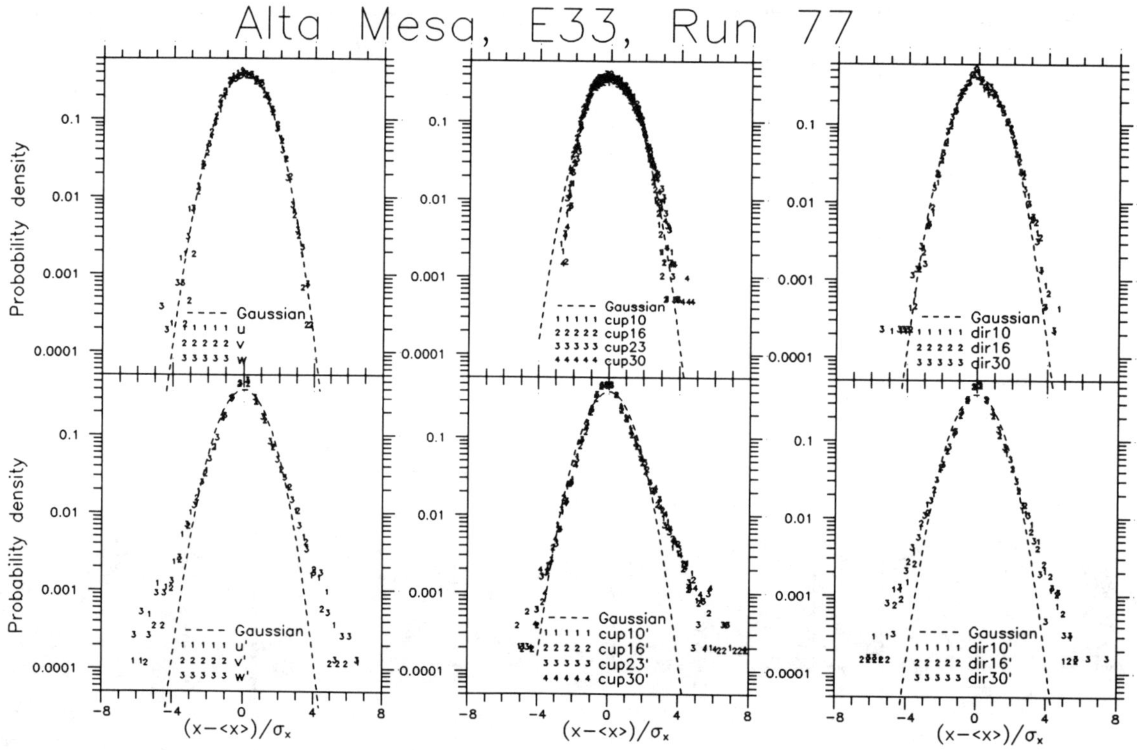

Figure 9 Alta Mesa. Probability distributions, from the left -sonic anemomenter, cupanemometers, winddirection. Top figures are the velocities, bottom figures are velocity differences.

6. ACKNOWLEDGEMENTS

The Tændpibe and the Nørrekær Enge measurements were supported by the CEC, the Alta Mesa measurements were supported by the Danish Ministry of Energy that also supplied funding for the Nørrekær Enge measurements.

REFERENCES

[1] Højstrup, J., I.Katic and P.Nørgård, 1988: Supervising and measuring at Tændpibe windfarm. CEC Euroforum New Energies conference, Saarbrücken, FRG.

[2] Højstrup, J., 1990: Turbulence measurements in a windfarm. British Wind Energy Association conference 12, Norwich, UK.

[3] Højstrup, J., 1990: Turbulence measurements in a windfarm. ECWEC90, Madrid, Spain.

[4] Højstrup, J. and P.Nørgård, 1990: Tændpibe windfarm measurements 1988. Risø M-2894. 44pp., Risø National Laboratory, Roskilde, Denmark.

[5] Højstrup J., 1991: Measurements of structural loads in the Nørrekær Enge Wind farm. Proceedings of the BWEA13, Swansea, UK, April 1991. Ed. by D.C.Quarton and V. C. Fenton.

[6] Højstrup J., Courtney M.S., and Sanderhoff P., 1991: Measurements of structural loads in the Nørrekær Enge Wind farm. Proceedings of the EWEC '91, Amsterdam, October 1991. Ed. by F.J.L.Van Hulle, P.T.Smulders, and J.B. Dragt.

[7] Højstrup, J., 1991: Probability structure of turbulence in a windfarm In: Proceedings from 20th meeting of experts - Wind characteristics of relevance for wind turbine design. Ed.: B.M.Pedersen, R.Windheim. Kernforshungsanlage Jülich. p.15-19.

[8] Pedersen, T.F., S.M.Petersen, K.Thomsen, P.H.Madsen, J.Højstrup, 1991: Loads for wind turbines in inhomogeneous terrain. Measurement report. Risø M-2922. 150 pp. Risø National Laboratory, Roskilde, Denmark.

[9] Courtney M.S., Frandsen S., and Sanderhoff P., 1991: Measurements in the Nørrekær Enge Wind farm. Proceedings of the BWEA14, Nottingham March 1992. Ed. by B. R. Clayton.

[10] Christensen, C.J., M.S.Courtney and J.Højstrup,1992: Turbulence and turbine dynamics in a wind farm. In: The potential of wind farms, conference proceedings, Herning, Denmark. paper D4.

[11] Højstrup, J., M.S.Courtney, C.J.Christensen, P.Sanderhoff,1993: Full scale measurements in wind-turbine arrays. Nørrekær Enge II. CEC/JOULE. Risø report I-684. Risø National Laboratory, Roskilde, Denmark.

[12] Dutton, J.A. and J.Højstrup, 1979: A probabilistic model of turbulent velocity fluctuations. Wind Characteristics and wind wnergy siting conference, Oregon, USA. Paciific Northwest Laboratory report PNL-3214. p.59-67.

APPLICATION OF A GENERAL PURPOSE NAVIER-STOKES CODE TO WIND TURBINE FLOW PROBLEMS

H.A. Madsen, F. Rasmussen
The Test Station for Wind Turbines, Risoe National Laboratory
Roskilde, DK-4000, Denmark

ABSTRACT: The complex nature of the aerodynamics of wind turbines requires the use of advanced, state-of-the-art computational tools. The fast growth of computing power on e.g. work stations combined with the development of new iterative solvers for large matrix systems has now made Navier Stokes calculations a realistic and efficient operational tool. At RISØ Navier-Stokes calculations have now been used for some time on different wind turbine flow problems. The present paper presents results on the 2D flow over an airfoil in stall and results on the steady and unsteady flow through an actuator disc which is equivalent to the dynamic inflow problem for wind turbines.

1. INTRODUCTION

In the on-going process of optimization and design of new wind turbine rotors the complex nature of the aerodynamics requires the use of advanced, state-of-the-art computational tools. So far, two zone models combining a surface singularity method with a boundary layer method have been widely used for the study of the flow past a single blade or a full rotor. In aeroelastic codes, the aerodynamic part is typically based on a blade element/momentum model or a vortex line/panel model.

However, in many flow problems related to wind turbines the assumptions introduced by using a full potential model or a two zone viscous/inviscid interaction model give uncertain results. As an example can be mentioned the flow past an airfoil in deep stall where the assumption of an inner viscous boundary layer and an outer potential flow is questionable.

The alternative to the above models is the use of a Navier-Stokes model. The very fast growth of the computing power on e.g. workstations combined with the development of new efficient iterative algorithms for the solution of big matrix systems has now made it possible to solve most 2D flow problems within a reasonable time as well as smaller or simplified 3D problems.

At Risø the general purpose 3D NS code FIDAP, based on the finite element discretization technique of the NS equations, has now been used for some time to study different wind turbine flow problems. Of the features facilitating the application on wind turbine aerodynamics can be mentioned; 2D/3D steady/transient analysis, fixed or rotating frame of reference and analysis of flow influenced by volume forces (actuator disc analysis).

The main application areas are:

a) stall and post-stall characteristics of airfoils
b) actuator disc steady/transient flow
c) interaction between stalled and non-stalled regions on a blade (3D problem)
d) tip flow
e) calculation of flow fields to be used as input in aeroacoustic calculations.

So far, results from the first two application areas have been obtained and will be presented in the paper.

2. THE MODEL

The unsteady, incompressible Navier-Stokes (NS) equations and the equation of continuity can be written on the form

$$\rho\left(\frac{\partial \boldsymbol{u}}{\partial t} + \boldsymbol{u}\cdot\nabla \boldsymbol{u}\right) = -\nabla p + \rho_o \boldsymbol{f} + \mu\nabla^2 \boldsymbol{u}$$

$$\nabla\cdot\boldsymbol{u} = 0$$

where **u**, **f** is the velocity vector and the body force, respectively.

The general purpose NS code FIDAP [1] based on the finite element discretization technique is used for solving the equations.

The turbulent stresses are simulated by the k-ϵ model combined with a law-of-the-wall model to simulate the viscous sub-layer.

In the following the variables have been made dimensionless with respect to a characteristic length L and a characteristic velocity U_0 for the actual problem. This means $x^* = x/L$, $\mathbf{u}^* = \mathbf{u}/U_0$, $t^* = t\,U_0/L$, $p_* = p/\rho U_0^2$, $\mathbf{f}^* = \mathbf{f}L/U_0^2$. However, the star which denote dimensionless quantities will not be shown in the following.

2. FLOW AROUND AN AIRFOIL

The NS calculations have been made for an 18 % thick airfoil developed by the National Technical University of Athens within the JOULE 1 project **AIRFOIL SECTION DESIGN AND ASSESSMENT**,[2]. The reason for choosing this airfoil is that it has been designed for application particularly on stall controlled rotors but also because pressure measurements are available for this airfoil. The pressure measurements were made by the University of Southampton during the above mentioned project, [2].

The Reynolds number (Re. no.) during the wind tunnel tests was around 1 million and the pressure measurements were carried out at two different rows A and B of pressure taps.

For the NS calculations, a C-type grid with around 11000 nodes was used, Fig. 1. The computed pressure distributions at

8 and 12 degrees incidence are shown in Figures 2 and 3 together with the experimental data. For comparison is also

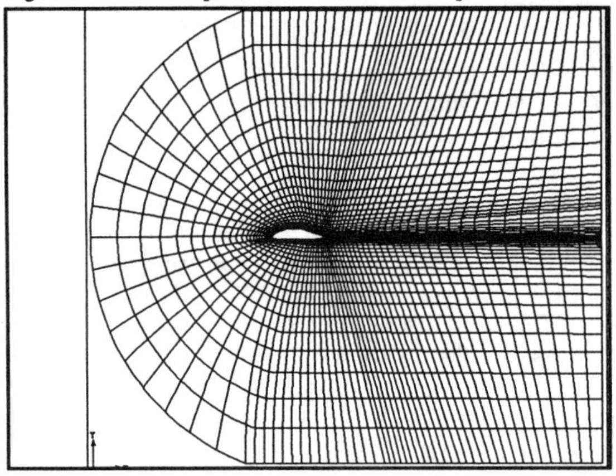

Figure 1. A C-type grid with about 11000 nodes is used for the NS calculations.

shown the computed pressure using the viscous/inviscid interaction code XFOIL, [3]. The inviscid flow in XFOIL is computed with a panel method and an integral method is used for the viscous flow.

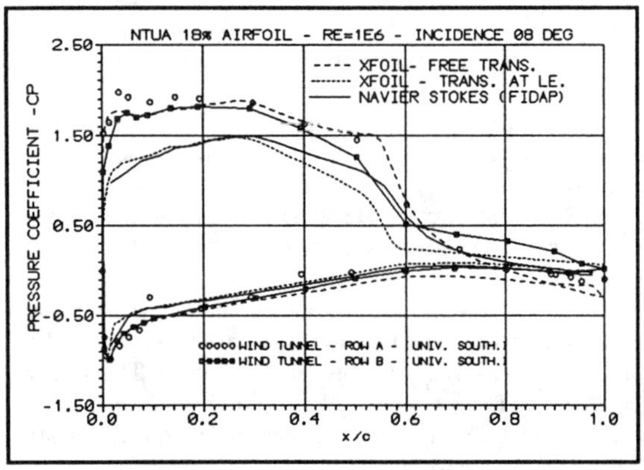

Figure 2 Comparison of pressure distributions at 8 deg. incidence with experiment.

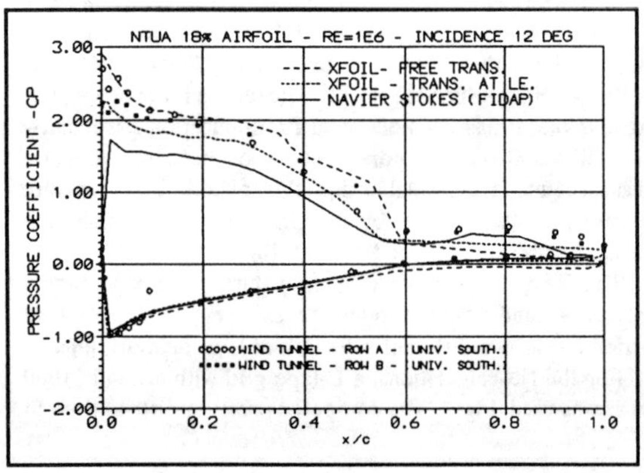

Figure 3. Comparison of computed pressure distributions at an incidence of 12 deg. with experiment.

Comparing now the NS pressure coefficient curves with experiment it can be seen that there is a considerable difference on the suction side whereas a good agreement is found on the pressure side. The main reason for this is that in the present version of the NS code only a full turbulent or a full laminar flow can be handled. However, for this particular airfoil there is a considerable laminar boundary layer on the front part of the aerofoil and the airfoil characteristics are affected very much if the boundary layer is triggered to be turbulent at the leading edge. Unfortunately, no measurements were carried out with a full turbulent boundary layer. In the XFOIL code the position of the transition point is based on the e^n method and it is seen that predictions from this code compare very well with experiment, both at an incidence of 8 as well as at 12 deg. It should be noted that 8 deg. incidence is just before C_{Lmax} is reached whereas at 12 deg. incidence there is separation on a considerable part of the airfoil.

To be able to evaluate better the NS predictions, the XFOIL code was also run with the position of the transition set at the leading edge. In Fig. 2 it is seen that there now is a much better agreement with the NS calculations and this also shows that the pressure distribution on this airfoil is influenced very much by the position of the transition point. At 12 deg., Fig. 3, the differences between the two predictions are bigger and it seems that the NS prediction does fully catch the very steep pressure gradient at the leading edge. However, this can probably be improved by refining the grid.

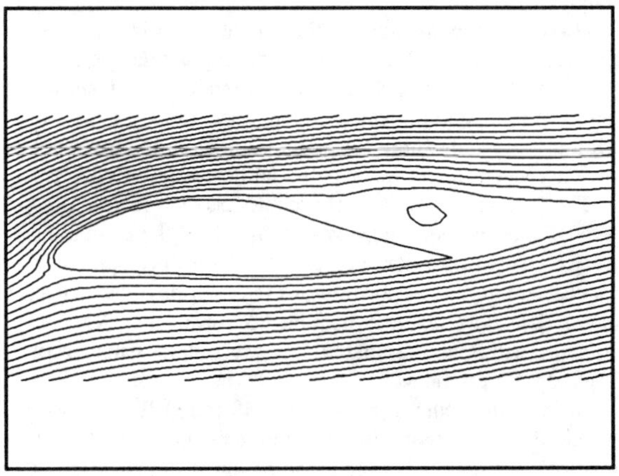

Figure 4. Streamlines at 12 deg. incidence predicted with the NS code.

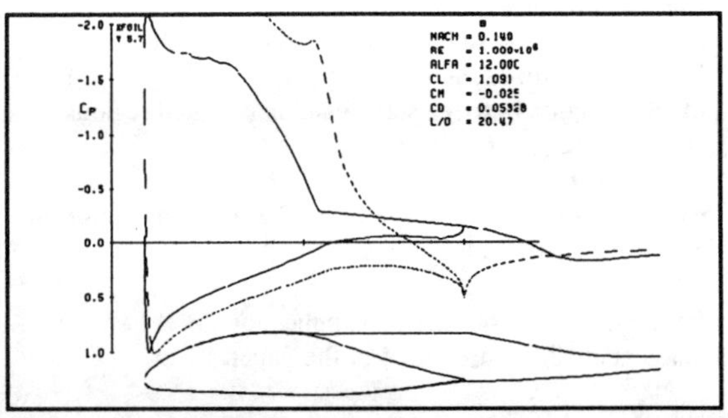

Figure 5. Pressure distribution and wake geometry predicted with XFOIL. Transition at LE. Incidence 12 deg.

In Fig. 4 is shown the streamlines predicted with the NS code at 12 deg. of incidence. It is seen that the flow separates on a considerable part of the airfoil. The same is predicted with XFOIL, Fig. 5. and the geometry of the wake compares very well with the NS prediction.

It can be concluded that for the present airfoil the XFOIL code predictions compare very well with experiment, even at massive separation. Good results have also been obtained for other airfoils of moderate thickness using the XFOIL code[4]. However, convergence problems have been encountered for thick airfoils and it is believed that the NS code will be particular suitable for such airfoils.

3 STEADY AND UNSTEADY ACTUATOR DISC FLOW

The actuator disc is the classical model of the propeller/turbine and combined with the annual stream tube momentum model (ASTM), this is the core in most of the aerodynamic and aeroelastic models used today. The ASTM model works fine for most applications but on the other hand, the assumptions about the flow conditions made in this model give some limitations for its use. The three most important assumptions are; 1) independence between the individual annual stream tubes; 2) steady flow and 3) potential flow (no viscous effects).

The reason for using the NS model on this flow problem is thus to highlight the importance of these assumptions and give basis for possible improvements of the ASTM model.

3.1 Steady Actuator Disc Flow

The actuator disc is modelled by specifying volume forces f_z in the streamwise direction over a short distance. In the present simulations f_z has been specified within one element column with a width of 0.1 R where R is the radius of the actuator disc and equal to 1 (dimensionless). An axisymmetric flow is assumed and the dimensions of the grid are 5R upstream, 8R downstream and 4R in lateral direction as shown in Fig. 6 where the bottom line is the rotor axis. These dimensions have been chosen in order to have a grid where the different variables have approached an almost constant value at the boundaries.

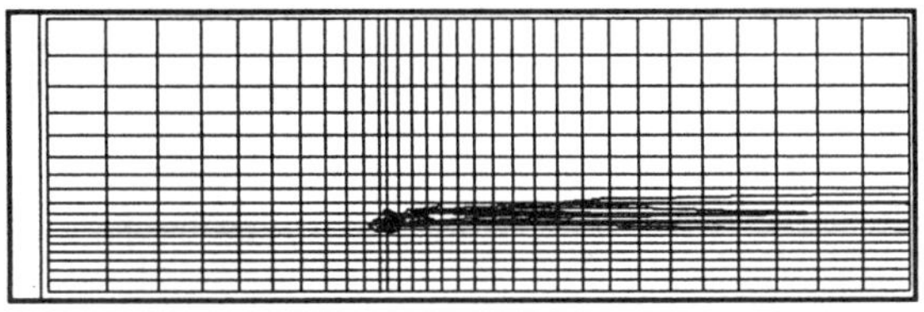

Figure 6. The grid used for the actuator disc simulations superimposed on a plot of vorticity contours.

The simulations have been run at a Re. no of 1000 and for laminar flow. This is considerably lower than the Re. no. for a typical full size rotor which will be around 10^7 when based on the rotor radius. The lower value was chosen in order to avoid convergence problems. However, calculations have been performed at higher Re. no. and it seems that this only has a minor influence on the flow field. Finally, the total number of grid points were 2331 with 63 in the streamwise direction and 37 in the radial direction. The CPU time for this simulation is a few minutes.

The result of a simulation with a constant loading on the disc corresponding to a rotor thrust coefficient of 0.9 is shown in Fig. 7. The parameter shown is the axial velocity component just downstream the disc as function of radius. The ASTM model gives a constant velocity of 0.67 for r<1 but it is seen that the NS code give a velocity that is decreasing towards the tip. The physics behind this is that a concentrated vortex sheet is shed at r=1 as can be seen in Fig. 6. The axial induced velocities from this sheet will be strongest close to r=1 and therefor we get a non-uniform velocity distribution. The reason that the ASTM model is not able to predict this is that with the strong gradient of the loading at the tip, the annual stream tubes cannot be treated completely independent as is assumed in the model.

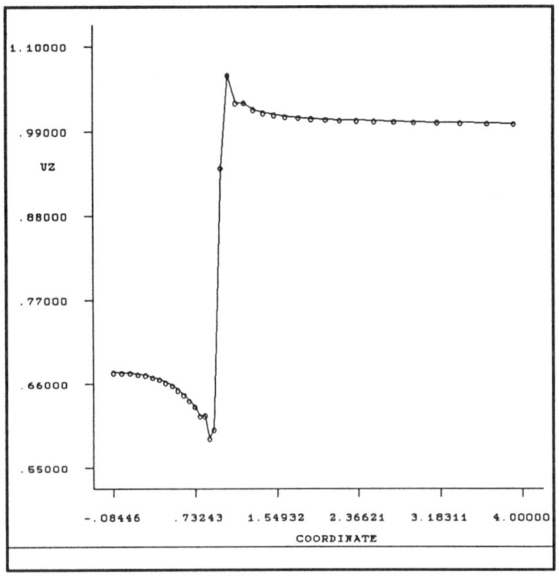

Figure 7. The variation of the axial velocity at the disc as function of radius and for a constant loading of CT=0.9

Now, for calculations on a real rotor the loading does not have the same discontinuity at the tip although the radial gradient can be rather big here. Therefor the differences between the two methods will be smaller but the above results indicate that the loading on the outer part of the blade can be overestimated by the ASTM model. The same tendency was found using an 2D Euler model for simulation of the flow through an actuator disc [5].

3.2 Unsteady Actuator Disc Flow

As already mentioned, one of the assumptions in the ASTM model is steady flow. However, wind turbine rotors operate in a highly unsteady environment due to several reasons. The inflow is turbulent, the rotor disc loading is varying and the alignment of the disc perpendicular to the flow is time dependent.

In the present paper, an example of simulating an unsteady rotor disc loading will be shown. This can occur on a turbine if the pitch is changed within a short time. This flow problem has been studied detailed within the JOULE 1 project, Dynamic Inflow [6].

The result of an unsteady simulation where the rotor disc thrust coefficient is changed in a step from 0.5 to 0.9 and later back from 0.9 to 0.5 is shown in Fig. 8. The axial velocity at the disc and close to the rotor axis (upper curve) and the same velocity component but close to the tip (lower curve) are shown as function of dimensionless time. Comparing the shape of the curves for increasing load and decreasing load, respectively, it is seen that they are different. Assuming an exponential shape, the time constant for increasing load is derived to 3.96 * R/U_0 and 2.56 * R/U_0 for decreasing load. The different time scales for an increasing load and for a decreasing load seems also to be found in the measurements of [6]. However, the above time constants are higher than what is measured and this led us to run calculations at higher Re. no. and it seems that for the unsteady flow, variations of the Re. no. in the interval above 1000 has some influence on the result. At 1 million the time constants became 3.37 * R/U_0 (decreasing load) and 2.92 * R/U_0 (increasing load).

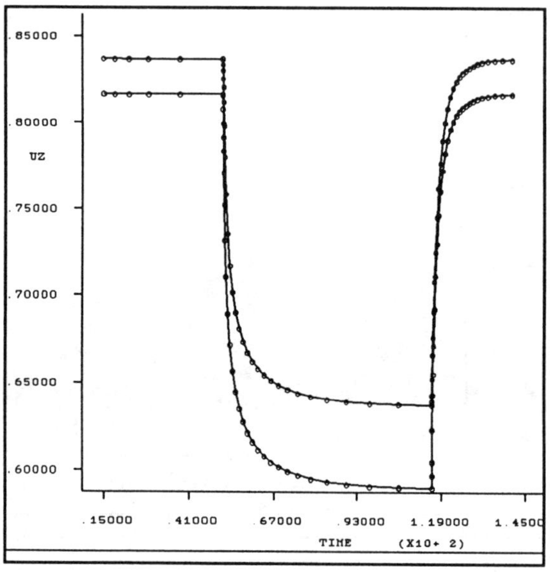

Figure 8 The axial velocity at the disc as function of time for a stepwise change in rotor thrust coefficient from 0.5 > 0.9 > 0.5. Upper curve close to rotor axis and lower curve close to the tip.

3.3 Steady, Turbulent Actuator Disc Flow

The third assumption mentioned above about the flow in the ASTM model was potential flow. For wind turbine array calculations it is well-known that the influence of turbulence is an important parameter but it has also been mentioned that turbulence can have some influence on the flow rate through a single wind turbine rotor [7].

The result of a calculation with turbulence at a rotor thrust coefficient of 0.9 is compared with a similar laminar calculation, Fig. 9. The axial velocity at $r = 0.2$ and as function of the streamwise coordinate is shown for the two cases. Including turbulence the velocity increases again at some distance downstream the disc whereas for the laminar flow the velocity approaches a steady value. It is also seen that there is a small difference between the velocities at the disc with lower value for the turbulent flow. This is different from what was found in another study [7] where the flowrate was slightly increased when the turbulent stresses were included. One explanation could be that in [7] the turbulent stresses were only included in the wake region.

4 CONCLUSION

The application of Navier Stokes solutions to different wind turbine flow problems has been shown. With a CPU time on

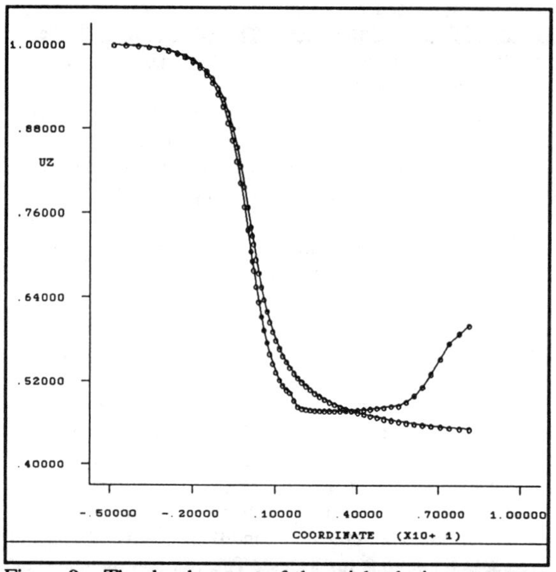

Figure 9 The development of the axial velocity component for a laminar and a turbulent calculation, respectively, at a rotor thrust coefficient of 0.9.

a RISC 6000 workstation ranging from a few minutes for small 2D problems to some hours for bigger 2D problems, NS calculations have been found to be an efficient and flexible tool for the study of complex flow problems. In particular, it seems that the NS calculations can be used as basis for improvement of simpler engineering codes.

REFERENCES

[1] FIDAP Users Manual, Revision 6.0, April 1991, Theoretical part.

[2] Hoadley, D. Wellicome, J.F.- University of Southampton; and Madsen, H.A. - Risø National Laboratory; and Papailiou, K.D. et. al. - National Technical University of Athens. "Airfoil Section Design and Assessment". Final Report on Contract JOUR 0079, CEC, January 1993.

[3] Drela M. "XFOIL: An Analysis Design System for Low Reynolds Number Airfoils", MIT Dept of Aeronautics and Astronautics, Cambridge, Massachusetts. (In: Low Reynolds Number Aerodynamics. Springer-Verlag, Lec. Notes in Eng. 54 1989).

[4] Filippone, A. and Madsen, H.A., "Implementation and Test of the XFOIL code for Airfoil Analysis and Design", Risø-R-644, March 1993.

[5] Madsen, H.A. "Some Results on an Integrated Rotor and Wake Model", Proceedings of the Third IEA Symposium on the Aerodynamics of Wind Turbines, Edited by K.F. McAnulty, June 1990.

[6] Snel, H. and Schepers, J.G. "Joint Investigation of Dynamic Inflow Effects and Implementation of an Engineering Method for Response Calculations", Proceedings Wind Energy R&D Contractor Meeting, Alghero, 3-5 June 1992, pp 181-209.

[7] Madsen, H.A. and Paulsen, U.S. "An Integrated Rotor and Wake Model Compared With Experiment", Proceedings of the European Community Wind Energy Conference,Madrid 10-14 September, 1990. pp. 269-273.

A 3D SHEAR LAYER PREDICTION METHOD FOR STEADY VISCOUS FLOWS AROUND HAWTs

F. Karagiannis, G. Simandirakis
P. Chaviaropoulos, K.D. Papailiou
National Technical University of Athens
Laboratory of Thermal Turbomachines
P.O. Box 64069, 157 10 Athens, Greece

ABSTRACT: The aim of the present work is the development of a 3D shear layer calculation method for the investigation of the viscous flow development along the blades of a Horizontal Axis Wind Turbine. The resulting calculation method is used to assess the radial movements of low energy material along the blade surfaces. The equations used are those of momentum and energy, integrated across the shear layer thickness is deficit form. The viscous calculation involves the solution of a system of 3 integral equations with three working variables chosen among the shear layer quantities. In order to obtain closure, a semi-empirical frame is constructed by employing two different velocity profiles in the streamwise and transverse directions. A second order accurate space marching Runge- Kutta method was chosen for the solution of the resulting system of differential equations, over the surface grid. The method has been validated in several 3D cases and good agreement between theory and experiment was observed. The study of the flow around a HAWT rotor blade is included in the results section.

INTRODUCTION

The numerical simulation of the flow field around a rotating HAWT rotor blade remains a difficult task even in the present days when CFD techniques are pretty matured and the computing power is strongly increased. Most difficulties rise from the modelling of the complicated flow patterns which are due to unsteady effects (triggered by the wind shear and yaw) and/or three-dimensional separation. It is known that the separation in a three-dimensional flow often leads to a formation of vortices in which it is very difficult to recognize a familiar boundary layer. The three-dimensionality adds a further difficulty to the laminar to turbulent transition modelling which is essential in the range of Reynolds numbers that HAWTs usually operate.

In contrast to other applications, like for example the analysis and design of turbomachines, the investigation of the wind turbines aerodynamics is still based on two-dimensional flow models. However, recent experiments performed in RISOE [1] and CRANFIELD I.T. [2] proved that three-dimensionality effects are too important to be ignored. Numerical experiments performed on rotor blades using a 3-D potential code [3] revealed no cross-flow effects (with the exception of the near-tip region). It is therefore believed that the 3-D structure of the flow field is mainly due to viscous effects. This observation motivated the authors to develop a viscous flow solver in order to assess the radial movements of low energy material along a HAWT blade surface. Assuming that there exists a main flow radial movements of low energy material along a HAWT blade surface. Assuming that there exists a main flow direction along which the diffusion is negligible (which is not true in off-design conditions) the Navier-Stokes equations may be reduced to the so-called boundary layer equations, which may be solved either in a differential or in an integral way. The integral approach has been selected here since it is generally much faster. The development of such integral methods for non-rotating shear layers is reported by Smith [4] and, more recently, by Cousteix [5]. All these methods use the integral momentum and entrainment equations to model the streamwise direction and the Johnston's [6] (or Mager) profile for the crossflow.

In the present method the governing equations have been developed at a second order level, as it was felt that they ought to be able to deal with separated flow. On the other hand, a non-orthogonal coordinate system was adopted (such as first used be P.D. Smith [4]), which was judged to be flexible enough, in order to accommodate the geometrical particularities of an horizontal axis wind turbine blade. Equations developed in a rotating frame of reference were used as a basis. Three integral equations were then developed, two momentum and one for the kinetic energy. Contrary to what has been the practice for other 3D published methods, the integral kinetic energy equation was used instead of the entrainment one. According to LTT/NTUA's experience better results were anticipated using it. Coriolis force and solid surface curvature effects upon the turbulent quantities were intended to be taken into account according to the LTT/NTUA practice [7]. On the other hand, the development of the equations was performed using contravariant velocity components. This particular formulation was intended, along with the use of CFD techniques for the numerical integration of the equations, to achieve a fast and numerically robust computational method. For the present case, the existing two-dimensional semi-empirical frame, which was developed for the benefit of the existing 2D counterpart of the presently targeted method was used. In the third dimension simple relations according to Johnston were used, which were found to be acceptable for engineering purposes. The resulting p.d.es are discretized with finite difference schemes and integrated with a second order accurate space marching Runge-Kutta scheme combined with a Newton solver. The method has been validated in several quasi and fully 3D cases including the NLR and ONERA-D infinite swept wings. Good agreement between theory and experiment was observed in all cases. Finally, the method has been applied to the Cranfield HAWT rotor blade under realistic flow conditions.

2. GOVERNING EQUATIONS AND DESCRIPTION OF THE GEOMETRY

The governing equations are the mass, momentum and energy conservation laws developed on a blade surface fitted coordinate system (see fig.1). Let (u^1,u^2) be the surface parametrization using curvilinear coordinates and (g_1,g_2,g_3) the covariant vector base. The surface metric can be then written as:

$$ds^{(2)}=h_1^{(2)}du^{1(2)}+h_2^{(2)}du^{2(2)}+du^{3(2)}+2g\,du^1du^2 \quad (1)$$

where

$$h_1=\sqrt{g_{11}} \quad , \quad h_2=\sqrt{g_{22}} \quad and \quad g=g_{12} \quad (2)$$

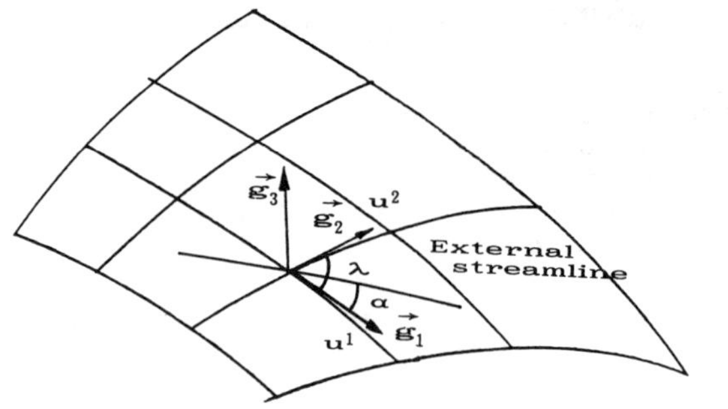

Fig. 1 : Presentation of the surface grid and coordinate system.

and g_{ij} are the metrics tensor components.
The Jacobian of the coordinate transformation takes the form

$$J = h_1^{(2)} h_2^{(2)} - g^{(2)} \tag{3}$$

where superscript (n) denotes n^{th} power. The velocity magnitude may be expressed in terms of the contravariant velocity components as

$$W^{(2)} = (W^1)^2 g_{11} + (W^2)^2 g_{22} + 2W^1 W^2 g_{12} + (W^3)^2 \tag{4}$$

The curvature tensor of the transformed domain which is needed in the present work, may be expressed in terms of the Christoffel symbols of the second kind (Γ_{ij}^k).

In order to develop the equations in their final form it is necessary to make use of the two-zone model concept. It is well known that it is possible to divide the flow field into a viscous part, near the solid boundaries and an inviscid (external) part, which covers the rest of the field. In order to make use of this concept, besides the Real Viscous Flow (RVF), it is necessary to assume the existence of the External Inviscid Flow (EIF) with the condition that at the end of the shear layer, all flow quantities are identical for both flows. Evidently, the (EIF) governing equations derive from the (RVF) ones if the viscous terms are neglected. In order to express the governing equations in deficit integral form one has to substract the (EIF) equations from the corresponding (RVF) and integrate the resulting deficit equations across the shear layer thickness. Defining the displacement (Δ_i), momentum (Θ_{ij}) and energy (Ω_j) thicknesses in the u^i,u^j (i,j=1,2) directions as,

$$\Delta_i = \frac{h_{i_w}}{J_w \rho_{e_w} W_{e_w}} \int (J\rho_e W_e^i - J\rho W^i)\, du^3 \tag{5}$$

$$\Theta_{ij} = \frac{h_{iw} h_{jw}}{J_w \rho_{e_w} W_{e_w}^{(2)}} \int J\rho W^j (W_e^i - W^i)\, du^3 \tag{6}$$

$$\Omega_j = \frac{h_{j_w}}{J_w \rho_{e_w} W_{e_w}^{(3)}} \int J\rho W^j (W_e^i W_e^i - W^i W_i)\, du^3 \tag{7}$$

where subscript (e) denotes external flow conditions (EIF) and subscript (w) solid wall conditions, the governing integral equations in deficit form read (for further details see ref.[8]).

Continuity equation

$$\frac{\partial}{\partial u^1}\left(\frac{J_w \rho_{e_w} W_{e_w}}{h_{1_w}} \Delta_1\right) + \frac{\partial}{\partial u^2}\left(\frac{J_w \rho_{e_w} W_{e_w}}{h_{2_w}} \Delta_2\right) = J_w \rho_{e_w} W_{e_w}^{\ 3} \tag{8}$$

Momentum equation u^i-component (i=1,2)

$$\frac{\partial}{\partial u^i}\left\{\frac{J_w \rho_{e_w} W_{e_w}^{(2)}}{2h_{iw}h_{jw}}\left[\Theta_{ij} + \Theta_{ji} + h_{iw}\frac{W_{e_w}^{\ i}}{W_{e_w}}\Delta_j + h_{jw}\frac{W_{e_w}^{\ j}}{W_{e_w}}\Delta_i\right] + P_{ij} - T_{ij}^{\ t}\right\} - J_w \rho_{e_w} W_{e_w}^{\ i} W_{e_w}^{\ 3} = F_e^i - \Gamma_{jk_w}^{\ \ i}\left\{\frac{J_w \rho_{e_w} W_{e_w}^{(2)}}{2h_{jw}h_{kw}}\left[\Theta_{jk} + \Theta_{kj} + h_{jw}\frac{W_{e_w}^{\ j}}{W_{e_w}}\Delta_k + h_{kw}\frac{W_{e_w}^{\ k}}{W_{e_w}}\Delta_j\right] + P_{jk} - T_{jk}^{\ t}\right\} - J_w \tau^{*i3} \tag{9}$$

(Summation on j and k; j,k=1,2).
The W_{ew}^3 term appearing in (9) is substituted from equation (8).

Energy equation

$$\frac{1}{2}\frac{\partial}{\partial u^1}\left(\frac{J_w \rho_{e_w} W_{e_w}^{(3)}}{h_{1_w}}\Omega_1\right) + \frac{1}{2}\frac{\partial}{\partial u^2}\left(\frac{J_w \rho_{e_w} W_{e_w}^{(3)}}{h_{2_w}}\Omega_2\right) = -\frac{1}{2}J_w \rho_{e_w} W_{e_w}\left[\frac{\Delta_1}{h_{1_w}}\frac{\partial}{\partial u^1}(W_{e_w}^{\ k} W_{e_{w_k}}) + \frac{\Delta_2}{h_{2_w}}\frac{\partial}{\partial u^2}(W_{e_w}^{\ k} W_{e_{w_k}})\right] - \bar{P}_1 - \bar{P}_2 - \bar{P}_3 - \bar{F}_1 - \bar{F}_2 - \bar{F}_3 + \int J\tau^{*31}\frac{\partial W_1}{\partial u^3}du^3 + \int J\tau^{*32}\frac{\partial W_2}{\partial u^3}du^3 - \int W_1 [A]\, du^3 - \int W_2 [B]\, du^3 - \int W_3 [C]\, du^3 \tag{10}$$

In the above equations the terms p_{ij}, T_{ij}^* and T_{ij}^t represent pressure and stresses integrals which are suitably modeled, representing the second order corrections. The p_i,F_i terms in the energy equation count for the external flow pressure influence and the external forces (Centrifugal and Coriolis).

The [A], [B] and [C] terms in the energy equation include the effect of turbulent stresses, W_i are the covariant velocity components and the τ^{*13}, τ^{*23} terms represent the wall shear stress values which are proportional to the skin friction coefficients in the u^1,u^2 directions, respectively.

3. THE ROTATION OF THE COORDINATES

In the general case the external flow direction (defining the streamwise flow) does not coincide with any of the coordinate directions, but forms an angle (α) with the u^1-direction (see fig.1, where (λ) is the angle between the coordinate lines). If (s_1,s_2) is the local streamwise coordinate system where s_2 represents the crossflow, the velocity components in the grid and streamwise systems may be interrelated through a rotation matrix M_{ij}, as

$$W_{sj} = W^i M_{ij} h_i \tag{11}$$

where the M_{ij} matrix elements are trigonometric functions of angles (α) and (λ).

The definition of the displacement (δ_i), momentum (θ_{ij}) and energy (ω_{ij}) thicknesses in the streamwise system, are similar to those of the coordinate system (eqs 5-7), where, now the streamwise and cross-flow velocity components substitute the contravariant velocities.

The (s_1,s_2) system integral thicknesses may be related to the (u^1,u^2) system thicknesses through the rotation matrix. The same holds for the wall shear stresses.

4. CONSTRUCTION OF THE 3-D SEMIEMPIRICAL FRAME

It is important to note that the physical flow model demands the reference direction coinciding with the external flow direction inside the shear layer. This direction is a privileged one in that the turbulent flow semi-empirical information, which is valid in the 2-D case, is applicable along it. In the present formulation the s_1 component of the shear layer velocity will be assumed to follow that of a 2-D shear layer and is described by the turbulent velocity profile of Kuhn and Nielsen [9]. The laminar part is described by the Hartree profile while a transition criterion based on Re_θ accounts for laminar to turbulent transition. The transverse profile is assumed to follow the Johnston's triangular law.

$$W_{s_2}=\tan\beta\, W_{s_1} \qquad \text{near the wall}$$
$$W_{s_2}=\tan\gamma\,[W_{e_v}-W_{s_1}] \qquad \text{over the rest part of the shear layer} \qquad (12)$$

where the so called limiting angle (β) is related to $\tan\gamma$ through the skin friction coefficient and the external flow Mach number.

The three independent variables chosen in the present work are the momentum thickness θ_{11} and the shape factor H for the streamwise direction and $\tan\gamma$ for the crossflow. The shape factor H is defined as

$$\bar{H}=\frac{\frac{1}{\rho_{e_v}W_{e_v}}\int \rho\,(W_{e_v}-W_s)\,du^3}{\theta_{11}} \qquad (13)$$

Any shear layer property may be then expressed in terms of $(\theta_{11},H,\tan\gamma)$ including the thicknesses $(\delta_i,\theta_{ij},\omega_{ij})$ and, consequently, the thicknesses $(\Delta_i,\Theta_{ij},\Omega_i)$. With than in mind, the governing equations may be written in symbolic form as

$$\frac{\partial\vec{F}(\vec{x})}{\partial u^1}+\frac{\partial\vec{G}(\vec{x})}{\partial u^2}=\vec{S}(\vec{x}) \qquad (14)$$

where, for example $F=(F_1,F_2,F_3)^T$ and $x=(\theta_{11},H,\tan\gamma)^T$.

5. NUMERICAL INTEGRATION

The numerical integration of the initial value problem (14) is performed using a second order accurate Runge-Kutta space marching scheme. Initial conditions are assigned to x at the flow inlet plane ($u^1=u^1_{min}$). During the Runge Kutta steps the following non-linear system has to be solved for x at each u^1=const. grid section

$$\vec{F}(u^1_{(i)},\vec{x})=F^* \qquad (15)$$

The above equation must be solved twice during each step of the Runge-Kutta Scheme and the values of $(\theta_{11},H,\tan\gamma)$ at a grid point are found after the computation of the flux vector F^*. Equation (15) is solved using a fast Newton iterative procedure in which the Jacobian $\partial F/\partial x$ has to be analytically known. The use of Johnston's profile in the transverse direction permits the analytical expression of the above Jacobian matrix. The space marching is done along the u^1-direction and the solution is found simultaneously for all the grid points along the u^2-direction. The discretization of the $\partial/\partial u^2$ term is performed using a centered or upwind scheme depending on the local characteristics of (14).

6. RESULTS AND DISCUSSION

The 3D sher layer code was validated against experiment for several infinite swept wing cases. Computational results are given here for the ONERA-D section infinite swept wing and the infinite swept wing of NLR. A rotating HAWT blade has been also studied under realistic flow conditions.

6.1 The Onera-D section infinite swept wing

This test case has been studied by Pailhas [10]. It concerns an infinite swept wing of 22.5° of sweep. The airfoil section of the model is an ONERA-D profile. A freestream velocity of 35m/s is used at a normal incidence of 8° and a chord of 200mm. The results shown in fig.(2) were obtained using the experimental velocity distribution. A generally good agreement is obtained, between theory and experiment. In the same figure the results obtained by PAILHAS are also presented.

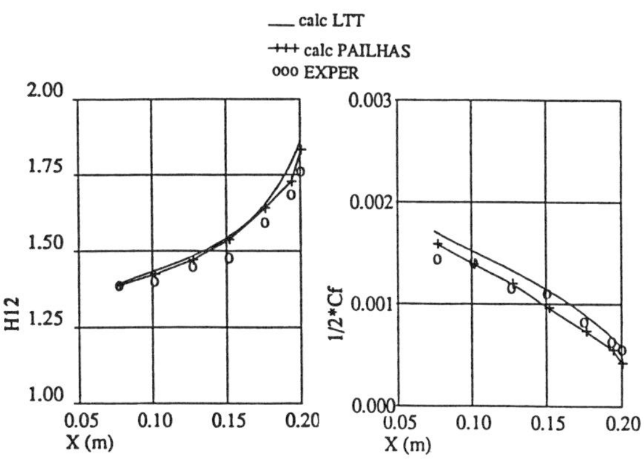

Fig. 2 : Numerical results versus measurements for the ONERA D infinite swept wing experiment

6.2 Hawt Rotor Blade

The 3-D shear layer code has been used to calculate the flow around the Cranfield 2.84 m diameter three blade rotor, the geometrical characteristics of which are described in detail in ref.[11]. This rotor has been selected among others because of its relevant simplicity but mainly because of the existence of full scale experimental measurements obtained by Hales and Bellia [2]. The experimental data base includes pressure measurements along specific cross cuts of a rotor blade in various combinations of wind speed and rotational speed, which could be used for the evaluation of the present results.

The external (inviscid) flow field around the Cranfield rotor has been computed with the full potential 3-D code of ref.[3]. The code run for several combinations of rotational speed and wind speed in order to produce the inviscid flow field which supports the viscous flow solver. From them the case with ω=334.3 rpm and wind speed 5 m/s (EXP 164) with zero yaw angle and pitch setting 8.0° was selected for the viscous flow computation. Results obtained by a 2-D and a 3-D potential code indicate that the 3-D (radial) effects are rather poor on the inviscid field. The comparison with the experimental results (which have been averaged in terms of their azimuthal fluctuations), however, was fair at 75% where the geometric incidence is 0.6° but very bad at 35% (geometric incidence 9.2°). The large discrepancies of the pressure field in the near hub region is probably due to massive 3-D separation in this region.

The computed shape factor H contours along the blade pressure and suction side are shown in figures 3a and b respectively. The H minimum and maximum pressure side values are 1.429 and 1.904 respectively, showing that the flow is attached there. The acceleration of the flow in the near leading edge region of the pressure side reduces H in that region. The H values increase gradually after the 50% blade chord line. The same patterns are observed on the suction side, as well, with the exception of the trailing edge region. In this region H increases rapidly because of the flow deceleration and turbulent separation appears at the hub at 40% of the chord (H=2.56). The separated region is then extended in the radial direction. Since the present code is unable to pass through severe separation, the suction side computation has been performed up to the onset of the separation bubble. The $\tan\gamma$ contours which characterize the development of the transversal profile are shown in figure 4. It can be shown that $\tan\gamma$ takes minimum values near separation.

Since no viscous-inviscid interaction procedure is employed, the above results are qualitatively meaningful but not quantitatively accurate. Nevertheless they demonstrate the mechanism of the separation bubble onset as well as the development of this bubble in the radial direction.

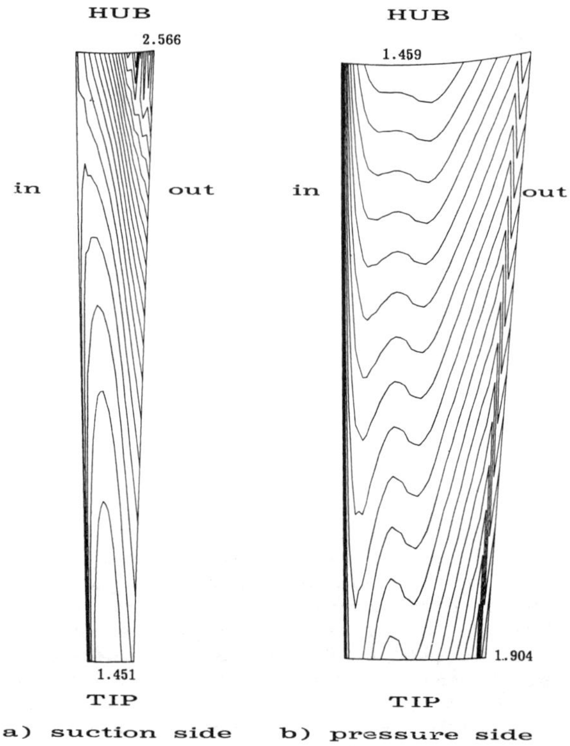

Fig. 3 : Shape factor (H) contours on the Cranfield HAWT blade
a) Suction side b) Pressure side

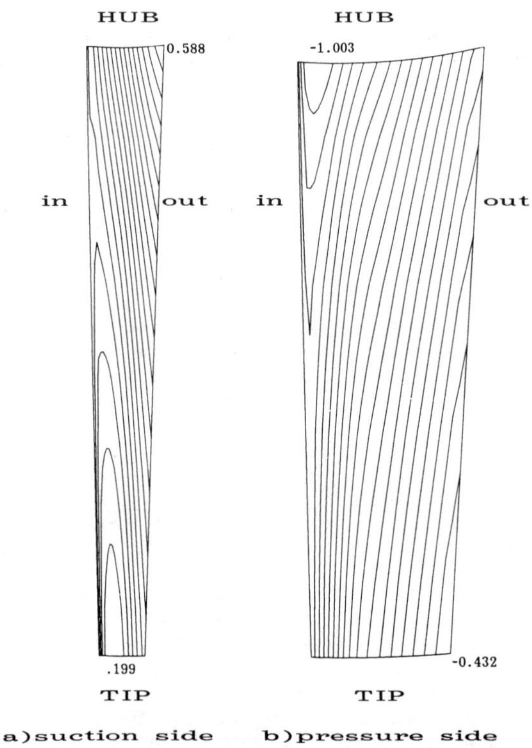

Fig. 4 : Transverse profile angle ($\tan\gamma$) contours on the Cranfield HAWT blade a) Suction side b) Pressure side

7. CONCLUSIONS

A 3-D integral shear layer solver has been developed for the investigation of the viscous effects on a rotating HAWT blade. The method solves the integral momentum and energy equations in deficit form on non-orthogonal surface grids. The 3-D semiempirical frame, employed for closure, is based on the 2-D Kuhn and Nielsen profile for the mainstream direction and the Johnston's triangular profile for the crossflow. Modern CFD techniques are used for both the development of the equations and their solution procedure. The method has been validated using infinite swept wing experimental data. The study of the flow around a HAWT rotor blade is also presented.

REFERENCES

[1] H.A. Madsen, "Measured Airfoil Characteristics of Three Blade Segments on 19 m HAWT Rotor", Progress report on Contract EN3W-0032-DK(B), RISOE National Laboratory, October 1989.

[2] J.M. Bellia, R.L. Hales, "Mobile Test Results for the Fully-Instrumented Rotor", Cranfield Institut of Technology, Final Report for CEC, DGXII, Contract No EN3W/0033/GB.

[3] B. Bouras, E. Bonataki, G. Kakouros, P. Chaviaropoulos, K.D. Papailiou, "Aerodynamic Investigation of Horizontal Axis Wind Energy Machine Rotors Using Advanced Computational Tools", Proc. of the EUROFORUM - New Energies Congress, Volume 3, 24-28 October 1988, Saarbrucken, Germany.

[4] P.D. Smith, "An Integral Prediction Method for Three-Dimensional Compressible Turbulent Boundary Layers", RAE Techn. Report, 72228, 1972.

[5] J. Cousteix, "Three-Dimensional Boundary Layers. Introduction to Calculation Methods", Paper # 1, AGARD R-741 (1986).

[6] J.P. Johnston, "On the Three-Dimensional Turbulent Boundary Layer Generated by Secondary Flow", Journal of Basic Engineering - Series D., Trans. ASME, Vol. 82, pp. 233-248 (1960).

[7] K.D. Papailiou, B. Bouras, "Arbitrary Blade Section Design Based on Viscous Considerations", von Karman Institute, Lecture Series on "Inverse Methods in Airfoil Design for Aeronautical and Turbomachinery Applications", VKI LS (1990), AGARD-R-780.

[8] F. Karagiannis, P. Chaviaropoulos, K.D. Papailiou, "The Development of a 3D Shear Layer Prediction Method, Adapted to the Steady Viscous Flow Along the Surfaces of a Rotating Horizontal Axis Wind Turbine Blade", Final Report prepared for the EEC Contract JOUR 0019-C(CD).

[9] J.D. Kuhn, J.N. Nielsen, "Prediction of Turbulent Separated Boundary Layers", AIAA No 73-663, July 1973.

[10] G. Pailhas, "Couche Limite et Sillage Turbulents for Dimensionnels - Champ Moyen -- Tensions de Reynolds", Memotre CNAM (1979).

[11] R.H. Hales, J.M. Bellia, "Aerodynamic Measurements on a Small Hawt Rotor in Axial and Yawed Flow", Proc. of European Community Wind Energy Conference, Madrit, Spain 10-14 Sept. 1990, pp. 213, 217.

SECTIONAL PREDICTION OF 3-D EFFECTS FOR STALLED FLOW ON ROTATING BLADES AND COMPARISON WITH MEASUREMENTS

H. Snel
Netherlands Energy Research Foundation,
P.O. Box 1,
1755 ZG Petten, NL

R. Houwink, J Bosschers and W.J. Piers
National Aerospace Laboratory NLR,
P.O. Box 90502,
1006 BM Amsterdam, NL.

G.J.W. van Bussel and A. Bruining
Delft University of Technology,
Faculty of Civil Engineering,
Institute for Wind Energy
Stevinweg 1,
2628 CN Delft, NL.

Summary.

A method is described to evaluate the first order effects of blade rotation on stall characteristics, by way of the solution of a simplified form of the 3D boundary layers equations. The local blade solidity, c/r, is shown to be the most important parameter for the deviations from the 2D stall behaviour. Results are matched with wind tunnel measurements of pressure distributions on rotating blades. Subsequently, the results are translated into a very simple correction formula, that can be used by designers to obtain an estimate of 3D profile characteristics, at different radial stations. Finally, results obtained by application of this formula are compared with measurements on the DUT test rotor, and with the power curve of a commercial wind turbine.

1. INTRODUCTION

Both for performance calculations of wind turbine rotors and for the aero-elastic analysis of the blades, aerodynamic forces are frequently calculated by strip theory. Lift and drag forces are then determined using two dimensional airfoil data, either obtained by separate calculations or from wind tunnel measurements at the appropriate Reynolds number. This procedure gives reasonable results for attached flow, but it is known to become unreliable for separated or partially separated flow conditions. One of the first researchers to address this problem was Himmelskamp [1], who noted the presence of very high lift coefficients on the inboard sections of a propeller blade. Much more theoretical and experimental work has been done since on 3D effects in flows over rotating blades. The main results are briefly, but by no means completely, reviewed in this introduction.

Although there may be some 3D potential flow effects, especially decambering (Woods [2]), these can be shown to be small for high aspect ratio blades, following the analysis of Sears [3] and Fogarty and Sears [4]. Also, Fogarty [5] showed 3D (i.e. cross flow) effects to be small for the attached laminar boundary layer and his arguments can be extended straightforwardly to the turbulent boundary layer case. He suggests that the observed deviations from 2D behaviour might occur in the separated flow region, to which his analysis does not extend. The same is suggested by McCroskey [6], on the basis of measurements and flow visualization on a rotating blade. McCroskey notes no appreciable change in the transition or separation location, but observes significant radial flow in separated flow regions, even in laminar separation bubbles. In a much more recent work [7], Sørensen numerically solved the 3D equations for the boundary layer on a rotating surface in strong interaction with the potential outer flow. His results are in good agreement with those mentioned before in that the position of the separation line along the blade is basically the same as for 2D predictions, but in regions with pronounced separation, there is a large difference between the lift coefficients calculated from the 2D and 3D flow fields respectively. Recent measurement on rotating wind turbine blades have shown the importance of these effects in stall, especially on inboard sections of the blade [8,9,10]. Finally, Narramore and Vermeland [11] recently published a numerical Navier Stokes study of the flow on a helicopter blade, also reaching much higher than 2D lift forces. However they predict effective postponement of stall on the rotating blade.

This investigation is funded by the Netherlands Agency for Energy and the Environment NOVEM.

A method that has been used by designers to correct the 2D airfoil characteristics in stall is the one proposed by Viterna and Corrigan [12]. However, in this method modification of the lift coefficient is based on measurements of the power output in stall, which contains only a small part of the lift, from which a large part of the profile drag must be subtracted. The method proposed in the present paper is based on direct comparison with measured normal forces, which have become available since.

The physical explanation of the differences between 2D stall and 3D separation on rotating blades, can be formulated as follows: significant radial flows only develop in separated flow regions, as will be discussed in the present paper. Flow towards the tip develops at the suction side and results in a Coriolis force in main flow direction, acting as a favourable pressure gradient. This decreases the displacement thickness of the separated boundary layer, so that there will be less decambering and higher lift coefficients.

Section 2 discusses the results of an order of magnitude analysis of the different terms in the 3D boundary layer equations for the rotating blade. This gives rise to a set of equations that is much simpler than the full 3D equations: the (non-linear) convective terms can be used in two dimensional form, while the important effects of centrifugal and Coriolis forces are retained. Solution of this set is much simpler than that of the full 3D equations, as in [7] or the 3D Navier Stokes equations as in [11]. Moreover, such a formulation is consistent with the strip approach used by the blade element momentum theory.

The results of the implementation of the method in a 2D strong interaction b.l. computational model are discussed in section 3. Calculated data are compared with measurements on a rotating turbine blade in the controlled environment of a large wind tunnel [9]. Possible causes for the discrepancies are discussed, and an empirical matching factor is determined. Results obtained with this factor are translated into a simple formula to obtain an estimate for the 3D profile characteristics in stall, on the basis of the 2D data and the local chord to radius ratio. Finally, the formula is applied to the configuration of [10] and results are compared with preliminary results of field measurement.

2. THE BOUNDARY LAYER EQUATIONS

2.1 Formulation of the equations

For the formulation of the equations, cylindrical coordinates rotating with the blade will be used, with the origin in the centre of rotation, as shown in figure 1. $\{u,v,w\}$ are the velocity components in directions $\{\theta,r,z\}$

respectively. Undisturbed free streamlines are circular arcs r = constant. The 3D boundary layer equations for stationary incompressible flow with respect to these co-ordinates can be formulated as follows (see e.g. [13]):

continuity:

$$\frac{1}{r}\frac{\partial u}{\partial \theta} + \frac{v}{r} + \frac{\partial v}{\partial r} + \frac{\partial w}{\partial z} = 0 \ , \quad (1)$$

θ-momentum (2):

$$\frac{u}{r}\frac{\partial u}{\partial \theta} + v\frac{\partial u}{\partial r} + w\frac{\partial u}{\partial z} =$$

$$- \frac{1}{\rho r}\frac{\partial p}{\partial \theta} + \frac{1}{\rho}\frac{\partial \tau_1}{\partial z} + 2\Omega v - \frac{uv}{r} \ ,$$

r-momentum (3):

$$\frac{u}{r}\frac{\partial v}{\partial \theta} + v\frac{\partial v}{\partial r} + w\frac{\partial v}{\partial z} =$$

$$- \frac{1}{\rho}\frac{\partial p}{\partial r} + \frac{1}{\rho}\frac{\partial \tau_2}{\partial z} + \frac{(u-\Omega r)^2}{r} \ .$$

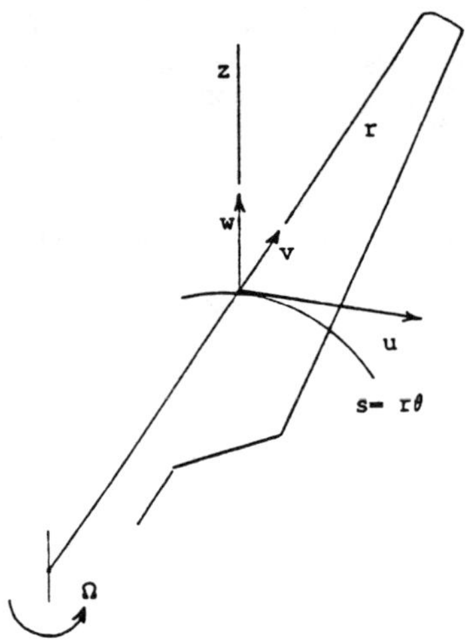

Figure 1. Coordinate system attached to rotating blade

It has been assumed that the surface on which the boundary layer develops is contained in the plane of rotation z=0; surface curvature in chord-wise direction has been neglected, with respect to the boundary layer thickness. Ω is the rotational speed (about the z-axis) with respect to an inertial system, while r is the distance to the centrum of rotation. τ_1 and τ_2 are shear stresses in θ and r direction respectively, which can be either laminar (viscous) or turbulent (Reynolds) stresses.

The momentum equations contain a number of 'inertia forces' type terms. The θ-momentum equation contains the Coriolis force term $2\Omega v$ and a co-ordinate curvature term uv/r. The r-momentum equation only contains the 'total' centrifugal force term resulting from the total circumpolar velocity $(u-\Omega r)$ in the inertial frame. This term is composed however of what formally are a co-ordinate curvature term (u^2/r), a Coriolis force component $(-2u\Omega r)$ and the 'formal' centrifugal force $\Omega^2 r$ resulting from the rotation of the co-ordinate system.

2.2 Discussion of the equations

In [14] an order of magnitude analysis of the different terms in the boundary layer equations is presented that enables the identification of important parameters. Only the results will be discussed here.

Geometrical scaling is done by means of the blade chord c for the θ dimension and by the value of radius r for the r-dimension. It should be kept in mind that, inherent to the boundary layer approximation, it has already be assumed that z scales with the boundary layer thickness δ, reason for which there is no z-momentum equation and the pressure p is approximated as independent of z.

For attached flow conditions, the following scaling rules apply:

$$u = O(\Omega r), \qquad v \div O(\Omega c), \qquad p - p_\infty = O(\rho\Omega^2 r^2) \quad (4)$$

if ρ is the air density. In the v scaling, it has been assumed that radial flow is driven by the radial pressure gradient and the centrifugal force, which are both $O(\rho\Omega^2 r)$.

Consequently:

- the v/r and $\partial v/\partial r$ terms in the continuity equation (1) are of $O(c/r)^2$ compared to the other two terms;
- all terms containing v in (2), viz: $v\partial u/\partial r$, uv/r and $2\Omega v$ are of $O(c/r)^2$ compared to the remaining terms;
- the leading terms in the r- momentum equation are of O(c/r) compared to the leading terms of the θ-momentum equation.

Hence to first order accuracy in c/r, the chord wise flow is not coupled with the radial flow and the situation is essentially two dimensional. If desired, the radial flow can be solved from (3) after substituting the solution of (2) for u. (3) is then a linear equation for v. This same result was obtained by Fogarty [5].

In a separated flow region, the situation is different however. Now the fluid is essentially transported with the blade, so that:

$$u \ll \Omega r \quad (5)$$

The chord wise pressure gradient is small compared to its value in attached flow, in the 2-D situation it would be essentially zero. It will then be assumed that the Coriolis force, possibly together with the pressure gradient and the shear stress, becomes the driving term for the chord wise acceleration. The radial accelerations are still governed by centrifugal forces and radial pressure gradients. As shown in [12], the following scaling rules are then valid:

$$u = O(\Omega r^{1/3}c^{2/3}),\ v = O(\Omega r^{2/3}c^{1/3}) \text{ and } v/u = O(r/c)^{1/3} \quad (6)$$

Note that now the chord wise velocity component u is small compared to the radial velocity component v for a high aspect ratio blade.

Introduction of all scaling rules into eq's (1), (2) and (3) gives the following results:

- the terms containing v in (1) are of $O(c/r)^{2/3}$ compared to the u and w derivatives.
- the radial convective acceleration terms in (2) and (3) are of $O(c/r)^{2/3}$ compared to the chord wise and z convective acceleration terms.
- as far as inertia forces are concerned in (2) and (3), only the Coriolis term in (2) and the $\Omega^2 r$ term in (3) are of the same order of magnitude as the leading terms.

Neglecting the $O(c/r)^{2/3}$ terms for a high aspect ratio blade, and in conjunction with the results for the attached flow regime, a set of equations is obtained that is valid for both attached and separated flow, to leading order in c/r. This set differs from (1), (2) and (3) in the neglect of the v-terms in the continuity equations, of the radial convective acceleration in the chord wise and radial momentum equations, and of the coordinate curvature term uv/r in the chord wise momentum equation.

The important aspect of these equations is that they are 2D in the convective terms, even the radial momentum equation. They can be solved by 2D b.l. methods, adding the Coriolis term to the chord wise momentum equation, and the additional radial momentum equation.

2.3 Implementation in the NLR ULTRAN-V program

As a first attempt to study the effects of the 3D terms described above, the equations have been implemented in an existing computer program, viz the NLR ULTRAN-V code. This program is based on the transonic small perturbation (TSP) modelling for the outer flow. The boundary layer equations are solved by an integral method, based on Green's lag entrainment model for the turbulent part. Transition to turbulence can be obtained as part of the solution based on the Granville criterion; it can also be fixed by the user at a certain position. If laminar separation is encountered, transition is enforced, so that no laminar separation bubbles are modelled. This in view of the fact that the calculated pressure gradient near the leading edge is not very realistic, because of the leading edge singularity inherent to small perturbation theory. The turbulent boundary layer equations are solved in strong interaction with the outer flow, so that realistic results can be obtained in modest separation. The method is described in more detail in [15].

Details of the implementation of the 3D extension in ULTRAN-V, including cross flow velocity profile assumptions for the additional momentum equation, are described in [14]. Here it will suffice to say that the integral equations, in terms of displacements thicknesses and momentum thicknesses, are cast in non dimensional form. The specific 3D terms then require an additional non dimensional parameter, which is c/r. For a given blade geometry, the sectional characteristics can be calculated at different stations along the span, by using the relevant value of c/r as input.

3. COMPARISON OF CALCULATED RESULTS WITH MEASUREMENT

3.1 FFA Wind tunnel measurements in CARDC

The modified code was used to perform computations of sectional air loads and pressure distributions for a number of sections of a HAT tested by FFA in the 12*16 m CARDC tunnel [9], having NACA 44xx sections. Calculations were carried out for c/r values of .37, .16 .09 and 0, pertaining to the r/R locations of 30%, 55%, 75% and to 2D conditions respectively. Measured results at these sections were published in [11]. For the calculations only the Ωr effect on the radial pressure gradient was accounted for, assuming no radial dependence of the pressure coefficient c_p. Results of this calculations was published and discussed in [14].

In fact, the calculated results qualitatively show the measured features of stall on a rotating blade very well, both in increased $c_{l,max}$ and in the shape of the pressure distribution. However, the effects are quantitatively exaggerated: measurements at a certain value of c/r agree well with calculations at a much lower value of c/r. To identify the possible causes for this discrepancy, a number of points could be addressed in future investigations:

- inclusion of the variation of pressure coefficient in radial direction in the $\partial p/\partial r$ term.
- accounting for the radial flow by including a source term to represent $\partial(rv)/\partial r$ in the right hand side of the continuity equation.
- likewise in the momentum equations by including transport terms as source terms in the rhs.
- implementation of the 3D terms in a 2D code without the limitations of the TSP approach, such as XFOIL [16]

In view of the qualitative agreement, for the time being a very pragmatic approach has been used. In order to match the calculated and measured lift curves for the FFA experiments, a matching constant of approximately 2.5 on the value of c/r is needed. That is, calculated results for c/(2.5 r) match well with measurements at c/r. Although this quite unsatisfactory from a modelling point of view, it can be of use to the designer who has to estimate power output and blade behaviour in stall. Figure 2 shows the agreement obtained by applying the matching factor.

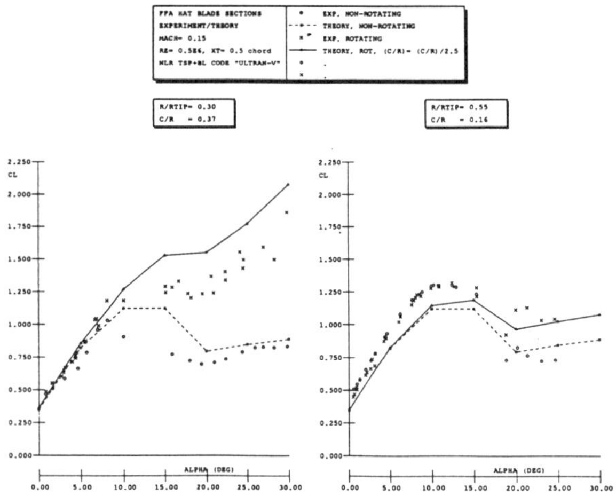

Figure 2. Result of matched calculations with ULTRAN-V compared to FFA CARDC measurements.

4. SIMPLE CORRECTION FROM 2D TO 3D AIRFOIL DATA

With the methods described above, 3D airfoil data can be estimated for different span wise locations with the aid of ULTRAN-V calculations. However, for the designer without access to the program, it would be very helpful if a simpler correction method could be used. In order to obtain such a method, a relatively good fit to the calculated results for the NACA 44xx section was found (by NLR) in the following way (see also figure 3).

Let $c_{l,lin} = dc_l/d\alpha \ (\alpha - \alpha_0)$

where α_0 is the angle of attack at zero lift, and $dc_l/d\alpha$ is slope of the lift curve. For the last quantity, the theoretical value of 2π/rad can be used, or a slightly lower value taken from 2D measurements at the appropriate Reynolds number. Next, let

$$\Delta c_l = c_{l,lin} - c_{l,2\text{-}D},$$

i.e the difference between the extrapolation of the linear lift relation and the measured 2-D lift coefficient. This quantity (a function of α) is used by a large number of semi-empirical methods for dynamic stall prediction. A good fit to the ULTRAN-V calculated 3D lift coefficient is obtained by:

$$c_{l,3\text{-}D} = c_{l,2\text{-}D} + 3\,(c/r)^2\,\Delta c_l \qquad (7)$$

Note that the multiplier of Δc_l should not be greater than 1, since then the inviscid limit is reached. In view of this, tanh(c/r) may be a good alternative for the multiplier. More excercices with the factor should be done to make a final

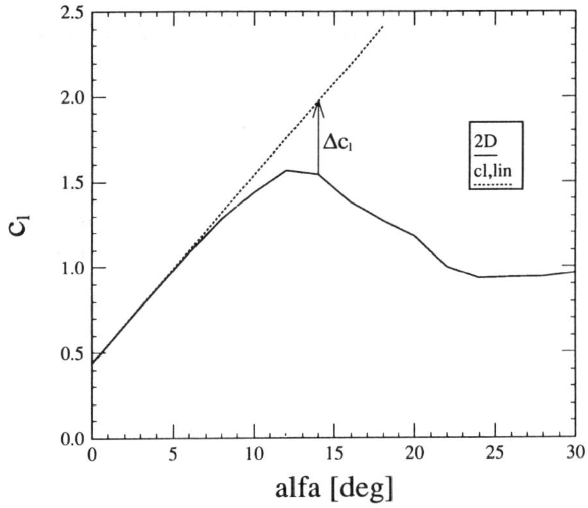

Figure 3. Definition of Δc_l.

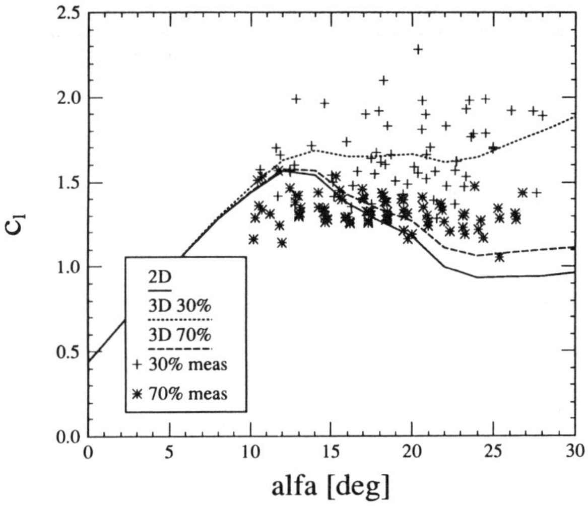

Figure 5. Comparison of results of correction formula with DUT field measurements.

choice.

In figure 4, the results of applying the formula to the NACA 4418 profile is compared with the result of the FFA CARDC measurements.

It should be emphasized that the result is of a preliminary nature, as it has been validated for the NACA 4418 profile only. It is quite possible that the multiplier of Δc_l is profile and Reynolds dependent. In general, one must also expect effects of the blade planform (taper and twist), which is not reflected in the formula. Nevertheless, application of the formula to a number of cases, shown below, show encouraging results, which have tempted the authors to propose the formula at this stage.

5. FURTHER COMPARISONS

5.1 The open air measurements of DUT

In order to establish the more general usefulness of the formula mentioned above, it was applied to the NLF 416 profile of the DUT test rotor, described in [10]. Results are shown in figure 5, where the calculated 3D lift coefficient for the 30% and 70% sections are compared with the 2D data. First measurements obtained at the open air experiments on at the DUT (described in detail in [10] although showing quite a lot of scatter, are also shown in the figure for the 30% section and the 70% section. The angle of attack is not a quantity that can be measured directly. For the results presented, it has been inferred from the measurements in the following way. The normal force on the section (chord) F_n can be measured accurately, while the incoming wind speed is measured upstream of the hub. The induced velocity u_i, as used in blade element momentum theory is related to the total axial force in the relevant 'annulus' through:

$$4 \pi \rho u_i (U_w - u_i) = F_{ax} = 2 F_n \cos\theta_p \qquad (8)$$

where θ_p indicates the blades pitch angle and the multiplication by 2 reflects the number of blades. With known 'induced velocity', the angle of attack and lift coefficient can be obtained.

Values derived from measurements for smaller angles of attack (< 10 degrees) are not shown in the figure. The flow conditions are then such that a 'turbulent wake' is reached and (8) does not give sensible results. No final conclusion about the agreement is possible, due to the large amount of scatter, but the results are certainly not in contradiction. More work is necessary to determine if the experimental scatter (which is also seen at higher tip speed ratios and hence smaller angles of attack) is due to lift hysteresus or other causes.

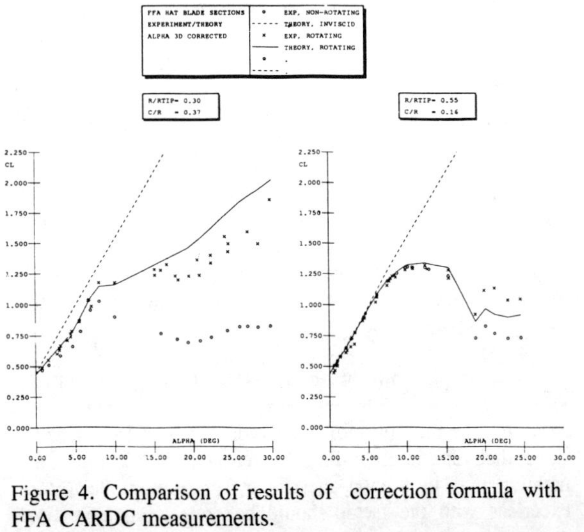

Figure 4. Comparison of results of correction formula with FFA CARDC measurements.

5.2 The Nordtank 300 kW wind turbine

To see the effects of the 3D corrections on the calculated performance curve of a real wind turbine, a further application was made to the Nordtank 300 kW, stall

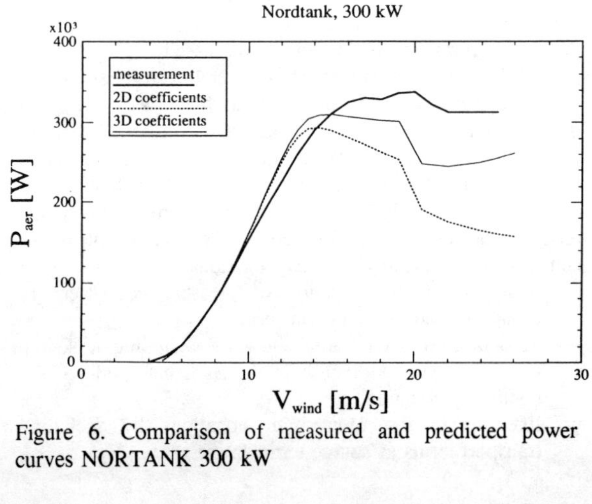

Figure 6. Comparison of measured and predicted power curves NORTANK 300 kW

regulated machine, using NACA 63-2xx profiles. Figure 6 compares the measured P-V curve with calculated ones, using 2D and 3D corrected profile data respectively. As can be seen, quite an improvement is obtained with the correction.

6. CONCLUSIONS

A method is presented to model the dominant radial flow effects on the aerodynamic characteristics in stall on a rotating blade. The method is implemented in an existing code, and calculational results are compared with measuremetns. Although only qualitative agreement is obtained by straightforward calculations, with the use of an empirical matching factor, quantitative agreement can be much improved. Finally, the results are fitted to a simple, algebraic method to obtain 3D lift coefficients for different locations on a wind turbine blade. Although the proposed method is not extensively validated, first results are promising.

REFERENCES

[1] H. Himmelskamp: Profile investigations on a rotating airscrew, Doctoral Thesis, Göttingen, 1945

[2] D.H. Wood, A three-dimensional analysis of stall-delay on a horizontal-axis wind turbine, Journal of Wind Engineering and Industrial Aerodynamics, 37, pp.1-14 (1991).

[3] W.R. Sears: Potential flow around a rotating cylindrical Blade. J.Aero.Sci., vol.17, no.3, pp.183-184, March 1950.

[4] L.E. Fogarty and W.R. Sears: Potential flow around a rotating, advancing cylindrical blade. J.Aero.Sci., vol.17, no.9, p.599, September 1950.

[5] L.E. Fogarty: The laminar boundary layer on a rotating blade. J.AeroSci., vol.18, no4., pp.247-252, September 1951.

[6] W.J. McCroskey: Measurements of boundary layer transition, separation and streamline direction on rotating blades. NASA TN D-6321, April 1971

[7] J.N. Sørensen: Prediction of three-dimensional stall on wind turbine blade suing three level, viscous-inviscid interaction model. Proceedings of the EWEC 1986, European Wind Energy Association Conference and Exhibition, Rome, October 7-9 1986.

[8] Bellia and Hales, Cranfield Institute of Technology

[9] G. Ronsten: Static pressure measurements in a rotating and a non-rotating 2.35 m wind turbine blade. Comparison with 2D calculations. Proceedings of the EWEC '91 conference, Amsterdam, October 1991.

[10] A. Bruining: Pressure measurements on a rotating wind turbine blade on the Open Air Rotor Research Facility of the Delft University of Technology. ECWEC 1993 Conference, Travemunde, March 1993.

[11] J.C. Narramore and R. Vermeland: Navier-Stokes Calculations of Inboard Stall Delay Due to Rotation.

[12] L.A. Viterna and R.D. Corrigan: Fixed Pitch Rotor Performance of Large Horizontal Axis Wind Turbines. DOE/NASA Workshop on Large Horizontal Axis Wind Turbines, July 1981.

[13] J.C. Cook and M.G. Hall: Boundary Layers in three dimensions. Progress in Aero.Sci, vol.2, p222-282, 1962

[14] H. Snel, R. Houwink and J. Piers: Sectional prediction of 3-D effects for separated flow on rotating blades.

[15] R. Houwink, J.A. van Egmond and P.A. van Gelder: Computation of viscous aerodynamic characteristics of 2-D airfoils for helicopter applications. NLR MP 88052 U (1988).

[16] M. Drela: XFOIL: An Analysis and Design System for Low Reynolds Number Airfoils. In: Low Reynolds Number Aerodynamics. Springer Verlag Lec. Notes in Eng. 54, 1989.

WAKE AND WIND FARM MODELLING

F.J. Verheij*, J.W. Cleijne*, S.G. Voutsinas (CRES, Greece),
S. Huberson (Limsi, France), A. Crespo (UPM, Spain),
D. Delaunay (CSTB, France) and G.J. Taylor (UK)

*TNO Environmental and Energy Research
P.O. Box 342, 7300 AH Apeldoorn, The Netherlands

ABSTRACT: This paper describes parts of the results of the CEC JOULE project Wake and Wind Farm Modelling.
A scheme for a new modular wind farm model has been designed. Turbine characteristics, single wake models and wake interaction models can be incorporated easily. The wind farm model is capable of handling wake models describing 3-dimensional wake effects. The wind farm model has been based on a sort/march technique and assumes that the wind field in a wind farm is only affected by upstream machines.
The modules calculate the wake in three different regions. The modules either exist of functional expressions or data bases. Turbulence characteristics and wake interaction have been incorporated in separate modules. The contents of the modules are derived from advanced wake models.
An evaluation of two existing gust models has resulted in a scheme for one combined wake gust model. The description of the gusts is based on known relationships expressed in non-dimensional parameters and will be validated with full-scale wind speed time series measured in the Dutch experimental wind farm at Sexbierum.

1 INTRODUCTION

In order to design economic clusters of wind turbines, engineers need to be able to predict accurately the efficiency of a wind farm. To ensure a long life time of the wind turbines a reliable prediction of the fatigue rate on the machines operating in a wind farm is also needed. This requires a good knowledge of the flow, and particularly the turbulence properties, in wind turbine wakes.

In this paper parts of the results of the CEC JOULE project Wake and Wind Farm Modelling are described. The project can be regarded as a continuation of previous CEC activities carried out in the concerted actions 'Wake Effects' and 'Gust Modelling'. The present project aimed at:

- improvement and extension of existing wake models (turbulence, wake interference etc.);
- development of an improved modular wind farm code;
- development of a wake gust model.

Chapter 2 deals with the first two items, as the last item is discussed in chapter 3. Conclusions are drawn in chapter 4. More detailed information can be found in the papers of Crespo & Hernández [4] and Voutsinas & Rados [14].

2 WIND FARM MODELLING

2.1 Existing wake models

There are three types of wake models which are used to develop an improved wind farm code. The three types of models are the eddy-viscosity model (Smith & Taylor [9]), the k-ε model (Crespo et al. [2]) and the vortex-particle model (Voutsinas et al. [11, 12]). These existing wake models have been improved and extended in order to increase the possibilities of the models. Most attention has been paid to the interaction of wakes, the modelling of the added turbulence in the near wake and to numerical improvements with respect to grid size and calculation time.

Eddy-viscosity model

Smith & Taylor [9] make a number of gross assumptions concerning the momentum transfer which takes place in the wake of a wind turbine operating in the wake of another machine. The wake of the second machine developes inside the wake of the first machine. Smith & Taylor determined a simple correction, which takes this effect into account. In this way they are able to predict the superposition of wake wind speeds properly.

This proposed method has been checked against numerical results from PHOENICS (k-ε model of UPM, see below) showing very good agreement.

K-ε model

Crespo [3, 4] has worked on the structure of the turbulence in the wake. Assuming anisotropy in the wake, similar to the atmospheric flow, the added turbulence intensity I_{add} can be defined as:

$$I_{add} = 1.04\,(k_{add})^{0.5} / U_\infty$$

k_{add} ($k_{add} = k - k_\infty$) being the added turbulence kinetic energy created in the wake.

In the near-wake region the kinetic turbulence energy and the dissipation rate ε can be estimated:

$$k_{add} = \Delta U^2 / 8 \quad \text{and} \quad \varepsilon = \varepsilon_\infty\,(k/k_\infty)^{1.21}$$

ΔU being the velocity deficit in the wake.

Un upper limit of I_{add} kan be calculated with known relations between ΔU and U_∞ expressed in the thrust coefficient of the turbine C_T:

$$I_{add} = 0.37\,(1 - (1 - C_T)^{0.5})$$

In figure 1 this analytical expression is compared with numerical results (x) and experiments in the TNO wind tunnel (O) and Nibe (+).

In the far wake region the relation for the added turbulence is used presented by Quarton [8]. UPM proposes minor adjustments of the coefficients.

From calculated results they found that there is an approximate equilibrium between energy production and dissipation.

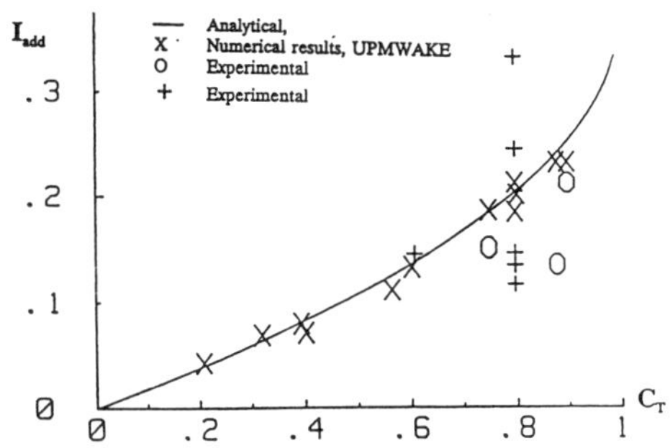

Figure 1 Maximum value of the added turbulence intensity in the near wake as a function of the thrust coefficient. Comparison of the analytical prediction with exoeriments and numerical calculations (UPMWAKE).

Vortex-particle model

A non-linear vortex method (code name: GENUVP/UPROP) using the vortex particle approximation of the wake has been developed jointly by CRES and LIMSI [13]. This new code takes into account the geometry of the rotor blades and calculates the profiles of wind velocity just behind the rotor and of the turbulence added by the rotor (see figures 2a+b).

GENUVP/UPROP has been coupled with AXI-NS to calculate the distribution of the various wind speed parameters in the near wake. AXI-NS solves the turbulent Navier-Stokes equations assuming axisymmetric flow conditions. Calculations have been made and have been compared with the single and double wake Nibe measurements [10]. The results show good agreement, both for the mean wind speed and for the turbulence intensity in the wakes (see figures 3a+b).

Validation tests have been performed for the energy interaction relation introduced by CRES [11].

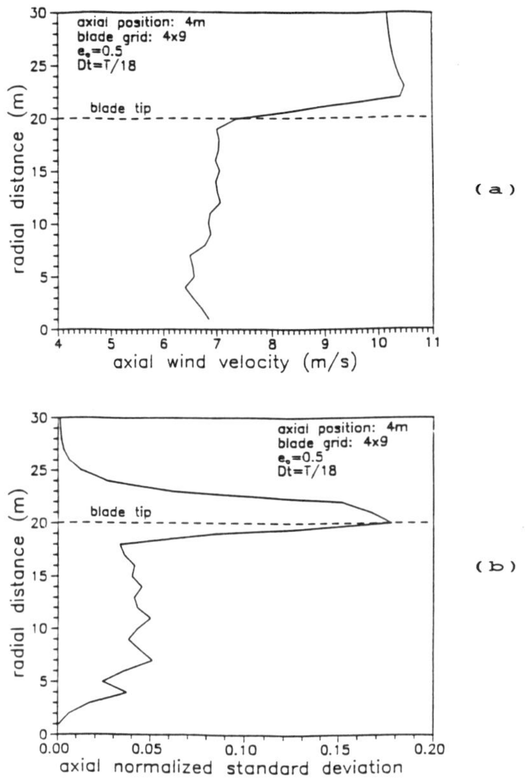

Figure 2 Axial velocity and turbulent velocity profile U_0 = 8.55 m/s.

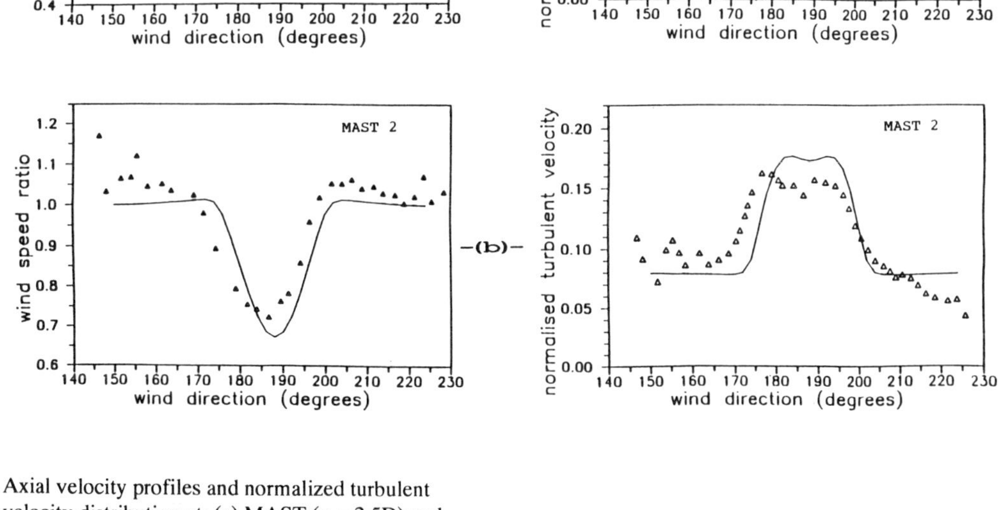

Figure 3 Axial velocity profiles and normalized turbulent velocity distribution at: (a) MAST (x = 2.5D) and (b) MAST 2 (x = 4D).

2.2 Turbulence characteristics

A model to estimate the evolution of turbulent spectra (PSD) in the wake of a wind turbine is being developed by Crespo [3]. The interaction of the atmospheric turbulence associated with the atmospheric boundary layer and the shear layer in the wake is non-linear and the production of turbulence is largely enhanced. Besides the turbulence kinetic energy is larger in the upper part of the ring shear layer in the wake. The UPM codes are able to predict these effects. Consequently, it is considered convenient to calculate k and ε from the wake models and then to obtain the spectra assuming an equilibrium -5/3 law. In this way the spectrum of the unperturbed basic flow is recovered when the effect of the wake becomes negligible.

Voutsinas [13] carried out an extensive parametric analysis in order to get a better understanding of the numerical aspects of the modelling of the turbulence characteristics in the rotor region (see paragraph 2.3).

TNO (Verheij [15]) analyzed PSD's for the Dynamic Loads in Wind Farms project. The data concern 4D and 7D single wake measurements in Sexbierum. They found that the PSD's can be described according to the von Kármán spectral equations [7]. To this end the local mean and variance of the wind speed as well as the local integral length scale of the turbulence are used to express the concerning parameters in non-dimensional expressions. An example of the comparison between measured and calculated PSD is given in figure 4. The data are measured at 4D on the centre-line of a single wake at U_0 = 7.7 m/s at hub height 35 m. The difference between the

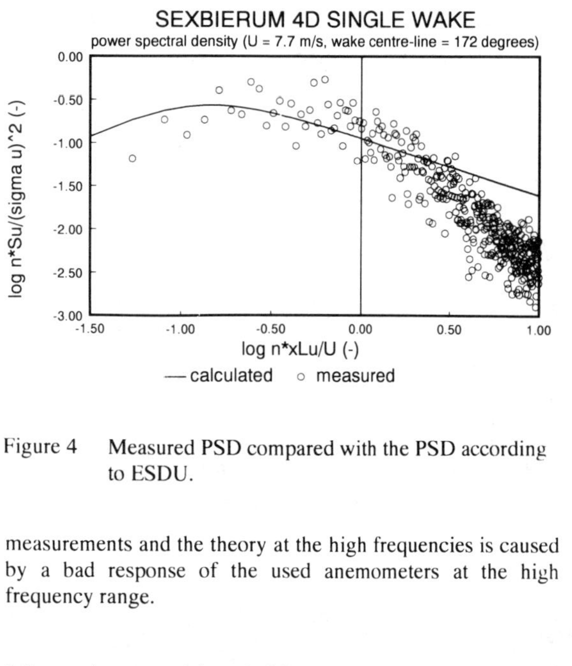

Figure 4 Measured PSD compared with the PSD according to ESDU.

measurements and the theory at the high frequencies is caused by a bad response of the used anemometers at the high frequency range.

2.3 A new modular wind farm model

A scheme for a new modular wind farm model has been designed by Cleijne [1]. Turbine characteristics, single wake models and wake interaction models can be incorporated easily. The wind farm model is capable of handling wake models describing 3D wake effects. The wind farm model has been based on a sort/march technique and assumes that the wind field in a wind farm is only affected by upstream machines.

The wake has been divided in three regions, i.e. the rotor region, the near-wake region and the far wake region. The modules calculating the wake in these regions either exist of functional expressions or data bases. The contents of the modules will be derived from the extended versions of the advanced wake models. In the first type of modules the expressions used for the improved kinematic wake models will be used. In the second type of modules the combined vortex-particle/k-ε model will be used to fill the data bases. Turbulence characteristics and wake interaction will be incorporated in separate modules.

For validation purposes experimental data is and will be used from Nibe, Sexbierum and Norrekaer Enge.

3 WAKE GUST MODELLING

An evaluation of two existing gust models from CSTB and TNO has resulted in a scheme for one combined model [5]. The gusts give a discrete description of the wind fluctuations. Important parameters are: the gust amplitude, the gust duration and its spatial dimension. Due to the rotation of the rotor blades, a wind turbine will experience so-called rotor effective gusts.

The new gust model will be extended with the description of gusts in wind turbine wakes. To this end full-scale wind speed time series, measured in the Dutch experimental wind farm at Sexbierum will be analyzed. The following information is required as an input:

– the characteristics of the site: wind climate, terrain roughness, position in the wake of the upstream wind turbine;
– the characteristics of the upstream wind turbine: rotor diameter, hub height, thrust coefficient and rotational speed;
– the characteristics of the turbulence in the wake.

output of the model will give overall distributions of gust durations and amplitudes as well as conditional distributions of the amplitude given a fixed duration. The distributions will be given for single-point and rotor effective gusts, both for the longitudinal and the lateral wind component. The distributions are expressed in non-dimensional parameters. The measurements of the Sexbierum wind data will be used to check if the known relationships can also be applied to wake situations.

An important parameter for the description of rotor effective gusts is the turbulence length scale. A literature survey of the description and measuremensts of the turbulence length scales in wind turbine wakes has been finished [6]. The attention has been focussed on single wakes, particularly in the range between twice and ten times the rotor diameter of the upstream machine. Very interesting results have been obtained by Smith and Taylor [9]. They have found a linear relationship

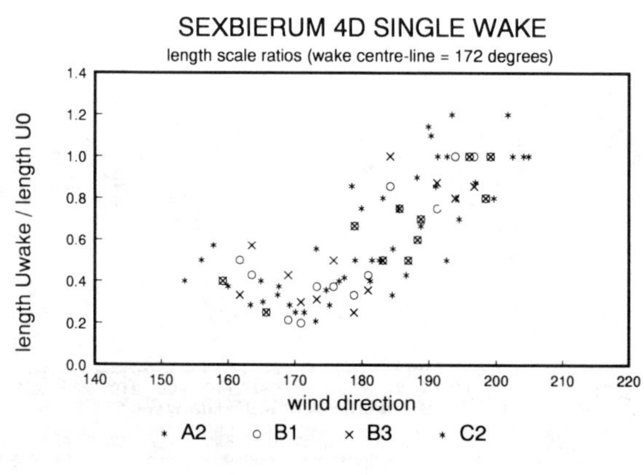

Figure 5 The ratio of the integral length scale of the turbulence between 4D wake situation and undisturbed flow situation.

between the turbulence length scale and the wake width, the latter to be derived from currently available wake models. This relationship will be compared with the Sexbierum wind farm data. From 4D single wake measurements in the Sexbierum wind farm [15] it can be shown (fig. 5) that the ratio between the length scale measured in the wake and that simultaneously measured in the undisturbed flow, strongly depends on the position in the wake.

4 CONCLUSIONS

Existing wake models have been improved and extended with respect to turbulence, ground effects and wake interference. Calculations agree well with measured results. The turbulence properties in the wake can be described in terms of the velocity deficit, the thrust coefficient and the added turbulence kinetic energy.

The modules for the new wind farm code will consist of functional relations or data bases. The data bases and the functional expressions will be extracted from calculations with the so-called advanced wake models.

The evaluation of existing gust models resulted in a scheme for a combined gust model. The known relationships of the gust amplitudes and gust durations will be applied to wake situations. The Sexbierum wind data will be used for validation purposes.

REFERENCES

[1] Cleijne, J.W.; The Development of a Wind Farm Code. Part A: Model Structure; TNO note 91-07430, Apeldoorn, 1991.

[2] Crespo, A. et al.; Numerical Modelling of Wind Turbine Wakes; Proc. EWEC'90, Madrid, 1990, pp. 166-170.

[3] Crespo, A.; Progress report for JOUR-0087 Wake and Wind Farm Modelling; Madrid, October 1992.

[4] Crespo, A. and Hernández, J.; Analytical correlations for the turbulence characteristics in the wake of wind turbines; Proc. ECWEC'93, Lübeck-Travemünde, 1993.

[5] Delaunay, D. and Verheij, F.J.; A Gust Model for Wind Turbine Wakes. Preliminary Analysis; CSTB/TNO, TNO report 92-173, Nantes (F) / Apeldoorn (NL), 1992.

[6] Delaunay, D.; Progress report for JOUR-0087 Wake and Wind Farm Modelling; Nantes, October 1992.

[7] ESDU; Characteristics of atmospheric turbulence near the ground. Part II: single point data for strong winds (neutral atmosphere); Engineering Sciences Data Item 85020, October 1985.

[8] Quarton, D.C.; Characterisation of Wind Turbine Wake Turbulence and its Implications on Wind Farm Spacing; Contractor report ETSU WN 5096, 1989.

[9] Smith, D. and Taylor, G.J.; Further Analysis of Turbine Wake Development and Interaction Data; BWEA '91 Proceedings, 1991.

[10] Taylor, G.J.; Wake measurements on the Nibe Wind Turbines in Denmark. Part 2: Data Collection and Analysis; Final report, CEC contract EN#W.0039.UK, 1990.

[11] Voutinas, S.G. et al.; On the Effect of the Rotor Geometry on the formation and the Development of its Wake; Proc. EWEC'91, Amsterdam, 1991, pp. 402-406.

[12] Voutsinas, S.G. et al.; Investigation of the Effect of the Initial Velocity Profile on the Wake Development of a Wind Turbine; Proc. EWEC'91, Amsterdam, 1991, pp. 407-411.

[13] Voutsinas, S.G. and Huberson, S.; Progress report for JOUR-0087 Wake and Wind Farm Modelling; Athens (GR) / Orsay (F), October 1992.

[14] Voutsinas, S.G. and Rados, K.G.; Wake effects in wind parks. A new modelling approach; Proc. ECWEC'93, Lübeck-Travemünde, 1993.

[15] Verheij, F.J.; Dynamic Loads in Wind Farms. Analysis of power spectral densities, coherence functions and turbulence length scales in single wakes. TNO, Apeldoorn, to be published 1993.

THE MEASUREMENT OF INFLOW VELOCITY AND ANGLE WITH A SPHERIC PROBE

A.J. Brand, J.G. Schepers and M. Späth
ECN - Renewable Energy
P.O. Box 1, 1755 ZG Petten, Netherlands

ABSTRACT: For a spheric probe with five measuring points on its surface, the magnitude and the direction of the inflow velocity is a function of three pressure coefficients. These inflow quantities can therefore be measured by measuring the pressure coefficients. A windtunnel test of such a probe reveals that there is a linear relation between the measured and the set value of the inflow angles, and the measured inflow velocity is in agreement with the set one. However, to reach this result the resolution of the pressure measurements must be sufficient and the effective position of the measuring points must be determined.

1 A VELOCITY PROBE FOR FULL-SCALE ROTOR AERODYNAMICS EXPERIMENTS

In order to determine experimentally the aerodynamic coefficients of an element of a full-scale rotating blade of a horizontal-axis wind turbine, it is needed to measure the magnitude and the direction of the relative wind velocity at the blade element. To achieve this goal, Paulsen [3] and Madsen [2] used a five-hole pitot tube. Unfortunately, their method can not universally be applied because it depends heavily on the value and the accuracy of an empirical constant. More importantly, it is erroneous (an inflow angle should be in the plane which contains a hole and the stagnation point, and not in the plane which contains a hole and a projection of the stagnation point), and consequently employs poor pressure coefficients.

In this paper we describe an alternative method to determine the inflow velocity and the inflow angle of the flow when the pressures are known in a number of points at the surface of a sphere. In the following we present the concept (§2), tests of its performance (§3), and an evaluation of our findings (§4).

2 THE CONCEPT OF A SPHERIC PROBE

If the Reynolds number of the flow around a sphere is sufficiently large (i.e. larger than a critical value of about 3 10^5), the theory of inviscid fluid flow can be applied to describe it. As a result, the pressure in any point at the surface of the sphere can be predicted as a function of the inflow velocity and the inflow angle. Inversely, if the pressure is known in a number of points at the surface of the sphere, the theoretical relations can be used to find the inflow velocity and the inflow angle. These inverse relations are the basics of the spheric-probe concept, and allow to avoid laborious calibration sessions involving an extensive range of inflow velocities and angles.

First, on basis of the streamfunction which describes the inviscid flow around a sphere [1, p. 452], we calculate the pressure p at the surface of the sphere as a function of the angle φ between a measuring point and the stagnation point. Applying Bernouilli's theorem yields that the pressure difference $p' = p - p_0$ (where p_0 is the reference pressure) obeys

$$p'(\varphi) = \frac{1}{8}\rho U_0^2(-5 + 9\cos^2\varphi), \quad (1)$$

with ρ the density of the fluid and U_0 the inflow velocity.

Next, we define the geometric base angle τ. This angle gives the position of five measuring points at the surface of the sphere (figure 1):

$$\begin{array}{ll} 1) & \phi_M = 0 \text{ and } \theta_M = \pi/2, \\ 2) & \phi_M = \tau \text{ and } \theta_M = \pi/2, \\ 3) & \phi_M = -\tau \text{ and } \theta_M = \pi/2, \\ 4) & \phi_M = 0 \text{ and } \theta_M = \pi/2 - \tau, \\ 5) & \phi_M = 0 \text{ and } \theta_M = \pi/2 + \tau; \end{array} \quad (2)$$

where ϕ_M and θ_M are the coordinates of a measuring point in the spherical coordinate system of the sphere (figure 1). If the position of the stagnation point is given by the azimuthal inflow angle (the 'azimuth') θ_S and the longitudal inflow angle (the 'longitude') ϕ_S (i.e. the coordinates of the stagnation point in the spherical coordinate system; see figure 1), by calculating the angle φ_i between the ith measuring point and the stagnation point, we find that the pressure difference p'_i is governed by

$$p'_i = \frac{1}{8}\rho U_0^2(-5 + 9\cos^2\varphi_i) \quad (3)$$

(i = 1, 2, 3, 4 and 5), with

$$\begin{aligned} \cos^2\varphi_1 &= \sin^2\theta_S \cos^2\phi_S, \\ \cos^2\varphi_2 &= \sin^2\theta_S\,((c^2 - s^2)\cos^2\phi_S + cs\sin 2\phi_S + s^2), \\ \cos^2\varphi_3 &= \sin^2\theta_S\,((c^2 - s^2)\cos^2\phi_S - cs\sin 2\phi_S + s^2), \\ \cos^2\varphi_4 &= c^2\sin^2\theta_S\cos^2\phi_S + cs\sin 2\theta_S\cos\phi_S + s^2\cos^2\theta_S, \\ \cos^2\varphi_5 &= c^2\sin^2\theta_S\cos^2\phi_S - cs\sin 2\theta_S\cos\phi_S + s^2\cos^2\theta_S; \end{aligned} \quad (4)$$

where $c \equiv \cos\tau$ and $s \equiv \sin\tau$.

Subsequently, we introduce three pressure coefficients:

$$K_m \equiv \frac{p'_5 + p'_4 - 2c^2 p'_1}{p'_3 + p'_2 - 2(c^2 - s^2)p'_1}, \quad (5)$$

$$L_m \equiv \frac{p'_1}{p'_3 + p'_2 - 2(c^2 - s^2)p'_1}, \quad (6)$$

$$M_m \equiv -3s^2 \frac{p'_1}{p'_5 + p'_4 + p'_3 + p'_2 - 2(2c^2 - s^2)p'_1}. \quad (7)$$

Inserting the 'theoretical' pressures p'_i into these coefficients yields their theoretical values:

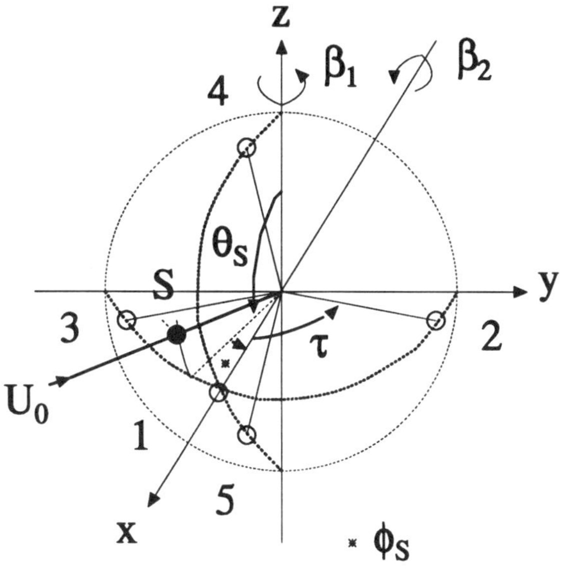

Figure 1: A schematic of the spheric probe. Indicated are the coordinate axes x, y and z; the stagnation point S; the inflow angles θ_S (azimuth) and ϕ_S (longitude); the inflow velocity U_0; the geometric base angle τ; the measuring points 1, 2, 3, 4 and 5; and the orientation angles β_1 and β_2.

$$K_s = \frac{5 - 9\cos^2\theta_S}{1 + 9\cos^2\theta_S}, \tag{8}$$

$$L_s = \frac{1}{2s^2}\frac{5 - 9\sin^2\theta_S\cos^2\phi_S}{1 + 9\cos^2\theta_S}, \tag{9}$$

$$M_s = \frac{1}{4}(-5 + 9\sin^2\theta_S\cos^2\phi_S). \tag{10}$$

In the following we will refer to K_s, L_s and M_s as the set values of the pressure coefficients (in contrast to the measured values K_m, L_m and M_m). It thus appears that K_m is a function of the azimuth only, whereas L_m and M_m depend on both the azimuth and the longitude.

By evaluating the position of the stagnation point with respect to the position of the 4th and the 5th measuring point, and inverting equation (8), we obtain the estimate θ_S^* for the azimuthal inflow angle:

$$\theta_S^* = \begin{cases} \theta_S^o & \text{if } p_4' > p_5' \\ \pi/2 & \text{if } p_4' = p_5' \\ \pi - \theta_S^o & \text{if } p_4' < p_5', \end{cases} \tag{11}$$

with

$$\theta_S^o = \arccos\frac{1}{3}\sqrt{\frac{5 - K_m}{1 + K_m}}. \tag{12}$$

Similarly, the estimate ϕ_S^* for the longitudal inflow angle is

$$\phi_S^* = \begin{cases} \phi_S^o & \text{if } p_2' > p_3' \\ 0 & \text{if } p_2' = p_3' \\ -\phi_S^o & \text{if } p_2' < p_3', \end{cases} \tag{13}$$

with

$$\phi_S^o = \arccos\sqrt{\frac{5(1 + K_m) - 12s^2 L_m}{2(2 + 5K_m)}}; \tag{14}$$

and the estimate U_0^* for the inflow velocity is

$$U_0^* = \sqrt{\frac{2}{\rho}\frac{p_1'}{M_m}}. \tag{15}$$

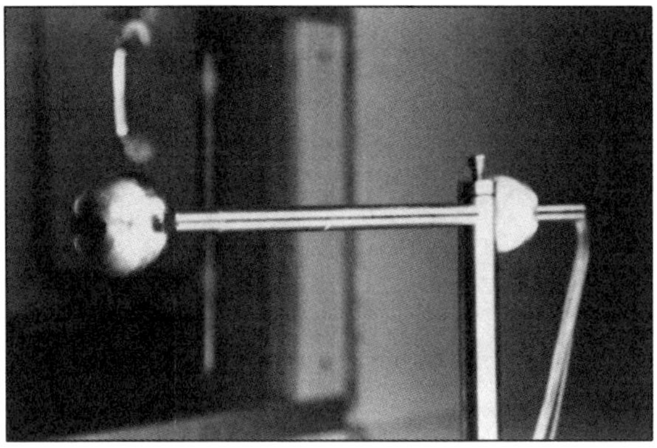

Figure 2: The spheric probe with $\tau = 45°$ in the test section of the windtunnel.

The equations (11) ... (15) express the inflow angles and the inflow velocity in terms of the measured pressures. Note that the estimate for the inflow velocity depends on neither the azimuth nor the longitude.

3 Test of concept and calibration

3.1 Windtunnel tests

In order to test the concept, we mounted a spheric probe with a diameter of 55 mm in the test section of a windtunnel (figure 2). A pressure sensor (Druck PDCR 22) with a range of 70 mbar, mounted in a digital pressure indicator (Druck DPI 100), was used to measure the difference between the pressure on the surface of the sphere and the pressure inside the windtunnel. A rotary pressure scanner (Scanivalve Type 48) was employed to connect the surface pressures to the pressure sensor. The absolute error of a pressure measurement was less than ±0.05 mbar. In §3.2 and §3.3 we will present separately the results for a spheric probe with a large (45°) and a small (15°) geometric base angle.

3.2 Large geometric base angle

The experiments on a spheric probe with $\tau = 45°$ were performed at three Reynolds numbers in the range 7.7 10^4 ... 1.4 10^5. As an example of the results, in figure 3 we present the non-dimensional pressures p_i'/q (where q is the stagnation pressure $\rho U_0^2/2$) as a function of the orientation (β_1, β_2) of the sphere (see figure 1), and for the Reynolds number 7.7 10^4. Figure 3 shows that, for this subcritical Reynolds number, the pressure at the surface of the sphere is in agreement with the one predicted by inviscid theory, provided that the angle between the flow and a measuring point is less than 45°. (This finding is similar to the one reported in the literature; see e.g. [4, p. 20]). As a consequence, the measured values of the pressure coefficients did not reach their theoretical values, which implied that the inverse relations could not be applied. On conclusion, the geometric base angle was too large because for the given layout always one measuring point is in the region of separated flow.

This windtunnel test also made it clear that, due to the non-ideal shape of the sphere, the position of a measuring point as 'seen' by the flow can be different from the geometric position. As a consequence, in practice, the base angle of the ith measuring point is $\tau_i = \tau_e + \varepsilon_i$ rather than $\tau_i = \tau$, where τ_e is the effective base angle and ε_i its error.

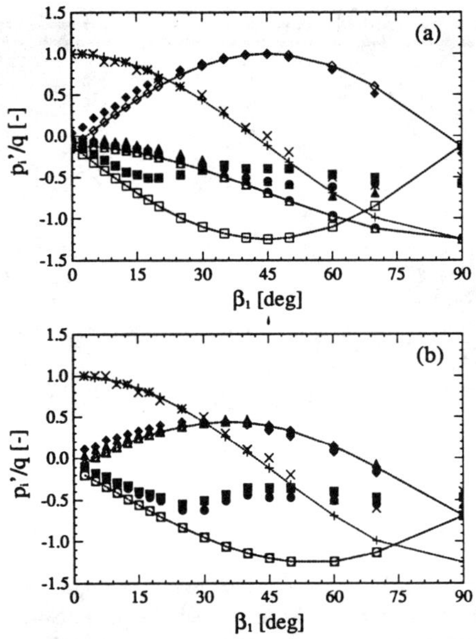

Figure 3: The non-dimensional pressures p_i'/q as a function of the sphere orientation β_1 for (a) $\beta_2 = 0°$ and (b) $\beta_2 = 45°$; spheric probe with $\tau = 45°$. The markers refer to the measuring points $i = 1$ ($\times, +$), 2 ($\square, \blacksquare$), 3 ($\diamond, \blacklozenge$), 4 ($\triangle, \blacktriangle$) and 5 ($\bigcirc, \bullet$); the first marker indicates the measured and the second the theoretical value. The Reynolds number was 7.7 10^4.

3.3 Small geometric base angle

Subsequently, a spheric probe with a geometric base angle of 15° was tested at a Reynolds number of 8.2 10^4. By rotating this probe about its x axis, and averaging the pressures p_i', we found the effective base angle $\tau_e = 12.5°$, and the errors $\varepsilon_2 = 0.5°$, $\varepsilon_3 = -\,0.5°$, $\varepsilon_4 = 0.5°$ and $\varepsilon_5 = -\,0.5°$.

The error in the position of the measuring points greatly affects the set values of the pressure coefficients. Specifically, by inserting $\cos\tau_i \approx \cos\tau_e - \varepsilon_i \sin\tau_e$ and $\sin\tau_i \approx \sin\tau_e + \varepsilon_i \cos\tau_e$ into the equations (4), and assuming $\varepsilon_3 \approx -\varepsilon_2$ and $\varepsilon_5 \approx -\varepsilon_4$, we obtain the compensated set values:

$$K_s = \frac{5(1-1.8\delta_5) - 9\cos^2\theta_S}{1 - 9\delta_3 + 9(1+\delta_3)\cos^2\theta_S}, \qquad (16)$$

$$L_s = \frac{1}{2s^2}\frac{5 - 9\sin^2\theta_S\cos^2\phi_S}{1 - 9\delta_3 + 9(1+\delta_3)\cos^2\theta_S}, \qquad (17)$$

$$M_s = \frac{1}{4}(-5 + 9\sin^2\theta_S\cos^2\phi_S). \qquad (18)$$

Here the compensation terms δ_3 and δ_5 depend on the actual position of the stagnation point:

$$\delta_3 = \frac{1}{2}(1 - \frac{c^2}{s^2})(\varepsilon_3 - \varepsilon_2)\sin 2\phi_S, \qquad (19)$$

$$\delta_5 = \frac{1}{2}(1 - \frac{c^2}{s^2})(\varepsilon_5 - \varepsilon_4)\sin 2\theta_S \cos\phi_S, \qquad (20)$$

where $c \equiv \cos\tau_e$ and $s \equiv \sin\tau_e$.

A test of the spheric probe with $\tau = 15°$ (Reynolds number: 8.2 10^4) showed that the measured and the theoretical values of the non-dimensional pressures (figure 4) and the pressure coefficients (figure 5) agree when the error in the position of each measuring point is compensated for. However, the absolute error in K_m, L_m and M_m (not shown) is large because the absolute error in the pressure measurements is relatively large (i.e. 1.7% of the stagnation pressure).

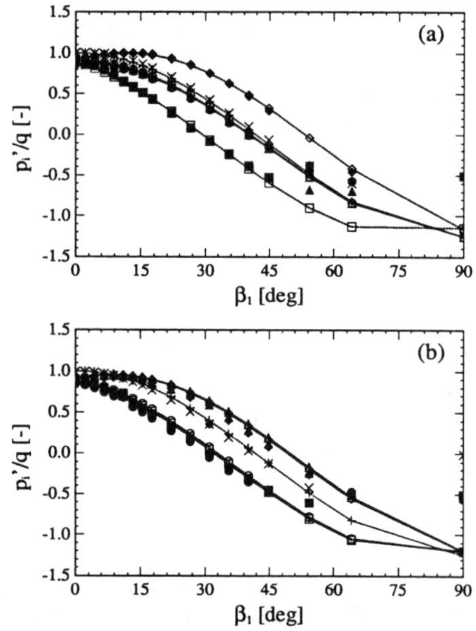

Figure 4: As figure 3, but for the spheric probe with $\tau = 15°$; the Reynolds number was 8.2 10^4.

By evaluating the position of the measuring points with respect to the stagnation point, and inverting the equations (16) ... (18), we find that the error in the position of the measuring points affects the estimates (12), (14) and (15) for the inflow quantities too. Specifically, if $\Delta_5 = 9q(\varepsilon_5 - \varepsilon_4)/4$ is the pressure offset in the xz plane of the spheric probe, the compensated estimate of the azimuthal inflow angle is:

$$\theta_S^* = \begin{cases} \theta_S^o & \text{if } p_4' > p_5' + \Delta_5 \\ \pi/2 & \text{if } p_4' = p_5' + \Delta_5 \\ \pi - \theta_S^o & \text{if } p_4' < p_5' + \Delta_5 \end{cases} \qquad (21)$$

with

$$\theta_S^o = \arccos\frac{1}{3}\sqrt{\frac{5(1-1.8\delta_5) - K_m(1-9\delta_3)}{1 + K_m(1+\delta_3)}}. \qquad (22)$$

Similarly, if $\Delta_3 = 9q(\varepsilon_3 - \varepsilon_2)/4$ is the pressure offset in the xy plane of the spheric probe, we find that the compensated estimate for the longitudal inflow angle is

$$\phi_S^* = \begin{cases} \phi_S^o & \text{if } p_2' > p_3' + \Delta_3 \\ 0 & \text{if } p_2' = p_3' + \Delta_3 \\ -\phi_S^o & \text{if } p_2' < p_3' + \Delta_3, \end{cases} \qquad (23)$$

where ϕ_S follows from

$$\cos^2\phi_S^o = \frac{5(1 + K_m(1+\delta_3)) - 12s^2 L_m(1 - 0.67\delta_3 - 1.5\delta_5)}{2(2(1+2.25\delta_5) + 5K_m)}. \qquad (24)$$

Finally, the compensated estimate for the inflow velocity is

$$U_0^* = \sqrt{\frac{2}{\rho}\frac{p_1'}{M_m}\frac{1}{1 + 1.5(\delta_5 + \delta_3')}}; \qquad (25)$$

with $\delta_3' = \delta_3 \sin^2\theta_S$. For a spheric probe with a non-ideal position of the measuring points, the equations (21) ... (25) express the inflow angles and the inflow velocity in terms of the measured pressures. Since the compensation terms δ_3, δ_3' and δ_5 depend on the inflow angles, these equations are to be solved iteratively.

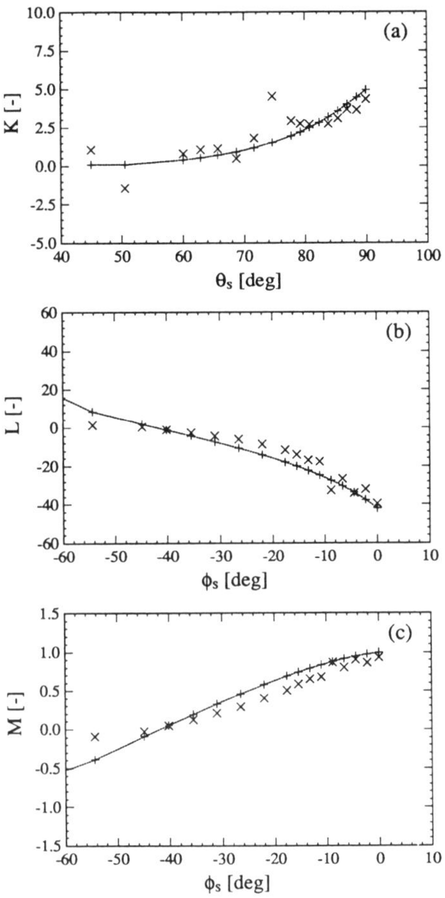

Figure 5: The measured (×) and the set values (+) of the pressure coefficients (a) K as a function of the azimuth θ_S, (b) L as a function of the longitude ϕ_S, and (c) M as a function of the longitude ϕ_S; spheric probe with $\tau = 15°$.

Solving the equations (21) ... (25) for the inflow quantities, we find that the estimated inflow velocity was 24.6 ± 1.9 m/s. Since the set inflow velocity is 22.2 ± 0.2 m/s, we conclude that the estimated inflow velocity is too large.

In figure 6 we present the compensated estimates of the azimuthal and the longitudal inflow angle on basis of the measured pressure coefficients K_m, L_m and M_m. These figures show that there is a linear relation between the measured and the set value of an inflow angle. However, in particular in the regions near $\theta_S = 90°$ and $\phi_S = 0°$, the difference between the measured and the set value is large.

The difference between the measured and the set value of an inflow quantity originates from the poor resolution of the present pressure measurements. For example, if an ideal probe has a geometric base angle $\tau = 15°$ and the stagnation point is at $(\theta_S, \phi_S) = (85°, 5°)$, for an inflow velocity of 22.5 m/s and an air density of 1.23 kg/m^3, the pressure readings are $p'_1 = 3.0$ mbar, $p'_2 = 2.8$ mbar, $p'_3 = 2.2$ mbar, $p'_4 = 2.8$ mbar and $p'_5 = 2.2$ mbar. This yields the pressure coefficients $K_m = 3.0$, $L_m = -15$ and $M_m = 0.76$, and, consequently, the estimates of the inflow quantities $\theta^*_S = 76°$, $\phi^*_S = 14°$, and $U^*_0 = 25.5$ m/s. If, on the other hand, the pressure measurements had three significant digits (rather than two), the pressure readings would have been $p'_1 = 2.97$ mbar, $p'_2 = 2.81$ mbar, $p'_3 = 2.22$ mbar, $p'_4 = 2.81$ mbar and $p'_5 = 2.22$ mbar. This would have given $K_m = 4.48$, $L_m = -26.0$ and $M_m = 0.953$, and, consequently, a better estimate of the inflow quantities: $\theta^*_S = 84°$, $\phi^*_S = 6°$, and $U^*_0 = 22.7$ m/s. We thus see that, if the measured inflow angles are to correspond to the set ones, the resolution of the pressure measurements must be sufficient.

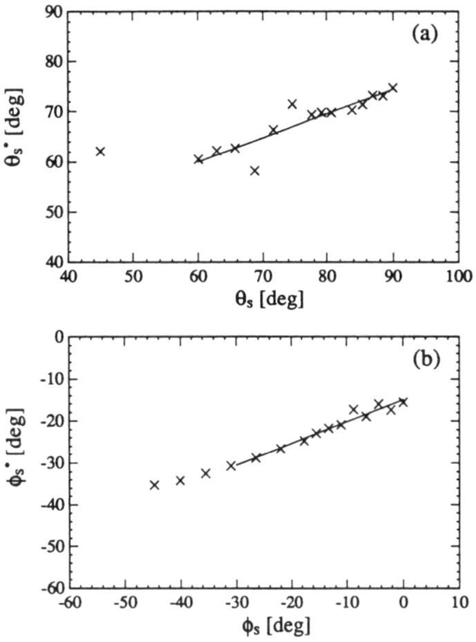

Figure 6: The estimated inflow angles (a) θ^*_S and (b) ϕ^*_S as a function of their set values; spheric probe with $\tau = 15°$.

4 Conclusion

The windtunnel tests have revealed that the geometric base angle of a spheric probe must be such that no measuring point is in the region of separated flow. For a spheric probe with a sufficiently small geometric base angle there is a linear relation between the measured and the set value of an inflow angle, provided that the error in the position of the measuring points is compensated for. In practice, these relations must be used as calibration lines, in particular when the resolution of the pressure measurements is poor (which usually occurs when the inflow velocity is small). As to the inflow velocity, the method gives too large a measured inflow velocity when the resolution is poor. Finally, we would like to emphasize that the geometric base angle only is a rough estimate of the position of a measuring point. This implies that for each spheric probe these positions must be determined experimentally.

Acknowledgements

This work is part of a project funded by the Netherlands Ministry of Economic Affairs. The windtunnel tests were performed at the Hogeschool Haarlem.

References

[1] G.K. Batchelor, 1967, *An introduction to fluid dynamics*, Cambridge University Press

[2] H.A. Madsen, 1991, *Aerodynamics of a horizontal-axis wind turbine in natural conditions*, Risø National Laboratory, Report Risø-M-2903

[3] U.S. Paulsen, 1990, *Calibration of rosemount M858 pitot tube sensor*, Risø National Laboratory, Report Risø-M-2878

[4] H. Schlichting, 1979, Boundary-layer theory, 7th edition, McGraw-Hill

FLOW ANALYSIS AROUND BLADE TIP OF HORIZONTAL AXIS WIND TURBINE WITH MIEVANE

Y. Shimizu, H. Imamura and T. Maeda
Department of Mechanical Engineering, Mie University
Kamihama-cho, Tsu, Mie, Japan

S. Matsumura
Mitsubishi Motor Co. Ltd., Wakayama Works, Japan

G.J.W. van Bussel
Institute for Wind Energy, Delft University of Technology
Stevinweg 1, The Netherlands

ABSTRACT: Power augmentation and velocity measurements in the wake of a HAWT with Mievane (a tip device on the main blade) are presented. The maximum C_p with Mievane is 15 percent larger than that without Mievane. Power augmentation caused by Mievane is mainly due to the reduction of tip vortex and the increase of the wake region by Mievane. The effects of Mievane are quantitatively verified by the velocity distributions around the tip of the main blade. The velocity distributions are measured by three-dimensional hot wire probes, which measures the axial, radial and tangential velocity components. The circulation distributions along main blade with Mievane and without Mievane are obtained from the measured velocity distributions. The circulation persists up to the tip of main blade due to the presence of Mievane.

1. INTRODUCTION

It has been demonstrated that the power of a wind turbine increases remarkably as a result of attaching the Mie type tipvane (Mievane) on the tip of the main blade [1],[2],[3]. The effects of the Mievane have been mainly determined as follows.

(1) The induced velocity in the rotor plane is decreased due to Mievane.
(2) The input energy available for the wind turbine rises due to the increase of the area of the captured stream tube.
(3) The rotational speed of the rotor can be controlled by changing the installation angle of the Mievane [4].

The effects (1) and (2) are qualitatively confirmed with the aid of flow visualization techniques such as the tuft grid method and the smoke flow method with vapor paraffin [2]. The exact quantitative measurement, however, is difficult, and not yet carried out to sufficient extend. This paper shows the axial, radial and tangential components of the velocity which are measured in the wake near the rotor plane and near the blade tip. The measurements are carried out using the three-dimensional hot wire probes, which have good performance for the signal response to the velocity fluctuation. The results of this experiment prove the above effects (1) and (2) due to the Mievane.

2. NOMENCLATURE

R : radius of wind turbine blade (m)
U : resultant velocity of u, v and w (m/s)
U_0 : wind velocity upwind of wind turbine (m/s)
u, v, w : velocity components of x, y and z direction, respectively (m/s)
Δu, Δw : amplitude of u and w due to blade passage, respectively (m/s)
x, y, z : axial, radial and tangential coordinate, respectively (m)
Γ : circulation strength around blade (m^2/s), Eq.(3)
λ : tip speed ratio, $R\,\omega/U_0$
ω : angular velocity of rotor revolution (rad/s)

3. EXPERIMENTAL APPARATUS AND PROCEDURE

3.1 Wind Tunnel and Wind Turbine

Figure 1 shows the experimental equipment. The wind tunnel has a outlet diameter of 1800 mm, and it is an open jet type. The wind tunnel reaches a maximum wind speed of 15 m/s and its turbulence intensity keeps below 1.8 percent. The test wind turbine has a diameter of 1158 mm. It has two blades which are twisted and tapered, and equipped with FX74-CL6-140 airfoil sections. The twist angle is given in Refs. [1] and [2].

The wind turbine is located at 1.2 m downwind of the outlet of wind tunnel. In order to minimize the influence of jet disturbance on the flow at the model rotor position, a free space of 4.6 m is taken downwind of the turbine.

Figure 2 shows a coordinate system used in this study. It is referred to a main blade set in the horizontal position. This coordinate system is fixed in the horizontal plane.

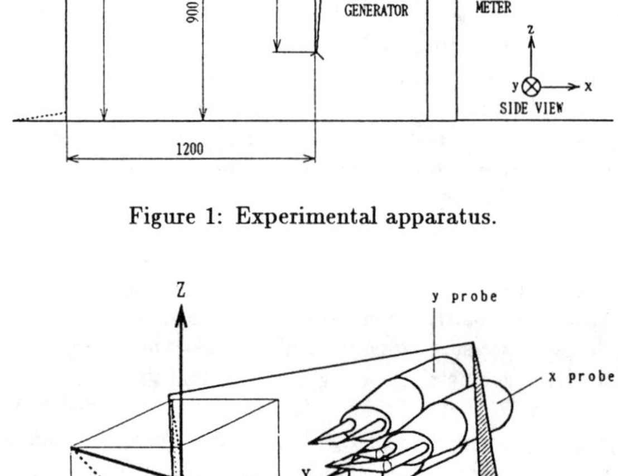

Figure 1: Experimental apparatus.

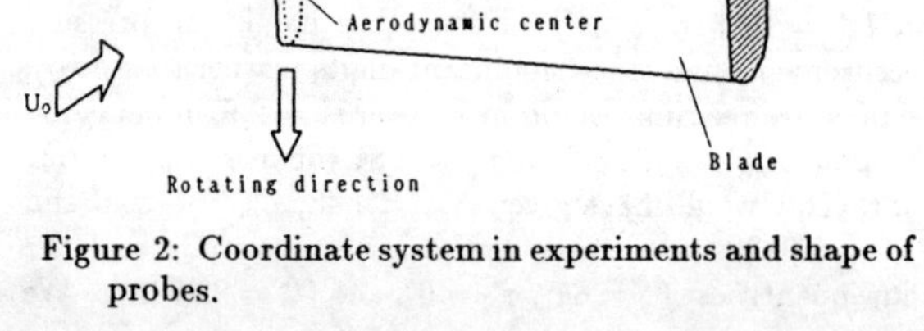

Figure 2: Coordinate system in experiments and shape of probes.

In this paper the V-type Mievane are used, as shown in Fig. 3, because it showed a higher efficiency in the previous studies.

Figure 4 shows the relation between the power coefficient C_p and the tip speed ratio λ. In the case of the turbine with Mievane, the power at the optimum efficiency point is improved 15 percent compared with the value without Mievane [3].

3.2 Hot Wire Probes

In order to measure the axial, radial and tangential components of the wind velocity, the three-dimensional hot wire probes are made as shown in Fig.2. The distances of the prongs are held at 3 mm. The hot wire is made by the tungsten filament of diameter 5 μm whose both end parts are plated with copper exclusive of the middle part of 1 mm.

3.3 Experimental Technique

The wind velocity downwind of the turbine is measured using three hot wire probes. The experiments are carried out with a fixed wind velocity U_0 = 5 m/s and a fixed tip speed ratio $\lambda = 5$. The optimum efficiency of the test turbine is obtained at $\lambda = 5$. In this condition, a Reynolds number of 1.3×10^5 which is referred to the chord length and the rotational speed at the blade tip is achieved. The results of the measurements with Mievane are compared with those without Mievane.

The azimuthal position of the blade is detected by a potentiometer.

All the output of three hot wire anemometers and the potentiometer are sampled by a small computer with an A/D-converter. The sampling frequency is 2.8 kHz, which corresponds to 3 degrees for the azimuth angle of the blade.

The output from the anemometers is sampled over 10 rotor revolutions. The sampled data are processed by the computer to the ensemble-averaged ones per 1 revolution.

4. RESULTS AND DISCUSSION

4.1 Ensemble-Averaged Velocity

Figures 5 shows the changes of all the components of the ensemble-averaged velocities in the wake after the rotor with Mievane at an axial position $x/R = 0.05$. The dotted, the solid and the chain line indicate the values at $y/R = 1.00$, 0.91 and 0.74, respectively. Three characteristics are found from figures as follows.

(1) The velocity components at arbitrary radial positions, $y/R < 1$, rapidly change at the moment when the blades pass the probes. The axial component u increases with the blade approaching the probes, and it rapidly drops just after the blade passage. After that, the component u increases again. The tangential component w rapidly increases at the moment of the blade passage. The azimuth angles for the rapid changes of the velocities ① is independent of the downwind position of the probes. The magnitude of the velocity fluctuations, Δu and Δw, decrease with increasing the distance from the rotor plane and with increasing the radial distance towards the blade tip. The velocity changes ① are caused by the velocity induced by the bound vorticity on the blade [5], [7]. The effects of the bound vorticity at the blade are always seen in the same azimuthal position.

(2) After the velocity fluctuations due to the bound vorticity ① are observed, all the velocity components, u, v, and w, are disturbed within a certain range of the azimuth angles, ② . The azimuth angles for these velocity fluctuation ② depend on the axial distance of the probes from the rotor plane. The disturbance ② appears at a later azimuth angle when the distance of the probes from the rotor plane is increased. The disturbance ② is caused by the vortex sheet released from the trailing edge of the blade. The vortex sheet itself enclosed in the turbulent boundary layer emanated from the rotor plane, so that the disturbance are observed for all the velocity components.

(3) In the area of the blade tip, $y/R = 1$, the increase in the velocity ③ is seen for all the velocity components, u, v and w, after the fluctuation due to the bound vorticity, ①. The velocity fluctuation ③ is not seen inside the rotor, $y/R < 1$, and it is limited within the area near the blade tip. The

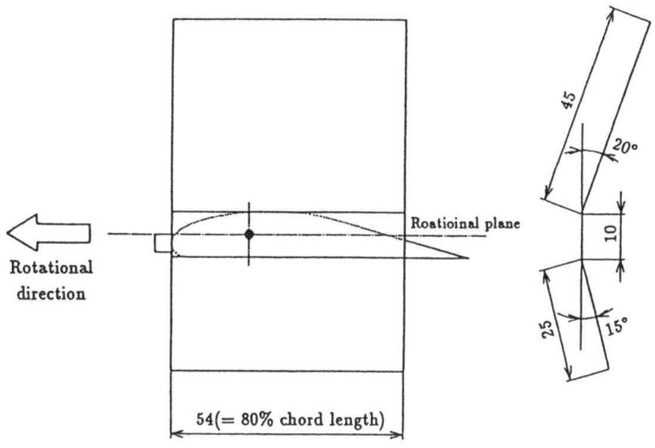

Figure 3: Configuration of V-type Mievane.

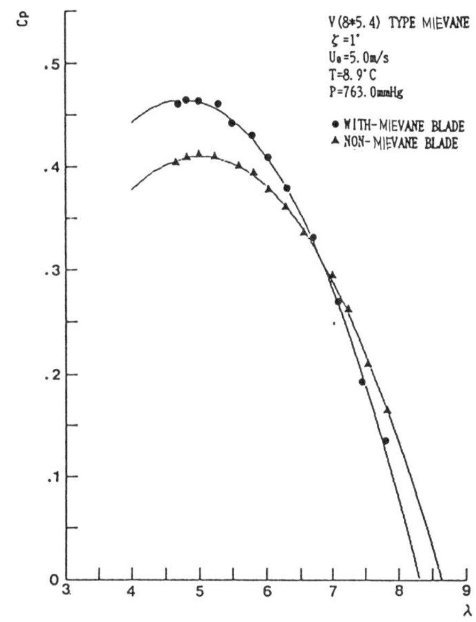

Figure 4: Power coefficient of model wind turbine rotor with and without Mievane.

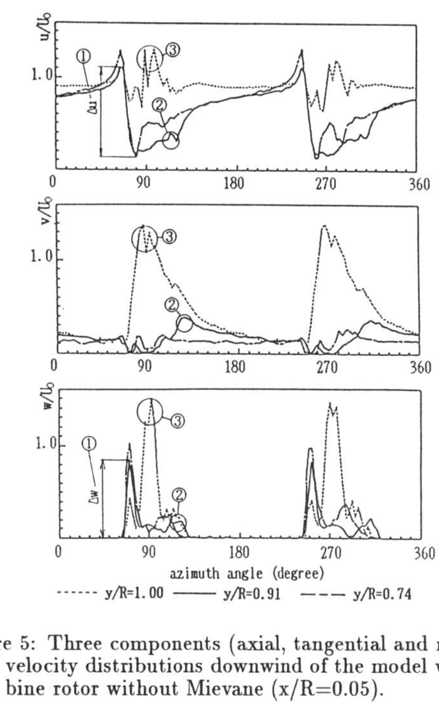

Figure 5: Three components (axial, tangential and radial) of velocity distributions downwind of the model wind turbine rotor without Mievane (x/R=0.05).

phenomenon of the fluctuation③is observed later at the more downstream axial position x/R = 0.12, which is similar to the characteristic (2). It seems that the velocity fluctuation③is caused by the tip vortex. Note that the velocity fluctuation due to the bound vorticity,①, takes a small value at the blade tip.

Figure 6 shows three velocity components of the wake downwind of the blade with Mievane. The measurements are carried out at the same position as shown in Fig.5. The local disturbance③caused by the tip vortex, which was observed in Fig.5, is absent, and the fluctuation①due to the blade passage is remarkably persistent. In particular, the velocity fluctuation③due to the tip vortex is seldom observed in the region of $0 < x/R < 0.073$ which corresponds to the downwind extension of the wake due to the Mievane. These phenomena are explained by the bound vorticity at the blade, which is transferred to the downwind side of the Mievane. A theoretical survey of this effect is given in [6].

Figure 7 shows the velocity components downwind of the Mievane. The solid line represents the velocity fluctuation just inside the Mievane, x/R = 0.09 and y/R = 1.02, and the dashed line shows just outside the Mievane, x/R = 0.09 and y/R = 1.03. It is seen that the velocity changes rapidly due to the vortex, ④, which is similar to the fluctuation③ in Fig.5. From this examination, it is confirmed that the position, at which the bound vortex is released, shifts toward the downstream edge of the Mievane.

In the case of the velocity measurements with Mievane, the two characteristics (1) and (2) mentioned above about the conventional rotor blade can be recognized.

4.2 Circulation

The velocity ū induced by a vortex is proportional to the circulation strength Γ and it is inversely proportional to the distance r from the rotor plane. If the vortex filament is assumed to be infinitely long, that is, two-dimensional, the induced velocity is given according to the Biot-Savart's laws by :

$$\bar{u} = \frac{\Gamma}{2\pi r}. \tag{1}$$

It can be shown in Ref. [7] that for both Δu and Δ w the following relation holds :

$$\Delta u, \Delta w = \frac{\Gamma}{2\pi x}. \tag{2}$$

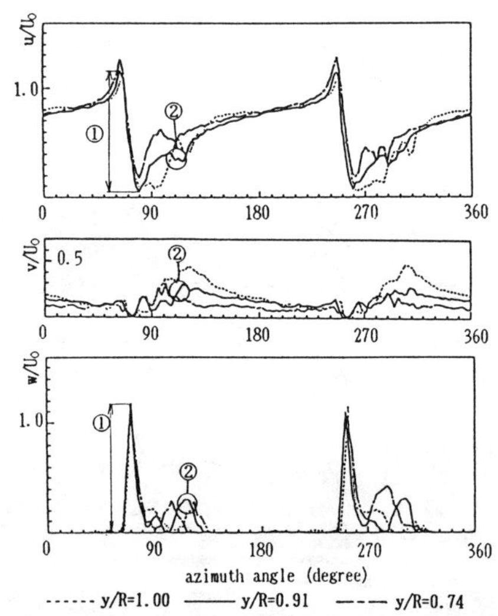

Figure 6: Three components (axial, tangential and radial) of velocity distributions downwind of the model wind turbine rotor with Mievane (x/R=0.05).

The circulation strength can be obtained by the detailed measurements of both the axial distance and the induced velocity. Equation (2) can be transformed into :

$$\Gamma = 2\pi\left(\frac{\partial(1/\Delta u)}{\partial x}\right)^{-1} \tag{3}$$

and is also valid for Δw. The use of Eq.(3) has the advantage that it is independent of the exact distance x between the probe and the bound vorticity at the rotor blade. This procedure has first been used in Ref.[7]. The circulation estimated with Eq.(3) using the measured velocity determined by the hot wire anemometer. One should be aware that the measured velocity includes the effects of the tip vortex and the existence of the Mievane. Figure 8 shows the distribution of the circulation strength Γ along the spanwise position. The marks ■ and □ show the values obtained from Δu, and ◆ and ◇ are obtained from Δw.

In the case of taking the Mievane off, the circulation strength Γ based on Δu is decreasing towards the blade tip. The value of the calculated circulation, however, does not completely vanish at the blade tip. This is caused by the assumption that the vortex filament is infinitely long.

For the situation with Mievane, the circulation inside the blade, $y/R < 0.65$, shows a similar tendency to those without Mievane. The circulation strength Γ for the blade with Mievane is, however, gradually increases towards the blade tip. The circulation takes a maximum value near the blade tip, and decreases in the proximity of the blade tip. It must be realized that the position and strength of the vortex filaments around the Mievane near the blade tip have an effect on the inflow at the blade. The real circulation strength Γ can not be determined so far since the procedure used here assumes two dimensional vorticity. By measurements in the wake downwind of the turbine as depicted in Figs.9 and 10, and the visualization shown in [2], it can be concluded that the vorticity is largely transformed from the blade tip to the Mievane.

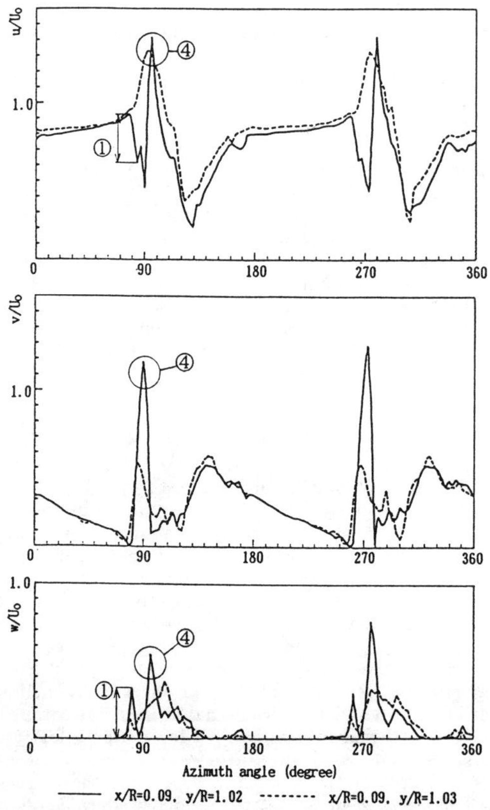

Figure 7: Three components (axial, tangential and radial) of velocity distributions just downwind of Mievane.

The circulation estimated by the velocity amplitude Δw decreases towards the blade tip both for the blade with Mievane and without Mievane. The circulation for the application of the Mievane takes quantitatively larger value than those without Mievane.

4.3 Velocity Distribution of Wake

Figures 9 and 10 show the distribution of the time-averaged axial velocity u measured at several downwind position with respect to the wind turbine. Figure 9 shows the results for the rotor without Mievane. Non-dimensional wind velocity u/U_0 outside the blade tip takes nearly the value 1.0, where the velocity inside the blade tip is reduced to about 70 percent on the average. The boundary of the wake and the uniform stream is extended radially outward according as the measurement position is moved downwind. For the rotor with Mievane, the velocity reduction has almost the same value as that without Mievane as shown in Fig.10. In the case of application of the Mievane, the region of the wake, however, is more extended radially outward than that without Mievane. The velocity of the wake increases gradually in the radial direction. The region of the wake extends largely in terms of the time-averaged value of the unsteady axial velocity.

Considering these results and the results of [2], it can be concluded that the vortex generated from the downwind edge of the Mievane is smaller than the tip vortex of the rotor without Mievane. There is a gradual leveling off of the bound vorticity over the downwind side of the Mievane.

5. CONCLUSIONS

(1) The Mievane has a suppressive effect on the formation of the tip vortex on the blade. The released position of the tip vortex is moves to the downwind edge of the Mievane. The discrete strength of the tip vortex is smaller than that without Mievane.

(2) It is seen that the circulation persists up to the blade tip due to the presence of the Mievane.

(3) The Mievane extends the diameter of the wake downwind of the wind turbine.

ACKNOWLEDGEMENTS

The authors would like to thank Mr. Nord-Jan. Vermeer, Delft University of Technology, for his advice and dissemination of knowledge with respect to the interpretation of velocity measurements in the wake of wind turbine model rotors.

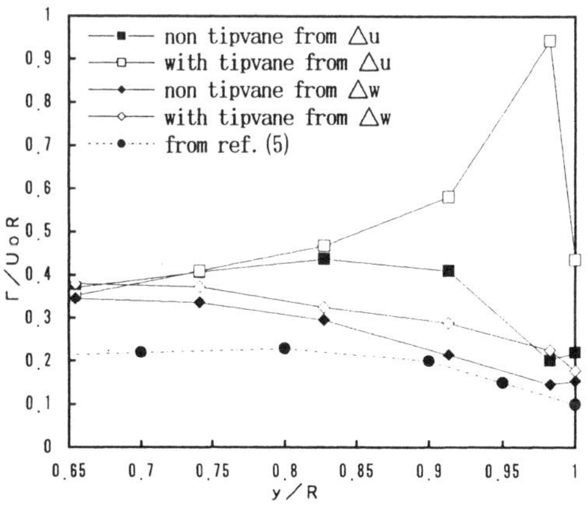

Figure 8: Spanwise distribution of circulation strength Γ.

REFERENCES

[1] SHIMIZU, Y., et al., " Power Augmentation Effects of Horizontal Axis Wind Turbine by Tip Vane (1st report, Relationships between turbine performance and configuration of tip vanes) ", *Trans. JSME*, Vol. B-522, 1990, pp. 495-501 (in Japanese).

[2] SHIMIZU, Y., et al., " Power Augmentation Effects of Horizontal Axis Wind Turbine by Tip Vane (2nd report, Flow Visualization around Blade with Tip Vane) ", *Trans.JSME*, Vol. B-522, 1990, pp. 502-507 (in Japanese).

[3] SHIMIZU, Y., et al., " Power Augmentation of Horizontal Axis Wind Turbine with Tip Vane ", *Trans. JSME*, Vol. B-543, 1991, pp. 3845-3850(in Japanese).

[4] SHIMIZU, Y., et al., " Rotation Speed Control of Horizontal Axis Wind Turbine by Tip Vane ", *Trans. JSME*, Vol. B-547, 1992, pp. 804-809 (in Japanese).

[5] TUTUI, Y., et al., " LDV Measurements of a Flow Field around a Wind Turbine ", *Trans. JSME*, Vol. B-505, 1988, pp. 2471-2477 (in Japanese).

[6] Van BUSSEL, G.J.W., " A momentum theory for winglet on horizontal axis wind turbine rotor and some comparison with experiments ", *Proceedings of 4th IEA Symposium on the Aerodynamics of Wind Turbines*, Rome, Itary, 1990.

[7] VERMEER, N. J. and Van BUSSEL, G. J. W., " Velocity measurements in the near wake of a model rotor and comparison with theoretical results ", *Proceedings of 15th European Rotational Forum*, Amsterdam, the Netherlands, 1989.

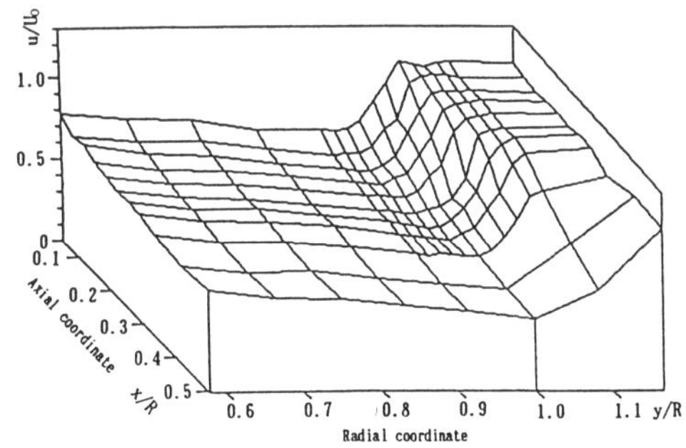

Figure 9: Distribution of axial time-averaged velocity without Mievane.

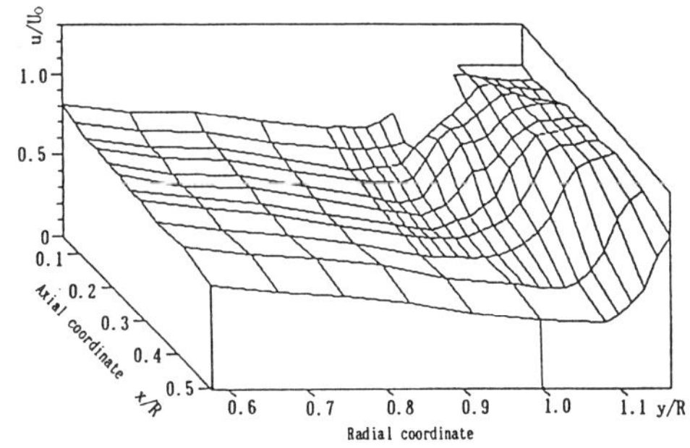

Figure 10: Distribution of axial time-averaged velocity with Mievane.

EXPERIMENTAL AND THEORETICAL INVESTIGATIONS OF UNSTEADY AERODYNAMIC FLOW ABOUT A WIND TURBINE

W.A.A.M. Bierbooms, G.J.W. van Bussel and L.J. Vermeer
Institute for Wind Energy, Faculty of Civil Engineering
Delft University of Technology
Stevinweg 1, 2628 CN Delft
The Netherlands

ABSTRACT: In the open jet wind tunnel, detailed velocity measurements have been performed in the near wake of a model rotor for yawed conditions. The use of hot wire equipment provides a fast and accurate measuring method. The passage of the blades results in a fast deterministic variation in the velocity, which is directly caused by the circulation distribution on the blade.

From the yaw measurements a hysteresis loop is obtained for the bound circulation as function of the angle of attack. Such a hysteresis loop can be used to identify the empirical coefficients for the particular aerofoil which are needed in the semi-empirical method of ONERA which accounts for unsteady profile aerodynamics.

The computer code PREDICDYN is based on the acceleration potential method and calculates the circulation distribution over the blade. A comparison will be made between the calculated circulation distribution and the circulation distribution reduced from the calculated wake velocities.

1. MEASUREMENTS

1.1 Equipment

The measurements have been performed in the open jet wind tunnel of the Institute for Wind Energy. The layout of this tunnel is described in [1]. The maximum velocity is 14.5 m/s. The degree of turbulence is 0.8 per cent and the maximum deviation from the average velocity is 1.6 per cent.

The model has been specially designed for aerodynamic measurements in the wind tunnel, see Figure 1. Specifications of the model:

Radius R	0.6 m
Number of blades	2
Aerofoil section	NACA 0012
Root cutout	30%
Chord c	0.08 m
Blade length	0.42 m
Blade twist	$\Theta(r/R) = (6 + \Theta_{tip}) - 6.67\ r/R^{\circ}$ for $0.3 \leq r/R \leq 0.9$
Replaceable tip	0.06 m span, no twist

The undisturbed tunnel wind velocity is measured with a Pitot tube connected to a Betz manometer. The Pitot tube is placed at a position representing the average velocity over the outlet area.

The velocities in the near wake are measured with constant temperature hot wire equipment; the necessary voltage to maintain the temperature of the hot wire is a measure for the local wind velocity perpendicular to the hot wire.

The probe position is set by a traversing system. In radial direction, this can be done by computer, with an absolute error of 0.2 mm. In axial direction, the adjustment is done by hand; the absolute error is 5 mm, but with an incremental error of only 0.2 mm.

A data acquisition system with a high-speed voltmeter is used to measure and store the signals of the hot wires and the azimuth counter.

1.2 Data acquisition

At each fixed position of the probe in the wake velocity samples were taken of the hot wires every 0.5° azimuth angle of the blades. In Figure 2, a typical plot is given of the

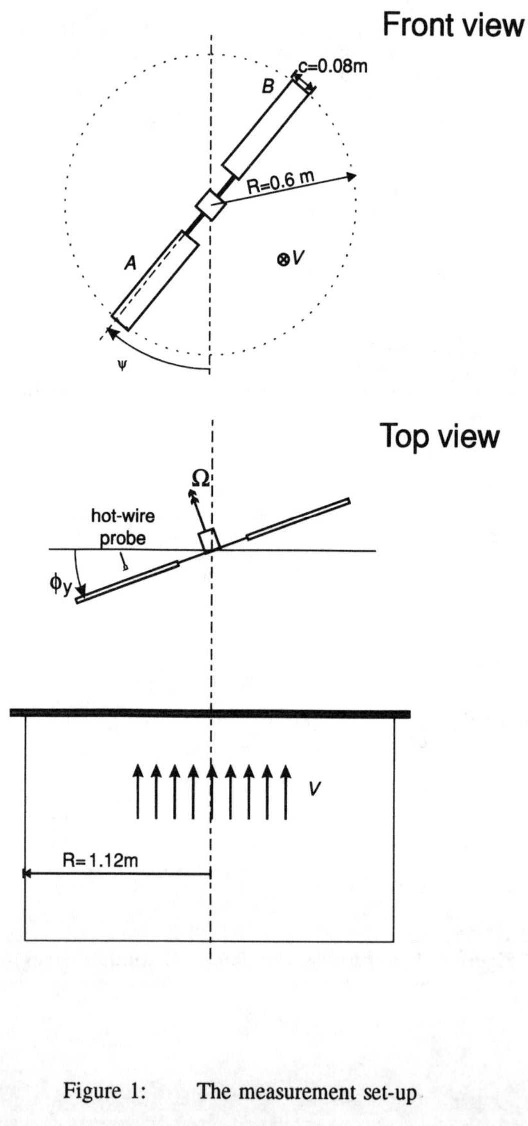

Figure 1: The measurement set-up

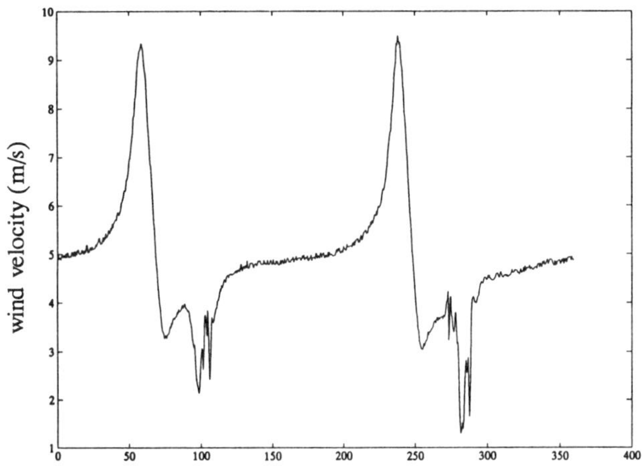

Figure 2: Near wake velocities of a rotor in yaw

velocity distribution over 1 revolution (position of the hot wire probe: axial distance x=0.03 m and azimuthal position ψ_{probe}=60°). The blade passages along the hot wire probe can clearly be distinguished at 60° (first blade) and 240° (second blade) by the deterministic fluctuations; the stochastic fluctuations around 100° and 280° are caused by the vortex sheet (consisting of trailing and shed vortices).

1.3 Data reduction

Representing the rotor blade with an infinite vortex line, it can mathematically be shown that, when moving along this vortex line, a fluctuation is induced by the vortex, with a peak-peak value ΔV proportional to the circulation strength Γ; also in case of yawed flow. When the absolute position of the rotor plane is not exactly known, the circulation can be calculated from the decrease of the peak-peak value with increasing axial distance:

$$\Gamma = 2\pi \left\{\frac{\partial(1/\Delta V)}{\partial x}\right\}^{-1}$$

In [2], this method has been shown to give promising results in an unyawed situation. A check has been made to verify the method also works with the rotor in yaw. In Figure 3, a plot is

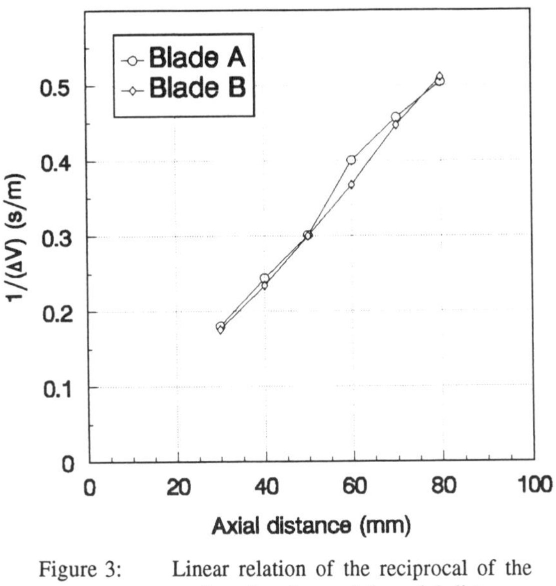

Figure 3: Linear relation of the reciprocal of the peak-peak value and the axial distance

shown of the reciprocal of the peak-peak values against axial distance for 30° yaw, giving a straight line. This means that the vortex sheet and tip and root vortices have only a minor contribution to the peak-peak value.

More details about the wind tunnel measurements can be found in [3].

2. UNSTEADY PROFILE AERODYNAMICS

2.1 Semi-empirical method of ONERA

Under yawed conditions, the blade will experience a varying angle of attack. If these fluctuations have high frequencies then a quasi-steady treatment of the aerodynamics (in other words: as a series of succeeding steady flow patterns) is no longer justified. This is an unsteady aerodynamic phenomena with length scale in the order of a chord (unsteady profile aerodynamics). In an unsteady situation the lift coefficient can amply exceed the maximum quasi-steady value and hysteresis will occur. A measure of how fast these fluctuations are is the so-called reduced frequency:

$$k = \frac{\omega \; 1/2 \; c}{U}$$

with

ω: the frequency of the variations in the angle of attack
U: the wind velocity (as seen by the aerofoil)

In case the fluctuations are caused by yawed flow, the frequency ω equals the rotor angular velocity Ω and U can be approximated by the rotational speed Ω r (with r the radial position). This means that the reduced frequency varies with the span position. In our wind tunnel experiments the reduced frequency was equal to 0.13 (at 50% span position).

For design purposes one has to fall back upon semi-empirical methods to account for unsteady profile aerodynamics. The method of ONERA (Office National d'Etudes et de Recherche Aerospatiales) is one of the existing semi-empirical methods. One of the starting points of the ONERA method is the fact that a linear system can be represented by a transfer function. Transformation to time domain results in a set of linear differential equations with constant coefficients. In case of an oscillating aerofoil, or an aerofoil experiencing a fluctuating wind, there is a linear relation between the aerodynamic forces and the changes in angle of attack (due to the blade oscillations and/or the fluctuating wind). The corresponding transfer function can be approximated by a simple fraction and in time domain a second-order differential equation for the bound circulation is obtained. The full differential equations for the bound circulation are given in [4] and [5].

Some of the coefficients in the differential equations depend on the particular aerofoil. They have to be determined for each aerofoil from dynamic wind tunnel measurements like the above mentioned yaw measurements. Up to now, these coefficients are available for typical helicopter aerofoils only.

2.2 Measurements

In this subparagraph the yaw measurements which are carried out will be discussed. The purpose of these preliminary experiments is to consider the possibility to produce the necessary data, with these kind of wind tunnel measurements, for identifying the aerofoil dependent coefficients in the semi-empirical method of ONERA.

The velocity measurements were performed for the following configuration: the pitch angle of the blade was set at Θ_{tip}=0°, the rotor was in a 30° yaw position, see Figure 1. The rotational speed was kept constant at f = 11.7 Hz, the tunnel velocity was set at V = 7.35 m/s, to match a tip speed ratio of λ = 6. In this way the Reynolds number at midspan is kept at Re = 130,000.

At a radial position r/R = 0.5, a "azimuthal traverse" is made with the hot wire probe from 0 to 360° with increments of 20°, for two axial positions: x = 0.03 and 0.06 m. The radial position is chosen to avoid interference with the root and tip vortices.

By measuring the velocities in the near wake at two axial positions an estimate of the bound circulation can be obtained, as explained in the first paragraph of this paper. The corresponding angles of attack can be calculated from the rotor angular velocity, the tunnel speed, the pitch angle, the yaw angle and the mean induced velocity. The latter follows from the mean velocity measured with the hot wires. From these data a hysteresis loop is constructed for the bound circulation Γ as function of the angle of attack α (Figure 4). This loop is compared with the steady bound circulation which follows from the steady 2-D lift coefficients C_l (from [6]) according to the relation:

$$\Gamma = 1/2\ c\ U\ C_l$$

The dotted line gives the calculations with the ONERA method for an aerofoil undergoing the same variations in the (tangential) wind velocity as a blade element (at 50% span) of the rotor in 30° yaw as in the wind tunnel measurements. For these calculations the values for a NACA 0012 profile are used which are given in reference [4] (for Re=3.10^6). Although the shape and position of the measured hysteresis loop differs a lot from the calculated one, the variations in the bound circulation in the measurements agrees reasonable well with the calculations. The difference in level between the measurements and calculation can be due to an error in the measurements or the data reduction; another explanation is that 3-D effects occur. For more information about 3-D effects see e.g. [7].

It can be concluded that these kind of wind tunnel experiments are in principle suited to determine the aerofoil dependent coefficients for the ONERA method with the aid of system identification techniques. However, the reproducibility and the inaccuracy of the measurements and the data reduction must be considered first.

In the future other experiments will be carried out for different situations: varying mean angle of attack, varying amplitude of angle of attack, etc. If more experience is gained the same measurements can be carried out for aerofoils which are now commonly used for wind turbines (e.g. NACA 4415).

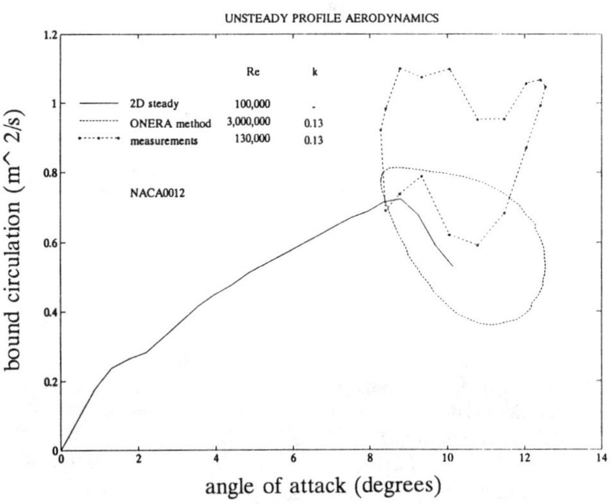

Figure 4: Hysteresis loop of the bound circulation

3. PREDICTION MODELS FOR ROTOR AERODYNAMICS

3.1 Asymptotic acceleration potential models

At the Institute for Wind Energy a number of aerodynamic load prediction models have been developed based upon asymptotic acceleration potential theory. Integration of accelerations experienced by particles of air determines the velocities in the rotor plane, or at any other spatial position. With the calculated (induction) velocities at the rotor blade the aerodynamic loads can be calculated. An introduction to the method applied to wind turbine aerodynamics can be found in [8]. The models have been implemented in various codes e.g. a performance prediction codes and flow structure codes (for comparison with velocity measurements in wind tunnel experiments).

3.2 Performance prediction codes

The two performance prediction codes that are currently operational are PREDICHAT1 and PREDICHAT2. The first code inhibits the acceleration potential equivalent of a prescribed wake model with an axially deformed wake. The second code is a free inflow code; the equivalent of a free wake model. Both codes are based upon inviscid aerodynamics. The viscous effects are however taken into account as an afterward correction of the inviscid result. For the corrections use is made of 2-D aerofoil characteristics.

PREDICHAT1 is very suitable in a design environment even though it implies discrete blade modelling (in contrast to e.g. blade element-momentum models). PREDICHAT2 is currently operational as a research tool within the Institute.

Both codes are restricted to a uniform (steady) perpendicular inflow of the rotor. For unsteady flow conditions where the unsteady phenomena have their effect on the scale of the rotor (dynamic inflow), such as coherent wind gusts, blade pitching and varying rotational speed, a dynamic version of PREDICHAT2 is developed. This code is designated PREDICDYN. In [9] the basic formulas and some calculated results are presented.

3.3 Circulation calculation in yawed flow.

Recently the computer code PREDICDYN is applied for the calculation of rotor loads in yawed flow conditions. This work is carried out in the framework of the CEC JOULE2 program on dynamic inflow, also reported in [10]. Here we restrict ourselves to comparison with wind tunnel measurements on a model rotor in yawed flow. The geometry of the model is given in section 1.1. The pitch angle of the blade was set at θ_{tip} = 4° in order to prevent the occurrence of large stall areas on the blade. The wind speed was V = 6 m/s and the rotational speed was f = 12 Hz. Figure 5 shows the calculated circulation distribution along the blade span as a function of the azimuth angle, for a yaw angle of 20°. In Figure 6 the measured circulation distribution is depicted. This circulation distribution is obtained from the peak-peak value in the measured velocities in the near wake of the model rotor, see also 1.4 and Figure 2.

It can be seen that the general trend in both figures is more or less similar. It must be pointed out that in the calculations no unsteady profile aerodynamics (as discussed in paragraph 2) are included. Some of the fluctuations in the measured results may have something to do with measurement accuracy. The accuracy of the circulation result depicted in Figure 6 is estimated to be about 15% of the average value. The velocity signal however is believed to be much more accurate. The measurement procedure provided a check of the velocity signal for several azimuthal positions, see [3], which showed that the accuracy of the peak-peak value can be determined within some 5%.

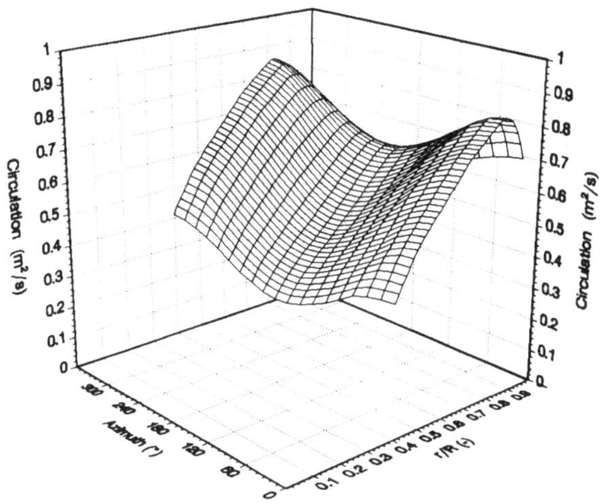

Figure 5: Calculated circulation distribution in yaw

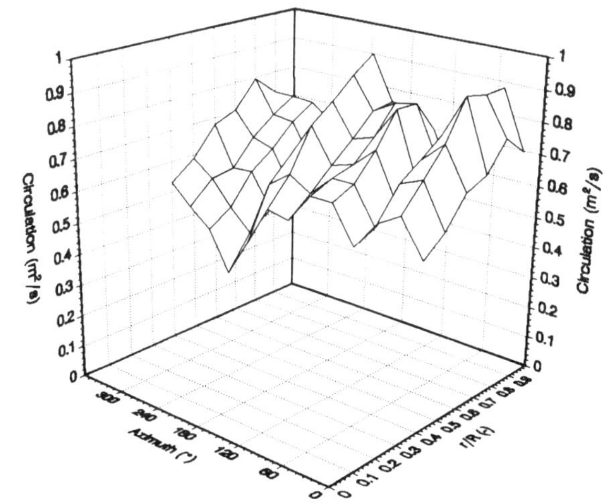

Figure 6: Measured circulation distribution in yaw

ACKNOWLEDGEMENT

We would like to thank Simon Toet for his assistance during the measurements.

REFERENCES

[1] N-J Vermeer, Local circulation on rotating wind turbine blades from velocity measurements in the wake of a model rotor, fourteenth BWEA Wind Energy Conference, Nottingham, 1992.

[2] N-J Vermeer, G.J.W. van Bussel, Velocity measurements in the near wake of a model rotor and comparison with theoretical results, European Community Wind Energy Conference, Madrid, 1990.

[3] L.J. Vermeer, J.K. de Boer, Wake structure of a rotor model in yaw, European Community Wind Energy Conference, Travemünde, 1993.

[4] D. Petot, Differential equation modelling of dynamic stall, Rech. Aerosp., No. 5, 1989.

[5] W.A.A.M. Bierbooms, A comparison between unsteady aerodynamic models, European Wind Energy Conference, Amsterdam, 1991.

[6] Unpublished NACA 0012 measurements, Low Speed Wind Tunnel Laboratory, Faculty of Aerospace Engineering, Delft University of Technology, The Netherlands, 1979.

[7] A. Bruining & W.A. Timmer, Airfoil characteristics of rotating wind turbine blades, European Wind Energy Conference, Amsterdam, 1991.

[8] G.J.W. van Bussel, The use of the asymptotic acceleration potential method for horizontal axis windturbine rotor aerodynamics, European Wind Energy Conference, Amsterdam, 1991.

[9] Y. Hasegawa, G.J.W van Bussel, Application of the asymptotic acceleration potential method for the calculation of dynamic inflow effects on horizontal axis wind turbines, The fourth International Symposium on transport phenomena and dynamics in Rotating Machinery (ISROMAC-4), Honolulu, 1992.

[10] H. Snel, J.G. Schepers, Investigation and modelling of dynamic inflow effects, European Community Wind Energy Conference, Travemünde, 1993.

NOTATION

c	blade chord	m
C_l	lift coefficient: $l/(½\rho V^2 c)$	-
E	voltage of hot wire bridge	V
f	speed of revolutions	Hz
k	reduced frequency	-
l	lift force per length	N/m
r	radial coordinate	m
R	rotor radius	m
V	velocity	m/s
x	axial coordinate	m
α	angle of attack	degrees
Γ	circulation	m^2/s
Θ	pitch angle	degrees
λ	tip speed ratio: $\Omega R/V$	-
ρ	air density	kg/m^3
ϕ_y	yaw angle	degrees
ψ	azimuth angle	degrees
Ω	rotational speed: $2\pi f$	rad/s

WIND TUNNEL RESULTS FOR A 25% THICK WIND TURBINE BLADE AIRFOIL

W.A. Timmer[1] and R.P.J.O.M. van Rooy[2]

[1]Institute for Wind Energy
Delft University of Technology, Stevinweg 1, 2628 CN Delft, The Netherlands
[2]Faculty of Aerospace Engineering
Delft University of Technology, Kluyverweg 1, 2629 HS Delft, The Netherlands

ABSTRACT: Measurements were performed in the low-speed low-turbulence wind tunnel of Delft University of Technology to determine the two-dimensional aerodynamic characteristics of a 25% thick airfoil designated DU 91-W2-250 in the Reynolds number range of 1.0 million to 3.0 million. Turbulators at the 5% chord station were used to trip the boundary-layer. Vortex generators and a gurney flap with a depth of one and two percent chord were applied to investigate the airfoil performance with lift-enhancing devices. The tests show that in general the design goals were met. The airfoil performance in the rough condition showed to be much better compared to earlier tests on a 25% thick representative of the NACA 63 series. Stall, however, appeared to be more abrupt than was expected on the basis of calculations. With vortex generators at the 20% chord station the peak lift coefficient could be raised from 1.37 to almost 2 at a Reynolds number of 2 million.

1. INTRODUCTION

Airfoils of the NACA six digit series are still widely used for application in horizontal axis wind turbines. Due to the combination of camberline and thickness distribution specific for these series the airfoils exhibit a decrease in performance with increasing relative thickness, especially with respect to roughness. This was experimentally verified during wind tunnel tests on the NACA $63_{(421)}$-425 airfoil (ref. 1). As a first step to find alternatives for thick airfoils of the NACA 63 and 64 series a 25% thick airfoil was designed within the frame work of the CEC JOULE I programme. The airfoil was designated DU 91-W2-250 and wind tunnel tested in July 1992

2. DESIGN OF THE AIRFOIL

Fig. 1 gives the inviscid pressure distributions of the airfoil. The design was done with the XFOIL code and the LSL-ADA (Airfoil Design and Analysis) code which is one of the simple codes available at the Low Speed Laboratory of the Faculty of Aerospace Engineering. Though the ADA code is based on the conformal mapping procedure and lacks a strong interaction scheme, it has proven to give a rather good prediction of transition location in the unstalled flow regime. More about the design of this airfoil, the design approach, the design tools and a comparison of calculation and wind tunnel results for several other airfoils have been reported in ref. 2. The design goals for this laminar airfoil were: a peak lift coefficient of about 1.5 and relatively smooth stall and insensitivity to roughness.

3. EXPERIMENTAL ARRANGEMENT

3.1 Model

The GRP model, built in Poland, had a chord of 0.60 m. and completely spanned the 1.85x1.25 octagonal test section. A total of 92 pressure orifices, 0.4 mm. in diameter, was located in upper and lower surface in staggered formation. The model had a trailing edge thickness of 0.6%c which enabled the installation of a pressure orifice right at the 100% chord station.

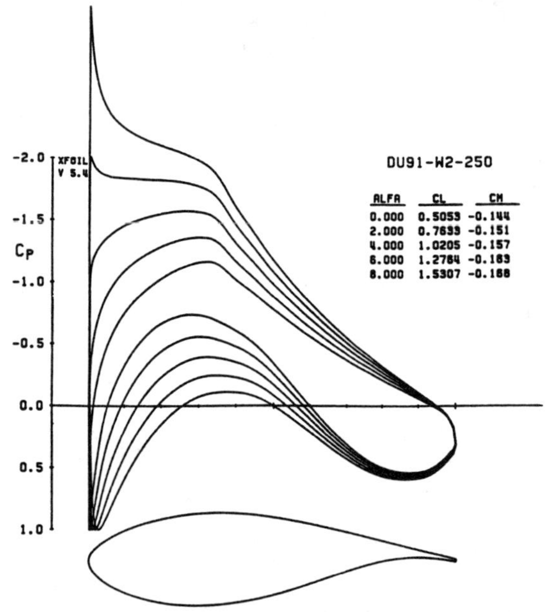

Fig. 1: Potential flow pressure distributions

3.2 Wind tunnel

The measurements were performed in the low-speed low-turbulence wind tunnel of the Faculty of Aerospace Engineering of Delft University of Technology. The wind tunnel is of the closed single-return type. The free-stream turbulence level in the 2.7 m. long test section varies from .02% at 10 m/s to 0.1% at 90 m/s.

3.3 Instrumentation

The 12 static and 50 total pressures in the wake were measured with a wake rake located half a chord length downstream from the trailing edge. All pressures were recorded by an automatically reading multi-tube liquid manometer. The data were reduced to pressure distributions and force coefficients and plotted on-line using the HP 9000/800 laboratory computer.

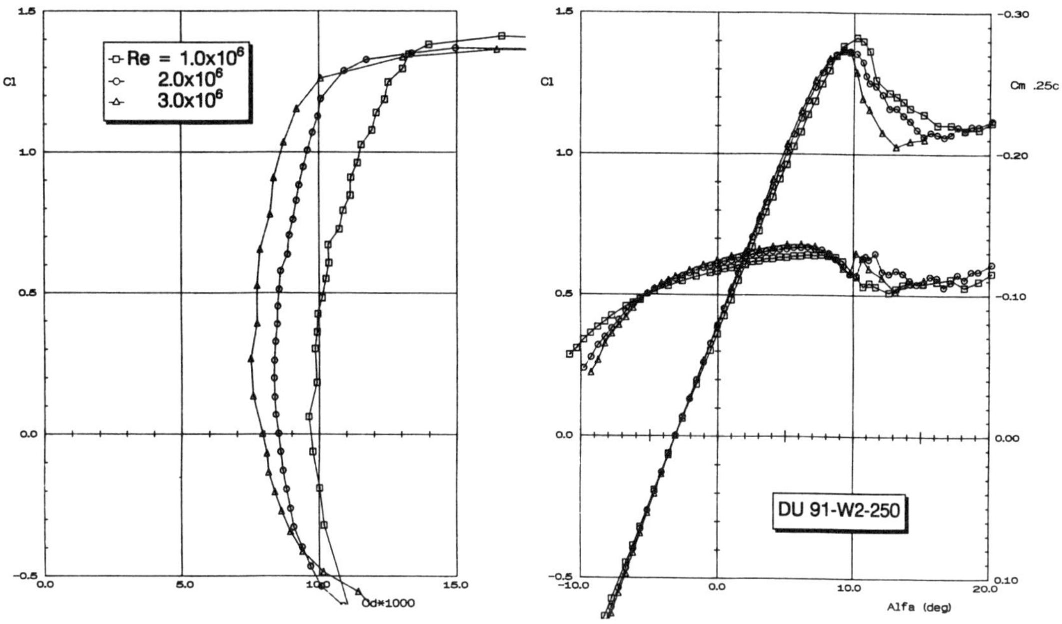

Fig. 2: Performance of DU 91-W2-250 for three Reynolds numbers

3.4 Wind tunnel wall corrections

Wind tunnel boundary corrections were made according to the method of Allen & Vincenti (ref.3). The effective height for the vertically mounted model is 1.656 m., leading to a blockage ratio of 0.36. In ref. 3 it is stated that the validity of the correction method at high lift is restricted to blockage ratios below 0.4.

4. RESULTS AND DISCUSSION

4.1 Free transition

Fig. 2 shows the results for three Reynolds numbers in the smooth surface configuration. Fig. 3 gives a comparison with XFOIL lift calculations for a Reynolds number of 3 millions. It appears that though the airfoil has a blunt trailing edge of 0.6%c the XFOIL calculations with a sharp trailing edge fit best to the measured lift curve. The transition locations on upper and lower surface were measured with a stethoscope. The trailing edge separation points were deduced from the measured pressure distributions. In general XFOIL tends to calculate a transition location which is too favourable, but in this case the measured lower surface transition locations and those of the upper surface up to a lift coefficient of appr. 1 compare very well with the predicted ones. However for lift coefficients higher than 1 the measured transition location on the upper surface deviates significantly from the prediction. This is believed to be a combined effect of a slight divergence of the model contour from the nominal coordinates between 20% and 30% of the chord and the calculated trailing edge separation starting too late. As a consequence of the latter the calculated suction peak on the nose is not lowered by a change of circulation due to separation, triggering transition in an earlier stage. The post stall lift curve is more abrupt than was expected on the basis of the calculations. Apparently the combination of the transition location and the local pressure gradient makes the turbulent boundary-layer grow faster than calculated.

4.2 Forced transition

Fig. 4 gives a comparison of the airfoil performance in the smooth and the rough condition at a Reynolds number of 3 million. To trip the boundary layer zig-zag tape turbulators with a thickness of 0.32 and 0.40 mm and tape with 0.15 mm bulges were consecutively located at the upper surface 5% chord station. The critical roughness height on that particular spot is about 0.15 mm, as calculated by the method of ref. 3. As can be seen the peak lift coefficient is somewhat affected by the roughness, a consequence of free transition not being in the nose region in the clean airfoil case. The results however compare favourably with earlier tests on a 25% thick representative of the NACA 63-4xx series (fig. 5).

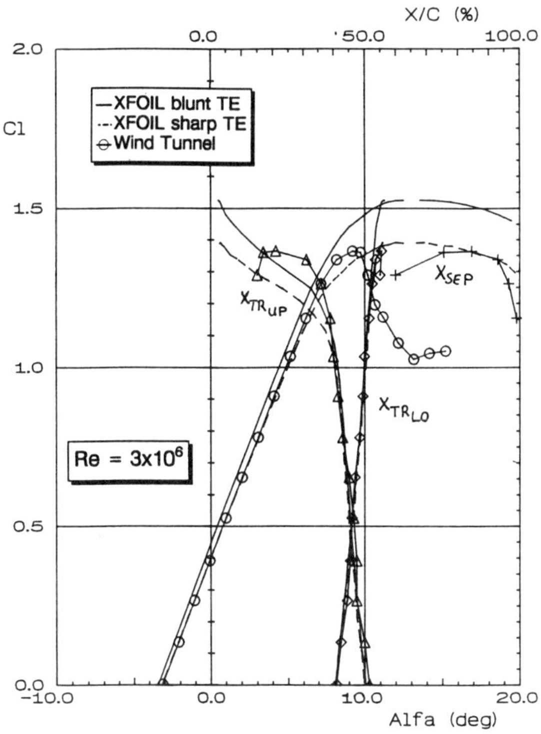

Fig. 3: Comparison of wind tunnel tests and calculations

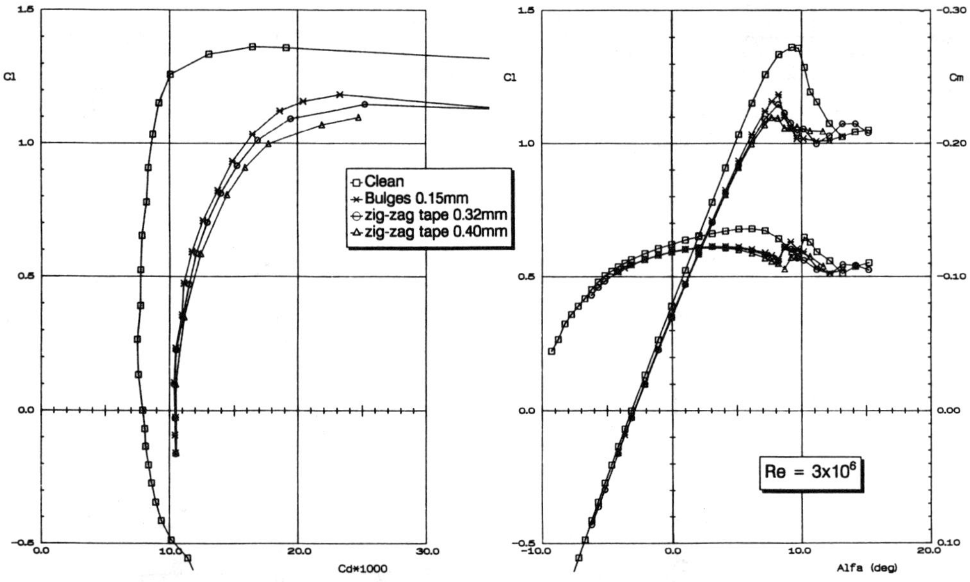

Fig.4: The influence of upper surface roughness at 5%c for Re=3x10^6

4.3 Vortex generators

In the rotating situation airfoils can exhibit higher peak lift coefficients and stall delay, the extent of which in general increases towards the root of the blade. Vortex generators extract momentum from the local free-stream and energize the boundary-layer, helping it to negociate the pressure rise towards the airfoil trailing edge just like span wise pressure gradients and Coriolis forces in the rotating situation can help to suppress the development of trailing edge separation. No doubt there are differences in both systems, but v.g. tests certainly can give a idea of what is possible. To investigate the potency of this airfoil, i.e. to what peak lift coefficient and angle-of-attack stall can be postponed (as if the airfoil were placed in the root for example) measurements have been performed with vortex generators installed at 20% and 30% chord (fig. 6). Wake rake drag measurements in span wise direction showed a regular wavy pattern with a wavelength equal to the spacing of the v.g. 's and an amplitude of about .0025. At 14^0 AOA the boundary layer separates in front of the vortex generators at 30%c, making them ineffective. With the vg 's installed at 20%c the peak lift coefficient almost reached 2.0 at α=17°

4.2 Gurney flaps

A Gurney flap is a flap of very small depth, usually 1 to 2% of the chord, attached perpendicular to the lower surface trailing edge of the airfoil. Several tests have shown that with the application of a gurney flap the airfoil performance can be improved. The benefits of a gurney flap - in the form of higher (peak) lift and a possible lower drag- will generally increase with increasing trailing-edge-included angle. The DU 91 airfoil here considered has a very small TE-angle and it was not expected that any drag reduction would be accomplished. However, due to the relatively simple means required it was decided to do some tests with an L-shaped

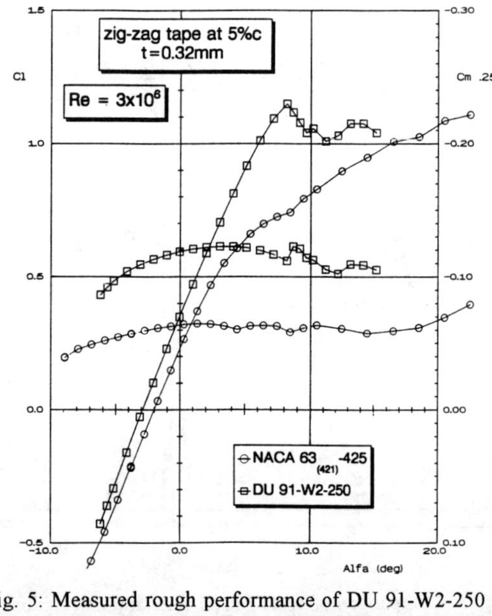

Fig. 5: Measured rough performance of DU 91-W2-250 and NACA 63$_{(421)}$-425, zig-zag tape 0.32 mm thick at 5%c

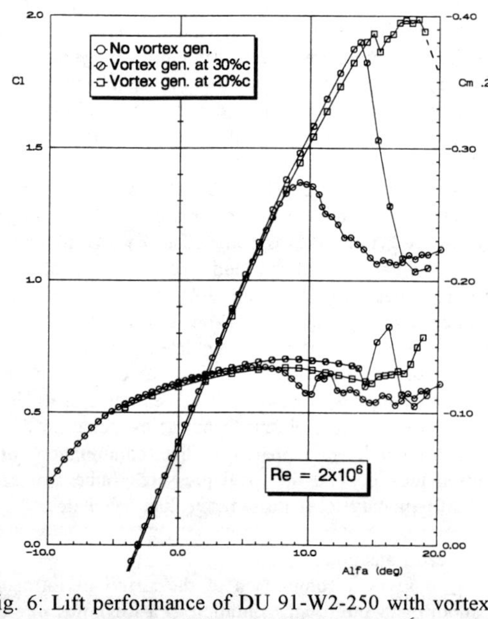

Fig. 6: Lift performance of DU 91-W2-250 with vortex generators at 20%c and 30%c, Re=2x10^6

metal strip attached to the lower surface trailing edge. Fig. 7 shows the pressure distribution of the airfoil at zero incidence and a Reynolds number of 1 million with and without a gurney flap of 12 mm (2%c) height. As a result of the flap the whole pressure distribution is blown up. At the trailing edge upper surface the flow is accelerated (higher negative-C_p). Due to the dead air region in front of the strip on the lower surface the local pressure increases. Fig. 8 gives the result of measurements at a Reynolds number of 2 millions with flap heights of 1 and 2%c compared to the no-flap case. The figure shows that indeed a drag reduction did not occur, at least not in the lower AOA-range. The maximum lift coefficient was raised from 1.37 to 1.59 for the 6mm high flap and to 1.73 for the 12mm high flap. The stall curve however becomes very sharp due to the higher upper surface negative-Cp peaks on the nose of the airfoil resulting in a rapid thickening of the boundary layer and a sudden shift of separation to the midchord position. The application of a 12 mm high serrated gurney flap with 90^0 wedges having the same frontal area as the 6mm high flap gave no significant increase in performance compared to the 6mm high flap case.

With both vortex generators at 20%c and a gurney flap of 12mm height a peak lift coefficient of 2.22 was measured.

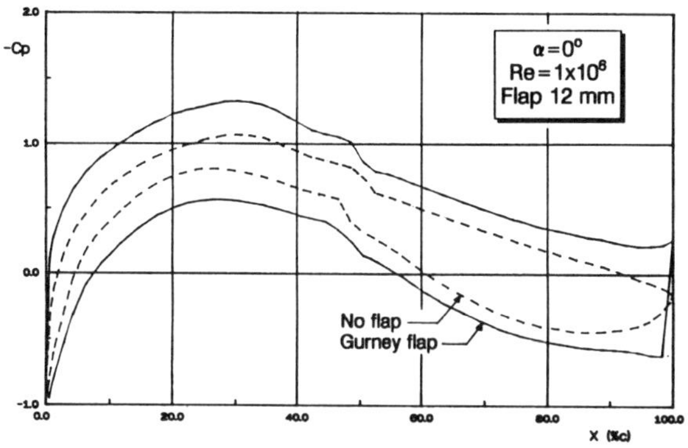

Fig. 7: Pressure distribution with and without a Gurney flap

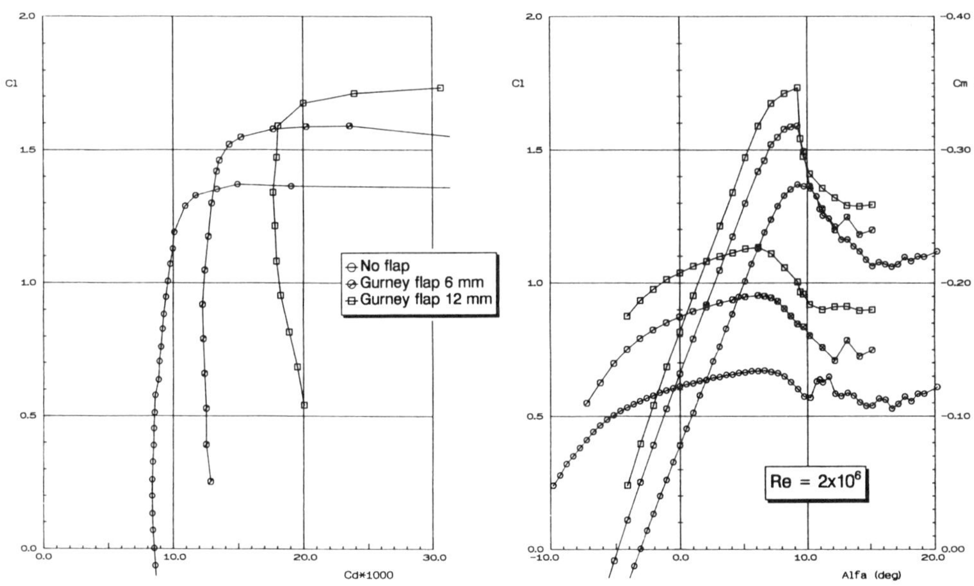

Fig. 8: The influence of Gurney flaps of 1 and 2%c height on the airfoil performance

5. CONCLUSIONS

A 25% thick airfoil designated DU 91-W2-250 has been designed and wind tunnel tested. Comparisons show that in general the design requirements have been met. The performance of the airfoil in the rough condition is much better compared to that of a more conventional evenly thick NACA 63 series airfoil. The post stall liftcurve however is too steep, which will be further investigated. The maximum lift potential of the airfoil in a fully three dimensional rotating situation is at least 2 as deduced from vortex generators measurements. Placed at 20% chord the vg's will raise the peak lift coefficient from 1.37 to appr. 2 at a Reynolds number of 2 millions. Possibly due to the small TE-included angle of this airfoil the application of a Gurney flap leads to an increase in the lift but also to higher drag at moderate angles-of-attack.

REFERENCES

[1] W.A. Timmer, Wecs blade airfoils - the NACA 63-4xx series. Proceedings of the European Community Wind Energy Conference 1990. Madrid, Spain.

[2] W.A. Timmer, R.P.J.O.M. van Rooy, Thick airfoils for HAWTs, Journal of Wind Engineering and Industrial Aerodynamics, vol 39, may 1992, Elsevier Science Publishers B.V. Amsterdam

[3] H.J. Allen, W.G. Vincenti, Wall interference in a two-dimensional-flow wind tunnel with consideration of the effect of compressibility. NACA Report 782, 1944

[4] A.E. Braslow, E.C. Knox, Simplified method for determination of critical height of distributed roughness particles for boundary-layer transition at mach numbers from 0 to 5. NACA Technical Note 4363, 1958

PREDICTION OF YAW LOADS ON A HORIZONTAL AXIS WIND TURBINE

S. Pesmajoglou and J.M.R. Graham
Department. of Aeronautics
Imperial College of Science, Technology and Medicine
London SW7 2BY, United Kingdom

ABSTRACT: The development of an unsteady vortex lattice method and its application to the flow over the blades and wake of horizontal axis wind turbines is presented. The model is modified to simulate the effect of stall along parts of the blades. Free stream turbulence is also simulated by using a statistical Gaussian turbulence model modified to take account of the effect of height and terrain roughness. Time histories of power, yaw moment and blade root bending moments are all calculated.

1. INTRODUCTION

Yaw loads induced on the blades and on the nacelles of horizontal axis wind turbines (HAWTs) account for a large number of fatigue failures of various components of free-yaw and driven-yaw systems. Problems arise with these loads because of the limited understanding of yaw dynamics and the inability to correctly predict the cyclic loads arising from misalignment of the rotor axis to the wind.

The basic approach for the prediction of aerodynamic performance of a HAWT is the commonly used modification of blade-element theory originally applied by Glauert [1] to a yawed propeller and extended by Wilson et al. [2] to wind turbines in order to predict the nature of stability in yaw. This method yields an almost neutral stability which does not agree with experimental results by Smulders et al. [3]. A prescribed wake model has been used by Zhong et al. [4] to estimate yawing moments by applying a simple moment balance procedure to the whole mass flow passing through the rotor disc. The wake skew angle is empirically modelled following observations from visualisation experiments.

Free-wake models, based on the steady vortex lattice method, were developed by Simoes et al. [5] and Rosen et al. [6] to investigate steady flow conditions for HAWTs and this method has been modified to account for separated flow, Simoes et al. [5]. Although both of these methods lead to results comparable to experimental values, they are not directly applicable to the case of a yawed rotor.

For the yawed case the resulting flow field is unsteady in all frames of reference and a fully unsteady flow computation is needed to predict the geometry of the wake and the distribution of aerodynamic loads and moments on the impulsively started rotor. This paper presents calculations using an unsteady vortex lattice method and attempts to establish the suitability of the unsteady vortex lattice technique for the turbine rotor.

2. COMPUTATIONAL FORMULATION

2.1 Unsteady vortex lattice method

The task of the computational method is to model the flow around arbitrary blade planforms, with camber, taper and twist, moving through an inviscid and incompressible fluid. The vorticity in the flow is restricted to a thin region around the lifting surface and its wake and the flow outside this region is assumed to be irrotational. Both the blades and the wake are represented by sheets of vorticity. For an inviscid fluid the Kelvin-Helmholtz theorem specifies continuity of pressure across the wake, hence all wake vorticity is transported with the local flow velocity. This condition generates a force-free wake and is used to obtain its position. The Kutta-Joukowski condition must be applied, in its unsteady flow form, specifying that all vorticity generated along the sharp trailing edges of the blades, where the Kutta-Joukowski condition is enforced, is shed into the wake.

For a rotor blade the vortex lattice method involves dividing the camber planform surface into a number of quadrilateral panels. An equivalent number of quadrilateral vortex rings, of initially zero vorticity, are then superimposed with their leading elements positioned on the quarter chord line of each panel. The shape of the wake is not specified a priori but is shed and allowed to convect and deform at each time step. Starting from rest, figure 1, the wake vorticity at t = 0 is set to zero.

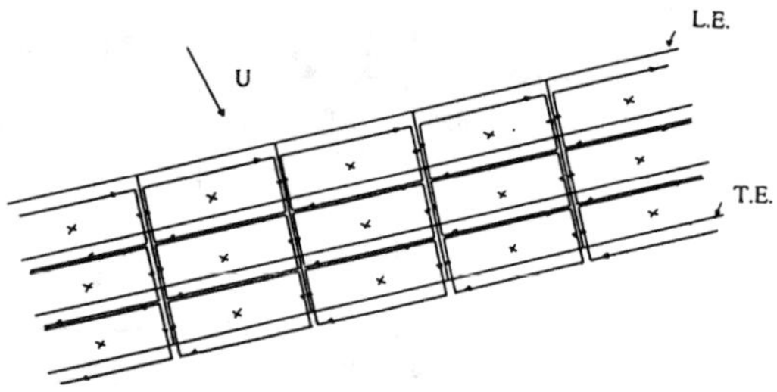

Figure 1. Example of uvlm (time = 0)

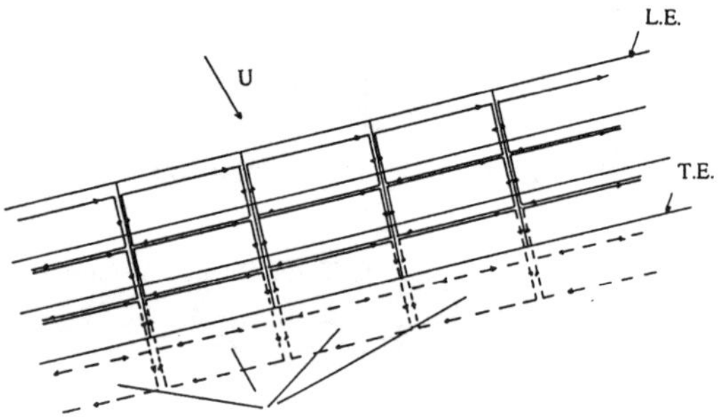

Figure 2. Example of uvlm (time = dt)

The flow tangency condition, which requires the normal component of velocity of the fluid relative to the lifting surface to be zero on the lifting surface, is applied at the control points that are located at the geometrical centre of each of the blade vortex rings. Applying the Biot-Savart law for the velocity induced by a straight line vortex filament in three-dimensional space this boundary condition leads to a number of simultaneous equations that can be written in a matrix form:

$A .\Gamma_b .n - \Omega .r .n + U.n = 0$ (1)

where A is the influence coefficient matrix for the blade,

Γ_b is the unknown vorticity on the blade,

U is the free stream velocity,

n are the direction cosines of the unit vector normal to corresponding panels,

Ω is the rotational velocity, and

r is the radius of the control point from the axis of rotation.

The unknown values of vorticity on the blade panels, are obtained by solving this system of linear equations. Consequently, when motion starts a bound vortex is formed along the trailing edge of the blade. The trailing edge panels are then shed to form the first row of wake panels, figure 2. Applying the unsteady Kutta-Joukowski condition along the sharp trailing edge the strength of the first row of the wake panels are set equal to that of the last row of the quadrilateral vortex rings of the blade. For all subsequent time steps (t > 0) equation (1) is modified to take account of the velocity Vw induced by the existing wake at the lifting surface control point:

$A.\Gamma_b.n + V_W.n - \Omega.r.n + U.n = 0$ (2)

The new values of the blade vorticity are calculated, the nodes of the existing wake panels are convected by the local velocity to new positions and a new row of panels is shed from the trailing edge into the wake. The procedure is repeated until an asymptotic state is reached.

The matrix A is calculated by considering the geometry of the total number of blades on the rotor disc and Vw is calculated by considering the complete wake geometry. For the solution of equations (1) and (2) a direct method, that involves the inversion of the matrix A, is used. In a body-fixed frame of reference A and n do not change as the control points move round the rotor disc. Hence, the inversion of A has to be performed only once at the beginning of the calculations.

For the present method a first order Euler integration (time stepping) formula has been employed to determine the new node positions from the local velocities evaluated at the start of the time step.

The forces on the blade are calculated at the midpoints of the vortex filaments of the camber surface by an application of the Bernoulli theorem. This leads to result similar to the Joukowski theorem that specifies :

$F = \rho V \times \Gamma + \rho\, d\Gamma/dt$

This is sufficiently accurate provided the reduced frequency of the blade chord, which is approximately equal to $\pi\sigma$, where σ is the local blade solidity, is small. From these forces the torque is calculated by taking the product :

$Q = F \times r$,

where r is the local radial distance from the centre of rotation and hence, the power is given by :

$P = Q.\Omega$

2.2 Computational efficiency

The efficiency of the computer code is improved by a) reducing the number of panels in the wake using an amalgamation procedure, and b) by "freezing" the wake after a specified number of turns.

For all presented results the first six rows of the wake panels are left unchanged and then every two rows, after the sixth, are amalgamated. This effectively halves the number of panels in the wake. The resultant circulation is calculated by

$$\Gamma_r = \frac{\sum_1^n A\Gamma}{\sum A}$$

where A is the area of each of the n panels to be amalgamated and Γ is the corresponding circulation value.

The effect of "part-freezing" the wake, which has been done for the results given here after one revolution is completed, is to neglect any further self induced effects. From this point on the wake is convected by the undisturbed stream only and hence no further wake expansion occurs after the first revolution. Although this might seem to introduce errors, the use of the "part-frozen" wake can be justified, since small changes in the wake far downstream of the rotor plane do not significantly affect the flow conditions on the blades.

For high aspect ratio blades it is known that the rolling up of the tip vortices takes place in a very narrow region of the span. This leads us to the use of a cosine distribution for the spanwise panels of the lattice, extending to a distance of a quarter of the spanwise panel length from the tips, Rubbert [7].

2.3 Separation model

Wind turbines operating at high winds often experience high angles of incidence over the whole or part of the span of their blades, figure 3.

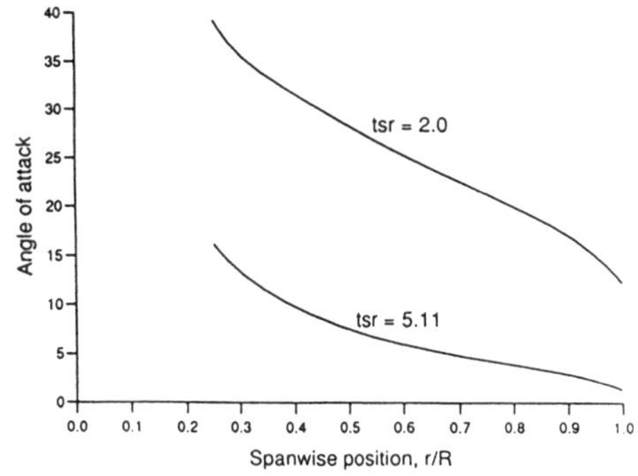

Figure 3. Relative angle of attack versus radial position

In order to represent separation on various spanwise sections a wake source model was used to simulate the flow over the blade. This formulation was based on the idea developed following the theory of Parkinson and Jandali [8] and involves the use of a number of sources, distributed over the surface of various blade stations. The strength of the sources is, in the present computations related to the behaviour of the drag for different angles of incidence. This model was implemented for the case of the HAWT and the results are presented in the next section. One of the limitations of this method is that it does not address the problem of rotor stall delay (the Himmelskamp effect). Large suction peaks have been measured near the leading edge of the inboard sections of rotors, at high angles of incidence, due to three dimensional effects in the flow (Brown and Graham [9]). As a result the predicted power coefficient may be lower than measured at low tip speed ratios.

2.3 Incident flow turbulence model

In order to study the effects of turbulence on the operation of HAWTs a statistical Gaussian turbulence model was used to represent the level of turbulence in the earth's lower boundary layer. This model was selected since the nature of the computational method enables the input of any arbitrary time history that can represent the structure of the incident flow field.

The turbulence model was modified in order to take account of the height of the wind turbine and of the terrain roughness. For all presented results the lateral component of the turbulent signal was used since it can be shown that (to a first order of accuracy) the yaw force is proportional to the lateral velocity component. An example of the signal used and its spectrum distribution is given in figures 4 and 5.

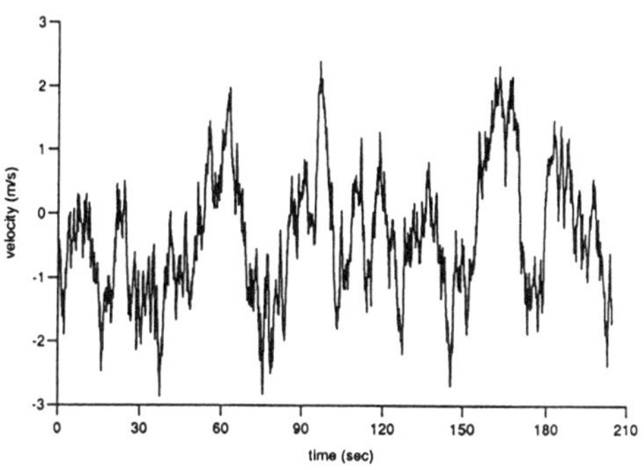

Figure 4. Lateral velocity turbulent signal

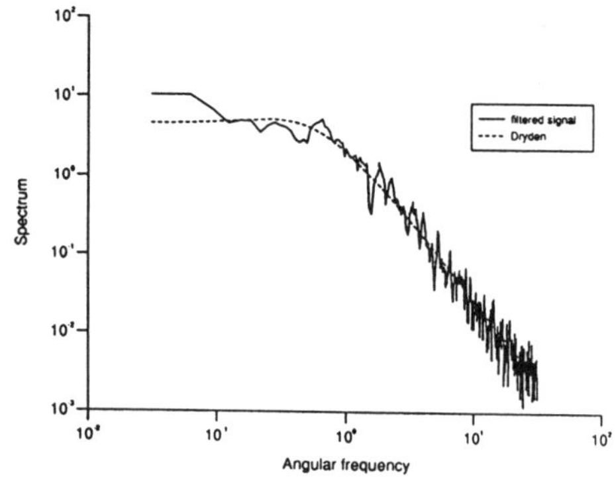

Figure 5. Spectrum of turbulent lateral velocity

The simulations presented in the next sections were done using an IRIS Indigo 4000 computer and the program took approximately 2.5 days of CPU time to produce a set of data equivalent to 60 seconds of flow time. This large running time makes the use of the present method prohibitively expensive for long time simulations. This led us to the use of an alternative formulation that is known as the indicial response technique, based on control theory for linear systems and usually applied in aircraft simulations. This technique uses the response of a system to a unit pulse input in order to calculate the response to an arbitrary input. For the present case the vortex lattice method was run until an asymptotic state was obtained and then a unit lateral velocity input was applied to obtain the response for the yawing moment of the rotor. This response was used to simulate the response to an arbitrary velocity input by time supeposition.

3. RESULTS

In order to validate the program results, a number of experiments were performed using a three-bladed model turbine that was developed by Noakes [10]. The major modification to the model was to replace the original cambered plate blades with three aerofoil section ones that were based on a 28m diameter Howden machine. The experiments took place in the 1.2mx1.3m wind tunnel of the Aeronautics Department of Imperial College. The measurements for the flap bending moment, from one of the blades, and the mean yaw moment were compared with predictions of the present method, figures 6 and 7.

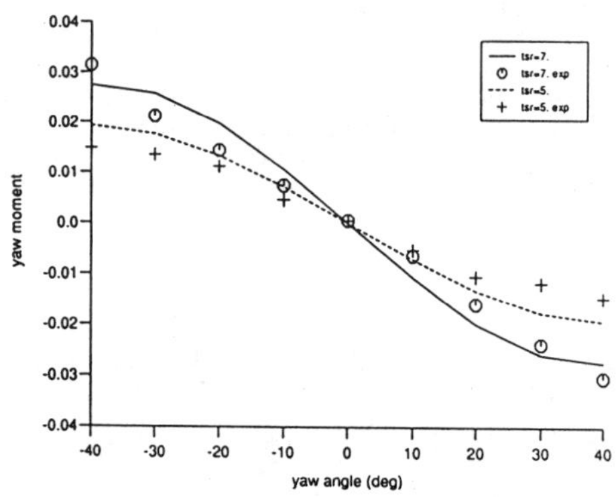

Figure 6. Mean yawing moment versus yaw angle

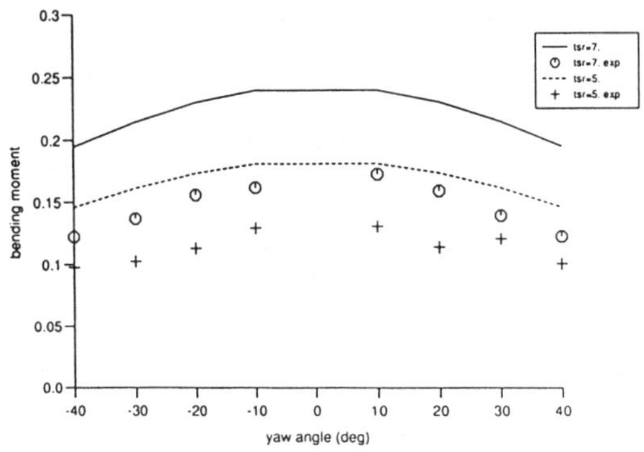

Figure 7. Mean flap bending moment versus yaw angle

As a second test the performance of a three-bladed Marlec wind turbine (3m diameter, rated at 1KW) was considered. The blades are tapered, untwisted, have a NACA4415 airfoil section and are set at a constant pitch angle of 6 degrees to the chord line.

For the case of the yawed rotor a set of results is presented for the rotor operating at a constant tip speed ratio of 5.11 at yaw angles in the range -40 and 40 degrees.

Figure 8 shows the shape of the wake, after the asymptotic state is reached for 4 angles of yaw. The wake after the first revolution follows the direction of the flow as expected due to freezing. However, deviations from this line can be seen in the first turn of the wake.

Figure 9 shows the variation of power coefficient at the selected angles of yaw. The value of Cp decreases as the absolute value of yaw angle increases following a relationship which is very close to $\cos^3\theta$ (where θ is the yaw angle). The results are in good agreement with both the predictions of the blade element theory, as modified by Ackerman [11] for the case of yaw, and also with the results by Smulders et al. [3] and Noakes et al. [10].

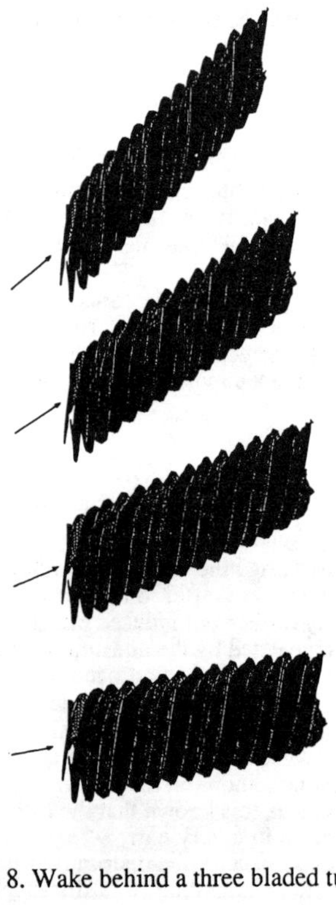

40 degrees

30 degrees

20 degrees

10 degrees

Figure 8. Wake behind a three bladed turbine at yaw.

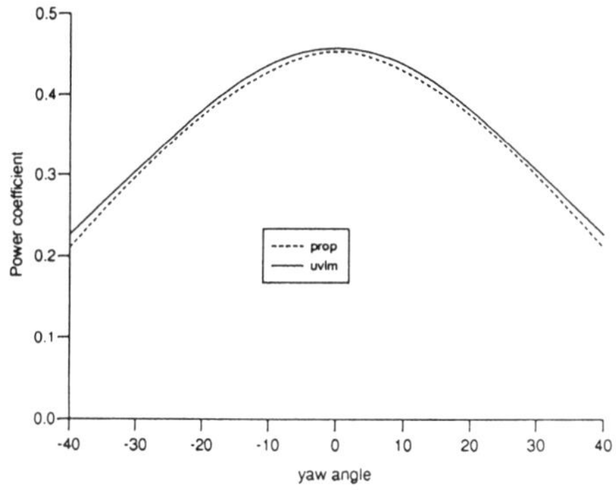

Figure 9 Power coefficient versus yaw angle

The input turbulence flow was applied on the model of a 55m diameter Howden machine. For the present results it is considered that the free stream direction is along the axis of the turbine and the lateral velocity is superimposed on the rotor disk. This in effect induces an unsteady yaw angle over a mean value of zero. The effect of the lateral component is mostly evident on the trace of the yawing moment of the wind turbine, figure 10. For the flap bending moment, figure 11, the turbulence contribution consists of a high frequency mode that is superimposed on the mean value due to the axial flow.

The results of the indicial method were compared against the predictions of the present method, figure 12. The running time for this method is a mere 1.5 seconds of CPU time (compared to 2.5 days!) and the agreement between the two signal is quite satisfactory. The small discrepancies can be corrected if a larger number of points in the system response to the pulse input is used (for this case only 200 time steps were considered). The initial differences are due to the establishment of the wake.

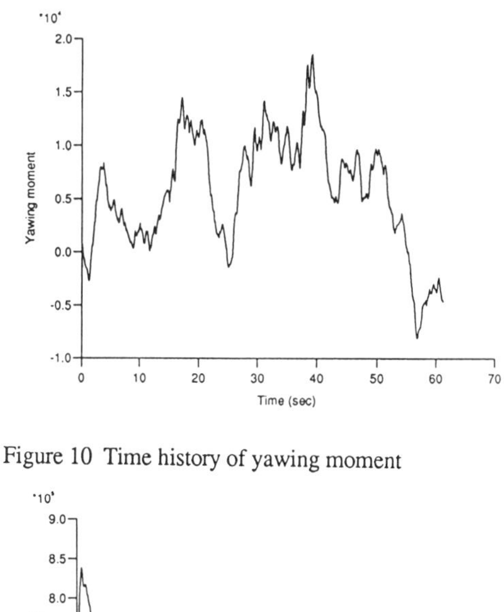

Figure 10 Time history of yawing moment

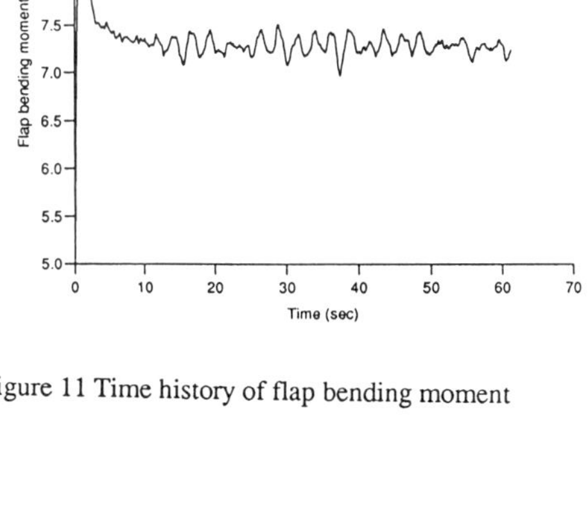

Figure 11 Time history of flap bending moment

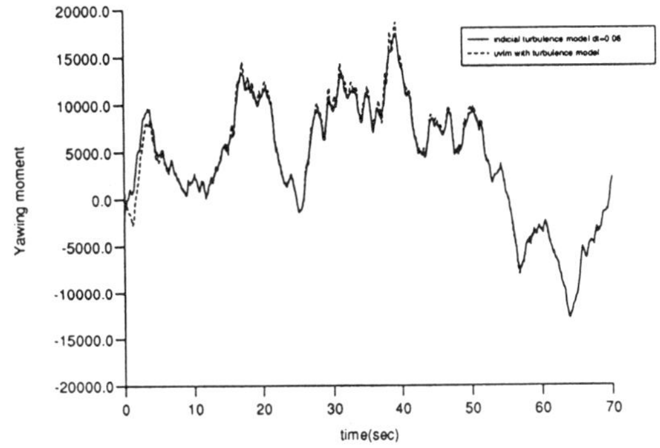

Figure 12 Indicial response to turbulent flow

4. CONCLUSIONS

Results have been presented for HAWTs in non-axial flow and turbulent conditions. The unsteady vortex lattice method works satisfactorily for almost all operating conditions. The indicial technique can be used very effectively in order to produce simulated data of the order of 1 hour of real flow time for statistical analysis of the loading of the yaw system.

ACKNOWLEDGMENTS

The work was supported by Wind-Harvester and ETSU (AEA Technology Ltd).

REFERENCES

[1] Glauert, H. "Miscellaneous Airscrew Problems", Aerodynamic Theory (ed. W.F. Durand), Div. L, Chapter XII, Dover Publ., Inc., N.Y. 1934

[2] Wilson R.E., Lissaman P.B.S., Walker S.N., "Aerodynamic Performance of Wind Turbines", Department of Mechanical Engineering, Oregon State University, ERDA/NSF/04014-76-1, June 1976

[3] Smulders P.T., Lenssen G., Leeuwen H. van, "Experiments with Wind Rotors in Yaw", Presented at : International Symposium on the Application of Fluid Mechanics and Heat Transfer to Energy and Environmental Problems, Patras, Greece, 1981

[4] Zhong S., Infield D.G., "Prediction of Wind Turbine Performance in Axial and non-Axial Flows by a Prescribed Wake Model", 13th BWEA Conference, 1991

[5] Simoes F.J., Graham J.M.R., "A Free Wake Model of the Wake of a Horizontal Axis Wind Turbine", 12th BWEA Conference, 1990

[6] Rosen A., Lavie I., Seginer A., "A General Free Wake Efficient Analysis of Horizontal Axis Wind Turbines", Wind Engineering, Vol. 14, No 6, 1990, pp. 362-373

[7] Rubbert P.E., "Theoretical Characteristics of Arbitrary Wings by a Non-Planar Vortex Lattice Method", Boeing Aircraft Co., Report D6-9244, February 1964

[8] Parkinson G.V. and Jandali T. "A Wake Source model for Bluff Body Potential Flow", JFM volume 40, 1970

[9] Brown C. J. and Graham J.M.R, "Investigation into the Effect of Turbine Rotor on the Maximum Lift Coefficient", Final report, Dept of Aeronautics, Imperial College, 1989

[10] Noakes J. and Sharpe D.J., "The measurements of Yaw Loads on a Model Upwind Horizontal Axis Wind Turbine", 13th BWEA Conference, 1991

[11] Ackerman M.C., "Yaw Modelling of Small Wind Turbines", EWEA Conference, Amsteram, 1991

THE FREE WAKE / HYBRID WAKE CODE ROVLM - A TOOL FOR AERODYNAMIC ANALYSIS OF WIND TURBINES

R. Bareiß, S. Wagner
Institut für Aerodynamik und Gasdynamik (IAG)
Universität Stuttgart, Pfaffenwaldring 21, 7000 Stuttgart 80
Tel. (49) 711 / 685-3401, Fax (49) 711 / 685-3438
Bundesrepublik Deutschland

ABSTRACT
The prediction of the flow field around a wind turbine is of essential interest. This is true especially for noise predictions, load calculations and interpretation of experiments regarding to the development of fast engineering tools. The objective of the study was to develop a fast simulation program that is able to predict the flow field at arbitrary points around a horizontal axis wind turbine without the time consuming task of space discretization. This information can be e.g. used for load calculations on the rotor. Due to its modular structure the code or parts of it can easily be extended even by other users or used within other simulation programs.

1. DESCRIPTION OF METHOD

The free wake code ROVLM (**RO**otor-Vortex-Lattice-Method) [1, 2, 3, 4] based on the Vortex-Lattice-Method (VLM) [5, 6] was developed to simulate the effects of vortex shedding and roll-up in the wake of a wind turbine. Based on the Laplace equation the rotor blade is covered with a set of quadrilateral singularity panels - here: doublets of constant strength - as in a classical panel methods (*figure 1*). The kinematic boundary condition is

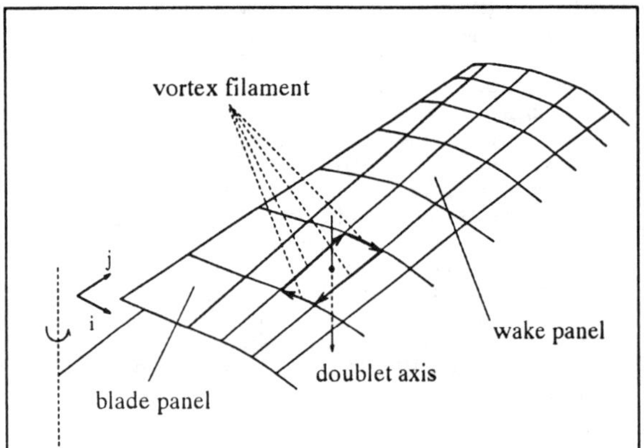

Figure 1: Discretization of blade and wake.

applied in the external Neumann form, i.e. vanishing normal velocity component in the panel center, to determine the unknown initial singularity distribution. Whereas the classical panel method is finished at this stage since it uses a prescribed wake to compute the complete induced velocities, in the present method the vorticity is shedded into a row of wake panels emmanating step by step from the separation line, here the trailing edge. The amount of shedded vorticity is determined by the Kutta condition. The wake itself now alters the induced velocities and therefore a new singularity distribution has to be calculated. This procedure is repeated for every time step, resulting in a growing wake. The deformation of the wake is calculated by a simple Euler update, i.e. multiplying the wake velocities with the time step. The simulation procedure is summerized in *figure 2*.

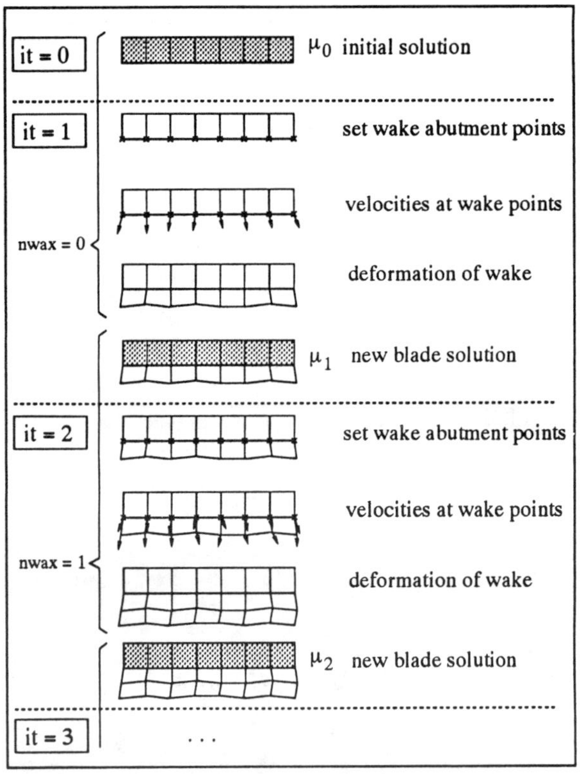

Figure 2: Simulation procedure.

The code can handle arbitrary flow conditions, e.g. gusts or yawed operation and several blade geometries, e.g. different blade tips. Apart from the discretization of the blades, the selection of a vortex core model and a time step no empirical input is required for a calculation. To reduce the high computational effort of a free wake calculation a hybrid wake model was developed [3] containing a simplified far wake transport.

The inviscid loads can be calculated by the Kutta-Joukowsky law and can be corrected with 2D profile tables to account for viscous effects [2]. With the help of the postprocessor FLOW velocities can be calculated at arbitrary field points (planes, cylinders, spheres, etc.). Therefore, the velocity inductions of the known singularities on the wake and on the blades are collected at the desired field points.

2. Results

The present method is applied in the CEC funded projects "Dynamic Inflow" (JOU2-CT92-0186) and "Development of An Aeroacoustic Tool For Noise Prediction of Wind Turbines" (JOU2-CT92-0148).

The following section summarizes some selected results generated during postprocessing with the program FLOW. *Figure 3* depicts the geometry of the rotor blades and the free wake of the two-bladed wind turbine UNIWEX (Universal Wind Turbine for Experiments) of Universtity of Stuttgart (ICA) at a wind speed of 7.44m/s and a rotational speed of 71.6rpm ($\lambda = 8.1$). The blades are discretized into 15 panels in spanwise and one in chordwise direction. The panel density is increased near the tip to resolve the roll-up of the tip vortex.

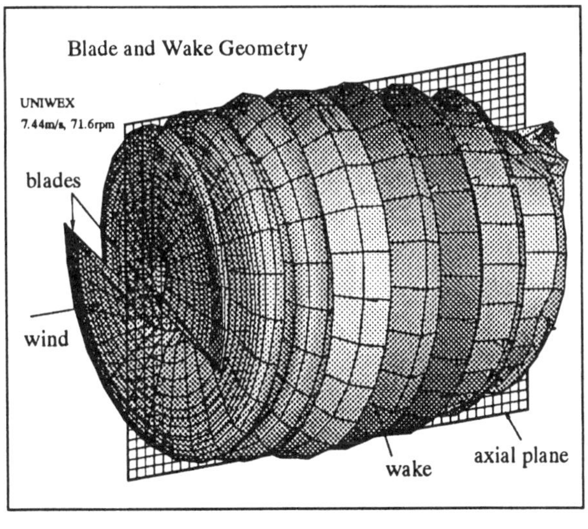

Figure 3: Blade and wake geometry of the UNIWEX turbine.

The axial induction factor (V_z/V_∞) in the axial plane (figure 3) is shown in *figure 4* after 4.25 revolutions. The cores of the root as well as of the tip vortices can be located. Downstream, the effect of the starting vortex is still visible indicated by a very low velocity level ($V_z/V_\infty \approx 0.3$) and the distorted flow field. The flow field can be mainly subdivided in three nearly cylindrical areas: (1) the region with an approximate radius of the root cut-out, where the velocities are higher than the freestream values (see below), (2) a part extending from root cut-out to the blade tip, where the freestream velocity is reduced to values of approximately $0.6 \cdots 0.8$ and (3) the region extending from the tip outboard with velocities close to the freestream level. The inductions near the root vortices

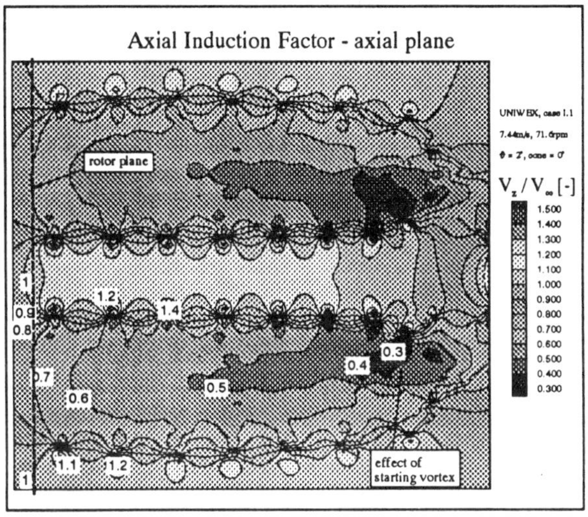

Figure 4: Axial induction factor in axial plane.

are higher because of the higher level of vorticity near the root compared to that at the tip. This is caused by the high angles of attack (> 20°) near the root area. Within the model the separation line is fixed at the trailing edge. In reality the flow will separate more outboard, which will result in lower levels of vorticity. An improvement of the root flow is expected through application of a separation model resulting in a variable separation line.

A typical calculated inflow pattern (V_z/V_∞) in a plane 3cm above the rotor plane is depicted in *figure 5* for a 20° yawed operation of the TU Delft wind tunnel model with a radius of 0.6m at a wind speed of 6m/s ($\lambda = 7.5$). The inflow velocities are calculated for one revolution consisting of 24 time steps and are averaged over this time period. The displacement of the wake in the negative y-direction due to the yaw velocity component results in a asymmetric distribution near the root whereas the tip region is nearly symmetrical. In the outboard region atop some slightly higher velocities compared to the freestream values can be observed.

Contours of the absolute lateral (V_y) velocity component are shown in *figure 6* for the same case as in figure 5. The maximum value is located downstream and is about 136% of the free stream component in lateral direction whereas the minimum value is reached upstream with 68%. This is a result of the lateral wake displacement too. Additionally, the stream lines - also inserted in the plot - support the interpretation that the flow is accelarated in the main area of the rotor plane.

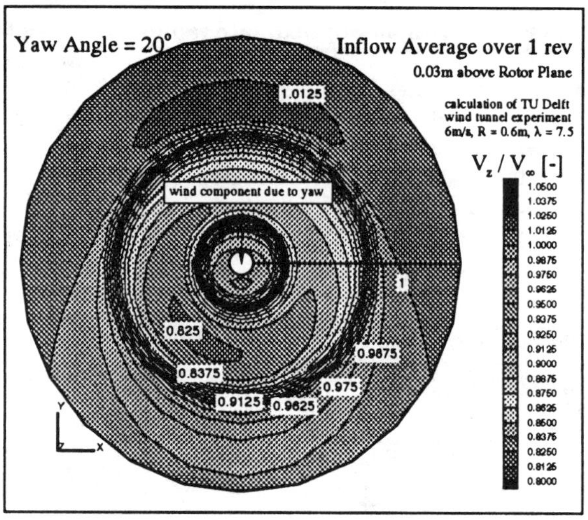

Figure 5: Time averaged inflow above rotor plane at 20° yaw.

The effect of a modified blade tip - expected as important for aerodynamic noise reduction [7] - is presented in *figure 7*. The rotor blade of the Debra is covered with 24 spanwise and 5 chordwise panels. A free wake calculation is performed with an azimuthal resolution of 5° for the original blade and a modified blade. In the latter case a secondary vortex can be observed leaving the kink at the trailing edge. The effect of the blade geometry modification on the vorticity distribution can be observed in four radial planes located in the near wake. The plane located close to the trailing edge is shown additionally in *figure 8*. The maximum value is reduced and the location of the new maximum seems to be shifted downstream away from the tip. The sensitivity of the spatial resolution as well as of the core model should be investigated before drawing further conclusions concerning tip

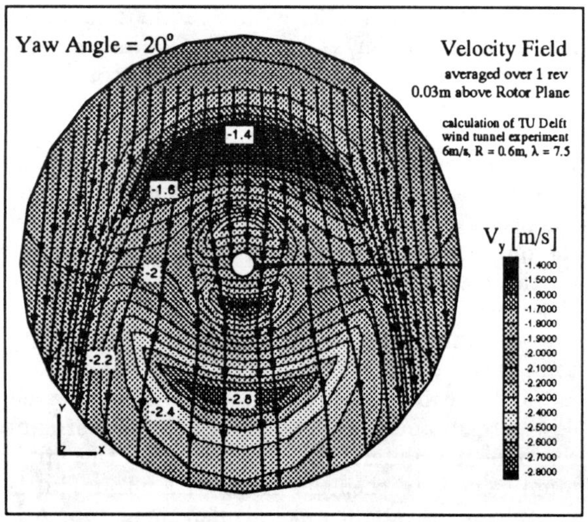

Figure 6: Time averaged velocity field above rotor plane at 20° yaw.

modifications.

3. Conclusion and Future Work

A method including vortex shedding and roll-up of the vortex sheet for the calculation of a wind turbine wake was presented. Quasi-steady loads on the rotor blades can be determined as well as the velocity field at arbitrary points in the flow field. This flow field information offers insight into the phenomena of vorticity convection around the turbine and together with the loads build a basis for understanding the underlying flow with regard e.g. to the development of a noise prediction tool.

Future work will include effects of tower / rotor interaction, blade thickness, partial pitch control and further investigation of noise relevant blade tip modifications.

4. References

[1] Röttgermann, A., Behr, R., Schöttl, Ch., Wagner, S. : *Calculation of Blade Vortex Interaction of Rotary Wings in Incompressible Flow by an Unsteady Vortex Lattice Method Including Free Wake Analysis*, Paper presented at the GAMM–Symposium, Kiel, Jan. 1991. Notes on Numerical Fluid Mechanics, Vol. 33: Numerical Techniques for Boundary Element Methods, (W. Hackbusch (Ed.)), Vieweg 1991, pp. 153–166.

[2] Bareiß, R.; Wagner, S. : *Load Calculations on Rotor Blades of a Wind Turbine*, Proceedings of the 5th IEA Symposium, University of Stuttgart, December 1991.

[3] Bareiß, R.; Wagner, S. : *A Hybrid Wake Model For Wind Turbines*, STAB Symposium Köln-Porz, 1992. to be published as a DGLR-Report.

[4] Bareiß, R.; Wagner, S. : *Aerodynamic Codes As A Basis For Aerodynamic Noise Prediction Of Horizontal Axis Windturbines*, to be presented at the International Conference on Noise and Vibration Control, St. Petersburg, Russia, May 31 - June 3, 1993.

[5] Behr, R.; Wagner, S. : *A Vortex-Lattice Method for the Calculation of Vortex Sheet Roll-up and Wing–Vortex Interaction*, Finite Approximations in Fluid Mechanics II, Notes on Numerical Fluid Mechanics, Vol. 25, pp. 1 – 14, Vieweg-Verlag, Braunschweig 1989.

[6] Behr, R.; Wagner, S. : *Application of a Low Order Panel Method to Slender Delta Wings at High Angles of Attack.* Paper presented at the Symposium of the International Association for Boundary Element Methods (IABEM 90), Universita degli Studi di Roma "La Sapienza", October 15–19, 1990. Boundary Integral Methods – Theory and Application, (L. Morino, R. Piva (Ed.s)). Springer 1991, pp. 105–114.

[7] Braun, K.A.: *Investigation Of Blade Tip Modifications For Acoustic Noise Reduction And Rotor Performance Improvement.* Proceedings of the CEC Wind Energy R & D Contractor Meeting, Alghero, 3-5 June 1992.

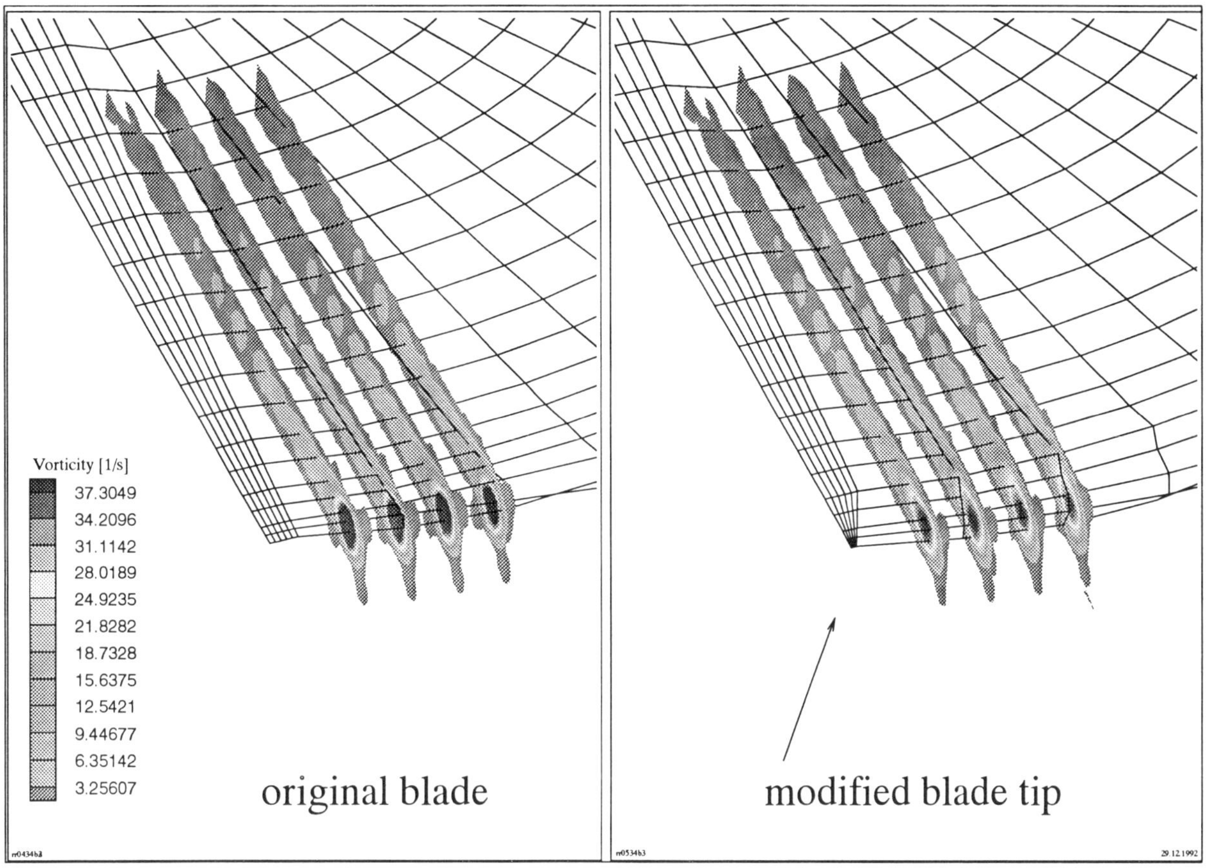

Figure 7: Influence of modified tip shape on vorticity in the wake (1).

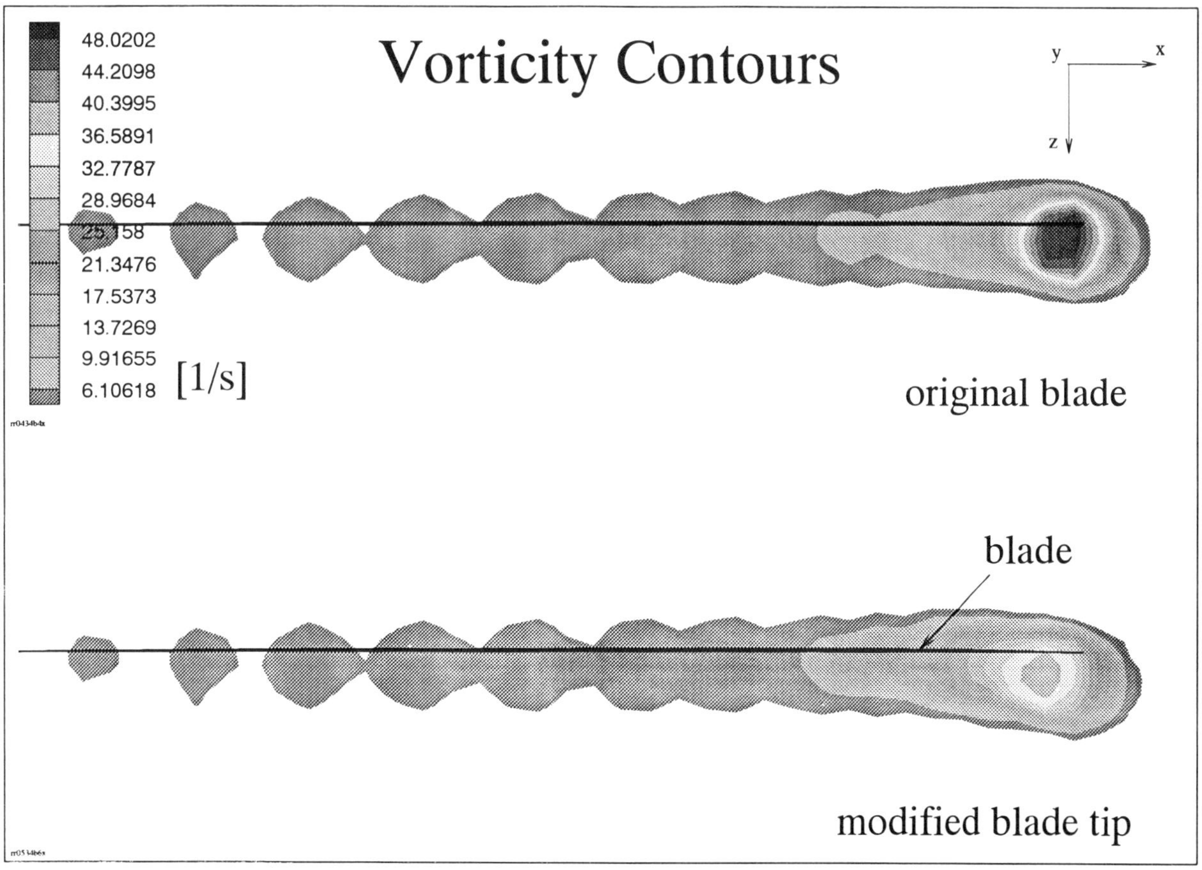

Figure 8: Influence of modified tip shape on vorticity in the wake (2).

DYNAMIC INFLOW EFFECTS AND VORTEX PARTICLE METHODS

Voutsinas S.G.[a], Belessis M.A.[a], Huberson S.[b]
[a] National Technical University of Athens, Fluids Section,
P.O. Box 64070, 15710 Zografou, Athens, Greece
[b] Universite du Havre, Laboratoire de Mecanique des Fluides,
Quai Frissard, B.P. 265, 76055 Le Havre Cedex, France

ABSTRACT: The modeling of the response of Wind Turbines to dynamic conditions, such as fast pitch angle changes, safety stops, sheared inflow is considered. The framework of analysis is defined by the assumptions that the fluid is incompressible and inviscid whereas the flow is unsteady and rotational (vortical). Under these assumptions the spatially distributed or free-vorticity representing the wakes of the blades will be concentrated forming free vortex sheets. The work uses a free-wake model by the name GENUVP (GENeral Unsteady Vortex Particle model). GENUVP is an unsteady code based on the vortex particle approximations for the free vorticity and vortex-lattice approximation for the flow induced by the blades. In the sequel GENUVP is presented together with some representative results for different cases of dynamic inflow conditions.

(this work is partially financed by the C.E.C. under the JOUR-0083)

1. INTRODUCTION

The response of an HAWT to dynamic inflow conditions is a special case of the aerodynamic performance problem of rotors, demanding the treatment of a three-dimensional and non-linear vortex or rotational flow. For a numerical analysis of the dynamic inflow effects, Vortex methods are among the most cost effective numerical models [1]. Based on the vortex particle approximation of the free vorticity and the boundary element approximation of the boundary conditions, a free-wake model by the name GENUVP (GENeral Unsteady Vortex Particle model) has been developed, that constitutes an improved version of the one presented in [2]. In the sequel, GENUVP is briefly described together with some representative results.

2. DESCRIPTION OF THE NUMERICAL MODEL

We consider the unsteady flow of an inviscid and incompressible fluid around a combination of N_B three-dimensional lifting surfaces $\boldsymbol{B_k}$ (i.e. the blades) with boundaries S_k, k=1(1)N_B that form the rotor of a wind turbine. Under the assumptions made for the fluid, the presence of lifting bodies imposes the introduction of the wakes S_{Wk}, k=1(1)N_B they generate, as active boundaries of the flow. Let,

$$\vec{x}_{Wk} \in S_{Wk}:\ \vec{x}_{Wk} = \vec{x}_{Wk}(\xi^1,\xi^2;t),\ \xi^1 \in [-1,1],\ \xi^2 \geq 0,\ t \geq 0 \quad (1)$$

denote a parametric representation of S_{Wk} shedded from the k-th lifting component of the configuration along its trailing and possibly its tip edges, i.e. the vorticity emission line. Considering S_{Wk} as a surface generated by the sequence of material lines leaving the emission line, $\vec{x}_{Wk}(\xi^1,\xi^2;t)$ is identified as the position at time t of a material element that was shedded at time ξ^2 and at the point along the emission line defined by ξ^1.

Using Helmholz's decomposition theorem and Green's representation theorem, the velocity field $\vec{u}(\vec{x};t)$ $\vec{x} \in D$, $t \geq 0$ takes the form:

$$\vec{u}(\vec{x};t) = \vec{U}_*(\vec{x};t) + \nabla\varphi(\vec{x};t), \quad \vec{x} \in D, \quad t \geq 0 \quad (2)$$

$$\varphi(\vec{x}_0;t) = -\sum_{k=1}^{N_B} \int_{S_k} \mu_k(\vec{x};t) \cdot \frac{\vec{\nu}(\vec{x}) \cdot (\vec{x}_0 - \vec{x})}{4\pi |\vec{x}_0 - \vec{x}|^3} dS(\vec{x}) - \sum_{k=1}^{N_B} \int_{S_{Wk}} \mu_{Wk}(\vec{x};t) \cdot \frac{\vec{\nu}_W(\vec{x};t) \cdot (\vec{x}_0 - \vec{x})}{4\pi |\vec{x}_0 - \vec{x}|^3} dS_W(\vec{x}) \quad (3)$$

where $\mu_k(.;t)$ and $\mu_{Wk}(.;t)$ denote the dipole distributions of the k-th thin lifting surface and its wake respectively. In (2) $\vec{U}_*(.;t)$ is a given div-free velocity field and $\varphi(.;t)$ the disturbance velocity potential. $\vec{U}_*(.;t)$ is set to include the inflow velocity $\vec{U}_\infty(.;t)$ as well as the velocity induced by the free vorticity; therefore it represents the rotational part of the velocity field.

In order to determine $\mu_k(.;t)$, $\mu_{Wk}(.;t)$ and $S_{Wk}(t)$ we dispose:

(a) The non-entry conditions on all the solid surfaces

$$\frac{\partial\varphi}{\partial\nu}(\vec{x};t) = \vec{\nu}(\vec{x};t) \cdot \left[\vec{U}_{Bk}(\vec{x};t) - \vec{U}_*(\vec{x};t)\right], \ \vec{x} \in S_k(t) \quad (4)$$

where $\vec{U}_{Bk}(.;t)$ denotes the rigid body velocity distribution of the k-th body.

(b) The zero pressure jump throughout the wakes, or equivalently Kelvin's theorem:

$$\frac{d_m \mu_{Wk}}{dt} = \frac{\partial \mu_{Wk}}{\partial t} + (\vec{U}_m \cdot \nabla_{Wk})\mu_{Wk} = 0 \quad (5)$$

Based on (1), (4) can be recast as:

$$\mu_{Wk}(\vec{x}_{Wk}(\xi^1,\xi^2;t);t) = \mu_{Wk}(\vec{x}_{Wk}(\xi^1,\xi^2;\xi^2);\xi^2) \quad (6)$$

stating that the dipole intensity carried by the material element ξ is equal to the value this element had when it was first shedded from the emission line of the body.

(c) The equations of motion for the wakes S_{Wk}, $k=1(1)N_B$,

$$\frac{d\vec{x}_{Wk}}{dt} = \vec{U}_*(\vec{x}_{Wk};t) + \nabla\varphi(\vec{x}_{Wk};t) \qquad (7)$$

In order to solve numerically the above equations, piecewize constant approximations are introduced for both $\mu_k(.;t)$ and $\mu_{Wk}(.;t)$. Let S_k^e, $e=1(1)E_k$ and S_{Wk}^e, $e=1(1)E_{Wk}$, denote the elements of S_k and S_{Wk}, $k=1(1)N_B$ with dipole intensities $\mu_k^e(t)$ and $\mu_{Wk}^e(t)$ respectively (E_k and E_{Wk} are the number of elements that form S_k respectively). In order to determine $\mu_k^e(t)$, $e=1(1)E_k$, $k=1(1)N_B$, the non-entry condition is applied to the centers $\vec{x}_c^e$ of the elements that form the solid boundaries S_k. As regards $\mu_{Wk}^e(t)$, the application of (6) to all the surface elements S_{Wk}^e of the wakes reduces the unknown degrees of freedom of $\mu_{Wk}(.;t)$ to the values of $\mu_{Wk}^e(t)$ along the corresponding emission lines only. On the other hand if (6) is applied along the emission line of S_k, the remaining unknowns of the $\mu_{Wk}(.;t)$ are set equal to the values $\mu_W^e(.;t)$ of the elements adjusted to this line.

Since the problem is time dependent, a time marching scheme is used, with time step Δt. According to the analysis given in the previous section, all information concerning the vortex sheets of the flow, is known from the previous steps, except of the *near part*, i.e. the part generated during the current time step. Consequently different approximations can be used for the near ("new" part) and the far ("old" part) region of the free vortex sheets. More specifically the vortex sheet assumption is retained only for the near region of every wake. On the contrary the rest, i.e. the "old" part, is transformed into free spatial vorticity, in the sense that a vortex particle approximation is introduced. Therefore, if $S_{Wk}^{\Delta}, S_{Wk}^{*}$, $k=1(1)N_B$ denote the near and far parts of the wake of the k-th lifting surface (Figure 1),

$$\varphi(\vec{x};t) = \Phi + \Phi_W^{\Delta} + \Phi_W^{*} \qquad (8)$$

where $\Phi_W^{\Delta}(.;t)$ and $\Phi_W^{*}(\cdot;t)$ denote the potentials induced by the near and far wakes respectively. As described in [3,4] there is an equivalence between dipole distributions and surface vorticity that permits to take $\nabla\Phi_W^{*}$ in a vortex particle form as follows:

$$\nabla\Phi_W^{*}(\vec{x};t) = \vec{U}_\omega(\vec{x};t) = \sum_{j\in J(t)} \frac{\vec{\Omega}_j(t)\wedge[\vec{x}-\vec{Z}_j(t)]}{4\pi\,|\vec{x}-Z_j|^3} \qquad (9)$$

where $\vec{\Omega}_j(t)$ and $\vec{Z}_j(t)$ denote the intensities and positions of the vortex particles obtained by integrating the surface vorticity carried by the wakes, whereas $J(t)$ denotes the index set for the vortex particles. Thus instead of calculating the geometry of the vortex sheets and the dipole distributions they carry, we follow the evolution of the vortex particles defined by the following dynamic equations:

$$\frac{d\vec{Z}_j(t)}{dt} = \vec{u}(\vec{Z}_j;t)\ , \quad j\in J(t) \qquad (10)$$

$$\frac{d\vec{\Omega}_j(t)}{dt} = \left(\vec{\Omega}_j(t)\cdot\nabla\right)\vec{u}(\vec{Z}_j;t) \qquad (11)$$

As the near parts still retain their character as vortex sheets their determination is different. Let $\vec{U}_{em}$ denotes the mean velocity at a point $\vec{X}_{em}$ along the vorticity emission line of a lifting body. The geometry of the near part of the corresponding wake S_{Wk} is determined kinematically through the following relation:

$$\vec{X}^{\Delta} = \vec{X}_{em} + \Delta t\cdot\vec{U}_{em} \qquad (12)$$

where $\vec{X}^{\Delta} - \vec{X}_{em}$ denotes the width of S_{Wm}^{Δ} in vectorial form.

Finally, the calculation of the unsteady loads on the rotor blades was based on the Bernoulli's equation:

$$-\frac{\Delta p_k}{\rho} = \frac{\partial\mu_k}{\partial t}(\vec{x};t) + (\vec{U}_m(\vec{x};t)\cdot\nabla)\mu_k(\vec{x};t) \qquad (13)$$

giving the pressure jump defined over S_k.

Due to the essentially inviscid character of the modelling adopted can only predict the "inviscid" part of the loading. The rest, i.e. the loads due to viscous effects must be superimposed. In this connection an approximate a-posteriori scheme, similar to the one used by the classical strip theory was applied. This scheme is based on the calculation of the radial distribution of the effective angle of attack. Next the drag coefficient C_D is estimated from given airfoil data. These drag forces are integrated over radial strips and thus an account of viscous effects is obtained.

3. RESULTS AND DISCUSSION

In the sequel, representative results are given for two pitch variation axisymmetric cases with velocities 7deg/s and 1deg/s, (Fig.2,3 and Fig.4,5 respectively), one pitch variation case with wind shear (Fig.6,7) and a safety stop case (Fig.8,9). The form of the vortex particle patterns in the wake is given in Figures 10 and 11 for a pitch variation case and a safety stop respectively.

The comparison between predictions and full-scale measurements show that the flow quantities that do not depend on viscous effects are well predicted both in global and in distributed form (e.g. the flapping moment). For flow quantities that depend strongly on viscous effects (e.g. the rotorshaft torque) though the predictions have the correct behaviour, their accuracy is not equally good. In this connection two improvements of GENUVP are planned: (a) the development of a more accurate scheme for the estimation of viscous effects,, and (b) the introduction of thickness effects.

REFERENCES

[1] A Leonard, "Computing three-dimensional incompressible flows with vortex filaments", Ann. Rev. Fluid Mech. (1985) Vol.17, 523.

[2] A Zervos, S Huberson, A Hemon, "Three-dimensional free wake calculation of wind turbine wakes" J. Wind Engineering and Industrial Aerodynamics (1988), Vol.27, 65.

[3] S G Voutsinas, "Theoretical and numerical analysis of three dimensional subsonic inviscid flows", PhD. Thesis, NTUA, Dept. Mech. Engineering (1990).

[4] A Zervos, S G Voutsinas, "Numerical simulation of three-dimensional vortical flows", NTUA Report (1991).

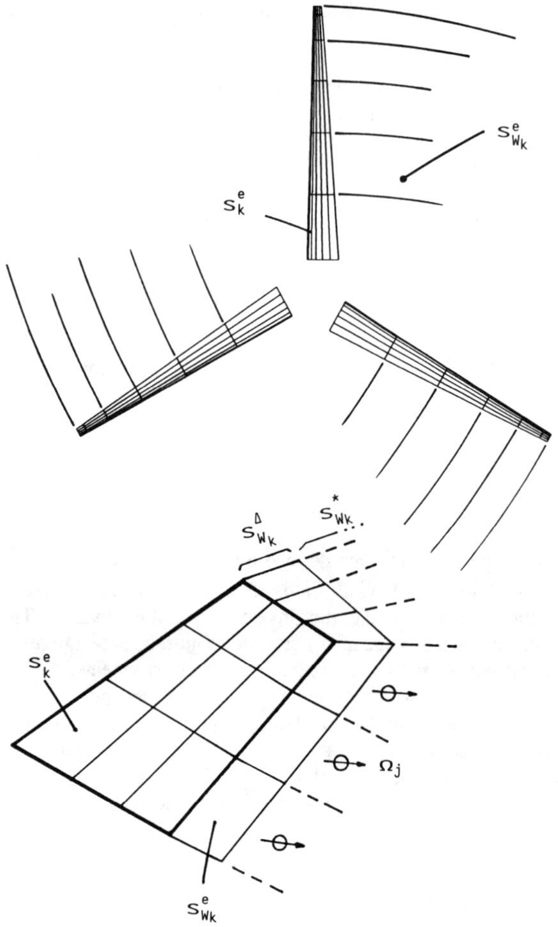

FIGURE 1:
The basic notations

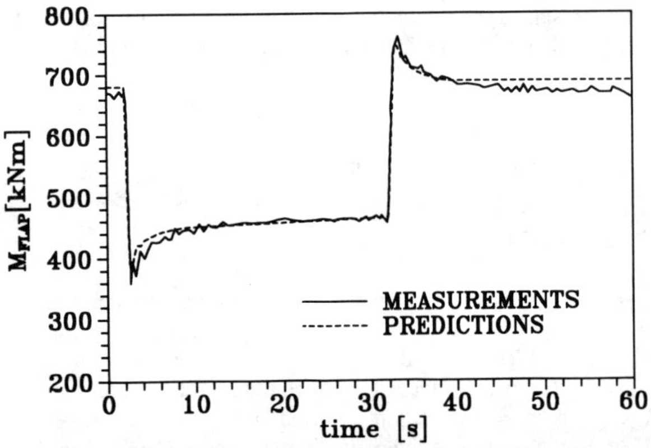

FIGURE 2:
Tjaereborg HAWT. Flapping moment at blade root during pitch variation (V_{WIND}=8.7m/s, pitch variation=3.6 deg., pitch variation velocity=7deg/s, step up at t=2.0s, step down at t=32.0s)

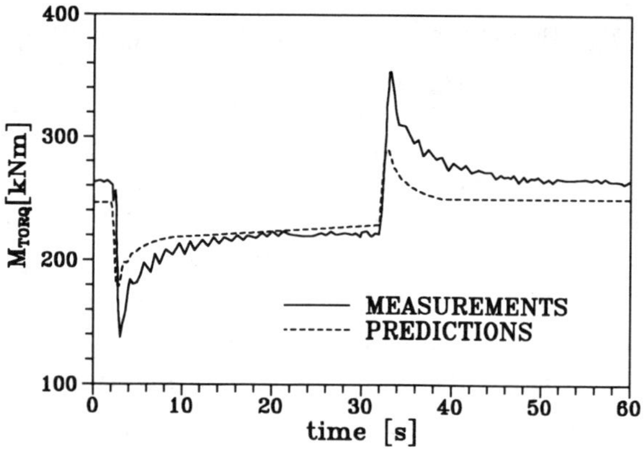

FIGURE 3:
Tjaereborg HAWT. Rotorshaft torque during pitch variation (V_{WIND}=8.7m/s, pitch variation=3.6 deg., pitch variation velocity=7deg/s, step up at t=2.0s, step down at t=32.0s)

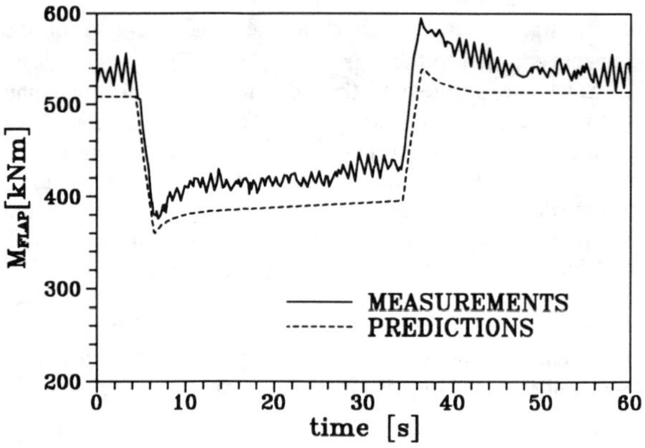

FIGURE 4:
Tjaereborg HAWT. Flapping moment at blade root during pitch variation (V_{WIND}=7.4m/s, pitch variation=2. deg., pitch variation velocity=1deg./s, step up at t=4.4s, step down at t=34.4s)

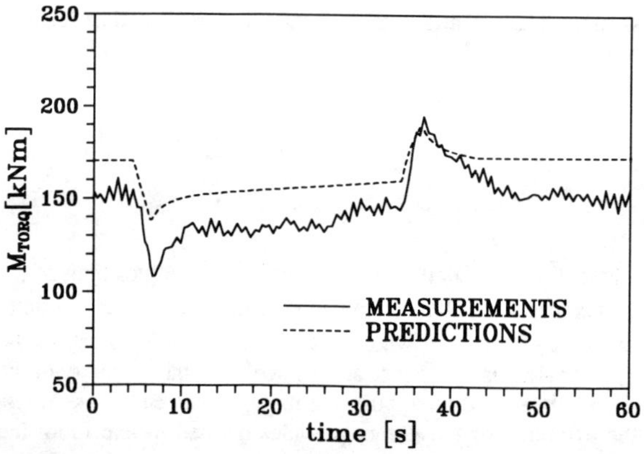

FIGURE 5:
Tjaereborg HAWT. Rotorshaft torque during pitch variation (V_{WIND}=7.4m/s, pitch variation=2. deg., pitch variation velocity=1deg./s, step up at t=4.4s, step down at t=34.4s)

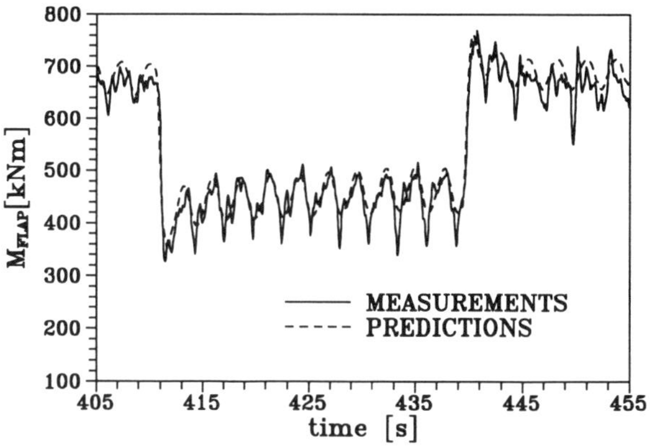

FIGURE 6:
Tjaereborg HAWT. Flapping moment at blade root during pitch variation (V_{WIND}=8.74m/s, wind shear according to power law with exponent = 0.11, pitch variation=3.7deg., step up at t=410.7s, step down at t=439.4s)

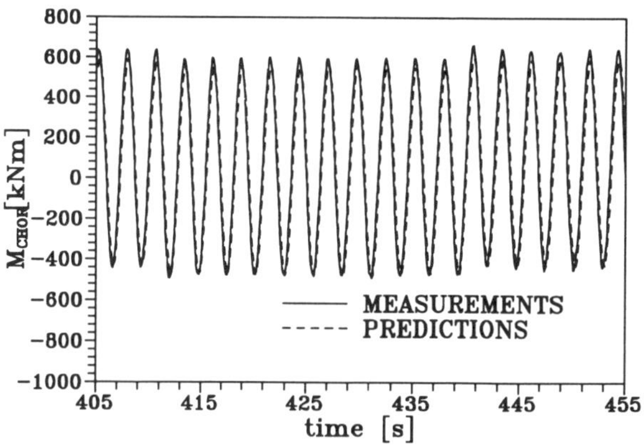

FIGURE 7:
Tjaereborg HAWT. Chordwise moment at blade root during pitch variation (V_{WIND}=8.74m/s, wind shear according to power law with exponent = 0.11, pitch variation=3.7deg., step up at t=410.7s, step down at t=439.4s)

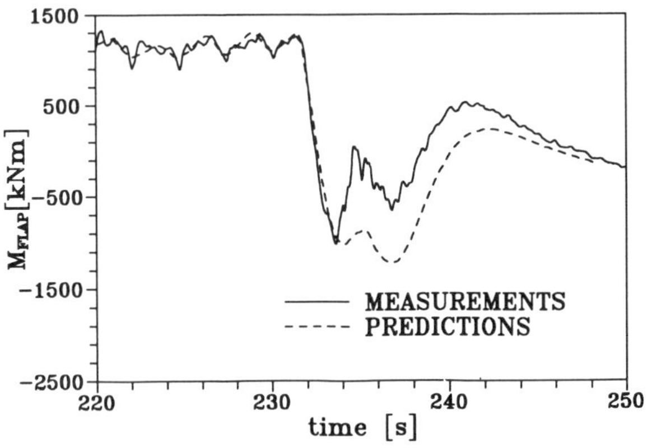

FIGURE 8:
Tjaereborg HAWT. Flapping moment at blade root during safety stop operation (V_{WIND}=7.09m/s, wind shear according to power law with exponent = 0.158, pitch variation=89deg.)

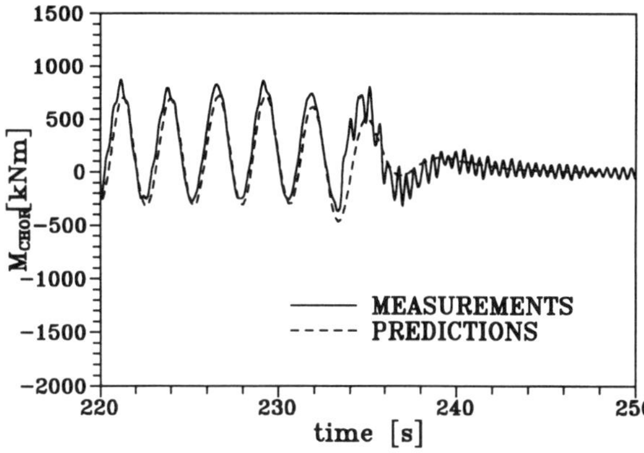

FIGURE 9:
Tjaereborg HAWT. Chordwise moment at blade root during safety stop operation (V_{WIND}=7.09m/s, wind shear according to power law with exponent = 0.158, pitch variation=89deg.)

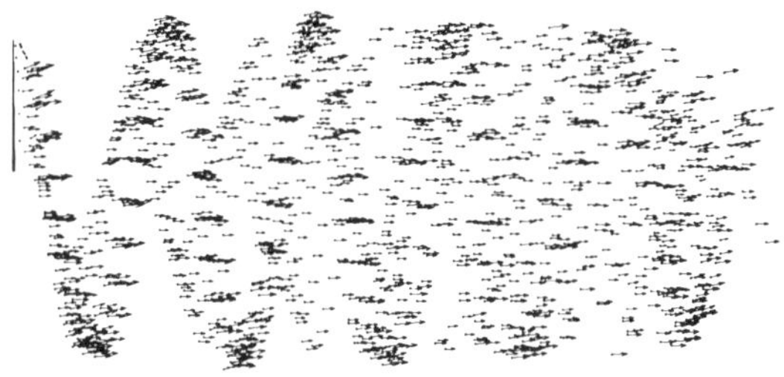

FIGURE 10:
Tjaereborg HAWT. Velocity vector field of the particles of the wake from one blade after pitch variation (V_{WIND}=12.5m/s, pitch variation=2.03deg.)

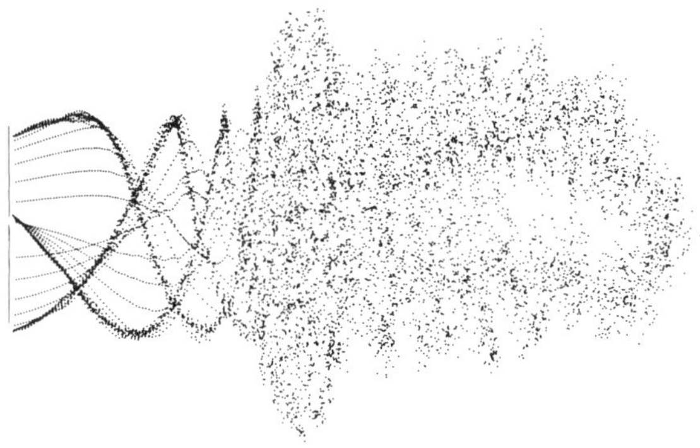

FIGURE 11:
Tjaereborg HAWT. View of the particles of the wake during safety stop operation (V_{WIND}=7.09m/s, wind shear according to power law with exponent = 0.158, pitch variation=89deg.)

WAKE STRUCTURE OF A ROTOR MODEL IN YAW

L.J. Vermeer[1] and J.K. de Boer[2]

[1]Institute for Wind Energy, Delft University of Technology
Stevinweg 1, 2628 CN Delft, The Netherlands
[2]Laboratory of Applied Physics, *University* of Groningen
Nijenborgh 18, 9747 AG Groningen, The Netherlands

ABSTRACT: Detailed velocity measurements have been performed in the open jet wind tunnel in the near wake of the rotor model. The wake velocities have been measured with hot-wire equipment, providing a very fast and accurate measuring method. For yaw angles 10°, 20° and 30° of the rotor, radial traverses are performed in 12 azimuthal directions at an axial distance of x/R = 0.05. At every position of the traverses, velocity samples are taken every 0.5° azimuth angle of the rotor. In previous experiments, it has been shown that the velocity signal is well determined. The passage of blades, vortex sheets and the influence of the tip vortex can be recognized clearly. As results will be presented: azimuthal distribution of the velocity at stationary probe positions, velocity distribution over the rotor plane for several yaw angles and the effect of yaw on the induction factor.

NOTATION

c	blade chord	m
E	voltage of hot-wire bridge	V
f	speed of revolutions	Hz
r	radial coordinate	m
R	rotor radius	m
V	velocity	m/s
V_0	undisturbed wind velocity	m/s
x	axial coordinate	m
Θ	pitch angle	°
λ	tip speed ratio: $\Omega R/V$	-
ϕ_y	yaw angle	°
ψ	azimuth angle	°
Ω	rotational speed: $2\pi f$	rad/s

1. INTRODUCTION

In the JOULE1 project Dynamic Inflow (JOUR 0083), load analysis have been made for step-wise wind fluctuations and for fast blade pitch excursions. The calculation results have been compared with field experiments from the Tjæreborg turbine in Denmark. As an extension of this program, wind tunnel experiments have been added. Besides the repetition of the above mentioned experiments in a conditioned environment, also measurements on a rotor model in yaw will be performed. In the underlying paper, these velocity measurements in the near wake of the rotor model in yaw will be presented.

2. EQUIPMENT

2.1 Wind tunnel

The measurements have been performed in the open jet wind tunnel of the Institute for Wind Energy. See Figure 1.

The tunnel consists of a flow channel of circular cross section (diameter 2.24 m), with a fan at the inlet, a flow straightener and gauzes. The flow straightener is made of thin walled aluminium, with a hexagonal honeycomb structure. There are three identical gauzes: one just after the flow straightener and two at a mutual distance of 0.2 m approximately 2 m before the outlet.

Figure 1: The rotor and the open jet wind tunnel.

The maximum velocity is 14.5 m/s. The degree of turbulence is 0.8% and the maximum deviation from the average velocity is 1.6%. The non-uniformity is radially distributed, so the "operational turbulence" is less then 0.4% (at λ = 7.5). This operational turbulence is a more representative quantity because a velocity deviation over the outlet area may be static, but is "felt" as a fluctuation by a rotating blade.

2.2 Model

The model has been specifically designed for aerodynamic measurements in the wind tunnel, see Figure 1. In the research, the emphasis is put on the rotor, which must therefore be well-defined and suitable to produce data for verification of numerical simulation programs. The aerofoil of the blades is not a common HAWT blade section, but it was chosen because of the availability of wind tunnel data over a wide Reynolds number range. The blades are manufactured on a numerical milling machine to obtain the highest feasible form accuracy.

Specifications of the model:

Radius	0.6 m
Number of blades	2
Aerofoil section	NACA 0012
Root cutout	30%
Chord	0.08 m
Blade length	0.42 m
Blade twist	$\Theta(r/R) = (6 + \Theta_{tip}) - 6.67r/R°$, for $0.3 \le r/R \le 0.9$
Replaceable tip	0.06 m span, no twist

The rotor is placed on an extended tube leaving 0.75 m free space behind the rotor plane.

Specifications of the test rig:

Hub height	2.33 m (= tunnel axis height)
Pitch angle	adjustable within 0.1°
Axial force meter	50 N, between hub and first bearing, transmission by sliprings
Torque meter	10 Nm
Azimuth counter	1 and 720 pulses/revolution
Motor/generator	1.5 kW, 4 quadrant operation, speed of revolutions adjustable to 16 Hz

2.3 Instrumentation

The undisturbed tunnel wind velocity is measured by means of a Pitot tube connected to a Betz manometer. The Pitot tube is placed at a position representative for the average velocity over the outlet area.

The velocities in the near wake are measured with constant temperature hot-wire equipment. The X-wire probe is calibrated using King's law: $E^2 = E_0^2 + BV^n$.

The probe position is set by a traversing system. In radial and azimuthal directions, this can be done by computer, with an absolute error of 0.2 mm. In axial direction, the adjustment is done by hand; the absolute error is 5 mm (the exact position of the rotor plane is difficult to detect!).

A data acquisition system, containing a high-speed voltmeter, a multiplexer and a counter, is used to measure and store the signals of the hot-wires and the azimuth counter. The maximum sample frequency of the voltmeter is 100 kHz, this is enough to keep track of the 720 samples per revolution for both wires at f = 16 Hz.

3. MEASUREMENTS

3.1 Configuration and conditions

In figure 2 a schematic overview is shown of the rotor and its position to the wind tunnel. Further, the conventions for the yaw angle and the azimuth angle are depicted. Seen from above, a positive yaw angle means that the rotor is rotated anti-clockwise. Looking downwind, the azimuthal angle increases as the rotor rotates clockwise. Pointing downwards, the azimuthal direction is zero. The axial distance is defined with respect to the rotor axis. So, the probe position is now defined by its radial, azimuthal and axial position and the position of the two blades by the azimuthal position of blade A.

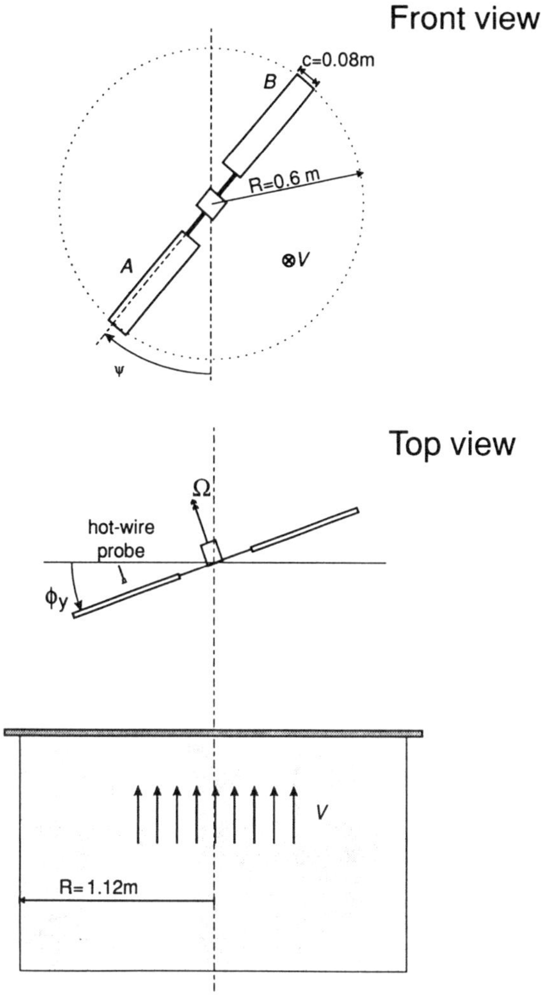

Figure 2: Schematic overview of test set-up with frame of reference.

In order to get an image of the wake velocities over the whole rotor plane, the hot-wire probe is positioned on 10 radial and 12 azimuthal positions. For the radial stations, the probe is placed from r/R = 0.4 to 1.1 with increments of 0.1, and additional stations at r/R = 0.95 and 1.05. The hub of the turbine prevents a closer approximation of the rotor axis, but this missing part represents only 12% of the rotor area. In azimuthal direction, the probe is placed from 0° to 360° with 30° increments.

The axial distance between the probe and the rotor area is kept constant at 3 cm (x/R = 0.05) with an absolute error of 5 mm.

During the measurements, the probe orientation is set with the plane of the wires parallel to the rotor plane, with the front wire in vertical position. In this way, the orientation of the front wire is unaffected by rotation in yaw direction, and it is most representative for measuring the total velocity.

For the rotor is operated in yaw, the angle of attack is varying over the azimuth position. To avoid stall on the blade, a rather high pitch angle of $\Theta_{tip} = 4°$ is chosen, which results in a low rotor loading.

During the measurements the rotational frequency of the rotor is kept constant at f = 12 Hz and the tunnel velocity at V = 6 m/s, which results in a tip speed ratio of $\lambda = 7.5$ and a Reynolds number at the tip of Re = 240.000.

The measurements are done for yaw angles ϕ_y = 10°, 20° and 30°. Additionally some control measurements are done in an unyawed situation.

3.2 Data acquisition

For each probe station, the velocity is measured during one revolution of the blades. During this revolution, 720 samples are taken from both hot-wires. This means that, every 0.5° azimuth angle of the blades the velocity is measured with a vertical and a horizontal hot-wire. This sampling frequency gives enough detail to recognize blade passages and vortex sheets easily.

With the current probe-traversing system only the left side of the rotor area can be reached and only one quarter at the time. The other half of the rotor plane is measured on the same side, but with a negative yaw angle. Control measurements showed that this is allowed. This way of traversing eliminates influence of possible tunnel differences between the two sides. Using an overlap between the different quarters at the four horizontal and vertical azimuthal traverses gives a way to control the reproducibility of the measurements. In this way, the limitations of the traversing system result in useful comparisons.

4. DATA REDUCTION

On every probe position the velocity is measured over one complete revolution of the rotor. As examples of typical measurements in a yawed situation, two azimuthal plots are given in figures 3 and 4. Both signals are taken at a 30° yawed rotor. The first is taken with the probe at r/R = 0.8 and Ψ = 0° and the second with the probe at r/R = 0.8 and Ψ = 180°.

In these measurements, the two passing blades are clearly visible as oscillations due to the bound blade vortices. The turbulence areas a little further on are from the vortex sheets of the blades. These vortex sheets contain trailing and shed vortices.

The difference between the two signals can be explained by different inflow directions. At Ψ = 0°, the blade is moving upwind and at Ψ = 180°, it is moving downwind. The induced velocities from the bound blade vortices together with a more or less constant wind component gives different velocities in both cases. With a vertical hot-wire, this results in the shown differences between the signals. Similar, any other probe positioning and orientation has its own typical signal shape.

Averaging a velocity trace over the whole revolution gives a value for the time averaged velocity at the corresponding point of the rotor plane. This can be done for all probe positions which gives a surface plot for the mean velocity over the whole rotor plane for different yaw angles.

Further reduction can be performed to get a value for the axial induction factor of the turbine (the decrease of the wind velocity due to the turbine). For that, the time averaged velocities are also averaged over the rotor area. The axial induction factor is the proportional decrease of this averaged value to the undisturbed wind velocity: $a=1-V_{av}/V_0$. By doing this for the different yaw angles, the influence of yaw on the induction factor becomes apparent.

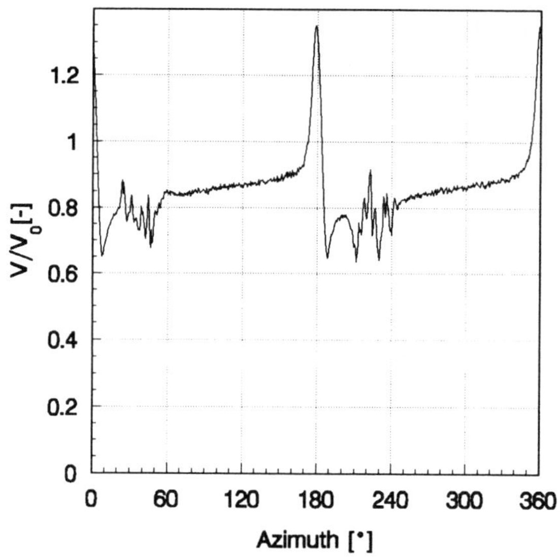

Figure 3: Measured velocity signal for ϕ_y = 30°, r/R = 0.8, Ψ = 0° measured with vertical hot-wire.

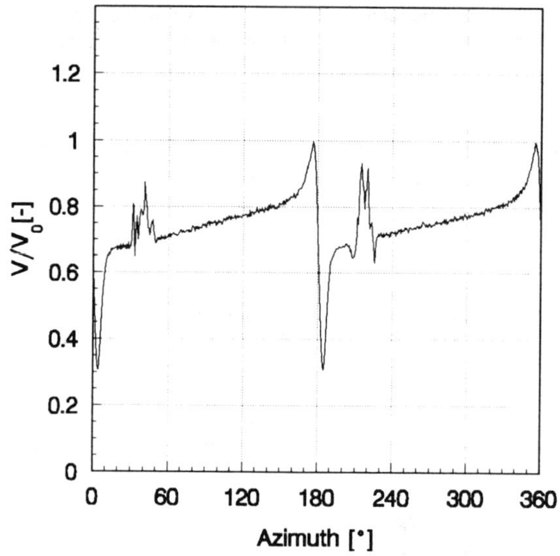

Figure 4: Measured velocity signal for ϕ_y = 30°, r/R = 0.8; Ψ = 180° measured with vertical hot-wire.

5. RESULTS

The surface plots of the time averaged wind velocity over the rotor plane for the different yaw angles are given in figure 5, 6 and 7. The surface plots represent the rotor area for r/R = 0.4 to 0.95. At r/R = 1.0, the radial position of the probe is very critical to be just inside or just outside the part that is slowed down by the turbine. Especially for a yawed turbine, the variance of the mean velocity for r/R = 1.0 over the azimuth angle, is merely due to crossing this border.

It is shown that for all positions of the probe the averaged velocity measured by the probe increases with the yaw angle. So, all parts of the rotor plane are working less effective when the yaw angle is increased.

Examining the figures from 10°, 20° to 30° yaw angle an increasing effect becomes apparent. In the surface plots a decrease in velocity is visible. This decrease is not located at the same azimuthal position for

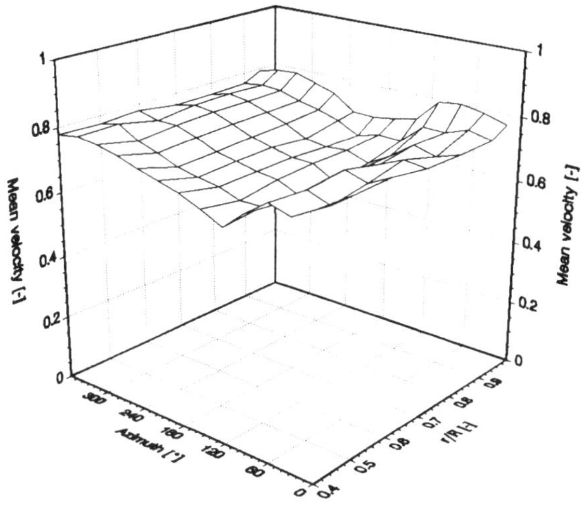

Figure 5: Mean velocity over the rotor plane for yaw angle ϕ_y = 10° measured with vertical hot-wire.

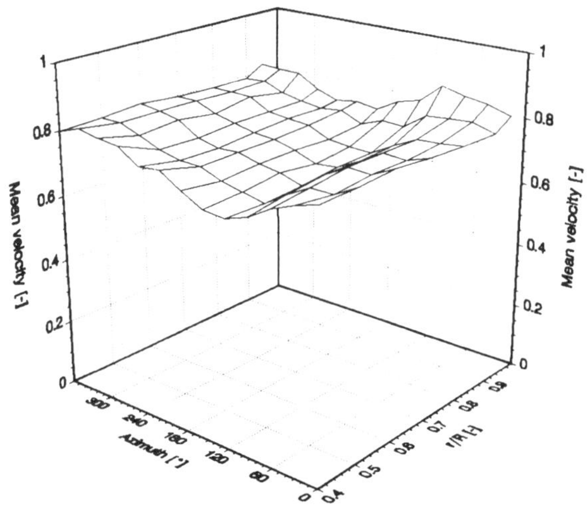

Figure 6: Mean velocity over the rotor plane for yaw angle ϕ_y = 20° measured with vertical hot-wire.

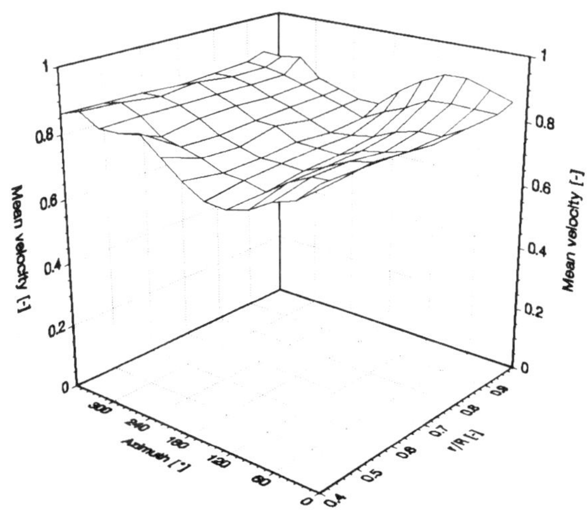

Figure 7: Mean velocity over the rotor plane for yaw angle ϕ_y = 30° measured with vertical hot-wire.

different radial stations, but varies from r/R = 0.4; Ψ = 120° to r/R = 0.95; Ψ = 180°. Apparently the rotor experiences the highest loading in this region. This effect is increasing with the yaw angle.

If the rotor is less effective locally everywhere over the whole rotor plane then it's obvious that this is also the case for the turbine as a whole. Therefore the induction factor decreases with the yaw angle. This is shown in figure 8.

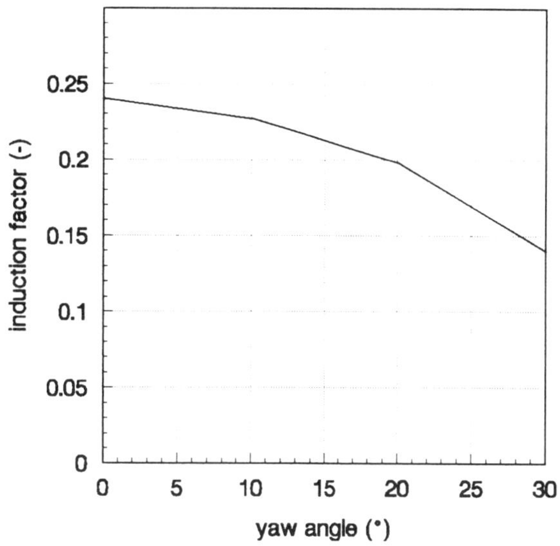

Figure 8: Induction factor of the rotor from averaged velocities over rotor area from r/R = 0.4 to r/R = 0.95.

The velocities are averaged over the rotor plane from r/R = 0.4 to 0.95, which represents 83% of the rotor plane to calculate the induction factor.

6. CONCLUSIONS

Throughout the rotor plane, the local effectivity of the rotor is decreasing with an increasing yaw angle.

As a result, the induction factor of a wind turbine, determined by averaging velocities over the rotor plane, decreases with the yaw angle.

In the used conventions, the rotor is working most effective in the region from r/R = 0.4; Ψ = 120° to r/R = 0.95; Ψ = 180°. In this region the wind is slowed down most.

ACKNOWLEDGEMENTS

We would like to thank Simon Toet for his helpful assistance during the measurements.

REFERENCES

[1] N-J Vermeer, Local circulation on rotating wind turbine blades from velocity measurements in the wake of a model rotor, Proceedings of the 1992 fourteenth BWEA Wind Energy Conference, Nottingham, 1992.

[2] N-J Vermeer, G.J.W. van Bussel, Velocity measurements in the near wake of a model rotor and comparison with theoretical results, ECWEC90 Proceedings, Madrid, 1990.

[3] J.K. de Boer, Wake structure of a rotor model in yaw, IW-93061R, Institute for Wind Energy, Delft University of Technology, 1993: to be published.

ANALYTICAL CORRELATIONS FOR TURBULENCE CHARACTERISTICS IN THE WAKES OF WIND TURBINES

A. Crespo and J. Hernández*

Mecánica de Fluidos, E.T.S.I. Industriales, Universidad Politécnica de Madrid (UPM).
José Gutiérrez Abascal, 2 - 28006 Madrid, Spain.
*Universidad de Murcia, Paseo Alfonso XIII, 48 - Cartagena, Spain.

ABSTRACT: A model to estimate the evolution of turbulence characteristics in wind-turbine wakes is being developed. Based on experimental results and on calculations from the code UPMWAKE combined with some analytical considerations, simple analytical expressions are proposed for the calculation of the turbulent kinetic energy, k, and its dissipation rate, ϵ. To obtain the turbulence spectra in the wake a classical law used for atmospheric turbulence is assumed, in which the height and friction velocities are estimated from the values of k and ϵ, so that when the effect of the wake is negligible the spectrum of the unperturbed basic flow is recovered.

1. INTRODUCTION

When a wind turbine is in the wake of another one the turbulence intensity incident over the machine is larger than in the absence of the wake, originating additional unsteady loads and fatigue damage that is of interest to study. The purpose of this work is to give simple and reliable expressions to calculate turbulence characteristics in wakes, that will be of use to the designer of the machines. Previous works of this type have been carried out in refs. [1] to [6]; however, the method proposed her is more complete, although some of the assumptions used are not entirely justified, and further research on this important problem is very necessary.

2. DEFINITIONS AND TURBULENCE ANISOTROPY

In the logarithmic layer of the atmospheric boundary layer, under neutral conditions, the standard deviations of the three velocity components can be related to the turbulence friction velocity by (Panofski and Dutton [7])

$$\sigma_{u\infty}=2.4u^*,\ \sigma_{v\infty}=1.9u^*,\ \sigma_{w\infty}=1.25u^*, \qquad (1)$$

where u is the velocity component in the average wind direction, w is the vertical component and v is perpendicular to both u and w. From Eq. (1) the turbulence kinetic energy is given by

$$k_\infty=5.47u^{*2}=0.95\sigma_{u\infty}^2, \qquad (2)$$

where subscript ∞ refers to ambient conditions unperturbed by the machine. The turbulence intensity I_∞ is defined as the standard deviation of the wind velocity component in the average wind direction divided by the wind velocity. From Eq. (2),

$$I_\infty=\sigma_{u\infty}/u_\infty=1.026k_\infty^{0.5}/u_\infty. \qquad (3)$$

The machine creates an additional turbulence kinetic energy in the wake that is denoted by

$$k_{add}=k-k_\infty. \qquad (4)$$

In this work it will be assumed that the anisotropy in the wake is similar to that in the basic atmospheric flow, so that, as in Eq. (2), $k=0.95\sigma_u^2$, and $I=\sigma_u/u_\infty=1.026k^{0.5}/u_\infty$. This assumption is validated by the experimental results of Cleijne [8]. We can define then an added turbulence intensity

$$I_{add}=(I^2-I_\infty^2)^{0.5}=1.026k_{add}^{0.5}/u_\infty. \qquad (5)$$

For the dissipation rate of the turbulence kinetic energy in the surface layer we have the classical expression

$$\epsilon_\infty=2.5u^{*3}/z, \qquad (6)$$

and the quantity created in the wake is

$$\epsilon_{add}=\epsilon-\epsilon_\infty. \qquad (7)$$

3. NEAR AND FAR WAKE.

As it is usually done, two regions are distinguished: the near wake (1D to 3D downstream), where an annular shear separating the inner core from the ambient flow is formed, and a far wake. The intermediate region has not been considered yet.

In the near wake, the highest turbulence kinetic energy is produced in the upper part of the annular shear layer, where the ambient shear and the shear due to the velocity defect add; conversely, in the lower part of the ring these two shears subtract and the maximum of k_{add} is lower. This situation is typically represented in Fig. 1, where experimental results of Hojstrup [3] for the turbulence intensity, normalized with the turbulence intensity of the basic flow, I/I_∞, are compared with the numerical predictions of UPMWAKE.

In the far wake the ring of local maximums of k_{add} reduces to a single point located above the turbine height.

The results of UPMWAKE show that the behavior of ϵ_{add} is qualitatively similar to that of k_{add}.

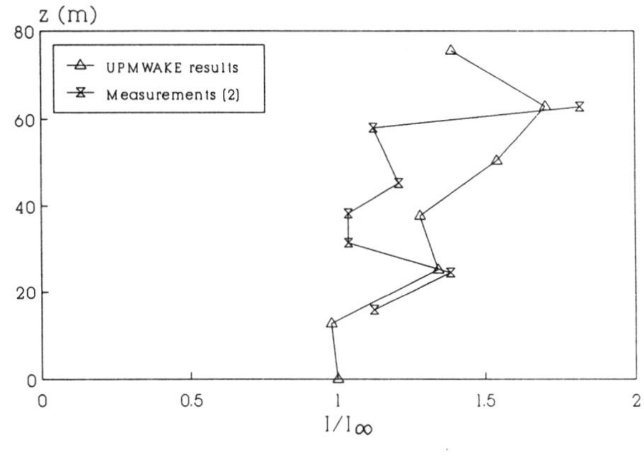

Figure 1: Vertical distribution of turbulence intensity normalized with its ambient value. Comparison of numerical (UPMWAKE) and experimental results, ref. [3], 1D.

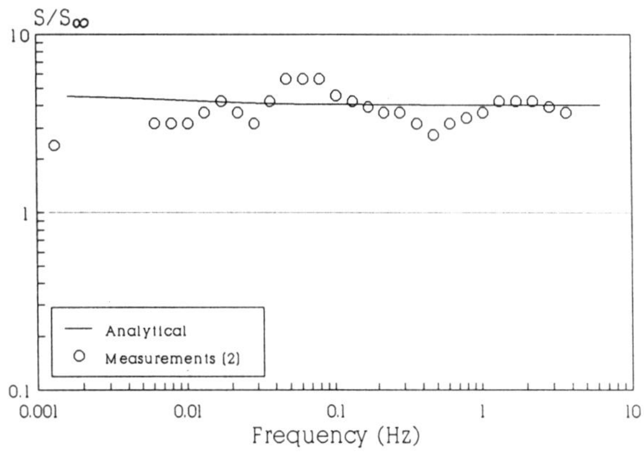

Figure 2: Turbulence spectra in the near wake, normalized with value at ∞. Nibe case 11 (ref. [2]), 1D downstream, z=64m

4. TURBULENCE SPECTRA

The unperturbed atmospheric flow has a typical spectrum for u (Panofski and Dutton [7]) of the form

$$nS_{u\infty}=19.2fk_{\infty}/(1+33f)^{5/3}, \qquad (8)$$

where f is a nondimensional frequency:

$$f=nz/u_{\infty}. \qquad (9)$$

It is assumed that the spectrum in the wake has the same form given by Eq. (8):

$$nS_{u}=19.2f'k/(1+33f')^{5/3}, \qquad (10)$$

where

$$f'=nz'/u_{\infty}, \quad z'=2.5(k/5,47)^{3/2}/\epsilon. \qquad (11)$$

The value of k_∞ is substituted by k, and the length scale, z, is substituted by a combination of k and ϵ that the dimensional analysis shows that is unique except for a multiplicative constant that is determined with the condition that for the unperturbed flow Eq. (8) is recovered. k and ϵ are obtained from the UPMWAKE model or from the expressions given below.

In Figs. 2 and 3 are shown spectra nondimensionalized with their unperturbed values at ∞; they respectively correspond to the upper and lower parts of the near wake (see Fig. 1). Experimental results corresponding to measurements of the Nibe machine are presented for comparison, Hojstrup [3]. Since k_{add} is positive, $S_u > S_{u\infty}$ for all frequencies. As it will be shown later, in the upper part of the near wake ϵ_{add} is proportional to $k_{add}^{1.42}$; as the exponent is smaller but quite close to 1.5, Eq. (11) shows that the modified typical length z' will be slightly larger than z. As S_u decreases with nz', the ratio S_u/S_∞ will decrease with n, as shown in Fig. 2. On the other hand, in the lower part of the near wake ϵ_{add} grows faster than $k_{add}^{1.5}$ and the opposite behavior occurs, as shown in Fig. 3. For the far wake, the results at the turbine height are shown in Fig. 4 (the spectrum at 4 diameters has been normalized with that at 7.5 D, because there were not available experimental results at ∞); the behavior of the nondimensionalized spectrum is similar to that in the upper part of the near wake, shown in Fig. 3.

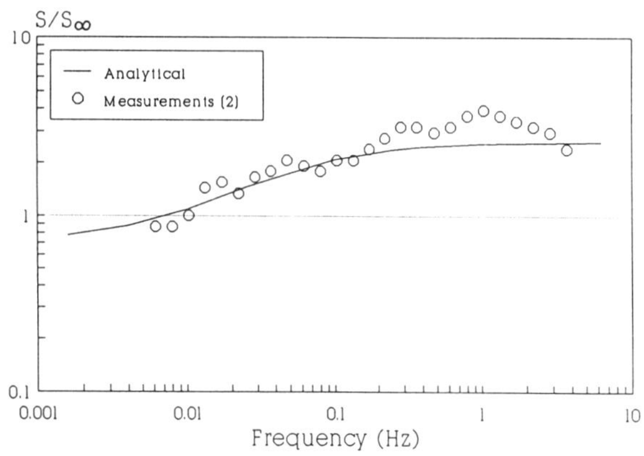

Figure 3: Turbulence spectra in the near wake, normalized with value at ∞. Nibe case 11 (ref. [2]), 1D downstream, z=25m

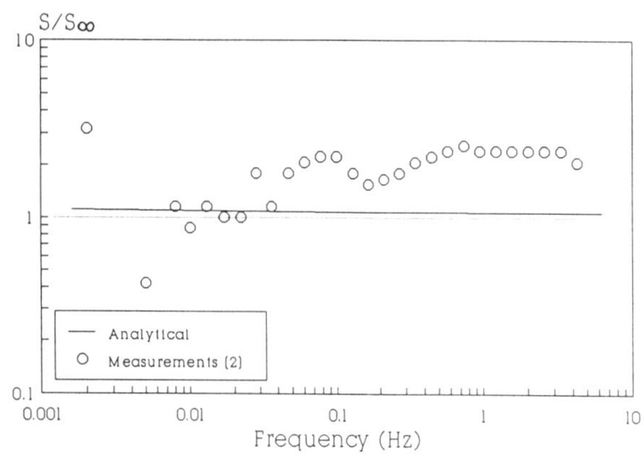

Figure 4: Turbulence spectra in the far wake, normalized with conditions at 7.5 D. Nibe case 10 (ref. [2]), 6D downstream, z=45m

5. CORRELATIONS FOR k AND ϵ

5.1 Near wake

The annular shear layer that is formed close to the turbine in the near wake can be treated locally as bidimensional, as shown schematically in Fig. 5. Using a boundary layer approximation, the equations of conservation of momentum in the main wind direction and of turbulent kinetic energy are

$$u\partial u/\partial x + v\partial u/\partial y = \partial(\nu_T \partial u/\partial y)/\partial y, \quad (12)$$

$$u\partial k/\partial x + v\partial k/\partial y = \partial(\nu_T \partial k/\partial y)/\partial y - \epsilon + \nu_T(\partial u/\partial y)^2. \quad (13)$$

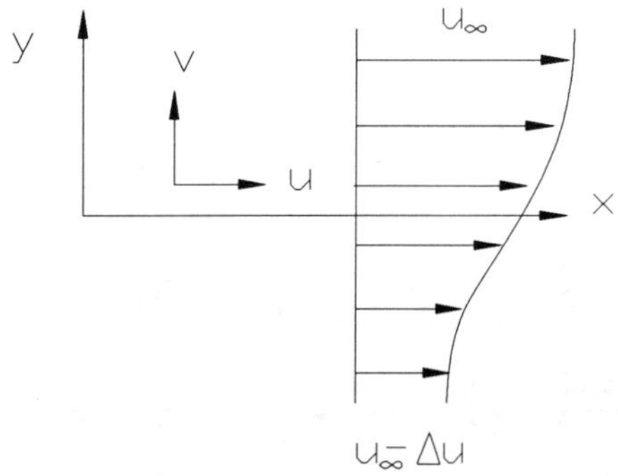

Figure 5: Schematic showing the shear layer in the near wake.

The notation is shown in Fig. 5 and ν_T is the eddy viscosity that is assumed to be equal to the eddy diffusivity. The last term of (13) corresponds to production of k by shear; near the origin the thickness of the shear layer is infinitely small and $(\partial u/\partial y)$ goes to infinity; however, if there is no ambient turbulence ν_T tends to zero near the origin and the production term is of the same order of the dissipation ϵ. In our case there is ambient turbulence, ν_T is expected to be finite near the origin, and the production term is much larger than ϵ. If ϵ is neglected in (13), it may be easily shown that

$$k_{add} = 0.5(u_\infty - u)(u - u_\infty + \Delta u), \quad (14)$$

satisfies (12) and (13) and k_{add} is zero at the borders of the shear layer. The maximum value of k_{add} corresponds to the averaged value of u, $u = u_\infty - \Delta u/2$:

$$k_{add\ max} = \Delta u^2/8. \quad (15)$$

Using Eq. (5) the maximum I_{add} can be written as:

$$I_{add\ max} = 0.725a = 0.362(1-(1-C_T)^{1/2}), \quad (16)$$

where a is the induced-velocity factor, $a = \Delta u/(2u_\infty)$, and C_T is the thrust coefficient, that according to the actuator disk theory is $C_T = 4a(1-a)$. In Fig. 6 a comparison is given of this simple expression with some experimental results of refs. 3 and 4 and values calculated numerically using UPMWAKE.

If an Eq. similar to (13) is written for conservation of ϵ, and the convective and production terms are considered dominant, it may be shown that $\epsilon = \epsilon_\infty(k/k_\infty)^{C\epsilon 1}$, where the coefficient $C_{\epsilon 1}$ multiplies the production term in the equation of ϵ, and is given the value 1.44 in the literature, although in UPMWAKE is 1.21 to adapt to the atmospheric conditions in the ambient basic flow. This result, however,

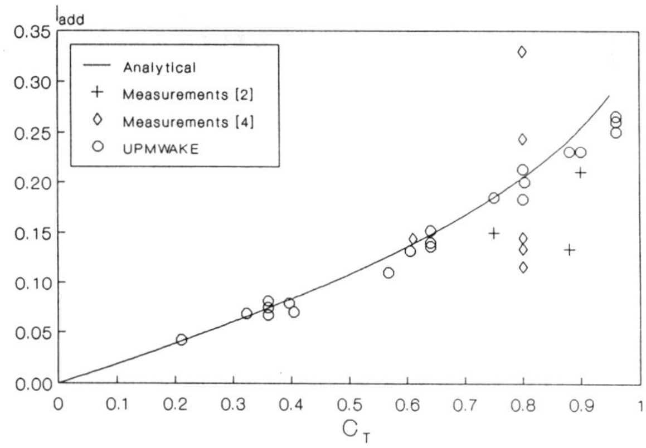

Figure 6: Maximum value of the added turbulence intensity in the near wake.

is not in good agreement with UPMWAKE results. For maximum k_{add} the exponent is larger than $C_{\epsilon 1}$, and the following correlation is found to be better:

$$\epsilon_{addmax}/(u_\infty^3/z) = 0.151(k_{addmax}/u_\infty^2)^{1.416} = 0.136 I_{addmax}^{2.832}. \quad (17)$$

This correlation is presented in Fig. 7 and is compared with UPMWAKE calculations.

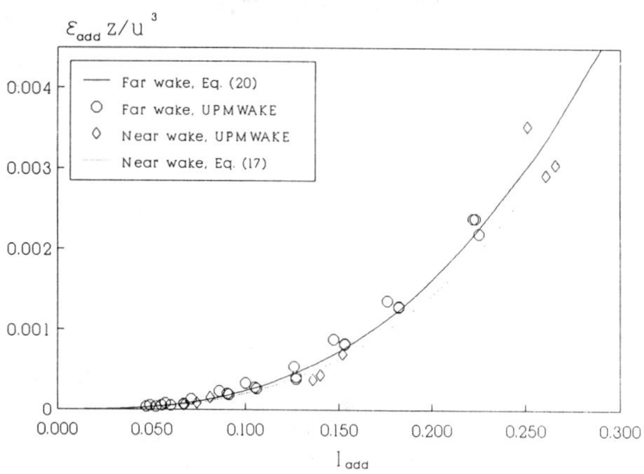

Figure 7: Correlation for ϵ_{addmax} as function of I_{addmax}.

5.2 Far wake.

For the far wake the following expression has been proposed in [1] for I_{add}:

$$I_{add} = 4.8 C_T^{0.7} I_\infty^{0.68} (X/X_N)^{-0.57}, \quad (18)$$

where X is the downstream distance and X_N is the length of the near wake as estimated by Vermeulen [6], and in appendix 2 of [1]. However, it has been found that, when comparing with some experiments and with the results of UPMWAKE, the exponents of X and I_∞ are somewhat high. An alternative expression that fits better the UPMWAKE results is

$$I_{addmax} = 0.73 a^{0.8325} I_\infty^{-0.0325} (X/D)^{-0.32}. \quad (19)$$

In Eq. (19) a has been used instead of C_T, but the behavior is roughly similar. In Eq. (18) I_{add} increases with I_∞, whereas in (19) it slightly decreases with it, this is due to diffusion by ambient turbulence. It has been considered more useful to nondimensionalize X with the diameter D instead of doing it with X_N. In Fig. 8 correlation (19) is compared with UPMWAKE calculations and with results of both field and wind tunnel experiments that have been collected in ref. [5]. Some of these experiments correspond to the centerline, where I_{add} is smaller than its maximum value; from UPMWAKE results it is estimated to be about 0.8 I_{addmax}, and this expression has also been presented in Fig. 8.

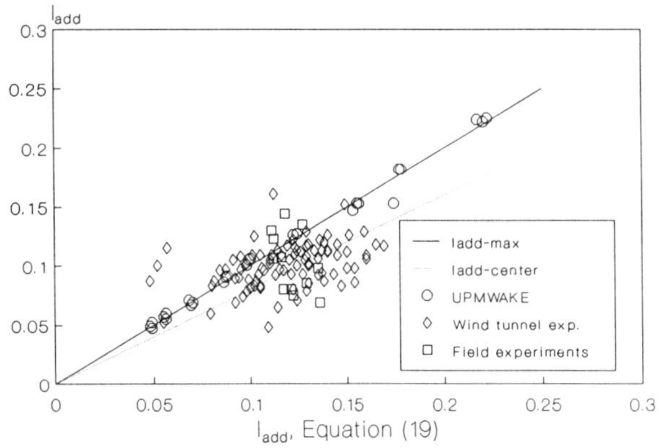

Figure 8: Proposed correlation for I_{addmax}. Comparison with experiments and UPMWAKE results.

For ϵ_{addmax} the results of UPMWAKE show that an expression very similar to that in the near wake, Eq. (17), is obtained,

$$\epsilon_{addmax}/(u_\infty^3/z) = 0.155(k_{addmax}/u_\infty^2)^{1.385} = 0.14 I_{addmax}^{2.77}. \quad (20)$$

this Eq. is compared with UPMWAKE results in Fig. 7.

ACKNOWLEDGMENT.

This work has been supported by the CEC through contract JOUR-0087 of the Joule programme.

REFERENCES

[1] D. C. Quarton. Final Report from Garrad Hassan and Partners to the Energy Technology Support Unit of the Department of Energy of the UK, contract ETSU WN 5096. Wake Turbulence Characterization. 1989.

[2] J. Hojstrup. Final Report on CEC contract no. EN3W.0039.UK(H1). Wake Measurements on the Nibe wind-Turbines in Denmark. Appendix 1. Nibe Wake 2: Data Report- Power Spectra. 1990.

[3] G. J. Taylor. Final Report on CEC contract no. EN3W.0039.UK(H1). Wake Measurements on the Nibe wind-Turbines in Denmark. Part 2. Data Collection and Analysis. 1990.

[4] U. Hogstrom et al. A Field Study of the Wake Behind a 2 MW Wind Turbine. Atmospheric Environment, vol. 22, nº 4, pp. 803-820. 1988.

[5] E. Luken The wind load of wind turbines in clusters. TNO report 89-160. May 1989

[6] P. Vermeulen. An experimental analysis of wind turbine wakes. Proc. 3rd Int. Symp. of Wind Energy systems. Copenhagen, August 1980.

[7] H. A. Panofski and J. A. Dutton. Atmospheric Turbulence. John Willey & Sons 1984.

[8] J. W. Cleijne Results of Sexbierum Wind Farm. Internal report of TNO 92-388. November 1992.

MODELIZATION OF WIND FARMS IN COMPLEX TERRAIN. APPLICATION TO THE MONTEAHUMADA WIND FARM

A. Crespo, F. Manuel, J.C. Grau and J. Hernández*

Mecánica de Fluidos, E.T.S.I. Industriales, Universidad Politécnica de Madrid.
José Gutiérrez Abascal, 2 - 28006 Madrid. Spain.
*Universidad de Murcia, Paseo Alfonso XIII, 48 - 30203 Cartagena, Spain.

ABSTRACT: The interaction between a wind-turbine wake and the topographic characteristics has been modelled by assuming linear superposition of the velocity perturbations created individually by both the wake and the terrain irregularities. Whereas the linear superposition of wake defects for turbines in a row overestimates the total defect, in the case of velocity perturbations due to topography and to a single wake this approach seems to work. Application has been made to the Monteahumada Wind Farm where velocity perturbations due to topography and the wind-turbine wakes are of a similar order. The characteristics of the wind-turbine wake are calculated with the code UPMWAKE and the topography effect is calculated with WASP.

1. INTRODUCTION

Usually wind farm models make the assumption that the terrain is flat and that the unperturbed wind velocity is uniform. This assumption is not correct in many cases of interest; as it is well known terrain irregularities can be used to enhance or concentrate wind power. For terrains that are moderately complex a simple procedure is to add linearly the velocity perturbations due to the terrain and the wakes; this procedure was applied to the Ampurdán wind farm [1, 2], and apparently the results were satisfactory. A similar procedure has been used in [3] to take into account the interaction of obstacle and turbine wakes in the Zeebrugge wind farm. Nevertheless, in all the previous cases there were simultaneously interactions of terrain and several wakes, creating some uncertainty about the validity of the results; as it is well known, the linear superposition of several wind turbine wakes overestimates the velocity defect, and some alternative procedures have been suggested in refs. [3] to [6] to take into account this interaction. Ref. [7] gives a procedure to take into account non-uniformity in wind velocity and the curvature of the streamlines in wind farms with small terrain irregularities, that is similar to the linear superposition; some sample calculations are made, but no experiments are presented to validate the method. In ref. [8] measurements in wind tunnel are made that show that the changes in the wake characteristics due to topography may be important. Ref. [9] studies theoretically the modification of aerodynamic and near-wake characteristics due to the terrain. Second order corrections to the linear superposition of terrain and wake effects are given in [10], that have been calculated using PHOENICS; it has been found that near the ground the terrain irregularity creates additional turbulent diffusion that diminishes the wake effect; on the other hand, above the apex of a hill the streamlines concentrate, thus enhancing wake defects.

In the Monteahumada wind farm there are only single wake interactions of an order of magnitude similar to those created by the terrain; this configuration is thus appropriate to examine the validity of the assumption of linear superposition of wake and terrain effects. On the other hand the data are scarce (only daily averages) and incomplete for some turbines; this complicates a direct comparison of interfering machines because in some cases is precisely the data of the upstream machine that is lacking, and because the daily energy production can not be obtained from the machine characteristics, unless some additional information is given about the daily distribution of winds. To avoid these problems, for each wind direction a reference machine has been chosen such that its velocity is closest to that in the measuring mast and its data are reliable; then, the theoretical power of the other machines is deduced from this reference machine by considering that all machines are equal and that the power only depends on the actual velocity incident on each machine.

This work shows that for a moderately irregular terrain the linear superposition of wake and terrain effects seems to work. It is of interest to notice that for the interaction of wakes, with perturbations of an order of magnitude similar to those presented here, this assumption is not valid in general.

2. DESCRIPTION OF MONTEAHUMADA WIND-FARM

The Monteahumada wind farm is located in the South of Spain near the Gibraltar Strait (Fig. 1). The farm is handled by the Spanish utilities "Sevillana de Electricidad" and ENDESA. There are seven machines, manufactured by the Spanish utility ENDESA, whose characteristics are as follows:

- 150 kW, rated power at 15 m/s.
- 3 blades.
- 46 rpm rotating speed
- Operate in the range 5 to 25 m/s.
- 20 m rotor diameter
- 21.8 m hub height.
- Thrust coefficient is 0.78 at 5 m/s and 0.4 at 15 m/s.

Prevailing winds are from E, 12 m/s average, and W, 6.7 m/s. The average value of wind velocity is 8.5 m/s. The position of the machines in the farm is indicated in Fig. 2. There is also a meteorological tower East of turbine 5. Wake effects are expected between machines 2 and 7, and between machines 1 and 6. Turbine 2 and the meteorological tower are located in a plateau, with valleys at E and W, and hills at N and S. Turbine 3 is at the highest location. The interfering machines are located approximately at the same level: turbine 6 is 3 m below turbine 1 and turbine 7 two meters above turbine 2.

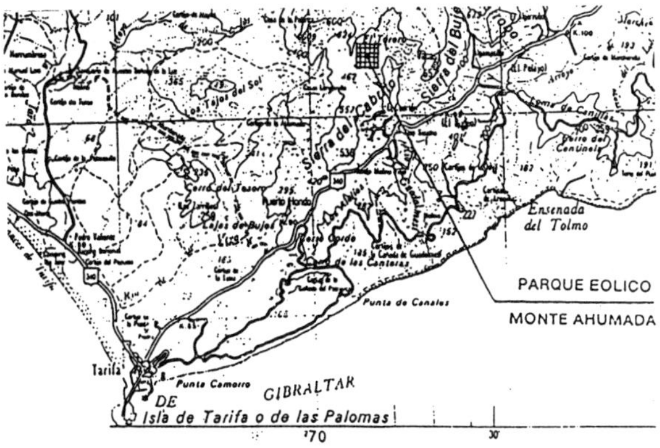

Figure 1. Map showing the location of the Monteahumada wind farm.

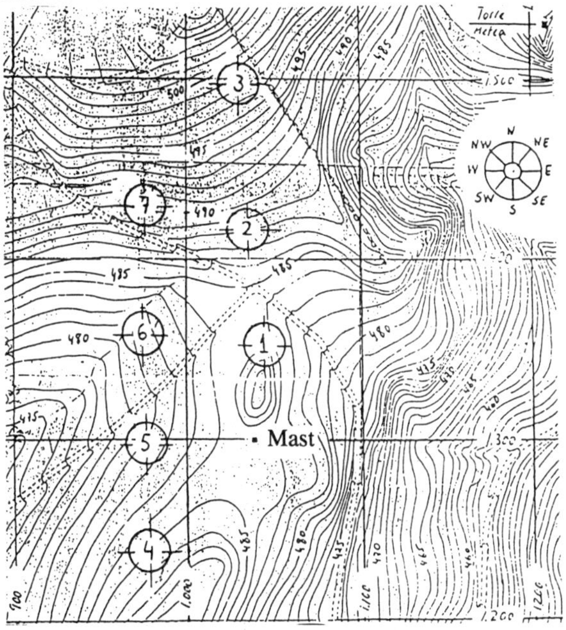

Figure 2. Schematic showing the Monteahumada wind farm.

3. MEASUREMENTS

The total production of energy is measured by direct reading at each machine; this is done regularly on a daily basis. However, as the winds have prevailing directions from East and West that last for periods of several days, the measurements can still be used to validate the calculations. The collected data correspond to a period of a year. The power-velocity curve of the machine given by the manufacturer would have to be averaged over a daily period; instead of doing that, the power-velocity of the machine to be used as a reference is estimated from the measurements made at the machine located at a position where there are no wake effects and where the velocity is closest to that in the meteorological mast.

At the mast tower, the wind velocity, direction, and temperature are measured at a height of 10 m. The wind characteristics are sampled every five seconds, and averages and standard deviations are calculated and stored every 10 minutes in a tape. There is a possibility of storing the whole sequence of measurements of a 10 minute period corresponding to some representative cases. Nevertheless, as the power is only measured on a daily basis, only the daily averages of velocity are of interest.

4. CALCULATIONS

4.1 Wake effects.

The velocity defects at each of the machines pairs (1-6 and 2-7) are presented in Table 1. They have been calculated using UPMWAKE. The surface roughness of the ground has been taken equal to 1 cm, and the Monin-Obukhov length is considered infinity. Machines 6 and 1 are right in the centerline of the wake of the other one for the prevailing winds, but it has been assumed that there are variations of the wind direction, whose standard deviation have been estimated to be 6º. The line connecting machines 2 and 7 is estimated to form 12º with the prevailing wind direction.

Table 1: Wake velocity defect as a function wind velocity

Wind velocity (m/s)	ΔU_w(1-6) (m/s)	ΔU_w(2-7) (m/s)
4	−0.68	−0.56
6	−1.08	−0.765
8	−1.36	−1.00
10	−1.44	−1.03
12	−1.305	−0.975
14	−1.31	−0.96
16	−1.305	−0.94
18	−1.155	−0.825

4.2 Topography effects.

The terrain effect has been estimated using the WASP code. In Table 2 the speed-up is given as a fraction of the velocity at the meteorological mast.

Table 2: Velocity variation due to topography, $\Delta U_o/U_{mast}$

Turbine	East	West
1	−0.029	−0.029
2	−0.059	−0.059
3	−0.076	−0.070
4	+0.035	+0.023
5	−0.024	−0.029
6	−0.059	−0.058
7	−0.059	−0.058

It may be observed that the topographic effect is very similar for W and E winds; this may be due to the symmetrical configuration of the ground. The wind velocity is highest at turbine 4 and lowest at turbine 3.

5. COMPARISON BETWEEN EXPERIMENTS AND CALCULATIONS.

The reference curve giving the power of the least perturbed machine as a function of the wind velocity at the mast is obtained from the experimental results of the machine that is expected to have a velocity closest to that of the mast. For E winds the reference machine is machine 5, and for W winds is machine 4; none of them have wake effects, and according to Table 2 they have the lowest absolute value of ΔU_o. Besides, the data for these two machines is complete and reliable. The power/mast-velocity curves of the other machines are deduced from this reference curve by considering that all machines are equal and that the power only depends on the actual velocity incident on each machine:

$$U = U_{mast} + \Delta U_o + \Delta U_w,$$

where

$$U_{mast} = U_5/0.976, \quad \text{for E winds,}$$
$$U_{mast} = U_4/1.023, \quad \text{for W winds.}$$

The estimated curves are then compared with experimental results; this comparison is shown in Figs. 3 to 8. In each figure the corresponding reference curve is given. Figs. 3, 4 and 5 correspond to Eastern winds and Figs. 6, 7 and 8 to Western winds. For E winds the data is more abundant and the range of values of the velocity is larger than for W winds.

In Figs. 3 and 6 are presented the results of the machines that are in the wake of another one; it may be observed that the agreement between prediction and measurements is in general acceptable. In Fig. 4, for E winds, are presented the results of machines 3 and 4 that are only affected by topography. The agreement is acceptable for machine 4 and the measured power in machine 3 is always smaller than the predicted power; this may be either because the influence of topography is larger than predicted or because the machine is malfunctioning (this discrepancy also happens for W winds, as may be observed in Fig. 7). In Fig. 5 are presented the results of the two machines that are least affected by topography with E winds. As it may be seen in Figs. 7 and 8, the powers measured in machines 3, 6 and 7 for W winds are smaller than the predicted ones; this may be due to the fact that the measurements for these machines in W winds were scarce and less reliable than for the others. As a matter of fact, a direct comparison of the powers of upstream and downstream machines (7-2 and 6-1) in W winds, will show a much smaller wake effect. The use of the reference machine proposed in this work obviates the uncertainties in the comparisons due to the topography or to deficient data.

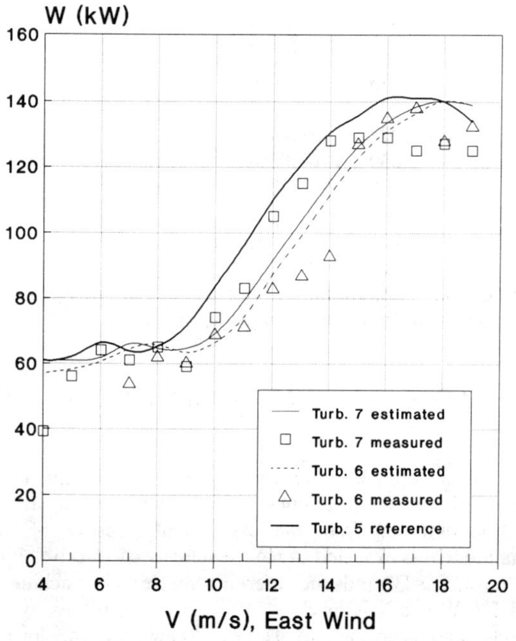

Figure 3. Power-velocity curves for machines 6 and 7. East wind.

Then, it may be concluded that the results of Figs. 3 and 6 show that the hypothesis made in this work of linear superposition of wake and topographic effects is acceptable.

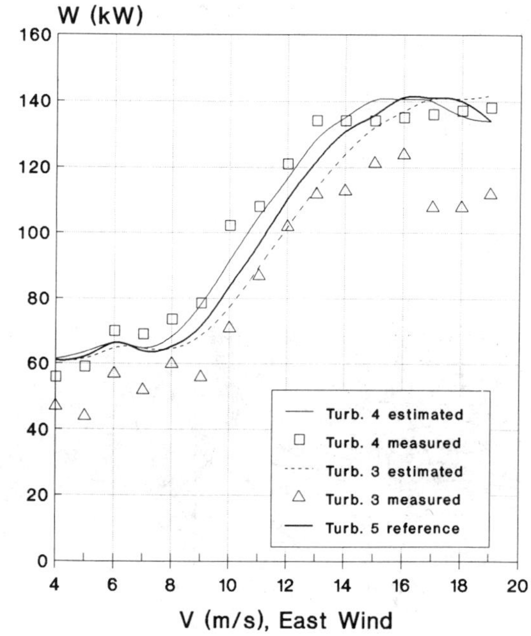

Figure 4. Power-velocity curves for machines 3 and 4. East wind.

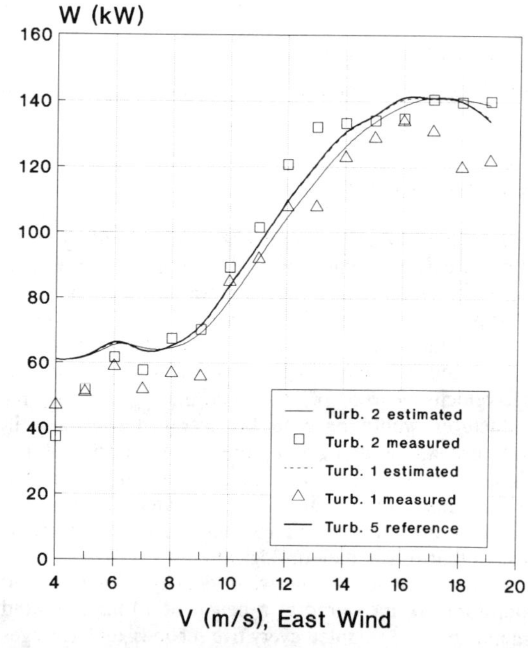

Figure 5. Power-velocity curves for machines 1 and 2. East wind.

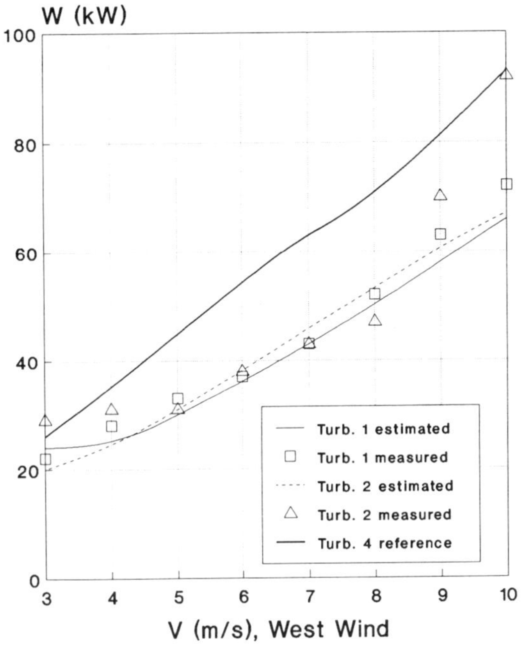

Figure 6. Power-velocity curves for machines 1 and 2. West wind.

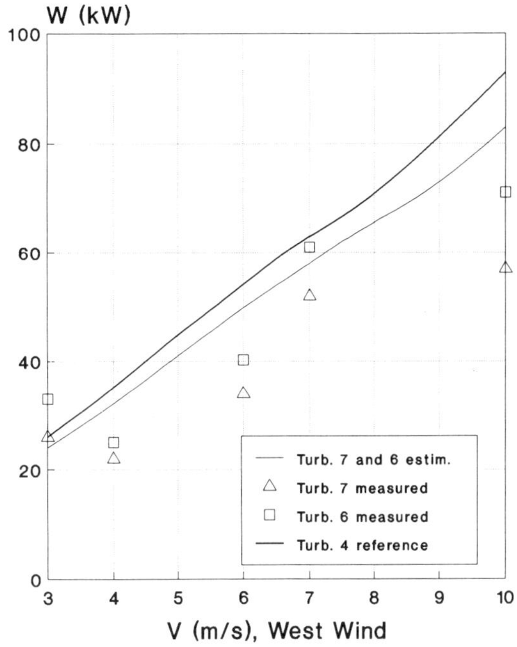

Figure 8. Power-velocity curves for machines 6 and 7. West wind.

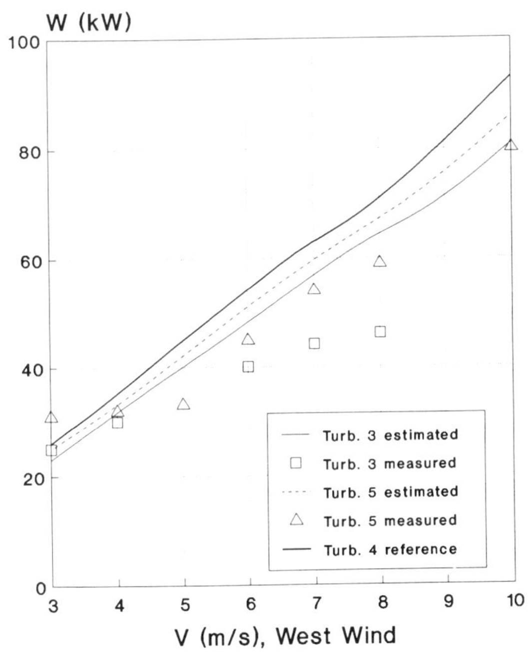

Figure 7. Power-velocity curves for machines 3 and 5. West wind.

ACKNOWLEDGMENTS

This work has been supported by the CEC through contract JOUR-0064 of the Joule programme. We are also very grateful to the utilities ENDESA and "Sevillana de Electricidad" for supplying the data of the Monteahumada wind farm, and to "RISE consults aps" for providing us with the results of WASP calculations.

REFERENCES

[1] A. Crespo and J. Hernández. "A numerical model of wind turbine wakes and wind farms." Proceedings of the EWEC ' 86, Rome, pp. 111 - 115. 1986

[2] A. Crespo, J. Hernández, E. Fraga, and C. Andreu. Final Report of CEC contract EN3W / 0020 / E(B). "Analysis of Wind Turbine Wakes." 1988.

[3] J. van Leuven. "The energetic effectiveness of a cluster of wind turbines." Thesis presented at the Universite Catholique de Louvain. Institut des Sciences Naturelles Appliquees. September 1992.

[4] I. Katic, J. Hojstrup, and N. O. Jensen. "A simple model for cluster efficiency." Proceedings of the EWEC ' 86, Rome, pp. 407 - 410. 1986.

[5] D. Smith and G. J. Taylor. "Further analysis of turbine wake development and interaction data." Proceedings of the 1991- 13th BWEA Wind Energy Conference. pp 325 - 331. 1991

[6] A. Crespo, F. Manuel, and J. Hernández. "Numerical modelling of wind turbine wakes." Proceedings of the ECWEC ' 90, Madrid, pp. 111 - 115. 1990

[7] S. G. Voutsinas, K. G. Rados, and A. Zervos. "The effect of the non-uniformity of the wind velocity field in the optimal design of wind parks." Proceedings of the ECWEC ' 90, Madrid, pp. 181 - 185. 1990.

[8] G. J. Taylor and D. Smith. "Wake measurements over complex terrain." Proceedings of the 1991- 13th BWEA Wind Energy Conference, pp 335 - 342. 1991.

[9] A. Hemon, S. Huberson, and A. Zervos. "Numerical study of wind turbine operation in complex terrain." Proceedings of the 1991- 13th BWEA Wind Energy Conference, pp 335 - 342. 1991.

[10] H. van Oort, P.H. van Gemert, and A. Crespo. Final Report of CEC contract EN3W / 0030 /NL. "Wind Farms in Complex Terrain." MT-TNO report 89 - 233. 1989.

WAKE EFFECTS IN WIND PARKS. A NEW MODELLING APPROACH

Voutsinas S.G.[a,b], Rados K.G.[a,b], Zervos A.[a,c]
[a] National Technical University of Athens
Fluids Section, P.O. Box 64070, 15710 Zografos, Athens, Greece
[b] Center for Renewable Energy Sources
19 Km Marathonos Ave., 190 09 Pikermi, Greece
[c] Commission of the European Communities
DG XII, Rue de la Loi 200, B-1049 Brussels, Belgium

ABSTRACT: In the present paper, wake effects are modelled by means of multi-parametric an approach based on the decomposition of the wake region into three distinct but interrelated parts: the rotor region where a vortex particle model is used, the near-wake region where the turbulent Navier-Stokes equations are solved and the far wake where explicit formulas are defined. This approach first developed in [1,2] is presented in its latest version which is capable of providing predictions for the velocity field and the turbulence characteristics in the wake. Representative results for the Nibe turbines are given. The comparison between predictions and measurements are regarded satisfactory. Finally a Wind Farm model based on the multi-parametric approach is also outlined.

(this work was partially financed by the C.E.C. under the JOUR-0087 project)

1. INTRODUCTION

Conceptually, the wake of a Wind Turbine (WT) is the rotational flow region just downstream the rotor plane. Within this region the wind inflow is reduced whereas the evolution of the pronounced shear layer originating from the tip region of the blades increases the flow fluctuations. Consequently, if a WT is placed in the wake of another one, it will not only operate at reduced inflow conditions, but it will also experience additional fluctuating loads. In order to model numerically these two effects we need: (a) a wake model capable of providing good predictions of the flow characteristics and in particular the turbulence properties within the wake, and (b) an interaction model that would take into account the effect of overlapping wakes.

Several models of varying complexity have been developed and validated. They can be classified according to their complexity in three categories: (a) the explicit or kinematic models, (b) the grid oriented implicit wake models based on solvers of the Navier Stokes equations, and (c) the implicit vortex models based on the solution of the vorticity transport equations. Due to their considerable computational requirements, the implicit models are not suitable for wind park analysis, at least in their present form. On the other hand all the existing explicit models are insufficient as they do not give estimates for the turbulence characteristics within the wake. Therefore, in order to define an improved wake model a compromise between the simple wake models and the more elaborate ones is necessary. In view of these remarks, the development of a *multi-parametric wake model* was initiated two years ago [1,2,3], based on the following guidelines:

> The wake is decomposed into three regions: the rotor region where a vortex particle model is used, the near-wake region where the turbulent Navier-Stokes equations are solved, and the far-wake region where explicit formulas are introduced based on similarity assumptions.

The first version of the multi-parametric model used in the rotor region a linear vortex model based on the lifting-line approximation of the rotor blades [1]. This model was found insufficient not only because the accuracy of the predictions for the radial velocity profiles was poor but also because it was difficult to retrieve estimates of the turbulence added by the rotor. The work conducted in [3] clearly revealed the necessity of a more careful turbulence modeling. Therefore a non-linear vortex model was used instead.

In the sequel this improved version of the multi-parametric model is presented together with results obtained for the Nibe turbines [4].

2. DESCRIPTION OF THE MODEL

Assuming the flow axi-symmetric, the multi-parametric model is defined by the following steps:

(a) Given the geometry of the rotor and the operational conditions (inflow velocity profile and rotation speed) the vortex particle model GENUVP (GENeral Unsteady Vortex Particle model) [6] is used to calculate the velocity distribution $\vec{u}$ as well as the contribution of the rotor to the turbulent characteristics of the flow. The latter corresponds to the velocity fluctuations due to the unsteady character of the flow. Let $\sigma_r(\vec{x})$ denote the standard deviation obtained at a fixed point $\vec{x}$ in space through time correlations of the unsteady velocity calculations. If N_T time steps are used, then:

$$\sigma_r^2(\vec{x}) = \frac{\sum_{n=1}^{N_T} [\, u(\vec{x};n\cdot\Delta t) - \overline{U}(\vec{x})\,]^2}{N_T} \tag{1}$$

where $\overline{U}(\vec{x})$ denotes the mean axial velocity at $\vec{x}$ given by the expression:

$$\overline{U}(\vec{x}) = \frac{\sum_{n=1}^{N_T} u(\vec{x};n\cdot\Delta t)}{N_T} \tag{2}$$

(b) Based on the above data, the inlet conditions for the near-wake region are defined. The velocity distribution used is the one given by the analysis of the rotor region whereas the inlet distributions of the turbulent kinetic energy k and the length scale L are modified in order to account for the specific ambient conditions.

Let σ^2 and σ_a^2 denote the total and ambient variance of the velocity respectively. Then,

$$\sigma^2 = \sigma_r^2 + \sigma_\alpha^2 \tag{3}$$

where according to [5]:

$$\sigma = 2.4 \cdot u^* , \qquad k = C_\mu^{-0.5} \cdot (u^*)^2 \cdot \phi_k \tag{4}$$

As regards L, for r > R (where R denotes the radius of the rotor) it is set equal to the ambient length scale L_α, whereas for r < R, local equilibrium between production and dissipation of k is used [7]. This assumptions was found to be the best choice according to the conclusions of a parametric analysis presented in [6].

(c) Solve the turbulent Navier-Stokes equations (AXI-NS code) [7] to obtain the velocity and turbulent intensity profiles in the near-wake region.
(d) Obtain through matching the parameters defining the far-field.
(e) Give the far-field profiles.

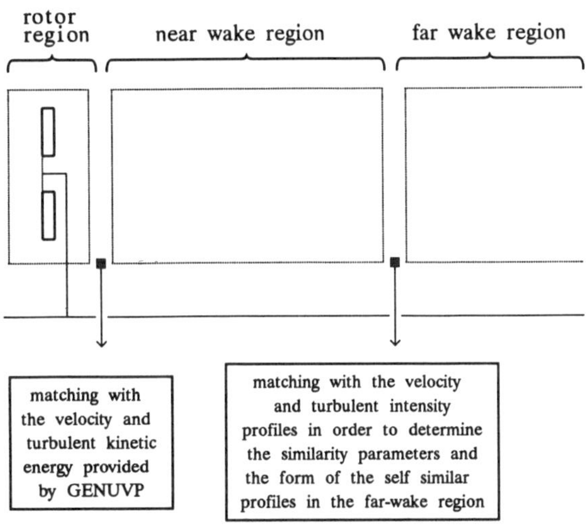

Figure 1: Schematic overview of the wake model

For the analysis of multiple wake situations, in [8] and within the context of a kinematic wake model, an explicit interaction formula was proposed. This formula is based on the equilibrium of energy losses and has the following form:

$$[\, U_w - u(\vec{x}) \,]^2 = \sum_{i=1}^{N_{WT}} [\, U_i - u_i(\vec{x}) \,]^2 \tag{5}$$

where, U_w denotes the ambient wind velocity, $\vec{x}$ the point of calculation, $u(\vec{x})$ the velocity at $\vec{x}$, N_{WT} the number of the wind turbines placed upstream of the point of evaluation $\vec{x}$, U_i the inflow velocity of the i-th wind turbine and $u_i(\vec{x})$ the velocity in the wake of the i-th wind turbine at point $\vec{x}$. The $u_i(\vec{x})$ velocity is provided by the single wake model for inflow wind velocity equal to U_i.

Finally for the analysis of Wind Farms, since the distances between the WTs are usually greater than four rotor diameters, only the far-wake profiles have to be defined. This remark is essential as it leads to the definition of a cost effective numerical tool for Wind Farm modeling. More specifically it suffices to complete a data base with input fields the type of the WT, the ambient turbulence and the inflow wind velocity. As output fields the data base will provide the data required for the definition of the far field. This data base together with the interaction formula will constitute the Wind Farm model.

In order to complete the data base, the combined (GENUVP+AXI-NS) code can be used. In this connection, it was found in [3] that the far-field characteristics can be obtained by a restricted number of complete runs.

3. RESULTS

3.1 The single wake results

In Figure 3 the axial velocity profile and the corresponding normalized standard deviation are given as calculated by the GENUVP code on a plane 4m downstream the rotor plane. The form of the profile as well as the order of magnitude can be regarded reasonable. The kinks of the curves are due to the particle approximation of the free-vorticity. It is noted that the experience gathered, showed that these "deficiencies" do not affect the numerical analysis of the near-wake region.

In Figures 4, 5 and 6 typical results are given corresponding to wind velocity at hub-height U = 8.55m/s and ambient turbulence intensity 9%. Figure 3 shows the variation of the centreline velocity deficit whereas Figures 5 and 6 give the axial velocity profiles and the normalized turbulent velocity distributions at x = 2.5D and x = 4D respectively. As a whole the results can be regarded satisfactory in spite of the axi-symmetric character of the approach. Although there are some deviations with respect to measurements, the form of the profiles for both the velocity and turbulence characteristics closely follow the behaviour experimentally observed. Since part of the inaccuracies could be attributed to three-dimensional effects there is need for a fully three-dimensional analysis in order to conclude on the validation of the model.

3.2 The double wake results

Setting two machines in tandem, at first the flowfield was predicted by means of a consecutive application of the combined GENUVP + AXI-NS code as described in Figure 2.

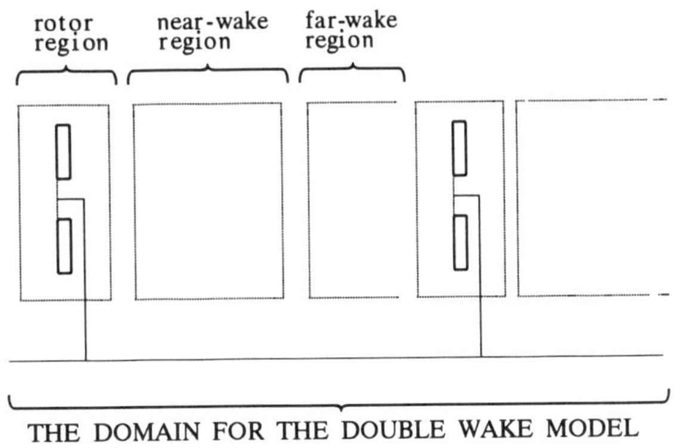

Figure 2: Schematic overview of the consecutive application of GENUVP + AXI-NS code

In Figure 7 the variation of the centreline velocity is compared to measurements. It is observed that the predictions are satisfactory.

Next, the validation of the energy formula (5) is examined for the explicit estimation of the interactive operation of the two wind turbines. The corresponding results are shown in Figure 8. More specifically, the solution obtained by the sequential application of the combined model (solid line) is compared with the estimate the energy relation gives (squares). In general the agreement is regarded satisfactory. The maximum error appears at x=2.5D downstream the second wind turbine. It is reminded that this was also the point with the maximum error for the isolated wind turbine. Though we cannot be sure, it appears that there is a problem as regards the predictions in the core of the second wind turbine.

5. DISCUSSION

Though further validation and possibly improvement of the proposed approach is required, the work done up to now leads to the conclusion that the combined (Vortex Particle + k-ϵ) model can constitute the numerical tool to be used for the completion of the data base required for the multi-parametric Wind Farm modeling.

It is noted that the whole analysis was based solely on the experimental results for the Nibe turbines. Thus in order to better support the aforementioned discussion additional measurements are needed. In this connection the modeling approach described in the present paper is currently validated in the case of the Sexbierum Wind Farm for which measurements were recently obtained.

Apart that, it is also noted that the whole analysis concerns the axi-symmetric case only. Therefore an additional validation step is required that would conclude on the limitations of this approximation through comparisons with fully three-dimensional predictions.

REFERENCES

[1] S G Voutsinas, K G Rados, A Zervos, J Wind Engineering and Industrial Aerodynamics (1992) 283.

[2] S G Voutsinas, J P Glekas, A Zervos, J Wind Engineering and Industrial Aerodynamics (1992) 293.

[3] S G Voutsinas, K G Rados, "Wake and Wind Farm Modeling", CRES contribution to the 1st Year Progress Report of JOU-0087 Project (1991).

[4] G J Taylor, Wake measurements on the Nibe wind-turbines in Denmark. Part 2: Data collection and analysis, Final Report, CEC Contract EN#W.0039.UK(H1).

[5] H Panofsky, J Datton, Atmospheric Turbulence, John Wiley & Sons (1984).

[6] S G Voutsinas, K G Rados, "Wake and Wind Farm Modeling", CRES contribution to the 2nd Year Progress Report of JOU-0087 Project (1992).

[7] J Hernandez, A Crespo, CHAM (1990), Vol.3, 330.

[8] S G Voutsinas, K G Rados, A Zervos, Wind Engineering Vol.14 (1990) 204.

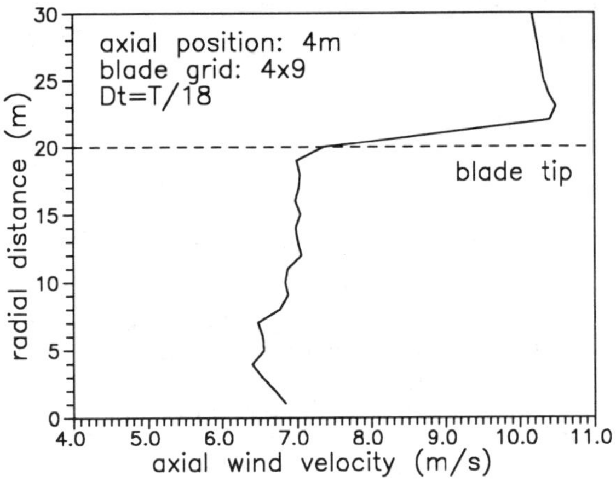

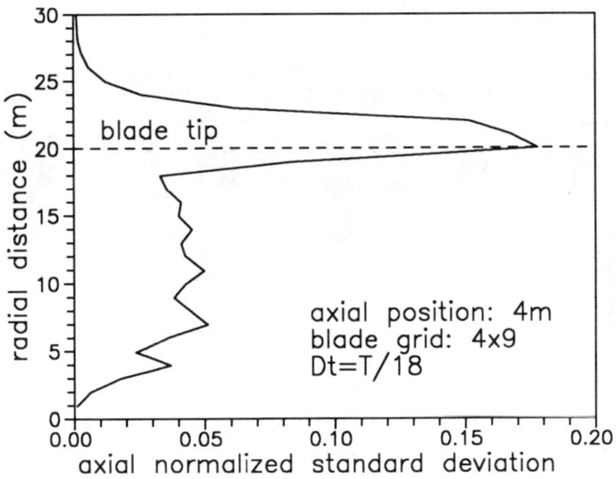

Figure 3: Axial and turbulent velocity profile as given by GENUVP numerical code

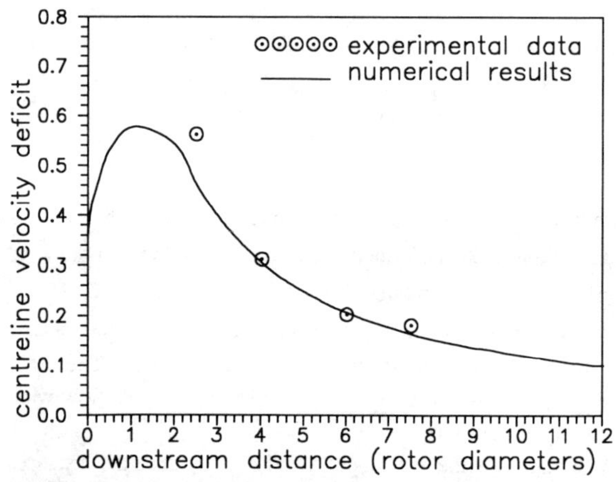

Figure 4: Variation of the centreline velocity deficit in the single wake of the Nibe B' wind turbine

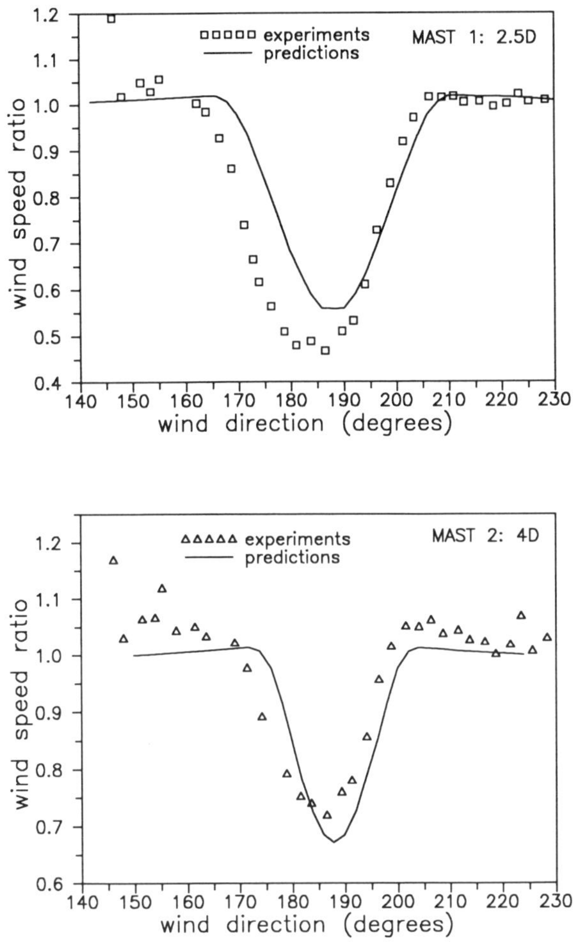

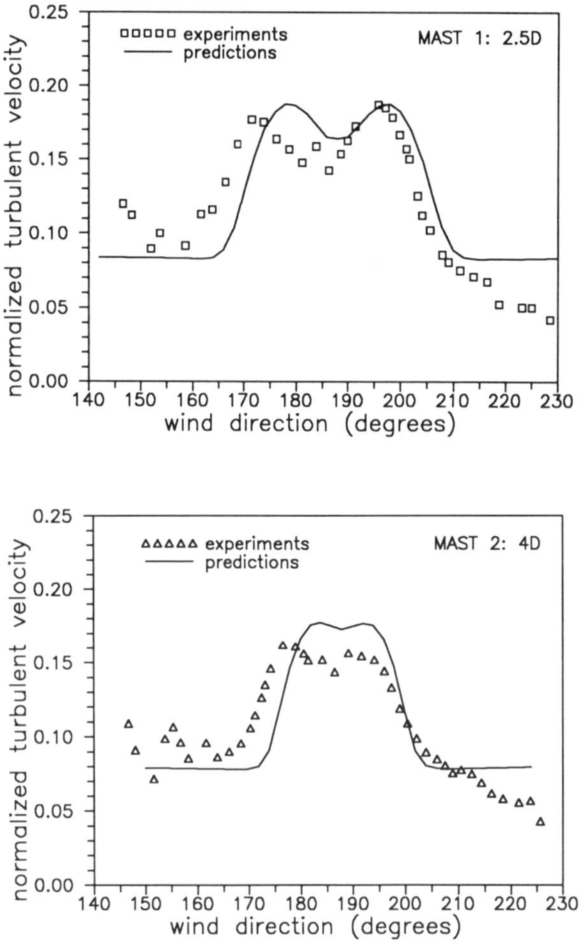

Figure 5: Non-dimensional axial velocity profiles in the single wake

Figure 6: Normalized turbulent velocity distribution in the single wake

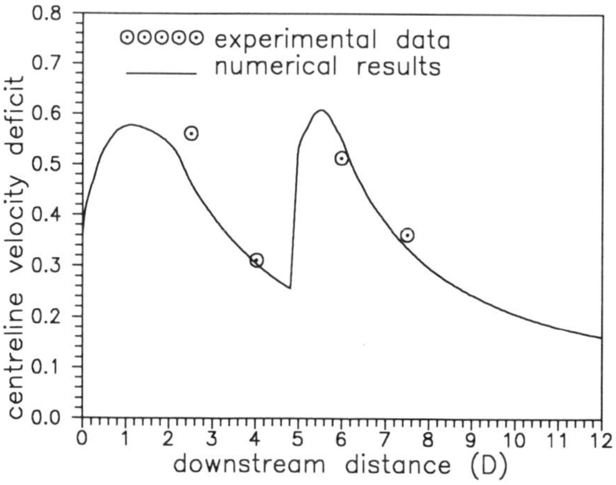

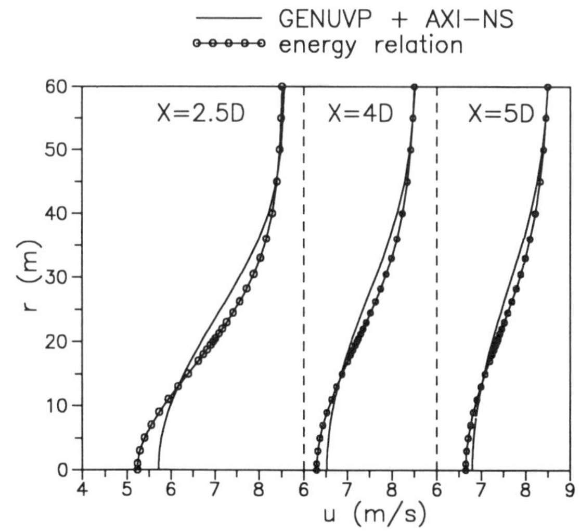

Figure 7: Variation of the centreline velocity deficit in the double wake of the two Nibe wind turbines operating in tandem

Figure 8: Validation of the explicit energy relation used for the double wake interaction of the two Nibe wind turbines operating in tandem

NUMERICAL WAKE-SIMULATION OF A HAWT CONSIDERING TOPOGRAPHY AND USING A MESOSCALE TURBULENCE MODEL

Günther, P.; Fallen, M.; Wolfanger, T.
Department of Fluid Mechanics, University of Kaiserslautern, Prof. Dr.-Ing. H. Rosenberg,
P.O. 3049, D-6750 Kaiserslautern, Tel: ..49/631/205-2301, Fax: ..49/631/205-2300

There are many highlands in Europe with a known wind potential. The understanding of a WTs wake is necessary to optimise the arrangement of WTs in wind parks to receive maximum energy yield. Numerical simulations seem to be a suitable way to obtain realistic and low-cost results.
In this paper some results calculated with the CFD-program PHOENICS are presented. The influence of a WT on the flow-field using a mesoscale turbulence model is shown for a plain area. To calculate the influence of a WT on top of a two-dimensional hill, it is necessary to enlarge the solution domain. The distribution of wind speed and turbulence quantities in front of the WT and the expansion of the turbine wake are affected by the hill.
The structure of the wake induced from the WT on the hill is slightly different from the wake appearing in a plain area. Comparisons to full-scale measurements are necessary to adapt the calculated results and transfer them to other problems.

1 Introduction

The knowledge of the wind-turbine-wake is very important for many aspects in wind-energy applications. Especially by the fact of energy production in wind parks, it is possible to optimise the arrangement of the WTs with respect to topography and to calculate the energy yield in dependence on wind speed, wind direction and variation of wind sped with teight.

In this paper some results calculated with the CFD-program PHOENICS are presented. The influence of wind-turbines on the flow-field is shown for a plain area and an area with a two-dimensional hill.

2 Some remarks on PHOENICS

PHOENICS is a Stokes solver developed by CHAM (U.K.). It solves the finite-volume equation of mass, momentum and energy conservation two- or three-dimensional. The grid may be cartesian, cylindrical polar, or body-fitted for complex domains. The linkage of mass- and momentum equations is treated by an improved version of the SIMPLE algorithm [1].

The calculations were carried out in use of PHOENICS Version 1.6.6 on a Silicon Graphics RISC-Workstation and a Siemens VP-100 vector-processor.

3 Initialisation of flow field and wind turbine

Considering the atmospheric boundary layer, some simplifications are made. Wind direction is constant with height, and wind speed is a function of surface roughness, friction velocity and heat flux on the ground. The atmospheric stability is described by the Monin-Obukhov length. Coriolis forces are neglected and turbulence is due to mechanical and thermal effects. The velocity-profile is logarithmic with nearly 15 m/s in 200 m height. The air-density is constant in the flow-field, the influence of the gravitational forces are simulated by the Boussinesq-Approximation [2]. The area covered by the wind turbine's propeller is approximated by a polygon, the withdrawal of energy from the wind is simulated by defining sinks in the domain.

4 Results

First, calculations are made on a plain area with a single wind turbine. The turbulence models implemented in PHOENICS by default does not cover any temperature influence in its source terms. In atmospheric simulations the influence of temperature cannot be neglected so for instance the standard k-ε model has to be modified. This is done by a turbulence model proposed from ETLING and DETERING [3], which refers to temperature influence by the gradient of potential temperature. This gradient of potential temperature is used to modify the source-terms of the equations of turbulent kinetic energy k and the dissipation rate of turbulent kinetic energy ε.

Fig. 1 shows the distributions of the mean wind speed divided by the unperturbed wind speed at hub height in two vertical planes. The location of these planes are 100 and 160 m behind the turbine. The wind speed ratio calculated with the standard k-ε model and the modified k-ε model in comparison to the incident flow and full-scale measurements on Nibe-turbines (rotor-diameter 40 m, hub height 45 m) in Denmark is marked.

Our calculations correspond satisfactory with full-scale measurements. The far wake is described well, but the velocity defect near the turbine is calculated too weak.

The standard k-ε model calculates only a small velocity defect.

In fig. 2 the wind speed in hub height is shown in relation to the distance from the WT. Even 8 rotor diameter downstream the wake is still perceptible.

Similar results met HERNANDEZ and CRESPO [4]. They also use the PHOENICS-code and their turbulence model is a modified LAUNDER and SPALDING k-ε model.

In a second step the domain is extended (10 x 20 x 40 cells, 100 x 200 x 2000 m) and the surface structure is taken into consideration. A primitive, quasi 2 dimensional hill has been introduced into the domain with a WT placed on top of the hill.

A contour plot of the wind speed in z-direction is shown in fig. 3. The downstream distance between the marked planes z=11 (WT plane) and z=24 is about 10 rotor diameter. In comparison to a site without hill the calculated wake is longer. Remarkable is the velocity development in the near wake and the influence of the WT on higher layers of the atmosphere.

5 Conclusions

Meteorological phenomena, especially the wake of WTs, are very complex. Numerical solutions of the time averaged finite-volume equations of mass, momentum and energy conservation can be found ony by using turbulence models. These turbulence models mostly contain parameters which must be adapted to the considered problem. In order to do this, it is absolutely necessary to compare the results of numerical solutions with full-scale measurements.

Solutions of differential equations are strongly influenced by boundary conditions. To receive realistic results for a WTs wake, it is necessary to introduce proper boundary conditions near the WT. This is only possible for a plain area, where the distribution of wind speed and turbulence quantities are known. In more complex terrain these distributions are not valid. By enlarging the domain and modelling a real surface, it is possible to calculate the distribution of wind speed and turbulence quantities directly. This is a way to avoid the problem of setting realistic boundary conditions in the direct surrounding of the WT.

References

[1] Patankar, S.V.: Numerical heat transfer and mass flow. Hemisphere Publications Corp., Washington, 1980.

[2] Pielke, R.: Mesoscale Meteorological Modelling. Academic Press, 1984

[3] Etling, D.; Detering, H.; Wamser, M.: Parametrisierung turbulenter Flüsse in mesoskaligen Simulationsmodellen. Meteorologische Rundschau 39, 178-188, 1986.

[4] Hernandez,; Crespo,: Parabolic and elliptic models of wind-turbine wakes. 4th International PHOENICS User Conference in Miami.

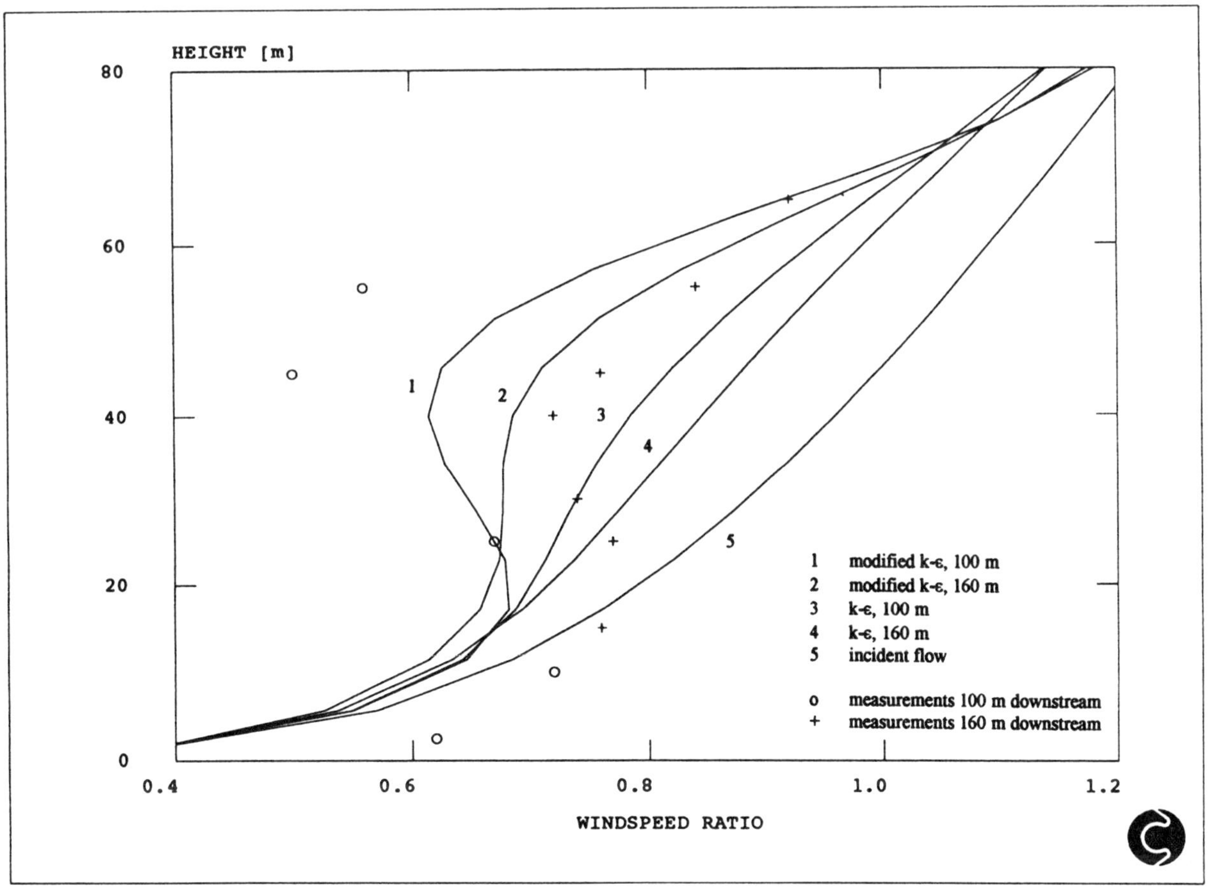

Fig. 1: Wind speed ratio 100 / 160 m behind turbine.
Comparison full scale measurements, k-ε, modified k-ε

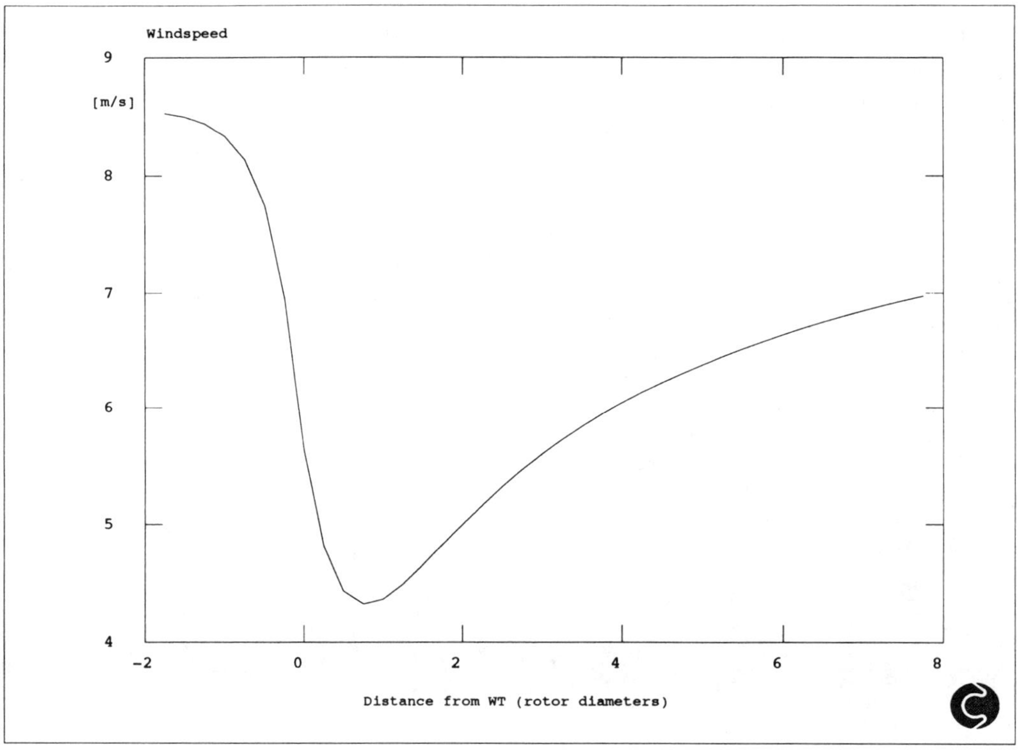

Fig. 2: Wind speed in m/s at hub height over distance from WT

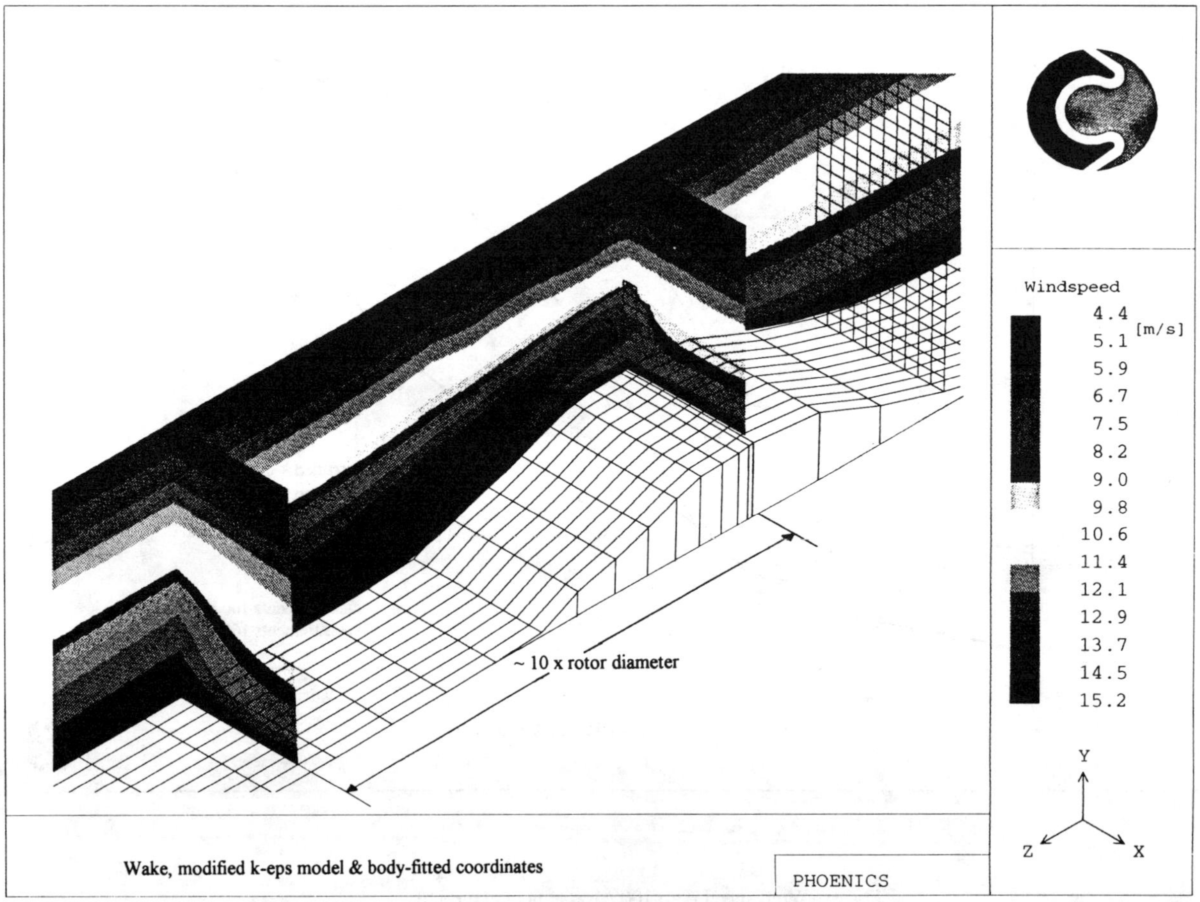

Fig. 3: Turbine on a 2-dimensional hill. Wind speed in z-direction in m/s

ATMOSPHERIC STABILITY INFLUENCE ON TURBINE WAKES

Mikael Magnusson
Department of Meteorology
Uppsala University
S-751 20 Uppsala, Sweden

ABSTRACT: In the present paper the effects of atmospheric stability on the velocity deficits behind wind turbines are studied. The general result is that the stability have a rather large influence on the deficits, not only regarding their magnitude but also the general behavior of the wakes. This may be of importance for load calculations on wind turbines situated down wind other turbines. An equation describing the velocity deficit during stable and unstable stratification is presented.

1. INTRODUCTION

The velocity deficit behind wind turbines have been studied during different atmospheric stabilities. Not many studies concerning the influence of stability in full scale measurements have been made [1]. The influence of the stability is assumed to be rather small or not existing. It is, however, here found that during different stability conditions the deficit can differ by almost a factor of 2 at 10D. The vertical and horizontal distribution of the wake for stable and unstable stratification are also different. The result of this is that the loads on wind turbines situated downwind other turbines will be quite different during different meteorological situations. The deficit as a function of averaging period and free wind speed interval have also been studied.

2. MEASUREMENTS

The measurements were made in a wind farm on the island Gotland in the Baltic sea. The farm is situated on a coastal area and the land is grasped by cattle. the location and configuration of the wind farm is presented in fig. 1.

The data set is collected during a period of two years, June 1990- July 1992. In the wind farm there are two meteorological towers, wind profile measurements in both. The temperature profile is measured in tower 1. In June 1991 the towers were extended from 42m to 54m, in order to be able to measure the total vertical extent of the wakes.

During the first period the wind profiles were measured at 7 levels (4, 10 18.4 24.4, 30.8, 35.5 40.7 m), and during the second period at 8 levels (10, 18.4, 24.4, 30.8, 35.5, 40.7, 47.3, 53.3 m), using anemometers developed at MIUU [2]. Before digitizing the signal at 1 Hz, the analog signal pass electronic filters with a half energy point of 0.5 Hz, in order to avoid aliasing. The temperature profile was measured with Pt500-sensors placed in ventilated and radiation shielded tubes at five levels (2.8, 10, 18.4, 30.8, 40.7 m), during the whole period. At the lowest level the absolute temperature is measured while two Pt-sensors is used to measure the temperature difference between adjacent levels higher up to get the profile [3]. The data from the instruments is evaluated as 10 min. mean values on line. and the 1 Hz data is stored on streamer tapes.

Mean values for the averaging periods, 1, 10 and 30 min., have been calculated from the 1 Hz data. For most of the studies only 1 min. mean values have been used, except when studying the effects of meandering.

Only wake data when the wind turbines produce more than 20 kW have been used. A free wind speed exceeding 12 m/s, from where the turbines starts to regulate, has been used as an upper limit.

Most of the results presented here are from measurements at 4.2D, as there are most data for this case. This is due to the configuration of the wind farm and the climatological wind direction distribution [4]. The deficit is defined as dU/U_f, where dU is the difference between the free wind speed, U_f, and the velocity behind the turbine.

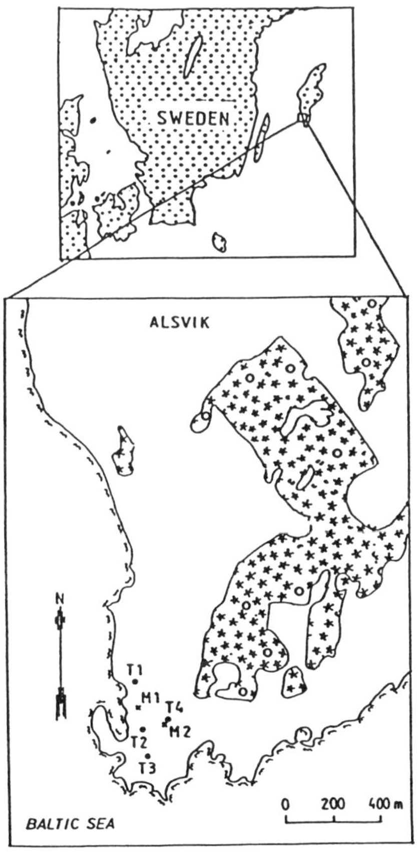

Fig. 1. Location and configuration of the wind farm.

3. ATMOSPHERIC STABILITY

Due to different meteorological situations the air can be categorized into three different stratification classes: stable, neutral and unstable.

To categorize the different stability situations the Richardson number has been used. It is defined:

$$Ri = \frac{g}{T_0} \frac{\frac{\partial \theta}{\partial z}}{\left(\frac{\partial \bar{u}}{\partial z}\right)^2} \qquad (1)$$

where:

$\frac{\partial \theta}{\partial z}$ = Potential temperature gradient

$\frac{\partial \bar{u}}{\partial z}$ = Velocity gradient

In this study Richardson number is calculated as a bulk number around 30 m, Using the levels 40.7 and 18.4 m.

The stratification classes is defined as unstable stratification for $Ri < 0$ and as stable stratification for $Ri > 0$

4. RESULTS

Fig. 2 shows the centre line deficit at hub height as a function of Richardson number for three distances. For $Ri < 0$, unstable cases, the deficit is approximately constant. Near $Ri = 0$, neutral stratification there is an increase (approximately at $Ri = -0.05$), for all distances. The increase is continued for $Ri > 0$, stable cases, reaching a leveling state above $Ri > 0.05$. This behavior is obvious for 4.2D and 9.6D, but not at 6.1D. This could be explained by the fact that there is only a small amount of data available at this distance, for $Ri > 0$.

Mean values of the centre line deficit at the three distances and for stable and unstable stratification are given in Table 1. We can see that the ratio between stable and unstable deficits increases with distance. This is an expected result, as during unstable conditions the air inside the wake is mixed more effectively with the surrounding air than for stable situations.

x/D	Unstable	Stable	Stable/Unstable
4.2	0.39	0.52	1.3
6.1	0.23	0.35	1.5
9.6	0.14	0.24	1.7

Table 1. Centre line deficit.

As a result of the above behavior, the measured data can be separated into three classes, unstable, neutral and stable stratification. Here the stable and unstable situations will be studied. Unstable data are those with $Ri < -0.05$, and stable those with $0.05 < Ri < 0.25$. The upper limit for the stable cases is chosen because above that number there are small amounts of data, and there are also uncertainties about the behavior of the atmospheric turbulence.

Fig. 3a-b and 4a-b show the horizontal and vertical distribution of the mean and standard deviation of the deficit for unstable and stable cases, at 4.2D. These four figures shows that it is not only the magnitude of the deficit that is different between stable and unstable cases. The shape of the wake is also different. The magnitude of the standard deviation is for the unstable case approximately independent of locations within the wake, while for the stable cases, it is larger in the centre of the wake and decreases towards the edges. This behavior may be explained by the fact that the interaction between the surrounding air and the air in the wake is less pronounced for stable situations. At the centre of the wake it is the behavior at the rotor produced turbulence that is most important.

To be able to study the shape of the wake, contour plots are presented in Fig. 5a-b. As can be seen from these figures the difference between the two cases is rather large. In the stable case the whole wake is 'tilted', while there is no such tendency for the unstable case. This may be a result of the fact that the mixing of the wake is so slow for the stable stratification, so the change of wind direction with height is affecting the wake.

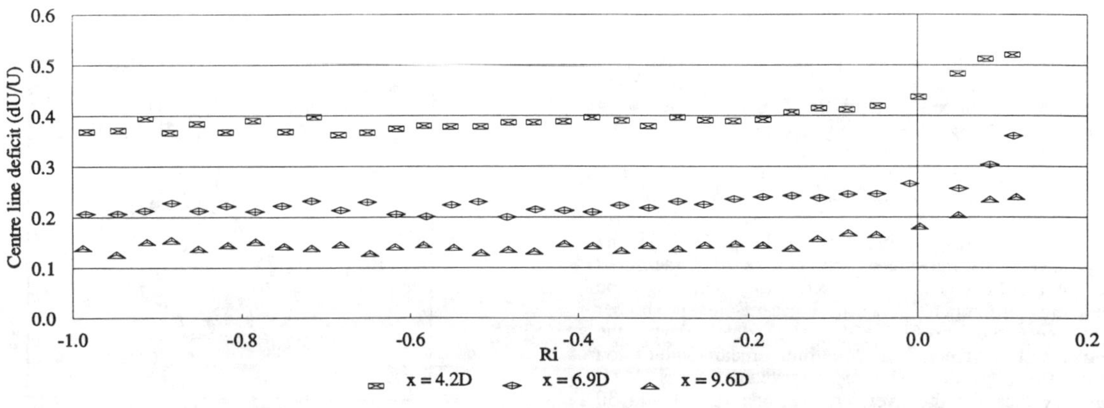

Fig. 2. Centre line deficit at hub height as a function of Richardson number, for three distances.

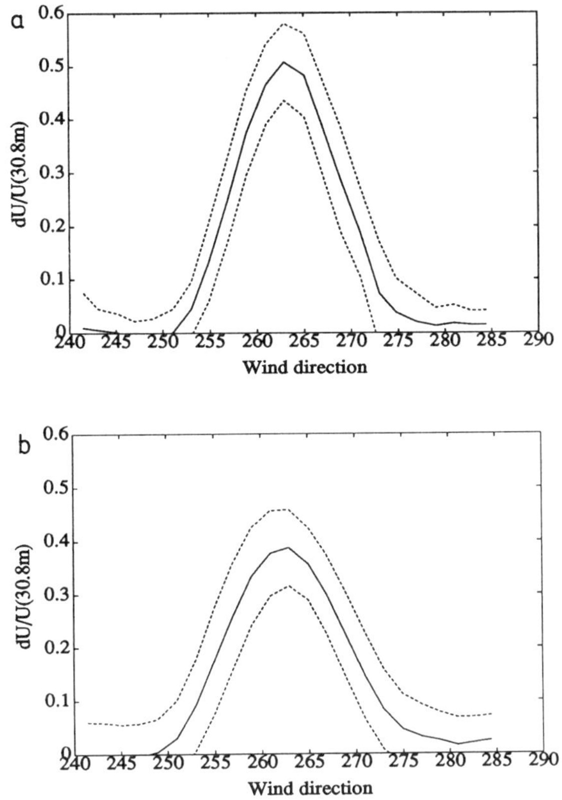

Fig. 3. Horizontal mean value and standard deviation profiles of the velocity deficit at hub height, 4.2D. a) Stable case. b) Unstable case.

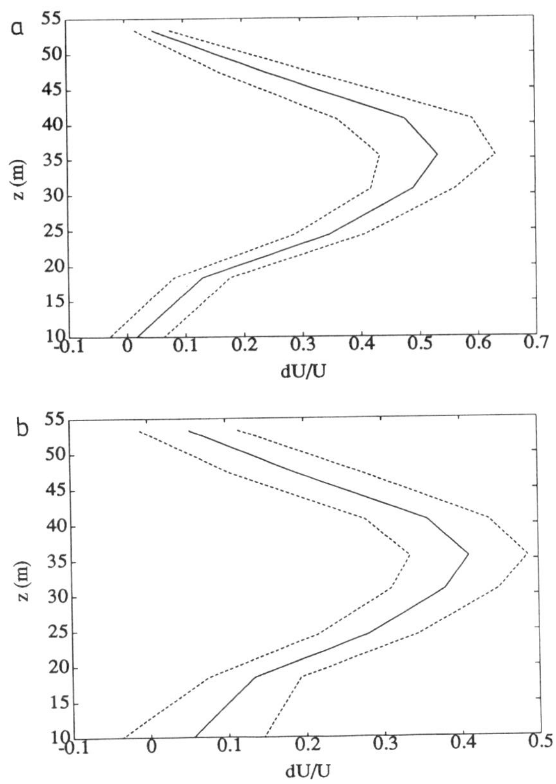

Fig. 4. Vertical mean value and standard deviation profiles of the velocity deficit at the centre line, 4.2D. a) Stable case. b) Unstable case.

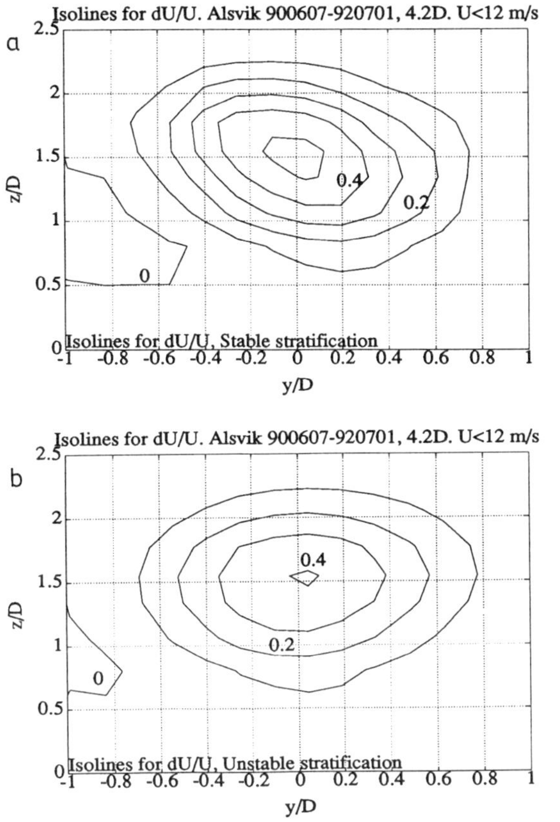

Fig. 5. Contour plots of the deficit as function of normalised horizontal and vertical extention, 4.2D. a) Stable case. b) Unstable case.

An attempt to numerically describe the shape of the wake for stable and unstable situations have been made, in analogy with dispersions of air pollution. The deficit is described by a Gaussian function, using mirroring at the surface, The velocity deficit can then be written:

$$dU/U(x,y,z)=A(x/D)\cdot\exp(-(y/D)^2/2\sigma_y^2)\cdot(\exp(-((z-z_c+B)/D)^2/2\sigma_z^2)+\exp(-((z+z_c+B)/D)^2/2\sigma_z^2) \qquad (2)$$

D is the rotor diameter,
The coefficients in equation (2) is dependent on the stability, see table 2.

Coeff.	Stable cases	Unstable cases
z_c	0.12ln(x/D)+1.32.	0.12ln(x/D)+1.32.
y_c	-0.32ln(x/D)+0.42	-0.154ln(x/D)+0.21
σ_y	$0.19\sqrt{x/D}$	$0.22\sqrt{x/D}$
σ_z	$0.21\sqrt{x/D}$	$0.24\sqrt{x/D}$
A(x/D)	$2.33(x/D)^{-1.053}$	$2.36(x/D)^{-1.235}$
B	$7\cdot(y-y_c)$	0

Table 2. The coefficients in Eq. (2).

Comparisons between measured and calculated vertical and horizontal profiles at 4.2D for the two stability classes is presented in Fig. 6a-b. From these figures it is seen that the shape of the profiles is well described by a Gaussian function. The reason why the calculated value of the deficit for the stable cases is somewhat smaller than the measured one, is that the equation that describes dU/U_f as a function of x/D, cf. Fig 7. is based on data from all three distances 4.2, 6.1 and 9.6D. The value at 6.1D reduces the magnitude, and as mentioned earlier there is not much data for the stable cases at 6.1D.

The empirical coefficients in eq. (2) is valid for this particular wind farm. The validity for other types of turbines is not investigated here, and there are very few data for comparison, as for the majority of full scale measurements there are no temperature profiles available, and thus not possible to analyse the stability effects.

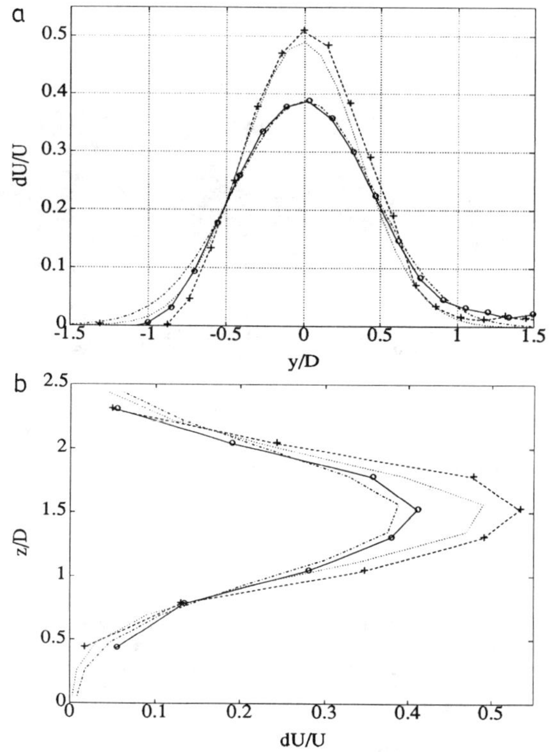

Fig. 6. Comparison betwen measured and calculated profiles, 4.2D. -o-: measured unstable, -.: calculated unstable, --+--: measured stable, ..: calculated stable. a) Horizontal profile at hub height. b) Vertical profile at centre line.

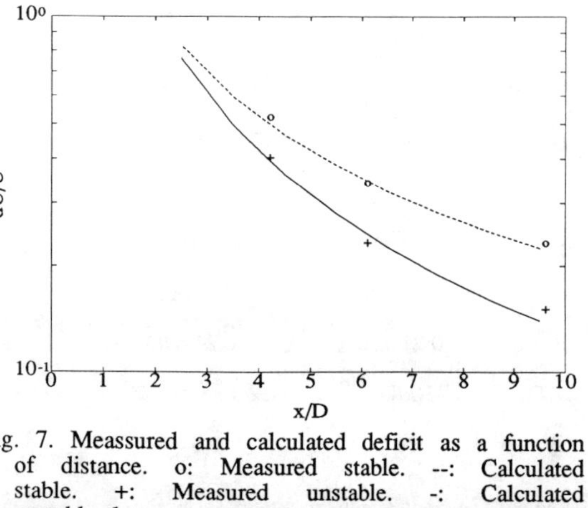

Fig. 7. Meassured and calculated deficit as a function of distance. o: Measured stable. --: Calculated stable. +: Measured unstable. -: Calculated unstable data.

Table 3 shows the deficit as a function of the averaging period. the size of the deficit is decreased as the averaging period is increased. This behavior can be explained by the fact that the whole wake is moved by large vortices of the same dimensions as the wake, meandering cf. [5]. Similar results have been presented cf. [6]. The dependency of free wind speed is also visible in table 3. The differences for different intervals is due to the changes of the C_T-value. For further investigation it is therefore necessary to separate the data not only according to stabilities but also into different wind speed intervals. The large difference between class III and IV is a result of regulation of the turbine.

Centre line deficit dU/U (%)

	4.2D		
Interval	1 min.	10 min.	30 min.
I 6<U≤8	41.0	39.3	35.8
II 8<U≤10	41.5	40.1	37.1
III 10<U≤12	38.7	38.2	35.0
IV 12 < U	31.1	29.5	28.4

Table3. Wake deficit dependence on averaging time for different free wind intervals.

5. CONCLUSIONS

The influence of the atmospheric stability on the flow behind wind turbines can be large. The relative difference between a stable and an unstable atmosphere is increased with increasing distance. Not only the magnitude of the velocity deficit varies with the stability, but the shape of the wake is also affected. This leads to different load situation during different stratifications, for downwind turbines.

Result from a simple numerical description of the wake deficit as a function of vertical and horizontal extensions, shows that the wake behavior can be treated similar to that for dispersion, and described as a Gaussian relation.

6. ACKNOWLEDGEMENT

This project is sponsored by NUTEK, Swedish Board for Industrial and Technical Development.

7. REFERENCES

1. HögströmU. Smedman A-S. Some additional data on full scale wind turbine wakes. Wind energy report from department of meteorology, Uppsala 1990.
2. BergströmH. and Lundin K. Climatological mean wind and turbulence measurements using a quality cup anemometer system. Proceedings of ECWEC' 93, Travemünde 1993.
3. Smedman A-S. Lundin K. Psychrometer systems for measurements of temperature and humidity profiles in the atmospheric boundary layer. Report No. 76, Dept. meteorology Uppsala Univ. Uppsala 1987
4. Magnusson M. The influence of atmospheric stability on turbine wakes. MIUU report. To be published.
5. U. Högström,D. N. Asimakopoulos, H. Kambezidis, C. G. Helmis and A. Smedman. A field study of the wake behind a 2 MW wind turbine. Atmospheric environment Vol. 22 no. 4 pp. 803-820. 1988.
6. Botta G. Giraldi V. A. Results of wake measurements at the Alta Nurra wind power station. ENEL 1990.

MODELLING IMPROVEMENT FOR WIND TURBINE WAKES

H. D. Kambezidis, M. A. Logothetis, B. E. Psiloglou, A. E. Tsangrassoulis

Atmospheric Research Team , Institute of Meteorology and Physics of the Atmospheric Environment , National Observatory of Athens, P.O. Box 20048, GR-11810 Athens, Greece

and A. G. Georgakopoulos

Centre for Renewable Energy Sources, 19th km Marathonos Ave. , GR-19009 Pikermi, Greece

ABSTRACT : This paper demonstrates the efficiency of the upstream wind logarithmic law if it is applied to wind turbine wake models. Comparison between some models against data is made.

1. INTRODUCTION

A Wind Energy Conversion System (WECS) is a device that extracts power from the wind in the Atmospheric Boundary Layer, usually near the ground. The flow downstream of such machines is markedly influenced by the extraction process itself. The resulting momentum deficit wakes behind the WECSs contain flows of lower wind velocities than the upstream values. Therefore the performance of other wind turbines located at distances downstream of such wakes (as in the case of an array of them) can be affected seriously by this slower flow.

The research (theoretical and experimental) on the wakes of WECSs is relatively recent. The first attempt of approaching the problem of wind turbines interaction is the report by Templin [1]. Later Lissaman [2] presented a two-dimensional model (called WINDS) for the evolution of the wake of a WECS over flat terrain based on the jet analogy [3]. Lissaman's model was modified by Vermenlen [4] who introduced the turbulence generated by the blades of the turbine and the supporting tower (MILLY code). Lissaman modified later his original code [5]. At MT-TNO a further development of MILLY code resulted in a more flexible version, called FARMS, containing additional input and output routines in order to handle different types of WECSs within a windfarm and facilitate the comparison of various alternative wind farm configurations. Lissaman et al. [6] have also presented a new version of WINDS for hilly terrain named AVENU but not yet fully tested against real data because of lack of relative measurements. Voutsinas et al. [7] have presented a two-region wake development.

2. APPLICATION OF THE WIND LOGARITHMIC LAW

Kambezidis et al. [8] have introduced the logarithmic wind profile that exists upstream of a WECS and shown that this application gives more realistic wind velocity deficits. Georgakopoulos et al. [9] have applied this technique to a case during NIBWAK '84 experiment [10].

In particular, the wind profile upstream of a wind turbine or any array of WECSs up to a height equal to the hub height plus the radius of the blades is introduced to the code. The profile is either known from local measurements or from wind field codes runs in the area. This procedure modifies the wind profiles in the wake accordingly.

3. RESULTS

Fig. 1 shows comparison of data during NIBWAK '84 with WINDS '82 and LOG WINDS '82 models; Fig. 2 is the same but MILLY and LOG MILLY codes are used. The prefix LOG refers to the application of the logarithmic law to the model. The tower data is referred to the upstream profile, while sodar one in the wake at a distance of 217 m downstream of the WECS. In both cases the logarithmic law gives more realistic wind velocities in the wake that go to zero at ground level. The discrepansy between the minimum of the sodar data and that of the models is due to the meandering effect that has been observed in wakes.

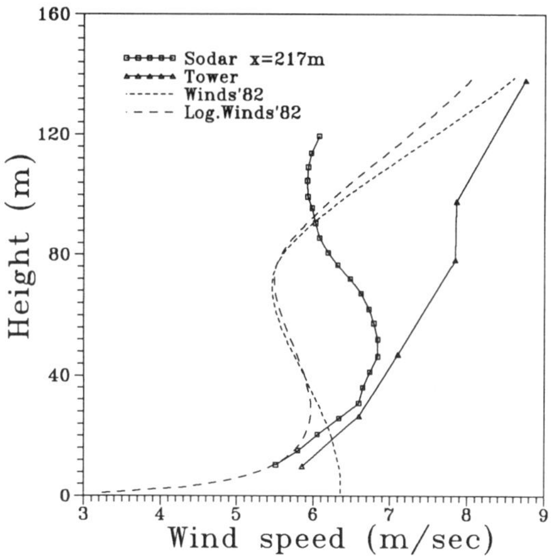

Figure 1 : Comparison of Winds '82 and Log Winds '82 codes with sodar data during NIBWAK '84 experiment. Tower data is for undisturbed air flow.

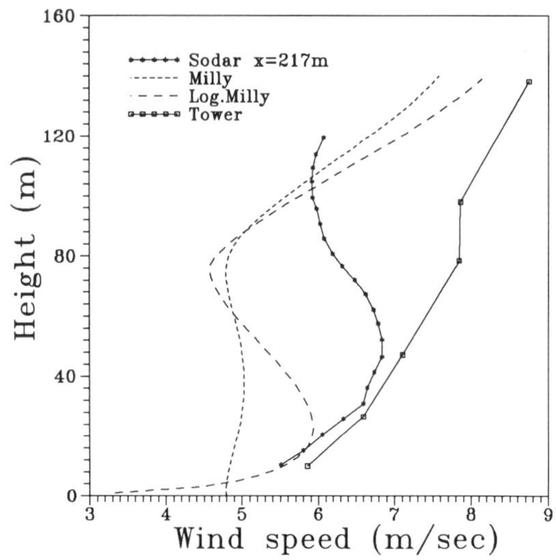

Figure 2: As in Fig. 1 but with Milly and Log Milly codes.

Fig. 3 gives the same comparison as Fig. 1 but in the case of the NIBE experiment [11] with WINDS '82 model and Fig. 4 with MILLY one. Tower 1 is in the wake. The improvement of both models after the application of the logarithmic law is impressing !

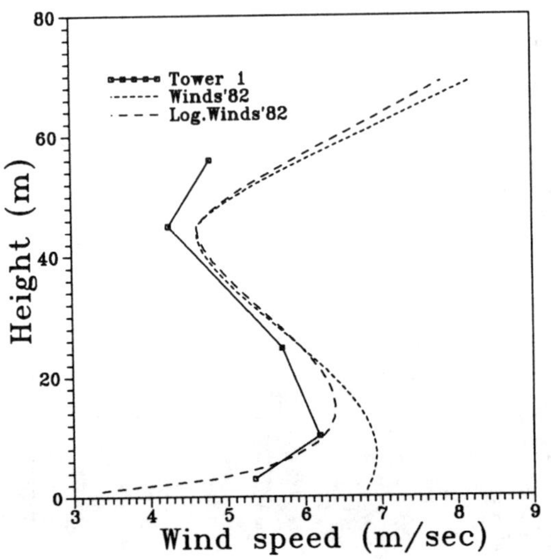

Figure 3: Comparison of Winds '82 and Log Winds '82 codes with Tower 1 data in the wake during NIBE experiment.

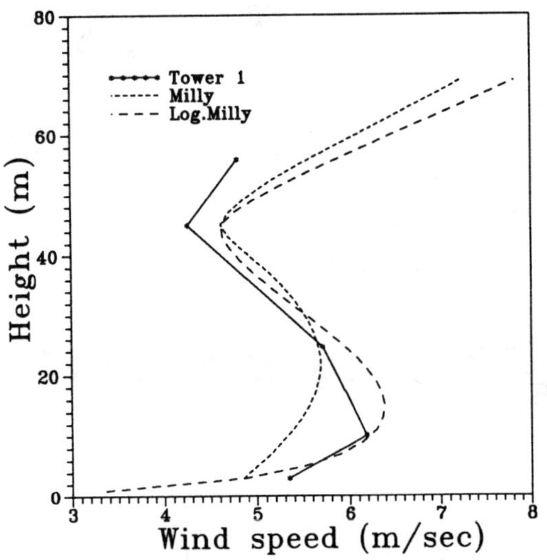

Figure 4: As in Fig. 3 but with Milly and Log Milly codes.

REFERENCES

[1] R. J. Templin, Technical Report LTR-LH 1711, Canada (1974).

[2] P. B. S. Lissaman, Proc. 17th Conf. , N. Orleans , USA (1979).

[3] G. N. Abramowich, MIT Press (1963).

[4] P. E. J. Vermeulen, Proc. 3rd Int. Symp. Wind-Energy Systems, Copenhagen (1980), 431.

[5] P. B. S. Lissaman, G. W. Gyatt, A. D. Zalay, Technical Report PNL-4183, USA, (1982).

[6] P. B. S. Lissaman, D. R. Foster, B. D. Hibbs, D. T. Lindberg, C. K. Thomson, J. H. Rumbaugh, Proc. Int. Conf. on Windfarms, Leeuwarden (1987).

[7] S. Voutsinas, K. Rados, A. Zervos, Wind Engineering, (1991).

[8] H. D. Kambezidis, A. P. Georgakopoulos, D. N. Asimakopoulos, Proc. EWEC, Madrid (1990) 196.

[9] A. P. Georgakopoulos, H. D. Kambezidis, D. N. Asimakopoulos, Proc. BWEA, Swansea, UK (1991).

[10] V. Högström, D. N. Asimakopoulos, H. D. Kambezidis, C. G. Helmis, A.S. Smedman, Atmospheric Environment 22, (1988) 803.

[11] G. J. Taylor, CEC Technical Report No EN3W.0039, (1990) 116.

AN EXPERIMENTAL INVESTIGATION OF WIND TURBINE WAKES USING PARTICLE IMAGE VELOCIMETRY

Whale, J. and Anderson, C.G.

Fluid Dynamics Unit
Department of Physics
University of Edinburgh
James Clerk Maxwell Building
The Kings Buildings
Mayfield Road
Edinburgh EH9 3JZ

ABSTRACT: Determination of the aerodynamic performance of horizontal-axis wind turbines (HAWT) is complicated by problems such as wake deformation effects and free stream turbulence. An accurate prediction of the wake structure is a dominant factor for reliable wind turbine aerodynamic prediction codes. The technique of Particle Image Velocimetry is used to provide detailed full-field data of the wake geometry in the immediate vicinity of the rotor of a reduced-scale wind turbine. The results strongly validate the PIV technique applied to studying wind turbine wakes. The potential for this method is significant and can be used to compare wake data from other experiments as well as wake geometries predicted by theoretical models.

1 Introduction

1.1 Background

Wind turbine performance is critically dependent on the geometry of the rotor wake. The wake structure determines the rotor inflow conditions and hence the forces on the blades. For design purposes it is essential to predict the blade loads accurately, to estimate rotor power output and structural stress. Thus, an accurate prediction of the wake structure is a dominant factor for reliable wind turbine aerodynamic prediction codes.

However, the lack of detailed experimental data in the wake of a turbine, and the difficulty of obtaining it, is widely appreciated. Full-scale visualisation experiments are difficult to perform and are limited by the problems of expense and non-repeatable conditions. Thus, though the investigations of wind turbine wakes are comparatively recent, many have involved simulators and model wind turbine generators. In the late 1970s, wind tunnel simulations began in Holland (at TNO Apeldoorn), Sweden (Aeronautical Research Institute) and in the UK (at CERL Leatherhead). The work was co-ordinated by the International Energy Agency (IEA) and concentrated on hot-wire measurements of flow over tipvane type simulators. A summary of the experimental results for models of size 200-300mm in diameter can be found in Vermuelen[1]. Green[2] later refined and developed the modelling techniques using laser doppler anemometry (LDA) on a modified model aircraft propeller of diameter 150mm.

The problems of poor signal to noise ratio and condensation of seeding in the wind tunnels experienced by LDA workers ([3], [4]) together with the desire to obtain simultaneous multipoint measurements with minimum disturbance to the flow caused by instrumentation, suggest the method of Particle Image Velocimetry (PIV) to studying flows of this type.

PIV is a non-intrusive velocity measurement technique which allows complete two-dimensional flow fields to be captured at a single instant. The basis of PIV is to stroboscopically illuminate a two-dimensional plane of flow containing small neutrally buoyant seeding particles by means of a sheet of pulsed light. A double (or multiple) exposure photograph of this plane is taken. The spacing between the images of each particle on the film gives the local velocity. This photograph is then analysed over a grid of points to determine the local flow velocities across the whole field.

The technique of PIV was introduced to the field of wind turbine aerodynamics by Smith et al.[5] who conducted tests on a Rutland wind turbine using pulsed lasers. The tests established the applicability and usefulness of the PIV technique as a velocimetry tool for wind turbines. However, the high slipstream velocities involved in wind tunnel testing meant that particles separating at the trailing edge of the blades were dispersed out of the light sheet causing problems of illuminating the wake structure. This limited the study to flow close to the blade and again illustrated the problems of seeding in wind tunnels.

In an attempt to overcome the problems of illumination and seeding density experienced by previous investigators in wind tunnel studies, some preliminary tests of a model turbine rig in water were conducted in the Fluid Dynamics Unit of the University of Edinburgh starting in December 1991. The PIV experiments were performed using a high powered continuous wave (CW) laser with the turbine placed in a water tank capable of generating currents. The blade passing frequency of the rotor and the upstream current were carefully arranged so that, Reynolds number apart, aerodynamic similarity existed between the model and a full-scale wind turbine.

1.2 Theory

The most common basis for HAWT performance prediction codes is a combined blade-element momentum theory. In high winds, however, where the rotor is partially or wholly stalled, this method often underpredicts rotor power. This discrepancy has been the subject of experimental and theoretical examination for some time. The theory is also deficient at very low windspeeds, when high blockage exists; empirical methods are usually employed in this region.

Vortex wake models attempt a more realistic representation of the wake, consistent with the notion of a helical vortex sheet shed from the trailing edge of each blade and convected downstream at the wake velocity. These trailing vortex filaments undergo self-induced distortions as well as being influenced by other filaments. The wake deforms into a vortex system comprising of an intense tip-vortex outer region and a weak diffused vortex sheet inner region (See Figure 1). In the flow recording process, the PIV illumination extracts behaviour in the wake corresponding to a cross-section of the helical vortex system.

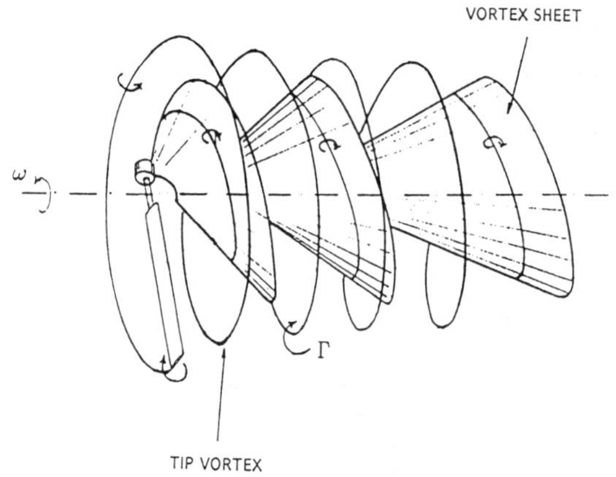

Figure 1. Schematic diagram of helical vortex system.

Many references can be cited for the development and application of vortex wake models (e.g. [6], [7]). There is a lack of consensus, however, in the way the vortex wake geometry should be represented, and the models are sensitive to the choice of assumed wake geometry. Inaccuracies must be expected as many of the features of the flow are not yet fully understood, or are simply unknown, due to the lack of quantitative experimental data on the wake. By providing full-field wake geometry data, PIV experiments provide the opportunity to validate some of the assumptions made in performance prediction codes.

2 Experiments

The experiments were carried out in a glass-based, two-dimensional wave flume equipped with a recirculating pump which allows a steady flow velocity to be established. The model turbine rig was placed across the tank subjecting the rotor to a uniform current. The turbine was driven by a electric motor/generator suspended on a frame above the water level. The rotor is located at the end of an inverted tube, or tower, which was perpendicular to the oncoming flow to ensure symmetric inflow conditions.

It is important that the mechanism of wake generation is accurately produced. As well as geometric similarity between the model and a full-scale machine, an appropriate range of tip speed ratios λ are used (ratio of blade rotation to freestream velocity) so that, notwithstanding the Reynolds number, kinematic similarity is assured. Motor speed was altered to achieve the desired tip speed ratios

$$\lambda = \frac{\omega R}{V_0},$$

where R is the radius of the rotor, ω is the blade passing frequency and V_0 is the freestream velocity.

The PIV set-up at Edinburgh is based on the scanning beam system of flow illumination[8]. The beam from the 15W CW Argon ion laser scans over a parabolic mirror which directs the beam vertically upwards through the base of the flume, illuminating a vertical cross-section of the flow. The water was seeded with conifer pollen of average diameter 70μm with concentrations maintained at a level to ensure a high density of particles on the resulting film record.

Two types of rotors were used in the experiments. In the preliminary investigations the turbine was a 2-blade model aeroplane propeller of 175mm diameter and 16mm in chord length at a blade span of 70%. The propeller was run in reverse so that its twist and chord distributions approximated those of a wind turbine rotor. Tip speed ratios of approximately 3-12 were used with Reynolds number based on tip chord in the range 6,000 - 35,000. Photographs of the flow were taken with a standard 35mm Nikon camera with a f2.8 50mm flat focus lens (e.g. Figure 2).

The second set of experiments involved some improvements in generating a more realistic wind turbine simulator and enhancing the quality of the measurement data. The propeller was not an ideal turbine model as its aerofoil section operated trailing edge first so it was replaced by flat-plate blades of the same diameter with varying thickness, camber and taper. The tower of the rig was streamlined with a foam plastic shroud in order to limit its wake interfering with the top half of the turbine wake structure, as was noticeable in early tests. Also, the nacelle of the model was moved out of the laser sheet so that just the tips of the blades passed through the sheet. A wide range of velocities were recorded in the previous experiments, from the undisturbed upstream flow to the region of almost stagnant fluid behind the rotor. Thus, for these experiments, an image shifting system was used to impose a translational velocity on the recorded image to increase dynamic range and eliminate directional ambiguity. The shift system was comprised of a Hasselblad 500 EL/M camera and 80mm lens mounted on a turntable and controlled by a microcomputer which rotated the turntable to a given shift speed and triggered the camera when it reached the correct position.

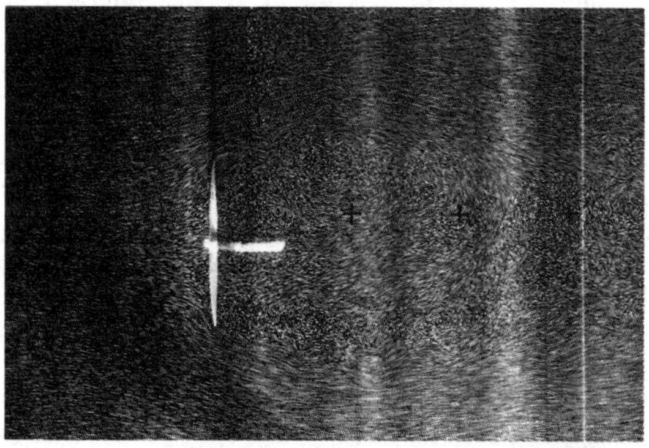

Figure 2. Photographic PIV record of flow past propeller.

The PIV negatives were analysed on a 32 bit microcomputer to yield 2D velocity vector maps. The film was interpreted point by point over a dense grid using methods of optical and digital analysis. This involves scanning a small part of the negative at a time with a probe laser to produce an interference pattern from the multiple particle images in that area. The interference fringes are measured and recorded in digital form and the data Fourier transformed to yield the particle velocities at that point. The whole negative is scanned in this way to build up a flow velocity map, which forms the basis of all subsequent analysis. Firstly the shift velocity was removed to reveal the actual flow pattern and then the mean horizontal component of the velocity was subtracted at all points in order to highlight the induced flow velocities. The velocity information can also be processed to produce vorticity maps. The contours of the plots join points of equal vorticity with vortex strengths represented by varying degrees of lightness.

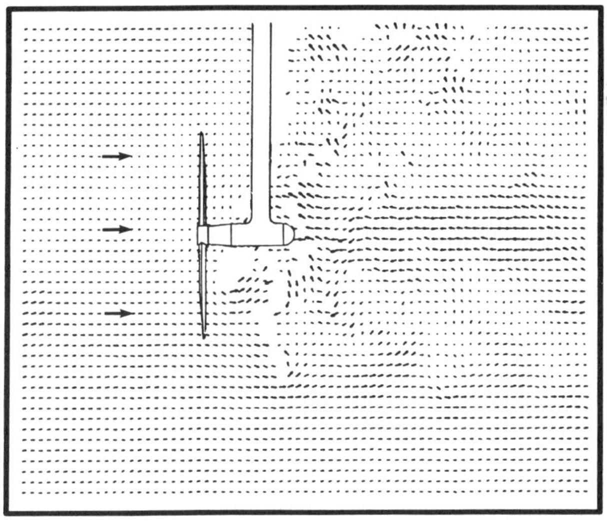

Figure 3. *PIV velocity vector map of propeller in windmill state.*

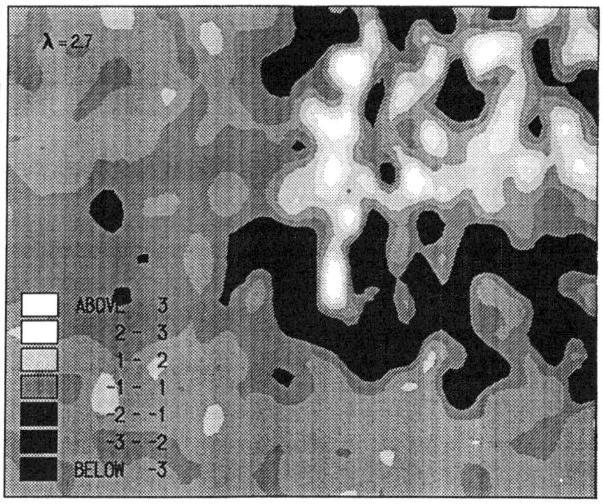

Figure 4. *PIV vorticity map corresponding to Figure 3.*

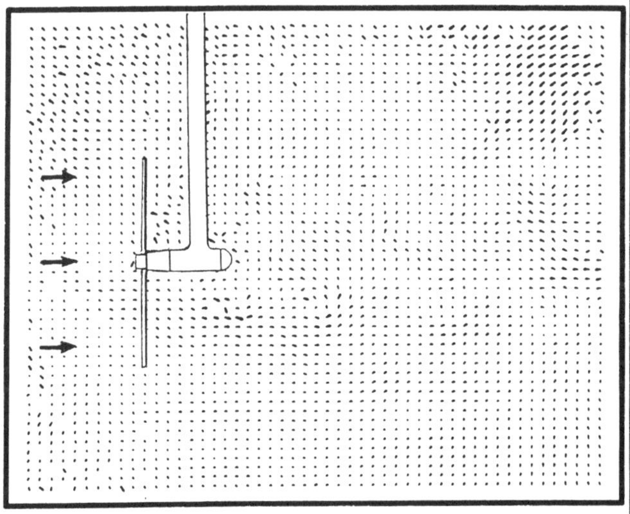

Figure 5. *PIV velocity vector map of flat-plate blade in windmill state.*

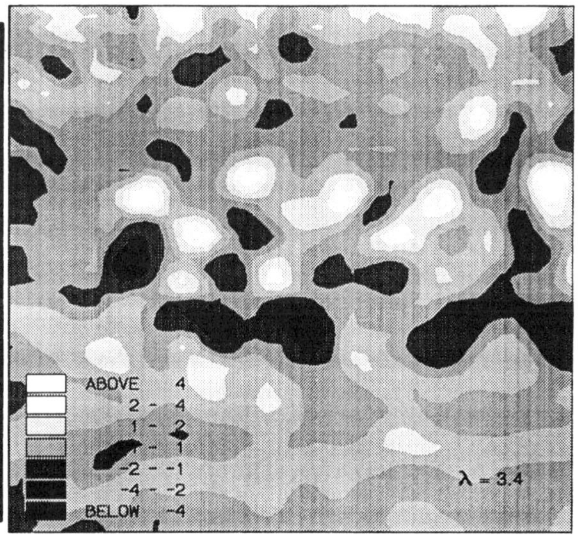

Figure 6. *PIV vorticity map corresponding to Figure 5.*

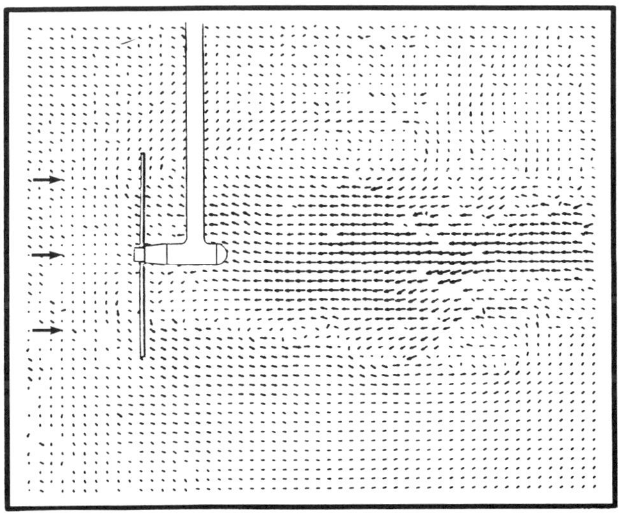

Figure 7. *PIV velocity vector map of flat-plate blade in turbulent wake state.*

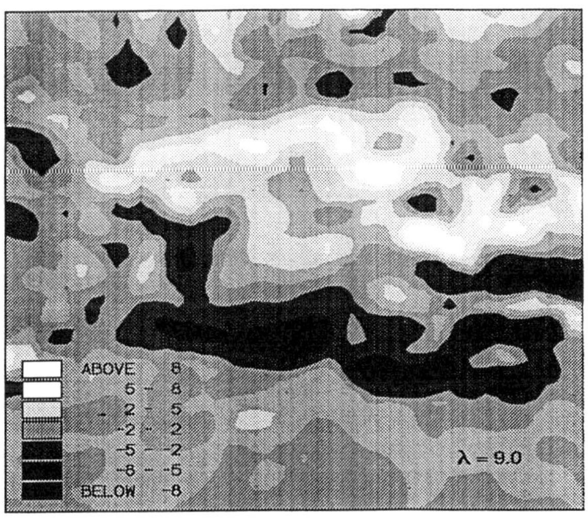

Figure 8. *PIV vorticity map corresponding to Figure 7.*

3 Experimental Results

At low tip speed ratios, the rotor is in the windmill state. The wake of the turbine contains an area of expanding flow with reduced velocity. This is verified by the PIV velocity map of the propeller at $\lambda = 2.7$ (See Figure 3) in that the velocity vectors in the wake point upstream after subtracting the mean horizontal velocity. In the lower half of the wake, angled strips of vorticity can be clearly seen; the expected pattern for the cross-section of a helical vortex sheet. Also seen from the velocity map is a sinusoidal pattern imposed on the boundary of the wake. This is consistent with the prescence of trailing tip vortices originating from the rotating blades and passing downstream.

Figure 4 displays the corresponding vorticity map of Figure 3. Note how the wake is roughly divided into regions of positive(light) and negative(dark) vorticity, lying on either side of the centreline. This is again as expected from the cross-section of a helical vortex. The appearance of two sets of trailing vortices in each half of the wake can be explained from the design of the model propeller which has a rapid change in chord and twist near the hub as well as a change in chord near the tip. The subsequent rate of change of circulation around these parts of the blade induces velocities which roll up the vortex sheet both at the tip and at the hub.

Figure 5 is a PIV velocity vector map for a flat-plate blade in the windmill state ($\lambda = 3.4$). The flat-plate had a thickness of 1.26mm, a camber of 6% and a linear taper with a chord length of 83% of the hub chord at the 70% span position. It can be seen from Figure 5 that the image shifting system was successful in resolving small velocities in the almost stagnant region of flow and produced a more complete vector map. Figure 6 is the vorticity contour plot corresponding to Figure 5. It shows the presence of the tip vortices highlighted by moving the nacelle out of the light sheet. Streamlining the tower also proved effective in reducing the inteference with the top half of the wake and produced a more symmetrical flow pattern.

Figures 7 and 8 are for the same flat-plate blade at a higher tip speed ratio ($\lambda = 9.0$). The rotor is in the turbulent wake state and the boundary of the wake is characterized by initial expansion followed by marked contraction. Figure 8 displays a large increase in the strength of areas of concentrated vorticity compared to Figures 4 and 6. This is because, as the tip ratio is increased, the wake becomes turbulent preserving the energy within the trailing vortices and producing a very strong vortex pattern.

4 Conclusions

The models clearly exhibited operating states characteristic of a wind turbine, namely the windmill state and the turbulent wake state. The windmill state is the intended state for wind turbine operation. However, turbulent wake states do occur during operation and much modelling needs to be done to have a complete understanding of the forces and moments on a rotor in this state.

Care must be taken when comparing the performance of a model wind turbine to that of a full scale machine. As pointed out by Galbraith et al.[9], interpretation of turbine performance requires reliable data on the characteristics of the aerofoil at the appropriate Reynolds number; the overall properties of the downstream wake, however, are less sensitive to Reynolds number and inferences can be made from bulk-flow phenomena.

The results strongly justify using the PIV technique to study flows of this type. PIV provides the means whereby detailed full-field data can be obtained and an experimental data base for wake velocities and structure established. The amount of high quality detailed data of the entire flow field obtained by the uncomplicated, quick and flexible approach of PIV compared favourably with the previous methods explored by investigators. The potential for this method is significant and can be used to compare wake data from other experiments as well as wake geometries predicted by numerical codes. In particular, it could provide information about those regions of flow where theoretical techniques give least satisfactory results, for example in stalled flows. This is a question of some importance to the wind turbine industry as current design methods for stall-regulated wind turbines are essentially empirical.

References

[1] P.E.J. Vermeulen. Studies of the wake structure of model wind turbine generators. *TNO Report 79-012904 Apeldoorn, Holland*, 1979.

[2] D.R.R. Green. Modelling large wind turbines and wakes. *PhD Thesis, Loughborough University*, 1985.

[3] M.B. Anderson, D.J. Milborrow, and J.N.. Ross. Performance and wake measurements on a 3m diameter HAWT : comparison of theory, wind tunnel and field test data. *ETSU contract E/5A/CON/1090/177/020 Cavendish*, 1982.

[4] J.N. Ross and J.F. Ainslie. Wake measurements in clusters of wind turbines using laser doppler anemometry. *Proceedings of the 3rd British Wind Energy Conference, Cranfield*, 1981.

[5] G.H. Smith, I. Grant, A. Liu, and D.G. Infield. Diagnostics of wind turbine aerodynamics by particle image velocimetry. *Proceedings of the 12th British Wind Energy Conference, Norwich*, 1990.

[6] A.A. Afjeh and T.G. Keith. A prescribed wake model for performance prediction of horizontal axis wind turbines. *European Wind Energy Assoc. Conference and Exhibition, Rome*, 1986.

[7] F.J. Simoes and J.M.R. Graham. A free vortex model of the wake of a horizontal axis wind turbine. *Proceedings of the 12th British Wind Energy Conference, Norwich*, 1990.

[8] C. Gray, C.A. Greated, D.R. McCluskey, and W.J. Easson. An analysis of the scanning beam PIV illumination system. *J.Phys :Measurement Science and Technology*, 2:717–24, 1991.

[9] R.A.McD. Galbraith, F.N. Coton, E. Saliveros, and G. Kokkodis. Aerofoil scale effects and the relevance to wind turbines. *Proceedings of the 9th British Wind Energy Conference, Edinburgh*, 1987.

Acknowledgements

The authors sincerely wish to thank Dr. D. Skyner and Mr. T. Bruce for their advice and assistance on matters relating to PIV and Dr. M.L. Jacobsen for assistance with laser beam alignment.

USE OF A MINI-SODAR FOR INVESTIGATION OF TURBULENCE IN A WIND FARM

A. Schomburg, V. Mellert, H.-P. Waldl, H. Klug*, U. de Witt*,

Dep. of Physics, Carl von Ossietzky Universität Oldenburg
D-2900 Oldenburg, P.O. Box 2503, phone: +49-441-798-3248 (fax: -3698)
* Deutsches Windenergie-Institut (DEWI)
D-2940 Wilhelmshaven, Ebertstraße 96, phone: +49-4421-4808-0 (fax: -43)

ABSTRACT: We present SODAR measurements of velocity and turbulence profiles of the wakeflow behind a 55 kW wind turbine sited in a straight line wind farm at the North Sea Coast. The wind farm consists of five turbines with a hub height of 22 m and a rotor diameter of 16 m.

1 SODAR PRINCIPLE

1.1 Determination of wind velocity and turbulence intensity

Remote sensing of the atmospheric boundary layer is easily done by acoustic waves. The SODAR (Sound Detection And Ranging) technique uses powerful sound pulses radiated into the atmosphere. The analysis of the backscattered signals gives information about wind velocity and turbulence. The measuring volume is determined by pulse length and beam width. Scattering occurs due to fluctuations of the acoustic refraction index, i.e. wind and temperature fluctuations. Since the scattering structure is moving with wind velocity, the scattered signal is dopplershifted. The doppler frequency distribution contains the information about the statistics of the velocity component in direction of the scatter. Evaluation of this doppler shift is performed by a spectral analysis of the scattered signal in comparison with the original signal. The choice of geometry between the transmitting antenna and the receiving microphone permits a vectorial decomposition of the wind vector, most preferably by using several microphone positions or several sound beams. In addition, the scattered signal gives information about turbulence. The fluctuations in wind velocity cause corresponding variations in the frequency shift. These frequency modulations lead to a spectral broadening of the scattered wave. The spectrum of the scattered signal consists of the background noise, which must be determined in each measurement separately, and of the spectral distribution of the scattered wave. After subtracting the noise, the moments of the distribution are calculated. The mean value directly gives the average wind velocity. The second moment incorporates the spectral broadening due to the finite window length of the spectral analysis and the variation of the wind velocity during the measuring interval.
Since the sound speed is much higher than the wind velocity, SODAR gives an instant representation of the velocity distribution inside the scattering volume. An analysis of the statistics of the Doppler sound thus yields a true mapping of the statistics of the scatterer. Compared to conventional mechanical or hot wire anemometers only little smoothing of turbulence parameters takes place.
By calculating the power spectrum of the scattered signal during a time interval of about 1 s no information about the spectral broadening is lost but even enhanced if the statistics are stationary during the averaging process.
The spectra depicted in fig. 1 and 2 are not only broadened because of turbulence but also because of the inevitable time windowing of the received scattering signal. In order to extract the turbulence level quantitavely out of those spectra it is therefore necessary to separate these two effects.
Fig. 3 is a sketch of a calibration experiment to serve this purpose. The lowpass filtered noise ($f_c \approx 5$ Hz) has a spectrum similar to that of atmospheric turbulence. The modulation of the SODAR pulse is simulated with a Voltage Controlled Oscillator. Further signal processing is done exactly as in the case of real SODAR measurements. The amplitude of the noise generator is a direct measure of the frequency variance of the modulating signal.
Fig. 4 shows the resulting broadening of the spectrum as a function of the standard deviation of the modulating frequency. The finite bandwidth even for an unmodulated sig-

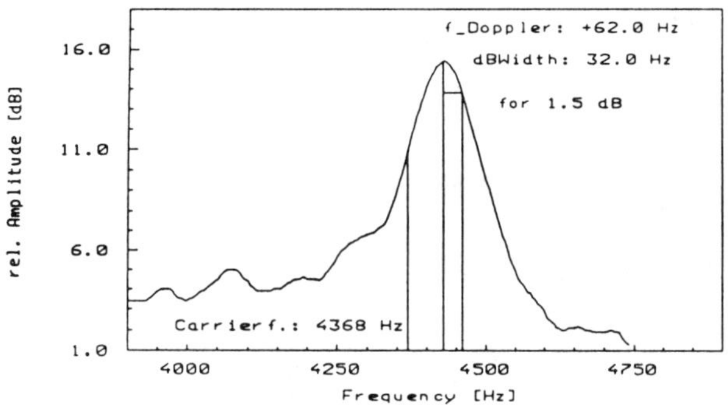

Fig. 1: *Spectrum of a backscattered signal from a height of 24 m (in the range of the hub height) in the wake of a wind turbine. Carrierfrequency, dopplershift and the 1.5 dB bandwidth are annotated, the length of the time window is 10 ms.*

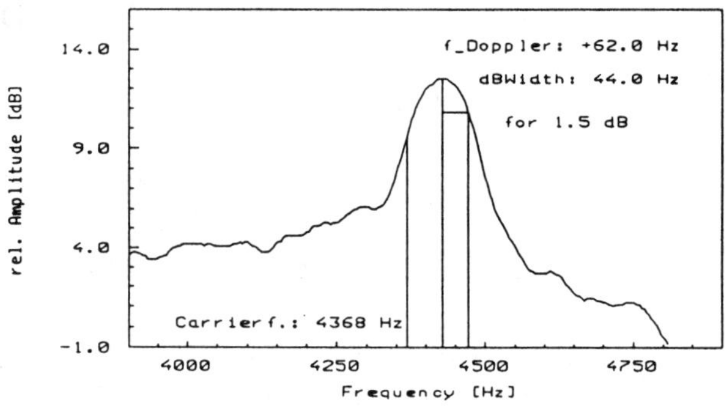

Fig. 2: *Spectrum of a signal from a height of approx. 27 m height (rotor range). The broadened spectrum indicates a higher turbulence level.*

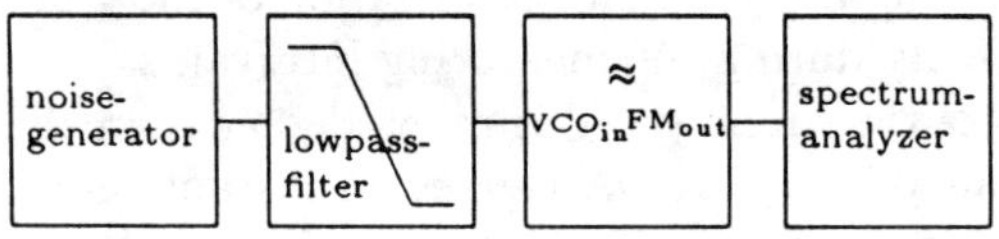

Fig. 3: *Calibration experiment to simulate the modulation of SODAR signals.*

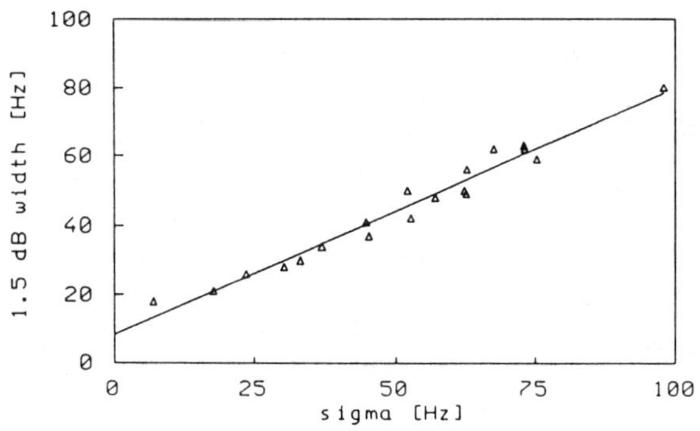

Fig. 4: *Broadening of the spectrum as a function of the standard deviation of the modulating frequency.*

nal is due to transient time windowing prior the FFT. With the help of this linear relationship between bandwidth and standard deviation of the modulating signal a quantitave evaluation of the turbulence level from spectral measurements is achieved.

1.2 Experimental Configuration

In this investigation a specially constructed SODAR is used which allows for easy handling with light weighted antennas. By using frequencies between 3 kHz and 7 kHz it is possible to scale down the antenna size accordingly. The 'Mini-SODAR' consists of 37 piezo loudspeakers mounted on a circular plate of 0.8 m diameter. Compared to conventional SODAR systems the higher operating frequency decreases the maximum range for remote sensing due to atmospheric absorption. However, for the intended application of measuring the wake behind wind turbines the sensitivity is sufficient. The backscattered acoustic pulses with a typical pulse length between 10 ms and 20 ms are received with a directional microphone yielding a vertical resolution between 1.5 m and 3 m. Pulse length and pulse repetition time (1 s typically) are variable. The acoustic power of 1.3 W is radiated within a narrow beam with a 3 dB width of ± 3.8°. The lower limit of the measurement height is about 5 m.

2 MEASUREMENTS

2.1 Measurement arrangement

The measurements were done in a wind farm situated a few hundred metres inland from the Northsea coast. The wind farm consists of five wind turbines (15 m rotor diameter, 55 kW rated power) arranged in a straight line from south-east to north-west with a spacing of four rotor diameters. Profiles of wind velocity and turbulence intensity were measured over the full height of the wind turbines at two distances (30 and 60 m) in the wake. During the measurement the mean wind direction was perpendicular to the linear configuration of the wind turbines. Fig. 5 is a sketch of the SODAR setup behind a wind turbine.

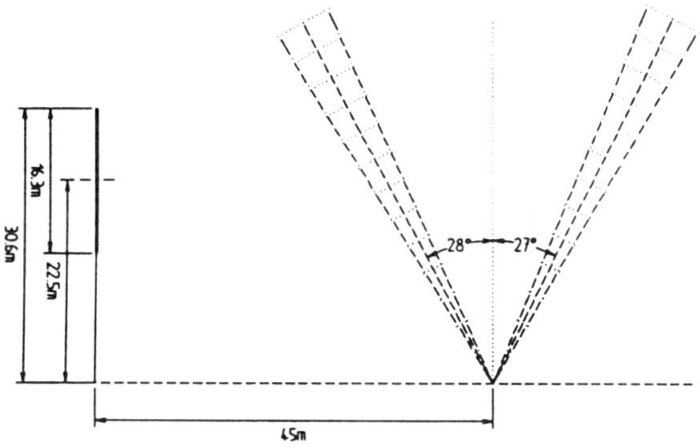

Fig. 5: *Side view of the SODAR setup behind a wind turbine. The antenna is placed at a distance of 45 m from the turbine (hub height 22 m, rotor diameter 16 m).*

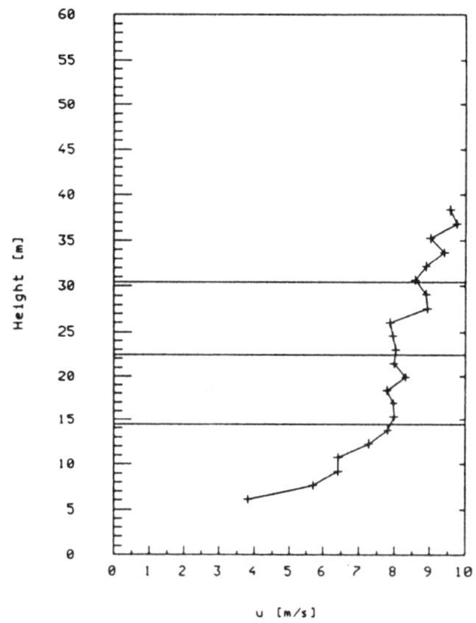

Fig. 6: *Wind profile of a SODAR measurement in the wake of a wind turbine (distance 60 m). The hub height and the rotor diameter are indicated by lines.*

2.2 Results of wake measurements

Fig. 6 shows the wind velocity profile at a distance of about four rotor diameters downstream. Since we are in the far field of the turbine the wind velocity profile shows no obvious structure. An unexpected high roughness length is remarkable. Fig. 7 shows the corresponding turbulence profile. The turbulence level $\sigma/\overline{u}$ is calculated from the standard deviation σ of the wind velocity (i.e. frequency fluctuation) scaled to the mean velocity $\overline{u}$ (Fig. 6). The turbulence decreases with height as expected. But certain structures may be notified, which appear more obvious just behind the turbine (Figs. 8 and 9). The wind profile in about 30 m distance (two rotor diameters) shows two pronounced dips, lying symmetrically in height around the hub. In these regions a significant increase of the turbulence can be observed. Taking into account that longer averaging times yield a smoothing of the meteorological fluctuations we are in good qualitative agreement with the results published in [1].

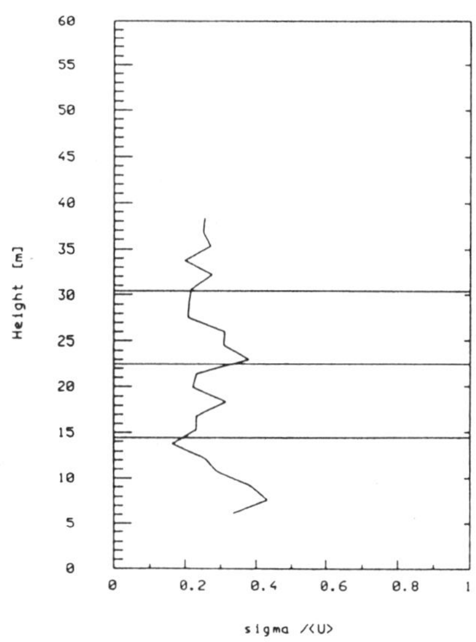

Fig. 7: *Turbulence profile corresponding to Fig. 6.*

2.3 Theoretical wake model compared with SODAR measurements

The analytical treatment of the wake behind wind turbines has been carried out by several authors ([2],[3]). Calculations with the model proposed by Ainslie are in good agreement with measurements of the wind velocity [4] (horizontal profile of the wind velocity). The model of Ainslie is based on the solution of the Reynolds equation. Fig. 10 is a plot of the vertical wind velocity profile as proposed by the Ainslie model in comparison with the SODAR measurement (Fig. 8). In addition

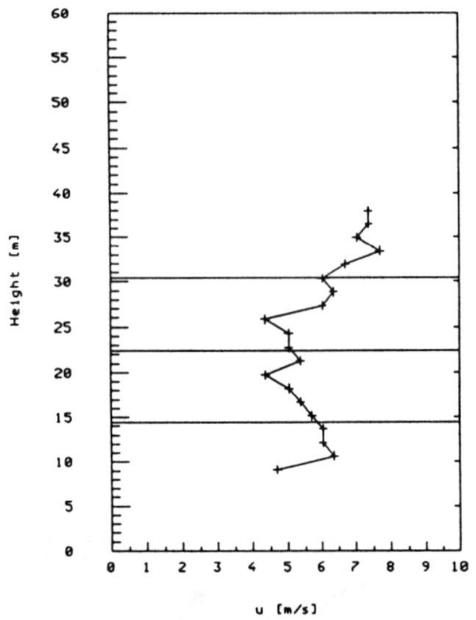

Fig. 8: *Wind profile measured in distance of 30 m. The decreasing wind speed in the region symmetrical around the hub is clearly visible.*

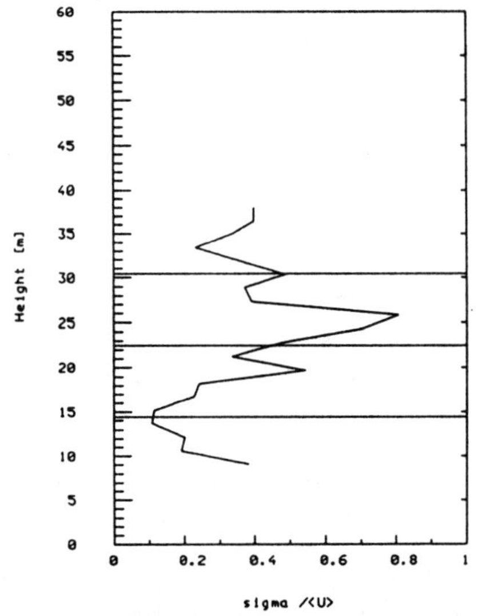

Fig. 9: *Turbulence profile corresponding to Fig. 8.*

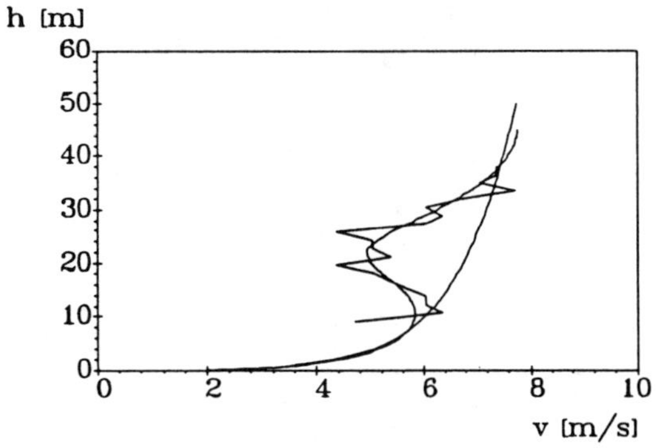

Fig. 9: *Profile of wind velocity from Fig. 8 compared with theoretical model (Ainslie). In addition the undisturbed logarithmic wind profile is shown.*

the undisturbed logarithmic windprofile is depicted.

3 CONCLUSIONS

The MINI SODAR proves to be an interesting measuring device for studying the wind field within wind farms. It can easily be mounted and neither does it disturb the velocity profiles nor does it raise the occurring difficulties when mounting a mast equipped with measuring sensors. True turbulence measurements are possible by the introduced advanced analysis of the Doppler frequency distribution.

REFERENCES

[1] U. Högström, D.N. Asimakopoulos, H. Kambezidis, C.G. Helmis, A. Smedman
A Field Study of the Wake behind a 2 MW Wind Turbine.
Atmospheric Environment, Vol 22, No 4, 803-820, 1988

[2] J.F. Ainslie
Calculating the Flowfield in the Wake of Wind Turbines.
Journal of Wind Engineering and Industrial Aerodynamics, Vol 27, 213-224, 1988

[3] P.B.S. Lissaman et al.
Numeric Modelling Sensitivity Analysis of the Performance of Wind Turbine Arrays.
Aerovironment Inc., Pasadena, California 1982

[4] H.G. Beyer, W. Schlez, W. Schmidt, H.P. Waldl
Messungen der Windgeschwindigkeit in der Nachlaufströmung eines Windenergiekonverters in einem Windpark.
8. Internationales Sonnenforum, Berlin 1992

DYNAMIC ANALYSIS OF GAMMA 60 WIND TURBINE GENERATOR AND CONTROL SYSTEM: DESIGN AND THEORETICAL AND EXPERIMENTAL VALIDATION

S. Avolio, S. Golinelli, G. Moleti, F. Morabito, C. Pierdominici (*);
V. Arcidiacono, S. Corsi, M. Pozzi (+); G. Botta (#)

(*) Alenia - Via Vitorchiano 81 - ROMA (Italy)
(+) ENEL Centro Ricerca di Automatica - MILANO (Italy)
(#) ENEL Centro Ricerche Elettriche - MILANO (Italy)

ABSTRACT: GAMMA 60 is an advanced two-bladed, 60 m-rotor, upwind Wind Turbine Generator, with a rated power of 1500 kW, developed by Alenia-WEST (formerly Aeritalia). The prototype has been developed with funds from ENEA, ENEL, and the DG XVII (DIRECTORATE GENERAL FOR ENERGY) of the European Community, as well as from Alenia-WEST. ENEL placed an order for the supply of the prototype, plus two further units, to be defined on the basis of experience gained on the first machine. The GAMMA 60 prototype was installed in the first half of 1992 at ENEL's Alta Nurra test site in Sardinia and, after the end of the ongoing commissioning activity, will be tested jointly by ENEA, ENEL, and the manufacturers.
The paper deals with the Gamma 60 WECS in terms of mechanical and aerodynamic model used for the dynamic analysis of the control system, giving a general description of the control loops, of the software packages developed for testing the control system performance, and of the global simulation carried out in collaboration between the manufacturer Alenia-West and the Automation Research Centre of ENEL. The paper also presents the approach, the tools and the results of the special shop tests performed in order to verify the adequacy of the system in terms of regulation and other performances.

GENERAL INTRODUCTION

Technical features of GAMMA 60 can be broken down into "basic" characteristics (upwind rotor and teetered hub), which were chosen from among well-proven solutions, and "advanced" characteristics (variable speed and power regulation through yaw control), which represent the innovative aspect of GAMMA 60 as compared with other large wind turbine generators.

These technical features have been implemented in order to reach two basic objectives: reduction of initial and maintenance costs and increasing of WTG performances (energy capture and fatigue life).

Operating speed varies over a broad range because rotor performance can thereby be kept at near-peak conditions, and because more power can be transmitted through the gearbox and generator without increasing torque or generator current. It also allows to increase the fatigue life.

Controlled yaw is used instead of pitch control for aerodynamic torque modulation. As compared with pitch control, this feature makes for a much simpler machine for purposes of both manufacture and maintenance, eliminating all control components from the rotating system, and thus facilitating access for inspection.

The structural design philosophy is that the structural components should be made relatively flexible. The "softness" of the rotor, tower, and drive system has the effect of attenuating both the harmonic and the transient stresses in the machine. This makes for reduction in weight, while retaining long, conservative fatigue life.

1. FOREWORD

The control system of a windturbine is, by its own nature, tightly linked to the "Windturbine System". For this reason the approach taken in the development of the control system of Gamma 60 has been the one of parallel development with the system.

Three areas can be identified:

- Regulation and operation algorithms and relevant modeling/analyses;
- Hardware and Software of Control and Supervision;
- Hardware and Software of Data Acquisition and Data Management.

This paper deals with the algorithms definition and validation; another paper [1] deals with hardware and software of the prototype.

2. OPERATING MODES

In order to understand the control system strategies and needs it is useful to give a short description of the GAMMA 60 operating modes. According to different ranges of wind speed, three different operating modes are followed, as can also be seen in Fig. 1.

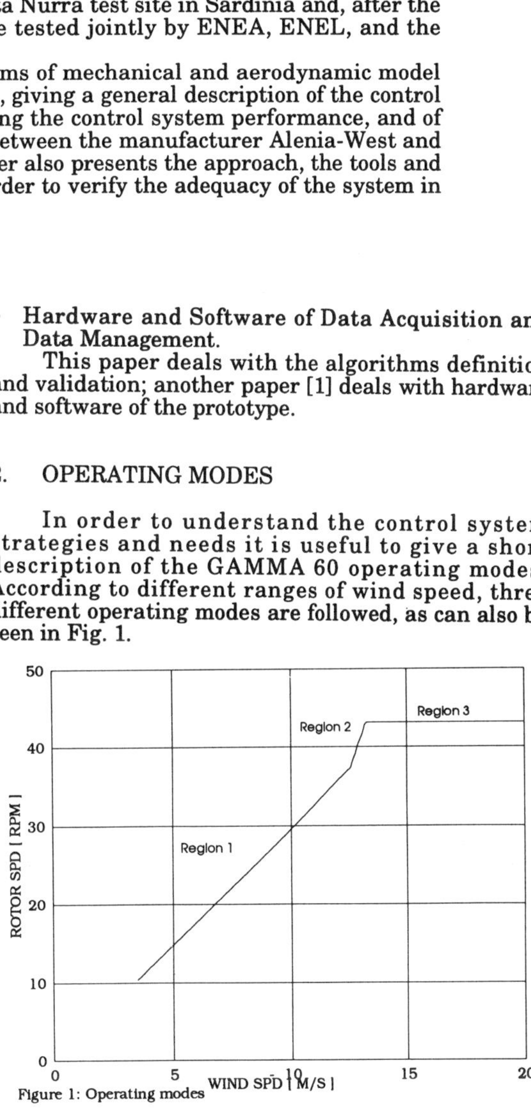

Figure 1: Operating modes

2.1 Operation with Constant Cp (Region 1)

This operating mode occurs for wind speeds (at hub height) ranging between 5 m/s and 12.5 m/s.

The WTG is started up by using the synchronous machine as a motor for the purpose when the wind exceeds 5 m/s (i.e. cut-in wind speed) and begins to feed useful power into the grid when the speed of the rotor reaches 15 r.p.m.

As wind speed increases, the control system manages the torque requested to the power converter so that the rotor speed also increases in proportion, in such a way that the power coefficient C_p of the rotor is always kept constant and equal to its maximum value.

For a wind of 12.5 m/s at the hub, a rotor speed of 35.5 r.p.m. and the rated torque of 380 kNm are reached.

2.2 Operation at Constant Torque (Region 2)

For wind speeds between 12.5 m/s and 13.5 m/s, the machine operates in constant torque mode.

As wind speed increases, the torque is kept constant at the rated value of 380 kNm, while the speed of the rotor grows linearly with the power, up to the rated value of 44 r.p.m., corresponding to a power supply to the grid of 1500 kW.

It is worth to say that the machine is designed to be able to extend this operating zone up to 55 r.p.m. with a wind of 14.6 m/s, corresponding to a power output of 1900 kW.

The nacelle is kept aligned with the wind in all this Region, as well as in Region 1.

2.3 Operation at Constant Power (Region 3)

For a wind of between 13.5 m/s and 27 m/s, the machine is operated at constant power.

As wind speed increases, the control system, via the yaw actuator, forces an appropriate disalignment of the nacelle in relation to the wind, in such a way as to keep power output constant. The combined action of this disalignment and of the regulation of the restraining torque opposed by the electric generator, which is kept constant at the nominal value, also causes rotor speed to remain constant and equal to 44 r.p.m.

When the wind exceeds 27 m/s (the cut-out wind speed) or drops below 5 m/s (cut-in wind speed), the windturbine is stopped by moving the nacelle 90° out of the wind.

3. MODELING

The control of the wind turbine generator requires two regulation loops: one, acting through the generator torque control, for maximizing the power ratio coefficient and limiting the mechanical torque, and the other, acting through the yaw angle control, for limiting the rotor speed and the electric power at high winds.

The early effort on the Gamma 60 program was devoted to defining the optimal torque and speed ratings for a 60 meter rotor and defining the basic system configuration.

The control system necessary to employ the broad range variable speed concept implemented in Gamma 60 must provide a means to vary the restraining torque on the rotor shaft and provide a means to control the aerodynamic torque produced by the rotor as required to hold the rotor at maximum speed.

A general schematic for the variable speed wind turbine control is shown in Fig. 2.

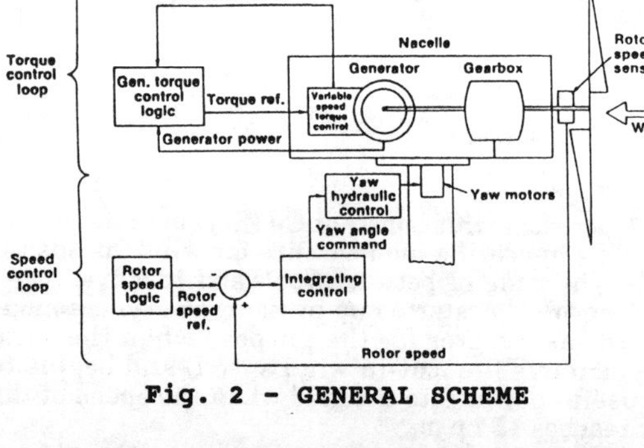

Fig. 2 - GENERAL SCHEME

At low wind speed, where the rotor is to be held at its performance peak, the generator torque is adjusted relative to the aerodynamic torque to manoeuvre the rotor speed, as required to maintain constant velocity ratio.

Once the gearbox torque limit is reached, additional power can still be put through the system in the form of increased rotor speed.

An electrohydraulic yaw system is used to limit the aerodynamic torque on the rotor. This system yaws the rotor out of the wind to maintain a torque balance with essentially constant generator torque and, thereby, holds the rotor speed at nominal speed limit.

The change of aerodynamic torque with yaw position is highly nonlinear. This and numerous other nonlinear features of the system have been incorporated in the computer dynamic simulation model in order to obtain a good evaluation of the system transient performance over the full range of operating condition, including large disturbances as loss of grid and various failure scenarios.

3.1 Rotor speed control

The control is designed to hold the rotor into the wind when the system is operating below rated rotor speed and power. In high winds, when the system is operating at rated power, the control modulates yaw heading to maintain nominal limit rotor speed.

3.2 Generator speed control

The digital generator speed controller modulates the electrical generator torque to control rotor speed in wind speeds below rated torque. The controller senses generator speed and power to control the power electronics, so that the generator torque is regulated.

4. SIMULATION

4.1 Control System design

Before the starting of Gamma 60 Program Hamilton Standard had developed analytical codes and tests in order to validate the feasibility of the variable speed yaw control operation.

The models and the code evolved during the definition of the various control algorithms, up to a stable (in Configuration Control sense) configuration.

The model of the windturbine and of the related control loops are implemented in a simulation code called "CODEW" (COntrol Dynamic Evaluation of Windturbines), through which the regulation loops dynamics, following various wind perturbations and critical events, have been studied.

Control algorithms were directly derived from CODEW, in order to assure the precise correspondance between as-analyzed and as-implemented.

During the design phase the configuration changes, both in terms of models and parameters, have then been controlled.

In order to assure the validity of the analyses, specific actions have been undertaken time by time:

- Detailed modeling of the hydraulic yaw system, also as a means to ensure compliance with requirements;
- Sensitivity analyses in CODEW to test influences of discrepancies between as-modeled and as-built;
- Identification tests in the shop, in order to confirm the match between the simulation model and the as-built system.

4.2 Control System validation

The Automation Research Centre of ENEL has performed as "Design Reviewer" of the performances of the Control System.

The study has been worked out with a new simulation code developed through a general purpose commercial dynamic simulation tool called EASY5 (Engineering Analysis SYstem).

Control system dynamic performances have been examined using a detailed mechanical and aerodynamic model, different from the one implemented in CODEW, defined in collaboration with the Electric Research Centre of ENEL, based upon the following assumptions:

- The windturbine is described by means of four rigid bodies (blades and hub, low speed shaft and gearbox, high speed shaft and electric generator, nacelle) connected together as to allow up to four degrees of freedom (teetering, torsion, shaft rotation, yaw motion).
- The rotor is described by two equal blades, with wind action in terms of aerotorque, lateral force, teetering torque and thrust. Curves of c_p, c_m and c_t as a function of the velocity ratio (lambda) were given by Alenia-West.
- Control and Operation strategies have been implemented in the model on the basis of Alenia-West specification documents; actuators (power converter and yaw hydraulics) models also came from Alenia-West. Some changes have been added concerning the critical speed avoidance logic and servovalves unlinearities compensation.

Control system dynamic performances have been examined in depth, in presence of small and large perturbations, in order to validate the design choices of the control loops structures and parameters, to test the achievement of the steady-state and dynamic requirements, to identify possible improvements to the control algorithms and parameters.

The study has fully confirmed the validity and the effectiveness of the designed control solutions. The main findings and considerations follow.

- The torque and generator speed control loop has shown a satisfactory behaviour, both in terms of static performances (steady-state error in various working points in Regions 1 and 2) and dynamic ones (speed of response and degree of stability).
- The power and rotor speed control loop (an integral regulator with derivative compensation) has shown adequate stability and speed, thus allowing for limited excursion of the electric power at high wind. Static and dynamic performances of the yaw positioning loop have been carefully considered, with the due attention to the influence of non-linearities in the servovalves and of hydraulic damping of yaw modes.
  The cyclic variation of the yaw inertia, small as a consequence of the teetering hub, has shown an even smaller influence on the dynamic response of the yaw loop and on the loads trasmitted to the actuation system;
  Also the maximum teeter angle has been shown to be within the limits foreseen by the ALENIA-WEST analysis.

5. ANALYSIS AND SIMULATION RESULTS

5.1 Design Analyses

In the development of the control strategies of Gamma 60 an huge amount of analyses and simulations has been made.

Among these:

- Loss of grid effects in terms of rotor overspeed;
- Yaw actuation failure effects in terms of rotor overspeed;
- Effects of wind speed and direction changes over power quality;
- Startup and shutdown procedures;
- Rotor speed avoidance logic.

Specific simulations have also been made in order to test the effect of possible differences between the models utilized for the analyses and the machine as-built. These analyses have shown a good behaviour even in case of significant differences.

Finally, small and large changes in wind speed and direction have been the basis of comparison between CODEW and EASY5 simulation.

5.2 Simulation results and comparisons

Simulation of the same cases are show for both Code W and Easy 5 in fig. 3 and 4.

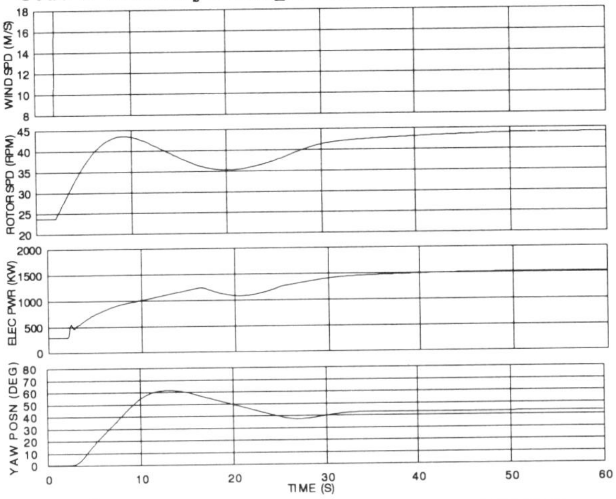

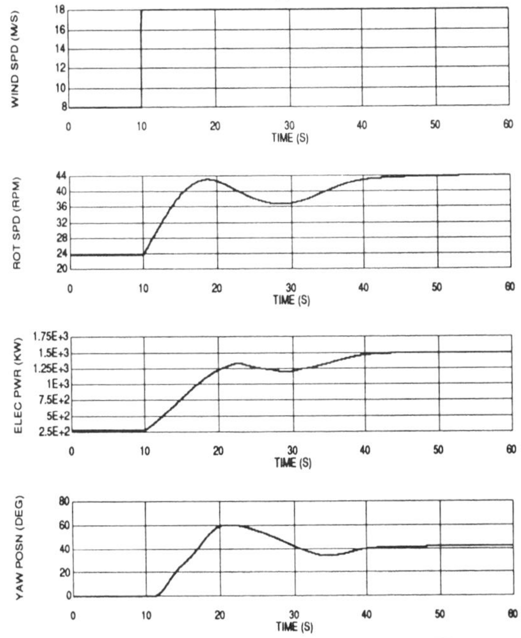

Fig. 4 - EASY 5 SIMULATION

6. EXPERIMENTAL VALIDATION

6.1 Activator and Shop Tests

The basic requirement of the shop tests was to have the machine operating in real time, in all possible operating conditions, in order to test the overall system, with special attention to the hydraulics, to the drive train, and to the Supervisory and Control hardware and software.

For this purpose a special simulator was conceived, made by a computer with I/O boards for interfaces with the SCADA system, and the relevant software was developed to simulate the wind speed and direction, together with other external condition, and the aerodynamics of the rotor. The simulator provides a torque reference signal driving the generator/converter as a motor, thus allowing speed variation similar to the ones with the blades and connection to the grid.

Sensors (and actuators) not physically present in the shop have been simulated by mean of signal generators.

The tests performed with this system have assured:

- The proper behaviour of the actuation devices (hydraulic and electrical);
- The proper interface of the Control and Data Acquisition System to sensors and actuators;
- The proper behaviour of the Data Acquisition;
- The validity of the modeling of the wind turbine (identification tests);
- The proper implementation of the regulation algorhythms;
- The proper implementation of the supervision strategies;
- The stability of the behaviour for extended operation (endurance tests).

The control system was tested on the Wind Turbine in the shop tests, in conditions with real response dynamics, including the hydraulic actuation. In all essential aspects the real sensors were active, the only simulation influencing performance of the control being the aerodynamic of the rotor, calculated in the special computer program called Activator.

This method of testing the control is vastly superior to anything that can be accomplished in a real wind at the test site. Results demonstrated that the Regulator was performing as predicted in the simulation studies and that the supervisory control performed as expected. Fig. 5 shows the results obtained in the same cases as the ones analyzed with Code W and Easy 5.

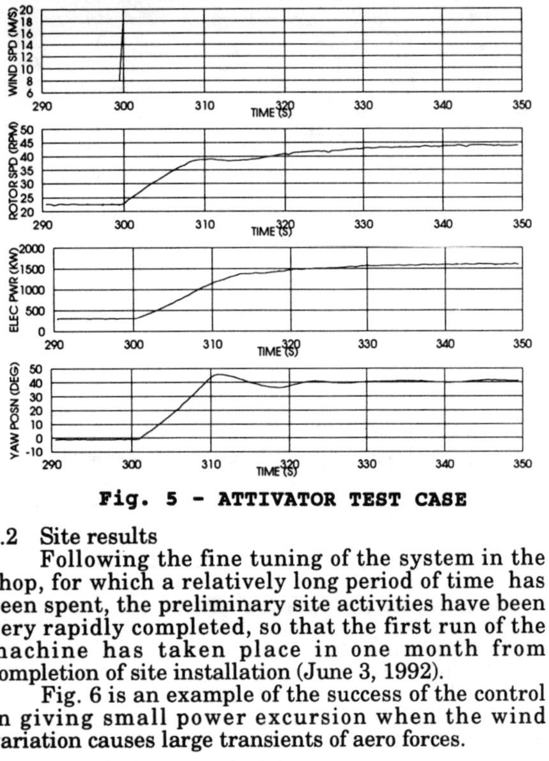

Fig. 5 - ATTIVATOR TEST CASE

6.2 Site results

Following the fine tuning of the system in the shop, for which a relatively long period of time has been spent, the preliminary site activities have been very rapidly completed, so that the first run of the machine has taken place in one month from completion of site installation (June 3, 1992).

Fig. 6 is an example of the success of the control in giving small power excursion when the wind variation causes large transients of aero forces.

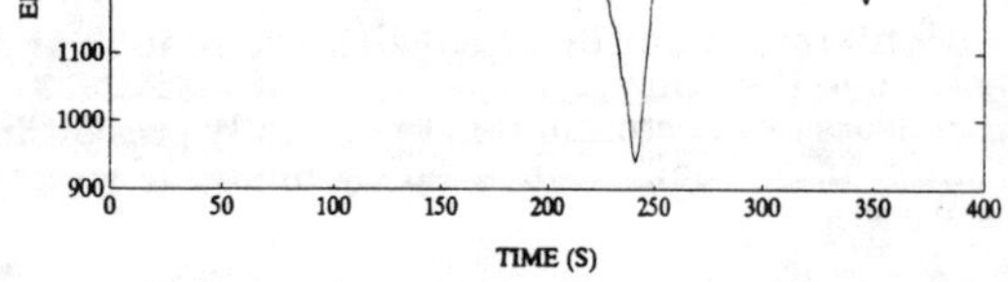

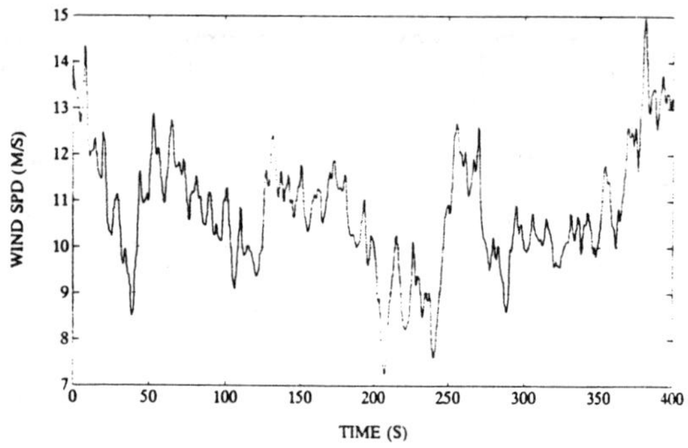

Fig. 6 - FIELD TEST RESULTS

7. CONCLUSIONS

An indepent study and evaluation of the Gamma 60 control system was conducted by ENEL-CRA in parallel with the ALENIA-WEST development and refinement of the sistem. The CRA investigation provided and expanded understanding of design and construction features and verified the good performance of the conrol. Based upon modeling with varying complexity, the CRA and WEST studies provided a clear focus for the fine-tuning and verification of control performance in the test program. Use of the activator to simulate real-time wind variations during the shop testing made it possible to evaluate the control performance during all the same cases that were used in the control development simulation. If was found that the conrol functioned almost exactly as predicted.

Testing of the control on the installed wind turbine does not permit quantitative evaluation comparable to that obtained in the shop with the activator because the actual, effective, wind is an unknown. The on-site testing therefore tends to be less interesting and significant than was the shop testing.

If can be concluded that the control performs wery well and as expected from the development simulation. Also, the investment in extensive simulated development has been paid back by the absense of problems and surprises during the test program, both in the shop and in field tests.

8. ACKNOWLEDGEMENTS

The Gamma 60 Project has seen the efforts of various bodies: Alenia-West and Hamilton Standard for System activities and blades; major suppliers included: Escher Wyss (Ravensburg-D and Schio-I) for design and manufacturing of hub, mechanics and actuation; Rexroth (D) for hydraulics; Thyssen (D) for the gearbox; Twiflex (GB) for the brake; Galtarossa (I) for steel structures and civil works; Ansaldo (I) for the electric subsystem; Gavazzi Space (I) for electronic apparata.

The internal team was made by: A. Aliberti, F. Anaclerio, S. Avolio, C. Calò, G. Caturano, G. Doman, S. Fugaro, S. Golinelli, P. Lecce, G. Moleti, F. Morabito, M. Musolino, G. Nettis, C. Palmari, C. Pierdominici, M. Rotondi, I. Sebastianelli, L. Serafini, A. Smarchi.

Thanks go also to all the other people that in different stages have contributed to the success of the program.

9. REFERENCES

[1] S.Avolio, G.Moleti, F.Morabito
Supervisory, control and data acquisition: when a prototype needs to have prototype components.
1993 European Community Wind Energy Conference Proceeding.

[2] S.Avolio, A.Aliberti, I.Sebastianelli
Design, analysis and testing: methods and procedures for large machines illustrated by Gamma 60.
1993 European Community Wind Energy Conference Proceeding.

UNIWEX - AN INTERNATIONAL RESEARCH PROJECT — HISTORY - RESULTS - SPIN-OFF - FUTURE

J. Argyris, K.A. Braun
Institute for Computer Applications, ICA
M. Müller
Institute for Statics and Dynamics, ISD
Both University of Stuttgart, Pfaffenwaldring 27, 7000 Stuttgart 80
Germany

ABSTRACT
The UNIWEX[1] project is based upon several previous projects and therefore represents one of several steps in a long-term effort. It should be seen in the context of its history which is briefly presented. A survey of the results obtained is given, mentioning both the direct results as hardware, software and intellectual outcome and those manifesting themselves as technology transfer to industry. Last but not least, the future integration of UNIWEX in European R&D projects is shown, demonstrating the usefulness of this tool and justifying the long development time and financial effort.

1. PURPOSE OF THE WORK

The first step necessary to increase the exploitation of wind energy is to arrive at more economical wind turbines. The overall aim of the UNIWEX project is to increase the technical reliability and the economy of horizontal axis wind turbines. This is possible by the improvement of the concepts of such machines. The large variety of concepts, however, can be investigated only by means of representative numerical and physical simulations. One outcome of the project is that these complementary simulation tools are applicable to a wide field of HAWT problems.

2. HISTORY OF PROJECTS

In order to reach the aforementioned goal, several R&D projects have been conducted since 1978. In what follows, a short survey is given of them, retaining, however, the stress on the UNIWEX project.

2.1 Large-Scale WECS

In 1978/1979, an IEA project "Investigation of rotor stressing and smoothness of operation of LS WECS" was conducted. The attempt was made to reduce all loads in the hub other than the torque by specific hub concepts. Numerical simulations for the isolated rotor, wind tunnel experiments with small models and exploratory experiments with a 7.2 m ∅ indoor model were performed.

2.2 Project OPTIWA I

The first phase of the BMFT project "Optimierung großer Windenergieanlagen" lasted from November 1979 to June 1982. A new computer-controlled indoor model with 7.6 m ∅ was built, based upon former experience. The servo-hydraulically activated fully articulated hub and the data handling were developed. On the side of the numerical simulation, software packages for the structural lay-out of the blade and for the aero-elastic investigation of the coupled tower-rotor-generator system were developed and tested.

2.3 Project OPTIWA II

In the second phase until January 1986, the rotor-blades and the tower were laid out and the turbine was simulated using the software developed before. The test wind turbine with 16 m ∅ was designed, constructed, commissioned and tested on the "Ulrich Hütter wind test field". First validations of parts of the software took place. The results are summarized in [1] and [2].

2.4 Project UNIWEX

This project entitled "Universelle Windturbine für Experimente" was in force from May 1988 to February 1992 and, in the overall framework of the above aims, had the topics: further development, validation and application of the hardware and software tools.

In the course of the UNIWEX project, the computer-controlled, two-bladed experimental wind turbine OPTIWA underwent a thorough redesign. Among other things, safety relevant components and the complete drive train were exchanged. The operational range w.r.t. to wind speed, r.p.m. and power output was enlarged. The measuring and control equipment was upgraded, now allowing the simulation of almost every two-bladed full pitch HAWT.

The second main effort within the project was the further development of the numerical simulation software using the Finite Element Method. One software system now allows the aero-elastic simulation of a wind turbine using a fast super-computer, giving answers about the stability behaviour, the displacements, loads, stresses in both the frequency and time domain (see fig. 1). By means of another, smaller software system, it is possible to lay out/optimize the structure of the fibre rotor blades w.r.t. stiffness and stresses.

For both the experimental and the numerical approach, it is mandatory to develop appropriate, user-friendly post-processing software in order to evaluate in finite time and represent in a readable manner the immense amount of data coming from the measurement system and the super-computer respectively.

[1] UNIWEX is a R&D project of the International Energy Agency IEA, and is conducted together with ECN-Holland and FFA-Sweden, the Institute for Computer Applications (ICA) of the University of Stuttgart being the Operating Agent. It was sponsered by Germany (BMFT), The Netherlands (NOVEM) and Sweden (NUTEK)

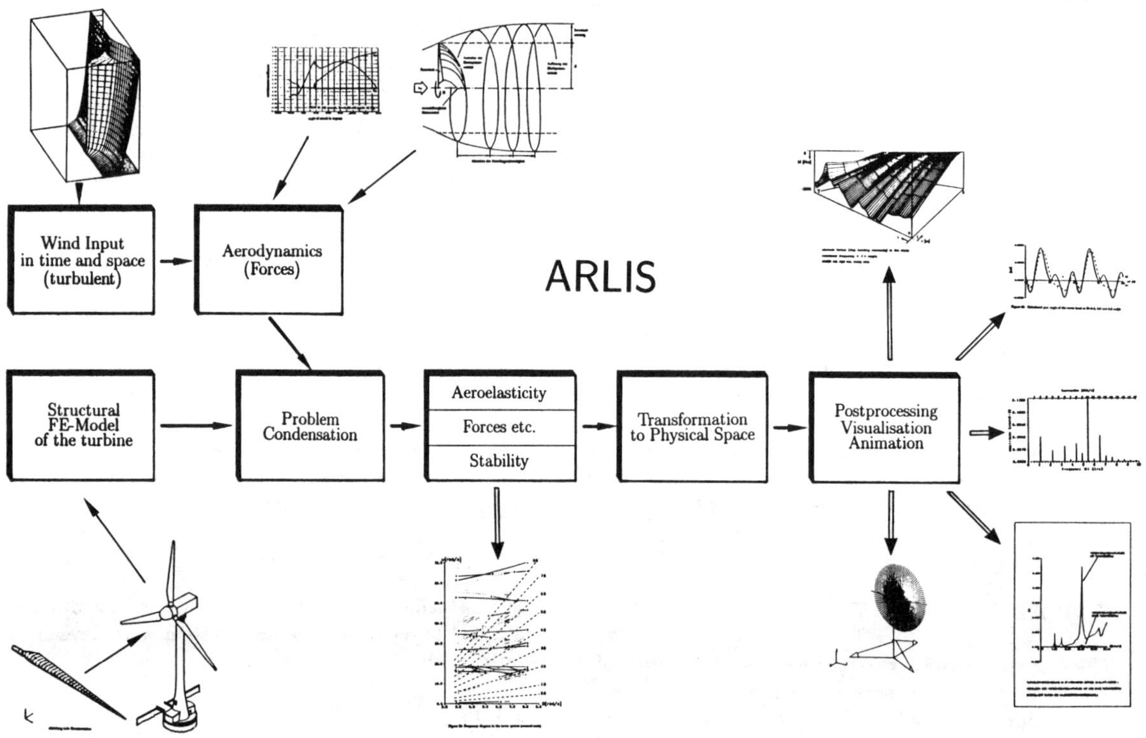

Figure 1: Flow chart of the aero-elastic simulation with ARLIS

3. DIRECT RESULTS

3.1 Validation of codes

For the validation of different computer codes, specific well-designed measurement campaigns were conducted and compared with the results of the simulations. Since the UNIWEX turbine has a fully articulated hub with adjustable stiffness and damping parameters around the flap- and lead-lag hinges, 'unusual' turbine configurations could also be operated, allowing the codes for such extreme cases to be checked as well. The possibility to vary r.p.m., pitching rate and the generator characteristics and the option for free yaw was of great value within this task.

The Dutch participants in the UNIWEX project checked three different aero-elastic simulation codes which are typically applied in the design and certification phase: the ECN code Phatas II which works in the time domain, uses an elastic blade model and non-linear analysis, the SPE code Flexlast also in the time domain simulating the blade elasticity by spring loaded hinges and the DUT code Duwecs for simulation of wind energy conversion systems with emphasis on the control system. For details, see [3] and [4].

The aero-elastic code ARLIS developed at ICA was checked and validated (see [5]) for three different hub concepts (rigid, teeter and individual flapping).

3.2 Simulation and comparison of hub concepts

All calculations were preceded by the re-idealisation of the whole turbine and a thorough model identification taking into account all hardware changes. Using the ARLIS code, the aforementioned hub concepts were simulated and compared covering the whole operational rpm-range of the UNIWEX turbine ($3 < \omega < 8$ [rad/s]). The stability characteristics and the structural answers at steady-state operation and some simulations for instationary wind situations were performed [6]. Only two results will be shown here:

- Flap displacements over Ω

In fig. 2, amplitudes and mean values of the flap-wise blade tip displacements are given versus Ω (rpm). An operation with $\lambda = 8$ was simulated, which means that $v_w \sim \Omega$. It can be seen from fig. 2 that the mean values for rigid and teetering hub are almost the same whereas for the amplitudes, teetering and flapping hub behave in the same way.

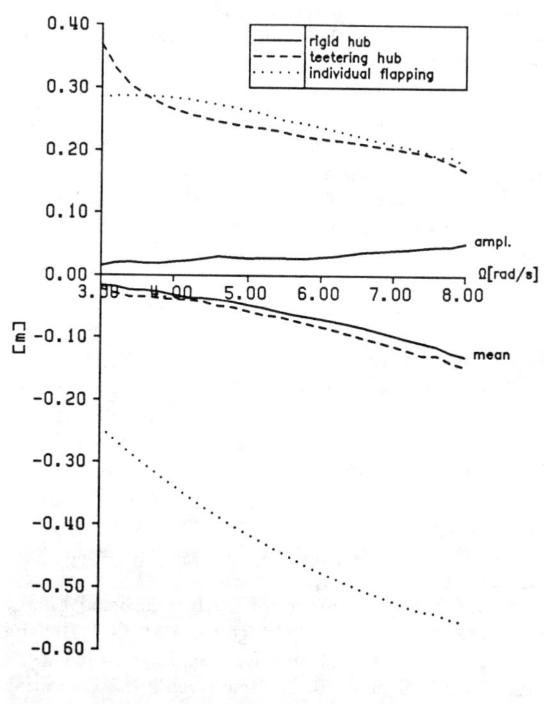

Figure 2: Flap-wise blade tip displacements (mean and amplitudes) over r.p.m.

• Flap-wise bending moments over one revolution

Fig. 3 shows the moments at radial position $r = 1$m. For the flapping blade, the bending moment is close to zero, as it should be. The teetering hub reacts with teeter excursions on asymmetric rotor loads (wind shear, tower wake) and therefore has considerably smaller load amplitudes than the rigid hub. The two answers on the tower wake can be seen at t/T≈ 25% and 75% whereas the rigid hub exhibits only one large answer.

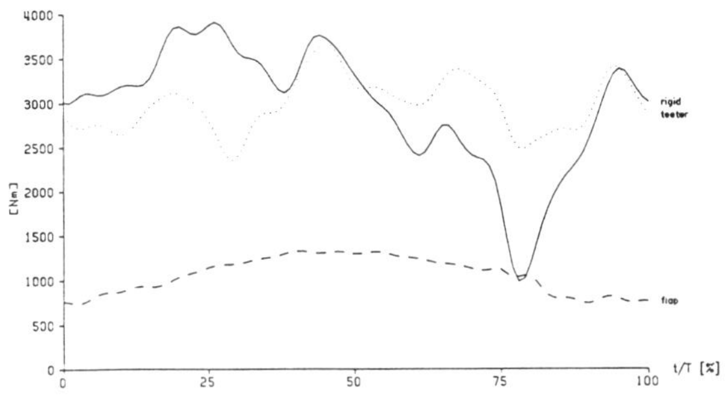

Figure 3: Flap-wise bending moment at $r = 1.0$ m ($\Omega = 7.4$ rad/s)

3.3 Experimental investigations

On the one hand, the UNIWEX turbine was used to validate the computer code ARLIS; on the other hand, the code was used to predict interesting operational regions and parameter choices for the wind turbine. Besides for the validation, the turbine, however, was extensively used for the experimental investigation of the operational behaviour of different rotor types, upwind and downwind configuration and with different generator characteristics. Load spectra were recorded and $C_p - \lambda$ diagrams established. Another field of interest was control strategies for const. r.p.m., const. torque, const. tip speed ratio λ or const. power using pitch, generator control or active yawing. Besides this, some aerodynamic effects in connection with stall, rain and icing were observed (see [7]).

4. SPIN-OFF

The spin-off of the research project can be seen on the one hand in the technology transfer to industry and on the other hand in the raising of new R&D topics.

4.1 Industrial projects

The expertise won within the UNIWEX project has been accepted and has led to several contracts for the numerical investigation of innovative/large HAWT's. Up to now, aero-elastic simulations have been performed for

- MONOPTEROS 400 (370 kW, 48 m ∅, MBB)
- AEROMAN 12.5/33 (33kW, 12.5 m ∅, MAN) plus a 3-bladed derivation
- WKA 60-II (1.2 MW, 60 m ∅, MAN)
- NAESUDDEN II (3 MW, 80 m ∅, KVAERNER/MBB)
- AEOLUS II (3 MW, 80 m ∅, MBB/KVAERNER)

(compare fig. 4)

4.2 New R&D topics

In the course of the project and the related industrial projects, numerous new questions arose, part of which became topics of further research, student and doctorate theses respectively. From the following incomplete list, it may be guessed how much inspiration spins off from projects like UNIWEX and which favourable influence they have on university education:

- Aero-elastic behaviour of a 1-bladed derivation of the AEROMAN.
- Engineering models for instationary induced velocities at HAWT's.
- Modelling of turbulent inflow to a HAWT.
- Induced velocities for a HAWT under yaw.
- Fatigue behaviour of the UNIWEX turbine operating in stall

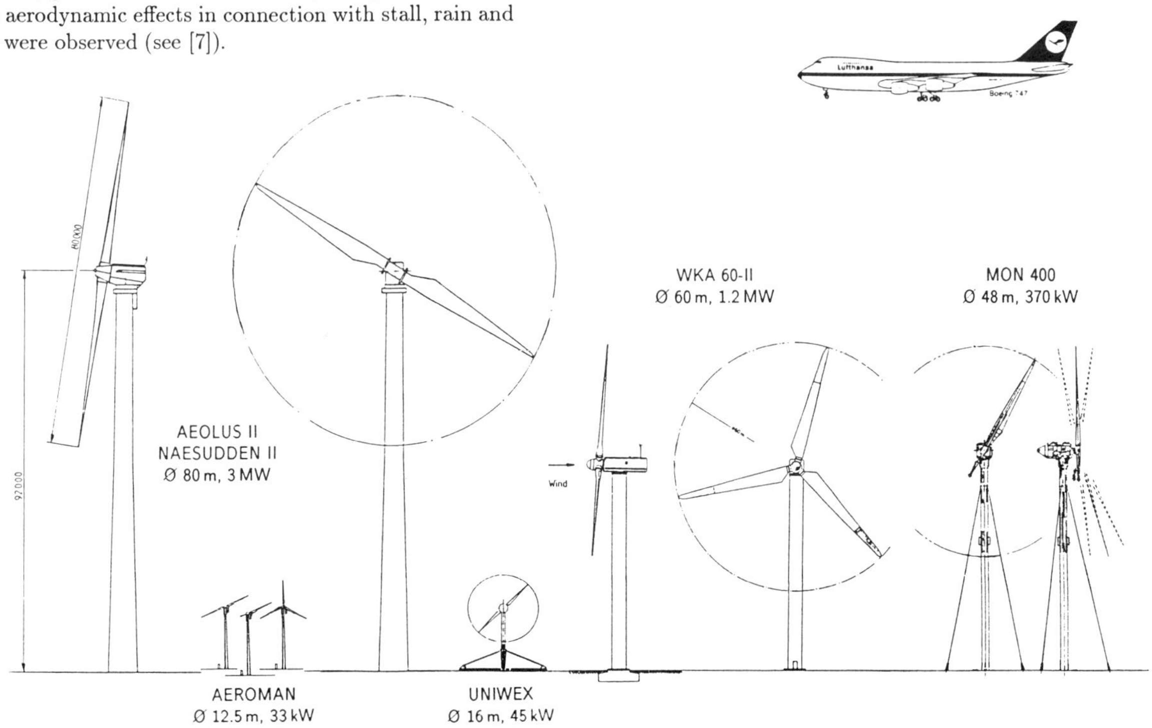

Figure 4: Wind turbines which were investigated at the ICA up to now using ARLIS

5. DIVERSIFICATION

The UNIWEX project was planned to shed more light on the structural and fatigue behaviour of wind turbines with special emphasis on hub concepts reducing 'unnecessary' loads. Due to its universal concept - which, by the way, caused a long development time - it is, however, also applicable to other research fields in connection with HAWT's. Having proven its usefulness as a testbed for specific fields of research, it became integrated in two new research projects of the European Commission dealing with the dynamic inflow to rotors and with the aero-acoustics of wind turbines.

5.1 Aero-acoustics of blade tips

Noise emission of modern (fast running) wind turbines and in particular the aero-acoustics has a direct impact on the environment which leads to problems with the acceptance of wind energy by the public. Under this general aspect a CEC-project "Investigation of Blade Tip Modifications for Acoustic Noise Reduction and Rotor Performance Improvement" was launched with the participation of German and Dutch participants. It is understood as the first phase of a multiphase project, in which empirical investigations are undertaken, putting emphasis on the noise shed from the blade tip region (see [8]). Several tip shapes are selected (comp. fig. 5) which will be tested on the UNIWEX turbine, using adopted measurement techniques. The UNIWEX turbine combines several favourable, features such as e.g. easy access to the rotor and the nacelle by the tilt mechanism of the tower, independent pitch setting for the two blades, variable r.p.m., which seem to make the turbine a good choice for the task.

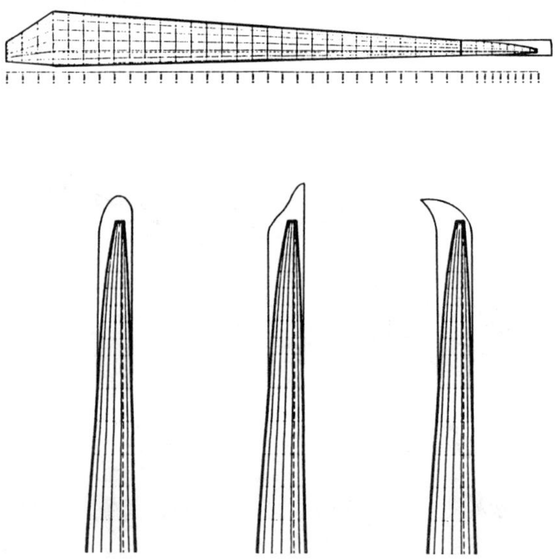

Figure 5: Planforms of tip shapes selected for aero-acoustic measurements with the UNIWEX turbine

5.2 Dynamic inflow II

In order to close the gap in diameter scale, the UNIWEX turbine was selected for the second phase of the CEC-project "Investigation and modelling of dynamic inflow effects" which is conducted under participation of Dutch, Danish, French, German, Greek and UK organisations. The turbine, being designed with aero-elastic investigations in mind, offers in an ideal way all prerequisites needed for the respective measurements. The pitching and yawing capabilities of the turbine will be of special interest. The aeroelastic simulation code ARLIS, as a typical representative of a lay-out and certification code, will be applied/adapted for operation under yawed conditions. In this context, the periodically induced velocities are of special interest.

6. CONCLUSION

The project UNIWEX together with its predecessor projects was, with intervening 'calms', in force for more than a decade. The outcome can be categorised in direct results in the form of actual hardware, software and data obtained both by measurements and numerical simulation. Besides this, a technology transfer to industry takes place and an impressive amount of scientific work was undertaken, for both of which the UNIWEX project was a necessary pre-requisite. Finally, the outcome in the above sense is accepted by new research projects, thus paying back the investments of the past years. Retrospectively, it can be concluded that the project was worth the intellectual and financial efforts.

7. REFERENCES

[1] J. Argyris, K.A. Braun: "OPTIWA - Optimierung großer Windenergieanlagen", Abschlußbericht Phase II, Teil I, Berechnung und Softwareentwicklung, ICA-Bericht Nr.17/I Stuttgart 1987

[2] J. Argyris, M. Müller: "OPTIWA - Optimierung großer Windenergieanlagen", Abschlußbericht Phase II, Entwicklung einer rechnergesteuerten Versuchs-Windturbine, ICA-Bericht Nr.17/II Stuttgart 1987

[3] H. Arnold, G. van Baars, P.M.M. Bongers, M. Müller, H. Schwartz: "Experiments and Validation of Dynamic Models linked with UNIWEX - First Results" European Wind Energy Conference, Amsterdam, EWEC'91 Proceedings, pp. 557-561

[4] A.J. Brand, H. Snel, N.O.T. Hansen, G.A.M. van Kuik, G.E. van Baars, P.M.M. Bongers: "Validating wind turbine design codes by analysing teeter, 'free' flap and 'free' lead/lag motions", ECWEC93 paper

[5] K.A. Braun, A. Finkel, A. Hopf, M. Søderberg: "Two Computers Codes for Aeroelastic Analysis of HAWT's, Comparison with Experiments" European Wind Energy Conference, Amsterdam, EWEC'91 Proceedings, pp. 562-566

[6] J. Argyris, K.A. Braun, A. Finkel: "Aeroelastische Simulation der UNIWEX Windturbine", Deutsche Windenergie-Konferenz 1992, Wilhelmshaven 28./29. Okt. 1992, Proceedings

[7] H. Arnold, H. Schwartz: "Ergebnisse aus experimentellen Parameter- und Konzeptstudien mit der Horizontalachsen-Windturbine UNIWEX", ibid.

[8] K.A. Braun: "Investigation of Blade Tip Modifications for Acoustic Noise Reduction and Power Performance Improvement", Proceedings, Wind Energy R&D Contractor Meeting (CEC, JOULE), Alghero 3-5 June 1992

INVESTIGATION OF YAW MISALIGNMENT AND TOWER FATIGUE LOADS, CAUSED BY WIND DIRECTION CHANGES AND BASED ON LONG TERM WIND MEASUREMENTS.

H. Ganander and H. Johansson
Teknikgruppen AB
Box 21, S 19121 Sollentuna
Sweden

ABSTRACT: Wind direction behaviour is not very well described in design codes, norms or even known, though important for the fatigue design of wind turbines. This conserns yaw error and corresponding loads on blades and nacelle as well as tower load spectrum. Since almost three years measurements have been carried out at the Alsvik site on the island of Gotland in the Baltic sea. Based on these measurements studies of yaw systems have been done. Results of these studies are presented in this paper, together with questions conserning fatigue design of the tower.

1. INTRODUCTION.

The objective of this study is to use some years of relatively high frequency wind measurements in order to contribute to the knowledge of the behaviour of wind direction changes, regarding consequencies of yaw misalignment and tower fatigue design. Designing codes normally says very little about yaw errors. In some cases a mean value and a standard deviation is mentioned. But these informations says nothing about frequency of the variations. Compared with knowledge about longitudinal wind speed behaviour, wind direction and/or lateral speeds, are much less known.

Since nearly three years now, there are measurements from the Alsvik site which have been used to evaluate the wind direction behaviour and used for studies of yaw error and tower fatigue loads. One of the objectives is to obtain a method to study yaw errors and tower fatigue loads influenced by wind direction behaviour. Other objectives are to obtain recommendations about yaw errors for design purposes and to indicate if it is conservative or not to use constant wind direction for thrust and bending moment reactions on the tower.

Two kinds of investigations will be presented. The first one concerns yaw misalignment due to different properties of the yaw system. The other one concerns fatigue load spectrum of the tower structure, not only due to wind speed variations, but also due to wind direction changes.

2. WIND MEASUREMENTS.

The measurements used up till now are based on what has been stored at the Alsvik site. At Alsvik two meteorological masts are equipped with 7-8 combined speed and direction sensors each. The sampling rate is 1 hz and the measurements have been going on since April 1990, with rather few interrupts. They covered originally 43 m vertically, but was extended to 53 m in mid 1991. This site is located near the shore line, which means that due to wind direction the topological conditions vary considerably [2]. West winds comes from the sea with low surface roughness induced turbulence. East wind travels over land and forests until they reach the site, of cause introducing much higher turbulence. For west winds the surface roughness has characteristic values in the wind direction itself. This may cause boundary layer variations, with consequences for turbulence and instant vertical gradients. Wind coming from north or south are alinged with the coast line. This means that the surface roughness varies across the wind direction (laterally) and this may create horisontal behaviour of the turbulence or instant horisontal gradient.

An other intreresting consequence of this locations is that temperature variations between land and sea varies a lot, one year around. Of that reason the masts are also equipped with temperature sensors in order to judge what meteorological situation is present, when evaluating behaviour of turbulence and wind directions. Are they neutral, stable or even unstable? It is noticable that most metērorological knowledge used in windturbine engineering assumes neutral situations.

These excellent good conditions for evaluating topological as well as meteorological phenomenas based on the long term measurements are at the moment also focused on other research projects, aimed at verifying simulated wind speed fields at mentioned different conditions.

For other planned investigations measurements at the Näsudden site also located on the island of Gotland started in April 1991. They consist of seven speed and three direction sensors along the 145 m high mast.

On the swedish west coast one 70 m high meteorological mast was errected mid 1992 at the Lyse site where one Bonus 450 kW and the NWP 400 kW turbine are located. This mast is equipped with seven combined speed and direction sensors, of the same MIUU-type as at Alsvik [1]. The idea is to finally store measured data on cheap Video-8 tapes which carries about 2000 Mb. Using compact facilities they may carry more than 5000 Mb.

Evaluation of wind direction behaviour may be made in different ways depending on actual purpose. There are two main purposes. One for alignment with the wind in order to produce as much energy as possible. The other is to reduce misalignments which creates load variations. But what is the best yaw position regaring the whole rotor area? Yawing too often may be restricted in active yaw systems. What about passive or free systems do they feel a 'better' wind direction?

There are lots of questions but few answers yet. In the following we will make use of the unique data set we have, in order to answer some of the questions.

3. YAW ERRORS.

What normally is ment by yaw error is the difference between the turbine yaw position and the measured wind direction at one point of the nacalle. An interesting question: is it possible to use the turbine itself to measure what could be the best estimate of the wind direction. Yaw errors that occur are results of how the wind direction is measured and several other parametrs

3.1 Method.

The main idea of studying this problem is to simulate what really happens. In this way most practical conditions may be involved. In more detail the yaw system of this study is described by the following quite general parameters:

- time delay constant of measured wind direction
- dead band of yaw error
- yaw speed

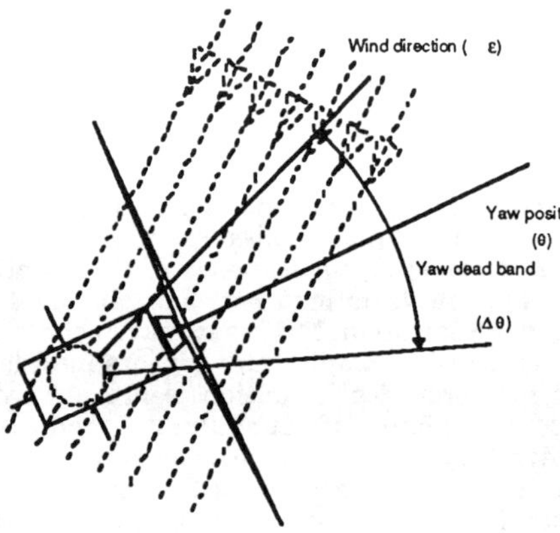

Fig 3.1 Yaw system definitions

A schematic exampel of time series of directions and angles of these class of yaw systems are shown in the figure 3.2 below.

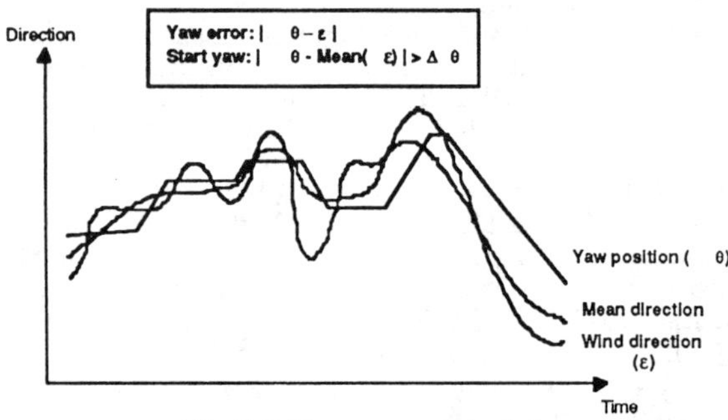

Fig.3.2 Yaw error principle

There are many ways of defining the real wind direction which is representatively felt by the turbine rotor. In above figures wind direction ε is what the nacelle sensor measures. The measurements carries several sensors vertically located within the rotor area. These are weighted in space and time in order be a more load representative value of the wind direction. The weighting procedure is aimed at being an estimate of the wind mass flow.

In this study the follwing values of the yaw system parameters are used:

- time delay constant	15 - 240 seconds
- dead band of yaw error	3 - 10 degrees
- yaw speed	2 degrees/second

For each set of yaw parameters the whole wind data set is used to simulate the yaw behaviour. The results of each simulation are time series, which are statistically evaluated in terms of:

- standard deviation of yaw error
- max yaw error
- number of yawings/hour

3.2 Wind conditions used.

As an example of yaw error analysis a 190 hour measurement is chosen. Mean wind speed was 8.5 m/s at height 30 m. Wind direction was distributed at a large part of the whole turn around. Turbulence intensity more than 10%.

3.3 Results.

Conserning the yaw error the conclusions are that standard deviation of yaw error seems (fig 3.3) to be linearly dependent on yaw dead band as well as a simple relationship with the time delay. Maximum values of the yaw error are nearly independent of the yaw dead band, see fig 3.4. One reason for that is that very fast wind direct changes occur. On the other hand the max error is stongly dependent on the averaging time. As foreseen (fig 3.5) the number of yawings per hour decreases with yaw dead band and averaging time.

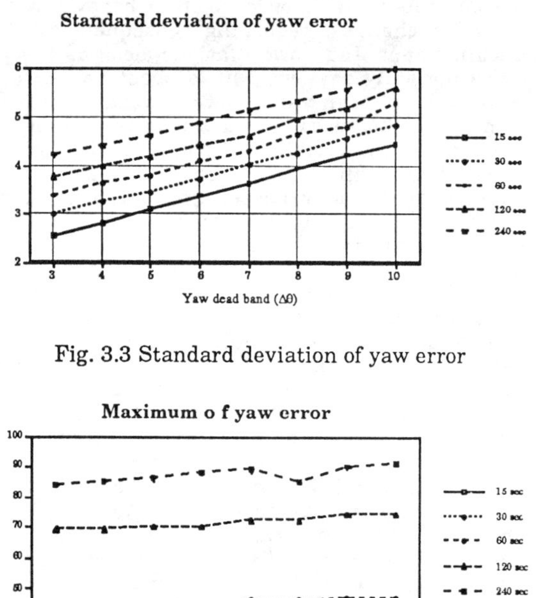

Fig. 3.3 Standard deviation of yaw error

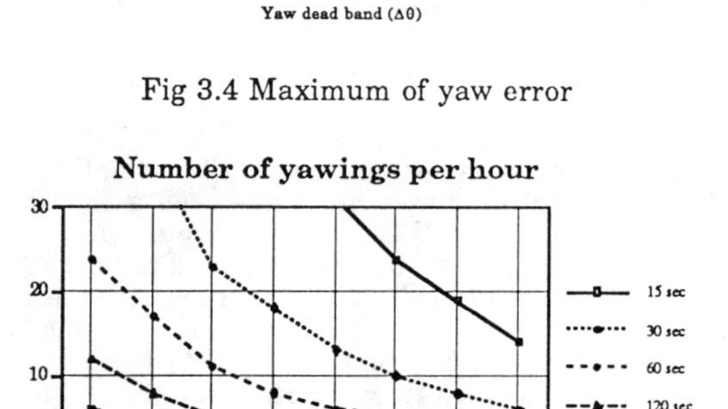

Fig 3.4 Maximum of yaw error

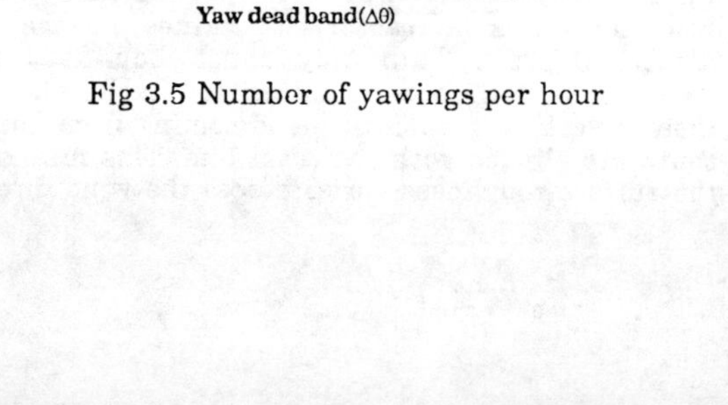

Fig 3.5 Number of yawings per hour

4. TOWER FATIGUE LOADS.

The normal way of calculating tower load spectrum is to regard wind speed variations incuding turbulence and dynamics as well as start and stops. These calculations normally assume constant wind direction. Is that conservative and how important is the thrust and torque moment behaviour of the turbine in terms of power control? Are there typical differences between power control methods as :

- pitch control
- stall controll
- yaw controll

How important is the turbine rotational speed concept? Lots of other questions may be raised but only a few of them are up till now studied, indicating how it can be done.

4.1 Method.

4.1.1 Tower loads and fatigue.

Main loads at tower top that influenses the fatigue conditions along the tower are two horisontal loads and moments, see fig 4.1 below.

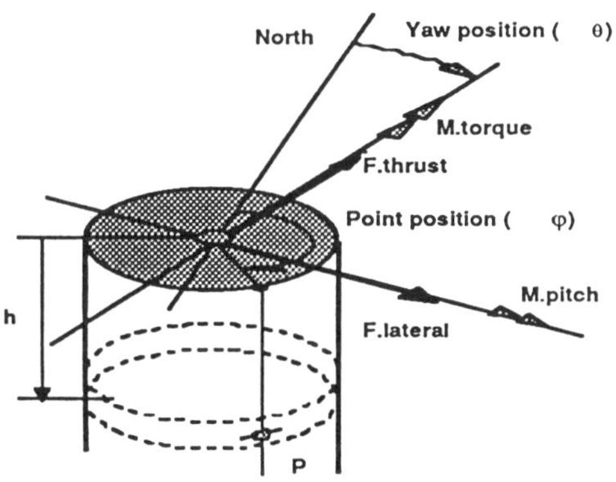

Fig 4.1 Main tower top loads to concider

The bending stress σ at point P in fig 4.1 is then

$$\sigma = \frac{(M1)}{I} * R*\cos(\varphi-\theta) + \frac{(M2)}{I} * R*\sin(\varphi-\theta) \qquad (1)$$

where
- I : section bending stiffness
- R : radius at height h from top
- h : vertical distance from top
- θ : yaw position relative north
- φ : point position relative north
- M.t: driving torqe
- F.t: thrust load
- M.p: pitch moment on tower
- F.l : lateral force
- M1 : =M.p+F.t*h
- M2 : =M.t+F.l*h

Here the dependence of the stress σ at P is shown as function of wind speed and direction, because driving torque (M.t) and thrust load (F.t) are strongly influnced by wind speed and power control and the yaw position follows wind direction ε. The loads M.p and F.l are normally more sensitive to turbulence (local gradients) and dynamics.

In most cases it is reasonable to neglect the vertical load and the yaw moment. Due to power control method following schematic behaviour of thrust load and driving torque may be more important to regard.

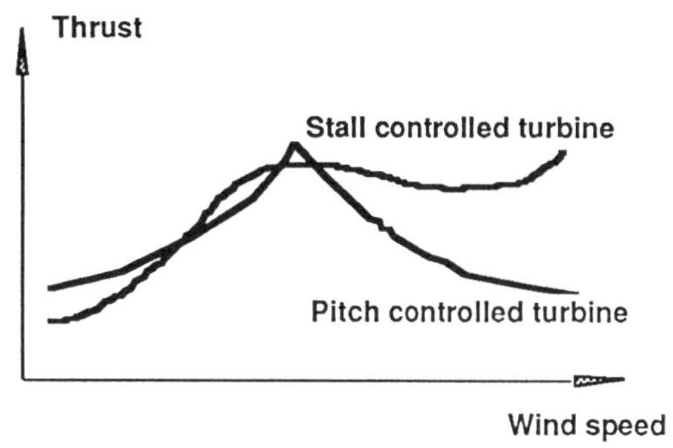

Fig 4.2 Thrust behaviour of pitch and stall turbines

Fatigue design of the tower means that all points of the tower shall be considered, along the tower as well as around it. Thus wind direction changes contribute to fatigue, like rotating bending does. The time series of the stress of individual points are thus influenced by wind speed variations as well direction changes. And finally these stress variations have to be evaluated according to the rain flow count (RFC) method, which is the method which is recommended for fatigue evaluation of varying loads and stresses.

4.1.2 Wind representation.

A method [3] has earlier been presented which describes evaluation of wind speed variations in a RFC sense and stored in a matrix for fatigue design use. In this case variations of not only the wind speed but also direction has to be regarded. An extended method has therefore been developped in order to take care of these two varying quantities at the same time. One way to solve this new increased problem could be to directly use measured wind speed and direction and use them for simulation of the tower loads at the top as function of time. A final step would then be to convert this loads to stresses at all points of interest of the tower.

To reduce computer time of this crude way the originally measured data was reduced to hold what is really needed to finally calculate RFC stress cycles of each actual point of the tower. It shoud be mentioned here that details of the yaw system behaviour, as presented in chapter 3, are not regarded here. It is assumed that the yaw system follows wind direction perfectly.

A major problem of evaluating wind direction changes is to handle changes around 360 degrees. Talking about RFC changes over a long period of time also raises the problem of how many rotations.

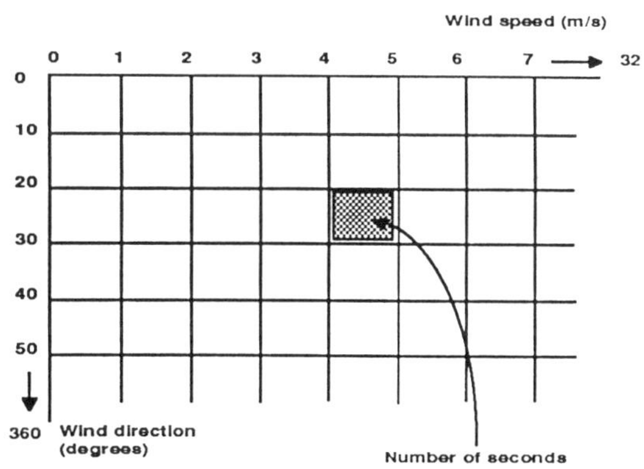

Fig 4.3 Wind speed and direction duration matrix

To have enough information of variations of measured wind speed and direction changes it was finally desided to store time duration at choosen speed and direction in a matrix, see fig 4.3 above.

The variations are evaluated in a way, which is shown in fig 4.4 below. Max and min point of the both speed and direction are identified and corresponding times are observed and stored on a file. This file has information about what changes and when did they occur, which are needed for further calculation of stress variations.

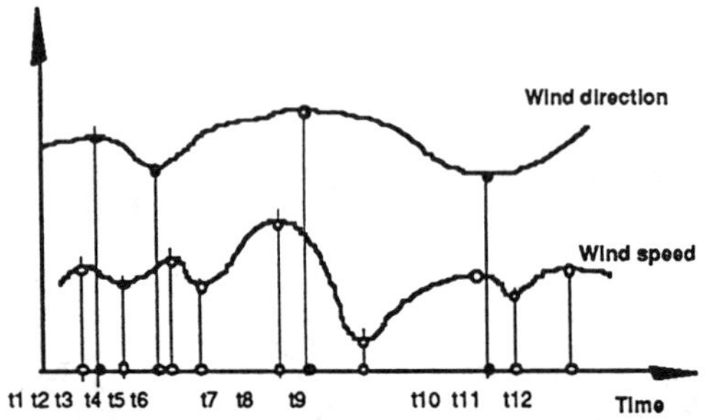

Fig 4.4 Speed and direction time series

4.1.3 Calculation of stress spectrum.

For a choosen turbine with a certain power control method the first step is to calculate mentioned tower top loads as function of wind speed, as shown in fig 4.2 above. In this study a simplified description of turbulence has been used, but stationary dynamics due to gradients and rotation are included. This means that the turbine behaviour in a nacelle fixed system is known as function of wind speed.

The first part of the stress variation contribution is a result of the duration matrix and calculated load - wind speed relations. It means that wind speed situations are spred out around the tower, but the stress at every point of the tower is calculated according to (1). Load variations of this first part are cause by turbulence and dynamics, and they are already RFC evaluated [4].

The other part representing variations according to fig.4.4 above is carried out in a more direct way. At instants where either the speed or the direction has a maximum or a minimum the stress at actual point is calculated according to (1). The final step of this second part is then to determine the RFC- cycles of stresses at actual points. This procedure is then repeated for each point of interest. The total RFC stress spectrum is finally made up of the duration result and the variation part together.

4.2 Calculations and results.

Preliminary calculations have been carried out in order to indicate how important the wind direction behavoiur may be for tower fatigue design of hawts. First six calculations are based on measured wind conditions during one week of April 1991. Required section modulus for sixteen points at three levels were calculated for pitch and stall control methods. This was done regarding wind direction changes and not doing it. The number of cycles were multipied with a factor to represent twenty years operation time. In table 4.1 the largest section modulus requirement at each level for the two cases with and without wind direction influence are relatively presented.

For the mid section level of the tower corresponding calculations are carried out based on measurements of one year duration.

Table 4.1 : Required section modulus regarding wind direction (Ww), relative not to regarding it, Wo.

Measurement	Power control	Level	Ww/Wo [%]
1 Week	Pitch	top	114
		mid	129
		bottom	129
	Stall	top	111
		mid	129
		bottom	129
1 Year	Pitch	mid	90
	Stall	mid	100

Results of these preliminary calculations based one one week wind data are that tower fatigue design regarding wind direction changes may increase dimension requirements with about 30%, independent of power control method used in these calculations. On the other hand looking at one year wind data the results are quite different. One major reason for this is that the wind direction distributions are quite different. During the one week measurement two dominating 180 degrees opposite conditions occurred, while the direction was reasonably even distributed during the one year measurement.

The main conclusion to draw is that characteristics of wind direction distributions may have influence on tower fatigue design and it is not allways conservative to assume constant wind direction. Further and more complete studies must be done, in order to draw more definite conclutions and recommendations how to regard wind direction changes. More measurements are very important ingrediencies of these kind of studies.

5. COMMENTS.

But finally it must be mentioned here that the climatological conditions at the Alsvik site has a constant wind direction over the year during 60-70 %. Further investigations of other locations are therefore planned to be carried out and it is assumed that local conditions may vary a lot from site to site. Questions about yaw controlled turbines is another interesting questions which will be studied in the near future.

6. REFERENCES

[1] K. Lundin, A. Smedman, U. Högström. A System for Wind and Turbulence Measurements in a Wind Farm. ECWEC in Madrid, 1990.

[2] J. Dahlberg, M. Poppen, S.E. Thor. Load/Fatigue effects on a Wind Turbine Generator in a Wind Fram. EWEC in Amsterdam,1991 .

[3] Hans Ganander. Wind Measurements from Alsvik. Results of two 190 Hours campaigns conserning gradient and wind direction. IEA-meeting of Experts, in Stockholm, March 7-8 1991.

[4] H. Ganander, H. Johansson. Fatigue Design by Using a Modified RFC Description of the Wind. AWEA in Honolulu, 1988.

DYNAMIC LOADS IN WIND FARMS

A J Tindal and A D Garrad Garrad Hassan & Partners Ltd., UK
G Schepers and B Bulder ECN, The Netherlands
H Hutting KEMA, The Netherlands
F Verheij TNO, The Netherlands

Abstract: This paper describes the Joule I project (JOUR 0084). Measured data from two wind farms in simple terrain - Sexbierum in the Netherlands and Nørrekær Enge II in Denmark have been used to validate and improve combined wind farm and machine behaviour models. A detailed study of wind conditions in the wake has also been investigated using wind turbine models in a wind tunnel. The main aims of the study are:

- To quantify the level of increase in loading of machines in wind farms.
- To estimate how the loading varies with spacing and position.
- To assess the maturity of the combined wind farm and machine behaviour models.

1 Introduction

This work has been funded by the CEC and the national programmes of the UK, through the DTI, and The Netherlands. In previous work funded by these sources, the description of wind turbine wakes and the dynamic response of the machines have been considered separately. The purpose of this project is to bring together those two areas of work into a single project, aimed at describing the dynamic response of a wind turbine in a wind farm. For this project it was decided that it was important to have very simple topography in order to remove its influence from the response. The Sexbierum site in northern Netherlands is very flat and it is used as the main source of data in this project. In addition to using full scale measurements wind tunnel measurements have also been made. The two experimental approaches are complemented by numerical models of the machine behaviour and the wind flow. Towards the end of the project the numerical tools developed and validated with data recorded at Sexbierum and in the wind tunnel have been applied to the Nørrekær Enge II wind farm. The Nørrekær Enge II comparison acts as a check on the general validity of the tools developed for Sexbierum.

2 Data sources

Measured data from three different sources has been used in this project. A summary is given below:

- **The Sexbierum wind farm in the Netherlands**
  18 Holec 300 kW wind turbines
  Variable speed
  Pitch regulated
  Steel blades
  One instrumented machine

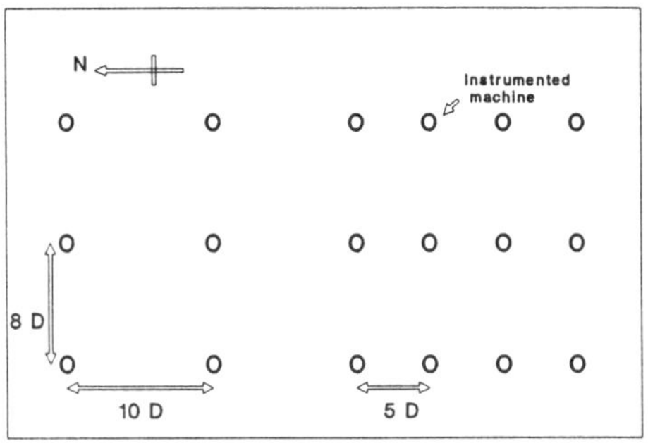

Figure 1 Layout of the Sexbierum wind farm

- **Wind tunnel tests carried out in UK and the Netherlands**
  Atmospheric boundary layer tunnel
  1:300 scale wind turbine models

- **Nørrekær Enge II wind farm in Denmark**
  42 Nordtank 300 kW wind turbines
  Fixed speed
  Stall regulated
  GRP blades
  Two instrumented machines

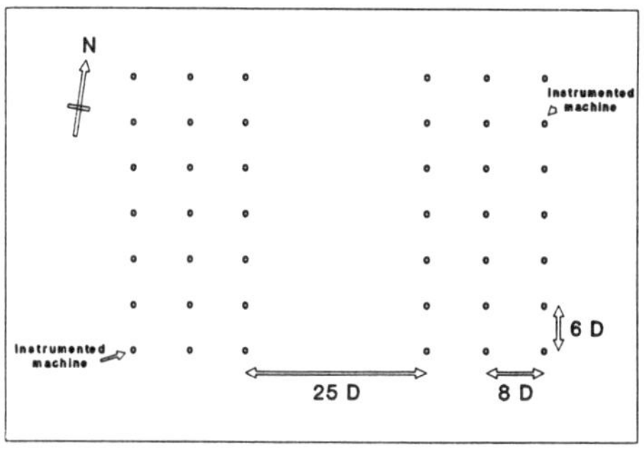

Figure 2 Layout of the Nørrekær Enge II wind farm

3 Wind structure in the wake

In order to simulate the applied loading of a wind turbine in a turbulent wind field it is necessary to use a model of the wind which has the correct representation of both the temporal and spatial structure of the longitudinal wind speed fluctuations. The wind simulation methods adopted by Garrad Hassan and ECN are based on that described by Veers [1]. The time series generated may have any user-specified auto-spectral density and coherence properties. For simulations of the operation of isolated wind turbines it is common to assume von Karman properties for the wind.

One of the aims of this project has been to investigate the structure of the wind in the wake in order to evaluate the wind input required to simulate satisfactorily the applied loading of a wind turbine operating in a wind farm. Figure 3 presents the auto-spectral density of the wind recorded in the free stream and 2.5 D behind an operational wind turbine on the machine centre line at hub height. ESDU spectra fitted to the measured data are also presented. The measurements were made in an atmospheric boundary layer wind tunnel using a 1:300 scale simulation. Experimental details are given in [2]. It is clear that the wind in the wake has the same structure as the ambient wind, there is, however, a marked reduction in the longitudinal turbulence length scale - $^{x}L_{u}$. Measurements were made at 2.5, 7.5 and 10 D downstream at a range of lateral positions at machine hub-height. The resulting length scale profiles are presented in Figure 4. The greatest reduction in length scale is seen at 2.5 D on the wake centre line where the length scale has reduced to a quarter of the free stream value. The length scale profile is generally seen to broaden and recover in similar manner to velocity profile. The results show the length scale at 10 D to be smaller than that at 7.5 D which is a surprising result. It was observed that the ambient length scale with which the 7.5 D results were normalised was approximately 10 % lower than the 2.5 D and 10 D results. If the 7.5 D results were normalised to the same ambient levels as 2.5 D and 10 D the trend described above would be reversed.

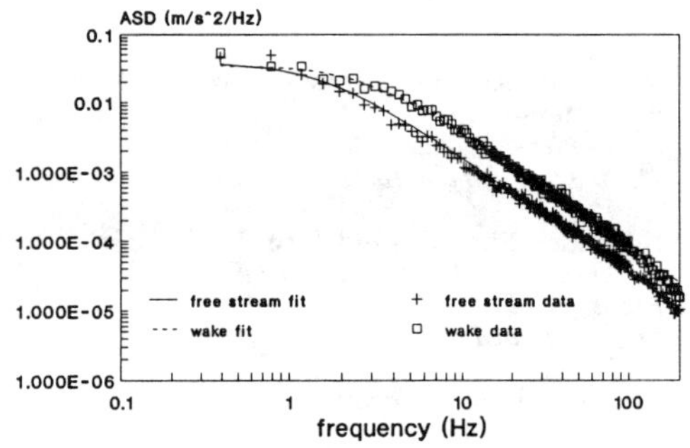

Figure 3 Spectrum of wind speed in free stream conditions and on the wake centre line at hub height 2.5 D behind the rotor

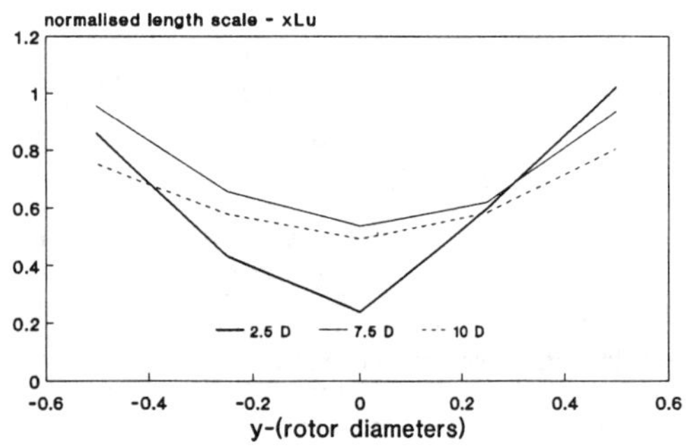

Figure 4 Cross wind profiles of turbulent length scale at hub height at 2.5, 7.5 and 10 D behind the rotor

Measurements made in the ambient wind and in the wake at downstream positions of 4 and 7 D have been recorded at Sexbierum. These data have been analysed by TNO in a fashion similar to that used for the wind tunnel data. The results have also shown that no significant alteration is seen in the structure of the wind in the wake except for change in length scale. Reduction in the longitudinal turbulence length scale to between 20 and 40 % of the free stream value on the wake centre line at hub height are seen at a downstream distance of 4 D. This result is broadly in agreement with the results from the wind tunnel, although, the estimate of length scale from the full scale data is more difficult than from the wind tunnel data.

The results from the two data sources described above lead to the important conclusion that the structure of the wind in a wake is well represented by a von Karman spectrum. In addition an indication is given of the likely variation of longitudinal length scale with wake position. This allows the applied loading of a wind turbine in a wind farm to be simulated with confidence using a von Karman model of the wind, with an appropriately reduced length scale.

A data base of mean and turbulent wind conditions in the 5 D and 8 D single and multiple wake situations which occur in the Sexbierum wind farm was generated. These conditions have also been replicated in the wind tunnel by TNO and GH. Replication of a part of the full scale wind farm in the wind tunnel allows more detailed wind measurements to be made than are practical in the field. Figure 5 presents surface plots of the wind speed at the second and third machine positions for a westerly wind direction at Sexbierum. These results were measured in the wind tunnel by removing the second and third turbines in turn and making wake traverses in their place. The results show that the centre line deficit at the third machine position is not significantly greater than at the second machine position. This observation is important and is further discussed in Section 4. The wake is, however, wider and extends to a greater height at the third machine position than at the second.

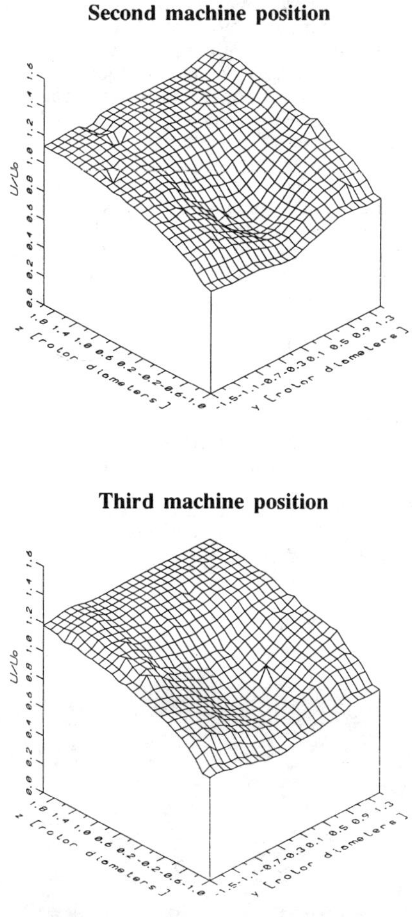

Figure 5 Surface plots of wind speed at 2nd and 3rd machine positions for 3 wind turbines with 5 D spacing

4 Dynamic loads and fatigue damage in wind farms

Accurate computer models of the wind turbines have been created. The ability of GH and ECN to predict the behaviour of the machines for free stream conditions has been confirmed through detailed analysis and comparison with measured data.

Measurements of individual machine power at Sexbierum and Nørrekær Enge II have been used to validate the wind farm models. Figure 6 presents the measured mean power for each wind turbine in the wind farm for two wind directions. Predictions from the GH proprietary code EVFARM are also included on the figure. Results from Sexbierum show similar trends and levels of agreement to those presented above. It is considered from these results that the ability of the code to predict the wind conditions within the wind farm is good. This allows the task of predicting wind farm machine loading to be tackled with confidence. The TNO code FARMS has also been shown to replicate measured wind farm behaviour well.

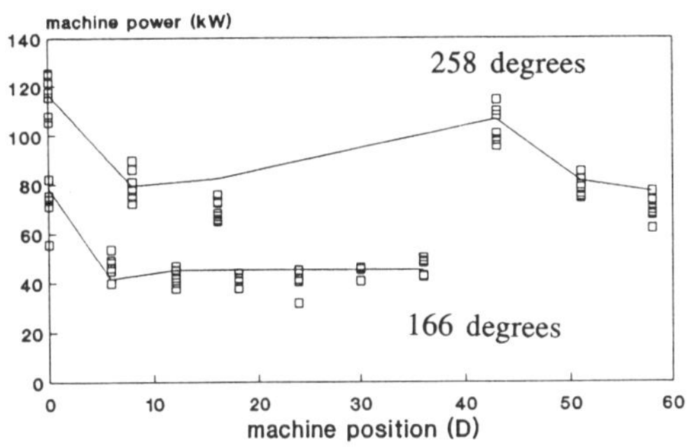

Figure 6 Measurements and predictions of individual mean machine power for two wind directions at Nørrekær Enge II

4.1 Campaign data

Two wake cases have been considered at the Sexbierum wind farm: the operation of the instrumented machine behind a wind turbine 8D upstream and behind a wind turbine 5D upstream. An example of the level of agreement achieved between GH and ECN predictions and measurements for an 8 D wake case is given in Figure 7 which presents the auto-spectral density of blade flatwise bending moment at a radial station of 3.5 m. The level of agreement of the predictions with measurements is similar to that achieved for free stream conditions. The data in Figure 7 is presented as just one example result. Flatwise and edgewise blade bending moment at two radial stations, slow speed shaft torque and tower bending moments have also been considered for a range of wake and wind speed conditions. In addition to single wake conditions a limited number of multiple wake conditions have been investigated.

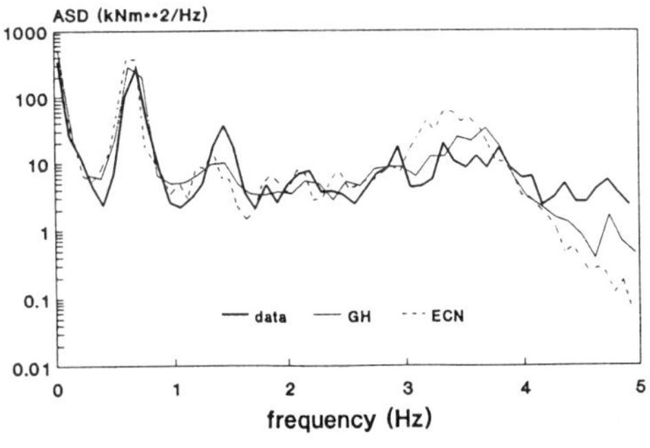

Figure 7 ASD of flatwise blade bending moment at 3.5 m for Holec WPS-30 during wake operation

Data from Nørrekær Enge II has been used towards the end of the project to confirm the results from Sexbierum. For machine loading only a westerly wind direction and one blade radial station have been considered. The presence of two instrumented machines in the wind farm allows the behaviour of a machine at the front of the wind farm to be compared directly with that of a machine at the back of the wind farm. An example result which shows the auto-spectral density of flatwise bending moment near the root is presented in Figure 8. The predictions agree well with measurements.

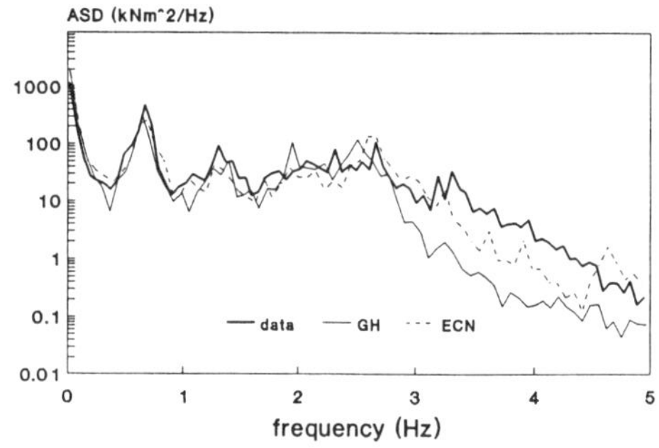

Figure 8 ASD of flatwise blade bending moment at the root for Nordtank 300 kW machine during wake operation

In order to assess the ability of GH and ECN to predict the loading of the wind turbines in free stream and wake conditions a parametric representation of the severity of loading is used. For this purpose the "1 Hz fatigue equivalent bending moment" is evaluated. This load is defined as the amplitude of sinusoidal bending moment which, if applied to the blade at 1 Hz, results in the same fatigue damage as the actual complex bending load time history. This approach allows simple comparisons to be made between predictions and measurements and different wind speeds and wake situations. Table 1 presents example comparisons of measured and predicted 1 Hz fatigue equivalent bending moment results for specific ten minute data campaigns recorded at Sexbierum and Nørrekær Enge II. The results indicate that GH and ECN can predict free stream and wake conditions in a wind farm in flat terrain with reasonable accuracy.

Campaign Description	Data	GH	ECN
Sexbierum			
Free stream 14.1 m/s	1.0	1.06	1.07
5 D single wake 9.3 m/s	1.0	1.07	0.95
8 D single wake 9.5 m/s	1.0	1.0	1.11
Nørrekær Enge II			
Free stream 9.8 m/s	1.0	0.93	1.0
8 D multiple wake 9.8 m/s	1.0	0.93	0.92

Table 1 1 Hz fatigue equivalent load predicted by GH and ECN for a range of measured campaign data sets normalised to the measured value - SN slope of 3 assumed.

4.2 Continuous data

Continuous load data has been recorded for the instrumented machine at Sexbierum. This has been binned automatically into one of six wind speed bins and four wake classes, and rainflow cycle counted. It provides a data base which allows the relative severity of different wind speed and wake conditions to be assessed. The cycle count data has been normalised to one hour's operation and the 1 Hz fatigue equivalent load has been calculated. The results are presented, normalised to the free stream class, in Figure 9. The cumulative load spectra for each wake and wind speed condition have been presented graphically by ECN in [3]. The most severe wake condition is the 5 D separation at all wind speeds. The wind speed at which the wake effects are at their greatest is the 10 to 12 m/s bin. The greatest increase in wake loading is just over 30 %.

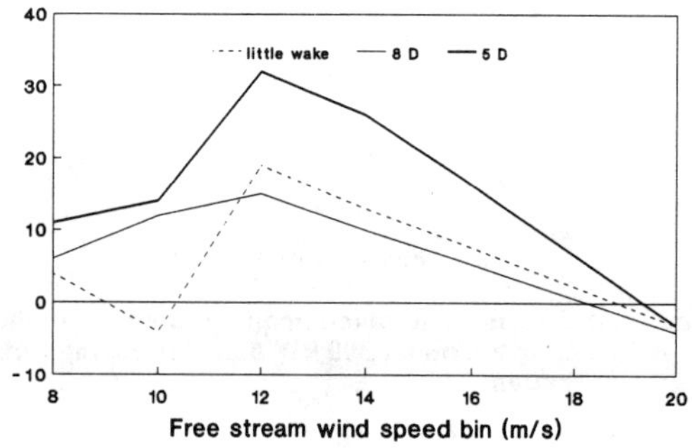

Figure 9 Percentage increase in 1 Hz fatigue equivalent load with wake and wind speed conditions measured at Sexbierum.

4.3 Fatigue loading within the wind farm

In order to assess the increase in loading in a wind farm the following approach has been used:

- The wind speed at each machine in the wind farm for all wind directions and wind speeds between cut-in and cut-out has been predicted.
- The loading experienced by each machine for the wind conditions evaluated above have been calculated.
- The total fatigue loading each machine in the wind farm will experience has been estimated by combining the above two predictions with the wind speed direction frequency for the site.
- The process is repeated for one isolated machine in place of the wind farm.

The results of this procedure allow the fatigue damage of each of the machines within the wind farm to be compared with the damage which an isolated machine on the same site would have seen. Clearly this task is onerous. Some simplifying assumptions have been made in order to make it practicable. Most notably the influence of partial wake immersion and changes in turbulence length scale have been neglected. Figure 10 presents 1 Hz fatigue equivalent flatwise blade load at 50 % span normalised to the isolated machine result. The greatest increase in fatigue loading is 9 %. The wind farm results have also been normalised to the severity of fatigue loading experienced by an isolated wind turbine operating the same wind regime but with a continuous turbulence level of 18 %. This figure is close to the 17 % assumed in the TC 88 standard [4]. The most severely loaded machine in the wind farm experiences a 1 Hz fatigue equivalent load which is 14 % lower than the isolated machine with 18 % turbulence. This indicates that if the WPS-30 were designed to this standard there is a little scope for closer machine spacing at Sexbierum wind farm without reducing the design life of the machines.

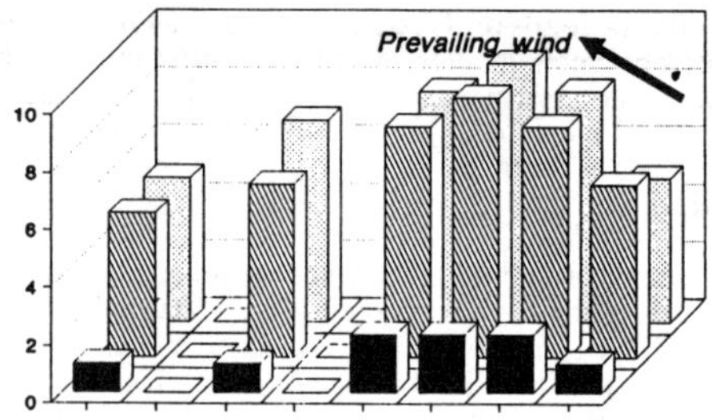

Figure 10 Predicted percentage increase in 1 Hz fatigue equivalent load for each machine in the Sexbierum wind farm

The above results are specific to the Sexbierum wind farm. From this project, however, results indicate that the reduction in energy capture and severity of machine loading deep within a large wind farm is not likely to greatly increased over that experienced by the machines in the second or third row. Machines within large, closely packed, wind farms would therefore be expected to experience greater increases in fatigue loading than found at Sexbierum due to the greater proportion of time spent in wake conditions rather than the greater severity of wake conditions themselves.

5 Conclusions

The project described is nearing completion. The following conclusions are drawn from the work.

- The structure of the wind in the wake is well represented by the von Karman spectrum. A significant reduction in turbulent length scale is seen in the wake.
- The ability of the combined wind farm and machine behaviour model to predict loading of wind turbines in wind farms in simple terrain is shown to be good.
- Measured results from Sexbierum indicate that the greatest increase in loading in the wind farm is approximately 30 %. This occurs for westerly or easterly winds of speed 10 to 12 m/s with machines at their closest spacing of 5 D
- Predictions indicate that over the lifetime of the wind farm the greatest increase in loading due to wind farm operation is likely to be in the order of 10 % for the Sexbierum wind farm. The increase in loading is seen to be sensitive to wind turbine spacing and position.

Multiple wakes and the effects of topography are still not well understood and will be addressed in the Joule II project "Dynamic loads in wind farms II".

6 References

1 Veers, P S "Three dimensional wind simulation" SAND88 -0152, Sandia National Laboratories, March 1988.

2 Hassan U "A wind tunnel investigation of the wake structure within small wind turbine arrays" Final report on Department of Energy contract E/5A/CON/5113/1890, June 1992.

3 Schepers J G and Bulder B H " Load prediction and evaluation of the WPS-30 wind turbine in the Sexbierum wind farm" Proceedings of EWEA conference, Denmark, September 1992

4 "Draft standard on the safety of wind turbine generators" IEC TC88 Issue 21 B

ACTIVE PITCH CONTROL: THE CONTROLLERS INFLUENCE ON PERFORMANCE AND LOADS OF A WIND TURBINE

H. Bindner, P.H. Madsen
Test Station for Wind Turbines, Risø National Laboratory
Kurt Andersen
Institute of Automatic Control Systems, Technical University of Denmark
Knud Erik Thomsen
VESTAS Wind Systems A/S

ABSTRACT: The paper presents a model suitable for controller design and analysis primarily for the design of pitch controllers. The scope of the model is to be able to predict the performance of a pitch controlled turbine and to give information on loads in the drive train. Results regarding model complexity and the impact of key parameter variation are presented.

1 INTRODUCTION

Recently a project called 'Active control of wind turbines' was initiated at Risø in collaboration with the wind turbine manufacturer, VESTAS, and the Technical University of Denmark. The purpose of the project is to study the use of active control schemes for wind turbines. Emphasis is on the trade off between maximizing energy capture and power quality while minimizing loads considering constraints especially in the control system. The project is connected to a measurement project on a pitch controlled turbine and initially pitch control is studied.

The first objective of the project has been to develop a model suitable for design and analysis of pitch controlled wind turbines. This model is presented below. The model is a linear state space model that includes important structural parts and detailed models of gearbox and generator.

The model is used to model a VESTAS WD34 400 kW pitch controlled wind turbine.To verify the model it is compared to measurements in both time and frequency domain.

Two PI controllers are designed to give an impression of the influence of the controllers impact on performance and loads. The difficulties of suppressing the wind fluctuations are illustrated.

Finally the effects of key parameter variations are discussed. This includes variation of shaft stiffness and generator slip and some results are presented regarding both performance and load are presented.

2 MODEL PRESENTATION

The scope of the model is to have a handy model for controller design and analysis. It has to represent the important phenomena in the relevant frequency range. It also has to be easy to use in the controller design phase. The important phenomena the presented model has to explain is the torque on the main shaft and the electrical power produced. The relevant frequency range is from 0.1 Hz to 10 Hz. The model presented here is a linear model linearized in a operating point. The model at present includes first tower bending mode, symmetric flapwise blade bending, suspended gearbox, induction generator, and rotational sampling and spatial filter on the wind input.

2.1 Component Models

The structural part of the model consists of the first tower bending mode, the symmetric flapwise blade bending and the fundamental in plane rotor mode. The flapwise bending of the blades is modelled as if the rotor is a ring with mass only at the edge. The individual bending of the blades is not modelled. The reason for including tower and flap bending is to represent the aerodynamical damping these phenomena cause on the aerodynamic torque.

A quite detailed model of a suspended gearbox, [1], is included. It models the inertia of the gear as well as the movement of the gearbox itself. It includes the induction generator inertia.

To study the impact of the dynamics of an induction generator a dynamic model has been developed, [2]. This model contains the dynamics of both the stator and the rotor. The main reason for the development of the model is to study whether such a detailed model is relevant for the design of pitch controllers. The model is linearized and different levels of modelling is studied.

The aerodynamics is also linearized at a operating point. It includes rotor thrust and torque and the aerodynamic damping. The model is driven by point wind and the model includes filters to model the spatial averaging, [3], and the rotational sampling of the rotor.

2.2 Total Model

The above mentioned component models is implemented in MATLAB, [4], and connected in SIMULINK. The model parameters are found from data from the WD34 turbine. They are found at two operating points, below and above rated wind speed. Values for above rated wind speed are used except for the time series simulation.

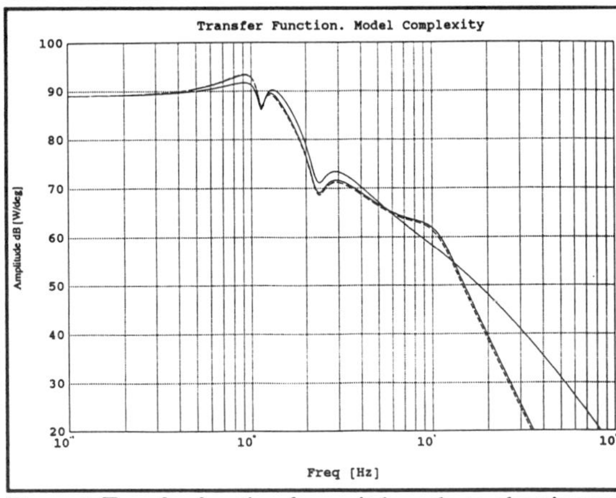

Figure 1 Transfer function from pitch angle to electric power, amplitude. 3 different levels of complexity. (Full model solid high peak, medium dashed, reduced solid low peak)

Since the drive train is fairly detailed modelled some model reductions are studied. On fig 1 and 2 the open loop transfer function from pitch angle to electrical power is shown for three different levels of drive train complexity. One including the full dynamics as mentioned above, one where gearbox movement and generator stator dynamic are neglected and one where gearbox movement all generator dynamic are neglected. The figures show that gearbox movement and stator dynamics can be safely neglected whereas the rotor dynamics of the generator has some role to play within the relevant frequency range. In the rest of the paper the medium complex model is used.

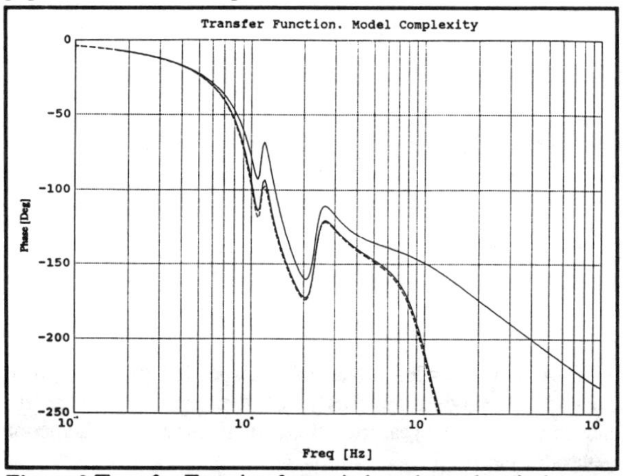

Figure 2 Transfer Function from pitch angle to electrical power, phase. 3 different levels of complexity. (Full model solid below, medium model dashed, reduced model solid above)

The first resonance peak on fig 1 is the first drive train eigenfrequency. This frequency is quite undamped. The second peak is the tower bending. This eigenfrequency is also undamped. Between these peaks there is zero in the transfer function. This is also due to the nacelle movement. At the higher frequencies the flapwise bending of the rotor is visible at 2.3 Hz and the generator rotor at approx. 10 Hz. On the phase plot it is seen that the zeros of the transfer function all are in the left half plan.

3 MODEL VERIFICATION

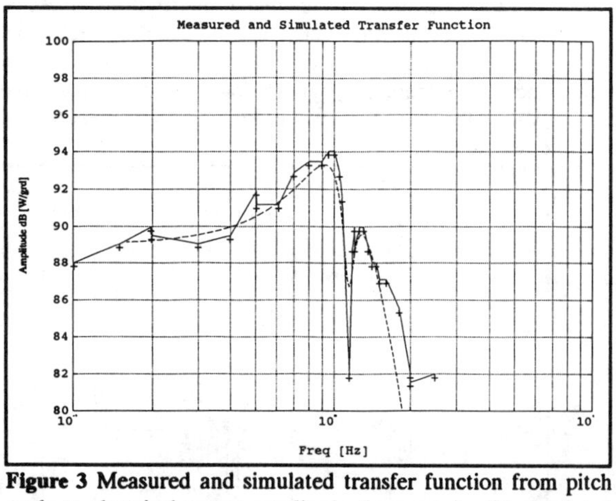

Figure 3 Measured and simulated transfer function from pitch angle to electrical power, amplitude. (measured solid, simulated dashed)

The use of a model depends on the degree of trust one can have in it, therefore a model has to be verified. A full verification is very difficult and comprehensive. To do some verification of the presented model two different things have been performed.The real turbine has been operated in open loop from pitch reference signal to measured electrical power and the frequency response has been recorded at several distinct frequencies. The other form of verification has been through time series comparison where both the time series and calculated power spectra are compared.

3.1 Transfer Function

The transfer function from pitch angle to electrical power has been measured. Fig 3 and 4 shows the bode plot of the measurements compared to the model. The frequency response was measured at distinct frequencies (the +'s on the figures) and with high resolution around the two peaks. The measurement conditions makes it difficult to measure at higher frequencies.

The shape of the amplitude of the transfer functions compare very well. The two peaks have the correct position and height. The small third peak on the measurements are caused by 3P excitation which is present only in the measurements.

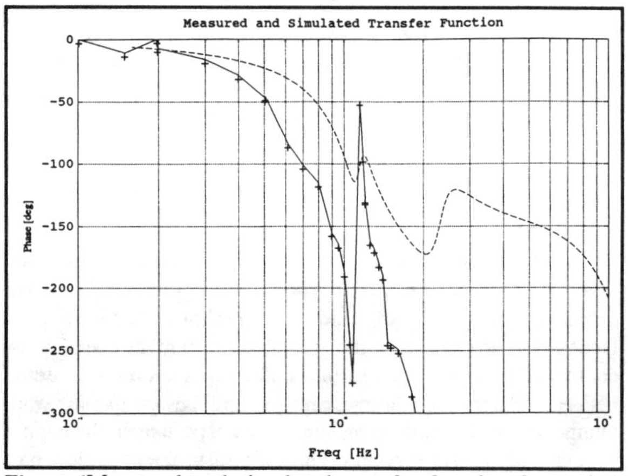

Figure 4 Measured and simulated transfer function from pitch angle to electrical power, phase. (measured solid, simulated dashed)

The measured phase of the transfer function does not correspond as well. The measured phase starts to fall much earlier than the simulated and it continues to fall at higher frequencies where the simulated is flatter. Some of the difference is because transducers are not modelled.

3.2 Output Spectra and Time Series

The model is compared to measured time series at wind speeds below rated where the turbine is operating in open loop. The parameters for lower wind speeds are used. The measured wind speed is driving the model through the spatial and rotational sampling filters. A 10 min time series with a mean wind speed of approx. 6.5 m/s where used for this comparison. The mean value of the measured time series has been subtracted since the model operates around an operating point.

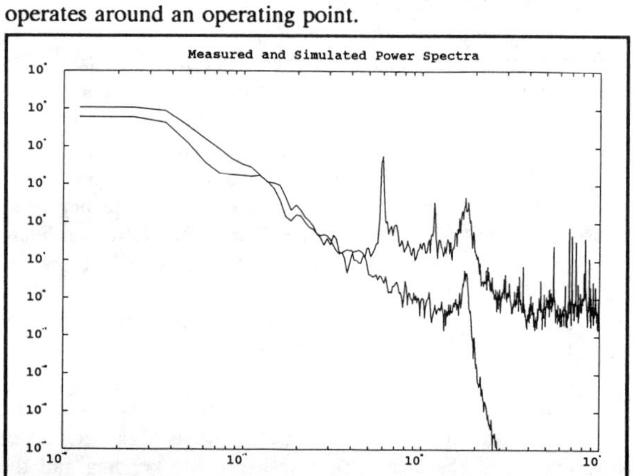

Figure 5 Measured and simulated power spectra of electrical power for a 10 min time series. (measured flat, simulated steep)

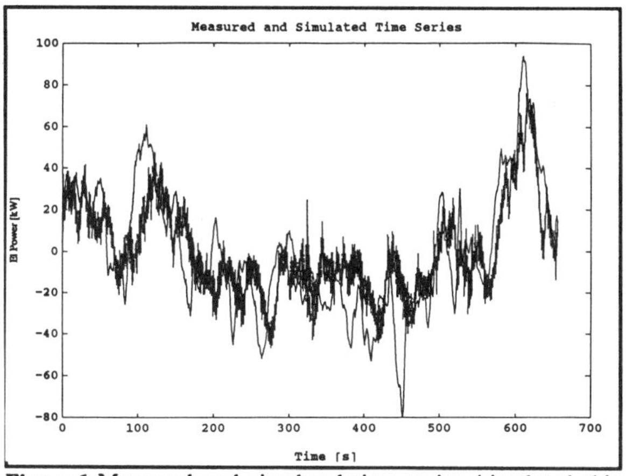

Figure 6 Measured and simulated time series (simulated thin curve)

On Fig 5 are the power spectra of both measured and simulated electrical power shown. They show the same overall behavior up to approx. 2 Hz. Above 2 Hz the simulated spectrum continues to fall. It is also seen that the models gain at low frequencies is higher than that of the real turbine and that the opposite is true at higher frequencies. The high gain at low frequencies is mostly due to the linearization of the aerodynamics. On the real turbine the gain at low wind speeds is highly unlinear and is decreasing when the wind speed is decreasing. The low gain at higher frequencies is caused by the spatial filter. The time series plots in fig 6 exhibits the same overall nature (the measurements contains 1P fluctuations in the power due to some imbalances. These effects are not modelled). The above mentioned discrepancies results in a higher variance in the simulated power than in the measured.

4 SIMPLE CONTROLLER IMPACT

4.1 Simple Design

The impact of the controller is illustrated by two PI designs. None of the designs pretends to be optimal they serve the purpose of illustration.

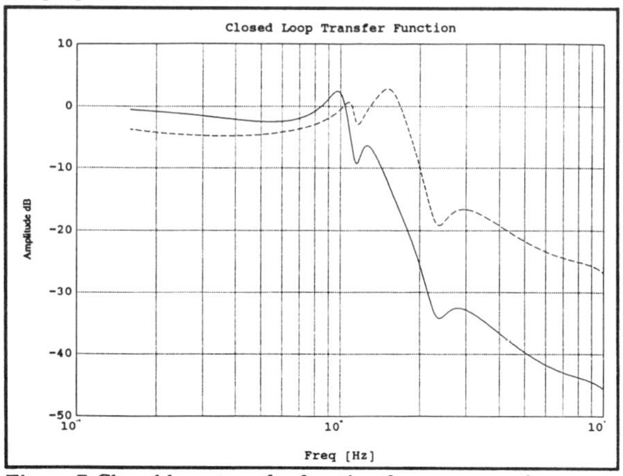

Figure 7 Closed loop transfer function from power reference to power (PI1 solid, PI2 dashed).

The controller parameters are P= 3.54e-5 and I=1s, PI1, respectively P=4e-6 and I=0.05s, PI2. This results in the two closed loop transfer functions (from power reference to power) in fig 7. They both have a pole relatively close to origo. The main difference is in the frequency range at 1 Hz. This difference is also illustrated on fig 8 which shows the closed loop transfer function from rotor wind speed to electrical power compared with the open loop transfer function. It is seen how PI1 amplifies the input at 1 Hz and how PI2 amplifies it at approx. 1.5 Hz. Since 3P is at 1.75 Hz this amplification could cause larger loads on the drive train even if the peak is lower than that of PI1. This movement of the two peaks is one that has to be considered when designing the controller.

4.2 Parameter Study

To give some indications on how the system changes behavior when some of the key parameters are changed the main shaft stiffness and generator rotor resistance are changed. The shaft stiffness is halved and the resistance is doubled. The open loop transfer function from pitch angle to electrical power is shown on fig 9. It is seen how the change in shaft stiffness results in a higher resonance peak at a lower frequency. This could be a desirable effect since the high frequency gain is lower. The change in generator rotor resistance (which is equivalent to a higher slip generator) also has a positive effect on the transfer function. The drive train resonance peak is much better damped

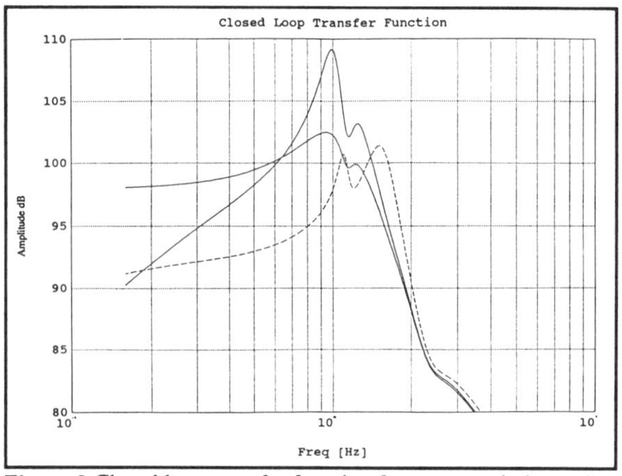

Figure 8 Closed loop transfer function from rotor wind speed to power. (Open Loop solid, PI1 solid high peak, PI2 dashed)

and the gain at 3P is lower than nominal gain.

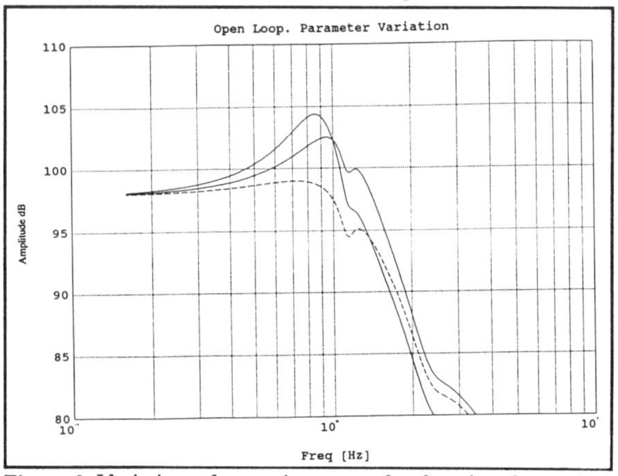

Figure 9 Variation of open loop transfer function from pitch angle to power when stiffness and resistance are changed. (Nominal System solid in the middle, Stiffness solid high peak, Resistance dashed)

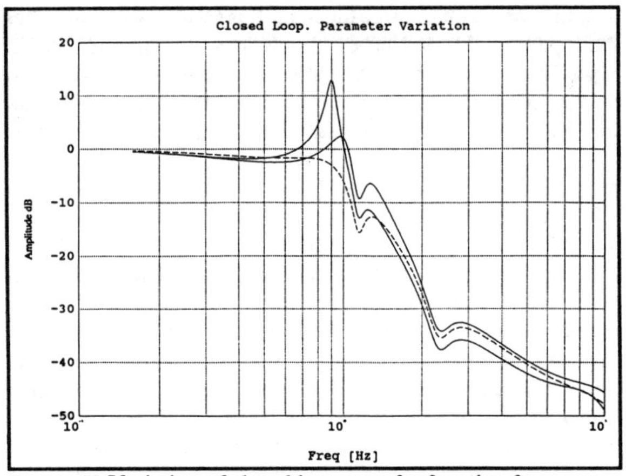

Figure 10 Variation of closed loop transfer function from power reference to power when stiffness and resistance are changed. (Nominal system middle, Stiffness solid high peak, Resistance dashed)

On fig 10 and 11 the closed loop transfer functions from power reference respectively rotor wind speed is shown together with the nominal values. Again it is seen how the gain at the second peak is reduced and that the system with high slip generator is better behaved. It is also seen that a high slip generator is a much better solution than a flexible main shaft.

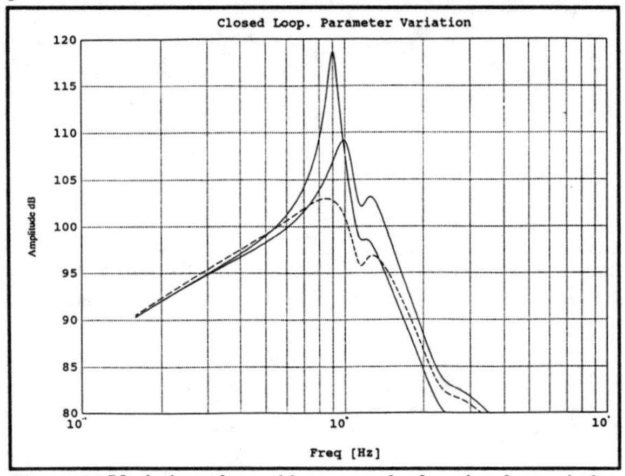

Figure 11 Variation of cosed lopp transfer function from wind to power when stiffness and resistance are changed. (Nominal solid middle, Stiffness solid high peak, Resistance dashed)

5 CONCLUSION

A linear model of a pitch controlled wind turbine has been presented. The complexity of the model is moderate and it is suitable for controller design. The model has been compared to measurements and the agreement between model and measurements are good as far as the shape of the amplitude of the transfer function from pitch angle to power. The measured phase shift is much faster than the simulated. Part of that is caused by the transducers. The time series simulation at low wind speeds show good agreement although especially the spatial filter needs some tuning.

Some model development will be done in the future, this includes sensor/actuator models. Also further measurement will do done in order to establish more accurate parameter values.

The papers shows that care has to be taken when designing controllers since a bad design will excite 3P more than necessary.The parameter variations show that much can be achieved when the right components are chosen. Especially the slip of the generator is important.

Future work will focus on a detailed study of controllers with emphasis on measuring the performance.

REFERENCES

[1] Lundsager, P., Christensen C.J. : 'Main Results from Risø's Wind-Diesel Programme 1984-1990' Risø-M-2906.

[2] Bindner, H : 'Active Control: The Connection Between Shaft Torque and Electrical Power' (in danish).

[3] Madsen, P.H., Frandsen, S.: 'Pitch Angle Control for Power Limitation', EWEC'89, Glasgow.

[4] Matlab User's Guide.

PROCEDURES TO DETERMINE THE FATIGUE LIFE FOR LARGE SIZE WIND TURBINES

H.J. van Grol and B.H. Bulder (Editors)
Netherlands Energy Research Foundation ECN
P.O. Box 1, 1755 ZG Petten, The Netherlands

ABSTRACT

The project "Reference procedure to establish fatigue stresses for large size wind turbines" (REFSTRESS) is performed within the framework of the European Wind Energy R&D programme JOULE-I and is carried out with the following partners:

- Centro de Investigaciones Energeticas Medioambientales y Tecnolgicas	[CIEMAT]	:	E. Soria Lascorz
- Garrad Hassan and Partners	[GH]	:	M.A. Wastling
- Netherlands Energy Research Foundation	[ECN]	:	B.H. Bulder, H.J. van Grol and C. Lindenburg
- Risø National Laboratory	[RNL]	:	J.T. Petersen
- Stork Product Engineering	[SPE]	:	M. Callanan and F. Hagg
- Technical University of Delft	[TUD]	:	D. van Delft
- Teknikgruppen AB	[TA]	:	H. Ganander
- Technical University of Denmark	[TUDk]	:	K.S. Hansen, B. Maribo Pedersen

The Tjaerborg 2 MW three bladed wind turbine at Esbjerg has been used as the reference machine. To arrive at the estimated life time a large number of steps have to be completed, viz.: establishment of design load conditions; description/modelling of wind input, in particular the stochastic part (turbulence); execution of load calculations, with proper interpretation of the criteria and on basis of sufficiently advanced aerodynamic and structural dynamic modelling; stress calculations; fatigue analyses, ending up with fatigue stress factors FSF and specific damage factor (SDF). Though not complete a large number of load cases have been considered: "production", "start & stops" and "idling" and two faulted conditions on yaw and pitch. Azimuthal binned values of load, load and stress spectra nd FSF-values have been calculated for two typical cross-sections of the rotor blade and at the front bearing of the rotor shaft. The results are presented in terms of: recommendations regarding design load conditions criteria; qualifications of methodologies, in particular of describing wind turbulence by either deterministic or stochastic methods; comparison of calculation results, on the basis of load and stress range and level crossing spectra and in particular FSF and SDF.

1. INTRODUCTION

General

The present work on "Reference procedures to establish fatigue stresses for large size wind turbines (REFSTRESS)", is part of the "Programme Joule 1989-1992; Wind Energy R&D" of the Commission of the European Communities. The Commission financed 50% of the estimate. The remainder was funded by the participating organizations and/or their national agencies.

Scope

A more substantial penetration of wind energy in the electricity production is to a large extent dependent on the ability to perform reliable design life calculations in particular of medium to large size wind turbines, viz. in the category of 0.5 ÷ 2.0 MW. This is especially true for the rotor and drive train which are designed for "safe life" and which are at the same time difficult, i.e. relatively very costly, to maintain, contrary to for instance helicopter or airplane structures which can be maintained indoor.
As meanwhile comparatively much effort has been spent on the development of national design criteria as well as on the harmonization of these criteria (see [1] up to [4] inclusive) and the determination of the maturity of models and codes for load calculations (see [5] through [8]) it is indeed appropriate to investigate the strength and weakness of design life predictions of modern large size wind turbines. The present REFSTRESS project serves as an embarkment of this investigation, attempting to cover most of the aspects from specifying load conditions down to the calculation of local stress fatigue margins; see figure I. It is further noted that from the early beginning of the work, wind farm aspects were not to be included as part of the project.

Objective

The objective of the REFSTRESS project is:
"To assess procedures and methodologies by means of which the design life of main components of wind turbines can be determined". In particular emphasis was put on:

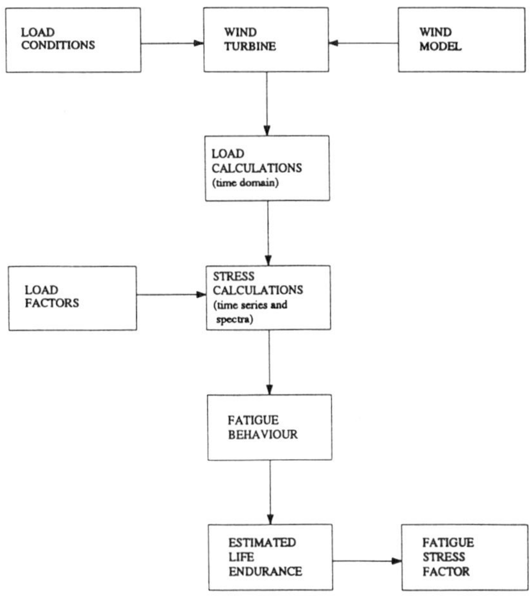

FIG. I: BREAKDOWN OF REFSTRESS ACTIVITIES

- An attempt to establish the relative importance of design load conditions, which has been specified in standards, on load and/or stress spectra;
- The effects of methodologies/procedures on load and fatigue stress spectra, provided a standard has been chosen;
- The search for an engineering measure to quantify the fatigue endurance margin.

2. WORK ORGANISATION, TASK AND OUTPUT

Work organisation

The following main types of work were distinguished and carried out:

- computational work : eight parties, see abstract;
- wind turbine data supply : one party, see section 5;
- technical management : one party, ECN.

Though there were eight separate parties, the computational work was carried out by all these parties except for TUD who joined ECN as the "joint NL party".

Tasks and outputs

The tasks and corresponding output, in which the proposed project was itemized, are summarized below.

Task I: Load conditions

- Inventory of formally specified load conditions for design life calculations.
- Selection of a limited number of, preferably nationally accepted, reference load conditions.

Task II: Wind input

- Inventory of wind models used for the fatigue design of medium to large size wind turbines.
- Selection of an appropriate wind model in order to obtain a basis for comparing results.

Task III: Wind turbine

- Selection of wind turbine.

Task IV: Load calculations

- Calculation of internal load spectra. Participants are free to perform these calculations on basis of a time and/or frequency domain methodology.
- Selection of internal reference load spectra for reasons of comparison.
- Establishing combinations of internal reference load spectra (CIRLO-spectra) to be used for stress calculations.

Task V: Stress calculations

- Stress calculations to be performed on basis of the slender beam theory approximation.
- Perform stress calculations for at least one rotor blade station and possibly for one other major component, like gear box, tower, etc.

Task VI: Fatigue damage model and design safety factor(s)

- Establish fatigue model, including fatigue data of concerned materials.
- Establish design safety factors.

Task VII: Design life analyses

- Estimation of "fatigue life" for the concerned components.

The actual execution of the work took place in accordance with the above mentioned, except for the following:

(i) The participants performed all their calculations with their own description of the wind (wind models), contrary to the intended item two of task I;

(ii) Ibid, but now with regard to the load spectra (or time series), contrary to the intended items two and three of task IV.

3. SELECTION OF WIND TURBINE

The main requirements for a candidate wind turbine were specified as:

- power : in the range of 1 to 2 MW;
- status : in operation;
- measurement : available.

Although the use of measurements was not required as mandatory in the firm proposal, it was decided to do so in order to be able to qualify the results to some extent. The survey concerned the following machines:

- DK : ESBJERG 1;
- E : AWEC-60;
- GB : HOWDEN 100/55 or WEG LS1;
- NL : NEWECS-45;
- S : MAGLARP or NASUDDEN.

In order to make a proper choice which wind turbine was most suited for the project, additional data was collected on basis of completed questionnaires, with emphasis on:

- measured quantities, in particular loads;
- operating and faulted conditions;
- accuracy of measured data;
- output formatting.

Finally it was unanimously decided to use the ESBJERG machine as the reference wind turbine for the REFSTRESS project. For the main characteristics of this wind turbine see table I.

TABLE I: MAIN CHARACTERISTICS OF ESBJERG 1 WIND TURBINE [Data supplied by TUDk]

- Rated electrical power		: 2.0 MW
- Rated shaft power		: 2.2 MW
- Synchronous rotor speed		: 21.93 rpm
- Rated power rotor speed		: 22.36 rpm
- Wind speed:	cut in	: 5.0m/s (λ_{in}= 13.96)
	rated	: 14.3m/s (λ_r= 5.0)
	cut out	: 25.0m/s (λ_{out}= 2.86)
- Number of blades		: three (upwind)
- Rotor diameter		: 61.1 m (incl. tip cap)
- Cone angle		: 0°
- Tilt angle		: 3°
- Power control		: full span´pitch
- Pitch angle:	operational	: 0-35°
	idling	: 55°
	stop	: 90°
- Hub :	height	: 61 m
	type	: rigid
	mass	: 42 ½ 10^3 kg
- Hub to tower center		: 6.76 m
- Tower diameter:	at h = 56 m	: 4.25 m
	at h = 28 m	: 4.75 m
	at h = 0 m	: 7.25 m
- Gear box:	support	: rigid
	ratio	: 68.4
- Generator:	type	: asynchronous
	speed	: 1500/1530 rpm
- Operation		: since 1988
- Frequencies:	blade flat	: 1.2 Hz (non-rotating)
	blade edge	: 2.3 Hz (non-rotating)
	tower bending	: 0.8 Hz
	shaft rotation	: 0.37 Hz (21.9/22.4 rpm)
- Blade:	material	: GFRP
	type	: tubular
	mass	: $9 \cdot 10^3$ kg (including steel hub connection)

4. PROCEDURE TO ACHIEVE THE OBJECTIVE

General

The work procedure to achieve the objective, apart from the selection of the wind turbine, was based on the following general approach:

(i) Mutual comparisons of load conditions and criteria;

(ii) Mutual comparisons of calculation results obtained by the participants;

(iii) Comparison of calculation results with measurements.

Specific

More specifically the following technical terms were chosen, along which the REFSTRESS project was carried out:

(iv) Questionnaire inquiry on design load conditions and criteria;

(v) Selection of a common standard to which the calculations should comply;

(vi) Calculated quantities to be compared with measurements should be loads or strongly related to loads, like $C_K(\lambda)$ and $C_T(\lambda)$;

(vii) Considered internal loads and stress spectra were, for reasons of comparison, of the following type:

- cumulative load range spectra;
- cumulative above level crossing peak spectra;
- cumulative below level crossing trough spectra.

(viii) Calculated and, if available, measured results refer to:
- rotor blade inner (near root: ≈9%) cross-section;
- rotor blade middle (≈51%) cross-section;
- rotor shaft cross-section at front bearing.

Note The choice of these locations was based on the following considerations:

Inner rotor blade cross-section:
* material: steel only;
* large loads;
* measured data available for comparison.

Middle rotor blade cross-section:
* material: GFRP only;
* critical fatigue load bearing capacity.

Rotor shaft at first bearing:
* important main component;
* critical fatigue load bearing capacity.

(ix) Questionnaire inquiries were performed with regard to the following items:
- wind input modelling;
- wind turbine modelling;
- applied load factors and effective fatique stress quantity.

Classification

The calculation rounds were arranged as:
A. Preliminary calculations;
B. Load calculations;
C. Fatigue calculations.

Preliminary calculations (set A)

The preliminary calculations served among others as job familiarization, in particular with respect to items such as:
- input transfer;
- output formatting;
- methods of comparison and presentation of results.

In order to reduce the number of different influences (parameters) through which results might be diffused the preliminary calculations are kept relatively straight-forward. Main assumption on the input side were:
- Load condition: "Normal-Production", limited to one wind speed interval above and one wind speed interval below V_{rated};
- Use of measured and modelled (unsteady) wind input data;
- Prescribed deterministic vertical wind shear in those cases where the wind input was simulated (modelled).

The set A calculations are, apart from job familiarization, focussed on:
- Qualification of calculated results by comparing them with measurements, through:
  * aerodynamic performance coefficients: $C_K(\lambda)$;
  * axial rotor trust coefficient: $C_T(\lambda)$;
  * pitch angle above V_{rated};
  * flat- and edgewise bending moment at rotor blade inner cross-section.
- Methodology assessment on basis of the analysis of azimuthal binned values and range, peak and trough spectra of different loads supported by wind turbine and wind model comparisons. The latter included the comparison of wind spectra as well.

Load calculations (set B)

As the actual project was aimed at the establishment of a "complete" load spectrum it took some time to find a reasonable balance between "complete" and "available budget". This was in particular the case for the following subjects:
- what standard(s) shall be applied for the load conditions;
- which load conditions shall be incorporated as generic, which shall only serve as examples and which ones shall be deleted altogether.

The outcome of the first question was that all participants should use the "Recommendation for a European Design Standard" [1] with the "Preliminary Regulations for the Certification of Wind Energy Conversion Systems" of the Germanischer Lloyd [3] as a back up. Regarding the extensiveness of the load conditions emphasis was put on the load condition "NORMAL". For the faulted conditions only an uncontrolled pitching and yawing case were to be considered as examples. Other load conditions were not considered at all. As to the main conditions assumed for the load spectra calculations, these were as follows:
- the average wind speed distribution is described by Weibull's distribution function;
- other aspects of the wind input modelling (or description) were free;
- structural flexibility and wind turbine control actions shall be accounted for.

Finally, the activities regarding the outcome of the load spectra calculations were focussed on:
- applicability and applied standard(s) on load conditions, among others by means of a questionnaire inquiry;
- determination of range, peak and trough spectra of loads in view of getting insight in the relative importance of different load conditions on the fatigue life.

Fatigue calculations (set C)

As the calculation of the local stress distribution in itself was not an objective of the REFSTRESS project, the slender beam theory was applied as being acceptable for the calculation of stresses. This is also the case for the material fatigue damage accumulation model and material fatigue properties. For this reason the linear damage summation rule of Pålmgren-Miner was applied straightforward. The set C calculations were focussed on the following items:
- stress spectra calculations in a number of pre-selected locations per cross-section;
- determination of relevant engineering fatigue quantities;
- comparison of these quantities in order to obtain:
  * a feeling for the degree of maturaty of the performed life time predictions based on the spread in calculation results;
  * insight in the relative importance of the different load conditions on the life time.

5. RESULTS ON LOAD CONDITIONS

General

(01) It was generally felt that a comparison of load spectra, merely brought about by different regulations, would show such large discrepancies, that a free choice on this matter was abandoned viz. less appropriate to cover in the REFSTRESS project.

(02) During the time that this part of the work was carried our (1991) no clear position on the application of mandatory standards could be established, even if the participants were grouped to their countries, viz.:

Denmark
There existed a national draft standard [4] used as a general approach (RNL and TUDk) but there was also preference for using the standard of the Germanische Lloyd [3], in particular by TUDk;

Spain
No formal national document existed, neither a firm position on the standard to be used (CIEMAT);

Sweden
No unique requirement was in force. The applied standard is the outcome of negotiations between manufacturers, purchasers and an approved "classification body";

The Netherlands
Official approach on basis of a national standard of CIWI [2] (SPE and ECN), but with exceptions on wind input modelling (ECN);
United Kingdom
No formal national document was in force. Preference for the standard of the Germanische Lloyd was expressed (GH).

Normal operating condition: "production"

(03) Four basic different ways were applied to describe/ model the wind input; see section 6 for further information.

(04) Lateral turbulence and misalignment gave rise to mixed up load cases. This can be avoided if the standards on load conditions more clearly require distinguished analysis for:
- stationary misalignment caused by kind of control;
- stochastic misalignment due to horizontal wind turbulence;
- misalignment as a result of improper system function (faulted condition);
- the proper act of yawing as such.

(05) Vertical turbulence and instationary vertical wind shear are not really accounted for.

(06) The information on grid failure was rather dis- appointing; ranging from no information to differences as much as a factor 5.

Normal operating conditions: "starts&stops" and "standstill"

(07) Quantification of the start & stop load spectra in terms of loads and number of cycles show differences due to:
- whether wind turbulence should be accounted for;
- the problem of the start & stops process, i.e. only at one, two or maybe three different wind speeds on the one hand or more complicated cases on the other hand, such as: start at V_1 (> V_{cut-in}), stop at $V_{cut-out}$, start at < $V_{cut-out}$ and stop at V_2 (< $V_{cut-out}$ but $\neq V_1$).

(08) Standstill calculations are mostly performed for $V > V_{cut-out}$ and sometimes only in view of strength implications and not so for fatigue.

Faulted conditions

(09) All faulted conditions are assumed to occur at $V \geq V_{rated}$.

(10) The classification of faulted conditions, and moreover load cases in general, as contributing to only strength or only fatigue, shows arbitrariness. Reason for this is the absence of a rationale on this matter. For further dis- cussion ECN proposed a recommendation; see below.

Estimated frequency of occurrence per year, υ_i	Design frequency of occurrence per design life of wind turbine	Load safety factor
$\upsilon_i<0.01$	0	not applicable
$0.01 \leq \upsilon_i \leq 1/L$ [1]	1 per case i	1.10
$\upsilon_i>1/L$	$\upsilon_i L$	... [2]
[1] L: life time	[2] According to adopted regulations	

(11) The frequency of occurrence of a failed system can be estimated by means of the "event tree" methodo- logy and "fault tree" analyses; see [7] and [9]. Only one of the participants (ECN) mentioned this possibility.

Installation and maintenance

(12) Only one participant (SPE) supplied some data on loading conditions related to installation and maintenance.

6. RESULTS OF PRELIMINARY CALCULATIONS

6.1. Summary of results on wind modelling

Turbulence modelling (approach)

Four different approaches have been used by the participants to discribe the stochastic wind behaviour viz.:
- Measured data base : CIEMAT
- Dutch Handbook Method : (ECN-CN), JNL and SPE
- Power density spectra approach : ECN, GH, RNL, SPE and TUDk
- Wind Data Matrix : TA

A survey of the turbulence characterization is given in table II.

TABLE II TURBULENCE CHARACTERIZATION

Aspect	CMT	ECN	JNL	GH	RNL	SPE	TA	TUDk
1 Applied method								
- emperical: data base	+	-	-	-	-	-	-	-
handbook	-	+[1]	+	-	-	+	-	-
data matrix	-	-	-	-	-	-	+	-
- power density spectra	-	+	-	+	+	(+)	-	+
2 Turbulence component								
- longitudinal (x)	+	+	+	+	+	+	+	+
- horizontal (y)	+[2]	+[1][2]	-	-	-	+[2]	-	-
- vertical (z)	-	-	-	-	-	-	-	-
3 Spatial dependency								
- none (flat)	n.a.	+[1]	+	n.a.	n.a.	+	n.a.	n.a.
- one dimensional (vert.)	+	n.a.	(+)	n.a.	n.a.	-	+	n.a.
- two dimensional	-	+	-	+	+	(+)	-	+
4 Turbulence statistics								
- homogeneous	n.a.	+	n.a.	+	+	(+)	n.a.	+
- stationary	n.a.	+	n.a.	+	+	(+)	n.a.	+
[1] ECN-CN [2] As misalignment								

Measured data base

The approach, applied by CIEMAT is empirical. Turbulence is not modelled but described by means of site specific 10 min time series for wind speed and direction at hub height. These data are used as coherent longitudinal and lateral wind fluctuations. As a matter of fact this method is only of limited value for design purposes.

Dutch Handbook Method [10]

The point of departure is to design by rule and not by analyses, in so far as it concerns the description of turbulence. Basically each wind speed interval of 2 m/s is divided in two main groups: one on gust duration and the other on gust amplitude. In turn each group is subdivided in three subcases. Thus per wind speed interval nine (3 * 3) deterministic gusts are prescribed in shape and frequency of occurrence. The method is empirical in the sense that it is tuned to measured results and backed-up with PDS calcula- tions. A basic disadvantage of the method is that resulting rotor blade loads on the rotor shaft are far from realistic.

Power density spectra method [11]

This method is semi empirical, based on measured power spectral densities of stochastic (turbulent) wind speeds; issued by ESDU (Engineering Sciences Data Unit). In so far the PSD method was used only the longitudinal turbulence component was modelled, however including the spatial dependency over the rotor plane. In other words also incoherent gust (eddy slicing effects of rotating rotor blades) are taken into account. This is implemented trough the spatial coherence function of the exponential type of ESDU. In practice only the following parameters need to be known for the turbulence quantification:
- average wind velocity at hub height;
- turbulence intensity at hub height;
- terrain roughness length.

The method is well suited for design calculations, viz. wind turbulence modelling. A practical shortcoming of the method is that low frequency gusts are not modelled, i.e. absent due to the low frequency cut-off of the PDS.

Wind data matrix [12]
The wind turbulence simulation developed by TA is basically an engineering approach. It is empirically directed towards obtaining essentials for the designer on gust amplitude and duration including the frequency of occurrence. The method is implemented along the so-called wind data matrix, abbreviated as WDM. Ongoing wind turbulence measurements (campaigns) at different sites constitute the data base. Presently only the longitudinal turbulence component is considered. The (spatial) turbulence distribution over the rotor plane is accounted for by non-stationary vertical one dimensional wind shear, being either symmetric with reference to the hub height or asymmetric. As a matter of fact the maturity of the method depends on the amount of collected data in the WDM, as in particular for design purposes site specifications might differ in turbulence intensities while at the same time information on low frequency gusts asks for long measuring campaigns.

Calculated wind speed range spectra
In the figs. II (case A) and III (case B) measured and calculated wind speed range spectra are depicted. Compared to the measured results fig. II (based on only one ≈ 10 min campaign) the general tendency is that all participants overestimate the turbulence,and that these quite differ mutually also. Looking to figure III again the calculated results show large differences. It can be summarized:

- The results based on measurements (CIEMAT) show an increasing absence of low cycle ranges with increasing wind speed. Evidently the used data base suffers from a lack of large gust campaigns. Note that it concerns coherent wind speed ranges;
- The DHM method (ECN & TUD and SPE) show in particular large low cycle ranges, which is conceivable in view of the inclusion of rare occurrence gusts. Also in this case the gusts are coherent;
- The results based on the PDS approach ECN, GH, RNL and TUDk) show less agreement than expected. Rather close is the agreement between ECN and TUDk (fig. III) apart at low cycles. The reason for this is that TUDk supplemented the PDS approach with "$V_{av.}$ - from - to" changes as well. Depicted wind speed ranges are point values;
- The WDM (TA) method, for which only data over the entire wind speed interval exist, seems to be unconservative, notwithstanding the fact that also in this case the wind speed ranges are of a coherent nature. This unconservatism is however not reflected in the corresponding load spectra.

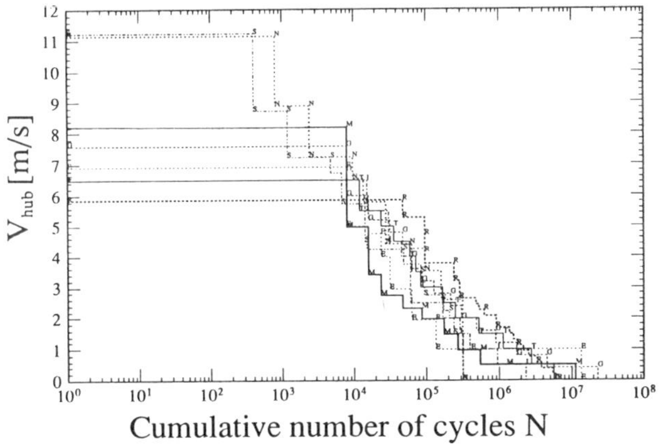

FIG. II (1/2) WIND SPEED RANGE SPECTRA PER CALENDAR YEAR AT V_{av} = 10 M/S

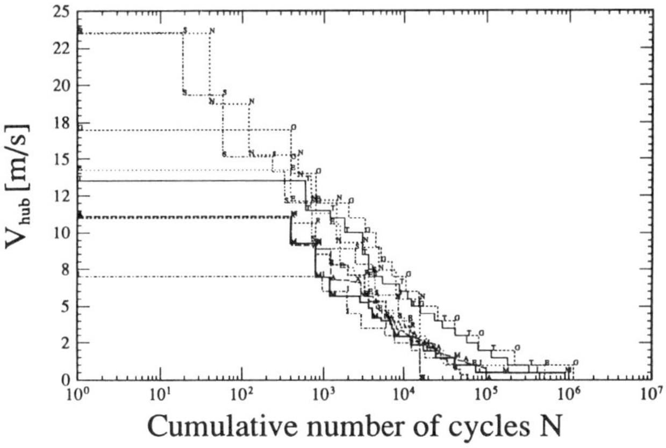

FIG. II (2/2) WIND SPEED RANGE SPECTRA PER CALENDAR YEAR AT V_{av} = 20 M/S

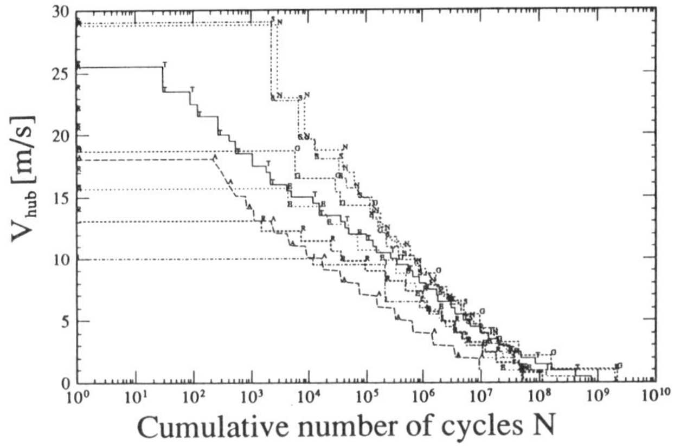

FIG. III WIND SPEED RANGE SPECTRA FOR LOADING CONDITION "PRODUCTION" PER LIFE TIME

6.2. Summary of results on wind turbine modelling

Comparing the questionnaire results with former results carried out within the framework of the "Wind Turbine Benchmark Exercise on Mechanical Loads" (WTBE-ML) [5], yields the following conclusions:

(i) The basic approach (conservation of axial and tangential momentum combined with blade element theory on uncoupled tubular flow) and modelling has not been changed since the completion of the WTBE-ML. This implies that all participants used the so-called "first generation rotor code". Or in other words no account, in terms of "routine" design calculations, has been given yet to such phenomena as:
  - three dimensional aerodynamic effects along the span of the rotor blade;
  - instationary aerodynamics, either wake effects (dynamic inflow through the non-stationary development of wakes; see [13]) or boundary layer thickness effects (delayed stall);

(ii) Contrary to what was the case during the execution of the WTBE-ML there were no participants who applied the frequency domain methodology. It is noted here that GH and RNL who also participated in the former benchmark exercise, used the frequency domain methodology at that time. As there is still some "competition" between the two methods, one may wonder why this is the case. The reason is in fact a mix of practical and theorectical considerations. For instance transients and stall effects can only be assessed on basis of the time domain methodology. Further the fact that only rotor blade loads can easier be handled by the frequency domain codes than loads on the nacelle or rotor shaft, is a plea for the time domain approach. The frequency domain approach however is a very fast tool for say sensitivity studies. In concluding the frequency domain methodology is used for scoping/pre-design studies while the time domain approach for final (accurate) design calculations.

(iii) The implementation of the number of structural deformation degrees of freedom still shows a large spread, maybe too large. Note that most participants applied the method of modal expansion. Nevertheless the calculated and measured natural frequencies showed good agreement, see table III.

TABLE III CALCULATED AND MEASURED NATURAL FREQUENCIES

Rotor blade	FLATWISE[1)]		EDGEWISE[1)]	
Participants	Ground mode[2)]	First harmonic[2)]	Ground mode[2)]	First harmonic[2)]
CIEMAT[3)]	1.18 (3.2 P)	3.48 (9.4P)	2.30 (6.2 P)	7.74 (20 P)
ECN	1.25 (3.4 P)	3.44 (9.3P)	2.45 (6.6 P)	8.20 (22 P)
GH	1.28 (3.5 P)	3.47 (9.4P)	2.33 (6.3 P)	-
RNL	1.31 (3.6 P)	-	2.31 (6.3 P)	-
SPE	1.38 (3.7 P)	-	-	-
TUDk	1.28 (3.5 P)	3.52 (9.5P)	2.34 (6.3 P)	-
TA[3)]	1.16 (3.1 P)	3.27 (8.9P)´	2.32 (6.3 P)	-
Measured[3)]	1.17 (3.2 P)	-	2.30 (6.2 P)	-
Tower	Measured: 0.81 Hz (2.2 P)			

[1)] Values in Hz; in parenthesis relative to rotational frequency (= 0.369 Hz)
[2)] Rotating value, unless stated otherwise
[3)] Non-rotating value

6.3. Aerodynamic power coefficient, axial rotor trust coefficient and rotor blade pitch angle

Aerodynamic power coefficient (see fig. IV)
By definition the agreement above V_{rated} (below λ_{rated}) must be perfect as $C_k(\lambda)$ follows from:

$$C_k(\lambda) = 3.36(\lambda/10)^3 \quad \left(= \frac{P_{aer.}(=2.18\ 10^6)}{\tfrac{1}{2}\rho(\Omega_r R)^3 A}\lambda^3 \right)$$

The spread in the results below V_{rated} (±8% in $C_k(\lambda)$ at λ = 12) must be expected to increase for increasing values of λ, because of the net results of the competing quantities $C_l \sin\varphi$ versus $C_d \cos\varphi$ and the increasing influence of the use of different turbulent wake state models; see also [5].
The poor agreement between calculated and measured C_k values is likely to be attributed to the increasing importance of inaccurately measured wind speeds and drive train losses.

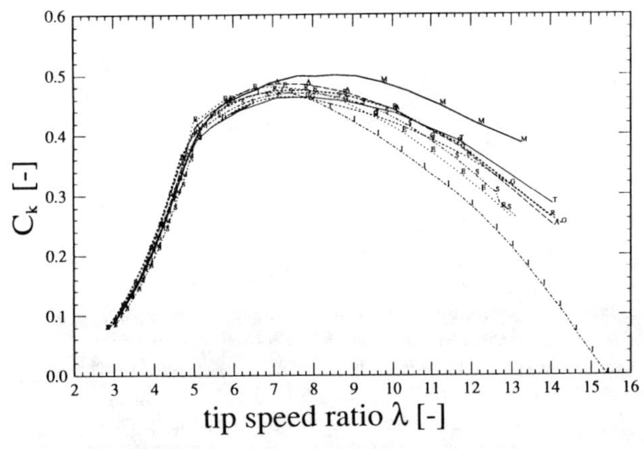

FIG. IV (1/2) AERODYNAMIC POWER COEFFICIENT

Axial rotor trust coefficient (see fig. IV)
For the same reason as mentioned above the agreement at low values of λ should be good and that at below V_{rated} less. From elementary rotor aerodynamics, $C_k = 4a(1-a)^2$ and $C_T = 4a(1-a)$ it follows that: $C_T \approx C_k\{1-\tfrac{1}{4}C_k(1+\tfrac{1}{2}C_k)\}^{-1}$. The calculated values are higher than the estimated because of the assumption that $a^1 = 0$ and $C_d = 0$.

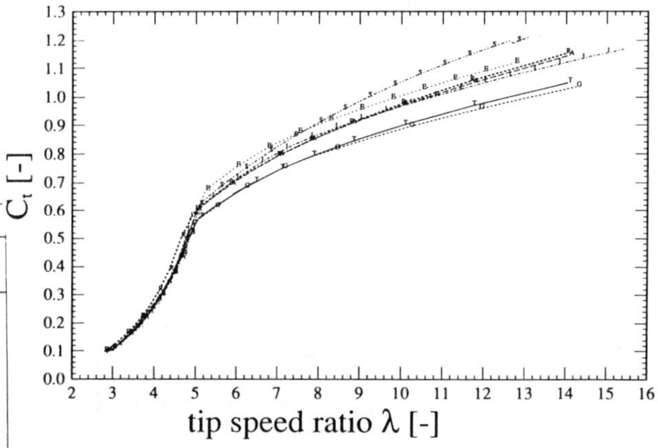

FIG. IV (2/2) AXIAL ROTOR TRUST COEFFICIENT

Rotor blade pitch angle (see fig. V)
The agreement between measured and calculated results is good: the spread in the pitch angle is ±1° approxi-mately. This might be a good reason for the manufacturer always to be able to adjust the pitch angle to ±1½ deg at the approval test. The decrease of the pitch angle with an increasing tip speed ratio is conceivable. However the rather complete linear dependency is not evident/straightforward.

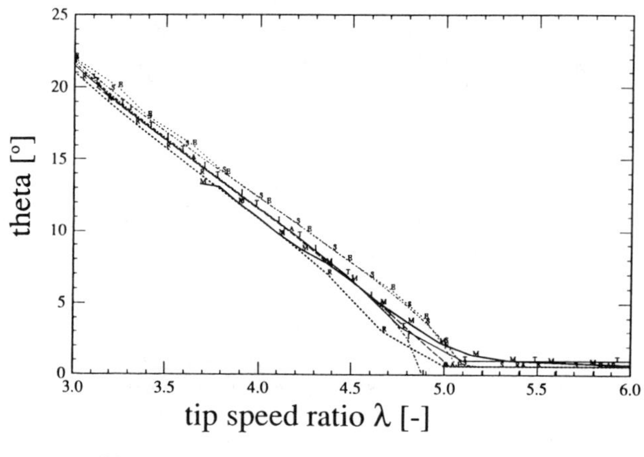

FIG. V PITCH ANGLE

6.4. Typical measured and calculated results

Electrical power range spectra below V_{rated} (fig. VI)
Differences between calculated and measurd results can be traced back as due to differences in the wind input; see fig. II. But the disagreement is striking.

Flatwise moment, binned and range spectra, below V_{rated}
The agreement, see fig. VII, is reasonable in particular if normalized to the same average wind speed. Within a chosen wind speed interval the average design wind speed may be different due to regulatory requirements. The spread in the averaged normalized flatwise moment is about ±6%, while that in the peak value (= minimum-average) is much larger: ranging from 109 kNm (lowest value of CIEMAT) and 145 kNm (highest value of ECN), or in other words showing a spread of ±15%.

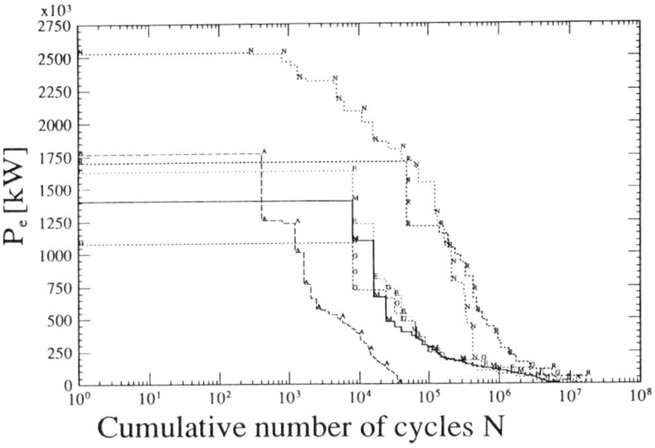

FIG. VI ELECTRICAL POWER RANGE SPECTRA PER DALENDAR YEAR AT V_{av} = 10 M/S

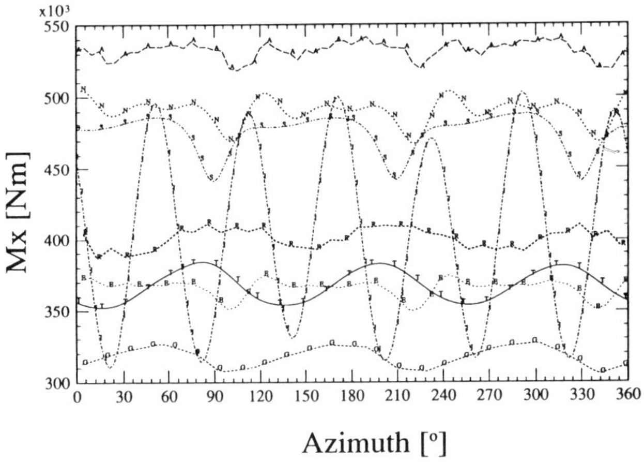

FIG. VIII (1/2) ROTOR SHAFT TORQUE AZIMUTHAL BINNED VALUES AT V_{av} = 10 M/S

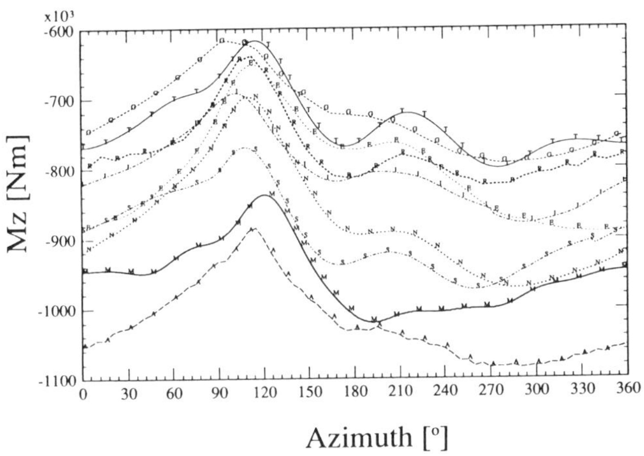

FIG. VII (1/2) FLATWISE MOMENT AZIMUTHAL BINNED VALUES AT V_{av} = M/S

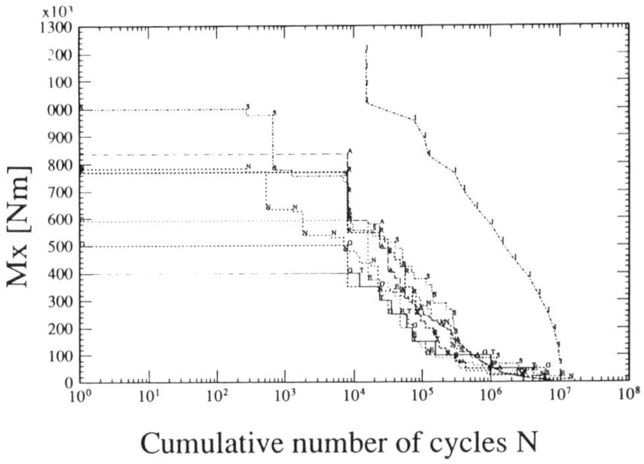

FIG. VIII (2/2) ROTOR SHAFT TORQUE RANGE SPECTRA; PER CAL.YR AT V_{av} = 10 M/S

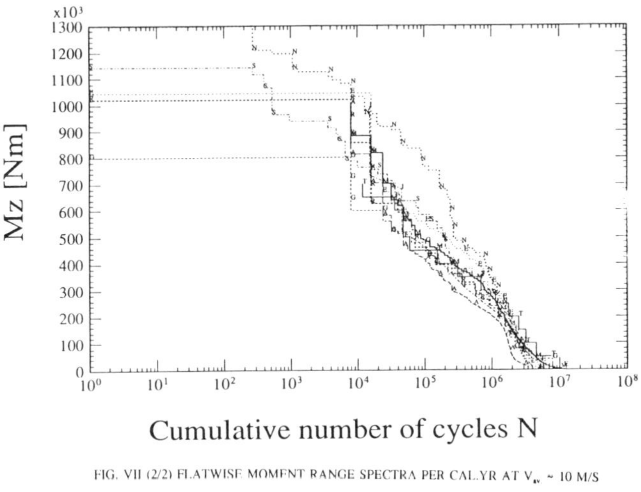

FIG. VII (2/2) FLATWISE MOMENT RANGE SPECTRA PER CAL.YR AT V_{av} ~ 10 M/S

Rotor shaft torque, binned and range spectra, below V_{rated} (fig. VIII)
The agreement is rather poor. The spread in the normalized binned average is about ±15%; ranging from 318 kNm (lowest value of GH) and 423 kNm (highest value of TA). Apart from these extremes the others comply with the spread of ±5% in $C_k(\lambda)$. The shape of the curves of the binned values clearly reflects the differences in drive train modelling; CIEMAT did not included the generator slip and RNL, TA and TUDk included one DOF for drive train flexibility (ECN did so for case B but not for A). The torque range spectra show large differences in particular at low cycles.

Pitch angle range spectra, above V_{rated} (fig. IX)
All calculated pitch angle responses strongly overpredict the actual wind turbine behaviour, which is rougly due to the corresponding modelled wind input (fig. II). Nevertheless the difference between JNL and SPE are mainly due to different dynamic control modelling.

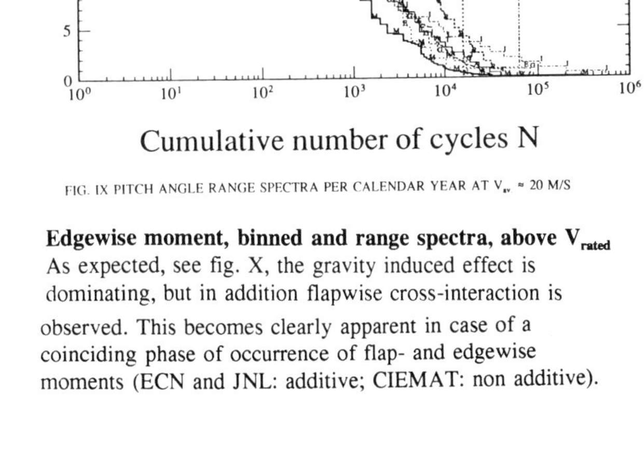

FIG. IX PITCH ANGLE RANGE SPECTRA PER CALENDAR YEAR AT V_{av} = 20 M/S

Edgewise moment, binned and range spectra, above V_{rated}
As expected, see fig. X, the gravity induced effect is dominating, but in addition flapwise cross-interaction is observed. This becomes clearly apparent in case of a coinciding phase of occurrence of flap- and edgewise moments (ECN and JNL: additive; CIEMAT: non additive).

This driblet-effect is also reflected in the edgewise range spectrum. The reason for this effect is that ECN and JNL did not include structural damping in their calculations.

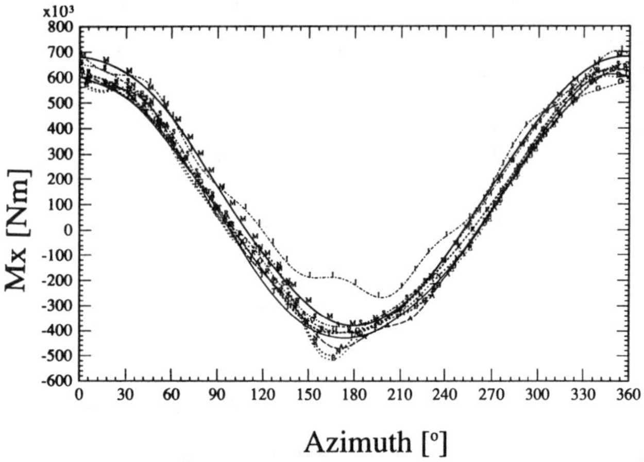

FIG. X (1/2) EDGEWISE MOMENT AZIMUTHAL BINNED VALUES AT $V_{av.}$ = 20 M/S

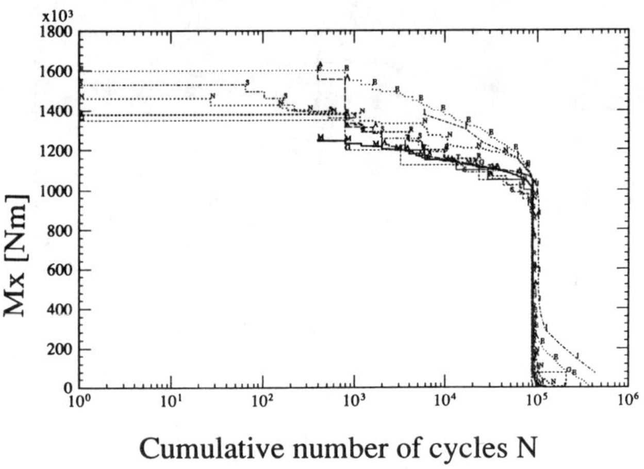

FIG. X (2/2) EDGEWISE MOMENT RANGE SPECTRA PER CALENDAR YEAR AT V_{rated} = 20 M/S

7. LOAD CALCULATIONS

General

Referring to section 4 under "load calculations" it suffices to mention that only a limited set of load conditions was considered. Thus the complete set of load conditions so as discussed in 5, was deliberated put aside. Applied were:

- Class I : "Normal operating conditions", with the subclasses: production, starts/stops and idling.
- Class II : "Faulted conditions", only for the purpose of demonstration: faulted pitch and faulted yaw.

The set of B calculations were structured as follows:

B.01 : Load condition I, production : $V_H < V_{rated}$;
B.02 : Load condition I, production : $V_H \geq V_{rated}$;
B.03 : Load condition I, start & stop : $V_H < V_{rated}$;
B.04 : Load condition I, start & stop : $V_H \geq V_{rated}$;
B.05 : Load condition I, idling : $V_H < V_{in}$;
B.06 : Load condition I, idling : $V_H > V_{out}$;
B.07 : Combination of (B.01) and (B.02);
B.08 : Combination of (B.01) up to and including (B.06);
B.09 : Load condition II, uncontrolled pitching;
B.10 : Load condition II, uncontrolled yawing;
B.11 : Combination of (B.08), (B.09) and (B.10).

The calculation matrix is given below as table IV.

TABLE IV:CALCULATION MATRIX FOR LOAD CONDITIONS (case B)

LOAD CONDITIONS			Rotor blade		Rotor shaft[3] cross-section
			Inner[1] cross-section	Outer[2] cross-section	
I	Normal	Production Start/stop Idling (parking) Standstill	x x x not considered	x x x not considered	x x x not considered
II	Faulted Conditions	Uncontr. pitching Uncontr. yawing Others	x x not considered	x x not considered	x x not considered
III	Transport, Installation, Maintenance, etc.		not considered	not considered	not considered
IV	Environmental Conditions (extremes)	Ice Birds Snow Other Hail	not considered	not considered	not considered

1) Inner cross-section [$RBCS_i$] : material: steel only r = 2,75 [m]
2) Outer cross-section [$RBCS_o$] : material GFRP only r = 15,46 [m]
3) Rotor shaft cross-section [RSCS] : material steel only and assumed to be positioned at foremost bearing (2.44 m behind rotor center).

Results

All participants were asked to classify the components in order to establish the partial safety factors for loads (PLSF). According to [1] the following classification is applicable:
- Good access & high failure consq.: class 1; PLSF = 1.35;
- Poor access & high failure consq. : class 2; PLSF = 1.20.

All participants responded (RNL and TUD excepted). They classified all three locations as class 1, however with one exception (TUDk) who classified $RBCS_i$ and RSCS as class 2. This indicates that the regulatory requirements on this matter are in need of a better specification of what is ment by good/poor accessibility.
The internal loads were all calculated as time series for each load case B. Both cases B.01 and B.02 were subsequently subdivided in 2 m/s wind speed intervals, except for TA. Nc further results are given here in view of the space needed therefore; see for the relative importance of methods and load cases on fatigue: the following section.

8. FATIGUE CALCULATIONS

General

For the basic approach see section 4 under "fatigue calculations". In addition the following is mentioned. The material fatigue data were supplied by TUDk. It suffices here to note that for $RBCS_i$, $RBCS_o$-shell and RSCS the number of cycles to failure (N) is only a function of the stress/strain range while for $RBCS_o$-spars both the mean and amplitude come into play. Concerning the effective fatigue quantity it turned out that all participants (no information from RNL and TUD) used the cross-sectional normal stress/strain (for steel and GRP) with the exception that for RSCS both CIEMAT and ECN used the von Mises stress equivalent (for ECN: $\sigma_{eff} = SC(\sigma^2+3\sigma^2)^{1/2}$ in which: $SC \equiv \sigma/|\sigma|$).

Fatigue stress factor

If the calculated life time of a component of a wind turbine differs from the design life time the question arises whether this difference is straightforward to interpret. This is not the case as the problem of estimating the life time is not a linear problem. Or to put things otherwise if two load or stress spectra intersect each other it is generally not immediately clear which spectra is most severe. In order to overcome this problem there are different modes of comparisons, viz.:
- life time factor;
- fatigue stress factor (FSF);
- equivalent one cycle per revolution fatigue stress;
- equivalent one cycle stress.

The use of the FSF-factor (proposed by ECN) proved to be a very effective engineering quantity. It is defined as the multiplication factor of the fatigue stress so that $L_{calc.} = L_{des.}$; thus:

$$L_{des}\sum \frac{n_{i,yr.}}{N(FSF\,\sigma_{f,i})} = 1$$

Note that if N is a function of $\sigma_{av.}$ and $\Delta\sigma_f$ than two FR-factors might be introduced one with regard to $\sigma_{av.}$ and the other with regard to $\Delta\sigma_f$, viz.:

$$L_{des}\sum \frac{n_{i.yr.}}{N(\overline{FSF}\sigma_{av,i},\ FSF\Delta\sigma_{f,i})} = 1$$

The fatigue stress factor is an effective engineering quantity to evaluate the amount of conservatism in terms of material thicknesses or errors in either the prevailing stress or the number of stress cycles. For thin shells for instance the FSF-value is a direct measure for the shell thickness margin. For example FSF = 1.30 means that the shell thickness can be reduced by about 30%. For a shaft however the FSF value is proportional to $diam.^3$; thus FSF = 1.30 implies that the diamter can be reduced by at most 10%.

Results

The calculated FSF at different locations for three concerned cross-sections are compiled in the tables V through VII. From these data it appears that the lowest value of the FSF show the following spread:

- rotor inner cross-section : ≈ 1.77 to 2.98
- rotor outer cross-section : ≈ 0.86 to 1.69
- rotor shaft : ≈ 0.83 to 1.20

TABLE V: FSF for rotor blade outer cross-section (material GFRP)

Location	IERC	GH	ECN	JNL	RNL	SPE	TA	TUDk
1	1.74	5.27	4.34	2.18	4.51	2.58	3.85	4.76
2	3.91	3.49	2.94	2.23	2.82	2.55	2.68	3.66
3	1.88	1.73	1.43	1.07	1.33	1.17	1.19	1.80
4	1.49	1.65	1.36	0.96	1.21	1.02	1.13	1.70
5 *	1.22	1.77	1.40	0.94	1.28	0.90	1.16	1.78
6	1.48	3.27	2.50	1.45	2.56	1.52	2.12	3.17
7	1.10	1.73	1.39	0.86	1.68	1.33	1.35	1.69
8	1.24	1.74	1.40	1.06	1.30	1.10	1.18	1.79
9	1.08	1.72	1.39	0.97	1.26	1.01	1.14	1.75
10	1.08	2.21	1.72	1.03	1.54	1.06	1.35	2.10

TABLE VI: FSF for rotor blade inner cross-section (material steel)

Location	IERC	GH	ECN	JNL	RNL	SPE	TA	TUDk
1	2.87	2.64	3.15	2.62	3.80	3.82	4.29	2.98
2	3.85	3.72	1.79	3.30	2.54	3.11	3.85	4.04
3	5.25	5.92	1.85	3.68	1.94	2.64	4.95	6.29
4 *	3.38	3.29	2.29	2.69	2.11	2.32	3.44	3.63
5	3.97	2.64	3.00	2.59	4.15	2.90	2.87	2.99
6	3.86	3.75	1.77	3.25	2.50	3.01	3.89	4.06
7	5.25	6.23	1.88	3.86	1.91	2.61	5.32	6.09
8	3.37	3.32	2.29	2.75	2.10	2.28	3.56	3.60

TABLE VII: FSF for rotor blade inner cross-section (material steel)

Location	IERC	GH	ECN	JNL	RNL	SPE	TA	TUDk
1	0.91	0.83	1.08	?	0.99	1.21	0.89	0.87
2	0.92	0.83	1.07	?	0.91	1.21	0.89	0.88
3 *	0.92	0.83	1.10	?	0.86	1.20	0.89	0.87
4	0.92	0.83	1.10	?	0.88	1.22	0.89	0.88
5	0.92	0.84	1.10	?	0.98	1.20	0.89	0.88

As a preliminary conclusion it may be said that the spread in FSF for the outer RBCS is too large (FSF = 1.3 ± 30%) and for both other cross-sections reasonable (both about ± 25%) and in anyway not larger than the margin implied by the application of the safety factor of 1.35. As an example of the relative importance of the different load cases and of the relative importance of damage related to energy production (SDF values) see below and [14] respectively.

* For location see end of paper

ROTOR BLADE INNER CROSS-SECTION; LOCATION 6											
Case	B.01	B.02	B.03	B.04	B.05	B.06	B.07	B.08	B.09	B.10	B.11
FRF	1.90	2.30	5.12	6.15	3.33	9.24	1.81	1.79	3.16	3.81	1.77

9. CONCLUSIONS AND RECOMMENDATIONS.

As the REFSTRESS project is not yet finished the following conclusions are of a preliminary nature.

(01) It is felt that if present available certification rules are applied to one and the same wind turbine the outcome of the estimated life will differ too much.

(02) The firm adoption of one European standard is emphasized; the use of a recommended standard is discouraged. The European standard might be an adopted existing standard, supplemented with specific national/site requirements.

(03) The state of the art of load spectra calculations regarding the normal load condition "production" is more developed than that of start & stops, idling and/or parking and certainly that of faulted and extreme conditions. In particular for the faulted conditions the event tree and fault tree analyses can be used, but were hardly mentioned.

(04) A clear rationale shall be adopted on whether a load case is to be included in fatigue calculations or not. A recommendation has been given as a start off.

(05) In order to reduce the change that a wrong (too low) value of the partial load safety factor is chosen, the regulatory requirements on the matter "poor versus good accessibility" must be improved.

(06) Compared to the maturity by means of which cross-sectional loads can be calculated, that regarding the modelling/description of turbulence seems to be lacking, in particular in the low cycle wind speed ranges. In fact three clearly different wind input approaches were applied, viz.: that based on the power density spectrum of the turbulence, a deterministic approach (Dutch Handbook Method) and finally that based on a continuous "growing" wind data matrix. Presently it is to early to draw firm conclusions on the best approach, though that based on PDS in itself has the best potential for further developments.

(07) In so far as measured data were available all calculated results were conservative, for example: C_k (λ) showing better performance than calculated; edgewise and flapwise range spectra showing smaller and for less frequent loads; ibid for pitch angle range and electrical power range spectra;

(08) Calculated natural frequencies show good agreement mutually as well as with measured data. Nevertheless dynamic responses of in particular the edgewise moment and rotorshaft torque may differ too much.

(09) Presently the normal cross-sectional stresses (strain) is used as the effective fatigue stress (strain). Hardly any account is given to the effect of shear stresses.

(10) During the execution of the project two engineering design quantities have been introduced, viz. the so-called fatigue stress factor (FSF) being directly related to the geometry of the concerned component. For example: FSF = 1.35 for a shell location means that the local shell thickness can be reduced by ≈ 30%.

(11) The overall spread in the minimum FSF values was large if not too large: ± 25% to ± 30%.

(12) For the considered loading conditions that of production below V_{rated} is most damaging and that of above V_{rated} thereafter. the faulted conditions on yaw and pitch are about of equal importance as that of "idling" below $V_{cut\text{-}in}$.

(13) In summary from the designers' point of view the "Reference Procedure to Establish Fatigue Stresses for Large Size Wind Turbines" shall be based on analyses. The analysis shall be structured on basis of load conditions (as is presently the case). The amount of detail of the calculated load spectra can best be validated by means of the so-called fatigue stress factor. For values close to one (1.0 ± 0.1) it must be considered whether sensitivity calculations have to be carried out.

10. REFERENCES

[1] Van Hulle, F.J.L. (editor) Recommendation for a European Wind Turbine Design Standard; Load cases and loads. Commission of the European Community Directorate General of Energy DG XVII. Final report, April 1991

[2] Stam, W.J. (editor) Regulations for the Type-Certification of Wind Turbines; Technical Criteria. ECN--91-001, February 1991

[3] Preliminary Regulations for the Certification of Wind Energy Conversion Systems. Germanischer Lloyd, July 1989

[4] Technical Basis for Type Licencing and Certification of Wind Turbines in Denmark (draft in Danish). Ministery of Energy, January 1991

[5] Van Grol, H.J., Snel, H., and Schepers, J.G. Wind Turbine Benchmark Exercise on Mechanical Loads; A state of the art report. ECN--91-030: Volume I (part A); Main body of the report, ECN--91-031: Volume I (part B); Annexes, Tables and Figures

[6] De Vries, O. Fluid Dynamic Aspects of Wind Energy Conversion AGARD-AG-242; July 1979

[7] Van Grol, H.J. Codes and Models in the Design of Horizontal Axis Wind Turbines. Euroforum New Energies; Proc. of an Int. Congress held at Saarbrücken, Germany, 24 ÷ 28 October 1988

[8] Matthies, H.G., Meier, H. and Schulz, H. Comparison of Standards for Wind Turbines: Overview. Euroforum New Energies; Proc. of an Int. Congress held at Saarbrücken, Germany, 24 ÷ 28 October 1988

[9] Rademakers, L.W.M.M., Seebregts, A.J. and van den Horn, B.A. Reliability in Wind Engineering. Eur.Com.Wind Energy Conf. and Exhibition: 8 ÷ 12 March 1993 (Travemünde, Germany).

[10] Verheij F.J., Föllings F.J., and Curvers A.P.W.M. Handboek Wind Data for Wind Turbine Design, version 3. TNO Institute for Environmental and Energy Technology, Apeldoorn, The Netherlands, January 1992.

[11] Veers, Paul, S. Three-Dimensional Wind Simulation. Sandia National Laboratory Report: SAND 88-0152. March 1988.

[12] Ganander, H. Calculation of Total Load Spectrum and Component Dimensions of WECS based on Wind Matrix and Simulation. Teknikgruppen AB (Sweden), IEA Expert Meeting; March 7th and 8th, 1991, Stockholm (Sweden).

[13] Snel, H. and Schepers, J.G. Investigation and Modelling of Dynamic Inflow. Eur.Com.Wind Energy Conf. and Exhibition: 8 ÷ 12 March 1993 (Travemünde, Germany).

[14] van Grol, H.J. and Bulder B.H. (editor) Reference Procedure to Establish Fatigue stresses for Large Size Wind Turbines "A start of the art report" ECN report (to be issued summer 1993).

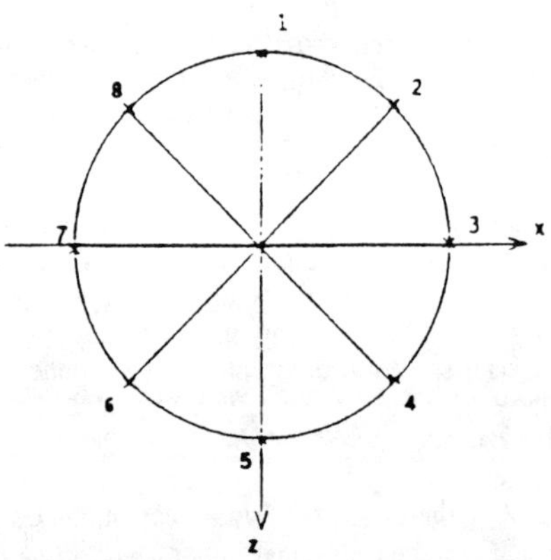

Rotor blade inner cross section

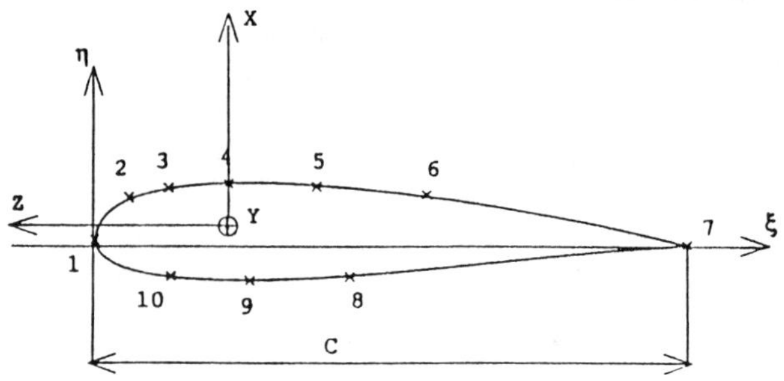

Rotor blade outer cross section

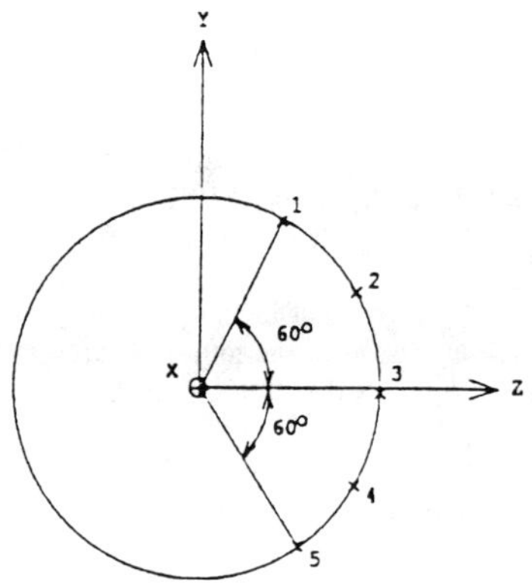

Rotor shaft at first bearing

MODELING A WIND TURBINE IN A TURBULENT WIND

Y. Sheinman and A. Rosen
Faculty of Aerospace Engineering
Technion - Israel Institute of Technology
Haifa 32000, Israel

ABSTRACT: Wind turbulence affects the wind turbine energy production and causes power fluctuations. During recent years the authors have been involved in a research on both aspects. Theoretical models to describe the influence are presented. These models exhibit good correlation with measured data. The models are easy to use and as an input they need the kind of data which is usually collected in most of the sites.

1. INTRODUCTION

Most of the performance analyses of wind turbines are based on averaging the wind behavior over relatively long periods of time (ten minutes or an hour are commonly used). Usually these analyses do not take into account the turbulent nature of the wind. The influence of turbulence has two main aspects, it affects the energy production and it causes power fluctuations.

Various researchers have pointed out the important influence that turbulence may have on the turbine performance. Different approaches have been presented in order to account for this influence. These approaches can be divided into two main groups:

(a) A detailed simulation of the wind turbine dynamic behavior when exposed to a turbulent wind. In this case a certain wind history is used as an input to the model. The turbine response is calculated by a tedious integration (with respect to time) of the system equations of motion. This approach requires a detailed information (high sampling rate) on the wind behavior at the site (wind speed and direction as functions of time), over long periods of time. This kind of information is rarely available since it requires complicated and expensive data acquisition techniques. Furthermore, the computations usually need large computing times.

(b) Simplified approaches that somehow account for the dynamic nature of the wind but do not take into account the dynamic characteristics of the turbine. These approaches are not accurate enough since it is the interaction between the wind and turbine dynamic characteristics that determines the system performance.

During the last few years the authors of the present paper have been involved in a continuous effort [1-6] to develop a new model for predicting the performance of a certain turbine at a certain site. This model presents an intermediate approach between the simple and complicated methods that were described above. While the dynamic characteristics of the turbine and wind, as well as their interaction, are taken into account, the input to the model includes integral information which is available on most sites.

The initial results of the new model were presented at the Amsterdam European Wind Energy Conference in October 1991 [4]. Other parts of the research are in various stages of publication (1-3, 5,6]. The purpose of the present paper is to describe briefly the progress of the research since the presentation of the initial results.

The paper is divided into two main parts. The first part discusses the influence of yaw errors on the average power and energy production of a turbine. The second part discusses the power fluctuations as a result of turbulence. The paper ends with conclusions which are based on the present results.

2. THE INFLUENCE OF YAW ERRORS ON THE PERFORMANCE

During the first stage of the research [3,4] it was assumed that the incoming wind direction coincides with the shaft direction, namely the yaw error is always zero. This assumption was adopted since simulations [1,2] confirmed previous indications in the literature that for turbines with active yaw the influence of yaw errors on the energy production is relatively small. Yet, in an attempt to increase the accuracy of the calculations the influence of yaw errors on the energy capture was recently added to the model.

In the past, most of the research on yaw error and yaw motion concentrated on their influence on the loads that act on the rotor (see for example [7,8]). When it comes to performance most of the analyses adopted the "cosine law" [9,10] which states that the output power is proportional to the third power of the cosine of the yaw error. Glasgow [11,12] investigated the yaw influence and concluded that the test data had indicated only fair agreement using the cosine correction to account for yaw effects on rotor power.

Detailed description of the present modeling of yaw errors' influence on performance appears in [5], here only a brief description will be presented.

In a similar manner to the analysis of the influence of variations in the wind speed, three different models of the influence of yaw-errors were considered [5]:

a) A "static" approach where it is assumed that the turbine characteristics (specifically efficiency) are not influenced by the yaw error, which is assumed to be small. This approach is analogous in fact to the cosine law.

b) A quasi-steady approach where dynamic variations of the wind speed component normal to the rotor disk, due to yaw errors, are taken into account, but the dynamic characteristics of the turbine are ignored.

c) An "accurate" dynamic simulation [1,2] where the dynamic characteristics of the turbine and their coupling with the dynamic variations in the wind direction, are taken into account.

Detailed investigation of the case of harmonic variations of the yaw error (assuming constant wind speed) indicated that the "static" analysis results in increasing errors in comparison with the "accurate" dynamic simulation. On the other hand, for practical purposes the differences between the results of the simulation and the quasi-steady approach are negligible. Thus the more simple quasi-steady approach can be used.

If $\delta_w(t)$ is the momentary value of the yaw error (see Fig. 1), then its average value, δ_{wc}, and standard deviation, σ_δ, are defined as:

$$\delta_{wc} = \frac{1}{T_a}\int_0^{T_a} \delta_w(t)\, dt \tag{1}$$

$$\sigma_\delta = \left\{\frac{1}{T_a}\int_0^{T_a}\left[\delta_w(t) - \delta_{wc}\right]^2 dt\right\}^{1/2} \tag{2}$$

where T_a is the averaging period of time.

The average output power according to the quasi- steady approach, in the case of a constant wind speed with arbitrary small yaw errors, is [5]:

$$\underline{P}_{qs}(\underline{V},\delta_{wc},\sigma_\delta) = \frac{1}{2}\rho A_r \underline{V}^3 C_p(\underline{V}) \left[1 - \frac{1}{2}\frac{C_{pP}(\underline{V})}{C_p(\underline{V})}\left[\sigma_\delta^2 + \delta_{wc}^2\right]\right] \tag{3}$$

where ρ is the air mass density, A_r is the rotor area, $\underline{V}$ the (constant) wind speed and $C_p(\underline{V})$ is the turbine power coefficient which is a function of the wind speed $\underline{V}$. Various functions which are associatted with $C_p(\underline{V})$ are defined below:

$$\begin{aligned} C_{pP}(\underline{V}) &= 3C_p(\underline{V}) + C_{p1}(\underline{V}) \\ C_{pM}(\underline{V}) &= 3C_p(\underline{V}) + 3C_{p1}(\underline{V}) + C_{p2}(\underline{V}) \\ C_{p1}(\underline{V}) &= C'_p(\underline{V})\,\underline{V} \\ C_{p2}(\underline{V}) &= C''_p(\underline{V})\,\underline{V}^2 \end{aligned} \tag{4}$$

An upper prime ( )' indicates differentiation with respect to $\underline{V}$.

In the general case there are arbitrary variations of the wind speed, as well as arbitrary variations in its direction relative to the rotor shaft. The wind speed V(t) is commonly described as:

$$V(t) = \underline{V} + u \tag{5}$$

where $\underline{V}$ is the average wind speed along T_a and u is the gust speed. The direction of the wind velocity relative to the rotor shaft projection onto a horizontal plane ($\tilde{z}_H$), is defined by the "shear" angle δ_s and the yaw error δ_w (see Fig. 1). The average output power along the time period T_a, in a "real" wind, will be (for more details see [5]):

$$\underline{P}_c(\underline{V},S_u,\delta_{wc},\delta_{sc},\sigma_\delta,\sigma_{\delta s}) = \frac{1}{2}\rho A\underline{V}^3 C_p(\underline{V}) \cdot \left[1 + I^2_{v,mod1}\frac{C_{pM}}{C_p} - \frac{1}{2}\frac{C_{pP}}{C_p}\left(\delta_{wc}^2 + \sigma_\delta^2\right) - \frac{1}{2}\frac{C_{pP}}{C_p}\left(\delta_{sc}^2 + \sigma_{\delta s}^2\right)\right] \tag{6}$$

The second term in the square brackets represents the influence of variations in the wind speed.

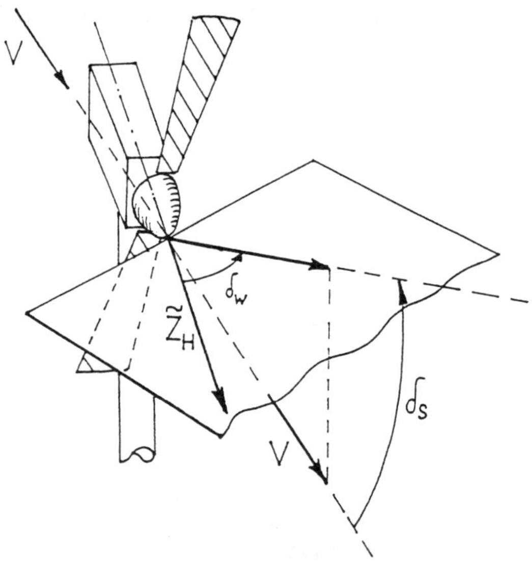

Fig. 1: Definition of the yaw and shear angles.

$I_{v,mod1}$ is the modified intensity of turbulence [3,4] which is a function of the average wind speed, the Power Spectral Density (PSD) function of the wind speed on the site $S_u(f)$, and the dynamic characteristics of the turbine. δ_{sc} and $\sigma_{\delta s}$ are the average and standard deviation, respectively, of the shear angle along the averaging period.

The calculation of the energy production of a certain turbine at a certain site was described in [4]. In order to account for the influence of yaw errors, the amount of energy which is lost due to these errors, $E_{\delta w}$, is defined:

$$E_{\delta w} = H\sum_{i=1}^{N} f(V_i)\, P_s(V_i) \sum_{k=1}^{L} f_{L,i}(\delta_{w,k})\, F(V_i,\delta_{w,k},\sigma_\delta) \tag{7}$$

H is the period of time (usually in hours) along which the energy production is calculated. N bins of average wind speed are defined. V_i is the representative wind speed of the ith bin. The frequency of occurrence of that bin is $f(V_i)$. Each wind speed bin is divided into L sub-bins. Each sub-bin represents a certain range of yaw errors. $\delta_{w,k}$ is the representative value of the yaw error of the kth bin. $f_{L,i}(\delta_{w,k})$ is the frequency of occurrence of yaw error values that belong to the kth sub-bin of yaw errors (of the ith speed bin). The function F in Eq. (7) is defined as [5]:

$$F(V_i,\delta_{w,k},\sigma_\delta) = -\frac{C_{pP}(V_i)}{C_p(V_i)}\left[\sigma_\delta^2(V_i,k) + \delta_{w,k}^2\right] \tag{8}$$

In order to validate the model use is made of a test data on the performance of a Vestas V25 wind turbine which is located at the Beit Yatir site. Detailed information was gathered along 153 hours and used previously [1-4] for validation purposes. Because of problems of measuring the wind direction, only 107 hours of the entire data base are used in order to calculate the influence of yaw errors.

Along these 107 hours the turbine actually produced 7988 kWhr of electric energy. If energy production is calculated while neglecting the yaw errors but taking into account the dynamic wind speed variations and their interaction with the turbine dynamic characteristics (E_c, see [4]), then an amount of 8199 kWhr is obtained. This presents an overestimation of 2.5% compared to the measured amount. If the yaw error is taken into account then the energy loss due to yaw-errors, $E_{\delta w}$, is equal to 161 kWhr, which is 2% of the total measured energy. Thus the predicted energy production (8038 kWhr) is only 0.6% higher than the amount of energy that was actually measured.

These results present the behavior of a certain turbine at a certain site, for a relatively short period of time. Yet it seems that they indicate that the yaw error influences on the turbine energy production are much smaller than the influences of variations in the wind speed. Nevertheless, taking the yaw errors into account while predicting the energy production seems to increase the accuracy of the results. Since these calculations do not cause any major complication in the analysis or data acquisition technique, it is desirable to include these influences in the calculations.

3. POWER FLUCTUATIONS

Turbulence results in power fluctuations. Power fluctuations are related to the quality of the output power and the fatigue life of the turbine system. The intensity of these power fluctuations, I_p, is defined as follows:

$$I_p = \sigma_p / \underline{P} \qquad (9)$$

where $\underline{P}$ is the average output power along a time period T_a, and the power standard deviation, σ_p, equals:

$$\sigma_p = \left\{ \frac{1}{T_a} \int_0^{T_a} \left[P(t) - \underline{P} \right]^2 dt \right\}^{1/2} \qquad (10)$$

It is expected that the intensity of power fluctuations will be a function of the intensity of turbulence, I_v,

$$I_v = \sigma_v / \underline{V} \qquad (11)$$

where the wind speed standard deviation σ_v is defined as (see Eq. (5)):

$$\sigma_v = \left[\frac{1}{T_a} \int_0^{T_a} u^2 dt \right]^{1/2} \qquad (12)$$

In addition it is also expected that the power fluctuations will be influenced by the dynamic characteristics of the turbine. An analysis [6] which is similar to the analysis that lead to the definition of the modified intensity of turbulence $I_{v,mod1}$ (which is associated with the average output power, see Section 2 above,) leads to the definition of another modified intensity of turbulence, $I_{v,mod2}$. This parameter is initially [6] defined for the case of harmonic gusts. Then, extension of the derivation leads to the definition of $I_{v,mod2}$ for a real wind. For that purpose, for a certain representative average wind speed, the frequency spectrum of the gusts is divided into N_s bins, where Δf is the "width" of each bin. If $S_u(f_i)$ is the frequency of the ith bin, then:

$$I_{v,mod2} = \left[\sum_{i=1}^{N_s} S_u(f_i)\, \Delta f\, \underline{G}_2^2(f_i, \underline{V}) \right]^{1/2} \Big/ \underline{V} \qquad (13)$$

$\underline{G}_2(f,\underline{V})$ is a function that describes the frequency response of the turbine to harmonic gusts [6]. As indicated it is a function of the frequency of the gust (f), and the average wind speed ($\underline{V}$). Moreover, in general, it is also a function of the gust amplitude, A. In Fig. 2 $G_2(f,\underline{V},A)$ is presented as a function of f, for $\underline{V}$ = 10 m/sec and three different amplitudes of gust (0.5, 1 and 1.5 m/sec). It is shown that the influence of the gust amplitude is relatively small, which is true for other values of $\underline{V}$. Thus it is possible to use a representative amplitude of the gust (1 m/sec is a typical value) and ignore the dependence on the amplitude (namely use $\underline{G}_2(f,V)$ instead of $G_2(f,V,A)$).

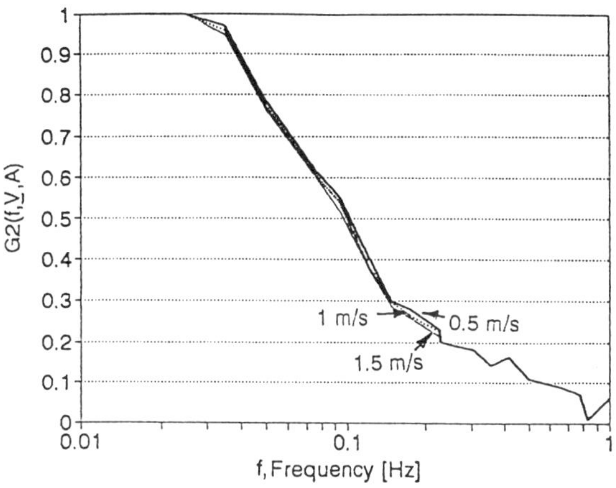

Fig. 2: The function $G_2(f,\underline{V},A)$ for an average wind speed $\underline{V}$=10 m/sec.

Analysis predicts an almost linear relation between the intensity of power fluctuations and $I_{v,mod2}$. In order to validate these results the data base of the Vestas V25 at the Beit-Yatir site was used. Representative examples are shown in Figs. 3-5, for $\underline{V}$ equal 9, 10 and 11 m/sec, respectively. It is shown that, as theory predicts, there is practically a linear relation between the measured intensity of power-fluctuations and the modified intensity of turbulence $I_{v,mod2}$. The quality of the linear relation reduces as the average wind speed is increased. More results are presented in [6].

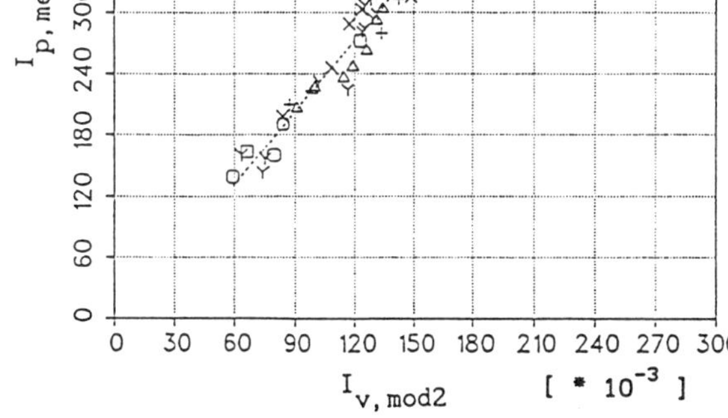

Fig. 3. $I_{p,mes}$ as a function of $I_{v,mod2}$. $\underline{V}$ = 9 m/sec.

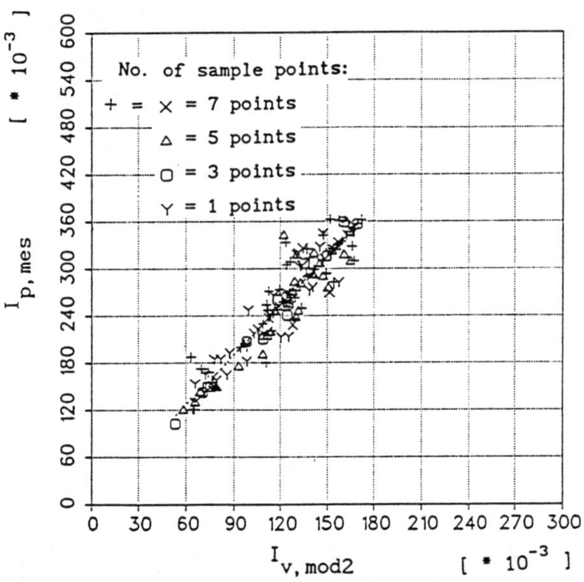

Fig. 4. $I_{p,mes}$ as a function of $I_{v,mod2}$. $\underline{V}$ = 10 m/sec.

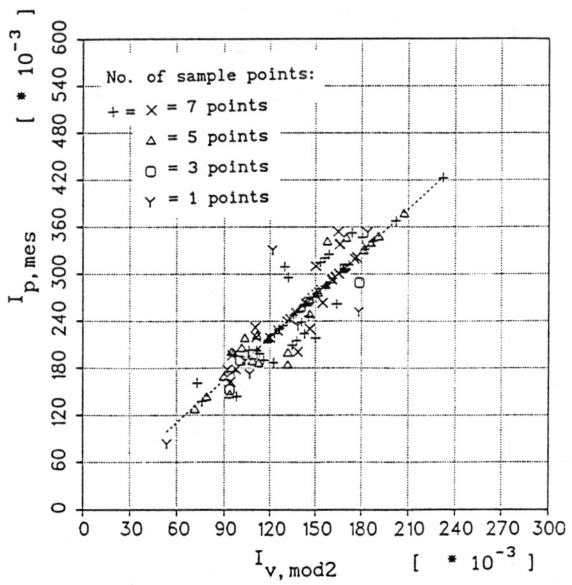

Fig. 5. $I_{p,mes}$ as a function of $I_{v,mod2}$. $\underline{V}$ = 11 m/sec.

4. CONCLUSIONS

Wind turbulence is comprised of dynamic variations of speed and direction. Turbulence affects the energy production of the turbine and results in output power fluctuations.

In the case of turbines having active yaw control, variations of the wind direction result in yaw errors. A method of calculating the influence of yaw errors on the energy production was presented. This method is relatively easy to apply. Although the influence of yaw errors on the energy production is relatively small, its inclusion in the analysis increases the accuracy of predicting the energy production of a certain turbine at a certain site.

The dynamic variations of the wind speed result in power fluctuations. These fluctuations are also influenced by the turbine dynamic characteristics. The intensity of power fluctuations is approximately linearly proportional to a modified intensity of turbulence that was defined in the paper. This parameter presents the coupling between the wind and turbine turbine dynamic characteristics.

ACKNOWLEDGEMENT

This research was supported by the Israel Ministry of Energy and Infrastructure.

REFERENCES

[1] Y Sheinman, A Rosen, "A Dynamic Model for Performance Calculations of Grid-Connected Horizontal Axis Wind Turbines: Part I - Description of the Model", Wind Engineering 15 (1991) 211-228

[2] A Rosen, Y Sheinman, "A Dynamic Model for Performance Calculations of Grid-Connected Horizontal Axis Wind Turbines: Part II - Validation", Wind Engineering 15 (1991) 229-239

[3] A Rosen, Y Sheinman, "The Average Output Power of a Wind Turbine in a Turbulent Wind", has been submitted for publication

[4] A Rosen, Y Sheinman, "A Dynamic Model of the Influence of Turbulence on the Power Output of a Wind Turbine", J Wind Engineering and Applied Aerodynamics 39 (1992) 329-341 (also presented at The Amsterdam European Wind Energy Conference, October 14-18, 1991)

[5] Y Sheinman, A Rosen, "The Influence of Yaw Errors on the Performance of Wind Turbines With Active Yaw Control", to be submitted for publication

[6] A Rosen, Y Sheinman, "The Power Fluctuations of a Wind Turbine", to be submitted for application

[7] A C Hansen, C Xudong, N J Siedschlag, "Yaw Dynamics of HAWT", First Annual Report, SERI Subcontract No. KL-6-05078-2, University of Utah, (1987)

[8] D M Eggelston, F D Stoddard, Wind Turbine Engineering Design, Van Nostrand Reinhold Co. New York (1987)

[9] P T Smulders, G Lenssen, H van Leeuvan, "Experiments with Wind Rotors in Yaw", Proceedings Int. Symp. of Fluid Mech. and Heat Transfer to Energy and Environmental Problems. Dept. of Physics, Eindhoven Univ. of Tech. The Netherlands, June (1981)

[10] L L Freris, Wind Energy Conversion Systems, Prentice Hall, The University Press, Cambridge (1980)

[11] J C Glasgow, R D Corrigan, R Miller, "The Effect of Yaw on Horizontal Axis Wind Turbine Loading and Performance", NASA TM-82778 (1981)

[12] J C Glasgow, R D Corrigan, R Miller, "Results of Free Yaw Tests of the MOD-0 100kW Turbine", Proceedings 6th Biennial Wind Energy Conference, Washington DC, (1981) 74-89.

ESTIMATES OF DYNAMIC WIND TURBINE LOADS FROM METEOROLOGICAL DATA

Ann-Sofi Smedman
Department of Meteorology
University of Uppsala
S-75120 Uppsala, Sweden

ABSTRACT: A very simple model has been constructed, which relates the variation in bending moment at the blade root in the flap wise direction to two meteorological parameters; turbulence intensity and wind shear. A time simulation model (VIDYN) has been used to separate the effect on loading from the two factors. The bending moment can thus be calculated from measurements of σ_u/U and $\Delta U/U$ which often are available from routine wind speed measurements on towers. The model can so far only be used for stiff turbines and for wind speeds below the rated speed.

1. INTRODUCTION

For large scale exploitation of wind energy it will be necessary to site wind turbines not only in flat coastal areas but also in rough inhomogeneous terrain. In order to maximize the use of costly land with a suitable wind resource wind turbines will also be put together in wind parks, where they will interact with each other.

All this is bound to create potential problems in terms of excessive loads on the turbines. The purpose of the present paper is to study the loads on the blades of stiff wind turbines and to explore the possibilities to express these loads in terms of the basic and relatively easily measured parameters; turbulence and vertical wind shear. The model can thus be used in siting investigations.

The natural wind is never strict stationary but always in a state of variation. These variations occur over an extremely wide range of scales which extends continuously from milliseconds to hours, days and even years. From a load point of view these variations can be divided into three categories: 1) oscillations with periods longer than about a minute which are bound to cause what can be treated as essentially a slowly varying steady load; 2) oscillations in the frequency range 1/minute to 1/second (see below); 3) very rapid fluctuations that are 'smeared out' over the rotor. The wind variations in the intermediate range are those most often referred to as 'turbulence' (although in a strict notation the high frequency parts of range 1 and the entire range 3 is also turbulence). Steady loads coming from frequency range 1 do not present much of a problem in terms of wind turbine design, while fluctuations in the intermediate range (2) are likely to cause *dynamic* loads on a typical full scale wind turbine.

The fact that the blades of the turbine are turning with constant speed of rotation results in what is referred to as rotational sampling. It is easy to see that due to shear, gravity, tower shadow etc. this will cause a cyclic load at the frequency of rotation, but it is perhaps not so evident that it also redistributes some of the turbulent energy of range 2 to this particular frequency. Thus the purely stochastic turbulent forcing causes what can be regarded as a cyclic load component.

In the future when performing meteorological site investigations it will not be sufficient only to map down mean wind condition. Turbulence structure and wind speed variation with height will probably be equally important. Variation of dynamic loads must thus be described as functions of simple meteorological parameters which can easily be measured or calculated from routine observations. The dynamic loads will of course affect all parts of a wind turbine but here we will concentrate on the blade loads.

An attempt is now made to relate the dynamic variation of flap wise bending moment of one blade of the 2 MW wind turbine at Näsudden in Sweden to statistical parameters such as turbulence intensity and mean wind shear.

2. THE NÄSUDDEN WIND TURBINE AND THE SITE

The wind turbine is a 2 MW upwind machine. It has two steel blades rotating at 25 rpm and the diameter of the rotor is 75 m. The hub is rigid with the primary shaft tilting 10° and the rotor has no cone angle. The height of the tower is 77 m and it is made of concrete. Cut-in wind speed is 7 m/s and at rated wind speed which is 12.5 m/s pitch control of the blades starts.

The Näsudden wind turbine is situated on a peninsula at the southwest coast of the island of Gotland in the Baltic. The ground is quite flat, with scattered juniper bushes. The distance from the machine to the shore line is 1500-2000 m for a wide sector from south over to northwest, with long unobstructed fetch over the open sea.

Wind speed and turbulence measurements have been taken on the 145 m meteorological tower at the site during several field experiments. It can be shown that the influence of the cross wind components (v,w) is small for a horizontal axis wind turbine during normal operational conditions. Thus the necessary information on the approach flow is usually assumed to be the variation in the longitudinal wind component (u).

3. CYCLIC BLADE LOADING.

As stated above fluctuations in the approach wind field are the main cause for flap wise blade loading of a horizontal axis wind turbine. The cyclic blade loads can be divided into a deterministic part and a stochastic part. For a turbine with constant rotation angular velocity loads which depend on the azimuthal position of the rotor become periodic while the turbine rotates. These deterministic loads are due to wind shear, tower interference, gravity and to the rotation such as centrifugal forces. The stochastic loads on the other hand depend on the turbulent fluctuations in the longitudinal wind component.

3.1 Wind shear

Most so far published full-scale investigations concerning blade and rotor loading are dealing with three bladed turbines with hub heights around 20-40 m For those rather small turbines the influence of the wind shear seems to be negligible. However, for large two bladed machines the wind shear can be as important as the turbulence for the cyclic loads.

3.2 Turbulence

Loads and structural responses of a wind turbine rotor are described in a rotating reference system and it is thus desirable to describe turbulence in the same system. Due to rotation of the wind turbine in the turbulent wind field, the turbulence seen from a point on a moving blade is altered. This is called 'rotational sampling'.

The effect of 'rotational sampling' on the spectrum of the longitudinal wind component is to redistribute energy from lower frequencies to the rotational frequency and its harmonics. Figure 1 shows an example of spectrum of the longitudinal wind component measured at hub height at Näsudden together with the spectrum of flap moment at the blade root. The u spectrum is normalized to give the best visual fit to the measured load spectrum at low frequencies. From n ~ 0.05 the redistribution of energy because of rotation is evident.

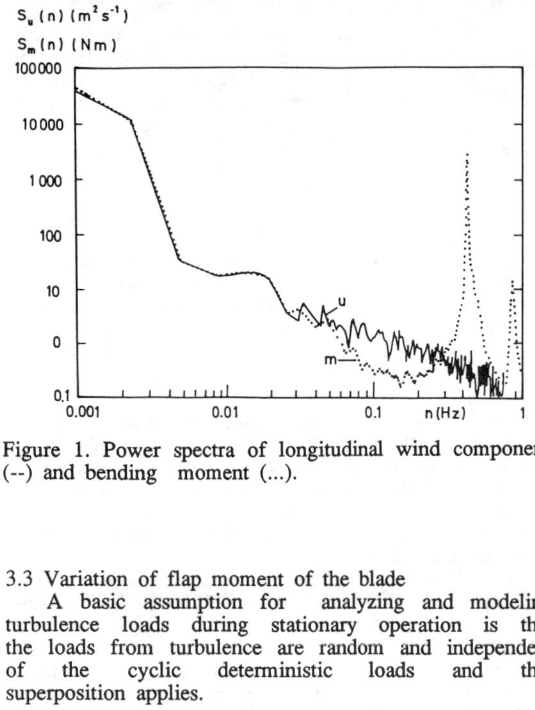

Figure 1. Power spectra of longitudinal wind component (--) and bending moment (...).

3.3 Variation of flap moment of the blade

A basic assumption for analyzing and modeling turbulence loads during stationary operation is that the loads from turbulence are random and independent of the cyclic deterministic loads and that superposition applies.

The variation of flap moment of a blade can be expressed

$$\sigma_m^2 = (\sigma_m^2)_{turb} + (\sigma_m^2)_{shear} + (\sigma_m^2)_{mean} \qquad (1)$$

where
turb = part of the variance depending on turbulence
shear = wind shear
mean = mean wind, gravity, tower shadow

To be able to separate the influence of the three different parts in (1) a simulation model has been run. The model VIDYN, which is described in the next Section, has been developed by Teknikgruppen AB and has been used during the evaluation of the two large wind turbine prototypes in Sweden situated in Maglarp and Näsudden.

4. THE VIDYN MODEL.

The VIDYN model is developed as a time simulation model and has proven to be a useful tool for the designer of the overall mechanical system and also of the control system. The model is capable of treating all kinds of turbines and nonlinearities such as variable rotational speed and aerodynamic stall behavior.

5. MODEL SIMULATIONS.

The VIDYN model has been run for three meteorological situations which we have found to be typical for the Näsudden site. For all three cases the wind is coming from the sea.

For each case first a simulation with a constant mean wind speed as input was performed. The second group of simulations were run with a constant wind shear and finally simulations were made with real time series.

In addition 32 runs with different mean wind shear have been carried out.

All simulations are made for wind speeds in the range 6 - 13 m/s i.e. with the turbine working without pitch control

6. RESULTS.

6.1 Simulations

To be able to interpret the results from the model runs power spectra have been calculated with a FFT routine of the simulated time series of the blade root bending moment in the flap wise direction.

Figure 2 shows the contribution to the cyclic variation of flap moment from the three parts in expression (1) for one specific run. From the Figure it is clear that the variation caused by effects such as gravity and tower shadow when the turbine is working in a constant mean wind field can be neglected in comparison with turbulence and wind shear.

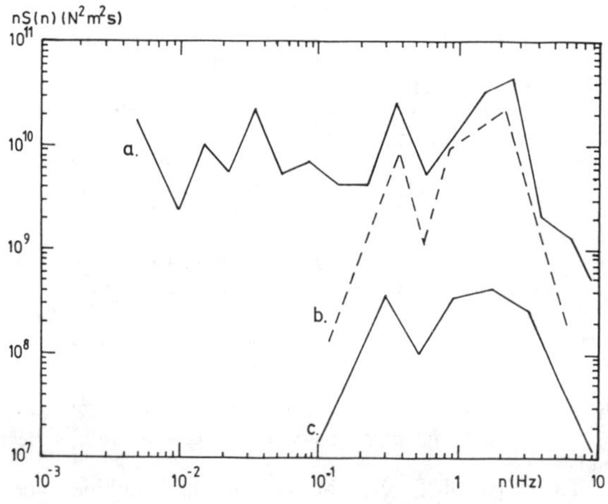

Figure 2. Power spectra of simulated bending moment. The three curves represent a) total variation of bending moment b) variation depending on wind shear and c) variation depending on tower, gravity etc.

The contribution caused by the vertical variation of wind speed can be calculated by using the 32 runs with different constant wind shear. The mean wind variation over the rotor expressed as ΔU (difference in wind speed over the rotor) should be the relevant parameter rather than the shear itself. The normalized velocity difference $\Delta U/U$ has been plotted against the normalized variation of flap moment σ_m / M in Figure 3.

The data are well represented by the following expression:

$$(\sigma_m / M)_{shear} = 0.7\ \Delta U / U \qquad (2)$$

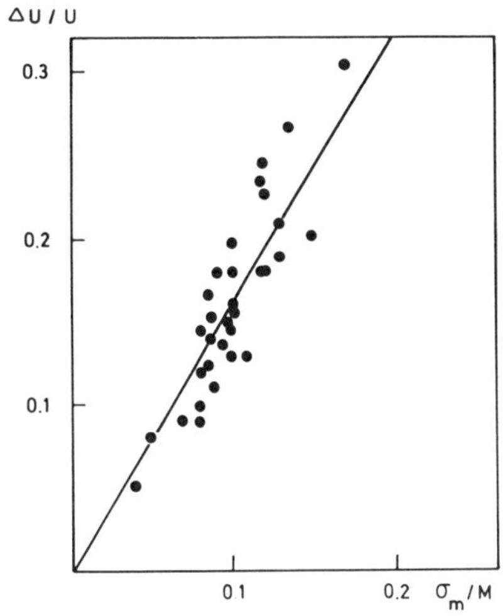

Figure 3. Simulated σ_m/M - values as a function of $\Delta U/U$ for 32 runs.

The first term in expression (1) represents the variation due to turbulence. The flap moment at the blade root can be written

$$M = Const.\ U^2 c_t$$

and for the Näsudden turbine c_t can approximately be said to vary as ~ 1/ U. Thus M ~ const · U
Differentiating yields

$\delta M/M=\delta U/U(6)$

$(\sigma_m / M)_{turb} \sim \sigma_u / U$

Expression (1) can then be written

$$(\sigma_m/M)^2=(\sigma_u/U)^2+(0.7\Delta U/U)^2 \qquad (3)$$

The first term on the right hand side of (3) gives the variation of bending moment caused by turbulence. As stated above turbulence at lower frequencies will exitate energy at higher frequencies because of rotation of the turbine. To find a simple way to handle it, we just suppose that the areas denoted 1 and 2 in Figure 4 are equal. The Figure shows simulated curves for the longitudinal wind component and bending moment, when the variation caused by wind shear has been subtracted. The u spectrum has been normalized to give the best fit to the spectrum of bending moment at low frequencies. The turbulence intensity in expression (3) will thus represent the total σ_u measured for that specific time period, in this case one minute.

To verify expression (3) five simulations for the first meteorological case have been used. For each run the normalized variation of flap moment has been calculated according to (3) and compared with the simulated one. The result, which is given in Table I, is very promising. The correlation coefficient between calculated and simulated values is as high as 0.99.

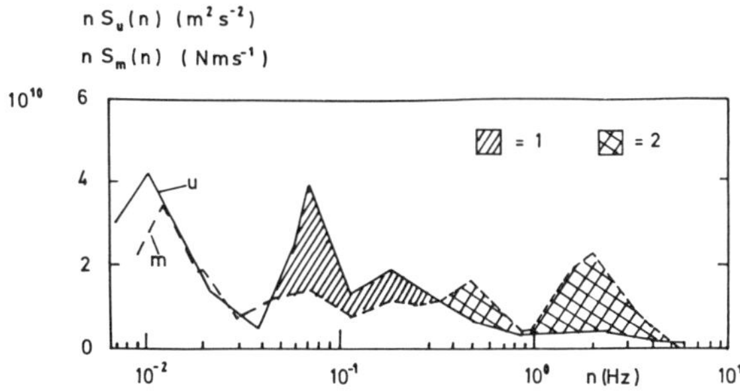

Figure 4. Simulated spectral curves of bending moment (m) and longitudinal wind component (u). The cyclic variation due to wind shear has been removed. As a first approximation area 1 ~ area 2 (see text for further explanation).

Table 1. Comparison between simulated and measured flap moments.

σ_u/U	$\Delta U/U$	$(\sigma_m/M)_{sim.}$	$(\sigma_m/M)_{cal.}$
0.11	0.07	0.12	0.10
0.30	0.14	0.21	0.24
0.15	0.14	0.17	0.17
0.14	0.05	0.12	0.11
0.05	0.06	0.08	0.07
MEAN		0.140	0.138

6.2 Measurements at Näsudden

An extensive measuring program has been carried out at Näsudden. Measurements comprise meteorological variables such as wind speed, direction and temperature but also values from sensors mounted on the turbine and the tower, for example strain gauges and accelerometers.

During one period from October 1983 to February 1985 (478 days) meteorological parameters were recorded as one minute averages. Wind speed was measured at six levels on the 145 m tower together with fast response Gill anemometers at three heights. The sensors on the turbine were not sampled continuously but in short (60 s) campaigns.

From the measurements comprising bending moment 24 campaigns have been selected. The criteria for choosing the data were that the turbine should be working without pitch regulation, there should be no start or stop procedure and no specific tests should be performed during the campaign. In Figure 5 the measured normalized bending moments are plotted against values calculated with expression (3). The mean value of measurements is 0.133 to be compared with 0.140 for the calculations. The correlation coefficient is 0.75. In view of the rather crude model and all simplifications the result in Figure 5 is encouraging.

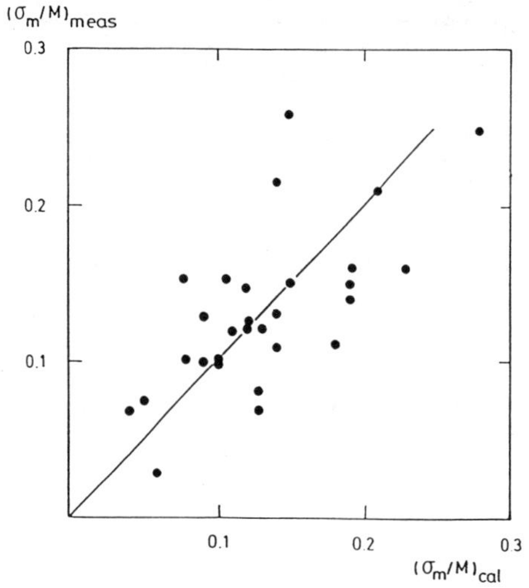

Figure 5. Measured values of σ_m/M as a function of calculated ones.

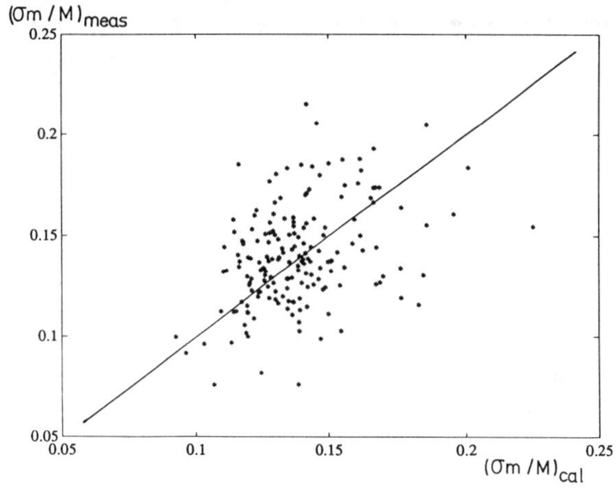

Figure 7. Measured values of σ_m/M as a function of calculated ones.

7. THE ALSVIK MEASUREMENTS

The Alsvik wind farm consists of four Danwin 23/180kW machines situated only 50 m from the shoreline on a very flat coastal strip on the west coast of the island of Gotland. The turbines are three-bladed, with a diameter of 23 m and a hub height of 30 m. They are stall-regulated and have a rated power of 180kW. Wind speed, direction and turbulence intensity are measured at 8 levels on two 42 m high towers. One turbine is also instrumentated with strain gauges and accelerometers.

Expression (3) has also been tested against measurements from Alsvik. The constant, which has the value 0.7 for the Näsudden data set, is now found to be around 1. This is an indication of the fact that wind shear is as important for blade load calculations for small wind turbines as it is for large.

Figure 6 and 7 show comparisons between measured and calculated flap moments. In Figure 6 time series during a 30 min period is given and in Figure 7 1 -min averages are calculated. The mean values are 0.138±0.020 for the measured flap moments and 0.140±0.025 for the calculated ones.

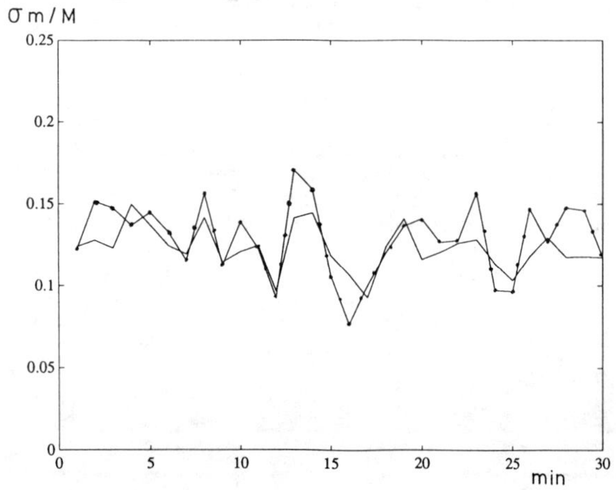

Figure 6. Time series of calculated flap moment (....) and measured flap moment (——).

8. CONCLUSIONS.

From a combination of meteorological and aerodynamic measurements and model simulations it has been possible to construct a very simple method to calculate normalized standard deviations of dynamic bending moment at the blade root in the flap wise direction of a wind turbine.

It has been shown that variation of flap moment σ_m/M depends mainly on two meteorological parameters; turbulence intensity and wind shear. The relative importance of the two seems to a certain degree to be turbine dependent. Both factors should, however, always be considered when performing load and fatigue calculations.

ACKNOWLEDGEMENTS.

The author would like to thank Mr H. Ganander and Mr H. Johansson at Teknikgruppen AB for running the model simulations. Many thanks also to Mr J-Å Dahlberg at the Aeronautical Research Institute of Sweden who carried out the evaluation of the meteorological tower data at Näsudden.The work has been sponsored by the National Energy Administration Sweden.

PRESENTATION OF THE HÖNÖ TEST MACHINE AND RESULTS OF MECHANICAL BEHAVIOUR IN STALL OPERATION

Hans Ganander
Teknikgruppen AB
Box 21, S 191 21 Sollentuna
Sweden

Anders Björck and Maria Poppen
The Aeronautical Research Institute of Sweden
Box 11021, S 161 11 Bromma
Sweden

ABSTRACT: In the Swedish wind energy R&D program questions were raised about the dynamical behaviour of a soft two bladed teetered turbine in stall operation. To get some indications of the answers it was decided to rebuild the existing 40 kW test machine at Hönö. Modification of the turbine included not only the rotor but also yaw and hydraulic systems. Demands on measured signals was now increased to include mechanical quantities to a much larger extent than before. Preliminary tests started in March 1992 and were continued during following autumn. Results from stall operation during this first test period are that loads are reasonably well predicted and that teeter motion does not show any instability.

1. INTRODUCTION.

In the Swedish wind energy R&D program questions were raised about the dynamical behaviour of a two bladed teetered turbine in stall operation at high wind speed. Stall experiences from Denmark based on three bladed relatively stiff designs are good, but what a about two bladed soft designs with variable rotational speed? To get some indications of the answers it was decided to rebuild the existing 40 kW test machine at Hönö. Measurements, evaluation and verification of calculations would be an important part of the project and in this way give apportunities to verify teoretical model as well as practical ideas from several years of R&D work.

This paper deals with mechanical aspects and experiencies of the project, while electrical and control issues are delt with in [1].

Characteristic parameters of the new turbine are:

- number of blades 2
- diameter 13.5 m
- aerodynamic profile NACA 63-200, FFA-W3-xxx
- pitch adjustment ± 1.5 degrees
- cone angle adjustment 0 - 5 degrees
- hub with or without teeter motion
- brakes at primary and secondary shafts
- yaw system with hydraulic damping devices
- variable rotational speed max 75 rpm
- rated power 35 kW
- rated wind speed 10 m/s
- rated tip speed ratio 8
- power control modes
- induction generator with frequency converter
- signals covering blade roots, nacelle, tower
- wind sensores at two masts

2. MECHANICAL DESIGN

2.1 Modifications.

Main modification activities of the wind turbine are first of all the new two bladed teetered rotor and the hub. Because of no pitch mechanism a primary shaft disc break was installed to fullfil the safety requirements in case of emergency. A completely new hydraulic system was also built in. This system is used for the primary shaft brake of rotor and generator shafts as well as yawing functions of the wind turbine.

2.2 Design loads.

As a first step of the project, preliminary loads were calculated in order to see to what extent the modified turbine would stand static and fatigue loads. A number of load cases were studied using the simulation program VIDYN. Calculated loads were used to check stresses of the main components such as blades, hub, nacelle and tower. Concerning blades this procedure is described more in detail in Chapter 3.

2.3 Hub and teeter design.

The teeter bearing is a conventional journal bearing. Mechanical stops limit the teeter angle to 8 degrees. At teeter angles larger than some degrees a progressive rubber spring introduces a teeter moment. The teeter motion can be locked, which makes it possible to run the machine with fixed hub. By mounting conical washes between the blade root flanges and the hub flanges, cone angles of 2.5 and 5 degrees may be introduced.

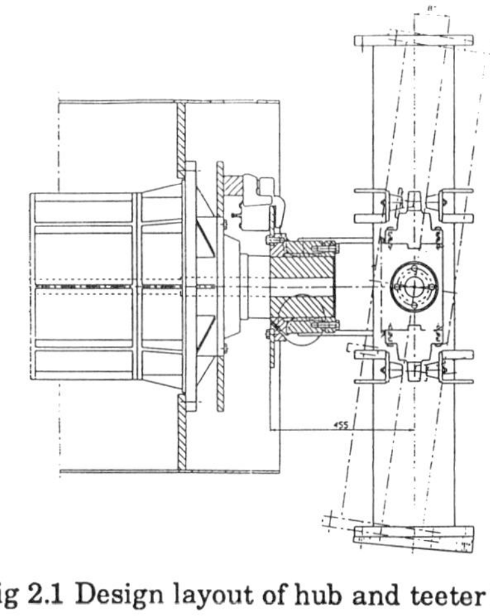

Fig 2.1 Design layout of hub and teeter hinge.

2.4 Primary shaft brake.

In the design phase it was of great importance to investigate the dynamic behaviour of emergency stop situations at a loss of grid or generator faults. At high wind speed, the rotational speed increases rapidly unless the breaking moment disaccelerates the rotor speed. To reduce dynamic overloads of such breaking events of specially the blades and the drive train, a damping device was introduced in the hydraulic breaking system.

2.5 Yaw system.

There are several possibilities to control the yaw position of the turbine. The first one is conventional. It calculates the yaw error and compares it with adjustable limits. A second method uses control valves of the hydraulic system and/or the self yaw-aligning property of up-wind turbines. Other methods are based on possibilities of measurement and control systems.

3. BLADES.

3.1 Design philosofy and main data.

The blades are designed to achieve aerodynamic and dynamic characteristics of the larger 35 meter diameter NWP 400 kW turbine. The Hönö turbine is thus much smaller so scale effects will inevitably be encountered. The blades are designed with constraints of an existing generator and gear-box. There is also the desire to get a lot of opportunities to measure in the stall regime. This desire and as high chord Reynolds numbers as possible led to a diameter of 13.5 meter and a rather low tip speed of 53 m/s.

The blades are made of solid wood, since this is a cheap manufacturing method for one-off blades. In order to get as good dynamic scaling properties as possible, the light weight wood Thuja Plicata is used. Main data of the blade and performance calculations as well as mass and stiffness distribution of the blade, is given in [2].

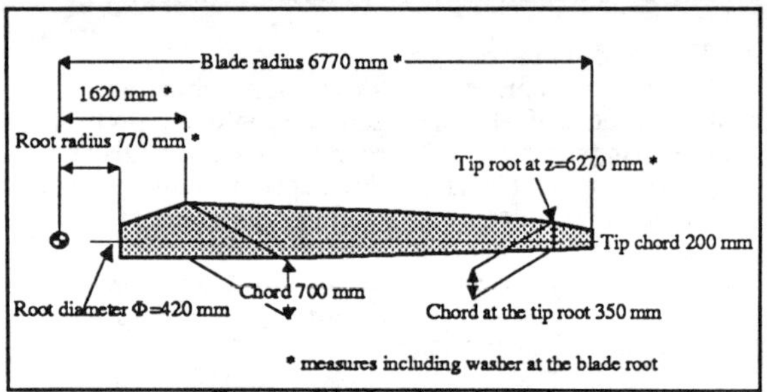

Fig 3.1 Blade plan form data.

The blades have removable blade tips in order to make it simple to exchange tips to study aerodynamic and acoustic effects of other tip forms.

3.2 Structural analysis.

3.2.1 Material Data

The material used for the Hönö blades, Thuja Plicata, is almost unknown with respect to strength, modulus and fatigue behaviour. Some limited testing had to be performed in order to get a possibility to make structure analyses. Four specimens were loaded statically in tension to failure. The load deflection curve was monitored for each specimen. The modulus of elasticity and the ultimate tensile strength were calculated [3]. The result is shown below.

Specimen number	UTS [MPa]	Modulus of elastic. [GPa]
1	39.8	5.89
2	36.7	6.38
3	38.7	5.72
4	40.5	6.80

Safety factors and statistical evaluation give the permitted stress under static loading of 8 MPa.

3.2.2 Fatigue Curve.

A fatigue curve had to be estimated. The base is the fatigue curve for pine and the static strength values for Thuja Plicata. A safety factor of 1.5 is used. The constant life diagram is shown in Figure 2.17. The triangular shape based on R = -1 loading is supposed to be a safe estimation as well.

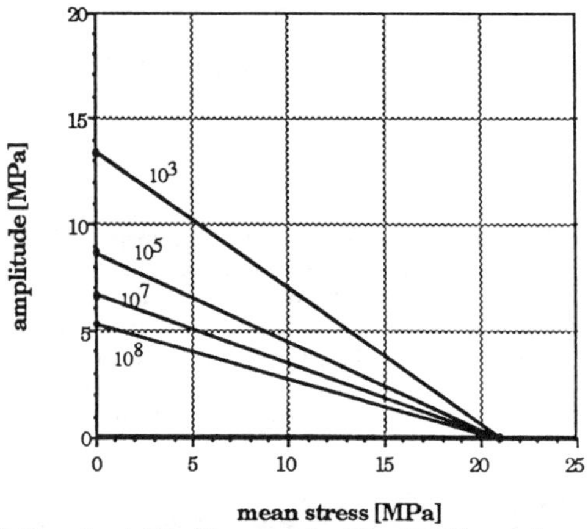

Fig 3.3 Constant life diagram for Thuja Plicata, based on static properties.

3.2.3 Stress Calculations.

Stress calculations are performed for the radial stations, r = 2, 3, 4, 5 and 6 meters. The program VIDYN was used to calculate the loads on the blade for several different operation modes [4]. Calculations are performed both for rigid and teetered hubs. The yawing spring were stiff as well as soft.

Table 3.1: Sumary of load cases for stress calculations.

Mode	Speed [m/s]	Shear [m/s]	Hub	Yaw
Normal operation	4-20	4	locked	stiff
-"-			teetered	stiff
Emergency break	10-15	4	teetered	stiff
Overspeed	10-20	4	teetered	stiff
Storm	50		teetered	stiff
Wind shear	4-20	6	locked	stiff
-"-	4-10	6	teetered	stiff
-"-	4-10	6	teetered	soft
Cross wind ±40	4-20	4	locked	stiff
-"-	4-10	4	locked	soft
-"-	4-10	4	teetered	soft

The highest static stress in flapwise direction is the over-speed load case at the wind speeds 20 and 15 m/s. The load for the over-speed load cases is taken as the load occurring after 4 seconds of free rotation. The crosswind load cases also give quite high stresses.The edgewise static stresses are below 1 Mpa and negligible.

Conserning fatigue loads the crosswind load cases of 10, 15 and 20 m/s from ±40 degrees causes most damage.

3.2.4 Full-scale Testing.

The two blades have been tested statically [5]. The test was performed to check the stress calculations and to find faults like bad bonding etc. The tested load case was the crosswind from ±40 degrees at a wind speed of 15 m/s. The blades were loaded with bags of sand. The deflection of the blades was measured during the stepwise increased loading, both at the trailing and leading edges to see if it was torsioned. The difference between the blades is small, probably within the accuracy of the measurement method. These deflection values are also used to estimate the equivalent Youngs modulas of the blade design.

4. MEASURING SYSTEM

4.1 System description.

The measuring system is designed to measure mechanical loads on the turbine, electrical power and meteorological data. The system consists of three data acquisition units, which convert the sensor signals into serial data format. During a measurement, the data are stored on the hard disk of the 386 personal computer that is connected to the data acquisition units (B) and (C) via two RS232 lines. Selected baud rate is 38400 baud.This is shown in fig 4.1.

Unit	Rate (samples/s)
(A)	46
(B)	0.8
(C)	23

When a measurement campaign is finished, the raw data stored on the hard disk can be processed and evaluated. Larger data files can be stored and copied on 8 mm video cassettes. When no testing is carried out at least wind measuremenst are performed continously.

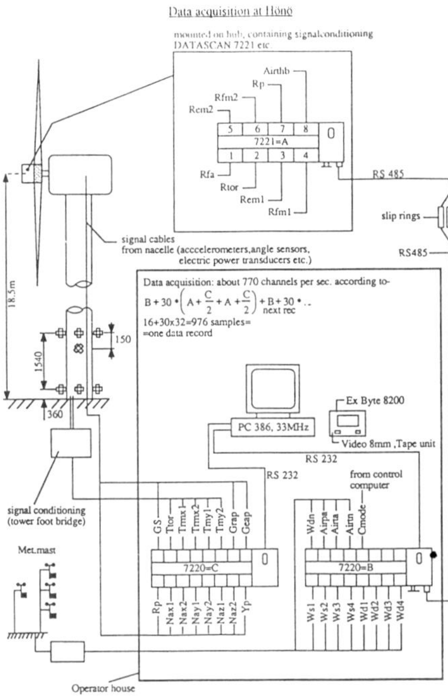

Fig 4.1 Measurement system

4.2 Signals.

One data acquisition unit (A) is mounted on the hub and rotates with the rotor. It collects bending moments in the blade roots, shaft torque and rotor teeter angle. The serial data are transferred to unit (B) via an RS485 line. Transactions on the RS485 line are packet-orientated.

The two remaining units are mounted in the operator house together with the personal computer. One of these (B) collects four wind speed and wind direction from the two masts. This unit also collects air pressure, air temperature and control mode of the system.

The six accelerometers in the nacelle are connected to unit (C). Strain gauges measure bending moments and torque in the tower foot. The signals are amplified and connected to unit (C). Unit (C) also collects rotor position, yaw position, generator speed and electrical power.

5. MEASUREMENTS AND EVALUATION.

The measurements have been carried out in two periods. It was of great interest to run the machine, as early as possible, in stall situations with the rotor in teeter mode. This was done in March-92 in order to get indications of the teeter stability and also in order to find out the recommendable value of the free teeter motion regarding security during further testing. The consequences of this first period were followed up in August, when the main part of the test program started. The first part of the test program was carried out until November, when teeter behaviour and loads in stall mode operation were measured and analizied.

5.1 Preliminary measurements in March and August 1992.

On the first occasion in March the wind speed varied between 12 and 16 m/s with a tip speed ratio down to nearly 3. The turbulence intensity was 15%. This run was about twenty minutes long. Early evaluation and comparison with calculations with the actual turbine and wind data show quite good agreement. At this first measurement, the free teeter angle was ± 1.6 degrees. The results are summarized below.

Quantity	Mean	Std	Max-Min
Medge.root	527	1469	6836
Mflap.root	3756	725	5634
Teeter (dgr)		2.30	6.06

Calculations with the VIDYN program based on measured wind signals as input give following results:

Quantity	Mean	Std	Max-Min
Medge.root	786	1578	6410
Mflap.root	4593	567	4991
Teeter (dgr)		2.31	6.82

In order to find out the influence of the free teeter angle, it is reduced to ± 0.5 degrees in the following calculation, with remainded wind input.

Quantity	Mean	Std	Max-Min
Medge.root	827	1562	6339
Mflap.root	4604	470	3535
Teeter (dgr)		1.41	4.81

Loads as well as teeter angle seems to be reduced when decreasing the free teeter angle from 1.5 to 0.5 degrees. In August-92 it was possible to perform measurements with the reduced free teeter angle at a reasonably strong wind speed 7-15 m/s, with 15% turbulence intensity. The rotational speed at operation was set to 41 rpm and the tip speed ratio was around 2.5. Preliminary evaluated blade-bending moments and teeter angles are:

Quantity	Mean	Std	Max-Min
Medge.root	806	1504	6200
Mflap.root	4304		3750
Teeter (dgr)		0.79	3.30

5.2 Measurements in August-November 1992

The intention is to study situations around the three tip speed ratios 3, 5 and 7 for varied wind and rotational speeds.

The measurements were made at different constant rotational speeds, during conditions with normally varying wind speed. The results are recorded on the video8 tapes. At the evaluation, periods of 5 minutes are classified according to average tip speed ratio during 5-minute periods. A catalogue of these measurements, where wind and turbine conditions as well as loads, teeter angle and accelerations, is made up of mean and standard deviation during the 5-minute periods.

Seven situations have been chosen in this presentation. Some wind characteristics of the seven chosen cases are found in Table 5.1 below:

Table 5.1:Wind and turbine conditions during evaluated 5 minute periods.

Wind [m/s] mean	std	Turb. intensity	Rpm	Tip
13.3	1.51	0.08	54	2.0
14.9	1.51	0.11	54	2.9
12.9	1.66	0.11	64	3.0
9.7	0.98	0.10	54	3.9
9.1	1.02	0.11	64	5.0
7.6	1.14	0.15	54	5.0
6.5	1.25	0.19	54	5.8

The varifying calculations are based on measured wind signals as wind input to the VIDYN calculations. Thus, the wind is known in three points located at the top, hub and bottom positions of the rotor plane. A presentation of some of the measured and calculated quantities are found in Table 5.2 and 5.3 below.

Table 5.2 Evaluated quantities in kNm and degrees.

Medge std	Mflap std	max	min	Δ	Teeter std
1.73	0.50	7.5	2.5	5.0	1.3
1.82	1.08	11.	4.0	7.0	2.2
1.78	1.04	11.	5.0	6.0	1.9
1.68	0.65	6.5	2.5	4.0	1.0
1.55	0.73	8.0	3.5	4.5	0.8
1.69	0.76	4.	-0.5	4.5	0.9
1.61	0.64	4.	0.0	4.	0.7

Table5.3 Calculated quantities in kNm and degrees.

Medge std	Mflap std	max	min	Δ	Teeter std
1.93	0.52	7.5	3.5	4.0	1.4
2.02	0.65	11.2	4.9	6.3	1.7
1.99	0.73	10.3	5.7	4.6	1.8
1.95	0.45	6.7	3.2	3.5	1.0
1.89	0.47	8.1	4.5	3.6	1.1
1.91	0.49	6.4	2.7	3.7	0.9
1.94	0.66	6.1	2.0	4.1	0.8

These results show that the calculated standard deviation of the blade root bending edge moment roughly is over-estimated by 10-15%, while the standard deviation of the blade root bending flap moment is under-estimated by 30%. Extreme values of the flap-bending moment (Mflap), which of course is very dependent on details about what really happened to the wind and the control of the machine, are reasonably well predicted.

Regarding the teeter angle it is found that small values are over-estimated by about 10%, while larger values are under-estimated by approximately 10-15%, which can be seen in Fig. 5.4 below.

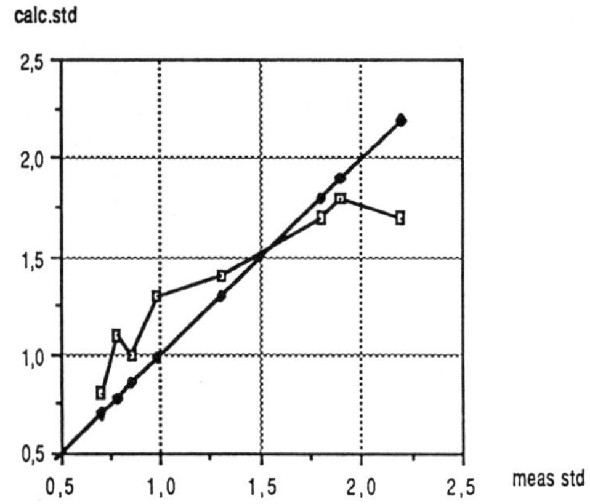

Fig 5.4 Comparison of evaluated and calculated standard deviation of the teeter angle.

Among those circumstances that cause differences between measured and calculated results the following uncertainties may be mentioned:

- wind speed and direction
- yaw stiffness
- yaw and drive train control

The accuracy of the evaluation is of great importance and should also be mentioned. The blade root moments are very dependent on the actual inching calibration measurements.

6. CONCLUSIONS.

- Loads are within calculated margins
- Teeter angle does not show any instability
- Comparisons between measured and calculated quantities are reasonable
- The benefit of reducing the free teeter motion (1.5 > 0.5) has been verified
- The Hönö test turbine is now suitable for several kinds of tests, theoretical verifications as well as practical wind turbine design issues.

7. REFERENCES.

[1] Ola Carlson et al " Stall Control with Variable Speed and an Induction Generator". ECWEC-93 in Trawemünde.

[2] Anders Björck and Maria Poppen, "Blade data for the Hönö2 blade", FFAP-A-959, The Aeronautical Research Institute of Sweden, to be published 1992.

[3] Maria Poppen, "Bedömning av hållfastheten hos Thuja Plicata bladen på nya Hönöaggregatet", FFAP-H-1143 in Swedish, The Aeronautical Research Institute of Sweden, 1991.

[4] Hans Ganander et al "Stall control with Variable Speed at Hönö2 Wind Turbine". Department of Electrical Machines and Power Electronics, Chalmers University of Technology, 1993.

[5] Maria Poppen, "Provbelastning av Hönöbladen", FFAP-A-928, in Swedish, The Aeronautical Research Institute of Sweden, 1991.

THEORETICAL AND EXPERIMENTAL INVESTIGATION ON A ONE-BLADED MEDIUM SIZE WT TO IMPROVE THE DESIGN CRITERIA

C.H.J. Stork
Riva Calzoni S.p.A.
Via Emilia Ponente 72
40133 Bologna
Italy

D.C. Quarton
Garrad Hassan & Partners Ltd.
The Coach House
Folleigh Lane Long Ashton
BS18 9JB Bristol
United Kingdom

S. Attanasi
ENEA FARE
Via Anguillarese 301
00060 S.Maria di Galeria RM
Italy

ABSTRACT: One aim of the work presented in this paper was to examine the sensitivity of the fatigue life predictions to assumed wind conditions, model complexity and adopted standards.
Beside this activity the possibility for flap system improvements was investigated. A common problem of a WTG with a teetering hub is that of maintaining flap angle stability under all possible conditions. For the single bladed M30 wind turbine the most critical condition is the emergency shut down transient which was analyzed by means of both computer predictions and on-site experiments.

1. INTRODUCTION

In 1990 Riva Calzoni (RC) started a research activity with two aims: to verify the design approach for fatigue predictions for the M30 prototype and to investigate possible flap system simplifications. The aims were obtained by both on-site measurements on ENEL's test field of Alta Nurra (Sardinia) and computer predictions carried out with different aeroelastic codes. This research activity, terminated in autumn '92, was partially supported by the EEC in the framework of the JOULE-program, and by ENEA. Furthermore ENEA and Garrad Hassan & Partners (GH&P) gave a technical contribution in measured data analysis and computer predictions.

2. PROJECT OVERVIEW

The main activities for the experimental and theoretical work of this project were the following ones:
- instrumental improvements
- site measurements
- definition and verification of compliance matrix
- dynamic tools validation
- fatigue life extimation
- flap system improvements

To obtain the experimental data for aeroelastic code validation, the data acquisition system had to be extended. The improvements consisted of the installation of further sensors, modifications in the signal conditioning and a new set-up of the recording devices. In November 1991 the data sets necessary for the research activities consisting of a 10 minute file at rated wind speed, 10 minutes at high wind speeds and a file containing the emergency shut down transient were recorded.

The aeroelastic code of GH&P, BLADED, was extended to model single bladed wind turbines and verified by means of the three mentioned records. The code, together with the compliance matrix, which transfers the hub loads and accelerations into nacelle frame stresses, provided damage rate predictions for single points on the nacelle structure. One "hot point" and a minor loaded node were analyzed as described in section 3.

The same code was used to study the flap angle behaviour during the emergency shut down transient under different conditions. The results could be verified at least qualitatively by on-site experiments. The investigations on the flap system are described in detail in section 4.

3. INVESTIGATION OF THE DESIGN APPROACH FOR FATIGUE

The fatigue life for two welds located on the nacelle structure was calculated by means of the aeroelastic codes of RC and GH&P with different assumptions for the wind conditions.

Furthermore GH&P's code was used to carry out a sensitivity study of the fatigue life predictions to:
- the number of wind speed bins used to represent the range of operating conditions;
- the turbulence intensity assumed for the site;
- the annual mean wind speed assumed for the site.

A draft of the IEC standard "Safety of Wind Turbine Generator Systems" [1] was used as a guide for the selection of site conditions and partial safety factors. All predictions were performed according to both Eurocode 3 and the British Standards BS 5400, Part 10.

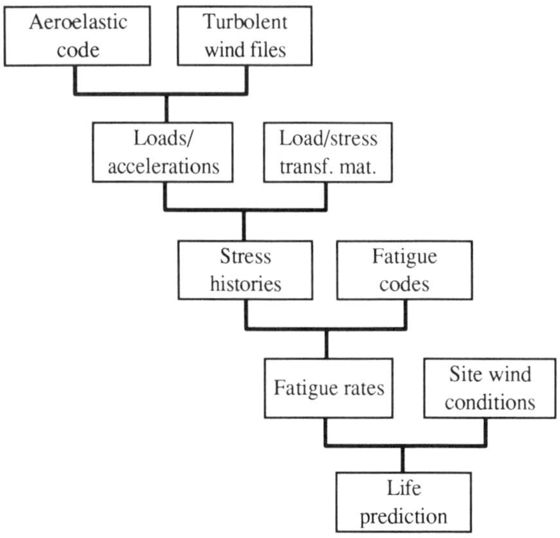

Figure1: Overview of fatigue life prediction method

Figure 1 is a schematic diagram showing the overall approach to the fatigue life predictions. For each wind speed bin, the calculation is based on a wind file characterized by a mean wind speed, the turbulence intensity and the turbulence length scale. The wind file supplied to the aeroelastic model is made up of a large number of longitudinal turbulent wind speed time histories arranged on a grid having dimensions larger than the rotor disk. A shear profile, yaw misalignment and tower shadow were superposed on the turbulent wind field.

The loads and accelerations of the hub center are calculated and converted into stresses by means of compliance matrices.

According to the adopted standard, Eurocode 3 or BS 5400, different stress parameters have to be calculated and subsequently reduced to a rainflow cycle count. The life fractions consumed by all constituent cycles are then summed linearly as determined by the relevant fatigue strength curves to predict the life fraction consumed by the complete sequence. The life fraction consumed during the design life is predicted by considering the number of occurrences of each sequence. This calculation was carried out on the basis of a Rayleigh wind speed distribution.

Calculations were performed for 3 sets of simulations as defined in Table 1. The wind speeds are defined by the WTG operation characteristics: 11 meters per second is the rated and 20 meters per second the cut out wind speed.

	mean wind speed [m/s]	turbulence intensity [%]	operational range [m/s]
Set 1	7.7	21	4.50 - 9.35
	11	21	9.35 - 12.65
	20	21	12.65 - 20.00
Set 2	7.7	15	4.50 - 9.35
	11	15	9.35 - 12.65
	20	15	12.65 - 20.00
Set 3	6	21	4.50 - 6.85
	7.7	21	6.85 - 8.35
	9	21	8.35 - 10.00
	11	21	10.00 - 12.50
	14	21	12.50 - 15.50
	17	21	15.50 - 18.50
	20	21	18.50 - 20.00

Table 1: Set definition for fatigue analysis

The Table 2 shows the results for a typical high loaded weld located on the nacelle structure. The normalized life time for the three sets of calculations and different annual mean wind speeds are presented in table. Results were given according to the two design standards used. The sensitivity of the fatigue life prediction to the number of wind speed bins is indicated by comparing the results of set 1 and set 3. According to the results based on 7 wind speed bins, the predicted life is significantly longer than that obtained using only three bins. Use of 7 bins increases the predicted life by a factor of about 3.3 for Eurocode 3 and 1.6 for BS 5400.

The results indicate that the life time is clearly sensitive to both the turbulence intensity and the annual mean wind speed. As expected, the fatigue life predictions based on the BS 5400 standard are shorter than those obtained with Eurocode 3. The difference is mainly caused by a lower S-n curve and the absence of a fatigue endurance limit in the case of BS 5400.

		Eurocode 3			BS 5400		
	Vave [m/s]	6	7	8	6	7	8
Set 1	3 bin, 21%	3.9	1.7	1.0	0.9	0.6	0.4
Set 2	3 bin, 15%	14.4	6.0	3.4	1.9	1.2	0.8
Set 3	7 bin, 21%	12.8	5.7	3.2	1.5	1.0	0.7

Table 2: Lifetime normalized with respect to set 1
8 meters per second annual mean wind speed and Eurocode 3

4. INVESTIGATION FOR FLAP SYSTEM IMPROVEMENTS

A common problem for wind turbines without rigid hubs is the control of teetering or flapping movements during start up, shut down and with the rotor parked. In the case of the M30 prototype this control is realized by a hydraulic system which permits free flapping during normal operation. A damping is introduced for start up, shut down and with the rotor parked to avoid excessive upwind flapping particularly during the emergency shut down, which is realized by passive blade pitching only.

The aim of the investigation was to fully understand the flap movement behaviour in order to simplify as much as possible the damping device.

As a first step, the aeroelastic codes were modified to improve the agreement between predictions and measurements for the emergency shut down transient. Both GH&P's and RC's codes are able to predict the flap transients quite well, but with some discrepancy on the final flap angle as described in [2].

The measured minimum flap angles are lower than the predicted ones and therefore more critical, since the blade gets nearer to the tower than according to calculations. Comparing the different computer simulation results, it can be seen that the minimum flap angle is independent of the damping coefficient as well as of the blade pitch change ramp. Both parameters influence mainly the flap angle velocity during the first transient after the shut down event.

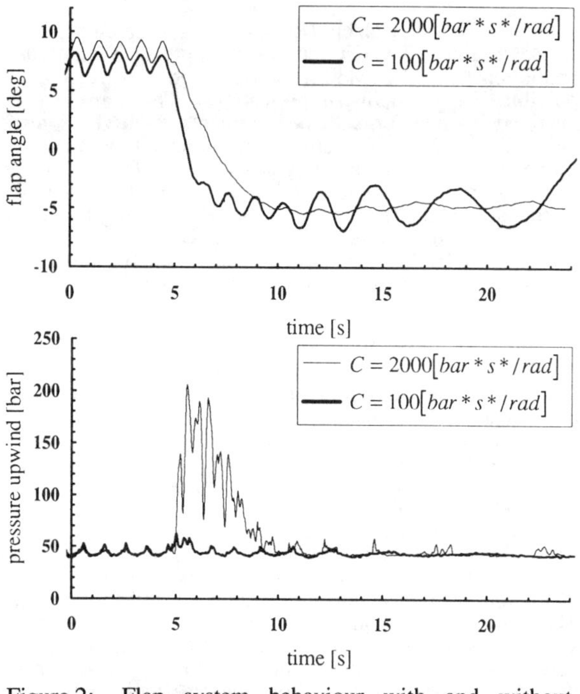

Figure 2: Flap system behaviour with and without damping

Following the computer calculations, the flap behaviour was also examined by on-site experiments. Figure 2 shows the flap angle and the upwind flap pressure for two cases. Without damping the blade is flapping very rapidly, at about 8 degrees per second, and moves from the mean operating flap angle of about 8 degrees to the minimum angle of about -5 degrees. The flap velocity with the damper active is much lower at about 3.8 degrees per second, but the minimum mean flap angle is unchanged. The pressure inside the damper increases up to 200 bar above the static pressure corresponding to a transmitted moment of 10 kNm. In addition to the higher flap velocity without hydraulic damping the larger flap fluctuation can be noted. This is because the aerodynamic damping decreases in a first approximation linearly with the rotation speed.

In addition to the test described above, the influence of other parameters such as nominal rotational speed, pitch speed ramp and nacelle yaw angle were examined.

The effect of the latter parameter is shown in Fig. 3 by the comparison of the flap dynamics for the nacelle correctly aligned into the wind and also misaligned by 30 degrees of yaw. During normal running the flap angle fluctuations have nearly the same amplitude. With decreasing rotation speed the flap angle fluctuation in the misaligned case is considerably greater than in the case of zero yaw misalignment.

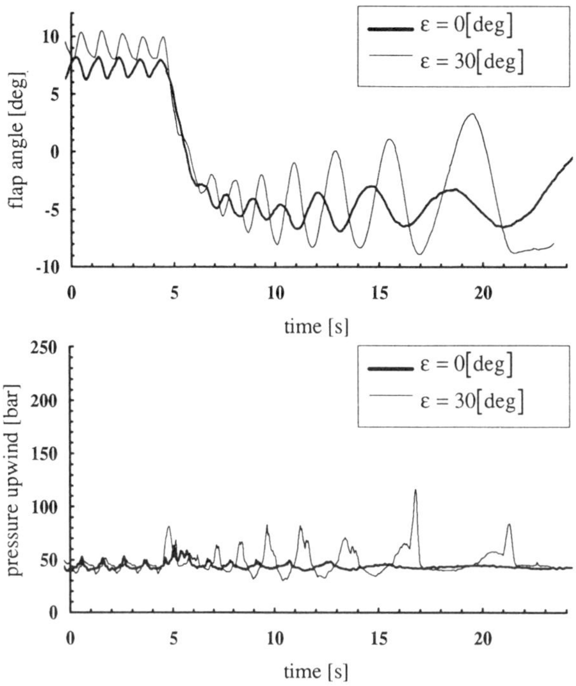

Figure 3: Flap system behaviour with nacelle correctly aligned into the wind and misaligned by 30 degrees

5. CONCLUSIONS

The project provided confidence in both the behaviour of the wind turbine and the validity of the aeroelastic codes. A high degree of accurancy for predicted loads is clearly essential when analysing fatigue damage rates. The fatigue life results were sensitive to all parameters considered in the study such as the number of wind speed bins, turbulence intensity and annual mean wind speed. In addition, considerable differences in lifetime were found by the use of the different standards as Eurocode 3 and BS 5400.

The other important result of the project is that the flap behaviour during emergency shut down is much less critical than expected even without any damping device. A significant simplification of the involved components will be the conseguence for future single-bladed wind turbines.

6. REFERENCES

[1] Safety of Wind Turbine Generator Systems
Draft of IEC TC 88 Secretariat 21
Report no. 92/26064 (1992) 1992

[2] C Stork, G Solmi, U Foli, L Pirazzi
Proceedings of Amsterdam EWEC'91
Van Hulle et al Ed (1991) 452

STALL INDUCED VIBRATIONS

Flemming Rasmussen, Jørgen Thirstrup Petersen
Risø National Laboratory, Denmark

Danny Winkelaar
ECN, Holland

Robert Rawlinson-Smith
Garrad Hassan and Partners, Bristol, Great Britain

ABSTRACT: The paper presents some of the results of a common effort between Risø, ECN and Garrad Hassan, supported by JOULE 1 (and for GH partly funded by the UK Department of Trade and Industry), with the aim to clarify the physical phenomena, identify the important parameters for the occurrence of stall induced vibrations and set up necessary restrictions and design guidelines for larger turbines.

1. INTRODUCTION

The possibility of stall induced vibrations as predicted by the aeroelastic codes in the early eighties has to some extent been hindering the development of large stall regulated wind turbines. The methods used to predict stalled behaviour have however, been based on two dimensional quasi-static aerodynamics, ignoring the effects of three dimensionality and unsteadiness.

The aims of the work in this joint project have been to clarify the physical phenomena involved, identify the important parameters for the occurrence of stall induced vibrations and set up restrictions for larger and possibly more flexible turbines. The program of work undertaken to achieve these aims has been based on theoretical modelling of four different stall regulated wind turbines.

2. STALL MODEL DEVELOPMENT AND VERIFICATION

An important precondition for obtaining realistic response predictions in stalled operation, is that a model for unsteady aerodynamics is incorporated into the aeorelastic prediction code. Each of the participants chose a different method for modelling dynamic stall, dependent upon the experience at the respective organizations. It was anticipated that the use of different models would aid in the understanding of the important aspects of the aerodynamic phenomena involved.

The ONERA-model [1,2,3] adopted by ECN describes the 2D unsteady non-dimensionalized lift and moment coefficients ϕ_F, ϕ_M as a set of two differential equations, which match the steady flow limit ϕ_F-$C_F(\alpha)$ and ϕ_M-$C_M(\alpha)$ and depend linearly on $\dot{\phi}_F$, $\dot{\phi}_M$ and $\dot{\alpha}$, where ($\dot{}$) stands for $\frac{b}{V}\frac{d}{dt}$:

$$\phi_F - C_F(\alpha)+\Lambda_F(\dot{\phi}_F,\dot{\phi}_M,\dot{\alpha})$$
$$\phi_M - C_M(\alpha)+\Lambda_M(\dot{\phi}_F,\dot{\phi}_M,\dot{\alpha})$$

The model is simplified by limiting the order of the differential equations and neglecting a number of terms, for instance, the coefficients coupling $\dot{\phi}_F$ and $\dot{\phi}_M$. The simplification cannot be derived from theoretical considerations and must therefore be validated empirically by comparison of model results with wind tunnel measurements. By analyzing (measured) transfer functions, which relate the variations of incidence to the unsteady terms, it was found that in the time domain the model should be described by a set of second order differential equations with variable coefficients. ONERA has identified the coefficients for a number of aerofoils, and there is also a set of "generally applicable" coefficients, which are used in the present study.

At Garrad Hassan the model adopted for inclusion of unsteady behaviour of aerofoils is that due to Beddoes [4]. It has been used successfully by Harris [5] and Galbraith [6] in the prediction of the behaviour of vertical axis wind turbines and by Hales and Garside [7] for calculation of unsteady periodic loading of horizontal axis wind turbines. The Beddoes model attempts to reproduce the behaviour of an aerofoil section in dynamic conditions through a combination of indicial aerodynamics and Kirchoff flow along with a number of empirically derived time lags associated with the physics of the flowfield.

The dynamic stall model implemented by Risø is based on simplifications to the Beddoes model performed by Øye [8]. It assumes that the dynamic value of the separation is lagging behind its static value, described by a time constant, which is proportional to the reduced time c/V. This time constant is determined from the best fit to experimental data, e.g. measured HAWT data.

For verification and parameter setting of the Risø model, predictions have been compared to measurements of

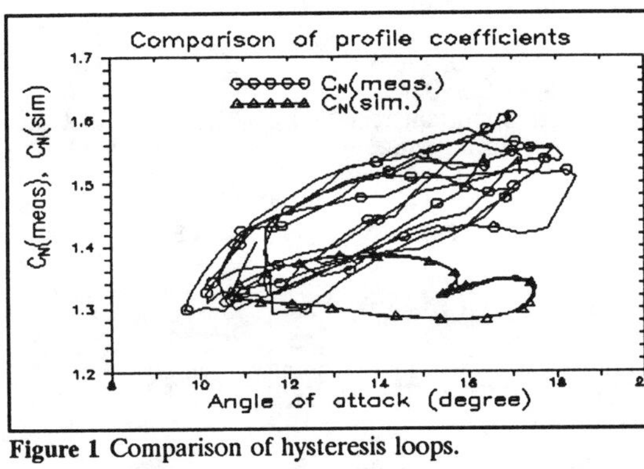

Figure 1 Comparison of hysteresis loops.

blade segment forces from the "HAWT-aerodynamics Concerted action" project (JOULE1 JOUR-0019-C). Such a comparison for the Risø model is shown in Fig. 1, which represents close to periodic variations from measurements in yawed flow. For a time constant of *3c/V* the loops are opening equally, but 3D effects in the measurements causes the average slope to deviate and this is not yet accounted for in the models. Comparisons with the other two models are not shown here, but they give in general similar results.

For comparison of the different aeroelastic codes the

modal damping using the same input data and steady aerodynamics is computed (for the WPX-20) and presented in Fig. 2. Obviously the two codes does not give identical results.

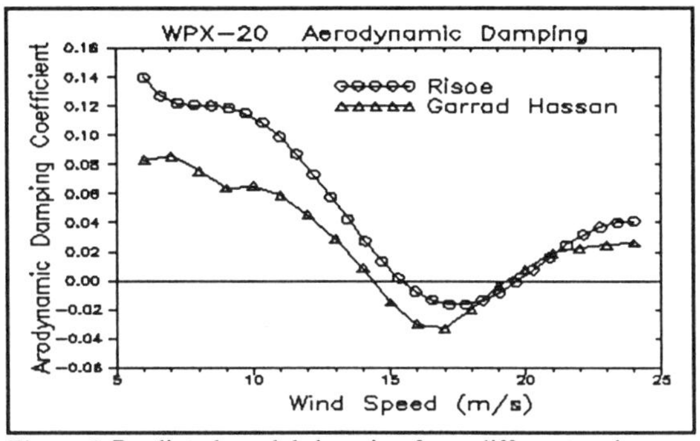

Figure 2 Predicted modal damping from different codes.

This is due to different assumptions and simplifications for the r models. However, the principal characteristics are in accordance, i.e. aerodynamic damping is negative, due to negative lift curve slope over an important part of the blade. This region of negative damping would suggest the possible occurrence of stall induced vibrations.

3. COMPARISON OF MEASUREMENTS AND PREDICTIONS FOR DIFFERENT TURBINES

In order to be able to investigate the problem of stall-induced vibrations within an appropriate wide range of situations, four quite different turbines, with respect to design philosophy and size, have been selected for measurements and modelling. These are the Nordex 250 kW, the ECN experimental turbine equipped with the commercial two-bladed 20 m diameter WPX-20 rotor, the Newinco 500 kW and the 630 kW Nibe A.

For the calculations the participants have been using their respective aeroelastic models, which are believed to provide an adequate modelling of the structural dynamics. The models all use quasi static aerodynamics in the basic version, but have been extended to include the effects of stall hysteresis. The modelled wind field includes simulated turbulence.

The measured and the calculated flapwise blade bending moments have been compared on basically three levels for representation of the response, and the comparison is primarily of a qualitative character, which intends to provide an indication of the extent to which the characteristics of the measured load history are represented in the predicted data.

Examples of representative time histories for two of the turbines in stall are shown in Fig. 3 and 4, respectively. The example calculations include the stall hysteresis models and show comparable characteristics. In general this is also the case for the waveform and the spectral density, although all participants have had some problems with the modelling of the WPX-20 turbine, which in the present test version has its fundamental blade flap wise frequency close to 3P. This is the only one of the investigated turbines, which showed response that might be interpreted as stall induced vibrations.

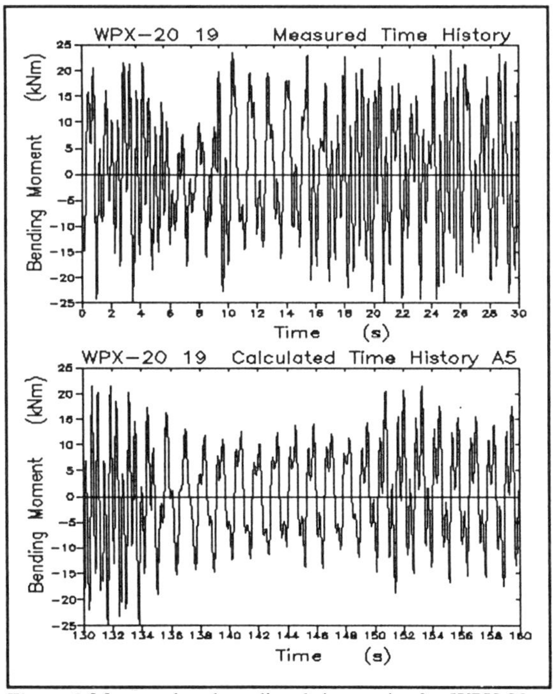

Figure 3 Measured and predicted time series for WPX-20 at 16.6 m/s.

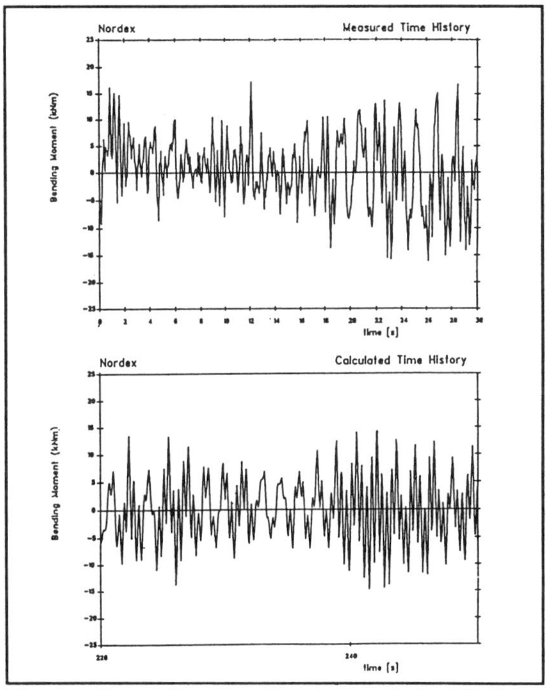

Figure 4 Measured and predicted time series for Nordex at 17.2 m/s.

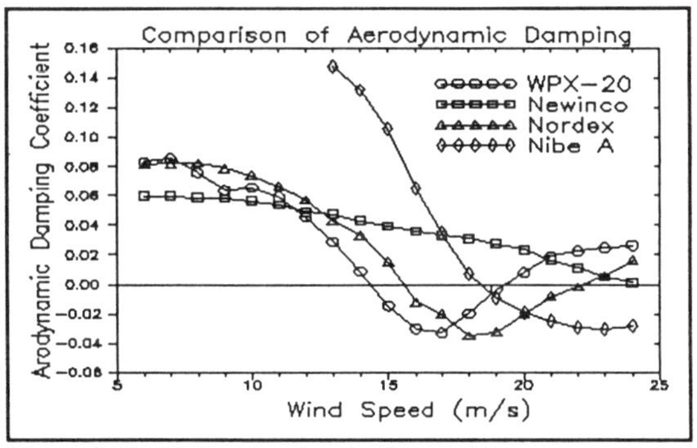

Figure 5 Modal damping for the four turbines.

In Fig. 5 the aerodynamic damping coefficient corresponding to the fundamental flapwise mode and static aerofoil data is shown for the four turbines. This is one of the most important characteristics of the turbine related to stall vibrations. A region of negative aerodynamic damping would suggest the possibility of divergent or limit cycle behaviour for this mode of vibration.

4. PARAMETRIC STUDY

In Chapter 3 it was revealed that the responses of the turbines were quite different in nature and were modelled with different success. In the modelling it is clear that the influence of some parameters is well predicted , while others are associated with some uncertainty. This has to do with the uncertainty on the assumptions and the simplifications in the aeroelastic models.

The aeroelastic models consist of three main parts: the input wind and turbulence part, the structural dynamics part and the unsteady aerodynamics interaction. It is believed that the aeroelastic models are by now so much developed and verified that the structural dynamics is quite well represented, which is an important precondition for using the codes in relation to wind turbine optimization and design.

One assumption in the turbulence models is that only the longitudinal component is taken into account. This has been a well established practice until now, but experience from the present work has indicated that the transversal components might be of importance as well.

The last part, the aerodynamic interaction contains unsteady aerodynamics in stall and represents the item where quite uncertain assumptions are likely to be taken, especially due to the lack of aerodynamic data.

The uncertainty on the parameters is different and the sensitivity of the turbine response to changes in these parameters is thus quite important with respect to design. This is treated further in this chapter in order to identify the important features determining the vibrational level in stall.

The aerodynamic design i.e. the chord, twist and profile distribution determines the distribution of the separated flow and thus the area of negative damping on the blade. The amount of local negative damping on the blade is directly related to the drop in lift coefficient after stall for the actual aerofoil section. As shown in Fig. 1 this drop is less for wind turbine operation due to 3D effects, which always means that using 2D data is conservative with respect to appearance of stall induced vibrations.

For chosen profile data and blade planform the twist distribution is also important for the stall distribution and a relatively small change in the blade tip setting also changes the aerodynamic damping. This is illustrated in Fig. 6, which shows the integrated aerodynamic damping as a function of wind speed for WPX-20 for three different blade settings.

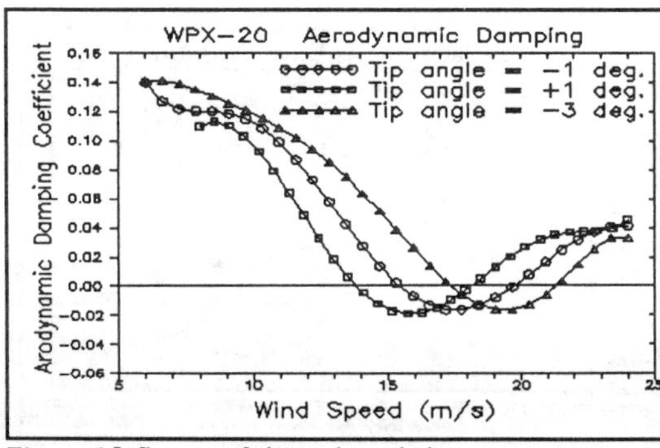

Figure 6 Influence of tip angle variation.

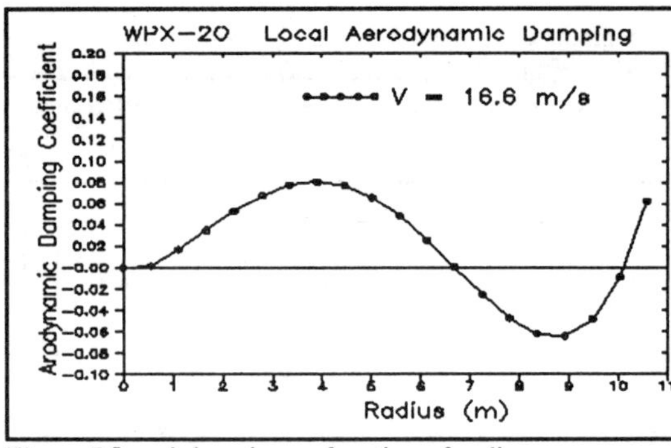

Figure 7 Local damping as function of radius.

The results of a more detailed investigation of the local aerodynamic damping at a certain wind speed are illustrated in Fig. 7 along the blade span. Typically at this wind speed the local aerodynamic damping is negative around the three quarter station of the blade, and positive closer to the root and at the tip.

The total integrated aerodynamic modal damping as represented in Fig. 6 is obtained by multiplication of the local aerodynamic damping in Fig. 7 by the mode shape of the blade and integrating. This means that also the mode shape is of importance for stall induced vibrations.

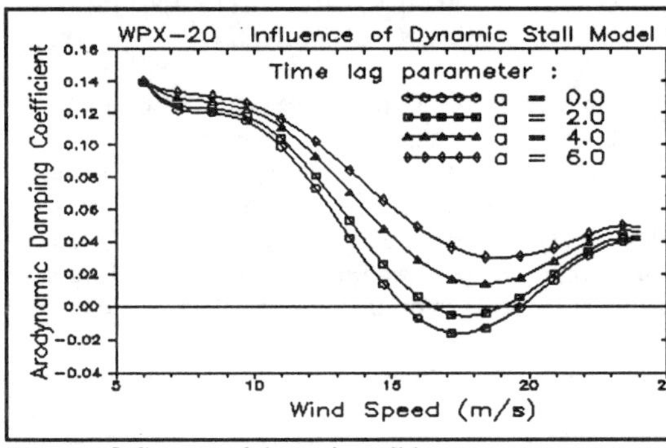

Figure 8 Influence of dynamic stall lag parameter on aerodynamic damping.

In order to illustrate the influence of the dynamic stall model on the presence of stall induced vibrations, the modal aerodynamic damping has been calculated for different parameter settings corresponding to different values of the time lag parameter in the dynamic stall models. The result is presented in Fig. 8

It is obvious that a time lag parameter slightly higher than 2 is necessary in order to create positive damping. This is quite well in accordance with Fig. 1, where a value of approximately 3 was obtained from comparison to measurements.

5. CONCLUDING REMARKS

The analysis of measurements and predictions on different turbines performed within this project has made it clear that determination of the vibrational level in stall for certain turbines is associated with some uncertainty. However, by use of unsteady aerodynamic models it has been possible to obtain quite realistic responses. It is clear that the structural dynamics is quite well represented in the aeroelastic codes, the turbulence reasonably well and that the unsteady and three dimensional aerodynamics involves most uncertainty. The principal influence of the last, however, is defined and always acts to increase aerodynamic damping relative to 2D steady assumptions.

REFERENCES

[1] Tran,C.T. and D.Petot. "Semi-empirical model for the dynamic stall of airfoils in view of the application of a helicopter blade in forward flight". In 6th European Rotorcraft and Powered-Lift Aircraft Forum, Bristol, UK, September 1980.

[2] Dat, R. and C.T.Tran. "Investigation of the stall flutter of an airfoil with semiemperical model of 2-D flow". Vertica, Vol. 7, 1983.

[3] Peters, D.A. "Toward a unified aerodynamic model for use in rotor blade stability analysis". In 40th Annual National Forum of the American Helicopter Society, Arlington, Virginia, USA, May 16-18 1984.

[4] Beddoes, T.S. "A synthesis of unsteady aerodynamic effects including stall hysteresis". Vertica, Vol. 1, 1976.

[5] Harris, A. "The Role of Unsteady Aerodynamics in Vertical Axis Wind Turbines". Recent Developments in the Aerodynamics of Wind Turbines. BWEA workshop, University of Nottingham, February 1990.

[6] Galbraith, R.A.McD, A.J.Niven and F.N.Coton. "Aspects of unsteady aerodynamics of wind turbines". Recent Developments in the Aerodynamics of Wind Turbines. BWEA workshop, University of Nottingham, February 1990.

[7] Hales, R.L. and A.J.Garside. "Design and Analysis Techniques for Wind Turbines in Unsteady Flow Environments". Final Report for DEn Research Contract Number E/5A/CON/5004/177/025, July 1985.

[8] Øye, S. "Dynamic Stall - Simulated as time lag of separation". Proc. of the 4th IEA Symposium on the Aerodynamics of Wind Turbines.
Ed. by K.F:McAnulty, Rome, Italy,January 1991.

RESULTS FROM WIND TUNNEL TESTS ON A Ø 5.35 M HAWT IN YAW OPERATION

Jan-Åke Dahlberg, Göran Ronsten & Eskil Ulén

FFA, The Aeronautical Research Institute of Sweden
P.O. Box 11021, S-161 11 Bromma, SWEDEN
Tel: +46 8 634 1000, Fax: +46 8 253481

INTRODUCTION

The method to reduce the power and/or the rotational speed by yawing the rotor disc out of wind has been widely and successfully used for small and medium sized wind turbines. There exists an interest to apply this power/overspeed control method on large wind turbines.

To minimise the risk and reduce the cost during the design phase and operation of a wind turbine using yaw control it is essential to use a validated computational tool. A prediction of the behaviour of a wind turbine concerning forces and movements will be trustworthy only if it can be shown that the method in use previously has been verified against experimental data.

There is however not enough experimental data available for a wind turbine operating in such well-defined situations as in a wind tunnel. Therefore, a wind turbine was designed, built and tested as a joint project between FFA and CARDC (China Aerodynamics Research and Development Centre).

Several tests with a 5.35 m diameter turbine were carried out in the CARDC large (test section 12 m x 16 m) low speed wind tunnel during March 1992. Emphasis was put on mapping the turbine load situation and teeter movements for different yaw operation modes. Rigid hub and teetered hub with different delta-3 angles were tested. The following operational modes were examined:

- Yaw rate tests, ±75 °. Yaw rates 1, 5, 10 and 15 °/sec.
- Test of active damping of the tower lateral resonance
- Test of a yaw controlled variable speed system with an induction generator.
- Yaw stability for a turbine with cone angle (0,3 and 6°)

The following parameters were measured:

- Blade root bending moments (flap and edge wise).
- Forces (3) and moments (3) in the top of the tower.
- Teeter angle.
- Rotational speed.
- Yaw angle
- Azimuth turbine angle.
- Accelerations in the tower top
- Wind speeds.
- Wind tunnel wall pressure distributions

YAW STABILITY

Rigid hubs with coning angles of 0°, 3° and 6° and teetered hubs with Δ-3 angles of 0° and 30° were tested in order to map the yaw stability. The yaw stability is determined by the turning moment around a vertical axis through the turbine plane of rotation and the side force that acts on the turbine rotor.

A distance between the yaw axis and the centre of the hub will create a lever arm for the side force and thereby cause a yaw moment. Figure 1 shows the side force gradient as function of tip speed ratio for the tested hub configurations. The gradient is determined by a linear fit applied to the side force within a ±20° yaw angle range around 0° yaw angle.

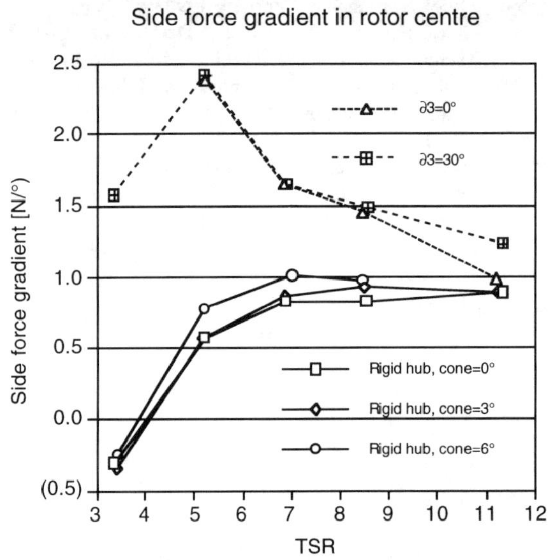

Figure 1: Side force gradient, in the rotor centre, for various hub configurations, as function of tip speed ratio.

A positive side force tends to turn the turbine out of wind. As can be seen in figure 1, the side force is de stabilising for all tested hub configurations. Figure 2 shows the yaw moment gradient as function of tip speed ratio for the tested hub configurations.

The rigid hubs create a stabilising yaw moment gradient at medium to high tip speed ratios while the teetered hubs give de stabilising yaw moments at all tip speed ratios. Maximum yaw stability was achieved using the rigid hub with a coning angle of three degrees. The test show that the turning moment is of great importance for the yaw stability and that the side force always is de stabilising. Yaw stability can be achieved by the using a rigid hub with a coning angle and by keeping the distance between the hub and the yaw axis as short as possible.

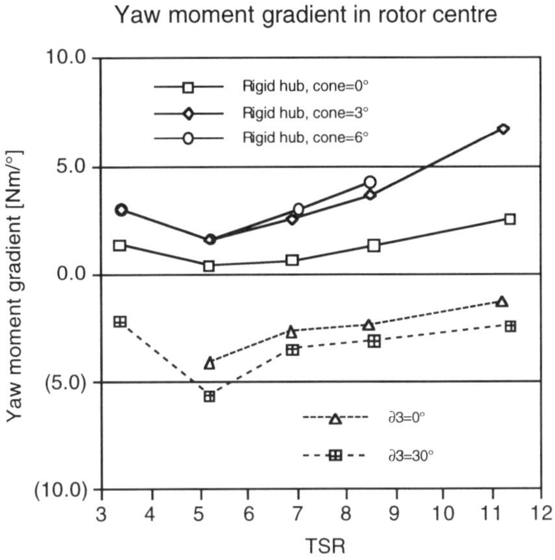

Figure 2: Yaw moment gradient, in the rotor centre, for various hub configurations, as function of tip speed ratio.

YAW CONTROL

The system is characterised by:

- two blades with fixed pitch
- variable and fixed speed
- rigid and teetered hub
- torque-regulated induction generator

The tests were performed with variable speed and yaw control and at fixed speed. Figure 3 shows a simplified diagram of the control system.

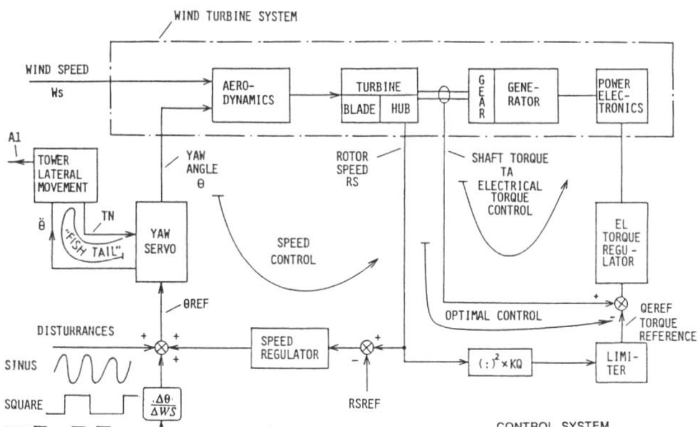

Figure 3: A simplified diagram of the control system.

Examples of wind speed and yaw disturbances

Wind speed and yaw angle disturbances were introduced. The figure below shows a wind speed variation, the resulting yaw angle, thrust force and teeter angle in a variable speed system with yaw control (RSref=25 rad/sec, see figure 3).

Figure 5: Test with a sinusoidal yaw disturbance at 240 RPM, λ=6. The plots show yaw rate, yaw angle, teeter angle and motor torque. →

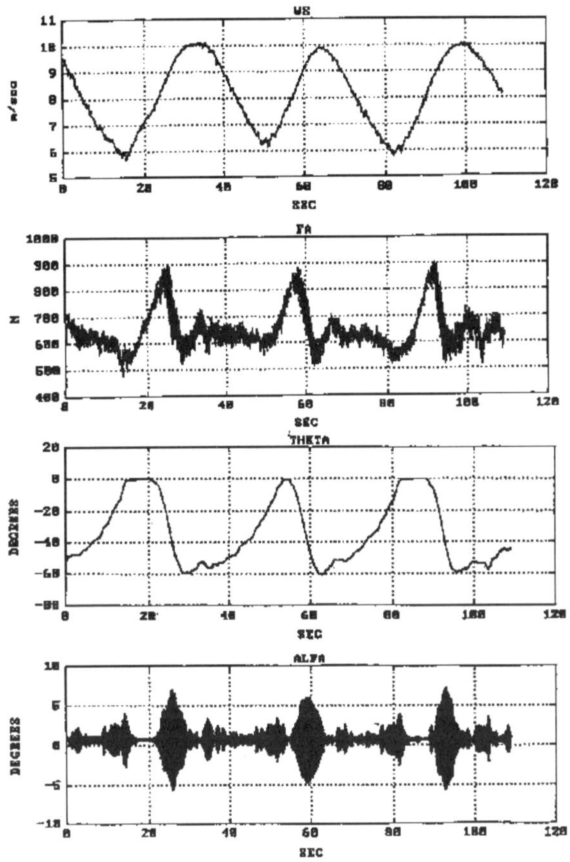

Figure 4: Wind speed variation. The graphs show wind speed, thrust force, yaw angle, and teeter angle.

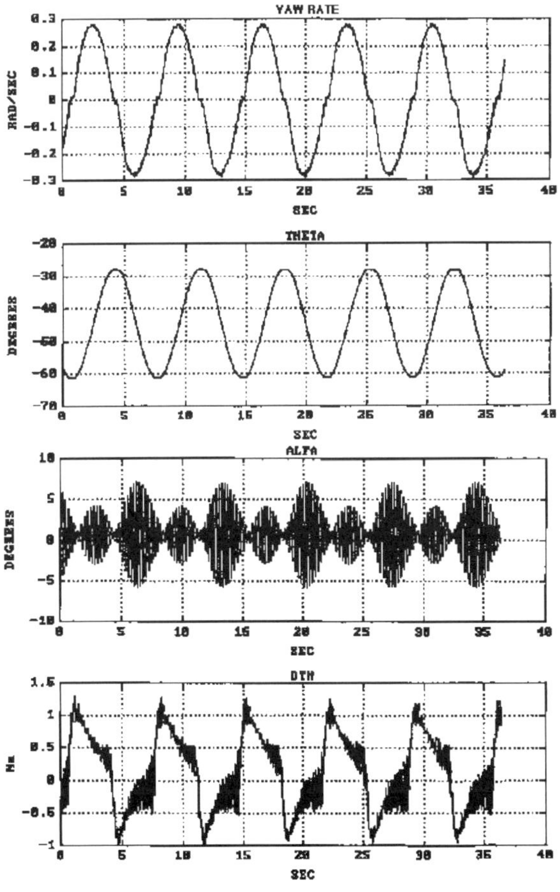

Active damping of tower lateral resonance

It is normally not possible to avoid a distance between the centre of gravity of the nacelle and the yaw axis. Rapid yaw movements can then excite the tower resonance. The loop is in fact closed, see figure 6.

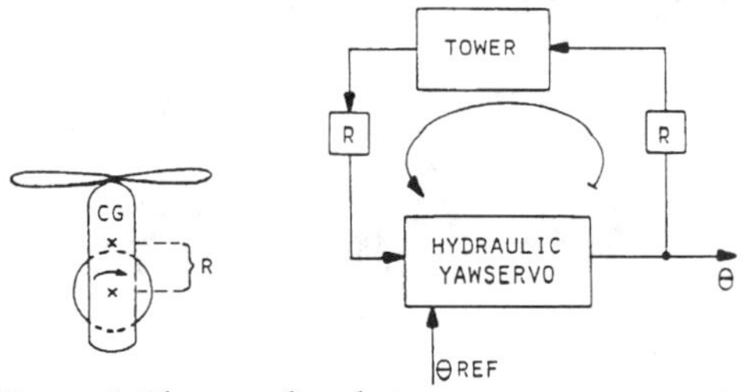

Figure 6: The coupling between yaw movements and the tower lateral resonance.

A damping can be introduced by means of a relatively big leakage in the servo motor. A better method is to make the leakage frequency dependent. The tower resonance at higher frequency will then be damped while the yaw servo performance is maintained. A combination of accumulators and orifices that are coupled in parallel over the motor or a small piston in combination with an orifice or possibly a membrane can be used.

Active damping by means of torque feedback in the yaw servo was used in our test. Figure 7 shows a simplified yaw servo diagram.

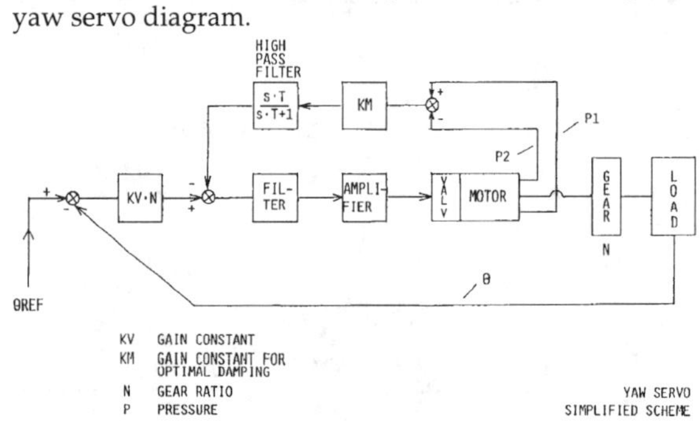

Figure 7: Simplified yaw servo scheme.

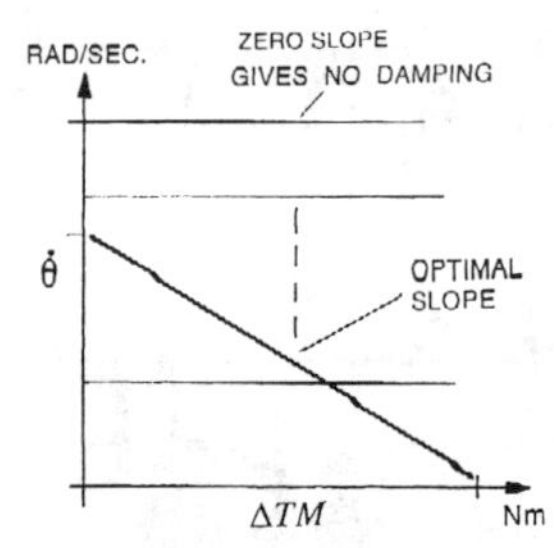

Figure 8: Linear torque-velocity diagram

A linear approach shows that the damping is dependent on the slope in the torque-velocity diagram. This slope is determined by the gain in the torque feedback. The magnitude and frequency behaviour of this gain can easily be changed.

A closer analysis shows that a certain gain is optimal in the sense that it produces maximum obtainable damping. However, in a valve controlled servo there is already a slope in the torque-velocity diagram. The torque feedback can then enable augmentation of the damping from a lower level to the optimal. Figure 9 shows an example of torque feedback and its effect on the tower lateral acceleration.

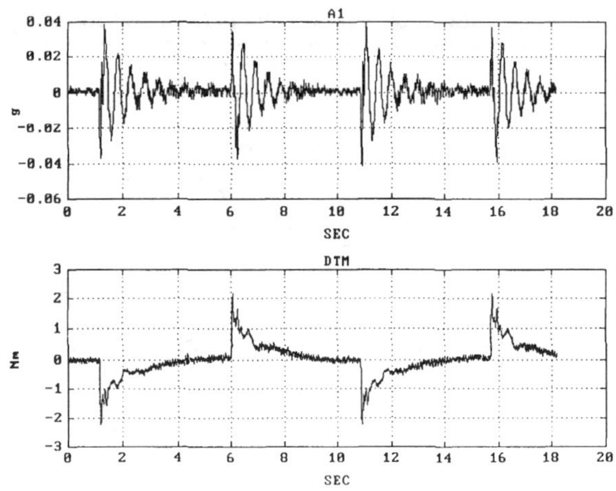

Figure 9: Acceleration and motor torque response to yaw angle steps.

TEETER ANGLE: Maximum amplitude of the teeter angle depends on the rotor- and the yaw rate. Figure 10 shows approximate values. There is a small bias <1°.

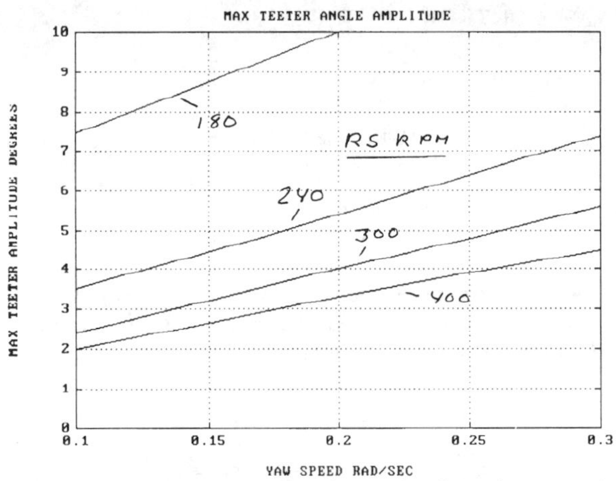

Figure 10: Maximum teeter angle amplitude as function of yaw rate and RPM.

BLOCKAGE CORRECTIONS

Any object inserted in a wind tunnel will cause the wind speed to change. A wind speed correction is therefore necessary to apply to a set of data if it is going to be compared with the results from calculations.

Wall interference velocity

A first step in the process of estimating the blockage is to calculate the wind tunnel wall induced velocities. For this purpose it was found necessary to measure the static pressures along the wind tunnel walls. Wall pressure measurements were carried out in the test section at 196 positions using 5 Scanivalve units connected to SETRA differential pressure transducers (±350 Pa). The total pressure from a Prandtl tube located at the start of the test section was used as the reference pressure.

Numerous methods are available for calculating wall induced velocities in a wind tunnel but few claim to be valid when the model is a wind turbine. One reason is that not enough test data are available and another is that the assumption of a small enough volume of irrotational flow is questionable in this case. A computer code

(WIN3LS) [1], was made available to FFA by NLR, 'National Aerospace laboratory', Holland. The method has not been validated for use on wind turbines. As the blockage corrections also are to be calculated by CARDC it was seen as an opportunity to compare these results with those from the method developed by Jiang [2].

An analysis of the wall-induced velocities from WIN3LS, as performed in [3], shows that the axial variation along the test section and the variation across the turbine are reasonable. It was assumed that the difference in wind tunnel wall pressure between the actual test case and the empty tunnel decreased linearly to zero from the end of the test section to a position located 25 m downstream.

Blockage correction

The maximum calculated wall interference velocity along the center-line was used when applying the ε–max. method as referenced in [4] with the exception that the horizontal buoyancy was accounted for using the expression $\Delta C_{T,Bo} = -1/4 * A/C * C_{T,U}^2$ instead of using the pressure gradient and the model volume.

The maximum power coefficient at yaw angle = 0° decreased from 0.58 to 0.48 and the corresponding tip speed ratio decreased from 8 to 7.5. The corrected power coefficients were found to agree well with 3rd order polynomials. This is shown in figure 11.

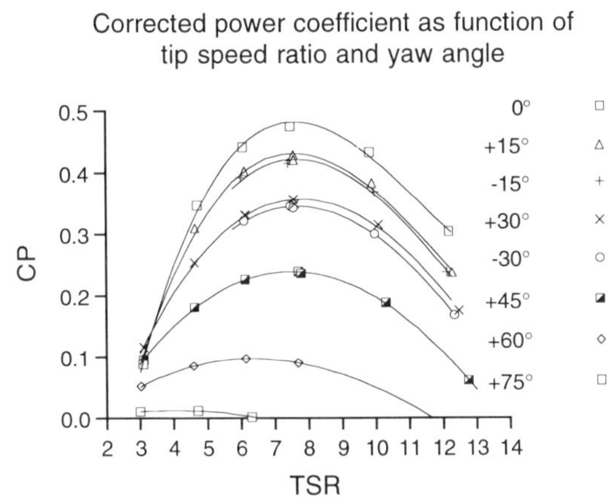

Figure 11: Power coefficients (corrected for blockage) as function of yaw angle and tip speed ratio.

The power coefficient for a turbine in stall is larger at a yaw angle of 30° then at 0°. The magnitude of the blockage correction is small for a tubine in stall.

The trust coefficient at yaw=0° and a tip speed ratio of 13.3 decreased from 1.4 to 1.1 and the corresponding tip speed ratio to 12.2. The best thrust force coefficient estimations were logarithmic at yaw=0° and ±15° while 2nd order polynomials were sufficient at larger yaw angles, figure 12.

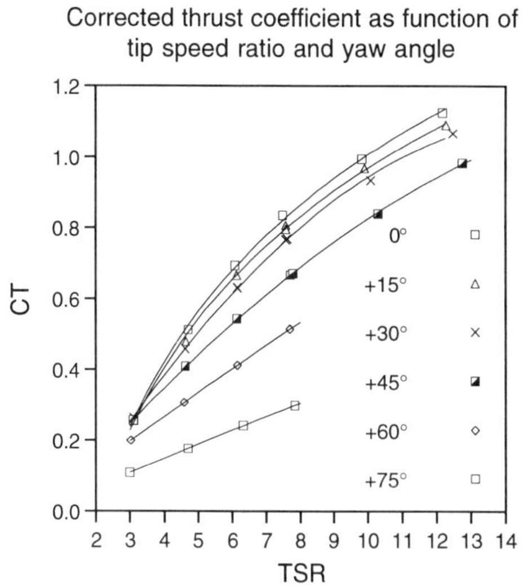

Figure 12: Thrust force coefficients (corrected for blockage) as function of yaw angle and tip speed ratio.

CONCLUSIONS

Yaw stability

Yaw stability can be achieved by using a rigid hub with a coning angle and by keeping the distance between the hub and the yaw axis as short as possible.

Control

Maximum teeter angle, damping of the tower resonance, the required yaw motor torque and disturbances around the yaw axis are of interest in a yaw controlled system.

The maximum amplitude of the teeter angle depends on the rotor- and the yaw rate.

Optimal damping of the tower lateral resonance can be achieved if torque feedback is applied in combination with a properly designed yaw servo. The bandwidth of the torque gain must be about four times higher than the tower resonance frequency.

Yaw motor torque was in our test mainly needed to overcome the non-linear friction.

The yaw torque disturbance depending on turbine rotation and yaw rate are small for a turbine with a teetered hub and cause no problem in regard to the torque feedback.

REFERENCES

[1] Labrujère, Th. E., " WIN3LS, A program for wall-interference assessment in a solid-wall wind tunnel", Memo AT-88-011 L, 1988

[2] Jiang Guiqing, "New advances in wind tunnel wall interference correction for pressure measurements", CARDC, ICAW 1991, 1991

[3] Samuelsson, I., "PM concerning the analysis of wall pressure data used in calculations of wind tunnel interference" (in Swedish), FFA, 1992

[4] Hacket, J. E. and Wilsden, D. J., "Estimation of tunnel blockage from wall pressure signatures: a review of recent work at Lockheed-Georgia", AIAA-78-828, 1978

Experimental and theoretical results of the UNIWEX wind turbine ‡

Gregor E. van Baars, Peter M.M. Bongers §, Herbert Schwartz, Hans Arnold ♯

Abstract

From theoretical modelling point of view the transfer function from blade pitch to flap moment is extremely interesting because it is mainly determined by aerodynamics and aeroelastics. These modelling aspects have not been properly validated in detail for full scale real life wind turbine systems. This is mainly due to the complexity of the wind field acting on the rotor. System identification can help to solve these complex validity questions. This paper provides two extensions to standard Prediction Error identification for the detailed dynamics observed from a real life flexible wind turbine system. Although the wind drives the wind turbine it still is considered as a disturbance. In this paper rough knowledge of the wind field is used to structure the stochastic part in the identification procedure. Some disturbances clearly relate to rotor position. A transformation is proposed and applied to the original data in such a way that every rotation consists of the same number of samples (while the time to complete a round varies). This clearly improves the identification results obtained from experimental data measured at the UNIWEX test turbine.

Keywords: system identification, application, wind power plants, model validation, prediction error methods, disturbance modelling, measurements

1 introduction

To achieve a significant contribution of wind generated power to public demand of electical power wind turbine systems have to gain efficiency, life time and reliability. It is expected that these demands can be satisfied if WECS are developed that allow *flexibility* in the different subsystems that compose the complete wind turbine system [5]. The dynamic behavior of a wind turbine system will be a mix of aerodynamics, structural dynamics, and electrical conversion phenomena. This mix will become stronger if these flexibilities are allowed in the different components of the wind turbine. This poses a more complicated *design* and *control* problem for a flexible wind turbine compared to conventional rigid wind turbines. To solve these design problems an accurate *model* describing the relevant dynamics of the complete wind turbine is necessary. In [5] an *integrated* dynamic model of a flexible wind turbine is presented which takes into account both the aero-elastic rotor parts as well as the rotating machinery. Models of this kind are scarce and only have been developed recently. For these models to produce reliable predictions they need to be *validated* with respect to their dynamics for different configurations. Some of these dynamic models have been partly validated by quasi-stationairy measurements. A model of a rigid wind turbine described in [6] is also validated with respect to its dynamics.

From theoretical modelling point of view the transfer function from blade pitch to flap moment is extremely interesting because it is mainly determined by aerodynamics and aeroelastics. These modelling aspects have not been properly validated in detail for full scale real life wind turbine systems.

The validity is pursued by application of system identification to experimental data obtained from the UNIWEX test turbine in Stuttgart, Germany.

The UNIWEX experimental wind turbine offers the possibility to emulate a large set of (flexible) wind turbine configurations [8]. The benefits of this experimental wind turbine lie in the fact that there are no hardware changes involved when switching between different configurations. A lot of different rotor systems have been measured at different wind speeds.

In general straightforward application of system identification to raw wind turbine data does not lead to acceptable models. See for results [1, 3, 9]. The results are for example sensitive to changes of model order which makes judgement upon the reliability of the identified models difficult.

Because the wind speed is not uniform over the rotor disc the relative wind speed acting on the blade changes during one revolution.

The three major causes for this are wind shear, tower shadow, and yawed flow. These cyclic disturbances are clearly related to the position of the rotor. Therefore the data which are sampled with fixed sampling frequency are transferred to data that are equidistant in rotor position. Incorporating a signal that represents the area that is influenced by the tower shadow improvements in identification results (blade pitch to flap moment) have been obtained and will be discussed in the paper.

The paper is organized as follows. The system identification in general and the application to wind turbine data is the subject of Section 2. Two extensions to the identification procedure are introduced in Section 3 and Section 4. Results obtained with the extended approach are given in Section 5. The paper end with conclusions in Section 6.

‡This research was supported by the Netherlands agency for energy and environment under grant 40.35-001.10

§Mech. Eng., Systems and Control Group, Delft University of Technology, Mekelweg 2, NL-2628 CD Delft, Tel +31 15 785617, e-mail baars@tudw03.tudelft.nl

♯Institute for Computer Applications, University of Stuttgart, Pfaffenwaldring 27, 7000 Stuttgart 80

2 identification of wind turbine dynamics

The topic of identifying wind turbine dynamics will now be discussed. The system identification procedure of the well known prediction error type [7] will be glanced at followed by a discussion how the wind turbine fits in this framework. In the sequel the abbreviation PEM will be used sometimes. The typically wind induced disturbances are characterized and their expected influence on the identifation problem are reflected upon. The identification is focussed on the transfer function between pitch angle and flap moment at the blade root. Two extensions to the identification procedure are proposed to overcome difficulties that have been encountered during identification of this transfer function.

2.1 prediction error identification

Usually the system identification procedure consists of the following steps:

1 design and perform identification experiment. specify input signal, sampling rate, duration.

2 treat the data to obtain a dataset suitable for identification. remove mean values, eliminate spikes, scale signals.

3 specify model structure.

4 specify model order. Decide how many parameters are used in the model structure.

5 choose criterion function.

6 estimate model.

7 evaluate model.

It is straightforward that this normally does not deliver a reliable model at the first attempt. As a start simple model structures are used. Some indication can be obtained of the relevant order that can be put in more complex models. The process of identification typically consists of iterations of trial and error. This will be referred to as the *standard* approach.

The usual system representation in black box system identification [7] is given in Fig. 1.

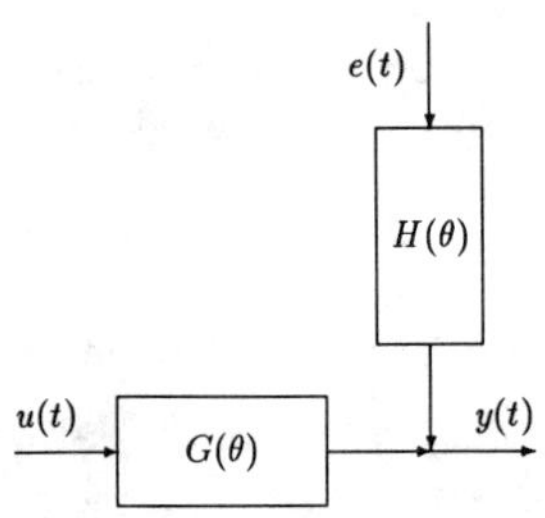

Fig. 1: black box identification model representation

The process of interest is the deterministic model part G. This part describes how the inputs $u(t)$ influence the output variables $y(t)$. All disturbances are modelled as filtered white noise $e(t)$. The transfer funtion H is therefore called the stochastic model part. This model represents a linear time invariant description of the system.

2.2 wind turbine identification

The wind turbine system can be represented by the following picture Fig. 2

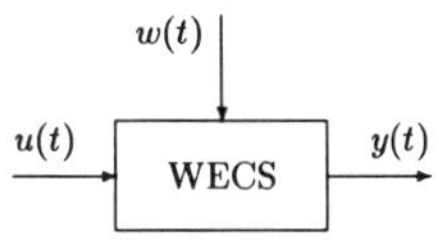

Fig. 2: wind turbine system representation

The wind $w(t)$ is not a contollable input to the system. Therefore it is considered as a disturbance or noise contribution to the output. The blade pitch angle θ is the input that is excited during the experiments and we want to identify the dynamical relation between the pitch angle and the observed outputs (rotor speed, rotor torque, flap bending moments at the blade root, etc.). The question is under what conditions the wind turbine system of Fig. 2 fits in the prediction error model structure of Fig. 1. The influence that the input has on the outputs depends on the point of operation in which the wind turbine system has been driven by the actual wind speed. PEM models are linear and can only be realistically applied if the average wind speed remains approximately constant during the experiment. Next it is assumed that the contribution of wind condition variations around this average can be characterized by the stochastic model part driven by unknown white noise.

2.3 disturbance considerations for wind turbine identification

The identification of wind turbine dynamics introduces some problems which do not occur in most other applications of system identification. First the role of the wind velocity on the identification process will be discussed in more detail.

1 point of operation
For almost every plant the point of operation can be controlled, but for a wind turbine to arrive in some prespecified operating condition the corresponding wind speed has to be present. In practice this means that for some measurements to be made one may have to wait weeks or longer.

2 system nonlinearity.
In combination with the natural fluctuation of wind velocity, the system runs through a section of a nonlinear range of operation. For identification a linear approximation of the dynamical relations between inputs and outputs needs to be realistic. Therefore only moderate fluctuations of wind speed and direction should occur. Because the wind is the variable that is beyond control the guaranty of gathering informative data can not be given beforehand. Always a trade off has to be made between restrictions on the wind conditions and the required amount of data to be able to estimate reliable models (quality against quantity).

3 periodic disturbances
These are introduced by the rotation of the turbine rotor and are not related to the applied inputs. The so called *tower shadow* affects the rotor both in down wind and down wind configuration (but stronger in down wind). Another aspect is *wind shear*. This is the phenomenon that the average wind speed varies with heigth. Since the blades rotate they feel on the average a variation of wind speed during a revolution. Therefore wind shear is also considered to be a disturbance.

4 turbulence of the wind.
Even if the average wind speed is uniform and constant over the rotor disc the actual wind speed is a function of place and time. This stochastic process generates noise on the output signals in addition to measurement noise.

2.4 remarks

Only if all conditions remain constant and rotor speed is constant some of these disturbances are periodic in time. Normally this will not be the case.

In most system identification applications the disturbances are of stochastic nature (e.g. measurement noise). The disturbances described above are not stochastic but have a more or less deterministic (but unknown) character. Clearly they do not fit in the class of PEM disturbances.

If the disturbances do not play a dominant role in the observed signals and the time scale of the system transfer function to be identified differs enough from the frequency range that is covered with noise standard identification performs well. All disturbances can be lumped into one stochastic noise contribution. Statisfactory results have been obtained with this approach for the identification of the transfer function from pitch angle to rotor speed and torque for a rigid configuration of the UNIWEX wind turbine [4, 1]. These models appeared to be a good basis for the design of a robust controller of the rotor speed by acting on the pitch of the blades [2] and worked satisfactory after implementation in practice.

2.5 identification of flap dynamics

As mentioned in Section 1 the transfer function from blade pitch angle to blade root flap moment is very interesting with validation of the theoretical model in mind. Also for control purposes it is essential to make reliable prediction of blade loads if the controller aims at minimization of fatigue in combination with power output optimization.

Standard identification did not produce satisfactory models. Models of different structure and different orders for both deterministic and stochastic model part have been calculated, but the differences between e.g. seperate models were too large to judge one model to be more reliable compared to another.

For the case of flap moment as output the disturbances discussed in the previous subsection have severe impact. Both tower shadow and shear are felt directly in flap direction. Therefore it is no surprise that this transfer function is diffucult to identify in a standard manner.

Although this strong influence on the flap moment signal the disturbance is in some sense structured. For example the tower shadow influence can be related to a specific region of the rotor disc. This structural knowledge is not used in the standard identification process yet. The idea, exploited in the sequel, is now to use this knowledge by a constructing a signal representing the tower shadow influenced area. This more or less artificial signal is used as extra input signal in the identification data set (multi input single output). The identification should now be able to relate the flap moment response to the tower shadow passage of the blades to this help signal. Thereby the parameters in the deterministic part can be fully utilized to fit the flap moment response related to pitch changes. The construction of this help signal is the next subject.

The second deviation from standard identification originates from the observation that not exactly periodic in time due to variations of rotor speed. If data can be transformed such that every revolution consists of the same number of samples the rotor speed variations are cancelled out and the repeating disturbances are exactly periodic. Hopefully this will improve the identification of the periodic disturbances and thereby the identification of the deterministic transfer function we desire to identify.

The paper continues with the discussion of these two suggested extensions can contribute to the identification accuracy. Results of first elaboration of the extended approach can be found in [9].

3 extending the noise model

In this section the extension of the noise model is discussed. First changes to the standard identification setting of Fig. 1 are evaluated. The construction of an artificial help signal describing the area where the tower shadow influences the blades is explained. Some remarks will be made about the implications of this extension.

3.1 extended identification setting

The identification model structure including the extra signal is depicted in Fig. 3

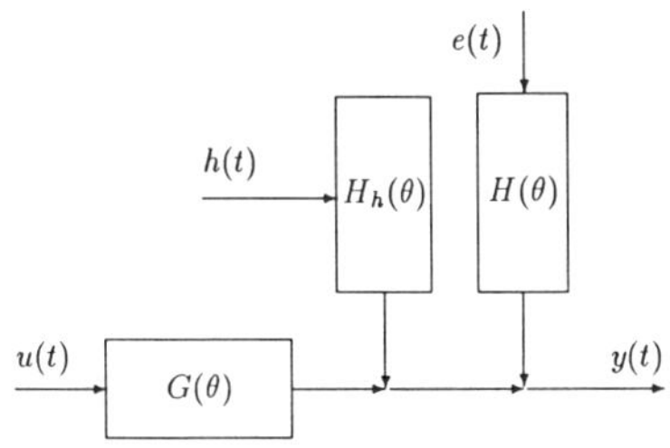

Fig. 3: extended model structure

The signal $h(t)$ is the help signal that represents the rough knowledge about some aspect of the disturbance on the system output. It passes through the transfer function $H_h(\theta)$ where the subscript h stands for *help*. The signal $h(t)$ is plugged into the data set as an extra input and the extended data set $Z_{ex}(t)$ is defined as :

$$Z_{ex}(t) = \left[\ y(t)\ \ \left[\ u(t)\ \ h(t)\ \right]\ \right]$$

Although $h(t)$ is treated as a deterministic input it is meant to model disturbances. All the noise and disturbance contributions to the data should be covered by $H_h h(t) + He(t)$.

Whether or not the noise model extension H_h has a physical interpretation is not of primary importance. In the first place it is introduced to make the identification of the deterministic transfer function G easier.

3.2 remarks

Some remarks have to be made about this proposed extension to the standard system identification model structure.

1 No guaranty can be given about the cuasality of the wake response. If the pulses are made too narrow the flap response may start before the pulse is present and no causal relation can be found. Therefore the width of the pulses needs to be chosen with care.

2 It may be expected that changes of average wind velocity generate different tower shadow responses while the height of the pulses is chosen fixed throughout the data set. In case this causes problems the pulse height can be chosen as a function of the actual wind speed at that specific tower passage. On the other hand this should not be a problem because large wind velocity variations would never be allowed in the data selection stage because of linearity requirements.

3 To cover the interaction of another blade passing the tower shadow on the blade we are modelling the same help signal can be used shifted over the corresponding difference of azimuth angle.

4 transformation of data

This section introduces a transformation of data from equidistant samples in time to equidistant samples of rotor position (called azimuth angle). Therefore a revolution consists of exactly the same number of samples while the original data set use slightly different number of samples due to small rotor speed variations. The benefits of this are twofold. In the first place the construction of the help signal $h(t)$ (Section 3) is easier because the pulse width and the number of samples between the pulses are both constant. In the second place it is expected that if the input signal and the disturbance contribution of the tower shadow are exactly periodic the identification of $H_h(\theta)$ performs better.

4.1 notation

To make discussion of the data transformation a bit more precise first some notation is introduced. Let Z_T^N denote the originally measured data consisting of N samples with time sample interval T. The transformed data set Z_ϕ^M consists of M samples with azimuth sample interval ϕ. This azimuth interval needs to be chosen such that the transformation of data conserves the information contained in the data as good as possible. Therefore M should be approximately equal to N. In addition to this the sample interval should fit an integer number of times in a revolution: $2 \cdot \pi = k \cdot \phi$ with $k \in \mathbb{N}$.

4.2 interpolation

The data points of the transformed set obviously exist in the neighborhood of the original data. These values will be determined by means of interpolation. The accuracy of the interpolation will depend on the "distance" between the time samples and the azimuth samples Fig. 4.

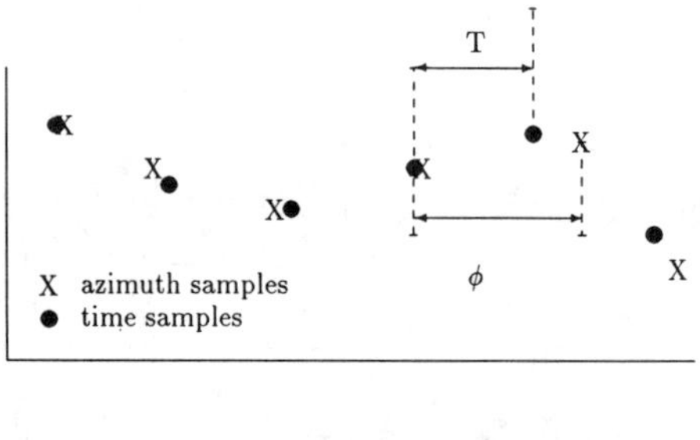

Fig. 4: original and transformed data

4.3 remarks

The transformation is likely to generate large interpolation inaccuracy in the case of strong and abrupt variations of rotational speed. This is not a serious problem because such variations were not allowed in the data selection stage in the first place.

The question rises if models in the azimuth domain have a physical interpretation. The interpretation becomes clear by the observation that for a constant rotor angular speed there is no essential difference between a sample in time or a sample in azimuth. Their mutual relation is a constant factor. Therefore the models can be seen as models describing the dynamics of interest as if the rotor speed is exactly fixed (perfectly usable for validation of the theoretical model in the corresponding point of operation).

5 results

Now results of the proposed approach for experime
are presented. The results have mainly been extracted from [9].

5.1 identification experiment

An experiment has been designed and performed with identical excitation of the pitch angle of both blades. All other inputs are kept constant. The input excitation was chosen to be a random sequence of steps between 2 and 7 degrees with clock frequency of $0.5Hz$ because this theoretically should be informative enough and it was relatively easy to implement at the test site. The data acquisition has been done with sample rate of approximately $48Hz$. Theoretically models up to $25Hz$ can be identified which is fast enough to cover most interesting dynamics. Besides the rotor speed, rotor shaft torque and generator pressure all rotor angles and bending moments have been measured.

5.2 data treatment

After removal of averages from the data no further data treatment was done. At this stage the data are transformed to azimuth domain. This is done before the help signal is constructed because in the azimuth domain the help signal is exactly periodic. The result of the data transformation is given in Fig. 5.

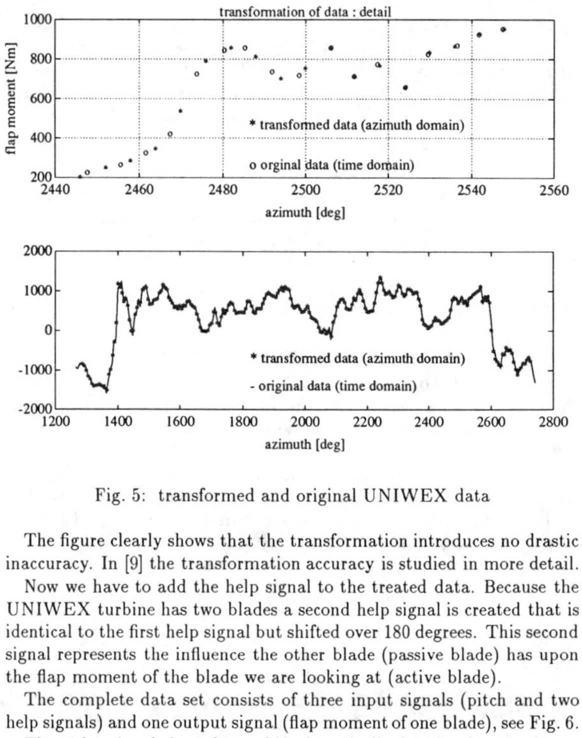

Fig. 5: transformed and original UNIWEX data

The figure clearly shows that the transformation introduces no drastic inaccuracy. In [9] the transformation accuracy is studied in more detail.

Now we have to add the help signal to the treated data. Because the UNIWEX turbine has two blades a second help signal is created that is identical to the first help signal but shifted over 180 degrees. This second signal represents the influence the other blade (passive blade) has upon the flap moment of the blade we are looking at (active blade).

The complete data set consists of three input signals (pitch and two help signals) and one output signal (flap moment of one blade), see Fig. 6.

The pulse signals have been shifted vertically for visual convenience. The sharp flap moment response at the tower shadow passage can be clearly located at the samples where the pulses at the input have been defined.

5.3 model identication

Now the identification results are discussed. Previous studies have demonstrated Box-Jenkins models to have the model structure that suits identification of wind turbine dynamics best. Also in this case models of this structure came up with the best results. In Fig. 7 (lower right plot) the overall model fit (summation of deterministic model contributions $Gu + H_{act}h_{act} + H_{pas}h_{pas}$) is compared with the flap moment data. The fit is remarkably good taking the large shadow influence into account.

In Fig. 7 also the seperate deterministic contributions have been plotted with their specific input signal (plots upper and lower left and upper right plot). The flap response to pitch seems to be realistic.

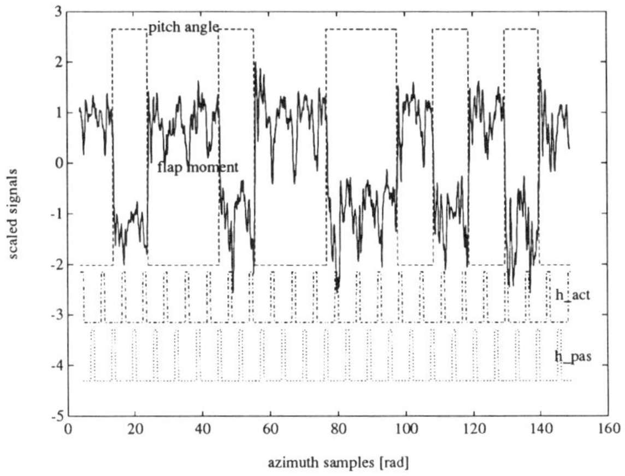

Fig. 6: complete identification data set

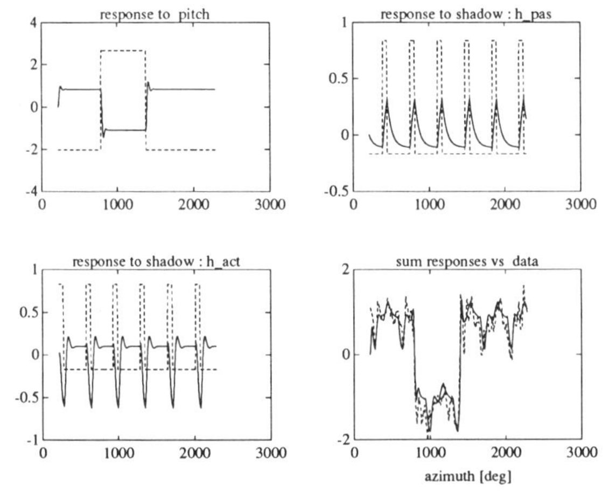

Fig. 7: identification result : sum of deterministic contributions vs data

Slight model order changes do not result in completely other models. Moreover the response can be physically understood. The same holds to some extent for the response to the two help signals. The sharpest response is felt by the active blade while the interaction of the passive blade is only moderate.

The results of this extended approach are clear improvementts compared with results of the standard procedure. The consistency of models estimated following the extended approach compares favourably with the standard results where every model seems to be equally good Illustration of the comparison between standard and extended approach is given in [9].

The power lies above all in the fact that by means of the help signals a lot of disturbance influence can be captured and thereby the identification algorithm is offered an opportunity to model the real relationship between pitch and flap moment whereas in the standard procedure the disturbances should be covered in a noise model part that is not properly utilized to model disturbances of this more or less deterministic but unknown type. Better experimental models make validation of theoretical models more precise. In [10] such a model is confronted with UNIWEX data and gives a remarkably good simulation fit to the data.

6 conclusions

The wind plays a typical role which makes the identification of wind turbine dynamics fundamentally different in some cases to most other systems. The character of some important disturbances does not fit well in the stochastic setting of prediction error modelling that normally suffices to model the noise part of the data. They are unknown but certainly not of stochastic nature. The more or less deterministic character is not used in the modelling process so far.

Although standard system identification has delivered acceptable models for some transfer functions (because the disturbances did not overwhelm the deterministic response that we wanted to identify) it did not work for the transfer function from blade pitch angle to blade root flap bending moment.

Physical insight initiated two extensions to the standard system identification procedure. Firstly the construction of a help signal representing the influence of the tower shadow on the passing blade is proposed. With this signal as extra input for the identification data set a major part of the tower shadow disturbance is captured. Secondly a transformation of data from samples in time to samples in azimuth angle is proposed to pronounce the periodicity of some disturbances and thereby simplify their elimination via the artificial help signal.

These extensions were heuristically tried out on real life measurements of the UNIWEX wind turbine and the approach turned out to give improved models with physically understandable model contributions to the overall fit with the flap moment data.

For the wind turbine system and particularly for the vaidation of the theoretical model the identification results have to be broadened over different points of operation and different configurations. This may open the possibility to say something about the relation of the identified dynamics between pitch and flap moment response as they were found in those different situations. As a spin off it may be possible to observe structural coherence between the identified tower shadow response models. This may turn out to be of importance with respect to validation of theoretical models describing the tower shadow phenomenon.

acknowledgements

The authors thank the ICA of the universtity of Stuttgart for the possibility to experiment with the UNIWEX turbine, and the people working at the test site for the pleasant coöperation.

References

[1] Baars G.E. van , *Experimental validation of the UNIWEX wind turbine* (in Dutch). Delft University of Technology, The Netherlands, 1991.

[2] Baars G.E. van, P.M.M. Bongers , "Wind turbine control design and implementation based on experimental models", in *proc. 31th IEEE Conf. on Decision and Control, Tucson Arizona, 16-18 december 1993,* pp.2454-2459, 1992.

[3] Baars G.E. van, P.M.M. Bongers , "Model Description of the UNIWEX Wind Turbine: Experimental and Theoretical Results", Delft University of Technology, The Netherlands, 1992.

[4] Bongers P.M.M., G.E. van Baars , "Experimental validation of a flexible wind turbine model", in *Proc. 30th IEEE Conf. on Decision and Control, Dec. 11-13, Brighton, UK,* pp.1660-1661, 1991.

[5] Bongers P.M.M., W.A.A.M. Bierbooms, Sj. Dijkstra, Th. van Holten , *An Integrated Dynamic Model of a Flexible Wind Turbine.* Delft University of Technology, The Netherlands, 1990.

[6] Bongers P.M.M., T.G. van Engelen, Sj. Dijkstra, Z.D.Q.P. Kock , "Optimal control of a wind turbine in full load", in *Proc. European Wind Energy Conference 1989, Glasgow, UK,* 1989.

[7] Ljung L. , *System Identification: theory for the user.* Prentice Hall,, Inc. Englewood Cliffs, New jersey, 1987.

[8] Müller M. , "Experimental investigations with the universal test wind turbine", in *Proc. European Wind energy Conf. 1989.,* Glasgow, Schotland, pp.832-836, 1989.

[9] Mosterd, H. , *experimental modelling of the UNIWEX turbine.* Delft University of Technology , The Netherlands, 1993.

[10] Bongers P.M.M., K. Wefelmeier, M. Barnsley , "Dynamic models of wind turbine systems for design purposes", in *Proc. Europ. Community Wind Energy Conference,* to appear, 1993.

INDEPENDENT BLADE CONTROL OF HAWT'S

W.E., Leithead, M.C.M., Rogers, P.R.D., Agius
University of Strathclyde, UK.

ABSTRACT : Many constant speed grid-connected horizontal axis wind turbines have the ability to vary the angle of pitch of their blades or part of their blades. This capability is frequently used to regulate the power above rated windspeed. The conventional approach is to measure the generated power and vary the angle of pitch of the blades, with all blades acting in unison, in response to the error between the generated power and the rated power. The purpose of this paper is to investigate the extent to which the above potential advantages of independent blade control, both on its own and in combination with power control, can be realised. Independent blade control has no real advantage over power control. The analysis is to a great extent independent of both the choice of variable on which the control acts, and the degree of correlation between the wind speed sensed by individual blades and the wind speed sensed by the complete rotor.

1. INTRODUCTION

Many constant speed grid-connected horizontal axis wind turbines have the ability to vary the angle of pitch of their blades or part of their blades. This capability is frequently used to regulate the power above rated windspeed. The conventional approach is to measure the generated power and vary the angle of pitch of the blades, with all blades acting in unison, in response to the error between the generated power and the rated power. The performance and design of these power control systems has been thoroughly investigated and is well understood [1].

An alternative approach to the control of wind turbines might be independent blade control whereby each blade has an independent actuator which is driven by a feedback loop in response to a separate measurement made locally at each blade. The angle of pitch of each blade can be adjusted individually in response to the load it experiences. Independent blade control might appear to have the following advantages:

(i) Since control is enacted directly on the rotor, control response to wind turbulence induced transients is faster than power control in which the control response occurs after the transients have propagated through the power-train.

(ii) The control can respond to transients, such as the cylcic loads at the blade passing frequency Ω, which can not be detected from the generated power and so cannot be regulated by power control.

The transients in both the above are induced by the aggregate influence of the wind-field on each blade.

The purpose of this paper is to investigate the extent to which the above potential advantages of independent blade control, both on its own and in combination with power control, can be realised.

2. ACTUATORS AND MEASUREMENTS

In independent blade control a measurement local to the blade of the aggregate influence of the wind-field is required ; that is, an indirect measurement is made of the aggregate windspeed, w. The required tip angle, G(w), at which power output would be rated power, is derived. It is proposed to cause the actual pitch angle of the blade, β_a, to follow G(w) locally at each tip thereby smoothing out the transient loads induced on the blade and so the wind turbine. The controller acts on the error (β_a - G(w)) to output a demanded pitch angle, β_d, which is fed to the actuator.

To enable an actuator to respond sufficiently rapidly to control demands, it is enclosed in a feedback loop. The gain of the forward path of the actuator is required to be high compared to the gain of the rest of the system. Accordingly, the difference between β_a and β_d is driven to zero over the frequency range relevant to control action. Hence, the actuator causes β_a to follow β_d.

Central to independent blade control is a suitable measurement [2] of the aggregate windfield affecting a blade. One such measurement is the aerodynamic pitching moment on the blade M_{aero}. Although, M_{aero} cannot be measured directly it can be inferred from a measurement of the moment applied to the blade by the actuator, M_{app} and a measurement of the actual pitch angle, β_a, which is already known from the feedback loop enclosing the actuator. The relationship of M_{aero} to M_{app} is

$$M_{aero} = J\ddot{\beta}_a - M_{app} \qquad (1)$$

where J is the blade inertia. M_{aero} is a function of both β_a and the v_0, the windspeed seen by the blade. The relationship may be linearised as in Fig. 1 where

$$K_{m2} = -\frac{\partial M_{aero}}{\partial \beta_a} \;;\; K_m = \frac{\partial M_{aero}}{\partial v_o} \qquad (2)$$

3. COMPARISON OF CONTROL SYSTEMS

A true comparison of the benefits of the two types of control action can only be made if similar blade angle accelerations and velocities are demanded of the actuator ; that is, any bias caused by one of the control actions being favoured by a more powerful actuator must be avoided. In the following investigation, individual blade control is expressed in terms similar to power control so that the two systems can be compared directly and combined as required.

Both control systems can be thought of as controlling perturbations of the torque on the drive-train about the desired operating point. The two control systems each have their own method of measuring these perturbations.

(i) *Power Control* : The power output of the machine is measured and compared with rated power and the error is used for control action.

(ii) *Independent blade control* : Measurements, local to a blade, are made in order to estimate the current wind speed. The error for control action is β_a-G(w).

In both cases the error is used to update the demanded tip angle, β_d, via a controller. Control action could be based on either of these controllers or on a combination of the two. The system dynamics consist of the dynamics of the actuator and the drive-train.

The two control systems are subject to different disturbances. The wind seen by an individual blade and that seen by the whole rotor are not identical. At low frequencies the wind speeds do not correlate exactly while at higher frequencies the strength of the spectral peaks induced by rotational sampling are expected to differ. The limiting factor on any control system is the saturation of the actuator by high frequency disturbances. Hence, if the high frequency spectral peaks on (β_a - G(w)), in comparison to those on the power error, are less prone to cause saturation of the actuator then there would be an advantage in using independent blade control.

(a) *Power Control*

It should be noted that whilst on-design at an operating point the derivatives of the aerodynamic torque, T, satisfy

$$\frac{dT}{d\beta_a} = \frac{\partial T}{\partial \beta_a} + \frac{\partial T}{\partial v_1} \cdot \frac{dv_1}{d\beta_a} = 0$$

and it follows that

$$\frac{\partial T}{\partial v_1} \Big/ \frac{\partial T}{\partial \beta_a} = -\frac{d\beta_a}{dv_1}$$

Hence, linearising the system for power control with error ($P-P_o$), it can be expressed as in Fig. 2. A is the actuator transfer function (β_d-β_a); c_1 is the power controller ; v_1 is the wind speed seen by the rotor ; D is the drive-train transfer function incorporating $\partial T/\partial\beta a$.

(b) *Independent blade control*

Linearising the system with error (β-G(ω)), it can be expressed as in Fig 3. C_2 is the independent blade controller ; v_o is the wind speed seen by the blade ; C is the actuator transfer function, $\delta\beta_d \rightarrow \delta(\beta_a$ - G(w)) ; B is the actuator transfer function, $\delta v_o \rightarrow \delta(\beta_a$ - G(w)).

The two control loops, δPower $\rightarrow$ δPower, are identical if the two controllers are related by

$$C_2 = \frac{AD}{C} C_1$$

Independent blade control requires to be combined with a coordinating power control loop to maintain rated power. For a control system which combines both power control and independent blade control the error signals should be combined as in Fig. 4 where $k_1 + k_2 = 1$ and k_1 is a low pass filter.

When the high frequency noise in individual blade control is less than in power control, the advantages of using a low pass filter are that the high frequency noise in the ($P-P_o$) signal is significantly reduced without affecting the phase shift round the control loop, and the filter allows ($P-P_o$) control action at the low frequencies where the wind seen by an individual blade is not the same as the wind seen by the rotor.

Whether independent blade control can be used to achieve better performance depends solely on the high frequency noise content of $\delta\beta_d$ and its effect on the actuator. The spectral peaks due to rotational sampling, in comparison to measurement noise, are the more important component of the high frequency noise. Therefore, the main indicator of comparative performance is the transfer function (A B)/C.

4. PERFORMANCE ASSESSMENT

In this section, a systematic analytic approach to the assessment of independent blade control is developed. The estimation of the pitch angle required for rated power for the current wind speed is investigated avoiding as many assumptions as possible about its form.

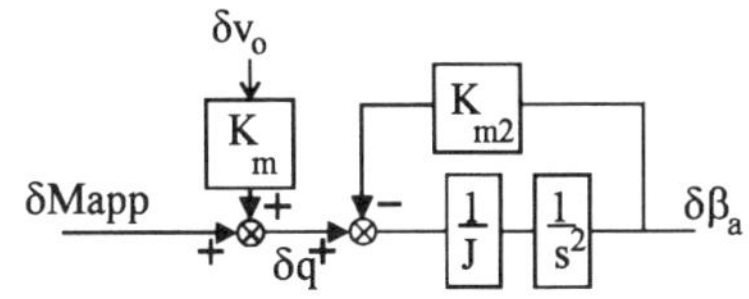

Figure 1 : Linearising actuator interaction with aerodynamics

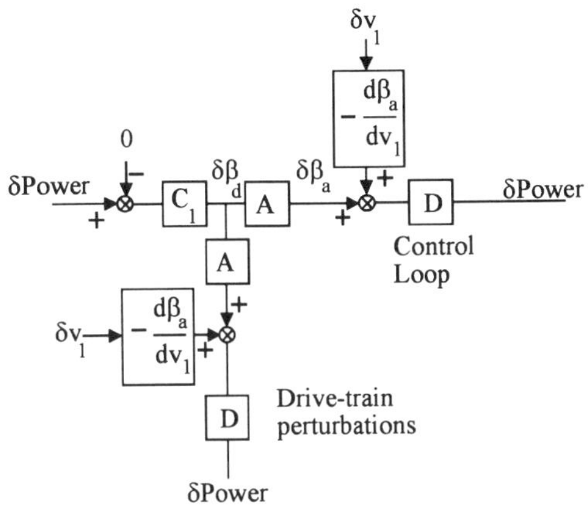

Figure 2 : Linearised system with power control

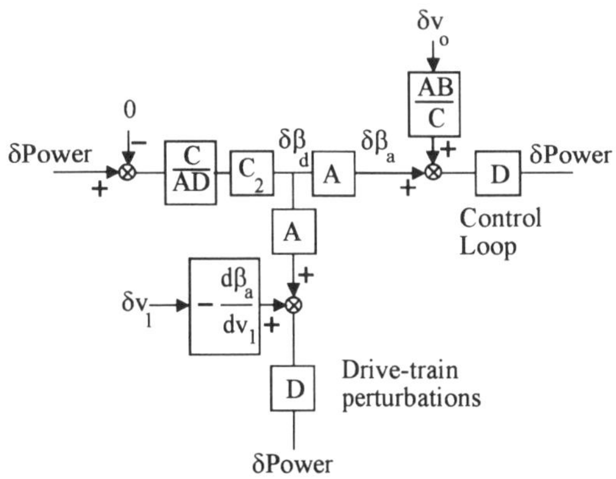

Figure 3 : Linearised system independent blade control

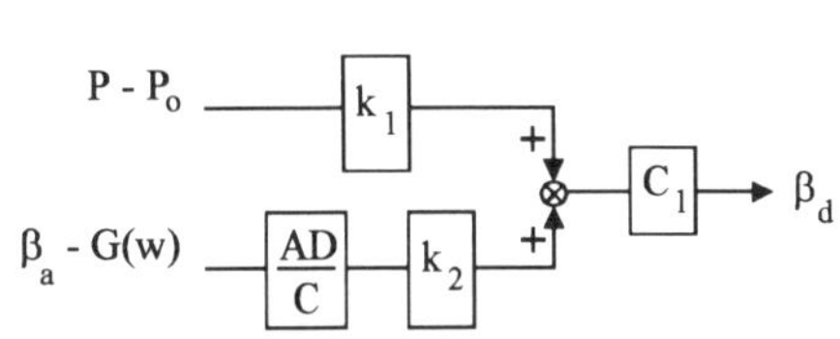

Figure 4 : Combined power and independent blade control

Two variables are available in the actuator ; namely, β_a the pitch angle, and Maero, the aerodynamic moment, using (1) and measurements of the tip angle acceleration, β_a, and the applied moment, Mapp. Hence, in general any estimate of the required pitch angle for rated power for current wind speed must be a function $G_2(\beta_a$, Maero) where Maero itself depends on the wind speed, v_o, and the blade angle, β_a. Since G_2 is the required pitch angle for the current wind speed, it should change only in response to a change of wind speed and not in response to a change in pitch angle - that is,

$$\frac{dG_2(\beta_a, Maero)}{d\beta_a} = \frac{\partial G_2}{\partial \beta_a} + \frac{\partial G_2}{\partial Maero}\frac{\partial Maero}{\partial \beta_a}(\beta_a, v_o) = 0$$

Hence,

$$\frac{\partial G_2}{\partial \beta_a}(\beta_a,\text{Maero}) = -\frac{\partial G_2}{\partial \text{Maero}}(\beta_a,\text{Maero})\,\frac{\partial \text{Maero}}{\partial \beta_a}(\beta_a,v_o)$$

The implications of this relationship are explored below.

4.1 Estimation of the Required Pitch Angle

One implication of (3) is that in the neighbourhood of an operating point, with β_a belonging to small interval enclosing that operating point,

$$G_2(\beta_a,\text{Maero}) = G(\text{Maero} - f(\beta_a))$$

where

$$\frac{df(\beta_a)}{d\beta_a} = \frac{\partial \text{Maero}}{\partial \beta_a} \qquad (3)$$

Hence, when close to on-design operating conditions, there exists a function $f(\beta_a)$, for all β_a such that the required pitch angle, β_r, for rated power for the current wind speed is

$$\beta_r \simeq G(\text{Maero} - f(\beta_a))$$

4.2 Investigation of the Transfer Function (A B/C)

The main indicator of performance for independent blade control is the transfer function (AB/C) where A is the transfer function from $\delta\beta_d$ to $\delta\beta_a$, and B is the transfer function from δv_o to $\delta(\beta_a - \beta_r)$ and C is the transfer function from $\delta\beta_d$ to $\delta(\beta_a - \beta_r)$.

Firstly, consider the transfer function C. Since,

$$\delta(\beta_a - G_2(\beta_a,\text{Maero})) = \delta\beta_a - \frac{\partial G_2}{\partial \beta_a}\delta\beta_a - \frac{\partial G_2}{\partial \text{Maero}}\delta\text{Maero} \qquad (4)$$

From (1) it follows that

$$C = \left(1 - \frac{\partial G_2}{\partial \beta_a}\right)A - J\frac{\partial G_2}{\partial \text{Maero}}T_{\beta_d\ddot{\beta}_a} + \frac{\partial G_2}{\partial \text{Maero}}T_{\beta_d\text{Mapp}}$$

where $T_{\beta_d\ddot{\beta}_a}$ is the transfer function from $\delta\beta_d$ to $\delta\ddot{\beta}_a$ and $T_{\beta_d\text{Mapp}}$ is the transfer function from $\delta\beta_d$ to δMapp. However, from Fig. 1 with δv_o equal to zero, it can be seen that

$$A = \frac{T_{\beta_d\ddot{\beta}_a}}{s^2}$$

and

$$A = \frac{T_{\beta_d\text{Mapp}}}{(Js^2 + K_{m2})}$$

Hence,

$$C = \left[1 + \left(\frac{\partial \text{Maero}}{\partial \beta_a} + K_{m2}\right)\frac{\partial G_2}{\partial \text{Maero}}\right]A$$

using (2) and the transfer functions C and A are equal. Secondly, consider the transfer function B. From (1) and (4)

$$B = \left(1 - \frac{\partial G_2}{\partial \beta_a}\right)T_{v_o\beta_a} - J\frac{\partial G_2}{\partial \text{Maero}}T_{v_o\ddot{\beta}_a} + \frac{\partial G_2}{\partial \text{Maero}}T_{v_o\text{Mapp}}$$

where $T_{v_o\beta_a}$ is the transfer function from δv_o to $\delta\beta_a$, $T_{v_o\ddot{\beta}_a}$ is the transfer function from δv_o to $\delta\ddot{\beta}_a$ and $T_{v_o\text{Mapp}}$ is the transfer function from δv_o to δMapp. However, from Fig. 1 and (2) it can be seen that

$$T_{v_oq} = T_{v_o\text{Mapp}} + K_m$$

where T_{v_oq} is the transfer function from δv_o to δq. In addition,

$$T_{v_o\beta_a} = \frac{1}{s^2}T_{v_o\ddot{\beta}_a}$$

and

$$T_{v_o\beta_a} = \frac{1}{(Js^2 + K_{m2})}T_{v_oq}$$

It follows that,

$$B = \left[1 - \left(\frac{\partial \text{Maero}}{\partial \beta_a} + K_{m2}\right)\frac{\partial G_2}{\partial \text{Maero}}\right]T_{v_o\beta_a} - K_m\frac{\partial G_2}{\partial \text{Maero}}$$

using (2). Hence,

$$B = T_{v_o\beta_a} - K_m\frac{\partial G_2}{\partial \text{Maero}}$$

Since δv_o is a disturbance in the actuator which has a high forward path gain (cf Section 2) and associated disturbance rejection properties, the transfer function $T_{v_o\beta_a}$ has very low gain at low frequency. Therefore, as s tends to zero the transfer function B tends to the constant - $K_m\,\partial G_2/\partial$Maero.

In sub-section 4.1, it was demonstrated, that when close to on-design operating conditions,

$$G_2(\beta_a, \text{Maero}) = G(\text{Maero} - f(\beta_a)) = G(\psi)$$

and thus,

$$\frac{\partial G_2}{\partial \text{Maero}} = \frac{dG(\psi)}{d\psi}$$

When on the on-design curve

$$\beta_a = G(\text{Maero} - f(\beta_a))$$

and

$$\frac{d\beta_a}{d\beta_a} = 1 = \frac{dG(\psi)}{d\psi}\left(\frac{d\text{Maero}}{d\beta_a} - \frac{df(\beta_a)}{d\beta_a}\right)$$

But

$$\frac{d\text{Maero}}{d\beta_a} = \frac{\partial \text{Maero}}{\partial \beta_a} + \frac{\partial \text{Maero}}{\partial v_o}\frac{dv_o}{d\beta_a}$$

and it follows that

$$\frac{dG(\psi)}{d\psi} = 1 \Big/ \left(\frac{\partial \text{Maero}}{\partial \beta_a} + \frac{\partial \text{Maero}}{\partial v_o}\frac{dv_o}{d\beta_a} - \frac{df(\beta_a)}{d\beta_a}\right)$$

$$= \frac{d\beta_a}{dv_o} \Big/ \frac{\partial \text{Maero}}{\partial v_o}$$

using (3). Hence,

$$-K_m \frac{\partial G_2}{\partial Maero} = -K_m \frac{d\beta_a}{dv_o} / \frac{\partial Maero}{\partial v_o} = -\frac{d\beta_a}{dv_o}$$

and

$$B = -\frac{d\beta_a}{dv_o} B^*$$

with the gain of the transfer function B^* tending to 1 as s tends to zero.

To enable the direct comparison of power control to independent blade control Fig. 3 can be amended with C_2 substituted by

$$C_2 = \frac{AD}{C} C_1 = DC_1.$$

The transfer function B^* is effectively 1 for the frequency range of interest. Hence, comparing Fig. 3 to Fig. 2 whether independent blade control has any inherent advantage over power control, depends solely on spectral density functions for δv_o and δv_1 ; that is, which has the least high frequency content and thus is least likely to saturate the actuator.

For independent blade control, the spectral density function of δv_o has rotationally induced spectral peaks at each integer multiple of the rotor angular velocity as sensed by one blade. For power control, the spectral density function for δv_1 has similar spectral peaks but sensed by the complete rotor. At low frequencies the contributions from each blade strongly correlate, with the result that only the spectral peaks at integer multiplies of n times the rotor angular velocity are present in the drive-train and so power control, where n is the number of blades. Although the magnitude of each remaining spectral peak is increased by a factor n, the content of the spectral density function is much reduced at low frequency since the magnitude of the individual blade contributions to the spectral peaks decrease rapidly as frequency increases. At higher frequency, the correlation across the wind-field which is the source of the spectral peaks, decreases. Consequently, the concentration of the spectral density function for power control at every n^{th} peak decreases with the intermediate frequencies becoming stronger. However, the magnitude of every n^{th} peak decreases from n times to $\sqrt{n}$ times that of the corresponding peaks of the spectral density function for a single blade. Hence, if no account is taken of spatial filtering then at very high frequencies the magnitude of the spectral density function for power control can be 5 dB greater for a 3 bladed machine than the magnitude of the spectral density function for independent blade control. Nevertheless, it is unlikely to outweigh the weaker magnitude of the spectral density function for power control at the low and middle frequency ranges. However, even if the 5 dB advantage is assumed for all the spectral peaks the bandwidth of the controller can only be made slightly greater. Hence, for the purpose of regulating drive-train transient loads independent blade control has no real advantage over power control. The control loops are subject to effectively the same level of middle to high frequency disturbances.

4.3 Assessment of Compliance Control

It has been established that power control, Fig. 2, is equivalent to independent blade control, Fig. 3, in that the transfer function descriptions of the control loops are identical provided the controllers, C_1 for power control and C_2 for independent blade control are related by

$$C_2 = D\, C_1$$

and in that the control loops are subject to disturbances of similar strengths. In addition, just as the power controller requires reasonable robustness margins to cater for the uncertainty in the description of the rotor aerodynamic torque, so the independent blade controller requires reasonable robustness margins to cater for the uncertainty in the description of the blade aerodynamic moment. Hence, even when the wind speed sensed by a blade is identical to that sensed by the complete rotor, independent blade control can have no inherent advantage over power control. Any controller, C_2, for independent blade control has a corresponding controller, C_1, for power control which achieves the same performance.

Although independent blade control has no inherent advantages over power control for the regulation of transient drive-train loads, it might be employed in an alternative role. One such alternative role might be to actively control the Ω loads on the blades. To actively regulate Ω the gain and phase of the controller need to be considerably increased at Ω which is greater than the usual controller crossover frequency. Accordingly, the gain of the controller must be raised considerably over the middle frequency range. Hence, in wind speeds close to rated the actuator activity is considerably increased over all frequencies except Ω at which frequency the activity is decreased. Hence, greater actuator activity would occur than for regulation of transient drive-train loads. Since the wind speed sensed by the blade differs from that sensed by the rotor, independent blade control is not appropriate on its own but should be combined with power control. They should be combined as in Fig. 4. The bandwidth of k_1 must be significantly less than Ω say $\Omega/3$, because only the independent blade control senses the Ω disturbance. The performance of the control system in terms of controlling the drive-train transient loads from $\Omega/3$ to Ω is degraded since it is not done solely in response to wind speed v_1 but a combination of v_1 and v_o. As for the design of controller on the power control loop, it is evident from the preceding that it should be the same as for power control on its own.

5. CONCLUSIONS

An analysis of the dynamics of independent blade control is presented in this paper. The main conclusions are as follows.

(i) *Independent blade control without power control*

For the purpose of regulating drive-train transient loads, independent blade control has no real advantage over power control.

(ii) *Independent blade control with power control*

For the purpose of actively controlling the Ω loads on the blades, independent blade control has the potential but greater actuator activity is experience than for power control. It is accompanied by a degradation in the regulation of drive-train transient loads between $\Omega/3$ and Ω.

The analysis of the dynamics of independent blade control which gives rise to the above conclusions, is to a great extent independent of both the choice of variable on which the control acts, and the degree of correlation between the wind speed sensed by individual blades and the wind speed sensed by the complete rotor.

ACKNOWLEDGEMENTS

The authors wish to thank the DTI (Department of Energy) for funding the work and for permission to publish this paper.

REFERENCES

1. Leithead, W.E., S.A. de la Salle, D. Reardon, *Wind turbine control systems modelling and control, Phase I and II*, University of Strathclyde Report prepared for D.En.

2. Anderson, C.G., T.J. Campbell, *Front-end aerodynamic control of HAWT's*, University of Edinburgh Report prepared for D.En.

CONTROL OF VARIABLE SPEED WIND TURBINES WITH SUPERPOSITION GEARBOX

P. Caselitz, R. Grebe, T. Krüger, K. Pischel*, A. Schoo*
Institut für Solare Energieversorgungstechnik (ISET)
Department of Control Engineering
D-3500 Kassel, Königstor 59
*Thyssen Industrie AG Henschel
D-3500 Kassel, Henschelplatz 1

The project was funded by the Bundesministerium für Forschung und Technologie.

ABSTRACT: In the paper a new concept for variable speed wind turbines is presented. In contrast to known concepts, which use electrical converters to decouple the generator speed from the grid frequency, this concept uses a superposition gearbox to control the wind turbine speed. This allows a directly grid-coupled synchronous generator to be employed which leads to a remarkable reduction of harmonic distortions and reactive power loads.
The paper explains the steady state and dynamic behaviour of a WEC with superposition gearbox, the controller design and the integration of the superposition gearbox into the drive train. Technical and cost aspects of the new concept are compared to other concepts for variable speed wind turbines.

1. Introduction

In comparison with fixed speed wind turbines variable speed wind turbines offer many advantages: The dynamic loads in the drive train are reduced and the electrical power output is smoothed. In addition, variable speed operation permits an increase of the gained energy as well as a reduction of the aerodynamic noise.

State of the art variable speed concepts are realized by feeding the generated power into the grid via a frequency converter. One of the disadvantages of these systems lies within their strong current harmonics.

Fig. 1 shows the scheme of a variable speed system based upon a superposition gearbox. Unlike an ordinary gearbox a superposition gearbox has a third shaft which permits the variation of the speed relation between the rotor and the generator. This way the generator speed can be kept constant despite the variable rotor speed and a directly grid-coupled synchronous generator can be employed.

The 'control shaft' is driven by a converter fed electrical machine, which transforms a maximum of 10 to 15% of the total power. Since the main part of the power is supplied by the synchronous machine the total harmonic distortion of the produced current is limited to an acceptable value.

2. Steady State Characteristics of a Wind Energy Converter with Superposition Gearbox

2.1 Rotational Speeds, Torques, and Power of an Ideal Superposition Gearbox

The relation between the shaft speeds of the ideal superposition gearbox in fig. 1 is given by [2]:

$$i_1\omega_R - i_2\omega_C - \omega_G = 0 \ , \qquad (1)$$

with i_1 and i_2 being constructive parameters. According to this relation the choice of two rotational speeds determines the rotational speed of the third shaft.

When using a directly grid-coupled synchronous generator the rotational speed ω_G is determined by the line frequency and by the generator's number of pole pairs ($\omega_G = \omega_{Go}$). In this case the rotor speed can be adjusted by the rotational speed of the control shaft:

$$\omega_R = \frac{1}{i_1}\omega_{G_o} + \frac{i_2}{i_1}\omega_C \ . \qquad (2)$$

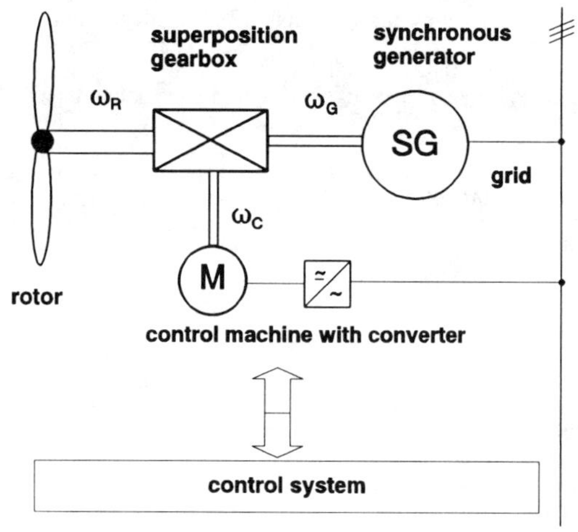

Fig. 1 Scheme of a wind energy converter with an electrically controlled superposition gearbox

A special case occurs when $\omega_C = 0$. Then the rotor speed is equal to the 'basic speed' ω_{R_o} of the wind turbine which depends entirely on the gear ratio i_1 and on the (constant) generator speed:

$$\omega_{R_o} = \frac{1}{i_1}\omega_{G_o} \ . \qquad (3)$$

The relations of the torques within the superposition gearbox shafts are:

$$\frac{M_R}{M_G} = i_1 \ , \quad \frac{M_C}{M_G} = i_2 \qquad (4)$$

which means that there is only <u>one</u> independent torque which determines the others two.

Equations (1) and (4) imply that the power within the control shaft is a function of the rotor speed

$$P_C = P_R\left(1 - \frac{\omega_{R_o}}{\omega_R}\right) \ . \qquad (5)$$

This means that the share of power converted by the control machine depends entirely on the relation between the rotor speed and the basic speed of the wind turbine. The corresponding equation for the generator power is:

$$P_G = P_R \frac{\omega_{R_0}}{\omega_R} . \qquad (6)$$

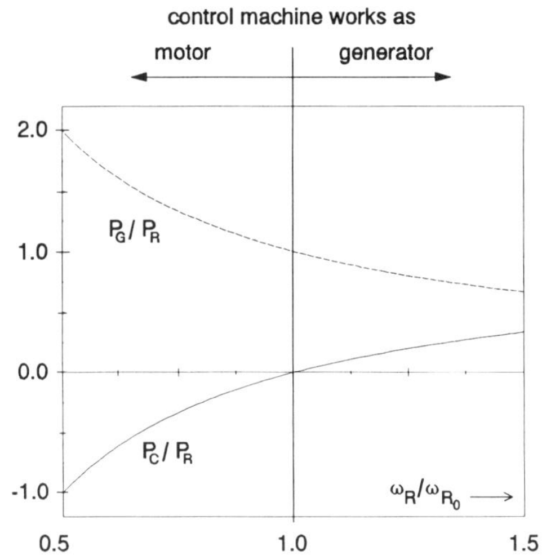

Fig. 2 P_C/P_R and P_G/P_R as a function of the normalized rotor speed

2.2 Projected Power of the Control Machine
Figure 2 shows the power flux within the system as a function of the normalized rotor speed. If the rotor speed is higher than the basic speed the control machine will work as a generator, i.e. it will act as a second generator feeding the grid parallel to the synchronous generator.

Else if the rotor speed is lower than the basic speed, the control machine will work as a motor. In that case the control drive receives power from the grid, which is transmitted to the drive train, and finally fed back into the grid by the generator. In that case the power of the generator is larger than that of the rotor. However, due to the circulating power the motor operating of the control machine leads to a lower efficiency of the overall system.

2.3 Choice of the Steady State Rotor Speed
According to eq. (4) the torque of the control machine will always be positive in steady state operation, because the rotor torque is positive, too:

$$M_C = \frac{i_2}{i_1} M_R . \qquad (7)$$

Hence the switching from motor to generator operation is effected through the change of the rotary direction of the control machine.

In both operating conditions the maximum variation of the rotor speed $\Delta\omega_{R\max}$ is limited by the maximum power $P_{C\max}$ of the control machine:

$$\frac{\Delta\omega_{R_{\max}}}{\omega_{R_0}} = \frac{1}{\frac{P_R}{P_{C_{\max}}} - 1} . \qquad (8)$$

The steady state rotor speed can in principle be chosen arbitrarily, as long as the play for dynamic speed deviations remains large enough. In fact it would be advantageous to have a stationary rotor speed at the basic speed in order to provide a maximum play for dynamic reactions.

However, the specifications of control drives do not permit this choice of the stationary rotor speed. With most types of electrical drives, the operating at very low speed should be avoided (because of the high pendulum moments etc.). Since the steady state motor operation should also be avoided because of its low total efficiency the rotor speed has to be chosen to be sufficiently far on the generator side.

The operating speed can also be chosen depending on the wind speed with regard to maximizing the rotor efficiency. In this case the turbine has to be operated at a low rotor speed, i.e. with the control machine in motor operation, when the wind speeds are low. When passing over to higher wind speeds the rotor speed is adjusted in the direction of an upper limit with the control machine working as a generator and a remaining play for dynamic reactions.

2.4 Control Machine Types
The control machine has to meet the following design criteria:
- high speed variations
- high torque even at low speeds
- short-term overloads
- protection against water and dust

These requirements can be met by using inverter fed asynchronous machines with a squirrel cage rotor. These machines are very tough and have a high degree of protection.

DC drives can be used as well but they need more maintenance and a higher constructive effort in order to reach the required degree of protection.

If the control machine is fed by a pulsed inverter the harmonic distortions can be significantly reduced.

3. Control of a Wind Energy Converter with a Superposition Gearbox
In order to test the dynamic properties of a wind energy converter with a superposition gearbox and a synchronous generator a mathematical model of the system was developed and digitally simulated. The results of the simulations were tested in the laboratory (80-kW experimental drive train).

3.1 The Mathematical Model of the Control Plant
Figure 3 shows the simulation model of the wind turbine, which consists of the rotor, the synchronous generator, the control machine (models described in [1]), and the superposition gearbox. The behaviour of the ideal superposition gearbox can be described by a linear equation system [3]:

$$\begin{bmatrix} \dot\omega_R \\ \dot\omega_G \\ \dot\omega_C \end{bmatrix} = G \begin{bmatrix} M_R \\ M_G \\ M_C \end{bmatrix} \qquad (9)$$

The matrix G depends on the gear ratios and on the efficiency of the superposition gearbox as well as on the rotor's and the control machine's inertia. The input of the system is the electrical torque of the control machine. This torque does influence the acceleration and speeds within the system and the shaft torques. The system is completely controllable via this input.

3.2 Controller Design

The control of the wind energy converter has to meet the following requirements:

1.) During steady state the rotor speed has to be adjusted to the reference value supplied by the supervisory control system
2.) The torques within the rotor shaft must be smoothed and the gradient of the electrical power output has to be limited
3.) The dynamic deviations of the rotor speed must be limited to permissible values.

A PI controller was designed to meet these criteria (see fig. 3). It uses the rotational speed of the control machine as input signal. Customary electrical drives include a PI speed control. However, the control parameters have to be changed in order to reach the control objectives in this context.

To evaluate the performance of the PI controller a state space controller was designed and tested in simulations. The state space controller, which uses the generator speed and its voltage angle as additional inputs, offers better performance especially with fast changes of the wind energy input. However, it's implementation is non-standard.

3.3 Simulations and Laboratory Results

In order to test the steady state and dynamic behaviour of the WEC with superposition gearbox an 80 kW experimental plant was built up.

The measurement results of the laboratory drive train are compared with simulations of a 500 kW unit. Fig. 4 and 5 show the dynamic response to a simulated wind gust and a wind course recorded in Travemünde in Nov. 1992.

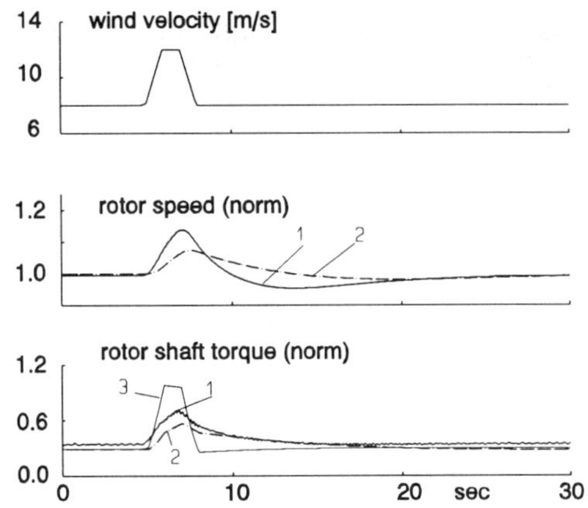

Fig. 4 Behaviour of the 80 kW experimental drive train (1) as response to a simulated wind gust in comparison with the simulation of a 500 kW plant (2). Curve (3) shows the normalized torque of the 500 kW rotor.

In both cases the evading rotor speed corresponds with a significant reduction of the rotor shaft torque peaks. Yet, the smoothing characteristic of the simulated 500 kW plant is stronger than that of the 80 kW drive train, even though its relative speed variation is smaller. This is caused by the rotor time constant which depends strongly on the rotor's inertia.

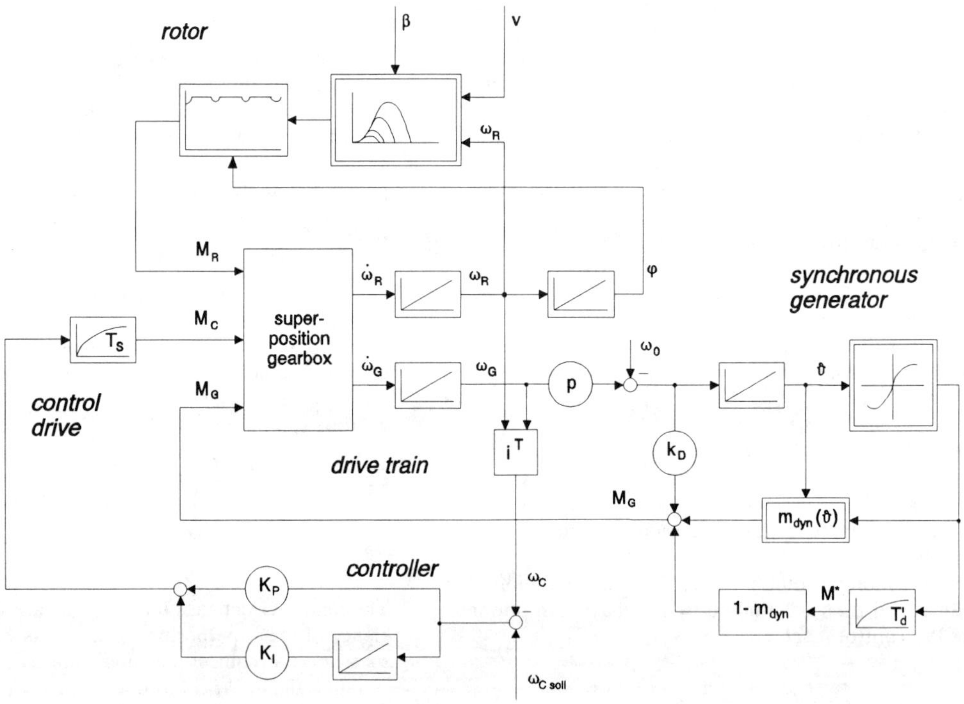

Fig. 3 Simulation model of a WEC with electrically controlled superposition gearbox and synchronous generator

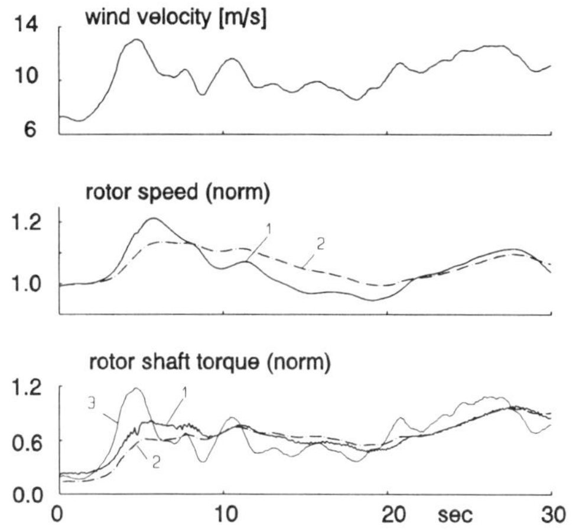

Fig. 5 Response of the 80 kW experimental drive train (1) to a measured wind course in comparison with the simulation of a 500 kW plant (2). Curve (3) shows the normalized torque of the 500 kW rotor.

4. Integration of the Superposition Gearbox into the Drive Train of a WEC

4.1 The Structural Placement inside the Drive Train

Superposition gearboxes for wind turbines consist of an actual superposition stage as well as of one or several adjustment stages in the direction of the rotor and the control machine. The superposition stage itself is a special kind of planetary gear. The use of a superposition stage within larger wind turbines, which are being equipped with planetary gearboxes anyway, does not represent an additional gearbox element, but a constructional variant only.

It is advisable to design the last stage in front of the generator as the superposition stage. A high speed level in this stage reduces the gear ratio (and thus the torques and mechanical losses) at the control shaft. The best solution for large WEC's with a co-axial drive train is to install the generator sided planetary stage as a three shaft gearbox. Within small and medium-sized WEC's, which are normally designed with spurwheel gearboxes, the fastest spurwheel stage has to be replaced by a planetary superposition stage.

4.2 The Physical Placement within the WEC

In general, gearboxes of larger wind turbines have to be specially constructed. This allows the construction to be adjusted to the physical conditions of the electrical components and auxiliary equipment while the WEC is being designed.

The required control machine is small enough to flange it directly onto the gearbox case. Depending on the size and the conception of the wind turbine nacelle, the control machine can be installed either next to or on top of the generator, but also on top of the gearbox case.

5. Comparison with Electrical Concepts for Variable Speed Operation

An important criterion for the assessment of a wind turbine with superposition gearbox are the costs in comparison with those of other variable speed turbine types with similar features. Their harmonic distortion and efficiency have to be taken into account likewise.

5.1 WEC with superposition gearbox

Depending on the construction type and on the rated power, the additional cost for a superposition gearbox in comparison with an ordinary gearbox is between 10 and 30%. The synchronous generator is available on the market with all control and protection devices at comparitively low cost. The control machine, too, is available on the market in the form of an ordinary drive. Its costs depend on the desired speed variability.

5.2 WEC with synchronous generator and DC link

While the synchronous generator is inexpensive, the frequency converter is comparatively expensive. Its strong harmonic distortion has to be kept away from the grid by the use of filters and its reactive power consumption has to be compensated. The frequency converter has a higher loss of power than the control drive of the superposition gearbox. As another disadvantage the frequency converters for synchronous generators with DC link are not mass-produced.

5.3 WEC with Doubly Fed Induction Generator

In regard to the effectivity, this system is the best of the three compared concepts. In all other respects, its characteristics are similar to those of a WEC with superposition gearbox. However, the control of the generator is complicated, and it is not mass-produced, too. The concept requires a slip ring induction generator which is about twice as expensive as a machine with squirrel cage rotor.

According to our estimations a WEC with superposition gearbox should cost approximately the same as a WEC with synchronous generator and DC link, but significantly less than a WEC with doubly fed induction generator.

6. Conclusions

The investigations have demonstrated that WEC's with a superposition gearbox possess good dynamic characteristics. The electrical system can be realized with standard components, and the construction of superposition gearboxes is not more complicated than that of ordinary gearboxes. According to rough estimations, this concept is indeed competitive, especially when certain grid voltage quality constraints have to be met in weak grids.

References

[1] Kleinkauf, W., Leonhard, W. u. a., "Betriebsverhalten von Windkraftanlagen", Band I und II, BMFT-Forschungsbericht T 84-154, Universität Gh Kassel, Fachgebiet Elektrische Energieversorgungssysteme, TU Braunschweig, Institut für Regelungstechnik, 1984.

[2] Dubbel, H., "Taschenbuch für den Maschinenbau", Springer Verlag, 16. Auflage, Berlin 1987

[3] Krüger, T., "Simulation und Regelung von drehzahlvariablen Windkraftanlagen mit Umlaufgetriebe", Diplomarbeit II, Universität Gh Kassel/ISET

[4] Caselitz, P., Grebe, R., Krüger, T., Pischel, K., Schoo, A., "Drehzahlvariable Windkraftanlagen mit elektrisch geregeltem Überlagerungsgetriebe", DEWEK 92, Wilhelmshaven

AN EXPERIMENTAL COMPARISON OF DIFFERENT WIND TURBINE CONTROL POLICIES.

A.D.Simmons. Consultant, Wind & Sun, The Howe, Watlington, Oxon, U.K.

L.L.Freris. Department of Electrical Engineering, Imperial College, London, U.K.

J.A.M.Bleijs. Department of Engineering, University of Leicester, Leicester, U.K.

D.G.Infield. Energy Research Unit, Rutherford Appleton Laboratory, Chilton, Didcot, Oxon, U.K.

Several theoretical studies of wind turbine control policies have been conducted around the world over the years. Conflicting conclusions have been drawn from these studies regarding the realisable operational advantages of the different control policies considered. Particular questions remain as to the possible advantages of variable speed over fixed speed operation.
A fully instrumented experimental wind turbine test rig, situated at RAL in the UK, has recently been enhanced through hardware and software development. This has allowed the direct experimental comparison of different wind turbine control policies which have been implemented on the same machine and under similar wind conditions.
Extensive comparative testing of the wind turbine has been carried out, to assess the relative merits of the different control strategies. Experimental results are presented, demonstrating the comparative effectiveness of these strategies. The experimental data is then introduced into a customised DUWECS model of the wind turbine to compare and validate the theoretical prediction against the experimental results.
The comparative advantages of fixed speed and variable speed operation are discussed, with particular regard to energy capture.

1. INTRODUCTION

The Imperial College Wind Turbine Generator (ICWTG), located at the Rutherford Appleton Laboratory (RAL) wind test site in Oxfordshire, was designed and built at Imperial College [1]. The ICWTG is a flexible experimental test bed which has been used for a number of research projects over the years, each investigating different aspects of wind energy conversion.

As reported in previous papers, the subject of this research project is the effect of various control policies on structural loadings and energy extraction.

This paper presents some initial data from the research machine, and introduces this data into a mathematical simulation model (DUWECS, references [2] & [3]).

2. EXPERIMENTAL RESULTS

The ICWTG has been operated over several hours in fixed speed and variable speed modes in similar wind conditions. The data collected has then been processed to produce simultaneous time traces of the wind speed, rotor speed, torque and power signals for fixed speed and variable speed operation. For ease of introduction into the computer simulation, only five minutes worth of data are considered here.

2.1 Fixed Speed Operation

Figures 1 to 3 show a 5 minute sample of time series data obtained while operating the wind turbine under fixed speed control. The fixed speed controller is able to keep the rotor speed to within 4.13% of the set point rotor speed of 36 rad/s, with a mean value of 36.00 rad/s.

2.2 Variable Speed Operation

Figures 4 to 6 show a 5 minute sample of time series data obtained while operating the wind turbine under variable speed control. The variable speed controller is able to keep the tip speed ratio to within 8.31% of the set point tip speed ratio of 8, with a mean tip speed ratio of 8.00.

2.3 Speed Control Comparison

Figures 7 and 8 show the binned torque and power versus wind speed curves for the experimental fixed speed and variable speed operation of the wind turbine. From these initial graphs it can be seen that the power-wind speed curve is higher in the variable speed case than in the fixed speed case, as expected. Similarly, the torque-wind speed curves show greater torque for variable speed than fixed speed.

3. DUWECS SIMULATION

The Delft University Wind Energy Conversion System was designed to enable the performance of time series simulation and analysis of flexible wind turbines see references [2] and [3].

The DUWECS model consists of the four distinct sub-models : rotor, transmission, tower and generator, with specific sub-model i/o interconnections defined through data files. To simulate and analyse a closed loop controlled wind turbine, controller state-space matrices are defined in the matrix processing environment of Matlab to be used by DUWECS.

Much work has been done to customise the basic DUWECS model to the Imperial College wind turbine. The rotor module supplied has been adapted to provide constant tip speed ratio control (as an alternative to constant rotor speed control) for variable speed operation. The generator module supplied has been re-written specifically for the permanent magnet generator used on the ICWTG (reference [1]).

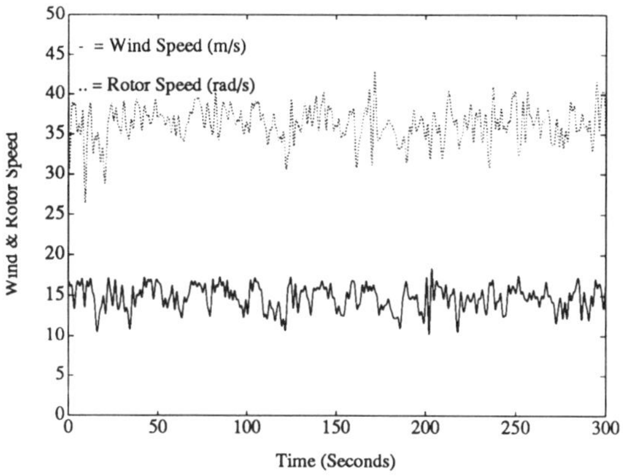

Figure 1 Experimental Fixed Speed Wind & Rotor Speed vs. time

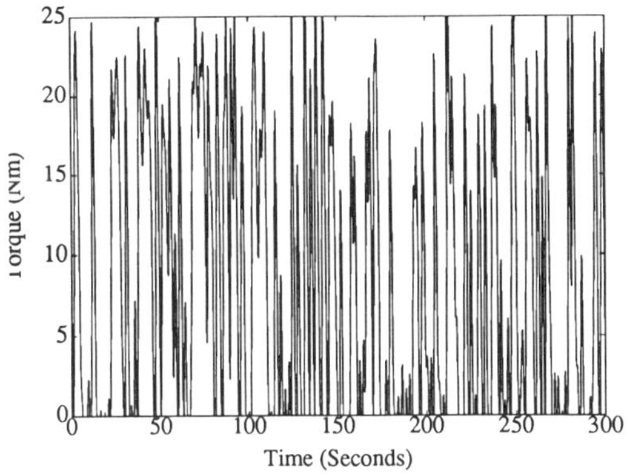

Figure 2 Experimental Fixed Speed Torque vs. time

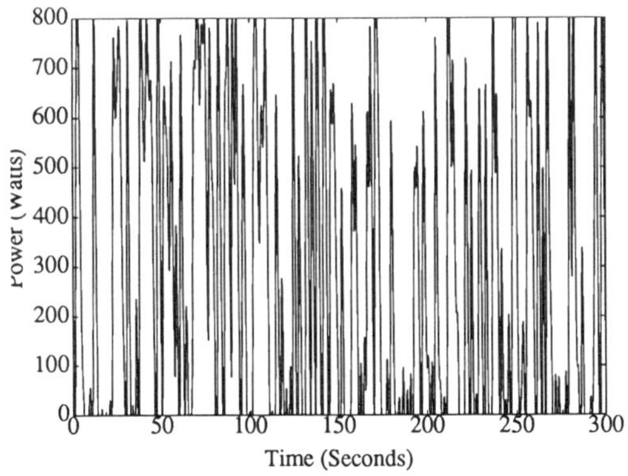

Figure 3 Experimental Fixed Speed Power vs. time

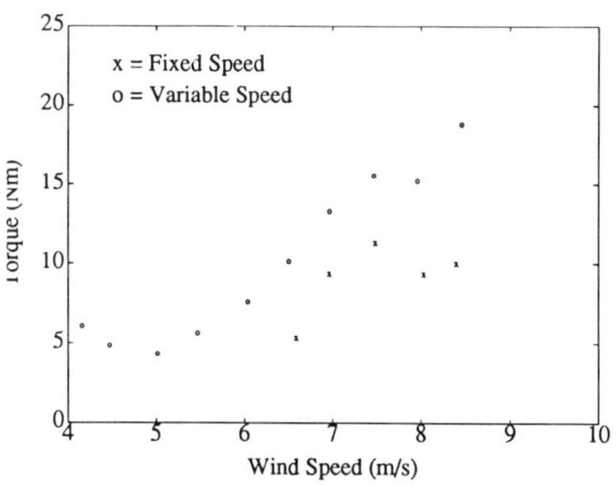

Figure 4 Experimental Speed Comparison Torque vs. Wind Speed

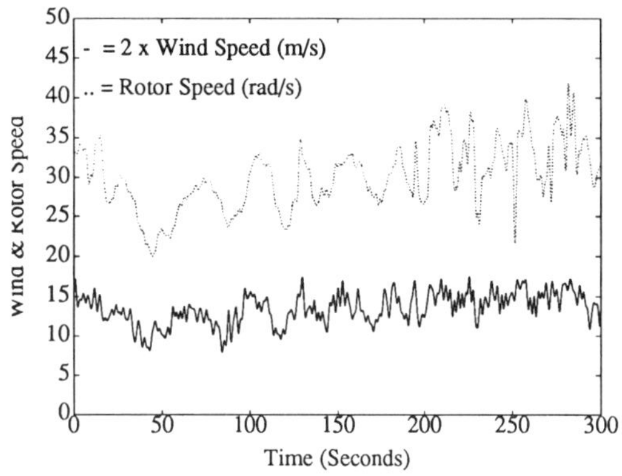

Figure 5 Experimental Variable Speed Wind & Rotor Speed vs. time

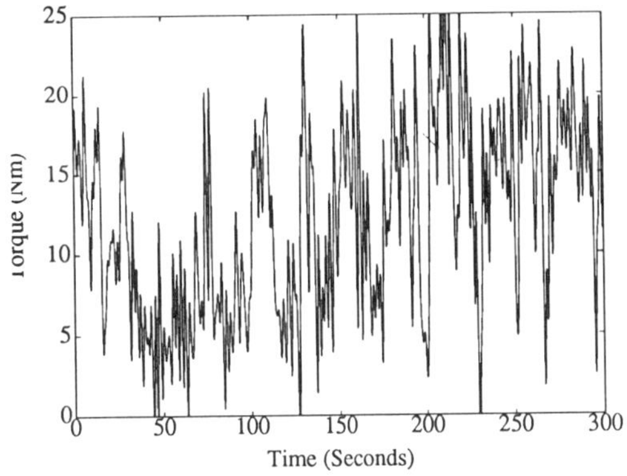

Figure 6 Experimental Variable Speed Torque vs. time

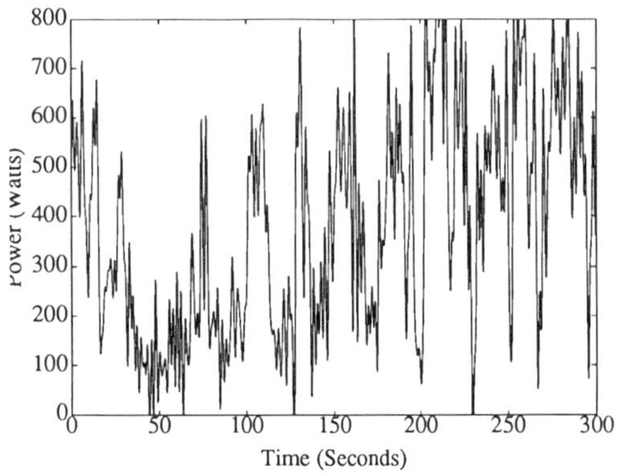

Figure 7 Experimental Variable Speed Power vs. time

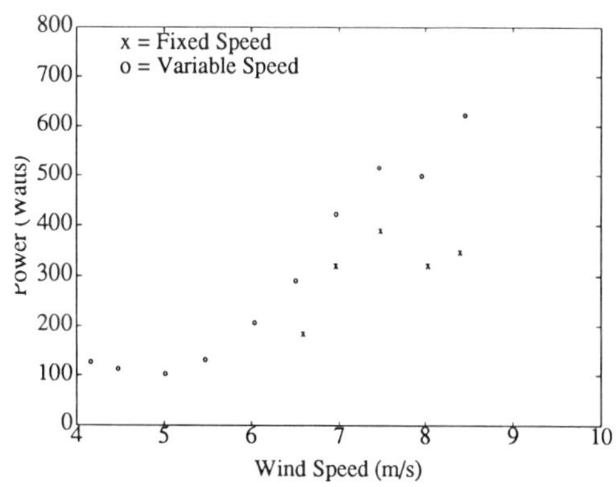

Figure 8 Experimental Speed Comparison Power vs. Wind Speed

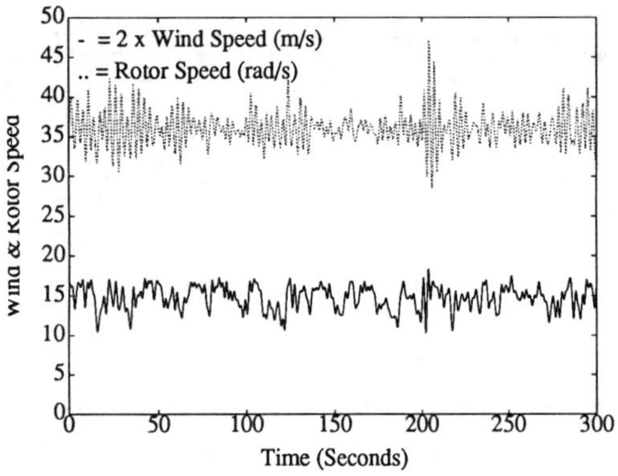

Figure 9 Simultated Fixed Speed Wind & Rotor Speed vs. time

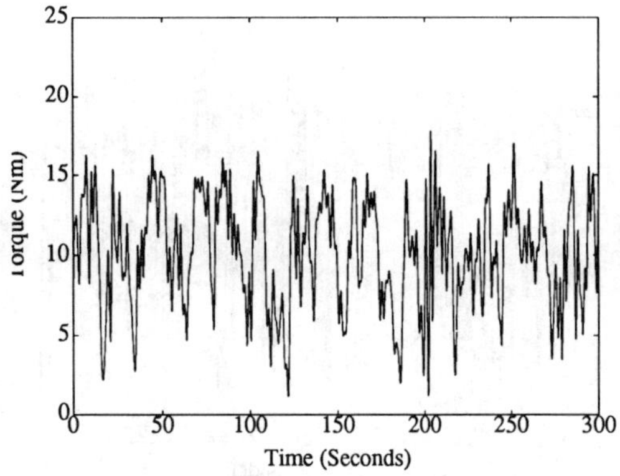

Figure 10 Simulated Fixed Speed Torque vs. time

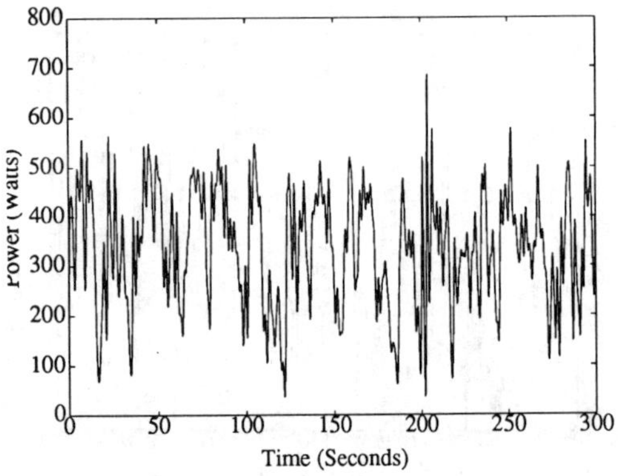

Figure 11 Simulated Fixed Speed Power vs. time

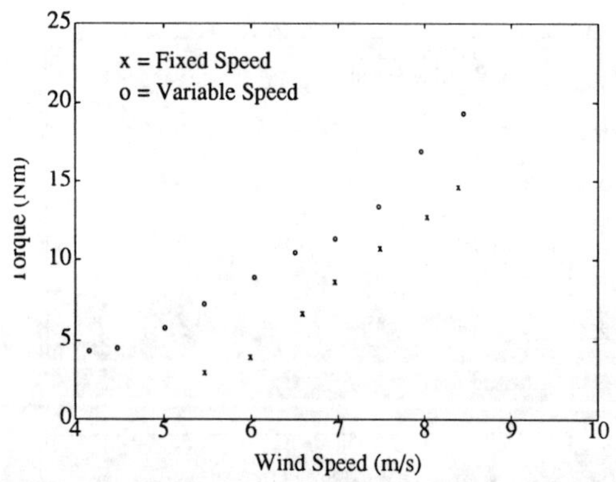

Figure 12 Simulated Speed Comparison Torque vs. Wind Speed

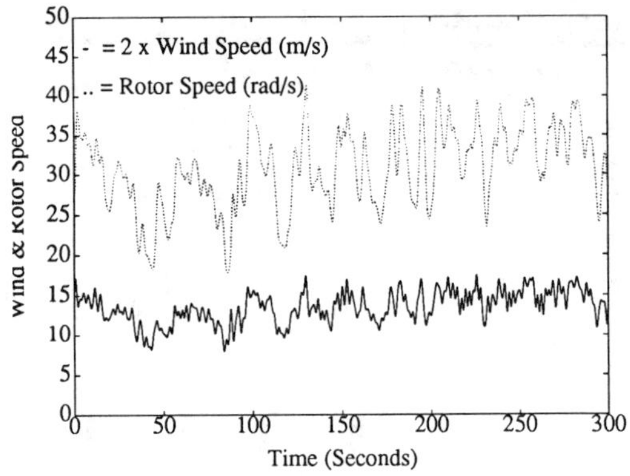

Figure 13 Simulated Variable Speed Wind & Rotor Speed vs. time

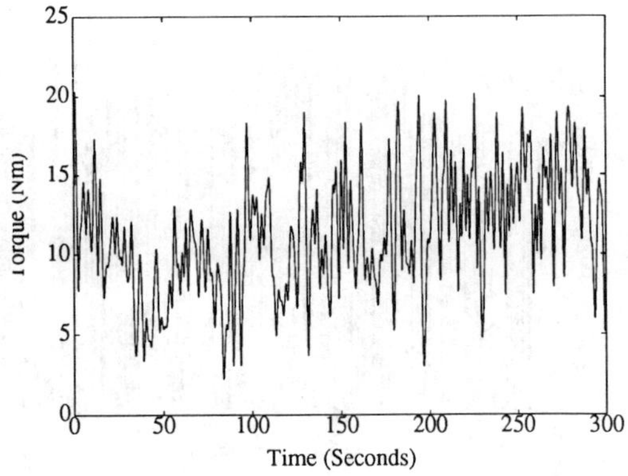

Figure 14 Simulated Variable Speed Torque vs. time

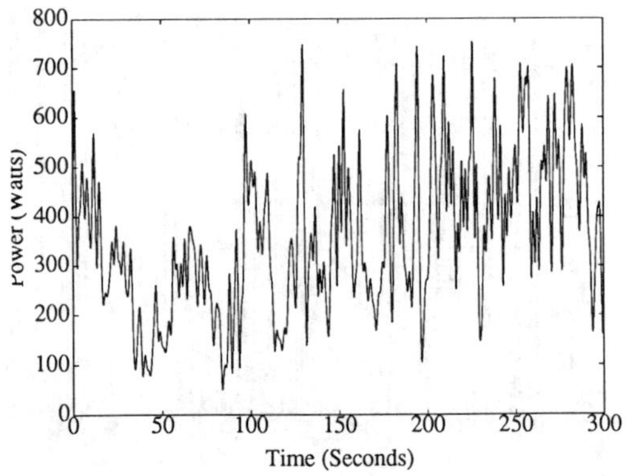

Figure 15 Simulated Variable Speed Power vs. time

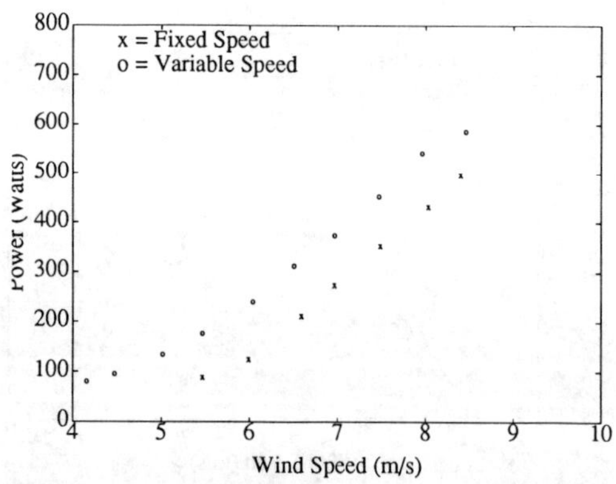

Figure 16 Simulated Speed Comparison Power vs. Wind Speed

Initial simulation runs, feeding the experimental wind speed data through the DUWECS model, produced greater power and torque readings than had been observed in the experimental data. To rectify this the Cp-λ curve was 'tuned' to bring the simulated power curve closer to the experimental power curve. Controller parameters were also 'tuned' from those used in the ICWTG controller to values that produced a stable response to a step change wind speed input for the computer model. Once this had been done a full simulation for fixed speed and variable speed operation was undertaken.

3.1 Fixed Speed Operation

Figures 9 to 11 show a 5 minute sample of time series data obtained from the DUWECS model from the introduction of the wind speed time series presented in section 2. for operation of the wind turbine under fixed speed control. The simulated fixed speed controller is able to keep the rotor speed to within 4.50% of the rotor speed set point of 36 rad/s with a mean value of 36.32 rad/s.

3.2 Variable Speed Operation

Figures 12 to 14 show a 5 minute sample of time series data obtained from the DUWECS model from the introduction of the wind speed time series presented in section 2. for operation of the wind turbine under variable speed control. The simulated variable speed controller is able to keep the tip speed ratio to within 8.86% of the tip speed ratio set point of 8 with a mean tip speed ratio of 7.85.

3.3 Speed Control Comparison

Figures 15 and 16 show the binned torque and power versus wind speed curves for the simulated fixed speed and variable speed operation of the wind turbine. From these simulation graphs, as with the experimental data, it can be seen that the power-wind speed and torque-wind speed curves are higher in the variable speed case than in the fixed speed case.

4. DISCUSSION

Further processing of the 5 minutes of data introduced into the DUWECS model has provided an indication of the energy capture of the wind turbine in the fixed speed and variable speed modes of operation. From the experimental data the energy capture advantage of variable speed over fixed speed has been calculated to be 6.90% based on the mechanical power. For the computer simulation the energy capture advantage of variable speed over fixed speed has been calculated to be 6.62%, based on mechanical power.

The computer simulation produces much smoother torque and power, with lower peak values, than are seen with the experimental data. Also the simulation as produces greater variations in speed for both fixed and variable speed operation. Much work is still needed to improve the accuracy of the model.

5. CONCLUSIONS

This paper has presented initial experimental and corresponding computer simulation data for fixed speed and variable speed operation of an experimental wind turbine.

Experimentally the wind turbine has been shown to operate under fixed or variable speed operation. Variable speed operation produces more, smoother power than fixed speed operation. Torque transients are also reduced with variable speed operation. Initial calculations have shown that there is a definite, although small, improvement in energy capture with variable speed compared to fixed speed operation.

The simulation conducted produced results similar to the real experimental data. However, the DUWECS model still needs further refinement.

6. ACKNOWLEDGEMENTS

The support of the Science and Engineering Research Council, U.K. and the Wind Energy Group, U.K. in the provision of a CASE studentship to carry out this research is acknowledged. Also acknowledged is the support of the Energy Research Unit at the S.E.R.C. Rutherford Appleton Laboratory, U.K. where all the experimental work was carried out. I also thank Peter Bongers for his help with getting the DUWECS modelling package operational with this particular configuration.

REFERENCES

[1] SIMMONS, A.D., FRERIS, L.L., BLEIJS, J.A.M. Comparison of energy capture and structural implications of various policies of controlling wind turbines, British Wind Energy Association annual conference, Swansea, Wales, 1991.

[2] BONGERS, P.M.M. Reference Guide Delft University Wind Energy Converter Simulation program, Delft University of Technology, The Netherlands, 1990.

[3] BONGERS, P.M., DIJKSTRA, S., BOSGRA, O. Modelling and Control of flexible wind turbines, European Community Wind Energy Conference, Madrid, Spain, 1990.

CONTROL OPTIONS FOR WIND ENERGY CONVERSION SYSTEMS ‡

Peter Bongers, Wilco Sturm§ , Ilya Kraan, Sjoerd Dijkstra, Okko Bosgra
Mech. Eng., Systems and control Group, Delft University of Technology,
Mekelweg 2, NL-2628 CD Delft, Tel. + 31 15 78 4703, email: bongers@tudw03.tudelft.nl

Abstract

The economic costs for electricity generated by wind power plants can be reduced by decreasing the construction costs, while increasing the lifetime and efficiency of the energy conversion. In an ideal situation choices with respect to the control system should be taken into account when a wind turbine system is optimized with respect to the above mentioned items. However in the present day to day practice some individual components are optimized, while the remaining part of the turbine is copied from existing ones. It is obvious that such a designed wind turbine will not be optimal. When turning to existing turbines or wind turbines on the drawing table, the control systems will become relevant on the site. Now the question will be to what extend a control system can improve the dynamic behavior of the turbine. In this paper we will discuss the contribution of control systems in order to achieve a better energy to cost ratio for a complete wind turbine system. In what way an optimum can be achieved will depend strongly on the applied wind turbine components, for example an asynchronous generator (almost constant speed wind turbine) or a synchronous generator with AC/DC/AC link (a variable speed wind turbine).

1 Introduction

To increase the operational life and energy output of the wind turbine without raising the construction costs flexible wind turbines can be realized with soft characteristics in all subsystems in order to reduce the internal stresses, and essentially allowing the application of lighter and less costly components [4].
The most important underlying motivation for wind energy research for large scale energy production is the aim to reduce the price of produced electrical energy. One single wind energy conversion system will then have:

low construction costs. Low construction costs can be achieved by using lighter, and thereby more flexible, components. Allowing more flexibility in the components also introduces the possibility of reducing internal stresses. However the dynamics of each component should be designed more carefully otherwise the internal stresses are magnified.

long lifetime. Fatigue loads have to be kept within acceptable margins in order to guarantee a long lifetime and thereby achieve an economic attractive system.

efficient energy conversion. Higher energy production out of the same amount of wind implies lower costs of the produced energy.

It is evident that each of these points more or less independently reduce the costs of the generated electrical energy.
The efficiency of wind energy conversion into electrical energy is of importance in partial load, where the maximal amount of the available wind energy has to be converted. This has to be achieved without an excess of mechanical loads, to maintain a reasonable operational life of the wind turbine. The application of a variable speed generator (like a synchronous generator with DC-link [17, 20]) enables a high efficiency energy conversion. In full load however the rated amount energy production must be maintained at minimal fatigue loads occurring in the wind turbine. By variable speed wind turbines variations of wind speed can partially be buffered variations in rotor speed.
The design of well controlled flexible wind turbines seems to be attractive for commercial applications. However such a design must be carried out very carefully otherwise the favorable soft characteristics will be lost or even reversed. Thus an accurate model describing of the relevant dynamics of the complete wind turbine is imperative in the design phase of the wind turbine (including the controller).
For our study of controller design for wind turbine systems, we assume that the wind turbine has been designed already. The two main type of wind turbine systems are studied in this paper are: a constant speed wind turbine and a variable speed wind turbine.

‡This research was supported by the CEC under grant JOUR-0110 and the Netherlands Agency for Energy and Environment under grant 24.301/5021
§Currently at the T.N.O Industrial research, P.O. Box 6033, 2600 JA Delft, The Netherlands

In order to evaluate the impact of the controllers on the wind turbine systems, a dynamical model of the wind turbines is a necessity. In Section 2 the integrated dynamic models of the different wind turbine systems are discussed. These models are implemented in the DUWECS package [3] which is used to obtain linear models and to simulate the non-linear dynamic behavior. The control systems are designed using $\mathcal{H}_\infty$ Control Theory [5, 6]. This implies that the controllers are designed on a linear state-space description of the non-linear wind turbine model. Because each wind turbine configuration has different control possibilities in Section 3 different control systems for each type of wind turbine system are designed. In Section 4 the dynamic models of Section 2 controlled by the controllers of Section 3 are used in the DUWECS program to obtain dynamic responses of the different controlled wind turbine configurations. Conclusions of this paper are given in Section 5.

2 Wind turbine configurations

The behavior of a variable speed wind turbine and a constant speed wind turbine are studied in this paper. The constant speed wind turbine has an asynchronous generator directly coupled to the public grid and allows an active pitch control. The variable speed turbine has a synchronous generator coupled with a direct current link to the public grid. Both wind turbine configurations fit into the integrated dynamic model model description of flexible wind turbines [4]. Each wind turbine, schematically depicted in Fig. 1, can be seen as a set of interacting submodels.

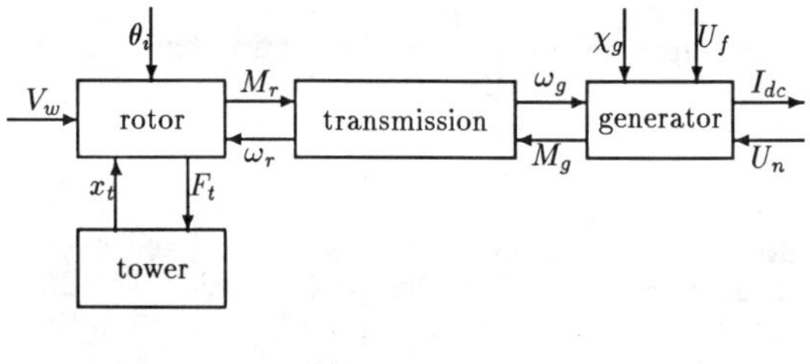

Fig. 1: modular structure of a wind turbine

The importance of the modular structure of Fig. 1 is the fact that each module is bilateral coupled with its environment. In the model description of the wind turbine configurations, the following building blocks are used:

rotor *(rot5a)* A two bladed rigid rotor with pitch, yaw and tilt freedom. The equations of motion are derived by using the method of Kane [12]. Each blade is divided into 9 sections with section dependent blade profile characteristics, corde, mass and twist. The 3D profile characteristics are obtained by applying the methods described in [22]. The aerodynamic forces for each blade section are calculated using methods described in [7], with tip-correction following [16], dynamic stall following [2] and a simple model of dynamic inflow [23]. The wind velocity, as felt by a blade section, is shaped with wind shear and tower shadow. The pitch angle of the blades can be used for control purposes.

tower *(tower1)* The dynamics of the tower are described by its lowest bending mode in axial direction.

transmission *(trans1)* The dynamics of the transmission are described by its lowest torsion mode.

two different type of generators are used:

synchronous machine *(irfgen)* The synchronous machine coupled by a thyristor rectifier to a DC-link is described in [11], a simple model of the excitation circuit is incorporated. The DC-link is coupled by an invertor to the public grid. The excitation voltage and the delay angle of the rectifier can be used for control purposes.

asynchronous machine *(asm)* A model of the asynchronous generator coupled directly to the grid is given in [18, 14]. There are no means to control this electrical system.

controller *(dicon0)* It is obvious that the specific controller parameters will depend on the wind turbine it has to control. Therefore a general purpose discrete time controller in state-space form is used.

The above mentioned mathematical models are implemented in DUWECS [3], which enables us to create different wind turbine configurations using the given building blocks. A variable speed wind turbine is created using *rot5a, tower1, unitra, irfgen, dicon0*, while a constant speed wind turbine is created using *rot5a, tower1, unitra, asm, dicon0.*

The physical parameters used in the building block are related to a 20kW rated wind turbine with blades of 8m. The validity of the mathematical model and the parameters, especially those used in the rotor module, are a topic of [1].

3 Controller synthesis

In this section the control design method, together with the specific choices to be made, will be discussed.

In order to achieve the control objectives, the pitch angle θ can be used for both configurations, where in the variable speed wind turbine also the delay angle α_g at the generator side can be used. For both configurations the produced current I_g and the generator speed ω_g are measured for control purposes.

For each wind turbine configuration one controller has been designed in one full load operating condition.

The control objectives are:

- low dynamic loads in the rotor shaft as well as in the blade roots, even under considerable measurement noise
- the feedback system has to be robust against uncertainties in the wind turbine model.
- the controller itself has to be stable, to make the implementation of the controller more acceptable.
- small control effort
- constant amount of produced electrical energy, in order to prevent off-design loads.

The main control objective, reduction of dynamical loads, cannot be achieved directly in a control synthesis procedure yet. For mechanical components the loads are mainly determined by the damping of the mechanical resonance frequencies. Therefore it is natural to apply an $\mathcal{H}_\infty$ control design method [1], by reasoning that reduction of the $\mathcal{H}_\infty$-norm [2] will increase the damping of the resonance frequencies. Thereby lower dynamical loads occur and a reduction in the construction costs can be achieved.

The control objectives are conflicting, for example it is impossible to independently control the objectives load reduction and constant energy production. However it is possible to achieve load reduction at the cost of a less constant amount of energy production. This means that in the control design a trade-off between the different control objectives is necessary. In order to achieve such a trade-off all design specifications are weighted in one single criterion. A different trade-off will be achieved by changing the specific weights. Next, consider for the following general feedback configuration of Fig. 2:

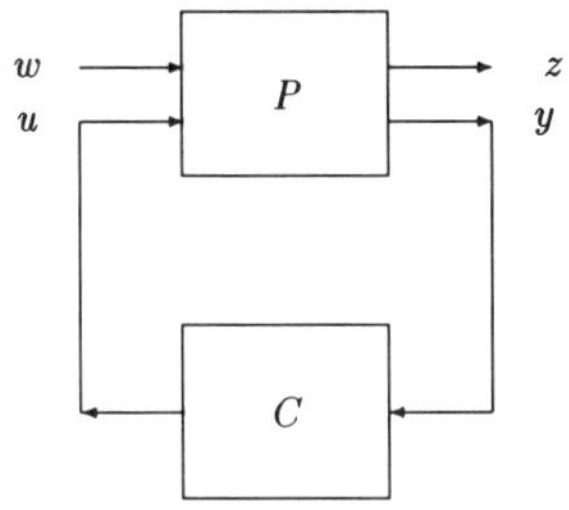

Fig. 2: feedback configuration

with

P	wind turbine	w	exogenous inputs
C	controller	z	controlled outputs
		u	control inputs
		y	measured outputs

The scheme of Fig. 2 can be written as:

$$\begin{pmatrix} z \\ y \end{pmatrix} = P \begin{pmatrix} w \\ u \end{pmatrix}$$

where P is partitioned as $\begin{bmatrix} P_{11} & P_{12} \\ P_{21} & P_{22} \end{bmatrix}$, with P_{ij} having the appropriate dimensions. The closed loop transfer function from w to z will be written as a lower linear fractional transformation on the plant and controller:

$$\mathcal{F}_l(P,C) = P_{11} + P_{12}C\,(I - P_{22}C)^{-1}\,P_{21}$$

The $\mathcal{H}_\infty$ optimal control problem is to find a controller C, such that $\mathcal{F}_l(P,C)$ is stable and $\|\mathcal{F}_l(P,C)\|_\infty$ is minimal.

Note that $\|\mathcal{F}_l(P,C)\|_\infty$ is the scalar criterion representing the trade-off of all weighted control specifications.

In [6, 10] it is shown that two Riccati equations need to be solved, in order to calculate the central controller. The order of the controller equals the order of the plant, including the order of the weighting functions.

Now in order to design a controller for wind turbine systems, such that the control objectives are satisfied, we need to translate the control objectives to properties that can be used in the $\mathcal{H}_\infty$-controller design. For this purpose the wind turbine model is extended with weightings, exogenous inputs and controlled outputs to obtain a standard plant suited for the control design, see Fig. 3 for the variable speed wind turbine.

The the control setup for both wind turbine configurations is the same. The fact that two control inputs are available for the variable speed wind turbine causes that the input uncertainty weightings are multivariable.

In the control setup it can be seen that the wind velocity V_w and the rotor shaft torque M_{as} are used. It should be noted that this information is not used as measured signals in the designed controller.

The weighting functions W_x are designed in such a way that the controllers will satisfy the objectives. For example the transfer function $w_1 \to z_1$ is the weighted rotor shaft torque due to filtered wind speed disturbances.

4 Results

In this section some simulation results are given for four wind turbine configurations, where the left part of Table 1 refer to the abbreviations used in the figures:

[1] Easy references on $\mathcal{H}_\infty$ control are for example [9, 15]

[2] For single input single output linear systems, the $\mathcal{H}_\infty$-norm is the maximum amplitude of the transfer function over all frequencies

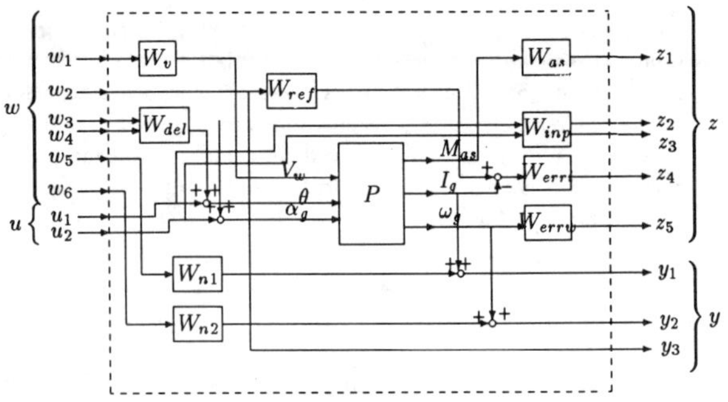

Fig. 3: weighted variable speed wind turbine model

ASM	uncontrolled constant speed wind turbine
ASM+control	controlled constant speed wind turbine
SMDC	uncontrolled variable speed wind turbine
SMDC+control	controlled variable speed wind turbine

Table 1: simulated wind turbine configurations

The simulations are performed using DUWECS [3], assuming an effective wind speed at rotor shaft height according to [8] including an extra wind shear of 30% and tower shadow.
The time history of the produced electrical power is given in Fig. 4.

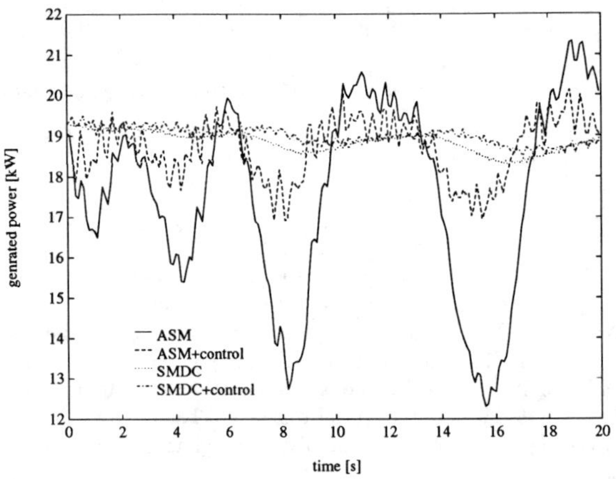

Fig. 4: Generated electrical power

For the constant speed wind turbine the power fluctuations are reduced by a factor of three by applying a controller. The power fluctuations for variable speed wind turbines are even in an uncontrolled mode small.
The main point of for modern control design is their capability to reduce loads. In Fig. 5 rainflow counting of the rotor shaft torque is given.
It can be seen in Fig. 5 that for both wind turbine configurations the application of the designed controller results in lower rotor shaft loads. The uncontrolled variable speed wind turbine is slightly smaller loads than controlled constant speed wind turbine, whereas the rotor shaft loading in the controlled variable speed wind turbine are almost vanished.
Finally in Fig. 6 the rainflow counting of the blade root flap moment are given.
It can be seen in Fig. 6 that in the constant speed wind turbine the application of control actions slightly increases the fatigue loading. Whereas the application of the designed control actions in the variable speed wind turbine reduced the fatigue loads in the blade root flap moments.
The fatigue loading presented depends heavily on the wind velocity fed into the simulation. Therefore, despite to the argumentation

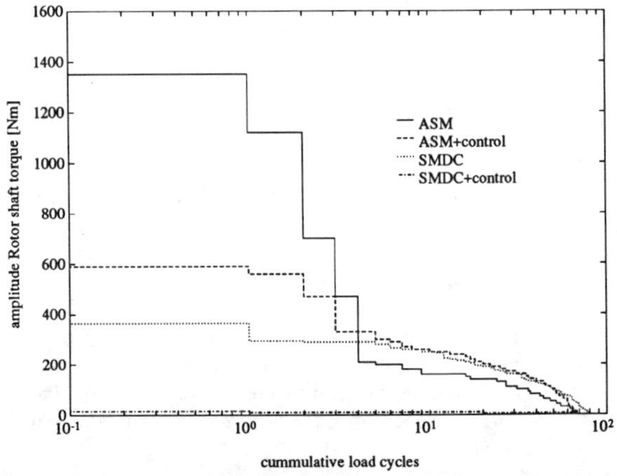

Fig. 5: Rainflow cycles rotor shaft torque

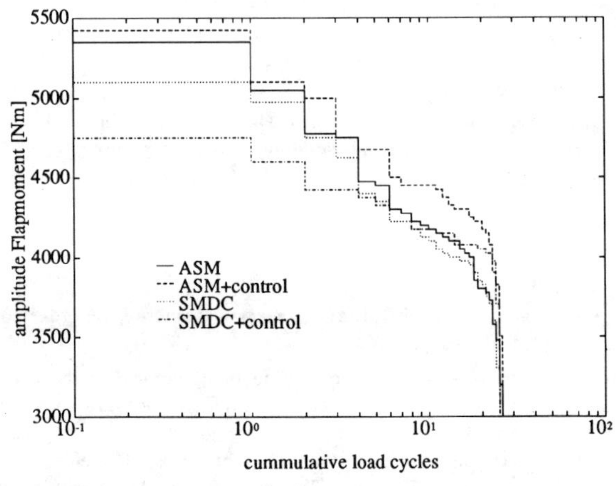

Fig. 6: Rainflow cycles blade root flap moments

of [8], one can say that a different wind characteristic should be used. One can investigate the effect of a different wind characteristic, using the transfer functions from wind velocity to rotor shaft torque (see Fig. 7).

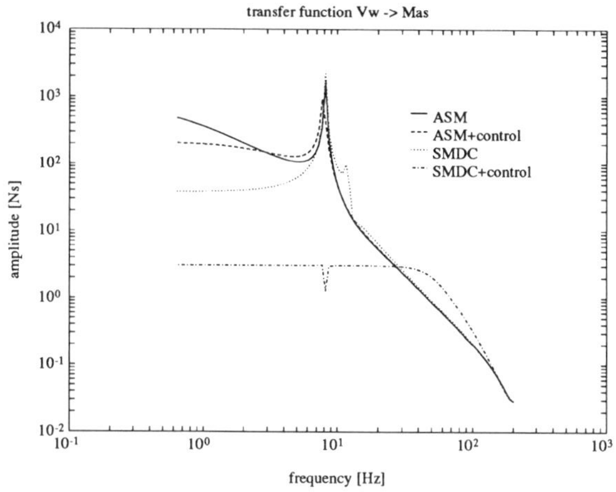

Fig. 7: Transfer functions wind velocity to rotor shaft torque

In Fig. 7 it can be seen that the application of a controller reduces the wind velocity variations on the rotor shaft torque for almost all frequencies. Because the controllers are designed to maintain approximately this level of performance, even when serious approximation errors in the model of the wind turbine occur, one may be confident that reduction of fatigue loading will be achieved for other shapes of the wind field.

Further aspects

The presented controllers are designed for one operating condition. A control strategy for the whole operating envelope of the wind turbine can be obtained by a smooth transition from one controller to the next controller, as described in [13]. By the application of this strategy, the results presented here can be generalized to the whole operating envelope of the wind turbine.

In the control synthesis procedure, implicitly, uncertainty descriptions are used. The structure singular value analysis procedure [5] provides a tool to analyze the different sources of uncertainty in more detail. Along the lines of [19] the impact of physical different sources of uncertainty can be evaluated. The necessity of further modeling the physical phenomena can be evaluated by this kind of studies.

For existing wind turbines, where no reliable theoretical model is available, it is possible to design controllers following the procedure outlined in [21]. In this procedure recursively: An experimental model is estimated on signals measured when the wind turbine is operating with a low performance controller; a higher performance controller is synthesized on the new model; This procedure is terminated when the control objectives are satisfied.

5 Conclusions

Although the simulation results of the previous section are obtained with specific wind turbine configurations (rated power, rotor diameter, wind shear, etc.) the conclusions can be extended to the more general difference between constant speed wind turbines and variable speed wind turbines.

Using a constant speed wind turbine it is possible to reduce the variations in produced power and rotor shaft loads at the expense of slightly increasing blade loads.

Using a more expensive variable speed wind turbine it is possible to keep the produced electrical energy constant and simultaneously reduce the mechanical loads.

References

[1] Baars G.E. van, P.M.M. Bongers, H. Schwartz, H. Arnold , "Experimental and theoretical results of the UNIWEX wind turbine", in *Proc. Europ. Community Wind Energy Conf.*, to appear, 1993.

[2] Beddoes T. , "Representation of airfoil behaviour", *Vertica*, vol.7, 1983.

[3] Bongers P.M.M. , *DUWECS Reference Guide, Delft University Wind Energy Conversion Simulation Program.* Delft University of Technology, The Netherlands, 1990.

[4] Bongers P.M.M., W.A.A.M. Bierbooms, Sj. Dijkstra, Th. van Holten , *An Integrated Dynamic Model of a Flexible Wind Turbine.* Delft University of Technology, The Netherlands, 1990.

[5] Doyle J.C. , "Analysis of feedback systems with structured uncertainties", *IEE proc., part D,* vol.129, pp.242–250, 1982.

[6] Doyle J.C., K. Glover, P.P. Khargonekar, B.A. Francis , "State–space solutions to standard H_2 and H_∞ problems", *IEEE Trans. Automat. Contr.,* vol.AC–34, pp.831–847, 1989.

[7] Eggleston, D.M., F.S. Stoddard , *Wind turbine engineering design.* Van Nostrand Reinhold Company, New York, 1987.

[8] Engelen, T.G. van, Bongers P.M.M., J.T.G. Pierik, G.E. van Baars , "Integral control for damping of mechanical resonance and protection against grid failures: results in a full load operating condition", in *Proc. Europ. Community Wind Energy Conf.,* to appear, 1993.

[9] Francis B.A. , *A course in $H\infty$ control theory.* Lecture notes in control an information sciences–88, Springer Verlag, Berlin, 1987.

[10] Glover K., J.C. Doyle , "State–Space formulae for all stabilizing controllers that satisfy an H_∞ norm bound and relations to risk sensitivity", *Syst. Control Lett.,* vol.1, pp.167–172, 1988.

[11] Hoeijmakers M.J. , *A possibility to incorporate saturation in the simple global model of a synchronuous machine with rectifier.* Eindhoven university of Technology, The Netherlands, 1989.

[12] Kane T.R., D.A. Levinson , *Dynamics, theory and applications.* McGraw–Hill series in mechanical engineering, 1985.

[13] Kraan I. , *Control design for a flexible wind turbine* (in Dutch). Delft University of Technology, TUD–WBMR–A–613, 1992.

[14] Leonhard W. , *Control of Electrical Drives.* Springer Verlag, Berlin, Germany, 1985.

[15] Maciejowski J.M. , *Multivariable feedback design.* Addison–Wesley Publishing Company, 1989.

[16] Prandtl L., O.G. Tietjens , *Applied Hydro– and Aeromechanics.* Dover Publications Inc., 1957.

[17] Raina, G., O.P. Malik , "Variable speed wind energy conversion using synchronous machine", *IEEE Trans. Aerospace and Electronic Systems,* vol.AES–21, pp.100–105, 1985.

[18] Schumacher W. , *Mikrorechner geregelter asynchronstellantrieb.* PhD Thesis, Technische Universitat Carolo–Wilhelmina, Braunschweig, 1985.

[19] Schutte D. , *Robustness analysis of flexible wind turbines* (in Dutch). Delft University of Technology, TUD–WBMR–A–613, 1993.

[20] Steinbuch M. , *Dynamic modelling and robust control of a wind energy conversion system.* PhD Thesis, Delft University of Technology, The Netherlands, 1989.

[21] Sturm W. , *Identificatie ten behoeve van regelaarontwerp voor een windturbine.* Delft University of Technology, The Netherlands, 1991.

[22] Viterna L., R. Corrigan , "Fixed pitch rotor performance of large horizontal axis wind turbines", in *DOE/NASA workshop on large horizontal axis wind turbines, Cleveland, USA.,* 1981.

[23] Øye S. , "Unsteady effects caused by pitch–angle changes", in *IEA Symposium – joint action on aerodynamics of wind turbines, London, Great Britain.,* 1986.

A CASE STUDY ON AN EASY DEFINABLE OPERATING HIGH SPEED CONTROL AND SAFETY SYSTEM (DOCS) PART I: DESCRIPTION DOCS & FIRST RESULTS

T.G. van Engelen , I.G. Kamphuis

Netherlands Energy Research Foundation (ECN)
Business Unit Renewable Energy, P.O. Box 1, 1755 ZG Petten, The Netherlands

Abstract

DOCS is an easy to use computer program package for control and safeguard of industrial processes with real-time data storage and message logging capabilities. It has been developed for advanced control in variable speed wind turbines. The main features of DOCS are the modular set-up and complete definability of control strategies. The user defined control algortihms can be layered on two levels for process protection and basic operation on one layer and advanced control on the other. Also they can be split-up into fast and slow parts, which can be allocated on different CPU-boards. The control can be partitioned to experiment phases like start-up, normal operation and shut-down. The main features of DOCS are shown through experiments on a computer simulator of a variable speed horizontal axis wind turbine with passive pitch control.

1 INTRODUCTION

The Definable Operating Control and Safety system (DOCS) has been developed[1] for experiments with advanced control in the Dutch research programme FLEXHAT [3] and the subsidiary project IRFLET [5, 2]. Laboratory scale results, from the IRFLET project, are reported in another contribution to this conference [2]; control objectives are are damping of mechanical resonance in the drive-train and power regulation. Before application of DOCS on the field scale FLEXHAT-testrig, the test-facility 25 m HAWT of ECN equipped with a free teetering passive pitch-controlled turbine rotor, a case study is performed on a real-time simulator of the 25 m HAWT for testing and demonstration purposes.

Up till now the 25 m HAWT is controlled by a PDP11/85 process-computer. With the applied control software [4], this system allows maximum cycle frequencies of only 4 to 8 Hz, while implementation of advanced control algorithms is possible with a great deal of restrictions. Nevertheless, the flexible set-up of this old control software formed the basis of DOCS: through analysis of the old and the desired new functionality the following requirements on DOCS were derived:

- implementation of multivariable feedback control (cycle time 0.01 to 0.02 s);
- implementation of process protection against extreme conditions and unexpected controller behaviour;
- basic process operation, like start-up, positioning and shut-down, well adjusted to advanced control;
- easy definable process signals, control algorithms and operational data, like setpoints, parameters and constraints;
- on-line user modifyability of parameters in the control algorithm by simple name reference;
- synchronized recording of system messages, user actions and algorithm data;
- multi-processor capability for parallel execution of fast and slow user code;
- distributed development capability: pre-testing of control algorithms should be enabled in a software development environment (e.g. work station with process simulator).
- portability: application on different computer platforms should be possible with minor changes.

[1] Financed as a separate project by the Netherlands Ministery of Economic Affairs.

Section 2 describes the developed computer program package and gives an overview of the definable items when applied to a particular industrial process. Performance, implementation and availability are also dealt with.

In section 3 the main features and process signals of the applied wind turbine simulator in the case study are mentioned, the control actions by DOCS are specified and first results of the case study are presented.

Conclusions and future work are given in section 4.

Appendix A outlines the DOCS application to the simulator.

2 PROGRAM DESCRIPTION

The executable tasks of DOCS set up a three-layer integrated control system. The lowest one, the SHELL-layer, is nearest to the computer hardware level and consists of fixed program code in which the two user programmable upper layers are embedded. The upper layers set up the actual control strategy, which is dealt with in 2.1. The SHELL performs (fig. 1):

- device I/O for process signals,
- checks on constraints,
- experiment phasing,
- user interfacing,
- message logging and data recording.

Before operation the SHELL is configured to a specific process and control strategy in a preceeding *definition* phase. Properties and constraints of process signals, operational data and experiment phases (turbine start-up, normal operation, shut-down) are read from input files. These data are mapped to the I/O-devices, user

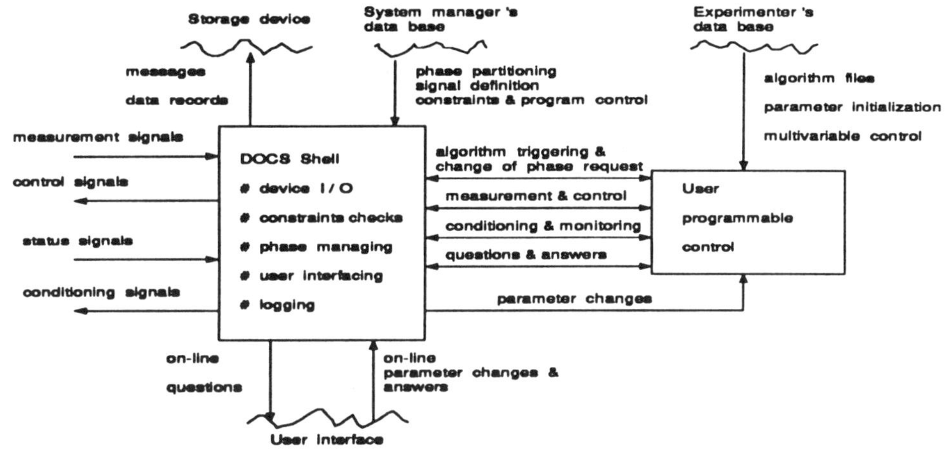

Figure 1: Functional layout of DOCS

interface and user program code. After definition, the SHELL triggers the appropriate, phase related, user code.

Through this configuration and embedding mechanism a user is able to implement any control strategy without detailed knowledge of the underlying hard- and software structure. The built-in capabilities of the SHELL and supporting software for the user program code enables a control strategy to

- measure and control,
- text based conditioning and status monitoring,
- react on constraint violations,
- jump from one phase to another,
- state on-line question/answer procedures with the user,
- apply on-line user modifyable operational data,
- communicate between parts of the user code.

2.1 User program code

The user code sets up an 'in four partitioned' control algorithm (fig. 2):

- two partitions are triggered with high cycle frequency and two with low frequency;
- one pair of fast and slow partitions (on the *machine* level) offers the other pair (on the *experiment* level) a process in which protective, basic operation and filtering of planned advanced control are incorporated.

Together with the process equipment and the SHELL, the fast and slow partitions on the machine level establish a *virtual machine*, which is to be operated on by those on the experiment level; the latter set up an *experiment execution system*. These two algorithm partition pairs set up respectively the VMS[2]- and EES[3]-algorithm and act on the *medium* and *upper* layer of DOCS. In accordance to this these layers are defined as the VMS- and EES-layer. On each layer, the fast and slow partition are able to communicate.

The control algorithm should also be structered according to the defined experiment phases. Therefore each of the four parts contains callable functions from the SHELL with user code related to a specific phase. Further refinement has been obtained by enabling definable subphases, *entries*, within a phase. The prescribed function names result from concatenation of phase and entry names. The VMS-algorithm *must* contain both slow and fast control functions for all defined phase and entries, while the EES-algorithm *may* contain control functions *only* for those phases, for which a flag has been set.

For each of the four algorithm parts a unique file must be supplied, while for each part any number of files may be linked to DOCS. An algorithm file gets a 'slow' or 'fast control meaning' by linking it to either the fast or slow control task. In the definition phase, a file quartet is selected through input file data. For this each *algorihm file* contains a unique identifier and a flag for application on either the VMS- or EES-layer. Further obliged file contents are:

- variable declarations for exchange of process signals, communication variables;
- function declarations for synchronized calls from the SHELL to the phase/entry related user code on *all 4* algorithm files;
- function definitions with data processing and program flow (phase jumping) statements; the names of process signals, operational data, phases/entries, etc. are used as variables;

The operational objectives[4] should be pursued from the appropriate algorithm parts. Consequently the user is forced by the supporting software to well-considered allocation of actions:

- The utilities for linear multivariable feedback are only available on the EES-layer, while all control signals *must be confirmed* on the VMS-layer.
- Conditioning and status monitoring capabilites as well as question/answer procedures are unique for the slow VMS-layer, while the (*results* are retrievable on the whole VMS-layer.
- Change of phase requests from the EES-layer will only be honoured if experiment level control is allowed in the new phase, while those from the VMS-layer are always prioritized.

Measurement signals are available in all algorithm parts, while on-line modifyable operational data are allowed.

[2]Virtual Machine's Software controlled part

[3]Experiment Execution System

[4]Objectives of all process operation actions, including protection, basic operation and advanced control.

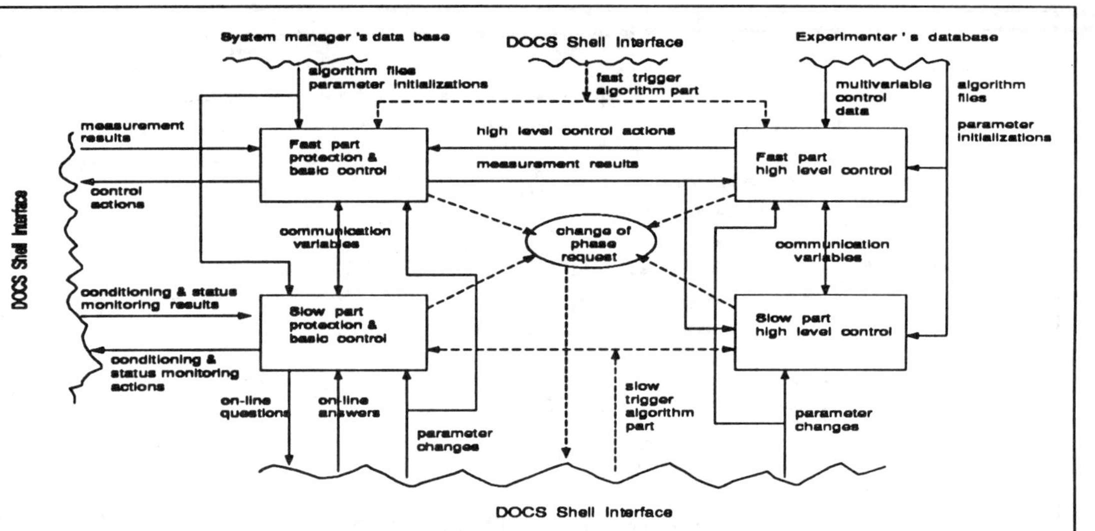

Figure 2: User programmable part of DOCS

2.2 Performance, implementation and availability

The minimum realizable slow and fast cycle times depend on the applied I/O-devices, computer platfrom, required I/O and algorithm complexity. DOCS has been programmed in standard C and consists of four executable tasks

- INP input processing task,
- QDT question display task,
- FCT fast control task,
- SCT slow control task.

Tasks FCT and SCT are built by the user through linking fast and slow algorithm files to the DOCS support libraries. The connection to the SHELL is established by including two header files and making variable declarations through predefined macro's. When DOCS is used in a developing/testing environment with simulated process, then also a simulator program module must be linked to FCT.

When starting DOCS, the executable tasks may be allocated over one to four processor boards, realized through a transparant intertask/interprocessor communication mechanism. Choosing two boards enables parallel processing for fast and slow control, while input and question handling will be time-shared with slow control. If high speed conditioning and monitoring is pursued next to advanced control then a 3-processor configuration will be preferred.

Implementations under UNIX[5] and OS9[6] are available. Porting DOCS to other computer platforms takes minor effort because a minimum number of, isolated, system calls is applied (e.g. *no* operating system specific flag- or event mechanisms).

The case study is performed with the UNIX-version on a SUN sparc 1+ workstation. The OS9-implementation (2 Motorola 68030 CPU's in a CompControl[7] modular VMEbus system) is used in the IRFLET-project, and will be used within the FLEXHAT-programme. Typical performance requirements are :

[5]UNIX is a trademark of Bell laboratories.
[6]OS9 is a trademark of Microware Systems Corp. and Motorola Inc.
[7]CompControl Int. B.V., P.O. Box 921 5600 AX Eindhoven, Holland

- control / measurement signals: 4 / 16,
- control cycle time: 0.01 to 0.02 s,
- conditioning / status signals: 30 / 48,
- conditioning cycle time: 0.2 to 0.5s,
- 2-input / 2-output high order multivariable control.

In summer '93 DOCS will be available with the next documentation:

- folder,
- system requirements and specifications,
- user guide and algorithm writer guide,
- technical documentation.

3 CASE STUDY

The case study demonstrates application to an industrial process. The process is a simulator of the 25 m HAWT test-facility at ECN, realized as a callable program module (later on DOCS will be applied to the 25 m HAWT itself). A *simulated* process was chosen because of the exactly known process properties, fast executable and reproducible experiments and completely controllable 'external' conditions. When periodically called to, the simulator program module behaves like a real machine: the core process (wind energy conversion) is conditional to the status of auxiliary systems, like power supplies, pumps and brakes, and the actions of a built-in primary safety system, while the full range of operation is covered.

DOCS is coupled to the simulator through user event setting and signal definitions on prescribed pseudo I/O-devices. The simulator program module, named RTSim, must include a header file with the same signal definitions. Also the header file may contain definitions of on-line modifyable *simulator* operational data The case study environment has the next properties:

- After phase definition, DOCS states a fixed number (=*value* simulator event variable) of calls per fast time cycle to module RTSim without interruption. This enables pseudo-continuous, (scaled) real-time, process simulation and allows more experiments within one session by 'looping' in the control algorihm.

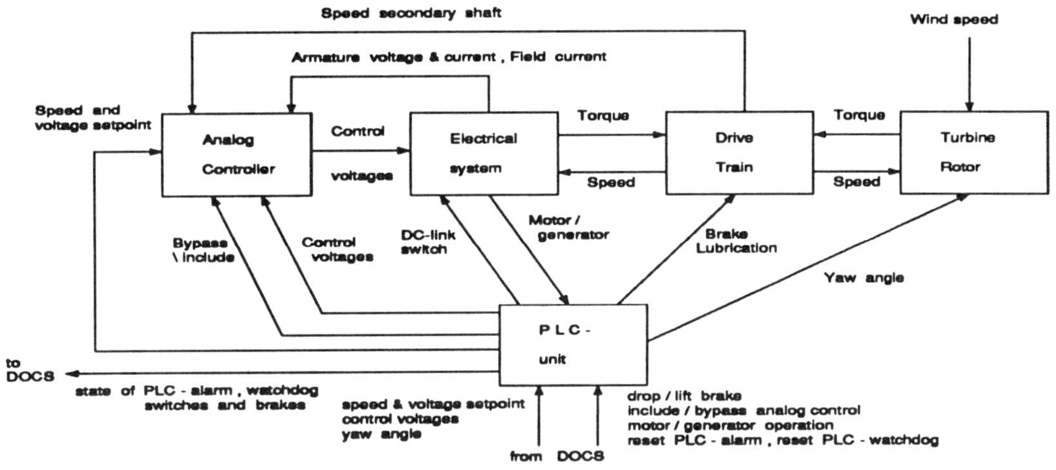

Figure 3: Layout of the wind turbine simulator

- Measurement and control signals are exchanged every fast cycle; conditioning, status signals and modified operational data every slow cycle.
- At feedback control in the fast algorithm part a time delay of one fast time cycle exists in the closed loop.

3.1 Simulator and process signals

The layout of the simulator is pictured in fig. 3. Initially, user supplied files with wind speed, - direction and grid voltage sequences are parsed. On-line, variations may be added to the file input data and failure events may be signaled to the power supplies and lubrication system of the drive-train. Simulator properties:

- Passive pitch controlled turbine rotor by centrifugal forces with active yaw facility. A conditioning and status signal exist for the power supply of the yaw facility; a control signal for yaw angle setting. The blade-tip pitch angle, rotor speed and yaw angle are measured.
- Flexible drive-train with slip coupling, brake on secondary shaft and gearbox with lubrication system (eigenfrequency drive-train ca. 4 Hz, fast oil temperature raising when lubrication fails). Conditioning and status signals exist for the gearbox lubrication system and brake. The deformation angle of the flexible shaft, speeds of primary and secondary shaft and gearbox oil temperature are measured.
- Variable speed electrical system for motor and generator operation. A DC-machine is grid-connected through a controllable invertor for power delivery and rectifier for field excitation. A fast safety switch is included in the armature current circuitry. Analog control equipment supplies the control voltages. Conditioning and status signals exist for the operation mode selector, power supply of the power electronics and fast safety switch. Armature and field voltage and current in the DC-machine are measured.
- Analog controller for rotor speed and voltage through invertor and rectifier control voltages; may be bypassed for direct control of power electronics by DOCS. Conditioning and status signals exist for the usage mode selector and power supply. Conditioning and status signals for the usage mode selector (include or bypass) and power supply. Control signals from DOCS are setpoints for rotor speed and voltage if analog control is included, else these are both the control voltages and setpoints; the latter are now used for including analog control again without large transients. The outputs of both control loops and the emitted control voltages are measured.
- PLC-unit for pirmary safeguarding and monitoring DOCS. Protective action (shut-down) is performed if DOCS omits cyclic watchdog resetting and in case of wrong status signals or first defence level constraint violation (rotor speed, -acceleration, armature current, oil temperature, wind speed). In case of protective action, conditioning and control signals from DOCS are blocked and the PLC-alarm is set. Conditioning and status signals for DOCS exist for the PLC-alarm system and watchdog; only *status* signals exist for the PLC-constraint checkers (reset automatically if DOCS resets the PLC-alarm).

3.2 Control strategy

The operational requirements to be met in the case study are:

- An experiment life cycle involves
  - motor start and generator operation;
  - generator operation without analog control, for
    - obtaining data records for model estimation (controller design),
    - advanced control (power regulation, resonance damping);
  - turbine shut-down by electrical braking and yawing.
- Pre-emptive turbine shut-down for prevention of protective action by the PLC-unit.
- Automatic experiment repetition when operation is allowed again after turbine shut-down, unless the operator stops the session.

The required data and algorithm files for embedding this control strategy in DOCS is described in appendix ??.

3.3 First results

The experiment course is outlined according to fig. 4. After 4

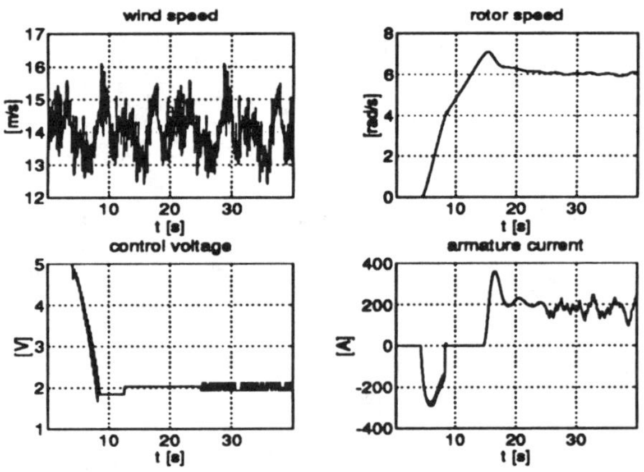

Figure 4: Experiment for system identification

seconds state the preconditioning state of the startup phase has been completed (starting the gearbox lubrication system is the time-critical factor, see table 3). From then the turbine is started in motor operation until the rotor speed amounts to 4 rad/s; the aramture current is negative. Switching into generator operation initially causes the armature current being 0 Ampere. When the rotor speed exceeds 6 rad/s, the analog controller is switched into bypass mode. After about 15 seconds the armature voltage exceeds the inverter voltage so that the armature current now becomes positive. After 25 seconds, phase StartUp is finished and phase Operation is entered for identification; a testsignal on the control voltage for the inverter is applied, which causes the high frequent variations in the armature current. The applied windspeed signal is designed such that, when passed through power coefficient data, a realistic driving torque is realized [2, 1].

4 CONCLUSIONS

Through a case study description, the complete definability of DOCS and the way to use it are showed.

Control definition on both input files and algorithm files leads to a minimum of programming effort, while through the supplied header files and other supporting software only elementary knowledge of the C-programming language is required.

The 2-time-scale/2-layer control algorithm with partitioning of functionality according to predefined experiment phases and entries enables flexible, surveyable and operational aim structured implementation of control strategies.

Future work

In the near future DOCS will be applied on the 25 m HAWT for resonance damping and power regulation.

ACKNOWLEDGEMENTS

Many thanks are due to Erik J. van der Leeden (CompControl), for his contributions in the conceptual phase of the design of DOCS, and Jan T.G. Pierik (ECN) for co-reading and providing illustrations.

References

[1] T.G. van Engelen. Miscellaneous usage of computer program TURBU-I: Spatial turbulence in wind input for power coefficient data. Intern report ECN-I–93-xxx (in preparation), ECN, 1993.

[2] T.G. van Engelen, P.M.M. Bongers, J.Th.G. Pierik, and G.E. van Baars. *Integral control for damping of mechanical resonance and protection against grid failures: Results in a full load operating condition.* In *Proceedings European Committee Wind Energy Conference 1993*, Travemunde, Germany.

[3] G.A.M. van Kuik, R.H.F. Tikkemeijer, and J.W.M Dekker. The development of advanced rotor systems. a survey of the flexhat program, including full scale results. In *Proceedings European Committee Wind Energy Conference 1990*, Madrid, Spain.

[4] F. Lekkerkerk. Bedieningsaanwijzingen voor automatisch bedrijf van de 25 m hawt. Technical Report FYS-WIND-88-17, ECN, May 1988.

[5] J.Th.G. Pierik, P.M.M. Bongers, T.G. van Engelen, and G.E. van Baars. Damping of mechanical resonance and protection against grid failure for variable speed wind turbines (irflet). Technical Report ECN-C–93-004, ECN, Petten, february 1993.

A DOCS APPLICATION

The application of DOCS to an industrial process generally involves the next steps:

- choosing computer hardware and tuning DOCS to it,
- making a phase/entry partitioning,
- definition of process signals on I/O-devices,
- definition of question/answer procedures,
- grouping conditioning and monitoring actions,
- writing VMS- and EES-algorithm files,
- writing VMS- and EES-run-time definition files.

These steps are dealt with below and highlighted with samples from the case study application.

Tuning to hardware. Usually DOCS will be tuned to the particular computer hardware and system software at delivery.

Allowed I/O devices are AD- and DA-boards for measurement and control signals and DDI-, and DDO-boards for measurement and status signals and control and conditioning signals. Size and conversion properties of the I/O-devices are defined in a separate I/O software package.

For the intertask/interprocessor communication utilities data about shared memory are defined in a message based communication package.

For allocation to CPU's and starting the exectuable tasks with the right environment variables (logging paths etc.) 'task start-up files' are provided.

The case study is performed on a SUN Sparc 1+ work station. In the simulation environment the (pseudo) I/O-devices are dynamically allocated on the shared memory, while measurement and control signals are treated as float variables and status and conditioning signals as logicals. No specific I/O-device definitions are required. The task start-up files e.g. automatically generate a window for each task, thus enabling source level debugging for each task.

Phase/entry partitioning. The system manager usually provides phase/entry partitionings. These are input files (extension .ped) with keywords for phases and nested entries and flags for control by the EES-algorithm and 'looping' (jumping back to a former phase). The order of the phases on the .ped-file determines jumps as forward or backward. The first defined phase/entry is entered after phase definition.

Below tabled phase/entry partitioning is used in the case study. The entry names are short form descriptions of the pursued operational aims.

phase	entries	EES	Looping
StartUp	Precondition, Yawalign, Motor, Generator, BypassAnalog	no	no
Operation	Identification, Control	yes	no
ShutDown	SlowingInc, SlowingByp, Parking, Postcondition	no	yes

Table 1: Phase/entry partioning in case study

Signal definition. The attributes of the process signals are parsed from a system manager's supplied configuration file (ext. cfg). In tables 2 and 3 sample signal definitions are given.

Measurement and control signals require the I/O-device (real or pseudo DA-/AD-board or real DDO-/DDI-board), (set of) line number(s), signal name, physical unit, calibration points and constraint definitions, consisting of upper/lower (time-derivative) limits and program flow definition.

Constraint violation may automatically induce a phase jump according to the *defined program flow*, which consists of a set of departing and destination states[8] for each signal.

device	PseudoAD
line	0
name	Vwind
unit	m/s
calibration {[V],[unit]}	{1, 0}, {5, 50}
value limits [unit]	0, 24
gradient limits [unit/s]	$-\infty$, 15
program flow {dpr,dst}	{StartUpPrecondition, ShutDownPostconditon} {StartUpYawalign, ShutDownParking} {StartUpMotor, ShutDownSlowingInc} {StartUpGenerator, ShutDownSlowingInc} {StartUpBypassAnalog, ShutDownSlowingByp} {OperationIdentifcation, ShutDownSlowingByp} {OperationControl, ShutDownSlowingByp}

Table 2: Measurement signal definition for wind speed

Conditioning and status signals require the I/O-device (real or pseudo DDO-/DDI-board), line number, target description, possible transitions and statuses, allowed transition times and initial transition.

Question/answer procedures. The system manager's supplied question/answer definition file (ext. .qad) links questions as displayed on the user interface to *variables* for questioning and answer retrieval in the VMS-algorithm. In the case study a question/answer procedure exists for decision about the actual turbine start-up in motor operation when the turbine has been aligned to the wind, i.e. at the end of state StartUpYawalign.

[8]State: phase/entry combination.

device	PseudoDDO
line	2
target	LubrSystemGearbox
transitions	PutOn, PutOff
transition times [s]	4, 0.5
initial transition	PutOff

Table 3: Conditioning signal definition for lubrication system

Conditioning and monitoring sequences. Conditioning and monitoring actions on more targets can be grouped on the system manager supplied sequence definition file (ext. .seq). A sequence is defined by a sequence name, to be used as *variable* in the VMS-algorithm for triggering, by action typing (conditioning and/or monitoring[9]) and action specifications (targets with transition and/or status). Sequences that involve conditioning can be executed in a parallel or serial way. A sample sequence definition is given in table 4.

name	PrepTurbine	
type	conditoning and monitoring	
execution	parallel	
target	action	status
PowSupplYawFacility	PutOn	On
PowSupplESystem	PutOn	On
PowSupplAnControl	PutOn	On
LubrSystem	PuOn	On

Table 4: Conditioning sequence definition for turbine preparation

Algorithm files . According to the phase/entry partionioning in table 1 the VMS-algorithm files *must* contain the functions StartUpPreCondition(), StartUpYawalign() to ShutDownPostCondition(); the EES-algorithm files only *may* contain functions OperationIdentification() and OperationControl(). A sample of the *template* fast VMS-algorithm file, with unique identifier 'CaseFastVMS' is listed below. As an illustration, a few variables are included in the template:

- Vwind: wind speed, measurement signal (MS);
- Nr1Set: setpoint rotor speed, control signal (CS);
- PowSupplyYaw: power supply yaw facility, validated conditioning target (VCT);
- PowSupplyESystem: power supply electrical system, validated conditioning target (VCT);
- PowSupplyAnControl: power supply analog controller, validated conditioning target (VCT);
- LubrSystem: gear box lubrication system, validated conditioning target (VCT);
- PrepTurbine: preparing sequence for turbine operation, validated conditioning sequence (VCDSEQ);
- Nr1MotGen: rotor speed for switch from motor to generator operation, run-time definable parameter (IP).

Comments in the file template are stated between /* and */:

```
/* Declarative part */
#include <DOCSTypes.h>
PMCLIST MS(Vwind), CS(Nr1Set), ... ENDLIST
IPLIST IP(Nr1MotGen) ... ENDLIST
LMCLIST VCT(LubrSystem), VCT(PowSupplyYaw),
```

[9]Combined conditioning and monitoring is defined as validated conditioning.

```
 VCT(PowSupplyESystem), VCT(PowSupplyAnControl),
  VCDSEQ(PrepTurbine) ... ENDLIST
... /* other variable declarations */

/* Function definition part */
StartUpPreCondition()
{  /* trigger turbine preparation */
   LMCTrig(PrepTurbine);
   if(LMCINQ(PrepTurbine)==LMCSize(PrepTurbine))
   { /* turbine prepared; go yawing */
     nexstate(StartUpYawlign);}
   else if(LMCINQ(PrepTurbine)<0)
   { /* time violation; stop experiment */
     nexstate(ShutDownPostCondition);}
   else {} /* turbine preparation busy */
   return();
}
... /* other function definitions */

/* Make connections with SHELL */
#include <DOCSInterfaces.h>
PMCLIST MS(Vwind), CS(NrlSet), ... ENDLIST
IPLIST IP(NrlMotGen) ... ENDLIST
LMCLIST VCT(LubrSystem), VCT(PowSupplyYaw),
 VCT(PowSupplyESystem), VCT(PowSupplyAnControl),
  VCDSEQ(PrepTurbine) ... ENDLIST
... /* other variable mappings to SHELL */
VMSCFLIST NEXTCF(StartUpPreCondition),
 NEXTCF(StartUpYawAlgin),
  ... /* other function mappings to SHELL */
   NEXTCF(ShutDownPostConditioning) ENDLIST
LINKDEF(CaseFastVMS, <auxiliary link data>
```

Run-time definition files. The run-time file for experiment level control is used for starting up DOCS. It contains the name of the run-time file for machine level control, identifiers for a fast and slow EES-algorithm file, storage path specifications, timing variables and parameter initializations. The VMS-run-time file contains the names of the auxiliary input files with the phase/entry partitioning, process signal definitions, question/answer procedures and monitoring and conditioning sequences. Furthermore it contains identifiers for a fast and slow VMS-algorithm file and parameter initializations.

INTEGRAL CONTROL FOR DAMPING OF MECHANICAL RESONANCE AND PROTECTION AGAINST GRID FAILURES: RESULTS IN A FULL LOAD OPERATING CONDITION ‡

T.G. van Engelen §, P.M.M. Bongers ♮, J.T.G. Pierik § G.E. van Baars ♮

Abstract

Variable speed operation of wind turbines enables reduction of electric power fluctuation and fatigue loads on rotor blades and drive-train. Besides, it allows tuning to local conditions and switching onto the grid with relative ease. As results of the IRFLET-project, experimentally verified solutions are presented for two major *problems* of variable speed wind turbines, viz. the low damped mechanical resonance of the drive-train and the commutation failure of the grid convertor resulting from a grid fault. The experiments are performed on the IRFLET-test-rig, a laboratory scale variable speed wind turbine system, consisting of an emulated turbine rotor with passive pitch control, a flexible shaft and a synchronous machine with a machine commutated rectifier and grid commutated invertor.

1 Introduction

In the Netherlands the research programme FLEXHAT is being executed. It aims at the development of low weight, flexible components for a next generation of cost effective megawatt-scale horizontal axis wind turbines [15]. The FLEXHAT concept includes variable speed operation for:

- reduction of fatigue loads on drive-train and rotor blades,
- reduction of fluctuations in electric power,
- passive pitch control by centrifugal forces,
- maximum net energy capture in partial load,
- on-site tuning to local conditions.

A well controlled variable speed electrical system is a necessary condition to meet these operation features. Such a system is designed, realized and partly tested in a separate project named IRFLET, which is the subject of this paper.

Based on previous research a brushless synchronous machine with DC-link was chosen for the electrical system [2]. The rectifier and invertor are respectively commutated by the synchronous machine and grid voltage. Such a system enables variable speed operation, has good control possibilities and can be switched easily onto the grid.

Two major problems in variable speed systems with DC-link are low damped mechanical resonance of the drive-train, typically in a range of 1 to 10 Hz, and the commutation failure of the invertor resulting from grid faults. These phenomena may respectively cause high fatigue loading of the drive-train and severe increase of the generator current, and possibly the shaft torques too.

On a theoretical level has been demonstrated that the aimed FLEXHAT operation features can be realized and the drive-train resonance can be damped critically by advanced controllers [6, 7, 14, 22]. Also is shown that the effects of a commutation failure can be suppressed by very fast changing the direct voltage at the rectifier side through the firing angle [13].

The electrical system and a process computer for control is incorporated in the IRFLET-test-rig (see Fig. 1), a 30 kW laboratory scale variable speed wind turbine system [20], for verification of the theoretical results. The applied computer programme for control [11] enables embedding of advanced control schemes and of protection algorithms with respective cycle frequencies of 100 and 1000 Hz. The remaining components of the test-rig are:

- turbine rotor simulator, set up by a controlled DC-machine with fast rotating flywheel;
- flexible shaft, causing a mechanical resonance of 3 Hz;
- gearbox between DC-machine and flexible shaft;
- gearbox between synchronous machine and flexible shaft.

Measurement devices are installed for the direct voltage at the invertor side, direct current and generator shaft torque and speed.

The scope of the work considered in this paper is restricted to one full load operating condition. Primary solutions to the control and safety problem will be demonstrated through experiments on the IRFLET-test-rig, more details can be found in [4, 21].

The layout of this paper is as follows: In section 2 the development of the controller is described. The successive subsections deal with definition of the control requirements and the controller development traject, consisting of modeling, controller synthesis, verification and evaluation of fatigue loads. Section 3 deals with the analysis and verified solution of the protection against grid faults and commutation failure. Section 4 states the conclusions.

2 Control development

The next requirements on the controller should be satisfied for realization of the FLEXHAT operation features and to deal with the mechanical resonance problem in full load conditions:

- the peak in the transfer function from wind speed to shaft torque should be reduced considerably. If the resonance peak is reduced to about 1.25 times the value at low frequencies, then a much smaller gearbox is allowed.
- average electrical power should be kept constant.
- only budget measurement devices should be applied, which implies no shaft torque feedback.

Because a passive pitch mechanism is present, the objectives have to be achieved by controlling the electromagnetic torque. According to now operating variable speed wind turbine systems, for example [3], first a PID controller from the firing angle of the rectifier to the direct current to achieve a bandwidth

‡This research was supported by the Netherlands Agency for Energy and Environment under grant 24.301/5021

§Netherlands Energy Research Foundation, Business Unit Renewable Energy, P.O. Box 1, 1755 ZG Petten, The Netherlands

♮Delft University of Technology, dept. of Mech. Eng., Systems and Control Group, Mekelweg 2, 2628 CD Delft, The Netherlands

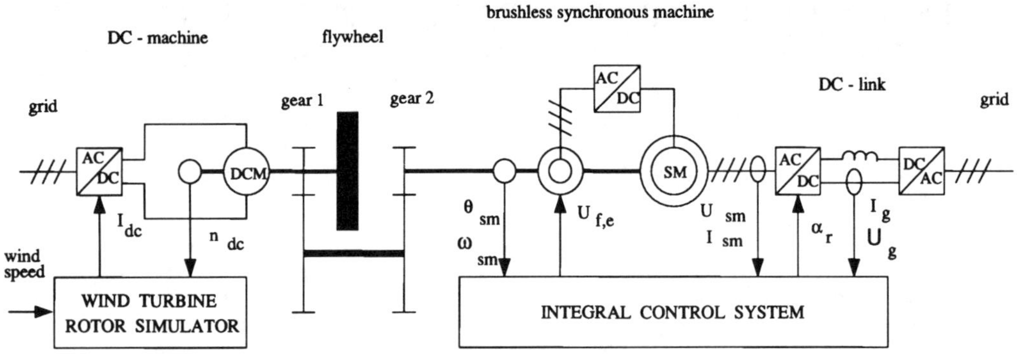

Fig. 1: The IRFLET test-rig

of approximate 10 Hz need to be designed. This would enable torque control by a cascade of PID controllers. Due to specific choices in the construction of the test-rig, the achievable bandwidth is limited to approximate 2 Hz. This observation implies that the control objectives are unachievable using PID controllers, hence controllers having more freedom than PID controllers have to be designed.

The control design methodology pursued in the sequel is hinged on the availability of an accurate linear dynamic model of the system to be controlled.

For this purpose the test-rig of Fig. 1 is schematized (Fig. 2). with[1]:

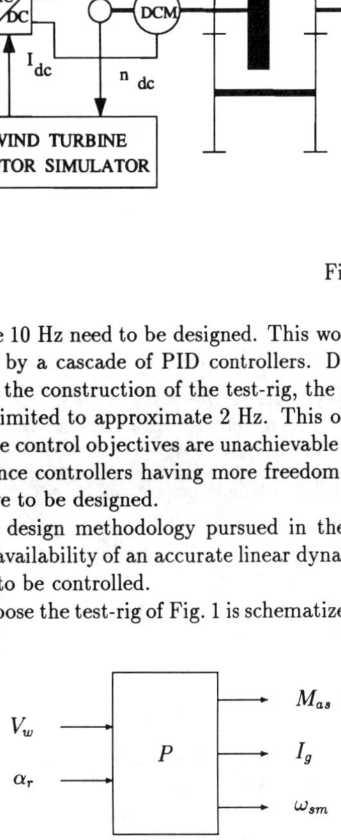

Fig. 2: Wind turbine

vw	V_w	wind speed	$[m/s]$
alf	α_r	delay angle rectifier	$[rad]$
Mas	M_{as}	rotor shaft torque	$[Nm]$
Ig	I_g	direct current	$[\ A\ ]$
omg	ω_{sm}	generator speed	$[rpm]$

The eventual controlled situation is schematically drawn in Fig. 3, where the controller uses only u and y to influence the transfer function from $w \to z$.

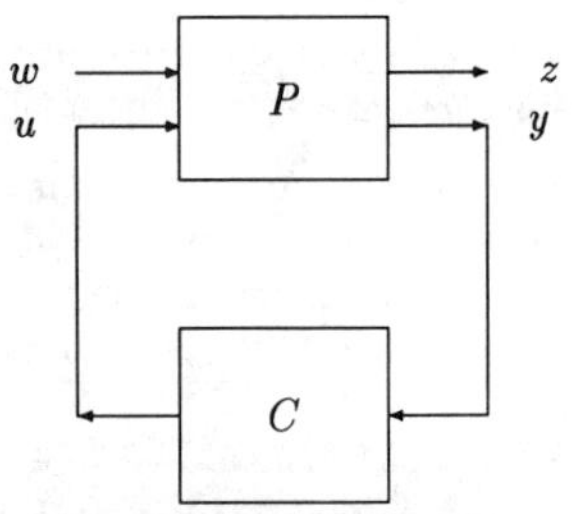

Fig. 3: feedback configuration

with

P	test-rig	w	exogenous inputs	V_w
C	controller	z	controlled outputs	$\begin{pmatrix} M_{as} \\ I_g \end{pmatrix}$
		u	control inputs	α_g
		y	measured outputs	$\begin{pmatrix} I_g \\ \omega_{sm} \end{pmatrix}$

The applied $\mathcal{H}_\infty$-design method enables the user to specify the desired transfer function from the exogenous inputs to the controlled outputs, see Fig. 3.

This implies that a linear dynamic model is required, describing the relations between (V_w, α_r) and $(M_{as}, I_g, \omega_{sm})$.

The controller development traject involves four steps, while each next step may lead to re-execution of former ones:

1 **Making a model.** A linear dynamic model will be derived for the considered operating condition. For obvious reasons the linear model is an approximation of the complex reality, hence also a description of the approximation error has to be derived. By lack of a verified theoretical model the linear dynamic model is obtained from experiments on the test-rig.

2 **Controller synthesis.** The controller will be designed using an $\mathcal{H}_\infty$ minimization technique, which enables the designer to deal with the required performance objectives and the knowledge about the existing uncertainty in the used model.

3 **Verification.** In this step the assumptions and approximations of the previous steps are evaluated. To this end the predicted closed loop operation of the controller and the linear dynamic model are compared to those obtained from experiments with the controller on the test-rig.

4 **Evaluation of fatigue loading.** The effect of the controller on fatigue loading is determined by comparing load cycles in the flexible shaft obtained from successive experiments with active and inactive control. To this end a wind signal is constructed such that the mechanical system is excitated in a similar way as a real wind turbine.

2.1 Experimental modeling

As mentioned before no reliable theoretical model is available yet. Therefore a linear dynamic model of the test-rig will be derived by application out of well known system identification techniques [17, 16] to data measured at the test-rig. In order to use the measured data for model estimation and control design, it has been reassured that the test-rig satisfies the following conditions:

- the data is measured at a sample frequency such that a controller can act on the dynamics of the test-rig over a

[1] The symbols in the first column refer to the labels used in the figures.

frequency range below 10 Hz. A sampling rate of 100 Hz has been chosen;

- the data is measured with proper anti-aliasing filters, otherwise frequencies above half the sampling rate disturb the low frequencies and accurate modeling is impossible.

Furthermore we assume that in a small area around one operating condition, the test-rig behaves linearly. Then the following model can be used to describe the test-rig in one operating condition:

$$\hat{y}_t = G(q^{-1}, \theta)u_t + H(q^{-1}, \theta)e_t \tag{1}$$

with:

$G(q^{-1}, \theta)$	deterministic transfer function
$H(q^{-1}, \theta)$	stochastic transfer function
θ	vector containing the model coefficients
q^{-1}	backward shift operator ($q^{-1}u_t = u_{t-1}$)
$\hat{y}_t$	predicted output at time t
u_t	input generated at time t
e_t	white noise at time t

When y_t denotes the measured output, the prediction error generated by model (1) is

$$\epsilon_t = y_t - \hat{y}_t$$

A quadratic cost function can be defined as the weighted mean square of the prediction error:

$$V_N(\theta, Z^N) = \frac{1}{N}\sum_{t=1}^{N}\frac{1}{2}L(q^{-1})\epsilon_t^2(\theta) \tag{2}$$

with $Z = \left[\begin{array}{cc} y_i & u_i \end{array}\right]$ the measured data vector, N the number of data points available, $L(q^{-1})$ a residual filter. The residual filter is used to specify the frequency range where a more accurate model estimate is desired, at the expense of a less accurate estimate outside this frequency range.

The identification problem is to determine the coefficients in (1), $\hat{\theta}_N$, in such a way that the difference between the model prediction and the data is minimized in the weighted least squares sense of (2). In order to estimate the transfer functions, the inputs are excited by test signals. For the delay angle a pseudo random binary noise PRBS with a cycle frequency of 5Hz has been chosen. This signal takes care of excitation of the test-rig dynamics over a broad frequency range. Although in a real wind turbine the wind velocity cannot be excited, we started the system identification in the most simple way: by exciting and measuring the wind velocity. This excitation has a flat spectrum below 10Hz and no frequency content above 10Hz. It is noticed that reliable models of the wind turbine can be estimated, even without user excitation of the wind velocity [1]. For convenience two separate experiments where executed, one for each input signal. The identification method is applied for each channel individually, therefore the parameter identification scheme is processed six times.

As an example of the model estimation results, the transfer function of the delay angle to the direct current is shown in Fig. 4. In this figure the empirical transfer function estimate (ETFE) [2] [16] and the estimated parametric model according to (1) is given. It can be seen that there exists a closed resemblance between the experimental models and the ETFE. This holds for both the amplitude as well as the phase.

2.2 Controller synthesis

The control design objectives, derived from the control requirements, modeling properties, basic operational requirements and implementation constraints are:

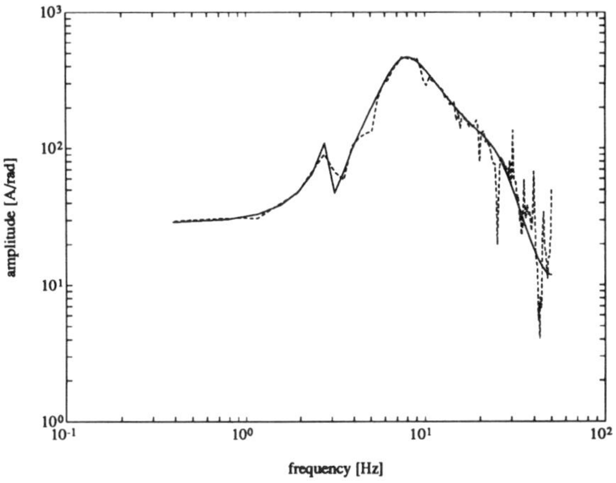

Fig. 4: transfer function `alf -> Ig`, (—) parametric model, (- -) ETFE

- peak reduction around 3 Hz in transfer function from wind speed to shaft torque to 1.25 times the value at lower frequencies;
- robustness against model uncertainties and measurement noise;
- low excursions of the delay angle;
- stable behavior of the controller itself;
- digital implementation with 100 Hz cycle frequency, 0.01 s time delay and a maximum controller order of 25.

The controller should meet these objectives for small variations around the considered operating point.

The main control objective is the reduction of rotor shaft torque variations. For the transmission, the service factor is mainly determined by the damping of the mechanical resonance frequencies. Therefore it is natural to apply an $\mathcal{H}_\infty$ control design method [3], by reasoning that reduction of the $\mathcal{H}_\infty$-norm [4] will increase the damping of the resonance frequencies. Thereby a smaller service factor is necessary and allow a reduction in the construction costs.

In order to design a controller for the test-rig, such that the control objectives are satisfied, the control objectives are translated to properties that can be used in the $\mathcal{H}_\infty$-controller design. Due to space limitations we skip the details of the design procedure and refer to [5, 7]. The result of the design is presented in Fig. 5. It can be seen in Fig. 5 that indeed the resonance peak is reduced by a factor of 10. Referring to the control objectives, the resonance peak is reduced to .67 times the value at 1 Hz.

This controller is said to satisfy the control objectives, thus we proceed by verifying its behavior at the test-rig.

2.3 Controller verification

In this section the predicted results of the previous section will be verified with measurements on the test-rig. For this purpose the controller has been implemented and controlling the test-rig. The same wind speed signal as in Section 2.1 has been applied. In Fig. 6 the experimental determined closed loop transfer function using an ETFE, and the predicted closed loop transfer function is given. Compared to Fig. 5 it can be seen that the maximum amplitude of rotor shaft torque variations are reduced by a factor of 7 in the frequency domain.

[2]An ETFE is a well conditioned spectral estimate of the transfer function and thereby based on the measured data only.

[3]Easy references on $\mathcal{H}_\infty$ control are for example [12, 18]

[4]For single input single output linear systems, the $\mathcal{H}_\infty$-norm is the maximum amplitude of the transfer function over all frequencies

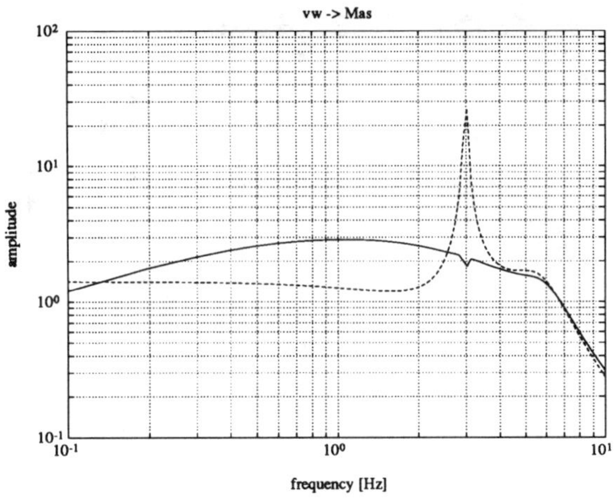

Fig. 5: design transfer functions `Vw -> Mas`, (—) no control, (- -) controlled

For the resonance frequency of the rotor shaft, interpreted as a second order system, the controller increases the damping from less than 0.05 to more than 0.5. The other signals: DC-current and generator speed are behaving well, the input used remains small.

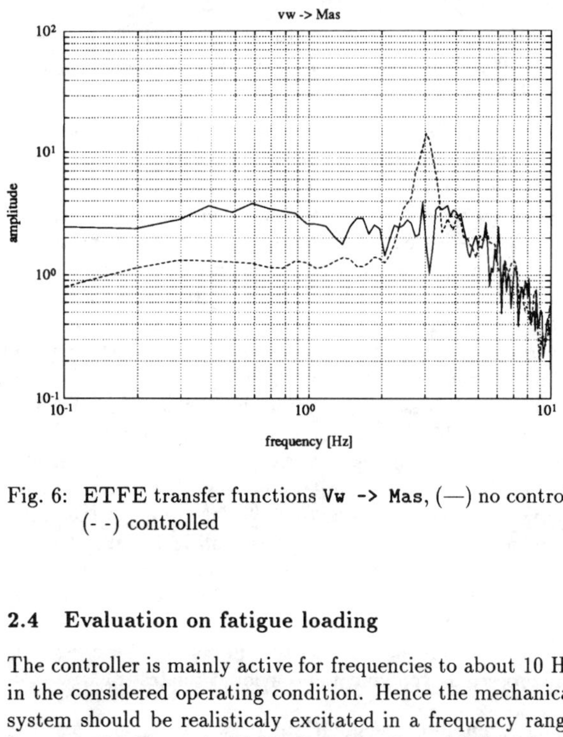

Fig. 6: ETFE transfer functions `Vw -> Mas`, (—) no control, (- -) controlled

2.4 Evaluation on fatigue loading

The controller is mainly active for frequencies to about 10 Hz in the considered operating condition. Hence the mechanical system should be realisticaly excitated in a frequency range from say 0.1 Hz up to 10 Hz for evaluation of fatigue loading. In general this would require real-time simulation of a turbulent wind field with turning rotor: the partial correlation of the turbulence in the rotor disc [9] and tower passing mostly enlarge the excitation near the eigenfrequency of the drive-train.

However the driving wind torque in the rotor simulator is just based on power coefficient data. Inputs are the measured rotor shaft speed and a user defined wind speed signal, while the pitch angle is calculated from the rotor speed [20]. Under certain assumptions a wind speed signal can be defined such for power coefficient data, that the excitation spectrum on the drive-train is equal to that obtained from real-time wind field simulation[10]. The main assumptions[5] are:

- linear transfer functions for all dynamic phenomena;
- homogeneous wind turbulence in rotor disc;
- rigid rotor blades in lead-lag direction;
- no impact of rotor speed fluctuations on windfield sampling.

Such an artificial wind signal has been generated by Fourier synthesis [19] of a power spectrum obtained with computer programme TURBU-I [11]. The rotor simulator does not include tower passing because the available power coefficient data are not azimuth dependent. Experiments with a duration of about three minutes have been performed successively with and without controller. Load spectra have been derived by sorting the load cycles in the shaft torque records according to amplitude ranges (Fig. 7). These load cycles were obtained by rainflow counting. To this computer programme LIFEHAT-I [8] was applied. It can roughly be stated that the fatigue loads on the drive-train are reduced by about 40%. This reduction will even be more when tower passing is taken into account.

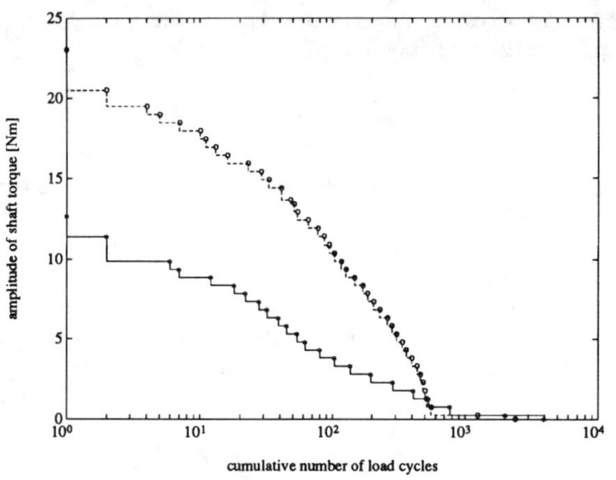

Fig. 7: Load spectra *with* (–,∗) and *without* (- -,o) control

3 Protection against grid faults and commutation failure

Figure 8 represents the DC-link consisting of a machine commutated controllable rectifier, a choke and a line commutated invertor. Through the short circuit of the line commutated convertor, for instance by a dip in grid voltage, the direct current will increase to 5 to 10 times the nominal value within 10 ms. Consequently the electromagnetic and shaft torque will also grow rapidly and the fuses may rupture. To prevent these undesired phenomena, a protection has been designed, which increases the firing angle of the rectifier when a potential short circuit is detected. This action decreases the direct voltage at the rectifier side. The detection procedure exist in measurement of the direct voltage at the invertor side and testing on lower bound excession. A value of 200 V or less activates the protection mechanism. A criterion using the direct current magnitude and rising rate did not guarantee fast enough action. The protection algorithm for 1000 Hz monitoring and control is implemented on the process computer.

[5] These are no essential drawbacks for comparative calculations in full load operation, when *not* governed by stall.

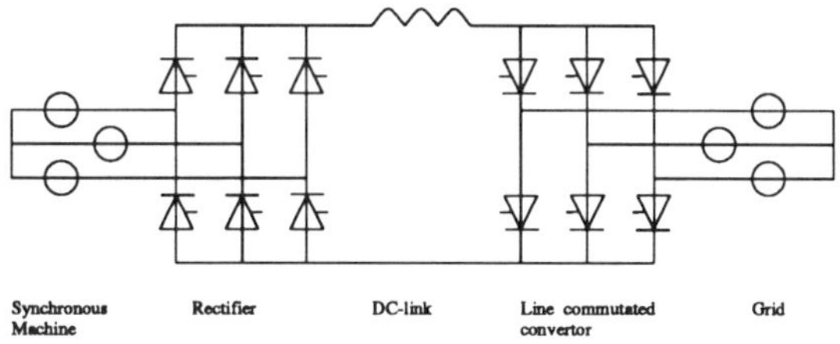

Fig. 8: Schematic representation of the DC-link.

Table 1: Faulty conditions without and with protective action; initial conditions: $I_{DC} = 35$ A, T = 81 Nm.

Case	Faulty condition	Protection	$I_{DC, max}$ [A]	$\overline{T}_{max}$ [Nm]
1 A	1 Phase voltage dip	no	800	210
		yes	170	85
1 B	1 Phase of the grid interrupted	no	800	200
		yes	170	85
1 C	Firing pulses of grid convertor interrupted	no	800	210
		yes	160	85
2 A	3 Phases of the grid shortcircuit	no	370	200
		yes	80	85
2 B	Firing angle of grid convertor 90 ^{0}	no	370	180
		yes	120	85
3 A	3 Phases of the grid interrupted	no	30	80
3 B	2 Phases of the grid shortcircuit	no	30	80

For the evaluation of the protection against grid faults and commutation failure a number of faulty conditions were identified and measurements were performed for these conditions with and without protective action. It was not the intention to examine all possible faults, faults of the most severe category (reversion of the DC-link voltage at the invertor) are included however. Table 1 lists the faulty conditions and gives the results in terms of maximum direct current and maximum shaft torque.

The cases 1A, 1B and 1C are the most severe. A short description of case 1A follows. For the cases 2A and 2B a short circuit over the two thyristors conducting at the moment of the fault occurs and the DC-link voltage at the invertor falls to zero. The maximum currents are less than half as high as for the cases 1A-C, the maximum torque is about the same. For the cases 3A and 3B no voltage reversal or short circuit occurs, the current is not commutated to the next set of thyristors, the ones conducting at the moment of the fault are quenched.

3.1 Case 1A: one phase voltage dip

A voltage dip in one grid phase was created using a variac, which causes a commutation failure. Measurements with and without protective action are shown in figure 9. The DC-link voltage at the invertor follows the line voltage of the grid, so the DC-link voltage changes polarity. After two periods of the grid voltage the current has produced enough heat to rupture the fuses in two phases of the grid, which are still connected to the DC-link. If no protective action is taken, the direct current increases from about 35 A to 800 A. The averaged maximum torque is about 210 Nm. A high-frequency resonance (not

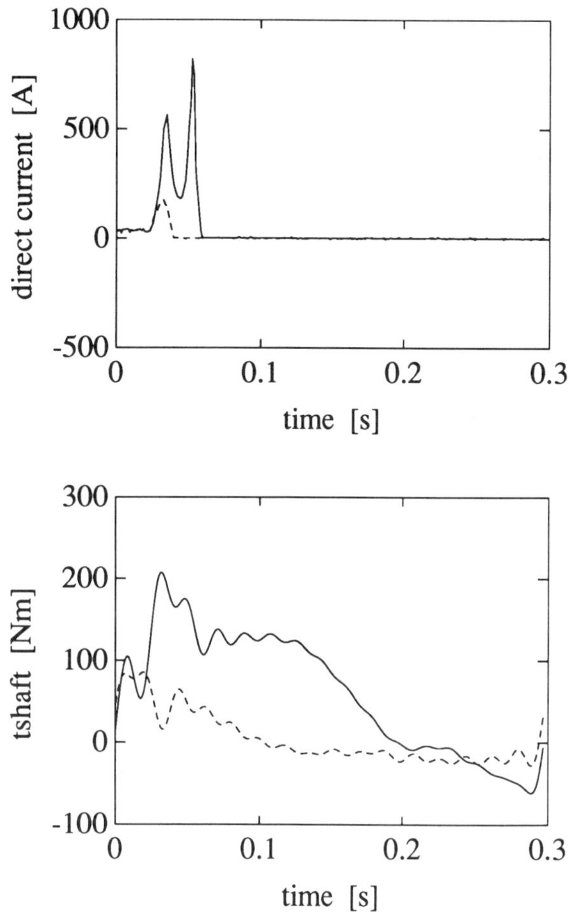

Fig. 9: One phase voltage dip *with* (- -) and *without* (—) protection.

shown in the figure), which is typical for the test-rig, but not for an actual turbine, was not taken into account.

In case the protection is active the direct current increases to 170 A in 10 ms followed by a decrease to zero. The averaged shaft torque is limited to about 80 Nm, which is the nominal torque. The polarity of the DC-link voltage changes, but the increase of the current is less than the increase in the unprotected case due to the reduced DC-link voltage at the rectifier. The fuses remain in tact.

4 Conclusions

Using experimental models, a controller has been designed through which the fatigue loads on the drive-train are reduced and the electric power is regulated in a laboratory environment.

The torque vibration at the mechanical eigenfrequency of 3 Hz has been reduced by the controller with a factor 7 (experimentally), while at frequencies below 2 Hz the impact of wind speed fluctuations on the torque in the flexible shaft is slightly larger on transfer function level. Notice that due to nonlinear behavior of the test-rig the measured fluctuations are lower when the controller is applied.

Although regulation of electric power has not been pursued in the control design, the electric power is regulated with a bandwidth of 0.5 Hz. At the mechanical eigenfrequency no severe power fluctuations occur.

The fatigue loading on the drive-train, determined from experiments with realistic excitation by the wind (except tower passing), has been reduced by the controller by ca. 40%.

A protection against grid faults and commutation failure of the invertor has been designed and evaluated for a number of faulty conditions. The maximum direct current after a fault has been reduced from about 800 A to about 170 A. The protection keeps the shaft torque at the rated torque.

Future parts of the IRFLET-project involve:

- verification of the physical knowledge based model of the electrical system;
- controller optimization in the considered full load condition by utilizing the exciter voltage for control;
- extension of control to all operating conditions;
- automatic transition from faulty to normal conditions.

Acknowledgements

The authors like to thank Mr. Baltus and Mr. de Haan of the Netherlands Energy Research Foundation as well as Mr. Hoeijmakers and Mr. Huisman of the Delft University of Technology for their assistance and stimulating discussions.

References

[1] Baars G.E. van, P.M.M. Bongers , "Model Description of the UNIWEX Wind Turbine: Experimental and Theoretical Results", Delft University of Technology, The Netherlands, In preparation, 1993.

[2] Baltus C.W.A., J.A.N. de Bonte, J.Th.G. Pierik, and G.P. Valter. , *Electrical conversion systems for flexhat-configurations* (in Dutch). ECN, Petten, Contract report ECN-C-90-040, 1990.

[3] Boer W.W. de, M.W. Mortier, P.A. Romar, G.J.F. Smithuis , "Measurement results of the wind turbines of the experimental wind farm (SEP) in Friesland", in *Wind Energy - Technology and Implementation, Proc. European Wind Energy Conf. Oct. 14-18, amsterdam, The Netherlands,* F.J.L. Van Hulle, P.T. Smulders, J. Dragt, pp.63-67, 1991.

[4] Bongers P.M.M. , "Control design and implementation for a wind turbine system", in *Proc. 2nd Europ. Contr. Conf. (to appear),* Groningen, The Netherlands, 1993.

[5] Bongers P.M.M., G.E. van Baars, Sj. Dijkstra , *Load reduction in a wind energy conversion system using an H_∞ controller.* Delft University of Technology, TUD-WBMR-N-422, Also submitted 2nd CCA, 1993.

[6] Bongers P.M.M., W.A.A.M. Bierbooms, Sj. Dijkstra, Th. van Holten , *An Integrated Dynamic Model of a Flexible Wind Turbine.* Delft University of Technology, The Netherlands, 1990.

[7] Bongers P.M.M., W. Sturm, I. Kraan, Sj. Dijkstra, O.H. Bosgra , "Control options for wind energy conversion systems", in *Proc. Europ. Community Wind Energy Conf.,* this issue, 1993.

[8] Bulder B.H. , *LIFEHAT-I, Theory and User's Manual.* ECN, Petten, in preparation, 1993.

[9] Dragt J.B. , *Load fluctuations and response of rotor systems in turbulent wind fields..* ECN, Petten, External report ECN-172, 1985.

[10] Engelen T.G. van , *Miscellaneous usage of computer program TURBU-I: Spatial turbulence in wind input for power coefficient data.* ECN, Petten, in preparation, 1993.

[11] Engelen T.G. van, T.D. Oei, H.C. Rieffe, J.B. Dragt. , *TURBU-I, computer program for turbulence load analysis of horizontal axis wind turbines: Theory and program description..* ECN, Petten, Internal report ECN-I-91-073, 1991.

[12] Francis B.A. , *A course in H∞ control theory.* Lecture notes in control an information sciences-88, Springer Verlag, Berlin, 1987.

[13] Haan S.W.H. de, J.T.G. Pierik, A.T. Veltman. , *Calculation method for torque transients in drive trains of variable speeds wind turbines after grid failure..* ECN, Petten, in preparation, 1993.

[14] Kraan I. , *Control design for a flexible wind turbine* (in Dutch). Delft University of Technology, TUD-WBMR-A-613, 1992.

[15] Kuik G.A.M. van, R.H.F. Tikkemeijer, J.W.M Dekker. , "The development of advanced rotor systems. a survey of the flexhat program, including full scale results.", in *Proceedings European Committee Wind Energy Conference,* Madrid, Spain., 1990.

[16] Ljung L. , *System Identification: theory for the user.* Prentice Hall Inc., Englewood Cliffs, New jersey, 1987.

[17] Ljung L., T. Söderström , *Theory and practice of recursive identification.* MIT Press Cambridge, MA, USA, 1983.

[18] Maciejowski J.M. , *Multivariable feedback design.* Addison-Wesley Publishing Company, 1989.

[19] Oscar D.S., T.L. Paez. , *Analysis of wind turbines on offshore support structures excited by random wind and random waves..* SANDIA, Technical Report SAND87-1689•UC-60, 1989.

[20] Pierik J.T.G., M.J. Hoeijmakers, J.M. Vleeshouwers, T.G. van Engelen, C.W.A. Baltus, A.T. Veltman, A. Warmenhoven , "A variable speed system with integral control for wind turbines (IRFLET): design of the testrig", in *Wind Energy - Technology and Implementation, Proc. European Wind Energy Conf. Oct. 14-18, Amsterdam, The Netherlands,* F.J.L. Van Hulle, P.T. Smulders, J. Dragt, pp.90-93, 1991.

[21] Pierik J.T.G., P.M.M. Bongers, T.G. van Engelen, G.E. van Baars. , *Damping of mechanical resonance and protection against grid failure for variable speed wind turbines (IRFLET).* ECN, Petten, Technical Report ECN-C-93-004, 1993.

[22] Steinbuch M. , *Dynamic modelling and robust control of a wind energy conversion system.* PhD Thesis, Delft University of Technology, The Netherlands, 1989.

Wind Tunnel Testing of a Blade Tip Actuator with Aerodynamic Moment Feedback

Jean-Baptiste Richon, Colin G. Anderson
Fluids Dynamics Unit
Physics Dept., King's Buildings
University of Edinburgh
Edinburgh, EH9 3JZ

ABSTRACT: On a large tip-controlled wind turbine, giving the tips a pitch compliant characteristic may bring significant reductions in magnitude of the cyclical loads seen by the blade roots and the drivetrain components. Pitch compliance at the tips can be achieved with aerodynamic moment feedback applied to a tip pitched behind its aerodynamic center. The Feasibility of the control technique is established in wind tunnel tests, and it is shown that at frequencies up to 3P, a pitch-compliant tip can provide a large amount of disturbance load attenuation, compared with a position-controlled tip.

1 Introduction

It is generally agreed that, to be commercially attractive, large wind turbines cannot afford to be over-engineered. This in turn means that control of the rotor input torque in high winds needs to be relatively accurate, to prevent excessive load transients acting on the structure. Whereas on smaller stall-regulated wind turbines high duty factors are applied to the drivetrain components, this becomes uneconomic at larger scale.

The commonest form of control on large HAWTs is pitch regulation. The electrical output power is measured and used as a feedback signal to drive the pitch actuators. To date, however, conventional pitch controlled machines have failed to show the level of output power accuracy required to allow drivetrain margins to be reduced.

A number of solutions to this problem are being sought, including improved closed-loop control algorithms [1] or the use of variable-speed drivetrains. In both cases, however, the pitch feedback loop is still used to convert an input demand to an output pitch angle.

An alternative approach is to control the aerodynamic load on the pitchable surface, rather than its position. This technique is known as *pitch compliance*. With a pitch compliant actuator the input signal is treated as a pitching moment demand, and the actuator output is aerodynamic moment, as opposed to position.

The perceived advantages of such a system are:

1. Individual response to local variations due to windshear, yaw error or localised gusting. This gives cyclic load alleviation on the pitch surface.
2. improved disturbance rejection properties for the power controller.
3. Reduced blade root fatigue loads.

The potential merits of this technique were highlighted in a number of earlier studies of pitch compliance applied to tip-controlled rotors [2][3]

The main objectives of the wind tunnel tests were:

1. To establish the feasibility of aerodynamic moment control.
2. To assess the load alleviation potential of compliance in comparison to a conventional position control system.

2 Description of the Experiments

2.1 Rig Design and Instrumentation

The function of the rig was to reproduce, at a scale of about one-third, the dynamic behaviour of a blade tip under compliant pitch control laws, with the possibility of a comparison with conventional angular position control. To simplify the interpretation of the experiments, the rig was designed for two-dimensional test aerofoils only.

A general view of the rig is given in Fig.1. The actuator was attached at one side of a rigid steel frame surrounding the wind tunnel control section. It was powered by a low inertia torque motor transmitting the torque to the test aerofoil through a belt-sector drive with adjustable gear ratio (Fig.2). A preload was applied to

Figure 1: General View of Test Rig

Figure 2: Detail of Belt-driven Servoactuator

each belt to remove any backlash. The test section, mounted on self-aligning bearings, was supported by a two-axis strain-gauge load cell at each side.

The test aerofoil sections, 1.1m chord by 0.3m span, were made of foam core sections cut using a computer-controlled hot-wire apparatus and assembled onto carbon fiber spars.

The rig instrumentation included transducers for the following measurements:

1. angular position,
2. angular acceleration,
3. applied torque,
4. lift and drag.

2.2 Pitch Control System

Different closed-loop control configurations had to be tested during the course of the investigation. For this reason, the control electronics had to provide a flexible environment, in which the transducers and the control circuitry could be easily changed or modified.

In order to demonstrate the advantages of compliant tip control, a comparison was necessary with conventional pitch angle control. Therefore the control system could be switched instantly from position to moment control mode.

Compliance was achieved by a combining applied torque and inverted acceleration into a composite feedback signal (Fig.3). This signal was calibrated so that it would be a good approximation to aerodynamic pitching moment over an operating frequency range of 10Hz. Again, additional pitch damping was needed to ensure adequate stability margins.

3 Experimental Results

All experiments were performed at windspeeds varying between 15 and 30m/s, i.e. Reynolds numbers in the range 300,000–600,000. The test aerofoils were pitched at 45% chord, a realistic pitch axis location for the full-scale device.

3.1 Unsteady Aerodynamics

This series of experiments was carried out in position control mode. Frequency sweep tests revealed that the aerodynamic pitching moment lagged behind pitch angle by a small but significant amount of phase at pitch

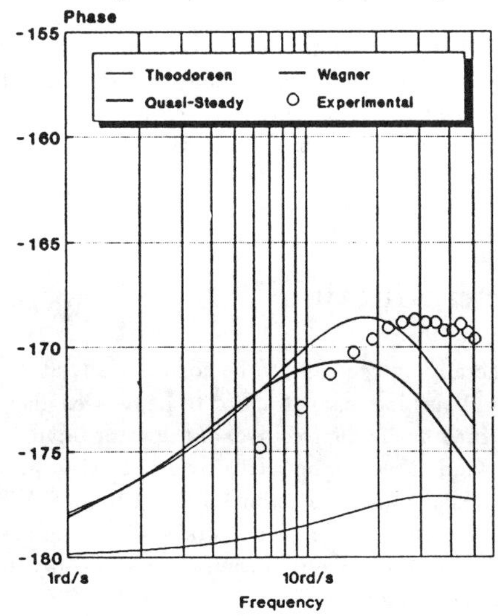

Figure 4: Phase Lag of Pitching Moment w.r.t. Tip Pitch Angle; Theory vs. Experiment

frequencies as low as 1 Hz (Fig.4). The magnitude and trend of the observed phase lag with frequency agreed well with results obtained from a 'Theodorsen' frequency-domain model, which uses an exact harmonic solution for potential, unsteady flows [5]. This phase lag corresponds to an aerodynamic damping force well in excess of that predicted by quasi-static aerodynamics, which actually assists control stability when the tip is flying.

The experiments did not reveal any occurence of negative aerodynamic damping forces over the range ow windspeeds and pitch frequencies investigated, confirming theoretical estimates that flutter-type instabilities were very unlikely in the tested configurations.

3.2 Feasibility of Compliant Control

The aerodynamic moment feedback loop was initially closed with low proportional gain. The controller compensation included integral action with a reset time of .1s, and a first-order low-pass filter rolling off at 10Hz to attenuate high frequency noise in the torque transducer signal. Damping was added to ensure adequate stability in all conditions of operation.

Frequency sweep tests were run in windspeeds up to 30m/s, and led to the following observations:

1. The aerofoil section showed a high degree of stability using an aerodynamic moment control loop based on measured acceleration and applied moment.

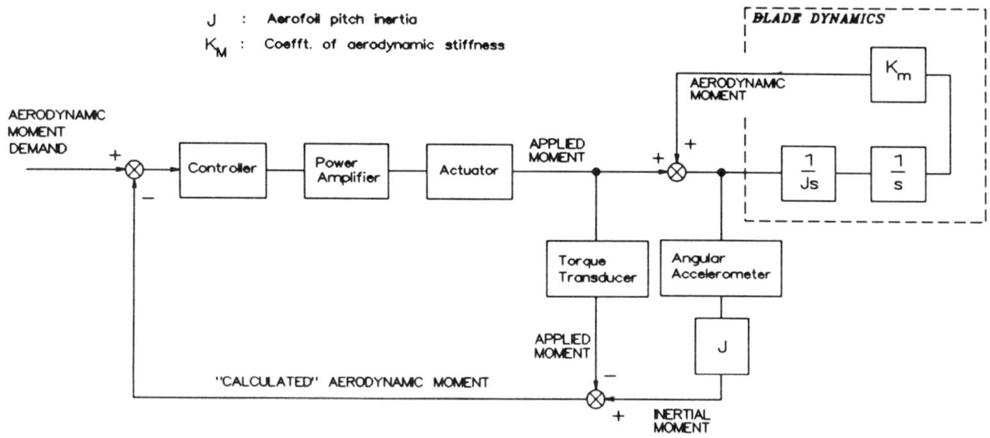

Figure 3: Block Diagram of a Compliant Tip Control Loop Based on Aerodynamic Moment Feedback

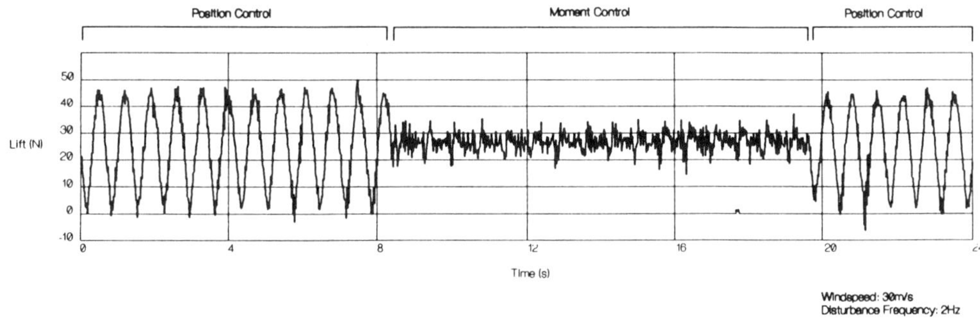

Figure 5: Response of a Tip Model to a Sinusoidal Airflow Disturbance in Position and Moment Control Modes

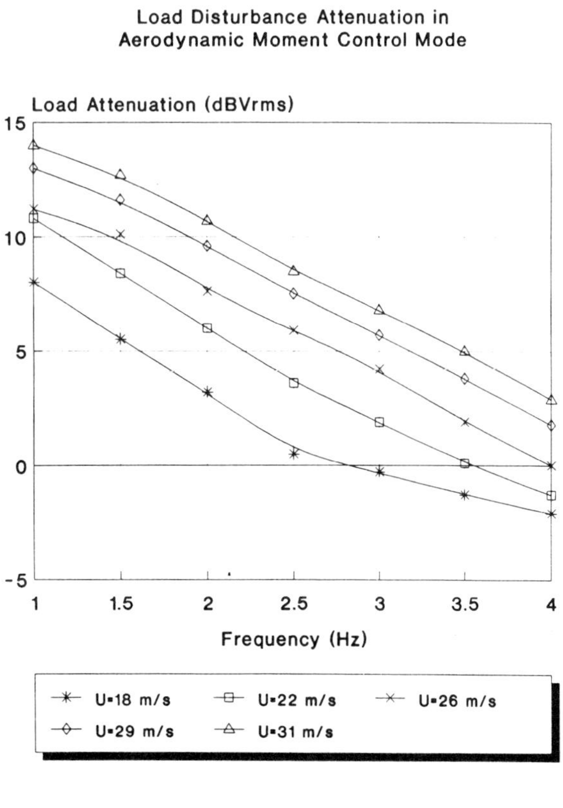

Figure 6

2. The system could only be operated at low gains and with marginal stability when the pitch acceleration signal was removed.

3. On removal of the actuating torque, the blade section deployed very rapidly into an air-braking position.

3.3 Flow Disturbance Rejection

Unlike a position-controlled tip, a moment-controlled tip will respond in pitch to a flow disturbance and attempt to keep the aerodynamic moment at the setpoint value. Lift variations induced by a flow disturbance on the test aerofoil should therefore be attenuated greatly (ideally: completely) in moment control mode, and very little in position control mode.

The lift response of the test section to a sinusoidal flow disturbance was recorded in aerodynamic moment and position control modes, for disturbance frequencies between 1 and 4Hz. The measured lift attenuation was initially a factor of 2 (3dB) only, but by optimisation of the high frequency noise filtering, it was possible to increase the moment feedback loop gain by an order of magnitude, and bring the lift attenuation up to a factor of 6 (15dB) at 1Hz. Figure 5 shows a comparison of the disturbance rejection ability for the two modes, and clearly demonstrates the load alleviation characteristic associated with compliance. At 31m/s, the measured load alleviation is above 10dB at frequencies up to 2Hz (Fig.6), which means that a compliant tip could filter out effectively load variations in the 1- to 3-P frequency range.

4 Conclusions

1. No major feasibility problem was encountered in applying aerodynamic moment feedback to a tip actuator rig in a wind tunnel environment.

2. Control stability was found to be good in all the conditions tested. Phase lags in lift due to unsteady aerodynamics have a positive effect on stability by increasing considerably the aerodynamic damping present in the system.

3. Significant cyclic blade load reductions can be expected when aerodynamic moment feedback is used at full scale. The wind tunnel measurements indicated that load attenuations of 10 to 15dB were achievable at frequencies up to 3P. However, the extent to which these measurements can be extrapolated is not known.

In its currently implemented version, compliant control requires the use of an angular accelerometer, a potentially fragile transducer for a system normally operating in harsh environments. However, the availability of reliable, cost-effective commercial designs for piezoresistive accelerometers makes this requirement tolerable in the full-scale control design.

5 Further Work

Single-degree-of-freedom aerodynamic instabilities are not an issue in normal operating conditions. However studies have shown [7] that pitch-flap coupling between the main blade and the tip may, in certain conditions, lead to a loss of aeroelastic stability in a pitch compliant tip actuator. This issue is now being addressed in a collaborative SERC project to investigate the aeroelastic stability and 3-D aerodynamics of the pitch compliant tip actuator.

Acknowledgements

This work was carried out under SERC grant No.26324.

References

[1] Leithead, W.E., de la Salle, S., and Reardon, D., "Wind Turbine Control Objectives and Design", *Procs. European Wind Energy Conference (ECWEC90),* Madrid, September 10-14th, 1990, pp. 510-515.

[2] Anderson, C.G., "The Use of Pitch Compliance for Improved Control Response on Tip-Controlled Horizontal Axis Wind Turbines", *2nd Progree Report to the Royal Society of Edinburgh,* University of Edinburgh, May 1989.

[3] Anderson, C.G., "Aerodynamic Sensitivity Analysis of the Howden HP330/33 Wind Turbine comparing Flying Tip and Full-Span Pitch Control Operation", *3rd Progress Report to the Royal Society of Edinburgh,* University of Edinburgh, January 1990.

[4] Anderson, C.G., Campbell , T.J., "'Front-End' Aerodynamic Control of Horizontal Axis Wind Turbines", Final Report under ETSU contract E/5A/CON/6037/2259, Sept.11, 1992, Ref.L300-186.

[5] Fung, Y.C.,*An Introduction to the Theory of Elasticity,* Galcit Aeronautical Series, John Wiley ans Sons, New York, 1955.

DESIGN RULES FOR MINIMIZATION OF ENERGY COST OF LARGE WECS

F.J. Föllings, INTRON, Institute for Material and Environmental Research B.V.
p.o. box 226, 3990 GA Houten, The Netherlands

H. Snel, ECN, Energy Research Foundation, p.o. box 1, 1755 ZG Petten, The Netherlands

G. Jenkins, University of Sunderland, Renewable Energy Centre, Langham Tower, Ryhope Road, Sunderland SR2, 7EE England

The energy output of a wind turbine is determined by the performance of the rotor, the design of the power control and the efficiency of the gear box and generator. Beside these machine features the distribution and annual mean wind speed of the wind determine the annual energy output. In general the shape of the wind speed distribution is given by a shape factor. The performance of the rotor is determined by the relation between power coefficient and tip speed ratio, the power control (depending on the type of control) by the "design" wind speed (rotor speed) and/or the "rated" wind speed (installed generator power). For the efficiency curve of the gear box and generator a shape factor is used.

The influence of these parameters on the annual energy output of the turbine has been investigated. It appears to be possible to give rather simple rules to maximize the energy output. These rules relate the design and/or rated wind speed to the annual mean wind speed. The control systems considered are:

- constant rotor speed and constant pitch angle (stall)
- constant rotor speed and variable pitch angle
- variable rotor speed and variable pitch angle.

Of the design parameters the design wind speed (rotor speed) and the rated wind speed (installed generator power) have a significant effect on the cost of the turbine. The necessary mass/size of nearly all the component of the turbine are to a certain degree affected by these parameters, especially the rotor, tower, low speed shaft, gear box, generator and transformer. With the cost model of the University of Sunderland, developed within the frame work of the JOULE Study on the Next Generation of Large Wind turbines, the influence of the mentioned parameters on the investment cost of the wind turbine have been investigated. The capital cost resulting from the investment cost form the major part of the energy cost. Based on the results of the analyses, design rules are given to minimize the energy cost of medium and large wind turbines.

1. GENERAL

A simple first order approximation of the relation between mass of wind turbine components and rotor diameter is a third power. If it is assumed that the cost of a turbine per unit of mass are more of less constant, the cost of a turbine will thus be proportional to the third power of the rotor diameter. The energy produced is proportional to the rotor surface, which is equivalent to the second power of the rotor diameter. This implies that the increase in the energy produced is less than the increase in turbine cost.

However, the turbine tower height in general also increases with rotor diameter. A higher tower results in an increase in average wind speed, so the energy produced per unit of rotor area will also increase. Because the wind speed is a logarithmic function of tower height and the cost increase a little less than with the third power (see [1]), it is likely that there will be an economic optimal range. This is illustrated in figure 1.

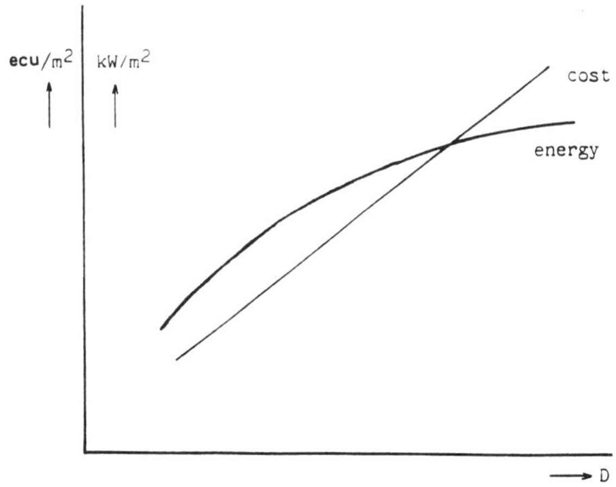

Figure 1. Wind turbine cost and energy yield per unit of rotor area.

The next two sections deals with the influence of relevant design parameters of wind turbines to the annual energy yield. With the presented approach dimensionless values can be determined for which the energy yield has a maximum.

2. ENERGY PRODUCTION OF A WIND TURBINE

The annual energy production of a wind turbine, that is connected to a grid, can be determined using both the power curve of a wind turbine and the wind speed probability distribution. The power characteristic of a wind turbine depends on various parameters. It can be calculated with the following equation:

$$P = \eta_t \, \eta_g \, 0{,}5\rho \, C_p \, V^3 \pi \, R^2 \qquad (1)$$

As is illustrated by equation (1) the generated power depends on three characteristics:

- the rotor power coefficient C_p (Ω R/V)
- the gearbox efficiency η_t (P/P_r)
- the generator efficiency η_g (P/P_r)

Figure 2 gives an example of each characteristic.

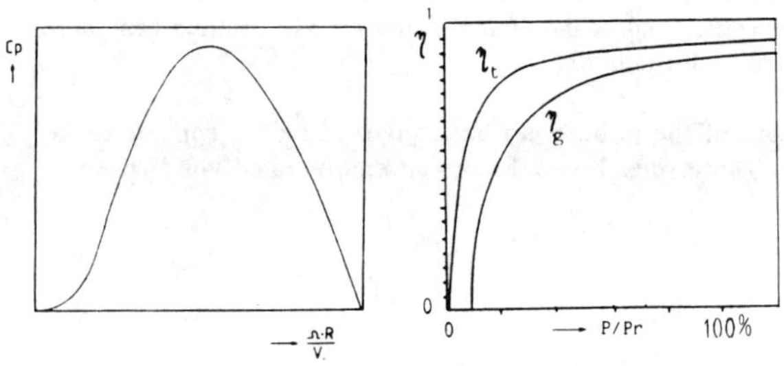

Figure 2. Examples of efficiency curves

The power control system of the turbine is also of importance. Three main control strategies can be distinguished:

1. variable rotor speed combined with a blade/tip control
2. constant rotor speed combined with a blade/tip control
3. constant rotor speed combined with a fixed pitch (stall control)

The last two control strategies have a (nearly) constant rotor speed over the full operational range of the rotor and above the rated wind speed the power output is limited by the mentioned systems/mechanisms.

The first control system permits rotor speed variations and above the rated wind speed the power output is limited by blade pitch/tip control. Figure 3 displays characteristic power curves as a function of wind speed for each control system.

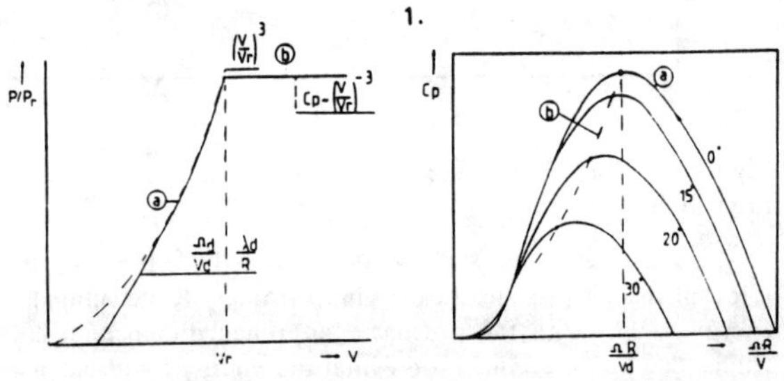

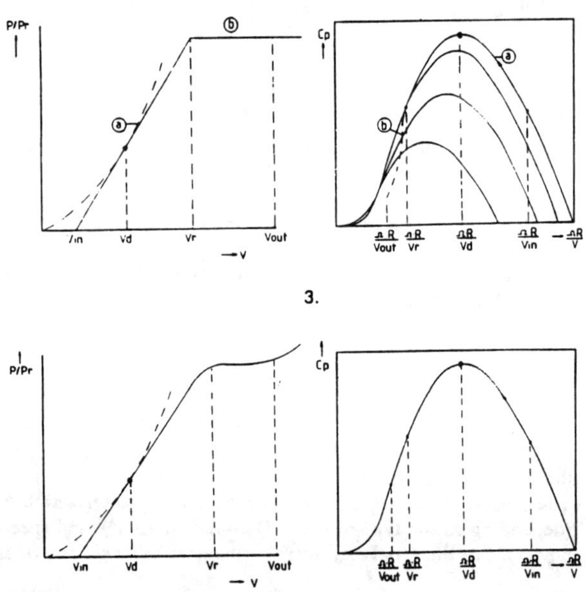

Figure 3. Characteristic power curves.

The probability function of wind, as characterized by the Weibull function, together with the power curve of the wind turbine makes it possible to calculate the energy output of a wind turbine.
With a computer code the dimensionless energy output (e_{syst}) can be calculated.

The code has to take into account all of the above mentioned parameters. It is made dimensionless with the following equation:

$$e_{syst} = \frac{\int \eta_t \, \eta_g \, 0{,}5 \, \rho \, C_p \, V^3 \pi \, R^2 \, dt}{\eta_{gr} \, \eta_{tr} \, 0{,}5 \, \rho \, C_{pd} \, V_h^3 \, \pi \, R^2 \, T} \qquad (2)$$

with: T = number of hours per year.

The actual energy output is compared to the energy output based on the mean annual wind speed at hub height V_h.

3. OPTIMIZATION OF ENERGY PRODUCTION

Firstly the influence of variations of the rated wind speed V_r and the design wind speed V_d (the latter wind speed is only applicable for turbines with a constant rotor speed) on the total energy output of a representative wind turbine is investigated. Figure 4 and 5 illustrate these variations for respectively a constant and a variable constant rotor speed turbine, both equipped with blade/tip regulation.

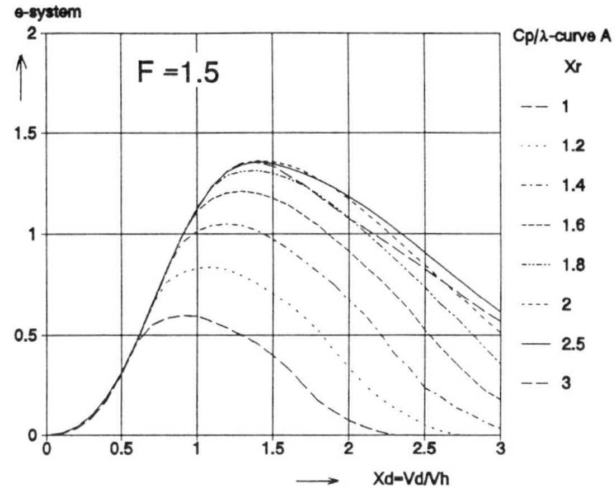

Figure 4. Relative energy output as a function of the dimensionless design wind speed and rated wind speed for a constant rotor speed with blade/tip control.

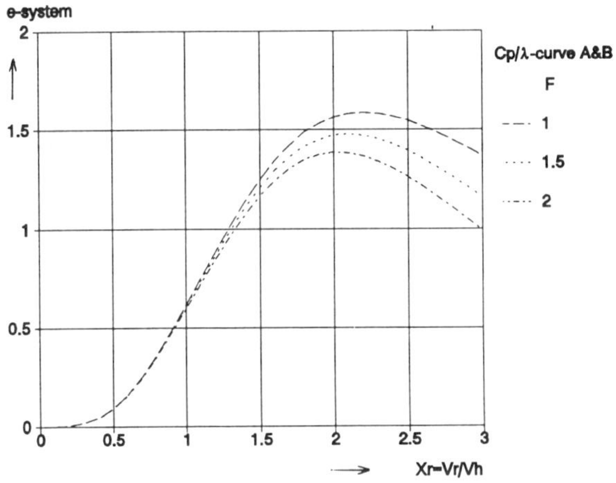

Figure 5. Relative energy output as a function of the dimensionless rated wind speed (with the service factor F of the gearbox) as parameter of a variable rotor speed turbine with blade / tip control.

From the latter figure it appears that the maximum energy output is obtained with a rated wind speed (V_r) of around 2 times the average annual wind speed at hub height. For a specific wind turbine of course the actual characteristics have to be used.

In case of a constant rotor speed turbine the optimum is also reached with a rated wind speed of around 2 times and a design wind speed of around 1.35 times the average annual wind speed at hub height. These values appear to be fairly sensitive for variations in turbine properties, such as the partial (fractional) load efficiency and the service factor F of the gearbox.

In case of stall controlled wind turbines the rated wind speed is not defined. Ideally the power curve of a stall controlled turbine is more or less equal to a constant rotor speed turbine with a blade/tip control, which implicates that also for a stall controlled turbine the power has to be limited above around $V_r = 2 \cdot V_h$.

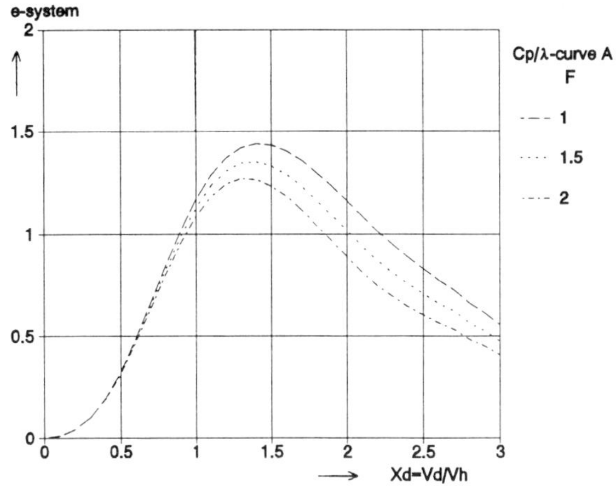

Figure 6. Relative energy output as a function of the dimension less design wind speed (with the service factor F of the gearbox as parameter) of a stall controlled rotor.

The relative energy output of a stall controlled rotor has also a maximum for a design wind speed of around $1.35 \cdot V_h$

4. OPTIMIZATION OF ENERGY PRODUCTION COST

The choices of dimensionless design wind speed and rated wind speed as discussed above are optimal with regard to energy yield. The chosen value for these wind speeds however affect the machine cost (in determining the nominal drive train torque, rotor and tower loads) and consequently energy production costs. The effects on machine cost are different for the three different control strategies. For this study, the Sunderland Wind Turbine Cost Model [2] was used to examine the effects in a coherent manner. It should be emphasized that the results, discussed in the present section, are only indicative; they should form a good starting point however for the optimisation of a specific design.

The following base values have been used for the calculations:

- three bladed 60 m diameter wind turbine
- annual mean wind speed at heub height: 7.7 m/s
- design tip speed ratio: 7.

For the constant rotor speed pitch control option X_d and X_r are independent design parameters. Their separate effects will be discussed qualitatively before interpreting the model results. The starting point is a fixed C_p-λ/λ_d curve and a fixed value of the design tip speed ratio λ_d.

* An increase in X_d implies an increase of the rotational speed, and at constant X_r, the nominal drive train torque is decreased. However, if X_d is not equal to its energy yield optimum, energy yield is decreased due to a sub optimal match of the rotor characteristics with the wind speed distribution.
* At constant $L = X_d/X_r$, rated power is proportional with X_r^3, because the tip speed ratio at V_r and hence the power coefficient remains unchanged. Both the energy yield and the drive train costs will increase with increasing X_r, but not at the same rate.
* For constant X_d and increasing X_r, λ_r will decrease and hence C_p at rated conditions will be lower. The installed power now increases less rapidly than X_r^3.

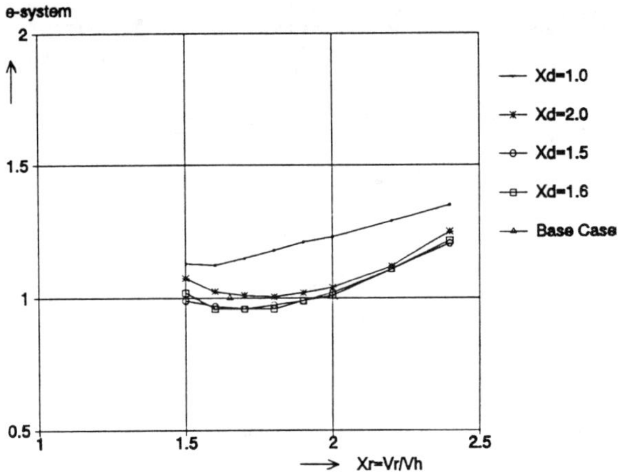

Figure 7. Relative energy cost of constant speed rotor as a function of the dimensionless rated wind speed and with the design wind speed as parameter.

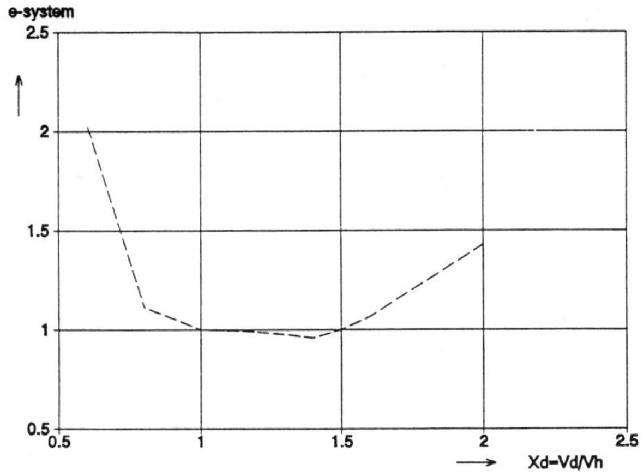

Figure 8. Relative energy cost of constant speed rotor and fixed pitch (stall control) as a function of dimensionless design wind speed.

Figure 7 shows one of the results of the model calculations. It can be seen that a minimum energy production cost is predicted for X_d approximately equal to 1.6 and X_r slightly higher than 1.6. Hence the optimal value X_d is suggested to be appreciably higher than its energy yield optimum of 1.35, and also than what is common practice (1.3 to 1.4). This is an effect of a higher rotational speed and hence a lighter drive train. The value of 1.6 to 1.7 for X_r is lower than its energy yield optimum (close to 2).

The high value of X_d suggested by the model calculation implies a high tip speed. For the example used this would be 86 m/s, possibly higher than advisable from the point of view of acoustic noise production. The calculations suggest a energy cost increase of approximately 3% when X_d is taken at 1.4, for a 75 m/s tip speed.

In case of a constant rotor speed and fixed pitch (stall control) the dimensionless design and rated wind speed are not independent quantities for a given C_p-λ curve. In fact, λ_d is the λ value where $dC_p/d\lambda$ equals zero, while λ_r is the λ value for which $d(C_pV^3)/d\lambda$ equals zero, or: $dC_p/d\lambda = 3\ C_p/\lambda$. Hence $X_r/X_d = \lambda_d/\lambda_r$ is fixed by the shape of the C_p-λ curve. The optimization is done for a specific shape of the C_p-λcurve. This implies that at increasing X_d the rotor speed increases proportionally and the installed (rated power) with the third power. Hence the drive train nominal torque will increase with X_d^2. In practice, the rotor designer may change the C_p-λ curve, by changing the blade planform or the pitch setting. This is not contemplated in here.

The model calculations again give a surprisingly high optimal value of X_d viz. 1.4, as shown in figure 8. This implies a tip speed, for the conditions considered here of 75 m/s, higher than what is commonly used for 3 bladed stall control turbines ($\approx$ 60 m/s). It would be interesting to study the influence of the specific C_p - λ curve on this result.

For the case of the variable speed turbine, X_d is not a design parameter, as the turbine operates at optimal λ for a range of wind speeds, usually up to and including rated wind speed. Hence $L = X_r/X_r$ is now equal to one. The rated (maximum) rotor speed varies linearly with X_r, rated power as X_r^3 and nominal torque as X_r^2 just as was the case with the stall control rotor, but now with $L = 1$.

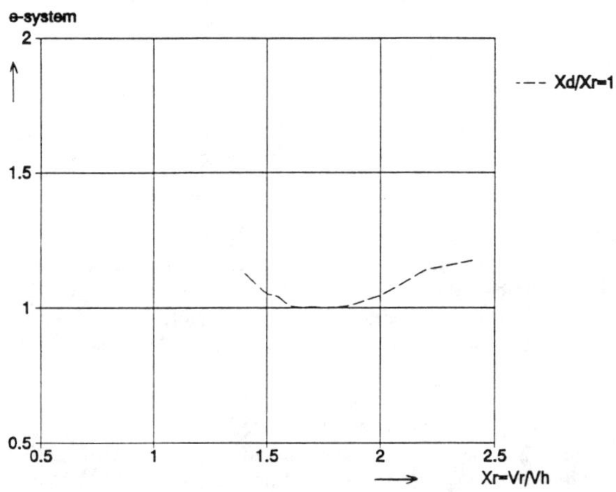

Figure 9. Relative energy cost of variable speed rotor as a function of dimensionless rated wind speed.

The model calculations now show a rather flat minimum for the energy production cost for $1.65 < x_r < 1.8$.

5. REFERENCES

[1] Economics of Large Stall Controlled Wind Turbines, F.J. Föllings, Intron report no. 90117, 1990 (in Dutch).

[2] Study on the Next Generation of Large Wind Turbines, ETAPLAN, ECN (SPE, Intron), University of Sunderland, DEFU, Commission of the European Communities DG XII (to be published).

A SOFTWARE FOR DESIGN AND CALCULATION OF WIND TURBINE COMPOSITE ROTOR BLADES (*)

B. Paluch
O.N.E.R.A. / I.M.F.L. - 5, Boulevard Paul Painlevé - 59000 Lille - France

ABSTRACT : Blades are one of the major componants of wind turbines and as such the blade manufacturer needs - using available technological solutions - to meet the specifications set by the wind turbine manufacturer. Although composite materials are quite flexible to use and lend themselves to the making of such structures, one needs an easy-to-use dimensioning tool to gradually, quickly and reliably refine the material arrangement of blades before mass production. Since the dimensioning procedure is iterative, the software relies on a database which includes the geometrical characteristics of the blades and the materials used to make them, and on a set of modular programs linked together through the files that make up the database. These modules mainly calculate the equivalent stiffnesses of the blade sections, the first natural frequencies at rest and the associated modeshapes, the forces acting on the blades, and the fatigue limit curves. This code was validated using an actual carbon-fiber blade. The comparison between the calculation and the tests relating to equivalent stiffnesses and natural frequencies gave good results.

1. INTRODUCTION

Beside the power train and the generator, the efficiency of the wind turbine is conditionned by the blades that make up the rotor. The blade manufacturer must therefore be able - on the basis of the given specifications - to quickly meet the demands of the wind turbine manufacturer, taking into account the technological solutions and materials available to make these blades. Although composite materials are quite flexible to use and lend themselves to making this type of structure, one needs an easy-to-use design and dimensioning tool that makes it possible to gradually, quickly and reliably refine the constructive dispositions of the blades before their mass production. It is with this purpose in mind that we conceived the software whose assumptions and calculation methods will be presented below, and which is mainly centered on the determination of the intrinsic blades properties, *i.e.* :

- their equivalent stiffnesses and mass properties,
- their vibratory behavior,
- and their fatigue behavior.

The dimensioning procedure being iterative, the sofware relies on :

- a database comprising the geometric and aerodynamic data of the airfoils, the shape of the blades, and the mechanical properties of the materials used to make them,
- and a set of modular programs linked together through the files that make up the database.

2. SOFTWARE ORGANIZATION

The wind turbine manufacturer's specifications generally include two facets: the first relates to the geometrical characteristics of the blades, whilst the second relates to the operating conditions of the wind turbine and the service strength of the blades. The first and part of the second are the starting point of the dimensioning. The other part of the secont facet assesses or not the constructive dispositions set by the blade manufacturer which defines - based on his experience and on the manufacturing technology used - the internal arrangement of the materials inside the blades. The software is structured around the database (Fig. 1) and includes the following modular programs :

- **Airfoil** : conditions the aerodynamic coordinates defining the airfoils of the blades, these points are used by the next program.
- **Section** : calculates the equivalent stiffnesses and mass properties of the blade sections.
- **Modal** : determines the first natural frequencies of the blades at rest and the associated modeshapes.
- **Loads** : calculates the distribution of forces under various blade operating conditions.
- **Fatigue** : calculates the fatigue limit curves of the blades.

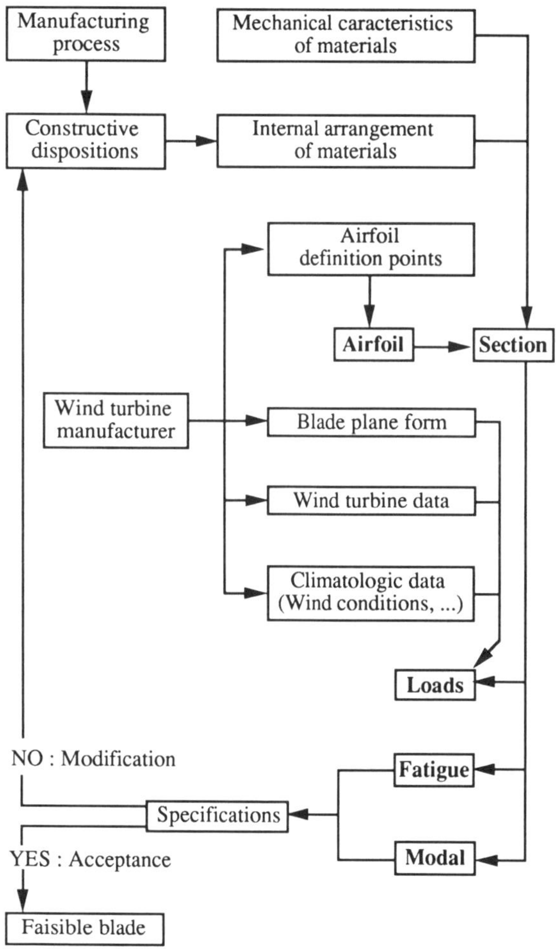

Fig. 1 : Software flow chart

Other less important programs help condition the data and handle their graphic presentation for quick analysis and interpretation of the results.

3 CALCULATION METHODS

3.1 Equivalent mechanical properties of the sections

The aerodynamic definition points of the profiles cannot generally be used as such for calculating the equivalent properties of the sections, they are stored in a profiles library in the form of non dimensional coordinates (normalized with respect to the chord). Each airfoil (Fig. 2a) is smoothed using cubic parametric splines [1] and then discretized (using the **Airfoil** module) by 100 equidistant points along the

(*) This work was supported by the french agency ADEME

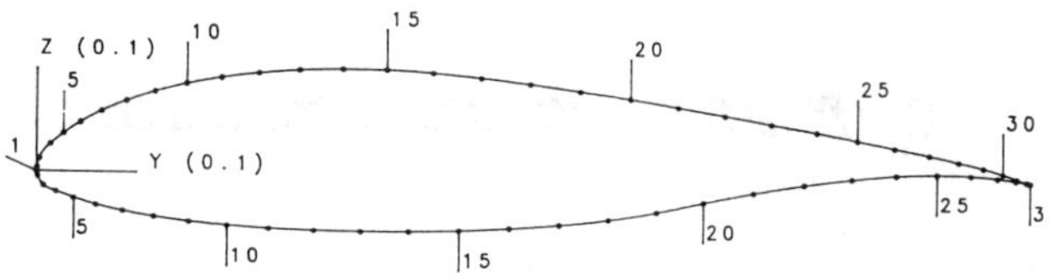

Fig. 2a : Airfoil aerodynamic definition points

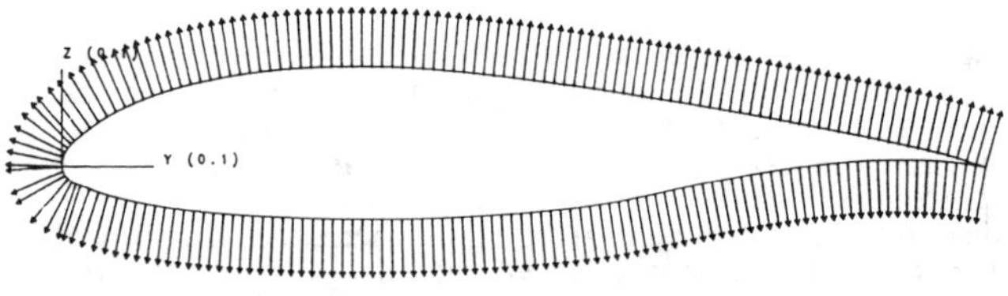

Fig. 2b : Airfoil structural definition points

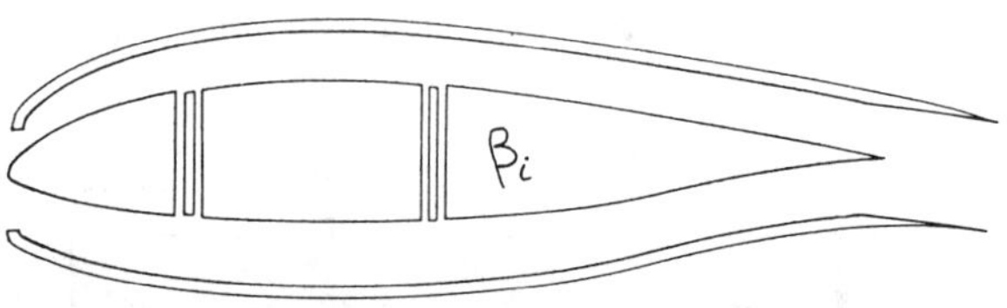

Fig. 2c: Sub-domains breaking down of a section

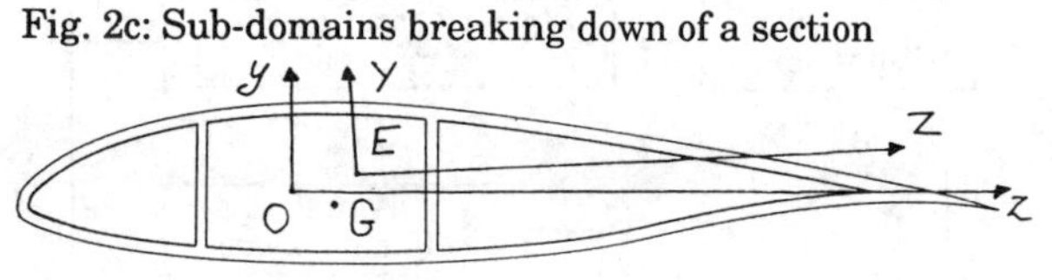

Fig. 2d : Typical blade section

curvilinear abscissa on the extrados and intrados faces. For each value along the abscissa the program calculates the coordinates of these new definition points - called structural points - as well as the coefficients of the normals oriented toward the inside of the airfoil (Fig. 2b). These normals are used to calculate the coordinates of the inside airfoil, *i.e.* the one derived from the outside airfoil based on the thickness of the layer of which it is made.

As blade results from the assembly of composite materials - and in some cases metals - of different types, let us call phase β_i a given assembly of materials. The coordinates (y_e,z_e) of the elastic center E of section are easily derived from the following formula :

$$(y_e,z_e) = \frac{1}{\langle ES \rangle} \sum_{i=1}^{N} E_i \int_{S_i} (y,z)\, dS \ , \ \langle ES \rangle = \sum_{i=1}^{N} E_i S_i$$

N, <ES> and E_i being respectively the number of phases, the equivalent stiffness in traction, and Young's modulus for phase i along the direction x normal to the plane of the section. In the reference system (E,y,z) associated with the elastic center, the values of the equivalent flexural stiffnesses are then :

$$(\langle EI_{yy} \rangle, \langle EI_{zz} \rangle, \langle EI_{yz} \rangle) = \sum_{i=1}^{N} E_i \int_{S_i} (z^2, y^2, yz)\, dS$$

The position of the main inertial reference system (E,Y,Z) is sought by making the term $\langle EI_{yz} \rangle$ equal to zero. The equivalent mass properties are calculated in the same way as above, replacing the equivalent modulus E_i by the equivalent density ρ_i . This yields the mass per unit length $\langle m_l \rangle$, the coordinates (y_g,z_g) of the center of gravity G, and the equivalent mass polar inertia, calculated with reference to G, and such that :

$$\langle \rho I_p \rangle = \sum_{i=1}^{N} \rho_i \int_{S_i} (y^2 + z^2)\, dS$$

As all the equivalent stiffnesses and mass properties - except the torsional stiffness - are summable quantities, it is possible to calculate them using a simple integration method. Each section is broken down into a series of adjacent sub-domains (Fig. 2c) constituted of a phase β_i, whose equivalent properties are precisely calculated by the **Section** program using the Gauss integration method [2]. The profile's skin and the section walls are made up of layers of composite materials, while the cores are made of low-density polystyrene used as filling. The thickness of the laminate material together with their equivalent modulus E_i and density ρ_i are automatically calculated from the thicknesses and the elastic and massic characteristics of each composite material layer making up the blade. A data file describes the arrangement of the various layers in each section, which help us to determine both the number of sub-domains, their shapes and materials distributions. The **Section** program is the core of the sofware since the equivalent properties it yields are used by practically all the other calculation modules.

The wind turbine blades made of carbon fibers - which provides for a non-neglible gain in weight by reason of the strong mechanical properties of the fibers - have a partionned structure, that is to say that the sections - which have thin walls separated by foam cores - are multiple-connected structures (Fig. 2d). As these foam cores have weak shearing strength properties, their contribution to the torsional stiffness <GJ> of the sections can be considered as negligible. It is then possible to calculate <GJ> using an analytical formulation based on the membrane analogy method derived from the classical elasticity theory assumptions [3]. When the blades have a much more complex structure, *i.e.* when they no longer verify the assumptions relating to thin walls, one needs to use a specific finite element program (not included in this software) which calculates the <GJ> values by determining the Saint-Venant warping function over the entire section [4].

3.2 Natural frequencies

It is sometimes necessary to know the first natural frequencies in order to ascertain whether they are located in critical frequency ranges with respect to the instability phenomena liable to occur under certain operating conditions of the wind turbine. As wind turbine blades are highly elongated structures - with an aspect ratio greater than 10 - they can be likened to beam structures. The **Modal** program therefore relies on the use of classical two nodes finite beam elements with six degrees of freedom per node, and calculates the first natural frequencies of the blades at rest (Fig. 3). Bernouilli's assumptions are used to model the mechanical behavior of the sections (the warping effects induced by transverse shearing are not taken into account). The first natural modes [Λ] and the associated modeshapes [Φ] are extracted using the subspace

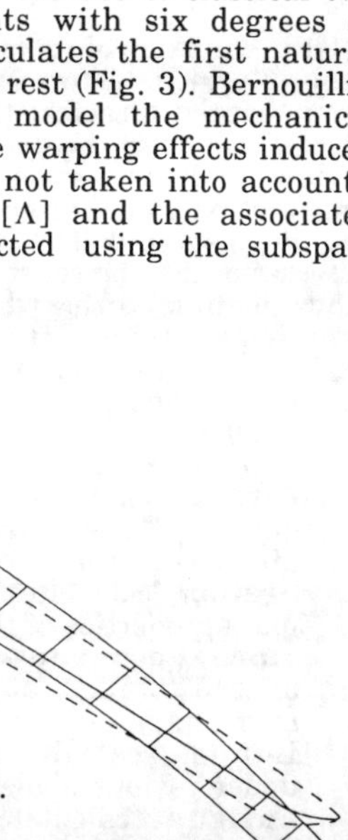

Fig.3 : 3 m carbon fiber blade.
Third modeshape (f=35 Hz)

algoritm [5] which consists in solving the system :

$$[K]\,[\Phi] = [M]\,[\Phi]\,[\Lambda]$$

[K] and [M] respectively represent the stiffness and mass matrices obtained from the equivalent stiffnesses and mass properties yielded by the **Section** program. The modeshapes are then interpolated by polynomial regression.

3.3 Calculation of the fatigue limit of the blades

Faced with the complex problem of estimating realistically the fatigue spectrum of the blades in service over very long periods of time - on the order of several decades - a statistic treatment of the data relating to the wind spectrum and to the frequency with which various operating modes occurs is needed. Although this type of approach tends to provide quite a faithful picture of reality, one can nevertheless, in a first step, handle the blade fatigue problem in a simpler way. Indeed, the method we present here consists rather in calculating the limit envelopes in dynamic moments of the blades, for static loads induced by a maximum wind regime. Among all the wind turbine operating cases, it is indeed possible to select the one that generates the highest static loads on the blades, based on which one determines the values of the maximum permissible dynamic flapping and lead-lag moments, over a given service period, that are representative of the blades' intrinsic fatigue behavior. These maximum dynamic moments can be compared to the dynamic moments calculated for different operating configurations, which makes it possible to calculate the safety margins and to locate the critical sections.

The loads being applied on the elastic center E of a section (Fig. 4) and in the main inertial frame include 6 components :

- N_x is the force perpendicular to the plane of the section principally generated by the centrifugal and gravitational forces.
- T_y and T_z, are the shearing forces.
- M_x is the torsional moment.
- M_y and M_z are the flapping and lead-lag moments.

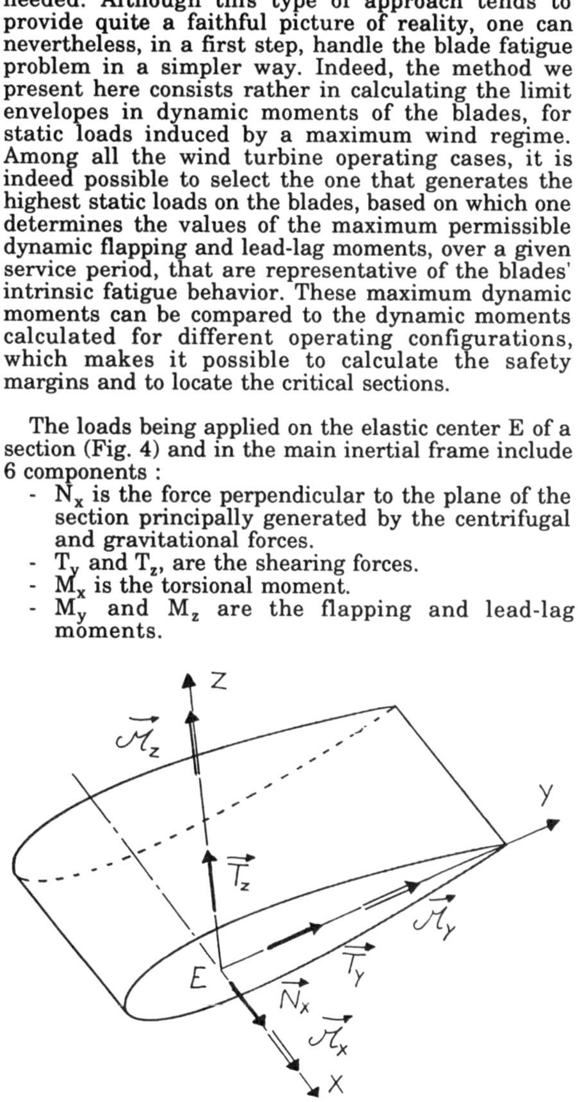

Fig. 4 : The 6 loads components acting on a blade section

When blades are partitionned off and made from carbon fiber-based composite materials, one can neglect the warping of section induced by the shearing forces T_y and T_z and by the torsional moment M_x because the blades are stiff in torsion and the phenomena caused by transversal shearing are negligible. As a result, the program does not take into account the shearing stresses when calculating fatigue strength. For each phase β_i the normal force N_x and the moments M_y and M_z generate a normal stress σ_x whose value is :

$$\sigma_{x_i} = \sigma^s_{x_i} + \sigma^d_{x_i} = E_i \left(\frac{N_x}{\langle ES\rangle} + z\frac{(M^s_y+M^d_y)}{\langle EI_{YY}\rangle} - y\frac{(M^s_z+M^d_z)}{\langle EI_{ZZ}\rangle} \right)$$

and broken down into a static part (s) and a dynamic part (d). As N_x varies very little, its dynamic component can be neglected compared to its static component. Fatigue is then generated by the dynamic components of the flapping and lead-lag moments.

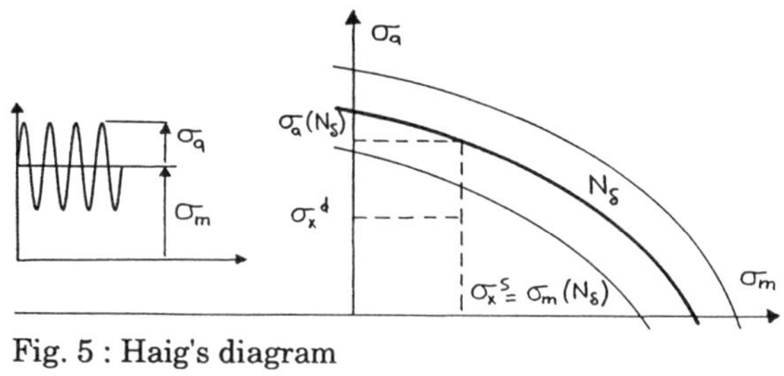

Fig. 5 : Haig's diagram

For a material or an assembly of materials with a given phase β_i, the Haig diagram (Fig. 5) can be used to plot - for different numbers of fatigue cycles N_c, for example 10^6 cycles - the evolution of the static stress $\sigma_m(N_c)$ as a function of the amplitude of the fatigue stress $\sigma_a(N_c)$ also called dynamic stress. Let there be a point P belonging to a phase β_i, for a number H_δ of hours corresponding to the operating case δ we will have for the number of cycles :

$$N_\delta = H_\delta\, \Omega\, 3600 / 2\pi$$

Ω being the rotational spedd of the rotor. Knowing $\sigma_m(N_\delta)=\sigma_x^s$, we can derive the value of the dynamic stress $\sigma_a(N_\delta)$ which must not be exceeded. In order that the material located at point P withstand the fatigue, it is necessary that :

$$\sqrt{(M_y^d)^2 + (M_z^d)^2} < \sqrt{(M_y^d)_\delta^2 + (M_z^d)_\delta^2} = (M_R^d)_\delta$$

$(M_y^d)_\delta$ and $(M_z^d)_\delta$ being the dynamic moments of flapping and lead-lag determined from $\sigma_a(N_\delta)$ and the Haig diagram knowing $\sigma_m(N_\delta)$, $(M_y^d)_\delta$ being the resulting moment. If we now express $(M_y^d)_\delta$ as a function of $(M_z^d)_\delta$, then :

$$\mathrm{tg}\,\phi = (M_y^d)_\delta \,/\, (M_z^d)_\delta$$

which yields for $\sigma_a(N_\delta)$:

$$\sigma_a(N_\delta) = (M_y^d)_\delta \left(\frac{z}{\langle EI_{YY}\rangle} - \frac{1}{\mathrm{tg}\,\phi}\frac{y}{\langle EI_{ZZ}\rangle} \right)$$

From the set of points P_k belonging to the phase β_i, we keep the point for which the resulting moment $(M_R^d)_\delta$ is minimum (Fig. 6). The resulting moments of each phase are also analysed in the same way for a number of cycles N_δ. The fatigue limit curve of a section is therefore the geometrical locus of the minima of the resulting moments which the section can withstand. The fatigue limit envelopes of the sections arc calculated by the **Fatigue** program.

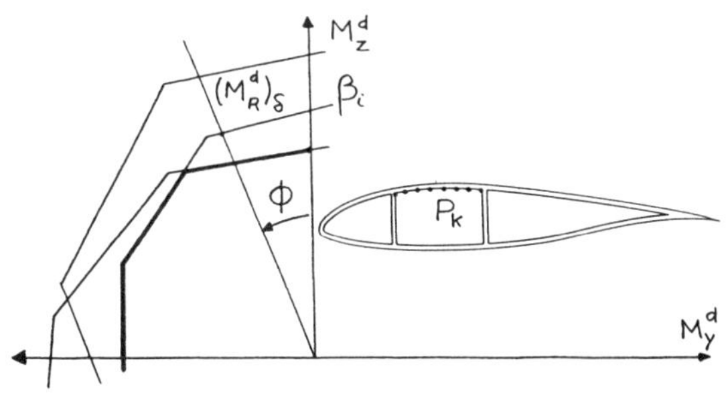

Fig. 6 : Determination of the section fatigue limit curve

3.4 Simplified calculation of the loads affecting the blades

The **Loads** program is first and foremost designed to yield a preliminary estimate of the forces affecting the blades assumed to be infinitely rigid (rigid rotor), the wind being stationary and the wind gradient being taken into account. For one blade the program calculates the gravity, centrifugal and aerodynamic forces and moments per unit length over a certain number of equidistant sections, graduated from the bladeroot to the tip end. The aerodynamic forces values are detrmined from the Glauert's theory [6], and the tower shadow is taken into account by considering that the wind flow around the tower arises from a potential. The evolutions of forces and moments is then obtained by integrating lineic forces an moments along the blade span, this method being reiterated for a certain number of azimuthal positions. Lastly, for each section where one whishes to calculate the static and dynamic components of the flapping and lead-lag moments, the program operates a break-up into Fourier series. The static and dynamic components of the moments are then used to determine the fatigue limit envelopes of the sections deemed most critical.

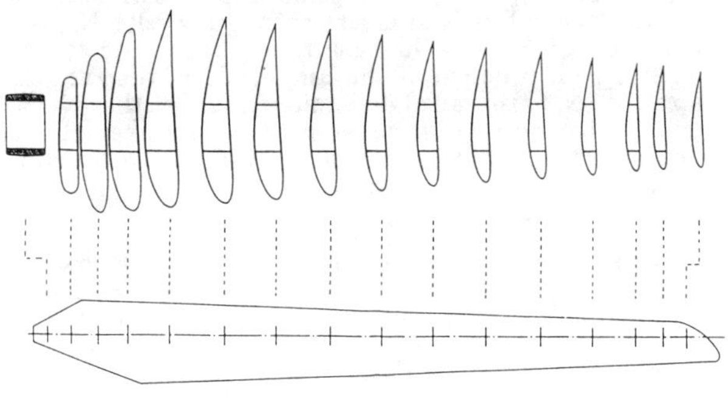

Fig. 7 : Sections of the 3 m carbon fiber blade

4. CALCULATION-TESTS COMPARISON

The comparison was made using the results yielded by the **Section** and **Modal** programs tested on a carbon fiber blade with a span of 3 meters and a relative thickness of the airfoil - belonging to the NACA 44xx series - which decreases linearly from the blade root(18%) to the tip end (12%) (Fig. 7). In order to asses the degree of accuracy of the method calculating the equivalent stiffnesses of the sections, we compared the flapping $<EI_{yy}>$ and torsional $<GJ>$ stiffnesses to values coming from the tests. These two quantities are indeed more easy to measure. After applying bending and torsional moments, the deflections were measured using a contact-free diode laser displacement sensor on 10 equidistant points 250 mm in span and located on the 25% axis (from R=480 mm to R=2730 mm) for $<EI_{yy}>$, and on 10 pairs of points for the $<GJ>$ (each pair corresponding to one point located on the leading edge and the another point on the trailing edge). In the first case the deflections were smoothed by polynomial regression, and the values of the deflection derivatives were thus calculated at each one of the 10 points in order to trace back the $<EI_{yy}>$. In the second case the differences in the displacements at each pairs of points yielded the values of the rotational angles which, after smoothing by polynomial regression, were used to calculate the unit torsion angles and then the values of the torsional stiffness. The difference between the calculated and measured $<EI_{yy}>$ and $<GJ>$ never exceed 5%. Another way to check the coherence of the calculated stiffnesses is to compare - at the same points as above - the measured deflections and the deflections calculated using a finite element method by taking as starting point the distributions of the $<EI_{yy}>$ and $<GJ>$ determined by the **Section** program. As in the case of the stiffnesses, the differences between deflections never exceed a few percent. A similar calculation was carried out on a 8 meters span carbon fiber blade, with a difference of less than 1% between the calculated and measured deflection at the tip end. The modal analysis of blade was carried out up to the sixth mode.

N°	Mode	Meas. f. Hz	Calc. f. Hz	Diff. %
1	Flapping	9	9.1	1
2	Lead-lag	18.7	22.3	19
3	Flapping	38.4	35	9
4	Flapping	82	83	1
5	Lead-lag	106.7	118.8	11
6	Torsion	110	109	1

Tab. 1 : Calculated and measured natural frequencies

Table 1 shows that the differences between measured and calculated natural frequencies relative to the flapping and torsion modes are low. The largest differences - occuring for the lead-lag mode - can be explained by the fact that the lead-lag clamping stiffness taken into account in the calculation is probably overestimated with respect to the actual stiffness. Except for the problem linked to the correct determination of the lead-lag boundary conditions, the software also yields good results in terms of predicting the first modes of the blade at rest.

5. CONCLUSION

The calculation modules we developped give access to the intrinsic blade properties and can be used to meet quickly a given set of specifications. The comparison of the calculated stiffnesses and natural frequencies and those measured on an average-size blade gave excellent results. However the correct modeling of the clamping conditions at the blade root will be required for the program to yield natural frequency values - particularly in lead-lag - that are representative of reality. The determination of the fatigue limit envelopes of the sections rests on a specific characterization of the composite materials fatigue behavior, this still needs to be achieved at this stage. At the same time the loads calculation - which will need to include information about rotor aeroelasticity, as for example in [7] - over periods of several decades can only be achieved through a statistic treatment in order to trace back - from the wind spectrum - the load spectrum. Finally, it should be noted that the use of this software on blades of different shapes will provide a valuable data base which will still need to be compared with experimental results. This continuing examination and comparison should help designers to better understand the mechanical behavior of the blades depending on the layer arrangement they use to make them. This will undoubtedly increase their experience and help them to identify and solve the main problems that arise at the design stage.

REFERENCES

[1] R.H. Bartels et al., B-splines, Hermès, 1987

[2] W.H. Press et al., Numerical recipes, Cambridge University Press, 1988

[3] L. Solomon, Linear elasticity, Masson, 1968

[4] B. Paluch, A finite element method for shear stresses calculation in composite blade models, 17th European Rotorcraft Forum, Berlin, Sept 24-27 1991, Paper N° 91-83

[5] K.J. Bathe, E.L. Wilson, Numerical methods in finite element analysis, Prentice Hall, 1976

[6] D.M. Eggleston, F.S. Stoddard, Wind turbine engineering design, Van Nostrand, 1990

[7] H. Snell, C. Lindenburg, Aeroelastic rotor system code for horizontal axis wind turbines : Phatas II, European Community Wind Energy Conference, 10-14 Sept. 1990, Madrid, pp. 284-290

DYNAMIC MODELS OF WIND TURBINE SYSTEMS FOR DESIGN PURPOSES ‡

Peter Bongers §, Klaus Wefelmeier ♯, Malcolm Barnsley ♭

Abstract

In this paper a dynamic model of the complete wind turbine, including control design, such that the behavior of various wind turbine configurations can be analyzed will be propagated. For this purpose the DUWECS code will be extended, both in the number of available models of wind turbine parts, as well as the accuracy of the models. This DUWECS code, which will become available at the end of this project, runs on a wide range of computers varying from PC-XT to DEC-workstations. Furthermore the dynamics of the submodels need to be verified by measurements. This paper will show results about: Dynamic building blocks of the wind turbine system, the DUWECS code, validation of an integrated dynamic model by measurements and finally results about robustly controlling wind turbine systems.

1 Introduction

The most important underlying motivation for wind energy research for large scale energy production is the aim to reduce the price of produced electrical energy. One single wind energy conversion system will then have:

low construction costs. Low construction costs can be achieved by using lighter, and thereby more flexible, components. Allowing more flexibility in the components also introduces the possibility of reducing internal stresses. However the dynamics of each component should be designed more carefully otherwise the internal stresses are magnified.

long lifetime. Fatigue loads have to be kept within acceptable margins in order to guarantee a long lifetime and thereby achieve an economic attractive system.

efficient energy conversion. Higher energy production out of the same amount of wind implies lower costs of the produced energy.

It is evident that each of these points more or less independently reduce the costs of the generated electrical energy. In order to achieve this aim the dynamic behavior of the complete wind turbine has to be optimized with respect to:

dynamics of the components. The choice of one component determines to a large extend the choices left open for the other components. To illustrate this, we will focus on two components: the generator/power electronics part and the rotor part.

Possible choices for electrical conversion are for example: an asynchronous machine, a synchronous machine with AC/DC/AC link or a doubly fed asynchronous machine.

The rotor system one can be chosen to have rigid blades, a teeter, lead-lag and/or flap flexible blades. Power limitation can for example be achieved by stall or partial span pitch control.

For example, the choice of a stall regulated rotor implies that the generator has to achieve a fixed rotational speed. Thereby the choice of electrical conversion is fixed.

control system design. The choice of the components determines active control possibilities. This possibility ranges from no control (stall regulated with fixed generator speed) to advanced efficient control pitch control with a variable speed generator). When control is possible the control design can have an enormous impact on the complete wind turbine behavior. A lot can be gained by adequate control design. Where a PI or PID controller fails in controlling power and blade loads, more advanced control design will be able to achieve both specifications.

Therefore dynamic models of the complete wind turbine are necessary to obtain an optimized wind turbine system. These integrated dynamic models are very useful in the design phase of new wind turbine systems, where the effects of different component choices need to be evaluated on the basis of their impact on the complete wind turbine behavior.

One of the very few computer programs capable of simulating the dynamic behavior of a *complete* wind turbine system is the DUWECS code. The user can easily change to different wind turbine configurations and evaluate the dynamics in a fast and user friendly environment.

In this paper a summary of results obtained during the JOUR 0110 DYNAMO research will be given, previous progress has been reported in [7, 8, 9]. This project is a continuation of a previous CEC funded research on *Integrated Dynamic Models of Flexible Wind Turbines* [12], which resulted in the computer code DUWECS [6].

Although a number of wind turbine models have been available [12], some aspects have been neglected, or have not been described in full detail. This paper describes some ways to tackle the observed discrepancies.

In Section 2 model extensions are discussed. First the rotor dynamics and aerodynamics are described in more detail, where the accuracy of the aerodynamic performance predictions can be increased by reliable 3D profile data, validated by wind tunnel measurements. A simple model of a synchronous machine with DC-link has been the only realistic generator model. Therefore it is sensible to describe other generator systems, like an asynchronous machine directly coupled with the grid or a doubly fed asynchronous machine.

The model developments are integrated in the complete dynamic model description of the DUWECS computer code, discussed in Section 2.

A model can be used for accurate predictions of the dynamic behavior only when it is validated with measurements on existing wind turbines. To this end some aspects of the integrated model are verified using wind turbine measurements in Section 4.

The impact of control design on the dynamic behavior of the complete wind turbine is investigated in Section 5. The impact of proper control system design is evaluated on generated power and fatigue load reduction.

Finally in Section 6 conclusions are drawn and recommendations for further research is discussed.

2 Wind turbine system building blocks

The main idea behind the modular structure of a wind turbine modeling, depicted in Fig. 1, is the flexibility to describe complete different wind turbines in a unified way. For this purpose interaction variables are used to assure a two-sided coupling between the wind turbine parts. Given a fixed, minimal, set of interaction variables the developed models can be replaced easily.

The following submodels have been added:

Rotor. A model consisting of flexible blades with flap, lead-lag and teeter freedom. The mechanical equations of motion are derived by using the method of Kane [17]. Each blade is divided into N sections with section dependent blade profile

‡This research was supported by the CEC under grant JOUR-0110

§Mech. Eng., Systems and control Group, Delft University of Technology, Mekelweg 2, NL-2628 CD Delft, Tel. + 31 15 78 4703, email: bongers@tudw03.tudelft.nl

♯Technical University of Braunschweig, Hans-Sommer straße 66, D-3300 Braunschweig

♭University of Southampton, Highfield, GB-Southampton SO9 5NH

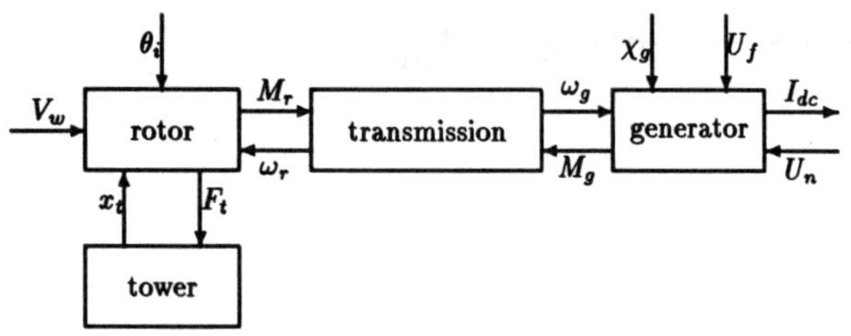

Fig. 1: Modular structure wind turbine model

characteristics, corde, mass and twist. Apart from the usual 2D profile characteristics, also 3D profile characteristics such as [25] or [4] can be used. The aerodynamic forces for each blade section are calculated using methods described in [15], with tip-correction following [24], dynamic stall following [5] and a simple model of dynamic inflow [26]. The wind velocity, as felt by a blade section, is shaped with wind shear and tower shadow. The used level of aerodynamic modeling is of the same level of complexity as the rotor computer programs in the CEC benchmark [16].

Generator. an asynchronous generator directly coupled to the grid, a doubly fed asynchronous generator, a synchronous generator with field excitation and AC/DC/AC link.

Controller. A framework to use multiple controllers has been developed. The transition between the different controllers can be obtained by instantaneous switching or is conditioned by Kalman filter gains.

3 DUWECS code

The mathematical model has to be implemented in software in order use the model for simulation studies. The following specifications where posed on the software:

- Simulate the non-linear, possibly controlled, flexible wind turbine behavior.
- Produce a linear state-space form of the non-linear wind turbine model, necessary to design and to optimize control strategies.
- Facilitate an easy exchange of wind turbine components, in order to simulate different wind turbine configurations.
- Possesses an user-friendly menu driven interface. For a specific simulation run the choices to be made by the user have to be presented in menu's and default values has to be available, while the log-file of a specific run can be used as a batch-file.
- Data exchange with standard programs for pre-processing or post-processing of the linear models or simulated data.
- The program has to be able to run on different computer systems, varying between MS-DOS, UNIX and VMS operating systems.

Although there are existing commercial available simulation packages as MATRIX$_x$ [21], MATLAB [20] or ACSL [1] these general packages do not satisfy the posed specification. Therefore a tailor made simulation package DUWECS has been developed. All the presented simulation results used in the sequel of this paper are obtained with the DUWECS code.

In order to illustrate one of the DUWECS code capabilities, the blade root flap moment is depicted in Fig. 2. In this figure the effect of applying flexibility in teeter, flap, lead/lag and lead/lag combined with flap direction is shown.

The modular structure has been set up for N modules. When only one wind turbine is described, as is the case here, nine modules are used. The number of used modules can be extended to incorporate for example wind-diesel systems or a complete wind farm consisting of different wind turbines connected to one weak grid.

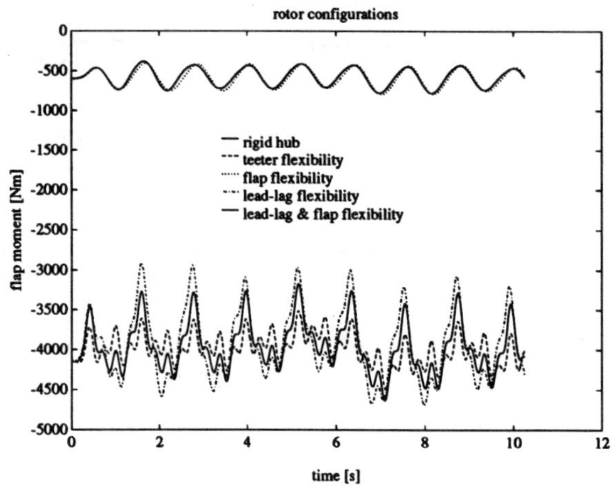

Fig. 2: Simulation different wind turbine configurations

4 Model validation

The wind turbine used for measurements is the UNIWEX wind turbine [23] of the university of Stuttgart. The first results have been obtained with a rigid rotor and validation of the flexible configurations is continuing.

This validation work on wind turbine models is mainly covered in [10, 2, 3, 22]. Here we present some figures to illustrate the approach and obtained results.

The advantage of this approach is that except the desired transfer functions also a model of the actual wind field, during the experiments, is calculated. The white noise generating the wind realization is not known, therefore the wind model can be used in frequency domain only. The estimation of individual transfer functions enables a more detailed validation of wind turbine models.

In Fig. 3 a validation example is given. It can be seen that in time domain the model predict only a part of the response. This is due to the absence of the estimated wind realization. In the lower part of Fig. 3 the wind model is added to the deterministic prediction, it can be seen that our experimental model describes the measured data very well.

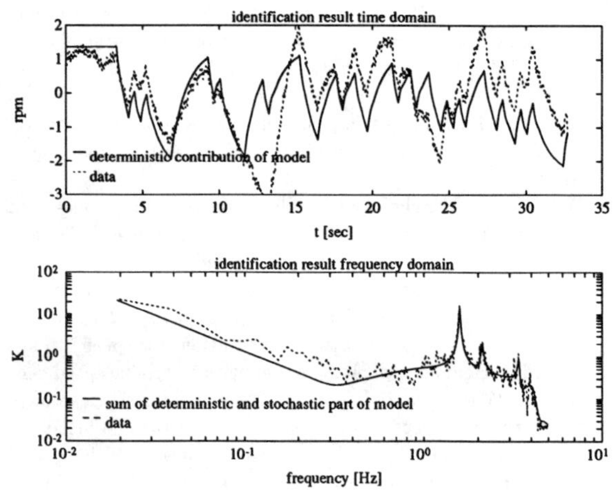

Fig. 3: Identification result in time domain and frequency domain

The validity of the used aerodynamic model description can be investigated using the blade root flapping moment. In Fig. 4 the DUWECS simulation is compared with measured data. It can be concluded by the resemblance between the DUWECS simulation and the measurement that the used aerodynamic model is satisfactory. The observed differences between the measurement and the simulation are due to: a) a simplified aerodynamic model; b) only mean wind conditions are used; c) at the UNIWEX the flap moments are also influenced by the controlled hydraulic cylinders.

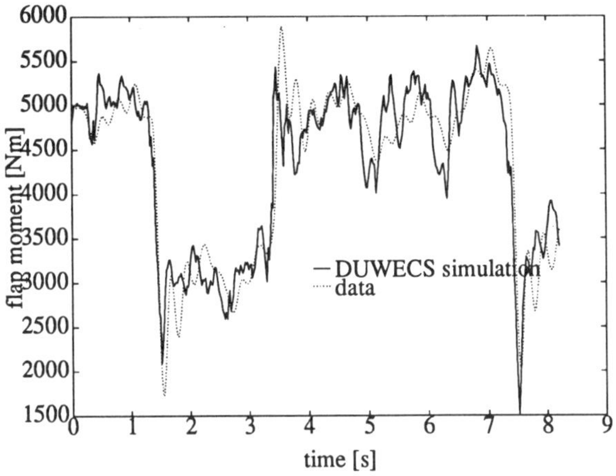

Fig. 4: Validation blade root flapping moment

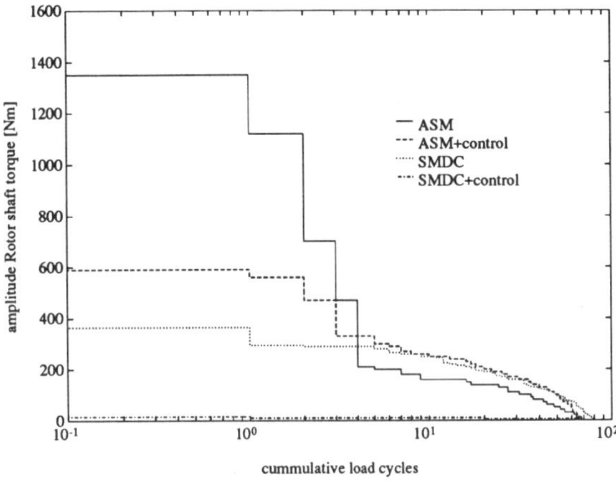

Fig. 5: Rainflow cycles rotor shaft torque for different wind turbine configurations

5 Impact of control design

The control design work is mainly covered in [11, 13, 14, 18]. This series of papers shows that by applying modern control techniques it is possible to reduce the blade root moments for certain wind turbine configurations. These modern control designs by far outperform the classical designs, like PI -or PID controllers, which are frequently used in wind turbines, whilst being almost as easy to implement.

In Fig. 5 the impact of proper designed controllers, see for example [14], on different wind turbine configurations around one operating point will be shown. To this end a constant speed wind turbine is compared with a variable speed wind turbine.

ASM	uncontrolled constant speed wind turbine
ASM+control	controlled constant speed wind turbine
SMDC	uncontrolled variable speed wind turbine
SMDC+control	controlled variable speed wind turbine

Table 1: simulated wind turbine configurations

It can be seen in Fig. 5, the abbreviations refer to Table 1, that for both wind turbine configurations the application of proper controllers results in lower rotor shaft loads. The uncontrolled variable speed wind turbine is slightly smaller loads than controlled constant speed wind turbine, whereas the rotor shaft loading in the controlled variable speed wind turbine are almost vanished. In the variable speed wind turbine, also the blade root bending moments are reduced.

In order achieve optimal control the wind turbine over whole operating envelope, it is necessary to use several controllers. Abrupt transition between the controllers cause large loads in the wind turbine. As described in [19] it is possible reduce these loads by applying a smoothing technique. This means that the states of the next controller to be actively controlling the wind turbine are conditioned using Kalman filtering. The advantage of the conditioning technique is demonstrated in Fig. 6.

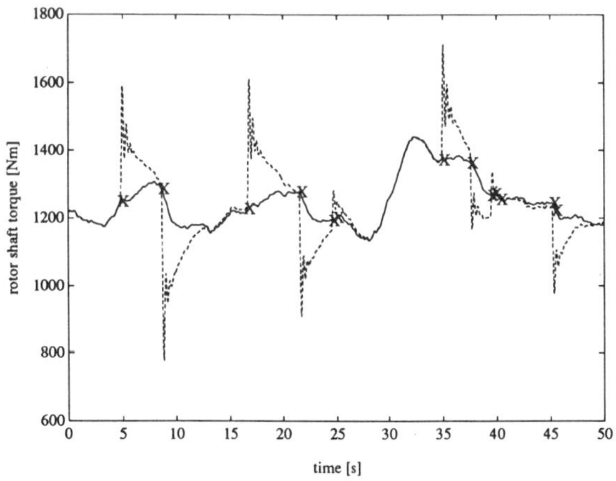

Fig. 6: Response of rotor shaft torque under noisy wind conditions, (—) conditioned controllers, (- -) non conditioned controllers

6 Conclusions and Further research

- The number of available models is increased with a doubly fed asynchronous generator, a detailed flexible rotor and a switching controller framework. The 2D aerodynamic profile coefficients are adapted to a 3D wind turbine environment.
- A rotor model, describing the UNIWEX turbine, has been validated by measurements. The applied description of the rotor dynamics and the aerodynamics is satisfactory.

- Significant blade root load reduction by application of well designed controllers has been shown. Even if the turbine is not exactly known those controllers comprise robustly stable controlled wind turbine. Furthermore the advantage of a conditioned switching between several controllers has been shown.

Further research

Therefore the main objective of further research is to significantly improve the complete wind turbine model by:

- Describing the 3D behavior of the aerodynamic model in more detail
- Modeling the synchronous machine with AC/DC/AC link and line interaction in more detail
- Developing a transparent link between fatigue life and control design methods
- Implementing those controllers on real life wind turbines
- Validating the complete model

References

[1] ACSL , *Reference Manual Rev. 4.0.* Mitchell and Gauthier Associates, Concord, MA, USA, 1986.

[2] Baars G.E. van, P.M.M. Bongers , "Wind turbine control design and implementation based on experimental models", 31th IEEE Conf. on Decision and Control, To appear, 1992.

[3] Baars G.E. van, P.M.M. Bongers, H. Schwartz, H. Arnold , "Experimental and theoretical results of the UNIWEX wind turbine", in *Proc. Europ. Community Wind Energy Conf.*, to appear, 1993.

[4] Barnsley M. , *Dynamic wind turbine models for desing purposes, 3D aerodynamic behaviour.* University of Southampton, Great Brittain, 1993.

[5] Beddoes T. , "Representation of airfoil behaviour", *Vertica*, vol.7, 1983.

[6] Bongers P.M.M. , *DUWECS Reference Guide, Delft University Wind Energy Conversion Simulation Program.* Delft University of Technology, The Netherlands, 1990.

[7] Bongers P.M.M. , *Dynamic models of wind turbine systems for design purposes.* Report on the CEC meeting on Aerodynamics, 11-12 october 1990, Athens, 1990.

[8] Bongers P.M.M. , "Dynamic models of wind turbine systems for design purposes, progress report I", in *Proc. Wind Energy R&D contractor meeting, Maastricht, 12-14 june, 1991*, pp.98-103, 1991.

[9] Bongers P.M.M. , "Dynamic models of wind turbine systems for design purposes, progress report II", in *Proc. Wind Energy R&D contractor meeting, Alghero, Italy, 3-5 june, 1992*, pp.145-154, 1992.

[10] Bongers P.M.M., G.E. van Baars , "Experimental validation of a flexible wind turbine model", in *Proc. 30th IEEE Conf. on Decision and Control, Dec. 11-13, Brighton, UK*, pp.1660-1661, 1991.

[11] Bongers P.M.M., G.E. van Baars, H. Huisman, S. Dijkstra , "Investigation of load reduction controllers for wind turbine systems with different electrical conversion systems", in *to appear at the IEA R&D Wind annex XI workshop on electrical systems for wind turbines*, Oct. 7-8 Goteborg, Sweden, 1991.

[12] Bongers P.M.M., W.A.A.M. Bierbooms, Sj. Dijkstra, Th. van Holten , *An Integrated Dynamic Model of a Flexible Wind Turbine.* Delft University of Technology, The Netherlands, 1990.

[13] Bongers P.M.M., Th. van Holten, Sj. Dijkstra , "Control of Wind turbine Systems", in *Wind Energy - Technology and Implementation, Proc. European Wind Energy Conf. Oct. 14-18, amsterdam, The Netherlands,* F.J.L. Van Hulle, P.T. Smulders, J. Dragt, pp.68-72, 1991.

[14] Bongers P.M.M., W. Sturm, I. Kraan, Sj. Dijkstra, O.H. Bosgra , "Control options for wind energy conversion systems", in *Proc. Europ. Community Wind Energy Conf.*, this issue, 1993.

[15] Eggleston, D.M., F.S. Stoddard , *Wind turbine engineering design.* Van Nostrand Reinhold Company, New York, 1987.

[16] Grol, H.J. van, H. Snel, J.G. Schepers , *Wind turbine benchmark exercise on mechanical loads, a state of the art report.* Netherland Energy Research Foundation, Petten, The Netherlands, 1991.

[17] Kane T.R., D.A. Levinson , *Dynamics, theory and applications.* McGraw-Hill series in mechanical engineering, 1985.

[18] Kraan I., P.M.M. Bongers , "Control of a wind turbine using several linear robust controllers" (in Dutch), Delft University of Technology, submitted 32nd CDC, 1993.

[19] Kraan I. , *Control design for a flexible wind turbine* (in Dutch). Delft University of Technology, TUD-WBMR-A-613, 1992.

[20] MATLAB , *Matlab User's guide.* The Mathworks Inc., South Natick, MA, USA, 1989.

[21] $MATRIX_x$, *7.0 Release notes.* Integrated Systems Inc., Santa Clara, CA., USA., 1988.

[22] Mosterd H. , *Experimental modelling of the UNIWEX turbine* (in Dutch). Delft University of Technology, TUD-WBMR-A-672, 1993.

[23] Müller M. , "Experimental investigations with the universal test wind turbine", in *Proc. European Wind energy Conf. 1989.*, Glasgow, Schotland, pp.832-836, 1989.

[24] Prandtl L., O.G. Tietjens , *Applied Hydro- and Aeromechanics.* Dover Publications Inc., 1957.

[25] Viterna L., R. Corrigan , "Fixed pitch rotor performance of large horizontal axis wind turbines", in *DOE/NASA workshop on large horizontal axis wind turbines, Cleveland, USA.*, 1981.

[26] Øye S. , "Unsteady effects caused by pitch-angle changes", in *IEA Symposium - joint action on aerodynamics of wind turbines, London, Great Britain.*, 1986.

VALIDATING WIND TURBINE DESIGN CODES BY ANALYSING TEETER, 'FREE' FLAP AND 'FREE' LEAD/LAG MOTIONS

A.J. Brand[1] and H. Snel
ECN - Renewable Energy
P.O. Box 1, 1755 ZG Petten, Netherlands

N.O.T. Hansen, G.A.M. van Kuik,
Stork Product Engineering SPE
P.O. Box 379, 1000 AJ Amsterdam, Netherlands

G.E. van Baars, P.M.M. Bongers
Mech. Eng. Systems and Control Group, Delft Univ. of Techn.
Mekelweg 2, 2628 CD Delft, Netherlands

ABSTRACT: The wind turbine design codes Flexlast, Phatas-II and Duwecs are validated by comparing their predictions with the measurements on the UNIWEX in the teeter, the 'free' flap and the 'free' lead/lag mode. The predicted behaviour agrees quite well with the measured behaviour for the teetered and the 'free' flapping configurations, but not for the 'free' lead/lag configuration.

1 CODE VALIDATION ON UNUSUAL ROTOR CONFIGURATIONS

Current aerodynamic and aeroelastic design codes are validated by comparing their predictions with experimental results. Since usually validations are confined to a stiff rotor with a fixed rotor speed, there is not much data available on a rotor which exploits one or more degrees of freedom (like a teeter hinge or a variable speed conversion system) to obtain load alleviation. If a code is validated on an unusual rotor configuration, it is tested under extreme conditions, and as a result code bugs, model errors or a poor understanding of the physics will manifest most clearly. Moreover, in this way it is possible to explore the benefits and disadvantages of the given rotor configuration.

The Dutch contribution to the UNIWEX[2] project involved the acquisition of data on a rotor which exploits many degrees of freedom, such as a variable rotor speed, teetering, individual free flap, free lead/lag and free yaw. The data from these experiments is stored in a database [1].

In this paper the results are reported on the validation of the ECN code Phatas-II, the SPE code Flexlast and the DUT code Duwecs on basis of experiments with a rotor in the teeter, free flap and free lead/lag mode (for high as well as low tip-speed ratios). First, we present an outline of the experiments and the modelling (§2). Next, we address the outcome of the experiments, the simulations and a simple model (§3), and an evaluation of our findings (§4).

2 OUTLINE OF THE EXPERIMENTS AND THE MODELLING

2.1 The experiments

The UNIWEX is a hydraulically activated and computer controlled two-bladed horizontal-axis wind turbine which is equipped with a great variety of sensors. In this context it is important to note that none of the employed rotor configurations is really 'free': since the teeter, the flap and the lead/lag motions are controlled by hydraulic actuators some stiffness and damping is introduced into the motions under investigation. In practice, however, the 'spring stiffness' was observed to be relatively small.

2.2 The modelling of 'free' teeter motion

In the dynamic module of Flexlast, the teetering rotor was modelled as a completely free teeter, i.e. without any spring stiffness. The measured wind speed and rotor speed were used as input to the aerodynamic module of Flexlast. Ideally, the rotor speed should have been the output of the simulation. However, since the braking torque applied during the experiment is not known, the actual rotor speed was used as input. For the simulation of the blade dynamics this procedure has little or no effect. As to the aerodynamics of Flexlast, 2D profile coefficients were used. More information on the Flexlast simulations is available in a separate report [5].

In the Phatas-II simulation of 'free' teeter motion the teetering degree of freedom was enabled with zero spring stiffness [7]. Also the flapwise and the lead/lag-wise elastic deformation degrees of freedom were enabled; the flapwise degree of freedom has a strong interaction with the teeter motion. To obtain a slow down of the rotor speed, a constant brake torque of 305 Nm was applied to the low-speed shaft such that the same range of rotor speeds was encountered as in the experiment. As input the wind speed measured at the hub, together with a mean wind shear according to a 1/7th-power law, was used. The tower-shadow model introduces excitation to the various degrees of freedom. The aerodynamic data was 2D too.

Duwecs is a simulation package for integrated dynamic models of (flexible) wind turbines [3]. The dynamic model of the UNIWEX was constructed by using the *rot5flta*, *unitra*, *tower0a*, *gener0* and *diconu0* modules of the module library. Using system identification techniques, dynamical aspects of these models are validated (see e.g. [2]). In this way it is also possible to analyse the dynamic behaviour of the wind turbine in more detail so that a comparison with the simulation models can be made. The teeter stiffness to be

[1]Corresponding author.

[2]UNIWEX (Universal Wind turbine for Experiments) is a R&D project of the International Energy Agency IEA, and is partly funded by the Netherlands Agency of Energy and Environment NOVEM.

applied in the simulation was determined by a least-square fit between the measured teeter moment and teeter angle. The components of the horizontal wind speed were computed from the wind speed and the wind direction as measured at hub height, and filtered by a 4th-order butterworth filter with a cut-off frequency of 5 Hz. To this signal wind shear and tower shadow were added.

2.3 The modelling of 'free' lead/lag motion

To derive a simple model for the lead/lag motion, the mass of each blade is concentrated in a point that moves in the rotor plane. The motion of each point (as determined by the aerodynamic force) is constrained by two massless rigid arms, which are connected to the hub by an eccentric lead/lag hinge. A Lagrangian approach is used to derive the equations of motion for the rotor azimuth ϕ and the two lead/lag angles ψ_1 and ψ_2. Inspection of these equations reveals that there are two key parameters in the model: the aerodynamical parameter (which is proportional to the wind speed, the lift coefficient and the blade size) and the gravitational parameter (which is proportional to the gravitational constant and inversely proportional to the blade size). On basis of the experimental data, the complexity of the equations of motion is reduced by retaining only the the second order, the first order and the gravitational terms. By examining separately the four cases $\phi \approx (i-1)\ 90°$ (i = 1, 2, 3 and 4), and introducing linear operators, the differential equations for the three degrees of freedom are derived. For the governing experimental conditions it is possible to solve these equations by standard methods and express their solutions as superpositions of goniometric and hyperbolic functions. A separate report presents a detailed description of this model [4]

3 THE OUTCOME OF THE EXPERIMENTS, THE SIMULATIONS AND THE MODEL

3.1 General

The measured and the simulated dynamic behaviour agree reasonably well for the teetering and the flapping configurations. Although the teeter angle and the flap angles were observed to increase significantly for low tip-speed ratios, during the measurements and in the simulations instabilities did not occur.

For the configurations with lead/lag freedom the 'stiffness and damping' approach did not lead to satisfactory simulation results. Probably, the very non-linear stiffness and damping (as applied by the hydraulic system) is not appropriately modelled in the codes. On the other hand, errors may be present in the until now untested lead/lag model. To explore this further, a simple model for free lead/lag motion was developed.

In the following we will present in greater detail the simulations of the teeter motion (§3.2) and the prediction of the 'free' lead/lag motion (§3.3).

3.2 'Free' teeter motion

3.2.1 General description of the experiment

A wind turbine configuration which has received attention recently, is a stall controlled rotor with a teetered hub (see e.g. [6]). For this configuration one of the points to address is teeter stability in stall. Since the teetering motion has a 1P eigenfrequency it is in resonance with the 1P excitation of e.g. the mean wind shear. The amplitude of the

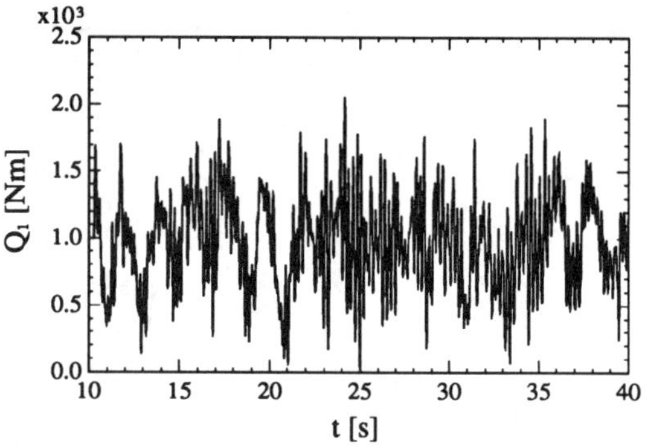

Figure 1: The measured flap moment Q_1 of blade 1 as a function of time; part of datafile 0867.

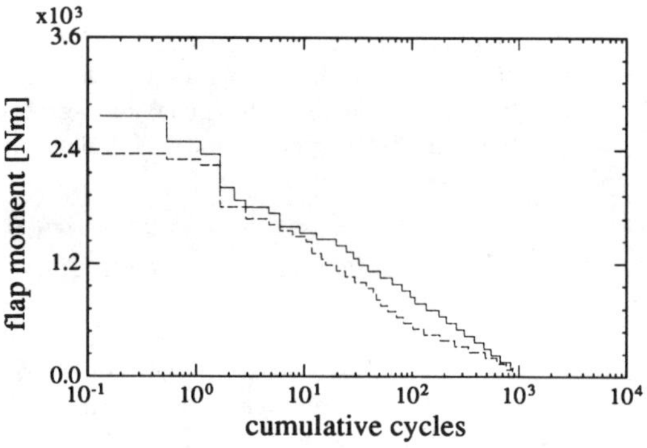

Figure 2: Measured (solid line) and simulated (Flexlast; dashed line) rainflow count of the flap moment of blade 1; result based on the first 145 s of the datafile.

teetering motion is only governed by aerodynamic damping, which becomes small (or even negative) in stall.

To study teetering behaviour, measurements were performed on the UNIWEX with the blade flapping control emulating a teetered hub (datafile 0867). By increasing the absorbed shaft torque, in a time lapse of 200 s the rotor speed was reduced from 30 rpm to approximately 10 rpm. At a wind speed of about 8.4 m/s the tip-speed ratio varied from 3.2 to 1.0, approximately. The blade-tip angle was maintained constant at 2°. Figure 1 shows a representative part of the flap bending moment of blade 1 at a radial position of 1.0 m. For t > 145 s (not shown) large peaks occurred when the mechanical limits were reached. This happened for tip-speed ratios of less than 2.

3.2.2 Flexlast

Figure 2 presents the measured and simulated rainflow counts of the flap bending moment for the first 145 s of the datafile. It appears that the agreement is very satisfactory.

Figure 3a shows the measured and the simulated flap angle of blade 1 from t = 10 to 40 s. It appears that the simulated flap angle is periodically too small or too large. This effect is caused by the 1P turbulence component: due to the rotating-turbulence model [8], this 1P component periodically increases and decreases the wind shear. To analyse this effect, the simulation was redone with an increased shear amplitude but without a turbulence model. As figure 3b shows, there is a better agreement between the measured and the simulated flap angle. As a side effect, this procedure

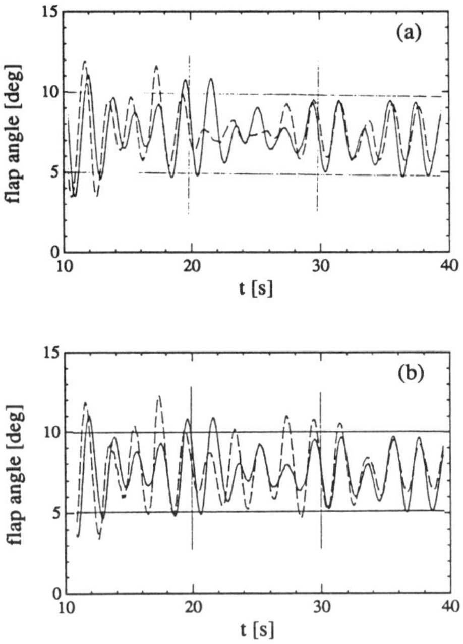

Figure 3: The measured (solid line) and simulated (Flexlast; dashed line) flap angle of blade 1 as a function of time: simulation with (a) a turbulence model and (b) an increased wind shear but without a turbulence model.

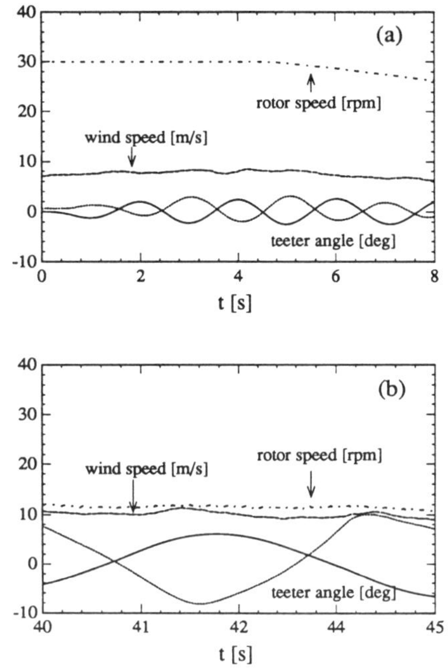

Figure 5: The teeter angle as a function of time for a tip-speed ratio of (a) 3 and (b) 1; measured (dotted line) and simulated (solid line) data. Also indicated are the wind speed and the rotor speed.

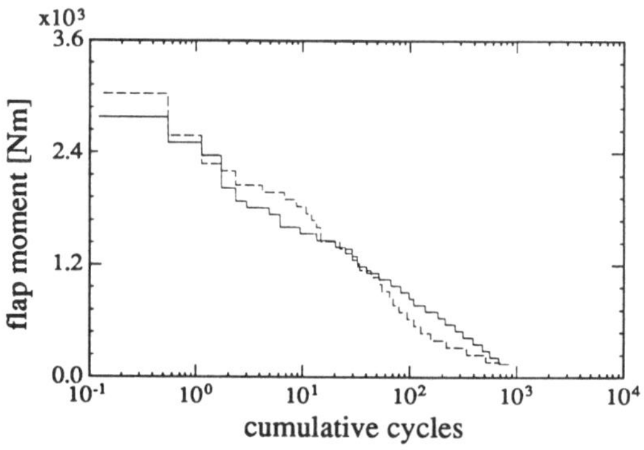

Figure 4: As figure 2; simulation with an increased wind shear but without a turbulence model.

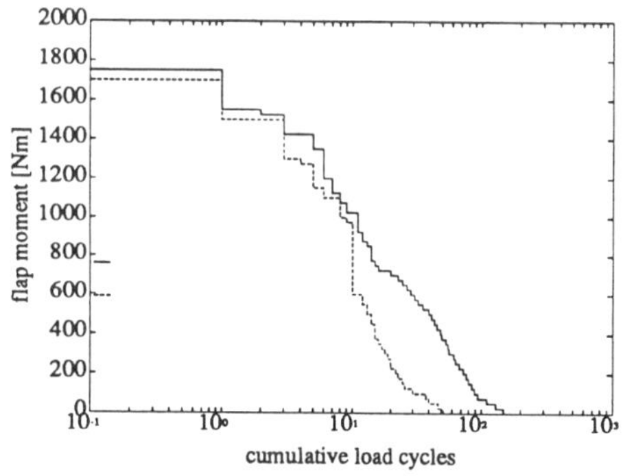

Figure 6: Measured (solid line) and simulated (Duwecs; dashed line) load spectrum of the blade root flap moment.

slightly improved the rainflow count of the simulated flap bending moment (figure 4).

On conclusion, the agreement between the results of the measurement and the Flexlast simulation is very satisfactory.

3.2.3 Phatas-II

Figures 5a and 5b present the measured and simulated teeter angle for a short time lapse, with a tip-speed ratio of 3 and 1, respectively. In these figures the wind speed and the rotor speed are indicated too. The measured teeter angle was determined as the difference between the instantaneous mean of the two flap angles and the cone angle, where the cone angle is the time average of the mean flap angle of the two blades.

Figures 5a and 5b show that until almost standstill no instabilities were encountered. Furthermore, both in the measurements and in the simulations large teeter amplitudes of approximately 10° occurred at a tip-speed ratio of 1. In view of the fact that 2D aerodynamic coefficients were used in the simulation, the agreement between the measured and the simulated teeter angle is quite reasonable.

It thus can be concluded that, for this case, Phatas-II gives a reliable simulation of the teeter dynamics.

3.2.4 Duwecs

In this section a glimpse of time-domain comparison of the Duwecs models with the measured data is discussed. Figure 6 presents the measured and simulated cumulative load cycles of the blade root flap moment. It appears that the simulated flap moments are smoother than the measured flap moments. This is due to the lack of 1) a coherent wind field in the simulation (by which the flap moment is expected to experience a large influence), and 2) a modelling of the servo valves at the flap hinges of the UNIWEX (by their way of achieving 'free' teeter motion it is expected that these valves cause significant flap moments).

Both the time and frequency domain results of the Duwecs simulation thus show a reasonable good agreement with the measured data.

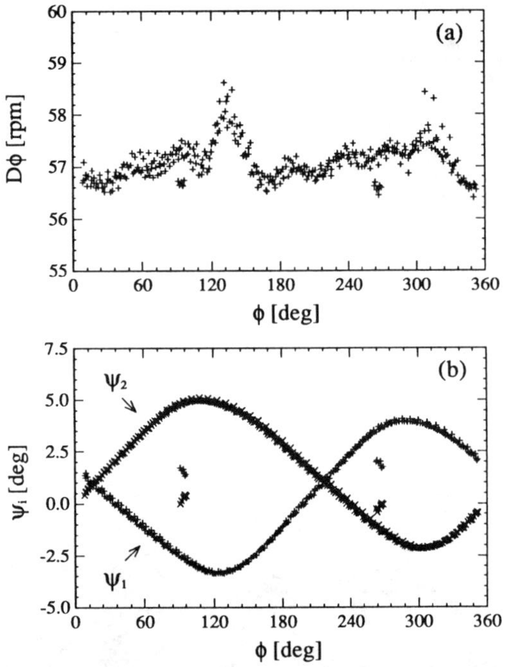

Figure 7: The measured 'free' lead/lag motion (datafile 0843): (a) the rotor speed $D\phi$ and (b) the lead/lag angles ψ_1 and ψ_2 as a function of the rotor azimuth ϕ.

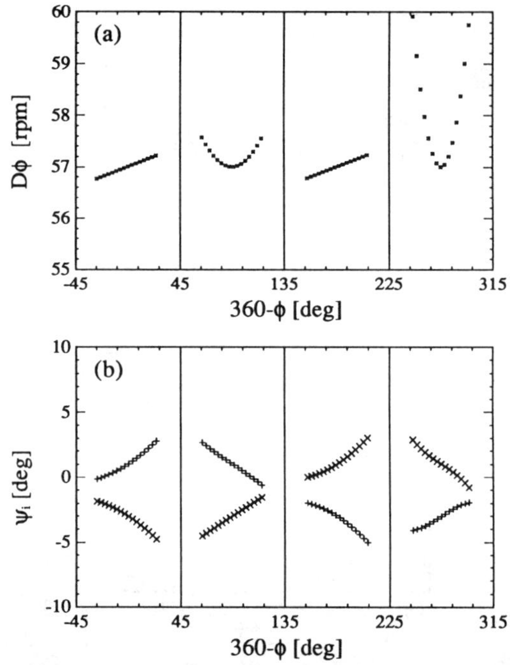

Figure 8: As figure 7 but for the predicted free lead/lag motion.

3.3 'Free' lead/lag motion

In the literature on horizontal-axis wind turbines no experience is documented on the application of blades with a marked flexibility in lead/lag direction. A feature of wind turbine lead/lag is the pronounced periodic loading by gravity, and the small aerodynamic damping. In practice, coupling of lead/lag eigenfrequencies with the eigenfrequencies of e.g. the flap degree of freedom may cause instabilities.

Figure 7 presents the main features of 'free' lead/lag motion on basis of datafile 0843: 1) The lead/lag angles are asymmetrical, 2) There is an offset in the zero crossing of each blade, and 3) Per revolution the displacement in lead/lag direction causes two peaks in the rotor speed. These results were obtained with a wind speed of about 5 m/s. Recall however that the lead/lag motion is not really free due to an unspecified amount of damping.

To gain some understanding of the observed lead/lag motion, it is compared with the predictions of the simple model. Figure 8 shows that the lead/lag motion introduces exponential growth rates into the rotor speed and that the lead/lag behaviour is not purely asymmetrical.

It thus appears that the differences between the conditions for the experiments and the model do not hamper a proper treatment of the main features of 'free' lead/lag motion, and that the model contains enough of the essential physics.

4 Conclusion

On conclusion, we have found that, for a wind turbine with the rotor in the teeter or in the 'free' flap mode, the agreement between the simulations and the measurements is very satisfactory. In particular, for low tip-speed ratios the rotor with a teetered hub experienced neither instabilities nor huge teeter angles. For the wind turbine with the rotor in the 'free' lead/lag mode, on the other hand, no satisfactory simulation results were obtained.

Acknowledgements

The authors would like to acknowledge the staff of the Institute for Computer Applications ICA of the University of Stuttgart for operating the UNIWEX.

References

[1] G.E. van Baars, 1992, *UNIWEX database user's guide*, Delft University of Technology

[2] G.E. van Baars, P.M.M. Bongers, H. Schwartz and H. Arnold, 1993, *Experimental and theoretical results of the UNIWEX wind turbine*, In: Proc. European Community Wind Energy Conference 1993, Lübeck-Travemünde, March 1993

[3] P.M.M. Bongers, 1990, *DUWECS Reference Guide, Delft University Wind Energy Conversion Simulation Program*, Delft University of Technology

[4] A.J. Brand, 1993, *A simple model for wind turbine free lead/lag motion*, (in preparation), Netherlands Energy Research Foundation ECN

[5] N.O.T. Hansen, 1992, *Handling of the UNIWEX datafiles to verify the wind turbine design program Flexlast, Part I and Part II*, Stork Product Engineering

[6] G.A.M. van Kuik and J.W.M. Dekker, 1991, *The FLEXHAT program: Technology development and testing of flexible rotor systems with fast passive pitch control*, In: Proc. European Wind Energy Conference 1991, Amsterdam, October 1991

[7] H. Snel and C. Lindenburg, 1992, *Phatas-II aeroelastic modelling*, ECN-C-92-027, Netherlands Energy Research Foundation ECN

[8] *Handboek Ontwerpwindgegevens Windturbines, Versie 3*, INTRON/TNO/ECN 1991

SUPERVISORY CONTROL OF A WIND PARK

K. Wefelmeier
Institut für Regelungstechnik, Technische Universität Braunschweig
Hans-Sommer-Straße 66, W-3300 Braunschweig, Germany

ABSTRACT: The paper shows the results gained from the simulation of a small wind park using a supervisory controller. It includes the model of a two-dimensional wind field, the rotor with steady state aerodynamic characteristics and a simplified model of the mechanical and the generator system. Variable speed generators with torque control reference inputs are mandatory for this operation. Further improvement is gained from additional wind speed information fed forward to the control system. These signals can either be used to smooth variations of power output or to reduce dynamic energy losses due to nonoptimal rotor/wind speed ratio. Maximum energy capture and minimum power fluctuations are found to be contradictory goals. Depending on the control scheme, either short term power variations can be significantly reduced, or the energy capture can be increased by up to 5% at low wind speeds.

1 INTRODUCTION

Most of todays windfarms represent just an accumulation of several turbines installed in one site operating almost independently of each other. A supervisory control of wind turbine generators in a wind park promises an improved performance of the system without significant extra costs. There are three different goals that can be set for the operation: smooth total power output, maximum energy capture or minimum torque variations of the mechanical system of the turbines. The basic effects are first demonstrated with a single turbine and then expanded for a wind park.

2 SIMULATION MODEL

The investigation of different control schemes requires a suitable model of the wind park. The wind speed signal is generated from a spectral distribution function by Fourier synthesis [1]. This single local value is extended to a two-dimensional function by introducing two distorting factors to the summation of the Fourier coefficients that are depending on the distance longitudinal and lateral to the wind direction [2]. This ensures that the correlation of the generated wind velocities decreases with their distance from the starting point. Wake effects are considered by equations similar to the Risø-model [3].

The mechanical system is simulated by a simple first order model with nonlinear static aerodynamic characteristics. As the time scale of wind induced power fluctuations is in the range of seconds, effects of tower shadow and oscillations of the drive train are neglected. A fixed speed generator system can be simulated by a simple asynchronous generator with a linear torque/speed characteristic. The variable speed generators are substituted by a first order lag representing the delays of power electronics and internal control. Figure 1 depicts the resulting block diagram of a single wind turbine.

3 OPTIMUM CONTROL OF A SINGLE WIND TURBINE GENERATOR

Maximum energy yield of a wind turbine demands the operation of the rotor at the peak of the aerodynamic c_p/λ characteristic, i.e. rotor speed should vary linearly with wind velocity. As the measurement of a representative wind speed signal is difficult, the use of a control characteristic based on the rotor speed to generate the torque demand of the generator was found to give satisfactory results. The control function can be derived from the power equation and gives a quadratic relation between torque and rotor speed that depends only on slowly changing parameters of the turbine. The characteristic shape factor may be adapted by a peak holding controller [4].

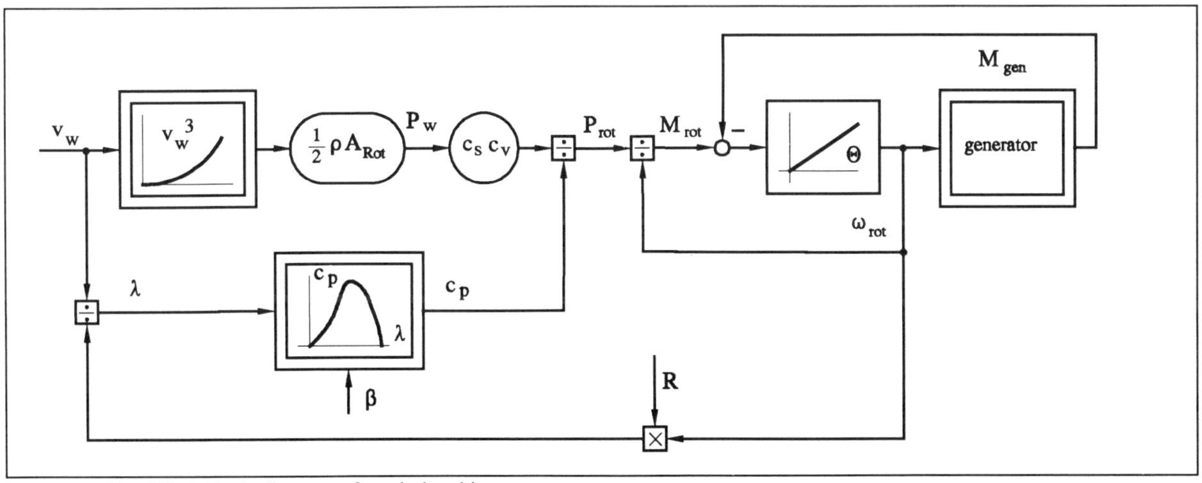

Figure 1 Simplified block diagram of a wind turbine

3.1 Speed Range Restrictions

The speed range of a wind turbine is usually restricted by centrifugal forces and aerodynamic noise at high speeds and tower resonances and low generator voltage at low speeds. Figure 2 depicts the remaining operating range of a wind turbine in the torque/speed plane. For maximum energy capture the turbine must be operated following the characteristic depicted in figure 2. Four different sections numerated I to IV are defined: I and III are characterised by a constant rotor speed at the bottom or top boundary. In these cases the machine will not operate at maximum c_p and the fixed speed of the generator causes high torque and power fluctuations. Part II will give optimum operation. In section IV of the characteristic the generator reaches rated torque where other mechanisms such as pitch control of the rotor blades or stall effects are used to keep the rotor speed within its permitted limits.

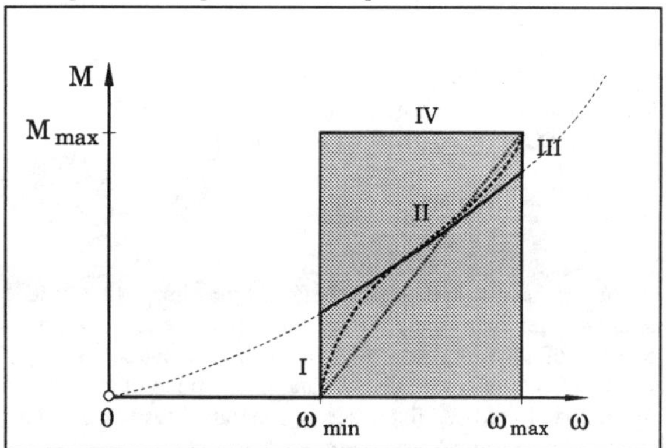

Figure 2 Optimum torque/ speed characteristic

The constant speed control in sections I and III can be avoided by using a simple linear function or an adapted parabolic characteristic (dotted and dashed lines in fig. 2), but additional energy losses have to be encountered. The effects depend on the control characteristic and the speed range of the generator. Under steady state conditions the energy capture will rise continuously with the speed range [5]. When using stochastic wind excitations no further improvement is achieved in the simulation above a speed range of 40% employing optimum control. The two other control characteristics even show a decrease of energy for a wider speed ranges caused by increased dynamic power losses at low wind velocities (figure 3).

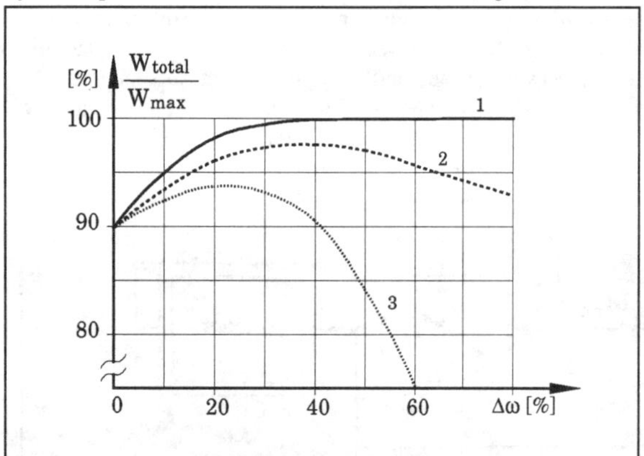

Figure 3 Energy capture versus speed range for different control characteristics

This loss is a result of the low torque that is available for the acceleration of the rotor at low wind speeds. As motor operation of the machine is usually excluded, only a slow speed adaption to changing wind velocities is possible resulting in an increased energy loss.

3.2 Feed Forward of a Wind Velocity Signal

As described above the variable speed wind turbine suffers some loss of energy capture due to nonoptimal operation at fast changes of wind velocity. This negative effect can be reduced by feeding forward a wind speed signal that can be gained from an anemometer or the output power of another wind turbine operating in wind direction of the affected turbine. The lead time of this signal allows an early adaption of rotor speed to wind velocity thus reducing dynamic energy losses. In the algorithm the wind information is smoothed by a low pass filter to give an estimation of the future velocity. Using the maximum c_p-value the optimum new rotor speed is calculated and feed forward to the speed controller of the generator. The results of the simulation are demonstrated in figure 4 for a gust excitation.

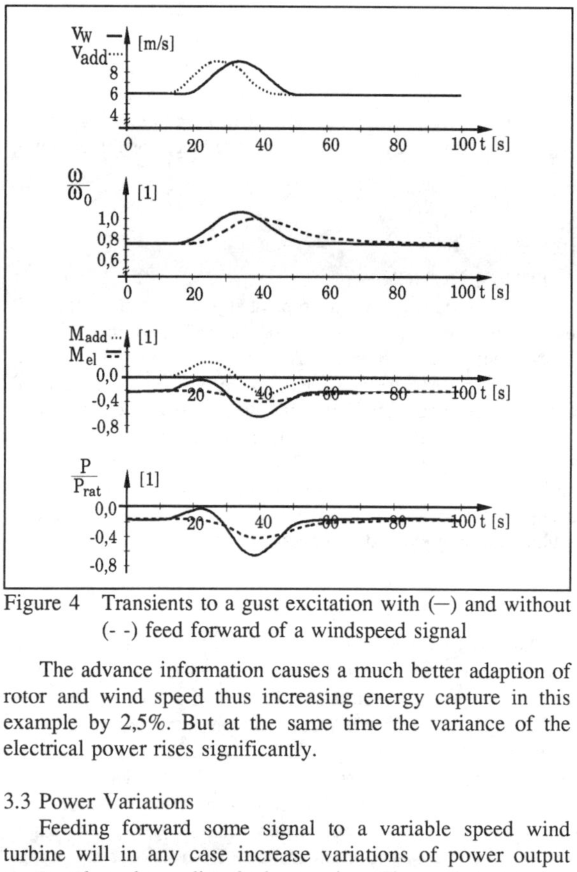

Figure 4 Transients to a gust excitation with (—) and without (- -) feed forward of a windspeed signal

The advance information causes a much better adaption of rotor and wind speed thus increasing energy capture in this example by 2,5%. But at the same time the variance of the electrical power rises significantly.

3.3 Power Variations

Feeding forward some signal to a variable speed wind turbine will in any case increase variations of power output compared to the undisturbed operation. Figure 5 depicts the standard deviation of the electrical power versus energy capture for the three different control characteristics of figure 2. It envisages the contradiction of high energy yield and smooth power output for a single wind turbine.

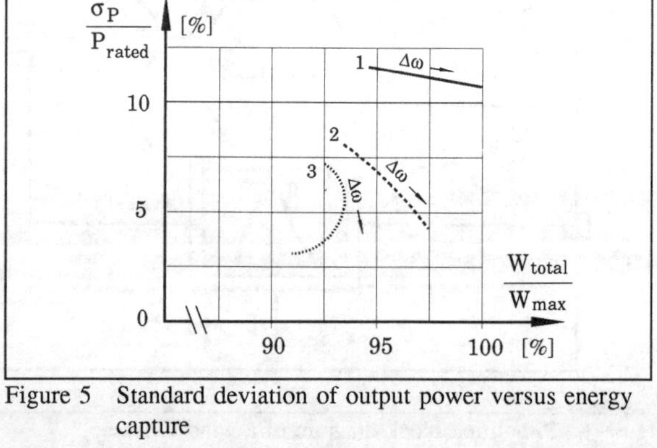

Figure 5 Standard deviation of output power versus energy capture

4 POWER CONTROL OF A WIND PARK

Parallel operation of generators in a wind park shows a smoother total power output compared to a single generator. The reason is that the wind speeds and the respective powers of different turbines are only partially correlated due to the stochastic nature of the wind excitation. This results in a compensation specially of high frequency power variations. If variable speed generators are available, they can be utilised for further smoothing of the total power output by a supervisory controller. In addition wind speed information can be used to reduce energy losses.

4.1 Feed Forward of a Wind Velocity Signal

Analogous to the single wind turbine generator wind speed information can also be used in a wind park. It is assumed that there are only a few signals available, so an algorithm is employed to select the most suitable signal depending on distance and wind direction. The simulation results show a small additional energy capture of 1% to a maximum of 5% at low wind speeds, which is a bit less compared to the single turbine. The variation of the power output signal rises by a minimum factor of 3.

4.2 Power Reference Level

Fast changes of the total power output can be reduced using a supervisory power controller. As the storage capacity of the rotors is limited to only a few seconds the main task of the control is to compensate fast power gradients. One problem is the generation of a suitable total power reference and its continuous adaption to varying wind velocities. The reference level is gained from the total power output of the whole wind park, smoothed by a low pass filter. Together with the actual power signal this acts like a delayed differentiator DT_1 in the feed back loop of the controller. As a result the control circuit will damp high frequency power variations but is insensitive to slow changes of wind velocity.

4.3 Power Splitting and Control Algorithm

The output signal M_{add} of the supervisory controller must be split and fed to the torque control loops of the individual variable speed turbines. There are two margins that have to be considered: none of the generators must leave its acceptable speed range and the additional torque should only be applied, when it moved the rotor speed to a higher c_p. These restrictions result quite often in a rejection of the additional torque signal thus opening the power control circuit which results in fast power transients.

This unsatisfactory behaviour may be dealt with by a modified approach that makes use of the rotor speed information of the generators. The velocity gives a indirect value of the local wind speed and the energy stored in the moving parts of the turbine. To avoid the problems mentioned above a small band is defined within the allowed speed range of the rotor and the generator is controlled to the centre of this band ω_c. This ensures an acceptance of either sign of torque.

The energy of the individual turbine is used to split the total torque output of the controller, so that every turbine is fed with a torque proportional to its potential of kinetic energy. To avoid operation out of the admissible speed range the additional torque amplitude is reduced near the absolute speed limits.

The problem of the power reference delay by the low pass filter is reduced by the feed back of the accumulated kinetic energy error of all variable speed wind turbines P_{fb}. This feedback signal represents the total energy, necessary to bring all turbines back to their centre band speed ω_c. It implies an additional excitation to the power control loop, but the dynamics of this error signal are quite slow being calculated from the rotor speeds of the turbines that are smoothed by the large inertias of the rotors (time constant approx. 10 s).

The blocks described so far are primarily used to compensate fast power variations. To reduce energy losses under steady state conditions the speed band for each individual turbine is continuously adapted to mean rotor speed that is influenced by the torque/speed characteristic of every turbine. Figure 6 depicts the block diagram of this supervisory control scheme, figure 7 gives an overview of the torque splitting and energy error calculating algorithm.

4.4 Results of Simulation

The operation of a wind park with this supervisory control shows a significant reduction of the power variations. Although no direct feed forward of a wind speed signal is provided the energy output of the park is slightly increased which seems to oppose the statements given for a single turbine in chapter 3. This can be explained by the indirect feed back of the rotor speeds. If for example a increasing wind velocity reaches the park, the power output of the first turbines will rise. The supervisory power controller tries to stabilize power output so the torque demands for all variable speed turbines are reduced causing an acceleration of the rotors. When the wind front reaches these turbines they are already adapted to the higher wind speed and operate with an improved c_p compared to the uncontrolled case.

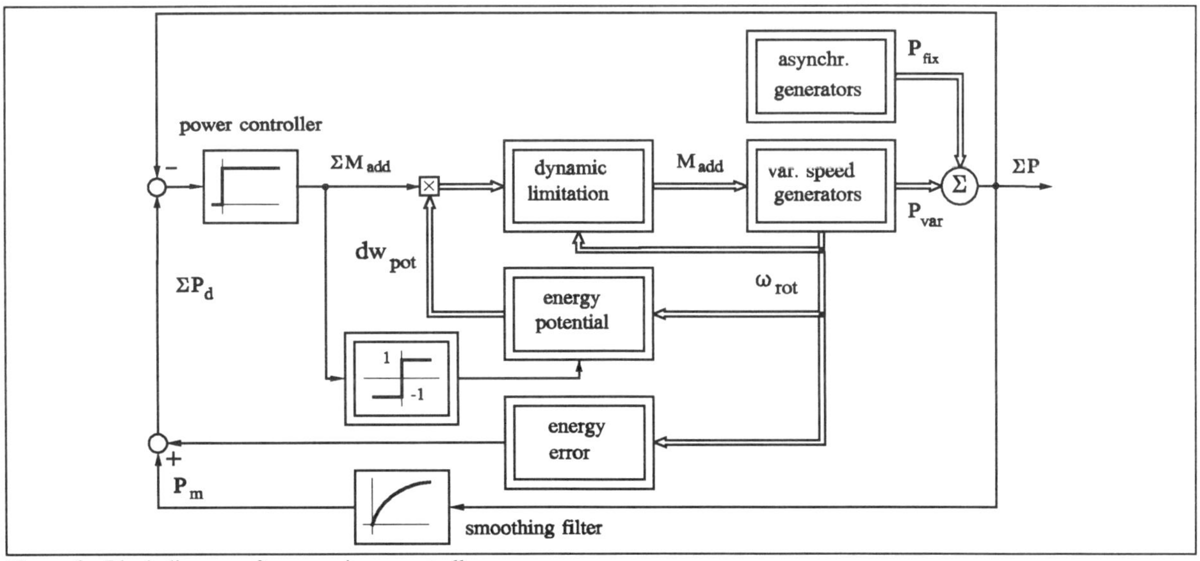

Figure 6 Block diagram of a supervisory controller

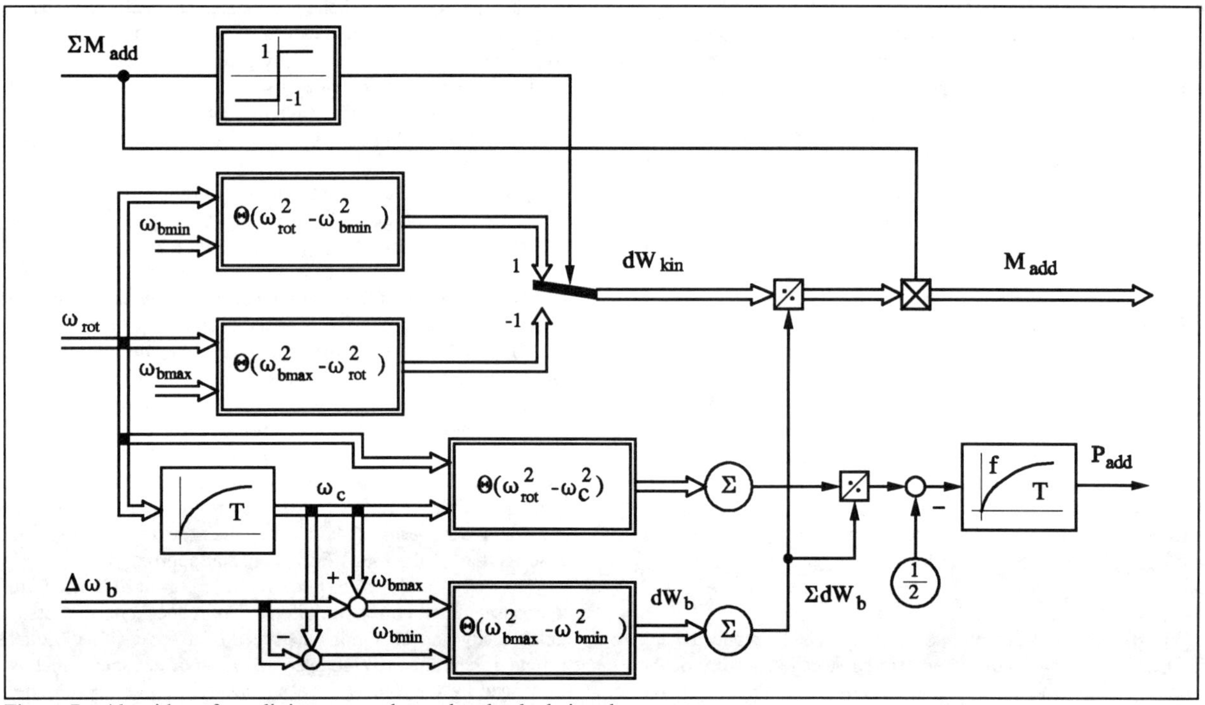

Figure 7 Algorithms for splitting torque demand and calculating the energy error

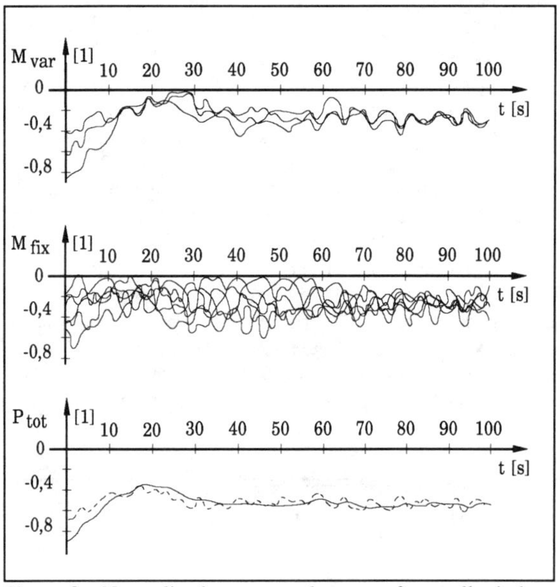

Figure 8 Normalized power total output of a small wind park with (—) and without (- -) supervisory control

This kind of control even allows a compensation of power fluctuations caused by fixed speed turbines. Figure 8 depicts the simulated power output of a windfarm of six fixed and three variable speed turbines all of equal power. The resulting output shows a significantly reduced power variation by approx. 40% although the total energy capture is slightly higher compared to the uncontrolled case. The efficiency was found to be almost independent of the wind direction and showed best performance at low wind speeds.

5 SUMMARY

The paper demonstrates with simulations the effects that can be gained from a supervisory controller for the variable speed wind turbines in a wind park. The total performance of the park can be improved to lower power fluctuations or to achieve a small additional energy capture at low wind speeds. These two goals are found to be contradictory but some acceptable compromise can be found. A combination with fixed speed generators is possible though at reduced performance. Additional slow torque variations of the variable speed turbines can not be avoided.

6 ACKNOWLEDGEMENTS

The work was part of a project founded by the German Federal Ministry of Research and Technology under contract no 032-8265-B. The author is responsible for the content of this publication.

7 REFERENCES

[1] Kaimal, Wynguard : Spectral Characteristics of Surface-Layer Turbulence, Quarterly Journal of the Royal Meteorological Society, Vol 98, S.563-589, 1972

[2] Wefelmeier K., Leonhard W., et. al.: Elektrische Energieversorgung mit Windkraftanlagen, BMFT final report, 1992

[3] Katić I., Højstrup J., Jensen N.O.: A Simple Model for Cluster Efficiency, Proceedings EWEC Rom 1986, H.S. Stephens & Associates, Bedford, 1986

[4] Ernst-Cathor J.: Drehzahlvariable Windenergieanlage mit Gleichstromzwischenkreis-Umrichter und Optimumsuchendem Regler, Dissertation, TU Braunschweig, 1987

[5] Atlas F.: Einfluß einer variablen Rotordrehzahl auf die Energieausbeute einer Windkraftanlage, Diplomarbeit, Institut für Regelungstechnik, TU Braunschweig, 1990

WIND AND WAVE LOADING OF OFFSHORE WIND TURBINES

D C Quarton and M A Wastling Garrad Hassan and Partners Ltd., UK
T E Schellin Germanischer Lloyd, Germany

ABSTRACT: Wind and wave loading of offshore wind turbines was considered as part of the Joule I study (JOUR 0072) of offshore wind energy in the European Community. The purpose was to compare the loading and response of machines in offshore and land based environments and hence determine the extent to which existing design guidelines should be modified for offshore applications. The paper describes methods used in the study and discusses results in terms of their implications for offshore wind turbine design.

1. INTRODUCTION

There have been a number of studies carried out, notably in the UK, Denmark, the Netherlands and Sweden, to investigate offshore wind energy. These studies have been in the form of outline design and costing and available resource assessment. More recently the Danes and the Swedes have taken this work one stage further with the construction of offshore prototypes; a wind farm of eleven Bonus 450kW machines located some 3km from the Danish coast at Vindeby and a 200kW WindWorld machine installed at 250m offshore of the harbour of Nogersund in the south of Sweden. These machines are sited in relatively shallow water.

The potential for offshore wind energy in Europe is of considerable interest to the European Commission. Germanischer Lloyd and Garrad Hassan were funded under the Joule I Programme to investigate this potential. In addition to EC funding, Germanischer Lloyd involvement in the project has been supported by BMFT and the Garrad Hassan work funded by ETSU for the UK Department of Trade and Industry. The project has had four main objectives:

- To establish the feasibility and potential for offshore wind energy in Europe;
- To review existing information on offshore construction, erection and maintenance with regard to offshore wind farms;
- To provide design guidelines for the design of offshore wind turbines;
- To investigate the impact of combined wind and wave loading on machine loads and dynamic response.

The work reported in this paper is concerned principally with the last of these objectives.

2. ANALYTICAL TOOLS

Garrad Hassan has invested considerable effort in the development of analytical tools which are capable of investigating the effects of both wind and wave loading of offshore wind turbines. This work has involved the extension of well validated in-house methods [1] for dealing with land based machines to enable analysis of wind turbine response due to hydrodynamic loading of the offshore support structure. Two separate and equally important methods of analysis have been developed as described below.

2.1 Time Domain Analysis

A model of the complete offshore wind turbine was developed based on a modal description of the coupled rotor-tower structural dynamics. The modal dynamics of the coupled system are formulated in the time domain as a series of differential equations. The rotor aerodynamics are solved using standard blade element theory, and the model also incorporates representations of the wind turbine power train and pitch control system.

The loading and behaviour of the wind turbine may be calculated in response to both periodic and turbulent aerodynamic loading of the rotor and wave loading of the support structure. The wind field supplied to the model is made up of a large number of turbulent wind speed time histories arranged on a grid having dimensions larger than the rotor disk. The turbulent wind speed time histories may have user-specified spectral and spatial correlation characteristics although for this project von Karman properties [2] were assumed. A mean wind speed, shear profile, yaw misalignment and tower shadow are superposed on the turbulent wind field.

Wave load time histories are applied to appropriate nodes on the support structure. Any form of wave load time history, including that due to a breaking wave, can be applied to the model.

The model, which represents a development of the Garrad Hassan code BLADED, was commissioned and used principally for analysis of the extreme wave loading of offshore wind turbines. A schematic indicating the various components of the time domain analysis is shown in Figure 1.

2.2 Frequency Domain Analysis

A standard approach to the loading analysis of offshore structures is to model the variation of wave height for a particular sea state as a spectrum and apply this to a frequency domain representation of the structural dynamics. This approach is particularly valuable in the context of fatigue analysis of offshore structures.

Over recent years Garrad Hassan have developed and validated a frequency domain treatment of the aerodynamic loading of wind turbines. This treatment, described in detail in [1], was extended for this project to enable wave loading spectra to be applied to a model of the support structure coupled in a simple way to the rotor dynamics. The approach is based on a linear representation, it does not permit a complicated model of the support structure dynamics and also requires wind and wave loading to be acting in the same direction. The importance of these simplifications was judged by comparison of predicted loading with results obtained from time domain analysis. With little loss of accuracy, the frequency domain method offers significant advantages in terms of easing and speeding fatigue loading calculations required for offshore wind turbines.

The frequency domain tool is an extension of the Garrad Hassan code TURBLOAD and is presented in schematic form in Figure 2.

3. WIND AND WAVE LOADING

The correlation of the wind and wave loads experienced by offshore wind turbines is clearly of importance to the machine designer. A number of methods have been investigated for combining these two forms of load and details of the approach taken in the present study are described below.

3.1 Fatigue Loads

Fatigue loading of the operational offshore wind turbine is affected by the combination of cyclic and random aerodynamic loads experienced by the rotor and by the random hydrodynamic loads acting on the support structure. The aerodynamic loads are associated with periodic variations in wind speed due to wind shear, tower shadow, yaw misalignment etc. and random variations associated with turbulence. For a frequency domain analysis of wind loads, the turbulent wind field is represented by an expression for the temporal and spatial cross correlation of longitudinal wind speed fluctuations. The cross correlation function, based on the von Karman model [2], is transformed to a frame of reference moving with the rotor blades.

There are many possible methods for calculating wave loads for fatigue assessment of offshore structures. The methods divide into the two broad categories of deterministic, regular wave analysis and spectral random wave analysis. For the purposes of the present study the spectral approach was adopted with the assumption of linear wave theory.

There are a number of wave spectral formulae in common use for offshore design analysis. The Pierson-Moskowitz [3] model was adopted for the present study and defines the auto-spectrum of wave height as follows:

$$S_{\eta\eta}(\omega) = \frac{4\pi^3 H_s^2}{T_z^4 \omega^5} \exp\left[-\frac{1}{\pi} \left[\frac{\omega T_z}{2\pi} \right]^{-4} \right]$$

The parameters H_s and T_z are known as the significant wave height and zero crossing period, respectively, and characterise the sea state. The circular wave frequency ω is in units of rad/s.

With the wave height spectral shape defined as indicated above, the problem remaining is one of correlating the significant wave height and zero crossing period for a particular sea state with the mean wind speed at the site. Neumann and Pierson [4] presented an empirical relationship between sea state parameters and mean wind speed as follows:

$$H_s = 0.0212\ U^2$$

$$T_z = 0.81\ (2\pi U/g)$$

Here U is the mean wind speed at the site.

These relationships allow the sea state to be correlated with the mean wind speed, and hence a series of fatigue load cases may be defined in terms of the appropriate combination of wind and wave loading of the wind turbine.

The wind shear, turbulence intensity and integral length scale of turbulence for a particular load case are related to the sea state through the surface roughness. Over land the roughness length is specific to the type of terrain and independent of wind speed. Over the sea, the roughness length is not directly related to the wave height although this might be thought to be a logical extension of the over land case, but rather relates to small wavelets or ripples on the surfaces of those waves. The wavelets or ripples depend on wind speed and, therefore, the roughness length at sea is itself a function of wind speed.

The wave load auto- and cross-spectra at specified nodal points on the wind turbine support structure were obtained from the wave height auto-spectrum by means of calculations of the form:

$$S_{FF}(\omega) = H_F(\omega)\ H_F^*(\omega)\ S_{\eta\eta}(\omega)$$

where S_{FF} is the one-sided wave load spectrum, H_F is the complex vector transfer function of wave load at a nodal point, H_F^* is the transposed complex conjugate of H_F, $S_{\eta\eta}$ is the wave height auto spectrum and ω the circular wave frequency in rad/s.

Transfer functions of the wave load at nodal points on the support structure were calculated on the basis of the Morison formula. This formula is generally accepted to be valid for predicting wave loads on hydrodynamically transparent structures. To calculate the force on a vertical pile in a wave field, Morison et al. [5] proposed an approach in which the horizontal force is expressed as the sum of a drag force and an inertia force. The drag force is non-linear and caused by viscous effects and the associated downstream wake. The linear inertia force results from the Froude-Krylov and hydrodynamic mass forces. Drag and inertia coefficients of C_D = 1.25 and C_M = 2.0 were assumed for the calculations in accordance with [6]. Wave particle velocities were calculated according to linear (Airy) small amplitude wave theory.

The horizontal wave loads acting on the support structure are not harmonic functions due to the effects of the non-linear drag term. However, for the purposes of the study, equivalent harmonic loads were obtained based on equalising the impulse over one wave period.

Accurate results can be expected with spectral wave analysis only for relatively small wave heights. Since waves of small to moderate heights predominate in the make-up of the collection of cyclic wave loadings, use of the spectral approach for the fatigue analysis was felt to be justified.

3.2 Extreme Loads

Support structures for offshore wind turbines will inevitably experience large wave loads as well as wind loads. The problem exists, therefore, of how to combine wind and wave loads to predict an overall extreme condition. Furthermore, there is uncertainty in how to calculate the extreme wave loading of the support structure.

To explore the sensitivity of the result to the calculation procedure and to the assumptions made in the analysis, extreme loads were calculated in a variety of different ways. Two basic approaches were adopted:

- A deterministic, design extreme wave was applied simultaneously with an extreme gust wind load. Various methods for predicting the magnitude of the extreme wave and extreme gust that can be applied at the same instance were considered.
- A design extreme sea state was combined with an extreme hourly mean wind speed and associated wind turbulence. The extreme response was predicted using both time and frequency domain calculations.

Both approaches were pursued with various levels of analytical sophistication. The impact of structural dynamic effects on the combined extreme loads was of particular interest.

4. EXAMPLE CALCULATIONS

The analytical methods described above were used to investigate the combined wind and wave loading of three reference offshore wind turbines. The purpose of the analysis was to compare the loading and response of machines in offshore and land based environments.

4.1 Reference Machines and Site Characteristics

The three reference wind turbines considered in the study were:

(a) The two-bladed, 100m diameter, 6MW machine designed by Wind Energy Group (WEG) Ltd as part of the Offshore Wind Energy Assessment undertaken for the UK Department of Energy. The machine design considered in the present work is that reported in [7]. The rotor is teetered and has full span pitchable blades constructed from wood/epoxy composite. The rotor and nacelle are supported on a cylindrical steel tower standing on a concrete caisson.

(b) A three bladed, 80m diameter, 3.8MW machine mounted on a tripod tower support structure designed by Dutch Heerema bv, refer [8]. The rotor blades were assumed to be manufactured from grp and are pitch regulated.

(c) A three bladed, stall regulated 35m diameter, 450kW machine mounted on a cylindrical concrete tower with a concrete caisson. This support structure was designed by Renewable Energy Systems (RES) Ltd as part of an offshore wind turbine installation feasibility study [9] funded by the UK Department of Energy.

For the purposes of the study the reference wind turbines (a) and (b) above were assumed to be located at offshore sites with the following characteristics:

- Annual mean wind speed 8.3m/s
- Rayleigh distribution of hourly mean wind speeds
- 50 year return 3s gust wind speed 47m/s
- Mean ambient turbulence intensity 10.8%
- Mean water level 25m

For wind turbine (c), the site characteristics assumed were those taken by RES Ltd for their feasibility study [9]. The main difference in this case was a reduced water depth of 12m.

4.2 Fatigue Loads

The time domain analysis provided by BLADED was compared with the frequency domain approach of TURBLOAD for a series of calculations involving both wind and wave loading of the WEG offshore wind turbine. Typical results of these calculations are presented in Figure 3 in terms of the flatwise bending moment at the blade root.

The blade load presented in Figure 3 is that corresponding to operation of the machine at a mean wind speed of 10m/s and in a turbulent wind field of turbulence intensity of 21% appropriate to wind farm operation. In the first case wave loading was not included in the calculation. The flatwise root moment auto spectrum is characterised by loading at low frequencies and, as a result of the rotor being teetered, at even harmonics of the rotational speed.

Also shown in Figure 3 is the flatwise root moment resulting from wave loading applied to the support structure. In this case all sources of applied wind load were omitted and only aerodynamic damping loads were retained. The wave loads are those associated with a sea state appropriate to a mean wind speed of 10m/s. In the case of the frequency domain analysis the wave loads are applied in spectral form to the model of the wind turbine, whereas in the case of the time domain calculations the wave load spectrum is first transformed into an equivalent time history. The blade bending moment resulting from the wave loading of the support structure is dominated by load at the tower fundamental natural frequency (0.35Hz) and is considerably smaller than that associated with wind loading.

An interesting feature of the results shown in Figure 3 is the spectral peak at 0.64Hz indicated in the BLADED prediction of blade load which is completely absent from the TURBLOAD results. This peak arises because of nodding of the tower head through a mechanism which is not modelled by the frequency domain analysis. With the exception of this spectral peak, the time and frequency domain results shown in Figure 3 are in good agreement providing confidence in the two analytical methods.

Figure 4 presents the bending moment at the tower base for wind loading only and wave loading only as described above. In the case of the tower base moment, the wind loads and wave loads are comparable in magnitude.

4.3 Extreme Loads

Wave load time histories for use in the extreme loading calculations were based on the Morison formula. In this case wave particle kinematics were obtained using stream function theory which gives reliable results for waves with heights close to the breaking limit. Wave forces on the support structure were assumed to act up to the calculated wave profile. The resulting horizontal forces on the wind turbine support structure were generally not harmonic functions.

Results of example load calculations are shown in Figures 5 and 6. These calculations were based on a deterministic design wave train applied in combination with turbulent wind loading. The figures show sample time histories and provide a qualitative comparison of the relative importance of wind and wave loading.

The calculations were based on a fault condition in which the machine was shut down and the blades set a pitch angle of zero degrees. The site conditions were as follows;

- hub height hourly mean wind speed 30m/s (1 year return value)
- hub height turbulence intensity 8.5%
- wave period 9s
- wave height 14m

An examination of the extreme loads indicated in Figures 5 and 6 suggests that at the blade root the impact of wind loading is considerably greater than that of wave loading. The extreme load at the tower base contains significant components of both wind and wave loading.

5. CONCLUDING REMARKS

Analytical methods based on both time and frequency domain approaches have been developed for calculation of the combined wind and wave loading of offshore wind turbines. These methods have been validated by means of mutual comparison of predicted results for wind and wave loading of a two-bladed, 100m diameter offshore machine.

A number of methods have been investigated for combining wind and wave loads for offshore machines. In the case of fatigue loads, the approach adopted is based on spectral random wave analysis in conjunction with a spectral model of wind turbulence. The correlation of the wind and wave loading involves use of an empirical relationship between sea state parameters and the mean wind speed. The combination of wind and wave loads to predict an overall extreme condition has been examined on the basis of both deterministic and spectral methods.

On the basis of results obtained in the study to date it appears that rotor fatigue and extreme loads are largely unaffected by hydrodynamic wave loading of the wind turbine support structure. Both wind and wave loads are, however, important for the design of the support structure itself.

REFERENCES

(1) Quarton D C, Wastling M A, Garrad A D and Hassan U. The calculation of wind turbine loads. A frequency or time domain problem? British Wind Energy Conference, Nottingham, March 1992.

(2) Engineering Sciences Data Unit. Characteristics of atmospheric turbulence near the ground. Part II: Single point data for strong winds, ESDU 74301, 1974.

(3) Pierson W J and Moskowitz L. A proposed spectral form of fully developed wind seas based on the similarity theory of SA Kitaigorodshii. J Geophys. RES., Vol. 69.

(4) Neumann G and Pierson W J. Principles of physical oceanography. Prentice-Hall, 1963.

(5) Morison J R, O'Brian M P, Johnson J W and Schaaf S A. The force exerted by surface waves on piles. Pet Trans AIME, TP 2846, 1950, 189.

(6) Department of Energy Guidance on Design and Construction of Fixed Offshore Structures, 4th Edition, 1990, London: HMSO.

(7) Simpson P B and Kuhn U. A re-appraisal of the cost of UK offshore wind energy. British Wind Energy Conference, Swansea, 1991.

(8) Van Iperen J. Duurzame Energie 2050 Windenergie Buitengaats. Project No. P-0096, Fugro-McClelland and Eng. bv, Leidschendam, 1987.

(9) Gardner P, Morgan C A, Noakes J anf Ffrench R G. Feasibility study for an offshore wind turbine. British Wind Energy Conference, Nottingham, March 1992.

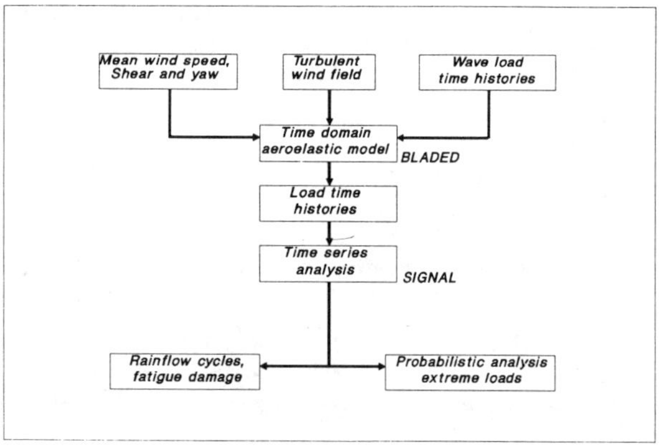

Figure 1 Time domain approach

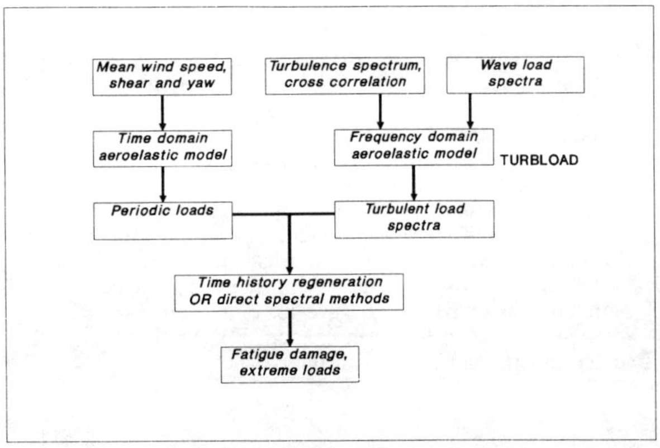

Figure 2 Frequency domain approach

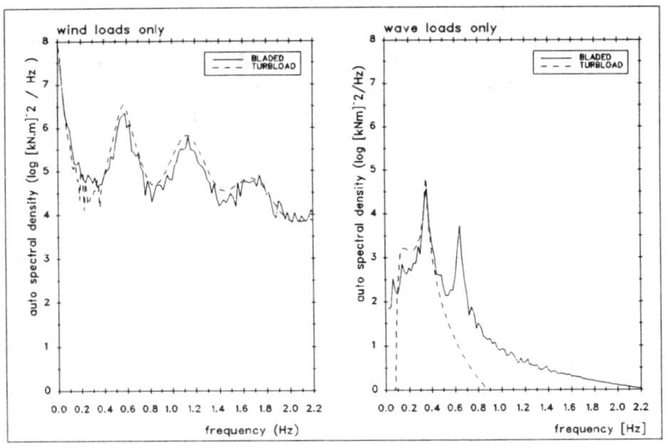

Figure 3 Fatigue loads – flapwise bending moment at blade root

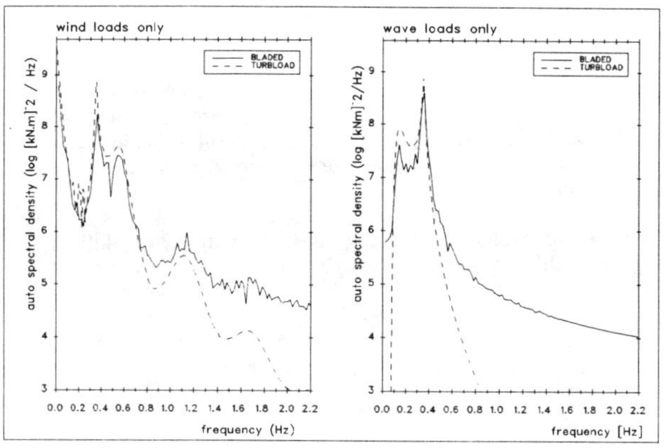

Figure 4 Fatigue loads – Bending moment at tower base

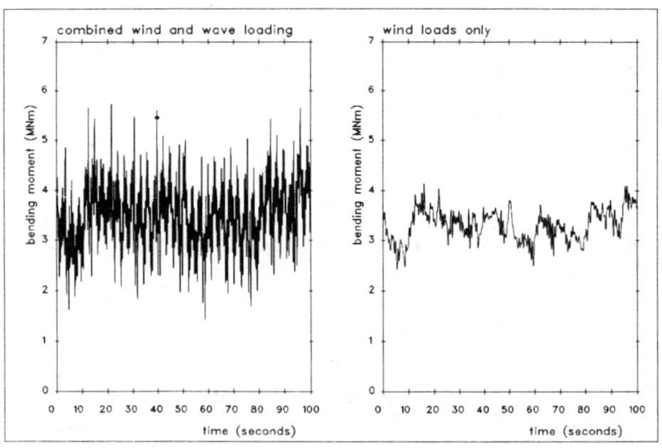

Figure 5 Extreme loads – flapwise bending moment at blade root

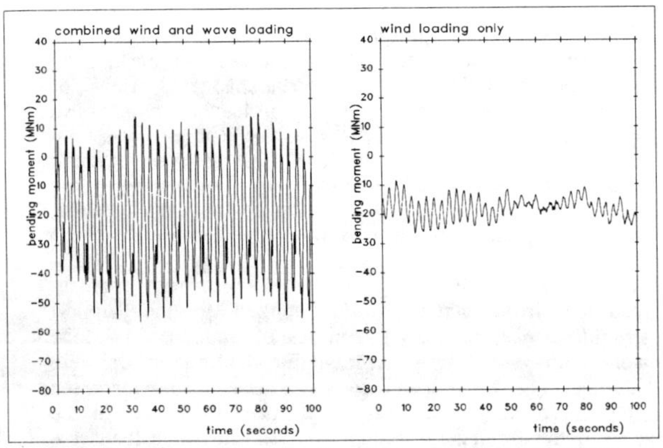

Figure 6 Extreme loads – Bending moment at tower base

AN EVALUATION OF MEASURED AND PREDICTED FATIGUE LOADS FOR THE TJÆREBORG WIND TURBINE.

Kurt S. Hansen, B. Maribo Pedersen, Stig Øye
Department of Fluid Mechanincs, Technical University of Denmark
DK-2800 Lyngby, Denmark

Peggy Friis
ELSAMPROJEKT A/S, DK-7000 Fredericia, Denmark

ABSTRACT: The Tjæreborg Wind Turbine 60m / 2MW is a prototype, built for demonstration and R&D purpose in the Danish wind energy programme. During the latest two years of operation an intensive measuring campaign has been carried through. The wind turbine duty cycles have been determined and based on these the cumulative spectra have been established for some of the most important structural loads like edgewise and flatwise bending moments, main shaft torque, tilt and yaw moments.

1. INTRODUCTION

1.1 The wind turbine

The Tjæreborg Wind Turbine is a horizontal axis-upwind turbine with a rotor diameter of 61.1m located in the western part of Jutland near Esbjerg. The rotor is connected to a 2 MW enlarged slip induction generator through a 3-stage planetary gearbox. Above rated wind speed, power is regulated by electrohydraulic full-span pitch control. The concrete tower supports the nacelle and the rotor at a height of 60 m (hub height) above ground. The rotor axis is tilted 3° with respect to the horizontal. The rotor runs at a constant speed of 22.3 rpm which gives a blade tip speed of 71 m/s.

Status for the wind turbine at the beginning of 1993 is an obtained availability of 79.3% during two years (1991-1992).

1.2 The measurement system

In a central computer 10-min averages of all channels (both structural and meteorological) are recorded and stored together with the standard deviation and max/min values for the 10-min period. These 10-min values are used to determine the duty cycles of the wind turbine.The data recordings for the fatigue evaluation are based on

i) campaign measurements which are recorded on command with a length of 10 minutes.
ii) capture matrix measurements recorded automatically with a duration of 3 min. These recorded timeseries more or less cover the whole operational range (wind speed interval [5 - 25 m/s] and turbulence intensity [0 - 25%]).

2. THE WIND TURBINE DUTY CYCLES

An operation period from the 1 of november 1990 to 31 of october 1992 has been monitored intensively and based on the operational statistics and the recordings from the measurement system [basically 10-min average values] the operational behaviour for this period is summarized in table 1.

Table 1 shows an obtained wind turbine availability of 79.3% and a data availability of 64% from the measurement system during the monitoring period.

Based on table 1 it is possible to estimate the duty cycles for the wind turbine eg. period with gridconnected operation, period with idling and number of start and stop sequences.

Table 1 : Recordings for the Tjæreborg Wind Turbine, based on two years of regular operation.

Operational statistics	
Turbine availability	13918 Hours
Gridconnected operation	9756 Hours
Idling	3740 Hours
Start/ stop sequences	422 Hours
Gross production	6636 MWh
Number of grid connections	2529
Recordings from measurement system	
Availability of meas. system	14788 Hours
Idling	2194 Hours
Generator online	6282 Hours
Generator online, ti[1] ≤ 5%	1048 Hours
Generator online, 5% <ti≤ 10%	3338 Hours
Generator online, 10% <ti≤ 15%	1647 Hours
Generator online, 15% <ti≤ 20%	228 Hours
Generator online, 20% < ti	21 Hours
Gross production	4322 Mwh

[1] ti = turbulence intensity [%]

This duty cycles determination has been done in accordance to the recommendations [1]. The estimated numbers of hours covering a whole year of operation with 100% wind turbine availability are given in table 2. This table express the operational duty cycles during the phase where the measurement programme has been in operation, but does not express the expected duty cycles for the turbine in regular operation [without operational measurement programme].

Table 2 express the wind speed distribution during gridconnected operation referring to 100% availability.

Table 2.: Annual operational hours and numbers based on 100 % availability

Modes	Wind Speed bin [m/s]												Sum
	0-5	5-7	7-9	9-11	11-13	13-15	15-17	17-19	19-21	21-23	23-25	> 25	
Idling [hours]	1677	527	48	30	15	11	7	17	19	8	1	0	2360
Start [numbers]	420	594	246	130	103	55	34	12	0	0	0	0	1594
Stop Normal [numbers]	787	261	186	145	101	41	25	25	15	1	0	0	1587
start/stop [Hours]	101	72	36	23	17	8	5	3	1	0	0	0	267
Operation [hours] 0% < ti ≤ 5%	4 ·	301	372	219	84	32	10	4	1	0	0	0	1027
Operation [hours] 5% < ti ≤ 10%	14 ·	779	995	627	394	222	152	73	5	2	1	0	3265
Operation [hours] 10% < ti ≤ 15%	9 ·	370	461	423	237	74	24	11	2	0	0	0	1611
Operation [hours] 15% < ti ≤ 20%	2 ·	53	69	56	31	8	3	1	0	0	0	0	223
Operation [hours] 20% < ti	1 ·	6 ·	6	5	2	1 ·	0	0	0	0	0	0	21
Operation [hours]	30 ·	1508	1903	1330	748	337	189	89	8	2	1	0	6146

· lack of measurements

3. SIMULATED LOADS

The simulated loads have been calculated by use of the aerolastic code FLEX4 developed at the Fluid Mechanics Department [2], which is a time domain model of the flexible wind turbine where the mean wind field over the rotor plane includes wind shear, yaw and tower shadow. The turbulent part of the wind is included in the model as timeseries of simulated turbulence. The simulated loads are identical to those used in the JOULE project REFSTRESS [3] and covers gridconnected operation in 10 wind speed bins ranging from 5-25 m/s, each with a constant turbulence intensity of 10%.

The length of the simulated timeseries are 200 sec. and these are in the fatigue evaluation weighted according to the operational hours for the turbulence intensity interval ranging from 5 - 15%

4. MEASURED LOADS

The measured loads are mainly recorded during a 2 year period starting from the beginning of 1991 to the end of 1992 and consist of 3 different types of timeseries:

i) Capture matrix timeseries with turbine grid connected, duration = 80 × 180 seconds ≈ 4 hours
ii) On command recorded timeseries with turbine grid connected, duration = 60 × 600 seconds ≈ 10 hours
iii) On command recorded timeseries, with turbine in idling mode, in start or stop sequences, duration = 38 × (150-600) seconds ≈ 4 hours

This means that the fatigue analysis is based on almost 18 hours of measurements recorded at highly different conditions.

All these timeseries are validated and characterized according to the duty cycles given in table 2, and the recorded timeseries fit into this table with only a small lack of data (indicated with small dots).

In this investigation the following sensors are analyzed:

1) Edgewise bending moment, fixed to blade, radius = 2.75 m, which is in-plane-bending moment (at pitch=0°).
2) Flatwise bending moment, fixed to blade, radius = 2.75 m, which is out-of-plane bending moment (at pitch=0°).
3) Torque in main shaft.
4) Tilt moment acting at the front bearing.
5) Yaw moment acting at the front bearing.

5. LOAD SPECTRA CALCULATIONS

Each of the simulated and measured timeseries is processed with a "rain flow" counting method and this results in a large number of Markov - matrices, one for each sensor and each timeseries (method described in [1]).

The cumulative spectrum for each of the main components are found by weighing and adding together the results from each of these Markov matrices (in this process all the load level information is lost).

5.1 Comparison of simulated and measured loads.

The measured data covering the turbulence interval ranging from 5 - 15 % and wind speed range from 5 - 25 m/s represent 30 × 4875 of grid connected hours without idling and start / stop sequences. These 30 × 4875 hours represent 79% of the expected hours with gridconnected operation during 30 years.

The simulated loads (10 × 200 seconds) represent 30 × 4875 hours of gridconnected operation with a constant turbulence intensity of 10% in the wind speed range from 5 to 25 m/s.

The cumulative simulated and measured load spectra are compared and shown in Figure 1 through 5.

The results in Figure 1 agree well as expected, as this load spectrum is dominated by gravity.

The measured flatwise bending moment, Figure 2 contains a small number of large ranges compared to the simulated, which can be explained by the variance in the turbulence intensity [5-15%] for the measured data compared to the simulated with a constant turbulence intensity of 10%.

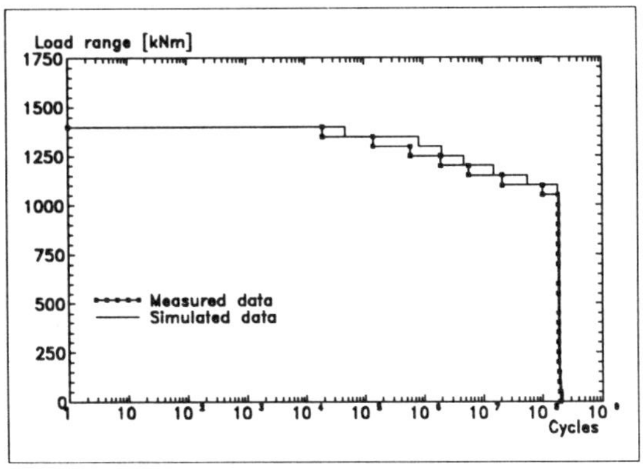

Figure 1 : Simulated and measured cumulative edgewise bending moment, r = 2.75 m

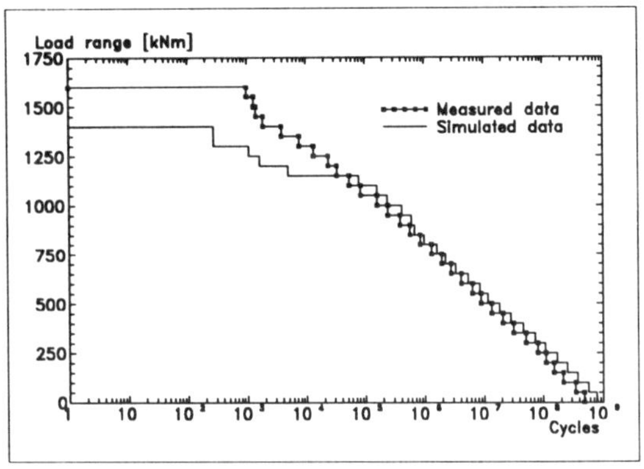

Figure 2 : Simulated and measured cumulative flatwise bending moment, r = 2.75 m

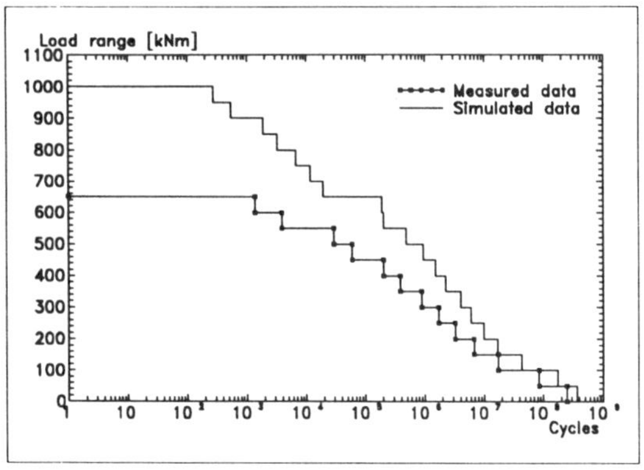

Figure 3 : Simulated and measured cumulative main shaft torque.

The simulated mains shaft torque, Figure 3 show larger load ranges compared to the measured. This difference can partly be explained by a difference in the assumed linear generator slip (2.0% at rated power) compared to the actual nonlinear generator slip effectively of 2.8% at rated power. The result is also affected by a conservative choice of turbulence length scale and coherence.

The calculated tilt and yaw moments are somewhat below the measured values for the higher load ranges, i.e. high wind speeds. This could be a result of the above mentioned "conservative" choice of coherence being nonconservative for these loads. Furthermore, the in-plane turbulence components were neglected, which means that no winddirection changes were simulated. The importance of this is expected to increase with the wind speed.

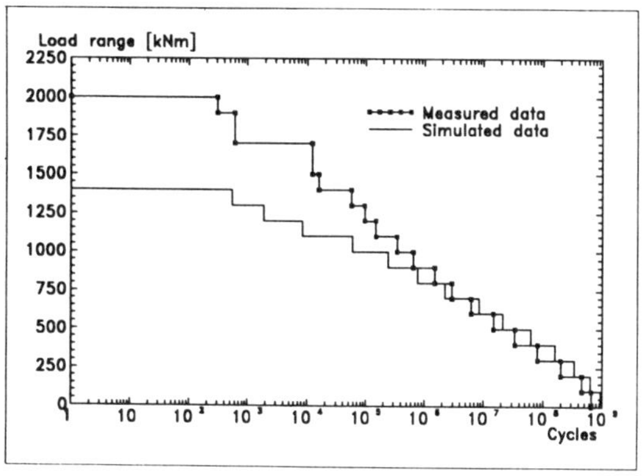

Figure 4 : Simulated and measured cumulative tilt moment acting at the front bearing.

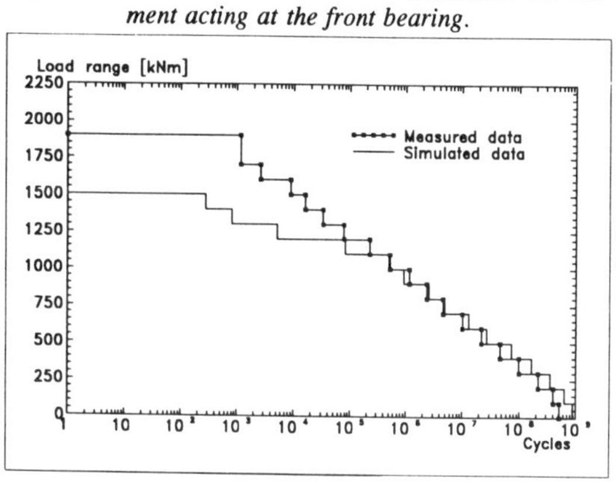

Figure 5 : Simulated and measured cumulative yaw moment acting at the front bearing

5.2 Evaluation of the measured loads.

The measured loads represent the area with gridconnected operation as defined in table 2 (= 30 x 6146 hours) (all wind speed ranges and all turbulence levels). The cumulative spectra are determined for gridconnected operation alone and gridconnected operation including idling and start / stop sequences.

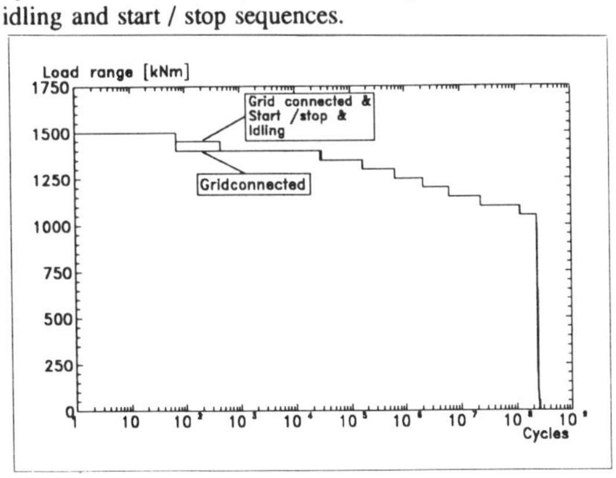

Figure 6 : Measured cumulative spectra for the edgewise bending moment, r = 2,75 m.

The start / stop sequences do not influence the cumulative spectrum for the edgewise bending moment.

The flatwise cumulative spectrum, Figure 7 shows a low number of larger ranges which are caused by the stop sequences distributed over the medium wind speed range from 12 - 18 m/s. The contribution from the idling mode (2360 hours/yr,pitch = 55°) is marked on Figure 7.

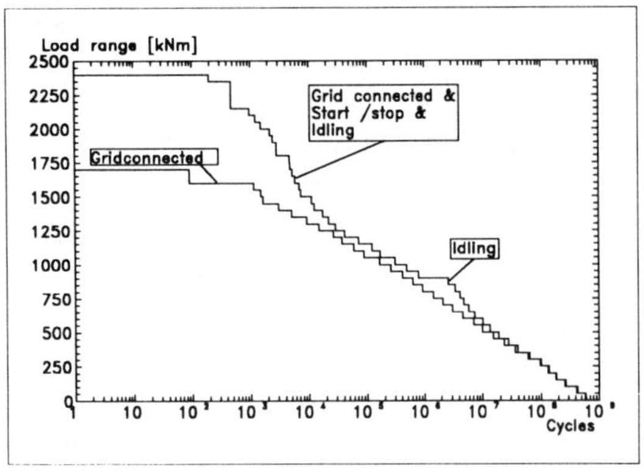

Figure 7 : Measured cumulative spectra for the flatwise bending moment, r = 2,75 m.

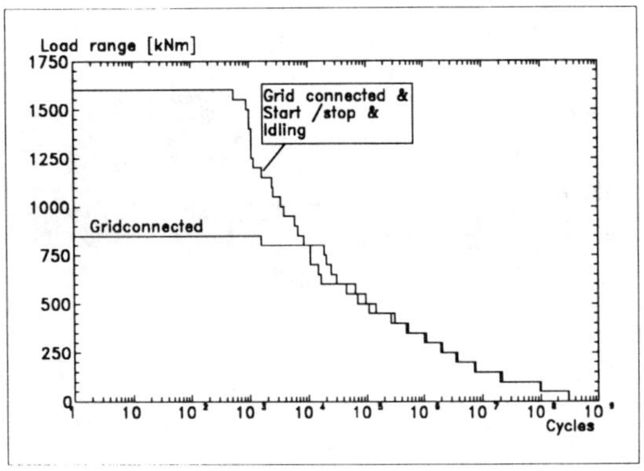

Figure 8 : Measured cumulative spectra for the main shaft torque.

The cumulative spectrum for the main shaft torque, Figure 8 shows again a few larger ranges, which are caused by the stop sequences.

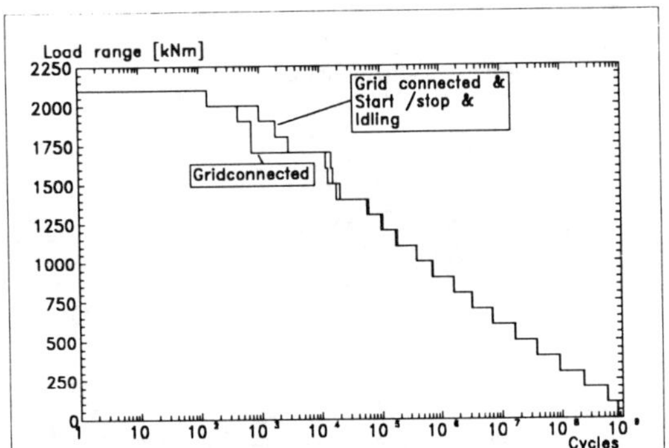

Figure 9 : Measured cumulative spectra for the tilt moment acting at the front bearing.

The cumulative spectra for both tilt and yaw moments, Figure 9 & 10 are not influenced by the idling mode or by the start / stop sequences.

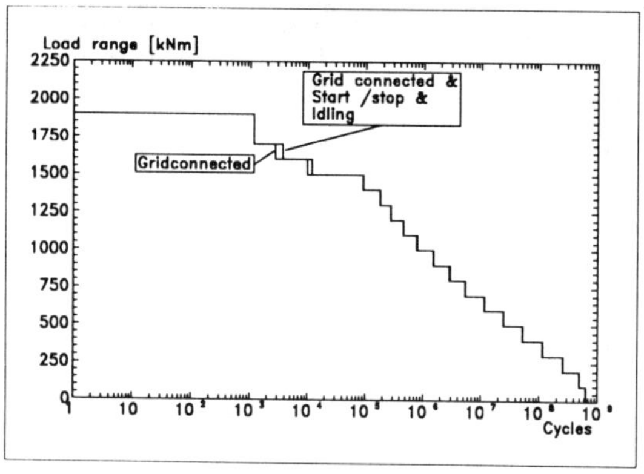

Figure 10 : Measured cumulative spectra for the yaw moment acting at the front bearing.

6. CONCLUDING REMARKS

The cumulative load spectra shown in the figures 6 - 10 must be considered as representative for the Tjæreborg Wind Turbine as they are based on more than 18 hours of measurements recorded during different conditions.

The large number of start and stop sequences in the whole wind speed range form 5 - 25 m/s are not unusual for at prototype wind turbine, whereas the start and stop sequences will be grouped around cut-in and cut-out wind speed for a commercial wind turbine.

The simulated loads used in the REFSTRESS project [3] are generally in good agreement with the measurements except for the main shaft torque, which overestimates the load spectra mainly because of the modelling of the generator slip and the turbulence length scale.

7. REFERENCES

[1] 3. Fatigue Loads, 2. ed. 1990
Recommended Practices for Wind Turbine Testing and Evaluation.
IEA Recommendation

[2] S.Øye (tech. Univ. of Denmark)
Simulation of loads on a Wind Turbine
20th Meeting of Experts - Wind Characteristics of Relevance for Wind Turbine Design
Stockholm, March 7-8, 1991

[3] Reference Procedures to establish fatigue stresses for Large Size Wind Turbines (REFSTRESS)
Programme Joule (1989-1992) JOUR/0083
by H.J. van Grol,
Netherlands Energy Research Foundation, ECN

TURBULENCE CHARACTERISTICS OF RELEVANCE TO WIND TURBINES IN COMPLEX TERRAIN

D. Founda[1], G. Glinou[2], A. Kotronaros[2], M. Petrakis[3],

D. Asimakopoulos[1] and I. Antoniou[2]

1. Univ. of Athens, Lab. of Meteorology, 33 Ippokratous Str., 10680 Athens
2. Centre of Renewable Energy Sources, 19Km Av.Marathon, 19009 Pikermi, GR
3. National Observatory of Athens, Lofos Nymphon, 11810 Athens, GR

Abstract: Turbulence measurements on top of a hill in Andros island (Greece) are presented. The area combines strong winds, ground complexity and surface roughness. Mean turbulence parameters are influenced by both topography and a wind park established on the hill. Power spectra were derived and compared with the Kaimal-Teunissen and the Von karman formulations. The latter formulation gave a better overall fit to our measurements. The study indicated that power spectra are affected by terrain characteristics and by the wind park presence.

1. INTRODUCTION

The structure of turbulence over flat, homogeneous surfaces, and under various atmospheric conditions is sufficiently understood nowadays, mainly through experimental campaigns, results of which are reported elsewhere. The paper published by Kaimal et al., (1972), based on data from the Kansas experiment, is a useful tool in the field. However, in recent years the efforts to describe the turbulent boundary layer have focused on more complex surfaces (varying topography and roughness). The study of these types of terrain is very important when dealing with problems of pollution transport, wind energy conversion systems siting, etc. The literature however, remains poor in this field. In this paper, some characteristics of the horizontal and vertical turbulence structure over the Kalivari hill are presented. Kalivari hill is located at the Northern part of Andros island (Greece) and is surrounded by complex terrain.

2. DESCRIPTION OF THE SITE

The site of Kalivari (Fig.1) in Andros island was chosen for the measurements campaign. The main criteria for this selection included high wind speeds, complex terrain with a well defined hill and a reliable historical record of anemological data, indicating mean wind speeds of about 9.5 m/s. The Kalivari site is the highest point (altitude: 335m) in the northernmost part of the island at a 2.5 Km distance from the sea. It is worth noting that the prevailing wind direction is N - NNE and the sea extends uninterrupted for over 50 Km. The slope of the hill is of the order of 15% and the surrounding terrains are very complex and rough with a roughness height of approximately 10-20 cm. A wind park of 1.5 MW is located and operating at the site since the Spring of 1992.

3. INSTRUMENTATION - DATA ACQUISITION

Four meteorological masts of 10m height were installed to the N,E,S and W arround the hill, equipped with vanes and cup anemometers. Three 30m masts were erected on the summit (M16) and also upwind (M15) and downwind (M17) of the hill. In Figure 1 the positions of the three 30m masts as well as the positions of the nearest wind turbines are shown. Finally a 20m mast (M18) was erected at a small peninsula at the southern coast of Andros in direct line about 12 Km south of the main mast. All three 30m towers at the main mast site were equipped as follows: a) at 3 levels (10,20 and 30m a.g.l.) with vanes and cup anemometers (NRG model # 200P and #40) with distance constant of 3m. b) at the 20m level with UVW propeller anemometers (Friedrichs and Co. model 4212.0). c) at the 30m level with sonic anemometers (Gill model MET-log). Additionally a matched pair of Pt-100 thermometers was placed at the central mast (M16) at two levels (10 and 30m) to measure ΔT. The 20m mast (M18) was instrumented at two levels (10 and 20m) with cups and vanes and with a 3D propeller at 20m. Sonics and 3D propellers were sampled at 4 Hz and 1Hz or 2Hz respectively for the turbulence measurement campaign. The data acquisition system used for the fast response instruments consisted of laptop personal computers equipped with A/D converters (Data Translation DT2814) and pulse counters (Advantech PCL730).

4. DATA ANALYSIS - RESULTS

4.1 Data analysis

Measurements have been recorded from 4 to 12 August 1992 on a 24hour base. The wind direction was in the sector [60° - 115°] (with 0° corresponding to the East), and the mean wind speed at the summit of the hill was within the range of 9m/s to 16m/s. Mean turbulence intensities standard deviations, and shear stresses have been evaluated for all the three masts on the hill and both heights (20 and 30m) by averaging 30 runs of one hour long each. These runs cover a variety of meteorological conditions and wind directions which remained considerably constant during the one hour experiment.

4.2 Mean turbulence parameters

In Table 1 , the three components of turbulence intensity and the friction velocity for the three 30m masts and both heights are shown. As seen from the table, there is a large dispersion of shear stress values for both heights.

Turbulence levels at 20m are higher than those at 30m. However, the value of the horizontal component σ_u/U at 30m of M16 is larger than the corresponding one at 20m. This can be attributed to the presence of the wind turbines (WT) which affect the flow at M16 for certain directions when the mast lies inside the wake. After a gross estimation it was found that M16 is under the influence of the WT for directions smaller than 80° and greater than 125° (from East). The presence of the WT causes increase in σ_u and decrease in the mean wind speed U. The combination of these two factors leads to

Table 1: Mean Turbulence intensities for all wind directions, for the three masts and both heights.

All Directions		σ_u/U	σ_v/U	σ_w/U	u_* m/s
z=30 m	M16	.135 ±.04	.104 ±.02	.079 ±.01	.427 ±.17
	M15	.128 ±.03	.112 ±.02	.085 ±.02	.456 ±.18
	M17	.187 ±.05	.140 ±.04	.114 ±.04	.367 ±.15
z=20 m	M16	.122 ±.03	.135 ±.05	.106 ±.03	.429 ±.22
	M15	.155 ±.04	.165 ±.06	.087 ±.02	.498 ±.26
	M17	.177 ±.04	.146 ±.07	.135 ±.03	.687 ±.21

significantly increased values of σ_u/U at 30m of M16. At the height of 20m of M15 an increased lateral component is observed in agreement with the observations over Lastours by Sacre et al [1].

The comparison between the three masts showed that M17 exhibits higher levels of turbulence than M15 and M16 for both heights and all components.

Refering to the upwind site (M15) and the summit of the hill (M16), a decrease of turbulence intensity values over the summit was expected, because of the larger increase of the mean flow (U) than that of σ_u, σ_v, σ_w [2]. The higher mean value of σ_u/U observed at 30m of M16 in relevance to the corresponding value of M15 is again attributed to some very high values of this component observed at this point, when M16 was under the influence of the WT.

Similar explanation is given for the increased mean value of the vertical component (σ_w/U) at 20m of M16 compared to that of M15. It can be concluded therefore, that when M16 is inside the wake of the WT, σ_u/U increases considerably at 30m (which corresponds to the WT hab height), while at lower heights (20m) the increase is more intense for the vertical component of turbulence intensity, σ_w/U.

The standard deviations σ_u, σ_v, σ_w at 30m of M16 were observed to vary noticeably with direction (Table 2).

Table 2: Variation of the standard deviations at 30m of M16 with wind direction sectors.

Dir (°)	σ_u (m/s)	σ_v (m/s)	σ_w (m/s)
60-70	1.91 ±.09	1.28 ±.11	1.05 ±.09
70-80	2.18 ±.20	1.22 ±.12	0.99 ±.11
80-95	1.38 ±.25	1.08 ±.16	0.82 ±.13
95-105	1.11 ±.22	1.02 ±.11	0.73 ±.09
105-115	1.50 ±.23	1.33 ±.13	0.97 ±.13

The increased values of the standard deviations in the sector [60° - 70°] can be again attributed to the influence of the WT. In the sector [70° - 80°] even larger values of the longitudinal component are observed. When the wind direction is inside this sector, M16 lies near the boundary of the wake where the perturbation is larger. The topography which exhibits large variations at these directions may also be responsible for such large values of turbulence.

For the other sectors, M16 lies outside the wake of the WT and the variations of the σ_u, σ_v, σ_w values, are attributed to the topography. The cross-sections of the topography which corresponds to these sectors are presented in Figure 2. As it can be seen, the terrain exhibits large scale irregularities for the sectors [80° - 95°] and [105° - 115°). The smallest values of σ_u, σ_v, σ_w are observed for the sector [95° - 105°]. The slope of the topography which corresponds to this sector is steeper (Fig.2) but the variations of the terrain are smaller.

As far as the topography is concerned it seems that it is the irregularities of the terrain and not the slope itself that increases turbulence at 30m. The opposite behaviour was observed at 20m where turbulence (especially the vertical component) increased with increasing slope. Therefore, smaller heights experience the effect of the slope mainly, while the complexity of the terrain is more important at higher levels.

4.3 Power Spectra

Power spectra calculated through a Fast Fourier Transform are presented for all wind components at 30m. They are compared to Kaimal and modified Kaimal formulations (eq. a,b,c) as well as to Von Karman expressions (eq. d,e,f) according to Teunissen [3].

$$\frac{nS_u(n)}{u_*^2} = \frac{105f}{(C+33f)^{5/3}} \quad \text{(a)}$$

$$\frac{nS_v(n)}{u_*^2} = \frac{17f}{(C+9.5f)^{5/3}} \quad \text{(b)}$$

$$\frac{nS_w(n)}{u_*^2} = \frac{2f}{(C+5.3f)^{5/3}} \quad \text{(c)}$$

f=nz/U

C=1 for Kaimal model

C=0.44, 0.38, 0.44 for eqs a,b,c respectively for modified Kaimal model

$$\frac{nS_u(n)}{\sigma_u^2} = \frac{4kL_u^x}{[1+70.7(kL_u^x)^2]^{5/6}} \quad \text{(d)}$$

$$\frac{nS_v(n)}{\sigma_v^2} = 4kL_v^x \frac{1+188.4(2kL_v^x)^2}{[1+70.7(2kL_v^x)^2]^{11/6}} \quad \text{(e)}$$

$$\frac{nS_w(n)}{\sigma_w^2} = 4kL_w^x \frac{1+188.4(2kL_w^x)^2}{[1+70.7(2kL_w^x)^2]^{11/6}} \quad \text{(f)}$$

where k=f/z.

The scales L_u^x, L_v^x, L_w^x were estimated by considering the frequency to which the energy maximum of our spectra corresponds and using Taylor's hypothesis.

In Fig.3 spectra of the three wind components at 30m of M15 are presented and compared with the models. It is obvious that the Kaimal and modified Kaimal expressions fail to represent the experimental curves for this case. This is because of the uncertainty involved in the estimation of the friction velocity (already indicated by other authors [4,5]) which is used for normalization in the Kaimal expressions and the fact that these expressions were derived exclusively over homogeneous and flat areas. On the contrary, the normalization by the variances in the Von Karman expressions and the use of the parameter L^x which is not dependent on terrain type, contribute to a good representation of our spectra by the Von Karman curves.

As for the horizontal components, a deficit of energy at some low frequencies is observed in agreement with Panofsky et al, [6]. It is remarkable that the spectra in Fig.3 are very similar to those observed over complex terrain at Great Dun Fell by Gallagher et al,[7]. They have observed the same energy deficit which they called "spectral lag".

In Fig.4 simultaneous spectra of the u-component at 30m of M16 and M15 and for a wind direction where M16 is influenced by the WT are presented. A shift of the energy maximum towards higher frequencies and a deficit of energy at lower frequencies is observed at M16, while the total energy is much higher than that at M15.

Finally, Fig.5 displays two spectra of the u-component at 30m of M16 which correspond to the same wind direction. The difference in the shapes of the two spectra can be understood considering the different atmospheric conditions to which the two curves correspond. The spectrum (1) was derived under moderately stable atmosphere (just before sunrise, moderate wind speed), while the spectrum (2) under neutral conditions (afternoon, high wind speed). The high energy at very low frequencies of the spectrum (1) may be due to wave activity. It is obvious that the spectrum (1) is influenced by the topography which strongly confirms the argument that under stable conditions the effect of topography becomes more important.

5. CONCLUSIONS

The presence of a WT seems to affect differently the horizontal and vertical component of the turbulence intensity with height.

At the height of 30 m turbulence increases with increasing complexity of the terrain, whilst at 20 m the local slope is responsible for the turbulence increase (especially for the vertical component).

The observed power spectra were satisfactorily described by the Von Karman formulation . Spectra are altered by the effect of the topography and the presence of the WT. The effect of topography becomes more important under stable conditions.

REFERENCES

[1] Sacre C. and Flori J.P., 1989: "Turbulence characteristics in complex terrain for wind turbines setting up" Proc. EWEC, Glasgow, UK, 523-526.

[2] Britter R.E., Hunt J.C.R. and Richards K.J., 1981: "Air flow over a two-dimensional hill: studies of velocity speed-up, roughness effects and turbulence", Q.J.R.Met.Soc. 107, 91-110.

[3] Teunissen H.W., 1980: "Structure of mean winds and turbulence in the planetary boundary layer over rural terrain" B.L.M,19,187-221.

[4] Bradley E.F., 1980: "An experimental study of the profiles of the wind speed, shearing stress and turbulence at the crest of a large hill" Q.J.R.Met.Soc. 106, 101-124.

[5] Kaimal J.C. and Hougen D.A., 1969: "Some errors in the measurement of Reynolds stress" J.of Appl.et.,8,460-462.

[6] Panofsky H.A., Larko D. Lipschutz R. and Stone G., 1982: "Spectra of velocity components over complex terrain" Q.J.R.Met.Soc. 108, 215-230.

[7] Gallagher M.W., Choularton T.W. and Hill M.K., 1988: "Some observations of the airflow over a large hill of moderate slope",B.L.M,42, 229-250.

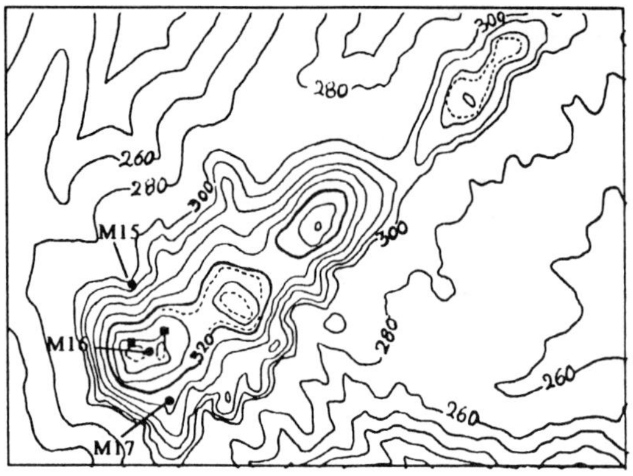

Figure 1: Topography of the site of Kalivari. (●):positions of the masts. (■):positions of the WT.

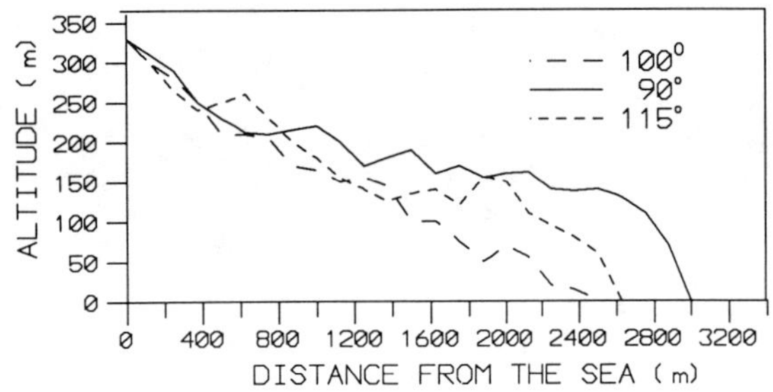

Figure 2: Topographic cross sections corresponding to different sectors of wind direction.

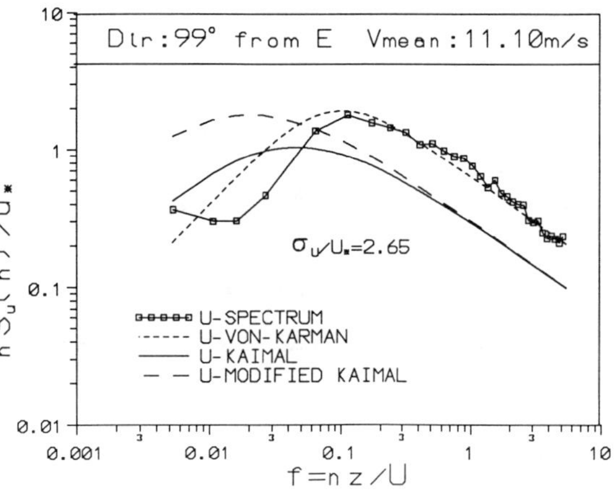

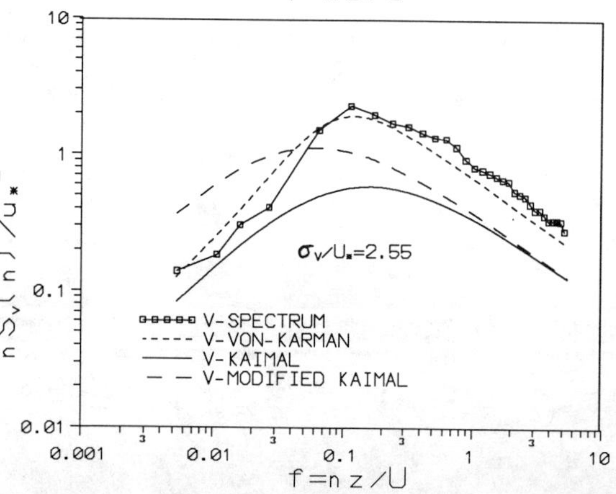

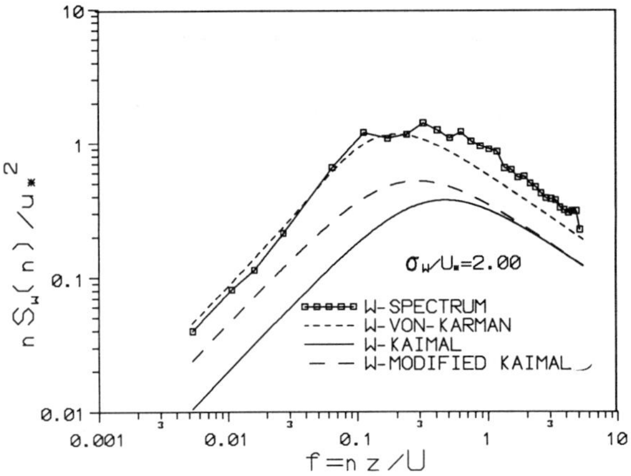

Figure 3: Normalized power spectra for the three wind components at 30m of M15. Comparison with Kaimal, modified Kaimal and Von Karman formulations.

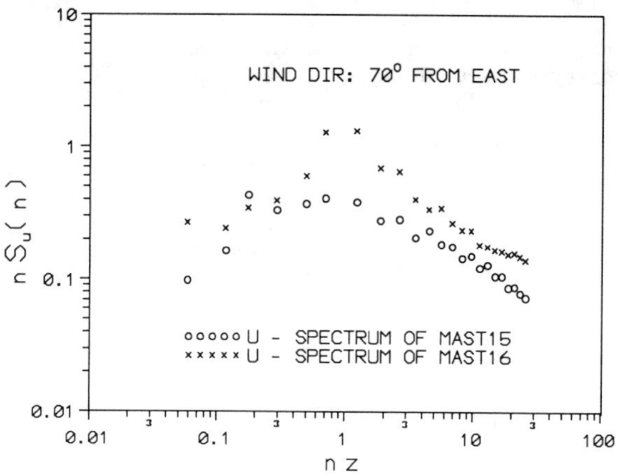

Figure 4: Power spectrum of the u-component at 30m of M15 and M16 when M16 is under the influence of the WT.

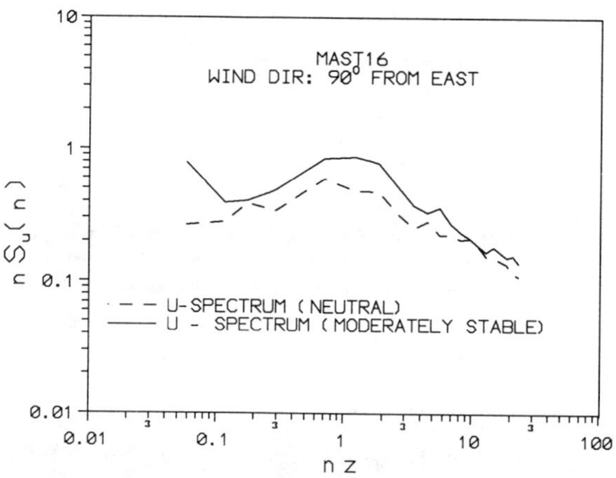

Figure 5: Power spectrum of the u-component at 30m of M16, for the same wind direction but under different atmospheric conditions.

WIND POWER ESTIMATION OVER COMPLEX TERRAIN

H.-T. Mengelkamp*, A. Raabe**
*GKSS Forschungszentrum Geesthacht, D-2054 Geesthacht, FRG
**University of Leipzig, FB Geophysik, D-7010 Leipzig, FRG

ABSTRACT: A numerical, non-hydrostatic, three-dimensional mesoscale model is used to prepare a wind power distribution for the complex terrain area around the Harz mountain. The frequency distribution of the geostrophic wind which drives the model is deduced from surface pressure measurements. This avoids use of possibly non-representative near-surface windspeed observations. Model results from a simulation of windspeed and wind power amplification over an isolated hill are compared with observations from a wind tunnel study. The agreement of the overall pattern and the amplification values is excellent.

1. INTRODUCTION

Wind power evaluations are usually based on long-term wind records at synoptic stations. Over relatively flat terrain with nearly homogeneous wind characteristics near-surface windspeed measurements may be representative for a wide area and simple interpolation methods may be adequate to derive a reasonable wind power distribution. Upper winds may be estimated with a certain degree of accuracy by well established logarithmic or exponential formulae. Over complex terrain, however, the spatial distribution of wind energy varies substantially over short distances. Measurements at specific, mostly exposed stations can be thought of as representative only for their immediate vicinity and the observational network is generally not dense enough for wind energy assessments on a regional scale.

On a local scale orographic particularities like isolated hills or a steep coast often exhibit an above average wind power potential. For optimal siting of a wind turbine a detailed study of the ambient airflow is necessary. In many practical cases a wind tunnel study or field experiment means too much effort.

Numerical models of different complexity have been used for wind energy investigations. They range from mass consistent models [8] over linearized spectral models [6] to hydrostatic mesoscale models [7]. In this paper the applicability of a non-hydrostatic mesoscale model for wind power investigations on a regional and local scale is demonstrated.

2. METHODOLOGY

In order to select and compare particular sites for wind power utilization and reasonably estimate the potential power output of a wind turbine information not only about windspeed u $[m/s]$ but on wind energy E $[W/m^2]$ is desirable. This is due to the non-linear relationship

$$E = \frac{1}{2}\,\rho\, u^3 \tag{1}$$

where ρ $[kg/m^3]$ is air density. It is apparent from (1) that wind power is most sensitive to changes in wind speed and its spatial and temporal averaging procedures [4]. The climatological mean of wind speed $\overline{u}$ at any particular location (x, y, z) is calculated from

$$\overline{u}(x,y,z) = \sum_{k=1}^{K}\sum_{n=1}^{N} f_{k,n}\left[u_{g_k}\,\frac{\vec{v}_n(x,y,z)}{u_{geo}}\right] \tag{2}$$

$f_{k,n}$ is the frequency distribution of the geostrophic wind for $k = 1...K$ geostrophic windspeed classes and $n = 1...N$ wind direction sectors. u_{g_k} $[m/s]$ is the mean geostrophic wind of class k, u_{geo} is the prescribed geostrophic wind at the upper boundary of the model domain and $\vec{v}_n(x,y,z) = \sqrt{u_n^2 + v_n^2}$ is the wind vector at position (x, y, z) with components (u_n, v_n) resulting from a simulation with u_{geo} from direction n. The mean wind energy flux $\overline{E}$ $[W/m^2]$ is calculated from

$$\overline{E} = \frac{1}{2}\,\rho \sum_{k=1}^{K}\sum_{n=1}^{N} f_{k,n}\left[u_{g_k}\,\frac{\vec{v}_n(x,y,z)}{u_{geo}}\right]^3 \tag{3}$$

Although apparent we should note that

$$\overline{E} = \frac{1}{2}\rho\,\overline{u^3} \neq \frac{1}{2}\rho\,\overline{u}^3 \qquad (4)$$

which stresses the difficulty to get reasonable accurate wind power information from windspeed maps. In equ. (2) and (3) the bar indicates an average over windspeed classes and wind directions. Equ. (4) also holds for spatial and temporal averages.

In equ. (2) and (3) the assumption is made that the relationship between u_{geo} and $\vec{v}_n$ does not depend on u_{geo}. This reduces the computational effort significantly because only 1 simulation must be run for each wind direction n (otherwise $K \times N$ simulations would be neccessary). The numerical model used is the three-dimensional, non-hydrostatic, and anelastic mesoscale model GESIMA (Geesthacht Simulation Model of the Atmosphere). A derivation of the dynamic equations is given in [1], the model including parameterization schemes for turbulent momentum fluxes and the surface enery balance is described in [5]. For this particular study thermal forcing was eliminated from the model because we are interested in a steady-state solution for flow over complex orography where dynamical effects dominate.

4. WIND POWER OVER THE HARZ AREA

The simulation domain covers an area of $90 \times 60\,km^2$ with a horizontal grid size of $1.5 \times 1.5\,km^2$ (Fig. 1). The more elevated area lies in the northwestern part with the Brocken beeing the highest point (1142 m). The roughness was subdivided into 5 classes with roughness lengths of 0.005 m (water), 0.2 m (farmland), 0.5 m (bushes), 0.8 m (forest) and 1.0 m for urban areas. The general characteristics are farmland at the environs and forests over the whole Harz area where heights exceed 400 m.

The frequency distribution of the geostrophic wind was taken from an analysis by Manier and Dietzer [3]. They analyzed pressure data from 60 synoptic stations for the years 1962 to 1966. By applying a second-order polynomial pressure maps were constructed by means of a least squares method from which the components of the geostrophic wind were calculated. Westerly winds dominate.

With a constant geostrophic wind simulations were run for each of 8 wind direction sectors until a quasi steady-state solution was reached.

The wind energy distribution (Fig. 2) shows characteristics as typical for flow from westerly directions. Areas with above average values are associated with regions of steep slopes in the luff of the Harz mountain, some isolated peaks and in particular the high elevation area around and west of the Brocken. The northeasterly fringe of the Harz, particularly the leeside of the Brocken, shows below average wind power. Mainly due to a sheltering effect large parts at the eastern side of the Harz or east of single hills exhibit a lower wind energy potential than the Harz surrounding. The higher roughness, though of minor importance, may also contribute to this effect.

5. FLOW OVER AN ISOLATED HILL

Natural or manmade orographic particularities like isolated hills can modify the wind field in a way that wind power utilization may be economic at certain places even in an area with an otherwise low wind power potential.

The simulation domain covers an area of $1400 \times 800\,m^2$, the horizontal grid size was $20 \times 20\,m^2$. The roughness length was 0.1 m throughout. One simulation was run with a geostrophic forcing of 10 m/s from west.

Fig. 3 shows a vertical cross section of the windspeed amplification in west-east direction through the center of the dump. The amplification factor is defined as p/p_i where p is windspeed at any particular point and p_i is its value of the undisturbed flow at the same height above ground. Not considering the small terrace at x=400 m two amplification maxima are found at the edges of the dump with the greater values at the windward side. Leeside divergence creates a region of amplification factors less than one. Fig. 4 taken from [2] shows the windspeed amplification around a dump with 1:3 slopes as measured in a wind tunnel. The agreement of the overall pattern as well as the amplification values with Fig. 3 is excellent.

ACKNOWLEDGEMENT

Gratitude is expressed to Ellen Heidelmann for digitizing the orographic maps. Discussions on the model dynamics with Hartmut Kapitza and Dieter Eppel are much appreciated.

REFERENCES

[1] Kapitza, H. and D. Eppel, Beitr. Phys. Atmosph. 65, (1992), 129

[2] Lohmeyer, A. and E. Plate, Arbeitshefte Ruhrgebiet, Ed.: Kommunalverband Ruhrgebiet, (1986), available from the author

[3] Manier, G. and B. Dietzer, Meteor. Rdsch. 32, (1979), 35

[4] Mengelkamp, H.-T., Int. J. Energy Res. 12, (1988), 113

[5] Mengelkamp, H.-T., Bound.-Layer Meteor. 57, (1991), 323

[6] Petersen, E.L., I. Troen, S. Frandsen and K. Heedegard, Risø National Laboratory, R-428, Roskilde, Denmark, (1981)

[7] Segal, M., Y. Mahrer and R. Pielke, Bound.-Layer Meteor. 22, (1982), 373

[8] Sherman, C.A., J. Appl. Meteor. 17, (1978), 312

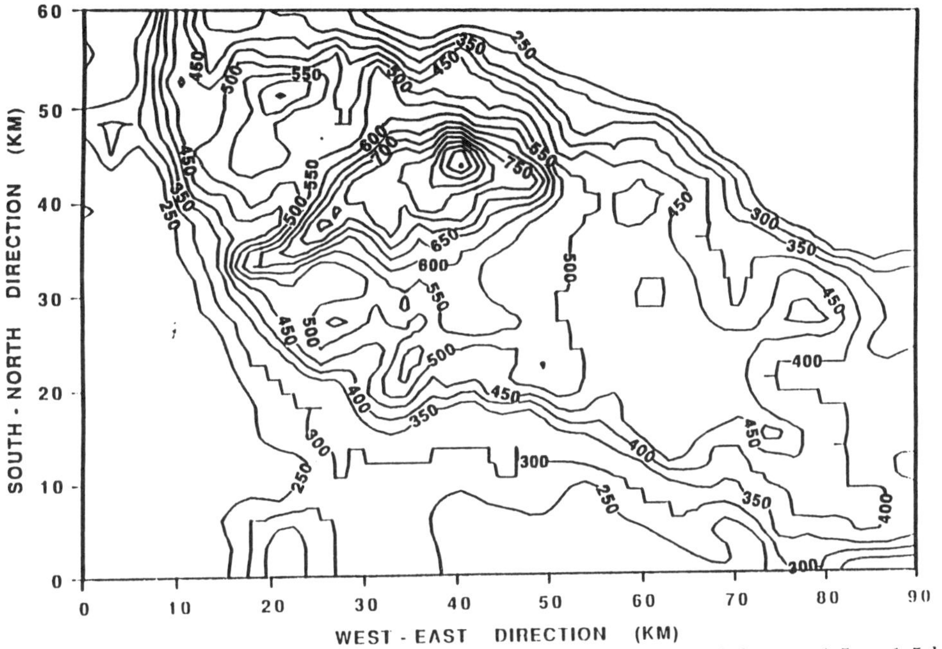

Fig. 1: Orography of the Harz area. The height isolines are interpolated from a 1.5 × 1.5 km^2 grid.

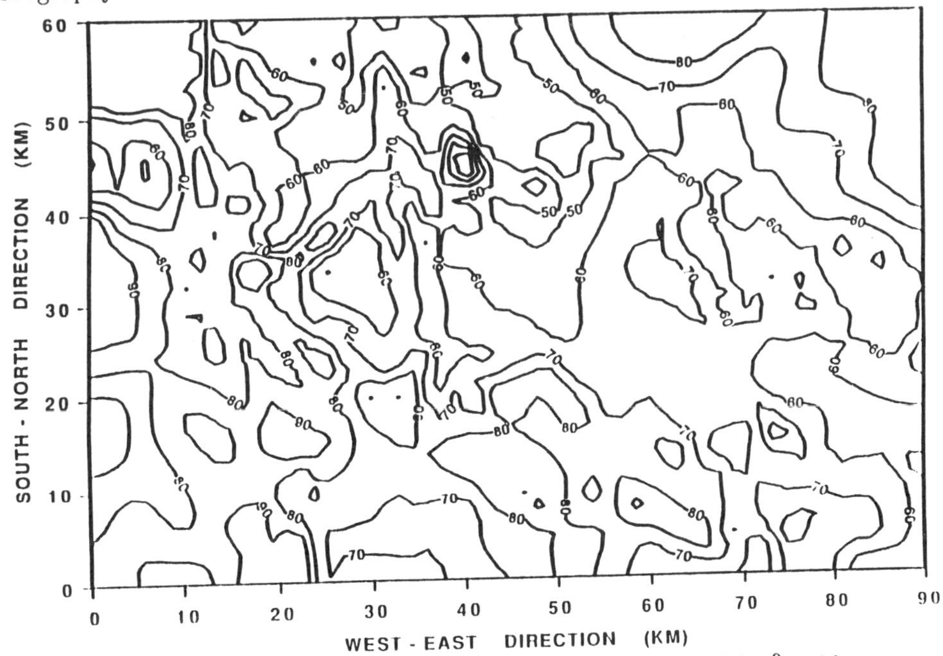

Fig. 2: Isolines of wind energy in W/m^2, interpolated from a 1.5 × 1.5 km^2 grid.

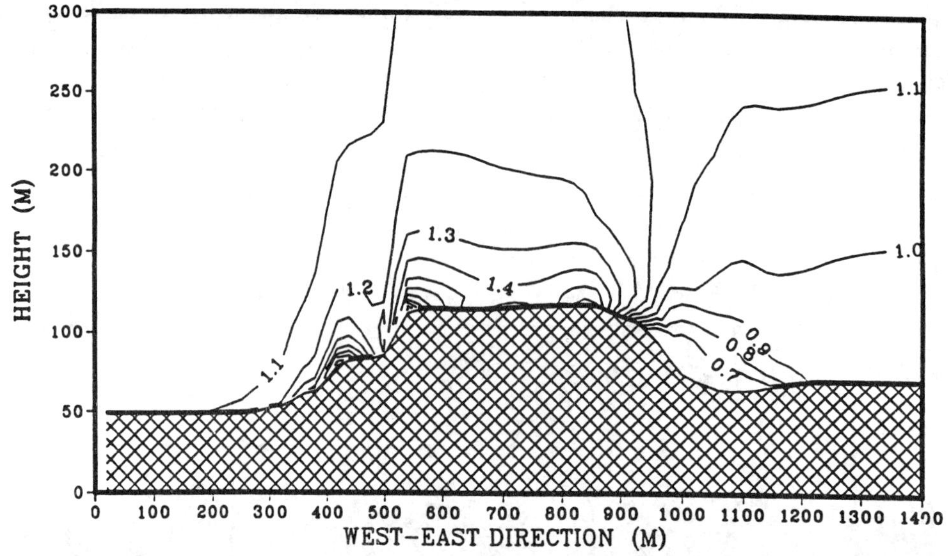

Fig. 3: Vertical cross section of the windspeed amplification factor. The wind comes from the left.

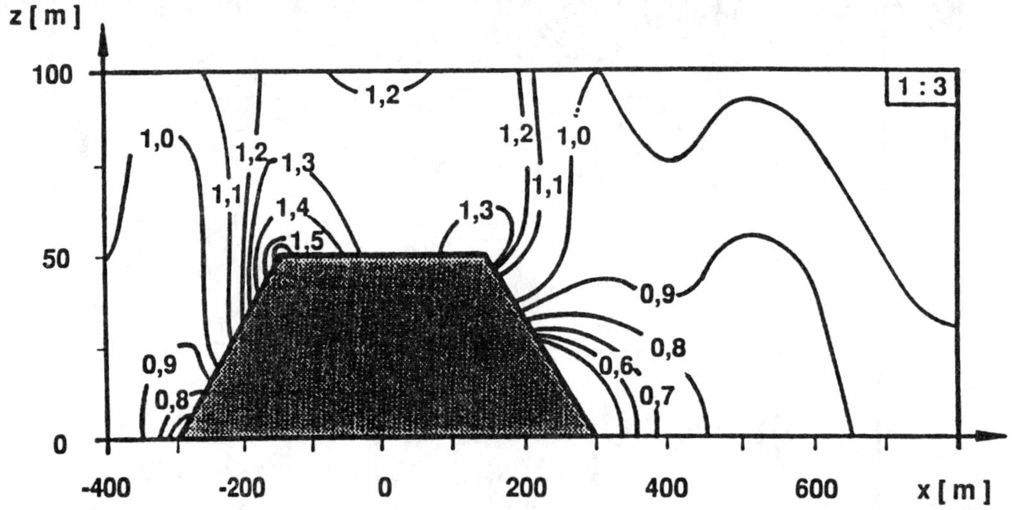

Fig. 4: as Fig. 3, but from a wind tunnel study. Taken from [2].

FLOW OVER COMPLEX TERRAIN ESTIMATED BY A GENERAL PURPOSE NAVIER-STOKES SOLVER.

L.K. Alm and T.A. Nygaard
Institutt for energiteknikk
P.O. Box 40 N-2007 Kjeller
Norway

A general purpose Navier-Stokes solver has been applied to atmospheric flow on microscale. For flat terrain the model produces the well known logarithmic profile in the surface layer and describes the effect of roughness change. At Askervein Hill, good agreement between model results and measurements is observed also behind the hilltop, where linearized models overpredict the wind speed.

1. INTRODUCTION

To get some practical experience with installation and operation of wind turbines, eight wind turbines have been erected in Norway with governmental support, in addition to two privately owned. All are sited along the coast in the western and northern part of the country, which is the most interesting part for wind energy. The terrain varies from small smooth islands leaving the wind nearly undisturbed, to areas with steep mountains some kilometres away from the site, separating the flow. To estimate the expected energy capture, simultaneous wind monitoring at the sites and weather stations with known long term wind data have been used [1].

The model WASP [8] developed by RISØ has also been applied. Through comparisons with the measurements, we have become increasingly aware of the limitations of linearized flow models as WASP.

If the use of wind monitoring could be reduced, the savings would be significant. This fact, and the exciting development of computer capacity/price has inspired us to try to take more of the complex physics of atmospheric flows into account when estimating local terrain effects.

The software package PHOENICS developed by Cham Ltd. has been used at Institutt for energiteknikk for some years, mainly for gas flow in tubes. It was therefore possible to test this package for atmospheric flow with moderate effort, as a starting point for development of a siting tool for norwegian conditions.

2 MODEL DESCRIPTION

2.1 Simplifications and assumptions

The model will mainly be used for comparison of mean wind speeds at hub height (30m-40m) at different sites.

If the sites are located a few kilometres apart, the following simplifications are proposed for dry neutral conditions:

- The density is a function of height above sea level only. It can be shown that for Mach numbers (velocity divided by speed of sound) below 0.3, the flow can be treated as incompressible, i. e. density variations will mainly be determined by hydrostatic pressure and temperature. In the initial test cases, the density has been assumed to be constant.

- The Coriolis force is neglected. Scale considerations support this assumption near the ground in the Prandl layer (up to say 50 meters), where shear stress, acceleration and pressure gradient terms are dominating. At higher elevations, the Coriolis force becomes relatively more important, and our solution containing no turning of the wind with height becomes physically unrealistic.

- The wind conditions averaged over a period much larger than the time scale of the locally produced turbulence vary slowly over the transportation time of a fluid particle across the domain. Time - averaged equations with steady boundary conditions can thereby be applied.

- The mixing is equal in all directions (isotropic turbulence) .The shear stress is approximated by the widely used k - ε model, with a wall function at the first node above ground. This implies that near the ground, a balance between production and dissipation of turbulent kinetic energy exist. At the remaining nodes, transport equations are solved for production and dissipation. This means that the flow will adjust slowly to a roughness change , from the ground and upwards, which is more realistic than the instantaneous change in shear stress observed when using simpler mixing length closures.

2.2 Equations

The assumptions above reduces the problem in the following way:

Time variation, Coriolis terms and gravity term are left out of the three momentum balances. The pressure is divided into a hydrostatic part (weight of the atmosphere above the point of interest), which with the assumptions above is a function of height above sea level only, and a a non-hydrostatic part. The gravity term and the hydrostatic term cancel, and only the non-hydrostatic pressure remain in the equations. The pressure in the solution should therefore be regarded as the perturbation from the hydrostatic pressure.

The energy balance can be omitted, because the density is prescribed. The dependent variables are then three velocity components, pressure perturbation, turbulent kinetic energy and dissipation .

In tensor notation the one phase steady mass balance equation reads

$$\frac{\partial}{\partial x_i}(\rho u_i) = 0 \qquad (1)$$

where ρ is the density and u_1, u_2, u_3 are the velocity components. In the initial test case with constant density, the density vanish from the mass balance. Stating that the divergence of the velocity should be zero, is generally sufficient for most low-Mach number flows.

Adopting the Boussinesq approximation for the Reynolds stresses, the steady momentum equation for the time averaged velocity component u_i can be written

$$u_j \frac{\partial u_i}{\partial x_j} - \frac{\partial}{\partial x_j}\left((\nu_T + \nu_l) \frac{\partial u_i}{\partial x_j} \right) + \frac{1}{\rho} \frac{\partial}{\partial x_i} p = 0 \quad (2)$$

where ν_T is the turbulent viscosity, ν_l is the molecular viscosity and p is the pressure perturbation.

The turbulent viscosity is determined by

$$\nu_T = C'_\mu k^2 / \varepsilon \quad (3)$$

together with transport equations for the turbulent kinetic energy k and the dissipation of turbulent energy ε

$$u_j \frac{\partial k}{\partial x_j} - \frac{\partial}{\partial x_j} \left(\frac{\nu_l + \nu_T}{\sigma_k} \frac{\partial k}{\partial x_j} \right) = P_k - \varepsilon \quad (4)$$

$$u_j \frac{\partial \varepsilon}{\partial x_j} - \frac{\partial}{\partial x_j} \left(\frac{\nu_l + \nu_T}{\sigma_\varepsilon} \frac{\partial \varepsilon}{\partial x_j} \right) = \frac{\varepsilon}{k} (C_{\varepsilon 1} P_k - C_{\varepsilon 2} \varepsilon) \quad (5)$$

σ_k and σ_ε are the turbulent Prantl numbers for k and ε respectively. The production P_k is given by

$$P_k = \nu_T \left(\frac{\partial u_i}{\partial x_j} + \frac{\partial u_j}{\partial x_i} \right) \frac{\partial u_i}{\partial x_j} \quad (6)$$

while the constants in the k-ε closure is set to

$$\begin{aligned} &C_{\mu'} = C_\mu \cdot C_D = 0.0324 \\ &\sigma_k = 1.0 \qquad \sigma_\varepsilon = 1.85 \\ &C_{\varepsilon 1} = 1.44 \quad C_{\varepsilon 2} = 1.92 \end{aligned} \quad (7)$$

Some of the constants are not the standard ones [9]. For the atmosphere, Panofsky and Dutton [3] stated that the proper value for $\sqrt{C_{\mu'}}$ should be 0.18.

To account for this Beljaars et al. [4] increase the turbulent Prantl number for epsilon to 1.85, according to the following equation

$$C_{\varepsilon 2} - C_{\varepsilon 1} = \kappa^2 / (\sqrt{C_{\mu'}} \cdot \sigma_\varepsilon) \quad (8)$$

κ is von Karman's constant (= 0.4).

The x_1 and x_3 axes are oriented horizontally, with x_3 along the main flow direction. x_2 is oriented vertically.

2.3 Boundary conditions

Steady boundary conditions are applied. The flow entering the domain is assumed to be in equilibrium with the flat surface, and the flow leaving the domain is assumed just to convect out. At the top of the boundary layer the vertical velocity is assumed to be zero. This gives problems similar to blockage effects in a wind tunnel experiment, and the terrain feature of interest must therefore be small compared to the total domain.

The boundary conditions at the first node above ground is determined by the logarithmic law of the wall

$$U^+ = \frac{1}{\kappa} \ln(E y^+) \quad (9)$$

where

$$y^+ = x_2 \frac{U_*}{\nu_l} \quad (10)$$

and

$$U^+ = \frac{U}{U_*}. \quad (11)$$

U is the wind speed parallell to the wall.

Let τ_w be the wall shear stress. The friction velocity U_* is then given by

$$\tau_w = \rho U_*^2 = \rho s U^2 \quad (12)$$

Combining (9) and (12) and defining $Re = U^+ y^+$, s can be calculated by the iteration

$$s = \left[\frac{\kappa}{\ln(Re \cdot E \cdot \sqrt{s})} \right]^2 \quad (13)$$

E is set to

$$\frac{29.7\ \nu_L}{h_r \cdot U_*} \qquad \text{for } \frac{U_* \cdot h_r}{\nu_L} > 100 \quad (14)$$

If the physical roughness height h_r is interpreted as $29.7 y_0$ (aerodynamic roughness), we recognize the usual logarithmic law for the Prandtl layer.

Let d be the distance from the wall to the nearest node. Assuming that production equals dissipation, the boundary conditions for k and ε near the ground are given by

$$k = \frac{U_*^2}{\sqrt{C_{\mu'}}} \quad (15)$$

$$\varepsilon = \frac{U_*^3}{\kappa d} \quad (16)$$

At the outlet the pressure perturbation is set to zero.

$$p = 0 \quad (17)$$

while for the other dependent variables there should be no change along the x_3-direction:

$$\frac{\partial u_1}{\partial x_3} = \frac{\partial u_2}{\partial x_3} = \frac{\partial u_3}{\partial x_3} = 0 \quad (18)$$

$$\frac{\partial k}{\partial x_3} = \frac{\partial \varepsilon}{\partial x_3} = 0 \quad (19)$$

This means that there is no diffusion out of the domain of momentum, dissipation or turbulence. Only mass is convected out.

The upper, left and right boundaries are impenetrable .i.e.

$$u_n = 0 \quad (20)$$

where u_n is the velocity component normal to the boundary.

2.4 Discretisation

The physical domain is divided into computational cells. The velocity components are represeneted at the cell interfaces, while the other dependent variables are represented at the cell centers (staggered grid).

The differential equations are integrated over the control volumes (cells) to form a set of finite difference equations. A so called hybrid scheme is applied to the convective terms, in order to minimize false diffusion and thereby numerical errors.

The finite difference equations are solved by an iterative solving technique: Semi Implicit Pressure Linked Equations (SIMPLE), [5].

The grid spacing is implemeted to increase upwards with an exponential rate while the grid is uniform in the horizontal directions.

3 SIMULATION RESULTS

3.1 Roughness change on a flat plain

The roughness change simulations were carried out to test whether the model could reproduce a wind profile over a plain with varying aerodynamic roughness.

The boundary conditions were chosen as described above, except for the conditions at the ground, where the aerodynamic roughess was set to 0.01m at the inlet, and abrupt changed to 0.1m one kilometer downwind. This could be similar to the roughness change at the coastline.

Figure 3.1 shows the velocity for different heights.

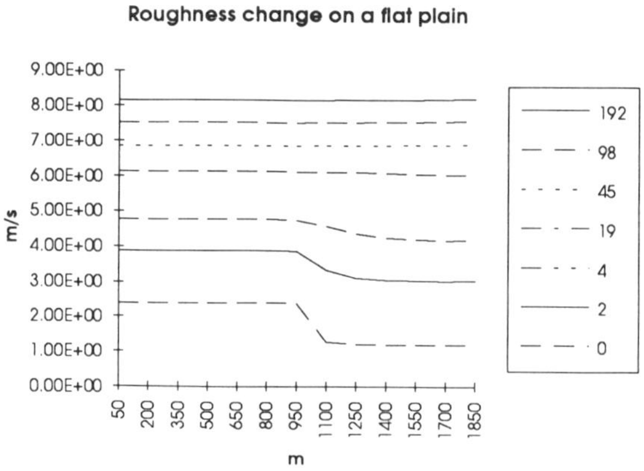

Figure 3.1: Horizontal velocity at different heights. The horizontal distance is 2000 meter, and the rougness change is positioned at 1000 meter.

The velocity located at 800m, 1100m and 1700m are shown in figure 3.2.

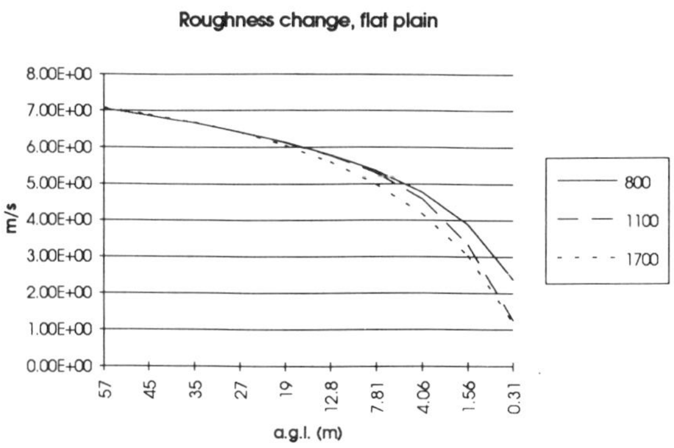

Figure 3.2: Velocity versus height for three wind profiles; at 800m, 1100m and 1700m. The roughness change is located at 1000m.

3.2 Comparison against the Askervein Hill experiments

Askervein Hill is a smooth hill at the west coast of the Hebrides (Scotland). It has been considered to be a good site for testing of wind models.

A series of wind measurements were carried out over Askervein Hill, during 1982-83, see [6]. Some of the results are reported in [7] and compared with the model MS3DJH/3. In [8] the measurements are compared with simulations with WASP (BZ) and MSFD. MSFD is described in [4]. Both WASP, MSFD and MS3DJH/3 performs well upwind the hill, but at the leeside the codes overestimate the wind speed.

Wind measurements with an upwind direction of 210 degrees was chosen for the comparison tests. Contour maps of the physical topography of Askervein Hill was obtained by a contouring facility in WASP. A coarse map taken from [8] was magnified and digitized. The contour lines were further processed by the code WGRID to form the coordinates of the lowest grid plane. WGRID creates a uniform grid spacing in the horizontal directions.

A separate module was developed to read the WGRID-file, create a vertical grid spacing and write out a 3D grid file in a PHOENICS compatible format.

Figure 3.3 shows the computational grid together with a contour map of Askervein Hill. The simulated normalised speed (local speed dived by upstream speed at same height above ground) at the three cross sections together with the measured values are shown in figure 3.5, 3.6 and 3.7.

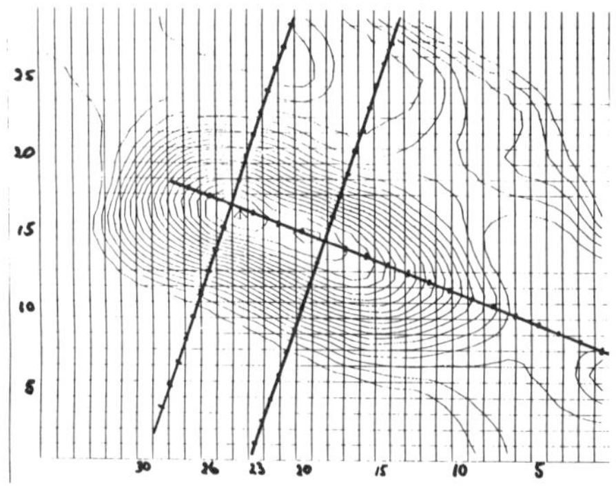

Figure 3.4: Computional grid and contour map of the 3D Askervein Hill model.

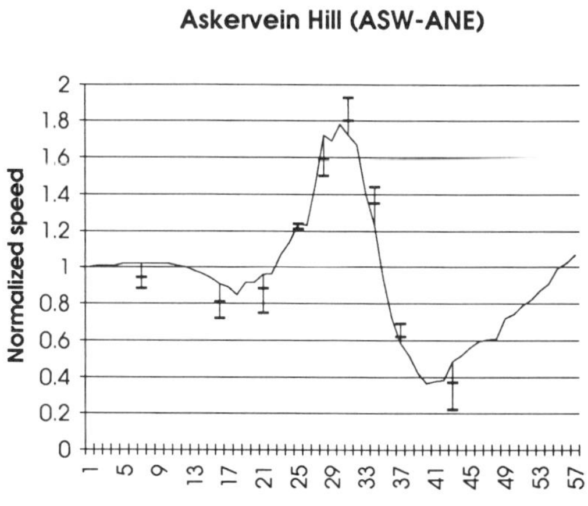

Figure 3.5 Normalised speed at 10 m a.g.l. at the SW-NE measurement line through the hilltop (HT).

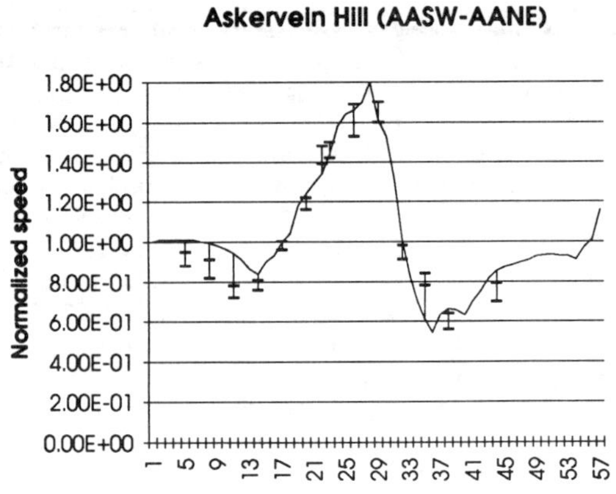

Figure 3.6 Normalised speed at 10 m a.g.l. at the SW-NE measurement line through the central point (CP).

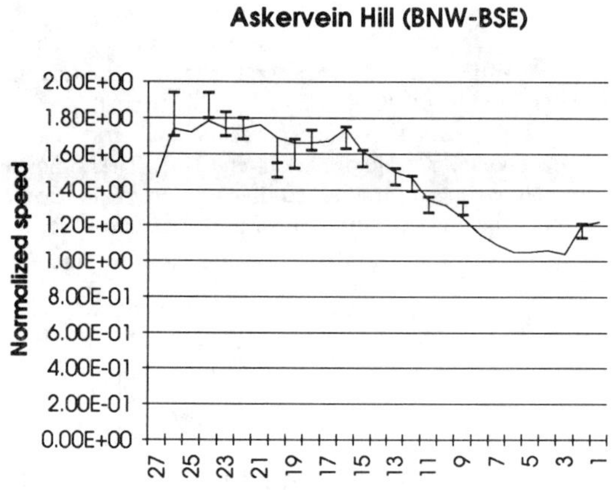

Figure 3.7 Normalised speed at 10 m a.g.l. at the NW-SE measurement line at the hill ridge.

There is good agreement between the simulations and the measurements at all three cross sections. Note the excellent results on the leeside.

Upstream the hilltop simulations overpredict the wind speed. At the hilltop, figure 3.5, the wind speed is underpredicted.

The main reason for the differences between the measured and the simulated wind speeds are considered to be the coarse contouration and the grid with long and thin cells causing numerical inaccuracy.

It was not possible to do grid refinements due to limitations in swapping space on the disc. The results should therefore be regarded as preliminary, until a sensitivity analysis on grid refinements have been performed, when a new swapping disc has been installed.

4 CONVERGENCE AND CPU-TIME

4.1 Convergence

Initially the simulations were run with inlet conditions (velocity profiles and turbulence parameters) calculated from simplified analytical solutions. These profiles turned out not to be in completely equilibrium with the computed solution, and the convergence was very poor.

The solution to this problem was to let the model itself compute the inlet conditions by simulating wind over a flat plain in parabolic mode. This gave a fully developed boundary layer, which then was used as inlet conditions for elliptic runs.

4.2 CPU time

The simulations were carried out on a workstation (HP 720). Typical computing time for the 3D Askervein Hill run (with grid size 36x58x40) is 75 CPU minutes. If a complete simulation should be done for 12 different wind directions and several wind speeds, the total CPU time is presently a problem for practical applications.

There may though be possibilities of making shortcuts reducing the total CPU time, and computer resources is not considered to be a limitation in a few years.

5 CONCLUSION

The Navier-Stokes solver PHOENICS has been used to model microscale atmospheric flow over terrain features. The initial test cases with prescribed density and neutral conditions has demonstrated the models ability to describe influence of roughness change and terrain elevation. The main advantage over simpler linearized models seems to be a better description of the flow behind the hill, where nonlinear effects may be important. When areas with multiple hills are considered, this could give a better representation of the flow field.

ACKNOWLEDGEMENTS

This work has been supported by Departement of Trade and Energy through the norwegian wind energy programme.

REFERENCES

[1] Nygaard T.A., 1992; "Estimating energy capture at potential wind turbine sites in Norway", Journal of Wind Engineering and Industrial Aerodynamics, 39, page 385-393. Elsevier Science Publishers B.V, Amsterdam.

[2] Spalding D.B. 1991; "The PHOENICS beginner's guide" / TR100. Concentration, Heat and Momentum Limited, London.

[3] Panofsky H.A. and Dutton J.A., 1984; Atmospheric Turbulence: "Models and Methods for Engineering Applications". New York, John Wiley and Sons.

[4] Beljaars, A.C.M. , Walmsley J.L. and Taylor P.A. 1986; "A mixed spectral finite-difference model for neutrally stratified boundary-layer flow over roughness changes and topography". Boundary Layer Meteorology 38 (1987) 273-303.

[5] Patankar S.V. 1980; "Numerical heat transfer and fluid flow". New York: Hemisphere Publishing Corporation

[6] Taylor, P.A. , Teunissen H.W. (1986); "The Askervein Hill project; overview and background data". Boundary Layer Meteorology 39 (1987) 15-39.

[7] Walmsley, J.L. and Salmon, J.R. , 1984; "A boundary layer model for wind flow over hills. Comparison of model results with Askervein 1983 data.". European Wind Energy Conferance and Exhibition, Hamburg Oct. 1984.

[8] Troen, I. and Lundtang Petersen, E. "Siting of wind turbines"; EC Wind Energy Conference 1988.

[9] Andersson I., 1988 "Introduction to turbulence modelling". Lecture notes in Turbulent Flow, NTH Trondheim.

NUMERICAL AND EXPERIMENTAL STUDIES IN COMPLEX LAND/SEA BOUNDARIES

Hans Bergström and Birgitta Källstrand
Department of Meteorology, Uppsala University
Box 516, S-751 20 Uppsala, Sweden

ABSTRACT: Pilot balloon measurements at a coastal site in the Baltic Sea, during a three month period in spring and early summer, showed that low level jets are a common phenomenon, at least during that part of the year when the water temperature is lower than the temperature over land. Comparing the observed wind profiles with those calculated with a simple wind model indicated that the increase in wind speed caused by the LLJs may be of the order of 34% at 125 m. It is shown that these jets may develop as an analogy in space to an inertial oscillation.
A higher order closure model was used to map the climatological mean wind field over southeastern Sweden and the southwestern Baltic Sea area off the Swedish south coast. Comparing the results with earlier ones using simpler models, not capable of modelling LLJs, revealed a 5% increase in mean wind speed over the sea. This could be interpreted as the climatological contribution from low level jets. As the model domain, however, did not include the land areas across the Baltic Sea, from where LLJs might also originate, this figure is probably at the lower limit.

1. INTRODUCTION

The existing wind measuring weather stations in southern Sweden are primarily located in the coastal area, and the horizontal distance between them is rather large, typically of the order of 50 km. Also, the measurements are usually confined to one height, 10 m. Consequently there is great need for models and additional measurements in order to get more detailed information regarding the horizontal and vertical variation of the wind.

For large scale exploitation of wind energy it will be necessary to site wind turbines not only in flat coastal areas but also further out at sea. The choice of the more expensive locations off shore depends, to a high degree, on the expected more favourable meteorological conditions over the sea, with higher mean wind speed and lower turbulence intensity. Besides very sparse measurements taken from lighthouses, practically no climatological observations exist over the Baltic Sea. The interest in off shore siting thus still more increases the need for numerical and experimental studies, not only in the vicinity of the coast but also far out at sea.

Furthermore, an often observed feature in the Baltic Sea area is a low level wind maximum (low level jet-LLJ) slightly above the boundary layer. Although this marine LLJ over the Baltic Sea has been observed and studied on several occasions during the last decade, see e.g. [1], no climatological mapping exists. The Baltic Sea jet is supposed to develop as relatively warm air flowing out over a coast is subject to frictional decoupling over a colder sea, and an inertial oscillation evolves in space during the subsequent transport over the water.

2. THE EXPERIMENTAL STUDY

To investigate the occurrence of LLJs, an extensive measuring programme was carried out during 910324-910614. Wind and temperature was measured on a 25 m high tower and pilot balloon measurements were made at Utlängan (see Figure 1). Except for one period (7-19 of April), and for days with low clouds or fog, there are one measuring occasion each day. These occasions are randomly distributed over the hours of the day. At every occasion three balloons were released and the mean of the three profiles have been used here. This resulted in 52 usable wind profiles, representing 52 days.

The pibal measurements were carried out in the following way: A balloon, filled with hydrogen gas, was released, and its movements followed through a theodolite - an instrument with a telescopic sight for measuring angles (horizontal and vertical angle). Readings of these angles were recorded on a cassette tape every 8th second, and later transferred to a PC to calculate profiles of wind speed and wind direction.

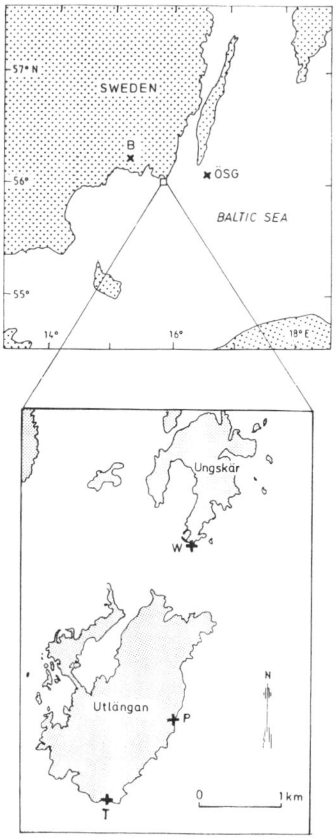

Figure 1: Map of the experimental area. The survey map shows the horizontal domain of the model used in Section 3. ÖSG=Ölands södra grund, B=Bredåkra, T=25 m tower, P=pibal site, W=synoptic weather station.

2.1 Experimental results

To get information about low-level jets from the pibal profiles, these were studied both numerically and by eye. There is no unique definition of a LLJ in the literature and different criteria have been used to identify LLJ:s in different investigations. Examples of used criteria are, cf. [2] and [3]: wind speeds greater than a specific value, the relation between the wind speed maximum and the geostrofic wind or the relation between the maximum wind speed and the wind speed above. Because the LLJs look quite different from case to case - with a more or less distinct maximum and a stronger or weaker absolute wind speed difference etc. - the relation between the wind speed maximum and the wind speed above was principally studied, but no "critical value" was chosen, cf. [4].

Examples of the measured profiles are plotted in Figure 2, and in 33 of them a more or less distinct LLJ was present. This indicates that LLJs are indeed a common phenomenon in this area, at least during spring and early summer. Since the wind direction appeared to be an important parameter, a partitioning of the profiles with LLJs in cases with and without a large wind direction change with height, was made. In 23 of the 33 LLJ cases there were only a small wind veer (type 1), while the remaining 10 profiles had a considerable wind direction shear (type 2). The type 2 jets are not likely to be caused by an inertial oscillation (cf. below), but by e.g. a thermal wind, a frontal zone, or a sea breeze circulation.

The added velocity from the LLJs was estimated by first interpolating the observed wind speed profile between 25 m and the heights above the jet. The increase of the wind speed at the height of the maximum of the LLJ, was then estimated by $(U_{max}-U_n)/U_n$, where U_{max} is the jet maximum wind speed and U_n is the interpolated wind speed at the height of the jet maximum.

This ratio for the LLJs of type 1 varied between 12% and 110%, with a mean value of 45%, while the mean value for the type 2 jets is 50%. Due to the limited number of profiles these figures must, however, be taken as a rough approximation. Note also, that it is not possible to claim that the wind speed at a specific height was increased with 45%, since the height of the wind speed maximum for the different profiles is not the same. It is 75 m in two cases and as high as 475 m for one case, even if the most frequent heights are around 100-300 m, with a mean value of 200 m.

A study of the frequency of low level jets for different wind direction sectors, was also done (see Table I). The climatologically most frequent wind directions in the area are around southwest, and the table shows that these directions are also the most common ones during LLJs, with 85% for SW and 70% for W. For these wind directions the air originates from Germany (SW) and southern Sweden (W), and has been transported over water for about 3-8 hours when it reaches Utlängan. This is in approximate agreement with the period of the presumed inertial oscillation, which have its maximum after about 6 hours. Consequently, this study seems to support the hypothesis that the inertial oscillation is an explanation of the LLJ:s. Also winds from southeast and east, although not very common, have a high frequency of LLJs, with 67% and 80% respectively.

Table I. Frequency of cases with low level jets (LLJ) for different wind direction sectors (wd, wind direction at 25 m). N = number of profiles and n = number of profiles with LLJ.

wd	N	NE	E	SE	S	SW	W	NW	All
N	2	7	5	3	3	20	10	2	52
n	0	2	4	2	0	17	7	1	33
n/N	0.0	0.29	0.80	0.67	0.0	0.85	0.70	0.50	0.63

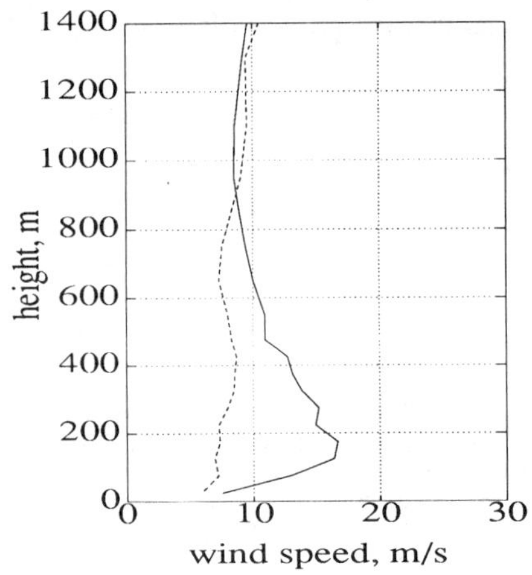

Figure 2: Examples of wind profiles with (—: 910406 17.15h) and without (---: 910510 16.15h) a low level jet.

To get further support for the spatial analogy to the inertial oscillation, the trajectories of the air over Utlängan was studied in more detail. The distance to the nearest upstream mainland was measured from maps, and the travel time to the measurement site was calculated. The observed ratio $(U_{max}-U_g)/U_{max}$ was plotted as a function of this travel time, and compared with a theoretical description of the inertial oscillation. This may be done starting out from an example in [2]. The solution of the inertial oscillation may be written as:

$$\begin{cases} u = U_g + F_u \cdot \sin(ft) - F_v \cdot \cos(ft) \\ v = F_u \cdot \cos(ft) + F_v \cdot \sin(ft) \end{cases}$$

where t is the time, f is the Coriolis parameter, which for the actual latitudes is approximately $1.2 \cdot 10^{-4}$ s^{-1}, u and v are the horizontal wind components. The coordinate system is chosen such that U_g represents the total geostrofic wind, and the parameters F_u and F_v are representing the departure of the wind from U_g at the start of the oscillation. The wind speed is then: $U = \sqrt{u^2 + v^2}$ and the ratio $(U - U_g) / U$ was calculated.

Comparing the theoretical curve with the LLJ cases of type 1, shows a good agreement between theory and observations, as can be seen in Figure 3. In general, the measurements agree quite well with the theoretical curve, although some points at around 4-5 hours are quite high. With regard to the uncertainty in the parameters, the result is surprisingly good.

A governing condition for the inertial oscillation to develop is the temperature difference between land and water, since it is supposed that the oscillation starts when relatively warm air is advected out over a colder sea. The interesting parameters in this case are, accordingly, the land temperature when and where the air leaves the mainland, and the water temperature off that coast. Figure 4 shows the relation between the water temperature and the land temperature for the LLJ cases of type 1. Except for one case, the land temperature is higher than - or in some cases almost equal to - the water temperature. Taking the uncertainties of the trajectory estimates into account, this result is also satisfactory and form another support for the hypothesis of an inertial oscillations in space.

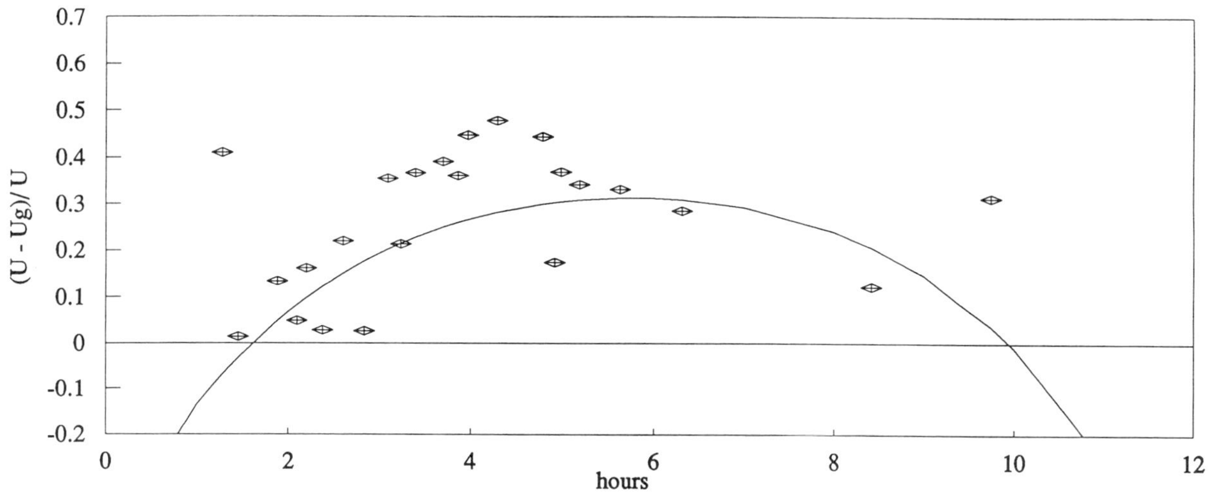

Figure 3: The ratio $(U-U_g)/U$ as function of time: i.e. for the observations (marks) travel time from upstream coast line, and for the theoretical curve (full line) time since onset of inertial oscillation.

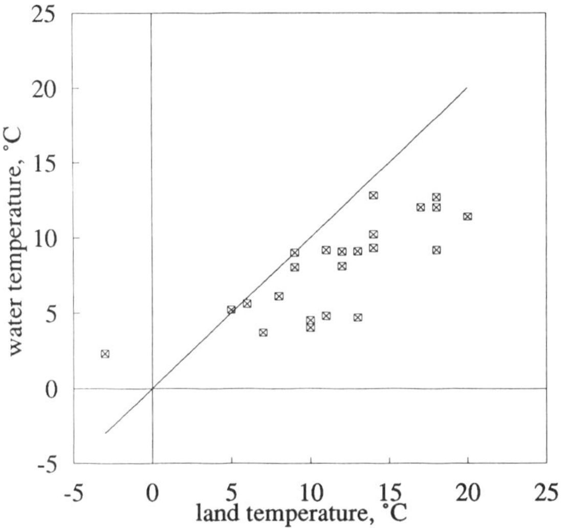

Figure 4: Water temperature versus air temperature over land at the upstream coastal passage (onset of inertial oscillation).

2.2 Comparison with simple model

The wind model used here is described in detail in [5]. With this model it is possible to determine the wind profile throughout the atmospheric boundary layer from a few, close to the ground, measured parameters. These parameters are the wind at one level and, to include the stability effects, the temperature at two levels. Alternatively the stability may be estimated from information about clouds (height and amount) in relation to wind speed, date, and time of the day. The model consists of two parts, one for the surface layer and one for the Ekman layer. It is possible to calculate the wind profile at a certain distance from a discontinuity in roughness and thermal stability, using the equilibrium profiles for the two surfaces. It is also possible to correct a measured transitional wind profile to the equilibrium profile for a surface before a discontinuity.

Since the model consequently is able to calculate the wind profile at the pibal measurement site, using data from the tower as input, it is possible to compare the observed wind speed and wind direction (from the pibal profiles) with the modelled values (see Table II). For 75 m and 125 m, the mean wind speed from the pibal profiles is 18% and 34% higher, respectively, than the modelled values. As the wind model is not capable of generating low level jets, this comparison - assuming that nothing else affects the difference - shows how much the LLJs affect the mean wind speed during these measurements (52 profiles). Still, it is important to remember that it is during spring and early summer the conditions for this phenomenon is most likely to occur, why the contribution of the LLJs to the mean wind speed for the whole year is probably smaller.

Table II: Utlängan, comparison between pibal profiles and wind speed calculated with the model, [m/s], (m=mean value, std=standard deviation).

	pibal profiles		modelled profiles	
height	m	std	m	std
75 m	9.6	4.9	8.1	4.3
125 m	11.4	6.2	8.5	4.9

3. THE NUMERICAL STUDY

To get a more detailed mapping regarding the horizontal variations of the climatological mean wind speed, a more sophisticated three-dimensional mesoscale higher order closure model of the atmospheric boundary layer has been used, [6] and [7]. The model includes terrain and is also capable of generating LLJs, which makes it possible to get an estimate of their contribution to the wind energy potential off shore.

This type of model is, however, quite computer time consuming, why the number of model runs must be kept at a minimum. The forcing parameters primarily affecting the wind field have to be identified and model runs made for a number of combinations of these. The most important parameters was here found to be: season, geostrophic wind speed and direction, stratification, and land/sea temperature difference. The modelled area is 340×340 km (see Figure 1) and contains 29×29 grid points. The grid is telescopic with a grid distance of 2 km in the central area and decreasing resolution towards the boundaries, in order to minimize influence from these. A total of 20 grid points in the vertical are log-linearly spaced between the roughness length and the model top at 3 km, in order to get a high resolution close to the surface. The terrain height has been taken from maps over the area and the roughness length has also been estimated from maps, judging the relative occurrence of forests, fields, and urban areas within each grid area.

The model was run for 288 different meteorological conditions and the resulting modelled wind fields were weighted together using climatological data from weather stations in the area. The details of this is described in [8]. A comparison between the modelled climatological mean wind and the observations at three weather stations showed good agreement, both as regards monthly and annual data. The annual results are given in Table III. It thus seems feasible to use a higher order model to resolve details of the wind field on the scale of a few kilometres, a resolution impossible to get from routine wind observations alone.

Table III. Comparison between observed and modelled annual mean wind speed (m/s) at Ölands södra grund (ÖSG), Ungskär (Ur), and Bredåkra (B). The locations are given in Figure 1.

	height (m)	model	observation
ÖSG	34	7.9	8.0
Ur	14	7.3	7.7
B	10	3.9	3.8

The resulting annual mean wind speed at the heights 15 m and 107 m are shown in Figures 5a-b. At the lower height the transition from higher wind speed over the sea to lower winds over land is rather sharp, and the strait (Kalmarsund) clearly influences the wind field. At higher levels the decrease in wind speed from sea to land becomes more of a transitional zone and the influence from the strait diminishes. It is also noteworthy that the model indicates wind speeds not much reduced, compared to those further out at sea, close to the southeastern corner of Sweden, which probably is a consequence of the complicated land/sea boundaries in the area.

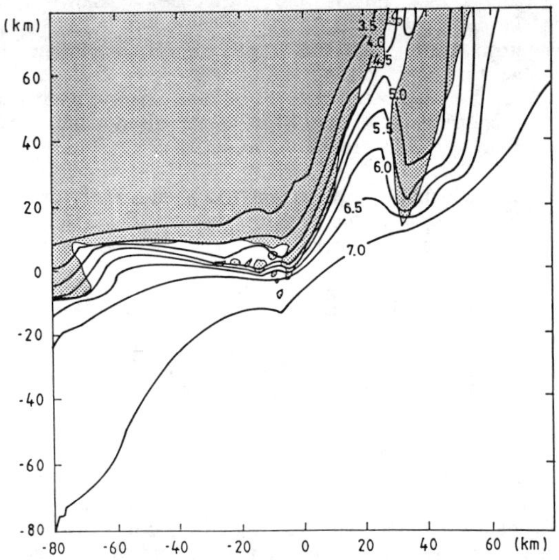

Figure 5a: Modelled annual mean wind speed at 15 m.

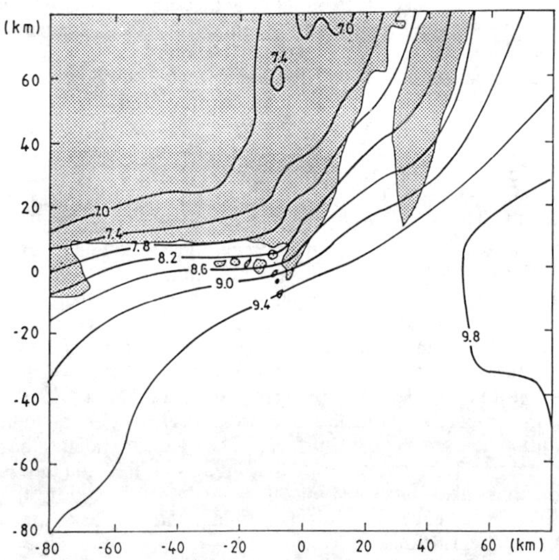

Figure 5b: Modelled annual mean wind speed at 107 m.

Looking at the mean wind speed over the open sea, it was found to be about 5% higher at 100 m height compared to earlier investigations with more simple models, not capable of generating LLJ's. This number is much smaller than found in the experiment, but since the occurrence of LLJ's is likely to be more frequent during spring and early summer than during other seasons, the increase in mean wind speed found in the previous section (around 20 to 30%) is probably to large on an annual basis. The modelled increase may be expected to give a more realistic climatological estimate of the contribution to the mean wind speed from the LLJ's, but more wind observations during longer periods of time throughout the height interval of the LLJ is needed to confirm this.

4. SUMMARY

The experimental study confirms that low level jets are indeed a common phenomenon in the Baltic Sea region, at least during spring and early summer when the water temperature is usually lower than the temperature over the surrounding land areas. By comparing the travel time from the nearest land area to the pibal site at Utlängan with the period of an inertial oscillation, it is made highly probable that the origin of the LLJ is an inertial oscillation which is onset as the air passes the coast line from a relatively warm land area out over a colder water surface.

The climatological contribution of the LLJ's to the wind energy potential is hard to judge from the pibal measurements alone, as they were made during the period of the year when LLJs are supposed to be most frequent. The higher order closure model results, however, suggest that they may increase the annual mean wind speed with at least 5% at 100 m, probably somewhat more as the model domain did not include Germany, Poland, and the Baltic states, from where LLJs may also originate.

Acknowledgement: The work was sponsored by NUTEK, Swedish Board for Industrial and Technical Development.

REFERENCES

[1] U. Högström, and A. Smedman, 1984: The wind regime in coastal areas with special reference to results obtained from the Swedish wind energy program. Boundary-Layer Meteorol., 33, 351-373.

[2] H. Kraus, J. Malcher, and E. Schaller, 1985: A nocturnal low level jet during PUKK. Boundary-Layer Meteorol., 31, 187-195.

[3] R., Stull, 1988: An introduction to boundary layer meteorology. Kluwer Academic Publishers.

[4] B. Källstrand, 1993: Low level jets in the Baltic Sea area. Wind Energy Report WE 93:2, Department of Meteorology, Uppsala University. To be published.

[5] H. Bergström, 1986: A simplified boundary layer wind model for practical applications. J. Clim. Appl. Meteorol., 25, 813-824.

[6] L. Enger, 1986: A higher order closure model applied to dispersion in a convective PBL. Atmos. Environment, 24A, 2431-2446.

[7] M. Tjernström, L. Enger, and A. Andrén, 1988: A three-dimensional numerical model for studies of atmospheric flows on the meso-γ-scale. J. Theor. Appl. Mech., 7, 167-194.

[8] H. Bergström, 1992: A climatological study of wind power potential in the Blekinge area using a meso-γ-scale higher order closure model. Wind Energy Report WE 92:01, Department of Meteorology, Uppsala University, 74 pp.

TESTS OF PREDICTION EFFECTIVENESS OF WIND ENERGY COMPUTER MODELS IN COMPLEX TERRAIN

M.Tombrou, Dept. of Applied Physics, University of Athens
D.P.Lalas, D.A.Tryfonopoulos, Lamda Technical Ltd.
J.Panourgias, Center for Renewable Energy Sources, Greece

ABSTRACT
Measurements in complex terrain in Andros, Greece are utilized to investigate the ability of four different operational mathematical models to predict the wind potential. The models used in the present study are, MS-MICRO, a Fourier decomposition model, MSFD a derivative of MS-MICRO to include turbulence effects, NOABL* an enhanced mass-consistent code, and a full 3D Navier Stokes solver which embodies the two-equation k-ε turbulence model. The comparison between the results obtained and the existing wind measurements reveals the differences of the various models. The wind speed predictions of all the models are in general satisfactory and it proves that they are useful engineering tools for the estimation of the available wind energy potential over complex terrain.

1. INTRODUCTION

The need for reliable prediction of the wind field over complex terrain has led during the last decade to the development of various computer codes, which differ both in complexity and accuracy.

The concept of the linear perturbation theory of Jackson and Hunt [1] and Mason and Sykes [2] gave rise to the development of several models, which have given satisfactory results in geometries where no flow reversal existed (J.L.Walmsley [3]).

A different approach to modelling the wind flow over complex terrain is the one of Sherman [4], where a non divergent wind field is produced via a perturbation potential field. Traci et al. [5] have developed a mass consistent model which was improved by D.P.Lalas [6].

A more complex approach is the 3D Navier-Stokes model (Tryfonopoulos et al. [7]) which is capable of predicting recirculating flows, but on the other hand is more computer time consuming than other simpler techniques.

In this paper four different models based on the aforementioned ideas are used and the obtained results are compared against existing wind measurements.

2. DESCRIPTION OF THE SITE

The meteorological masts have been erected in the northern edge of the island at terrain heights which vary from 230m to 330m a.m.s.l. As can be seen from Figure 1 the five masts are located in a cross pattern and have been assigned code numbers (11, 12, 13, 14, and 16), with mast No 16 being the central one. Measurements from an additional mast (No 18) which is located at the southern sea-coast have also been collected. From Figure 1 where the terrain contours have been plotted it can be seen that the site of the five masts is complex, with large terrain slopes whose sign changes from positive to negative within a few kilometers. All the masts are 10m high, with the exception of the No 16 and No 18 which are 30m and 20m respectively, having the ability of taking measurements at three levels (i.e. 10m a.g.l., 20m a.g.l., and 30m a.g.l.). Five NRG cup anemometers are used for the wind speed measurements. The mean wind speed and the relevant standard deviation are recorded every 10 mins from 30 sec measurements.

3. THE NUMERICAL MODELS USED

The NOABL* is an enhanced mass-consistent model which produces a three dimensional, terrain dependent, divergence free wind field. Substantial improvements of the original code [5] have been made by D.P. Lalas [6] to include more physical information. Thus different type of terrain (i.e. sea, land, e.t.c.) can be represented, the atmospheric stratification can be taken in to account in a more suitable way, and an algorithm for the parameterization of the planetary boundary layer is included in the code. Also the code has been enriched with a telescoping procedure (M.Tombrou and D.P.Lalas [8]), enabling the focussing on special areas of interest.

The 3D Navier Stokes model solves the momentum equations along the three coordinate directions coupled with the continuity equation for the incompressible fluid. A coordinate transformation is first made so that the complex domain is transformed to a unit cube and then the equations are rewritten in the new coordinate system [7]. The two-equation k-ε model after modifications appropriate for atmospheric flows [9] is adopted for the turbulence closure, as it proves to be the best compromise between accuracy and computational cost.

MSFD is a linear mixed spectral finite difference model for the study of neutrally stratified atmospheric surface layer over complex terrain. The governing equations are transformed from physical to spectral space using fast Fourier transform in the horizontal direction, with finite difference methods being used in the vertical one. Numerical solutions are calculated iteratively in spectral variables and then the inverse transform is applied to obtain the solution in physical variables. MSFD model is considered to be superior to similar models based on the approximate analytic theory of Jackson and Hunt [1], which they are restricted by their use of simple mixing length theory and constant advection velocities within the inner and outer layers.

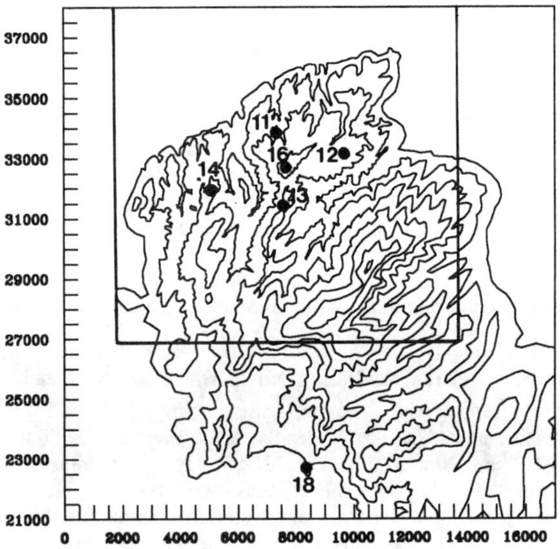

Figure 1: Terrain contours of the northern Andros.

The NOABL* and 3D Navier Stokes codes were used to simulate the flow field over the topography shown in Figure 1, while MS-MICRO and MSFD were applied to the smaller square region in Figure 1, with center the mast No 16. The models were run with both coarse (500m horizontal resolution) and fine (140m horizontal resolution) grids with the exception of the 3D Navier Stokes solver which was run only with the coarse resolution due to limited computational time available. Neutral northern wind conditions were simulated, since the measurements show that this is the prevailing wind direction in Andros. The results of the various applications are presented in the next section.

4. RESULTS AND DISCUSSION

In Figures 2, 3, and 4 the resulting wind speed distributions over the complex topography from NOABL*, 3D Navier Stokes, and MSFD models are presented at about 10m a.g.l. for the 500m x 500m resolution. For the shake of comparison the normalized wind speed contours with respect to the inlet wind speed have been plotted. The gross features of the wind fields from the different models are compare favorably between them. Areas of flow acceleration and deceleration coincide. The main difference is the lower overspeed in the high altitude regions predicted by the NOABL* compared to two other models. The different performance of each one of the models is more obvious in Figure 5, where the wind speed variation along the South-North cross section passing through the mast No 16 site is

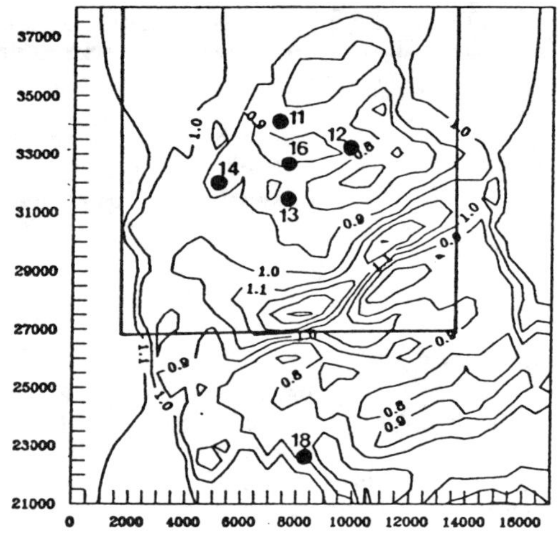

Figure 2: Normalized wind speed contours at 10m a.g.l. from NOABL* (500m x 500m).

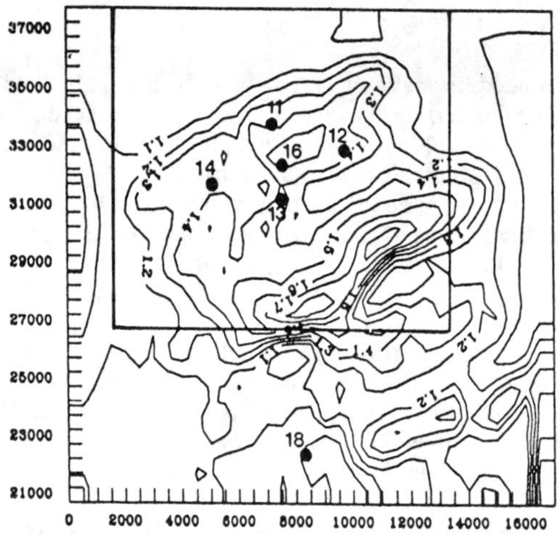

Figure 3: Normalized wind speed contours at 10m a.g.l. from K-ε model (500m x 500m).

Figure 4: Normalized wind speed contours at 10m a.g.l. from MSFD model (500m x 500m).

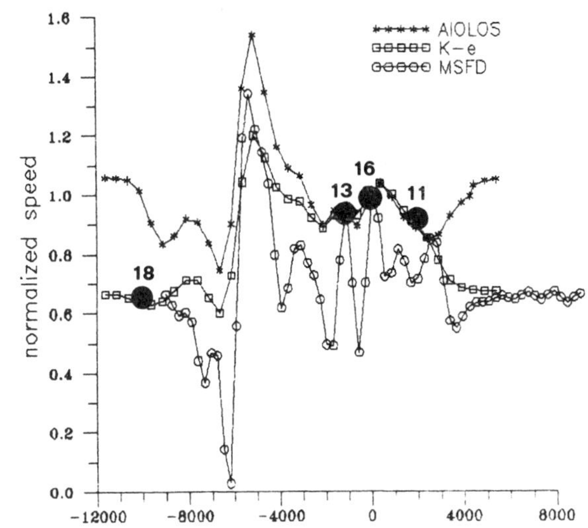

Figure 5: Wind speed variation from MSFD, NOABL*(AIOLOS) and k-ε model (500m x 500m).

presented. The wind speed is normalized, with the value of the speed at mast No 16, which lies on the top of an open hill. Negative values of the distance imply the locations downwind of the mast No 16 which is assumed to be located at zero coordinate value. The existing mean wind speed values over 2 hours at four masts are also presented in the same graph. The resulting curves from NOABL* and 3D Navier Stokes models appear to have the same trend, although the wind speed seems to be rather overestimated by the mass consistent code. On the other hand the MSFD curve is very different from the other two, while there exist a trend towards underestimation of the wind speed. The intense variations of the wind speed resulting from MSFD are partly due to the different way of introducing the topography information in the model, which allows a more detailed description of the terrain, and partly due to the use of Fourier transform. The measurements are in general agreement with the predictions of all the models, the greatest differences

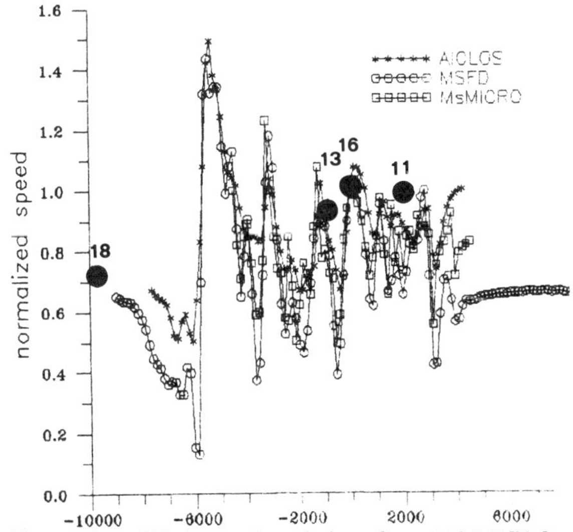

Figure 6: Wind speed variation from MS-MICRO, MSFD, and NOABL*(AIOLOS) (140m x 140m).

being observed in the case of NOABL* results.

Figure 6 presents the wind speed variation resulting from the mass consistent code, the MS-MICRO and the MSFD model along the previous cross section, but for the fine horizontal resolution (140m x 140m). The curves from MSFD and MS-MICRO are in close agreement, while the NOABL* results in less intense wind speed variations. Again the general agreement with the measured values is good, although no improvement of the results, due to the finer grid used, is apparent.

Table 1 summarizes the results of all the test cases and displays the comparison with the measurements. All values are normalized with respect to the wind speed at the site of mast No 16. The important feature is that the measured values differ between them by a maximum factor of 13%, with the exception of mast No 18, where a substantial flow deceleration is present. Also the predicted normalized values vary in the range from 0.89 to 1.0 in agreement with the experiments.

Mast No	11	12	13	14	16	18
Measured	0.94	0.87	0.90	0.93	1.0	0.71
MS-MICRO (fine)	0.93	0.99	0.97	0.97	1.0	-
MSFD (coarse)	0.85	1.0	0.97	0.94	1.0	-
MSFD (fine)	0.89	0.96	0.89	0.93	1.0	-
NOABL* (coarse)	0.89	1.01	0.98	0.9	1.0	-
K-ε	0.93	0.94	0.92	0.90	1.0	0.60

Table 1: Comparison of normalized wind speeds at 10m a.g.l.

As mentioned above two of the models used, namely 3D Navier Stokes and MSFD are capable of producing information about the turbulent characteristics of the wind field. Figures 7 and 8 present the distribution of the turbulent kinetic energy (TKE) at 10m a.g.l. from the two models respectively. The general features (i.e. regions of maxima and minima) are similar from the two models, but the magnitude of the TKE are very different. These differences as well as the unrealistic negative values of TKE resulting from MSFD should be attributed to the failure of the local equilibrium boundary conditions (with the shear stress proportional to TKE) in steep topographies. However in the vicinity of the measuring masts, where the terrain slopes are smaller, the predicted TKE values at 10m a.g.l. from the two models are of the same order, as Figure 9, where the vertical profiles of TKE at the masts sites are presented, implies. The predicted turbulence intensities vary from 6.5% to 7% and practically there are no difference between them. Mean turbulence intensity values from the measurements confirm this similarity and also report values around 11%, which ensure the reliability of the results obtained, taking into account the coarse grid resolution (500m x 500m) used.

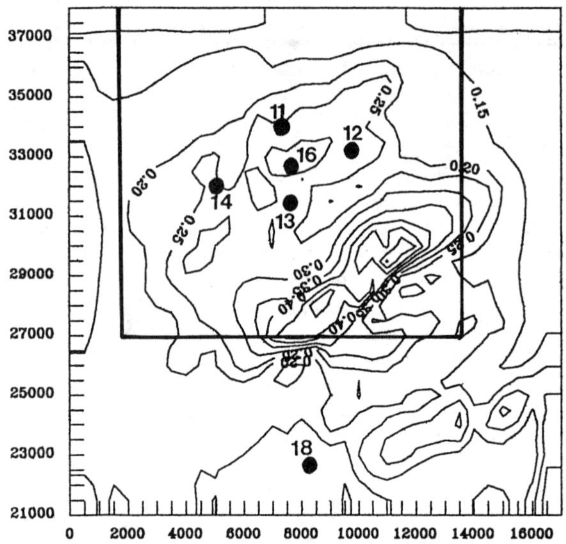

Figure 7: Turbulent kinetic energy at 10m a.g.l. from k-ε model. (5.5m/s upwind wind speed).

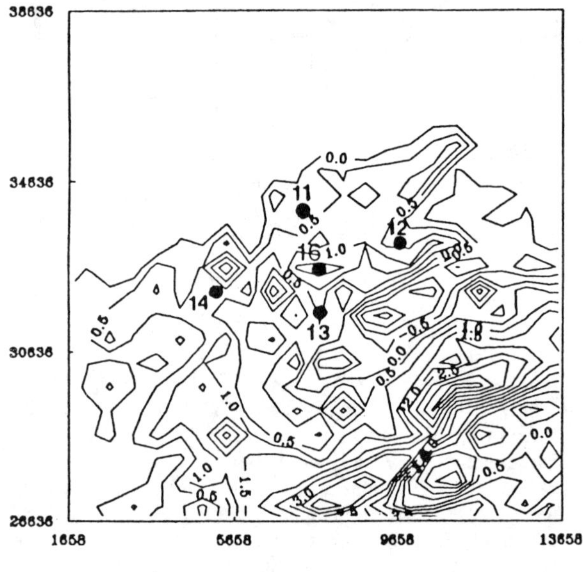

Figure 8: Turbulent kinetic energy at 10m a.g.l. from MSFD model. (10 m/s upwind wind speed at 10m a.g.l.).

5. CONCLUSIONS

Four different operational wind flow models were used in this paper and their results were compared between them and against the available wind measurements. All the models reproduced satisfactorily the gross features of the flow field, while differences were observed between them especially in the regions of decelerated flow. MSFD and MS-MICRO tend to underestimate the wind speed in the upslope regions and to overestimate it in the regions of accelerated flow, while NOABL * and 3D Navier-Stokes resulted in a more smooth variation of the wind speed. As far as the turbulent characteristics are concerned, the TKE values resulting from MSFD are much greater than the ones from the 3D Navier-Stokes model in regions with great terrain slopes. On the contrary in less steep terrain the values from the two models are comparable.

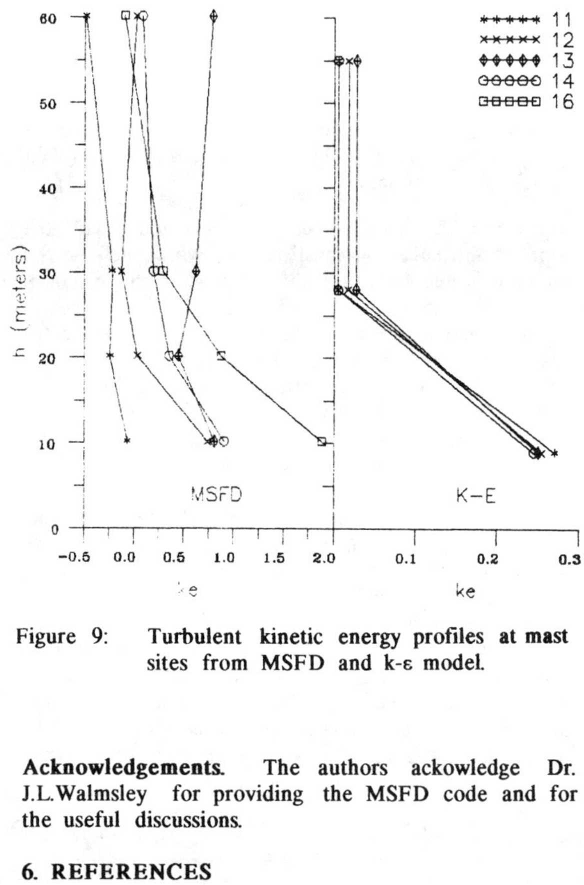

Figure 9: Turbulent kinetic energy profiles at mast sites from MSFD and k-ε model.

Acknowledgements. The authors ackowledge Dr. J.L.Walmsley for providing the MSFD code and for the useful discussions.

6. REFERENCES

[1] P.S. Jackson and J.C.R. Hunt, Q.J.R. Met. Soc., 101, (1975), 929-955.
[2] P.J. Mason, R.I. Sykes, Q.J.R. Met. Soc., 105, (1979), 383-395.
[3] J.L.Walmsley, P.A. Taylor and T. Keith, Bound. Layer Met., 36, 1986, 157-186.
[4] C.A. Sherman , J. Appl. Met., 17, 1978, 312-319.
[5] R.M.Traci, G.T. Phillips, P.C. Patnaik and B.E. Freeman, Report RLO/2440-11, U.S. Dept. Of Energy, 1977.
[6] D.P. Lalas, J. Solar Energy, 3, 43-67, 1985.
[7] D.A. Tryfonopoulos, J.P. Glekas, G.C. Bergeles, Wind Engineering, Vol. 13, No 6, 1989, 324-337.
[8] M. Tombrou and D.P.Lalas, Proc. 1990 ECWEC, 66-70.
[9] D.A. Tryfonopoulos, PhD, National Technical University of Athens, 1992.

WIND MEASUREMENT AND MODELLING IN MOUNTAINOUS REGIONS OF PORTUGAL. PRELIMINARY RESULTS

A. Restivo
Dept. of Mechanical Engineering
Faculty of Engineering, University of Porto
R. dos Bragas, 4099 Porto Codex, Portugal

E. L. Petersen
Department of Meteorology and Wind Energy
Risø National Laboratory
4000 Roskilde, Denmark

ABSTRACT: The paper presents preliminary results of the Portuguese participation in "Wind Measurement and Modelling in Complex Terrain", a multi-partner project supported by the European Community within the JOULE Programme. The acitivities described here were carried out by the University of Porto, Faculty of Engineering.

The research programme aims at improving the present knowledge of the wind resource in regions of Portugal with complex topography. Upon the selection of relevant areas to be studied in detail, a total of ten meteorological masts were installed and are in operation since July 1991. The wind data collected until now has been processed and summary results are presented, comprising the mean wind speed and power density (at 10 m a.g.l.), the maximum recorded 10 min average and the maximum gust (3 sec. average).

The report includes detailed information on the sites that are being studied, an updated summary of the recorded wind characteristics for all stations and sample preliminary results of the application of WAsP.

1. INTRODUCTION

It is known that the wind resources vary significantly in local scales imposed by the characteristics of ground topography, namely in mountainous areas. Among other regions of Europe, such topographic effects are encountered in the north of Portugal, where peak altitudes in excess of 1000 m occur at distances of only 30 km from the sea. In addition, the coast is oriented in a nearly north-south direction, the prevailing winds come from west and the mountains are cut by deep valleys of several rivers that flow towards the Atlantic.

In a country where the modest wind potential in plain areas [1] has not motivated much interest in the exploitation of wind energy, the identification of sites with local wind concentration may be of primary importance.

However, estimating the wind energy potencial at any point in the mountains of Portugal generally poses severe difficulties [1], that stem from the complexity of ground just mentioned and from insufficient meteorological data. It is generally recognized that only the collection of more detailed data on the wind in candidate sites will provide the necessary evidence of their energy potential. However, a local study of the wind characteristics in complex terrain may produce results far beyond the goal of fulfilling the need for wind data. Wind in mountainous areas remains a severe test case for current wind prediction capabilities, and its measurement may also contribute to the subsequent development of more powerful models for use in other

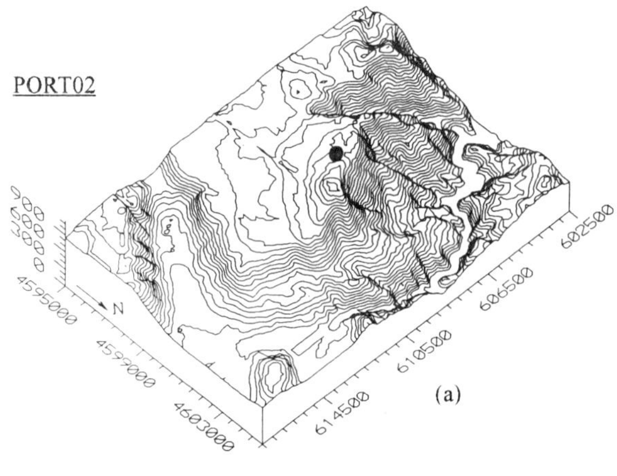

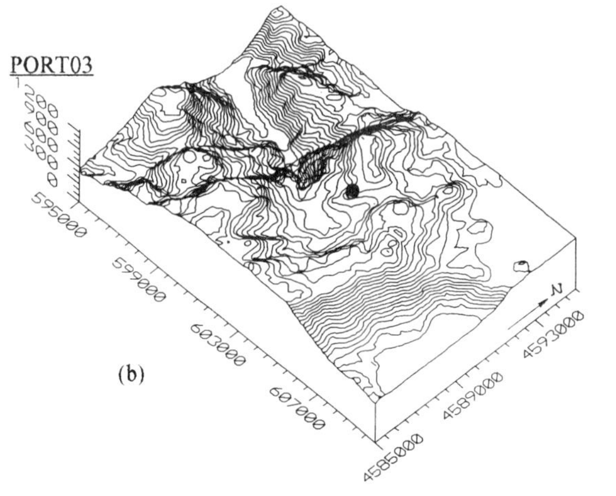

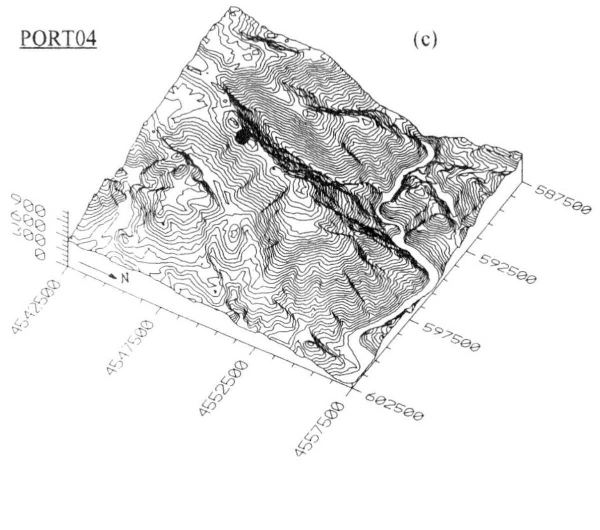

Figure 1 (a), (b) and (c). Digitized maps and locations of stations PORT02, PORT03 and PORT04.

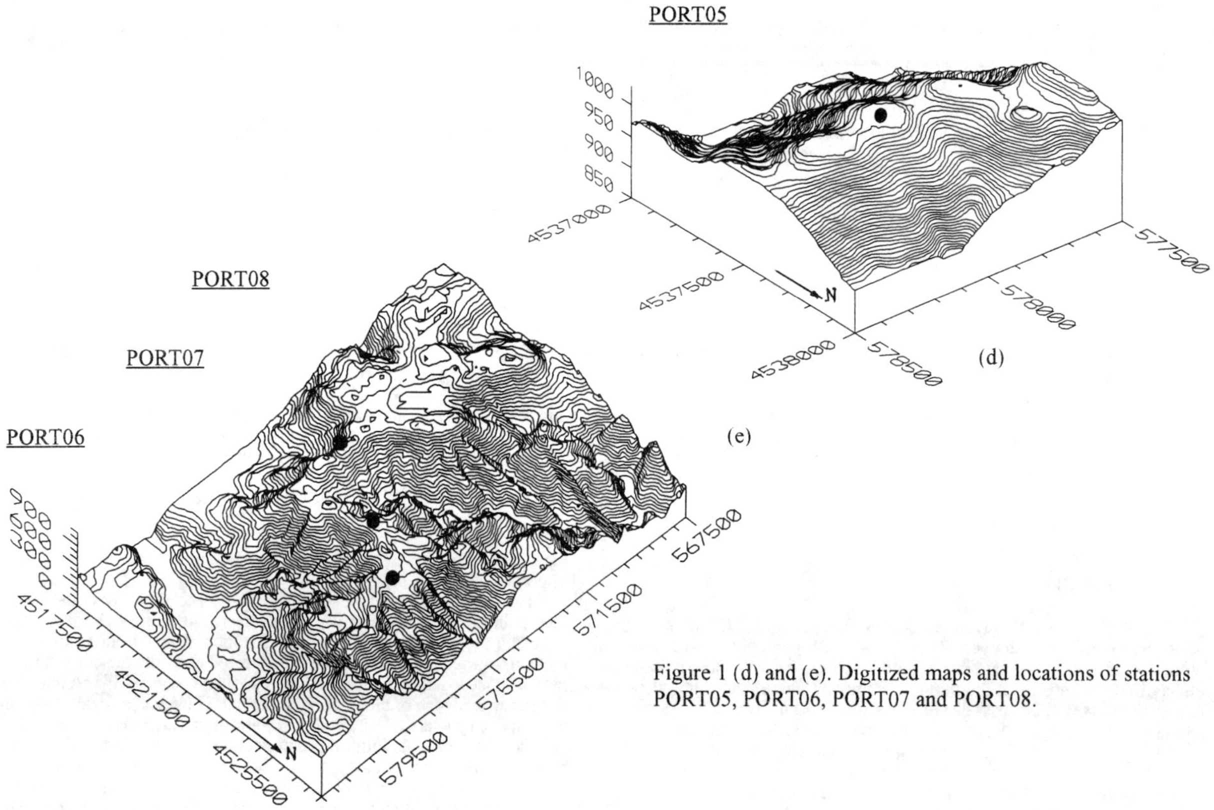

Figure 1 (d) and (e). Digitized maps and locations of stations PORT05, PORT06, PORT07 and PORT08.

areas of complex terrain.

With this motivation, the University of Porto joined a multi-partner European project, with the title "Wind Measurement and Modelling in Complex Terrain", supported by the EEC within the JOULE Programme. The project deals with the collection of wind data and the development of modelling tools in regions of Europe with complex topography, see reference [2]. Details of the portuguese participation in the project are presented in the following section, and then subsequent sections contain preliminary results of the measurements and of the application of WAsP to the measured data.

2. PROJECT OBJECTIVES AND PROGRAMME.

In general, the project points towards the identification of sites in Portugal with favourable characteristics for the utilization of wind energy, associated with local concentration effects determined by ground topography. The central objective is to obtain good quality data on such concentration effects for regions of complex terrain as exist in the north and centre of Portugal and, through its analysis, check and eventually adjust current wind energy distribution models. In particular, the wind modelling tools behind the European Wind Atlas will be employed and its accuracy tested.

The Department of Meteorology and Wind Energy, Risø National Laboratory, in Denmark, is responsible for the coordination of the entire project, which involves participations from Finland, France, Greece, Ireland, Italy, Portugal and Spain.

3. AREAS UNDER STUDY AND MEASUREMENT PROCEDURE

The region of Portugal being studied is the northernmost part of the country, at latitudes above 40°, where the highest mountains are concentrated.In this region, zones with potencial concentration efffects were first selected through analysis of maps. In general, land higher than 900m has scarce vegetation or exposed, rocky soil, providing open areas that may in principle be suitable for wind energy utilization. It formed the basis for the present study.

Each zone was then inspected locally. Some of the candidate points had to be discarded because they were too far from any road or near the limits of natural reserves.

The final selection of measurement points is detailed in reference [3] and identified in table 1. Locations 2 to 10 are in areas of complex topography, in the mountains of Alvão (two points), Montemuro - Meadas (two points), Arada - S. Macário (three points) and Freita (two points). Location number 1 is on flat terrain and was selected to represent, as much as possible, the wind conditions that would prevail over the studied region, in the absence of topographic concentration effects. For that purpose, the mast was installed in very open ground near the coast, in the largest flat zone available in this region. The nearest mountain is Freita, about 30 km towards East.

Table 1. Locations of measurement stations (UTM 29, m).

Station	Name	Longitude	Latitude	Altitude
PORT01	Murtosa	527375	4510125	0
PORT02	Trandeiras	608664	4599803	1100
PORT03	Alvadia	603204	4590208	1058
PORT04	Lamego	595046	4549625	1056
PORT05	Aveloso	577992	4537508	1020
PORT06	Pena	577146	4525019	932
PORT07	Drave	575598	4523085	982
PORT08	Coelheira	573770	4520487	1057
PORT09	Arouca	563421	4525458	1082
PORT10	Castanheira	562053	4522395	1012

This preliminary choice was guided by the application of WAsP to predict the wind potential in each zone, based on the wind data for Porto contained in the European Wind Atlas.

Where possible, preference was given to well defined orographic features, namely valleys of rivers and hill tops potentially capable of producing large orographic effects.

Some locations of measurement points are visible in figures 1.a to 1.e. They are graphical representations of digitzed terrain data, produced for each station from topographic maps. The data was taken from maps in 1:25000 scale and transferred into computer files under operation of WAsP.

In each site a 10m mast was erected, composed of a lattice tower with steel cables, anchored to concrete foundations produced on site. The installation was completed in May 1991.

The measuring and recording equipment was supplied and installed by Risø National Laboratory. Each station comprises a cup anemometer (Risø 70), a direction sensor (Aanderaa Instruments 2750), a sensor scanner board (Aanderaa Instruments 3010), a wind gust board, a data storage unit (Aanderaa Instruments 2990), a battery and interconnecting cables.

The complete system was in operation at the end of June 1991.

4. PRELIMINARY RESULTS OF MEASUREMENTS

The wind data collected until December 1992 has been processed and results were compiled in table 2. For each station the table presents the mean wind speed and power density (at 10 m a.g.l.), the maximum 10 min average and gust (3 sec. average). The wind roses for all stations (frequency of occurrence, in 12 sectors) are presented in figure 2.

Wind speed and power density are indicated in table 2 for two different 12 month periods. Power density has been corrected for the decrease in air density at 1000m (a value of 1.11 kgm^{-3} was adopted).

Table 2. Measured results. July 1991 - December 1992.

STATION Data		PORT 01	PORT 02	PORT 03	PORT 04	PORT 05	PORT 06	PORT 07	PORT 08	PORT 09	PORT 10
Mean velocity (ms^{-1})	(1)	4.1	6.1	6.5	6.7	6.1	4.7	5.4	6.1	5.8	(5)
Power density (Wm^{-2})	(1)	113	222	366	400	306	122	187	266	226	(5)
Mean velocity (ms^{-1})	(2)	4.4	6.0	6.5	(4)	6.2	4.6	5.4	6.1	5.8	5.6
Power density (Wm^{-2})	(2)	127	226	369	(4)	324	126	196	282	254	210
Max 10 mn aver. (ms^{-1})	(3)	19.1	23.9	26.9	26.2	32.1	23.4	27.7	27.9	32.6	22.4
Max 3 sec gust (ms^{-1})	(3)	33.5	35.0	35.2	33.4	43.0	34.2	37.2	39.8	41.8	35.1

Notes:
(1)-Period: July 1991 to June 1992.
(2)-Period: January 1992 to December 1992.
(3)-Period: July 1991 to December 1992 (all available data).
(4)-Operation was suspended from December 1991.
(5)-Operation was interrupted between August 1991 and January 1992.

5. APPLICATION OF WAsP TO THE MEASURED DATA

The data collected from the stations has been employed in several applications of WAsP. Preliminary tests were performed to assess the sensitivity of predictions to the refinement of the topographical description of the terrain supplied as input to the programme. It was concluded that, in general, an area of 4 km radius centred at the prediction site was sufficient, but the details of the topography near each site are critical. In obtaining the present results, the terrain was represented by digitized contour lines in 50 m steps, and lines in 10 m steps over the nearest 500 m from each site. In addition, detailed contour lines were provided as necessary within the first 50 m from each site to ensure that the grids used by WAsP accurately follow the local features of the terrain. A detailed map constructed for the area of station 05 is shown in figure 1.d.

The terrain roughness had to be prescribed on the basis of visual examination. It was found that a uniform roughness of 0.05 m is probably a reasonable estimate for most of the sites in the mountainous areas under study, and that value was adopted.. Sample tests with values between 0.03 and 0.10 m produced only minor changes in the computed results.

Table 3. WAsP predictions using stations PORT02 and PORT03 .

	Measured (1992)	Predictions PORT02	from station PORT03
Applic. site:	V (ms^{-1}) E (Wm^{-2})	V (ms^{-1}) E (Wm^{-2})	V (ms^{-1}) E (Wm^{-2})
Station PORT02	6.0 226	5.9 227	6.5 437
Station PORT03	6.5 369	6.6 369	6.5 366

WAsP was employed firstly to build a wind atlas from each station, and then test the ability to predict each station from its own atlas and from the atlas of another station. The atlas were produced based on the measured data por 1992, summarized in table 2.

Sample results are shown in table 3, for stations PORT02 and PORT03, and in table 4 for stations 6, 7 and 8, in terms of mean wind

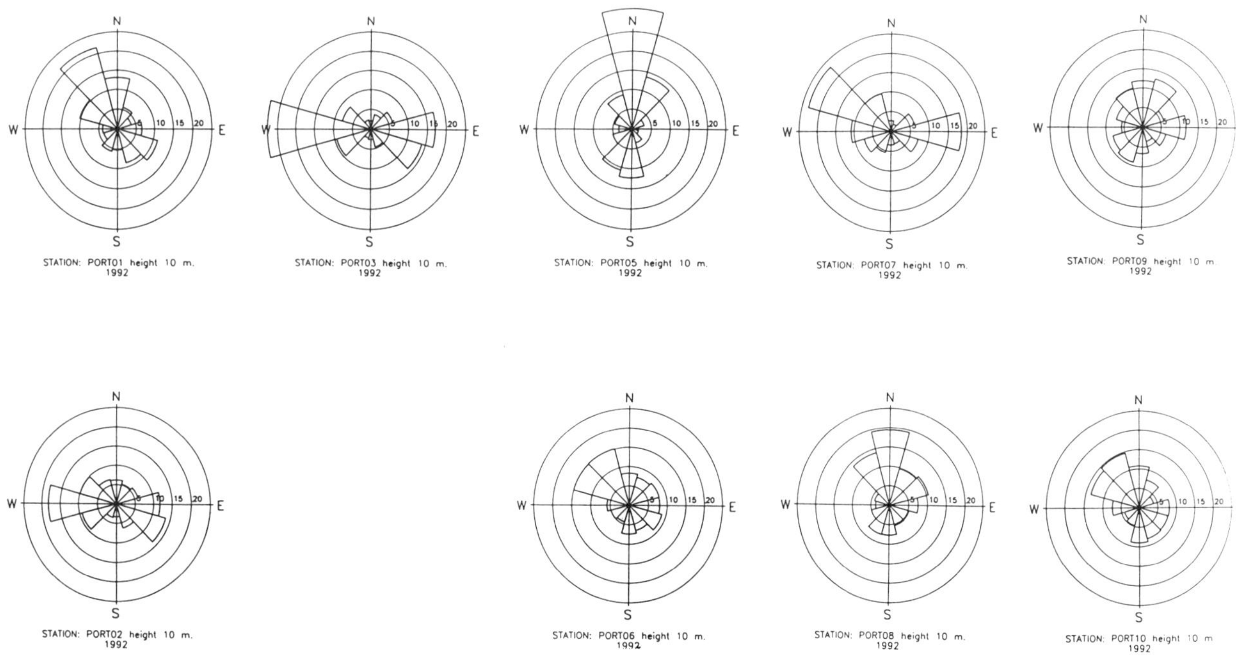

Figure 2. Measured frequency distributions of wind direction. Data collected in 1992, stations PORT01 - PORT10.

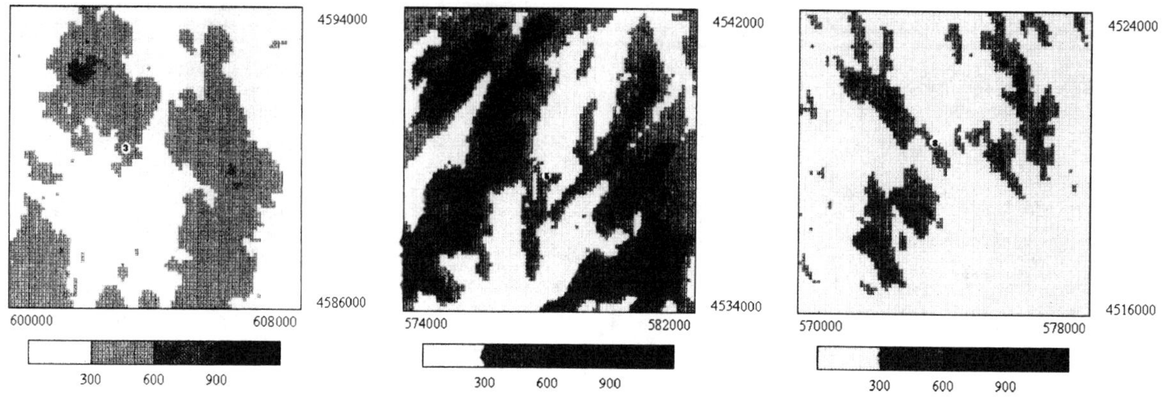

(a) (b) (c)

Figure 3. Wind resource maps for the areas of stations (a)-PORT03; (b)-PORT05; (c)-PORT08. WAsP prediction of power density in Wm^{-2}, at 50 m a.g.l.

speed and power density (again referred to a value of 1.11 kgm^{-3} for air density, to account for the effect of altitude).

Table 4. WAsP predictions using stations PORT06, PORT07 and PORT08 .

	Measured (1992)	Predictions from station PORT06	PORT07	PORT08
Applic. site	V (ms^{-1}) E (Wm^{-2})	V (ms^{-1}) E (Wm^{-2})	V (ms^{-1}) E (Wm^{-2})	V (ms^{-1}) E (Wm^{-2})
Station PORT06	4.6 126	4.5 126	4.4 119	6.1 360
Station PORT07	5.4 196	5.4 215	5.3 195	7.3 634
Station PORT08	6.1 282	5.1 200	4.6 172	6.0 282

In spite of the limitations suggested by these tables, the programme was subsequently used to construct detailed wind resource maps for the areas under study. In each case, the "RESOURCEFILE" utility of WAsP was employed to calculate the power density at the standard height of 50 m on a regular grid of 100 m spacing, covering an area of 8 x 8 km around a station. The results were converted into wind resource maps, using a correspondence between calculated values and a series of grey shades. Sample maps, for stations PORT03, PORT05 and PORT08 are contained in figures 3.a, 3.b and 3.c.

6. DISCUSSION AND CONCLUSIONS

The wind data collected to date confirms that there are sites with good wind potential in the areas under study, that can be explored by adequate siting. The concentration effects are important, as seen by comparing the measured power densities in flat terrain (station PORT01) and in most of the other stations: a magnification factor of up to 3 has been observed in the best cases.

It is also worth noting that the highest wind energy levels are being recorded at the stations further inland, about 100 km away from the Atlantic coast, suggesting that other mountain ridges far from the sea may provide favourable wind conditions.

The extension of the measurement period is not yet representative of long term climatological data, as seen by comparing the average values listed in table 1 for the first and the latest 12 months of measurement. No attempt was made to estimate the probable errors of the data due to climatic fluctuations.

The intercomparisons presented in tables 3 and 4 indicate that, even within distances of few kms, the application of WAsP may encounter difficulties in areas of very complex orography. Further insight is gained from the inspection of the sectorwise frequency distributions in figure 2. The windroses for stations PORT03 and PORT05, for example, show that the wind in both cases is markedly directed by the orientations (E-W and N-S, respectively) of local valleys, visible in figures 1.b and 1.d, and the strong corrections needed must be adequately modelled.

Nevertheless, the resource maps in figures 3.a, 3.b and 3.c have the practical value of guiding the selection of the best sites in each area. They also suggest that some of the stations are not located in the most favourable points, an information that would have been most valuable when the project started. The general level of wind power density at 50 m a.g.l. may be less outstanding than the measured data collected at 10 m a.g.l. could suggest, because the overspeeding produced at hill tops is most effective closer to the ground. Still, the best locations exhibit a predicted power density of the order of 1 kWm^{-2}, a value that goes well with a wind turbine.

ACKNOWLEDGEMENTS

The finantial support provided by the European Comunity, through Contract JOUR-0067-C (JOULE Programme), was essential to perform the work reported. From January 1993, additional support for continuation of the present work has been granted to FEUP through the STRIDE Programme (Contract STRDA/C/ENR/143/92). Both are gratefully acknowledged.

REFERENCES

[1] I. Troen and E. L. Petersen, 1989, European Wind Atlas. Risoe National Laboratory, Roskilde, Denmark.

[2] E. L. Petersen, 1990, Wind Resources in Complex Terrain, Measurements and Models. Proceedings of ECWEC'90 Wind Energy Conference in Madrid, September, pp.61-65.

[3] A. Restivo, 1991, Resource Assessment in Regions of Portugal with Complex Terrain. Wind Energy: Technology and Implementation (Proceedings of Amsterdam EWEC'91), pp.797-801.

WIND MEASUREMENT AND MODELLING IN IRELAND

Rick Watson
Dept. of Electronic and Electrical Engineering
University College Dublin, Dublin 4, Ireland

ABSTRACT: The wind measurements being carried out in Ireland are detailed along with results for 1992. Some preliminary results from the application of WAsP to the results at two stations are presented. These results are considered promising given the lack of high resolution digitised contour data available. The priorities for further work are outlined.

1. INTRODUCTION

There are three wind measurement campaigns being carried out in Ireland at present. One is being carried out by University College Dublin (UCD) under a JOULE project, a second is being carried out by the Electricity Supply Board (ESB) and the third is being carried out by a private firm Arigna Fuels Ltd. (AFL). There has been considerable cooperation between the different campaigns in order to avoid overlap and to optimise the benefits to the different participants.

This cooperation has taken the form of agreement on the choice of

- ♦ sites — to obtain the best spread over the wind rich areas which are chiefly along the western seaboard
- ♦ of data loggers and wind measuring instruments — NRG loggers and instruments are being used
- ♦ of logging interval — ten minutes at most stations, one minute at some
- ♦ of heights above ground level at which the measurements of windspeed (mean and standard deviation) and wind direction are taken — in most cases ten metres and thirty metres and in some cases thirty metres and fifty metres
- ♦ of logging period — to the end of 1993 for most stations.

In addition it has been agreed to share the wind data collected and also to share detailed site description information including contour maps, roughness maps and location and dimensions of obstacles.

In this paper the objectives of the combined project are outlined, the current status of all three campaigns is reported upon and summary results for the year 1992 at fourteen stations are presented. Preliminary results of applications of the WA^S^P tool to the data collected at some of the stations are also presented and the plans for further work are set out.

2. OBJECTIVES OF THE PROJECT.

The objectives of the project are:

- ♦ to identify and investigate the wind resource at the selected sites.
- ♦ to evaluate the suitability of WA^S^P as a tool for wind resource assessment in the complex terrains typical of the west of Ireland.
- ♦ to investigate the effects of spatial diversity of the instantaneous wind resource on two scales:
    - firstly on the scale of the national grid by collecting synchronous time series of ten minute means of windspeed and wind direction at dispersed sites with separations of the order of hundreds of kilometres.
    - secondly on the scale of a single windfarm by collecting synchronous time series of one minute means of windspeed and direction at sites with separations of the order of a few kilometres.

The objectives were to be realised originally by measuring ten minute mean windspeeds at four sites dispersed by hundreds of kilometres and by measuring one minute mean windspeeds at four sites within a few kilometres of each other in one area. However the number of stations included in the monitoring programme has been greatly increased due to the informal cooperation with the wind site assessment study being carried out by the ESB and with the investigation of the upland inland area of Arigna being carried out by the firm AFL.

3. THE STATUS OF THE WIND MEASUREMENT CAMPAIGNS.

The locations of the monitoring stations are indicated by numbered circles on the map in Figure 1. Five general areas are covered i.e. Donegal in the north, Mayo in the west, the Shannon area in the midwest, the Arigna area in the northwest and Wexford in the southeast. As may be seen the full network of stations provides reasonably good coverage of the country's wind rich areas.

The details of the stations which are or have been monitored (sixteen in all) are listed in Table 1. Each station has been assigned a number — used to identify the station in Figure (1) — and a three letter code e.g. CMY is Cahermurphy in Co.Clare.

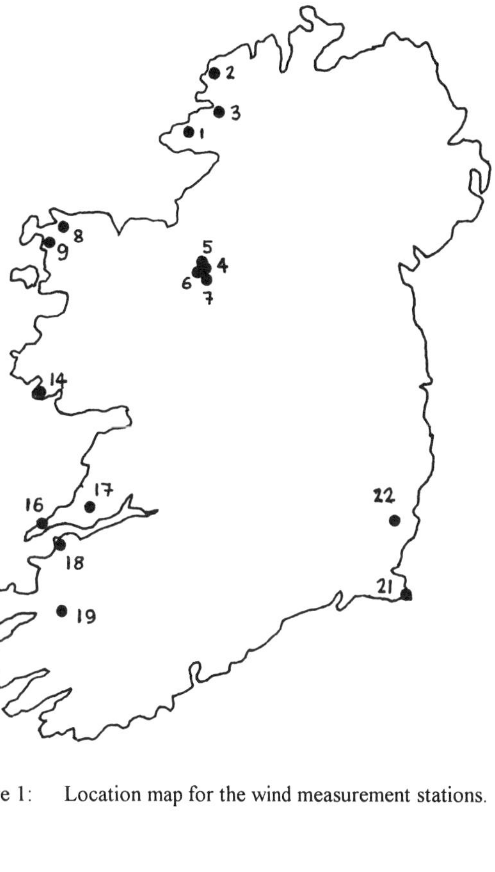

Figure 1: Location map for the wind measurement stations.

No.	Code	Name	Location	Operator	Start	Mast	Logger	Instruments	Altitude	Interval
1	MSG	Mulmusog	Co.Donegal	UCD	May 92	TE	NRG9600	u50,d50,u30	350 m	10 min
2	CDY	Clady	Co.Donegal	UCD	Sep 91	ESB	NRG9600	u30,d30,u10,d10	60 m	10 min
3	DGE	Dungloe	Co.Donegal	UCD	Apr 92	TE	NRG9600	u50,u30,d30	20 m	10 min
16	KAG	Knocknagarhoon	Co.Clare	UCD	Dec 91	ANSO	NRG9600	u30,d30,u10,d10	121 m	10 min
18	KAN	Knockanore	Co.Kerry	UCD	Dec 91	ANSO	NRG9600	u30,d30,u10,d10	264 m	10 min
22	DGD	Drumgoold	Co.Wexford	UCD	Apr 92	TE	NRG9600	u50,d50,u30	70 m	10 min
6	KRU	Kilronan Upper	Co.Roscommon	ESB	Dec 91	TT	NRG9600	u30,d30,u10	320 m	10 min
8	GMY	Glenamoy	Co.Mayo	ESB	May 92	TT	NRG9600	u30,d30,u10	37 m	10 min
9	RTH	Rath Hill	Co.Mayo	ESB	May 92	TT	NRG9600	u30,d30,u10	59 m	10 min
17	CMY	Cahermurphy	Co.Clare	ESB	Jan 92	TE	NRG9600	u30,d30,u10	175 m	10 min
19	EGM	Eagle Mountain	Co.Kerry	ESB	Dec 91	TT	NRG9600	u30,d30,u10	400 m	10 min
4	AG1	Altagowlan 1	Co.Roscommon	AFL	Sep 91	TT	NRG9600	u30,d30,u10	400 m	10 min
7	AG2	Altagowlan 2	Co.Roscommon	AFL	Aug 92	TT	NRG9600	u30,d30	400 m	10 min
5	KRL	Kilronan Lower	Co.Roscommon	AFL	Sep 91	TT	NRG9600	u30,d30,u10	220 m	10 min
21	CSP	Carnsore Point	Co.Wexford	ESB	Jan 93	TT	NRG9600	u30,d30,u10	10 m	10 min
14	MHD	Mace Head	Co.Galway	UCG	Sep 91	UCG	PC	u20,d20	10 m	1 min

Table 1: Wind monitoring station details.

The information given in Table 1 includes the name and location and operator of the station, the date monitoring began, the type of mast being used (TT is an NRG talltower, TE is a Telecom Eireann communication mast, ESB is an ESB communication mast, ANSO is an Air Navigation communication mast), the type of logger in use, the altitude above sea level of the station and the logging interval being used, the location and type of measuring instruments e.g. for CMY u30, d30, u10 indicates windspeed is being measured at 30 m and 10 m a.g.l. and wind direction is being measured at 30 m a.g.l. In addition to the 10 minute mean windspeed and the wind direction the standard deviation of the windspeed is also recorded every ten minutes which will allow calculations of turbulence intensity to be carried out. Each time a data chip on a logger is exchanged the maximum two second gust since the last datachip exchange is also recorded.

The installation program for the 10 min measurements is almost complete (one further station is planned). There are fourteen stations operating at this time, thirteen at ten minute logging interval and one, the MHD station at a one minute logging interval — MHD is an atmospheric monitoring station operated by University College Galway (UCG).

Logging has been discontinued at two of the Arigna stations AG1 and KRL and the masts and instruments were relocated to alternative sites. The installation of four one minute stations planned for the Mayo area has been delayed but it is hoped to have them installed over the next few months.

4 WIND MEASUREMENT RESULTS FOR 1992.

The raw data from all the monitoring stations is analysed by UCD. The data is assembled into weekly and monthly files ready for further processing. The ten minute time series can be plotted on a daily, weekly or monthly time frame. This proves very useful in checking for inconsistencies in the data. The data quality is further checked by comparing it with summarised Meteorological Service data for the same period.

Initial summary statistical analysis on the data collected during 1992 has been carried out. The windspeed data has been analysed on a monthly basis in terms of monthly data availability and monthly mean windspeed. The data for fourteen stations operating has been analysed in this way (the MHD data although available has not yet been analysed).

The windspeed data monthly percentage availability at 30 m a.g.l. for fourteen stations in 1992 is summarised in Table 2. As can be seen there are only a limited number of stations where a full years data is available. At six of the stations logging only started in mid year while at two others logging was discontinued (as indicated by the letter d). There were serious gaps in data at KRU due to a faulty anemometer and at MSG where an anemometer was destroyed by a lightning strike and the logger was also damaged. There was also faulty instruments at CMY in the latter months while at EGM problems in the earlier part of the year were solved and operation has been excellent since then.

STATION	CDY	DGE	MSG	KAG	KAN	DGD	RHL	GMY	AG1	AG2	KRL	KRU	CMY	EGM
jan	100			100	100				87		42.2	32		61.9
feb	100			100	98.6				99.1		100	100	97.8	97.9
mar	100			100	100				97.7		98.1	98.1	99.9	75.5
apr	100			100	100				89.8		99.8	94.3	84.7	77.6
may	100	100		100	100	100			100		59.6	65.2	85	0
jun	81.2	81.2	81.1	81.6	82.3	94.7	100	100	d		d	4.1	100	27.1
jul	100	100	100	100	100	100	100	100	d		d	0	100	100
aug	100	99	100	100	100	100	100	81.6	d	100	d	0	100	100
sep	98.8	100	98.2	94.6	94.6	93.3	100	19.2	d	100	d	42.4	46.2	100
oct	100	100	100	87.1	87.3	96.6	100	98.9	d	100	d	0	0	100
nov	80.9	19.7	19.5	97.7	96.2	100	99.9	52.3	d	100	d	0	0	100
dec	78.9	46.6	0	95.2	100	100	72.4	nc	d	nc	d	nc	nc	nc
1992	95	81	71.3	96.4	96.6	98.1	96	75	94.7	100	79.9	39.7	71.4	76.4

Table 2: Mean monthly windspeed data availabilities [%] at 30 m a.g.l. for fourteen stations in 1992.

STATION	CDY	DGE	MSG	KAG	KAN	DGD	RHL	GMY	AG1	AG2	KRL	KRU	CMY	EGM	WIND INDEX
jan	6.9			7.8	8.4				8.6		8	10.3		10.1	0.80
feb	9.1			10.2	10.9				10.8		7.9	8.8	8.9	9.7	1.09
mar	9			11.3	12.3				11.7		9.2	10.3	10.5	9.9	1.32
apr	7.5			9.7	10.6				9.8		7.8	8.5	8.7	8.8	1.23
may	7.6	6.5		8.8	9.7	5.1			9.1		8	8.8	9		1.17
jun	5.2	4	6.4	5.5	5.9	3.9	6.4	5.8	d		d	7.1	5.7	4.8	0.88
jul	6.3	5.3	6.9	6.8	7.4	4.7	7.9	6.6	d		d	m	6.5	6.7	1.09
aug	6.7	6.1	7.6	8.6	9.4	5.5	9.2	7.7	d	9.1	d	m	7.7	7.9	1.31
sep	6.6	5.7	8.1	8.2	8.6	5.5	8.8	9.9	d	8.6	d	5.7	9.6	7.9	1.05
oct	5.9	4.6	7.5	8.8	8.9	5.5	7.9	6.7	d	7.8	d	m	m	6.9	0.95
nov	9.2	8.6	9.9	10.7	11.7	6.7	12	8.9	d	11.8	d	m	m	9.9	1.17
dec	6.5	5.3	m	9.2	9.3	5.4	9.9	nc	d	nc	d	nc	nc	nc	0.92
1992	7.2	5.5	7.4	8.9	9.5	5.3	8.8	7.1	10	9.3	8.2	8.9	8.2	8.4	1.08

Table 3: Mean monthly windspeeds [m/s] at 30 m a.g.l for fourteen stations and wind index for 1992.

The mean monthly windspeeds (for the time for which data was available) at 30 m a.g.l. for the fourteen stations is summarised in Table 3. Missing data is indicated by the letter m. For some of the entries the mean indicated is just for part of the month as can be seen by consulting the data availabilities in Table 2. The data for December 1992 had not yet been obtained for GMY, KRU, CMY, AG2 and EGM as indicated by the letters nc.

In addition a wind index for each month is listed in the last column. This index is an indication of how windy a particular month was. It has been calculated by taking the ratio of the average of all the coastal meteorological service stations mean monthly windspeeds measured at 10 m a.g.l. in 1992 to the average of their longterm (1962-1984) mean monthly windspeeds for each month. As may be seen the year as a whole was about 8% above the average.

The highest mean windspeed for a station with a full years data has been recorded at KAN where the mean for the year was 9.5 m/s at 30 m a.g.l.. The KAN station is located on a well exposed smooth hill (altitude 264 m) close to the coast south of the Shannon estuary. The KAG station is located on the other side of the Shannon estuary again on a well exposed smooth hill (altitude 121 m) and shows a mean windspeed of 8.9 m/s at 30 m a.g.l.. Two other stations are located in the same region i.e. CMY and EGM. They are located about 25 km inland also on well exposed high ground (altitudes 175 m and 400 m respectively) and they show lower mean winds, 8.2 and 8.4 m/s (note however that the data availability at these two stations was 76.4 % and 71.4 % respectively).

	KAG 10	KAG 30	KAN 10	KAN 30
KAG 10	7.8	7.6	8.4	8.4
KAG 30	9.1	8.9	9.7	9.8
KAN 10	8.4	8.2	8.6	8.7
KAN 30	9.2	9.0	9.4	9.5

Table 4: Mean windspeeds for 1992 at the stations down along the first column, (as predicted by stations along the top row)

5 PRELIMINARY WA^{S}P RESULTS.

WA^{S}P has been applied to the results for one year (1992) at two stations KAN and KAG located on either side of the Shannon estuary. The data availability for the two stations for this period was excellent. The availability of high resolution digitised contour data is however limited at present.

The Ordnance Survey is at this time in the process of producing a full set of 1:50000 scale maps from digital data however these new maps do not yet cover the areas of interest in this project. The Ordnance Survey expect to finish their work by the mid 1990's. It was therefore necessary to use the WA^{S}P facility to digitise the available half inch map (scale 1:126720). On this map the contour resolution is 100 ft (30.5 m) and even greater above 1000 ft (305 m). Roughness maps were also prepared using the WA^{S}P facilities.

Sector	KAN10 predicted			KAN10 measured		
°	A	k	f	A	k	f
0	6.8	1.61	6.2	8.9	1.95	6.7
30	8.1	1.62	5.0	8.6	2.12	6.6
60	7.2	1.81	3.3	6.1	1.89	4.3
90	6.2	2.06	3.8	6.0	2.0	2.9
120	6.1	1.85	4.3	7.3	2.18	3.3
150	6.3	1.85	5.4	7.5	1.97	6.1
180	9.2	2.06	9.3	9.3	1.89	10.6
210	12.5	2.37	16.4	11.1	2.36	16.3
240	11.8	2.35	17.0	11.7	2.49	13.8
270	10.2	2.47	13.5	11.1	2.49	12.7
300	8.8	2.28	9.0	10.7	2.53	9.5
330	7.3	1.88	6.9	9.1	1.98	7.5
Total	9.5	1.90	100.0	9.8	2.05	100.0

Table 5: Comparison between KAN station 10 m level as predicted by KAG 10 m level and as actually measured.

The results for the mean windspeed for 1992 are listed in Table 4. The stations across the first row are the predicting stations, the results across from column one are the predicted values. As may be seen when the KAG stations are predicting the results are reasably good but the KAN stations overpredict the KAG stations.

The results when the KAG 10 m station is predicting are shown in more detail in Table 5 and Table 6. The results for the weibull A parameter and the frequency of occurance for KAN 30 m as predicted by KAG 10 m are shown plotted against direction sector in Figures 3 and 4.

As can be seen from these Tables and Graphs the results are quite promising. It is planned that more detailed digitised data for the two hilltops in question (KAN and KAG) will be purchased shortly and superimposed on the regional WA^{S}P maps. More detailed roughness maps will also be made up and it is then hoped that the promise of the present results will be improved upon.

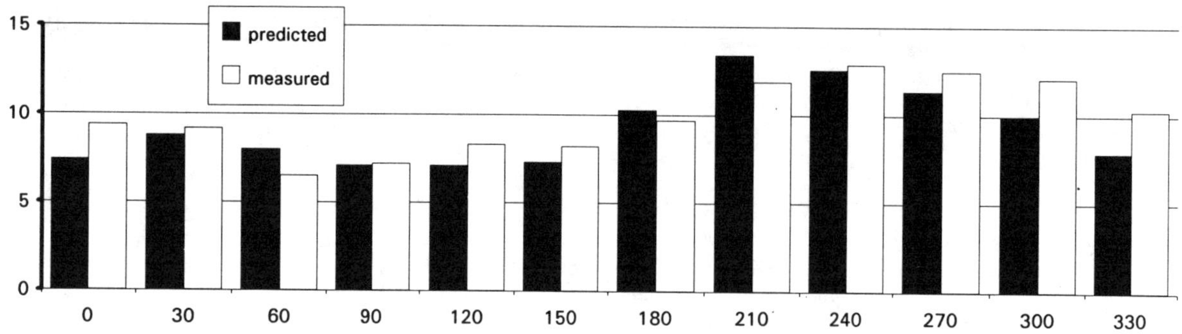

Figure 3: Comparison of the sectorwise KAN 30 m weibull parameter [m/s] as predicted by KAG 10 m and the measured value.

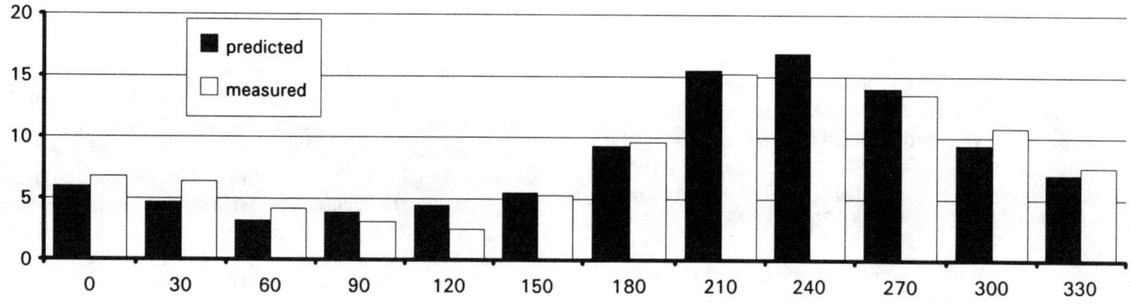

Figure 4: Comparison of the sectorwise KAN 30 m frequency of occurance [%] as predicted by KAG 10 m and the measured value.

Sector	KAN30 predicted			KAN 30 measured		
°	A	k	f	A	k	f
0	7.4	1.76	6.0	9.4	2.00	6.8
30	8.8	1.83	4.7	9.2	2.14	6.4
60	8.0	2.03	3.2	6.5	1.92	4.2
90	7.1	2.27	3.9	7.2	2.31	3.1
120	7.1	2.03	4.5	8.3	2.12	2.5
150	7.3	2.03	5.5	8.2	1.86	5.3
180	10.3	2.26	9.3	9.7	1.92	9.6
210	13.4	2.52	15.5	11.9	2.43	15.2
240	12.6	2.49	16.9	12.9	2.44	15.0
270	11.4	2.55	14.0	12.5	2.59	13.5
300	10.0	2.39	9.4	12.1	2.61	10.8
330	7.9	1.94	7.0	10.3	2.02	7.6
Total	10.4	2.04	100.0	10.8	2.12	100.0

Table 6: Comparison between KAN station 30 m level as predicted by KAG 10 m level and as actually measured.

6. CONCLUSIONS AND FURTHER WORK.

The measurement campaigns have been reasonably successful to date with the exception of a number of stations where the data availability has been poor. It is hoped to improve on the availability during the remainder of the project. Measurements will continue at least until the end of 1993.

The first applications of WA^{S}P have proved encouraging if the already mentioned constraint of the availability of only poor resolution digitised contour data is taken into account. This situation will be rectifed at some of the stations by the purchase of high resolution surveys.

The priorities for further work over the next year are:

- to complete the installation progamme of the one minute measurement stations and one further 10 minute station.
- to assemble more detailed descriptions of all the sites including digitised contour data (with high resolution data in the vicinity of the stations), terrain roughness estimations and location of obstacles.
- to apply WA^{S}P to the results obtained by extrapolating from one station to the other and comparing the WA^{S}P results with those measured.
- to carry out a more detailed statistical analysis of the results and in particular the analysis from the point of view of spatial diversity of the wind resource.

7. ACKNOWLEDGEMENTS.

The financial support of the Commission of the European Communities under the JOULE programme and the cooperation of the Electricity Supply Board and Arigna Fuels Ltd and University College Galway in allowing access to their wind data is gratefully acknowledged.

WIND RESOURCES, PART II: CALCULATIONAL METHODS

N.G. Mortensen, E.L. Petersen and L. Landberg
Meteorology and Wind Energy Department
Risø National Laboratory
Roskilde, Denmark

ABSTRACT: The Wind Atlas Analysis and Application Program represents one of several calculational methods or tools that are used routinely in the analysis of wind data and estimation of wind resources. Based on the experience gathered at Risø it is concluded that the program works very well in relatively uncomplicated topography. Less is known about the performance in mountainous terrain, but it is concluded that high-quality wind and topographical data, as well as great attention to detail, are a prerequisite to the succesful application of such models in complex terrain. A case study, using data from an ongoing measuring campaign in very complex terrain in Northern Portugal, is used to illustrate the capabilities and limitations of current models.

1 INTRODUCTION

During the last half a dozen years numerical flow models and other advanced computational tools have become widespread in the wind energy community. One such collection of computer tools is the Wind Atlas Analysis and Application Program (WA^{S}P) which was developed during the investigation leading to the European Wind Atlas. As the name indicates the program is used mainly for two purposes:

1. Establishing of regional wind climatologies, ie wind atlases
2. Micro siting, ie determination of optimum wind turbine sites in specific areas.

Over the years, WA^{S}P has been distributed to several hundred institutions and companies in more than 50 countries around the world. The application of standard tools and common guidelines in the analysis of wind data and wind resource estimation has led to a common frame of reference and a much needed standardization. This paper is a brief report on the status of WA^{S}P, including some observations on the proper use of the program. The limitations of current models in complex terrain are illustrated by a recent example from Portugal and possible future initiatives are touched upon.

2 REGIONAL WIND CLIMATOLOGIES

The European Wind Atlas, or more precisely the EC Wind Atlas, was produced using WA^{S}P and published in 1989. The experience gained from using the regional climatologies is discussed in an accompanying paper (E.L. Petersen: The European Wind Climatology). During the following years wind atlases were produced and published in Algeria, Jordan, Finland, Sweden, Switzerland and Western Australia. Other countries are currently undertaking work using WA^{S}P, which eventually will lead to national atlases in eg Egypt, Syria, Libya, Turkey, Bulgaria, Poland and Israel. Limited regional studies have been performed in India, China, Mongolia, Somalia, Cabo Verde, Chile, Australia and USA.

As discussed in the first part of this paper the reliability of the calculated regional wind climatologies depends strongly on the quality of the data base, the complexity of the topography and the climate of the region. Hence, any published wind atlas should contain a discussion on the representativeness of the statistics based on a proper verification. There are usually three steps in the verification procedure: 1) letting the stations in the data base predict each other, 2) comparisons with measurements from tall masts and 3) comparisons with other types of available information such as voluntary ship reports or power production statistics from wind turbines. As an example, the Swedish Wind Atlas uses the verification to rank the stations of representativeness and actually dismisses a few of them. It is found that especially in Southern Sweden the reliability of the wind atlas statistics is high. On the other hand, in Finland it is found that the quality of the data base is generally not sufficient due to the location of most of the stations in very sheltered surroundings.

3 MICRO SITING

WA^{S}P has been used for siting a large number of wind turbines, single or in wind farms. For countries where a national wind atlas exists the common siting procedure is to select the proper wind atlas statistics and further to acquire a set of meteorological data from the nearest meteorological station. By means of WA^{S}P both data sets are then made representative for the site in question. Ideally, the power production estimates calculated for the two sets of statistics should be close; otherwise it is necessary to look deep into the circumstances under which the two data sets originated – certainly not a trivial matter. In the more general case no wind atlas statistics exist, but data have been collected 'in the neighbourhood' for a few years; often just one year or even shorter. Due to the seasonal and interannual variations of the wind climate use of data series shorter than 3 years should be avoided; the use of short series may add considerably to the uncertainty of the mean power production estimates.

In order to check the mean power production estimates of WA^{S}P it is necessary to pick a number of wind turbines that have produced reasonably continuous over several years. This has been done in Denmark and Southern Sweden with the conclusion that WA^{S}P performs very well in this region of relatively uncomplicated topography. For most of the wind turbines over the world, however, it is necessary to wait some years before sufficient statistics have been collected.

A number of test studies have been performed over the years, largely based on data from two experimental meteorological masts, in order to check whether WA^{S}P was able to predict the measurements from one mast using the data from the other mast. Some of these studies have had rather negative conclusions, but in each case a close look revealed that one or more of the following major errors were committed:

1. the two time series were very short, often only a few months long and sometimes not even covering the same period of time.
2. the orography was not described with sufficient accuracy and detail, especially close to the mast.
3. the position of one, or both, of the masts was entered wrongly to WA^{S}P which make the orography calculations and corrections meaningless.
4. the wind speed and direction series contained serious errors.
5. the author did not understand how to use the program.

There are, however, clear limitations to the use of WA^{S}P: On the micro scale very steep slopes with separating flows are not treated correctly, and on the large scales the dynamics are too simplified by the neglect of the large scale stratification, and also in relation to local, thermally driven flow systems. It should be mentioned, however, that the problems related to large scale flows are alleviated if input data come from a station not too far from the point of application relative to the horizontal scale of the important flow systems.

One of the objectives of the JOULE project "Wind measurement and modelling in complex terrain" is to establish a data base of high-quality wind measurements that can be used in the evaluation and further development of existing models. The project has now come so far as to allow preliminary investigations of this nature.

4 A CASE STUDY IN MOUNTAINOUS TERRAIN

As part of the University of Porto's contribution to the project mentioned above, ten meteorological masts have been erected in the northernmost part of Portugal. A detailed description of the stations and their surroundings, the instrumentation used and the data obtained so far is given by Restivo and Petersen (1993, this volume).

For the purpose of this investigation six stations were selected, representing three different degrees of topographical complexity along a 50-km long profile: from the Atlantic coast in the W to the more than 1000-m high mountains of Arada-S. Macário in the E. In the following we will refer to the stations by the numbers 01, 10, 09, 08, 07, and 06 (going from W to E); a detailed description of the stations, including their names and coordinates, are given by Restivo and Petersen (ibid).

4.1 Station characteristics

Station 01 is situated close to the Atlantic coast in a flat and level area, and the terrain surface close to the station consists of water, farmland and marsh. There are no sheltering obstacles close to the station. Hence, the areas of different roughness are by far the most important topographic influence on the wind measurements.

Stations 10 and 09 are situated approx. 34 km ENE of station 01, in the mountains of Freita. The altitudes of the two stations are 1012 and 1082 m a.s.l, respectively, and the distance between the stations is 3 km. Here the roughness is much more uniform than at station 1, but the height variations around the stations give rise to pronounced orographic effects.

Stations 08, 07 and 06 are situated approx. 14 km further inland, or about 40 km from the Atlantic coast. The three stations are situated a few km from each other in very complex, mountainous terrain, see Restivo and Petersen (ibid). The altitudes of the stations are 1057, 982 and 932 m a.s.l, respectively. Detailed maps of the surroundings of each station are shown in Fig. 1 and some simple orographic characteristics are listed in Tab. I.

Table I: Differences in altitude between the lowest and highest level (Δz) and standard deviations of grid point heights (σ_z) within unit areas of 64 km^2 and 1 km^2 around the six stations. The grid size ($\Delta x = \Delta y$) is 100 m and 10 m, respectively.

	8×8 km^2		1×1 km^2	
#	Δz	σ_z	Δz	σ_z
01	$O(1)$	$O(1)$	$O(1)$	$O(1)$
10	900	187	100	16
09	800	200	92	19
08	718	197	248	51
07	800	178	257	57
06	750	183	260	45

Digital height contour maps covering an area of 8×8 km^2 around each site was used as input to the flow model of WA^{S}P. The altitude differences listed in Tab. I for these maps correspond approximately to the *relative relief* given in the European Wind Atlas. It appears, that neither this measure, nor the standard deviation of 100-m grid point heights seem to differentiate the five mountain stations. The same measures, however, calculated for areas of 1×1 km^2 around each site, show clear differences between the three groups of stations.

4.2 Model calculations and input data

The model used for the wind climate estimations below is version 4.0 of WA^{S}P, which is the latest version of the model that was used for the European Wind Atlas (Troen and Petersen, 1989). The application of this version is described by Mortensen et al. (1993a, 1993b).

The input data to WA^{S}P consist of wind measurements and a detailed topographical description of the surroundings of each station. The instrumentation used for measuring the wind speed and direction is described by Restivo and Petersen (ibid). For the purpose of this study the time-series covering one full year (1992) were used.

The orographic information is contained in digital maps giving the coordinates of height contour lines. In this case, each map covers an area of approx. 8×8 km^2, with a contour interval of 50 m. However, within an area of 1×1 km^2 centered around the stations the contour interval is 10 m. The station levels have been marked with additional height contours in order to get as detailed and accurate a description of the terrain as possible.

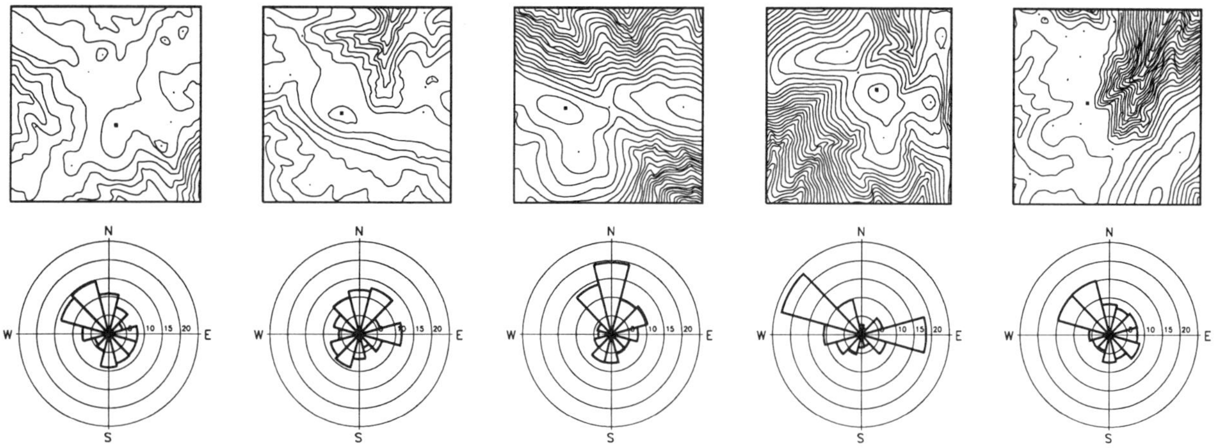

Figure 1: a) Detailed 1×1 km^2 maps of the surroundings of stations 10, 09, 08, 07 and 06 (upper panel). Left to right corresponds to W–E. The contour interval is 10 m. b) Sector-wise distribution of the winds measured at the five stations (lower panel).

The surface roughness at the mountain stations were assumed uniform with a z_0 of 0.05 or 0.03 m. The roughness lengths were estimated from photographs and maps. From the same material it was concluded that neither of the six stations are likely to be influenced by sheltering obstacles. – The roughness classification for station 01 was entered to WA^{S}P as a *roughness map*, ie a digital map of roughness change lines (Mortensen et al, 1993b).

4.3 Results

Using the wind data and topographical information described above, wind atlas data sets were generated for the six stations. These data sets were subsequently used together with the topographical information to predict the wind climate at 10 m a.g.l. at the other five stations. The results are presented below in a so-called score scheme (Troen and Petersen, 1989), see Tab. II. The first column contains the predicted stations and the first row the number of the predictor stations. The diagonal line of bold face figures thus indicates the stations predicting themselves; these values are close to those actually measured, which are listed in the last column.

Table II: Measured and estimated mean wind speeds (ms^{-1}) and mean wind power densities (Wm^{-2}) at 10 m a.g.l. for the six stations. The power densities have not been corrected for the influence of different air densities.

	01	10	09	08	07	06	Meas.
PORT01	**4.2**	4.6	4.6	4.4	3.4	3.4	4.4
	118	137	148	140	58	58	127
PORT10	5.4	**5.5**	5.5	5.2	4.0	4.3	5.6
	262	**227**	233	209	96	115	231
PORT09	5.3	6.0	**5.7**	5.7	4.1	4.3	5.8
	242	321	**272**	293	125	124	279
PORT08	6.7	6.8	6.5	**6.0**	4.6	5.1	6.1
	569	516	398	**313**	191	216	310
PORT07	6.3	7.3	7.2	7.2	**5.3**	5.3	5.4
	354	579	592	653	**216**	229	216
PORT06	5.4	6.2	6.1	6.1	4.4	**4.5**	4.6
	233	354	337	384	136	**139**	139

As one might expect, the picture emerging from Tab. II is not as simple and unambiguous as is often seen in cases of less complex terrain.

Stations 01, 10 and 09 compare fairly well, the differences in mean wind speed and power density being mostly within 10% and 15%, respectively.

The comparisons become difficult, however, when we look at stations 08, 07 and 06. Station 08 predicts stations 01, 10 and 09 reasonably well, but is not predicted well by the same stations. The prediction by station 09 is the closest, possibly because of the orographic resemblance to station 08.

Stations 07 and 06 predict significantly lower mean values at the other four stations than is actually measured. And, conversely, the predictions of 07 and 06 by these stations are significantly higher than the wind climate observed. However, it is noteworthy, that the two stations 07 and 06 compare very well. The reasons for this, as well as for the discrepancies between 07 and 06 and the other stations are not yet clear, but are likely related to the orographic characteristics of these sites – which the model is obviously not able to correct for. Therefore, these stations can be considered representative only of their specific locations and nearby surroundings with similar orographic characteristics.

Finally, it should be mentioned that several different roughness classifications have been employed during the station intercomparisons. It was found, that even though the predictions change somewhat with different roughness descriptions, the overall picture remained essentially the same.

4.1 Comparison with the European Wind Atlas

As mentioned above, only one full years worth of data has been collected so far, hence a comparison with the stations in the European Wind Atlas must be preliminary. The mean wind speeds and mean energy densities for the six stations, at 50 m a.g.l. over roughness class 1 is given in Tab. III.

Table III: Estimated mean wind speeds (ms^{-1}) and mean wind power densities (Wm^{-2}) 50 m a.g.l. over a uniform, flat surface of z_0 = 0.03 m (roughness class 1).

Station	01	10	09	08	07	06
Wind speed	5.1	5.7	5.5	5.4	4.1	4.2
Density	162	209	204	213	80	87

For comparison, the near-coastal stations of the Wind Atlas, Porto and Viana do Castelo, are reported as having mean wind speeds and energy densities of (5.5 ms^{-1}, 198 Wm^{-2}) and (4.7 ms^{-1}, 221 Wm^{-2}), respectively. Coimbra, 60 km to the S and 40 km inland, is reported as (4.8 ms^{-1}, 225 Wm^{-2}). A tentative conclusion is that these (admittedly sparse) observations does not seem to change the picture given in the European Wind Atlas much.

5 CONCLUSION

Comparisons of wind data measured at meteorological stations, as well as actual power productions measured over several years in Denmark and Southern Sweden, have shown that WA^{S}P performs very well in relatively uncomplicated topography. A similar data base of high-quality wind and/or power production measurements is not yet readily available for most areas of complex terrain. Consequently, less is known on the performance of common computational tools in such areas. Furthermore, the application of programs such as WA^{S}P in complex terrain, require great attention to the details of the wind measurements and topographical input in order to get meaningful results.

However, as indicated by the case study above, there are limitations to the use of programs like WA^{S}P. It has long been recognized that in order to do realistic wind climate assessments in mountainous areas there is a need for a solid dynamical model as well. Work along this line has recently been initiated, starting out with the Karlsruhe Mesoscale Model (KAMM). Several goals have to be accomplished:

1. The models are extremely computer time consuming therefore they have to be simplified – if possible – considering the use in relation to wind energy.

2. Establishing methods for the use of such models, which are fundamentally prognostic, for diagnostic wind climate analysis.

The difference between the models in WA^{S}P and KAMM is that the model in WA^{S}P is a local model which can handle very detailed information of the topography out to approximately ten kilometres from the sites. In contrast, the mesoscale models calculate the wind resources over several hundreds of kilometres in squares of one to five square kilometres. Whether the goals are met will be tested by a massive measurement programme in mountainous terrain; employing more than 30 instrumented masts of 30 m height. The masts are currently being put into operation in Crete, Central Italy, Northern Portugal and Southern Germany.

Yet another type of model is necessary for siting in mountainous areas, ie a model which can predict where separating flow occurs. An attempt is currently being made to construct such a model based on the governing equations for the turbulence and using the k-ε closure scheme.

Hence, it is foreseen that future wind climate analysis and siting in mountainous terrain will require the use of all three models mentioned above.

ACKNOWLEDGEMENTS

Prof. A. Restivo, University of Porto, provided the digital maps and other information necessary for the case study. The work was supported by the Commission of the European Communities, DG XII, under contract No. JOUR/0067.

REFERENCES

[1] Mortensen, N.G., L. Landberg, I. Troen and E.L. Petersen (1993a). Wind Atlas Analysis and Application Program. Vol. 1: Getting Started. Risø–I–666(EN)(v.1). Risø National Laboratory, Roskilde. 30 p.

[2] Mortensen, N.G., L. Landberg, I. Troen and E.L. Petersen (1993b). Wind Atlas Analysis and Application Program. Vol. 2: User's Guide. Risø–I–666(EN)(v.2). Risø National Laboratory, Roskilde. 133 p.

[3] Petersen, E.L. (1993). Wind resources, Part I: The European wind climatology. This volume.

[4] Troen, I. and E.L. Petersen (1989). *European Wind Atlas.* ISBN 87–550–1482–8. Risø National Laboratory, Roskilde. 656 p.

[5] Restivo, A. and E.L. Petersen (1993). Wind measurement and modelling in mountainous regions of Portugal. Preliminary results. This volume.

Assessment of Wind Potential and Area Available for Wind Farms at The Lower Saxonian Coast

Thomas Pahlke, Gerhard J. Gerdes, Armin Keuper

Deutsches Windenergie-Institut, Ebertstr. 96, D-2940 Wilhelmshaven, FRG
Tel.: ++49/4421/4808-0, Fax: ++49/4421/4808-43

On behalf of the Ministry of Environment of Lower Saxony the German Wind Energy Institute (DEWI) carried out a resource assessment for the area and determined wind potential in 10 rural districts and cities at the coastal region of lower Saxony. The procedures to determine suitable areas, wind speeds and possible energy output are explained.

1. Introduction

Due to recently improved electricity tariff systems and to state subsidy programmes wind energy can now be used with an economical benefit. For this reason the local authorities are confronted with a flood of applications for planning and building permissions for single wind energy converters and wind farms. The district administrations being responsible for environmental planning, nature and landscape protection have to decide on the applications. The fact that a large scale use of wind energy has not been considered in the environmental plans and lack of experience led to some uncertainties in the decision making process.
Additionally the aim of the environmental ministry of Lower Saxony to install wind turbines with the total power of 1000MW by the year 2000 requires environmental planning of wind energy use.

Thus, the environmental ministry of Lower Saxony has commissioned DEWI to assess the wind potential and the areas available for wind farms of 10 Lower Saxonian districts (Figure 1) located near the coast of the North Sea.

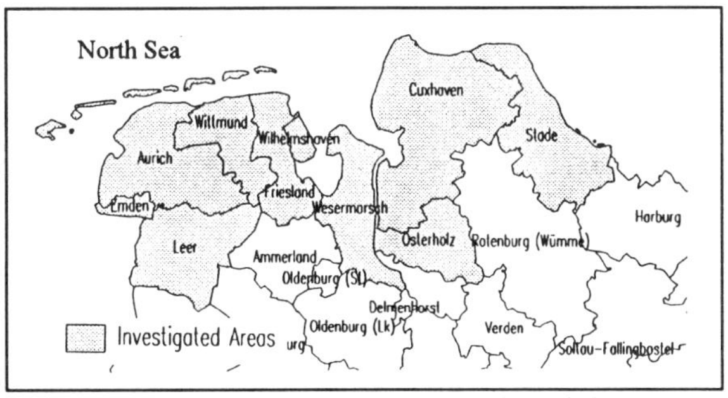

Figure 1: Map of 10 districts and cities of the coastal region of Lower Saxony which were under investigation.

2. Purpose of the study

Therefore this study's objective was to determine areas available for wind parks, to calculate wind speeds at heights of 30m, 40m and 60m and to evaluate the possible energy output of typical wind turbines (500KW, 1MW). The results of this study, maps with areas of calculated wind and energy potential, shall be the basis for planning and installation of wind energy converters for the concerned districts and cities.

3. Selection of wind potential areas

The selection of areas suitable for wind energy has been carried out in cooperation with the concerned local authorities. Guidelines were created in cooperation with the environmental ministry of Lower Saxony. The exclusion of areas not available for the use of wind energy was concentrated on areas reserved for nature protection, military installations, and residential areas. Distances to determine wind potential areas may be subdivided into three categories:

Residential areas:
For these areas the distances are determined under consideration of the sound pressure level fixed in "TA Lärm":

Residential areas	500m
Single houses	300m

Special areas:

Military installations	outer protective zone
Nature protectorate	200m
Forest	100m
Main dikes	200m
Other dikes	100m
standing water areas above 1/2 ha extent, rivers	200m

Traffic roads, supply, communication:

Motor way	40m
ship canals	40m
Railways	40m
Airports	boundary of prohibited building
High voltage transmission lines	50m
Transmitter	50m

In general distances to traffic roads etc. were only taken into account at the borders of the potential areas. In case of concrete planning it is always possible to ensure the prescribed distances by means of variation of the wind park geometry. Turbine to turbine distances have to be larger than 200 meters.

The following limits for sound pressure levels are given in the immission protection act of the FRG "TA Lärm":

	daytime	**night time**
Industrial areas	70 dB(A)	**70dB(A)**
Most parts industry	65 dB(A)	**50dB(A)**
Mixed Areas	60 dB(A)	**45dB(A)**
Most parts residential	55 dB(A)	**40dB(A)**
Residential areas	50 dB(A)	**35dB(A)**

In fact the sound pressure levels for night time were taken into account when calculating the distances to residential areas and single buildings. For larger wind parks the distances had to be increased. Figure 2 shows typically suitable areas for wind parks as results of the above mentioned distances.

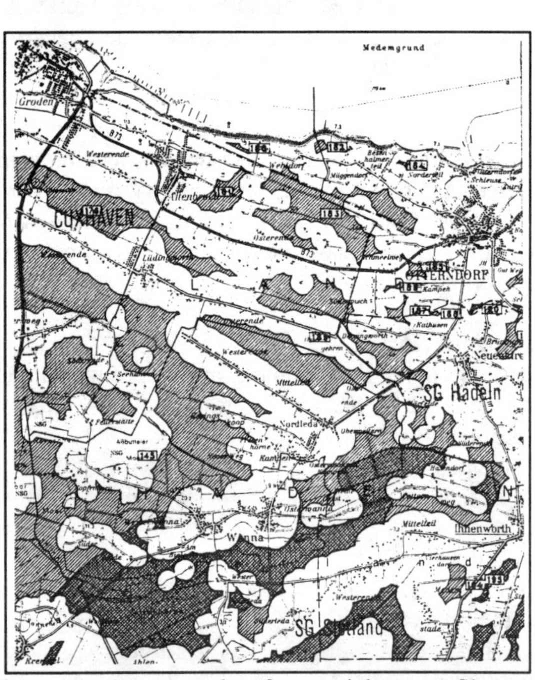

Figure 2: An example of potential areas. Size and shape of the areas are the results of the distance rules, for example distance circles around single houses.

These potential areas are only preliminary selections of the theoretically possible areas. For concrete realisation of wind parks the authorities of the local districts will have to make a further selection.

4. Wind potential calculations

Wind speed calculations were performed with the European wind atlas method (Wind Atlas Analysis and Application Program, WASP, Risø National Laboratory, Roskilde, Denmark). To use the European wind atlas method relevant areas were digitised as roughness maps. Because of overall calculations single obstacles could not be taken into account. In addition the influence of obstacles with heights well below 40 meters will be very small. The wind climate data of the following reference stations were taken into account: Eelde, Bremen and Hamburg.

The roughness map was verified in comparison with wind expertises carried out in form of site seeing and tabular registration of environmental roughness. This roughness map and the local wind climate data were the foundation for the calculation of velocity contour maps of the lower saxonian coastal area (Figure 4, 5).

5. Energy output calculations

In order to compute energy output an exemplary power curve of an 500 kW wind turbine was created based on power curves of several wind turbine manufacturers. In addition the power output of future 1 MW wind turbines was estimated. The advantage of 1 MW wind turbines compared to 500 kW turbines is less area consumption per kW power. In wind parks with constant spacing (measured in rotor diameters) the area needed increases only to 170% if power is doubled. The 1 MW power curve was derived under consideration of the logarithmic wind speed profile from the 500 kW wind turbine.

	WTG 500 kW	**WTG 1 MW**
Hub height	: 40m	: 60m
Diameter	: 40m	: 52m
Sound press. level	: 100 dB(A)	: 103 dB(A)
Control	: Pitch	: Pitch

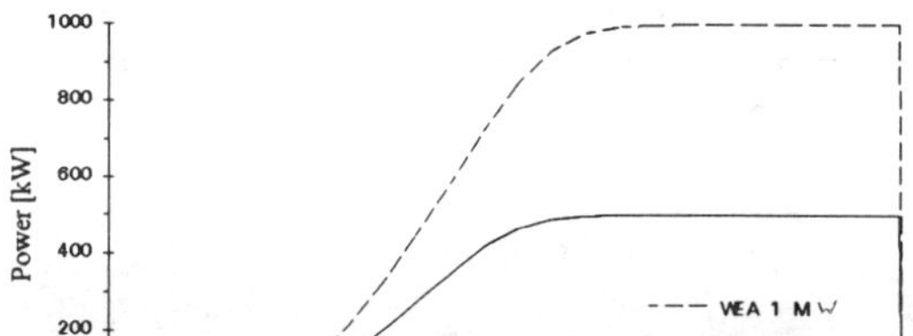

Figure 3: Power curves of the 500kW und 1 MW wind turbine generators. The 500 kW power curve is based on power curves of turbines of different manufacturers.

Figure 4: Digitised roughness map of the Lower Saxonian coastal area. Shown lines are separating areas of different roughness.

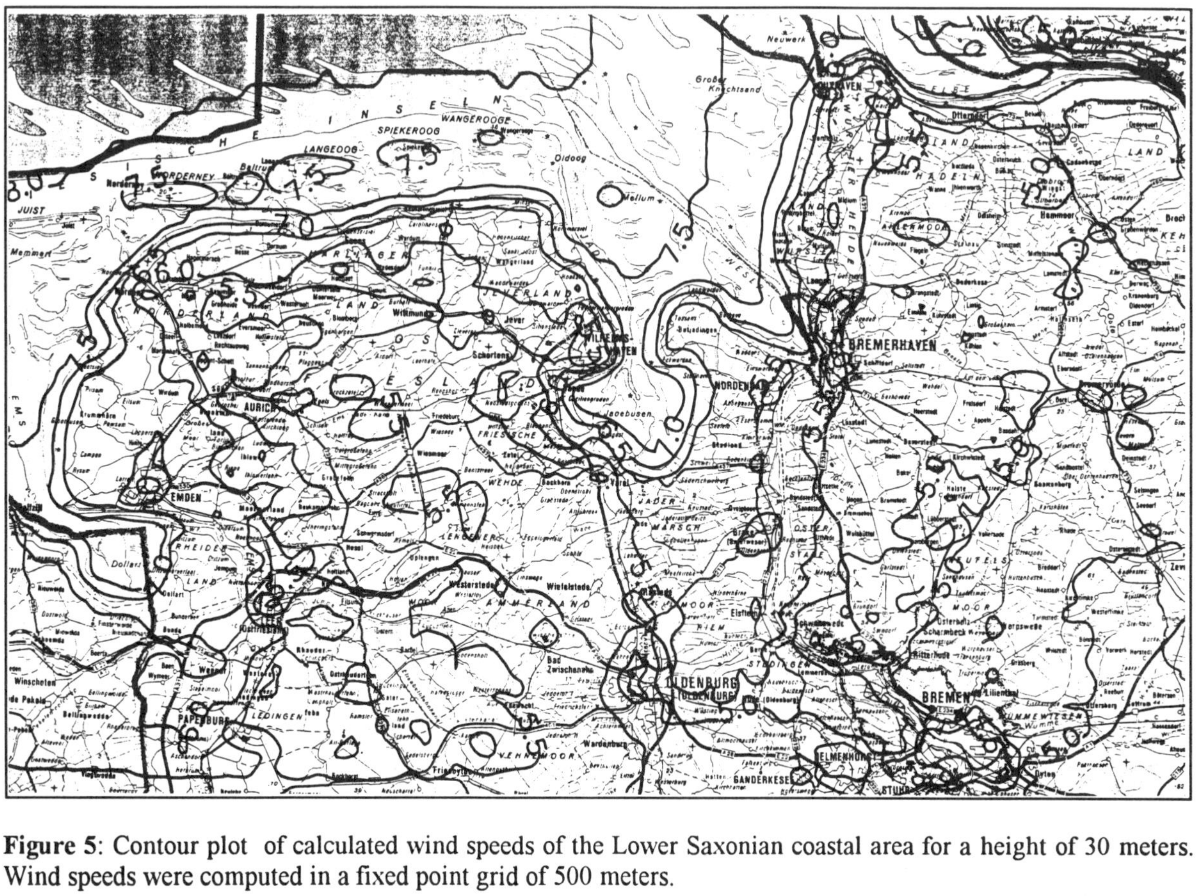

Figure 5: Contour plot of calculated wind speeds of the Lower Saxonian coastal area for a height of 30 meters. Wind speeds were computed in a fixed point grid of 500 meters.

To determine the distances of wind turbine generators in wind parks extensive calculations with different grid spacings were carried out. The Park efficiency depends on grid spacing and especially with smaller distances shadowing effects decrease wind speed and energy output.
Wind park configurations with smaller distances (for example 4D / 4D) are producing more energy due to a larger number of wind turbines that can be integrated into the park. However park efficiency decreases. Additionally, life time of wind turbines working in the wakes of other leading machines may be shortened. The higher turbulence in wind turbine wakes at small distances is leading to higher loads.
In case of concrete realisation of wind parks the sound pressure levels fixed in "TA Lärm" have to be considered. Calculations show that in case of narrow spacing (4D/4D) the number of possible wind turbine sites has to be reduced in such a way that the resulting number of wind turbines are only insignificantly higher than in case of 6D/6D spacing.
In some cases site spacing with larger distance in the prevailing wind direction may lead to better results but for overall calculations it does not seem to be reasonable.
Because of good exploitation of the possible areas and taking into account lifetime and efficiency of wind turbine generators, further calculations were done with a turbine spacing of 6D / 6D.

6. Wind park layout

Park efficiencies were computed assuming site specific wind distributions with the program "PARK" (Risø National Laboratory, Denmark). Wind park calculations could not be performed for the whole set of potential areas. Therefore typical areas were chosen with the following criteria:

- Distance from the coast line
- Neighbourhood of towns
- Environs roughness
- Size of the area

In the following example an optimisation of a wind park is performed in order to comply with the prescribed sound pressure levels. Figure 6 shows an area covered with 500 kW turbines with a spacing of 6D / 6D. The grey levels are showing 5 dB(A) sound pressure level graduations. The example of figure 6 leads to inadmissible levels higher than 40 dB(A) at some residential areas. The prescribed levels could only be met if the number of turbines is reduced from 147 to 92 (figure 7). The Park efficiency increases about 2%.

7. Results

The results of this study are exploitation maps with potential areas, wind speeds and energy output. These may be the foundation for planning boards to declare parts of these areas suitable for wind energy use. Publication rights are reserved by the Ministry of Environment of Lower Saxony.

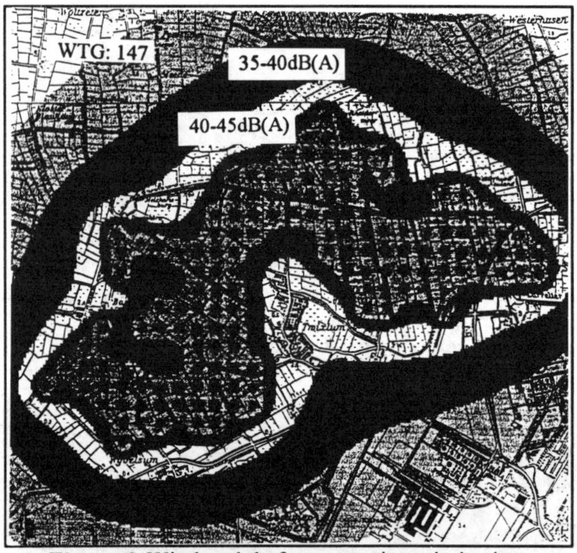

Figure 6: Wind park before acoustic optimisation. Number of turbines: 147, spacing 6D / 6D.

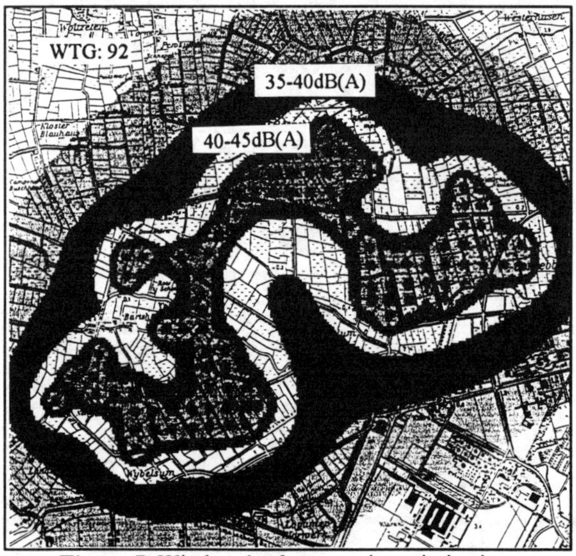

Figure 7: Wind park after acousic optimisation. Number of turbines: 92, spacing 6D / 6D

References

[1] Niedersächsisches Innenministerium, *Empfehlungen zur Standortsicherung und raumordnerischen Beurteilung von Windenergieanlagen*, Bek. d. MI vom 3.7.1991.

[2] Niedersächsisches Innenministerium, *Landes-Raumordnungsprogramm Niedersachsen*, Entwurf, Mai 1992.

[3] I. Troen, N.G. Mortensen und E. L. Petersen, *WASP - Wind Atlas Analysis and Application Programme*, Risø National Laboratory, Roskilde, Denmark

BESTIMMUNG VON GEBIETEN GLEICHER MITTLERER WINDGESCHWINDIGKEITEN IM KOMPLEXEN GELÄNDE AM BEISPIEL DER SCHWÄBISCHEN ALB

J. Albiger[1], A. Böhringer[2], M. Kaltschmitt[3], H. Müh[2]
[1]Institut für Energiewirtschaft und Rationelle Energieanwendung (IER), Universität Stuttgart
Heßbrühlstr. 49a, Tel.: 0711/78061-0, W-7000 Stuttgart 80
[2]Energie-Versorgung Schwaben AG (EVS)
Kriegsbergstr. 32, Tel.: 0711/128-0, W-7000 Stuttgart 80
[3]Kuratorium für Technik und Bauwesen in der Landwirtschaft (KTBL)
Bartningstr. 49, Tel.: 06151/7001-0, W-6100 Darmstadt

Im Mittelgebirge wie der Schwäbischen Alb lassen sich geeignete Flächen für eine Installation von Windkraftanlagen nur umständlich finden und beurteilen. Durch die Topographie und den Bewuchs werden sowohl die Flächen eingeschränkt als auch die Windverhältnisse geprägt. Die vorliegende Untersuchung beschreibt ein Verfahren, das es erlaubt, mit Windmessungen an wenigen Standorten, einem topographischen Geländemodell und Landnutzungsdaten Gebiete gleicher mittlerer Windgeschwindigkeiten und somit auch das Windpotential zu ermitteln. Hierzu wird im ersten Schritt aus ca. einjährigen Windmessungen durch Korrelation mit Windmessdaten des Deutschen Wetterdienstes auf die langjährigen Windverhältnisse geschlossen. Für den nächsten Schritt wird ein entwickeltes Modell eingesetzt, das unter Berücksichtigung der Topographie und der Rauhigkeit in Abhängigkeit von der Windrichtung die Windgeschwindigkeit für jeden Ort im Untersuchungsgebiet ermittelt. Auf diesen beiden Schritten aufbauend können nun Gebiete gleicher mittlerer Windgeschwindigkeiten und geeignete Standorte ausgewiesen werden. In diesem Beitrag wird zunächst die Methotik vorgestellt und anschließend für ein Fallbeispiel angewandt.

1 Zielsetzung

Ziel des Beitrages ist es, eine Methode vorzustellen, die auf der Basis gemessener Windgeschwindigkeiten und unter Berücksichtigung kleinräumiger Effekte eine Bestimmung von Gebieten gleicher mittlerer Windgeschwindigkeiten und daraus letztlich eine Identifikation optimaler Konverterstandorte ermöglicht.

Zunächst müssen die langjährigen mittleren Windverhältnisse für einzelne Referenzstandorte aus gemessenen Winddaten bestimmt werden. Dazu wurde ein Windmeßprogramm aufgelegt [1]. Anschließend werden diese gemessenen Winddaten mit den Meßergebnissen der Stationen des Deutschen Wetterdienstes korreliert. Damit können die langjährigen Mittelwerte in Abhängigkeit der Windrichtung ermittelt werden. Sie lassen sich anschließend durch ein Windmodell auf die jeweiligen Umgebungen um die entsprechenden Meßstationen übertragen, wenn standortspezifische Gegebenheiten, die die Windverhältnisse entscheidend beeinflussen, berücksichtigt werden. Dies ist auf der Grundlage geeigneter Ansätze aus der Strömungslehre und einer räumlich hoch aufgelösten Modellierung des Geländes möglich. Somit können Gebiete gleicher mittlerer Windgeschwindigkeiten innerhalb der untersuchten Gebietsfläche ermittelt und letztlich durch eine Verknüpfung mit Landnutzungsdaten in windtechnische Flächenpotentiale überführt werden.

Vor dem Hintergrund dieser grundsätzlichen Vorgehensweise ist es das Ziel dieses Beitrags, die Methotik zu beschreiben. Die computerunterstützte Methode wird anschließend am Beispiel eines bestimmten Gebietes auf der Schwäbischen Alb angewandt und erste Ergebnisse dargestellt.

2 Windmeßprogramm

Im Zuge eines Meßprogramms der Energie-Versorgung Schwaben AG (EVS) [1] werden an verschiedenen Standorten auf der Schwäbischen Alb für 6 bis 12 Monate Windmessungen durchgeführt. Daraus können die langjährigen mittleren Windverhältnisse bestimmt werden, wenn die Stundenmittel der Windgeschwindigkeiten in Abhängigkeit der Windrichtung mit denen der nahegelegenen Stationen des Deutschen Wetterdienstes korreliert werden. Diese "langjährigen" Mittelwerte, die näherungsweise unabhängig von den kurzzeitigen Windverhältnissen während der Meßperioden sind, sind typisch für den jeweiligen Standort.

Abbildung 1 zeigt die Lage der Meßstationen in Abhängigkeit des Rechts- und Hochwertes des Gauß-Krüger-Koordinatensystems. Der Hochwert gibt den Abstand zum Äquator und der Rechtswert - versehen mit einem Offset von 3 500 000 m - den Abstand zum 9°-Meridian an.

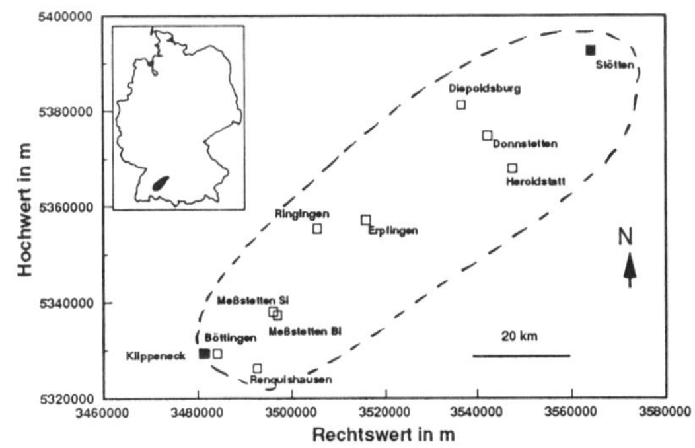

Abb. 1: Geographische Lage der Meßstationen

2.1 Bestimmung langjähriger Mittelwerte

Die Windverhältnisse an den verschiedenen Meßstationen werden nur für begrenzte und unterschiedliche Zeitperioden gemessen. Sie sind aber weder untereinander vergleichbar noch repräsentativ hinsichtlich der langjährigen mittleren Windverhältnisse. Es ist notwendig, die aus diesen Kurzzeitmessungen erhaltenen Winddaten in die zu erwartenden langjährigen Mittelwerte zu überführen. Dazu werden die arithmetischen Durchschnittswerte für die in der jeweiligen Meßperiode erfaßten Windgeschwindigkeiten ermittelt und ins Verhältnis gesetzt zu den Mittelwerten der für die gleiche Zeitperiode an den Stationen des Deutschen Wetterdienstes gemessenen Winddaten. Dieser Zusammenhang wird für jede Himmelsrichtung getrennt ermittelt. Dieses Verhältnis kann anschaulich dahingehend gedeutet werden, daß für die entsprechende Zeitperiode und Richtung die Windgeschwindigkeit an dem Meßstandort um dieses Verhältnis größer bzw. kleiner ist als an der Bezugsstation. Unter der Annahme, daß dieses Verhältnis unabhängig von der gewählten Zeitperiode und näherungsweise konstant ist, können die langjährigen Windverhältnisse an der Meßstation abgeschätzt werden, da die langjährigen Windverhältnisse an den Wetterstationen des Deutschen Wetterdienstes bekannt sind [2]. Nach dieser Korrelation, die für jede Windrichtung getrennt durchgeführt werden muß, kann für jede Meßstation eine mittlere Windverteilung mit den jeweiligen Windgeschwindigkeiten angegeben werden. Daraus läßt sich dann der langjährige Mittelwert der Windgeschwindigkeit der Station errechnen.

Tabelle 1: Mittlere langjährige Windgeschwindigkeiten an Standorten auf der Schwäbischen Alb

	Mittlere Windgeschwindigkeit	Höhe über NN
Meßstationen (EVS-Meßprogramm)		
Böttingen	4,3 m/s	993 m
Diepoldsburg	4,0 m/s	805 m
Donnstetten	4,6 m/s	838 m
Erpfingen	3,9 m/s	800 m
Heroldstatt	4,3 m/s	816 m
Meßstetten-Blumersberg	4,0 m/s	993 m
Meßstetten-Sickersberg	4,0 m/s	978 m
Renquishausen	3,7 m/s	885 m
Ringingen	3,2 m/s	822 m
Wetterstationen (Deutscher Wetterdienst)		
Stötten	4,5 m/s	734 m
Klippeneck	3,7 m/s	973 m

Anhand dieser Vorgehensweise werden für die unterschiedlichen Meßstandorte, von denen Windmessdaten vorliegen (vgl. [1]), die langjährigen Windgeschwindigkeitsmittel auf der Basis der Wetterstation Stötten berechnet. Sie sind in Tabelle 1 zusammengestellt. Demnach liegen an den untersuchten Standorten die mittleren Geschwindigkeiten der bewegten Luftmassen zwischen 3,2 und 4,6 m/s.

2.2 Analyse und Interpretation der Ergebnisse

Ein die Windverhältnisse auf der Schwäbischen Alb mitbestimmendes großräumiges Strömungshindernis ist der westlich gelegene Schwarzwald; bei Westwinden liegt das Untersuchungsgebiet im Windschatten des durch eine geodätische Höhe von bis zu 1 500 m reichenden Mittelgebirges. Der Abschattungseffekt wird deutlich, wenn die mittleren Windgeschwindigkeiten aus Westen (Winkelsegment von Nordwest bis Südwest) in Abhängigkeit des Rechtswertes der jeweiligen Meßstation aufgetragen werden (vgl. Abbildung 2); dabei wird mit zunehmendem Rechtswert des Gauß-Krüger-Koordinatensystems der Abstand zum Strömungshindernis "Schwarzwald" größer. Dies führt zu einem Anstieg der jahresmittleren Windgeschwindigkeiten. Die schraffierte Fläche kennzeichnet die Bandbreite, in der die Windgeschwindigkeiten liegen. Obwohl die Meßstationen Diepoldsburg, Donnstetten und Heroldstatt kleinere geodätische Höhen aufweisen als z. B. Klippeneck oder die Meßstationen in Meßstetten, werden auf der östlichen Alb höhere mittlere Windgeschwindigkeiten aus westlicher Richtung als im westlichen Teil dieses Mittelgebirges gemessen.

Da der Westwind mit einer durchschnittlich höheren Windgeschwindigkeit weht und im Vergleich zu den anderen Windrichtungen mit einer größeren Häufigkeit auftritt, zeigt sich der Abschattungseffekt des Schwarzwaldes auch bei der richtungsunabhängigen jahresmittleren Windgeschwindigkeit. Trotz kleinerer geodätischer Höhen im Osten der Alb sind die mittleren Windgeschwindigkeiten höher als auf der höhergelegenen Westalb.

Abbildung 2 zeigt aber auch deutlich, daß die Windgeschwindigkeiten an den Meßstationen, die teilweise nur einige Kilometer auseinanderliegen und sich auch in ihrer Höhe über Normalnull kaum unterscheiden, unterschiedlich sind. Dies liegt - neben anderen Faktoren - am Einfluß der Topographie und der Oberflächenstruktur der jeweiligen Umgebung um die Meßstation auf die Strömungsverhältnisse. Durch mechanische und thermische Rauhigkeit werden Turbulenz und Windprofil verändert. Die Topographie kann die Strömung kanalisieren und beschleunigen, oder aber auch blockieren und abbremsen.

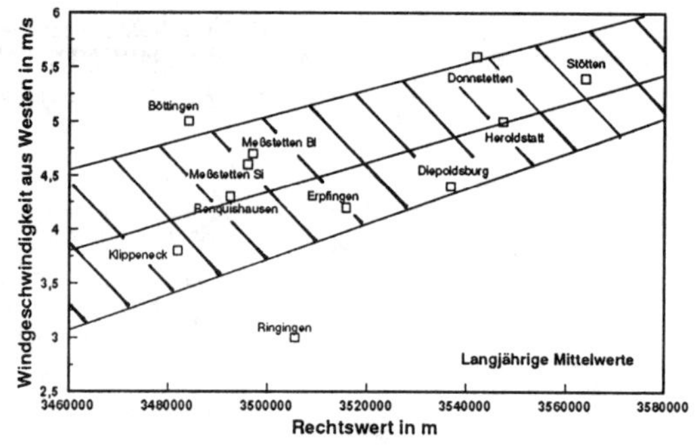

Abb. 2: Mittlere Windgeschwindigkeit aus Westen in Abhängigkeit des Rechtswerts; die schraffierte Fläche stellt die Bandbreite der Windgeschwindigkeiten dar

3 Methodisches Vorgehen

Um aus den Windverhältnissen an einem Standort auf die Windverhältnisse in der Umgebung schließen zu können, ist es notwendig, standort- und oberflächenspezifische Effekte, die die Windverhältnisse beeinflussen, zu quantifizieren. Dazu muß die Topographie und die Oberflächenstruktur in einer hohen räumlichen Auflösung abgebildet werden. Dann sind Ansätze zu formulieren, die den Zusammenhang zwischen der Geländeform und den Windströmungen beschreiben. Erst dann können die Windverhältnisse unter Berücksichtigung der geodätischen Höhe auf die nähere und weitere Umgebung transformiert werden. Dies muß für die unterschiedlichen Windrichtungen getrennt erfolgen, um großräumige Effekte gut abbilden zu können.

3.1 Modellierung der Topographie

Für die Modellierung der Topographie des Untersuchungsgebietes wird das Digitale Höhenmodell Baden-Württemberg zugrunde gelegt [3]. Bei diesem Höhenmodell wird die Gebietsfläche des Bundeslandes in quadratische Rasterelemente mit einer Kantenlänge von 50 m aufgeteilt. Für jedes Element wird die geodätische Höhe über Normalnull in Metern angegeben. Die geographische Lage eines Quadrates ist durch den Rechts- und den Hochwert im Gauß-Krüger-Koordiantensystem festgelegt. Damit ist es mit einem entsprechenden Programmsystem möglich, die Oberfläche des Untersuchungsgebietes in dieser hohen räumlichen Auflösung darzustellen. Im Umkehrschluß sind aber auch die Voraussetzungen geschaffen, die kleinräumigen bzw. lokalen Einflüsse auf das Windenergieangebot abzubilden und bei der Simulation der richtungsabhängigen Windgeschwindigkeiten zu berücksichtigen.

3.2 Modellierung der Oberflächenbeschaffenheit

Neben den topographischen Gegebenheiten entscheidet die Struktur und die Beschaffenheit der Oberfläche ebenfalls wesentlich über die Windverhältnisse. Zur realistischen Modellierung der Erdoberfläche werden Satellitenaufnahmen zugrunde gelegt. Dazu werden zwei sich überlappende Szenen von LANDSAT-TM Satellitenaufnahmen aus verschiedenen Umläufen ausgewertet. Anhand von sogenannten Trainingsgebieten, von denen genaue Landnutzungs- bzw. Bodendaten vorliegen, werden den spektralen Werten, wie sie der Satellit in einer Auflösung von 30 x 30 m liefert, die wesentlichen Landnutzungsarten, wie Wald, besiedeltes Gebiet und landwirtschaftlich genutzte Flächen zugeordnet. Anschließend können die Satellitendaten in die Information zur Oberflächenbeschaffenheit transformiert werden. Dabei muß jedoch beachtet werden, daß die Grenze zwischen den einzelnen Landnutzungsinformationen z. T. fließend ist. In Wirklichkeit existieren mehr als drei Landnutzungsarten. Dies führt zu teilweise unsinnigen Ursprungsdaten. Um solche Unstimmigkeiten zu korrigiren, bearbeitet ein Fehlersuchalgorithmus vor der Weiterverarbeitung die transformierten Landnutzungsdaten.

Danach müssen die klassifizierten und berichtigten Landnutzungsdaten aufgrund der Schräglage der Satellitenaufnahme durch eine Paßpunktentzerrung und eine anschließende geometrische Transformation in das Gauß-Krüger-Koordinatensystem überführt werden. Anschließend erfolgt noch eine Informationsverdichtung auf eine räumliche Auflösung von 50 x 50 m, damit eine vergleichbare Datenstruktur wie bei dem Digitalen Höhenmodell gegeben ist. Abschließend werden die derart ermittelten Daten im gleichen Format abgelegt wie die des Höhenmodells; dadurch ist - infolge des gleichen Aufbaus der derart ermittelten Daten - eine einheitliche Zugriffsmöglichkeit geschaffen.

3.3 Modellierung der kleinräumigen Windverhältnisse

Der Einfluß der in hoher örtlicher Auflösung vorliegenden Topographie und Oberflächenbeschaffenheit des untersuchten Gebietes auf die Windverhältnisse kann mit Hilfe von Ansätzen aus der Strömungslehre beschrieben und damit abgebildet werden.

Zunächst wird von einer zweidimensionalen Beschreibung des Geländes ausgegangen. Für einen bestimmten Winkel in der horizontalen Ebene wird eine vertikale Schnittebene angenommen. Dieser Querschnitt im Gelände ist eine zweidimensionale Konturlinie, deren Höhenverlauf bzw. Merkmale hinsichtlich Landnutzung die Datengrundlage für die Ansätze bilden. Anhand der Ergebnisse aus den Ansätzen kann dieser Konturlinie ein Verstärkungsfaktor bzw. Verminderungsfaktor für die Windgeschwindigkeit in Abhängigkeit der jeweiligen Topographie und Oberflächenrauhigkeit zugeordnet werden. Die Schnittebene wird nun sukzessive um eine vertikale Achse im Untersuchungspunkt gedreht. Dabei entstehen neue Konturlinien, die analog ausgewertet werden. Für eine Windrichtung (die Windrose besteht hier aus acht Windrichtungssektoren) werden die Faktoren der in diesem Sektor liegenden Konturlinien gemittelt. Dieser Mittelwert ist dann die charakteristische Größe für den untersuchten Richtungssektor.

Um den Einfluß der Topographie abzuschätzen, wird von einer stationären inkompressiblen zweidimensionalen Strömung ausgegangen, die sich aus einer Grundströmung und einem durch die topographischen Strukturen verursachten Anteil zusammensetzt. Unter Zugrundelegung der Theorie der dünnen Tragfläche [4] und vereinfachenden Annahmen (reibungsfreie Strömung, die immer an der Oberfläche anliegt; Verhältnis der Höhe zu horizontalen Ausdehnungen ist klein) kann ein Zusammenhang zwischen der Geländeform, hier der Form der Konturlinie, und der Geschwindigkeitsänderung angegeben werden.

Zusätzlich wird die Struktur der Landoberfläche berücksichtigt. Bei dem daraus resultierenden Rauhigkeitsmodell werden die klassifizierten Landnutzungsdaten verwendet, die auf der jeweiligen Konturlinie gespeichert sind. Für die verschiedenen Oberflächennutzungsarten werden die im europäischen Windatlas [5] ausgewiesenen Rauhigkeitslängen zugrunde gelegt. Die Rauhigkeitslängen der windaufwärts gelegenen Elemente der Konturlinie werden gemittelt, wobei durch eine Gewichtungsfunktion die Werte der näher am Untersuchungspunkt liegenden Elemente stärker bei der Mittelung berücksichtigt werden. Bezieht man die gemessene Windgeschwindigkeit auf eine Geschwindigkeit, die sich bei einer recht glatten Oberfläche (Rauhigkeitslänge = 0,03 m) in Bodennähe einstellen würde, erhält man ein Maß für die Geschwindigkeitsänderung durch die gegebene Rauhigkeit.

Diesem auf dieser Basis entwickelten Programmsystem - das die beschriebene Vorgehensweise und die zugrunde liegenden Daten zusammen mit einer graphischen Ausgabe zu einem lauffähigen Modell verbindet - liegen eine Reihe vereinfachender Annahmen zugrunde; dies liegt darin begründet, daß immer ein Kompromiß zwischen dem Aufwand bzw. der benötigten Rechenzeit und der erzielbaren Genauigkeit gefunden werden muß. Deshalb müssen die Modellergebnisse in einem weiteren Schritt durch einen Vergleich simulierter zu noch nicht im Modell implementierter Windmessdaten überprüft und ggf. das Modell besser den tatsächlichen Gegebenheiten angepaßt werden.

4 Ergebnisse der Modellierung

Die in Kapitel 3 vorgestellte Vorgehensweise wird exemplarisch für ein bestimmtes Gebiet auf der Schwäbischen Alb angewendet. Das ausgewählte 4 x 4 km^2 große Untersuchungsgebiet rund um den Römerstein liegt ca. 2 km westlich zur Meßstation Donnstetten (vgl. Abbildung 1). Es ist hinsichtlich Bewuchs bzw. Bebauung durch ein für die Schwäbische Alb typisches Erscheinungsbild gekennzeichnet. Das Untersuchungsgebiet ist in Abbildung 3 in einer dreidimensionaler Darstellung zu sehen. Die vertikale Achse, d. h. die Höhe über Normalnull, ist - um die Struktur des Geländes hervorzuheben - gestreckt.

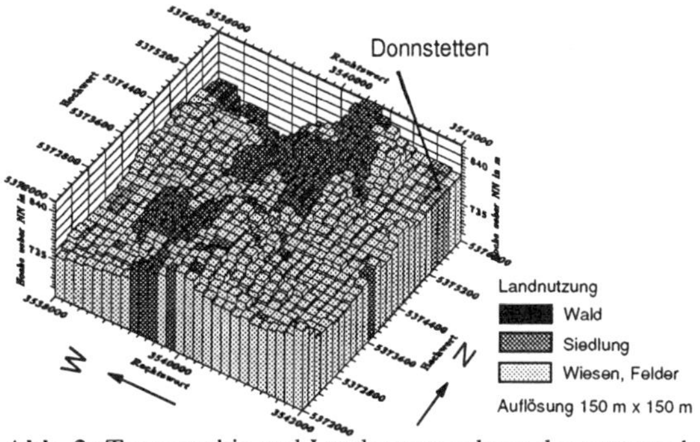

Abb. 3: Topographie und Landnutzungsdaten des untersuchten Gebietes "Römerstein"

Die Landnutzung wird als Grauwert über die einzelnen Rasterelemente gelegt. Aus Gründen der einfacheren Darstellbarkeit werden die 50 x 50 m Raster zu Elementen mit einer Kantenlänge von 150 m zusammengefaßt. Durch die unterschiedliche Höhe wird die Geländeform beschrieben; in der nordwestlichen Ecke des Gebietes ist das Lenninger Tal erkennen, das zu einer bewaldeten Hügelkette ansteigt, die im mittleren Teil des Bildes mit dem bewaldteten Römerstein (872 m) die höchste Erhebung erreicht (in der Darstellung liegt durch die grobe Rasterung die Höhe bei ca. 860m). Die gegenwärtige Landnutzung geht aus den unterschiedlichen Grauschattierungen der einzelnen Säulen hervor; dunkle Schattierungen kennzeichnen dabei die Landnutzungen Wald bzw. Siedlung. Bei einem Vergleich mit topographischen Landkarten zeigt sich eine gute Übereinstimmung mit den visualisierten Computerdaten.

Für jedes Rasterelement in der untersuchten Gebietsfläche wird unter Berücksichtigung der kleinräumigen Windverhältnisse und einer für jede Windrichtung getrennt durchgeführten Interpolation das langjährige Jahresmittel abgeschätzt (vgl. Abbildung 4). Erwartungsgemäß ist das Tal in

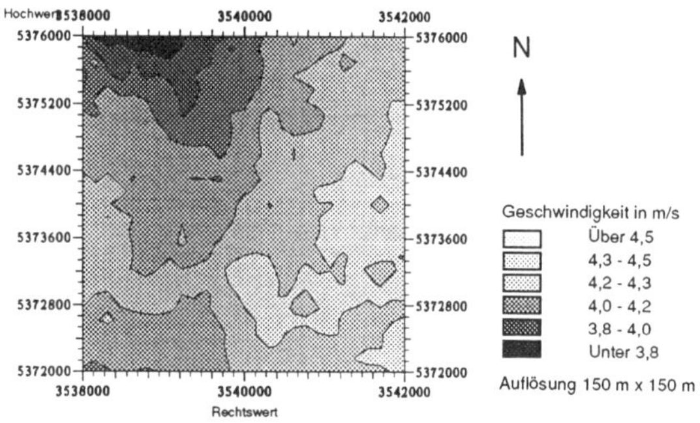

Abb. 4: Simulierte mittlere Windgeschwindigkeiten auf der untersuchten Gebietsfläche

der rechten oberen Ecke durch geringe mittlere Windgeschwindigkeiten gekennzeichnet. Obwohl die bewaldete Hügelkette die höchste Erhebung im Bild ist, liegen die Windgeschwindigkeiten wegen der hohen Oberflächenrauhigkeit des Waldes nur bei ca. 4.3 m/s. Die höchsten mittleren Windgeschwindigkeiten sind dort anzutreffen, wo das Gebiet frei anströmbar ist und geringe Rauhigkeiten aufweist.

Aufbauend auf dieser derart bestimmten Windgeschwindigkeitsverteilung innerhalb der untersuchten Gebietsfläche lassen sich durch eine Verknüpfung mit den Landnutzungsdaten die windtechnischen Flächenpotentiale bestimmen. D. h., für verschiedene Geschwindigkeitsbereiche können die Flächen ermittelt werden, die für eine Installation von Windkraftanlagen geeignet sind. Rund 37 % der gesamten exemplarisch untersuchten Gebietsfläche ist bewaldet bzw. besiedelt. Diese Gebiete sind für eine Installation von Windkonvertern nicht geeignet. Die verbleibende Fläche ist - aufgrund der Lage des untersuchten Gebietes auf der Hochfläche der Schwäbischen Alb - durch mittlere Windgeschwindigkeiten zwischen 4,0 und 4,7 m/s gekennzeichnet. Mittlere Geschwindigkeiten zwischen 4,0 und 4,3 m/s sind dabei auf rund 17 % der verbleibenden Fläche, zwischen 4,3 und 4,6 m/s auf rund 42 % und von mehr als 4,6 m/s auf ca. 4 % der für eine Windkraftanlageninstallation geeigneten Gebietsflächen gegeben.

Der Vergleich mit Windmessdaten des untersuchten Gebietes verdeutlicht, daß der gewählte Ansatz - trotz der gemachten Vereinfachungen - zu sinnvollen und realitätsnahen Ergebnissen führt.

Anhand dieser Informationen können potentielle Windkraftanlagenstandorte, d. h. Flächen mit hohem Windenergieangebot, identifiziert werden.

5 Schlußbetrachtung

Ziel dieses Beitrages ist es, Methoden darzustellen, die es ermöglichen, örtlich hoch aufgelöste Gebiete mit gleichem-Windgeschwindigkeitsmittel zu identifizieren und darauf aufbauend günstige Windkraftanlagenstandorte zu bestimmen. Das Verfahren wurde am Beispiel der Schwäbischen Alb entwickelt und soll auch auf andere Mittelgebirge übertragbar sein.

Zur Berücksichtigung kleinräumiger topographischer Gegebenheiten wurde dazu die Topographie in einer Auflösung von 50 x 50 m abgebildet. Mit der gleichen räumlichen Auflösung wurden aus LANDSAT-TM Satellitenaufnahmen die Landnutzungsarten ermittelt. Auf der Basis dieser Daten und der mittleren Windverhältnisse in einem dichten Stationsnetz wurde ein Simulationsmodell entwickelt, das es ermöglicht, für jeden beliebigen Standort innerhalb der betrachteten Gebietsfläche die jahresmittleren Windverhältnisse in Abhängigkeit von der Windrichtung zu berechnen. Unter der Annahme, daß eine Installation von Windkonvertern auf besiedelten und bewaldeten Gebieten auszuschließen ist, können die Flächenpotentiale, die zur Errichtung von Windkraftanlagen geeignet sind, bestimmt werden. Daraus lassen sich dann bezüglich des Windenergieangebots optimale Standorte ermitteln.

Diese Vorgehensweise wurde auf einem begrenzten Gebiet auf der Schwäbischen Alb exemplarisch angewendet. In diesem Gebiet, das ein für diese Region typisches Erscheinungsbild hinsichtlich Topographie und Landnutzung hat, liegen die simulierten Windgeschwindigkeiten zwischen 3,4 und 4,7 m/s. Diese Bandbreite korrespondiert auch mit den langjährigen Mittelwerten der auf der Schwäbischen Alb gemessenen Windgeschwindigkeiten. Der Anteil der Flächen, der davon für eine Aufstellung von Windkraftanlagen geeignet und durch ein hohes Windgeschwindigkeitsmittel (über 4,6 m/s) gekennzeichnet ist, ist jedoch nur gering und liegt im exemplarisch untersuchten Gebiet bei ca. 4 %.

Die entwickelte Methode kann nun auch auf das gesamte Gebiet der Schwäbischen Alb übertragen werden. Damit können für dieses Mittelgebirge die gesamten windtechnischen Flächenpotentiale abgeschätzt werden. Darüber hinaus ist eine von den örtlichen Gegebenheiten abhängige Zuordnung der Flächenpotentiale für z. B. einen Verwaltungsbezirk möglich. Die vorgestellte Methodik kann also nicht nur zur Abschätzung von Gesamtpotentialen, sondern auch zur räumlich hoch aufgelösten Potentialflächenanalyse verwendet werden. Außerdem ist die Identifikation von optimalen Windkraftanlagenstandorten möglich.

Literatur

[1] Böhringer, A.
Auswahl und Beurteilung von geeigneten Standorten für Windkraftanlagen auf der Schwäbischen Alb
7. Internationales Sonnenforum 1990, 9. - 12. 10. 1990, Frankfurt, Bundesrepublik Deutschland, Tagungsbericht Band 3, S. 1789 - 1795

[2] Christoffer, J.; Ulbricht-Eissing, M.
Die bodennahen Windverhältnisse in der Bundesrepublik Deutschland
Berichte des Deutschen Wetterdienstes Nr. 147
Selbstverlag des Deutschen Wetterdienstes, Offenbach a. M., 1989, 2. vollständig neu bearbeitete Auflage

[3] Landesvermessungsamt Baden-Württemberg (Hrsg.)
Digitales Höhenmodell Baden-Württemberg
Landesvermessungsamt Baden-Württemberg, Stuttgart, 1990

[4] Karamcheti, K.
Principles of ideal-fluid aerodynamics
John Wiley and Sons, New York, 1966

[5] Troen, I.; Petersen, E. L.
Europäischer Windatlas
Riso National Laboratory, Roskilde, Dänemark, 1990

WINDENERGY POTENTIAL IN THE PRE-ALPINE REGION OF UPPER BAVARIA MEASUREMENTS AT EIGHT SITES IN THE AREA SUPPLIED BY THE ISAR-AMPERWERKE AG

Dipl.-Ing. Gerhard Pfeiffer,Dipl.-Ing. (FH) Herbert Reiner, Dipl.-Ing. Helmut Stadtmüller
Isar-Amperwerke AG, Postfach 37 02 20, 8000 München 2

Isar-Amperwerke AG has been measuring wind speed and direction continuously at several sites in the pre-alpine region of Upper Bavarian Alps for some years. The recorded results indicate that the exploitation of the wind is economically unfeasible, even at the high alpine site. The frequency distribution at all sites differs significantly from the Rayleigh distribution. Therefore wind measurements should be carried out before the erection of a wind generator in the inland.

1. Isar-Ampwerke AG

As a regional electricity utility company, Isar-Ampwerke AG (IAW) serves almost all of Upper Bavaria but the city of Munich. Against the background of rising environmental awareness environment-friendly electricity generation and rational energy application are gaining importance. Therefore IAW has been investigating the feasibility of exploiting renewable energy sources in their region of supply. To minimize the risk of investing large amounts of money in windmills at unsuitable sites, the most significant wind parameters have been collected at several sites for the last couple of years.

2. Selecting the sites for the measurements

A wind potential of a mean wind speed above 5 m/s — as in the coastal regions — can only be expected at a few sites in Upper Bavaria. Thus IAW has been erecting measuring units in co-operation with the German Weather Service, Wetteramt München. The most important criteria for the selection of a specific site was the expected wind potential, which had to be as high as possible. The wind potential is significantly affected by the shape of the terrain. Another criteria was a short distance to the public grid, so that the connection of a wind converter to the public grid could be done at reasonable costs. At last, the sites had to be accessible by car. The remote site 'Tölzer Hütte', however, located at 1 835m height in the Karwendel mountain range, had been selected on suggestion by the German Alpen Verein (DAV), section Bad Tölz, to investigate the conditions for a stand alone grid based on wind power.

3. Data recording and processing

The measurements are carried out at all eight sites with a wind measuring unit in a height of ten meters according to general accepted standards. Recorded are the mean wind speed, the direction and the maximum gust within each ten-minute interval. Via computer programs these data are being analysed regarding specific characteristics such as mean wind speed and its frequency distribution, direction of the wind and its frequency distribution, maximum gust, etc. Furthermore, the exploitable energy potential is determined in connection with the power curves of various commercial wind turbines. Some of the results and findings are presented.

4. Average wind speed

The recorded mean wind speeds at the eight sites range from 2.2 m/s at the site 'Winklmoosalm' to 3.4 m/s at the 'Tölzer Hütte' (**Figure 1**). As a result the mean wind speeds at the investigated sites fall clearly short of the wind speed of 4 m/s which is the generally accepted minimum for an economically promising exploitation, whereas the mean wind speed at coastal areas ranges about 5 m/s. At the investigated isolated grid at the 'Tölzer Hütte' a wind speed of 3.1 m/s was found for the time from the middle of May to Mid-October, when the hut is open to the public. During the winter the hut is hardly accessible and closed except for one winter refuge room. Only during this period, however, a larger wind potential is given at 3.7 m/s wind speed. Thus an exploitation of the wind energy at this site is inadvisable, which is emphasised by the evaluation of the frequency distribution.

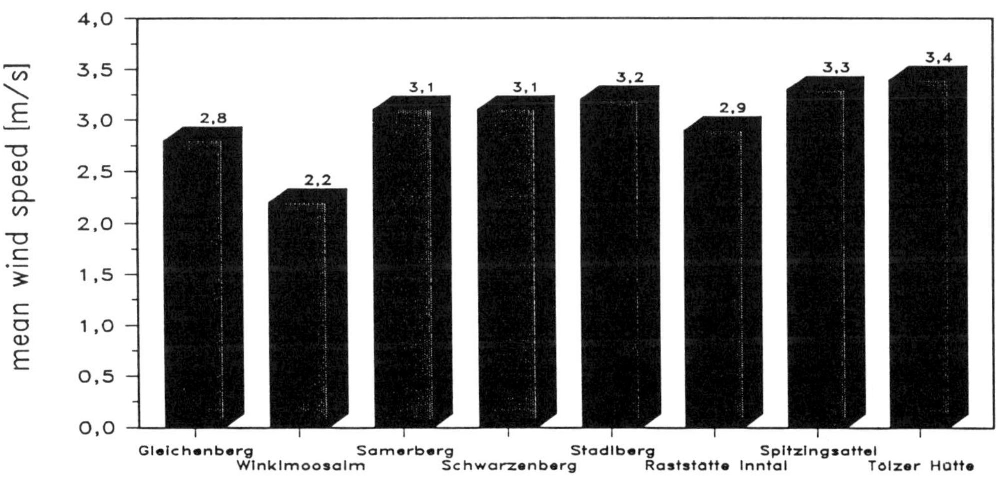

Figure 1: Mean wind speeds of the sites

5. Frequency distribution of the wind speed

An important factor for the evaluation of the exploitable energy is the frequency distribution of the wind speed. This will be demonstrated exemplary for the site 'Schwarzenberg' (**Figure 2**).

In the graph the following parameters are shown:

— recorded frequency distribution, determined from the mean wind speed in each ten minute interval, as well as the Weibull distribution determined for the site. The k-factor of 1.42 ensues from both the given mean wind speed and the frequency distribution.

— As a comparison, the Rayleigh distribution, a special form of the Weibull distribution with a k-factor of 2, based on the recorded mean wind speed determined for the site 'Schwarzenberg' is also shown.The Rayleigh distribution is typical for coastal sites only.

The comparison of Weibull and Rayleigh distribution based on the same mean wind speed shows a clear left hand shift of the recorded frequency distribution.

— Lower wind speeds occur more often so that the cut-in wind speed of wind turbines (above 3 m/s) is reached less frequently as in the Rayleigh distribution.

— Shares of higher wind speeds are of positive influence towards the possible energetic yield so that the cinetic energy is higher than it could be expected according to the Rayleigh distribution.

It is therefore essential for the evaluation of the suitability of a certain site to determine its frequency distribution along with the mean wind speed. The knowledge of the frequency distribution allows to take into consideration the influence of the terrain on the wind resources. For the site 'Schwarzenberg' the given distribution indicates that more energy is contained by the wind than it could be expected according to a regular Rayleigh-distribution.

The k-factors for the sites in the survey range about 1.2 to 1.8 i.e. below the factor of the Rayleigh distribution.

6. Survey of the direction of the wind

The predominant direction of the wind was south-westerly for all eight sites but one. At the exceptional site 'Raststätte Inntal' a southerly fall wind with relatively high wind speed, the so-called 'Erler Wind', occurs daily from 0.00 to 11.00 a.m. similar to the well-known winds at Lake Garda in Italy.

7. Energy assessment

Based on the power curves of the various wind turbines and the recorded wind speeds both the attainable energy output and, to compare the different turbines, the plant load factors based on a full year are calculated. The nominal power of the turbines taken into consideration, which are first and foremost designed for a cut-in wind speed of about 3 m/s, ranges from 45 kW to 300 kW. The full load hours given in **Figure 3** are referring to the machine most appropriate for each specific site. However, only wind turbines with recorded power curves have been considered. The maximum of full load hours amounted to more than 1000 at the site 'Raststätte Inntal' compared to 2 000 to 2 500 hours in coastal areas.

To cross-check the plausability of the results, the overall efficiency of all examined turbines was determined from the recorded wind potential and the theoretically attainable yield. The efficiency consists of the c_p-value (ratio of the mechanic energy that can be attained from the wind to the energy contained in the wind) which is given for modern windturbines at 42%, and the mechanic-electric energy ratio of about 85% for well-designed turbines. The overall efficiency therefore ranges from 35% to 40%.

However, in some cases the overall efficiencies calculated from the recorded data exceeded 50%. This results from determining power curves by mathematical simulation rather than by actual measurements. It is therefore recommended to always check the origin of the power curve.

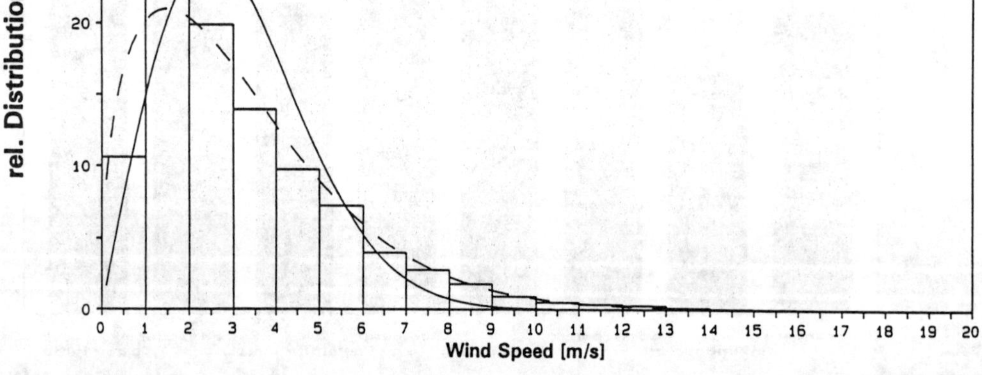

Figure 2: Frequency distribution of wind speed

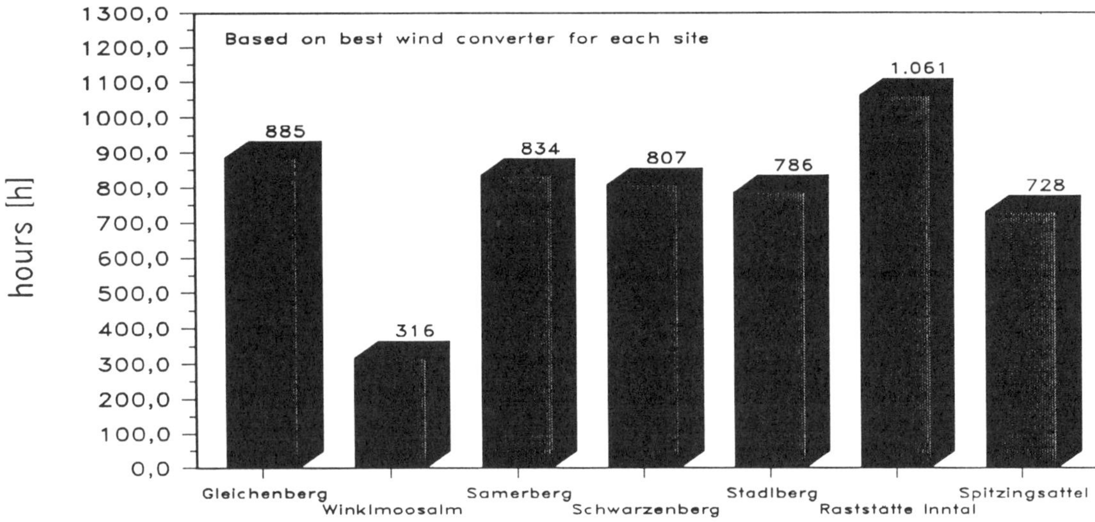

Figure 3: Full load hours

8. Economical assessment

The economical assessment showed that the generating costs of the wind energy at the site 'Raststätte Inntal' (which showed the most full load hours) were about 40 Pf/kWh in the best case. This assessment was carried out under the customary estimates for life-span (20 years for a modern wind turbine), rate of interest (8% p.a.) and investment for the turbine, including foundation and transmission to the grid. Under these assumptions probably no wind turbine can operate economically in Upper Bavaria, especialy not at the surveyed sites.

9. Conclusion

The measurements indicate that for the selection, respectively erection, of a wind turbine in the inland the essential parameter for the evaluation of the exploitable energy is the frequency distribution. It is thus not sufficient to base the decision for such an investment on the recorded mean wind speed of a meteorological station nearby. An actual onsite measurement for at least one year is strictly recommended.

ESTIMATING THE WIND RESOURCE IN THE COMMONWEALTH OF INDEPENDENT STATES: NUMERICAL PROCEDURES AND STANDARDS

V J Nikolaev, Y A Grintsevich, L V Ponomarkenko, A N Baklanova, A S Tremsin
State Scientific Research Center, TSAGI, 17 Radio Str., Moscow 107005 USSR

Abstract: In this paper the results of development of reliable techniques for estimation of wind power resources (WPR) and the wind-systems (WPS) efficiency of the CIS territory are covered. This technique is used as the basis for the State Standard development.

In the draft, the standard domestic and international experience of estimation of WPR and WPS efficiency are generalised; the coordination and regulation of the available techniques were carried out and the common approaches to WPR and WPS efficiency estimation were developed.

The majority of problems, connected with the development, commissioning and effective exploitation of wind power stations (WPS) are concerned with the increase of energy efficiency of the system:- wind power station and wind. Efficiency is determined by two factors: availability of technically satisfactory WPS and use of the wind energy source with optimal parameters. There are two classes of task, which are important at the planning, developing, commissioning and exploitation of some of the WPS. The first class task deals with the type of WPS which would be more actively and economically effective at the usage of the winds with known parameters. For the second task the reverse problem is considered, where it would be better to place WPS with fixed working parameters to obtain the maximum energy. The second class of problem is the determination of energy productivity of different types of WPS in one or another region of the country.

The wind parameter set, which determines the efficiency of wind energy involves up to ten characteristics which are different for different regions: daily, annual or several years dependence of average velocity and direction of the wind, its vertical distribution at the surface closed layer, repetition of the velocity and direction by the graduation and their uninterrupted duration at a given range and also the maximum velocity and acceleration at the gust. Meteorological and aerological probe data from the network of stations are the basis for wind energy characteristics investigation at the surface closed layer. The number of stations in the network may be up to 3.2 thousand over USSR region. The variety of developed and suggested WPS in the world and the soviet market is greater than several dozen now. The labour intensive multivariant and joint analysis of energetic characteristic of the WPS and the wind's parameters is necessary for solving many actual wind energy problems. Therefore the amount of dual variants, which need to be considered, may exceed 10,000. It is obvious that solving this task can be achieved only on the basis of automatic systems, containing databases for the wind parameters at a given region and the working characteristics of WPS of known schemes.

The technology of automatic analysis and question solving mentioned above, is developed at TSAGI. This technology is based upon modern methods of wind power parameters and WPS productivity simulation.

The adequacy of evaluation of wind power potential is determined by accuracy of the probability and time structure of the wind in our models. Pointed models are currently created, primarily based on data obtained from the meteorological and aerological stations network or on the limited measurements or other non-systematic investigations.

With time position distribution investigation of wind power potential, and empirical modelling of wind conditions at a given region, the principal questions are:

1. Determination, for a given region, of values of wind power potential on the basis of a meteorological station network.

2. The problem of recovery of the limited data, obtained from the network of meteorological stations. However, for many stations, only prevailing directions and monthly average velocities of the wind are available from the reference books.

3. The problem of insufficient frequency of meteorological observations from the station network (as a rule - four times a day and insufficiently long date sets).

4. Possibility of interpolation of various wind parameters in inter-station space, taking into account orographical and other local peculiarities.

5. Reliable data extrapolation of surface meteorological observations on altitude.

Progress in solving the first and last problem has been made in this paper by using observations from aerological stations network. The second and fourth problems are addressed by using a range of developed methods, by means of statistical modelling or the base of maximum complete usage of meteorological information, known for given regions. The most reliable results here were obtained by using the local wind density distribution function.

Maximum efficiency and comprehension of the modelled meteorological information is provided by using an automised database called "FLUGER".

"FLUGER" is an original information search system developed by the authors using IMB PC compatible computers. It can be applied to solving tasks, which concern the modelling and calculation of wind field and wind energetic potential, and also the efficiency WPS evaluation at a specified region and location in the USSR.

The information base of the system includes a statistical summary of many years' data of atmosphere probes on the network of meteorological (greater than 3000) and aerological (up to 200) stations making the most complete and reliable reference books regarding the climatic and atmospheric parameters in the territory of USSR. It contains single station information for annual and daily velocity behaviour and direction for: altitude profile, scaler wind velocity distribution functions, air density data, exposure height of station and also for the period of observation. The system gives information about the main tendencies and peculiarities of the wind regime at a location of interest and also makes recommendations about wind parameters and wind power potential modelling.

The database system provides a choice of interesting regions and stations on the functional map and gives the nearest representative station selected by the user. The Helman model is used for information data over altitudes. Parameters of this model, determined by the closed representative aerological stations take account of wind and orographical peculiarities of searched point.

The values of the most probable wind conditions at the interstation gaps are determined by means of physical statistical modelling using meteorological data from near the investigated point stations and by means of numerical models, which allow the space time wind structure and relief peculiarities to be taken into account.

The monthly annual maximum values of wind energy potential can be calculated from a created wind model for a given local region. The probabilities of wind energy potential values, at a given velocity and time range can also be calculated from this model. The system is an effective tool for wind mesoscale space and time analysis at a specified USSR region and also it can be used as teaching material.

The data base includes information from technical and working parameters of several dozen different WPS, which are under serial production or are under development.

At a specific local region, wind model and working (power) wind energy characteristic, evaluation of monthly and annual electro-energy yield, coefficient of nominal power usage and probability of working WPS at a different power regime can be performed. As a result of such investigations, the choice of the most effective WPS type can be made for a region under consideration, including foreign regions.

There is a possibility to extend the database and to include new information blocks.

The required information can be represented as a file, which can be printed or visualized on the screen in a numerical or graphical form. This information can also be further processed by some programs.

The system requires disk memory resources of PC (up to 10 MB) and provides high performance and maximum simplicity and convenience for user.

Control of this system supports the interactive dialogue region by means of menu test and graphical explanations.

The "FLUGER" system can be adopted for any IBM PC compatible computer. The presence of hard disk is not necessary, although it would be preferable, but this system can be loaded from floppy disk.

The calculated technology allows maximum completeness to solve the following tasks.

1. The evaluation of wind resources at a given point in the USSR region. The construction of the local statistical models for the most probable wind characteristic on the basis of many years' observations: annual, daily and several years' of changing velocity and direction, height, distribution of the velocity, the probability of calm and wind velocities at a given range, impenduosity and the other characteristics, which is determined by taking into account the height of the location and its relief.

2. The evaluation of energy efficiency of the different type WPS, with known or expected characteristics, at the points given by customers (annual and monthly average yields, the nominal power usage coefficient, probability of calm and storm standing idle time and also work of WPS at different power regimes).

3. Effective planning of wind prospecting at proposed places for WPS regimes the optimal point choice for their placing. The program, and techniques of undertaking such prospecting, have to be formulated after complete processing of available regime meteorological information and using all possibile physical and numerical modelling.

This technology may be of great practical interest for WPS projects in definite regions by means of complete usage of the information for regional peculiarities of the wind, while choosing calculation parameters and working characteristics of WPS.

The benefit of a developed system is undoubtedly to join the regimes of network load.

The possibility of reliable splitting of USSR territory into appropriate regions by wind energy parameters may be of great practical interest in the planning and connection of networks of certification stations.

The presence of such networks could allow specific WPS schemes, at conditions close to proposed regions for WPS to be established.

In the authors' opinion, the system "FLUGER" and realised in it technology of solving wind energy tasks may find useful application in the practice of design, energy, construction and economic organisations.

One of the conditions for successful development and formation of domestic wind power engineering is the presence of reliable techniques for estimation of wind power resources (WPR) and the systems wind (WPS) efficiency of the CIS territory. Acceptance of such techniques, as the standard norm, arouses interest both of potential consumers of wind power production and of manufacturers. For the consumers, these techniques are important during the development of the power supply strategy and the choice of WPS types, most effective for specific wind conditions and power loads. For the manufacturers, they are required for this production competitiveness increase.

In this paper the results of the State Standard developments are given. This work was carried out by the authors in accordance with the order of Ministry of Fuel and Power of Russia.

In this draft the standard domestic and international experience of estimation of WPR and WPS efficiency are generalised; the coordination and regulation of the available techniques were carried out and the common approaches to WPR and WPS efficiency estimation were developed.

WINDMILL OPERATION PARAMETER CALCULATION TECHNIQUE

If the wind distribution function is known at rotor axis height H, it is possible to estimate some windmill operation parameters;

- established power use coefficient, P nom,

$$K = \frac{P}{P_{nom}}$$

- windmill mean power (kWt):

- mean generation for the time period T(kWt/hour)

$$W = P_{nom} * K * T$$

all year round $W = 8760 * P_{nom} * K$

- hours of the established power use, P_{nom}/year:

$$t = 8760 * K$$

- the proportion of time that the windmill is operational within the power range (P_1, P_2) (equivalent to the wind within the corresponding velocity range (V_1, V_2) where: $P_1 = P(V_1)$ $P_2 = P(V_2)$

$$F(V, V_o) = \int_{V_o}^{V} f(v)\, dv$$

- idle time proportion corresponds to the wind proportion within the velocity interval (0, V_{min}). Corresponding value, represented by the mean number of hours per year: $8760 * F(V_o, V)$

- mean continuous duration of operation (idle time) of a windmill corresponds to the mean continuous duration of the wind velocity higher (lower) than the starting one $t(v > V_{min})$. It can be estimated by indirect method using the equation:

$$t(v > V_{min} = 0.7\ V_{mean} = 0.09\ F(v > V_{min}) - 0.06$$

where $f(v > V_{min}) = \int_{V_{min}}^{V_{max}} f(v)\, dv$

for idle time:

$$t\ (v < V_{min}) = t(v > V_{min}) \frac{1 - F(v > V_{min})}{F(v > V_{min})}$$

- the mean number of periods with continuous duration $t(v > V_{min})$ per year

$$n = \frac{f(v) > V_{min})}{t(v > V_{min})} * 8760$$

All the above calculations are based on knowledge of wind distribution f(V), below the problems of incomplete acceptable information concerning the distribution and methods of rectifying this are considered.

WIND DISTRIBUTION ESTIMATION TECHNIQUE ACCORDING TO THE WIND MEAN VELOCITY

To calculate the windmill operation parameters, information for the database concerning the wind distribution at 10m height (distribution is represented in the form of an histogram on the basis of the wind duration frequencies within 15 different intervals of velocity) is used. In the case of insufficient observations (less than 10 years) the database contains information concerning only the wind mean velocity at height 10m, the program uses the following technique of distribution, according to wind mean velocity.

The tabulated distribution is used after static processing of all the distributions from the database (Table 1).

For arbitrary value of the mean velocity V, situated between the velocities V_j and V_{j+1} of the Table, the wind distribution frequencies can be found by linear approximation:

$$f_i = f_i^{\,j} + \frac{(f_{j+1} - f_j)}{(v_{j+1} - v_j)} (v - v_j)$$

The choice of a distribution interpolation method according to the mean velocity was needed due to the lack of analytical function, reliability approximating the distribution. Tabulated distribution was obtained by formal statistical methods minimizing the error of prediction. In this case, the mean error of the wind occurrence frequency definition within any velocity interval does not exceed 10 - 20 %.

In the case of, for example, Weibull approximating distribution, error of definition of reiteration frequencies within low and especially high velocities reaches 50 - 100 %.

In spite of this, Weibull distribution is acceptable for a rough approximation of empirical distribution to calculate the integral characteristics of the windmill operation, (eg of the annual generation). In this case, the mean error does not exceed 5 - 10 %.

However, if the parameter calculation on the basis of the wind reiteration within the narrow range of velocities (such calculation requires more exact knowledge of distribution), the tabulated distribution gives more accurate results.

HEIGHT H WIND DISTRIBUTION ESTIMATION TECHNIQUE ACCORDING TO THE WIND DISTRIBUTION AT HEIGHT 10 M

To obtain the required estimates for the rotor axis placement at height 10 H, it is necessary to estimate the wind distribution at height H according to available distribution at height 10m.

Wind profile within the near-ground layer represents, as a rule, the function which increases together with the height. This function can be approximated by the degree of dependence:

$$\frac{V}{V_o} = \left\{ \frac{H}{H_o} \right\}^m$$

The mean statistical dependence of degree index m on the wind velocity, used at calculation, is given in Table:

v	1.5	3.5	5.5	7.5	9.5	11.5	13.5
m	.42	.31	.25	.21	.18	.16	.15
v	15.5	17.5	20.5	24.5	28.5	34.5	40.5
m	.14	.13	.13	.13	.12	.12	.12

The dependence is obtained by providing an average according to the great number of the wind profiles. In spite of the fact that sometimes the individual profiles may considerably differ from the characteristic ones (calculated using the table), because we calculate windmill parameters average for the durable period of time, it will not lead us to substantial errors.

Table 1

v	0-1	2-3	4-5	6-7	8-9	10-11	12-13	14-15	16-17	18-20	21-24	25-29	30-34	35-39	>40
.0	100.	.0	.0	.0	.0	.0	.0	.0	.0	.0	.0	.0	.0	.0	.0
1.4	77.1	15.5	5.2	1.5	.4	.2	.1	.0	.0	.0	.0	.0	.0	.0	.0
2.0	58.4	24.4	10.7	4.0	1.4	.4	.3	.2	.1	.1	.0	.0	.0	.0	.0
3.0	37.9	28.5	18.6	8.8	3.4	1.1	.9	.4	.3	.1	.0	.0	.0	.0	.0
4.0	24.9	27.6	23.1	13.0	5.9	2.2	1.8	.7	.6	.2	.0	.0	.0	.0	.0
4.9	18.2	23.4	23.2	16.0	8.7	4.4	3.1	1.3	1.0	.6	.1	.0	.0	.0	.0
6.0	14.6	18.5	20.6	16.2	11.4	7.3	4.8	2.6	2.0	1.5	.3	.1	.1	.0	.0
7.0	12.0	14.2	18.5	15.8	13.1	10.2	6.5	3.8	2.8	2.1	.5	.3	.1	.1	.0
8.1	10.0	11.8	16.0	15.0	13.2	11.2	7.4	4.9	4.0	4.0	1.3	.7	.4	.1	.0
9.3	8.4	8.5	13.5	14.2	12.7	12.1	8.8	6.3	5.8	5.8	2.0	1.0	.7	.2	.0
10.1	7.4	7.0	11.3	13.3	12.1	12.6	10.0	7.3	6.7	7.2	2.6	1.3	.9	.3	.0
11.1	6.4	6.5	9.5	11.0	11.5	12.7	10.0	8.0	8.0	8.5	4.0	2.0	1.3	.6	.0
12.0	5.6	6.0	8.3	9.6	10.2	12.3	10.2	8.5	9.2	10.0	4.8	2.8	1.6	.8	.1
12.7	5.2	5.5	7.2	8.7	9.8	11.8	9.8	8.8	9.6	11.0	5.8	3.5	2.2	1.0	.1
13.0	5.0	5.5	6.1	6.8	8.8	11.0	9.0	8.0	9.5	11.8	7.0	4.0	3.0	1.3	.2
14.5	5.0	5.5	5.7	6.2	8.0	10.0	8.0	7.5	9.0	12.6	10.0	7.0	3.6	1.6	.3

Form of a function m(v) is typical for the layer 10-100 m of smooth and open locality (as aerological data showed), there are some variants to this function ie for relief of different forms and roughness. Since in our database we have no information concerning these parameters, we have to use the mean statistic dependence m(v), which leads to qualitatively correct results, but for the complex forms of relief, the estimates on it lead to an error (which equals zero for H=10 m), which increases together with the height (which reaches 20-30 % for H=30 m).

A mathematical method of distribution lifting to height H was realised as follows:

Let us consider the wind velocity interval (v_i, v_{ih}) for 10 m height. The wind occurrence frequency within this interval is f. It is necessary to have characteristic profile and velocity at height H and wind velocity at 10m, which equals v. In this case it is found on the basis of degree dependence:

$$v_i^H = v_i \left(\frac{H}{10} \right)^m ; \text{ similarly: } v_{i+1}^{10} = \left(\frac{H}{10} \right) V_{i+1}$$

It is affirmed that the wind reiteration frequency within the velocity interval (v_i, v_{i+1}) at H height equals the wind frequency within the interval(v_i, v_{i+1}) at 10m height. As it was mentioned above, if we think that the wind profile is strictly restricted to the degree of dependence with the index m(v), which was prescribed by somesome function, then such a statement does not lead to considerable errors at statistic estimations.

At program realisation, we firstly obtain the new values of velocity gradation intervals, for example:

height	gradations				
	0-1	2-3	4-5		...
10m	0.	1.5	3.5	5.5	...
25m	0.	2.2	4.65	6.92	...

According to our affirmation, reiteration frequencies within the new intervals are the same as at 10 m height. We now have to bring the obtained histogram to the standard form with the standard intervals of velocity:

$$f_1^H = 0.68\, f_1^{10}$$

$$f_2^H = 0.32\, f_1^{10} + 0.53\, f_2^{10}$$

$$f_3^H = 0.47\, f_1^{10} + 0.37\, f_2^{2} \quad \text{etc.}$$

REAL WIND DISTRIBUTION APPROXIMATION BY WIEBULL DISTRIBUTION

Wind structure data currently available, makes it possible to affirm that the wind distribution empirical functions within the near-ground layer of atmosphere can be described by some known functions for statistic theory of functions.

However, on the basis of our experience of turbine-wind double-component system modelling for the wind parameter qualitative estimates, the following Wiebull distributions are suggested:

$$f(v) = \frac{\gamma}{\beta} \left(\frac{V}{\beta}\right)^{\gamma-1} \exp\left[-\left(\frac{V}{\beta}\right)^{\gamma} \right]$$

Where β and γ - distribution parameters.

They can usually be found by equating the 1st and 2nd moments of empirical distribution to the corresponding moments of Weilbull distribution. We can obtain an equation system of which it is possible to calculate β and γ numerically:

$$V_{mean} = \beta \times \Gamma^{-1} \left(\frac{\gamma+1}{\gamma}\right) \qquad (1)$$

where - V_{mean} - mean velocity the first moment of empirical distribution.

Γ (x) - gamma - function

Since in equation (1) the coefficient $1/\Gamma(1/\gamma + 1) = 0.87 - 0.92$ within the range, which is interesting as it is possible to express via V m clearly by:

$$\beta = 1.12 \text{ V m} \qquad (4)$$

It does not lead to substantial error but it considerably simplifies the estimates because we have to find only the parameter from equation (2).

Thus the resultant Wiebull distribution can suitably approximate the real distribution.

The checks show that Kolmogorov criterion with 5% is satisfied by all available distributions. In other cases, despite the discrepancies, a Wiebull distribution can be used for some estimates.

We do not observe the simple dependence of β on the mean velocity V_{mean}. For the same value of V_{mean}, β can vary from 1.0 to 2.0 for different meteorological stations (at height 10m).

γ increases together with height and when it is higher it reaches 2.0 (in this case the Weibull distribution becomes Maxwell distribution).

At the ground at 10m height β in mean equals 1.2 - 1.7 βs dependence on period of averaging of observations is observed. So, for weathercock measurements of the wind with the averaging period of 2 min, v, as a rule, is lower than it is at anemometric ones with period of 10 min.

We have presented above the development perspective for wind power energy in some regions of USSR, including Aserbaidgan. The experience of working with the system shows that it allows considerable increase of velocity, completeness and reliability of typical calculation of wind energy potential and efficacy of WPS on the territory of USSR.

This technique is used as the basis for the State Standard development.

The draft of the standard includes several sections:

- introduction, containing the analysis of the available information and defining the tasks and the field of this standard application;
- the general idea of WPR's and possibilities of their effective use;
- the main appropriateness and peculiarities of space-time distributions of WPR's on the territory of Russia and CIS;
- analytical methods of WPS and estimates;
- experimental methods of wind power parameter estimates;
- ecological estimates of WPR and WPS efficiency estimates regarding the local conditions;
- appendices including necessary explanations, maps, tables and examples of calculations.

Today, the main task of scientific and technical approbation of this draft consists in bringing it into line with the available foreign and international standards of ISO, MAC etc.

ADVANCED POWER INTERFACE AND CONTROLLER FOR VARIABLE SPEED WIND TURBINES

A.T. Veltman, S.W.H. de Haan,
C.F.A. Frumau, E. Wildenbeest
Netherlands Energy Research Foundation ECN
Business Unit Renewable Energy
Postbus 1, 1755 ZG Petten
The Netherlands
Tel: (+31) 2246 4113, Fax: (+31) 2246 3214
E-mail: veltman@ecn.nl

G.A. Smith
Leicester University
Department of Engineering
University Road
Leicester LE1 7RH
United Kingdom
Tel: (+44) 533 522 535
Fax: (+44) 533 522 619

D.G. Infield, D.A. Simmons
Rutherford Appleton Laboratories
Energy Research Unit
Chilton, Didcot
Oxfordshire, OX 14 0QX
Tel: (+44) 235 821 900
Fax: (+44) 235 446 863
E-mail: dgi@uk.ac.rl.ib

ABSTRACT: A 30 kVA AC/DC/AC converter for variable speed wind turbines has been designed and built. Tests have been carried out using a wind turbine simulating test rig, equipped with a synchronous generator. Efficiencies and grid behaviour have been measured under laboratory conditions and at different power levels. Also a wind turbine controller, controlling the turbine power by the converter input current under different wind turbine control regimes has been developed and is tested on the wind turbine simulating test rig.

1. INTRODUCTION

The objective of the work described here is the development of a 30 kW advanced AC/DC/AC converter system and an integral control unit combined with a synchronous generator, to realise a rugged generator system for variable speed wind turbines which is very appropriate to the mains and has a high energy yield.

Expected achievements were to be obtained by applying:
- variable speed wind turbine systems as (opposed to fixed speed wind turbines),
- advanced converters on both generator and grid side instead of conventional line commutated converters,
- a synchronous generator (preferably excitated by permanent magnets).

The advantages of variable speed wind turbines to be mentioned are: increased energy output due to operation in a closer vicinity of the optimal C_p-λ curves, reduced stresses in the drive train due to the flywheel capabilities of a turbine rotor loaded with constant torque rather than rotating with constant speed, and the possibility of rotating at reduced speed - hence less acoustic noise generation, especially in crowded regions and at night.

Further benefits that can be expected by applying the advanced AC/DC/AC converter aimed at, consist of a much improved behaviour of the mains current with respect to its displacement factor, reducing importantly the active power to be supplied by the utility grid. Further, a lower harmonic distortion, leading with conventional converters to disturbances of the grid voltage shape, which should be sinusoïdal. Also mentionable are the reduced energy losses in the generator, due to more appropriate power angles and a lower harmonic currents content, as well as a better controllability of the torque, eliminating the large torque pulsations which would otherwise be induced by grid voltage disturbances.

The method of approach consisted of the completion of the following tasks:
- the development of a 30 kW advanced AC/DC/AC power electronic converter,
- the development of an integral control system, implemented on a personal computer,
- integration of the 30 kW converter with the integrated controller on a wind turbine simulating test rig for evaluation of the electrical conversion system under closely defined laboratory (wind as well as electrical) conditions.

2. AC/DC/AC CONVERTER DESIGN

2.1 General

As outlined in the introduction, the AC/DC/AC converter design should provide:
- a high degree of controllability of the generator power to reduce torque fluctuations induced by for instance grid voltage disturbances, .
- highly sinusoïdal mains, and preferably also machine, currents, e.g. low current harmonics content,
- power factor close to one,
- high efficiency.

For the variable speed operation, one of the main objectives of the power-electronic interface is to allow the synchronous generator to operate over at least a 3:1 range of speed depending on wind condition and return good quality power to the power system. The original concept envisaged identical machine side and network side reversible rectifiers with the former being synchronised to the machine emf's and the latter to the supply voltage. The reversible rectifiers would then control the machine and network current both in waveform quality and in power factor.

Reversible rectifiers are essentially rectifiers in which the individual diodes have a controlled switch connected across them. Insulated Gate Bi-Polar Transistors, IGBT's, were selected as they are an integrated diode-switch with low on-state losses and a high-frequency switching capability. Such a rectifier, when properly controlled, principally exhibits only high frequency current components with very low energy content and a fundamental component with zero displacement angle.

2.2 Input converter design

Originally one converter was designed at a 3 kW power level for validation of the proposed system. Unfortunately technical problems associated with first generation IGBT's and EMC problems in the circuit due to the fast switching transients greatly extended this phase of the work and this together with restrictions in time led to the simplifying solution for the input converter as shown in figure 1 for the 30 kW version.

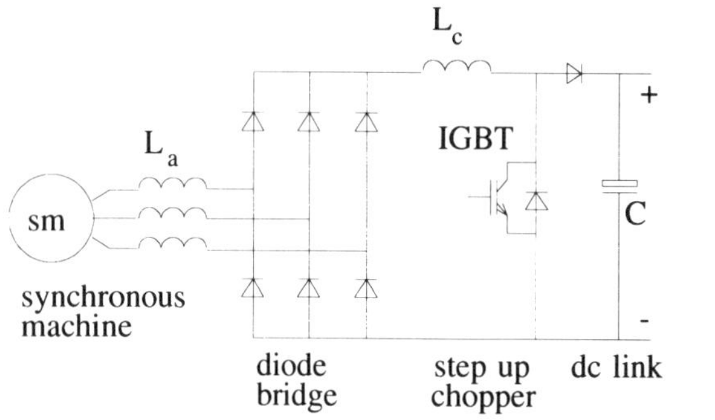

Figure 1: Configuration of the input converter.

This figure 1 shows a diode bridge rectifier on the input side (no reactive currents), followed by a step up chopper to adequately control the input current, hence power.

2.3 Output converter design

Mindful of extension up to very high powers requiring connnection to a high voltage line through a network transformer, the output was based on three identical 'H' bridge rectifiers each connected to an insulated secondary winding of the network transformer as shown in figure 2.

A current reference sinewave is generated in phase with

the supply emf's and the H bridge IGBT's are switched to force the current to the supply to be constrained to lie within

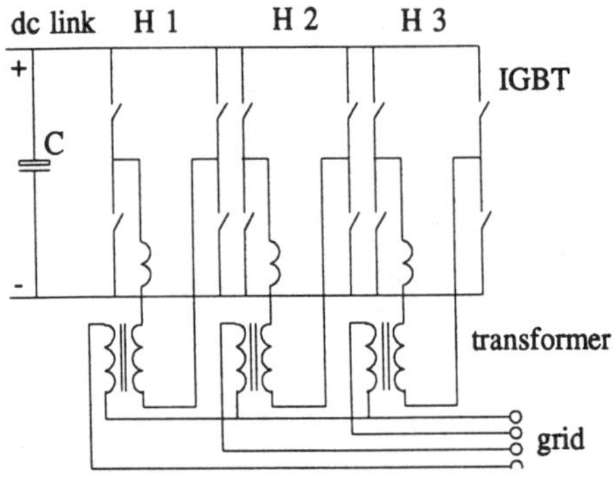

Figure 2: Grid side converter set up

a preset boundary of that sinewave. This is known as hysteresis control and results in switching frequencies that vary across the sinewave period and which are dependent on the supply voltage and series reactance. Thus a broad band of frequencies are present, avoiding a high energy density at one particular frequency.

2.4 Protection and converter control

Protection of the 'H' bridge IGBT's include gate circuits preventing a dc link short circuit and fitted with saturation voltage detectors. The latter will turn off the IGBT's if excessive on-state voltage, due to excessive currents is observed. In addition the dc link has been provided with excessive rate and magnitude current detectors to again turn off all devices. The link reactor limits the rate and reduces reverse recovery losses in the diodes. A fuse in series with the reactor gives ultimate protection with the parallel diode ensuring a return path for any stored energy in the output circuit. For correct operation the H bridge diodes must be reverse biased hence it is necessary to ensure a minimum voltage on the DC link capacitor C in figure 1 and 2. To achieve this when the generator voltage is changing with operating speed, the step-up chopper is used. Also this one is operated by hysteresis control at the output of the rectifier diode bridge. Where the subtransient quadrature axis reactance of the synchronous machine will prevent the generators' synchronous reactance from appearing on the input side of the rectifier during the transient states of the chopper IGBT, the switching frequency is controlled by a single reactor on the dc side. Thus there are two current control systems in operation: one for machine currents and the other for network current. The latter being synchronised at unity power factor to the network actually controls the output power. Both control loops are separated by the dc link capacitor to obtain as few interaction as possible of both control loops. The effect of controlling the machine side current is precisely known as the power factor is constant. Providing an appropriate relationship is programmed in the PC this current can be used to control generator torque. Where input chopper current is controlled by the system computer, the output current must be controlled internally by either voltage underspill or overspill to maintain the link capacitor voltage within prescribed limits.

Fault shutdowns will always be indicated and advise the master control computer for shutdown or can directly shutdown the power converter and apply dynamic braking resistors.

3. OPERATION OF AC/DC/AC DRIVE WITH SMALL PITCH CONTROLLED WIND TURBINE

3.1 Experimental arrangement

A two bladed, downwind, 3.5 m diameter wind turbine with active pitch control was used to demonstrate the control and operation of the AC/DC/AC drive to achieve variable speed operation. The turbine, which has a maximum power output slightly in excess of 1 kW is fully described in [1].

It should be pointed out that the original intention of the project was to test a specifically developed 3 kW AC/DC/AC drive on this wind turbine. Technical difficulties prevented the completion of this smaller drive and a modified design was constructed at 30 kW rating for later application in the Netherlands. Despite the size miss-match it was agreed to gain early experience in the operation of the drive by running it with the small turbine.

Output from the permanent magnet generator of the wind turbine is fed to the AC/DC/AC drive with a resistive dump load connnected in the case of grid failure and emergency shut down. The drive is connected to the three phase grid via a transformer as described. Control is performed by an industrial PC with suitable I/O boards. Data collection is handled by another PC.

3.2 Control algorithm for variable speed operation

All control operations including run up, shut down and protection are handled by the PC. Of most concern here is the algorithm implemented when the wind turbine is below rated power. Above rated power pitch control is used to limit power but below rated power variable speed operation with the aim of maximising energy capture is provided. This is achieved by controlling the AC/DC/AC drive. The PC provides a current demand signal to the drive which is used to control the DC current delivered by the bridge rectifier on the machine side of the DC link. The power delivered by the drive to the grid is controlled internally in relation to the DC link voltage.

Turbine control is specified in a feed forward manner in that the load to be applied to the permanent magnet generator is calculated as giving maximum C_p under steady state conditions; i.e. the load is set proportional to the measured wind turbine rotor speed cubed. The current demand signal is calculated from equation

$$I_d = 3k_u\sqrt{\frac{1}{2}\iota^2 - \frac{1}{2}\sqrt{\iota^4 - 4\frac{(P_{AC}/3)^2}{\rho^2}}}$$

describing the drive and permanent magnet generator [2] and in which I_d is the chopper command current. k_u is a constant and ι and ρ are functions of generator rpm and/or power.

3.3 Experimental results

Difficulties were experienced in running up the turbine: The drive was set up with a fixed minimum current from the chopper. For a appropriately sized wind turbine this would not have mattered but in this case the load applied to the turbine irrespective of the control demand signal, was a significant portion of the rated output of the turbine. As a result the system could only be run in very high winds and run up was achieved by temporarily disconnecting the turbine from the drive.

Figures 3 and 4 show the response of the system over 500 seconds during which the wind speed varied between 4 ms^{-1} and 16 ms^{-1}

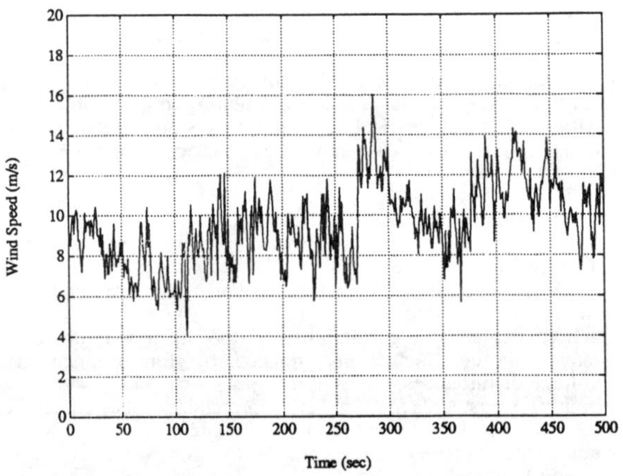

Figure 3. Wind speeds between 6 ms^{-1} and 16 ms^{-1}.

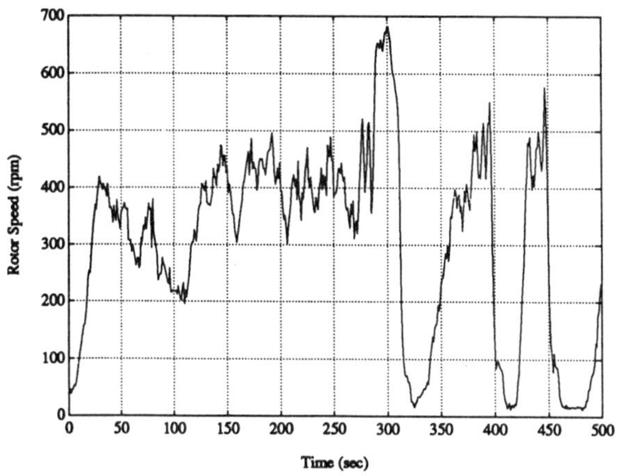

Figure 4: Rotor speed during the same interval as in figure 3.

Up to about 300 seconds the drive is being adjusted to successfully keep the rotor speed in the range of 200 rpm to 500 rpm. A dramatic gust then pushes the rotor into overspeed despite blade pitch action, at which time the controller sets the drive current demand to zero and applies the dump load. The generator voltage rises and as a result of the minimum drive current the DC link voltage is pushed up, producing a increase in drive output power. After this event the controller makes two attempts to run the turbine up but due to the wind turbulence overspeed and shut down occurs.

4. CONVERTER MEASUREMENTS

4.1 Electrical measurements

After the construction of the 30 kW converter and after the controller measurements on the 1 kW wind turbine were finished, the converter has been tested on a 30 kW wind turbine simulating test rig consisting of a 150 kW DC drive with controllable torque, speed and power, a control computer and power measurement and analysis equipment.

The efficiency of the converter has been determined by separately measuring the chopper bridge rectifier and the inverter bridge including the transformer. Afterwards, losses in the transformer have been measured. The results are shown in **figure 6** as a function of AC output power.

This measurement shows a overall system efficiency of approximately 82 % at 5 kW and 7.5 kW and of 88 % at 10 kW and above.

Where the input current harmonics will mainly be determined by the diode bridge input rectifier with its well known relative harmonic values $u_n = 1/n$, $n = 2k \pm 1$, $k \in \mathbf{N}$, and a part of the heavily filtered chopper current harmonics, the output current harmonics are measured and shown in **figure 5**.

It is clearly visible that not a single harmonic exceeds a value of 2 %. Table 1 shows total harmonic distortions measured up to 20 kW. As a result of the EMC problems also mentioned in section 2.2, it turned out that powers exceeding 20 kW were not extractable from the converter.

Table 1: Total bridge current output harmonics

Power [kW]	Total converter output current harmonic distortion percentage of fundamental component [%]
5	5.9
10	6.0
15	6.3
20	7.4

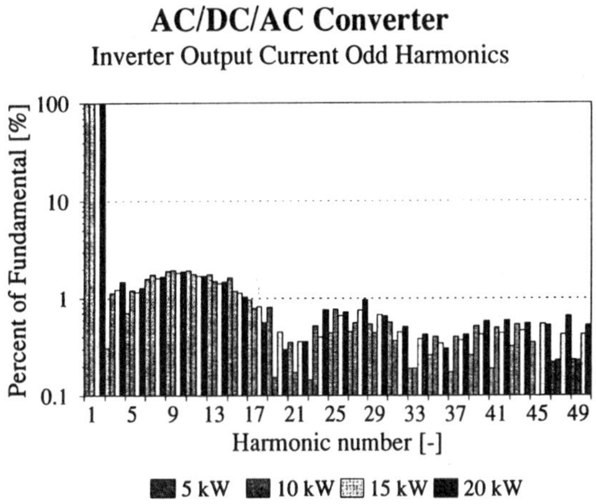

Figure 5: Converter output current odd harmonics

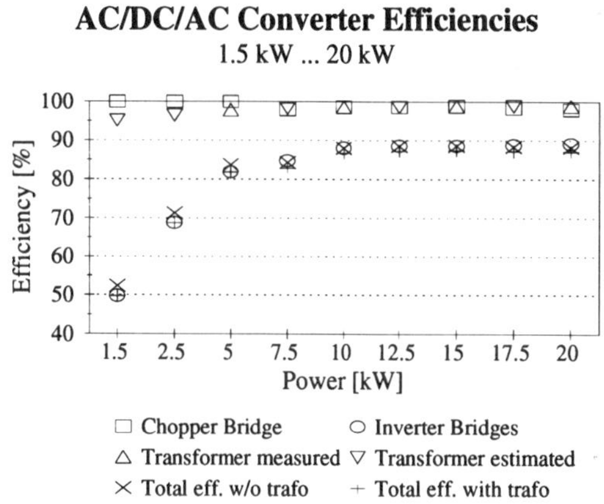

Figure 6: Efficiency measurements on the converter

The displacement angle between inverter phase current and voltage shows a good behaviour with respect to reactive current demand by the inverter. The cosine of this angle, cos(**φ**), is shown in figure 7 and shows excellent behaviour over a wide power range.

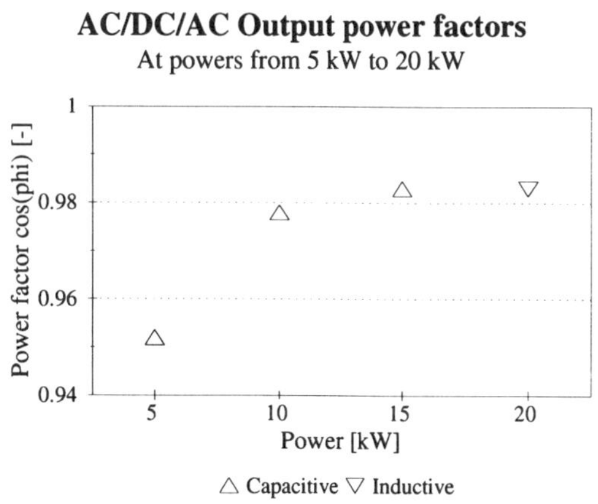

Figure 7: Displacement angle of the output converter

At 5 kW the cos(φ) measures 0.95 capacitive and grows towards one until approximately 17.5 kW, at which point the value can be estimated as one. Above this power level, the character of the displacement angle changes from slightly capacitive to slightly inductive. Note that this value is reached without any reactive current compensation circuitry as would be normally used with conventional, i.e. line commutated converters.

Further measurements have been made on the response of the output power due to a step in the input current demand. As the chopper frequency is approximately 0.7 kHz and the DC link inductor current control forces the input current to follow within 1/(0.7 kHz), i.e. within 1.4 ms, input power can be considered to react within 2 ms. As the efficiency of the input chopper is very at this frequency (see figure 6), higher switching frequencies up to several kHz seem possible, increasing linearly the response speed. Figure 8 shows the respons of the link voltage (U_d) and the output current in one phase (I_R) due to a step in the chopper input current demand (I_{ch}) from 11 A to 23 A.

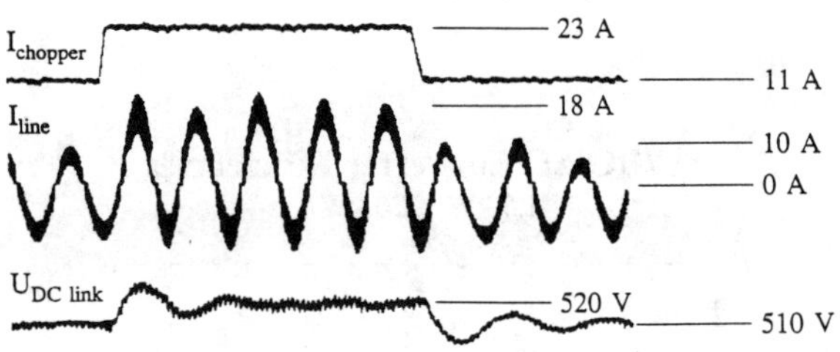

Figure 8: DC link voltage and converter output current responses due to a step change in input current demand. Each AC current cycle takes 20 ms.

Due to the increased input current, the DC link capacitor charges up, producing a voltage overspill which causes the current demand signal of the output converter to increase, hence output current increases, stabilising the voltage overspill to a new value, corresponding to the increased power level.

4.2 Wind turbine simulator measurements

After completion of the electrical measurements on the AC/DC/AC converter tests have been carried out using a wind turbine simulating test rig. This rig incorporates a microcomputer for simulation of a wind turbines' behaviour during pre-programmed wind conditions. In this test rig also the rotational mass and shaft torsional flexibility of a large wind turbine have been incorporated by a flywheel and a flexible shaft respectively, resulting in a dominating 3 Hz system eigenfrequency [3]. In this set-up the response of the converter, controller and wind turbine simulator to a cosine shaped wind gust has been measured. The wind speed profile was:

$$v_w = 8 \text{ ms}^{-1}, \qquad t \le t_0 \;;$$

$$v_w = 10 - 2\cos\left[\frac{\pi}{4}(t - t_0)\right] \text{ ms}^{-1}, \qquad t_0 \le t \le t_0 + 4 \text{ s} \;;$$

$$v_w = 12 \text{ ms}^{-1}, \qquad t + 4 \text{ s} \le t \le t + 9 \text{ s} \;;$$

$$v_w = 10 + 2\cos\left[\frac{\pi}{4}(t - t_0 - 9)\right] \text{ ms}^{-1}, \quad t_0 + 9 \text{ s} \le t \le t_0 + 13 \text{ s} \;;$$

$$v_w = 8 \text{ ms}^{-1}, \qquad t \ge t_0 + 13 \text{ s}.$$

As a response to this gust, which is shown in figure 9, the generator speed increases from 75 rpm to 82.5 rpm. For matching the Cp-λ curve of the wind turbine with the generator and converter characteristics, the control was programmed to increase power from 2.5 kW at 76 rpm to 20 kW at 85 rpm. It can be observed that after the turbine speed increases due to the higher wind speed, the integrated controller reacts with a increase in the demand of the input current $I_{chopper}$, followed by a increase in torque proportional to $n_{turbine}$ - 75 [rpm] as programmed - mind that the relative change in $n_{turbine}$ is low, as well as generator voltage, so with a first order approximation power as well as torque will

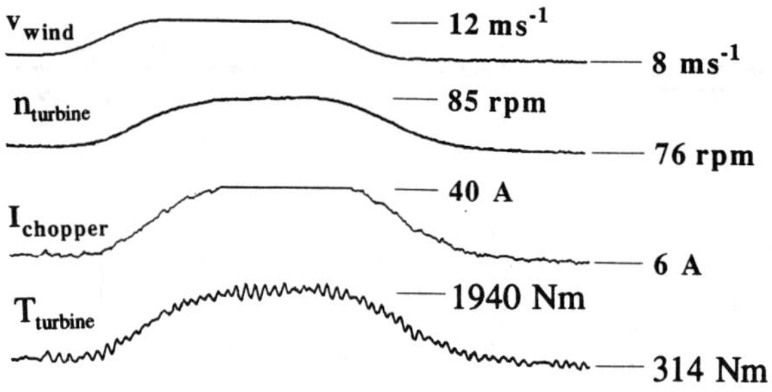

Figure 9: Response of the converter and integrated controller to a wind turbine simulator excited with a wind gust from 8 to 12 ms^{-1}

behave proportional to $I_{chopper}$ hence to $n_{turbine}$ - 75 [rpm].

Also note the 3 Hz component in the torque signal which is caused by rotational vibrations in the drive train due to the flexibility of the drive and the additional degree of freedom introduced by decoupling the generator speed from the mains frequency by the converter.

5. CONCLUSIONS

It turns out that the converter technique adopted is adequate with respect to the following project aims:

- good quality of power delivered to the grid in terms of displacement angle and harmonic distortion,
- high speed controllability of the wind turbines' power hence torque.

The controller developed for this project enables the turbine to follow the quasi-static, pre-programmed optimal power-rpm curves.

The converter techniques investigated are now being adopted commercially based upon the original concept of two three-phase reversible rectifiers and are available for powers up to 650 kW.

6. REFERENCES

[1] A D Simmons, L L Freris, J A M Bleijs, Comparison of energy capture and structural implications of various policies of controlling wind turbines, Proc. BWEA 13, Swanea, 1991.

[2] A T Veltman, S W H de Haan, Power factor and power control of the boost rectifier and the chopper controlled rectifier, Technical report ECN-C--92-054, ECN, Petten, August 1992.

[3] J T G Pierik, P M M Bongers, T G van Engelen, G E van Baars, Damping of mechanical resonance and protection against grid failure for variable speed wind turbines (IRFLET) Technical report ECN-C--93-004, ECN, Petten, January 1993.

Programmed Pole Generator for Wind Turbines

D Levy
University of Limerick
Limerick, Ireland

Abstract:
The programmed pole generator is an electrical machine based on ceramic magnet programming technology; continuous printing of the desired number of poles on the rotor allows fixed output frequency independent of speed. The rotor of the machine is normally made of ceramic magnet (Aniosotropic Ferrite pieces) with no slip rings or commutators. The machine can work in two modes of operation, "self excitation" to feed power to an isolated grid and "grid excitation" to feed power to the national grid. The stator of the machine consists of an output winding and an exciter winding. The latter characterizes this machine and makes it different from others. Theoretically, the machine is useful as a generator over a wide speed range.

1. Introduction

The machine invented in 1970 by J.F. Roesel Jr. [1:17], is basically, a synchronous machine with continuous variation of the number of poles to match the rotor speed and therefore, it is in fact a "synchronous machine" for a wide range of speed [2].

The machine can be regarded as a "high power tape recorder" with its output frequency always equal to the recorded frequency during the recording stage and therefore, it is independent of the tape speed (within a definite range of speed). Thus, the machine consists of two electrical circuits. An exciter (writing head) winding, and an output (read) winding.

The machine is available today up to 60kw and is used for NIPs (Non-Interruptive Power systems) [9] in the U.S.

2. Machine Description

The output frequency of any synchronous machine is given by the angular speed divided by the number of pole pairs. All conventional synchronous electrical machines have an even number of poles determined by physical windings. This forces the output frequency to vary with the rotational speed. The Roesel generator is different in that the number of poles is varied continuously and is inversely proportional to the speed, so that the output frequency is maintained at constant value.

The basic diagram of the Roesel generator is shown in Fig. 1. The stator, with its windings connected to the grid (or to an external load), is located on the inside of the generator. The rotor, which contains the field poles, rotates on the outside of the stator. The stator contains an excitation coil wrapped around the exciter head in addition to the usual output windings. The mutual inductance between the excitation coil and the output windings is practically equal to zero in order to prevent direct power feeding from the exciter to the output winding. In this way, the two coils are practically decoupled from each other. The rotor is built in two layers, with the outer layer being high-permeability laminated generator steel and the inner layer being a hard magnetizable material such as barium ferrite. Ferrites typically do not have the mechanical strength characteristics of steel so this design helps to maintain mechanical integrity by having the steel carry the magnetizing and centrifugal forces. The ferrite would have to be much stronger if the rotor were inside the stator. A reference frequency (or line frequency) is applied to the excitation coil and magnetizes a pole on the rotor as it turns around the stator. This is called "writing" a pole. The pole then induces a voltage in the stator windings at the same frequency. The output frequency then has the same precision as the input frequency, independent of shaft speed at steady state (zero acceleration).

As the rotor speed increases, the circumferential length of the poles increases and fewer of them are written around the circumference of the rotor. As the rotor speed decreases, the length of the poles shortens, so, more of them are written around the periphery of the rotor. There will be an even-pole synchronous speed where an even number of equal-length poles are spaced uniformly around the rotor. At the other extreme, there will be an odd-pole synchronous speed, where an odd number of equal-length poles are equally spaced around the rotor. Between these extremes there will be fractional poles in the vicinity of the excitor head as poles are being partially rewritten. At the even-pole synchronous speed, the poles remain in the same position from one revolution to the next, so no rewriting of poles actually takes place. There will not be rotor hysteresis loss in this case, since the rotor iron magnetization does not change with time. At the odd-pole synchronous speed, however, every positive pole is being replaced with a negative pole during each revolution, so rotor hysteresis losses will be a maximum at this speed.

The inventor manufactured the machine with double bearings, that is, the rotor (Ferrite) rotating outside the stator in order to prevent the magnetizing and centrifugal forces from detaching the Ferrite pieces from its place (Fig.1). However, in the present design, in order to simplify the mechanical construction, the rotor was built inside the stator as a normal machine. The stator laminations are shown in Fig. 2. The figure shows the exciter head with its coil space (occupies 22% of the stator). This is not an optimal design, but for a large machine, the exciter will occupy much less space (as in Fig. 1 for example).

This study was sponsored by the Commission of the European Communities, Directorate General for Science, Research and Development under Jour program on Wind Energy

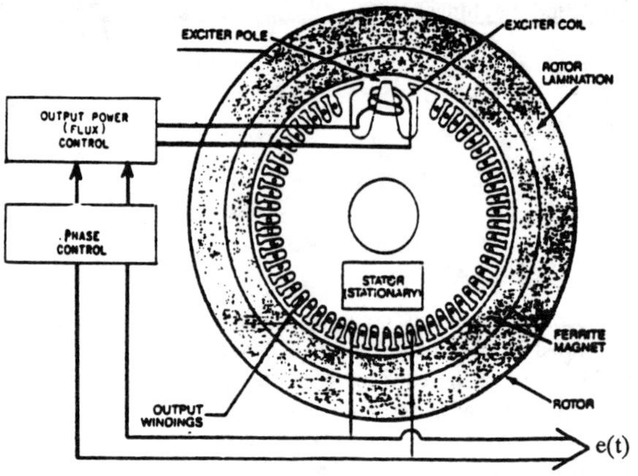

Fig 1 The Roesel Generator CONSTANT FREQUENCY OUTPUT

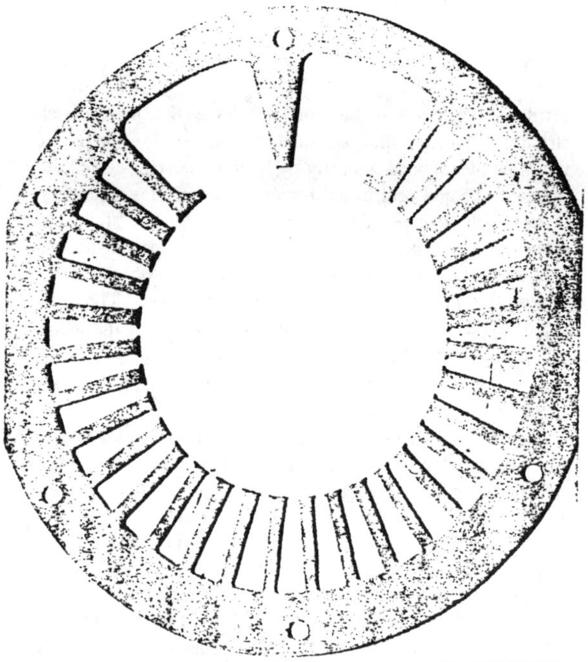

FIG. 2 PROGRAMMED POLE GENERATOR LAMINATIONS (REDUCED IN SIZE)

3. Analysis

The electromechanical and electrical equations of the machine are defined by two sets of nonlinear differential equations [16,17]:

$$J \frac{d\Omega}{dt} = T_s - T \quad (1)$$

$$\frac{d\,Li}{dt} = e(t) - iR - N\frac{d\phi}{dt} \quad (2)$$

$$T = -\frac{1}{\Omega} i.(N \frac{d\phi}{dt}), \qquad \Omega \neq 0 \quad (3)$$

where

J - Total rotor inertia

Ω - Machine speed
T_s - Turbine torque
T - Machine torque ($T = T_s$ at steady state)
L - Output winding inductance
i - Output or grid current
e(t) - Grid voltage
R - Output winding resistance
$N\frac{d\phi}{dt}$ - Open circuit output voltage (a function of speed)
e(t) - and T_s are independent variables (input to the machine)
i,Ω - are dependant variables, they are functions of time which should be derived from the solution of (1):(3)

The open circuit voltage $N\frac{d\phi}{dt}$ was calculated in [17] and is shown in Appendix 1 in a simple form for a winding with pitch factor and distribution factor equal to one. The solution of 1:3 will provide information on machine behaviour such as torque/speed, electrical power/speed, mechanical power/speed, reactive power/speed and efficiency/speed characteristics.

Transient state and steady state analysis of the machine was made [17] and some of the results will be summarised in this paper.

4. Machine Characteristics

The basic theory of operation is given in Appendix 1. However, due to the winding factor, (in particular the pitch factor), the output voltage is not constant but speed dependent as shown in Fig. 3 and Fig. 5.
These figures show the open circuit output voltage for different windings, when the machine is used to drive a passive load (isolated grid). For this application, the output voltage needs to be regulated. Fig. 4 and Fig. 6 show the output power fed to the grid when the machine is used in parallel with the national grid system (active load). The output voltage in this case need not be regulated since the machine acts as a current source and the output voltage is constant and equal to the grid voltage.

Most of the machine secrets lie in efficient writing of the poles on the rotor. Machine analysis was made by assuming the poles are ideally written on the rotor. Fig. 3

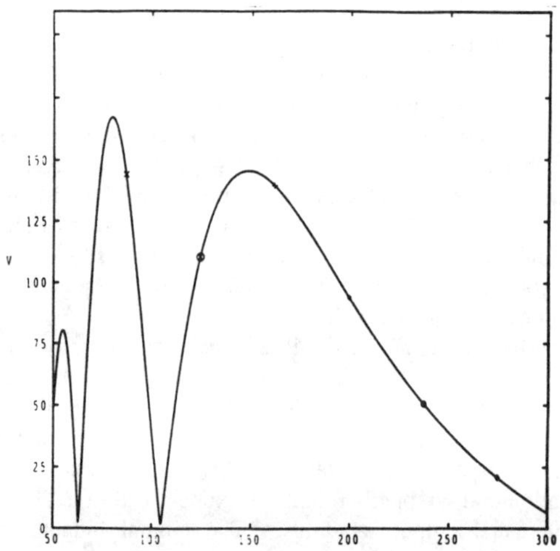

FIG.3 OPEN CIRCUIT OUTPUT VOLTAGE PASSIVE MODE OF OPERATION (PITCH = 100 DEGREES)

and Fig. 4 show the open circuit output voltage and the output power fed to the grid when the output winding is connected across constant grid voltage and the stator of the machine wound as a 4 pole winding (coil pitch about 100 degrees).

The output power was plotted against the speed taking the phase between the exciter and the output winding as a parameter. This phase needs to be adjusted continuously with the speed in order to increase the operating range of

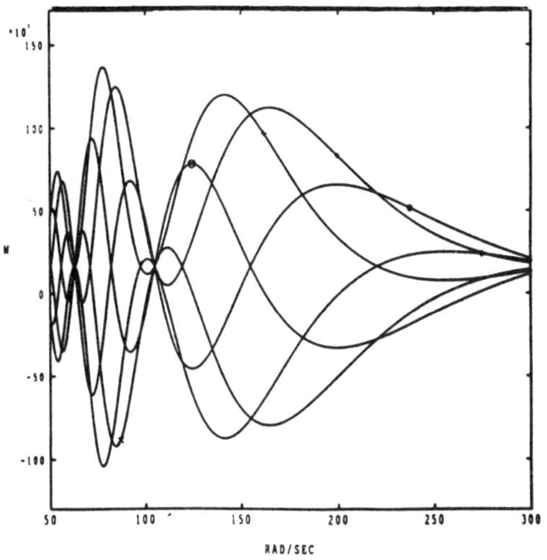

FIG.4 POWER FED TO THE GRID (PITCH = 100 DEGREES) (NEGATIVE = GENERATING , POSITIVE = MOTORING) ACTIVE MODE OF OPERATION

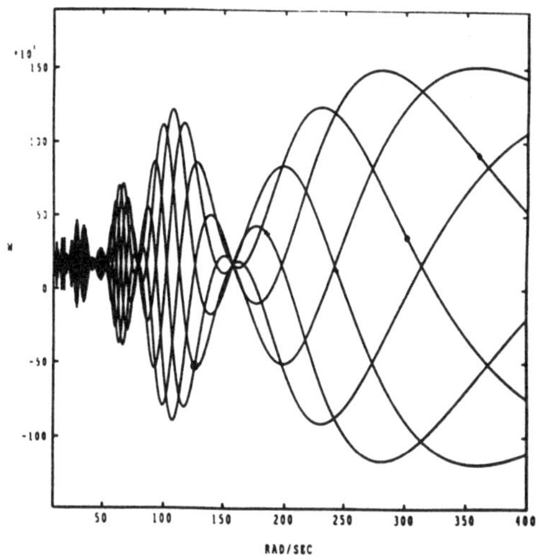

FIG.6 POWER FED TO THE GRID (PITCH = 180 DEGREES) (NEGATIVE = GENERATING , POSITIVE = MOTORING) ACTIVE MODE OF OPERATION

the machine as much as possible. Useful power can be fed to the grid within the speed range 120:180 rad/sec (ratio up to 50%).

Fig. 5 and 6 show the machine characteristics for a two pole stator winding. The coil pitch in this case is 180 degrees and we can see a big improvement in the characteristics. If the exciter phase is continuously adjusted, then the speed range would be from 160 to more than 400 rad/sec or ratio up to 250%.

This type of winding and the phase relationship between the output voltage and the exciter voltage are today under study, together with efficient methods for writing the poles on the rotor.

In active load operation parallel with the national grid, the power factor is lagging in the generator mode except at synchronous speed when the power factor is leading. In that case, the machine becomes a synchronous generator and the exciter can be disconnected. Fig. 4 and 6 shows the output power can also be positive. In that case, the machine runs as a motor with a lagging power factor.

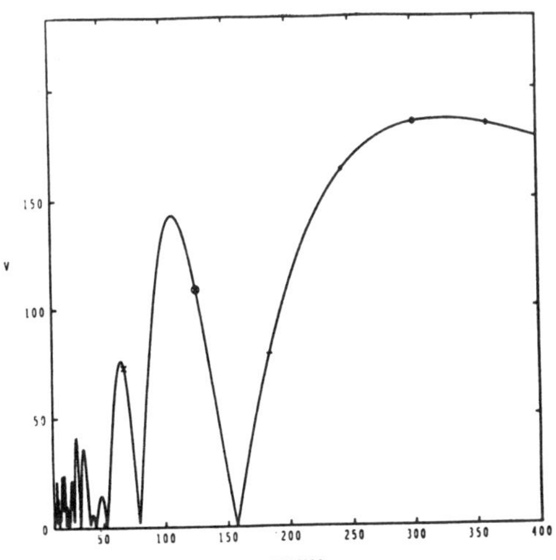

FIG.5 OPEN CIRCUIT OUTPUT VOLTAGE PASSIVE MODE OF OPERATION (PITCH = 180 DEGREES)

5. Acknowledgement

The author wishes to acknowledge the Commission of the European Communities, Directorate General for Science and Development for his support in this study as part of the Joule program on Wind Energy.

6. Conclusions

The programmed pole machine is a brushless machine without rotating windings which has the unique ability to decouple rotating field speed and frequency of its enerating voltage. The machine was designed and developed in the University of Limerick for output power up to 2kW, single phase, 50Hz. Theoretical analysis was carried out together with simulation, and compared with measurements from the machine itself. The main objective is the study of optimal design and performance of the machine, in particular, efficient "writing" of the poles on the rotor.

In parallel to the national grid operation (active load mode), automatic adjustment of the phase between the exciter voltage and the output voltage as a function of speed will provide a machine which will be optimally adapted to the wind turbine characteristic i.e. the wind power can always be adapted to the generator characteristic for optimal operation at a broad speed range.

If the above objectives are achieved, then the wind turbine will not need to run anymore at a constant speed nor will electronic power processing and conversion units need to be used in order to run the turbine at a variable speed.

In the passive load mode of operation (isolated grid), the output voltage varies with speed (Fig. 3 and 5) and the output voltage regulation can be achieved by regulating the flux density of the exciter (variation of the exciter voltage in order to keep the output voltage constant)

Annex 1

Difference between Alternator (Synchronous) Generator and the Programmed Pole (Roesel) Generator.

The open circuit output voltage of a machine, is given by:

$$Voltage = N\frac{d\varphi}{dt} \qquad (1)$$

where

$$\varphi = \varphi_0 \sin p\, \Omega\, t \ldots\ldots\ldots\ldots (2)$$

is the flux of the machine
p = No of poles pairs of the rotating magnet (rotor)
Ω = speed
t = time
For an alternator
From (1) and (2)
The output voltage = Np Ω Cos p Ω t
Both voltage and frequency are function of the speed Ω.
For Programmed Pole (Roesel) Machine

$$p = \frac{w}{\Omega}$$

where
w = grid angular frequency
voltage = Nw Cos wt
that is constant frequency
independent on speed

Annex II
Date of the Programmed Pole Generator:
Speed range (theoretical) = 160 : 400 [rad/sec] (1530 : 3800 r.p.m.)
Lower operating speeds at narrow ranges are not considered
Grid voltage : 110 : 220 [V] single phase
Exciter (or grid frequency) = 50 Hz
Output power : up to 2kW
No of slots for the output winding = 20
No of slots for the feedback winding (deliver power to drive the exciter) = 8
Output winding pitch = 180 degrees
Feedback winding pitch = 80 degrees
Rotor length = 150mm
Rotor diameter = 55mm
Air gap = 1mm
Ferrite thickness = 5mm
Ferrite type = Perimag 8A measured flux density inside the windings = 0.1 approx. [Web/m^2]
No of conductors inside each slot = 37
Output winding resistance = 1.5 [ohm]
Output winding inductance = 34.4[mH]

References :

1. Roesel J.F., Jr., Electric Power Generator, U.S. Patent No. 3,521,149, July 21, 1970.
2. Johnson G.L., Wind Energy Systems, Prentice Hall, N.J. 1985 p.p. 271 : 275.
3. Ott. R.R., R.J. Barber and J.F. Roesel, Jr The Roesel Generator : A Unique Variable -Speed, Constant - Frequency Generator, IEEE 75CH0964 - 7MAG.
4. Herman, L.R. : The Roesel Generator, Constant Frequency with Variable Speed, "Paper A 76035 - 6, IEEE Power Engineering Society Winter Meeting, New York, Jan. 25 - 30, 1976.
5. Lindsley, E.F., "New Alternator Delivers 60 - Cycle Power At Any Speed", Popular Science; Vol. 203; No. 1, p.p. 38 : 40, July 1973.
6. Barber R.J., Ott R.R. and Morash R.T., "Line Isolation for Industrial Control Systems" Proceedings of the Milwaukee Symposium on Automatic Controls, March 1974.
7. Marshall O.W., Morash R.T. and Barber R.J. "Independent Energy Systems for Better Efficiency" 9th Intersociety Energy Conversion Engineering Conference Proceedings, American Society of Mechanical Engineers, August 1974.
8. Morash R.T., Ott R.R., Barber R.J., and Marshall O.W. "Precise Power for Industrial Controls " Milwankee Symposium on Automatic Control, April 1975.
9. Data-sheet, Model 35-60, 35KVA NIPS, Precise Power Corporation, Bradenton, Florida 1986.
10. Bak, J.D., Variable-pole generator supplies constant - frequency output, Design News, East Coast, U.S.A., April 21, 1986.
11. Roesel J.F.Jr, Constant Frequency Motor Generator Set with only one Rotor, U.S.A., Patent 3931535, Jan 6 1976.
12. Roesel, J.F. Jr, Non-interruptible Power Supply Systems, U.S. Patent 4,168,459, September 18, 1979.
13. Morash R.J., Barber, R.J., Roesel, J.F. Jr, Variable Speed A.C. Motor, A Continuous Pole Printing Technology, Proceedings of the 18th Intersociety Energy Conversion Engineering Conference, Orlando, Florida, Aug 21-26, 1983, p.p. 374-9, Vol. 1 AICHE, N.Y., USA.
14. Roesel J.F.Jr., Variable Speed A.C. Motor, U.S. Patent 4,227,136, Oct 7, 1980.
15. Morash R.T., Flywheel Delivers Precise 60Hz Over Wide Speed Range, 1980 Flywheel Technology Symposium Proceedings, Oct. 1980, Scottsdale, AZ, USA.
16. Levy, D., Characteristic of the Roesel Machine used for Small Scale Hydro/Wind Electric Power Generation, Proc. ASCE Waterpower 87 Conference, Portland, Oregan USA, August 1987, p.p. 1645 - 1654.
17. Levy, D., Analysis of the Roesel Machine used for Small Scale Hydro/Wind or Sea Wave Electric Power Generation and Storage, Internal Report, Department of Electronic & Computer Engineering, University of Limerick, Ireland, 1978, not for publication at this stage.

VARIABLE SPEED DIRECT-DRIVEN PM-GENERATOR WITH A PWM CONTROLLED CURRENT SOURCE INVERTER

M. Alatalo, M. Sc (E.E) and T. Svensson, Ph. D
Department of Electrical Machines and Power Electronics
Chalmers University of Technology
S-412 96 Göteborg
Sweden

ABSTRACT:An attempt is made to design a direct-driven PM-generator by utilizing modern permanent magnets (NdFeB) and an axial field approach in a wind mill application. The construction of a PM-generator is uncomplicated. The stator winding, on both sides of the rotor disc, is fixed to the stator core without slots (air gap winding). Cogging, noise and iron losses are decreased. The power density is improved by increasing the number of poles. A 4.7 kW, 12 pole axial field PM-generator with NdFeB magnets is experimentally examined. The efficiency is about 92 %. Calculations on a 200 kW generator is performed. The radius is 1 m, the number of poles are 100 and the weight of the active material is 850 kg. The magnets dominate the cost of the generator but the total cost, including the frequency converter, is competitive compared to an asynchronous generator and frequency converter. It is suggested that the frequency converter should be based on a forced commutate current source inverter (CSI). The CSI is comparable to a voltage source inverter (VSI) with regard to the PWM technique and control of active and reactive power but the CSI has two main advantages. One is the capacitors, incorporated at the AC-side of the CSI, which make the current fed to the grid sinusoidal. The second one is the step-up ability, i.e. the CSI does not need any step-up dc-converter which is in contrast to the VSI.

1. INTRODUCTION

In wind mills feeding power to the grid, the armature of the generator is either connected directly to the grid or a frequency converter is utilized, transforming the frequency of the generator to the frequency of the grid. In the first case, the fluctuating power of the wind, including the influence of the tower, is directly imposed to the grid. If the grid is weak compared to the power of the generator, the stability of the grid is influenced. Two speed generators adapt approximately wind and generator speed. In the second case, the frequency converter is able to separate the value of the wind power absorbed by the turbine and the electrical power fed to the grid. The difference is stored as variable speed of the rotating masses. The stability of the grid is not affected and a forced commutated frequency converter can be utilized to compensate reactive power and voltage stabilization. Some negative phenomenon occur as a consequence of the converter, i.e. increased cost of the plant and harmonics of the current feeding the grid. It is of great interest to find solutions that reduce these effects and the aim is also to reduce the number of components in the electromechanical drive system. A generator directly driven by the turbine should exclude the gear-box. It is known that the weight of a low-speed generator is high. However, the increase of the weight will be reduced if the number of poles and the radius is increased by choosing the rated frequence of the armature current higher than the normal one of 50 Hz.

This discussion leads to the choice of a direct-driven permanent-magnet generator. Permanent magnets with very high energy density, amorphous metals and iron powder materials are new interesting options for the constructor of electrical machines.

Transversal flux and axial flux generators are not new principles but new materials make them competitive to ordinary radial flux generators [1], [2]. The design of the armature winding and the stator core is simplified by the use of these magnets. An airgap winding eliminates the need of slots of the stator core.

The choice of the frequency converter in a wind mill application is made considering that the current fed to the grid should not have harmonics. In the choice between line-commutated converters and forced commutated converters, the cost is still an advantage for line-commutated converters. However, the cost of the AC-filter is much higher because of the higher current harmonics and the reactive power consumption.

In the choice between voltage source (VSI) and current source (CSI) forced commutate converters it should be noted that the AC-filter is incorporated in the CSI. In a VSI, the capacitor is stabilizing the DC-voltage and has no influence on the AC current. Furthermore, the CSI does not need a step-up DC-converter in contrast to the VSI. Therefore, a frequency converter based on a CSI is a reasonable choice in a wind mill.

2. TEST SET-UP

The test set-up for the laboratory experiments is outlined in Figure 1.It contains a 12-pole PM-generator, driven by a DC-machine, and a frequency converter based on a CSI. The harmonics of the PWM-controlled current are cancelled by the LC-filter and the current fed to the grid is sinusoidal.

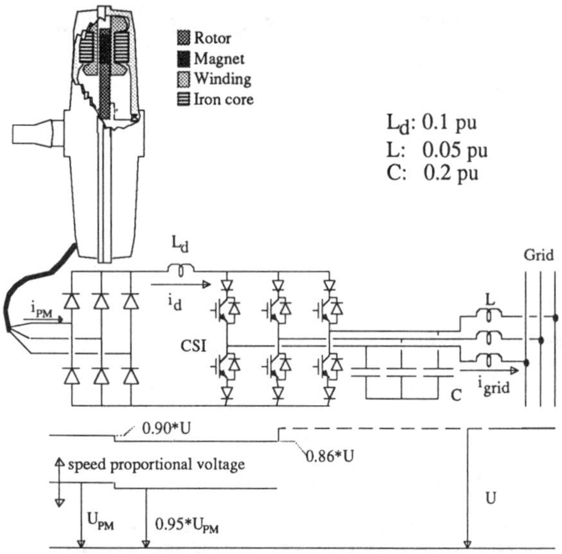

Figure 1: Test set-up for laboratory experiments.

2.1 The generator

A 12-pole PM-generator has been designed and built. The rated power is 4.7 kW (= 1 pu) at 1000 rpm. The generator has a slotless stator onto which a three-plane winding is glued as shown by the picture in Figure 2. The winding is embedded in epoxy resin, with relatively high thermal conductivity. A picture of the PM-rotor is shown in Figure 3. NdFeB magnets are glued to an aluminium disk.

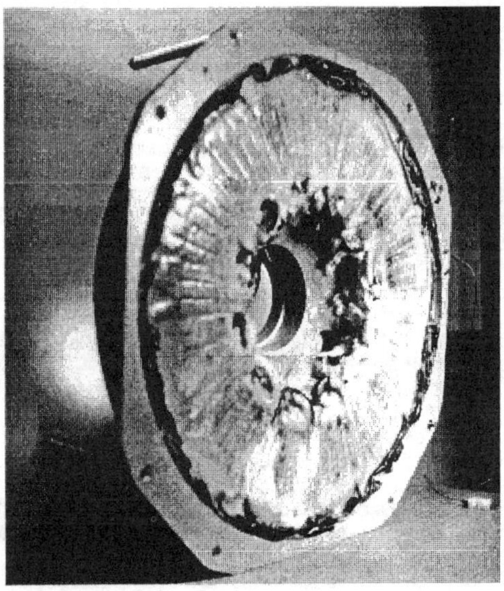

Figure 2: One half of the generator.

Table 1. Generator data

Component	Radius (m)	Thickn. (m)	Weight (kg)	Losses (W)
Iron	0.148	0.030	18	30
Magnet	-	0.015	3.5	20
Copper		0.016	7.5	370
Alu				50

Figure 3: The rotor of the test machine.

The remanent flux density of a NdFeB-magnet is B_r=1.2 T and the coercive force is H_c=900 kA/m. The thickness of the armature winding is 8 mm on both sides of the rotor, i.e. the rotor rotates between two stators. The ratio of inner to outer radius of the stator core is 0.7.

The power produced to a resistive load is measured at 1000 rpm. With an acceptable heating of the stator, the rated phase current is chosen to 6.5 A. At this current the power of the generator is 4700 W and the efficiency is 91.6 %.

2.2 The frequency converter.

The output power of the generator, i.e. approximately the power supplied to the grid, is controlled by the CSI. The switching frequency is 3.5 kHz using IGBT switches. The DC-current is controlled by the CSI by means of PWM-control. The CSI feeds the grid capacitors with 7 kHz current pulses which are sinusoidally distributed. The value of the capacitors is chosen in order to reduce the ripple of the capacitor voltage to a maximum of 2.5 % with respect to 7 kHz current pulses. Therefore, the 7 kHz ripple of the grid current is negligible. The power factor is controllable but the CSI is optimally utilized if the power factor is equal to one.

The control system is divided into two parts. The PWM-control of the CSI is the main part which controls the value of the power supplied to the grid. This part of the control system also distributes sinusoidal three-phase PWM-current.

The ripple of the rectified current generates non-harmonic current components of the current supplied to the grid, especially when the ripple frequency is near to the resonant frequency of the LC filter. Therefore, an active two-quadrant DC-filter is suggested, however, not yet tested. The stability is controlled by a separate control system

2.3 Experimental results.

In order to simulate the wind mill application in the tests, the power load of the generator is held proportional to the cube of the speed as shown in Figure 4.

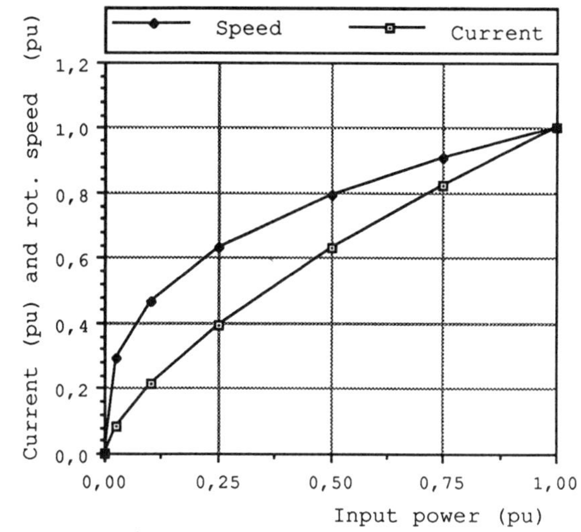

Figure 4. Current and rotational speed.

Measurements are performed on the conditions of Figure 4. The losses and the efficiency of the generator and of the frequency converter are outlined in Figure 5 and Figure 6.

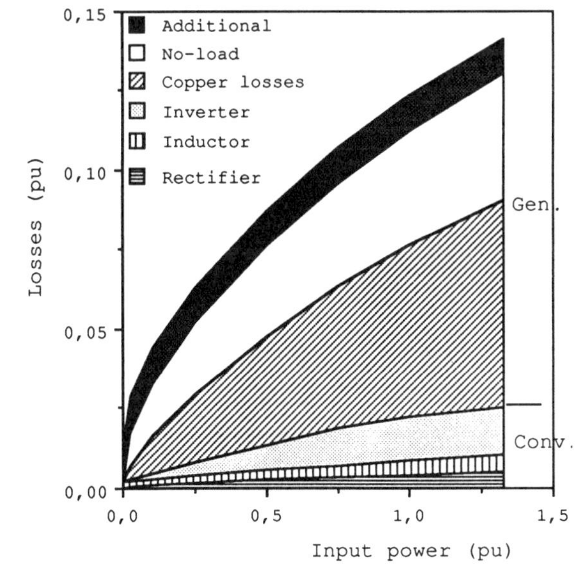

Figure 5: Losses in different parts of the generator.

The wave shape of the current supplied to the grid is presented in Figure 7. This test was performed with an extra intermediate choke in order to avoid ripple of the DC-current.

An active DC-filter will soon be tested.

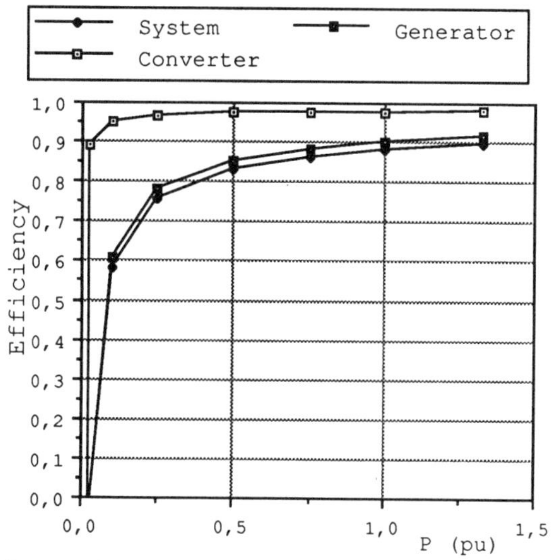

Figure 6. Efficiency

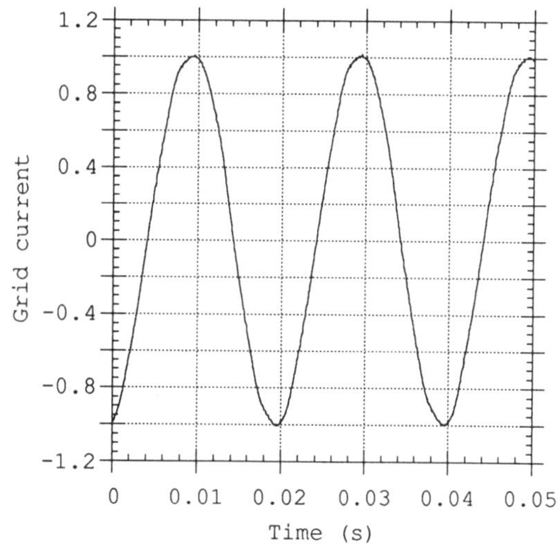

Figure 7. Grid current.

3. A DESIGN EXAMPLE

A generator and converter are dimensioned to a wind turbine of 25.5 m. The generator is constructed with a so called toroidal winding, [3,4,5,6]. The winding is closed on the counter part of the stator core, i.e. both sides of the core are used.

When constructing the generator there are two contradictory demands. The cost should be low and the efficiency should be high. The efficiency is easily increased using more copper and magnets, but this is expensive. The rotational speed is depending on the maximum tip speed of the turbine and this affects the generator weight. A tip speed of 120 m/s, corresponding to a single-bladed WEC, and a 140-pole generator render a generator weight of 550 kg. With a more moderate tip speed, 100 m/s, the generator weight is 850 kg. The nominal rotational speed is 75 rpm.

At 11.6 m/s the turbine power is 235 kW and the torque is 29.4 kNm. The generator is out-lined in Figure 8.

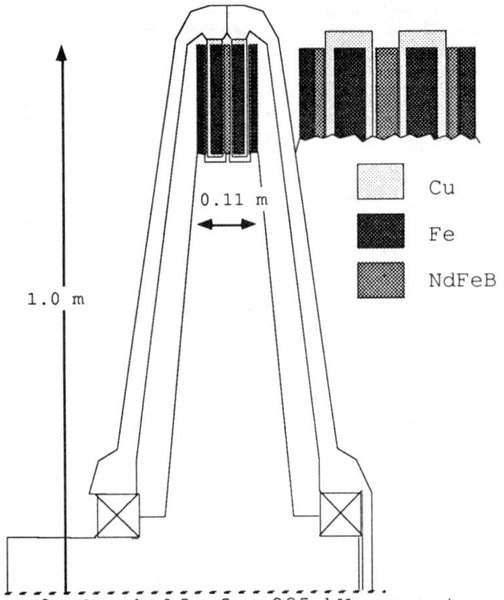

Figure 8: One half of a 235 kW generator.

In order to achieve the torque demands the generator is constructed with two stators sandwiched between three rotor discs. The outer radius of the core is 1.0 m and the number of poles is 100. The data for the generator are:

Table 2: Generator data, 235 kW

Component	Radius (m)	Thickn. (m)	Weight (kg)	Losses (kW)
Iron	1.0	0.010	551	0.69
Magnet	-	0.007	159	0.20
Copper		0.005	147	10.3

A part of the copper losses is depending on eddy currents in the conductor. The losses due to eddy currents can be calculated according to (1), [5].

$$P_{cueddy}=\frac{\omega^2}{\rho}d^2B_1^2\frac{1}{36} \qquad (1)$$

P_{cueddy} $\left(\frac{W}{m^3}\right)$ losses in copper details

due to eddy currents

$\omega=2\pi f$ (rad/s)

d diameter of the copper strand (m)

B_1 air gap flux density (T)

The actual power of the system is for the time being the maximum capacity of a single IGBT-converter. The calculation is based on 400 A/1200 V IGBTs, which are considered capable to handle a phase-to--phase voltage of 610 V and phase current of 180 A. The voltage is not a standard voltage but higher voltage IGBT could in the future handle 690 V.

The power produced with the direct-driven generator and CSI is shown in Figure 9. It is assumed that the power limitation is made with turning the turbine blades.

The efficiency of the whole electric system is shown in Figure 10.

The annual produced energy is calculated using the wind speed distribution measured at Näsudden. The direct-driven generator together with the CSI produce 13 % more energy than a geared asynchronous generator controlled by a voltage stiff PWM-converter.

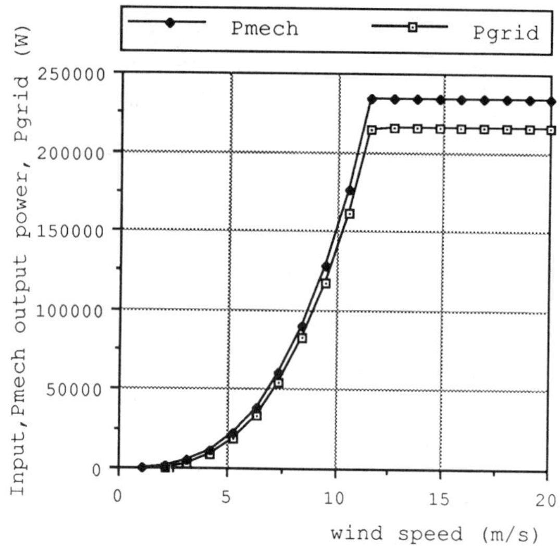

Figure 9. Input and output power.

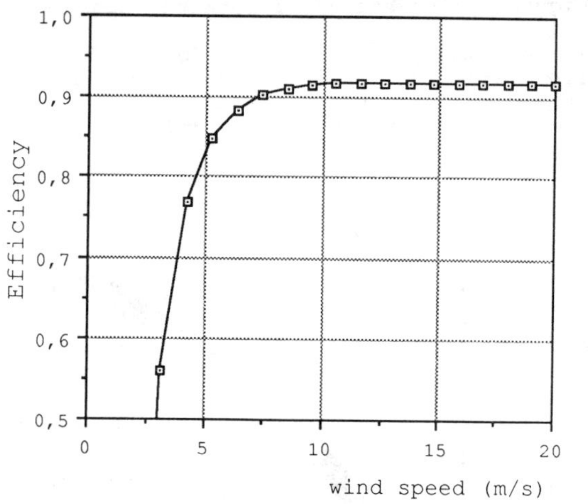

Figure 10. Efficiency of the electric system.

REFERENCES

[1] Weh, Hoffmann, Landrath, Mosebach, Poschadel, 'Directly Driven Permanent Magnet Excited Synchronous Generator for Variable Speed Operation' ,ECWEC 1988 Herning Denmark

[2] Weh H., Mayer R., 'Synchronous Machine with Permanent Magnet Excitation As a Flexible Propulsion Motor with High Efficiency', ICEM Lausanne 1984

[3] Spooner E., Williamson,'PERMANENT-MAGNET GENERATORS FOR WIND POWER APPLICATIONS',ICEM 1992 Manchester

[4]Stiebler Manfred,'A Permanent-Magnet Toroid Wind generator',ICEM 1992 Manchester

[5]Spooner E., Chalmers B.J.,'Toroidally-Wound, Slotless, Axial-Flux, Permanent-Magnet, Brushless-DC Motors', ICEM 1988

[6] Di Napoli A., Caricchi F., Crescimbini F., Noia G.,'Design Criteria of a Low-Speed Axial-Flux PM Synchronous Machine',SM-100 Zürich, 1991

MULTIPOLE, PERMANENT MAGNET GENERATOR STUDIES AND PLANNED PROTOTYPES

P. Rasmussen, N. Vilsboell
Folkecenter for Renewable Energy
P.O. Box 208
DK-7760 Hurup Thy
Denmark

A. Larsen
Berendsen Teknik A/S
Telefonvej 6
DK-2860 Soeborg
Denmark

1. INTRODUCTION

Permanent magnet generators (PMGs) and motors (PMMs) will become more common in widespread applications. For wind power in particular the multipole PMG has become common for small machines with power nominal to a few kilowatt. Those small machines are extremely simple with a minimum of wear parts, often only two bearings in the generator.

The idea behind PMG applications for windturbines is to avoid the step up gearbox and only having a multipole generator at the same speed as the windturbine rotor. This idea is well-known and proven from waterturbines using multipole synchronous generators in megawatt range.

The physical condition is ideal for this solution in spite of a constant relation between the blade tip speed (nominal 50 to 70 m/s) and the PMG airgap speed (nominal 3 to 5 m/s) independent of the sizes. It means, that the airgap diameter of the PMG is from 4% to 10% of the windturbine rotor diameter.

The PMG is a synchronous machine without possibility to regulate the rotor magnetization field and the output voltage. Use of permanent magnets makes the machine extremely simple related to the ordinary, traditional synchronous machine with DC excitation of the rotor field.

The market for permanent magnets offers traditional magnets as ferrite and alnico and strong new magnets of rare earth materials. The price is still in favor of a traditional magnet as the ferrite, however, use of big numbers might change the demand to rare earth magnets due to reductions in prices.

2. DESIGN METHODS

The choice of methods is conservative in order to minimize risks and surprise from a prototype. Furthermore the choices are based on practical experiences from PMGs already realized in the range from 1 kW to 20 kW.

The mechanical construction depends on more or less integration in the support structure of the windturbine nacelle. By choosing the stator as the inner part and the rotor as the outer part of the PMG (figure 1), the electromagnetic part can be adapted to all proposals as shown in the design figures 5 to 9. The PMG may be placed integrated in the hub as a sophisticated solution or as a single component fixed to the nacelle bottom plate as a traditional solution.

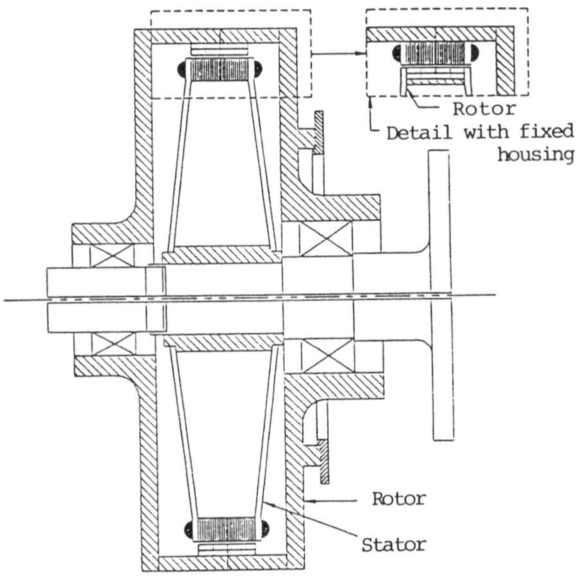

Figure 1: View of the PM generator.

The electromagnetic construction of the PMG is conventional with stator of laminated magnetic iron with coils placed in sloths. The rotor has salient poles with poleshoes and permanent magnets placed in between the poles instead of the traditional DC excitation coils (figure 2 next page).

By placing the magnets in between the poles and not with surface direct to the airgap, it is possible to amplify the flux density to a convenient value for the airgap flux with limitation from saturation in the teeth of the stator and in part of poleshoe.

An important experience from the past is a test, that showed too high airgap reactance also expressed zigzag

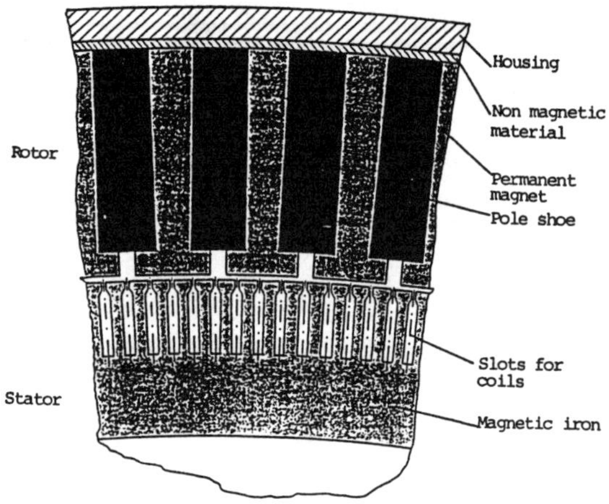

Figure 2: Details of the stator and rotor showing some pole divisions. The permanent magnet material is ferrite.

reactance. By a multipole generator it is a factor that has to be taken into consideration and calculation. The literature about airgap reactance is poor, but an important test (ref. 1) of a prototype took place at the Technical University of Copenhagen in 1982 at a 20 kW PMG with 58 poles. The reactance turned out to be much higher than expected, and the reactance was recalculated with the airgap reactance taken into consideration. The conclusion was, that the number of sloths had to increased from 120 to 168 in order to reduce the reactance sufficiently. Due to a mechanical failure a new stator with 168 sloths was never realized.

In a study project (ref. 2) the calculations with traditional equation electric and magnetic conditions were compared to measured values of three PMGs of a nominal power of 1.5 kW, 10 kW and 20 kW respectively. It showed high accordance between theory and practice.

As the ratio between blade tip speed and PMG airgap speed is constant, the circuit distance of the airgap is proportional to the blade tip circuit and to the number of poles in the PMG. It means, that a pole division has nearly a constant distance independent of the generator size. In practice it is between 30 mm and 50 mm for the range from 1 kW to 500 kW. The number of slots are 3 to 4 per pole division. This physical dimensions give the limits to a practical size for manufacturing and mounting of coils of the electromagnetic parts. For smaller dimensions the frequency of the generated AC voltage could be increased to more than 50-60 Hz, but it is not convenient of practical reasons.

3. TEST OPERATION OF A PMG

A test run of a 10 kW PMG linked to the grid through a converter (figure 3) for variable speed showed good performances, although the PMG itself had high reactance (voltage drop) and poor efficiency.

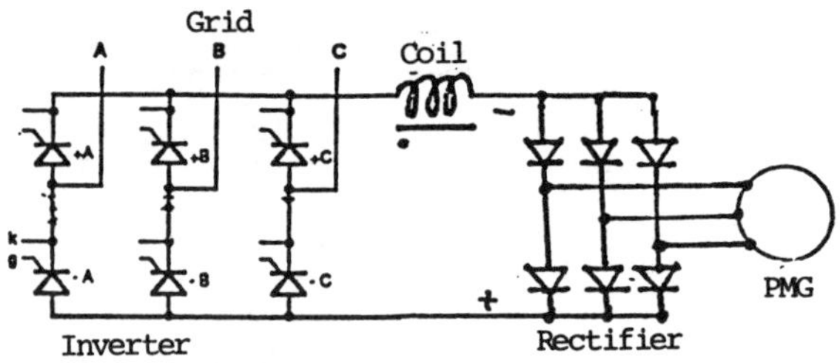

Figure 3: Test set-up for a converter to provide variable speed at the PMG and feed the energy to the grid.

The measured power points (figure 4) for the PMG nearly fit the ideal power curve for a windturbine, but does not fit completely due to high no load voltage of the PMG. By means of simple modifications the voltage drop could have been lowered considerably to fit the windturbine perfectly.

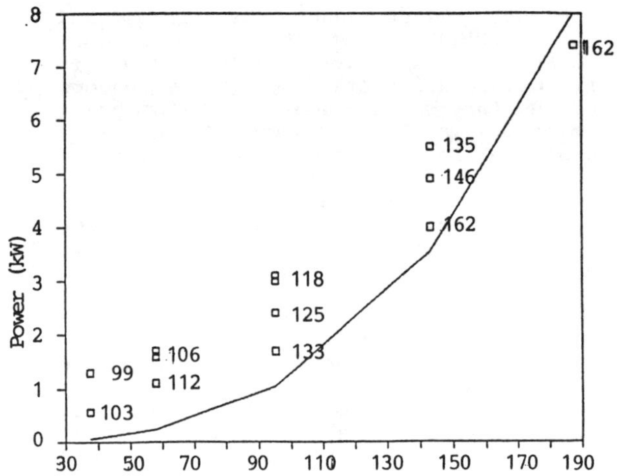

Figure 4: Measured points for power versus rotational speed characteristic for the PMG. The numbers at the points are the firing angle for the thyristorinverter. The lower power limitation is the nominal current of the PMG. The full line indicates an ideal power characteristic for an 8 kW windturbine.

4. WORK IN PROGRESS WITH PROTOTYPES

Two prototypes are prepared and outlined with the following data:

Power:	20 kW	100 kW
Rotational speed:	66 rpm	46 rpm
Number of poles:	90	130
Frequency, max:	50 Hz	50 Hz
Phases:	3	3
Voltage, max.	230/400Volt	400 Volt

The 20 kW PMG is a multi-purpose machine for stand alone, converter and grid connection operation, while the 100 kW PMG is for converter operation only. The multi-purpose machine has a limited voltage variation within 210 Volt and 240 Volt, good sine wave voltage curve and possibility for 1 phase 230 Volt and 3 phases 400 Volt.

For the time being finite element method (FEM) computer programs are installed and training are going on. New fully graphic supported programs are not powerful enough, however, the old strong programs originally used for main frame computers have all performances wanted, but are not special user-friendly. It has not been possible to have FEM results for this presentation.

The two prototypes are subsidized with fund from the EC Joule II programme and from the Danish Council for Renewable Energy, Ministry of Energy.

REFERENCES

[1] Reactance for a 58 Poles Generator (In Danish). Prof. Ulrik Krabbe. November 1982. Technical University of Denmark.

[2] Studyproject: Multipole Permanent Magnet Generator for Wind Power etc. Peter Rasmussen and Niels Vilsboell. (In Danish with English summary) January 1992. Folkecenter for Renewable Energy.

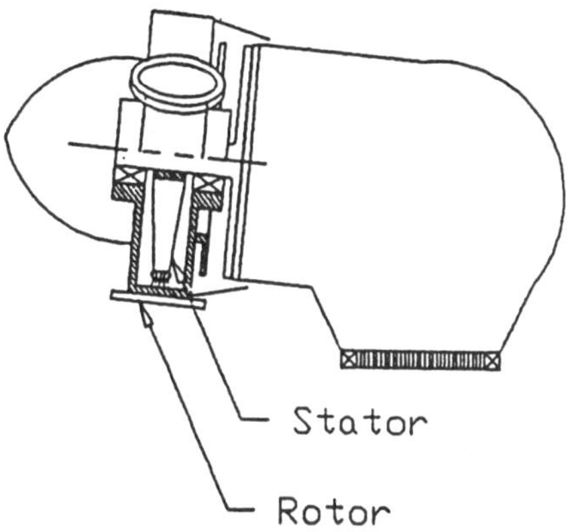

Figure 5: Design 1. Generator has been made with the house as the rotating part and is at the same time made as hub so that the blades can be mounted directly on the generator house. Bearings and stator together with the rotating parts have been designed to transfer of all rotor powers. The advantage of total integration is minimum consumption of materials and processing.

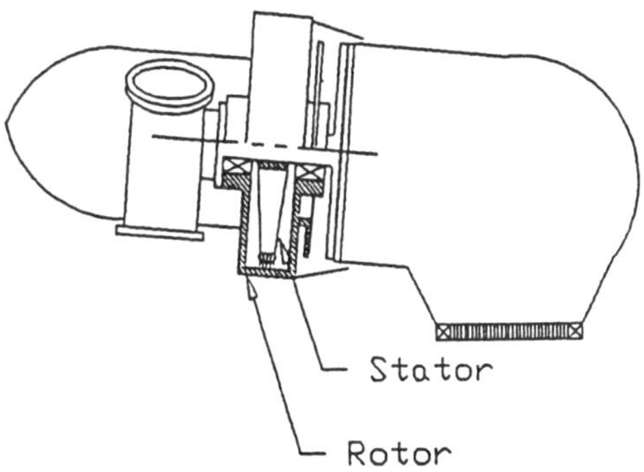

Figure 6: Design 2. Generator has been made with the house as the rotating part of the generator, rotor and stator are intended for transfer of all rotor loads. The advantage is that the hub can be mounted directly on the rotor so that only 2 bearings are required.

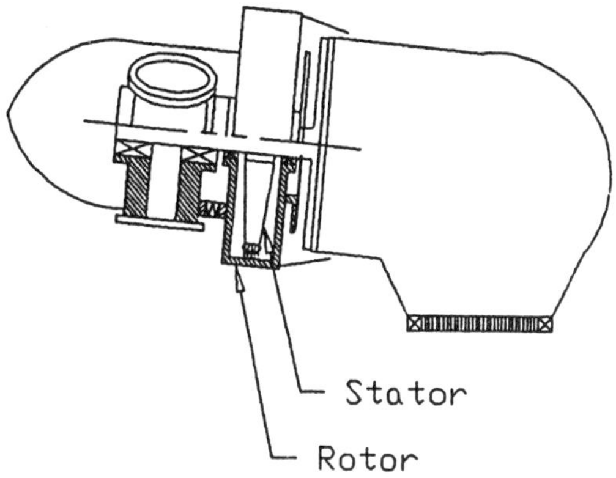

Figure 7: Design 3. Generator has been made with the house as the rotating part of the generator and stator is intended for transfer of own weight and torsion. The advantage is that it is possible to carry out a flexible coupling between hub and generator where both are embedded to the same main shaft.

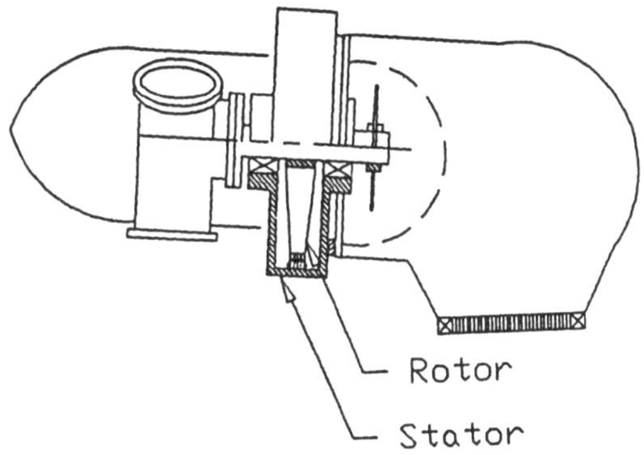

Figure 8: Design 4. The generator is made with the house as stator with built-in flange for direct mounting on the nacelle, and the rotor shaft is intended for direct mounting of hub and thereby for transfer of all rotor loads. The advantage is a simple and compact building and only 2 bearings are used for the whole transmission.

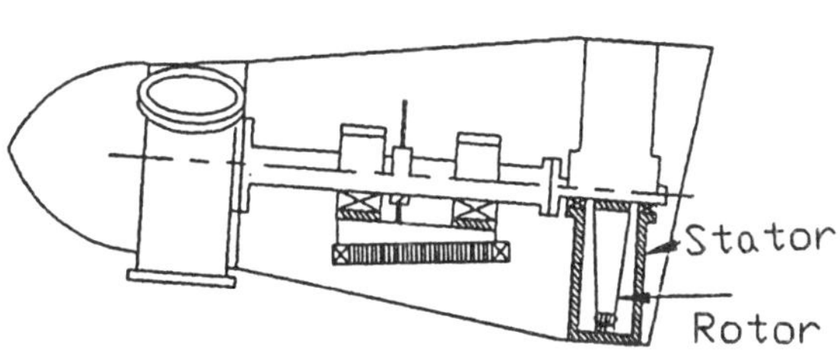

Figure 9: Design 5. Generator is made with the house as stator and the rotor shaft is made for transfer of own weight and torsion moment. The advantage is a good distribution of the weight, the hub and generation being placed on each side of the tower and a short distance between the main shaft and the yawing which minimizes the yawing moments and the influence of the air gap is minimized.

POWER CONVERSION FOR THE CONNECTION TO THE MEDIUM-VOLTAGE GRID OF A VARIABLE SPEED WIND GENERATOR

S.Avolio, C.Palmari
ALENIA
Via Vitorchiano, 81
00189 ROMA (Italy)

C.Pincella, P.Zanotti
ENEL S.p.A.- CREl
Via Volta, 1
20093 Cologno M.se
MILANO- Italy

O.Honorati
Università di Roma
"La Sapienza"
Via Eudossiana, 18
00184 ROMA (Italy)

C.Torri
ANSALDO
INDUSTRIA
Vilale Sarca, 336
20126 MILANO (Italia)

The use of medium-large size wind-generators for wind energy conversion into electric energy requires the solution of problems related to the maximum exploitation of the available wind energy, but it also involves considerations of their economicity and reliability.
Such an approach suggests the adoption of an electric generator operating at variable speed together with a static converter, which allows to decouple the generator output frequency from the constant frequency grid.

The electric system presented in this paper is an essential part of the wind generator project denominated GAMMA 60. This project has been developed by an industrial consortium (of which Alenia is file-leader) for ENEA and ENEL, and considered the construction of a prototype which has been installed in the ENEL test station in Alta Nurra, Sardinia, Italy.

This prototype compounds the experimental wind turbine, a synchronous generator at variable speed, and a naturally commutated static conversion system (rectifier - intermediate DC bus - inverter) which supplies the ENEL's medium-voltage AC system.

Because of the experimental peculiarity of the system, one crucial aspect relies on testing the electric subsystem in order to verify the requested performances.
With this aim, the performances of the generation-conversion system have been evaluated (including AC fault transients) during the following phases: tests of the electric subsystem off-site and during the first period of the overall system operation.
The paper describes the characteristics of the technical solution adopted for the power conversion as well as the obtained performances, with particular reference to its interactions with the ENEL grid.

1. INTRODUCTION

A wind turbine generator is defined as a system for producing electric power, consisting basically of a wind-powered motor (generally with 2 or 3 blades) coupled to an electric generator, almost always via an overdrive.

The aerodynamic torque from the motor (blades) is proportional to the square of the wind speed, random in character and frequently subject to sudden and wide fluctuations. This makes the control of a wind turbine very different from that of an hydraulic or steam turbine, in which the torque can easily be regulated by using the water distributor or the steam inlet valves, respectively. On a wind turbine it is only possible to reduce the torque as compared with the maximum value which would otherwise be obtained at that particular speed.

Most of the wind turbines so far built have practically constant speed, since they use an a.c. generator, directly connected to the distribution grid, which determines its speed of rotation.

The GAMMA 60 project, on the other hand, enables the unit to operate with a broad range of variable speed, which enables the wind turbine to produce a higher energy output - by exploiting the wind more efficiently - with a simpler overall design (using fixed-pitch blades) thus resulting in a reduction in the unit cost versus the power produced.

To this end use is made of a power electronic converter, which also makes it possible to adapt the electric power gen erated by the wind turbine to the frequency of the ENEL grid.

The rated power output of the GAMMA 60 experimental project is 1.5 MW.

2. DUTY REQUIREMENTS OF THE CONVERSION SYSTEM

The design and basic operation of the GAMMA 60 wind turbine have been already described in previous papers [1,2]. Here the typical 'modes of operation' of the GAMMA 60 generator, each one at different wind speed range, are recalled:

a) Starting of the machine as a motor (with the turbine not aligned with the wind).
b) Running at constant power coefficient Cp (peak value) for speeds betweeen 5 and 12.5 m/s.
c) Running at constant torque (rated value) for wind speeds between 12.5 and 13.5 m/s.
d) Running at constant power output (rated value) for wind speeds between 13.5 and 27 m/s.

The range of speed for the above duty goes from about 15 up to 44 rpm.The rotor speed can be controlled through a proper modulation of the load torque, in this case the electrical generator one, without any torque choking, except when exceeding the torque rated value[1].

. For each wind speed there is a 'best' rotor speed which maximize the generator power output. This speed is taken as a reference point in the rotor speed control loop and then is translated in a reference torque in the load torque inner control loop.

An adjustable speed drive has therefore been selected to match the above requirements, thus permitting to comply with the needs of the electrical network and the maximum energy capture. The action of the drive is a smooth and continuous regulation of the load torque with the electrical machine running both as a generator and as a motor.

3. THE GAMMA 60 ELECTRIC CONVERSION SYSTEM

On the basis of the above mentioned requirements the design of the electric system configuration, which links the

[1]This situation is typical of the d) operating mode, where a constant output is is mantained through yaw control of the wind machine

wind turbine to the distribution electrical grid, has been made. A number of technical solutions were taken into account, all of them including an electronic power converter. In the end two of them were found both quite acceptable.

The first was basically a squirrel cage converter-fed asynchronous machine. The frequency converter, a current-source type, consisted in a line-side line-commutated converter with a d.c. link between it and the machine-side self-commutated converter.

In the second solution a high speed (1800 rpm) synchronous machine with coaxial brushless-type exciter was fed by a convertor quite similar to the previous described, apart from the fact that both rectifier and inverter were line-commutated.

Finally the converter-fed synchronous machine solution was preferred, not only due to the good efficiency and reliability of its power electronics but also because of the more favourable cost/power ratio, at least for a drive of the GAMMA 60 rating.

As regards the electric grid (absorption of reactive power and injection of currents harmonics), this system behaves exactly like one consisting of a d.c. machine and a naturally-commutated converter, operating as an inverter.

4. DESCRIPTION OF THE ELECTRIC SYSTEM

The electrical layout of the wind turbine generator GAMMA 60 is shown in Fig. 1, where the main power components of the conversion system are included:

a) A brushless-type synchronous generator.
b) An electronic power converter, with intermediate d.c. link.
c) A static exciter.
d) A step-up transformer (line side).
e) A power factor correction/filtering bank.

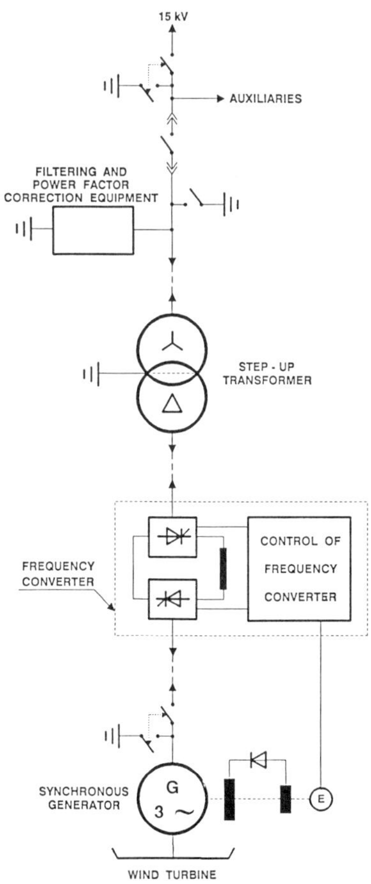

Fig. 1 - General diagram of the power circuits of the GAMMA 60 wind turbine generator

All of them, but the filter, which will be discussed later, are shortly presented hereafter.

4.1 Synchronous Generator

In order to satisfy all of the service conditions a 4 poles, 60 hZ, 2225 kVA $\cos\varphi=0.9$ synchronous generator was chosen. It's a self-ventilated, closed machine, with a constant torque operating range between 1200 and 1800 rpm.

Through a special brushless exciter the field in the machine can be produced also when the rotor is locked, thus allowing the motor starting of the wind turbine. A particular care in fixing the rated voltage of the machine was paid. In fact the project value (1200 V) has been selected with regard to the convertor rating to get the best overall design of the conversion system.

4.2 Frequency Converter

4.2.1 General Features

The frequency convertor is basically set up as two six pulse line-commutated thyristor bridges linked through a smoothing reactor. Thanks to the choice of the 1200 V voltage value of the synchronous generator, in each arm of the converter bridges a single large-size thyristor (today easily available on the semiconductors market) is enough to insure the necessary degree of reliability for the converter and finally, to the system.

Here are some of the typical data for each bridge:

Thyristors:	VDRM/VRRM	5200 V
	Silicon Diameter	82 mm
	Cooling	forced air, open-circuit
Line-to-line Voltage (Uvo)	1450 V ± 10%	(valve side)
D.C. Rated Current (Idn)	1325 A	(30% overload for 1 min. every 30 min.)

4.2.2 Convertor Protection and Control

A double-level protection criterium is active in the convertor:

* On-line monitoring of the main electrical quantities and of special service conditions. This function is typical of the convertor control system.
* High-level inherent safety components on the bridges, together with suitable protective devices.

As far as the first protection criterium is concerned, the convertor control system (Fig. 2) has the following basic tasks:

- to set the current at a value which agrees with the demanded torque.
- to operate the generator at virtually costant power factor, thus mantaining a torque proportional to the current.

These actions take place in co-ordination with the static exciter control, which is demanded to keep a constant voltage/frequency ratio (constant flux in the generator) over the whole range of speeds of the wind turbine generator.

The convertor control system also allows the so-called 'pulsed operation' of the convertor during the starting of the wind turbine. In this mode of operation the flux of the synchronous machine is reversed and the latter works as a motor.

Lastly, the convertor control system acts as a rapid protection system for the bridges during anomalous conditions by blocking the firing pulses. These conditions can be both inner and external to the convertor, such as:

- Overcurrents.
- Unbalance between the alternating currents on the generator and the transformer sides.

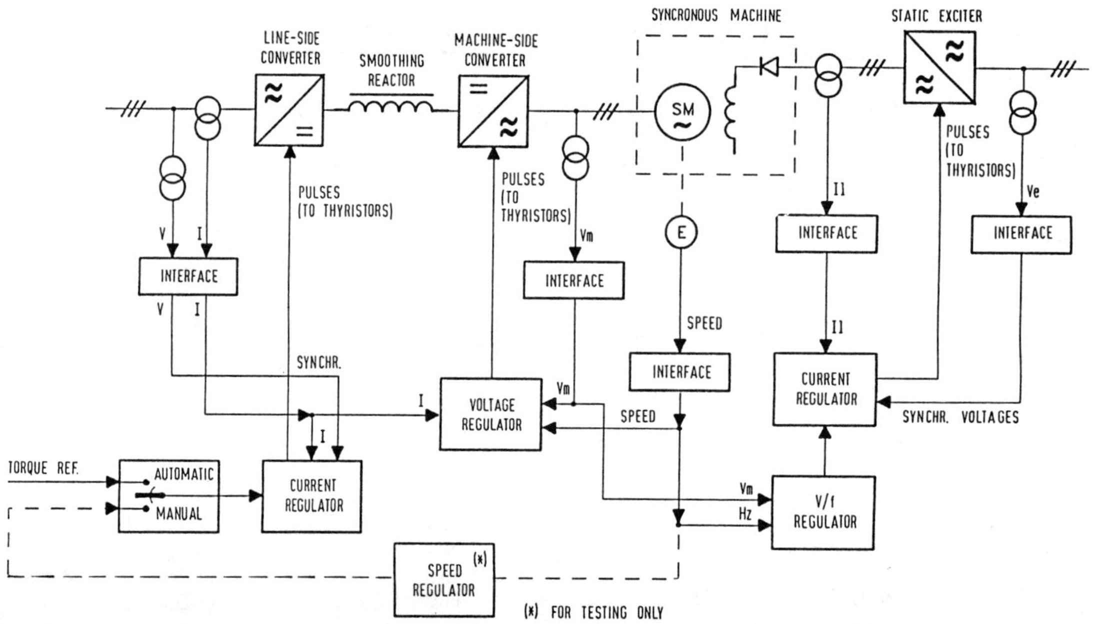

Fig. 2 - GAMMA 60 wind turbine generator - Schematic of the static converter control system

- Absence of grid or generator voltage.
- Thermal overload in the smoothing reactor.

In the second protection criterium can be especially included:

- Zinc-oxide arresters on the low voltage windings of the transformer (phase-to-phase).
- self-firing (emergency only) of the thyristors at a voltage value greater than 80% VDRM.
- The thyristor Fault Suppression Capability.

No fuse was installed because of the above actions.

4.3 Static Exciter

The static exciter consists of an a.c.\ a.c. electronic voltage regulator feeding the brushless device through a 6 pulse rectifier.

4.4 Step-up Transformer

An isolation transformer has been provided between the frequency converter and the ENEL 15 kV network. Between the two windinds of the 15/1,45 ±10%kV, 2715 kVA machine an electrostatic screen has been set up. The transformer is an outdoor ONAN cooled type.

5. CONNECTION OF THE WIND TURBINE GENERATOR TO THE ENEL GRID

The GAMMA 60 wind turbine generator will be connected to the ENEL 15 kV network through a 11 km line from the Porto Torres 150/15 kV primary substation.

Given the experimental nature of the installation, it was deemed preferable not to feed other users from this line, which is therefore substantially a dedicated one, unless called upon to supply, under emergency conditions, the auxiliary services of the nearby Fiume Santo thermal power plant.

5.1 Reactive Power Consumption and Harmonic Emission of the Convertor

As already reported, the generator runs at constant V/f ratio on most of its range of operation (which means a constant flux in the machine). The rotor speed is therefore linear with the supplied voltage/frequency. In this condition there is also a linearity between the generator output current and its torque.

In chapter 2 the 'modes of operation' were defined for the wind turbine generator.

Slowing down in speed from mode c) towards mode b) constant torque operation is active at first, where the reactive power absorbed in the line-side convertor increases with the reduction of the rotor speed (in fact the apparent power exchanged is constant). A further speed reduction switches the convertor in mode b), in which the torque is strongly decreasing. As a consequence, a reduction of the reactive power absorption on the network, even in presence of a lower and lower power factor, is given.

Table I shows the calculated power (both active and reactive) exchanged with the grid for a number of operating points.

5.2 Voltage Variation

The value of the short circuit power at the Porto Torres 15 kV bus (primary substation), which is basically determined by the reactance of the medium voltage line supplying the wind turbine conversion system, is about 40 MVA.

With reference to this short-circuit power and taking into account the active and reactive power supplied and absorbed, respectively, by the convertor (see the previous section) it is not

TABLE I - Active power fed into and reactive power absorbed from the grid by the gamma 60 wind turbine generator

SYNCHRONOUS GENERATOR		MARGIN ANGLE OF THE FREQUENCY CONVERTER	15 KV GRID BUS	
SPEED (rpm)	OUTPUT POWER (kW)	(°)	ACTIVE POWER (kW)	REACTIVE POWER (kVAr)
1800	2000	34.1	1927	1577
1350	1500	50.6	1438	2023
1200	1324	55.4	1265	2122
1000	650	63.7	605	1334
495	86.4	78.1	75	375

expected that there will be any significant variations in the supply voltage caused by operation of the wind turbine.

Therefore it was decided not to install any power factor correction device, apart from filters to limit the convertor harmonic emission.

5.3 Harmonic Control & Filter Rating

The maximum voltage distortion, both as individual harmonic and THD, for the GAMMA 60 generating group was chosen on the base of criteria suggested by CIGRE [3]. The fact that the GAMMA 60 feeder was both relatively long and a dedicated one prompted to the imposition of different admissible voltage distortion levels:

- Low levels at the Porto Torres primary sub-station bus-bars, valid for the generality of the loads.
- Higher levels, quite similar to the one usually accepted for industrial users, at the GAMMA 60 connection point.

In fixing the levels admissible in the primary sub-station, account was taken of the existing distortion (assessed by means of measurements), thus leaving available to the GAMMA 60 convertor a 60% of the existing margin between the values measured and the maximum tolerable limits suggested by CIGRE. As a result of the above procedure the admitted distortion limits of Table II came out.

TABLE II - GAMMA 60 Project
Admitted Voltage Distortion Limits

HARMONIC	INDIVIDUAL HARMONIC DISTORTION	
	PORTO TORRES	GAMMA 60
3rd	2.5-3.%	6%
5th	2.5-3.%	8%
7th	2.5%	7%
11th	1.5%	5%
13th	1.3%	4.5%
TOTAL HARMONIC DISTORTION (THD)	4%	10%

To secure the observance of the above limits a tuned filter (between the 4th and 5th harmonic), to be installed on the GAMMA 60 at the 15 kV side of the isolation transformer, was deemed satisfactory. In the filter design possible grid resonances were taken into account. The filter, which has a quality factor Q=30, should produce 1.5 MVAr.

6 THE TEST PROGRAM

Because of the experimental nature of the system, one crucial aspect relies on testing the electric subsystem in order to verify the requested performances.

6.1 Factory Tests

Two different series of tests were performed on the GAMMA 60 electric conversion subsystem:

- At single power assemblies or component level, type and routine tests (shop tests). These tests have been especially done on the frequency converter, in agreement with the existing Standards, where possible.
- Operational tests of the whole drive train at the manufacturer bench, including the line-side transformer and the actual control system of the convertor.

The expected behaviour of each power component at the specific duties was verified under service conditions since the first series of tests. However, due to the random nature of the power source (i.e., the wind), and therefore the difficulty to perform extensive field testing, the need of further investigation in order to check the correct operation, both steady-state and transient, of the whole drive was deemed necessary.

The drive test bench, set up at the manufacturer's factory, is shown in Fig. 3. It can be seen that the wind turbine action was replaced by a d.c. machine driving the synchronous generator.
In the test schedule the following main tests and checks were included:

* Load test
* Temperature rise test
* Starting of the drive (pulsed operation)
* Commutation from motor to generator operation
* System response to a sudden variation
* Transfer of a switching/ligthning impulse between the transformer windings
* Measurements of system efficiency and power factor
* Measurements of the harmonic currents
* Checking of correct thyristor self-firing intervention

In particular it was verified the system response to a 5% frequency step (from 33 to 30 hZ). The consequent fast torque variation took place in about 50 ms, as expected. Satisfactory was also the behaviour of the system when passing from motor to generator (reverting the torque), which was achieved through a smooth current regulation on the d.c. link.

The load/thermal tests were run at different speeds. A short report on the measured efficiency of the frequency converter and of the overall electrical conversion system is given in Table III, while comparative data on the measured current harmonics together with the expected 'typical' ones of the convertor, injected at full load (1500 kW) on the 15 kV ENEL line, are shown in Table IV.

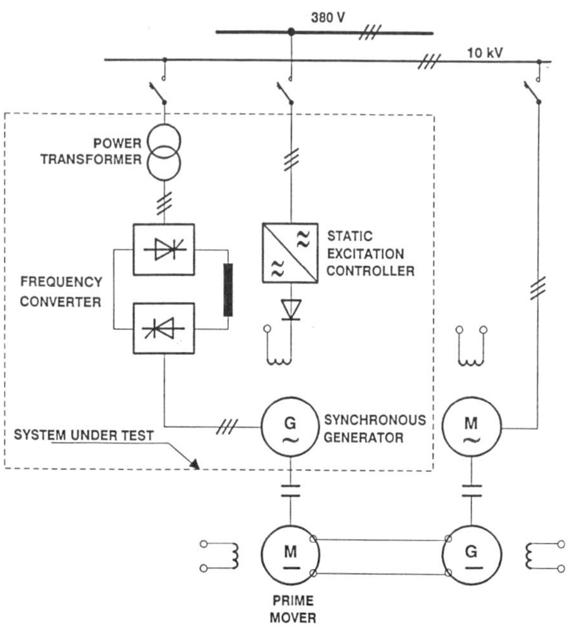

Fig. 3 - Layout of machinery during shop tests on the electrical system of the GAMMA 60 wind turbine generator

TABLE III - Summary of the gamma 60 electrical drive efficiencies at various loads, measured at the test bench

Generator Speed [R.P.M.]	Power at the shaft [kW]	Convertor Efficiency p.u.	Drive Efficiency p.u.	Output Power at the 15 kV network [kW]
493	121.3	0.865	0.672	81.5
660	266	0.94	0.795	211.5
862	598	0.97	0.86	514.2
1100	1132	0.976	0.871	986
1202	1426	0.983	0.881	1256.3
1304	1496	0.976	0.876	1310.5
1488	1680	0.980	0.885	1487

TABLE IV – Harmonic currents emitted by the convertor, as by design and measures during the load test at the manufacturer factory

Harmonic n.	Amplitude [A]	
	Design	Test (*)
5	19	17,8
7	13	11,3
11	7	7,05
13	6	6,7
17	3	4,5
19	2,5	4

(*) at full load, on the medium voltage side of the input transformer

During the bench testing a resetting and implementation of the logics of the supervision system (which has the task of insuring a correct and safe operation of the drive in every duty condition) was also possible, thus saving a heavier job on the windmill installation site.

The test results showed an operation of the drive as predicted in each load situation.

6.2 Field Tests for the Analysis of the Interactions with the ENEL MV Grid

The field tests to be performed will be devoted to investigate the following items:

- Efficiency of the whole electric system.
- Harmonic impact on the ENEL a.c. network.
- Voltage variations on the GAMMA 60 MV bus-bars.
- Behaviour of the electric system in presence of a.c. faults on the MV ENEL grid.

Particular attention has been paid in the design stage to the last item, with the goal to reduce as much as possible the sensitivity of GAMMA 60 convertor to voltage dips.

To this aim, automatic restart procedures will be adopted in presence of voltage dips caused by faults on others MV lines connected to the primary substation of Porto Torres. In this way the effect of the aforesaid events, that are an important part of the total number of expected voltage dips, will be limited to a short temporary stop of the convertor (< 1s), with negligible impact on the overall plant behaviour.

The following field tests will be performed to check the expected behaviour of the GAMMA 60 electric system in presence of a.c. faults:

- Single phase or poliphase faults on other MV lines connected to Porto Torres.
- Single phase or poliphase faults on the GAMMA 60 line; these events will cause the permanent stop of the convertor and, therefore, of the overall plant.

7 CONCLUSIONS

A broad-variable speed, wind turbine machine like the GAMMA 60 should offer, with respect to the traditional systems, good inherent quality project together with a competitive energy/cost ratio.

From the point of view of the electrical conversion system the choice of a variable speed drive like the converter-fed synchronous machine, well known in a number of industrial applications in its motor basic configuration, should grant both a good reliability and the design expected performances.

The impact of the electronic power converter in terms of reactive power absorption and harmonic currents emission on the ENEL network have been evaluated.

Special care was paid in each step of the work progress to verify not only the convertor ratings but also the overall drive behaviour in factory just before its installation on site.

Finally, a field program aiming at investigating interfacing aspects between the machine and the actual ENEL grid has been boosted, along with the first service experiences.

REFERENCES

[1] D. Coiante, U. Foli, E. Sesto, A. Taschini, S. Avolio, F. Zappala
"GAMMA 60 1.5 MW Wind Turbine Generator", EWEC'89 Conference Proceedings.
Glasgow, UK, July 1989.

[2] S. Avolio, C. Calò, U. Foli, L. Rubbi, C. Casale, E. Sesto
"GAMMA 60 1.5 MW Wind Turbine Generator", Proceedings of the European Community Wind. Energy Conference and Exhibition, Madrid, Spain, September 1990.

[3] WG CIGRE/CIRED CC02 "Equipment producing harmonics and conditions governing their connection to the main power supply"
ELECTRA, N. 123, March 1989

HYBRID SYSTEMS - BURG AUF FEHMARN AND PELLWORM

Dipl.-Ing. Gert Nimz,
SCHLESWAG AG, Rendsburg
Dipl.-Ing. Hans-Jürgen Lowalt,
Telefunken SystemTechnik GmbH, Wedel

ABSTRACT: By means of hybrid plants with an offer of primary energy decoupled in a time dependent way, periods can be reduced during which no energy is available. Important areas have been further developed in technical respect by means of which the outputs have been increased and the harmonic waves as well as the wattless power have been reduced.

1 REGENERATIVE SOURCES OF ENERGY FOR THE POWER SUPPLY OF A SEWAGE WORKS

1.1 Plant description

Due to the favorable location for wind and solar energy, a new energy supply conception for the sewage works was projected within in the framework of the extension plans. The base load supply from the utility (SCHLESWAG AG) was extended via the combination of a solar generator of 140 kWp, a wind energy converter of 250 kW and a combined heating and power station (BHKW) of 30 kWel.

With a capacity of 140 kWp, the solar generator converts solar energy directly into electric energy. Different module types of multicrystalline silicon are employed. All components - from the cell to the module up to the generator - have been manufactured by Telefunken SystemTechnik.

The wind system is the type HSW 250 developed by Husum shipyard.

The MOEWE system (modular energy supply system for heat and electricity), type WF 3, developed by Daimler-Benz AG, is integrated in the complete system to use the accumulated sewage gas.

The storage is a biogas reservoir of 300m3 capacity. This reservoir serves as a short-time energy buffer.

This project is unique in Europe in its constellation sun, wind and biogas.

The meteorological characteristics of Fehmarn are a mild, maritime, bracing climate and relatively low rainfall (about 520 mm a year, compared with country's average of 800 mm), furthermore pleasant southwest winds (with speeds of 4 m/s and more) and a high number of sunshine hours (about 1830, which is about 400 hours more than in Essen, for example). The average global radiation, i. e. the radiation intensity from direct and indirect sunlight, is about 3.0 kWh/m2 a day. Therefore, the island excellently fulfills the requirements for the economic use of regenerative energy sources such as wind and solar energy. Furthermore, Fehmarn is suitable for a pilot plant with the topic "Regenerative Energy Systems" due to the following reasons:

- Good adaptability between PV energy generation and consumption, as the sewage treatment plant is operated in a summer tourist area (consumption summer/winter abt. 2.5 : 1)
- Due to the internal inertia of the storage, the energy consumption "sewage treatment plant" enables the load to be adapted to the available regenerative energy to a certain extent.
- Due to the use of a biogas driven methane motor in the total system, a part of the energy to be supplied can be controlled in respect of its time of use.

The purpose of the project is to adapt the existing single components (sun, wind, biogas) to a system programme in such a way that a flexible total system with high reliability and availability results.

An operation management system has been installed for the energy management, into which a data storage station has also been integrated. The amount of energy which is fed into the public mains at times of surplus regenerative energy supply and the amount which is withdrawn from it are balanced separately. The operation management has been organised in such a way that the sewage plant consumer units (pumps, screw conveyors, agitating machines, ventilators, etc.) are combined in functional groups. The functional groups which are not used, can be switched off. Thus, the energy supply of the regenerative energy sources can be adapted very well to the energy requirement of the consumer units.

As the output of the solar generator depends on the amount of sunshine, this must also be considered at energy conversion. By automatically switching on and off both invertors, the plant is operated at its highest efficiency.

1.2

Technical description of the total system

The block diagramme shows a summary of the combined action of the main plant components (fig. 1):

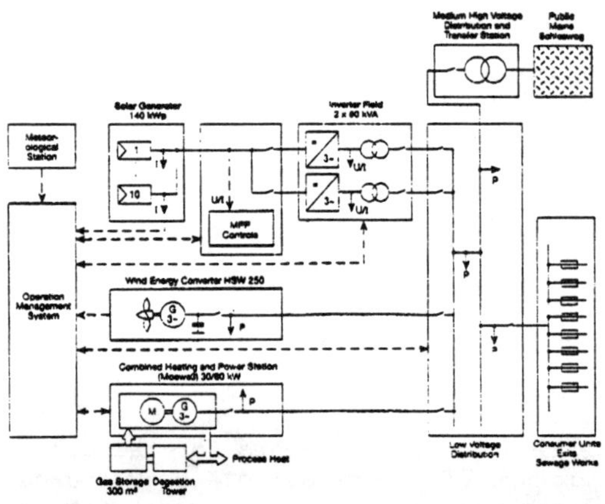

Block Diagramme fig 1

1.3 Technical data of the plant components

1.3.1 Photovoltaic Generator

The solar generator consists of 3,840 modules and is divided into 11 rows spaced at 7m.

Direct current is converted into three-phase current 380 V, 50 Hz by means of two pulse inverters of 80 kVA each. The two inverters are parallely operated in half-load mode so that the inverters are always operated in the range of full load = highest efficiency.

Technical data:

Modular solar generator:	PQ 10/40 3,840 pcs
(multicrystalline)	in various designs
Tilt angle:	45°
Nominal direct voltage:	410 V DC
Nominal current AM 1.5:	370 A

1.3.2 Wind energy converter - WEC

The wind system type HSW 250 consists of the main components: pipe tower, wings, rotor hub, generator and controls.

A change-pole, three-phase, asychronous generator has been installed to convert the wind energy into electric energy. It is located directly behind the mechanical load transmission elements of the wind turbine in the WEC gondola.

Technical data:

Tower:	steel
Tower height:	27.3m, accessible
Wing material:	GFK
Rotor diameter:	25 m
No. of blades:	3
Generator construction:	asynchronous, change-pole
Nominal capacity:	80 /250 kW
Rpm:	1,000/1,500
Nominal voltage:	380 V
Frequency:	50 Hz

1.3.3.Combined Heating and Power Plant (BHKW)

The energy supply system MOEWE (Modular Energy Supply System for Heat and Electricity), Type WE 3, is essentially composed of an combustion engine and a synchronous generator for generating electric energy and at the same time utilizing the heat of cooling water, exhaust gas and oil. The combustion engine is driven by biogas from the putrefaction processes.

Technical Data:

Daimler-Benz engine:	M 102
Generator type:	B 182
Construction:	synchronous
Nominal output:	45 kVA
Rpm:	3,000
Mains voltage:	380 V
Frequency:	50 Hz

1.4 Network influence

The network fault power at the connecting point to the 11 kV grid is 61 MVA. The 11/0.4 kV transformer has a power of 400 kVA.

In November 1989, the feeding of solar current via the pulse inverter was analysed in the range of current as well as the range of voltage:

Meas. No.	Oper. state	K-factor
181	Mains voltage Op. with WEC without inv.	0.0076
132	Phase current of 1 inv. Op. with 2 inv. and WEC VDC = 425, IDC = 117 A	0.1616

WEC = Wind energy converter; inv. = inverter

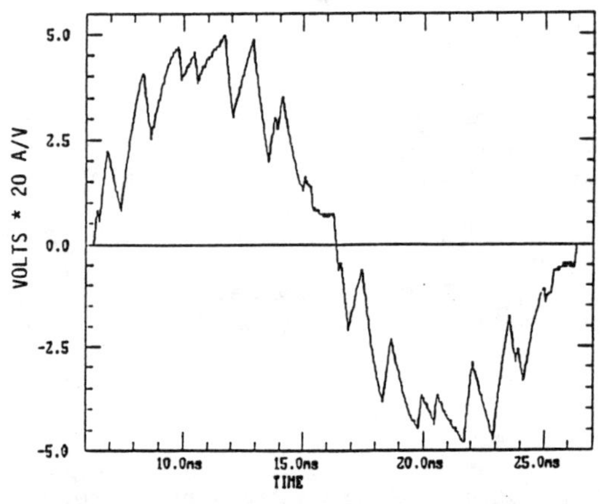

Fig. 2 : Phase current of an inverter (Characteristic No. 132)

1.5 Energetic balance

In the period 09/89 to 03.03.93, the different components supplied the following energy amounts:

PV	423,950 kWh
WEC	1,471,579 kWh
BHKW	103,528 kWh
	1,999,057 kWh

Additionally, the SCHLESWAG bought 374,371 kWh. 1,220,036 kWh were supplied to the SCHLESWAG.

About 172 % of the sewage plant needs were generated, i. e. a percentage of 68 % was supplied on time. The remaining 32 % were supplied by the utility grid.

2 SOLAR ELECTRIC SUPPLY FOR PELLWORM

2.1 Plant description

In spring 1980, the European Commission invited tenders for the development and construction of solar-electric plants. A total of 17 plants with a capacity of 30 kW up to a plant with 300 kW - Pellworm - which is the largest by far were ordered. The operative plant of Pellworm started operation in mid-year 1983.

This solar-electric power plant with a dc peak power of 300 kWp independently supplied the recreation centre of the North Sea island Pellworm. Surplus energy was fed into the network of the SCHLESWAG AG.

The direct current generated by the solar generator was converted by self-commutated inverters for the independent supply of the recreation centre and by a line-commutated inverter for network supply. For power supply during night or during spells of bad weather, a battery of 6,000 Ah with 346 V was available. A supply from the grid was kept for emergency supply.

The energy was supplied by a 20 kV system from the solar generator to the consumer, the recreation centre.

For the higher ranking control tasks, a microprocessor system was used in order to coordinate the entire energy flux.

The seasonal energy demand of the recreation centre is perfectly adjusted to the sun energy available in the different seasons. The consumers need most energy in summer and are not working during most of the winter months.

2.2 Technical description of the total system

The solar generator itself, consisting of 17,568 modules, is divided into 21 subgroups switchable according to operation. All groups which are controlled via a process computer are either part of the dc bus bar I or II (fig. 3):

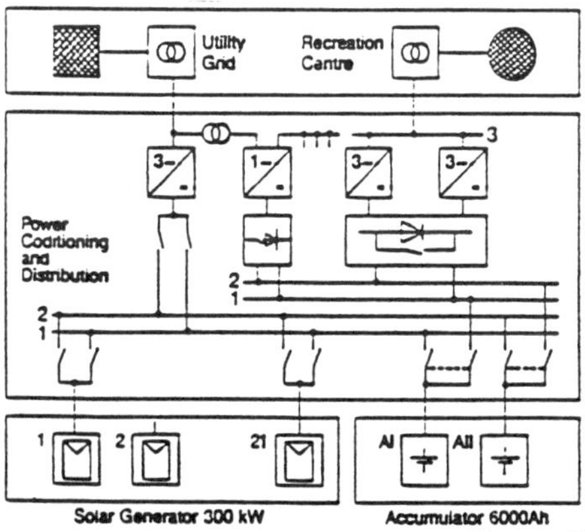

Block Diagramme Solar System Pellworm I Fig. 3

Each bus bar was provided with one half of the battery of 3,000 Ah each. One half was used as a consumer buffer battery while the other was charged by the solar generator. Thus, the batteries were operated as gentle as possible because the charging process could be maintained until a battery half was completely charged - without any discharging processes in the meantime.

The use of a charging controller was not necessary due to the switchable solar generator groups. When the threshold voltage at begin of gassing was reached, the charging current was limited to a permissible value by means of switching over generator groups to the consumer bus bar.

In the case of an energy offer higher than the energy demand of the recreation centre, the missing energy was taken from the buffer battery and as many solar generator groups as possible were switched over to the buffer battery bus bar without interrupting the charging current of the other battery unit.

In the case of an energy offer higher than the demand, the system voltage was kept near the voltage in the MPP by limiting the charging current of the buffer battery by switching over generator groups; furthermore, a higher charging current was applied at the battery to be charged. After completed charging, the consumer was connected to this battery and the previous buffer battery was charged.

If both battery halfs were completely charged, and in the case of an energy offer higher than the energy demand, the surplus energy was fed into the public network. By means of this, no solar generator groups had to be switched off due to operating states.

2.3 Technical data of the components

2.3.1. Photovoltaic power plant

- Structure area: 16,500 m2
- Module area: 4,500 m2
- Cell area: 3,500 m2
- Modular solar generator: PQ 1/20; (multicrystalline) 17,568 pcs.
- Tilt angle: 40°
- Row distance: 7 m
- Nominal direct current: 346 V DC
- Nominal current AM 1.5: 1,000 A

2.3.2. Inverter

A line-commutated inverter of 450 kVA was used for network supply.

Two self-commutated inverters of 75 kVA each were used for the independent energy supply of the recreation centre. The second inverter was additionally phased according to the energy needed.

2.3.3 Energy storing

The recorded mean values of insolation in the North Sea area of Schleswig-Holstein from 1972 to 1980 showed that the solar generator might offer less energy than required by the recreation centre for five days running. Considering these weather conditions also, the battery was dimensioned to 6,000 Ah. Then, it was operated to a discharge depth of 70 % when completely charged to 2.45 MWh.

In 1990, the service life of the battery ended. Early in 1991, the complete battery set was dismounted and disposed of. Since this time, the old solar plant has been completely out of operation.

2.4 Network influence

As the plant was not the propriety of the SCHLESWAG AG during the period of island operation (from November 1989), the SCHLESWAG AG did not carry out any measurements regarding harmonic waves. Due to the analogy of the inverters, the above mentioned data for Fehmarn and the publications as to Kobern-Gondorf can be considered (Ref.3).

2.5 Energetic balance

It becomes very clear in the graphic chart (fig.4) how large the losses of this independent PV battery system were. The losses occurred in the battery and in the self-commutated inverters which were only loaded by some kW during night.

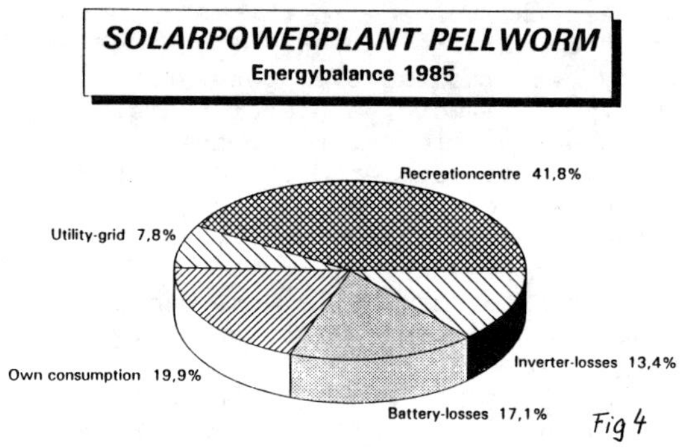

Fig 4

2.6 Enlargements

In May 1989, 3 WECs of 33 kW were additionally installed.

Technical data of wind energy converters:

Tower:	steel
Tower height:	14.5 m
Wing material:	GFK
Rotor diameter:	12.5 m
No. of blades:	2 pcs.
Generator construction:	asynchronous
Nominal capacity:	33 kW
Rpm:	1,500
Nominal voltage:	380 V
Frequency:	50 Hz

The construction of a plant with additional 300 kWp of latest solar technology was started on August 15th of 1991.

The system has been completed by a 300 kW wind energy converter which feeds energy into the 20 kW grid via a 12 pulse inverter. Thus, 900 kW of inverter capacity are connected to the network spur of the island of pellworm Sk = 41 MVA. Fig.5+6

2.7 Data of the components

2.7.1 Photovoltaic power plant

Technical data:

- Structure area:	9,500 m2
- Module area:	3,000 m2
- Cell area:	2,240 m2
- Modular solar generator: (multicrystalline)	PQ 1/40; HRD 50 6,048 pcs.
- Tilt angle:	30°
- Row distance:	6 m
- Nominal direct current:	418 V DC
- Nominal current AM 1.5:	720 A

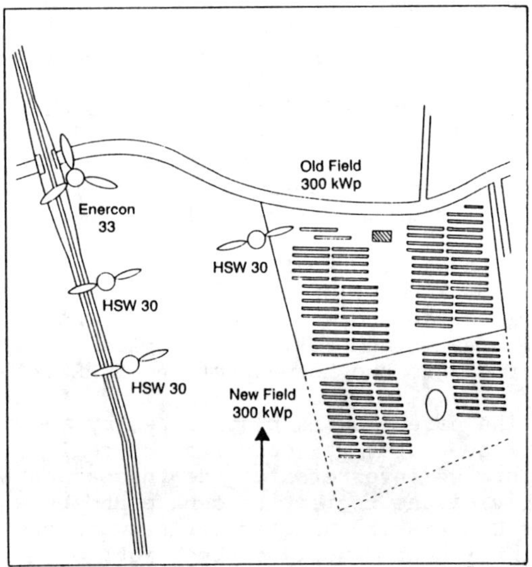

Map of the new Hybrid Plant Fig. 5

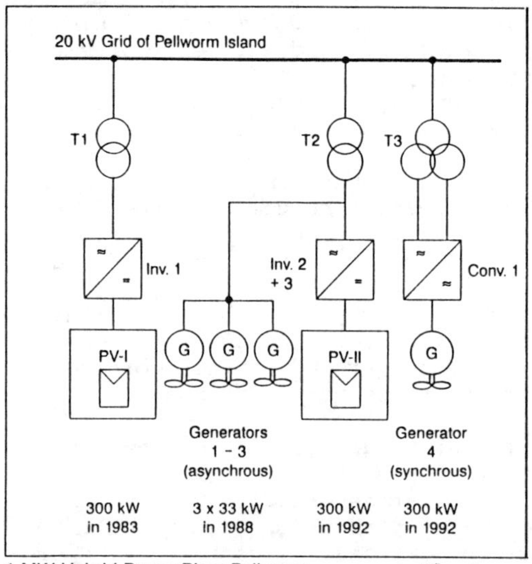

1 MW Hybrid Power Plant Pellworm Fig.6

With an efficiency of the solar modules of 12 %, the cell area for a peak power of 300 kW has only a size of 2,420 m2 compared to 16,500 m2 of the first construction phase. The structure area was reduced from 16,500 m2 to 9,500 m2. Furthermore, the tilt angle and the row distance were reduced. The community of Pellworm makes the site next to the current plant available for extending the solar plant.

2.7.2 Inverter

In order to keep the network influence of the inverters as small as possible, the new PV plant was connected to the network via two pulse modulated inverters with high frequency.

A distribution of 2 * 70 kW and 2 * 90 kW has been chosen according to the different solar field sizes.

Some of the advantages of the modular design are given in the following:

- Improvement of grid supply due to low single hysteresis of the network modulating current.
- Beat of all modulating currents has a compensating effect.
- Increased operating safety due to smaller modules.
- Reduced transport due to prefabricated wall.
- If one module fails, operation can be continued with reduced output.

The advantage of this design in contrast to the use of thyristors is the nearly sinusoidal output current being nearly in phase with the network voltage. No wattless current compensation is required. A 6 pulse thyristor inverter generates square currents which are not in phase with the network voltage due to the small firing angle.

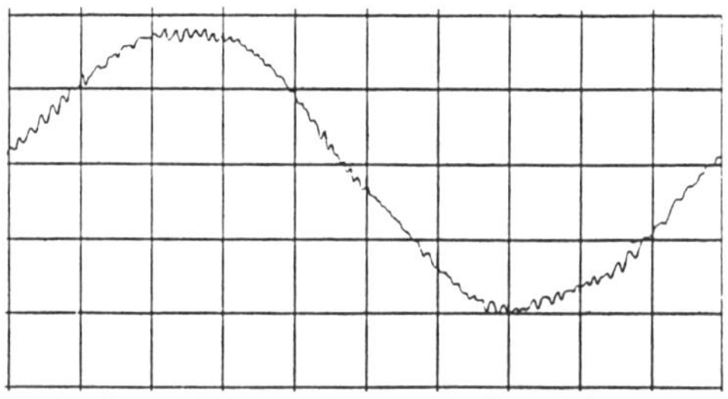

Fig. 7 : Phase current of an inverter
I = 170 A

A compensation system must provide the required capacitive wattless power (switchable or controlable in steps in the case of large units). Compensation systems themselves influence the natural frequencies of the network causing unfavorable effects or troubles at ripple control. Currently, the costs and the low electric strength of about 1,200 V are disadvantages of the IGBT.

A MPP controller is superimposed for the operation at the solar field.

2.7.3 Wind energy converter - Data of Enercon 332

Technical data:

Tower:	reinforced concrete moulded by centrifugal action
Tower height:	13.5 m
Wing material:	GFK
Rotor diameter:	32 m
No. of blades:	3
Generator construction:	synchronous
Nominal capacity:	300 kW
Rpm:	variable
Network coupling:	12 pulse inverter
Nominal voltage:	380 V
Frequency:	50 Hz

2.8 Prospects

The energy generation extrapolated to a year on the island of Pellworm which provides of a lot of sunshine and wind holds out a prospect of interesting energetic operating results (fig. 8). The results obtained late in 1992 and early in 1993 are also shown.

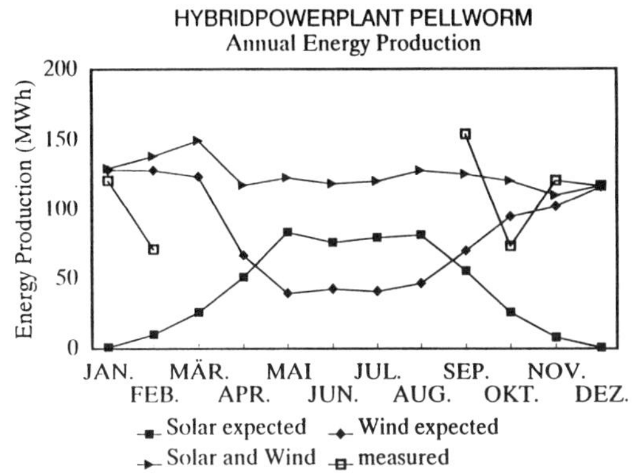

The result is mainly the output of the WEC. The new solar field coud be taken in automatic operation with MPP control just a short time ago. The first results are positive.

The old solar field is not in operation, ample overhauling of cabling is to be done: cables and connectors have been damaged by the UV radiation. The inverter must also be retrofitted.

By means of hybrid plants with an offer of primary energy decoupled in a time dependent way,periods can be reduced during which no energy is available.Important areas have been further developed in technical respect by means of which the outputs have been increased and the harmonic waves as well as the wattless power have been reduced.

References:

Ref. (1) Telefunken SystemTechnik GmbH, Neue Technologien: "Sonne, Wind und Biogas - Kläranlage Fehmarn"

Ref. (2) Telefunken SystemTechnik GmbH, Neue Technologien: "Solarelektrische Energieversorgung für Pellworm"

Ref. (3) Dipl-Ing. Rolf Hotopp: "Auf dem Weg zum besten Konzept für Photovoltaik-Wechselrichter, Teil 2: Netzrückwirkungen" etz Vol. 111 (1990), No. 7/8.

Higher electrical efficiency with variable speed

Anders Grauers, M.Sc.
Chalmers University of Technology
Dept. of Electrical machines and Power electronics
S-412 96 Gothenburg
Tel. int.+46 31 772 16 31
Fax int. +46 31 772 16 30

ABSTRACT: This article present a general view of the losses in the generator, frequency converter and gear box both at constant speed and variable speed operation. Theoretical calculations show that a high-efficient variable speed generator and converter system can have lower total losses than a directly grid-connected generator at constant speed. In comparison with a two-speed generator the efficiency is about the same, a little higher on low wind speed sites and a little lower on high wind speed sites. The generator and converter loss model are verified with measurements on two small systems. The system described is a synchronous generator with a diode rectifier and a thyristor inverter. The losses of an induction generator system are similar, except for the converter efficiency which generally is a little lower.

1. GENERATOR LOSSES

1.1 Grid-connected generator

In the grid connected generator the losses can not be controlled. The generator speed and magnetic flux are determined by the grid. As a result, the friction losses and core losses are constant. The stator and field-winding losses only depend on the produced power. Producing only a few percent of the rated power, the losses of the grid-connected generator still remain to about 1/3 of the losses at full load. By using a two–speed generator the losses can, however, be reduced when the output power is less than 30 % of the rated power.

1.2 Variable speed generator

In the variable speed generator system the converter allows the speed and magnetic flux of the generator to change. The speed is adjusted to keep the turbine efficiency at its maximum, but still the magnetization can be controlled freely. The magnetization control can be used to reduce core losses and field-winding losses by reducing the magnetic flux of the generator. But at a specific power the reduction of the flux reduces the voltage and leads to higher generator current and higher resistance losses. The two types of losses must be balanced to keep maximum efficiency. The total losses of the variable speed generator can be much lower than those of a grid-connected generator for low input powers. At rated power and rated speed the losses are a little higher in the variable speed generator due to harmonics in current and voltage. Also the bearing friction and windage losses are reduced by the generator speed. They are, however, a small part of the total losses. If the losses of a variable-speed generator is compared with the ones of a two–speed generator, the losses are still lower at low output power but the difference is much less.

When the converter allows the generator speed to change there is a possibility to run the generator at higher speed than what is possible with a grid-connected generator. Since ordinary generators are designed for a frequency up to 60 Hz at least 20 % overspeed can be used compared to 50 Hz operation. This reduces the losses even more and also makes it possible to use the full 50 Hz ratings of the generator even with non sinusoidal current and voltage.

The efficiency of the generator both grid-connected and at variable speed with maximized efficiency can be seen in Figure 2. A curve of a two-speed generator is also plotted. The option of higher speed at rated power is used in the variable speed generator .

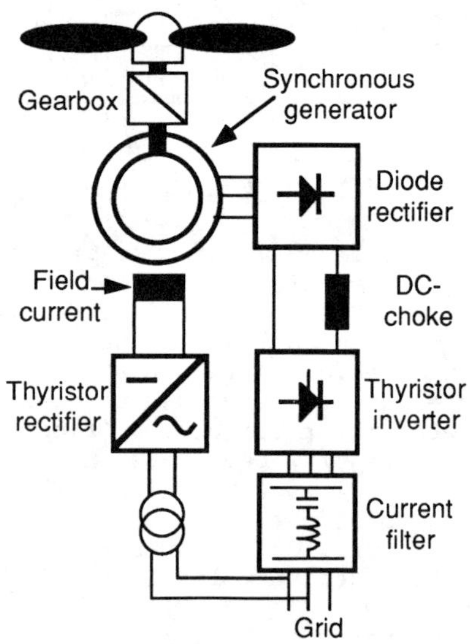

Figure 1: The variable speed system used in these calculations.

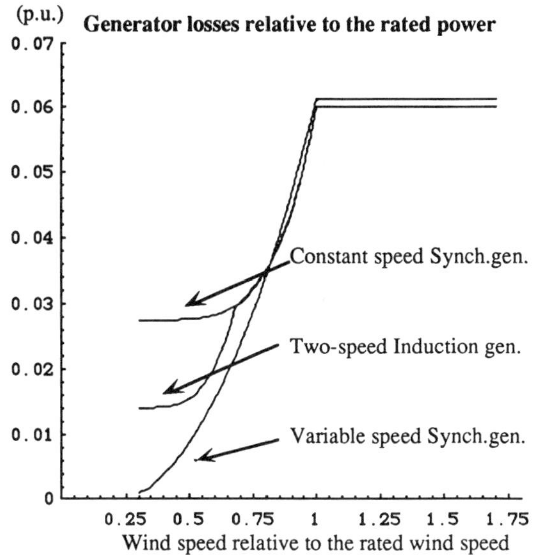

Figure 2: Losses in the generator at constant speed, variable speed, and two-speed operation.

2. CONVERTER LOSSES

The losses in the frequency converter are due to the voltage drop of the rectifier diodes, resistive loss of the dc-filter, voltage drop of the inverter thyristors and power for the control circuits and cooling fans. If the voltage is reduced at a specific power the converter current must increase. Increasing current leads to increasing losses both due to semiconductor voltage drop and resistance. That implies that by reducing generator core losses one increases converter losses. Since the generator losses are larger than the converter losses, the total losses still decreases with lower voltage on the generator at low wind speeds. The efficiency of an efficient 250 kW converter at rated power is about 98 %. At variable speed operation with low magnetization the efficiency is lower at low wind speed, about 95 % at 20 % of the rated power.

Apart from the direct converter losses the converter also causes extra losses in the generator. The diode bridge leads to non sinusoidal generator currents and generator voltage, which causes more resistive losses and more core losses than sinusoidal currents do.

3. THE GEARBOX LOSSES

About 75 % of the gear losses, the gear mesh losses, have the same percentage of the input power at all speeds. However, the friction losses of the bearings and the windage and churning losses, are reduced at variable speed operation. They are reduced at least linearly with the shaft speed. The efficiency at the rated power of a 250 kW gearbox is about 97 %. At 20 % of the rated power the efficiency at the rated speed is 94.5 %, and at the speed of a variable speed turbine it is 96 %. (The gearbox losses have not been measured.)

4. TOTAL LOSSES

The total losses of the gearbox and generator system at constant speed operation both for a one-speed and two-speed generator and at optimized variable speed operation can be seen in Figure 3. In Figure 4 the probability of different wind speeds is plotted. By looking at the probability of different wind speeds it can be seen that the losses at rated wind speed is not as important as the losses at a wind speed of about half the rated wind speed.

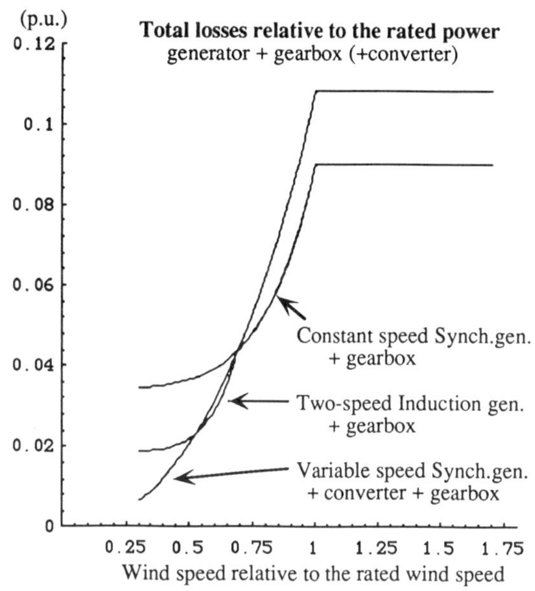

Figure 3: Total losses in generator, gearbox and, for the variable speed system, converter.

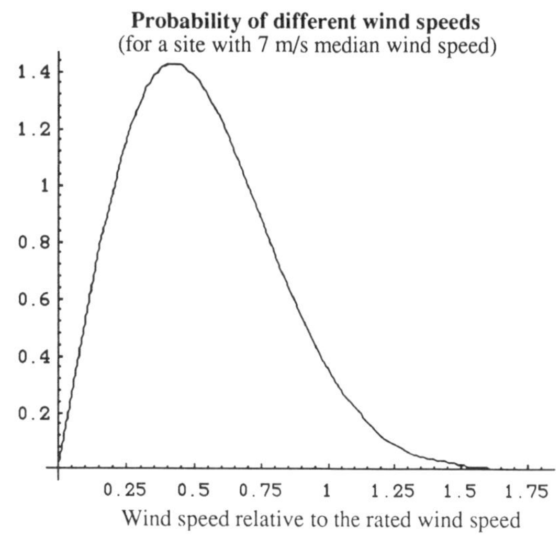

Figure 4: Probability curve for different wind speeds on a site with a median wind speed of 7 m/s (rated wind speed is 14 m/s).

5. ANNUAL AVERAGE EFFICIENCY

The annual average efficiency is calculated as the annual average power loss divided by the annual average power input from the turbine. The annual losses are calculated from the loss function in Figure 3 multiplied by the probability function in Figure 4 and integrated over wind speeds from the start of the system to the stop. The figures calculated are the average power loss for a year in percent of the rated turbine power. The annual energy production is calculated in the same way, by creating a function of captured turbine power at different wind speeds, and then performing the same integration with the probability curve to get the annual average power production in percent of the rated turbine power.

6. RESULTS

The calculations of generator system average losses and gearbox average losses together with calculations of the average turbine power are carried out for a 250 kW wind turbine. The losses of the generator system have also been measured for a 15 kW induction generator system and a 40 kW synchronous generator system. The average efficiency at sites with different median wind speed is presented in Table 1.

7 CONCLUSIONS

As can be seen in Table 1 a variable speed wind turbine can have higher electrical efficiency than one with constant speed. When comparing an optimized variable speed system and a two-speed generator it shows that they can have almost the same electrical efficiency. The variable speed system is a little more efficient on low wind sites and a little less efficient on high wind sites. In practice a variable speed system can not be perfectly optimized in all parts. Therefore it is probably realistic to expect some percent lower efficiency in a well designed variable speed system.

Table 1: The annual average efficiency at different median wind speeds of the different generator system with and without the gearbox losses. The converter losses are included in the variable speed system efficiency. The difference in turbine efficiency is not included in the figures.

Median wind speed m/s	5.6	6.3	7.0	7.7	8.4
(measured losses)	Annual average efficiency (%)				
15 kW Ind.Gen.					
Variable speed	72.1	74.3	75.8	76.8	77.6
Constant speed	70.4	73.3	75.1	76.6	77.4
40 kW Sync.Gen.					
Variable speed	82.2	84.4	85.8	86.7	87.3
Constant speed	79.9	84.1	86.8	88.5	89.6
(theoretical)	Annual average efficiency (%)				
250kW Sync.Gen.					
Variable speed	88.5	89.5	90.1	90.6	90.9
(incl. gear losses)	(83.6)	(85.0)	(86.0)	(86.7)	(87.2)
250kW Sync.Gen.					
Constant speed	82.7	86.2	88.4	89.8	90.7
(incl. gear losses)	(76.3)	(80.9)	(83.7)	(85.5)	(86.7)
250/75kW two-speed Ind.Gen.					
Constant speed	87.8	89.8	90.9	91.6	92.1
(incl. gear losses)	(82.7)	(85.1)	(86.7)	(87.7)	(88.4)

Estimation of Rotor Voltage Vector on the DEIM used in WECS

M. T. EL-HAGRY M. N. ESKANDER

ELECTRONICS RESEARCH INSTITUTE

National Research Center Dokki Cairo

EGYPT

Abstract

The aim of this work is to generate the rotor voltage of A SRIM that is going to be used as a generator in wind energy conversion systems with speed range extending from the sub synchronous range to double the synchronous speeds at the most efficient operating conditions. Hence the machine is to be operated at its rated electric torque which is realised by having both rotor and stator currents magnitudes constant and that the angle between the rotor current and stator current is kept constant irrespective of the rotor speed.

INTRODUCTION

The slip ring induction machine has many applications as double excited induction machine (DEIM). The DEIM may be used as a variable speed drive with wide range of operating speeds (from standstill to double the synchronous speed). It can be used as a generator at sub- and super-synchronous speeds when the applied voltage is of the suitable magnitude and phase angle. This DEIM performance is proved in previous research by the authors [1], which leads to higher power generation when used in wind energy conversion systems (WECS).

In spite of the advantages of compactness, high torque and power capabilities, as well as wide speed range of operation of the DEIM, a major problem appears when dealing with practical applications. This problem is that of generating a voltage signal with slip frequency in synchronism with the rotor voltage to be used in controlling the firing pulses of the inverter connected to the rotor circuit. Previous methods proposed for implementing rotor voltage used either specially designed machine with search coil, or a pilot machine having the same design as the main induction machine [2], or by manual adjustment [3]. The search coil would sense the resultant field rather than the rotor field, including the effect of all switching devices used, which add noises at very high levels. Pilot machines requires additional cost and complicated control circuits.

Modern digital control methods [4] could be useful in detecting the rotor voltage magnitude and phase in the DEIM. Such methods are now applied for controlling the squirrel cage induction machine as a variable speed drive. However, to the authors knowledge, no proposals for detecting the rotor voltage vector by digital control methods have been presented [5-7].

In this paper a scheme is proposed for the determination of the rotor voltage vector for the double excited slip ring induction machine used in WECS, i.e., with its rotor mechanically coupled to the wind turbine rotor. The stator of the machine is connected to the main grid, while its rotor circuit is connected via two bridge converters and a dc link to the main grid [1]. The rotor voltage is varied in magnitude and phase angle by changing the firing angles of both bridges via a closed loop control system following the control strategy of operating the DEIM as a generator with machine current and torque limited by the rated values to allow maximum possible power generation. The advantage of this control strategy is that it leads to the stable operation of the DEIM all over the considered speed range. Otherwise the DEIM will be unstable due to the presence of negative damping torque [8]

The proposed method is based on the theory of the extended Kalman filter [9], since the state space model that describes the system is nonlinear. The chosen states are the synchronously rotating d and q components of the rotor current, and the mechanical speed, in addition to the d component of the stator current which is taken as a reference. The control parameters taken are the mechanical torque, and the two components of the rotor voltage. The measured parameters are the mechanical speed, the stator current, and the dc link current. The d-component of the synchronously rotating frame is assumed to coincide with the stator current in order to simplify the equations.

The aim of this work is to generate the two components of the rotor voltage that will result in an optimum operation of the SRIM as a generator. It has been found that operating the machine at its rated values (including stator current and voltage, rotor current, and electric torque) may be considered an optimal operation from the point of view of generation.

PROBLEM FORMULATION

The system equations of the SRIM for synchronously rotating frame that coincides with the direct component of the stator current are given by

$$\dot{i}_{ds} = a_1 i_{ds} + a_2 i_{dr} + a_3 \omega_m i_{qr} - b_1 V_{ds} + b_2 V_{dr} \quad (1)$$

$$\dot{i}_{dr} = a_4 i_{ds} + a_5 i_{dr} + \omega_s i_{qr} + a_6 \omega_m i_{qr} + b_3 V_{ds} + b_4 V_{dr} \quad (2)$$

$$\dot{i}_{qr} = a_7 \omega_m i_{ds} - \omega_s i_{dr} + a_8 \omega_m i_{dr} + a_9 i_{qr} + b_s V_{qr} \quad (3)$$

$$\dot{\omega}_m = a_{10} \omega_m + a_{11} i_{ds} i_{qr} + b_6 T_l \quad (4)$$

where

$a_1 = r_s L_r / Z$, $a_2 = -r_r M/Z$, $a_3 = -M L_r / Z$, $b_1 = -L_r / Z$

$b_2 = M/Z$, $a_4 = -r_s M/Z$, $a_5 = r_r L_s / Z$, $a_6 = L_r L_s / Z$

$b_3 = M/Z$, $b_4 = -L_s / Z$, $a_7 = -M L_s / Z$, $a_8 = -L_s L_r / Z$

$a_9 = r_s L_s / Z$, $b_5 = -L_s / Z$, $a_{10} = -K_d / J$

$a_{11} = -M n_p^2 / J$, $b_6 = -n_p / J$, $Z = M^2 - L_s L_r$

The three control parameters that are shown in these equations are the direct and quadrant components of the rotor voltage in addition to the input mechanical torque T_l. The input mechanical torque (having negative sign) is the actual energy that is to be converted to the electric form and hence it has to be used as efficient as possible. The other two parameters are the rotor voltage components and they are functions of the mechanical speed of the machine only when we considered both rotor and stator currents are constants and that the relative angle between them is constant too to satisfy the condition of constant electric torque operation. Stator voltage components are constant as far as stator and rotor currents are constant and they aren't functions of the mechanical speed of the rotor. The steady state equations of the components of the stator and rotor voltages are given by the following set of equations after equating the derivative component with zero.

$$V_{ds} = r_s I_{ds} - \omega_s M\, I_{qr} \tag{5}$$

$$V_{qs} = \omega_s L_s I_{ds} + \omega_s M\, I_{dr} \tag{6}$$

$$V_{dr} = r_r I_{dr} - \omega_s L_r I_{qr} + \omega_m L_r I_{qr} \tag{7}$$

$$V_{qr} = (\omega_s - \omega_m)(M\, I_{ds} + L_r I_{dr}) + r_r I_{qr} \tag{8}$$

where I_{ds} is the rated stator current and

$$I_{qr} = -\, T_e / M n_p I_{ds} \tag{9}$$

$$I_{dr}^2 = I_r^{\,2} - I_{qr}^{\,2} \tag{10}$$

T_e is the electric torque of the machine and it is negative value in case of generation, while I_r is the rated rotor current. I_{dr} has negative polarity.

Equations (1) to (4) can be written in the form of state matrix vector form as follows:

$$\dot{x} = f(x,u) \tag{11}$$

where x are the state variables and *u* are the control parameters. Expanding equation (11) using taylor series about the nominal value and neglecting all terms except those of the first order then we obtain will end up with the following state equations after substituting for V_{ds}, V_{qs}, V_{dr} and V_q from equations (5) to (8)given by the following matrix equation

$$\delta\dot{x} = \partial f(x,U)/\partial x\big|_{x}^{*}\ \delta x \ + \ \partial f(x,U)/\partial U\big|_{U}^{*}\ \delta U \tag{12}$$

$$\dot{X} = F(t)\ X + G(t)\ u$$

$$\dot{X} = \delta\dot{x} = \dot{x} - \dot{x}^*,\ u = \delta U\ ,\ F(t) = \partial f/\partial x,\quad G(t) = \partial f/\partial U$$

equation (12) represent the linear perturbation equation which describe the motion of the actual system relative to the nominal. Equation (12) is reduced to the linear difference equation as follows

$$X(t_1) = \phi\ (t_1,t_o)\ X(t_o) + \Gamma(t_1,t_o)\ u(t_o) \tag{13}$$

where $d\phi(t_1,t_o)/dt = F(t)\phi(t_1,t_o)$

$$\Gamma(t_1,t_o) = \int_{to}^{t1} \phi(t_1,\tau)\ G(\tau)\ d\tau$$

$$X_1(k) = i_{ds}(k) - i_{ds}(k-1),$$

$$X_2(k) = i_{dr}(k) - i_{dr}(k-1),$$

$$X_3(k) = i_{qr}(k) - i_{qr}(k-1),$$

and $X_4(k) = \omega_m(k) - \omega_m(k-1)$.

equation (13) can be rewritten in the following form

$$X(k) = \phi(k,k-1)\ X(k-1) + f(k-1) + w(k-1) \tag{14}$$

where $\phi(k,k-1) = I + h\ F[x(k-1)]$

and $f(k-1) = h\ G\ \delta T_l$, $\quad \delta T_l = T_l(k) - T_l(k-1)$

The noise process w(k) is assumed to be gaussian with zero mean and co variance Q(k).

$$\phi(k,k-1) = \begin{bmatrix} O_{11} & O_{12} & O_{13} & O_{14} \\ O_{21} & O_{22} & O_{23} & O_{24} \\ O_{31} & O_{32} & O_{33} & O_{34} \\ O_{41} & O_{42} & O_{43} & O_{43} \end{bmatrix}_{\text{at } k-1}$$

where the components of the matrix at k-1 are given by

$O_{11} = 1 + hr_sL_r/Z$, $O_{12} = -hr_rM/Z$, $O_{13} = -hML_r\hat{x}_4^{k-1}/Z$

$O_{14} = -hML_r\hat{x}_3^{k-1} + hMc_2/Z$, $O_{21} = -hr_rM/Z$,

$O_{22} = 1 + hr_rL_s/Z$, $O_{23} = h\omega_s + hL_sL_r\hat{x}_4^{k-1}/Z$,

$O_{24} = hL_sL_r\hat{x}_3^{k-1}/Z - hL_sC_2/Z$, $O_{31} = -hML_s\hat{x}_4^{k-1}/Z$,

$O_{32} = -h\omega_s - hL_sL_r\hat{x}_4^{k-1}/Z$, $O_{33} = 1 + hr_rL_s/Z$,

$O_{34} = -hML_s\hat{x}_1^{k-1}/Z - hL_sL_r\hat{x}_2^{k-1}/Z$, $O_{41} = -hMn_p^{\,2}\hat{x}_3^{k-1}/J$,

$O_{42} = 0$, $O_{43} = -hMn_p^{\,2}\hat{x}_1^{k-1}/J$, $O_{44} = 1 - hK_d/J$

The available measurements that are related to the states are stator current, dc current, and the mechanical rotor speed. The dc current in the dc link is related to the direct and quadrant components of the current in the rotor nonlinearly. The measurement equations are also linearized as follows

$$i_{ds}(k) = i_{ds}(k) + v_1(k) \tag{15}$$

$$i_{dc}(k) = C\ i_r(k) = C\ [i_{dr}^{\,2}(k) + i_{qr}^{\,2}(k)]^{1/2} \tag{16}$$

$$\omega_m(k) = \omega_m(k) + v_3(k) \tag{17}$$

The linearized equations are

$$Y(k) = H(k)\ x(k) + V(k) \tag{18}$$

where

$$H(k) = \begin{bmatrix} 1 & 0 & 0 & 0 \\ 0 & 0 & 0 & 1 \\ 0 & \dfrac{C\,\hat{x}_2^{k-1}}{\hat{i}_r^{k-1}} & \dfrac{C\,\hat{x}_3^{k-1}}{\hat{i}_r^{k-1}} & 0 \end{bmatrix}$$

where $\hat{i}_r^{k-1} = \sqrt{\hat{x}_2^{k-1\,2} + \hat{x}_3^{k-1\,2}}$

From equations (14) and (18) the predicted values of the state vector x is given by

$$\hat{X}'_k = \phi\ (k,k-1)\ \hat{X}(k-1) + f(k-1) \tag{19}$$

The estimated $\hat{x}$ value at time k is given by

$$\hat{X}(k) = \hat{X}'(k) + K_k[y_k - H_k\hat{X}'(k)] \tag{20}$$

where K is the gain matrix and is given by

$$K_k = P'_k\ H_k^T[H_kP'_k\ H_k^T + R_k]^{-1} \tag{21}$$

$$P' = \phi(k,k-1)\ P_{k-1}\phi^T(k,k-1) \tag{22}$$

where P_k is the estimation error covariance matrix and P_k is given by

$$P_k = P'_k - K_kH_kP'_k \tag{23}$$

Usually estimation problems using kalman

filters is used for estimating non-measurable quantities, however it is used here to estimate the value of the components of the machine currents in addition to the rotor speed. The rotor speed is a measurable quantity and it is the only quantity used in calculating the rotor voltage components that satisfies the operation at the rated values of the machine as given by equations (7) and (8) and the more closer to the actual state the more accurate the control will be. Measuring the rotor current components is not straight forward and subject to many uncertainties. But the correction signal to be used in the control of firing should rely on the accurate components of the rotor current to estimate the value of the components of the rotor voltage Fig. 2.

SYSTEM DESCRIPTION

The system as shown in Fig 1 comprises two converters CS and CR separated by a dc link. BR is connected to the rotor of the machine while CS to utility grid. The specification of the machine under consideration is given in Appendix A. The machine is to be operated at its rated values in the generating mode as follows:

I_{ds} = 5.66 Amp, I_{qs} = 0.0 Amp, V_{ds} = -265.62 V,

V_{qs} = 109.51 V, I_{dr} = -5.04 Amp I_{qr} =2.56 Amp,

at T_e = -10 Nm.

This condition keeps the torque angle constant, and keeps the machine in the generating mode and hence it should supply its rated power from the stator side irrespective of the rotor speed. However the power taped from the rotor side is really dependent on the rotor speed where the rotor circuit is consuming energy from the grid in the sub- synchronous range and is delivering power to the utility grid in the super-synchronous speeds. The converters in the rotor circuit would control the components of the rotor voltage, to fix the rotor and stator currents at their rated value and hence keep the torque angle fixed. Both CS and CR controls the rotor voltage magnitude, frequency and phase through out the whole range under consideration as shown from Fig. 2 in the super synchronous range the direct component of the rotor voltage is positive, whereas the direct component of the current is negative which means that the machine is generating from the rotor as well, however at the positive slip values the direct component of the rotor voltage is negative, while the direct component of the rotor current is unchanged and hence the rotor circuit is consuming energy and the system is generating from the stator side only. Hence in the range $0.6 \geq s \geq S_r$ (where S_r is the rated slip) CS function as a converter while CR as an inverter while in the speed range $-S_r \geq s \geq -1.0$ CS and CR swap their functions.

The expected output of the generator, when it follows exactly the control strategy is optimum where as shown in Fig. 3 the power factor of the generated power is exceeding 0.9 in the slip range $-1.0 \leq S \leq 0.5$ with generated power amounts to double the rated power of the machine at slip equal to -1.0 and equal to 40% of the rating of the machine at 0.5 slip. For our machine we are generating in the speed range $3000 \geq \omega_m \geq 600$ rpm without yawing the nacelle. With a gear ratio equal to 25 this will be equivalent to a turbine rotor speed ranges from 24 RPM to 120 RPM which is quite a good range comparing with the conventional induction generator with its very narrow speed range and that its speed regulation is completely dependent on tilting the nacelle. At double the synchronous speed the machine is delivering double its rated output which is actually achieved at its slip frequency as shown in the Fig. 4. However, Operating the machine at an electric torque lower than its rated value at the same rated stator voltage is possible but with sacrificing the high power factor which is a major advantage of such a system beside its wide speed operating range, Fig.2 shows the block diagram of the control system that generates the firing circuit control voltage δV_r and $\delta\theta_r$. The Extended kalman filter (EKF) (19)-(23) is fed by the measured signal together with the torque signals and the initial conditions. The rotor speed is fed to the optimal operation box to generate the corresponding optimal rotor voltage components V^*_{dr} and V^*_{qr} and the optimal stator voltage components V^*_{ds} and V^*_{qs}, these rectangular components are transformed to the polar form V^*_r and θ^*_r and θ^*_s. The output of the EKF are the states are fed to the estimation operation box to calculate the estimated components of the rotor and stator voltages, those are in turns transformed to the polar coordinates and results with $\hat{V}_r$, $\hat{\theta}_r$, and $\hat{\theta}_s$. The estimated values are compared with the optimal values and the comparison outputs δV_r and $\delta\theta_r$ are fed to the firing control block after transformed to the rotor coordinates and to the supply coordinated .

Conclusion

It is clear from Fig. 5 , that the optimal control, operating the machine at its rated torque, means that the two components of the rotor voltage have to change linearly with the rotor speed, and that the direct component of the rotor current possesses zero value at the negative rated slip value at which the power generation equals the rated power of the machine tapped from the stator side only where, practically, the rotor voltage magnitude is almost zero. This is equivalent to conventional induction generation. Moreover the power factor is better than 0.9 for slip values less than 0.57 and drop severely above this value. The phase difference between the rotor and stator voltages, Fig. 6, is changing from first quadrant in the sub- synchronous range to the forth quadrant in the super- synchronous range, which confirm that CR has to work as inverter above the negative value of the slip to deliver power to the rotor while CS is a controlled rectifier. below the negative rated slip CR is a converter, while CS as an inverter.

APPENDIX A

Machine Specifications

A three, 50 Hz, wound rotor induction machine with the following characteristics and measured parameters has been used as the basis for all quantitative calculations:

Rated power : 1.5 KW
Number of poles: 4
Rated voltage: 220
Moment of inertia: 1.5 Kg.mt^2
Damping factor: 0.03 N-M/rad/sec.
Turns ratio N_s/N_r : 1
r_s = 2.131 Ω
r_r = 2.553 Ω
L_s = 0.246 H
L_r = 0.245 H

$M = 0.23$ H
$T_e = 10$ N-M
Rated phase stator current = 4 Amp (rms)
Rated phase rotor current = 4 Amp (rms)

REFERENCES

[1] Double fed induction machine used in wind electric energy conversion system, Ph.D. Thesis, Faculty of Eng., Menoufia Univ. Egypt.

[2] M.G.Uoannidou, J.A.Tegopoulos, "Performance of a double fed induction motor with controlled rotor voltage magnitude and phase angle", IEEE Trans. on EC, Vol. EC-2, No. 2, June 87.

[3] Gordon A Smith, "A current source inverter in the secondary circuit of a wound rotor induction motor provides sub- and super-synchronous operation", IEEE Trans IA, Vol. IA-17, No.4, july/aug. 1981

[4] G.F. Franklin, J.D. Powel M.L. Workman, Digital Control of Dynamic Systems, Addison - wesley publishing company, 1990.

[5] D.J.Atkinson , P.P. Acronley and J.W.Finch, "Application of estimation technique in vector controlled induction motor devices", IEE Japan IAS'91, Annual conference on industrial application society, pp 358-363.

[6] A.Delli Aquila, L.Salvatore, "Flux estimation by kalman filter in inverter fed induction motor", , IEE Japan IAS'91, Annual conference on industrial application society pp 388-393.

[7] J.W. Finch, D.J. Atkinson, P.D. Acornley, "Scaler to vector : General principles of modern induction motor control" , IEE Japan IAS'91, Annual conference on industrial application society pp 364-369.

[8] M.T.El-Hagry, A.A.Mohamed, M.N.Eskander, "stabilization of double excited induction machine used in WECS in two modes of operation" Mansoura Engineering Journal (MEJ) Vol. 15 No. 2, Dec.1990

[9] H.W. Sorenson, Advance in control systems, Theory and application Vol.3 Kalman filtering technique, pp 219-289

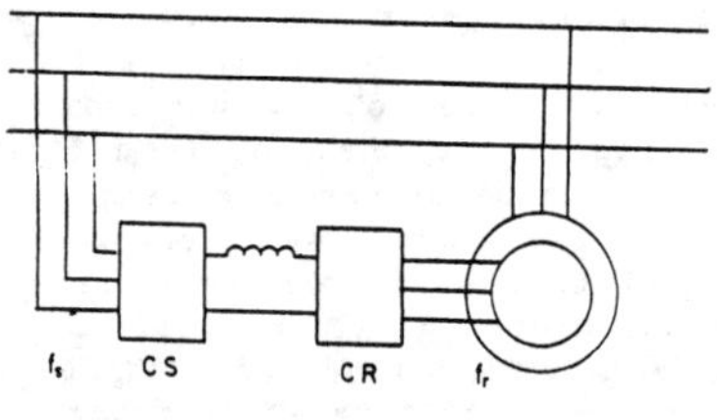

Fig. 2 The DEIM Scheme

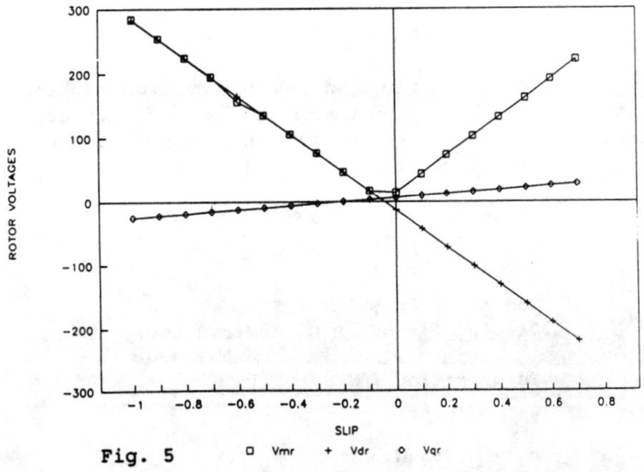

Fig. 5 □ Vmr + Vdr ◇ Vqr

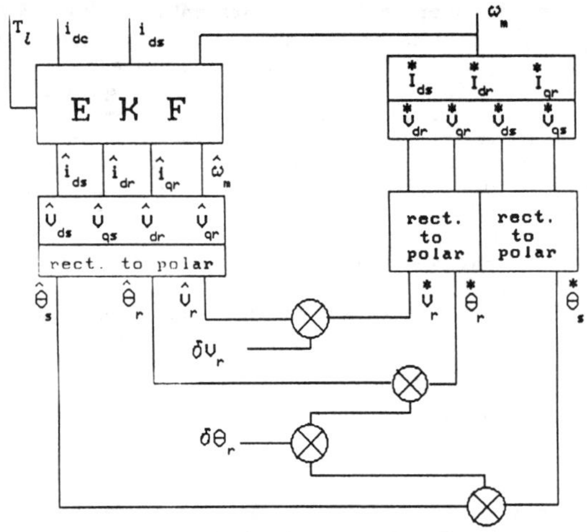

δV_r and $\delta\theta_r$ are two error signals to be fed to the firing control circuit of CS and CR.

Fig. 2- Block diagram of the generation of rotor voltage error signals

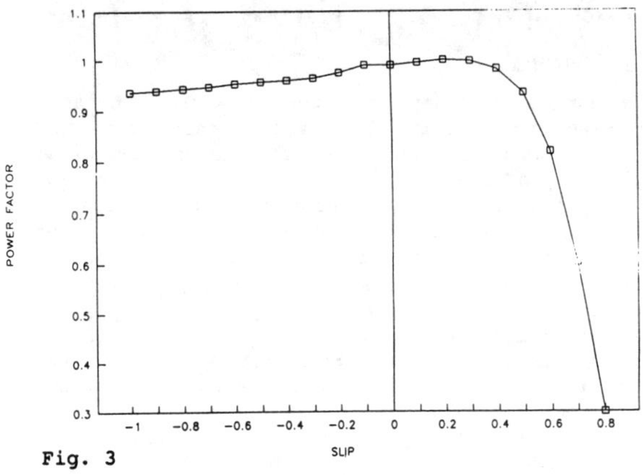

Fig. 3

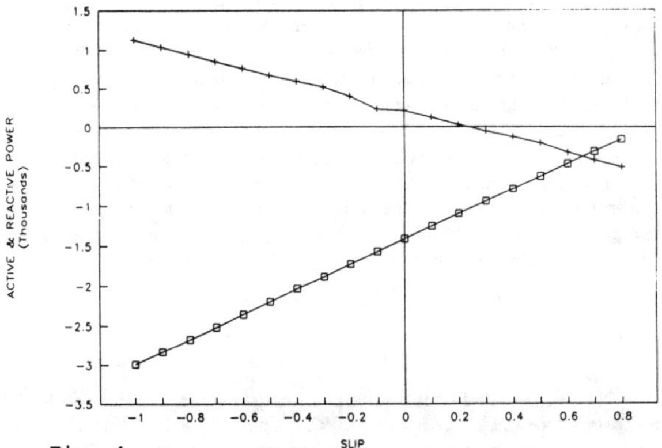

Fig. 4 □ ACTIVE POWER + REACTIVE POWER

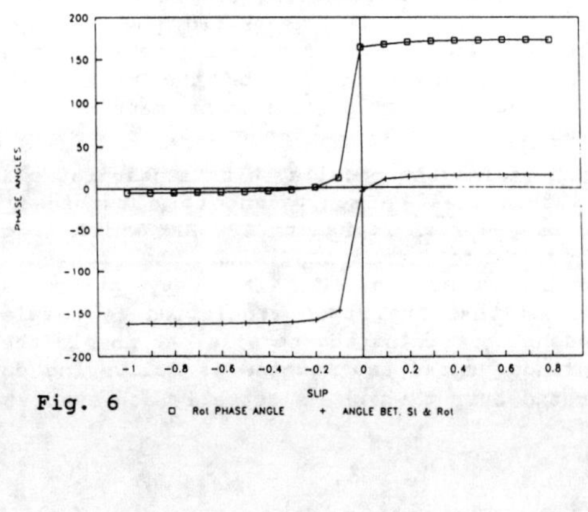

Fig. 6 □ Rot PHASE ANGLE + ANGLE BET. St & Rot

WIND RESOURCES PART I, THE EUROPEAN WIND CLIMATOLOGY

Erik L. Petersen
Risø National Laboratory, Denmark

1 Summary

The European Wind Atlas was produced with the purpose of establishing the meteorological basis for the assessment of the wind resources in the European Community. The comprehensive guidelines and computational tools which were generated together with the Atlas have constituted the basis for a number of similar studies outside the Community thereby making the calculated power production from wind turbines comparable from one country to another. A standardization seriously desired and needed by the wind energy community.

The Atlas depicts through maps and tables the regional distribution of the wind resources over land and a recent study has complemented this by presenting an offshore wind resource map. It appears from the Atlas that areas potentially suitable for wind energy applications are dispersed throughout all the countries of the European Community. It is the general experience that the Atlas provides a reliable picture of the overall and general distribution of the wind resources in most of the European Community. In regions with complicated topography and for which only sparse statistics exist the estimation of wind resources is very uncertain. To improve the situation in some of these regions the Commission has initiated measurements under the Joule 1 program in Northern Portugal, Spain, Italy, Greece and Ireland. Most of the measurements started in the Summer of 1991 and will run for two years. The overall impression of the intermediate results is they will not change the picture given in the Atlas much. However, they will add significantly to our understanding of the local distribution of the wind resources in mountainous terrain and thereby help to improve the model for flow over hills.

Recently a very comprehensive study has been undertaken by Deutsche Wetterdienst - Seewetteramt, Hamburg: "Wind and wave conditions in 55 European coastal sea areas, determined from weather and wave observations of voluntary commercial ships". The statistics presented in the study can be compared directly with the statistics in the Atlas maps. The comparison comes out very good, from the Northern Seas to the Aegean Sea there are very few discrepancies.

A careful comparison between the Wind Atlas predictions and the ship statistics can be used to rank a number of near-coastal stations in the Wind Atlas. It remains to be mentioned that the interpretation of the ship statistics is not without problems - on the contrary - but that will be discussed elsewhere.

2 The wind resources of the European Community

The wind resources over land. The map on Fig.1 from the Atlas depicts the regional distribution of the wind resources over land. It appears from the map that areas potentially suitable for wind energy applications are dispersed throughout all the countries of the European Community. It is the general experience that the Atlas provides a reliable picture of the overall and general distribution of the wind resources in most of the European Community. In regions with complicated topography and for which only sparse statistics exist the estimation of wind resources is very uncertain. To improve the situation in some of these regions the Commission has initiated measurements under the Joule 1 program in Northern Portugal, Spain, Italy, Greece and Ireland. Most of the measurements started in the Summer of 1991 and will run for two years. The overall impression of the intermediate results is they will not change the picture given in Fig.1 much. However, they will add significantly to our understanding of the local distribution of the wind resources in mountainous terrain and thereby help to improve the model for flow over hills. The Commission has further taken the initiative to supplement this model by one or more mesoscale models. The difference is that the present model is a very local model which can handle very detailed information of the topography out to approximately ten kilometers from the sites. In contrast the mesoscale models calculates the wind resource over several hundreds of kilometers in squares of one to five square kilometers. The models are currently being developed and they will be tested by a massive measurement programme where more than 30 instrumented mast of 30 meters height are currently being put into operation on Crete, Central Italy and Northern Portugal.

The wind resources offshore. An offshore position is in this context understood to be ten kilometers or more away from the nearest coast out to sea. The 15 colored wind resource maps in the Atlas are only colored over land such as the map on Fig.1. But this is only for reason of clarity, the statistics in the Atlas combined with some common knowledge of the European wind climate allows one straight away to construct an offshore wind resources map as shown in Fig.2. First of all, the basis idea behind the methodology of the Atlas is to present for each of the 220 stations used, regional representative wind statistics. By regional representative is meant that the derived statistics can be used out to a distance of approximately 100 km from the station. The experience tells that this hypothesis works well in non-mountainous terrain. The derived regional statistics is for each station presented for four terrain conditions - also called roughness classes - and one of them is open sea. And because more than 100 of the Atlas stations are either very close to the coast or even at sea it is straight forward to use the Atlas for estimating offshore wind resources. In addition, each of the maps, like Fig.1, contains a column giving the conditions at open sea. In the next section the resources portrayed by the map will be verified against the statistics in the Atlas and results from some recent studies.

The wind resources in the coastal region. When the air moves from sea to land areas or vice versa, two effects are of major importance for wind resource climatologies, namely: the change of surface roughness and thermal surface properties. Well away from the coast, the wind climate is either of the maritime or inland type, but in between it is a mixture. The width of the coastal zone varies with climate and topography. The Wind Atlas assumes a width of 10 km on either side of the coast.

Figures 3 and 4 from Ref.1 show the change of mean energy density at a height of 50 meters calculated for two different coastal zones. The change is given as function of distance from the coast. A North Sea coast in the Netherlands and a Mediterranean coast in France are taken as examples. Generally, the shape of the curves depends on the geographical position and orientation of the coast as well as the roughness class and orography of the coastal terrain. The conditions of a specific coast must in general be calculated by an appropriate siting procedure (Ref.2).

In the first example from a North Sea coast in the Netherlands it is seen that the change to open-sea conditions takes place over approximately 10 km whereas the change to land conditions occurs more abruptly, that is over approximately 4 km. The second example shows the conditions at the French Mediterranean coastline in a region strongly influenced by the Mistral. The two coastlines are perpendicular and parallel to the Mistral, respectively.

3 Verification of the offshore and coastal wind resource estimates

The roughness of a particular surface area is determined by the size and distribution of the roughness elements it contains; for land surfaces these are typically vegetation, built-up areas and the soil surface. The roughness length of surfaces covered by vegetation may vary with the wind speed. For example, the bending of stalks by the wind can change the form of the surface. A similar phenomenon occurs for water waves where both the height and the form of the waves are dependent on wind speed. From dimensional arguments, the well known Charnock's relation (see Ref.1) can be found. The relation gives the surface roughness as a function of the surface stress and the gravitational acceleration.

In the Atlas it was attempted to use both Charnock's relation and a fixed value for the roughness of water areas. It turned out that a fixed value of 0.0002 m gave results as good as the relation for the moderate to high wind speeds of interest to the Wind Atlas, hence all statistics in the Wind Atlas are obtained with this value. It should be noted that over some water areas the roughness can vary strongly with the time of the day and from season to season under the influence of phenomena as tidal waters and ice cover.

The coastal zone. It is the general belief that the roughness over water in the coastal zone is higher than at the open sea. The reason should be that the wave field is never in equilibrium with the wind field and therefore extracts more energy from the wind. In the Atlas, however the same roughness is used for all water surfaces the justification being the verification against measured wind statistics. Of special importance was the validation against high meteorological masts placed very close to the sea. For each mast the data from the lowest measurement level have been used to predict wind distributions at the higher levels. Tables 1 to 3 list Weibull parameters and wind power density for measured and predicted wind distributions for the Portuguese Ferrel mast, the Swedish Nasudden mast and the Danish Sprogø mast. It is obvious that the models ability to explain the variation of the wind statistics with height is quite good. Hence the use of a roughness length of 0.0002 m is not contradicted. The mast data also provides a test of the skill of the model to handel the modifications of the wind profile due to the different stability conditions over land and sea.

The major part of the verification in the Atlas is performed by letting the surface stations predicting each others wind statistics. And because half of the stations are located in the coastal zone, the success of the model validation heavily depends of is ability to model local variations in this zone. The conclusion that can be drawn from the validations in the Atlas is that under reasonable moderate conditions the model performs very well. When the conditions are complex such as in some Mediterranean areas with strong thermal differences in the horizontal combined with steep reliefs then the model results might prove less reliable.

The open sea. The Wind Atlas for the North Sea and the Norwegian Sea (Ref.3) provides an overall picture of the wind conditions in the Northern waters of Europe. A picture which is in good accordance with that given by Figures 1 and 2. This can be seen from Fig.5 extracted from Ref.3: The average wind field in about 1000 m above sea level. The lines drawn are based on computed averages over a 27 years period from 1957 to 1981 for grid points with a spacing of 150 km. Four wind values per day are obtained from analyzing more than 39000 weather maps.

It is quite obvious that very few data sets exists from fixed installations at open sea. Those which can be found are often measured at production platforms, they are difficult to access, they are usually infested with large gaps of missing data and the influence of the platform on the wind flow to the anemometers are considerable and difficult to correct for. Nevertheless that's what is available and Ref.6 presents an analysis of eight years of data (1982-1989) from the platform K13 situated 100 km from the Dutch coast. At a height of 75 m a mean value of 9.0 m/s is found which compares well with Fig.2. The corresponding Weibull parameters are A = 10.25 m/s and k = 2.04; it is found that the Weibull distribution gives a good fit to the data. The average roughness length is found to be close to 0.0002 m/s.

Ref.7 gives a mean wind speed of 9.5 m/s measured over almost 4 years at 50 meters height at a gas platform located 50 km South of Roches Point (Irish Coast). Again, this value compares well with Fig.2.

A very comprehensive study has been undertaken by Deutsche Wetterdienst - Seewetteramt, Hamburg: "Wind and wave conditions in 55 european coastal sea areas, determined from weather and wave observations of voluntary commercials ships" (Ref.6). A large number of ships carry out meteorological and oceanographic observations along the route of the ships at fixed 3-hourly time intervals. The observations start as soon as a ship is out of the harbour; the determination of the wind speed is done by an estimation of the wind force according to the appearance of the sea surface and following the well known Beaufort scale. Numerous studies have been performed over the years in order to establish the correct conversion from the Beaufort scale to meters per second. In short, for the purpose of this paper it can be noted that a generally accepted scale exists and that the resulting wind speeds in m/s are considered representative for the winds 25 meters over the sea surface. Hence the statistics presented in Seewetteramt study can be compared directly with Fig.2. The comparison comes out very good, from the Northern seas to the Aegean Sea there are very few discrepancies. When offshore predictions from single station from the Wind Atlas are compared with statistics from the ships large discrepancies are found, especially for Portugal and Spain. But also very good agreements are found, for example where the Mistral meets over the Mediterranean Sea the Atlas station Istres predicts an energy density of 828 Wm-2 and the ships statistics 822 wm-2. Other examples of good agreement are Roches Point, Ireland (658 653), Rønne, Denmark (527 518), Middelkerke, Belgium (566 599), Brest, France (616 656), Faro, Portugal (334 387), Barcelona, Spain (243 282).

A careful comparison between the Wind Atlas predictions and the ship statistics could well lead to a subset of stations to be recommended and therefore it is the intention to do so. It remains to be mentioned that the interpretation of the ship statistics is not without problems - on the contrary - but that will be discussed elsewhere.

4 Conclusion

The European Wind Atlas was published in 1989, but work has been carried out since then for improving the knowledge og the European wind climatology. Comprehensive measurements have been initiated in several EC countries in mountainous terrain with a twofold purpose: to help the understanding and the modelling skill of the local distribution of the wind resources in such terrain and to gain general climatological knowledge. Preliminary analysis of some of the measurements leaves the impression that the overall distribution as given by Fig.1 will not be changed very much.

The distribution of the European wind resources over open water seems to be known to a sufficient degree of accuracy. It is however a question whether the open sea resources will ever be utilized. The coastal zone on the contrary attracts much more attention, but in these areas it is of much importance for accurate resource estimation that reliable wind statistics are utilized and that the siting procedure is capable off handling the influence on the wind profile of the land-water transitions and the topography of the coastal area.

5. References

[1] Mortensen, N.G., L. Landberg, E.L. Petersen, and I. Troen, 1992: *Wind Atlas Analysis and Application Program (WA^SP) User's Guide*. Risø National Laboratory, Denmark.

[2] Troen, I. and E.L. Petersen, 1989: *European Wind Atlas*. Published for the Commission of the European Communities by Risø National Laboratory. 656pp.

[3] Børresen, J.A., 1987: *Wind Atlas for the North Sea and the Norwegian Sea*. Norwegian University Press, The Norwegian Meteorological Institute. 183pp.

[4] Report TNO Environmental and Energy Research, 1991: *Description of the North Sea Wind climate for offshore wind energy applications*. 95pp.

[5] Burke, L., 1988: *EC Wind Energy Atlas: An Assessment of the Irish Results*. Work group document. 14pp.

[6] Schmidt, H., J. Püttker, 1991. *Wind and wave conditions in 55 European coastal areas, determined from weather and wave observations of voluntary commercial ships*. Report from Deutscher Wetterdienst to Germanischer Lloyd. 30pp.

Table 1: Ferrel mast, Portugal. The data cover the two years 1977-78. Located at the coast 10 km northeast of Cabo Carvoeiro. The mast was situated 3.5 km northeast of the village of Ferrel. The distance to the sea is 300 m to the northwest. To the southeast the landscape is undulating and covered by vegetation. Close to the mast the terrain consists of sand dunes.

Height	Measured			Predicted		
z	A	k	E	A	k	E
10 m	5.8	1.83	171	5.8	1.83	171
30 m	7.2	2.09	289	7.0	1.94	286
100 m	8.6	2.33	450	8.4	2.04	467

Table 2: Näsudden mast, Sweden. Data cover the years 1980-85. The mast is located on the small peninsula Näsudden on the island of Gotland. The distance to the Baltic sea is a little more than one kilometre in directions from south to southwest. Near the mast the terrain consists of low dense bushes and scattered trees. Further away from the mast the terrain is hilly and forested.

Height	Measured				Predicted			
z	M	A	k	E	M	A	k	E
10 m	5.1	5.8	1.99	157	5.1	5.7	1.96	154
38 m	7.0	7.8	2.02	381	7.0	7.9	2.10	385
54 m	7.5	8.4	2.10	465	7.7	8.6	2.13	493
75 m	8.0	9.0	2.10	555	8.1	9.1	2.13	585
96 m	8.4	9.5	2.12	647	8.4	9.5	2.12	654
120 m	8.6	9.7	2.13	692	8.8	9.9	2.10	750
145 m	9.1	10.3	2.15	818	9.1	10.3	2.08	845

Table 3: Sprogø mast, Denmark. Data cover 4 years (1983-88). The mast is situated on the small island of Sprogø in the middle of Storebælt (Great Belt). The mast stands on a long narrow spit of land only 50 m wide and extending 300 m east of the island proper. Except for the 240°-270° sector, the approach to the mast is over several kilometres of open water. The distance to the islands of Sjælland to the east and Fyn to the west is aproximately 10 km.

Height	Measured			Predicted		
z	A	k	E	A	k	E
8 m	7.1	2.02	289	7.0	2.01	282
18 m	7.6	2.07	342	7.8	2.14	357
55 m	8.9	2.26	513	8.9	2.33	499
68 m	9.2	2.31	547	9.2	2.31	555

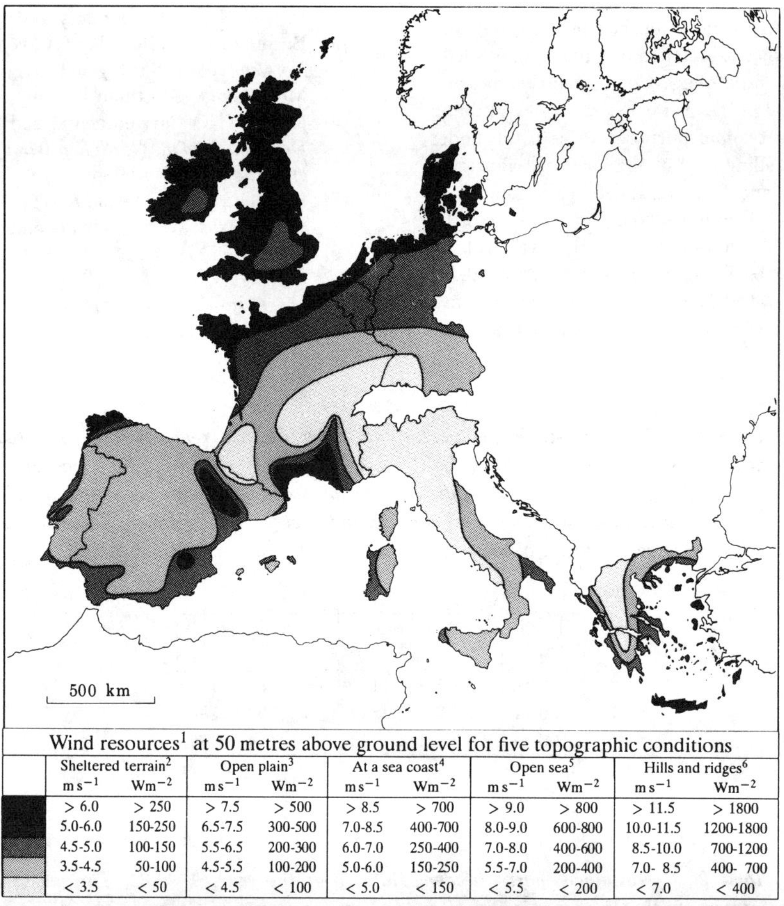

Wind resources[1] at 50 metres above ground level for five topographic conditions

	Sheltered terrain[2]		Open plain[3]		At a sea coast[4]		Open sea[5]		Hills and ridges[6]	
	$m\,s^{-1}$	Wm^{-2}	$m\,s^{-1}$	Wm^{-2}	$m\,s^{-1}$	Wm^{-2}	$m\,s^{-1}$	Wm^{-2}	$m\,s^{-1}$	Wm^{-2}
	> 6.0	> 250	> 7.5	> 500	> 8.5	> 700	> 9.0	> 800	> 11.5	> 1800
	5.0-6.0	150-250	6.5-7.5	300-500	7.0-8.5	400-700	8.0-9.0	600-800	10.0-11.5	1200-1800
	4.5-5.0	100-150	5.5-6.5	200-300	6.0-7.0	250-400	7.0-8.0	400-600	8.5-10.0	700-1200
	3.5-4.5	50-100	4.5-5.5	100-200	5.0-6.0	150-250	5.5-7.0	200-400	7.0- 8.5	400- 700
	< 3.5	< 50	< 4.5	< 100	< 5.0	< 150	< 5.5	< 200	< 7.0	< 400

1. The resources refer to the power present in the wind. A wind turbine can utilize between 20 and 30% of the available resource. The resources are calculated for an air density of 1.23 kg m^{-3}, corresponding to standard sea level pressure and a temperature of 15°C. Air density decreases with height but up to 1000 m a.s.l. the resulting reduction of the power densities is less than 10%, see Table B.1 in Appendix B.
2. Urban districts, forest and farm land with many windbreaks (roughness class 3).
3. Open landscapes with few windbreaks (roughness class 1). In general, the most favourable inland sites on level land are found here.
4. The classes pertain to a straight coastline, a uniform wind rose and a land surface with few windbreaks (roughness class 1). Resources will be higher, and closer to open sea values, if winds from the sea occur more frequently, i.e. the wind rose is not uniform and/or the land protrudes into the sea. Conversely, resources will generally be smaller, and closer to land values, if winds from land occur more frequently.
5. More than 10 km offshore (roughness class 0).
6. The classes correspond to 50% overspeeding and were calculated for a site on the summit of a single axisymmetric hill with a height of 400 metres and a base diameter of 4 km. The overspeeding depends on the height, length and specific setting of the hill.

Figure 1: Distribution of wind resources in Europe. By means of the legend the available wind energy at a height of 50 meters can be estimated for five topographicconditions, (from Ref. 1).

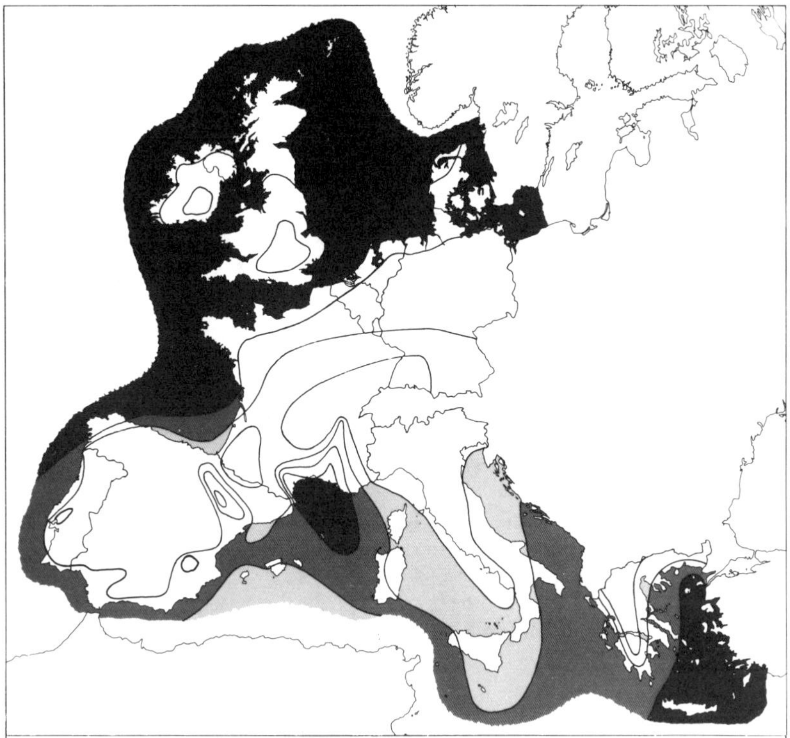

Wind resources over open sea (more than 10 km offshore) for five standard heights										
	10 m		25 m		50 m		100 m		200 m	
	$m\,s^{-1}$	Wm^{-2}	$m\,s^{-1}$	Wm^{-2}	$m\,s^{-1}$	Wm^{-2}	$m\,s^{-1}$	Wm^{-2}	$m\,s^{-1}$	Wm^{-2}
	> 8.0	> 600	> 8.5	> 700	> 9.0	> 800	> 10.0	> 1100	> 11.0	> 1500
	7.0-8.0	350-600	7.5-8.5	450-700	8.0-9.0	600-800	8.5-10.0	650-1100	9.5-11.0	900-1500
	6.0-7.0	250-300	6.5-7.5	300-450	7.0-8.0	400-600	7.5-8.5	450-650	8.0-9.5	600-900
	4.5-6.0	100-250	5.0-6.5	150-300	5.5-7.0	200-400	6.0-7.5	250-450	6.5-8.0	300-600
	< 4.5	< 100	< 5.0	< 150	< 5.5	< 200	< 6.0	< 250	< 6.5	< 300

Figure 2: Distribution of wind resources in Europe over open sea. The table is calculated by means of figures 4.2 and 4.3 in Ref. 1.

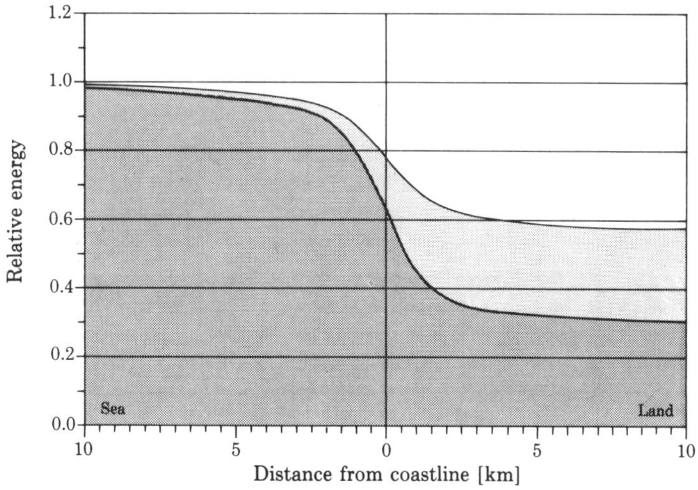

Figure 3: Mean energy density at a height of 50 m calculated as a function of the distance to the North Sea coast in the Netherlands for two different land roughnesses. The coast is oriented SW-NE and the terrain is of roughness class 1 (upper line) and roughness class 3 (lower line), respectively, (from Ref. 1.).

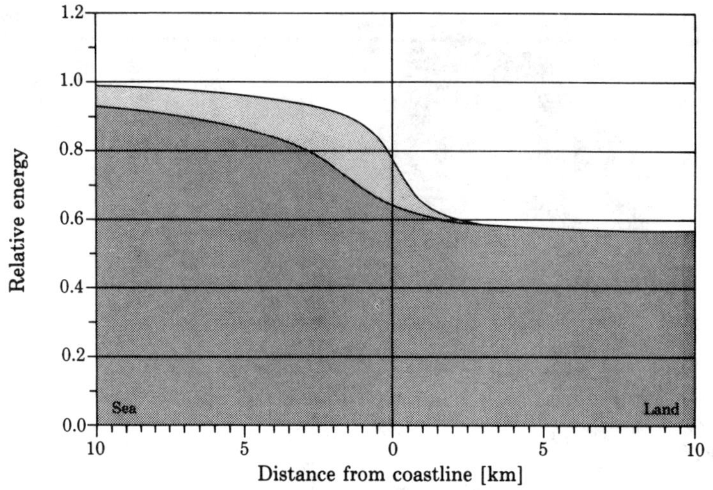

Figure 4: Mean energy density at a height of 50 m calculated for two coastlines as a function of the distances to the Mediterranean Sea in France. the two costalines are situated in the region influenced by the Mistral; one is perpendicular (lower line) the other parallel (upper line) to the direction of the Mistral. The terrain is of roughness class 1, (from Ref 1.)

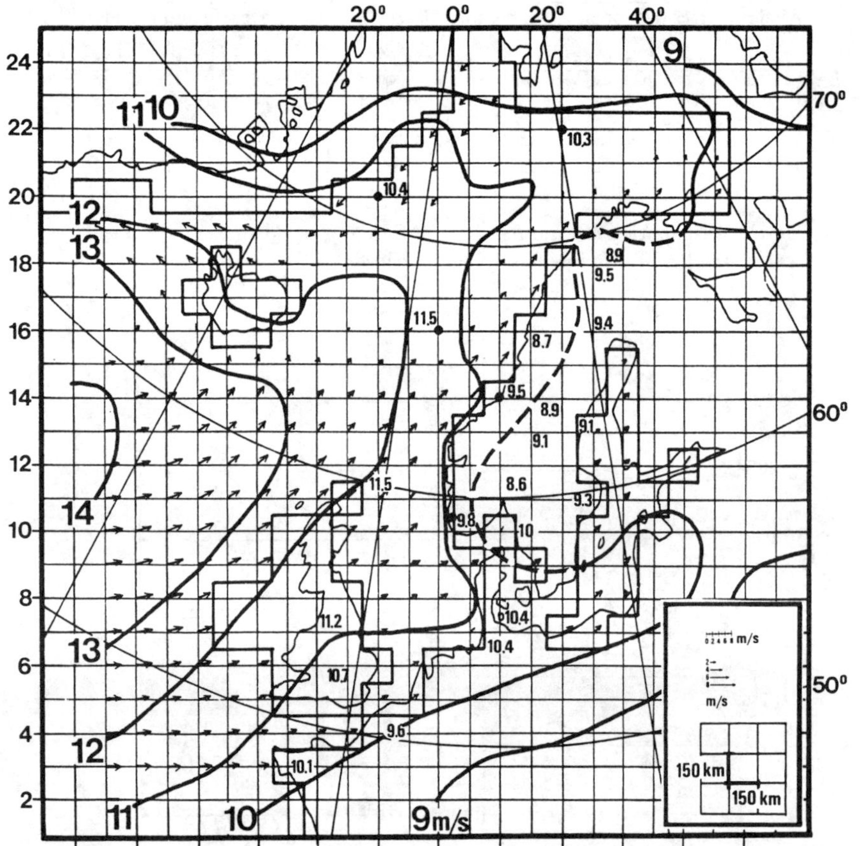

Figure 5: The average wind field in about 1000 m above sea level. The lines drawn are based on computed averages over a 27 year period from 1957 to 1981 for all grid points. Four wind values are available per day. The inserted figures are averages from measurements at the 850 mb level made by radiosondes (Ref. 3).

WIND AND WAVE CONDITIONS IN 55 COASTAL SEA AREAS OF THE EUROPEAN COMMUNITY DETERMINED FROM WEATHER OBSERVATIONS OF VOLUNTARY SHIPS

Heiner Schmidt

Deutscher Wetterdienst
-Seewetteramt-
Postfach 30 11 90
D 2000 Hamburg 36

ABSTRACT: Wind speed and wave height estimations from ships were used to calculate the mean wind speed and the mean potential wind power production rate as well as the extreme hourly mean wind and wave height with a mean recurrence period of 50 years. All wind parameters are given for a height of 25 m above sea level. Some problems in using ship data are discussed.

1. INTRODUCTION

The Germanische Lloyd ordered an expertise from the Seewetteramt (marine weather office) of Deutscher Wetterdienst about some statistical wind and wave parameters determined from marine weather observations. It will be used for an investigation on the possibilities of wind energy utilization in European coastal sea areas.

For this purpose the long range average and the frequency distribution of the wind speed had to be determined for the calculation of the wind power potential and the mean power production rate. Additional knowledge is required about mean and extreme wind and wave conditions for the constructional design of the supporting off-shore platforms.

2. USING SHIP OBSERVATIONS

The Seewetteramt keeps a large archive of nearly 50 millions of data sets containing marine meteorological observations from all parts of the world. These are voluntary observations (mostly done by the nautical officers) from all kinds of ships. They are carried out more or less routinely. The observations are coded according to the international regulations of the World Meteorological Organisation (WMO).

The data sets consist of meteorological parameters (weather state, air pressure, wind speed and direction, air and water temperature and others), of sea state parameters (wave height and period), and of organisational data (date, time, position). Some problems with naval wind and wave data will be discussed in the following.

2.1 Wind velocity

In most cases the determination of the wind speed is done by an estimation of the wind force according to the appearance of the sea surface, i.e. the wind action upon the sea. The so called "Beaufort" forces are estimated, which are discernible by well-defined different sea surface states. A direct estimation of the wind speed (by feeling) without sight of the sea surface is impossible.

In contrast wind measurements are very problematic, we therefore excluded them for this expertise. (Problems are: unknown disturbance by the ship's body; partly incorrect, insufficient or omitted vector subtraction of ship's course and speed; unknown measuring height)

According to the WMO regulations wind values are not coded and transmitted in Beaufort, but as speeds in metric units. Observers are held only to encode the so called "Beaufort equivalent values", average wind speeds for the Beaufort classes, but in reality they also set wind values in between. But (even if interpolated) the wind speed values always result from the nonlinear relation between wind speed and Beaufort forces, assigned by WMO in 1949. During the preceding 20 years it became obvious, that the WMO scale is slightly incorrect.

On the average the wind speed increases with the height above sea surface. It thus depends on height, whereas the wind force (the wind action at 0 m height) does not. For the assignment of speeds to forces therefore the reference height is essencial. The WMO scale was designed for a height of 10 m.

For this expertise we used a relation developed by KAUFELD [4]. A reanalysis resulted in a continuous transformation of estimated wind speed values into true wind speeds at 25 m height above sea level, see fig.1.

For each coastal sea area the frequency distributions of the original estimated wind speeds in the range of 0 to 33 m/s were counted. The boundaries of the wind speed classes were then transformed into true wind speeds at 25 m height using the above mentioned transformations. Afterwards the distributions were converted to empirical probabilities of exceedance by normalising with the total amount of observations.

2.2 Wave heights

The meteorological data sets also contain information about wave height and period. They are given in two categories, one for the wind waves, the other for the swell. Both are clearly discernible by their definition, but in practice often they are not. We took both entries as two different and independent observations (events), which seems to enhance the statistical stability. The frequency distributions of wave heights are treated like those of the wind speed as described before.

The wave height reported and coded after the WMO regulations is the so called

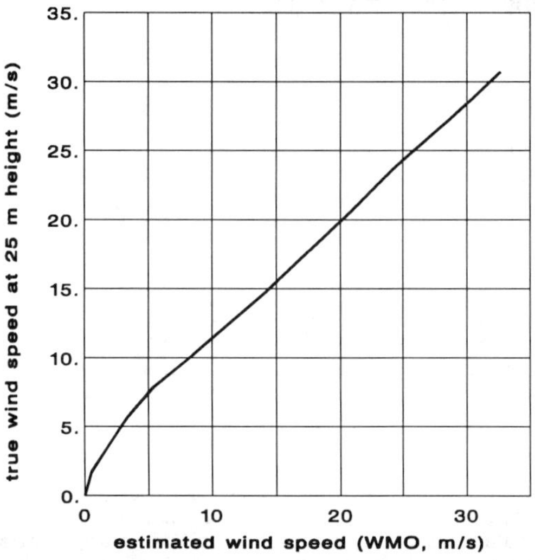

Figure 1: True wind speeds at 25 m height above sea level versus estimated wind speeds (Beaufort estimates encoded to m/s according to the WMO scale) from ship observations

significant wave height $H_{1/3}$, which is the average of the upper third of all waves. The estimation of $H_{1/3}$ is feasible als long as roll and pitch of the ship is small. But from a wind wave height of about 6 m on (which is connected to a typical wave period of 8 seconds and a wave length of 100 m) the ship's movements change drastically. The estimation of $H_{1/3}$ becomes more difficult, and for subjective reasons the observers with increasing wave height tend to estimate the height of peak waves, H_{max}, instead of $H_{1/3}$. Peak waves are always higher than the significant wave heights, a typical factor is 2. For extreme waves this factor may be even larger.

2.3 Approximation by the Weibull distribution

The empirical distributions of wind speed and wave height were approximated for each coastal sea area by a two-dimensional Weibull distribution. With the inverse function the wind speeds or wave heights are easily calculated from the probability of exceedance. In this way the wind speeds and wave heights with 50 years recurrence period were calculated (see below).

Frequency distributions of the wind speed at sea normally cannot exactly be described by a Weibull distribution. The empirical distributions are slightly curved, such, that the wind speeds belonging to small and high probabilities of exceedance are a bit greater than for a Weibull distribution. For this reason the Weibull distributions were only fitted in a limited wind speed range.

This range was chosen to be different for wind energy calculations (range 3-20 m/s) and for extreme value estimation (range 7-26 m/s). The speed range above 26 m/s was excluded from the fitting routine. The randomness of these rare events may otherwise lead to unrealistic results. It must be emphazised however, that these data have **not** been excluded as errors. They were counted as valid events while calculating the frequency of exceedance.

For the above mentioned reasons the Weibull approximation of wave height distributions was only performed in the range between H=0.5 m and about 6 m (probability 0.01). The 1 percent of highest waves (with their excessive heights) thus only determine the approximation by their number, but not by their height.

2.4 Estimation of extreme values

In order to be able to determine an extreme wind speed or wave height with an average return period of 50 years empirically, a continuous time series of all events (or the annual extremes) of a time period of at least 150 years would be required. Such time series do not exist at sea.

Unfortunately even for ship observations from 40 years none of the other well known methods of extreme value estimation can be applied. The reason is the method of sampling, which is extremely incoherent in time. Annual extrema cannot be deduced. In fact, our data can not at all be regarded as continuous time series.

Additionally, the application of extreme value analysis is further restricted by the fact, that the highest class of wind force, Beaufort 12, is an open class. Any wind speeds higher than its lower boundary (30.7 m/s) cannot really be distinguished by estimation. And at sea Beaufort 12 is more frequent than once per year almost everywhere.

Consequently we had to use another approximative solution. The empirical frequency distributions of wind and waves in the 55 coastal sea areas were determined from the observations of 1950-1990. Due to the strong variations of the total numbers in different areas and due to the above mentioned discontinuity in time the distributions may only be interpreted as empirical probabilities. But generally we assume the empirical distributions to be sufficient samples for the determination of the main statistical properties of the underlying stochastic process "wind" (or "waves").

After approximation by a proper distribution function (model) they can be used in a limited way for an extrapolation down to relatively low probabilities. This implies the additional assumption, that the extreme values are generated by the same process as the bulk of the values. For the European coasts this assumption seems to be valid.

The calculation of wind or wave values with an average recurrence of once per 50 years from a probability function requires the definition of one event. In the case of the wind speed (which results from an estimation of the wind force) we take the average time needed for the adjustment of the sea surface to a different Beaufort force as one event. This time is approximately one hour. We therefore interprete the estimated wind values of our archive as hourly averages.

A year now consists of 8766 events (hours). To get the wind with an average return period of 50 years the fitted probability functions have to be extrapolated down to a probability of 1/(50*8766). The conversion of a low probability of exceedance into a long return period is only valid under certain assumptions. The most restricting one is, that the wind must not have any long time trend. (We are not certain about this).

The development and decay of waves deserves a longer time than the adjustment of the sea surface to a new Beaufort force. We therefore define a "significant wave height event" by a time interval of 6 hours. The annual number of events is now 8766/6=1461.

3. RESULTS

Some of the results are presented in maps within the following text. Along the coasts of the European Community 55 coastal sea areas were defined. Each area is a group of at least two 1-degr-squares. The arrangement of the coastal sea areas takes care of known or supposed borders of different climatological or geographical influences. The areas were chosen to be relatively large where the data are few.

3.1 Mean wind speed

All results for wind parameters given in the following are valid for a height of 25 m above sea level. The annual mean wind speed (fig. 2) is insufficient for an estimation of wind power potential, but it gives an impression of the general wind conditions. It can easily be compared with other investigations.

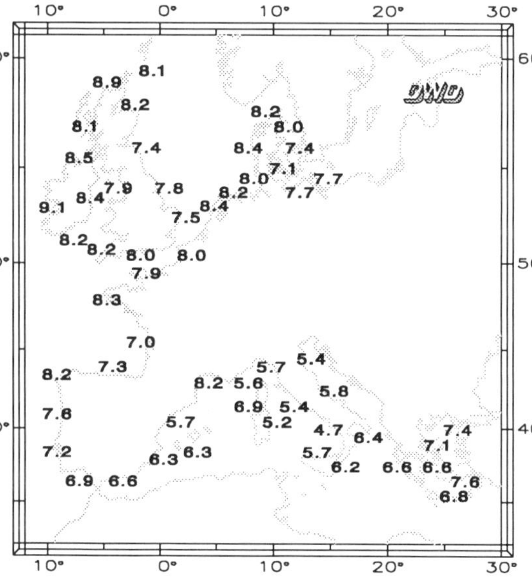

Figure 2: Mean wind speed (m/s) in coastal sea areas of the European Community. Height 25 m above sea level.

The highest wind speeds are found west of Ireland and north of Scotland. Speeds are only slightly lower near the capes of the continental European coasts. Lowest values are found in the shielded areas of the Mediterranean Sea. Rather windy is the Aegean Sea, as is well known.

The highest wind speeds of the Mediterranean Sea occur in the Golfe du Lion. The reason is a twofold jet-like enhancement of the wind between the Pyrennees, the Massif Central and the Alpes. Both wind systems flow into the Golfe du Lion and run out between Sardinia and the Baleares.

3.2 Wind energy

For this purpose a Weibull distribution was fitted in the wind speed range of 3-20 m/s. From its shape and scale parameters the wind energy flowing through an area of 1 m^2 perpendicular to the wind was calculated. Fig. 3 gives the mean (potential) power production rate (W/m^2). The regional distribution is similar to that of the mean wind speed.

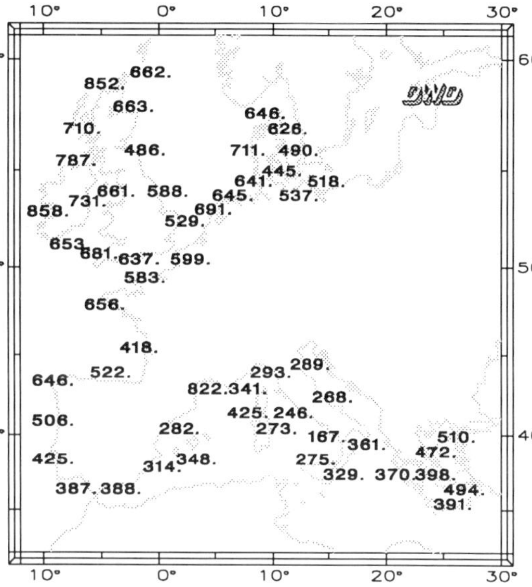

Figure 3: Mean power production rate (W/m^2) in coastal sea areas of the European Community. Height 25 m above sea level.

The wind power potential does not depend upon machine specific properties of planned wind converters. The calculation of the real power production requires a new integration taking into account the speed dependancy of the machine's efficiency factor and the real rotor area.

3.3 Wind speeds with 50 years return period

In this case the empirical distributions were fitted in the speed range of 7-26 m/s using the Weibull distribution for extrapolation as described above. The hourly wind values with a mean return period of 50 years are shown in fig.4.

The values for the Mediterranean areas are only slightly lower than those for the north west coasts of Europe. The reason is an enhanced "flatness" of the Mediterranean wind speed distributions, which is described by low shape parameters of the Weibull distribution, and results from a relative increase of high wind speeds. This is only partly compensated by lower scale parameters. We regard this general result as climatologically reasonable.

The statistical security is estimated to be about ±10-15% for areas with more than 50,000 observations. In areas with low numbers (<7,000) the security decreases drastically. The results of these areas (Aegean Sea (west), Adriatic Sea (north), Sicily (north side), Bay of Biscay (south), Kiel Bight and southerly Baltic) should not be used, and better be replaced with physical care from neighbouring values.

The wind speed results can be compared with other investigations in some regions,

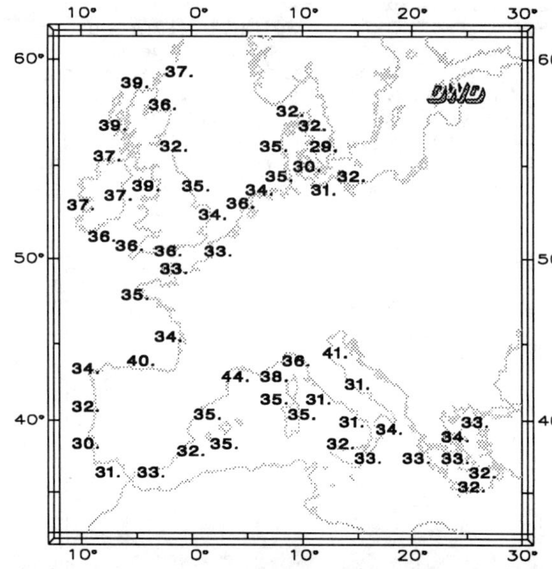

Figure 4: Hourly mean wind (m/s) with a mean recurrence period of 50 years in coastal sea areas of the European Community. Height 25 m above sea.

these are for the German Bight: [1] and [3], for the whole North Sea: [2], and for the Netherlands: [6]. The "European Wind Atlas" [5] can be used for a comparison, too. Its results for roughness class 0 ("open sea") and a height of 25 m from coastal stations can be compared to ours, provided that the average coastal distance of the observations in our sea areas is large enough (>20 km). The agreement is generally very good.

3.4 Wave heights with 50 years recurrence period

All wave heights in this expertise are significant wave heights ($H_{1/3}$). The determination procedures have been explained above. The Weibull distribution was fitted in the wave height range from 0.5 to about 6 m. Similar to the case of the wind in the Mediterranean Sea there are smaller scale parameters (and mean wave heights) and smaller shape parameters. This combination again leads to relatively high extreme significant wave heights with 50 years return period which are shown in figure 5.

For some coastal sea areas the meaning of the resulting significant wave heights with 50 years return period is somewhat limited by the following: The ordered size of this expertise did not include any investigation about directional dependancy of wind or waves. There are sea areas where the prevailing wind with high speeds is directed from land to sea. The coastal distance (the fetch) will generally be sufficient for a development of significant wave heights in the Weibull fitting range of 0.5-6 m, but in some cases be insufficient for higher waves as extrapolated for a recurrence of 50 years.

An example for the limitation is the Golfe du Lion where we calculated 13 m wave height. The prevailing wind direction is north west. The distance between the average position of the observations and the coast is 80-100 km. From empirical relations between wind speed, fetch and wave height (e.g. [7]) with this short fetch no significant wave height higher than 10 m can be deduced, even if the wind blows with 40 m/s over a longer time. The calculated height of 13 m thus cannot occur.

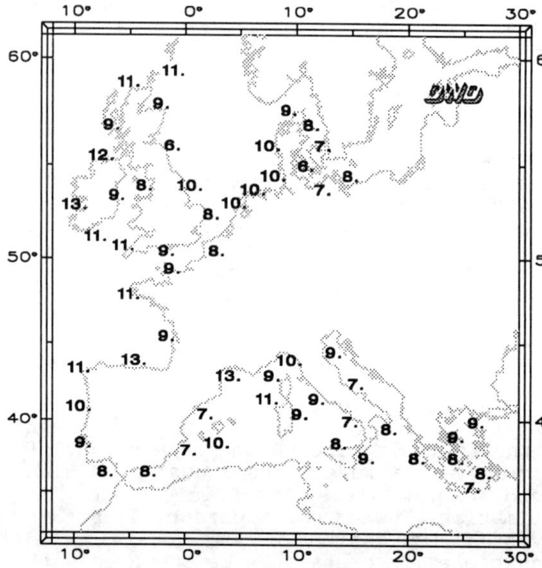

Figure 5: Significant wave heights (m) with a mean recurrence period of 50 years in coastal sea areas of the European Community.

REFERENCES

[1] Christoffer, J. and M. Ulbricht-Eissing (1989): Die bodennahen Windverhältnisse in der Bundesrepublik Deutschland. Berichte des Deutschen Wetterdienstes Nr. 147; Offenbach am Main.

[2] Børresen, J.A. (1987): Wind Atlas for the North Sea and the Norwegian Sea. Norwegian University Press; Oslo.

[3] Duensing G. et al. (1985): Meteorologische Untersuchung über Stürme an der deutschen Nordseeküste. Deutscher Wetterdienst, Einzelveröffentl. des Seewetteramtes Nr. 108; Hamburg.

[4] Kaufeld, L. (1981): The Development of a new Beaufort Equivalent Scale. Meteorologische Rundschau, **34**, 17-23; Gebr. Bornträger, Berlin, Stuttgart.

[5] Troen, I. and E.L. Petersen (1988): European Wind Atlas. Risö National Laboratory; Roskilde.

[6] Wieringa, J. and P.J. Rijkoort (1983): Windklimaat van Nederland. KNMI, Staatsuitgeverij; Den Haag.

[7] WMO (1988): Guide to Wave Analysis and Forecasting. World Meteorological Organization (WMO) publ. No. 702; Geneva.

EXPERIENCES OF USING WAsP AND ON-SITE-MEASUREMENTS FOR SITING A WIND FARM

H Holttinen and E Peltola
VTT Technical Research Centre of Finland
Lab. of electrical and automation eng.
PO Box 34
SF-02151 Espoo FINLAND

ABSTRACT The WAsP-method together with the Finnish Wind Atlas and on-site measurements have been used to assess the wind energy resources on 6 sites on the western coast of Finland. The WAsP-method seems to give reasonably accurate results on the case sites. Using the monthly mean wind speeds in the wind atlas and mean temperatures, estimates for the production each month have been made, too. The temperature correction of the power curve of a wind turbine increases the annual production estimate by about 5 % in coastal Finland.

1. INTRODUCTION

A correct production estimate for a single wind turbine or a wind farm is of vital importance for the project developer. In case of a utility it is also a new element beside a common cost breakdown of the investment. As a reliable estimate of the production is known, the developer is able to calculate, whether the project is worth the investment or not.

The WAsP-method is probably the most commonly used tool for siting wind turbines. The WAsP-method uses the European Wind Atlas prepared for European Community and similar national atlases such as the Finnish Wind Atlas [1] as its database.

The WAsP-method as well as the Finnish Wind Atlas is originally designed for calculating average annual wind conditions and wind energy production based on them. The value of wind energy depends, however, much on the seasonal as well as the diurnal variations of the production and the consumption and their coincidence. The seasonal variation of the production and the accuracy of the estimates in general can be checked by on-site measurements.

2. THE SITES

This paper is based on the experience from the cases given in Table 1. The sites are all located on the western coast of Finland as shown in Fig. 1.

The Korsnäs project has already been developed thus giving an opportunity to make comparisons between monthly production estimates and real production.

In the case of Kokkola, one proposed site was studied for Kokkola Energy Works. The project is so far postponed but may be developed later although the site is not as good as the the other presented cases. In this case, much attention was paid to the comparison of results given by different wind atlas files.

The site in Kemi is located near the coast of a larger island. The meteorological station is located about 2 km S-SW from the site giving thus a good base for the calculations. On-site measurements have been started in early 1993 and the turbines are to be constructed by summer -93.

The three sites in the distributing area of Revon Sähkö distributing utility were chosen based on an inquiry made on the local municipalities. Additional measurements were started because there is a gap between existing meteorological stations in the area. The sites of Kalajoki, Siikajoki and Hailuoto are to be constructed in 1993.

Table 1: The studied sites.

SITE or PROJECT	Scope of assessments	Present status of the project
KORSNÄS	One site, annual and monthly energy production	4 * 200 kW operational since November/1991
KOKKOLA	One site, annual production	Postponed
KEMI	One site, annual and monthly production	3 * 300 kW to be constructed in 1993
REVON SÄHKÖ LTD	1. Annual and monthly production for 3 sites 2. Measurements compared to WAsP	6 * 300 kW to be constructed in 1993 (Kalajoki, Siikajoki, Hailuoto)

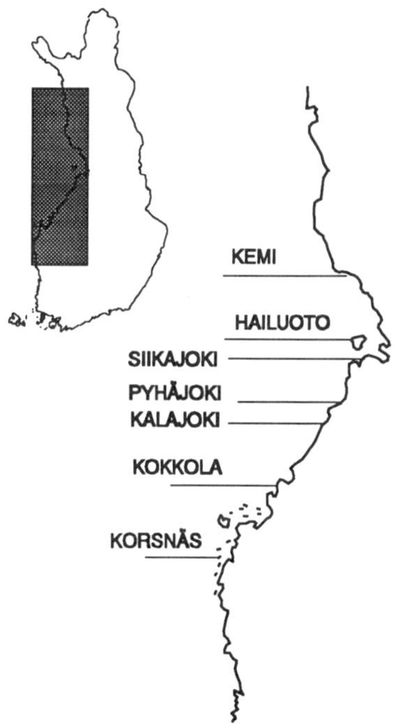

Figure 1. The cases used as examples.

3. ESTIMATING MONTHLY PRODUCTION USING WAsP-MODEL

WAsP-model uses three kinds of input for average wind speed and yearly production estimate : a wind atlas with the average wind speed distribution parameters, a site description with roughness, obstacle and orography descriptions, and additionally a power curve for the annual production estimate.

To attain the monthly production estimates the wind atlas and the power curve files were scaled to achieve 12 monthly files. The roughness file could be changed, too, to take into account the differences in the terrain - iced up sea has a different roughness compared to open sea, and the woods are also different wintertime with no leaves on trees etc. - but this has not been done for the time being.

Two auxiliary programmes for PC have been developed to have the monthly production estimates in as a simple way as possible with WAsP. Their principles are explained in 3.1 and 3.2.

3.1 Monthly wind atlas files

The model uses the wind speed statistics as distributions. If we assume a Weibull distribution for wind speed, the whole data can be described with two parameters A and k as

$$P(\bar{v} > \bar{v}_x) = e^{-(\frac{\bar{v}_x}{A})^k} \quad (1)$$

One wind atlas file is comprised of Weibull parameters for different heights and roughnesses. These wind atlas parameters were scaled with monthly mean wind speeds to attain monthly wind atlas files.

Only the scale parameters A were changed while the shape parameter k in (1) as well as the wind direction distribution were kept as such. Thus the differences in monthly wind direction distributions were not taken into account in this rather simple procedure, which in some cases may result to some errors in results.

For every month, the new Weibull parameters A' were produced as the product of old parameters A (for the whole year) and the ratio of the month's mean wind speed and the annual mean wind speed

$$A' = A \frac{\bar{v}_{month}}{\bar{v}_{year}} \quad (2)$$

This resulted in 12 different wind atlas files, one for each month.

3.2 The temperature dependence of power curves

To correct the power curve according to the mean air density of each month the method presented in IEA Recommended practices [2] for calculating the power curve in standard air density at sea level of 1.225 kg/m^3 was used in reversed direction. Thus, the power values of the power curve were changed according to

$$P_{month} = P_{STD} \frac{\bar{\rho}_{month}}{1.225} \quad (3)$$

The monthly mean air density values ρ_{month} are calculated as

$$\bar{\rho}_{month} = 1.225 \frac{288.15}{\bar{T}_{month}} \frac{\bar{B}_{month}}{1013.3} \quad (4)$$

In figure 2 the power curves corresponding the standard conditions (lower curve) and mean temperature of -12 °C (January in Kemi) are shown for 300 kW stall regulated wind turbine. The difference between the two curves is about 10 %, giving a signinficant contribution to the energy output.

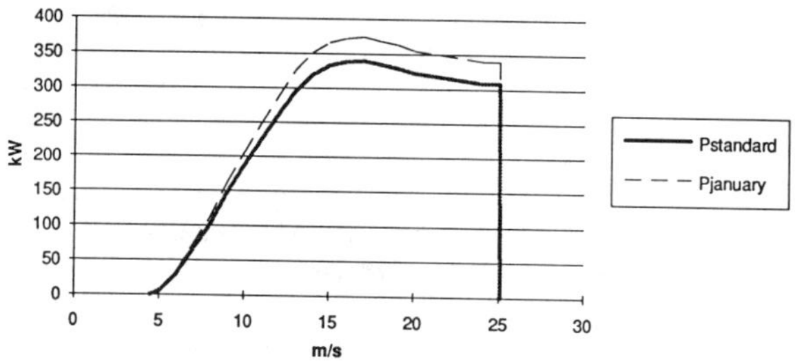

Figure 2. The power curve in standard conditions (+15°C, 1013 mbar) compared to power curve in the average temperature of January in Kemi (-12°C, 1011 mbar).

For pitch regulated wind turbines the correction (2) applies only to power levels less than 70 % of rated power. For power levels above 70 % of rated power the correction should be made to wind speed instead of power values, and the correction ratio should be cube root of the ratio of the standard air density and monthly air density [2].

3.3 Results

To obtain the monthly distribution of wind power production the WAsP-model is run 12 times using the files for wind atlas and power curves calculated for each month. The 12 results from the model are results for the whole year, so the results must then be reduced to one twelfth of the WAsP-results to apply for each month.

Using the altered wind atlas files doesn't change the overall yearly production estimate (difference is less than 1 %), whereas using the altered power curves affects the production estimate - the colder the climate, the greater air densities and thus production.

The air density corrections for sites in Middle Finland (see Fig 1, the sites are between 62° and 65° latitude), and the correction gives 4 - 6 % greater annual production depending on how far North we go (annual mean temperatures between +1° - +3°C).

The effect of the correction is shown for Kemi and Kalajoki in figures 3 and 4 . The monthly WAsP-runs using the standard power curve but 12 scaled wind atlas files gives the shape of the seasonal variation of wind power production without altering the overall yearly production estimate. When using additionally 12 monthly power curve files the production estimates for each month become greater. The correction is negligible for the summer months, when the average temperature is between +12 - +15 °C, but is about 10 % for winter months with average temperature about -10°C.

3.4 Comparison with measured production

Korsnäs windfarm, 4 units 200 kW each, was erected in November 1991. The average monthly production estimate and production in 1992 is shown in figure 5.

The production in 1992 has been 1640 MWh. Estimated power production was 1600 MWh (availability and wake effect together 93 %). Even though the estimate for the annual production was correct within 5 %, at least for the first year, the monthly values have varied from the estimate as much as -50 to + 50 %.

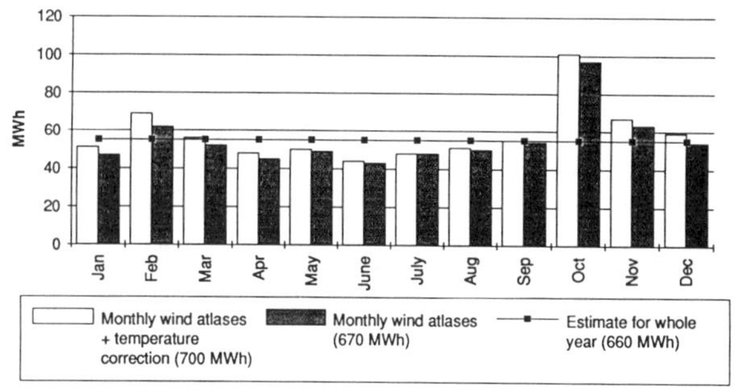

Figure 3. The monthly average wind power production estimated in Kemi Ajos.

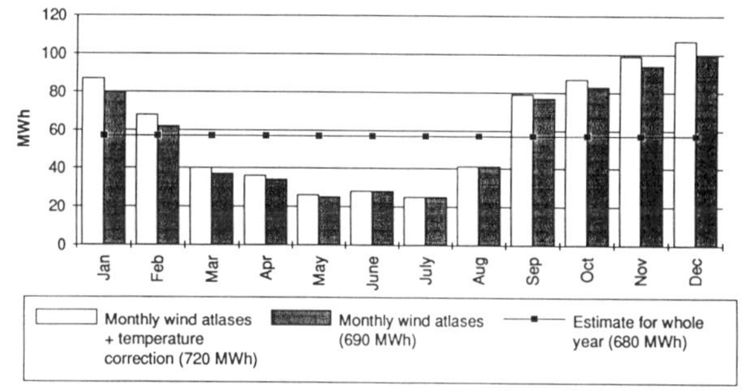

Figure 4. The monthly average wind power production estimated in Kalajoki Rahja.

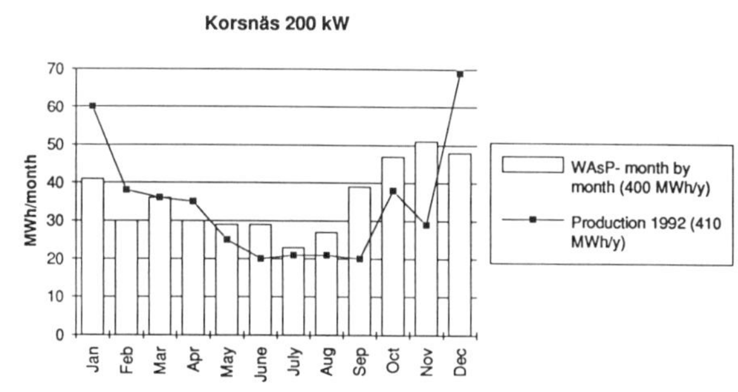

Figure 5. The monthly wind power production in Korsnäs versus WAsP-estimate.

4. CHOOSING THE WIND ATLAS FILE

In some cases a meteorological station to be used as wind atlas input cannot be found near the site. It is sometimes necessary to choose the wind atlas used from two or three ones, each for example 30 - 50 km away. Two wind atlas files from close-by stations should, in principle, give the same result in WAsP-model, because the meteorological data is processed when producing a wind atlas file by modelling of he surrounding terrain. In practise, however, this is not always the case.

In Kokkola, our next example, we had to choose from 3 stations, one an airport station (Kruunupyy about 20 km South) and two stations situated on small islands quite far from the coast (Mässkär about 30 km South-West and Ulkokalla 55 km North). The different atlas files gave annual energy production of 380 MWh (Mässkär), 400 MWh (Kruunupyy) and 420 MWh (Ulkokalla) for a 200 kW turbine.

To investigate closer which one to choose we looked at the distribution of wind direction at each of the stations - this is something that is not modified, when a wind atlas file is produced from the measured wind data. Our site was on the coast, totally open to the North and having the town Kokkola about 5 km to the South. The distributions of wind direction at the meteorological stations are shown in Figures 6.

Without any measurements it is difficult to say which of these wind direction distributions would be the closest to what is actually found at the site. The lowest estimation from Mässkär seems to result from a wind direction distribution that is not favourable to our site, reducing the amount of energy available from the North.

Finnish Meteorological Institute has made some corrections to the Finnish Wind Atlas files at the end of 1992. The files where run using WAsP version 3.0 (instead of WAsP 2.0 as they were before). Also some systematic errors (especially found in the inland stations' data) were corrected. The new files give somewhat different results. Now the wind atlases Kruunupyy and Mässkär give approximately same yearly production estimate (350 MWh) and Ulkokalla still an estimate of more than 10 % greater energy production. In this case it is quite easy to choose the best estimate, as Ulkokalla is situated furthest away.

5 ON-SITE MEASUREMENTS AND WAsP-MODEL

In Pyhäjoki, located in the area of Revon Sähkö Utility, measurements were carried out for 11 months from July 1991 to June 1992. The measuring site was chosen to be in the middle of Utility's long coastline between the existing meteorological stations, Ulkokalla 35 km South-West and Hailuoto 70 km North.

The monthly average wind speeds and wind direction distributions from the measuring period are shown in figure 7.

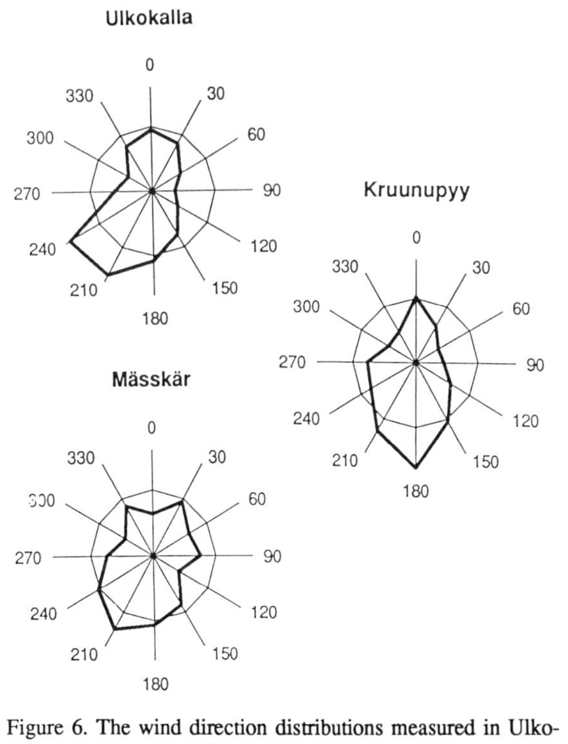

Figure 6. The wind direction distributions measured in Ulkokalla (a), Kruunupyy (b) and Mässkär (c).

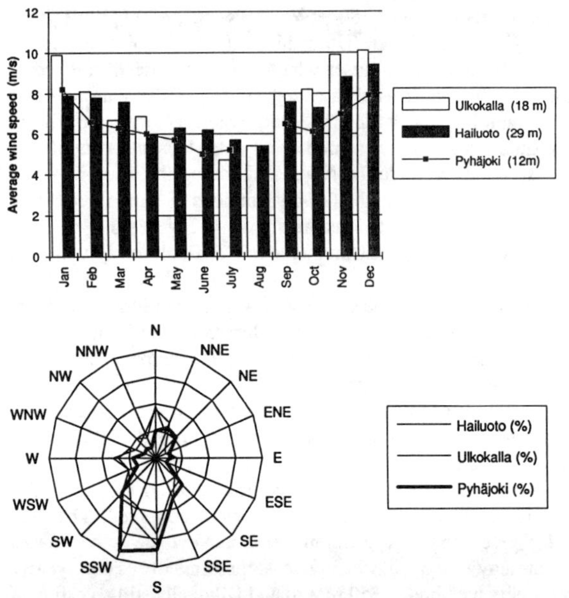

Figure 7. The measured monthly average wind speeds and wind rose from Pyhäjoki (continuous measurements, 10 min averages) compared to meteorological stations' wind speed and direction from the same period (measurements every 3 hours).

The difference between the wind direction distributions is not very great. The South - South-West direction is more frequent on our measuring site, whereas North and West directions somewhat less frequent.

The wind power production of a 200 kW turbine was calculated from the 10 minute average wind speeds measured, and compared with the estimates of WAsP. Monthly average wind speeds and monthly power production are shown in figures 8 and 9. In terms of the annual energy production the difference between the WAsP-results from the two stations is small. Also they are rather close to the result of the measurements. The closer station Ulkokalla seems to predict the seasonal variation better for Pyhäjoki.

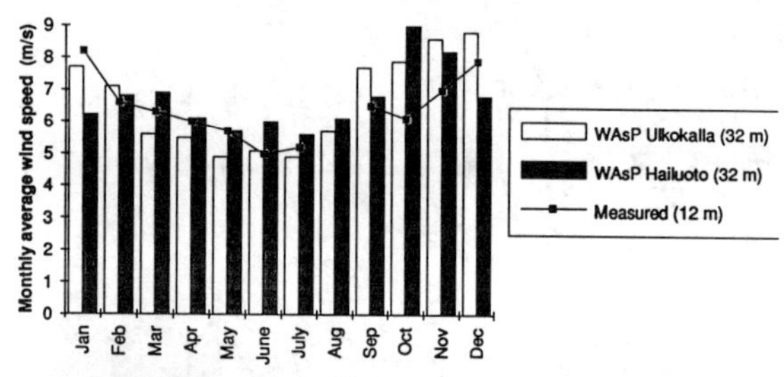

Figure 8. The measured monthly wind speeds in Pyhäjoki compared with estimated monthly wind speeds using two meteorological stations as input.

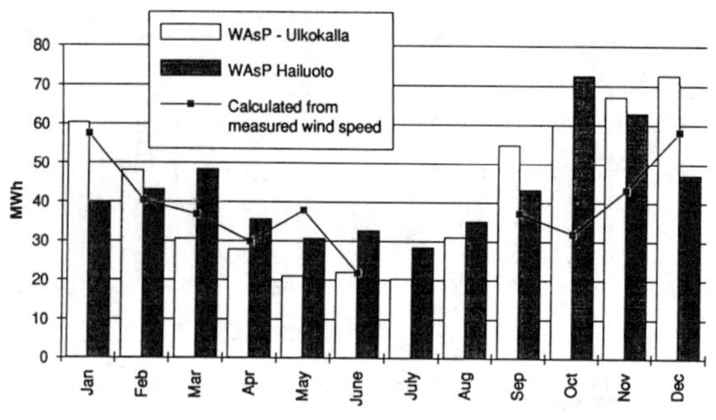

Figure 9. The monthly wind power production calculated from measured wind speeds in Pyhäjoki compared with estimated monthly production using two meteorological stations as input.

6. CONCLUSIONS

Resource assessments using both the WAsP-method alone and together with measurements have been made during recent years. Especially the utilities would appreciate the assessment of the seasonal variation of wind electricity production. To make such estimates the long term monthly mean wind speeds have been used to scale windatlas input files of WAsP. Since the power curves measured in +15°C are actually valid for only one month a year in Finland, temperature corrections for power curves have also been made.

In some cases it has been difficult to choose the wind atlas file because no meteorological stations have been close, or when the wind roses from different stations differ considerably (case Kokkola). WAsP results and on-site wind measurements have given comparable results in Korsnäs and Pyhäjoki, one of the Revon Sähkö Ltd sites. Production of Korsnäs wind farm during the first year was well within 5 % from the WAsP estimate.

Careful application of the WAsP-method seems to give satisfactory results in coastal regions of Finland. On-site measurements don't necessarily make the estimates more accurate, but are required in more complex terrain (e.g. archipelago and mountains).

REFERENCES

[1] Tammelin, B., Suomen Tuuliatlas (Finnish Wind Atlas), Finnish Meteorological Institute, 1991. (in Finnish)

[2] Frandsen, S. and Pedersen, B.M. (ed.), Recommended practises for wind turbine testing 1. Power performance testing, International Energy Agency, 1990.

SHORT-TERM PREDICTION OF LOCAL WIND CONDITIONS

L. Landberg
Dept. of Meteorology and Wind Energy
Risø National Laboratory
DK-4000 Roskilde
Denmark

S.J. Watson, J.A. Halliday
Energy Research Unit
Rutherford Appleton Laboratory
Chilton, Didcot, OXON, OX11 0QX
United Kingdom

J.U. Jørgensen
Danish Meteorological Institute
Lyngbyvej 100
DK-2100 Copenhagen Ø
Denmark

ABSTRACT In this paper it is shown that for a large scale electricity supply grid with a significant penetration of wind power, fossil fuel savings could be increased by up to 25% by using 'intelligent' forecasts of wind energy production. Wind speed forecasts are made using numerical weather prediction models combined with either a physical or a statistical model taking local microscale effects into account. The forecasts are used in conjunction with a time series model which simulates the scheduling of power plant in the England and Wales electricity supply grid. Results are shown in the form of fossil fuel savings for different forecasting methods and, in addition, show the benefits of using wind speed forecasting errors to help plan conventional plant scheduling.

1 INTRODUCTION

With the ever increasing number of wind farms becoming grid connected throughout Europe, there is a need to assess how significant amounts of wind power could be integrated to the best advantage into large scale national grid networks. This paper assesses the use of wind speed forecasting in the integration of wind power into a large scale grid.

The first part of this paper describes the formulation of the wind power predictions. Numerical Weather Prediction (NWP) computer model forecasts are used as the basis for formulating these predictions. These forecasts are adapted to specific sites by the application of either a physical siting model (WAsP) or a statistical technique (Model Output Statistics (MOS)).

The second part of this paper describes the application of these forecasts to the prediction of wind power in a grid scheduling simulation model (the Reading University/Rutherford Appleton Laboratory National Grid Model).

2 THE NWP MODELS

In this study two Numerical Weather Prediction (NWP) models have been used:

The U.K. Meteorological Office mesoscale model: This model has a grid point spacing of 15 km and is run at midday and midnight producing hourly forecasts up to 18 hours ahead ($t + 18$). The mesoscale model covers the whole of the United Kingdom, the Republic of Ireland, and some of continental Europe. Site forecasts are made by taking a weighted average of the three nearest grid points [4].

The High Resolution Limited Area Model (HIRLAM): used by the Danish Meteorological Institute. This model has a grid point spacing of 51 km and produces forecasts at midday and midnight in 3-hourly steps up to 36 hours ahead [5]. HIRLAM covers all of Europe, most of the Atlantic (including Greenland) and parts of Northern America. Forecasts for a given location are derived by bi-linear interpolation of the four nearest grid points.

It is obvious that the forecasts produced by the NWP models cannot take account of local site conditions and climatology which are on a scale of less than ~30 km and ~100 km for the two models respectively. The forecasts are made site specific by the two methods described below.

3 WAsP

WAsP (Wind Atlas Analysis and Application Program), [6,7], is a model that corrects for the influence of local effects. The local effects are:

- Shelter from obstacles, such as houses, rows of trees, etc.
- The effect of roughness and roughness changes, eg the change from water to land.
- The effect of the orography, eg the speed-up at hill tops.

WAsP is used to calculate these corrections and the result is applied to the free-stream wind in the surface layer.

To calculate the free-stream wind, the geostrophic drag law

$$G = \frac{u_*}{\kappa}\sqrt{\left[\ln\left(\frac{u_*}{fz_0}\right) - A\right]^2 + B^2} \qquad (1)$$

and the logarithmic wind profile

$$u(z) = \frac{u_*}{\kappa}\ln\left(\frac{z}{z_0}\right) \qquad (2)$$

are used on the output from the NWP model. These two equations transform the geostrophic wind (*i.e.* the wind outside the atmospheric boundary layer), G, to the free-stream wind at the surface, $u(z)$. u_* is the friction velocity, linking the geostrophic wind to the wind at the surface, κ is the Von Karman constant (=0.4), f the Coriolis parameter, z_0 the aerodynamic roughness of the surface, and A and B are constants, equal to 1.8 and 4.5, respectively. The direction of the wind at the surface, α, is calculated from

$$\tan\alpha = \frac{-B}{\ln\left(\frac{u_*}{fz_0}\right) - A} \qquad (3)$$

4 THE MOS MODEL

The MOS technique is described in ref. [9,8] and will only be briefly mentioned here.

Hourly observed and forecast wind speed values were compared for several sites at 10 m above ground level for one year's worth of data. An Analysis of Variance (ANOVA) technique was used to assess for which parameters the forecasting error varied systematically. The significant parameters were found to be time of day, season, forecast wind direction and forecast precipitation. For each combination of levels of these parameters *e.g.* daytime, winter, east and rain the ratio of the mean observed to mean forecast wind was calculated and this ratio was used to 'correct' future forecasts. This method gives a regression of the mesoscale forecast wind speed onto the MOS predicted wind speed where the regression coefficient (the ratio of mean observed to mean forecast wind speed) is dependent on the combination of four parameter levels i, j, k, l:

$$\hat{w}_{ijkl} = r_{ijkl} \times w_{ijkl} \quad (4)$$

where $\hat{w}_{ijkl}$ is the MOS predicted wind speed, w_{ijkl} is the mesoscale forecast wind speed and r_{ijkl} is the regression coefficient. Each set of regression coefficients r_{ijkl} must be calculated for each site as the prevailing topography, climate *etc.* will be different for each site. These forecasts will be known as MESO/MOS forecasts and will be referred to as 'enhanced' NWP forecasts. In addition, the standard deviation of the forecasting error was calculated for each combination of level parameters after having corrected each forecast in the original data set using the regression coefficients.

5 FORECAST COMPARISONS

To assess the forecast improvement from a wind power point of view, each observed and verifying forecast wind speed was converted into wind power using the power curve of a Vestas V27 (225 kW) wind turbine. For each verifying observation and forecast the power error or 'penalty' was calculated and the mean and standard deviation 'power penalty' deduced from a year's worth of data. Table I shows these mean and standard deviation power penalties for ten sites comparing the 'raw' mesoscale forecasts with the MESO/MOS forecasts for the period April 1989 to March 1990 (using statistics from 1988/89). Table I shows that there is a reduction in absolute mean power error for the MOS regressed forecasts. The standard deviation error shows only a small reduction for most of the sites which seems to reflect the stochastic nature of the wind and the finite time resolution of the mesoscale model. The standard deviation of the forecasting error is important from an integration point of view as will be shown later in the paper.

Figure 1 shows a comparison of r.m.s. wind speed forecasting error as a function of forecast lead-time for HIRLAM/WA^{s}P, MESO/MOS and persistence. Persistence is the simplest *a priori* form of forecasting which assumes that the forecast of a parameter at time $t + x$ is the same value as it was at time t. This is a good approximation for low values of x but becomes less accurate as x becomes larger. This is reflected in Figure 1. R.m.s. values in this figure are calculated for the period December 1990 to November 1991 for nine U.K. sites: Benbecula, Birmingham, Blackpool, Bournemouth, Eskdalemuir, London, Manchester, Valley and Wick. Note that HIRLAM forecasts are three hourly but for the purposes of comparison with the mesoscale model, forecasts were interpolated to obtain intermediate values. Figure 1 shows that persistence is better than MESO/MOS for up to $t+2$ and is better than the HIRLAM/WA^{s}P for up to $t+4$. Beyond these lead times the two enhanced NWP forecasts perform significantly better than persistence. It is also apparent from Figure 1 that the enhanced NWP forecasts have a fairly constant accuracy over all lead times up to $t+18$ but MESO/MOS performs significantly better than HIRLAM/WA^{s}P. Recent investigations have shown that this reflects the higher spatial resolution of the mesoscale model.

Table II shows a comparison of the mean and standard deviation power penalties for the nine sites for the two different forecasting methods. Table II clearly shows, once again, the smaller standard deviation error for MESO/MOS than for HIRLAM/WA^{s}P. It is noticeable that HIRLAM/WA^{s}P forecasts have a slightly smaller mean error on average than MESO/MOS forecasts, though this varies widely from station to station and is not significant.

Table I: Mean and standard deviation 'power penalties' in kW for the period of data FY 89/90

	Forecast Method			
	MESO		MESO/MOS	
Site	*Mean*	*S.D.*	*Mean*	*S.D.*
Aberporth	-1.19	94.56	4.57	84.72
Belfast	14.43	95.40	11.59	99.71
Benbecula	10.59	81.10	2.96	78.22
Birmingham	25.89	89.73	6.30	83.79
Blackpool	27.38	84.19	15.81	79.36
Cairngorm	7.18	132.90	14.35	100.84
Coltishall	20.25	97.57	4.85	88.36
Ronaldsway	10.94	89.48	-1.23	83.11
St. Mawgan	12.69	74.14	10.84	79.96
Wick	19.32	94.64	2.22	86.54
Mean	14.75	35.89	7.23	33.44

6 THE NGM

6.1 Model Description

The National Grid Model (NGM) [1,2,3] is an hourly simulation computer model of the electrical load demand on the England and Wales National Grid which is met by scheduling the available system plant in a merit order. The model plans the start-up and shut-down of various plant using actual hourly load data supplied by the U.K. utilities. Wind power is predicted using either persistence, NWP forecasting or 'perfect' forecasting. To start up a steam turbine (500 MW) from cold is assumed to require 8 hours. At any time a certain amount of 'spinning reserve' *i.e.* part loaded steam turbine plant that can quickly be brought up to full power, is carried on line to meet any unforeseen surges in load demand or shortfalls of wind power generation. The spinning reserve consists of two parts: SR1 which is a fixed fraction of the predicted demand and SR2 which can either be a fixed fraction of the predicted wind power or proportional to the standard deviation of the wind power forecasting error. The fractions SR1 and SR2 are varied in order to minimise the fossil fuel costs for the specified simulation period with the *important* constraint that there are *no losses-of-load.*

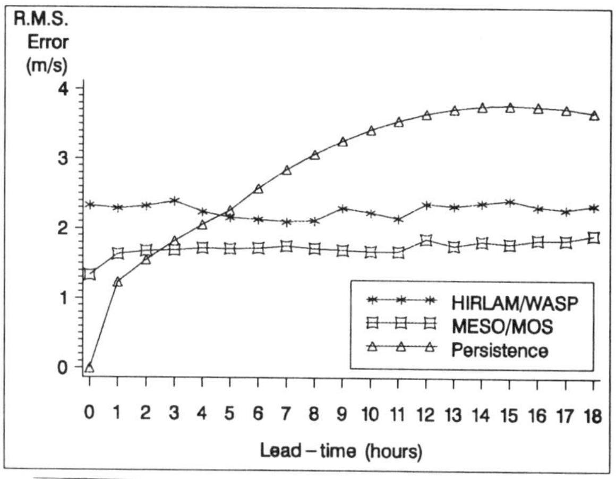

Figure 1: R.m.s. forecasting errors for three different wind speed forecasting methods

At each specific hour the actual load is simulated by multiplying the load data which was used as the prediction by a randomly distributed gaussian function with mean 1 and standard deviation 0.015 which reflects the actual uncertainty of prediction quoted by the utilities. The actual wind power at each specified hour is calculated from observed wind speed values at the chosen sites.

6.2 NGM Results

The results show fossil fuel savings as a function of wind power penetration for different forecasting methods. The England and Wales grid is assumed to have a conventional plant Declared Net Capacity (DNC) of 51 GW. The results show fuel savings for wind power total installed capacities of 5 GW, 10 GW, 20 GW, 30 GW and 40 GW corresponding to penetrations of 9%, 16%, 28%, 37% and 44% respectively. Wind power is assumed to be installed in equal quantities at ten sites throughout the U.K.: Aberporth, Belfast, Benbecula, Birmingham, Blackpool, Cairngorm, Coltishall, Ronaldsway, St. Mawgan and Wick. A spread of sites was used as previous studies have shown that dispersing wind turbines over a wide area helps to reduce the variability of wind power [3]. Again, observed and forecast power is calculated using a Vestas V27 (225 kW) power curve. Each wind speed (observed or forecast) is extrapolated to hub height (31.5 m) taking into account site roughness [7]. The standard deviation of the forecast wind speed errors are used to calculate the standard deviation of the forecast power errors. Fossil fuel prices are those as of 1989/90. Savings are calculated by taking the difference between the total minimised fossil fuel costs for a grid with no wind power installed and subtracting the total minimised fossil fuel costs for a given amount of installed wind power capacity.

Table II: Mean and standard deviation 'power penalties' in kW for the period of data Dec 90 to Nov 91

	Forecast Method			
	MESO/MOS		HIRL/WAsP	
Site	*Mean*	*S.D.*	*Mean*	*S.D.*
Benbecula	8.84	73.26	7.04	84.18
Birmingham	5.90	84.60	6.07	89.57
Blackpool	37.72	86.18	34.28	90.85
Bournemouth	4.25	88.16	2.81	98.25
Eskdalemuir	-5.92	91.25	-14.06	100.53
London	5.04	80.08	6.80	94.37
Manchester	7.69	85.80	-0.73	89.99
Valley	-8.55	74.84	-16.34	85.24
Wick	6.40	81.19	2.57	87.57
Mean	6.82	34.27	3.16	44.19

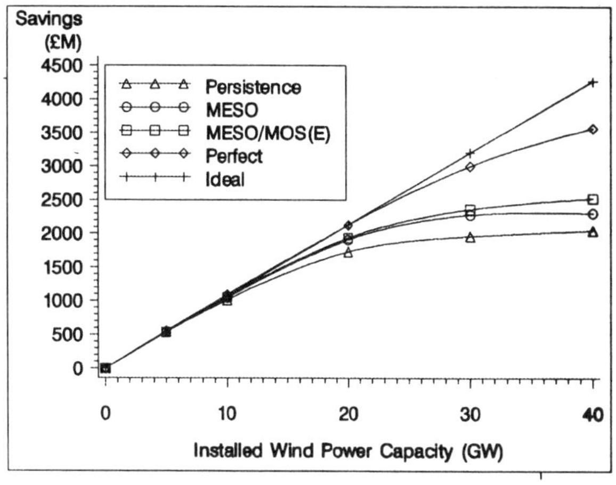

Figure 2: Fossil fuel savings for the England and Wales grid against wind power total installed capacity using different wind speed forecasting methods (FY 89/90).

Figure 2 shows fossil fuel savings for the financial year 1989/90 (FY 89/90). MOS regression coefficients are calculated using 1988/89 statistics. Ideal savings represent the savings that would be made by installing increasing amounts of wind power if there were no 'operating' penalty *i.e.* wind power has no effect on the running of the grid, and if there were no 'discarding' penalty *i.e.* all available wind power can be absorbed by the grid. The graph clearly shows the benefit of using NWP mesoscale forecasting instead of persistence at medium penetrations (>10 GW). Using MESO/MOS forecasts in conjunction with standard deviation errors to plan spinning reserve (referred to henceforth as MESO/MOS(E) forecasts) increases fossil fuel savings at high penetrations (>30 GW). Perfect forecasting yields less than ideal savings at penetrations of greater than 20 GW because of:

1. The assumed inability to shutdown nuclear plant at times of low demand and hence the discarding of some wind power
2. The intermittent nature of wind power meaning that more conventional plant 'stops' and 'starts' are required

Results from a simulation of the financial year 1988/89 (FY 88/89) using 1988/89 forecast data and 1989/90 forecast

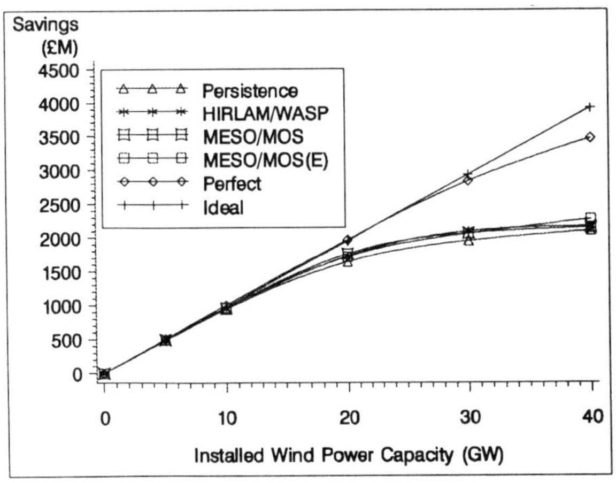

Figure 3: Fossil fuel savings for the England and Wales grid against wind power total installed capacity using different wind speed forecasting methods (December 1990 to November 1991).

MOS statistics yields a similar trend in relative savings as a function of forecasting method.

Figure 3 shows a comparison between MESO/MOS forecasts, MESO/MOS(E) forecasts and HIRLAM/WA^{S}P forecasts. This was produced for the period of concurrent data December 1990 to November 1991 using the nine chosen coincident stations. The plot seems to indicate little difference between the MESO/MOS and the HIRLAM/WA^{S}P forecasts. In addition both these forecasting methods fair little better than persistence. MESO/MOS(E) forecasts do show a significant improvement on persistence at high penetrations but not as substantial as for the two financial year periods FY 88/89 and FY 89/90.

When understanding these results several caveats should be borne in mind:

- An optimisation over one year is very susceptible to one poor forecast. If one particular forecast severely overestimates the observed wind power then a disproportionately large amount of spinning reserve needs to be carried throughout the year to prevent a loss-of-load occurring.
- Optimisations over several years would be required to produce error bands on relative savings by forecasting method.
- The inclusion of capital costs in addition to fossil fuel costs may change the relative savings.
- The model cannot take account of factors such as transmission line costs, grid synchronisation and stability, distances between consumers and generators *etc.*

An attempt has been made to address the first point by looking at the fuel savings for different forecasting methods at a given penetration allowing a number of losses-of-load to occur in each simulation period. Looking at fossil fuel savings as a function of increasing loss-of-load for high (40 GW) penetration shows little increase in savings for FY 88/89 and FY 89/90 for either persistence, MESO/MOS or MESO/MOS(E) for up to three losses-of-load. However, Figure 4 which compares persistence, MESO/MOS, HIRLAM/WA^{S}P and MESO/MOS(E) for Dec. 1990 to Nov. 1991 shows a sharp increase in savings for MESO/MOS and MESO/MOS(E) between zero and one loss-of-load and for MESO/MOS between one and two losses-of-load. Further investigation revealed that these losses-of-load were occurring when the wind speed over the country picked up over a number of hours resulting in the furling of all wind turbines at one specific hour and hence no wind power for this hour. This is an example of one or two uncharacteristically poor forecasts adversely affecting the results.

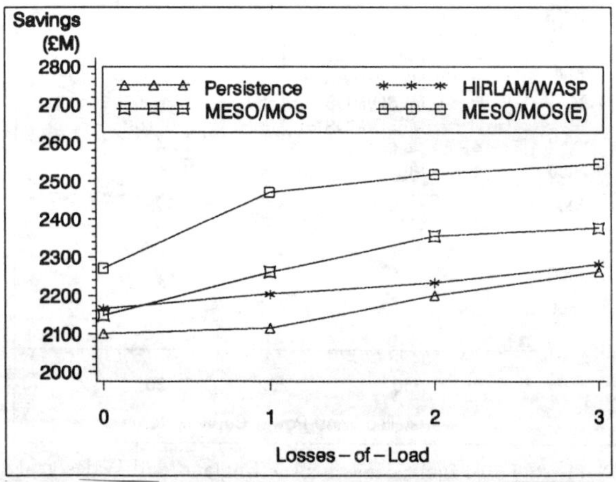

Figure 4: Fossil fuel savings for different forecasting methods as a function of permitted losses-of-load (December 1990 to November 1991).

7 CONCLUSIONS

It has been shown that MOS can be used to reduce mean forecasting error due to systematic discrepancies in an NWP model. Comparison between HIRLAM/WA^{S}P and MESO/MOS has shown that the MESO/MOS forecasts have a lower r.m.s. error for lead-times up to $t+18$. This reflects the higher spatial resolution (but smaller area covered) of the mesoscale model. However, comparing these two types of forecasts using the National Grid Model shows that the large difference in r.m.s. error is not necessarily so strongly reflected in the prediction of wind power for integration in a grid. This is due to the one or two poor forecasts resulting in the need for a significantly larger amount of spinning reserve to be carried to cover these prediction shortfalls.

Furthermore, the inclusion of MOS errors in the planning of spinning reserve does appear to be significant in reducing fuel costs particularly at high penetrations (>30 GW). Although such high penetrations are unlikely to be achieved for a country such as the U.K. as a whole, penetrations on a localised scale could well approach levels for which such forecasting techniques would be beneficial. In conclusion, although the National Grid Model can only address certain aspects of wind power integration in a large scale grid, the results indicate that NWP wind power forecasting systems can produce fossil fuel savings of between 15% and 25% better than persistence when used to help plan plant scheduling. The results for perfect wind power forecasting show that the potential savings could be much higher if future NWP wind speed forecasting accuracy could be improved. However, the understanding of forecasting errors has been shown to be at least as important in increasing these savings given that perfect wind forecasting is never likely to be achieved.

8 ACKNOWLEDGEMENTS

The authors would like to thank the Commission of the European Communities for the funding of this work (contract JOUR-0091-C(MB)) under DG XII and the JOULE programme of work in the field of non-nuclear energy. We would also like to thank Mr. Gil Ross of the U.K. Met. Office for advice and help in obtaining and analysing forecast data. Finally, we would like to thank Mr. Robert Paynter for his help in converting wind speed data into wind power data and for help in modifying the National Grid Model.

[1] Bossanyi, E.A., 1983 *Wind Engineering* **7**, 233–246.
[2] Bossanyi, E.A. and J.A. Halliday, 1983 Proc. 5th British Wind Energy Association Conference, Cambridge University Press, 62–74.
[3] Halliday, J.A., 1988 PhD. thesis, University of Strathclyde. Published as Rutherford Appleton Laboratory internal report RAL T 075.
[4] Golding B.W., 1990 *Met. Mag.* **119**, 81–96.
[5] Machenhauer, B. (editor), 1989: *The HIRLAM Final Report.* HIRLAM Technical Report No. 5, Danish meteorological Institute, Denmark.
[6] Mortensen, N.G., L. Landberg, I. Troen and E.L. Petersen, 1993 *Wind Atlas Analysis and Application Program (WA^{S}P), User's Guide.* Risø National Laboratory, Denmark.
[7] Troen I. and E.L. Petersen, 1989 *The European Wind Atlas.*
[8] Glahn H.R. and D.A. Lowry, 1972 *J. Appl. Met.* **11** 1203–1211
[9] Watson, S.J., J.A. Halliday, and L. Landberg, 1992 Proc. 14th British Wind Energy Association Conference, Mechanical Engineering Publications Ltd, London, 291-297.

DEVELOPMENT OF THE MEASURE-CORRELATE-PREDICT STRATEGY FOR SITE ASSESSMENT

A. Derrick
National Wind Turbine Centre
NEL, East Kilbride, UK, G75 0QU

ABSTRACT: The measure-correlate-predict technique is widely used for estimating the long term wind regime at potential wind farm sites. This paper highlights some of the findings of a study into the accuracy and sensitivity of the technique when applied to typical UK sites. The conclusions serve as guidelines for applying the technique in the UK, indicating the likely length of measurement campaigns required, suggested methods for processing the measured data and the level of prediction accuracy that can be expected.

1. INTRODUCTION

Economic viability of any proposed development is the primary concern of all wind farm developers. One of the major factors influencing the viability is the long-term wind regime at the specific site. Various methods are available(1) to allow the best sites to be identified and estimates of the long-term wind regime to be made, all of which are based on extrapolating a long-term measured wind regime at some reference site to the specific wind farm location using sophisticated computer flow models or some other technique. Despite their sophistication, the variability of the topography and the climate in the U.K. places severe demands on computer flow models and ultimately measurements must be taken at the proposed wind farm site to validate initial computer predictions. The technique used for final validation invariably takes the form of a methodology known as measure-correlate-predict(MCP).

2. MEASURE-CORRELATE-PREDICT(MCP)

MCP techniques are based on extrapolating the results of short-term wind speed measurements to the long-term by relating the short-term wind speeds measured at the prediction site to those at a long-term reference site. The methods of relating the wind speeds at the reference and prediction sites have included:

- fitting continuous functions to relate, normally on a directional basis, the short-term measured wind speed at the prediction site to that at the reference site.
- binning the data according to the wind speed at the reference site and determining wind speed factors for each bin to yield the prediction site wind speeds.
- determining relationships between the Weibull parameters fitted to measured data at both the reference and prediction sites(2).

Using speed-up factors for each wind speed bin could be made to work but would require very long measurement campaigns to ensure full coverage of the wind speed range for all directions. Other researchers have shown that short-term Weibull parameters(2) do not provide a very reliable method of predicting wind speed relationships and that regression analyses using simple continuous functions are much more successful. Continuous functions have been used in this study to relate the short-term measured data mainly because regression analysis lends itself to objective statistical confidence analyses and, as the functions are continuous, by definition, gaps in the measured data are objectively filled.

The standard MCP technique requires that simultaneous short-term time series data are available for both the prediction and reference sites. The data are then binned according to the reference direction(twelve 30 degree bins are used here) and simple functions determined relating the prediction site wind speed to that at the reference site. The twelve wind speed relationships are then applied to the appropriate wind direction sector long-term wind speed distribution measured at the reference site. The long-term reference dual parameter distribution of wind speed and direction is transposed to a similar dual parameter distribution table with the magnitude of the wind speed bins modified according to the fitted function. Hence a predicted long-term mean wind speed can be calculated. This is the technique in its simplest form, however it is subject to many variables and can be enhanced considerably as will be described.

3. ANALYSES

This study has analysed the sensitivity of the prediction technique to the processing applied to the measured data and to the methods used in the correlation stage. Specific attention was given to:

- the length of measurement campaigns
- the influence of individual pressure systems
- reference/prediction site separation
- removal of short-term wind speed variations(less than diurnal)
- removal of diurnal variations
- incorporation of objective confidence analyses
- low wind speed data
- directional veer predictions

Two pairs of sites were available for which short-term measured time series and long-term measured wind speed distribution tables were available at both reference and prediction sites, hence allowing long-term predictions to be compared directly with long-term measurements. Numerous other pairs of sites were analysed for which there were short-term time series data available for both sites but for which long-term wind speed distributions were available for only one of the two sites. Although, in such cases, long-term predictions could not be compared with measurements they did provide sites at a range of separations, highlighting the likely variations in short-term correlation quality. Specific aspects of the analyses under the most significant of the above headings are considered in the following sections.

3.1. Objectivity

Experience of carrying out correlations of short-term wind speed data at many sites had shown that there could be considerable scatter in the data which would have some bearing on the confidence in the long-term predictions. Figures 1(a) and 1(b) show typical short-term hourly mean wind speed

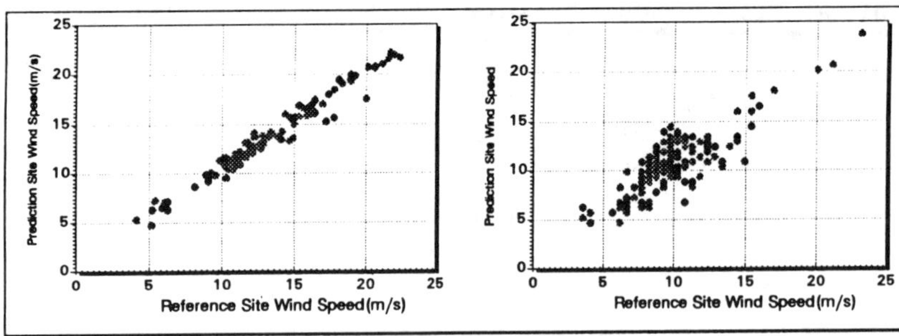

(a) Site Separation 1.5km (b) Site Separation 50km

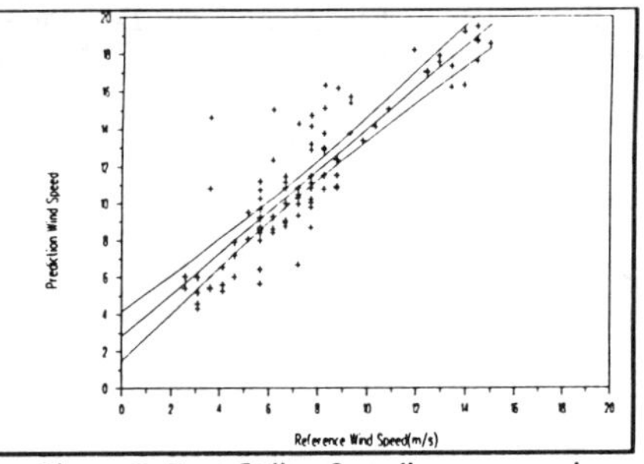

Figure 2: Best fit line from linear regression with 99% limits of confidence.

Figure 1: Examples of short-term wind speed correlations

correlations for a particular direction sector at two different pairs of sites separated by 1.5km and 50km respectively.

The resulting uncertainty could not be quantified by the direct application of continuous correlation functions alone. However by restricting the correlation functions to a simple line or power law, standard linear regression techniques could be used to determine the best-fit continuous function and the standard deviations of the regressed values. For example, a best-fit line with equation:

$$y = mx + c$$

where y is the prediction site wind speed and x is the reference site wind speed has a standard deviation for each predicted y given by:

$$\sigma(y) = \sqrt{x^2 \operatorname{var}(m) + \operatorname{var}(c) + 2x\operatorname{cov}(m,c)}$$

When combined with the regressed line, this yields the best-fit line with confidence limits in terms of the standard deviation as shown in the example of figure 2.

The best-fit function and standard deviation relationship for each direction sector are then used to determine the mean wind speed, Y, and standard deviation, $\sigma(Y)$, for each sector using the appropriate long-term wind speed distribution, p(y). Hence,

$$Y = \Sigma_i p_i y_i$$

and

$$\sigma(Y) = \sqrt{\Sigma_i (p_i \sigma(y)_i)^2 + \Sigma_i \Sigma_j 2\, p_i p_j \operatorname{cov}(y_i, y_j)}$$

The covariance term is defined by,

$$\operatorname{cov}(y_i, y_j) = x_i x_j \operatorname{var}(m) + (x_i + x_j)\operatorname{cov}(m,c) + \operatorname{var}(c)$$

and is required as the y values all derive from the same regressed line and are therefore not independent. Combining the results for each of the direction sectors, the overall mean wind speed and standard deviation are,

$$V = \Sigma_k P_k Y_k$$

$$\sigma(V) = \sqrt{\Sigma_k (P_k\, \sigma(Y)_k)^2}$$

In this case the Y values are independent as there is a different regressed line for each direction sector, hence there is no covariance term.

For this analysis to be applicable, the reference wind speed must be taken as the x-ordinate and the prediction site wind speed as the y-ordinate. Hence, the resulting predicted wind speed is obtained objectively along with a measure of the uncertainty. However, it should be noted that the resulting uncertainty is a function of the quality of the short-term correlations; the short-term correlations themselves may not be representative of the long-term, especially if a very short measurement campaign is used. For reasons to be considered later, it is generally the case that correlation quality will decrease as the length of measurement campaign is increased although prediction accuracy will increase.

3.2. Sources of Uncertainty

Irrespective of the site separation, the topography between and around the reference and prediction sites is usually fixed in the short-term, therefore the major source of scatter and hence uncertainty in the wind speed correlations resulting from topographic effects is likely to be turbulence arising from flow separation and blockage. As such effects are pseudo-random they are best filtered out by averaging the time series data at both reference and prediction sites in the short-term, hence improving the correlation quality. Whereas a straightforward average would tend to remove short-term fluctuations, it could also diminish any apparent trends by reducing the overall number of data points available for correlation. This situation was avoided somewhat by employing a running average scheme. A progressively more severe filtering effect could be obtained by using the output from the first stage of the running average as input to a second stage and so on. This averaging scheme was analysed in the frequency domain(3) to reveal the cut-off characteristics shown in figure 3.

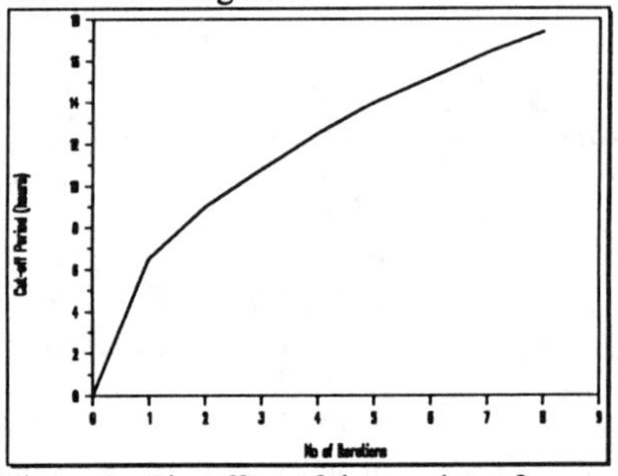

Figure 3: The effect of the number of running average iterations on the cut-off period

As the relationship shown in figure 3 becomes asymptotic with increasing number of iterations there seemed little point in applying more than five iterations (corresponding to a cut-off period of about 17 hours) of the averaging scheme to any particular data set. Whereas it had been thought that the averaging scheme would significantly reduce scatter and reveal a clear wind speed relationship for each particular direction sector, the actual result was quite surprising. Figures 4(a)-(c) show examples of a 30 degree direction sector correlation using 0, 1, and 5 iterations of the running average scheme for a pair of sites separated by about 40km. The data were measured over the period of a month.

It was surprising to note that although there was an obvious reduction in the random nature of the data, the overall envelope enclosing the data did not significantly reduce. In fact, the envelope enclosing the data points was of similar extent to that of the original hourly data(figure 4(a)). The filtering had removed some short term effects, but the overall correlation variance was dominated by effects of period longer than 17 hours. Inspection of figure 5 shows that, over the period of the

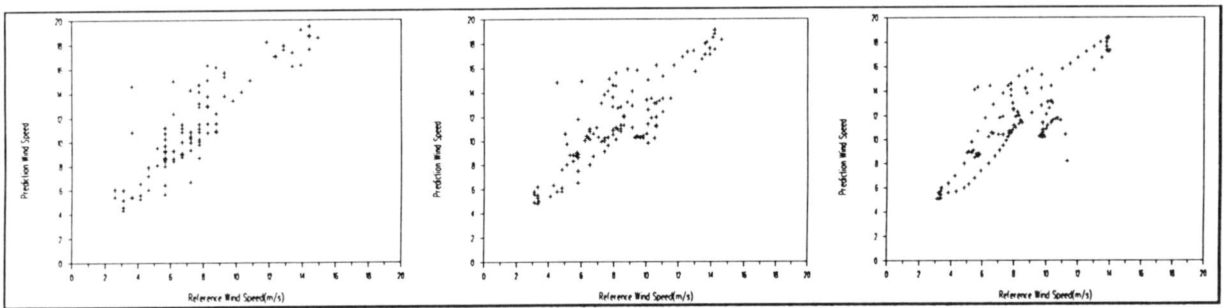

Figures 4(a)-(c): Wind speed correlations using unsmoothed hourly data and smoothed data(0, 1 and 5 iteration running averages).

month, on the occasions when wind direction returned to this particular direction sector, the wind speed relationships achieved were nearly always different. Two particular 'runs' consisting of data points which are consecutive in time are highlighted as 'A' and 'B' in figure 5. Run 'A' seems, intuitively, to be normal whereas run 'B' represents a decidedly abnormal wind speed relationship. Several other examples of 'normal' and 'abnormal' wind speed relationships can be readily identified in the figure.

Isobaric charts corresponding to the two particular days when runs 'A' and 'B' occurred were inspected in an attempt to identify the source of such widely varying wind speed relationships. These charts are reproduced in figures 6(a)-(b). The reference site is on the south west coast of Scotland and the prediction site is some 40km inland to the north east.

These figures show that even over relatively short distances isobaric curvature has a significant effect on the wind speed relationships between two sites. In figure 6(a), corresponding to the normal correlation, the isobars over the reference and prediction sites are relatively straight and parallel whereas in figure 6(b), the isobaric curvature is high. Therefore in the vicinity of the sites the curvature and hence wind direction would probably have changed significantly over the course of the day as the pressure system and fronts crossed the country. For any given wind direction at the reference site, the wind direction at the prediction site could experience significant variation, hence local topographic and roughness effects would not be consistent for any particular reference direction sector. This would be more of a problem in areas of complex terrain. In severe cases where sites are separated by a relatively long distance, the winds at the two sites could conceivably be driven by different pressure systems separated by a front. This may be what happened in the example of figure 6(b).

The significance of this to the UK is highlighted by an analysis which surveyed the UK synoptic charts for every day from 1868-1967(4) and classified the weather type according to direction and cyclonic or anti-cyclonic nature. Table 1 summarises the average yearly findings.

Weather Type	Category	Occurrence(%)
Changeable Westerly	W	25.7
North Westerly	NW	4.7
Northerly	N	7.4
Easterly	E	7.6
Southerly	S	8.6
Anticyclonic	A	24.9
Cyclonic	C	17.5
Unclassified	-	3.6

Table 1:Average Occurrence of Weather Categories for the UK 1861-1967(4).

The highest winds are generally associated with category W which is also characterised by secondary depressions and cold fronts crossing the UK(4). By definition category C results in complex depressions crossing the UK mainland. Categories C and W are particularly prevalent and it can be assumed that the proximity of depression centres and fronts to the UK will increase the chance of high isobaric curvature over the mainland and the ensuing wind speed characteristics described previously. The variability in the short-term correlations caused by pressure system effects must also be carried through to the long-term predictions in terms of uncertainty and indeed this is the case with the objective correlation procedures described previously. This is not something that is easily achieved using complex computer flow models which usually assume linear transposition of geostrophic winds from the reference to the prediction sites.

By filtering out all of the wind speed disturbances of period less than 17 hours, pressure system effects have been revealed, however the diurnal variability caused by thermal effects still remain. Application of a numerical implementation of a Butterworth filter did allow diurnal trends to be conveniently removed from the short-term measured time series data. However, although it was possible to identify changes in the correlations caused by the diurnal variability, as shown in the example of figure 7, the effect was by no means consistent.

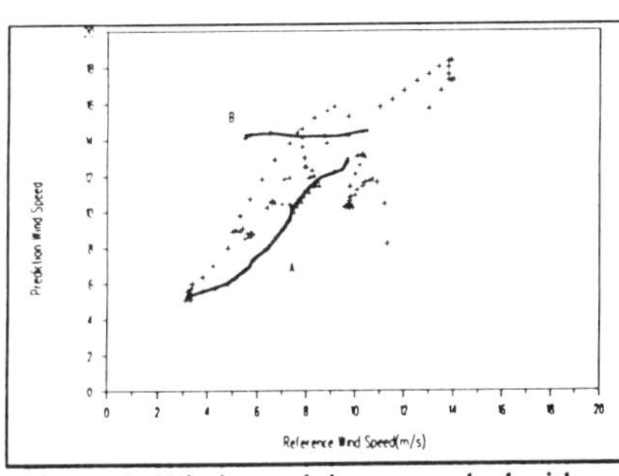

Figure 5: Wind speed data smoothed with a five iteration running average with normal(A) and abnormal(B) runs highlighted.

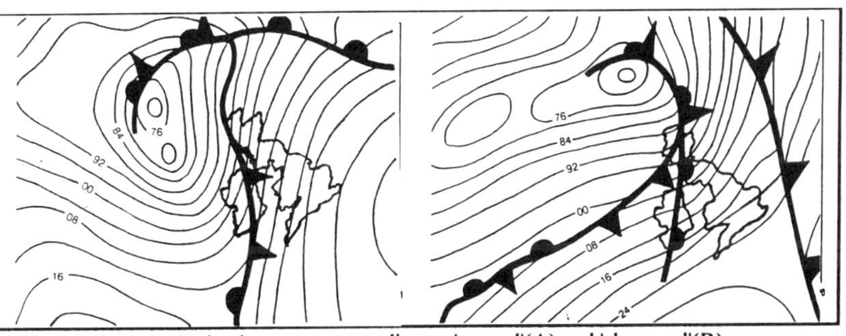

Figure 6(a)-(b): Isobaric charts corresponding to 'normal'(A) and 'abnormal'(B) correlations in figure 5

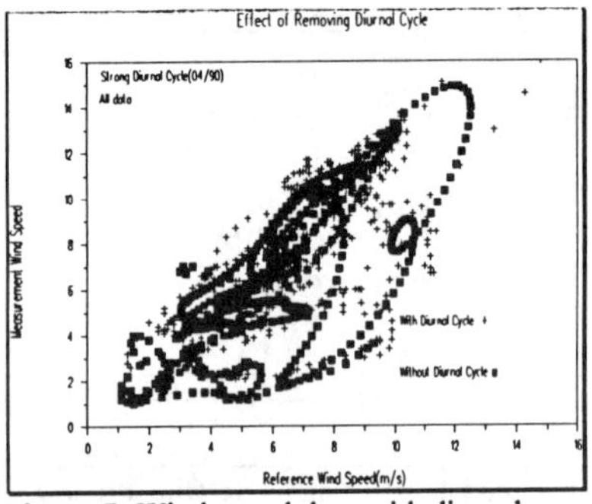

Figure 7: Wind speed data with diurnal trend, superimposed on the same data with the diurnal trend removed.

The major problem arises from not knowing if the prediction site diurnal trend determined from the short-term measurements is typical of the long-term. This is especially so in the UK where the trends are less consistent than in mainland Europe. Therefore, in this study, it was decided not to remove the diurnal trends from the short-term time series. Rather, the trends were retained to add to the overall uncertainty associated with the correlations and hence the long-term predictions.

3.3. The Duration of the Measurement Campaign

Investigations into the effect of the length of the measurement campaign on prediction accuracy were tackled by comparing long-term measurements with long-term predictions based on short-term measurement campaigns of duration from 1-12 months and also by extracting examples of short-term time series of fixed duration from different time periods. Tables 2(a) and 2(b) show examples of the effect on prediction accuracy of increasing the length of the measurement campaigns for two pairs of sites.

3.4. Veer Predictions

A similar MCP strategy was applied to the directional data to determine the long-term directional distribution of wind speed at the prediction site. As with the wind speed, directional relationships between sites separated by some distance are not only influenced by the topography but also by pressure system effects. However, figures 8(a)-(b), which represent two different years of measured data for a particular pair of reference and prediction sites, show that an underlying site characteristic can be identified, although the trends do exhibit varying degrees of scatter.

A polynomial may be fitted to these data relating reference direction to veer with the requirement that it gives continuity at the 0/360 degree cross-over point. The polynomial is then applied to the long-term reference directional distribution to give a prediction of the directional distribution of wind at the prediction site. The results of the wind speed and directional predictions may then be combined. Figures 9(a) and 9(b) compare examples of the predicted and measured long-term(15 year) directional and mean wind speed distributions based on a 12 month measurement campaign for a pair of sites separated by some 50km. The predictions of the direction are certainly not as impressive as the predictions of wind speed, mainly because the scatter in the directional data is considerably more extensive than that in the wind speed data. This is ignored by the continuous polynomial representation of the veer, so although the general trend of the prediction is correct, the details are less accurate.

No of Months	1	2	3	4	5	6	7	8	9	10	11	12
Mean Wind Speed (m/s)	8.1	8.7	8.6	8.6	8.5	8.6	8.4	8.4	8.4	8.5	8.5	8.5
Wind Speed Error (%)	-2.4	4.8	3.6	3.6	2.4	3.6	1.2	1.2	1.2	2.4	2.4	2.4
Energy Yield Error(%)	-5.3	9.4	6.2	4.4	5.5	6.2	1.1	0.8	0.7	1.1	1.1	0.4

Table 2(a) Effect of length of measurement campaign on 15 year predicted wind speed and energy yield: case 1(40km)

No of Months	1	2	3	4	5	6	7	8	9	10	11	12
Mean Wind Speed (m/s)	6.1	6.5	6.4	6.6	6.7	6.8	6.8	6.8	6.9	6.9	7.0	6.9
Wind Speed Error (%)	-6.1	0.0	-1.5	1.5	3.1	4.6	4.6	4.6	6.2	6.2	7.7	6.2
Energy Yield Error(%)	-22	-3.7	-12	-8.6	-8.9	-4.0	-3.1	-3.1	1.7	4.4	5.3	4.9

Table 2(b) Effect of length of measurement campaign on 15 year predicted wind speed and energy yield: case 2(50km)

Both tables show that, although by chance it may be possible to obtain a very accurate prediction after only a few months measurement, the predicted energy yield error tends to reduce with increasing length of measurement campaign. This is mainly a reflection on the pressure system effects noted previously as it is possible to experience pressure systems typical of the long-term after only a very short measurement period. However, without the benefit of the prediction site long-term measured data used in this study for comparison, there would be no way of knowing this at the time. Therefore, it seems that the only way to determine if the measurement campaign is long enough is to repeat the MCP procedure at monthly intervals until such time as the predictions converge. The inevitable result is that measurement campaigns of 8 months minimum duration are likely to be required if a broad range of seasonal pressure system effects are to be experienced.

Wind turbine mean energy yields have been more reliably predicted than wind speed because correlations are generally very poor in low wind speeds, resulting in poor predictions of this wind speed range(<4m/s). However, as the low winds are outwith the operational range of typical turbines, the influence on the energy yield predictions is negligible.

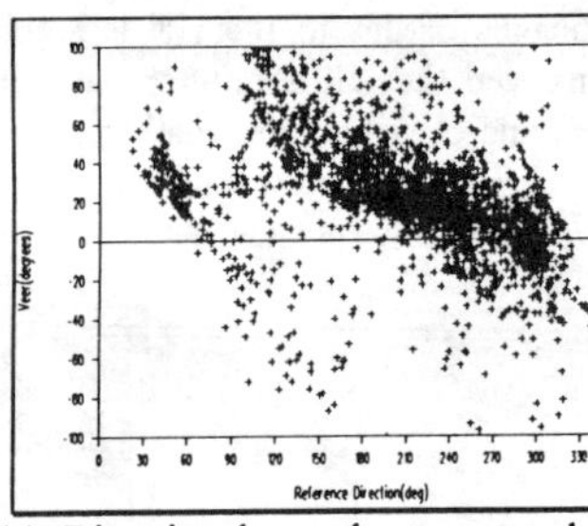

Figure 8(a): Directional veer data measured during 1989

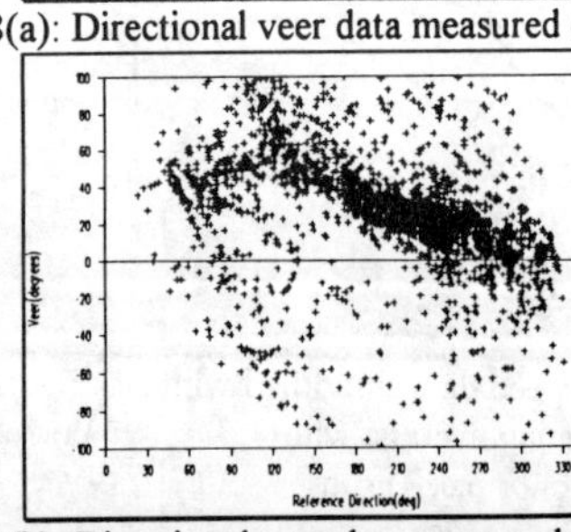

Figure 8(b): Directional veer data measured during 1990

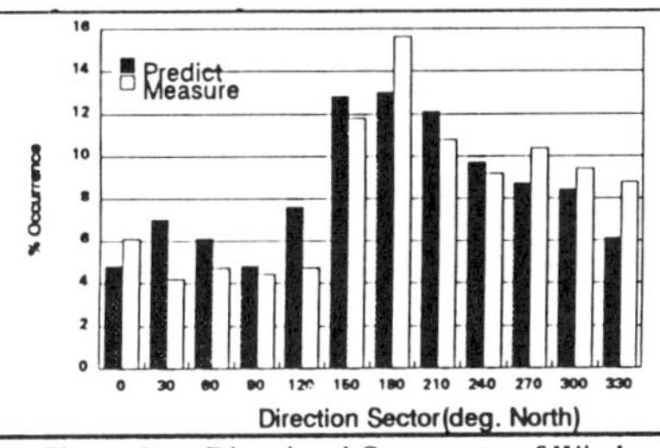

Figure 9(a): Directional Occurrence of Wind

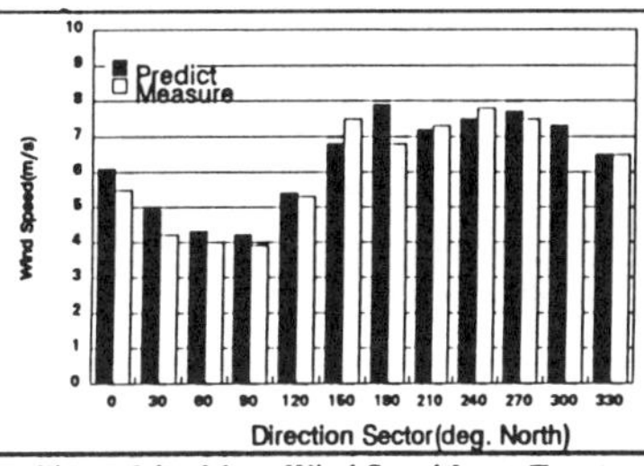

Figure 9(b): Mean Wind Speed(Long-Term)

Figures 9(a)-(b): Comparison of long-term predictions and measurements

4. CONCLUSIONS

The Measure-Correlate-Predict analysis has proved to be an entirely methodical and acceptably accurate means of determining the long-term wind regime at potential wind turbine or wind farm sites. Throughout the analyses of many combinations of sites and measurement periods, numerous factors have emerged as being of importance and similarly the significance of others has reduced. Considering each of the three MCP stages in turn, the main conclusions concerning the short-term measurement campaign are:

(a) The length of the measurement campaign must be such that a broad range of typical weather systems are experienced, giving full directional coverage and a wide range of wind speeds within each direction sector. These conditions can be objectively satisfied by a repeated application of the MCP analysis at monthly intervals until such time as the final predictions converge. However application of this technique during this study tends to confirm the common sense conclusion that measurements during all seasons should be obtained, dictating that measurements should ideally last for a minimum of 8 months with the emphasis on the generally windier Winter months(UK climate).

(b) The further apart sites are the greater the influence of pressure system curvature on the wind speed relationships and hence the longer the measurement campaign required to generalise these relationships to approximate the long-term.

Considering the conclusions of the correlation investigations:

(c) As the ultimate aim is to make a long-term prediction, random short-term effects will generally degrade the relevance to the long-term of the short-term wind speed relationships. Therefore all of the hourly time series data should be filtered to remove very short-term effects(less than diurnal).

(d) It is recommended that the diurnal trend should not be removed from the data, as it is a genuine uncertainty in the short-term to long-term relationships and as such is accounted for in the prediction uncertainty analysis.

(e) To maintain objectivity in the full MCP procedure, the prediction site data should be regressed onto the reference Meteorological Office data. That is for a best-fit line, $y = mx + c$, x is the reference site data and y is the prediction site data.

(f) The reference and prediction site wind speed data should be binned according to reference site direction and the wind speed correlations carried out on the data within each direction bin.

(g) Low wind speed data at both the reference and prediction sites should be discarded before carrying out the correlations as the erratic behaviour of the wind vane in such conditions can spoil the wind speed correlations within each directional bin.

(h) Wind direction relationships are best determined by fitting a continuous function to relate the short term veer and reference direction. Low wind speed data(less than 4m/s) should be discarded as the meandering behaviour of wind vanes in light winds can totally obscure any high wind speed directional trends.

Finally, the prediction stage of the MCP analysis resulted in the following conclusions:

(i) Predictions of long-term wind speed of better than 6 per cent accuracy have been consistently achieved using the MCP techniques described in this report and it is believed that this could be reliably repeated at wind farm sites with good potential.

(j) All of the predictions have been based on reference sites on the West coast of the UK. Hence the reference data have been relatively free from disturbance in the prevailing wind direction. The technique still requires verification at sites well inland or to the East of the UK where there could be considerable topographic effects at both the reference and prediction sites.

(k) This study has clarified many issues surrounding what is a very simple technique. However, it has been shown that with careful application very accurate predictions can be obtained which are of direct use to wind energy developers.

5. ACKNOWLEDGEMENTS

Most of the findings of this paper are based on work funded by the UK Department of Trade and Industry and the assistance of their project officer, Dr Henry Parkinson, in supporting this project and supplying data is gratefully acknowledged. Ray Hunter of NWTC also provided much valuable feedback and input to the investigations.

6. REFERENCES

1. Barton, S.M., Methods used by the UK National Wind Turbine Centre for site assessment, NEL Report NWTC/2/89, National Engineering Laboratory, East Kilbride, 1989.
2. Uitzinger, M.J., Verkuijlen, E., van Mierlo, B., On the impracticability of Weibull statistics in the comparison of wind speed time series, Wind Engineering Vol 13,(4) 207.
3. Derrick, A., Development of the measure-correlate-predict strategy for site assessment, Proceedings of 14th BWEA conference, Nottingham 1992, MEP.
4. Chandler, T.J.(ed.), Gregory, S.(ed.), The climate of the British Isles, Longman 1976.

PERFORMANCE AND LIMITATIONS OF WIND SIMULATION MODELS IN NW OF SPAIN

R. Zubiaur, F. Martin

CIEMAT. Av. Complutense, 22 MADRID - 28040 SPAIN
Tel.: 34-1-346 60 41 Fax: 34-1-346 60 37

ABSTRACT: In the North West of the Iberian Peninsula (Galicia), several projects for the evaluation of wind resources are carried out in order to install wind farms. For this work several simulation models are used, which have serious difficulties because of this region's characteristics: complex orography, the "corner" effect on the prevailing winds, coastal area, influence of the seasonal deviations from the polar front. Also, wind measurements at three other locations between 50 and 100 Kms far, are used. The results obtained are compared with the measurements which are simultaneously carried out, although at the moment, the agreement between model's results and the measurements are not satisfactory enough, specially in the case of directional distributions.
It has been specially analyzed the sensitivity of the WASP's results versus the meteorological model's parameters and it has been found that some of them have a great influence, and so, very different results have been obtained from the same wind data and topography. Nevertheless the adequate selection of these parameters is not inmediate in order to decide which are the better values for the model's most apropiate use.

1. INTRODUCTION

Northwest area of the Iberian Peninsula (Galicia Coast) is one of the most promising zones for wind energy exploitation, due to the existing high wind conditions. There are already installed in this region some wind power plants (AWEC 60, 1.2 MW; Cabo Villano Wind Farm, 3.6 MW; Estaca de Bares, 0.36 MW) (figure 1), and, which is more important, the concerted plannings for this area call for the installation of a number of wind farms with a global capacity in the range of several hundred magawatts for the coming years.

At the same time, there are in course some programs for wind resources assessement in the area, as a previous step for site selection: enlargement of Cabo Villano Wind Farm (15 MW), Barbanza Wind Farm (20 MW), and other projects like "Wind Measurement and Modelling in Complex Terrain", in the context of the EC JOULE Program.

For those activities in wind resources evaluation and site selection, beside the measuring campaings, several simulation models are being applied: methods based on mass conservation, methods based on correlation-extrapolation and WASP model.

Due to the particular characteristics of the region (very complex terrain, "corner" effect on the prevailing winds, coastal area, influence of the seasonal oscillations of polar front, etc), the application of models for local wind conditons assessement present serious difficulties when comparing simulation results and measurements.

Available measurements for the data analysis come from: local meteorological observatories (La Coruna and Santiago de Compostela, previously used in the European Wind Atlas), two stations at Cabo Villano site, three other at Barbanza site and four stations corresponding to the JOULE Program (figure 2).

2. MODELS APPLICATION

2.1. Mass consistent model.

A mass consistent model, MATHEW [1], has been used for a mesoscale range evaluation of resources in Galicia. In this kind of codes the advantages are:
- consideration of topography.
- integration of all available measurements.
- softness of particular measuring errors introduced for each station.
- consistency of results with measurements of all stations.

The main disadvantage found out in the application to this region is the low spatial resolution in the topography discretization (4 Km), due to the size of the area to evaluate (200x200 km²) and to the limits of code capacity. Other limitation is the shortage of useful measuring points. That makes that the microscale effects of the topography, specially important at Galicia, are filtered. The improvement of resolution implies the reduction of measuring points.

Using this model, a global assessement of the whole area was made (figure 3). In a broad scheme, the results are not too far from those contained in the European Wind Atlas, but it is observed the filtering of orographycally singular points. In the case of Malpica site, a small hill 182 m maximum height, 1 Km² surface, the topography effect disappears in a 4x4 km² cell.

The applicability of mass consistent models, in the case of large areas with few measuring stations, is limited to the cualitative global wind resource assessement.

In small areas with higher number of stations, these models are used and good results in local wind resources evaluation have been obtained [2].

2.2. Correlation-Extrapolation Methods

In site evaluation for wind farms installation it is necesary to carry out some local measurement campaings. Then estimation of wind conditions have been made using correlation-extrapolation techniques. For this region, the effect of polar front displacement between summer and winter seasons must be considered in the application of such techniques and taken into account separately.

The obtained correlations are in general very good, with correlation coefficients higher than 0.8, between observatories tens of kilometers far away, and so, the extrapolations may be considered reliable (figure 4)

2.3. WASP Model

As a first step in the model application, a previous action was made in order to evaluate different sites based on observatories included in the European Wind Atlas [4]: La Coruna and Santiago de Compostela. As an exemple, the results obtained for Malpica site (figure 5), at distances of 37 and 60 Kms respectively, are in the following table:

Refer. Obs.	Malpica (estimation at 10 m agl)	
La Coruna	8.7 m/s	1001 W/m^2
Santiago de C.	7.4 m/s	537 W/m^2

From the disparity of the displayed results, there are serious doubt on what observatory is more adequated for their use in the wind resources estimation of the zone, moreover when the yearly averages at the reference sites are 3.9 and 3.6 m/s respectively wich are not considered as high wind conditions. In order to include some high wind potential reference values, measurements from the JOULE Program located at high wind areas, have been used.

In the adequate use of WASP, it is esential the verification of results obtained coming from a particular station, with measurements carried out in another station at the same region.

An uncertainty in the model application is presented when considering the validity of physical parameters included in the PARAMETER FILE, specially those related to local characteristics, like the high and intensity of thermal inversion, and the high from which wind speed can be considered as constant (parameters #6, #7 and #28 of WASP.PAR).

In Galicia region there are not available measurements of those parameters. Because of that, a sensitivity analysis of results has been made in order to see what values are the best to reproduce the measurements. As an exemple, the results obtained with 2 months of continue and simultaneous measurements in winter conditons at two sites in complex topography, Malpica and Barbanza (25x10 km^2, 690 m height, figure 6).

The average wind conditions measured at the two sites (period: November and December, 1992), are:

Malpica: 8.6 m/s; 822 W/m^2
Barbanza: 6.7 m/s; 456 W/m^2

Using Malpica as the reference, "atlas" with different values of parameters #6, #7 and #28 have been obtained. A low influence on the results from the parameter #28 has been found. From these "atlas" the values corresponding to Barbanza site have been predicted (table 1).

As can be observed, the best results for those winter period (1% error in energy) are reached up with the parameter 7 (softness of inversion) = 1.0. (The physical meanning in this case is that there are not thermal inversions in the area in this period).

In this sense, works carried out in other zones [5] have obtained different figures in the adequated parameter values, associated to different climatological conditions.

Beside this, it is already known that there are other limitations in the use of WASP, as separating flows are not treated correctly [6].

3. CONCLUSIONS

In the wind resources assessement and wind farms site selection activities now in course at the Galicia region, different types of wind estimation models have been applied:

- Mass consistent model, in order to evaluate the region at a mesoscale level.
- Correlation-extrapolation models.
- WASP model, in order to introduce with a high resolution the effects of topography.

The first model do not permit the detailed analysis of local effects, due to the low density of the stations network.

For the application of correlation-extrapolation techniques in this region it is neccesary to take into account the seasonal differences derived from general atmospheric circulation. The results obtained have been very reliable.

In the application of WASP it's important to select adequately the reference observatory. It is recomended the verification of results by using simultaneous measurements from several stations. That permits to choose the best values of phisical parameters in the PARAMETER FILE (#6, #7), which have a great influence on the results.

REFERENCES

[1] C. Sherman. MATHEW - A Mass Consistent Model For Wind Fields over Complex Terrain. J. Appl. Meteor. 1978.

[2] P. Moreno, R. Zubiaur. Wind Resources on the Southern Coast of the Iberian Peninsula. ECWEC 90.

[3] I. Troen. WASP USER's GUIDE. 1988.

[4] European Wind Atlas. 1989.

[5] E. L. Petersen, B. Tammelin, R. Hyvonen. Wind Resources Assessment in Complex Terrain in the Far North. Boreas 13,2. 1992.

[6] S.E. Larsen, N.O. Jensen. Flow around and over the ridge of Pedreiras as seen by WASP and in reality. Riso. 1992.

PREDICTED WIND CONDITIONS IN BARBANZA FROM MALPICA AT 10 m agl (m/s ; W/m^2)

HEIGHT OF INVERSION (parameter 6)	SOFTNESS PARAMETER (parameter 7) 1.0	0.8	0.5
800 m	7.2 ; 461	7.8 ; 582	8.6 ; 803
1000 m	7.2 ; 461	7.6 ; 549	8.3 ; 709

Table 1. Results from WASP.

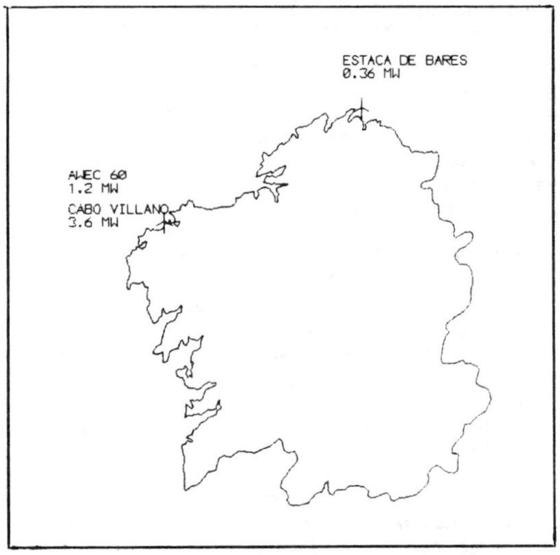

Figure 1.- Location of wind power plants at Galicia.

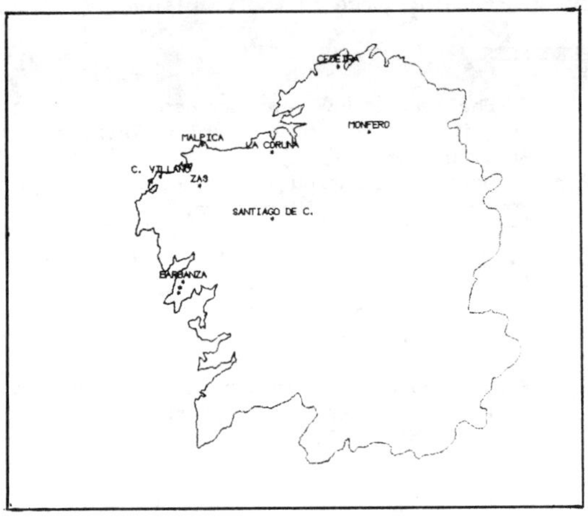

Figure 2.- Wind measurement stations.

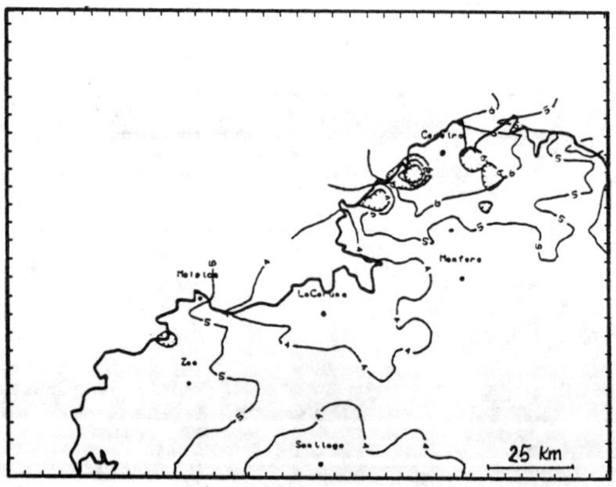

Figure 3.- Results from MATHEW. Annual isovents map (m/s at 10 m. agl).

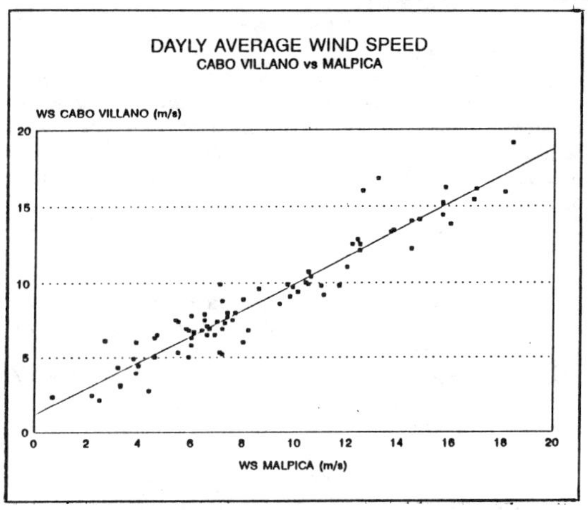

Figure 4.- Wind speed correlation.

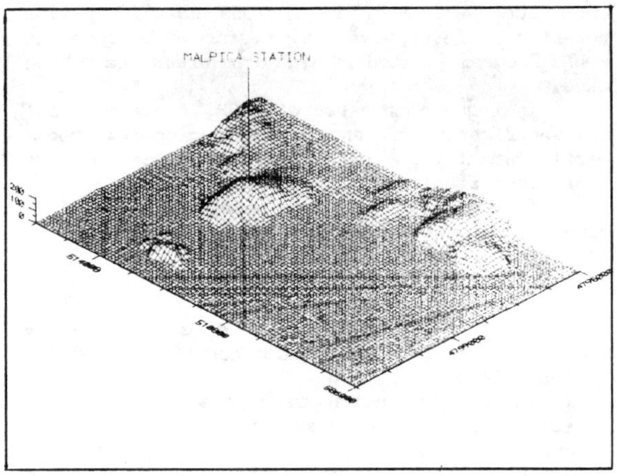

Figure 5.- View of Malpica Topography from NW (area 7 x 10 Km^2, cells 0.1 x 0.1 Km^2).

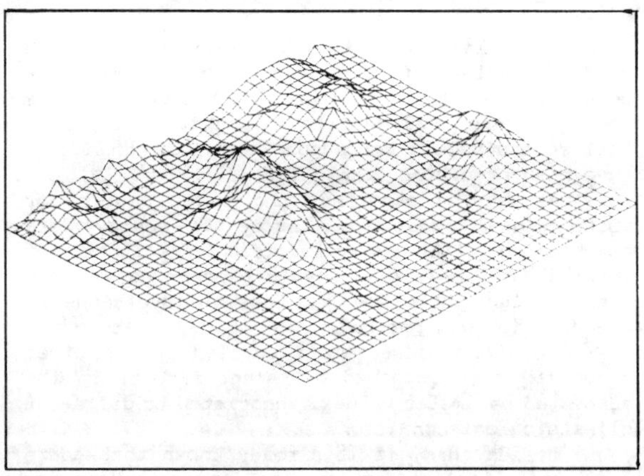

Figure 6.- View of Barbanza Topography from SW (area 35 x 40 Km^2, cells 1 x 1 Km^2).

WIND ENERGY IN THE CZECH REPUBLIC

J. F. Rimal, RNDr., Ph.D.
Czech Technical University of Prague
Faculty of Civil Engineering
7 Thakurova
166 29 Prague 6

ABSTRACT: The factors determining the possibilities of exploitation of wind energy in the Czech Republic are discussed. The approximate estimate of the wind potential for the territory of the Czech Republic, is based on data from ten localities obtained by measurements during a period of five years. This paper on exploitation of wind energy presents an answer to three basic questions:

a) time distribution of wind energy during the day, month, year and in course of the entire period of measurements

b) the estimate of the wind potential for the entire territory of the Czech Republic with the aid of the defined concepts - theoretical wind potential, technical wind potential, economically usable wind potential

c) delimitation of suitable regions with favourable wind conditions in the Czech Republic

1. INTRODUCTION

The input for the analysis of the wind potential in the Czech Republic are wind data from 10 meteorological stations (Fig. 1) in the period from 1976 to 1980 obtained from the Czech Hydrometeorological Bureau.

1. Primda	754 m a.s.l.	2. Churanov	1122 m a.s.l
3. Praha	376	4. Milesovka	833
5. Svratouch	737	6. Cervena hora	750
7. Brno	238	8. Znojmo	334
9. Luka	510	10. Lysa hora	1324

Figure 1: Meteorological stations in the Czech Republic provading wind data.

For the technical use of wind energy it is assumed necessary to build the wind park - turbine groups. Individual wind turbines properly situated at the given locality will be energetically connected to help one another and integrate total energetic output interconnected with contemporary energetic network. The choice of these localities and optimum distribution of wind turbines inside the chosen areas is the principal task of wind energy exploitation in future.

The most suitable locality should have the following meteorological characteristics:

- high annual mean wind velocities, the most important determining factor for the annual energy yield per wind power plant,
- rare occurrence of extreme turbulence, i.e. on the average a low shear of wind direction and velocities as prerequisite for disturbance-free operation of the plant,
- dominating main wind directions to minimize the area required for a multi-unit plant.

Besides of meteorological characteristics, various other criteria play essential roles in the ultimate choice of a location for the utilization of wind power. These are as follows:

- the available transport network and the possibility of integration into the existing grid,
- economic factors (e.g., the price of land),
- legal rules e.g., nature conservation laws and safety regulations for flights, populated areas and the immediately affected population, especially farmers,
- negative influences upon the environment such as noise, landscape disturbance and the disturbance of radio and television reception.

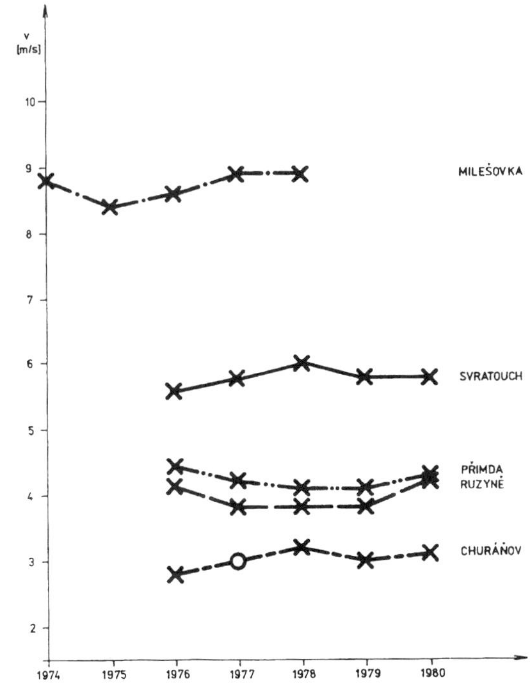

Figure 2: Annual mean wind velocities.

2. EVALUTIONS OF WIND DATA

2.1 Annual Mean Wind Velocities

The annual mean wind velocities have been obtained from various meteorological stations at the same height - 10 m above the ground. Their values are shown in Fig. 2 and Fig. 3. At the level of 10 m above ground a great influence of the relief and the existence of turbulence can be observed. The measurements made at higher levels are characterized by smaller pulsations of wind velocity and represent a real contribution to the calculated wind energy potential. The level from 50 to 100 m is supposed to be optimal from the wind turbine construction point of view.

To obtain the information on wind velocity, as indicated the well-known Hellmann formula had been used

$$v(h_2) = v(h_1)\left[\frac{h_2}{h_1}\right]^{\alpha}$$

where h_1 : height of measurement

h_2 : height of interest

$v(h_1)$: velocity at height h_1

$v(h_2)$: velocity at height h_2

α : altitute exponent

Annual and monthly average wind velocities have been computed for the heights of 50 and 100 m above the ground.

Localities Churanov, Cervena, Znojmo, Turany have a low annual and monthly average velocities corresponding to minimum input velocity for the action of wind turbines.

The highest velocities are in the region of Milesovka and Lysa, both in mountain areas. High values have been obtained at the localities Luka and Primda. The best conditions are at the locality Svratouch where for the height of 50 m above the ground, the annual mean velocities throughout the 5 year period are between 8 and 9 m/s and for the height of 100 and between 10 and 11 m/s.

Three last localities Luka, Primda, Svratouch have at the height of 100 m above ground annual average velocities higher than the minimum input velocity required for the action of big wind turbines. The Svratouch area seems to be the optimum one.

2.2 Monthly Mean Wind Velocities

In Figs. 4 and 5 the monthly mean velocities for the localities Svratouch and Luka confirm the general fact that the heighest wind velocities occur during winter seasons and the lowest ones during summer seasons.

The social requirements of maximum energetic sources during winter seasons are very important and should have an essential influence on decisions in the field of exploitation of the wind energy for common interests. In comparison with the solar energy which has its maximum during summer seasons, the wind energy is potentially superior. In a suitable combination with the solar energetic sources or as a supplement of the classical energetic sources it could help to cover the energetic peak-loads during winter seasons.

2.3 Extreme Wind Velocity

The operating safety of the wind power plants requires the design of the constructions elements and of the whole system to correspond to the least favourable actually occuring momentary wind fields. The extreme values of forces and moments at blades and towers which are subject to rapid temporal and spatial change can be established only by simultaneous measurements of wind velocities, accelerations and direction in the total area. As long as such measurements are not available, the design of wind power plants as regards operating safety must be determined by the existing extreme values of hourly or ten-minute mean measurements at the standard height of measurements. Figs. 5 and 6 give the maximum annual mean wind velocities.

2.4 Potential of Wind Pover

Theoretical wind potential is defined as the energy which has come to the "unit area" during one year or one month. It can be computed from the following equation

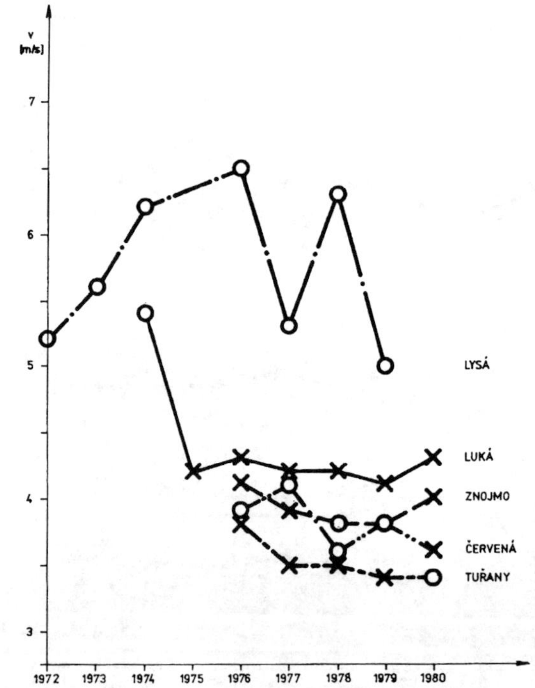

Figure 3: Annual mean wind velocities.

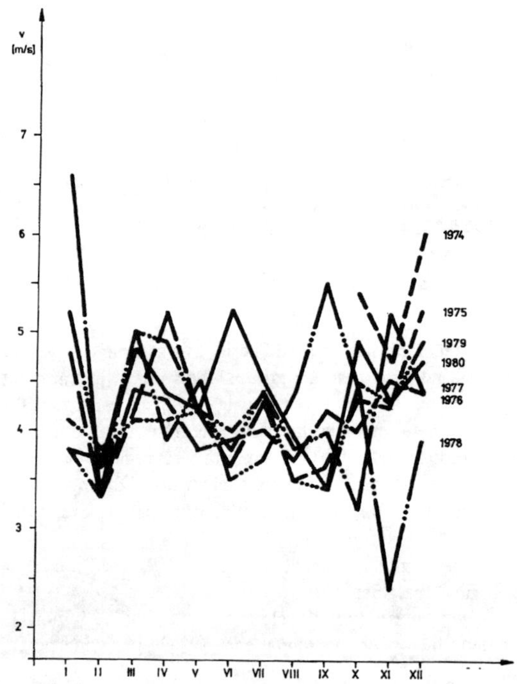

Figure 5: Monthly mean velocities for the location Luka.

$$W_p = \frac{1}{2}\rho.v^3 \ /W.m^{-2}/$$

where ρ : is density of air
v : wind velocity

The technical potential is then about 34 per cent of the theoretical wind potential. Now, let us make an approximate estimate of the wind energy potential for the territory of the Czech Republic from data obtained at 10 measuring localities during the 5 year period using monthly potentials.

For a reasonable estimate of wind balance the extreme high measurements obtained at mountain localities Milesovka and Lysa and the extreme low ones from Churanov will not be considered.

The mean theoretical potential is then:

height /m/	potential /w.m^{-2}/
10	51
50	154
100	258

Recalculating for the territory of the Czech Republic during one year yields the following data on theoretical and technical wind potential:

Theoretical potential			
10 m	34 TWh	during one year	= 122 PJ
50 m	104 TWh		= 374 PJ
100 m	194 TWh		= 698 PJ
Technical potential			
10 m	11 TWh	during one year	= 40 PJ
50 m	35 TWh		= 126 PJ
100 m	66 TWh		= 238 PJ

For the analysis of the reality and the correction limits the official governmental data on technically usable part of wind energy of the total FRG territory, i.e. 220 TWh a year, has been used. It corresponds to 792 PJ a year for the territory of about 245000 km^2. Thus from our point of view under conditions of our energetical estimate we can assume that FRG has (with the exclusion of coastal regions) roughly similar wind conditions as our territory. In proportion, the territory of Czech Republic with its 78000 km^2 can have an estimated technical potential of 70 TWh a year, what is surprisingly identical to the amount of 66 TWh calculated for one year at the height of 100 m above the ground.

In spite of the approximation in the wind energy calculation, it must be admited that great potential reserves exist in the wind energy. Its exploitation is limited only by technical development of the wind aggregates.

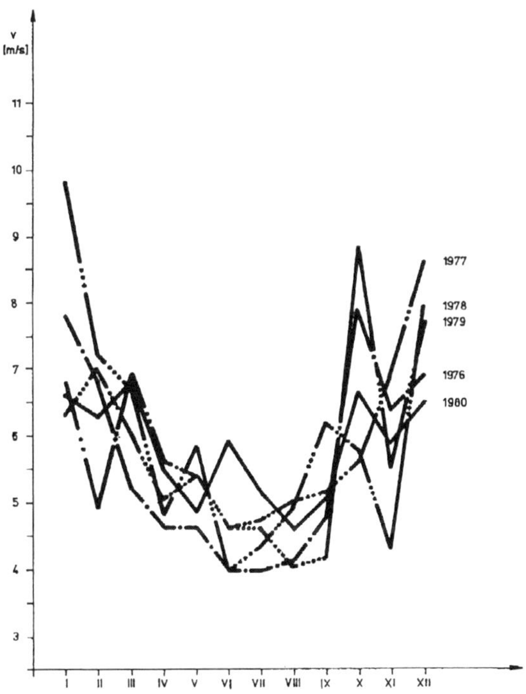

Figure 4: Monthly mean velocities for the location Svratouch.

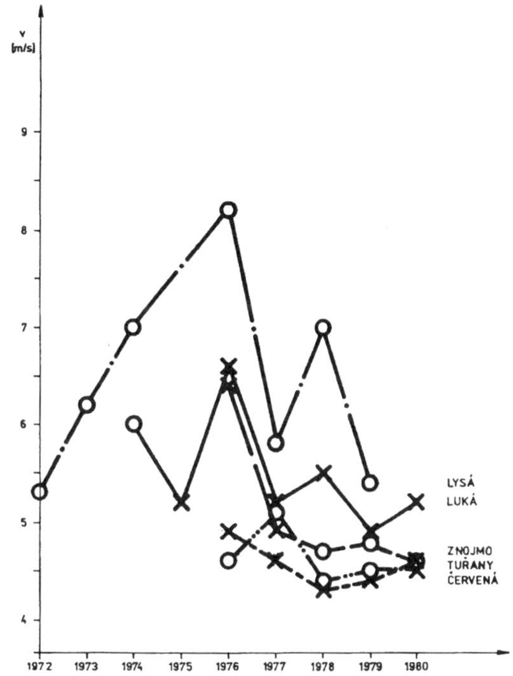

Figure 6: Annual maximum mean wind velocities.

3. SUMMARY

3.1 The characteristic data on wind conditions in the Czech Republic on the basis of data from 10 measuring localities during the five years period 1976-1980 have been used as the base for calculating the wind energy available.

3.2 The estimate of energy potential for the Czech territory is as follows:

Theoretical potential			
10 m	34 TWh	during one year	= 122 PJ
50 m	104 TWh		= 374 PJ
100 m	194 TWh		= 698 PJ
Technical potential			
10 m	11 TWh	during one year	= 40 PJ
50 m	35 TWh		= 126 PJ
100 m	66 TWh		= 238 PJ

Taking into account the working height of wind turbines between 50 and 150 m, the wind energy compared with the hydroenergetic potential in the Czech Republic:

- theoretical potential 47 PJ a year
- technical potencial 11 PJ a year

seems to be very favourable even after considering all circumstances of calculation of very changeable wind "field".

3.3 The best wind conditions from 10 chosen localities are in Svratouch locality where for the height of 50 m above the ground the mean velocities during the five years period are in the

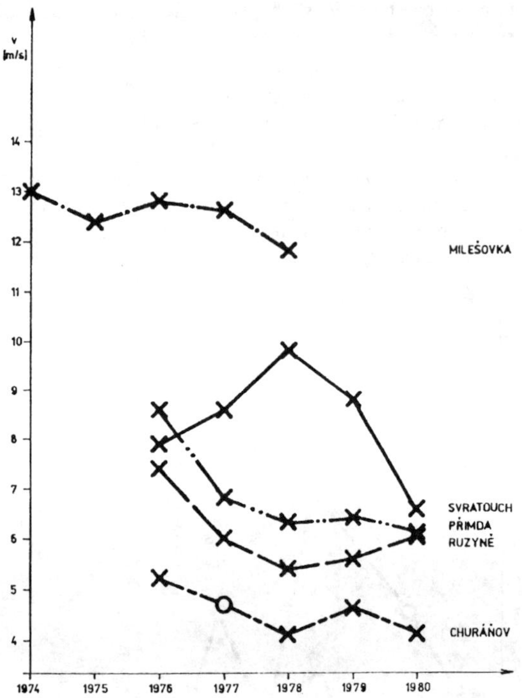

Figure 7: Annual maximum mean wind velocities.

interval between 8 and 9 m/s and for the height of 100 m between 10 and 11 m/s. These values are within the technical requirements of wind turbines.

3.4 The wind energy has its maximum energetic effectivity during winter seasons and its minimum during summer seasons. Compared with the solar energy which has its maximum during summer seasons the wind energy is potentially superior.

REFERENCES

[1] L. Jarass, L. Hoffmann, A. Jarass, G. Obermair, Wind Energy, Springer-Verlag, Berlin (1981)
[2] J. M Sherman, M. S. Gresham, D. L. Fergason, Wind System Life Cycle Cost Analysis, Rockwell International Corporation, U.S. Department of Energy (1983)
[3] The Potential for Wind Energy in Developing Countries, U.S. Department of Energy (1988)

"Determination of the hourly wind speed field over complex terrain in Southern Uruguay"

C. López[1] , J. Cataldo[2]

[1] Centro de Cálculo, [2] Instituto de Mecánica de los Fluidos e Ingeniería Ambiental

Facultad de Ingeniería, J.H. y Reissig 565, CP11300, Montevideo, Uruguay.

ABSTRACT: Numerical and physical modelling techniques are applied jointly to deduce the velocity wind field up to the meteorogical microscale in Southern Uruguay. The calculated wind velocities in some points are verified against measurements taken with cup anemometers.

1. Introduction.

Wind climate determination is the common problem in wind power evaluation programs. The wind velocity long time series are known in stations of the weather net. The weather stations are usually located in sites with little interest for installation of wind power conversion systems (WECS). In Uruguay as in other places, the sites of principal interest for wind power exploitation are over complex terrain zones where measurements of wind velocity do not exist. Druyan,L., 1985 suggest a summary of methods to deduce the wind velocity field. Such methods can be classified in subjective and objective methods. The first one, justified where very scarce data are available, use the observation of tree deformation or other forest indicators to deduce the wind velocity. Between the objective methods on site direct measuring for short periods (Barros et.al.,1983), physical modelling when some data are known near to the place of interest (Neal,D.,1979) and extrapolation of routine meteorological data by mass consistent models (Endlich, et.al., 1982), cab be mentioned as well as another method mentioned in Tombrou,M.,et.al., 1990.
The methodology applied to estimate the hourly wind field in Southern Uruguay up to the meteorological microscale uses a combination of numerical and physical modelling. As will be explained later the numerical model is a mass consistent code to solve the mean wind velocity field at 5 to 30 km side mesh and the physical model allows determining the mean velocity and the turbulence characteristics of the flow over the complex terrain at meteorological microscale.
The field measurements with which wind mean velocity and direction used for comparision with model results were obtained, will be described.

2. Numerical model.Method description.

The numerical model used is a mass consistent code. This model uses as data the hourly mean wind velocity time series obtained in the weather stations located in the volume to be studied. This volume (Fig. 1) has a 200x200km base and its height is equal to the one of the atmospheric boundary layer (500 to 1500m).
Essentially, the methodology is similar to the one used in Sherman, 1978. A first guess of the mean wind field is obtained by a weighting mean of the known mean velocities, where the weighting factors are proportional to the inverse of the distance between the site and the weather station. The mean velocity field is calculated as the one which makes the integral square difference minimum between the first guess and the mean velocity field and verifying

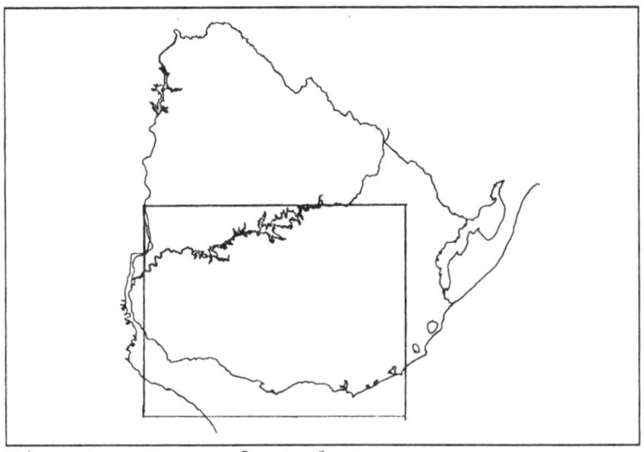

Fig. 1 - Zone of study

the equation of continuity. The equation obtained is a Poisson equation which is solved using the finite element method. López, 1992 points out details of the methodology and compares calculation and theory for potential flows.
The data of the velocity vectors are expressed in an orthogonal frame of reference

$$V_i = V_m + \sum_{i=1}^{i=n} a_{i,j} \cdot e_j$$

where V_i is the known velocity vector at the time i, V_m is the time mean of the velocity vectors, e_j are the reference vectors and $a_{i,j}$ are the coordinates of the vector V_i-V_m in the e_j base. The aforementioned calculations are applied to the vectors V_m and e_j. In this case the frame of reference is composed by the eigenvectors of the covariance matrix of the data (e_j vectors). The results obtained applying the methodology to the three or four eigenvectors with higher eigenvalues shows little difference with the results obtained for all the eigenvectors. Table 1 shows the mean velocity values calculated using the three eigenvectors with higher eigenvalues and all eigenvectors at different sites.
Fig. 2 shows the mean velocity field obtained at 30m over the terrain in the grid nodes with 15km mesh size.

3. Physical model.

The physical modelling was used to solve the wind velocity field up to the meteorological microscale. Thus, it was necessary to simulate the Atmospheric Boundary - Layer flow (ABL) and the terrain.

The ABL was modelled for "high wind"

Site	N⁰ of vectors 3	10
C. del Toro	26.6	27.2
Caracoles	27.5	27.8
Animas	24.0	24.6
José Ignacio	24.0	24.6
Aigua	23.0	23.4

Table 1 - Calculated mean velocity values.

conditions, which meant a turbulent, steady, thermically neutral and aerodynamically rough flow. The obtained mean velocity profile was adjusted to a

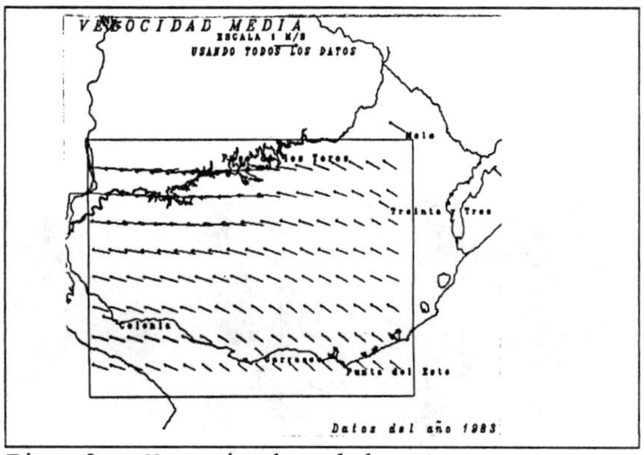

Fig. 2 - Numerical model out put.

logarithmic law up to heigth z_l (height of the inertial sub-layer) and to power law over that zone (wake sub-layer). The intensity of turbulence profile was compared to the results reported by Harris, 1969 and E.S.D.U., 1985. The length scale of the turbulence profile were obtained from the last reference while Counihan, 1975, was used to compare the results. From the match between measured spectral peak and the von Karman spectral peak, the geometrical scale (1/6000) used to compare all the aforementioned magnitudes was obtained.
The characteristic roughness length in the zone to be studied is 5cm and the heigth of the boundary-layer was initially estimated in 600m.
The method reported first in Counihan, 1969 and later in Robins, 1979 for ABL simulation was used. The physical modelling was made in a wind tunnel with a 1.2x1.6x3.6m test section. The measurements of velocity were made with a TSI hot-wire anemometer. The mean velocity was calculated from the anemometer indication and the calibration curve. The mean square root and the power spectral density of the anemometer signal was obtained from a spectra analyzer Hewlett-Packard 3582A. The hot-wire probe was positioned using a four freedom degrees robot. Fig. 3, 4 and 5 show the mean velocity profile, intensity of turbulence profile and the longitudinal scale length of the turbulence profile. It is wortwhile to points out that the turbulence characteristics did not vary after eight heights of simulated ABL, as is reported in Robins, 1979, but the mean velocity profile changed up to 12 heights of ABL simulated downstream from the ABL simulation system. Cataldo, 1992 gives more details about this physical simulation.
The terrain was modelled to scale 1/6000 as was deduced before. The models were

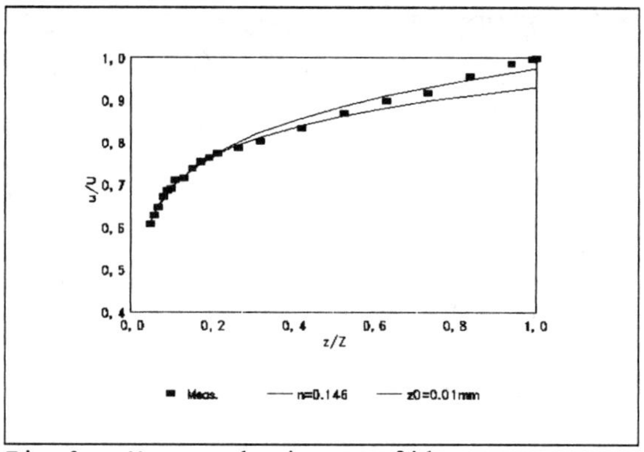

Fig.3 - Mean velocity profile.

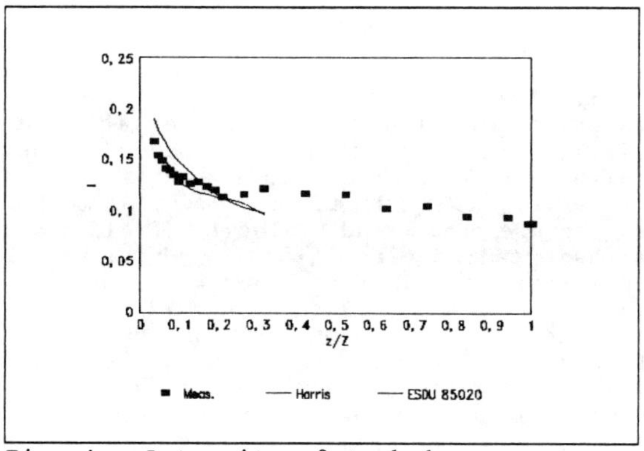

Fig. 4 - Intensity of turbulence.

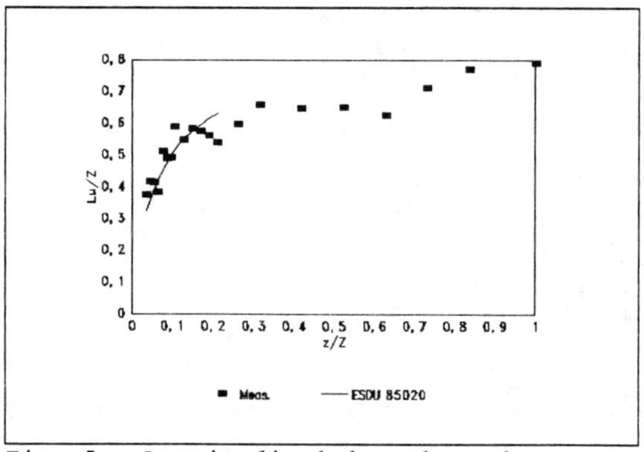

Fig. 5 - Longitudinal length scale.

constructed following the method described in Neal, 1979 and they were located where the flow in the wind tunnel was able to be considered self-preserving. The measures were made over a set of points of the model, where the existence of a wind farm is assumed. The local speed-up and the intensity of turbulence were deduced at 30m (at the prototype scale) over each point.

4. Field measurements.

The field measurements used in this paper were undertaken as part of the URU/87/028 program and they were obtained from cup anemometer and vane B.A.P.T. mounted on 10m towers. The on-site measuring was made for short periods as part of the overall methodology reported in Barros,et. al., 1983. Basically, the measurements were undertaken for three month periods. The towers were deployed in several sites in Southern Uruguay. Some towers were located in homologous sites to the model selected points. These points were identified as Cerro del Toro, Sierra de Caracoles, Sierra de Animas and José Ignacio.

5. Calculation vs. measurements comparisons. Results and discussion.

Applying the speed-up factors deduced from the physical model to the wind velocity time series obtained from the numerical modelling the wind velocity time series was obtained at the sites where there are field measurements. Thus, several statistical parameters of these time series were obtained at the different sites, specially for the interval of time with field measurements. The same statistical parameters were calculated for the field measurements.
Table 2 shows the calculated and measured mean velocity values. The first one was obtained for one year period while the measurements were taken during three to six months.

Site	Mean cal. vel. km/h	Mean meas. vel. (m/s)	Dif. (%)
C. del Toro	27.2	25.8	5.3
S. de Animas	24.6	24.5	0.4
S. de Caracoles	27.8	29.5	-5.5
José Ignacio	24.6	24.5	0.2

Table 2 - Mean velocities.

A maximun difference of 5.5% can be observed. The change of sign in the difference can be assigned to different seasonal mean velocity value.
Figs. 6, 7, 8 and 9 show the probability density function calculated and measured for the Cerro del Toro, Animas, José Ignacio and Sierra de Caracoles sites. In the two first cases the calculated and measured probability functions present the peak at the same velocity value, while in José Ignacio site the measured peak is lower than the calculated one and in Sierra de Caracoles site the opposite is founded. In all cases a calculated function smoother than the measured one is observed, which to the elimination of experimental errors from the calculations can be assigned.
Table 3 shows the energy calculated with the wind velocity time series calculated and measured at the considered points using the perfomance curve of a real wind turbine.

Site	Cal. (kw.h/year)	Meas. (kw.h/year)
C. del Toro	773508	853400
S. de Caracoles	724452	1004772
José Ignacio	572904	551442

Tabla 2 - Calculated energy.

7. Conclusions.

The described method provides a relatively fast and low cost tool that provides a good estimation of the wind mean velocity in complex terrain regions of interest in the choice of locations for WECS. This can be deduced from the preliminary results presented in this paper. The mean velocity values deduced (Table 2) from calculated and measured time series show a low difference, but in some cases high enough to estimate the energy (see Table 3). However, it is a good method to make comparisons between different sites (compare Tables 2 and 3) where more intensive studies must to be done. The method also allow us to deduce the turbulence characteristics of the sites necessary to find out structural problems and control system operation prior of the installation.
Two difficulties should be pointed out about this method: 1)the results are highly dependent on the data quality, but following the principal components method used in this paper the quality can be studied; and 2)the

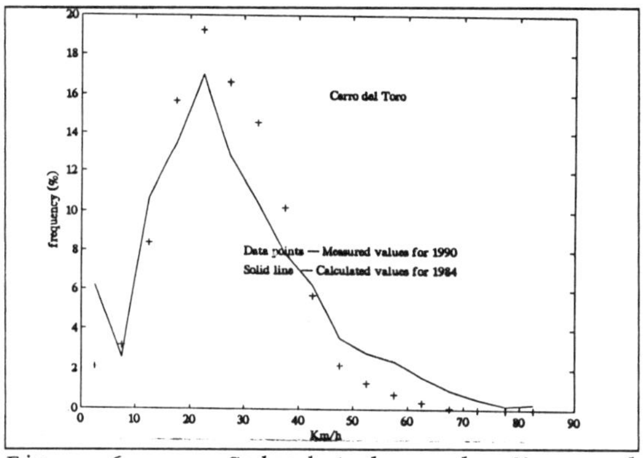

Fig. 6 - Calculated and Measured Probability in Cerro del Toro site.

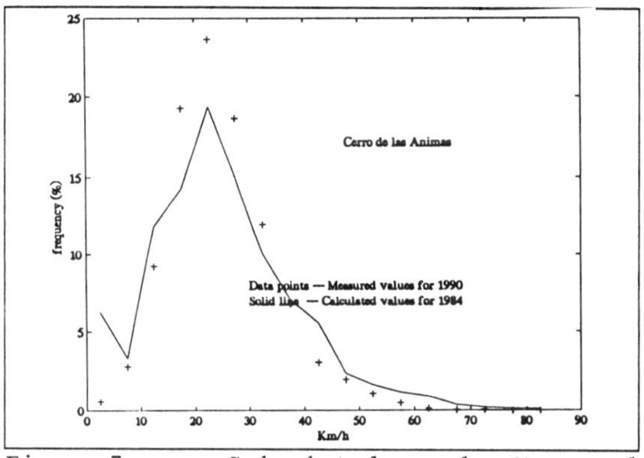

Fig. 7 - Calculated and Measured probability in Animas site.

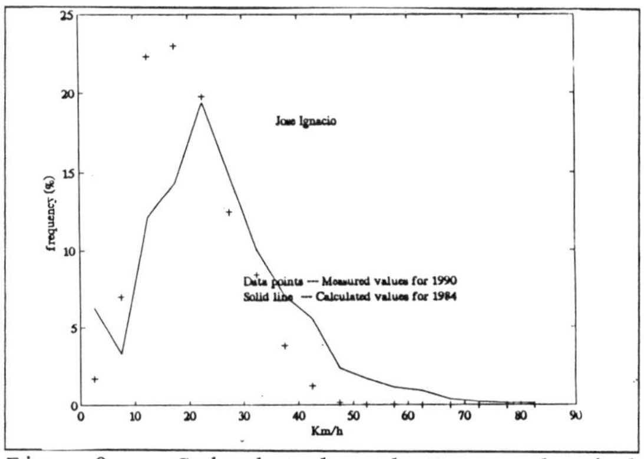

Fig. 8 - Calculated and Measured wind velocity probability density in José Ignacio site.

kind of terrain can limit the test

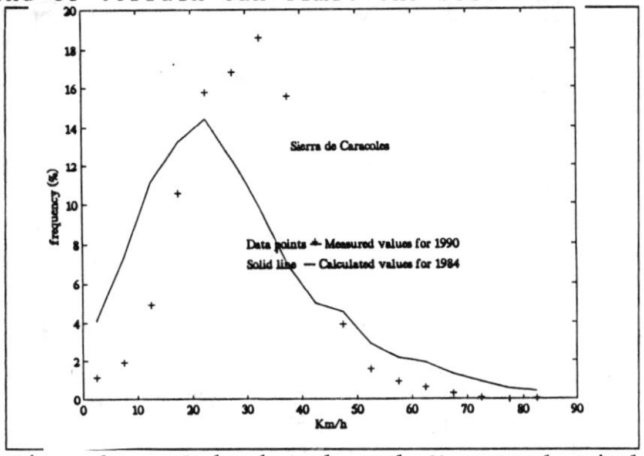

Fig. 9 - Calculated and Measured wind velocity probability density in Sierra de Caracoles site.

directions of the models, which may imply that the speed-up factors for those directions must be assumed.
This method provides a fast and low cost tool to make a first relatively conservative estimation of the available wind power in extensive zones of complex terrain.

8. References.

Barros,V. and Rodríguez Seró, J. "Measurement strategies: use of short observation records for estimating the annual wind variations", Centro Nacional Patagónico, Conicet, Argentina, 1983.

Cambilargiu, E "Primer mapa isoviento del Uruguay", Boletín de la Facultad de Ingeniería y Agrimensura, vol. VIII, N^{0}11, Montevideo, 1963.

Cataldo,J. and Farell,C. "Comparación entre simulaciones físicas de flujos tipo Capa Límite Atmosférica a escalas 1/500 y 1/6250", XV Congreso Latinoamericano de Hidráulica, Cartagena, 1992. (In spanish, with english abstract)

Counihan, J. "An improved method of simulating an atmospheric boundary-layer in a wind tunnel", Atmos. Env.,vol.3, 197-214, 1969.

Counihan, J. "Adiabatic atmospheric boudary-layers: A review and analysis of data from the period 1880-1972", Atmos. Env., vol. 9, 871-905, 1975.

Druyan,L. "Wind climate srudies for WECS siting", J. of Climatology, vol. 5, 95-104, 1985.

Endlich,R.M., Ludwig,F.L., Bhumralkar,C.M. and Estoque,M.A."A diagnostic model for estimating winds at potential sites for wind turbines", J. Appl. Meteorol., 21, 1441-1454, 1982.

E.S.D.U. "Characteristics of atmospheric turbulence near the ground. Part II: single point data for strong winds (neutral atmosphere)", Item Number 85020, 1985.

Harris,I. "Measurements of wind structure at heigths up to 598ft above ground level", Symposium on Wind Effects on Buildings and Structures Organized by Loughborough University of Technology National Physical Laboratory Royal Aeronautical Society, 1969.

López,C.L. "Estimación del campo de velocidades a partir de un número insuficiente de datos", 1ª Reunión del Grupo de Trabajo sobre Hidromecánica, División Latinoamerica de la IAHR, Salto Grande, Marzo, 1992. (In spanish, with english abstract)

Neal,D. Wind flow and structure over Gebbies Pass, New Zealand: a comparision between wind tunnel simulation and field measurements", Ph.D. thesis, University of Canterbury, Christchurch, New Zealand, 1979.

Robins, A.G. "The development and structure of simulated neutrally stable atmospheric boundary-layers", J. of Industrial Aerodynamics, vol. 4, 71-100, 1979.

Sherman,C. "A mass-consistent model for wind fields over complex terrain", J. Appl. Meteorol., 17, 312-319, 1978.

Tombrou,M. and Lalas, D.P. " A telescoping procedure for local wind energy potential assessment", European Community Wind Energy Conference, Madrid, 1990.

CLIMATOLOGICAL MEAN WIND AND TURBULENCE MEASUREMENTS USING A HIGH QUALITY CUP ANEMOMETER SYSTEM

Hans Bergström and Knut Lundin
Department of Meteorology, Uppsala University
Box 516, S-751 20 Uppsala, Sweden

ABSTRACT: A combined cup anemometer/wind vane instrument, developed at the Department of Meteorology in Uppsala, is presented. The instrument has a frequency response of about 1 Hz, making it suitable not only for mean wind measurements but also for studies of atmospheric turbulence. It is robust enough to be used for climatological purposes. Comparisons with data from a hot-film anemometer show good agreement, both as regards standard deviations and the spectral decomposition of the turbulent wind signal.
The cup anemometer/wind vane instrument is currently used at two sites within the Swedish wind energy research programme. These measurements are shortly described, and a few examples of the results are also given.

1. INTRODUCTION

Wind measurements with high temporal resolution during long periods of time are so far only available from a limited number of sites, mainly due to the high cost of the instrumental equipment, e.g. sonic anemometers.

This type of data is, however, of great interest in wind energy research, e.g. when estimating load and fatigue on wind turbines, caused by the atmospheric turbulence. Also when investigating the extreme wind speed distribution for short averaging times, long time high resolution data is needed. This is also the case when studying rapid changes in wind speed and direction or to get climatological data on added wake turbulence.

At the Department of Meteorology in Uppsala a combined *C*up *A*nemometer/*Wi*nd *V*ane (CAWiV) instrument has been developed. The instrument has a distance constant of about 1 m, making it possible to measure the turbulent fluctuations of the two horizontal wind components up to a frequency of 1 Hz. This is sufficient for wind energy purposes, and thus at a rather low cost makes it possible to take high quality measurements of the atmospheric turbulence structure.

2. TECHNICAL DESCRIPTION

A schematic picture of the instrument and a block diagram illustrating its components is shown in Figure 1. The wind speed is measured with a three-cup anemometer of the pulse type. The radius of the rotor is 4 cm, and the cups are conical in shape, made of plastic, and has a weight of only 10 g. The cups are of the same shape as a larger version, which has been extensively tested in many field experiments, and it has been demonstrated [1] that the over-speeding of this larger anemometer is negligible.

A 40-slot disc connected to the shaft of the anemometer chops an infrared beam. The pulses are detected by an infrared sensitive photo transistor, and the pulse frequency is fed into an F/V converter, where the frequency is transformed to an analog voltage signal. This voltage is then a direct measure of the wind speed.

The wind direction part of the instrument consists of a potentiometer, sensing the position of a high performance wind vane, made of balsa wood. The size of the wind vane is 8×16 cm.

The complete instrument thus makes it possible to measure the two horizontal wind components with a high frequency resolution. Before storing data on cassette tapes at a sampling frequency of 1 Hz, in order to avoid aliasing the analog signals are passed through electronic filters with their half energy points at 0.5 Hz.

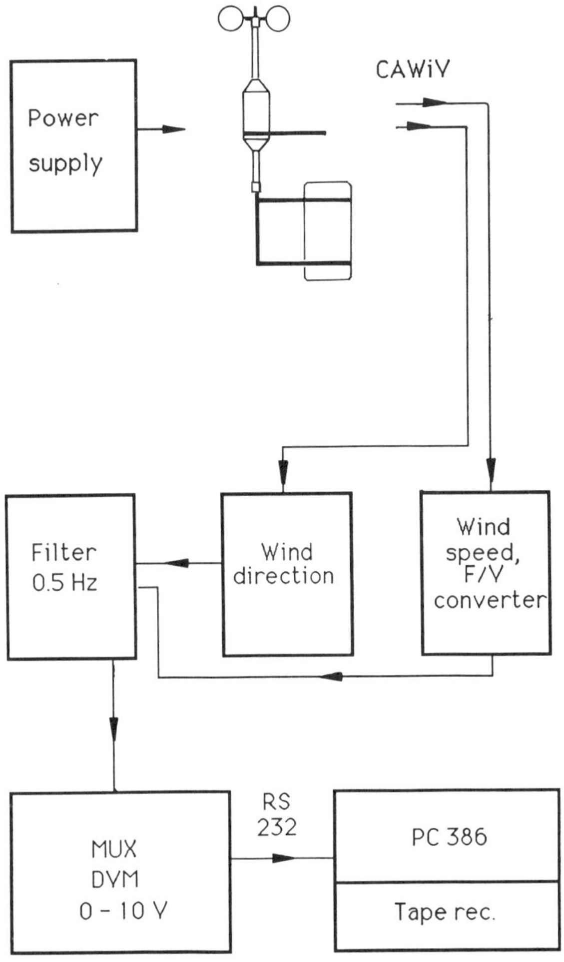

Figure 1: Schematic picture and block diagram of the cup anemometer/wind vane system.

Before taken into use, 16 anemometers of this type were calibrated in a wind tunnel. These calibrations showed that the mechanical parts of the anemometer (bearing, shaft, slot disc) gave practically identical wind response when compared using the same cups. Comparisons of the different cups also showed good agreement, all of them being within ±1.5% in measured wind speed.

3. FIELD PERFORMANCE TESTS

Data from the Alsvik site (cf. below) sampled with a frequency of 1 Hz was compared with turbulence data taken with a hot-film instrument, [2], sampled at 20 Hz. The comparison shows that the CAWiV instrument has negligible over-speeding, and is capable of measuring the variance of the longitudinal and the lateral wind components with high accuracy, as can be seen in Figures 2a-b. At 40 m the cup anemometer instrument catches 93% and 95% of the standard deviations of the longitudinal and lateral wind components respectively, compared to measurements with the hot film instrument. The remaining 5-7% are the contributions from frequencies between 1 Hz and 20 Hz. The same result was arrived at in [3], regarding a mini-sonde with a cup anemometer of the same type as used here. The losses can easily be corrected for by assuming the inertial subrange -2/3-slope to be valid above 0.5 Hz, and integrating over the higher frequencies to estimate the losses.

Spectral densities of the longitudinal and lateral wind components were determined using the FFT technique. Also spectra from the CAWiV instrument compare well with those measured with the hot film instrument. This can be seen in Figures 3a-b, showing the longitudinal and lateral spectral densities at 31 m during a 2 hour period. The curves follow each other closely at all frequencies.

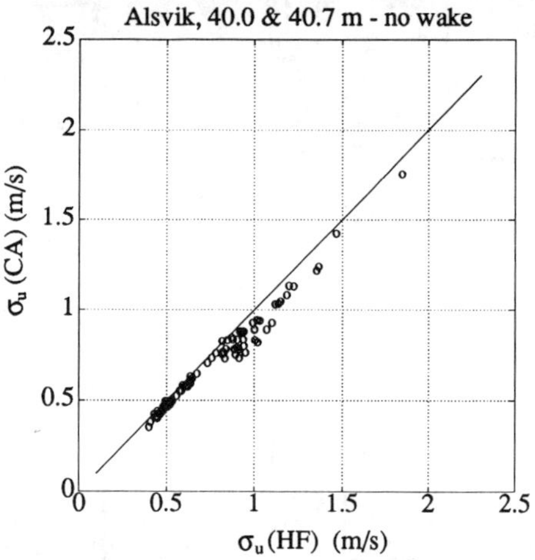

Figure 2a: Relation between standard deviation of longitudinal wind component measured with the cup anemometer (CA) and with the hot film instrument (HF).

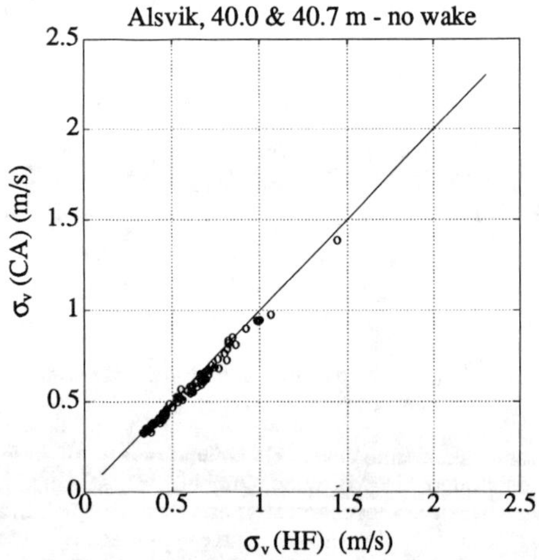

Figure 2b: Same as in Figure 2a but for the lateral wind component.

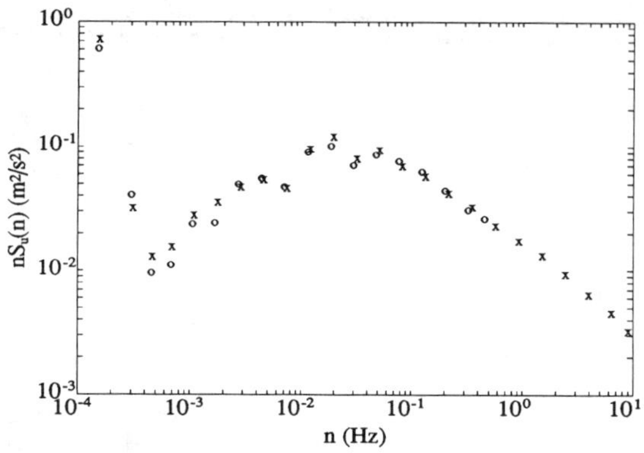

Figure 3a: Comparison between spectrum of the longitudinal wind component measured with the cup anemometer system (o) and with the hot film instrument (x).

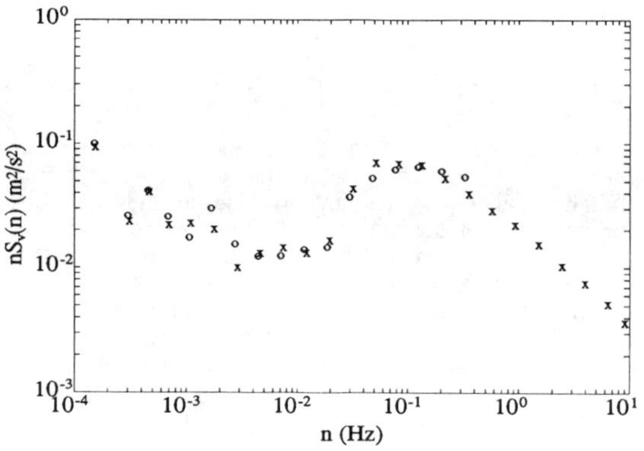

Figure 3b: Same as in Figure 3a but for the lateral wind component.

4. EXAMPLES OF APPLICATION

The CAWiV instrument is presently used at two of the Swedish wind turbine sites, Alsvik on the island of Gotland in the Baltic Sea, and Lyse at the Swedish west coast, see Figure 4.

4.1 Alsvik

Measurements at Alsvik have been taken since April 1990. Two towers (M1 and M2 in Figure 4) are since June 1991 equipped with anemometers at 8 levels, (10, 18, 24, 31, 36, 41, 47, and 53 m), which are sampled with 1 Hz and stored on data cartridge tapes. Before June 1991 instruments were mounted at 7 heights from 3 to 41 m. The measurements also include temperature at five levels (3, 10, 18, 31, and 41 m) in tower M1.

The Alsvik measurements are primarily intended for the study of wind turbine wakes at a number of distances from 1 to 9.6 rotor diameters downwind of the four 23 m diameter, 180 kW turbines located at this site, 50-200 m from the shore-line (T1-T4 in Figure 4). But they also give valuable information on the structure of the atmospheric turbulence, suitable for load and fatigue estimates, e.g. RFC (rain flow count) statistics of the wind speed and direction, [3] and [4].

The site is very flat and the coastal area surrounding the turbines is mainly grassland, occasionally partly covered with small ponds when the water level of the Baltic Sea is high, making it difficult to accurately determine the exact location of the coast line. To the east of the site, towards the interior of the island, there are some forested areas with mainly 5-10 m high pines.

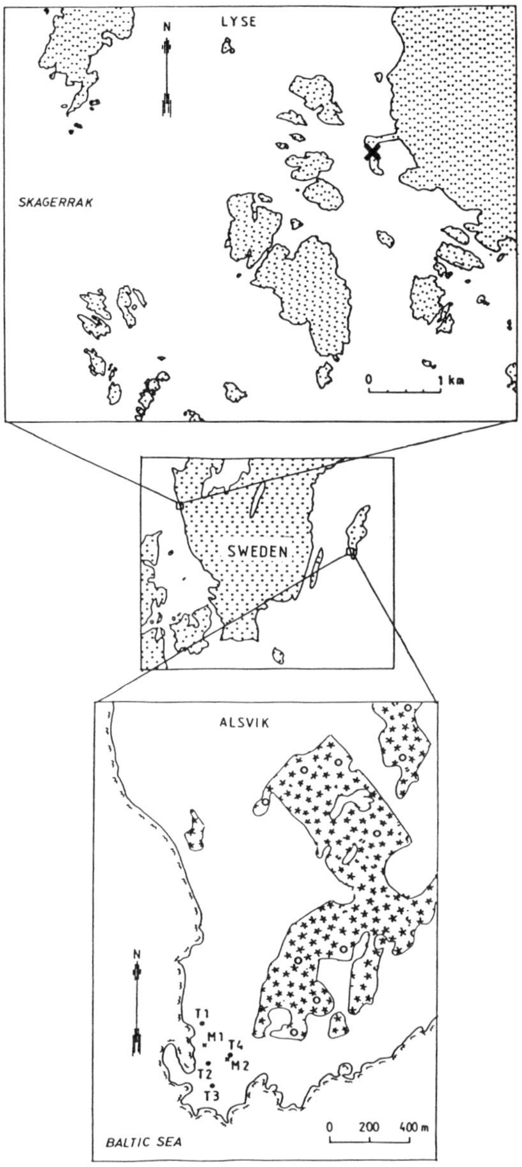

Figure 4: Detailed maps of the two experimental sites. Lyse: tower marked **x**. Alsvik: towers marked M1 and M2, wind turbines marked T1-T4.

The location of towers and turbines allows for free flow measurements in either of the two towers from all directions except for a 36° sector around NNE and a 19° sector around ESE. But as winds from these sectors climatologically have a low probability, we could expect to get undisturbed measurements about 90% of the time.

The climatological influence from the four wind turbines upon the turbulence characteristics could e.g. be illustrated by comparing the cumulative distribution of turbulence intensity, σ_u/U (where σ_u=standard deviation of the wind, and U=mean wind speed), in the free flow with that at tower M2. This is shown in Figure 5, where we can see that the contribution from the four wakes on a 'climatological' basis is clear but not very large. The median value increases from 0.08 to 0.10 although the added turbulence from a single wake is much more pronounced.

An example of measurements on tower M2 of the wake from turbine T2 (distance 4.2D) is given in Figure 6, showing the added wake turbulence intensity expressed as parts of the free flow value. Thus 1 unit here means that the added part equals the free flow value, 2 units that it equals two times the free flow value. It is obvious that the added turbulence intensity is of the order once or twice the value in the undisturbed flow,

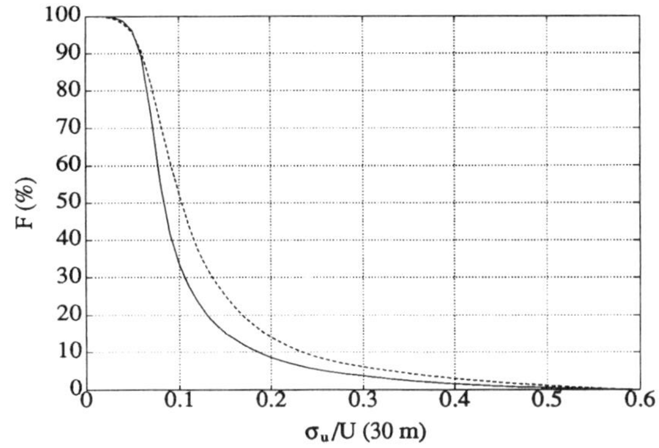

Figure 5: Cumulative frequency distribution of turbulence intensity at Alsvik during a one year period 1991-92; —: free flow data, ---: data from tower M2 alone, affected by wakes from turbines T1-T4.

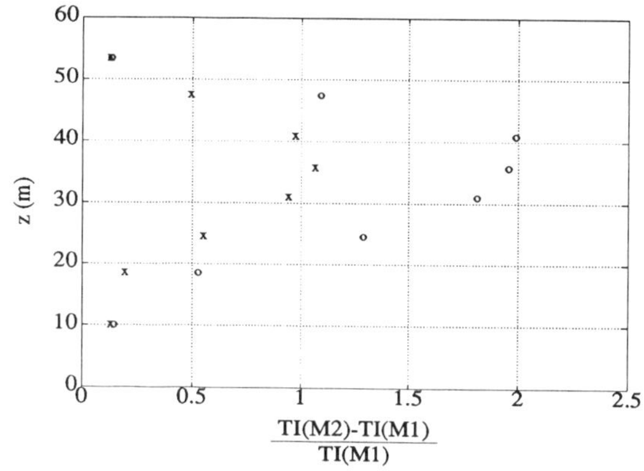

Figure 6: Vertical profiles of wake added turbulence intensity from turbine T2 at tower M2 (4.2D); o: stable stratification, x: unstable stratification.

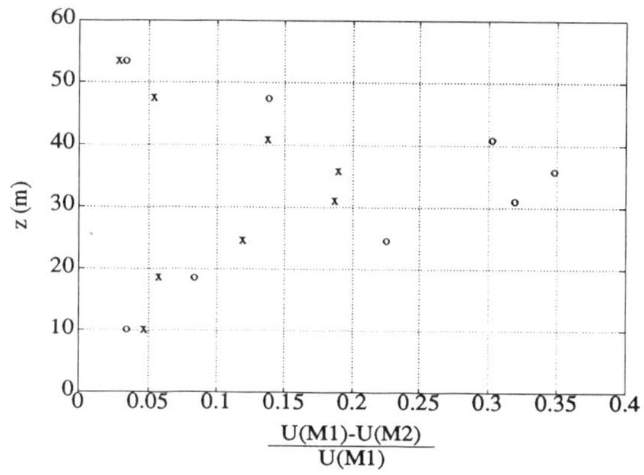

Figure 7: Vertical profiles of velocity deficit from turbine T2 at tower M2 (4.2D); o: stable stratification, x: unstable stratification.

and depends to a high degree on the atmospheric stability. During unstable conditions the ambient turbulence is larger than with stable stratification, but the added turbulence is on the other hand larger in stable conditions, so that the wake centre line turbulence intensity is about the same for stable and unstable stratification.

Also the velocity deficit is highly dependent on atmospheric stability, as can be seen in Figure 7, showing that for stable stratification the deficit could be almost twice that found in unstable conditions. A more detailed analysis of the wake data from Alsvik is presented in [5] and [6].

4.2 Lyse

Since the autumn of 1992 similar 1 Hz tower measurements are taken at the Lyse wind turbine site, using the same type of instruments. To be able to get undisturbed measurements from all directions, anemometers have been mounted at 7 heights (10, 23, 31, 40, 48, 57, and 64 m) on the northwestern side of the tower, and at 5 heights (32, 40, 49, 57, and 65 m) on the southeastern side. Temperature is measured at 5 heights (2, 10, 23, 40, and 64 m), and relative humidity and air pressure at 2 m.

The terrain at Lyse is, in contrast to the very flat and rather homogeneous Alsvik site, much more heterogeneous. The coastal area to the east is rocky with peaks reaching 50 m a.s.l. within the nearest 1-2 km. Nearby islands to the west and southwest are 30 to 40 m high. Both the islands and the nearby mainland rocks have only sparse vegetation, with patches of grass and herbs, and only very few bushes and trees. Thus the roughness of the surface could be expected to be rather small, except for the heterogeneous rocky landscape itself.

A comparison between the turbulence at Lyse and Alsvik reveals that with winds from the sea sector, the more heterogeneous terrain at Lyse increases the turbulence intensity at 30 m from 0.08 at Alsvik to around 0.12, see Figure 8. The small kink in the Lyse profile at about 35 m is at the top of the internal boundary layer which builds up over the 500-600 m wide bay to the west of the tower.

With easterly winds, i.e. from land, the turbulence intensities at Lyse are of the same order of magnitude as those at Alsvik above about 30 m, indicating that the turbulence intensity above a heterogeneous hilly terrain with sparse vegetation may be approximately the same as over flat, partly forested, terrain. At lower heights, the small bay east of the tower at Lyse reduces the turbulence intensity to values well below those at Alsvik at the 10 m level. It must be noted, however, that the amount of data from Lyse included in this comparison only covers a period of about two months, and consequently the exact numbers may be altered when more data is available.

Finally it is interesting to note that the neighbouring topographic features at Lyse alters the expected spectral characteristics of turbulence so that spectra do not scale with height in the ordinary manner. This can be seen in Figures 9a-b, where examples of u-spectra from the heights 32 m, 49 m, and 65 m are plotted for the sea- and land sectors respectively. The spectral curves fall close together with only minor deviations between the levels. In an 'ideal' boundary layer the frequency of the spectral peak should vary by about a factor of two between the 32 m and the 65 m level.

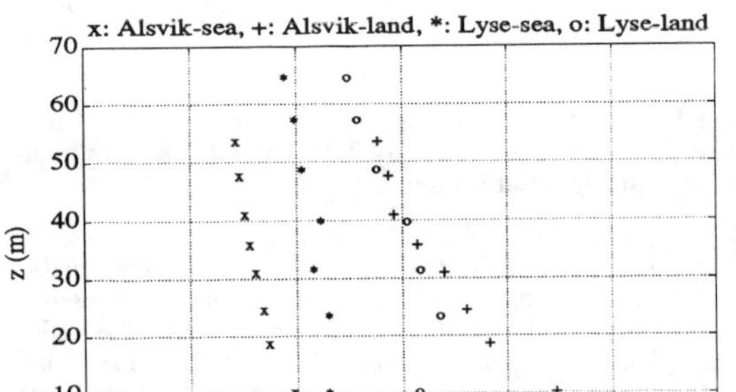

Figure 8: Comparison between vertical profiles of turbulence intensity at Alsvik and Lyse for two wind direction sectors, sea and land (i.e. west and east).

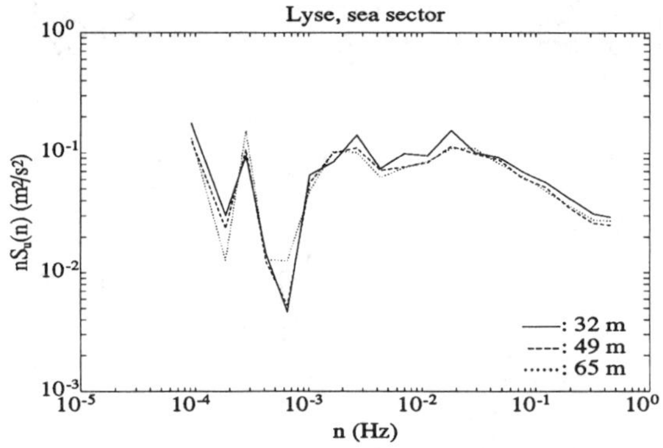

Figure 9a: Examples of spectra of the longitudinal wind component at three heights at Lyse, wind from the sea.

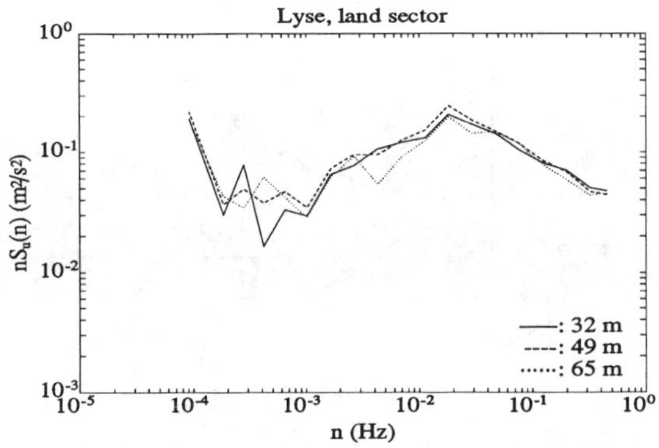

Figure 9b: Same as Figure 9a but wind from land.

5. SUMMARY

The field test reveals excellent performance of the CAWiV instrument. It truthfully records fluctuations in the two horizontal wind components of all scales up to the frequency limit 0.5 Hz set by the sampling frequency 1 Hz, and thus illustrates the possibility of using a relatively low cost instrument for high quality wind and turbulence measurements. With a global calibration curve for all instruments, the accuracy of the measured wind speed will be ±1.5%. Using individual calibrations for each instrument this small uncertainty could of course be further reduced.

Acknowledgement: The work was sponsored by NUTEK, Swedish Board for Industrial and Technical Development.

REFERENCES

[1] U. Högström,1988: Non-dimensional wind and temperature profiles in the atmospheric surface layer. Boundary-Layer Meteorol., 42, 55-78.

[2] U. Högström, 1982: A critical evaluation of a turbulence instrument. J. Appl. Meteorol., 5, 1838-1844.

[3] H. Bergström, 1991: Rain flow count and gust modeling of turbulent winds in the atmospheric boundary layer. Wind Energy Report WE 91:1, Department of Meteorology, Uppsala University, 29 pp.

[4] H. Bergström, H. Ganander, and H. Johansson, 1992: Wind description for designing Wecs, based on RFC evaluated wind measurements. Proceedings of Amsterdam EWEC'91, 762-766.

[5] M. Magnusson, 1993: Atmospheric stability influence on turbine wakes. Proceedings of the Travemünde ECWEC'93.

[6] M. Magnusson, 1993: Influence of atmospheric stability on wind turbine wakes. Wind Energy Report WE93:1, Department of Meteorology, Uppsala University. To be published.

GEOSTROPHIC WIND SPEEDS OVER THE NORTH SEA — A COMPARISON OF DIFFERENT CALCULATION METHODS

J.P. Coelingh and A.J.M. van Wijk

Utrecht University, Department of Science, Technology and Society
Padualaan 14, 3584 CH Utrecht, The Netherlands
telephone +31+30+537621/7600; telefax +31+30+537601

Summary — In the near future, placing wind turbines offshore in the North Sea may become a reality. In order to assess the technical and economic feasibility, it is necessary to have sufficiently detailed knowledge of the wind climate. This also includes knowledge of the geostrophic wind field, because the geostrophic wind speed is an important parameter for modelling wind speeds near the surface. In this paper, calculation methods for the geostrophic wind speed have been compared. A database of the K13 platform, located in the North Sea 100 km off the Dutch coast, has been used. It is found that, on average, calculation methods assuming neutral conditions of the atmosphere give the same results as calculation methods incorporating stability effects. Furthermore, the mean geostrophic wind speed at K13 over the period 1985–1990 was found to be between 11.6 m/s and 11.9 m/s, while the measured wind speed (at a height of 75 m above MSL) was 9.8 m/s. This is about 15% higher than values reported in the literature using pressure data for the calculation of the geostrophic wind speed. Comparison with four meteorological stations on land shows that the period of 6 years considered here, can be regarded as fairly representative for the climatological (30-year) mean wind speed.

1 Introduction

In The Netherlands, the government goal for wind energy is to have 1000 MW installed in the year 2000, and 2000 MW in the year 2010. Of these 2000 MW 200 MW should be placed offshore, e.g. in the North Sea. When placing wind turbines offshore, the generally higher mean wind speeds have to balance the higher construction, operation and maintenance costs. Therefore, knowledge of the wind climate of the North Sea has to be obtained. The geostrophic wind field is an important variable for describing the wind climate over large areas. It can be used for calculating the wind speed and the energy production of a wind turbine at a specific site. One of the characteristics of the geostrophic wind field is that it is not influenced by the surface underneath (land or sea). This means, it can be used in methodologies in which wind speed information of a measuring site is transformed to other locations nearby.

The aim of this paper is to calculate the geostrophic wind on an hourly basis, using five more or less different methods. Two methods do not take stability conditions into account, three do. A comparison of these methods will be made. It is expected that stability conditions do not play such an important role as they do on land [1,2]. For the calculations a database has been used of the gas production platform K13, which is one of the meteorological stations in the Measuring Network North Sea [3]. It will be shown that the limited period of six years of which data were available, gives a good representation of the climatological mean. Furthermore, the resulting mean geostrophic wind speeds calculated with these five calculation methods based on near-surface measurements, will be compared to available results obtained from upper-air pressure data.

For wind energy purposes, more detailed information than just a long-term mean wind speed is needed, such as interannual variations, the annual course, the diurnal course, the frequency distribution, the distribution by direction, extremes and stability conditions. These quantities can be calculated with the methods used in this paper.

2 Theory

2.1 Introduction

The geostrophic wind is defined as the result of the balancing of the pressure gradient force and the Coriolis force (caused by the earth's rotation). At this height the influence of the earth's surface is negligible, so there is no friction. This is the reason why the geostrophic wind is also known as the 'free' wind. In the absence of horizontal temperature gradients, and if the isobars are straight lines, the geostrophic wind approaches the wind above the planetary boundary layer (PBL), which is the wind calculated by the models used in this paper.

Theoretically, it is possible to relate the two horizontal components (x, y) of the geostrophic wind speed u_{geo} and v_{geo} to the pressure p (see e.g. [4]), resulting in

$$u_{geo} = -\frac{1}{\rho f_{Cor}} \frac{\partial p}{\partial y} \tag{1}$$

and

$$v_{geo} = \frac{1}{\rho f_{Cor}} \frac{\partial p}{\partial x} \tag{2}$$

where f_{Cor} is the Coriolis parameter
ρ is the air density.

Several investigations have been carried out to calculate the geostrophic wind field over the North Sea using pressure data. These had been collected on a 6- or 12-hourly basis with radiosondes [5,6]. The method used by Børresen [5] makes use of expressions (1) and (2).

The North Sea is divided into gridsquares with a length of 150 km. The pressure difference of adjacent gridsquares divided by the gridlength gives an approximation of $\frac{\partial p}{\partial x}$ and $\frac{\partial p}{\partial y}$. This results in a map showing the mean geostrophic wind field over the North Sea. The method described by Moore [6] also uses pressure data. This results in a formula to calculate the geostrophic wind speed as function of the position only.

The geostrophic wind speed can also be calculated from near-surface measurements. There are several methods, of which five have been used in this paper (see Table 1). In the following sections these models will be described.

Table 1: The five different methods to calculate geostrophic wind speeds from near-surface measurements.

Method 1	Neutral scaling to the PBL height
Method 2	Diabatic scaling to the PBL height
Method 3	Neutral geostrophic resistance law
Method 4	Diabatic geostrophic resistance law
Method 5	Two-layer method by Bergström

2.2 Neutral and diabatic scaling to the height of the planetary boundary layer

It is assumed that Monin-Obukhov similarity theory, using the logarithmic wind profile, is valid up to the top of the planetary boundary layer [8]. In neutral conditions (Method 1), with the wind speed U known at one particular height z, the Charnock relation for the roughness length z_0 (3) and the logarithmic wind profile (4) form a set of two equations with two unknowns, which can be solved

$$z_0 = \alpha \frac{u_*^2}{g} \tag{3}$$

where g is the acceleration due to gravity
u_* is the friction velocity
α is a constant

and

$$U_z = \frac{u_*}{\kappa} \ln(\frac{z}{z_0}) \tag{4}$$

where κ is the von Kármán constant.

The next step is to calculate the PBL height h for neutral conditions, as described in [8]. Finally, the geostrophic wind speed U_{geo} can be calculated as the wind speed at the top of the PBL by inserting $z = h$ into (4).

Method 2, incorporating stability corrections, is analogous to Method 1. Instead of the logarithmic wind profile, the diabatic wind profile (5) is being used (see e.g. [8])

$$U_z = \frac{u_*}{\kappa}(\ln(\frac{z}{z_0}) - \Psi(\frac{z}{L})) \tag{5}$$

where Ψ is a function accounting for stability
L is the Obukhov length.

This means the calculation of an extra parameter, the Obukhov length L. In this case a set of three equations with three unknowns is obtained, which can also be solved [8]. Different expressions for h are used for neutral or unstable conditions, and for stable conditions [8]. The last step to obtain the geostrophic wind speed is to insert $z = h$ into (5).

2.3 Neutral and diabatic geostrophic resistance law

Method 3, the neutral geostrophic resistance law, has been described in [9]. When using it over sea, the only input required is the wind speed at one particular height. First u_* and z_0 have to be calculated from (4) and (3) as described in Method 1. Then the geostrophic wind speed is calculated according to

$$U_{geo} = \frac{u_*}{\kappa}\sqrt{\left[A - \ln(\frac{u_*}{z_0 f_{Cor}})\right]^2 + B^2} \tag{6}$$

with the two constants $A = 1.8$ and $B = 4.5$.

Method 4 is very similar to Method 3, and has been used in the European Wind Atlas [7]. Here stability conditions are taken into account by making the constants A and B into functions of the stability parameter $\mu \equiv \frac{\kappa u_*}{f_{Cor} L}$. Calculation of u_*, L and z_0 is similar as in Method 2, with some small alterations. The geostrophic wind speed is calculated according to (6), replacing A and B by $A(\mu)$ and $B(\mu)$, respectively.

2.4 Two-layer method by Bergström

In this model two layers are defined [10]: a surface layer with height z_s, in which Monin-Obukhov similarity theory can be applied, and an Ekman layer starting at height z_s upwards, in which the classical Ekman approach can be applied. Neglecting the thermal wind components, continuity at $z = z_s$ for both the wind speed and the shear stress, and assuming that the wind direction is constant in the surface layer, results in expressions for the scalar geostrophic wind speed (U_{geo}) and for the angle of the geostrophic wind speed vector relative to the wind direction in the surface layer (α_{geo}). The height of the surface layer and the wind speed at this height have to be calculated iteratively from two expressions. The values of L, u_* and z_0 follow from applying Monin-Obukhov theory exactly as described in Method 2.

3 The K13 database

3.1 General description

The K13 database contained reliable data of the period 1985–1990. The location of the K13 platform is shown in Fig. 1. The distance to the Dutch coast is about 100 km.

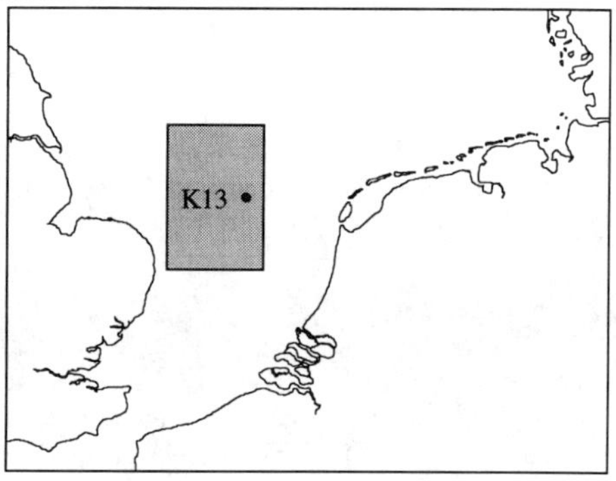

Figure 1: Location of the K13 platform in the North Sea, the rectangle indicates the area of which the seawater temperatures have been taken.

The wind speed was measured at a height of 75 m above Mean Sea Level (MSL), the air temperature at a height of 15 m above MSL [3]. Seawater temperatures were obtained from Voluntary Observings Ships from the rectangular area indicated in Fig. 1. During this period only 3.6% of the hours had to be excluded from analysis because data for either air temperature or wind speed were missing. Because a fitting procedure was used to obtain the seawater temperatures, these were never missing [1,2]. If the measured wind speed was 3 m/s or less (occurring 6.6% of all hours), no Monin-Obukhov similarity theory was applied. In these cases the wind profile was assumed to be uniform with height, the geostrophic wind speed was equal to the measured wind speed.

3.2 Climatology

Results of the analysis of the K13 database have been reported in earlier publications [1,2]. However, closer study has revealed that the wind speed data of the years 1982–1984 may not be reliable. Therefore, at this moment only the six years 1985–1990 are considered. In order to compare the mean wind speed of this 6-year period to a long-term (30-year) average, the following procedure was applied. KNMI (Royal Netherlands Meteorological Office) publishes monthly weather reports since a number of years. For the wind speed five stations have been selected for publication, that give an overall view of the wind speed in The Netherlands. Of four of these stations (De Bilt, Beek, Eelde and Vlissingen), the long-term mean wind speeds are also known. In 1992 these climatological means have been updated to the period 1961–1990 [11]. By taking the monthly means of these stations, and dividing these by the long-term average, the relative variations of the years under consideration can be determined. From this analysis it follows that, for these four stations, the mean wind speed over the period 1985–1990 is 99% of the mean wind speed over the period 1961–1990. Therefore, it is concluded that data from the period 1985–1990 represent the climatological mean wind speed quite well.

4 Results

4.1 Mean wind speeds and interannual variations

In Table 2 the annual mean measured wind speed and the annual mean geostrophic wind speeds, for each of the five different methods are shown. In the last column the mean geostrophic wind speeds are shown for the whole period 1985–1990.

Table 2: Interannual variations of the annual mean measured wind speed and the annual mean geostrophic wind speeds, calculated according to five different methods.

Year:	1985	1986	1987	1988	1989	1990	Mean
	Measured wind speed [m/s]						
	9.2	10.4	9.2	10.2	9.6	10.6	9.8
	Geostrophic wind speed [m/s]						
Method 1	11.2	12.5	10.9	12.2	11.3	12.7	11.8
Method 2	11.0	12.2	10.6	11.9	11.4	12.6	11.6
Method 3	11.3	12.5	10.9	12.2	11.4	12.8	11.9
Method 4	11.2	12.4	10.8	12.1	11.5	12.8	11.8
Method 5	11.3	12.5	10.9	12.2	11.6	12.9	11.9

It can be seen that the interannual variations deviate from the overall mean wind speed between -9% in 1987 and +8% in 1990. The results show further that the five models used to calculate geostrophic wind speeds (as indicated in Table 1), yield very similar results, The five means over the period 1985–1990 are within a range of 0.3 m/s.

4.2 Comparison with other results

As mentioned earlier, pressure-based results are also available of the geostrophic wind speeds over the North Sea. Børresen [5] used pressure data of a 27-year period to calculate the geostrophic wind field over the entire North Sea. The grid used here is rather coarse (150 km × 150 km). The value of the mean geostrophic wind speed at the location of the K13 platform read of the resulting map is about 10.5 m/s. Moore [6] analysed 10 years of pressure data resulting in a location-dependent formula to calculate the mean geostrophic wind speed over the North Sea. For the location of the K13 platform this formula results in a value of about 10.2 m/s. It can be seen that these values agree well with each other, but that the mean geostrophic wind speeds shown in Table 2 are much higher (11.6–11.9 m/s). Moreover, the measured wind speed at 75 m height of 9.8 m/s even approaches the values of the geostrophic wind speeds.

4.3 Detailed results

4.3.1 Diurnal courses

As shown in earlier publications [1,2], there is no diurnal course in the measured wind speed, nor in the seawater or air temperature. Therefore, it is expected that the geostrophic wind speeds also show no diurnal course. Calculations indeed show that the variations during the hours of the day are negligible. Therefore, it is concluded that there is no diurnal course in the geostrophic wind speeds.

4.3.2 Annual courses

In Fig. 2 the annual courses of the measured wind speed and the geostrophic wind speeds are shown, for each of the five different calculation methods.

It can be seen that the minimum wind speed is reached in the month June, while the maximum is reached in February. There is not much difference between the results of the different calculation methods.

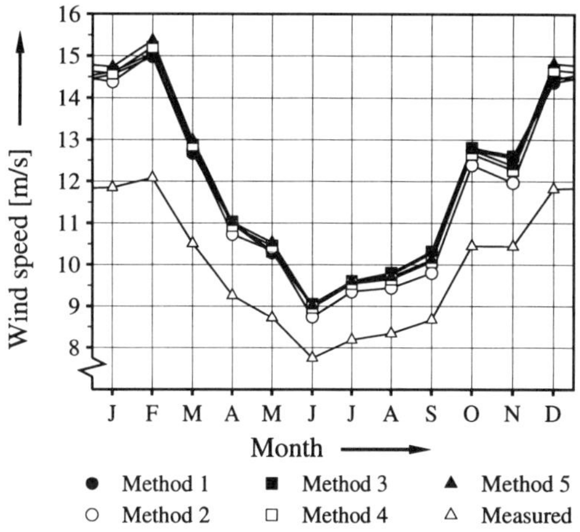

Figure 2: Annual courses of the measured wind speed and the calculated geostrophic wind speeds.

It can be seen that there is not much difference in the results of the five different methods. Another conclusion is that the differences between the Weibull means and the actual mean wind speeds are only 0.2–0.4 m/s, which indicates that the Weibull distribution function describes the frequency distributions of the geostrophic wind speeds well.

Table 3: Weibull parameters of the measured wind speed and the geostrophic wind speeds, calculated according to five different methods.

Weibull parameters:	k-factor	a-factor [m/s]	mean [m/s]
	Measured wind speed		
	2.11	11.1	9.9
	Geostrophic wind speed		
Method 1	1.88	13.0	11.5
Method 2	1.87	12.7	11.2
Method 3	1.86	13.1	11.7
Method 4	1.87	13.0	11.6
Method 5	1.86	13.1	11.6

4.3.4 Distribution by wind direction

The wind directions have been transformed from tens of degrees into 12 sectors of 30°, where sector 1 represents the sector around the north (345°–15°), sector 2 the sector from 15° to 45°, etc. In Figs. 3–4, the frequency of occurrence and the mean geostrophic wind speed by wind sector are shown, respectively.

It can be seen that the frequencies of the sectors 7–10 (south to west) are the highest (together some 50%), while the mean wind speeds in these sectors are also the highest. In terms of wind energy capture, this means that these four sectors would provide far more than half of the total energy production if a wind turbine would be placed at such an offshore site.

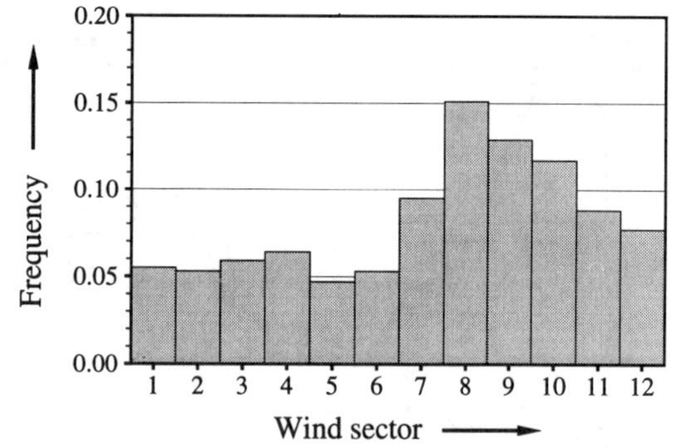

Figure 3: Frequency of occurrence by wind sector.

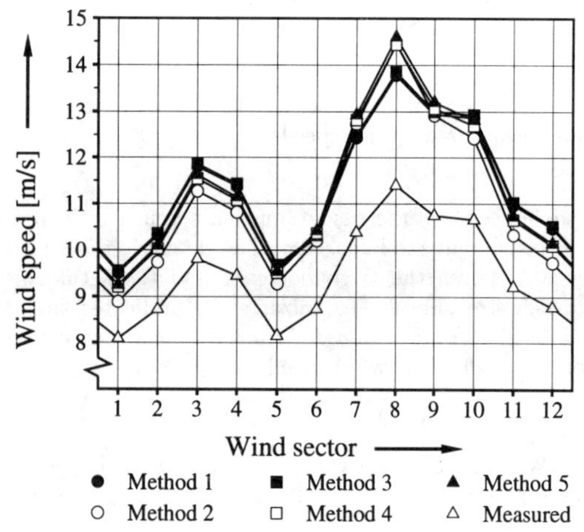

Figure 4: Measured wind speed and calculated geostrophic wind speeds by wind sector.

5 Discussion

The results of this study show that the five models (as indicated in Table 1) used to calculate geostrophic wind speeds from near-surface measurements yield very similar results. Although stability plays an important role, it appears that, on average, the neutral approach is sufficient for calculation of the mean geostrophic wind speeds. In [1,2] it was shown that at K13 (very) unstable conditions ($L < 0$) prevail. Conditions are (very) stable in only 30% of all hours. As unstable conditions imply small corrections with regard to the diabatic wind profile compared to stable conditions, these effects appear to cancel out in the mean geostrophic wind speed.

Comparison of the values found for the mean wind speeds with pressure-based results show that they are much higher than both the value of 10.5 m/s derived from [5], and the value of 10.2 m/s derived from [6]. A possible explanation for these differences may be that the geostrophic wind obtained from pressure observations is not compatible with the geostrophic wind calculated from near-surface measurements. The pressure-based geostrophic wind speed only approaches the wind speed at the top of the PBL in ideal conditions.

The mean geostrophic wind speeds found at K13 for the period 1986–1990 ranged from 11.6 to 11.9 m/s, depending on the calculation method used. It has been shown that, although the period analysed is rather short for climatological purposes, this period can be regarded as fairly representative for the climatological mean.

The details needed for wind energy applications in the analysis of geostrophic wind speeds, can be obtained from models using near-surface measurements. The interannual variations are between 10.6-10.9 m/s in 1987 and 12.6-12.9 m/s in 1990. The mean annual course shows variations between 8.7-9.1 m/s in June to 15.0-15.4 m/s in February. There is, however, no diurnal course. The four sectors from south to west show the highest wind speeds, and also the highest frequency of occurrence. The Weibull distribution function describes the observed frequency distributions well, with values for k ranging from 1.86 to 1.88.

Acknowledgements

Hereby NOVEM (Netherlands Agency for Energy and the Environment) is thanked for its financial support to make this project possible. Furthermore, KNMI (Royal Netherlands Meteorological Institute) is thanked for its contribution by supplying the data and further assistance in the analysis, especially dr. B. Holtslag. Finally, R. Engelen is thanked for his contribution to the computer programming.

References

[1] J.W. Cleijne, J.P. Coelingh, A.J.M. van Wijk, Description of the North Sea wind climate for offshore wind energy applications, 91-327, TNO Institute of Environmental and Energy Technology, Apeldoorn, The Netherlands, 1991.

[2] J.P. Coelingh, A.J.M. van Wijk, J.W. Cleijne, R. Pleune, Description of the North Sea wind climate for wind energy applications, Journal of Wind Engineering and Industrial Aerodynamics, **39**, pp. 221–232, 1992.

[3] H.C.L.J. Witteveen, H. van Rees, J.W. Hillebrand, Meetnet Noordzee — Overzicht lokaties en sensoren (*in Dutch*), Rijkswaterstaat, Direktie Noordzee, Hoek van Holland, The Netherlands, 1989.

[4] H.A. Panofsky, J.A. Dutton Atmospheric turbulence — Models and methods for engineering applications J. Wiley & Sons, New York, 1984.

[5] J.A. Børresen, Wind Atlas for the North Sea and the Norwegian Sea, Norwegian University Press/The Norwegian Meteorological Institute, Kristiansand, Norway, 1987.

[6] D.J. Moore, 10 to 100 m winds calculated from 900 mb data, Proceedings 4th BWEA Wind Energy Conference, BHRA Fluid Engineering, Cranfield, UK, pp. 197–207, 1982.

[7] E.L. Petersen, I. Troen, The European Wind Atlas, Department of Meteorology and Wind Energy, Risø National Laboratory, Roskilde, Denmark, 1989.

[8] A.J.M. van Wijk, A.C.M. Beljaars, A.A.M. Holtslag, W.C. Turkenburg, Evaluation of stability corrections in wind speed profiles over the North Sea, Journal of Wind Engineering and Industrial Aerodynamics, **33**, pp. 551–566, 1990.

[9] J. Wieringa, Roughness-dependent geographical interpolation of surface wind speed averages, Quart. J. R. Met. Soc., **112**, pp. 867–889, 1986.

[10] H. Bergström, A simplified boundary layer wind model for practical application, Journal of Climate and Applied Meteorology, **25**, pp. 813–824, 1986.

[11] KNMI (Royal Netherlands Meteorological Institute, Klimatologische gegevens van Nederlands stations — Normalen en extreme waarden van 15 hoofdstations voor het tijdvak 1961–1990 (*in Dutch*), 150-27, KNMI, De Bilt, The Netherlands, 1992.

Modelling the coherence of wind field fluctuations

H.G. Beyer, J. Handwerker, J. Luther, R. Steinberger-Willms
Department of Physics, Renewable Energy Group,
Carl von Ossietzky University of Oldenburg,
D(W)-2900 Oldenburg, FRG, P.O. Box 2503,
Tel. ++49-441-798-3544, Fax ++49-441-798-3201

Abstract

The fluctuations of wind power production caused by wind speed fluctuations may cause problems in local supply grids with high wind energy penetration. These fluctuations are reduced when several WTG sites are located in one utility grid. The reduction of fluctuations (normalized to the installed capacity) depends on the spatial structure of the wind field which can be characterised by the coherence. In this paper we establish a model for the coherence in time scales between some minutes and one hour and for distances between 200 m and 3 km.

We describe the coherence as an exponential function of the frequency. The decay parameter is shown to be a linear function of the distance where the gradient depends on the thermal stratification and the wind direction.

1 Introduction

Due to the variability of wind speed, the power output of wind turbine generators (WTGs) shows strong fluctuations. When WTGs are connected to a utility grid, fluctuations with periods below one hour are of interest since grid interference problems caused by wind power fluctuations have to be considered. In grids with a high proportion of renewable energy (e.g. stand-alone systems) the utilisability of the wind potential may be affected by power fluctuations.

It may be expected, though, that the fluctuations in the lumped power output of several geographically dispersed WTGs will be reduced due to the weak correlation of the fluctuations with periods shorter than one hour. Our work aims at a quantitative description of the smoothing effect due to geographical dispersion with the help of wind field coherence models.

Up to now several investigations analysed the correlation of fluctuations of wind speed dealing with distances up to some 300 m and time scales of some seconds and some minutes (e.g. [2], [7], [8], [11], [12]) or distances between 25 km and 150 km [1] and time scales of some hours up to days. In this paper mesoscale distances between 200 m to 3 km are considered. The time scale ranges from minutes to one hour. Since the data base in this time resolution was very sparse, we conducted own measurements.

2 Measurements

We operated a network of 5 meteorological stations for the recording of wind speed data over a span of approximately one year. The location was on an open terrain less than 50 km inland from the northwest German coast. Each station consists of a mast with a hight of 32 m. Data of wind speed and temperature are measured at heights of 10m and 32m a.g.l. Wind direction was recorded at 32 m. The time resolution was 10 or 60sec.

We split the measuring campaign in two phases. During the first phase the masts were sited as shown in Fig. 1. Distances varied between 210 m and 850 m. During the second phase we increased the distances between the masts by a factor of 4, conserving the geometry. Distances then varied between 850 m and 3 km.

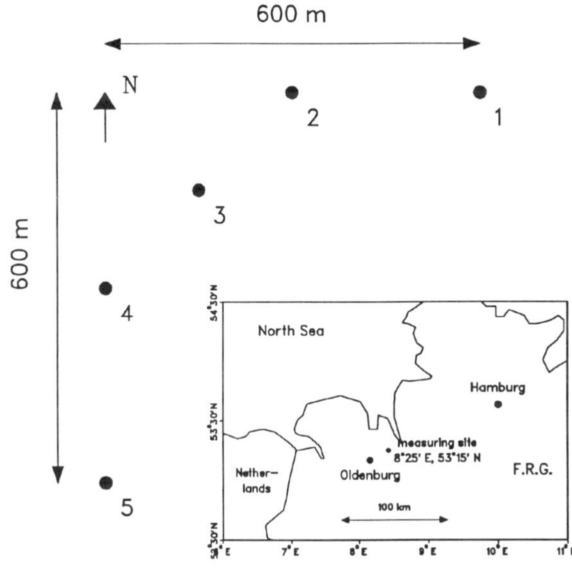

Figure 1: Measurement setup of the meteorological data recording stations in the first phase of the project and geographical location. In a second phase the distances between sites were increased whilst preserving the geometry of the setup.

3 Mathematical description of the statistical wind field structure

Temporal and spatial wind field fluctuations may be characterised by the (cross)correlation function or the (cross) spectral power density.

The cross-correlation of two random processes $v_i(t), v_j(t)$, defined by

$$R_{ij}(\tau) = \lim_{T\to\infty} \frac{1}{T} \int_0^{\infty} v_i(t)\ v_j(t+\tau)\ dt,$$

compares the fluctuations of the two processes, taken at time delay τ.

The Fourier transform of the cross-correlation is called cross spectral power density:

$$S_{ij}(f) = \int_{-\infty}^{\infty} R_{ij}(\tau)\ e^{-i2\pi\tau f}\ d\tau$$

where f denotes the frequency. In the case of identical indices $i = j$ the autocorrelation and the spectral power density are obtained. The integral of the spectral power density equals the variance σ_v^2 of the fluctuations of $v_i(t)$:

$$\sigma_v^2 = \int_{-\infty}^{\infty} S_{ii}(f)\ df$$

The spectral power density may be interpreted as an analysis of the variance in the frequency domain. For $i \neq j$ the cross spectral power density may be rewritten as

$$S_{ij}(f) = \sqrt{S_i(f) S_j(f)}\ \gamma_{ij}(f)\ e^{-i\theta_{ij}(f)},$$

where the coherence

$$\gamma_{ij}^2(f) = \frac{|S_{ij}(f)|^2}{S_{ii}(f)\ S_{jj}(f)}$$

describes the coupling of the fluctuations and $\theta_{ij}(f)$ is the phase lag of the two time series.

The spectral power density of the process defined by the sum of N processes may be calculated by

$$S_\Sigma(f) = \sum_{i=1}^{N} \sum_{j=1}^{N} S_{ij}(f).$$

Assuming that all single point spectra are identical ($S_{ii} = S$) and neglecting the phases of the cross spectra one may rewrite the spectral power density of the sum of processes as

$$\begin{aligned} S_\Sigma(f) &= S(f) \sum_{i=1}^{N} \left(1 + 2 \sum_{j=i+1}^{N} \gamma_{ij}(f)\right) \\ &= S(f) \cdot H(f) \cdot N^2, \end{aligned}$$

where $H(f)$ is a filter function. This gives a first estimate of the relative reduction of fluctuations.

For detailed analysis of the fluctuations of wind power on a untility grid the power output of each single WTG has to be calculated. Thus time series of the wind input with realistic spectrum and with realistic coherences are necessary. On the basis of models for the spectrum and the coherence this type of data sets may be generated using the Shinozuka method [9].

Models for spectra are given in e.g. [4], [5], [6], [3]. For the type of calculations sketched out above a model for the coherence is needed in addition.

A widely applied model is given by

$$\gamma^2(f) = e^{-af} \qquad (1)$$

first proposed by [2]. The decay paramter a separates the strongly and the weakly correlated fluctuations: Fluctuations with durations much longer than a are strongly correlated whilst those with durations much shorter than the decay parameter are weakly correlated.

Many publications show that for *small* distances (i.e. smaller than about 300 m) the decay parameter can be estimated by $a = b_0 d/v$ where d is the distance between the to stations and v is the average wind speed. The factor b_0 differs with the wind direction relative to the measuring sites and the thermal stratification of the lower atmospheric boundary layer.

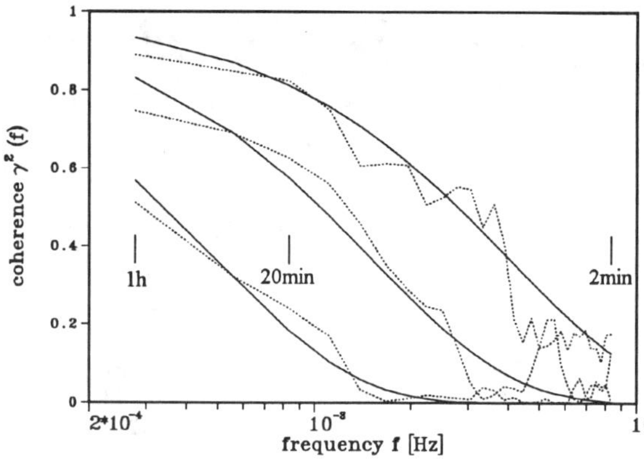

Figure 2: Comparison of three measured coherence functions (dashed) and the exponential model (solid). The coherences are measured at three different situations with distances of 210 m, 1400 m and 2100 m (from the top)

4 Calculation of the decay parameter

Our aim is to develope a parametrisation of the decay parameter a for mesoscale distances (here 200 m - 3 km). The coherence data forming the basis of this analysis were obtained in the following way:

Using wind speed data at a height of 32 m we first identified time sections with steady state weather situations of a duration of 6 hours. A steady state was defined by small variations of wind speed, wind direction and thermal stratification. The thermal stratification was estimated by using the difference of the temperatures measured at 10 m and 32 m. To avoid problems with harmonics of frequencies smaller than 1/1h we first removed wind speed fluctuations with periods longer than one hour by filtering the time series using the filter $h(\tau)$:

$$u(t) = \int_{-\infty}^{\infty} h(\tau) v(t-\tau) d\tau$$

where

$$h(\tau) = \begin{cases} \delta(\tau) - \dfrac{\sin(2\pi f_0 \tau)}{\Phi \cdot \pi\tau} & \forall \quad |\tau| \leq \frac{1}{2f_0} \\ 0 & \text{else} \end{cases}$$

with $f_0 = 1/1\text{h}$ and $\Phi = 1.179$. This value ensures that the integral over the filter $h(\tau)$ equals zero.

We devided the 6 hour period into 31 parts each with a length of 1 hour. Consecutive periods have an overlap of 50 minutes. For each part we calculated the cross spectral power density and the spectral power densities. Then these functions were averaged over the 31 parts. The average (cross) spectral power densities were used to calculate the coherence.

For each 6 hour period we got 10 coherences (all possible pairs given with 5 masts). Each coherence was associated with an average wind speed, average wind direction, thermal stratification and distance between the two masts. We calculated the decay parameter for each single coherence by a least-square-method (see Fig. 2). Comparing the decay parameters determined for similar situations showed that the scatter of the decay parameter is considerable. So we sorted each measurement into a class defined by a distance range, wind speed range, a range of wind direction relative to the mast alignment and the thermal stratification. All the coherences in one class were averaged for estimating a typical decay parameter of this class.

This procedure resultet in 1500 classes (10 distances, 10 wind speeds, 5 wind directions and 3 thermal stratifications). The classes of the thermal stratification were defined in the following way: A neutral (or adiabatic) stratification was given by a temperature difference $T_{32} - T_{10}$ between -0.3 K and -0.1 K. A lower value defined an instable stratification and a higher value a stable one. It has to be noted that we did not find any strongly instable situations.

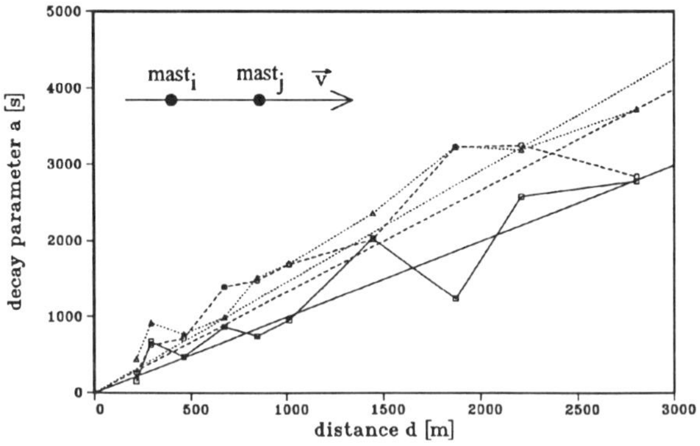

Figure 3: The dependence of the decay parameter on the distance shown for longitudinal wind flow. The straight lines represent the linear model. (Solid: stable stratification, dashed: neutral, dotted: instable)

A dependence of the decay parameter a on the wind speed v could only be detected at the smallest distances and for longitudinal wind flow. For larger distances the model

$$a = a_0 \, d \tag{2}$$

gives a good approximation for the decay parameter. As an example Fig. 3 shows the comparison of the linear model with the measured decay parameters for longitudinal wind flow. Regarding the small differences that were found between neutral and instable situations it has to be taken into account that only weakly instable stratification occured during the measurements.

The dependence of a_0 on the wind direction is shown in Fig. 4 for the different thermal stratifications. The values shown in this plot, together with equations (1) and (2) provide the model for the mesoscale coherence.

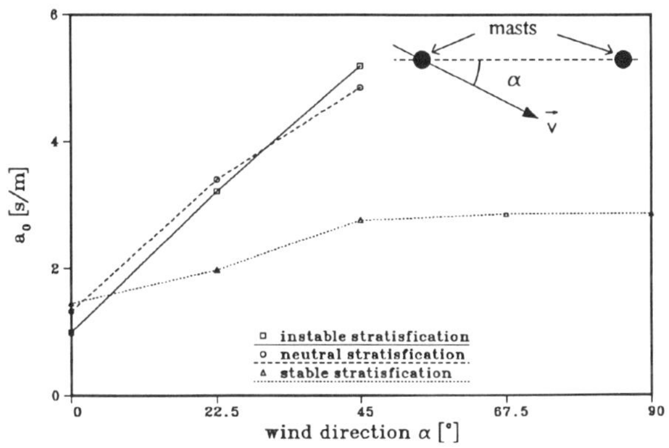

Figure 4: The factor a_0 as a function of wind direction. $\alpha = 0°$ is longitudinal, $\alpha = 90°$ lateral wind flow. (Solid: stable, dashed: neutral, dotted: instable stratification)

The model is not able to describe precisely the coherence for an individual time series. Instead it should represent the average coherence for all sequences characterised by a given meteorological class as described above. Thus discrepancies of individual coherences to the class average may occur.

To test the applicability of the model we calculated the filter function $H(f)$ from the model for a given situation and compared it with the measured one. This comparison is shown in Figure 5.

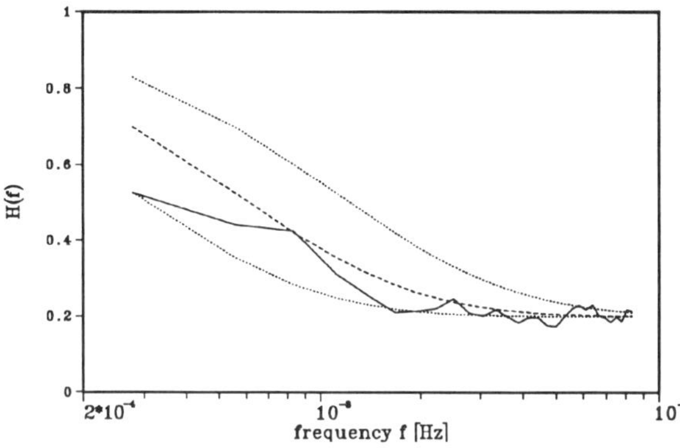

Figure 5: Comparison of the measured filter and the model. The solid line is calculated directly from the measurements of June 10, 1991 23.00 to June 11, 5.00. The dashed line is calculated with our model. The dotted lines describe the filter function if all decay parameters are estimated too low or too high by a factor of 2.

5 Conclusion and outlook

On the basis of the exponential model for the coherence and the linear model for the decay parameter it is possible to calculate the fluctuation of the lumped power output of wind turbines, sited with distances below 3 km using Shinozuka's method for synthesizing time series with fixed coherences [9]. It should be noted that these calculations are only valid as long as the mutual interference of the WTGs caused by wake effects can be neglected. It should be possible to show that the fluctuations of the lumped power output of wind turbines at frequencies of 1/10min and higher will be reduced by a factor of $1/N$ where N is the number of installed wind turbines in a cluster with distances up to some kilometers.

Two further measurement phases are concerned with the evaluation of coherences of wind speed at separations up to 400 km. On the base of these data the models will be refined for larger distances.

6 Acknowledgements

This research was partly supported by the "German Bundesministerium für Forschung und Technologie."

References

[1] H.G. Beyer, J. Luther, R. Steinberger-Willms: *Fluctuations in the Combined Power Output from Geographically Distributed Grid Coupled Wind Energy Conversion Systems - An Analysis in the Frequency Domain,* Wind Engineering **14** (1990) 179-192

[2] A.G. Davenport: *The Spectrum of Horizontal Gustiness Near the Ground in High Winds,* Quarterly Journal of the Royal Meteorological Society **87** (1961) 194-211

[3] N.O. Jensen, J.E. Simpson: *Atmospheric Turbulence,* Engineering Meteorology, E. Plate (edt.), Amsterdam 1982

[4] J.C. Kaimal, J.C. Wyngaard, Y. Izumi, O.R. Coté: *Spectral characteristics of Surface-Layer Turbulence,* Quart. J. R. Met. Soc. **98** (1972) 563-589

[5] J.C. Kaimal: *Turbulence Spektra, Length Scales and Structure Parameters in the Stable Surface Layer,* Boundary Layer Meteorology **4** (1973) 289-309

[6] T. von Kármán *Progress in the Statistical Theory of Turbulence,* Proc. National Acad. Sci. **34** (1948) 530-539

[7] L. Kristensen, N.O. Jensen: *Lateral Coherence in Isotropic Turbulence and in the Natural Wind,* Boundary Layer Meteorology **17** (1979) 353-373

[8] L. Kristensen, H.A. Panofsky, S.D. Smith: *Lateral Coherence of Longitudinal Wind Components in Strong Winds,* Boundary Layer Meteorology **21** (1981) 199-205

[9] M. Shinozuka: *Simulation of Multivariate and Multidimensional Random Processes,* J. Acoustical Soc. of America **49** (1971) 357-368

[10] R. Steinberger-Willms: *Untersuchung der Fluktuationen der Leistungsabgabe von räumlich ausgedehnten Wind- und Solarenergie-Konvertersystemen in Hinblick auf deren Einbindung in elektrische Versorgungsnetze,* Dissertation, Oldenburg 1993

[11] H. W. Teunissen: *Structure of Mean Winds and Turbulence in the Planetary Boundary Layer over Rural Terrain,* Boundary Layer Metrology **19** (1980) 187-221

[12] S. Wacongne, A. Babiano: *Space-Time Turbulent Characteristics in the Atmospheric Surface Layer,* Boundary Layer Meteorology, **24** (1982) 429-450

M7 WIND PUMPING SYSTEMS SUITABLE FOR IRRIGATION ON THE COAST RIO GRANDE DO SUL (BRAZIL)

G.Baldi, D.Bedard, M.Cavagioni,
Riva Calzoni SPA, Italy

D.Saddhu G.C.Occacia Federal University
of Rio Grande do Sul, Brazil

ABSTRACT

The aims of the present project, are the acquisition of market information and the identification of small-sized wind plants for the Brazilian market. These activities are carried out by Riva Calzoni in cooperation with the Universidade Federal do Rio Grande do Sul (UFRGS) and through EEC financial support (Technology Exchanges and Transfer with Third Party Countries), to create the conditions to promote local manufacturing of wind energy conversion (WEC) systems for water pumping applications.

1) INTRODUCTION

Since 1978 Riva Calzoni has been working in the wind energy field to develop small, medium and large size machines. For the developing country market, Riva Calzoni will promote local manufacturing, through the creation of joint ventures.

PROMEC a team of the Mechanical Engineering Department of URFGS, has built prototypes of wind powered systems for irrigation purposes and has achieved a deep knowledge on the local applications of wind energy.

2) SITE IDENTIFICATION

Two small independent pilot plants are being installed to irrigate small plots of rice "ad hoc", but capable of supplying realistic data on rice production during this period. With the aim of developing a product for the Brazil market, 2 sites on the coast of Rio Grande do Sul (Brazil) have been selected on these basis:

- -good average wind speed
- -under 200 km from Porto Alegre to facilitate maintenance and installation
- -accessibility to site with heavy machines
- -water pumping application for rice cultivation and other use to keep the system operational during the whole year
- -active cooperation of the owner

The first site is Sao Simao at 200 km from Porto Alegre and has a head of about 1.5m. The site has good accessibility. The second site is Granja Pimenta at 65km from Porto Alegre with an acceptable road access. This site has 25 ha rice field, 30 ha grass for cattle, 5 ha for corn and a fish pond. The water head is between 2-3m.

Two anemometers have been installed since September 91. Average wind speed for the first site of Sao-Simao is 4,9 m/s. The second site of Granja Pimenta has an average wind speed of 5.8 m/s.

3) WIND PUMPING TECHNOLOGY

The M7 wind turbine characteristics, power curve and yield energy are given in Table-1 and Fig.1-2.

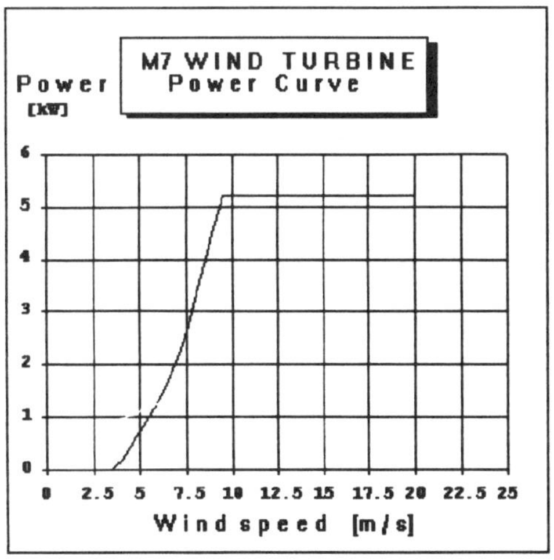

Fig.1: M7 Power Curve

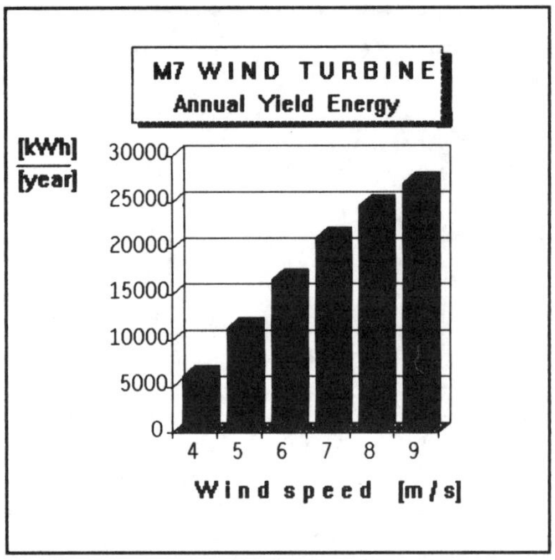

Fig.2: M7 Yield Energy

The best technology available for shallow wind pumping has been evaluated. The electric transmission has been chosen to run the asynchronous pump motor under variable frequency. In this way, batteries and inverter are not necessary. Higher performance and a cheaper overall cost, are also achievable.

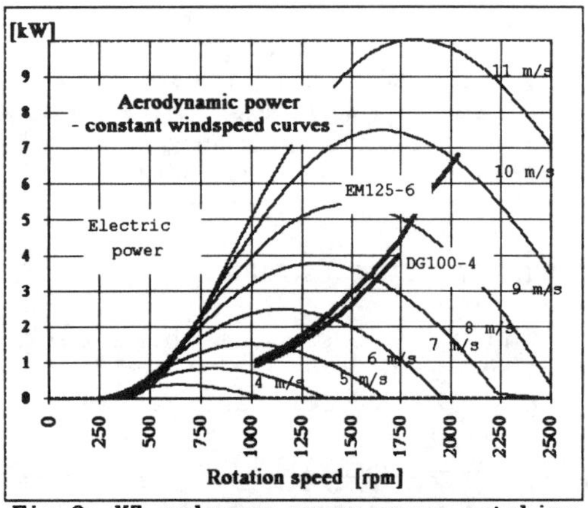

Fig.3: M7 and pump power curve matching

The pump manufacturer was selected according to product performance, reliability and participation to the experiment cost. VITEX pumps from MAJMAR, were selected. The blade propeller pump is another possible solution under evaluation.

The VITEX pump is patented for a new kind of high efficiency impeller which is optimized for lower water head than normal centrifugal pumps.

The first site with a 1.5 m water head, could be equipped with a propeller pump or a VITEX EM-125/6 and the second site with a 2-3 m water head, will get a VITEX DG-100/4 submerged pump. The submerged pump will be suspended by a chain over a platform.

Pumping testing were done at the MAJMAR laboratory facilities. Using a 15 kW inverter and a 4 kW-380 Vac motor on the VITEX pump results were as follow:

Pump Type	Freq. (Hz)	Q (m3/h)	Head (m)	Gen. (Watt)
DG100/4	26	76	2.0	725
DG100/4	35	108	2.0	1050
DG100/4	50	154	2.0	2700
DG100/4	55	170	2.0	3450
EM125/6	24	120	1.4	720
EM125/6	35	146	1.4	900
EM125/6	50	206	1.4	2400
EM125/6	55	226	1.4	3150

At nominal wind speed (9m/s) the M7 is producing at 50 Hz but under gust condition the frequency could climb up to 55 Hz for a short time. The pump start at 35 Hz but when rotating the pump still work until the frequency drop to 25 Hz.

Under computer simulation , we obtain that the pumping system should start pumping at a wind speed around 5 m/s. The predicted output flow rate should reached 1274 and 1070 m3/day respectively at the 3m and 1.5m water head sites.

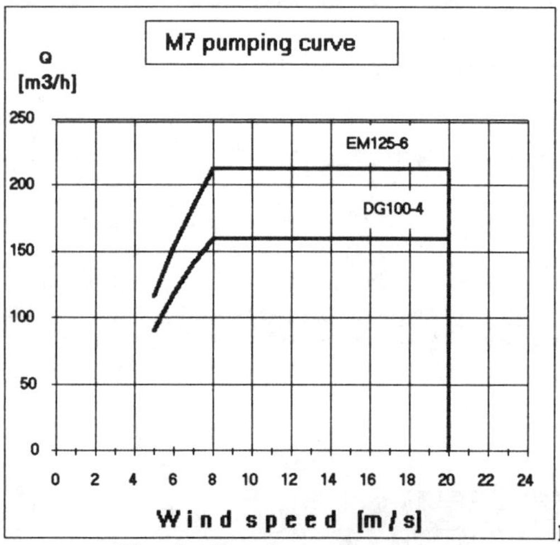

Fi

Fig.4: Pumping power curve

The predicted water flow rate for the 2 wind pumping systems is given in Fig.3-4.

During the pump rig testing, we are also taking confidence in the M7 wind turbine, installed at our 1300m test site (Fig-5), which up to now, has shown good results under very turbulent and hard wind conditions .

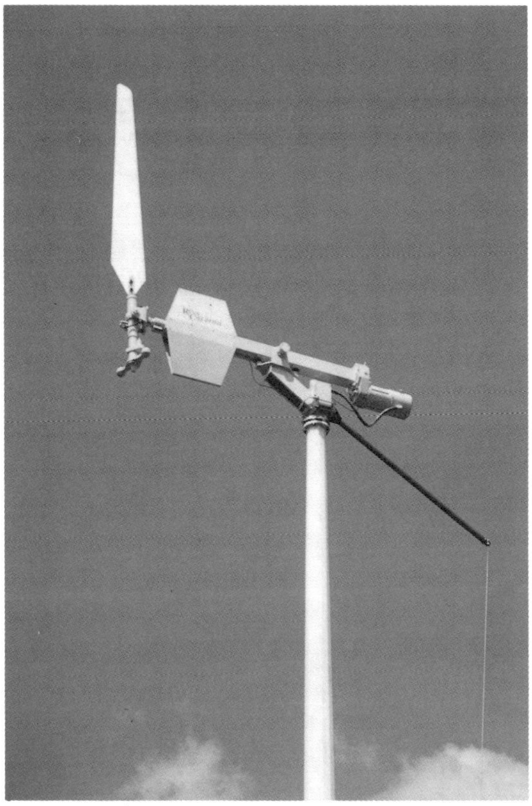

Fig.5: M7 Wind turbine at Mt Cavallo

4) MARKET STUDY

The market is mainly directed to individual users and farms located far from the electric grid, for domestic applications such as powering of small machinery, water pumping, etc.

All rice field in the Rio Grande del Sul state are irrigated by gravity 30% and the rest by lift pumps. With the gradual electric grid extension, electric motors are replacing the diesel engine. For now 34% of engine are electric and 36% use fuel for nearly 228,000 hp of diesel engine in the field.

It is unrealistic to cover the whole rice growing region with the electric grid since much of it is sparsely populated and has no local industry.

Rice growers are subsidised through a 30% discount rate. Even if the lines are already overloaded, specially in summer, rice growers are lobbying for the electric grid expansion.

Rice cultivation is concentrated mainly in the south and west of the state. The main pump requirement is large discharge at low head. Nearly 90% of pumps are operating at head under 8m and the upper limit could reach 18m.

There are already many American multibladed windmills used mainly for domestic water supply but they are too small for irrigation which could absorb on average 300 m3/ha per day of water.

Some preliminary surveys gave positive results on the possibility of replacing the thermal units presently used for rice irrigation with wind systems on the coast of RGS. Favourable conditions have been particularly noted as follows:

- Irrigation used for flooding rice fields allows the possibility of accumulating energy at a tolerated level (maximum and minimum).
- The wind data show that during summer period when the irrigation is required , its energy is at the highest.

Rice irrigation is carried out on a large scale using large generators (50-100 kW or more). Replacement or integration of present irrigation systems requires the use of medium-size wind powered generators.

Given the financial outlay involved and the risk taken by the farmer, medium size machines can only be proposed after demonstration on a small scale. This is why the project was realized with the 5.2 kW M7 wind turbine.

Diesel pumping cost is compared with wind powered generators, to establish the competitiveness and consequent market size for the industrial initiative as a later joint venture.

The average power demand in this area is about .7 kW/ha. Motor pumping is estimated around $.06/m3 which contribute to a mere 15-33% of the $1350/ha for rice crop in Brazil. Division of the irrigation system in RGS is:

- about 25% of mechanical type (rural 10-12.5%)
- about 35% of electric type
- about 40% of fall and flood type (Dam type)

Only mechanical pumping has higher cost than wind pumping and then could be replaced (cost between 800-900$/ha). Along these pumping systems, there is a prospect of 2170 plants (around 80 ha by wind pumping system) which could be converted to wind pumping. Realistic figure show that above 20% of it could be replaced by wind pumping within 10 years.

5) CONCLUSION AND FUTURE WORK

Installation and experimentation of the 2 wind plants in real condition should be ready to start at the end of spring 1993. The next irrigation campaign will start in August 1993. The demonstration of the 2 wind plants will include:

-Optimize the starting behaviour and pumping at low frequency through generator excitation alteration or by using 2 different pump size combination

-Propeller pump testing.

-Delivery of the two wind plants with accessories

-Experimental testing of the M7 for pumping by PROMEC.

-Demonstration of the wind plants operation in a realistic situation.

-To verify after that the wind plants are operating that they are adequate with respect to the demand and to study the technical assistance needed by the users.

The pumping technology used on the 2 small wind plant is however scale reproducible on a medium size 250 kW wind turbine as our M30. The M30 is one of the few medium size wind turbines available for non grid connected application.

After testing of the 2 demonstration plants, we are planning to go on with this project aas follows:

-Wind generators manufacturing with local components.

-Involvement of local plant engineering experts necessary to activate the market of wind generators.

TABLE. I

SUMMARY OF THE TECHNICAL FEATURES OF THE M7 WIND ENERGY CONVERTER

ROTOR	
No. of blades:	1
Diameter:	6.5 m
Rotor position:	downwind
Hub type:	teetering hub
Pitch angle range:	2° -> 65°
BLADE	
Chord length:	0.175-0.347 m
Material type:	CRP (Epoxy)
Section profile:	NACA 44XX serie
TOWER	
Height:	12 m to hub
Shape:	polygonal conical
Diameter:	0.4- .2 m
Material:	painted steel
CONTROL	
Pitching System:	mechanical/hydraulic
Yawing System:	passive
Parking:	blade in vane position free flap
DRIVE TRAIN	
Gear:	1 stage/parallel
Gear ratio:	1 : 5.13
Generator type:	synchronous
Generator power:	5.2 kW
OPERATION	
Wind speed :	9 m/s at rated power
Operating range:	4 - 20 m/s
Survival wind gust:	50 m/s

CONTROL OF A LARGE WIND-PUMPING SYSTEM
Laboratory Results - Energy Optimization

M. Könemund (TU Braunschweig);
Institut für Regelungstechnik, Hans-Sommer-Straße 66, 3300 **Braunschweig**, Germany,
☎: (49) 531 / 391 - 3826, Telefax: (49) 531 / 391 - 5194

Storing energy poses a major issue with renewable sources like wind or solar. An installation (wind: 500 kW, solar: 60 kW) in combination with circular water pumps will be realized at the site of the pumped storage power plant (Hamburger Electricitäts Werke AG) in Geesthacht, close to the city of Hamburg, Germany. The controls of a grid independently operated wind-pumping system and its realization in a laboratory prototype setup are presented. Energy and efficiency aspects as well as controller design problems are discussed. An electrical connection between wind-generator and pump-motors has been projected calling for electrical control actuators. For laboratory purposes electronic switching devices of external rotor resistors have been realised proving their effectiveness for control while also guaranteeing simplicity because of slip power chopping. A second approach involving power electronics inverters in the stator circuit has been investigated from a energy and flexibility in terms of control strategy point of view.

1 INTRODUCTION AND GENERAL ASPECTS

Fig. 1 sketches the setting of the wind pumping system as a part of the pumped storage power plant in Geesthacht.This article focuses on concepts and realisation of the control systems and operating system for the pumps.

Notation:
For the context of this article ***Pump [1]*** is used as a synonym for the KSB OMEGA 150-250 A centrifugal water pump (P = 320 kW at 2100/min, $H_{geo.+Losses}$= 110 m, Q = 215 l/s), ***Pump [2]*** is used for the KSB OMEGA 300-560 B (P = 500 kW at 2100 kW, $H_{geo.+Losses}$= 100 m, Q = 450 l/s).

Characteristics of the plant:
1. Pumping head varying from 76 m to 92 m,
2. Variable frequency operation of turbine and pump system;
3. Energy transmission and conversion:
   Windturbine - synchronous generator - two (or three) induction motors - centrifugal water pumps;
4. The electrical coupling between generator and motors offers freedom in placement of different parts of the system thus providing best conditions for the turbine (high wind speeds on the rim of the reservoir) and pumps (high heads resulting in good storage conditions).

A number of restrictions partly from technical constraints (a) partly from representative premises (b) were assigned. Latter ones result from the plant's character as a demonstration project showing the feasibility of this kind of plant and size for renewable energy storage purposes to potential clients and to the interested public:

(a) · The operating range is limited from 51 Hz (structural mechanics of the turbine, i.e. resonant frequency of the tower) to 69 Hz (noise problems) thus requiring rather rigid control;

(b) · Simple electronic control actuator with pulse width modulated slip ring resistors for the pump induction motors illustrating a possible application in remote areas with limited access to maintenance services;
· Grid independent operation which demonstrates the new approach taken for a plant of this scale.

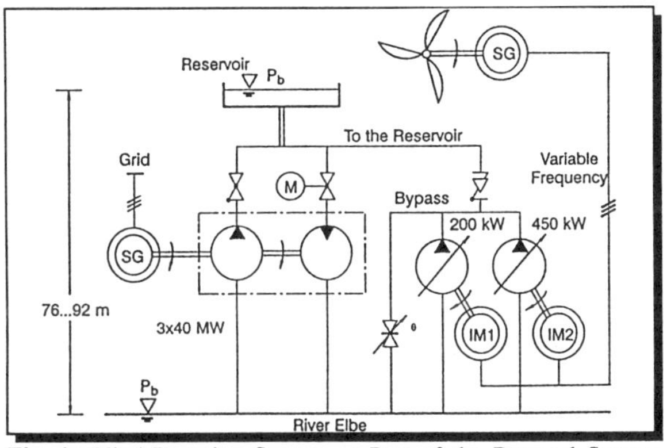

Fig. 1: Wind-Pumping System as Part of the Pumped Storage Power Plant at Geesthacht

During the process of design of this project a number of these points have changed specifically because a different turbine has to be chosen. The major control strategies and operating conditions still apply because performance of the turbine and centrifugal water pumps are fixed by the location of the plant.

(bb) · More sophisticated power electronics involving inverters with rated power of 500 kW will be installed because the prices for semiconductor devices have decreased. This solution gives greater flexibility in speed control of all parts and promises better efficiency as pumps and turbine can be run at their specific best points.

2 ESTIMATED ANNUAL ENERGY GAIN

From previous investigations /1/ the annual energy gain

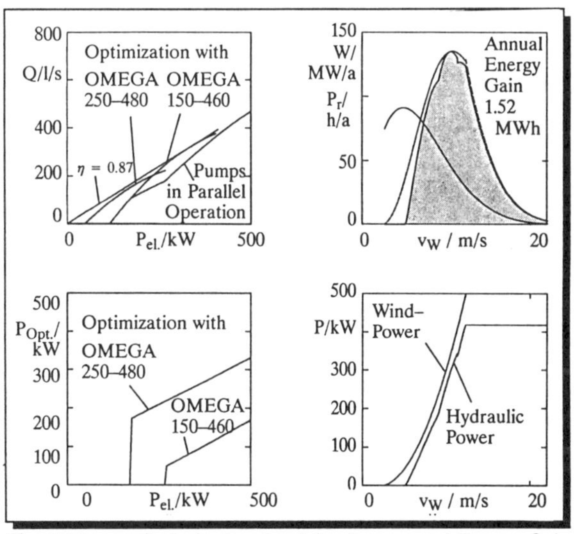

Fig. 2: Energy Optimization Applying Incremental Energy Gain Principle: U(pper)L(eft): Flow vs. Electrical Power; UR: Energy vs. Wind-Velocity; LL: Optimal Power Distribution; LR: Power (Aerodyn.,Hydr.) vs. Wind Velocity

for the direct three phase electrical coupling solution applied to a MBB Monopteros 50 WEC (P_{rated} = 640 kW, Pumps [1] and [2]) was estimated to be about 1150 MWh.

An analysis /2/,/3/ of the solution incorporating power electronics inverters and a ENERCON 40 WEC (P_{rated} = 500 kW) is shown in Fig. 2 where water pumps (KSB OMEGA 250-480, P = 450 kW; Pump [1]) are chosen to match the energy production of this low wind WEC. The annual energy gain for a pumping head of 92 m results in 1517 MWh which can be even slightly improved to 1551 MWh when adding a third small pump (KSB OMEGA 100-375, P = 85 kW) thus profiting from very low wind speeds and reduced losses due to low efficiency in switching transition from one pump to another.

3 Electrical Coupling

Two solutions for the electrical coupling between wind-generator and pump-motors have been investigated:

1. A *direct electrical three phase coupling (so called "Electrical Shaft")* between the wind-generator and the pump-motors was adopted in the beginning of the project. Variable resistors at the slip rings of the induction machines serving as electrical control actuators (electrical efficiency 70 % .. 92 %, Section 7).
2. Electrical energy transmisson including *power electronic converters* (one rectifier and two or three inverters, electrical efficiency 92 %, Manufacturer data).

Although the measures are different the control actions for the two solutions (1) and (2) are the same ones:

I. Limitation of pump speeds to their respective range of operation (HIGH: Net Positive Suction Head (NPSH) boundary; Measure from which protection of the pump against cavitation (wear) can be estimated in particular in high speed operation, LOW: realization of flow against head of reservoir). Therefore a speed control for the pumps is required.

II. Provision of switching hysteresis in order to avoid frequent acceleration and deceleration of all mechanical parts. This is possible with both solutions by smooth distribution of power delivered from the turbine to the pumps in operation.

III. Realization of frequency control by means of adequate load to the generator depending on energy (wind) input which is generated by the pump system.

From numerical simulation and laboratory experiment it can be seen that variation of the slip requires little electrical power but it results in large variation of hydraulic power. Since centrifugal water pumps are very rugged devices it looks quite promising to use them to control frequency (Sections 7.2 and 7.3) rather than rely purely on pitch angle control of the turbine.

4 System Performance

The system performance is governed by turbine and water pumps which can be seen from Fig. 3. Ideal pairing of aerodynamical and hydraulic fluid machines applies only partially. Due to high heads no pumping through the pressure pipes is possible for a wide range of speeds. Hence the feasible range of operation is restricted to a small speed band which even does not cover the whole limited range allowed for the WEC MON 50 (see Fig. 3).

Considering the constraints for the pumps as well as for the turbine (resonant frequency for low speeds, noise problems for high speeds) the initial approach to let the system find its own equilibrium without any control activity required could not be pursued. Pump characteristics for variable speed pumps include pumping heads $H = f(Q, \omega)$, power $P = f(Q, \omega)$, $NPSH_{required} = f(Q, \omega)$, and $NPSH_{available} = f(Q, H_{Suction})$.

Controls with regard to voltage control of the synchronous generator (including the speed dependent reference value, no-load startup procedure) and switching procedure between pumps have been shown previously /4/.

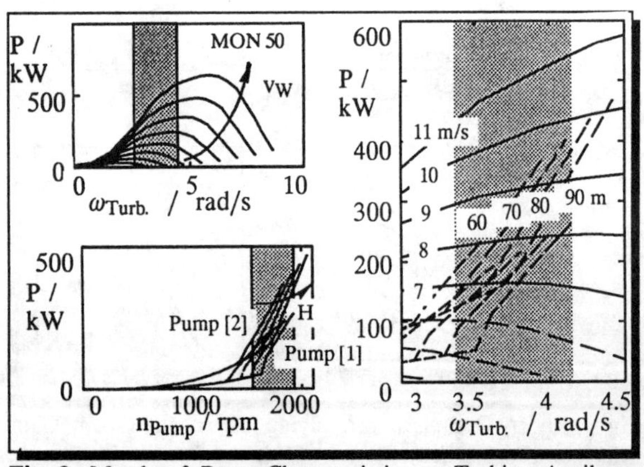

Fig. 3: Match of Pump-Characteristics to Turbine Attributes (UL: Turbine $P = f(\omega, v_W)$; LL: Pumps $P = f(n, H_{geo})$; R: Match on the operating range of the turbine)

5 Operating System

The major task of the operating system comprises decisions on operation of the different pumps (and their combination) and the generation of appropriate reference values for the controllers (Fig. 5). The main operating points can be characterised as follows:

1. So called "BP12": Operation of pump [2] as frequency controller and pump [1] with high efficiency.
   Speed P-controller for pump [2].
2. "BP2"/"BP1": Operation of pump [2]/[1] with protective PI-speed control.
   On enter from "BP12"/"BP21" dynamic reintialisation of speed controller.
3. "BP21": Operation of pump [1] as frequency controller and pump [2] with high efficiency. Higher flows than in "BP12", less actuator power due to lower input power of pump [1].
   Speed P-controller for pump [1].

Realization of these tasks can be transfered to structured analysis programming tools (Computer Aided Software Engineering, CASE-tools) in order to make programming more efficient and independent from specific hardware platforms and even programming structures.

6 Laboratory Setup

In laboratory experiments the control system and the operating system are tested. Three sets of machines each including one DC-machine and one three phase machine (synchronous generator, induction machine respectively) are used to represent the wind-pumping system on a scaled-down basis (10 kW).

The three phase machines are interconnected - corresponding to the real system - without direct connection to the grid. The energy flow is characterised by Fig. 6 indicating that the energy distribution in point 1 follows from the pump load models as well as from control actions and associated performance of the induction machines in points 2a, 2b. The 4-quadrant converters im-pose a Microcomputer propor-

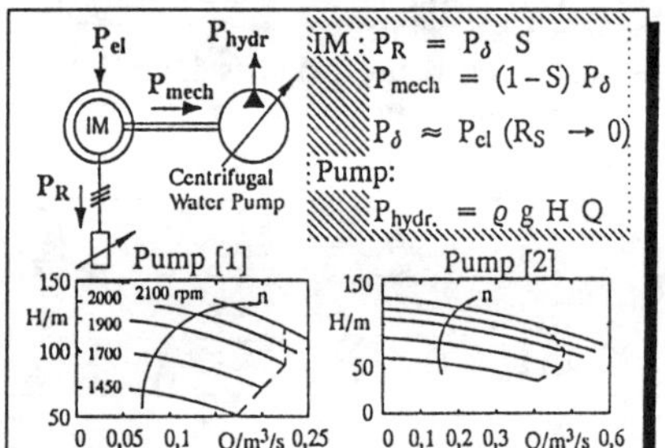

Fig. 4: Characteristics of Centrifugal Water Pump [1] and Pump [2]

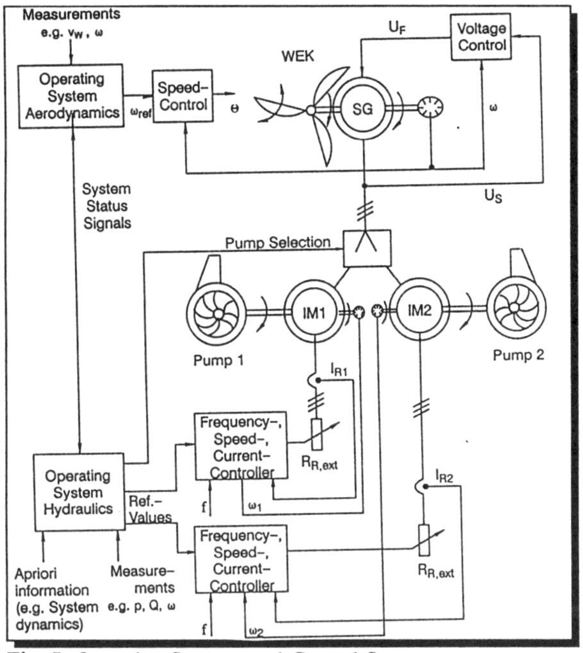

Fig. 5: Operating Systems and Control Systems

tional torque (fluid machine simulation) signal to the armature current of the DC machines. The variable rotor resistors on the slip ring induction machines are realized by 1 kHz pulse-width-modulated chopper circuits shortcutting fixed valued external resistors according Fig. 7. For laboratory purposes $R_{external} = 100 \cdot R_R = 10\ \Omega$ were chosen and proved effective in particular when low loads are required for frequency control in gusty wind conditions.

7 Results from Laboratory Experiments

Experiments indicate that frequency control with centrifugal pump [1] offers a better overall efficiency (63 %; pump [2]: 45 %) while the pump [2] gives higher actuator power thus more rigid control. The electrical efficiency results to 70 % to 93 % for both cases.

7.1 Wind regime

Different synthetic wind velocity as energy input to the wind-pumping system have been chosen for test and controller design purposes (Ramps of wind velocity between 9.5 m/s to 11.5 m/s applying wind gradients of ±0.4 m/s² and ±0.13 m/s²). They correspond to the transition from partial to full power

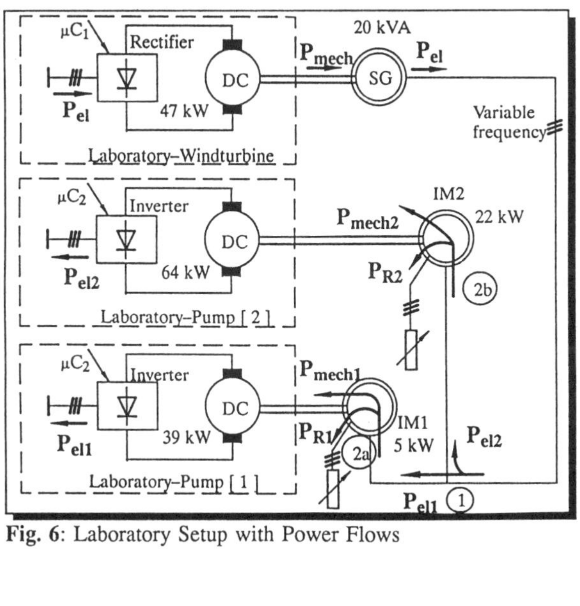

Fig. 6: Laboratory Setup with Power Flows

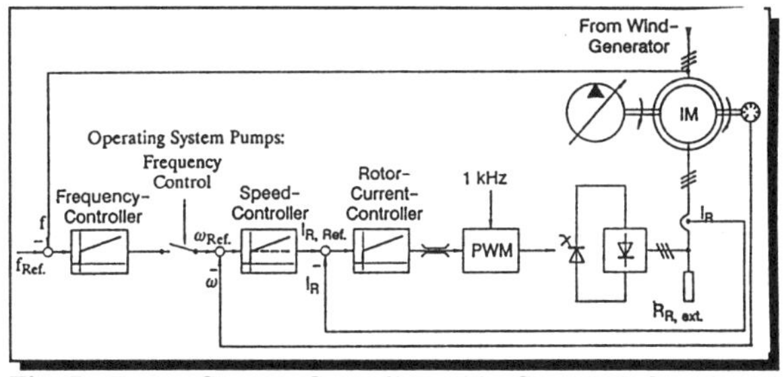

Fig. 7: Rotor Current Controller with Cascaded Speed and Frequency Controller

(and speed) where the frequency controller of the pump system becomes effective.

7.2 Frequency Control Using Pump [1]

From Fig. 8 the results at the "BP21"-operating point (see Section 5 for definition) with two very different control regimes can be seen.

- Frequency control of the wind-pumping system by loading of the generator when wind velocity rises (B_1 - A_2).
- Speed control by the underlying controller for pump [1] when cavitation is about to occur thus (securing safe operating but also) limiting the actuator range (A_1 - B_1, A_2 - B_2).

The overall efficiency of η = 62 % is considerably higher (Tab. 1) than for "BP12" (η = 38% .. 55%, Tab. 2) even though the pump [1] is more restricted in power and speed. Offering more control potential the complementory control "BP12" will be investigated in the following sections.

7.3 Frequency Control Using Pump [2]

In this section results from "BP12" (Section 5) are evaluated. From Fig. 9 the hydraulic flow Q_2 can be seen, which is quite small for this pump because it is operating far away from its design point of best efficiency (see Fig. 4). It becomes evident that the maximum power is required when accelerating the controlling pump. This phase of increasing windspeed should be anticipated as early as possible by the pump system in order to avoid unnecessary stresses by dynamic controller actions.

7.4 Change of Controller Settings

In this section an approach to adjust the controller settings to the performance of the system in "BP12" (Section 5) is presented.

The speed controller reaches its lower boundary for the period when wind speeds are low and the controlling pump runs with minimum speed (point A in Fig. 11). This boundary was chosen with respect to the integral gain of the current controller. In point A (Fig. 10) the frequency of the turbine reaches its reference value and the fast PI-frequency controller becomes effective. It results in some oscillations produced by the P-

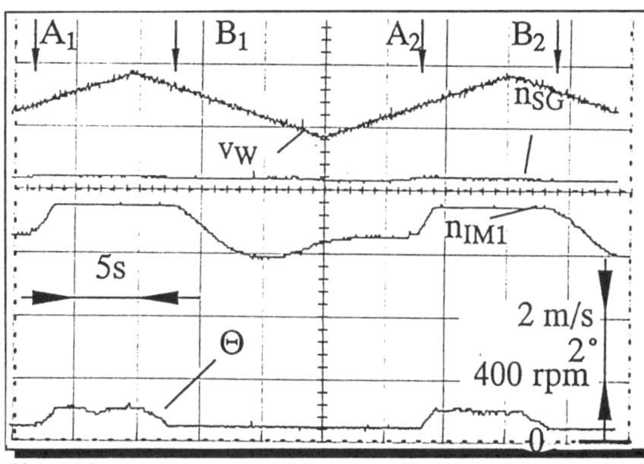

Fig. 8: System Response using the **smaller** Water Pump [1] as Actuator for Frequency Control. Plots: *Wind-Velocity, Speed of Turbine and Pump [1], Pitch Angle*

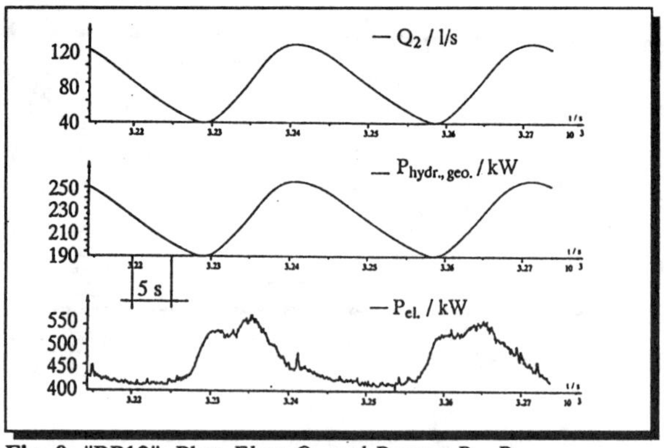

Fig. 9: "BP12". Plot: *Flow Q_2 and Powers P_{el}, $P_{hydr.}$*

speed controller which in turn produces maximum output, thus maximum acceleration of pump [2]. In point B the frequency of the turbine drops below its set point again. A second control action is sometimes required as seen in point C when the load shedding of pump [2] becomes too big and the turbine results in renewed acceleration. This time the speed controller output does not need to interfere because the pump is already at high speed. Nevertheless attention has to be brought to this second activity in terms of the underlying current controller and actuator which has to deal with high voltages when at this speed (frequency) level the chopping switch has to be closed and opened again.

8 Perspectives and Conclusion

The realization of the wind-pumping system in Geesthacht is now scheduled for late 1993. In a subsequent measurement and evaluation period of the plant energy gain aspects as well as control dynamics and operating system questions will be investigated. For these purposes different projects with the laboratory prototype plant are under way employing a high performance microcomputer (VMEbus-based rtVAX300 processor single board computer, TI TMS 320C30 coprocessor AD-board) for measurement and control tasks. A study on the application of structured analysis with real time extension methods (SA/RT) for the design of the operating system will be performed taking advantage of the smooth implementation of these software development tools for the real time computer system as well as for numerical simulation.

The presented results show a promising approach to control of a grid independent operated wind-pumping system. Laboratory results give an indication how simple implementation can be achieved with high efficiency. The methode includes cascaded control of rotor currents, pump speeds and system frequency where frequency regulation of the pumps interact with pitch angle control of the turbine. The combination of these still independently designed manually tuned controllers assure a good overall performance of the wind pumping system.

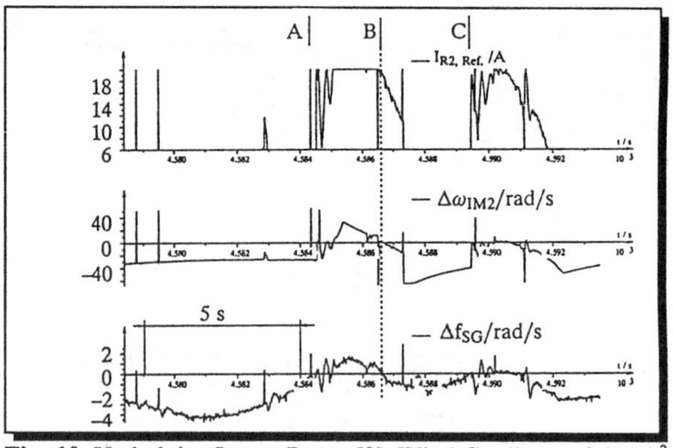

Fig. 10: Underlying Loops Pump [2], Wind Gradients +0.3 m/s^2 and -0.14 m/s^2, PI-Settings: V_R: 5.0 → 7.0; T_I: 0.2 s → 0.05 s; Plot: $I_{R,Ref.}$, $\Delta\omega_{ASM\,2}$, Δf

References:

/1/ Mickeler, S.; Preiß, A.: Pumpenkennlinien und Jahresenergien, Internal Report MBB, 1991, Hoykenkamp

/2/ Leonhard, W.: Regelung in der elektrischen Energieversorgung, Teubner, Stuttgart, 1980

/3/ Könemund, M.: Wind-Pumpensystems Geesthacht: Auswahl der Kreiselpumpen für maximalen Energieertrag, 1992, Institut für Regelungstechnik, Braunschw., Internal Report

/4/ Van Hulle, F. et al. (ed.): Wind Energy Technology and Implementation, Proceedings of Amsterdam EWEC'91: Könemund, M.; Leonhard, W.; Mickeler, S.; Newi, G.: Control of a Large Wind-Pumping System; Elsevier, 1991, Amsterdam

Partners in the project are the Hamburger Electricitätswerke AG, ENERCON GmbH, Aurich, and the Institut für Regelungstechnik, Braunschweig, Germany. It is supported by the Federal Ministry for Research and Technology.

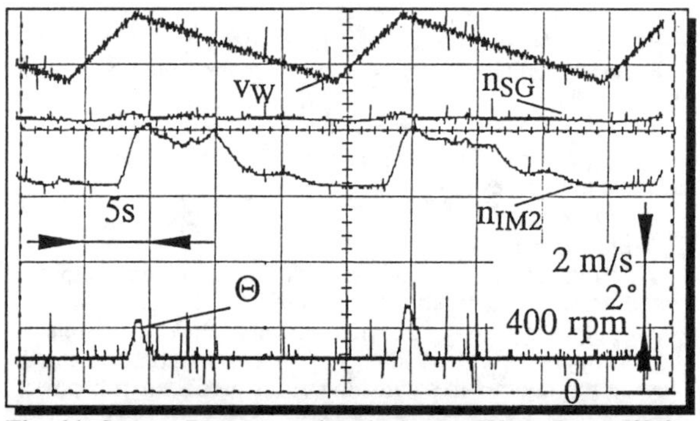

Fig. 11: System Response using the **larger** Water Pump [2] for Frequency Control. Plots: *Wind-Velocity, Speed of Turbine and Pump [1], Pitch Angle*

Elec.Power P_{el} / kW	Hydr.Power $P_{hydr.}$ / kW	Flow[1] Q_1 / l/s	Flow[2] Q_2 / l/s	Efficiency η / %	Diss.Power P_{R1} / kW	Diss.Pow. P_{R2} / kW	Elec.Efficiency $\eta_{el.,1}$ / %	Elec.Efficiency $\eta_{el.,2}$ / %
550	250	200	120	45	30	22	87	93
440	190	200	40	43	30	63	87	70

Tab. 1: Efficiency Values for Frequency Control Employing Pump [2] ("BP12")

Elec.Power P_{el} / kW	Hydr.Power $P_{hydr.}$ / kW	Flow[1] Q_1 / l/s	Flow[2] Q_2 / l/s	Efficiency η / %	Diss.Power P_{R1} / kW	Diss.Pow. P_{R2} / kW	Elec.Efficiency $\eta_{el.,1}$ / %	Elec.Efficiency $\eta_{el.,2}$ / %
520	320	150	250	63	66	21	70	93
410	250	60	250	62	14	21	87	93

Tab. 2: Efficiency Values for Frequency Control Employing Pump [1] ("BP21")

Layout of wind/solar-hybrid systems for the autonomous supply of small electrical loads

Hans Georg Beyer, Hansjörg Gabler, Klaus Nottebaum

Dept. of Physics, Carl von Ossietzky Universität Oldenburg

D-2900 Oldenburg, Germany

Abstract

Small electrical loads at remote locations may be supplied by a combination of a wind generator and PV-modules. We present a method for the identification of an optimal system configuration. Systems, that reach the same level of supply reliability are compared by the figure 'investment costs'. The supply reliability of a given system configuration it determined using the meteorological conditions of the prospected site. The system performance over one year is evaluated by a time step simulation model. This model is validated by comparision with the measured performance of an experimental system. The meteorological input data are synthesized according to the climatic data given in the European Radiation Atlas and the European Wind Atlas.

Introduction

Autonomous systems on the basis of wind and solar energy are an option for the supply of small electrical loads at remote locations (telecommunication installations, alpine huts or data logging stations for environmental parameters). For the system configuration different layout options may be regarded:

- supply by wind power
- supply by a photovoltaic (PV) generator
- combined use of PV and wind power

The preferable configuration depends mainly on the climatic conditions at the considered site. As for a large number of regions the seasonal pattern of wind and solar energy flows are mutual supplements, the combination of wind and solar sources has in many cases advantages. In those cases the level of reliability of supply of hybrid systems could be gained in monovalent systems only by an unreasonable increase of the size of converters or storage. In the following we will present the evaluation of optimal system configurations for various european regions. These calculations are based on a simulation model that is validated against operational data of an experimental system.

Layout and evaluation of an experimental system

To evaluate the performance of hybrid systems an experimental system was set up and operated at Oldenburg University. It is designed for the supply of a data logging station with a daily energy consumption of 240 Wh (see fig.1). A PV-generator consisting of two modules with 40W peak power each and a wind turbine with a nameplate rating of 50W deliver the energy. As storage unit a 12V/45 Ah lead acid battery is used.

The system was operated for several months and extensively monitored. The performance data gained were used to test a simulation model of the system.

The model was implemented in the blockdiagram orientated simulation system INSEL developped at Oldenburg University. Taking values of wind speed, solar irradiance and load the time pattern of battery charging and discharging currents, battery voltage and state of charge are calculated by time step simulation [1]. In accordance with the time resolution of the measured data the calculations are done with a time step of 15minutes.

Fig.2 gives a comparision of the measured and calculated time traces of battery current, battery voltage and state of charge for a period of 11 days. Here the "experimental" curve for the state of charge of the battery was gained by an ampere hour balance based on the measured battery current in connection with a model for the ampere hour efficiency and a recalibration procedure at full charge conditions. These data prove that the simulation model gives a good estimation of the state of the battery. This is the key figure for the estimation of the long term performance of the system in view of the supply reliability.

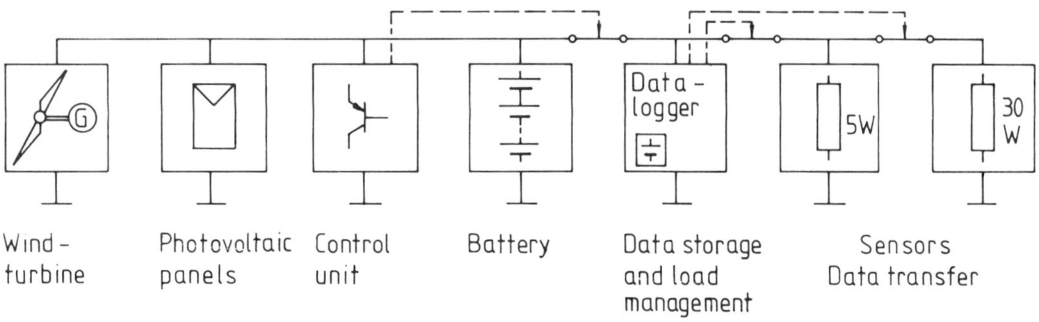

Fig.1
Structure of the experimental system operated at Oldenburg University. The power is supplied by PV-modules and a wind generator. Two switchable loads and the data logging unit have an average energy consumption of 240 Wh/day. The battery (12V/45Ah) is protected by a control unit.

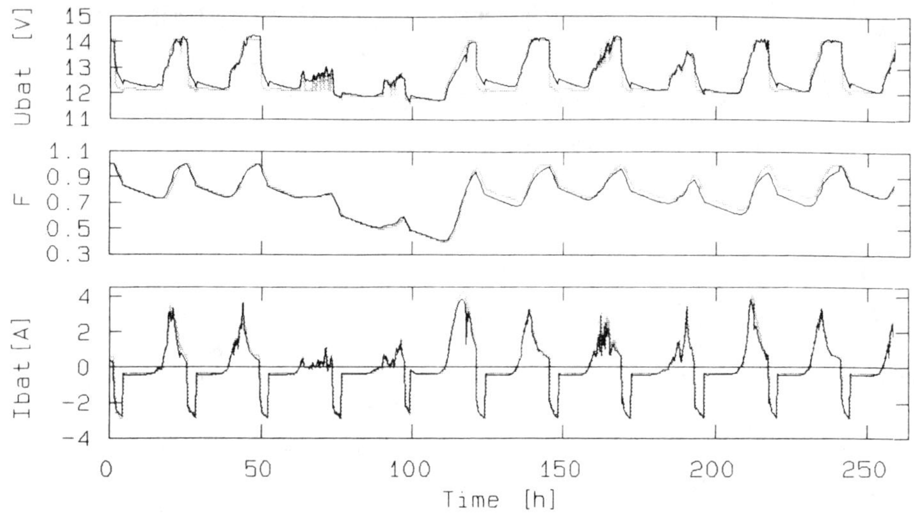

Fig.2
Measured and simulated time series of battery voltage (upper curves), state of charge (middle curves) and battery current (lower curves). The measured time traces are given by the bold lines; the dotted lines refer to the simulation.

The analysis of the annual system performance at various sites

The simulation model enabels us to evaluate the annual performance of systems with different configurations and at various sites. If hourly time series of wind speed, solar radiation and the time pattern of the load are available the overall reliability of supply may be determined. As a measure of supply reliability we choose the "loss of load probability" (LLP) of the system. The LLP is defined here by the the relative number of hours in which the load can not be supplied due to battery depletion.

The aim of these calculations is the identification of optimal system configurations (pure wind, pure solar or hybrid) for various sites in Europe. This is done under the following constraints:

- the average load to be covered is 20 W
- the battery size is 12V/200Ah (sufficient to supply the load for 5 days)
- the LLP of the system should be less than or equal 1%

For the system configuration the following options are regarded:

- variations of the number of PV-modules (40 W_{peak} each)
- variation of the wind turbine size (nameplate rating 50W, 100 W, 250W and 600W)

The sizes of the wind turbines refer to types available on the market. All turbines are installed on masts with a hight of 15m.

For this task hourly time series of solar irradiance and wind speed must be available for the sites inspected. As measured time series are not available for a large number of sites we use input data generated synthetically on the basis of informations given in the European Wind Atlas [2] and the European Solar Radiation Atlas [3].

Generation of synthetical input data

The European Solar Radiation Atlas contains informations on monthly mean radiation sums for a large number of stations. These data are input to a procedure that generates solar irradiance time series with an hourly resolution. The procedure is composed of methods described by Gordon and Reddy [4], Graham et al. [5] and Aguiar and Collares-Pereira [6]. With these methods data for the irradiance on horizontal surfaces are synthesized. As PV-generators are normally installed with a certain inclination towards the equator the irradiance data sets have to be transformed accordingly. Here it is assumed that the PV-modules are set up with an inclination equal to the latitude of the site under consideration.

The European Wind Atlas (EWA) was mainly created to back up a method for an estimation of the wind energy potential at arbitrary sites in Europe. For this purpose wind data from meteorological stations in the countries of the European Community have been analysed. From these base station data the necessary input sets for the estimation of the wind energy potential at sites in the respective geographical region have been produced. The estimation of the energy potential can be done with the WAsP method [7]. For the application of the method the site must be characterized by the surface conditions of its surroundings - parametrisized by a characteristic length (roughness length) - and the prospected height above ground.

Besides the data necessary for the wind potential estimation the EWA gives additional characteristics of the wind climate of the base stations. These are: the mean daily evolution of the wind speed for each month and, as a measure for the variability of the wind speed, the annual evolution of the mean cubed wind speed. The temporal characteristics of the wind speed fluctuations are given by their spectral power densities. These informations form the input to the procedure used in this work for the generation of synthetical sets of hourly wind speed time series [8]. As the EWA gives the characteristics of the time series measured at the base stations, the synthetical series reflect the respective mean wind speeds and energy potentials. To take into account the different conditions at the prospected site a scaling may be applied to the generated data sets. This can be done by multiplying the series by the ratio of the annual mean wind speed at the prospected site and the mean wind speed at the base station. The mean wind speed at the sites can be evaluated using the WAsP method. For this purpose the sites must be characterized by a roughness class. In the following two types of sites with roughness class 2 (moderatly open terrain) and 3 (obstructed terrain) are taken into account.

The irradiance and wind time series are generated independently. Thus mutual correlations of both data sets are only existing in form of the seasonal trends represented by the mean annual pattern.

The annual performance of combined systems

In this section we will give an example of the procedure for the identification of optimal system configurations. This is done for the region Hamburg (northern Germany) and a site with roughness class 2.

In the first step it has to be determined which system configurations can fullfill the requirement of a LLP of 1%. The LLP of a given configuration is determined by an annual time step simulation using the synthetical input data. Fig.3 shows the annual time pattern of the power production from a wind turbine (100 W) and a PV-generator (2 modules à 40 W) at the site Hamburg.

A comparision of the LLP's for different configurations is given in fig.4. The LLP is given as a function of the number of PV-modules for various sizes of the wind turbine. As we are interested in LLP values of about 1% the differences in the system performance are more clearly visible in a logarithmic presentation. The LLP = 1% is indicated by a value of -log(LLP) = 2.0. We find annual LLP's of less than 1% for systems with: wind turbines bigger than 250W and no PV-power, wind turbines of 100W and 2 PV-modules or wind turbines of 50 W and 12 PV-modules. For the site Alghero (Sardinia,Italy) the following configurations give the required LLP (lower plot in fig. 4): wind turbines with 600W and 1 PV-module, wind turbines with 250W and 2 PV-modules, wind turbines with 100W and 3 PV-modules, wind turbines with 50W and 5 PV-modules or 8 PV-modules.

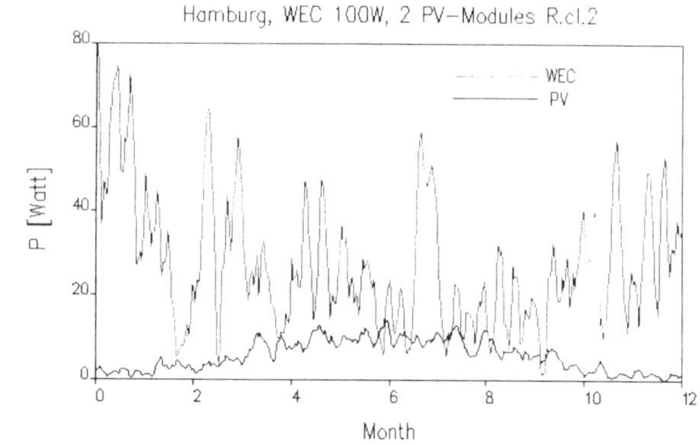

Fig.3
Annual time pattern of the power output of a wind generator (100 W) and a PV-generator (80 W) (bold curve) for the site Hamburg. The curves give 5 day moving average values.

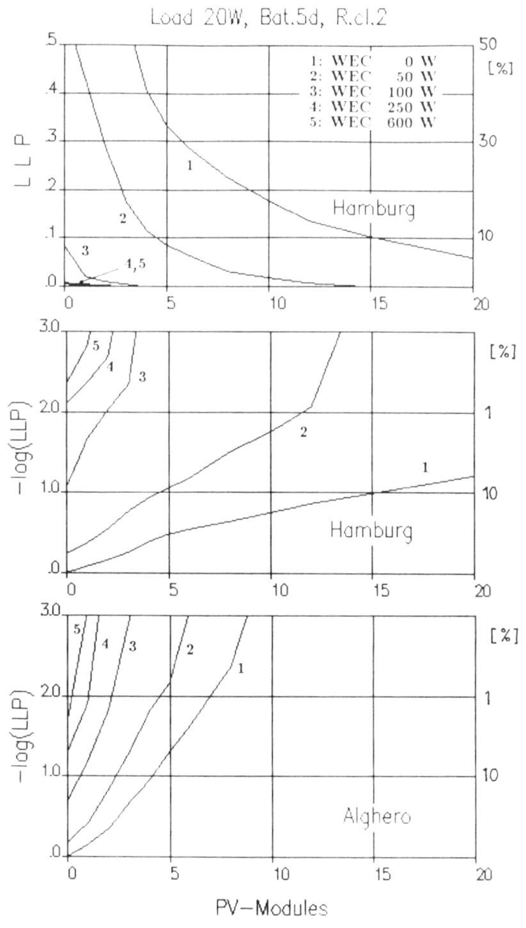

Fig.4
Loss of load probability for systems with different configurations. The value of LLP is given as a function of the number of PV-modules for wind generators of different sizes. In the middle and lower plot a logarithmic presentation is given. The value -log(LLP) = 2.0 indicates a supply reliability of 99%.

System optimization

The calculations described above will identify different system configurations with the same level of supply reliability for every site under consideration. These system configurations may be ranked according to their investment costs. The following prices for the generators are assumed:

- 1 PV-module 40 W: 500 ECU
- wind turbine 100W: 2500 ECU
- wind turbine 250W: 3750 ECU

The battery and balance of system costs are assumed to be the same for all configurations.

For 26 sites in Europe configurations with the lowest investment costs have been determined. These configurations are indicated in the maps given in fig. 5. The two maps refer to sites in terrain of roughness class 3 and 2. It can be seen that the use of wind turbines for these small systems is mostly preferable in the regions around the north sea. This holds for both type of wind conditions. For the mediterranean region the combined use of wind and solar at sheltered sites (roughness class 3) is only recommendable in the case of very good regional wind climates (e.g. Istres). Even for less obstructed sites (class 2) the pure PV-system still gives the optimum in regions with very poor wind conditions (e.g. Pisa) or very good solar conditions (e.g. Almeria).

Conclusions

We have presented a method for the optimization of small wind/solar hybrid systems. The results indicate that the combination of both power sources is in general preferable at sites in northern Europe. In the Mediterrenean areas the irradiance conditions are more favourable; wind generators compete with PV only at a very limited number of sites.

The calculations presented are based on data generated for a single year. The year to year variability in the wind speed and insolation time patterns - which may to some extend be reproduced in the simulated time series - should increase the generation capacity necessary to guarantee a fixed level of annual reliability.

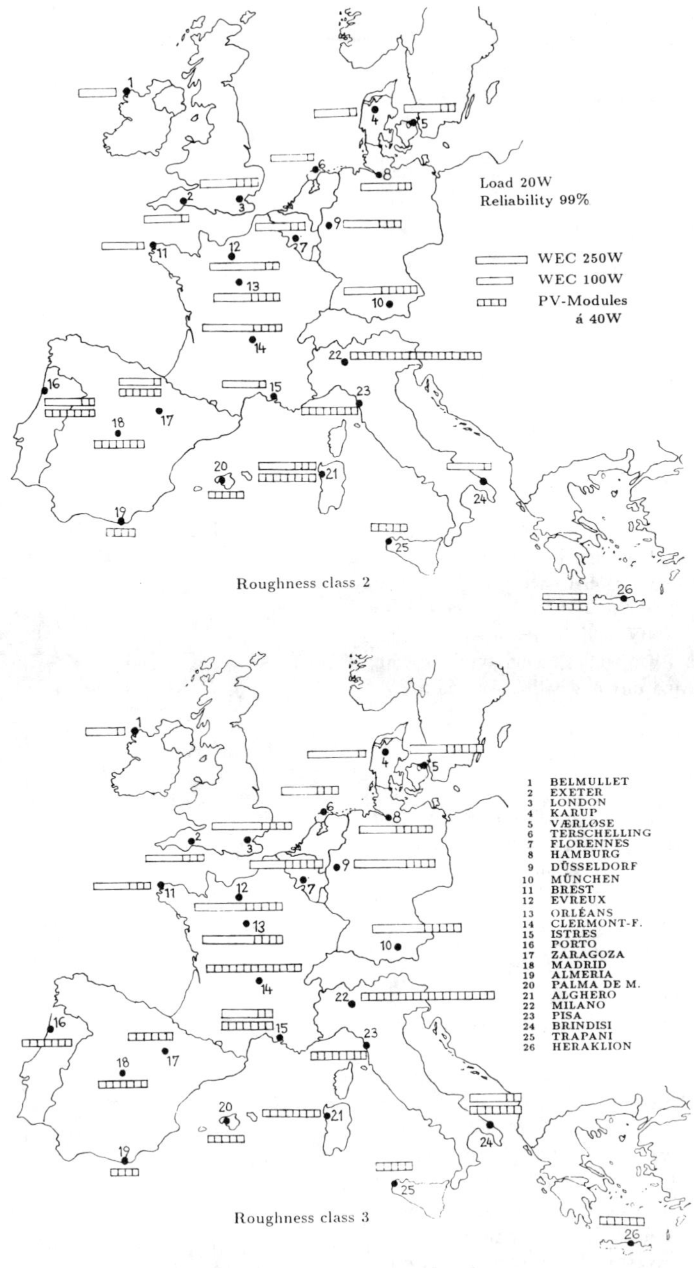

Fig.5
The optimal system configuration with respect to investment costs. The two maps refer to sites with roughness class 2 (upper map) and roughness class 3 (lower map). The bars next to the base stations indicate the system layout for the respective region. The area of the bars is proportional to the investment costs. Different configurations leading to the same investment costs are indicated only in case of a competition of monovalent and hybrid systems.

References

[1] Klaus Nottebaum, Umweltmeßstation mit Energieversorgung aus Sonne und Wind - Auslegung für Europäische Standorte, Diploma-Thesis, Dept. of Physics, Carl von Ossietzky Universität Oldenburg, 1993

[2] Commision of the European Communities: European Wind Atlas. Risø National Laboratory 1989

[3] Commision of the European Community, Atlas über die Solarstrahlung Europas, Verlag TÜV Rheinland 1984

[4] J.M.Gordon, T.A.Reddy, Time series analysis of daily horizontal solar radiation, Solar Energy, 41,1988

[5] V.A.Graham, K.G.T. Hollands, T.E. Unny, A time series model for Kt with application to global synthetic weather generation, Solar Energy 40, 83-92, 1988

[6] R. Aguiar, M.Collares-Pereira, TAG: a time dependent, autoregressive model for generating synthetic hourly radiation, Solar Energy, 49, 1992

[7] I.Troen, N.G. Mortensen, E.L. Petersen, WASP Wind Atlas Analysis and Application Programme. Users Guide, Risø National Laboratory 1988

[8] H.G.Beyer, Synthesis of annual wind speed time series on the basis of European Wind Atlas data, Internal Report, Dept. of Physics, Carl von Ossietzky Universität Oldenburg, 1993

TEST AND OPTIMIZATION OF WIND/SOLAR HYBRID SYSTEMS FOR AUTONOMOUS ENERGY SUPPLY OF SMALL ELECTRICAL CONSUMERS

M. Schubert*, B. Sundermann, J. Twele*
TU Berlin, Institut für Luft- und Raumfahrt, Sekr. F5, Marchstr. 12, W-1000 Berlin 10, Germany
*ATLANTIS gGmbH - Windkraftanlagen, Glogauerstr. 19, W-1000 Berlin 36, Germany

At locations with respective climatic conditions the coupling of PV systems with small wind turbines can lower energy costs and, at the same time, increase the securitiy of a minimum supply.
A wind turbine intended for integration in a hybrid system must show as favourable operation conditions as the durable and almost maintenance-free PV panels. A major component to meet this requirement is the safety system. It was designed as a thrust controlled rotor tilt into a horizontal "helicopter" position. A simple analytical theory has been developed for the description of the tilt governor, which was consequently improved with a numerical investigation. Both results had been checked with experimental data.
Another focus for the design of the wind-driven battery charger was an optimal matching of the electrical system components: wind and solar generators, battery storage, load. For this purpose a computer simulation with time step integration was executed in order to analyse the system behaviour of the entire hybrid plant and thus optimize it for different parameters such as storage capacity, load or metereological data. A comparison of the results from measurements and computer simulation shows good correspondance.

1. INTRODUCTION

For the grid independent energy supply of small electrical consumers, photovoltaic (PV) systems have a well established market. As a design without moving parts they are appreciated as simple to understand, easy to install, reliable and almost maintenance-free. There is also an increasing market for grid connected PV systems in the private sector.

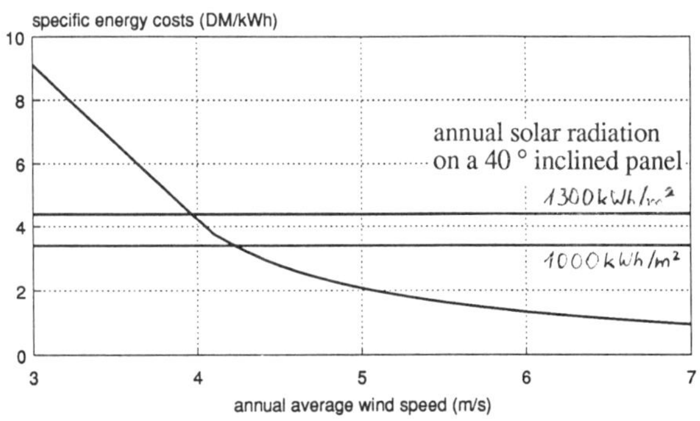

Fig. 1: Comparison of specific energy costs from wind and solar generators (300W installed capacity, roof mounted incl. charge controller, 10% interest rate, 10 years investment return)

Given enough wind at the location, small wind turbines are a cheaper alternative (fig. 1). The combination of a solar with a wind generator may therefore lower energy costs and at the same increase the security of a minimum supply. This advantage of combined wind/solar hybrid systems was first theoretically proven for different climatic areas on the bases of their respective metereological data [1]. Especially in Germany wind and solar energy complement one another in an ideal manner.

But a more detailed analysis is necessary to assure an uninterrupted energy supply for autonomous consumers such as measuring posts in remote areas. A computer simulation of the energy balance of a hybrid system with constant electrical load (e.g. from measurement equipment to collect environmental and/or climatic data) will be introduced in chapter 3.

A wind turbine intended for integration in a hybrid system must show as favourable operational conditions as the PV panels. A key component to meet this requirement is the power regulation and safety system of the rotor.

2. TILT GOVERNOR AS SAFETY SYSTEM

Active safety systems are to expensive for small wind turbines. An apparently simple option for a passive, self-governing power regulation is the use of the rotor thrust. Until a well defined wind speed the rotor stays in a vertical plane. Then with further increasing wind speed it continuesly tilts into a horizontal "helicopter" plane. The rotor thrust is balanced out by a counter weight (fig. 2a).

A simple analytical theory has been developed for the description of the tilt governor. It assumes a reduced wind speed

$$v' = v * \cos \alpha$$

on the rotor determined by the component parallel to the rotor axis tilted with α [2, 3]. Following this theory constant power output is achieved when the characteristic of the tilt governor keeps the reduced wind speed v' constantly on the level v_m for maximal allowed power output:

$$\frac{v}{v_m} = \frac{1}{\cos \alpha}$$

This required control characteristic for the tilt governor was confirmed by experimental data gathered during measurements of full scale wind turbines (1.5 and 2 m diameter) on a test car (fig. 2b) [4].

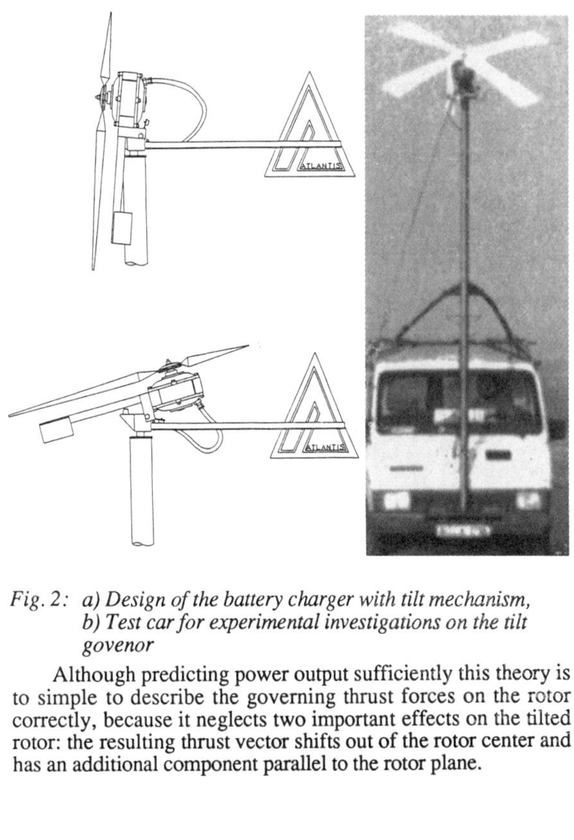

Fig. 2: a) Design of the battery charger with tilt mechanism, b) Test car for experimental investigations on the tilt govenor

Although predicting power output sufficiently this theory is to simple to describe the governing thrust forces on the rotor correctly, because it neglects two important effects on the tilted rotor: the resulting thrust vector shifts out of the rotor center and has an additional component parallel to the rotor plane.

The theoretical description of the tilt governor was consequently improved with a numerical investigation of the aerodynamic and inertia forces on the tilting rotor. The computer programme is based on blade element theory.

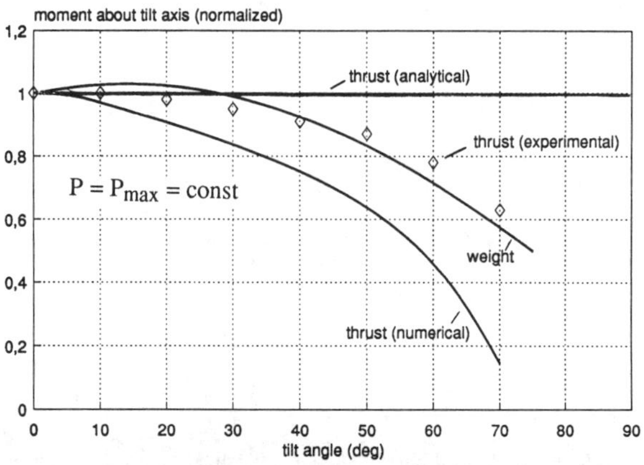

Fig. 3: Comparison of theoretical and experimental moments about the tilt axis normalized with the moment at beginning of tilt movement

Both theoretical results were compared with experimental data from full scale measurements on the test car (fig. 3). These measurements confirm that the analytical theory does not model the tilt governor correctly whereas the numerical description turns out to be sufficient for a first go.

Fig. 2a shows a design of the tilt governor as realized for wind turbines with 300 and 600 W nominal output (at 10 m/s wind speed; 1.5 and 2 m rotor diameter respectively). During the development of the tilt governor, the following design rules turned out to be important for its safe function:

1) Use stiff blades. The flapwise bending of a blade is often coupled with a torsion of the outer blade section. This yields a (dynamic) change of the aerodynamic forces on the blade which are essential for the governing characteristic of the tilt mechanism.

2) Do not go for optimum tip speed ratio. The tilted rotor tends to stall easyly (again a dynamic process). Best results had been obtained when the load keeps the rotor on a tip speed ratio above the optimum value for a factor of 1.2 ··· 1.3.

3) Use large geometric dimensions between rotor and tilt axis. The tilting movement is superimposed by a whole range of dynamic effects (e.g. gyroscopic forces). Influence of these effects should be kept small in relation to the governing forces in order to prevent instability. This was achieved with a large lever arm for the thrust.

Measured data from a 600 W / 2 m wind turbine on a water level monitoring post in the north sea prove that the tilt governor is a suitable feature under practical conditions (fig. 4) [5].

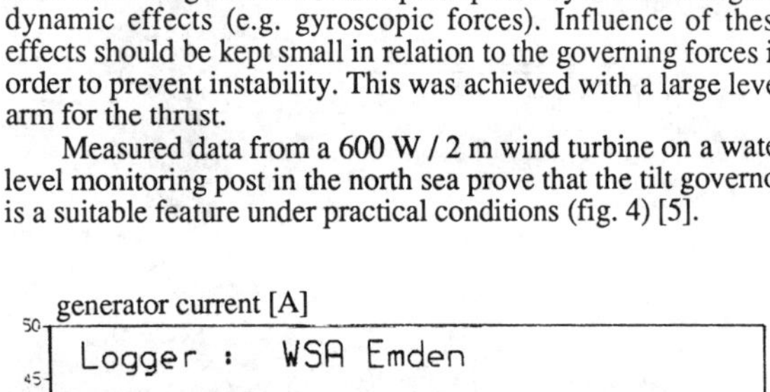

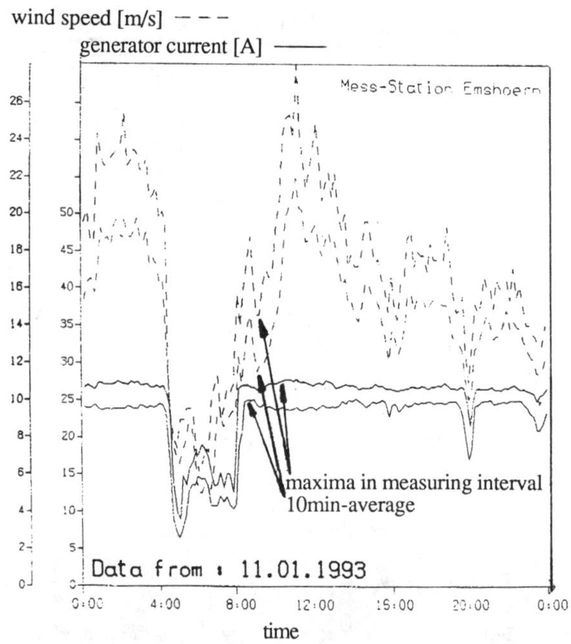

Fig. 4: a) Power characteristics (below left) and b) time record of generator current and wind speed measured on a 2 m wind turbine (system voltage: 28,6 V)

3. SYSTEM MONITORING AND MEASUREMENT PROGRAM

Wind/solar hybrid systems were set up at four sites in Germany: At the Kaiser Wilhelm Koog (KWK) wind test field of the German Lloyd, in Munich, in Mainz and at the wind test field of the Technical University Berlin, Berlin Gatow. The systems were equipped with a ATLANTIS WB 15 battery charger (300 W nominal power output), four Siemens SM 53 solar panels (212 W nominal power output) and equipment to measure temperature, system voltage, power output of wind and solar generator, wind speed and power consumption of load and dump load. The battery capacity was 300 Ah. The data were used to monitor the system performance and to compare it to the results of the simulation.

3.1 Time Step Simulation

In order to allow a matching of the system components to different site conditions and to optimize the energy supply a model for numerical simulation was developed. The system was discribed with a time dependent differential equation which was solved by numerical integration with the simulation program TRNSYS (TRaNsient SYstem Simulation). It was originally developed to design energy supply, heating and cooling systems in general, particulary with regenerative energy. Therefore it uses routines for modelling electronical and electrical devices like regulators, battery storages and rectifiers.

To generate time series of weather data the TRNSYS weather data generator was used and an own routine was added to obtain Weibull distributed wind speed time series. Input for this module are the statistical Weibull parameters A (scale factor) and k (form factor) or the mean wind speed $\bar{v}$ (in this case k is set to the default value 2). Using the reverse function of the Weibull distribution windspeed, data in 10 minute time steps are generated and sorted out to match typical day- or monthlike pattern of the middle European wind regime: eg. lower wind velocities in summer, higher in winter. In a second mode a higer time step resolution of 2 sec. is possible by using a Shinozuka method which is eg. useful to calculate the wind speed - power curve of a wind energy converter [6].

Additional routines were created to compute the power output of the WEC and the photovoltaic systems. The complete structure is shown in fig.5. Time step size for all calculations was 10 minutes, according to the wind data generator.

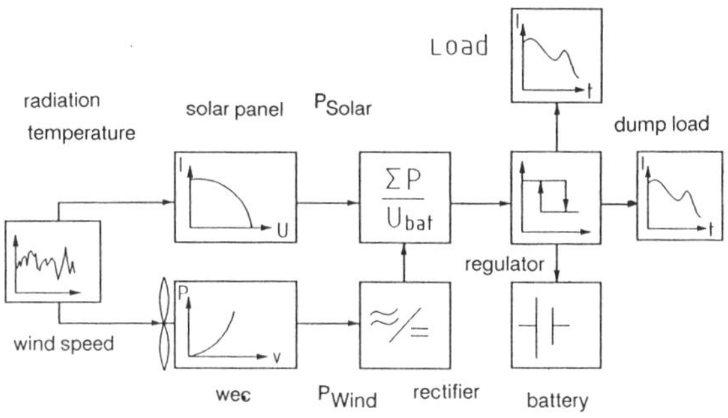

Fig. 5: Block diagram wind/solar hybrid system

In order to prove the simulation model the correspondence, of measured and calculated data was tested. Especially important was the performance of the battery and the regulator model because they are directly related to the problem of minimal power supply security. Fig. 6 shows a good correspondence in an example from Munich: For 400 hours in November 91 measured and calculated current at the consumer are compared.

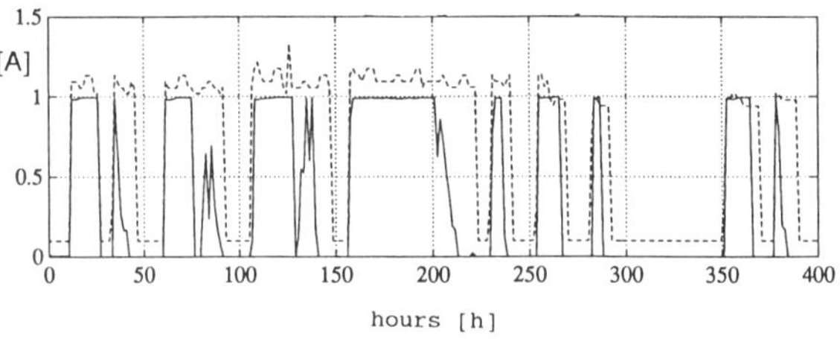

Fig. 6: Comparison of measured and calculated currents consumed by load
—— calculated current
---- measured current (to increase clarity, zero current is numericaly set to 0.1 A)

It was easy to supply the load of 25 - 30 Watt in spring and summer on all sites but not in autumn and winter in Munich, Berlin and Mainz. So it was studied, under which conditions power supply security could be raised. Tab. 1 shows the produced electrical energy in October and November 1991 in Berlin and KWK. Time series from these sites and months were used in the following examples.

	Solar Energy		Wind Energy	
	October	November	October	November
Berlin	17.2	5.6	10.1	10.3
KWK	7.9	4.4	37.7	59.9

Tab. 1: Amount of produced energy in kWh, Berlin and KWK October and November 1991,[7]

3.2 Some Results of Simulation

In Fig.7 the results of a simulation with different values for Weibull parameters, loads and number of solar panels are plotted. At a state of 0.1 the battery is "empty" and the load is separated from the system, at a state > 1 further loading of the battery is suspended and the spare power converted to heat in the dump load. A 100% security of power supply in this example would have been possible by reducing the load or adding more solar panels.

The measured mean wind speed in Berlin in October and November 1991 was very low compared to the average values from statistical data, so a simulation was executed using Weibull parameters from the European Wind Atlas [8]: With the parameters A = 5.3 m/s, k = 2.18 for October and A = 6.1 m/s, k = 2.29 for November the calculated wind energy output raised to 39.3 kWh and 49 kWh.

Under this conditions a higher rate of supply security would have been also possible, fig 7.

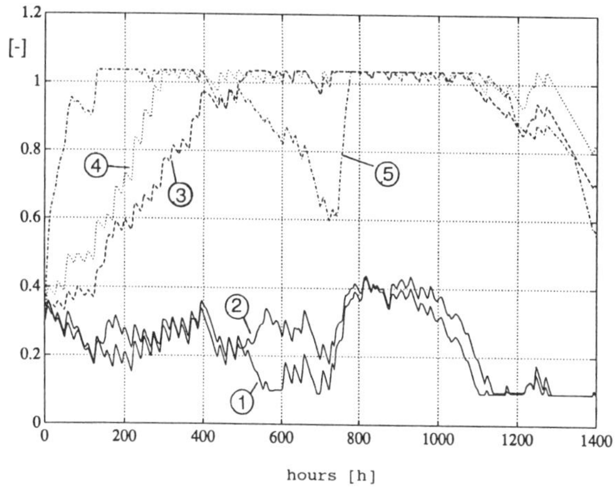

Fig. 7: State of battery charge October - November in Berlin Battery capacity 300 Ah
1) Measured data input 2) 6 solar panels SM 53
3) 8 Solar panels SM 53, 4) Load reduced to 15 Watt
5) Wind speed input with new Weibull parameters

The following figures show the calculated battery behaviour in October and November 1991 at KWK. In the opposite to Berlin it was no problem to supply a minimum power of 30 W. The calculations were executed to demonstrate possibilities of varying load and battery capacity. In fig. 9 the load (40 W) is the same as in fig. 8 (dotted curve), but only a 300 Ah battery was used for calculation. Fig. 10 shows the hybrid system.

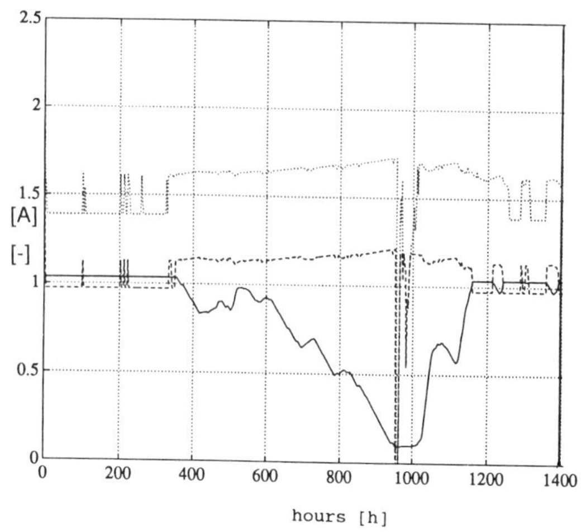

Fig.8: Calculated current consumed by load and state of the battery charge
—— battery charge, 30 W, 300 Ah
---- current consumed by load, 30 W , 300 Ah
..... current consumed by load , 40 W, 500 Ah

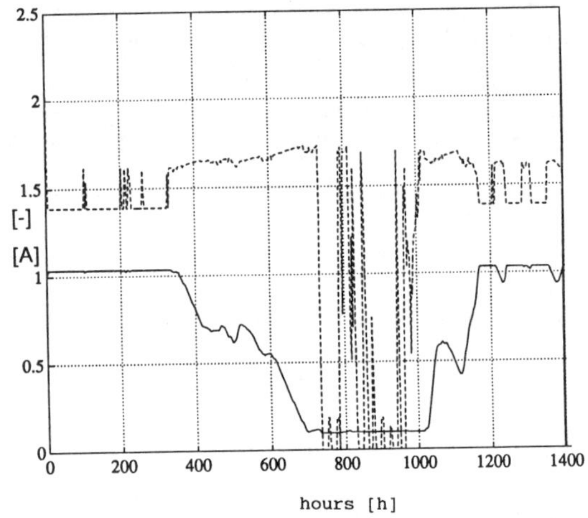

Fig.9: Calculated current consumed by load and state of the battery charge, load = 40 W, Battery 300 Ah

— battery charge

---- current consumed by load

Fig. 10: Wind/solar hybrid system with tripod tubular tower on Kaiser Wilhelm Koog (300 W wind, 212 W solar, 300 Ah battery storage, 30 W permanent load)

4. REFERENCES

[1] J. Twele: Hybrides Energieversorgungssystem für autonome Anwendungen mit kontinuierlichem Leistungsbedarf; 7. Internationales Sonnenforum; Frankfurt 1990.

[2] R. Gasch (Editor): Windkraftanlagen; Teubner Studienbücher - Maschinenbau; Stuttgart 1991.

[3] J. Twele et al.: Leistungsregelung kleiner Windkraftanlagen durch schubabhängiges Kippen um eine horizontale Achse; 8. Internationales Sonnenforum; Berlin 1992.

[4] M. Schubert et al.: Entwurf und Erprobung einer Leistungsregelung für Windkraftanlagen im Leistungsbereich unter 2 kW durch Kippen des Rotors in "Helikopterstellung"; Tagungsband zur Deutschen Windenergie Konferenz; Wilhelmshaven 1992.

[5] Data supplied by Zentralstelle für Solartechnik, Hilden.

[6] J. Knissel: Erstellung eines Rechnerprogramms zur Simulation des Anlageverhaltens von Windkraftanlagen, diploma thesis, Technical University Berlin , 1991

[7] Research project "Mikroklimameßstation, Energieversorgungssystem", final report, MAN Technologie/Ins. of Aeroengineering Berlin, Munich/Berlin 1992

[8] European Wind Atlas, Risø National Laboratory, Roskilde 1990

THEORETICAL AND EXPERIMENTAL STUDY OF VAWT STRAIGHT BLADED DARRIEUS ROTOR

F. Kamissoko, L. De Wilde R. Derdelinckx and Ch. Hirsch

Vrije Universiteit Brussel, Pleinlaan 2, 1050 Brussel, Belgium
Tel. 02/641.23.91, Fax.02/641.28.80

ABSTRACT

The wind turbine "STRO-V 8" is a Darrieus rotor with straight blades coupled to two double acting pumps (type STORK) with piston diameters of 0.1 m and 0.13 m.

The field measurement of the wind turbine-pump system comprises wind speeds, pump flow rate, rotor torque and r.p.m. Rotor power and power coefficients are derived from these quantities.

Theoretical calculations are obtained from the computer code VAWTAP (Vertical Axis Wind Turbine Applications). The code is based on the double-multiple stream tube theory including dynamic stall effects and has been used to predict the aerodynamic loads on the rotor and performance optimization of a Darrieus wind turbine pump system.

The paper will show a comparison between experimental data and the results obtained from the calculation program.

1. DESCRIPTION OF THE WIND TURBINE

The wind turbine system consists of a 8 m diameter Darrieus wind turbine driving two piston pumps through a speed-reducing gear box. A gear ratio of 1 has been choosen in order to obtain a sufficiently high speed for an efficient operation of the pump. To protect the rotor from overspeed, a spring system, attached to the blades, controls the blade pitch setting and thus the aerodynamic performance by centrifugal forcing.

A schematic view of the system is given in figure 1.

1.1 Characteristics of the "STRO-V8"

- Straight bladed vertical axis wind turbine
- 3 NACA 0015 profiled blades in reinforced polyester
- Rotor dimensions : 8x 6 m (diameter, height)
- Tower construction : rigid, 6 post galvanized steel (hexagonal base)
- Hub height : 6.5 m (centre of rotor)
- Power and overspeed control : aerodynamic, centrifugally activated pitch control
- Application : water pumping and/or electricity production
- Rated power : 8 kW (mechanical)
- Rated wind speed : 10 m/s
- Cut-in wind speed : 3.5 m/s
- Survival : 50 m/s
- Weight : 2000 kg
- Design life : all components of the wind turbine : 20 years minimum.

1.2 Characteristics of piston pumps

- Type of pump : horizontal axis piston pump
- Total manometrical head at site : 11 m
- Maximum total manometrical head : 60 m
- Transmission ratio : 1/1
- Pump SP1000 : 0.100x0.170 m (diameter of piston, stroke), maximum capacity of 10 m^3/h
- Pump SP2000 : 0.130x0.170 m (diameter of piston, stroke), maximum capacity of 20 m^3/h

2. THEORETICAL STUDY OF DARRIEUS WIND TURBINE COUPLED TO TWO PISTON PUMPS

The computer code VAWTAP (Vertical Axis Wind Turbine Applications) is a program developed at the Department of Fluid Mechanics (VUB).

It is used to predict the aerodynamic loads on the rotor and to calculate the performance characteristics of a Darrieus wind turbine with straight blades coupled to two piston pumps (double acting pump).

The program is based on the Double Multiple Stream Tube (D.M.S.T.) theory [3]. Dynamic stall effects [2], according to the Boeing-Vertol approach, are included. The following secondary effects are included in the code as well :

- effect of centrifugal forces on blade pitching
- flow curvature effect
- effect of finite aspect ratio
- power losses due to the blade struts
- power losses due to friction forces
- streamtube expansion effect.

3. EXPERIMENTAL STUDY OF DARRIEUS WIND TURBINE INSTALLED AT THE POLDER OF WILLEBROEK (BELGIUM)

3.1 Location of Willebroek

The test field is in an open terrain, covered by grass and bordered by a row of dense trees and a lake. Height of the trees is about 10 m.

Photographs of Darrieus wind turbine driving two pumps are given in figure 2.

In the dominant south-west in direction no obstacles are found.

The wind regime of the site is appropriate for a good operation of the Darrieus rotor.

3.2 Field measurement procedure and location of instrumentation

Both long term and short term measurements have been carried out. The long term measurement procedure consisted of taking 10 minute averages of wind speed, water quantity, torque and rotor speed. To this end a DT 100I Datataker connected to an Apple Macintosh computer is used. The wind speed is measured by an anemometer, mounted on a mast at 10 m height and placed at a distance of 10 m from the wind turbine. Rotor speed and torque are measured by a torque meter attached to the rotor axis. The pumping rate is measured with a water meter (diameter 80 mm).

For the short term measurements a PCM data-acquisition system is connected to an Apple Mac II computer with Labview II software.

3.3 Results and discussions.

In order to compare the results to data from literature and to calculated data, an averaging period of 10 minutes has been choosen. Tests data are elaborated by means of a bin sort procedure.

Dynamic stall is included in the calculations. Figure 3 shows characteristics of mechanical power speed curves compared dynamic stall included and no dynamic stall. This figure shows that at high wind speeds the power values are lower, when dynamic stall effect is not considered in the calculations. The difference between dynamic stall included and no dynamic stall at maximum power is 25%.

Figure 4 shows a combination of characteristic power speed curves of the wind pump system for different wind speeds compared with experimental data. It is observed from this figure that there is a close agreement between the prediction and the measured data. From calculations of the power curve of a wind rotor/displacement pump combination is normally linear. The intersections with the curve of the mechanical power required for the pump are the operating points of the wind turbine. For a wind speed of 6 m/s a power of 400 W at 35 r.p.m. is found.

Figures 5 and 6 show the calculation results of the power coefficient versus tip speed ratio curves for different pitch angles. The optimum of rotor performance is obtained at pitch angle of 5°.

Figure 7 shows the power coefficient versus tip speed ratio for a blade pitching of 5°. From this figure. It is seen that the comparison between calculated values of power coefficients and the measurements is more than satisfactory. The spread in the experimental data are due to the variations of rotor speed and the resulting pitch angle variation during the data acquisition period. The optimum predicted C_p values are equal to 0.25 for a tip speed ratio of λ=1.8 dynamic stall included and to 0.19 for a tip speed ratio of λ=2 no dynamic stall. The average C_p value measured of 0.24 for λ=1.75.

The flow rate (m^3/h) versus rotor speed curves of the predictions and measurements are shown in figure 8. The experimental point of first delivery is around 15 r.p.m. The figure shows a good agreement of the predictions with experimental data for rotor speeds above 30 r.p.m. The discrepancies between the prediction and measurement at low r.p.m. are due to the cyclic variation of the pump torque and to the losses produced by the pumping system. In a decreasing wind speed the wind turbine operates at very low wind speeds. Flow rate maximum values are respectively 9 m^3/h (measured) and 10 m^3/h (calculated). The average wind speed is ranging between 4 m/s and 5 m/s. The maximum measured rotor speed at site during the measurement period is approximately 46 r.p.m. and the power required for the wind turbine pump system is around 600 W.

4. CONCLUSIONS

The wind turbine "STRO-V8", coupled to a SP1000 and a SP2000 pump at a gear ratio of 1 and for a total manometrical head of 11 m, starts up at about 3 m/s. The centrifugally controlled safety system protects the turbine from overspeed and limits the rotational speed to approximately 60 rpm.

The comparison between the experimental results and calculated values shows that the VAWTAP code can be used with reasonable accuracy to determine the performance characteristics of a Darrieus rotor with straight blades.

5. REFERENCES

[1] E. LYSEN. "Introduction to Wind Energy" CWD, P.O. Box 85, CWD 2nd, 1983 Amersfoort, Netherlands,

[2] D. BULTEEL. "Centrifugaal Beveiliging Van Een Verticale As Wind Turbine" Thesis : 1988 Department of Fluid Mechanics "Vrije Universiteit Brussel" (VUB), (Belgium)

[3] I. PARASCHIVOIU. "Double Multiple Stream Tube Model For Studying Vertical Axis Wind Turbine" Journal of Propulsion and Power, Vol.4, No.4, July-August, 1988

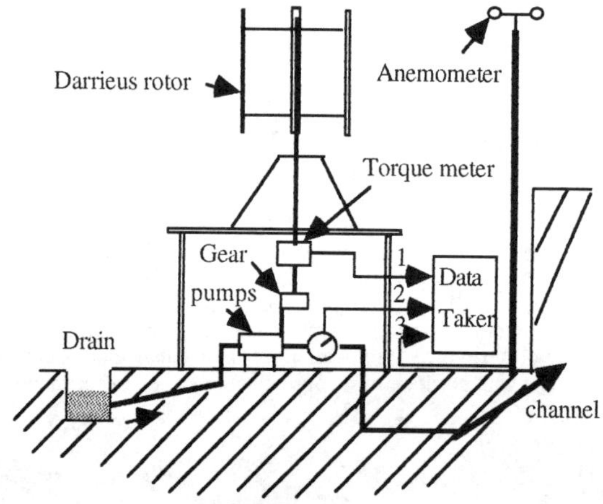

Site of Willebroek

Measurements signals :

1 : Torque (Nm) and r.p.m.
2 : Flow of water (m3/h)
3 : Wind speed (m/s)

Figure 1 : A schematic view of the wind turbine pump system

Figure 2a : Photograph of the wind turbine "STRO-V8"

Figure 2b : Photograph of piston pumps

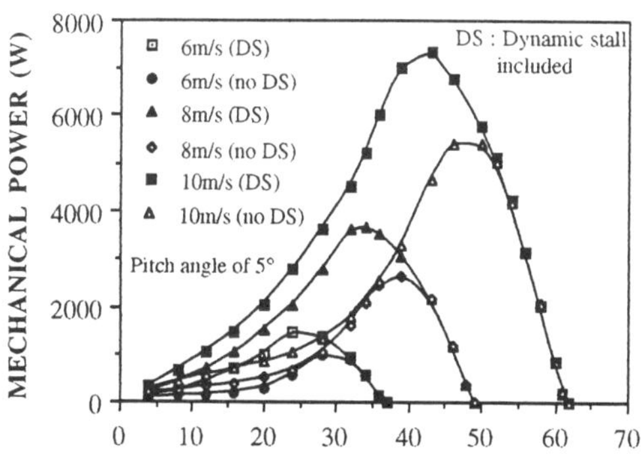

Figure 3 : Comparison of mechanical power versus rpm (dynamic stall included and no dynamic stall)

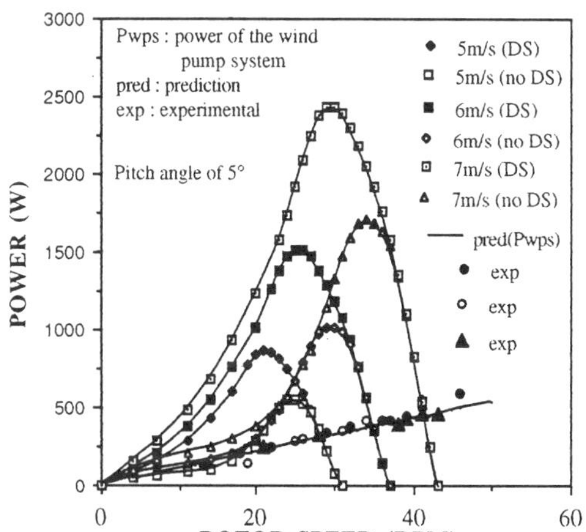

Figure 4 : Comparison of experimental and predicted mechanical power required for the pump The parabolic curves represent the performance of the rotor for different wind speeds.

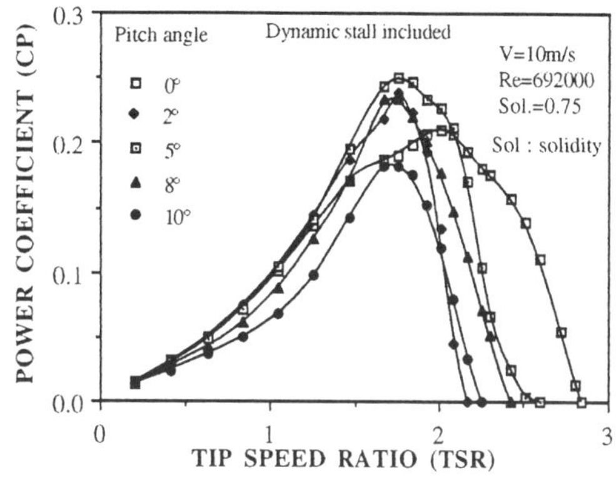

Figure 5 : Power coefficients versus tip speed ratios at different pitch angles

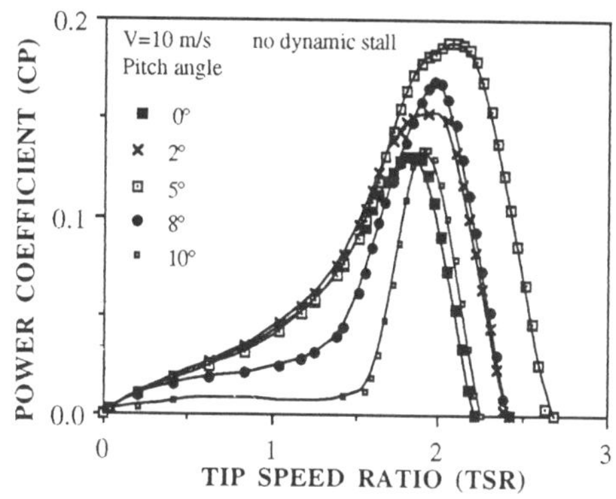

Figure 6 : Power coefficients versus tip speed ratios at different pitch angle

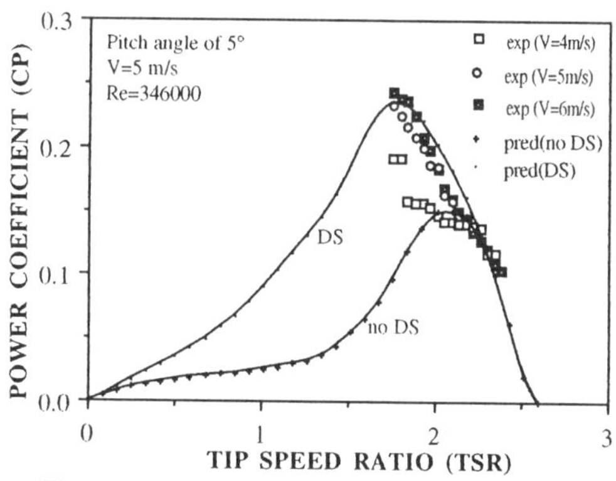

Figure 7 : Comparison of experimental and predicted power coefficient versus TSR

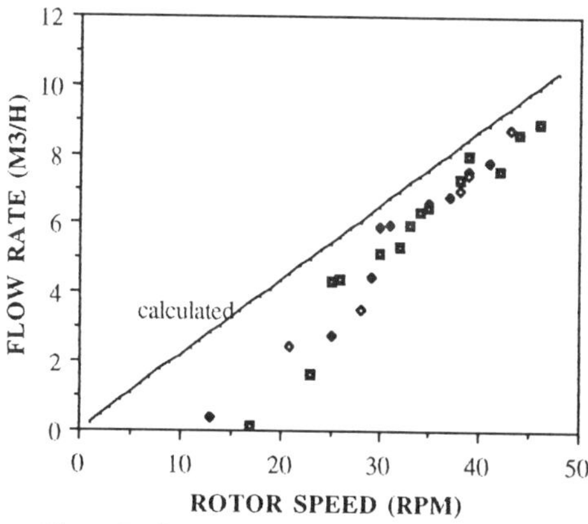

Figure 8 : Comparison of experimental and predicted flow rate versus rotor speed

THE INVENTUS 6 STAND-ALONE SYSTEM

Dipl. Ing. Christof Erban

Hoofdstraat 40, 6461 Kerkrade
Netherland

The INVENTUS 6 stand-alone system is a horizontal axis wind turbine (hawt) with a rotor diameter of 6m and a rated power output of 5 kW at a windspeed of 10m/s. It is a further development of the INVENTUS 6 grid parallel hawt which is performing successfully for about 5 years in the german inland. The INVENTUS 6 stand-alone system is distinguished by capabilities that are extraordinary for this size of hawt. Power regulation by passive pitch of the rotorblades, inherent security system due to the overspeed controlsystem within the rotor, automatical aerodynamical counterbalancing of the rotor. It furthermore provides three optional operation modes in it's stand alone version.

1. System description

1.1. The Inventus 6 turbine

The Inventus 6 is a four blade horizontal axis wind turbine, that was originally specifically designed for inland operation whose specific demands:

- high starting torque at low windspeeds,
- high power output at low windspeeds,
- low noiseemission in urban areas
- capability to handle the large windspeed gradient and the high turbulence in the swept rotor area due to the comparatively low tower

were taken into consideration during the lay out of the rotor. The patented[1] INVENTUS rotor concept was developed. This concept contains three features within one system:

1.1.1. Power control using the passive pitch moment of the rotorblade.

The windspeed-related thrustforces upon the rotorblade cause a pitching moment of the rotorblade due to the fact that the resulting pressure point does not match with the axis of the rotorblade (picture 1). The pitching moment is regulated by stressing a spring using a special computer designed lever-system. Since the thrustforces upon the rotorblade, the torque moment and the power output of the hawt are by aerodynamical laws related quantities, not only the pitching moment, but the torque moment and therefore the power output are regulated. This principle of power regulation provides three advantages compared to the active pitch regulation:

- no external energy is needed;
- variable gradients of windspeeds cause a variable pitching rate, that means a heavy gust causes a high speed reply of power regulation (no high power peaks with the corresponding overload of the transmission and the generator, and therefore less emergency stopps caused by overload), whereas a small change of the windspeed causes a low speed reply of power regulation (less power oscillation caused by an overreacting power regulation);
- the power regulation works within the rotor. The stress of the transmission and the generator is kept at a low level.

1.1.2. Overspeed control by the r.p.m. related centrifugal forces.

Since the rotorblade is frontwards inclined out of the plane of rotation, and the center of gravity of the rotorblade does not match the rotorbladeaxis, a r.p.m related momentum towards the feathering position is created. This momentum causes a featherwise motion of the blades if a predefined r. p. m. level is reached. This rotor behaviour does not depend on the load of the system. Even while running without any loads an excessive acceleration is prevented. Therefore the rotor itself is an inherent security system.

1.1.3. Automatical aerodynamical counter balancing of the rotorblades

Each of the rotorblades is connected to the power regulating spring mechanism, which permits a pairwise motion of the opposite rotorblades. This ability of pairwise motion does not interfere the motion of the rotorblades that arises during power regulation. The rotor is counterbalanced by pairwise adapting the angle of attack of the opposite rotorblades as long as the acting forces are balanced. Thus a reduction of the bending moments of the rotoraxis can be achieved. Even an operation in the relatively turbulent area close to the ground with its high windspeed gradients is possible.

1.2. The operational control system (ocs)

The operation system consists of four major parts: the digital r.p.m. counter; the programmable control unit; the output stage and dump loads. Depending on the users demands the utilities e.g. battery charger, the waterpump, heating devices, 400V/220V power devices or 12V/24V consumers are to be added (picture 2). The operational control system provides several options of priority connections and operation modes. The standart priority connection is defined as: battery charging up to a definable charging current and additional connection of other utilities depending on the power that is provided by the INVENTUS 6. Furthermore each of the utilities can be driven alone to satisfy specific demands e.g. maximal battery loading, waterpumping, heating, hot water production or 400V/220V output.

The INVENTUS 6 stand alone system provides three different operation modes with different ranges of frequency and/or voltage. The operation modes are achieved by controlled dumpload steering of the INVENTUS 6. These modes are

- the generation of electricity with constant frequency/voltage (50Hz/400V ±4%) as long as the power consumption of the connected users does not exceed the power output of the hawt. In case of persisting mismatch between the power consumption and the power output, the users are disconnected. As soon as sufficient power is available the non hazardous users are reconnected automatically, whereas hazardous consumers e.g. circular saw/welding transformers are only reconnectable by hand;

- the generation of electricity with maximum power output of the hawt. This operation mode yields to a power output according to the line of the maximum power coefficient c_p of the rotor, independent of the actual power consumption of the connected utilties. Surplus power is to be used in dump loads. This mode produces a variable frequency/voltage that can be used for heating purposes, battery charging or other DC consumers.
- the waterpumping mode. An electrical industrial deepwell-waterpump that accomplishes the demands (delivery head/amount of water) of the users is connected to the system. By adapting the generation of electricity in frequency/voltage according to the demands of the required waterpump, a very high system efficiency is achieved [2].

The integration of photovoltaic generators or battery charging wind turbines within the INVENTUS 6 stand alone system is possible. The substitution of the wind energy by a diesel driven generator as back up system in times of continuous dead calm is also possible.

2. Aspekts of the INVENTUS 6 stand alone system layout

There are a lot of different arguments to build a stand-alone system the one or the other way. To understand the "philosophy" that the INVENTUS 6 stand-alone system is based upon, a couple of aspects are discussed furtheron.

A hawt of this size (6m rotordiameter; 13 or 19 m tower height and 5kw rated power output) is preferable used in developing countries, since in most places of industrial countries, electricity provided by a public net is easily available. The usage of hawt in developing countries however arises a lot of specific demands, concerning technology, price, transportation, errection, operation and maintanance of the hawt.

The major components of the INVENTUS 6 stand-alone system are the 4 blade rotor, the nacelle, the guyed, divisible tower and the ocs. Each of the components is portable by hand, since none of the components exceed 6m in length or 120kg in weight. The assembling of the hawt takes place on the ground, using a support for the tower. The errection of the hawt is achieved by hand-winch and the special errecting devices. For maintanance the tower can be laid down again. No special tools are required.

To provide a wide range of possible usage of the windenergy, a synchronous generator is connected to the rotor. It delivers a frequency/voltage that depends on the revolutions of the rotor. Using dump loads for controlled load management, the three operation modes mentioned above are possible. The usage of electricity as the means of energy transport provides furthermore the possibility to connect users with different demands and different locations (up to 500m) at the same time to the hawt.

Due to the fact, that most of the stand-alone systems are waterpumping windwills, a high value on special research of the waterpump mode of the INVENTUS 6 was set.

An electrical industrial deepwell waterpump is connected to the hawt for maximum reliability. The performance of this type of pump normally takes place at a rated frequency and a rated voltage. Its pumping efficiency is dependent on its rate of flow of water. Since the power output of the hawt also varies with the rate of flow of the air in a similar way as the waterpump and furthermore the percentual annual energy output differs with the windspeed, it is obvious to relate the maximum pumping efficiency with the windspeed that provides the highest annual energy output (picture 3). Unfortunately running an electrical powered deepwell waterpump under these conditions leads to an operation far away from the rated frequency and voltage. Converting the rated power of a given electrical deepwell waterpump according to the laws of similarity of non rated frequency/ voltage/ flow rate and delivery head, it is possible to choose the best pump for each given delivery head and flow rate.

Picture 4 shows the flow rate, the transformation efficiency and the system efficiency over the windspeed.

3. Test results

The INVENTUS 6 (waterpumping mode / delivery head 50m) was tested for one year by the German Lloyd[3] on their windtestsite Kaiser Wihelm Koog. The test contained aspects like handling, setting up the system, security, power output, availability, actual loads of stressed parts during operation and in special circumstances, operation troubles. The tested parameters were windspeed, winddirection, effective power, actual and added up water output, pressure behind the pump, voltage, frequency, bending and torque moment of the rotor axis, flatwise and chordwise bending moment of the rotorblade, rotorpostition, nacelle acceleration and tower bending moments with a resolution of 0.1 s. The testresults were within the range of the predictions. No major troubles appeared. The annual water output of the INVENTUS stand alone system in 1991 was more than 2500000 l.

Picture 5 shows the results of a short-time-messurement of a switching on operation of the waterpump at approximately rated windspeed. The switching on operation takes place at second no. 66. Before this time only the static toque moment of the rotor with pulled brake appears. Between second no. 66 and 88 the rotor first moderate, then very fast accelerates. At second no. 85 (the appropriate frequency is reached) the waterpump is automatically connected by the ocs. The peak in torque moment of the rotor appears at the same time. The value of maximal appearing torque moment does not exceed the calculated values. At second no. 100 rated power and rated waterflow are reached.

It is obvious that the rate of waterflow maintains allmost constant at reated level, although changes of the windspeed with a high frequency appear.

The system was certified by the German Lloyd.

References:

1. Patent DE 3628626 C2, dated 8.12.88 "Windpowercontrol"
2. Erban, C.: lay out and optimization of a winddrven waterpump system. Study at the RWTH Aachen, Germany, 25.11.1989
3. Rohden, R.; Krabbe, N.: selection, verification and test of small windenergyconverters for developing countries. Interim report on the wind energyconverter INVENTUS 5, WE24/91, German Lloyd, 1991.

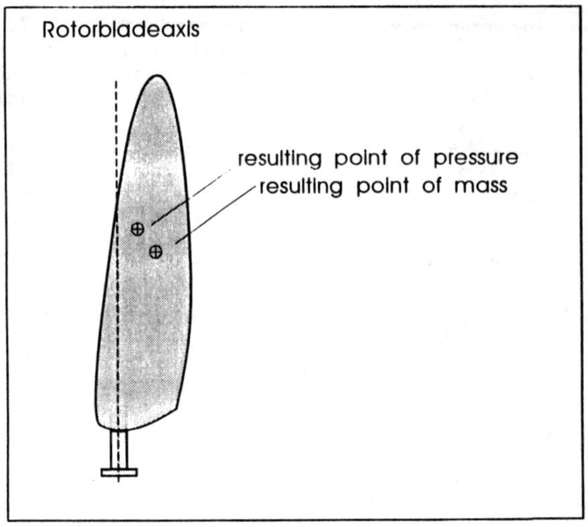

Picture 1: Missmatch of rotorbladeaxis and resulting point of pressure/mass

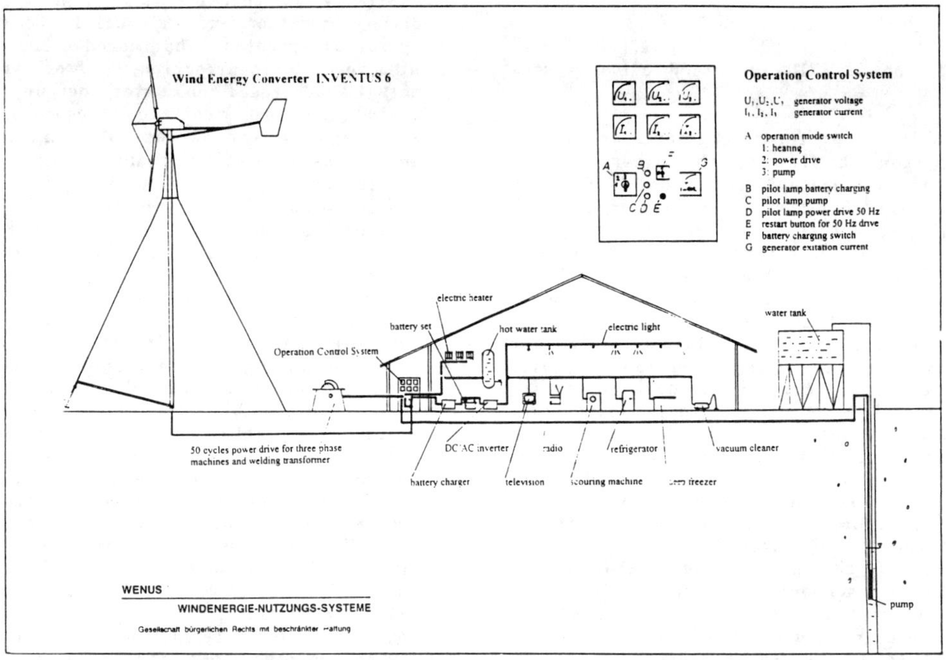

Picture 2: The INVENTUS 6 stand-alone system

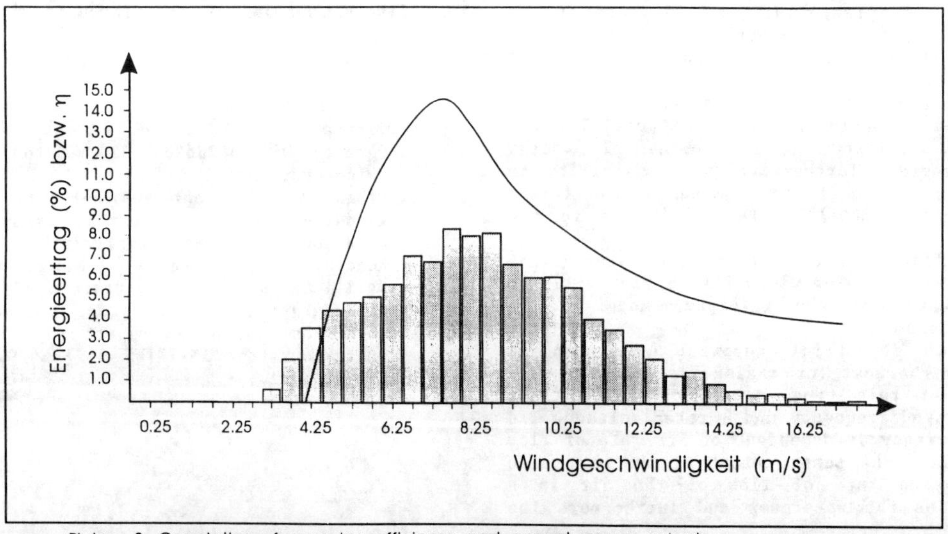

Picture 3: Correlation of pumping efficiency and annual power output

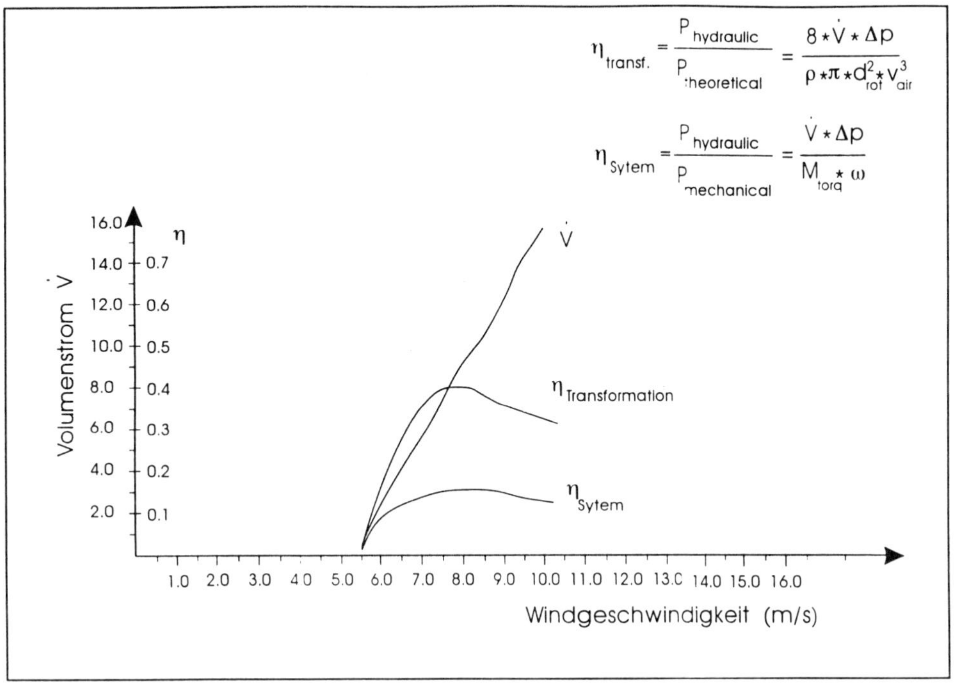

Picture 4: Flow rate, transformation and system efficiency versus windspeed

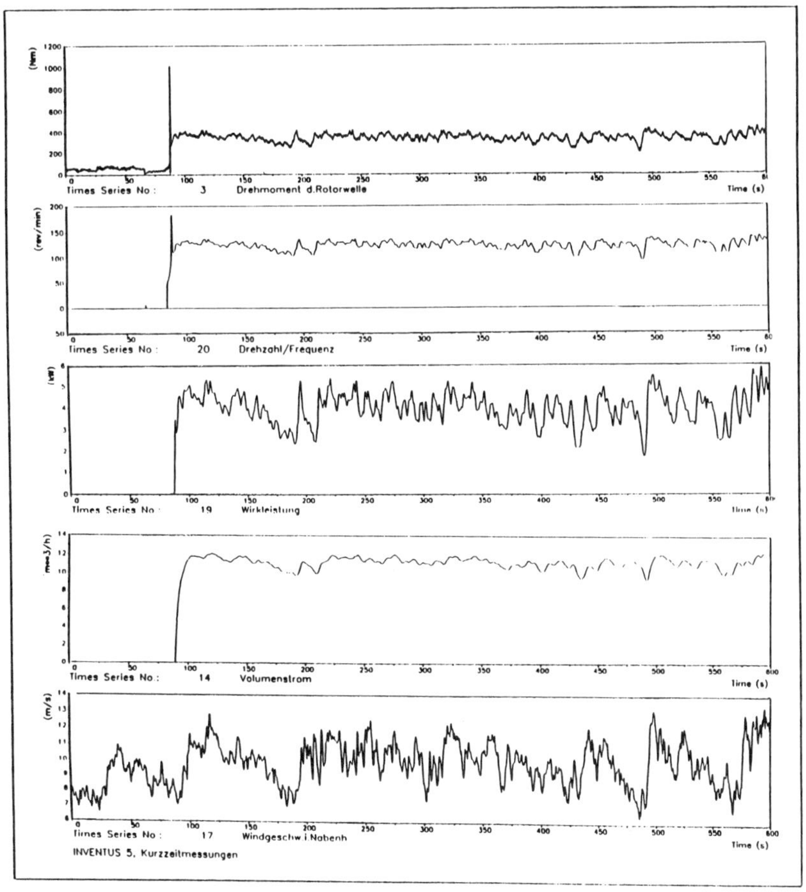

Picture 5: Short time messurement of torque moment, r.p.m./frequency, power output, flow rate and windspeed[3]

FIELD TEST OF A SMALL WECS IN ANTARCTICA

S. Kimura*1, Y. Ichikawa*2, K. Ishizawa*3, T. Takanaga*4 & K. Suzuki*5

*1: Kanagawa Inst. of Tech., 1030 Shimo-Ogino, Atsugi, Kanagawa, Japan 243-02
*2: Sumitomo Precision Products, 1-10 Fuso, Amagasaki, Hyogo, Japan 660
*3: National Inst. of Polar Research, 1-9-10 Kaga, Itabashi, Tokyo, Japan 173
*4: Kandenko Co. Ltd., 4-8-33 Shibaura, Minato, Tokyo, Japan 108
*5: Tokai Univ., 1117 Kita-Kaname, Hiratsuka, Kanagawa, Japan 259-19

ABSTRACT: Field test for a WECS in Antarctica has been carried out throughout almost one year at the small camp by the Japanese Antarctic Research Expedition in order to investigate the newly developed turbine system and confirm its reliability accordingly. Though this was the 16th attempt for utilization of a wind turbine system in Antarctica, it was the first system accompanied with complete measuring activities. Wind velocity and direction, rotor revolutions and electric characteristics of the circuit were measured successively by the automatic data acquisition system using a personal computer, and stored into floppy disks as occasion demands.

1. INTRODUCTION

Field test has been carried out at Asuka observation camp in Antarctica by the staff of Japanese Antarctic Research Expedition in 1991. She was located in 71.5 degrees South and 24.1 degrees East, and 155km away from the coast near which the observation ship has been always anchored, which is shown in Figure 1. Though 15 attempts have been so far done for producing electricity at isolated stations, little information useful for analysis could be obtained unfortunately. As stated later, the wind condition is very satisfactory for operating a WECS. In addition, the energy circumstance surrounding the Japanese observation stations is rather severe since the observation ship transporting fuel plies only once a year. Hence, a demand for the alternative energy, such as the natural energy resources, has been high since the first operation. Consequently, the project was started in 1988 and the research activities including the system design and manufacturing have been implemented. The turbine system was brought into the camp from the observation ship on sleighs pulled by snow vehicles at the end of 1990, and subsequently set up.

2. WIND TURBINE SYSTEM

Figure 2 shows the entire view of the turbine and the wind velocimeter just after the erection on the ice field near the warehouse at the edge of the camp. The rotor rotational plane is faced to the direction of ESE from which the prevailing wind blows. The 2 bladed GFRP rotor is 2m in diameter. The blades with pitch angle fixed at 30 degrees and no yaw control mechanism are installed. The stainless steel protector is applied 10% in chord from the leading edge to avoid erosion due to collision of inflow ice particles in blizzard. The hub height is 6m from the ground, which can make the rotor operate away from the high dense layer of floating ice particle. A permanent magnet generator is employed so that the turbine system could be operated stand alone because of small size of grid. The rotor is supported by the generator, that is to say , the rotor is connected directly to the generator with no multiplying gear. The rotor hub and the generator were completely covered with the GFRP nacelle in order to prevent them from penetration of ice particles as well. Rechargeable batteries and resistance loads were installed and connected consecutively to the circuit to

Table 1. Specifications of wind turbine system

ITEM	SUB-ITEMS	SPECIFICATIONS
WIND VELOCITY	CUT-IN RATED CUT-OUT MAXIMUM	10.0 m/s 20.0 m/s none 60.0 m/s
ROTOR	ARRANGEMENT DIAMETER RATED REVOLUTIONS ALLOWED MAX. REV. REV. DIRECTION	UP-WIND 2.0 m 300 rpm 1500 rpm COUNTERCLOCKWISE (SEEN FROM WINDWARD)
BLADE	NO. OF BLADE LENGTH MATERIAL WING SECTION	2 0.8 m GFRP (FILLED WITH URETHAN FOAM) NACA 44 SERIES
GENERATOR	TYPE MAGNET MATERIAL NO. OF POLES MAX. OUTPUT MAX. REVOLUTIONS MAX. FREQUENCY	3-PHASE PERMANET MAGNET, BRUSHLESS 2-17-SAMARIUM-COBALT 12 5 kW 1500 rpm 150 Hz
TOWER	TYPE HUB HIEGHT	STEEL MONO-POLE (4 PARTITIONS) 6.0 m
CONTROL	TYPE PITCH CONTROL YAW CONTROL	AUTOMATIC LOAD REGULATION FIXED (REMOTELY ADJUSTABLE) FIXED (MANUALLY ADJUSTABLE)
WEIGHT	NACELLE UNIT TOWER	180 kg 450 kg
BASE	TYPE CONFIGURATION MATERIAL	BURYING IN SNOW ROUND SHAPE (3m IN DIAMETER) WOOD (REINFORCED BY ANGLE IRONS)

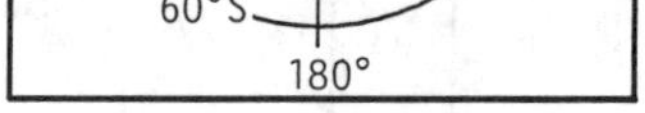

Figure 1. Antarctica and Japanese stations

Figure 2. Wind turbine and anemometer at Asuka camp

absorb the power from the generators the revolutions change. The turbine tower was supported by the pedestal which was buried in the ice/snow surface layer. Specifications of the turbine are summarized in Table 1. As shown in Figure 2, the wind velocimeter was set up beside the wind turbine. Both the wind velocimeter and the turbine were standing parallel to the primary wind direction so that the wake from the turbine would not affect the anemometer.

The data acquisition for the wind velocity and direction, and the system performance was implemented according to the program as indicated in Figure 3. Though additional test as to acquisition of strain on the blade and the tower was sometime carried out separately from the data acquisition for the rotor performance, this would be explained some other time. All data were stored automatically into the hard disk controlled by the personal computer, and transferred to floppy disks as occasion arises. The output controller, several sets of load and the measuring instruments were placed in a warehouse. Because of the snow drift, the warehouse was approximately 2m beneath the snow surface, which fortunately kept the ambient temperatures of the instruments moderate rather than on the surface.

3. ANALYSIS OF METEOROLOGICAL DATA

3-1. Wind Velocity

Figure 4 shows the monthly wind velocity change in 1991. The annual mean velocity was arithmetically calculated as 13m/s, which is enough high for turbine operation compared with the ordinary value. The wind velocity duration can be seen in Figure 5, together with the cumulative and the energy. From the standpoint of utilization of wind energy, the energy would be the most convenient measure for determination of the operative wind velocity. A turbine system should be, therefore, designed as it can extract mainly the energy from the wind

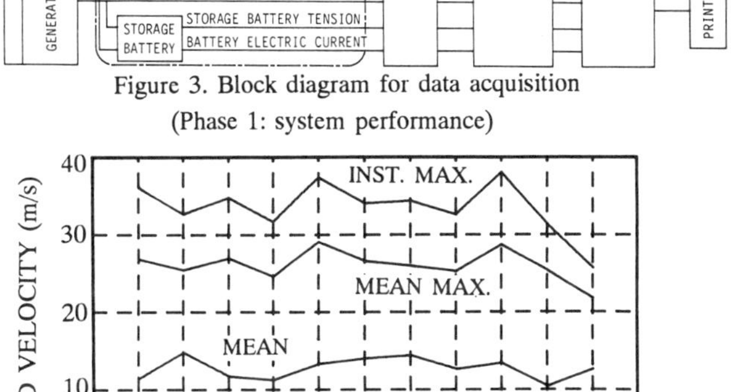

Figure 3. Block diagram for data acquisition (Phase 1: system performance)

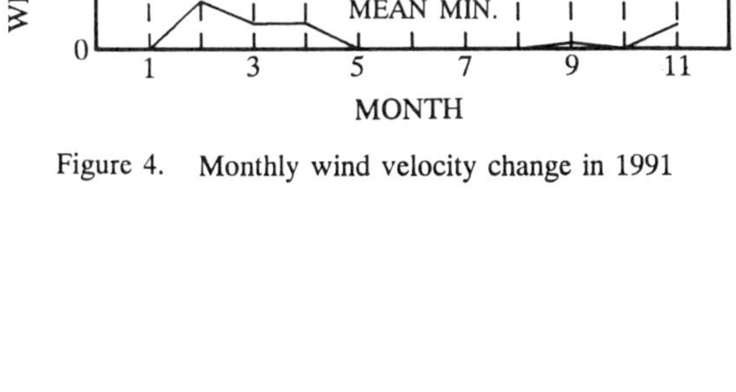

Figure 4. Monthly wind velocity change in 1991

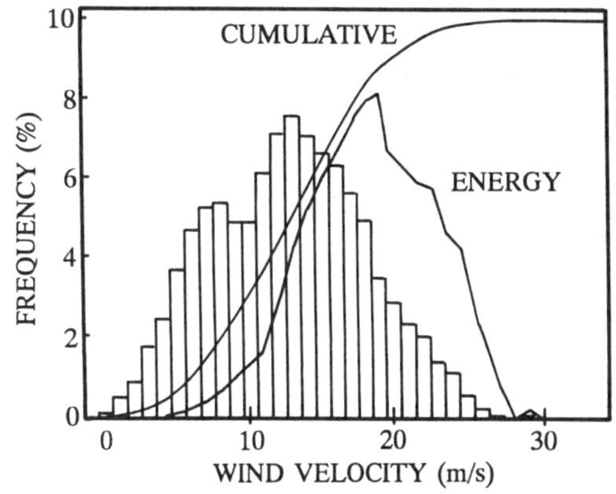

Figure 5. Wind velocity appearance frequency

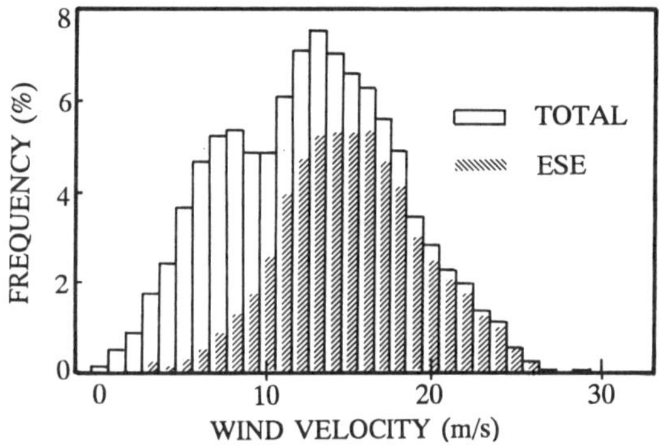

Figure 6. Appearance of ESE wind in velocity

around 20m/s at the camp, and the turbine system had been actually designed in this manner. Figure 6 shows the appearance of wind velocity especially restricted in the direction of ESE to which the rotor faces. It can be seen that in higher velocity region the ESE wind appears at the high rate to the respective total frequency. It is easily deduced that no use of yawing control mechanism was very proper to the turbine for that region of Antarctica.

The wind velocity fluctuation for the duration of 512 seconds are depicted by auto-correlation and power spectrum in Figure 7 (a), (b) and (c). Since the revolutional response of the turbine rotor of this size can be mentioned to be up to several seconds at most, the duration would be appropriate. In the expression of power spectrum, the prevailing frequency can be read at 0.016Hz, which means that the turbine can be operated without any delay under such fluctuated wind.

3-2. Wind direction

Figure 8 shows the wind direction duration for 5 years from 1987 to 1991. Because the location of the camp in the katabatic wind climate, ESE wind has been prevailing in measurement of 5 years' duration. In particular, that feature was remarkable in higher wind velocity range of more than 10m/s, which can be seen in Figure 9. The same deduction of the propriety of no installation of a yawing mechanism could be derived as well as analysis of the wind velocity in Figure 6.

The instantaneous fluctuation of wind direction is seen in Figure 10. Data were acquired for the same duration as the analysis of wind velocity. The distinguished prevailing frequency can not be found. It fluctuated very rapidly with relatively small deviated angles to the mean. Though the unsteadiness in the rotor aerodynamic must be evaluated in some sense, it would be easily deduced that no yawing control has to be considered.

3–3. Temperature

Figure 11 shows the temperature change in 1991. Low temperatures affect the function of lubricant in the generator and the strength of material, and the conversion efficiency from the mechanical power to the electrical one of the generator. The extreme low temperature of the year of 1991 was approximately −43 degrees centigrade. Since the low temperature proof materials up to −60 degrees centigrade were employed, no obvious problem attributable to low temperatures was recognized in the test.

4. SYSTEM PERFORMANCE

4–1. Performance

As stated above, the measurement of the system characteristics has been carried out continuously, and all the acquired data have been stored into the hard disk once and transferred to floppy disks as occasion demanded. On the other hand, the materials which could be transported from the camp to the main station by sleighs were strictly restricted. Hence, all acquired data could not be provided for analysis.

The power converted by the generator was plotted to the wind velocity with every 1m/s when the instantaneous wind direction was only in the range of 22.5 degrees represented by ESE in Figure 12. Over the wind velocity of 13m/s the extracted power was increasing linearly with some sort of extent as the wind velocity increased. Though the rotor could rotate under the condition of the wind velocity not exceeding

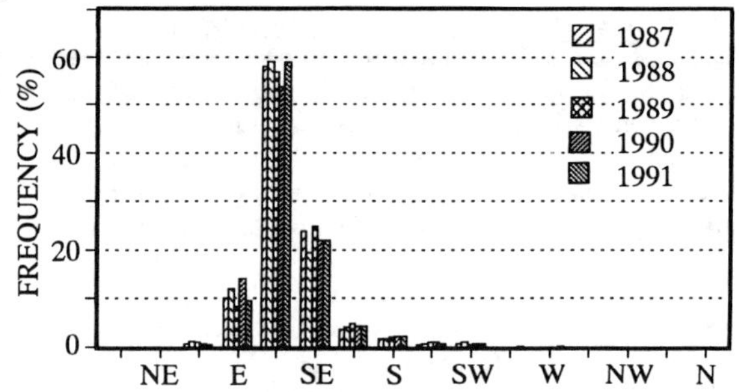

Figure 8. Wind direction appearance frequency

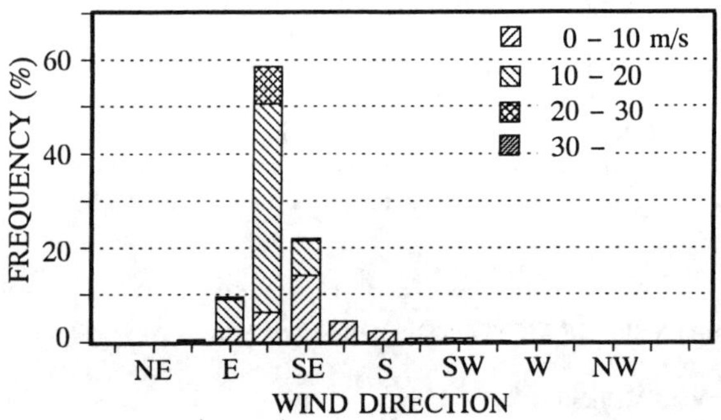

Figure 9. Appearance of wind direction classified by wind velocity

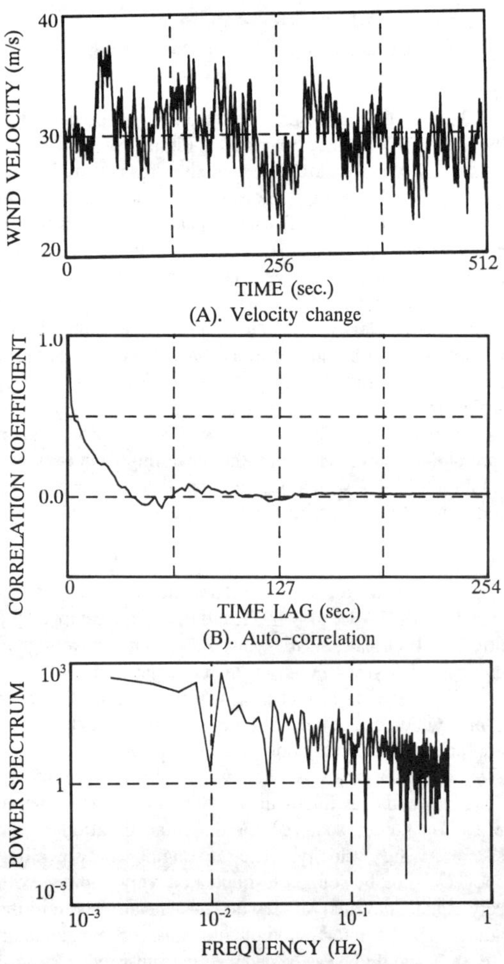

(A). Velocity change

(B). Auto-correlation

(C). Power spectrum

Figure 7. Characteristics of wind in velocity

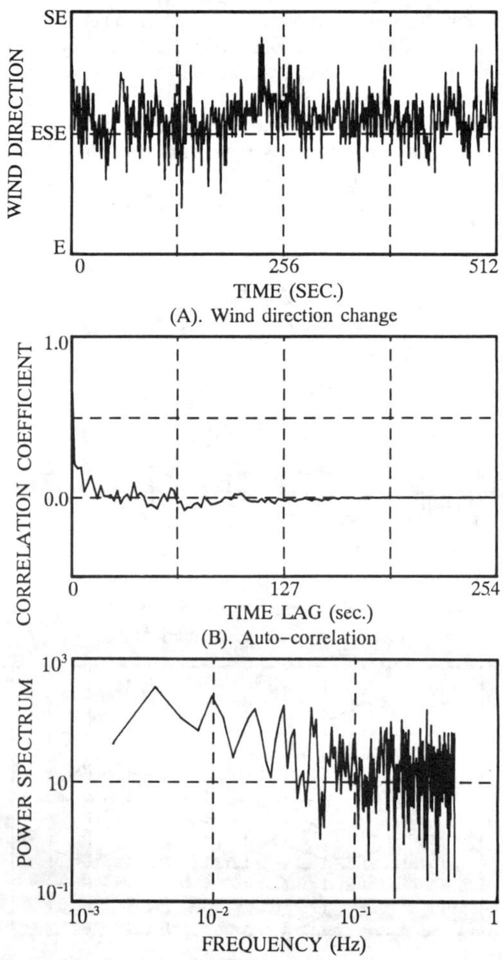

(A). Wind direction change

(B). Auto-correlation

(C). Power spectrum

Figure 10. Characteristic of wind in direction

approximate 12m/s, power producing could be scarcely expected from the generator. The system performance based on analysis of the test results was arranged by the method of BIN and indicated in Figure 13 in which the analytical performance was drawn together with it. The analytical performance was calculated with using the results of the wind tunnel test for the rotor aerodynamic characteristics and the results of the low temperature test for the generator and the output controller. It turned out that both are well coincident each other, and consequently the analytical one can be adapted for the further or future investigation. Using this performance, the annual power output can be estimated as approximately 1400kWh employing the annual wind velocity duration.

4-2. Rotor revolution

The relation between the rotor revolution and the wind velocity change is shown in Figure 14 in the expression as the cross-correlation. Though the fluctuation of the rotor revolution depends on that of the wind velocity and that of the wind direction, the former would be considered to be dominant. From the analysis of cross-correlation, it is understood that the rotor could follow the variation of the wind velocity.

Numerical investigation of the rotor motion was performed by solving the equation of motion governed by $I_R \dot{\omega} = Q_R - Q_L$, where I_R, ω, Q_R and Q_L denote the rotor inertia moment about the rotational axis, the revolutional speed, the rotor producing torque and the generator torque respectively. The rotor torque is the function of the rotational speed, yawing angle and unsteadiness due to fluctuation of the inflow angle or the wind velocity, and the generator torque the function of the rotational speed, the load rate, the temperature and the mode of increasing or decreasing revolution. Furthermore, (•) expresses the differentiation by time. The comparison is shown in Figure 15. It turns out that if the numerical solution be right, the system worked mostly as designed, and alternatively the numerical analysis would be applied properly for the future analysis about the rotor motion.

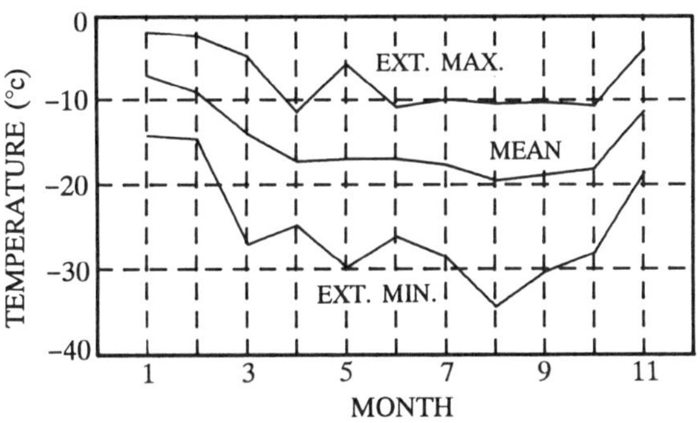

Figure 11. Monthly temperature change in 1991

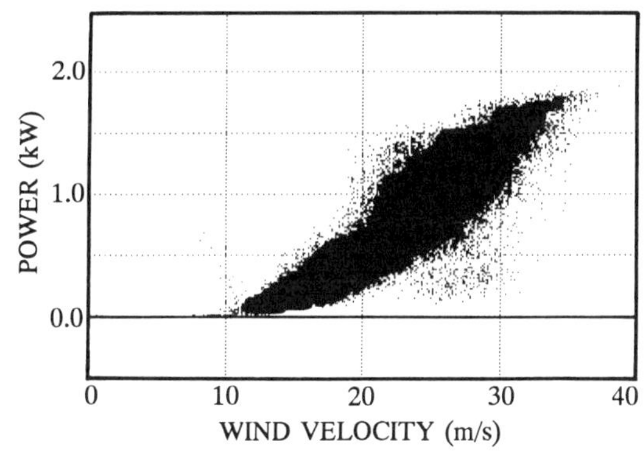

Figure 12. Power output from system

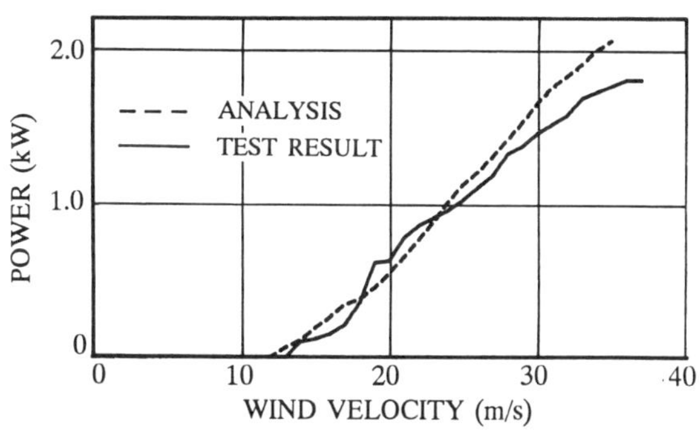

Figure 13. Performance of wind turbine system

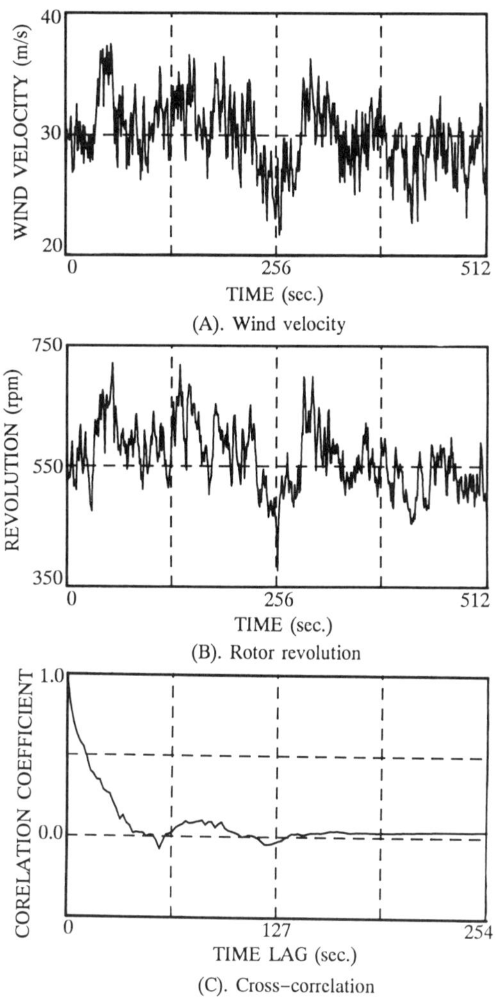

Figure 14. Relation between rotor revolution and wind velocity

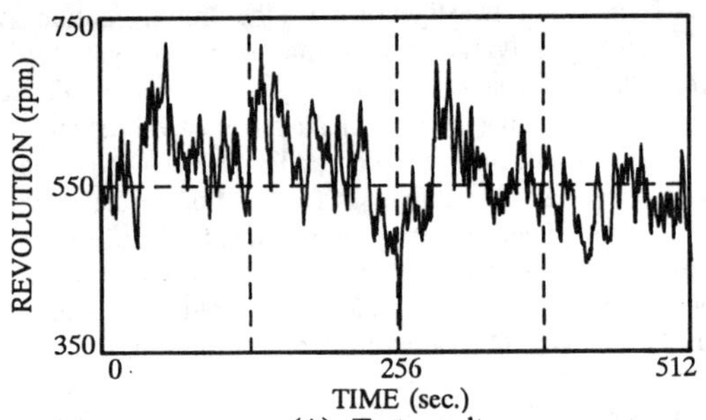

(A). Test result

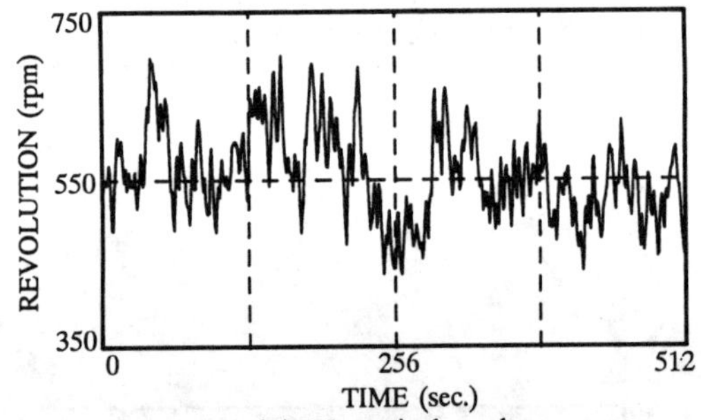

(B). Numerical result

Figure 15. Comparison of numerical estimation with result for rotor motion

5. MISCELLANEOUS

5-1. Erosion of rotor blades

Some blades were eroded due to the collision of ice particles in the previous attempts. In the new turbine system, no scratch on the blades could be found probably for the reason of application of the leading edge protector and the enough clearance between the rotor and the ground. If the rotor of the GFRP blade be necessary to be mounted on the relatively low tower, the leading edge protector should be applied.

5-2. Effect of static electricity

Because of the very low *absolute* humidity, it was reported that static electricity of high intensity was generated and frequently had much influence on the electrical measurement. A discharging device was equipped on the aft nacelle in which the revolution detector and the slipring to transmit signals from the strain gauges on the blade were attached to the generator. No evaluation on the device could be made since no operation without such a device was performed. However, because of no report for malfunction of the revolution detector and no extraordinary signal from the strain gauges was recorded, it certainly worked well throughout a test, and should be surely equipped in the future utilization of a turbine in Antarctica.

6. CONCLUDING REMARKS

Field test for a small WECS at a small camp in Antarctica was accomplished. It could be confirmed that the turbine system worked well as expected even under very sever condition. Hence much information about operating a WECS in a cold region was obtained. Moreover the meteorological data were acquired as well, which will be able to be offered to the future utilization of WECSs in Antarctica.

7. ACKNOWLEDGEMENT

The authors are pleased to acknowledge the laborious undertaking of the research staff at the Asuka camp. They transported materials by pulling sleighs, assembled and erected the wind turbine system, and contributed to measurement under very severe conditions in Antarctica. The authors are indebted to the staff for the successful experiment.

8. REFERENCES

[1]. Kimura, S., et al: Operation program for wind turbine in Antarctica, Proc. of EWEC'91, 1991

[2]. Bisplinghoff, et al: Aeroelasticity, Addison-Welsly, 1957

[3]. Hino, M.: Spectrum analysis, Asakura, 1979 (in Japanese)

KADAM EL-JABAL PILOT PROJECT FOR ELECTRICITY GENERATION

EL-OSTA, W. B., KALIFA, Y. M. & EL-AMIN, K
Centre For Solar Energy Studies
P. O. Box 12932, TRIPOLI-LIBYA

ABSTRACT: Conventional power plants supply electricity to various areas in Libya, except some remote ones, to which stand alone wind turbines could be the best solution. This paper presents the possibility of using wind energy in such areas. Kadam El-Jabal project is assumed to be the first project of its kind, in which a 10 kw wind turbine supplies electricity to a small laboratory used as a control room. The annual energy output from the WECS was estimated to be 19.01 MWh/year, using WASP software. The Economical feasibility of the project was investigated.

1. INTRODUCTION

Few Libyans live in remote areas, where the cost of electricity generated using conventional fuels (diesel engines) is high. Obviously, conventional power plants have an adverse effect on the environmental air quality and consequently on the global climate balance and human life. Libya, in addition to oil and natural gas, is blessed with vast renewable energy sources such as Solar and Wind energy. Libya should be prepared for the post fossil fuel era, and start developing its renewable energy sources. These sources are freely available, eventhough, they are restricted to a certain times, e.g. wind power is limited by season, time of day and physical features of the area (topography) which may obstruct the flow of the air. Wind power could contribute to a substantial percentage of the electrical energy demand, especially on the eastern parts of the country. However, substantional study is required to assess the wind potential in order to contribute to the total power generation. Although some studies had been conducted [1,2], developing a wind atlas is a must for a serious and wide applications of wind energy in Libya.

2. THE SITE INVESTIGATION

Kadam El-Jabal project is located in Wadi El-Hira, about 60km south of Tripoli. It is a fairly high land, crossed by valleys and hills as shown in (fig. 1). The WECS was installed on the crest of a small hill of about 260m a.s.l. The latitude and longitude of the site are 32° 45' North and 13° East respectively. The topography of the site is uneven with different sized rocks and vegetation, like alfalfa, on both, the top and the valley. The neighboring regions are flat, crossed by some dry valleys, and some forest trees of about 2 meters heigh.

3. WIND TURBINE DESCRIPTION AND INSTALLATION

The wind energy conversion system (WECS) was imported from LMW wind energy b.v. (the Netherlands). It has a low speed permanent alternator. The output of the alternator is a variable voltage, variable frequency, 3-phase AC. The AC is converted into DC through a voltage control system to be fed to a bank of batteries in order to supply the control room with its electric energy demand. The rated power of the WECS is 10kw. The detailed description is shown in (Table 1).

The wind turbine was installed by late June 1991 and became operational on July 1st, 1991. By the end of August, there was a failure in the mechanical brake, which was fixed immediately, and by September 19, 1991, the wind turbine completly stopped due to a failure in the generator. The company was reported on the progress of the operation and it was asked for a subtitute for the generator, but its staff still working to sort out the failure in order to develop a better product.

Table 1 : Wind Turbine Description

Alternator	- Permanent magnet alternator, 3-phase, 10kw rated power
Blades	- Rotor diameter 7 meter. 3-blades upwind composed of Fiber glass reinforced plastic pultrusions
Output Control	- The output alternator is near sinsoidal, variable voltage, variable frequency, three phase AC, which is not compatible with the battery charging. The VCS-10 (voltage control system) is used to interface the alternator to 120VDC battery bank (20 batteries of 6V each, conncted in series).
Tower	- It consists of 3-galvanized steel pipes, with total height of 24 m,
Rotor speed	- 60-350 RPM

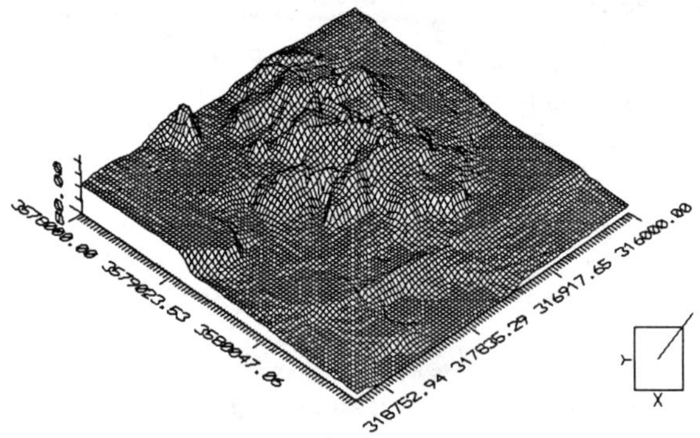

Figure 1: Kadam El-Jabal WECS site

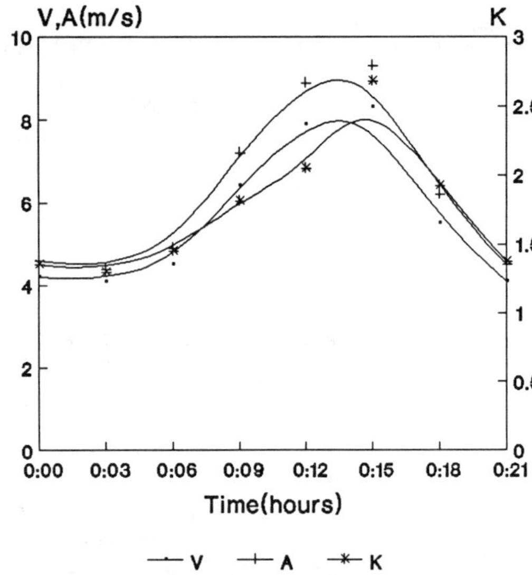

Figure 2: Diurnal Pattern for wind speed Weibull scale and shape parameter for (1979-1988)

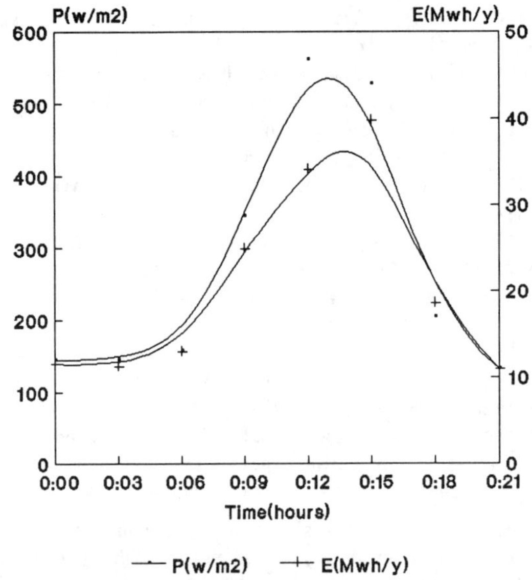

Figure 3: Diurnal Pattern for the site available powr and energy expected for WECS (1979-1988)

Table 2: Yearly Wind Data for the Site

Year	V(m/s)	P(W/m²)	E(Mwh/y)	C.F.(%)	C(m/s)	K
79	6.1	325	22.57	25.7	6.8	1.66
80	6.1	324	22.96	26.2	6.9	1.7
81	5.9	314	21.12	24	6.5	1.57
82	5.6	261	19.79	22.6	6.3	1.63
83	5.4	167	13.22	15.1	4.9	1.4
84	5.7	270	20.44	23.3	6.4	1.66
85	5.6	254	19.82	22.6	6.3	1.67
86	5.9	298	21.58	24.6	6.6	1.66
87	5.1	218	16.66	19	5.6	1.51
88	5.9	291	21.67	24.7	6.6	1.7
Ave.	5.7	272	20.13	23	6.3	1.62

4. ANALYSIS OF WIND DATA

Wind Data from Tripoli Airport meteorological station formed the primary data base. These data included wind speed and direction of 3-hours readings for a period of 10-years (1979-1988) in conjunction with topographic information and site roughness and obstacles as well as the power curve of the WECS. All these information was analysed using WASP software (developed by Risoe). The analysis showed that the average wind speed at the site, at the WECS hup height, is 5.7 m/s, the available site power is 272 W/sq.m, the average Weibull scale parameter is 6.3 m/s and the average Weibull shape parameter is 1.62, the expected energy from the WECS is 20.13 Mwh/y and the expected average capacity factor is 23%.

(Fig. 2) shows the diurnal pattern for the wind speed, Weibull scale and shape parameters and (fig. 3) illustrates the diurnal pattern for the site available power and the expected energy output from the WECS at the site. The monthly average wind speed as well as the monthly average Weibull scale and shape parameters are presented in (fig. 4), while the monthly average site available power and the expected energy from the WECS are illustrated in (fig. 5). The yearly average values for the above parameters are tabulated in (Table 2).

5. ECONOMIC EVALUATION OF THE PROJECT

The detailed cost of the WECS and its installation is summarized in (Table 3), which indicate that the installaion cost is fairly high compared to the total cost. This is due to many factors such as the availability of crames and their high rental cost as well as the high labour cost. The cost of the electricity generated by the WECS was estimated as 0.077 L.D/kwh (0.25 US $/kwh) assuming a life time of 25years for the wind turbine, an intrest rate of 5.5%, an inflation rate of 5% and the operation and maintenace cost of 2% of the capital investment.

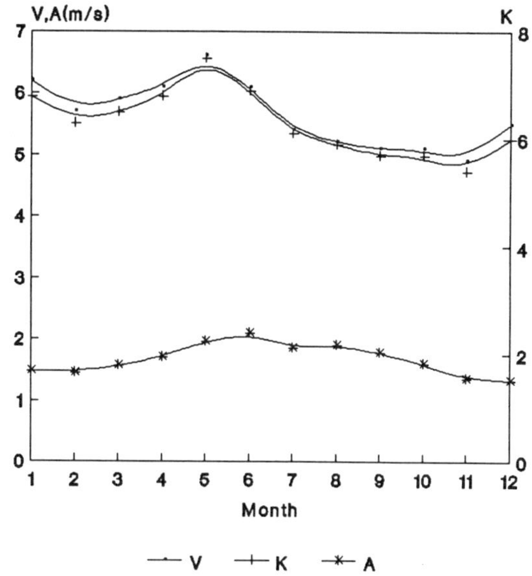

Figure 4: Monthly Average wind speed, Weibull scale and shape parameter (1979-1988)

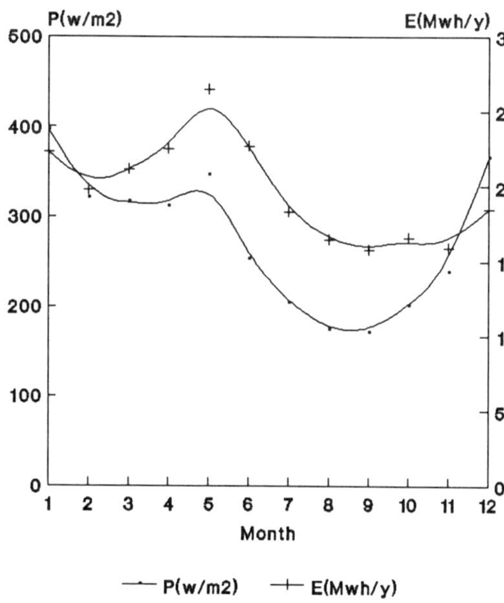

Figure 5: Monthly Ave. available power and energy expected for WECS(1979-1988)

6. CONCLUSIONS

Although the first project was not successful, but it did not discourage the staff in the wind energy department at CSES. So other programs are still in process. A wind farm is planned to be established in Zwara (west of Tripoli), for demonstration purposes and in order to convence decession markers with the value of wind energy in saving fuel and saving the environment. Also, it is expected that a wind atlas will be developed for the assessment of wind resources in Libya.

The previous investigations of the wind potential and siting studies favour the wind utilization as an alternative for electricity generation and for oil saving especially on the east coast.

Table 3: Detailed cost of WECS.

Unit	cost (LD)
-Wind turbine LMW 10/7.	3628.8
-Voltage control MP 85A-120V.	730.3
-Guyed mast of 24m hight.	1414
-50m electrical cable.	180.3
-Set of batteries, total 20 PCS.	1234.65
-Inverter ATLAS 120/1200W.	469.1
-Output 220V, 50HZ.	
-Transportation from Roterdam Habrbour - Libya.	287.9
- Sum.	7845
Installation cost.	12600
Total system cost.	20455.9

REFERENCES :

1. El-Osta, W.B., Kalifa, Y., Elhadi, M., Taher, A. and Hawass, A., "WIND ENERGY POTENTIAL IN LIBYA", The National Solar Energy Conf. Solar 90, ISES Annual Conf., March 19-22, 1990, Austin, Texas, American Solar Enrgy Socity Inc. 1990.
2. El-Osta, W.B. and Taher A., "Forcast Wind Energy Potential In Libya", Proceedings of EWEC 91, Oct. 14-17,1991, Amesterdam, The Netherlands.

DESIGN AND TESTING OF A SMALL-SIZE WIND-PHOTOVOLTAIC SYSTEM PROTOTYPE

F. Caricchi, F. Crescimbini, O. Honorati, E. Santini
Electrical Eng. Dept. - University of Rome "La Sapienza"
Via Eudossiana, 18 - 00184 Rome (Italy)
ph. +39 - 6 - 44585552; fax +39 - 6 - 4883235

ABSTRACT: This paper presents a newly-conceived configuration for stand alone power plants which generate electric power from wind and photovoltaic energy. The main features of the proposed configuration are a gearless axial-flux PM wind generator and a double-input single-output power electronics interface. Details concerning the design of these system components are given, as well as the paper discusses a control strategy which allows the maximum exploitation of the system energy sources. Finally, the paper reports the experimental results achieved from testing a small-size prototype.

1. INTRODUCTION

There is a "stand alone" market all over the world, and it is widely acknowledged that this market is potentially vast. Many developing countries will probably venture into wind and photovoltaic energy by introducing small systems for water pumping and battery charging. There are also uses for stand alone systems in the developed world where power is required at a costly distance from the grid, such as in the case of power supply systems for telecommunication apparatus located in mountain regions.

In rural districts the level of energy consumption per capita is very low. The basic energy needs are for pumping water, electricity for hospitals, lighting, cooling, and telecommunications. In vast rural regions these needs cannot be met by centralised energy production systems because the cost of transmitting energy to sparsely populated locations is prohibitive. There is therefore a growing demand for decentralised power systems, which can be arranged by using diesel generator sets together with photovoltaic and wind energy systems.

Stand-alone small-size wind and photovoltaic energy systems have a good potential for becoming economically feasible and fulfilling needs over the short term. At the same time such generating systems provide an excellent platform for the introduction of renewables, encourage public acceptance of the technology and lead to the development of local expertise for the adoption of larger wind and photovoltaic energy systems.

European technology should be developed to meet this perspective demand. Technical developments are required before this market can be readily exploited in conjunction with diesel power generation sets, but already there is a market for a relevant number of battery charging units per year. Smaller wind machines than those used at present on the "grid connected" market will be in demand, and therefore machine developments to satisfy this market are quite distinct from the other more conventional markets.

In concern of the subject discussed above, this paper presents a newly-conceived configuration for stand alone power plants which use wind and photovoltaic energy for electricity generation purpose. The layout of this autonomous generating system compounds a wind turbine-driven PM generator and a photovoltaic array as power generating units. Both these generators supply a single power electronics interface, which connects the generating units to the utility circuit and operates in order to achieve the best exploitation of the two renewable sources.

One innovative concept of the power plant configuration under investigation is the use of an axial-flux PM generator. In comparison with conventional generators such a new type of electrical machine is particularly suitable for low-speed high-torque applications, since it can be designed with a torque/weight ratio as high as the machine number of pair of poles, whereas the machine efficiency decreases slightly with an increase of the machine poles. Hence, axial-flux PM generators can be designed to be driven directly by a wind turbine, and the step-up gearbox generally employed in conventional wind energy conversion systems can be avoided thereby.

Another system feature is given by the topology of the output power conditioner. The layout of such a power converter results from the connection in series of two switch-mode step-up d.c.-d.c. converters, so that the power conditioner is able to manage at the same time the power coming from both the wind generator and the photovoltaic array, while provides the power supply of both the user load and the battery storage.

In the following this paper discusses a system control strategy which allows the maximum exploitation of the renewable energy resources. Finally, the design characteristics of a small-size power plant prototype are given together with the preliminary results taken from testing the system prototype components.

2. GENERATING SYSTEM LAYOUT

In one sense the wind and solar energy have complementary characteristics. Sunny days are generally characterized by a poor wind activity, whereas wind speeds above the average are observed often during cloudy days and the night. The use of both the solar and wind resources of a site to generate power results thereby in an improvement of the average availability of energy, and in a significant number of cases the needs of electricity users located in rural areas can be met by means of small-size power plants which use wind and photovoltaic energy together with a limited amount of energy storage. Hence, such an approach suggests the development of integrated generating systems, which consist basically of a wind generator and a photovoltaic array operating in the same power circuit. Both these generating units feed a power conditioner, which is devoted to control suitably the power flow and supply the battery system.

Fig. 1 shows a configuration for such stand-alone power plants which uses a wind turbine-driven PM generator and a photovoltaic array as power generating units. The wind generator supplies a three-phase diode rectifier, and altogether this system produces a variable d.c. output power which depends on the wind speed and the turbine rotor speed. The photovoltaic array is made up by means of a large number of solar cells connected in series and parallel, and this d.c. generator delivers a variable output power depending on the sunlight intensity and the cell temperature. Both these generators supply a power electronics interface, which connects the generating units to the utility circuit and operates in order to achieve the best exploitation of the two renewable sources as discussed later on.

2.1 Axial-Flux PM Generator

The wind conversion system shown in Fig. 1 uses an axial-flux PM generator directly driven by the wind turbine. If compared with conventional generators designed for low-speed applications axial-flux PM generators exhibit torque/weight ratio as high as the machine number of poles, whereas the machine efficiency results still comparable with that one of conventional high-speed machines having the same rated power [1]. Hence, such a new type of generators can be designed with a number of poles as high as required to drive directly the machine by a wind turbine, and the step-up gearbox generally employed in wind energy conversion systems can be avoided thereby [2].

Axial-flux PM generators have a robust and compact structure which results from a relatively easy construction. This electrical machine has a slotless strip-wound toroidal core, which supports a three-phase concentrated winding and is positioned between two rotors made up by means of discs of steel. Each rotor disc supports Nd-Fe-B permanent magnets and is connected rigidly to the machine mechanical shaft. The PMs drive the magnetic flux across the two annular air-gaps into the stator, and the current flowing through each coil of the stator winding interacts with the rotor flux producing the machine electromagnetic torque.

In comparison with conventional machines, one further advantage of the axial-flux generators is that a higher exploitation of the active materials can be achieved, since the two rotating discs act naturally as fans, and thereby the rotor structure can be designed suitably in order to remove the heating due to the copper and iron losses in the machine stator.

2.2 Photovoltaic Array

The photovoltaic cells produce a d.c. voltage when they are exposed to sunlight, and their i-v characteristic at a given sunlight intensity and temperature consists basically of a constant-voltage segment and a constant-current segment. The current is limited as the cell is short-circuited, whereas the intersection of the i-v characteristic with the voltage axis gives the cell open-circuit voltage.

The maximum power condition occurs at the knee of the i-v characteristic where the two segments meet. In order to achieve a high efficiency in the energy conversion, the cell should be operated at the maximum power point. Ideally, a pure d.c. current should be drawn from the photovoltaic array during the power generation process, though the reduction in delivered power is not very large even in presence of a fair amount of ripple current due to the power conditioner modes of operation.

2.3 Power Electronics Interface

One innovative concept of the system configuration shown in Fig. 1 is the topology of the power conditioner, which is arranged by the connection in series of two switch-mode step-up d.c.-d.c. converters. Hence, the resulting converter topology has a single output from a double input, and thereby this power conditioner is able to manage at the same time the power coming from both the wind generator and the photovoltaic array, while provides the utility circuit supply.

The modes of operation of the power conditioner are controlled by means of two power semiconductor switches (i.e. the IGBTs S_w and S_p shown in Fig. 1), which operate at a fixed switching frequency. Therefore, the power flow due to each generating unit can be regulated at the desired level by adjusting the duty cycle of both the switches, so that for given characteristics of the two renewable sources (i.e. the sunlight intensity and the wind speed) the system control strategy determines the power conditioner output which meets continuously the power required by the user load and the storage system.

3. SYSTEM CONTROL STRATEGY

Whenever the user requires the peak load and the battery storage is fully discharged, it is desiderable that the system control strategy operates in order to maximize continuously the energy extracted from both the wind and solar resources. Such an approach requires variable speed

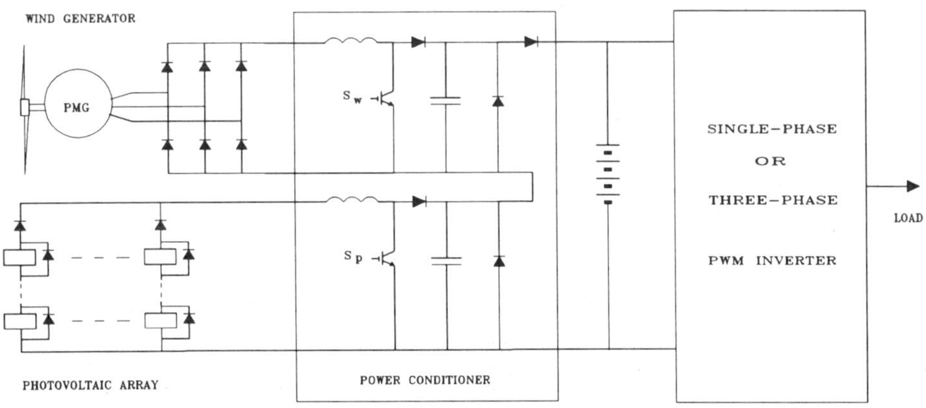

Figure 1. Wind-photovoltaic stand-alone power plant

operations of the wind generator, since the operating point at which the wind turbine delivers the maximum power changes accordingly with the wind speed. Therefore, the duty cycle of the switch S_w has to be regulated accordingly with the wind speed in order to control the current, and thereby the torque, of the wind generator, so that the wind turbine rotates at the speed which corresponds to the maximum power delivery.

To ensure that the photovoltaic array keeps on the maximum power point, the current flowing in the solar cells has to be controlled suitably by means of the duty cycle of the switch S_p. To this purpose a perturb-and-adjust technique can be used, where at regular intervals (i.e. once every few seconds) the amount of current drawn is perturbed and the resulting output power is observed. If an increased current results in a higher power, it is further increased until the power output begins to decline. On the other hand, if an increase in current results in less power than before, then the current is decreased until the power output stops increasing and begins to go down.

If the user and the battery storage require altogether less power than the maximum power achievable from the renewable sources, the system control strategy provides for a suitable reduction of the wind generator output power. For a given wind speed the current of the wind generator is regulated thereby at the level which permits that the overall power generated by means of the photovoltaic array and the wind turbine meets the power required by the utility circuit. As a consequence of that, the machine speed increases and the wind turbine operating point moves far from the point of maximum power delivery.

This procedure can be carried out till the wind turbine reaches its cut-off speed, beyond which the wind generator is shot down by means of its mechanical braking system. Whenever this fact occurs, the wind generator output voltage goes to zero, and the diode put in parallel at the output terminals of the relative step-up converter determines the exclusion of the generating unit from the utility power circuit. Afterward, the system operations rely only on the power generated by the photovoltaic array, and further reductions of the system output power required by light load conditions have to be accomplished by moving the operating point of the solar cells along the constant-voltage segment of the cells i-v characteristic.

4. PROTOTYPE DESIGN AND TESTING

A 2 kW prototype of the wind-photovoltaic power plant shown in Fig. 1 is going to be set up on the roof of the Engineering Faculty building, which is located on the top of a hill dominating the Colosseum. The generating system prototype will be set up by using a 1 kW photovoltaic array arranged by means of monocrystalline silicon cells, while the wind energy conversion system will compound a 1 kW axial-flux PM machine driven by a three-blades wind turbine. The system components have been constructed already, and preliminary laboratory tests have been carried out to verify the operating characteristics of the wind generator and the power conditioner.

4.1 Wind generator prototype

The design of the wind generator prototype was carried out by using a CAD-CAE procedure [3] devoted to the optimization of geometrical and electrical parameters, such as the machine weight, sizing and efficiency. Table 1 reports the main characteristics of the axial-flux PM generator prototype resulting from the machine optimum design. Laboratory tests were arranged in order to evaluate significant machine characteristics, such as the EMF waveform, the no-load losses, the stator winding losses and the machine efficiency for a number of rotor speeds and load conditions.

Machine tests at no-load considered open stator winding terminals and the machine shaft driven by means of a dynamometric unit. Fig. 2 shows the waveform of the machine E.M.F., whereas Fig. 3 indicates the peak value of the open circuit voltage vs. the machine output frequency. During the course of the no-load tests the machine losses were evaluated by measuring the input torque at a given rotor speed. Although the machine no-load losses include a number of loss mechanisms, such as iron core losses, stator winding eddy-current losses and frictional losses, the measured values were quite small, and this result is in accordance with the low value of the average flux density (i.e. about 0.5 T) in the machine airgap.

Machine load conditions were experimented by using the generator prototype to feed a diode-bridge rectifier, which supplies a variable d.c. load through an inductive filter. By considering different values of both the rotor speed and the d.c. load current, several running conditions of the machine prototype were experimented in order to achieve data and calculate curves of the machine efficiency vs. the output peak current. To this purpose the generator input torque was measured at given values of the rotor speed and the rectifier output current. Hence, by taking into account the machine no-load losses corresponding to the given rotor speed, such tests allowed the drawing of the machine efficiency curves shown in Fig. 4.

Table 1: Prototype design characteristics.

Number of poles	24	
Magnet thickness	4.3	mm
Peak airgap flux density	0.6	T
Number of phases	3	
Machine outer diameter	0.24	m
Machine inner diameter	0.16	m
Machine axial lenght	43	mm
Rotors weight (without PMs)	4.3	kg
Toroidal core weight	2.4	kg
Permanent Magnets weight	0.95	kg
Stator winding weight	0.96	kg
Rating speed	350	rpm
Peak EMF at the rating speed	40	V
Rating torque	32	Nm
Torque/weight	3.7	Nm/kg

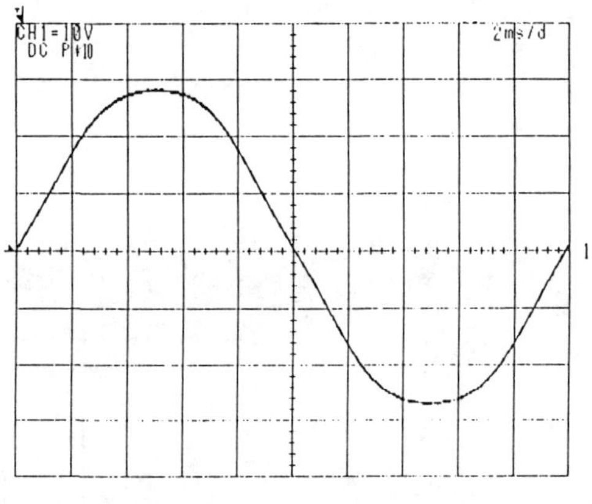

Figure 2. Wind generator prototype EMF waveform

4.2 Power conditioner prototype

The design of the power conditioner benefited of an accurate investigation of the converter modes of operation. Such an analysis was carried out by a computer simulation which takes into account suitable models of the wind and photovoltaic generators. Each one of the two step-up converters was modelled by means of its LC elements and a switching function representing the converter switching at fixed frequency and variable duty-cycle. A PI regulator acting on the value of the duty-cycle was simulated, in order to control the output current of the relative generator accordingly with the system control strategy discussed above.

A prototype of the power conditioner was arranged by using IGBTs power modules rated 600 V, 25 A. The experimental tests were carried out by using a photovoltaic array set up on the laboratory

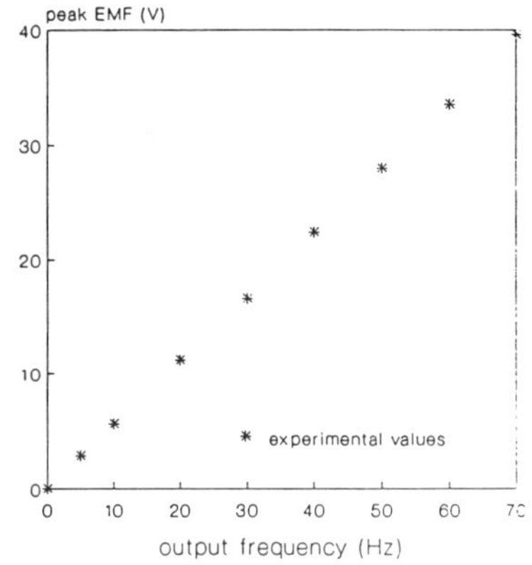

Figure 3. EMF peak value vs. output frequency

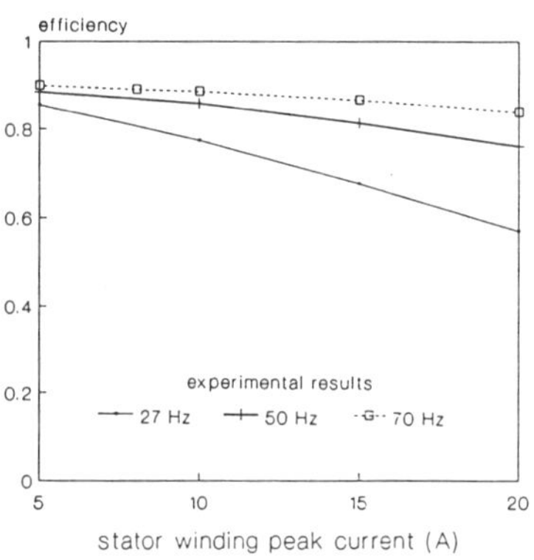

Figure 4. Wind generator efficiency

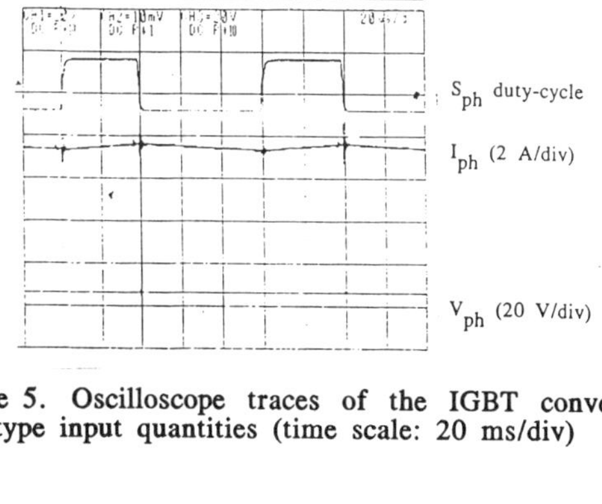

Figure 5. Oscilloscope traces of the IGBT converter prototype input quantities (time scale: 20 ms/div)

roof and the axial-flux PM generator prototype driven by a PM d.c. motor drive. The power conditioner prototype was operated at a 10 kHz switching frequency resulting from the comparison between a triangular carrier wave and a d.c. voltage signal proportional to the duty-cycle value commanded by a PI regulator. The utility circuit was arranged by using a battery and a passive d.c. load connected in parallel. Figure 5 shows oscilloscope traces of the power conditioner input quantities related to a converter output of 100 V, 6.5 A.

5. CONCLUSIONS

The development of stand alone integrated generating systems which use wind and photovoltaic energy has a potentially vast market in the developing countries, since these small-size power plants are an important option for meeting the basic electricity needs of millions of people who live in isolated rural areas of the world. In the developed countries wind-photovoltaic systems can find application where power is required at a costly distance from the grid, such as in the case of the power supply systems devoted to telecommunication apparatus located in inaccessible mountain regions.

This paper has discussed a configuration for integrated wind and photovoltaic energy power plants which utilizes innovative concepts for both the wind generator and the power conditioner. The advantages resulting from such a configuration can be summarised as follows:

- no gearbox in the wind conversion system;
- maximum exploitation of the renewable resouces of the site achieved by means of a suitable control of the output power of each generating unit;
- minimum energy storage required because the two energy sources are complementary inherently;
- the wind generator characterized by an high compactness and robustness, easy construction, high torque/weight ratio, and efficiency comparable with that one of conventional machine designed for high-speed applications;
- the power conditioner has a double-input and a single-output, so that it can handle the same time the power generated by the wind and the photovoltaic generators while provides for the control of the system operating conditions;
- easy connection in parallel with diesel generator sets or mini hydro generators, which may be required in the case of a poor availability of the renewables in order to achieve the continuous supply of the user load.

REFERENCES

[1] A. Di Napoli, F. Caricchi, F. Crescimbini, G. Noia, "Design Criteria of a Low-Speed Axial-Flux PM Synchronous Machine", Proceedings of the International Conference on Evolution and Modern Aspects of Synchronous Machines SM'100, Zurich (Switzerland), August 25-27, 1991, pp. 834-839.

[2] O. Honorati, F. Caricchi, F. Crescimbini, G. Noia, "Gear-less Wind Energy Conversion System using an Axial-Flux PM Synchronous Machine", Proceedings of the EWEC'91, Amsterdam, October 14 - 18,1991, pp. 814-818.

[3] F. Caricchi, F. Crescimbini, A. Di Napoli, E. Santini, "Optimum CAD-CAE Design of Axial-Flux Permanent Magnets Motors", Proceedings of the International Conference on Electrical Machines 1992, 15-17 September 1992, Manchester, U.K., pp. 637-641.

COST OF DESALTED WATER THROUGH R.O., ON A LARGE-SCALE, BY MEANS WIND ENERGY

J.A.Carta and R.Calero

University of Las Palmas de G.C.
High Technical School of Industrial Engineering
Edificio Departamental de Ingenierías. Tafira Baja s/n
35017 Las Palmas - Spain

ABSTRACT:The purpose of this paper is to expound a methodology that makes possible to know the economical viability of wind energy applications to sea-water desalination on a large-scale. That aspect could be crucial for the future development of many areas of the Earth and particularly in Canary Islands, Where the desalted water demands increases quickly, with costs subordinated to the prices evolution of conventional fuels and even with possible problems in their supply.
This work has been developed since 1988 in the Mechanical Engineering Department of Las Palmas University and has been financed by the State Energy Office of Canarian Government.

1 METHOD OF APPROACH

The methodology that are proposed in order to get the predicted objectives it can become in the following steps:

1.1 Knowledge of desalination potential.

The unsalted water flow that can be obtained from wind energy in a predetermined wind area depend, on the one hand by electric energies generated from commercial wind turbines in that area, and on the other hand, the configuration type and desalination wind park-plant system working.Among the different alternatives, viable technically, in order to connect wind park to desalination plants, in this paper only one of them is taken into consideration [1][2]:"Wind park composed by synchronous wind turbines, isolated from the local electrical grid, electrically connected to a reverse osmosis plant of variable capacity (independent desalination modules that can be connected as disconnected at will in order to adapt to energetic variation), and working at constant regime of flow and pressure" (Figure 1).

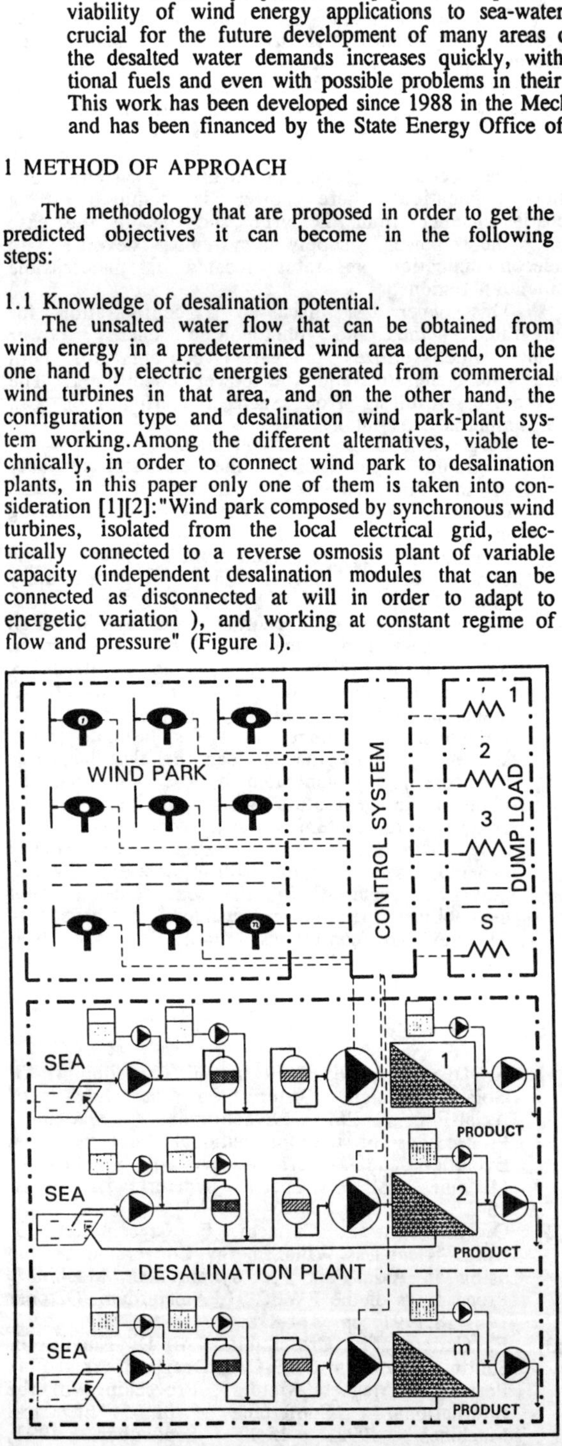

Figure 1

Obviously, in a predetermined wind park, the annual unsalted water that can be obtained depends among other factors of the desalination modules nominal capacity (Membranes/Module). So, when the modules capacities are lower, the park energetic exploitations are bigger and therefore higher annual desalted flows (Figure 2).

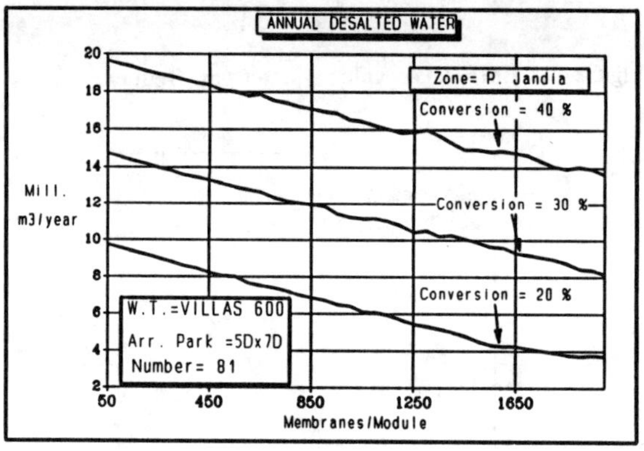

Figure 2

However, in compensation, when the modules capacities are lower, bigger number of this ones are necessaries in order to cover the energetic offer and the percentage is higher with short annual use. In this context and don't taking into consideration, at the moment, questions of economic nature, it's reasonable to think if the system operative or efficacy were measure in term of number of installed modules, when the less number of modules make the desalination plant up more favored will be its efficacy. Hence, as conclusion is taken out the advantage to analyze the flow percentages, with respect to the maximum, that are obtained when modules not very used are eliminated (Figure 3).

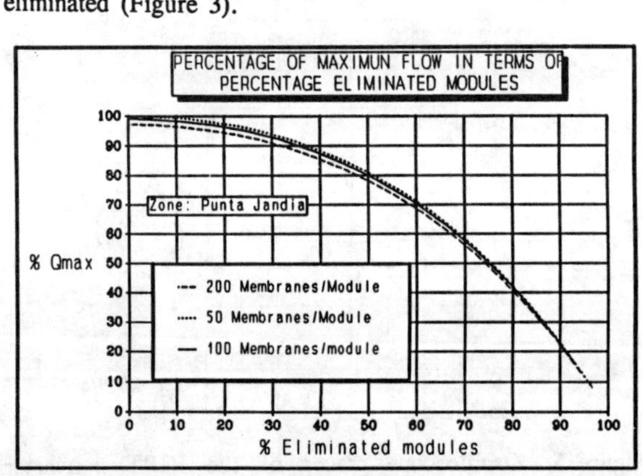

Figure 3

1.2 Definition of the economical model

At literature about selection of inversions multiples criterion to establish the decision to invert can be obtained. Developing this work we have inclined by chronologic methods (Accumulate Present Value "APV" and the Internal Rate Return "IRR"), which take into consideration the chronology of cash flow and they use a capitalization law or discount in order to homogenize the reliable and received quantities at different time.

1.3 Definition of the economical parameters

All reply to the different questions that are raised about cost and income generated when sea water is desalted from wind energy, it's implicit the previous knowledge about a set of economic parameters which determine the costs. Obviously, such set will be composed by very different parameters, that why in order to make easy the analysis its classification by three influence groups is proposed (Figure 4).

Parameters associated with wind parks	
Investment	Operating costs
Lot	Personnel
Wind turbines	Maintenance
Building site	Operation
Installations	Rent
Engineering	

Parameters associated with desalination plants	
Investment	Operating costs
Lot	Personnel
Desalination plants	Maintenance
Building site	Operation
Installations	Membranes replacement
Engineering	

Parameters associated with the economical model
Inflation
Insurance
Reduction of taxes
Technical depreciation
Subvention
Credit
Tax

Figure 4

1.4 Determination of desalted water cost

At this stage proceed from two different point of view to determine the generated cost when sea water is desalted from wind energy:

• When the installed capacity of the binomial park-plant is the maximum capacity. That is to say, the purpose would be to save the major possible quantity of conventional energy, even though the m³ cost of unsalted water not be the minimum.

To determinate these costs is only necessary to replace the entry variables to the developed economic model the corresponding values to the park-plant system which provide the maximum water production (Less modules capacity)

• When installed capacity in the above mentioned binomial make minimum the m³ cost of produced water.

The situation of not very use that take place in desalination modules when the maximum flow is looked for, it contribute, more or less in proportion to the m³ cost of unsalted water increase, it depends as it's presumable the use degree; therefore it'll essential for imposed purpose performance to analyse the influence that in the m³ cost of unsalted water has to eliminate a determinated number of modules, and that not only for the module capacity that provide the maximum annual flow but those ones "size" that has more capacity and they provide less annual flow, since "a priori" it'd be risky to eliminate any of them (Figure 5).

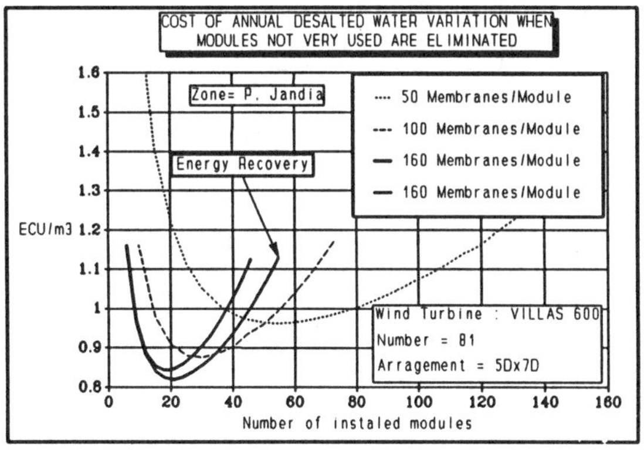

Figure 5

From so many similar chart realization like the showed one in figure 5 as wind turbines have been implicated in this study, it is possible to determine:

▪ Unsalted water cost performance when the modules capacity is varied.

▪ What is the plant "size" that make minimum the m³ unsalted water cost.

From the minimum unsalted water cost superposition for each type of wind turbine (Figure 6) it'll possible to deduce the minimum absolute cost of the produced water in a pre determinate wind area.

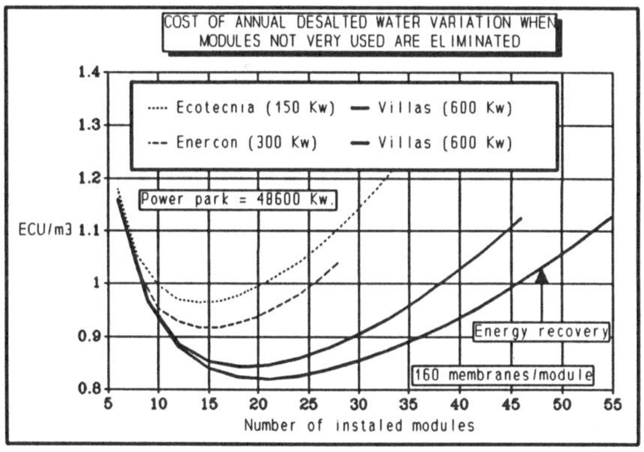

Figure 6

1.5 Cost sensibility

The purpose of this section analyse the influence that in the minimum cost per m³ of desalted water has the variation of those parameters more outstanding implicated to determinate it (Figure 7 y 8).

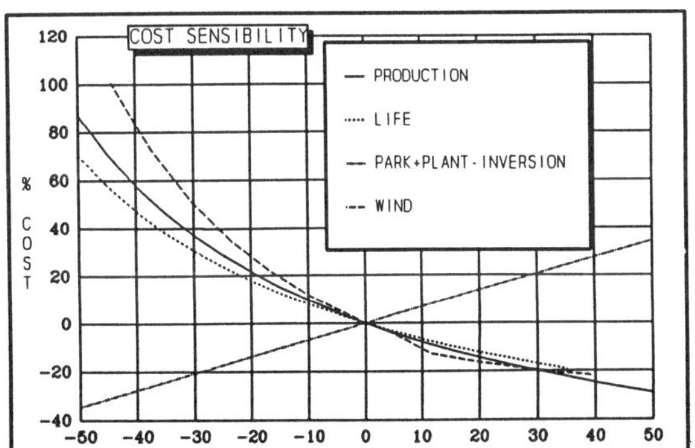

Figure 7

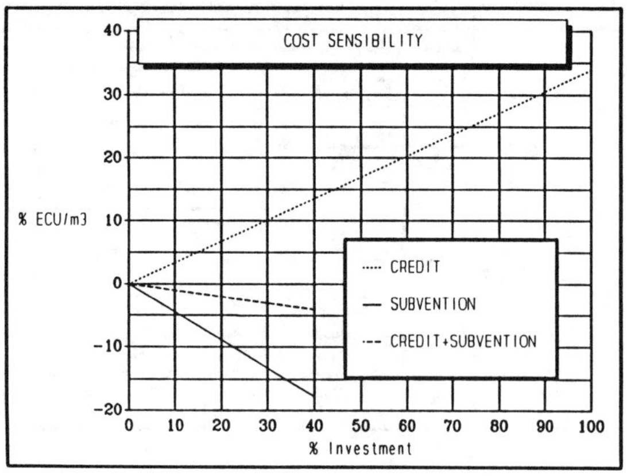

Figure 8

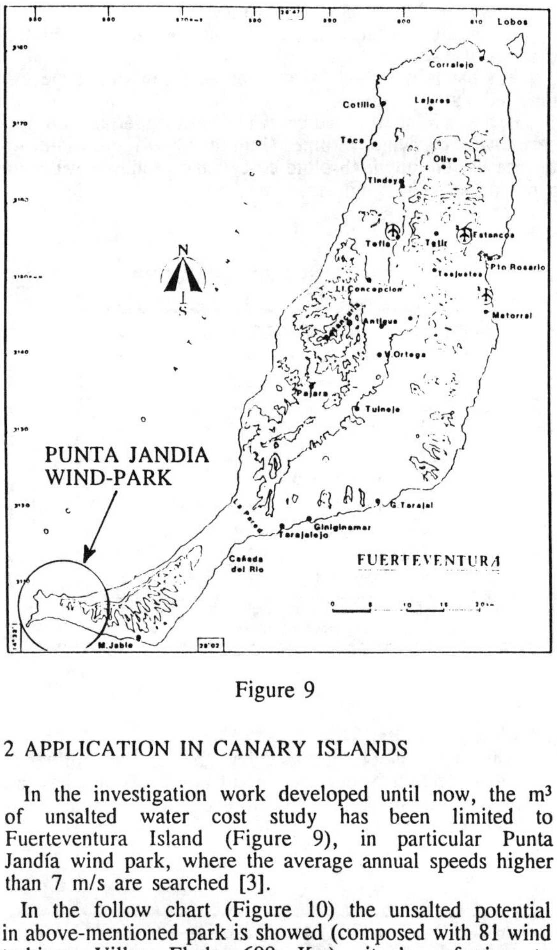

Figure 9

2 APPLICATION IN CANARY ISLANDS

In the investigation work developed until now, the m³ of unsalted water cost study has been limited to Fuerteventura Island (Figure 9), in particular Punta Jandía wind park, where the average annual speeds higher than 7 m/s are searched [3].

In the follow chart (Figure 10) the unsalted potential in above-mentioned park is showed (composed with 81 wind turbines Villas Floda 600 Kw), it 's referring to desalination modules of reverse osmosis composed each one with 50 membranes (1150 m³/day nominal conditions). So in figure 10 the numbers of wind turbines evolution in function of the demand of unsalted water in Fuerteventura predicted evolution.

To calculate the m³ of water have been taken into consideration all associated cost to the acquisition and exploitation of the wind park and modular desalination plant, an 30% credit from the total inversion is supposed, 15% annual interest, to give it back in 10 years. The rest of the capital is a direct contribution and not any subvention has been suppose.

So to the 81 wind turbines 600 Kw connected to 147 reverse osmosis modules with 50 membrane each ones, the produced flow is 19 Hm³, and resultant cost is 1.34 ECU/m³. In this park-plant conditions, the use factor on

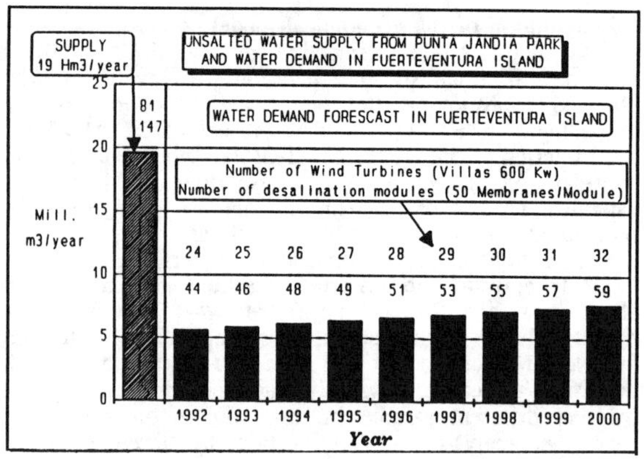

Figure 10

the first one is 0.3 and on the second one is 0.41.

Given that on the one hand the plant cost come down when the capacity in each module increase, besides, the set became cheaper too when the modules not very used are eliminated, the minimum cost 0.817 ECU/m³ is obtained when 21 modules with 160 membranes each ones are used, connecting to the same wind park. In this conditions, the plant use factor increases to 0.73, whereas the park factor decreases to 0.21. However, the desalination potential decreases become only 16 Hm³.

In short, at new conditions, flow comes down approximately 30% meanwhile the cost downs come 40%, from 1.34 ECU a 0.817 ECU/m³. The last cost, in Fuerteventura island case, is sensibility lower to the obtained one at present in a conventional system.

3 SIGNIFICANT RESULTS

- In the considered alternatives exist an optimum combination park-plant that minimize the cost of drinkable water.
- The volume and cost of desalted water that can be obtained by the connection of desalination plants with wind parks are highly sensitive to the local regime of wind and in a minor way, to the possible existence of an energy recovery system.
- Because of the influence of the several park-plant system components cost, the plant capacity has to be increased, even though the park capacity is decreased.
- In the isolated systems, the costs decrease as the number of membranes which compose each desalination module as well as the elimination of infrequently used modules.
- The desalted water costs obtained with the systems connected to the electrical grid are less than the ones produced by isolated systems.
- Even through the theoretical and experimental replies found in this work we think they establish the longed-for "green light" to break away the use a large-scale with an unpolluted wind energy to desalt sea water, the direct path toward the total use of this application with wind energy we think it must begin with the decision to an experiment a short-scale, the obtained results in the above theoretical stage, specially, the different elements performance that take part in a "modulate" plant in order to the future extended experiences that can be realized progress on solid bases where they can lean

In this sense we have to indicate that it is in the stage of project the connection of a small wind park (composed by 3 or 4 synchronous wind turbine), in isolated working, with some modules of reverse osmosis desalination, in constant working, and with a nominal capacity of about 50 cubic meters per day each module.

REFERENCES

[1] Optimization technical-economic of the desalination system worked by wind energy. J.A. Carta and R. Calero, European Wind Energy Conference and Exhibtion, Madrid, Spain, September 1990.

[2] Fundamentos teóricos de la desalación de agua de mar con energía eólica. J.A. Carta and R. Calero, Symposium Internacional de Desalación y Reuso del Agua Canagua 92. Gran Canaria, Spain, 1992.

[3] Canaries Wind 2000 Project. R.Calero and J.A. Carta, European Community Wind Energy Conference, Lübeck-Travemünde, Germany, March 1993.

CONNECTION OF WIND POWER GENERATING FACILITIES TO THE POWER DISTRIBUTION SYSTEM AND POTENTIAL EFFECTS ON THE SYSTEM

Hans-J. Fiss
Wolfgang Weidemann
SCHLESWAG
Kieler Straße 19
D-2370 Rendsburg

Karl-Heinz-Weck
Fritz Weinel
Forschungsgemeinschaft für Hochspannungs- und Hochstromtechnik e.V.
(FGH)
Hallenweg
D-6800 Mannheim-Rheinau

SUMMARY
The increasing installation of wind power plants in the supply area of SCHLESWAG, Germany, makes the considerations of possible supply voltage disturbances during the planning of the connection of such plants to the system necessary. This reports deals with measurement results of the parameters relevant for such disturbances obtained from generators with ratings between 25 kW and 330 kW. From these measurements and from commonly adopted recommendations connection conditions are elaborated, which are today applied by SCHLESWAG in the approval of wind power plants in their supply area.

1. Introduction

The coastal regions of Schleswig-Holstein, including the islands in the North Sea and the Baltic, are among the regions with the highest wind probabilities in the Federal Republic of Germany. Legislative initiatives by the Federal Government and the State of Schleswig-Holstein have given rise to a rapid increase in the number of wind power generating facilities connected to the public power supply system. Within the area supplied by SCHLESWAG, there were wind power generators with a total capacity of about 80 MW in service as standalone units of in windparks by the end of 1992, with a further 550 MW plus in planning.

Since the wind power yield is a function of wind speed to the power of three, the operators of wind power generators endeavour to install their generating facilities as close to the coast as possible. This goal conflicts directly with the power distribution system conditions as they actually exist in SCHLESWAG 's supply region. SCHLESWAG 's medium voltage distribution network is designed to ensure a reliable power supply in this region of relatively low power density, and because of its low transmission capacity and low short circuit power it is not capable of absorbing arbitrarily high wind power levels into the existing system without adverse effects on the reliable supply of power to the connected loads.

Since 1989 measurements in the "Windpark Westküste" were carried out and the results were interpreted in therms of their probable implications for the power distribution network. These wind power plants are listed in Table 1 as the upper four. Since then, the political and economic development led towards larger power units, mainly operated by private investors. At these power levels, the low voltage distribution network is unsuitable as a connection point. They are connected local distribution transformers directly to the 10 or 20 kV-distribution networks. Therefore, similar measurements were carried out in four commercially-operated wind farms using wind power generators in the 150-300 kW power rating range shown in Table 1 as the lower four.

2. Measurement results

In the course of the measurement period, it proved expedient to subdivide the range of voltage disturbances caused by the variations in wind power input into four categories:

- Power changes causing slow voltage fluctuations
The power change is defined as the change in active power and reactive power integrated over a time interval of 1 minute.

Type	Number of generators	Total power kW	kind
Aeroman-125/30	20	600	asynchron
electromat-25	5	125	synchron
ENERCON 16	5	275	synchron
Adler 25	2	330	asynchron
HSW 250	4	1000	asynchron
Vestas V25-200	5	1000	asynchron
AN Bonus 150/30	7	1050	asynchron
ENERCON 32	10	3300	synchron

Table 1 : Investigated wind power generators

Generator	Rated power kW	Max. power / Rat. power	power factor
HSW 250	250	1.21	0.8
Vestas V25	200	1.01	0.9
AN Bonus 150	150	1.05	0.95
ENERCON 32	330	1.00	0.95

Table 2 : Power characteristics of the wind generators

- Power fluctuations
The power fluctuation is defined as the fluctuation of the power and reactive over a time interval of 1 minute.

- Light flicker
The possible light flicker is calculated from the current and an assumed system impedance by means of a computer-simulated flicker meter and is expressed as the P_{st}-value.

- Current harmonics
Simultaneous measurements of the harmonic voltage made it possible to distinguish between harmonics generators and harmonics absorbers.

The following is a selection of the most interesting aspects discovered in the course of the analysis of the measurement results. Further information will be published in the reference.

- Powers and power factors
The active powers produced by the various types of generator may at times lie significantly above

their rated power output. The compilation given in Table 2 for the larger generators shows that, though in the case of most generators the maximum active power output is only slightly higher than their rated output, there are some instances, in which the active power output can be considerably higher than rated. The power factors also vary within very wide limits, with the result that the apparent power of some generators, which is responsible for system disturbances can be up to 60 % higher than their rated output.

- Power changes
Depending on the prevailing winds, the power output from the wind power generating facilities may change within a short time by practically the full rated power output of the plant.

- Power fluctuations
The power generated by a wind power plant varies in line with changes in wind speed. Fig. 1 shows examples of the active power of three generators over a time interval of 1 minute. High-amplitude power fluctuations with a slow rate of rise can be clearly distinguished from low-amplitude, periodically recurring fluctuations attributable to local wind conditions at the generator. These correspond to the frequency with which, in this example, one of the three blades of the rotor enters the wind shadow behind the mast which, in the case of the generators studied, gives rise to power fluctuations with frequencies between 2 and 2,2 Hz. The high-amplitude, slow power fluctuations occur in all generator types, the periodic fluctuations in asyncronous generators only.

When a number of generators are operated in parallel in the same wind park, the relative power fluctuatuions are reduced. In the case of the higher-rated generators, the relative powe fluctution in reduced in line with a stochastic superposition following an exponential function of $1/\sqrt{n}$, n being the number of generators involved. This function applies both to the wider-amplitude power fluctuations due to wind gusts and to the periodic fluctuations.

- Light flicker
Depending on the inherent impedance of the supply system, the fluctuating power input can give rise to system flicker. As examples, Fig. 1 shows also the P_{st} -values for the three generators for the short circuit power of the system equal to 25-times the rated power of the generators. The appreciably higher value for the Vestas as compared to the AN Bonus is attributable to the fact that this generator accounts for a considerably larger portion of the power fluctuation caused by the rotor blade frequency of 2.2 Hz. Without this component, both generators would cause about the same light flicker. For the synchron generator ENERCON the influence of the rotor blade frequency is negligible and the P_{st} -value ist small. As already described for the power fluctuations, the flicker induced by multiple generators operating in parallel is superimposed in accordance with stochastic functions, i.e. given n identical generators operating in parallel and identical ratios between the network short circuit power and the total generator rated power, light flicker is reduced according to the function $1/\sqrt{n}$.

- Switching operations
The asynchronous generators HSW 250, Vestas V25 and AN Bonus 150/30 studied in the second measurement period were equipped with converter-controlled starting to limit the maximum current to 2 to 3 times the rated current. Despite this limitation, significant system flicker was observed to occur due to generator start-up and in particular due to the rapid power changes from full power to zero power when the generators have to be braked because of excessive wind strengths.

- Harmonics
Wind power plants with synchronous generators and line-commutated power converters feed harmonic currents into the receiving power distribution system. To illustrate the peculiarities observed in the course of the harmonic measurements, Fig. 2 shows the harmonics spectrum of the current generated by an ENERCON 32 unit. This generator has a 12-pulse converter, a series resonant circuit filter for the 11th and 13th harmonics, and is connected to the medium voltage distribution network via a transformer. In addition to the harmonics of the orders typical of power converters, harmonics of even orders and those divisible by three occur on an noteworthy scale. Inter-harmonics are also identifiable.
The reason for the occurence of harmonics not typical of power converters is the use of conventional converters with analog firing angle control whose accuracy is not sufficient to avoid these harmonics. Furthermore, parallel resonances occur between the series resonant circuit filter and the terminal transformer, amplifiying the harmonic currents such as here in the region of the 9th harmonic. Inter-harmonics arise due to inconstancy of the firing angle or due to incomplete smoothing of the direct current in the d.c. part allowing the frequency imposed by the synchronous generator to pass through to the a.c. side.

3. Connection to the power distribution system

3.1 Criteria for connection

Since at the time of the study there were no defined conditions for connection of wind power plants to the power distribution system and even now conditions have been defined only for connection to the low voltage distribution system, it was first necessary to specify criteria for ensuring the required voltage quality. As wind power may be high, these criteria are applicable to the total wind power of a system part.

- Power change
The voltage change caused by systematic changes in the power fed by wind power plants into the power distribution system must be so small, that the low voltage at the consumer remains within the tolerances prescribed in IEC-Standard 38. This ist ful filled for all systems and load conditions if the voltage increase caused by the wind plant is lower than 4 %.

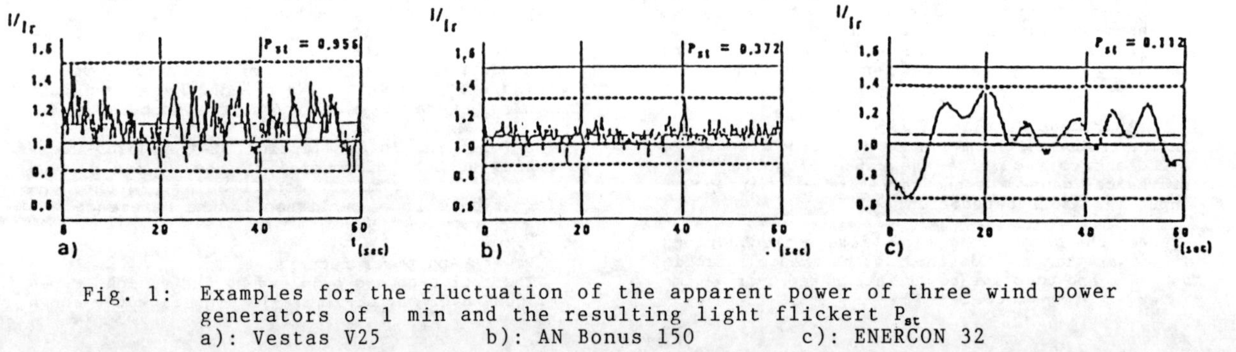

Fig. 1: Examples for the fluctuation of the apparent power of three wind power generators of 1 min and the resulting light flickert P_{st}
a): Vestas V25 b): AN Bonus 150 c): ENERCON 32

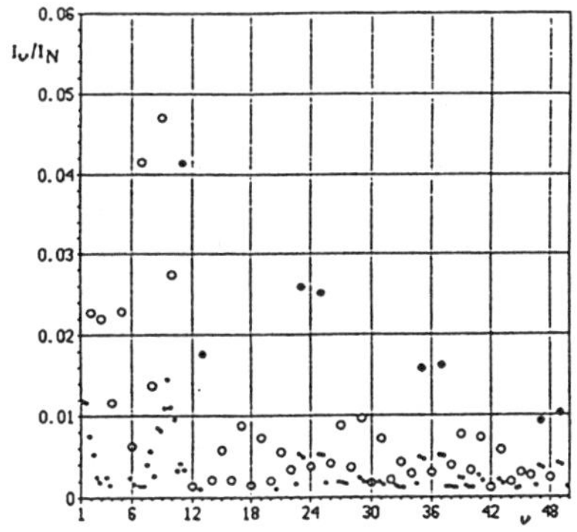

Fig. 2: Relative current harmonics dependent on the order for a synchron generator ENERCON 32 with filter for the 11./13. order

● : Converter typical orders
O : Converter untypical orders
• : Interharmonics

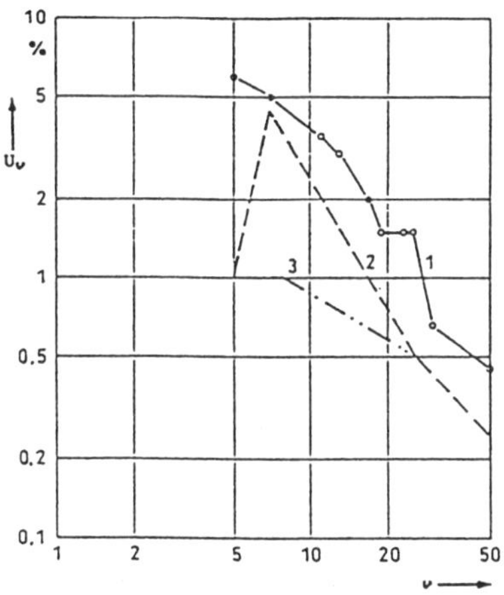

Fig. 3 : Permissible voltage harmonics
1 : Permissible levels in the low voltage system acc. to IEC 1000-2-2
2 : Permissible levels in the medium voltage system
3 : Permissible levels calculated form current and short circuit power in the medium voltage system including parallel resonance

- Power fluctuations
The limits specified in the VDEW recommendations, 2 % for the medium voltage distribution system and 3 % for the low voltage distribution system, are recommended als limit criteria for the voltage fluctuations.

- Light flicker
The criteria is that the so-called P_{st}-values in the supply voltage is below the disturbance level of $P_{st} = 1$. The acutal light flicker will be lower at least by a factor of 0,9

- Switching operations
SCHLESWAG specifies that the starting currend must be limited to twice the rated current.

- Harmonics
The basic criterion established is that the levels specified in IEC 1000-2-2 must not be exceeded in the low voltage system. For the specification of connection criteria, however, it is important to remember that the harmonic voltage occuring in the low voltage distribution system are not explicitly known, but have to be estimated from the harmonic currents input from the wind power generating facility. It must be taken into account, therefore, that at higher frequecies resonances may cause considerably higher voltage harmonics than those estimated form the short circuit power. This applies especially if, as is almost certainly the case, there are capacitive compensation facilities in the low voltage system. The following steps to arrive at a connection condition are used:

1. The limits to IEC 1000-2-2, Fig. 3, Curve 1, apply to the maximum allowable voltage harmonics in the low voltage system.

2. The possibility of resonances in the low voltage system means that the harmonic voltages calculated from the short circuit resistance must by lower, and Fig. 3, Curve 2 is obtained. This curve also takes into account the permissible load of capacitors.
Fig. 3, Curve 2 also gives the maximum allowable harmonics in the distribution system.

3. Especially at high cable portion, parallel resonances may occur in the distribution system, where the connected loads act as the only damping. Fig. 3, Curve 3, makes allowance for the harmonic voltage increase, above the levels calculated from the short circuit.

Similar principles apply to the orders of harmonics not typical for DC/AC converters. ,

3.2 Connection conditions

Load flow calculations and harmonics calculation programs can be used to assist in transferring the connection criteria into connection conditions. This method has, however, proven too complicated for practical applications. For this reason, simplifications were introduced to make the connections conditions applicable even without numerial calculations.

From the criterien for the voltage change, the wind power connectable to the medium voltage system is obtained als follows:

$$\sum S_{WKA} \left(\frac{1}{S_{KE}} - \frac{1}{S_{Kss}} \right) \leq \frac{1}{33}$$

S_{WKA} : Wind generator power
S_{KE} : System short circuit power at the connection point
S_{KSS} : Short circuit power at the transformer station busbar

From voltage fluctuations and light flicker the following conditions are obtained:

$$\sqrt{\sum \left(\frac{P_{WKA}}{S_{KE}} \right)^2} \leq \frac{1}{25}$$

$$\sqrt{\sum \left(\frac{P_{WKA}}{S_{KE}} P_{st} \right)^2} \leq \frac{1}{25}$$

P_{st}: Light flicker of the generator calculated from the current at short circuit power equals 25 times the generator rated power

Comparison of the connecting conditions derived from the voltage fluctuations and the system flicker with those derived from the voltage change shows that allowance need by made for these only if the total possible power is supplied by only a few generators. Given the power ratings of the units commonly in use today, power fluctuations ans system flicker will be of significance only when connecting wind power generating to the low voltage system, while as a rule they can be neglected for the medium voltage system.

The connection conditions deriving from the maximum allowable harmonic voltages apply only for synchronus generators with d.c. circuit and converters. They are dependent on the generator design, and especially on the filter technology used, and the following applies only to the ENERCON 32 that was included in the study. Table 3 shows the necessary ratios between system short circuit power and generator power, as determined for three different generators of this type. It is evident that in all cases it is the high orders that are responsible for the system fault level required.

For this reason, additional filters were installed at one of the generators to reduce the higher-order harmonic currents. The required short circuit power decreased for the orders typical, but that the non-typical orders resulting from the firing inaccuracies still require high short circuit power values because of the lower allowable voltage levels. These values, however, also decrease, if more generators are operating in parallel, as shown in Table 4.

Order	Generator No.	$\frac{S_{KE}}{S_{WKA}}$	Critical order
odd. not divisible by 3	1	175	49
	2	140	47
	3	175	37, 49
even	1	150	26, 28
	2	100	28
	3	300	39
odd. divisible by 3	1	250	45
	2	250	45
	3	300	39

Table 3 : Necessary short circuit power S_{KE} related to generator power S_{WKA} for 3 different generators ENERCON 32 with filter for 11. / 13. harmonic.

Order	Number of Generator	Location	$\frac{S_{KE}}{\Sigma S_{WKA}}$	Critical order
odd. not divisible by 3	4	EP	65	23
		SS	70	23
	10	EP	45	23, 37
		SS	65	11
even	4	EP	250	20
		SS	300	20
	10	EP	55	20
		SS	65	20
odd. divisible by 3	4	EP	150	21
		SS	240	21
	10	EP	35	21
		SS	55	21

Table 4 : Necessary short circuit power related to wind park power ΣS_{WKA} for wind parks with ENERCON 32 with additional filters.
EP : considered location is connection point
SS : considered location is busbar

For the harmonics of the orders divisible by two and by three there is a noticeable reduction when several generators are operated in parallel. If thyristor gate control ist improved, for instance by means of digital firing angle control the harmonics not typical orders still occur. In this case parallel operation of several wind power generators reduces the harmonic currents only slightly as compared to those from a single generator.

Further reduction of the harmonics typical of power converters is possible only by introduction of precisely-tuned series resonant circuit filters.

If this development works is consistantly continued, it may be possible to raise the allowable wind energy input from synchronous generators with d.c. converters to about 1/40 of the short circuit power at the point of connection from the value of 1/65 applied today. These factors, however, are applied only to the wind generator ENERCON, for which measurement results are available. For generators, for which no information on the produced current harmonics is available a ratio of 1/250 is applied.

4. Summary and consequences

The measurements performed between 1989 and 1991 on wind power plants in the SCHLESWAG power supply region the assessment of possible effects of such installations on the voltage of the power system. In the case of wind power plants with asynchronous generators connected to the medium voltage system, the change in voltage resulting form the slow changes in power fed into the system ist the deciding connection condition. Owing to its high short circuit power, the medium voltage system can sustain a total wind power which, at present, can be attained only ba a large number of generators operated in parallel. The mutually compensatory effects of this mean that voltage fluctuations and and light flicker caused by fluctuations in power remain within the allowable limits.

When wind power generating facilities are connected to the low voltage system, the maximum wind power input deriving from the maximum allowable voltage change can be delivered by a single generator. Allowance must also be made for the connecting conditions deriving form the maximum allowable voltage fluctuations and light flicker. Depending on the type of generator, the light flicker occurring at full power, and especially during starting and braking of the generators, may be the crucial factor limiting the connectable wind power capacity.

For wind power generating facilities comprising synchronous generators with converters of state-of-the-art design, it is the voltage harmonics that decide the connectable wind power.

SCHLESWAG always uses the connection conditions described here in its applications for installation of any new wind power generating facility. Since the wind power plant connection capacity of the existing medium voltage distribution system has already largely been exhausted, planning for further large-scale wind power plants is concentrating on the construction of special "wind power cable systems". In the lead-up to these planning activities, SCHLESWAG has assisted the town and country planning efforts of the state government and the development planning of the local authorities in the regions with promising wind conditions by identifying the transformer substation areas that can be developed for installation of wind power generating installations at comparatively low cost. In the 20 kV distribution system, cable lengths of about 8 km, centered on the transformer substations, are found to be viable while fully satisfying the recommended connection conditions. Provided the external conditions, e.g. etc., permit, the intention is to change the land use plans to provide for "wind power reserves" within the areas served by the transformer found to be suitable.

As a further development towards system compatible wind power generation the certification of wind generators regarding their possible supply voltage disturbance is in discussion.

REFERENCES

H.-J. Fiß e.a.: Netzrückwirkungen und Netzanbindung von Windenergieanlagen,
Published in : Elektrizitätswirtschaft Heft 22, 1992

ECONOMICALLY OPTIMIZED DESIGN OF WIND FARMS.

Kristiansen, J., Hansen, J.C.H.

ELSAMPROJEKT A/S
Kraftværksvej 53
DK-7000 Fredericia
Denmark
Fax: + 45 75 56 44 77
Tel: + 45 75 56 44 11

Summary.

The electrical layout of the power collecting system in wind farms has been investigated with regard to electrical losses. The losses have been evaluated to obtain the total lifetime costs, including capitalized losses and equipment costs. Experience gained by the investigations has shown that the electrical layout for a wind farm on a specific site is highly dependable on local conditions such as prices for selected equipment, interest rate, energy price, soil compound regarding excavation etc. and wind distribution. The amount of electrical losses has an important influence in choice of solution. Though, the alternative with the lowest electrical losses is not necessarily the optimum choice when the lay-outs are evaluated with regard to lifetime costs including capitalized electrical losses. Whether a low-voltage solution or a high-voltage solution is the economical optimum solution is highly dependable on the parameters used in the evaluation and the size of the wind turbines.

1. INTRODUCTION.

The increasing size of wind turbines used in wind farms makes it profitable to evaluate the layout of the electrical power collecting system, not only based upon voltage variations, but also with regard to the electrical losses. The amount of losses together with costs for the installed equipment are investigated in order to achieve the minimum lifetime costs for a wind farm.
For that purpose a design tool has been developed. WIFAGO (WInd FArm Grid Optimizing) is a design tool for optimizing the electrical power collecting system in the wind farm.
On the basis of general information about the wind farm and wind statistics etc., the tool gives solutions to the economically optimum choice of equipment. The wind farm is optimized during several interactions making it possible to compare different configurations. All cables and transformers etc. represented are divided into groups of identical configurations to reduce the number of calculations to be performed.

This paper presents investigations made and considerations, regarding the above. The investigations are based upon evaluations and comparisons of a proposed wind farm in Australia.
The wind farm consists of 32 300 kW wind turbines equipped with induction generators. The induction generators are partly compensated for their reactive power consumption.

2. OPTIMIZATION.

2.1 Optimization method.

The wind farm is optimized during several interactions making it possible to compare different configurations in order to achieve the minimum lifetime costs for a wind farm.
The wind farm is divided into groups represented by an equal number of wind turbines and identical lines and transformers. Each transformer or cable within the group is separately optimized. This reduces the number of calculations to be performed.
With a given wind speed distribution for the area and the power curve for the wind turbines, a load duration curve is calculated, allowing the energy losses in the power collecting system to be calculated.

The losses in all components are capitalized using the NPV-method (Net Present Value). This gives a figure for comparison of the future costs. With a given interest rate, energy price and lifetime, the NPV of the losses is calculated. The total lifetime costs are the sum of equipment costs and NPV losses of the wind farm lifetime.

2.2 Optimization procedures.

The optimization is made by finding the equipment with the lowest combined equipment price and cost of capitalized lifetime losses.
Before optimization of the electrical layout, the micro-siting and location of the wind turbines have been completed by parameters such as general wind direction, topography and wake effects.

Calculating the load distribution, the basic data for the generator and the wind conditions must be defined. The generator is described by the nominal voltage, maximum active and reactive power output and the power curve for the wind turbine.

The local wind conditions are described by the Weibul A and C parameters which are used to calculate the wind distribution.

Wind speed frequency W_f:

$$W_f=\frac{C}{A}e^{((C-1)\ln(\frac{WS}{A}))}e^{(-e^{(C\ln(\frac{WS}{A}))})}$$

WS - Wind speed
A - Weibul A scale factor
C - Weibul C shape.

The annual number of hours during which a certain wind speed occur is:

$W_t = W_f$ 8760 hours

In figure 1, the wind distribution for the specific site is shown. The Weibul parameters A and C are 8 and 1.9 respectively.

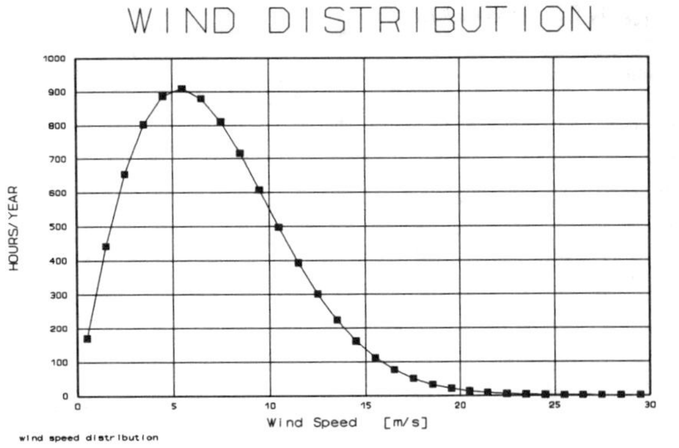

Figure 1. Wind distribution for site.

Combining the wind turbine power curve with the wind distribution curve, giving the generator load curve, the power production and the losses can be calculated as:

$$\Delta P = \frac{r}{U_n^2}\,(P^2+Q^2)$$

where P and Q are the power transmitted.

In the NPV-calculations the losses are multiplied with the energy price. The energy price can be fixed or change over the years.

The NPV-function calculates the Net Present Value with the formula:

$$NPV = \sum_{i=1}^{n} \frac{c_i}{(1+interest)^i}$$

n expected wind farm lifetime in years or duration with fixed energy price.
interest interest rate
c_i cost for yearly losses.

When the Net Present Value of the losses is known, the lifetime costs are calculated as the sum of costs for NPV-losses and the equipment costs. Foundation costs have to be included in transformer costs, and costs for excavation, reconditioning underground installations and backfilling have to be included in the cable costs.

The voltage increase in cables and transformers are calculated using the approximate formula for evaluating the voltage variations in the wind farm. These calculations are only performed to observe the voltage increase at the generators at full production. If more accurate calculations are needed, load flow calculations have to be performed.The voltage increase are calculated using the approximate formula:

$$\Delta U = \frac{-P_{max}r + Q_{max}x}{U_n^2}$$

U_n Nominal voltage
P_{max} Maximum active power transmitted
Q_{max} Maximum reactive power transmitted
r Resistance
x Reactance.

2.3 Parameters influencing the WIFAGO model and the electrical layout.

In order to design and evaluate the different layout solutions, it is essential to identify the parameters influencing the layout. For the optimization model data has to be collected and evaluated regarding their influence on the layout. Examples of topics to be considered are mentioned below:

- Material prices: Cables, transformers, switchgear etc.
- Soil compound regarding costs of excavation for cable trenches and foundations.
- Topography and roads on site regarding cabling.
- Local wages regarding the electrical works.

To determine the capitalized lifetime costs of the losses in the power collecting system, the following information is needed:

- Development in energy price
- Interest rate
- Wind speed distribution
- Expected lifetime
- Equipment characteristics

Deviations and uncertainties in the above data have highly varied influence on the layout. To minimize the total uncertainty, the topics of major importance have to be as accurate as possible.
To illustrate the WIFAGO model sensitivity, some of the above-mentioned parameters and their influence on the modelled layout are shown in figure 2. The values in figure 2 are relative and show the change in total lifetime costs for the group evaluated and are based upon a certain change in the parameter specified. The impact on the lifetime costs expressed in percentages depends on the extent of equipment included in the calculations.

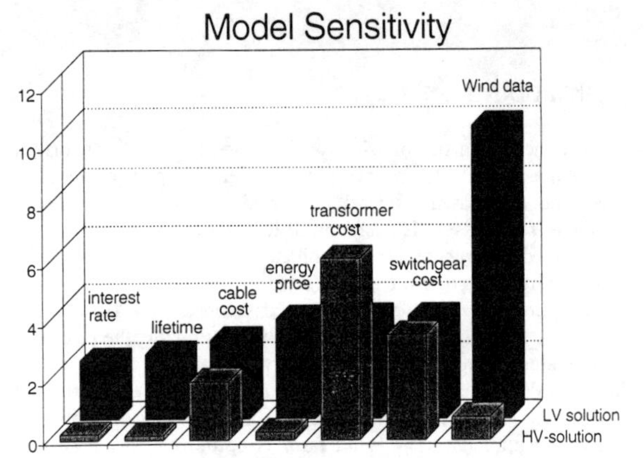

Figure 2. The parameters' relative influence on costs for the modelled layout.

As seen in figure 2 a certain deviation in data for wind distribution causes a large change in the evaluation of a LV/HV-solution due to the influence which the wind data has on calculating the capitalized costs of the losses. In general, the data used for calculating the capitalized costs of the losses are important for evaluating a LV-solution. Evaluating the HV-solution, the equipment costs, such as costs of transformers, cables and switchgear, are parameters of major importance to the total lifetime costs. The data used for calculating the costs of capitalized losses are of minor importance.

3. DESIGN OF WIND FARM.

3.1 Wind farm data.

Examples of design considerations are given for illustrating some of the aspects mentioned above. A considered wind farm in Australia consists of 33 300 kW stall controlled wind turbines equipped with induction generators with a nominal voltage of 690 V. The induction generators are partly compensated for their reactive power consumption. The voltage level of the power collecting system is 22 kV and 66 kV for the grid connection. The wind turbines are located in groups of 3 and 4 wind turbines with an approx. mutual distance of 200 m.

3.2 Reactive power compensation.

Due to regulations and claims from the local utility the wind farm are compensated for the reactive power demand at idle. Each wind turbine is equipped with a capacity bank with a capacity equal to the reactive power demand on the wind turbine generator at idle. The capacity bank is connected to the grid when the wind turbine is connected. This secures a cos φ between 0.97 and 1.0 for all load cases and no production of reactive power out of the wind farm. Only solutions with this compensation are considered.

In figure 3, the power curve for the 300 kW wind turbine is shown. Both active, reactive power consumption and compensated reactive power consumption is shown.

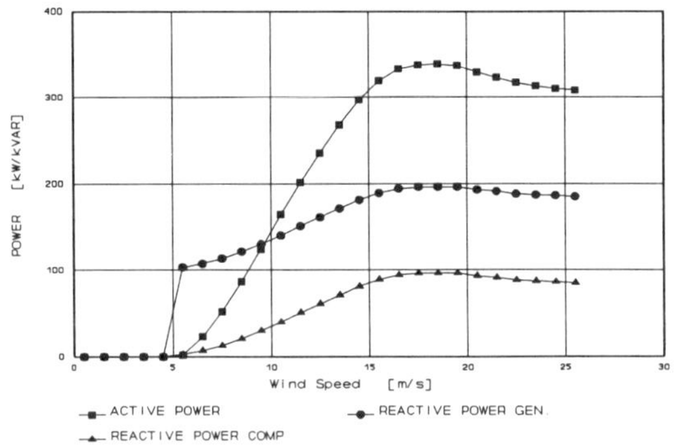

Figure 3. Power curve characteristic for wind turbine.

3.3 Grouping the wind farm.

As the groups of wind turbines in the wind farm are almost identical regarding number of turbines and mutual distance, the layout is determined on the basis of evaluating one representative group. The design of the power collecting system from these groups to the grid connection is the same with no regard to the solutions within the groups. This reduces the number of calculations to be performed.

A LV/HV-solution and a HV-solution are investigated to determine the most economical solution and to emphasize some of the parameters influencing the layout.

The costs used to evaluate the layouts are based on prices for cables inclusive of excavating, laying and backfilling, transformers, switchgear, housings, foundations and losses calculated and collected for this specific study. The capitalized losses are calculated on basis of 7% interest rate, an energy price of 0.05 ECU/kWh and 20 lifetime. Only costs for the representative group of 4 wind turbines have been investigated. Additional costs have to be included for the total wind farm costs.

To determine the transformer location, a group of 4 wind turbines is chosen. The 4 wind turbines are, via a LV-switchgear, connected to a single step-up transformer by LV-cables. On the HV-side of the transformer, a ring main unit is used for looping to and from other groups. The transformer, switchgear and ring main unit are located in a common housing at turbine no. 6.

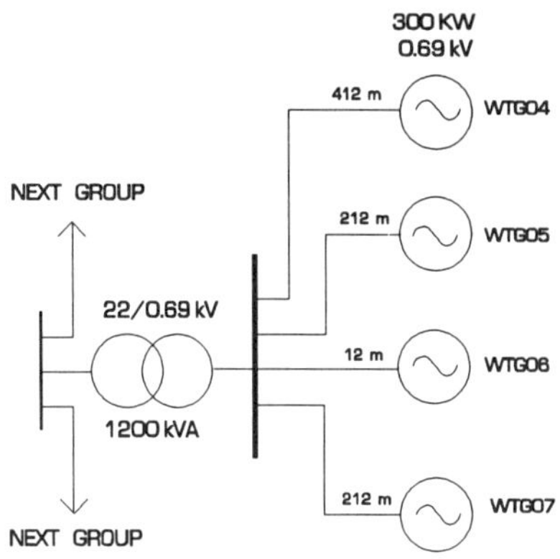

Figure 4. Representative group with LV/HV layout.

In table 1 below, a calculation of the total lifetime costs for a representative LV/HV group is shown, including costs for equipment and capitalized losses calculated by the NPV-method. All LV-cables are 3*240 mm^2 aluminium cable.

All figures in ECU	Equip-ment	NPV Losses	Total
LV-cables	25.404	16.234	41.638
HV-cables			
Transformer	12.000	6.533	18.533
HV-switchgear	6.667		6.667
LV-switchgear	6.667		6.667
Housing/found.	13.333		13.333
	64.071	22.767	86.839

Table 1. Costs for locating 1 transformer for each 4 turbines.

In the HV-solution a step-up transformer is located at each wind turbine and a HV-ring main unit is located at turbine no. 6. In this solution the LV-cables are as short as possible in order to reduce the losses.

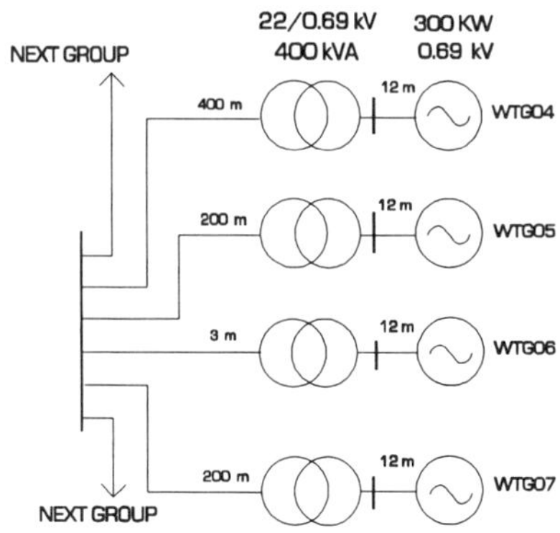

Figure 5. HV-solution with transformer at each wind turbine.

Evaluation of the total lifetime costs for a HV-solution gives:

All figures in ECU	Equip-ment	Losses	Total
LV-cables	1.333	874	2.207
HV-cables	19.806	44	19.850
Transformer	22.400	7.467	29.867
HV-switchgear	13.333		13.333
LV-switchgear			
Housing/found.	16.000		16.000
	72.872	8.385	81.257

Table 2. Costs for locating the transformers next to the turbines.

In the HV-solution the equipment costs are higher than in the LV/HV-solution and the capitalized losses are less than for the LV/HV-solution. Calculations for this representative group show that in this particular location of the wind turbines a HV-solution has the lowest lifetime costs.

If some of the parameters affecting the evaluation of the layout deviate from the values used in the above calculations, the result may be different. If for example the wind turbines are located with a mutual distance of 100 m instead of the 200 m used in the above, the total lifetime costs for a LV-solution are slightly less than for a HV-solution.

3.4 Optimizing a cable.

All cables in the power collecting system are optimized with regard to capitalized losses and cable costs. As an example, a cable is chosen in the wind farm. The cable has a length of 700 m and loops from 4 groups with a total of 14 wind turbines to the main panel in the central control building. In the library in the WIFAGO model, 5 different 22 kV high-voltage cables are accessible for the optimization: 25, 50, 95, 150 and 240 mm² aluminium, all cables with 3 sector-formed conductors and a copper screen. With the load duration curve calculated on the basis of wind duration, power curve for the wind turbines connected and availability factor etc.

The total lifetime costs for the cables are stated in table 3.

Dimension mm²	NPV Life-time losses ECU	Cable costs ECU	Total costs ECU
3*25+16	22853	13347	36200
3*50+16	12207	14933	27141
3*95+25	6094	18293	24387
3*150+25	3961	22307	31838
3*240+35	2434	29400	31834

Table 3. Lifetime costs for a 700 m cable with a load of 14 wind turbines.

As seen in table 3 the best choice of cable regarding lifetime costs, including capitalized losses is a 3*95+16 mm² cable, showing that the alternative with the lowest losses or the lowest equipment costs is not necessarily the best choice.
In figure 6 it is illustrated how the choice of cables is influenced by the interest rate and energy price, and in figure 7 the total lifetime costs, including cable costs and capitalized losses, are shown. The dotted lines indicate the choice made on basis of 7% interest rate and an energy price of 0.05 ECU/kWh.

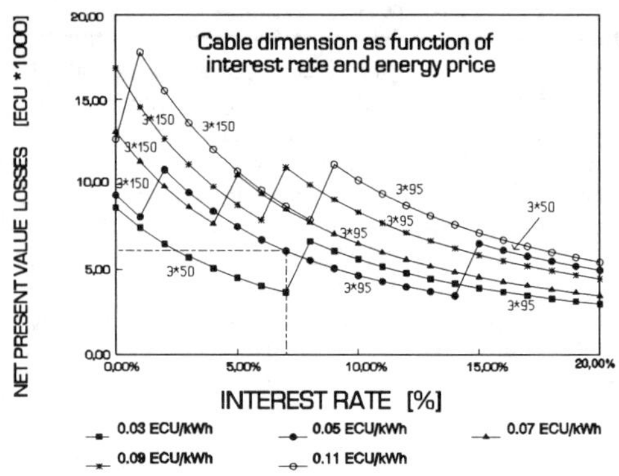

Figure 6. Choice of cable dimension as a function of interest rate and energy price (700 m cable, load of 14 wind turbines.

Conclusions.

The electrical layout of the power collecting system in wind farms has been investigated with regard to electrical losses. The losses have been evaluated to obtain the total lifetime costs, including capitalized losses and equipment costs.
Experience gained by the investigations has shown that the electrical layout for a wind farm on a specific site is highly dependable on local conditions, such as prices for selected equipment, interest rate, energy price, soil compound regarding excavation, foundation and wind distribution data.
The investigations show that the amount of losses has an important influence on the choice of solution. The alternative with the lowest losses is not necessarily the best choice when the layouts are evaluated with regard to lifetime costs. The parameters of major importance for the calculation of the capitalized losses are cable length of LV-cables, the energy price and wind distribution data.

Parameters which are in favour of a HV-solution are: large mutual distance between (large) wind turbines, high energy price and low transformer prices. In favour of a LV-solution are: short mutual distance between (small) wind turbines, low energy price and high transformer prices. A general conclusion is that it is normally an economical solution to locate transformers as close as possible to the wind turbines. Whether a low or a high voltage solution is the economical optimum is highly dependable on the parameters used in the evaluation and the size of wind turbines. Therefore, it has to be evaluated for each individual case.
No specific criteria can be given for cabling due to the influence from local conditions. The general rule should be to make the power flow (cabling) in the general direction of the grid connecting point. This should be incorporated in the routing of access roads, as the cables are usually following these.

REFERENCES

[1] Jensen, U.S. a.o.
European Wind Power Study
EEC DGXII, JUR-0041-DK

FULL SCALE MEASUREMENTS IN WIND TURBINE ARRAYS

G J Taylor National Power, Cleeve Road, Leatherhead KT22 7SE, U K
H Cleijne MT-TNO, P O Box 342, 7300 AH Apeldoorn, Netherlands
A Crespo UPM, Jose Gutierrez Abascal 2, 28006 Madrid, Spain
J Højstrup RISØ National Laboratory, DK-4000 Roskilde, Denmark
H Hutting KEMA, Utrechtseweg 310, Arnhem 6812 AR, Netherlands
J van Leuven RUCA, Middelheimlaan 1, Antwerp B2020, Belgium

ABSTRACT: This paper presents the results of a series of full scale measurements on wind turbine arrays located in a variety of environments and configurations. The measurements provide information on the perturbed velocity field within the arrays resulting from turbine wake interactions, the associated power losses and also the enhanced dynamic structural loads experienced by individual turbines within the wind farms.

1. INTRODUCTION:

It has long been recognized that turbines within wind farms will experience reduced wind speeds and increased turbulence levels resulting from the wakes of upstream machines. This will result not only in a diminished energy capture but also in the turbines within the array experiencing increased dynamic loads (and thus a potentially reduced fatigue life) compared with isolated turbines.

Whilst the structure of turbine wakes and the associated power losses have been the subject of theoretical and wind tunnel studies, few detailed full scale data are available to corroborate the resulting predictions. In addition, very little information is available on turbulence levels within wind farms or on the resulting dynamic loads and turbine design codes are thus often based on atmospheric turbulence levels.

This paper presents the results of a series of full scale measurements designed to address these issues. Data are available from four wind farms located in differing topography:

- Sexbierum (Netherlands) - flat, uniform site.
- Nørrekær Enge (Denmark) - flat, uniform coastal site.
- Zeebrugge (Belgium) - linear array with obstacle effects.
- Monteahumada (Spain) - complex, (near mountainous) terrain.

Long term statistics of wind speeds in and around the array and of turbine power output have been collected at all sites. In addition, detailed time series measurements of turbulence and of the structural loads experienced by selected turbines have been made at two sites. Measurements have been made over a period of more than two years.

The data are also being used within other CEC funded projects to develop improved array efficiency models and to investigate further dynamic loading within wind farms.

The results presented here provide only an overview of the wind speed and power output measurements. Analyses of dynamic load and fatigue data [1] and the turbulence structure within a wind farm [2] are available elsewhere.

2. RESULTS

2.1 Nørrekær Enge:

This wind farm comprises 42 x 300kW turbines located on a flat coastal site in N. Jutland (Fig. 1). Wind measurements are made at two masts (M1 & M2) and detailed structual load measurements are available from two nearby turbines (A1 & F6).

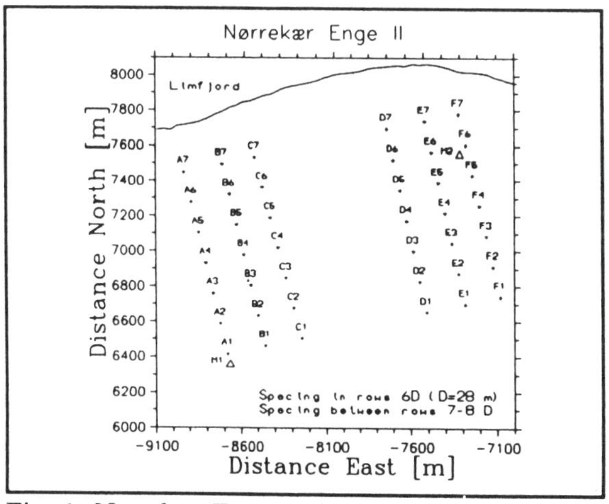

Fig. 1: Nørrekær Enge - schematic layout

Figs. 2 & 3 show the power output of individual machines in row A relative to the mean for the row. For the wind directions depicted, these machines should all be operating in an undisturbed flow free of any wake effects. Even though the site was expected to be free of terrain effects, significant power output perturbations are apparent which are similar in magnitude to those expected from wake interactions. (A comparison of these data with those from other rows confirms that the observed effects do not arise from variations in performance characteristics between machines).

For incident winds within the sector 255-285° (Fig. 2), the upwind fetch lies predominantly over water (Limfjord), with the lower numbered turbines being furthest from the coast. The observed large (17%), monotonic power variations seen in Fig. 2 appear to result from the lower numbered turbines being located in a more developed and deeper internal boundary layer, corresponding to an increased surface roughness over land. This results in a reduced mean wind speed over the rotor disc and lower power production. Similar effects are apparent in Fig. 3, but in this case the influence of a small group of buildings 400-800m upstream of the array also significantly reduces the power output of the central machines.

An attempt has been made to account for the effects of the coast and nearby buildings using the RISØ WasP and PARK models (Fig. 2 & 3), yielding a reasonable description of the observed behaviour.

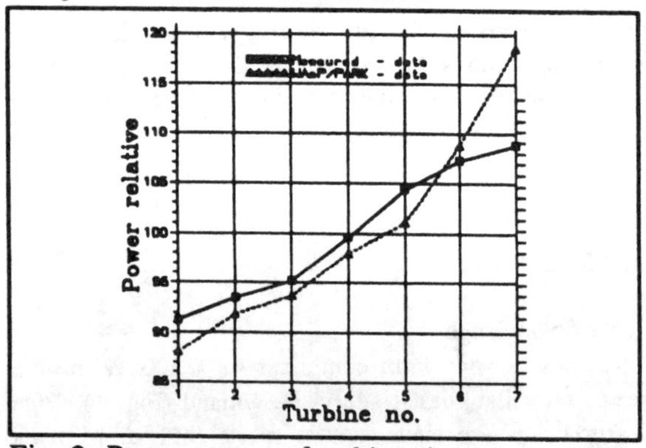

Fig. 2: Power output of turbines in row A - (255 - 285o)

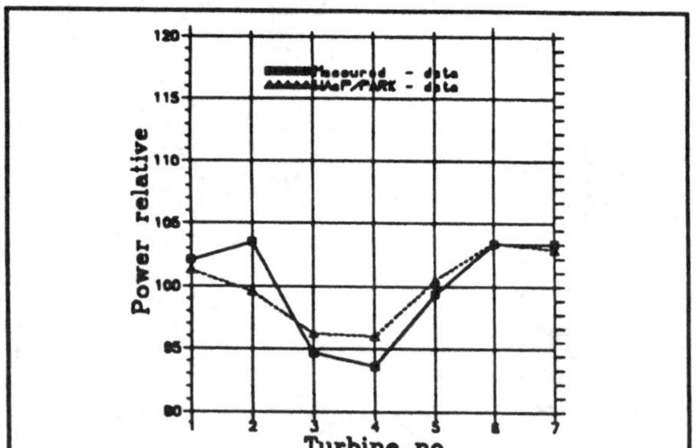

Fig. 3: Power output of turbines in row A - (225-255o)

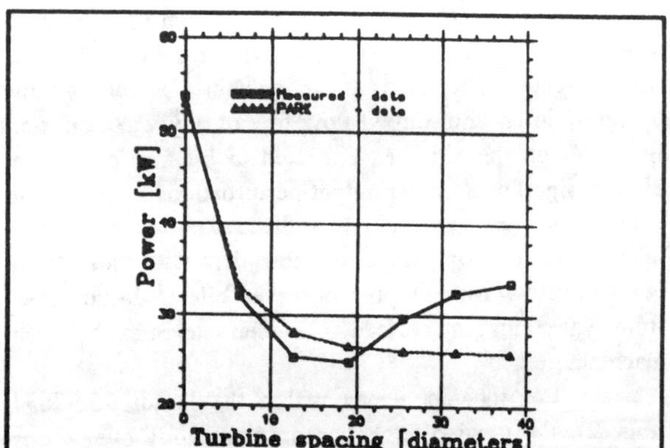

Fig. 4: Power output of turbines in row A (166o)

Fig. 4 shows the power output of members of a line of turbines separated by 6D, with the wind directed along the inter machine axis. Significant power losses are apparent for the second and third machines. Further downstream, however, the machine output recovers again. This phenomenon probably results from the more developed shear layers on the periphery of multiple wake regions (compared with a single wake) yielding an increased turbulent momentum flux and more rapid wake recovery. Similar behaviour, dependent on machine separation, has been suggested by wind tunnel studies [3]. It should be noted, however, that this is not consistent with the predictions of simple array efficiency models which result in a monotonic decay of turbine output downstream as shown by the behaviour calculated by the PARK model.

2.2 Sexbierum:

The wind farm consists of 18x30m turbines located on an irregular 6x3 rectangular grid in flat uniform terrain. Seven fixed meteorological masts are located around the array and further detailed wind measurements are available from 3 transportable masts equipped with 3-d anemometers. A single turbine is also instrumented for structural measurements.

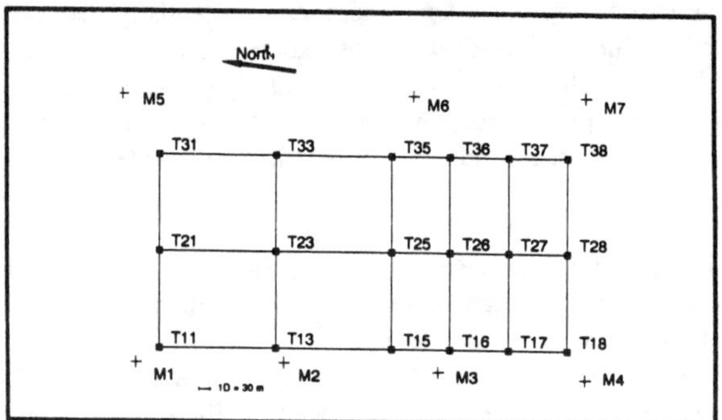

Fig. 5: Sexbierum - schematic layout

Details of the measured turbulence structure within a wake at various distances downstream of a single turbine are shown in Figs. 6, 7 & 8.

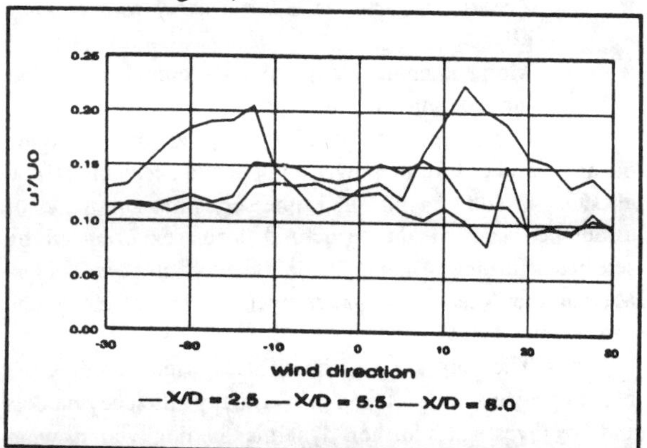

Fig. 6: Turbulence intensity (u'/U_0) in single wake at 23m height for various downstream distances

Close behind the turbine, the turbulence intensity measured across the wake at hub height (Fig. 6) shows the expected double peak structure associated with the well developed shear layers on the wake periphery. Further downstream these peaks slowly decay and spread as the wake disperses and the velocity gradients relax.

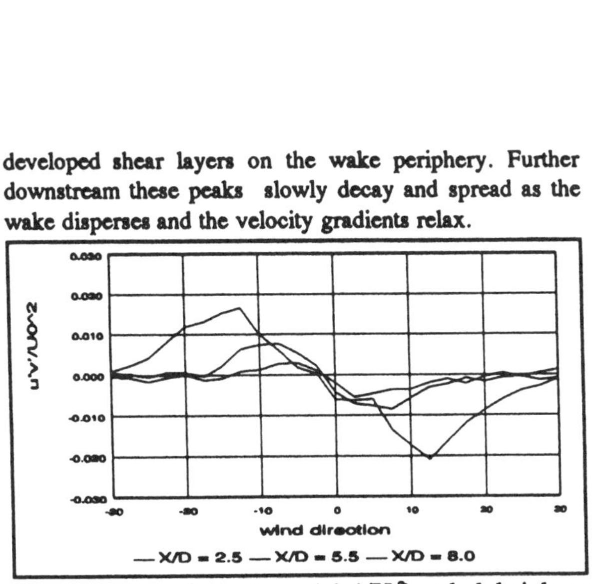

Fig. 7: Lateral shear stress ($u'v'/U_0^2$) at hub height at various downstream stations

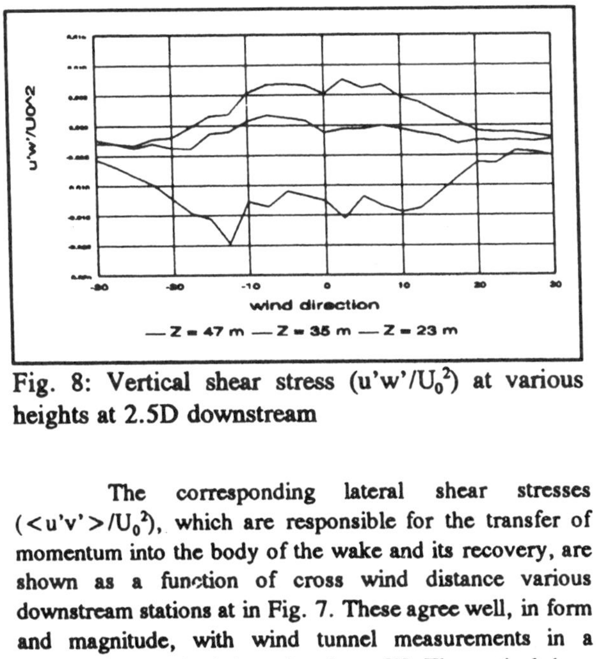

Fig. 8: Vertical shear stress ($u'w'/U_0^2$) at various heights at 2.5D downstream

The corresponding lateral shear stresses ($<u'v'>/U_0^2$), which are responsible for the transfer of momentum into the body of the wake and its recovery, are shown as a function of cross wind distance various downstream stations at in Fig. 7. These agree well, in form and magnitude, with wind tunnel measurements in a simulated atmospheric boundary layer [3]. The vertical shear stresses ($<u'w'>/U_0^2$) measured at various heights at 2.5D downstream are shown in Fig. 8. Once again these are in good agreement with wind tunnel measurements.

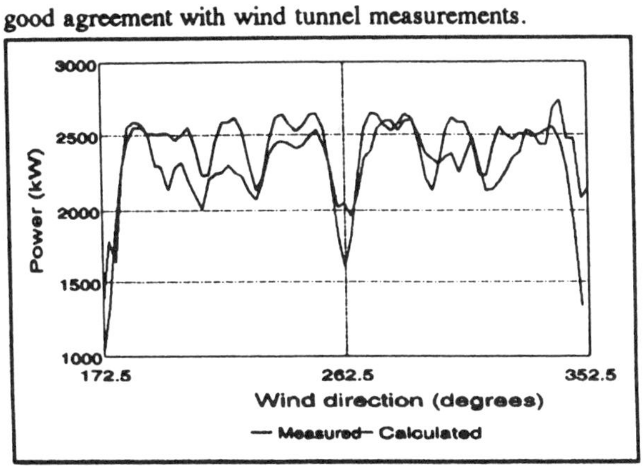

Fig. 9: Calculated and measured wind farm output ($10 < u < 11$ ms^{-1})

The measured array performance as a function of wind direction is shown in Fig. 8, where the results are compared with the predictions of the TNO FARMS model.

This model utilises a simple kinematic model of the turbine wake and uses the linear addition of individual wake deficits to estimate the mean velocity field in multiple wake regions. Although the model reproduces the general features of the observed behaviour quite well, it is clear that it significantly overestimates the velocity and power deficits in cases where the turbines are aligned. Indeed similar behaviour is found here to that noted in Fig. 4 above, with the power losses tending to saturate quite quickly on passing down a row of turbines.

2.3 Monteahumada:

The main part of the wind farm consists of 7x150kW horizontal axis turbines located in a valley in very irregular terrain with mountains nearby and allows the interaction of wakes and flow over topographical features to be examined.

A number of studies based either on wind tunnel tests [4] or numerical calculations [5] suggest that local topography may significantly influence wake development. van Oort et al. [5], for instance, have investigated the corrections to the simple superposition of wake and terrain perturbations using PHOENICS and find that, near the ground, terrain features may lead to additional turbulent diffusion, thus reducing wake deficits. Over the peak of a hill, on the other hand, the converging streamlines may act to concentrate and increase the perturbations.

This array provides an opportunity to study the combined effect of topography on wake and flow development since the terrain induced perturbations are expected to be of a similar magnitude to typical wake deficits. The velocity perturbations due to topography over the area occupied by the machines have been analysed using WasP and are found to be, typically, about 10%.

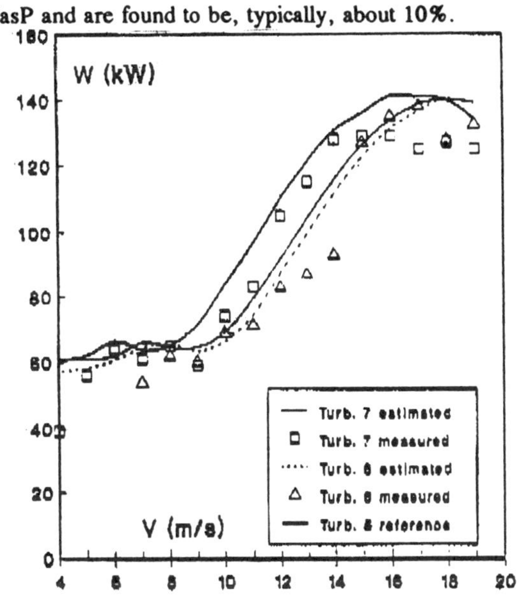

Fig. 10: Power curves for turbines 6 & 7 (E. winds)

Fig. 10 shows the power output curves for two turbines (7 & 8) when operating in Easterly winds, each in the wake of a single upstream turbine. The mean incident wind speed has been deduced from the power output of the upstream machine using a measured performance curve corrected for topographic effects (turbine 5). The mean velocity at the downstream machine has then been determined by simply adding the calculated wake deficit and the additional terrain induced perturbations between the two turbines. Reasonable agreement is found between the measured and calculated behaviour, suggesting that it may be possible to combine the effects of wakes and terrain in a straightforward way. Further details are supplied in Ref. 6.

2.4 Zeebrugge:

The Zeebrugge wind farm contains 21x200kW

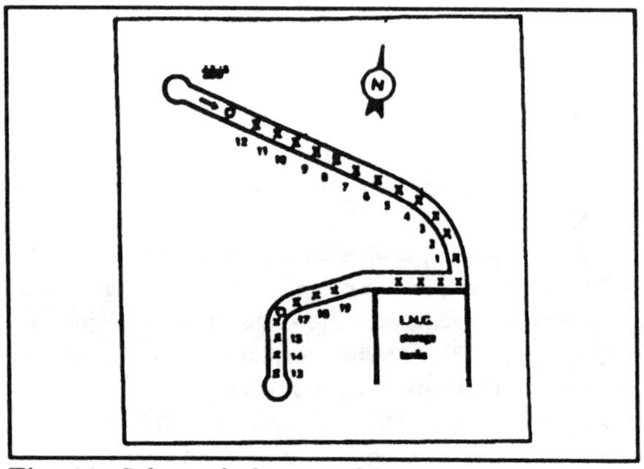

Fig. 11: Schematic layout of Zeebrugge wind farm.

turbines and two smaller machines located along the harbour breakwater.

A variety of perturbations to the local flow may be observed here including those arising from the harbour wall itself.

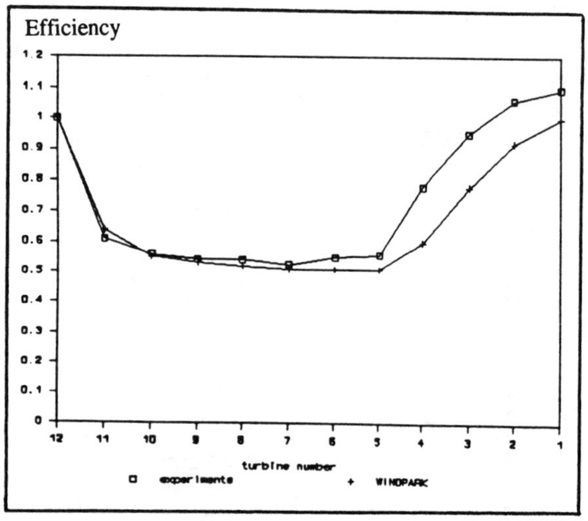

Fig. 12: Turbine efficiency - wind along breakwater

The output of the individual turbines along the outer harbour wall corresponding to the wind direction lying directly along the inter-machine axis is presented in Fig. 11. Once again, it is immediately apparent that the wake losses saturate rather quickly. Also shown are the predictions of an array efficiency model, WINDPARK. In this case it is assumed that the velocity deficits resulting from multiple wake interactions are dominated by the deficit of the immediate upstream machine. This provides a quite reasonable description of the observed behaviour.

3. CONCLUSIONS:

A significant body of data describing the wind field and turbulence structure within wind farms and the associated power output has been accumulated. The limited overview presented here suggests that:

(i) Even in apparently flat uniform terrain, flow disturbances arising from small clusters of buildings or local surface roughness changes may produce significant perturbations to the performance of turbines in the array. Such effects need to be incorporated in future models.

(ii) The turbine power losses encountered within an array are more complex than suggested by current simple models. In the cases studied, turbine wake losses appear to saturate 2 or 3 machines into the array and, for smaller separations, may actually decrease on passing down a row of turbines.

(iii) The turbulence structure found within wakes is consistent with that determined from wind tunnel studies.

(iv) It appears that a reasonable prediction of the combined effects of terrain and wake losses may be obtained from a simple linear superposition of the corresponding wind speed perturbations.

4. ACKNOWLEDGEMENTS:

This work has been part funded by the CEC under Contract JOUR 0064.

5. REFERENCES:

1. Christensen C.J., Courtney M.S., Højstrup J., 1992, "Turbulence and turbine dynamics in a wind farm", International Conference on Wind Farms, Herning, Denmark.

2. Højstrup J., Courtney M.S., 1993, "Turbulence in windfarms", this conference.

3. Smith D., 1991, "Further analysis of turbine wake development and interaction data", 13th BWEA Conference, Swansea, UK.

4. Taylor G.J., Smith D., 1991,"Wake measurements over complex terrain", 13th BWEA Conference, Swansea, U.K.

5. van Oort H., van Gemart P.H., Crespo A., 1989,"Wind farms in complex terrain", TNO Report 89-233 on CEC Contract EN3W/0030/NL.

6. Crespo A., Manuel F., Grau J.C., Hernandez J., 1993, "Modelization of wind farms in complex terrain - Application to the Monteahumada wind farm", this conference.

EARLY EXPERIENCE IN UK WIND FARMS

D Lindley°, P Musgrove°, J Armstrong+, M Hitner+

° National Wind Power Ltd., Riverside House, Meadowbank, Furlong Road, Bourne End, Buckinghamshire, UK

+ Wind Energy Group Ltd., Greenford House, 309 Ruislip Road East, Greenford, Middlesex, UK

ABSTRACT: The introduction of a Non Fossil Fuel Obligation (NFFO) when the UK Electricity Supply Industry was privatised in 1990 has led to the construction of over 15 wind farms in the past 18 months. The first was completed in December 1991 and by December 1993 the installed wind farm capacity will exceed 120 MW. National Wind Power (NWP) has been responsible for the development, financing, construction and operation of four wind farms with a total capacity of 25 MW. One of these Cemmaes (7.2 MW) is described in some detail. It was the subject of the first Public Inquiry into a wind farm planning application, and the successful outcome of this has paved the way for many subsequent wind farms. Cemmaes wind farm has also been the first to be constructed on a difficult upland site that is representative of many British wind farm sites. The turnkey construction of the Cemmaes wind farm was undertaken for NWP by Wind Energy Group (WEG), using 24 WEG MS3 turbines, each 35m diameter and 300 kW rated. Construction and early operational experiences are described. The wind farm commenced generation over a month ahead of schedule, and has been operating at high availability since the beginning of 1993. Details of the arrangements for the next round of the NFFO are now awaited by everyone who is involved in wind energy in the UK: these details will - it is expected - be made public in mid-1993 with power purchase contracts being signed in 1994. This third NFFO round will encourage wind farming in Scotland and Northern Ireland as well as providing further support for wind farming in England and Wales.

1. INTRODUCTION

The UK Wind Energy Resource has for some years been recognised as the best in Europe, but the recent report of the Renewable Energy Advisory Group (REAG) gives a revised assessment of the UK wind resource which is even larger than previous estimates. REAG was set up in the UK in late 1991, chaired initially by the Minister with responsibility for Renewable Energy, Mr Moynihan, to review the potential and prospects within the UK of the various renewable energy options. The REAG report, published [1] in December 1992, concluded that the 'accessible' potential for wind energy from sites on land in the UK was over 300 TWh/yr; the wind energy potential slightly exceeds the present UK total electricity consumption, and is far larger than the assessed accessible potential of any other renewable energy source in the UK.

The REAG report also estimates what contribution could be made by the various renewable energy technologies in the UK by the year 2005, after allowing for constraints such as integration into the electricity network, the success rate in achieving planning consents and the economics of renewable energy technologies relative to the more conventional generation options; it concludes that wind farms in the UK could provide about 30 TWh/yr by the year 2005 at a cost of about 4 pence/kWh (about 0./05 ECU/kWh). The report proceeds to recommend that the UK government should underpin support for renewable energy projects in the UK to a level that would correspond to the installation of up to 2000 MW of wind farm capacity by the year 2000; however the government's response to the recommendations made in the REAG report will not be known until the third quarter of 1993.

Though the UK wind energy resource is large, the UK has also benefitted from the large reserves of oil and gas that lie beneath the North Sea. Through the 1980s and for a few years more the UK has been a net energy exporter; there has therefore been less political motivation in the UK than elsewhere in Europe to stimulate the application of renewable energy technologies. Growing concern over the environmental consequences of conventional methods of generating electricity, e.g. acid rain and its effects and the climate changes that are likely to result from current levels of carbon dioxide emissions, have however ensured a continuing UK renewable energy research and development programme and sustained public interest in the potential of renewable energy sources. And, fortuitously, the privatisation of the UK electricity supply industry in April 1990 provided a mechanism whereby the applications of renewable energy technologies could be encouraged; this mechanism is the Non Fossil Fuel Obligation, NFFO.

2. THE NON FOSSIL FUEL OBLIGATION

The privatisation of the electricity supply industry in the UK in 1990 resulted, in England and Wales, in the conventional generation capacity of the former CEGB passing into the ownership of two public companies, National Power plc and Powergen plc; nuclear power stations are now owned by Nuclear Electric plc, which remains a state-owned company. Electricity distribution at the local level is the responsibility of twelve Regional Electricity Companies (RECs), who collectively own the high voltage transmission network. The RECs purchase most of the power they need from National Power, Powergen and Nuclear Electric, though there is a growing number of independent power producers who have financed the construction of new, gas-burning power stations. Many of the RECs have investments in these new gas-fired power stations.

The privatisation legislation empowered the Government to specify a series of Non Fossil Fuel Obligations, i.e. obligations on the RECs to purchase at a premium price the output from specified power stations that do not burn fossil fuel. The original purpose of the Non Fossil Fuel Obligation (NFFO) was to ensure that the RECs continued to buy - at a premium price - the output from nuclear power stations, with the extra cost being recovered by a levy on all consumers which adds approximately 10% to their electricity costs. However the Government decided in 1989 that it would also use this mechanism, the NFFO, to support a number of renewable energy projects. It has done this, so far, by issuing two renewable energy NFFO orders, one in September 1990 and the second in November 1991. Details of the arrangements for the third NFFO for renewable energy projects will, it is anticipated, be announced in the third quarter of 1993 though the actual, third, NFFO order won't be issued until 1994. In 1990 and 1991 the NFFO arrangements were only applied to England and Wales. The third NFFO, now awaited, will also be applied in Scotland and Northern Ireland so as to encourage renewable energy projects in these countries as well as in England and Wales.

In 1990 and in 1991 proposers of renewable energy projects had to provide technical and financial details of the schemes they proposed, and bid competitively to secure power purchase contracts. For wind farms this meant that would be developers had to supply details of the sites selected, evidence of the site wind speed, details of the number of turbines and turbine type proposed, details of the grid connection, evidence of progress with the planning application, and evidence that the wind farm could be financed. Developers had finally to bid the premium price at which they would be willing to supply electricity, given that this premium price would only be paid until December 1998. The 1990 NFFO order led to power purchase contracts being signed in September 1990 for nine wind projects with a nameplate capacity of 28.4 MW. The 1991 NFFO order led to power purchase contracts being signed for a further forty-nine wind projects with a nameplate capacity of 192 MW.

3. THE 1990 NFFO WIND FARMS

Though the 1990 NFFO resulted in power purchase contracts for nine wind projects, three of these were for individual wind turbines and one was for a group of four different experimental machines at Carmarthen Bay in South Wales. The remaining five projects were wind farms, whose current status (February 1993) is as specified below.

(1) Delabole, Cornwall: 10 Vestas 400 kW turbines, each 35m diameter. Developed by Windelectric Ltd. Operational since December 1991. Capacity 4.0 MW

(2) Cemmaes, mid-Wales: 24 WEG 300 kW turbines, each 33m diameter. Developed by National Wind Power Ltd. Operational since November 1992. Capacity 7.2 MW

(3) Chelker, Yorkshire: 4 WEG 300 kW turbines each 33m diameter. Developed by Yorkshire Water. Operational since December 1992. Capacity 1.2 MW

(4) Ovenden, Yorkshire: 23 Vestas 400 kW turbines, each 35m diameter. Developed by Yorkshire Wind Power (jointly owned by Yorkshire Water and Yorkshire Electricity, the local REC). Under construction with completion due in Summer 1993. Capacity 9.2 MW

(5) Kirkby Moor, Cumbria: 12 Vestas 400 kW turbines, each 35m diameter. Developed by National Wind Power. Under construction with completion due in Summer 1993. Capacity 4.8 MW

Only the first of these, Delabole, had been granted planning consent before September 1990, when the power purchase contracts were signed. This planning consent was subject to detailed agreement over noise levels which took some months to finalise; financing arrangements also required prolonged negotiation, but the decision by National Power to make a substantial investment in the project helped to resolve this issue. Construction commenced in June 1991 on the relatively flat wind farm site, and the facility was operational by the end of the year - Britain's first wind farm.

The other four wind farms had all to secure planning consent subsequent to the September 1990 signing of their power purchase agreements. Though in each case the local planning authority (the District Council) was willing to give planning consent all four projects were delayed whilst wider planning issues relating to the environmental impact of wind farms were reviewed. The financial stimuli given in the UK to encourage wind farming is set by Government at the lowest level that will achieve the desired result. This means that very windy sites must be selected; these are invariably exposed, and usually in rural areas of considerable landscape value. Cemmaes, Ovenden and Kirkby Moor wind farms were therefore all required to go to Public Inquiries; Chelker, with just four turbines, was not required to go to Public Inquiry but was nonetheless delayed for many months whilst the Department of the Environment was reviewing whether or not it should be subject to a Public Inquiry.

Cemmaes was the first proposed wind farm to go to a Public Inquiry in the UK, in April 1991, and the Inquiry issues were discussed in the paper presented by Musgrove and Lindley [2] at the 1991 Amsterdam Wind Energy Conference. The main concern voiced by objectors was that the wind turbines would be unacceptably visually intrusive in an area of considerable landscape value; moreover the wind farm would be visible from the nearby Snowdonia National Park, which some objectors considered to be unacceptable. The very thorough Inspector's Report [3] reviewed all the evidence presented during the eight days of the Inquiry, and concluded very clearly in favour of the wind farm. This established a very important precedent, and allowed National Wind Power to proceed and place orders for construction in January 1992. This was a full twelve months later than had been expected when the power purchase contract was signed in September 1990, so the Inquiry had cost the project over £2 million in lost revenue as well as the direct cost to the developers of over £300,000 in legal fees and other charges. Notwithstanding this delay National Wind Power succeeded in raising the finance for this windfarm with a project loan provided by Barclays Bank and equity provided by National Wind Power's parent companies, National Power and Taylor Woodrow. Cemmaes wind farm has been a very important pioneering project in the UK. As indicated above it was the first wind farm to go to Public Inquiry, and the successful outcome of this has helped subsequent wind farms secure their planning consents. It was also the first wind farm to be sited on an elevated location that is quite different from typical European and American wind farm sites, but is representative of many sites in the UK. Details of its construction and some early operating experiences are given in section 4.

Returning to the other 1990 NFFO wind farm projects, Kirkby Moor went to Public Inquiry in June 1991, before the Cemmaes Inquiry decision was announced in September 1991. In addition to the issue of visual intrusion, Kirkby Moor wind farm faced opposition because it had been designated a Site of Special Scientific Interest (SSSI) - after the wind farm planning application had been submitted. The moor contains a large expanse of upland heather, which in many parts of Britain has suffered from overgrazing due to the financial pressures on hill farmers (financial pressures which wind farming may help alleviate in the future, to the benefit of the local ecology). The Inquiry Inspector accepted that damage to the heather would be minimal, but recommended rejection on visual grounds; the Secretary of State for the Environment decided, however, that the clean energy contributed by the wind farm outweighed the visual concerns, and he gave the project his consent in March 1992. Detailed planning conditions took some months to finalise, and National Wind Power agreed a very detailed Construction Method Statement with English Nature, so as to ensure that any damage to the habitat is minimised. Construction started at Kirkby Moor in January 1993 and completion is due in August 1993.

Ovenden Wind Farm went to Public Inquiry in December 1991, and so was able to benefit from the positive outcome of the Cemmaes Inquiry. It did however also involve a planning issue that had not previously arisen, as Ovenden Moor is in a Green Belt, i.e. a rural area, close to a large conurbation, which is subject to greater than normal planning constraints. The Inquiry decision was given in mid-1992 and was again in favour of the wind farm proceeding; construction commenced in Summer 1992, with completion scheduled for Summer 1993.

All five wind farms that secured power purchase contracts in September 1990 are therefore either already in operation or are under construction, a remarkable 100% planning success rate despite the fact that three projects had to go to Public Inquiry.

4. CEMMAES WIND FARM

As has been indicated above Cemmaes wind farm has been an important pioneering project in the development of wind farming in the UK. Not only was it the first wind farm to go to Public Inquiry, where all the environmental issues for and against were reviewed in great detail, and where planning consent was granted even though the site is in attractive surroundings and close to a National Park: it is also an upland hill top site that is very different from what is typical in Europe and the USA, but is very representative of many UK sites.

Cemmaes wind farm is located approximately 12 km east and 5 km north of Machynlleth, in mid-Wales, on the Mynydd y Cemais ridge which runs for approximately 5 km parallel and to the east of the River Dyfi; to the north and west, on the other side of the Dyfi valley, is Snowdonia National Park. The prevailing westerly winds are funnelled along the valley and encounter the 400m high ridge, accelerating up its sides which slope at up to 45°. It is one of the most rugged sites yet developed in Europe, and presents particular difficulties both to the wind turbine designer and to the windfarm constructor.

Figure 1: Cemmaes site-access road.

Wind Energy Group Ltd. (WEG) was awarded the contract for the turnkey construction of this windfarm, including the supply of twenty-four 300 kW MS3 wind turbines, by National Wind Power in January 1992. The MS3 is a 33m diameter, two-bladed and teetered upwind machine which was prototyped in 1988 and has since seen operation in a range of locations from Altamont Pass in the USA to Sicily. The wood epoxy composite blades are manufactured at WEG's factory in Southampton and the nacelles are assembled in Cardiff.

Construction work at Cemmaes started in Spring 1992. The first tasks were the widening and strengthening of the narrow, steep, minor public road which provides access to the bottom of the ridge, and the construction of the site track up to the ridge itself. The public road had to be altered to take the 12m long tower sections and 16m long blades, mounted on their low loaders, and the 3.1m wide erection crane. The steepness of the road, with a stream alongside, and the need to maintain key environmental features added to the difficulties of the job. The on-site track had to be built into a hillside of up to 45° slope. This resulted in three hairpin bends (see Fig. 1) and a gradient in places of 1 in 4. The local stone used as hardcore had a tendency to break down into a slurry, and as a result most of the low-loader trucks carrying towers or blades had to be towed up the hill by tractor.

On top of the ridge the main problem was boggy ground conditions in some stretches. In one case a turbine foundation had to be installed where the peat depth was 3m, and this required an additional 170 cubic metres of concrete. It was, incidentally, impossible to drive concrete lorries up the steep site track, and premixed concrete had to be ferried up the hill in purpose-built skips. In places where the peat gave way to rock, cable trenches had to be blasted into the rock to allow burial to the specified depth.

By Summer 1992 the foundations and civil works were in place, and tower deliveries started in July. Nacelles were delivered from Cardiff and installed between August and October. An earlier decision to use a mobile crane was reversed when it was seen that a crawler crane, cheap enough to keep on site throughout the erection phase, would give more flexibility in planning the lifts. Completion of the grid connection was delayed until October, and this delayed the commissioning of the wind turbines. One of the unexpected problems this caused was that lack of power for the electrical cabinet heaters allowed condensation to occur in the damp, exposed conditions where the tower bases were often in low cloud. In spite of this, first generation took place in late October and all 24 turbines were operational in time for the formal opening of the windfarm by the Rt Hon David Hunt, Secretary of State for Wales, on 25 November 1992.

During early December, the turbines were being tuned up for operational duty under the control of a SCADA system located in the nearby substation. In late December and January, the site was hit by a series of severe storms, with all turbines recording winds in excess of 40 m/s and more than half in excess of 45 m/s. Some turbines also saw highly skewed winds, possibly due to topographical effects.

WEG's experience is that the teetered 2-bladed wind turbine is better able to cope with turbulent conditions than a fixed hub machine. However, shutting down the rotor in storm conditions can increase teeter movements and this needs careful design and control. The MS3 has a highly effective teeter damper system, but in the conditions experienced at Cemmaes it was found necessary to adjust the shutdown procedure so as to reduce the time spent by the rotor at certain rotational speeds during shutdown. These modifications took until late January to complete. In spite of this, the windfarm has operated since the beginning of the year with high availability, and continues to do so. The completed wind farm is shown in Fig. 2.

The performance of the Cemmaes wind farm will be monitored in detail by National Wind Power as part of a programme jointly funded between NWP and the Energy Division of the Department of Trade and Industry. This co-funded multi-million pound programme will look at many aspects of wind farm operation, technical and environmental, and will be centred on the three wind farms (Cemmaes, Llangwyryfon and Cold Northcott) that were given grant assistance by the former UK Department of Energy. Llangwyryfon and Cold Northcott wind farms secured power purchase contracts in the 1991 NFFO round, see section 5, and are now nearing completion.

5. THE 1991 NFFO WIND FARMS

As stated in section 2, the 1991 NFFO resulted in power purchase contracts being signed in November 1991 for forty-nine wind projects, with a total installed capacity of 192 MW. As with the 1990 NFFO the power purchase contracts only guarantee a premium price until December 1998; thereafter the price reverts to the 'Pool' price. The Pool price is currently low, i.e. less than 3p/kWh (0.038 ECU/kWh) and its magnitude after 1998 is uncertain. Wind farms must therefore be financed such that their capital cost is fully recovered, and some modest return provided to shareholders, before the end of 1998. For the 1990 NFFO wind farms this requirement led to a premium price of about 9p/kWh (about 0.11 ECU/kWh) being paid for the period to 1998. For the 1991 NFFO the premium price is available for one year less (since the 1998 termination date is unchanged) and this led to the premium price being set at 11p/kWh (about 0.14 ECU/kWh). Those of the wind farm projects completed or under construction by early 1993 include:

(1) Carland Cross, Cornwall: 15 Vestas 400 kW
Operational September 1992 Capacity 6.0 MW
(2) Rhyd-y-Groes, Anglesey: 24 Bonus 300 kW
Operational November 1992 Capacity 7.2 MW
(3) Blyth Harbour, Northumberland: 9 HMZ 300 kW
Operational December 1992 Capacity 2.7 MW
(4) Penrhyddlan, mid Wales: 43 Mitsubishi 300 kW
Operational December 1992 Capacity 12.9 MW
(5) Coal Clough, Lancashire 24 Vestas 400 kW
Operational January 1993 Capacity 9.6 MW
(6) Great Orton, Cumbria: 10 Carter 300 kW
Operational February 1993 Capacity 3.0 MW
(7) Goonhilly, Cornwall: 14 Vestas 400 kW
Completion due Spring 1993 Capacity 5.6 MW
(8) Llidiartywaun, mid Wales 60 Mitsubishi 300 kW
Completion due Spring 1993 Capacity 18.0 MW
(9) Cold Northcott, Cornwall: 21 WEG 300kW + 1 WEG 400kW. Completion due Spring 1993
Capacity 6.7 MW
(10) Llangwyryfon, mid Wales 20 WEG 300kW
Completion due Spring 1993 Capacity 6.0 MW

Several other projects are close to the commencement of construction and by the end of 1993 at least 120 MW of wind farm capacity will be operational in the UK from approximately 20 wind farms. Given that Britain's first wind farm only commenced operation in December 1991 this represents very substantial progress.

Most UK wind farms have only been operational for a few months, so it is too early to draw any definitive conclusions as to how the public perceives them. Preliminary indications are that most people are relaxed about the visual appearance of wind farms in scenic rural areas; many comment favourably and some - inevitably - comment unfavourably. The latter would seem to be a small minority.

There has been greater concern about noise. Though general practice has been to site turbines at least 350 metres from any dwelling, and planning authorities specify that noise levels should not exceed 40-45 dBA, there have been noise complaints from people living in the vicinity of some windfarms (involving turbines made by at least three leading manufacturers). Generally the problem has been caused by tonal noise emissions and the manufacturers concerned are making strenuous efforts to minimise these.

6. CONCLUSIONS

Britain's first 4 MW wind farm commenced operation in December 1991. By the end of 1992 the operational wind farm capacity totalled 52 MW and by the end of 1993 the installed capacity will exceed 120 MW. It is encouraging to see that at last Britain is starting to harness its quite exceptional wind resource, and the public reaction to the wind farms already in operation has generally been very positive. Cemmaes wind farm, illustrated in the paper, is the first of many wind farms to be built on the difficult upland sites that will be typical of many British installations.

Applications in Britain have been encouraged by the Non Fossil Fuel Obligation (NFFO), which pays a premium price for a limited period for electricity generated from power stations that do not burn fossil fuels. NFFO power purchase contracts signed in 1990 and 1991 have led to the construction of the 120 MW of wind farm capacity referred to above, but no further NFFO contracts will be signed until 1994.

So far all the wind farm construction activity has been in England and Wales, mostly at elevated locations where average wind speeds exceed 7.5 m/s. The 1994 NFFO will cover wind farms in Scotland and Northern Ireland, as well as in England and Wales. Details of the arrangements for the 1994 NFFO round will be announced in mid-1993, and are eagerly awaited.

Figure 2: Cemmaes wind farm.

REFERENCES

[1] Renewable Energy Advisory Group: Report to the President of the Board of Trade. November 1992. Energy Paper no. 60. HMSO, London.

[2] P Musgrove, D Lindley. Wind Farm Developments in the UK. Amsterdam Wind Energy Conference EWEC '91. Proceedings vl pp 914-918. Elsevier.

[3] Report by Mr D Sheers, Inquiry Inspector, on an application by Wind Energy Group for 24 wind turbine generators at Mynydd y Cemmaes. Welsh Office. June 1991.

INITIAL EXPERIENCE WITH THE NORDIC WINDPOWER 400 KW PROTOTYPE

Bruno Bergkvist and Staffan Engström
Nordic Windpower AB, Box 968, S-191 29 Sollentuna, Sweden
Tel +46-8-754 96 20 — Fax +46-8-754 99 47

Göran Dalén
Vattenfall AB, S-162 87 Vällingby, Sweden
Tel +46-8-739 74 43 — Fax +46-8-739 58 82

Abstract

In August 1992 the prototype of the Nordic Windpower 400 kW wind turbine was erected at Vattenfall's Windpower Station close to Lysekil in the Wouth West of Sweden. A conventional three-bladed Danish wind turbine (Bonus 450 kW) has been installed at the same site, allowing comparisons regarding performance etc. The purpose of the project is to develop and evaluate new wind power technology. An advanced data acquisition system and a 66 meter met tower will be utilized for the evaluation program, which will be carried on until the end of 1994.

The Nordic 400 has a 35 meter diameter two-bladed, fixed pitch rotor that combines stall-control with a variable speed electrical system and a teetering hub. The soft steel-shell tower gives a hub height of 40 meters. Emergency braking is performed by the rotatable blade tips, backed up by a mechanical brake on the secondary shaft.

A main feature of the soft dynamics is that practically all internally generated dynamic forces have been eliminated. The teetering hub and variable rotor speed reduce the aerodynamically induced fatigue. These features and the simple design have reduced the system weight with about 40 percent compared with a conventional three-bladed wind turbine, which gives the design a potential for high competitiveness.

The commissioning and early testing of the prototype have so far confirmed that the design fulfils the requirements. In general the concept works as predicted. No teeter instabilities have been detected.

1. Introduction

For many years wind power activities in Sweden were almost entirely directed towards the development of large wind turbines. This work continues, with the 3 MW Näsudden II unit currently under commissioning.

A new development line was initiated some years ago with the aim to find a design with low weight, an uncomplicated design and a cost effective power control. In August 1990 Vattenfall decided to erect two wind turbines in the 400 - 500 kW range. One prototype, using the concept described above, and one conventional wind turbine.

There are two main ideas behind the decision to initiate these projects. First, wind turbines in this size range are currently economically competitive, i e there is a market today. Furthermore, the solutions used in the prototype should be possible to implement in larger wind turbines. This means that we believe in the long-term competitiveness of large wind turbines.

The development of the Nordic 400 is supported by Vattenfall AB, by the Swedish Board for Industrial and Technical Development (NUTEK) and by the Swedish Energy Development Corporation (SEU). Nordic Windpower AB is owned by consultant company ÅF-Industriteknik and manufacturer Hägglunds Component, the latter being a member of the Asea Group.

2. The site

For a long time there had been a wish to demonstrate wind power in the distribution area of the regional division Vattenfall Västsverige (=Western Sweden). A suitable site was found on the coast close to the city of Lysekil. The area is known for tourism and for a today declining fishing. The Brofjorden oil refinery nearby is a single large industrial installation. Thus the area is quite sensitive for environmental disturbance.

The Lyse Windpower Station, which is owned by Vattenfall, is situated on an artifical island created around two islets, Basteviksholmarna, connected with the shore by a road built on top of a pier. Besides the wind turbines, the island is used as a harbour for some some small fishing and pleasure boats. The two turbines are situated on the islets, so the foundations could be anchored in solid rock. The distance to the nearest dwellings is about 400 meters.

The site is equipped with a 66 meter meteorology mast and a small building which holds an exhibition and facilities for the evaluation program.

3. The reference wind turbine

The reference turbine was purchased through a normal tendering procedure. A Bonus 450 kW Mk II 36 meter diameter turbine was installed in June 1992. During the rest of 1992 it produced

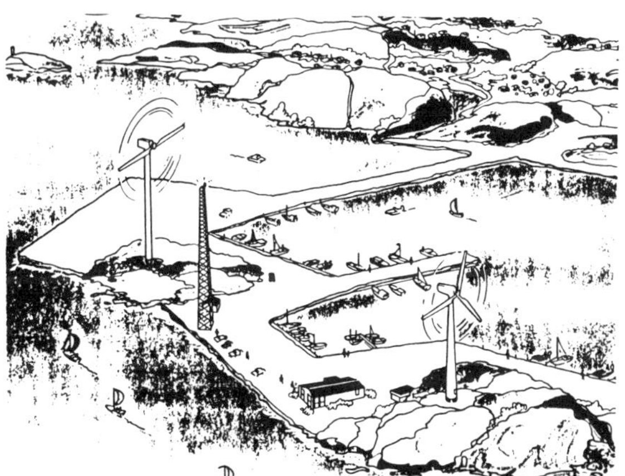

Fig 1. The Lyse Wind Power Station with Nordic 400 (two-bladed) and Bonus wind turbines.

501 MWh. Except for the problems related to the power quality in the local grid mentioned further down, it has worked according to the expectations.

4. The design of the 400 kW turbine

The Nordic 400 kW turbine has a 35 meter diameter two-bladed upwind rotor with teeter hinge. The rotor is supported by the main rotor bearings integrated in the planetary gear box. Teeter bearings and teeter stops are elastomeric. Both gear box and the induction generator are flanged to the tubular machine bed, see Fig. 2.

The blades are made of GRP and manufactured by LM Glasfiber in Denmark. The airfoils utilized are NACA 63.2xx on the outer parts of the blades and FFA-W3 on the inner parts. The rotable tips serve as emergency brakes and are kept aligned with the blades during normal operation by pressure being applied on hydraulic cylinders in the hub region. A small hydraulic unit is situated in the hub. Thus the activation of the tips is achieved passively, by relieving the hydraulic pressure. Electric power and signals are transformed over slip rings. Emergency braking is performed by the blade tips assisted by a mechanical brake on the secondary shaft capable of delivering nominal torque. At all operational conditions an over-run condition can be avoided by either the two blade tips or one blade tip and the mechanical brake.

The turbine is run at variable speed at wind speeds below rated and stall-regulated at a nearly constant RPM at higher wind speeds. The purposes of the variable speed system are multiple: to achieve a soft drive train, to increase energy yield and decrease noise disturbance at low wind speeds, to allow adjustment of power level according to air pressure, temperature, turbulence level etc and finally to facilitate the commissioning. The power electronics system of an AC-DC-AC grid commutated type is situated in a separate container building on ground level. Power is transformed from the nacelle through flexible cables.

The nacelle is made of GRP and high enough to allow a man full standing height.

Yawing is performed by a flexibly mounted geared electrical servo motor. Thus the turbine is also soft in yaw.

The soft tower is a welded steel shell construction. A vertical ladder inside the tower provides access to the nacelle.

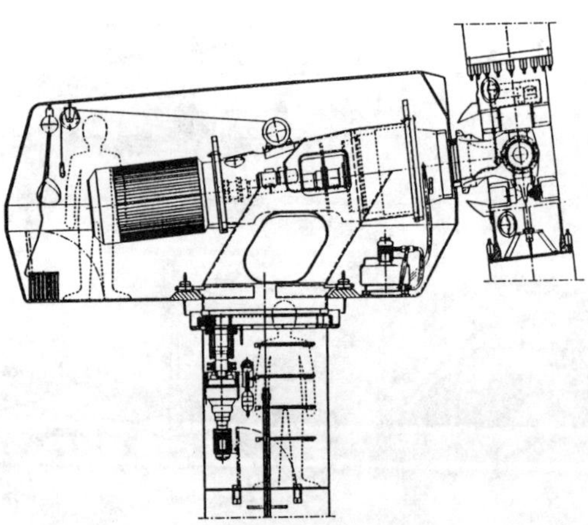

Fig. 2. The Nordic 400 machinery.

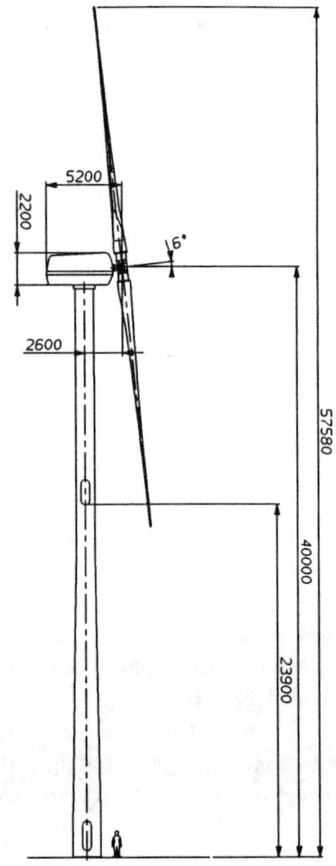

Fig. 3. The Nordic 400 wind turbine.

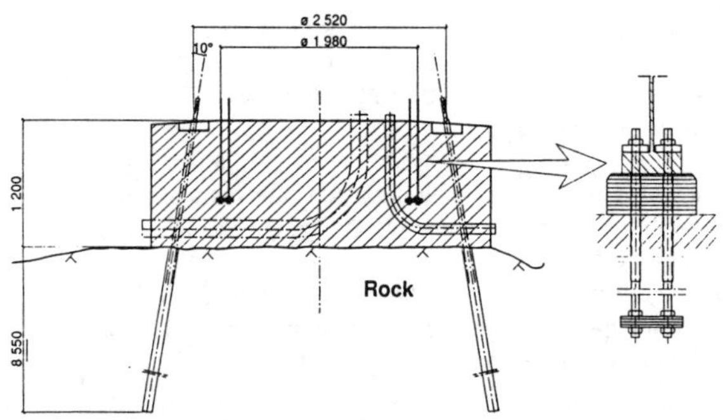

Fig. 4. The rock-anchored foundation of the Nordic 400.

The foundation is secured to the solid rock by 8 meter prestressed steel bars. The tower is fixed to the foundation by the use of 120 bolts.

Generally the turbine is all soft: teeter hinge, soft shaft, soft yaw and a soft tower. To achieve the objective - that short term wind speed variations due to wind shear and turbulence shall not increase the loading of the turbine - the softness however has to be connected with damping and elasticity. To find the right combination, that does relieve stress without trigging any dynamic phenomena, required a high number of computer simulations, but once found, the "recipe" is quite robust and unsensitive to variations in the parameters.

During the design period the risque of getting teeter instabilities in stall at high wind speeds was anticipated. According to the conventional aerodynamic two-dimensional theory instabilities would occur, but nowhere in the literature such instabilities had been reported. To investigate the phenomenas a scaled-down turbine was tested on the Chalmers-Hönö 12 meter diameter test turbine[1]. No instabilities were detected on that turbine, and so far no instabilities have occurred on the real turbine either.

The soft and simple design has resulted in a low weight, about 40% less than a conventional three-bladed wind turbine.

The design is certificated by Det Norske Veritas.

5. Construction phase

The tower was transported by rail from the Hägglunds factories in Northern Sweden in one piece to a place close to the Lyse wind power station, whereas the complete nacelle and the blades arrived by truck. The erection was performed with a mobile crane and with no problems.

6. Commissioning

The commissioning so far has revealed that the concept works as predicted and fulfills the requirements. It has been greatly facilitated by the variable RPM system, which made it possible to make first runs without wind and then gradually increase the RPMs allowed when running in wind conditions.

One peculiar problem encountered during the commissioning was that the Bonus turbine refused to start when the Nordic was operating. This was found to be due to harmonics created by the power electronics from the variable RPM system. Both turbines are connected to the same transformer, which formed a much smaller low voltage grid than in the normal applications for this industrial drive equipment. The problem was solved by installing some reactors in the Nordic electrical system.

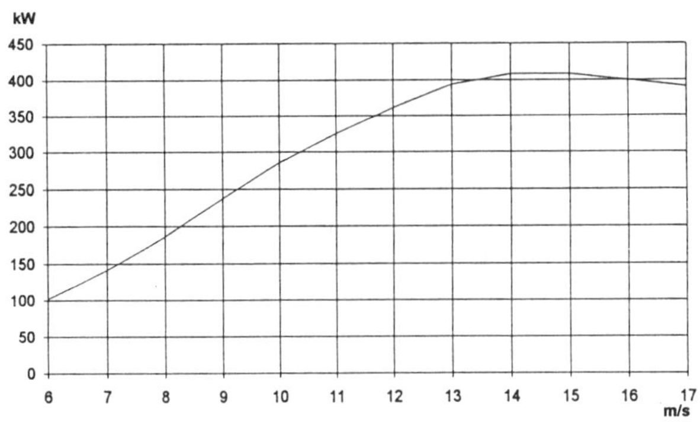

Fig. 5. A preliminary power curve for the Nordic 400.

In January 1993 there was a storm with steady wind speeds of more than 30 m/s. This revealed weak points in some details, which made it necessary to exchange the wind sensor stand and the bolts securing the nacelle.

A 500 hour acceptance test is planned to take place early spring 1993.

7. Evaluation programme

A two year evaluation program will be performed. For both turbines availability, maintenance requirements, electricity production, power quality and noise will be investigated, whereas a more thorough technical evaluation will be performed for the Nordic turbine. It will primarily cover the dynamic behaviour and stress situation in comparison with the calculated response.

8. Further development

The further development of the Nordic 400 concept is performed along two lines. One is to develop a marketable series production version of the turbine, starting with an initial serie of around five machines. The other is to further develop the concept to the 1 MW size, which is today performed in a project sponsored mainly by Vattenfall, with additional funding from the CEC as well as from Swedish sources.

Fig. 6. The Lyse Windpower Station Data Acquisition System.

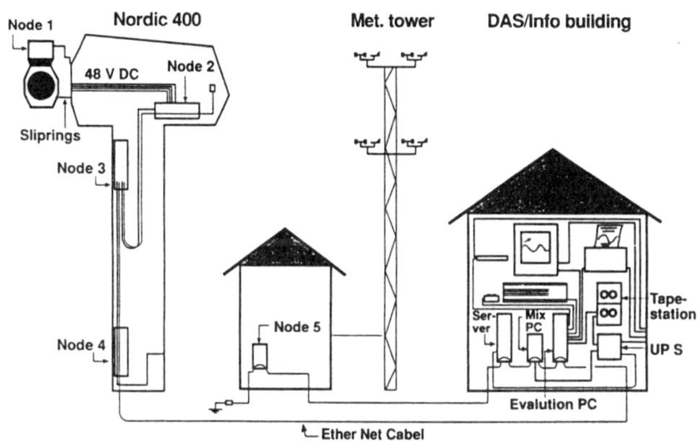

9. References

[1]H Ganander et al: Stall Control with Variable Speed at the Hönö 2 Wind Turbine. Department of Electrical Machines and Power Electronics. Chalmers University of Technology. 1993.

Main data

Rotor	
Number of blades	2
Diameter	35,2 m
Rotor speed	20-41 RPM
Tip speed	37-75 m/s
Orientation	upwind
Type of hub	teeter
Airfoil	NACA 63.2xx/FFA-W3
Tower	
Material	steel shell
Hub height	40,0 m
Diameter top/base	1,3/1,9 m
Machinery	
Gear box	2-stage planetary 1:40 with integral turbine shaft bearings
Generator	induction
Voltage	690 V
Conversion system	AC-DC-AC grid commutated
Performance	
Rated power	400 kW
Cut i	5 m/s
Rated	13 m/s
Cut-out	25 m/s
Design wind speed (2 s)	53 m/s
Weights	
Nacelle with blades	13 500 kg
Tower	15 000 kg
Total excl. foundation	28 000 kg

The Scientific Measurement and Evaluation Programme in the German "250 MW Wind" Programme

C. Enßlin, G. Füller, B. Hahn, M. Hoppe-Kilpper, K. Rohrig

Institut für Solare Energieversorgungstechnik (ISET), Kassel
D-3500 Kassel, Königstor 59, Tel. +49 (0) 561/7294-0

Abstract

The "Scientific Measurement and Evaluation Programme" (WMEP) is accompanying the German "250 MW Wind" Programme in order to obtain statistically relevant data from the practical application of wind energy converters (WECs) in Germany. The congress paper summarizes the development of WMEP from 1990 up to now. Evaluations from more than 10,000 months of WEC supervision are presented under the topics WEC performance and failure statistics. This includes results from data acquisition by the WMEP remote measurement network. Furthermore WMEP's new information system WISY is introduced.

1 Introduction

In June 1989 the "250 MW Wind" programme was presented by the German Ministry for Research and Technology (BMFT) scheduling installation of WECs with a total capacity of 250 MW within five years. Each of these approximately 1250 WECs (mean rated power of 200 kW assumed) is supervised for a period of ten years by WMEP.

As the operational behaviour of all subsidized WECs is to be closely monitored, maintenance and repair measures as well as energy production data are recorded by so called logbook records and evaluated at the Institut für Solare Energieversorgungstechnik (ISET). Up to now records from more than 10,000 months of WEC operation have been evaluated, which considerably broadened the statistical basis for estimation of the performance and economy of WEC models operated in Germany.

For measurement data acquisition of WEC's operational characteristics, a remote measurement network is being installed transferring high resolution wind data and WEC power data from up to 450 selected WECs to the WMEP control center.

This supplements the logbook evaluation in certain essential aspects and the comparison of WEC performance is rendered possible.

2 Project status, spring 1993

By the end of February 1993 a total of 742 WECs representing an installed capacity of 121 MW, were operating under the supervision of WMEP. The corresponding mean rated power of 163 kW has only slightly risen since the beginning of the evaluation. In the coming months however the erection of considerably larger WEC models is expected in the 500 kW rated power range.

The overall statistics show that 94 percent of the supervised WECs are operating in the four following federal states:

Schleswig-Holstein	281 WECs (58.4 MW)
Lower Saxony	246 WECs (39.4 MW)
North Rhine-Westphalia	127 WECs (10.7 MW)
Mecklenburg-Vorpommern	30 WECs (5.5 MW)

Especially has to be notified that in the eastern federal state Mecklenburg-Vorpommern a total of 5.5 MW has been erected in the space of a few month.

The past years showed a large increase of wind farm installations. Meanwhile approximately 60 percent of the installed WEC capacity in WMEP are operating in wind farms. This figure is based on the most general definition, regarding groups of at least two WECs at one location as wind farms.

The majority of the large wind farms (more than 1 MW installed capacity) is operated by commercial companies, followed by utilities and private owners, who operate a considerable share of the smaller wind farms.

Fig. 1 shows the model characteristics of the supervised WECs, listed according to blade number, power control,

		Dec.'90	Dec.'91	Dec.'92
Number of WECs in WMEP		111	369	719
Blade number	2-Blade	21%	12%	13%
	3-Blade	79%	88%	87%
power control	stall	40%	54%	56%
	pitch	60%	46%	44%
Generator	synchronous	31%	30%	33%
	asynchronous	69%	70%	67%
speed control	variable	48%	39%	39%
	constant	52%	61%	61%

Fig. 1: WEC models in WMEP (frequency of model characteristics)

generator and speed control.

The distribution shows that the following technical characteristics are still prevelant:

- o 3-bladed WECs (87 %)
- o stall-regulated (56 %)
- o asynchronous generator (67 %)
- o constant rotational speed (61 %).

These characteristics are typical mostly to WECs of Danish origin and two of the important German manufactures (HSW, Tacke).

Enercon as another important German manufacturer is represented with 11 % of the total WEC number (models E 16, E 17, E 18), 14 % (models E 32, E 33) respectively. Like the German manufacturer Krogmann (share of 5 %) and the Dutch manufacturer Lagerwey (share of 8 %), this company prefers a variable rotational speed concept.

The shown trend is of course influenced by the quota of subsidized WEC models in the "250 MW Wind" programme.

By the end of 1992 the WMEP remote measurement network included more than 100 data acquisition units (DAU) sending daily wind and power data to the control center at ISET. The choice of connected WECs guarantees a broad selection of different WEC locations and models. Today the data pool of the control center incompasses more than 29,000 daily DAU files, as well as energy production and failure data from more than 10,000 months of operation. This data will be made accessible to external users by means of special data base called WISY (see Chapter 5).

3 WEC-Performance

WEC operators participating in the WMEP programme are obliged to document monthly energy meter readings for evaluation at ISET. They are used for the calculation of power output and demand, capacity factors, for single WECs as well as groups, and also for calculation of the BMFT subsidies.

Fig. 2 gives an overview of the annual energy production of WECs supervised in WMEP. This figure shows that almost 200 million kWh of electrical energy were generated by WECs under the WMEP programme in 1992.

Part of the WMEP evaluation is the calculation of average capacity factors for Federal States and rated power classes. In 1990 the results ranged from 5 percent (North Rhine-Westphalia) to 28 percent. In 1992 the largest mean capacity factor (32 %) was attained in the Federal State of Schleswig-Holstein in the 281-359 kW class. Within this rated power class individual converters attained annual production rates of more than 1,000,000 kWh corresponding to about 3000 full-load hours. These high values indicate a higher mean annual wind velocity in comparison to the previous years.

	1990	1991	1992
Number of WECs in WMEP	111	369	678
Total months of supervision	648	3005	6880
Installed capacity [MW]	16,2	61	110
Energy production [Mill. kWh]	14,9	83	194

Fig.2: WEC energy production under WMEP

Using data from the remote measurement network (power output and wind velocity) some initial deductions can be made about the control and supervision characteristics of the converters.

Fig. 3 shows 10 min mean values of a specifically difined power charactersitic (power divided by rotor area and free wind speed) drawn over windspeed measured in hub-height. These data values were acquired for three pitch-controlled converters over a time period of 1,5 month (see fig. 3 a, b, c). It can be seen, that these WECs start production at wind speeds of 3 - 3,5 m/s. The maximum specific power values are achieved within a broad range of approximately 4 to 10 m/s on the converter shown in fig. 3 a. Beyond approximately 10 m/s the figure shows the typical power control

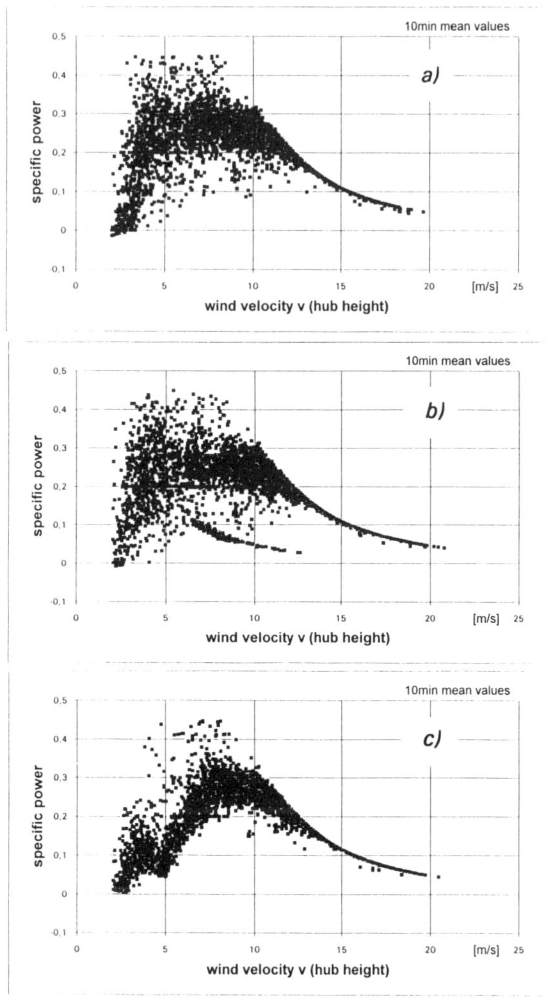

Fig. 3 a, b, c: Power characteristics of selected WECs

behaviour of pitch-controlled machines. Fig. 3 b shows a visible deviation from the expected behaviour according to power regulation between 6 - 12 m/s. The converter, illustrated in fig. 3 c, shows a power output reduction between 4 - 5 m/s.

Such evaluations are repeated at regular intervals (e.g. every half a year) and give valuable information about the long-time behaviour of the supervised WEC. These evaluations can also be used by converter manufacturers in order to optimize the converters under varying conditions at different locations (e.g. at coastal or inland sites).

Furthermore special evaluations of the short term behaviour and performance of WECs are possible by use of the optional raw data acquisition at 10 Hz. By means of a spectral analysis of the power signal, e.g. failures of WECs can be detected.

4 WMEP Failure Statistics

Participants in the WMEP are obliged to notify ISET in each instance of maintenance or repair to their WECs. In 1992 a total of nearly 1500 maintenance/repair reports for 449 WECs were sent in for evaluation. These reports were evaluated regarding failure causes, resulting malfunctions, repair measures and costs.

Fig. 4 shows a summary of the 1992 failure statistics. The overall technical availability was calculated at 97,5 percent, whereas the rated power classes partially show large deviations from this value. The 1-70 kW class only achieved a technical availibility of 95 percent. The highest technical availibility was achieved by the 141-210 kW class (99 percent) and 14 WECs of the 421 - 490 kW class (99,5 percent).

Documented failure causes show, as in previous years, WECs being sensitive to control system failures (30 percent). This corresponds to the distribution of repaired or exchanged components, which is "Microprocessor" (17 percent).

Examination of costs of maintenance and repair measures shows that the highest costs average approximately 20,000 DEM are for the relatively rare failure cause "Lightning Strike", whereas the frequent cause of failure "Control System" only results in average repair costs of 5,000 DEM. Fig. 4 shows the evaluated average repair costs per repair measure being liable to pay costs for each cause of failure.

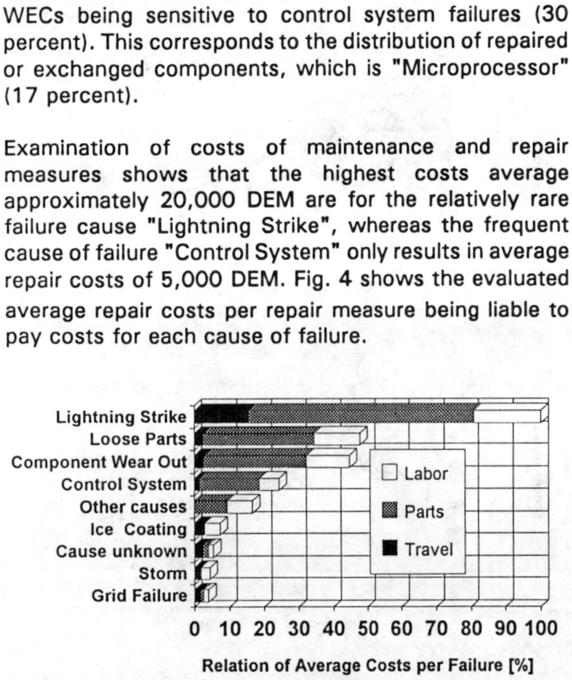

Fig. 5: Costs per failure cause 1992

The WMEP Annual Report 1992, which is to be issued in spring 1993, will contain a detailed account of WEC O&M costs (repair, maintenance contract, insurance, site rent, etc.), as well as the revenue incentives from energy sale to utilities and saved by private use of generated energy.

A number of spectacular WEC damages by lightning strike led to a short study based on the WMEP data base system, the results of which are discribed in the following.
Between January 1992 and February 1993 these failure cause was documented 65 times, a total of approx. 720 WECs being supervised by the end of this period. Only nine cases led to considerable damages by destroyed rotor blades, which caused repair costs of up to 150,000 DEM. For the most recent cases only estimates are available.

Overvoltage or black-outs after lightning strikes led on the other hand to considerable downtimes, although the damage was limited to protective devices.

Repair cost for lightning damage could be significantly lowered by avoiding or reducing blade damage and avoiding prolonged downtime after minor damages.
Approximately 80 percent of the WEC locations hit by lightning strike are on inland locations, 2/3 of which are found in exposed, i.e. mountainous, inland regions. Most of the affected inland WECs have enlarged towers.

Failure Statistics 1992	
Recieved reports	1477
WECs with reports	449
Repaired or exchanged components	1372
Documented downtime (hrs)	124.623
Relative downtime	2,5%
WECs in WMEP	719
Failure cause total	**1196**
Storm	110
Grid	101
Lightning Strike	56
Ice coating	34
Controlsystem failure	329
Component wear out or defect	258
Loose Parts	51
Other cause	143
Unknown cause	114
Failure effects total	**1217**
Overspeed	87
Overload	21
Noise	74
Vibrations	70
Reduced Power output	59
Cause of following failures	40
WEC standstill	718
Other effects	148
Repair Measures	
Faultless functions without repair after	
Reset	340
Change of control parameters	58
Repaired or exchanged components total	**1372**
Hub	48
Rotor blades	63
Generator	97
Other electrical parts	337
Sensors	193
Microprocessor	234
Gearbox	58
Brake	47
Other (Drive Train)	28
Hydraulic System	121
Yaw System	112
WEC Structure/ Nacelle	34

Fig. 4: WMEP Failure Statistics 1992

5 Wind Energy Information System WISY

The new Wind Energy Information System WISY makes an online access to the WMEP data possible. This provides a platform for supplying data and information about the 250 MW programme. The user groups addressed are at present WEC manufacturers, WEC owners, utilities, other research institutes and specialist publications.

Data and information retrieval is possible via terminal masks or via download of data files.
The main characteristics of WISY are:
- quick retrieval of current data,
- direct computer-aided processing possible.

Technical requirements for communication with WISY are a Personal Computer, standard modem and standard telecommunication software. A hierarchically structured menu system supports the user during the application of terminal masks. These masks show pre-processed and compressed WMEP data. At present the following masks are available:

1. WEC operation records on daily/monthly basis, with the following quantities (among others):
   -Mean power and wind speed
   -Power output and demand
   -Number of grid connections
   -Grid failures
   -Duration of part/full load
   -Mean power
   -Capacity factor

2. Failure Statistics
   -Failure cause
   -Malfunction
   -Repair measure

3. O&M costs
   -Repair, maintenance
   -Maintenance contracts
   -Site rent
   -Miscellaneous.

In addition, further general information masks about the current status of the WMEP programme (Fig. 6) are available.

WISY access rights are restricted according to data protection regulations and special user rights, which have to be agreed also by BMFT. E.g. access of WEC manufacturers should be limited to information on converters produced by their own respective company, WEC owners, respectively, to inform about their own WECs.

6 Outlook

The past year was dominated by building up the WMEP remote measurement network, by implementing software tools for data processing at ISET and by achieving a standard evaluation of logbook information. Now the technical conditions are made up for enlarged WMEP evaluations and putting data at other users' disposal.

The newly installed platform for WMEP data transfer called WISY will be further developed according to its users' special requirements. Any request from potential users of the system is welcome.

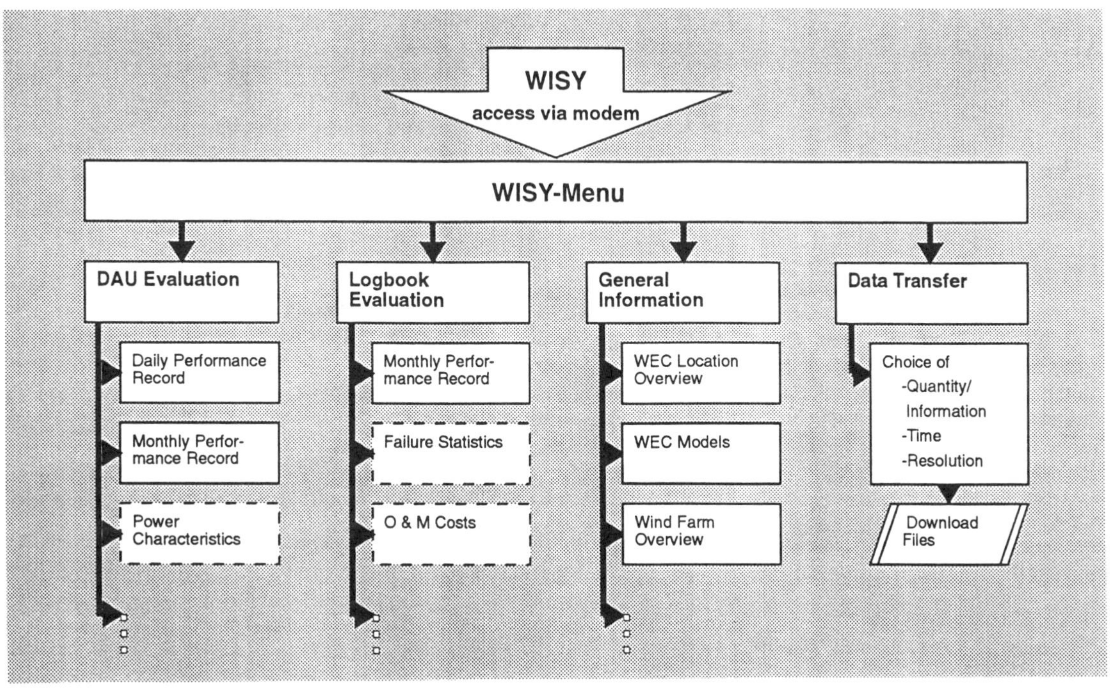

Fig. 6: Wind Energy Information System WISY

Evaluation and Short-Time-Forecast of WEC-Power within the power grid of SCHLESWAG AG

U. Papke
SCHLESWAG AG
Postfach 260
2370 Rendsburg
Germany

A. Petersen
WINDTEST KWK GmbH
Sommerdeich 14 b
2222 Kaiser-Wilhelm-Koog
Germany

V. Köhne
WINDTEST KWK GmbH
Sommerdeich 14 b
2222 Kaiser-Wilhelm-Koog
Germany

ABSTRACT: Because of the increasing number of WEC's connected to the power grid of 'SCHLESWAG AG', utility in Schleswig-Holstein, it was necessary to provide more information about the varying power fed into the grid from WEC's. Therefore SCHLESWAG AG, in cooperation with WINDTEST KWK GmbH, installed measurements of meteorological data on several locations spread all over Schleswig-Holstein. Based on this data and with the knowledge of the locational factors of all WEC's, which are connected to the utilities power grid, a personal computer calculates the power, which is fed into the grid by WEC's each minute. Furthermore it is possible to get a power-forecast for the next hour. The measuring system and the method of power calculation are described. Some results and a look to future developments are shown.

1 INTRODUCTION

In Germany wind energy concentrates in the coastal regions in the north, especially in the states of Schleswig-Holstein and Lower Saxony. The utilities in this regions have to integrate the power, which is generated by Wind Energy Converters (WECs), into their power grid. The capacity of windpower, which is integrated in the grid of SCHLESWAG AG, increased in the last few years. This is shown in figure 1. If the installed

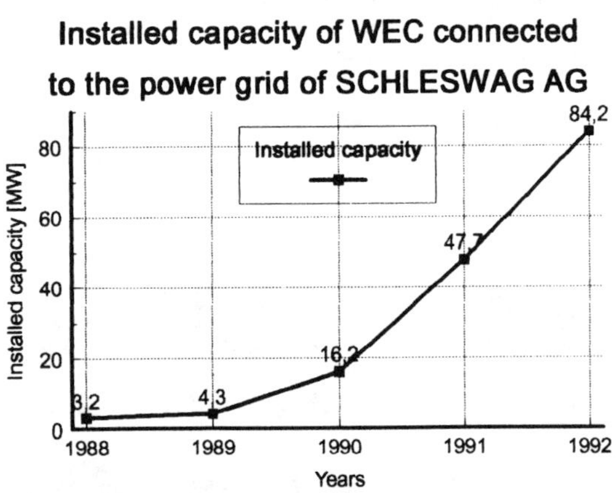

Figure 1: Installed capacity of WEC connected to the power grid of SCHLESWAG AG

power of WECs becomes a notable portion of the total grid power this can affect the economics of the distribution grid. To prevent this the power distributor of SCHLESWAG AG needs information about the power output fed into the grid from WEC. The mostly needed informations are the aktually produced power output. and the prediction of the power output for the next hour. As it is to expencive to take up the power output of each WEC, SCHLESWAG AG decided to estimate the power output with a computer program, based on measured meteorological data [1].

2 MEASURING SYSTEM

Meteorological data are collected at several places in Schleswig-Holstein (see figure 2). At each station data logger registers input signals like:

- wind speed in two different heights,
- wind direction,
- temperature,
- air pressure,

and if possible

- power output of nearby windfarms.

If power output measurement of some windfarms are possible, the calculated values can be verified for this windfarms. This would be a possibility to scrutinize the calculation and if necassary to improve the calculation methods.

Each data logger works independent from the total measuring system. The loggers are reducing the data to mean values of a period of one minute. Thus, each minute of the day is represented by one value. Furthermore the loggers comprise the data into statistics.

A personal computer at the headoffice of SCHLESWAG AG in Rendsburg communicates with the measuring stations via modem to transfer the measured data to the headoffice. For saving time it is possible to use several modems at the same time. The communication network of SCHLESWAG AG is used as well as the communication network of Telekom. An important task

Figure 2: Location of the measuring stations

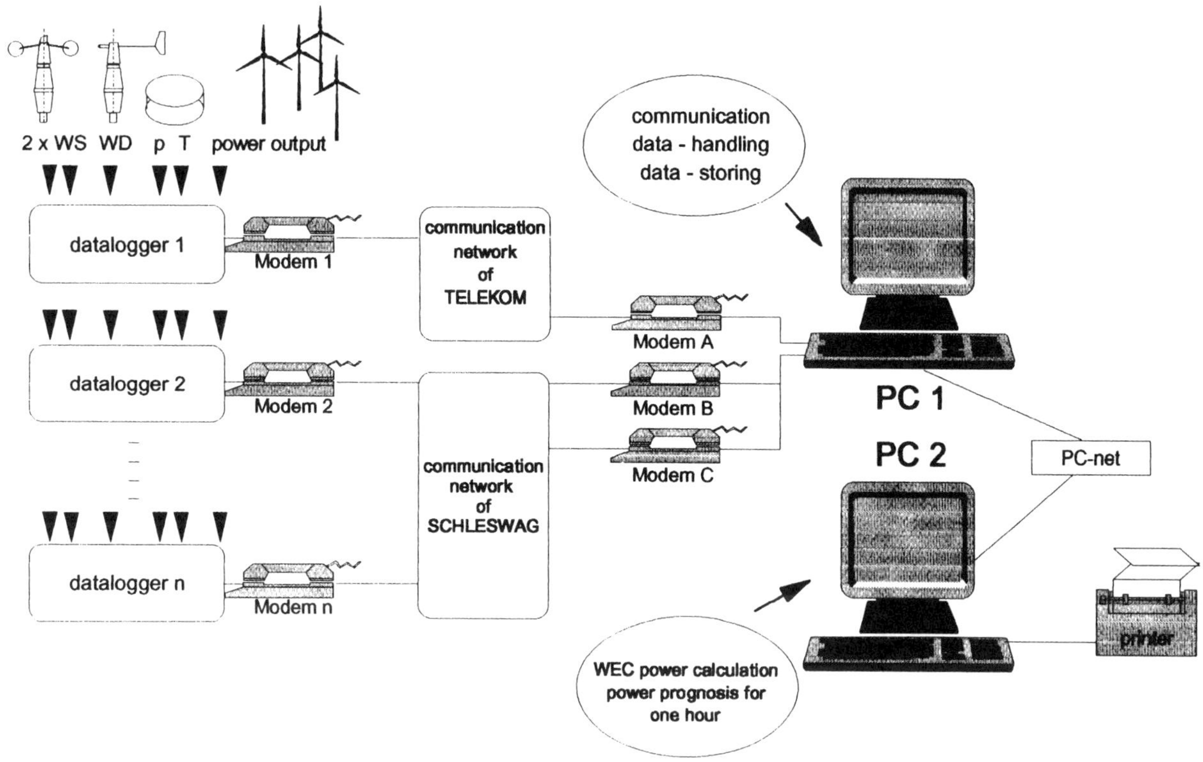

Figure 3: Measuring system

of the PC is to call each station automatically once a day to store and update the data.

A second PC is working together with the first one in a network. Its only task is the calculation and the prediction of the power output. An illustration of the measuring system shows figure 3.

3 POWER CALCULATION

To calculate the power output of WECs it is necessary to get information about every WEC. Above the technical data, e.c. rated power, hubheight or rotor diameter, information about the locational factors of the WEC are needed. In particular the roughness at the locations has to be determined. To get a relation between the measured windspeed and the power output of the different WECs, a power curve is necessary for each different type. All data is stored in a data base, which is updated monthly. In December 1992 the data base contained information about 393 WECs.

The configuration of the measuring stations and the locational factors at the measuring sites are stored. This includes information about the measuring height and used channels as well as the kind of input signals.

Together with the measured meteorological data a program named 'PELWIN' (**P**rogramm zur **E**valuierung von **L**eistung aus **Win**dkraftanlagen) which was created by WINDTEST for this special task, is able to calculate the power output for each minute. Furthermore it is possible to get a forecast of power output for the next hour (see figure 4).

Because the power output is determined from the power curve, the wind speed at hub height of the WEC is needed. The wind speed is measured at several stations all over Schleswig-Holstein. To calculate the windspeed at the specific location of each WEC a model was created. It interpolates the data from all stations depending on their distance to the position of the WEC. The method is insensitive to measuring failures.

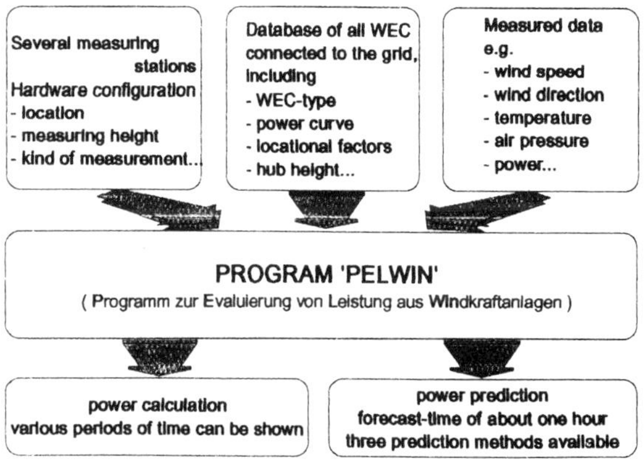

Figure 4: Data basis for power output calculation

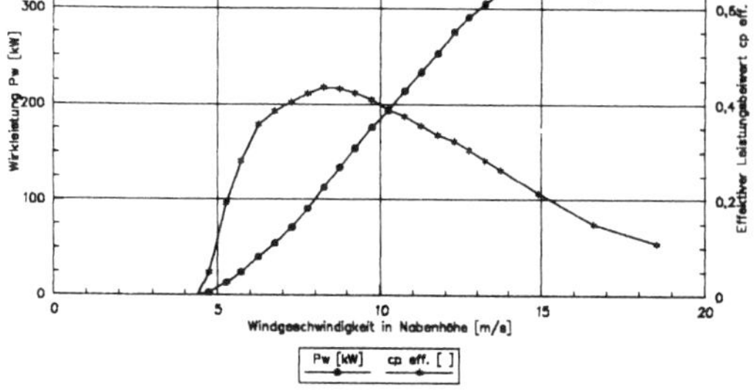

Figure 5: Typical power curve, measured by WINDTEST

The power curves should be measured according to the 'IEA: Recommended practices for wind turbine testing and evaluation' [2] (see figure 5). These curves are corrected to a reference air density of 1.225 kg/m³. For usage in the calculation the measurements have to be applied to the actual air density. The actual air density is calculated from the measured temperature and the air pressure and taken into account.

The power of each WEC is added and stored. As the WECs are numbered it is possible to determine the power output of windfarms or special regions.

A continuous calculation for each minute is possible, for the meteorological data is updated every day. The 'Windows'-standard allows a comparison of different periods in several windows simultaneously. All evaluations can be printed out immediately. An elapse of the calculated power output of all WECs, which are connected to the power grid of SCHLESWAG AG, is shown in figure 6.

4 POWER PREDICTION

Three different methods are available for the prediction of the WEC power output in the next hour. These models are named the statistical model, the translatorical model and the meteorological model.

The statistical model is a trend analysis based on the meteorological data of the last three hours. The windspeed distribution influences on the curvation of the predicted elapse of power output.

For fast weather changes the translatorical model is designed. It takes up the weather situation at a certain time and to change the local position of the measuring stations each minute depending on wind speed and wind direction. In interaction with the midel of power calculation the translatorical model enables it to integrate weather changes into the forecast (see figure 7).

Air movement is caused by differences in air pressure. Knowing about the correlation between this differences and the Coriolisforce is enabling to calculate the windvelocity and the winddirection. Thus a prediction for the next hour can be made. This is the idea of the meteorological model.

All models have individual advantages and disadvantages. The discussion of the three methods resulted in a combination of the first and the second method. Whether adding the third method to the combination shows any advantages will be tested in the next few month. An example of a prediction is shown in figure 8.

5 RESULTS

With this measuring system the wind power fed into the grid of SCHLESWAG AG can be calculated. For a period of one hour the output power can be predicted. With the program PELWIN all stored data can be displayed on screen and printed out. The data from different periods can be compared (PELWIN uses MDI-technology). The calculated values can be compared with measured data of windfarms, so a scrutinization is possible. Output power of different regions can be displayed and compared.

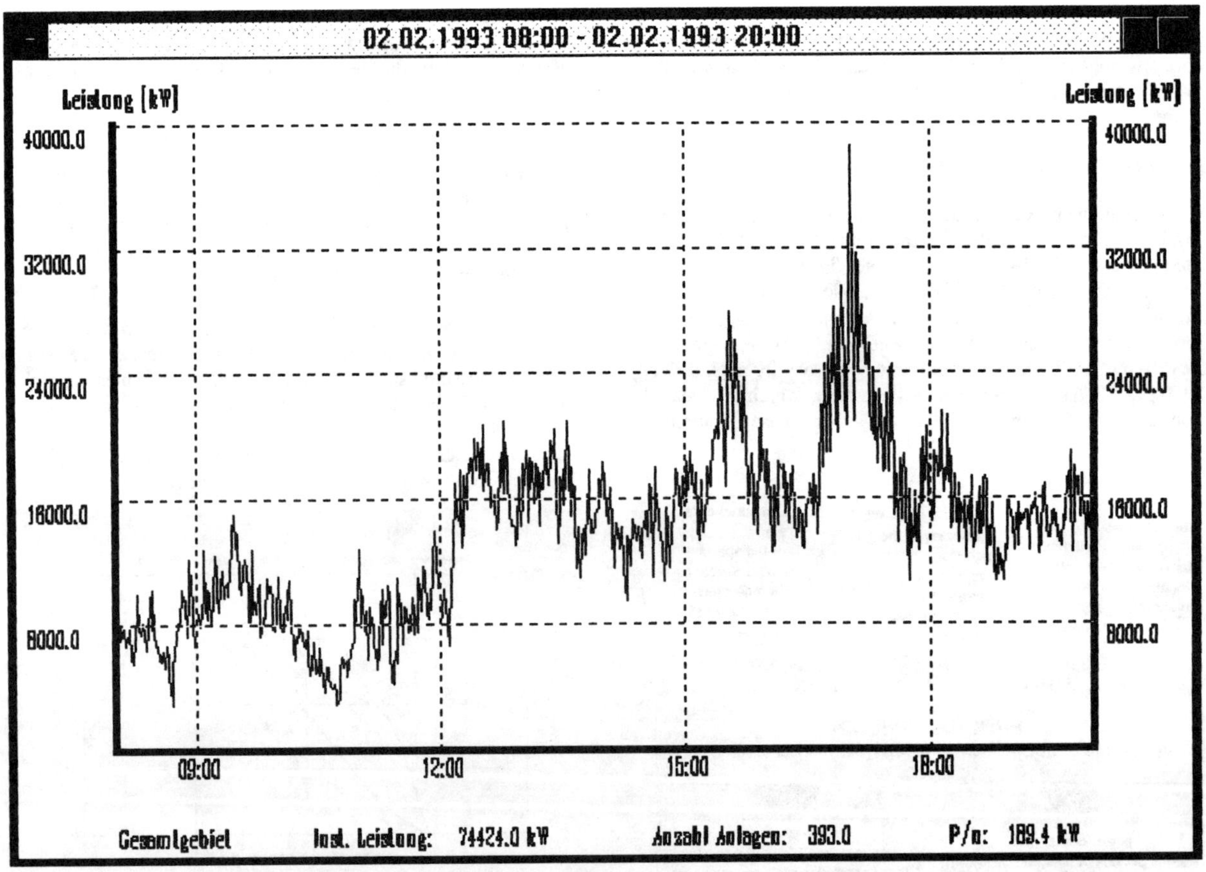

Figure 6: Example of the calculated power output from all WECs, which are connected to the grid

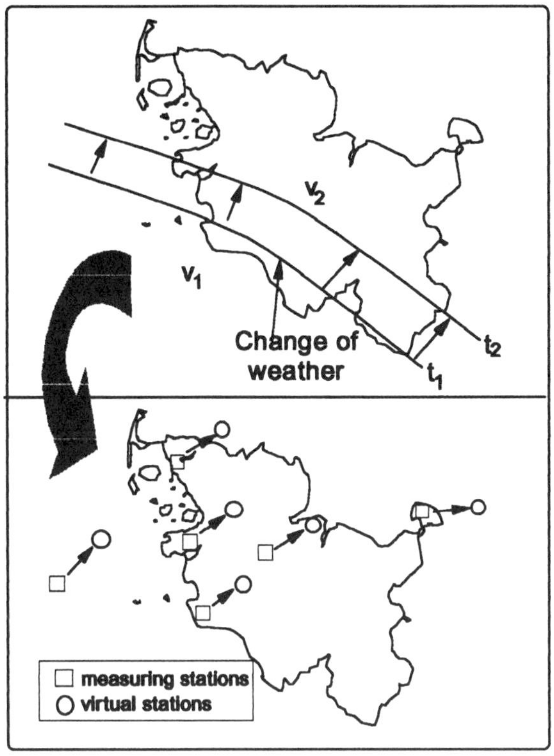

Figure 7: Translatorical model

In the further test period the most important aim is to decide which prediction models are useful. Another target is to get a framework of measuring stations in which the maximum distance between the stations is at the most 50 km. The number and the distribution of the measuring stations have an major impact on the accuracy of the power calculation. In the other way the increasing mass of data is slowing the calculation down. A compromize is to be found between both aims. The calculation of a prediction takes at the moment approximately 10 Minutes. To reduce this time, efforts will be made to accelerate the data transfer from the stations to the headquarter. Then it will be possible to calculate a prediction in a few minutes.

6 REFERENCES

[1] Volker Köhne: 'Erstellen eines Systems zur Bestimmung der aktuellen Leistung aus Windkraftanlagen im Netz der SCHLESWAG AG und zur Prognose der unmittelbar bevorstehenden Leistung auf Basis eines Windmeßnetzes', Diplomarbeit an der Universität Hannover, Institut für Elektrische Energieversorgung, Prof. Dr.-Ing. habil. B.R. Oswald, 1992

[2] N.N.: IEA Recommendated Practices for Wind Turbine Testing and Evaluation, 1. Power Performance Testing, 2. Edition 1990

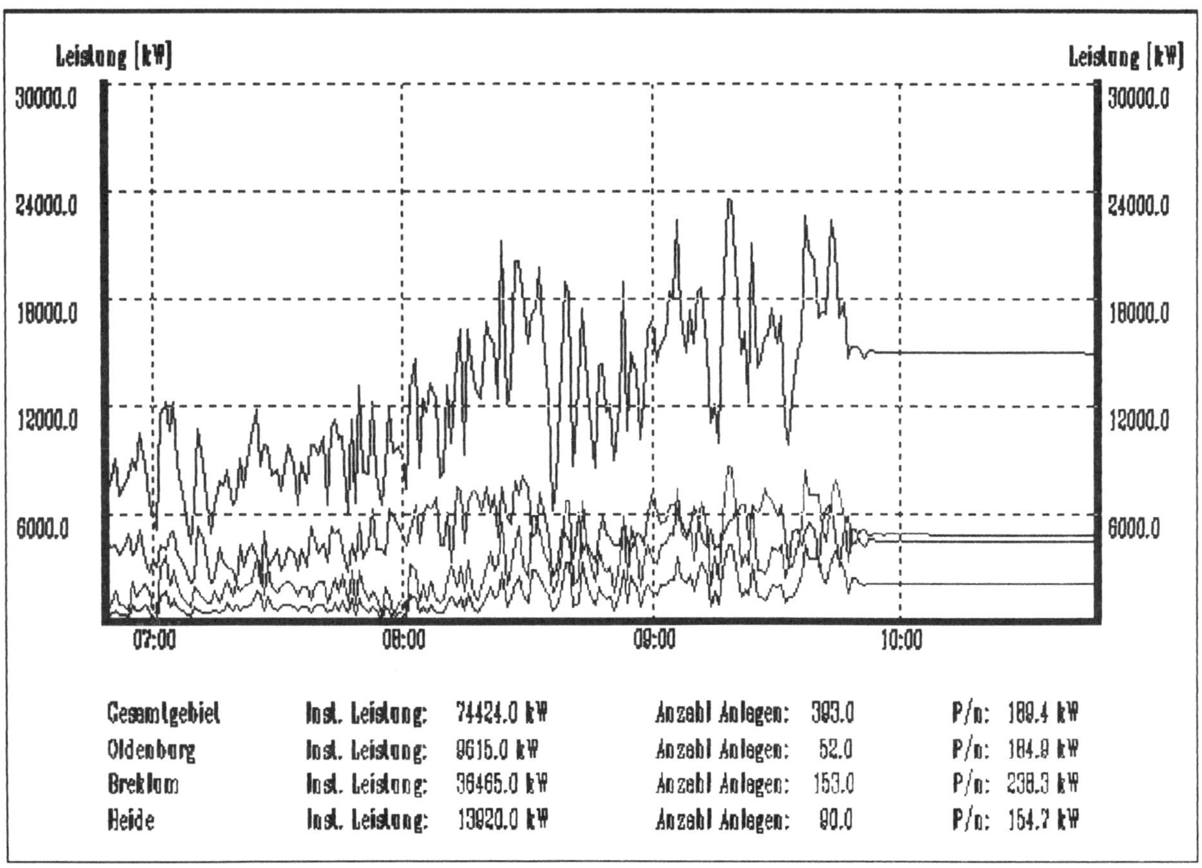

Figure 8: Example for a forecast of one hour

A 10 SEC. FORECAST OF WIND TURBINE OUTPUT WITH NEURAL NETWORKS

John O. G. Tande **Lars Landberg**

Dept. of Meteorology and Wind Energy
Risø National Laboratory
DK-4000 Roskilde, Denmark
(Phone: +45 42371212 - Fax: +45 46755619)

ABSTRACT. In this paper an experiment for obtaining a 10 second forecast of the one second average output wind turbine power is described. Measured time series of one second averages of corresponding values of wind speed, wind direction and output wind turbine power are considered, and two different neural network multilayer backpropagtion network (MBPN) models for providing a continuous 10 second forecast of the one second average output power are described and evaluated.
The models are tested on real time series, and the results are presented and analyzed. The results of applying the different MBPN models are evaluated against each other and against a simple persistence model assuming the output power to remain constant within the forecast period. The paper indicates the usability of neural networks for short-term forecasting of wind power, and it concludes with a ranking of the different models in terms of their estimated confidence level.

1 Introduction

1.1 Background

World wide many power systems exists which may show economic prospects for introducing wind energy to high penetration levels, i.e. especially those in remote areas with high operation costs and good wind conditions. In the most simple high wind energy penetration systems without any energy storage or load management options, the conventional power supply system may be operated with a large spinning reserve for managing possible sudden drops in the wind power production. In fact, the spinning capacity may be as large as without any wind power production even during high wind periods. This is unfortunate both because it reduces the fuel efficiency and because the load on the conventional units may get too low causing operation problems or forcing dissipation of energy for keeping the load above the acceptable minimum level. However, if an appropriate wind power output forecast could be obtained, this forecast could be used in the continuous operation planning of the conventional power supply system for reducing the spinning reserve and obtaining substantial additional fuel savings. The "appropriate" wind power output forecast to be used in the continuous operation planning would depend on the conventional power supply system size and characteristics, i.e. in particular the start up time of the generating units which can be some seconds for diesel generator sets in the range up say 500 kW and several hours for larger coal fired units.

1.2 Scope of works

In this paper recorded time series with one second average output power of a single wind turbine is considered, and three models for providing a continuous 10 second forecast of the one second average output power are described and evaluated.

The three models assumed are the persistence model and two neural network models. Basically, this paper considers the usability of neural networks for providing a continuous 10 seconds forecast, whereas the persistence model is included mainly for the sake of comparison with the neural network models.

There exists p.t. no proved rules for the choice of the dimensions of a neural network, so the selection is solely based on trial and error. The dimensions of the assumed neural network models are more or less arbitrary selected, and then the neural networks are trained and tested without further modifications. Thus, the neural network models presented in this paper should not be taken as anything else than a first attempt to be further modified and made better in possible later efforts.

This paper does not include general theory on neural networks and time series forecasting. These topics are treated in [1], [2] and [3].

2 Methodology

2.1 Experimental setup

Time series with corresponding values of one second averages of wind speed, wind direction and wind turbine output power were provided. The assumed time series are from the department's test site measurements on a Windane 34 wind turbine (400 kW, activ pitch control). The measurement setup was as sketched in figure 1.

2.2 Train and test time series

The train and test time series of the wind turbine output power assumed for this experiment are shown

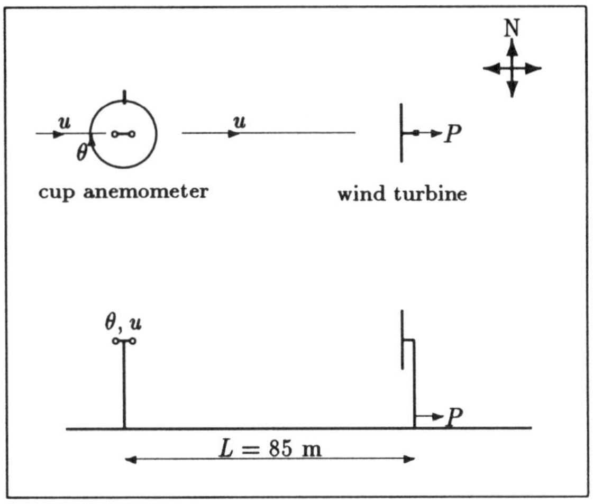

Figure 1: *Wind turbine test site setup of cup anemometer and wind turbine.*

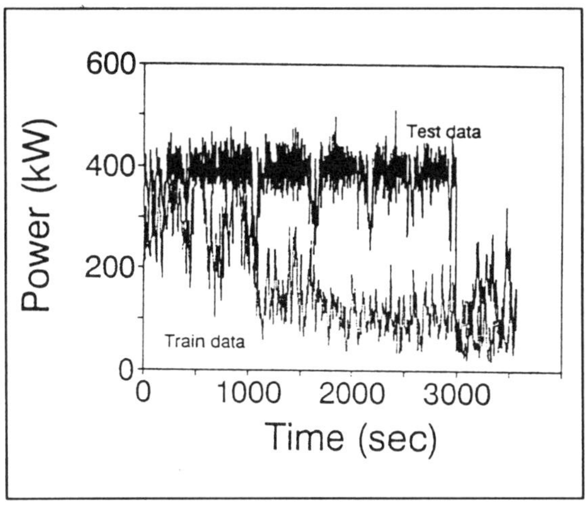

Figure 2: *Assumed test and train time series of wind turbine output power.*

in figure 2. The series were constructed by combining two actual measured time series, i.e. the two measured time series (A and B) were split into four series (A1, A2, B1 and B2) and the train and test series were put together as A1+B1 and A2+B2.

The characteristics of the assumed time series for this experiment are summarized in table 1.

Table 1: *Summary statistics for assumed test and train data (3569 points).*

Test data			
	Wind speed m/s	Wind dir. deg.	Output kW
Minimum	4.6	213	21
Maximum	16.4	304	512
Mean	12.0	281	342
STD	2.33	18.6	103
Correlation between wind speed and output			0.83
Train data			
Minimum	3.9	209	19
Maximum	14.5	307	435
Mean	8.4	252	167
STD	2.0	19.5	96.8
Correlation between wind speed and output			0.72

2.3 Assumed models

2.3.1 Persistence model

The "persistence model" is the most simple model for wind power production forecasting. The persistence model assumes the wind power production, $P(t)$, to remain constant in the forecast period, ΔT:

$$P(t + \Delta T)_{per} = P(t)$$

2.3.2 Neural network model #1

The idea of the "neural network model #1" applied in this experiment is to include measured up-stream wind speed, wind direction and wind turbine output power as network inputs. The wind speed and direction is included for the network to utilize parts of the frozen turbulence theory, whereas the wind turbine output power, P(t), is included as persistence may provide a sound and simple forecast. Thus, the model seeks to utilize parts of both the persistence and the frozen turbulence theory.

As the wind direction of both the test and train data for this experiment are reasonable constant, it is for simplicity not included as an input to the neural network. Thus, the neural network inputs are in this experiment limited to the wind speed and output power samples as specified in fig 3.

The neural network assumed is a Multilayer Backpropagtion Network (MBPN) with 8 input neurons, 5 hidden layer neurons and 1 output neuron as specified in fig. 3. Each neuron in the input layer is connected to all neurons in the second layer, and each neuron in the second layer is connected to the output neuron. In addition, for including a linear relation between the input and output, each input neuron is also connected directly to the output neuron.

The assumed network was trained with the train data set for 2001 iterations arriving at a root sum square error of 3.7 kW.

2.3.3 Neural network model #2

The "neural network model #2" applied in this experiment assumes time series of measured wind turbine output power as input. The idea and hope is that the network should learn to make an accurate forecast from the "historic" wind turbine output trend.

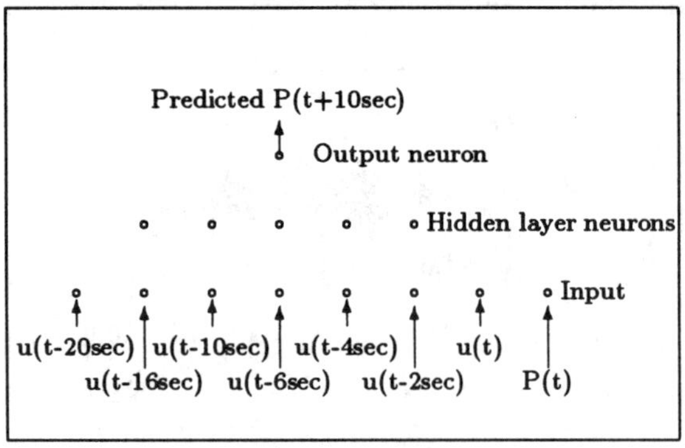

Figure 3: *Assumed neural network layout. Note that each neuron in the input layer is connected to all neurons in the second layer as well as to the output neuron, and each neuron in the second layer is connected to the output neuron.*

The neural network assumed is a Multilayer Backpropagtion Network (MBPN) with 5 input neurons, 3 hidden layer neurons and 1 output neuron as specified in fig 4. The network neurons are connected to each other as for the neural network #1.

The assumed network was trained with the train data set for 2001 iterations arriving at a root sum square error of 4.0 kW.

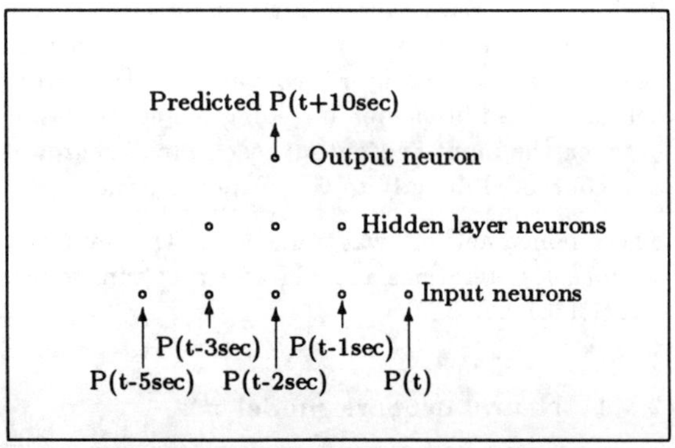

Figure 4: *Assumed neural network layout. Note that each neuron in the input layer is connected to all neurons in the second layer as well as to the output neuron, and each neuron in the second layer is connected to the output neuron.*

3 Results

Table 2: *Summary statistics for output models vs desired output power for the test data time series (3569 points).*

Persistence model			
	Model out kW	Desired out kW	Error kW
Minimum	21	21	-231
Maximum	512	512	375
Mean	342	341	0.78
STD	103	104	45
Correlation coefficient			0.91
Best lin. fit for XY plot of pers. out and desired out			
Intercept (kW)	29	Slope	0.9
Neural net model #1			
Minimum	53	21	-236
Maximum	486	512	351
Mean	333	341	-8.4
STD	96	104	43
Correlation coefficient			0.91
Best lin. fit for XY plot of neural out and desired out			
Intercept (kW)	11	Slope	1.0
Neural net model #2			
Minimum	40	21	-226
Maximum	429	512	351
Mean	320	341	-21.0
STD	89	104	43
Correlation coefficient			0.91
Best lin. fit for XY plot of neural out and desired out			
Intercept (kW)	3.0	Slope	1.0

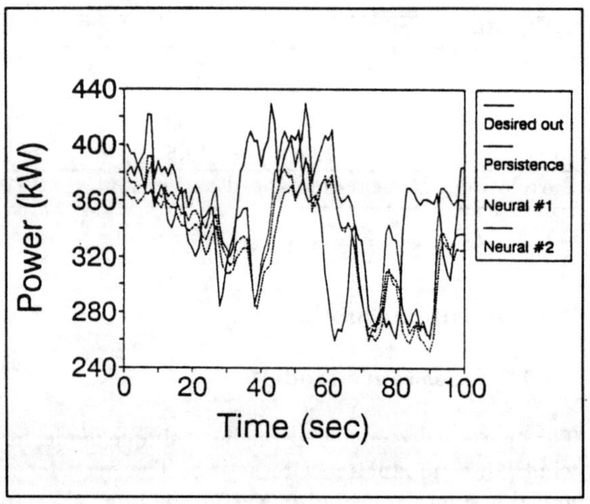

Figure 5: *Detail of time series plot of test data actual power, $P(t+10sec)$, and model outputs, $P(t+10sec)_{mod}$.*

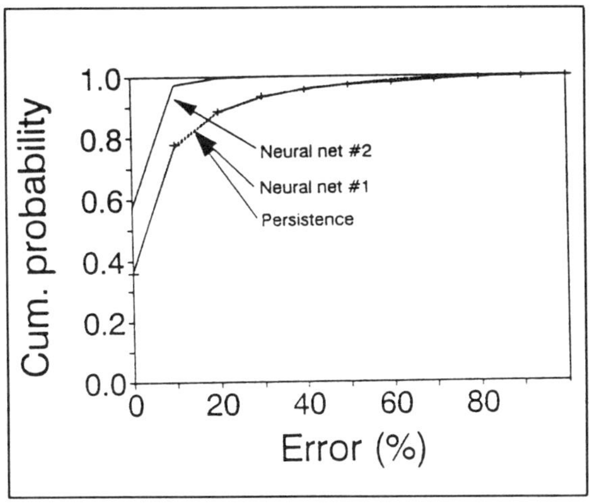

Figure 6: *Cumulative probability distribution of forecast estimate error, $e = |P_{for} - P_{desired}|$ in percent of the peak maximum wind turbine output power (700 kW).*

4 Conclusion

The test results of the three models include some summary statistics, time series plots and distribution of forecast estimate error. The results do not come out one-sided, and care should be taken while ranking the models.

The summary statistics in table 2 indicate that all three models may give errors in the same range. The persistence model naturally has a very low average error as it always will, whereas the two neural net models provides some higher average error. Neglecting the mean error as a measure suitable for forecasting models ranking, the summary statistics then may indicate that the tree models perform more or less the same.

The time series plots in figure 5 illustrates that the output of two neural network models tend to follow the output of the persistence model.

The forecast confidence level for the three models may be read from figure 6. Assuming a 95% confidence level, the test for the neural network model #2 gives a forecast estimate error not exceeding 10% of the peak maximum wind turbine output (700 kW), whereas the estimate for the neural network model #1 and the persistence model may be as high as 37%. Using this as the ranking criteria, the neural network model #2 may be ranked as the winner, whereas the neural network model #1 and the persistence model may share the second place. Indeed, the models must be further tested and evaluated before any recommendations of their usability can be stated.

There exists p.t. no proved rules for the choice of the dimensions of a neural network, so the selection must be solely based on trial and error. As both of the assumed networks represent a first attempt only, many other network layouts can be argued and could have been selected. The neural networks may thus be further modified in search for better performance. A possible continuation of the efforts may thus include more modifications of the tested network layouts and dimensions. New neural networks inputs could also be assumed. These could be time history wind power data, running (~ 5 sec.) average wind power and maybe the output of a traditional forecast model such as the ARIMA (*a*uto*r*egressive *i*ntegrated *m*oving *a*verage).

References

[1] Landberg, L., 1992: *Neural networks utilized for the prediction of a geophysical time series, A comparison.* In Danish. Risø–R–656(DA). Risø National Laboratory, Roskilde, Denmark, 76 pp.

[2] Lapedes, A. and R. Farber, 1987: *Nonlinear Signal Processing Using Neural Networks: Prediction and System Modelling.* Preprint LA-UR-87-2662. Submitted to: Proceedings of IEEE.

[3] Rumelhart, D.E., J.E. McClelland and the PDP Research Group, 1986: *Parallel Distributed Processing. Explorations in the Microstructure of Cognition.* Volume 1: Fundations. MIT Press.

INTEGRATION EINER WINDTECHNISCHEN STROMERZEUGUNG IN DEN KONVENTIONELLEN KRAFTWERKSPARK

Manfred Fischedick*, Martin Kaltschmitt**
*Institut für Energiewirtschaft und Rationelle Energieanwendung (IER), Universität Stuttgart
Heßbrühlstr. 49a, Tel.: 0711/78061-0, W-7000 Stuttgart 80
**Kuratorium für Technik und Bauwesen in der Landwirtschaft (KTBL),
Bartningstr. 49, Tel.: 06151/7001-0, W-6100 Darmstadt

ABSTRACT: Zur Schonung der fossilen Reserven und Ressourcen und zur Verminderung der Kohlendioxidemissionen wird in zunehmenden Maße ein verstärkter Einsatz erneuerbarer Energieträger gefordert. Dies gilt insbesondere auch für den Bereich der Stromerzeugung. Der Gewinnung elektrischer Energie aus Windenergie kommt dabei eine Schlüsselrolle zu, da sie eine der wenigen regenerativen Optionen ist, die an günstigen Standorten derzeit bereits wirtschaftlich sein kann. Darüber hinaus ist sie innerhalb der Gebietsgrenzen der Bundesrepublik Deutschland durch ein technisches Potential in einer energiewirtschaftlich relevanten Größenordnung gekennzeichnet, das auch großtechnisch erschlossen werden könnte. Eine verstärkte Nutzung der Windenergie würde aber zwangsläufig zu einem gemischten Strombereitstellungssystem aus konventionellen und regenerativen Konversionsanlagen führen, mit neuen Anforderungen an den konventionellen Kraftwerkspark. Vor diesem Hintergrund ist es das Ziel dieser Ausarbeitung, die Folgen einer großtechnischen Stromerzeugung aus Windkraft auf den konventionellen Kraftwerkspark und die Fahrweise der einzelnen Kraftwerke zu untersuchen und zu quantifizieren.

1 EINLEITUNG

Eine großtechnische Nutzung der technischen Potentiale der Windenergie zur Stromerzeugung würde - zumindest für eine Übergangszeit - zu einem konventionell-regenerativem Mischsystem der Strombereitstellung führen, bei dem die Struktur des konventionellen Kraftwerksparks im wesentlichen der heutigen entspricht. Aufgrund der z. T. starken Variationen unterworfenen Elektrizitätsgewinnung aus Windkraft können dann bisher nicht bzw. nicht in diesem Ausmaß gekannte technische Probleme im bzw. Anforderungen an den konventionellen Kraftwerkspark bei der Erfüllung des im deutschen Energiewirtschaftsgesetz festgelegten Versorgungsauftrages (d. h. zu jeder Zeit eine ausreichend sichere Stromversorgung zu gewährleisten) auftreten. In diesem Kontext ist es das Ziel dieses Beitrags, beispielhaft für ein Versorgungsunternehmen die mit der Integration einer Windstromerzeugung in das konventionelle Versorgungssystem verbundenen Folgen zu identifizieren und zu quantfizieren; darauf aufbauend werden mögliche Lösungsansätze diskutiert.

2 SIMULATION DER WINDTECHNISCHEN STROMERZEUGUNG

Vor der Quantifizierung der Folgen einer windtechnischen Stromerzeugung muß zunächst die technisch mögliche Elektrizitätsgewinnung analysiert werden. Innerhalb eines bestimmten zu betrachtenden Gebietes ist dabei, aufgrund des lokal sehr verschiedenartigen Windenergieangebots, eine örtlich hoch aufgelöste Untersuchung des meteorologischen Energieangebotes notwendig (vgl. /3, 4/). Hierzu muß - aufgrund nur begrenzt vorliegender Windgeschwindigkeitszeitreihen - eine methodische Vorgehensweise entwickelt werden, die eine Simulation des - durch die meteorologische Großwetterlage korrelierten - Windenergieangebots für jeden potentiellen Windkraftanlagenstandort und zu jeder Stunde des Jahres ermöglicht (vgl. /1/). Ausgehend von diesem Windenergieangebot können dann mögliche Windkonverterstandorte unter Berücksichtigung der lokal unterschiedlichen restriktiven Parameter identifiziert werden (vgl. /1/).

Auf der Basis der Bestimmung potentieller Konverterstandorte und der räumlich und zeitlich hoch aufgelösten Bestimmung des Windenergieangebots kann nun eine Methode zur Anwendung kommen, die - unter der Vorgabe einer gewünschten zu installierenden Leistung - die Simulation der korrespondierenden Windstromerzeugung für ein zu betrachtendes Gebiet für jede Stunde des Jahres ermöglicht. Bei dieser Simulation werden neben den Anlagenkenngrößen auch die jahreszeitlich und ortsabhängig unterschiedlichen Standortcharakteristiken berücksichtigt. Zur Erzielung realistischer Ergebnisse wird unterstellt, daß jeweils diejenigen Windkonverterstandorte zuerst genutzt werden, die unter den vorliegenden ökonomischen Randbedingungen durch die günstigsten Stromgestehungskosten gekennzeichnet sind /2/.

3 STROMNACHFRAGE UND DEREN DECKUNG

Die Stromnachfrage im Netz der öffentlichen Versorgung muß entsprechend dem deutschen Energiewirtschaftsgesetz jederzeit sicher und kostengünstig gedeckt werden. Vor diesem Hintergrund wird im folgenden eine Vorgehensweise dargestellt, die eine Simulation der Nachfrage nach elektrischer Energie und der in Koppelproduktion zu erzeugenden Fernwärme ermöglicht. Darauf aufbauend wird ein methodischer Ansatz für die Bestimmung der kostenoptimalen Fahrweise des Kraftwerksparks diskutiert, der die Dekkung dieser Nachfrage realsiert. Um dabei den zeitlichen Schwankungen des Windenergieangebotes gerecht zu werden, ist eine zeitlich hoch aufgelöste Betrachtungsweise erforderlich.

3.1 Simulation der Nachfrage

Für die Stromverbrauchssektoren der Haushalte und der Kleinverbraucher kann die stundenmittlere Nachfrage nach elektrischer Energie im Jahresverlauf, aufbauend auf vorliegenden Untersuchungen, in denen typische Ganglinien für diese Verbrauchergruppen für verschiedene Tage im Jahr gemessen wurden, auf der Basis meteorologischer und weiterer Kenngrößen (z. B. saisonale Abhängigkeiten), simuliert werden. Mit einer ähnlichen Vorgehensweise kann auch das zeitliche Nachfrageverhalten des Elektrowärmemarktes (d. h. der Speicherheizungen) abgebildet werden; es orientiert sich im wesentlichen an den Außentemperaturen. Die Differenz zwischen der Summe der simulierten Nachfrage der Haushalte, der Kleinverbraucher und des Elektrowärmemarktes und der für bestimmte Wochen vorliegenden gemessenen Stammlast des betrachtetetn Gebietes ergibt die verbleibende Nachfrage der sonstigen, hauptsächlich industriellen Kunden. Sie kann anschließend - unter Berücksichtigung der sie bestimmenden Größen (u. a. Nettoproduktionsindices) - auf den verbleibenden Rest des Jahres extrapoliert werden. Aus der Summe der stundenmittleren Stromnachfrage der Haushalte, der Kleinverbraucher, des Elektrowärmemarktes und der sonstigen Nachfrager ergibt sich damit die gesamte Stammlast im Jahresverlauf. Nach Berücksichtigung der

Netzverluste ermittelt sich daraus die entsprechende Netzeinspeisung der Konversionsanlagen des betrachteten Versorgungsgebietes. Mit einer ähnlichen Vorgehensweise kann auch die in Koppelproduktion produzierte Fernwärmenetzeinspeisung simuliert werden.

3.2 Kraftwerkseinsatzoptimierung

Die Stromerzeugung der dargebotsabhängigen Optionen - hierzu zählt neben der windtechnischen Elektrizitätsgewinnung auch die Stromerzeugung aus Wasserkraft (vgl. /1/) - kann nicht oder nur in einem geringen Maße beeinflußt bzw. an die Nachfrage im Netz angepaßt werden. Im Gegensatz dazu sind die thermischen Kraftwerke eines Kraftwerksparks durch eine Reihe von Freiheitsgraden bei der Erfüllung ihres Versorgungsauftrages gekennzeichnet. Da im Energiewirtschaftsgesetz auch eine kostengünstige Elektrizitätsversorgung festgeschrieben ist, müssen diese Freiheitsgrade im Hinblick auf minimale Stromgestehungskosten optimiert werden. Aus diesem Zusammenhang heraus wird hier die Fahrweise der konventionellen Kraftwerke mit einem Kraftwerkseinsatzoptimierungsmodell /2/ bestimmt.; d. h. für jede Stunde des Jahres wird hier die Fahrweise des Kraftwerksparks ermittel, die letztlich über das gesamte Jahr betrachtet zu minimalen Kosten der Strombereitstellung führt.

Die Bestimmung der Fahrweise der konventionellen Kraftwerke ist dabei aus modell- und rechentechnischen Gründen nicht für jede Stunde des Jahres möglich. Deshalb werden aus den Jahresganglinien der vom konventionellen Park zu deckenden Strom- und Fernwärmenachfrage durch ein Verfahren der automatischen Klassifikation zunächst Gruppen von Wochenganglinien mit vergleichbaren Charakteristiken ermittelt und zu Clustern zusammengefaßt (z. B. alle Hochsommerwochen). Aus jeder solchen Gruppe wird dann eine konkrete und als repräsentativ anzusehende Woche ausgewählt, die dem Durchschnitt aller diesem Cluster zugeordneten Wochen am ehesten entspricht. Damit liegen letztlich Wochenganglinien der stundenmittleren Nachfrage nach Strom und Wärme vor, die für einen definierten Zeitraum des Jahres stehen und als Grundlage für die Ermittlung der Fahrweise des Kraftwerksparks dienen können.

Zur Ermittlung der kostenminimalen Betriebsweise der einzelnen Blöcke muß der gesamte Anlagenpark, der für die Deckung der Strom- und Wärmeachfrage auf der untersuchten Gebietsfläche derzeit verfügbar ist bzw. u. U. in Zukunft vorhanden sein wird, mit seinen anlagenbeschreibenden Kenngrößen definiert werden (vgl. Abbildung 1). Dabei wird unterschieden zwischen Kraftwerken, die nur elektrische Energie liefern, Heizkraftwerken, die in Koppelproduktion Strom und Wärme bereitstellen, Heizwerken, die die Fernwärmespitzenlast decken, sowie Speicher- bzw. Pumpspeicherkraftwerken zur Spitzenlaststromerzeugung. Zusätzlich dazu sind eine Reihe weiterer energiewirtschaftlich relevanter Größen vorzugeben (z. B. die zu verstromenden Primärenergiekontingente, die verfügbaren bzw. nicht einsetzbaren Anlagen laut Revisionsplan).

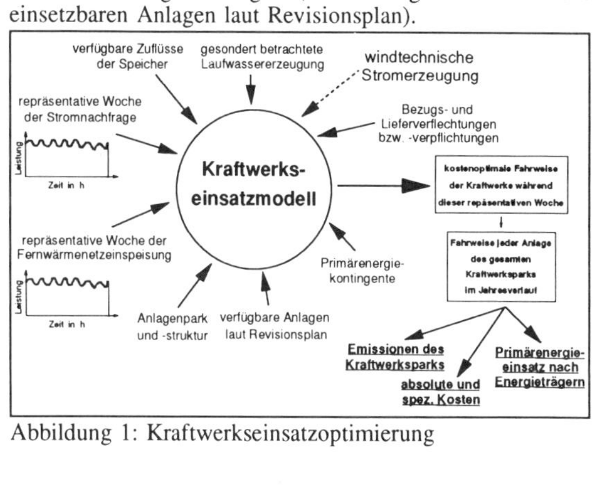

Abbildung 1: Kraftwerkseinsatzoptimierung

Das verwendete Optimierungsprogramm ermöglicht eine interaktive Abbildung des betrachteten Kraftwerkssystems, indem menügesteuert die wesentlichen Kenngrößen abgefragt und in ein lineares Gleichungssystem überführt werden. Dazu wird die gesamte Erzeugungsstruktur in einzelne Elemente unterteilt, die letztendlich, unter Berücksichtigung des Systemverhaltens und dem Zusammenspiel mit anderen Komponenten, das Gesamtsystem bilden. Das diesem System korrespondierende mathematische Modell wird anschließend durch eine lineare Optimierung nach dem Simplex-Algorithmus mit dem Ziel der Minimierung der variablen Erzeugungskosten gelöst. Diese Ermittlung der kostenoptimalen Fahrweise des Kraftwerksparks erfolgt dabei für jede der als repräsentativ ermittelten Wochen. Eine entsprechende Wichtung der für die einzelnen Wochen erzielten Ergebnisse liefert die gewünschten Jahresergebnisse. Letztlich können die erzielten anlagenspezifischen Ergebnisse u. a. bezüglich der Emissionen, des Primärenergieeinsatzes oder der absoluten und spezifischen Kosten ausgewertet werden.

4 FALLSTUDIE

Die Analyse der Auswirkungen einer Einbindung einer windtechnischen Stromerzeugung in den konventionellen Kraftwerkspark und damit die Umsetzung der beschriebenen Vorgehensweise wird im folgenden am Beispiel einer Fallstudie realisiert. Da Speicher- und Pumpspeicherkraftwerke im Rahmen der Einbindung eines fluktuierenden Energieangebotes eine besondere Rolle spielen können, wird im Rahmen der Fallstudie ein Versorgungsunternehmen betrachtet, das über einen relativ hohen Anteil an Pumpspeicherkraftwerksleistung verfügt. Im folgenden wird zunächst der Kraftwerkspark des ausgewählten Versorgungsunternehmens kurz dargestellt. Anschließend werden unter Umsetzung der in Kapitel 2 dargestellten Vorgehensweise drei Nutzungsstrategien einer Windstromerzeugung im betrachteten Versorgungsgebiet entwickelt. Aufbauend hierauf werden die Auswirkungen einer Einbindung dieser regenerativen Stromerzeugung auf den Lastgang diskutiert. Anschließend wird analysiert, inwieweit die Stromerzeugung aus Windkraft zu einer Veränderung der Fahrweise des konventionellen Kraftwerksparks führt; auch werden Möglichkeiten untersucht, den Kraftwerkspark optimal an das fluktuierende Energieangebot anzupassen.

4.1 Rand- und Rahmenbedingungen

Der konventionelle Kraftwerkspark des betrachteten Unternehmens besteht im wesentlichen aus Laufwasserkraftanlagen, Grundlastkraftwerken auf nuklearer Basis, fossilen Mittel- und Spitzenlastkraftwerken sowie Pumpspeicheranlagen (vgl. Tabelle 1). Darüber hinaus bestehen verschiedene Bezugs- und Verkaufsverträge.

Tabelle 1: Kraftwerkspark des untersuchten Unternehmens

Wasserkraft		112,0 MW
Grundlastkraftwerke		1 540,0 MW
fossile Mittellastkraftwerke		1 580,0 MW
fossile Spitzenlastkraftwerke		130,0 MW
Pumpspeicher (Turbine)		1 140,0 MW
Summe		4 502,0 MW
Windkraft	konventionell:	0,0 MW
	Strategie I:	39,0 MW
	Strategie II:	310,0 MW
	Strategie III:	1 100,0 MW

In diesen Anlagenpark soll eine windtechnische Stromerzeugung integriert werden. Dabei stellen hier neben wirtschaftlichen Gesichtspunkten auch technische Aspekte begrenzende Faktoren dar. Entsprechend der in Kapitel 2 angesprochenen Vorgehensweise ermittelt sich für das betrachtete Versorgungsgebiet ein windtechnisches Stromerzeugungspotential von etwa 1 470 GWh_{el}/a. Ausgehend von diesem maximalen Potential, sind drei Strategien zur Implementierung einer windtechnischen Stromerzeugung in den konventionellen Kraftwerkspark definiert worden. Innerhalb einer Strategie I wird ein moderater Ausbau der Windenergienutzung unterstellt; es wird hier von einer zu installierenden Leistung von 39 MW ausgegangen. Die Strategie II ist durch eine Installation von ca. 310 MW an Windkonverterleistung gekennzeichnet. In der Strategie III wird letztlich von einer vollständigen Ausschöpfung des technischen Potentials ausgegangen; dementsprechend sind rund 1 100 MW an Windkraftleistung in den konventionellen Kraftwerkspark einzubinden.

4.2 Auswirkungen auf den Lastverlauf

In Folge der Integration einer angebotsorientierten und damit nur in engen Grenzen steuerbaren Windstromerzeugung in den bestehenden Kraftwerkspark, kommt es zu einer Beeinflussung der durch die konventionellen Kraftwerke noch zu deckenden Stromnachfrage. Abbildung 2 zeigt die sich einstellenden Veränderungen am Beispiel einer Winterwoche für die Strategie III. Die Abbildung macht zunächst noch einmal deutlich, daß das windtechnische Stromangebot durch sehr große Energieangebotsunterschiede gekennzeichnet ist. Während zu Beginn der hier betrachteten Woche noch eine nennenswerte Stromerzeugung realisiert werden kann, ist die windtechnische Elektrizitätsgewinnung gegen Ende der dargestellten Woche sehr gering. Diese Schwankungen im Energieangebot führen insgesamt zu einer deutlichen Verungleichmäßigung der Restlast und stellen damit die sich für das konventionelle Stromerzeugungssystem einstellenden Veränderungen deutlich dar. Dies gilt insbesondere für die durch ein vergleichsweise hohes Windenergieangebot gekennzeichneten Wintermonate.

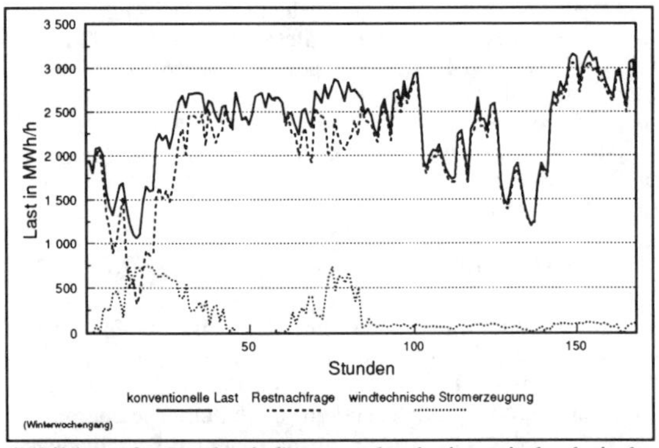

Abbildung 2: Lastbeeinflussung durch eine windtechnische Stromerzeugung

Zusätzlich zu den Fluktuationen des windtechnischen Energieangebotes im Stundenbereich, treten auch im Substundenbereich z. T. extreme Schwankungen auf; sie müssen ebenfalls durch das konventionelle Stromerzeugungssystem ausgeglichen werden. Während die Nachfrage- und Angebotsschwankungen im Subminutenbereich durch die speziellen Regelsysteme des Verbundnetzes (Primärregelung) ausgeregelt werden, sind die Fluktuationen im Minutenbereich durch die einzelnen Unternehmen selber auszugleichen. Dafür kommen i. allg. schnell startbare Kraftwerke, wie z. B. Gasturbinen oder Pumpspeicherkraftwerke, zum Einsatz. Für kurzfristigere Schwankungen ist das Vorhalten einer rotierenden Reserve, d. h. die nicht vollständige Auslastung der in Betrieb befindlichen Kraftwerke, möglich.

Bei einzelnen Konvertern kann es zu Leistungsschwankungen im Kurzzeitbereich in der Größenordnung von bis zu 75 % bezogen auf die installierte Leistung innerhalb von einer Minute kommen. Aufgrund der geringen Energiedichte der Windkraft ist bei einer großtechnischen Windstromerzeugung aber von einer weiträumigen Verteilung der Konverter auszugehen. Aufgrund des zunehmend unkorrelierteren Windenergieangebots mit zunehmenden räumlichen Abstand der Konverter, ist dann aber einer Dämpfung dieser Leistungsschwankungen bzw. eine Vergleichmäßigung des Energieangebotes zu erwarten; die Leistungsschwankungen dürften sich im Kurzzeitbereich im Verbund von n Windenergiekonvertern auf einer Gebietsfläche von etwa 1 000 km^2 näherungsweise um den Faktor $1/\sqrt{n}$ reduzieren (vgl. /5/). Werden die für dieses Beispielunternehmen vorhandenen Möglichkeiten zum Ausgleich kurzfristiger Angebotsschwankungen zusammengefaßt, so wird deutlich, daß die im Kraftwerkspark verfügbare Gasturbinen- und Pumpspeicherkraftwerksleistung etwa in gleichen Größenordnung wie das technische Windenergiepotential liegt. Für das betrachtete Unternehmen bedeutet dies, daß Schwankungen des windtechnischen Stromangebotes im Kurzzeitbereich durch den bestehenden Kraftwerkspark, auch bei vollständiger Ausschöpfung der windtechnischen Stromerzeugungspotentiale, mit hoher Wahrscheinlichkeit ausgeglichen werden können.

4.3 Auswirkungen auf die Fahrweise des Kraftwerksparks

Der Kraftwerkspark des betrachteten Unternehmens setzt sich aus Laufwasserkraftwerken, thermischen Kraftwerken und Pumpspeicherkraftwerken zusammen (vgl. Tabelle 1). Die wassertechnische Stromerzeugung ist dabei der Grundlaststromerzeugung zuzuordnen. Dies ist zum einen auf die relativ geringen variablen Stromgestehungskosten zurückzuführen, zum anderen kann Laufwasserkraftwerken eine gesicherte Leistung zugewiesen werden. Die verfügbaren Pumpspeicherkraftwerke dienen derzeit vorwiegend als saisonale Speicher und damit zum Ausgleich zwischen den nachfrageärmeren Sommermonaten und den nachfragestärkeren Wintermonaten. Sie werden primär für die Spitzenlast-, in Zeiten hohen Zuflusses (d. h. Schneeschmelze) z. T. auch für die Mittellaststromerzeugung und zur Abdeckung des Reservebedarfs eingesetzt.

Die thermischen Kraftwerke realisieren die Grund-, Mittel- und verbleibende Spitzenlaststromerzeugung; diese Anlagen sind durch z. T. deutlich unterschiedliche Stromerzeugungskosten gekennzeichnet und werden im Regelfall entsprechend der anfallenden (variablen) Kosten eingesetzt. Dementsprechend werden die im Kraftwerkspark verfügbaren Kernkraftwerke in der Grundlast, die Steinkohlenkraftwerke in der Mittellast und die gas- oder ölgefeuerten Anlagen für die Spitzenlast- oder Reservestromerzeugung eingesetzt. Z. T. sind jedoch Abweichungen von dieser Einsatzreihenfolge erforderlich; z. B. kann eine Brennstoffkontigentierung zu einer höheren oder geringeren Auslastung bestimmter Kraftwerke führen.

Tabelle 2 zeigt die mit der beschriebenen Vorgehensweise ermittelten Ergebnisse für die Fahrweise des betrachteten Kraftwerkspark. Dabei wurde neben den verschiedenen windtechnischen Nutzungsstrategien auch eine rein konventionelle Fahrweise (Referenz) untersucht; letztere stellt eine Bezugsbasis dar und entspricht der derzeitigen Anlagenbetriebsweise des Kraftwerksparks.

Bei den beiden ersten Nutzungsstrategien der Windenergie zeigen sich im untersuchten Kraftwerkspark nur geringe Veränderungen in der Fahrweise der einzelnen Anlagen (vgl. Tabelle 2). Dies ist im wesentlichen auf den im Vergleich zu der gesamten im Jahresverlauf zur Verfügung zu stellenden elektrischen Arbeit nur geringen Anteil der windtechnischen Stromerzeugung zurückzuführen. Demgegenüber kommt es zu deutlicheren Auswirkungen der Einbindung des fluktuierenden Energieangebotes auf die Fahrweise des konventionellen Kraftwerksparks, wenn das gesamte windtechnische Stromerzeugungspotential ausgeschöpft wird. Dies betrifft im wesentlichen den Bereich der Mittel- und Spitzenlaststromerzeugung, während sich innerhalb der Grundlaststromerzeugung nur sehr geringe Ände-

Tabelle 2: Stromerzeugung der konventionellen Kraftwerksklassen (in GWh) in Abhängigkeit der windtechnischen Stromerzeugung

	Referenz	S I	S II	S III
Laufwasser	505	505	505	505
Grundlastkraftwerke	11 092	11 092	11 092	11 070
fossile Mittel- & Spitzenlastanl. (Vollaststunden)	4 769 (2 794)	4 699 (2 752)	4 363 (2 556)	3 389 (1 985)
PSW - Zulauf[1]	1 671	1 671	1 671	1 671
PSW - Pumpe[2]	99	107	142	274
Vertragssaldo	- 721	- 721	- 721	- 721
Wind (Anteil Arbeit[3]) (Anteil Leistung[4])		73 (0,4 %) (0,8 %)	422 (2,4 %) (6,9 %)	1 469 (8,1 %) (24,4 %)
Summe	17 316	17 319	17 332	17 383

1 Pumpspeicherkraftwerke - natürlicher Zulauf
2 Pumpspeicherkraftwerke - Arbeit der Pumpen
3 Anteil der Arbeit aus Wind an der gesamten elektrischen Arbeit
4 Anteil der Leistung aus Wind an der gesamten installierten konventionellen Kraftwerksleistung

rungen einstellen (vgl. Tabelle 2). Dies ist darauf zurückzuführen, daß nur ein geringer Anteil der windtechnischen Stromerzeugung (etwa ein Sechstel bis ein Siebtel) im Bereich der Grundlaststromerzeugung liegt und diese elektrische Energie aus wirtschaftlichen Gesichtspunkten zudem in den zur Verfügung stehenden Pumpspeicherkraftwerken zwischengespeichert werden kann. Dieser Strom wird dann zu späteren Zeitpunkten, d. h. in Zeiten einer höheren Stromnachfrage, in der Mittel- oder Spitzenlast wieder in das Netz eingespeist. Der nur geringe Anteil der windtechnischen Stromerzeugung im Grundlastbereich liegt dabei u. a. darin begründet, daß aufgrund des ausgeprägten Jahresganges des Windenergieangebotes der größere Teil der windtechnischen Stromerzeugung im nachfragestärkeren Winterhalbjahr liegt; hier liegt das Lastniveau aber i. allg. deutlich oberhalb der zur Verfügung stehenden Grundlastleistung.

Damit substituiert der größte Anteil der windtechnischen Stromerzeugung direkt oder indirekt konventionellen Strom der Mittel- oder Spitzenlast. Hierdurch sinkt die Auslastung der Mittel- und Spitzenlastkraftwerke; gleichermaßen nimmt aber auch die Starthäufigkeit dieser Kraftwerke z. T. deutlich zu. Aufgrund des größeren Windenergieangebotes tritt dieser Effekt verstärkt im Winterhalbjahr auf. Innerhalb dieser Fallstudie sinkt bei einer vollständigen Ausschöpfung des windtechnischen Stromerzeugungspotentials (Strategie III) die Auslastung der fossilen Mittel- und Spitzenlastkraftwerke um etwa 22 % ab; gleichzeitig steigt die Starthäufigkeit dieser Kraftwerke um den Faktor drei. Ebenso ist festzustellen, daß, in Folge der Einbindung des erneuerbaren Stromangebotes, vermehrt fossile Kraftwerke eingesetzt werden, die durch eine vergleichsweise geringe installierte Leistung gekennzeichnet sind und darüber hinaus schnell einsetzbar sind (kurze Anfahrzeiten). Bei diesen Kraftwerken handelt es sich z. T. um die älteren Kraftwerke im Anlagenpark, die gegenüber den größeren und moderneren Anlagen durch einen schlechteren Wirkungsgrad gekennzeichnet sind. Diese kommen dann nur noch in einem geringeren Maße zum Einsatz, da mit der Einbindung der Windstromerzeugung die verbleibende Nachfrage zunehmend nur noch geringfügig oberhalb der zur Verfügung stehenden Grundlastleistung liegt. Dementsprechend nehmen auch die Stillstandszeiten dieser Anlagen zu.

Wird das konventionell-regenerative Mischsystem der Stromerzeugung des untersuchten Unternehmens aus energiewirtschaftlicher Sicht betrachtet, bedeutet die Tatsache, daß das erneuerbare Stromangebot zum überwiegenden Anteil Strom der Mittel- bzw. Spitzenlast substituiert, daß eine kostengerechte Einspeisevergütung des erneuerbaren Energieangebotes - unter Zugrundelegung eines nur geringen Kapazitätseffektes der windtechnischen Stromerzeugung /8/ - sich an den variablen Kosten der Mittel- und Spitzenlaststromerzeugung orientiern müßte.

5 SCHLUßBETRACHTUNG

Ziel der vorliegenden Untersuchung ist die Analyse der Folgen einer Integration einer windtechnischen Stromerzeugung in den konventionellen Kraftwerkspark. Dazu wurde zunächst eine methodische Vorgehensweise entwickelt, die die Bearbeitung dieser Problemstellung in zeitlich und örtlich hoher Auflösung ermöglicht und auf ein Fallbeispiel angewendet. Diese Fallstudie hat deutlich gemacht, daß es bei einer nennenswerten Einbindung einer windtechnischen Stromerzeugung in den konventionellen Kraftwerkspark zu einer Veränderung der Fahrweise des Kraftwerksparks kommen kann. Dabei ersetzt die windtechnische Stromerzeugung im wesentlichen Strom in der Mittel- und Spitzenlast. In windstarken oder lastschwachen Zeiten kann das erneuerbare Stromangebot jedoch auch bis in den Grundlastbereich hineinreichen. Stehen im Kraftwerkspark ausreichende Speicherkapazitäten zur Verfügung, wird diese, die Grundlast verdrängende windtechnische Stromerzeugung, zwischengespeichert und in Zeiten höherer Stromnachfrage wieder in das Netz zurückgespeist. Dies führt zu einer fast vollständigen Ausnutzung des erneuerbaren Energieangebotes im Mittel- und Spitzenlastbereich. Dementsprechend nimmt die Auslastung der größeren fossilen Mittellastkraftwerke ab und deren Stillstandszeiten zu. Dagegen nimmt die Starthäufigkeit kleinerer und schnell verfügbarer Kraftwerke zu.

Eine optimale Ausnutzung einer windtechnischen Stromerzeugung ist demnach dann gegeben, wenn im Kraftwerkspark sowohl ausreichende Speicherkapazitäten, als auch kleine Kraftwerke vorhanden sind, die flexibel einsetzbar und darüber hinaus für einen Ausgleich der Leistungsschwankungen im Minutenbereich geeignet sind. Dies gilt in ähnlicher Weise für eine verstärkte Integration einer photovoltaischen Stromerzeugung in den konventionellen Kraftwerkspark (vgl. /6, 7/). Soll es daher zukünftig zu einer verstärkten Nutzung der ernuerbaren Stromerzeugungsoptionen kommen, müssen die zuvor genannten Aspekte bei der zukünftigen Ausgestaltung des konventionellen Kraftwerksparks entsprechend berücksichtigt werden.

Literatur

/1/ Kaltschmitt, M.; Voß, A.
Integration einer Stromerzeugung aus Windkraft und Solarstrahlung in das Elektrizitätsversorgungssystem von Baden-Württemberg
Studie im Auftrag der Stiftung Energieforschung Baden-Württemberg
2. Zwischenbericht, Institut für Energiewirtschaft und Rationelle Energieanwendung, Universität Stuttgart, Oktober 1991

/2/ Hanselmann, M.; Friedrich, R.; Hönes, R.; Schaal, D.
Tageseinsatzoptimierung mit dem Programmsystem PROFAKO
Elektrizitätswirtschaft 91(1992), 9, S. 515 - 522

/3/ Kaltschmitt, M.; Wiese, A.
Potentiale und Kosten regenerativer Energieträger in Baden-Württemberg; Forschungsbericht des Instituts für Energiewirtschaft und Rationelle Energieanwendung, Band 11, Universität Stuttgart, April 1992

/4/ Kaltschmitt, M.
Möglichkeiten und Grenzen einer Stromerzeugung aus Windkraft und Solarstrahlung am Beispiel Baden-Württembergs; Forschungsbericht des Instituts für Energiewirtschaft und Rationelle Energieanwendung, Band 7, Universität Stuttgart, Dezember 1990

/5/ Beyer, H.G.; Luther, J.; Steinberger-Wilms, R.
Power Fluctuations From Geographically Diverse, Grid Coupled Wind Energy Conversion Systems, Proceedings European Wind Energy Conference, Glasgow, 1989, p. 306/310

/6/ Fischedick, M.; Kaltschmitt, M.
Integration of a photovoltaic electricity production into the conventional power plant system; 11 th European Photovoltaic Solar Energy Conference and Exhibition, Montreaux, Oktober 1992

/7/ Fischedick, M.; Kaltschmitt, M.
Integration einer windtechnischen und photovoltaischen Stromerzeugung in den konventionellen Kraftwerkspark; 8. Internationales Sonnenforum, Berlin, Juni 1992

/8/ Kaltschmitt, M.; Voß, A.
Kapazitätseffekte einer Stromerzeugung aus Windkraft und Solarstrahlung; Elektrizitätswirtschaft 90(1991), 8, S. 365 - 371

IMPACT OF HIGH WIND PENETRATION ON THE POWER SYSTEM OF LARGE ISLANDS

M. Papadopoulos P. Malatestas S. Papathanassiou N. Bilios

National Technical University of Athens
Department of Electrical Engineering
42 Patission Str. - 10682 Athens - Greece
Fax 0030.1.3605604

ABSTRACT : The impact of high wind penetration on the power systems of large or medium size islands is investigated, where an important percentage of the load will be covered by dispersed wind parks. The method of analysis and the main results for two study cases are presented : The first concerns Crete, the largest Greek island (300 MW peak), where 60 MW of wind power are scheduled for the next years. The second concerns the island of Hios (25 MW peak), where three scenarios of wind penetration levels have been considered, corresponding to 3.4 MW, 9.1 MW and 12.7 MW total installed wind power respectively. It is concluded that a high wind penetration can be achieved, under the reservation that the wind power rate of change is forecasted with reasonable accuracy and the controls of the system are properly selected. A monitoring system for all the power units seems to be of primary importance for the safe operation of the power system.

1. INTRODUCTION

Wind turbines are usually connected to large power systems, where only the local distribution network is influenced by their connection. A number of wind turbines has also been installed in small islands with diesel power systems or in isolated autonomous wind-diesel systems. The problems arising from the connection of wind turbines to such power systems have been extensively analysed by many researchers and utilities, as well as by the authors, [1,2].

Currently, the perspective of high wind penetration in relatively large systems, where an important percentage of the load will be covered by dispersed large wind parks, is seriously considered. This is the case in Crete, the largest Greek island, where more than 60 MW of wind power are scheduled to be installed until the year 2000. Taking into account that the estimated maximum/minimum demand for 1995 is 300/100 MW, it is clear that the system might occasionally operate with a wind power penetration approximately equal to 60% of its load.

Other cases of interest are medium size islands, such as Hios, where penetrations even higher than 60% are economically justifiable, due to the favourable wind conditions. In Hios, 3.4 MW of wind turbines have already been installed, the maximum/minimum demand for 1992 being 23/6 MW respectively. Three scenarios of wind penetration have been considered, corresponding to 3.4 MW, 9.1 MW and 12.7 MW of installed wind power.

A description of the method of analysis applied to the study cases of Crete and Hios and some of the results obtained are presented and discussed.

2. METHOD OF ANALYSIS

2.1 The computer program

The analysis of the repercussions from the connection of wind turbines on the operation of the power systems was performed with a suitable computer simulation program, presented in [2], with certain modifications and additions. In this program standard power system analysis methods are used, with emphasis to the proper simulation of the loads and the representation of the wind turbines, which are treated as "special loads". Special techniques for the aggregation of dispersed loads and wind turbines are also used.

2.2 Characteristics of the disturbances

The proper selection of the disturbances to be simulated is crucial for the correct assessment of the behaviour of the system. The expected disturbances are due either to wind power changes or to faults in the system.

2.2.1 Wind Power Changes

The output power of a wind turbine can be represented by a relatively slow changing component, P_{av}, plus a fast fluctuating component, P_{fl}. The calculations of the frequency and voltage deviations are usually made considering only the changes of the "average" power, P_{av}. However, flickering problems may arise for the voltage in some cases, because of the fast component P_{fl}. These fast voltage fluctuations can be seperately calculated and compared with the regulations limits.

The rate of change of the "average" wind power obviously depends on the length of the time intervals considered. From the point of view of the repercussions on the power system, the time intervals of interest depend upon the characteristics of the conventional units. We believe that two cases deserve special attention:

(a) Changes of the wind power in time intervals of the order of some seconds (up to one minute), which correspond to the reaction time of the control systems of the units already in operation. (Fast variations).
(b) Changes in intervals of the order of several minutes (up to one hour), which are related with the possibility of starting a new unit. (Slow variations).

The frequency of appearance of these disturbances is apparently of great significance. Indicative values for the rate of the above changes are shown in the Figures 1 and 2.

In Fig.1 the frequency of the output power changes of one wind park, as a percentage of its installed capacity, is shown. Fig.2 refers to the case of three wind parks, approximately 40 Km away from each other. These curves are based on simultaneous wind speed measurements of 1 minute sample length, made by PPC (Public Power Corporation of Greece) at three of the Cyclades islands. It can be seen, for example, that there is a 2% probability of losing wind power equal to 21.5% or more of the installed capacity in 1 min, if one wind park is in operation. In the case of the three wind parks, the same probability corresponds to a lower level of wind power loss in 1 min (15% or more).

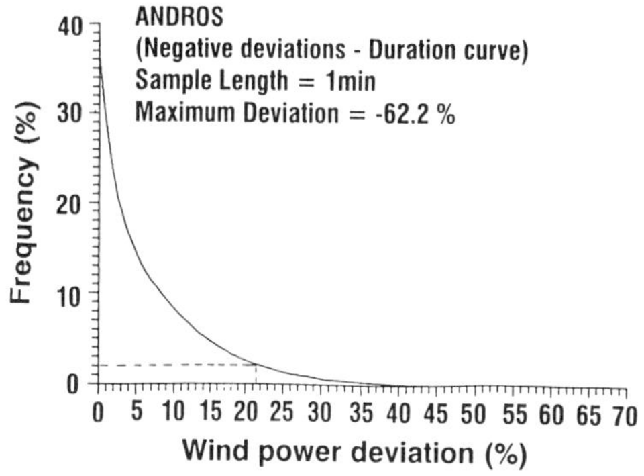

Figure 1. Wind power deviations in 1 minute time intervals - One wind park.

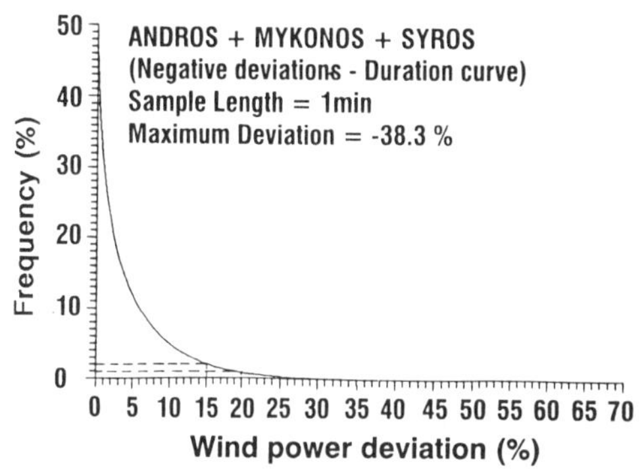

Figure 2. Wind power deviations in 1 minute time intervals - Three wind parks.

The selected fast and slow wind power changes for the two study cases are presented in sections 3 and 4.

2.2.2 Failures of the system.

In power system studies, the behaviour of the system in case of short circuits and loss of generation is usually considered. The impact of wind power loss could be safely assessed by considering that the biggest wind park is suddenly disconnected. The investigation of the behaviour of the wind parks and particularly their impact on the stability of the system during short circuits has to be analysed, as well.

3. CASE STUDY 1: CRETE

The power system of Crete consists of two power stations, interconnected with a 150 KV transmission system, eleven 150/15 KV substations and an extended 15 KV distribution network. The installed conventional power units are the following:

231 MW	Steam units
134 MW	Gas turbines
49 MW	Diesel units
132 MW	Combined cycle (under construction)
------	
546 MW	Total

Nine wind parks are scheduled by PPC for the next seven years, that is a total installed capacity of 61 MW. Half of this capacity is to be installed in one wind park and the rest in six wind parks of about 5 MW each. Apart from this, a considerable number of dispersed private wind turbines is expect to be connected to the distribution network in the next few years.

The behaviour of the system was analysed under certain, rather severe, wind power variations, selected according to the information of section 2.2. Further research is, however, necessary on the modelling of the wind power changes and the probability of their occurence, based on data and measurements obtained at the sites of the wind parks.

For the time being, the power system of Crete is not equipped with an Automatic Generation Control (AGC) system. Such a system is planned for the next years and the expected development of wind power generation should be taken into account for its design. Thus, the analysis has considered only the response of the automatic regulators of each power unit, which act for a period of some tens of seconds after the disturbance ("primary control").

The operating conditions of the system at 07.00h of 24/1/91 were selected as a typical case. The total generated power at that specific hour was 93 MW and the max/min demand of the day 211/76 MW. By projecting these figures to 1995, the following total load and produced wind power were considered:

Conventional power	104.78 MW	66.4%
Wind power (7 wind parks)	51.10 MW	33.6%
	----------	-----
Total load (11 substations)	157.88 MW	100.0%

For the above operating conditions, the generated power by the conventional units is given in Table 1, as calculated by the load flow module of the Dynamic Analysis program, for two alternative spinning reserve policies (high and low spinning reserve).

Table 1. Generated power from the conventional units, before the disturbance.

Units	Rating (MW)	Generated Power	
		High / Low Spinning Reserve (MW)	(MVAR)
Steam A	38.8	18.0/30.0	14.0/22.5
Steam B	75.0	35.0/60.0	18.5/37.2
Gas Turb.	60.0	30.0/ -	25.0/ -
Diesel	49.2	22.6/16.0	12.0/13.1
Total	223.0	105.6/106.0	69.5/72.8

The behaviour of the system, operating with high or low spinning reserve, has been simulated for sudden of gradual wind power loss of different magnitudes (10, 20 and 30 MW) and rates (from 1% up to 5% per second). Some of the results of the simulations are shown in the following figures.

3.1 Wind power decrease of 20 MW.

In Fig.3.1(a) and (b) the frequency and voltage changes are shown, after the sudden loss of 20 MW of wind power from the 30 MW wind park, when the system operates under high spinning reserve. The active power output of the conventional units is shown in Fig.3.1(c). The corresponding variations of the mechanical power of the conventional units can be seen in Fig.3.1(d). The difference in the response of the steam turbines from the gas turbines and the diesel units is clear, regarding their capability to take over load.

It is remarkable that, if the system operates under the low spinning reserve policy, indicated in Table 1, then the same disturbance results in loss of synchronism for the system. The same happens in the case of a sudden disconnection of the whole 30 MW wind park, for both spinning reserve policies.

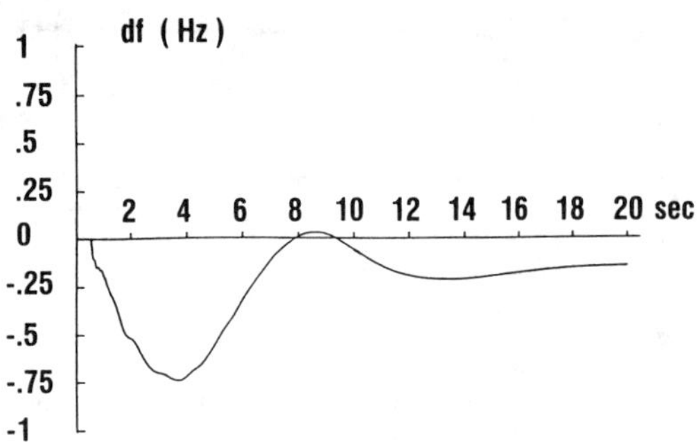

(a)

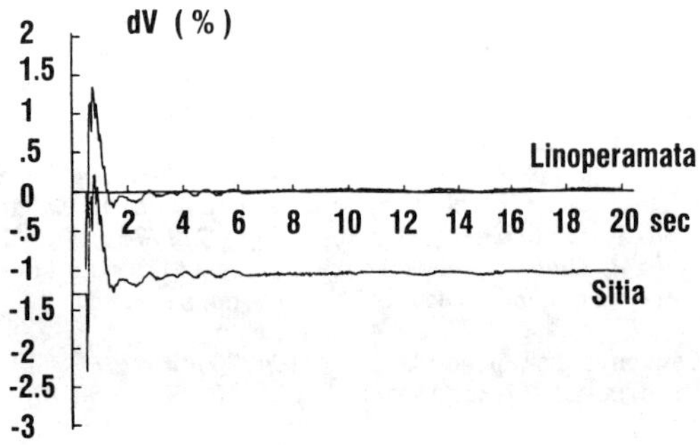

(b)

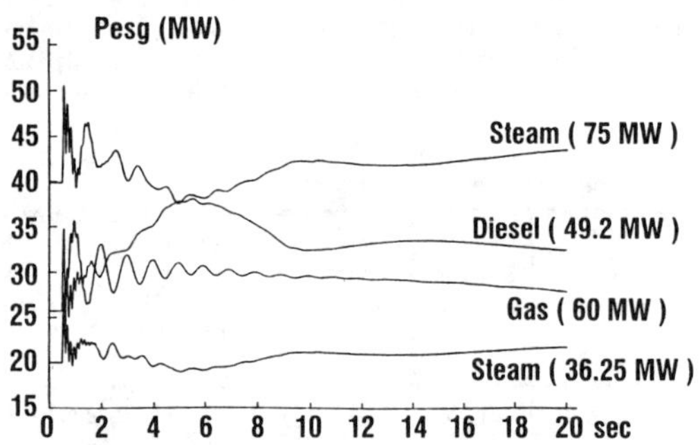

(c)

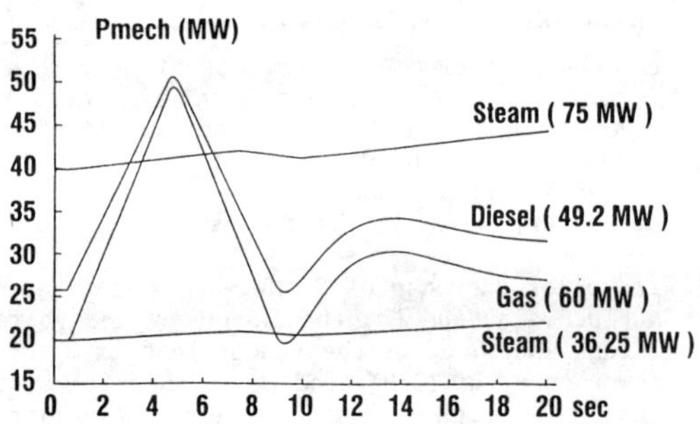

(d)

Figure 3.1 Frequency, voltage, active power and mechanical power variations, after step disconnection of 20 MW of wind power.

If the disturbance of the 20 MW wind power reduction is considered to take place gradually in 20 sec, then the system remains stable, even under the low spinning reserve conditions. The frequency and mechanical power changes for this disturbance are shown in Fig.3.2(a) and (b), respectively.

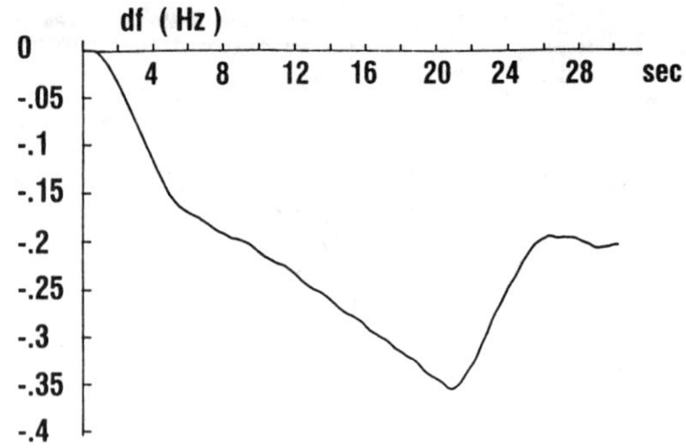

(a)

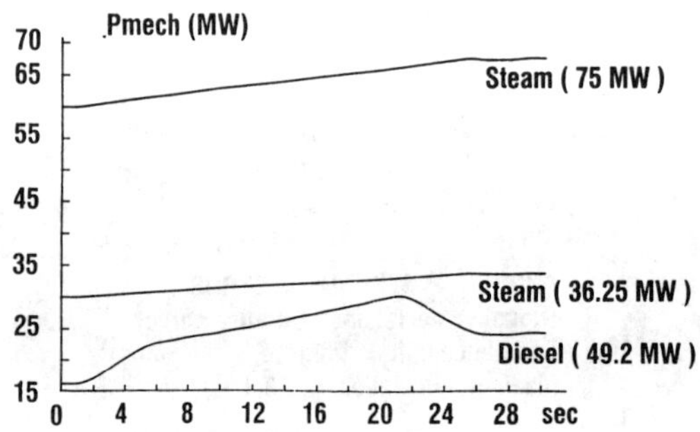

(b)

Figure 3.2 Frequency and mechanical power variations after gradual loss of 20 MW of wind power in 20 sec.

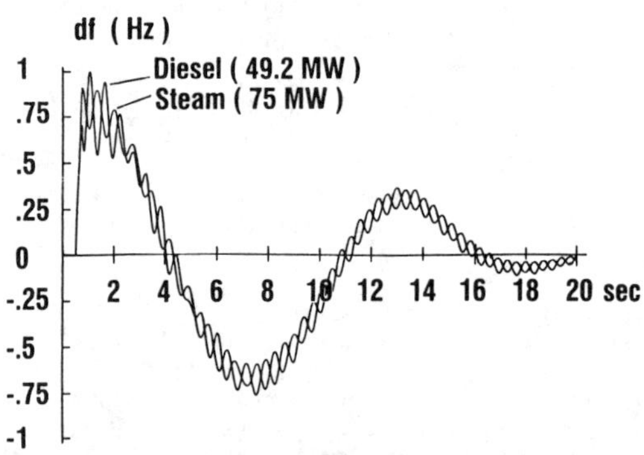

(a)

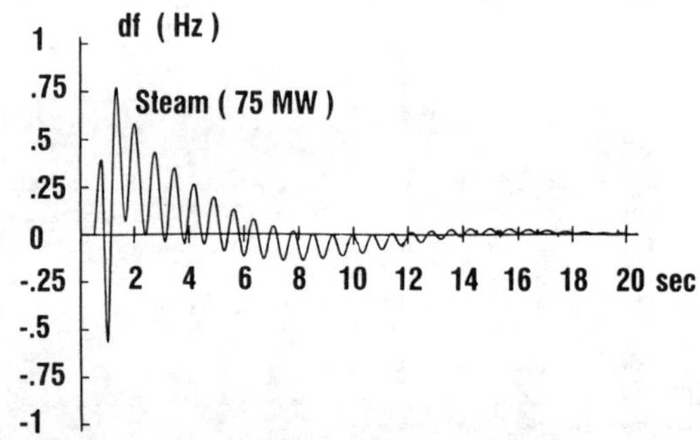

(b)

Figure 3.3 Frequency of the system in the case of three phase short circuit. (a) Without wind turbines (b) With wind turbines

3.2 Short circuits

The main question to be answered is how the transient behaviour of the power system in case of faults and particularly its transient stability limits are affected by the connection of the wind turbines.

In Fig.3.3(a) and (b) the frequency of the system is shown, with and without the wind turbines resp., in the case of a three phase short circuit near the Sitia substation, which is eliminated in 0.2 sec. It is evident that the system undergoes a more serious transient when the wind turbines are not connected.

4. CASE STUDY 2: HIOS

The power system of Hios island includes a Diesel power station with 6 units (G1:2x4.0MW, G2:3x5.85MW, G3:1x11.0MW) and a 20 KV distribution network. One 10x100 KW wind park is already in operation and another 8x300 KW one is in commissioning.

The presented results come from a study carried out in collaboration with Rutherford Appleton Laboratory (RAL) and PPC. Its objective was the development of an assessment tool for the determination of the technically and economically acceptable wind power penetration in medium size islands. Hios island was chosen as a study case. Three wind penetration scenarios were considered and the operation of the system during one year was simulated, in order to calculate the fuel savings for each scenario (for this purpose the Logistic Model of RAL was used). Then, the dynamic behaviour of the system was simulated under certain critical operating conditions, selected from the output of the Logistic Model.

In Fig.4.1 the operation of the system is shown for a high wind day, under the conditions of Scenario 3 (four wind parks with a total installed capacity of 12.7 MW). The points indicated as Cases 1, 2 and 3 were selected for the analysis of the dynamic behaviour of the system, considering disturbances similar to the ones of the previous study case.

In Fig. 4.2(a), (b) and (c), the frequency, the diesel output power and the voltage at the power station are shown, for a linear wind power reduction from 8.9 MW to 5.9 MW in 10 sec, if the system operates as in Case 1 of Fig.4.1. It can be seen that the system remains stable and the frequency and voltage deviations are within the acceptable limits. In case, however, of a step disconnection of the same amount of wind power, the system goes out of step.

5. CONCLUSIONS

High wind penetration levels in large islands can be achieved, provided that their power systems are properly designed to withstand exceptional cases of fast wind variations. Nevertheless, more research has to be carried out concerning the determination of the impact of dispersed wind parks and the frequency of occurrence of marginal wind variations. In any case, the development of Dispatching Centers for the monitoring and control of the whole system, based on short term wind forecasting, seems to be essential for the good operation of the power systems.

REFERENCES

[1] M.Papadopoulos, P.Malatestas, B.Papadias, "Penetration of wind turbines in islands with diesel power stations" - EWEC 1988.

[2] M.Papadopoulos, P.Malatestas, N.Hatziargyriou, "Simulation and analysis of small and medium size power systems containing wind turbines" - Paper presented at IEEE/PES 1991 Winter Meeting.

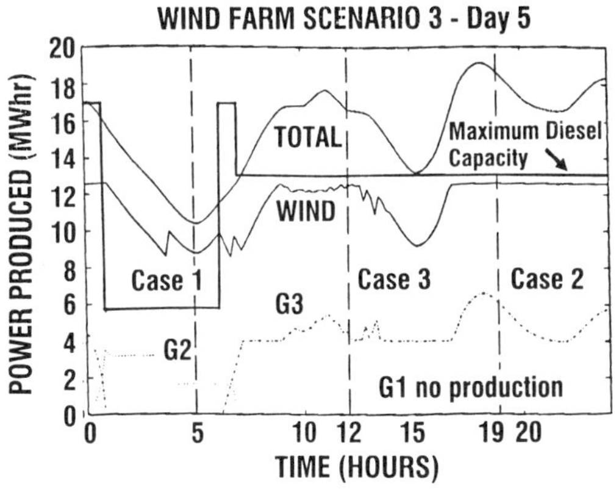

Figure 4.1 Operation of the system during a high wind day, under the high penetration of scenario 3.

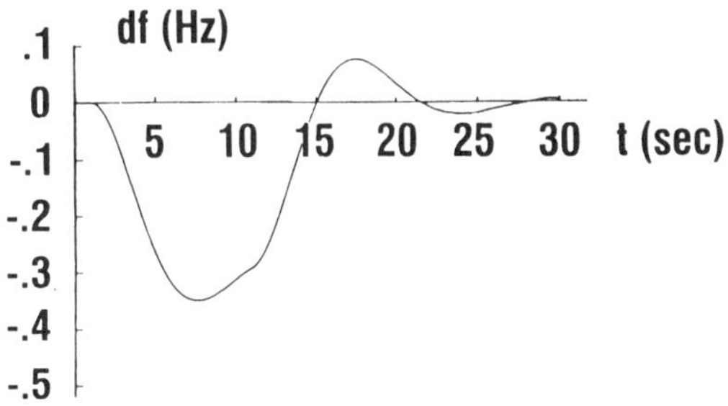

(a)

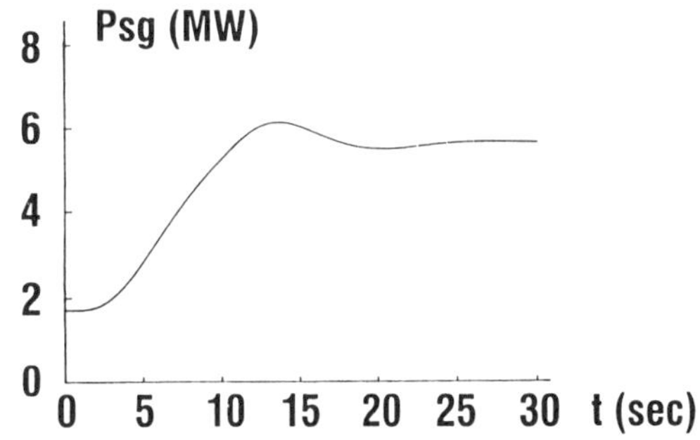

(b)

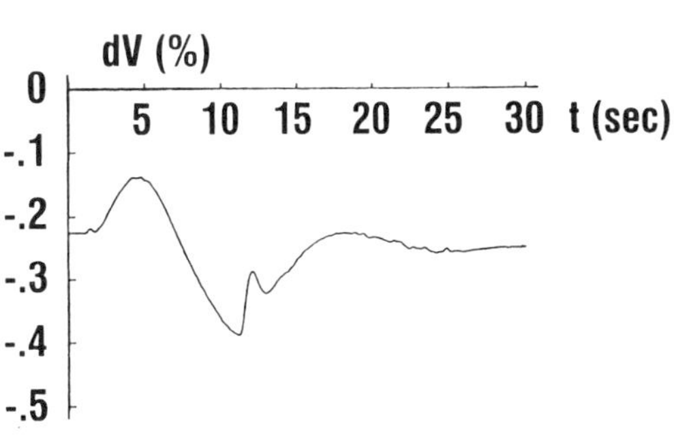

(c)

Figure 4.2 Frequency, diesel output power and voltage after a linear wind power reduction of 3 MW in 10 sec.

Modelling of wind power forecast uncertainty

Dr. Lennart Söder
Department of Electric Power Systems
Royal Institute of Technology
S-100 44 Stockholm, Sweden

Abstract With a large-scale introduction of wind power, the power system operation will be influenced since extra uncertainties are introduced. To study this a more detailed analysis of system operation is required. This analysis has to consider wind speed forecast uncertainties, system load forecast uncertainties and the possibilities of the non-wind power stations to follow the total forecasted and unforecasted variations caused by the system load, wind power and forced outages of all generating units. This paper suggest a method to simulate realistic forecasts of total produced wind power to be used in simulation of the daily operation of the power system. These simulations can be used to estimate the value of better wind speed forecasts and/or the integration cost of wind power.

1 Introduction

With a large scale introduction of wind power it can be assumed that several wind speed forecasts will be available for several sites. A simulation of forecasts of total produced wind power has to be based on realistic estimations of wind speed forecast uncertainty, correlation between forecast errors and knowledge about systematic wind speed forecast errors.

2 Simulation of wind speed forecasts

The method used here is to produce realistic wind speed forecasts from real wind speed time series and simulated realistic wind speed forecast errors. The wind speed forecast errors are divided into one part v_{si} concerning the systematic error which is dependent on the wind speed level and a random part X_i. With real wind speed $v_i(t)$ at site i for hour t, a corresponding wind speed forecast $V_{Fi}(t)$ can be simulated as

$$V_{Fi}(t) = v_i(t) + v_{si}(v_i(t)) + X_i(t) \qquad (2.1)$$

How to simulate $X_i(t)$ is treated in this chapter while v_{si} is treated in appendix C.

2.1 Simulation of one wind speed forecast

In Sweden the possibilities to perform good wind speed forecasts to be used for wind power forecasts have been investigated [2]. The result can be presented as the minimal possible RMSE that can be obtained for different forecast periods using different forecasting methods, see figure 2.1.

A common method to simulate time series is the *autoregressive* method, see appendix B. To include

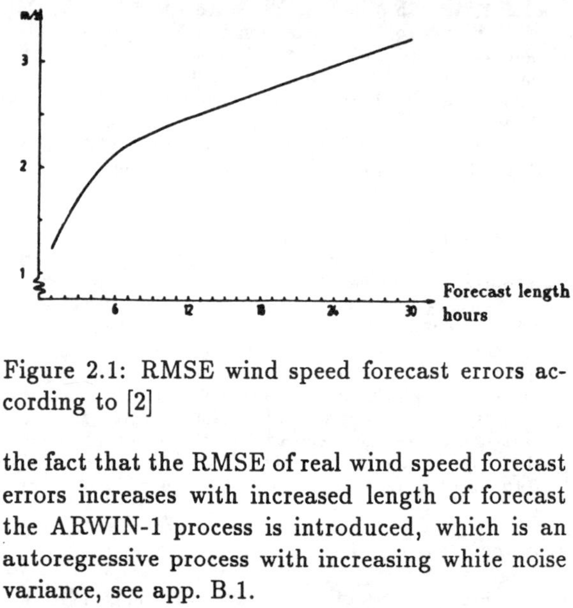

Figure 2.1: RMSE wind speed forecast errors according to [2]

the fact that the RMSE of real wind speed forecast errors increases with increased length of forecast the ARWIN-1 process is introduced, which is an autoregressive process with increasing white noise variance, see app. B.1.

The parameters of an ARWIN-1 model can be estimated by fitting the RMSE-curve of an ARWIN-1 model to the real RMSE:s according to figure 2.1 [1]. The result is shown in table 2.1.

ARWIN-1 param.	$\hat{\Phi}_1$	$\hat{\sigma}_{ZX}$	$\hat{\sigma}_{ZXa}$
value :	0.8073	1.1638	0.2800

Table 2.1: ARWIN-1 parameters of wind speed forecasts according to [2]

2.2 Simulation of simultaneously performed wind speed forecast

In section 2.1 the wind speed forecast was modelled with an ARWIN-1 process. Assume now that several wind power units are located within a limited area. Assume also that the wind speed is forecasted and measured at each site with a wind power unit. The corresponding time series of the wind speed forecast errors have a dependence since the same unforecasted weather situation affects the wind speed (and the wind speed forecast error) at several sites, see appendix A.

The model used here is to simulate the wind speed forecast errors with a Multivariate AutoRegressive process of first order With Increasing white Noise variance, a MARWIN-1 process, according to appendix B.2. Assume two sites having a dependence between wind speed forecast errors according to the mean correlation between six sites in southern Sweden shown in figure A.2. Assume also that at both sites the forecasts have an RMSE according to the forecasts shown in figure 2.1 corresponding to ARWIN-1 parameters shown in table 2.1. With these ARWIN-1 parameters the MARWIN-1 parameters can be estimated by fitting a curve of correlations of two ARWIN-1 processes to a curve of real correlations [1]. The result, i.e. the parameters of the MARWIN-1 process, are shown in table 2.2. In figure 2.2 the mean correlation between the six sites in southern Sweden from figure A.2 are shown together with the correlation of the MARWIN-1 process with parameters according to table 2.2.

parameter	value
$\hat{\Phi}_1$	0.8073
$\hat{\alpha}$	0.3155
$\hat{\sigma}_1$	0.8408
$\hat{\sigma}_3$	0.2621
$\hat{\sigma}_2$	0.8047
$\hat{\sigma}_4$	0.0984

Table 2.2: MARWIN-1 parameters of two correlated ARWIN-1 processes

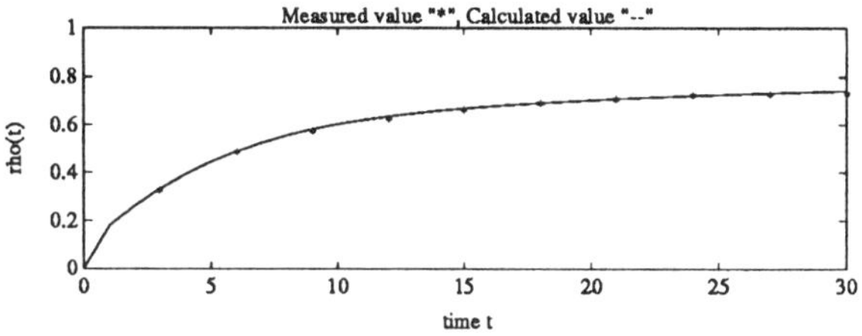

Figure 2.2: Measured correlation and MARWIN-1 process correlation

3 Simulation of wind power forecasts

In this chapter the simulated wind speed forecast errors in chapter 2 will be used to simulate wind power forecasts. The Näsudden II unit will be used to illustrate the methodology of how to simulate wind power forecasts. Figure 3.1 shows the power production at different wind speeds of this wind power unit.

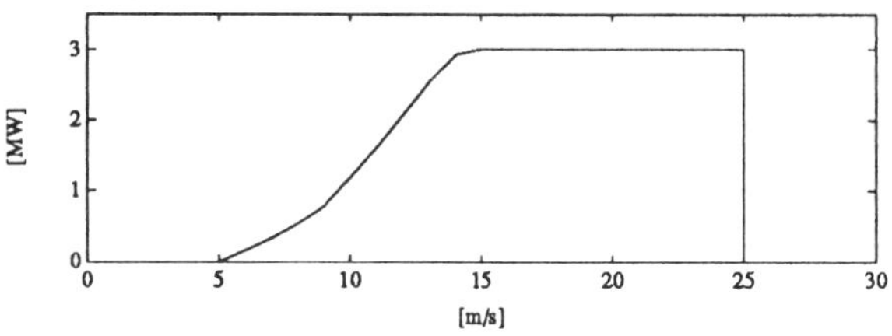

Figure 3.1: Wind speed - power transfer function $P_N(\cdot)$ of Näsudden II

3.1 Simulation of forecasted power from one wind power unit

The forecasted power for one wind power unit at hour t, P_{F1t}, can be simulated as

$$P_{F1t} = P_N\left(V_{Fi}(t)\right) \tag{3.1}$$

In figure 3.2 the power from one Näsudden II wind power unit with wind speeds from Maglarp at Period 1 (=February 5, 1980), is shown as a straight line. The figure also shows 10 examples of simulated wind power forecast, calculated with eq. 3.1, where the wind speed forecast calculations uses eqs. 2.1, B.3, C.1 and table 2.1.

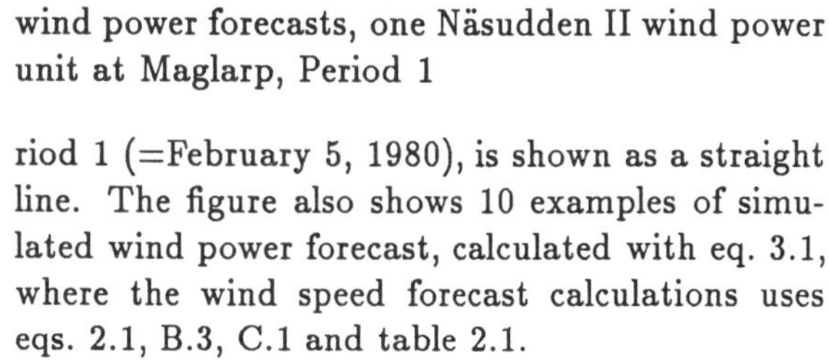

Figure 3.2: Real wind power '-', and 10 simulated wind power forecasts, one Näsudden II wind power unit at Maglarp, Period 1

3.2 Simulation of forecasted power from several wind power units

The method of how to simulate the forecast of power from several wind power units will here be illustrated with an installation of 400 Näsudden II units (= 1200 MW) installed in the coastal areas of Skåne in southern Sweden. It can be approximated that the mean correlation between wind speed forecast errors of the 400 units is about the same as the mean correlation between wind speed forecast errors at the six sites in figure A.1.

The forecasted power from the 400 units can now be estimated as

$$P_{F400t} = \sum_{i=1}^{10} 40 \cdot P_N\left(V_{Fi}(t)\right) \qquad (3.2)$$

where all X_{it}:s are simulated as a MARWIN-1 process with parameters according to table 2.2. The measured wind speeds at the six sites in figure A.2 have been used for the 10 sites with wind farms.

In figure 3.3 the real power from the 400 Näsudden II wind power units is shown as a

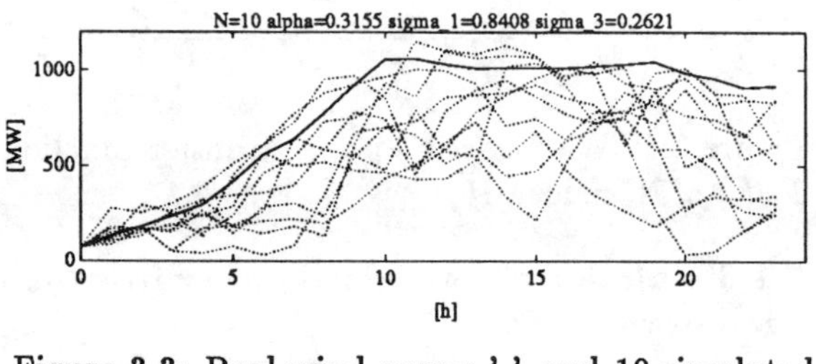

Figure 3.3: Real wind power '-', and 10 simulated wind power forecasts, 400 Näsudden II wind power units, Period 1

straight line. The figure also shows 10 examples of forecasts of total wind power calculated with eqs. 3.2, 2.1, B.4, C.1 and table 2.2.

In figure 3.4 the RMSE of the total wind power forecast error at this specific day is shown.

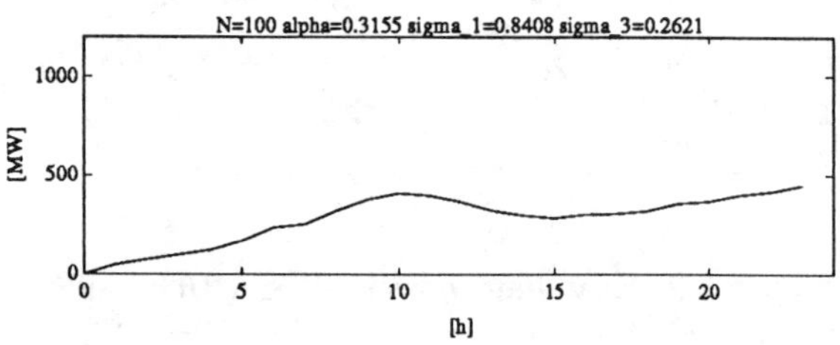

Figure 3.4: RMSE of 100 simulated time series of total wind power

A Wind speed forecast error correlations in Sweden

To determine the uncertainty of total produced wind power it is necessary to have an estimation of the correlation between the forecast errors at different sites.

To obtain an estimation of the correlation between wind speed forecast errors several performed persistance forecasts have been used. Simultaneous wind speed time series have been available for the year 1980 for the six sites in the southern part of Sweden, see figure A.1 .

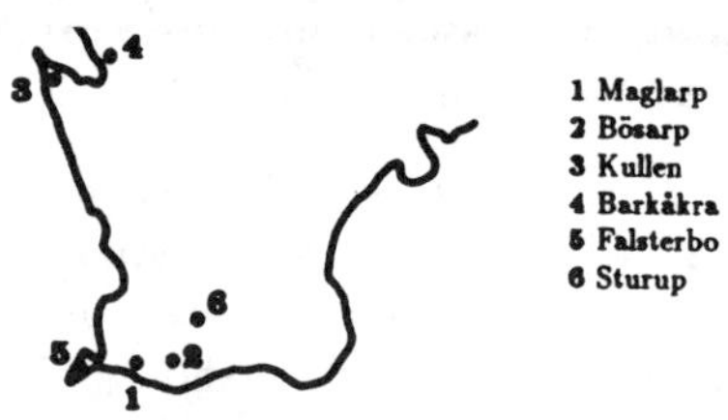

Figure A.1: Six sites with wind speed measurements in southern Sweden

In figure A.2 the mean correlation for different pairs of sites in figure A.1 are shown as a function of forecast length.

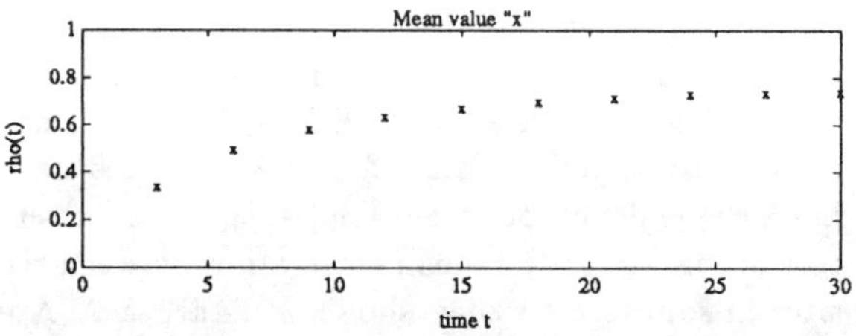

Figure A.2: Mean value of correlations between the 15 sites in figure A.1

B Autoregressive processes

A stochastic model which is commonly used in the representation of certain practically occuring time series is the so-called *autoregressive* model. In this model, the current value of the process is expressed as a finite, linear aggregate of *previous values of the process* and a white noise process Z_{Xi} [1].

The first-order autoregressive process, AR-1, can be written as

$$X_t = \Phi_X X_{t-1} + Z_{Xt} \qquad (B.1)$$

where $|\Phi_X| < 1$ and $\sigma^2(Z_{Xt}) = \sigma^2_{ZX}$

B.1 First-order AutoRegressive process With Increasing white Noise variance, ARWIN-1

For the ordinary AR-1 process in equation B.1 the white noise that was added in each step was assumed to have a constant variance independent of the time. Assume now that the variance of the white noise is increasing linearly with time i.e.

$$\sigma^2_{ZXi} = \sigma^2_{ZX} + i \cdot \sigma^2_{ZXa} \qquad (B.2)$$

where both σ^2_{ZX} and σ^2_{ZXa} are constant which implies that σ^2_{ZXi} increases linearly with time. This implies that eq. B.1 can be written as

$$X_t = \Phi_X X_{t-1} + Z_{Xt} + \sqrt{t} Z_{Xta} \qquad (B.3)$$

where Z_{Xt} and Z_{Xta} are independent white noise processes having variances σ^2_{ZX} and σ^2_{ZXa} respectively.

B.2 Multivariate AutoRegressive process of first order With Increasing white Noise variance, MARWIN-1

Assume first two correlated ARWIN-1 processes with the same ARWIN-1 parameters according to eq. B.3 :

$$\begin{aligned} X_{1t} &= \Phi_X X_{1(t-1)} + Z_{X1t} + \sqrt{t} Z_{X1ta} \\ X_{2t} &= \Phi_X X_{2(t-1)} + Z_{X2t} + \sqrt{t} Z_{X2ta} \end{aligned} \quad \text{(B.4)}$$

Z_{X1t} and Z_{X2t} are correlated as well as Z_{X1ta} and Z_{X2ta}. The correlation between these two processes must behave as real correlations according to appendix A. To obtain this, the four added noises can be defined as functions of independent white noises as :

$$\begin{aligned} Z_{X1t} &= \sqrt{1-e^{-\alpha t}} Z_{X0t} + \\ &+ \sqrt{e^{-\alpha t}} Z_{X1tb} + Z_{X1tc} \\ Z_{X2t} &= \sqrt{1-e^{-\alpha t}} Z_{X0t} + \\ &+ \sqrt{e^{-\alpha t}} Z_{X2tb} + Z_{X2tc} \\ Z_{X1ta} &= Z_{X0ta} + Z_{X1td} \\ Z_{X2ta} &= Z_{X0ta} + Z_{X2td} \end{aligned} \quad \text{(B.5)}$$

where the variances of the white noises on the right hand side of eq. B.5 are :

$$\begin{aligned} \sigma^2(Z_{X0t}) &= \sigma^2(Z_{X1tb}) = \sigma^2(Z_{X2tb}) = \sigma_1^2 \\ \sigma^2(Z_{X1tc}) &= \sigma^2(Z_{X2tc}) = \sigma_2^2 \\ \sigma^2(Z_{X0ta}) &= \sigma_3^2 \\ \sigma^2(Z_{X1td}) &= \sigma^2(Z_{X2td}) = \sigma_4^2 \end{aligned} \quad \text{(B.6)}$$

In figure B.1 the variances of the two processes and the correlation coefficient are shown.

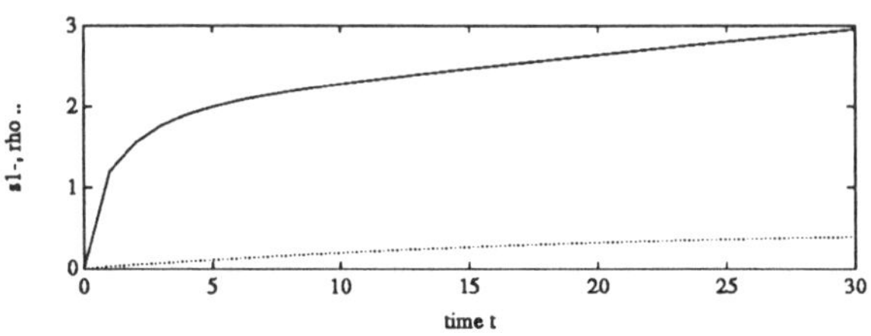

Figure B.1: Standard deviations and correlations of two ARWIN-1 processes $\Phi_{X1} = \Phi_{X2} = 0.8$, $\sigma_1 = 0.4$, $\sigma_2 = 1.1$, $\sigma_3 = 0.2$, $\sigma_4 = 0.15$, $\alpha = 0.04$

C Wind speed forecasts at increasing or decreasing wind conditions

To obtain realistic wind speed forecasts it is necessary to take into account systematic errors for certain situations. With e.g. a persistance forecast, an increasing trend is by definition impossible to forecast.

Assume that the real wind speed increases with 5-7 m/s within 9 hours from the hour when the forecast is performed. For 88 studied forecasts [1] this is the case for 15 forecasts. In figure C.1 the forecast errors of these forecasts are shown. As

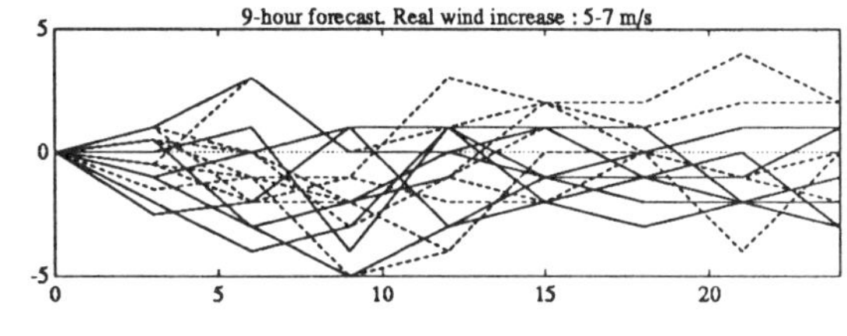

Figure C.1: Wind speed forecast errors for 15 days with increasing real wind

shown in the figure the forecast errors for the 9-hour forecast (=15.00 GMT) are mainly negative, i.e. the increase of real wind speed is underestimated. The mean value of the 15 forecast errors at 15.00 GMT is -1.87 m/s. It should be noted that this value is much better than -6 m/s which is the value for persistance forecasts.

Corresponding values for other forecast lengths and increase or decrease of real wind speeds (v_{ch}), can be calculated for mentioned 88 forecasts [1]. The mean values of all systematic errors (v_s) are shown in figure C.2. The figure also shows a linear

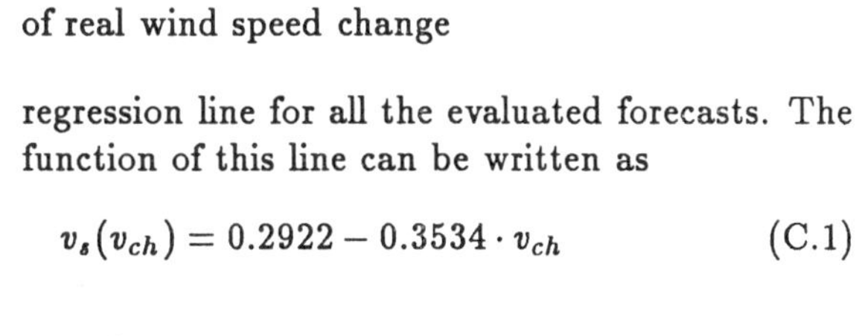

Figure C.2: Systematic forecast error as function of real wind speed change

regression line for all the evaluated forecasts. The function of this line can be written as

$$v_s(v_{ch}) = 0.2922 - 0.3534 \cdot v_{ch} \quad \text{(C.1)}$$

References

[1] Söder L., *Modelling of wind power forecast uncertainty* Royal Institute of Technology, Stockholm, Sweden, (1993)

[2] Törnevik H., Olsson E. and Gollvik S., *Forecast of wind in the layer between 50 and 150 meters*, Swedish Meteorological and Hydrological Institute, Promis-report, nr 4 (1985), (in Swedish)

GRID CONNECTION OF WIND ENERGY CONVERTERS

S. Heier, W. Kleinkauf

Universität Gh-Kassel
Institut für Elektrische Energietechnik
Wilhelmshöher Allee 73
3500 Kassel

Institut für Solare Energieversorgungstechnik e.V.
(ISET)
Königstor 59
3500 Kassel

Summary

Test results from differently designed wind energy plants have shown the existence of certain problems. These problems are presented in this paper and the solutions evaluated. The main points of consideration are wind energy plants with asynchronous generators and plants with grid commutated inverters. The insufficient quality of current which is fed into the grid results in high losses in the capacity of the grid connecting points which seriously restricts the spread and use of wind energy.

This paper also considers more sucessful ways of integrating wind energy plants and improving their grid compatibility. It is here that both the design of wind farms and the capacity of grid connecting points play an important role. Test results are presented which show how systems with a dc interconnecting circuit may significantly reduce their harmonics by using special control measures. Methods are described for the reduction of both distortion harmonics and harmonics of a certain order. Furthermore, this paper also describes the possible ways of improving the current quality by using a combination of harmonics-afflicted converter plants with asynchronous generator systems.

1 Introduction

Wind energy plants are mostly used in parallel grid operation. It is possible to use them in single operation or in the configuration of a wind farm.

The ways of connecting wind energy converters (WEC) with grids do not only depend on grid connecting conditions and the capacity of the grid connecting points, but also on the electrical characteristics of the plant. As well as the wind dependent fluctuations of the input power, loads are also especially problematic, which are related to the reactive power rate and the harmonics generated by WECs.

2 Technical aspects (of the plant)

With regard to the electrical supply and to the grid influences, the main components of a wind energy converter are: the turbine, the mechanical gear drive with generator, the electrical power transfer, and the control and supervisory systems which coordinate the whole converter.

Turbines of plants with a power limit of approx. 500 kW are mostly designed with fixed rotor blades. The input power is limited by the so-called stall operation. Larger units, but also some smaller ones, have blades which can change their pitch in longitudinal axis. The power input from the wind can thus be varied by changing the blade pitch angle, which is achieved by using an adjustable drive motor (e.g. hydraulic, electric motor). Thus, limiting the power is always possible at such plants by using control and supervisory systems.

For mechanic-electrical energy conversion of wind energy converters, three phase machines are used because of their stable construction. These generators are principally divided into asynchronous and synchronous generators with direct grid connection as well as converter coupling. Furthermore, asynchronous generators can be classified as cage rotor or slip-ring machines. Finally, synchronous machines can be divided into two groups: generators with a permanent magnet excitation or generators with an electrical excitation unit. The latter may be subdivided into self-excited, brushless or slip-ring versions. Mostly, asynchronous generators with direct grid coupling are used. Systems with a synchronous machine, a rectifier, a dc interconnected circuit and a grid commutated inverter allow the turbine rotating speed to be decoupled from the grid frequency. These are becoming more and more important because they reduce the strains on the mechanical parts and the electrical power variations.

3 Grid Influences

The integration of wind energy converters in the mains produces a number of interactions. The following points should be observed:

- general compatibility
- voltage variations with possible flicker effects
- voltage asymmetries
- harmonics
- interharmonics
- disturbance emissions.

Large grid interactions are to be expected in weak grids where wind energy contributes much of the power. However, practical tests have shown, that is possible for the wind energy share to be as high as 100 % of the feed-in power when the grid and its components are specially designed. By using measures which are able to influence the characteristics of the grid, possible grid influences may be avoided by integrating a phase shifter, a battery storage with converter, a grid filter, a compensation unit and a grid controller (see fig. 1). Thus it is even possible to integrate photovoltaic systems with an inverter feed-in.

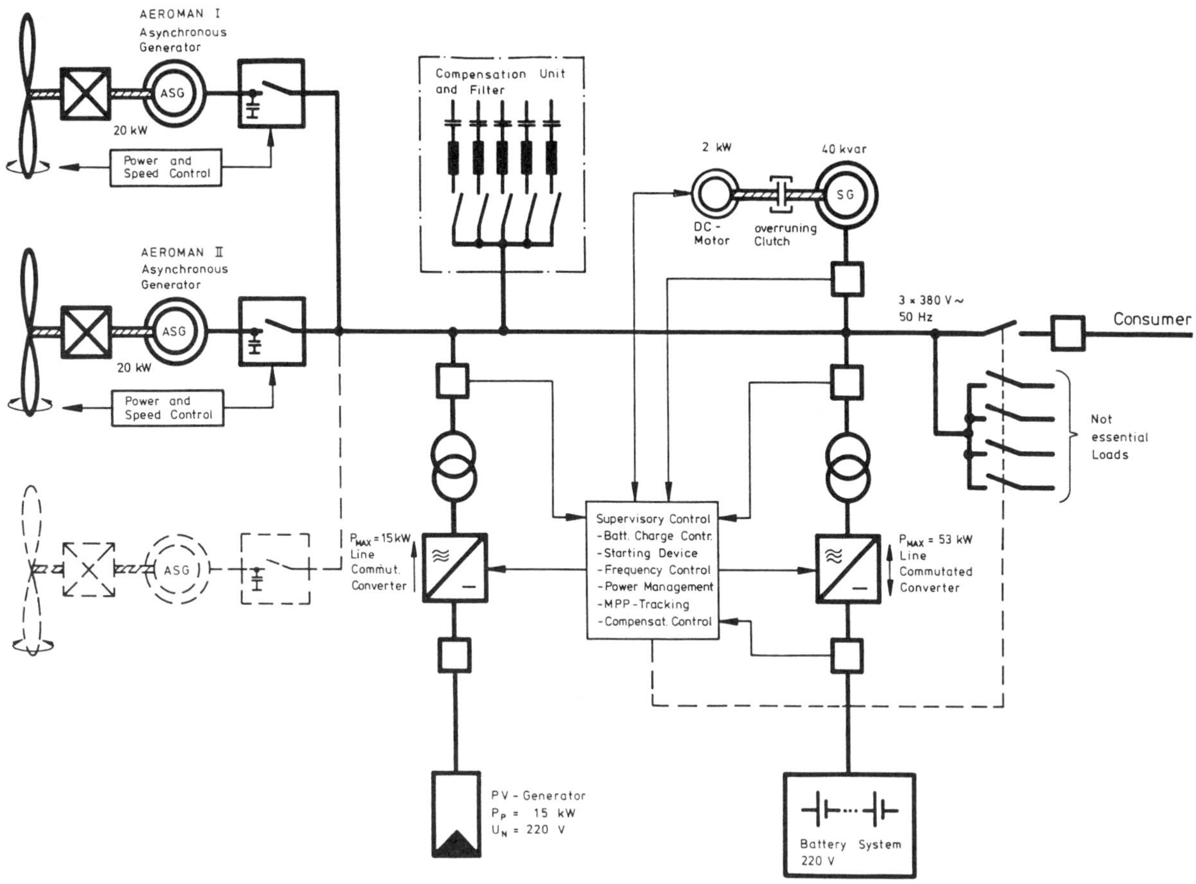

Fig. 1: Electrical energy supply with a wind-photovoltaic-system and battery storage (village supply in Jurf, Jordan)

Figure 2 clearly shows the effects of a filter and compensation unit of the supply system from figure 1. Figure 2a) and 2b) show the voltage (L_1) in both the frequency and time domains at approximately half of the rated load. Fig. 2a) presents the results with the filter unit turned off, and fig. 2b) with the filter unit in operation. It is not of central importance whether the harmonics are produced by the consumers, the photovoltaic plant, the battery storage or the wind energy converters.

In fig. 2a) the voltage spectrum shows a broad range of harmonics which are reduced in fig. 2b) to only a few typical harmonics as found in 6 pulse converters. The waveform in fig. 2b) is almost sinusoidal which shows the function of the filter unit.

In small grids, where wind energy makes up a large share, the use of a replacement energy source which has a short-time power control (e.g. battery with inverter) is the main problem for achieving a stable grid, because of the changing wind energy supply and the varying load behaviour of the consumers. In contrast to this, interconnected grids have a relatively low wind energy content which means that the harmonics are the dominating grid influence.

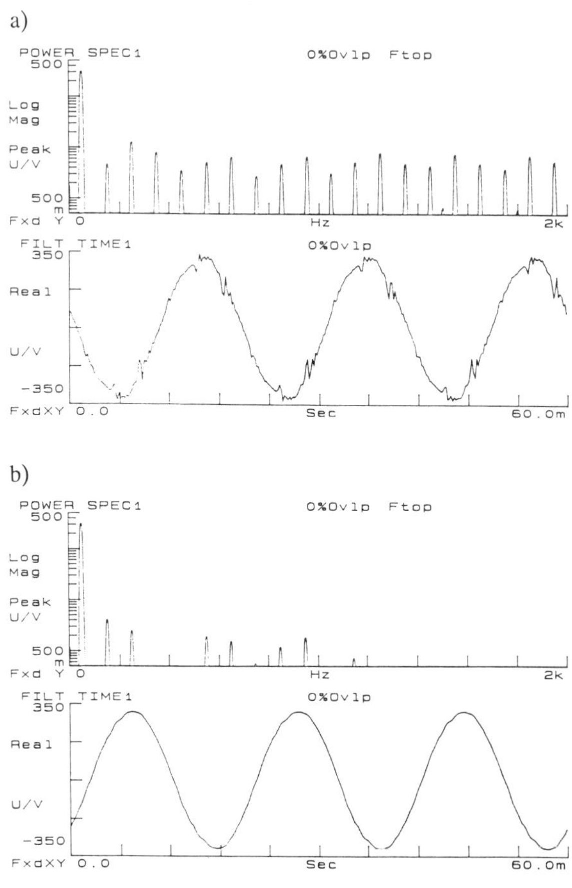

Fig. 2: Frequency spectrum and time waveform of the voltage for the village supply
a) without b) with filter and compensation unit

4 Harmonics

Different converter systems react differently at various grid connections. Increasing numbers of asynchronous generators which are directly connected to the grid generally do not show an increase in grid influences. As measurements have shown, harmonics and interharmonics are mostly reduced.

In contrast, an increasing number of (wind energy) plants with converter feed-in, results in a higher harmonic factor. It is here that grid commutated 6-pulse inverters generate stronger grid influences at the 5th and 7th harmonic compared to a 12-pulse inverter.

By combining wind energy plants which feed into the grid via a converter with directly grid coupled asynchronous generators, it is found that due to the increased short circuit power and the filtering effects of both the generator and compensator reactances, there is a significant reduction in grid influences (see fig. 3a, b).

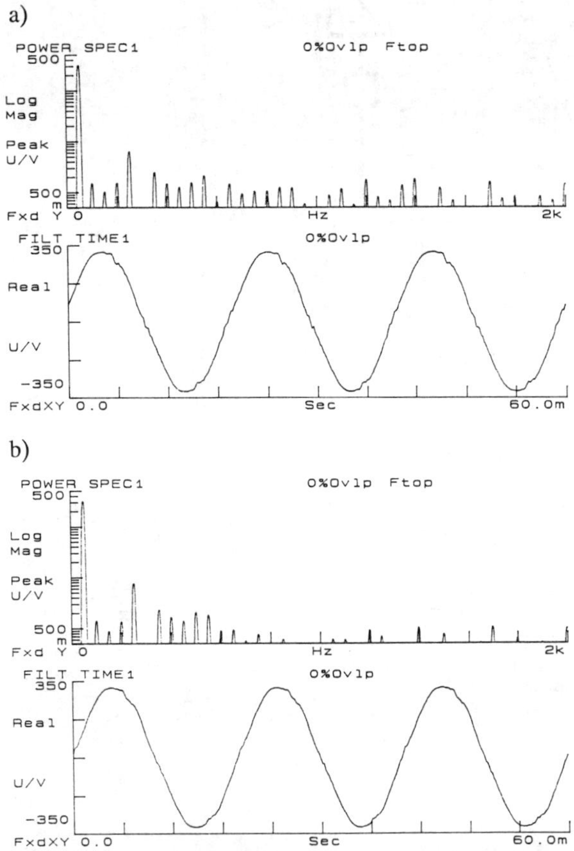

Fig. 3: Frequency spectrum and time waveform of the voltage at
a) 275 kW feed-in power via 6-pulse inverter
b) 320 kW feed-in via 6-pulse inverter and 320 kW feed-in via directly grid coupled asynchronous generator

For the feed-in of power from big wind energy converters into weak grids like the existing grid branches in windy regions on the coast, large grid influences have to be expected. An example for this is the WKA 60 II with 1.2 MW rated power, connected to the Marne grid branche at Kaiser-Wilhelm-Koog with approx. 50 MVA short circuit power of the grid.

The WKA 60 II feeds the electrical energy into the grid via a converter with controlled rectifier, a dc interconnected circuit and a grid commutated 6-pulse bridge circuit inverter. The synchronous generator is electrically excited, which allows a change in the load angle and thus a change in the reactive power supply for the machine controlled rectifier.

Due to the situation of the grid in Kaiser-Wilhelm-Koog, the setting up of the 1.2 MW-plant has produced grid influences of new dimensions. During the test trials, in spite of the fact that grid filters had been installed to use the conventional method of being acceptor circuits for the 5th and 7th harmonics, high harmonic levels were found at an existing grid resonance point around the 25th to 29th ordinal numbers. This phenomenon can be attributed more to the constellation of the grid than to the influences of the wind energy converter itself, which means that the plant would operate well if it were connected to a grid with a larger short circuit power. Furthermore, this example clearly shows the limitations of filter units.

In order to adhere to the energy supply companies' regulations on compatibility levels at grid connecting points, the whole electricity supply system has been specially equipped with suitably designed filters. Such a procedure is not only very work intensive, but also requires a great deal of exact information about of the grid and the plant.

A new method, which will have to be examined in more detail, concerns the installation of several wind energy converters with line commutated inverter feed-in, where the plant inverters work in the grid with different firing angles. With two plants it is possible to set the firing angles of the two inverters by selecting the voltage of the interconnected circuits or transformers. If the two systems are then staggered by a time period of half a cycle the critical 29th harmonic which is of interest here will be erased (see fig. 4).

Here it is assumed that with a constant dc voltage converter can operate normally up to a voltage break-down of 15 %, which corresponds to a rated firing angle of α_{1N} = 138°. Thus, according to fig. 5 the rated firing angle α_{2N} = 131.79° must be for the inverter 2 to achieve the lowest amplitude of the 29th harmonic. The amplitude spectrum of the superimposed grid current (calculated value) clearly shows a large reduction in all harmonics between the 11th and 49th ordinal number (see fig. 6). Therefore, this method which uses a precise system layout is able to reduce the grid influences and save on additional construction expenses. This not only applies during rated operation but also there is only a partial load.

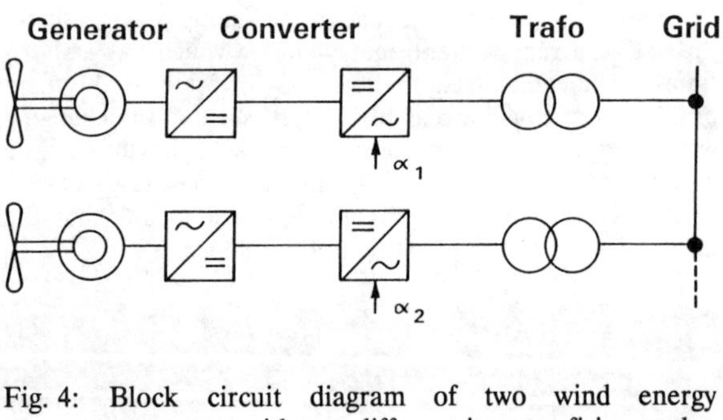

Fig. 4: Block circuit diagram of two wind energy converters with two different inverter firing angles

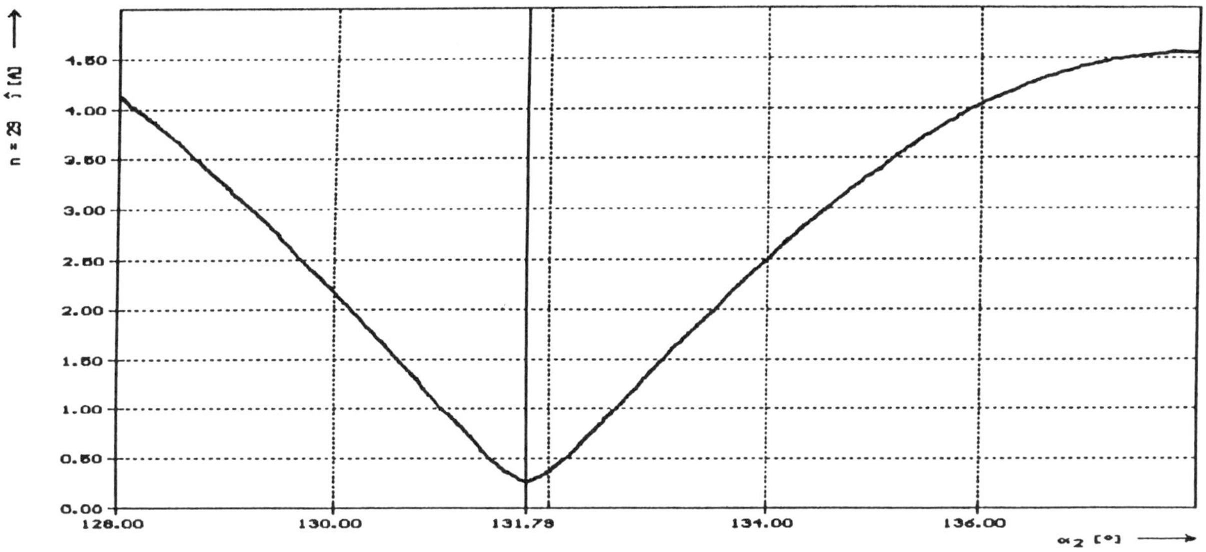

Fig. 5: Current amplitude of the 29th harmonic depending on the firing angle α_2

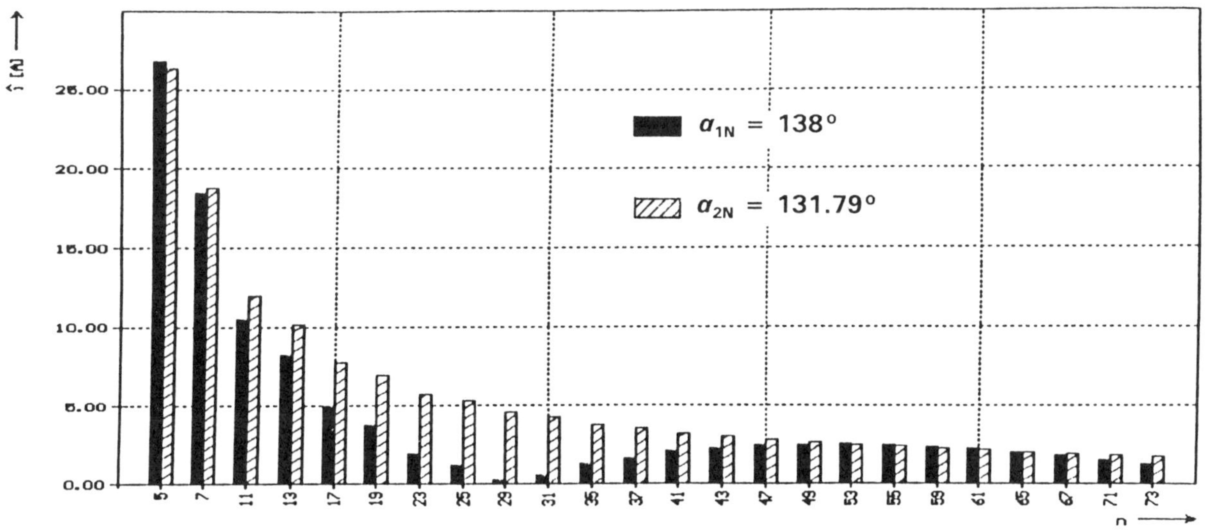

Fig. 6: Amplitude spectrum of the superimposed grid circuit current at α_{1N} = 138° and α_{2N} = 131.79°

Bibliography:

[1] Heier, S.; Kleinkauf, W. et al.: Betriebsverhalten von Windenergieanlagen
- Regelung und dynamisches Verhalten
- mechanisch-elektrische Energiewandler. BMFT - Forschungsbericht T 84-154, ISSN 0340-7608, 1984

[2] Heier, S.: Grid Influence by Wind Energy Converters. International Energy Agency, Expertmeeting Göteborg 1991

[3] Stromversorgungseinrichtung mit mindestens zwei Stromrichtern. Pat. ang. 1992 Kennzeichen P 42 32 356.8

APPLICATION OF A MESO-SCALE GRID INTEGRATION MODEL TO CHIOS ISLAND, GREECE

A G DUTTON, J A HALLIDAY
Energy Research Unit
Rutherford Appleton Laboratory
Chilton
Didcot
Oxon OX11 0QX
UK

ABSTRACT

The application of a time series meso-scale grid integration model to the Greek island of Chios is described. The characteristics of the conventional diesel power station are accurately modelled so that the plant mix and annual fuel consumption are well simulated. The importance of characterising the diesel power station correctly is stressed and in this respect the inclusion of an algorithm to model maintenance scheduling was found to be essential. Using this model, three scenarios of wind penetration were explored. The potential fuel savings and required changes to operating strategy of the conventional power station are considered.

1. INTRODUCTION

This paper summarises the findings from a recent CEC-funded project [1] to apply a meso-scale grid integration model to the Greek island, Chios. The time series model, which simulates operation of an island power station with conventional diesel, gas and steam turbine power generation before and after the introduction of wind power generation, was originally developed as part of a Scottish Hydro Electric/University of Strathclyde/RAL project to examine the wind potential of Shetland [2]. The model was extensively tested and validated for the "no wind" case. Although a full validation for an island with significant wind penetration is still required, the model provides a basis on which to draw preliminary conclusions about the long term impact of wind power generation on the operation of a given grid.

Several improvements have been made to the model for the current project. These improvements include the addition of an algorithm to model the effect of maintenance on plant scheduling, extended graphics facilities, and minor adjustments to the scheduling algorithm for conventional plant.

The model is a simulation of power station operation and in no way attempts to optimize any of the parameters concerned with operation of the power station. It is essential that an extensive and accurate description of the plant and its operation is obtained. Experience shows that this must be obtained from the operators themselves, since design guidelines are rarely followed to the letter.

The model includes a range of different forecasting techniques for both load and wind forecasting. Detailed studies of the benefits of improved wind and load forecasting have been made using the same grid integration model [3,4] but these techniques have not been used in the current study.

2. DESCRIPTION OF POWER STATION AND POTENTIAL WIND FARM SITES

The Chios island power station comprises six diesel generator sets with a total installed nameplate capacity of 36.5 MWe. In practice, these units are considered to have a total capacity of 29.0 MWe. There are three different makes of diesel corresponding to the three different sizes (see table I).

The characteristics of the diesels are described in more detail in the final report [1], but an example of the load ramp characteristic for the largest diesel is shown in figure 1.

The wind turbines were considered to be distributed between three wind farms and various scenarios of increasing wind penetration were defined (see table II). Since the intention of the model is to assess the potential of a given grid to absorb wind power, it was not considered necessary to model each individual wind turbine. Instead, the turbines at each farm were grouped into a single 'mega' turbine seeing a single wind speed and with an appropriately scaled power curve. This strategy can lead to large oscillations in the available wind power when the speed at a given farm exceeds the furling speed. In the model, the whole wind farm is lost in one timestep, whereas in reality it is expected that the turbines will be furled at different times. This leads to a more severe condition than might normally be the case. It does not, of course, negate the need to examine the short term dynamics of the grid, but that is beyond the scope of this paper.

Table I : Actual rating of diesel generator sets in Chios power station

Diesel type	Group	Rating	Number
4.0 MWe Fiat	G1	3.5 MW	2
5.85 MWe GMT	G2	4.0 MW	3
11.0 MWe Cegielski	G3	10.0 MW	1

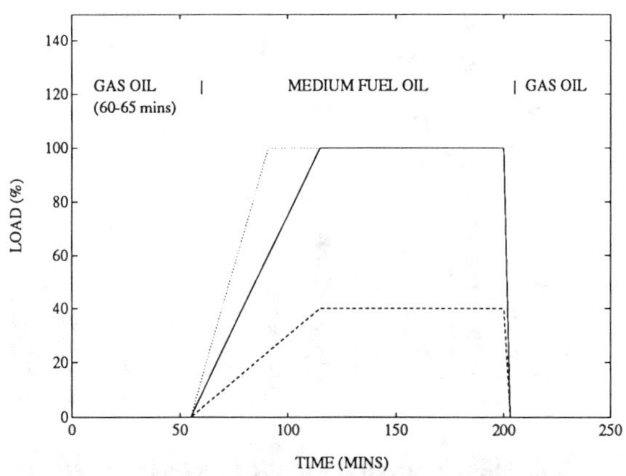

Figure 1 : Load ramp/deramp characteristics of the 11.0 MW diesel unit showing part load limit (dashed line), hot start characteristics (dotted line) and regimes of fuel use from the moment a decision was made to start the unit

Table II : Wind penetration scenarios for Chios island

Scenario 1 Total capacity 3.4 MW	Region A : 10 x 100 kW Region B : 8 x 300 kW
Scenario 2 Total capacity 9.1 MW	Region A : 10 x 100 kW Region B : 8 x 300 kW Region C : 19 x 300 kW
Scenario 3 Total capacity 12.7 MW	Region A : 10 x 100 kW Region B : 8 x 300 kW Region C : 31 x 300 kW

3. MODEL VALIDATION : SIMULATION OF POWER STATION OPERATION WITHOUT WIND

Two types of model validation were carried out for the "no wind" case. The first was to ascertain that the frequency of stops and starts matched that occurring in practice, that the correct size of diesel unit to start and stop was selected, and that decisions were made at the correct time. This was carried out using a week (17/3/1991-23/3/1991) of 10-minute demand data. The second validation was carried out using hourly demand data to compare the net fuel consumption and total generation by each of type of plant during a full calendar year (1990). In the second case the hourly demand data was interpolated to 10-minute values with the same statistical distribution about the hourly values as found during the analysis of the week of 10-minute data. For both types of validation the basic timestep, decision time interval, and type of load forecast were varied to find the optimum values for simulating actual operation of the power station.

Spinning reserve is defined in terms of load margin and wind margin. Even without wind, a certain load margin must be maintained in order to cover unexpected fluctuations in the demand. In the model this is expressed as a fraction of the total load at the current timestep. Similarly, an extra spinning reserve to cover drops in the wind is defined in terms of the wind margin - a fraction of the total wind power at the current timestep.

The model simulates actual decision making within the power station at a decision interval chosen by the user (typically 30 minutes). At these times predictions of demand and available wind power are made for a given number of decision intervals ahead. Predictions of both variables may currently be made on the basis of persistence (the current value is assumed to remain constant throughout the decision interval) or perfect (the future values are known to the decision algorithm) forecasts. Predictions of demand may also be made based on the values pertaining the previous day or week, with options of scaling by a factor reflecting the change in load between days or weeks, as appropriate, and/or including a randomisation factor. On the basis of whether there is an expected surplus or deficit, decisions are made whether to stop or start diesel generators. Since the predictions are only made at the decision intervals and wind farms are lumped under a single wind regime, it is possible that, even with 'perfect' forecasting, large oscillations in wind power (such as when a whole wind farm is furled) can give rise to loss of load events (ie power cuts).

3.1. Simulation of short term diesel power station operation

Table III shows the effect of load forecasting technique and load margin on the number of diesel starts predicted by the model compared with the number which actually occurred. Note that the number of diesel starts (and, in fact, the fuel consumption) increases gradually as the load margin is increased. This is the price that must be paid for increased security of supply. Forecasting based on load values the previous day does not appear to confer any benefit, but this is not true since, in fact, the single loss of load event occurs on the first day when no previous data is available and persistence forecasting has had to be used.

Table III : Effect of load forecasting technique on net fuel consumption/diesel starts

Load forecast technique	Load margin	Diesel starts	Comments
Actual operation		20	
Perfect	0.0	15	No loss of load events
Persistence	0.0	15	1 loss of load event
Persistence	0.3	28	No loss of load events
Past day/same hour scaled for change in load between days	0.0	19	1 loss of load event
Past day/same hour scaled for change in load between days	0.1	25	1 loss of load event

3.2. Simulation of long term (1 year) diesel power station operation

A comparison of the plant mix over a full year using load forecasting based on the past day, same hour scaled by a factor reflecting the change in load between days (load margin 0.1) is shown in table IV. It can be seen that the plant mix used by the model is very similar to that used in actual operation. All the wind penetration studies reported below are carried out using this load forecast technique.

Table IV : Power output during 1990 (MWhr)

Unit sizes	Actual power output (MWhr)	Modelled power output (MWhr)
4.0 MW	8,031	6,276
5.85 MW	28,917	27,814
11.0 MW	54,236	57,514
TOTAL	91,184	91,604
Fuel used (GJ)	807,602	809,992

It was also found that if maintenance is not included in the model then the power output of the 11.0 MW unit is some 20,000 MWhr greater, since it is used as base load plant whenever available. Since it is the most efficient unit, this, in turn, reduces the net fuel consumption by some 50,000 GJ (6%). Since, high wind penetrations have a similar effect of displacing the 11.0 MW unit in favour of the smaller units, the fuel savings due to wind power are often reduced accordingly.

4. WIND PENETRATION STUDIES

4.1. Effect of wind margin

Once the conventional power plant had been satisfactorily

Table V : Effect of wind penetration on total power station fuel consumption for simulations with load prediction based on the same time the previous day (scaled by a factor reflecting the change in load between days)/persistence wind forecasting

Scenario	Wind margin	Wind energy generated (MWhr)	Wind energy spilt (MWhr)	Total fuel consumption (GJ) (Loss of load events)	Fuel saved (GJ)	Fuel saved (GJ/MW) per unit of installed wind power
NO WIND				809,992 (1)		
Scenario 1 3.4 MW total capacity	0.0	17,022	55.5	656,632 (6)	153,360	45,106
	0.4	16,931	146.4	656,685 (1)	153,307	45,090
Scenario 2 9.1 MW total capacity	0.0	35,073	2,785	525,120 (71)	284,872	31,305
	0.6	31,243	6,615	532,064 (1)	277,928	30,542
Scenario 3 12.7 MW total capacity	0.0	41,720	9,262	467,929 (176)	342,063	26,934
	0.2	41,018	9,965	476,663 (68)	333,329	26,246
	0.4	37,365	13,618	493,164 (20)	316,828	24,947
	0.6	34,819	16,164	501,691 (2)	308,301	24,276

modelled, the wind farms were introduced to the model. Table V shows the simulated wind power generation for the three scenarios defined in section 2. At high wind penetration, if a large number of loss of load events are to be avoided, then it is necessary to run at high wind margin.

Figures 2a and 2b show a summary of the daily power production figures compared to the demand for scenario 3 with a wind margin of 0.6. At this wind margin the large 11.0 MW unit (G3) is run continuously, except when it is off-line for maintenance (as can be seen in figure 2b).

This has a major effect on the projected fuel savings to the extent that scenario 1 has savings of 45,100 GJ of fuel per installed MW, whereas scenario 3 (at a wind margin of 0.6) has only 24,300 GJ of fuel per installed MW (and only 26,900 GJ fuel per installed MW at a wind margin of 0.0). Table VI shows the variation in plant mix as the wind margin is increased. Note that, as well as reducing the number of loss of load events, operating at a high wind margin can actually reduce the number of diesel starts (1050 compared with 1228 starts in the no wind case), albeit with the penalty of running the diesels at low load factor. Thus high wind penetration does not necessarily imply high cycling of existing conventional plant.

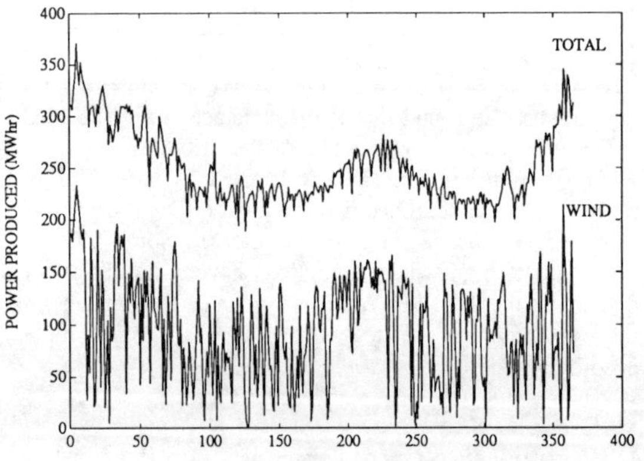

Figure 2a : Wind farm scenario 3 : simulation of wind power production - previous day load forecasting (load margin = 0.1) and persistence wind forecasting (wind margin = 0.6)

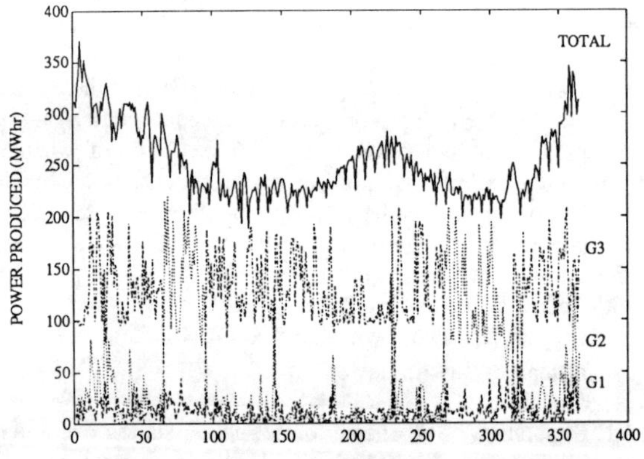

Figure 2b : Wind farm scenario 3 : simulation of power plant operation - previous day load forecasting (load margin = 0.1) and persistence wind forecasting (wind margin = 0.6)

Table VI : Wind farm scenario 3 : effect of wind margin on simulated power station operation with load prediction based on the same time the previous day (scaled by a factor reflecting the change in load between days)/persistence wind forecasting

Parameter / Wind margin	0.0	0.2	0.4	0.6
4.0 MW units (MWhr)	8,178	6,300	5,851	4,298
5.85 MW units (MWhr)	20,139	22,566	18,426	15,653
11.0 MW unit (MWhr)	21,427	21,657	29,956	36,834
Wind power (MWhr)	41,720	41,018	37,365	34,819
Total generation (MWhr)	91,464	91,541	91,598	91,604
Number of diesel starts	1420	1472	1176	1050
Wind power spilt (MWhr) (Number of events)	9,262 (18,509)	9,965 (20,582)	13,618 (22,738)	16,174 (24,336)
Loss of load (MWhr) (Number of events)	258 (176)	107 (68)	21 (20)	1.4 (2)

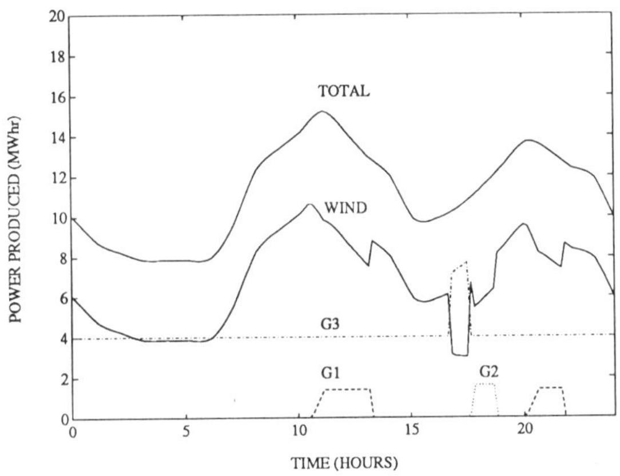

Figure 3 : Wind farm scenario 3 : simulation detail of day 214 showing power plant operation with wind margin = 0.6

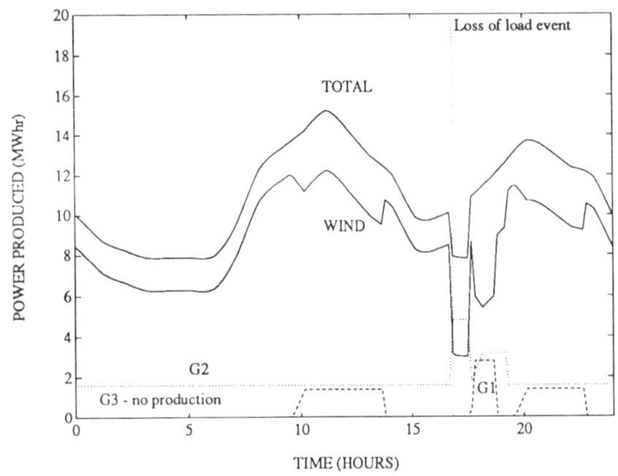

Figure 4 : Wind farm scenario 3 : simulation detail of day 214 showing power plant operation with wind margin = 0.2

4.2. Detailed simulation of a day with high wind penetration

Figure 3 shows an example of a day with high wind power penetration (scenario 3 - wind margin = 0.6). It is seen that at this wind margin the large 11.0 MW unit (G3) is run continuously at its part load limit and excess wind power is being 'spilt'. Operation of the smaller diesel sets is needed to maintain the required spinning reserve during times of peak demand. The most interesting feature is a sudden dip in the wind at about 17.00 hrs, which is easily covered by the 11.0 MW unit (G3).

Consider, next, a simulation of the same day for an operational strategy to maintain a wind margin of only 0.2 (figure 4). The 11.0 MW unit is not run and the required spinning reserve is carried by the smaller units. An increased penetration of wind power is obtained, but when the sudden lull in the wind power occurs there is insufficient reserve to prevent a loss of load event.

5. CONCLUSIONS

The inclusion of diesel maintenance scheduling in the grid integration model has enabled very close simulation of the conventional power station operation on Chios Island in terms of fuel use and distribution of load between the diesel power generators. Thus the model provides a good basis on which to predict the likely impact of wind power generation on the long term operation of the grid. It has been found that at low wind penetrations (scenario 1 = 11.7% of actual rated diesel capacity) very little change is needed in the operating strategy of the conventional plant and considerable fuel savings can be made (19% of annual consumption). However, at high wind penetrations (scenario 3 = 44% of actual rated diesel capacity) considerable extra spinning reserve is required to avoid loss of load events and the diesels need to be operated at or close to their part load limit. This reduces the anticipated fuel savings to only a further 19% of annual consumption on top of the low penetration case.

REFERENCES

[1] Halliday, J.A., Dutton, A.G., 'Adaptation of logistic model to the Chios Island grid' in 'Wind Integration into Diesel Grids: Development of an Assessment Tool', Final Report, CEC Contract No. WE 053/88 HE, October 1992

[2] Halliday, J.A., et al, 'Simulation of wind integration for the Shetland medium-scale diesel grid', Proc. of ECWEC '88, Herning, Denmark, 1988, p. 518-523

[3] Fellows, A., Hill, D.C., 'Wind and load forecasting for integration of wind power into a meso-scale electricity grid', Proc. of ECWEC '90, Madrid, 1990, p. 636-640

[4] Hill, D.C., et al, 'Integration of wind power into a meso-scale electricity grid', Proc. of the 13th BWEA Wind Energy Conference, Swansea, 1991, MEP, p. 289-296

THE DEVELOPMENT AND APPLICATION OF A MODEL FOR POWER OUTPUT FLUCTUATIONS IN A WIND PARK

A.I. Estanqueiro*, R. Aguiar*, J. Gil Saraiva*
Rui M.G. Castro, J.M. Ferreira de Jesus****

* INETI/DER, Az. Lameiros, 1699 Lisboa Codex, PORTUGAL
Tel - 351 1 7162712 (ext-2725); Fax - 351 1 7160901; email - der@donald.lneti.pt
** INTERG / IST - Instituto da Energia / Instituto Superior Técnico, Technical University of Lisbon
Av. Rovisco Pais, 1096 Lisboa Codex, PORTUGAL
Tel - 351 1 808257; Fax - 351 1 8484235; email - rui@interg.pt

ABSTRACT: The study of the wind park's electric power output fluctuation effect on the a.c. System voltage and frequency regulation are a necessity when the connection of a Wind Park to the existing utility grid, specially a weak a.c. System, is intended.

The aim of this work in a global perspective is to analyse the electric power produced by a park in terms of stability of the grid and "quality" of the output. In this paper a non-linear model of a two WECS wind park, together with a model to generate correlated wind time series', is used to simulate the power output fluctuations. The time series' of the individual power generated by each WECS and the park's output series are analysed both in frequency and time domain.

1 - INTRODUCTION

The effect of WECS electric power output fluctuations in the existing a.c. system is a matter of great concern, since the utility distribution normal procedures concerning operation and planning do not generally take into account for independent power production [1,2].

This situation can be particularly serious for high levels of wind penetration or in weak grids, this being the case of Portugal where windy sites are sparsely populated, and in most of the cases have a very weak grid.

Therefore, the grid connection of wind parks with a considerable amount of installed power may pose some problems, not only to the utility grid management, but also to the wind park owner, since its production capacity may be limited by the utility due to stability problems or extreme voltage deviations.

The lack of an accurate tool which allows to assess the interaction between the wind parks and the utility grid leads to a peculiar situation: neither the utility is able to properly account for the perturbations induced by the wind parks in their system, nor the independent producers are able to predict the impact that the utility normal practises have in their installations.

Those are some aspects why an accurate model to simulate wind parks connection to the existing a.c. system is currently being developed. The ultimate aim of this work is to build a tool, to be used both by the utilities and by the wind park investors, which will allow an assessment of the impacts resulting from the connection of wind parks to the grid in selected windy locals.

In this first step only a two-turbine wind park was considered. Nevertheless, a considerable amount of work was spent in order to correctly model the spatial correlated wind in the different locations, the interaction between the turbine/generator groups inside the park and the utility grid near the wind park that can be disturbed by its performance.

2 - WIND PARK'S MODEL

If a dynamic time dependent model of a wind park is intended, some aspects have to be addressed, namely:

i) a spatial and time wind correlation method;
ii) a time-domain method to simulate the WECS and the electrical system;
iii) the computation of the electrical interaction between generators and transformers inside the park.

A wind model to generate correlated wind time series was developed to perform these studies [3]. The situation addressed here corresponds to a two side by side WECS wind park - thus wind turbines' wakes are not modelled.

The wind park's model is based in a dynamic WECS model developed previously. This WECS model results from the integration of a wind turbine's characteristic equation that accounts for the time variation of the shaft rotational speed and an existing induction generator model - developed in due time to apply to small hydro power plants [4] - that was adapted to perform the WECS typical behaviour with time variable torque [5,6].

The generator's model accounts for saturation effects of the induction generators thus including its non linear working range. The interaction between the two turbine/generator groups inside the power plant and the conditions at the interconnection busbar were established in order to model the electric power output.

2.1 - Wind Models

To achieve a dynamic model of a wind park, the knowledge of the simultaneous wind velocity time series, at the locations of the various turbines, is essential. This means not just using average wind velocities as input, but also the superimposed turbulence. This is required by the existence of phenomena associated with turbulence effects on the blade's loads, so on the turbine's power output, and even on the turbine's responses to electrical grid faults.

A preliminary wind model was developed, based on the so called "Shinozuka method", i.e. the generation of wind speed time series from the inverse FFT of a complex spectrum, whose real part is a wind power spectrum and the imaginary part is a random phase spectrum.

The current model is based on i) the Taylor hypothesis of "frozen" homogeneous turbulence; ii) the assumption that each size class of eddies is represented in the power spectrum by the energy at the frequency $f = L/v$, where L is the eddy size (along wind) and v the mean wind speed; and iii) the assumption that eddies with size less than the distance H between two certain sites will be too small to affect both of them at the same time - and therefore will bring a null average contribution to the cross-correlation.

Within these hypothesis, the Shinozuka method is used to generate cross-correlated wind speed series for any two places H apart taking the phase spectrum as equal for both places up to f'=H/v, and random thereon [3]; the well-known "Davenport spectrum" shape was used as the power spectrum for both places with common average speeds at 10 m high of 7 m/s, and a roughness coefficient of 0.008.

Samples from the models stochastic wind series where taken as appropriate when yielding cross-correlations close to those estimated with exponential fits to field observations, with decay lengths of 200 m along-wind, and 50 m across-wind.

2.2 - WECS Model

To simulate the wind turbine performance the already mentioned time dependent model that uses as input the wind's velocity instantaneous value was used [5,6]. To describe the torque's characteristic of the turbine's rotor a characteristic equation that accounts for the shaft's rotational speed variation is used:

$$T_M(\Omega, V_0) = A(V_0)\Omega + B(V_0)$$

where $A(V_0)$ e $B(V_0)$ are wind's velocity dependent parameters, and Ω is the rotor's shaft rotational speed. This equation was found by applying numerical interpolation to the rotor's performance simulation results with an improved version of the PROPshaft code.

It has to be referred that the turbine's model doesn't include the aeroelastic effects on the blades neither the disk averaging effect. Thus the high frequency turbulence (aprox. over 1 - 2 Hz) present on the wind time series that is not filtered by the shaft and the generator will be present in the power output time series.

2.3 - Electric Equipment Models

Models have been developed to describe the behaviour of the different subsystems that constitute the electrical part of the system studied [4]. All models have been developed in the synchronous reference frame to avoid undesired dependence of rotor position.

The prediction of induction generator steady-state and transient performance requires proper account of saturation effects, thus a detailed model based on Von der Embse circuit theory was developed [7].

Each generator has its own reactive power compensation system, which has been sized in order that the admittance of the capacitor bank equals the slope of the linear part of the no-load magnetisation characteristic of the induction generator. This size was chosen as a result of previous studies in this domain [8] and complies to the Portuguese legislation.

The local loads were modelled as constant active and reactive power for the computation of initial conditions and as a constant impedance in the simulation studies. Both transformers, the feeder and the a.c. system were modelled also as constant impedance.

In order to obtain the initial conditions for the set of differential equations which accurately describe the system, a numerical model taking as input only the system parameters and the instantaneous wind speed was built. This technique enables the computation of the initial steady-state operating point through a modified power flow which takes into account the electro-mechanical characteristics of the turbine/generator groups to evaluate the rotor speed of each machine.

The WECS overall model was first presented in [5,6] and is now implemented in a software package named INDUSAT, which allows to perform several studies either in steady-state or transient conditions [9].

3 - TIME AND SPECTRAL ANALYSIS OF THE POWER OUTPUT FLUCTUATION SIMULATIONS

The wind park's model was applied to the topology represented in Fig.1. The system is basically composed by two side by side WECS equipped with squirrel cage induction generators, with reactive power compensation.

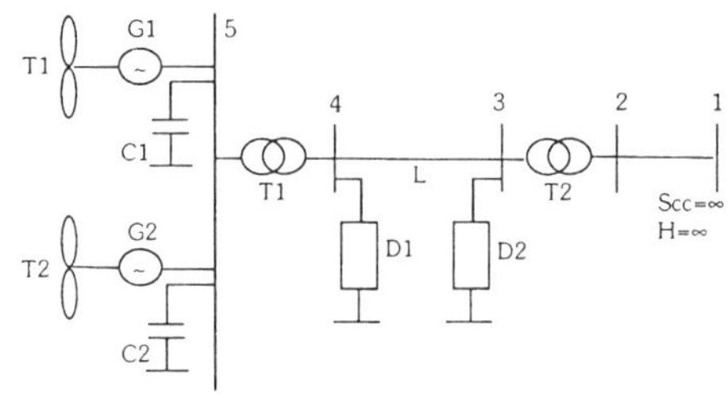

Fig. 1 - Grid's topology.

The generators are connected to a feeder through a step-up transformer, which feeds a local load. The interconnection substation is represented by another step-up transformer, which also has a local load connected to its low voltage side. Characteristics of the WECS and the grid simulated in this work can be found in reference [10].

The short-circuit power at the interconnection point was assumed to be 50 times the wind park rated power, thus the existing a.c. system is represented by a reactance with the appropriate value.

Using the wind park's model described above, some simulations were performed in order to obtain results in time domain. Moreover other type of studies have been carried out to assess some statistical characteristics of the power output (cross-correlation and standard deviation) as well as to examine the frequency domain results (Power Spectral Density - PSD) of the park's output.

The time domain output power series presented correspond to wind series with a cross correlation factor equal to 0.33. This value was selected as an example since it correspond to the typical cross wind distance for this kind of turbines (diameter -12 m).

Fig. 2, 3 and 4 show some results obtained in time domain. Fig. 2 depicts an sample of the correlated wind time series' (crosscor=0.33).

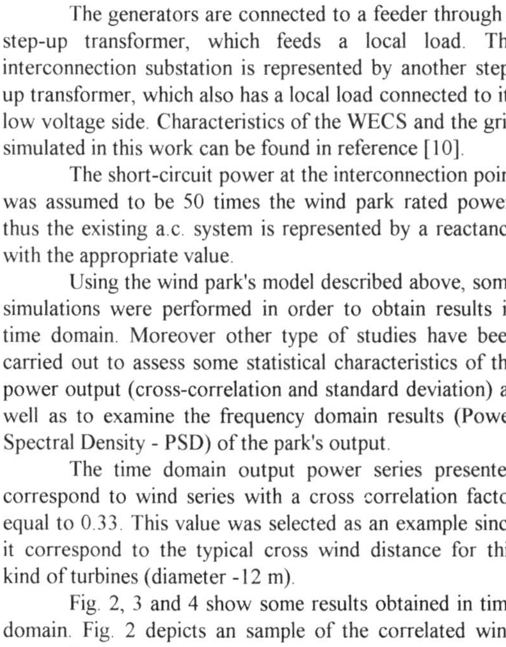

Fig. 2 - Wind times series (crosscor=0.33).

Fig. 3 and 4 display the power time series for each turbine (Fig.3) and the wind park's power output (Fig. 4).

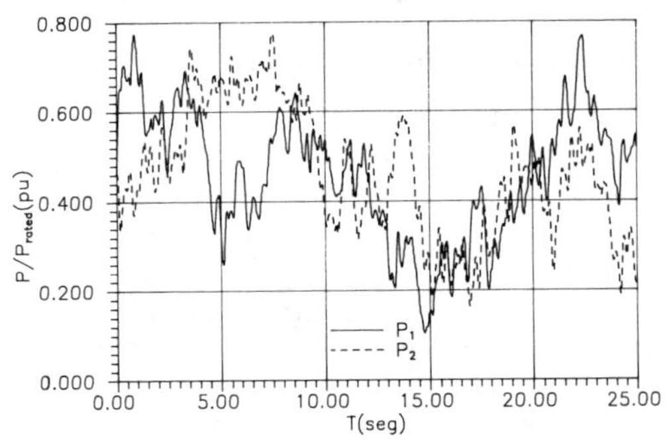

Fig. 3 - Turbine's power output (wind's crosscor=0.33).

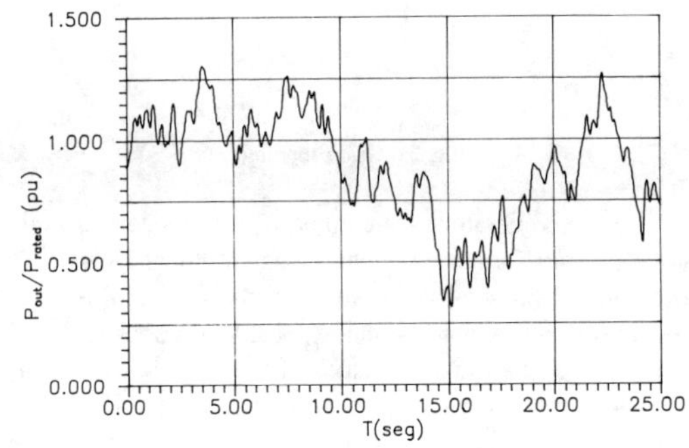

Fig. 4 - Park's power output (wind's crosscor=0.33).

Figures 5 and 6 deal with the statistical properties of the power produced by the wind park. In Fig. 5 it is represented the individual power series cross correlation factor between the two turbines against the cross correlation factor of the input wind time series.

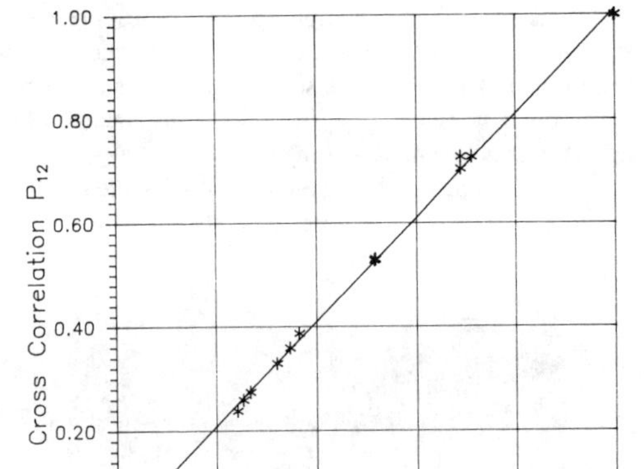

Fig. 5 - Park's power output crosscor vs. wind's crosscor.

In Fig. 6 the evolution of park's power output standard deviation versus the wind's cross correlation factor is plotted.

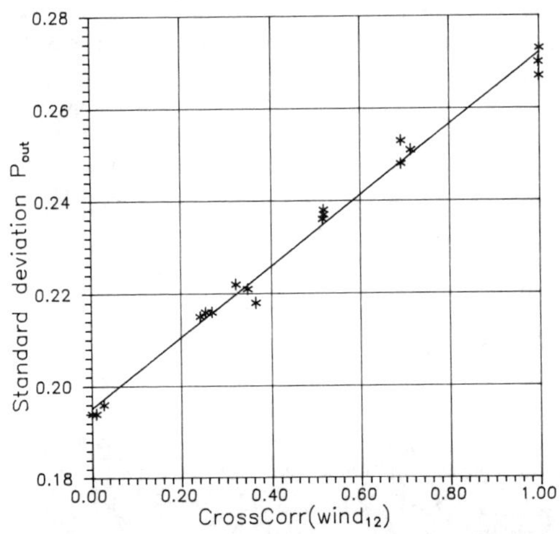

Fig. 6 - Park's power output standard deviation vs. wind's crosscor.

The power spectral density (PSD) of the park power output fluctuations' is plotted in Fig. 7. For each wind correlation factor the average PSD of the samples is computed and is presented. It may be seen that the lower curve corresponds to zero correlation and the upper curve to correlation factor equal to one (low frequency range).

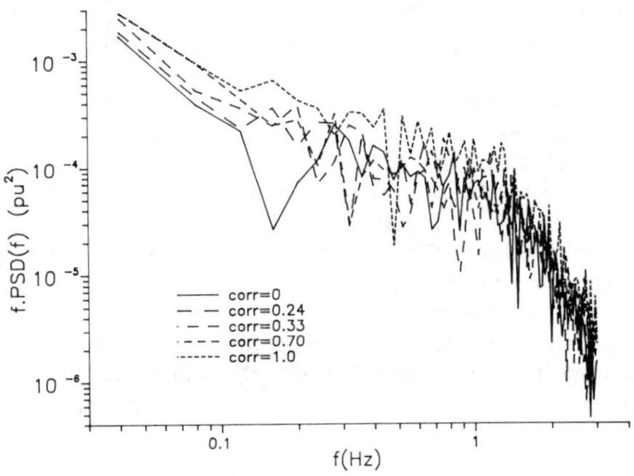

Fig. 7 - PSD of Park's Power Output fluctuations. The lower curve corresponds to wind's crosscor=0 and the upper curve to wind's crosscor=1.

Figure 8 corresponds to the power spectral densities of one sample of the power outputs of each turbine (P1, P2) and the park (Pout), for wind time series' cross correlation factor equal to 0.33.

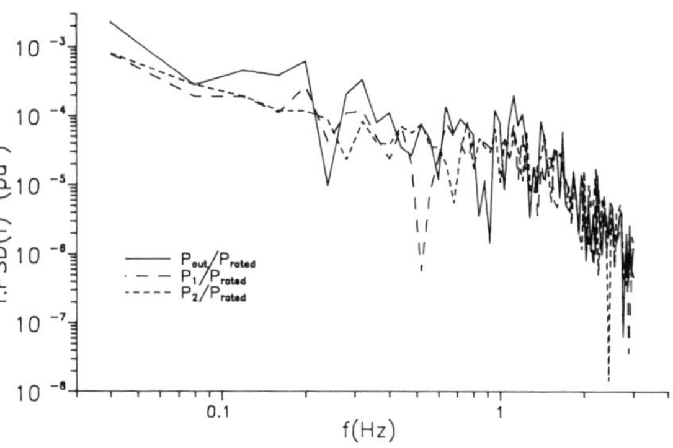

Fig. 8 - PSD of park's power output (solid line) and PSD of each turbine's power output (P1-dotted line, P2-dashed line).

The number of samples presented is not yet large enough to be representative of a stochastic process. However as shown in Fig. 5 and 6 the results dispersion is quite low.

Analysis of Fig. 5 shows a linear dependence between the two quantities in spite of the simulation model being non linear. In what concerns Fig. 6, it clearly shows an increase of the standard deviation with the wind's cross correlation factor.

Analysis of Figures 6 and 7 allows to conclude that an increase in the wind's cross correlation factor corresponds to a larger fluctuation in both the voltage and power output of the wind park. This result is in agreement with results pointed out by several authors [1, 2,11].

The N turbine smoothing effect is less obvious than expected in figure 8, which may partially be explained by the low number of turbines involved.

4 - CONCLUSIONS

The results presented agree quite well with results previously published on this subject (e.g. Beyer et al [1] and G. McNemey and R. Richardson [11]). The smoothing ("cancellation" of the turbine's individual high frequency fluctuations) in the park's power output fluctuations when the wind is uncorrelated is illustrated by Fig. 7 and 8. However attention must be paid to the space scale associated with this phenomena, since in this case only two turbines are involved and the distances between turbines are smaller (the frequency "cancellation" limit being higher as a consequence) than those of references [1,11].

Results show a dependence - that seems to be linear from the short number of simulated samples - on the wind's cross correlation factor, though a non linear model wind park is used.

The analysis of PSD of the park's output (Fig. 7 and 8) in the low frequency range brings up the idea that mean and standard deviation analysis should be complemented with both frequency and absolute value of extreme occurrences.

ACKNOWLEDGEMENTS

The authors want to thank to Dr. P. Christiansen by the valuable suggestions to the application of the wind model and to the procedure analysis of the stochastic output.

REFERENCES

[1] Beyer, H.G. et al; "Power Fluctuations from Geographically Diverse, Grid Coupled Wind Energy Conversion Systems; EWEC '89 Proc.; pp 306-310; Glasgow, 1989.

[2] Peltola, E. et al; "Simulations and Measurements on the Effect of a Wind Farm on the Local Grid", Paper D6 , EWEA Special Topic Conference '92 on the Potential of wind Farms, 1992

[3] Aguiar, R. et al "Uma abordagem espectral à Geração de Séries Correlacionadas de Vento" to be presented in "Congresso Ibérico de Energia Solar", Lisbon, April, 1993.

[4] Rui M.G. Castro, "Induction Generators: It's Use in Small Hydro Plants" (in Port.), MSc Thesis, IST., Lisbon, November 1988.

[5] Estanqueiro, A. I.; "Horizontal Axis Wind Turbines - A Behaviour's Model"., (in Port.) MSc Thesis, IST, Lisboa, April, 1991.

[6] Estanqueiro, A.I.; Ferreira de Jesus, J.M.; Gil Saraiva, J.A.; "WECS Unsteady Output Simulation", EWEC'91 Proc., pp 618-622. Amsterdam, 1991

[7] Ferreira de Jesus, J.M.; "A Model for Saturation in Induction Machines", IEEE Transactions on Energy Conversion,Vol EC-3, 1988

[8] Rui M.G. Castro, A. Eugénio Gomes, J.M. Ferreira de Jesus, "The Influence of Reactive Power Compensation in the Transient Behaviour of an Induction Generator", Proceedings IFAC, Seoul, August 1989.

[9] Rui M.G. Castro, J.M. Ferreira de Jesus, "INDUSAT Basic Reference Guide V1.0 - Software Package for WECS Simulation", Instituto Superior Técnico, Lisbon, January 1993.

[10] Estanqueiro, A.I, Ferreira de Jesus, J.M:, Lalos, D.; "Transient Behaviour of a WEC under Different Load Conditions", presented in "The potential for Small & Medium sized Wind Energy Applications in Mediterranean Countries" , Rhodes, Greece, 1992.

[11] McNemey, G. and Richardson. R., "The Statistical Smoothing of Power Delivered to Utilities by Multiple Wind Turbines"; IEEE/PES Winter Meeting paper 92 WM 024-0 EC,1992.

[12] Lipman; N.H. et al., "Fluctuations in the output of wind turbine clusters". Wind Engineering Vol. 4, no. 1, 1980

[13] Taylor, G. J.; "Power Quality Measurements on a 24 m Windmill", IEE Proceedings, Vol. 134, Pt. A, nº 5, May, 1987.

[14] Smedman, A.; "Effect of Wind Speed Fluctuations on the Power Output of the Wind turbine in Kalkugnen, Sweden", 4th International Symposium on Wind Energy systems, Sept., 1982

[15] Beyer, H. G. et al; "Fluctuations in the combined Power Output from Geographically Distributed Grid Coupled Wind Energy Conversion Systems - An analysis in the Frequency Domain", Wind Engineering Vol.14, nº3, 1990.

[16] W. E. Feero, W. B. Gish, "Overvoltages Caused by DSG Operation: Synchronous and Induction Generators", IEEE Transactions on Power Delivery, January 1986.

[17] R. Dugan, D. Rizy, "Electric Distribution Protection Problems Associated with the Interconnection of Small Dispersed Generation Devices", IEEE Transactions PAS, Vol. PAS-103, Nº6, June 1984.

OPERATIONAL TOOL FOR SHORT TERM PREDICTION MODEL OF ENERGY PRODUCTION IN WIND POWER PLANTS AT TARIFA (SPAIN)

F. Martin*, R. Zubiaur*, P. Moreno**, S. Rodriguez**, M. Cabre***, M. Casanova***, A. Hormigo****, M. Alonso*****

* CIEMAT, Av. Complutense, 22; Madrid-28040, SPAIN
Tel.: 341-1-346 60 41, Fax: 341-1-346 60 37
** PMSA, Sector Embarcaciones Bloque 4-Esc. 17-1'D. Tres Cantos. Colmenar Viejo; Madrid-28770, SPAIN
Tel.: 341-1-803 25 34, Fax: 341-1-803 24 89
*** ECOTECNIA, Dem"stenes, 6; Barcelona-08026, SPAIN
TEL.: 330 78 60, Fax: 411 23 45
**** C. SEVILLANA E, Av. de la Borbolla, n' 5; Sevilla-41004, SPAIN
Tel.: 41 73 11, Fax: 41 21 28
***** ENDESA, Pr!ncipe de Vergara, 187; Madrid-28002, SPAIN
Tel.: 341-1-563 10 42, Fax: 341-1-563 09 23R. Zubiaur, F. Martin

ABSTRACT: The aim of this project (carried out under agreement between local utilities,wind farms managers, and public and private wind energy research centres) is to develop an operational tool for short term forecasting of local wind conditions, and consequently of the energy production, in the wind power plants installed at Tarifa (Spain), taking into account the climatological situation, the effect of the wind regime at local scale, and the response of the parameters of performance of wind plants.

In the Tarifa area, located at the Gibraltar Strait, there are now running several wind farms with a total installed capacity great than 40 Mw, because of the well known high wind energy potential.The wind conditions are characterized by the marked presence of the two stable prevailing directions, Eastern and Westernlies, as a response to concrete situations clearly identified of the general climatology.

The overall circunstances (well identified climatological regime, wind data availability and the installed and planned wind power capacity) are favourable conditions to carry out this project. At the current stage of wind energy development in Tarifa, it is very useful to have available this operational tool, because of the advantages on operation and maintenance programe, the produced energy management and prices, and also on the plant.

1. INTRODUCTION

The high level of wind energy development and the degree of exploitation reached up in Spain, besides the perspectives of increasing the installed capacity, make neccesary to have available an operational tool for short term forecasting of local wind conditions, in order to optimize the management of energy production.

At the high wind potential areas generally the wind regime maintains relatively stable patterns of behaviour, and it is possible to identifie some characteristic synoptic situations that permit to stablish predictions at short-term with higher level of reliability.

In the Tarifa area, located at the Gibraltar Strait, there are now running several wind farms with a total installed capacity great than 40 Mw, because of the well known high wind energy potential.The wind conditions are characterized by the marked presence of the two stable prevailing directions, Eastern and Westernlies, as a response to concrete situations clearly identified of the general climatology.

The overall circunstances (well identified climatological regime, wind data availability and the installed and planned wind power capacity) are favourable conditions to carry out this project. At the current stage of wind energy development in Tarifa, it is very useful to have available this operational tool, because of the advantages on operation and maintenance programe, the produced energy management and prices, and also on the plant.

This subject involves several aspects related to the wind energy exploitation (the wind resource characterization, the operation of wind plants, the technical and economical management of the produced energy, etc.), and consequently, it has been accorded a collaboration agreement between publics and privates research and development centers with activities in the field of wind energy exploitation in Spain.

2. SCOPE OF THE PROJECT

- Objective: To develop a model for short-term prediction (3-6 days) of the energy production in wind power plants at Tarifa (Spain), taking into account the climatological situation, the repercussion on the local scale wind regime and the response of wind farms performance parameters. The final aim is to implement an operational tool in order to improve the wind power plants energy management.
- Participants:
  CIEMAT (Public Research Centre)
  PMSA (Private Research Centre)
  ECOTECNIA (Wind machines manufacturer)
  CSE (Regional Utility)
  ENDESA (Public Utility)

Tasks

1. Local climatology analysis.
2. Local wind regime analysis.
3. Wind power plants operation parametrization.
4. Evolution models definition.
5. Prediction model development.
6. Model implementation and operation.
7. Data acquisition and validation.

8. Results and conclusions.
Duration: 3 years: 1 year for model definition; 2 years for operation and validation.

3. METHODOLOGY

The first approach to the problem consists in the identification and characterization of relevant synoptic situations at the Gibraltar Strait Area. Up to seven tipical synoptic patterns were identified, and their associated local wind behaviour was studied, and also their frequency, persistency and evolution.

This characterization lead to a better knowledge of the physical phenomenon that appears in the Gibraltar Strait, which origins the two prevailing winds: Levante (East) and Poniente (West), the former being the most important because of the energy contents.

"Levante" situations are mainly influenced by the phenomenon called orographic "dipole" [1]. This phenomenon is a mesoscale structure that consists in a overpressure at windward side and a depression at leeward side of a mountains ridge, respect to the macroscale general flow. In the case of Gibraltar Strait, a gap in the mountains bow along the edge of Alboran Sea, for eastern wind situations, an overpressure at the Mediterranean side and a low at the Cadiz Gulf are present [2]. The consequence of it is the wind channelling through the strait in eastern conditions, with the wind direction perpendicular to the isobars.

In order to evaluate the validity of the hypothesis of the orographic dipole, correlations between measured pressure gradients at both sides of the Strait have been analysed. It has been found that, effectively, the pressure gradient defines the situations of each type (Eastern and Western) of wind conditions at the Strait zone.

Pressure measurements at Malaga Airport have been used as representatives of the Mediterranean side, and Jerez have been used for the Atlantic side pressures reference. Even with the limitations associated to the ground based measurements, the results are encouraging.

Once identified the pressure as the determinant variable for wind estimation in the area, the methodology to apply is based on the use of the routine forecastings of the pressure field at ground level obtained by the limited area models (LAM) running at the National Meteorological Institute (INM), and then adapting the results, with a MOS technique, up to microscale resolution [3].

Predictions coming from operative LAM model at the INM are made each 6 hours and, by the moment, have 2 days of temporal projection. This timespan, by now, is relatively short for wind power plants requirements, but by other hand the reliability of routine forecasting is continuosly improved and the timescale of predictions is also going to be increased.

At this time, analysis and validation of pressure field results of LAM model, and comparisons with ground measurements, are being carried out.

REFERENCES

[1] A. Jansa Clar. Notas sobre analisis meteorologico mesoscalar en niveles atmosfericos bajos. INM 1990.

[2] E. M. Garcia et al. Contribucion al estudio y prediccion de los vientos en el Estrecho de Gibraltar.

[3] R. Diaz Pavon, B. Orfila. Sistema Operativo de Analisis y Prediccion Numerica (LAM) en el Instituto Nacional de Meteorologia. Rev. Fisica de la Tierra, 3. 1991.

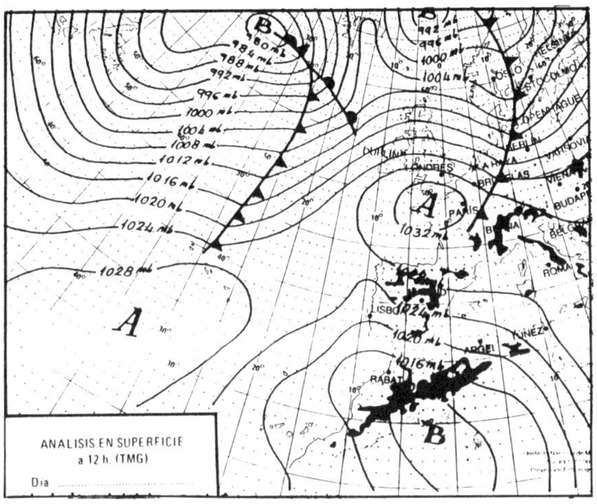

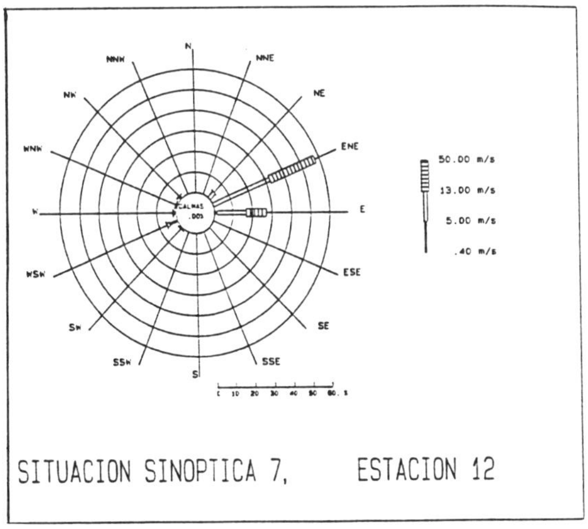

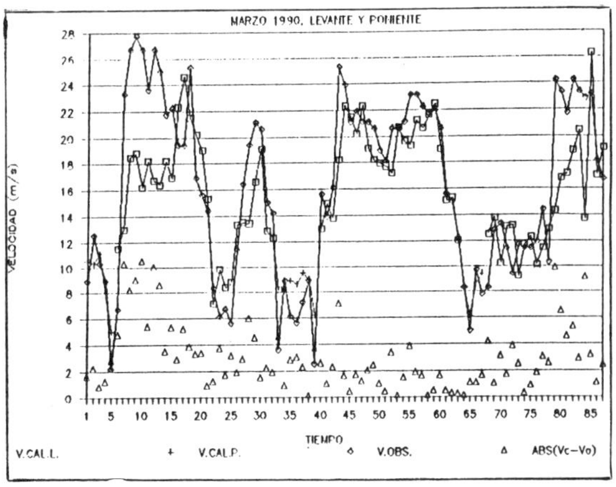

USE OF EXPERT SYSTEM IN WIND TURBINE APPLICATIONS

Lj.Pilić-Rabadan, Z.Milas, D.Stipaničev, R.Vujčić, M.Jelavić, B.Klarin, M.Sansević

University of Split,
Faculty of Electrical Engineering, Mechanical Engineering and Naval Architecture,
Rudera Boškovića bb, 58000 Split,
Republic of Croatia

ABSTRACT: The paper deals with the layout of expert system (ES) for wind energy utilization and its application to specific project. ES-clarification procedure is developed to cope with most of the user's requests (accounting for project limitations and goals) and specifying thereby the ES inference engine reasoning to match WPS to the given site. Hierarchical multilevel addressing provides efficient manipulation with data bases containing wind energy data in Croatia and commercially available wind turbine generators (WTG) data respectively.
The prospect of wind energy utilization on the Middle Adriatic island of Lastovo i.e., contribution of the wind power station Lastovo in electric power supply to the mainland grid and the local community, is predicted by applying E.S. inference engine. WPS-installed power is determined on account of comprehensive analysis which contains specific investment and energy cost optimisation criteria, as well as WPS matching to the island's grid.

INTRODUCTION

An computer program has been developed in order to study the phenomena affecting the power of the horizontal axis wind turbine [1]. It is necessary, among other things, to analyze the influence of aerodynamic characteristics of the given airfoils, the induced speeds and the design speed on the wind turbine power characteristics [2]. We also studied the wind potentials of our side of the Adriatic coast. Comparing these potentials to those of the European coast, we concluded that the Adriatic coast potentials could be used as a support to the Croatian power system, which does not supply enough energy to meet the needs of Croatia [3].
Some of our investigations have proved the possibility of obtaining wind power at a reasonable price, with regard to prices in European countries [4],[5].The research results were used to study optimum turbine requirements for the given potentials.
An expert system (ES) has been developed to integrate our investigations regarding locations (potentials, areas, sites) and wind turbine optimization (WTG) to the commercial WTG data base designed by European designers in order to offer the users optimum solutions according to their demands and priorities.

2. EXPERT SYSTEM DESIGN

When designing WPS for the given operating location with the characteristic wind potential, it is important to decide which WTG best suits the given location. Wind turbine optimization is a complex process since the WT blades should be made so as to have acceptable aerodynamic and structure characteristics .
An ES is used as a design tool with the purpose of discovering the corresponding WTG for the given location (the designed wind speed), according to the user's needs.
Fig. 1 shows the design of an expert system. Its main parts are three knowledge bases (KB): a wind potential (location) KB, a commercial WTGS KB, and a KB with parameters affecting geometry and power characteristics for the primary simulation WT design.

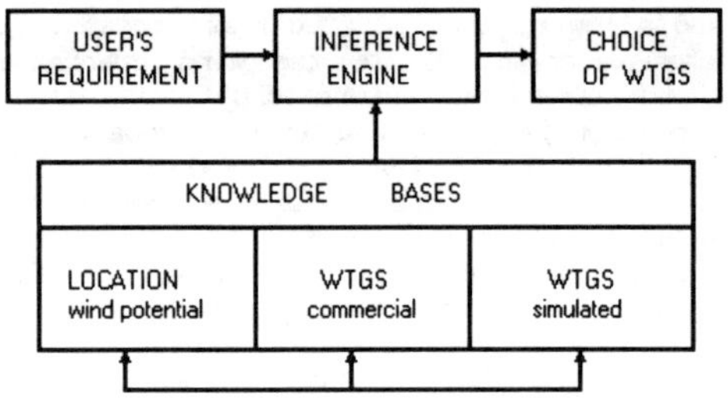

Fig. 1 Design of expert system

ES is of open structure type, and its data bases can be examined and updated, particularly with the characteristic parameters of aerodynamic airfoils, important for the optimum WT simulation.

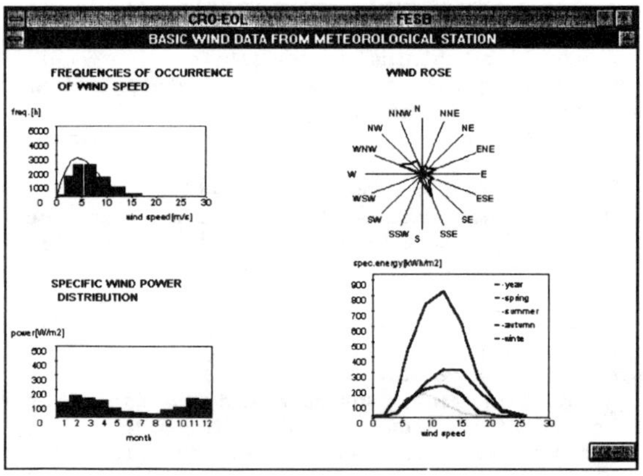

Fig. 2 Basic wind data from meteor. stations

Inference mechanism is the central part of an ES. Axial WTGS are suggested for the given location as a result of inferencing from users' demands.User interface meets a wide range of users' demands. A user can demand for a choice of best locations with regard to potential, distance from the road, from the electric network, from the nearest habitation or port, etc. KB-s dealing with potentials of different locations are made by processing data from the location data base,

Fig.2. In a selection process, the best locations with the greatest number of points in all parameters are chosen (Fig.3).
A user may have demands regarding the power level at the given location and/or the requirements for the power generation per year.

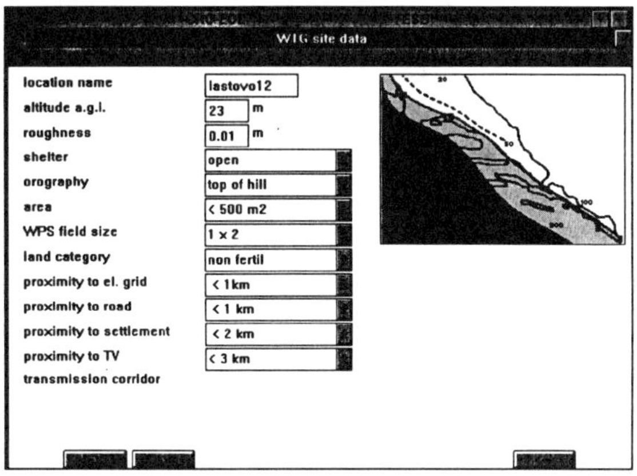

Fig. 3 WPS site data

The designed wind speed for the given locations is calculated. This wind speed and the airfoil characteristics are the basic parameters for optimum turbine simulation. Turbines obtained by simulation are characterised by the same number of blades, same rotation diameter and speed, and the same nominal power from the commercial WTGS short list. From the commercial WTGS data base, a KB on the possible power generation in the given locations is formed. A selection of WTGS from the commercial WTGS data base is performed in a number of steps with regard to limits set by users. This selection may be limited by parameters of maximum installed power and/or dimensions (D and H); by the number of units, the price per USD/kW power unit, and by the total price of the installed power. In this way, according to the given criteria, and taking into account the user's priority demands, two or three commercial WTGS are selected for each location (Fig. 4).
Inference mechanism for the final WTGS selection operates in two steps. The selection criteria are: the least difference between power generated of commercial WTGS on the given location and the power of simulation WTGS (Fig. 5); and the lowest price of USD/kWh generating power unit.

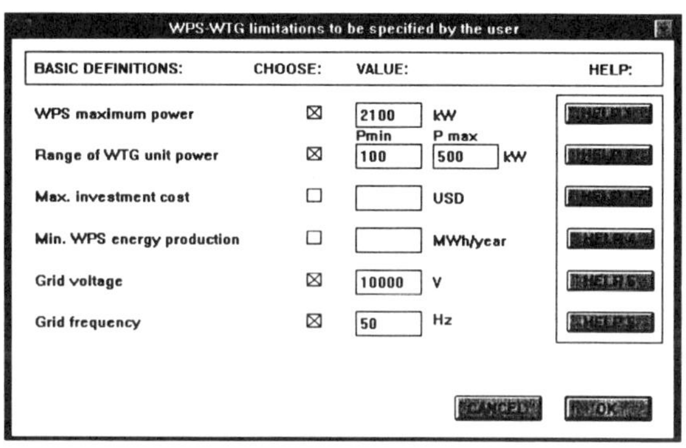

Fig. 4 WPS-WTG limitations

The user can give priority to one of these two criteria. The given mechanism implementing the choice out of a number of commercial units is suitable for quick and simple accomplishment of the given aim.

In the decision making process the criterion for comparing the commercial WTGS with the simulation one gives preference to modern solutions. The logic of some other inference and selection mechanisms is the subject of our further investigation.

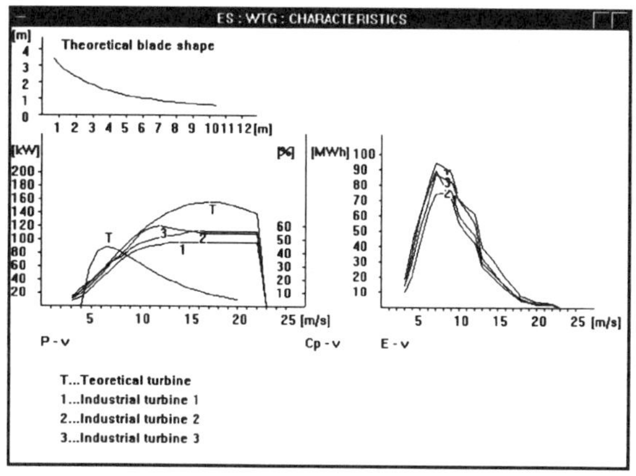

Fig. 5a WTG characteristics

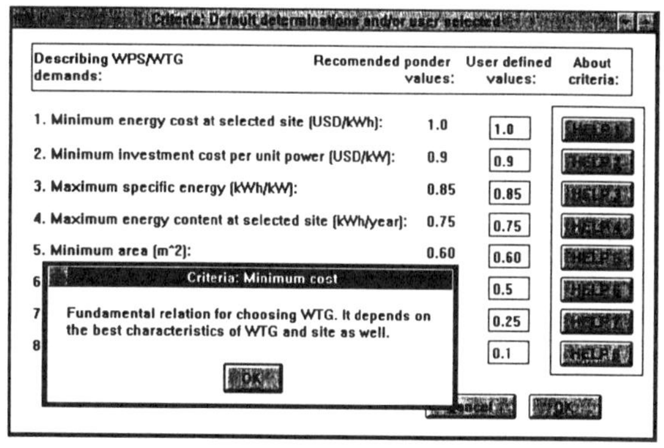

Fig. 5b Definition of WTG selecting criteria

3. WPS LASTOVO PROJECT

The project of WPS on the island of Lastovo was in many occasions a starting point in developing ES, particularly in establishing its user interface (UI) structure [8]. After completion of UI with other conceivable project situations, WPS Lastovo project was a test of the ES inference engine effectiveness. Emphasis will therefore be put on those part of the project which contributed to the ES development.

3.1. WPS Lastovo Project Parameters

Work on ES was preceded by the studies [5,7] dealing with general aspect of the wind energy utilization in coastal area of Croatia. The choice of the island of Lastovo for the WPS carrier to check the prospects more realistically was made due to the following:

- wind observation site measurements place the island among the few best WPS sites in Croatia with average wind speed of v_a=6.2 m/s and e_w=2.4 MWh/m^2 specific wind energy in an avarage year.
- unobstructed wind inflow to the island from almost all directions except the NE due to the island of Korchula located 11 km away from the Lastovo island (Fig.3). This makes the prediction of the wind energy on most of the Lastovo island sites (Fig.6), much easier.

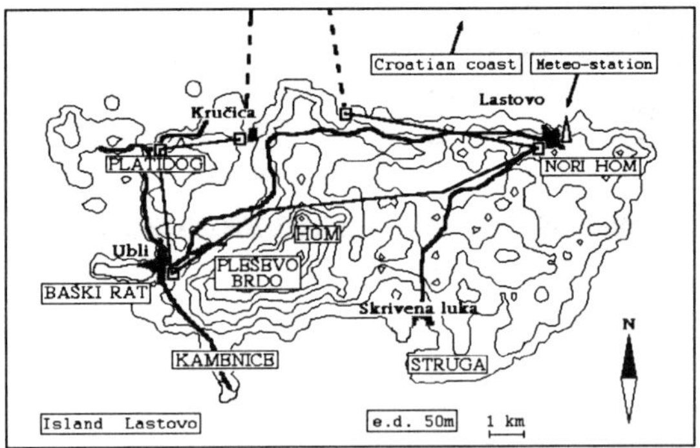

Fig.6. Contour map of Lastovo island

- island's topography is moderately complex-hilly (average peak altitude of 150-200 m a.s.l. and ~0.3 slope) with dominating hill Plesevo located along the southern island coast. Plateaus are scattered and sheltered among the hills at low altitudes. Due to the low tree-high bush Mediterranean vegetation roughness of all the hillsides is uniform and gradually deminishes towards the hill ridges
-most of the population resides in two settlements (Lastovo, Ubli) providing large areas for the WPS site.

Area (km^2)	53
El. energy consump.	900 MWh in 1988.
Peak load (MW)	$P_{C\ MAX}$ = 1 (1988.)
Grid capacity (MW)	3 (10/0.4 kV)
Submarine cable:	1 2
capacity (MW)	2.1 2.8
length (km)	16 16

Tab.1.Island Lastovo data

When establishing the project goals two different approaches were considered:
U)WPS as an integral part of the utility grid of Croatia subject to corresponding criteria in power engineering i.e., maximum energy production with satisfactory economical efficiency. Island grid facilities play an important part in WPS Lastovo planning. They are limited with the capacity of submarine cables supplying el. energy from the mainland grid via the island Korchula transformer station (Tab.1). Three U(tility) projects alternatives were conside-red with respect to the additional investment cost shown in Tab.2.

	$P_{WPS,max}$ (MW)	Investment prerequisites	Cost mill.USD
U1	$\leq P_{SC1}$=2.1	None	0
U2	$\leq P_{SC1}+P_{SC2}$ =4.9	Repair of cable 2.	~50% C_{RC} =0.2
U3	$\leq P_{ts}$=13	Laying down new submarine cable	100% C_{SC} =0.8

Tab.2. Project alternative goals

C) WPS operating as the island community power plant adjusted to the community power-energy demand and substituting supply from conventional power plants. This approach strongly depends on the community development plans predicting the peak power demand up to 4 MW until 2000. It will be considered in the next project phase because of extensive overlapping analysis of the available wind energy with community energy demand.

3.2. WPS Siting

Raw wind data were available as hourly averages in 1979.-1982. period measured at the island anemograph station. The wind observation site description (Tab.3) indicates the minor speed-up in the measuring results.

Altitude (m)	186 m a.s.l
Location	on the top of a steep hill
Terrain	open in all directions, close to the station the ground is devoid of trees and bushes
Anemometer place	7 m above the roof of a 8 m high meteo-station building

Tab.3. Wind observation site data

However most of the island orography is very much alike the environment of the anemograph site. This allows for the wind data to be allocated straightforwardly to the WPS sites except for those of Plesevo and Platidog with distinctively long ridges. Fig.2,7. illustrate the wind data proccessing results that can be summarized as follows:
- wind density probability function is almost equal to Rayleigh distribution (k~2)
- app 70% of the wind energy is available through winter/autumn and falls within 7-17 m/s wind speed range.
-dominant, open sea, southern(SE-SW) winds (scirocco) characterized by their persistence; weak Etesian (NW sector) winds blowing through summer.

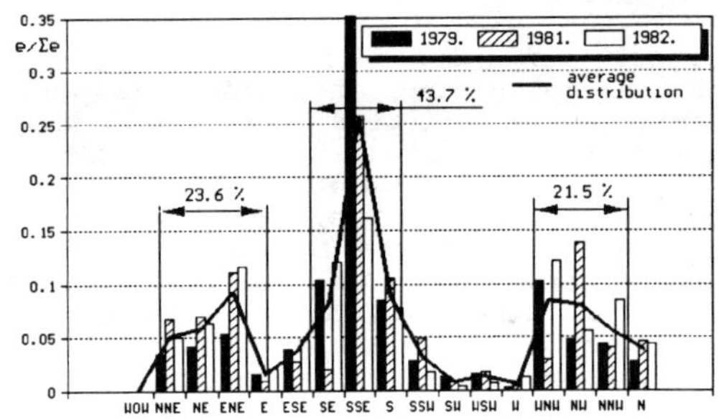

Fig.7 Wind energy distrib. by wind directions

Position of the WPS sites close to the anemograph removed the need for modelling wind on a large geographic scale (between remote anem. sites). However, medium term measeurements (Fig.8) do not provide quite a reliable basis for predicting averages corresponding to WTG service life. This triggered the idea of implementing a routine in ES in order to mutually compare wind data of near-by anem. sites in real time. Short term averages can be thereby extended to longer periods.

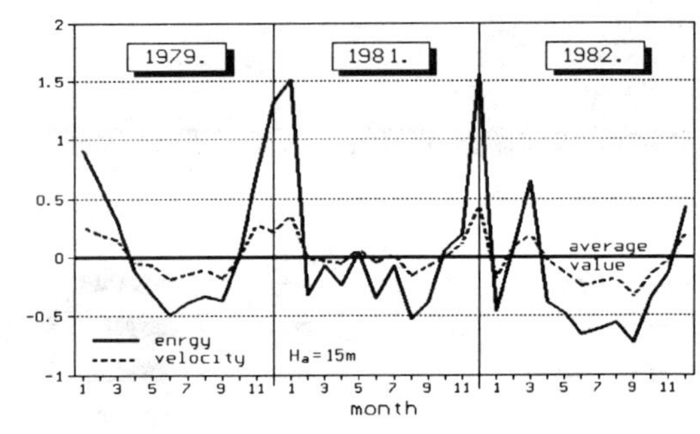

Fig.8 Wind speed-wind energy relative deviation in real time

Exact corelation, experimental or numerical, of the on-site wind data and those of the observation site was not available. It was substituted with orography benefit analysis based on simplified assessment of the WPS site position with respect to the wind flow obstructions and prevailing wind directions. This kind of partial modelling of the wind adjustment to the site orography is efficient when selecting some of the most favourable sites or eliminating the poor ones.

Similar benefit analysis was applied to the WPS field size by identifying areas with moderate terrain slope and WPS proximity to the electric/road network on the island. Characteristics of the selected locations are shown in Tab.4.

	Location-altitude m a.s.l.	L_f m	L_{CG} m	L_{CR} m
I	Plesevo 400-300	1750	1675	1825
II	Platidog 200-160	1000	500	975
III	Struga 90-80	1300	3582	440

Tab.4. Selected WPS sites data

Terrain configuration provides WTG spacing along the hill ridges resulting in single row arrangement. Evidently such a WPS field layout favours large machines that can easily overcome high local terrain slope and achieve high WPS installed power with consequently lower fixed WPS cost per unit power.
Locations of Plesevo and Platidog are distinguished by their long ridges stretching accros the southern wind direction, the latter being surpassed by Plesevo for 200 m at the distance of 2.5 km. Calculation of the relative speed-up [9] due to hill and shelter effect exhibits their marginal influence. However the hill effect results are to be questioned since they strongly depend on roughness parameter value (z_0=0.05 m as taken from [7]) and unprecise height a.g. definition.
Various WTG types, with corresponding data available from ES, were compared with regard to the capacity factor C_f, the mean power coefficient C_{pe} and the specific energy production in an average year as illustrated in Fig.9.

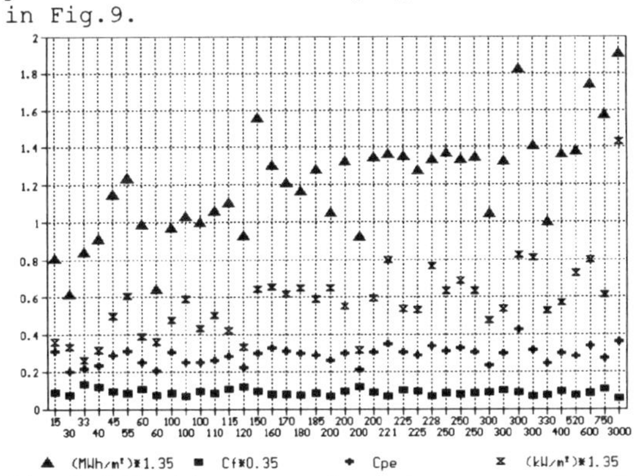

Fig.9 Effectiveness of various commercial WTG at Lastovo island

Tab.5. presents the results of the final WTG selection that suits best to the utility-site constraints and provides max. WPS energy/min. spec. energy cost. It was assumed for simplicity for each WPS sites to be spaced with single WTG type.

		U1			U2
		I	II	III	I+II+III
P_{WPS}	MW	2.1	2.08	2.0	4.08
E_{WPS}	MWh	5181	3945	4756	11843
C_{WPS}	10^6 USD	3.63	2.87	3.61	8.3
ce	USc/kWh	6.56	6.82	7.04	6.56

Tab.5 WPS effectiveness

WPS capital investment calculation included up to date WTG manufacturers prices and the cost of: transport and assembly; civil engineering (foundation + access road); electric enginering (WPS grid + transformers + high voltage connecting grid).

Specific energy cost was calculated on the basis of 20 years annuity period and 7% interest rate.

CONCLUSIONS

WPS Lastovo specific energy cost is close to the economical price of el. energy from conventional power plants in Croatia.
Project should be extended to the locations on the SE Lastovo coast as well as to the nearby island. This particulary refers to the island Vis which has many common features with the Lastovo island.
Refinement of ES prediction of on-site wind energy is required to provide more reliable planning of WPS contribution to the utility grid of Croatia.
Designing of the optimal simulated WTG for the given site provides efficient evaluation of the possible commercial WTG improvements. Simulated WTG can achieve 21% higher energy output than commercial WTG of the same size selected by ES.

ACKNOWLEDGEMENTS

This work has been supported by the Croatian Ministery of Science through grant nr.2-08-196.
Following the suggestion of the Conf. Technical Committee two papers: 'Expert system inference engine' by Lj.Pilic-Rabadan, D.Stipanicev, M.Sansevic and 'WPS Lastovo project' by Z.Milas, R.Vujcic, B.Klarin, submmitted for preview were combined.

REFERENCES

[1] Lj.Pilic-Rabadan,M.Jelavic:'Performance of wind turbines by computer aided analysis and some results of wind energy potential on the Adriatic coast',Sonnenforum,Berlin,1984.
[2] Lj.Pilic-Rabadan,M.Jelavic:'Analysis of propeler wind turbine operation, the influence of radial and axial vortexes',Strojarstvo 28(2), Zagreb,1986.
[3] Lj.Pilic-Rabadan, M.Sansevic:'Energija vjetra u elektroenergetskom sustavu zemlje',Congress proc. Energy and Environment,Opatija,1992.
[4] D.Lindley:'The commercialisation of wind energy',New Energ.Congr., Saarbrücken,1988.
[5] Lj.Pilic-Rabadan,Z.Milas et all.:'Supply of electric energy to Adriatic Islands by WPS', Conf.proc.,Supply of remote Adr.islands,Split,1989.
[6] Lj.Pilic-Rabadan,Z.Milas,D.Stipanicev:'Basic of expert system for wind energy utilization',SONT proc.,Opatija,1992.
[7] D.Poje, B.Cividini:'Wind energy potential in Croatia',Solar Energy Vol.41(6),1988.
[8] Z.Milas,R.Vujcic,B.Klarin:'Optimal WTG adjustment to specific site',Congr.proc. Energy and Environment,Opatija,1992.
[9] I.Troen,E.L.Petersen: 'European wind atlas', Risø National Lab.,Roskilde,1989.

VORTEILHAFTE BAUSTELLEN ZUR VERWERTUNG DER WINDENERGIE IN DER BANATER GEBIRGE AUS RUMANIEN

I.Preda x), F. Gyulai xx)
x) Technische Universität Timişoara
xx) Vorschungszentrum für Windenergie Timişoara
Bd. Mihai Viteazul 1, RO - 19oo Timişoara
Tel.+4o-96-11233o/383 Telefax +4o-96-19o321

ZUSAMMENFASSUNG:Windenergie scheint auch in Rumänien eine günstige Ersatzenergie zu bilden. Im Folgenden unterstützt man, durch einige synthetische Daten, die Ausbildung der Windfarms im Westen des Landes.

Die Banater Gebirge bieten in ihrer Gesamtheit günstige Bedingungen zur Verwertung der Windenergie in der Süd-West Zone Rumäniens, verdankend einerseits der Windregimen und anderseits der morphologischen Strukturen des Bodens.

Unser Studium begründet sich auf eine vorgehende Untersuchung von INMH Bukarest, in Landkarten mitgeteilt, in dem man Isolinien gleicher Dauerwerte der verschiedenen Schwellen von der Windgeschwindigkeit (2, 4, 6 ... m/s) eingezeichnet wurden. Nachlegend, jede Landkarte jener Geschwindigkeitstreppe zugeschrieben, stellt einem "Dauerberg" dar; setzt man dann diese Oberflachen übereinander und durchstecht sie mittels lotrechte Geraden in verschiedenen Orten, so bekommt man die nützlichen Daten für die Konstruktion der zugehörigen Verteilungsfunktionen.

In einem Rechenprogramm haben wir

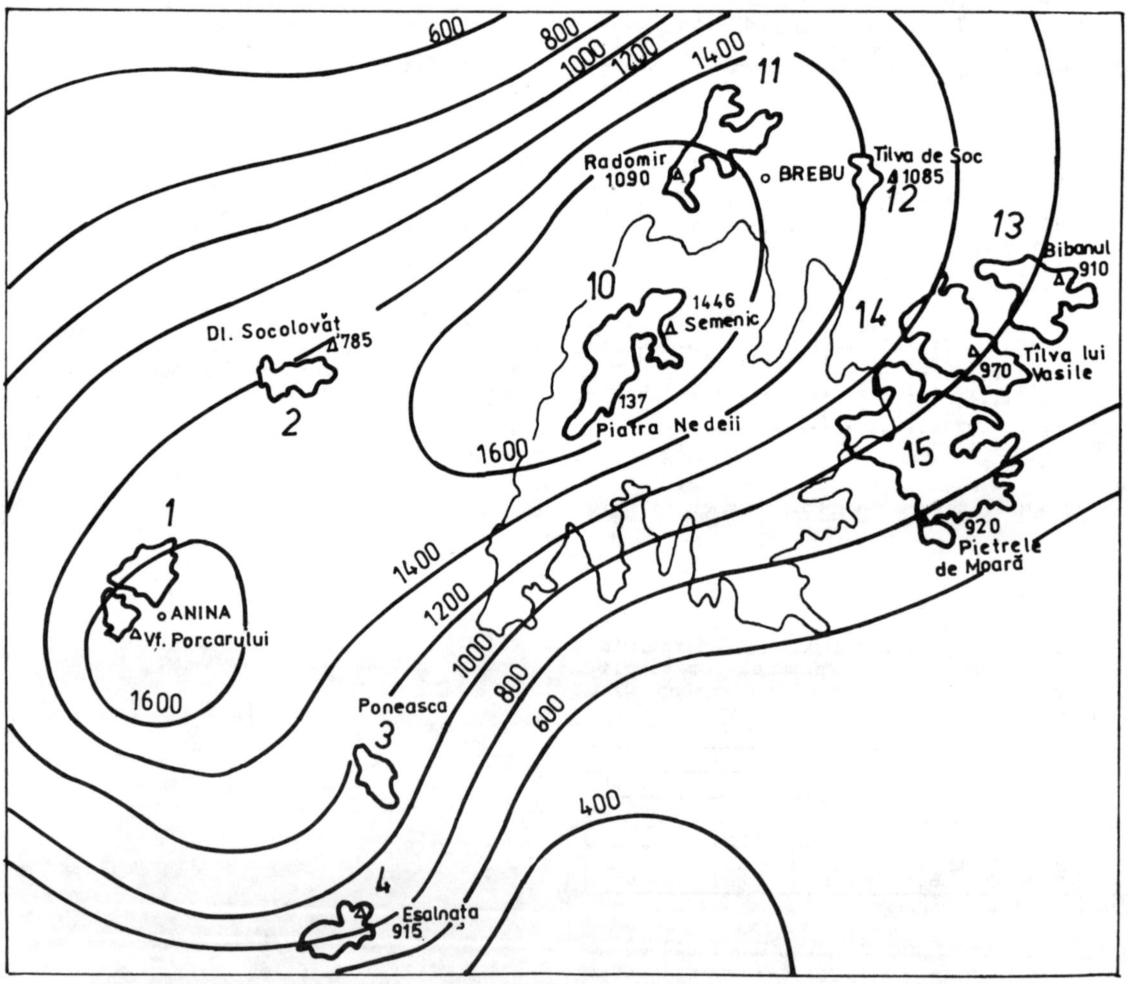

Abb.1: Lokalisierung von Windfarms in der Banater Gebirge Rumäniens. (Die aufgeschriebenen Kurven in der Landkarte bezeichnen Linien gleichen Brutto-Potentials KWh/m^{2}u.Jahr).

Weibull-sche Funktionen benützt, dreiparametrischer Art; wir konnten dabei die drei Beiwerte mittels empirischen Daten ausrechnen.

Zusätzlich erhalteten wir durch ein Integrationsverfahren, für verschiedenen geographischen Punkten, die gesuchten Energiequoten. In den annähernden Schätzungen, nahmen wir an dass die spezifische Masse der Luft gleichmässig in der Zone verteilt sei; also = 1,1 kg/m^3.

Nach der Bearbeitung des so erhaltenen Potentialberges, haben wir, wie die Landkarte in Abb.1 zeigt, Linien Gleicher windenergetischer Angebote aufgeschrieben.

In dem man den Brutto-Windpotenţial in Rechnung nimmt, ist sie in KWh/m^{2}u. Jahr ausgedrückt, also pro Einheitsflache die durch die Laufschaufeln der Windturbine gefegt wird.

Zusätzlich, weil man bei der Verwertung der Windenergie durch Windfarms rechnerisch beweisen kann dass die Energieproduktion der verschiedenen Baustellen unabhänglich von dem Laufrad-Durchmesser der Windturbine ist, kann man auch deren erzeugte Energie abschätzen.

Natürlicher Weise, die Baustellen der Wind-Farms bedingten, unter anderen die Auswahl der Zonen höherer bruttto Windpotentials (höher als 1000 KWh/m^2.u.Jahr, also pro Einheitsfläche die durch die Laufschaufeln der Windturbine gefegt wird), wie auch die Auswahl der Zugeschriebenen günstigen Eigenheiten der Baustellungen und Betriebsbedingungen der technischen Ausrüstungen, die den nachfolgenden Kriterien entsprechen: Austellung des Terains gegen beherrschende Winde, ihre gleichförmige Struktur (kleine Abhänge, mässige Niveaudifferenzen), nicht in anderen oekonomischen Tätigkeiten eingezogen, zugänglichkeit, usw.

Die Studien brachten im Vorschein 15 Baustellen, deren technische Charakteristiken in die folgende Zahlentafel gegeben sind :

Zahlentafel 1 :

No.	Ort	Höhenlage m	Oberflache km^2	Potential KWh/m^2.u.Jahr	inst. Leistung MW	erzeugte Energie GWh/Jahr
1.	Anina (Vf.Porcarului)	800	4,4	1600	88	208
2.	Caraşova (Dl.Socolovăţ)	800	2,8	1400	50	116
3.	Poneasca	600	1,2	1200	19	43
4.	Esalnaţa	800	2,8	800	28	66
5.	Pleşina (Est)	800	4	400	-	-
6.	Moldoviţa (Est)	600	2	700	20	41
7.	Gîrnic (Tîlva Pietroasă)					
8.	Gîrnic	600	7,2	700	72	150
9.	Sfînta Elena	400	7,2	800	72	171
10.	Complex Semenic	1200	10,4	1600	260	496
11.	Gărîna (Radomir)	1000	8	1600	160	381
12.	Brebu (Tîlva de Soc)	1000	1,2	1400	21	50
13.	Bibanul	800	6,8	1000	95	202
14.	Tîlva lui Vasile	800	8,8	1000	123	262
15.	Pietrele de Moară	800	20,8	1000	291	620

Wesentlich wurden Studien durchgeführt, einerseits um die Verteilung des wahrscheinlichen jährlichen brutto Potentials in dem Gebiet, und anderseits um die typischen Charakteristiken der Windturbinen, die den günstigsten Baustellen nach gegebenen Kriterien zugeschrieben wurden, zu erhalten.

Aus den 15 ausgewählten Baustellen die eine Oberfläche von 87,6 km^2 belegen und für eine instal. Leistung von 1300 MW eine wahrscheinliche Produktion von 2806 GWh/Jahr bieten, wurden nur 9 vorgeschrieben, die die Schwelle des brutto Potentials von 1000 kWh/m^2.u.Jahr überschreiten. Daren belegte Oberfläche ist zu 64,4 km^2 (74 %) geschätzt, hat eine gesamte installierte Leistung von 1107 MW und bietet eine wahrscheinliche Energieproduktion von 2380 GWh/Jahr.

Einige der wichtigsten Baustellen sind mit der selben Ordnungs nummer in der Landkarte Abb. 1 eingeschrieben.

Bibliographie

1. Preda I. Metodologia estimării potenţialului aeroenergetic, Conferinţa Naţională de Energetică, Bucureşti, 1983

2. Preda I. Asupra exploatării agregatelor aeroelectrice în formaţie, Conferinţa de Maşini Hidraulice, Timişoara, 1985

3. INMH Bucureşti Studiul climatologic privind potenţialul energetic eolian în teritoriul judeţului Caraş-Severin

4. Preda I., Gyulai F. Preliminary Studies on Locations Identification for Wind Farms in the Banat Mountains, Energetic National Conference Neptun, June 1992, Romania.

FIRST DEMONSTRATIVE WIND FARM IN ROMANIA

eng. G.Voicu, eng. M.Tudor
ICEMENERG (Energy Research and Modernising Institute)
8, Bd. Energeticienilor, 79619 Bucharest 3, Romania,
Fax (400) 129315

ABSTRACT: The Romanian wind potential is estimated at several tens of Millions kWh/year. There are two zones of interest for industrial applications, each of them concentrating about 25% of hole potential the high area of the Carpathian mountains and the Black Sea coast (Danube Delta and continental platform included). The feasibility studies which have been previously carried out hance identified sites, where several hundred of MW can be installed under favourable conditions. The specialists forecast the generation of 1% from the Romanian electricity demand by years 2005-2010. The site on the Semenic mountains (providing the installation of 310 turbines of 300 kW each and electricity 206.7 billions kWh) has been chose for the erection of first demonstrative wind farm in Romania. The wind farm will be fitted with 8-12 units, having the same rated power and the electrical power output will deliver 20 kW to the available network. It is expected that the works for the wind farm erections will be stated in 1993 and will be finished in 1995. Estimated output power is of about 6 millions kWh/year. The studies carried out for this wind farm pointed out certain surprising, but favourable ecological aspects. Thus biologist state that the turbines location in the Semenic plateau expected for the final stage will result in increasing the zonal humidity by wind speed reduction, which will improve the vegetal and animal conservation safety for the existing ecological micro-system.

1. INTRODUCTION

Previous studies proved that Romania disposed of a large wind potential, estimated to be over 3000 kW, the high mountainous area having an over 25% harnessable technical potential.

One of the mountainous sites taken into account as early as 1985, due to its favourable logistic conditions, was the plateau of the Semenic mountainous. The studies drawn up by ICEMENERG and its collaborators made their contribution in identifying possibilities to erect a 93 MW wind farm.

The Semenic plateau, at a height of 145o m, having a flat-spread shape, over a NV-SE axis, covers some Km2 of barren area. Fir tree forests come into sight below 1360 m. In the middle of the plateau here is a little mountainous touristic resort, connected by an update road, to the urban centres (lying at) the mountain foot. Wind conditions are favourable, the average speed being 5.9 m/s at a standard height of 10 m (Weibull parameters A= 6,6 m/s, K= 1.67).

The easy access, the wide plateau, the good wind conditions and the relative moderate climatic mountainous conditions, the availability of several facilities (access road, 20 kW electric line etc.) made possible the instalment in Semenic of specially mountainous designed experimental units.

At present 4 wind turbines are under various erection and montage stages; they are research models or industrial pilots of 300 kW rated power, the first of then being in operation since December 1989.

2. OBJECTIVES OF A DEMONSTRATIVE POWER PLANT

In order to carry out a steady programme of industrial applications, besides the equipment availability (Domestic or foreign one), an win operating experience within a first experimental-demonstrative power plant is also required.

Such a power plant supposed the availability of a sufficient number of units within the harnessed areal, meant to enable the rapid meeting of 3 major objective:

- comparing the performances of various unit types (or various generations);
- operating behaviour analyzes of the turbines located in identical conditions;
- investigation of units inter-influence and direct estimation of generating energy, under the severe mountainous climate conditions.

3. WIND FARM DEMONSTRATIVE IN THE SEMENIC PLATEAU

The present available works and facilities in Semenic can make-up a sufficient logistic basis for ensuring the development at the level of an experimental plant.

Thus, the development of new roads was not considered as necessary and the power plant installed capacity was restricted at the level of the maximum transmission capacity of the 20 kW electric line in the area.

Obviously, the purpose was to integrate the 4 units - which are in various accomplishment stages - into the harnessing diagram; therefore, units location will be made on two parallel rows, starting from the available turbines.

Power estimating computations have proved that, with no modifications of the 20 kW network, maximum 12 turbines x 300 kW can be installed on Semenic, being able to directly supply several consumers in the area with maximum 800 kVA; the remaining port of energy is taken over by the public network.

A 6/20 kV connection electric substation will be achieved in the weighting centre on the two turbine rows.

The overall area covered by the demonstrative plant is of 26,000 m2; to organize the civil work, platforms of 15,000 m2.

The two windgenerator rows will represent, as a matter of fact, a very part of the final harnessing structure, the 12 wind generators making part of the 310 turbines to be installed in the Semenic plateau.

4. ECONOMIC-FINANCIAL SURVEY

4.1. Equipment cost

A 300 kW serially manufactured turbine has been taken into account.

The offers of the following companies have been considered: C.M.Bocsa (Romania), Micon (DK), HMZ Wind Master (Belgium) and Bonus (DK).

The lowest domestic prices at the western factories are of about $ 800/kW, to which the international transportation express of about $50/tm should be added.

The Romanian company offers turbines of the same power level at $740/kW and, since the factory is at about 50 km from the site, the transportation cost will fall to a great extent (quality quarrantees are however different).

4.2. Cost of the Erection-Montage works

The harnessing diagram takes into consideration the arrangement of turbines rows at 6 D (distance between 2 turbines of 2.5 D) perpendicularly on the dominant wind speed which, in the plateau, is NNV-SSE.

The scheme of power plant connection to the network provided, "in the inner side" of the plant, the connection of 4-5 units at a 0.4/20 kV transformer-equipped site, with availability at the 20 kV local grid.

The costs of the erection-montage works represent about 47% of the turbine costs.

The local investments for the demonstrative stage are of about $3 mil. (8 + 4 turbines).

4.3. Annual expenses

4.3.1. The distributions are, according to the Romanian regulations, 3.5%/year of the turbine cost.

4.3.2. The operation - maintenance expenses have been estimated at 1% of the annual cost of the initial capital.

4.3.3. The labour force expenses, totalizing 0.8 man/MWi, are of $2.7/kWi - taking into account the present salaries in Romania.

4.4. Energy Cost

4.4.1.Mean Electricity production per power plants of 2.400 kWh/kW (considering the natural unavailabilities, the incidents and failure estimated at 8%)

4.4.2. Energy cost, resulted from the above mentioned expenses, also considering the variable operating expenses ($0.125/MWh) is of 5.64 cents/kWh.

Mention should be made that in 1992 the most expensive energy import made by Romania (from Ukraina) did not exceed 5.1 cents/kWh.

The rate of the updated incomes, which is the ratio between the incomes resulted from energy sells and the expenses made, lead to a superunitary value what proves that the investment is good.

4.5. Accomplishment graph

The investment is expected to be approved in March 1993.

The power plant erection begins in May 1994 with the foundation accomplishment (at 14 months forme the equipment order) and the montage of the 8 new turbines will end in October 1994 (i.e. no erection-montage works can be done during winter).

Starting from January 1995, the 12 turbines are expected to have a production of 8.5 mil kWh/year.

Fig. 1 The 300 kW turbine. The research model carried out by the Technical University of Timisoara, commissioned in December

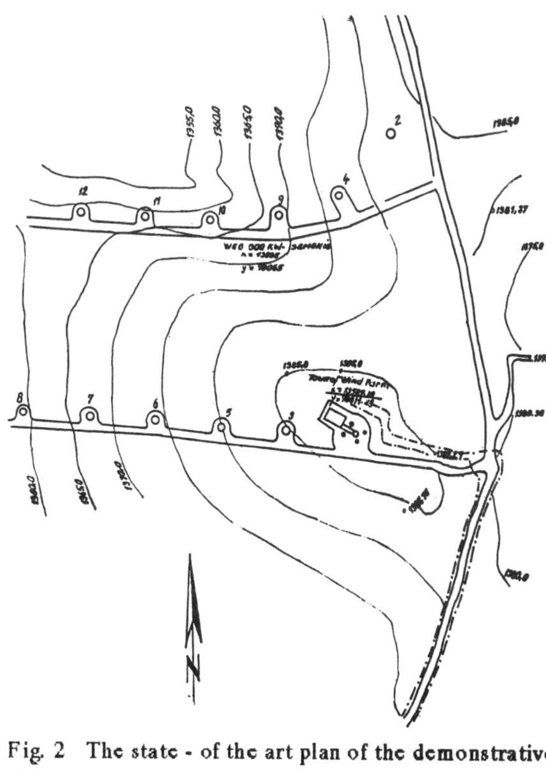

Fig. 2 The state - of the art plan of the demonstrative plant.

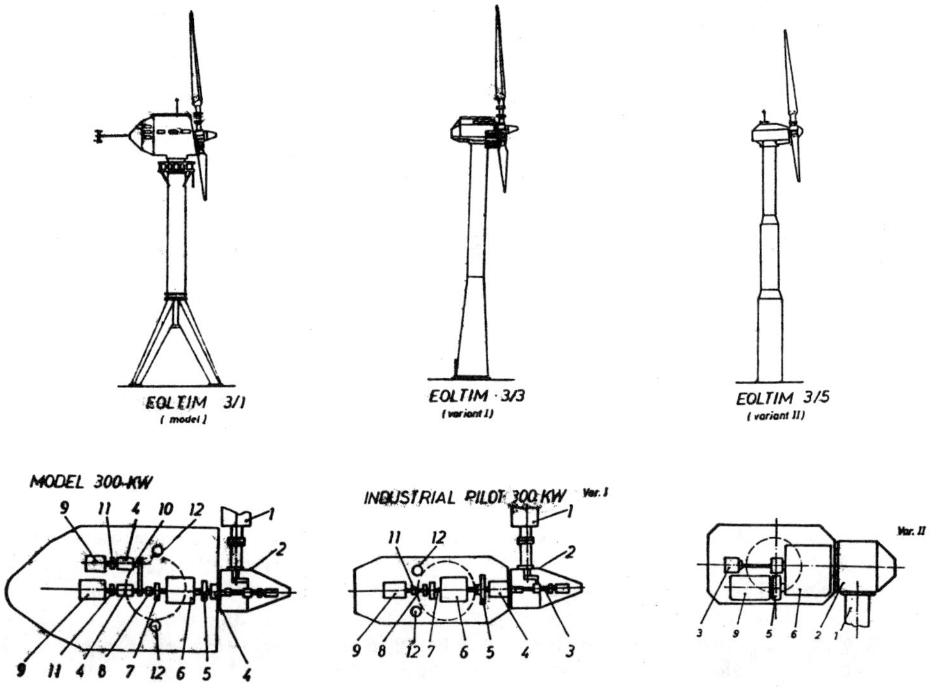

Fig. 3 Development of the 300 kW turbines at HIDROTIM S.A.- Timisoara.

Fig. 4 The Semenic plateau - sight to the Semenic (1400 m alt). To the left: the road to the touristic resort.

FIRST EXPERIENCE RESULTS ON GRID CONNECTED M30 WEC OF ITALIAN INDEPENDENT PRODUCERS

G.Baldi, D.Bedard, D.Comand, E.Dalpane
Riva Calzoni
Via Emilia Ponente #72
40133 Bologna, Italy

ABSTRACT: The paper presents the particular case of independent electric producers in Italy by wind energy. Single blade converters based on a light and flexible design are particularly suitable for operating in difficult environment with hard access and high turbulence, typical of mountainous sites, which are the prevalent windy sites of interest in Italy.

1. WIND TURBINE DESCRIPTION

Riva Calzoni is the only company in the world which has developed at an industrial level the advanced single-blade technology. The production of wind energy converters is characterized by innovative machines with high energy output and high reliability. At present it includes the M7 (5.2 kW) in the small size and the M30 (250 kW) in the medium size wind generator.

The M30-A is a medium size Wind Energy Converter (WEC) with 33m of diameter and a 250 kW rated power. The machine is especially designed for optimum production in low wind regimes with a second wind generator of 55 kW. A high wind version, the M30-S2 rated 350kW with one generator only, a new blade and rotor design, would be ready by late 1993. The principal characteristics of the M7, M30-A and M30-S2, are given in Table 1.

The M30-A can rightly be considered a second-generation WEC for its highly innovative features, including:

- A single-piece, tip to counterweight, CRP/GRP single blade structure which thus reduces load at critical points of blade connection where the blade structure is connected to the supporting structure.
- A teetering hub with an active fail-safe pitch control mechanism which leads to a considerable reduction of design loads.
- A 90Á pitch, and unbraked rotor with braked yawing, which ensure low loads and stability in parking position.
- An electrical system, which in the standard version includes two in-line asynchronous generators with 6 and 4 poles and a nominal power of 55 and 250 kW (previously 200 kW), respectively. This system has been designed for direct grid connection to large networks and is particularly suited for windfarm applications.
- An "ultrasoft" design, 33 m high tapered tower, which maximizes reduction of component stressing.

In addition to these technologically innovative features, a careful attention has also been devoted to ensuring WEC's maximum safety both during operation and maintenance.

2. DEVELOPMENT OF WIND PLANTS

Along request from potential customers, many sites have been studied in Italy to verify the feasibility of developing a wind farms. These studies are part of a complete package that Riva Calzoni offer to his customers, within the wind farm development project.

The first step concerns the evaluation of the wind potential through site ground inspection and site monitoring for a 1-2 year period. If data are available from an existing station in the area, the monitoring period can be reduced. After, if it is found that the wind potential is acceptable (wind speed in the range 5.5-7 m/s has been found in many sites mainly located in the Mountain Appennines), the project will start with Riva Calzoni assistance for all the necessary actions as: wind farm project, financing, authorizations, construction, commissioning and maintenance.

3. WIND PLANTS INSTALLED

The first M30 prototype unit was installed at the ENEL wind test site of Alta Nurra (oct. 89, Sardinia, Italy) and is in operation since more than 3 years. Also a variable speed M30 prototype, is located at Marienhafe (dec. 89, Gemeinde Brookmerland, Germany).

After two years testing of both prototypes, the first commercial units were installed and put in operation in 1992, this allowed to achieve axperience in solving the different siting difficulties and various problems encountered by independent produ-cers, related mainly to road access, authorizations and grid con-nection. Problems related with road access concerned site locali-zation in mountainous area, where main windy site are located. They are characterized by step curve, high slope and rough road.

Problems related with authorizations concerned the large number of public authorities involved to get a consensus. A complex, long and variable request procedure is necessary at each public level.

Problems encountered with grid connection are technicals and administrative. Each ENEL compartment has special regulations which vary from zone to zone. Also, indepentent measurements of the energy supplied is requested by the utility and by the state tax office. The definition of the medium voltage design, technical specifications, is very complicate and the construction is expensive and difficult, due to the complex laud to be crossed.

As examples, we report three typical installations, where the mentioned problems were successfully solved.

a) Tocco da Casauria

The site is located in a narrow valley between mountains which reach up to 2000 m. A micro-climatic effect caused by a large plain in the centre of Abruzzo (Italy) and the Adriatic sea, causes an almost constant high wind speed from 21 pm to 11 am every day. The 2 wind turbines are installed on a small ridge perpendicular to the valley stream at 341 m on sea level. The site showed a wind speed of 6 m/s averaged over a 2 year monitoring period and very high turbulence.

Using VALOREN (EEC) financial support, and additional contributions obtained by ABRUZZO Region and ENEA two M30 have been installed for the municipality of Tocco da Casauria.

This was the first commercial M30 plant to start operation, so allowing the Tocco Commune to be self sufficient in energy and to sell the extra energy production to the utility. The wind turbine shows a very good behaviour to the high turbulence of the site. The wind turbine shape fits pleasantly with the adjacent landscape corresponding to a pre-Appenine area of recent industrialization (Fig-1).

The civil work at Tocco was particularly difficult. The access road was too narrow and had to be enlarged, requesting all necessary authorizations to do it. A medium tension line of about o.4 km was built in a very steep terrain.

Fig.1: M30 Wind Turbines at Tocco da Casauria

b) San Simone

Under the VALOREN (EEC) program, one M30 has been installed for the "Consorzio di Bonifica della Sardegna Centrale" (Commune of Posada, Sardinia). The WEC is on a small hill a few kilometers from the sea (Fig-2). It was ordered by a consortium to help reduce its high electric bill for irrigation water pumping. The grid connected wind turbine work in combination with a storage water pumping unit of 700 kW. This unit feeds the irrigation system of Posada.

The wind turbine permit to supply a minor part of the electric consumption during the summer time, when the need of irrigation is very high and the wind average is low. During the other windier seasons when irrigation need is lighter or inexistent and only drinking water is needed, the wind generator permit to stop the big pumping unit. The extra energy production is also sold to the utility.

For this plant also, the mounting was mainly delayed by the difficult road access. The 1 km access road had to tight curve and steep curve (30%) for the articulate truck and crane. The interconnection were easy in this case by using the existing medium voltage board of the pumping station. Neverless, 1 km long electric line was built.

c) Collarmele

With a Demonstration EEC project financial support, one M30A has been installed at Collarmele in Abruzzo (It) for a private company. This plant will permit to test the M30 behaviour in mountainous surroundings (over 1000m, Fig-3).

The site is a hilly plain between mountain which reach over 2000m. The site being at few kilometres from Tocco da Casauria, was selected after few months of wind data survey since average wind speed was in the same range and a good correlation was found.

After a 9 month survey period, the site did show an average wind speed of 5.53 m/s. The site with some snow fall and cold weather will procure interesting data on the M30-A behaviour in extreme weather conditions.

Fig.2: M30 Wind Turbine at San Simone

4. FUTURE PERSPECTIVE

It appears that many new sites with average wind speed in the range of 5.5-7 m/s, will be available as a result of wind surveys presently ongoing. There are good perspectives to install small size wind farm of 2-5 MW during next year, mainly in small quite windier areas.

The Appenine area and the mountainous zone of Centrale Sardegna, are the most favourable site for wind energy development in Italy. Wind energy is quite good and problems related with landscape approval are lower than on the sea side.

The mono-blade M30 is probably the best suitable wind turbine for such sites. This is because the lower weight (around 40% less than a conventional three-blade machine of similar swept area) permit a more economical installation. Transportation is easier and small size cranes can be used in difficult access sites, where larger ones could not be used. The flexible of M30, allows a long life in high turbulence winds and optimizes the wind energy yield during wind gusts quite common in such site.

Figure 3: Collarmele Landscape with the M30-A Wind Turbine in background.

DESCRIPTION	M 7	M 30 A	M 30 S2
ROTOR			
No. of blades:	1	1	1
Diameter:	6.5 m	33 m	33 m
Rotor position:	downwind	downwind	downwind
Hub type:	teetering hub	teetering hub	teetering hub
Pitch angle range:	2° -> 65°	0° -> 90°	0° -> 90°
BLADE			
Chord length:	0.175 - 0.347 m	0.26 - 1.77 m	0.60 - 2.06 m
Material type:	CRP (Epoxy)	CRP/GRP (Epoxy)	CRP/GRP (Epoxy)
Section profile:	NACA 44XX series	FX84-W series	FX84-W series
TOWER			
Height: (from base to hub)	12 m	33 m	33 m
Shape:	polygonal conical	polygonal conical	polygonal conical
Diameter:	0.4 - 0.2 m	1.6 - 1.1 m	1.6 - 1.1 m
Material:	painted steel	painted/galvanized steel	painted/galvanized steel
CONTROL			
Pitching System:	mechanical/hydraulic	-computerized with hydraulic actuators	-computerized with hydraulic actuators
Yawing System:	passive	active	active
Parking:	blade in vane position	blade in vane position	blade in vane position
	free flap	flap locked	flap locked
DRIVE TRAIN			
Gear:	1 stage/parallel shaft	2 stage / parallel shaft	2 stage / parallel shaft
Gear ratio:	1 : 5.13	1 : 25.6	1 : 28.3
Manual brake:	disc, optional	disc (maintenance)	disc (maintenance)
Generator type:	synchronous	asynchronous	asynchronous
Generator number:	1	2 (high+low wind speed)	1
Generator power:	5.2 kW	250-55 kW	350 kW
OPERATION			
Wind speed: at rated power	9 m/s	12 m/s	13 m/s
Operating range:	4 - 20 m/s	4 - 25 m/s	4 - 25 m/s
Survival wind gust:	50 m/s	70 m/s	70 m/s

Table 1: A summary of the technical features of the M7 and M30 wind energy converters

STALL CONTROL WITH VARIABLE SPEED AND AN INDUCTION GENERATOR

S. Tsiolis M.Sc. and O. Carlson Ph.D.
Department of Electrical Machines and Power Electronics
Chalmers University of Technology
S-412 96 Göteborg
Sweden

E. Ulén and A. Björck, M.Sc.
The Aeronautical Research Institute of Sweden
Box 110 21
S-161 11 Bromma
Sweden

ABSTRACT: The aim of the work is to investigate the operation of a wind turbine with variable speed, stall control and a teeter hinge. The rotor of the wind turbine has a diameter of 13.5 meter. The blades are made of solid wood and have a mixture of NACA 63-200 and FFA-W3-xxx profiles. An electrical system with an induction generator and a frequency converter has been designed, built and tested in the laboratory and set into operation in the wind turbine. This system makes it possible to operate the wind turbine with variable speed and to control the torque in the drive train. From the control point of view the turbine is unstable in the stall region and the torque control of an induction generator requires special attention. Measurements are carried out on the rotor, in the nacelle and in the tower and results from operation in stall mode show the possibility to operate with a soft rotor speed control system.

1. INTRODUCTION

In the Swedish wind energy R&D program questions were raised about the dynamical behaviour of a two-bladed teetered turbine in stall operation at high wind speed. Experiences from Denmark based on three-bladed relatively stiff designs are good, but what about two-bladed soft designs? Power control of a variable-speed machine in stall operation is also of great interest to study. The total idea was that these ingredients would be of major importance in order to reduce the cost of electric energy produced by wind turbines. To get some indications it was decided to rebuild the existing three-bladed full-span pitch test machine at Hönö and use it for test operation. Measurements, evaluation and verification of calculations are also important parts of the project. This project has the ambition to include results and ideas, theoretical ones as well as practical, from several years of R&D work within the Swedish wind energy program. [1]. The mechanical aspects of this work are also presented in [2]. Characteristic parameters of the new turbine are listed below:

- number of blades 2
- diameter 13.5 m
- aerodynamic profile NACA 63-200, FFA-W3-xxx
- pitch adjustment (manually) ± 1.5 degrees
- cone angle adjustment (manually) 0, 2.5, 5 degrees
- hub with or without teeter motion (man. adjusted) max 8 deg.
- brakes at primary and secondary shafts
- yaw system with hydraulic damping devices
- variable rotational speed max 75 rpm
- rated power 35 kW
- rated wind speed 11 m/s
- rated tip speed ratio 8
- power control modes
- induction generator with frequency converter
- mechanical signals covering blade roots, nacelle and tower
- wind measurements from two masts with 4 (3+1) combined speed and direction sensors

2. HUB AND TEETER DESIGN

The hub is a welded cylindrical tube structure between the two blade flanges. The teeter bearing is a conventional journal bearing. Mechanical stops limit the teeter angle to 8 degrees. At teeter angles larger than 0.5 to 1.5 degrees a progressive rubber spring introduces a teeter moment. The teeter motion can be locked and thus make it possible to run the machine with fixed hub. By mounting conical washes between the blade root flanges and the hub flanges, cone angles of 2.5 and 5 degrees may be introduced.

3. BLADES

The blades are designed to achieve aerodynamic and dynamic characteristics that as much as possible resemble the characteristics of the larger Nordic Wind Power turbine NWP 400 with a diameter of 35 meter.

The Hönö turbine is much smaller than the NWP 400 turbine so that scale effects will inevitably be encountered. The blades are designed with constraints put up by the use of an existing generator, gear-box and other existing equipment and structure together with the wish for cheap manufacturing.

There is also the desire to get a lot of opportunities to measure in the stall regime. This desire together with the wish to get as high chord of Reynolds numbers as possible led to a diameter of 13.5 meter and a rather low tip speed for a two-bladed turbine of maximum 50 m/s. With a low tip speed, low tip speed ratios and the stall regime will occur at low winds and increase the chances to get many occasions for measuring campaigns in the stall regime.

The blades are made of solid wood since this is a cheap manufacturing method for one-off blades. In order to get as good dynamic scaling properties as possible the lightweight wood Thuja Plicata is used.

NACA 63-2xx airfoils were chosen for the part of the blade with t/c less than 18 % and FFA-W3-xxx airfoils are used for the inner part with t/c larger than 21 %. For the part with a relative thickness between 18 and 21 %, interpolated airfoils are used. Figure 3.1 shows the distribution of airfoils along the blade.

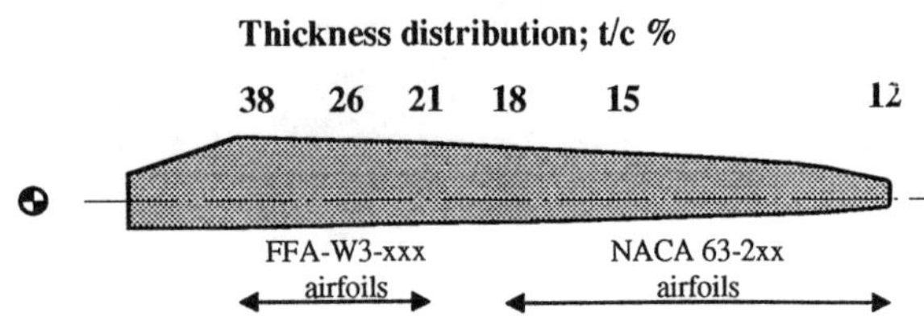

Figure 3.1: Thickness distribution and types of airfoils for the blades.

3.1 Aerodynamic characteristics

The power characteristics have been estimated by blade element/momentum calculations. As discussed and shown in [1] the characteristics will vary with the rotational speed, since the characteristics are very sensitive to Reynolds number, specially around peak power and Clmax for the airfoils.

Figure 3.2 shows calculations of the power at a constant RPM of 70 RPM and some limited measurements. The figure shows 1 min average values of the mechanical power on the main shaft and the bin-average curve together with calculations. Measurements and calculations show very good agreement. The calculations are made with data for Reynolds number of 1 million and smooth airfoils with a Clmax of 1.22 for the NACA 63-215 airfoil. The marked drop in power over 11 m/s is shown both in measurements and calculations.

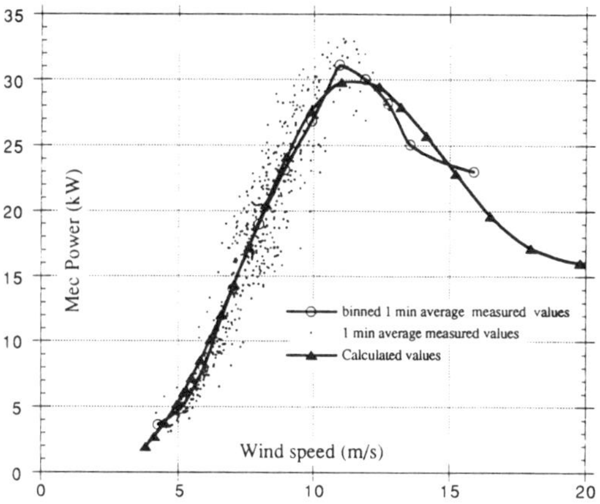

Figure 3.2: Power at 70 RPM.

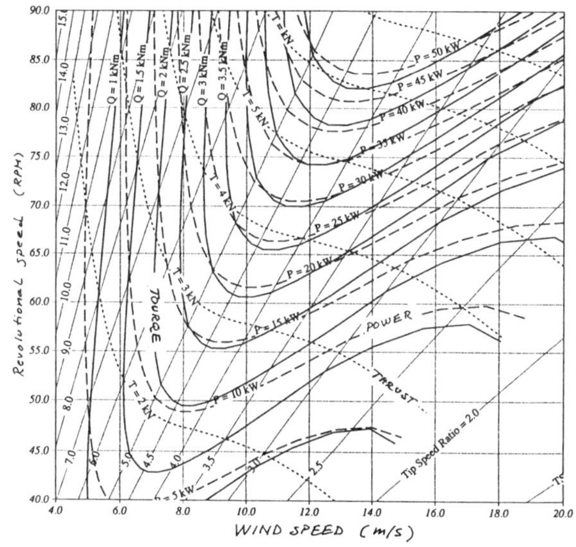

Figure 3.3: Lawson-Tacncred nomogram for the Hönö turbine based on Cp and CT curves from calculations.

Figure 3.3 shows power, torque and thrust-curves in a so-called LT-nomogram. This nomogram which is constructed only from a Cp-lambda, a CT-lambda curve, a value for the density of air and the turbine diameter, is very useful for e.g. control strategy considerations.

4. ELECTRICAL SYSTEM

The electrical system consists of one induction generator and three converter units and is designed to manage 40 kW power.

The left part of the converter block in Figure 4.1 shows the ac-dc converter with power transistors as current valves. With this converter it is possible to have a full control of the torque in the speed range of 0.1-1.5 times the nominal frequency. The converter is controlled by the PWM (Pulse Width Modulation) in such a way that the lower harmonics are cancelled. Reactive current to the generator is provided by the parallel capacitor.

The middle one of the converter units is a two-quadrant dc-dc converter, which means that the power flow is bi-directional, based on a dead-beat control strategy [3]. The aim of the dc-dc converter is to provide a regulated value of the voltage, DCV, by controlling the inductor current, DCC, so that the value of DCV stays proportional to the frequency of the generator.

The output converter unit converts the dc power to ac for an interconnection to the utility grid. This unit is actually a double thyristor line-commutated converter. This type of converter requires reactive power and causes current harmonics to the power line. These harmonics can be eliminated by filters. The power factor for the electrical system is 0.9.

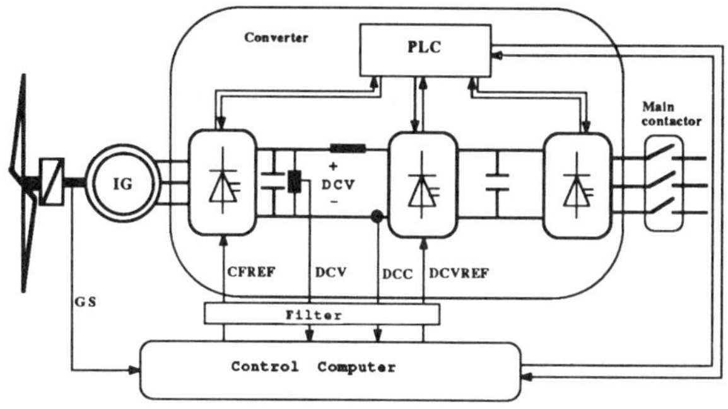

Figure 4.1: Block diagram of the electrical system.

The torque control is carried out in the control computer of the wind turbine. The torque on the generator shaft, GTORE, is calculated from the DCC, the DCV, the generator speed, GS, and the losses in the generator and the converter. The output control unit from the computer is the frequency of the converter voltage, CFREF.

The electrical system with torque control has been simulated in a computer. The model of the electrical machine and the converters is valid for dynamic conditions. In the simulation model the inertias of the wind turbine rotor, J_T, and of the generator, J_G, are used. In the simulations (Fig. 4.2) a step is applied to the aerodynamical torque, RTORA, while the reference value, GTEREF, of the electrical torque stays at zero.

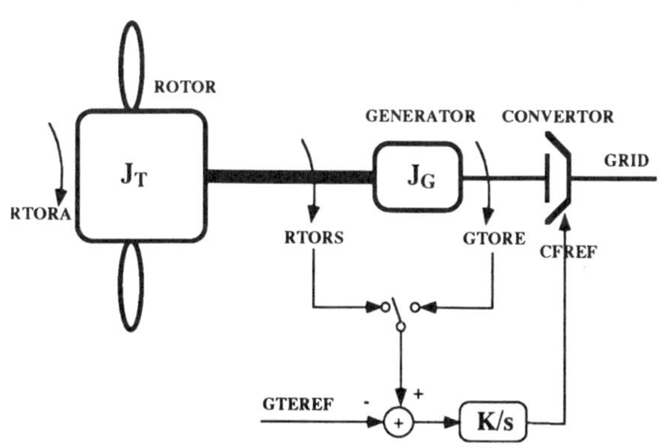

Figure 4.2: Electrical system simulations.

Curve 2 (Fig 4.3) shows the shaft torque, RTORS, with the feedback of the shaft torque, and curve 1 shows the shaft torque with the feedback of the electrical torque, GTORE. When the step is applied, the rotor starts to accelerate and the torque in the drive train only increases by 20 % of the applied torque in curve 2 and 33 % of the step value in curve 1.

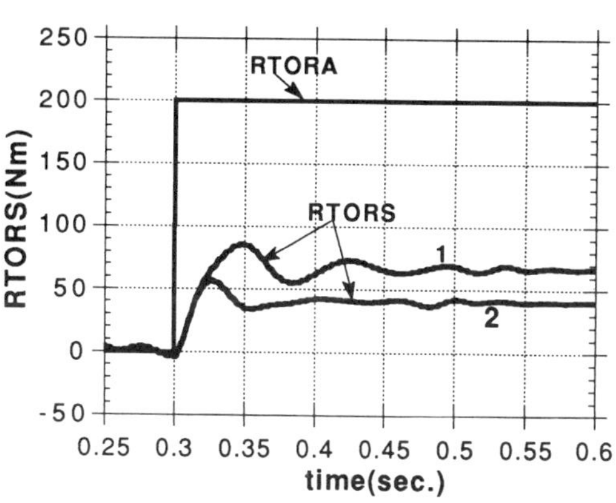

Figure 4.3: Electrical system simulation results.

5. CONTROL

From a control point of view the system is characterized by:
- fixed pitch - torque-regulated induction generator
- variable speed - stall-regulated blades

The aerodynamic performance is shown in Figure 3.3 by the so-called LT-diagram.

5.1 Generator torque control

The main goal of the control is to make the electrical torque, GTORE, dependent on the reference, GTEREF, and as much as possible independent of the generator speed, GS. A rather simple model can be used, if small deviations around a working point are considered (Fig 5.1).

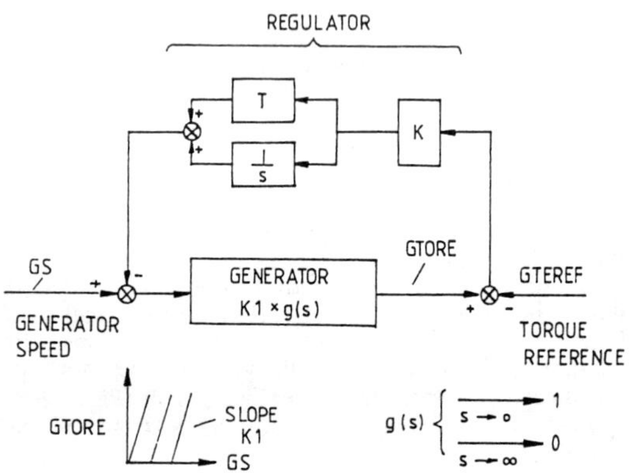

Figure 5.1: Model for the electrical control.

If the loop gain is high (K x K1 >> 1), the following equation will be valid for frequencies < the bandwidth

$$\Delta \text{GTORE} / \Delta \text{GS} \approx s/K * 1/(s\,T+1)$$

The drive train behaviour will be very dependent on the generator control (Fig. 5.2).

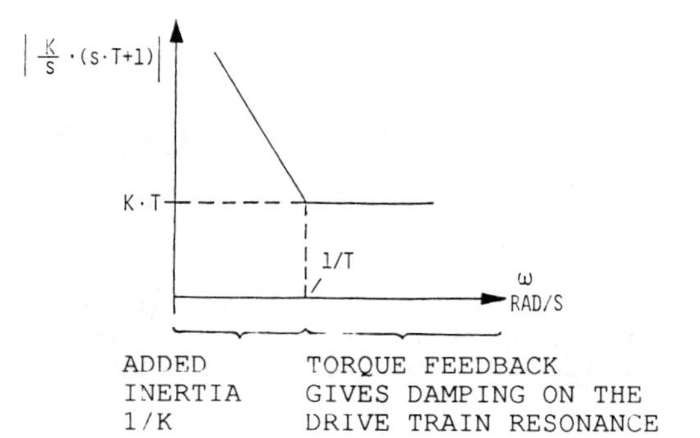

Figure 5.2: BODE-diagram for the feedback operator.

5.2 The wind turbine system

For the wind turbine system a laboratory test has been performed at the CTH-laboratory. The aerodynamic torque, RTORA, has been simulated by means of a DC-machine (Fig. 5.3). Figure 5.4 shows some results of the test.

A rather low gain was obtained because of dead time in the converter.

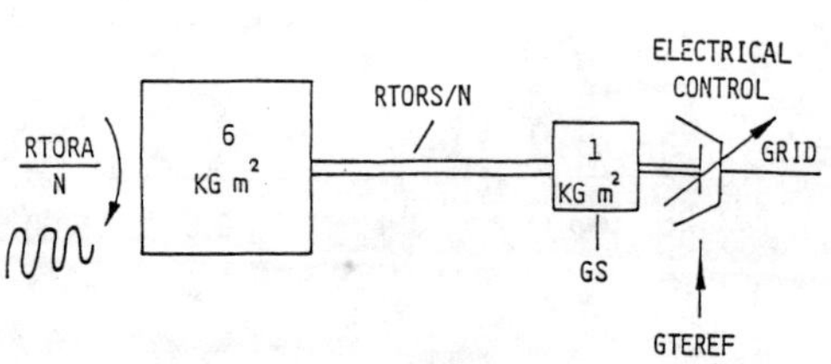

Figure 5.3: Block diagram of the electrical set-up in the laboratory.

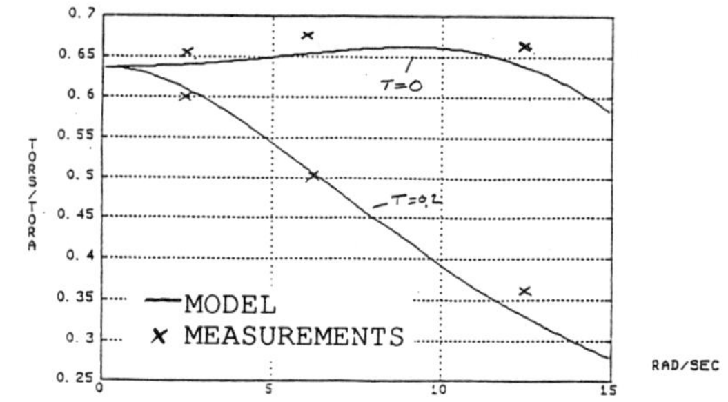

Figure 5.4: Bode diagram with input torque to shaft torque.

5.3 Main control

The driving aerodynamic torque, RTORA, on the rotor is:

$$\text{RTORA} = 0.5 * \rho * A * R * Cq(\lambda) * WS * WS$$

where ρ is the air density, A is the rotor area
R is the rotor radius, WS is the wind speed

Figure 5.5 shows the $Cq(\lambda)$ - curve.

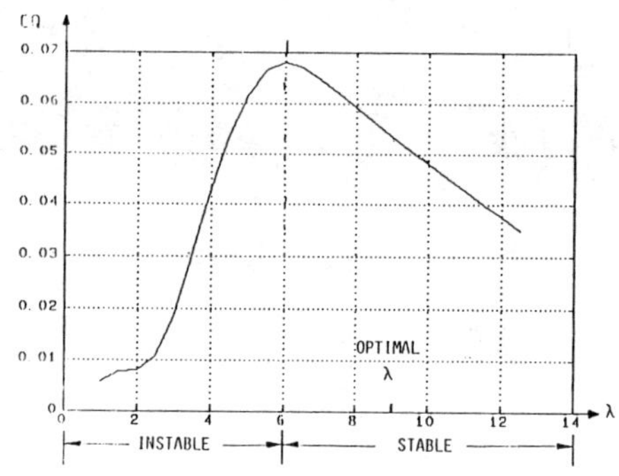

Figure 5.5: The $C_q(\lambda)$ curve.

The behaviour can be studied more in detail in the LT - diagram (Fig. 3.3). Three different modes can be defined depending on the wind speed.

0 - 6	m/sec	optimal mode(PPR-mode)
6 - 13	m/sec	intermediate mode
13 - 20	m/sec	stall mode

5.4 Optimal Control

The torque loop is chosen, (Fig 5.6) [4].

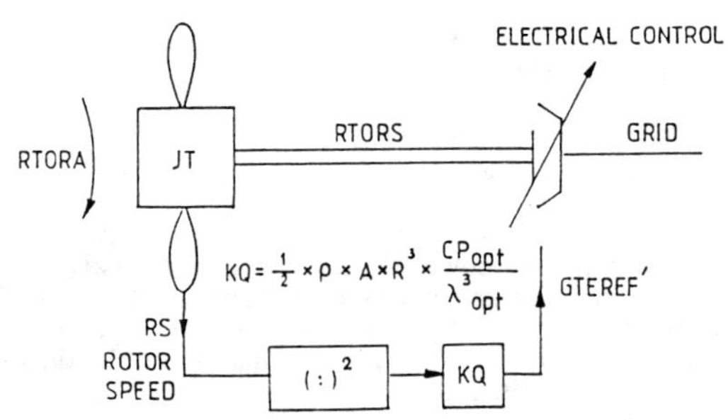

Figure 5.6: Torque loop for optimal control.

5.5 Speed Control

A rather soft speed control is used in the intermediate and stall modes. In order to stabilize the system for $\lambda < 6$ a feedforward from RS to GTORE is implemented. The following equation is valid:

$$\Delta \text{RTORA}/\Delta \text{RS} = 0.5 * \rho * A * R * R * \Delta C_q(\lambda)/\Delta\lambda * WS$$

where Δ means small deviations of RS around the reference value. Figure 5.7 shows $\Delta C_q/\Delta\lambda$ as a function of λ.

Figure 5.8 shows a block diagram of the regulator for speed control.

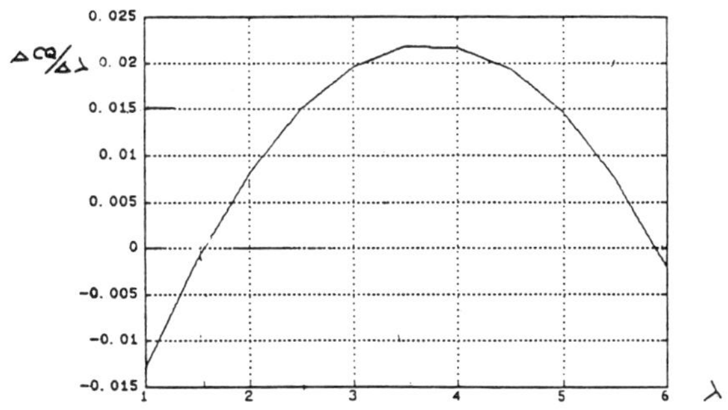

Figure 5.7: $\Delta C_q/\Delta\lambda$ as a function of λ.

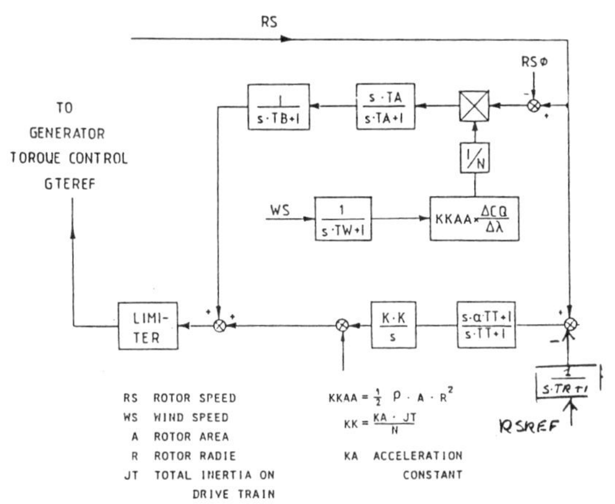

Figure 5.8: Block diagram of the speed control.

The feedforward action gives a normal inertia load for all frequencies. Without feedforward it will be an increasing negative phase shift at lower frequencies (the time constant

$$JT / \frac{\Delta\,\mathrm{RTORA}}{\Delta\,\mathrm{RS}}$$ will interfere).

6. MEASURING SYSTEM

The measuring system is designed to measure mainly mechanical loads on the turbine, electrical power and meteorological data.

The system consists of three data acquisition units, which convert the sensor signals into a serial data format. During a measurement, the data are stored on the hard disk of the 386 personal computer. Lager data files can be stored on 8 mm video cassettes.

One data acquisition unit (A) is mounted on the hub. It collects bending moments in the blade roots, shaft torque and rotor teeter angle.

The two remaining units are situated in the operator house together with the personal computer. Unit (B) collects wind speed and wind direction from four anemometers mounted on two masts. This unit also collects air pressure, air temperature and control mode.

The six accelerometers in the nacelle and the strain gauges in the tower foot are connected to unit (C). Unit (C) also collects rotor position, yaw position, generator speed and electrical power.

Measurement rate:	Unit	Rate (samples/s)
	(A)	46
	(B)	0.8
	(C)	23

7. CONCLUSIONS

The wind turbine is well equipped for tests and evaluations of aerodynamic and mechanical loads, control and electrical systems.

The main modifications are new wooden blades, teeter motion of the rotor, increased measurement system of mechanical quantities, variable hydraulic yaw system, power electronics, and control routines.

In low wind speeds, the turbine operates well with optimal tip speed ratio. In intermediate and stall regions the speed control system is stable and operates well. With a loose control of the speed, the system is stable and the wind disturbance interferes less on the drive train.

The blades are built in a cheap way for one-off blades in solid wood of the lightweight wood Thuja Plicata. The first measurements at 70 RPM are in good agreement with the calculations.

8. FUTURE WORK

The future work can be split up in different parts as suggested below:

Evaluation of the present profile by power curves for different constant speeds and with the turbine working in auto mode with speed and power control. After this evaluation it should be interesting to design blades with a new profile to increase the stall and minimize the thrust. The blades have removable blade tips in order to make it simple to exchange tips to study acoustic noise or aerodynamic effects of the other tip forms.

Tests will be made in order to find out if the electrical control can be improved by a more sophisticated control with higher gain.

Tests to minimize the teeter movement by a control of the yaw position will be carried out.

The stability of the yaw will be tested with different cone angles, rigid hub and a soft control.

Results from the different studies can be used to form a concept of a soft, simple and economic wind turbine.

9. ACKNOWLEDGEMENT

In a project like this, which includes many types of work, e.g. mechanical design, electrical control and mechanical fabrication, a lot of people are involved. The authors would like to thank all the colleagues involved in this project for their invaluable effort and good co-operation and especially Magnus Ellsén and Asterios Kastoris for their enthusiastic work with the measuring system and the operation of the wind turbine.

The financial support given by the Swedish National Board for Industrial and Technical Development and by Vattenfall AB is gratefully acknowledged.

REFERENCES

[1] Ganander H., et al."Stall Control with Variable Speed at the Hönö 2 Wind Turbine.", Department of Electrical Machines and Power Electronics, Chalmers University of Technology, 1993.

[2] Ganander H., et al. "Presentation of the Hönö test machine and results of mechanical behaviour in stall operation.", ECWEC '93, Travemunde.

[3] Al-Hosini F., "Design of function regulators for dc converters based on a dead-beat control strategy.", Technical report No 75 L, Chalmers University of Technology, 1989.

[4] Ulén E. and Dalén G. "Some control problems with a broad range variable speed for horizontal axis wind turbines.", FFA TN 1990-37, The Aeronautical Research Institute of Sweden, 1991.

AUTHOR INDEX